CHAPTER ESSAYS, CONNECTION MODUL... AND TALKING ABOUT SCIENCE MODUL...

CH 1	Essay: A Big-Billed Bird Rebounds
1.9	Biology is connected to our lives in many ways
CH 2	Essay: Nature's Chemical Language
2.2	Trace elements are common additives to food and water
2.5	Radioactive isotopes can help or harm us
2.16	Acid precipitation threatens the environment
CH 3	Essay: Got Lactose?
3.6	How sweet is sweet?
3.10	Anabolic steroids and related substances pose health risks
3.15	Linus Pauling contributed to our understanding of the chemistry of life
CH 4	Essay: The Art of Looking at Cells
4.11	Abnormal lysosomes can cause fatal diseases
CH 5	Essay: Cool "Fires" Attract Mates and Meals
5.9	Many poisons, pesticides, and drugs are enzyme inhibitors
5.20	Faulty membranes can overload the blood with cholesterol
CH 6	Essay: How Is a Marathoner Different from a Sprinter?
6.4	The human body uses energy from ATP for all its activities
6.11	Certain poisons interrupt critical events in cellular respiration
CH 7	Essay: Plant Power
7.13	Photosynthesis moderates global warming
7.14	Mario Molina talks about Earth's protective ozone layer
CH 8	Essay: Rain Forest Rescue
8.10	Growing out of control, cancer cells produce malignant tumors
8.20	An extra copy of chromosome 21 causes Down syndrome
8.22	Abnormal numbers of sex chromosomes do not usually affect survival
8.23	Alterations of chromosome structure can cause birth defects and cancer
CH 9	Essay: Purebreds and Mutts-A Difference of Heredity
9.8	Genetic traits in humans can be tracked through family pedigrees
9.9	Many inherited disorders in humans are controlled by a single gene
9.10	New technologies can provide insight into one's genetic legacy
9.17	Genetic testing can detect disease-causing alleles
9.24	Sex-linked disorders affect mostly males
CH 10	Essay: Sabotage Inside Our Cells
10.18	Many viruses cause disease in animals
10.19	Plant viruses are serious agricultural pests
10.20	Emerging viruses threaten human health
CH 11	Essay: To Clone or Not to Clone
11.11	Reproductive cloning has valuable applications, but human reproductive cloning raises major ethical issues
11.12	Therapeutic cloning can produce stem cells with great medical potential
11.19	Mary-Claire King discusses mutations that cause breast cancer
11.20	Avoiding carcinogens can reduce the risk of cancer
CH 12	Essay: DNA and Crime Scene Investigations
12.6	Recombinant cells and organisms can mass-produce gene products
12.7	DNA technology is changing the pharmaceutical industry and medicine
12.9	DNA microarrays test for the expression of many genes at once
12.12	DNA technology is used in courts of law
12.13	Gene therapy may someday help treat a variety of diseases
12.15	The Human Genome Project is an ambitious application of DNA technology
12.17	The science of genomics compares whole genomes
12.18	Genetically modified organisms are transforming agriculture
12.19	Could GM organisms harm human health or the environment?
12.20	Genomics researcher Eric Lander discusses the Human Genome Project
CH 13	Essay: Clown, Fool, or Simply Well Adapted?
13.5	Scientists can observe natural selection in action
13.8	The Hardy-Weinberg equation is useful in public health science
13.10	Endangered species often have reduced variation
13.13	The evolution of antibiotic resistance in bacteria is a serious public health concern
CH 14	Essay: Mosquito Mystery
14.7	Polyploid plants clothe and feed us
14.9	Peter and Rosemary Grant study the evolution of Darwin's finches
CH 15	Essay: Are Birds Really Dinosaurs with Feathers?
15.4	Tectonic trauma imperils local life
CH 16	Essay: How Ancient Bacteria Changed the World
16.3	Stanley Miller's experiments showed that organic molecules could have arisen on a lifeless Earth
16.14	Some bacteria cause disease
16.15	Bacteria can be used as biological weapons
16.16	Prokaryotes help recycle chemicals and clean up the environment
CH 17	Essay: Plants and Fungi—A Beneficial Partnership
17.12	Agriculture is based almost entirely on angiosperms
17.14	Plant diversity is a nonrenewable resource
17.19	Parasitic fungi harm plants and animals
17.22	Fungi have enormous ecological benefits and practical uses
CH 18	Essay: What Am I?
18.23	Humans threaten animal diversity by introducing non-native species
CH 19	Essay: How Are We Related to the Neanderthals?
19.7	Human skin colors reflect adaptations to varying amounts of sunlight

(continued on next page)

(continued from previous page)

19.8   A genetic difference helped humans start speaking

CH 20   Essay: Climbing the Walls

20.8   Artificial tissues have medical uses

20.11   New imaging technology reveals the inner body

CH 21   Essay: Getting Their Fill of Krill

21.7   The Heimlich maneuver can save lives

21.10   Bacterial infections can cause ulcers

21.17   Vegetarians must be sure to obtain all eight essential amino acids

21.20   Do you need to take vitamin and mineral supplements?

21.21   What do food labels tell us?

21.22   Obesity is a human health problem

21.23   What are the health risks and benefits of fad diets?

21.24   Diet can influence cardiovascular disease and cancer

CH 22   Essay: Surviving in Thin Air

22.6   Smoking is a deadly assault on our respiratory system

22.11   The human fetus exchanges gases with the mother's bloodstream

CH 23   Essay: How Does Gravity Affect Blood Circulation?

23.8   What is a heart attack?

23.10   Measuring blood pressure can reveal cardiovascular problems

23.14   Too few or too many red blood cells can be unhealthy

23.16   Stem cells offer a potential cure for blood cell diseases

CH 24   Essay: An AIDS Uproar

24.10   Monoclonal antibodies are powerful tools in the lab and clinic

24.12   HIV destroys helper T cells, compromising the body's defenses

24.16   Malfunction or failure of the immune system causes disease

24.17   Allergies are overreactions to certain environmental antigens

CH 25   Essay: Let Sleeping Bears Lie

25.5   Do we need to drink eight glasses of water each day?

25.8   Alcohol consumption can damage the liver

25.12   Kidney dialysis can be a lifesaver

CH 26   Essay: Testosterone and Male Aggression: Is There a Link?

26.8   Diabetes is a common endocrine disorder

26.10   Glucocorticoids offer relief from pain, but not without serious risks

CH 27   Essay: Baby Bonanza

27.7   Sexual activity can transmit disease

27.8   Contraception can prevent unwanted pregnancy

27.19   Reproductive technology increases our reproductive options

CH 28   Essay: Can an Injured Spinal Cord Be Fixed?

28.9   Many drugs act at chemical synapses

28.17   Injuries and brain operations provide insight into brain function

28.20   Changes in brain physiology can produce neurological disorders

CH 29   Essay: An Animal's Senses Guide Its Movement

29.7   Artificial lenses or surgery can correct focusing problems

29.11   What causes motion sickness?

29.13   Our sense of taste may change as we age

CH 30   Essay: Elephants Do the "Groucho Gait"

30.5   Broken bones can heal themselves

30.6   Weak, brittle bones are a serious health problem, even in young people

30.11   Athletic training increases strength and endurance

CH 31   Essay: A Gentle Giant

31.1   Plant scientist Natasha Raikhel studies the *Arabidopsis* plant as a model biological system

31.15   Asexual reproduction is a mainstay of modern agriculture

CH 32   Essay: Plants That Clean Up Poisons

32.7   You can diagnose some nutrient deficiencies in your own plants

32.9   Soil conservation is essential to human life

32.10   Organic farmers must follow ecological principles

32.11   Agricultural research is improving the yields and nutritional values of crops

CH 33   Essay: What Are the Health Benefits of Soy?

33.8   Plant hormones have many agricultural uses

33.13   Joanne Chory studies the effects of light and hormones in the model plant *Arabidopsis*

33.15   Plant biochemist Eloy Rodriguez studies how animals use defensive chemicals made by plants

CH 34   Essay: A Mysterious Giant of the Deep

34.3   Environmental problems reveal the limits of the biosphere

34.18   Ecologist Ariel Lugo studies tropical forests in Puerto Rico

CH 35   Essay: Leaping Herds of Herbivores

35.6   Imprinting poses problems and opportunities for conservation programs

35.19   Behavioral biologist Jane Goodall discusses dominance hierarchies and reconciliation behavior in chimpanzees

35.22   Both genes and culture contribute to human social behavior

35.23   Edward O. Wilson promoted the field of sociobiology and is a leading conservation activist

CH 36   Essay: The Spread of Shakespeare's Starlings

36.8   Principles of population ecology have practical applications

36.9   Human population growth has started to slow after centuries of exponential increase

CH 37   Essay: Dining In

37.8   Fire specialist Max Moritz discusses the role of fire in ecosystems

37.14   A production pyramid explains why meat is a luxury for humans

37.20   Ecosystem alteration can upset chemical cycling

37.21   David Schindler talks about the effects of nutrients on freshwater ecosystems

CH 38   Essay: Saving the Tiger

38.4   Pollution of the environment compounds our impact on other species

38.5   Rapid global warming could alter the entire biosphere

38.10   The Yellowstone to Yukon Conservation Initiative seeks to preserve biodiversity by connecting protected areas

38.12   The Kissimmee River project is a case study in restoration ecology

Log on.

Tune in.

Succeed.

Your steps to success.

STEP 1: Register

All you need to get started is a valid email address and the access code below. To register, simply:

1. Go to www.campbellbiology.com
2. Click the appropriate book cover.
   Cover must match the textbook edition being used for your class.
3. Click **"Register"** under **"First-Time User?"**
4. Leave **"No, I Am a New User"** selected.
5. Using a coin, scratch off the silver coating below to reveal your access code.
   Do not use a knife or other sharp object, which can damage the code.
6. Enter your access code in lowercase or uppercase, without the dashes.
7. Follow the on-screen instructions to complete registration.
   During registration, you will establish a personal login name and password to use for logging into the website. You will also be sent a registration confirmation email that contains your login name and password.

Your Access Code is:

                                                  *

Note: If there is no silver foil covering the access code, it may already have been redeemed, and therefore may no longer be valid. In that case, you can purchase access online using a major credit card. To do so, go to www.campbellbiology.com, click the cover of your textbook, click **"Buy Now,"** and follow the on-screen instructions.

STEP 2: Log in

1. Go to www.campbellbiology.com and click the appropriate book cover.
2. Under **"Established User?"** enter the login name and password that you created during registration. *If unsure of this information, refer to your registration confirmation email.*
3. Click **"Log In."**

STEP 3: (Optional) Join a class

Instructors have the option of creating an online class for you to use with this website. If your instructor decides to do this, you'll need to complete the following steps using the Class ID your instructor provides you. By "joining a class," you enable your instructor to view the scored results of your work on the website in his or her online gradebook.

To join a class:

1. Log into the website. For instructions, see "STEP 2: Log in."
2. Click **"Join a Class"** near the top left.
3. Enter your instructor's **"Class ID"** and then click **"Next."**
4. At the Confirm Class page you will see your instructor's name and class information. If this information is correct, click Next.
5. Click **"Enter Class Now"** from the Class Confirmation page.

- *To confirm your enrollment in the class, check for your instructor and class name at the top right of the page. You will be sent a class enrollment confirmation email.*
- *As you complete quizzes on the website from now through the class end date, your results will post to your instructor's gradebook, in addition to appearing in your personal view of the Results Reporter.*

To log into the class later, follow the instructions under "STEP 2: Log in."

Got technical questions?

Visit http://247.aw.com. Email technical support is available 24/7.
Call 1-800-677-6337. Phone support is available 8:00am – 8:00pm Eastern, Monday-Friday and 5:00pm - 12:00am (midnight) Eastern, Sunday.

SITE REQUIREMENTS

For the latest updates on Site Requirements, go to www.campbellbiology.com, choose your text cover, and click Site Reqs.

WINDOWS
366 MHz CPU
Windows 98/2000/XP
64 MB RAM
1024 x 768 screen resolution, thousands of colors
Browser: Internet Explorer 5.0/5.5/6.0; Netscape 7.01
Plug-Ins: Shockwave Player 8; Flash Player 7; QuickTime 6
Internet Connection: 56k modem minimum

MACINTOSH
266 MHz PowerPC
OS 9.2/10.2.4/10.3.2
64 MB RAM
1024 x 768 screen resolution, thousands of colors
Browser: Internet Explorer 5.2; Netscape 6.2.3/7.01
Plug-Ins: Shockwave Player 8; Flash Player 7; QuickTime 6
Internet Connection: 56k modem minimum

Register and log in

Join a class

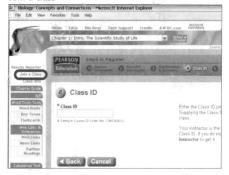

Important: Please read the Subscription and End-User License agreement, accessible from the book website's login page, before using the Campbell Biology website and CD-ROM. By using the website or CD-ROM, you indicate that you have read, understood, and accepted the terms of this agreement.

Features of the Website and Student CD-ROM for
Biology: Concepts & Connections, Fifth Edition

Chapter Guide
The Chapter Guide organizes media under central concepts that correlate directly to the text.

Activities
Explore approximately 200 activities, including animations, interactive review exercises, and videos. Test your knowledge with Activities Quizzes that can be e-mailed to instructors.

Thinking as a Scientist
Perform 56 virtual investigations that develop scientific thinking skills such as data collection, analysis, and communication of results via e-mailable worksheets. With a subscription to Biology Labs On-Line, link to 12 more extensive virtual labs and complete assignments that can be e-mailed.

You Decide
These activities will help you learn how to interpret research data and make informed decisions by examining topics such as low fat versus low carb diets and global warming.

Quizzes
Assess your understanding with over 3,600 multiple-choice questions. Each chapter includes a Pre-Test to diagnose current knowledge, an Activities Quiz with graphics that tests understanding of the media Activities in the chapter, and a comprehensive Chapter Quiz (50 questions per chapter on average). Each quiz has hints, immediate feedback, grading, and e-mailable results. New to the fifth edition are Key Concept Quizzes that ask students to synthesize information found over multiple modules.

Cumulative Quizzes
Create customized tests on multiple chapters at once, by choosing the chapters and number of questions. Each question provides a hint and reference to the relevant module(s) in the text.

Word Study Tools
Use electronic Flash Cards to test your knowledge of the key terms and definitions for each chapter or multiple chapters. Includes audio pronunciations for selected terms. Key Terms allow you to study the list of key terms for each chapter, looking up definitions and hearing selected audio pronunciations. Word roots are also provided to improve vocabulary skills. The Glossary provides definitions of boldface terms and selected audio pronunciations.

Art and Videos
View all the art from the textbook and a collection of 85 videos.

Web Links and References
Organized by chapter, Web Links and References allow you to extend your knowledge through links to relevant websites, access news links on recent developments in biology, and consult an archive of relevant biology news articles and lists of further readings for each chapter.

BIOLOGY

CONCEPTS & CONNECTIONS

Fifth Edition

Neil A. Campbell
University of California, Riverside

Jane B. Reece
Berkeley, California

Martha R. Taylor
Cornell University

Eric J. Simon
New England College

PEARSON

Benjamin Cummings

San Francisco Boston New York
Cape Town Hong Kong London Madrid Mexico City
Montreal Munich Paris Singapore Sydney Tokyo Toronto

Editor-In-Chief: *Beth Wilbur*
Acquisitions Editor: *Chalon Bridges*
Developmental Director: *Deborah Gale*
Editorial Project Manager: *Ginnie Simione Jutson*
Developmental Editors: *Evelyn Dahlgren, Kim Johnson Krummel, Pat Burner, John Burner, Beth N. Winickoff, Alice E. Fugate, Robin Fox*
Art Director: *Blakeley Kim*
Assistant Editor: *Nora Lally-Graves*
Managing Editor, Production: *Erin Gregg*
Production Supervisor: *Vivian McDougal*
Developmental Artists: *Hilair Chism, Quade and Emiko Paul*
Illustrations: *Imagineering Media Services and Blakeley Kim*
Photo Editor: *Donna Kalal*
Copyeditor: *Janet Greenblatt*

Proofreader: *Pete Shanks*
Indexer: *Deborah E. Patton*
Marketing Managers: *Jeff Hester and Josh Frost*
Media Producer: *Christopher Delgado*
Media Project Manager: *Brienn Buchanan*
Media Editors: *Alexandra Fellowes, Amy Austin, Karen Gulliver*
Permissions Editor: *Sue Ewing*
Production Management and Art Coordination: *Jon Peck, Dovetail Publishing Services*
Compositor: *Dovetail Publishing Services*
Text Design: *Carolyn Deacy*
Cover Design: *Yvo Riezebos*
Manufacturing Buyer: *Pam Augspurger*
Cover Printer: *Phoenix Color*
Printer: *RR Donnelly*

On the cover: Photograph of a brown pelican (*Pelecanus occidentalis*) © The Image Bank.

Credits continue in the appendices.

ISBN 10: 0-321-51244-8
ISBN 13: 978-0-321-51244-4

Library of Congress Cataloging-in-Publication Data

Biology : concepts and connections / Neil A. Campbell ... [et al.].-- 5th ed.
  p. cm.
 Includes bibliographical references and index.
 1. Biology. I. Campbell, Neil A., 1946–2004

QH308.2.B56448 2005
570--dc22

                               2004020141

PEARSON

Benjamin Cummings

Benjamin Cummings
1301 Sansome St.
San Francisco, CA 94111
www.aw.com/bc

1 2 3 4 5 6 7 8 9 10—DOW—08 07

About the Authors

Neil A. Campbell

Jane B. Reece

Martha R. Taylor

Eric J. Simon

Neil A. Campbell taught general biology for over 30 years and with Dr. Reece recently coauthored *Biology*, Seventh Edition, the most widely used text for biology majors. His enthusiasm for sharing the fun of science with students stemmed from his own undergraduate experience. He began at Long Beach State College as a history major, but switched to zoology after general education requirements "forced" him to take a science course. Following a B.S. from Long Beach, he earned an M.A. in zoology from UCLA and a Ph.D. in plant biology from the University of California, Riverside. He published numerous articles on how certain desert plants thrive in salty soil and how the sensitive plant (*Mimosa*) and other legumes move their leaves. His diverse teaching experiences included courses for non-biology majors at Cornell University, Pomona College, and San Bernardino Valley College, where he received the first Outstanding Professor Award in 1986. For many years, Dr. Campbell was a visiting scholar in the Department of Botany and Plant Sciences at UC Riverside, which recognized him as the university's Distinguished Alumnus for 2001. In addition to *Biology*, Seventh Edition, he coauthored *Essential Biology* (First and Second Editions) and *Essential Biology with Physiology*.

Jane B. Reece has worked in biology publishing since 1978, when she joined the editorial staff of Benjamin Cummings. Her education includes an A.B. in biology from Harvard University (where she was initially a philosophy major), an M.S. in microbiology from Rutgers University, and a Ph.D. in bacteriology from the University of California, Berkeley. At UC Berkeley and later as a postdoctoral fellow in genetics at Stanford University, her research focused on genetic recombination in bacteria. Dr. Reece taught biology at Middlesex County College (New Jersey) and Queensborough Community College (New York). During her 12 years as an editor at Benjamin Cummings, she played a major role in a number of successful textbooks. Subsequently, she was a coauthor of *The World of the Cell*, Third Edition, with W. M. Becker and M. F. Poenie. With Dr. Campbell, she coauthored *Biology*, Seventh Edition, *Essential Biology* (First and Second Editions), and *Essential Biology with Physiology*.

Martha R. Taylor has been teaching biology for over 20 years. As an undergraduate at Gettysburg College, she considered math, French, and religion as possible majors before settling on biology and receiving her B.A. in that subject. She then went on to earn her M.S. and Ph.D. in science education from Cornell University. She was assistant director of the Office of Instructional Support at Cornell for seven years. She has taught introductory biology for both majors and nonmajors at Cornell University for many years and is currently a visiting lecturer in Cornell's introductory biology laboratory course. Based on her experiences working with students from high school and community college through university, in both classrooms and tutorials, Dr. Taylor is committed to helping students create their own knowledge of and appreciation for biology. She has been the author of the Student Study Guide for all seven editions of *Biology*, by Drs. Campbell and Reece.

Eric J. Simon is an assistant professor of biology at New England College in Henniker, New Hampshire. He teaches introductory biology to both science majors and non-science majors, as well as upper-level courses in genetics, microbiology, and molecular biology. Dr. Simon received a B.A. in biology and computer science and an M.A. in biology from Wesleyan University and a Ph.D. in biochemistry from Harvard University. Currently, he is working toward an M.S.Ed. in educational psychology. His research focuses on innovative ways to use technology to improve teaching and learning in the science classroom, particularly for non-science majors. Dr. Simon is a coauthor of *Essential Biology*, Second Edition, and *Essential Biology with Physiology* with Drs. Campbell and Reece.

Neil A. Campbell died just after the manuscript for this edition was completed. He is deeply missed by his many friends and colleagues at Benjamin Cummings and throughout the biology community.

Introduce yourself to the chapter.

Find out where you're going.
Use the *chapter outline* to preview the chapter.

CHAPTER 20

Climbing the Walls

SPIDERMAN IS A FAMILIAR CHARACTER well known for his ability to climb walls, but few vertebrates have similar talents. One exception is the gecko, a small lizard commonly found in the tropics. Although perhaps not skilled at fighting crime or rescuing those in danger, geckos have no trouble walking up walls and even across ceilings. How do they do it? Several hypotheses, including either a sticky adhesive or suction cups on their toes, have turned out to be wrong. Instead, the explanation relates to hairs, called setae, on the gecko's toes. These hairs are made of the protein keratin, just like our own hair.

The micrographs on the opposite page reveal the microscopic structure of setae. They are arranged in rows, and each seta ends in many split ends called spatulae, which have flattened tips. It took a multidisciplinary team of biologists and engineers to work out how the setae stick to surfaces with enough strength to support the animal's weight. In a recent study using the Tokay gecko (*Gekko gecko*), engineers designed an apparatus to measure the force of attraction between individual setae and the surface they touched—a difficult task because of the microscopic size of setae. This force of attraction turned out to be ten times greater than had been predicted.

But what is causing this attraction? The researchers attribute it to attractions between molecules at the tips of the spatulae and molecules making up the surface. Even uncharged molecules have regions that temporarily carry charges, and a region of positive charge on one molecule will be attracted to a region of negative

THE HIERARCHY OF STRUCTURAL ORGANIZATION IN AN ANIMAL

20.1 Structure fits function in the animal body
20.2 Animal structure has a hierarchy
20.3 Tissues are groups of cells with a common structure and function
20.4 Epithelial tissue covers the body and lines its organs and cavities
20.5 Connective tissue binds and supports other tissues
20.6 Muscle tissue functions in movement
20.7 Nervous tissue forms a communication network
20.8 Artificial tissues have medical uses
20.9 Organs are made up of tissues
20.10 Organ systems work together to perform life's functions
20.11 New imaging technology reveals the inner body

EXCHANGES WITH THE EXTERNAL ENVIRONMENT

20.12 Structural adaptations enhance exchange between animals and their environment
20.13 Animals regulate their internal environment
20.14 Homeostasis depends on negative feedback

Focus on what's most important.

Get the big picture.
Look for the *main headings* (blue bars) that organize the chapter into major sections.

Understand biology one concept at a time.
Each module features a *central concept,* announced in its heading.

Use both text and figures as you study.
The figures illuminate the text and vice versa. Text and figures are always together—so you'll never have to turn a page to find what you need.

ANIMAL CLONING

11.10 Nuclear transplantation can be used to clone animals

As mentioned in Module 11.3, cloning provides strong evidence that differentiated cells retain their full genetic potential. Animal cloning, including the cases described in the chapter introduction, is achieved through a procedure called **nuclear transplantation** (Figure 11.10). First performed in the 1950s on frog embryos, nuclear transplantation involves replacing the nucleus of an egg cell or a zygote with the nucleus of an adult somatic cell. The egg cell may then begin to divide. About 5 days later, repeated cell divisions form a blastocyst, a ball of cells. At this point, the blastocyst may be used for different purposes, as indicated by the branching in Figure 11.10.

If the animal to be cloned is a mammal, further development requires implanting the blastocyst into the uterus of a surrogate mother (Figure 11.10, upper branch). The resulting animal will be genetically identical to the donor of the nucleus—a "clone" of the donor. This type of cloning, which results in birth of a new individual, is called **reproductive cloning**. Scottish researcher Ian Wilmut and his colleagues

used reproductive cloning to produce the celebrated sheep Dolly (see introduction), using a fully differentiated mammary cell of an adult ewe as the source of the nucleus.

In a different cloning procedure (Figure 11.10, lower branch), **embryonic stem cells (ES cells)** are harvested from the blastocyst. In nature, embryonic stem cells give rise to all the different kinds of specialized cells of the body. In the laboratory, embryonic stem cells are easily grown in culture, where, given the right conditions, they can perpetuate themselves indefinitely. When the major aim is to produce embryonic stem cells for therapeutic treatments, the process is called **therapeutic cloning**. In the next two modules, we will discuss applications of reproductive and therapeutic cloning.

? What are the intended products of reproductive cloning and therapeutic cloning?

■ Reproductive cloning seeks the production of new individuals. Therapeutic cloning seeks the harvest of embryonic stem cells.

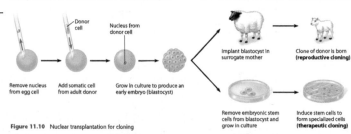

Figure 11.10 Nuclear transplantation for cloning

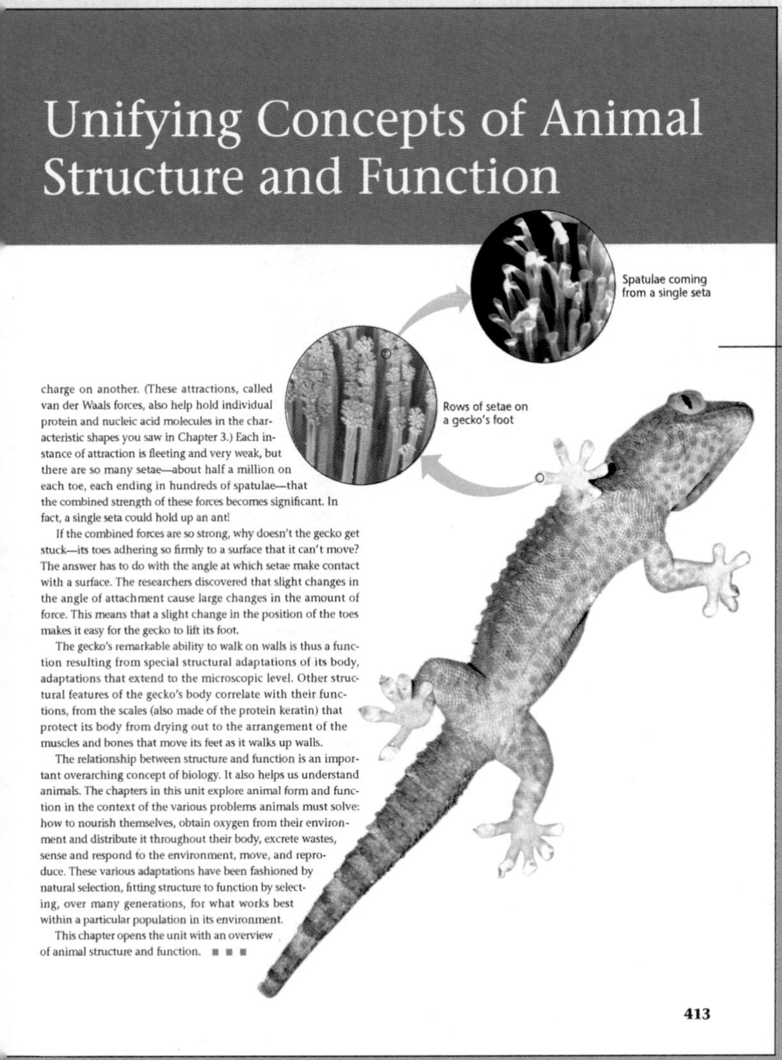

Unifying Concepts of Animal Structure and Function

Spatulae coming from a single seta

Rows of setae on a gecko's foot

charge on another. (These attractions, called van der Waals forces, also help hold individual protein and nucleic acid molecules in the characteristic shapes you saw in Chapter 3.) Each instance of attraction is fleeting and very weak, but there are so many setae—about half a million on each toe, each ending in hundreds of spatulae—that the combined strength of these forces becomes significant. In fact, a single seta could hold up an ant!

If the combined forces are so strong, why doesn't the gecko get stuck—its toes adhering so firmly to a surface that it can't move? The answer has to do with the angle at which setae make contact with a surface. The researchers discovered that slight changes in the angle of attachment cause large changes in the amount of force. This means that a slight change in the position of the toes makes it easy for the gecko to lift its foot.

The gecko's remarkable ability to walk on walls is thus a function resulting from special structural adaptations of its body, adaptations that extend to the microscopic level. Other structural features of the gecko's body correlate with their functions, from the scales (also made of the protein keratin) that protect its body from drying out to the arrangement of the muscles and bones that move its feet as it walks up walls.

The relationship between structure and function is an important overarching concept of biology. It also helps us understand animals. The chapters in this unit explore animal form and function in the context of the various problems animals must solve: how to nourish themselves, obtain oxygen from their environment and distribute it throughout their body, excrete wastes, sense and respond to the environment, move, and reproduce. These various adaptations have been fashioned by natural selection, fitting structure to function by selecting, over many generations, for what works best within a particular population in its environment.

This chapter opens the unit with an overview of animal structure and function. ■ ■ ■

413

Discover.
The *opening essays* introduce the chapter topic through stories that will pique your curiosity.

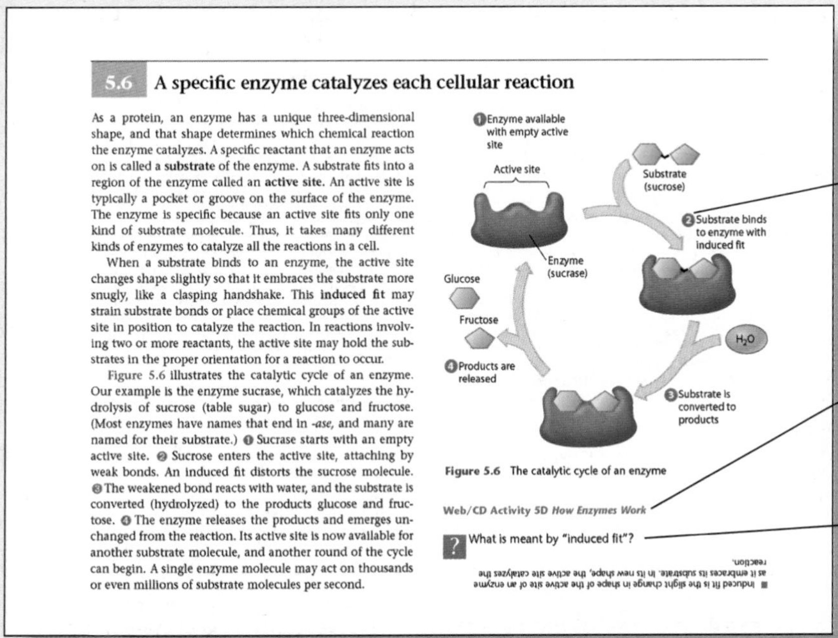

5.6 A specific enzyme catalyzes each cellular reaction

As a protein, an enzyme has a unique three-dimensional shape, and that shape determines which chemical reaction the enzyme catalyzes. A specific reactant that an enzyme acts on is called a **substrate** of the enzyme. A substrate fits into a region of the enzyme called an **active site**. An active site is typically a pocket or groove on the surface of the enzyme. The enzyme is specific because an active site fits only one kind of substrate molecule. Thus, it takes many different kinds of enzymes to catalyze all the reactions in a cell.

When a substrate binds to an enzyme, the active site changes shape slightly so that it embraces the substrate more snugly, like a clasping handshake. This **induced fit** may strain substrate bonds or place chemical groups of the active site in position to catalyze the reaction. In reactions involving two or more reactants, the active site may hold the substrates in the proper orientation for a reaction to occur.

Figure 5.6 illustrates the catalytic cycle of an enzyme. Our example is the enzyme sucrase, which catalyzes the hydrolysis of sucrose (table sugar) to glucose and fructose. (Most enzymes have names that end in *-ase*, and many are named for their substrate.) ❶ Sucrase starts with an empty active site. ❷ Sucrose enters the active site, attaching by weak bonds. An induced fit distorts the sucrose molecule. ❸ The weakened bond reacts with water, and the substrate is converted (hydrolyzed) to the products glucose and fructose. ❹ The enzyme releases the products and emerges unchanged from the reaction. Its active site is now available for another substrate molecule, and another round of the cycle can begin. A single enzyme molecule may act on thousands or even millions of substrate molecules per second.

❶ Enzyme available with empty active site

Active site

Substrate (sucrose)

❷ Substrate binds to enzyme with induced fit

Enzyme (sucrase)

Glucose

Fructose

H_2O

❹ Products are released

❸ Substrate is converted to products

Figure 5.6 The catalytic cycle of an enzyme

Web/CD Activity 5D *How Enzymes Work*

? What is meant by "induced fit"?

■ Induced fit is the slight change in shape of the active site of an enzyme as it embraces its substrate. In its new shape, the active site catalyzes the reaction.

Never got lost.
Figures describing a process take you through a series of *numbered steps* keyed to explanations in the text.

Interact.
Media references direct you to related activities and investigations on the CD-ROM and website.

Test yourself.
Get immediate feedback with a *checkpoint question* at the end of each module.

Learn about biology in your world.

Make a connection.
Connection modules relate biology to your life and interests—from blood doping to stem cell research to fad diets.

CONNECTION

23.14 Too few or too many red blood cells can be unhealthy

Adequate numbers of red blood cells (Figure 23.14) are essential for healthy body function. After circulating in blood for 3 or 4 months, red blood cells are broken down and their molecules recycled. Much of the iron removed from the hemoglobin is returned to the bone marrow, where new red blood cells are formed at the amazing rate of 2 million per second.

An abnormally low amount of hemoglobin or a low number of red blood cells is a condition called **anemia.** An anemic person feels constantly tired and is often susceptible to infections because the body cells do not get enough oxygen. Anemia can result from a variety of factors, including excessive blood loss, vitamin or mineral deficiencies, and certain cancers. Iron deficiency is the most common cause. Women are more likely to develop iron deficiency than men because of blood loss during menstruation. Pregnant women generally benefit from iron supplements to support the developing fetus and placenta.

The production of red blood cells in the bone marrow is controlled by a negative-feedback mechanism that is sensitive to the amount of oxygen reaching the tissues via the blood. If the tissues are not receiving enough oxygen, the kidneys produce a hormone called **erythropoietin (EPO)** that stimulates the bone marrow to produce more red blood cells. Patients on kidney dialysis often have very low red blood cell counts because their kidneys do not produce enough erythropoietin. Genetically engineered EPO has significantly helped these patients.

One of the physiological adaptations of individuals who live at high altitudes, where oxygen levels are low, is the production of more red blood cells. Many athletes train at high altitudes to benefit from this effect. But other athletes take more drastic and illegal measures to increase the oxygen-carrying capacity of their blood and improve their performance. Injecting synthetic EPO can increase normal red blood cell volume from 45% to as much as 65%. Other athletes seek an unfair advantage by blood doping—withdrawing and storing their red blood cells and then reinjecting them before a competition. One way that athletic commissions test for cheaters is by measuring the percentage of red blood cells in the blood volume. Athletic organizations also test for EPO-like chemicals; several athletes who tested positive in the 2002 Winter Olympics were stripped of their medals.

But there can be even more serious consequences. In some athletes, a combination of dehydration from a long race and blood already thickened by an increased number of red blood cells has led to severe medical problems, such as clotting, stroke, heart failure, and even death.

? Why might increasing the number of red blood cells result in greater endurance and speed?

■ The additional red blood cells increase the oxygen-carrying capacity of blood and thus the oxygen supply to working muscles.

Colorized SEM 3,400×

Figure 23.14
Human red blood cells

Meet the people behind the science.
Hear scientists discuss their careers, their research, and their thoughts about the future in *Talking About Science* modules.

TALKING ABOUT SCIENCE

12.20 Genomics researcher Eric Lander discusses the Human Genome Project

Dr. Eric Lander (Figure 12.20) is the founding director of the Broad Institute of MIT and Harvard, which uses genomics to develop new tools and approaches to understanding and treating disease. Previously, he supervised a team that played a leading role in the Human Genome Project (HGP). Although he is now a leader in the field of genomics, Dr. Lander's study of biology began much later in life than for most scientists. In fact, he taught economics at Harvard Business School before turning to biology full-time in 1990. In a recent interview, Dr. Lander discussed his unusual career path:

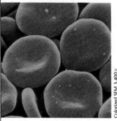

In high school, I took biology, but I loved math, and I was a math major in college. I went on to get my Ph.D. in mathematics but decided I didn't want to be a pure mathematician. One day, my brother suggested I might be interested in the coding theory of the brain and sent me some papers on mathematical neurobiology. I realized that to understand them, I had to learn something about cellular neurobiology. This required me to study cell biology. Next came molecular biology, and finally, I really had to know genetics. So one thing led to another—and here I am still learning genetics!

Figure 12.20 Eric Lander

Dr. Lander's interest in applying math to biology led him to the field of genetic analysis at a time when the HGP was starting. Dr. Lander describes the importance of the project:

One of the major surprises of the HGP was the number of human genes. Dr. Lander describes this finding:

My best guess today is that there are 20,000 to 25,000 protein-coding genes. One of the most surprising findings of the Human Genome Project was a gene count much lower than people had expected based on the total size of the human genome; not so many years ago, textbooks gave 100,000 as the likely number of human genes. When we got the rough draft of the sequence, we thought there might be as many as 40,000 genes, but we soon learned that many of these were actually pseudogenes, which are defective, nonfunctional copies of true genes. The gene count has been falling and falling.

In addition to the number of genes, the HGP has yielded many other unexpected results. Dr. Lander thus sees the completion of the HGP not as an end, but as the beginning of a deeper exploration:

The extent of our ignorance became clear from comparing the human genome with the mouse genome. When we lined up the two, we found that about 5% of the human genome showed strong similarity to that of the mouse, indicating strong evolutionary conservation since the last common ancestor of mouse and human. But only about a third of the 5% could be accounted for by known genes and regulatory sequences, leaving a lot more that evolution "cares about" than we can explain today. That's what I love about genomics. We're learning that there are vast tracts of biology we have missed. It's as if we suddenly could look at the whole Earth and see that, golly, there are several continents we hadn't known about! Genomics is revealing huge territories for the next generation of young scientists to explore.

Feel confident going into the test.

Review the main points.
Reviewing the Concepts helps you do just that, with helpful diagrams and references back to the text.

Connect the chapter's key concepts.
Connecting the Concepts activities test your ability to link topics from different modules and include active mapping, labeling, and categorizing exercises.

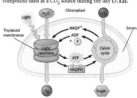
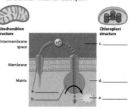

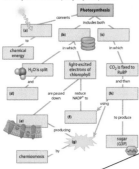

Prepare for the test using the multiple-choice questions that appear in the *Testing your Knowledge* section.

Test your understanding with questions that ask you to describe, compare, and explain. If you can restate a concept in your own words, you've probably learned it.

Get involved.
Applying the Concepts questions help you connect biology to your life and society.

Preface

Neil Campbell was the visionary educator behind the first edition of this groundbreaking text and instrumental in many of the improvements in succeeding editions. His death in October 2004 coincided with the completion of the manuscript for this fifth edition. In this Preface, we summarize Neil's vision as refined and extended in this edition.

The aim of this book has always been to engage students from a wide variety of majors in the wonders of the living world. Most of these students will not become biologists themselves, but their lives will be touched by biology every day. Today, understanding the concepts of biology and their connections to our lives is more important than ever. Whether we're concerned with our own health or the health of our planet, a familiarity with biology is essential. This basic knowledge and an appreciation for how science works have become elements of good citizenship in an era when informed evaluations of health issues, environmental problems, and applications of new technology are critical.

The "connections" to which the title of this book refers include many such practical applications of biology—and go beyond them. Biology has important connections with the other natural sciences and with the humanities and social sciences as well. And the study of life has no coherence without an understanding of the connections between the different areas of biology and an appreciation of the grand unifying theme of evolution. From its first edition, the hallmarks of this book have included an emphasis on connections within biology and between biology and other fields. In this fifth edition, we have increased the emphasis on connections to our everyday world while also doing more to help students connect the concepts of biology.

We could not have hoped to meet our ambitious goals for this book without extensive discussions with teaching colleagues throughout the world and feedback from many of the hundreds of instructors and hundreds of thousands of students who have used our earlier editions. We have been gratified by their enthusiastic responses and have paid close attention to their thoughtful suggestions for improvement. For this edition, we set out to create a book that would be an even more effective tool for learning biology. In addition, we worked to ensure that the book would integrate smoothly with the rich program of supporting materials on the CD-ROM and website that accompany the book.

How can we help students learn—and enjoy—biology? How can we help instructors teach biology? Our responses to these questions are reflected in the teaching strategies we bring to the book. Below we describe our main strategies as they are embodied in *Biology: Concepts & Connections*, Fifth Edition.

Focus Students on the Main Ideas of Biology

Biology is a vast subject that gets bigger every year, but an introductory biology course is still only one or two terms long. In that brief time, we explore all of life, from molecules to ecosystems, while also trying to share the excitement of important research breakthroughs. For beginning students confronting this avalanche of information, it can seem as important to memorize all the scientific terms and facts as it is to master and apply the major ideas. This situation changes, however, when students acquire a framework of key biological concepts into which they can fit the many new things they learn. It is this framework of concepts that will serve them long after they have forgotten specific facts and terms.

Concept Modules *Biology: Concepts & Connections* was the first introductory biology textbook to use concept modules to help students recognize and focus on the main ideas of each chapter. The heading of each module is not simply a topic title but a carefully crafted statement of a biological concept. Printed in large type, each concept heading serves as a focal point for a module, and all of the module's text and illustrations converge on that concept with explanation and, often, evidence. For example, "Sensory receptors convert stimulus energy to action potentials" announces a key concept in Chapter 29 (Module 29.2). The text and illustrations introduce the general principles of sensory reception and transduction, using the human sense of taste as an example. In this and other modules, we integrate the words and pictures to an unprecedented degree: The text walks the student through the illustrations, just as an instructor would do in class. In teaching a sequential process, such as the functioning of the receptor cells of the taste buds (Figure 29.2A), we number the steps in the text to correspond to numbered steps in the figure. The synergy between a module's verbal and graphic components transforms the concept heading into an idea with meaning to the student. In this edition, the illustration program has been extensively revamped to better serve both visual and nonvisual learners.

Integrated Media Printed in blue immediately following the text in many modules are one or more references to interactive Activities and Thinking as a Scientist investigations to be found on the student CD-ROM and website.

Checkpoint Questions At the end of every module is a checkpoint question. These questions encourage students to test themselves as they proceed through a chapter. Some questions simply ask the student to restate the main concept or a corollary; others test understanding of the supporting evidence or ask the student to connect the concept to another concept in the book; still others require the student to carry out a calculation using information in the module. Feedback is provided on the spot: The answer is printed upside down beneath the question. These questions are intended to make students think about the material they are studying and to build their confidence. In summary, each module provides everything a student needs to master a concept.

Help Students Connect the Concepts

How do we help students see the connections between concepts? First, we group the modules under prominent main headings, printed on blue bars, which form an overarching framework for the chapter. Students first see these main headings and their subordinate concepts and connections in the outline at the beginning of the chapter. Within the chapter, we make frequent use of overviews and reviews, and explanatory transitions at the beginnings or ends of modules help the modules flow as a continuous story. At the end of the chapter, the main headings organize the chapter summary, Reviewing the Concepts. Following the summary, a new section called Connecting the Concepts helps students tie the chapter's concepts together for themselves. This section includes concept maps for students to complete and other graphic activities.

Relate Biological Concepts to Everyday Life

Connection Modules Students are more motivated to study biology when they can connect it to their own lives and interests—for example, to health issues, economic problems, environmental quality, ethical controversies, and social responsibility. In this edition, light tan Connection banners mark the numerous application modules that go beyond the core biological concepts. You can preview the Connection module headings on the front endpapers of the book.

Introductory Essays Most of the illustrated essays that open the chapters discuss topics that relate to readers' lives and interests. For example, Chapter 3, "The Molecules of Cells," opens with a new essay on lactose intolerance. Other introductory essays continue our earlier tradition of featuring a nonhuman organism and describing how it is adapted to its environment. One example is our new essay for Chapter 1, which focuses on pelicans. Our hope is that essays like this one will nurture students' appreciation for biological diversity.

Applying the Concepts Many of the questions in Applying the Concepts, found at the end of every chapter, encourage students to use the concepts they have learned in thinking about various social and environmental issues.

Adapt This Book to Fit Your Course

Though a biology textbook's table of contents must be linear, biology itself is more like a web of related concepts without a single starting point or prescribed path. Courses can navigate this network starting with molecules, with ecology, or somewhere in between, and most courses omit some topics. *Biology: Concepts & Connections* is uniquely suited to serve this variety of courses. The seven units of the book are largely self-contained, and in a number of the units, chapters can be assigned in a different order without much loss of coherence. Moreover, the modular format of the chapters makes it easy to omit or to relocate modules within a syllabus.

Relate Biological Concepts to Evolution

The history of life on Earth goes back more than 3.5 billion years, and this past is the key to the present diversity of organisms. As the unifying theme of this textbook, evolution elevates biology from a collection of facts to a coherent study of changing life on a changing planet. In *Biology: Concepts & Connections*, students study the structure, function, and behavior of organisms in an evolutionary context. And throughout the book, students learn to view the unity and diversity of life—the similarities and differences among organisms—as the dual consequences of descent with modification. Our enhanced coverage of evolution in Chapter 1 and Unit III of this edition strengthens this unifying theme.

Help Students Learn the Process of Science

A biology course should familiarize students with the scientific process—in particular, with the posing and testing of hypotheses. With an improved introduction to the process of science in Chapter 1, students will be better equipped to appreciate the many examples throughout the book of how scientific concepts emerge from observations and experimental evidence. The book also puts human faces on science with Talking About Science modules.

Many of the questions in Applying the Concepts, at the end of each chapter, give students personal practice with science as a process. In addition, the CD-ROM and website provide numerous interactive You Decide and Thinking as a Scientist investigations. In fact, this book will work best for students who participate actively in learning about biological concepts and their applications.

■ ■ ■

Introductory biology is the only science course that many students will take during their college years. Long after today's students have forgotten most of the specific content of their biology course, they will be left with general impressions and attitudes about science and scientists. We hope that this new edition of *Biology: Concepts & Connections* helps make those impressions positive and supports the instructor's goals for sharing the fun of biology. To help us produce an even better text in the next edition, please send your comments and suggestions to us in care of our editor, Chalon Bridges, at Benjamin Cummings, 1301 Sansome Street, San Francisco, CA 94111. You can also send e-mail to chalon.bridges@aw.com.

Jane Reece, Martha Taylor, and Eric Simon

New to this edition

Chapter 1, "Biology: Exploring Life," has a new opening essay on the brown pelican, pictured on the cover of this book. A new module (1.3) highlights cells as the structural and functional units of life and introduces the concepts of emergent properties and biological systems. Module 1.8, on hypothesis-based science, more fully develops the everyday (broken flashlight) example and now includes research on mimicry in king snakes.

Unit I, The Life of the Cell, the book's introduction to the basic chemistry, structure, and energetics of cells, has benefited from a number of new chapter-opening essays, art improvements, and reordering and refocusing of many modules. Chapter 2 includes a new Connection module on trace elements as common additives to food and water. Hydrogen bonds are now covered in a separate module (2.10), as a prelude to the major section on water's life-supporting properties. The new opening essay in Chapter 3 uses the topic of lactose intolerance to introduce organic molecules. Chapter 4 includes a better visual and descriptive comparison of prokaryotic and eukaryotic cells. In Chapter 6, the modules on oxidation-reduction reactions and ATP generation have been combined and restructured. The presentation of fermentation (6.13) has been heavily revised. Chapter 7 has a new opening essay on energy plantations, and the photosynthesis art has been restructured. A mechanical analogy has been added to Module 7.9 to help explain electron flow in the light reactions.

Unit II, Cellular Reproduction and Genetics, incorporates important recent advances in the field. New chapter introductions discuss the rescue of extremely rare plant species (Chapter 8), the cloning of endangered animals (Chapter 11), and the use of DNA in forensic investigations (Chapter 12). Chapter 8 incorporates advances in our understanding of binary fission (8.3), the G_0 stage of the cell cycle (8.9), the causes of Down syndrome (8.20), and sex chromosome nondisjunction syndromes (8.22). Chapter 9 contains a heavily revised module on options for fetal genetic testing (9.10). The discussion of non-Mendelian genetics has been reorganized and includes a new module discussing the role of the environment (9.16). The final section of Chapter 10, now called "Microbial Genetics," includes coverage of both viruses (with an update on emerging viruses, such as SARS) and bacteria (10.22 and 10.23, formerly in Chapter 12). In Chapter 11, a new section called "Animal Cloning" highlights the differences between reproductive and therapeutic cloning. Chapter 12 has been largely revised, as reflected in its new title, "DNA Technology and Genomics." Modules on applications of DNA technology are now located immediately after the modules that present the corresponding methods. A new module (12.17) presents recent advances in genomics and proteomics. A new Talking About Science module (12.20) features a discussion of the Human Genome Project with Eric Lander.

Unit III, Concepts of Evolution, has been extensively reorganized and updated. In Chapter 13, Module 13.2 ("Darwin proposed natural selection as the mechanism of evolution") now follows directly from the account of Darwin's travels in Module 13.1. Module 13.4 refines and expands the discussion of the molecular evidence of evolution and includes a new figure on comparative embryology. Module 13.12, on sources of genetic variation, now includes recombination in bacteria and viruses. The revised opening essay for Chapter 14 describes new research on the mosquito species that spreads West Nile virus. A new Module 14.1 distinguishes microevolution and macroevolution. Material on macroevolution has been moved from Chapter 15 and expanded into a new three-module section at the end of Chapter 14. Module 14.11 includes a figure on the range of complexity of eyes in molluscs to illustrate the concept of gradual refinement in the evolution of complex structures. Module 14.12, on the role of changes in developmental genes in evolution, includes a discussion of homeotic genes. In Chapter 15, the modules on phylogeny and systematics are heavily revised, with updated information on molecular systematics. Module 15.8 explains how cladistic analysis includes birds in the reptilian clade, a classification now adopted by this book.

Unit IV, The Evolution of Biological Diversity, has been revised within an updated phylogenetic framework. In Chapter 16, the discussion of hypotheses about the origin of life now includes protobionts and their nutrition (16.6). Several modules describing general prokaryote characteristics now precede expanded coverage of various groups of archaea and bacteria. The revised protist section follows a tentative phylogeny of protist clades. The coverage of plants in Chapter 17 now begins with their evolution from green algae (17.1) and adaptations for life on land (17.2). The section on fungi in Chapter 17 has been rearranged and expanded to cover basic aspects of fungal life cycles and reproduction, a tentative phylogeny, and details of the life cycles of two fungal groups. A new module (17.21) presents leaf-cutting ants as an example of a mutualistic relationship between fungi and animals. In Chapter 18, two new modules (18.3 and 18.4) provide an overview of animal body plans and how they have been used to infer phylogenetic relationships. Another new module (18.15) presents a cladogram of the chordates. Lampreys are now covered in a separate module (18.16), and the coverage of chondrichthyans, ray-finned fishes, and lobe-fins has been updated (18.17). The characteristics and major groups of mammals are more fully described in Module 18.21. Chapter 19 has been significantly revised to reflect more current understanding of primate and human evolution. Two new Connection modules discuss the evolution of human skin color and human language.

Unit V, Animals: Form and Function, has been enhanced with improved art and more relevant connec-

tions to everyday life. New chapter introductions discuss the controversy surrounding AIDS vaccine trials (Chapter 24) and the increased use of fertility drugs (Chapter 27) as well as research on whether elephants really run (Chapter 30). New Connection modules discuss artificial tissues (Chapter 20); the Heimlich maneuver, vitamin requirements, obesity and evolution, and fad diets (all in Chapter 21); anemia and blood doping (Chapter 23); hydration needs ("Do we need to drink eight glasses of water each day?") and alcohol's effects on the liver (both in Chapter 25); the potential abuse of human growth hormone by athletes (Chapter 26); mental illnesses and other neurological disorders (Chapter 28); changes in taste perception with age (Chapter 29); and osteoporosis (Chapter 30). In addition, the presentation of the concepts underlying the applications has been improved. For example, in Chapter 24, "The Immune System," a new discussion of helper T cells (24.11) prepares the students for a new Connection module offering a detailed description of how HIV affects immunity (24.12). And the revised discussion of clonal selection (24.7) has a large new figure to help clarify this difficult topic for students. In Chapter 27, the topic of hormonal regulation of the menstrual cycle has been reworked to provide better text/art integration (27.5). Chapter 28 has a more current organization of the vertebrate peripheral nervous system (28.12).

Unit VI, Plants: Form and Function, opens with a new Talking About Science module featuring plant biologist Natasha Raikhel. The discussion of plant anatomy contains new connections to what students actually see in their food (31.4 and 31.12). A new summary figure (31.6) helps students visualize the organization of plant tissue systems throughout a plant body and its individual organs.

In Chapter 32, a new section ("Plant Nutrition and Symbiosis") has improved coverage of symbiotic relationships. Chapter 33 has an updated discussion of the role of hormones in apical dominance (33.4).

Unit VII, Ecology, has been updated with current ecological data, fresh photos, and new and revised Talking About Science and Connection modules. The chapter on animal behavior, which is now Chapter 35, opens with an essay on impalas and includes two new Connection modules (imprinting of whooping cranes on ultralight aircraft and a human mate choice study). Module 35.3 includes new research on monogamous prairie voles in discussing the influence of genes and environment on behavior. New modules include 35.10, on social learning, and 35.13, on mating systems and parental care as they relate to the needs of offspring and the certainty of paternity. In Chapter 36, Modules 36.9 and 36.10, on human population growth, have new material on demographic differences between developing and developed countries. Chapter 37, on communities and ecosystems, includes a new module on herbivory (37.5) and one on primary production (37.12). Module 37.2 includes a new figure illustrating resource partitioning in *Anolis* lizards, and Module 37.15 has a figure presenting a general model of nutrient cycling in ecosystems. Max Moritz describes the role of fire in some ecosystems in a new Talking About Science module. Chapter 38 begins with a new essay on tiger conservation in Myanmar and updated information on the biodiversity crisis in Module 38.1. Module 38.4 has new material on biomagnification, dead zones, and mercury pollution. Two new Connection modules describe the Yellowstone to Yukon Conservation Initiative and the Kissimmee River Restoration Project.

Acknowledgments

Biology: Concepts & Connections, Fifth Edition, results from the combined efforts of many people, and the authors wish to extend heartfelt thanks to all those who contributed to this and previous editions. Our work on this edition was shaped by input from the biologists acknowledged in the Fifth Edition reviewer list on p. xiii, who shared with us their experiences teaching introductory biology and provided specific suggestions for improving the book. In particular, we would like to thank Scott Meissner, Cornell University; Kevin Padian, University of California at Berkeley; and Ed Zalisko, Blackburn College. The unsolicited comments and suggestions we received from other biologists and from biology students were also very helpful. In addition, this book has benefited in countless ways from the stimulating contacts we had with numerous biologists during the recent preparation of the larger text, *Biology*, Seventh Edition. We are fortunate to be part of a truly global community that is dedicated to excellence in biology education.

Eric Simon would like to thank the many colleagues who offered advice and contributed their knowledge to improve the text, including Deb Dunlop (Unit VI) and Mark Sugalski (Unit V) of New England College, Kevin McMahon of the New Hampshire State Police Forensics Laboratory (Chapter 12), Jim Newcomb of Georgia State University (Chapter 28), Marshall Simon (Chapter 28, including the data for Figure 28.20A), the staff at the National Tropical Botanical Garden (Chapter 8 opening essay), Jamey Barone (Chapters 12 and 27), and Natalia Plotnikova. Special thanks to Amanda Marsh for helpful suggestions and edits.

The superb publishing team for this edition was headed up by editor-in-chief Beth Wilbur and acquistions editor Chalon Bridges. We cannot thank them enough for their unstinting efforts on behalf of the book and for their commitment to excellence in biology education. Directing the project on a daily basis was our esteemed editorial project manager, Ginnie Simione Jutson. We enormously appreciate her energy and commitment to this project. We are similarly grateful to our exceptionally talented developmental team of Evelyn Dahlgren and Kim Johnson Krummel, as well as developmental editors Pat Burner, John Burner, Beth Winickoff, Alice Fugate, and Robin Fox; we thank them for their thoroughness, creativity, and good humor, and the book is far better than it would have been without their efforts. Thanks also to assistant editor Nora Lally-Graves, who helped coordinate the supplements and many other details. We wish to express our appreciation to executive producer Lauren Fogel, developmental director Deborah Gale, editorial director Frank Ruggirello, and president Linda Davis for their ongoing support.

This book and all the other components of the teaching package are both attractive and pedagogically effective in large part because of the hard work and creativity of the production professionals on our team. We wish to thank Vivian McDougal and Erin Gregg, of Benjamin Cummings, and Jon Peck and Joan Keyes, of Dovetail Publishing Services. They were a pleasure to work with, as was our copyeditor, Janet Greenblatt. We are also grateful to proofreader Pete Shanks, indexer Deborah Patton, permissions editor Sue Ewing, and manufacturing buyer Pam Augspurger.

For users of this book, the illustrations and photos are as important as the prose. One of the major goals of this new edition was a major revamping of the art program. This effort was facilitated by Evelyn Dahlgren and Kim Johnson Krummel, and its success rests on the work of art director Blake Kim and developmental artists Hilair Chism and Emi and Quade Paul, as well as photo editor Donna Kalal. In addition, we love the interior design by Carolyn Deacy and the cover design by Yvo Riezebos; both add to the inviting new look of the book.

As always, this new edition seeks to enhance the connections to students' lives outside the biology classroom. For helping us reach this goal, we are particularly indebted to science journalist April Lynch for suggesting, researching, and writing drafts for several new and revised chapter introductions and Connection modules.

For creating the supplementary materials that support this book, thanks go to David Reid and Ed Zalisko, who prepared the new Instructor's Guide to Text and Media; further thanks to Ed for his work on the Test Bank. Jane Brundage and Joan Keyes coordinated the production of these materials, as well as the excellent Student Study Guide by Richard Liebaert. Chris Romero and John Hammett prepared the PowerPoint Lectures. Playing key roles in the development and production of the electronic supplements were media producer Christopher Delgado, media project manager Brienn Buchanan, media editors Alexandra Fellowes, Amy Austin, and Karen Gulliver, and Web developers Andrew Corbett, Steve Wright, Jim Hufford, and Sarah Young-Dualan. Thank you, one and all!

The members of the Addison Wesley/Benjamin Cummings sales group and the Benjamin Cummings marketing department—in particular, Josh Frost and Jeff Hester—have continued to help us connect with biology instructors and their teaching needs. We thank them for all their hard work and enthusiastic support.

Finally, we are deeply grateful to our families and friends for their support, encouragement, and patience.

Neil Campbell, Jane Reece, Martha Taylor, and Eric Simon

Reviewers

Reviewers of the Fifth Edition

Shylaja Akkaraju, *Bronx Community College*
Felix Akojie, *Paducah Community College*
Bert Atsma, *Union County College*
Mark Barnby, *Ohlone College*
Gail Baughman, *Mira Costa College*
Lisa Bellows, *North Central Texas College*
Robert Boyd, *Auburn University*
Agnello Braganza, *Chabot College*
James Bray, *Blackburn College*
Peggy Brickman, *University of Georgia*
Carole Browne, *Wake Forest University*
Hillary Cressey, *George Mason University*
Michael Davis, *Central Connecticut State University*
Jean DeSaix, *University of North Carolina at Chapel Hill*
Jamin Eisenbach, *Eastern Michigan University*
Gerald Esch, *Wake Forest University*
Shannon Kuchel Fehlberg, *Colorado Christian University*
Dennis Forsythe, *The Citadel*
Sandi Gardner, *Triton College*
Richard Groover, *J. Sargeant Reynolds Community College*
David Harbster, *Paradise Valley Community College*
Chris Haynes, *Shelton State Community College*
Janet Haynes, *Long Island University*
Catherine Hurlbut, *Florida Community College*
Hinrich Kaiser, *Victor Valley College*
Klaus Kalthoff, *University of Texas at Austin*
Jennifer Katcher, *Pima Community College*
Marlene Kayne, *The College of New Jersey*
Kevin Krown, *San Diego State University*
Vic Landrum, *Washburn University*
Tom Lehman, *Morgan Community College*
Harvey Liftin, *Broward Community College*
Colleen McNamara, *Albuquerque TVI Community College*
Scott Meissner, *Cornell University*
Joseph Mendelson, *Utah State University*
V. Christine Minor, *Clemson University*
James Moné, *Millersville University*
Steven Oliver, *Worcester State College*
Kevin Padian, *University of California at Berkeley*
Kathleen Pelkki, *Saginaw Valley State University*
Gary Peterson, *South Dakota State University*
Nirmala Prabhu, *Edison Community College*
Rongsun Pu, *Kean University*
Bob Ratterman, *Jamestown Community College*
David Reid, *Blackburn College*
Stephen Reinbold, *Longview Community College*
John Rinehart, *Eastern Oregon University*
Laura Ritt, *Burlington County College*
Lynn Rivers, *Henry Ford Community College*
Jennifer Roberts, *Lewis University*
Lynette Rushton, *South Puget Sound Community College*
Fred Schindler, *Indian Hills Community College*
Fayla Schwartz, *Everett Community College*
Michele Shuster, *New Mexico State University*
Phil Snider, *University of Houston*
Ralph Sorensen, *Gettysburg College*
David Stanton, *Saginaw Valley State University*
Amanda Starnes, *Emory University*
Mark Sugalski, *New England College*

Christopher Tabit, *State University of West Georgia*
Kenneth Thomas, *Hillsborough Community College*
Laura Thurlow, *Jackson Community College*
Anne Tokazewski, *Burlington County College*
Nancy Tress, *University of Pittsburgh at Titusville*
Michael Twaddle, *University of Toledo*
Rukmani Viswanath, *Laredo Community College*
Patricia Walsh, *University of Delaware*
Harrington Wells, *University of Tulsa*
Mary Jo Witz, *Monroe Community College*
William Yurkiewicz, *Millersville University*
Edward J. Zalisko, *Blackburn College*

Reviewers of Previous Editions

Michael Abbott, *Westminster College*
Daryl Adams, *Mankato State University*
Dawn Adrian Adams, *Baylor University*
Olushola Adeyeye, *Duquesne University*
Dan Alex, *Chabot College*
Sylvester Allred, *Northern Arizona University*
Jane Aloi-Horlings, *Saddleback College*
Loren Ammerman, *University of Texas at Arlington*
Dennis Anderson, *Oklahoma City Community College*
Marjay Anderson, *Howard University*
Gail Baker, *LaGuardia Community College*
Mark Barnby, *Ohlone College*
Chris Barnhart, *University of San Diego*
Stephen Barnhart, *Santa Rosa Junior College*
William Barstow, *University of Georgia*
Kirk A. Bartholomew, *Central Connecticut State University*
Jane Beiswenger, *University of Wyoming*
Tania Beliz, *College of San Mateo*
Ernest Benfield, *Virginia Polytechnic Institute*
Rudi Berkelhamer, *University of California, Irvine*
Harry Bernheim, *Tufts University*
Richard Bliss, *Yuba College*
Lawrence Blumer, *Morehouse College*
Mehdi Borhan, *Johnson County Community College*
Kathleen Bossy, *Bryant College*
William Bowen, *University of Arkansas at Little Rock*
Robert Boyd, *Auburn University*
Bradford Boyer, *State University of New York, Suffolk County Community College*
Paul Boyer, *University of Wisconsin–Parkside*
William Bradshaw, *Brigham Young University*
Agnello Braganza, *Chabot College*
Chris Brinegar, *San Jose State University*
Becky Brown-Watson, *Santa Rosa Junior College*
Charles Brown, *Santa Rosa Junior College*
Virginia Buckner, *Johnson County Community College*
Joseph C. Bundy, Jr., *University of North Carolina at Greensboro*
Warren Buss, *University of Northern Colorado*
Michael Bucher, *College of San Mateo*
Linda Butler, *University of Texas at Austin*
Jerry Button, *Portland Community College*
Carolee Caffrey, *University of California, Los Angeles*
George Cain, *University of Iowa*
James Cappuccino, *Rockland Community College*

M. Carabelli, *Broward Community College*
Cathryn Cates, *Tyler Junior College*
Russell Centanni, *Boise State University*
Van Christman, *Ricks College*
Ruth Chesnut, *Eastern Illinois University*
Vic Chow, *San Francisco City College*
Mary Colavito-Shepanski, *Santa Monica College*
Bob Cowling, *Ouachita Technical College*
Don Cox, *Miami University*
Robert Creek, *Western Kentucky University*
Norma Criley, *Illinois Wesleyan University*
Judy Daniels, *Monroe Community College*
Lewis Deaton, *University of Louisiana–Lafayette*
Lawrence DeFilippi, *Lurleen B. Wallace College*
James Dekloe, *Solano Community College*
Loren Denney, *Southwest Missouri State University*
Jean DeSaix, *University of North Carolina at Chapel Hill*
Veronique Delesalle, *Gettysburg College*
Mary Dettman, *Seminole Community College*
Kathleen Diamond, *College of San Mateo*
Alfred Diboll, *Macon College*
Jean Dickey, *Clemson University*
Stephen Dina, *St. Louis University*
Robert P. Donaldson, *George Washington University*
Gary Donnermeyer, *Iowa Central Community College*
Charles Duggins, *University of South Carolina*
Susan Dunford, *University of Cincinnati*
Betty Eidemiller, *Lamar University*
Norman Ellstrand, *University of California, Riverside*
Thomas Emmel, *University of Florida*
Cindy Erwin, *City College of San Francisco*
David Essar, *Winona State University*
Cory Etchberger, *Longview Community College*
Nancy Eyster-Smith, *Bentley College*
Laurie Faber, *Grand Rapids Community College*
Terence Farrell, *Stetson University*
Jerry Feldman, *University of California, Santa Cruz*
Eugene Fenster, *Longview Community College*
Dino Fiabane, *Community College of Philadelphia*
Kathleen Fisher, *San Diego State University*
Dennis Forsythe, *The Citadel Military College of
    South Carolina*
Robert Frankis, *College of Charleston*
James French, *Rutgers University*
Bernard Frye, *University of Texas at Arlington*
Anne Galbraith, *University of Wisconsin–LaCrosse*
Robert Galbraith, *Crafton Hills College*
Rosa Gambier, *State University of New York, Suffolk County
    Community College*
George Garcia, *University of Texas at Austin*
Gail Gasparich, *Towson University*
Shelley Gaudia, *Lane Community College*
Douglas Gayou, *University of Missouri–Columbia*
Robert Gendron, *Indiana University of Pennsylvania*
Rebecca German, *University of Cincinnati*
Grant Gerrish, *University of Hawaii*
Frank Gilliam, *Marshall University*
Patricia Glas, *The Citadel Military College of South Carolina*
David Glenn-Lewin, *Wichita State University*
Robert Grammer, *Belmont University*
Peggy Green, *Broward Community College*
Miriam L. Greenberg, *Wayne State University*

Sylvia Greer, *City University of New York*
Dana Griffin, *University of Florida*
Peggy Guthrie, *University of Central Oklahoma*
Maggie Haag, *University of Alberta*
Richard Haas, *California State University, Fresno*
Martin Hahn, *William Paterson College*
Leah Haimo, *University of California, Riverside*
James Hampton, *Salt Lake Community College*
Blanche Haning, *North Carolina State University*
Richard Hanke, *Rose State College*
Laszlo Hanzely, *Northern Illinois University*
Jim Harris, *Utah Valley Community College*
Mary Harris, *Louisiana State University*
Jean Helgeson, *Collin County Community College*
Ira Herskowitz, *University of California, San Francisco*
Paul Hertz, *Barnard College*
Margeret Hicks, *David Lipscomb University*
Jean Higgins-Fonda, *Prince George's Community College*
Phyllis Hirsch, *East Los Angeles College*
William Hixon, *St. Ambrose University*
Carl Hoagstrom, *Ohio Northern University*
Kim Hodgson, *Longwood College*
John Holt, *Michigan State University*
Laura Hoopes, *Occidental College*
Lauren Howard, *Norwich University*
Robert Howe, *Suffolk University*
George Hudock, *Indiana University*
Michael Hudecki, *State University of New York, Buffalo*
Kris Hueftle, *Pensacola Junior College*
Charles Ide, *Tulane University*
Mark Ikeda, *San Bernardino Valley College*
Georgia Ineichen, *Hinds Community College*
Charles Jacobs, *Henry Ford Community College*
Fred James, *Presbyterian College*
Ursula Jander, *Washburn University*
Alan Jaworski, *University of Georgia*
R. Jensen, *Saint Mary's College*
Russell Johnson, *Ricks College*
Florence Juillerat, *Indiana University–Purdue University
    at Indianapolis*
Tracy Kahn, *University of California, Riverside*
Tom Kantz, *California State University, Sacramento*
Marlene Kayne, *The College of New Jersey*
Judy Kaufman, *Monroe Community College*
Mahlon Kelly, *University of Virginia*
Kenneth Kerrick, *University of Pittsburgh at
    Johnstown*
Joyce Kille-Marino, *College of Charleston*
Joanne Kilpatrick, *Auburn University, Montgomery*
Stephen Kilpatrick, *University of Pittsburgh at Johnstown*
Lee Kirkpatrick, *Glendale Community College*
Peter Kish, *Southwestern Oklahoma State University*
Robert Koch, *California State University, Fullerton*
Eliot Krause, *Seton Hall University*
Mary Rose Lamb, *University of Puget Sound*
Carmine Lanciani, *University of Florida*
Deborah Langsam, *University of North Carolina at Charlotte*
Geneen Lannom, *University of Central Oklahoma*
Brenda Latham, *Merced College*
Steven Lebsack, *Linn-Benton Community College*
Karen Lee, *University of Pittsburgh at Johnstown*
Richard Liebaert, *Linn-Benton Community College*

Kevin Lien, *Portland Community College*
Harvey Liftin, *Broward Community College*
Ivo Lindauer, *University of Northern Colorado*
William Lindsay, *Monterey Peninsula College*
Kirsten Lindstrom, *Santa Rosa Junior College*
Melanie Loo, *California State University, Sacramento*
Dave Loring, *Johnson County Community College*
James Mack, *Monmouth University*
David Magrane, *Morehead State University*
Joan Maloof, *Salisbury State University*
Joseph Marshall, *West Virginia University*
Presley Martin, *Drexel University*
William McComas, *University of Iowa*
Steven McCullagh, *Kennesaw State College*
Mitchell McGinnis, *North Seattle Community College*
James McGivern, *Gannon University*
Timothy Metz, *Campbell University*
Iain Miller, *University of Cincinnati*
Robert Miller, *University of Dubuque*
Brad Mogen, *University of Wisconsin–River Falls*
Richard Mortensen, *Albion College*
Henry Mulcahy, *Suffolk University*
Christopher Murphy, *James Madison University*
James Nivison, *Mid Michigan Community College*
Peter Nordloh, *Southeastern Community College*
Stephen Novak, *Boise State University*
Bette Nybakken, *Hartnell College*
Michael O'Donnell, *Trinity College*
Karen Olmstead, *University of South Dakota*
Steven O'Neal, *Southwestern Oklahoma State University*
Lowell Orr, *Kent State University*
William Outlaw, *Florida State University*
Kay Pauling, *Foothill College*
Debra Pearce, *Northern Kentucky University*
David Pearson, *Bucknell University*
Patricia Pearson, *Western Kentucky University*
Andrew Penniman, *Georgia Perimeter College*
Margaret Peterson, *Concordia Lutheran College*
Russell L. Peterson, *Indiana University of Pennsylvania*
Paula Piehl, *Potomac State College*
Ben Pierce, *Baylor University*
Barbara Pleasants, *Iowa State University*
Kathryn Podwall, *Nassau Community College*
Judith Pottmeyer, *Columbia Basin College*
Donald Potts, *University of California, Santa Cruz*
James Pru, *Belleville Area College*
Charles Pumpuni, *Northern Virginia Community College*
Rebecca Pyles, *East Tennessee State University*
Jill Raymond, *Rock Valley College*
Brian Reeder, *Morehead State University*
Bruce Reid, *Kean College*
Stephen Reinbold, *Longview Community College*
Michael Renfroe, *James Madison University*
Douglas Reynolds, *Central Washington University*
Fred Rhoades, *Western Washington University*

Bruce Robart, *University of Pittsburgh at Johnstown*
Laurel Roberts, *University of Pittsburgh*
Duane Rohlfing, *University of South Carolina*
Jeanette Rollinger, *College of the Sequoias*
Steven Roof, *Fairmont State College*
Jim Rosowski, *University of Nebraska*
Stephen Rothstein, *University of California, Santa Barbara*
Donald Roush, *University of North Alabama*
Lynette Rushton, *South Puget Sound Community College*
Linda Sabatino, *State University of New York, Suffolk County Community College*
Douglas Schamel, *University of Alaska, Fairbanks*
Douglas Schelhaas, *University of Mary*
Robert Schoch, *Boston University*
Brian Scholtens, *College of Charleston*
Julie Schroer, *Bismarck State College*
Judy Shea, *Kutztown University of Pennsylvania*
Thomas Shellberg, *Henry Ford Community College*
Lisa Shimeld, *Crafton Hills College*
Brian Shmaefsky, *Kingwood College*
Mark Shotwell, *Slippery Rock University*
Jane Shoup, *Purdue University*
Linda Simpson, *University of North Carolina at Charlotte*
Gary Smith, *Tarrant County Junior College*
Gary Sojka, *Bucknell University*
Ralph Sorensen, *Gettysburg College*
David Stanton, *Saginaw Valley State University*
John Stolz, *Duquesne University*
Ross Strayer, *Washtenaw Community College*
Donald Streuble, *Idaho State University*
Gerald Summers, *University of Missouri–Columbia*
Marshall Sundberg, *Louisiana State University*
David Tauck, *Santa Clara University*
Hilda Taylor, *Acadia University*
Kathy Thompson, *Louisiana State University*
Bruce Tomlinson, *State University of New York, Fredonia*
Nancy Tress, *University of Pittsburgh at Titusville*
Donald Trisel, *Fairmont State College*
Kimberly Turk, *Mitchell Community College*
Leslie Vander Molen, *Humboldt State University*
John Vaughan, *Georgetown College*
Ann Vernon, *St. Charles County Community College*
Mary Beth Voltura, *State University of New York, Cortland*
Jerry Waldvogel, *Clemson University*
Robert Wallace, *Ripon College*
James Wee, *Loyola University*
Larry Williams, *University of Houston*
Neil Woffinden, *University of Pittsburgh at Johnstown*
Mark Wygoda, *McNeese State University*
William Yurkiewicz, *Millersville University of Pennsylvania*
Ray S. Williams, *Appalachian State University*
Sandra Winicur, *Indiana University, South Bend*
Patrick Woolley, *East Central College*
Tony Yates, *Seminole State College*
Uko Zylstra, *Calvin College*

Supplements for the Instructor

LECTURES

Campbell Media Manager for Biology: Concepts & Connections, Fifth Edition
(0-8053-7110-9)

The Campbell Media Manager combines all the instructor and student media for *Biology: Concepts & Connections* into one chapter-by-chapter resource and includes an Image Presentation Library, PowerPoint® presentations, the Introductory Biology Lecture Launcher (Discovery Channel) Videos, lecture outlines, test questions, student media activities and investigations, active lecture questions for use with clickers, and more.

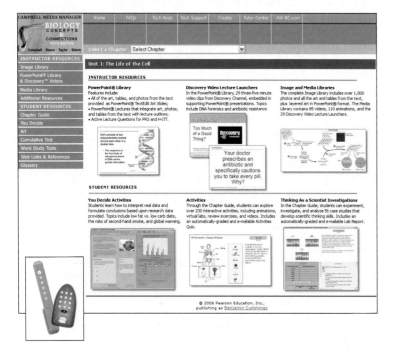

Instructor's Guide to Text and Media, Fifth Edition
(0-8053-7161-3)

This comprehensive guide provides chapter-by-chapter references to all the media resources available to instructors and students plus a list of transparency acetates. The guide also includes objectives, key terms, word roots, lecture outlines, and class activities. A separate chapter offers suggestions for effective uses of technology in teaching introductory biology.

Transparency Acetates
(0-8053-7162-1)

More than 650 full-color acetates include all illustrations and tables from the text, many of which incorporate photographs. Selected figures illustrating key concepts are broken down into layers for step-by-step lecture presentation.

ASSESSMENT

Printed Test Bank, Fifth Edition
(0-8053-7164-8)

Edward J. Zalisko, *Blackburn College*
Thoroughly revised and updated, the test bank includes an optional section with questions that test students on the Web/CD Activities. The test bank is also available in the instructor section of CourseCompass™, Blackboard, and WebCT.

Computerized Test Bank, Fifth Edition
(0-8053-7163-X)

Edward J. Zalisko, *Blackburn College*
Computerized version of the printed test bank.

COURSE MANAGEMENT

CourseCompass™, Fifth Edition
(0-8053-7111-7)

CourseCompass™ combines the strength of Benjamin Cummings content with state-of-the-art eLearning tools. CourseCompass™ is a nationally hosted, dynamic, interactive online course management system powered by Blackboard, leaders in the development of Internet-based learning tools. This easy-to-use and customizable program enables instructors to tailor content and functionality to meet individual course needs. Every CourseCompass™ course includes a range of preloaded content, such as testing and assessment question pools, chapter-level objectives, chapter summaries, photos, illustrations, videos, animations, and web activities—all designed to help students master core course objectives.
URL: http://cms.aw.com/coursecompass

CourseCompass™ Standalone Access Code Card for Biology: Concepts & Connections, Fifth Edition
(0-8053-6822-1)

WebCT, Fifth Edition
(0-8053-7115-X)
URL: http://cms.aw.com/webct

Blackboard Premium, Fifth Edition
(0-8053-7114-1)
URL: http://cms.aw.com/blackboard

Blackboard, Fifth Edition
(0-8053-7113-3)
URL: http://cms.aw.com/blackboard

LABS

Annotated Instructor's Edition for Laboratory Investigations for Biology, Second Edition

(0-8053-6792-6)

Jean L. Dickey, *Clemson University*

The Annotated Instructor's Edition includes margin notes with instructor overviews, time requirements, helpful hints, and suggestions for extending or supplementing labs; answers to questions in the Student Edition; and suggestions for adapting the labs to a 2-hour period.

Preparation Guide for Laboratory Investigations for Biology, Second Edition

(0-8053-6771-3)

Jean L. Dickey, *Clemson University*

This guide provides complete instructions on how to order materials and plan, set up, and run labs smoothly.

Symbiosis Lab Authoring Kit

Build a customized lab manual, choosing the labs you want, importing artwork from our graphics library, and even adding your own notes, syllabi, or other material.

Instructor's Lab Manual for BiologyLabs On-Line

(0-8053-7018-8)

This printed manual provides the individual and group assignments from the Student Lab Manual for BiologyLabs On-Line with suggested answers. Includes a subscription to BiologyLabs On-Line.
URL: http://www.biologylabsonline.com

ADJUNCTS

Adjunct Support Center

Part-time faculty will find instructors with more than 30 years combined experience available by phone, e-mail, and fax to offer advice on everything from syllabi to classroom strategies.
URL: http://www.aw-bc.com/tutorcenter/biology-adjunct.html

Supplements for the Student

STUDY TOOLS

Student CD-ROM and Website for Biology: Concepts & Connections, Fifth Edition (www.campbellbiology.com)

The student CD-ROM and website included with each book contain seven new You Decide activities and a customized testing feature, plus approximately 200 Activities, 56 Thinking as a Scientist investigations, 30 Connections, Flashcards, Word Roots, Key Terms linked to the glossary, over 3,600 quiz questions (a Pre-Test, Activities Quiz, Chapter Quiz, and new Key Concept Quizzes for each chapter), and a Glossary with pronunciations. Responses to questions can be e-mailed. In addition, the website provides access to all the art from the book, with labels and without labels, 85 videos, the Tutor Center, Web Links, News Links, News Archives, Further Readings, Instructor Resources, and Syllabus Manager. The CD-ROM and website are included with new books. Students who buy a used book may purchase the standalone CD-ROM and website or purchase a subscription to the website at www.campbell-biology.com. The website content is also available in CourseCompass™, Blackboard, and WebCT.

Student Study Guide for Biology: Concepts & Connections, Fifth Edition (0-8053-7116-8)

Richard Liebaert, *Linn-Benton Community College*
Students can master key concepts and earn a better grade with the thought-provoking exercises found in this study guide. A wide range of questions and activities help students test their understanding of biology. The study guide also includes references to student media activities on the Student CD-ROM and website.

Study Card for Biology: Concepts & Connections, Fifth Edition (0-8053-7117-6)

This fold-out quick-reference card provides students with an overview of the entire book, helping them see the connections between topics and understand the big picture.

Standalone Access Code Student Tutor Center, First Edition (0-201-72170-8)

This service provides students one-on-one tutoring in four different ways—phone, fax, e-mail, and interactive Web. Qualified college instructors are available to answer students' questions during evening hours and on weekends. Free when bundled with the text.
URL: http://www.aw-bc.com/tutorcenter

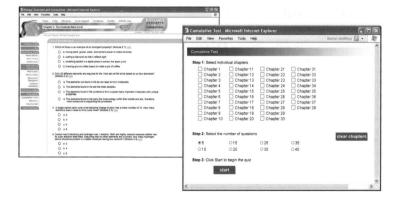

CURRENT TOPICS IN BIOLOGY

Current Issues in Biology, Volume 2 (0-8053-7108-7)
Current Issues in Biology, Volume 1 (0-8053-7507-4)

Exclusive to Benjamin Cummings and free when bundled with Neil Campbell's nonmajors biology texts, these accessible, timely, and relevant articles from *Scientific American* magazine cover current topics in biology, such as performance-enhancing drugs, global warming, schizophrenia, stem cell research, and more. Questions included at the end of each article make them easy to assign.

Understanding the Human Genome Project,
Second Edition (ISBN 0-8053-4877-8)

Michael A. Palladino, *Monmouth University*

Stem Cells and Cloning
(0-8053-4864-6)

David A. Prentice, *Indiana State University*

Biological Terrorism
(0-8053-4868-9)

Steve Goodwin and Randall Phillis,
University of Massachusetts, Amherst

Biology of Cancer
(0-8053-4867-0)

Randall Phillis and Steve Goodwin,
University of Massachusetts, Amherst

HIV and AIDS
(0-8053-3956-6)

Michael A. Palladino, *Monmouth University*

Emerging Infectious Diseases
(0-8053-3955-8)

Michael A. Palladino, *Monmouth University*

LABS

Laboratory Investigations for Biology,
Second Edition (0-8053-6789-6)

Jean L. Dickey, *Clemson University*
An investigative approach actively involves students in the process of scientific discovery by allowing them to make observations, devise techniques, and draw conclusions. Twenty carefully chosen laboratory topics encourage students to use their critical thinking skills to solve problems using the scientific method.

Student Lab Manual for BiologyLabs On-Line,
First Edition (0-8053-7017-X)

This printed student manual provides background information, instructions, assignments, and group assignments for BiologyLabs On-Line. Includes a subscription to Biology-Labs On-Line.

Detailed Contents

1 BIOLOGY: EXPLORING LIFE 1

Essay **A Big-Billed Bird Rebounds** 1

The Scope of Biology 2

1.1 Life's levels of organization define the scope of biology 2

1.2 Living organisms and their environments form interconnecting webs 3

1.3 Cells are the structural and functional units of life 4

Evolution, Unity, and Diversity 4

1.4 The unity of life: All forms of life have common features 4

1.5 The diversity of life can be arranged into three domains 6

1.6 Evolution explains the unity and diversity of life 8

The Process of Science 9

1.7 Scientists use two main approaches to learn about nature 9

1.8 With hypothesis-based science, we pose and test hypotheses 10

Biology and Everyday Life 12

1.9 *Connection* Biology is connected to our lives in many ways 12

Chapter Review 13

UNIT I

The Life of the Cell

2 THE CHEMICAL BASIS OF LIFE 16

Essay **Nature's Chemical Language** 16

Elements, Atoms, and Molecules 18

2.1 Living organisms are composed of about 25 chemical elements 18

2.2 *Connection* Trace elements are common additives to food and water 18

2.3 Elements can combine to form compounds 19

2.4 Atoms consist of protons, neutrons, and electrons 20

2.5 *Connection* Radioactive isotopes can help or harm us 21

2.6 Electron arrangement determines the chemical properties of an atom 22

2.7 Ionic bonds are attractions between ions of opposite charge 22

2.8 Covalent bonds join atoms into molecules through electron sharing 23

2.9 Unequal electron sharing creates polar molecules 24

2.10 Hydrogen bonds are weak bonds important in the chemistry of life 24

Water's Life-Supporting Properties 25

2.11 Hydrogen bonds make liquid water cohesive 25

2.12 Water's hydrogen bonds moderate temperature 25

2.13 Ice is less dense than liquid water 26

2.14 Water is the solvent of life 26

2.15 The chemistry of life is sensitive to acidic and basic conditions 27

2.16 *Connection* Acid precipitation threatens the environment 28

Chemical Reactions 29

2.17 Chemical reactions change the composition of matter 29

Chapter Review 30

---

3 THE MOLECULES OF CELLS 32

Essay **Got Lactose?** 32

Introduction to Organic Compounds 34

3.1 Life's molecular diversity is based on the properties of carbon 34

3.2 Functional groups help determine the properties of organic compounds 35

3.3 Cells make a huge number of large molecules from a small set of small molecules 36

Carbohydrates 37

3.4 Monosaccharides are the simplest carbohydrates 37

3.5 Cells link two single sugars to form disaccharides 38

3.6 *Connection* How sweet is sweet? 38

3.7 Polysaccharides are long chains of sugar units 39

Lipids 40

3.8 Fats are lipids that are mostly energy-storage molecules 40

3.9 Phospholipids, waxes, and steroids are lipids with a variety of functions 41

3.10 **Connection** Anabolic steroids pose health risks 41

Proteins 42

3.11 Proteins are essential to the structures and activities of life 42

3.12 Proteins are made from amino acids linked by peptide bonds 42

3.13 A protein's specific shape determines its function 43

3.14 A protein's shape depends on four levels of structure 44

3.15 **Talking About Science** Linus Pauling contributed to our understanding of the chemistry of life 46

Nucleic Acids 47

3.16 Nucleic acids are information-rich polymers of nucleotides 47

Chapter Review 48

4 A TOUR OF THE CELL 50

Essay The Art of Looking at Cells 50

Introduction to the Cell 52

4.1 Microscopes provide windows to the world of the cell 52

4.2 Most cells are microscopic 54

4.3 Prokaryotic cells are structurally simpler than eukaryotic cells 55

4.4 Eukaryotic cells are partitioned into functional compartments 56

Organelles of the Endomembrane System 58

4.5 The nucleus is the cell's genetic control center 58

4.6 Overview: Many cell organelles are connected through the endomembrane system 58

4.7 Smooth endoplasmic reticulum has a variety of functions 58

4.8 Rough endoplasmic reticulum makes membrane and proteins 59

4.9 The Golgi apparatus finishes, sorts, and ships cell products 60

4.10 Lysosomes are digestive compartments within a cell 60

4.11 **Connection** Abnormal lysosomes can cause fatal diseases 61

4.12 Vacuoles function in the general maintenance of the cell 62

4.13 A review of the endomembrane system 62

Energy-Converting Organelles 63

4.14 Chloroplasts convert solar energy to chemical energy 63

4.15 Mitochondria harvest chemical energy from food 63

The Cytoskeleton and Related Structures 64

4.16 The cell's internal skeleton helps organize its structure and activities 64

4.17 Cilia and flagella move when microtubules bend 65

Cell Surfaces and Junctions 66

4.18 Cell surfaces protect, support, and join cells 66

Functional Categories of Organelles 67

4.19 Eukaryotic organelles comprise four functional categories 67

Chapter Review 68

5 THE WORKING CELL 70

Essay Cool "Fires" Attract Mates and Meals 70

Energy and the Cell 72

5.1 Energy is the capacity to perform work 72

5.2 Two laws govern energy transformations 73

5.3 Chemical reactions either store or release energy 74

5.4 ATP shuttles chemical energy and drives cellular work 75

How Enzymes Function 76

5.5 Enzymes speed up the cell's chemical reactions by lowering energy barriers 76

5.6 A specific enzyme catalyzes each cellular reaction 77

5.7 The cellular environment affects enzyme activity 77

5.8 Enzyme inhibitors block enzyme action 78

5.9 **Connection** Many poisons, pesticides, and drugs are enzyme inhibitors 78

Membrane Structure and Function 79

5.10 Membranes organize the chemical activities of cells 79

5.11 Membrane phospholipids form a bilayer 79

5.12 The membrane is a fluid mosaic of phospholipids and proteins 80

5.13 Proteins make the membrane a mosaic of function 80

5.14 Passive transport is diffusion across a membrane 81

5.15 Transport proteins may facilitate diffusion across membranes 82

5.16 Osmosis is the diffusion of water across a membrane 82

5.17 Water balance between cells and their surroundings is crucial to organisms 83

5.18 Cells expend energy for active transport 84

5.19 Exocytosis and endocytosis transport large molecules 84

5.20 *Connection* Faulty membranes can overload the blood with cholesterol 85

5.21 Chloroplasts and mitochondria make energy available for cellular work 86

Chapter Review 86

6 HOW CELLS HARVEST CHEMICAL ENERGY 88

Essay How Is a Marathoner Different from a Sprinter? 88

Introduction to Cellular Respiration 90

6.1 Photosynthesis and cellular respiration provide energy for life 90

6.2 Breathing supplies oxygen to our cells and removes carbon dioxide 90

6.3 Cellular respiration banks energy in ATP molecules 91

6.4 *Connection* The human body uses energy from ATP for all its activities 91

6.5 Cells tap energy from electrons "falling" from organic fuels to oxygen 92

Stages of Cellular Respiration and Fermentation 93

6.6 Overview: Cellular respiration occurs in three main stages 93

6.7 Glycolysis harvests chemical energy by oxidizing glucose to pyruvate 94

6.8 Pyruvate is chemically groomed for the citric acid cycle 96

6.9 The citric acid cycle completes the oxidation of organic fuel, generating many NADH and $FADH_2$ molecules 96

6.10 Most ATP production occurs by oxidative phosphorylation 98

6.11 *Connection* Certain poisons interrupt critical events in cellular respiration 99

6.12 Review: Each molecule of glucose yields many molecules of ATP 100

6.13 Fermentation is an anaerobic alternative to cellular respiration 101

Interconnections Between Molecular Breakdown and Synthesis 102

6.14 Cells use many kinds of organic molecules as fuel for cellular respiration 102

6.15 Food molecules provide raw materials for biosynthesis 103

6.16 The fuel for respiration ultimately comes from photosynthesis 103

Chapter Review 104

7 PHOTOSYNTHESIS: USING LIGHT TO MAKE FOOD 106

Essay Plant Power 106

An Overview of Photosynthesis 108

7.1 Autotrophs are the producers of the biosphere 108

7.2 Photosynthesis occurs in chloroplasts 109

7.3 Plants produce O_2 gas by splitting water 110

7.4 Photosynthesis is a redox process, as is cellular respiration 110

7.5 Overview: Photosynthesis occurs in two stages linked by ATP and NADPH 111

The Light Reactions: Converting Solar Energy to Chemical Energy 112

7.6 Visible radiation drives the light reactions 112

7.7 Photosystems capture solar power 113

7.8 In the light reactions, electron transport chains generate ATP and NADPH 114

7.9 Chemiosmosis powers ATP synthesis in the light reactions 115

The Calvin Cycle: Converting CO_2 to Sugars 116

7.10 ATP and NADPH power sugar synthesis in the Calvin cycle 116

Photosynthesis Reviewed and Extended 117

7.11 Review: Photosynthesis uses light energy to make food molecules 117

7.12 C_4 and CAM plants have special adaptations that save water 118

Photosynthesis, Solar Radiation, and Earth's Atmosphere 119

7.13 *Connection* Photosynthesis moderates global warming 119

7.14 *Talking About Science* Mario Molina talks about Earth's protective ozone layer 120

Chapter Review 121

UNIT II

Cellular Reproduction and Genetics

8 THE CELLULAR BASIS OF REPRODUCTION AND INHERITANCE 124

Essay Rain Forest Rescue   124

Connections Between Cell Division and Reproduction   126
8.1   Like begets like, more or less   126
8.2   Cells arise only from preexisting cells   127
8.3   Prokaryotes reproduce by binary fission   127

The Eukaryotic Cell Cycle and Mitosis   128
8.4   The large, complex chromosomes of eukaryotes duplicate with each cell division   128
8.5   The cell cycle multiplies cells   129
8.6   Cell division is a continuum of dynamic changes   130
8.7   Cytokinesis differs for plant and animal cells   132
8.8   Anchorage, cell density, and chemical growth factors affect cell division   133
8.9   Growth factors signal the cell cycle control system   134
8.10   *Connection* Growing out of control, cancer cells produce malignant tumors   135
8.11   Review of the functions of mitosis: Growth, cell replacement, and asexual reproduction   136

Meiosis and Crossing Over   136
8.12   Chromosomes are matched in homologous pairs   136
8.13   Gametes have a single set of chromosomes   137
8.14   Meiosis reduces the chromosome number from diploid to haploid   138
8.15   Review: A comparison of mitosis and meiosis   140
8.16   Independent orientation of chromosomes in meiosis and random fertilization lead to varied offspring   141
8.17   Homologous chromosomes carry different versions of genes   142
8.18   Crossing over further increases genetic variability   142

Alterations of Chromosome Number and Structure   144
8.19   A karyotype is a photographic inventory of an individual's chromosomes   144
8.20   *Connection* An extra copy of chromosome 21 causes Down syndrome   145
8.21   Accidents during meiosis can alter chromosome number   146
8.22   *Connection* Abnormal numbers of sex chromosomes do not usually affect survival   147
8.23   *Connection* Alterations of chromosome structure can cause birth defects and cancer   148
Chapter Review   149

9 PATTERNS OF INHERITANCE 152

Essay Purebreds and Mutts—A Difference of Heredity   152

Mendel's Laws   154
9.1   The science of genetics has ancient roots   154
9.2   Experimental genetics began in an abbey garden   154
9.3   Mendel's law of segregation describes the inheritance of a single characteristic   156
9.4   Homologous chromosomes bear the two alleles for each characteristic   157
9.5   The law of independent assortment is revealed by tracking two characteristics at once   158
9.6   Geneticists use the testcross to determine unknown genotypes   159
9.7   Mendel's laws reflect the rules of probability   160
9.8   *Connection* Genetic traits in humans can be tracked through family pedigrees   161
9.9   *Connection* Many inherited disorders in humans are controlled by a single gene   162
9.10   *Connection* New technologies can provide insight into one's genetic legacy   164

Variations on Mendel's Laws   166
9.11   The relationship of genotype to phenotype is rarely simple   166
9.12   Incomplete dominance results in intermediate phenotypes   166
9.13   Many genes have more than two alleles in the population   167
9.14   A single gene may affect many phenotypic characteristics   168
9.15   A single characteristic may be influenced by many genes   169
9.16   The environment affects many characteristics   170
9.17   *Connection* Genetic testing can detect disease-causing alleles   170

The Chromosomal Basis of Inheritance   171
9.18   Chromosome behavior accounts for Mendel's laws   171
9.19   Genes on the same chromosome tend to be inherited together   172
9.20   Crossing over produces new combinations of alleles   172

9.21 Geneticists use crossover data to map genes 174

Sex Chromosomes and Sex-Linked Genes 175

9.22 Chromosomes determine sex in many species 175

9.23 Sex-linked genes exhibit a unique pattern of inheritance 176

9.24 *Connection* Sex-linked disorders affect mostly males 177

Chapter Review 178

10 MOLECULAR BIOLOGY OF THE GENE 180

Essay Sabotage Inside Our Cells 180

The Structure of the Genetic Material 182

10.1 Experiments showed that DNA is the genetic material 182

10.2 DNA and RNA are polymers of nucleotides 184

10.3 DNA is a double-stranded helix 186

DNA Replication 188

10.4 DNA replication depends on specific base pairing 188

10.5 DNA replication: A closer look 189

The Flow of Genetic Information from DNA to RNA to Protein 190

10.6 The DNA genotype is expressed as proteins, which provide the molecular basis for phenotypic traits 190

10.7 Genetic information written in codons is translated into amino acid sequences 191

10.8 The genetic code is the Rosetta stone of life 192

10.9 Transcription produces genetic messages in the form of RNA 193

10.10 Eukaryotic RNA is processed before leaving the nucleus 194

10.11 Transfer RNA molecules serve as interpreters during translation 194

10.12 Ribosomes build polypeptides 196

10.13 An initiation codon marks the start of an mRNA message 196

10.14 Elongation adds amino acids to the polypeptide chain until a stop codon terminates translation 197

10.15 Review: The flow of genetic information in the cell is DNA → RNA → protein 198

10.16 Mutations can change the meaning of genes 199

Microbial Genetics 200

10.17 Viral DNA may become part of the host chromosome 200

10.18 *Connection* Many viruses cause disease in animals 201

10.19 *Connection* Plant viruses are serious agricultural pests 202

10.20 *Connection* Emerging viruses threaten human health 202

10.21 The AIDS virus makes DNA on an RNA template 203

10.22 Bacteria can transfer DNA in three ways 204

10.23 Bacterial plasmids can serve as carriers for gene transfer 205

Chapter Review 206

11 THE CONTROL OF GENE EXPRESSION 208

Essay To Clone or Not to Clone? 208

Gene Regulation 210

11.1 Proteins interacting with DNA turn prokaryotic genes on or off in response to environmental changes 210

11.2 Differentiation yields a variety of cell types, each expressing a different combination of genes 212

11.3 Differentiated cells may retain all of their genetic potential 212

11.4 DNA packing in eukaryotic chromosomes helps regulate gene expression 213

11.5 In female mammals, one X chromosome is inactive in each cell 214

11.6 Complex assemblies of proteins control eukaryotic transcription 214

11.7 Eukaryotic RNA may be spliced in more than one way 215

11.8 Translation and later stages of gene expression are also subject to regulation 216

11.9 Review: Multiple mechanisms regulate gene expression in eukaryotes 217

Animal Cloning 218

11.10 Nuclear transplantation can be used to clone animals 218

11.11 *Connection* Reproductive cloning has valuable applications, but human reproductive cloning raises ethical issues 218

11.12 *Connection* Therapeutic cloning can produce stem cells with great medical potential 219

The Genetic Control of Embryonic Development 220

11.13 Cascades of gene expression and cell-to-cell signaling direct the development of an animal 220

11.14 Signal transduction pathways convert messages received at the cell surface to responses within the cell 221

11.15 Key developmental genes are very ancient 222

The Genetic Basis of Cancer 222

11.16 Cancer results from mutations in genes that control cell division 222

11.17 Oncogene proteins and faulty tumor-suppressor proteins can interfere with normal signal transduction pathways 224

11.18 Multiple genetic changes underlie the development of cancer 225

11.19 *Talking About Science* Mary-Claire King discusses mutations that cause breast cancer 226

11.20 *Connection* Avoiding carcinogens can reduce the risk of cancer 227

Chapter Review 228

12 DNA TECHNOLOGY AND GENOMICS 230

Essay DNA and Crime Scene Investigations 230

Bacterial Plasmids and Gene Cloning 232

12.1 Plasmids are used to customize bacteria: An overview 232

12.2 Enzymes are used to "cut and paste" DNA 233

12.3 Genes can be cloned in recombinant plasmids: A closer look 234

12.4 Cloned genes can be stored in genomic libraries 235

12.5 Reverse transcriptase helps make genes for cloning 235

12.6 *Connection* Recombinant cells and organisms can mass-produce gene products 236

12.7 *Connection* DNA technology is changing the pharmaceutical industry and medicine 237

Restriction Fragment Analysis and DNA Fingerprinting 238

12.8 Nucleic acid probes identify clones carrying specific genes 238

12.9 *Connection* DNA microarrays test for the expression of many genes at once 238

12.10 Gel electrophoresis sorts DNA molecules by size 239

12.11 Restriction fragment length polymorphisms can be used to detect differences in DNA sequences 240

12.12 *Connection* DNA technology is used in courts of law 242

12.13 *Connection* Gene therapy may someday help treat a variety of diseases 243

12.14 The PCR method is used to amplify DNA sequences 244

Genomics 244

12.15 *Connection* The Human Genome Project is an ambitious application of DNA technology 244

12.16 Most of the human genome does not consist of genes 245

12.17 *Connection* The science of genomics compares whole genomes 246

Genetically Modified Organisms 247

12.18 *Connection* Genetically modified organisms are transforming agriculture 247

12.19 *Connection* Could GM organisms harm human health or the environment? 248

12.20 *Talking About Science* Genomics researcher Eric Lander discusses the Human Genome Project 249

Chapter Review 249

Concepts of Evolution

13 HOW POPULATIONS EVOLVE 254

Essay Clown, Fool, or Simply Well Adapted? 254

Darwin's Theory of Evolution 256

13.1 A sea voyage helped Darwin frame his theory of evolution 256

13.2 Darwin proposed natural selection as the mechanism of evolution 258

13.3 The study of fossils provides strong evidence for evolution 260

13.4 A mass of other evidence reinforces the evolutionary view of life 262

13.5 *Connection* Scientists can observe natural selection in action 264

Population Genetics and The Modern Synthesis 265

13.6 Populations are the units of evolution 265

13.7 The gene pool of a nonevolving population remains constant over the generations 266

13.8 *Connection* The Hardy-Weinberg equation is useful in public health science 267

13.9   In addition to natural selection, genetic drift and gene flow can contribute to evolution   268

13.10   *Connection* Endangered species often have reduced variation   269

Variation and Natural Selection   270

13.11   Variation is extensive in most populations   270

13.12   Mutation and sexual recombination generate variation   270

13.13   *Connection* The evolution of antibiotic resistance in bacteria is a serious public health concern   272

13.14   Diploidy and balancing selection preserve variation   272

13.15   The perpetuation of genes defines evolutionary fitness   273

13.16   Natural selection can alter variation in a population in three ways   274

13.17   Sexual selection may produce sexual dimorphism   275

13.18   Natural selection cannot fashion perfect organisms   275

Chapter Review   276

14   THE ORIGIN OF SPECIES 278

Essay Mosquito Mystery   278

14.1   The origin of species is the source of biological diversity   280

Concepts of Species   280

14.2   What is a species?   280

14.3   Reproductive barriers keep species separate   282

Mechanisms of Speciation   284

14.4   Geographic isolation can lead to speciation   284

14.5   Reproductive barriers may evolve as populations diverge   285

14.6   New species can arise within the same geographic area as the parent species   286

14.7   *Connection* Polyploid plants clothe and feed us   287

14.8   Adaptive radiation may occur in new or newly vacated habitats   288

14.9   *Talking About Science* Peter and Rosemary Grant study the evolution of Darwin's finches   289

14.10   The tempo of speciation can appear steady or jumpy   290

Macroevolution   291

14.11   Evolutionary novelties may arise in several ways   291

14.12   Genes that control development are important in evolution   292

14.13   Evolutionary trends do not mean that evolution is goal directed   293

Chapter Review   294

15   TRACING EVOLUTIONARY HISTORY 296

Essay Are Birds Really Dinosaurs with Feathers?   296

Macroevolution and Earth's History   298

15.1   The fossil record chronicles macroevolution   298

15.2   The actual ages of rocks and fossils mark geologic time   299

15.3   Continental drift has played a major role in macroevolution   300

15.4   *Connection* Tectonic trauma imperils local life   302

15.5   Mass extinctions were followed by diversification of life-forms   302

Phylogeny and Systematics   304

15.6   Phylogenies are based on homologies in fossils and living organisms   304

15.7   Systematics connects classification with evolutionary history   304

15.8   Cladograms are diagrams based on shared characters among species   306

15.9   Molecular biology is a powerful tool in systematics   308

15.10   Arranging life into kingdoms is a work in progress   310

Chapter Review   311

UNIT IV

The Evolution of Biological Diversity

16   THE ORIGIN AND EVOLUTION OF MICROBIAL LIFE: PROKARYOTES AND PROTISTS 314

Essay How Ancient Bacteria Changed the World   314

Early Earth and the Origin of Life   316

16.1   Life began on a young Earth   316

16.2   How did life originate?   318

16.3   *Talking About Science* Stanley Miller's experiments showed that organic molecules could have arisen on a lifeless Earth   318

16.4   The first polymers may have formed on hot rocks or clay   320

16.5   The first genetic material and enzymes may both have been RNA   320

16.6 Membrane-enclosed molecular cooperatives may have preceded the first cells 321

Prokaryotes 322

16.7 Prokaryotes have inhabited Earth for billions of years 322

16.8 Bacteria and archaea are the two main branches of prokaryotic evolution 322

16.9 Prokaryotes come in a variety of shapes 323

16.10 Various structural features contribute to the success of prokaryotes 324

16.11 Prokaryotes obtain nourishment in a variety of ways 326

16.12 Archaea thrive in extreme environments-and in other habitats 327

16.13 Bacteria include a diverse assemblage of prokaryotes 328

16.14 *Connection* Some bacteria cause disease 329

16.15 *Connection* Bacteria can be used as biological weapons 330

16.16 *Connection* Prokaryotes help recycle chemicals and clean up the environment 330

Protists 332

16.17 The eukaryotic cell probably originated as a community of prokaryotes 332

16.18 Protists are an extremely diverse assortment of eukaryotes 333

16.19 A tentative phylogeny of eukaryotes includes multiple clades of protists 334

16.20 Diplomonads and euglenozoans include some flagellated parasites 334

16.21 Alveolates have sacs beneath the plasma membrane and include dinoflagellates, apicomplexans, and ciliates 335

16.22 Stramenopiles are named for their "hairy" flagella and include the water molds, diatoms, and brown algae 336

16.23 Amoebozoans have pseudopodia and include amoebas and slime molds 337

16.24 Red algae and green algae are the closest relatives of land plants 338

16.25 Multicellularity evolved several times in eukaryotes 339

Chapter Review 340

17 PLANTS, FUNGI, AND THE COLONIZATION OF LAND 342

Essay **Plants and Fungi-A Beneficial Partnership** 342

Plant Evolution and Diversity 344

17.1 Plants evolved from green algae 344

17.2 Plants have adaptations for life on land 344

17.3 Plant diversity reflects the evolutionary history of the plant kingdom 346

Alternation of Generations and Plant Life Cycles 348

17.4 Haploid and diploid generations alternate in plant life cycles 348

17.5 Mosses have a dominant gametophyte 348

17.6 Ferns, like most plants, have a dominant sporophyte 349

17.7 Seedless plants dominated vast "coal forests" 350

17.8 A pine tree is a sporophyte with tiny gametophytes in its cones 350

17.9 The flower is the centerpiece of angiosperm reproduction 352

17.10 The angiosperm plant is a sporophyte with gametophytes in its flowers 352

17.11 The structure of a fruit reflects its function in seed dispersal 354

17.12 *Connection* Agriculture is based almost entirely on angiosperms 354

17.13 Interactions with animals have profoundly influenced angiosperm evolution 355

17.14 *Connection* Plant diversity is a nonrenewable resource 356

Fungi 357

17.15 Fungi absorb food after digesting it outside their bodies 357

17.16 Fungi produce spores in both asexual and sexual life cycles 358

17.17 Fungi can be classified into five groups 358

17.18 Fungal groups differ in their life cycles and reproductive structures 360

17.19 *Connection* Parasitic fungi harm plants and animals 361

17.20 Lichens consist of fungi living mutualistically with photosynthetic organisms 362

17.21 Fungi also form mutualistic relationships with animals 362

17.22 *Connection* Fungi have enormous ecological benefits and practical uses 363

Chapter Review 364

18 THE EVOLUTION OF ANIMAL DIVERSITY 366

Essay What Am I? 366

Animal Evolution and Diversity 368
18.1 What is an animal? 368
18.2 The ancestor of animals was probably a colonial, flagellated protist 369
18.3 Animals can be characterized by basic features of their "body plan" 370
18.4 The body plans of animals can be used to build phylogenetic trees 371

Invertebrates 372
18.5 Sponges have a relatively simple, porous body 372
18.6 Cnidarians are radial animals with tentacles and stinging cells 373
18.7 Flatworms are the simplest bilateral animals 374
18.8 Nematodes have a pseudocoelom and a complete digestive tract 375
18.9 Diverse molluscs are variations on a common body plan 376
18.10 Annelids are segmented worms 378
18.11 Arthropods are segmented animals with jointed appendages and an exoskeleton 380
18.12 Insects are the most diverse group of organisms 382
18.13 Echinoderms have spiny skin, an endoskeleton, and a water vascular system for movement 384
18.14 Our own phylum, Chordata, is distinguished by four features 385

Vertebrates 386
18.15 Derived characters define the major clades of chordates 386
18.16 Lampreys are vertebrates that lack hinged jaws 387
18.17 Jawed vertebrates with gills and paired fins include sharks, ray-finned fishes, and lobe-fins 388
18.18 Amphibians were the first tetrapods-vertebrates with two pairs of limbs 389
18.19 Reptiles are amniotes-tetrapods with a terrestrially adapted egg 390
18.20 Birds are feathered reptiles with adaptations for flight 391
18.21 Mammals are amniotes that have hair and produce milk 392

Animal Phylogeny and Diversity Revisited 393
18.22 An animal phylogenetic tree is a work in progress 393
18.23 *Connection* Humans threaten animal diversity by introducing non-native species 394
Chapter Review 395

19 HUMAN EVOLUTION 398

Essay How Are We Related to the Neanderthals? 398

Primate Diversity 400
19.1 The human story begins with our primate heritage 400
19.2 Hominoids include humans and four other groups of apes 402

Hominid Evolution 403
19.3 The human branch of the primate tree is only a few million years old 403
19.4 Upright posture evolved well before an enlarged brain in hominids 404
19.5 Larger brains and reduced sexual dimorphism mark the evolution of *Homo* 404
19.6 When and where did *Homo sapiens* arise? 405
19.7 *Connection* Human skin colors reflect adaptations to varying amounts of sunlight 406
19.8 *Connection* A genetic difference helped humans start speaking 406

Our Cultural History and Its Consequences 407
19.9 Culture gives humans enormous power to change the environment 407
19.10 Scavenging, gathering, and hunting were the earliest human endeavors 407
19.11 Agriculture was a major development in human history 408
19.12 Development of complex tools affects human culture and the world 408
Chapter Review 409

UNIT V

Animals: Form and Function

20 UNIFYING CONCEPTS OF ANIMAL STRUCTURE AND FUNCTION 412

Essay Climbing the Walls 412

The Hierarchy of Structural Organization in an Animal 414
20.1 Structure fits function in the animal body 414
20.2 Animal structure has a hierarchy 415

18 THE EVOLUTION OF ANIMAL DIVERSITY 366

Essay What Am I? 366

Animal Evolution and Diversity 368

18.1 What is an animal? 368

18.2 The ancestor of animals was probably a colonial, flagellated protist 369

18.3 Animals can be characterized by basic features of their "body plan" 370

18.4 The body plans of animals can be used to build phylogenetic trees 371

Invertebrates 372

18.5 Sponges have a relatively simple, porous body 372

18.6 Cnidarians are radial animals with tentacles and stinging cells 373

18.7 Flatworms are the simplest bilateral animals 374

18.8 Nematodes have a pseudocoelom and a complete digestive tract 375

18.9 Diverse molluscs are variations on a common body plan 376

18.10 Annelids are segmented worms 378

18.11 Arthropods are segmented animals with jointed appendages and an exoskeleton 380

18.12 Insects are the most diverse group of organisms 382

18.13 Echinoderms have spiny skin, an endoskeleton, and a water vascular system for movement 384

18.14 Our own phylum, Chordata, is distinguished by four features 385

Vertebrates 386

18.15 Derived characters define the major clades of chordates 386

18.16 Lampreys are vertebrates that lack hinged jaws 387

18.17 Jawed vertebrates with gills and paired fins include sharks, ray-finned fishes, and lobe-fins 388

18.18 Amphibians were the first tetrapods-vertebrates with two pairs of limbs 389

18.19 Reptiles are amniotes-tetrapods with a terrestrially adapted egg 390

18.20 Birds are feathered reptiles with adaptations for flight 391

18.21 Mammals are amniotes that have hair and produce milk 392

Animal Phylogeny and Diversity Revisited 393

18.22 An animal phylogenetic tree is a work in progress 393

18.23 *Connection* Humans threaten animal diversity by introducing non-native species 394

Chapter Review 395

19 HUMAN EVOLUTION 398

Essay How Are We Related to the Neanderthals? 398

Primate Diversity 400

19.1 The human story begins with our primate heritage 400

19.2 Hominoids include humans and four other groups of apes 402

Hominid Evolution 403

19.3 The human branch of the primate tree is only a few million years old 403

19.4 Upright posture evolved well before an enlarged brain in hominids 404

19.5 Larger brains and reduced sexual dimorphism mark the evolution of *Homo* 404

19.6 When and where did *Homo sapiens* arise? 405

19.7 *Connection* Human skin colors reflect adaptations to varying amounts of sunlight 406

19.8 *Connection* A genetic difference helped humans start speaking 406

Our Cultural History and Its Consequences 407

19.9 Culture gives humans enormous power to change the environment 407

19.10 Scavenging, gathering, and hunting were the earliest human endeavors 407

19.11 Agriculture was a major development in human history 408

19.12 Development of complex tools affects human culture and the world 408

Chapter Review 409

UNIT V

Animals: Form and Function

20 UNIFYING CONCEPTS OF ANIMAL STRUCTURE AND FUNCTION 412

Essay Climbing the Walls 412

The Hierarchy of Structural Organization in an Animal 414

20.1 Structure fits function in the animal body 414

20.2 Animal structure has a hierarchy 415

20.3 Tissues are groups of cells with a common structure and function 415

20.4 Epithelial tissue covers the body and lines its organs and cavities 416

20.5 Connective tissue binds and supports other tissues 417

20.6 Muscle tissue functions in movement 418

20.7 Nervous tissue forms a communication network 418

20.8 *Connection* Artificial tissues have medical uses 419

20.9 Organs are made up of tissues 419

20.10 Organ systems work together to perform life's functions 420

20.11 *Connection* New imaging technology reveals the inner body 422

Exchanges with the External Environment 424

20.12 Structural adaptations enhance exchange between animals and their environment 424

20.13 Animals regulate their internal environment 425

20.14 Homeostasis depends on negative feedback 426

Chapter Review 426

Diets and Digestive Adaptations 440

21.13 Adaptations of vertebrate digestive systems reflect diet 440

Nutrition 442

21.14 Overview: A healthy diet satisfies three needs 442

21.15 Chemical energy powers the body 442

21.16 An animal's diet must supply essential nutrients 443

21.17 *Connection* Vegetarians must be sure to obtain all eight essential amino acids 443

21.18 A healthy diet includes 13 vitamins 444

21.19 Essential minerals are required for many body functions 445

21.20 *Connection* Do you need to take vitamin and mineral supplements? 446

21.21 *Connection* What do food labels tell us? 446

21.22 *Connection* Obesity is a human health problem 447

21.23 *Connection* What are the health risks and benefits of fad diets? 448

21.24 *Connection* Diet can influence cardiovascular disease and cancer 449

Chapter Review 450

21 NUTRITION AND DIGESTION 428

Essay Getting Their Fill of Krill 428

Obtaining and Processing Food 430

21.1 Animals ingest their food in a variety of ways 430

21.2 Overview: Food processing occurs in four stages 431

21.3 Digestion occurs in specialized compartments 432

Human Digestive System 433

21.4 The human digestive system consists of an alimentary canal and accessory glands 433

21.5 Digestion begins in the oral cavity 434

21.6 The food and breathing passages both open into the pharynx 434

21.7 *Connection* The Heimlich maneuver can save lives 435

21.8 The esophagus squeezes food along to the stomach by peristalsis 435

21.9 The stomach stores food and breaks it down with acid and enzymes 436

21.10 *Connection* Bacterial infections can cause ulcers 437

21.11 The small intestine is the major organ of chemical digestion and nutrient absorption 438

21.12 The large intestine reclaims water and compacts the feces 440

22 GAS EXCHANGE 452

Essay Surviving in Thin Air 452

Mechanisms of Gas Exchange 454

22.1 Overview: Gas exchange involves breathing, transport of gases, and exchange of gases with tissue cells 454

22.2 Animals exchange O_2 and CO_2 across moist body surfaces 454

22.3 Gills are adapted for gas exchange in aquatic environments 456

22.4 The tracheal system of insects provides direct exchange between the air and body cells 457

22.5 Terrestrial vertebrates have lungs 458

22.6 *Connection* Smoking is a deadly assault on our respiratory system 459

22.7 Breathing ventilates the lungs 460

22.8 Breathing is automatically controlled 461

Transport of Gases in the Body 462

22.9 Blood transports respiratory gases 462

22.10 Hemoglobin carries O_2 and helps transport CO_2 and buffer the blood 462

22.11 *Connection* The human fetus exchanges gases with the mother's bloodstream 463

Chapter Review 464

23 CIRCULATION 466

Essay How Does Gravity Affect Blood Circulation? 466

23.1 The circulatory system connects with all body tissues 468

Mechanisms of Internal Transport 468

23.2 Several types of internal transport have evolved in animals 468

23.3 Vertebrate cardiovascular systems reflect evolution 470

The Mammalian Cardiovascular System 471

23.4 The human heart and cardiovascular system are typical of mammals 471

23.5 The structure of blood vessels fits their functions 472

23.6 The heart contracts and relaxes rhythmically 472

23.7 The pacemaker sets the tempo of the heartbeat 473

23.8 *Connection* What is a heart attack? 474

23.9 Blood exerts pressure on vessel walls 475

23.10 *Connection* Measuring blood pressure can reveal cardiovascular problems 476

23.11 Smooth muscle controls the distribution of blood 477

23.12 Capillaries allow the transfer of substances through their walls 478

Structure and Function of Blood 479

23.13 Blood consists of red and white blood cells suspended in plasma 479

23.14 *Connection* Too few or too many red blood cells can be unhealthy 480

23.15 Blood clots plug leaks when blood vessels are injured 480

23.16 *Connection* Stem cells offer a potential cure for blood cell diseases 481

Chapter Review 482

24 THE IMMUNE SYSTEM 484

Essay An AIDS Uproar 484

Innate Defenses Against Infection 486

24.1 Innate defenses against infection include the skin and mucous membranes, phagocytic cells, and antimicrobial proteins 486

24.2 The inflammatory response mobilizes nonspecific defense forces 487

24.3 The lymphatic system becomes a crucial battleground during infection 488

Acquired Immunity 489

24.4 The immune response counters specific invaders 489

24.5 Lymphocytes mount a dual defense 490

24.6 Antigens have specific regions where antibodies bind to them 491

24.7 Clonal selection musters defensive forces against specific antigens 492

24.8 Antibodies are the weapons of humoral immunity 494

24.9 Antibodies mark antigens for elimination 495

24.10 *Connection* Monoclonal antibodies are powerful tools in the lab and clinic 496

24.11 Helper T cells stimulate humoral and cell-mediated immunity 497

24.12 *Connection* HIV destroys helper T cells, compromising the body's defenses 498

24.13 Cytotoxic T cells destroy infected body cells 499

24.14 Cytotoxic T cells may help prevent cancer 499

24.15 The immune system depends on our molecular fingerprints 500

Disorders of the Immune System 500

24.16 *Connection* Malfunction or failure of the immune system causes disease 500

24.17 *Connection* Allergies are overreactions to certain environmental antigens 501

Chapter Review 502

25 CONTROL OF THE INTERNAL ENVIRONMENT 504

Essay Let Sleeping Bears Lie 504

Thermoregulation 506

25.1 Heat is gained or lost in four ways 506

25.2 Thermoregulation involves adaptations that balance heat gain and loss 506

25.3 Reducing metabolic rate and body temperature saves energy 507

Osmoregulation and Excretion 508

25.4 Osmoregulation: Animals balance the gain and loss of water and solutes 508

25.5 *Connection* Do we need to drink eight glasses of water each day? 509

25.6 Animals must dispose of nitrogenous wastes 510

25.7 The liver performs many functions, inclduing the production of urea 511

25.8 *Connection* Alcohol consumption can damage the liver 511

25.9 The excretory system plays several major roles in homeostasis 512

25.10 Overview: The key processes of the excretory system are filtration, reabsorption, secretion, and excretion 513

25.11 From blood filtrate to urine: A closer look   514

25.12 *Connection* Kidney dialysis can be a lifesaver   515

Chapter Review   516

26 CHEMICAL REGULATION 518

Essay Testosterone and Male Aggression: Is There a Link?   518

The Nature of Chemical Regulation   520

26.1 Chemical signals coordinate body functions   520

26.2 Hormones affect target cells by two main signaling mechanisms   521

The Vertebrate Endocrine System   522

26.3 Overview: The vertebrate endocrine system   522

26.4 The hypothalamus, closely tied to the pituitary, connects the nervous and endocrine systems   524

Hormones and Homeostasis   526

26.5 The thyroid regulates development and metabolism   526

26.6 Hormones from the thyroid and the parathyroids maintain calcium homeostasis   526

26.7 Pancreatic hormones regulate blood glucose levels   528

26.8 *Connection* Diabetes is a common endocrine disorder   529

26.9 The adrenal glands mobilize responses to stress   530

26.10 *Connection* Glucocorticoids offer relief from pain, but not without serious risks   531

26.11 The gonads secrete sex hormones   532

Chapter Review   532

27 REPRODUCTION AND EMBRYONIC DEVELOPMENT 534

Essay Baby Bonanza   534

Asexual and Sexual Reproduction   536

27.1 Sexual and asexual reproduction are both common among animals   536

Human Reproduction   538

27.2 Reproductive anatomy of the human female   538

27.3 Reproductive anatomy of the human male   540

27.4 The formation of sperm and ova requires meiosis   542

27.5 Hormones synchronize cyclic changes in the ovary and uterus   544

27.6 The human sexual response occurs in four phases   546

27.7 *Connection* Sexual activity can transmit disease   546

27.8 *Connection* Contraception can prevent unwanted pregnancy   547

Principles of Embryonic Development   548

27.9 Fertilization results in a zygote and triggers embryonic development   548

27.10 Cleavage produces a ball of cells from the zygote   550

27.11 Gastrulation produces a three-layered embryo   550

27.12 Organs start to form after gastrulation   552

27.13 Changes in cell shape, cell migration, and programmed cell death give form to the developing animal   554

27.14 Embryonic induction initiates organ formation   554

27.15 Pattern formation organizes the animal body   555

Human Development   556

27.16 The embryo and placenta take shape during the first month of pregnancy   556

27.17 Human development from conception to birth is divided into three trimesters   558

27.18 Childbirth is hormonally induced and occurs in three stages   560

27.19 *Connection* Reproductive technology increases our reproductive options   561

Chapter Review   562

28 NERVOUS SYSTEMS 564

Essay Can an Injured Spinal Cord Be Fixed?   564

Nervous System Structure and Function   566

28.1 Nervous systems receive sensory input, interpret it, and send out appropriate commands   566

28.2 Neurons are the functional units of nervous systems   567

Nerve Signals and Their Transmission   568

28.3 A neuron maintains a membrane potential across its membrane   568

28.4 A nerve signal begins as a change in the membrane potential   568

28.5 The action potential propagates itself along the neuron   570

28.6 Neurons communicate at synapses   571

28.7 Chemical synapses make complex information processing possible   572

28.8 A variety of small molecules function as neurotransmitters   572

28.9 *Connection* Many drugs act at chemical synapses 573

An Overview of Animal Nervous Systems 574

28.10 Nervous system organization usually correlates with body symmetry 574

28.11 Vertebrate nervous systems are highly centralized and cephalized 575

28.12 The peripheral nervous system of vertebrates is a functional hierarchy 576

28.13 Opposing actions of sympathetic and parasympathetic neurons regulate the internal environment 576

28.14 The vertebrate brain develops from three anterior bulges of the neural tube 578

The Human Brain 578

28.15 The structure of a living supercomputer: The human brain 578

28.16 The cerebral cortex is a mosaic of specialized, interactive regions 580

28.17 *Connection* Injuries and brain operations provide insight into brain function 581

28.18 Several parts of the brain regulate sleep and arousal 582

28.19 The limbic system is involved in emotions, memory, and learning 583

28.20 *Connection* Changes in brain physiology can produce neurological disorders 584

Chapter Review 586

29 THE SENSES 588

Essay An Animal's Senses Guide Its Movement 588

29.1 Sensory inputs become sensations and perceptions in the brain 590

Sensory Reception 590

29.2 Sensory receptors convert stimulus energy to action potentials 590

29.3 Specialized sensory receptors detect five categories of stimuli 592

Vision 594

29.4 Several types of eyes have evolved among invertebrates 594

29.5 Vertebrates have single-lens eyes 594

29.6 To focus, a lens changes position or shape 595

29.7 *Connection* Artificial lenses or surgery can correct focusing problems 596

29.8 Our photoreceptors are rods and cones 597

Hearing and Balance 598

29.9 The ear converts air pressure waves to action potentials that are perceived as sound 598

29.10 The inner ear houses our organs of balance 600

29.11 *Connection* What causes motion sickness? 600

Taste and Smell 601

29.12 Taste and odor receptors detect chemicals present in solution or air 601

29.13 *Connection* Our sense of taste may change as we age 601

29.14 Review: The central nervous system couples stimulus with response 602

Chapter Review 602

30 HOW ANIMALS MOVE 604

Essay Elephants Do the "Groucho Gait" 604

Movement and Locomotion 606

30.1 Diverse means of animal locomotion have evolved 606

Skeletal Support 608

30.2 Skeletons function in support, movement, and protection 608

30.3 The human skeleton is a unique variation on an ancient theme 610

30.4 Bones are complex living organs 611

30.5 *Connection* Broken bones can heal themselves 612

30.6 *Connection* Weak, brittle bones are a serious health problem, even in young people 612

Muscle Contraction and Movement 613

30.7 The skeleton and muscles interact in movement 613

30.8 Each muscle cell has its own contractile apparatus 614

30.9 A muscle contracts when thin filaments slide across thick filaments 614

30.10 Motor neurons stimulate muscle contraction 616

30.11 *Connection* Athletic training increases strength and endurance 617

30.12 The structure-function theme underlies all the parts and activities of an animal 618

Chapter Review 619

UNIT VI

Plants: Form and Function

31 PLANT STRUCTURE, REPRODUCTION, AND DEVELOPMENT 622

Essay A Gentle Giant 622

31.1 *Talking About Science* Plant scientist Natasha Raikhel studies the *Arabidopsis* plant as a model biological system 624

Plant Structure and Function 625

31.2 The two main groups of angiosperms are the monocots and the dicots 625

31.3 A typical plant body consists of roots and shoots 626

31.4 Many plants have modified roots, stems, and leaves 627

31.5 Plant cells and tissues are diverse in structure and function 628

31.6 Three tissue systems make up the plant body 630

Plant Growth 632

31.7 Primary growth lengthens roots and shoots 632

31.8 Secondary growth increases the girth of woody plants 634

Reproduction of Flowering Plants 636

31.9 Overview: The sexual life cycle of a flowering plant 636

31.10 The development of pollen and ovules culminates in fertilization 636

31.11 The ovule develops into a seed 638

31.12 The ovary develops into a fruit 639

31.13 Seed germination continues the life cycle 640

31.14 Asexual reproduction produces plant clones 641

31.15 *Connection* Asexual reproduction is a mainstay of modern agriculture 642

Chapter Review 642

32 PLANT NUTRITION AND TRANSPORT 644

Essay Plants That Clean Up Poisons 644

The Uptake and Transport of Plant Nutrients 646

32.1 Plants acquire their nutrients from soil and air 646

32.2 The plasma membranes of root cells control solute uptake 647

32.3 Transpiration pulls water up xylem vessels 648

32.4 Guard cells control transpiration 649

32.5 Phloem transports sugars 650

Plant Nutrients and the Soil 652

32.6 Plant health depends on a complete diet of essential inorganic nutrients 652

32.7 *Connection* You can diagnose some nutrient deficiencies in your own plants 653

32.8 Fertile soil supports plant growth 654

32.9 *Connection* Soil conservation is essential to human life 655

32.10 *Connection* Organic farmers must follow ecological principles 656

32.11 *Connection* Agricultural research is improving the yields and nutritional values of crops 656

Plant Nutrition and Symbiosis 657

32.12 Fungi help most plants absorb nutrients from the soil 657

32.13 Most plants depend on bacteria to supply nitrogen 658

32.14 Legumes and certain other plants house nitrogen-fixing bacteria 658

32.15 The plant kingdom includes parasites and carnivores 659

Chapter Review 660

33 CONTROL SYSTEMS IN PLANTS 662

Essay What Are the Health Benefits of Soy? 662

Plant Hormones 664

33.1 Experiments on how plants turn toward light led to the discovery of a plant hormone 664

33.2 Five major types of hormones regulate plant growth and development 666

33.3 Auxin stimulates the elongation of cells in young shoots 666

33.4 Cytokinins stimulate cell division 668

33.5 Gibberellins affect stem elongation and have numerous other effects 668

33.6 Abscisic acid inhibits many plant processes 669

33.7 Ethylene triggers fruit ripening and other aging processes 670

33.8 *Connection* Plant hormones have many agricultural uses 671

Growth Responses and Biological Rhythms in Plants 672

33.9 Tropisms orient plant growth toward or away from environmental stimuli 672

33.10 Plants have internal clocks 673

UNIT VI

Plants: Form and Function

31 PLANT STRUCTURE, REPRODUCTION, AND DEVELOPMENT 622

Essay A Gentle Giant 622

31.1 *Talking About Science* Plant scientist Natasha Raikhel studies the *Arabidopsis* plant as a model biological system 624

Plant Structure and Function 625

31.2 The two main groups of angiosperms are the monocots and the dicots 625

31.3 A typical plant body consists of roots and shoots 626

31.4 Many plants have modified roots, stems, and leaves 627

31.5 Plant cells and tissues are diverse in structure and function 628

31.6 Three tissue systems make up the plant body 630

Plant Growth 632

31.7 Primary growth lengthens roots and shoots 632

31.8 Secondary growth increases the girth of woody plants 634

Reproduction of Flowering Plants 636

31.9 Overview: The sexual life cycle of a flowering plant 636

31.10 The development of pollen and ovules culminates in fertilization 636

31.11 The ovule develops into a seed 638

31.12 The ovary develops into a fruit 639

31.13 Seed germination continues the life cycle 640

31.14 Asexual reproduction produces plant clones 641

31.15 *Connection* Asexual reproduction is a mainstay of modern agriculture 642

Chapter Review 642

32 PLANT NUTRITION AND TRANSPORT 644

Essay Plants That Clean Up Poisons 644

The Uptake and Transport of Plant Nutrients 646

32.1 Plants acquire their nutrients from soil and air 646

32.2 The plasma membranes of root cells control solute uptake 647

32.3 Transpiration pulls water up xylem vessels 648

32.4 Guard cells control transpiration 649

32.5 Phloem transports sugars 650

Plant Nutrients and the Soil 652

32.6 Plant health depends on a complete diet of essential inorganic nutrients 652

32.7 *Connection* You can diagnose some nutrient deficiencies in your own plants 653

32.8 Fertile soil supports plant growth 654

32.9 *Connection* Soil conservation is essential to human life 655

32.10 *Connection* Organic farmers must follow ecological principles 656

32.11 *Connection* Agricultural research is improving the yields and nutritional values of crops 656

Plant Nutrition and Symbiosis 657

32.12 Fungi help most plants absorb nutrients from the soil 657

32.13 Most plants depend on bacteria to supply nitrogen 658

32.14 Legumes and certain other plants house nitrogen-fixing bacteria 658

32.15 The plant kingdom includes parasites and carnivores 659

Chapter Review 660

33 CONTROL SYSTEMS IN PLANTS 662

Essay What Are the Health Benefits of Soy? 662

Plant Hormones 664

33.1 Experiments on how plants turn toward light led to the discovery of a plant hormone 664

33.2 Five major types of hormones regulate plant growth and development 666

33.3 Auxin stimulates the elongation of cells in young shoots 666

33.4 Cytokinins stimulate cell division 668

33.5 Gibberellins affect stem elongation and have numerous other effects 668

33.6 Abscisic acid inhibits many plant processes 669

33.7 Ethylene triggers fruit ripening and other aging processes 670

33.8 *Connection* Plant hormones have many agricultural uses 671

Growth Responses and Biological Rhythms in Plants 672

33.9 Tropisms orient plant growth toward or away from environmental stimuli 672

33.10 Plants have internal clocks 673

33.11 Plants mark the seasons by measuring photoperiod 674

33.12 Phytochrome is a light detector that may help set the biological clock 675

33.13 *Talking About Science* Joanne Chory studies the effects of light and hormones in the model plant *Arabidopsis* 676

Plant Defenses 676

33.14 Defenses against herbivores and infectious microbes have evolved in plants 676

33.15 *Talking About Science* Plant biochemist Eloy Rodriguez studies how animals use defensive chemicals made by plants 678

Chapter Review 678

34.14 Temperate grasslands include the North American prairie 696

34.15 Broadleaf trees dominate temperate forests 697

34.16 Coniferous forests are often dominated by a few species of trees 698

34.17 Long, bitter-cold winters characterize the tundra 698

34.18 *Talking About Science* Ecologist Ariel Lugo studies tropical forests in Puerto Rico 699

Chapter Review 700

UNIT VII

Ecology

34 BIOSPHERE: AN INTRODUCTION TO EARTH'S DIVERSE ENVIRONMENTS 682

Essay A Mysterious Giant of the Deep 682

34.1 Ecologists study how organisms interact with their environment at several levels 684

The Biosphere 684

34.2 The biosphere is the total of all of Earth's ecosystems 684

34.3 *Connection* Environmental problems reveal the limits of the biosphere 685

34.4 Physical and chemical factors influence life in the biosphere 686

34.5 Organisms are adapted to abiotic and biotic factors by natural selection 687

34.6 Regional climate influences the distribution of biological communities 688

Aquatic Biomes 690

34.7 Oceans occupy most of Earth's surface 690

34.8 Freshwater biomes include lakes, ponds, rivers, streams, and wetlands 692

Terrestrial Biomes 693

34.9 Terrestrial biomes reflect regional variations in climate 693

34.10 Tropical forests cluster near the equator 694

34.11 Savannas are grasslands with scattered trees 694

34.12 Deserts are defined by their dryness 695

34.13 Spiny shrubs dominate the chaparral 696

35 BEHAVIORAL ADAPTATIONS TO THE ENVIRONMENT 702

Essay Leaping Herds of Herbivores 702

The Scientific Study of Behavior 704

35.1 Behavioral ecologists ask both proximate and ultimate questions 704

35.2 Early behaviorists used experiments to study fixed action patterns 705

35.3 Behavior is the result of both genes and environmental factors 706

Learning 707

35.4 Learning ranges from simple behavioral changes to complex problem solving 707

35.5 Imprinting is learning that involves innate behavior and experience 708

35.6 *Connection* Imprinting poses problems and opportunities for conservation programs 709

35.7 Animal movement may be a simple response to stimuli or involve spatial learning 710

35.8 Movements of animals may depend on internal maps 711

35.9 Animals may learn to associate a stimulus or behavior with a response 712

35.10 Social learning involves observation and imitation of others 712

35.11 Problem-solving behavior relies on cognition 713

Foraging and Mating Behaviors 714

35.12 Behavioral ecologists use cost-benefit analysis in studying foraging 714

35.13 Mating behaviors enhance reproductive success 715

35.14 Mating behavior often involves elaborate courtship rituals 716

Social Behavior and Sociobiology 717

35.15 Sociobiology places social behavior in an evolutionary context 717

35.16 Territorial behavior parcels space and resources 717

35.17 Rituals involving agonistic behavior often resolve confrontations between competitors 718

35.18 Dominance hierarchies are maintained by agonistic behavior 718

35.19 *Talking About Science* Behavioral biologist Jane Goodall discusses dominance hierarchies and reconciliation behavior in chimpanzees 719

35.20 Social behavior requires communication between animals 720

35.21 Altruistic acts can often be explained by the concept of inclusive fitness 721

35.22 *Connection* Both genes and culture contribute to human social behavior 722

35.23 *Talking About Science* Edward O. Wilson promoted the field of sociobiology and is a leading conservation activist 723

Chapter Review 724

36 POPULATION DYNAMICS 726

Essay The Spread of Shakespeare's Starlings 726

36.1 Population ecology studies how and why populations change 728

Population Structure and Dynamics 728

36.2 Density and dispersion patterns are important population variables 728

36.3 Life tables track mortality and survivorship in populations 729

36.4 Idealized models help us understand population growth 730

36.5 Multiple factors may limit population growth 732

36.6 Some populations have "boom-and-bust" cycles 733

Life Histories and Their Evolution 734

36.7 Evolution shapes life histories 734

36.8 *Connection* Principles of population ecology have practical applications 735

The Human Population 736

36.9 *Connection* Human population growth has started to slow after centuries of exponential increase 736

36.10 Birth and death rates and age structure affect population growth 738

Chapter Review 740

37 COMMUNITIES AND ECOSYSTEMS 742

Essay Dining In 742

Structural Features of Communities 744

37.1 A community includes all the organisms inhabiting a particular area 744

37.2 Competition may occur when a shared resource is limited 745

37.3 Predation leads to diverse adaptations in both predator and prey 746

37.4 Predation can maintain diversity in a community 747

37.5 Herbivores and the plants they eat have various adaptations 748

37.6 Symbiotic relationships help structure communities 748

37.7 Disturbance is a prominent feature of most communities 750

37.8 *Talking About Science* Fire specialist Max Moritz discusses the role of fire in ecosystems 751

37.9 Trophic structure is a key factor in community dynamics 752

37.10 Food chains interconnect, forming food webs 753

Ecosystem Structure and Dynamics 754

37.11 Ecosystem ecology emphasizes energy flow and chemical cycling 754

37.12 Primary production sets the energy budget for ecosystems 754

37.13 Energy supply limits the length of food chains 755

37.14 *Connection* A production pyramid explains why meat is a luxury for humans 756

37.15 Chemicals are recycled between organic matter and abiotic reservoirs 756

37.16 Water moves through the biosphere in a global cycle 757

37.17 The carbon cycle depends on photosynthesis and respiration 758

37.18 The nitrogen cycle relies heavily on bacteria 758

37.19 The phosphorus cycle depends on the weathering of rock 759

Ecosystem Alteration 760

37.20 *Connection* Ecosystem alteration can upset chemical cycling 760

37.21 *Talking About Science* David Schindler talks about the effects of nutrients on freshwater ecosystems 761

Chapter Review 762

38 CONSERVATION BIOLOGY 764

Essay Saving the Tiger 764

The Biodiversity Crisis: An Overview 766

38.1 Human activities threaten Earth's biodiversity 766

38.2 Biodiversity is vital to human welfare 767

38.3 Habitat destruction, introduced species, and overexploitation are the major threats to biodiversity 768

38.4 *Connection* Pollution of the environment compounds our impact on other species 769

38.5 *Connection* Rapid global warming could alter the entire biosphere 770

Conservation of Populations and Species 772

38.6 Two ways to study endangered populations are the small-population approach and the declining-population approach 772

38.7 Identifying critical habitat factors can guide conservation efforts 773

Managing and Restoring Ecosystems 774

38.8 Sustaining ecosystems and landscapes is a conservation priority 774

38.9 Protected areas are established to slow the loss of biodiversity 775

38.10 *Connection* The Yellowstone to Yukon Conservation Initiative seeks to preserve biodiversity by connecting protected areas 776

38.11 The study of how to restore degraded habitats is a developing science 778

38.12 *Connection* The Kissimmee River project is a case study in restoration ecology 779

38.13 Zoned reserves are an attempt to reverse ecosystem disruption 780

38.14 Sustainable development is an ultimate goal 781

Chapter Review 782

APPENDIX 1   Metric Conversion Table

APPENDIX 2   The Amino Acids of Proteins

APPENDIX 3   Chapter Review Answers

APPENDIX 4   Credits

GLOSSARY

INDEX

CHAPTER 1

A Big-Billed Bird Rebounds

ALONG THE COAST OF FLORIDA, a skilled fisher glides high above the waves. With the slow, powerful beating of its wings, it scans the bright blue water below, checking for any shadow or movement that betrays the presence of fish. Suddenly, the hunter spots a glistening school of silversides and morphs from flier into diver. Wings folded, the bird hurtles into the waves, surfacing an instant later with a round bulge beneath its bill—a fleshy pouch, now fully expanded and stuffed with fish.

This adept sea feeder is the brown pelican, shown on these two pages and on the cover of this book. Brown pelicans (*Pelecanus occidentalis*) have probably soared along the coasts of the Americas for millions of years. They belong to a group of animals called Aves, a group believed to have evolved from an ancient feathered dinosaur about 200 million years ago. Seven types, or species, of pelicans are now found worldwide. Brown pelicans are the smallest of the seven, but that doesn't make them petite.

Large, gregarious, and cartoonish, brown pelicans typically weigh 3–4 kilograms (kg; 8–10 pounds) and spread their wings up to 2.3 meters (m) wide.[1] Their feathers sprout in swaths of chestnut, white, yellow, and blackish brown. Brown pelicans feed primarily on fish, and lots of it, with adults needing almost 2 kg a day. To haul in that big a catch, the birds rely on excellent eyesight and diving skills, spotting fish from as high as 18 m above the water. After a steep dive, they scoop liters of water and fish into their expandable pouch. A pelican with a successful catch tilts its long bill to drain out the water and then throws back its head to swallow its prey.

These fishing skills reflect how well suited these birds are to their ocean habitat. The brown pelican is the only pelican that is

---

THE SCOPE OF BIOLOGY

1.1 Life's levels of organization define the scope of biology
1.2 Living organisms and their environments form interconnecting webs
1.3 Cells are the structural and functional units of life

EVOLUTION, UNITY, AND DIVERSITY

1.4 The unity of life: All forms of life have common features
1.5 The diversity of life can be arranged into three domains
1.6 Evolution explains the unity and diversity of life

THE PROCESS OF SCIENCE

1.7 Scientists use two main approaches to learn about nature
1.8 With hypothesis-based science, we pose and test hypotheses

BIOLOGY AND EVERYDAY LIFE

1.9 Biology is connected to our lives in many ways

---

[1] You can review metric measurements in Appendix 1.

Biology: Exploring Life

a plunge diver, and it can hit the water with such force that even fish 2 m below the surface are stunned, allowing for an easy catch. The birds are equally adept at making nests in trees or on the ground—important flexibility in coastal areas, where sand dunes are sometimes more plentiful than tree branches.

Brown pelicans are part of the web of life in their environment. Coastal islands from Louisiana to Texas and California are sometimes packed with noisy pelican nests, where the birds congregate to rear chicks and guard against predators. Through their hunting, pelicans help regulate the size of fish populations. A pelican's comfort on the waves also makes it an ideal scavenger. Small fishing boats seeking to toss away unwanted fish parts without polluting the water don't have to look any farther than the hungry pelican waiting hopefully nearby.

But pelicans' proximity to people has also meant trouble for the species. Human development in coastal areas has reduced the amount of land and vegetation available for pelican nests. Some fishermen have viewed the birds as hungry competition rather than trash collectors and have slaughtered thousands of the birds. In the past, some pelicans were hunted for their feathers.

The pelican's taste for fish almost led to its disappearance

In the middle of the 20th century, the pelican's taste for fish almost led to its disappearance. Heavy human use of pesticides such as DDT and endrin filled coastal waters with pesticide run-off, which worked its way up through the food web to accumulate in fish and the birds that ate them. The pesticides warped the development of pelican eggs, thinning shells so badly that protective pelican parents often crushed the eggs they were incubating.

The result came close to a pelican wipeout. The birds first disappeared entirely from Louisiana and then vanished from other coastal areas as well. In 1970, U.S. officials declared the once-plentiful brown pelican an endangered species. Biologists argued that only a ban on pesticides such as DDT would save the birds, and in 1972, such a ban was enacted in the United States.

Pelicans responded with a revival that commends the combination of biological research and government action. On Anacapa Island, off the coast of California, worried biologists counted only one successful young pelican from a population of 552 breeding pairs in 1970. Today Anacapa Island is home base for more than 4,000 breeding pairs. The birds are once again thriving in many coastal areas. But as their numbers grow, they still sometimes face problems when they encounter humans. The young pelican shown on the cover of this book, for example, was photographed at a bird sanctuary in Florida, where biologists care for birds injured by boat propellers or fishing line.

In its connections to the environment and other organisms, the brown pelican sets the stage for this chapter's introduction to **biology**, the scientific study of life. Let's begin with a panoramic view of the scope of biology. ■ ■ ■

1.1 Life's levels of organization define the scope of biology

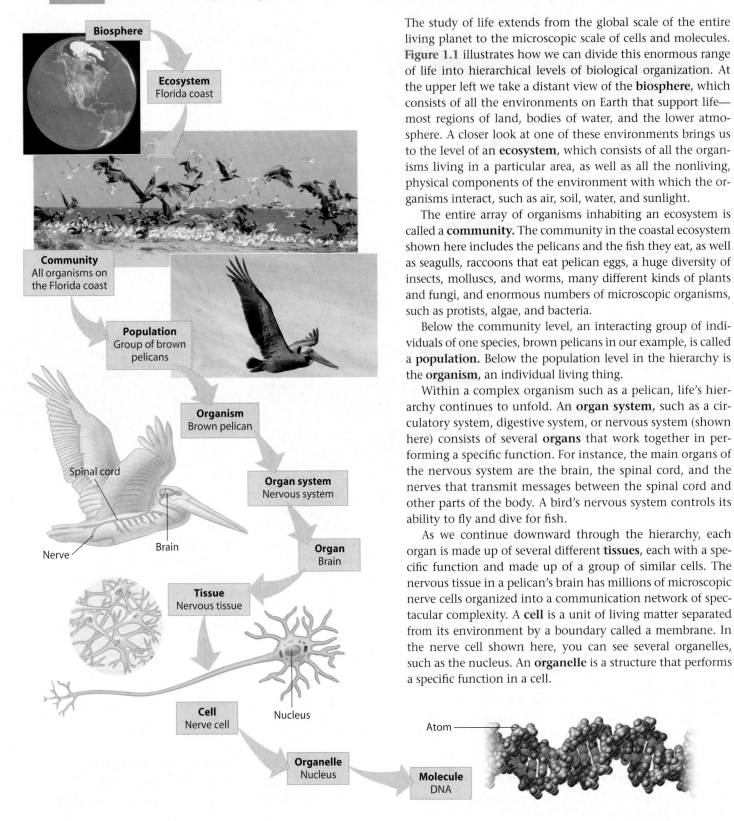

The study of life extends from the global scale of the entire living planet to the microscopic scale of cells and molecules. **Figure 1.1** illustrates how we can divide this enormous range of life into hierarchical levels of biological organization. At the upper left we take a distant view of the **biosphere**, which consists of all the environments on Earth that support life—most regions of land, bodies of water, and the lower atmosphere. A closer look at one of these environments brings us to the level of an **ecosystem**, which consists of all the organisms living in a particular area, as well as all the nonliving, physical components of the environment with which the organisms interact, such as air, soil, water, and sunlight.

The entire array of organisms inhabiting an ecosystem is called a **community.** The community in the coastal ecosystem shown here includes the pelicans and the fish they eat, as well as seagulls, raccoons that eat pelican eggs, a huge diversity of insects, molluscs, and worms, many different kinds of plants and fungi, and enormous numbers of microscopic organisms, such as protists, algae, and bacteria.

Below the community level, an interacting group of individuals of one species, brown pelicans in our example, is called a **population.** Below the population level in the hierarchy is the **organism**, an individual living thing.

Within a complex organism such as a pelican, life's hierarchy continues to unfold. An **organ system**, such as a circulatory system, digestive system, or nervous system (shown here) consists of several **organs** that work together in performing a specific function. For instance, the main organs of the nervous system are the brain, the spinal cord, and the nerves that transmit messages between the spinal cord and other parts of the body. A bird's nervous system controls its ability to fly and dive for fish.

As we continue downward through the hierarchy, each organ is made up of several different **tissues**, each with a specific function and made up of a group of similar cells. The nervous tissue in a pelican's brain has millions of microscopic nerve cells organized into a communication network of spectacular complexity. A **cell** is a unit of living matter separated from its environment by a boundary called a membrane. In the nerve cell shown here, you can see several organelles, such as the nucleus. An **organelle** is a structure that performs a specific function in a cell.

Figure 1.1 Life's hierarchy of organization

Finally, we reach the level of molecules in the hierarchy. A **molecule** is a cluster of atoms held together by chemical bonds. We show as our example DNA (deoxyribonucleic acid). DNA molecules provide the blueprint for constructing the organism's other important molecules and transmit this information from parents to offspring. In the computer graphic of a small section of DNA at the bottom of Figure 1.1, each of the spheres represents an **atom**, the smallest particle of ordinary matter.

As we discuss in later chapters, life's hierarchy builds from molecules to the biosphere. It takes many molecules to build organelles and make a cell, many cells to make a tissue, several kinds of tissues to make an organ, and so on. From the interactions within the biosphere to the molecular machinery within cells, biologists investigate life at its many levels. In the next two modules, we take a closer look at two biological levels near opposite ends of the scope of life: ecosystems and cells.

Web/CD Activity 1A *The Levels of Life Card Game*

? Which of the following levels of biological organization includes all others in the list: cell, molecule, organ, tissue?

■ Organ

1.2 Living organisms and their environments form interconnecting webs

At the level of the ecosystem, interactions of organisms with the living and nonliving components of the environment make up a complex web of relationships. Figure 1.2 is a simplified picture of such a web. Plants and other photosynthetic organisms are the **producers** that provide the food for a typical ecosystem. In photosynthesis, plants trap energy from sunlight and use carbon dioxide (CO_2) from the air, along with water (H_2O) from the soil, to make sugar molecules. Plants also absorb mineral nutrients from the soil and incorporate them into molecules. The **consumers** of the ecosystem eat plants and other animals. They take in oxygen (O_2) from the air and return CO_2. Their wastes return other chemicals to the environment.

Another vital part of the ecosystem includes the bacteria, fungi, and small animals in the soil that decompose the remains of dead organisms. These **decomposers** act as recyclers, changing the complex dead matter into simple mineral nutrients that plants can use. Thus, the web of relationships among plants, animals, microorganisms, and the physical environment gives the ecosystem its structure.

The dynamics of an ecosystem include two major processes—the recycling of chemical nutrients and the flow of energy. These processes are illustrated in Figure 1.2. The most basic chemicals necessary for life—the atoms making up carbon dioxide, oxygen, water, and various minerals—flow from the air and soil to plants, to animals and decomposers, and back to the air and soil. As shown by the blue arrows in the figure, chemical nutrients cycle within an ecosystem's structural web. By contrast, an ecosystem gains and loses energy constantly. Energy flows into the ecosystem when plants and other photosynthesizers trap light energy from the sun (yellow arrow) and convert it to the chemical energy of sugars and other complex molecules. Chemical energy, in the form of food molecules, is then passed through the ecosystem's web, powering each organism in turn (orange arrow). In the process of these energy conversions between and within organisms, the energy is eventually converted to heat, which is then lost from the system (red arrow). In contrast to chemical nutrients, which recycle within an ecosystem, energy flows through an ecosystem, entering as light and exiting as heat.

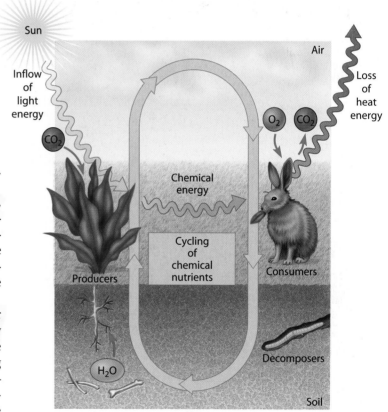

Figure 1.2 The web of interactions in an ecosystem

? Explain how the photosynthesis of plants functions in both the cycling of chemical nutrients and the flow of energy in an ecosystem.

■ Photosynthesis uses light to convert CO_2 and H_2O to energy-rich food, making it the pathway by which both chemical nutrients and energy become available to most organisms.

1.3   Cells are the structural and functional units of life

The cell has a special place in the hierarchy of biological organization. It is the lowest level of structure that can perform all activities required for life. A cell can regulate its internal environment, take in and use energy, respond to the environment around it, and develop and maintain its complex organization. The ability of cells to give rise to new cells is the basis for all reproduction and for the growth and repair of multicellular organisms. Your every movement and thought are based on the activities of muscle cells and nerve cells. Even a global process such as the cycling of carbon, a chemical element essential to life, is the result of cellular activities, including the photosynthesis of plant cells and the cellular respiration of nearly all cells, a process that breaks down sugar for energy and releases CO_2.

The properties of life that arise from the structural level of a cell illustrate another important theme of biology, called **emergent properties.** The familiar saying that "the whole is greater than the sum of its parts" captures this idea. Take another look at the levels of life in Figure 1.1. With each step upward in the hierarchy of biological order, novel properties emerge that result from the arrangement and interactions of the component parts.

Such a combination of components forms a more complex organization called a **system.** Cells are examples of biological systems, as are organisms and ecosystems. Systems and their emergent properties are not unique to life. Consider a box of bicycle parts. When all of the individual parts are properly assembled, the result is a mechanical system you can use for exercise or transportation. The emergent properties of life, however, are particularly challenging to study because of the unrivaled complexity of biological systems. An important focus of biology today is the study of the behavior of whole, integrated systems, ranging from the functioning of the biosphere to the complex molecular machinery of a cell.

Cells are the structural and functional units of all life. We can distinguish two kinds of cells, illustrated in **Figure 1.3.** The **prokaryotic cell** is much simpler and usually much smaller than the eukaryotic cell. The cells of the microorganisms we commonly call bacteria are prokaryotic. Forms of life such as plants, animals, and fungi are composed of eukaryotic cells. In contrast to a prokaryotic cell, a **eukaryotic cell** is subdivided by internal membranes into many different functional compartments, or organelles, including the nucleus that houses the cell's DNA.

Despite their differences, prokaryotic and eukaryotic cells share many characteristics. For example, every cell is enclosed by a membrane that regulates the passage of materials between the cell and its surroundings. And all cells use DNA to code their genetic information. In the next module, we explore how DNA unifies all of life.

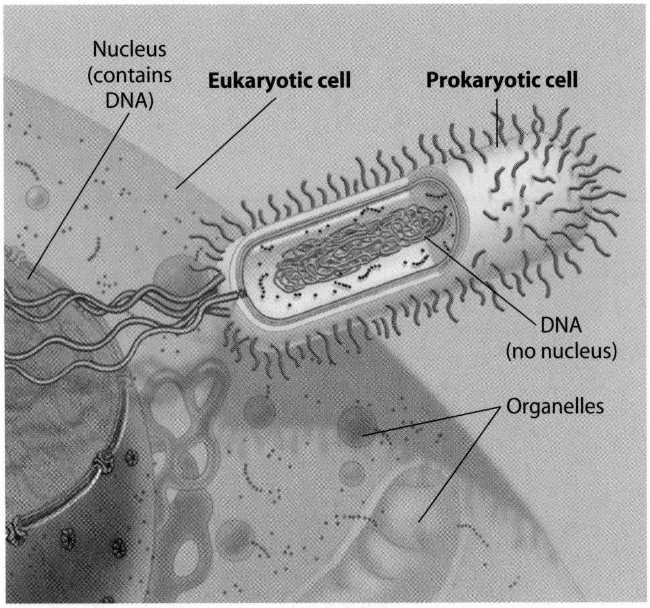

Figure 1.3 Comparison of prokaryotic and eukaryotic cells. (Cells are shown approximately 25,000 times their real size.)

? Explain why cells are considered the basic units of life.

■ They are the lowest level in the hierarchy of biological organization at which the properties of life emerge.

1.4   The unity of life: All forms of life have common features

The foundation for the unity of life is the genetic information in DNA molecules. The molecular structure of DNA accounts for its information-rich nature. Each DNA molecule is made up of two long chains coiled together into what is called a double helix (see Figure 1.1). The chains are made up of four kinds of chemical building blocks. **Figure 1.4A** diagrams a short section of one of the chains and indicates the four different building blocks with letter abbreviations of their names.

The way DNA encodes a cell's information is analogous to the way we arrange letters of the alphabet into precise sequences with specific meanings. The word *rat*, for example, conjures up an image of a rodent; *tar* and *art,* which contain the same letters, mean very different things. We can think of the four building blocks as the alphabet of inheritance. Specific sequential arrangements of these four chemical letters encode precise information in genes, which are typically

---

4   **CHAPTER 1** *Biology: Exploring Life*

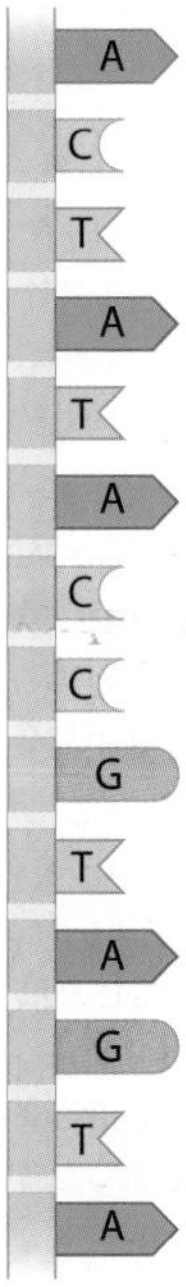

Figure 1.4A One chain of a DNA molecule, its message written in the order of the four building blocks labeled A, T, C, and G

A
C
T
A
T
A
C
C
G
T
A
G
T
A

hundreds or thousands of "letters" long. One gene in a bacterial cell may be translated as "Build a purple pigment." A particular human gene may mean "Make the hormone insulin." All forms of life employ essentially the same genetic code, which has made it possible to engineer cells that produce proteins normally found only in some other organism. Thus, bacteria can be used to produce insulin by inserting a gene for human insulin into bacterial cells.

Life occurs in a vast diversity of forms. This diversity stems from differences in DNA sequences—in other words, from variations on the common theme of storing genetic information in DNA. Bacteria and humans are different because they have different genes with different DNA sequences. But both sets of instructions are written in the same language.

Let's list some of the other properties that are common to all organisms. (1) *Order.* All living things exhibit complex organization, as seen in the highly ordered structure of

the sunflower shown in Figure 1.4B. (2) *Regulation.* The environment outside an organism may change markedly, but mechanisms regulate an organism's internal environment, maintaining it within limits that sustain life. For example, regulation of the amount of blood flowing through the blood vessels in the large ears of the jackrabbit in Figure 1.4C helps maintain a constant body temperature by adjusting heat exchange with the air. (3) *Growth and development.* Inherited information carried by genes controls an organism's pattern of growth and development. (4) *Energy utilization.* Organisms take in energy and transform it to perform all of life's activities. (5) *Response to the environment.* All organisms respond to environmental stimuli. The Venus flytrap in Figure 1.4D closed its trap rapidly in response to the environmental stimulus of a cricket landing on it. (6) *Reproduction.* Organisms reproduce their own kind. Figure 1.4E shows a panda with its baby. (7) *Evolution.* Reproduction underlies the capacity of species to change (evolve) over time. Evolutionary change has been a central, unifying feature of life since it arose about 4 billion years ago.

In the next module, we see how biologists attempt to organize the incredible diversity of life.

? What is the chemical basis for all of life's kinship?

■ DNA as the genetic material

Figure 1.4B Order

Figure 1.4C Regulation

Figure 1.4D Response to the environment

Figure 1.4E Reproduction

1.5 The diversity of life can be arranged into three domains

We can think of biology's enormous scope as having two dimensions. The "vertical" dimension, which we examined in Module 1.1, is the size scale that stretches from molecules to the biosphere. But biology's scope also has a "horizontal" dimension, spanning across the great diversity of organisms existing now and over life's long history.

Diversity is a hallmark of life. Biologists have so far identified and named about 1.8 million **species,** the term used for a particular type of organism, such as *Pelecanus occidentalis,* the brown pelican. Researchers identify thousands of additional species each year. Estimates of the total number of species range from 10 million to over 200 million. How do we make sense of this much diversity?

There seems to be a human tendency to group diverse items according to similarities. Grouping species that are similar is natural for us. We may speak of bears or butterflies, though we recognize that each group includes many different species. We may even sort groups into broader categories, such as mammals and insects. **Taxonomy,** the branch of biology that names and classifies species, arranges them into a hierarchy of broader and broader groups, which you will learn about in Chapter 15.

Until the last decade, most biologists adopted a taxonomic scheme that divided all of the diversity of life into five **kingdoms.** But new methods, such as comparison of DNA sequences, have led to an ongoing reevaluation of the number and boundaries of kingdoms. New classification schemes have been proposed that range from six kingdoms to dozens of kingdoms. As that debate continues, however, most biologists now agree that the kingdoms of life can be organized into three overarching groups called **domains.**

The organisms in the photographs on these two pages represent the three domains. Figures 1.5A and 1.5B show examples of domain **Bacteria** and domain **Archaea,** both of which consist of **prokaryotes,** organisms with prokaryotic cells. Most prokaryotes are unicellular and microscopic. Prokaryotes are the most widespread of all living organisms.

In the five-kingdom system, bacteria and archaea were combined in a single kingdom because they both have prokaryotic cells. But newer evidence from comparisons of DNA and other molecules suggests that they represent two very distinct branches of life, as you'll learn in Chapter 16.

All the **eukaryotes,** organisms with eukaryotic cells, are now grouped into the various kingdoms of domain **Eukarya** (Figure 1.5C). As you learned in Module 1.3, eukaryotic cells have a nucleus and other internal structures called organelles.

Most aquatic or moist habitats support members of domain Eukarya called protists. Protists are a diverse collection of mostly single-celled organisms. Some protists, including those commonly called algae, make their own food molecules by the process of photosynthesis. Another collection of protists, commonly called protozoans, are animal-like in that they eat other organisms, including algae and prokaryotes. The photo of protists in Figure 1.5C shows a number of different protists in a drop of water viewed with a microscope. The large, irregular, bluish cell in the center is an amoeba (a protozoan), and the smaller cells are mostly single-celled algae. Also present are multicellular algae (the rodlike organisms), which are considered protists because of their similarities to single-celled algae. In the five-kingdom scheme, protists were classified in a single kingdom, but evidence from molecular studies now indicates that they are more diverse than any other group of eukaryotes. The recent trend has been to split the protists into several kingdoms, although biologists have not yet reached agreement on the exact number.

The three remaining kingdoms within Eukarya contain organisms that are mostly multicellular. These three kingdoms are distinguished partly by their modes of nutrition. Kingdom Plantae consists of plants, which produce their own food by photosynthesis and have cells with rigid walls made of cellulose. The photo representing kingdom Plantae in Figure 1.5C is a tropical bromeliad, a plant native to the Americas.

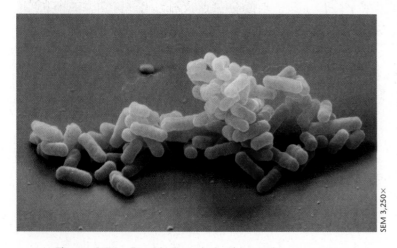

Figure 1.5A Domain Bacteria (prokaryotes)

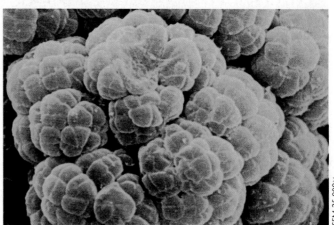

Figure 1.5B Domain Archaea (prokaryotes)

Kingdom Fungi, represented by the mushrooms in Figure 1.5C, is a diverse group that includes molds, yeasts, and mushrooms. Fungi are mostly decomposers. They break down the remains of dead organisms and organic wastes, such as leaf litter and animal feces, and then absorb the nutrients into their cells.

Representing the kingdom Animalia (animals), the sloth in Figure 1.5C resides in the trees of American rain forests. Animals obtain food by ingestion, which means they eat other organisms. Most animals are motile and are made of cells that lack rigid walls. The sloth is a slow-moving animal that spends most of its time hanging upside down eating leaves.

There are actually members of three kingdoms in the sloth photo. The sloth is clinging to a tree (kingdom Plantae), and the greenish tinge in the animal's hair is a luxuriant growth of photosynthetic prokaryotes (domain Bacteria). This photograph exemplifies a theme reflected in our book's title: connections between living things. The sloth depends on trees for food and shelter; the tree uses nutrients from the decomposition of the sloth's feces; the prokaryotes gain access to the sunlight necessary for photosynthesis by living on the sloth; and the sloth is camouflaged from predators by its green coat.

Life's diversity and its interconnectedness are evident almost everywhere. We have looked at life's unity and surveyed its diversity. In the next module, we describe how evolution explains both the unity and the diversity of life.

Web/CD Activity 1B *Classification Schemes*

? To which of the three domains of life do we belong?

■ Eukarya

275×

Protists (multiple kingdoms)

Kingdom Fungi

Kingdom Animalia

Kingdom Plantae

Figure 1.5C
Domain Eukarya (eukaryotes)

1.6 Evolution explains the unity and diversity of life

Figure 1.6A
Charles Darwin in 1859

In November 1859, British biologist Charles Robert Darwin (**Figure 1.6A**) published one of the most important and controversial books ever written. Entitled *On the Origin of Species by Means of Natural Selection,* Darwin's book was an immediate bestseller and soon made his name almost synonymous with the concept of evolution. Darwin stands out in history with people like Newton and Einstein, scientists who synthesized ideas with great explanatory power. Such comprehensive ideas are called **theories.**

The Origin of Species articulated two main points. First, Darwin presented evidence in support of the evolutionary view that species living today descended from ancestral species. Darwin called this evolutionary history "descent with modification." It was an insightful phrase, as it captured both the unity (descent from a common ancestor) and diversity (modification) of life.

Darwin's second point was to propose a mechanism for evolution, a mechanism he called **natural selection.** Darwin synthesized this concept from two observations that by themselves were neither profound nor original. Others had the pieces of the puzzle, but Darwin saw how they fit together. He inferred natural selection by connecting two readily observable features of life:

OBSERVATION #1: **Individual variation.** Individuals in a population vary in many heritable traits.

OBSERVATION #2: **Overproduction and competition.** A population of any species has the potential to produce far more offspring than will survive to produce offspring of their own. With more individuals than the environment can support, competition is inevitable.

INFERENCE: **Unequal reproductive success.** From these two observations, Darwin inferred that individuals are unequal in their likelihood of surviving and reproducing. Those individuals with heritable traits best suited to the environment will leave the greatest number of healthy, fertile offspring.

It is this unequal reproductive success that Darwin called natural selection. And the product of natural selection is **evolutionary adaptation,** the accumulation of favorable variations in a population over time.

Figure 1.6B uses a simple example to show how natural selection works. ❶ An imaginary beetle population has colonized an area where the soil has been blackened by a recent brush fire. Initially, the population varies extensively in the inherited coloration of individuals, from very light gray to charcoal. ❷ A predatory bird eats the beetles it sees most eas-

ily, the light-colored ones. This selective predation favors the survival and reproductive success of the darker beetles compared to the lighter ones. ❸ The surviving beetles have reproduced. The population is now quite different from the original one, as natural selection has produced a change in the frequencies of the different-colored beetles.

Our example also illustrates that natural selection is not a creative process, but an editing mechanism. It can only select among the variations that are present in the population. We can summarize natural selection as follows: *Natural selection occurs as heritable variations are exposed to environmental factors that favor the reproductive success of some individuals over others.* In our example, birds are the relevant environmental factor.

Darwin realized that numerous small changes in populations caused by natural selection could eventually lead to major alterations of species. He proposed that new species could evolve as a result of the gradual accumulation of changes over long periods of time. We will explore evolution and natural selection in more detail in Chapters 13 and 14.

We see the exquisite results of natural selection in every kind of organism. Each species has its own special set of evolutionary adaptations that evolved by means of natural selection. In **Figure 1.6C**, we see the pangolin, a mammal that lives in East African rain forests. One of its main adaptations, its tough body armor of overlapping scales, protects it from most predators. The pangolin also has an unusually long

❶ Population with varied inherited traits

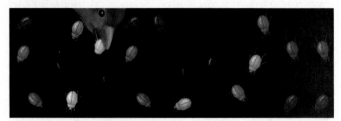

❷ Elimination of individuals with certain traits

❸ Reproduction of survivors

Figure 1.6B   Natural selection

tongue, which it uses to prod termites and ants out of their nests. Another mammal shown in Figure 1.6C, the killer whale (orca), is adapted for life at sea. It breathes air through nostrils on the top of its head and keeps in touch with its companions by emitting clicking sounds into the water. Orcas use sound echoes to detect obstacles and schools of fish or other prey. The pangolin's armor and the orca's echolocating ability arose over many, many generations as individuals with heritable traits that made them best adapted to their environment had the greatest reproductive success.

Evolution is biology's core theme—the one idea that makes sense of all we know about life. Next we explore some of the methods scientists use to learn about life.

Web/CD Thinking as a Scientist *How Do Environmental Changes Affect a Population?*

? How does natural selection enable a population of organisms to adapt to its environment?

■ On average, those individuals with heritable traits best suited to the local environment produce the greatest number of offspring that survive and reproduce. This increases the frequency of those traits in the population.

Killer whale

Pangolin

Figure 1.6C Examples of adaptations to the environment

THE PROCESS OF SCIENCE

1.7 Scientists use two main approaches to learn about nature

The word *science* is derived from a Latin verb meaning "to know." Science is a way of knowing. It stems from our curiosity about ourselves and the world around us. Science seeks natural causes for natural phenomena. This criterion limits the scope of science to the study of structures and processes that we can observe and measure, either directly or indirectly with the help of tools, such as microscopes, that extend our senses. This dependence on observations that other people can confirm demystifies nature and distinguishes science from belief in the supernatural. Science can neither prove nor disprove that angels, ghosts, or deities, either benevolent or evil, cause rainbows, storms, illnesses, and cures—for such explanations are outside the bounds of science.

Biology blends two main scientific approaches: discovery science, which is mostly about *describing* nature, and hypothesis-based science, which is mostly about *explaining* nature. Most research combines these two forms of inquiry.

Discovery Science Verifiable observations and measurements are the data of discovery science. In biology, discovery science describes life at its many levels, from the biosphere down to cells and molecules. An example is the sequencing of the human genome. While this research involves complicated methods and instruments, it is essentially just a detailed dissection and description of human DNA.

Discovery science can lead to important conclusions based on a type of logic called *inductive reasoning*. This kind of reasoning derives general principles from a large number of specific observations. "All organisms are made of cells" is an

inductive conclusion based on the discovery of cells in every microscopic biological specimen observed by biologists over two centuries of time. The careful observations of discovery science and the inductive conclusions they sometimes produce are fundamental to our understanding of nature.

Hypothesis-Based Science The observations of discovery science stimulate inquiring minds to ask questions and seek explanations. Such inquiry usually involves the proposing and testing of hypotheses. A **hypothesis** is a tentative answer to some question—an explanation on trial. A good hypothesis leads to predictions that scientists can test by recording additional observations or by designing experiments.

Deductive reasoning is the logic used in hypothesis-based science to come up with ways to test hypotheses. In deduction, the reasoning flows from the general to the specific. From general premises, we extrapolate to the specific results we should expect if the premises are true. If all organisms are made of cells (premise 1), and humans are organisms (premise 2), then humans are composed of cells (deduction about a specific case). This deduction is a prediction that can be tested by examining human tissues with a microscope.

In the next module, we'll use some additional examples to examine the methods of science more closely.

? What is the difference between discovery science and hypothesis-based science?

■ In the first, scientists observe and describe objects and phenomena; in the second, they propose hypotheses, make deductions, and test predictions.

1.8 With hypothesis-based science, we pose and test hypotheses

We will explore the elements of hypothesis-based science with the help of two scientific investigations, one from everyday life and one from a research project on snakes. Both of these case studies illustrate the "scientific method," a process of inquiry that involves observations, questions, hypotheses, predictions, and tests of predictions.

A Case Study from Everyday Life   We all use hypotheses in solving everyday problems. Let's say, for example, that your flashlight fails during a camp-out. That's an observation. The question is obvious: Why doesn't the flashlight work?

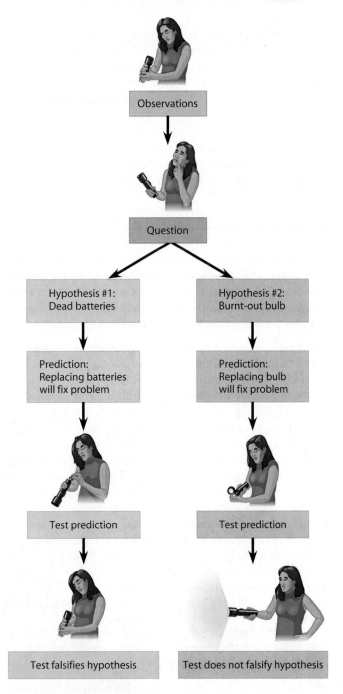

Figure 1.8A   The hypothesis-driven scientific method

Two reasonable hypotheses based on past experience are that either the batteries in the flashlight are dead or the bulb is burnt out. **Figure 1.8A** diagrams this campgound inquiry.

Now we're ready to use deductive reasoning. In the process of science, deduction usually takes the form of predictions of experimental results or observations we should expect *if* a particular hypothesis is correct. We then test our predictions by carrying out the appropriate experiments or observations to see whether or not the results are expected. This deductive testing uses "*If . . . then*" logic. For example, *if* the dead-batteries hypothesis is correct and I replace the batteries with new ones, *then* the flashlight should work. If the flashlight still does not work, we can test our other hypothesis by replacing the flashlight bulb.

The flashlight example illustrates two important properties of scientific hypotheses. First, a hypothesis must be *testable*—there must be some way to check its validity. Second, a hypothesis must be *falsifiable*—there must be some observation or experiment that could show that it is not true. As shown on the left in Figure 1.8A, the hypothesis that dead batteries are the sole cause of the problem was falsified by replacing the batteries with new ones. As shown on the right, the burnt-out-bulb hypothesis is the more likely explanation. Notice that testing supports a hypothesis not by proving that it is correct, but by not eliminating it through falsification. Perhaps the bulb was simply loose and the new bulb was inserted correctly. No amount of experimental testing can *prove* a hypothesis beyond a shadow of doubt, because it is impossible to exhaust all alternative hypotheses. A hypothesis gains credibility by surviving various attempts to falsify it.

A Case Study of Hypothesis-Based Science   One way to learn more about how hypothesis-based science works is to examine a case study from actual scientific research.

The story begins with a set of observations and generalizations from discovery science. Many poisonous animals are brightly colored, often with distinctive patterns. This so-called warning coloration apparently says "dangerous species" to potential predators. But there are also mimics. These imposters look like poisonous species but are actually harmless. A question that follows from these observations is: What is the function of such mimicry? A reasonable hypothesis is that such deception is an evolutionary adaptation that reduces the harmless animal's risk of being eaten.

In 2001, biologists David and Karin Pfennig, along with William Harcombe, one of their students, designed an elegant set of field experiments to test the hypothesis that mimics benefit because predators confuse them with the harmful species. A poisonous snake called the eastern coral snake has warning coloration: bold, alternating rings of red, yellow, and black **(Figure 1.8B)**. Predators rarely attack these snakes. The predators do not learn this avoidance behavior by trial and error; a first encounter with a coral snake would usually be deadly. A predator making that mistake will not pass its genes on to any more offspring. Natural selection has apparently

Figure 1.8B Eastern coral snake (poisonous)

Figure 1.8C Scarlet king snake (nonpoisonous)

increased the frequency of predators that inherit an instinctive ability to recognize the coloration of the coral snake.

A nonpoisonous snake named the scarlet king snake mimics the ringed coloration of the coral snake (Figure 1.8C). Both types of snakes live in North and South Carolina, but the king snakes' geographic range also extends into regions where no coral snakes are found.

The geographic distribution of these snakes made it possible for the researchers to test a key prediction of the mimicry hypothesis: Mimicry should help protect king snakes from predators, but only in regions where coral snakes also live. Avoiding snakes with warning coloration is an adaptation of predator populations that evolved in areas where the poisonous coral snakes are present. The mimicry hypothesis predicts that predators adapted to the warning coloration of coral snakes will attack king snakes less frequently than will predators in areas where coral snakes are absent.

To test this prediction, Harcombe made hundreds of artificial snakes out of wire covered with a claylike substance called plasticine. He made two versions of fake snakes: an "experimental group" with the red, black, and yellow ring pattern of king snakes and a "control group" of plain brown artificial snakes as a basis of comparison.

The researchers placed equal numbers of the two types of artificial snakes in field sites throughout North and South Carolina, including the region where coral snakes are absent. After four weeks, they retrieved the fake snakes and recorded how many had been attacked by looking for bite or claw marks. The most common predators were foxes, coyotes, and raccoons, but black bears also attacked some of the artificial snakes (Figure 1.8D).

The data fit the key prediction of the mimicry hypothesis. Compared to the brown artificial snakes, the ringed snakes were attacked by predators less frequently only in field sites within the geographic range of the poisonous coral snakes. The bar graph in Figure 1.8E summarizes the results.

This case study provides an example of a **controlled experiment**, one that is designed to compare an experimental group (the artificial king snakes, in this case study) with a control group (the artificial brown snakes). Ideally, the experimental and control groups differ only in the one factor the experiment is designed to test—in our example, the effect of the snakes' coloration on the behavior of predators. Without the control group, the researchers would not have been able to rule out the number of predators in the different test areas as the cause of the different number of attacks on the artificial king snakes. The clever experimental design left coloration as the only factor that could account for the low predation rate on the artificial king snakes placed within the range of coral snakes.

Web/CD Thinking as a Scientist *The Process of Science: How Does Acid Precipitation Affect Trees?*

? Why is it difficult to draw a conclusion from an experiment that is not controlled?

■ Without a control, you don't know if the experimental outcome is due to the variable you are trying to test or to some other variable.

Figure 1.8D Artificial king snake (left); artificial brown snake that has been attacked (right)

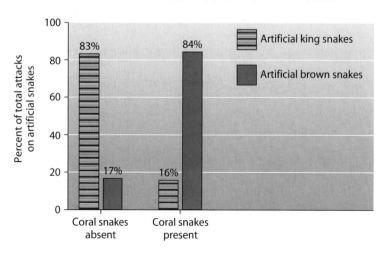

Figure 1.8E Results of mimicry experiment

1.9   Biology is connected to our lives in many ways

Endangered species, genetically modified crops, global warming, air and water pollution, the cloning of embryos, nutrition controversies, emerging diseases, medical advances—is there ever a day that we don't see several of these issues featured in the news? These topics and many more have biological underpinnings. Biology, the science of life, has an enormous impact on our everyday life (Figure 1.9).

Most of these issues of science and society also involve technology. Science and technology are interdependent, but their basic goals differ. The goal of science is to understand natural phenomena. In contrast, the goal of **technology** is generally to apply scientific knowledge for some specific purpose. Biologists and other scientists often speak of "discoveries," while engineers and other technologists more often speak of "inventions." The beneficiaries of those inventions also include scientists, who put new technology to work in their research. And scientific discoveries often lead to new technologies.

Figure 1.9   Biology in the news

The potent combination of science and technology has dramatic effects on society. For example, discovery of the structure of DNA by James D. Watson and Francis Crick some 50 years ago and subsequent achievements in DNA science have led to the many technologies of DNA engineering that are transforming many fields, including medicine, agriculture, and forensics (DNA fingerprinting, for example). Perhaps Watson and Crick envisioned their discovery as someday leading to important applications, but it is unlikely that they could have predicted exactly what those applications would be.

The directions that technology takes depend less on the curiosity that drives basic science than it does on the current needs and wants of people and on the social environment of the times. Debates about technology center more on "should we do it" than "can we do it." Among the difficult choices society must make about technology is deciding under what circumstances it is acceptable to use DNA technology to see if people have genes for hereditary diseases. Should such tests always be voluntary, or are there any circumstances when genetic testing should be mandatory? Should insurance companies or employers have access to the information, as they do for many other types of personal health data?

Technology has improved our standard of living in many ways, but not without consequences. Technology that keeps people healthier has enabled the Earth's population to grow more than tenfold in the past three centuries, to double to over 6 billion in just the past 40 years. The environmental effects of this growth can be devastating. Global warming, toxic wastes, acid rain, deforestation, nuclear accidents, and extinction of species are just some of the repercussions of more and more people wielding more and more technology. Science can help us identify such problems and provide insight into what course of action may prevent further damage. But solutions to these problems have as much to do with politics, economics, and cultural values as with science and technology. Now that science and technology have become such powerful aspects of society, every thoughtful citizen has a responsibility to develop a reasonable amount of scientific literacy. The crucial science-technology-society relationship is a theme that adds to the significance of any biology course.

Biology—from the molecular level to the level of the biosphere—is directly connected to our everyday lives. We hope this book will help you develop an appreciation for the science of life and help you apply that understanding to evaluating issues ranging from your personal health to the well-being of the whole world. Biology offers us a deeper understanding of ourselves and our planet and a chance to more fully appreciate life in all its diversity.

? How are science and technology related?

■ Technology is the application of scientific knowledge; new technologies increase the ability of scientists to search for new knowledge.

Reviewing the Concepts

The Scope of Biology (1.1–1.3)

Life's structural hierarchy defines the scope of biology, the scientific study of life. The biosphere is made up of all of Earth's ecosystems. An ecosystem consists of all the organisms living in a particular area, as well as the nonliving environmental components. All the living organisms in an ecosystem make up a community. A localized group of individuals of a species make up a population. An individual living entity is an organism. The hierarchy continues downward with organ systems, organs, tissues, cells, organelles, and molecules **(1.1)**.

Ecosystems are characterized by the cycling of chemical nutrients from the atmosphere and soil to producers to consumers to decomposers and back to the environment. Energy flows one-way through an ecosystem from the sun to producers to consumers and exits as heat **(1.2)**.

A cell is the basic unit of life. New properties emerge from the complex organization of a system, such as a cell. Eukaryotic cells contain membrane-enclosed organelles, including a DNA-containing nucleus. Prokaryotic cells lack such organelles **(1.3)**.

Evolution, Unity, and Diversity (1.4–1.6)

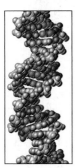

DNA is the genetic information for constructing the molecules that make up cells and organisms. Each species' genetic instructions are coded in the sequences of the four building blocks making up DNA's two helically coiled chains. All organisms share a set of common features: ordered structures, regulation of internal conditions, growth and development, energy use, response to environmental stimuli, and the ability to reproduce and evolve **(1.4)**.

Three domains. Organisms are grouped (classified) into the prokaryotic domains Bacteria and Archaea and the eukaryotic domain Eukarya, which includes protists (protozoans and algae, falling into multiple kingdoms) and the kingdoms Fungi, Plantae, and Animalia **(1.5)**.

Evolution and natural selection. Evolution explains the unity and diversity of life. Charles Darwin synthesized the theory of evolution by natural selection. Natural selection is an editing mechanism that occurs when populations of organisms, having inherited variations, are exposed to environmental factors that favor the reproductive success of some individuals over others. All organisms have adaptations that have evolved by means of natural selection **(1.6)**.

Observations

Individual variation	
	Inference
	Natural selection: unequal reproductive success
Overproduction of offspring	

The Process of Science (1.7–1.8)

Scientific inquiry. In discovery science, scientists describe some aspect of the world and use inductive reasoning to draw general conclusions. In hypothesis-based science, they attempt to explain observations by testing hypotheses **(1.7)**.

Hypothesis-based science involves observations, questions, hypotheses as tentative answers to questions, deductions leading to predictions, and then tests of the predictions to see if the hypotheses are falsifiable. Deductive reasoning is used: *If* a hypothesis is correct *and* we test one of its predictions (by performing an experiment or making observations), *then* a particular outcome will occur. In experiments designed to test hypotheses, the use of control groups and experimental groups helps to control variables **(1.8)**.

Biology and Everyday Life (1.9)

Biology is connected to many important issues in our lives. Technological advances stem from scientific research. The science-technology-society relationship is an important aspect of a biology course **(1.9)**.

Connecting the Concepts

1. Biology can be described as having both a vertical scale and a horizontal scale. Explain what that means.

2. Complete the following map organizing some of biology's major concepts.

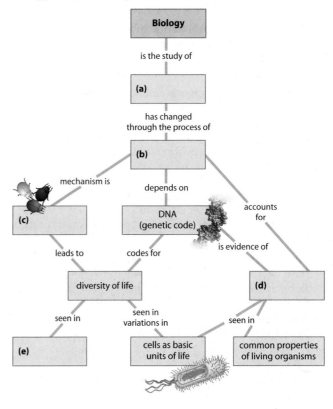

Testing Your Knowledge

Multiple Choice

3. Which of the following best describes the logic of the scientific process?
   a. If I generate a testable hypothesis, tests and observations will support it.
   b. If my prediction is correct, it will lead to a testable hypothesis.
   c. If my observations are accurate, they will support my hypothesis.
   d. If my hypothesis is correct, I can expect certain test results.
   e. If my tests are right, they will prove my hypothesis.

4. Amoebas and bacteria are grouped into different domains because
   a. amoebas eat bacteria.
   b. bacteria are not made of cells.
   c. bacterial cells lack a nucleus.
   d. bacteria decompose amoebas.
   e. amoebas are photosynthetic.

5. A biologist studying interactions among the protists in an ecosystem could *not* be working at which level in life's hierarchy? (*Explain your answer.*)
   a. the population level   d. the organism level
   b. the molecular level   e. the organ level
   c. the community level

6. Which of the following questions is outside the realm of science?
   a. Which organisms play the most important role in energy input to a rain forest canopy?
   b. What percentage of music majors take a biology course?
   c. What is the physical nature of the universe?
   d. What is the nature of the supernatural?
   e. What is the historical basis for the division of Earth's human population into ethnic groups?

7. Which of the following statements best distinguishes hypotheses from theories in science?
   a. Theories are hypotheses that have been proved.
   b. Hypotheses are tentative guesses; theories are correct answers to questions about nature.
   c. Hypotheses usually are narrow in scope; theories have broad explanatory power.
   d. Hypotheses and theories are different terms for essentially the same thing in science.
   e. Theories cannot be falsified; hypotheses are expected to be falsified.

8. The organisms in your backyard include trees, shrubs, grass, ants, mushrooms, birds, spiders, beetles, flies, and bacteria. Together, all these organisms make up
   a. an ecosystem.   d. an experimental group.
   b. a community.   e. both a and b.
   c. a population.

9. The core idea that makes sense of all of biology is
   a. the process of science.
   b. the correlation of function with structure.
   c. deductive reasoning.
   d. evolution.
   e. unity in diversity.

Describing, Comparing, and Explaining

10. In an ecosystem, how is the movement of energy similar to that of chemical nutrients, and how is it different?

11. Explain the role of heritable variations in Darwin's theory of natural selection.

12. Explain what is meant by this statement: The scientific process is not a rigid method.

13. Contrast technology with science. Give an example of each to illustrate the difference.

14. Explain what is meant by this statement: Natural selection is an editing mechanism rather than a creative process.

Applying the Concepts

15. The graph below shows the results of an experiment in which mice learned to run through a maze.

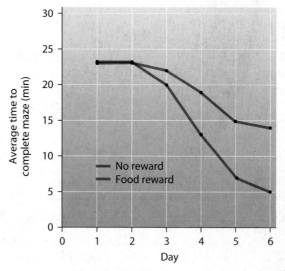

a. State the hypothesis and prediction that you think this experiment tested.
b. Which was the control group and which the experimental? Why was a control group needed?
c. List some variables that must have been controlled so as not to affect the results.
d. Do the data support the hypothesis? Explain.

16. In an experiment similar to the mimicry experiment described in Module 1.8, a researcher found that there were more predator attacks on artificial king snakes in areas with coral snakes than in areas outside the range of coral snakes. From this the researcher concluded that the mimicry hypothesis is false. Do you think this conclusion is justified? Why or why not?

17. The news media and popular magazines frequently report stories that are connected to biology. In the next 24 hours, record all the ones you hear or read about in three different sources, and briefly describe the biological connections you perceive in each story.

Answers to all questions can be found in Appendix 3.

For study help and Activities, go to campbellbiology.com or the student CD-ROM.

UNIT I

The Life of the Cell

2 THE CHEMICAL BASIS OF LIFE

3 THE MOLECULES OF CELLS

4 A TOUR OF THE CELL

5 THE WORKING CELL

6 HOW CELLS HARVEST CHEMICAL ENERGY

7 PHOTOSYNTHESIS: USING LIGHT TO MAKE FOOD

CHAPTER 2

ELEMENTS, ATOMS, AND MOLECULES

2.1   Living organisms are composed of about 25 chemical elements
2.2   Trace elements are common additives to food and water
2.3   Elements can combine to form compounds
2.4   Atoms consist of protons, neutrons, and electrons
2.5   Radioactive isotopes can help or harm us
2.6   Electron arrangement determines the chemical properties of an atom
2.7   Ionic bonds are attractions between ions of opposite charge
2.8   Covalent bonds join atoms into molecules through electron sharing
2.9   Unequal electron sharing creates polar molecules
2.10 Hydrogen bonds are weak bonds important in the chemistry of life

WATER'S LIFE-SUPPORTING PROPERTIES

2.11 Hydrogen bonds make liquid water cohesive
2.12 Water's hydrogen bonds moderate temperature
2.13 Ice is less dense than liquid water
2.14 Water is the solvent of life
2.15 The chemistry of life is sensitive to acidic and basic conditions
2.16 Acid precipitation threatens the environment

CHEMICAL REACTIONS

2.17 Chemical reactions change the composition of matter

Rattlebox moths mating

Nature's Chemical Language

THE TWO BALLS OF FROTH ON THE RATTLEBOX MOTH in the photo on the facing page contain a noxious chemical, one that is particularly distasteful to the spiders that prey on this insect. Officially called *Utetheisa ornatrix,* this moth is a native of central Florida. Its common name comes from the rattlebox plant (*Crotalaria mucronata*), the source of the moth's defensive chemical.

We know about the role of this chemical, a so-called alkaloid, from the work of Thomas Eisner, of Cornell University. Eisner is a pioneer of chemical ecology, which he defines as the study of the chemical language of nature. In particular, he studies how insects interact via chemical messages with each other and also with plants—messages ranging from "stay away!" to "come mate with me." Eisner's research has yielded insights into animal behavior, ecology, and evolution.

Eisner's interest in chemical communication emerged while he was at graduate school in the 1950s. It was the heyday of insect hormones. The discoveries then being made about hormones, chemicals that serve as signals *within* an organism, suggested to him that there might be chemicals that signal *between* organisms. Arriving at Cornell as a young professor, he pursued this idea. In an interview with us, he described what he has learned:

> It turns out that chemical signaling is everywhere. In fact, I've stopped thinking of air as air; I think of it as a carrier of mes-

The Chemical Basis of Life

Thomas Eisner

A rattlebox moth releasing defensive chemicals

sages. When I see a meadow filled with insects and other animals browsing on the vegetation, I think of the perfumes that attract pollinators to the flowers, as well as the repellent chemicals produced by plants that discourage a butterfly from laying her eggs or a caterpillar from feeding on leaves. These repellent chemicals are a defensive strategy for the plant; the attractive floral scents are a reproductive strategy. Meanwhile, the insects are interacting with each other. The ants are repelled by substances produced by beetles. Another insect [the rattlebox moth] procures defensive substances from plants it eats as a caterpillar; later, as an adult, it bestows the chemicals on its own offspring. And these and many other insects use chemicals to attract mates. It's all chemical!

In the case of the rattlebox moth, mating is especially chemical. Eisner found that while both male and female caterpillars contain a store of the defensive chemical from the plants they've eaten, the female moth receives an extra dose at mating. During the 8- or 9-hour copulation, shown in the photograph to the left, the male passes a large mass of sperm, nutrients, and alkaloid to the female, supplying additional protection for her and for their offspring, who carry some protection even as embryos. As Eisner put it:

Only a human bridegroom would buy life insurance for his bride. This classy moth gives a gift she can really use—a life assurance policy, if you will—that keeps paying off every time her life is in danger.

The alkaloid even seems to play a part in mate selection. During the courtship dance, the male moth releases into the air puffs of a chemical derived from the alkaloid; the female, sensing this chemical, can assess how much alkaloid he has.

Do we humans communicate chemically? Eisner thinks so:

Humans don't have a chemical that attracts males to ovulating females. On the contrary, the human female seems to be programmed to hide ovulation, chemically as well as anatomically. In this way, she induces the male—who can't be sure when she is fertile—to remain in attendance for long periods. (So what we call love has a subtle biological basis!) But clearly some kinds of chemical signaling go on between men and women. For instance, chemicals in the armpit of a male can apparently regularize a female companion's ovulatory cycle.

Chemicals play many more roles in life than signaling, of course: They are the very stuff making up our bodies, those of other organisms, and the physical environment. This first unit of chapters will take you from atoms to cells and explain their most basic chemical activities. This first chapter focuses on atoms and molecules, where it all begins. ■ ■ ■

2.1 Living organisms are composed of about 25 chemical elements

Why does a biology textbook begin with a chapter on chemistry? To learn about life, you will study the structures and functions of living organisms. In the process, you will travel through all the levels of the hierarchy of biological structure described in Module 1.1, from molecules to ecosystems. At the base of this hierarchy are elements, atoms, and molecules. You will see that the properties of life emerge from the arrangement of its chemical parts into higher and higher levels of biological organization.

Living organisms are composed of **matter**, which is anything that occupies space and has mass. (In everyday language, we can think of mass as an object's weight.) Matter, in forms as diverse as rock, water, air, and living organisms, is composed of chemical elements. An **element** is a substance that cannot be broken down to other substances by ordinary chemical means. Today, chemists recognize 92 elements occurring in nature; gold, copper, carbon, and oxygen are some examples. Each element has a symbol, the first letter or two of its English, Latin, or German name. For instance, the symbol for sodium, Na, is from the Latin word *natrium*; the symbol O stands for the English word *oxygen*.

Life requires about 25 chemical elements. As you can see in **Table 2.1**, four of these—oxygen (O), carbon (C), hydrogen (H), and nitrogen (N)—make up about 96% of the human body, as well as most other living organisms. These four elements are the main ingredients of biological molecules such as proteins, sugars, and fats. Calcium (Ca), phosphorus (P), potassium (K), sulfur (S), sodium (Na), chlorine (Cl), and magnesium (Mg) account for most of the remaining 4% of the human body. These elements are involved in such important functions as bone formation, nerve signaling, and DNA synthesis.

TABLE 2.1	CHEMICAL COMPOSITION OF THE HUMAN BODY	
Symbol	Element	Percentage of Human Body Weight
O	Oxygen	65.0 ⎫
C	Carbon	18.5 ⎪ 96.3
H	Hydrogen	9.5 ⎬
N	Nitrogen	3.3 ⎭
Ca	Calcium	1.5
P	Phosphorus	1.0
K	Potassium	0.4
S	Sulfur	0.3
Na	Sodium	0.2
Cl	Chlorine	0.2
Mg	Magnesium	0.1

Trace elements (less than 0.01%): boron (B), chromium (Cr), cobalt (Co), copper (Cu), fluorine (F), iodine (I), iron (Fe), manganese (Mn), molybdenum (Mo), selenium (Se), silicon (Si), tin (Sn), vanadium (V), and zinc (Zn).

The **trace elements** listed at the bottom of the table are essential, but only in minute quantities. We explore the importance of trace elements to your health next.

Web/CD Activity 2A *The Levels of Life Card Game*

Web/CD Thinking as a Scientist *Connection: How Are Space Rocks Analyzed for Signs of Life?*

? Which four chemical elements are most abundant in living matter?

■ Oxygen, carbon, hydrogen, and nitrogen

2.2 Trace elements are common additives to food and water

Some trace elements, such as iron (Fe), are needed by all forms of life. Iron makes up only 0.004% of your body mass but is vital for energy processing and for transporting oxygen in your blood. Other trace elements, such as iodine (I), are required only by certain species. The average human needs about 0.15 milligram (mg) iodine each day. Iodine is an essential ingredient of a hormone produced by the thyroid gland, which is located in the neck. An iodine deficiency in the diet causes the thyroid gland to grow to abnormal size, a condition called goiter. The goiter of the woman shown in **Figure 2.2A** can probably be reversed by iodine supplements. Adding iodine to table salt has reduced the incidence of goiter in many countries. Unfortunately, iodized salt is not available or affordable everywhere, and goiter still affects many thousands of people in developing nations.

Iodine is just one example of a trace element added to food or water to improve health. For more than 50 years, the American Dental

Figure 2.2A Goiter in a Malayan woman

Association has supported fluoridation of community drinking water supplies as a public health measure. Fluoride is a form of the element fluorine (F), an element in Earth's crust that is found in small amounts in all water sources. In many areas, fluoride is added as part of the municipal water treatment process to raise levels to a concentration that can reduce tooth decay.

Chemicals are added to food to help preserve the food, make it more nutritious, or simply make it look better. Look at the nutrition facts label from the side of the cereal box in Figure 2.2B to see a familiar example of how foods are fortified with mineral elements. Iron, for example, is a trace element commonly added to foods. (You can actually see the iron that has been added to a fortified cereal by crushing the cereal and then stirring a magnet through it.) Also note that the nutrition facts label lists numerous vitamins that are added to improve the nutritional value of the cereal. For instance, the cereal in this example supplies 10% of the recommended daily value for vitamin A. Vitamins consist of more than one element and are examples of compounds, which we consider next.

? In addition to iron, what other trace elements have been added to the cereal in Figure 2.2B?

■ Copper and zinc

Nutrition Facts
Serving Size ¾ cup (30g)
Servings Per Container About 11

Amount Per Serving	Whole Grain Total	with ½ cup skim milk
Calories	110	150
Calories from Fat	10	10

	% Daily Value**	
Total Fat 1g*	1%	1%
Saturated Fat 0g	0%	0%
Polyunsaturated Fat 0g		
Monounsaturated Fat 0g		
Cholesterol 0mg	0%	1%
Sodium 190mg	8%	11%
Potassium 90mg	3%	8%
Total Carbohydrate 23g	8%	10%
Dietary Fiber 3g	10%	10%
Sugars 5g		
Other Carbohydrate 15g		
Protein 2g		

Vitamin A	10%	15%
Vitamin C	100%	100%
Calcium	100%	110%
Iron	100%	100%
Vitamin D	10%	25%
Vitamin E	100%	100%
Thiamin	100%	100%
Riboflavin	100%	110%
Niacin	100%	100%
Vitamin B₆	100%	100%
Folic Acid	100%	100%
Vitamin B₁₂	100%	110%
Pantothenic Acid	100%	100%
Phosphorus	8%	20%
Magnesium	6%	10%
Zinc	100%	100%
Copper	4%	4%

Figure 2.2B Nutrition facts from a fortified cereal

2.3 Elements can combine to form compounds

Vitamin A is a **compound,** a substance containing two or more elements in a fixed ratio. Compounds are much more common than pure elements. In fact, few elements exist in a pure state in nature. Many compounds consist of only two elements; for instance, table salt (sodium chloride, NaCl) has equal parts of the elements sodium and chlorine. Pure sodium is a metal and pure chlorine is a poisonous gas. Chemically combined, however, they form a common seasoning (Figure 2.3). This example shows the emergence of novel properties with a higher level of structural organization.

Most of the compounds in living organisms contain at least three or four different elements, mainly carbon, hydrogen, oxygen, and nitrogen. Vitamin A, for example, is formed of just carbon, hydrogen, and oxygen. Proteins are compounds containing carbon, hydrogen, oxygen, nitrogen, and a small amount of sulfur. Different arrangements of the atoms of these elements determine unique properties for each compound.

The smallest unit of an element is an atom. Let's explore the structure of an atom, and then see how this structure determines the chemical properties of elements.

? Why are sodium chloride and vitamin A both classified as compounds?

■ They both contain more than one element.

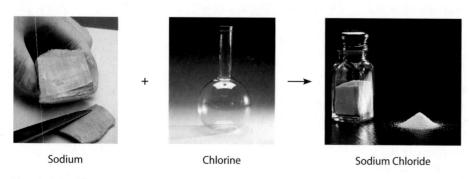

Sodium + Chlorine → Sodium Chloride

Figure 2.3 The emergent properties of a compound

2.4 Atoms consist of protons, neutrons, and electrons

Each element consists of one kind of atom, which is different from the atoms of any other element. An **atom**, named from a Greek word meaning "indivisible," is the smallest unit of matter that still retains the properties of an element. It would take about a million atoms to stretch across the period printed at the end of this sentence.

Subatomic Particles Physicists have split the atom into more than a hundred types of subatomic particles. However, we need to consider only three. A **proton** is a subatomic particle with a single positive electrical charge (+). An **electron** is a subatomic particle with a single negative electrical charge (−). A third type of subatomic particle, the **neutron**, is electrically neutral (has no electrical charge).

Figure 2.4A shows two very simple models of an atom of the element helium (He), the "lighter-than-air" gas that can make balloons rise. Notice that two neutrons and two protons are tightly packed in the atom's central core, or **nucleus.** Two electrons orbit the nucleus at nearly the speed of light. The attraction between the negatively charged electrons and the positively charged protons keeps the electrons near the nucleus. The left-hand model shows the number of electrons in the atom. The right-hand model, slightly more realistic, shows a spherical cloud of negative charge created by the rapidly orbiting electrons around the nucleus. Neither model is drawn to scale. In real atoms, the electrons are very much smaller than the protons and neutrons, and the electron cloud is much bigger compared to the nucleus. Imagine that this atom was the size of the baseball stadium of the New York Yankees: The nucleus would be the size of a fly in center field, and the electrons would be like two tiny gnats buzzing all around the stadium.

Differences in Elements Elements differ in the number of subatomic particles in their atoms. All atoms of a particular element have the same unique number of protons. In other words, the number of protons defines an element. This number is the element's **atomic number.** Thus, an atom of helium, with 2 protons, has an atomic number of 2. Carbon, with 6 protons, has an atomic number of 6 (Figure 2.4B). Note that in these atoms, the atomic number is also the number of electrons. Unless otherwise indicated, an atom has an equal number of protons and electrons and thus its net electrical charge is 0 (zero).

An atom's **mass number** is the sum of the protons and neutrons in its nucleus. For helium, the mass number is 4; for the carbon shown in Figure 2.4B, it is 12. The mass of a proton and the mass of a neutron are almost identical. An electron has only about $\frac{1}{2,000}$ the mass of a proton. So an atom's **atomic mass** (or weight) is approximately equal to its mass number, the sum of its protons and neutrons.

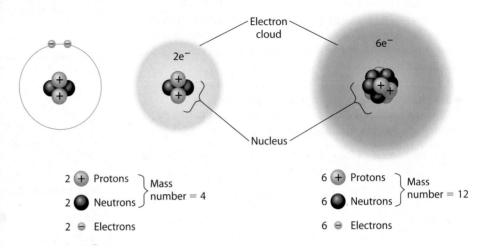

2 ⊕ Protons	} Mass number = 4	
2 ● Neutrons		
2 ⊖ Electrons		

Figure 2.4A Helium atom

6 ⊕ Protons	} Mass number = 12	
6 ● Neutrons		
6 ⊖ Electrons		

Figure 2.4B Carbon atom

Isotopes Although all atoms of an element have the same atomic number, some atoms of a given element may differ in mass number. The different **isotopes** of an element have the same numbers of protons and electrons and behave identically in chemical reactions, but they have different numbers of neutrons. Table 2.4 shows the numbers of subatomic particles in the three isotopes of carbon. Carbon-12 (usually written ^{12}C), with 6 neutrons, makes up about 99% of all naturally occurring carbon. Most of the other 1% consists of carbon-13 (^{13}C), with 7 neutrons. A third isotope, carbon-14 (^{14}C), with 8 neutrons, occurs in minute quantities. Notice that all three isotopes have 6 protons—otherwise, they would not be carbon.

Both ^{12}C and ^{13}C are stable isotopes, meaning their nuclei remain permanently intact. The isotope ^{14}C, on the other hand, is unstable, or radioactive. A **radioactive isotope** is one in which the nucleus decays spontaneously, giving off particles and energy. Radioactive isotopes pose serious risks to living organisms, but they also have many uses in biological research and medicine, as we see next.

TABLE 2.4	ISOTOPES OF CARBON		
	Carbon-12	**Carbon-13**	**Carbon-14**
Protons	6	6	6
Neutrons	6	7	8
Electrons	6	6	6

Web/CD Activity 2B *Structure of the Atomic Nucleus*

? A nitrogen atom has 7 protons, and the most common isotope of nitrogen has 7 neutrons. A radioactive isotope of nitrogen has 9 neutrons. What is the atomic number and mass number of this radioactive nitrogen?

■ Atomic number = 7; mass number = 16

2.5 Radioactive isotopes can help or harm us

Living cells cannot distinguish between isotopes of the same element. Consequently, organisms take up and use compounds containing radioactive isotopes in the usual way. Because radioactivity is easily detected, radioactive isotopes are useful as tracers—biological spies, in effect—for monitoring the fate of atoms in living organisms.

Basic Research Biologists often use radioactive tracers to follow molecules as they undergo chemical changes in an organism. For example, plant researchers have used them to study photosynthesis, the light-requiring process by which plants take in carbon dioxide (CO_2) from the air and use it to make sugar molecules. Researchers used CO_2 containing the radioactive isotope ^{14}C to trace the molecules plants make in the process. They found that the ^{14}C appeared first in one compound, then in several others, before finally showing up in the sugar glucose. These results showed the chemical route by which plants make glucose from CO_2.

Medical Diagnosis Radioactive tracers are also used in medicine. Certain kidney disorders, for example, are diagnosed by injecting a radioactive chemical into a patient's blood and then measuring the amount of radioactive material passed in the urine. In most diagnostic uses of radioactive tracers, the patient receives only a tiny amount of an isotope that decays completely in minutes or hours.

Radioactive tracers are often used for diagnosis in combination with sophisticated imaging instruments. Figure 2.5A shows a patient being examined by a technique called PET (positron-emission tomography), which can detect locations of intense chemical activity in the body. The patient is first injected with an isotope that emits subatomic particles called positrons. The positrons collide with electrons made available by chemical reactions in the body. A PET scanner detects energy released by these collisions and maps as "hot spots" the regions of an organ that are most chemically acitve.

Cancerous tissue has more chemical activity than other areas of the body. Figure 2.5B shows a PET image of the upper body of a cancer patient. (The patient is lying on his back; note the whitish vertebrae.) The bright yellow spot is cancerous tissue in the patient's throat. PET is also useful for diagnosing certain heart disorders and for basic research on the brain (see Module 20.11).

Dangers Though radioactive isotopes have many beneficial uses, uncontrolled exposure to them can harm living organisms by damaging molecules, especially DNA. The particles and energy thrown off by radioactive atoms can break chemical bonds and also cause abnormal bonds to form. The explosion of a nuclear reactor at Chernobyl, Ukraine, in 1986 released large amounts of radioactive isotopes into the environment, killing 30 people within a few weeks. The survivors have suffered increased rates of thyroid cancer and increased rates of birth defects in their children, and thousands may be at increased risk of future cancers. A

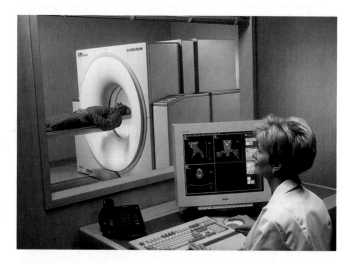

Figure 2.5A PET scanner

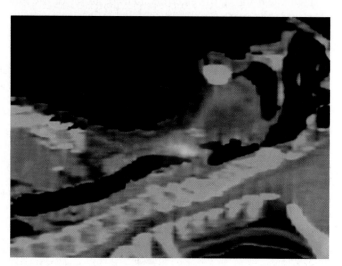

Figure 2.5B PET image of throat cancer

1999 nuclear accident at the Tokaimura nuclear power plant, about 150 kilometers (km) upwind from Tokyo, exposed 46 workers to high doses of radiation, two of whom died from their exposure. Many thousands of people in nearby towns are being monitored long-term for possible effects.

Natural sources of radiation can also pose a threat. Radon, a radioactive gas, may be a cause of lung cancer. Radon can contaminate buildings in regions where underlying rocks naturally contain uranium, a radioactive element. Homeowners can buy a radon detector or hire a company to test their home to ensure that radon levels are safe.

? Why are radioactive isotopes useful as tracers in research on the chemistry of life?

■ Organisms incorporate radioactive isotopes of an element into their molecules just as they do the nonradioactive isotopes, and researchers can detect the presence of the radioactive isotopes.

2.6 Electron arrangement determines the chemical properties of an atom

Of the subatomic particles we have discussed, it is mainly electrons that determine how an atom behaves when it encounters other atoms. Electrons vary in the amount of energy they possess. The farther an electron is from the nucleus, the greater its energy. Electrons in an atom occur only at certain energy levels, called **electron shells.** Depending on their atomic number, atoms may have one, two, or more electron shells.

Each shell can accommodate up to a specific number of electrons. The innermost shell is full with only 2 electrons. So in atoms with more than 2 electrons, the remainder are found in shells farther from the nucleus. For the main elements of life, the outermost (highest-energy) shell can hold up to 8 electrons (four pairs). It is the number of electrons in the outermost shell that determines the chemical properties of an atom. Atoms whose outer shells are not full (have unpaired electrons) tend to interact with other atoms—that is, to participate in chemical reactions.

Figure 2.6 shows the electron shells of atoms of the four elements that are the main components of biological molecules. Small dashed circles represent the unfilled "spaces" in the outer electron shells. Because the outer shells of all four atoms are incomplete, all these atoms react readily with other atoms. The hydrogen atom has only 1 electron in its single electron shell, which can accommodate 2 electrons. Atoms of carbon, nitrogen, and oxygen also are highly reactive because their outer shells, which can hold 8 electrons, are incomplete. In contrast, the helium atom that you saw in Figure 2.4A has a single, first-level shell that is full with 2 electrons. As a result, helium is chemically inert (unreactive).

How do chemical interactions between atoms enable them to fill their outer electron shells? When two atoms with incomplete outer shells react, each atom either shares, donates, or receives outer electrons, so that both partners end up with completed outer shells. These interactions usually result in atoms staying close together, held by attractions called **chemical bonds.** In the next two modules, we look at two important types of chemical bonds.

Web/CD Activity 2C *Electron Arrangement*

Web/CD Activity 2D *Build an Atom*

? Sodium has an atomic number of 11. How many electron shells does a sodium atom have, and how many electrons are in the outermost shell?

■ Three electron shells; one electron in the outermost shell

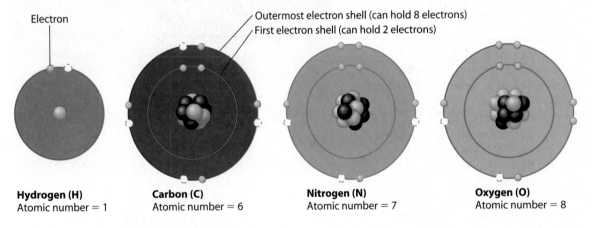

Electron

Outermost electron shell (can hold 8 electrons)
First electron shell (can hold 2 electrons)

Hydrogen (H)
Atomic number = 1

Carbon (C)
Atomic number = 6

Nitrogen (N)
Atomic number = 7

Oxygen (O)
Atomic number = 8

Figure 2.6 Atoms of four elements most common in living matter, all with incomplete outer electron shells

2.7 Ionic bonds are attractions between ions of opposite charge

At the top of the next page, Figure 2.7A shows how a sodium atom and a chlorine atom form the compound sodium chloride (NaCl). Notice that sodium has only 1 electron in its outer shell, whereas chlorine has 7. When these atoms interact, the chlorine atom strips sodium's outer electron away. In doing so, chlorine fills its outer shell with 8 electrons. Sodium, in losing 1 electron, ends up with only two shells, the outer shell having a full set of 8 electrons.

Since electrons are negatively charged particles, the electron transfer between the two atoms moves one unit of negative charge from sodium to chlorine. Sodium, with 11 protons but now only 10 electrons, has a net electrical charge of +1. Chlorine, having gained an extra electron, now has a net electrical charge of −1. In each case, an atom has become what is called an ion. An **ion** is an atom or molecule with an electrical charge resulting from a gain or loss of one or more electrons. As you can see in Figure 2.7A, the ion formed from chlorine is called a chloride ion. Two ions with opposite charges attract each other; when the attraction holds them together, it is called an **ionic bond.** The resulting compound, in this case NaCl, is electrically neutral.

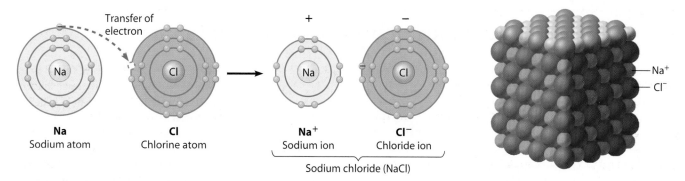

Figure 2.7A Formation of an ionic bond, producing sodium chloride

Figure 2.7B A crystal of sodium chloride

Sodium chloride is a familiar type of **salt**, a synonym for an ionic compound. Salts often exist as crystals in nature. **Figure 2.7B** shows the atoms in a crystal of sodium chloride. An NaCl crystal can be of any size (there is no fixed number of ions), but sodium and chloride ions are always present in a 1:1 ratio. The ratio of ions differs with the kind of salt.

Web/CD Activity 2E *Ionic Bonds*

? Explain what holds together the atoms in a crystal of table salt (NaCl).

■ Opposite charges attract. The positively charged sodium ions (Na^+) and the negatively charged chloride ions (Cl^-) are held together by ionic bonds, the attractions between oppositely charged ions.

2.8 Covalent bonds join atoms into molecules through electron sharing

The second kind of strong chemical bond is the **covalent bond**, in which two atoms *share* one or more pairs of outer-shell electrons. Two or more atoms held together by covalent bonds form a **molecule**. For example, a covalent bond connects the two hydrogen atoms in the molecule H_2, a common gas in the atmosphere. **Table 2.8** shows three ways to represent this molecule. The symbol H_2, called the molecular formula, merely tells you that the molecule consists of two atoms of hydrogen. The electron configuration diagram shows that the atoms share 2 electrons; as a result, both atoms fill their outer (only) shells. The third column shows a structural formula. The line between the hydrogen atoms stands for the single covalent bond formed by the sharing of one pair of electrons. In an O_2 molecule, the two oxygen atoms share two pairs of electrons, forming a **double bond**. In the structural formula, the double bond is indicated by a pair of lines between the oxygen atoms.

The number of covalent bonds an atom can form is equal to the number of additional electrons needed to fill its outer shell. This number is often called the valence, or bonding capacity, of an atom. Looking back at Figure 2.6, we see that H can form one bond; O can form two; N, three; and C, four.

H_2 and O_2 are molecules, but because they are composed of only one element, they are not compounds. An example of a molecule that is a compound is methane (CH_4), a common gas produced by certain bacteria. As you can see in the table, each of the four hydrogen atoms in this molecule shares one pair of electrons with the single carbon atom. Water is also a compound, and we look more closely at the covalent bonds in a water molecule next.

Web/CD Activity 2F *Covalent Bonds*

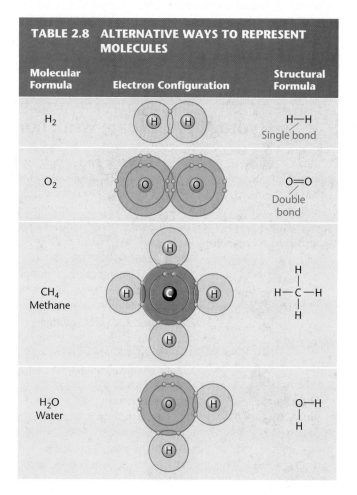

TABLE 2.8 ALTERNATIVE WAYS TO REPRESENT MOLECULES

Molecular Formula	Electron Configuration	Structural Formula
H_2		H—H Single bond
O_2		O=O Double bond
CH_4 Methane		H—C—H (with H above and below)
H_2O Water		O—H (with H below)

? What is chemically nonsensical about this structure?

H—C≡C—H

■ Each carbon atom has only three covalent bonds instead of the required four.

2.9 Unequal electron sharing creates polar molecules

A water molecule (H_2O), as shown in Figure 2.9, consists of two hydrogen atoms covalently bonded to a single oxygen atom. Atoms in a covalently bonded molecule are in a constant tug-of-war for the shared electrons of their covalent bonds. An atom's attraction for its electrons, including shared electrons, is called its **electronegativity**. The more electronegative an atom, the more strongly it pulls shared electrons toward its nucleus. In molecules of only one element, such as O_2 and H_2, the two identical atoms exert an equal pull on the electrons. The covalent bonds in such molecules are said to be **nonpolar** because the electrons are shared equally between the atoms. Some compounds—for instance, methane (CH_4)—also have nonpolar bonds. The atoms in such compounds are not substantially different in electronegativity.

In contrast to O_2, H_2, and CH_4, water is composed of atoms with very different electronegativities. Oxygen is one of the most electronegative of the elements. As indicated by the arrows in the figure, O attracts the shared electrons in H_2O much more strongly than does H, so that the shared electrons

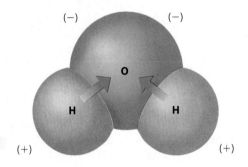

Figure 2.9 A water molecule

spend more time near the O atom than near the H atoms. This unequal sharing of electrons produces what is called a **polar covalent bond.** In a polar covalent bond, the pulling of shared electrons closer to the more electronegative atom makes that atom partially negative and the other atom partially positive. Thus, in H_2O, the O atom actually has a slight negative charge and each H atom a slight positive charge. Because of its V shape and its polar covalent bonds, water is a **polar molecule**—that is, it has an unequal distribution of charges. It is slightly negative at the O end of the molecule (apex of the V) and slightly positive at each of the two H ends.

Web/CD Activity 2G *Nonpolar and Polar Molecules*

? Why is it unlikely that two neighboring water molecules would be arranged like this?

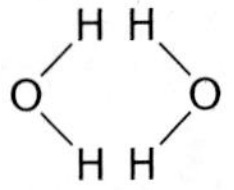

■ The hydrogen atoms of one molecule, with their partial positive charge, would repel the hydrogen atoms of the adjacent molecule.

2.10 Hydrogen bonds are weak bonds important in the chemistry of life

In living organisms, most of the strongest chemical bonds are covalent, linking atoms to form a cell's molecules. But crucial to the functioning of a cell are weaker bonds within and between molecules.

When an H atom is part of a polar covalent bond, its partial positive charge allows it to share attractions with other electronegative atoms, such as O or N. These weak but important bonds are best illustrated with water molecules, as shown in Figure 2.10. The charged regions on each water molecule are electrically attracted to oppositely charged regions on neighboring molecules. Because the positively charged region in this special type of bond is always a hydrogen atom, the bond is called a **hydrogen bond.** As Figure 2.10 shows, the negative (O) pole on each water molecule can form hydrogen bonds (dotted lines) to two H atoms. Thus, each H_2O molecule can hydrogen-bond to as many as four partners.

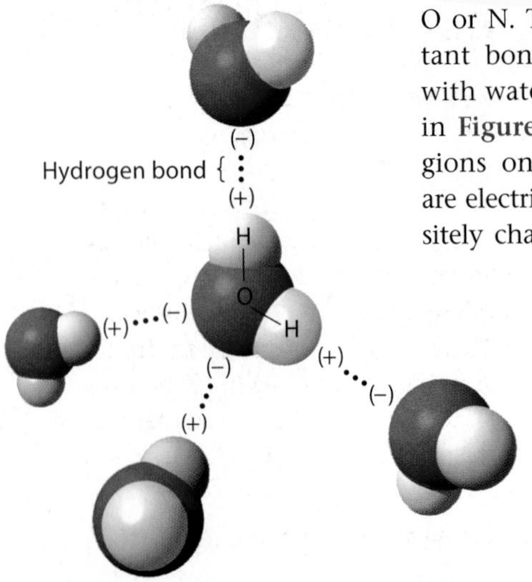

You will learn in Chapter 3 how hydrogen bonds help to create a protein's shape (and thus its function) and hold the two strands of a DNA molecule together. Indeed, these weak bonds are involved in many of the activities of a cell.

In the next few modules, we explore how water's polarity and hydrogen bonds give it unusual properties. Like no other common substance on Earth, water exists in nature in all three physical states: solid, liquid, and gas (water vapor). When astronomers search for signs of extraterrestrial life on distant planets, they look for evidence of water (such as that recently found on Mars). The unique properties of water and its abundance make it possible for life to thrive on Earth.

Web/CD Activity 2H *Water's Polarity and Hydrogen Bonding*

? What enables neighboring water molecules to hydrogen-bond to one another?

Hydrogen bond {

■ The molecules are polar, with the negative end (oxygen end) of one molecule attracted to the positive end (hydrogen end) of its neighbor.

Figure 2.10 Hydrogen bonds between water molecules

2.11 Hydrogen bonds make liquid water cohesive

Hydrogen bonds between molecules of liquid water last for only a few trillionths of a second, yet at any instant, many of the molecules are hydrogen-bonded to others. This tendency of molecules to stick together, called **cohesion**, is much stronger for water than for most other liquids. The cohesion of water is important in the living world. Trees, for example, depend on cohesion to help transport water and nutrients from their roots to their leaves. The evaporation of water from a leaf exerts a pulling force on water within the veins of the leaf. Because of cohesion, the force is relayed through the veins all the way down to the roots. As a result, water rises against the force of gravity.

Related to cohesion is **surface tension,** a measure of how difficult it is to stretch or break the surface of a liquid. Hydrogen bonds give water unusually high surface tension, making it behave as though it were coated with an invisible film. The insect in Figure 2.11, called a water strider, takes advantage of the high surface tension of water. It "strides" across ponds without breaking the surface.

Figure 2.11 The surface tension of water allows a water strider to walk on water

Web/CD Activity 2I *Cohesion of Water*

? In a tall tree, water in thin tubes within the trunk is pulled upward by evaporation from the leaves. What keeps the water molecules at the bottom of the tree moving?

■ Hydrogen bonds hold neighboring water molecules together in the liquid water; this cohesion helps the molecules move against the downward pull of gravity.

2.12 Water's hydrogen bonds moderate temperature

If you have ever burned your finger on a metal pot while waiting for the water in it to boil, you know that water heats up much more slowly than metal. In fact, because of hydrogen bonding, water has a better ability to resist temperature change than most other substances. Because of this property, Earth's giant water supply moderates temperatures, keeping them within limits that permit life.

Temperature and heat are related, but different. A swimmer crossing San Francisco Bay has a higher temperature than the water, but the bay contains far more heat because of its immense volume. **Heat** is the amount of energy associated with the movement of atoms and molecules in a body of matter. **Temperature** measures the intensity of heat—that is, the *average* speed of molecules rather than the *total* amount of heat energy in a body of matter.

When water is heated, the heat energy first disrupts hydrogen bonds and then makes water molecules move faster. Because heat is absorbed as the bonds break, water absorbs and stores a large amount of heat while warming up only a few degrees. Conversely, when water is cooled, more hydrogen bonds form. Heat energy is released when the bonds form, slowing the cooling process.

A large body of water can store a huge amount of heat from the sun during warm periods. At cooler times, heat given off from the gradually cooling water can warm the air. That's why coastal areas generally have milder climates than inland regions. Water's resistance to temperature change also stabilizes ocean temperatures, creating a favorable environment for marine life. And at 66% of your body weight, water helps moderate your internal temperature.

Hydrogen bonds also decrease water's tendency to evaporate, or vaporize. Liquids vaporize into a gas when some of their molecules move fast enough to overcome the attractions that keep the molecules close together. Water must absorb an unusually large amount of heat in order to vaporize because its hydrogen bonds tend to hold the molecules in place, giving water a high boiling point (100°C).

Another way water moderates temperatures is by evaporative cooling. When a substance evaporates, the surface of the liquid remaining behind cools down as the molecules with the greatest energy (the "hottest" ones) leave. It's as if the ten fastest runners on the track team left school, lowering the average speed of the remaining team. Evaporative cooling helps prevent land-dwelling organisms from overheating. Evaporation from a plant's leaves keeps them from becoming too warm in the sun, just as sweating helps to dissipate our excess body heat (**Figure 2.12**). On a much larger scale, the evaporation of surface waters cools tropical seas.

Figure 2.12
Evaporative cooling

? Explain the popular adage "It's not the heat, it's the humidity."

■ High humidity hampers cooling by slowing the evaporation of sweat.

2.13 Ice is less dense than liquid water

Water exists in nature as a gas (water vapor), liquid, and solid. Unlike most substances, water is less dense as a solid than as a liquid. Again, this unusual property is due to hydrogen bonds.

As water freezes, each molecule forms stable hydrogen bonds with four neighbors, holding them at "arm's length" and creating a three-dimensional crystal. In Figure 2.13, compare the spaciously arranged molecules in the ice crystal with the more tightly packed molecules in the liquid water. The ice crystal has fewer molecules than an equal volume of liquid water. Therefore, ice is less dense and floats on top of liquid water. If ice sank, then eventually all ponds, lakes, and even oceans would freeze solid. Instead, floating ice insulates the water below from colder air above, allowing life to persist under the frozen surface.

? Explain how the freezing of water can crack boulders.

■ Water expands as it freezes because the water molecules become spaced farther apart in forming ice crystals. When there is water in a crevice of a boulder, expansion of the water due to freezing may crack the rock.

Figure 2.13
Water molecule arrangement in ice versus water

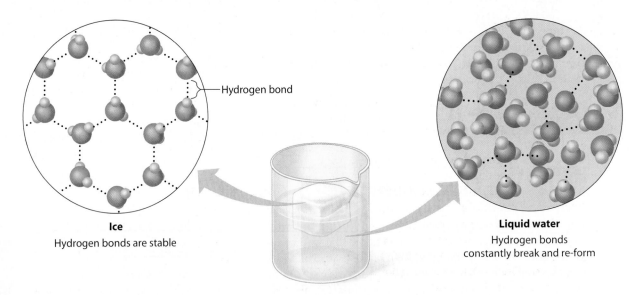

Hydrogen bond

Ice
Hydrogen bonds are stable

Liquid water
Hydrogen bonds constantly break and re-form

2.14 Water is the solvent of life

A **solution** is a liquid consisting of a uniform mixture of two or more substances. The dissolving agent is the **solvent**, and a substance that is dissolved is a **solute**. When water is the solvent, the result is called an **aqueous solution** (from the Latin *aqua*, water). As the solvent inside all cells, in blood, and in plant sap, water dissolves an enormous variety of solutes necessary for life.

Water's versatility as a solvent results from the polarity of its molecules. Figure 2.14 shows how a crystal of table salt dissolves in water. The sodium and chloride ions at the surface of the crystal have affinities for different parts of the water molecules. The positive sodium ions (Na^+, yellow) attract the electronegative regions (oxygen, red) of the water molecules. The negative chloride ions (Cl^-, green) attract the positively charged hydrogen regions (gray). As a result, H_2O molecules surround and separate individual sodium and chloride ions, dissolving the crystal in the process.

A compound doesn't need to be ionic to dissolve in water. Polar molecules such as sugars and many proteins dissolve as they become surrounded by polar water molecules.

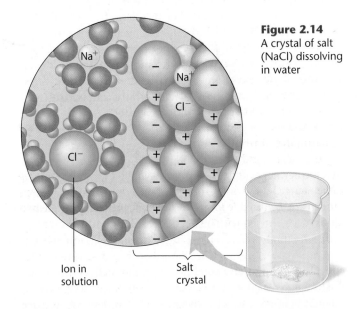

Figure 2.14
A crystal of salt (NaCl) dissolving in water

Ion in solution

Salt crystal

? Why are blood and most other biological fluids classified as aqueous solutions?

■ The solvent is water.

2.15 The chemistry of life is sensitive to acidic and basic conditions

In aqueous solutions, a very small percentage of the water molecules actually break apart (dissociate) into ions. The ions formed are called hydrogen ions (H^+) and hydroxide ions (OH^-). Hydrogen and hydroxide ions are very reactive, and the proper balance of these ions is critical for the proper functioning of chemical processes within an organism.

Some chemical compounds contribute additional H^+ to an aqueous solution, while others remove H^+ from it. A compound that donates hydrogen ions to solutions is called an **acid**. One example of a strong acid is hydrochloric acid (HCl), the acid in your stomach. In solution, HCl dissociates completely into H^+ and Cl^-. An acidic solution has a higher concentration of H^+ than OH^-.

A **base** is a compound that accepts hydrogen ions and removes them from solution. Some bases, such as sodium hydroxide (NaOH), do this by donating OH^-; the OH^- combines with H^+ to form H_2O, thus reducing the H^+ concentration. The more basic a solution, the higher its OH^- concentration and the lower its H^+ concentration.

We use the **pH scale** to describe how acidic or basic a solution is (pH stands for potential of hydrogen). As shown in Figure 2.15, the scale ranges from 0 (most acidic) to 14 (most basic). Each pH unit represents a tenfold change in the concentration of H^+ in a solution. For example, lemon juice at pH 2 has 10 times more H^+ than an equal amount of grapefruit juice at pH 3 and 100 times more H^+ than tomato juice at pH 4.

Pure water and aqueous solutions that are neither acidic nor basic are said to be neutral; they have a pH of 7. They do contain some hydrogen and hydroxide ions, but the concentrations of the two kinds of ions are equal. The pH of the solution inside most living cells is close to 7. Even a slight change in pH can be harmful because the molecules in cells are extremely sensitive to H^+ and OH^- concentrations. However, biological fluids contain **buffers**, substances that resist changes in pH by accepting H^+ when it is in excess and donating H^+ when it is depleted. For example, buffers normally maintain the pH of human blood very close to 7.4. A person cannot survive for more than a few minutes if the blood pH drops to 7 or rises to 7.8.

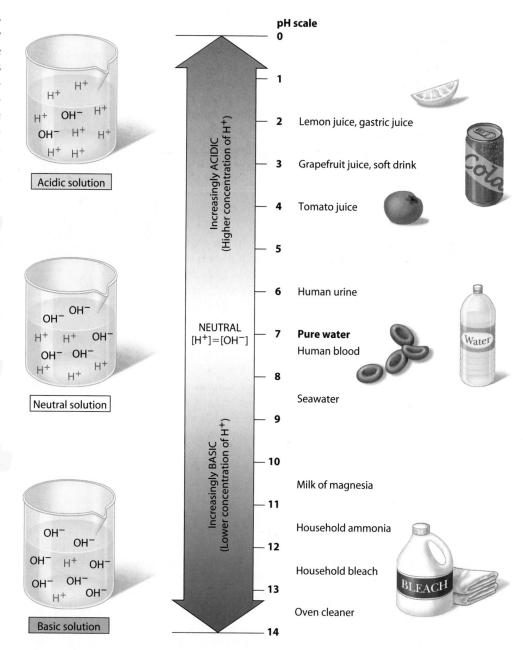

Figure 2.15 The pH scale

The environmental damage caused by an unfavorable pH is discussed in the next module on acid precipitation.

Web/CD Activity 2J *Acids, Bases, and pH*

? Compared to a basic solution at pH 9, the same volume of an acidic solution at pH 4 has ___ times more hydrogen ions (H^+).

100,000 ■

2.16 Acid precipitation threatens the environment

Imagine arriving for a long-awaited vacation at a mountain lake only to discover that since your last visit a few years ago, all fish and other forms of life in the lake have perished because of increased acidity of the water. Over the past quarter century, thousands of lakes in North America, Europe, and Asia have suffered that fate. This problem is due primarily to **acid precipitation**, defined as rain, snow or fog with a pH lower (more acidic) than 5.6, the pH of uncontaminated rain.

Acid precipitation results mainly from the presence in the air of sulfur oxides and nitrogen oxides, air-polluting compounds composed of oxygen combined with sulfur or nitrogen. These oxides react with water vapor in the air to form sulfuric and nitric acids, which fall to earth in rain or snow. Rain with a pH between 2 and 3—more acidic than vinegar—has been recorded in the eastern United States. Acid fog of pH 1.7, approaching that of the digestive juices in the human stomach, has been recorded downwind from Los Angeles.

Sulfur and nitrogen oxides arise mostly from the burning of fossil fuels (coal, oil, and gas) in factories and automobiles. Electrical power plants that burn coal produce more of these pollutants than any other single source. Ironically, the tall smokestacks built to reduce local pollution by dispersing factory exhaust help spread airborne acids. Winds carry the pollutants away, and acid rain may fall thousands of miles from industrial centers.

The effect of acid in lakes and streams is most pronounced in the spring, as snow begins to melt. The surface snow melts first, drains down, and sends much of the acid that has accumulated over the winter into lakes and streams all at once. Early meltwater often has a pH as low as 3, and this acid surge hits when fish and other forms of aquatic life are producing eggs and young, which are especially vulnerable to acidic conditions. Strong acidity can break down the molecules of living organisms or interfere with the essential chemical processes of life.

Figure 2.16B Damage to a statue in Germany between 1908 and 1968

There is evidence that acid precipitation has taken a toll on some North American forests and is contributing to the decline of European forests (Figure 2.16A). Acid precipitation falling on land washes away certain mineral ions, such as calcium and magnesium, that ordinarily help buffer the soil solution and are essential nutrients for plant growth. At the same time, minerals such as aluminum reach toxic concentrations. However, careful studies over the past decade seem to show that the vast majority of North American forests are not suffering substantially from acid precipitation. In cities, meanwhile, the corrosive effect of acid in the air is obvious in the worn surfaces of buildings and statues, as you can see in Figure 2.16B.

Many questions remain. We do not know for sure what the long-term effects of acid precipitation may be on plants and soils. Nor do we know much about the effects of airborne acids on terrestrial animals, including humans.

As with most environmental issues, there are no easy solutions to the acid precipitation problem. There is some hopeful news, however. In the United States, Canada, and Europe, emissions of sulfur oxides have declined significantly in recent decades, causing a decrease in acid precipitation. Laws that require reductions in emissions are thus already helping to alleviate the problem. Continued progress can come from the actions of educated citizens who understand the crucial role that water plays in the fitness of the environment for life on Earth.

Web/CD Thinking as a Scientist *Connection: How Does Acid Precipitation Affect Trees?*

? What is the relationship between fossil fuel consumption and acid precipitation?

Figure 2.16A The effects of acid precipitation on a fir forest in the Czech Republic

■ Compounds in the exhaust react with water vapor in the atmosphere to form sulfuric and nitric acids.

2.17 Chemical reactions change the composition of matter

The basic chemistry of life has an overriding theme: The structure of atoms and molecules determines the way they behave. As we have seen, the chemical properties of an atom are determined by the number and arrangement of its subatomic particles, particularly its electrons. Other properties emerge when atoms combine to form molecules and when molecules interact to form more complex substances, such as liquid water. Water is a good example, because its unusual properties sustain all life on Earth.

Hydrogen and oxygen can react to form water:

$$2\,H_2 + O_2 \rightarrow 2\,H_2O$$

This is a **chemical reaction**, a process leading to changes in the composition of matter. In this particular case, two molecules of hydrogen ($2\,H_2$) react with one molecule of oxygen (O_2) to give two molecules of water ($2\,H_2O$). The arrow indicates the conversion of the starting materials, called the **reactants** (H_2 and O_2), to the resulting **product** (H_2O). Notice that the same *numbers* of hydrogen and oxygen atoms appear on the left and right sides of the arrow, although they are grouped differently. Chemical reactions do not create or destroy matter; they only rearrange it in various ways. Chemical reactions involve the making and breaking of chemical bonds. In the example above, the covalent bonds holding hydrogen atoms together in H_2 and those holding oxygen atoms together in O_2 are broken, and new bonds are formed to yield the H_2O product molecules (**Figure 2.17A**).

Organisms cannot make water from H_2 and O_2, but they do carry out a great number of chemical reactions that re-arrange matter in significant ways. Let's examine one that relates to human nutrition. Much of the color of yellow and orange fruits and vegetables comes from molecules of beta-carotene and related compounds. Shown in **Figure 2.17B**, beta-carotene is also abundant in many green vegetables and is sold in concentrated form as a food supplement. This substance is important in our diet as a source of vitamin A, which is essential for normal vision and healthy skin. You saw in Module 2.2 that vitamin A can be added to cereals as a nutritional supplement. But our cells can also make vitamin A from beta-carotene in the following way:

$$\underset{\text{Beta-carotene}}{C_{40}H_{56}} + O_2 + 4\,H \rightarrow \underset{\text{Vitamin A}}{2\,C_{20}H_{30}O}$$

Although there are many atoms in beta-carotene and vitamin A, the chemical reaction that converts one to the other is essentially a simple one. Two molecules of vitamin A are made from each beta-carotene molecule by splitting the beta-carotene molecule in half. Notice that beta-carotene has 40 carbon (C) atoms, whereas each vitamin A molecule has 20 carbons. The red arrow in Figure 2.17B shows where beta-carotene is split. The other reactants are a molecule of O_2 and 4 H atoms contributed by other molecules in the cell. If you count up all the atoms, you will see that the same number of each type appears on each side of the reaction.

The conversion of beta-carotene to vitamin A is only one example of the thousands of chemical reactions routinely carried out in living cells. Like most of these reactions, our example involved compounds of the element carbon. We look at the carbon compounds of cells in more detail in Chapter 3.

? Fill in the blanks with the correct numbers in the following chemical process:

$$C_6H_{12}O_6 + \_\,O_2 \rightarrow \_\,CO_2 + \_\,H_2O$$

■ $C_6H_{12}O_6 + 6\,O_2 \rightarrow 6\,CO_2 + 6\,H_2O$

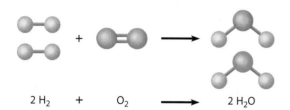

Figure 2.17A Breaking and making of bonds in a chemical reaction

$$2\,H_2 \quad + \quad O_2 \longrightarrow 2\,H_2O$$

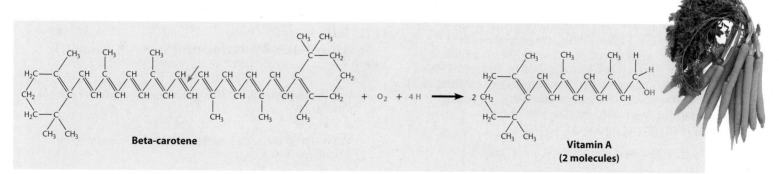

Figure 2.17B Chemical reaction converting beta-carotene to vitamin A

CHAPTER REVIEW

Reviewing the Concepts

Elements, Atoms, and Molecules (2.1–2.10)

Elements. Of the 25 chemical elements essential to life, carbon, hydrogen, oxygen, and nitrogen make up the bulk of living matter **(2.1)**. Trace elements are essential to health and may be added to food or water **(2.2)**. Chemical elements combine in fixed ratios to form compounds **(2.3)**.

 Atoms. Atoms, the smallest unit of an element, consist of protons and neutrons (found in the nucleus) and electrons (arranged in electron shells). Atoms of an element with varying numbers of neutrons are called isotopes **(2.4)**. Though potentially harmful, radioactive isotopes can be useful in medicine and science **(2.5)**.

Chemical bonds. An atom whose outermost electron shell is not full tends to interact with other atoms and gain, lose, or share electrons, resulting in attractions called chemical bonds **(2.6)**. Electron gain and loss create charged atoms, called ions. Oppositely charged ions attract one another in ionic bonds **(2.7)**. In covalent bonds, atoms complete their outer electron shells by sharing electrons **(2.8)**.

Polarity. A molecule is nonpolar when its covalently bonded atoms share electrons equally. In a polar molecule, such as water, electrons are not shared equally **(2.9)**. The slightly positively-charged H atoms in one water molecule may be attracted to the partial negative charge of neighboring O or N atoms, forming weak but important hydrogen bonds. Polarity and hydrogen bonds give water some unique properties **(2.10)**.

Water's Life-Supporting Properties (2.11–2.16)

Hydrogen bonds make water molecules cohesive, creating surface tension and allowing water to move from plant roots to leaves **(2.11)**. Water's ability to store heat moderates body temperature and climate. It takes a lot of energy to disrupt hydrogen bonds, so water is able to absorb a great deal of heat energy without a large increase in temperature. As water cools, heat is released as hydrogen bonds form. A water molecule takes energy with it when it evaporates, leading to evaporative cooling **(2.12)**. Hydrogen bonds hold molecules in ice farther apart than in liquid water. Floating ice protects lakes and oceans from freezing solid **(2.13)**.

Solutions. Polar or charged solutes dissolve when water molecules surround them, forming aqueous solutions **(2.14)**. A compound that releases H^+ in solution is an acid, and one that accepts H^+ is a base. Acidity is measured on the pH scale, from 0 (most acidic) to 14 (most basic). The pH of most cells is close to 7 (neutral) and kept that way by buffers, substances that resist pH change **(2.15)**. Some ecosystems are threatened by acid precipitation, which is formed when air pollutants from burning fossil fuels combine with water vapor in the air to form sulfuric and nitric acids **(2.16)**.

Chemical Reactions (2.17)

Matter is rearranged in chemical reactions as bonds are broken and formed to convert reactants to products **(2.17)**.

Connecting the Concepts

1. Fill in the blanks in this concept map to help you tie together the key concepts concerning elements, atoms, and molecules.

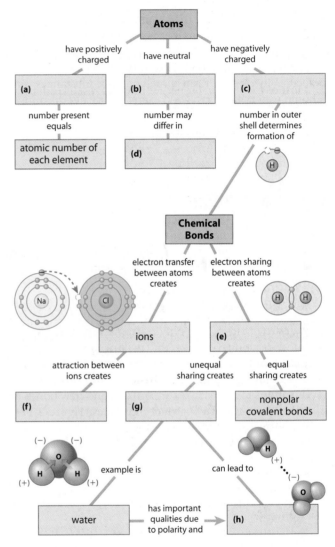

2. On a separate piece of paper, create a concept map to organize your understanding of the life-supporting properties of water. A sample map is in the answer section, but the value of this exercise is in the thinking and integrating you must do to create your own map.

Testing Your Knowledge

Multiple Choice

3. Your body contains the smallest amount of which of the following?
   a. nitrogen
   b. phosphorus
   c. carbon
   d. oxygen
   e. hydrogen

4. Changing the ____ would change it into an atom of a different element.
   a. number of electrons surrounding the nucleus of an atom
   b. number of bonds formed by an atom
   c. number of protons in the nucleus of an atom
   d. electrical charge of an atom
   e. number of neutrons in the nucleus of an atom

5. A solution at pH 6 contains ____ than the same amount of solution at pH 8.
   a. 2 times more H^+
   b. 4 times more H^+
   c. 100 times more H^+
   d. 4 times less H^+
   e. 100 times less H^+

6. Most of the unique properties of water result from the fact that water molecules
   a. are very small.
   b. are held together by covalent bonds.
   c. easily separate from one another.
   d. are constantly in motion.
   e. are polar and form hydrogen bonds.

7. A sulfur atom has 6 electrons in its outer shell. As a result, it forms ____ covalent bonds with other atoms. (*Explain your answer.*)
   a. 2
   b. 3
   c. 4
   d. 6
   e. 8

8. A can of cola consists mostly of sugar dissolved in water, with some carbon dioxide gas that makes it fizzy and makes the pH less than 7. In chemical terms, you could say that cola is an aqueous solution, where water is the ____, sugar is a ____, and carbon dioxide makes the solution ____.
   a. solvent . . . solute . . . basic
   b. solute . . . solvent . . . basic
   c. solvent . . . solute . . . acidic
   d. solute . . . solvent . . . acidic
   e. not enough information to say

9. Which of the following is *not* a chemical reaction?
   a. Sugar and oxygen gas combine to form carbon dioxide and water.
   b. Sodium metal and chlorine gas unite to form sodium chloride.
   c. Hydrogen gas combines with oxygen gas to form liquid water.
   d. Solid ice melts to form liquid water.
   e. Sulfur dioxide and water vapor join to form sulfuric acid.

True/False (*Change false statements to make them true.*)

10. Table salt, water, and carbon are compounds.

11. The smallest particle of an element is a molecule.

12. A bathtub full of lukewarm water may hold more heat than a teakettle full of boiling water.

13. If the atoms in a molecule share electrons equally, the molecule is said to be nonpolar.

14. Ice floats because water molecules in ice are more tightly packed than in liquid water.

15. Atoms in a water molecule are held together by the sharing of electrons.

16. Most acid precipitation results from the presence of pollutants from aerosol cans and air conditioners.

17. An atom that has gained or lost electrons is called an ion.

Describing, Comparing, and Explaining

18. Make a sketch that shows how water molecules hydrogen-bond with one another. Why do water molecules form hydrogen bonds? What unique properties of water result from water's tendency to form hydrogen bonds?

19. Describe two ways in which the water in your body helps stabilize your body temperature.

20. Compare covalent and ionic bonds.

21. What is an acid? A base? How is the acidity of a solution described?

Applying the Concepts

22. The diagram below shows the arrangement of electrons around the nucleus of a fluorine atom (left) and a potassium atom (right). What kind of bond do you think would form between these two atoms?

Fluorine atom   Potassium atom

23. Animals obtain energy through a series of chemical reactions in which sugar ($C_6H_{12}O_6$) and oxygen gas (O_2) are reactants. This process produces water (H_2O) and carbon dioxide (CO_2) as waste products. How might you use a radioactive isotope to find out whether the oxygen in CO_2 comes from sugar or oxygen gas?

24. One solution to the problem of acid precipitation is to use nuclear energy to produce electricity. Development of nuclear power in the United States virtually stopped after the accident at the Three Mile Island power plant in Pennsylvania in 1979. But proponents of nuclear power contend that accelerated development of nuclear energy is the answer to U.S. energy needs and tightened pollution standards. Besides reducing acid precipitation, what are other potential benefits of nuclear power? What are its possible costs and dangers? Do you think we ought to pursue the development of nuclear power? Why or why not? If a power plant were to be built near your home, would you prefer it to be a nuclear or coal-fired plant? Why?

Answers to all questions can be found in Appendix 3.

For study help and Activities, go to campbellbiology.com or the student CD-ROM.

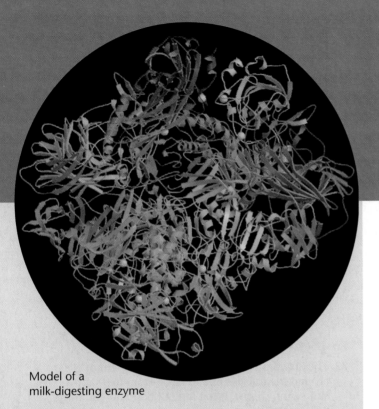

Model of a
milk-digesting enzyme

INTRODUCTION TO ORGANIC COMPOUNDS

3.1 Life's molecular diversity is based on the properties of carbon

3.2 Functional groups help determine the properties of organic compounds

3.3 Cells make a huge number of large molecules from a small set of small molecules

CARBOHYDRATES

3.4 Monosaccharides are the simplest carbohydrates

3.5 Cells link two single sugars to form disaccharides

3.6 How sweet is sweet?

3.7 Polysaccharides are long chains of sugar units

LIPIDS

3.8 Fats are lipids that are mostly energy-storage molecules

3.9 Phospholipids, waxes, and steroids are lipids with a variety of functions

3.10 Anabolic steroids and related substances pose health risks

PROTEINS

3.11 Proteins are essential to the structures and activities of life

3.12 Proteins are made from amino acids linked by peptide bonds

3.13 A protein's specific shape determines its function

3.14 A protein's shape depends on four levels of structure

3.15 Linus Pauling contributed to our understanding of the chemistry of life

NUCLEIC ACIDS

3.16 Nucleic acids are information-rich polymers of nucleotides

CHAPTER 3

Got Lactose?

IS A BIG GLASS OF MILK a way to a healthy diet—or an upset stomach? For much of the world's population, the answer is often the latter. Most of the world's people cannot easily digest milk-based foods. Those who can happily devour ice cream and cheese, scientists have found, are in the minority, benefiting from a lucky connection between human evolution and the major molecules that shape life.

Milk and other dairy products have long been recognized as highly nutritious foods, rich in protein and minerals necessary for good teeth and strong bones. But for millions of people, those health benefits come with a heavy dose of digestive discomfort. Such people suffer from lactose intolerance, or the inability to properly break down lactose, the main sugar found in milk.

For those with lactose intolerance, the problem starts once lactose passes through the stomach and enters the small intestine. There, to absorb the sugar, digestive cells need to secrete an enzyme called lactase, which is neces-

Lactose intolerance is actually the human norm

sary for the breakdown of lactose. (An enzyme, such as the one shown above, is a protein that speeds up a specific chemical reaction.) Those with lactose intolerance, however, produce insufficient amounts of the enzyme. Without enough lactase, lactose cannot be properly digested or absorbed, leading to uncomfortable symptoms that include nausea, cramps, and bloating.

Scientists who study the problem have known for many years that the condition is closely tied to age and heritage. When it comes to milk, most people are born equal. Nearly everyone produces enough lactase in infancy to digest the lactose in breast milk and dairy products. Thus, milk—high in protein, fats, and sugars—provides excellent nourishment for infants. But after the

The Molecules of Cells

age of 2, lactase levels start to decline in most of the world's populations, as if the lactase-producing mechanism in the body is being slowly shut down. In the United States, researchers have noted, as many as 75% of African-Americans and Native Americans and 90% of Asian-Americans are lactase-deficient once they reach their teenage years. People of European descent are the only group with relatively low levels of lactose intolerance; only about 15% are believed to suffer from the problem.

Researchers suspected a genetic link, but it took a study involving people from around the world to confirm one. In 2002, U.S. and Finnish scientists completed a study of the genes of 196 lactose-intolerant adults of African, Asian, and European descent. They determined that lactose intolerance is actually the human norm. It is "lactose tolerance" that represents a relatively recent mutation in the human genome.

The ability to make lactase into adulthood is concentrated in people of northern European descent, and the researchers speculated that lactose tolerance became widespread among this group because it offered a survival advantage. A look at a map quickly reveals what that advantage might have been. In northern Europe's relatively cold climate, only one harvest is possible a year. For earlier humans, that meant nonplant sources of food often had to be found. With milk and other dairy products at hand year-round, the lactose tolerance mutation would help keep people alive through a long, cold winter.

What works for this lactose-tolerant minority, however, can mean more complicated eating for the rest of the world. There is no treatment for the underlying cause of lactase deficiency, and lactose is now used widely in everything from bottled salad dressings and lunch meat to prescription drugs. But the condition's uncomfortable symptoms can be controlled through diet. People with lactose intolerance must watch what they eat and how much milk-based food they consume. Some avoid foods that contain lactose altogether, and doctors and nutritionists advise such people to take calcium supplements and vitamins. Others turn to substitutes: In many Asian cultures, for example, beverages have long been made from soy or rice instead of milk. Other foods are prepared from milk that has been pre-treated with lactase. Lactase, in pill form, can also be taken along with food to ease digestion.

Lactose intolerance, with its interplay between genes and milk sugar, illustrates the importance of biological molecules to the functioning of living cells and to human health. In people who easily digest milk, lactose, a sugar, is broken down by lactase, a protein, which is produced by a gene, made of DNA, a nucleic acid. If the gene for lactase production is not active, lactase is not present. And the presence of lactase can mean the difference between delight and discomfort when someone contemplates an ice cream sundae. Such molecular interactions, repeated in countless variations, drive all biological processes. In this chapter, we'll further explore the structure and function of these and other molecules that are essential to life. ■ ■ ■

3.1   Life's molecular diversity is based on the properties of carbon

Almost all the molecules a cell makes are composed of carbon atoms bonded to one another and to atoms of other elements. Carbon is unparalleled in its ability to form large, diverse molecules. Next to water, compounds containing carbon are the most common substances in living organisms.

Carbon-based molecules are called **organic compounds.** Well over 2 million organic compounds are known, and chemists continue to identify and make new ones. As we discussed in Chapter 2, an element's chemical properties are determined by the electrons in the outermost shell of its atoms. A carbon atom has 4 outer electrons in a shell that holds 8. Carbon completes its outer shell by sharing electrons with other atoms in four covalent bonds (see Module 2.8). Thus, each carbon atom is a connecting point from which a molecule can branch in up to four directions.

At the top of **Figure 3.1** are three illustrations of methane, one of the simplest organic molecules. The structural formula shows that covalent bonds link four hydrogen atoms to the carbon atom. Each of the four lines in the formula represents a pair of shared electrons. The two models show that methane is three-dimensional, with the space-filling version portraying its overall shape more accurately. The ball-and-stick model indicates the angles of the molecule's covalent bonds (the gray "sticks"). The four hydrogen atoms of methane are at the corners of an imaginary tetrahedron (an object with four triangular sides). This tetrahedral shape occurs wherever a carbon atom participates in four single bonds. Different bond angles occur where carbon atoms form double bonds. Large organic molecules can have very elaborate shapes. And the shape of a molecule usually helps determine its function.

Methane and other compounds composed of only carbon and hydrogen are called **hydrocarbons.** The figure illustrates some variations in hydrocarbon structure. Carbon atoms, with attached hydrogens, can bond together in chains of various lengths to form compounds such as ethane or propane, a gas used as household fuel. The chain of carbon atoms in an organic molecule is called a **carbon skeleton.** Carbon skeletons can be unbranched, as in butane, or branched, as in isobutane. Carbon skeletons may also include double bonds, as in 1-butene and 2-butene. These two compounds have the same molecular formula, C_4H_8, but differ in the position of their double bond. Compounds with the same formula but different structures are called **isomers.** Their different shapes result in unique properties. Cyclohexane and benzene, the only hydrocarbons in the figure that are liquids rather than gases, are examples of carbon skeletons arranged in rings.

A general characteristic of all hydrocarbons is that they are nonpolar molecules, owing to their nonpolar C—H bonds.

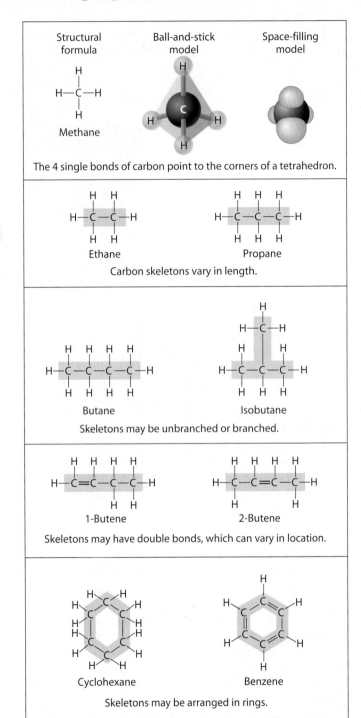

Figure 3.1   Variations in carbon skeletons

[?] Why do isomers, which have the same formula (same numbers of atoms), have different properties?

■ Isomers have different structures, or shapes, and the shape of a molecule usually helps determine the way it functions—how it interacts with other molecules.

3.2 Functional groups help determine the properties of organic compounds

The unique properties of an organic compound depend not only on the size and shape of its carbon skeleton but also on the atoms that are attached to the skeleton. In an organic molecule, the groups of atoms that usually participate in chemical reactions are called **functional groups.**

Table 3.2 illustrates five functional groups important in the chemistry of life. All of these functional groups are polar, because their oxygen or nitrogen atoms exert a strong pull on shared electrons. This polarity tends to make the compounds containing these groups **hydrophilic** (water-loving) and therefore soluble in water—a necessary condition for their roles in water-based life.

A **hydroxyl group** consists of a hydrogen atom bonded to an oxygen atom, which in turn is bonded to the carbon skeleton of a molecule. Ethanol, shown in the table, and other organic compounds containing hydroxyl groups are called alcohols.

A **carbonyl group** is a carbon atom linked by a double bond to an oxygen atom. If the carbon atom of the carbonyl group is at the end of a carbon skeleton, the compound is called an aldehyde. A compound whose carbonyl group is within a carbon chain is called a ketone. Sugars typically contain a carbonyl group and several hydroxyl groups.

A **carboxyl group** consists of a carbon double-bonded to an oxygen and also bonded to a hydroxyl group. The carboxyl group acts as an acid by contributing an H^+ to a solution (see Module 2.15) and becoming ionized. Compounds with carboxyl groups are called **carboxylic acids.** Acetic acid, shown in the table, gives vinegar its sour taste.

An **amino group** is composed of a nitrogen atom bonded to two hydrogen atoms. It acts as a base by picking up an H^+ from a solution. Organic compounds with an amino group are called **amines.** Amino acids, the building blocks of proteins, contain both a carboxyl and an amino group. Under cellular conditions, both groups are usually ionized.

A **phosphate group** consists of a phophorus atom bonded to four oxygen atoms. It is usually ionized and attached to the carbon skeleton by one of its oxygen atoms. This structure is abbreviated as Ⓟ in this text. Compounds with phosphate groups are called organic phosphates and are often involved in energy transfers. Three phosphate groups are found in the energy compound ATP, or adenosine triphosphate.

Figure 3.2 shows compounds that are female and male sex hormones in humans and other vertebrates. They differ mainly in the functional groups attached to their carbon skeletons. The different actions of these two molecules on many targets in the body help produce the contrasting features of females and males. Thus, sexual differences have their biological basis in variations in functional groups.

Web/CD Activity 3B *Functional Groups*

? Identify which of the following functional groups do *not* contain carbon: carboxyl, hydroxyl, amino, carbonyl.

■ The amino and hydroxyl groups

TABLE 3.2 FUNCTIONAL GROUPS OF ORGANIC COMPOUNDS

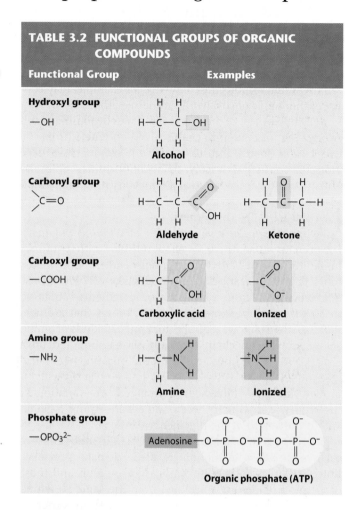

Functional Group — Examples

Hydroxyl group —OH / Alcohol

Carbonyl group >C=O / Aldehyde / Ketone

Carboxyl group —COOH / Carboxylic acid / Ionized

Amino group —NH₂ / Amine / Ionized

Phosphate group —OPO₃²⁻ / Adenosine / Organic phosphate (ATP)

Female lion

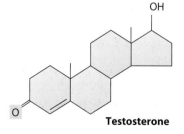

Male lion

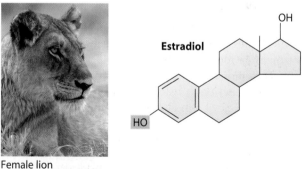

Estradiol

Testosterone

Figure 3.2 Differences in the functional groups of male and female sex hormones (carbons and their attached hydrogens omitted)

3.3 Cells make a huge number of large molecules from a small set of small molecules

The four main classes of large biological molecules are carbohydrates, lipids, proteins, and nucleic acids. On a molecular scale, many of these molecules are gigantic; in fact, biologists call them **macromolecules.** Proteins, for example, may consist of thousands of covalently connected atoms.

Cells make most of their large molecules by joining smaller organic molecules into chains called **polymers** (from the Greek *polys,* many, and *meros,* part). A polymer is a large molecule consisting of many identical or similar molecular units strung together, much as a train consists of many individual cars. The units that serve as the building blocks of polymers are called **monomers.**

Living cells make a vast number of different polymers. For proteins alone, there are about a trillion different kinds in nature, and the variety is potentially endless. One of life's most remarkable features is that a cell makes all of its diverse macromolecules from a small list of ingredients—about 40 to 50 common monomers and a few others that are rare. Proteins, for example, are built from only 20 kinds of amino acids, arranged in chains typically several hundred units long. DNA is built from just four kinds of monomers called nucleotides. As we will see, the key to the great diversity of protein and DNA molecules is arrangement—variation in the sequence in which the monomers are strung together and in the length of the polymer.

The variety in polymers accounts for the uniqueness of each organism. The monomers used to make polymers, however, are essentially universal. Your proteins and those of a tree or spider are assembled from the same 20 amino acids; the amino acids are just arranged in different sequences. Life has a simple yet elegant molecular logic: Small molecules common to all organisms are ordered into macromolecules, which vary from species to species and even from individual to individual.

Cells link monomers together to form polymers by a **dehydration reaction,** a reaction that removes a molecule of water. As you can see in **Figure 3.3A**, unlinked monomers have both hydroxyl groups (—OH) and hydrogen atoms (H). For each monomer added to a chain, a water molecule (H_2O) is removed. Notice that two monomers contribute to the H_2O molecule, one monomer (the one at the right end of the short polymer in this example) losing a hydroxyl group and the other monomer losing a hydrogen atom. As this occurs, a new covalent bond forms, linking the two monomers. Dehydration reactions are the same regardless of the type of polymer the cell is producing.

Cells not only make macromolecules, but also have to break them down. For example, food that an organism ingests is often in the form of macromolecules. To digest these substances and make their monomers usable, a cell carries out **hydrolysis.** Essentially the reverse of a dehydration reaction, hydrolysis means to break (*lyse*) with water (*hydro-*), and cells break bonds between monomers by adding water to them, as **Figure 3.3B** shows. In the process,

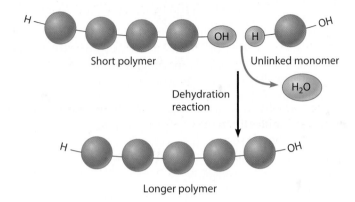

Figure 3.3A   Building a polymer chain

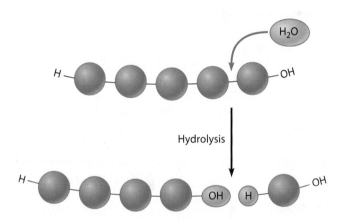

Figure 3.3B   Breaking a polymer chain

a hydrogen joins to one monomer, and a hydroxyl group joins to the adjacent monomer. The lactose-intolerant individuals you learned about in the chapter introduction are unable to hydrolyze such a bond in lactose because they lack the enzyme lactase. Both dehydration reactions and hydrolysis require the help of enzymes.

In the remainder of the chapter, we examine each of the four classes of large molecules in more detail. We will see that polysaccharides, proteins, and nucleic acids are all polymers assembled by dehydration reactions from their respective monomers. Lipids are generally smaller than polysaccharides, proteins, and nucleic acids, and they are not truly polymers. However, some of the largest lipids, the fats, are formed by dehydration reactions from several smaller molecules.

Web/CD Activity 3C *Making and Breaking Polymers*

? How many molecules of water are needed to completely hydrolyze a polymer that is 100 monomers long?

3.4   Monosaccharides are the simplest carbohydrates

The name **carbohydrate** refers to a class of molecules ranging from the small sugar molecules dissolved in soft drinks to large polysaccharides, such as the starch molecules we consume in pasta and potatoes.

The carbohydrate monomers (single-unit sugars) are **monosaccharides** (from the Greek *monos-*, single, and *sacchar*, sugar). The honey shown in **Figure 3.4A** consists mainly of monosaccharides called glucose and fructose. These and other single-unit sugars can be hooked together by a dehydration reaction to form more complex sugars and polysaccharides.

Monosaccharides generally have molecular formulas that are some multiple of CH_2O. For example, the formula for glucose, a common monosaccharide of central importance in the chemistry of life, is $C_6H_{12}O_6$. Its molecular structure, shown in **Figure 3.4B**, has the two trademarks of a sugar: a number of hydroxyl groups (—OH) and a carbonyl group ($>C=O$). The hydroxyl groups make a sugar an alcohol, and the carbonyl group, depending on its location, makes it either an aldehyde or a ketone. As you see in Figure 3.4B, glucose is an aldehyde and fructose is a ketone.

If you count the numbers of different atoms in the fructose molecule in Figure 3.4B, you will find that its molecular formula is $C_6H_{12}O_6$, identical to that of glucose. Thus, glucose and fructose are isomers; they differ only in the arrangement of their atoms (in this case, the positions of the carbonyl groups, highlighted in pink). Seemingly minor differences like this give isomers different properties, such as how they react with other molecules. In this case, the differences also make fructose taste considerably sweeter than glucose.

While the carbon skeletons of both glucose and fructose are six carbon atoms long, other monosaccharides have three to seven carbon atoms. The five-carbon sugars, called pentoses, and the six-carbon sugars, called hexoses, are among the most common. Note that many names for sugars end in the suffix *-ose*.

It is convenient to draw sugars as if their carbon skeletons were linear, but in aqueous solutions, most monosaccharides form rings, as shown for glucose in **Figure 3.4C**. To form the glucose ring, carbon 1 bonds to the oxygen attached to carbon 5. As shown in the middle representation, the ring diagram of glucose and other sugars may be abbreviated by not showing the carbon atoms at the corners of the ring. Also, the bonds in the ring are often drawn with varied thickness, indicating that the ring is a relatively flat structure with atoms and functional groups, such as —OH, extending above and below it. The simplified ring symbol on the right is often used in this book to represent glucose.

Monosaccharides, particularly glucose, are the main fuel molecules for cellular work. Cells also use the carbon skeletons of monosaccharides as raw material for manufacturing other kinds of organic molecules, including amino acids. Monosaccharides that cells do not use immediately are usually incorporated into disaccharides and polysaccharides, as described next.

Web/CD Activity 3D *Models of Glucose*

? Write the formula for a monosaccharide that has three carbons.

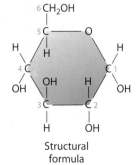

Figure 3.4A   Bees with honey, a mixture of two monosaccharides

$C_3H_6O_3$ ∎

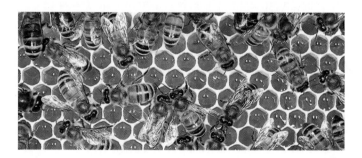

Glucose   Fructose

Figure 3.4B   Structures of glucose and fructose

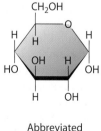

Structural formula   Abbreviated structure

Simplified structure

Figure 3.4C   Three representations of the ring form of glucose

3.5 Cells link two single sugars to form disaccharides

Cells construct a **disaccharide** from two monosaccharides by a dehydration reaction. **Figure 3.5** shows how maltose, also called malt sugar, forms from two glucose monomers. One monomer gives up a hydroxyl group and the other gives up a hydrogen atom from a hydroxyl group. As H_2O forms, an oxygen atom is left, linking the two monomers. Maltose, which is common in germinating seeds, is used in making beer, malted milk shakes, and malted milk ball candy.

The most common disaccharide is sucrose, which is made of a glucose monomer linked to a fructose monomer. The main carbohydrate in plant sap, sucrose nourishes all the parts of the plant. We extract it from the stems of sugarcane or the roots of sugar beets to use as table sugar.

? Lactose, the sugar found in milk, is a disaccharide formed by a dehydration reaction joining glucose and galactose. The formula for both these monosaccharides is $C_6H_{12}O_6$. What is the formula for lactose?

■ $C_{12}H_{22}O_{11}$

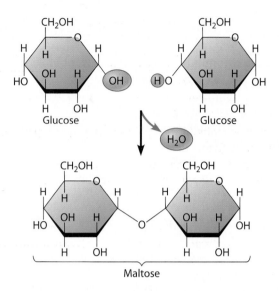

Figure 3.5 Disaccharide formation

3.6 How sweet is sweet?

The human tongue detects five basic taste qualities: the familiar bitter, salty, sour, and sweet and the newly identified umami, a meat-like taste. The taste we describe as sweet has been a familiar and beloved sensation throughout human history. You may be surprised to learn, however, that sugars are not the only substances perceived as sweet; in fact, scientists are discovering that a wide range of chemically different substances can trigger this one simple sensation.

We perceive sweetness when molecules of a substance attach to "sweet" taste receptors on our tongue, triggering a message to the brain, such as "Hmmm . . . sweet!" Many different kinds of molecules can bind to our "sweet" taste receptors, each causing a similar message to be sent. So the glucose and fructose in honey taste sweet—but so does the laboratory-produced compound called aspartame (sold under the trade names Equal and NutraSweet), which is made not of sugars but of two amino acids linked together (a structure you'll see in Module 3.12).

In attempting to quantify the sweetness of a new compound, scientists compare it with table sugar (sucrose), the gold standard of sweeteners. In the lab, volunteers compare the tastes of sucrose solutions with the tastes of solutions containing a test sweetener. For example, only one-fourth the concentration of fructose is needed to attain the sweetness of sucrose, so fructose is rated as four times sweeter than sucrose.

The chemical structure of a compound determines its shape, which in turn determines how well it fits into a taste receptor. Compounds that bind more tightly to "sweet" taste receptors send stronger "sweet" messages to the brain. Because natural sugars vary in their structure, they also vary in their shape and our perception of their sweetness. Some artificial sweeteners are far sweeter than sucrose, probably because these molecules fit more snugly into our "sweet" taste receptors than natural sugars. Neotame, a new artificial sweetener that received FDA approval in 2002, has been rated as 8,000 times sweeter than sucrose (Table 3.6).

Even when adjusted for strength of sweetness, however, sweet substances do not all taste exactly the same. This is because some of them, especially among the sugar substitutes, also bind to other kinds of taste receptors on the tongue. For example, a sweetener may have a bitter aftertaste because it binds to "bitter" receptors as well as "sweet" receptors.

? What makes a chemical taste sweet?

■ It has a shape that binds to "sweet" taste receptors on the tongue, causing a "sweet" message to be sent to the brain.

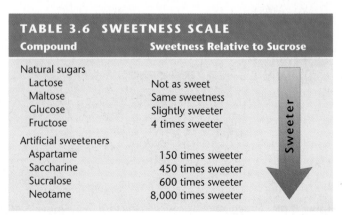

TABLE 3.6 SWEETNESS SCALE	
Compound	**Sweetness Relative to Sucrose**
Natural sugars	
Lactose	Not as sweet
Maltose	Same sweetness
Glucose	Slightly sweeter
Fructose	4 times sweeter
Artificial sweeteners	
Aspartame	150 times sweeter
Saccharine	450 times sweeter
Sucralose	600 times sweeter
Neotame	8,000 times sweeter

Polysaccharides are long chains of sugar units

Polysaccharides are polymers of monosaccharides linked together by dehydration reactions. Some polysaccharides are storage molecules, which cells break down as needed to obtain sugar. **Starch**, a storage polysaccharide in the roots and other tissues of plants, consists entirely of glucose monomers, as shown in **Figure 3.7**. Starch molecules coil into a helical shape because of the angles of the bonds joining their glucose units. A starch helix may be unbranched (as shown in the figure) or branched.

Plant cells often contain starch granules that serve as sugar stockpiles. Plant cells, like animal cells, need sugar for energy and as raw material for building other molecules. They break starch down by hydrolyzing the bonds between the glucose monomers. Humans and most other animals are able to use plant starch as food by hydrolyzing it within their digestive systems. Potatoes and grains, such as wheat, corn, and rice, are the major sources of starch in the human diet.

Animals store excess sugar in the form of another glucose polysaccharide, called **glycogen.** Glycogen is more highly branched than starch. Most of our glycogen is stored as granules in our liver and muscle cells, which hydrolyze the glycogen to release glucose when it is needed.

Some polysaccharides serve as structural compounds. **Cellulose**, the most abundant organic compound on Earth, forms cable-like fibrils in the tough walls that enclose plant cells. Cellulose resembles starch and glycogen in being a polymer of glucose, but its glucose monomers are linked together in a different orientation; they form an unbranched rod, rather than a coil. Arranged parallel to each other, cellulose molecules are joined by hydrogen bonds, forming part of a fibril. In wood, layers of cellulose fibrils combine with other polymers, making a material strong enough to support trees hundreds of feet high.

Unlike the glucose linkages in starch and glycogen, those in cellulose cannot be hydrolyzed by most animals. Therefore, cellulose is not a nutrient for humans, although it does help to keep our digestive system healthy. The cellulose that passes unchanged through our digestive tract is commonly known as "insoluble fiber." Most fresh fruits, vegetables, and grains are rich in fiber. Animals that do derive nutrition from cellulose, such as cows and termites, have cellulose-hydrolyzing microorganisms inhabiting their digestive tracts.

Almost all carbohydrates are hydrophilic owing to the many hydroxyl groups attached to their sugar monomers (see Figure 3.4C). Thus, cotton bath towels, which are mostly cellulose, are quite water absorbent. Next we look at a class of macromolecules that are not soluble in water.

Web/CD Activity 3E *Carbohydrates*

? Compare and contrast starch and cellulose, two plant polysaccharides.

■ Both molecules are polymers of glucose, but the glucose monomers are arranged differently. Starch functions mainly for sugar storage. Cellulose is a structural polysaccharide that is the main material of cell walls.

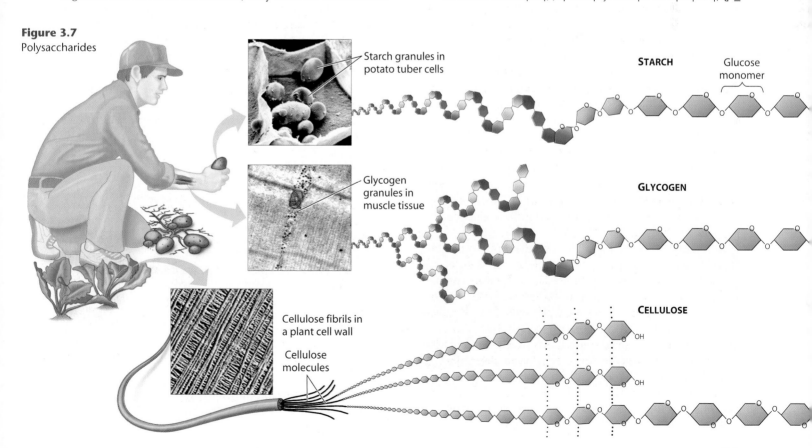

Figure 3.7
Polysaccharides

Starch granules in potato tuber cells

STARCH

Glucose monomer

Glycogen granules in muscle tissue

GLYCOGEN

Cellulose fibrils in a plant cell wall

Cellulose molecules

CELLULOSE

3.8 Fats are lipids that are mostly energy-storage molecules

Lipids are diverse compounds that consist mainly of carbon and hydrogen atoms linked by nonpolar covalent bonds. Being mostly nonpolar, lipid molecules are not attracted to water molecules. In contrast to carbohydrates and most other biological molecules, lipids are **hydrophobic** (water-fearing), You can see the effect of this chemical difference in a bottle of salad dressing: The oil (a type of lipid) separates from the vinegar (which is mostly water). Other oils make the feathers in **Figure 3.8A** repel water, which you see as beads on the surface. By keeping feathers from absorbing water, oils help waterfowl such as ducks stay afloat.

Oils are a liquid fat. A **fat** is a large lipid made from two kinds of smaller molecules: glycerol and fatty acids. Shown at the top in **Figure 3.8B**, glycerol is an alcohol with three carbons, each bearing a hydroxyl group. A fatty acid consists of a carboxyl group and a hydrocarbon chain with about 15 carbon atoms. The carbons in the chain are linked to each other and to hydrogen atoms by nonpolar covalent bonds, making the hydrocarbon chain hydrophobic. The main function of fats is energy storage. A gram of fat stores more than twice as much energy as a gram of a polysaccharide such as starch. This compact energy storage enables a mobile animal, such as a duck or a human, to get around much better than if the animal had to lug its stored energy around in the bulkier form of a carbohydrate. In addition to storing energy, fatty tissue cushions vital organs and insulates the body.

Figure 3.8B shows how one fatty acid molecule can link to a glycerol molecule by a dehydration reaction. **Figure 3.8C** shows a fat molecule, which consists of three fatty acids linked to one glycerol molecule. A synonym for "fat" is thus **triglyceride**, a term you may see on food labels or on medical tests for fat in the blood. The three fatty acids in a fat are often of different kinds, as in our example.

As shown in the third fatty acid in Figure 3.8C, some fatty acids contain double bonds, which cause kinks in the carbon chain. Double bonds prevent a carbon skeleton from bond-

ing to the maximum number of hydrogen atoms. Fatty acids and fats with double bonds are said to be **unsaturated**—that is, having less than the maximum number of hydrogens. Fats with the maximum number of hydrogens are said to be **saturated.** The kinks in unsaturated fats prevent the molecules from packing tightly together and solidifying at room temperature. Corn oil, olive oil, and other vegetable oils are unsaturated fats. When you see "hydrogenated vegetable oils" on a margarine label, it means that unsaturated fats have been converted to saturated fats by adding hydrogen. Unfortunately, hydrogenation also creates *trans* fats, a form of fat that recent research associates with heart disease.

Most plant fats are unsaturated oils, whereas most animal fats are saturated. This is why butter and lard are solids at room temperature. Diets rich in saturated fats and *trans* fats may contribute to cardiovascular disease by promoting a condition called atherosclerosis. In this condition, lipid-containing deposits called plaques build up within the walls of blood vessels, reducing blood flow.

> **?** On a food package, what does "unsaturated fats" mean?

■ Unsaturated fats are fats with some double bonds in the hydrocarbons of their fatty acids. These fats have fewer hydrogen atoms than they would without the double bonds.

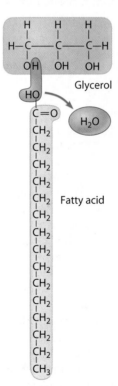

Figure 3.8B A dehydration reaction linking a fatty acid to glycerol

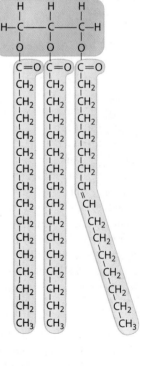

Figure 3.8C A fat molecule

Figure 3.8A Water beading on the naturally oily coating of feathers

3.9 Phospholipids, waxes, and steroids are lipids with a variety of functions

Fats are only one type of lipid important in living organisms. Other lipids are major constituents of cell membranes or perform such vital functions as protecting body surfaces and regulating cellular and body functions.

Phospholipids, a major component of cell membranes, are structurally similar to fats, but they contain the element phosphorus and have only two fatty acids instead of three. Phospholipids are very important biological molecules, as we will see when we discuss cell membranes in Chapter 5.

Waxes consist of one fatty acid linked to an alcohol. They are more hydrophobic than fats, and this characteristic makes waxes effective natural coatings for fruits such as apples and pears. Many animals, especially insects, also have waxy coats that help keep them from drying out.

Steroids are lipids whose carbon skeleton is bent to form four fused rings, as shown in the structural diagram of cholesterol in **Figure 3.9**. All steroids have the same ring pattern: three six-sided rings and one five-sided ring. (The diagram omits the carbons making up the rings and most of the chain and also the hydrogens attached to these carbons.) Cholesterol is a common substance in animal cell mem-

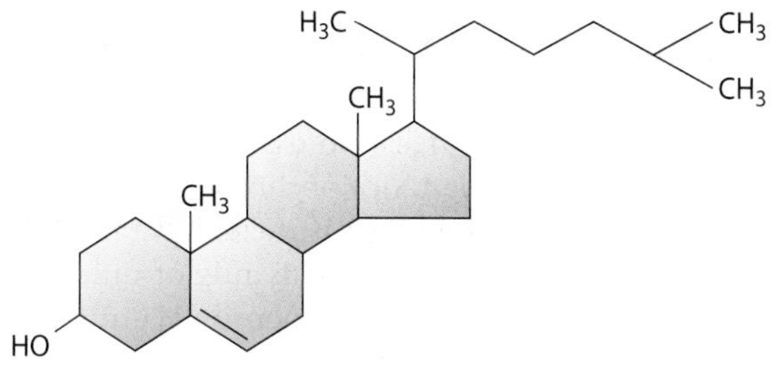

Figure 3.9 Cholesterol, a steroid

branes, and animal cells also use it as a starting material for making other steroids, including the female and male sex hormones. Too much cholesterol in the blood may contribute to atherosclerosis.

Web/CD Activity 3F *Lipids*

> **?** Human sex hormones belong to what family of lipids?

■ Steroids

CONNECTION

3.10 Anabolic steroids pose health risks

Anabolic steroids are synthetic variants of the male hormone testosterone. Testosterone causes a general buildup in muscle and bone mass in males during puberty and maintains masculine traits throughout life. Because anabolic steroids structurally resemble testosterone, they also mimic some of its effects. (Anabolism is the building of substances by the body.) As prescription drugs, anabolic steroids are used to treat general anemia and diseases that destroy body muscle. However, some individuals abuse these drugs, with serious consequences. Overdosing may cause violent mood swings ("steroid rage") and deep depression. The liver may be damaged, leading to cancer. The use of anabolic steroids can also alter cholesterol levels and lead to high blood pressure, increasing the risk of cardiovascular problems. Use of these drugs often makes the body reduce its output of natural male sex hormones, which can cause shrunken testicles, reduced sex drive, infertility, and breast enlargement in men. Use in women has been linked to menstrual cycle disruption and development of masculine characteristics. In teens, bones may stop growing, stunting growth.

Despite the risks associated with steroid use, some athletes use steroids to gain a competitive edge. Meanwhile, sports organizations ban their use, implement drug testing, and penalize violators in an effort to counter an environment in which athletes feel they have to take performance-enhancing drugs just to stay competitive.

In 2003, the discovery of a new designer steroid rocked the sports world. THG, short for tetrahydrogestrinone, is a drug modified to avoid detection in ordinary drug testing. The drug was discovered when a track coach mailed a syringe containing a sample of THG to the U.S. Anti-Doping Agency. With that sample, the agency was able to develop a test that has revealed the substance's use among track and field athletes as well as professional football players, and the International Olympic Committee has begun retesting frozen urine samples from the 2002 winter games. The U.S. Food and Drug Administration declared THG an illegal steroid, and athletic governing bodies from the Olympics to swimming, soccer, football, and baseball organizations now prohibit its use. In 2004, a British sprinter became the first athlete to be penalized for its use, with a permanent exclusion from the Olympics following his positive test for THG. Unscrupulous chemists, as well as dishonest trainers, coaches, and athletes, will continue to try to find ways to cheat to obtain a competitive advantage, while sports authorities struggle to keep athletics fair and protect the health of the athletes.

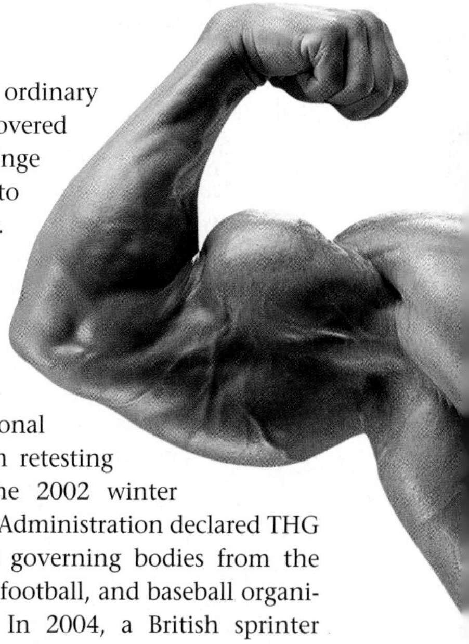

> **?** What kind of carbon ring structure would you find in an anabolic steroid?

■ Four fused carbon rings: three six-sided and one five-sided, as in all steroids

3.11 Proteins are essential to the structures and activities of life

The name *protein*, from the Greek word *proteios,* meaning "first place," suggests the importance of this class of macromolecules. A **protein** is a polymer constructed from amino acid monomers. Each of our many thousands of different kinds of proteins has a unique three-dimensional structure that corresponds to a specific function. Proteins are important to the structures of cells and organisms and participate in everything they do.

Probably the most important role for proteins is as **enzymes**, the chemical catalysts that speed and regulate virtually all chemical reactions in cells. Lactase, which you read about in the chapter introduction, is just one of the thousands of different enzymes produced by body cells.

In Figure 3.11, you can see examples of two other types of proteins. Structural proteins are found in hair and the fibers that make up tendons and ligaments. Muscles contain contractile proteins.

Other types of proteins include defensive proteins, such as the antibodies of the immune system, and signal proteins, such as many of the hormones and other messengers that help coordinate body activities by communicating between cells. Hemoglobin in red blood cells is a transport protein that delivers O_2 to working muscles. Different transport proteins move sugar molecules into cells for energy. Other proteins are storage proteins, such as ovalbumin, the protein of egg white, which serves as a source of amino acids for developing embryos. Milk proteins provide amino acids for baby mammals, and plants have storage proteins for the developing embryos in their seeds.

As we see next, this huge diversity of proteins is based on just 20 building blocks.

Web/CD Activity 3G *Protein Functions*

? Which of the following is not a protein: hemoglobin, cholesterol, an enzyme, an antibody?

Figure 3.11
Structural and contractile proteins

■ Cholesterol

3.12 Proteins are made from amino acids linked by peptide bonds

Of all of life's molecules, proteins are the most elaborate and the most diverse in structure and function. Protein diversity is based on differing arrangements of a common set of just 20 amino acids. **Amino acids** all have an amino group and a carboxyl group (which makes it an acid, hence the name *amino acid*). As you can see in the general structure shown in Figure 3.12A, both of these functional groups are covalently bonded to a central carbon atom, called the alpha carbon. Also bonded to the alpha carbon is a hydrogen atom and a chemical group symbolized by the letter R. The R group is the variable part of an amino acid. In the simplest amino acid (glycine), the R group is just a hydrogen atom. In others, such as those in Figure 3.12B, the R group consists of one or more carbon atoms with various functional groups attached. The structure of the R group determines the specific properties of each of the 20 amino acids in proteins (see Appendix 2).

The amino acids in Figure 3.12B represent two main types, hydrophobic and hydrophilic. Leucine (abbreviated Leu) is an example of an amino acid whose R group is non-

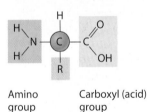

Amino group   Carboxyl (acid) group

Figure 3.12A General structure of an amino acid

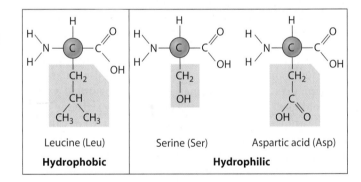

Leucine (Leu) **Hydrophobic**   Serine (Ser)   Aspartic acid (Asp) **Hydrophilic**

Figure 3.12B Examples of amino acids

polar and hydrophobic. Serine (Ser), with a hydroxyl group in its R group, is an example of an amino acid with a polar, hydrophilic R group. Aspartic Acid (Asp) is acidic and negatively charged at the pH of a cell. Other amino acids have basic R groups and are positively charged. Polar and charged amino acids help proteins dissolve in the aqueous solutions inside cells.

Now that we have examined amino acids, let's see how they are linked to form polymers. Cells join amino acids together in a dehydration reaction that links the carboxyl

group of one amino acid to the amino group of the next amino acid as a water molecule is removed (Figure 3.12C). The resulting covalent linkage is called a **peptide bond**. The product of the reaction shown in the figure is called a *dipeptide*, because it was made from *two* amino acids. Additional amino acids can be added by the same process to form a chain of amino acids, a **polypeptide**. To release amino acids from the polypeptide by hydrolysis, a molecule of H_2O must be added back to each peptide bond.

How is it possible to make thousands of different kinds of proteins from just 20 amino acids? The answer has to do with sequence. You know that thousands of English words can be made by varying the sequence of 26 letters. Although the protein alphabet is slightly smaller (just 20 "letters"), the "words" are much longer. Most polypeptides are at least 100 amino acids in length; some are a thousand or more. Each polypeptide has a unique sequence of amino acids. But a long polypeptide chain of specific sequence is not the same as a protein, any more than a long strand of yarn is the same as a sweater. A functioning protein is one or more polypeptide chains precisely twisted, folded, and coiled into a unique three-dimensional shape.

? In what way is the production of a dipeptide similar to the production of a disaccharide?

■ In both cases, the monomers are joined by a dehydration reaction.

Figure 3.12C
Peptide bond formation

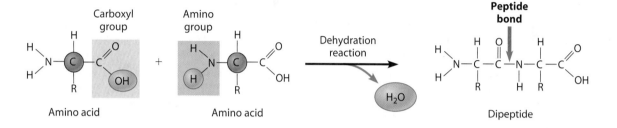

Amino acid + Amino acid →(Dehydration reaction / H_2O)→ Dipeptide

Carboxyl group · Amino group · Peptide bond

3.13   A protein's specific shape determines its function

Figure 3.13A shows a ribbon model of the protein lysozyme, an enzyme found in our tears and white blood cells. Lysozyme consists of one long polypeptide, represented by the ribbon. Roughly spherical, lysozyme's general shape is called globular. This overall shape is more apparent in **Figure 3.13B**, a space-filling model of lysozyme. Most enzymes and other proteins are globular in shape. Structural proteins, such as those making up hair and tendons, are typically long and thin—fibrous. General shape is one thing; specific shape is another. The coils and twists of lysozyme's polypeptide ribbon appear haphazard, but they represent the molecule's specific, three-dimensional shape, and this shape is what determines its specific function. Nearly all proteins must recognize and bind to some other molecule in order to function. Lysozyme, for example, can destroy bacterial cells, but first it must bind to specific molecules on the bacterial cell surface. Lysozyme's specific shape enables it to recognize and attach to its molecular target, which fits into the groove you see on the right in the figures.

The dependence of protein function on a protein's specific shape becomes clear when proteins are altered. In a process called **denaturation**, polypeptide chains unravel, losing their specific shape and, as a result, their function. Changes in salt concentration and pH can denature many proteins, as can excessive heat. For example, visualize what happens when you fry an egg. Heat quickly denatures the clear proteins surrounding the yolk, making them solid, white, and opaque. One of the reasons why extremely high fevers are so dangerous is that some proteins in the body become denatured and cannot function.

Given the proper cellular environment, a newly synthesized polypeptide chain spontaneously folds into its functional shape. We examine the four levels of a protein's structure next.

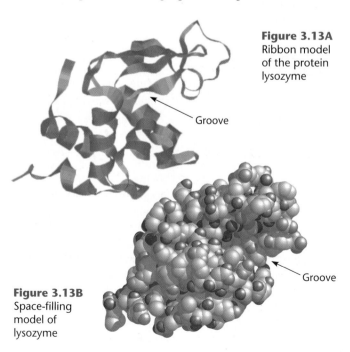

Figure 3.13A
Ribbon model of the protein lysozyme

Groove

Figure 3.13B
Space-filling model of lysozyme

Groove

? Why does a denatured protein no longer function normally?

■ The function of each protein is a consequence of its specific shape, which is lost when a protein denatures.

3.14 A protein's shape depends on four levels of structure

Primary structure The **primary structure** of a protein is its unique sequence of amino acids. As an example, let's consider transthyretin, an important transport protein found in our blood. It is a globular molecule whose specific shape enables it to transport two key chemicals throughout the body, vitamin A and a hormone from the thyroid gland. A complete molecule of transthyretin has four polypeptide chains, each made up of 127 amino acids. **Figure 3.14A** shows one of these chains partially unraveled for a closer look at its primary structure. The three-letter abbreviations represent amino acids. One of the 20 amino acids occupies each of the 127 positions along the chain.

For this or any other protein to perform its specific function, it must have the correct collection of amino acids arranged in a precise order. The primary structure of a protein is determined by inherited genetic information. Even a slight change in a protein's primary structure may affect its overall shape and its ability to function. For instance, a single amino acid change in hemoglobin, the oxygen-carrying blood protein, causes sickle-cell disease, a serious blood disorder.

Secondary structure In the second level of protein structure, parts of the polypeptide coil or fold into local patterns called **secondary structure**. Coiling of a polypeptide chain results in a secondary structure called an **alpha helix;** a certain kind of folding leads to a **pleated sheet.** Both of these patterns are maintained by regularly spaced hydrogen bonds between the hydrogens of the $>$N—H groups and the oxygens of the $>$C$=$O groups of neighboring peptide bonds along the polypeptide chain. The hydrogen bonds are represented in **Figure 3.14B** by a row of dots. Because the R groups of the amino acids are not involved in the hydrogen bonds forming these secondary structures, they are omitted from the diagrams.

Transthyretin has only one alpha-helix region (see **Figure 3.14C**). In contrast, many fibrous proteins, such as the structural protein of hair, have the alpha-helix structure over most of their length. Pleated sheets make up the core of many globular proteins, as is the case for transthyretin. Pleated sheets also dominate some fibrous proteins, including the silk protein of a spider's web, shown to the left. The teamwork of so many hydrogen bonds makes each silk fiber of a web as strong as steel. Potential uses of spider silk proteins include surgical thread, fishing line, and bulletproof vests. Industrial researchers have spliced spider genes into goats and harvested silk protein from their milk in the effort to commercially produce spider silk proteins.

Tertiary structure The term **tertiary structure** refers to the overall, three-dimensional shape of a polypeptide. As already mentioned, most tertiary structures can be roughly described as either globular or fibrous. As you can see in Figure 3.14C, a transthyretin polypeptide has a generally globular shape, which results from the compact combination of an alpha helix and several pleated-sheet regions. The indentations and bulges arising from its particular arrangement of coils and folds give the polypeptide the specific shape that fits it to its function.

Tertiary structure generally results from interactions among the R groups of the amino acids making up the polypeptide. For example, globular proteins found in aqueous solution, such as transthyretin, are folded so that the hydrophobic R groups are on the inside of the molecule and the hydrophilic groups on the outside, exposed to water. In addition to the clustering of hydrophobic groups, hydrogen bonding between polar side chains and ionic bonding of some of the charged R groups help maintain the tertiary structure. A protein's shape may be reinforced further by covalent bonds called disulfide bridges.

Quaternary structure Many proteins consist of two or more polypeptide chains, or subunits. Such proteins have a **quaternary structure**, resulting from the association of the subunits. **Figure 3.14D** shows a complete transthyretin molecule with its four identical globular subunits.

Another example of a protein with quarternary structure is collagen, shown to the right. Collagen is a fibrous protein with helical subunits intertwined into a larger triple helix. This arrangement gives the long fibers great strength, suited to their function as the girders of connective tissue in skin, bone, tendons, and ligaments. (Collagen accounts for 40% of the protein in a human body.)

Many other proteins have subunits that are different from one another. For example, the oxygen-transporting molecule hemoglobin has four subunits of two distinct types (see Figure 22.10B). Both types of subunits consist primarily of alpha-helical secondary structure. Each subunit also has a nonprotein component, called a heme, with an iron atom that binds oxygen.

Polypeptide chain

Collagen

Web/CD Activity 3H *Protein Structure*

? A genetic mutation can change the primary structure of a protein. How can this destroy the protein's function?

■ Primary structure, the amino acid sequence, affects the secondary structure, which affects the tertiary structure, which affects the quaternary structure (if any). In short, the amino acid sequence affects the shape of the protein, and the function of a protein depends on its shape.

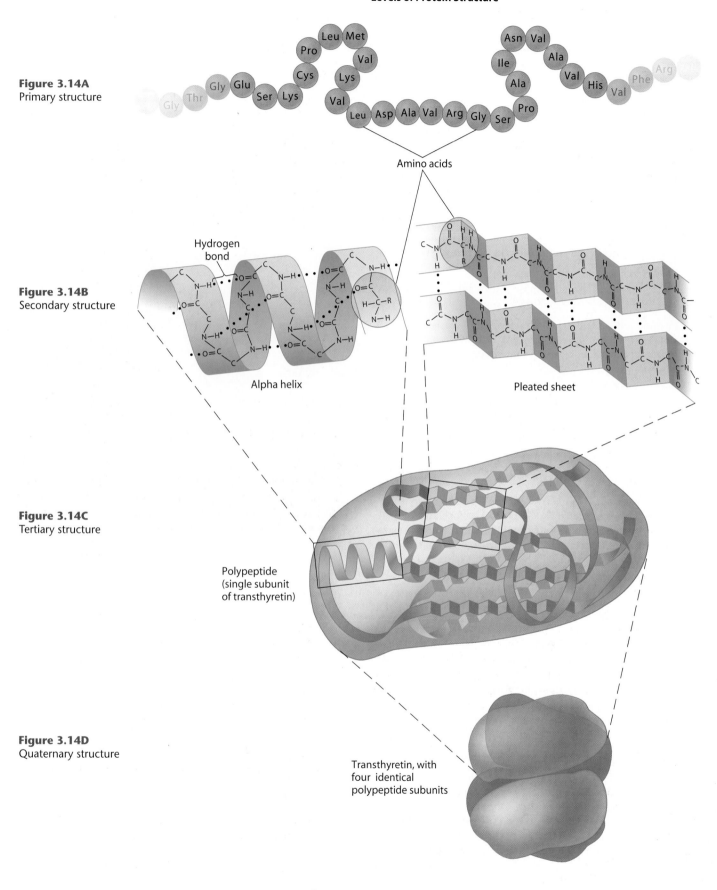

Figure 3.14A
Primary structure

Amino acids

Figure 3.14B
Secondary structure

Hydrogen bond

Alpha helix

Pleated sheet

Figure 3.14C
Tertiary structure

Polypeptide
(single subunit
of transthyretin)

Figure 3.14D
Quaternary structure

Transthyretin, with
four identical
polypeptide subunits

3.15 Linus Pauling contributed to our understanding of the chemistry of life

Linus Pauling, who died in 1994, was one of the giants of 20th-century science. Toward the end of his 93 years, Pauling was most often associated with his controversial belief that large doses of vitamin C can help prevent the common cold, cancer, and other diseases. Earlier in his career, Pauling made extraordinary contributions to both science and human affairs.

Driven by the desire "to understand the world," Pauling started out in chemistry and physics. He published a series of papers on chemical bonding that eventually led to a Nobel Prize in Chemistry in 1954. By that time, Pauling was also studying biological molecules. In an interview a number of years ago, we asked him about the value of studying molecules for understanding life, and he responded by talking about his work on hemoglobin:

> Life is too complicated to permit a complete understanding through the study of whole organisms. Only by simplifying the problem—breaking it down into a multitude of individual problems—can you get the answers. In 1935, Dr. Charles Coryell and I made our discovery about how oxygen molecules are attached to the iron atoms of hemoglobin, not by getting a cow and putting it through our magnetic apparatus, but by getting some blood from the cow and studying this blood and the hemoglobin from it. . . .
>
> Of course, the study of different parts of an organism leads to the question, Do these parts interact? Can we learn more about the living organism by putting two parts together to see to what extent the properties of the combination are different from those of the two separated parts? This approach permits further progress in understanding the organism. And yet I, myself, have confidence that all of the properties of living organisms could ultimately be discovered by the process of attempting to reduce the organism . . . to a combination of the different parts: essentially, the molecules that make up the organism.

Besides finding out how hemoglobin carries oxygen, Pauling discovered how an abnormal hemoglobin molecule causes sickle-cell disease. And it was Pauling who first described the two fundamental secondary structures of proteins, the alpha helix and the pleated sheet (**Figure 3.15**). His recounting of that work reveals more about his scientific attitudes and efforts:

> I began trying to find the structure of proteins in 1937, and didn't succeed. So I began working with my collaborators to determine the three-dimensional structures of amino acids and simple peptides. In 1937, no one had yet determined such structures. . . . And then in 1948 I found the alpha-helix and pleated-sheet structures in proteins. . . . I'm surprised that nobody else had done this job in the 11 years that intervened—in a sense, surprised that I hadn't done it in '37, when my ideas were all the right ones. I just hadn't worked hard enough.

Pauling's efforts were not limited to science. He also became the scientific community's leading advocate for halting the testing of nuclear weapons, resulting in the

Figure 3.15 Linus Pauling with a model of the alpha helix in 1948

accusation that he was a Communist and the revocation of his passport. Nevertheless, in 1963 he received the Nobel Peace Prize for helping produce a ban on nuclear testing. Pauling is the only person who has ever received two unshared Nobel Prizes.

The cancellation of Pauling's passport in 1952 kept him from working directly with scientists in Europe. This may have prevented Pauling from earning a third Nobel Prize, for at that time he was hot on the trail of the structure of DNA. But his triple-helix model was wrong, and it was James Watson and Francis Crick, working in England, who came up with the correct solution—a double helix, which we describe in the next module.

> **?** A decade after being awarded a Nobel Prize in Chemistry, Pauling received the Nobel Peace Prize for _____.

■ his efforts leading to a ban on nuclear testing

3.16 Nucleic acids are information-rich polymers of nucleotides

The **nucleic acids** are polymers that can serve as the blueprints for proteins. There are two types: **deoxyribonucleic acid (DNA)** and **ribonucleic acid (RNA)**. The genetic material that organisms inherit from their parents consists of DNA. **Genes** are specific stretches of a DNA molecule that program the amino acid sequences (primary structure) of proteins. And remember that amino acid sequences determine the specific three-dimensional structures and therefore the functions of proteins. Thus, by determining the primary structure of proteins, DNA controls the life of the cell and the organism.

DNA does not put its genetic information to work directly. It works through an intermediary—RNA. DNA's information is transcribed into RNA, which is then translated into the primary structure of proteins. We return to this chain of command and the functions of DNA and RNA later in the book. Here, we just want to introduce the structure of nucleic acids.

The monomers that make up nucleic acids are called **nucleotides.** As indicated in **Figure 3.16A**, each nucleotide has three parts. One part is a five-carbon sugar (blue); DNA has the sugar deoxyribose, whereas RNA has a closely related sugar called ribose. Linked to one end of the sugar in both types of nucleic acid is a phosphate group (yellow). At the other end of the sugar is a **nitrogenous base.** DNA has the nitrogenous bases adenine (A), thymine (T), cytosine (C), and guanine (G). RNA also has A, C, and G, but instead of thymine, it has uracil (U). Like the one in Figure 3.16A, all nitrogenous bases contain nitrogen.

Like polysaccharides and polypeptides, a nucleic acid polymer—a polynucleotide—forms from its monomers by dehydration reactions. In this process, the phosphate group of one nucleotide bonds to the sugar of the next monomer. The result is a repeating sugar-phosphate backbone in the polymer, as represented by the blue and yellow ribbon in Figure 3.16B.

RNA usually consists of a single polynucleotide strand, but DNA is a **double helix,** in which two polynucleotides wrap around each other (**Figure 3.16C**). The nitrogenous

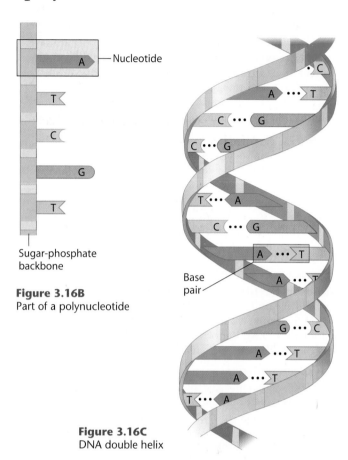

Figure 3.16B
Part of a polynucleotide

Figure 3.16C
DNA double helix

bases protrude from the two sugar-phosphate backbones into the center of the helix. There they always pair up as shown: A pairs with T, and C pairs with G.

The two DNA chains are held in a double helix by hydrogen bonds (dotted lines) between their paired bases. Most DNA molecules are very long, with thousands or even millions of base pairs. One long DNA molecule may contain many genes, each a specific series of hundreds or thousands of nucleotides along one of the polynucleotide strands. The specific sequence of nucleotides in a gene is the information that programs the primary structure of a protein.

This introduction to nucleic acids concludes our look at the four major classes of biological molecules. In the next chapter, we set the scene of molecular action—the cell.

Web/CD Activity 3I *Nucleic Acid Structure*

Web/CD Thinking as a Scientist *Connection: What Factors Determine the Effectiveness of Drugs?*

? In a double helix, a region along one DNA strand has this sequence of nitrogenous bases: TAGGCCT. What is the base sequence along the other strand of the double helix?

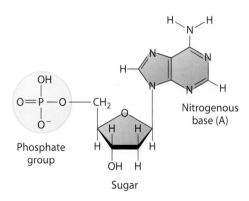

Figure 3.16A A nucleotide

■ ATCCGGA

Reviewing the Concepts

Introduction to Organic Compounds (Introduction–3.3)

Carbon. Carbon's ability to bond with four other atoms is the basis for building large and diverse organic compounds. Hydrocarbons are composed of only carbon and hydrogen. Isomers have the same molecular formula but different structures (**3.1**). Functional groups are particular groupings of atoms that give organic molecules specific properties (**3.2**).

Polymers. A huge number of different polymers are built from a small number of monomers (**3.3**).

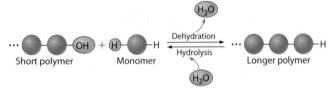

Carbohydrates (3.4–3.7)

Monosaccharides. A monosaccharide has a formula that is a multiple of (CH_2O) and contains hydroxyl groups and a carbonyl group. Two monosaccharides join to form a disaccharide (**3.4–3.5**). Various molecules taste sweet because they bind to "sweet" taste receptors on the tongue (**3.6**).

Polysaccharides are polymers of monosaccharides, such as glucose. Starch and glycogen are storage polysaccharides; cellulose is structural, found in plant cell walls (**3.7**).

Lipids (3.8–3.10)

Lipids, diverse compounds composed largely of carbon and hydrogen, are grouped together because they are hydrophobic. Fats, also called triglycerides, are energy-storage molecules consisting of glycerol linked to three fatty acids (**3.8**). Other lipids include phospholipids (found in cell membranes), waxes, and steroids (**3.9**). Use of anabolic steroids can cause serious health problems (**3.10**).

Proteins (3.11–3.15)

Proteins are involved in almost all of a cell's activities; as enzymes, they regulate chemical reactions (**3.11**). Protein diversity is based on different sequences of amino acids, monomers that contain an amino group, a carboxyl group, an H, and an R group, all attached to a central carbon. The R groups distinguish 20 different amino acids, each with specific properties. Amino acids are linked together by peptide bonds to form polypeptides (**3.12**). A protein consists of one or more polypeptide chains folded into a unique shape that determines the protein's function (**3.13**).

Protein structure. A protein's primary structure is the specific sequence of amino acids forming its polypeptide chains. Its secondary structure is the coiling or folding of the chain, stabilized by hydrogen bonds. Tertiary structure is the overall three-dimensional shape of a polypeptide, resulting from interactions between R groups. Proteins made of more than one polypeptide have quaternary structure (**3.14**). Linus Pauling made important contributions to our understanding of protein structure and function (**3.15**).

Nucleic Acids (3.16)

Nucleic acids—DNA and RNA—serve as the blueprints for proteins and thus control the life of a cell. The monomers of nucleic acids are nucleotides, composed of a sugar, a phosphate group, and a nitrogenous base. DNA is a double helix; RNA is a single-stranded polynucleotide (**3.16**).

Connecting the Concepts

1. The incredible diversity and complexity of life is staggering. Yet the molecular logic of life is simple and elegant: Small molecules common to all organisms are ordered into unique macromolecules. Explain why carbon is central to this diversity of organic molecules. How do carbon skeletons, functional groups, monomers, and polymers relate to this molecular logic of life?

2. There are four classes of organic molecules that are essential to all living organisms. Complete the following table to help organize your understanding of the structures and functions of these macromolecules.

Classes of macromolecules and their monomers	Functions	Examples
Carbohydrates Monosaccharides	Energy for cell, raw material	a. _____
	b. _____	Starch, glycogen
	Plant structure	c. _____
Lipids (don't form polymers) Glycerol Fatty acid Components of a fat molecule	Energy storage	d. _____
	e. _____	Phospholipids
	Hormones	f. _____
Proteins g. _____ h. _____ i. _____ Amino acid	j. _____	Lactase
	Structure	k. _____
	l. _____	Muscles
	Transport	m. _____
	Communication	Signal proteins
	n. _____	Antibodies
	Storage	Egg albumin
Nucleic Acids o. _____ p. _____ q. _____ Nucleotide	Heredity	r. _____
	s. _____	DNA and RNA

Testing Your Knowledge

Multiple Choice

3. A glucose molecule is to starch as (*Explain your answer.*)
   a. a steroid is to a lipid.
   b. a protein is to an amino acid.
   c. a nucleic acid is to a polypeptide.
   d. a nucleotide is to a nucleic acid.
   e. an amino acid is to a nucleic acid.

4. What makes a fatty acid an acid?
   a. It does not dissolve in water.
   b. It is capable of bonding with other molecules to form a fat.
   c. It has a carboxyl group that donates a hydrogen ion to a solution.
   d. It contains only two oxygen atoms.
   e. It is a polymer made of many smaller subunits.

5. Where in the tertiary structure of a water-soluble protein would you most likely find a hydrophobic amino acid R group?
   a. at both ends of the polypeptide chain
   b. on the outside, next to the water
   c. covalently bonded to another R group
   d. on the inside, away from water
   e. hydrogen-bonded to nearby amino acids

6. Cows can derive nutrients from cellulose because
   a. they produce enzymes that recognize the shape of the glucose-glucose bonds and hydrolyze them.
   b. they chew and rechew their cud to break down cellulose fibers.
   c. one of their stomachs contains microorganisms that can hydrolyze the bonds of cellulose.
   d. their intestinal tract contains termites, which produce enzymes to hydrolyze cellulose.
   e. they can convert cellulose to starch and then hydrolyze starch to glucose.

7. A shortage of phosphorus in the soil would make it especially difficult for a plant to manufacture
   a. DNA.   d. fatty acids.
   b. proteins.   e. sucrose.
   c. cellulose.

8. Lipids differ from other macromolecules in that they
   a. are much larger.
   b. are not truly polymers.
   c. do not have specific shapes.
   d. are nonpolar and therefore hydrophilic.
   e. contain nitrogen atoms.

9. Which functional group (or groups) is polar and tends to make organic compounds hydrophilic?
   a. carbonyl   d. carboxyl
   b. amino   e. all of the above
   c. hydroxyl

10. Unsaturated fats
    a. are more common in animals than in plants.
    b. have fewer fatty acid molecules per fat molecule.
    c. are associated with greater health risks than are saturated fats.
    d. have double bonds in their fatty acid chains.
    e. are usually solid at room temperature.

Describing, Comparing, and Explaining

11. List four different kinds of lipids and briefly describe their functions.

12. Explain why heat, pH changes, and other environmental changes can interfere with a protein's function.

13. How can a cell make many different kinds of protein out of only 20 amino acids? Of the myriad possibilities, how does the cell "know" which proteins to make?

14. Briefly describe the various functions performed by proteins in a cell.

15. Explain how DNA controls the functions of a cell.

16. Sucrose is broken down in your intestine to the monosaccharides glucose and fructose, which are then absorbed into your blood. What is the name of this type of reaction? Using this diagram of sucrose, show how this would occur.

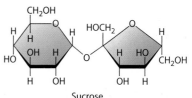

Sucrose

17. Circle and name the functional groups in this organic molecule. What type of compound is this? For which class of macromolecules is it a monomer?

Applying the Concepts

18. Linus Pauling believed that large doses of vitamin C can help prevent the common cold, cancer, and other diseases. Imagine you have been given a research grant by the National Institutes of Health to evaluate Pauling's claims. How would you go about setting up an experimental study to determine whether vitamin C can prevent colds? How would you evaluate your results?

19. Enzymes usually function best at an optimal pH and temperature. The following graph shows the effectiveness of two enzymes at various temperatures.

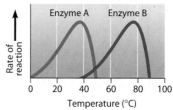

    a. At which temperature does enzyme A perform best? Enzyme B?
    b. One of these enzymes is found in humans and the other in thermophilic (heat-loving) bacteria. Which enzyme would you predict comes from which organism?
    c. From what you know about enzyme structure, explain why the rate of the reaction catalyzed by enzyme A slows down at temperatures above 40°C.

Answers to all questions can be found in Appendix 3.

For study help and Activities, go to campbellbiology.com or the student CD-ROM.

CHAPTER 4

INTRODUCTION TO THE CELL

4.1 Microscopes provide windows to the world of the cell
4.2 Most cells are microscopic
4.3 Prokaryotic cells are structurally simpler than eukaryotic cells
4.4 Eukaryotic cells are partitioned into functional compartments

ORGANELLES OF THE ENDOMEMBRANE SYSTEM

4.5 The nucleus is the cell's genetic control center
4.6 Overview: Many cell organelles are connected through the endomembrane system
4.7 Smooth endoplasmic reticulum has a variety of functions
4.8 Rough endoplasmic reticulum makes membrane and proteins
4.9 The Golgi apparatus finishes, sorts, and ships cell products
4.10 Lysosomes are digestive compartments within a cell
4.11 Abnormal lysosomes can cause fatal diseases
4.12 Vacuoles function in the general maintenance of the cell
4.13 A review of the endomembrane system

ENERGY-CONVERTING ORGANELLES

4.14 Chloroplasts convert solar energy to chemical energy
4.15 Mitochondria harvest chemical energy from food

THE CYTOSKELETON AND RELATED STRUCTURES

4.16 The cell's internal skeleton helps organize its structure and activities
4.17 Cilia and flagella move when microtubules bend

CELL SURFACES AND JUNCTIONS

4.18 Cell surfaces protect, support, and join cells

FUNCTIONAL CATEGORIES OF ORGANELLES

4.19 Eukaryotic organelles comprise four functional categories

The Art of Looking at Cells

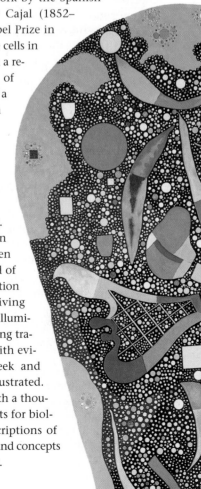

THE IMAGE BELOW, from a work by the Russian-born Wassily Kandinsky (1866–1944), was painted toward the end of the artist's career. During this period, he produced a number of works that show what appear to be cellular forms. In fact, there is evidence that these forms were indeed based on cells, for Kandinsky owned and studied a number of biology books, marking pages that contained drawings of cells. Apparently, he used these scientific illustrations as the starting point for his fanciful paintings. On the opposite page you see a work by the Spanish anatomist Santiago Ramón y Cajal (1852–1934), winner of the 1906 Nobel Prize in Medicine. This drawing of nerve cells in the retina of the eye appeared in a research paper on the structure of these cells. Each cell consists of a round part, the cell body, with one or more long projections extending from it. As this beautiful drawing indicates, Cajal was trained as an artist. Some argue that his research was well received in part because his drawings were so eye-catching.

These two images reveal an important relationship between art and biology, the most visual of the sciences: Artists find inspiration in the visual richness of the living world, and biologists use art to illuminate their findings. There is a long tradition of art serving biology, with evidence that some ancient Greek and Roman texts on plants were illustrated. The old adage "A picture is worth a thousand words" succinctly accounts for biology's dependence on art. Descriptions of anatomy and even of processes and concepts can often benefit from drawings.

A Tour of the Cell

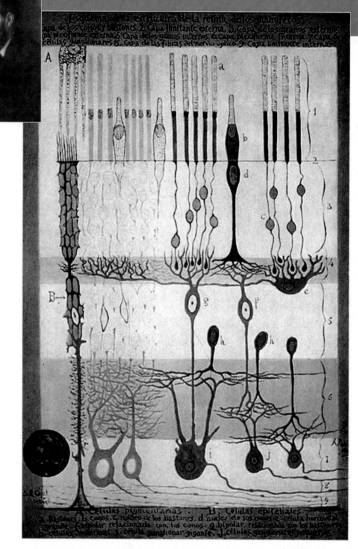

The first scientists to peer through microscopes certainly appreciated the power of illustration to describe a world not visible to the naked eye. When reporting his findings to the Royal Society of London, the Dutch microscopist Anton van Leeuwenhoek (1632–1723) included drawings with his descriptions of the tiny "animalcules" he saw in rainwater and other liquids. In England, Robert Hooke (1635–1703) produced a book, *Micrographia,* about his explorations of the microscopic world. His book featured over 60 illustrations, including a sketch of a magnified slice of cork (from the bark of an oak tree). Hooke compared the structures he saw in the cork to "little rooms"—*cellulae* in Latin—and the term *cells* stuck.

Since the days of Leeuwenhoek and Hooke, improved microscopes have vastly expanded our view of life at the cellular and subcellular levels, and photography and electronic imaging now enable biologists to capture microscope images directly. However, there is still room for the artist in helping us understand the microscopic world. You will see in this chapter, for example, that the use of different colors in drawings makes it easier to distinguish between the various cellular structures called *organelles* ("little organs"), most of which are actually colorless. You will also notice that drawings are often paired with **micrographs** of cell structures. (A micrograph is a photograph taken through a microscope.) This pairing provides the best of both worlds: The micrograph shows a structure as a biologist sees it, and the drawing helps you understand the micrograph by emphasizing specific details. And micrographs

Nerve cells in the retina of the eye as drawn by Santiago Ramón y Cajal (inset photo above).

themselves may also be an art form, as you can see in the examples in this book. There are even contests for outstanding images, such as the Nikon Small World Competition and *Nature's* Cell of the Month Image Competition.

This chapter focuses on the cellular structures that microscopes have revealed and describes how they work together in a living cell. Here and throughout your study of biology, you will continue to see examples of the relationship between art and science. Study the art in this book carefully, for it will illuminate your study of life. ■ ■ ■

4.1 Microscopes provide windows to the world of the cell

Our understanding of nature often parallels the invention and refinement of instruments that extend human senses to new limits. Before microscopes were first used in the seventeenth century, no one knew for certain that living organisms were composed of cells. The first microscopes, like the ones you may use in a biology laboratory, were light microscopes. A **light microscope (LM)** (Figure 4.1A) works by passing visible light through a specimen, such as a microorganism or a thin slice of animal or plant tissue. Glass lenses in the microscope bend the light to magnify the image of the specimen and project the image into the viewer's eye or onto photographic film or a video screen. **Magnification** is the increase in the apparent size of an object. Figure 4.1B shows a protist called *Euglena*. The notation "LM 1,000×" printed along the right edge of this micrograph is one we use throughout this book. It tells you that the photograph was taken through a light microscope and that this image is about 1,000 times the actual size of the organism. (This *Euglena* is about $\frac{2}{100}$ millimeters (0.02 mm) in length.) Athough this image could be magnified further, additional magnification would not allow us to see more detail. Indeed, at higher magnification, the image would begin to blur. Light microscopes can effectively magnify objects only about 1,000 times.

Thus, another important factor in microscopy (the use of a microscope) is **resolution,** a measure of the clarity of an image. Resolution is the ability of an optical instrument to show two close objects as separate. For example, what looks to your unaided eye like a single star in the sky may be resolved as two stars with the help of a telescope. Any optical device—be it an eye, a telescope, or a microscope—has a limit to its resolution. The human eye cannot resolve details finer than $\frac{1}{10}$ mm. The light microscope can resolve objects as small as 0.2 micrometer (μm), about the size of the smallest bacterium. But no matter how many times its image of such a bacterium is magnified, the light microscope cannot show the details of this small cell's internal structure. (The μ in the abbreviation for micrometer is the Greek letter mu.

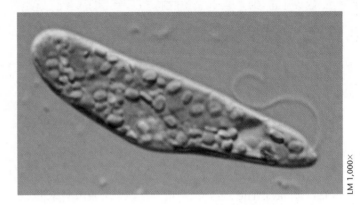

Figure 4.1B Light micrograph of a protist, *Euglena*

Table 4.1 gives the most common units of length that biologists use. As you can see, they are metric, so conversions are easy.)

From the year 1665, when Robert Hooke discovered cells, until the middle of the twentieth century, biologists had only light microscopes for viewing cells. But they discovered a great deal, including microorganisms, the cells composing animal and plant tissues, and also some of the structures within cells. By the mid-1800s, these discoveries led to the **cell theory,** which states that all living things are composed of cells and that all cells come from other cells.

Our knowledge of cell structure took a giant leap forward as biologists began using the electron microscope in the 1950s. Instead of light, the **electron microscope (EM)** uses a beam of electrons. The EM has a much greater resolution than the light microscope. Under special conditions, the most powerful EMs can detect individual atoms. In general, modern EMs can distinguish objects as small as about 2 nanometers (nm), a hundredfold improvement over the light microscope. This high resolution has allowed biologists to explore cellular ultrastructure, the complex internal anatomy of a cell. The highest-power electron micrographs you will see in this book have magnifications of about 100,000 times.

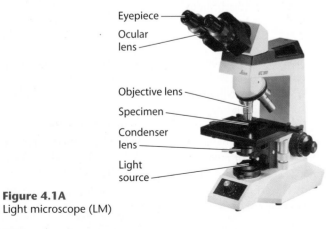

Eyepiece

Ocular
lens

Objective lens

Specimen

Condenser
lens

Light
source

Figure 4.1A
Light microscope (LM)

TABLE 4.1 MEASUREMENT EQUIVALENTS
1 meter (m) = 10^0 m = 39.37 inches
1 centimeter (cm) = 10^{-2} m (1/100 m) = 0.4 inch
1 millimeter (mm) = 10^{-3} m (1/1,000 m) = 1/10 cm
1 micrometer (μm) = 10^{-6} m (1/1,000,000 m)
1 nanometer (nm) = 10^{-9} m (1/1,000,000,000 m)
1 meter = 10^2 cm = 10^3 mm = 10^6 μm = 10^9 nm

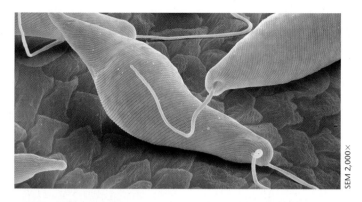

Figure 4.1C Scanning electron micrograph of *Euglena*

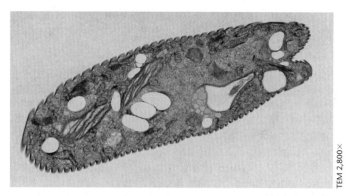

Figure 4.1D Transmission electron micrograph of *Euglena*

Figures 4.1C and 4.1D show images produced by two kinds of electron microscopes. Biologists use the **scanning electron microscope (SEM)** to study the detailed architecture of cell *surfaces*. The SEM uses an electron beam to scan the surface of a cell or group of cells that has been coated with a thin film of metal. When the surface is hit by the beam, it emits electrons. The electrons are detected by a device that then translates their pattern into an image projected onto a video screen. The scanning electron micrograph in Figure 4.1C highlights the *Euglena's* flagellum, a projection it uses for movement. Many structural details of cell surfaces have been discovered using the SEM. As you can see, the SEM produces images that look three-dimensional.

The **transmission electron microscope (TEM)** is used to study the details of internal cell structure. Specimens are cut into extremely thin sections and stained with atoms of heavy metals such as gold. The TEM aims an electron beam through a section, just as a light microscope aims a beam of light through a specimen. However, instead of lenses made of glass, the TEM uses electromagnets as lenses, as do all electron microscopes. The electromagnets bend the electron beam to magnify and focus an image onto a viewing screen or photographic film. The micrograph in Figure 4.1D shows internal details of *Euglena* as seen with the TEM. SEMs and TEMs are initially black and white, but are often artificially colorized to highlight or clarify structural features.

Electron microscopes have truly revolutionized the study of cells and cell organelles. Nonetheless, they have not replaced the light microscope. One problem with electron microscopes is that they cannot be used to study living specimens because the specimen must be held in a vacuum chamber; that is, all the air and liquid must be removed. For a biologist studying a living process, such as the movement of *Euglena*, a light microscope equipped with a video camera might be more suitable than either an SEM or a TEM. Thus, the light microscope remains a useful tool, especially for studying living cells.

There are different types of light microscopy, each using different techniques to enhance contrast and selectively highlight cellular components. **Figure 4.1E** shows a small worm, as seen using differential interference-contrast microscopy. This optical technique amplifies differences in density while still allowing living cells to be examined. **Figure 4.1F** is a striking example of fluorescence and confocal microscopy. In this technique, specific molecules are tagged

Figure 4.1E Micrograph produced by differential interference-contrast microscopy of a nematode (a kind of worm)

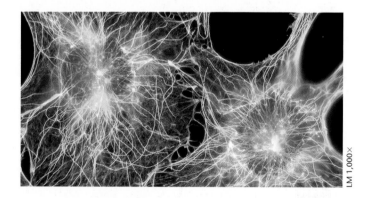

Figure 4.1F Micrograph produced by fluorescence confocal microscopy, showing cancer cells

with fluorescent dyes, and lasers and special optics bring a very thin section of the cell into focus. This beautiful image of cancer cells in culture was one of the Nikon Small World Competition winners, illustrating again the connection between art and science discussed in the chapter introduction.

Web/CD Activity 4A *Metric System Review*

? Which type of microscope would you use to study (a) the changes in shape of a living human white blood cell; (b) the finest details of surface texture of a human hair; (c) the detailed structure of an organelle in a human liver cell?

■ (a) light microscope; (b) scanning electron microscope; (c) transmission electron microscope

4.2 Most cells are microscopic

There is a reason that our knowledge of cells depended on the development of microscopes: most cells cannot be seen with the unaided eye. **Figure 4.2A** shows the size range of cells compared with objects both larger and smaller. The smallest cells are bacteria called mycoplasmas, with diameters as small as 0.2 μm. Some of the bulkiest cells are bird eggs, and the longest human cells are certain muscle and nerve cells. Most cells lie between these extremes, in the range indicated by the yellow area in the figure. The scale is logarithmic, with the length labels on the left ascending in powers of ten, to accommodate the range of sizes shown.

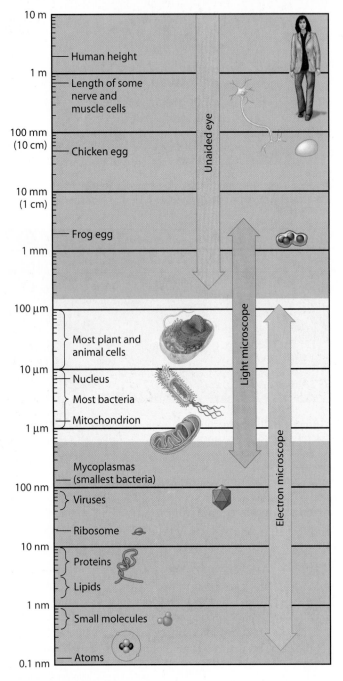

Figure 4.2A The sizes of cells and related objects

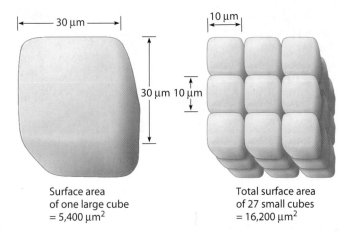

Surface area of one large cube = 5,400 μm^2

Total surface area of 27 small cubes = 16,200 μm^2

Figure 4.2B Effect of cell size on surface area

Thus, most plant and animal cells, with diameters ranging from 10 to 100 μm, are ten times larger than most bacteria.

The logistics of carrying out a cell's functions sets limits on cell size. At minimum, a cell must be able to house enough DNA, protein molecules, and internal structures to survive and reproduce. The maximum size of a cell is influenced by its requirement for enough surface area to obtain adequate nutrients and oxygen from the environment and dispose of wastes. Large cells have more surface area than small cells, but large cells have much less surface area *relative to their volume* than small cells of the same shape.

Figure 4.2B illustrates the relationship of surface area to volume using cube-shaped cells. The figure shows 1 large cell and 27 small ones. The total volume is the same in both cases:

Volume = 30 μm $\times$ 30 μm $\times$ 30 μm = 27,000 μm^3

In contrast to the total volume, the total surface areas are very different. Because a cube has six sides, its surface area is six times the area of one side. The surface areas of the cubes are as follows:

Area of large cube = 6 $\times$ (30 μm $\times$ 30 μm) = 5,400 μm^2

Area of small cube = 6 $\times$ (10 μm $\times$ 10 μm) = 600 μm^2

For all 27 of the small cubes, the total surface area is 27 $\times$ 600 μm^2, which equals 16,200 μm^2—three times the surface area of the large cube.

Thus, we see that a large cell has a much smaller surface area relative to its volume than smaller cells have. The need for a surface area large enough to service a cell's volume helps explain the microscopic size of most cells.

In the next module, we look briefly at the smallest cells—the prokaryotic cells of domains Bacteria and Archaea.

Web/CD Thinking as a Scientist *Connection: What Is the Size and Scale of Our World?*

? Red blood cells, which transport O$_2$, are among the smallest of human cells. Explain one functional advantage of their small size.

■ The larger area of plasma membrane of such small cells provides more surface area across which oxygen can diffuse.

Prokaryotic cells are structurally simpler than eukaryotic cells

Two kinds of structurally different cells have evolved over time. Bacteria and archaea consist of **prokaryotic cells**, whereas all other forms of life (protists, fungi, plants, and animals) are composed of **eukaryotic cells.**

All cells have several basic features in common. They are all bounded by a membrane, called a **plasma membrane.** All cells have genes made of DNA as their hereditary material. And all cells contain **ribosomes**, tiny structures that make proteins according to instructions from the genes. Eukaryotic cells are distinguished by having a membrane-enclosed nucleus, which houses most of their DNA.

The electron micrographs in **Figure 4.3A** contrast the size and complexity of prokaryotic and eukaryotic cells. It takes an electron microscope to clearly see the structural details of any cell, and this is especially true of prokaryotic cells because they are so small. Most prokaryotic cells range from 1 to 10 μm in length, averaging about one-tenth the size of a typical eukaryotic cell. A prokaryotic cell lacks a nucleus (its name comes from the Greek *pro,* before, and *karyon,* kernel, referring to the nucleus). The DNA of a prokaryotic cell is coiled into a **nucleoid** (nucleus-like) **region**, as seen in Figure 4.3A, but in contrast to the situation in eukaryotic cells, no membrane surrounds the DNA. Note the prominent nucleus in the eukaryotic cell in Figure 4.3A and the many other types of structures.

The cutaway diagram in **Figure 4.3B** reveals the structure of a generalized prokaryotic cell. Outside the plasma membrane (gray) of most prokaryotes is a fairly rigid, chemically complex **cell wall** (orange). The wall protects the cell and helps maintain its shape. In some prokaryotes, another layer, a sticky outer coat called a **capsule** (yellow), surrounds the cell wall and further protects the cell surface. Capsules also help glue prokaryotes to surfaces, such as sticks and rocks in fast-flowing streams or tissues within the human body. In addition to outer coats, some prokaryotes have surface projections. Short projections called **pili** (singular, *pilus*) help attach prokaryotes to surfaces. Longer projections called **prokaryotic flagella** (singular, *flagellum*) propel the prokaryotic cell through its liquid environment.

Prokaryotes will be described in more detail in Chapter 16. Eukaryotic cells are the main focus of this chapter.

Web/CD Activity 4B *Prokaryotic Cell Structure and Function*

? How is the nucleoid region of a prokaryotic cell unlike the nucleus of a eukaryotic cell?

■ There is no membrane enclosing the DNA of the nucleoid region.

Prokaryotic cell

Nucleoid region

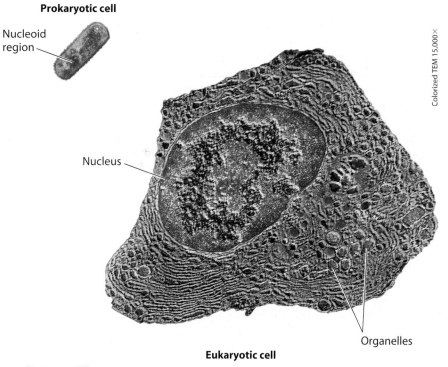

Nucleus

Organelles

Eukaryotic cell

Figure 4.3A Contrasting the size and complexity of prokaryotic and eukaryotic cells

Colorized TEM 15,000×

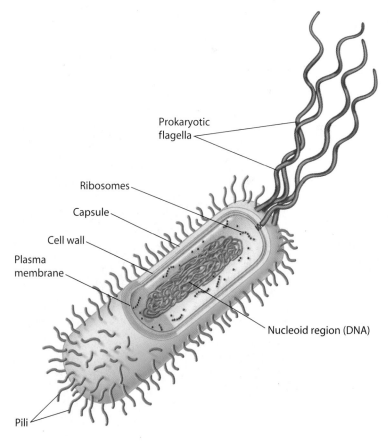

Prokaryotic flagella

Ribosomes

Capsule

Cell wall

Plasma membrane

Nucleoid region (DNA)

Pili

Figure 4.3B A prokaryotic cell

Eukaryotic cells are partitioned into functional compartments

All eukaryotic cells (Greek *eu,* true, and *karyon,* kernel)—whether from animals, plants, protists, or fungi—are fundamentally similar to one another and profoundly different from prokaryotic cells. Let's look at an animal cell and a plant cell as representatives of the eukaryotes.

Figure 4.4A is a diagram of an idealized animal cell, showing the details visible with the transmission electron microscope. As you saw in the micrograph in Figure 4.3A, eukaryotic cells are much more complex than prokaryotic cells. Besides the presence of a nucleus in the eukaryotic cell, the most obvious difference is the variety of structures in the **cytoplasm,** the fluid-filled region between the nucleus and the plasma membrane. (The term *cytoplasm* is also used for the interior of a prokaryotic cell.) These structures, along with the nucleus, are the **organelles,** and each type has a specific function in the cell. Notice that most eukaryotic organelles are compartments bounded by membranes; in the figure, the names of these "membranous organelles" and the plasma membrane are underlined. In essence, the internal membranes of a eukaryotic cell partition it into compartments. As mentioned in the chapter introduction, in this text we color-code the various organelles for easier identification, but keep in mind that most of them are actually colorless.

Many of the chemical activities of cells—activities known collectively as **cellular metabolism**—occur in the fluid-filled spaces within membranous organelles. These spaces are important as sites where specific chemical conditions are maintained, conditions that vary from one organelle to another and that favor the metabolic processes occurring in each kind of organelle. For example, while the endoplasmic reticulum is engaged in making steroid hormones, neighboring peroxisomes may be making hydrogen peroxide (H_2O_2) as a poisonous by-product of their activities. But because the H_2O_2 is confined within peroxisomes, where it is quickly converted to H_2O by resident enzymes, the hormones and the rest of the cell are protected from destruction.

Another benefit of internal membranes is that they greatly increase a eukaryotic cell's total membrane area. A typical eukaryotic cell, with a diameter about ten times greater than that of a typical prokaryotic cell, has a thousand times the

Figure 4.4A An animal cell

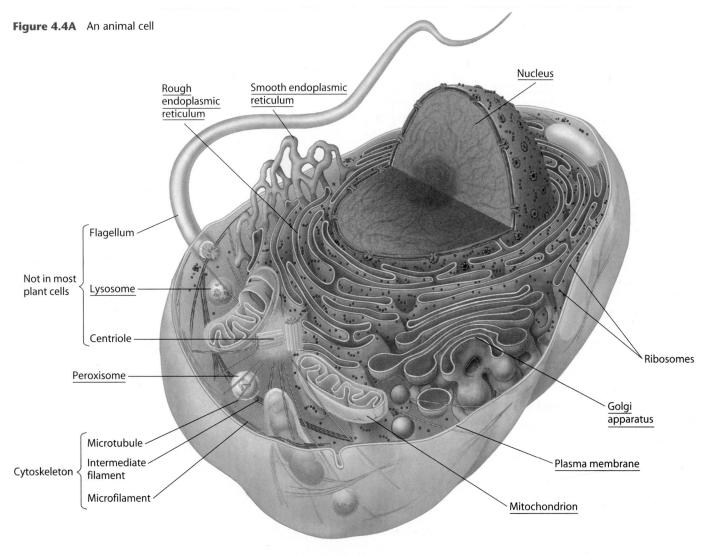

cytoplasmic volume but only a hundred times the plasma membrane area of the prokaryotic cell. In eukaryotic cells, internal (organelle) membranes provide the surfaces where many important metabolic processes occur; in fact, many enzymatic proteins essential for metabolic processes are built into organelle membranes.

Almost all of the membranous organelles appearing in Figure 4.4A are also present in plant cells. As you can see in Figure 4.4A, there are a few exceptions, the most obvious one being the flagellum. Among the plants, only the sperm cells of a few species have flagella. (The eukaryotic flagellum is different from the prokaryotic flagellum in both structure and operation, as you will learn later.)

A plant cell (Figure 4.4B) has some structures that an animal cell lacks. For example, a plant cell has a rigid, rather thick cell wall (as do the cells of fungi and many protists). Cell walls protect cells and help maintain their shape. Chemically different from prokaryotic cell walls, plant cell walls contain the polysaccharide cellulose. An important organelle found in plant cells but not in animal cells is the chloroplast, where photosynthesis occurs. (Chloroplasts are also found in some protists.) Unique to plant cells is a large central vacuole, a compartment that stores water and a variety of chemicals.

Although we have emphasized membranous organelles, eukaryotic cells contain nonmembranous structures as well (those with labels not underlined). Among them are the centriole and cytoskeleton, both of which consist of protein tubes called microtubules. Also, you can see by the many brown dots in both figures that ribosomes, the sites of protein synthesis, occur throughout the cytoplasm, as they do in prokaryotic cells. In addition to ribosomes in the cytoplasmic fluid, eukaryotic cells have many ribosomes attached to parts of the endoplasmic reticulum (making it "rough") and to the outside of the nucleus. Ribosome structure is presented in Chapter 10. In the remaining modules of this chapter, we explore the structure and functions of the other organelles of eukaryotic cells, starting with the nucleus.

Web/CD Activity 4C *Comparing Prokaryotic and Eukaryotic Cells*

Web/CD Activity 4D *Build an Animal Cell and a Plant Cell*

[?] Which of the following organelles does not belong in the list: mitochondrion, chloroplast, ribosome, lysosome, peroxisome? Why?

■ Ribosome, because it is the only organelle in the list that is not bounded by a membrane

Figure 4.4B A plant cell

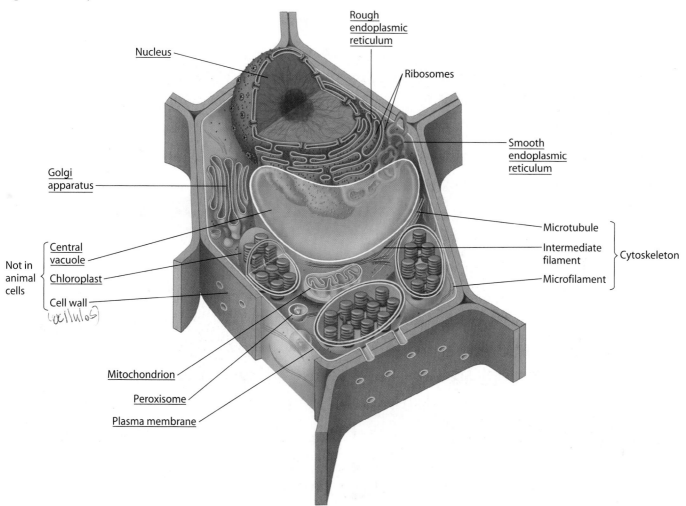

4.5 The nucleus is the cell's genetic control center

The **nucleus** contains DNA, the genetic material of a cell, and controls the cell's activities by directing protein synthesis. Nuclear DNA is attached to proteins, forming very long fibers called **chromatin** (the purple threads in **Figure 4.5**). Each fiber is a **chromosome**. Prior to cell division, the DNA is copied. During cell division, the chromatin fibers coil up, forming thicker structures that are individually visible with a light microscope.

Enclosing the nucleus is a **nuclear envelope**, a double membrane perforated with pores that control the flow of materials into and out of the nucleus. The **nucleolus** is a prominent structure in the nucleus. Building blocks of ribosomes are produced here and exit the nucleus via the pores.

Following instructions in the DNA, the nucleus synthesizes RNA. A type called messenger RNA moves through the pores to the cytoplasm and is translated there by ribosomes into the amino acid sequences of proteins. Protein synthesis will be explored in more detail in Chapter 10.

? What are the main functions of the nucleus?

■ To contain DNA and pass it on to daughter cells in cell division, to build ribosomes, to copy DNA instructions into RNA

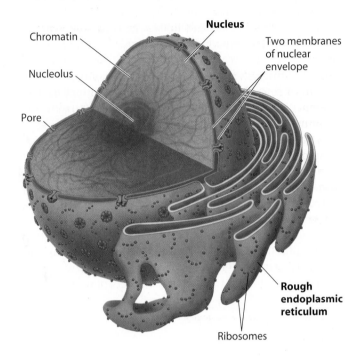

Figure 4.5 The nucleus and rough endoplasmic reticulum

4.6 Overview: Many cell organelles are connected through the endomembrane system

We now focus on eukaryotic organelles that are formed of interrelated membranes (Modules 4.6–4.13). Some of these membranes are physically connected and some are related by the transfer of membrane segments by tiny **vesicles** (sacs made of membrane). Collectively, these organelles constitute the **endomembrane system.** Many of these organelles work together in the synthesis, storage, and export of molecules.

An extensive network of flattened sacs and tubes called the **endoplasmic reticulum (ER)** is the prime example of the direct interrelatedness of parts of the endomembrane system. (The term *endoplasmic reticulum* comes from Greek words meaning "network within the cell.") As we discuss next, there are two regions of ER, rough ER and smooth ER,

which differ in structure and function. The membranes that form them, however, are continuous. Membranes of the rough ER are also continuous with the nuclear envelope, as you can see in Figure 4.5. The tubules and sacs of the ER enclose an interior space that is separate from the cytoplasmic fluid. Dividing the cell into separate compartments is a major function of the endomembrane system.

? Which structure includes all others in the list: rough ER, smooth ER, endomembrane system, nuclear envelope?

■ Endomembrane system

4.7 Smooth endoplasmic reticulum has a variety of functions

As **Figure 4.7** on the facing page indicates, **smooth endoplasmic reticulum**, or **smooth ER**, is a network of interconnected tubules, called *smooth* because it lacks attached ribosomes. Much of the activity of smooth ER results from enzymes embedded in its membrane. One of the most important functions of smooth ER is the synthesis of lipids, including fatty acids, phospholipids, and steroids. Different

types of cells make particular kinds of lipids. In mammals, for example, cells of the ovaries and testes synthesize the steroid sex hormones; they are rich in smooth ER.

Our liver cells also have large amounts of smooth ER, with additional kinds of functions. Certain enzymes in the smooth ER of liver cells help process drugs and other potentially harmful substances. The drugs detoxified by these

enzymes include, among others, sedatives such as barbiturates, and alcohol.

Undesirable complications can result when liver cells respond to drugs. As the cells are exposed to such chemicals, the amounts of smooth ER and its detoxifying enzymes increase, thereby increasing the body's tolerance to the drugs. This means that higher and higher doses of a drug are required to achieve a particular effect, such as sedation. Another complication is that detoxifying enzymes often cannot distinguish among related chemicals. As a result, the growth of smooth ER in response to one drug can increase tolerance to other drugs, including important medicines. Barbiturate use, for example, can decrease the effectiveness of certain antibiotics.

Smooth ER has yet another function, the storage of calcium ions. In muscle tissue, these are necessary for contraction. When a nerve signal stimulates a muscle cell, calcium ions leak from the smooth ER into the cytoplasmic fluid, where they trigger contraction of the cell.

? What are three functions of smooth ER?

■ Lipid synthesis; destruction of toxic substances (in liver cells); regulation of muscle contraction by uptake and release of calcium

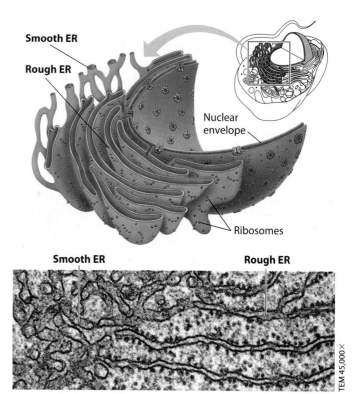

Figure 4.7 Smooth and rough endoplasmic reticulum

4.8 Rough endoplasmic reticulum makes membrane and proteins

The "rough" in **rough endoplasmic reticulum**, or **rough ER**, refers to the appearance of this organelle in electron micrographs. As Figure 4.7 shows, ribosomes stud the membranes. Rough ER has two main functions. One is to make more membrane. Some of the proteins made by ER ribosomes are inserted into the membrane, as are phospholipids made by ER enzymes. As a result, the rough ER membrane enlarges, and some of it is transported to other organelles.

The other major function of rough ER is to modify proteins that will be transported to other organelles or secreted by the cell. An example of a **secretory protein** is insulin, a hormone secreted by certain cells in the pancreas. Ribosomes attached to the rough ER synthesize the hormone's polypeptide, which is assembled into a functional protein inside the ER.

Figure 4.8 shows the synthesis, modification, and packaging of a secretory protein. ❶ As the polypeptide is synthesized by an attached ribosome, it passes into the ER, where it folds into its three-dimensional shape. ❷ Short chains of sugars are often linked to the polypeptide, making the molecule a **glycoprotein** (*glyco-* means "sugar"). ❸ When the molecule is ready for export from the ER, the ER packages it in a **transport vesicle.** ❹ This vesicle buds off from the ER membrane. The protein now travels to the Golgi apparatus (described in the next module) for further processing. From there, a transport vesicle containing the finished molecule will make its way to the plasma membrane and release its contents from the cell.

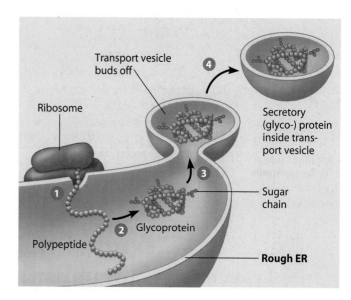

Figure 4.8 Synthesis and packaging of a secretory protein by the rough ER

Web/CD Activity 4E *Overview of Protein Synthesis*

? What makes rough ER rough?

■ Ribosomes attached to the membranes

4.9 The Golgi apparatus finishes, sorts, and ships cell products

The Golgi apparatus was named after Italian biologist and physician Camillo Golgi, whose career spanned the turn of the twentieth century. Using the light microscope, Golgi and his contemporaries discovered this membranous organelle and a number of others in animal and plant cells. The electron microscope has revealed that the Golgi apparatus consists of flattened sacs looking like a stack of pita bread. As you can see in **Figure 4.9**, the sacs are not interconnected like ER sacs. A cell may contain only a few Golgi stacks or hundreds. The number of Golgi stacks correlates with how active the cell is in secreting proteins—a multistep process that, as we have just seen, is initiated in the rough ER.

The **Golgi apparatus** performs several functions in close partnership with the ER. Serving as a molecular warehouse and finishing factory, a Golgi apparatus receives and modifies substances manufactured by the ER. One side of a Golgi stack serves as a receiving dock for transport vesicles produced by the ER. When a Golgi receives transport vesicles containing glycoprotein molecules, for instance, it takes in the materials and then modifies them chemically. One function of this chemical modification seems to be to mark and sort the molecules into different batches for different destinations. Molecules may be moved from sac to sac in the Golgi by transport vesicles, or, according to recent research, entire sacs may move from the receiving to the shipping side, modifying their protein cargo as they go. The shipping side of the Golgi stack serves as a depot from which finished secretory products, packaged in transport vesicles, move to the plasma membrane for export from the cell. Alternatively, finished products may become part of the plasma membrane itself or part of another organelle, such as a lysosome, which we discuss next.

? What is the relationship of the Golgi apparatus to the ER in a protein-secreting cell?

■ The Golgi receives transport vesicles that bud from the ER and that contain proteins synthesized by ribosomes attached to the ER. The Golgi finishes processing the proteins and then dispatches transport vesicles that secrete the proteins to the outside of the cell.

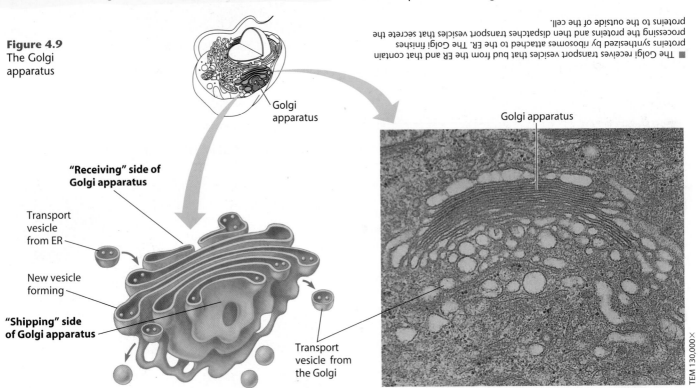

Figure 4.9
The Golgi apparatus

Golgi apparatus

"Receiving" side of Golgi apparatus

Transport vesicle from ER

New vesicle forming

"Shipping" side of Golgi apparatus

Transport vesicle from the Golgi

Golgi apparatus

TEM 130,000×

4.10 Lysosomes are digestive compartments within a cell

A fourth component of the endomembrane system, the lysosome, is produced in animal cells by the rough ER and the Golgi apparatus. The name **lysosome** is derived from two Greek words meaning "breakdown body," and lysosomes consist of digestive (hydrolytic) enzymes enclosed in a membranous sac. The formation and functions of lysosomes are diagrammed in **Figure 4.10A** on the next page. ❶ The rough ER packages the enzymes into transport vesicles. The Golgi apparatus chemically refines the enzymes and ❷ releases mature lysosomes. Lysosomes illustrate the main theme of eukaryotic cell structure—compartmentalization. The lysosomal membrane encloses a compartment where digestive enzymes are provided with an acidic environment, safely isolated from the rest of the cytoplasm.

Lysosomes have several types of digestive functions, as **Figure 4.10A** shows. ❸ Many protists engulf food particles into tiny cytoplasmic sacs called food vacuoles. ❹ Lysosomes fuse with the food vacuoles and then break down

the food, releasing nutrients to the cell. Our white blood cells ingest bacteria into vacuoles, and lysosomal enzymes emptied into these vacuoles rupture the bacterial cell walls (Figure 4.10B). Moreover, lysosomes serve as recycling centers for cells. Damaged organelles or small amounts of cell fluid become enclosed in a membrane vesicle. ❺ A lysosome fuses with such a vesicle and dismantles its contents, making organic molecules available for reuse (Figure 4.10C). With the help of lysosomes, a cell continually renews itself. A human liver cell, for example, recycles half of

its macromolecules each week. Lysosomes also play vital roles in embryonic development. For instance, lysosomal enzymes help destroy cells of the webbing that joins the fingers of early human embryos.

? When lysosomes were first discovered, they were sometimes called "suicide capsules." In what way does that nickname fit?

■ If a cell's lysosomes break open, the hydrolytic enzymes released may kill the cell.

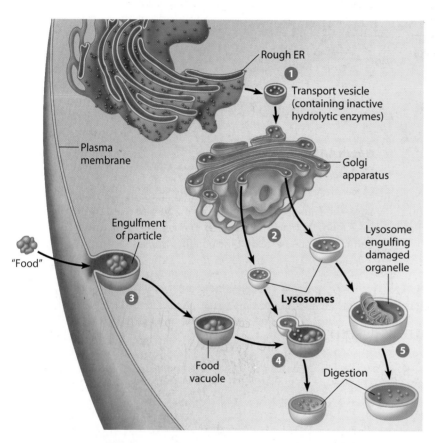

Figure 4.10A   Lysosome formation and functions

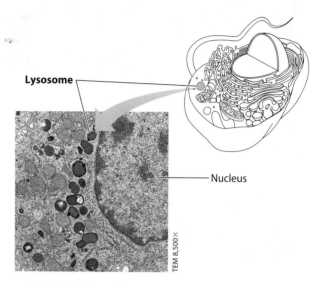

Figure 4.10B   Lysosomes in a white blood cell

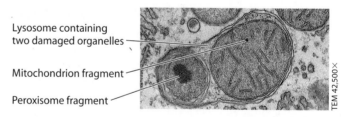

Figure 4.10C   Lysosome breaking down damaged organelles

4.11   Abnormal lysosomes can cause fatal diseases

1. Degrade lipids
2. Devistation of muscle-cell

The importance of lysosomes to cell function and human health is made strikingly clear by the serious hereditary disorders called lysosomal storage diseases. A person afflicted with a lysosomal storage disease is missing one or more of the hydrolytic enzymes of the lysosome. The abnormal lysosomes become engorged with indigestible substances, which eventually interfere with other cellular functions.

Most of these diseases are fatal in early childhood. In Pompe's disease, harmful amounts of the polysaccharide glycogen accumulate in muscle and liver cells because lysosomes lack a glycogen-digesting enzyme. Tay-Sachs disease ravages the nervous system. In this disorder, lysosomes lack

an enzyme needed to break down a lipid abundant in nerve cell membranes. Nerve cells in the brain are damaged as their lysosomes swell with undigested lipids. Fortunately, storage diseases are rare in the general population. For Tay-Sachs disease, carriers of the abnormal gene that causes it can be identified before they decide whether to have children.

? How can defective lysosomes result in excess accumulation of a particular compound in a cell?

■ If the lysosomes lack an enzyme needed to hydrolyze the compound, the cell will accumulate an excess of the compound.

4.12 Vacuoles function in the general maintenance of the cell

Like lysosomes, **vacuoles** are membranous sacs, generally larger than vesicles, that belong to the endomembrane system. Vacuoles come in different shapes and sizes and have a variety of functions. In Module 4.10, we saw that the food vacuole functions in collaboration with a lysosome. Here, in Figure 4.12A, we see a plant cell's **central vacuole**, which can function as a large lysosome. The central vacuole also helps the plant cell grow in size by absorbing water and enlarging, and it can store vital chemicals or waste products as well. Central vacuoles in flower petals may contain pigments that attract pollinating insects. Other central vacuoles contain poisons that protect against plant-eating animals.

Figure 4.12B shows a very different kind of vacuole in the protist *Paramecium*. Notice the two contractile vacuoles, looking somewhat like wheel hubs with radiating spokes. The "spokes" collect excess water from the cell, and the hub expels it to the outside. This function is necessary for freshwater protists because they constantly take up water from their environment. Without a way to get rid of the excess water, the cell fluid would become too dilute to support life, and eventually the cell would swell and burst. Thus, the contractile vacuole is vital in maintaining the cell's internal environment.

Figure 4.12A Central vacuole in a plant cell

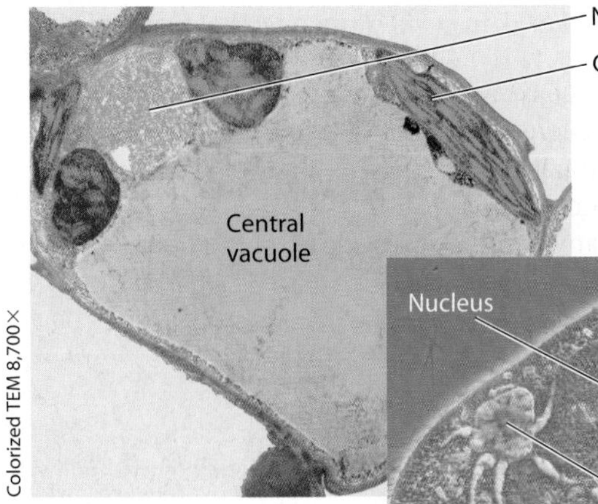

Nucleus

Chloroplast

Central vacuole

Colorized TEM 8,700×

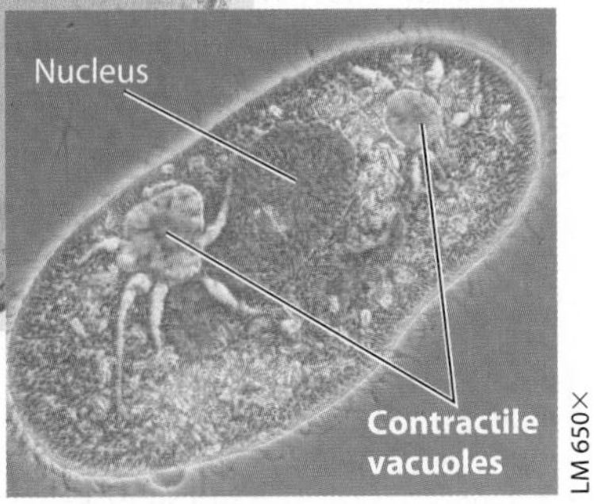

Nucleus

Contractile vacuoles

LM 650×

Figure 4.12B Contractile vacuoles in *Paramecium*

? The *Paramecium* cell in Figure 4.12B is about 0.08 mm long. Estimate the diameter in micrometers of the larger contractile vacuole. (Don't include the "spokes.")

■ 10–15 μm

4.13 A review of the endomembrane system

Figure 4.13 summarizes the relationships among the major organelles of the endomembrane system (all shown in gray). You can see the direct *structural* connections between the nuclear envelope, rough ER, and smooth ER. The red arrows show the *functional* connections within the endomembrane system, as transport vesicles travel from the ER to the Golgi and from there to other destinations. Some vesicles develop into lysosomes and vacuoles.

Other transport vesicles fuse with the plasma membrane and contribute their membranes to it. The blue color inside the endomembrane organelles and outside the cell highlights the fact that an ER product can get out of the cell without ever actually crossing through a membrane. A transport vesicle containing the product fuses with the plasma membrane and releases the product to the outside.

Next, we look at two organelles that are not part of the endomembrane system—the chloroplast and the mitochondrion. Both contain some DNA and ribosomes and make some of their own proteins—indications of their origin as bacteria that established residence in ancestral eukaryotic cells (see Module 16.17). Chloroplasts and mitochondria are energy processors: They convert energy to forms that living cells can use.

[handwritten: if not connected physically than they are functually.]

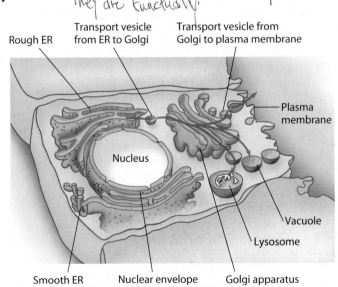

Rough ER   Transport vesicle from ER to Golgi   Transport vesicle from Golgi to plasma membrane

Plasma membrane

Nucleus

Vacuole

Lysosome

Smooth ER   Nuclear envelope   Golgi apparatus

Figure 4.13 Connections among the organelles of the endomembrane system

? How do transport vesicles help tie together the endomembrane system?

■ Transport vesicles move membranes and substances they enclose between other components of the endomembrane system.

4.14 Chloroplasts convert solar energy to chemical energy

Most of the living world runs on the energy provided by photosynthesis, the conversion of light energy from the sun to the chemical energy of sugar molecules. **Chloroplasts** are the photosynthesizing organelles of all photosynthetic eukaryotes. The chloroplast's solar power system is much more successful than anything yet produced by human ingenuity.

Befitting an organelle that carries out complex, multistep processes, internal membranes partition the chloroplast into three major compartments (Figure 4.14). The narrow intermembrane space, between the outer and inner membranes of the chloroplast, is one compartment. A second, the space enclosed by the inner membrane, contains a thick fluid called

stroma and a network of tubules and interconnected hollow disks formed of membranes. The space inside the tubules and disks constitutes a third compartment. Notice that the disks occur in stacks, each called a **granum** (plural, *grana*). The grana are the chloroplast's solar power packs—the sites where chlorophyll actually traps solar energy. As we will discuss in Chapter 7, each part of the chloroplast plays a particular role in converting solar energy to chemical energy.

? What does photosynthesis accomplish?

■ The conversion of light energy to chemical energy stored in sugar molecules

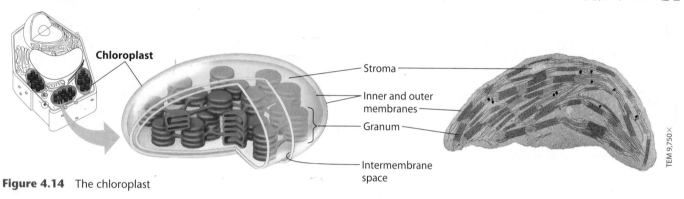

Figure 4.14   The chloroplast

4.15 Mitochondria harvest chemical energy from food

Mitochondria (singular, *mitochondrion*) are organelles that carry out cellular respiration in nearly all eukaryotic cells, converting the chemical energy of foods such as sugars to the chemical energy of a molecule called ATP (adenosine triphosphate). ATP is the main energy source for cellular work.

The mitochondrion's structure suits its function. Like the chloroplast, the mitochondrion is enclosed by two membranes (Figure 4.15). However, the mitochondrion has only two compartments. The **intermembrane space** forms one fluid-filled compartment. The inner membrane encloses the second compartment, containing a fluid called the **mitochondrial matrix**. Many of the chemical reactions of cellular respiration occur in the matrix. The inner membrane is highly folded, and enzyme molecules that make ATP are embedded in it. The folds, called **cristae**, increase the membrane's surface area, enhancing the mitochondrion's ability to produce ATP. We will discuss the role of mitochondria in cellular respiration in more detail in Chapter 6.

Web/CD Activity 4G *Build a Chloroplast and a Mitochondrion*

? What is cellular respiration?

■ A process that converts the chemical energy of sugars and other molecules to chemical energy in the form of ATP

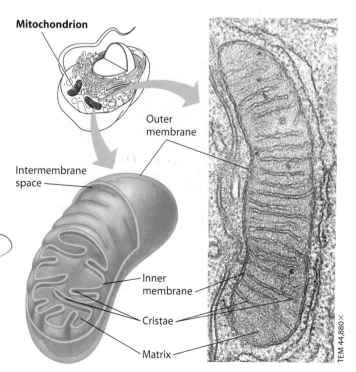

Figure 4.15   The mitochondrion

4.16 The cell's internal skeleton helps organize its structure and activities

Eukaryotic cells contain a meshwork of protein fibers, collectively called the **cytoskeleton**, extending throughout the cytoplasm of a cell. These fibers provide structural support (a skeleton) and are involved in various types of cell movement. Mounting evidence also suggests that they help regulate cellular activities by mechanically transmitting signals from the cell's surface to its interior.

Three main kinds of fibers make up the cytoskeleton: microfilaments, the thinnest type of fiber; microtubules, the thickest; and intermediate filaments, in between in thickness. **Figure 4.16** shows micrographs of three cells, each stained with a fluorescent dye that selectively highlights one of these three types of fibers.

Microfilaments, also called actin filaments, are solid rods composed mainly of globular proteins called actin, arranged in a twisted double chain (bottom left of Figure 4.16). Microfilaments form a three-dimensional network just inside the plasma membrane that helps support the cell's shape. Actin filaments often interact with other kinds of protein filaments to make cells contract. As we will see in Chapter 30, actin filaments and thicker filaments made of the motor protein myosin interact to cause contraction of muscle cells. Localized contractions brought about by actin and myosin are involved in the amoeboid (crawling) movement of the protist *Amoeba* and certain of our white blood cells. Actin filaments also help these cells change shape and move by assembling (adding subunits) at one end while disassembling (losing subunits) at the other.

Intermediate filaments are a varied group. They are made of fibrous proteins rather than globular ones and have a ropelike structure. Intermediate filaments serve mainly for reinforcing cell shape and anchoring certain organelles. For instance, the nucleus is often held in place by a cage of intermediate filaments.

Microtubules are straight, hollow tubes composed of globular proteins called tubulins. As shown in Figure 4.16, microtubules elongate by adding subunits consisting of tubulin pairs. They are readily disassembled in a reverse manner, and the tubulin subunits can then be reused. Microtubules that provide rigidity and shape in one area may disassemble and then reassemble elsewhere in the cell.

Other important functions of microtubules are to provide anchorage for organelles and to act as tracks for organelle movement within the cytoplasm. Organelles equipped with motor proteins may "walk" along microtubule tracks. For example, a lysosome might move along a microtubule to reach a food vacuole. Microtubules also guide the movement of chromosomes when cells divide, and as we see next, they are the main structural components of cilia and flagella.

? Which component of the cytoskeleton is most important in (a) holding the nucleus in place within the cell; (b) guiding transport vesicles from the Golgi to the plasma membrane; (c) contracting muscle cells?

■ (a) intermediate filaments; (b) microtubules; (c) microfilaments

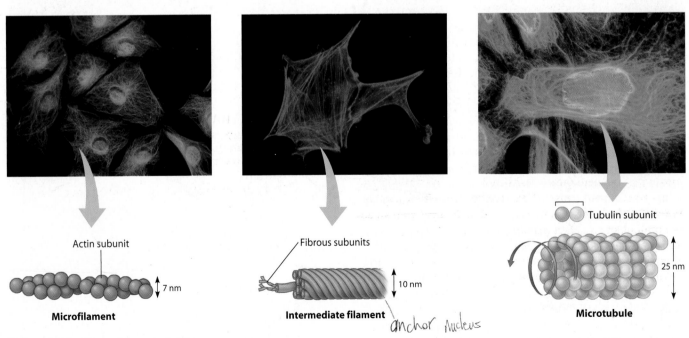

Actin subunit

7 nm

Microfilament

Fibrous subunits

10 nm

Intermediate filament

Tubulin subunit

25 nm

Microtubule

Figure 4.16 Fibers of the cytoskeleton

4.17 Cilia and flagella move when microtubules bend

The role of the cytoskeleton in movement is clearly seen in eukaryotic flagella and cilia, the locomotor appendages that protrude from certain cells. The short, numerous appendages that propel protists such as *Paramecium* (see Figure 4.12B) are called **cilia** (singular, *cilium*). Longer, generally less numerous appendages on protists such as *Euglena* (see Figure 4.1C) are called **flagella**. Some cells of multicellular organisms also have cilia or flagella. For example, **Figure 4.17A** shows cilia on cells lining the human windpipe. In this case, the cilia sweep mucus containing trapped debris out of our lungs. Most animals and some plants have flagellated sperm. A flagellum, shown in **Figure 4.17B**, propels the cell by an undulating whiplike motion. In contrast, cilia work more like the coordinated oars of a rowing team.

Though different in length and beating pattern, flagella and cilia have a common structure and mechanism of movement (**Figure 4.17C**). Both flagella and cilia are composed of microtubules wrapped in an extension of the plasma membrane. A ring of nine microtubule doublets surrounds a central pair of microtubules. This arrangement, found in nearly all eukaryotic flagella and cilia, is called the 9 + 2 pattern.

The microtubule assembly extends into an anchoring structure called a **basal body**, which has a pattern of nine microtubule triplets. When a cilium or flagellum begins to grow, the basal body may act as a foundation for microtubule assembly from tubulin subunits. Basal bodies are identical in structure to **centrioles** (shown in Figure 4.4A). We will encounter centrioles again in Chapter 8, when we discuss cell division.

How does this microtubule assembly produce the bending movement of these organelles? Bending involves motor proteins called dynein arms that are attached to each microtubule doublet. Using energy from ATP, the dynein arms grab an adjacent doublet and exert a sliding force as they start to "walk" along it. The arms then release and reattach a little farther along the doublet. The doublets are held together by crosslinking proteins (not illustrated) and radial spokes extending to the central microtubules; if they were not held in place, the walking action would make the doublets slide past each other. Instead, the movements of the dynein arms cause the microtubules (and consequently the flagellum or cilium) to bend.

Figure 4.17A Cilia on cells lining the respiratory tract

Colorized SEM 4,100×

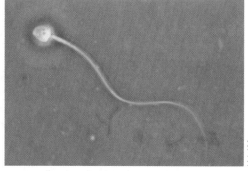

Figure 4.17B Undulating flagellum on a sperm cell

LM 600×

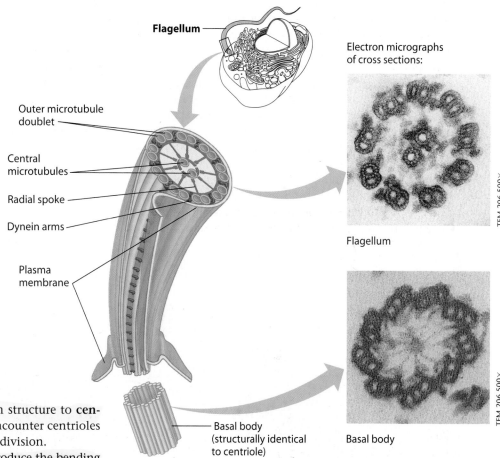

Figure 4.17C Structure of a eukaryotic flagellum or cilium

Flagellum

Outer microtubule doublet

Central microtubules

Radial spoke

Dynein arms

Plasma membrane

Basal body (structurally identical to centriole)

Electron micrographs of cross sections:

Flagellum

TEM 206,500×

Basal body

TEM 206,500×

Web/CD Activity 4H *Cilia and Flagella*

? How do cilia and flagella bend?

■ Dynein arms, powered by ATP, move neighboring doublets of microtubules relative to one another. Because they are anchored within the organelle, the doublets bend instead of sliding past one another.

4.18 Cell surfaces protect, support, and join cells

You might guess that the delicate plasma membrane alone could not handle all the challenges of the environment outside a cell. In fact, most cells have additional surface coverings surrounding the plasma membrane. We introduced the cell walls and capsules of prokaryotes in Module 4.3. Because most prokaryotes exist as single cells or as loose aggregates of cells, their surface coverings interact mainly with noncellular surroundings. In contrast, most eukaryotes are composed of many cells, which are organized into a single, functional organism.

In plants, rigid cell walls not only protect the cells but provide the skeletal support that keeps plants upright on land. Typically 10 to 100 times thicker than the plasma membrane, plant cell walls consist of fibers of the polysaccharide cellulose embedded in a matrix of other polysaccharides and proteins. This tough, fibers-in-a-matrix construction resembles that of fiberglass, also noted for its strength. **Figure 4.18A** shows how the walls of plant cells are arranged. Notice that the cell walls are multilayered. Between the walls of adjacent cells is a layer of sticky polysaccharides (dark brown) that glues the cells together. Rigid molecules called lignin strengthen the cell walls of many plants. These strong cell walls are the main component of wood.

Despite their thickness, plant cell walls do not totally isolate the cells from each other. To function in a coordinated way as part of a tissue, the cells must have cell junctions, structures that connect them to one another. As Figure 4.18A shows, numerous **plasmodesmata** (singular, *plasmodesma*), channels between adjacent plant cells, form a circulatory and communication system connecting the cells in plant tissues. Notice that the plasma membrane and the cytoplasmic fluid of the cells extend through the plasmodesmata, so that water and other small molecules can readily pass from cell to cell. Through plasmodesmata, the cells of a plant tissue share water, nourishment, and chemical messages.

Animal cells lack rigid cell walls, but most of them secrete and are embedded in a sticky layer of glycoproteins, the **extracellular matrix** (Figure 4.18B). This layer helps hold cells together in tissues and can have protective and supportive functions, too. Researchers have discovered that the extracellular matrix helps regulate cell behavior, probably through contact with proteins in the plasma membrane, which in turn are in contact with fibers of the cytoskeleton (see Figure 5.12).

Adjacent cells in many animal tissues also connect by cell junctions; there are three general types. **Tight junctions** bind cells very tightly together, forming a leakproof sheet. Such a sheet of tissue lines the digestive tract, preventing the contents from leaking into surrounding tissues. **Anchoring junctions** rivet cells together with cytoskeletal fibers, forming strong sheets. Anchoring junctions are common in tissues subject to stretching or mechanical stress, such as skin and heart muscle. **Gap junctions** are channels similar in function to the plasmodesmata of plants; they allow small molecules to flow between neighboring cells. For example, the flow of ions through gap junctions in the cells of heart muscle coordinates their contraction. These junctions are especially common in animal embryos, where chemical communication between cells is essential for development.

Web/CD Activity 4I *Cell Junctions*

> ? How is a plant or animal tissue different from just an aggregate of similar cells?

■ Cell junctions integrate the cells of a tissue.

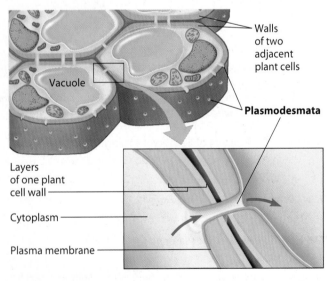

Figure 4.18A   Plant cell walls and cell junctions

Walls of two adjacent plant cells

Vacuole

Plasmodesmata

Layers of one plant cell wall

Cytoplasm

Plasma membrane

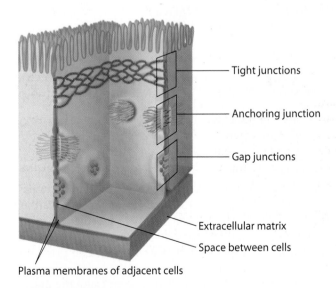

Figure 4.18B   Animal cell surfaces and cell junctions

Tight junctions

Anchoring junction

Gap junctions

Extracellular matrix

Space between cells

Plasma membranes of adjacent cells

4.19 Eukaryotic organelles comprise four functional categories

We have introduced many important cell structures in this chapter. To provide a framework for this information and reinforce the theme that structure is correlated with function, we can group the eukaryotic cell organelles into four categories by general function, as shown in Table 4.19.

The first category is manufacturing. Here we include not only the synthesis of molecules, but also their transport within the cell. The second category includes three organelles that break down and recycle materials that are harmful or no longer needed. (Vacuoles are included here, although, being multifunctional, they do not fit neatly into any one of our categories.) The third category contains the two energy-processing organelles. The fourth category is support, movement, and intercellular communication. These three functions are related because for movement to occur, there must be some sort of rigid support against which force can be applied. And when a supporting structure forms the cell's outer boundary, it is necessarily involved in the cell's communication with its neighbors.

Within each of the four categories, a structural similarity underlies the general function of the organelles. In the first category, manufacturing depends heavily on a network of metabolically active membranes. In the second category, all the organelles listed are composed of single membranous sacs, inside of which materials can be safely broken down. In the third category, expanses of metabolically active membranes and intermembrane compartments within the organelles make it possible for chloroplasts and mitochondria to perform complex energy conversions that power the cell. Even in the diverse fourth category, there is a common structural theme in the various fibers involved in the functioning of most of these organelles.

We can summarize further by emphasizing that these four categories of organelles form an integrated team and that properties of life at the cellular level emerge from the coordinated functions of the team members. We will see many examples of the integrated actions of cellular organelles in later chapters.

All life-forms on our planet share the fundamental features of (1) consisting of cells, each enclosed by a membrane that

TABLE 4.19	EUKARYOTIC ORGANELLES AND THEIR FUNCTIONS
General Function: Manufacturing	
Nucleus	DNA synthesis; RNA synthesis; assembly of ribosomal subunits (in nucleoli)
Ribosomes	Polypeptide (protein) synthesis
Rough ER	Synthesis of membrane proteins, secretory proteins, and hydrolytic enzymes; formation of transport vesicles
Smooth ER	Lipid synthesis; carbohydrate metabolism in liver cells; detoxification in liver cells; calcium ion storage
Golgi apparatus	Modification, temporary storage, and transport of macromolecules; formation of lysosomes and transport vesicles
General Function: Breakdown	
Lysosomes (in animal cells and some protists)	Digestion of nutrients, bacteria, and damaged organelles; destruction of certain cells during embryonic development
Peroxisomes	Diverse metabolic processes, with breakdown of H_2O_2 by-product
Vacuoles	Digestion (like lysosomes); storage of chemicals; cell enlargement; water balance
General Function: Energy Processing	
Chloroplasts (in plants and some protists)	Conversion of light energy to chemical energy of sugars
Mitochondria	Conversion of chemical energy of food to chemical energy of ATP
General Functions: Support, Movement, and Communication Between Cells	
Cytoskeleton (including cilia, flagella, and centrioles in animal cells)	Maintenance of cell shape; anchorage for organelles; movement of organelles within cells; cell movement; mechanical transmission of signals from exterior of cell to interior
Cell walls (in plants, fungi, and some protists)	Maintenance of cell shape and skeletal support; surface protection; binding of cells in tissues
Extracellular matrix (in animals)	Binding of cells in tissues; surface protection; regulation of cellular activities
Cell junctions	Communication between cells; binding of cells in tissues

maintains internal conditions very different from the surroundings; (2) having DNA as the genetic material; and (3) carrying out metabolism, which involves the interconversion of different forms of energy and of chemical materials. We expand on the subjects of membranes and metabolism in Chapter 5.

Web/CD Activity 4J *Review: Animal Cell Structure and Function*

Web/CD Activity 4K *Review: Plant Cell Structure and Function*

Web/CD Thinking as a Scientist *Connection: How Are Space Rocks Analyzed for Signs of Life?*

? How do mitochondria, smooth ER, and the cytoskeleton all contribute to the contraction of a muscle cell?

■ Mitochondria supply energy in the form of ATP. The smooth ER helps regulate contraction by the uptake and release of calcium. Microfilaments function as the actual contractile apparatus.

Reviewing the Concepts

Introduction to the Cell (4.1–4.4)

Microscopy. The light microscope can magnify cells, both living and preserved, up to 1,000 times. The greater magnification and resolution of electron microscopes reveals the ultrastructure of cells **(4.1)**. The microscopic size of most cells ensures a sufficient surface area across which nutrients and wastes can move to service the cell volume **(4.2)**.

Prokaryotic and eukaryotic cells. All cells have a plasma membrane, DNA, ribosomes, and cytoplasm. Prokaryotic cells—bacteria and archaea—are small, relatively simple cells that do not have a membrane-bounded nucleus **(4.3)**. All other forms of life are composed of more complex eukaryotic cells, distinguished by the presence of a true nucleus. Membranes form the boundaries of many eukaryotic organelles, compartmentalizing the interior of the cell and facilitating a variety of metabolic activities **(4.4)**.

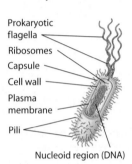

Prokaryotic flagella
Ribosomes
Capsule
Cell wall
Plasma membrane
Pili
Nucleoid region (DNA)

Organelles of the Endomembrane System (4.5–4.13)

Nucleus. Surrounded by a porous nuclear envelope, the nucleus contains the DNA that carries the cell's hereditary blueprint and directs its activities **(4.5)**.

Endomembrane system is a collection of membranous organelles that manufacture and distribute cell products **(4.6)**. The various organelles of the endomembrane system are interconnected structurally and functionally **(4.13)**.

Endoplasmic reticulum (ER) is a membranous network of tubes and sacs. **Smooth ER** synthesizes lipids, processes toxins and drugs in liver cells, and stores and releases calcium ions in muscle cells **(4.7)**. **Rough ER** manufactures membranes, and ribosomes on its surface produce proteins that are secreted, inserted into membranes, or transported in vesicles to other organelles **(4.8)**.

Golgi apparatus consists of stacks of membranous sacs that modify ER products, then ship them to other organelles or to the cell surface **(4.9)**.

Lysosomes are sacs of enzymes that function in digestion and recycling within the cell **(4.10–4.11)**.

Vacuoles. Plant cells contain a large central vacuole, with lysosomal, storage, and growth functions. Some protists have contractile vacuoles **(4.12)**.

Energy-Converting Organelles (4.14–4.15)

Chloroplasts are present in plants and some protists, converting solar energy to chemical energy in sugars **(4.14)**.

Mitochondria carry out cellular respiration, using the energy in food to make ATP for cellular work **(4.15)**.

The Cytoskeleton and Related Structures (4.16–4.17)

Cytoskeleton is a structural network of protein fibers. Microfilaments of actin enable cells to change shape and move. Intermediate filaments reinforce the cell and anchor certain organelles. Microtubules give the cell rigidity and act as tracks for organelle movement **(4.16)**. Eukaryotic cilia and flagella are locomotor appendages made of microtubules in a "9 + 2" arrangement **(4.17)**.

Cell Surfaces and Junctions (4.18)

Plant cells are supported by rigid cell walls made largely of cellulose. Plasmodesmata are connecting channels between plant cells.

Animal cells. The extracellular matrix of animal cells consists mainly of glycoproteins. Tight junctions bind cells to form leakproof sheets. Anchoring junctions rivet cells into strong tissues. Gap junctions allow substances to flow from cell to cell.

Functional Categories of Organelles (4.19)

Eukaryotic organelles fall into four functional groups: (1) manufacturing; (2) breakdown; (3) energy processing; and (4) support, movement, and communication between cells.

Connecting the Concepts

1. Label the structures in this diagram of an animal cell. Review the functions of each of these organelles.

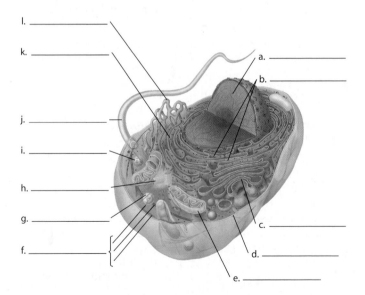

l.
k.
j.
i.
h.
g.
f.
a.
b.
c.
d.
e.

2. List three structures found in animal cells but not in plant cells.

3. List three structures found in plant cells but not in animal cells.

Testing Your Knowledge

4. The ultrastructure of a chloroplast could be best studied using a
   a. light microscope.
   b. telescope.
   c. scanning electron microscope.
   d. transmission electron microscope.
   e. light microscope and fluorescent dyes.

5. The cells of an ant and a horse are, on average, the same small size; a horse just has more of them. What is the main advantage of small cell size?
   a. Small cells are less likely to burst than large cells.
   b. A small cell has a larger plasma membrane surface area than does a large cell.
   c. Small cells can better take up sufficient nutrients and oxygen to provide for their needs.
   d. It takes less energy to make an organism out of small cells.
   e. Small cells require less oxygen than do large cells.

6. Which of the following clues would tell you whether a cell is prokaryotic or eukaryotic?
   a. the presence or absence of a rigid cell wall
   b. whether or not the cell is partitioned by internal membranes
   c. the presence or absence of ribosomes
   d. whether or not the cell carries out cellular metabolism
   e. whether or not the cell contains DNA

7. Which of the following structures is *not* directly involved in cell support or movement?
   a. microfilament   d. lysosome
   b. flagellum   e. cell wall
   c. microtubule

   Choose from the following cells for questions 8–12:
   a. muscle cell in thigh of long-distance runner
   b. pancreatic cell that secretes digestive enzymes
   c. ovarian cell that produces the steroid hormone estrogen
   d. cell in tissue layer lining digestive tract
   e. white blood cell that engulfs bacteria

8. In which cell would you find the most lysosomes?

9. In which cell would you find the most mitochondria?

10. In which cell would you find the most smooth ER?

11. In which cell would you find the most rough ER?

12. In which cell would you find the most tight junctions?

13. A type of cell called a lymphocyte makes proteins that are exported from the cell. You can track the path of these proteins within the cell by labeling them with radioactive isotopes. Which of the following might be the path of a protein from the site where its polypeptides are made to its export?
    a. chloroplast . . Golgi . . lysosomes . . plasma membrane
    b. Golgi . . rough ER . . smooth ER . . transport vesicle
    c. rough ER . . Golgi . . transport vesicle . . plasma membrane
    d. smooth ER . . Golgi . . lysosome . . plasma membrane
    e. nucleus . . Golgi . . rough ER . . plasma membrane

Describing, Comparing, and Explaining

14. What three cellular components are shared by prokaryotic and eukaryotic cells?

15. Briefly describe the three kinds of junctions that can connect animal cells, and compare their functions.

16. What general function do the chloroplast and mitochondrion have in common? How are their functions different?

17. How does a eukaryotic cell benefit from its internal membranes?

18. Describe two different ways in which the motion of cilia can function in organisms.

19. Explain how a protein inside the ER can be exported from the cell without ever crossing a membrane.

20. Is this statement true or false? "Animal cells have mitochondria; plant cells have chloroplasts." Explain.

Applying the Concepts

21. Imagine a spherical cell with a radius of 10 μm. What is the cell's surface area in μm^2? Its volume, in μm^3? What is the ratio of surface area to volume for this cell? Now do the same calculations for a second cell, this one with a radius of 20 μm. Compare the surface-to-volume ratios of the two cells. How is this comparison significant to the functioning of cells? (*Note:* For a sphere of radius *r*, surface area = $4\pi r^2$ and volume = $\frac{4}{3}\pi r^3$. The value of π is 3.14.)

22. The ciliated lining of your respiratory system sweeps mucus with trapped debris out of your lungs. Tobacco smoke, whether inhaled first or secondhand, irritates these cells, inhibiting or destroying the cilia. This interferes with the normal cleansing mechanisms and allows more toxin-laden smoke particles to reach the lungs. Human sperm rely on flagella to travel up the female reproductive tract. Interestingly, some men with a certain type of hereditary sterility also suffer from respiratory problems. Propose an explanation for this linkage of health problems.

23. Doctors at UCLA removed John Moore's spleen, standard treatment for his type of leukemia, and the disease did not recur. Researchers kept the spleen cells alive in a nutrient medium. They found that some of the cells produced a blood protein that showed promise as a treatment for cancer and AIDS. The researchers patented the cells. Moore sued, claiming a share in profits from any products derived from his cells. The U.S. Supreme Court ruled against Moore, stating that his suit "threatens to destroy the economic incentive to conduct important medical research." Moore argued that the ruling left patients "vulnerable to exploitation at the hands of the state." Do you think Moore was treated fairly? Is there anything else you would like to know about this case that might help you decide?

Answers to all questions can be found in Appendix 3.

For study help and Activities, go to campbellbiology.com or the student CD-ROM.

CHAPTER 5

A firefly flashing for a mate

ENERGY AND THE CELL

5.1 Energy is the capacity to perform work
5.2 Two laws govern energy transformations
5.3 Chemical reactions either store or release energy
5.4 ATP shuttles chemical energy and drives cellular work

HOW ENZYMES FUNCTION

5.5 Enzymes speed up the cell's chemical reactions by lowering energy barriers
5.6 A specific enzyme catalyzes each cellular reaction
5.7 The cellular environment affects enzyme activity
5.8 Enzyme inhibitors block enzyme action
5.9 Many poisons, pesticides, and drugs are enzyme inhibitors

MEMBRANE STRUCTURE AND FUNCTION

5.10 Membranes organize the chemical activities of cells
5.11 Membrane phospholipids form a bilayer
5.12 The membrane is a fluid mosaic of phospholipids and proteins
5.13 Proteins make the membrane a mosaic of function
5.14 Passive transport is diffusion across a membrane
5.15 Transport proteins may facilitate diffusion across membranes
5.16 Osmosis is the diffusion of water across a membrane
5.17 Water balance between cells and their surroundings is crucial to organisms
5.18 Cells expend energy for active transport
5.19 Exocytosis and endocytosis transport large molecules
5.20 Faulty membranes can overload the blood with cholesterol
5.21 Chloroplasts and mitochondria make energy available for cellular work

Cool "Fires" Attract Mates and Meals

BRIGHT YELLOW FLASHES in a dark field—we could be almost anywhere in the eastern or central United States. The light display comes from insects commonly known as fireflies or lightning bugs. Males on the wing do most of the flashing. The females are perched on leaves close to the ground.

Fireflies use light to send signals to potential mates, instead of using chemical signals like most other insects. When a female sees flashes of light from a male of her species, she reacts with flashes of her own. If the male sees her flashes, he automatically gives another display and flies in the female's direction. Members of both sexes are responding to particular patterns of light flashes characteristic of their species.

Each of the 2,000 or so species of fireflies has its own way to signal a mate

Mating occurs after the female's display leads a male to her, and most females stop flashing after they mate. But in a few species, a mated female will continue to flash, using a pattern that attracts males of *other* firefly species. A veritable *femme fatale*, she waits until the male gets close, then grabs him and eats him. In the photograph on the opposite page, the firefly clinging to the leaf is a female dining on a luckless male of another species.

Each of the 2,000 or so species of fireflies has its own way to signal a mate. Some flash more often than others or during

The Working Cell

A female firefly devouring a male of another species

different hours, while other species give fewer but longer flashes. Many species produce light of a characteristic color: yellow, bluish green, or reddish. In areas where fireflies congregate—often on lawns, golf courses, or open meadows—you can usually see several different species signaling. Because the insects respond instinctively to specific flash patterns, you can even attract males to artificial light if you flash it a certain way.

From a biological standpoint, fireflies are poorly named. They are beetles, not flies, and their light is almost cold, not fiery. In fact, in emitting light, they give off only about one hundred-thousandth of the amount of heat that would be produced by a candle flame of equal brightness.

What happens in a firefly that makes this light? The light comes from a set of chemical reactions that occur in light-producing organs at the rear of the insect. Light-emitting cells in these organs contain an acidic substance called luciferin and an enzyme called luciferase (both from the Latin term for "light bearer"). In the presence of oxygen (O_2) and chemical energy

from ATP molecules, luciferase catalyzes the conversion of luciferin to a molecule that emits light.

The luciferin-luciferase system is one example of how living cells put energy to work by means of enzyme-controlled chemical reactions. Light is a form of energy, and the firefly's ability to make light energy from chemical energy is an example of life's dependence on energy conversions.

Many of the enzymes that control a cell's chemical reactions, including the firefly's luciferase, are located in membranes. Membranes thus serve as sites where chemical reactions can occur in an orderly manner. Also, as we mentioned in Chapter 4, membranes control the passage of crucial substances into and out of cells, and they partition the eukaryotic cell into useful compartments. In fact, the firefly could not produce light, nor could any cell or organism survive, without membranes. Everything that happens in the firefly's light organs has some relation to this chapter's subject: how working cells use energy, enzymes, and membranes. ■ ■ ■

5.1 Energy is the capacity to perform work

eg. strach has potential energy

The title of this chapter is "The Working Cell." But just what type of work does a cell do? A cell is a chemical factory in which thousands of reactions occur within a microscopic space. Some of these reactions assemble monomers into polymers that can be used to build the elaborate structures you learned about in Chapter 4. Other reactions break down sugars and other fuels to release energy. The cell uses this energy to manufacture cellular parts and products, to transport materials across membranes, and to move cellular structures or even the whole cell. Life requires energy, and cells are energy transformers. To understand how the living cell works, you must have a basic knowledge of energy.

Energy is defined as the capacity to perform work. Work is performed when an object is moved against an opposing force, such as gravity or friction. Put another way, energy is the capacity to rearrange matter. There are two basic forms of energy: kinetic energy and potential energy.

Kinetic energy is the energy of motion (Figure 5.1A). Moving objects can perform work by transferring motion to other matter. The contraction of leg muscles pushes bicycle pedals, moving the bike and its rider up a hill. **Heat**, or thermal energy, is a form of kinetic energy associated with the movement of molecules or atoms in a body of matter. Light is

eg. muscle contraction.

another kind of kinetic energy that can be harnessed to perform work, such as powering photosynthesis in green plants. **Potential energy**, the second form of energy, is stored energy that an object possesses as a result of its location or structure. A cyclist at the top of a hill has potential energy as a result of elevation (Figure 5.1B).

Likewise, the negatively charged electrons of an atom have potential energy owing to their positions in electron shells at a distance from the positively charged nucleus. The molecules in a living cell have potential energy as a result of the arrangement of their atoms; this potential energy is the energy that is available to do the work of the cell. **Chemical energy**, the potential energy of molecules, is the most important type of energy for living organisms. And just as the potential energy of the cyclist can be converted to kinetic energy in the speedy ride downhill (Figure 5.1C), the chemical energy of molecules can be released to power the work of the cell. Life depends on the fact that energy can be converted from one form to another. Next we consider some basic principles of energy transformation.

? How can an object at rest have energy?

■ It can have potential energy as a result of its location.

Figure 5.1B Potential energy

Figure 5.1A Kinetic energy

Figure 5.1C Potential energy being converted to kinetic energy

Two laws govern energy transformations

Thermodynamics is the study of energy transformations that occur in a collection of matter. Scientists use the word *system* for the matter under study and refer to the rest of the universe—everything outside the system—as the *surroundings*. A system can be an electric power plant, a single cell, or the entire planet. A living organism is an open system; that is, it exchanges both energy and matter with its surroundings. The firefly in the chapter introduction, for instance, takes in food and oxygen and releases heat, light, and chemical waste products, such as carbon dioxide.

The First Law of Thermodynamics According to the **first law of thermodynamics**, also known as the law of energy conservation, the total amount of energy in the universe is constant. Energy can be transferred and transformed, but it cannot be created or destroyed. An electric company does not manufacture energy; it merely converts it to a form that is convenient to use. A light bulb does not make light; it converts electricity to light energy. By converting sunlight to chemical energy, a green plant acts as an energy transformer, not an energy producer. The firefly converts some of the chemical energy of its food to kinetic energy when it flies and to light energy when it flashes.

The Second Law of Thermodynamics If energy cannot be destroyed, then why can't organisms simply recycle their energy over and over again? It turns out that during every energy transfer or transformation, some energy becomes unusable—unavailable to do work. In most energy transformations, some energy is converted to heat, the energy associated with random molecular motion. Cells cannot use this heat to perform work; it is lost to the surroundings. When muscle cells convert the chemical energy of food molecules to kinetic energy, more than half of the energy is lost as heat. Sweating and evaporative cooling help release this excess heat to the surroundings (**Figure 5.2A**).

Heat is a disordered form of energy, and its release makes the universe more random, more disorganized. The amount of disorder in a system is called **entropy.** The **second law of thermodynamics** states that energy conversions reduce the order of the universe and increase its entropy.

Let's take a closer look at how living organisms increase the amount of disorder in the universe. Energy flows into Earth's ecosystems as sunlight; photosynthesis converts light energy into chemical energy; cells transfer chemical energy between molecules in chemical reactions and convert chemical energy to kinetic energy. At every step along the way, energy is lost to the surroundings as heat.

If the universe is becoming more disordered, how can we account for biological order? A cell creates ordered structures from less organized starting materials. For example, amino acids are ordered into the specific sequences of polypeptide chains. However, this increase in order, which corresponds to a decrease in entropy, is accomplished at the expense of ordered forms of energy taken in from the surroundings. As shown in **Figure 5.2B**, cells extract the chemical energy of glucose and other fuels and return heat and the lower-energy molecules of carbon dioxide and water to their surroundings. In a thermodynamic sense, a cell or an organism is an island of low entropy in an increasingly random universe.

As we will see in the modules that follow, the work of cells is powered by the potential energy contained in molecules.

Web/CD Activity 5A *Energy Transformations*

? Describe the energy transformations that occur when you climb to the top of a stairway.

■ You convert the chemical energy of food to the kinetic energy of your upward climb. At the top of the stairs, some of the energy has been stored as potential energy owing to your higher elevation. The rest has been converted to heat.

Figure 5.2A Energy lost as heat during the conversion of chemical energy to kinetic energy

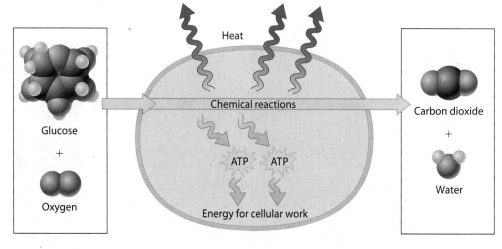

Heat

Chemical reactions

Glucose
+
Oxygen

ATP   ATP

Energy for cellular work

Carbon dioxide
+
Water

Figure 5.2B Energy transformations in a cell

Chemical reactions either store or release energy

Chemical reactions, including those that occur in cells, are of two types. One type, called **endergonic reactions**, requires a net input of energy (endergonic means "energy-in"). Endergonic reactions yield products that are rich in potential energy. As you can see in **Figure 5.3A**, an endergonic reaction starts out with reactant molecules that contain relatively little potential energy. Energy is absorbed from the surroundings as the reaction occurs, so that the products of an endergonic reaction store more energy than the reactants did. The energy is actually stored in the covalent bonds of the product molecules. And as the graph shows, the amount of additional energy stored in the products equals the difference in potential energy between the reactants and the products.

Photosynthesis, the process whereby plant cells make sugar, is one example of a strongly endergonic process. Photosynthesis starts with energy-poor reactants (carbon dioxide and water molecules) and, using energy absorbed from sunlight, produces energy-rich sugar molecules.

Some other chemical processes are exergonic. An **exergonic reaction** is a chemical reaction that releases energy (exergonic means "energy-out"). As indicated in **Figure 5.3B**, an exergonic reaction begins with reactants whose covalent bonds contain more energy than those in the products. The reaction releases to the surroundings an amount of energy equal to the difference in potential energy between the reactants and the products.

As an example of an exergonic reaction, consider what happens when wood burns. One of the major components of wood is cellulose, a large carbohydrate composed of many glucose monomers. Each glucose monomer is rich in potential energy. The burning of wood releases the potential energy of glucose as heat and light. Carbon dioxide and water are the products of the reaction.

Burning is one way to release energy from chemicals. Cells release energy by means of a different exergonic process, called cellular respiration. **Cellular respiration** is a chemical process that uses oxygen to convert the chemical energy stored in fuel molecules to a form of chemical energy that the cell can use to perform work. Burning and cellular respiration are alike in being exergonic. They differ in that burning is essentially a one-step process that releases all of a substance's energy at once. Cellular respiration, on the other hand, involves many steps, each a separate chemical reaction; you can think of it as a "slow burn." Some of the energy released from glucose by cellular respiration escapes as heat, but a substantial amount of released energy is converted to the chemical energy of ATP. Cells use ATP as an immediate source of energy.

Every working cell in every organism carries out thousands of endergonic and exergonic reactions, the sum of which is known as **cellular metabolism.** In a firefly, for instance, the light display discussed earlier is exergonic. In this case, light energy is released when reactants are converted to product molecules containing less energy than the reactants. In addition to generating light, fireflies, like all animals, must find, eat, and digest food; escape predators; grow; and reproduce. All these activities require energy, which is obtained from sugar and other food molecules by the exergonic reactions of cellular respiration. Cells then use that energy in endergonic reactions to make molecules and do the work of the cell. **Energy coupling**—the use of energy released from exergonic reactions to drive essential endergonic reactions—is a crucial ability of all cells. ATP molecules are the key to energy coupling. In the next module, we explore the structure and function of ATP.

Web/CD Activity 5B *Chemical Reactions and ATP*

? Cellular respiration is an exergonic process. Remembering that energy must be conserved, what becomes of the energy extracted from food during cellular respiration?

■ Some of it is stored in ATP molecules; the rest is released as heat.

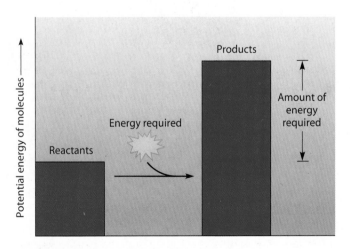

Figure 5.3A Endergonic reaction

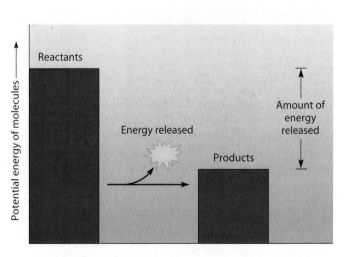

Figure 5.3B Exergonic reaction

5.4 ATP shuttles chemical energy and drives cellular work

ATP powers nearly all forms of cellular work, from the generation of light by fireflies to the movements of muscle cells that enable you to pedal a bicycle. Let's see how the structure of ATP fits its function as an energy shuttle.

ATP stands for adenosine triphosphate. As shown in **Figure 5.4A,** adenosine consists of adenine, a nitrogenous base, and ribose, a five-carbon sugar; triphosphate is a chain of three phosphate groups (each symbolized by ⓟ). All three phosphate groups are negatively charged. These like charges are crowded together, and their mutual repulsion contributes to the potential energy stored in ATP. The triphosphate chain of ATP is the chemical equivalent of a compressed spring.

Thus, the bonds connecting the phosphate groups of ATP are unstable and can readily be broken by hydrolysis. Notice in Figure 5.4A that when the bond to the third group breaks, the phosphate group leaves ATP—which becomes ADP (adenosine diphosphate)—and energy is released.

The hydrolysis of ATP is an exergonic reaction; that is, it releases energy. How does the cell couple this reaction to an endergonic one? It does so by transferring a phosphate group from ATP to some other molecule. This transfer is called **phosphorylation,** and most cellular work depends on ATP energizing molecules by phosphorylating them.

There are three main types of cellular work: chemical, mechanical, and transport. As shown in **Figure 5.4B,** ATP drives all three types of work. In chemical work, the phosphorylation of reactant molecules drives the endergonic synthesis of product molecules. In an example of mechanical work, the transfer of phosphate groups to special motor proteins in muscle cells causes the proteins to change shape and pull on actin filaments, causing the cells to contract. ATP drives transport work by phosphorylating certain membrane proteins. We discuss the ATP-driven transport of materials across membranes in Module 5.18.

Work can be sustained because ATP is a renewable resource that cells regenerate. **Figure 5.4C** shows the ATP cycle. Energy released in exergonic reactions, such as glucose breakdown during cellular respiration, is used to regenerate ATP from ADP. In this process, a phosphate group is bonded to ADP—the ADP is phosphorylated. This is an endergonic (energy-storing) reaction. A cell at work uses ATP continuously, and the ATP cycle runs at an astonishing pace. In fact, a working cell may consume and regenerate 10 million ATP molecules each second.

But even with a constant supply of energy, few metabolic reactions would occur without the assistance of enzymes. We explore these biological catalysts next.

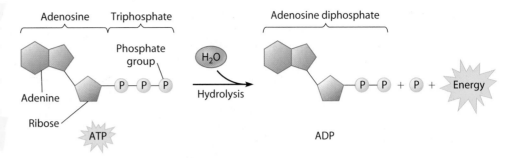

Figure 5.4A ATP structure and hydrolysis

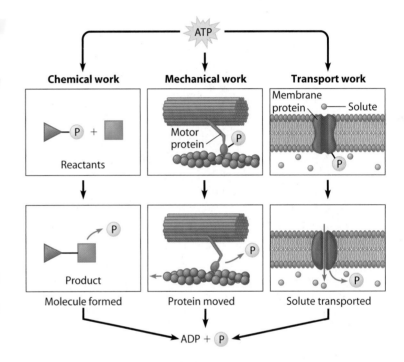

Figure 5.4B How ATP powers cellular work

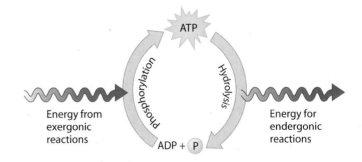

Figure 5.4C The ATP cycle

? Explain how ATP transfers energy from exergonic to endergonic processes in the cell.

■ By phosphorylation: Exergonic processes phosphorylate ADP to form ATP. ATP transfers energy to endergonic processes by phosphorylating other molecules.

5.5 Enzymes speed up the cell's chemical reactions by lowering energy barriers

Proteins, DNA, carbohydrates, phospholipids—most of the cell's molecules are rich in potential energy. What prevents these molecules from spontaneously breaking down into simpler, less energetic molecules? There is an energy barrier that must be overcome before a chemical reaction can begin. Energy must be absorbed to contort or weaken bonds in reactant molecules so that they can break and new bonds can form. This energy barrier is the amount of energy, called the **energy of activation (E$_A$)**, that reactants must absorb to become activated and start a chemical reaction.

Figure 5.5A illustrates the E$_A$ concept with an analogy involving Mexican jumping beans. These are seeds of certain desert shrubs, each containing an insect larva whose wriggling makes the bean jump. Jumping beans in the left-hand chamber of container 1 represent reactant molecules in a chemical reaction; those in the right-hand chamber represent product molecules. The partition symbolizes the E$_A$ of the chemical reaction. Notice that the level of the left-hand chamber is higher than that of the right-hand chamber. Because they have a higher position, beans on the left have more potential energy than those on the right. But even though the beans that reach the lower chamber end up with less energy, they first had to have extra energy to make it over the partition.

Jumping beans vary in how much energy they have. Those with the most actively wriggling larvae jump the most often and the highest. At any particular instant, only a few beans may jump high enough to clear the partition between the two chambers and become "products." Thus, it may take a very long time for a significant number of reactant beans to reach the right-hand chamber.

Now we have a dilemma. Most of the essential reactions of metabolism must occur quickly and precisely for a cell to survive. How can the specific reactions that a cell requires get over that energy barrier? One way to speed reactions is to add heat; even jumping beans jump higher when you warm them up. But heating a cell would speed all reactions, not just the necessary ones, and too much heat would denature proteins and kill a cell.

The solution lies in the special capabilities of enzymes. An **enzyme** is a protein molecule that functions as a biological catalyst, increasing the rate of a reaction without itself being changed into a different molecule. An enzyme does not add energy to a cellular reaction; it speeds up a reaction by lowering the E$_A$ barrier. Without enzymes, most metabolic reactions would occur too slowly to sustain life.

In Figure 5.5A, container 2 symbolizes the same chemical reaction as container 1, but catalyzed by an enzyme. The enzyme, in effect, pushes down the partition, making it possible for beans with less energy to clear it. As a result, in a given period of time, more beans will end up in the right-hand product chamber than would without the enzyme.

In **Figure 5.5B**, a graph shows the effect of an enzyme on the reaction it catalyzes. The black curve represents the course of the reaction without an enzyme; the E$_A$ barrier is higher than in the reaction with an enzyme (red curve). Notice that the net change in energy from start to finish, analogous to the difference in heights of the two chambers in the jumping-bean model, is the same for both curves.

In catalyzing metabolic reactions in cells, enzymes are essential to life. Next we consider how enzymes lower the energy of activation for the specific reactions they catalyze.

> **?** Explain why an enzyme cannot change an endergonic reaction into an exergonic one.

■ Although an enzyme speeds a reaction by lowering the energy of activation, it has no effect on the relative energy content of products versus reactants.

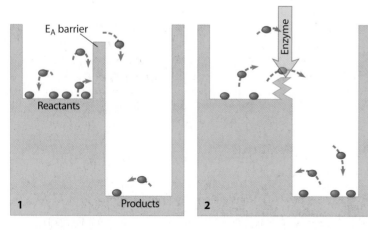

Figure 5.5A Jumping-bean analogy for energy of activation (E$_A$) and the role of enzymes

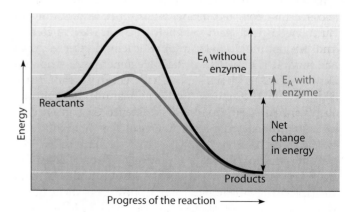

Figure 5.5B The effect of an enzyme on E$_A$

5.6 A specific enzyme catalyzes each cellular reaction

As a protein, an enzyme has a unique three-dimensional shape, and that shape determines which chemical reaction the enzyme catalyzes. A specific reactant that an enzyme acts on is called a **substrate** of the enzyme. A substrate fits into a region of the enzyme called an **active site.** An active site is typically a pocket or groove on the surface of the enzyme. The enzyme is specific because an active site fits only one kind of substrate molecule. Thus, it takes many different kinds of enzymes to catalyze all the reactions in a cell.

When a substrate binds to an enzyme, the active site changes shape slightly so that it embraces the substrate more snugly, like a clasping handshake. This **induced fit** may strain substrate bonds or place chemical groups of the active site in position to catalyze the reaction. In reactions involving two or more reactants, the active site may hold the substrates in the proper orientation for a reaction to occur.

Figure 5.6 illustrates the catalytic cycle of an enzyme. Our example is the enzyme sucrase, which catalyzes the hydrolysis of sucrose (table sugar) to glucose and fructose. (Most enzymes have names that end in -ase, and many are named for their substrate.) ❶ Sucrase starts with an empty active site. ❷ Sucrose enters the active site, attaching by weak bonds. An induced fit distorts the sucrose molecule. ❸ The weakened bond reacts with water, and the substrate is converted (hydrolyzed) to the products glucose and fructose. ❹ The enzyme releases the products and emerges unchanged from the reaction. Its active site is now available for another substrate molecule, and another round of the cycle can begin. A single enzyme molecule may act on thousands or even millions of substrate molecules per second.

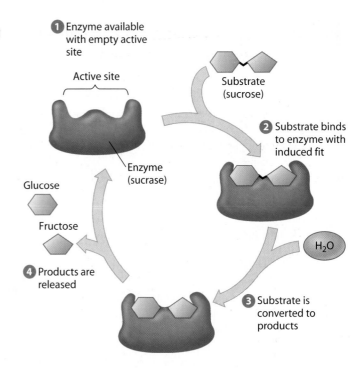

- ❶ Enzyme available with empty active site
- Active site
- Enzyme (sucrase)
- Glucose
- Fructose
- ❹ Products are released
- Substrate (sucrose)
- ❷ Substrate binds to enzyme with induced fit
- H_2O
- ❸ Substrate is converted to products

Figure 5.6 The catalytic cycle of an enzyme

Web/CD Activity 5D *How Enzymes Work*

? What is meant by "induced fit"?

■ Induced fit is the slight change in shape of the active site of an enzyme as it embraces its substrate. In its new shape, the active site catalyzes the reaction.

5.7 The cellular environment affects enzyme activity

As with all proteins, an enzyme's structure and shape are essential to its function, and its shape is affected by changes in its environment. For every enzyme, there are conditions under which it is most effective. Temperature, for instance, affects molecular motion, and an enzyme's optimal temperature produces the highest rate of contact between reactant molecules and the enzyme's active site. Higher temperatures denature the enzyme, altering its specific three-dimensional shape and destroying its function. Most human enzymes work best at 35–40°C (95–104°F), close to our normal body temperature. Bacteria that live in hot springs, however, contain enzymes with optimal temperatures of 70°C (158°F) or higher.

Salt concentration and pH also influence enzyme activity. Few enzymes can tolerate extremely salty solutions because the salt ions interfere with some of the chemical bonds that maintain protein structure. The same is true of the extra hydrogen ions present at very low pH. The optimal pH for most enzymes is near neutrality, in the range of 6–8. Outside this range, enzyme action, and thus the normal chemical

functioning of cells, may be impaired. In some locations, entire lakes are threatened by acid precipitation, which can make the water so acidic that the enzymes of aquatic organisms are affected (see Module 2.15).

Many enzymes will not function without nonprotein helpers called **cofactors.** Cofactors may be inorganic substances, such as the ions of zinc, iron, or copper. If the cofactor is an organic molecule, it is called a **coenzyme.** Most coenzymes are made from vitamins or are vitamins themselves. For example, vitamin B_6 is a coenzyme required by enzymes involved in converting one amino acid to another.

Web/CD Thinking as a Scientist *How Is the Rate of Enzyme Catalysis Measured?*

? A few human enzymes work best at very low pH, about 2. Where in the body do you think these enzymes are located?

■ In the stomach

5.8 Enzyme inhibitors block enzyme action

A chemical that interferes with an enzyme's activity is called an inhibitor. If an inhibitor attaches to the enzyme by covalent bonds, the inhibition is usually irreversible. Toxins and poisons are often irreversible inhibitors. Inhibition is reversible when weak bonds, such as hydrogen bonds, bind inhibitor and enzyme.

There are two types of enzyme inhibitors. A **competitive inhibitor** resembles the enzyme's normal substrate and competes with the substrate for the active site on the enzyme. As shown in the lower left of Figure 5.8, a competitive inhibitor reduces the productivity of an enzyme by blocking substrates from entering the active site. This type of inhibition can be overcome by increasing the concentration of substrate molecules, making it more likely that a substrate molecule will be nearby when an active site becomes vacant.

A **noncompetitive inhibitor** does not enter the active site. Instead, it binds to the enzyme somewhere else, and its binding changes the shape of the enzyme so that the active site no longer fits the substrate (lower right of figure).

Inhibitors are not always harmful; in fact, enzyme inhibition is an important mechanism for regulating cell metabolism. Most of the chemical reactions of a cell are organized into metabolic pathways in which a specific molecule is altered in a series of steps, each catalyzed by an enzyme, to form a certain product. If a cell is producing more of that product than it needs, the product may act as an inhibitor of one of the enzymes of the pathway. This sort of inhibition, whereby a metabolic reaction is blocked by its products, is called **feedback inhibition**, and it is one of the most important mechanisms that regulate metabolism.

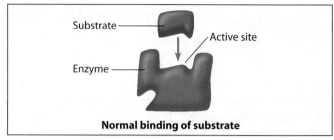

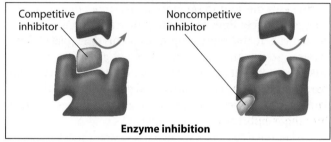

Figure 5.8 How inhibitors interfere with substrate binding

Even ATP can produce feedback inhibition. When a cell's supply of ATP exceeds demand, ATP can noncompetitively inhibit enzymes that catalyze certain early steps of the pathways driving ATP synthesis.

 What is the advantage of feedback inhibition to a cell?

■ It prevents the cell from wasting valuable resources by synthesizing more of a particular product than is necessary.

5.9 Many poisons, pesticides, and drugs are enzyme inhibitors

People have developed and used enzyme inhibitors as both constructive and destructive agents—as pesticides and beneficial drugs, but also as deadly poisons for use in warfare. When an inhibitor, especially an irreversible one, prevents an enzyme from catalyzing a crucial metabolic reaction, an organism may be poisoned.

Cyanide is a poison that inhibits an enzyme involved with the production of ATP during cellular respiration. Nerve gases such as Sarin, which was released by terrorists in the Tokyo subway in 1995, are small molecules that bind covalently to an amino acid in the active site of acetylcholinesterase. This enzyme is vital to the transmission of nerve impulses, and its inhibition leads to rapid paralysis of vital functions and death.

Pesticides such as malathion and parathion are toxic to insects because they also irreversibly inhibit the enzyme acetylcholinesterase. These agents can also be toxic to other animals, including humans, though doses that kill insects are not generally harmful to people.

Many antibiotics work by inhibiting enzymes—in this case, enzymes that are essential to the survival of disease-causing bacteria. Penicillin, for example, inhibits an enzyme that bacteria use in making cell walls. Because humans lack this enzyme, we are not harmed by the drug.

Ibuprofen and aspirin act as inhibitors of enzymes involved in inducing pain. Protease inhibitors are HIV drugs that target a key viral enzyme. And many cancer drugs are inhibitors of enzymes that promote cell division.

The anti-enzyme actions of many toxins, pesticides, and drugs help us appreciate how important enzymes are in living cells. Let's now turn to a discussion of membranes, which help organize the enzyme-catalyzed reactions that make up the cell's metabolism.

 How does the antibiotic penicillin work?

■ It inhibits a bacterial enzyme that functions in cell wall production.

5.10 Membranes organize the chemical activities of cells

So many metabolic reactions occur simultaneously in a cell that utter chaos would result if the cell were not highly organized and able to control its metabolic reactions precisely. Some of this organization comes from teams of enzymes that function like assembly lines. A product from one enzyme-catalyzed reaction becomes the substrate for a neighboring enzyme, and so on, until a final product is made. For a cell's assembly lines to operate, the right enzymes have to be present at the right time and in the right place. Membranes provide the structural basis for metabolic order. In eukaryotes, membranes form most of the cell's organelles, partitioning the cell into functional compartments that contain specific enzymes. And many enzymes are built right into the membranes of these organelles.

For all cells, the plasma membrane forms a boundary between the living cell and its surroundings and controls the traffic of molecules into and out of the cell. The plasma membrane takes up substances the cell needs and disposes of the cell's wastes. Like all the membranes of a cell, the plasma membrane exhibits **selective permeability;** that is,

it allows some substances to cross more easily than others.

For a structure that separates life from nonlife, the plasma membrane is amazingly thin. It would take a stack of over 8,000 to equal the thickness of this page. **Figure 5.10** displays a cross section of part of the plasma membrane of a red blood cell. Magnified 200,000 times, the membrane appears as two dark bands separated by a thicker light zone.

The key to how a membrane functions is its structure, which we explore next.

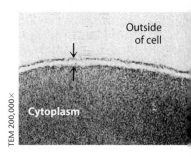

Outside of cell

TEM 200,000×

Cytoplasm

Figure 5.10
Plasma membrane
in cross section

? How do membranes organize the chemical activities of cells?

■ Membranes form organelles that contain enzymes in solution; many enzymes are embedded in cell membranes; the plasma membrane controls the passage of chemicals into and out of the cell.

5.11 Membrane phospholipids form a bilayer

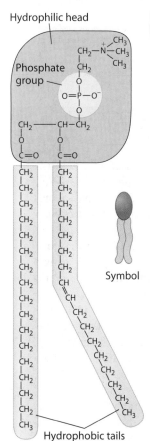

Hydrophilic head

Phosphate group

CH_3
$CH_2-N^+-CH_3$
CH_3
CH_2
$O=P-O^-$
CH_2—$CH-CH_2$
$C=O$ $C=O$
CH_2 ... Hydrophobic tails

Symbol

Figure 5.11A Phospholipid molecule

Lipids, mainly phospholipids, are the main structural components of membranes. A phospholipid is structured much like the fat molecules you learned about in Module 3.8, but it has a phosphate group and only two fatty acids instead of three. The structural formula in **Figure 5.11A** highlights the two parts of such a molecule, which have opposite interactions with water. The head (gray), with its charged phosphate group, is hydrophilic. The fatty acid double tail (gold) is nonpolar and hydrophobic. Thus, the tail end of a phospholipid is pushed away by water, while the head is attracted to water. Also shown is the phospholipid symbol this book uses in depicting membranes.

The structure of phospholipid molecules is well suited to their role in membranes. Phospholipids form a two-layer sheet called a phospholipid bilayer (**Figure 5.11B**). Their hydrophilic

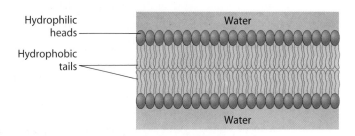

Hydrophilic heads

Hydrophobic tails

Water

Water

Figure 5.11B Phospholipid bilayer

heads face outward, exposed to the water on both sides of a membrane, and their hydrophobic tails point inward, mingling together and shielded from water.

The hydrophobic interior of the bilayer is one reason membranes are selectively permeable. Nonpolar, hydrophobic molecules are soluble in lipids and can easily pass through membranes. In contrast, polar molecules and ions are not soluble in lipids. Whether these hydrophilic molecules pass through the membrane depends on protein molecules in the phospholipid bilayer. In fact, much of a membrane's structure and function depends on membrane proteins, which we examine in the next module.

? Why do phospholipids tend to organize into a bilayer in an aqueous environment?

■ This structure shields the hydrophobic tails of the phospholipids from water, while exposing the hydrophilic heads to water.

5.12 The membrane is a fluid mosaic of phospholipids and proteins

Figure 5.12 shows the structure of the plasma membrane. Like other cellular membranes, it is commonly described as a **fluid mosaic.** The word *mosaic* denotes a surface made of small fragments, like pieces of colored tile cemented together in a mosaic floor or picture. A membrane is a "mosaic" in having diverse protein molecules embedded in a framework of phospholipids. The membrane mosaic is "fluid" in that most of the individual proteins and phospholipid molecules can drift laterally in the membrane. Helping to support the delicate membrane, some of the proteins are linked to the cytoskeleton and to fibers of the extracellular matrix, as shown in this drawing of an animal cell's plasma membrane.

Notice the kinked tails of many of the phospholipids. Resulting from double bonds in unsaturated fatty acid tails, the kinks make the membrane more fluid by keeping adjacent phospholipids from packing tightly together. In animal cells, the steroid cholesterol (see Module 3.9), wedged into the bilayer, helps stabilize the phospholipids at body temperature (about 37°C for humans) but also helps keep the membrane fluid at lower temperatures. In a cell, the phospholipid bilayer remains about as fluid as salad oil at room temperature.

The outside surface of the plasma membrane has carbohydrates (chains of sugars, shown here in green) bonded to proteins and lipids in the membrane. A protein with attached sugars is called a glycoprotein, whereas a lipid with sugars is called a glycolipid. The carbohydrate chains vary from species to species, from one individual to another in the same species, and even from one cell type to another in a single individual. Many function as cell identification tags that are recognized by other cells. This ability to distinguish among different cells is crucial to life. It allows cells in an embryo to sort themselves into tissues and organs. It also enables cells of the immune system to recognize and reject foreign cells, such as infectious bacteria. Next we look at some other functions of membrane proteins.

Web/CD Activity 5E *Membrane Structure*

[?] Why are cellular membranes described as a fluid mosaic?

■ Diverse proteins (a "mosaic") float in a "fluid" phospholipid bilayer.

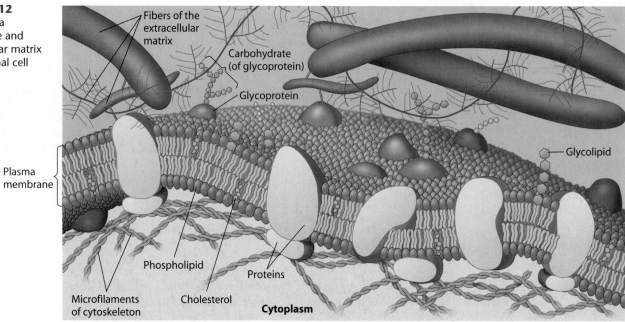

Figure 5.12
The plasma membrane and extracellular matrix of an animal cell

Fibers of the extracellular matrix

Carbohydrate (of glycoprotein)

Glycoprotein

Glycolipid

Plasma membrane

Phospholipid

Proteins

Microfilaments of cytoskeleton

Cholesterol

Cytoplasm

5.13 Proteins make the membrane a mosaic of function

The word *mosaic* as applied to a membrane can refer not only to the positioning of proteins in the phospholipid bilayer but also to the varied activities of these proteins. In fact, proteins perform most of the functions of a membrane. More than 50 different kinds of proteins have been found in the plasma membrane of human red blood cells so far. Different types of cells contain different sets of membrane proteins, and the various membranes within a cell each have unique proteins.

We have already mentioned several functions of membrane proteins, including attaching the membrane to the cytoskeleton and external fibers, providing identification tags, and forming junctions between adjacent cells (see Module 4.18).

Many membrane proteins are enzymes, which may function in catalytic teams for molecular assembly lines (Figure 5.13A). Other proteins function as **receptors** for chemical messengers from other cells (Figure 5.13B). A receptor pro-

tein has a shape that fits the shape of a specific messenger, such as a hormone, just as an enzyme fits its substrate. Often the binding of the messenger to the receptor triggers a chain reaction involving other proteins, which relay the message to a molecule that performs a specific activity inside the cell. This message-transfer process is called **signal transduction.**

Finally, an important function of membrane proteins is to help move substances across the membrane (Figure 5.13C). Although small, nonpolar molecules such as O_2 and CO_2 pass freely through the phospholipid bilayer, many essential molecules, such as glucose and even water, need assistance from proteins to enter or leave the cell. In the next several modules, we look at two types of molecular transport: passive transport, with no energy cost to the cell, and active transport, in which the cell expends energy to move substances across its membranes.

Web/CD Activity 5F *Signal Transduction*

Web/CD Activity 5G *Selective Permeability of Membranes*

Web/CD Thinking as a Scientist *How Do Cells Communicate with Each Other?*

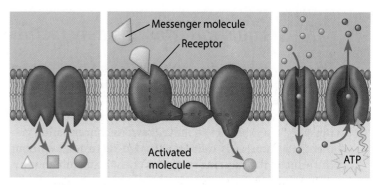

Figure 5.13A
Enzyme activity

Figure 5.13B
Signal transduction

Figure 5.13C
Transport

? The hormone epinephrine can cause a liver cell to hydrolyze its stored glycogen and release sugar without the hormone even entering the cell. Explain.

■ Epinephrine binds to a receptor on the liver cell surface, activating a signal transduction pathway inside the cell that leads to sugar release.

5.14 Passive transport is diffusion across a membrane

Diffusion is the tendency for particles of any kind to spread out evenly in an available space, moving from where they are more concentrated to regions where they are less concentrated. Diffusion requires no work; it results from the random motion (kinetic energy) of atoms and molecules.

The figures here will help you to visualize diffusion across a membrane. **Figure 5.14A** shows a solution of green dye separated from pure water by a membrane that is permeable to the dye molecules. Although each dye molecule moves randomly, there will be a *net* movement from the side of the membrane where the molecules are more concentrated to the side where they are less concentrated. Eventually, the solutions on both sides have equal concentrations of dye. Put another way, the dye diffuses down its **concentration gradient** until equilibrium is reached. At equilibrium, molecules continue to move back and forth, but there is no *net* change in concentration on either side of the membrane. Figure 5.14B illustrates the important point that two or more substances diffuse independently of each other; that is, each diffuses down its own concentration gradient.

Because a cell does not perform work when molecules diffuse across its membrane, the diffusion of a substance across a biological membrane is called **passive transport.** Passive transport is extremely important to all cells. In our lungs, for example, passive transport down concentration gradients is the sole means by which oxygen (O_2), essential for metabolism, enters red blood cells and carbon dioxide (CO_2), a metabolic waste, passes out of them.

Both O_2 and CO_2 are small, nonpolar molecules that diffuse easily across the phospholipid bilayer of a membrane. But can ions and polar molecules also move by passive transport into or out of a cell? They can if they are moving down their

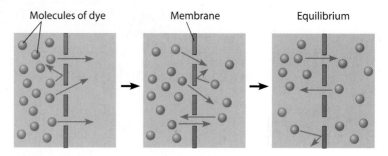

Figure 5.14A   Passive transport of one type of molecule

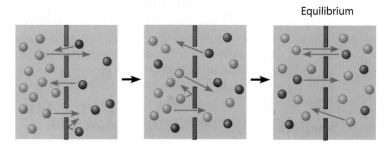

Figure 5.14B   Passive transport of two types of molecules

concentration gradients and if they have transport proteins to provide a pathway across the membrane, as we see next.

Web/CD Activity 5H *Diffusion*

? Why is diffusion across a membrane called passive transport?

■ The cell does not expend energy to transport substances that are diffusing down their concentration gradients.

5.15 Transport proteins may facilitate diffusion across membranes

Numerous substances that do not diffuse freely across membranes because of their size, polarity, or charge can move across a membrane with the help of specific transport proteins in the membrane. When one of these proteins makes it possible for a substance to move down its concentration gradient, the process is called **facilitated diffusion**. Without the protein, the substance does not cross the membrane or it diffuses across it too slowly to be useful to the cell. Facilitated diffusion is a type of passive transport because it does not require energy. As in all passive transport, the driving force is the concentration gradient.

Figure 5.15 shows a common type of transport protein, which provides a pore, or tunnel, for the passage of a solute. Another type of transport protein binds its passenger, changes shape, and releases its passenger on the other side. In both cases, the transport protein is specific for the substance it helps move across the membrane. The rate of facilitated diffusion depends on the number of transport protein molecules for a particular substance present in the membrane.

Substances that use facilitated diffusion for crossing cell membranes include a number of sugars, amino acids, ions—and even water. The water molecule is very small, but because it is polar (see Module 2.9), its unaided diffusion through a membrane's hydrophobic interior is slow. Rapid

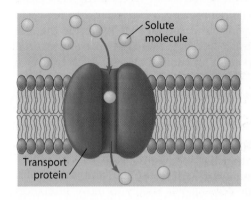

Figure 5.15
Transport protein providing a pore for solute passage

diffusion of water into and out of cells is made possible by transport proteins called aquaporins. One of the recipients of the 2003 Nobel Prize in Chemistry was the scientist who discovered these water channels.

Web/CD Activity 5I *Facilitated Diffusion*

? How do transport proteins contribute to a membrane's selective permeability?

5.16 Osmosis is the diffusion of water across a membrane

One of the most important substances that crosses membranes by passive transport is water. In the next module, we consider the critical balance of water between a cell and its environment. But first let's explore a physical model of the diffusion of water molecules across a selectively permeable membrane, a process called **osmosis.**

The top of **Figure 5.16** shows what happens if a membrane permeable to water but not to a solute (such as glucose) separates two solutions with different concentrations of solute. The solution on the right side has a higher concentration of solute than that on the left. Remember that water can cross this membrane; the solute cannot. As you can see, water crosses the membrane until the solute concentrations (molecules per milliliter of solution) are equal on both sides.

In the close-up view at the bottom of Figure 5.16, you can see what happens at the molecular level. Clusters of polar water molecules form weak bonds with solute molecules, so that fewer water molecules are free to diffuse across the membrane. The less concentrated solution on the left, with fewer solute molecules, has more free water molecules. There is a net movement of water down its own concentration gradient, from the solution with the lower solute concentration to that with the higher solute concentration. The result is the difference in water levels you see in Figure 5.16.

Here we show only one type of solute, but the same net movement of water would occur no matter how many kinds

hypo — lower [E]
hyper — higher [E]

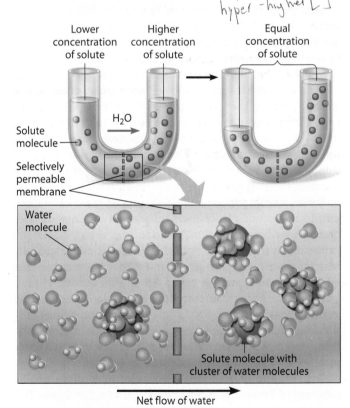

Figure 5.16 Osmosis

of solutes were present. The direction of osmosis is determined by the difference in *total* solute concentration, not by the nature of the solutes. Now let's apply osmosis to living cells.

? Indicate the direction of net water movement between two solutions—a 0.5% sucrose solution and a 2% sucrose solution—separated by a membrane not permeable to sucrose.

■ From the 0.5% sucrose solution (lower solute concentration) to the 2% sucrose solution (higher solute concentration)

5.17 Water balance between cells and their surroundings is crucial to organisms

Biologists use a special vocabulary to describe the relationship between a cell and its surroundings with regard to the movement of water. The term **tonicity** describes the tendency of a cell in a given solution to lose or gain water.

Figure 5.17 illustrates how the principles of osmosis and tonicity apply to cells. When an animal cell, such as the red blood cell shown in part 1 of the figure, is immersed in a solution that is **isotonic** to the cell (*iso*, same, and *tonos*, tension), the cell's volume remains constant. The solute concentration of a cell and its isotonic environment are essentially equal, and the cell gains water at the same rate that it loses it. Red blood cells are transported in the isotonic plasma of the blood. The cells of most animals are bathed in an extracellular fluid that is isotonic to the cells. Many marine animals, such as sea stars and crabs, are isotonic to seawater.

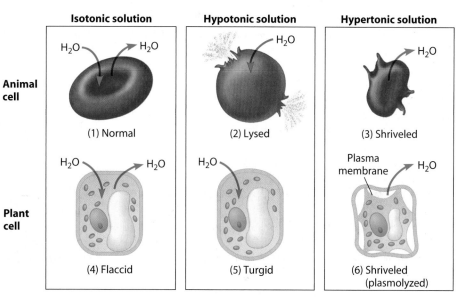

Figure 5.17 How animal and plant cells behave in different solutions

Part 2 of the figure shows what can happen to an animal cell in a **hypotonic** solution (*hypo*, below), a solution with a solute concentration lower than that of the cell. The cell gains water, swells, and may pop (lyse) like an overfilled balloon. Part 3 shows the opposite case—an animal cell in a **hypertonic** solution (*hyper*, above), a solution with a higher solute concentration. The cell shrivels and can die from water loss.

For an animal to survive in a hypertonic or hypotonic environment, it must have a way to prevent excessive uptake or excessive loss of water. The control of water balance is called **osmoregulation.** For example, a freshwater fish, which lives in a hypotonic environment, has kidneys and gills that work constantly to prevent an excessive buildup of water in the body. (We discuss osmoregulation further in Chapter 25.)

Water balance problems are somewhat different for plant cells because of their rigid cell walls. A plant cell immersed in an isotonic solution (part 4) is flaccid, and nonwoody plants, such as most houseplants, wilt in this situation. In contrast, a plant cell is turgid, and plants are healthiest, in a hypotonic environment (part 5). To become turgid, a plant cell needs a net inflow of water. Although the somewhat

elastic cell wall expands a bit, the pressure it exerts prevents the cell from taking in too much water and bursting, as an animal cell would in this environment. Part 6 shows that in a hypertonic environment, a plant cell is no better off than an animal cell. As a plant cell loses water, it shrivels, and its plasma membrane pulls away from the cell wall. This process is called **plasmolysis.** The walled cells of bacteria and fungi also plasmolyze in hypertonic environments.

The movement of water into and out of cells is always by passive transport, although it may be greatly speeded up by the presence of aquaporins. But not all transport proteins simply facilitate diffusion. Some use cellular energy to actively transport substances against their concentration gradients—a topic we consider next.

Web/CD Activity 5J *Osmosis and Water Balance in Cells*

Web/CD Thinking as a Scientist *How Does Osmosis Affect Cells?*

? Explain the function of the contractile vacuoles in the *Paramecium* cell in Figure 4.12B in terms of what you have learned about water balance in cells.

■ The pond water in which *Paramecium* lives is hypotonic to the cell, and thus there is a constant net osmosis of water into the cell. The contractile vacuoles expel this excess water, preventing the cell from bursting.

5.18 Cells expend energy for active transport

In contrast to passive transport, **active transport** requires that a cell expend energy to move molecules across a membrane *against* the solute's concentration gradient—that is, toward the side where it is more concentrated. ATP supplies the energy for most active transport.

Figure 5.18 shows a simple model of an active transport system that moves a solute out of the cell against its concentration gradient. ❶ The process begins when solute on the cytoplasmic side of the plasma membrane attaches to a specific binding site on the transport protein. ❷ ATP then phosphorylates the transport protein, ❸ causing it to change shape in such a way that the solute is released on the other side of the membrane. ❹ Subsequently, the phosphate group detaches, and the transport protein returns to its original shape, ready for a new round of active transport. (As indicated in the figure, two solute ions or molecules are usually transported together.)

Active transport allows a cell to maintain concentrations of small molecules that are different from the concentrations of its surroundings. For example, an animal cell has a higher concentration of potassium ions (K^+) and a lower concentration of sodium ions (Na^+) than the solution outside the cell. The generation of nerve signals depends on these concentration differences, and the sodium-potassium pump helps cells maintain these steep gradients by shuttling Na^+ and K^+ across the membrane.

Web/CD Activity 5K *Active Transport*

? Cells actively transport Ca^{2+} out of the cell. Is calcium more concentrated inside or outside the cell? Explain.

■ Outside the cell. If the cell uses active transport, it is moving calcium against its concentration gradient.

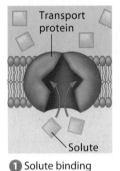 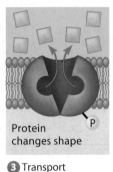 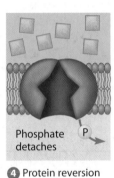

Figure 5.18 Active transport of a solute across a membrane

❶ Solute binding ❷ Phosphorylation ❸ Transport ❹ Protein reversion

5.19 Exocytosis and endocytosis transport large molecules } used for bigger molecules such as protein transfer

So far, we've focused on how water and small solutes enter and leave cells by moving through the plasma membrane. The story is different for large molecules such as proteins.

As illustrated in **Figure 5.19A**, a cell uses the process of **exocytosis** (from the Greek *exo*, outside, and *kytos*, cell) to export bulky materials. In the first step of this process, a membrane-enclosed vesicle filled with macromolecules (purple) moves to the plasma membrane. Once there, the vesicle fuses with the plasma membrane, and the vesicle's

contents spill out of the cell. When we weep, for instance, cells in our tear glands use exocytosis to export a salty solution containing proteins. In another example, certain cells in the pancreas manufacture the hormone insulin and secrete it into the bloodstream by exocytosis.

Figure 5.19B shows a transport process that is basically the opposite of exocytosis. In **endocytosis** (*endo*, inside), a cell takes in macromolecules or other particles by forming vesicles or vacuoles from its plasma membrane. **Figure 5.19C** shows

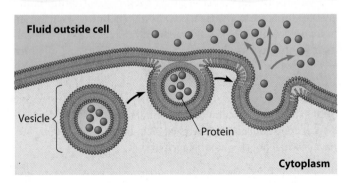

Figure 5.19A Exocytosis

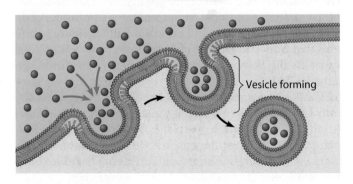

Figure 5.19B Endocytosis

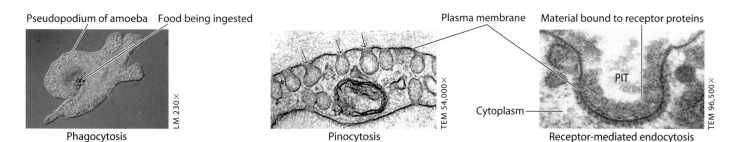

Figure 5.19C Three kinds of endocytosis

three kinds of endocytosis. The left micrograph shows an amoeba taking in a food particle, an example of **phagocytosis**, or "cellular eating." The amoeba engulfs its prey by wrapping extensions, called pseudopodia, around it and packaging it within a vacuole. As Module 4.10 described, the vacuole then fuses with a lysosome, and the lysosome's hydrolytic enzymes digest the contents.

The center micrograph shows pinocytosis, or "cellular drinking." The cell is taking droplets of fluid from its surroundings into tiny vesicles (arrows). Pinocytosis is not specific; it takes in any and all solutes dissolved in the droplets.

In contrast to pinocytosis, a third type of endocytosis, called receptor-mediated endocytosis, is highly specific. The micrograph on the right shows one stage of the process. The plasma membrane has indented to form a pit. The pit is lined with receptor proteins that have picked up particular molecules from the surroundings. The pit will pinch closed to form a vesicle that will carry the molecules into the cytoplasm.

Web/CD Activity 5L *Exocytosis and Endocytosis*

? Explain how a protein-secreting cell can synthesize and secrete its product without the protein ever having to cross a membrane (see Modules 4.8 and 4.13).

■ From the time the protein is made by ribosomes on rough endoplasmic reticulum (ER), it is enclosed within membranes, first in the ER interior, then within the Golgi, and finally in transport vesicles, which fuse with the plasma membrane, releasing the proteins by exocytosis.

CONNECTION

5.20 Faulty membranes can overload the blood with cholesterol

Few molecules make the news as often as cholesterol does. Cholesterol is notorious as a possible factor in heart disease, yet it is essential for the normal functioning of all our cells. As mentioned earlier, cholesterol is a component of membranes as well as a starting material for making other steroids, such as sex hormones. Problems tend to arise only when we have too much cholesterol in our blood.

In most of us, cells take in cholesterol from the blood by receptor-mediated endocytosis. Consequently, we have only small amounts of the substance in our blood. Cholesterol circulates in the blood mainly in particles called low-density lipoproteins, or LDLs. As **Figure 5.20** shows, an LDL is a globule of cholesterol (and other lipids) surrounded by a single layer of phospholipids in which proteins are embedded. One of the LDL proteins (pink) fits a specific type of receptor protein (purple) on cell membranes. Normally, body cells take up LDLs from the blood by receptor-mediated endocytosis using these LDL receptor proteins.

Worldwide, about one in 500 human babies inherits a disease called **hypercholesterolemia**, which is characterized by more than twice the normal level of cholesterol in the blood. In severe cases (about one in a million people), there are no functional LDL receptors, and cholesterol levels are even higher. People with the milder form of the disease have functional receptors but in low numbers. In either case, LDLs tend to accumulate in the blood. High levels of LDLs in the blood

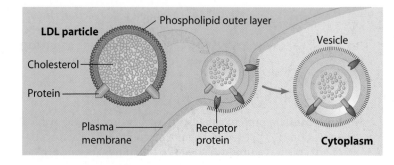

Figure 5.20 A cell using receptor-mediated endocytosis to take up an LDL

are life-threatening because the LDLs can deposit cholesterol in the lining of blood vessels, causing the lining to bulge inward and reduce blood flow. The blockage of blood vessels that nourish the heart leads to heart disease. Individuals with severe hypercholesterolemia may die from heart disease in early childhood. Milder cases are also dangerous, but they can be treated with medications and a low-fat diet. We discuss the genetics of hypercholesterolemia in Module 9.12.

? What type of endocytosis enables body cells to take in cholesterol?

■ Receptor-mediated endocytosis

5.21 Chloroplasts and mitochondria make energy available for cellular work

This chapter's main topics are energy, enzymes, and membranes. Two organelles in the cell, the chloroplast and the mitochondrion, are composed of membranes with enzyme assembly lines. Both organelles play central roles in harvesting energy and making it available for cellular work.

Nearly all the chemical energy that organisms use comes ultimately from sunlight. In Chapter 7, we'll explore how photosynthesis converts light energy to chemical energy in the chloroplast. Mitochondria, present in almost all eukaryotes, use the chemical energy stored in glucose to make ATP.

In the next chapter, we discuss how cellular respiration supplies cells with ATP.

Web/CD Activity 5M *Build a Chemical Cycling System*

 Energy comes into ecosystems as sunlight and leaves as heat. Relate this to the second law of thermodynamics.

■ In each energy transfer, some energy is lost to the disorganized energy of heat.

Chapter Review

Reviewing the Concepts

Energy and the Cell (5.1–5.4)

Energy is the capacity to perform work. Kinetic energy is the energy of motion. Potential energy is energy stored in the location or structure of matter **(5.1)**. According to the laws of thermodynamics, energy can change form, but cannot be created or destroyed. Energy transformations increase disorder, or entropy, and some energy is lost as heat **(5.2)**.

Chemical Reactions. Endergonic reactions require energy and yield products rich in potential energy. Exergonic reactions release energy. Energy coupling uses exergonic reactions to drive endergonic ones. Metabolism includes all of a cell's chemical reactions **(5.3)**. ATP drives endergonic reactions by phosphorylation, transferring a phosphate group to make molecules more reactive **(5.4)**.

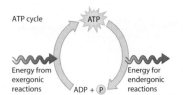

How Enzymes Function (5.5–5.9)

Enzymes are protein catalysts that decrease the energy of activation (E_A) needed to begin a reaction **(5.5)**. Each type of enzyme has a unique active site that binds specifically with its substrate **(5.6)**. Temperature and pH influence enzyme activity. Some enzymes require cofactors, such as metal ions or organic coenzymes **(5.7)**.

Inhibitors. A competitive inhibitor competes with the substrate for the active site. A noncompetitive inhibitor alters an enzyme's function by changing its shape. Feedback inhibition helps regulate metabolism **(5.8)**. Poisons, pesticides, and drugs may inhibit enzymes **(5.9)**.

Membrane Structure and Function (5.10–5.21)

Membranes provide structural order for metabolism. The selectively permeable plasma membrane controls the flow of substances into and out of a cell **(5.10)**. A phospholipid has a hydrophilic head and two hydrophobic tails. A membrane is a fluid mosaic, with protein molecules embedded in a phospholipid bilayer **(5.11–5.12)**. Functions of membrane proteins include enzymatic activity, signal transduction,

transport, and attachment to the cytoskeleton and extracellular matrix **(5.13)**.

Diffusion and Passive Transport. In passive transport, substances diffuse (move from higher to lower concentration) across membranes without work by the cell **(5.14)**. In facilitated diffusion, transport proteins provide passage across membranes for ions and polar molecules **(5.15)**. Osmosis is the diffusion of water across a membrane from a solution of lower solute concentration to one of higher solute concentration **(5.16)**. Osmosis causes cells to shrink in a hypertonic solution and swell in a hypotonic solution. In isotonic solutions, animal cells are normal but plant cells are limp. The control of water balance is called osmoregulation **(5.17)**.

Active Transport. Transport proteins can move solutes against a concentration gradient; this active transport requires energy **(5.18)**. In moving large molecules across a membrane, a vesicle may fuse with the membrane and expel its contents (exocytosis), or the membrane may fold inward, enclosing material from the outside **(5.19)**. Harmful levels of cholesterol can accumulate in the blood if plasma membranes lack cholesterol receptors **(5.20)**.

Enzymes and membranes are central to processes that make energy available to cells **(5.21)**.

Connecting the Concepts

1. Label the parts of this diagram illustrating the catalytic cycle of an enzyme.

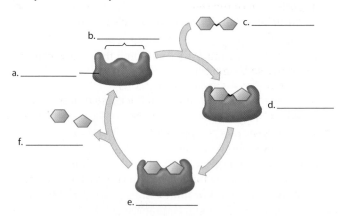

2. Fill in the following concept map to review the processes by which molecules move across membranes.

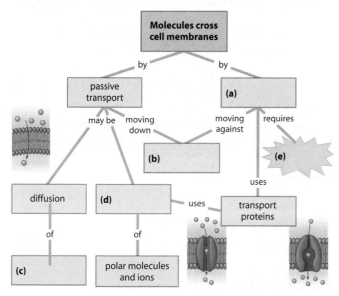

Testing Your Knowledge

Multiple Choice

3. Consider the following: chemical bonds in the gasoline in a car's gas tank and the movement of the car along the road; a climber at the top of a hill and the hike she took to get there. The first parts of these situations illustrate _____, and the second parts illustrate _____.
   a. the first law of thermodynamics . . . the second law
   b. kinetic energy . . . potential energy
   c. an exergonic reaction . . . an endergonic reaction
   d. potential energy . . . kinetic energy
   e. the second law of thermodynamics . . . the first law

4. Which best describes the structure of a cell membrane?
   a. proteins between two layers of phospholipid
   b. proteins embedded in two layers of phospholipid
   c. a layer of protein coating a layer of phospholipid
   d. phospholipids between two layers of protein
   e. cholesterol embedded in two layers of phospholipid

5. A freshwater *Paramecium* is placed in salt water. Which of the following events would most likely occur?
   a. an increase in the action of its contractile vacuole
   b. swelling of the cell until it becomes turgid
   c. swelling of the cell until it bursts
   d. shriveling of the cell
   e. diffusion of salt ions out of the cell

6. The sodium concentration in a cell is ten times less than the concentration in the surrounding fluid. How can the cell move sodium out of the cell? (*Explain.*)
   a. passive transport   d. osmosis
   b. diffusion   e. any of these processes
   c. active transport

7. The synthesis of ATP from ADP and ⓟ
   a. is an exergonic process.
   b. involves the hydrolysis of a phosphate bond.
   c. transfers a phosphate, priming a protein to do work.
   d. stores energy in a form that can drive cellular work.
   e. releases energy.

Describing, Comparing, and Explaining

8. How do the two laws of thermodynamics apply to living organisms?

9. What are the main types of cellular work? How does ATP provide the energy for this work?

10. Why is the barrier of activation energy beneficial for organic molecules? Explain how enzymes lower E_A.

11. How do the components and structure of cell membranes relate to the functions of membranes?

Applying the Concepts

12. Explain how each of the following food preservation methods would interfere with a microbe's enzyme activity and ability to break down food: canning (heating), freezing, pickling (soaking in acetic acid), salting.

13. A biologist performed two series of experiments on lactase, the enzyme that hydrolyzes lactose to glucose and galactose. First, she made up 10% lactose solutions containing different concentrations of enzyme and measured the rate at which galactose was produced (grams of galactose per minute). Results of these experiments are shown in Table A below. In the second series of experiments (Table B), she prepared 2% enzyme solutions containing different concentrations of lactose and again measured the rate of galactose production.

Table A: Rate and Enzyme Concentration					
Lactose concentration	10%	10%	10%	10%	10%
Enzyme concentration	0%	1%	2%	4%	8%
Reaction rate	0	25	50	100	200

Table B: Rate and Substrate Concentration					
Lactose concentration	0%	5%	10%	20%	30%
Enzyme concentration	2%	2%	2%	2%	2%
Reaction rate	0	25	50	65	65

   a. Graph and explain the relationship between the reaction rate and the enzyme concentration.
   b. Graph and explain the relationship between the reaction rate and the substrate concentration. How and why did the results of the two experiments differ?

14. Lead acts as an enzyme inhibitor and can interfere with the development of the nervous system. One manufacturer of lead-acid batteries instituted a "fetal protection policy" that banned female employees of childbearing age from working in areas where they might be exposed to high levels of lead. Women were involuntarily transferred to lower-paying jobs in lower-risk areas. A group of employees challenged the policy, claiming that it deprived women of job opportunities available to men. The U.S. Supreme Court ruled the policy illegal. What rights and responsibilities of employers, employees, and government agencies are in conflict in this situation? Who should determine what makes a safe environment?

Answers to all questions can be found in Appendix 3.

For study help and Activities, go to campbellbiology.com or the student CD-ROM.

CHAPTER 6

How Is a Marathoner Different from a Sprinter?

INTRODUCTION TO CELLULAR RESPIRATION

6.1 Photosynthesis and cellular respiration provide energy for life
6.2 Breathing supplies oxygen to our cells and removes carbon dioxide
6.3 Cellular respiration banks energy in ATP molecules
6.4 The human body uses energy from ATP for all its activities
6.5 Cells tap energy from electrons "falling" from organic fuels to oxygen

STAGES OF CELLULAR RESPIRATION AND FERMENTATION

6.6 Overview: Cellular respiration occurs in three main stages
6.7 Glycolysis harvests chemical energy by oxidizing glucose to pyruvate
6.8 Pyruvate is chemically groomed for the citric acid cycle
6.9 The citric acid cycle completes the oxidation of organic fuel, generating many NADH and $FADH_2$ molecules
6.10 Most ATP production occurs by oxidative phosphorylation
6.11 Certain poisons interrupt critical events in cellular respiration
6.12 Review: Each molecule of glucose yields many molecules of ATP
6.13 Fermentation is an anaerobic alternative to cellular respiration

INTERCONNECTIONS BETWEEN MOLECULAR BREAKDOWN AND SYNTHESIS

6.14 Cells use many kinds of organic molecules as fuel for cellular respiration
6.15 Food molecules provide raw materials for biosynthesis
6.16 The fuel for respiration ultimately comes from photosynthesis

ATHLETES WHO PARTICIPATE in track competitions usually have a favorite event in which they excel. For some runners, this event may be a sprint, a short race of only 100 or 200 m. For others, it may be a race of 1,500, 5,000, or even 10,000 m. It is unusual to find a runner who competes equally well in both 100-m and 10,000-m races; runners just seem to feel more comfortable running races of particular lengths. But why?

Could it be that runners' bodies "tell" them which races are best for them? There are indications that this is indeed the case. The muscles that move our legs contain two main types of muscle fibers, called slow and fast muscle fibers. Slow muscle fibers (also called "slow-twitch" fibers) are muscle cells that can sustain repeated, long contractions but don't generate a lot of quick power for the body. They perform better in endurance exercises, like long-distance running, which require slow, steady muscle activity. Fast muscle fibers ("fast-twitch" fibers) are cells that can contract more quickly and powerfully than slow fibers but fatigue much more easily; they function best for short bursts of intense activity, like weight lifting or sprinting.

All human muscles contain both slow and fast fibers, but muscles differ in the percentage of each. The percentage of each fiber type in a particular muscle also varies from person to person. For example, in the quadriceps muscles of the legs, most marathon runners have about 80% slow fibers, whereas sprinters have about 60% fast fibers. These differences, which are genetically determined, undoubtedly help account for our differing athletic capabilities. Training can't usually turn one kind of runner into another!

But what makes these two types of muscle fibers perform so differently? An important part of the answer is that they use different processes for making ATP (see Module 5.4), the molecule that supplies the energy for muscle contraction. While

How Cells Harvest Chemical Energy

both types of muscle cells break down sugar (chiefly glucose) to make chemical energy available for ATP production, slow fibers do it *aerobically,* using oxygen (O_2), while fast fibers work *anaerobically*, without oxygen.

A closer look at the differing structures of these two kinds of muscle cells helps explain their differing functions. The structure of slow fibers supports their aerobic function. The cells have many mitochondria, the organelles where aerobic ATP production occurs. And slow fibers contain many molecules of myoglobin, a red protein related to hemoglobin that, like hemoglobin, is a carrier of O_2 molecules. The myoglobin gives slow muscle fibers a reddish color. You've undoubtedly seen this color in the "dark meat" of cooked turkeys, whose leg muscles are composed mostly of myoglobin-rich slow fibers. The aerobic harvesting of energy from sugar by muscle cells (or other cells) is called **cellular respiration.** This process yields carbon dioxide (CO_2), water (H_2O), and a large amount of ATP—perfect for sustaining long muscle contractions.

Fast muscle fibers, on the other hand, have a structure that allows for a quicker energy-harvesting process that doesn't require O_2 but produces much less ATP per glucose molecule. Fast fibers are thicker than slow ones, have fewer mitochondria, and have much less myoglobin, making them pale in color. (Turkey breast muscles, specialized for quick bursts of flight, consist of fast fibers—"white meat.") The thickness of fast fibers enhances their power. But when oxygen is not available, they can't completely break down glucose to CO_2. Instead, they produce lactate, a compound that can build up in muscles and make them ache and fatigue. This is why fast muscle fibers are best at supplying short bursts of power. Anaerobic ATP production in our muscles is only effective for a minute or so.

Muscle cells are not the only cells in our body that break down sugars and other food molecules for ATP production. Nearly all our cells harvest chemical energy from food—as do the cells of all other organisms, eukaryotic and prokaryotic alike. Cells of most eukaryotic organisms function like slow muscle fibers in that they carry out the aerobic process of cellular respiration. Before learning more about how cellular respiration powers the work of cells, let's take a step back and consider the bigger picture of energy flow through an ecosystem. In other words, where does the energy for cellular work originate? ■ ■ ■

The New York City marathon

6.1 Photosynthesis and cellular respiration provide energy for life

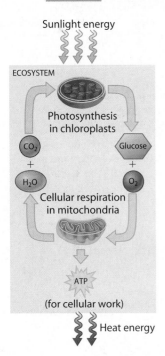

Sunlight energy

ECOSYSTEM

Photosynthesis in chloroplasts

CO_2 + H_2O

Glucose + O_2

Cellular respiration in mitochondria

ATP

(for cellular work)

Heat energy

Figure 6.1 The connection between photosynthesis and cellular respiration

Life requires energy. A cell uses energy to build and maintain its structure, transport materials, manufacture products, move, grow, and reproduce. This energy ultimately comes from the sun. Through the process of photosynthesis, green plants, algae, and photosynthetic protists and bacteria convert light energy into chemical energy. Photosynthesis is the topic of Chapter 7. But a brief overview here of the relationship between cellular respiration and photosynthesis will illustrate how these two processes, working together, provide energy for life.

As **Figure 6.1** shows, in photosynthesis in chloroplasts, the energy of sunlight is used to rearrange the atoms of CO_2 and H_2O to produce glucose and O_2. In cellular respiration in mitochondria, O_2 is consumed as glucose is broken down to CO_2 and H_2O; the cell captures the energy released in ATP.

This figure also shows that in these energy conversions, some energy is lost as heat. Life on Earth is solar powered, and energy makes a one-way trip through an ecosystem. Chemicals, however, are recycled. The CO_2 and H_2O released by cellular respiration are converted through photosynthesis to glucose and O_2, which are then used in respiration.

Before we take a closer look at how our cells harvest energy through cellular respiration, let's consider the connection between breathing and respiration.

? What is misleading about the following statement? "Plants perform photosynthesis and animals perform cellular respiration."

■ It implies that cellular respiration does not occur in plants. In fact, almost all eukaryotic cells use cellular respiration to obtain energy.

6.2 Breathing supplies oxygen to our cells and removes carbon dioxide

We often use the word *respiration* as a synonym for "breathing," the meaning of its Latin root. In this sense, respiration refers to an exchange of gases: An organism obtains O_2 from its environment and releases CO_2 as a waste product. Biologists also define respiration as the aerobic harvesting of energy from food molecules by cells. This process is called cellular respiration to distinguish it from breathing.

Breathing and cellular respiration are closely related. As the gymnast in **Figure 6.2** goes through her routine, her lungs take up O_2 from the air and pass it to her bloodstream. The bloodstream carries the O_2 to her muscle cells. Mitochondria in the muscle cells use the O_2 in cellular respiration, harvesting energy from glucose and other organic molecules to generate ATP, which the cells then use to contract. Simultaneous contraction of thousands of cells, precisely controlled by the nervous system, makes the gymnast's body move. The gymnast's bloodstream and lungs also perform the vital function of disposing of the CO_2 waste produced by cellular respiration.

? How is your breathing related to your cellular respiration?

■ In breathing, CO_2 and O_2 are exchanged between your lungs and the air. In cellular respiration, cells use O_2 to break down fuel, releasing CO_2 as a waste.

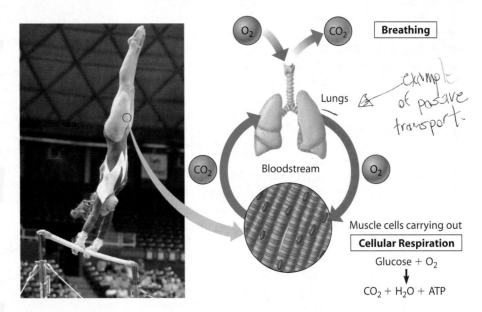

Breathing

O_2 → CO_2

Lungs ← *example of passive transport.*

Bloodstream

CO_2 O_2

Muscle cells carrying out

Cellular Respiration

Glucose + O_2
↓
CO_2 + H_2O + ATP

Figure 6.2 The connection between breathing and cellular respiration

6.3 Cellular respiration banks energy in ATP molecules

As the gymnast example implies, oxygen usage is only a means to an end. Generating ATP for cellular work is the fundamental function of cellular respiration.

$$C_6H_{12}O_6 \ + \ 6 \ O_2 \longrightarrow 6 \ CO_2 \ + \ 6 \ H_2O \ + \ ATPs$$

Glucose   Oxygen gas   Carbon   Water   Energy
                                             dioxide

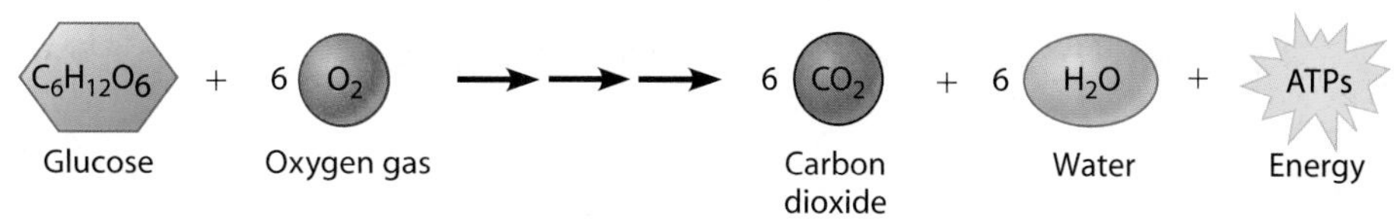

Figure 6.3 Summary equation for cellular respiration

The balanced chemical equation in **Figure 6.3** summarizes cellular respiration as carried out by cells that use O_2 in harvesting energy from the sugar glucose. Throughout this chapter, we use glucose as a convenient representative food molecule, although cells also "burn" many other organic molecules in cellular respiration. The summary equation tells us that the atoms of the starting (reactant) molecules glucose and O_2 regroup to form the products CO_2 and H_2O. In the process, glucose releases chemical-bond energy, which the cell stores (or "banks") in the chemical bonds of ATP (see Module 5.4). The series of arrows in Figure 6.3 indicates that cellular respiration consists of many steps, not just a single chemical reaction.

Cellular respiration can produce up to 38 ATP molecules for each glucose molecule, representing about 40% of the energy in glucose. The rest of the energy is released as heat. This may seem inefficient, but it compares very well with the efficiency of most energy-conversion systems. The average automobile engine, for instance, is able to convert only about 25% of the energy in gasoline to the kinetic energy of movement.

How great are the energy needs of a cell? A working muscle cell spends and must regenerate 10 million molecules of ATP per second! Let's consider the energy requirements for various human activities next.

? Why are sweating and other body-cooling mechanisms necessary during vigorous exercise?

■ The demand for ATP is supported by an increased rate of cellular respiration, but about 60% of the energy from food produces body heat instead of ATP.

CONNECTION

6.4 The human body uses energy from ATP for all its activities

Our body cells require a continuous supply of energy just to stay alive—to keep the heart pumping blood, to breathe, to maintain body temperature, and to digest food. These and other life-sustaining activities use as much as 75% of the energy a person takes in as food during a typical day. At any particular time, whether you are sleeping or active, most of your cells are busy with cellular respiration, producing ATP just to maintain your body.

Above and beyond the energy we need for body maintenance, the ATP made during cellular respiration provides energy for voluntary activities. Table 6.4 shows the amount of energy it takes to perform some of these activities. The energy units are kilocalories (kcal), commonly referred to simply as "calories" in nontechnical sources. The values shown do not include the energy the body consumes for its basic life-sustaining activities. Thus, sleeping or lying quietly does not consume any energy above the energy used in maintenance.

The U.S. National Academy of Sciences estimates that the average adult human needs to take in food that provides about 2,200 kcal of energy per day. This is just an estimate of the total amount of energy a person of average weight expends, or "burns," in both maintenance and voluntary activity.

TABLE 6.4	ENERGY CONSUMED BY VARIOUS ACTIVITIES (IN KCAL)
Activity	**Kcal Consumed per Hour by a 67.5-kg (150-lb) Person***
Bicycling (racing)	514
Bicycling (slowly)	170
Dancing (slow)	202
Dancing (fast)	599
Eating	28
Gymnastics	186
Laboratory work	73
Running (7 min/mi)	865
Sitting (writing)	28
Sleeping or lying still	0
Standing (relaxed)	32
Swimming (2 mph)	535
Walking (3 mph)	158
Walking (4 mph)	231

* Not including kcal needed for body maintenance

? Walking at 3 mph, how far would you have to travel to "burn off" the equivalent of an extra slice of pizza, which has about 475 kcal? How long would that take?

■ About 9 miles; about 3 hours (Now you understand why it is said that the most effective exercise for losing weight is pushing weight away from the table!)

Cells tap energy from electrons "falling" from organic fuels to oxygen

Just how *do* our cells extract energy from organic molecules? The energy available to a cell is contained in the arrangement of electrons in the chemical bonds that hold an organic molecule like glucose together. During cellular respiration, electrons are transferred to oxygen as the carbon-hydrogen bonds of glucose are broken and the hydrogen-oxygen bonds of water form. Oxygen attracts electrons very strongly, and an electron loses potential energy when it "falls" to oxygen. If you burn a cube of sugar, this electron "fall" happens very rapidly, releasing energy in the form of heat and light. Cellular respiration is a more controlled "fall" of electrons—more like stepping down an energy staircase, with energy released in small amounts that can be stored in ATP.

In the cellular respiration equation in **Figure 6.5A**, you cannot see any electron transfers. What you do see are changes in hydrogen atom distribution. Glucose loses hydrogen atoms as it is converted to carbon dioxide; simultaneously, molecular oxygen (O_2) gains hydrogen atoms in being converted to water. These hydrogen movements represent electron transfers because, as we saw in Chapter 2, each hydrogen atom consists of an electron and a proton.

The movement of electrons from one molecule to another is an oxidation-reduction reaction, or **redox reaction** for short. In a redox reaction, the loss of electrons from one substance is called **oxidation**, and the addition of electrons to another substance is called **reduction.** A molecule is said to become oxidized when it loses one or more electrons and reduced when it gains one or more electrons. Because an electron transfer requires both a donor and an acceptor, oxidation and reduction always go together. You can see the overall results of the redox reactions of cellular respiration in Figure 6.5A: Glucose loses electrons (in H atoms) and becomes oxidized, while O_2 gains electrons (in H atoms) and becomes reduced. The electrons lose potential energy along the way, and energy is released.

Two key players in the process of oxidizing glucose are an enzyme called a **dehydrogenase** and a coenzyme called NAD^+. **NAD^+** (nicotinamide adenine dinucleotide) is an organic molecule that cells make from the vitamin niacin and use to shuttle electrons in redox reactions. The top equation in **Figure 6.5B** depicts the oxidation of an organic molecule. We show only its three carbons (gray balls) and a few of its other atoms. Dehydrogenase strips two hydrogen atoms from this molecule. Simultaneously, as shown in the lower equation, NAD^+ picks up the two electrons (one of which

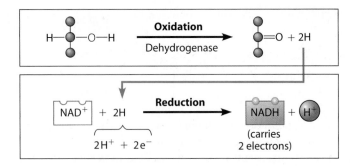

Figure 6.5B A pair of redox reactions, occurring simultaneously

neutralizes its positive charge) and is reduced to NADH. One proton (H^+) is released.

In the energy staircase analogy of electrons "falling" from glucose to oxygen, the transfer of electrons from an organic molecule to NADH represents the first step. **Figure 6.5C** shows NADH delivering these electrons to the rest of the staircase—an **electron transport chain.** The steps in the chain are electron carrier molecules, shown here as purple ovals, built into the inner membrane of a mitochondrion. At the bottom of the staircase is O_2, the final electron acceptor.

The electron transport chain involves a series of redox reactions in which electrons pass from carrier to carrier down to oxygen. The redox steps in the staircase release energy in amounts small enough to be used by the cell to make ATP.

With an understanding of this basic mechanism of electron transfer and energy release, we can now explore cellular respiration in more detail.

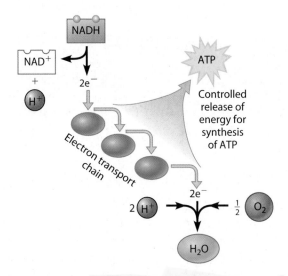

Figure 6.5C In cellular respiration, electrons "fall" down an energy staircase and finally reduce O_2

? What chemical characteristic of the element oxygen accounts for its function in cellular respiration?

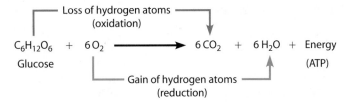

Figure 6.5A Rearrangement of hydrogen atoms (with their electrons) in cellular respiration

■ Oxygen is very electronegative (see Module 2.8), meaning that it is very powerful in pulling electrons from other elements.

6.6 Overview: Cellular respiration occurs in three main stages

Cellular respiration consists of a sequence of steps, which, for study purposes, can be divided into three main stages. Figure 6.6 below gives an overview of the three stages and shows where they occur in the cell.

Stage 1: Glycolysis (shown with an aqua background throughout the chapter) occurs in the cytoplasmic fluid of the cell, that is, outside the organelles. Glycolysis begins respiration by breaking glucose into two molecules of a compound called pyruvate.

Stage 2: The **citric acid cycle** (light salmon color), which takes place within the mitochondria, completes the breakdown of glucose by decomposing a derivative of pyruvate to carbon dioxide. As suggested by the smaller ATP symbols in the diagram, the cell makes a small amount of ATP during glycolysis and the citric acid cycle. The main function of glycolysis and the citric acid cycle, however, is to supply the third stage of respiration with electrons (gold arrows).

Stage 3: Oxidative phosphorylation (purple) involves the electron transport chain (which you just learned about in Module 6.5) and a process known as chemiosmosis. NADH and a related electron carrier, FADH$_2$ (flavin adenine dinucleotide), shuttle electrons to the electron transport chain. Most of the ATP produced by cellular respiration is generated by oxidative phosphorylation, which uses the energy released by the downhill flow of electrons from NADH and FADH$_2$ to O$_2$ to phosphorylate ADP. (Recall from Module 5.4

that cells generate ATP by phosphorylation—that is, by adding a phosphate group to ADP.)

What couples the electron transport chain to ATP synthesis? As the electron transport chain passes electrons down the energy staircase, it also pumps hydrogen ions (H$^+$) across the inner mitochondrial membrane. The result is a concentration gradient of H$^+$ across the membrane. In **chemiosmosis**, the potential energy of this concentration gradient is used to make ATP. The concentration gradient drives the diffusion of H$^+$ through **ATP synthases**, protein complexes built into the inner membrane that synthesize ATP. The details of this process are explored in Module 6.10. In 1978, British biochemist Peter Mitchell was awarded the Nobel Prize for developing the theory of chemiosmosis.

The small amount of ATP produced in glycolysis and the citric acid cycle is made by substrate-level phosphorylation, a process we discuss in more detail in the next module. In the next several modules, we look more closely at the three stages of cellular respiration and the two mechanisms of ATP synthesis.

Web/CD Activity 6A *Overview of Cellular Respiration*

? Of the three main stages of cellular respiration represented in Figure 6.6, which one uses oxygen to extract chemical energy from organic compounds?

■ Oxidative phosphorylation, using the electron transport chain and chemiosmosis

Figure 6.6
An overview of cellular respiration

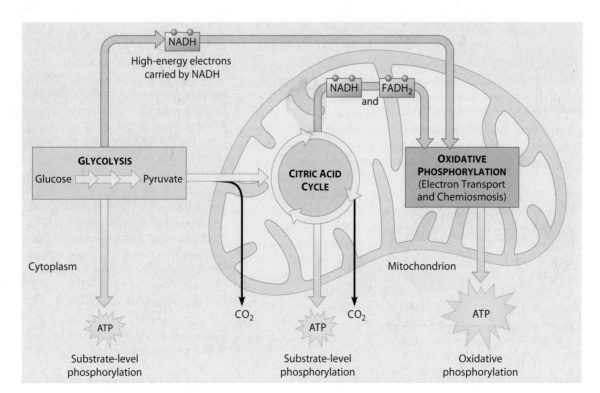

Glycolysis harvests chemical energy by oxidizing glucose to pyruvate

Now that we have introduced the major players and processes, it's time to focus on the individual stages of cellular respiration. The term for the first stage, *glycolysis*, means "splitting of sugar" (*glyco*, sweet, and *lysis*, split), and that's exactly what happens during this phase.

Figure 6.7A gives an overview of glycolysis in terms of input and output. Glycolysis begins with a single molecule of glucose and concludes with two molecules of another organic compound, pyruvate. The gray balls represent the carbon atoms in each molecule; glucose has six, and these same six end up in the two molecules of pyruvate (three in each). The straight arrow from glucose to pyruvate represents nine chemical steps, each catalyzed by its own enzyme. As these reactions occur, the cell reduces two molecules of NAD^+, forming two molecules of NADH, and produces two molecules of ATP by substrate-level phosphorylation.

In **substrate-level phosphorylation** (Figure 6.7B), an enzyme transfers a phosphate group from a substrate molecule directly to ADP, forming ATP. This process produces a small amount of ATP in both glycolysis and the citric acid cycle.

Thus, the energy extracted from glucose during glycolysis is banked in a combination of ATP and NADH. The cell can use the energy in ATP immediately, but for it to use the energy in NADH, electrons from NADH must pass down the electron transport chain in the mitochondrial membrane.

And the pyruvate still holds most of the energy of glucose; these molecules will be oxidized in the citric acid cycle.

Glycolysis is the universal energy-harvesting process of life. If we looked inside a bacterial cell, inside one of our own body cells, or inside virtually any other living cell, we would find the metabolic machinery of glycolysis in full swing. Because glycolysis occurs universally, does not require oxygen, and does not occur in a membrane-bounded organelle, it is thought to be an ancient metabolic system. In fact, what we call glycolysis today may be very similar to the process some of the first cells on Earth used to make ATP. Let's take a closer look at this venerable and vital process.

Figure 6.7C shows all the organic compounds that form in the nine chemical reactions of glycolysis. Commentary on the left highlights the main features of the reactions. As in Figures 6.7A and 6.7B, the gray balls represent the carbon atoms in each of the compounds named on the right. The compounds that form between the initial reactant, glucose, and the final product, pyruvate, are called **intermediates.** Each chemical step leads to the next one. For instance, the intermediate glucose-6-phosphate is the product of step 1 and the reactant for step 2. Similarly, fructose-6-phosphate is the product of step 2 and the reactant for step 3. Also essential are the specific enzymes that catalyze each of the chemical steps; however, to keep the figures simple, we have not included enzymes in the diagrams.

As indicated in Figure 6.7C, the steps of glycolysis can be grouped into two main phases. Steps ❶–❹, the first phase, are preparatory and *consume* energy. In this phase, ATP is used to energize a glucose molecule, which is then split into two small sugars that are now primed to release some energy. Steps ❺–❾, the second phase, yield an *energy payoff* for the cell. During these steps, two NADH molecules are produced for each initial glucose molecule, and four ATP molecules are generated. Since the preparatory steps use two molecules of ATP, *the net gain to the cell is two ATP molecules for each glucose molecule that enters glycolysis.*

These two ATP molecules from glycolysis account for only 5% of the energy that a cell can harvest from a glucose molecule. The two NADH molecules generated during step 5 account for another 16%, but their stored energy is not available for use in the absence of O_2. Some organisms—yeasts and certain bacteria, for instance—can satisfy their energy needs with the ATP produced by glycolysis alone. Most organisms, however, have far higher energy demands. The stages of cellular respiration that follow glycolysis release much more energy. In the next two modules, we see what happens in most organisms after glycolysis forms pyruvate.

Web/CD Activity 6B *Glycolysis*

? For each glucose molecule processed, what are the net molecular products of glycolysis?

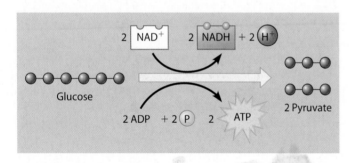

Figure 6.7A   An overview of glycolysis

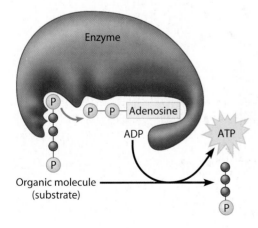

Figure 6.7B   Substrate-level phosphorylation

Two molecules of pyruvate, two molecules of ATP, and two molecules of NADH

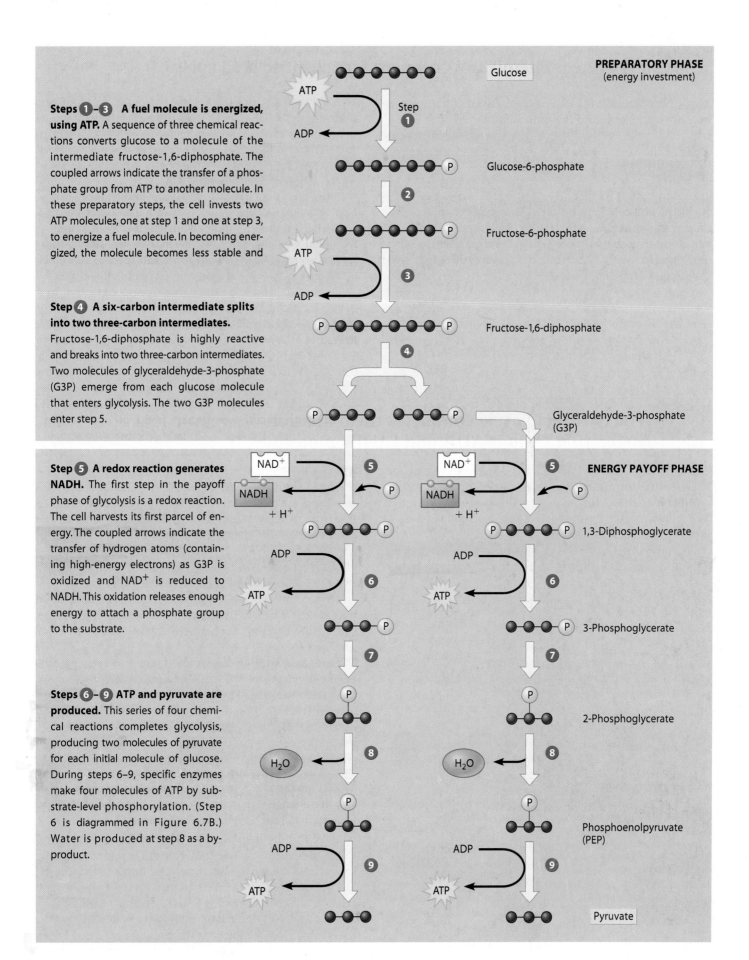

Steps ①–③ A fuel molecule is energized, using ATP. A sequence of three chemical reactions converts glucose to a molecule of the intermediate fructose-1,6-diphosphate. The coupled arrows indicate the transfer of a phosphate group from ATP to another molecule. In these preparatory steps, the cell invests two ATP molecules, one at step 1 and one at step 3, to energize a fuel molecule. In becoming energized, the molecule becomes less stable and

Step ④ A six-carbon intermediate splits into two three-carbon intermediates.
Fructose-1,6-diphosphate is highly reactive and breaks into two three-carbon intermediates. Two molecules of glyceraldehyde-3-phosphate (G3P) emerge from each glucose molecule that enters glycolysis. The two G3P molecules enter step 5.

Step ⑤ A redox reaction generates NADH. The first step in the payoff phase of glycolysis is a redox reaction. The cell harvests its first parcel of energy. The coupled arrows indicate the transfer of hydrogen atoms (containing high-energy electrons) as G3P is oxidized and NAD⁺ is reduced to NADH. This oxidation releases enough energy to attach a phosphate group to the substrate.

Steps ⑥–⑨ ATP and pyruvate are produced. This series of four chemical reactions completes glycolysis, producing two molecules of pyruvate for each initial molecule of glucose. During steps 6–9, specific enzymes make four molecules of ATP by substrate-level phosphorylation. (Step 6 is diagrammed in Figure 6.7B.) Water is produced at step 8 as a byproduct.

PREPARATORY PHASE
(energy investment)

Glucose

Glucose-6-phosphate

Fructose-6-phosphate

Fructose-1,6-diphosphate

Glyceraldehyde-3-phosphate
(G3P)

ENERGY PAYOFF PHASE

1,3-Diphosphoglycerate

3-Phosphoglycerate

2-Phosphoglycerate

Phosphoenolpyruvate
(PEP)

Pyruvate

Figure 6.7C Details of glycolysis

6.8 Pyruvate is chemically groomed for the citric acid cycle

As pyruvate forms at the end of glycolysis, it is transported from the cytoplasm into the mitochondria, the sites of the citric acid cycle. Pyruvate itself does not enter the citric acid cycle. As shown in Figure 6.8, it first undergoes some major chemical "grooming." A large, multienzyme complex cat-alyzes three reactions: ❶ A carbon atom is removed from pyruvate and released in CO_2; ❷ the two-carbon compound remaining is oxidized while a molecule of NAD^+ is reduced to NADH; and ❸ a compound called coenzyme A, derived from a B vitamin, joins with the two-carbon group to form a molecule called acetyl coenzyme A.

These grooming steps—a chemical "haircut and condition-ing" of pyruvate—set up the second major stage of cellular respiration. Acetyl coenzyme A, abbreviated **acetyl CoA**, is a high-energy fuel molecule for the citric acid cycle. For each molecule of glucose that entered glycolysis, two molecules of acetyl CoA are produced and enter the citric acid cycle.

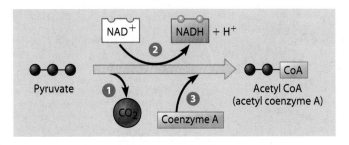

Figure 6.8 The conversion of pyruvate to acetyl CoA

? Which molecule in Figure 6.8 has been reduced?

■ NAD⁺ is reduced to NADH.

6.9 The citric acid cycle completes the oxidation of organic fuel, generating many NADH and FADH₂ molecules

The citric acid cycle is often called the Krebs cycle in honor of Hans Krebs, the German-British researcher who worked out much of this cyclic phase of cellular respiration in the 1930s. We present an overview figure first, followed by a more detailed look at this cycle.

As shown in Figure 6.9A, only the two-carbon acetyl part of the acetyl CoA molecule actually participates in the citric acid cycle. Coenzyme A helps the acetyl group enter the cycle and then splits off and is recycled. Not shown in this figure are the multiple steps that follow, each catalyzed by a specific enzyme located in the mitochondrial matrix. The acetyl group joins a four-carbon molecule. The result-ing six-carbon molecule is processed through a series of redox reactions, two carbon atoms are removed as CO_2, and the four-carbon molecule is regenerated. This regener-ation accounts for the word *cycle*; the six-carbon com-pound first formed in the cycle is citrate, the ionized (negatively charged) form of citric acid. Hence the name *citric acid cycle*.

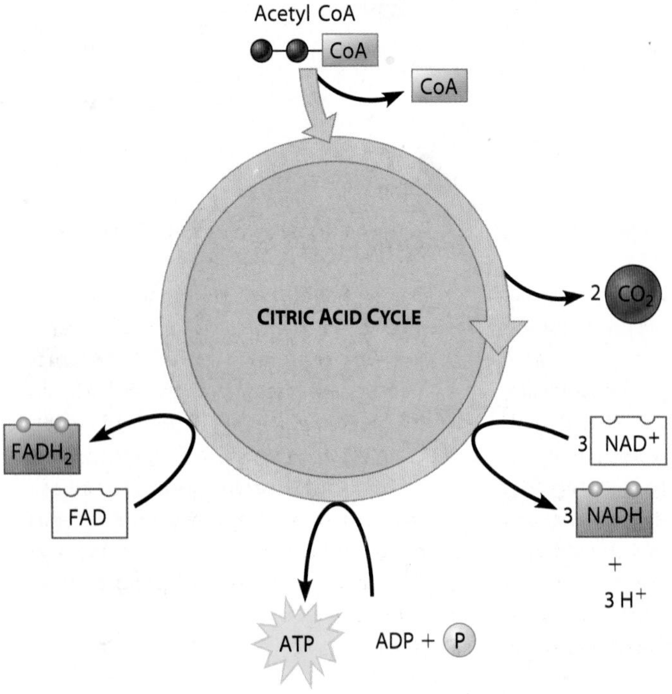

Compared with glycolysis, the citric acid cycle pays big energy dividends to the cell. Each turn of the cycle makes one ATP molecule by substrate-level phosphorylation (shown at the bottom of Figure 6.9A). It also produces four other energy-rich molecules: three NADH molecules and one molecule of a similar electron carrier, $FADH_2$. Since the citric acid cycle processes two molecules of acetyl CoA for each initial molecule of glucose, the overall yield per mole-cule of glucose is 2 ATP, 6 NADH, and 2 $FADH_2$. This yield is considerably more than the 2 ATP plus 2 NADH produced by glycolysis alone.

Overall, how many energy-rich molecules has the cell gained by processing one molecule of glucose through gly-colysis and the citric acid cycle? Up to this point, the cell has gained a total of 4 ATP (from substrate-level phospho-rylation), 10 NADH, and 2 $FADH_2$. Still, for the cell to be able to put to use the energy banked in NADH and $FADH_2$, these molecules must shuttle their high-energy electrons to the electron transport chain. There the energy from the ox-idation of organic fuels can be used for the oxidative phos-

Figure 6.9A An overview of the citric acid cycle

phorylation of ADP to ATP. Before we look at how oxidative phosphorylation works, you may want to examine the inner workings of the citric acid cycle in Figure 6.9B, below.

Web/CD Activity 6C *The Citric Acid Cycle*

❓ What is the total number of NADH molecules generated during the complete breakdown of one glucose molecule to six carbon dioxide molecules? (*Hint:* Combine the outputs of Modules 6.7–6.9.)

■ 10 NADH: 2 from glycolysis; 2 from the grooming of pyruvate; 6 from the citric acid cycle. (Did you remember to double the output due to the sugar-splitting step of glycolysis?)

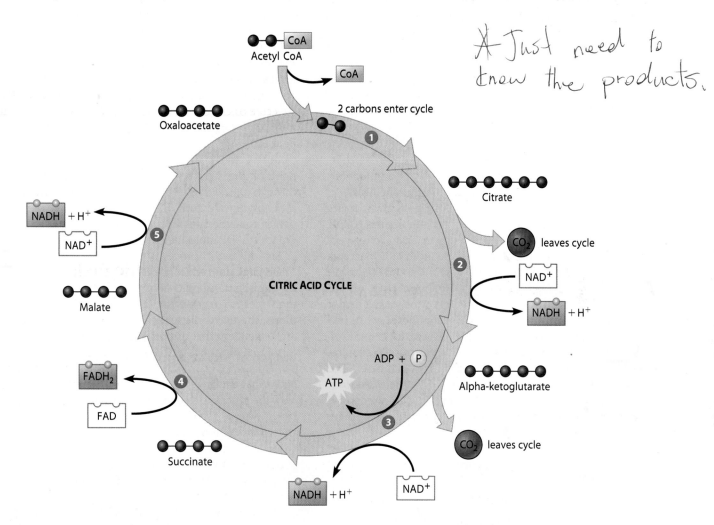

Just need to know the products.

Figure 6.9B Details of the citric acid cycle

Step ①

Acetyl CoA stokes the furnace.

A turn of the citric acid cycle begins (top center) as enzymes strip the CoA portion from acetyl CoA and combine the remaining two-carbon acetyl group with oxaloacetate (top left) already present in the mitochondrion. The product of this reaction is the six-carbon molecule citrate. Citrate is the ionized form of citric acid. All the acid compounds in this cycle exist in the cell in their ionized form, hence the suffix *ate.*

Steps ② and ③

NADH, ATP, and CO_2 are generated during redox reactions.

Successive redox reactions harvest some of the energy of the acetyl group by stripping hydrogen atoms from organic acid intermediates (such as alpha-ketoglutarate) and producing energy-laden NADH molecules. The redox reactions dispose of two carbon atoms that came from oxaloacetate. The carbons are completely oxidized and released as two molecules of CO_2. Energy is also harvested by substrate-level phosphorylation of ADP to produce ATP. A four-carbon compound called succinate emerges at the end of step 3.

Steps ④ and ⑤

Redox reactions generate $FADH_2$ and NADH.

Enzymes rearrange chemical bonds, eventually completing the citric acid cycle by regenerating oxaloacetate. Concurrently, the electron carriers FAD and NAD^+ are reduced to $FADH_2$ and NADH, respectively. One turn of the citric acid cycle is completed with the conversion of a molecule of malate to oxaloacetate. This compound is then ready to start the next turn of the cycle by accepting another acetyl group from acetyl CoA.

6.10 Most ATP production occurs by oxidative phosphorylation

The final stage of cellular respiration is oxidative phosphorylation, which involves the electron transport chain and chemiosmosis. Oxidative phosphorylation is a clear illustration of structure fitting function: The spatial arrangement of electron carriers built into a membrane makes it possible for the mitochondrion to use the chemical energy released by redox reactions to create an H^+ gradient and then use the energy stored in that gradient to drive ATP synthesis.

Figure 6.10 expands on our earlier discussion of oxidative phosphorylation in Module 6.6. It shows that the electron transport chain is built into the inner membrane of the mitochondrion. The folds (cristae) of this membrane enlarge its surface area, providing space for thousands of copies of the electron transport chain and many ATP synthase complexes. With all these ATP-making "machines," a mitochondrion can produce many ATP molecules simultaneously.

Starting on the left in Figure 6.10, the gold arrow traces the path of electron flow from the shuttle molecules NADH and $FADH_2$ through the electron transport chain to oxygen, the final electron acceptor. Each oxygen atom $(\frac{1}{2} O_2)$ accepts two electrons from the chain and picks up two hydrogen ions from the surrounding solution to form H_2O, one of the final products of cellular respiration. Most of the carrier molecules reside in three main protein complexes, while two mobile carriers transport electrons between the complexes.

All of the carriers bind and release electrons in redox reactions, passing electrons down the "energy staircase." The protein complexes shown here use the energy released from the electron transfers to actively transport H^+ across the membrane, from where they are less concentrated to where they are more concentrated. The green vertical arrows indicate that the hydrogen ions are transported from the matrix of the mitochondrion (its innermost compartment) into the mitochondrion's intermembrane space.

The resulting H^+ gradient stores potential energy, much the way a dam stores energy by holding back the elevated water behind it. The energy stored by a dam can be harnessed to do work (such as generating electricity) when the water is allowed to rush downhill, turning giant wheels called turbines. Similarly, the ATP synthases built into the inner mitochondrial membrane act like miniature turbines. The hydrogen ions tend to be driven back across the membrane by the energy of their concentration gradient. However, the membrane is not permeable to hydrogen ions, and they can only cross back through a channel in the ATP synthase, as shown on the far right of the figure. Hydrogen ions rush back "downhill" through an ATP synthase, spinning a component of the complex, just as water turns the turbines in a dam. The rotation activates catalytic sites in the synthase that attach phosphate groups to ADP molecules to generate ATP.

So why is this process called oxidative phosphorylation? The energy derived from the *oxidation*-reduction reactions of the electron transport chain that transfer electrons from organic molecules to oxygen is used to *phosphorylate* ADP. By chemiosmosis, the exergonic reactions of electron transport produce an H^+ gradient that drives the endergonic synthesis of ATP.

Web/CD Activity 6D *Electron Transport and Chemiosmosis*

> What effect would an absence of oxygen (O_2) have on the process illustrated in Figure 6.10?

■ There would be no ATP produced. Without oxygen to "pull" electrons down the electron transport chain, the energy stored in NADH cannot be harnessed for ATP synthesis.

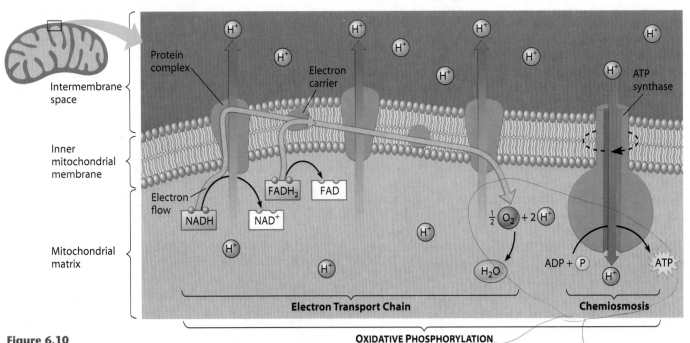

Figure 6.10
Oxidative phosphorylation, using electron transport and chemiosmosis in the mitochondrion

6.11 Certain poisons interrupt critical events in cellular respiration

A number of poisons produce their deadly effects by interfering with some of the events we have just discussed. **Figure 6.11** shows the places where three different categories of poisons obstruct cellular respiration.

Poisons in one category block the electron transport chain. A substance called rotenone, for instance, binds tightly with one of the electron carrier molecules in the first protein complex, preventing electrons from passing to the next carrier molecule. Rotenone is often used to kill pest insects and fish. By blocking the electron transport chain near its start and thus preventing ATP synthesis, rotenone literally starves an organism's cells of energy. Two other electron transport blockers, cyanide and carbon monoxide, bind with an electron carrier in the third protein complex. Here they block the passage of electrons to oxygen. This blockage is like turning off a faucet; electrons cease to flow through the "pipe." The result is the same as that of rotenone: No H^+ gradient is generated, and no ATP is made.

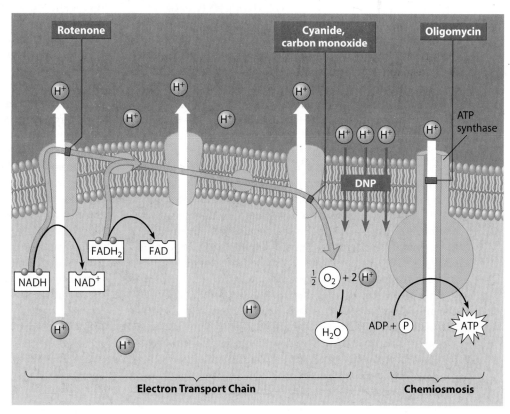

Figure 6.11   The effects of five poisons on the electron transport chain and chemiosmosis

A second kind of respiratory poison inhibits ATP synthase. On the right side of the figure, the antibiotic oligomycin blocks the passage of H^+ through the channel in ATP synthase. Oligomycin is used on the skin to combat fungal infections. It kills fungal cells by preventing them from using the potential energy of the H^+ gradient to make ATP. (Because the drug cannot get into the living skin cells, they are protected from its effects.)

A third kind of poison, collectively called uncouplers, makes the membrane of the mitochondrion leaky to hydrogen ions. Electron transport continues, but ATP cannot be made because leakage of H^+ through the membrane abolishes the H^+ gradient. Cells continue to consume oxygen, often at a higher than normal rate, but to no avail, for they cannot make any ATP through oxidative phosphorylation.

One uncoupler, dinitrophenol (DNP), is highly toxic to humans. DNP poisoning produces an enormous increase in metabolic rate, profuse sweating as the body attempts to dissipate excess heat, collapse, and then death. For a short time in the 1940s, some physicians prescribed DNP in low doses as weight loss pills, but fatalities soon made it clear that there were far safer ways to lose weight. When DNP is present, all steps of cellular respiration except chemiosmosis

continue to run, consuming fuel molecules, even though almost all the energy is lost as heat.

Poisons do have some good points. In addition to being useful as pesticides or antibiotics, toxins may be used in biochemical research. Discovering exactly what toxic substances do to the cell's respiratory machinery has, in many cases, helped biochemists understand how the machinery works. The effects of uncouplers, for example, made it clear that ATP synthesis is a complicated activity involving the distinct, but related, processes of electron transport and generation of a membrane H^+ gradient.

The function of cellular respiration is to generate ATP for cellular work. Without energy, a cell cannot live. We review cellular respiration and the essential production of ATP in the next module

? The poison DNP causes what one biochemist calls "mitochondrial wheel-spinning." Explain this metaphor.

■ Like an automobile wasting fuel by spinning its wheels and going nowhere, a mitochondrion poisoned with DNP consumes fuel and powers electron transport but makes no ATP. The DNP destroys the H^+ gradient required for ATP synthesis.

6.12 Review: Each molecule of glucose yields many molecules of ATP

Now that we have looked at all stages of cellular respiration, let's review what the cell accomplishes by oxidizing a molecule of glucose. Figure 6.12 puts all the stages together and indicates where they occur in the cell. At the bottom of the figure is a tally of ATP molecules, showing the potential energy payoff for a typical working cell. If you wish to refer back to earlier modules, this diagram summarizes glycolysis (Module 6.7), the chemical grooming of pyruvate (Module 6.8), the citric acid cycle (Module 6.9), and oxidative phosphorylation (Module 6.10).

Starting on the left, glycolysis, occurring in the cytoplasmic fluid, and the citric acid cycle, occurring in the mitochondrial matrix, contribute a net total of 4 ATP per glucose molecule by substrate-level phosphorylation. The cell harvests much more energy than this from the carrier molecules NADH and $FADH_2$, which are produced by glycolysis, the grooming of pyruvate, and the citric acid cycle. The energy of the electrons they carry is used to make an estimated 34 molecules of ATP using the electron transport chain and chemiosmosis in oxidative phosphorylation.

Let's see where the numbers in the diagram come from. Our model assumes that each NADH that transfers a pair of high-energy electrons from a food molecule to the electron transport chain contributes enough to the mitochondrion's H^+ gradient to generate 3 ATP. Another assumption is that each $FADH_2$ molecule yields only 2 ATP because it contributes its electrons later in the electron transport chain (see Figure 6.10). These numbers are maximums; the actual amounts may vary. Also, as indicated in this diagram, the cell uses a shuttle mechanism to pass the electrons from NADH produced in glycolysis into the mitochondrion. Depending on the shuttle used, either NAD^+ or FAD picks up the electrons (so the yield of ATP by oxidative phosphorylation may be 32 or 34 ATP). The total net yield of ATP molecules per glucose molecule has a theoretical maximum of about 38.

More important than the actual numbers of ATP molecules is the point that a cell can harvest a great deal of energy from glucose—up to about 40% of the molecule's potential energy. Because most of the ATP generated by cellular respiration results from oxidative phosphorylation, the ATP yield depends heavily on an adequate supply of oxygen (O_2) to the cell. Without oxygen to function as the final electron acceptor in the electron transport chain, chemiosmosis ceases, and cells die from energy starvation. However, as we discussed in the chapter introduction, muscle cells may continue to function for a while without oxygen. We look at how some cells and organisms can oxidize organic fuel and generate ATP *without* oxygen next.

Web/CD Thinking as a Scientist *How Is the Rate of Cellular Respiration Measured?*

? What would a cell's net ATP yield per glucose molecule be in the presence of the poison DNP? (See Module 6.11.)

■ 4 ATP, all from substrate-level phosphorylation

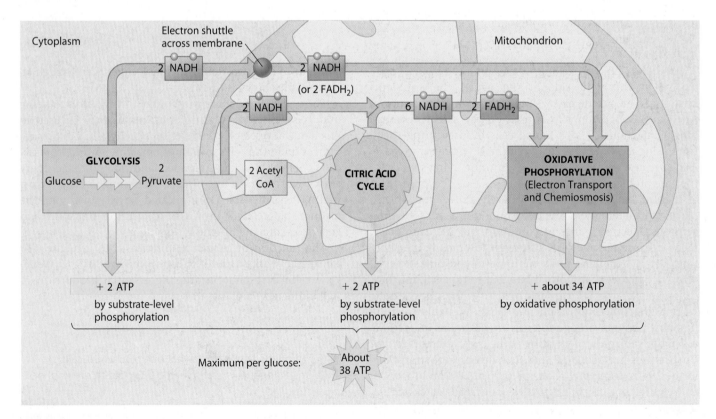

Figure 6.12 A tally of the ATP yield from cellular respiration

6.13 Fermentation is an anaerobic alternative to cellular respiration

The metabolic pathway that generates ATP during fermentation is glycolysis, the same pathway that functions in the first stage of cellular respiration. Remember that glycolysis uses no O_2; it simply generates a net gain of 2 ATP while oxidizing glucose to two molecules of pyruvate and reducing NAD^+ to NADH. The yield of 2 ATP is certainly a lot less than the 38 ATP per glucose generated during aerobic respiration, but it is enough to keep your muscles contracting for a short while when the need for ATP outpaces the delivery of O_2 via the bloodstream. And many microorganisms supply all their energy needs with the 2 ATP yield of glycolysis.

There is more to fermentation, however, than just glycolysis. To harvest food energy by glycolysis, NAD^+ must be present as an electron acceptor. This is no problem under aerobic conditions, because the cell regenerates its pool of NAD^+ when NADH drops its electrons down the electron transport chain. Fermentation provides an anaerobic step that recycles NADH back to NAD^+.

Your muscle cells, a few other cell types, and certain bacteria regenerate NAD^+ by a process called **lactic acid fermentation,** illustrated in **Figure 6.13A.** You can see that NADH is oxidized to NAD^+ as pyruvate is reduced to lactate (the ionized form of lactic acid). The lactate that builds up in muscle cells during strenuous exercise is carried in the blood to the liver, where it is converted back to pyruvate. The dairy industry uses lactic acid fermentation by bacteria to make cheese and yogurt. Other types of microbial fermentation turn soybeans into soy sauce and cabbage into sauerkraut.

Alcohol fermentation is a type of fermentation that people have made use of for thousands of years in brewing, winemaking, and baking. Yeasts are single-celled fungi that normally use aerobic respiration to process their food. But they are also able to survive in anaerobic environments when there is plenty of glucose to keep glycolysis operating. Yeasts and certain bacteria recycle their NADH to NAD^+ while converting pyruvate to CO_2 and ethanol (ethyl alcohol). The CO_2 provides the bubbles in beer and champagne. Bubbles of CO_2 generated by baker's yeast cause bread dough to rise.

During alcohol fermentation, shown in **Figure 6.13B,** the oxidation of NADH recharges the cell with a supply of NAD^+ that keeps glycolysis working. Ethanol, the two-carbon product, is toxic to the organisms that produce it. Yeasts release their alcohol wastes to their surroundings. When yeasts are confined in a wine vat, they die when the alcohol concentration reaches 14%.

Unlike muscle cells and yeasts, many bacteria that live in stagnant ponds and deep in the soil are **strict anaerobes,** meaning they require anaerobic conditions and are poisoned by oxygen. Yeasts and many other bacteria are facultative anaerobes. A **facultative anaerobe** can make ATP either by fermentation or by oxidative phosphorylation, depending on whether O_2 is available.

For a facultative anaerobe, pyruvate is a fork in the metabolic road. If oxygen is available, the organism will always use the more efficient process—aerobic respiration. Thus, to make

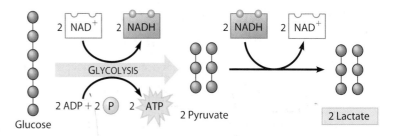

Figure 6.13A Lactic acid fermentation

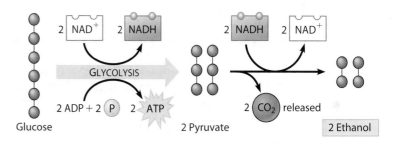

Figure 6.13B Alcohol fermentation

Figure 6.13C Fermentation vats for wine

wine and beer, yeasts must be grown anaerobically so that they will ferment sugars and produce ethanol. For this reason, the large fermentation vats in **Figure 6.13C** are equipped with one-way gas valves that vent off excess CO_2 but keep air out.

Web/CD Activity 6E *Fermentation*

? A glucose-fed yeast cell is moved from an aerobic environment to an anaerobic one. If the cell continues to generate ATP at the same rate, how will its rate of glucose consumption change in the anaerobic environment?

■ The cell must consume glucose at a rate about 19 times the consumption rate in the aerobic environment (2 ATP by fermentation vs. 38 ATP by cellular respiration).

6.14 Cells use many kinds of organic molecules as fuel for cellular respiration

Throughout this chapter, we have spoken of glucose as the fuel for cellular respiration. But free glucose molecules are not common in our diet. We obtain most of our calories as fats, proteins, sucrose and other disaccharide sugars, and starch (a polysaccharide). You consume all these types of molecules when you eat a bag of peanuts, for instance.

Figure 6.14 illustrates how the cell uses three main kinds of food molecules to make ATP. A cell can funnel a wide range of carbohydrates (polysaccharides and sugars) into glycolysis, as shown by the green arrows in the diagram. For example, enzymes in our digestive tract hydrolyze starch to glucose, which is then broken down by glycolysis and the citric acid cycle. Similarly, glycogen, the polysaccharide stored in our liver and muscle cells, can be hydrolyzed to glucose to serve as fuel between meals.

Proteins (purple arrows) can be used for fuel, but first they must be digested to their constituent amino acids. Typically, a cell will use most of the amino acids to make its own proteins, but enzymes will convert excess amino acids to intermediates of glycolysis or the citric acid cycle, and their energy is then harvested by cellular respiration. Dur-

ing the conversion, the amino groups are stripped off and later disposed of in urine.

Fats make excellent cellular fuel because they contain many hydrogen atoms and thus many energy-rich electrons. As the diagram shows (tan arrows), the cell first hydrolyzes fats to glycerol and fatty acids. It then converts the glycerol to glyceraldehyde-3-phosphate (G3P), one of the intermediates in glycolysis. The fatty acids are broken into two-carbon fragments that enter the citric acid cycle as acetyl CoA. Processed this way, a gram of fat yields more than twice as much ATP as a gram of starch. Because so many calories are stockpiled in each gram of fat, a person must expend a large amount of energy to burn fat stored in the body. This explains why it is so difficult for a dieter to lose excess fat.

? Animals store most of their energy reserves as fats, not as polysaccharides. What is the advantage of this mode of storage for an animal?

■ Because most animals are mobile, they benefit from a compact form of energy storage. A gram of fat stores about twice as much energy as a gram of carbohydrate.

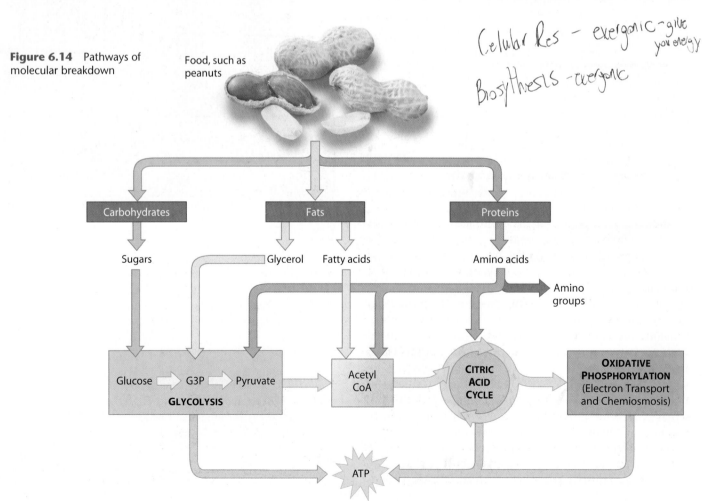

Figure 6.14 Pathways of molecular breakdown

Food, such as peanuts

Cellular Res – exergonic – gives you energy

Biosynthesis – exergonic

6.15 Food molecules provide raw materials for biosynthesis

Not all food molecules are destined to be oxidized as fuel for making ATP. Food also provides the raw materials a cell uses for biosynthesis, to make its own molecules for repair and growth. Some raw materials from food, such as amino acids, can be incorporated directly into an organism's molecules. However, our cells can make molecules that are not present in food by using some of the intermediate compounds of glycolysis and the citric acid cycle.

Figure 6.15 outlines biosynthetic pathways by which cells make three classes of macromolecules, starting with some of the small organic molecules produced in glycolysis and the citric acid cycle. These pathways consume ATP, rather than generate it. Notice that some pathways appear to be the exact reverse of those in Figure 6.14. However, the details of the pathways can be different. In particular, glucose synthesis (pink box) is *not* the exact reverse of glycolysis, although some of the key intermediates—such as G3P—are the same in the two processes.

Thus, we see an important distinction as well as clear connections between two aspects of metabolism that are central to life: the energy-harvesting process of cellular respiration and the biosynthetic pathways used to construct all parts of the cell.

> **?** Explain how one can gain weight and store fat even when on a low-fat, carbohydrate-rich diet. (*Hint:* Look for G3P and acetyl CoA in Figures 6.14 and 6.15.)

■ If caloric intake is excessive, whatever its form, body cells use metabolic pathways to convert the excess to fat. Fats are made from glycerol and fatty acids, which are made from G3P and acetyl CoA, respectively. Acetyl CoA and G3P are both produced from the oxidation of carbohydrates.

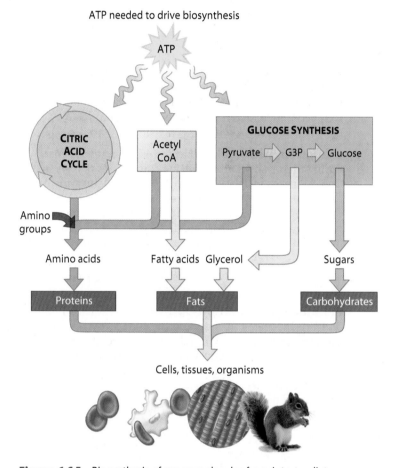

Figure 6.15 Biosynthesis of macromolecules from intermediates in cellular respiration

6.16 The fuel for respiration ultimately comes from photosynthesis

A giant panda eating its favorite food, a bamboo shoot, is an appropriate ending to this chapter and a good starting point for the next one (**Figure 6.16**). Almost entirely a vegetarian, a panda can get all the energy and nutrients it needs by eating a few armloads of bamboo leaves and shoots daily. The panda's digestive tract breaks the plant material down into fuel molecules, and its cells harvest energy from the molecules using cellular respiration.

The cells of all living organisms—those of pandas and bamboo plants included—have the ability to harvest energy from the breakdown of organic fuel molecules. When the breakdown process is cellular respiration, the atoms of the starting materials end up in CO_2 and H_2O. In contrast, the ability to make organic molecules from CO_2 and H_2O is not universal. Giant pandas—in fact, all animals—lack this ability, but plants have it. Animal cells can only convert the energy in the chemical bonds of organic molecules to other chemical forms, but plant cells can actually produce organic molecules from inorganic ones using the energy of sunlight. This process, photosynthesis, is the subject of Chapter 7.

> **?** The chemical ingredients from which a plant makes food by photosynthesis are _____ and _____.

■ carbon dioxide . . . water

Figure 6.16
Photosynthesis creates the plant material (bamboo) that nourishes a panda

CHAPTER REVIEW

Reviewing the Concepts

Introduction to Cellular Respiration (6.1–6.5)

Cellular respiration makes ATP. O_2 is consumed during the oxidation of glucose to CO_2 and H_2O. Photosynthesis uses solar energy to produce glucose and O_2 from CO_2 and H_2O **(6.1)**. Breathing provides for the exchange of O_2 and CO_2 between an organism and its environment **(6.2)**. ATP powers almost all cellular and body activities **(6.3–6.4)**.

Oxidation/Reduction. Electrons lose potential energy during their transfer from organic compounds to oxygen. Dehydrogenase removes electrons (in hydrogen atoms) from fuel molecules (oxidation) and transfers them to NAD^+ (reduction). NADH passes electrons to an electron transport chain. As electrons "fall" from carrier to carrier and finally to O_2, energy is released in small quantities **(6.5)**.

Stages of Cellular Respiration and Fermentation (6.6–6.13)

Cellular respiration produces ATP. A small amount of ATP is made in glycolysis and the citric acid cycle. In oxidative phosphorylation, cells use the energy released by "falling" electrons to pump H^+ across a membrane. In a mechanism called chemiosmosis, the energy of the H^+ gradient is harnessed to make ATP **(6.6)**. Cellular respiration can be divided into three stages.

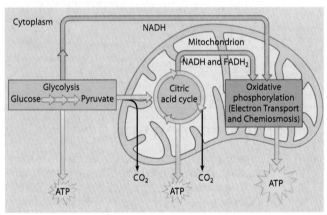

Glycolysis. ATP is used to prime a glucose molecule, which is split in two. These three-carbon intermediates are oxidized and converted to two molecules of pyruvate, yielding a net of 2 ATP and 2 NADH. ATP is formed by substrate-level phosphorylation, in which a phosphate group is transferred from an organic molecule to ADP **(6.7)**.

Citric Acid Cycle. Prior to the citric acid cycle, enzymes process pyruvate, releasing CO_2 and producing NADH and acetyl CoA **(6.8)**. In the citric acid cycle, the two-carbon acetyl group is added to a four-carbon compound, forming citrate, which is degraded back to the starting four-carbon compound. For each turn of the cycle, 2 CO_2 is released; the energy yield is 1 ATP, 3 NADH, and 1 $FADH_2$ **(6.9)**.

Oxidative Phosphorylation. Electrons from NADH and $FADH_2$ travel down the electron transport chain to oxygen, which picks up H^+ to form water. Energy released by these redox reactions is used to pump H^+ into the space between the membranes of the mitochondrion. In chemiosmosis, the H^+ diffuses back across the inner membrane (down its concentration gradient) through ATP synthase complexes, driving the synthesis of ATP **(6.10)**. Various poisons can block the movement of electrons, block the flow of H^+ through ATP synthase, or allow H^+ to leak through the membrane **(6.11)**. Oxidative phosphorylation, using electron transport and chemiosmosis, produces up to 38 ATP molecules for every glucose molecule that enters cellular respiration **(6.12)**.

Fermentation. Under anaerobic conditions, muscle cells, yeasts, and certain bacteria produce small amounts of ATP by glycolysis. NAD^+ is recycled from NADH as pyruvate is converted to lactate (lactic acid fermentation) or alcohol and CO_2 (alcohol fermentation) **(6.13)**.

Interconnections Between Molecular Breakdown and Synthesis (6.14–6.16)

Food as fuel. Carbohydrates, fats, and proteins can all fuel cellular respiration when they are converted to molecules that enter glycolysis or the citric acid cycle **(6.14)**.

Biosynthesis. Cells use some food molecules and intermediates from glycolysis and the citric acid cycle as raw materials. Biosynthesis consumes ATP **(6.15)**. All organisms can harvest energy from organic molecules. Plants (but not animals) can also make organic molecules from inorganic sources by the process of photosynthesis **(6.16)**.

Connecting the Concepts

1. Fill in the blanks in this summary map to help you review the key concepts of cellular respiration. (*Hint:* Most boxes are color-coded to match the text art.)

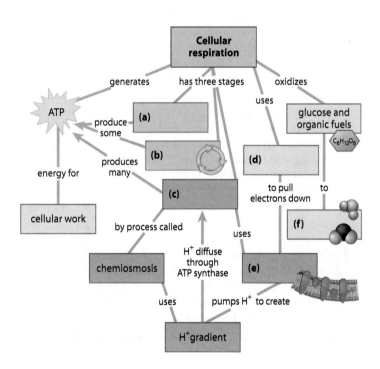

Testing your Knowledge

Multiple Choice

2. What is the role of oxygen in cellular respiration?
   a. It is reduced in glycolysis as glucose is oxidized.
   b. It provides electrons to the electron transport chain.
   c. It combines with the carbon removed during the citric acid cycle to form CO_2.
   d. It is required for the production of heat and light.
   e. It is the final electron acceptor for the electron transport chain.

3. When the poison cyanide blocks the electron transport chain, glycolysis and the citric acid cycle soon grind to a halt as well. Why do you think they stop?
   a. They both run out of ATP.
   b. Unused O_2 interferes with glycolysis and the citric acid cycle.
   c. They run out of NAD^+ and FAD.
   d. Electrons are no longer available from the electron transport chain.
   e. They run out of ADP.

4. A biochemist wanted to study how various substances were used in cellular respiration. In one experiment, he allowed a mouse to breathe air containing O_2 "labeled" by a particular isotope of oxygen (a procedure harmless to the mouse). In the mouse, the labeled oxygen atoms first showed up in
   a. ATP.
   b. glucose, $C_6H_{12}O_6$.
   c. NADH.
   d. carbon dioxide, CO_2.
   e. water, H_2O.

5. In glycolysis, ____ is oxidized and ____ is reduced.
   a. NAD^+ . . . glucose
   b. glucose . . . oxygen
   c. ATP . . . ADP
   d. glucose . . . NAD^+
   e. ADP . . . ATP

6. Which of the following is the most immediate source of energy for making most of the ATP in your cells?
   a. the reduction of oxygen
   b. the transfer of phosphates from intermediate substrates to ADP
   c. the diffusion of hydrogen ions across a membrane
   d. the splitting of glucose into two molecules of pyruvate
   e. the movement of electrons along the electron transport chain

7. Which of the following conversions represents a reduction reaction?
   a. pyruvate → acetyl CoA
   b. pyruvate → lactate
   c. glucose → pyruvate
   d. NADH + H^+ → NAD^+ + 2 H
   e. $C_6H_{12}O_6$ → 6 CO_2

Describing, Comparing, and Explaining

8. Which of the three stages of cellular respiration is considered the most ancient? Explain your answer.

9. Explain in terms of cellular respiration why we need oxygen and why we exhale carbon dioxide.

10. What is the main function of fermentation?

11. Compare and contrast fermentation as it occurs in human muscle cells and as it occurs in yeast cells.

12. Explain how your body can convert excess carbohydrates in the diet to fats. Can excess carbohydrates be converted to protein? What else must be supplied?

Applying the Concepts

13. An average adult human requires 2,200 kcal of energy per day. Suppose your diet provides an average of 2,300 kcal per day. How many hours per week would you have to walk to burn off the extra calories? Swim? Run? (See Table 6.4.)

14. Your body makes NAD^+ and FAD from two B vitamins, niacin and riboflavin. The recommended dietary allowance for niacin is 20 mg daily and for riboflavin, 1.7 mg. These amounts are thousands of times less than the amount of glucose your body needs each day to fuel its energy needs. How many NAD^+ and FAD molecules are needed for the breakdown of each glucose molecule? Why do you think your daily requirement for these substances is so small?

15. In a detail of the citric acid cycle not shown in Figure 6.9B, succinate is converted to a compound called fumarate, with the release of H^+. You are studying this reaction using a suspension of bean cell mitochondria and a blue dye that loses its color as it takes up H^+. You know from previous experiments that the higher the concentration of succinate, the more rapid the decolorization of the blue dye. You set up reaction mixtures with mitochondria, dye, and three different concentrations of succinate (0.1 mg/L, 0.2 mg/L, and 0.3 mg/L). Which of the following graphs represents the results you would expect, and why?

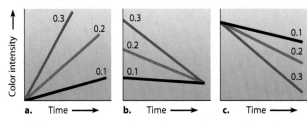

16. The consumption of alcohol by a pregnant woman can cause a complex of birth defects called fetal alcohol syndrome (FAS). Symptoms of FAS include head and facial irregularities, heart defects, mental retardation, and behavioral problems such as hyperactivity. The U.S. Surgeon General's Office recommends that pregnant women abstain from drinking alcohol, and the government has mandated that a warning label be placed on liquor bottles: "Women should not drink alcoholic beverages during pregnancy because of the risk of birth defects." Imagine you are a server in a restaurant. An obviously pregnant woman orders a strawberry daiquiri. How would you respond? Is it the woman's right to make those decisions about her unborn child's health? Do you bear any responsibility in the matter? Is a restaurant responsible for monitoring the health habits of its customers?

Answers to all questions can be found in Appendix 3.

For study help and Activities, go to campbellbiology.com or the student CD-ROM.

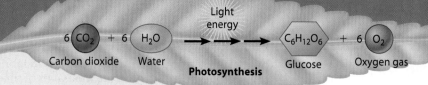

CHAPTER 7

The summary equation for photosynthesis

AN OVERVIEW OF PHOTOSYNTHESIS

7.1   Autotrophs are the producers of the biosphere
7.2   Photosynthesis occurs in chloroplasts
7.3   Plants produce O_2 gas by splitting water
7.4   Photosynthesis is a redox process, as is cellular respiration
7.5   Overview: Photosynthesis occurs in two stages linked by ATP and NADPH

THE LIGHT REACTIONS: CONVERTING SOLAR ENERGY TO CHEMICAL ENERGY

7.6   Visible radiation drives the light reactions
7.7   Photosystems capture solar power
7.8   In the light reactions, electron transport chains generate ATP and NADPH
7.9   Chemiosmosis powers ATP synthesis in the light reactions

THE CALVIN CYCLE: CONVERTING CO_2 TO SUGARS

7.10   ATP and NADPH power sugar synthesis in the Calvin cycle

PHOTOSYNTHESIS REVIEWED AND EXTENDED

7.11   Review: Photosynthesis uses light energy to make food molecules
7.12 C_4 and CAM plants have special adaptations that save water

PHOTOSYNTHESIS, SOLAR RADIATION, AND EARTH'S ATMOSPHERE

7.13   Photosynthesis moderates global warming
7.14   Mario Molina talks about Earth's protective ozone layer

Plant Power

WITH EVERY NEW CAR, DVD player, and wireless phone, the world's demand for energy grows. In an era when most energy is generated from pollution-laden fossil fuels, more power often means more air pollution, more acid precipitation, and more of the "greenhouse gases" that contribute to global warming. Is there a better energy solution, one that could somehow produce power without harming our atmosphere? Some scientists believe they've found such a solution—and if you can see green through a nearby window, their answer might literally be in sight.

Energy researchers from Brazil to upstate New York have seized on the potential of plant power, turning to trees and other green vegetation to create "energy plantations" as a promising fuel source. In many ways, this solution draws from an ancient one, relying on one of the oldest energy pathways on the planet—photosynthesis. In this process, green plants, algae, and certain bacteria use light energy to make sugar from carbon dioxide and water. During the process, they also produce the oxygen we need to breathe and store an energy source that could help solve the human power problem.

As you learned in Chapter 6, both carbon dioxide and water are waste products of cellular respiration. Plants take in this cellular "exhaust," absorbing CO_2 through their leaves and pulling up water through their roots. Their leaves also absorb the solar energy of sunlight. In a sort of molecular shuffling driven by this energy, six molecules of carbon dioxide and six molecules of water are converted to one molecule of the sugar glucose and six molecules of oxygen gas. The summary equation for photosynthesis is shown at the top of this page.

Photosynthesis happens on a microscopic level, but carried out repeatedly in plants around the world, it is responsible for an enormous amount of product. By converting the energy of sunlight to chemical energy, Earth's plants and other photosynthetic organisms make about 160 billion metric tons (176 billion tons) of carbohydrates each year. All of the food consumed by people can be traced back to photosynthetic plants—either eaten directly or fed to animals that are then consumed. And for most of human history, burning plant material has been a major source of heat, light, and cooking fuel. Its replacement by fossil fuels is relatively recent and centered in the developed world.

Photosynthesis: Using Light to Make Food

Now, as fossil fuel supplies dwindle and pollution accumulates, scientists are once again eyeing plants and working to make them a more efficient energy source. At energy plantations in Europe, South America, and upstate New York, certain types of fast-growing trees, such as willows, have shown promise as fuel to grow and burn in power generation facilities. Willows grow quickly and resprout once they are cut, limiting the amount of new planting, herbicides, and pesticides required to produce each generation of trees. They also pack more fuel and power than most native trees: Willow crops should be able to produce about 5–8 dry tons of wood per acre each year and generate between 84 and 134 BTU (units of heat energy) per acre each year, according to researchers in New York. In comparison, natural forests produce only about 0.5–1.0 ton of wood per acre and generate about 8–19 BTU each year.

Fuel from energy plantations is not only a renewable energy source; it can also be environmentally safer than fossil fuels. The products of burned wood do not include the sulfur contaminants released by fossil fuels, which contribute to acid rain. To be sure,

burning wood fuel releases carbon dioxide into the air. But as energy plantation crops are quickly regrown, their photosynthesizing removes carbon dioxide from the atmosphere even as recently harvested trees release the gas through a power plant's smokestack. In contrast, the burning of fossil fuels, which come from the remains of ancient organisms, releases carbon dioxide that was removed from the atmosphere through photosynthesis hundreds of millions of years ago.

Today, "biomass energy" accounts for only about 4% of all energy consumed in the United States. But the idea has drawn enough support to be studied or tried in many countries and may play a role in meeting future energy needs. In this chapter, you'll learn more about how plants convert solar energy to chemical energy through the essential process of photosynthesis. ■ ■ ■

Rows of young willow trees growing on an energy plantation in upstate New York

7.1   Autotrophs are the producers of the biosphere

Figure 7.1A   Forest plants

Figure 7.1B   Wheat field

Figure 7.1C   Kelp, a large alga

Figure 7.1D   Cyanobacteria (photosynthetic bacteria)

Plants are **autotrophs** (meaning "self-feeders" in Greek) in that they make their own food and thus sustain themselves without eating other organisms or even organic molecules. The chloroplasts of plant cells capture light energy that has traveled 150 million kilometers from the sun and convert it to chemical energy that is stored in sugar and other organic molecules.

Plants make their own organic molecules and are the ultimate source of organic molecules for almost all other organisms. They are often referred to as the **producers** of the biosphere (the global ecosystem) because they produce its food supply. Actually, plants are not the only producers in this sense; algae, certain other protists, and some prokaryotes also make food molecules from carbon dioxide, water, and other inorganic materials. All organisms that produce organic molecules from inorganic molecules using the energy of light are called **photoautotrophs.**

The photographs on this page illustrate some of the diversity among photoautotrophs. On land, plants, such as those in the forest scene in **Figure 7.1A** and the wheat field

in **Figure 7.1B**, are the predominant producers. In aquatic environments, algae and photosynthetic bacteria are the main food producers. **Figure 7.1C** shows kelp, a large alga that forms extensive underwater "forests" off the coast of California. **Figure 7.1D** shows prokaryotic cyanobacteria, abundant and important producers in freshwater and marine ecosystems.

Plants, algae, and photosynthetic bacteria all use light energy to drive the synthesis of organic molecules from carbon dioxide and water. In plants and algae, this process goes on in the cellular organelles called chloroplasts. In this chapter, we focus on photosynthesis in plants, and the next module is an overview of the location and structure of plant chloroplasts.

? Although they are "self-feeders," autotrophs are not totally self-sufficient. What do they require from the environment in order to synthesize sugar?

■ Light, carbon dioxide, and water. (Plants also require soil minerals, as you'll learn in Chapter 32.)

7.2 Photosynthesis occurs in chloroplasts

All green parts of a plant have chloroplasts in their cells and can carry out photosynthesis. In most plants, however, the leaves have the most chloroplasts (about half a million per square millimeter of leaf surface) and are the major sites of photosynthesis. Their green color is from chlorophyll, a light-absorbing pigment in the chloroplasts that plays a central role in converting solar energy to chemical energy.

Figure 7.2 zooms in on a leaf to show the actual sites of photosynthesis. The top center drawing is a cross section (slice) of a leaf as it would look under a light microscope. Chloroplasts are concentrated in the cells of the **mesophyll,** the green tissue in the interior of the leaf. Carbon dioxide enters the leaf, and oxygen exits, by way of tiny pores called **stomata** (singular, *stoma,* meaning "mouth"). Water absorbed by the roots is delivered to the leaves in veins.

As you can see in the light micrograph of a single cell (upper right), each mesophyll cell has numerous chloroplasts. The bottom electron micrograph and drawing show the structures in a single chloroplast. Membranes in the chloroplast form the framework where many of the reactions of photosynthesis occur, just as mitochondrial membranes do for the energy-harvesting machinery we discussed in Chapter 6. Like the mitochondrion, the chloroplast has an outer membrane and an inner membrane, with an intermembrane space between them. The chloroplast's inner membrane encloses a second compartment, which is filled with a thick fluid called **stroma.** The stroma is where sugars are made from carbon dioxide and water. Suspended in the stroma is a system of interconnected membranous sacs, called **thylakoids,** which enclose a third chloroplast compartment, called the thylakoid space. In some places, thylakoids are concentrated in stacks called **grana** (singular, *granum*). Built into the thylakoid membranes are the chlorophyll molecules that capture light energy. The thylakoid membranes also house much of the machinery that converts light energy to chemical energy.

Later in the chapter, we examine the function of these structures in more detail. But first, let's look more closely at the general equation for photosynthesis.

Web/CD Activity 7A *The Sites of Photosynthesis*

? Chloroplast is to _____ as _____ is to cellular respiration.

■ photosynthesis . . . mitochondrion

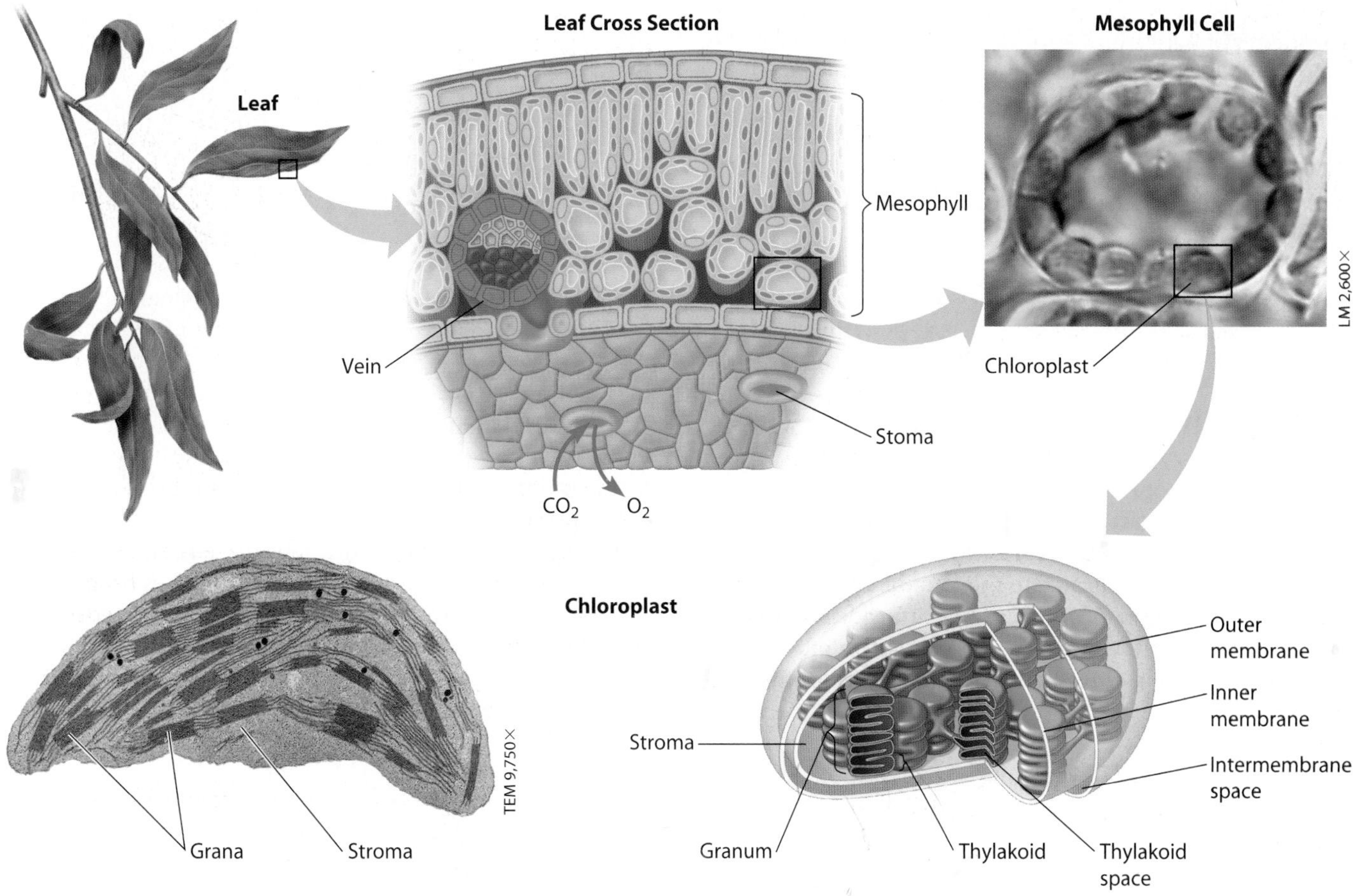

Figure 7.2 The location and structure of chloroplasts

7.3 Plants produce O_2 gas by splitting water

The leaves of plants that live in lakes and ponds are often covered with bubbles like the ones in **Figure 7.3A**. The bubbles are oxygen gas (O_2) produced during photosynthesis.

The basic equation for photosynthesis was determined in the 1800s, and most scientists assumed that plants produce O_2 by extracting it from CO_2. In the 1950s, scientists tested this hypothesis by using a heavy isotope of oxygen, ^{18}O, to follow the fate of oxygen atoms during photosynthesis. (This was one of the first uses of isotopes as tracers in biological research. To review isotopes and their use as tracers, see Module 2.5.) In the photosynthesis equations in **Figure 7.3B**, the red type denotes ^{18}O. The equations here are written in a slightly more detailed form than the summary photosynthesis equation you often see. These show that water is actually both a reactant and a product in the reaction.

In experiment 1, a plant given carbon dioxide containing ^{18}O gave off no labeled (^{18}O-containing) oxygen gas. But in experiment 2, a plant given water containing ^{18}O did produce labeled O_2. These experiments showed that the O_2 produced during photosynthesis comes from water and not from CO_2. It takes two water (H_2O) molecules to make each molecule of O_2.

Knowing where the O_2 comes from gives us a hint of what else happens during photosynthesis. Additional experiments have revealed that the oxygen atoms in CO_2 and the hydrogens in the reactant H_2O molecules end up in the sugar molecule and in water that is formed anew. **Figure 7.3C** summarizes the fates of all the atoms that start out in the reactant molecules of photosynthesis.

> **?** What step in cellular respiration reverses the water-splitting step of photosynthesis? (*Hint:* Review Module 6.10.)

■ The formation of water when electrons and hydrogen ions are added to oxygen at the "bottom" of the electron transport chain

Figure 7.3A Oxygen bubbles on the leaves of an aquatic plant

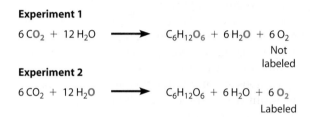

Experiment 1

$$6\,CO_2 \ + \ 12\,H_2O \longrightarrow C_6H_{12}O_6 \ + \ 6\,H_2O \ + \ 6\,O_2$$
Not labeled

Experiment 2

$$6\,CO_2 \ + \ 12\,H_2O \longrightarrow C_6H_{12}O_6 \ + \ 6\,H_2O \ + \ 6\,O_2$$
Labeled

Figure 7.3B Experiments tracking the oxygen atoms in photosynthesis

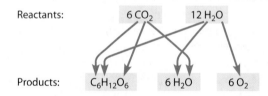

Reactants: $6\,CO_2$ $12\,H_2O$

Products: $C_6H_{12}O_6$ $6\,H_2O$ $6\,O_2$

Figure 7.3C Fates of all the atoms in photosynthesis

7.4 Photosynthesis is a redox process, as is cellular respiration

What actually happens when photosynthesis converts CO_2 and water into sugars and O_2? Photosynthesis is a redox (oxidation-reduction) process, just as cellular respiration is (see Module 6.5). As indicated in the summary photosynthesis equation (**Figure 7.4A**), when water molecules are split apart, yielding O_2, they are actually oxidized; that is, they lose electrons, along with hydrogen ions (H^+). Meanwhile,

CO_2 is reduced to sugar as electrons and hydrogen ions are added to it. Oxidation and reduction go hand in hand.

Now move from the food-producing equation for photosynthesis to the energy-releasing equation for cellular respiration (**Figure 7.4B**). Overall, cellular respiration harvests energy stored in a glucose molecule by oxidizing the sugar and reducing O_2 to H_2O. This process involves a number of

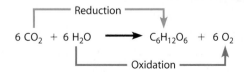

Reduction

$$6\,CO_2 \ + \ 6\,H_2O \longrightarrow C_6H_{12}O_6 \ + \ 6\,O_2$$

Oxidation

Figure 7.4A Photosynthesis (uses light energy)

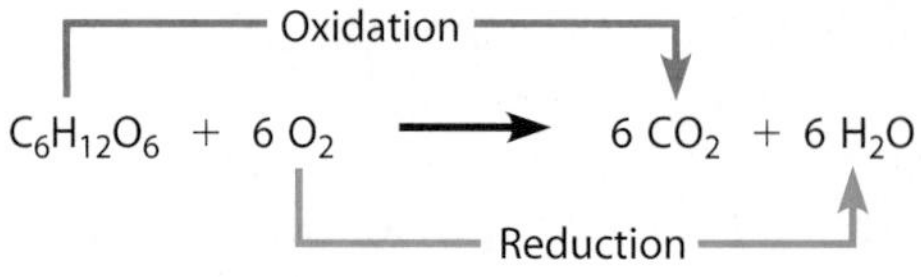

Oxidation

$$C_6H_{12}O_6 \ + \ 6\,O_2 \longrightarrow 6\,CO_2 \ + \ 6\,H_2O$$

Reduction

Figure 7.4B Cellular respiration (releases chemical energy)

energy-releasing redox reactions, with electrons losing potential energy as they travel down an energy hill from sugar to O_2. Along the way, the mitochondrion uses some of the energy to synthesize ATP, as we saw in Chapter 6.

In contrast, the food-producing redox reactions of photosynthesis involve an uphill climb. As water is oxidized and CO_2 is reduced during photosynthesis, electrons gain energy by being boosted up an energy hill. The light energy captured by chlorophyll molecules in the chloroplast provides the boost for the electrons. Photosynthesis converts the light energy to chemical energy and stores it in sugar molecules.

> **?** Which redox process, photosynthesis or cellular respiration, is endergonic? (*Hint:* See Module 5.3.)

7.5 Overview: Photosynthesis occurs in two stages linked by ATP and NADPH

The equation for photosynthesis is a simple summary of a very complex process. Actually, photosynthesis occurs in two stages, each with multiple steps. Figure 7.5 shows the inputs and outputs of the two stages and how the stages are related.

The **light reactions** (left in figure) include the steps that convert light energy to chemical energy and produce O_2 gas as a waste product. The light reactions occur in the thylakoid membranes. Light energy absorbed by chlorophyll molecules built into the membranes is used to make ATP from ADP and phosphate. It is also used to drive a transfer of electrons from water to $NADP^+$, an electron carrier similar to the NAD^+ that carries electrons in cellular respiration. Enzymes reduce $NADP^+$ to NADPH by adding a pair of light-excited electrons along with an H^+. This reaction temporarily stores the energized electrons. The electrons originally came from water; it is in the light reactions that water is split and O_2 is released.

In summary, the light reactions of photosynthesis are the steps that absorb solar energy and convert it to chemical energy stored in ATP and NADPH. Notice that these reactions produce no sugar; sugar is not made until the Calvin cycle, the second stage of photosynthesis.

The **Calvin cycle** occurs in the stroma of the chloroplast (Figure 7.5, right). It is a cyclic series of reactions that assembles sugar molecules using CO_2 and the energy-containing products of the light reactions. This second stage of photosynthesis is named for American biochemist and Nobel laureate Melvin Calvin. In the 1940s, Calvin and his colleagues traced the path of carbon in the cycle, using the radioactive isotope ^{14}C to label the carbon in CO_2. The incorporation of carbon from CO_2 into organic compounds, shown in the figure as CO_2 entering the Calvin cycle, is called **carbon fixation.** After carbon fixation, enzymes of the cycle make sugars by further reducing the carbon compounds.

As the figure suggests, it is NADPH produced by the light reactions that provides the high-energy electrons for reducing carbon in the Calvin cycle. And ATP from the light reactions provides chemical energy that powers several of the

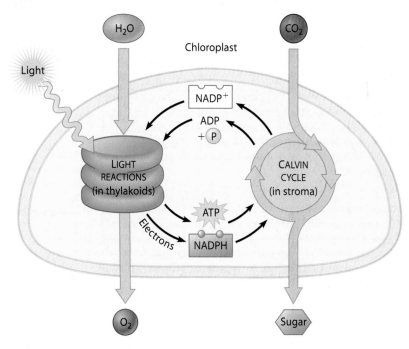

Figure 7.5
An overview of photosynthesis

steps of the Calvin cycle. The Calvin cycle does not require light directly. However, in most plants, the Calvin cycle runs during daytime, when the light reactions power the cycle's sugar assembly line by supplying it with NADPH and ATP.

The word *photosynthesis* capsulizes the two stages. *Photo-,* from the Greek word for light, refers to the light reactions; *synthesis,* meaning "putting together," refers to sugar construction by the Calvin cycle. In the next several modules, we look at these two stages in more detail. But first, let's consider some of the properties of light, the energy source that powers photosynthesis.

Web/CD Activity 7B *Overview of Photosynthesis*

> **?** For chloroplasts to produce sugar from carbon dioxide in the dark, they would require an artificial supply of
> _____ and _____.

7.6 Visible radiation drives the light reactions

What exactly do we mean when we say that photosynthesis is powered by light energy from the sun? Sunlight is a type of energy called radiation, or **electromagnetic energy.** Electromagnetic energy travels in space as rhythmic waves analogous to those made by a pebble dropped in a puddle of water. The distance between the crests of two adjacent waves is called a **wavelength.**

Figure 7.6A shows the electromagnetic spectrum, the full range of electromagnetic wavelengths from the very short gamma rays to the very long-wavelength radio waves. As you can see in the figure, visible light—the radiation your eyes see as different colors—is only a small fraction of the spectrum. It consists of wavelengths from about 380 nm to about 750 nm. Shorter wavelengths have more energy than longer ones. In fact, wavelengths that are shorter than those of visible light have enough energy to damage organic molecules such as proteins and nucleic acids. This is why ultraviolet (UV) radiation in sunlight can cause sunburns and skin cancer.

Figure 7.6B shows what happens to visible light in the chloroplast. Light-absorbing molecules called pigments, built into the thylakoid membranes, absorb some wavelengths of light and reflect or transmit other wavelengths. We do not see the absorbed wavelengths; their energy has been absorbed by pigment molecules. What we see when we look at a leaf are the green wavelengths that the pigments transmit and reflect.

Different pigments absorb light of different wavelengths, and chloroplasts contain several kinds of pigments. One, chlorophyll *a*, absorbs mainly blue-violet and red light. Chlorophyll *a* participates directly in the light reactions. It looks grass-green because it reflects mainly green light. A

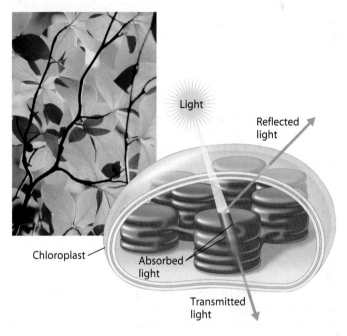

Light
Reflected light
Chloroplast
Absorbed light
Transmitted light

Figure 7.6B The interaction of light with a chloroplast

very similar molecule, chlorophyll *b*, absorbs mainly blue and orange light and reflects (appears) yellow-green. Chlorophyll *b* broadens the range of light that a plant can use by conveying absorbed energy to chlorophyll *a*, which then puts the energy to work in the light reactions.

Chloroplasts also contain a family of yellow-orange pigments called carotenoids, which absorb mainly blue-green light. Some may pass energy to chlorophyll *a*, as chlorophyll *b* does. Other carotenoids have a protective function: They absorb and dissipate excessive light energy that would otherwise damage chlorophyll. (Similar carotenoids may help protect our eyes from very bright light.)

The theory of light as waves explains most of light's properties. However, light also behaves as discrete packets of energy called photons. A **photon** is a fixed quantity of light energy, and, as you have just learned, the shorter the wavelength, the greater the energy. Each type of pigment absorbs certain wavelengths of light because it is able to absorb the specific amounts of energy in those photons. Let's see what happens when a pigment molecule such as chlorophyll absorbs a photon of light.

Web/CD Activity 7C *Light Energy and Pigments*

Web/CD Thinking as a Scientist *How Does Paper Chromatography Separate Plant Pigments?*

? What color of light is least effective in driving photosynthesis?

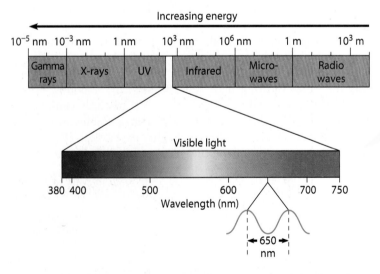

Increasing energy

10^{-5} nm 10^{-3} nm 1 nm 10^3 nm 10^6 nm 1 m 10^3 m

| Gamma rays | X-rays | UV | Infrared | Micro-waves | Radio waves |

Visible light

380 400   500   600   700 750
Wavelength (nm)

650 nm

Figure 7.6A The electromagnetic spectrum and the wavelengths of visible light

7.7　Photosystems capture solar power

When a pigment molecule absorbs a photon, one of the pigment's electrons jumps to an energy level farther from the nucleus, where it has more potential energy, and we say that the electron has been raised from a ground state to an excited state. The excited state is very unstable. Generally, when isolated pigment molecules absorb light, their excited electrons drop back down to the ground state in a billionth of a second, releasing their excess energy as heat. This conversion of light energy to heat is what makes a black car so hot on a sunny day (black pigments absorb all wavelengths of light).

Some isolated pigments, including chlorophyll, emit light as well as heat after absorbing photons. We can demonstrate this phenomenon in the laboratory with a chlorophyll solution, as shown in Figure 7.7A. When illuminated, the chlorophyll emits photons that produce a reddish afterglow called fluorescence. Figure 7.7B illustrates what happens in fluorescence: A light-excited electron falls back to the ground state, emitting its energy as heat and light. But chlorophyll behaves very differently in isolation than it does in an intact chloroplast. In its native habitat of the thylakoid membrane, chlorophyll passes its excited electron to a neighboring molecule. The objects outlined by dashed lines in Figure 7.7B represent molecules in the membrane that are associated with chlorophyll and can capture the excited electron before it drops back to the ground state.

In the thylakoid membrane, chlorophyll molecules are organized along with other pigments and proteins into clusters called photosystems (Figure 7.7C). A **photosystem** consists of a number of light-harvesting complexes surrounding a reaction center. These complexes have chlorophyll *a*, chloro-phyll *b*, and carotenoid pigments that function collectively as a light-gathering antenna. The pigments absorb photons and pass the energy from molecule to molecule (thin yellow arrows) until it reaches the reaction center. The **reaction center** is a protein complex that contains a chlorophyll *a* molecule and a molecule called the primary electron acceptor. The primary electron acceptor captures a light-excited electron from the reaction-center chlorophyll molecule and passes it to an electron transport chain.

Two types of photosystems have been identified, and they cooperate in the light reactions. They are referred to as photosystem I and photosystem II, in order of their discovery, although photosystem II actually functions first in the sequence of steps that make up the light reactions. Each photosystem has a characteristic reaction center. In photosystem II, the chlorophyll *a* molecule of the reaction center is called P680 because the light it absorbs best is red light with a wavelength of 680 nm. The reaction-center chlorophyll of photosystem I is called P700 because the wavelength of light it absorbs best is 700 nm (in the far-red part of the spectrum). Let's now see how the two photosystems work together in the light reactions to generate ATP and NADPH.

> ❓ Compared to a solution of isolated chlorophyll, why do intact chloroplasts release less heat and fluorescence when illuminated?

■ In the chloroplasts, the light-excited electrons are trapped by a primary electron acceptor rather than immediately giving up all their energy as heat and light.

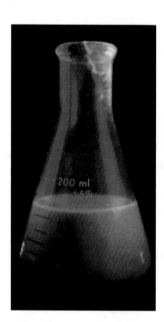

Figure 7.7A Fluorescence of isolated chlorophyll in solution

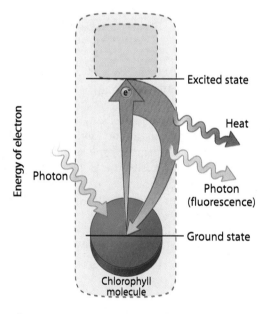

Figure 7.7B Light-excited chlorophyll in isolation

Energy of electron

Excited state

Heat

Photon

Photon (fluorescence)

Ground state

Chlorophyll molecule

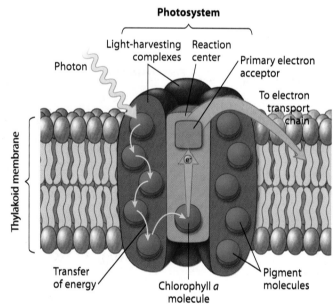

Figure 7.7C Light-excited chlorophyll embedded in a photosystem

Photosystem

Light-harvesting complexes　Reaction center

Photon

Primary electron acceptor

To electron transport chain

Thylakoid membrane

Transfer of energy

Chlorophyll *a* molecule

Pigment molecules

7.8 In the light reactions, electron transport chains generate ATP and NADPH

In the light reactions, light energy is transformed into the chemical energy of ATP and NADPH. In this process, electrons removed from water molecules pass from photosystem II to photosystem I to NADP⁺. Between the two photosystems, the electrons move down an electron transport chain (similar to the one in cellular respiration) and provide energy for ATP production.

Let's follow the flow of electrons (represented by gold arrows) in **Figure 7.8A**, which shows the two photosystems embedded in a thylakoid membrane. ❶ A pigment molecule in a light-harvesting complex absorbs a photon of light. The energy is passed to other pigment molecules and finally to the reaction center of Photosystem II, where it excites an electron of chlorophyll P680 to a higher energy level. ❷ This electron is captured by the primary electron acceptor. ❸ Water is split, and its electrons are supplied one by one to P680, replacing those lost to the primary electron acceptor. The oxygen atom combines with an oxygen from another split water molecule to form a molecule of O_2.

❹ Each photoexcited electron passes from photosystem II to photosystem I via an electron transport chain. The exergonic "fall" of electrons provides energy for the synthesis of ATP. ❺ Meanwhile, light energy excites an electron of chlorophyll P700 in the reaction center of photosystem I. The primary electron acceptor captures the excited electron and an electron from the bottom of the electron transport chain replaces the lost electron in P700. ❻ The excited electron of photosystem I is passed through a short electron transport chain to NADP⁺, reducing it to NADPH.

Figure 7.8B provides a mechanical analogy to help you understand how the two photosystems cooperate in generating ATP and NADPH. The input of light energy, represented by the large yellow mallets, boosts electrons in both

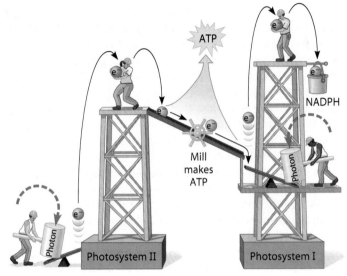

Figure 7.8B A mechanical analogy of the light reactions

photosystems up to the excited state. The electrons are caught by the primary electron acceptor on top of the platform in each photosystem. Photosystem II passes the excited electrons through an ATP mill. Photosystem I hands its excited electrons off to reduce NADP⁺ to NADPH.

NADPH, ATP, and O_2 are the products of the light reactions. Next we look in more detail at how ATP is formed.

❓ Tracing the light reactions in Figure 7.8A, there is a flow of electrons from _____ molecules to _____, which is reduced to _____, the source of electrons for sugar synthesis in the _____ cycle.

■ water . . . NADP⁺ . . . NADPH . . . Calvin

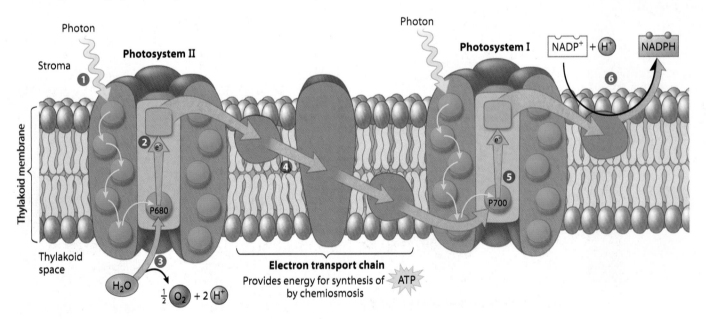

Figure 7.8A Electron flow in the light reactions of photosynthesis

7.9 Chemiosmosis powers ATP synthesis in the light reactions

You first encountered chemiosmosis in Modules 6.6 and 6.10, as the mechanism of oxidative phosphorylation (ATP formation) in a mitochondrion. Chemiosmosis is also the mechanism that generates ATP in a chloroplast. Recall that the process of chemiosmosis drives ATP synthesis using the potential energy of a concentration gradient of hydrogen ions across a membrane. The gradient is created when an electron transport chain pumps hydrogen ions across a membrane as it passes electrons down the chain.

Figure 7.9 illustrates the relationship between chloroplast structure and function in the light reactions of photosynthesis. As in Figure 7.8A, we show the two photosystems and electron transport chains, all located within the thylakoid membrane of a chloroplast. Here you can see that as photoexcited electrons are passed down the electron transport chain connecting the two photosystems, hydrogen ions (H^+) are pumped across the membrane from the stroma into the thylakoid space. This generates a concentration gradient across the membrane.

The flask-shaped structure on the right in the figure represents an ATP synthase complex, which is like the one we saw in the mitochondrion. The energy of the concentration gradient drives H^+ back across the membrane through ATP synthase. As you learned in Module 6.10, ATP synthase couples the flow of H^+ to the phosphorylation of ADP. In photosynthesis, this chemiosmotic production of ATP is called **photophosphorylation** because the initial energy input is light energy.

How does photophosphorylation compare with oxidative phosphorylation? In cellular respiration, the high-energy electrons passed down the electron transport chain come from the oxidation of food molecules. In photosynthesis, light energy is used to drive electrons to the top of the transport chain. Mitochondria transfer chemical energy from food to ATP; chloroplasts transform light energy into the chemical energy of ATP.

Notice that in the light-driven flow of electrons through the two photosystems, the final electron acceptor is $NADP^+$, not O_2 as in cellular respiration. Electrons do not end up at a low energy level in H_2O, as they do in respiration. Instead, they are stored at a high state of potential energy in NADPH. The ATP and NADPH produced during the light reactions are used in the next stage of photosynthesis, the Calvin cycle. Module 7.10 describes how that cycle makes sugar.

Web/CD Activity 7D *The Light Reactions*

? What is the advantage of the light reactions producing NADPH and ATP on the stroma side of the thylakoid membrane?

■ The Calvin cycle, which consumes the NADPH and ATP, occurs in the stroma.

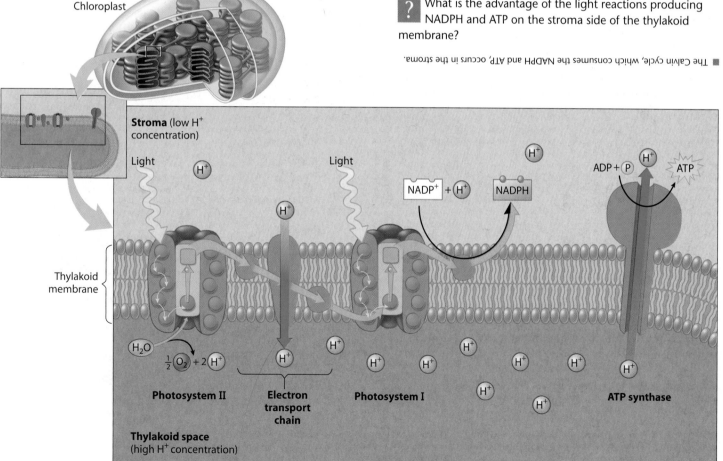

Figure 7.9 The production of ATP by chemiosmosis in photosynthesis

7.10 ATP and NADPH power sugar synthesis in the Calvin cycle

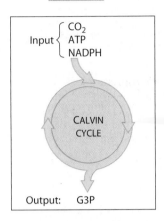

Figure 7.10A An overview of the Calvin cycle

The Calvin cycle functions like a sugar factory within a chloroplast. As **Figure 7.10A** shows, inputs to this all-important food-making process are CO_2 (from the air) and ATP and NADPH (both generated by the light reactions). Using carbon from CO_2, energy from ATP, and high-energy electrons from NADPH, the Calvin cycle constructs an energy-rich, three-carbon sugar, glyceraldehyde-3-phosphate (G3P). A plant cell can use G3P to make glucose and other organic molecules as needed. (You already met G3P in glycolysis: It is the three-carbon sugar formed by the splitting of glucose.)

Figure 7.10B presents the details of the Calvin cycle. It is called a cycle because, like the citric acid cycle in cellular respiration, the starting material is regenerated as the process occurs. In this case, the starting material is ribulose bisphosphate (RuBP). ❶ In the carbon fixation step, the enzyme rubisco attaches CO_2 to RuBP. ❷ In the next step, a reduction, NADPH reduces the organic acid 3-PGA to G3P with the assistance of ATP. To make a molecule of G3P, the cycle must incorporate the carbon atoms from three molecules of CO_2. The cycle actually incorporates one carbon at a time, but we show it starting with three CO_2 molecules so that we end up with a complete G3P molecule.

For this to be a cycle, RuBP must be regenerated. ❸ For every three CO_2 molecules fixed, one G3P molecule leaves the cycle as product, and the remaining five G3P molecules are rearranged, ❹ using energy from ATP, to regenerate three molecules of RuBP. Review the steps of the Calvin cycle below.

Web/CD Activity 7E *The Calvin Cycle*

? To synthesize one glucose molecule, the Calvin cycle must turn six times. In doing so, it uses ____ molecules of CO_2, ____ molecules of ATP, and ____ molecules of NADPH.

■ 6 . . . 18 . . . 12

Step ❶ Carbon fixation. An enzyme called rubisco combines CO_2 with a five-carbon sugar called ribulose bisphosphate (abbreviated RuBP). The unstable product splits into two molecules of the three-carbon organic acid, 3-phosphoglyceric acid (3-PGA). For three CO_2 entering, six 3-PGA result.

Step ❷ Reduction. Two chemical reactions (indicated by the two blue arrows) consume energy from six molecules of ATP and oxidize six molecules of NADPH. Six molecules of 3-PGA are reduced, producing six molecules of the energy-rich three-carbon sugar, G3P.

Step ❸ Release of one molecule of G3P. Five of the G3Ps from step 2 remain in the cycle. The single molecule of G3P you see leaving the cycle is the net product of photosynthesis. A plant cell uses two G3P molecules to make one molecule of glucose.

Step ❹ Regeneration of RuBP. A series of chemical reactions uses energy from ATP to rearrange the atoms in the five G3P molecules (15 carbons total), forming three RuBP molecules (15 carbons). These can start another turn of the cycle.

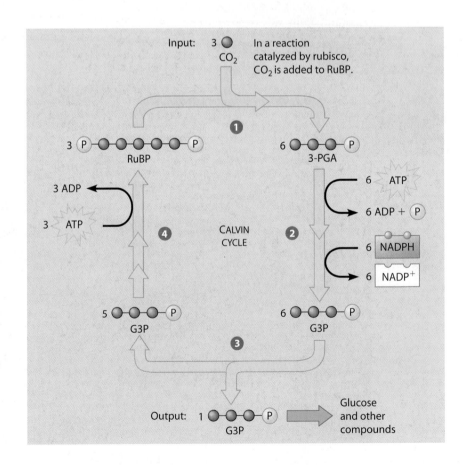

Figure 7.10B Details of the Calvin cycle

7.11 Review: Photosynthesis uses light energy to make food molecules

As we have discussed, most of the living world depends on the food-making machinery of photosynthesis. Figure 7.11 summarizes the two stages of this vital process and reviews where they occur in the chloroplast.

Starting on the left in the diagram, you see a summary of the light reactions, which occur in the thylakoid membranes (green). Two photosystems in the membranes capture solar energy, using it to energize electrons. Simultaneously, water is split, and O_2 is released. The photosystems transfer energized electrons to electron transport chains, where energy is harvested and used to make the high-energy molecules NADPH and ATP.

The chloroplast's sugar factory is the Calvin cycle, the second stage of photosynthesis. In the stroma, enzymes of the cycle combine CO_2 with RuBP and produce G3P. Sugar molecules made from G3P serve as a plant's own food supply. Plants use sugars as fuel for cellular respiration and as starting material for making other organic molecules, such as the structural molecule cellulose. Most plants make considerably more sugar than they need. They stockpile the excess sugar as starch, storing it in roots, tubers, and fruits.

Plants (and other photosynthesizers) not only feed themselves, but are also the ultimate source of food for virtually all other organisms. Humans and other animals make none of their own food and are totally dependent on the organic matter made by photosynthesizers. Even the energy we acquire when we eat meat was originally captured by photosynthesis. The energy in a hamburger, for instance, came from sunlight that was originally converted to a chemical form in chloroplasts in the cells of the grasses eaten by cattle. Photosynthesis is the ultimate source of all the food we eat and the oxygen we breathe.

This review of photosynthesis is an appropriate place to reflect on the metabolic ground we have covered in this chapter and the previous one. In Chapter 6, we saw that virtually all organisms, plants included, use cellular respiration to obtain the energy they need from fuel molecules such as glucose. We followed the chemical pathways of glycolysis and the citric acid cycle, which break glucose down and release energy from it. We have now come full circle, seeing how plants trap sunlight energy and use it to make glucose from the raw materials carbon dioxide and water.

In tracing glucose synthesis and its breakdown, we have also seen that cells use several of the same mechanisms—electron transport, redox reactions, and chemiosmosis—in energy storage (photosynthesis) and energy harvest (cellular respiration).

Web/CD Thinking as a Scientist *How Is the Rate of Photosynthesis Measured?*

? Explain why a poison that inhibits an enzyme of the Calvin cycle will also inhibit the light reactions.

■ The light reactions require ADP and NADP⁺, which are not recycled from ATP and NADPH when the Calvin cycle stops.

Figure 7.11
A summary of the chemical processes of photosynthesis

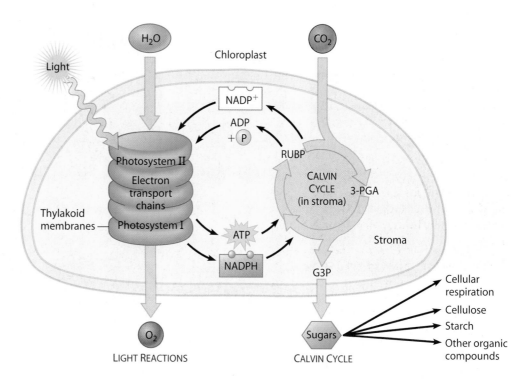

7.12 C₄ and CAM plants have special adaptations that save water

Plants in which the Calvin cycle uses CO_2 directly from the air are called **C₃ plants** because the first organic compound produced is the three-carbon compound 3-PGA (step 1 in Figure 7.10B). C₃ plants are common and widely distributed, and some of them, such as soybeans, oats, wheat, and rice, are important in agriculture. One of the problems that farmers face in growing C₃ plants, however, is that dry weather can reduce the rate of photosynthesis and decrease crop productivity.

Closing stomata on a hot, dry day is an adaptation that reduces water loss, but it also prevents CO_2 from entering the leaf and O_2 from leaving. As a result, CO_2 levels can get very low in the leaf, while O_2 from the light reactions builds up. When this happens, rubisco, the first enzyme of the Calvin cycle, adds O_2 instead of CO_2 to RuBP. A two-carbon product of this reaction is then broken down by the cell to CO_2 and H_2O. This process is called **photorespiration.** Unlike photosynthesis, photorespiration yields no sugar, and unlike cellular respiration, it produces no ATP. This seemingly wasteful process can drain away as much as 50% of the carbon fixed by the Calvin cycle.

In contrast to C₃ plants, **C₄ plants** have special adaptations that prevent photorespiration. When the weather is hot and dry, a C₄ plant keeps its stomata closed most of the time, thus conserving water. At the same time, it continues making sugars by photosynthesis, using the pathway shown in the left side of **Figure 7.12**. A C₄ plant has an enzyme that first fixes carbon into a four-carbon (4-C) compound. This enzyme has a high affinity for CO_2 and can continue to fix carbon even when the CO_2 concentration in the leaf is low. The four-carbon compound acts as a carbon shuttle; it transfers CO_2 to nearby cells called bundle-sheath cells, which are packed around the veins of the leaf. The CO_2 concentration in these cells remains high enough for the Calvin cycle to make sugars and avoid photorespiration. Corn and sugarcane are examples of agriculturally important C₄ plants. C₄ photosynthesis is advantageous in hot, dry climates, which is where C₄ plants evolved and thrive today.

A second photosynthetic adaptation to arid conditions has evolved in pineapples, many cacti, and succulent (water-storing) plants, such as ice plants and jade plants. Collectively called **CAM plants**, these species are adapted to very dry climates. A CAM plant (right side of Figure 7.12) conserves water by opening its stomata and admitting CO_2 only at night. When CO_2 enters the leaves, it is fixed into a four-carbon compound, as in C₄ plants. The four-carbon compound in a CAM plant banks CO_2 at night and releases it to the Calvin cycle during the day. This keeps photosynthesis operating during the day, even though the leaf's stomata remain closed. CAM stands for crassulacean acid metabolism, after the plant family Crassulaceae, in which this water-saving adaptation was first discovered.

In C₄ plants, carbon fixation and the Calvin cycle occur in different types of cells. In CAM plants, these processes occur in the same cells, but at different times. Keep in mind that CAM, C₄, and C₃ plants all eventually use the Calvin cycle to make sugar from CO_2.

Web/CD Activity 7F *Photosynthesis in Dry Climates*

? How would you expect the relative abundance of C₃ versus C₄ and CAM species to change in a geographic region where the climate becomes much hotter and drier?

■ C₄ and CAM species would replace many of the C₃ species.

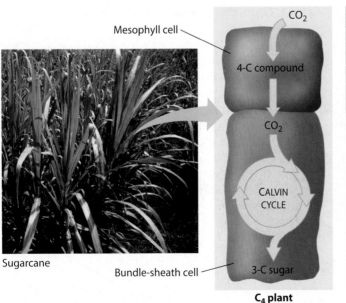

Mesophyll cell

CO_2

4-C compound

CO_2

CALVIN CYCLE

Sugarcane

Bundle-sheath cell

3-C sugar

C₄ plant

CO_2 Night

4-C compound

CO_2

CALVIN CYCLE

3-C sugar   Day

CAM plant

Pineapple

Figure 7.12 Comparison of photosynthesis in C₄ and CAM plants

7.13 Photosynthesis moderates global warming

The greenhouse in **Figure 7.13A** is used to grow plants where the weather outside is too cold. The glass or plastic walls of a greenhouse allow solar radiation to pass through. The sunlight heats the soil, which in turn warms the air. The walls trap the warm air, raising the temperature inside.

An analogous process, commonly called the **greenhouse effect**, operates on a global scale (**Figure 7.13B**). Solar radiation reaching Earth's atmosphere includes ultraviolet radiation and visible light. As we discuss in the next module, the ozone layer filters out most of the damaging UV. Visible light passes through and is absorbed by the planet's surface, warming it. Heat is radiated by the warmed planet, and these longer, infrared wavelengths are absorbed by gases in the atmosphere, which then radiate some of the heat back to Earth. This natural heating effect is highly beneficial. Without it, Earth would be much colder and much less hospitable to life.

The gases in the atmosphere that absorb heat radiation are called greenhouse gases. Some occur naturally, such as water vapor, carbon dioxide (CO_2), and methane (CH_4), while others are synthetic, such as chlorofluorocarbons (CFCs). Human activities are adding to the levels of these gases.

Carbon dioxide is one of the most important greenhouse gases. You have just learned that CO_2 is a raw material for photosynthesis and a waste product of cellular respiration. These two processes, taking place in microscopic chloroplasts and mitochondria, keep carbon cycling between inorganic and organic compounds on a global scale. Photosynthetic organisms absorb billions of tons of CO_2 each year. Most of that fixed carbon returns to the atmosphere via cellular respiration, the action of decomposers, and fires. But much of it remains locked in large tracts of forests and undecomposed organisms. And large amounts of carbon are in long-term storage in fossil fuels buried deep under Earth's surface.

Before 1850, carbon dioxide was estimated to make up less than 0.03% of the air we breathe. Since the start of the Industrial Revolution, the atmospheric concentration of CO_2 has increased about 30%, mostly due to the combustion of carbon-based fossil fuels, such as coal, oil, and gasoline.

Increasing concentrations of greenhouse gases have been linked to **global warming**, a slow but steady rise in Earth's surface temperature. Predicted changes of just a few degrees over the next 50 years may have dramatic and wide-ranging consequences. These would likely include melting of polar ice, rising sea levels, extreme weather patterns, droughts, and the spread of tropical diseases.

Unfortunately, the rise in atmospheric CO_2 levels during the last century coincided with widespread deforestation, which aggravated the global warming problem by reducing an effective CO_2 sink. As forests are cleared for lumber or agriculture and population growth increases the demand for

Figure 7.13A Orchid plants growing in a greenhouse

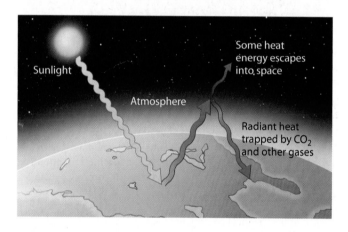

Figure 7.13B CO_2 in the atmosphere and global warming

fossil fuels, CO_2 levels will continue to rise. We discuss global warming in more detail in Module 38.5.

Can photosynthesis help further to mitigate this increase in atmospheric CO_2? As you read in the chapter introduction, "energy plantations" hold out the promise of a cleaner, renewable fuel source. Slowing the destruction of forests will sustain their photosynthetic and carbon-storing contributions. Taking a lesson from plants, we can explore technologies that utilize solar energy for some of our energy needs. Almost all life on Earth depends on the ability of plants and other photosynthetic organisms to convert light energy to the chemical energy of food molecules. Their contribution to life on Earth may also come to include increased removal of CO_2 from the atmosphere.

? Explain the greenhouse effect.

■ Sunlight warms Earth's surface, which radiates heat to the atmosphere. CO_2 and other greenhouse gases absorb and radiate some heat back to Earth.

7.14 Mario Molina talks about Earth's protective ozone layer

Figure 7.14A Mario Molina

As the process of photosynthesis consumes CO_2, it produces the O_2 on which plants, animals, and most other organisms depend for cellular respiration. This O_2 has another benefit: High in the atmosphere it is converted to ozone (O_3), which plays an important protective role for life on Earth. Among the scientists who study the ozone layer is Mario Molina of MIT (Figure 7.14A), who in 1995 shared a Nobel Prize for his research on how certain pollutants are damaging that layer. In an interview, Dr. Molina explained why the ozone layer is important:

The ozone layer shields the Earth's surface from powerful ultraviolet radiation that comes from the sun. This UV radiation is harmful to organisms, including humans. For example, UV radiation causes sunburn, and skin cancer can be a cumulative result of exposure. There is also evidence that UV can damage crops. Certain developing animals, such as the larvae of fish, seem to be particularly sensitive.

He described how ozone forms and how it is destroyed:

The ozone forms when high-energy solar radiation breaks apart O_2 molecules and frees oxygen atoms. These then react with unbroken O_2 molecules. The result is ozone, which has three oxygen atoms (O_3). So ozone is continuously forming by the action of sunlight on the atmosphere. This is balanced by continuous destruction of the ozone molecules when they react with other chemical compounds that are naturally present in the atmosphere. Humans have disrupted that balance by releasing certain industrial chemicals that hasten this destruction.

The research that won Dr. Molina the Nobel Prize dealt with the destruction of ozone by one particular class of industrial chemicals, called chlorofluorocarbons, or CFCs (mentioned as greenhouse gases in Module 7.13). In 1974, when Dr. Molina and his colleagues first published their CFC-ozone–depletion hypothesis, CFCs were used in large amounts as refrigerants, as propellants in spray cans, as solvents, and in the process for making plastic foams:

We predicted that the continuous release of CFCs would damage the protective ozone layer. CFCs are very stable compounds, and this stability allows them to make it up to the ozone layer, which is about 15 miles above the surface of the Earth. There solar radiation converts them to very reactive chemicals called free radicals, which then destroy the ozone.

How did scientists and the general public react to the alarming possibility that the ozone layer was thinning?

At first, there was very little reaction because not many people were aware of the importance of invisible things like the ozone

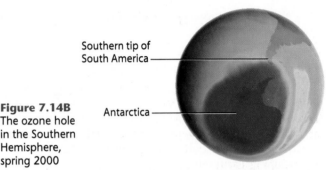

Southern tip of South America

Antarctica

Figure 7.14B
The ozone hole in the Southern Hemisphere, spring 2000

layer and UV radiation. Experts in our field quickly realized that the prediction of the CFC-ozone depletion theory was something to worry about, but many other scientists were skeptical, which is natural for scientists. Then, as we and others began doing experiments to test the idea, the evidence became strong, and more and more people, including politicians, became concerned about ozone depletion. Then, in 1985, scientists documented a drastic depletion of the ozone layer over Antarctica—an ozone hole.

Figure 7.14B shows the "ozone hole" in dark blue. This thinning of the ozone layer appears every spring. Is anything being done to fight this problem?

An international agreement was reached, called the Montreal Protocol, that required a complete phasing out of CFC production by developed countries by 1996. So that's already happened. These countries now use other refrigerants, which are destroyed in the atmosphere before they reach the ozone layer. There is a grace period for developing countries . . . [but] CFCs will be phased out everywhere within the next decade. Because CFCs are so stable, however, we predict that the ozone layer won't recover until the middle of the twenty-first century.

We asked Dr. Molina what he has learned since 1974 about the interface between science and politics:

One lesson is that science is not always something that politicians care very much about. Often, supporting science is considered a luxury, though actually it is a good investment. Another challenge is that many issues related to science and technology are long-term issues that require patience and long-term commitment.

Whether an environmental problem involves CFCs or carbon dioxide emissions, the scientific research is often complicated, and solutions are often complex and expensive. The connections between science, technology, and society, so clearly exemplified by the work of Mario Molina, are a major theme of this book. This theme will come up again in the next unit, on cellular reproduction and genetics.

? Where does the ozone layer in Earth's atmosphere come from?

Photosynthesis releases O_2. High in the atmosphere, radiation from the sun converts some of the O_2 to ozone, O_3.

Reviewing the Concepts

An Overview of Photosynthesis (Introduction–7.5)

Photosynthesis is summarized by the following equation:

$$6\ CO_2 + 6\ H_2O + \text{Light energy} \rightarrow C_6H_{12}O_6 + 6\ O_2$$

Plants, algae, and some bacteria are photoautotrophs, the producers of food consumed by virtually all organisms **(Introduction–7.1)**. In plants, photosynthesis occurs primarily in the leaves—within chloroplasts, which contain stroma and stacks of thylakoids called grana **(7.2)**.

The Two Stages of Photosynthesis. The O_2 liberated by photosynthesis comes from H_2O **(7.3)**. In photosynthesis, H_2O is oxidized and CO_2 is reduced **(7.4)**. The light reactions convert light energy to chemical energy and produce O_2. The Calvin cycle assembles sugar molecules from CO_2 using ATP and NADPH from the light reactions **(7.5)**.

The Light Reactions: Converting Solar Energy to Chemical Energy (7.6–7.9)

Photosystems. Certain wavelengths of visible light, absorbed by pigments, drive photosynthesis **(7.6)**. Thylakoid membranes contain multiple photosystems, each consisting of light-harvesting complexes of pigments and a reaction center with a primary electron acceptor that receives excited electrons from a reaction-center chlorophyll *a* **(7.7)**.

The Light Reactions. Two photosystems absorb photons and transfer energy to chlorophyll P680 and P700. Their excited electrons are passed from primary electron acceptors to electron transport chains. Electrons shuttle from photosystem II to I, providing energy to make ATP. Electrons from photosystem I reduce $NADP^+$ to NADPH. Photosystem II regains electrons by splitting water, releasing O_2 **(7.8)**.

Photophosphorylation. ATP is synthesized by chemiosmosis. The electron transport chain pumps H^+ into the thylakoid space. The diffusion of H^+ back across the membrane through ATP synthase powers the phosphorylation of ADP to produce ATP (photophosphorylation) **(7.9)**.

The Calvin Cycle: Converting CO_2 to Sugars (7.10)

The Calvin cycle, occurring in the stroma, consists of carbon fixation, reduction, release of G3P, and regeneration of RuBP. Using carbon from CO_2, electrons from NADPH, and energy from ATP, the cycle constructs G3P, which is used to build glucose and other organic molecules **(7.10)**.

Photosynthesis Reviewed and Extended (7.11–7.12)

A summary diagram of the two stages of photosynthesis is shown at the top of the next column **(7.11)**.

C_4 and CAM plants. In C_3 plants, a drop in CO_2 and rise in O_2 when stomata close on hot, dry days divert the Calvin cycle to photorespiration. C_4 plants first fix CO_2 into a four-carbon compound that provides CO_2 to the Calvin cycle.

CAM plants open stomata at night, making a four-carbon compound used as a CO_2 source during the day **(7.12)**.

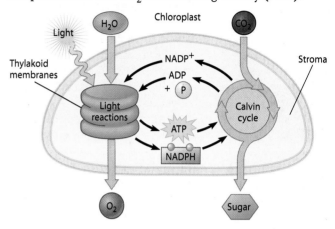

Photosynthesis, Solar Radiation, and Earth's Atmosphere (7.13–7.14)

Excess CO_2 is contributing to global warming. Photosynthesis, which removes CO_2 from the atmosphere, moderates this warming **(7.13)**. Solar radiation converts O_2 high in the atmosphere to ozone (O_3), which shields organisms from damaging UV radiation. Industrial chemicals called CFCs have caused dangerous thinning of the ozone layer, but international restrictions on CFC use is allowing recovery **(7.14)**.

Connecting the Concepts

1. This diagram compares the chemiosmotic synthesis of ATP in mitochondria and chloroplasts. In both cases, indicate which side of the membrane has the higher H^+ concentration. Then label the structures involved and the locations within the chloroplast.

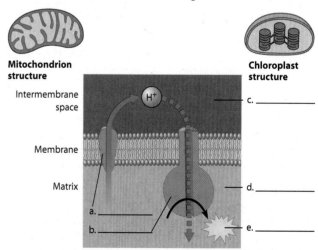

2. Continue your comparison of chemiosmosis and electron transport in mitochondria and chloroplasts. In each case,
   a. Where do the electrons come from?
   b. Where do the electrons get their energy?
   c. What picks up the electrons at the end of the chain?
   d. How is the energy given up by the electrons used?

3. Complete this summary map of photosynthesis.

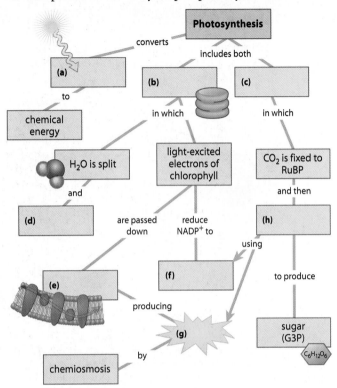

Testing your Knowledge

4. Photosynthesis consumes ____ and produces ____.
   a. O_2 ... H_2O
   b. H_2O ... CO_2
   c. CO_2 ... chlorophyll
   d. H_2O ... O_2
   e. glucose ... O_2

5. Which of the following are produced by reactions that take place in the thylakoids and consumed by reactions in the stroma?
   a. CO_2 and H_2O
   b. $NADP^+$ and ADP
   c. ATP and NADPH
   d. ATP, NADPH, and O_2
   e. CO_2 and ATP

6. In photosynthesis, ____ is oxidized and ____ is reduced.
   a. glucose ... oxygen
   b. carbon dioxide ... water
   c. water ... carbon dioxide
   d. glucose ... carbon dioxide
   e. water ... oxygen

7. Why is it difficult for most plants to carry out photosynthesis in very hot, dry environments such as deserts?
   a. The light is too intense and overpowers all the pigment molecules.
   b. The closing of stomata keeps CO_2 from entering and O_2 from leaving the plant.
   c. They must rely on photorespiration to make ATP.
   d. Global warming is intensified in a desert environment.
   e. CO_2 builds up in the leaves, blocking carbon fixation.

8. When light strikes chlorophyll molecules, they lose electrons, which are ultimately replaced by
   a. splitting water.
   b. breaking down ATP.
   c. oxidizing NADPH.
   d. fixing carbon.
   e. oxidizing glucose.

9. What is the role of $NADP^+$ in photosynthesis?
   a. It assists chlorophyll in capturing light.
   b. It acts as the primary electron acceptor for the photosystems.
   c. As part of the electron transport chain, it helps to synthesize ATP.
   d. It assists photosystem II in the splitting of water.
   e. It carries electrons to the Calvin cycle.

10. The reactions of the Calvin cycle are not directly dependent on light, but they usually do not occur at night. Why? *(Explain your answer.)*
    a. It is often too cold at night for these reactions to take place.
    b. Carbon dioxide concentrations decrease at night.
    c. The Calvin cycle depends on products of the light reactions.
    d. Plants usually close their stomata at night.
    e. Most plants do not make four-carbon compounds, which they would need for the Calvin cycle at night.

11. How many "turns" of the Calvin cycle are required to produce one molecule of glucose? (Assume one CO_2 is fixed in each turn of the cycle.)
    a. 1   b. 2   c. 3   d. 6   e. 12

Describing, Comparing, and Explaining

12. What are the major inputs and outputs of the two stages of photosynthesis?

13. What do plants do with the sugar they produce in photosynthesis?

Applying the Concepts

14. Most experts now agree that global warming is occurring, and in response to its potential threat, a number of countries have made a commitment to reduce CO_2 emissions significantly. However, some countries oppose taking strong action at this time. Several reasons are cited: First, a few experts think that the apparent warming trend may be just a random fluctuation in temperature. Second, if the temperature increase is real, it has yet to be shown that it is caused by increased CO_2. Some people also believe that it would be difficult to cut CO_2 emissions without sacrificing economic growth. Do you think we should have more evidence before taking action? Or is it better to play it safe and act now to reduce CO_2 emissions? What are the possible costs and benefits of each of these two strategies?

15. The use of biomass energy (see chapter introduction) avoids many of the problems associated with gathering, refining, transporting, and burning fossil fuels. Yet biomass energy is not without its own set of problems. What challenges do you think would arise from a large-scale conversion to biomass energy? How do these challenges compare with those encountered with fossil fuels? Which set of challenges do you think is more likely to be eventually overcome? Do you think any one type of energy has more benefits and fewer costs than the others? Which one, and why?

Answers to all questions can be found in Appendix 3.

For study help and Activities, go to campbellbiology.com or the student CD-ROM.

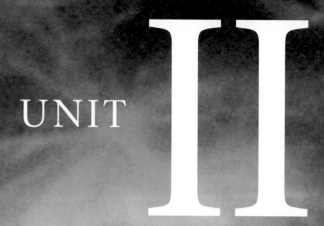

UNIT II

Cellular Reproduction and Genetics

8 THE CELLULAR BASIS OF REPRODUCTION AND INHERITANCE

9 PATTERNS OF INHERITANCE

10 MOLECULAR BIOLOGY OF THE GENE

11 THE CONTROL OF GENE EXPRESSION

12 DNA TECHNOLOGY AND GENOMICS

CHAPTER 8

CONNECTIONS BETWEEN CELL DIVISION AND REPRODUCTION

8.1   Like begets like, more or less
8.2   Cells arise only from preexisting cells
8.3   Prokaryotes reproduce by binary fission

THE EUKARYOTIC CELL CYCLE AND MITOSIS

8.4   The large, complex chromosomes of eukaryotes duplicate with each cell division
8.5   The cell cycle multiplies cells
8.6   Cell division is a continuum of dynamic changes
8.7   Cytokinesis differs for plant and animal cells
8.8   Anchorage, cell density, and chemical growth factors affect cell division
8.9   Growth factors signal the cell cycle control system
8.10 Growing out of control, cancer cells produce malignant tumors
8.11 Review of the functions of mitosis: Growth, cell replacement, and asexual reproduction

MEIOSIS AND CROSSING OVER

8.12 Chromosomes are matched in homologous pairs
8.13 Gametes have a single set of chromosomes
8.14 Meiosis reduces the chromosome number from diploid to haploid
8.15 Review: A comparison of mitosis and meiosis
8.16 Independent orientation of chromosomes in meiosis and random fertilization lead to varied offspring
8.17 Homologous chromosomes carry different versions of genes
8.18 Crossing over further increases genetic variability

ALTERATIONS OF CHROMOSOME NUMBER AND STRUCTURE

8.19 A karyotype is a photographic inventory of an individual's chromosomes
8.20 An extra copy of chromosome 21 causes Down syndrome
8.21 Accidents during meiosis can alter chromosome number
8.22 Abnormal numbers of sex chromosomes do not usually affect survival
8.23 Alterations of chromosome structure can cause birth defects and cancer

Rain Forest Rescue

DEEP IN A HAWAIIAN RAIN FOREST, the thrumming of a helicopter cuts through the warm tropical air. The craft touches down in a clearing and a pair of rescuers jump out, one of them carrying a precious bundle. After a 20-minute hike through dense foliage, they reach their target: the last known wild *Cyanea kuhihewa* plant (pictured at right). The rescuers now stand in the middle of a life-or-death drama, armed with the pollen of a species on the very brink of extinction.

Despite its tiny size, Hawaii is home to over 300 endangered or threatened species, including more endangered plants than any other state in the U.S. Of these, 11 plant species have fewer than five representatives growing anywhere in the wild. At the time of this rescue attempt in 2002, the small shrub *Cyanea kuhihewa* was among the rarest of the rare, with only a single known plant alive in nature.

Hawaii is home to more endangered plants than any other state

So when this lone wild *C. kuhihewa* bloomed, scientists from the National Tropical Botanical Garden on Kauai saw an opportunity. Their goal was to promote **sexual reproduction**, the reproductive process that involves **fertilization**, the union of a sperm and an egg. Using a fine brush, the botanical rescuers transferred sperm-carrying pollen from a garden-grown plant onto the egg-containing wild bloom. If sperm and egg join, the fertilized egg may soon divide into two cells. If development proceeds, the resulting cells will continue to divide and eventually form an embryo and then a seed. Upon germination of the seed, the embryo may further develop into a juvenile and later into an adult plant.

Development, from a fertilized egg to a new adult organism, is one phase of a multicellular organism's **life cycle**, the sequence of stages leading from the adults of one generation to the adults of the next. The other phase of the life cycle is reproduction, the formation of new individuals from preexisting ones. The reproductive phase of the life cycle entails the creation of offspring carrying genetic information, as DNA, from their parents. Sperm and egg each carry one set of genetic information—one copy of the organism's **genome**. Thus, the offspring of sexual reproduction

The Cellular Basis of Reproduction and Inheritance

Cyanea kuhihewa

inherit traits from two parents. In the case of *C. kuhihewa*, sexual reproduction could provide desperately needed genetic diversity.

Unfortunately, the *C. kuhihewa* fertilization attempt failed. Even worse, the last remaining wild specimen died in 2003. But hope remains for *C. kuhihewa*. The garden's botanists are trying **asexual reproduction,** the production of offspring by a single parent, without the participation of sperm and egg. Some organisms that normally reproduce sexually, such as many plants, can be induced to reproduce asexually in the laboratory. Indeed, work at Kauai's botanical garden has produced several new *C. kuhihewa* plants through asexual reproduction, using stem tissue snipped from wild plants in the 1990s.

These examples of sexual and asexual reproduction illustrate the main point of this

chapter: Cell division is at the heart of organismal reproduction. Plants created from cuttings result from repeated cell divisions, as does the more familiar development of a multicellular organism from a fertilized egg. And eggs and sperm themselves result from cell division of a special kind. In this chapter, we discuss the two main types of cell division and their functions in organisms. ■ ■ ■

Kauai, Hawaii

8.1 Like begets like, more or less

The ability of organisms to reproduce their own kind is the one characteristic that best distinguishes living things from nonliving matter. Only amoebas produce more amoebas, only people make more people, and only maple trees produce more maple trees. These simple facts of life have been recognized for thousands of years and are summarized by the age-old saying "Like begets like."

In a strict sense, "Like begets like" applies only to asexual reproduction, such as the single-celled amoeba dividing in Figure 8.1A. The amoeba's **chromosomes**, the structures that contain most of the organism's DNA, have been duplicated, and identical chromosomes have been allocated to opposite sides of the parent cell. When the parent cell divides, the resulting two daughter amoebas will be genetically identical to each other and to the original parent. (Biologists traditionally use the word "daughter" in this context only to indicate offspring, not to imply gender.) In asexual reproduction, there is one simple principle of inheritance: The lone parent and each of its offspring have identical genes.

The photograph of the family in Figure 8.1B makes the point that in a *sexually* reproducing species, like does not precisely beget like. Offspring produced by sexual reproduction generally resemble their parents more closely than they resemble unrelated individuals of the same species, but they are not identical to their parents or to each other. Each offspring inherits a unique combination of genes from its two parents, and this one-and-only set of genes programs a unique combination of traits. As a result, sexual reproduction can produce great variation among offspring. Notice in the photo that despite the family resemblances, each family member has a unique appearance. You probably resemble

Figure 8.1B Sexual reproduction produces offspring with unique combinations of genes.

your parents more closely than a stranger, but you do not look exactly like either parent or any of your siblings (unless you are an identical twin).

Long before anyone knew about genes, chromosomes, or the underlying principles of inheritance, people recognized that individuals of sexually reproducing species are highly varied. Furthermore, people learned to develop domestic breeds of plants and animals by controlling sexual reproduction. A domestic breed displays particular traits from among the great variety of traits found in the species as a whole. For example, though all domestic dogs belong to a single species, each breed of dog exhibits much less variability than the species as a whole. In producing each kind of domestic dog, breeders selected from a varied population of dogs certain individuals with specific traits and allowed these individuals to mate and produce offspring. The ancestry of a dog breed such as the German shepherd can be traced back many generations, during which breeders reduced variability in the breed by rigidly selecting for mating only those dogs with specific traits. In a sense, selective breeding is an attempt to make like beget like more than it does in nature.

? In terms of the "like begets like" adage, contrast asexual with sexual reproduction.

■ Asexual reproduction produces genetically identical offspring that inherit all their DNA from a single parent. The offspring of sexual reproduction show a family resemblance, but siblings vary because they inherit different combinations of genes from the two parents.

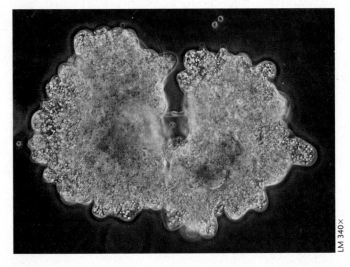

Figure 8.1A An amoeba producing genetically identical offspring through asexual reproduction

LM 340×

8.2 Cells arise only from preexisting cells

In 1858, German physician Rudolf Virchow stated an important biological principle: "Where a cell exists, there must have been a preexisting cell." Like many significant ideas now taken for granted, Virchow's principle (which he summarized as "Every cell from a cell") is both simple and profound. It tells us that the perpetuation of life, including all aspects of reproduction and inheritance, is based on the reproduction of cells, or **cell division.**

Cell division plays several important roles in the lives of organisms. In unicellular organisms, cell division can reproduce an entire organism. Cell division on a larger scale allows some multicellular organisms to reproduce asexually (such as plants that grow from cuttings). Among sexually reproducing organisms, cell division is the basis of sperm and egg formation. Cell division also enables sexually reproduc-ing organisms to develop from a single cell—the fertilized egg, or zygote—into an adult organism. After an organism is fully grown, cell division continues to function in renewal and repair, replacing cells that die from normal wear and tear or accidents.

So far, we have discussed only the division of eukaryotic cells, and we will emphasize them in this chapter. Prokaryotes, however, also illustrate Virchow's principle, and in the next module, we look briefly at prokaryotic cell division.

? Starting with a fertilized egg cell, a series of five cell divisions would produce an early embryo with how many cells?

■ *32 cells*

8.3 Prokaryotes reproduce by binary fission

Prokaryotes (bacteria and archaea) reproduce by a type of cell division called **binary fission** ("dividing in half"). In typical prokaryotes, most genes are carried on a circular DNA molecule that, with associated proteins, constitutes the organism's single chromosome. Although prokaryotic

Prokaryotic chromosomes

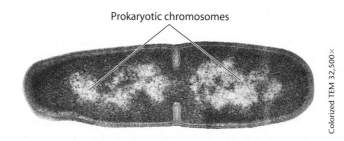

Figure 8.3B Electron micrograph of a dividing bacterium

chromosomes are much smaller than those of eukaryotes, replicating them in an orderly fashion and distributing the copies equally to two daughter bacteria is still a formidable task. Consider, for example, that when stretched out, the chromosome of the bacterium *Escherichia coli* is about 500 times longer than the cell itself. Accurately duplicating this molecule when it is coiled and packed inside the cell is no small achievement.

Figure 8.3A illustrates binary fission in a prokaryote. ❶ As the chromosome is duplicating, one copy moves toward the opposite end of the cell. ❷ Meanwhile, as chromosome duplication progresses, the cell elongates. ❸ When chromosome duplication is complete and the bacterium has reached about twice its initial size, the plasma membrane grows inward, dividing the parent cell into two daughter cells. **Figure 8.3B** is an electron micrograph of a dividing bacterium at a stage similar to step 3 in Figure 8.3A.

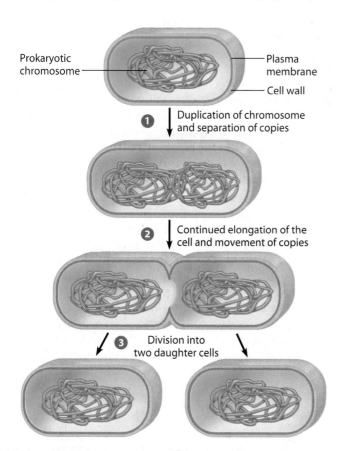

Figure 8.3A Binary fission of a prokaryotic cell

Prokaryotic chromosome — Plasma membrane — Cell wall

❶ Duplication of chromosome and separation of copies

❷ Continued elongation of the cell and movement of copies

❸ Division into two daughter cells

? Why is binary fission classified as asexual reproduction?

■ *Because the genetically identical offspring inherit their DNA from a single parent.*

8.4 The large, complex chromosomes of eukaryotes duplicate with each cell division

Eukaryotic cells are more complex and generally much larger than prokaryotic cells, and they have many more genes. Human cells, for example, carry about 30,000 genes, versus about 3,000 for a typical bacterium. Almost all the genes in the cells of humans, and in all other eukaryotes, are found in the cell nucleus, grouped into multiple chromosomes. (The exceptions include genes on the small DNA molecules of mitochondria.)

Most of the time, chromosomes exist as a diffuse mass of long, thin fibers. This material, called **chromatin**, is a combination of DNA and protein molecules. As a cell prepares to divide, its chromatin coils up, forming compact, distinct chromosomes. In this state, the chromosomes become visible under the light microscope. **Figure 8.4A** is a micrograph of a plant cell that is about to divide; each dark thread is an individual chromosome. In fact, chromosomes get their name (from the Greek *chromo,* colored, and *somes,* bodies) from their attraction for certain stains used in microscopy.

Like a prokaryotic chromosome, each eukaryotic chromosome contains one long DNA molecule bearing hundreds or thousands of genes and, attached to the DNA, a number of protein molecules. However, the eukaryotic chromosome has a much more complex structure than the prokaryotic chromosome. The eukaryotic chromosome includes many more protein molecules, which help maintain the chromosome structure and control the activity of its genes. The number of chromosomes in a eukaryotic cell depends on the species. For example, human body cells generally have 46 chromosomes, while the body cells of a dog have 78.

Well before a eukaryotic cell begins to divide, it duplicates all of its chromosomes. The DNA molecule of each chromosome is copied, and new protein molecules attach as needed.

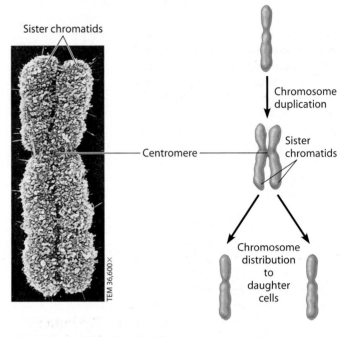

Figure 8.4B Electron micrograph of a duplicated chromosome

Figure 8.4C Chromosome duplication and distribution

The result is that each chromosome now consists of two copies called **sister chromatids**, which contain identical copies of the DNA molecule. **Figure 8.4B** is an electron micrograph of a human chromosome that has duplicated. The two chromatids are joined together especially tightly at a narrow "waist" called the **centromere.** The fuzzy appearance of the chromosome comes from the intricate twists and folds of its two chromatin fibers.

Figure 8.4C is a simple diagram making the point that when the cell divides, the sister chromatids of a duplicated chromosome separate from each other. Once separated from its sister, each chromatid is called a chromosome, and it is identical to the chromosome the cell started with. One of the new chromosomes goes to one daughter cell, and the other goes to the other daughter cell. In this way, each daughter cell receives a complete and identical set of chromosomes. In humans, for example, a typical dividing cell has 46 duplicated chromosomes (or 92 chromatids), and each of the two daughter cells that results from it has 46 single chromosomes.

? When does a chromosome consist of two identical chromatids?

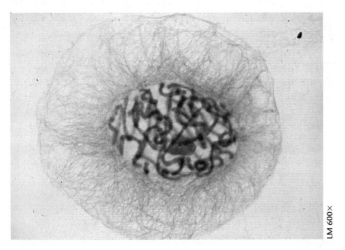

Figure 8.4A A plant cell (from an African blood lily) just before division

■ When the cell is preparing to divide and has duplicated its chromosomes, but before the duplicates actually separate

8.5 The cell cycle multiplies cells

How do chromosome duplication and cell division fit into the life of a cell—and the life of an organism? As you already know, cell division is essential to life. Cell division is the basis of reproduction for every organism. It enables a multicellular organism to grow to adult size. It also replaces worn-out or damaged cells, keeping the total cell number in a mature individual relatively constant. In your own body, for example, millions of cells must divide every second to maintain the total number of about 100 trillion cells. Some cells divide once a day, others less often, and highly specialized cells, such as our mature muscle cells, not at all.

The process of cell division is a key component of the **cell cycle**, an ordered sequence of events that extends from the time a cell is first formed from a dividing parent cell until its own division into two cells. The cell cycle con-

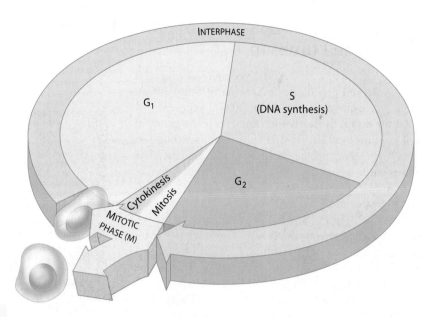

Figure 8.5 The eukaryotic cell cycle

sists of two broad stages: a growing stage (called interphase), during which the cell roughly doubles everything in its cytoplasm and precisely duplicates its chromosomal DNA, and the actual cell division (called the mitotic phase).

As **Figure 8.5** indicates, most of the cell cycle is spent in **interphase.** This is a time when a cell's metabolic activity is very high and the cell performs its various functions within the organism. Moreover, a cell in interphase increases its supply of proteins, creates more cytoplasmic organelles (such as mitochondria and ribosomes), and grows in size. Additionally, the chromosomes duplicate during this period. Typically, interphase lasts for at least 90% of the total time required for the cell cycle.

Interphase can be divided into three subphases: the G_1 phase ("first gap"), the S phase, and the G_2 phase ("second gap"). During all three subphases, the cell grows. However, chromosomes are duplicated only during the S phase. S stands for synthesis of DNA—also known as DNA replication—a process that will be discussed in Chapter 10. At the beginning of the S phase, each chromosome is single. At the end of this phase, after DNA replication, the chromosomes are double, each consisting of two sister chromatids. To summarize interphase, a cell grows (G_1), continues to grow as it copies its chromosomes (S), and then grows more as it completes preparations for cell division (G_2).

The **mitotic phase (M phase)**, the part of the cell cycle when the cell actually divides, accounts for only about 10% of the total time required for the cell cycle. The mitotic phase is divided into two stages, called mitosis and cytokinesis, although the second stage begins before the first one ends. In **mitosis** (light yellow area in the figure), the nucleus and its contents, including the duplicated chromosomes, divide and

are evenly distributed to form two daughter nuclei. During **cytokinesis**, the cytoplasm is divided in two. The combination of mitosis and cytokinesis produces two genetically identical daughter cells, each with a single nucleus, surrounding cytoplasm, and plasma membrane. Each newly produced daughter cell may then proceed through G_1 and repeat the cycle.

Mitosis is unique to eukaryotes and may be an evolutionary solution to the problem of allocating identical copies of a large amount of genetic material, in a number of separate chromosomes, to two daughter cells. Mitosis is a remarkably accurate mechanism. Experiments with yeast, for example, indicate that an error in chromosome distribution occurs only once in about 100,000 cell divisions.

A living cell viewed through a light microscope undergoes dramatic changes in appearance during the mitotic phase. During interphase, the cell's individual chromosomes are not distinguishable because they are in the form of loosely packed chromatin fibers. With the onset of mitosis, however, striking changes are visible in the chromosomes and other structures, as we see in the next module.

Web/CD Activity 8A *The Cell Cycle*

? A researcher treats cells with a chemical that prevents DNA synthesis from starting. This treatment would trap the cells in which part of the cell cycle?

G_1 ■

8.6 Cell division is a continuum of dynamic changes

The light micrographs in **Figure 8.6** show the cell cycle for an animal cell—in this case, from a newt. Interphase is included, but the emphasis is on the dramatic changes that occur during cell division, the mitotic phase. Mitosis is a continuum of changes, but biologists distinguish five main stages: **prophase**, **prometaphase**, **metaphase**, **anaphase**, and **telophase**. The drawings in Figure 8.6 show details not visible in the micrographs. For simplicity, only four chromosomes are drawn.

The chromosomes are the stars of the mitotic drama, and their movements depend on the **mitotic spindle**, a football-

Figure 8.6 The stages of cell division

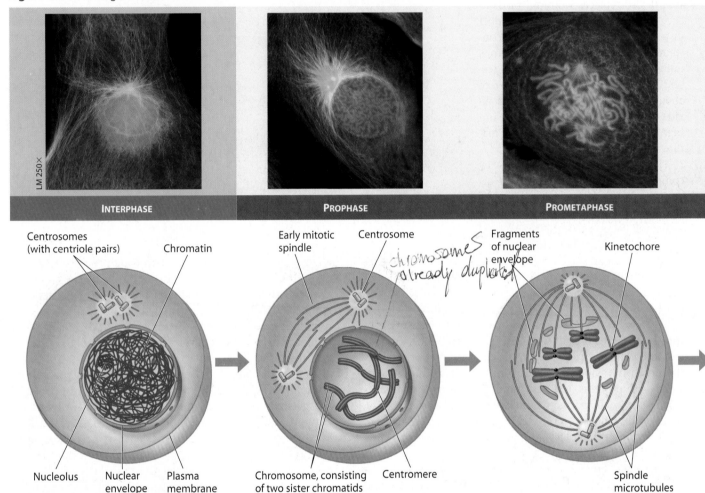

Centrosomes (with centriole pairs) Chromatin

Nucleolus Nuclear envelope Plasma membrane

INTERPHASE

Early mitotic spindle Centrosome chromosomes already duplicated Fragments of nuclear envelope Kinetochore

Chromosome, consisting of two sister chromatids Centromere

PROPHASE

Spindle microtubules

PROMETAPHASE

Interphase Interphase is the period of cell growth when the cell synthesizes new molecules and organelles. At the point shown here, late interphase (G$_2$), the cell looks much the same as it does throughout interphase. Nonetheless, by the G$_2$ stage the cell has doubled much of its earlier contents and the cytoplasm contains two centrosomes. Within the nucleus, the chromosomes are duplicated, but they cannot be distinguished individually because they are still in the form of loosely packed chromatin. The nucleus also contains one or more nucleoli, an indication that the cell is actively making proteins. Nucleoli are where the parts of ribosomes are assembled before export to the cytoplasm (see Chapter 10).

Prophase During prophase, changes occur in both the nucleus and the cytoplasm. Within the nucleus, the chromatin fibers become more tightly coiled and folded, forming discrete chromosomes that can be seen with the light microscope. The nucleoli disappear. Each duplicated chromosome appears as two identical sister chromatids joined together, with a narrow "waist" at the centromere. In the cytoplasm, the mitotic spindle begins to form as microtubules rapidly grow out from the centrosomes, which begin to move away from each other.

Prometaphase The nuclear envelope breaks into fragments and disappears. Microtubules emerging from the centrosomes at the poles (ends) of the spindle reach the chromosomes, now highly condensed. At the centromere region, each sister chromatid has a protein structure called a kinetochore (shown as a black dot). Some of the spindle microtubules attach to the kinetochores, throwing the chromosomes into agitated motion. Other spindle microtubules make contact with microtubules coming from the opposite pole. Forces exerted by protein "motors" associated with spindle microtubules move the chromosomes toward the center of the cell.

*sister chromatids only exist after interfase.

shaped structure of microtubules that guides the separation of the two sets of daughter chromosomes. The spindle microtubules emerge from two **centrosomes**, clouds of cytoplasmic material that in animal cells contain centrioles. (Centrosomes are also known as *microtubule-organizing centers*, a term describing their function.) The role of centrioles in cell division is a mystery; destroying them experimentally does not interfere with normal spindle formation, and plant cells lack them entirely.

Web/CD Thinking as a Scientist *How Much Time Do Cells Spend in Each Phase of Mitosis?*

? An organism called a plasmodial slime mold is one large cytoplasmic mass with many nuclei. Explain how such a "megacell" could form.

■ Mitosis occurs repeatedly without cytokinesis.

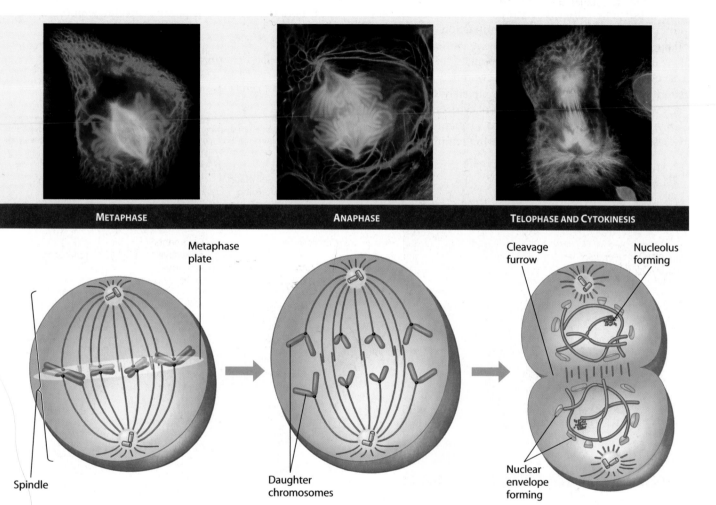

METAPHASE　　**ANAPHASE**　　**TELOPHASE AND CYTOKINESIS**

Metaphase At metaphase, the mitotic spindle is fully formed, with its poles at opposite ends of the cell. The chromosomes convene on the metaphase plate, an imaginary plane equidistant between the two poles of the spindle. The centromeres of all the chromosomes are lined up on the metaphase plate. For each chromosome, the kinetochores of the two sister chromatids face opposite poles of the spindle. The microtubules attached to a particular chromatid all come from one pole of the spindle, and those attached to its sister chromatid come from the opposite pole.

Anaphase Anaphase begins when the two centromeres of each chromosome come apart, separating the sister chromatids. Once separate, each sister chromatid is considered a full-fledged (daughter) chromosome. Motor proteins of the kinetochores, powered by ATP, "walk" the daughter chromosomes centromere-first along the microtubules toward opposite poles of the cell. As this happens, the spindle microtubules attached to the kinetochores shorten. However, the spindle microtubules not attached to chromosomes lengthen. The poles are moved farther apart, elongating the cell. Anaphase is over when equivalent—and complete—collections of chromosomes have reached the two poles of the cell.

Telophase Telophase is roughly the reverse of prophase. The cell elongation that started in anaphase continues. Daughter nuclei appear at the two poles of the cell as nuclear envelopes form around the chromosomes. Meanwhile, the chromatin fiber of each chromosome uncoils, and nucleoli reappear. At the end of telophase, the mitotic spindle disappears. Mitosis, the equal division of one nucleus into two genetically identical daughter nuclei, is now finished.

Cytokinesis Cytokinesis, the division of the cytoplasm, usually occurs along with telophase, with two daughter cells completely separating soon after the end of mitosis. In animal cells, cytokinesis involves a cleavage furrow, which pinches the cell in two.

Cytokinesis, or division of the cell into two, typically begins during telophase, although it may begin in late anaphase. In animal cells, cytokinesis occurs by a process known as cleavage. As shown in Figure 8.7A, the first sign of cleavage is the appearance of a **cleavage furrow,** a shallow groove in the cell surface. At the site of the furrow, the cytoplasm has a ring of microfilaments made of actin associated with molecules of the protein myosin. (Actin and myosin are the same proteins responsible for muscle contraction—see Module 30.8.) When the actin microfilaments interact wth the myosin, the ring contracts, like the pulling of drawstrings. The cleavage furrow deepens and eventually pinches the parent cell in two, producing two completely separate cells, each with its own nucleus and share of cytoplasm.

Cytokinesis is markedly different in plant cells, which possess cell walls (Figure 8.7B). During telophase, vesicles containing cell wall material (tan in figure) collect at the middle of the parent cell. The vesicles fuse, forming a membranous **cell plate.** The cell plate grows outward, accumulating more cell wall materials as more vesicles fuse with it. Eventually, the membrane surrounding the cell plate fuses with the plasma membrane, and the cell plate's contents join the parental cell wall. The result is two daughter cells, each bounded by its own plasma membrane and cell wall.

Web/CD Activity 8B *Mitosis and Cytokinesis Animation*

Web/CD Activity 8C *Mitosis and Cytokinesis Video*

?   Contrast cytokinesis in animals with cytokinesis in plants.

■ In animals, cytokinesis involves a cleavage furrow in which contracting microfilaments pinch the cell in two. In plants, it involves formation of a cell plate, a fusion of vesicles that forms new membrane and walls between the cells.

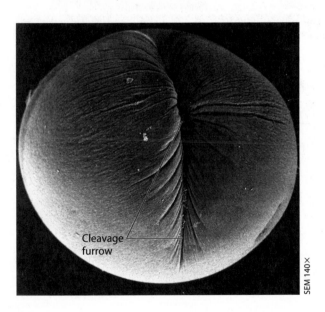

SEM 140×

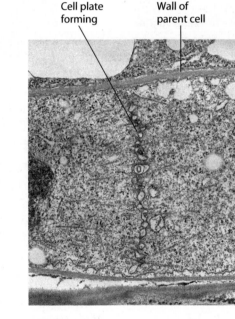

Cell plate forming   Wall of parent cell   Daughter nucleus

TEM 7,500×

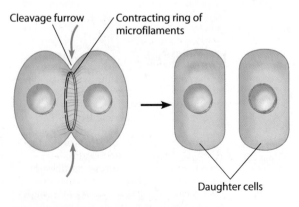

Cleavage furrow   Contracting ring of microfilaments

Daughter cells

Figure 8.7A   Cleavage of an animal cell

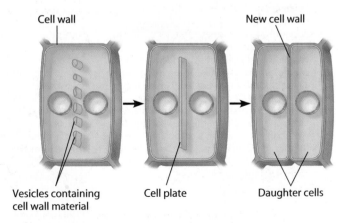

Cell wall   New cell wall

Vesicles containing cell wall material   Cell plate   Daughter cells

Figure 8.7B   Cell plate formation in a plant cell

8.8 Anchorage, cell density, and chemical growth factors affect cell division

For a plant or an animal to grow and develop normally and to maintain its tissues once full grown, it must be able to control the timing of cell division in different parts of its body. For example, in the adult human, skin cells and the cells lining the digestive tract divide frequently throughout life, replacing cells that are constantly being abraded and sloughed off. In contrast, cells in the human liver usually do not divide unless the liver is damaged. Cell division in this case repairs wounds.

Many questions about how cell division is controlled are still unanswered, but biologists have learned a great deal by studying cells grown in lab cultures. They have learned, for instance, that most animal cells exhibit **anchorage dependence**; they must be in contact with a solid surface—such as the inside of a culture dish or the extracellular matrix of a tissue—to divide.

Scientists have also found that animal cells growing on the surface of a dish multiply to form a single layer and usually stop dividing when they touch one another (Figure 8.8A). This phenomenon is called **density-dependent inhibition**, since crowded cells stop dividing. If some cells are removed, those bordering the open space begin dividing again and continue until the vacancy is filled.

Clearing a space in a cell culture is analogous to cutting your skin. When you cut yourself, skin cells all around the cut immediately begin dividing, and healing occurs as cells fill in the gap. The cells stop dividing when they encounter other cells.

What actually causes density-dependent inhibition? Studies of cultured cells suggest that it is primarily an inadequate supply of certain proteins, called growth factors, rather than physical contact with other cells, that is responsible for inhibiting cell division. A **growth factor** is a protein secreted by certain body cells that stimulates other cells to divide. Most cells require growth factors in order to begin dividing, and they stop dividing when they run out of these substances. For cells in culture, growth factors are provided in the nutrient medium. Apparently, density-dependent inhibition occurs when the cells become so crowded that each cell uses up the supply of growth factors in its immediate environment. As indicated in Figure 8.8B, flooding a layer of nondividing cells with a more concentrated solution of growth factors stimulates the cells to grow to a greater density than they would otherwise. The cells are still in a single layer but are smaller and more numerous.

Density-dependent inhibition mediated by the availability of growth factors is probably an important regulatory mechanism in the body's tissues. It may help keep cell populations at optimal levels in the tissues. How exactly do growth factors work? We pursue answers to this question in the next module.

? Compared to a control culture, the cells in an experimental culture are fewer but much larger when they cover the dish surface and stop growing. What is a reasonable hypothesis for this difference?

■ The experimental culture is deficient in one or more growth factors.

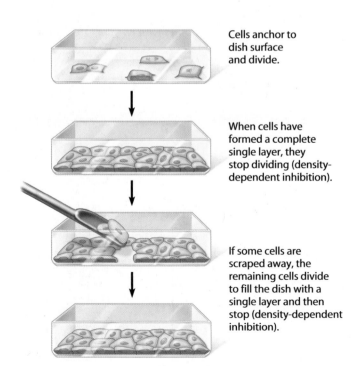

Figure 8.8A An experiment demonstrating density-dependent inhibition, using animal cells grown in culture

Cells anchor to dish surface and divide.

When cells have formed a complete single layer, they stop dividing (density-dependent inhibition).

If some cells are scraped away, the remaining cells divide to fill the dish with a single layer and then stop (density-dependent inhibition).

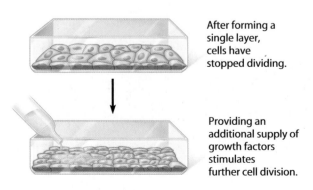

Figure 8.8B An experiment demonstrating the effect of growth factors on the division of cultured animal cells

After forming a single layer, cells have stopped dividing.

Providing an additional supply of growth factors stimulates further cell division.

8.9   Growth factors signal the cell cycle control system

In a living animal, most cells are anchored in a fixed position and bathed in a solution of nutrients supplied by the blood, yet they usually do not divide unless they are signaled by other cells to do so. Growth factors are the main signals, and their role in promoting cell division leads us back to our earlier discussion of the cell cycle.

The sequential events of the cell cycle, represented by the circle of flat blocks in Figure 8.9A, are directed by a distinct cell cycle control system, represented by the knob in the center. Analogous to the control device of an automatic washing machine, the **cell cycle control system** is a cyclically operating set of molecules in the cell that both triggers and coordinates key events in the cell cycle. The cell cycle is not like a row of falling dominoes, with each event causing the next one in line. Within the M phase, for example, metaphase does not automatically lead to anaphase. Instead, proteins of the cell cycle control system must trigger the separation of sister chromatids that marks the start of anaphase.

A checkpoint in the cell cycle is a critical control point where stop and go-ahead signals can regulate the cycle. (The signals are transmitted within the cell by signal transduction pathways—see Figure 5.13B.) Animal cells generally have built-in stop signals that halt the cell cycle at checkpoints until overridden by go-ahead signals.

The red barriers in Figure 8.9A represent three major checkpoints in the cell cycle: during the G_1 and G_2 subphases of interphase and in the M phase. Intracellular signals detected by the control system tell it whether key cellular processes up to each point have been completed and thus whether or not the cell cycle should proceed past that point. The control system also receives messages from outside the cell, indicating both general environmental conditions and the presence of specific signal molecules from other cells. When the cell cycle control system gets a go-ahead signal

at the G_1 checkpoint, for example, the cell soon enters the S phase of the cell cycle.

For many cells, the G_1 checkpoint seems to be the most important. If a cell receives a go-ahead signal—for example, from a growth factor—at the G_1 checkpoint, it will usually complete its cycle and divide. But if it does not receive a go-ahead signal at G_1, it will exit the cell cycle, switching into a nondividing state called the G_0 phase. For instance, nondividing nerve cells and muscle cells in your body are in the G_0 phase.

Figure 8.9B shows a simplified model for how a growth factor might affect the cell cycle control system at the G_1 checkpoint. A cell that responds to a growth factor has molecules of a specific receptor protein in its plasma membrane. Binding of the growth factor to the receptor triggers a signal transduction pathway in the cell, a pathway that in this case leads to cell division. The "signals" are changes that a molecule induces in the next molecule in the pathway. Via a series of relay molecules, a signal finally reaches the cell cycle control system and overrides the brakes that otherwise prevent progress of the cell cycle. In Figure 8.9B, the cell cycle is set off from the cell in a separate diagram because, unlike the components of the timer in a washing machine, the proteins making up the control system in the cell are not actually located together in one place.

Research on the control of the cell cycle is one of the hottest areas in biology today. This research is leading to a better understanding of cancer, which we discuss next.

?   At which of the three checkpoints described in this module do the chromosomes exist as duplicated sisters chromatids?

■ G_2 and M

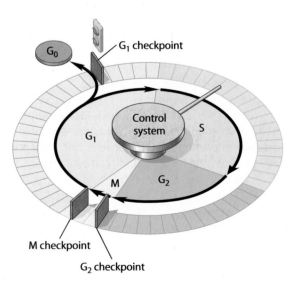

Figure 8.9A   Mechanical model for the cell cycle control system

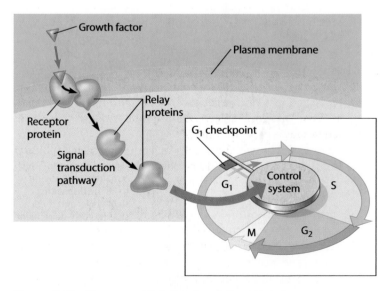

Figure 8.9B   How a growth factor signals the cell cycle control system

8.10 Growing out of control, cancer cells produce malignant tumors

Cancer, which currently claims the lives of one out of every five people in the United States and other developed nations, is a disease of the cell cycle. **Cancer cells** do not respond normally to the cell cycle control system; they divide excessively and can invade other tissues of the body. If unchecked, cancer cells may continue to grow until they kill the organism.

The abnormal behavior of cancer cells begins when a single cell undergoes transformation, a process that converts a normal cell to a cancer cell. The body's immune system normally recognizes a transformed cell as abnormal and destroys it. However, if the cell evades destruction, it may proliferate to form a **tumor**, an abnormally growing mass of body cells. If the abnormal cells remain at the original site, the lump is called a **benign tumor**. Benign tumors can cause problems if they grow in and disrupt certain organs, such as the brain, but often they can be completely removed by surgery.

In contrast, a **malignant tumor** can spread into neighboring tissues and other parts of the body, displacing normal tissue and interrupting organ function as it goes (Figure 8.10). Cancer cells may separate from the original tumor or secrete signal molecules that cause blood vessels to grow toward the tumor. A few tumor cells may then enter the blood and lymph vessels of the circulatory system and move to other parts of the body, where they may proliferate and form new tumors. This spread of cancer cells via the circulatory system beyond their original site is called **metastasis**. An individual with a malignant tumor is said to have cancer.

Cancers are named according to the organ or tissue in which they originate (for example, liver cancer starts in liver tissue) and are grouped into four categories. **Carcinomas** are cancers that originate in the external or internal coverings of the body, such as the skin or the lining of the intestine. **Sarcomas** arise in tissues that support the body, such as bone and muscle. Cancers of blood-forming tissues, such as bone marrow, spleen, and lymph nodes, are called **leukemias** and **lymphomas.**

From studying cancer cells in culture, researchers have learned that these cells do not heed the normal signals that regulate the cell cycle. For example, cancer cells do not exhibit density-dependent inhibition; they continue to divide even at high densities, piling up on one another. Many cancer cells have defective cell cycle control systems that proceed past checkpoints even in the absence of growth factors. Other cancer cells themselves synthesize growth factors that make them divide continuously. If cancer cells do stop dividing, they seem to do so at random points in the cell cycle, rather than at the normal cell cycle checkpoints. Moreover,

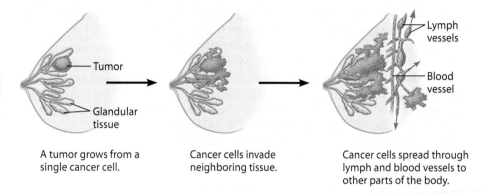

A tumor grows from a single cancer cell.

Cancer cells invade neighboring tissue.

Cancer cells spread through lymph and blood vessels to other parts of the body.

Figure 8.10 Growth and metastasis of a malignant (cancerous) tumor of the breast

in culture, cancer cells are "immortal"; they can go on dividing indefinitely, as long as they have a supply of nutrients (whereas normal mammalian cells divide only about 20 to 50 times before they stop).

Luckily, many tumors can be successfully treated. A tumor that appears to be localized may be removed surgically. Alternatively, it can be treated with high-energy radiation, which damages DNA in cancer cells much more than it does in normal cells, apparently because cancer cells have lost the ability to repair such damage. However, there is sometimes enough damage to normal body cells to produce bad side effects. For example, damage to cells of the ovaries or testes can lead to sterility.

To treat widespread or metastatic tumors, chemotherapy is used. During periodic chemotherapy treatments, drugs are administered that disrupt specific steps in the cell cycle. For example, the drug paclitaxel (trade name Taxol) "freezes" the mitotic spindle, which stops actively dividing cells from proceeding past metaphase. (Interestingly, Taxol was originally discovered in the bark of the Pacific yew tree, found mainly in the northwestern United States.) Vinblastin, a chemotherapeutic drug first obtained from the periwinkle plant (found in the rain forests of Madagascar), prevents the spindle from forming in the first place.

The side effects of chemotherapy are due to the drugs' effects on normal cells that rapidly divide. For example, nausea results from chemotherapy's effects on intestinal cells, hair loss from effects on hair follicle cells, and susceptibility to infection from effects on immune cell production.

We will return to the topic of cancer in Chapter 11, after studying the structure and function of genes. We will see that cancer results from changes in genes for proteins that control cell division.

? What is metastasis?

■ Metastasis is the spread of cancer cells via the circulatory system from their original site of formation to sites in the body.

8.11 Review of the functions of mitosis: Growth, cell replacement, and asexual reproduction

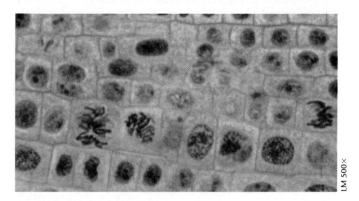

Figure 8.11A Growth (in an onion root)

The three micrographs in Figures 8.11A–8.11C summarize the roles that mitotic cell division plays in the lives of multicellular organisms. **Figure 8.11A** shows some of the cells from the tip of a rapidly growing onion plant root. Notice the large number of cells whose nuclei are in various stages of mitosis. Cell division in the root tip produces new cells, which elongate to bring about growth of the root.

Figure 8.11B shows a dividing bone marrow cell. Mitotic cell division within the red marrow of your body's bones (particularly within your ribs, vertebrae, breastbone, and pelvis) continuously creates new blood cells that replace older ones. Similar processes replace cells throughout your body. For example, dividing cells within your epidermis continuously replace dead cells that slough off the surface of your skin.

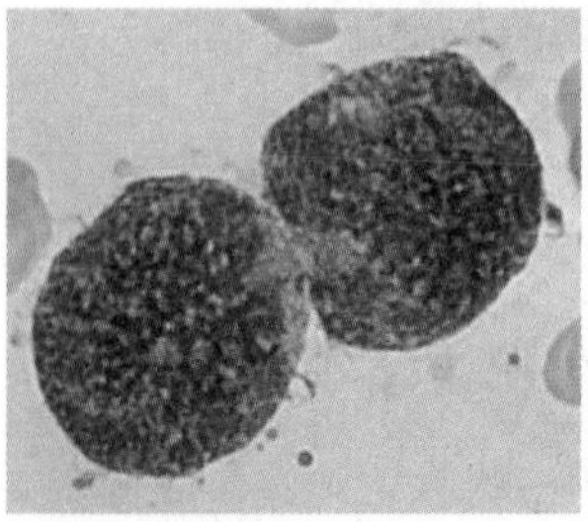

Figure 8.11C is a micrograph of a hydra, a common inhabitant of freshwater lakes. A hydra is a tiny

Figure 8.11B
Cell replacement (in bone marrow)

multicellular animal that reproduces by either sexual or asexual means. This individual is reproducing asexually by budding. A bud starts out as a mass of mitotically dividing cells growing on the side of the parent. The bud develops into a small hydra like the one in the photo here. Eventually, the offspring detaches from the parent and takes up life on its own. The offspring is literally a "chip off the old block," being genetically identical to (a clone of) its parent.

In all three of these cases—the growing onion root, replacement of blood cells, and budding hydra—the new cells have exactly the same number and types of chromosomes as the parent cells because of the way duplicated chromosomes divide during mitosis. Mitosis makes it possible for organisms to grow, regenerate and repair tissues, and reproduce asexually by producing cells that carry the same genes as the parent cells.

If we examine the cells of any individual organism, we see that almost all of them contain the same number and types of chromosomes. Likewise, if we examine cells from different individuals of any one species, we see that they have the same number and types of chromosomes. In the next module, we take a closer look at how chromosomes are organized in cells.

Figure 8.11C Asexual reproduction (of a hydra)

? If a human skin cell with 46 chromosomes divides by mitosis, each daughter cell will have __ chromosomes.

9ㄴ ■

MEIOSIS AND CROSSING OVER

8.12 Chromosomes are matched in homologous pairs

In humans, a typical body cell, called a **somatic cell**, has 46 chromosomes. If we use a microscope to examine human chromosomes in metaphase of mitosis, we see that each duplicated chromosome has a twin that is identical in length and centromere position. Altogether, we see 23 such matched pairs of duplicated chromosomes. Other species have different numbers of chromosomes, but these, too, are usually matched pairs. Moreover, when treated with special

dyes, the chromosomes of a pair display matching staining patterns. **Figure 8.12** illustrates one pair of metaphase chromosomes, with the staining pattern represented by colored stripes. Notice that each chromosome consists of two sister chromatids joined at the centromere. The two chromosomes composing a pair are called **homologous chromosomes** (or homologues) because they both carry genes controlling the same inherited characteristics. For example, if a gene that

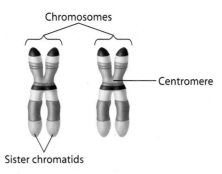

Figure 8.12
A homologous pair of chromosomes

Chromosomes

Centromere

Sister chromatids

determines whether a person has freckles is located at a particular place, or **locus** (plural, *loci*), on one chromosome—within the narrow orange band in our drawing, for instance—then the other chromosome of the homologous pair also has a gene for freckles at that locus. (However, the two homologues may have different versions of the freckles gene, perhaps one that promotes freckles and one that does not.)

The two distinct chromosomes called X and Y are an important exception to the general pattern of homologous chromosomes. Human females have a homologous pair of X chromosomes (XX), but males have one X and one Y chromosome (XY). Only small parts of the X and Y are homologous; most of the genes carried on the X chromosome do not have counterparts on the tiny Y, and the Y chromosome has genes lacking on the X. Because they determine an individual's sex, the X and Y chromosomes are called **sex chromosomes** (although they carry genes that perform other functions as well). The other 22 pairs of chromosomes are called **autosomes**.

For both autosomes and sex chromosomes, we inherit one chromosome of each pair from our mother and the other from our father, as discussed in the next module.

? What is the explanation for our having two of each type of chromosome?

■ *We inherit one of each kind from each parent.*

8.13 Gametes have a single set of chromosomes

Having two sets of chromosomes, one inherited from each parent, is a key factor in the human life cycle (outlined in **Figure 8.13**) and in the life cycles of all other species that reproduce sexually. (We examine other life cycles in Unit IV.)

Any cell with two homologous sets of chromosomes is called a **diploid cell**, and the total number of chromosomes is called the diploid number (abbreviated $2n$). For humans,

the diploid number is 46; that is, $2n = 46$. Humans are said to be diploid organisms because almost all our cells are diploid. The exceptions are the egg and sperm cells, collectively known as **gametes.** Each gamete has a single set of chromosomes: 22 autosomes plus a single sex chromosome, either X or Y. A cell with a single chromosome set is called a **haploid cell.** For humans, the haploid number (abbreviated n) is 23; that is, $n = 23$.

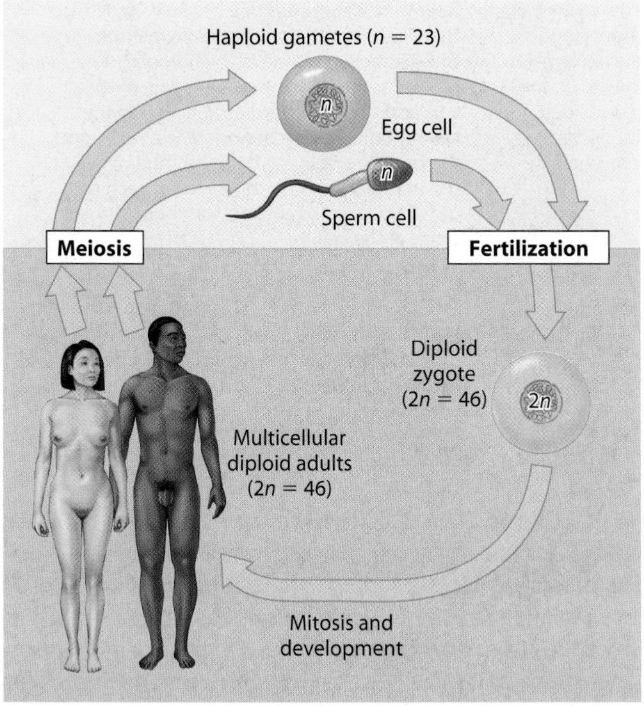

In humans, sexual intercourse allows a haploid sperm cell from the father to reach and fuse with a haploid egg cell of the mother in the process of fertilization. The resulting fertilized egg, called a **zygote**, is diploid. It has two haploid sets of chromosomes: one set from the mother and a homologous set from the father. The life cycle is completed as a sexually mature adult develops from the zygote. Mitotic cell division ensures that all somatic cells of the human body receive copies of all of the zygote's 46 chromosomes.

All sexual life cycles, including our own, involve an alternation of diploid and haploid stages. Having haploid gametes keeps the chromosome number from doubling in each generation. Gametes are made by a special sort of cell division called meiosis, which occurs only in reproductive organs (ovaries and testes in animals). Whereas mitosis produces daughter cells with the same numbers of chromosomes as the parent cell, meiosis reduces the chromosome number by half. We turn to meiosis next.

Web/CD Activity 8D *Asexual and Sexual Life Cycles*

? _____ is to somatic cells as haploid is to _____.

Haploid gametes ($n = 23$)

Egg cell

Sperm cell

Meiosis

Fertilization

Diploid zygote ($2n = 46$)

Multicellular diploid adults ($2n = 46$)

Mitosis and development

Figure 8.13 The human life cycle

■ *Diploid . . . gametes*

8.14 Meiosis reduces the chromosome number from diploid to haploid

Meiosis is a type of cell division that produces haploid gametes in diploid organisms. Many of the stages of meiosis closely resemble corresponding stages in mitosis. Meiosis, like mitosis, is preceded by the replication of chromosomes. However, this single replication is followed by two consecutive cell divisions, called meiosis I and meiosis II. These divisions result

Figure 8.14 The stages of meiosis

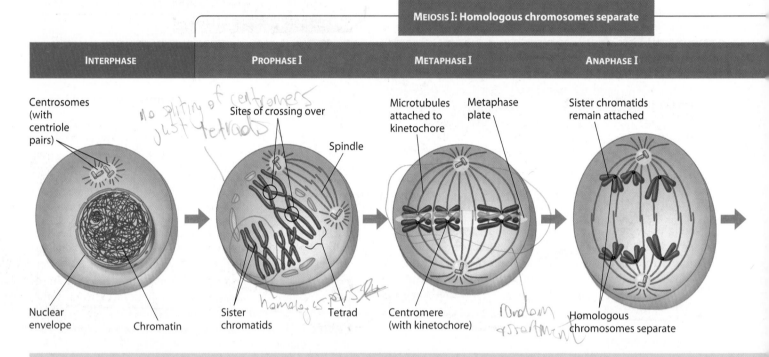

MEIOSIS I: Homologous chromosomes separate

INTERPHASE — PROPHASE I — METAPHASE I — ANAPHASE I

Centrosomes (with centriole pairs)
Nuclear envelope
Chromatin
Sites of crossing over
Spindle
Sister chromatids
Tetrad
Microtubules attached to kinetochore
Metaphase plate
Centromere (with kinetochore)
Sister chromatids remain attached
Homologous chromosomes separate

Handwritten annotations: "No splitting of centromeres just tetrads", "homolog. US P/S R4", "random assortment"

Interphase
Like mitosis, meiosis is preceded by an interphase, during which the chromosomes duplicate. At the end of this interphase, each chromosome consists of two genetically identical sister chromatids attached together. But at this stage the chromosomes are not yet visible under the microscope except as a mass of chromatin. The cell's centrosome has also duplicated by the end of this interphase.

Prophase I
Prophase I is the most complex phase of meiosis and typically occupies over 90% of the time required for meiotic cell division. Early in this phase, the chromatin coils up, so that individual chromosomes become visible with the microscope. In a process called synapsis, homologous chromosomes, each composed of two sister chromatids, come together as pairs. The resulting structure, consisting of four chromatids, is called a tetrad. During synapsis, chromatids of homologous chromosomes exchange segments in a process called crossing over. Because the versions of the genes on a chromosome (or one of its chromatids) may be different from those on its homologue, crossing over rearranges genetic information. As we will discuss in Module 8.18, the genetic shuffling produced by crossing over can make an important contribution to the genetic variability resulting from sexual reproduction.

As prophase I continues, the chromosomes condense further as the nucleoli disappear. Now the centrosomes move away from each other, and a spindle starts to form between them. The nuclear envelope breaks into fragments, and the chromosome tetrads, captured by spindle microtubules, are moved toward the center of the cell.

Metaphase I
At metaphase I, the chromosome tetrads are aligned on the metaphase plate, midway between the two poles of the spindle. Each chromosome is condensed and thick, with its sister chromatids still attached at their centromeres. Spindle microtubules are attached to kinetochores at the centromeres. In each tetrad, the homologous chromosomes are held together at sites of crossing over. Notice that, for each tetrad, the spindle microtubules attached to one of the homologous chromosomes come from one pole of the cell, and the microtubules attached to the other homologous chromosome come from the opposite pole. With this arrangement, the homologous chromosomes of each tetrad are poised to move toward opposite poles of the cell.

Anaphase I
Like anaphase of mitosis, anaphase I of meiosis is marked by the migration of chromosomes toward the two poles of the cell. In contrast to mitosis, however, the sister chromatids making up each doubled chromosome remain attached at their centromeres. Only the tetrads (pairs of homologous chromosomes) split up. Thus, in the drawing you see three still-doubled chromosomes (the haploid number) moving toward each spindle pole. If this were anaphase of mitosis, you would see six daughter chromosomes moving toward each pole.

in four daughter cells (rather than the two daughter cells of mitosis), each with a single haploid set of chromosomes. Thus, meiosis produces daughter cells with only half as many chromosomes as the parent cell. The drawings in Figure 8.14 show the two meiotic divisions for an animal cell with a diploid number of 6.

? A cell has the haploid number of chromosomes, but each chromosome has two chromatids. The chromosomes are arranged at the center of the spindle. What is the meiotic stage?

■ Metaphase II

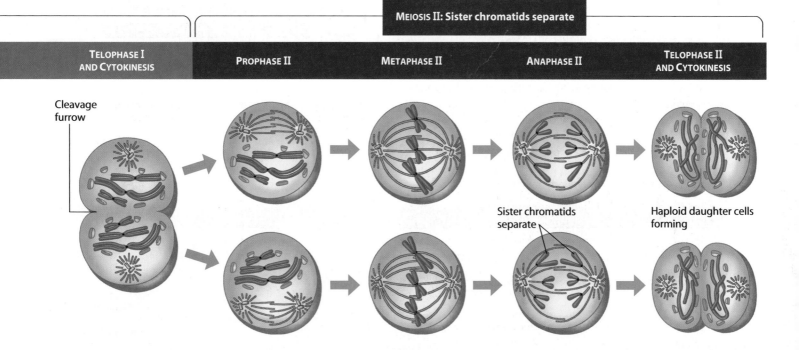

MEIOSIS II: Sister chromatids separate			
TELOPHASE I AND CYTOKINESIS	**PROPHASE II**	**METAPHASE II**	**ANAPHASE II**	**TELOPHASE II AND CYTOKINESIS**

Cleavage furrow

Sister chromatids separate

Haploid daughter cells forming

Telophase I and Cytokinesis

In telophase I, the chromosomes arrive at the poles of the cell. When the chromosomes finish their journey, each pole of the cell has a haploid chromosome set, although each chromosome is still in duplicate form at this point. In other words, each chromosome still consists of two sister chromatids. Usually, cytokinesis occurs along with telophase I, and two haploid daughter cells are formed.

Following telophase I in some organisms, the chromosomes uncoil and the nuclear envelope re-forms, and there is an interphase before meiosis II begins. In other species, daughter cells produced in the first meiotic division immediately begin preparation for the second meiotic division. In either case, no chromosome duplication occurs between telophase I and the onset of meiosis II.

Meiosis II

In organisms having an interphase after meiosis I, the chromosomes condense again and the nuclear envelope breaks down during prophase II. In any case, meiosis II is essentially the same as mitosis. The important difference is that meiosis II starts with a haploid cell.

During prophase II, a spindle forms and moves the chromosomes toward the middle of the cell. During metaphase II, the chromosomes are aligned on the metaphase plate as they are in mitosis, with the kinetochores of the sister chromatids of each chromosome pointing toward opposite poles. In anaphase II, the centromeres of sister chromatids finally separate, and the sister chromatids of each pair, now individual daughter chromosomes, move toward opposite poles of the cell. In telophase II, nuclei form at the cell poles, and cytokinesis occurs at the same time. There are now four daughter cells, each with the haploid number of (single) chromosomes.

8.15 Review: A comparison of mitosis and meiosis

We have now described the two ways that cells of eukaryotic organisms divide. Mitosis, which provides for growth, tissue repair, and asexual reproduction, produces daughter cells genetically identical to the parent cell. Meiosis, needed for sexual reproduction, yields haploid daughter cells—cells with one member of each homologous chromosome pair.

For both mitosis and meiosis, the chromosomes replicate only once, in the preceding interphase. Mitosis involves one division of the nucleus, and it is usually accompanied by cytokinesis, producing two diploid cells. Meiosis entails two nuclear and cytoplasmic divisions, yielding four haploid cells.

Figure 8.15 compares mitosis and meiosis, tracing these two processes for a diploid parent cell with four chromosomes. Homologous chromosomes are those matching in size.

All the events unique to meiosis occur during meiosis I. In prophase I, duplicated homologous chromosomes pair to form tetrads, and crossing over occurs between homologous (nonsister) chromatids. In metaphase I, tetrads (not individual chromosomes) are aligned at the metaphase plate. During anaphase I, sister chromatids of each chromosome stay together and go to the same pole of the cell as homologous pairs of chromosomes separate. At the end of meiosis I, there are two haploid cells, but each chromosome still has two sister chromatids.

Meiosis II is virtually identical to mitosis and separates sister chromatids. But unlike mitosis, each daughter cell produced by meiosis II has only a *haploid* set of chromosomes.

Web/CD Activity 8E *Meiosis Animation*

? Explain how mitosis conserves chromosome number while meiosis reduces the number from diploid to haploid.

■ In mitosis, a single replication of chromosomes is followed by one division of the cell. In meiosis, homologous chromosomes separate in the first of two cell divisions; after the second division, each new cell ends up with just a single haploid set.

Figure 8.15 Comparison of mitosis and meiosis

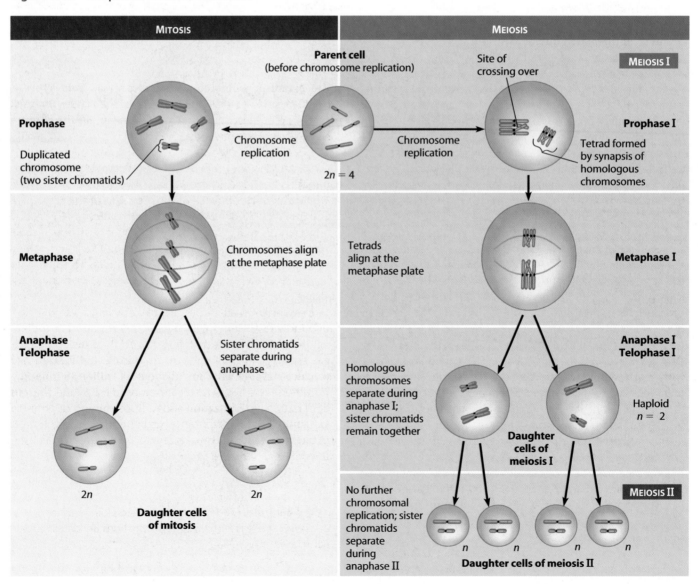

140 **UNIT II** *Cellular Reproduction and Genetics*

8.16 Independent orientation of chromosomes in meiosis and random fertilization lead to varied offspring

As we discussed in Module 8.1, off-spring that result from sexual reproduction are highly varied; they are genetically different from their parents and from one another. How do we account for this genetic variation? As you will learn in later chapters, mutations are the original source of genetic diversity. These changes in an organism's DNA create different versions of genes. Once these differences arise, reshuffling of the different versions during sexual reproduction produces genetic variation. When we discuss natural selection and evolution in Unit III, we will see that this genetic variety in offspring is the raw material for natural selection.

Figure 8.16 illustrates one way in which the process of meiosis contributes to genetic differences in gametes. The figure shows how the arrangement of homologous chromosome pairs at metaphase of meiosis I affects the resulting gametes. Here we use color to distinguish chromosomes that originated from the organism's two parents (red for maternal and blue for paternal). The colors highlight the important fact that each maternal chromosome differs genetically from its paternal homologue. In fact, each chromosome you received from your mother carries many genes that are different versions from those on the homologous chromosome you received from your father.

The orientation of the homologous pairs of chromosomes (tetrads) at metaphase I—whether the maternal or paternal chromosome is closer to a given pole—is as random as the flip of a coin. Thus, there is a 50% chance that a particular daughter cell will get the maternal (red) chromosome of a certain homologous pair and a 50% chance that it will receive the paternal (blue) chromosome. In this example, there are two possible ways that the two tetrads can align during metaphase I. In possibility 1, the tetrads are oriented with both red chromosomes on the same side of the metaphase plate. Therefore, the gametes produced in possibility 1 can each have only red *or* blue chromosomes (bottom row, combinations 1 and 2).

In possibility 2, the tetrads are oriented differently, with one red and one blue chromosome on each side of the metaphase I plate. This arrangement produces gametes that each have one red and one blue chromosome. Furthermore, half the gametes have a big blue chromosome and a small red one (combination 3), and half have a big red one and a small blue one (combination 4).

So we see that for this example, a total of four chromosome combinations is possible in the gametes, and in fact

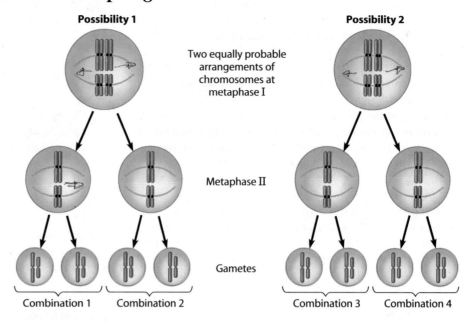

Figure 8.16 Results of the independent orientation of chromosomes at metaphase I

the organism will produce gametes of all four types in roughly equal quantities. For a species with more than two pairs of chromosomes, such as the human, *all* the chromosome pairs orient independently at metaphase I. (Chromosomes X and Y behave as a homologous pair in meiosis.)

For any species, the total number of combinations of chromosomes that meiosis can package into gametes is 2^n, where n is the haploid number. For the organism in this figure, $n = 2$, so the number of chromosome combinations is 2^2, or 4. For a human ($n = 23$), there are 2^{23}, or about 8 million possible chromosome combinations! This means that each gamete you produce contains one of roughly 8 million possible combinations of chromosomes inherited from your mother and father.

How many possibilities are there when a gamete from one individual unites with a gamete from another individual in fertilization? In humans, the random fusion of a single sperm with a single ovum during fertilization will produce a zygote with any of about 64 trillion (8 million × 8 million) combinations of chromosomes! While the random nature of fertilization adds a huge amount of potential variability to the offspring of sexual reproduction, there is in fact even more variety created during meiosis, as we see in the next two modules.

? A particular species of worm has a diploid number of 10. How many chromosomal combinations are possible for gametes formed by meiosis?

■ 32; 2n = 10, so n = 5 and 2ⁿ = 32

8.17 Homologous chromosomes carry different versions of genes

So far, we have focused on genetic variability in gametes and zygotes at the whole-chromosome level. We have yet to discuss the actual genetic information—the genes—contained in the chromosomes of gametes and zygotes. The question we need to answer now is this: What is the significance of the independent orientation of metaphase chromosomes at the level of genes?

Let's take a simple example, the single tetrad in Figure 8.17A. The letters on the homologous chromosomes represent genes. Recall that homologous chromosomes have genes for the same characteristic at corresponding loci. Our example involves hypothetical genes controlling the appearance of mice. *C* and *c* are different versions of a gene for one characteristic, coat color; *E* and *e* are different versions of a gene for another characteristic, eye color. (As you'll learn in later chapters, different versions of a gene contain slightly different nucleotide sequences in the chromosomal DNA.)

Let's say that *C* represents the gene for the brownish color of the mouse on the left in Figure 8.17B and that *c* represents

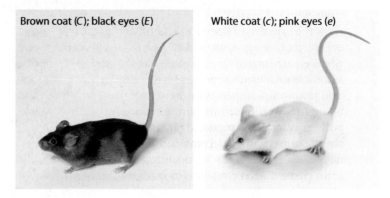

Brown coat (*C*); black eyes (*E*)   White coat (*c*); pink eyes (*e*)

Figure 8.17B   Coat-color and eye-color traits in mice

the gene for white coat color (mouse on right). In the chromosome diagram, notice that *C* is at the same locus on the red homologue as *c* is on the blue one. Likewise, gene *E* (for black eyes) is at the same locus as *e* (pink eyes).

The fact that homologous chromosomes can bear two different kinds of genetic information for the same characteristic (for instance, coat color) is what really makes gametes—and therefore offspring—different from one another. In our example, a gamete carrying a red chromosome would have genes specifying brownish coat color and black eye color, while a gamete with the homologous blue chromosome would have genes for white coat and pink eyes. Thus, we see how a tetrad with genes shown for only two characteristics can yield two genetically different kinds of gametes. In the next module, we go a step further and see how this same tetrad can actually yield *four* different kinds of gametes.

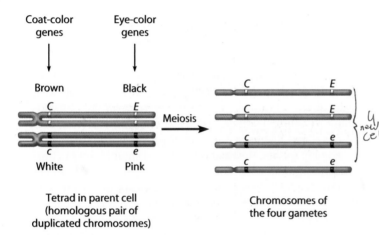

Coat-color genes   Eye-color genes

Brown   Black

Meiosis

White   Pink

Tetrad in parent cell
(homologous pair of duplicated chromosomes)

Chromosomes of the four gametes

Figure 8.17A   Differing genetic information on homologous chromosomes

?   In the tetrad of Figure 8.17A, use labels to distinguish the homologous pair of chromosomes from sister chromatids.

Sister chromatids
Homologous pair of chromosomes
Sister chromatids

8.18 Crossing over further increases genetic variability

Crossing over is an exchange of corresponding segments between two homologous chromosomes. The micrograph and drawing in Figure 8.18A show the results of crossing over between two homologous chromosomes during prophase I of meiosis. The chromosomes are a tetrad—four chromatids, with each pair of sister chromatids joined at their centromeres. The sites of crossing over appear as X-shaped regions; each is called a **chiasma** (Greek for "cross"). A chiasma (plural, *chiasmata*) is a place where two homologous (nonsister) chromatids are attached to each other. Figure 8.18B illustrates how crossing over can produce new combinations of genes, using as examples the hypothetical mouse genes mentioned in the previous module.

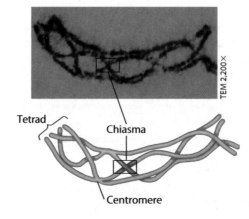

TEM 2,200×

Tetrad

Chiasma

Centromere

Figure 8.18A
Chiasmata

Crossing over begins very early in prophase I of meiosis. At that time, homologous chromosomes are paired along their lengths. Each gene on one homologue is aligned precisely with the corresponding gene on the other homologue.

At the top of the figure is a tetrad with coat-color (*C, c*) and eye-color (*E, e*) genes labeled. ❶ The DNA molecules of two nonsister chromatids—one maternal (red) and one paternal (blue)—break at the same place. ❷ Immediately, the two broken chromatids join together in a new way (red to blue and blue to red). In effect, the two homologous segments trade places, or cross over, producing hybrid (red/blue and blue/red) chromosomes with new combinations of maternal and paternal genes. ❸ When the homologous chromosomes separate in anaphase I, each contains a new segment originating from its homologue. ❹ Finally, in meiosis II, the sister chromatids separate, each going to a different gamete.

In this example, if there were no crossing over, meiosis could produce only two genetic types of gametes. These would be the ones ending up with the "parental" types of chromosomes, carrying either genes *C* and *E* or genes *c* and *e*. These are the same two kinds of gametes we saw in Figure 8.17A. With crossing over, two other types of gametes can result. One of these carries genes *C* and *e,* and the other carries genes *c* and *E*. Chromosomes with these combinations of genes would not exist if not for crossing over; thus, they are called "recombinant" because they result from **genetic recombination**, the production of gene combinations different from those carried by the original chromosomes.

In meiosis in humans, an average of one to three crossover events occur per chromosome pair. Thus, if you were to examine a chromosome from one of your gametes, you would most likely find that it is not exactly like any one of your own chromosomes. Rather, it is probably a patchwork of segments derived from a pair of homologous chromosomes, cut and pasted together to form a chromosome with a unique combination of genes.

We have now examined three sources of genetic variability in sexually reproducing organisms: crossing over during prophase I of meiosis, independent orientation of chromosomes at metaphase I, and random fertilization. When we take up molecular genetics in Chapter 10, we will see yet another source of variability—mutations, or rare changes in the DNA of genes. The different versions of genes that homologous chromosomes may have at each locus originally arise from mutations, so mutations are ultimately responsible for genetic diversity in living organisms.

Our discussion of meiosis to this point has focused on the process as it normally and "correctly" occurs. In the next, and last, major section of the chapter, we consider some of the consequences of errors in the process.

Web/CD Activity 8F *Origins of Genetic Variation*

Web/CD Thinking as a Scientist *How Is Crossing Over Measured in the Fungus* **Sordaria?**

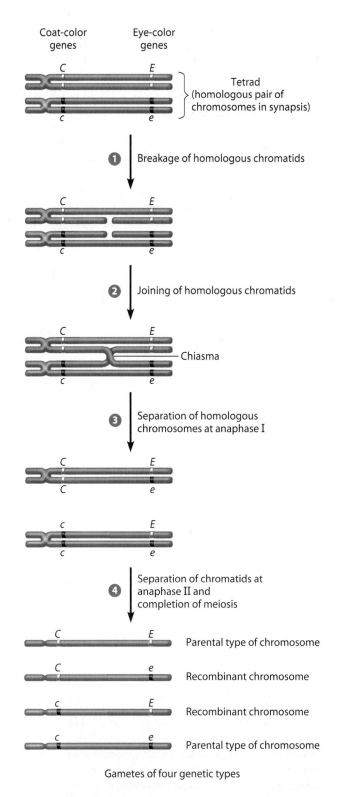

Figure 8.18B How crossing over leads to genetic recombination

⬛ Describe how the two processes that you have learned about over the last several modules account for the genetic variation among gametes formed by meiosis.

❓ (1) Crossing over creates recombinant chromosomes having a combination of genes that were originally on different, though homologous, chromosomes. (2) Homologous chromosome pairs are oriented randomly at metaphase of meiosis I (see Figure 8.16).

8.19 A karyotype is a photographic inventory of an individual's chromosomes

Errors in meiosis can lead to gametes containing chromosomes in abnormal numbers or with major alterations in their structures. Fertilization involving these sorts of abnormal gametes results in offspring with chromosomal abnormalities. Such conditions can be readily detected in a **karyotype**, an ordered display of magnified images of an individual's chromosomes arranged in pairs, starting with the longest. A karyotype shows the chromosomes condensed and doubled, as they appear in metaphase of mitosis.

To prepare a karyotype, medical scientists often use lymphocytes, a type of white blood cell. A blood sample is treated with a chemical that stimulates mitosis. After growing in culture for several days, the cells are treated with another chemical to arrest mitosis at metaphase, when the chromosomes, each consisting of two joined sister chromatids, are most highly condensed. Figure 8.19 outlines the steps in one method for the preparation of a karyotype from a blood sample.

The photograph in step 5 shows the karyotype of a normal human male. The 46 chromosomes of a single diploid cell are arranged in 23 homologous pairs: autosomes numbered from 1 to 22 and one pair of sex chromosomes (X and Y in this case). The chromosomes have been stained to reveal band patterns, which are helpful in differentiating the chromosomes and in detecting structural abnormalities. Among the alterations detected by karyotyping is trisomy 21, the basis of Down syndrome, which we discuss next.

? How would the karyotype of a human female differ from the male karyotype in Figure 8.19?

■ Instead of an XY combination for the sex chromosomes, there would be a homologous pair of X chromosomes (XX).

Figure 8.19 Preparation of a karyotype from a blood sample

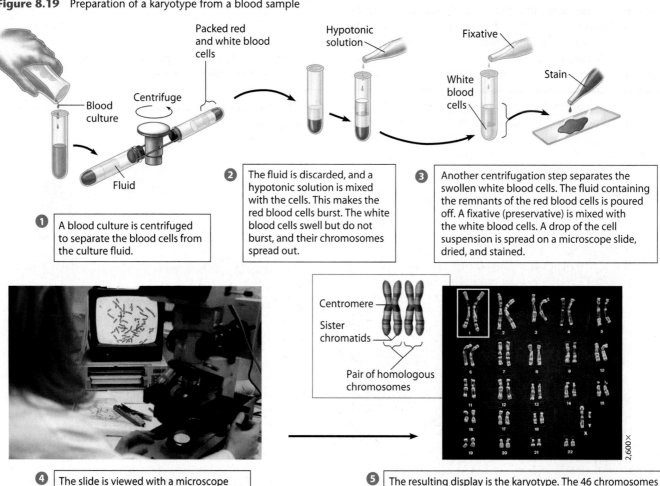

Packed red and white blood cells

Hypotonic solution

Fixative

Blood culture

Centrifuge

White blood cells

Stain

Fluid

❶ A blood culture is centrifuged to separate the blood cells from the culture fluid.

❷ The fluid is discarded, and a hypotonic solution is mixed with the cells. This makes the red blood cells burst. The white blood cells swell but do not burst, and their chromosomes spread out.

❸ Another centrifugation step separates the swollen white blood cells. The fluid containing the remnants of the red blood cells is poured off. A fixative (preservative) is mixed with the white blood cells. A drop of the cell suspension is spread on a microscope slide, dried, and stained.

Centromere

Sister chromatids

Pair of homologous chromosomes

❹ The slide is viewed with a microscope equipped with a digital camera. A photograph of the chromosomes is entered into a computer, which electronically arranges them by size and shape.

❺ The resulting display is the karyotype. The 46 chromosomes here include 22 pairs of autosomes and 2 sex chromosomes, X and Y. Although difficult to discern in the karyotype, each of the chromosomes consists of two sister chromatids lying very close together (see diagram).

8.20 An extra copy of chromosome 21 causes Down syndrome

The karyotype in Figure 8.19 shows the normal human complement of 23 pairs of chromosomes. The karyotype in **Figure 8.20A** is different; notice that there are three number 21 chromosomes, making 47 chromosomes in total. This condition is called **trisomy 21.**

In most cases, a human embryo with an abnormal number of chromosomes is spontaneously aborted (miscarried) long before birth. But some aberrations in chromosome number, including trisomy 21, appear to upset the genetic balance less drastically, and individuals carrying them can survive. These individuals have a characteristic set of symptoms, called a syndrome. A person with an extra copy of chromosome 21, for instance, has a condition called **Down syndrome** (named after John Langdon Down, who characterized the syndrome in 1866).

Trisomy 21 is the most common chromosome number abnormality. Affecting about one out of every 700 children born, it is the most common serious birth defect in the United States. Chromosome 21 is one of our smallest chromosomes, but an extra copy produces a number of effects. Down syndrome (**Figure 8.20B**) includes characteristic facial features, notably a round face, a skin fold at the inner corner of the eye, a flattened nose bridge, and small, irregular teeth, as well as short stature, heart defects, and susceptibility to respiratory infections, leukemia, and Alzheimer's disease.

People with Down syndrome usually have a life span shorter than normal. They also exhibit varying degrees of mental retardation. However, some individuals with the syndrome live to middle age or beyond, and many are socially adept and able to hold jobs. A few women with Down syndrome have had children, though most people with the syndrome are sexually underdeveloped and sterile. Half the eggs produced by a woman with Down syndrome will have the extra chromosome 21, so there is a 50% chance that she will transmit the syndrome to her child.

Figure 8.20B A child with Down syndrome

As indicated in **Figure 8.20C**, the incidence of Down syndrome in the offspring of normal parents increases markedly with the age of the mother. Down syndrome strikes less than 0.05% of children (fewer than one in 2,000) born to women under age 30. The risk climbs to 1.25% for mothers in their early 30s and is even higher for older mothers. Because of this relatively high risk, pregnant women over 35 are candidates for fetal testing for trisomy 21 and other chromosomal abnormalities (see Module 9.10).

What causes trisomy 21? We address that question in the next module.

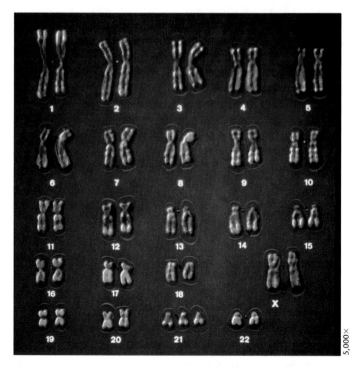

5,000×

Figure 8.20A A karyotype for trisomy 21 (Down syndrome)

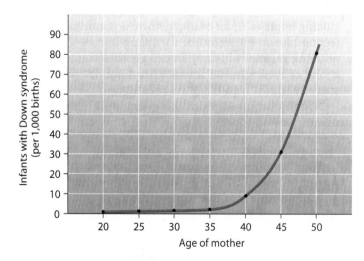

Figure 8.20C Maternal age and incidence of Down syndrome

? For mothers of age 47, the risk of having a baby with Down syndrome is about __ per thousand births, or _____ %.

■ 40 4

8.20 An extra copy of chromosome 21 causes Down syndrome

The karyotype in Figure 8.19 shows the normal human complement of 23 pairs of chromosomes. The karyotype in **Figure 8.20A** is different; notice that there are three number 21 chromosomes, making 47 chromosomes in total. This condition is called **trisomy 21.**

In most cases, a human embryo with an abnormal number of chromosomes is spontaneously aborted (miscarried) long before birth. But some aberrations in chromosome number, including trisomy 21, appear to upset the genetic balance less drastically, and individuals carrying them can survive. These individuals have a characteristic set of symptoms, called a syndrome. A person with an extra copy of chromosome 21, for instance, has a condition called **Down syndrome** (named after John Langdon Down, who characterized the syndrome in 1866).

Trisomy 21 is the most common chromosome number abnormality. Affecting about one out of every 700 children born, it is the most common serious birth defect in the United States. Chromosome 21 is one of our smallest chromosomes, but an extra copy produces a number of effects. Down syndrome (**Figure 8.20B**) includes characteristic facial features, notably a round face, a skin fold at the inner corner of the eye, a flattened nose bridge, and small, irregular teeth, as well as short stature, heart defects, and susceptibility to respiratory infections, leukemia, and Alzheimer's disease.

People with Down syndrome usually have a life span shorter than normal. They also exhibit varying degrees of mental retardation. However, some individuals with the syndrome live to middle age or beyond, and many are socially adept and able to hold jobs. A few women with Down syndrome have had children, though most people with the syndrome are sexually underdeveloped and sterile. Half the eggs produced by a woman with Down syndrome will have the extra chromosome 21, so there is a 50% chance that she will transmit the syndrome to her child.

Figure 8.20B A child with Down syndrome

As indicated in **Figure 8.20C**, the incidence of Down syndrome in the offspring of normal parents increases markedly with the age of the mother. Down syndrome strikes less than 0.05% of children (fewer than one in 2,000) born to women under age 30. The risk climbs to 1.25% for mothers in their early 30s and is even higher for older mothers. Because of this relatively high risk, pregnant women over 35 are candidates for fetal testing for trisomy 21 and other chromosomal abnormalities (see Module 9.10).

What causes trisomy 21? We address that question in the next module.

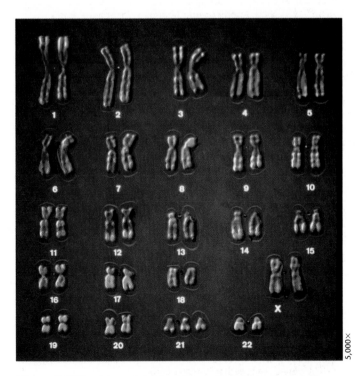

Figure 8.20A A karyotype for trisomy 21 (Down syndrome)

Figure 8.20C Maternal age and incidence of Down syndrome

? For mothers of age 47, the risk of having a baby with Down syndrome is about __ per thousand births, or _____ %.

■ 40 . . . 4

8.21 Accidents during meiosis can alter chromosome number

Meiosis occurs repeatedly in a human lifetime as the testes or ovaries produce gametes. Almost always, the meiotic spindle distributes chromosomes to daughter cells without error. But there is an occasional mishap, called a **nondisjunction**, in which the members of a chromosome pair fail to separate. **Figures 8.21A** and **8.21B** illustrate two ways that nondisjunction can occur. For simplicity, we use a hypothetical organism whose diploid chromosome number is 4. In both figures, the cell at the top is diploid ($2n = 4$), with two pairs of homologous chromosomes undergoing anaphase of meiosis I.

Sometimes, as in Figure 8.21A, a pair of homologous chromosomes does not separate during meiosis I. In this case, even though the rest of meiosis occurs normally, all the resulting gametes end up with abnormal numbers of chromosomes. Two of the gametes have three chromosomes; the other two gametes have only one chromosome each. In Figure 8.21B, meiosis I is normal, but one pair of sister chromatids fails to move apart during meiosis II. In this case, two of the resulting gametes are abnormal; the other two gametes are normal.

Figure 8.21C shows an example of what can happen when an abnormal gamete produced by nondisjunction unites with a normal gamete in fertilization. Here, an egg cell with two copies of one of its chromosomes (a total of $n + 1$ chromosomes) is fertilized by a normal sperm cell (n). The resulting zygote has an extra chromosome (a total of $2n + 1$ chromosomes). Mitosis will then transmit the anomaly to all embryonic cells. If this were a real organism and it survived, it would have an abnormal karyotype and probably a syndrome of disorders caused by the abnormal number of genes.

Nondisjunction can lead to an abnormal chromosome number in any sexually reproducing diploid organism, including humans. If, for example, there is nondisjunction affecting human chromosome 21 during meiosis I, half the resulting gametes will carry an extra chromosome 21. Then if one of these gametes unites with a normal gamete, trisomy 21 (Down syndrome) will result.

Nondisjunction explains how abnormal chromosome numbers come about, but what causes nondisjunction in the first place? We do not yet know the answer, nor do we fully understand why offspring with trisomy 21 are more likely to be born as a woman ages. We do know, however, that meiosis begins in a woman's ovaries before she is born but is not completed until years later, at the time of an ovulation (see Module 27.4). Because only one egg cell usually matures each month, a cell might remain arrested in the mid-meiosis state for decades. Some research points to an age-dependent error in one of the checkpoints that coordinates the process of meiosis. Nondisjunction can also affect chromosomes other than 21, as we see next.

? Explain how nondisjunction could result in a diploid gamete.

■ A diploid gamete would result if the nondisjunction affected all the chromosomes during one of the meiotic divisions.

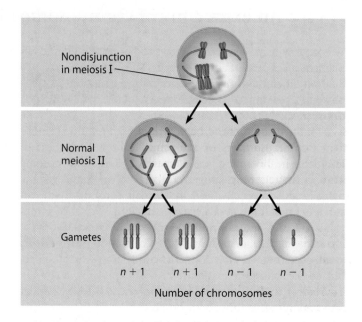

Figure 8.21A Nondisjunction in meiosis I

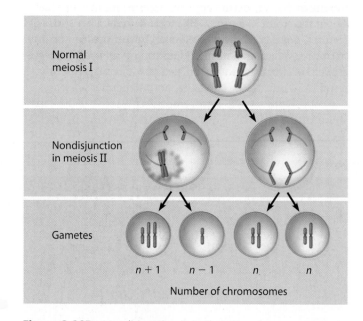

Figure 8.21B Nondisjunction in meiosis II

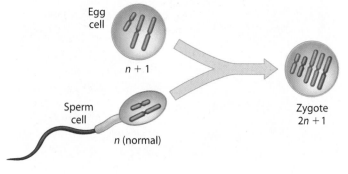

Figure 8.21C Fertilization after nondisjunction in the mother

8.22 Abnormal numbers of sex chromosomes do not usually affect survival

Nondisjunction in meiosis does not affect just autosomes, such as chromosome 21. It can also lead to abnormal numbers of sex chromosomes, X and Y. Unusual numbers of sex chromosomes seem to upset the genetic balance less than unusual numbers of autosomes. This may be because the Y chromosome carries relatively few genes. A peculiarity of X chromosomes in humans and other mammals also helps an individual tolerate unusual numbers of X chromosomes: In mammals, the cells usually operate with only one functioning X chromosome because extra copies of the chromosome become inactivated in each cell (see Module 11.7).

TABLE 8.22	ABNORMALITIES OF SEX CHROMOSOME NUMBER IN HUMANS		
Sex Chromosomes	Syndrome	Origin of Nondisjunction	Frequency in Population
XXY	Klinefelter syndrome (male)	Meiosis in egg or sperm formation	$\frac{1}{2,000}$
XYY	None (normal male)	Meiosis in sperm formation	$\frac{1}{2,000}$
XXX	None (normal female)	Meiosis in egg or sperm formation	$\frac{1}{1,000}$
XO	Turner syndrome (female)	Meiosis in egg or sperm formation	$\frac{1}{5,000}$

Table 8.22 lists the most common human sex chromosome abnormalities. An extra X chromosome in a male, making him XXY, occurs approximately once in every 2,000 live births (once in every 1,000 male births). Men with this disorder, called *Klinefelter syndrome,* have male sex organs, but the testes are abnormally small and the individual is sterile. The syndrome often includes breast enlargement and other female body characteristics (Figure 8.22A). The person is usually of normal intelligence. Klinefelter syndrome is also found in individuals with more than one additional sex chromosome, such as XXYY, XXXY, or XXXXY. These abnormal numbers of sex chromosomes probably result from multiple nondisjunctions; such men are more likely to be mentally retarded than XY or XXY individuals.

Human males with an extra Y chromosome (XYY) do not have any well-defined syndrome, although they tend to be taller than average. Females with an extra X chromosome (XXX) cannot be distinguished from XX females except by karyotype.

Females who are lacking an X chromosome are designated XO; the O indicates the absence of a second sex chromosome. These women have *Turner syndrome.* They have a characteristic appearance, including short stature and often a web of skin extending between the neck and the shoulders (Figure 8.22B). Women with Turner syndrome are sterile because their sex organs do not fully mature at adolescence. If left untreated, girls with Turner syndrome have poor development of breasts and other secondary sexual characteristics. Artificial administration of estrogen can alleviate these symptoms. Girls with Turner syndrome have normal intelligence. The XO condition is the sole known case where having only 45 chromosomes is not fatal in humans.

The sex chromosome abnormalities described here illustrate the crucial role of the Y chromosome in determining sex. In general, a single Y chromosome is enough to produce "maleness," even in combination with several X chromosomes. The absence of a Y chromosome yields "femaleness."

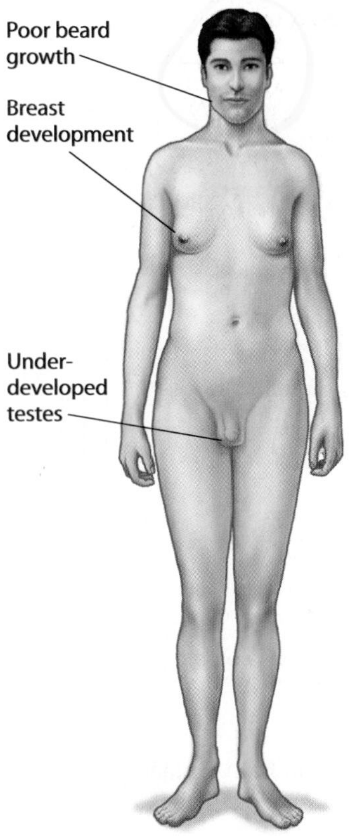

Poor beard growth
Breast development
Underdeveloped testes

Characteristic facial features
Web of skin
Constriction of aorta
Poor breast development
Underdeveloped ovaries

Figure 8.22A A man with Klinefelter syndrome (XXY)

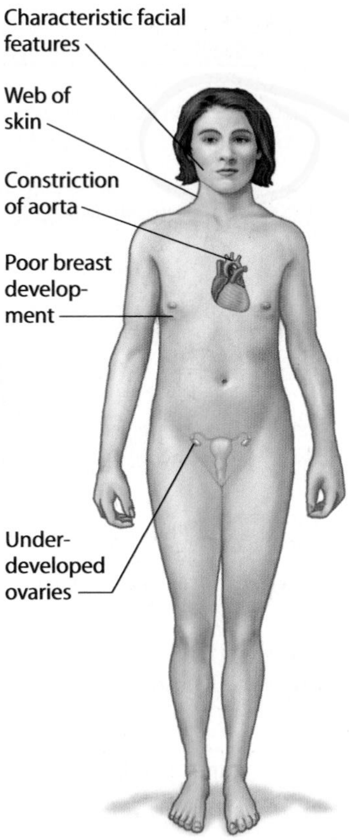

Figure 8.22B A woman with Turner syndrome (XO)

? What is the *total* number of chromosomes you would expect to find in the karyotype of a female with Turner syndrome?

8.23 Alterations of chromosome structure can cause birth defects and cancer

Even if all chromosomes are present in normal numbers, abnormalities in chromosome structure may cause disorders. Breakage of a chromosome can lead to a variety of rearrangements affecting the genes of that chromosome. Figure 8.23A shows three types of rearrangement. (The pink arrows indicate chromosome breaks.) If a fragment of a chromosome is lost, the remaining chromosome will then have a **deletion**. If a fragment from one chromosome joins to a sister chromatid or homologous chromosome, it will produce a **duplication**. If a fragment reattaches to the original chromosome but in the reverse direction, an **inversion** results.

Inversions are less likely than deletions or duplications to produce harmful effects, because in inversions all genes are still present in their normal number. Many deletions in human chromsomes, however, cause serious physical and mental problems. One example is a specific deletion in chromosome 5 that causes *cri du chat* ("cry of the cat") syndrome. A child born with this syndrome is mentally retarded, has a small head with unusual facial features, and has a cry that sounds like the mewing of a distressed cat. Such individuals usually die in infancy or early childhood.

Another type of chromosomal change is chromosomal **translocation**, the attachment of a chromosomal fragment to a nonhomologous chromosome. Figure 8.23B shows a translocation that is reciprocal; that is, two nonhomologous chromosomes exchange segments. Like inversions, translocations may or may not be harmful. Some people with Down syndrome have only part of a third chromosome 21; as the result of a translocation, it is attached to another (nonhomologous) chromosome.

Whereas chromosomal changes present in sperm or egg can cause congenital disorders, such changes in a somatic cell may contribute to the development of cancer. For example, a chromosomal translocation in somatic cells in the bone marrow is associated with *chronic myelogenous leukemia* (*CML*). CML is one of the most common of the leukemias, the cancers affecting cells that give rise to white blood cells (leukocytes). In the cancerous cells of most CML patients, a part of chromosome 22 has switched places with a small fragment from a tip of chromosome 9 (Figure 8.23C). This reciprocal translocation activates a gene that leads to leukemia. The chromosome ending up with the activated cancer-causing gene is called the "Philadelphia chromosome," after the city where it was discovered.

Because the chromosomal changes in cancer are usually confined to somatic cells, cancer is not usually inherited. We'll return to cancer in Chapter 11. In the next chapter, Chapter 9, we continue our study of genetic principles, looking first at the historical development of the science of genetics and then at the rules governing the way traits are passed from parents to offspring.

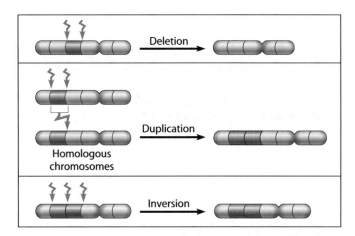

Figure 8.23A Alterations of chromosome structure involving one chromosome or a homologous pair

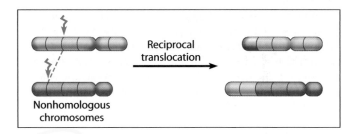

Figure 8.23B Chromosomal translocation between nonhomologous chromosomes

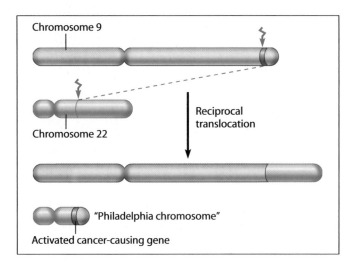

Figure 8.23C The translocation associated with chronic myelogenous leukemia

? How is reciprocal translocation different from normal crossing over?

■ Reciprocal translocation swaps chromosome segments between nonhomologous chromosomes. Crossing over normally exchanges corresponding segments between homologous chromosomes.

Reviewing the Concepts

Connections Between Cell Division and Reproduction (Introduction–8.3)

Cell division is at the heart of the reproduction of cells and organisms, because cells come only from preexisting cells. Some organisms reproduce asexually, and their offspring are all genetic copies of the parent and of each other. Others reproduce sexually, creating a variety of offspring, each with a unique combination of traits **(Introduction–8.2).** Prokaryotic cells reproduce asexually by cell division. As the cell replicates its single chromosome, the copies move apart; the growing membrane then divides the cell **(8.3).**

The Eukaryotic Cell Cycle and Mitosis (8.4–8.11)

Chromosomes. A eukaryotic cell has many more genes than a prokaryotic cell, and they are grouped into multiple chromosomes in the nucleus. Each chromosome contains a very long DNA molecule associated with proteins. Individual chromosomes are visible only when the cell is in the process of dividing; otherwise, they are in the form of thin, loosely packed chromatin fibers. Before a cell starts dividing, the chromosomes replicate, producing sister chromatids (containing identical DNA) joined together at the centromere. Cell division involves the separation of sister chromatids and results in two daughter cells, each containing a complete and identical set of chromosomes **(8.4).**

The eukaryotic cell cycle (8.5):

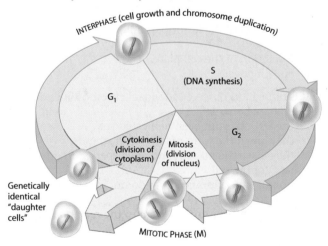

Mitosis distributes duplicated chromosomes into two daughter nuclei. After the chromosomes coil up, a mitotic spindle made of microtubules moves them to the middle of the cell. The sister chromatids then separate and move to opposite poles of the cell, where two new nuclei form. Cytokinesis, in which the cell divides in two, overlaps the end of mitosis **(8.6).** In animals, cytokinesis occurs by a constriction of the cell (cleaveage). In plants, a membranous cell plate splits the cell in two **(8.7).**

Cell cycle control. Most animal cells divide only when stimulated, and some not at all. In laboratory cultures, most normal cells divide only when attached to a surface. They continue dividing until they touch one another. Growth factors are proteins secreted by cells that stimulate other cells to divide **(8.8).** A set of proteins within the cell controls the cell cycle. Signals affecting critical checkpoints in the cell cycle determine whether a cell will go through the complete cycle and divide. The binding of growth factors to specific receptors on the plasma membrane is usually necessary for cell division **(8.9).**

Cancer cells divide excessively to form masses called tumors. Malignant tumors can invade other tissues. Radiation and chemotherapy are effective as cancer treatments because they interfere with cell division **(8.10).** When the cell cycle operates normally, mitotic cell division functions in growth, replacement of damaged and lost cells, and asexual reproduction **(8.11).**

Meiosis and Crossing Over (8.12–8.18)

Homologous chromosomes. The somatic (body) cells of each species contain a specific number of chromosomes; for example, human cells have 46, making up 23 pairs (two sets) of homologous chromosomes. The chromosomes of a homologous pair carry genes for the same characteristics at the same place, or locus **(8.12).** Cells with two sets of homologous chromosomes are said to be diploid. Gametes—eggs and sperm—are haploid cells with a single set of chromosomes. Sexual life cycles involve the alternation of haploid and diploid stages **(8.13):**

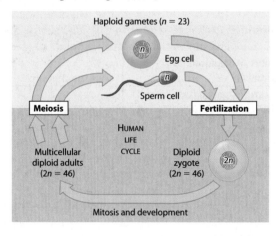

Meiosis, like mitosis, is preceded by chromosome duplication, but in meiosis the cell divides twice to form four daughter cells. The first division, meiosis I, starts with synapsis, the pairing of homologous chromosomes. In crossing over, homologous chromosomes exchange corresponding segments. Meiosis I separates each homologous pair and produces two daughter cells, each with one set of chromosomes. Meiosis II is essentially the same as mitosis: In each of the cells, the sister chromatids of each chromosome separate; the result is a total of four haploid cells **(8.14).** Figure 8.15 reviews and compares mitosis and meiosis **(8.15).**

Genetic variation. Each chromosome of a homologous pair differs at many points from the other member of the pair. Random arrangements of chromosome pairs at

metaphase I of meiosis leads to many different combinations of chromosomes in eggs and sperm. Random fertilization of eggs by sperm greatly increases this variation **(8.16)**. The differences between homologous chromosomes are based on the fact that they can bear different versions of a gene at corresponding loci **(8.17)**. Genetic recombination, which results from crossing over during prophase I of meiosis, increases variation still further **(8.18)**.

Alterations of Chromosome Number and Structure (8.19–8.23)

Chromosome number.   A karyotype is an ordered arrangement of images of a cell's chromosomes **(8.19)**. A person may have an abnormal number of chromosomes, which causes problems. Down syndrome is caused by trisomy 21, an extra copy of chromosome 21 **(8.20)**. An abnormal chromosome count is the result of nondisjunction, the failure of a homologous pair of chromosomes to separate during meiosis I or of sister chromatids to separate during meiosis II **(8.21)**. Nondisjunction can also produce gametes with extra or missing sex chromosomes, leading to varying degrees of malfunction in humans but not usually affecting survival **(8.22)**. Chromosome breakage can lead to rearrangements—deletions, duplications, inversions, and translocations—that can produce genetic disorders or, if the changes occur in somatic cells, cancer **(8.23)**.

Connecting the Concepts

1. Complete the following table to compare mitosis and meiosis.

	Mitosis	Meiosis
Number of chromosomal duplications		
Number of cell divisions		
Number of daughter cells produced		
Number of chromosomes in daughter cells		
How chromosomes line up during metaphase		
Genetic relationship of daughter cells to parent cell		
Functions performed in the human body		

Testing Your Knowledge

Multiple Choice

2. If an intestinal cell in a grasshopper contains 24 chromosomes, a grasshopper sperm cell contains ____ chromosomes.
   a. 3   d. 24
   b. 6   e. 48
   c. 12

3. Which of the following phases of mitosis is essentially the opposite of prophase in terms of nuclear changes?
   a. telophase   d. interphase
   b. metaphase   e. anaphase
   c. S phase

4. A biochemist measured the amount of DNA in cells growing in the laboratory and found that the quantity of DNA in a cell doubled
   a. between prophase and anaphase of mitosis.
   b. between the G_1 and G_2 phases of the cell cycle.
   c. during the M phase of the cell cycle.
   d. between prophase I and prophase II of meiosis.
   e. between anaphase and telophase of mitosis.

5. Which of the following is *not* a function of mitosis in humans?
   a. repair of wounds
   b. growth
   c. production of gametes from diploid cells
   d. replacement of lost or damaged cells
   e. multiplication of somatic cells

6. A micrograph of a dividing cell from a mouse showed 19 chromosomes, each consisting of two sister chromatids. During which of the following stages of cell division could this picture have been taken? (*Explain your answer.*)
   a. prophase of mitosis
   b. telophase II of meiosis
   c. prophase I of meiosis
   d. anaphase of mitosis
   e. prophase II of meiosis

7. Cytochalasin B is a chemical that disrupts microfilament formation. This chemical would interfere with
   a. DNA replication.
   b. formation of the mitotic spindle.
   c. cleavage.
   d. formation of the cell plate.
   e. crossing over.

8. It is difficult to observe individual chromosomes during interphase because
   a. the DNA has not been replicated yet.
   b. they are in the form of long, thin strands.
   c. they leave the nucleus and are dispersed to other parts of the cell.
   d. homologous chromosomes do not pair up until division starts.
   e. the spindle must move them to the metaphase plate before they become visible.

9. A fruit fly somatic cell contains 8 chromosomes. This means that ____ different combinations of chromosomes are possible in its gametes.
   a. 4   d. 32
   b. 8   e. 64
   c. 16

10. If a fragment of a chromosome breaks off and then reattaches to the original chromosome but in the reverse direction, the resulting chromosomal abnormality is called
   a. a deletion.   d. a nondisjunction.
   b. an inversion.   e. a reciprocal translocation.
   c. a translocation.

metaphase I of meiosis leads to many different combinations of chromosomes in eggs and sperm. Random fertilization of eggs by sperm greatly increases this variation **(8.16)**. The differences between homologous chromosomes are based on the fact that they can bear different versions of a gene at corresponding loci **(8.17)**. Genetic recombination, which results from crossing over during prophase I of meiosis, increases variation still further **(8.18)**.

Alterations of Chromosome Number and
Structure (8.19–8.23)
Chromosome number.   A karyotype is an ordered arrangement of images of a cell's chromosomes **(8.19)**. A person may have an abnormal number of chromosomes, which causes problems. Down syndrome is caused by trisomy 21, an extra copy of chromosome 21 **(8.20)**. An abnormal chromosome count is the result of nondisjunction, the failure of a homologous pair of chromosomes to separate during meiosis I or of sister chromatids to separate during meiosis II **(8.21)**. Nondisjunction can also produce gametes with extra or missing sex chromosomes, leading to varying degrees of malfunction in humans but not usually affecting survival **(8.22)**. Chromosome breakage can lead to rearrangements—deletions, duplications, inversions, and translocations—that can produce genetic disorders or, if the changes occur in somatic cells, cancer **(8.23)**.

Connecting the Concepts

1. Complete the following table to compare mitosis and meiosis.

	Mitosis	Meiosis
Number of chromosomal duplications		
Number of cell divisions		
Number of daughter cells produced		
Number of chromosomes in daughter cells		
How chromosomes line up during metaphase		
Genetic relationship of daughter cells to parent cell		
Functions performed in the human body		

Testing Your Knowledge

Multiple Choice

2. If an intestinal cell in a grasshopper contains 24 chromosomes, a grasshopper sperm cell contains ____ chromosomes.
   a. 3   d. 24
   b. 6   e. 48
   c. 12

3. Which of the following phases of mitosis is essentially the opposite of prophase in terms of nuclear changes?
   a. telophase   d. interphase
   b. metaphase   e. anaphase
   c. S phase

4. A biochemist measured the amount of DNA in cells growing in the laboratory and found that the quantity of DNA in a cell doubled
   a. between prophase and anaphase of mitosis.
   b. between the G_1 and G_2 phases of the cell cycle.
   c. during the M phase of the cell cycle.
   d. between prophase I and prophase II of meiosis.
   e. between anaphase and telophase of mitosis.

5. Which of the following is *not* a function of mitosis in humans?
   a. repair of wounds
   b. growth
   c. production of gametes from diploid cells
   d. replacement of lost or damaged cells
   e. multiplication of somatic cells

6. A micrograph of a dividing cell from a mouse showed 19 chromosomes, each consisting of two sister chromatids. During which of the following stages of cell division could this picture have been taken? (*Explain your answer.*)
   a. prophase of mitosis
   b. telophase II of meiosis
   c. prophase I of meiosis
   d. anaphase of mitosis
   e. prophase II of meiosis

7. Cytochalasin B is a chemical that disrupts microfilament formation. This chemical would interfere with
   a. DNA replication.
   b. formation of the mitotic spindle.
   c. cleavage.
   d. formation of the cell plate.
   e. crossing over.

8. It is difficult to observe individual chromosomes during interphase because
   a. the DNA has not been replicated yet.
   b. they are in the form of long, thin strands.
   c. they leave the nucleus and are dispersed to other parts of the cell.
   d. homologous chromosomes do not pair up until division starts.
   e. the spindle must move them to the metaphase plate before they become visible.

9. A fruit fly somatic cell contains 8 chromosomes. This means that ____ different combinations of chromosomes are possible in its gametes.
   a. 4   d. 32
   b. 8   e. 64
   c. 16

10. If a fragment of a chromosome breaks off and then reattaches to the original chromosome but in the reverse direction, the resulting chromosomal abnormality is called
    a. a deletion.   d. a nondisjunction.
    b. an inversion.   e. a reciprocal translocation.
    c. a translocation.

11. Why are individuals with an extra chromosome 21, which causes Down syndrome, more numerous than individuals with an extra chromosome 3 or chromosome 16?
    a. There are probably more genes on chromosome 21 than on the others.
    b. Chromosome 21 is a sex chromosome and chromosomes 3 and 16 are not.
    c. Down syndrome is not more common, just more serious.
    d. Extra copies of the other chromosomes are probably fatal.
    e. Nondisjunction of chromosome 21 probably occurs more frequently.

Describing, Comparing, and Explaining

12. Briefly describe how three different processes that occur during a sexual life cycle increase the genetic diversity of offspring.

13. In the light micrograph below of dividing cells near the tip of an onion root, identify a cell in interphase, prophase, metaphase, anaphase, and telophase. Describe the major events occurring at each stage.

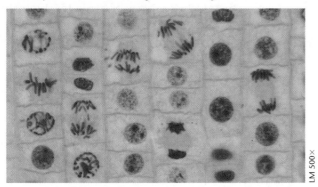

LM 500×

14. Discuss the factors that control the division of eukaryotic cells grown in the laboratory. Cancer cells are easier to grow in the lab than other cells. Why do you suppose this is the case?

15. Compare cytokinesis in plant and animal cells.

16. Sketch a cell with three pairs of chromosomes undergoing meiosis, and show how nondisjunction can result in the production of gametes with extra or missing chromosomes.

Applying the Concepts

17. Suppose you read in the newspaper that a genetic engineering laboratory has developed a procedure for fusing two gametes from the same person (two eggs or two sperm) to form a zygote. The article mentions that an early step in the procedure prevents crossing over from occurring during the formation of the gametes in the donor's body. The researchers are in the process of determining the genetic makeup of one of their new zygotes. Which of the following predictions do you think they would make? Justify your choice, and explain why you rejected each of the other choices.
    a. The zygote would have 46 chromosomes, all of which came from the gamete donor (its one parent), so the zygote would be genetically identical to the gamete donor.
    b. The zygote *could* be genetically identical to the gamete donor, but it is much more likely that it would have an unpredictable mixture of chromosomes from the gamete donor's parents.
    c. The zygote would not be genetically identical to the gamete donor, but it would be genetically identical to one of the donor's parents.
    d. The zygote would not be genetically identical to the gamete donor, but it would be genetically identical to one of the donor's grandparents.

18. Bacteria are able to divide on a much faster schedule than eukaryotic cells. Some bacteria can divide every 20 minutes, while the minimum time required by eukaryotic cells in a rapidly developing embryo is about once per hour, and most cells divide much less often than that. State several testable hypotheses explaining why bacteria can divide at a faster rate than eukaryotic cells.

19. Red blood cells, which carry oxygen to body tissues, live for only about 120 days. Replacement cells are produced by cell division in bone marrow. How many cell divisions must occur each second in your bone marrow just to replace red blood cells? Here is some information to use in calculating your answer: There are about 5 million red blood cells per cubic millimeter (mm^3) of blood. An average adult has about 5 L (5,000 cm^3) of blood. (*Hints:* What is the total number of red blood cells in the body? What fraction of them must be replaced each day if all are replaced in 120 days?)

20. A mule is the offspring of a horse and a donkey. A donkey sperm contains 31 chromosomes and a horse egg cell 32 chromosomes, so the zygote contains a total of 63 chromosomes. The zygote develops normally. The combined set of chromosomes is not a problem in mitosis, and the mule combines some of the best characteristics of horses and donkeys. However, a mule is sterile; meiosis cannot occur normally in its testes (or ovaries). Explain why mitosis is normal in cells containing both horse and donkey chromosomes but the mixed set of chromosomes interferes with meiosis.

21. Every year about a million Americans are diagnosed as having cancer. This means that about 75 million Americans now living will eventually have cancer, and one in five will die of the disease. There are many kinds of cancers and many causes of the disease. For example, smoking causes most lung cancers. Overexposure to ultraviolet rays in sunlight causes most skin cancers. There is evidence that a high-fat, low-fiber diet is a factor in breast, colon, and prostate cancers. And agents in the workplace such as asbestos and vinyl chloride are also implicated as causes of cancer. Hundreds of millions of dollars are spent each year in the search for effective treatments for cancer; far less money is spent preventing cancer. Why might this be the case? What kinds of lifestyle changes could we make to help prevent cancer? What kinds of prevention programs could be initiated or strengthened to encourage these changes? What factors might impede such changes and programs? Should we devote more of our resources to treating cancer or preventing it? Why?

Answers to all questions can be found in Appendix 3.

For study help and Activities, go to campbellbiology.com or the student CD-ROM.

CHAPTER 9

MENDEL'S LAWS

9.1   The science of genetics has ancient roots
9.2   Experimental genetics began in an abbey garden
9.3   Mendel's law of segregation describes the inheritance of a single characteristic
9.4   Homologous chromosomes bear the two alleles for each characteristic
9.5   The law of independent assortment is revealed by tracking two characteristics at once
9.6   Geneticists use the testcross to determine unknown genotypes
9.7   Mendel's laws reflect the rules of probability
9.8   Genetic traits in humans can be tracked through family pedigrees
9.9   Many inherited disorders in humans are controlled by a single gene
9.10 New technologies can provide insight into one's genetic legacy

VARIATIONS ON MENDEL'S LAWS

9.11 The relationship of genotype to phenotype is rarely simple
9.12 Incomplete dominance results in intermediate phenotypes
9.13 Many genes have more than two alleles in the population
9.14 A single gene may affect many phenotypic characteristics
9.15 A single characteristic may be influenced by many genes
9.16 The environment affects many characteristics
9.17 Genetic testing can detect disease-causing alleles

THE CHROMOSOMAL BASIS OF INHERITANCE

9.18 Chromosome behavior accounts for Mendel's laws
9.19 Genes on the same chromosome tend to be inherited together
9.20 Crossing over produces new combinations of alleles
9.21 Geneticists use crossover data to map genes

SEX CHROMOSOMES AND SEX-LINKED GENES

9.22 Chromosomes determine sex in many species
9.23 Sex-linked genes exhibit a unique pattern of inheritance
9.24 Sex-linked disorders affect mostly males

Purebreds and Mutts— A Difference of Heredity

THE BEAGLE PUPPIES SHOWN RUNNING HERE will grow up to look very much like their parents. It would be surprising if they didn't because the parents of this litter are purebred beagles. The parents, grandparents, and great grandparents of these puppies were all beagles with very similar genetic makeups. For the puppies on the facing page, however, the outcome is harder to predict. Their parents were mixtures of several breeds. As a result of this diverse genetic background, the final sizes, markings, and builds of these puppies, as well as their behavioral traits, will be more varied. But whether dogs are purebred or mutts, we can ultimately explain their innate traits by using **genetics**, the science of heredity.

Because their patterns of inheritance are relatively simple, purebred, or true-breeding, individuals are important in genetic research. As you will see in this chapter, Gregor Mendel, the founder of genetics, used purebred pea plants in his research. He interbred true-breeding plants that differed from one another in only one well-defined characteristic, such as the color of their pods. From looking at the offspring plants, he was able to deduce the basic rules of inheritance.

Today, genetic researchers often focus on more complex cases involving less well defined characteristics. And dogs are

Patterns of Inheritance

genome project is yielding a wealth of data that can be used in comparing the DNA of humans and dogs.

Biologists are also studying certain dog breeds to shed light on the relationship between genetic makeup and behavior. Dogs of different breeds tend to have different temperaments. Pit bulls and rottweilers, for instance, are more likely to be aggressive than some other breeds because they were bred as guard dogs: In generation after generation, the most aggressive dogs were selected for mating with each other. In fact, selective breeding is the basis for the many dog breeds that exist today. It's estimated that until about 14,000 years ago, dogs were very much like wolves, the species from which dogs evolved. It was at this time that dogs began to move with people into more permanent settlements, which were often geographically isolated. As a result, different populations of dogs became isolated from one another and eventually became inbred. At the same time, different groups of people selected dogs for different traits, depending on their needs. Herders selected dogs that were good at controlling flocks of animals, producing breeds such as the border collie. Hunters had developed breeds of dogs, such as the Labrador retriever, that were good at retrieving wounded prey. Such dogs were bred to be less aggressive than some other breeds because hunters don't want their dogs to eat the quarry!

> **It's estimated that until about 14,000 years ago, dogs were very much like wolves**

Genetics isn't everything: A dog's level of aggression and other behavioral characteristics are influenced not only by its genes but also by its environment and care. The same is turning out to be true of human behavioral characteristics. Traits such as shyness have some genetic basis but can also be amplified or reduced by upbringing.

In this chapter, we examine the rules that govern how inherited characteristics are passed from parents to offspring. We will look at several kinds of inheritance patterns and see how to predict the ratios of offspring with particular traits. Most importantly, we will uncover a basic biological concept—that the behavior of chromosomes during gamete formation and fertilization, discussed in Chapter 8, accounts for the patterns of inheritance we observe. ■ ■ ■

turning out to be useful for some of this work. Because of inbreeding, purebreds often suffer from serious genetic defects (you'll learn why in this chapter). For instance, a type of hereditary blindness called progressive retinal atrophy (PRA) is common among Labrador retrievers, spaniels, and several other breeds. Studies of such defects may be useful not only for veterinary medicine but also for human medicine because some 60% of the genetic disorders in dogs are similar to human genetic diseases. Examples include skeletal and heart malformations and nervous system defects. Moreover, in 2003, researchers sequenced the complete genome of a dog. This canine

A litter of puppies of mixed ancestry

9.1 The science of genetics has ancient roots

Attempts to explain inheritance date back at least to ancient Greece. Hippocrates, known as the father of medicine, suggested an explanation called pangenesis. According to this idea, particles called pangenes travel from each part of an organism's body to the eggs or sperm and are then passed to the next generation; moreover, changes that occur in various parts of the body during an organism's life are passed on in this way. The Greek philosopher Aristotle rejected this idea as simplistic, saying that what is inherited is the potential to produce body features, rather than particles of the features themselves.

Actually, pangenesis proves incorrect on several counts. The reproductive cells are not composed of particles from somatic (body) cells, and changes in somatic cells do not influence eggs and sperm. For instance, no matter how much you enlarge your biceps by lifting weights, muscle cells in your arms do not transmit genetic information to your gametes, and your offspring will not be changed by your weight-lifting efforts. This may seem like common sense today, but the pangenesis hypothesis and the idea that traits acquired during an individual's lifetime are passed on to offspring prevailed well into the 19th century.

By observing inheritance patterns in ornamental plants, biologists of the early 19th century established that offspring inherit traits from both parents. The favored explanation of inheritance then became the "blending" hypothesis, the idea that the hereditary materials contributed by the male and female parents mix in forming the offspring in the way that blue and yellow paints blend to make green. For example, according to this hypothesis, after the genetic information for the colors of black and chocolate brown Labrador retrievers is blended, the colors should be as inseparable as paint pigments. But this is not what happens: Instead, the offspring of a true-breeding black Lab and a true-breeding brown one will all be black, but some of the dogs in the next generation will be brown. The blending hypothesis was finally rejected because it does not explain how traits that disappear in one generation can reappear in later ones.

? Horse breeders sometimes speak of "mixing the bloodlines" of two pedigrees. How does this language relate back to the "blending" model of inheritance?

■ It implies that offspring are a blend of two parents, as in a liquid mixture.

9.2 Experimental genetics began in an abbey garden

The modern science of genetics began in the 1860s, when an Augustinian monk named Gregor Mendel deduced the fundamental principles of genetics by breeding garden peas (Figure 9.2A). Mendel lived and worked in an abbey in Brunn, Austria (now Brno, in the Czech Republic). Strongly influenced by his study of physics, mathematics, and chemistry at the University of Vienna, his research was both experimentally and mathematically rigorous, and these qualities were largely responsible for his success.

In a paper published in 1866, Mendel correctly argued that parents pass on to their offspring discrete heritable factors. (It is interesting to note that Mendel's publications came just seven years after Darwin's 1859 publication of *The Origin of Species*, making the 1860s a banner decade in the development of modern biology.) In his paper, Mendel stressed that the heritable factors (today called genes) retain their individuality generation after generation. In other words, genes are like marbles of different colors: Just as marbles retain their colors permanently and do not blend, no matter how they are mixed, genes permanently retain their identities.

Mendel probably chose to study garden peas because he was familiar with them from his rural upbringing, they were easy to grow, and they came in many readily distinguishable varieties. Perhaps most importantly, Mendel was able to exercise strict control over pea plant matings. As Figure 9.2B shows, the petals of the pea flower almost completely enclose the reproductive organs: the stamens and carpel. (For better visibility, the drawing omits one of the petals.) Consequently, pea plants usually **self-fertilize** in nature. That is, sperm-carrying pollen grains released from the stamens land on the egg-containing carpel of the same flower. Mendel could ensure self-fertilization by covering a flower with a small bag so that

Figure 9.2A Gregor Mendel in his garden

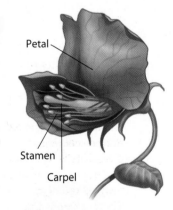

Figure 9.2B Anatomy of a garden pea flower

Petal
Stamen
Carpel

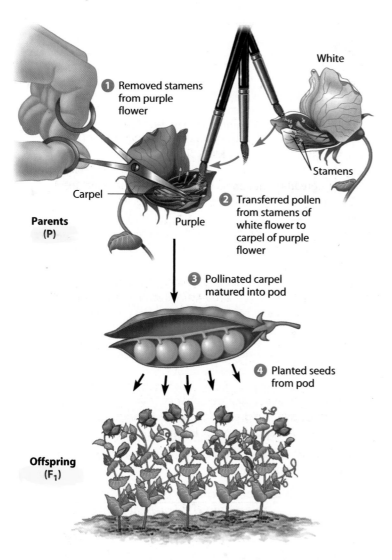

Figure 9.2C Mendel's technique for cross-fertilization of pea plants

① Removed stamens from purple flower

White

② Transferred pollen from stamens of white flower to carpel of purple flower

Stamens

Carpel

Parents (P)

Purple

③ Pollinated carpel matured into pod

④ Planted seeds from pod

Offspring (F₁)

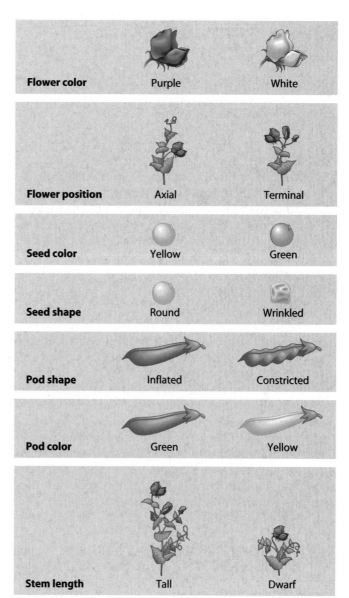

Figure 9.2D The seven pea characteristics studied by Mendel

Flower color	Purple	White
Flower position	Axial	Terminal
Seed color	Yellow	Green
Seed shape	Round	Wrinkled
Pod shape	Inflated	Constricted
Pod color	Green	Yellow
Stem length	Tall	Dwarf

no pollen from another plant could reach the carpel. When he wanted **cross-fertilization** (fertilization of one plant by pollen from a different plant), he used the method shown in Figure 9.2C. ❶ He prevented self-fertilization by cutting off the immature stamens of a plant before they produced pollen. ❷ To cross-fertilize the stamenless flower, he dusted its carpel with pollen from another plant. After pollination, ❸ the carpel developed into a pod, containing seeds (peas) that ❹ he planted. The seeds grew into offspring plants. Through these methods, Mendel could always be sure of the parentage of new plants.

Mendel's success was due not only to his experimental approach and choice of organism but also to his selection of characteristics to study. He chose to observe seven characteristics, each of which occurs in two distinct forms (Figure 9.2D). Mendel worked with his plants until he was sure he had **true-breeding** varieties—that is, varieties for which self-fertilization produced offspring all identical to the parent. For instance, he identified a purple-flowered variety that, when self-fertilized, produced offspring plants that all had purple flowers.

Now Mendel was ready to ask what would happen when he crossed his different true-breeding varieties with each other. For example, what offspring would result if plants with purple flowers and plants with white flowers were cross-fertilized? In the language of plant and animal breeders and geneticists, the offspring of two different varieties are called **hybrids**, and the cross-fertilization itself is referred to as a hybridization, or simply a **cross**. The true-breeding parental plants are called the **P generation** (P for parental), and their hybrid offspring are the **F₁ generation** (F for filial, from the Latin word for "son"). When F₁ plants self-fertilize or fertilize each other, their offspring are the **F₂ generation**. We turn to Mendel's results next.

? What is a "true-breeding" variety of pea plant?

■ A variety that, when self-fertilized, produces offspring identical to one another and to the parent.

9.3 Mendel's law of segregation describes the inheritance of a single characteristic

Mendel performed many experiments in which he tracked the inheritance of characteristics that occur in two forms, such as flower color. The results led him to formulate several ideas about inheritance. Let's look at some of his experiments and follow the reasoning that led to his hypotheses.

Figure 9.3A starts with a cross between a pea plant with purple flowers and one with white flowers. This is called a **monohybrid cross** because the parent plants differ in only one characteristic. Mendel observed that F_1 plants produced by these two true-breeding parents all had purple flowers; they were not a lighter purple, as predicted by the blending hypothesis. Was the white-flowered plant's genetic contribution to the hybrids lost? By mating the F_1 plants, Mendel found the answer to be no. Out of 929 F_2 plants, Mendel found that 705 (about $\frac{3}{4}$) had purple flowers and 224 (about $\frac{1}{4}$) had white flowers, a ratio of about three plants with purple flowers to every one with white flowers in the F_2 generation. Mendel reasoned that the heritable factor for white flowers did not disappear in the F_1 plants, but that only the purple-flower factor was affecting F_1 flower color. He also deduced that the F_1 plants must have carried two factors for the flower-color characteristic, one for purple and one for white.

Mendel observed these same patterns of inheritance for six other pea plant characteristics (see Figure 9.2D). From these results, Mendel developed four hypotheses, which we describe here using modern terminology (such as "gene" instead of "heritable factor"):

1. *There are alternative forms of genes that account for variations in inherited chacteristics.* For example, the gene for flower color in pea plants exists in two forms, one for purple and the other for white. The alternative versions of a gene are now called **alleles.**

2. *For each characteristic, an organism inherits two alleles, one from each parent. These alleles may be the same or different.* An organism that has two identical alleles for a gene is said to be **homozygous** for that gene (and is called a homozygote). An organism that has two different alleles for a gene is said to be **heterozygous** for that gene (and is called a heterozygote).

3. *If the two alleles of an inherited pair differ, then one determines the organism's appearance and is called the **dominant allele**; the other has no noticeable effect on the organism's appearance and is called the **recessive allele.*** We use uppercase letters to represent dominant alleles and lowercase letters to represent recessive alleles.

4. *A sperm or egg carries only one allele for each inherited trait because allele pairs separate (segregate) from each other during the production of gametes.* This statement is now known as the **law of segregation.** When sperm and egg unite at fertilization, each contributes its allele, restoring the paired condition in the offspring.

Figure 9.3B shows how Mendel explained the results given in Figure 9.3A. In this example, *P* represents the dominant

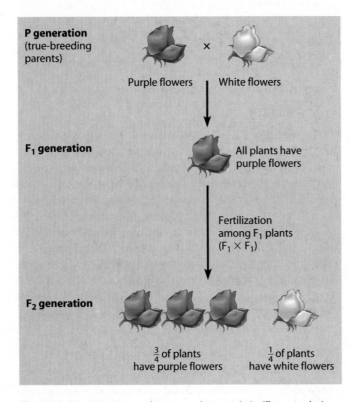

Figure 9.3A Crosses tracking one characteristic (flower color)

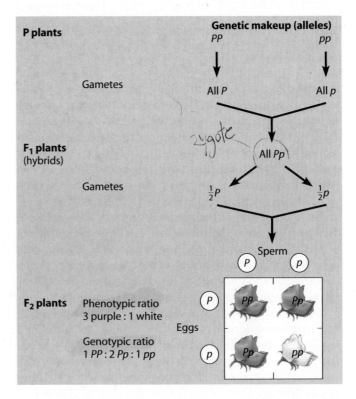

Figure 9.3B Explanation of the crosses in Figure 9.3A

allele (for purple flowers) and *p* stands for the recessive allele (for white flowers). At the top in Figure 9.3B, you see the alleles carried by the parental plants. Both plants were true-breeding, and Mendel's first two hypotheses propose that one parental variety had two alleles for purple flowers (*PP*) while the other had two alleles for white flowers (*pp*).

Consistent with hypothesis 4, the gametes of Mendel's parental plants each carried one allele; thus, the parental gametes in Figure 9.3B are either *P* or *p*. As a result of fertilization, the F_1 hybrids each inherited one allele for purple flowers and one for white. Hypothesis 3 explains why all of the F_1 hybrids (*Pp*) had purple flowers: The dominant *P* allele has its full effect in the heterozygote, while the recessive *p* allele has no effect on flower color.

Mendel's hypotheses also explain the 3:1 ratio ($\frac{3}{4}$ purple flowers to $\frac{1}{4}$ white flowers) in the F_2 generation. Because the F_1 hybrids are *Pp*, they make gametes *P* and *p* in equal numbers. The bottom diagram in Figure 9.3B, called a **Punnett square**, shows the four possible combinations of gametes.

The Punnett square shows the proportions of F_2 plants predicted by Mendel's hypotheses. If a sperm carrying allele *P* fertilizes an egg carrying allele *P*, the offspring (*PP*) will produce purple flowers. Mendel's hypotheses predict that this combination will occur in $\frac{1}{4}$ of the offspring. As shown in the Punnett square, the hypotheses also predict that $\frac{2}{4}$ of the offspring will inherit one *P* allele and one *p* allele. These offspring (*Pp*) will all have purple flowers because *P* is dominant. The remaining $\frac{1}{4}$ of the F_2 plants will inherit two *p* alleles and will have white flowers.

Because an organism's appearance does not always reveal its genetic composition, geneticists distinguish between an organism's expressed, or physical, traits, called its **phenotype** (such as purple or white flowers), and its genetic makeup, its **genotype** (in this example, *PP*, *Pp*, or *pp*). So now we see that Figure 9.3A shows the phenotypes and Figure 9.3B the genotypes in our sample crosses. For the F_2 plants, the ratio of plants with purple flowers to those with white flowers (3:1) is called the phenotypic ratio. The genotypic ratio, as shown by the Punnett square, is 1 *PP* : 2 *Pp* : 1 *pp*.

Mendel found that each of the seven characteristics he studied exhibited the same inheritance pattern: One parental trait disappeared in the F_1 generation, only to reappear in one-fourth of the F_2 offspring. The mechanism underlying this inheritance pattern is stated by Mendel's law of segregation: *Pairs of alleles segregate (separate) during gamete formation.* The fusion of gametes at fertilization creates allele pairs once again. Research since Mendel's time has established that the law of segregation applies to all sexually reproducing organisms, including humans. We'll return to Mendel and his experiments with pea plants in Module 9.5. But first let's see how some of the concepts we discussed in Chapter 8 fit with what we have said about genetics so far.

Web/CD Activity 9A *Monohybrid Cross*

? How can two plants with different genotypes for a particular inherited characteristic be identical in phenotype?

■ One could be homozygous for the dominant allele, while the other is heterozygous.

9.4 Homologous chromosomes bear the two alleles for each characteristic

Figure 9.4 shows a pair of homologous chromosomes (homologues)—chromosomes that carry genes controlling the same inherited characteristics. Recall from Chapter 8 that every diploid cell, whether from pea plant or human, has two sets of homologous chromosomes. One set comes from the organism's female parent, the other from the male parent.

The labeled bands on the chromosomes in the figure represent a few gene loci, specific locations of genes along the chromosome. The matching colors of corresponding loci on the two homologues highlight the fact that homologous chromosomes have genes for the same characteristics located at the same positions along their lengths. However, as the uppercase and lowercase letters next to the loci indicate, two homologous chromosomes may bear either the same alleles or different ones. Thus, we see the connection between Mendel's laws and homologous chromosomes: *Alleles (alternative forms) of a gene reside at the same locus on homologous chromosomes.*

The diagram here also serves as a review of some of the genetic terms we have encountered to this point. We will return to the chromosomal basis of inheritance in more detail beginning with Module 9.18.

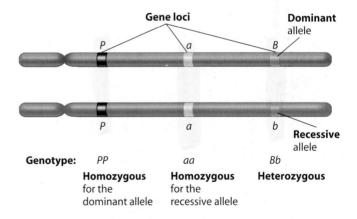

Figure 9.4 Homologous chromosomes

? An individual is heterozygous, *Gg*, for a genetic locus. According to the law of segregation, each gamete formed by this individual will have *either* the *G* allele *or* the *g* allele. Recalling what you learned about meiosis in Chapter 8, explain the physical basis for this segregation of alleles.

■ The *G* and *g* alleles are located on homologous chromosomes, which separate during meiosis and are packaged in separate gametes.

The law of independent assortment is revealed by tracking two characteristics at once

Two of the seven characteristics Mendel studied were seed shape and seed color. Mendel's seeds were either round or wrinkled in shape, and they were either yellow or green in color. From monohybrid crosses, Mendel knew that the allele for round shape (designated *R*) was dominant to the allele for wrinkled shape (*r*) and that the allele for yellow seed color (*Y*) was dominant to the allele for green seed color (*y*). What would result from a mating of parental varieties differing in two characteristics—a **dihybrid cross**? Mendel crossed homozygous plants having round yellow seeds (genotype *RRYY*) with plants having wrinkled green seeds (*rryy*). As shown in **Figure 9.5A**, the union of *RY* and *ry* gametes yielded hybrids heterozygous for both characteristics (*RrYy*)—that is, *dihybrids*. As we would expect, all of these offspring, the F₁ generation, had round yellow seeds. But were the two characteristics transmitted from parents to offspring as a package, or was each characteristic inherited independently of the other?

The question was answered when Mendel allowed fertilization to occur among the F₁ plants. If the genes for the two characteristics were inherited together, as shown on the left in the figure, then the F₁ hybrids would produce only the same two kinds (genotypes) of gametes—*RY* and *ry*—that they received from their parents. This hypothesis predicts that the phenotypic ratio of the F₂ generation will be 3:1 (three plants with round yellow seeds for every one with

wrinkled green seeds), as in the left Punnett square. If, however, the two seed characteristics segregated independently, then the F₁ generation would produce four gamete genotypes—*RY*, *rY*, *Ry*, and *ry*—in equal quantities. The Punnett square on the right shows all possible combinations of alleles that can result in the F₂ generation from the union of four kinds of sperm with four kinds of eggs. This Punnett square shows that there are nine different genotypes in the F₂. However, there are only four phenotypes, with a ratio of 9:3:3:1.

The Punnett square on the right also reveals that a dihybrid cross is equivalent to two monohybrid crosses occurring simultaneously. From the 9:3:3:1 ratio, we can see that there are 12 plants with round seeds to 4 with wrinkled seeds and 12 yellow-seeded plants to 4 green-seeded ones. These 12:4 ratios each reduce to 3:1, which is the F₂ ratio for a monohybrid cross. Mendel tried his seven pea characteristics in various dihybrid combinations and always obtained data close to the predicted 9:3:3:1 ratio. These results supported the hypothesis that *each pair of alleles segregates independently of the other pairs of alleles during gamete formation.* This is called Mendel's **law of independent assortment.**

Figure 9.5B shows how this law applies to the inheritance of two hereditary characteristics in Labrador retrievers: black versus chocolate coat color and normal vision versus the eye disorder called progressive retinal atrophy (PRA). As you

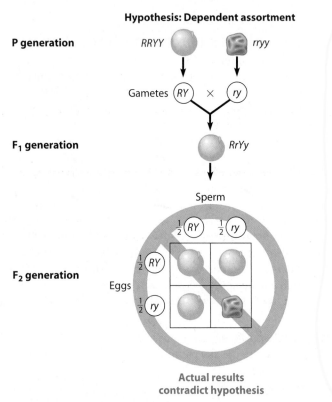

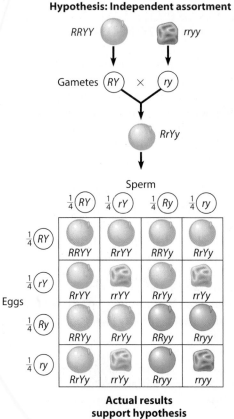

$\frac{9}{16}$	Yellow round
$\frac{3}{16}$	Green round
$\frac{3}{16}$	Yellow wrinkled
$\frac{1}{16}$	Green wrinkled

Figure 9.5A Two hypotheses for segregation in a dihybrid cross

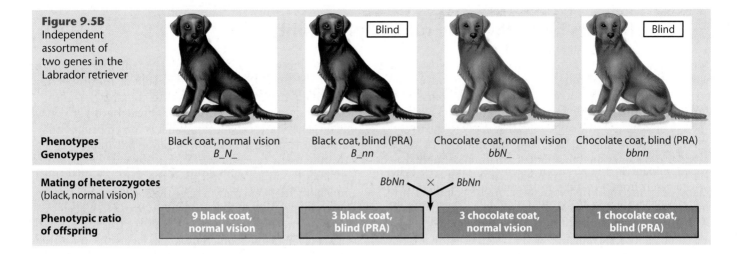

Figure 9.5B Independent assortment of two genes in the Labrador retriever

Phenotypes	Black coat, normal vision	Black coat, blind (PRA)	Chocolate coat, normal vision	Chocolate coat, blind (PRA)
Genotypes	*B_N_*	*B_nn*	*bbN_*	*bbnn*

Mating of heterozygotes (black, normal vision)

BbNn × *BbNn*

Phenotypic ratio of offspring

9 black coat, normal vision	3 black coat, blind (PRA)	3 chocolate coat, normal vision	1 chocolate coat, blind (PRA)

would expect, these characteristics are controlled by separate genes. Black Labs have at least one copy of an allele called *B*, which gives their hairs densely packed granules of a dark pigment. The *B* allele is dominant to *b*, which leads to a less tightly packed distribution of pigment granules. As a result, the coats of dogs with genotype *bb* are chocolate in color. The allele that causes PRA, called *n*, is recessive to allele *N*, which is necessary for normal vision. Thus, only dogs of genotype *nn* become blind from PRA. (In the figure, blanks in the genotypes indicate a second allele of either sort. And if you're wondering about yellow Labs, their color is controlled by a different gene altogether.)

The lower part of Figure 9.5B shows what happens when we mate two heterozygous Labs, both of genotype *BbNn*. The F_2 phenotypic ratio will be 9 black dogs with normal eyes to 3 black with PRA to 3 chocolate with normal eyes to 1 chocolate with PRA. These results are analogous to the results in Figure 9.5A and demonstrate that the *B* and *N* genes are inherited independently.

? Predict the phenotypes of offspring obtained by mating a black Lab homozygous for both coat color and normal eyes with a chocolate Lab that is blind from PRA.

■ All offspring would be black with normal eyes (*BBNN* × *bbnn* → *BbNn*).

9.6 Geneticists use the testcross to determine unknown genotypes

Suppose you have a Labrador retriever with a chocolate coat. Referring to Figure 9.5B, you can tell that its genotype must be *bb*, the only combination of alleles that produces the chocolate-coat phenotype. But what if you had a black Lab? It could have one of two possible genotypes—*BB* or *Bb*—and

there is no way to tell which is correct by looking at the dog. To determine your dog's genotype, you need to perform a **testcross**, a mating between an individual of unknown genotype (your dog) and a homozygous recessive (*bb*) individual—in this case, a chocolate Lab.

Figure 9.6 shows the offspring that could result from such a mating. If, as shown on the left, the black parent's genotype is *BB*, we would expect all the offspring to be black because a cross between genotypes *BB* and *bb* can produce only *Bb* offspring. On the other hand, if the black parent is *Bb*, we would expect both black (*Bb*) and chocolate (*bb*) offspring. Thus, the appearance of the offspring reveals the original black dog's genotype.

Mendel used testcrosses to determine whether he had true-breeding varieties of plants. The testcross continues to be an important tool of geneticists for determining genotypes.

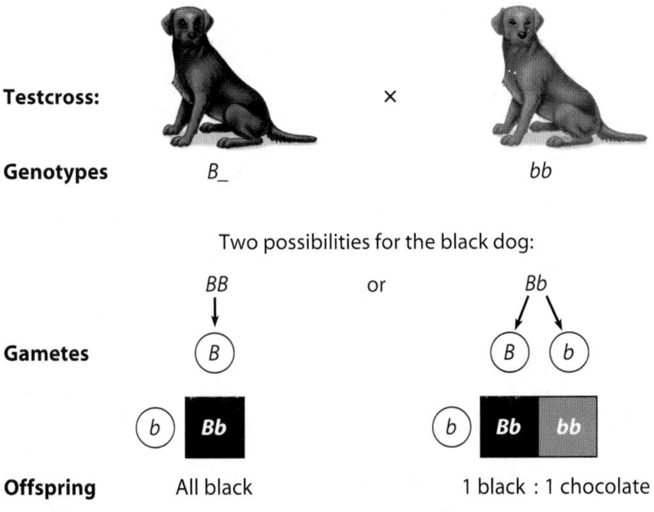

Web/CD Activity 9B *Dihybrid Cross*

? You use a testcross to determine the genotype of a Lab with normal eyes. Half of the offspring of the testcross are normal and half develop PRA. What is the genotype of the normal parent?

Testcross: *B_* × *bb*

Genotypes *B_* *bb*

Two possibilities for the black dog:

BB or *Bb*

Gametes *B* | *B* *b*

| *b* | **Bb** | | *b* | **Bb** | **bb** |

Offspring All black | 1 black : 1 chocolate

Figure 9.6 Using a testcross to determine genotype

Mendel's laws reflect the rules of probability

Mendel's strong background in mathematics served him well in his studies of inheritance. He understood, for instance, that the segregation of allele pairs during gamete formation and the re-forming of pairs at fertilization obey the rules of probability—the same rules that apply to the tossing of coins, the rolling of dice, and the drawing of cards. Mendel also appreciated the statistical nature of inheritance. He knew that he needed to obtain large samples—count many offspring from his crosses—before he could begin to interpret inheritance patterns.

Let's see how the rules of probability apply to inheritance. The probability scale ranges from 0 to 1. An event that is certain to occur has a probability of 1, while an event that is certain not to occur has a probability of 0. The probabilities of all possible outcomes for an event must add up to 1. With a coin, the chance of tossing heads is $\frac{1}{2}$, and the chance of tossing tails is $\frac{1}{2}$. In a standard deck of 52 playing cards, the chance of drawing a jack of diamonds is $\frac{1}{52}$, and the chance of drawing any card other than the jack of diamonds is $\frac{51}{52}$.

An important lesson we can learn from coin tossing is that for each and every toss of the coin, the probability of heads is $\frac{1}{2}$. In other words, the outcome of any particular toss is unaffected by what has happened on previous attempts. Each toss is an *independent event*.

If two coins are tossed simultaneously, the outcome for each coin is an independent event, unaffected by the other coin. What is the chance that both coins will land heads up? The probability of such a *compound event* is the product of the separate probabilities of the independent events—for the coins, $\frac{1}{2} \times \frac{1}{2} = \frac{1}{4}$. This is called the **rule of multiplication**, and it holds true for genetics as well as coin tosses. Figure 9.7 represents a cross between F_1 Labrador retrievers that have the *Bb* genotype for coat color. What is the probability that a particular F_2 dog will have the *bb* genotype? To produce a *bb* offspring, both egg and sperm must carry the *b* allele. The probability that an egg will have the *b* allele is $\frac{1}{2}$, and the probability that a sperm will have the *b* allele is also $\frac{1}{2}$. By the rule of multiplication, the probability that two *b* alleles will come together at fertilization is $\frac{1}{2} \times \frac{1}{2} = \frac{1}{4}$. This is exactly the answer given by the Punnett square in Figure 9.7. If we know the genotypes of the parents, we can predict the probability for any genotype among the offspring.

Now consider the probability that an F_2 Lab will be heterozygous for the coat-color gene. As Figure 9.7 shows, there are two ways in which F_1 gametes can combine to produce a heterozygous offspring. The dominant (*B*) allele can come from the egg and the recessive (*b*) allele from the sperm, or vice versa. The probability that an event can occur in two or more alternative ways is the *sum* of the separate probabilities of the different ways; this is known as the **rule of addition**. Using this rule, we can calculate the probability of an F_2 heterozygote as $\frac{1}{4} + \frac{1}{4} = \frac{1}{2}$.

By applying the rules of probability to segregation and independent assortment, we can solve some rather complex genetics problems. For instance, we can predict the results of trihybrid crosses, in which three different characteristics are involved. Consider a cross between two organisms that both have the genotype *AaBbCc*. What is the probability that an offspring from this cross will be a recessive homozygote for all three genes (*aabbcc*)? Since each allele pair assorts independently, we can treat this trihybrid cross as three separate monohybrid crosses:

Aa × *Aa:* Probability of *aa* offspring $= \frac{1}{4}$

Bb × *Bb:* Probability of *bb* offspring $= \frac{1}{4}$

Cc × *Cc:* Probability of *cc* offspring $= \frac{1}{4}$

Because the segregation of each allele pair is an independent event, we use the rule of multiplication to calculate the probability that the offspring will be *aabbcc*:

$\frac{1}{4} aa \times \frac{1}{4} bb \times \frac{1}{4} cc = \frac{1}{64}$

We could reach the same conclusion by constructing a 64-section Punnett square, but that would take a lot of time and a lot of space!

Web/CD Activity 9C *Gregor's Garden*

? A plant of genotype *AABbCC* is crossed with an *AaBbCc* plant. What is the probability of an offspring having the genotype *AABBCC*?

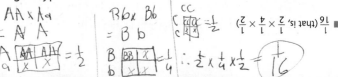

$\frac{1}{16}$ (that is, $\frac{1}{1} \times \frac{1}{4} \times \frac{2}{1}$)

F₁ genotypes

Bb male
↓
Formation of sperm

Bb female
↓
Formation of eggs

$\frac{1}{2}$ **B** $\frac{1}{2}$ **b**

$\frac{1}{2}$ **B**   **BB** $\frac{1}{4}$ **Bb** $\frac{1}{4}$

F₂ genotypes

$\frac{1}{2}$ **b**   **bB** $\frac{1}{4}$ **bb** $\frac{1}{4}$

Figure 9.7 Segregation and fertilization as chance events

9.8 Genetic traits in humans can be tracked through family pedigrees

Mendel's laws apply to the inheritance of many human traits. **Figure 9.8A** illustrates alternative forms of three human characteristics that are each thought to be determined by simple dominant-recessive inheritance at one gene locus. (The genetic basis of many other human characteristics—such as eye and hair color—are not well understood.) If we call the dominant allele of any such gene *A,* the dominant phenotype results from either the homozygous genotype *AA* or the heterozygous genotype *Aa.* Recessive phenotypes always result from the homozygous genotype *aa.* In genetics, the word *dominant* does not imply that a phenotype is either normal or more common than a recessive phenotype; wild-type traits (those prevailing in nature) are not necessarily specified by dominant alleles. In genetics, dominance means that a heterozygote (*Aa*), carrying only a single copy of a dominant allele, displays the dominant phenotype. By contrast, the phenotype of the corresponding recessive allele is seen only in a homozygote (*aa*). Recessive traits are often more common in the population than dominant ones. For example, the absence of freckles is more common than their presence.

How do we know how particular human traits are inherited? Since we obviously cannot control human matings, geneticists must analyze the results of matings that have already occurred. First, the geneticist collects as much information as possible about a family's history for the trait of interest. The next step is to assemble this information into a family tree—the family **pedigree.** Finally, to analyze the pedigree, the geneticist applies logic and the Mendelian laws you have learned.

Let's apply this approach to the example in **Figure 9.8B**, which shows part of a real pedigree from a family that lived on Martha's Vineyard, an island off the coast of Massachusetts, where a particular kind of inherited deafness was once prevalent. The letter *D* stands for the hearing allele, and *d* symbolizes the recessive allele for deafness. In the pedigree, squares represent males, circles represent females, and colored symbols indicate deafness. The earliest generation studied is at the top of the pedigree. Notice that deafness did not appear in this generation and that it showed up in only two of the seven children in the third generation.

By applying Mendel's laws, we can deduce that the deafness allele is recessive because that is the only way that

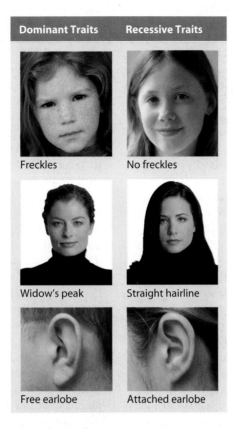

Dominant Traits	Recessive Traits
Freckles	No freckles
Widow's peak	Straight hairline
Free earlobe	Attached earlobe

Figure 9.8A Examples of inherited traits in humans

Jonathan Lambert could be deaf while both his parents were not. We can therefore label all the deaf individuals in the pedigree as homozygous recessive (*dd*). Mendel's laws also enable us to deduce the genotypes that are shown for most of the people in the pedigree. For example, both of Jonathan's parents must have carried a *d* allele (which they passed on to him) along with the *D* allele that gave them normal hearing. Two of Jonathan's children were deaf (*dd*), so his wife, who had normal hearing, must also have carried a *d* allele. Likewise, all of the couple's normal children must have been heterozygous (*Dd*). People who have one copy of the allele for a recessive disorder and do not exhibit symptoms are called **carriers** of the disorder.

What are the genotypes of Elizabeth Eddy's parents and Jonathan's sister Abigail? These three people had normal hearing, so they must have carried at least one *D* allele. And at least one of Elizabeth's parents must have had a *d* allele. But more than that we cannot say without additional information.

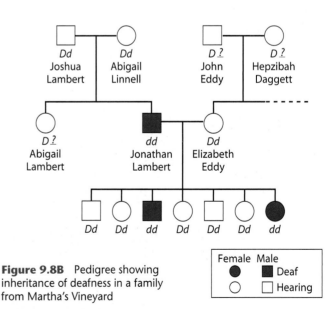

Figure 9.8B Pedigree showing inheritance of deafness in a family from Martha's Vineyard

? A man and a woman who are both carriers of the Martha's Vineyard deafness allele have had three children who are not deaf. If the couple has a fourth child, what is the probability that the child will be deaf?

¼ ■

9.9 Many inherited disorders in humans are controlled by a single gene

The hereditary deafness of Martha's Vineyard is just one of the thousands of human genetic disorders currently known to be inherited as dominant or recessive traits controlled by a single gene locus; ten more examples are listed in Table 9.9. These disorders show simple inheritance patterns like the traits Mendel studied in pea plants. The genes discussed in this module are all located on autosomes (chromosomes other than the sex chromosomes X and Y; see Module 8.12).

Recessive Disorders Most human genetic disorders are recessive. They range in severity from relatively mild, such as albinism (lack of pigmentation), to life-threatening, such as cystic fibrosis. Most people who have recessive disorders are born to normal parents who are both heterozygotes—that is, who are carriers of the recessive allele for the disorder but are phenotypically normal.

Using Mendel's laws, we can predict the fraction of affected offspring likely to result from a marriage between two carriers. Consider the example in Figure 9.8B. Suppose one of Jonathan Lambert's hearing sons (*Dd*) married a hearing woman whose pedigree indicated that her genotype was also *Dd*. What is the probability that they would have a deaf child? As the Punnett square in **Figure 9.9A** shows, each child of two carriers has a $\frac{1}{4}$ chance of inheriting two recessive alleles. Thus, we can say that about one-fourth of the children of this marriage are likely to be deaf. We can also say that a hearing ("normal") child from such a marriage has a $\frac{2}{3}$ chance of being a carrier (that is, 2 out of 3 of the offspring with the hearing phenotype are likely to be carriers). We can apply this same method of pedigree analysis and prediction to any genetic trait controlled by a single gene locus.

The most common lethal genetic disease in the United States is **cystic fibrosis.** Though the disease affects only about one in 17,000 African-Americans and about one in 90,000 Asian-Americans, it occurs in approximately one out of every 2,500 Caucasian (European ancestry) births. The cystic fibrosis allele is recessive and is carried by about one in every 25 Caucasians. A person with two copies of this allele has cystic fibrosis, which is characterized by an excessive secretion of very thick mucus from the lungs, pancreas, and other organs. This mucus can interfere with breathing, digestion, and liver function and makes the person vulnerable to recurrent bacterial infections. Untreated, most children with cystic fibrosis die before their fifth birthday. Although there is no cure for this fatal disase, a special diet, antibiotics to prevent infection, frequent pounding of the chest and back to clear the lungs, and other treatments can prolong life to adulthood.

Like cystic fibrosis, most genetic disorders are not evenly distributed across all ethnic groups. Such uneven distribution is the result of prolonged geographic isolation of certain populations. For example, the isolated lives of the Martha's Vineyard inhabitants between 1700 and 1900 led to frequent marriage between close relatives. Consequently, the frequency of deafness remained high, and the deafness allele was rarely transmitted to outsiders.

With the increased mobility in most societies today, it is relatively unlikely that two carriers of a rare, harmful allele will meet and mate. However, the probability increases greatly if close relatives marry and have children. People with recent common ancestors are more likely to carry the same recessive alleles than are unrelated people. Therefore, a mating of close relatives, called **inbreeding**, is more likely to produce offspring homozygous for recessive traits. Such effects can be observed in many types of inbred animals. For example, dogs that have been inbred for appearance may have serious genetic disorders, such as weak hip joints, eye problems, or undesirable behaviors. The detrimental effects of inbreeding are also seen in some endangered species, such as cheetahs (see Module 13.10).

Geneticists debate the extent to which human inbreeding increases the risk of inherited diseases. Many harmful mutations have such severe effects that a homozygous embryo spontaneously aborts long before birth. Still, most societies have taboos and laws forbidding marriages between close relatives. These rules may have arisen out of the observation that stillbirths and birth defects are more common when parents are closely related.

Dominant Disorders Although many harmful alleles are recessive, a number of human disorders are due to dominant alleles. Some are nonlethal conditions, such as extra fingers and toes or webbed fingers and toes.

One serious disorder caused by a dominant allele is **achondroplasia**, a form of dwarfism. In people with this

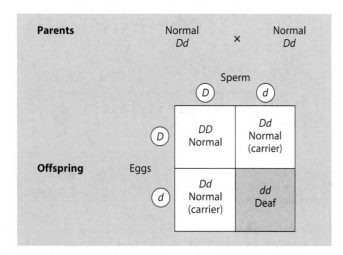

Figure 9.9A Offspring produced by parents who are both carriers for a recessive disorder

disorder, the head and torso of the body develop normally, but the arms and legs are short. (**Figure 9.9B** shows the late actor David Rappaport.) About one out of every 25,000 people has achondroplasia. The homozygous dominant genotype causes death of the embryo, and therefore only heterozygotes, individuals with a single copy of the defective allele, have this disorder. (This also means that a person with achondroplasia has a 50% chance of passing the condition on to any children.) Therefore, all those who do not have achondroplasia, more than 99.99% of the population, are homozygous for the recessive allele. This example makes it clear that a dominant allele is not necessarily more common in a population than the corresponding recessive allele.

Dominant alleles that cause lethal diseases are, in fact, much less common than lethal recessives. One reason for this difference is that the dominant lethal allele cannot be carried by heterozygotes without affecting them. Many lethal dominant alleles result from mutations in a sperm or egg that subsequently kill the embryo. And if the afflicted individual is born but does not survive long enough to reproduce, he or she will not pass on the lethal allele to future generations. This is in contrast to lethal recessive mutations, which are perpetuated from generation to generation by healthy (unaffected) heterozygous carriers.

A lethal dominant allele can escape elimination, however, if it does not cause death until a relatively advanced age. One such example is the allele that causes **Huntington's disease**, a degenerative disorder of the nervous system that usually does not begin until middle age. As the disease progresses, it causes uncontrollable movements in all parts of the body. Loss of brain cells leads to memory loss and impaired judgment and contributes to depression. Diminished motor skills eventually prevent swallowing and speaking. Death usually ensues 10 to 20 years after the onset of symptoms. By the time the symptoms of Huntington's disease become evident, the afflicted individual may have had children, each of whom has a 50% chance of having received the dominant allele for the disorder. This example makes it clear that the dominant allele is not necessarily "better" than the corresponding recessive allele.

Figure 9.9B Achondroplasia, caused by a dominant allele

? Peter is a 30-year-old man whose father died of Huntington's disease. Neither Peter's mother nor a much older sister, who is 48 years old, show any signs of Huntington's. What is the probability that Peter has inherited Huntington's disease?

$\frac{1}{2}$ ∎

TABLE 9.9 SOME AUTOSOMAL DISORDERS IN HUMANS

Disorder	Major Symptoms	Incidence	Comments
Recessive disorders			
Albinism	Lack of pigment in skin, hair, and eyes	$\frac{1}{22,000}$	Prone to skin cancer
Cystic fibrosis	Excess mucus in lungs, digestive tract, liver; increased susceptibility to infections; death in early childhood unless treated	$\frac{1}{2,500}$ Caucasians	See Modules 9.9 and 12.11
Galactosemia	Accumulation of galactose in tissues; mental retardation; eye and liver damage	$\frac{1}{100,000}$	Treated by eliminating galactose from diet
Phenylketonuria (PKU)	Accumulation of phenylalanine in blood; lack of normal skin pigment; mental retardation	$\frac{1}{10,000}$ in U.S. and Europe	See Module 9.10
Sickle-cell disease (homozygous)	Sickled red blood cells; damage to many tissues	$\frac{1}{400}$ African-Americans	Alleles are codominant; see Modules 9.13 and 9.14
Tay-Sachs disease	Lipid accumulation in brain cells; mental deficiency; blindness; death in childhood	$\frac{1}{3,500}$ Jews from central Europe	See Module 4.11
Dominant disorders			
Achondroplasia	Dwarfism	$\frac{1}{25,000}$	See Module 9.9
Alzheimer's disease (one type)	Mental deterioration; usually strikes late in life	Not known	
Huntington's disease	Mental deterioration and uncontrollable movements; strikes in middle age	$\frac{1}{25,000}$	See Modules 9.9 and 12.11
Hypercholesterolemia	Excess cholesterol in blood; heart disease	$\frac{1}{500}$ are heterozygous	Incomplete dominance; see Module 9.12

9.10 New technologies can provide insight into one's genetic legacy

Some prospective parents are aware that they have an increased risk of having a baby with a genetic disease. For example, many pregnant women over age 35 know that they have a heightened risk of bearing children with Down syndrome (see Modules 8.19 and 8.20), and some couples are aware that a certain genetic disease runs in their families. These potential parents may want to learn more about their baby's genetic legacy. Modern technologies offer ways to obtain such information before conception, during pregnancy, and after birth.

Identifying Carriers   Because most children with recessive disorders are born to healthy parents, the genetic risk for many diseases is determined by whether the prospective parents are carriers of the recessive allele. For an increasing number of genetic disorders, tests are available that can distinguish between individuals who have no disease-causing alleles and those who are heterozygous carriers. For example, tests are available that can identify carriers of the alleles for Tay-Sachs disease, sickle-cell disease, and one form of

cystic fibrosis. Such information can inform decisions about whether to have a child.

Fetal Testing   Several technologies are available for detecting harmful genetic conditions in the fetus. Genetic testing before birth requires the collection of fetal cells. In a procedure called **amniocentesis**, performed between weeks 14 and 20 of pregnancy, a physician carefully inserts a needle into the mother's uterus, avoiding the fetus (**Figure 9.10A**, left). The physician extracts about 10 milliliters (2 teaspoonful) of the amniotic fluid that bathes the developing fetus. Tests for genetic disorders can be performed on fetal cells (mostly sloughed off skin) present in the fluid. Before testing, these cells are usually cultured in the laboratory for several weeks. By then, enough dividing cells can be harvested to allow karyotyping and the detection of chromosomal abnormalities such as Down syndrome. Biochemical tests can also be performed on the cultured cells, revealing conditions such as Tay-Sachs disease (see Table 9.9). Additional information can

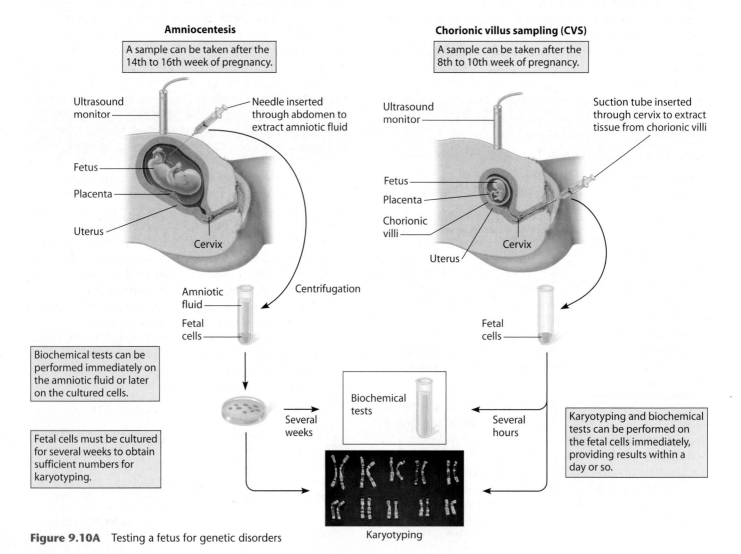

Figure 9.10A   Testing a fetus for genetic disorders

come from DNA testing (discussed further in Module 9.17 and Chapter 12.)

In another procedure, **chorionic villus sampling (CVS)**, a physician extracts a tiny sample of chorionic villus tissue from the placenta, the organ that carries nourishment and wastes between the fetus and the mother. The tissue can be obtained using a narrow, flexible tube inserted through the mother's vagina and cervix into the uterus (see Figure 9.10A, right). Results of karyotyping and some biochemical tests are available within 24 hours. The speed of CVS is an advantage over amniocentesis. Another advantage is that CVS can be performed as early as the 8th to 10th week of pregnancy.

Unfortunately, both amniocentesis and CVS pose some risk of complications, such as maternal bleeding, miscarriage, or premature birth. Complication rates for amniocentesis and CVS are about 1% and 2%, respectively. Because of the risks, these procedures are usually reserved for situations in which the risk of a genetic disorder is reasonably high.

Blood tests on the mother at 15 to 20 weeks of pregnancy can help identify fetuses at risk for certain birth defects—and thus candidates for further testing that may require more invasive procedures (such as amniocentesis). The most widely used blood test measures the mother's blood level of alpha-fetoprotein (AFP), a protein produced by the fetus. High levels of AFP may result from Down syndrome or neural tube defects in the fetus. (The neural tube is an embryonic structure that develops into the brain and spinal cord.) For a more complete risk profile, a woman's doctor may order a "triple screen test," which measures AFP as well as estriol and human chorionic gonadotropin (hCG), hormones produced by the placenta. Abnormal levels of these substances in the maternal blood may also point to a risk of Down syndrome.

Fetal Imaging

Other techniques enable a physician to examine a fetus directly for anatomical deformities. The most widely used such procedure is **ultrasound imaging,** which uses sound waves to produce a picture of the fetus. **Figure 9.10B** shows an ultrasound scanner, which emits high-frequency sounds, beyond the range of hearing. When the sound waves bounce off the fetus, the echoes produce an image on the monitor. The inset image in Figure 9.10B shows a fetus at about 18 weeks. Ultrasound imaging is noninvasive (no foreign objects are inserted into the mother's body) and has no known risk. In another imaging method, fetoscopy, a needle-thin tube containing a viewing scope and fiber optics is inserted into the uterus. Fetoscopy can provide highly detailed images of the fetus but, unlike ultrasound, carries risk of complications.

Newborn Screening

Some genetic disorders can be detected at birth by simple tests that are now routinely performed in most hospitals in the United States. One common screening program is for phenylketonuria (PKU), a recessively inherited disorder that occurs in about one out of every 10,000 births in the United States. Children with this disease cannot properly break down the amino acid phenylalanine, which may lead to mental retardation. However, if the deficiency is detected in the newborn, a special diet low

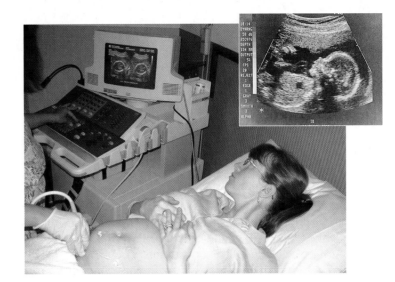

Figure 9.10B Ultrasound scanning of a fetus

in phenylalanine can usually prevent retardation. Unfortunately, very few other known genetic disorders are treatable at the present time.

Ethical Considerations

New technologies such as fetal imaging and testing raise new ethical questions. Consider the tests for identifying carriers of recessive diseases. Such information may enable people with family histories of genetic disorders to make informed decisions about having children. But these new methods for genetic screening pose problems, too. If confidentiality is breached, will carriers be stigmatized? For example, will they be denied health or life insurance, even though they themselves are healthy? Will misinformed employers equate "carrier" with disease? And will sufficient genetic counseling be available to help a large number of individuals understand their test results?

Couples at risk for conceiving children with genetic disorders may now learn a great deal about their unborn children. In particular, CVS gives parents a chance to become informed while the fetus is still quite young. What is to be done with such information? If fetal tests reveal a serious disorder that cannot be helped by surgery or therapy, the parents must choose between terminating the pregnancy and preparing themselves for a baby with severe problems. Identifying a genetic disease early can give families time to prepare—emotionally, medically, and financially.

Advances in biotechnology offer possibilities for reducing human suffering, but not before key ethical issues are resolved. The dilemmas posed by human genetics reinforce one of this book's themes: the immense social implications of biology.

? Review the genetic basis of Down syndrome in Module 8.20. (a) In what circumstances would fetal testing be particularly advisable for detecting a fetus with Down syndrome? (b) How would the genotype for Down syndrome reveal itself in a karyotype of a fetal cell?

■ (a) If the mother is in her late 30s or older; (b) there would be three copies of chromosome 21, instead of two.

9.11 The relationship of genotype to phenotype is rarely simple

Mendel's two laws explain inheritance in terms of discrete factors—genes—that are passed along from generation to generation according to simple rules of probability. Mendel's laws are valid for all sexually reproducing organisms, including garden peas, beagles, and human beings. But just as the basic rules of musical harmony cannot account for all the rich sounds of a symphony, Mendel's laws stop short of explaining some patterns of genetic inheritance. In fact, for most sexually reproducing organisms, cases where Mendel's laws can strictly account for the patterns of inheritance are relatively rare. More often, the inheritance patterns are more complex, as we will see in the next five modules.

? If a characteristic does not follow a Mendelian pattern, does that mean the characteristic is not inherited? Explain your answer.

■ No. Many inherited characteristics do not follow a simple Mendelian pattern in the relationship between genotype and phenotype.

9.12 Incomplete dominance results in intermediate phenotypes

The F_1 offspring of Mendel's pea crosses always looked like one of the two parental varieties. In this situation of **complete dominance**, the dominant allele had the same phenotypic effect whether present in one or two copies. But for some characteristics, the F_1 hybrids have an appearance *in between* the phenotypes of the two parental varieties, an effect called **incomplete dominance**. For instance, as **Figure 9.12A** illustrates, when red snapdragons are crossed with white snapdragons, all the F_1 hybrids have pink flowers. This third phenotype results from flowers of the heterozygote having less red pigment than the red homozygotes.

Incomplete dominance does *not* support the blending hypothesis, which would predict that the red and white traits could never be retrieved from the pink hybrids. As the Punnett square at the bottom of Figure 9.12A shows, the F_2 offspring appear in a phenotypic ratio of 1 red to 2 pink to 1 white, because the red and white alleles segregate during gamete formation in the pink F_1 hybrids. In incomplete dominance, the phenotypes of heterozygotes differ from the two homozygous varieties, and the genotypic ratio and the phenotypic ratio are both 1:2:1 in the F_2 generation.

We also see examples of incomplete dominance in humans. One case involves a recessive allele (h) that can cause *hypercholesterolemia*, dangerously high levels of cholesterol in the blood (see Module 5.20). Normal individuals are HH. Heterozygotes (Hh; about one in 500 people) have blood cholesterol levels about twice normal. They are unusually prone to atherosclerosis, the blockage of arteries by cholesterol buildup in artery walls, and they may have heart attacks from blocked

P generation

Red
RR × White
rr

Gametes R r

F₁ generation

Pink
Rr

Gametes $\frac{1}{2}$ R $\frac{1}{2}$ r

Sperm
$\frac{1}{2}$ R $\frac{1}{2}$ r

	$\frac{1}{2}$ R	$\frac{1}{2}$ r
$\frac{1}{2}$ R	Red RR	Pink rR
$\frac{1}{2}$ r	Pink Rr	White rr

F₂ generation Eggs

Figure 9.12A Incomplete dominance in snapdragon color

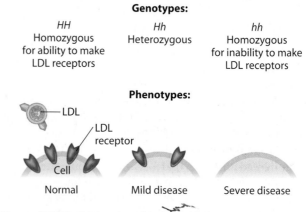

Genotypes:

| HH Homozygous for ability to make LDL receptors | Hh Heterozygous | hh Homozygous for inability to make LDL receptors |

Phenotypes:

LDL — LDL receptor — Cell

Normal · Mild disease · Severe disease

Figure 9.12B Incomplete dominance in human hypercholesterolemia

heart arteries by their mid-30s. Hypercholesterolemia is even more serious in homozygous individuals (*hh*; about one in a million people). Homozygotes have about five times the normal amount of blood cholesterol and may have heart attacks as early as age 2.

Figure 9.12B illustrates the molecular basis for hypercholesterolemia. The dominant allele, which normal individuals carry in duplicate (*HH*), specifies a cell-surface protein called an LDL receptor. LDLs, or low-density lipoproteins, are cholesterol-containing particles in the blood. The LDL receptors pick up LDL particles from the blood and promote their uptake by cells that break down the cholesterol. This process helps prevent the accumulation of cholesterol in arteries. Without the receptors, lethal levels of LDL build up in the blood. Heterozygotes (*Hh*) have only half the normal number of LDL receptors, and homozygous recessives (*hh*) have none.

Web/CD Activity 9D *Incomplete Dominance*

? Why is a testcross unnecessary to determine whether a snapdragon with red flowers is homozygous or heterozygous?

■ Because the homozygotes and heterozygotes differ in phenotype: red flowers for the dominant homozygote and pink flowers for the heterozygote

---

Many genes have more than two alleles in the population

So far, we have discussed inheritance patterns involving only two alleles per gene. But most genes can be found in populations in more than two forms, known as multiple alleles. Although any particular individual carries, at most, two different alleles for a particular gene, in cases of multiple alleles, more than two possible alleles exist in the wider population.

For instance, the **ABO blood group** phenotype in humans involves three alleles of a single gene. These three alleles, in various combinations, produce four phenotypes: A person's blood group may be either O, A, B, or AB. These letters refer to two carbohydrates, designated A and B, that may be found on the surface of red blood cells. A person's red blood cells may have carbohydrate A (type A blood), carbohydrate B (type B), both (type AB), or neither (type O). Matching compatible blood groups is critical for safe blood transfusions. If a donor's blood cells have a carbohydrate (A or B) that is foreign to the recipient, then the recipient's immune system produces blood proteins called antibodies (see Module 24.8) that bind specifically to the foreign carbohydrates and cause the donor blood cells to clump together, potentially killing the recipient. The clumping reaction is also the basis of a blood-typing test performed in the laboratory. **Figure 9.13** shows which combinations of blood groups result in clumping.

The four blood groups result from various combinations of the three different alleles, symbolized as I^A (for the enzyme that adds carbohydrate A to red blood cells), I^B (for B), and i (for neither A nor B). Each person inherits one of these alleles from each parent. Because there are three alleles, there are six possible genotypes, as listed in the figure. Both the I^A and I^B alleles are dominant to the i allele. Thus, I^AI^A and I^Ai people have type A blood, and I^BI^B and I^Bi people have type B. Recessive homozygotes, ii, have type O blood because their blood cells have neither the A nor the B carbohydrate. The I^A and I^B alleles are **codominant**; both alleles are expressed in heterozygous individuals (I^AI^B), who have type AB blood. Note that codominance (the expression of both alleles) is different from incomplete dominance (the expression of one intermediate trait).

? Maria has type O blood and her sister has type AB blood. The girls know that both of their maternal grandparents are type A. What are the genotypes of the girls' parents?

■ Their mother is I^Ai; their father is I^Bi.

Blood Group (Phenotype)	Genotypes	Antibodies Present in Blood	Reaction When Blood from Groups Below Is Mixed with Antibodies from Groups at Left			
			O	A	B	AB
O	*ii*	Anti-A Anti-B				
A	I^AI^A or I^Ai	Anti-B				
B	I^BI^B or I^Bi	Anti-A				
AB	I^AI^B	—				

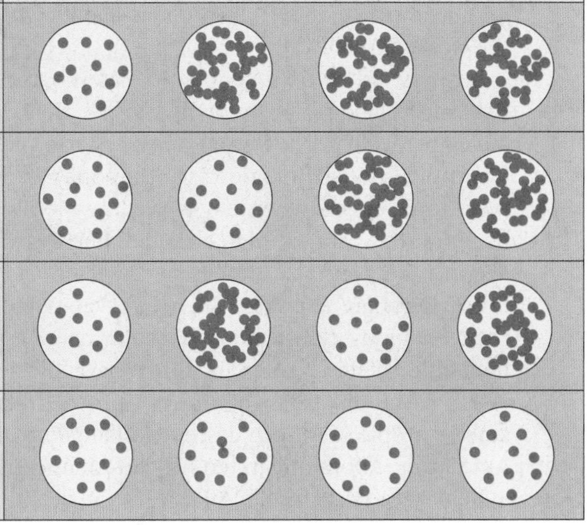

Figure 9.13 Multiple alleles for the ABO blood groups

9.14 A single gene may affect many phenotypic characteristics

All of our genetic examples to this point have been cases in which each gene specified only one hereditary characteristic. Most genes, however, influence multiple characteristics, a property called **pleiotropy** (from the Greek *pleion,* more).

An example of pleiotropy in humans is sickle-cell disease, a disorder characterized by the diverse symptoms shown in **Figure 9.14.** All of these possible phenotypic effects result from the action of a single kind of allele when it is present on both homologous chromosomes. The direct effect of the sickle-cell allele is to make red blood cells produce abnormal hemoglobin molecules. These molecules tend to link together and crystallize, especially when the oxygen content of the blood is lower than usual because of high altitude, overexertion, or respiratory ailments. As the hemoglobin crystallizes, the normally disk-shaped red blood cells deform to a sickle shape with jagged edges, as shown in the micrograph. Sickled cells are destroyed rapidly by the body, and the destruction of these cells may seriously lower the individual's red cell count, causing anemia and general weakening of the body. Also, because of their angular shape, sickled cells do not flow smoothly in the blood and tend to accumulate and clog tiny blood vessels. Blood flow to body parts is reduced, resulting in periodic fever, severe pain, and damage to various organs, including the heart, brain, and kidneys. Sickled cells also accumulate in the spleen, damaging it. Blood transfusions and certain drugs may relieve some of the symptoms, but there is no cure, and sickle-cell disease kills about 100,000 people in the world annually.

In most cases, only people who are homozygous for the sickle-cell allele suffer from the disease. Heterozygotes, who have one sickle-cell allele and one nonsickle allele, are usually healthy, although in rare cases they may experience some pleiotropic effects when oxygen in the blood is severely reduced, such as at very high altitudes. These effects may occur because the nonsickle and sickle-cell alleles are codominant: Both alleles are expressed in heterozygous individuals, and their red blood cells contain both normal and abnormal hemoglobin. Heterozygotes are said to have "sickle-cell trait."

Sickle-cell disease is the most common inherited disorder among people of African descent, striking one in 400 African-Americans. About one in ten African-Americans is a carrier—a heterozygote. Among Americans of other ancestry, the sickle-cell allele is extremely rare.

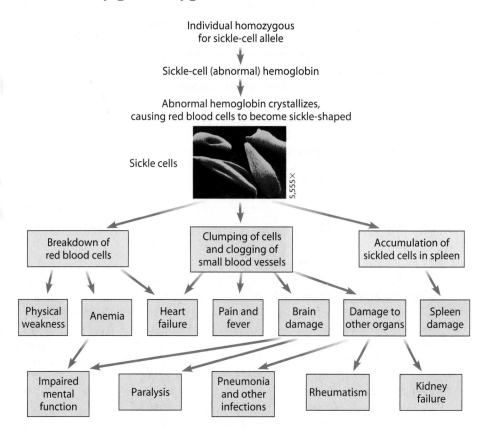

Figure 9.14 Sickle-cell disease, multiple effects of a single human gene

One in ten is an unusually high frequency of carriers for an allele with such harmful effects in homozygotes. We might expect that the frequency of the sickle-cell allele in the population would be much lower because many homozygotes die before passing their genes to the next generation. The high frequency appears to be a vestige of the roots of African-Americans. Sickle-cell disease is most common in tropical Africa, where the deadly disease malaria is also prevalent. The protistan parasite that causes malaria spends part of its life cycle inside red blood cells. When it enters those of a person with the sickle-cell allele, it triggers sickling. The body destroys most of the sickled cells, and the parasite does not grow well in those that remain. Consequently, sickle-cell carriers are resistant to malaria, and in many parts of Africa, they live longer and have more offspring than noncarriers who are exposed to malaria. In this way, malaria has kept the frequency of the sickle-cell allele relatively high in much of the African continent. To put it in evolutionary terms, as long as malaria is a danger, individuals with the sickle-cell allele have a selective advantage.

> ? How does sickle-cell disease exemplify the concept of pleiotropy?

■ Homozygosity for the sickle-cell allele causes abnormal hemoglobin, and the impact of the abnormal hemoglobin on the shape of red blood cells leads to a cascade of symptoms in multiple organs of the body.

9.15 A single characteristic may be influenced by many genes

Mendel studied genetic characteristics that could be classified on an either-or basis, such as purple or white flower color. However, many characteristics, such as human skin color and height, vary in a population along a continuum. Many such features result from **polygenic inheritance**, the additive effects of two or more genes on a single phenotypic characteristic. (This is the converse of pleiotropy, in which a single gene affects several characteristics.)

Let's consider a hypothetical example. Assume that the continuous variation in human skin color is controlled by three genes that are inherited separately, like Mendel's pea genes. (Actually, genetic evidence indicates that *at least* three genes control this characteristic.) The "dark-skin" allele for each gene (*A, B,* or *C*) contributes one "unit" of darkness to the phenotype and is incompletely dominant to the other allele (*a, b,* or *c*). A person who is *AABBCC* would be very dark, whereas an *aabbcc* individual would be very light. An *AaBbCc* person (resulting, for example, from a mating between an *AABBCC* person and an *aabbcc* person) would have skin of an intermediate shade. Because the alleles have an additive effect, the genotype *AaBbCc* would produce the same skin color as any other genotype with just three dark-skin alleles, such as *AABbcc*.

The Punnett square in **Figure 9.15** shows all possible genotypes of offspring from a mating of two triple heterozygotes (the F$_1$ generation here). The row of squares below the Punnett square shows the seven skin-color phenotypes that would theoretically result from this mating. The seven bars in the graph at the bottom of the figure depict the relative numbers of each of the phenotypes in the F$_2$ generation. In real human populations, skin color has even more variations than shown in the figure, in part for reasons we discuss in the next module.

Up to this point in the chapter, we have presented four types of inheritance patterns that are extensions of Mendel's laws of inheritance: incomplete dominance, codominance, pleiotropy, and polygenic inheritance. It is important to realize that these patterns are extensions of Mendel's model, rather than exceptions to it. From Mendel's garden pea experiments came data supporting a particulate theory of inheritance, with the particles (genes) being transmitted according to the same rules of chance that govern the tossing of coins. The particulate theory holds true for all inheritance patterns. In the next module, we consider another important source of deviation from Mendel's standard model: the effect of the environment.

? Based on the skin-color model in Figure 9.15, an *AaBbcc* individual would be indistinguishable in phenotype from which of the following individuals: *AAbbcc, aaBBcc, AabbCc, Aabbcc,* or *aaBbCc*?

■ All except *Aabbcc*

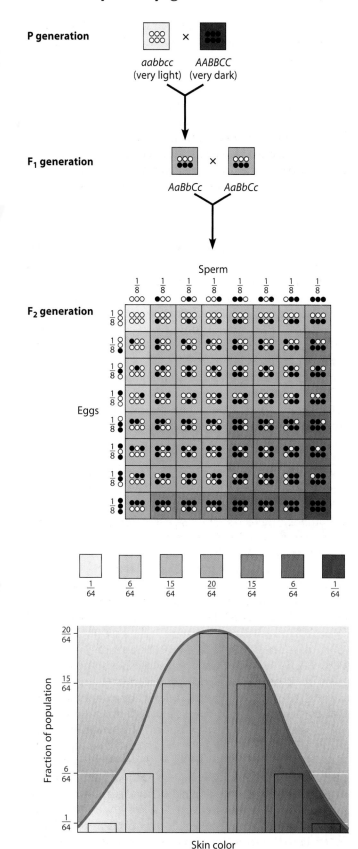

Figure 9.15 A model for polygenic inheritance of skin color

9.16 The environment affects many characteristics

In the previous module, we saw how a set of three hypothetical human skin-color genes could produce seven different skin-color phenotypes. If we examine a real human population for the skin-color phenotype, we would see more shades than just seven. The true range might be similar to the entire spectrum of color under the bell-shaped curve in Figure 9.15. In fact, no matter how carefully we characterize the genes for skin color, a purely genetic description will always be incomplete. This is because some intermediate shades of skin color result from the effects of environmental factors, such as exposure to the sun.

Many characteristics result from a combination of heredity and environment. For example, although a single tree is locked into its inherited genotype, its leaves vary in size, shape, and color, depending on exposure to wind and sun and the tree's nutritional state. For humans, nutrition influences height; exercise alters build; sun-tanning darkens the skin (**Figure 9.16**); and experience improves performance on intelligence tests. As geneticists learn more and more about our genes, it is becoming clear that many human phenotypes—such as risk of heart disease and cancer and susceptibility to alcoholism and schizophrenia—are influenced by both genes and environment.

Whether human characteristics are more influenced by genes or by the environment—nature or nurture—is a very old and hotly contested debate. For some characters, such as the ABO blood group, a given genotype mandates a very specific phenotype. In contrast, a person's blood count of red and white cells varies quite a bit, depending on such factors as the altitude, the customary level of physical activity, and the presence of infectious agents.

Figure 9.16 Environmental factors such as exercise and sun exposure may produce varying phenotypes

It is important to realize that the individual features of any organism arise from a combination of genetic and environmental factors. Simply spending time with identical twins will convince anyone that environment, and not just genes, affects a person's traits. However, there is an important difference between these two sources of variation: Only genetic influences are inherited. Any effects of the environment are not passed on to the next generation.

? Why was Mendel able to ignore environmental influences on his pea plants ?

■ The characteristics he chose for study were all entirely genetically determined.

9.17 Genetic testing can detect disease-causing alleles

The growing field of genetic testing (also called genetic screening) offers the chance to check for alleles associated with many hereditary diseases. Such genetic testing is rapidly becoming a significant component of health care.

For people who have a family history but no symptoms of a genetic disorder, a predictive test can help determine a person's risk for developing the disorder in the future. For example, DNA testing for FAP (familial adenomatous polyposis), a condition that almost always leads to colon cancer, can alert young people at risk to seek early medical care. And DNA tests for defective alleles of the genes *BRCA1* and *BRCA2* can reveal that a woman is at increased risk for developing breast cancer. (However, note that in neither case is a defect in a single gene sufficient to cause cancer.)

The growing number of genetic tests available to the public has raised concern about how they are used. Geneticists stress that patients seeking genetic testing should receive counseling both before and after—to clarify their family history, to explain the test, and to help them cope with the test results. Some patients may need to rethink their life plans or seek special medical care. Others may be relieved to learn that they do not carry the allele for a feared genetic disease.

As genetic testing becomes more widespread, geneticists are working to make sure that it does not cause more problems than it solves. Noted breast cancer researcher Mary-Claire King (see Module 11.19) cautions: "Testing is useful only if it is presented in such a way that the person understands what the limitations of the tests are and what the results mean. Another critical principle is one of social justice: We need to respect absolutely the rights of the individual. A person's genetic background should have no bearing at all on his or her ability to obtain health insurance, for example. For every one of us is predisposed to something."

? The presence of certain alleles indicates a predisposition to cancer but cannot indicate with certainty that the person will get cancer. Why?

■ Environmental influences and other genes may play an important role.

9.18 Chromosome behavior accounts for Mendel's laws

Mendel published his results in 1866, but not until long after he died did biologists understand the significance of his work. Cell biologists worked out the processes of mitosis and meiosis in the late 1800s (see Chapter 8 to review these processes). Then, around 1900, researchers began to notice parallels between the behavior of chromosomes and the behavior of Mendel's heritable factors. One of biology's most important concepts—the **chromosome theory of inheritance**—was emerging. The chromosome theory states that genes occupy specific loci (positions) on chromosomes and it is the chromosomes that undergo segregation and independent assortment during meiosis. Thus, it is the behavior of chromosomes during meiosis and fertilization that accounts for inheritance patterns.

We can see the chromosomal basis of Mendel's laws by following the fates of two genes during meiosis and fertilization in pea plants. In **Figure 9.18**, we picture the genes for seed shape (alleles *R* and *r*) and seed color (*Y* and *y*) as black bars on different chromosomes. We start with the F_1 generation, in which all individuals have the *RrYy* genotype. To simplify the diagram, we show only two of the seven pairs of pea chromosomes and three of the stages of meiosis: metaphase I, anaphase I, and metaphase II.

To see the chromosomal basis of the law of segregation, let's follow just the homologous pair of long chromosomes, the

ones carrying *R* and *r*, taking either the left or the right branch from the F_1 cell. Whichever arrangement the chromosomes assume at metaphase I, the two alleles *segregate* as the homologous chromosomes separate in anaphase I. And at the end of meiosis II, a single long chromosome ends up in each of the gametes. Fertilization then recombines the two alleles at random, resulting in F_2 offspring that are $\frac{1}{4}$ *RR*, $\frac{1}{4}$ *rr*, and $\frac{1}{2}$ *Rr*. The ratio of round to wrinkled phenotypes is thus 3:1 (12 round to 4 wrinkled), the ratio Mendel observed (see Figure 9.3A).

To see the chromosomal basis of the law of independent assortment, follow both the long and short (nonhomologous) chromosomes through the figure below. Two alternative, equally likely arrangements of tetrads can occur at metaphase I. The nonhomologous chromosomes (and their genes) assort independently, leading to four gamete genotypes. Random fertilization leads to the 9:3:3:1 phenotypic ratio in the F_2 generation, as you saw in Figure 9.5A.

? Which of Mendel's laws have their physical basis in the following phases of meiosis? (a) the orientation of homologous chromosome pairs in metaphase I; (b) the separation of homologues in anaphase I.

■ (a) the law of independent assortment; (b) the law of segregation

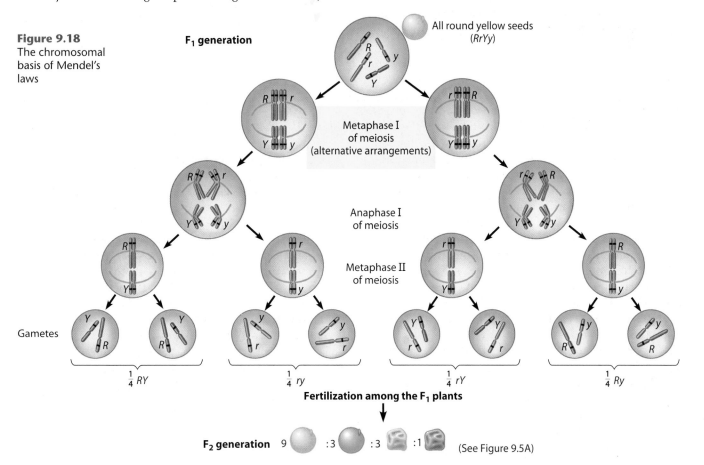

Figure 9.18
The chromosomal basis of Mendel's laws

F₂ generation 9 :3 :3 :1 (See Figure 9.5A)

Genes on the same chromosome tend to be inherited together

In 1908, British biologists William Bateson and Reginald Punnett (originator of the Punnett square) observed an inheritance pattern that seemed inconsistent with Mendelian laws. Bateson and Punnett were working with two characteristics in sweet peas, flower color and pollen shape. They crossed doubly heterozygous plants (*PpLl*) that exhibited the dominant traits: purple flowers (expression of the *P* allele) and long pollen grains (expression of the *L* allele). The corresponding recessive traits are red flowers (in *pp* plants) and round pollen (in *ll* plants).

The top part of **Figure 9.19** illustrates Bateson and Punnett's experiment. When they looked at just one of the two characteristics (that is, either cross *Pp* × *Pp* or cross *Ll* × *Ll*), they found that the dominant and recessive alleles segregated, producing a phenotypic ratio of approximately 3:1 for the offspring, in agreement with Mendel's law of segregation. However, when the biologists combined their data for the two characteristics, they did not see the predicted 9:3:3:1 ratio. Instead, as shown in the table, they found a disproportionately large number of plants with just two of the predicted phenotypes: purple long (almost 75% of the total) and red round (about 14%). The other two phenotypes (purple round and red long) were found in far fewer numbers than expected. What can account for these results?

The number of genes in a cell is far greater than the number of chromosomes; in fact, each chromosome has hundreds or thousands of genes. Genes located close together on the same chromosome tend to be inherited together and are called **linked genes**. Linked genes generally do not follow Mendel's law of independent assortment.

As shown in the "Explanation" in the figure, the sweet-pea genes for flower color and pollen shape are located on the same chromosome. Thus, meiosis in the heterozygous (*PpLl*) sweet-pea plant yields mostly two genotypes of gametes (*PL* and *pl*) rather than equal numbers of the four types of gametes that would result if the flower-color and pollen-shape genes were not linked. The large numbers of plants with purple long and red round traits in the Bateson-Punnett experiment resulted from fertilization among the *PL* and *pl* gametes. But what about the smaller numbers of plants with purple round and red long traits? As you will see in the next module, the phenomenon of crossing over accounts for these offspring.

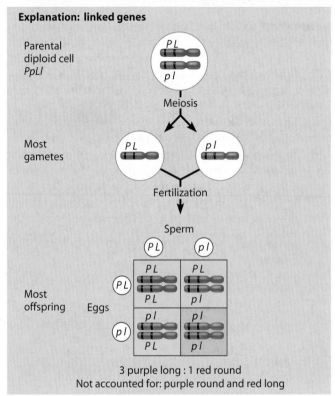

Experiment

Purple flower

PpLl × *PpLl*

Long pollen

Phenotypes	Observed offspring	Prediction (9:3:3:1)
Purple long	284	215
Purple round	21	71
Red long	21	71
Red round	55	24

Explanation: linked genes

Parental diploid cell *PpLl*

Meiosis

Most gametes

Fertilization

Sperm

Most offspring   Eggs

3 purple long : 1 red round
Not accounted for: purple round and red long

Figure 9.19 Experiment involving linked genes in the sweet pea

? What are linked genes?

Genes that tend to be inherited together because their loci are close together on the same chromosome

---

Crossing over produces new combinations of alleles

In Module 8.18, we saw that during meiosis, crossing over between homologous chromosomes produces new combinations of alleles in gametes. **Figure 9.20A** reviews this process, showing that two linked genes can give rise to four different gamete genotypes. Gametes with genotypes *AB* and *ab* carry parental-type chromosomes that have not been altered by crossing over. In contrast, gametes with genotypes *Ab* and *aB*

are recombinant gametes. The exchange of chromosome segments during crossing over has produced new combinations of alleles. We can now understand the results of the Bateson-Punnett experiment presented in the previous module: The small fraction of offspring with the recombinant phenotype must have resulted from fertilization involving recombinant gametes.

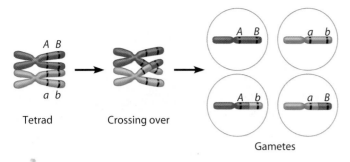

Figure 9.20A Review: Production of recombinant gametes

The discovery of how crossing over creates gamete diversity confirmed the relationship between chromosome behavior and heredity. Some of the most important early studies of crossing over were performed in the laboratory of American embryologist Thomas Hunt Morgan in the early 1900s. Morgan and his colleagues used the fruit fly *Drosophila melanogaster* in many of their experiments (**Figure 9.20B**).

Figure 9.20B
Drosophila melanogaster

Often seen flying around ripe fruit, *Drosophila* is a good research animal for genetic studies because it can be easily and inexpensively grown and can produce several generations in a matter of months.

Figure 9.20C shows one of Morgan's experiments, a cross between a wild-type fruit fly (gray body and long wings) and a fly with a black body and undeveloped, or vestigial, wings. Morgan knew the genotypes of these flies from previous studies. Here we use the following gene symbols:

G = gray body (dominant)

g = black body (recessive)

L = long wings (dominant)

l = vestigial wings (recessive)

In mating a gray fly with long wings (genotype $GgLl$) with a black fly with vestigial wings (genotype $ggll$), Morgan performed a testcross (see Module 9.6). If the genes were not linked, then independent assortment would produce offspring in a phenotypic ratio of 1:1:1:1 ($\frac{1}{4}$ gray body, long wings; $\frac{1}{4}$ black body, vestigial wings; $\frac{1}{4}$ gray body, vestigial wings; and $\frac{1}{4}$ black body, long wings). But because these genes were linked, Morgan obtained the results shown in Figure 9.20C: Most of the offspring had parental phenotypes, but 17% of the offspring flies were recombinants. The percentage of recombinants is called the **recombination frequency.**

When Morgan first obtained these results, he did not know about crossing over. To explain the ratio of offspring, he hypothesized that the genes were linked and that some mechanism occasionally broke the linkage. Tests of the hypothesis proved him correct, establishing that crossing over was the mechanism that "breaks linkages" between genes.

The lower part of Figure 9.20C explains Morgan's results in terms of crossing over. A crossover between chromatids of homologous chromosomes in parent $GgLl$ broke linkages between the G and L alleles and between the g and l alleles, forming the recombinant chromosomes Gl and gL. Later steps in meiosis distributed the recombinant chromosomes to gametes, and random fertilization produced the four kinds of offspring Morgan observed.

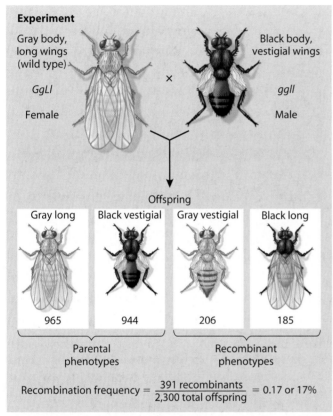

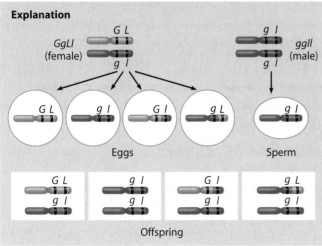

Figure 9.20C Fruit fly experiment demonstrating the role of crossing over in inheritance

? Return to the data in Figure 9.19. What is the recombination frequency for the flower-color and pollen-length genes?

■ 11%, or $\frac{42}{381}$

9.21 Geneticists use crossover data to map genes

Working with *Drosophila*, T. H. Morgan and his students greatly advanced our understanding of genetics. In the photo in Figure 9.21A, Morgan (back row, far right), several students, and a skeleton are celebrating the return of Alfred H. Sturtevant (left foreground) from World War I military service. One of Sturtevant's major contributions to genetics was an approach for using crossover data to map gene loci. Sturtevant started by assuming that the chance of crossing over is approximately equal at all points along a chromosome. He then hypothesized that the farther apart two genes are on a chromosome, the higher the probability that a crossover will occur between them. His reasoning was elegantly simple: The greater the distance between two genes, the more points there are between them where crossing over can occur. With this principle in mind, Sturtevant began using recombination data from fruit fly crosses to assign to genes relative positions on chromosomes—that is, to *map* genes.

Figure 9.21B represents a part of the chromosome that carries the linked genes for black body (*g*) and vestigial wings (*l*) that we described in Module 9.20. This same chromosome also carries a gene that has a recessive allele (we'll call it *c*) determining cinnabar eye color, a brighter red than the wild-type color. Figure 9.21B shows the actual crossover (recombination) frequencies between these alleles, taken two at a time: 17% between the *g* and *l* alleles, 9% between *g* and *c*, and 9.5% between *l* and *c*. Sturtevant reasoned that these values represent the relative distances between the genes. Because the crossover frequencies between *g* and *c* and between *l* and *c* are approximately half that between *g* and *l*, gene *c* must lie roughly midway between *g* and *l*. Thus, the sequence of these genes on one of the fruit fly chromosomes must be *g-c-l* (or the equivalent *l-c-g*).

Years later it was learned that Sturtevant's assumption that crossovers are equally likely at all points on a chromosome was not exactly correct. Still, his method of mapping genes worked, and it proved extremely valuable in establishing the relative positions of many other fruit fly genes. Eventually,

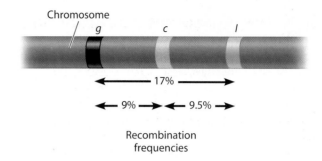

Figure 9.21B Mapping genes from crossover data

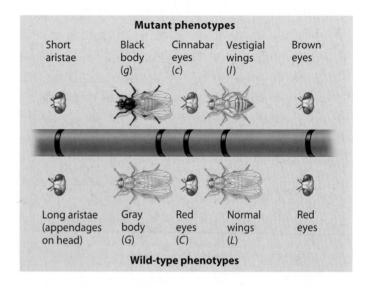

Figure 9.21C A partial genetic map of a fruit fly chromosome

enough data were accumulated to reveal that *Drosophila* has four groups of genes, corresponding to its four pairs of chromosomes. Figure 9.21C is a genetic map showing just five of the gene loci on part of one chromosome: the loci labeled *g*, *c*, and *l* and two others. Notice that eye color is a characteristic affected by more than one gene. Here we see the cinnabar-eye and brown-eye genes; still other eye-color genes are found elsewhere (see Module 9.23). For all these genes, however, the wild-type allele specifies red eyes.

Today, with DNA technology, geneticists can determine the actual distances in nucleotides between linked genes. The new genetic maps confirm the relative positions established by Sturtevant's mapping method.

Web/CD Activity 9E *Linked Genes and Crossing Over*

? You design *Drosophila* crosses to provide recombination data for a gene not included in Figure 9.21C. The gene has recombination frequencies of 5% with the vestigial-wing (*l*) locus and 5% with the cinnabar-eye (*c*) locus. Where is it located on the chromosome?

Figure 9.21A A party in Morgan's fly room

■ About halfway between the vestigial and cinnabar loci

9.22 Chromosomes determine sex in many species

Many animals, including fruit flies and all mammals, have a pair of **sex chromosomes**, designated X and Y, that determine an individual's sex. **Figure 9.22A** reviews what you learned in Chapter 8 about sex determination in humans. Individuals with one X chromosome and one Y chromosome are males; XX individuals are females. Human males and females both have 44 autosomes (nonsex chromosomes). As a result of chromosome segregation during meiosis, each gamete contains one sex chromosome and a haploid set of autosomes (22 in humans). All eggs contain a single X chromosome. Of the sperm cells, half contain an X chromosome and half contain a Y chromosome. An offspring's sex depends on whether the sperm cell that fertilizes the egg bears an X or a Y. In the fruit fly's X-Y system, sex is determined primarily by the number of X chromosomes, although the Y chromosome is essential for sperm formation.

The genetic basis of sex determination in humans is not yet completely understood, but one gene on the Y chromosome plays a crucial role. This gene, discovered by a British research team in 1990, is called *SRY* (for sex-determining region of Y) and triggers testis development. In the absence of *SRY*, an individual develops ovaries rather than testes. *SRY* codes for proteins that regulate other genes on the Y chromosome. These genes in turn produce proteins necessary for normal testis development.

The X-Y system is only one of several sex-determining systems. For example, grasshoppers, roaches, and some other insects have an X-O system, in which O stands for the absence of a sex chromosome (**Figure 9.22B**). Females have two X chromosomes (XX); males have only one sex chromosome (XO). Males produce two classes of sperm (X-bearing and lacking any sex chromosome), and sperm cells determine the sex of the offspring at fertilization.

In contrast to the X-Y and X-O systems, *eggs* determine sex in certain fishes, butterflies, and birds (**Figure 9.22C**). The sex chromosomes in these animals are designated Z and W. Males have the genotype ZZ; females are ZW. In this system, sex is determined by whether the egg carries a Z or a W.

Some organisms lack sex chromosomes altogether. In most ants and bees, sex is determined by chromosome *number*, rather than by sex chromosomes (**Figure 9.22D**). Females develop from fertilized eggs and thus are diploid. Males develop from unfertilized eggs—they are fatherless—and are haploid.

Most animals have two separate sexes; that is, individuals are either male or female. Many plant species have sperm-bearing and egg-bearing flowers borne on different individuals. Some plant species, such as date palms and marijuana, have the X-Y system of sex determination; others, such as the wild strawberry, have the Z-W system. However, most plant species and some animal species have individuals that produce both sperm and eggs. In such species, all individuals have the same complement of chromosomes.

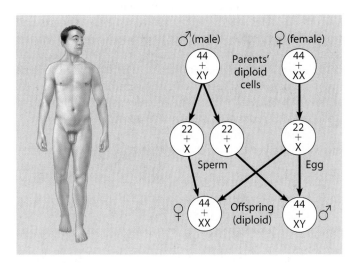

Figure 9.22A   The X-Y system

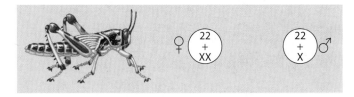

Figure 9.22B   The X-O system

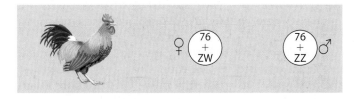

Figure 9.22C   The Z-W system

Figure 9.22D   Sex determination by chromosome number

? At the moment of conception in humans, what determines the sex of the offspring?

■ Whether the egg is fertilized by a sperm bearing an X chromosome (producing a female offspring) or by a sperm with a Y chromosome (producing a male)

Sex-linked genes exhibit a unique pattern of inheritance

Besides bearing genes that determine sex, the so-called sex chromosomes also contain genes for characteristics unrelated to femaleness or maleness. A gene located on either sex chromosome is called a **sex-linked gene**, although in humans the term has historically referred specifically to a gene on the X chromosome. (Be careful not to confuse the term *sex-linked gene*, which refers to a gene on a sex chromosome, with the term *linked genes*, which refers to genes on the same chromosome that tend to be inherited together.)

The figures here illustrate inheritance patterns for white eye color in the fruit fly, an X-linked recessive trait. Wild-type fruit flies have red eyes; white eyes are very rare (Figure 9.23A). We use the uppercase letter *R* for the dominant, wild-type, red-eye allele and *r* for the recessive, white-eye allele. Because these alleles are carried on the X chromosome, we show them as superscripts to the letter X. Thus, red-eyed male fruit flies have the genotype $X^R Y$; white-eyed males are $X^r Y$. The Y chromosome does not have a gene locus for eye color; therefore, the male's phenotype results entirely from his single X-linked gene. In the female, $X^R X^R$ and $X^R X^r$ flies have red eyes, and $X^r X^r$ flies have white eyes.

A white-eyed male ($X^r Y$) will transmit his X^r to all of his female offspring, but to none of his male offspring. This is because his daughters, in order to be female, must inherit his X chromosome, but his sons must inherit his Y chromosome. As shown in Figure 9.23B, when the female parent is a dominant homozygote ($X^R X^R$) and the male parent is $X^r Y$, all the offspring have red eyes, but the female offspring are all carriers of the allele for white eyes ($X^R X^r$). When those offspring are bred to each other, the classical 3:1 phenotypic ratio of red eyes to white eyes appears among the offspring (Figure 9.23C). However, there is a twist: The white-eyed trait shows up only in males. All the females have red eyes, whereas half the males have red eyes and half have white eyes. All females inherit at least one dominant allele (from their father); half of them are homozygous dominant, whereas the other half are heterozygous carriers, like their mother. Among the males, half of them inherit the recessive allele their mother was carrying, producing the white-eye phenotype.

Because the white-eye allele is recessive, a female will have white eyes only if she receives that allele on both X chromosomes. For example, if a heterozygous female mates with a white-eyed male, there is a 50% chance that each offspring will have white eyes (resulting from genotype $X^r X^r$ or $X^r Y$), regardless of sex (Figure 9.23D). Daughters with red eyes are heterozygotes, whereas red-eyed male offspring completely lack the recessive allele.

Web/CD Activity 9F *Sex-Linked Genes*

Web/CD Thinking as a Scientist *How Is the Chi-Square Test Used in Genetic Analysis?*

> [?] A white-eyed female *Drosophila* is mated with a red-eyed (wild-type) male. What result do you predict for the numerous offspring?

■ All female offspring will be red-eyed but heterozygous ($X^R X^r$); all male offspring will be white-eyed ($X^r Y$).

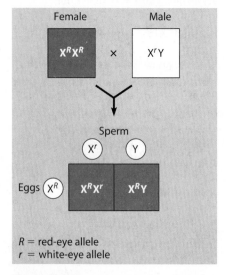

Figure 9.23A Fruit fly eye color, a sex-linked characteristic

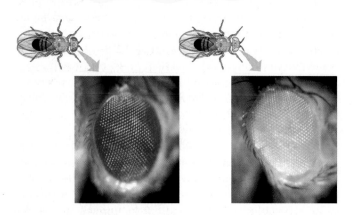

R = red-eye allele
r = white-eye allele

Figure 9.23B Homozygous, red-eyed female × white-eyed male

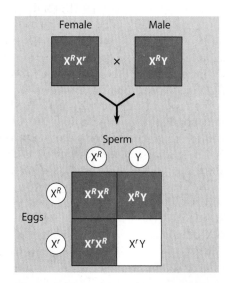

Figure 9.23C Heterozygous female × red-eyed male

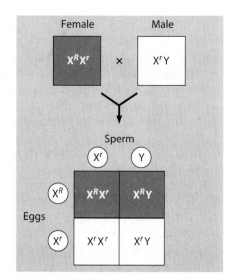

Figure 9.23D Heterozygous female × white-eyed male

9.24 Sex-linked disorders affect mostly males

Fruit fly genetics has taught us much about human inheritance. A number of human conditions result from sex-linked (X-linked) recessive alleles that are inherited in the same way as the white-eye trait in fruit flies. The fruit fly model also shows us why recessive sex-linked traits are expressed much more frequently in men than in women. Like a male fruit fly, if a man inherits only one sex-linked recessive allele—from his mother—the allele will be expressed. In contrast, a woman has to inherit two such alleles—one from each parent—in order to exhibit the trait.

Red-green color blindness is a common sex-linked disorder characterized by a malfunction of light-sensitive cells in the eyes. It is actually a class of disorders, involving several X-linked genes. A person with normal color vision can see more than 150 colors. In contrast, someone with red-green color blindness can see fewer than 25. For some affected people, red hues appear gray; others see gray instead of green; still others are green-weak or red-weak, tending to confuse shades of these colors. Mostly males are affected, but heterozygous females have some defects. (If you have red-green color blindness, you probably cannot see the numeral 7 in **Figure 9.24A.**)

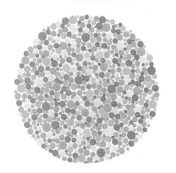

Figure 9.24A A test for red-green color blindness

Hemophilia is a sex-linked recessive trait with a long, well-documented history. Hemophiliacs bleed excessively when injured because they lack one or more of the proteins required for blood clotting. The most seriously affected individuals may bleed to death after relatively minor bruises or cuts.

A high incidence of hemophilia has plagued the royal families of Europe. The first royal hemophiliac seems to have been a son of Queen Victoria (1819–1901) of England. It is likely that the hemophilia allele arose through a mutation in one of the gametes of Victoria's mother or father, making Victoria a carrier of the deadly allele. Hemophilia was eventually introduced into the royal families of Prussia, Russia, and Spain through the marriages of two of Victoria's daughters who were carriers. Thus, the age-old practice of strengthening international alliances by marriage effectively spread hemophilia through the royal families of several nations. The photograph in **Figure 9.24B** shows Queen Victoria's granddaughter Alexandra, her husband Nicholas, who was the last czar of Russia, their daughters, and their son Alexis. The pedigree uses half-colored symbols to represent heterozygous carriers of the hemophilia allele. As you can see in the pedigree, Alexandra, like her mother and grandmother, was a carrier, and Alexis had the disease.

Another sex-linked recessive disorder is **Duchenne muscular dystrophy,** a condition characterized by a progressive weakening of the muscles and loss of coordination. Almost

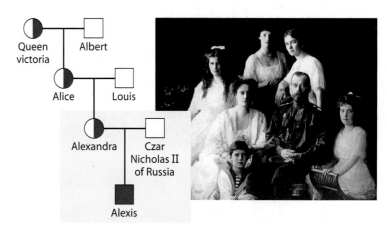

Figure 9.24B Hemophilia in the royal family of Russia

all cases are males, and the first symptoms appear in early childhood, when the child begins to have difficulty standing up. He is inevitably wheelchair-bound by age 12. Eventually, he becomes severely wasted, and normal breathing becomes difficult. Affected individuals rarely live past their early 20s.

For such a severe disease, Duchenne muscular dystrophy is relatively common. In the general U.S. population, about one in 3,500 male babies is affected, and the disease is even more common in some inbred populations. In one Amish community in Indiana, for instance, one out of every 100 males is born with the disease.

With the help of DNA technology (discussed in Chapter 12), the defective gene that causes Duchenne muscular dystrophy has been mapped at a particular point on the X chromosome. The gene's wild-type allele codes for a protein called dystrophin, which is present in normal muscle but missing in Duchenne patients.

The discovery of sex-linked genes and their pattern of inheritance in fruit flies and humans was one of many breakthroughs in understanding how genes are passed from one generation to the next. During the first half of the 20th century, geneticists rediscovered Mendel's work, reinterpreted his laws in light of chromosomal behavior during meiosis, and firmly established the chromosome theory of inheritance. The chromosome theory set the stage for another explosion of experimental work in the second half of the 20th century. This work was in molecular genetics, an area we explore in the next three chapters.

? Neither Tom nor Sue has Duchenne muscular dystrophy, but their first son does. If the couple has a second child, what is the probability that he or she will also have the disease?

■ $\frac{1}{4}$ ($\frac{1}{2}$ chance of a male child × $\frac{1}{2}$ chance that he will inherit the mutant X)

CHAPTER REVIEW

Reviewing the Concepts

Mendel's Laws (Introduction–9.10)

Mendelian genetics. The historical roots of genetics, the science of heredity, date back to ancient attempts at selective breeding **(Introduction–9.1)**. Modern genetics began with Gregor Mendel's quantitative experiments. Mendel crossed pea plants and traced traits from generation to generation. He hypothesized that there are alternative forms of genes, the units that determine heritable traits **(9.2)**. Mendel's law of segregation predicts that alleles separate in gametes **(9.3–9.4)**:

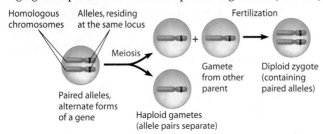

When the two alleles of a gene in a diploid individual are different, the dominant allele determines the inherited trait, whereas the recessive allele has no effect. Mendel's law of independent assortment states that the alleles of a pair segregate independently of other allele pairs during gamete formation **(9.5)**.

Mendelian crosses. The offspring of a testcross, a mating between an individual of unknown genotype and a homozygous recessive individual, can reveal the unknown's genotype **(9.6)**. Inheritance follows the rules of probability. The rule of multiplication calculates the probability of two independent events. The rule of addition calculates the probability of an event that can occur in alternative ways **(9.7)**.

Human genetics. The inheritance of many human traits follows Mendel's laws. Family pedigrees can be used to determine individual genotypes **(9.8)**. Many inherited disorders in humans are controlled by a single gene **(9.9)**. New technologies—including carrier screening, fetal testing, fetal imaging, and newborn screening—can provide insight for reproductive decisions but create new ethical dilemmas **(9.10)**.

Variations on Mendel's Laws (9.11–9.17)

Mendel's laws extended. Mendel's laws are valid for all sexually reproducing species, but genotype often does not dictate phenotype in the simple way his laws describe:

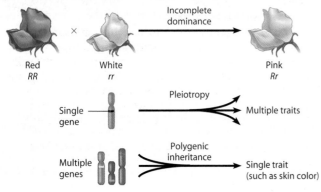

In a population, there may be multiple alleles for a characteristic, such as the three alleles for the ABO blood group. The alleles determining the A and B blood factors are codominant; that is, both are expressed in a heterozygote **(9.11–9.15)**. Many traits are affected, in varying degrees, by both genetic and environmental factors **(9.16)**. Predictive genetic testing may inform people of their risk for developing genetic diseases **(9.17)**.

The Chromosomal Basis of Inheritance (9.18–9.21)

Genes and chromosomes. Genes are located on chromosomes, whose behavior during meiosis and fertilization accounts for inheritance patterns **(9.18)**. Certain genes are linked; they tend to be inherited together because they reside close together on the same chromosome **(9.19)**. Crossing over can separate linked alleles, producing gametes with recombinant chromosomes **(9.20)**. Recombination frequencies can be used to map the relative positions of genes on chromosomes **(9.21)**.

Sex Chromosomes and Sex-Linked Genes (9.22–9.24)

Sex chromosomes determine sex in many species. In mammals, a male is XY, while a female is XX. The Y chromosome has genes for the development of testes, whereas an absence of the Y allows ovaries to develop. Other systems of sex determination exist in other animals and plants **(9.22)**. All genes on the sex chromosomes are said to be sex-linked. However, the X chromosome carries many genes unrelated to sex **(9.23)**. Most sex-linked human disorders are due to recessive alleles and are seen mostly in males. A male receiving a single X-linked recessive allele from his mother will have the disorder; a female has to receive the allele from both parents to be affected **(9.24)**.

Connecting the Concepts

1. Complete this concept map to help you review some key concepts of genetics.

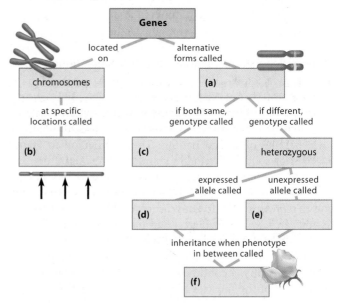

Testing Your Knowledge

Multiple Choice

2. Edward was found to be heterozygous (*Ss*) for sickle-cell trait. The alleles represented by the letters *S* and *s* are
   a. on the X and Y chromosomes.
   b. linked.
   c. on homologous chromosomes.
   d. both present in each of Edward's sperm cells.
   e. on the same chromosome but far apart.

3. Whether an allele is dominant or recessive depends on
   a. how common the allele is, relative to other alleles.
   b. whether it is inherited from the mother or the father.
   c. which chromosome it is on.
   d. whether it or another allele determines the phenotype when both are present.
   e. whether or not it is linked to other genes.

4. Two fruit flies with eyes of the usual red color are crossed, and their offspring are as follows: 77 red-eyed males, 71 ruby-eyed males, 152 red-eyed females. The allele for ruby eyes is
   a. autosomal (carried on an autosome) and dominant.
   b. autosomal and recessive.
   c. sex-linked and dominant.
   d. sex-linked and recessive.
   e. impossible to determine without more information.

5. In some of his experiments, Mendel studied the inheritance patterns of two characteristics at once—flower color and pod color, for example. He did this to find out
   a. whether genes for the two characteristics are inherited together or separately.
   b. how many genes are responsible for determining a characteristic.
   c. whether genes are on chromosomes.
   d. the distance between genes on a chromosome.
   e. how many different genes a pea plant has.

6. A man who has type B blood and a woman who has type A blood could have children of which of the following phenotypes?
   a. A or B only   d. A, B, or O
   b. AB only   e. A, B, AB, or O
   c. AB or O

Additional Genetics Problems

7. Why do more men than women have color blindness?

8. In fruit flies, the genes for wing shape and body stripes are linked. In a fly whose genotype is *WwSs*, *W* is linked to *S*, and *w* is linked to *s*. Show how this fly can produce gametes containing four different combinations of alleles. Which are parental-type gametes? Which are recombinant gametes? How are the recombinants produced?

9. Adult height in humans is at least partially hereditary; tall parents tend to have tall children. But humans come in a range of sizes, not just tall and short. Explain an extension of Mendel's model that could produce the hereditary variation in human height.

10. Tim and Jan both have freckles (see Module 9.8), but their son Mike does not. Show with a Punnett square how this is possible. If Tim and Jan have two more children, what is the probability that both will have freckles?

11. Both Tim and Jan (problem 10) have a widow's peak (see Module 9.8), but Mike has a straight hairline. What are their genotypes? What is the probability that Tim and Jan's next child will have freckles and a straight hairline?

12. In rabbits, black hair depends on a dominant allele, *B*, and brown hair on a recessive allele, *b*. Short hair is due to a dominant allele, *S*, and long hair to a recessive allele, *s*. If a true-breeding black, short-haired male is mated with a brown, long-haired female, describe their offspring. What will be the genotypes of the offspring? If two of these F_1 rabbits are mated, what phenotypes would you expect among their offspring? In what proportions?

13. A fruit fly with a gray body and red eyes (genotype *BbPp*) is mated with a fly having a black body and purple eyes (genotype *bbpp*). What offspring, in what proportions, would you expect if the body-color and eye-color genes are on different chromosomes (unlinked)? When this mating is actually carried out, most of the offspring look like the parents, but 3% have gray body and purple eyes, and 3% have black body and red eyes. Are these genes linked or unlinked? What is the recombination frequency?

14. A series of matings shows that the recombination frequency between the black-body gene (problem 13) and the gene for dumpy (shortened) wings is 36%. The recombination frequency between purple eyes and dumpy wings is 41%. What is the sequence of these three genes on the chromosome?

15. A couple are both phenotypically normal, but their son suffers from hemophilia, a sex-linked recessive disorder. What fraction of their children are likely to suffer from hemophilia? What fraction are likely to be carriers?

16. Heather was not able to see the numeral 7 in Figure 9.24A and was surprised to discover she suffered from red-green color blindness. She told her biology professor, who said, "Your father is color-blind too, right?" How did her professor know this? Why did her professor not say the same thing to the color-blind males in the class?

Applying the Concepts

17. In 1981, a stray black cat with unusual rounded, curled-back ears was adopted by a family in Lakewood, California. Hundreds of descendants of this cat have since been born, and cat fanciers hope to develop the "curl" cat into a show breed. The curl allele is apparently dominant and autosomal (carried on an autosome).  Suppose you owned the first curl cat and wanted to breed it to develop a true-breeding variety. Describe tests that would determine whether the curl gene is dominant or recessive and whether it is autosomal or sex-linked. Explain why you think your tests would be conclusive. Describe a test to determine that a cat is true-breeding.

Answers to all questions can be found in Appendix 3.

For study help and Activities, go to campbellbiology.com or the student CD-ROM.

CHAPTER 10

THE STRUCTURE OF THE GENETIC MATERIAL

10.1 Experiments showed that DNA is the genetic material
10.2 DNA and RNA are polymers of nucleotides
10.3 DNA is a double-stranded helix

DNA REPLICATION

10.4 DNA replication depends on specific base pairing
10.5 DNA replication: A closer look

THE FLOW OF GENETIC INFORMATION FROM DNA TO RNA TO PROTEIN

10.6 The DNA genotype is expressed as proteins, which provide the molecular basis for phenotypic traits
10.7 Genetic information written in codons is translated into amino acid sequences
10.8 The genetic code is the Rosetta stone of life
10.9 Transcription produces genetic messages in the form of RNA
10.10 Eukaryotic RNA is processed before leaving the nucleus
10.11 Transfer RNA molecules serve as interpreters during translation
10.12 Ribosomes build polypeptides
10.13 An initiation codon marks the start of an mRNA message
10.14 Elongation adds amino acids to the polypeptide chain until a stop codon terminates translation
10.15 Review: The flow of genetic information in the cell is DNA → RNA → protein
10.16 Mutations can change the meaning of genes

MICROBIAL GENETICS

10.17 Viral DNA may become part of the host chromosome
10.18 Many viruses cause disease in animals
10.19 Plant viruses are serious agricultural pests
10.20 Emerging viruses threaten human health
10.21 The AIDS virus makes DNA on an RNA template
10.22 Bacteria can transfer DNA in three ways
10.23 Bacterial plasmids can serve as carriers for gene transfer

Sabotage Inside Our Cells

A SABOTEUR DRIFTS STEALTHILY toward his target, a vital factory. Stopped at the perimeter by a guard, the intruder presents counterfeit identification to gain entry. Once inside, he surveys the scene and makes a quick decision: The time is not yet ripe for sabotage. So he lies low and waits silently, undetected, until he receives the go signal. The intruder now acts quickly, hijacking the factory

Molecular Biology of the Gene

Viral protein —

machinery and diverting production to his own diabolical ends. The factory, controlled by the saboteur, now manufactures replicas of the saboteur! When these duplicates are ready, they break their way out, destroying the factory as they exit. With the ruins behind them, they move off silently in search of new targets.

The scenario just described is played out millions of times each year. What is it? Industrial sabotage? Military espionage? In fact, this story describes cellular damage by a herpesvirus, the type of virus that causes cold sores, genital herpes (the most common sexually transmitted disease), chicken pox, and a number of other diseases. For genital herpes alone, usually caused by the herpesvirus called herpes simplex 2, as many as 500,000 Americans may be newly infected each year; the number who acquire the cold-sore virus, herpes simplex 1, is even higher. The colorized TEM at the right shows a herpesvirus magnified 500,000 times.

Once a person is infected with a herpesvirus, the virus remains permanently latent in the body

Viruses share some of the characteristics of living organisms, such as genetic material in the form of nucleic acid, packaged within a highly organized structure. A virus is generally not considered alive, however, because it is not cellular and cannot reproduce on its own. A virus is simply nucleic acid wrapped in a coat of protein and, for herpesviruses and some other animal viruses, a membranous envelope. While a herpesvirus is fairly large as viruses go—about 200 nm across—its diameter is less than 1/100 that of a typical human cell. Just about all a herpesvirus or any other virus can do is infect a host. It is the host that provides most of the tools and raw materials for viral multiplication.

Once in the body, a herpesvirus tumbles along until it finds a suitable target cell, recognized when protein molecules on the outside of the virus (see micrograph at right) fit into receptor molecules on the surface of the cell. Not perceiving the threat, the cell takes in the virus. Once inside the cell, the DNA of the herpesvirus enters the nucleus. In the nuclei of certain nerve cells, the viral DNA can remain dormant for long periods of time, until activated by a signal such as cellular stress. When activated, the viral DNA hijacks the cell's own molecules and organelles to produce new copies of the virus. Virus production eventually results in destruction of host cells—causing the sores that are characteristic of herpes diseases. The released viruses can then infect other cells.

Once a person is infected with a herpesvirus, the virus remains permanently latent (dormant) in the body, its DNA integrated into the chromosomes of nerve cells. Over 75% of American adults are thought to carry herpes simplex 1, and over 20% herpes simplex 2, although many people never develop symptoms. Herpesviruses are somewhat unusual in being able to remain latent inside our cells. Other viruses with this ability include HIV, the virus that causes AIDS.

Because viruses are much simpler than cells, they are relatively easy to study on the molecular level, far easier than Mendel's peas or Morgan's fruit flies. For this reason, we owe our first glimpses of the functions of DNA, the molecule that controls hereditary traits, to the study of viruses.

This chapter is about the DNA molecule and how it serves as the basis of heredity—the subject of **molecular biology.** Here we explore the structure of DNA, how it replicates (the molecular basis of why offspring resemble their parents), how it controls the cell by directing RNA and protein synthesis, and how it can change. We also look at viruses that infect bacteria, animals, and plants. We end with an examination of bacterial genetics. To start the chapter, we recount the story of how we know that DNA is the genetic material, a story in which a virus played a major role. ■ ■ ■

10.1 Experiments showed that DNA is the genetic material

Today, even schoolchildren have heard of DNA, and scientists routinely manipulate DNA in the laboratory and use it to change the heritable characteristics of cells. Early in the 20th century, however, the identification of the molecules of inheritance posed a major challenge. Biologists knew that genes are located on chromosomes. Therefore, the two chemical components of chromosomes—DNA and protein—were the candidates for the genetic material. Until the 1940s, the case for proteins seemed stronger because proteins appeared to be more structurally complex and functionally specific. Biologists finally established the role of DNA in heredity through studies involving bacteria and the viruses that infect them.

We can trace the discovery of the genetic role of DNA back to 1928. British medical officer Frederick Griffith was studying two strains of a bacterium: a pathogenic (disease-causing) strain that causes pneumonia and a harmless strain. Griffith was surprised to find that when he killed the pathogenic bacteria and then mixed the bacterial remains with living harmless bacteria, some living bacterial cells were converted to the disease-causing form. Furthermore, all of the descendants of the transformed bacteria inherited the newly acquired ability to cause disease. Clearly, some chemical component of the dead bacteria could act as a "transforming factor" that brought about a heritable change.

Most biologists doubted that DNA could be Griffith's transforming factor, primarily because so little was known about DNA. However, in 1952, American biologists Alfred Hershey and Martha Chase performed a very convincing set of experiments. They showed that DNA is the genetic material of a virus called T2, which infects the bacterium *Escherichia coli* (*E. coli*). Bacterial viruses are called **bacteriophages** ("bacteria-eaters"), or **phages** for short. Figure 10.1A shows the structure of phage T2, which consists solely of DNA (blue) and protein (yellow). Resembling a lunar landing craft, T2 has a DNA-containing head and a hollow tail with six jointed fibers extending from it. The fibers attach to the surface of a susceptible bacterium. Hershey and Chase knew that T2 could reprogram its host cell to produce new phages, but they did not know which component—DNA or protein—was responsible for this ability.

Hershey and Chase found the answer by devising an experiment to determine what kinds of molecules the phage transferred to *E. coli* during infection. Their experiment used only a few relatively simple tools: chemicals containing radioactive isotopes (see Module 2.5); a radioactivity detector; a kitchen blender; and a centrifuge, a device that spins test tubes to separate particles of different weights. (These are still basic tools of molecular biology.)

Hershey and Chase used different radioactive isotopes to label the DNA and protein in T2. First, they grew T2 with *E. coli* in a solution containing radioactive sulfur (yellow in Figure 10.1B). Protein contains sulfur but DNA does not, so as

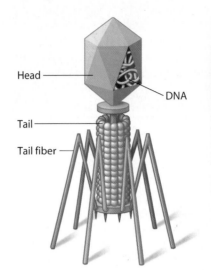

Head

DNA

Tail

Tail fiber

300,000×

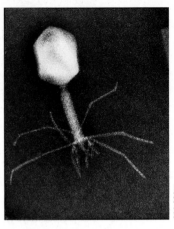

Figure 10.1A   Phage T2

new phages were made, the radioactive sulfur atoms were incorporated only into their proteins. The researchers grew a separate batch of phages in a solution containing radioactive phosphorus (green). Because nearly all the phage's phosphorus is in DNA, this labeled only the phage DNA.

Armed with the two batches of labeled T2, Hershey and Chase were ready to perform the experiment outlined in Figure 10.1B. ❶ They allowed the two batches of T2 to infect separate samples of nonradioactive bacteria. ❷ Shortly after the onset of infection, they agitated the cultures in a blender to shake loose any parts of the phages that remained outside the bacterial cells. ❸ They then spun the mixtures in a centrifuge. The cells were deposited as a pellet at the bottom of the centrifuge tubes, but phages and parts of phages, being lighter, remained suspended in the liquid. ❹ The researchers then measured the radioactivity in the pellet and the liquid.

Hershey and Chase found that when the bacteria had been infected with T2 phages containing labeled protein, the radioactivity ended up mainly in the liquid, which contained phages but not bacteria. This result suggested that the phage protein did not enter the cells. But when the bacteria had been infected with phages whose DNA was tagged, then most of the radioactivity was in the pellet, made up of bacteria. When these bacteria were returned to liquid growth medium, the bacterial cells were soon destroyed, lysing (breaking open) and releasing new phages that contained radioactive phosphorus in their DNA but no radioactive sulfur in their proteins.

Hershey and Chase concluded that T2 injects its DNA into the host cell, leaving virtually all its protein outside (as shown in Figure 10.1B). More importantly, they demonstrated that it is the injected DNA molecules that cause the

cells to produce additional phage DNA and proteins—indeed, new complete phages. This indicated that the DNA contained the instructions for making phages. **Figure 10.1C** outlines the reproductive cycle for phage T2 as we now understand it.

The Hershey-Chase results, added to earlier evidence, convinced most scientists that DNA is the hereditary material. What happened next was one of the most celebrated quests in the history of science: the effort to figure out the structure of DNA and how this structure enables the molecule to store genetic information and transmit it from parents to offspring.

Web/CD Activity 10A *The Hershey-Chase Experiment*

Web/CD Activity 10B *Phage T2 Reproductive Cycle*

? What convinced Hershey and Chase that DNA, rather than protein, is the genetic material of phage T2?

■ Radioactively labeled phage DNA, but not labeled protein, entered the host cell during infection and directed the synthesis of new viruses.

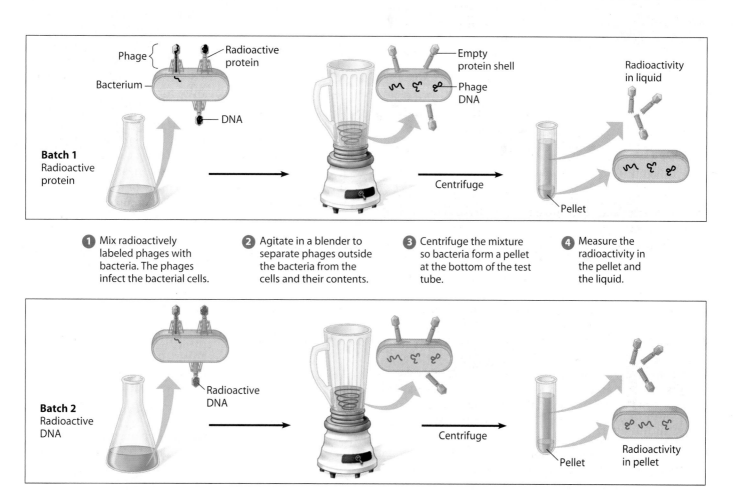

Batch 1 Radioactive protein

Batch 2 Radioactive DNA

1 Mix radioactively labeled phages with bacteria. The phages infect the bacterial cells.

2 Agitate in a blender to separate phages outside the bacteria from the cells and their contents.

3 Centrifuge the mixture so bacteria form a pellet at the bottom of the test tube.

4 Measure the radioactivity in the pellet and the liquid.

Figure 10.1B The Hershey-Chase experiment

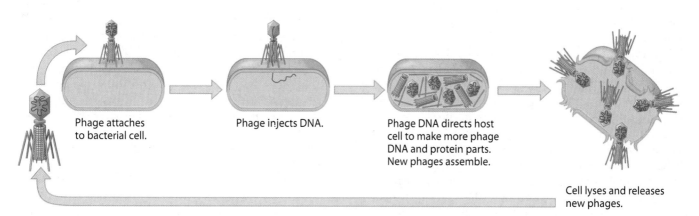

Phage attaches to bacterial cell.

Phage injects DNA.

Phage DNA directs host cell to make more phage DNA and protein parts. New phages assemble.

Cell lyses and releases new phages.

Figure 10.1C Phage reproductive cycle

10.2 DNA and RNA are polymers of nucleotides

By the time Hershey and Chase performed their experiments, a good deal was already known about DNA. Scientists had identified all its atoms and knew how they were covalently bonded to one another. What was not understood was the specific arrangement of parts that gave DNA its unique properties—the capacity to store genetic information, copy it, and pass it from generation to generation. However, only one year after Hershey and Chase published their results, scientists figured out DNA's three-dimensional structure and the basic strategy of how it works. We will examine that momentous discovery in Module 10.3. First, let's look at the underlying chemical structure of DNA and its chemical cousin RNA.

Recall from Module 3.20 that DNA and RNA are nucleic acids, which consist of long chains (polymers) of chemical units (monomers) called **nucleotides.** A very simple diagram of such a polymer, or **polynucleotide,** is shown on the far left in **Figure 10.2A.** This chain shows one arrangement of the four types of nucleotides that make up DNA. Each type of DNA nucleotide has a different nitrogenous base: adenine (A), cytosine (C), thymine (T), or guanine (G). Because nucleotides can occur in a polynucleotide in any sequence and polynucleotides vary in length from long to very long, the number of possible polynucleotides is enormous.

Looking more closely at our polynucleotide, we see in the center of Figure 10.2A that each nucleotide consists of three components: a nitrogenous base (in DNA, A, C, T, or G), a sugar (blue), and a phosphate group (yellow). The nucleotides are joined to one another by covalent bonds between the sugar of one nucleotide and the phosphate of the next. This results in a **sugar-phosphate backbone,** a repeating pattern of sugar-phosphate-sugar-phosphate. The nitrogenous bases are arranged as appendages all along this backbone.

Examining a single nucleotide in even more detail (on the right in Figure 10.2A), we note the chemical structure of its three components. The phosphate group has a phosphorus atom (P) at its center and is the source of the *acid* in nucleic acid. The sugar has five carbon atoms (shown in red here for emphasis)—four in its ring and one extending above the ring. The ring also includes an oxygen atom. The sugar is called deoxyribose because, compared with the sugar ribose, it is missing an oxygen atom. (Notice that the C atom in the lower right corner of the ring is bonded to an H atom instead of to an —OH group, as it is in ribose; see Figure 10.2C.) The full name for DNA is deoxyribonucleic acid, with the "nucleic" part coming from DNA's location in the nuclei of eukaryotic cells. The nitrogenous base (thymine, in our example) has a ring consisting of nitrogen

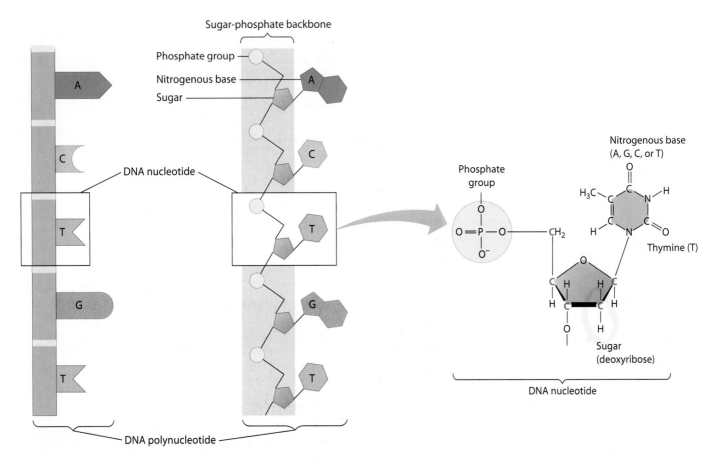

Figure 10.2A   DNA polynucleotide

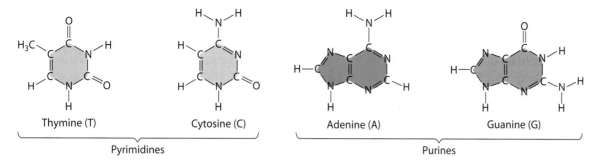

Figure 10.2B Nitrogenous bases of DNA

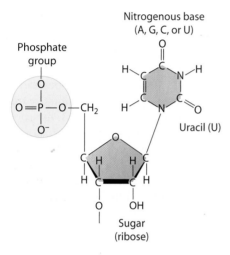

Figure 10.2C An RNA nucleotide

Key
- Hydrogen atom
- Carbon atom
- Nitrogen atom
- Oxygen atom
- Phosphorus atom

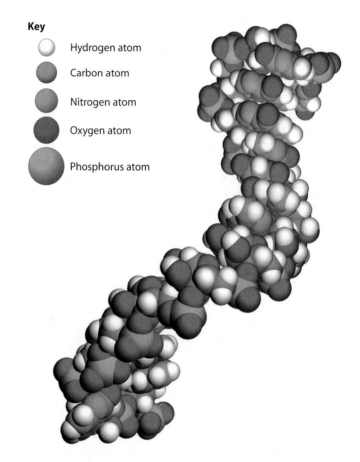

Figure 10.2D Part of an RNA polynucleotide

and carbon atoms with various functional groups attached. In contrast to the acidic phosphate group, nitrogenous bases are basic (hence their name).

The four nucleotides found in DNA differ only in their nitrogenous bases. Figure 10.2B shows the structures of DNA's four nitrogenous bases. At this point, the structural details are not as important as the fact that the bases are of two types. **Thymine (T)** and **cytosine (C)** are single-ring structures called *pyrimidines*. **Adenine (A)** and **guanine (G)** are larger, double-ring structures called *purines*. The one-letter abbreviations can be used for either the bases alone or for the nucleotides containing them.

What about RNA? As its name—ribonucleic acid—implies, its sugar is ribose rather than deoxyribose. Notice the ribose in the RNA nucleotide in Figure 10.2C; unlike deoxyribose, the sugar ring has an —OH group attached to the C atom at its lower right corner. Another difference between RNA and DNA is that instead of thymine, RNA has a nitrogenous base called **uracil (U)**. (You can see the structure of uracil in Figure 10.2C; it is very similar to thymine.) Except for the presence of ribose and uracil, an RNA polynucleotide chain is identical to a DNA polynucleotide chain. Figure 10.2D is a computer

graphic of a piece of RNA polynucleotide about 20 nucleotides long. The orange color of the phosphorus atoms at the center of the phosphate groups makes it easy to spot the sugar-phosphate backbone.

? Compare and contrast DNA with RNA.

■ Both are polymers of nucleotides. A nucleotide consists of a sugar + a nitrogenous base + a phosphate group. In RNA, the sugar is ribose; in DNA, it is deoxyribose. Both RNA and DNA have the bases A, G, and C; for a fourth base, DNA has T and RNA has U.

10.3 DNA is a double-stranded helix

After most biologists became convinced that DNA was the genetic material, a race was on to determine how the structure of this molecule could account for its role in heredity. By the beginning of the 1950s, the arrangement of covalent bonds in a nucleic acid polymer was well established, and researchers focused on discovering the three-dimensional structure of DNA. Among the scientists working on the problem were Linus Pauling, in California, and Maurice Wilkins and Rosalind Franklin (**Figure 10.3A**), in London. First to the finish line, however, were two scientists who were relatively unknown—American James D. Watson and Englishman Francis Crick (**Figure 10.3B**).

Figure 10.3A
Rosalind Franklin

The brief but celebrated partnership that solved the puzzle of DNA structure began soon after the 23-year-old Watson journeyed to Cambridge University, where Crick was studying protein structure with a technique called X-ray crystallography. While visiting the laboratory of Maurice Wilkins at King's College in London, Watson saw an X-ray crystallographic image of DNA produced by Wilkins's colleague Rosalind Franklin. Just a glance at the image enabled Watson to deduce the basic shape of DNA to be a helix with a uniform diameter of 2 nm, with its nitrogenous bases stacked about one-third of a nanometer apart. (For comparison, recall that the plasma membrane of a cell is about 8 nm thick.) The diameter of the helix suggested that it was made up of two polynucleotide strands. The presence of two strands accounts for the now-familiar term **double helix.**

Watson and Crick began trying to construct a double helix that would conform both to Franklin's data and to what was then known about the chemistry of DNA. Franklin had concluded that the sugar-phosphate backbones must be on the outside of the double helix, forcing the nitrogenous bases to swivel to the interior of the molecule. But how were the bases arranged in the interior of the double helix?

At first, Watson and Crick imagined that the bases paired like with like—for example, A with A, and C with C. But that kind of pairing did not fit the X-ray data, which suggested that the DNA molecule has a *uniform* diameter. An AA pair would be almost twice as wide as a CC pair, causing bulges in the molecule. It soon became apparent that a double-ringed base (purine) must always be paired with a single-ringed base (pyrimidine) on the opposite strand. Moreover, Watson and Crick realized that the individual structures of the bases dictated the pairings even more specifically. Each base has chemical side groups that can best form hydrogen bonds with one appropriate partner (to review the hydrogen bond, see Module 2.10). Adenine can best form hydrogen bonds with thymine, and guanine with cytosine. In the biologist's shorthand, A pairs with T, and G pairs with C. A is also said to be "complementary" to T, and G to C.

Watson and Crick's pairing scheme not only fit what was known about the physical attributes and chemical bonding of DNA, but also explained some data obtained several years earlier by American biochemist Erwin Chargaff. Chargaff had discovered that the amount of adenine in the DNA of any one species was equal to the amount of thymine and that the amount of guanine was equal to that of cytosine. Chargaff's rules, as they are called, are explained by the fact that A on one of DNA's polynucleotide chains always pairs with T on the other polynucleotide chain, and G on one chain pairs only with C on the other chain.

You can picture the model of the DNA double helix proposed by Watson and Crick as a twisted rope ladder with wooden rungs (**Figure 10.3C**). The side ropes are the equivalent of the sugar-phosphate backbones, and the rungs represent pairs of nitrogenous bases joined by hydrogen bonds.

Figure 10.3D shows three representations of the double helix. The shapes of the base symbols in the ribbonlike diagram on the left indicate the bases' complementarity. In the center is an atomic-level version showing four base pairs, with the helix untwisted and the hydrogen bonds specified by dotted lines; you can see that the two sugar-phosphate backbones of the double helix are oriented in opposite directions. (Notice that the sugars on the two strands are upside down with respect to each other.) On the right is a computer graphic showing every atom of part of a double helix. The

Figure 10.3B Watson and Crick in 1953 with their model of the DNA double helix

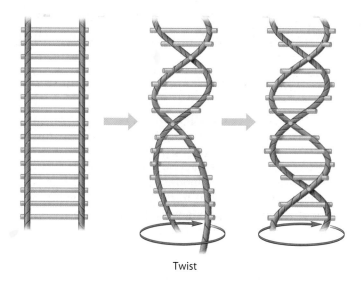

Twist

Figure 10.3C A rope-ladder model for the double helix

atoms that compose the deoxyribose sugars are shown as bright blue, phosphate groups as yellow, and nitrogenous bases as shades of green and orange.

Although the Watson-Crick base-pairing rules dictate the side-by-side combinations of nitrogenous bases that form the rungs of the double helix, they place no restrictions on the *sequence* of nucleotides along the length of a DNA strand. In fact, the sequence of bases can vary in countless ways, and each gene has a unique order of nucleotides, or base sequence.

In April 1953, Watson and Crick shook the scientific world with a succinct paper explaining their molecular model for DNA in the journal *Nature*. In 1962, Watson, Crick, and Wilkins received the Nobel Prize for their work. (Rosalind Franklin probably would have received the prize as well, but for her death from cancer in 1958; Nobel Prizes are never awarded posthumously.) Few milestones in the history of biology have had as broad an impact as the discovery of the double helix, with its AT and CG base pairing.

The Watson-Crick model gave new meaning to the words *genes* and *chromosomes*—and to the chromosome theory of inheritance (see Module 9.18). With a complete picture of DNA, we can see that the genetic information in a chromosome must be encoded in the nucleotide sequence of the molecule. One powerful aspect of the Watson-Crick model is that the structure of DNA suggests a molecular explanation for genetic inheritance, as we see in the next module.

Web/CD Activity 10C *DNA and RNA Structure*

Web/CD Activity 10D *DNA Double Helix*

? Along one strand of a double helix is the nucleotide sequence GGCATAGGT. What is the complementary sequence for the other DNA strand?

■ CCGTATCCA

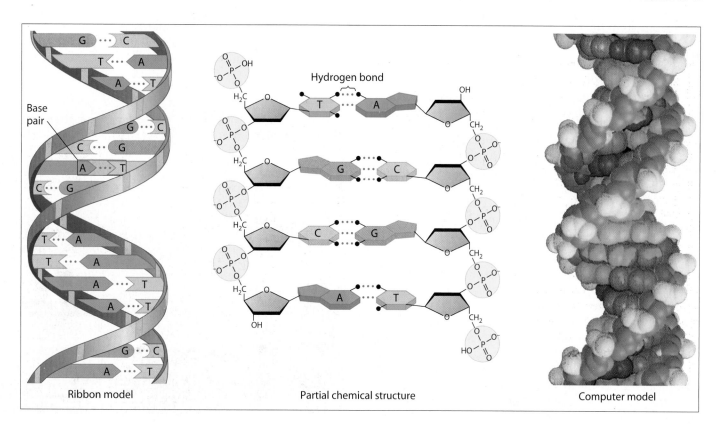

Ribbon model   Partial chemical structure   Computer model

Base pair

Hydrogen bond

Figure 10.3D Three representations of DNA

10.4 DNA replication depends on specific base pairing

One of biology's overarching themes—the relationship between structure and function—is evident in the double helix. The idea that there is specific pairing of bases in DNA was the flash of inspiration that led Watson and Crick to the correct structure of the double helix. At the same time, they saw the functional significance of the base-pairing rules. They ended their classic paper with this statement: "It has not escaped our notice that the specific pairing we have postulated immediately suggests a possible copying mechanism for the genetic material."

The logic behind the Watson-Crick proposal for how DNA is copied—by specific pairing of complementary bases—is quite simple. You can see this by covering one of the strands in the parental DNA molecule in Figure 10.4A with a piece of paper. You can determine the sequence of bases in the covered strand by applying the base-pairing rules to the unmasked strand: A pairs with T, G with C. Watson and Crick predicted that a cell applies the same rules when copying its genes. Figure 10.4A illustrates the template hypothesis for DNA replication. First, the two strands of parental DNA separate, and each becomes a template for the assembly of a complementary strand from a supply of free nucleotides. The nucleotides line up one at a time along the template strand in accordance with the base-pairing rules. Enzymes link the nucleotides to form the new DNA strands. The completed new molecules, identical to the parental molecule, are known as daughter DNA. The copying mechanism is analogous to using a photographic negative to make a positive image, which can in turn be used to make another negative, and so on.

Watson and Crick's model predicts that when a double helix replicates, each of the two daughter molecules will have one old strand, which was part of the parental molecule, and one newly made strand. This model for DNA replication is known as the **semiconservative model** because half of the parental molecule is maintained (conserved) in each daughter molecule. The semiconservative model of replication was confirmed by experiments performed in the 1950s.

Although the general mechanism of DNA replication is conceptually simple, the actual process involves complex

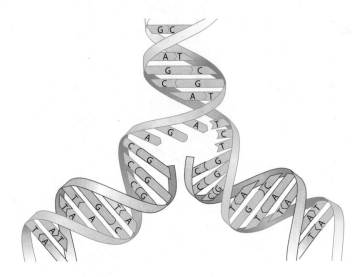

Figure 10.4B Untwisting and replication of DNA

biochemical gymnastics. Some of the complexity arises from the fact that the helical DNA molecule must untwist as it replicates and must copy its two strands roughly simultaneously (Figure 10.4B). Another challenge is the speed of the process. Nucleotides are added at a rate of about 50 per second in mammals and 500 per second in bacteria. We take a closer look at the mechanisms of DNA replication in the next module.

Web/CD Thinking as a Scientist *What Is the Correct Model for DNA Replication?*

? How does complementary base pairing make possible the replication of DNA?

■ When the two strands of the double helix separate, each serves as a "mold" upon which nucleotides can be arranged by specific base pairing into new complementary strands.

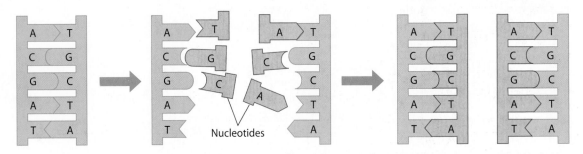

Figure 10.4A
A template model for DNA replication

Parental molecule of DNA

Both parental strands serve as templates

Nucleotides

Two identical daughter molecules of DNA

10.5 DNA replication: A closer look

DNA replication begins at specific sites on the double helix, called origins of replication, where proteins that start the process attach to the DNA and separate the strands. As shown in **Figure 10.5A**, replication then proceeds in both directions, creating what are called replication "bubbles." The parental DNA strands (blue) open up as daughter strands (gray) elongate on both sides of each bubble. The DNA molecule of a eukaryotic chromosome has many origins where replication can start simultaneously, shortening the total time needed for the process. Thus, thousands of bubbles can be present at once. Eventually, all the bubbles merge, yielding two completed daughter DNA molecules.

Figure 10.5B shows the molecular building blocks of a tiny segment of DNA, reminding us that the DNA's sugar-phosphate backbones run in opposite directions. Notice that each strand has a 3′ ("three-prime") end and a 5′ end. The primed numbers refer to the carbon atoms of the nucleotide sugars. At one end of each DNA strand, the sugar's 3′ carbon atom is attached to an —OH group; at the other end, the sugar's 5′ carbon has a phosphate group.

The opposite orientation of the strands is important in DNA replication. The enzymes that link DNA nucleotides to a growing daughter strand, called **DNA polymerases**, add nucleotides only to the 3′ end of the strand, never to the 5′ end. Thus, a daughter DNA strand can only grow in the 5′ → 3′ direction. You see the consequences of this enzyme specificity in **Figure 10.5C**. The forked structure represents one side of a replication bubble. One of the daughter strands (shown in gray) can be synthesized in one continuous piece by a DNA polymerase working toward the forking point of the parental DNA. However, to make the other daughter strand, polymerase molecules must work outward from the forking point. This new strand is synthesized in short pieces as the fork opens up. Another enzyme, called **DNA ligase**, then links (ligates) the pieces together into a single DNA strand.

Altogether, DNA replication requires the cooperation of more than a dozen enzymes and other proteins. The process is not only fast but also amazingly accurate; typically, only about

one DNA nucleotide per billion is incorrectly paired. In addition to their roles in linking nucleotides together, DNA polymerases carry out a proofreading step that quickly removes nucleotides that have base-paired incorrectly during replication. DNA polymerases and DNA ligase are also involved in repairing DNA damaged by harmful radiation (such as ultraviolet light and X-rays) or toxic chemicals in the environment.

DNA replication ensures that all the somatic cells in a multicellular organism carry the same genetic information. It is also the means by which genetic instructions are copied for the next generation of the organism. In the next module, we begin to pursue the connection between DNA instructions and an organism's phenotypic traits.

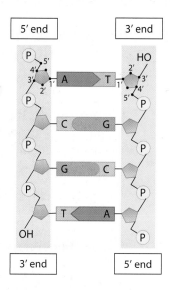

Figure 10.5B The opposite orientations of DNA strands

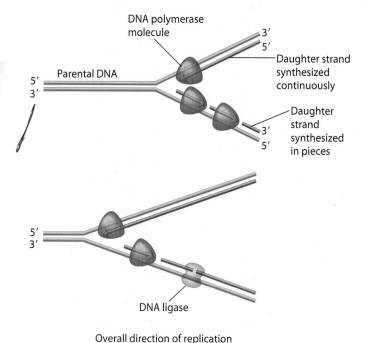

Figure 10.5C How daughter DNA strands are synthesized

Web/CD Activity 10E *DNA Replication*

? What is the function of DNA polymerase in DNA replication?

■ This enzyme covalently connects the nucleotides to form a new strand as the nucleotides line up along an existing strand according to the base-pairing rules.

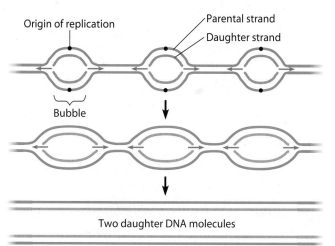

Figure 10.5A Multiple "bubbles" in replicating DNA

10.6 The DNA genotype is expressed as proteins, which provide the molecular basis for phenotypic traits

With our knowledge of DNA, we can now define genotype and phenotype more precisely than we did in Chapter 9. An organism's genotype, its genetic makeup, is the heritable information contained in its DNA. The phenotype is the organism's specific traits. So what is the molecular connection between genotype and phenotype?

The answer is that the DNA inherited by an organism specifies traits by dictating the synthesis of proteins. In other words, proteins are the links between genotype and phenotype. However, a gene does not build a protein directly. Rather, a gene dispatches instructions in the form of RNA, which in turn programs protein synthesis. This central concept in biology is summarized in **Figure 10.6A.** The chain of command is from DNA in the nucleus of the cell (purple area) to RNA to protein synthesis in the cytoplasm (tan area). The two main stages are **transcription,** the transfer of genetic information from DNA into an RNA molecule, and **translation,** the transfer of the information in the RNA into a protein. In the next nine modules, we will explore the steps in this flow of molecular information from gene to protein.

The relationship between genes and proteins was first proposed in 1909, when English physician Archibald Garrod suggested that genes dictate phenotypes through enzymes, the proteins that catalyze chemical processes in the cell. Garrod's idea came from his observations of inherited diseases. He hypothesized that an inherited disease reflects a person's inability to make a particular enzyme, and he referred to such diseases as "inborn errors of metabolism." He gave as one example the hereditary condition called alkaptonuria, in which the urine appears black because it contains a chemical called alkapton. Garrod reasoned that normal individuals have an enzyme that breaks down alkapton, whereas alkaptonuric individuals cannot make the enzyme. Garrod's hypothesis was ahead of its time, but research conducted decades later proved him right. In the intervening years, biochemists accumulated evidence that cells make and break down biologically important molecules via metabolic pathways, as in the synthesis of an amino acid or the breakdown of a sugar. As we described in Unit I, each step in a metabolic pathway is catalyzed by a specific enzyme. Therefore, individuals lacking one of the enzymes for a pathway are unable to complete the pathway.

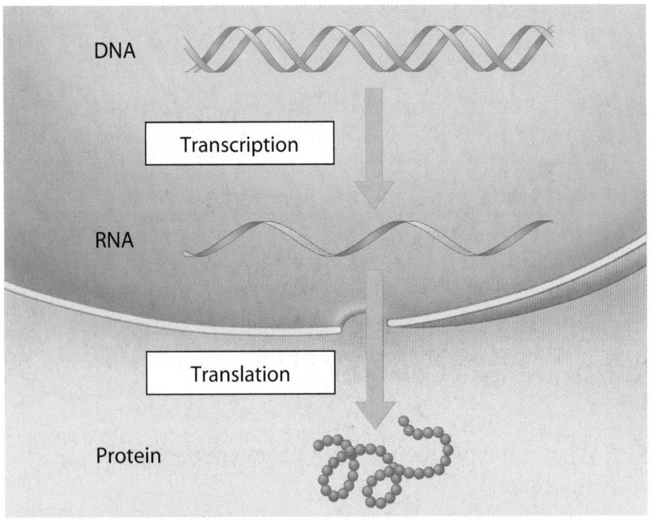

The major breakthrough in demonstrating the relationship between genes and enzymes came in the 1940s from the work of American geneticists George Beadle and Edward Tatum with the bread mold *Neurospora crassa* (Figure 10.6B). Beadle and Tatum studied strains of the mold that were unable to grow on a simple growth medium. Each of these so-called nutritional mutants turned out to lack an enzyme in a metabolic pathway that produced some molecule the mold needed, such as an amino acid. Beadle and Tatum also showed that each mutant was defective in a single gene. This result suggested the one gene–one enzyme hypothesis: that the function of a gene is to dictate the production of a specific enzyme.

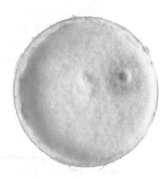

Figure 10.6B *Neurospora crassa* growing in a culture dish

The one gene–one enzyme hypothesis has been amply confirmed, but with some important modifications. First it was extended beyond enzymes to include *all* types of proteins. For example, keratin, the structural protein of your hair, and the hormone insulin are two examples of proteins that are not enzymes. So biologists began to think in terms of one gene–one *protein*. However, many proteins are made from two or more polypeptide chains (see Module 3.14), with each polypeptide specified by its own gene. For example, hemoglobin, the oxygen-transporting protein in your blood, is built from two kinds of polypeptides, encoded by two different genes. Thus, Beadle and Tatum's hypothesis has come to be restated as the one gene–one *polypeptide* hypothesis.

Web/CD Activity 10F *Overview of Protein Synthesis*

Web/CD Thinking as a Scientist *How Are Nutritional Mutations Identified?*

> **?** In the information flow from DNA to protein, what are the functions of transcription and translation?

DNA

Transcription

RNA

Translation

Protein

Figure 10.6A Flow of genetic information in a eukaryotic cell

■ Transcription is the transfer of information from DNA to RNA. Translation is the use of the RNA as information for making a protein.

10.7 Genetic information written in codons is translated into amino acid sequences

Genes provide the instructions for making specific proteins. But a gene does not build a protein directly. The bridge between DNA and protein synthesis is the nucleic acid RNA: DNA is transcribed into RNA, which is then translated into protein. Put another way, cells are governed by a molecular chain of command: DNA → RNA → protein.

Transcription and translation are linguistic terms, and it is useful to think of nucleic acids and proteins as having languages. To understand how genetic information passes from genotype to phenotype, we need to see how the chemical language of DNA is translated into the different chemical language of proteins.

What, exactly, is the language of nucleic acids? Both DNA and RNA are polymers made of nucleotide monomers. In DNA, there are four types of nucleotides, which differ in their nitrogenous bases (A, T, C, and G). The same is true for RNA, although it has the base U instead of T.

Figure 10.7 focuses on a small region of one of the genes (gene 3, shown in light blue) carried by a DNA molecule. DNA's language is written as a linear sequence of nucleotide bases on a polynucleotide, a sequence such as the one you see on the enlarged DNA strand in the figure. Specific sequences of bases, each with a beginning and an end, make up the genes on a DNA strand. A typical gene consists of hundreds or thousands of nucleotides in a specific sequence.

The pink strand underneath the enlarged DNA region represents the results of transcription: an RNA molecule. The process is called transcription because the nucleic acid language of DNA has been rewritten (transcribed) as a sequence of bases on RNA; the language is still that of nucleic

acids. Notice that the nucleotide bases on the RNA molecule are complementary to those on the DNA strand. As we will see in Module 10.9, this is because the RNA was synthesized using the DNA as a template.

The purple chain represents the results of translation, the conversion of the nucleic acid language into the polypeptide language (recall that proteins consist of one or more polypeptides). Like nucleic acids, polypeptides are polymers, but the monomers that make them up are the 20 amino acids common to all organisms. Again, the language is written in a linear sequence, and the sequence of nucleotides of the RNA molecule dictates the sequence of amino acids of the polypeptide. The RNA acts as a messenger carrying genetic information from DNA.

During translation, there is a change in language from the nucleotide sequence of the RNA into the amino acid sequence of the polypeptide. The brackets below the RNA indicate how genetic information is coded in nucleic acids. Notice that each bracket encloses *three* nucleotides on RNA. Recall that there are only four different kinds of nucleotides in DNA (A, G, C, T) and RNA (A, G, C, U). In translation, these four must somehow specify 20 amino acids. If each nucleotide base specified one amino acid, only 4 of the 20 amino acids could be accounted for. What if the language consisted of two-letter code words? If we read the bases of a gene two at a time, AG, for example, could specify one amino acid, while AT could designate a different amino acid. However, when the 4 bases are taken in doublets, there are only 16 (that is, 4^2) possible arrangements—still not enough to specify all 20 amino acids.

Triplets of bases are the smallest "words" of uniform length that can specify all the amino acids. Suppose each code word in DNA consists of a triplet, with each arrangement of three consecutive bases specifying an amino acid. Then there can be 64 (that is, 4^3) possible code words—more than enough to specify the 20 amino acids. Indeed, there are enough triplets to allow more than one coding for each amino acid. For example, the base triplets AAT and AAC both code for the same amino acid.

Experiments have verified that the flow of information from gene to protein is based on a **triplet code:** The genetic instructions for the amino acid sequence of a polypeptide chain are written in DNA and RNA as a series of three-base words, called **codons.** Notice in the figure that three-base codons in the DNA are transcribed into complementary three-base codons in the RNA, and then the RNA codons are translated into amino acids that form a polypeptide. We turn to the codons themselves in the next module.

? A particular protein is 100 amino acids long. In the gene for this protein, how many nucleotides are necessary to code for this protein?

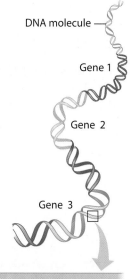

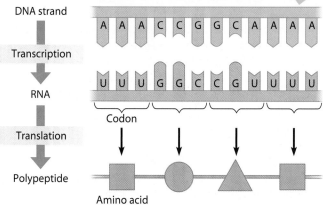

DNA molecule
Gene 1
Gene 2
Gene 3

DNA strand

Transcription

RNA

Translation

Polypeptide

A A A C C G G C A A A A

U U U G G C C G U U U U

Codon

Amino acid

Figure 10.7 Transcription and translation of codons

10.8 The genetic code is the Rosetta stone of life

In 1799, a large stone tablet was found in Rosetta, Egypt, carrying the same lengthy inscription in three ancient languages: Egyptian hieroglyphics, Egyptian script, and Greek. This stone provided the key that enabled scholars to crack the previously indecipherable hieroglyphic code.

In cracking the genetic code, scientists wrote their own Rosetta stone. It was based on information gathered from a series of elegant experiments that disclosed the amino acid translations of each of the nucleotide-triplet code words. The first codon was deciphered in 1961 by American biochemist Marshall Nirenberg. He synthesized an artificial RNA molecule by linking together identical RNA nucleotides having uracil as their base. No matter where this message started or stopped, it could contain only one type of triplet codon: UUU. Nirenberg added this "poly U" to a test-tube mixture containing ribosomes and the other ingredients required for polypeptide synthesis. This mixture translated the poly U into a polypeptide containing a single kind of amino acid, phenylalanine. Thus, Nirenberg learned that the RNA codon UUU specifies the amino acid phenylalanine (Phe). By variations on this method, the amino acids specified by all the codons were soon determined.

The **genetic code** is the set of rules giving the correspondence between codons in RNA and amino acids in proteins. As **Figure 10.8A** shows, 61 of the 64 codons code for amino acids. The triplet AUG has a dual function: It codes for the amino acid methionine (Met) and also can provide a signal for the start of a polypeptide chain. Three of the other codons (in white boxes in the figure) do not designate amino acids. They are the stop codons that mark the end of translation.

Notice in Figure 10.8A that there is redundancy in the code but no ambiguity. For example, although codons UUU and UUC both specify phenylalanine (redundancy), neither of them ever represents any other amino acid (no ambiguity). The codons in the figure are the triplets found in RNA. They have a straightforward, complementary relationship to the codons in DNA. The nucleotides making up the codons occur in a linear order along the DNA and RNA, with no gaps or "punctuation" separating the codons.

As an exercise in translating the genetic code, consider the 12-nucleotide segment of DNA in **Figure 10.8B**. Let's read this as a series of triplets. Using the base-pairing rules (with U in RNA instead of T), we see that the RNA codon corresponding to the first transcribed DNA triplet, TAC, is AUG. AUG says, "Place Met as the first amino acid in the polypeptide." The second DNA triplet, TTC, dictates RNA codon AAG, which designates lysine (Lys) as the second amino acid. We continue until we reach the stop codon.

The genetic code is nearly universal, shared by organisms from the simplest bacteria to the most complex plants and animals. In experiments, bacteria can translate human genetic messages, and human cells can translate bacterial RNA. A language shared by all living things must have evolved early enough in the history of life to be present in the common ancestors of all modern organisms. A shared genetic vocabulary is a reminder of the kinship that connects all life on Earth.

? Translate the mRNA sequence CCAUUUACG into the corresponding amino acid sequence.

■ Pro-Phe-Thr

Second base				
	U	**C**	**A**	**G**
U	UUU / UUC — Phe UUA / UUG — Leu	UCU / UCC / UCA / UCG — Ser	UAU / UAC — Tyr UAA Stop / UAG Stop	UGU / UGC — Cys UGA Stop / UGG Trp	U C A G
C	CUU / CUC / CUA / CUG — Leu	CCU / CCC / CCA / CCG — Pro	CAU / CAC — His CAA / CAG — Gln	CGU / CGC / CGA / CGG — Arg	U C A G
A	AUU / AUC / AUA — Ile AUG — Met or start	ACU / ACC / ACA / ACG — Thr	AAU / AAC — Asn AAA / AAG — Lys	AGU / AGC — Ser AGA / AGG — Arg	U C A G
G	GUU / GUC / GUA / GUG — Val	GCU / GCC / GCA / GCG — Ala	GAU / GAC — Asp GAA / GAG — Glu	GGU / GGC / GGA / GGG — Gly	U C A G

First base (left side) — Third base (right side)

Figure 10.8A Dictionary of the genetic code (RNA codons)

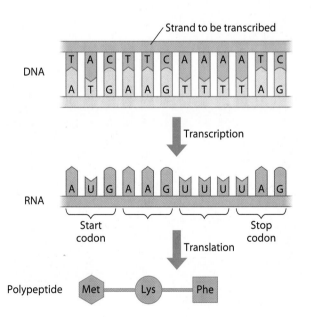

Figure 10.8B Deciphering the genetic information in DNA

Transcription produces genetic messages in the form of RNA

Transcription, the transfer of genetic information from DNA to RNA, occurs in the eukaryotic cell nucleus, as we indicated in Figure 10.6A. An RNA molecule is transcribed from a DNA template by a process that resembles the synthesis of a DNA strand during DNA replication. **Figure 10.9A** is a close-up view of this process. As with replication, the two DNA strands must first separate at the place where the process will start. In transcription, however, only one of the DNA strands serves as a template for the newly forming molecule. The nucleotides that make up the new RNA molecule take their places one at a time along the DNA template strand by forming hydrogen bonds with the nucleotide bases there. Notice that the RNA nucleotides follow the same base-pairing rules that govern DNA replication, except that U, rather than T, pairs with A. The RNA nucleotides are linked by the transcription enzyme **RNA polymerase**, symbolized in the figure by the large gray shape in the background.

Figure 10.9B is an overview of the transcription of an entire prokaryotic gene. (We focus on prokaryotes here; eukaryotic transcription is a similar process but somewhat more complex.) Specific sequences of nucleotides along the DNA mark where transcription of a gene begins and ends. The "start transcribing" signal is a nucleotide sequence called a **promoter**. A promoter is a specific binding site for RNA polymerase and determines which of the two strands of the DNA double helix is used as the template in transcription.

❶ The first phase of transcription, called initiation, is the attachment of RNA polymerase to the promoter and the start of RNA synthesis. ❷ During a second phase of transcription, the RNA elongates. As RNA synthesis continues, the RNA strand peels away from its DNA template, allowing the two separated DNA strands to come back together in the region already transcribed. ❸ Finally, in the third phase, termination, the RNA polymerase reaches a sequence of bases in the DNA template called a **terminator.** This sequence signals the end of the gene; at that point, the polymerase molecule detaches from the RNA molecule and the gene.

In addition to producing RNA that encodes amino acid sequences, transcription makes two other kinds of RNA that are involved in building polypeptides. We discuss these three kinds of RNA in the next three modules.

Web/CD Activity 10G *Transcription*

? **What is a promoter?**

■ A promoter is a specific nucleotide sequence at the start of a gene where RNA polymerase attaches and begins transcription.

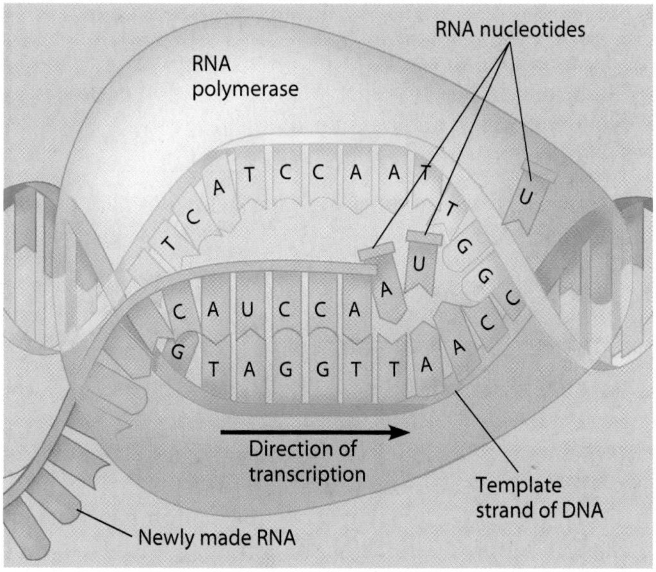

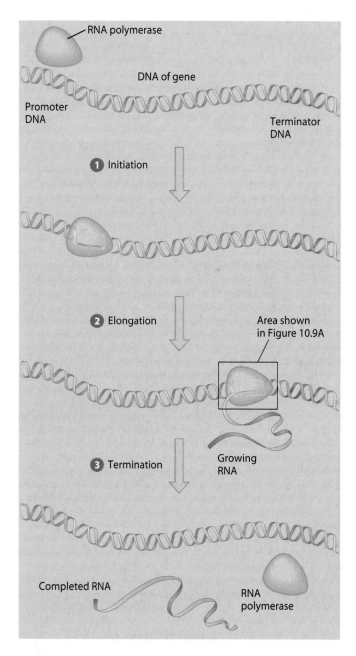

Figure 10.9B Transcription of a gene

Figure 10.9A A close-up view of transcription

10.10 Eukaryotic RNA is processed before leaving the nucleus

The kind of RNA that encodes amino acid sequences is called **messenger RNA (mRNA)** because it conveys genetic information from DNA to the translation machinery of the cell. Messenger RNA is transcribed from DNA, and the message in the mRNA is then translated into polypeptides. In prokaryotic cells, which lack a nucleus, transcription and translation occur in the same place (the cytoplasm). In eukaryotic cells, however, mRNA molecules and other RNA molecules required for translation must exit the nucleus via the nuclear pores and enter the cytoplasm, where the machinery for polypeptide synthesis is located.

Before leaving the nucleus as mRNA, eukaryotic transcripts are modified, or processed, in several ways. One kind of RNA processing is the addition of extra nucleotides to the ends of the RNA transcript (Figure 10.10). These additions include a small cap (a single G nucleotide) at one end and a long tail (a chain of 50 to 250 A's) at the other end. The cap and tail facilitate the export of the mRNA from the nucleus, protect the mRNA from attack by cellular enzymes, and help ribosomes bind to the mRNA. The cap and tail themselves are not translated into protein.

Eukaryotes require an additional type of RNA processing because, in most protein-coding genes, the DNA sequence that codes for the polypeptides is not continuous. Most genes of plants and animals include internal noncoding regions called **introns** (for "intervening sequences"). The coding regions—the parts of a gene that are expressed as amino acids—are called **exons**. As Figure 10.10 shows, both exons (darker color) and introns (lighter color) are transcribed from DNA into RNA. However, before the RNA leaves the nucleus, the introns are removed, and the exons are joined to produce an mRNA molecule with a continuous coding sequence. (The short noncoding regions just inside the cap and tail are considered parts of the first and last exons.) This cutting-and-pasting process is called **RNA splicing**. In most cases, RNA splicing is catalyzed by a complex of proteins and small RNA molecules, but sometimes the RNA transcript itself catalyzes the process. In other words, RNA can sometimes act as an enzyme that removes its own introns! As we will see in the next chapter (Module 11.7), RNA splicing also provides a means to produce multiple polypeptides from a single gene.

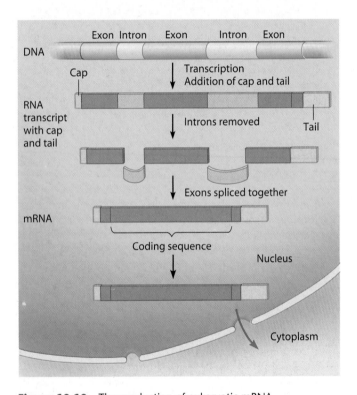

? Explain why many eukaryotic genes are longer than the mRNA that leaves the nucleus.

These genes have introns, noncoding sequences of nucleotides that are spliced out of the RNA transcripts.

■   ■   ■

We are now ready to see how the translation process works. Translation of mRNA into protein involves more complicated machinery than transcription, including:

- Transfer RNA, another kind of RNA molecule
- Ribosomes, the organelles where translation occurs
- Enzymes and a number of protein "factors"
- Sources of chemical energy, such as ATP

In the next two modules, we take a closer look at transfer RNA and ribosomes.

Figure 10.10 The production of eukaryotic mRNA

10.11 Transfer RNA molecules serve as interpreters during translation

Translation of any language requires an interpreter, someone who can recognize the words of one language and convert them to another. Translation of a message carried in mRNA into the amino acid language of proteins also requires an interpreter. To convert the three-letter words (codons) of nucleic acids to the one-letter, amino acid words of proteins, a cell employs a molecular interpreter, a special type of RNA called **transfer RNA (tRNA)**.

A cell that is ready to carry out translation has in its cytoplasm a supply of amino acids, either obtained from food or made from other chemicals. The amino acids themselves cannot recognize the codons in the mRNA. The amino acid

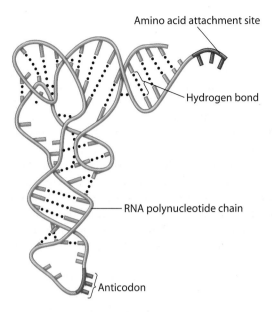

Figure 10.11A The structure of tRNA

Labels on Figure 10.11A: Amino acid attachment site; Hydrogen bond; RNA polynucleotide chain; Anticodon

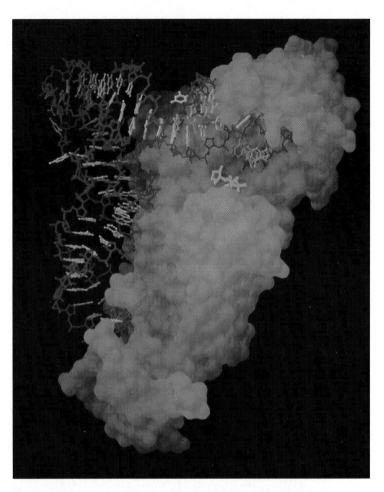

Figure 10.11C A molecule of tRNA binding to an enzyme molecule (blue)

tryptophan, for example, is no more attracted by codons for tryptophan than by any other codons. It is up to the cell's molecular interpreters, tRNA molecules, to match amino acids to the appropriate codons to form the new polypeptide. To perform this task, tRNA molecules must carry out two functions: (1) picking up the appropriate amino acids and (2) recognizing the appropriate codons in the mRNA. The unique structure of tRNA molecules enables them to perform both tasks.

As shown in **Figure 10.11A**, a tRNA molecule is made of a single strand of RNA—one polynucleotide chain—consisting of about 80 nucleotides. By twisting and folding upon itself, tRNA forms several double-stranded regions in which short stretches of RNA base-pair with other stretches. A single-stranded loop at one end of the folded molecule contains a special triplet of bases called an **anticodon.** The anticodon triplet is complementary to a codon triplet on mRNA. During translation, the anticodon on tRNA recognizes a particular codon on mRNA by using base-pairing rules. At the other end of the tRNA molecule is a site where an amino acid can attach.

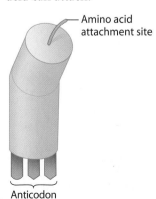

Labels on Figure 10.11B: Amino acid attachment site; Anticodon

Figure 10.11B A simplified model of tRNA

In the modules that follow, in which we trace the process of translation, we represent tRNA with the simplified shape that is shown in **Figure 10.11B**. This symbol emphasizes two parts of the molecule—the anticodon and the amino acid attachment site—that give tRNA its ability to match a particular nucleic acid word (codon) with its corresponding protein word (amino acid). Although all tRNA molecules

are similar, there is a slightly different variety of tRNA for each amino acid.

Each amino acid is joined to the correct tRNA by a specific enzyme. There is a family of 20 versions of these enzymes, one enzyme for each amino acid. Each enzyme specifically binds one type of amino acid to all tRNA molecules that code for that amino acid, using a molecule of ATP as energy to drive the reaction. The resulting amino acid–tRNA complex can then furnish its amino acid to a growing polypeptide chain, a process that we describe in Module 10.12.

The computer graphic in **Figure 10.11C** shows a tRNA molecule (red and yellow) and an ATP molecule (green) bound to the enzyme molecule (blue). In this picture, you can see the proportional sizes of these three molecules. The amino acid that would attach to the tRNA is not shown; it would be less than half the size of the ATP.

? What is an anticodon, and what is its function?

■ It is the base triplet of a tRNA molecule that couples the tRNA to a complementary codon in the mRNA. This is a key step in translating mRNA to polypeptide.

10.12 Ribosomes build polypeptides

We have now looked at many of the components a cell needs to carry out translation: instructions in the form of mRNA molecules, tRNA to interpret the instructions, a supply of amino acids, enzymes for attaching amino acids to tRNA, and ATP for energy. The final components are the ribosomes, organelles in the cytoplasm that coordinate the functioning of the mRNA and tRNA and actually make polypeptides.

A ribosome consists of two subunits, each made up of proteins and a kind of RNA called **ribosomal RNA (rRNA)**. In **Figure 10.12A**, you can see the actual shapes and relative sizes of the ribosomal subunits. You can also see where mRNA, tRNA, and the growing polypeptide are located during translation.

The ribosomes of prokaryotes and eukaryotes are very similar in function, but those of eukaryotes are slightly larger and are different in composition. The differences are medically significant. Certain antibiotic drugs can inactivate prokaryotic ribosomes while leaving eukaryotic ribosomes unaffected. These drugs, such as tetracycline and streptomycin, are used to combat bacterial infections.

The simplified drawings in Figures 10.12B and 10.12C indicate how tRNA anticodons and mRNA codons fit together on ribosomes. As **Figure 10.12B** shows, each ribosome has a binding site for mRNA and also two binding sites for tRNA. **Figure 10.12C** shows tRNA molecules occupying these two sites. The subunits of the ribosome act like a vise, holding the tRNA and mRNA molecules close together, allowing the amino acids carried by the tRNA molecules to be connected into a polypeptide chain. In the next two modules, we examine the steps of translation in detail.

? How does a ribosome function in protein synthesis?

■ A ribosome holds mRNA and tRNAs together and connects amino acids from the tRNAs to the growing polypeptide chain.

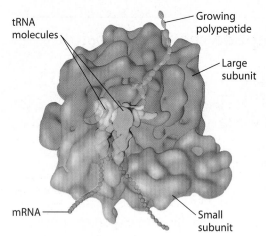

Figure 10.12A The true shape of a functioning ribosome

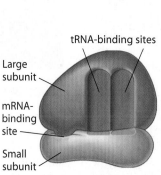

Figure 10.12B Binding sites of a ribosome

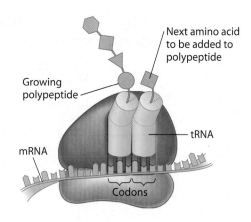

Figure 10.12C A ribosome with occupied binding sites

10.13 An initiation codon marks the start of an mRNA message

Translation can be divided into the same three phases as transcription: initiation, elongation, and termination. The process of polypeptide initiation brings together the mRNA, a tRNA bearing the first amino acid, and the two subunits of a ribosome.

As indicated in **Figure 10.13A**, an mRNA molecule transcribed from DNA is longer than the genetic message it carries. A sequence of nucleotides (light pink) at either end of the molecule is not part of the message but helps the mRNA bind to the ribosome. The role of the initiation process is to establish exactly where translation will begin, ensuring that the mRNA codons are translated into the correct sequence of amino acids.

Initiation occurs in two steps (**Figure 10.13B**). ❶ An mRNA molecule binds to a small ribosomal subunit. A special

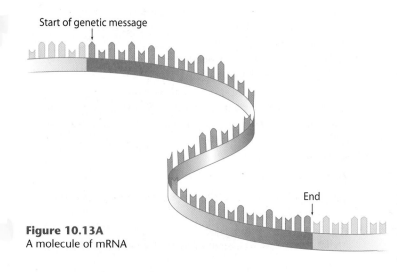

Figure 10.13A
A molecule of mRNA

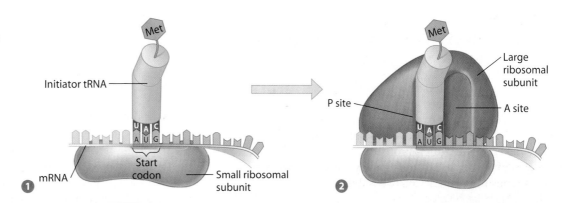

Figure 10.13B
The initiation of translation

Labels on figure: Initiator tRNA · Met · mRNA · Start codon · Small ribosomal subunit · ❶ · Large ribosomal subunit · P site · A site · Met · ❷

initiator tRNA binds to the specific codon, called the **start codon**, where translation is to begin on the mRNA molecule. The initiator tRNA carries the amino acid methionine (Met); its anticodon, UAC, binds to the start codon, AUG. ❷ Next, a large ribosomal subunit binds to the small one, creating a functional ribosome. The initiator tRNA fits into one of the two tRNA-binding sites on the ribosome. This site, called the

P site, will hold the growing polypeptide. The other tRNA-binding site, called the **A site**, is vacant and ready for the next amino-acid-bearing tRNA.

? What would happen if a genetic mutation changed a start codon to some other codon?

■ Any messenger RNA transcribed from the mutated gene would be nonfunctional because ribosomes could not initiate translation correctly.

10.14 Elongation adds amino acids to the polypeptide chain until a stop codon terminates translation

Once initiation is complete, amino acids are added one by one to the first amino acid. Each addition occurs in a three-step elongation process (**Figure 10.14**; the small green arrows indicate movement):

❶ **Codon recognition.** The anticodon of an incoming tRNA molecule, carrying its amino acid, pairs with the mRNA codon in the A site of the ribosome.

❷ **Peptide bond formation.** The polypeptide separates from the tRNA to which it was bound (the one in the P site) and attaches by a peptide bond to the amino acid carried by the tRNA in the A site. The ribosome catalyzes formation of the bond. Thus, one more amino acid is added to the chain.

❸ **Translocation.** The P site tRNA now leaves the ribosome, and the ribosome translocates (moves) the tRNA in the A site, with its attached polypeptide, to the P site. The codon and anticodon remain bonded, and the mRNA and tRNA move as a unit. This movement brings into the A site the next mRNA codon to be translated, and the process can start again with step 1.

Elongation continues until a **stop codon** reaches the ribosome's A site. Stop codons—UAA, UAG, and UGA—do not code for amino acids but instead act as signals to stop translation. This is the termination stage of translation. The completed polypeptide is released from the last tRNA and exits the ribosome, which then splits into its separate subunits.

Web/CD Activity 10H *Translation*

? What happens as a tRNA passes through the A and P binding sites on the ribosome?

■ In the A site, its amino acid receives the growing polypeptide from the tRNA that precedes it. In the P site, it gives up the polypeptide to the tRNA that follows it.

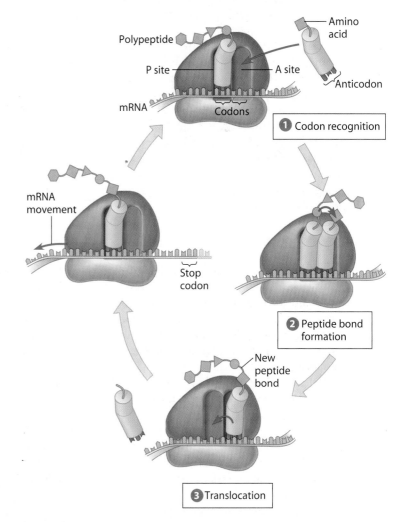

Labels on figure: Polypeptide · Amino acid · P site · A site · Anticodon · mRNA · Codons · ❶ Codon recognition · mRNA movement · Stop codon · ❷ Peptide bond formation · New peptide bond · ❸ Translocation

Figure 10.14 Polypeptide elongation

10.15 Review: The flow of genetic information in the cell is DNA → RNA → protein

Figure 10.15 summarizes the main stages in the flow of genetic information from DNA to RNA to protein. ❶ In transcription (DNA → RNA), the RNA is synthesized on a DNA template. In eukaryotic cells, transcription occurs in the nucleus, and the messenger RNA must travel from the nucleus to the cytoplasm.

❷–❺ Translation (RNA → protein) can be divided into four steps, all of which occur in the cytoplasm. When the polypeptide is complete, the two ribosomal subunits come apart, and the tRNA and mRNA are released (not shown in this figure). Translation is rapid; a single ribosome can make an average-sized polypeptide in less than a minute. Typically, an mRNA molecule is translated simultaneously by a number of ribosomes. Once the start codon emerges from the first ribosome, a second ribosome can attach to it; thus, several ribosomes may trail along on the same mRNA molecule.

Each polypeptide coils and folds, assuming a three-dimensional shape, its tertiary structure. Several polypeptides may come together, forming a protein with quaternary structure (see Module 3.14).

What is the overall significance of transcription and translation? These are the processes whereby genes control the structures and activities of cells, or, more broadly, the way the genotype produces the phenotype. The chain of command originates with the information in a gene, a specific linear sequence of nucleotides in DNA. The gene serves as a template, dictating transcription of a complementary sequence of nucleotides in mRNA. In turn, mRNA dictates the linear sequence in which amino acids appear in a specific polypeptide. Finally, the proteins that form from the polypeptides determine the appearance and the capabilities of the cell and organism.

> ❓ Which of the following molecules or structures does not participate directly in translation: ribosomes, transfer RNA, messenger RNA, DNA, ATP, enzymes?
>
> ■ DNA

Figure 10.15 Summary of transcription and translation

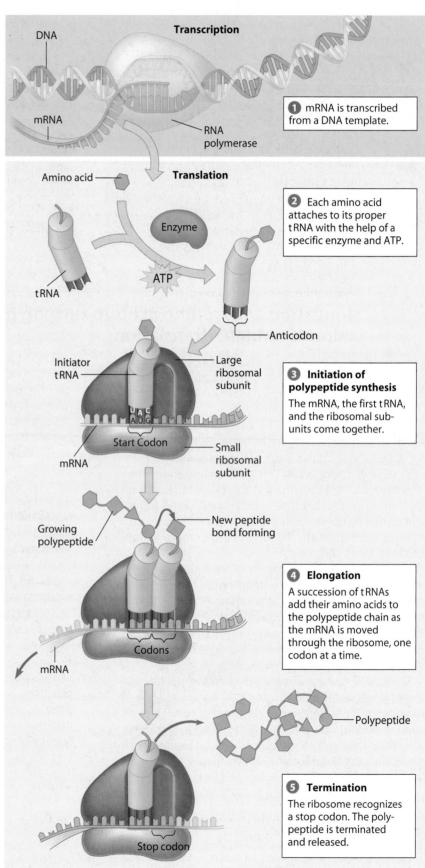

Transcription

DNA

mRNA

RNA polymerase

❶ mRNA is transcribed from a DNA template.

Translation

Amino acid

Enzyme

ATP

tRNA

Anticodon

Initiator tRNA

Large ribosomal subunit

U A C
A U G

Start Codon

Small ribosomal subunit

mRNA

❷ Each amino acid attaches to its proper tRNA with the help of a specific enzyme and ATP.

❸ **Initiation of polypeptide synthesis**
The mRNA, the first tRNA, and the ribosomal subunits come together.

Growing polypeptide

New peptide bond forming

Codons

mRNA

❹ **Elongation**
A succession of tRNAs add their amino acids to the polypeptide chain as the mRNA is moved through the ribosome, one codon at a time.

Polypeptide

Stop codon

❺ **Termination**
The ribosome recognizes a stop codon. The polypeptide is terminated and released.

10.16 Mutations can change the meaning of genes

Since discovering how genes are translated into proteins, scientists have been able to describe many inherited traits in molecular terms. For instance, when a child is born with sickle-cell disease (see Module 9.14), the condition can be traced back through a difference in a protein to one tiny change in a gene. In one of the two kinds of polypeptides in the hemoglobin protein, the sickle-cell child has a single different amino acid. This difference is caused by the change of a single nucleotide in the coding strand of DNA (Figure 10.16A). In the double helix, one base *pair* is changed.

We now know that the alternative alleles of many genes result from changes in single base pairs in DNA. Any change in the nucleotide sequence of DNA is called a **mutation**. Mutations can involve large regions of a chromosome or just a single nucleotide pair, as in sickle-cell disease. Here we consider how mutations involving only one or a few nucleotide pairs can affect gene translation.

Mutations within a gene can be divided into two general categories: base substitutions and base insertions or deletions (Figure 10.16B). A base substitution is the replacement of one nucleotide with another. Occasionally, a base substitution leads to an improved protein that enhances the success of the mutant organism and its descendants. Much more often, though, mutations are harmful.

In the second row in Figure 10.16B, A replaces G in the fourth codon of the mRNA. Depending on how a base substitution is translated, it can result in no change in the protein, in an insignificant change, or in a change that might significantly affect the organism. It is because of the redundancy of the genetic code that some substitution mutations have no effect. For example, if a mutation causes an mRNA codon to change from GAA to GAG, no change in the protein product would result, because GAA and GAG both code for the same amino acid (Glu). Other changes of a single nucleotide may alter an amino acid but have little effect on the function of the protein. But as in Figure 10.16A, base substitutions may cause changes in a protein that prevent it from functioning normally. And if a base substitution changes an amino-acid codon to a stop codon, a shortened, probably nonfunctional polypeptide will result.

Mutations involving the insertion or deletion of one or more nucleotides in a gene often have disastrous effects. Because mRNA is read as a series of nucleotide triplets (codons) during translation, adding or subtracting nucleotides may alter the **reading frame** (triplet grouping) of the message. All the nucleotides that are "downstream" of the insertion or deletion will be regrouped into different codons (Figure 10.16B, bottom). The result will most likely be a nonfunctional polypeptide.

The production of mutations, called **mutagenesis,** can occur in a number of ways. Errors that occur during DNA replication or recombination are called spontaneous mutations. Another source of mutation is a physical or chemical agent, called a **mutagen.** The most common physical mutagen in nature is high-energy radiation, such as X-rays and ultraviolet light. Chemical mutagens fall into several categories. One type, for example, consists of chemicals that are similar to normal DNA bases but that pair incorrectly.

Although mutations are often harmful, they are also extremely useful, both in nature and in the laboratory. It is because of mutations that there is such a rich diversity of genes in the living world, a diversity that makes evolution by natural selection possible. Mutations are also essential tools for geneticists. Whether naturally occurring (as in Mendel's peas) or created in the laboratory (Morgan used X-rays to make most of his fruit fly mutants), mutations create the different alleles needed for genetic research.

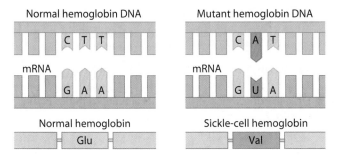

Figure 10.16A The molecular basis of sickle-cell disease

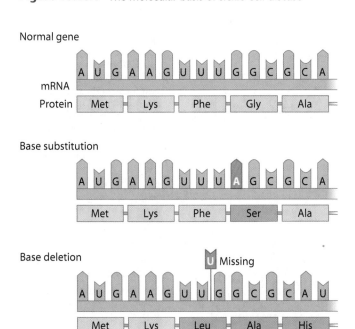

Figure 10.16B Types of mutations and their effects

Web/CD Thinking as a Scientist *Connection: How Do You Diagnose a Genetic Disorder?*

? What is the molecular basis of sickle-cell disease?

■ A single base difference in a hemoglobin gene results in an amino acid substitution in the protein, altering the structure and behavior of the hemoglobin.

10.17 Viral DNA may become part of the host chromosome

As we discussed earlier, bacteria and viruses served as models in experiments that uncovered the molecular details of heredity. Now let's take a closer look at viruses, focusing on the relationship between viral structure and the processes of nucleic acid replication, transcription, and translation. (Viruses, while not considered alive, are often grouped with microscopic organisms under the general term "microbes"; hence, this section is called "microbial genetics.")

In a sense, viruses are nothing more than "genes in a box": infectious particles consisting of nucleic acid enclosed in a protein coat called a **capsid** and, in some cases, a membrane envelope. Viruses are parasites that can reproduce only inside cells. In fact, the host cell provides most of the components necessary for replicating, transcribing, and translating the viral nucleic acid.

In Module 10.1, we described the reproductive cycle of phage T2. This sort of cycle is called a **lytic cycle** because it results in the lysis (breaking open) of the host cell and the release of the viruses that were produced within the cell. Some phages can also reproduce by an alternative route called the lysogenic cycle. During a **lysogenic cycle**, viral DNA replication occurs without destroying the host cell.

In **Figure 10.17**, you see the two kinds of cycles for a phage called lambda that infects *E. coli*. Both cycles begin when the phage DNA ❶ enters the bacterium and ❷ forms a circle. The DNA then embarks on one of the two pathways. In the lytic cycle (left), ❸ lambda's DNA immediately turns

the cell into a virus-producing factory, and ❹ the cell soon lyses and releases its viral products. In the lysogenic cycle, however, ❺ the DNA is inserted by genetic recombination into the bacterial chromosome. Once inserted, the phage DNA is referred to as a **prophage**, and most of its genes are inactive. ❻ Every time the *E. coli* cell prepares to divide, it replicates the phage DNA along with its own and passes the copies on to daughter cells. A single infected cell can quickly give rise to a large population of bacteria carrying the virus in prophage form. The lysogenic cycle enables viruses to propagate without killing the host cells on which they depend. The prophages may remain in the bacterial cells indefinitely. ❼ Occasionally, however, an environmental signal such as radiation or a certain chemical triggers a switchover from the lysogenic cycle to the lytic cycle.

Sometimes the few prophage genes active in a lysogenic bacterium can cause medical problems. For example, the bacteria that cause diphtheria, botulism, and scarlet fever would be harmless to humans if it were not for the prophage genes they carry. Certain of these genes direct the bacteria to produce the toxins responsible for making people ill. In the next module, we will explore animal viruses.

Web/CD Activity 10I *Phage Lysogenic and Lytic Cycles*

? Describe one way a virus can perpetuate its genes without destroying its host cell.

■ Some viruses can insert their DNA into a chromosome of the host cell, which replicates the viral genes when it replicates its own DNA prior to cell division.

Figure 10.17
Two types of phage reproductive cycles

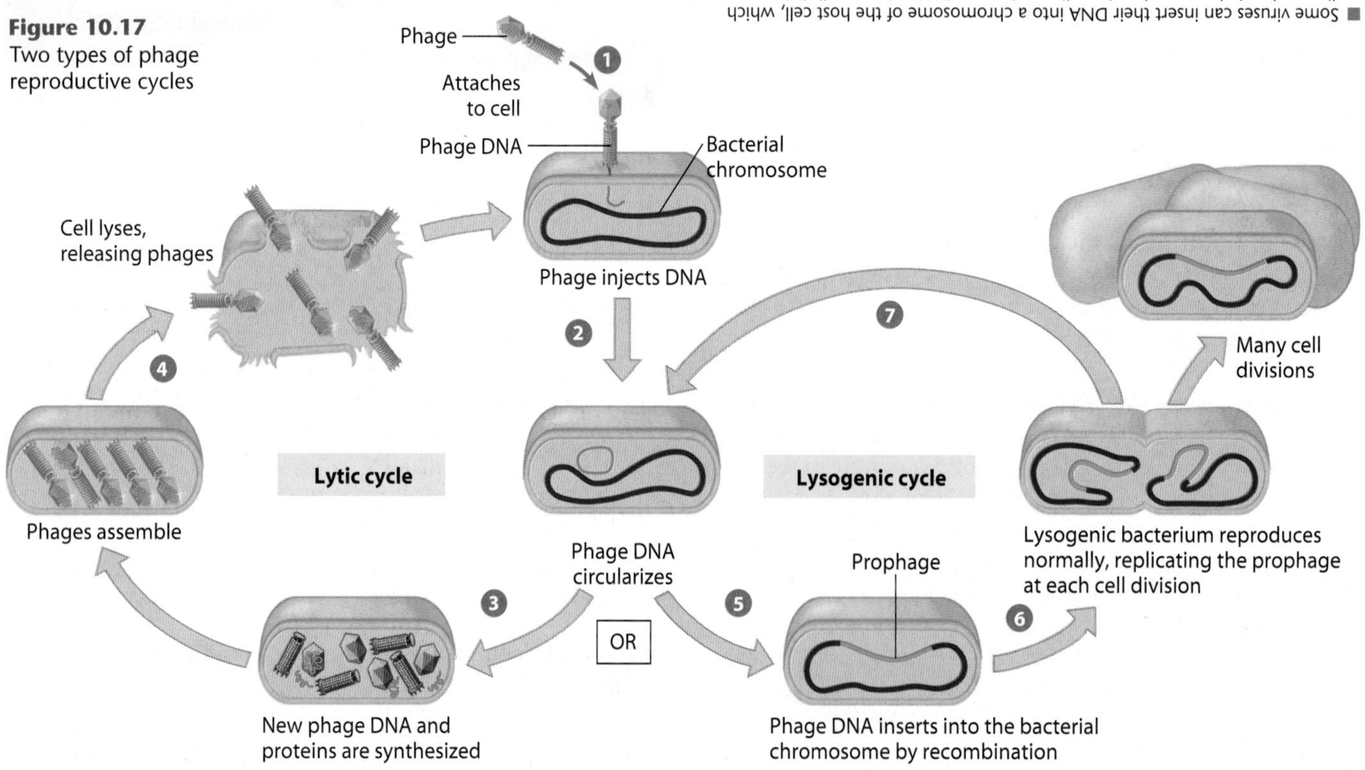

Phage

Attaches to cell

Phage DNA

Bacterial chromosome

Phage injects DNA

Cell lyses, releasing phages

Lytic cycle

Lysogenic cycle

Phages assemble

Phage DNA circularizes

OR

Prophage

Many cell divisions

Lysogenic bacterium reproduces normally, replicating the prophage at each cell division

New phage DNA and proteins are synthesized

Phage DNA inserts into the bacterial chromosome by recombination

10.18 Many viruses cause disease in animals

Viruses that infect animal cells are common causes of disease. We have all suffered from viral infections. **Figure 10.18A**

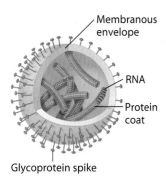

Figure 10.18A
An influenza virus

shows the structure of an influenza (flu) virus. This virus, like many that infect animal cells, has a membranous outer envelope and projecting spikes of glycoprotein (protein with attached sugars). The envelope helps the virus enter and leave the host cell. Like many viruses, flu viruses have RNA rather than DNA as their genetic material; the RNA of flu viruses is actually in eight pieces, each wrapped in protein. Other RNA viruses include those that cause the common cold, measles, and mumps, as well as ones that cause more serious human diseases, such as AIDS and polio. Examples of diseases caused by DNA viruses are hepatitis, chicken pox, and herpes infections.

Figure 10.18B shows the reproductive cycle of an enveloped RNA virus (the mumps virus). When the virus contacts a host cell, the glycoprotein spikes attach to receptor proteins on the cell's plasma membrane. The envelope fuses with the cell's membrane, allowing the protein-coated RNA to ❶ enter the cytoplasm. ❷ Enzymes then remove the protein coat. ❸ An enzyme that entered the cell as part of the virus uses the virus's RNA genome as a template for making complementary strands of RNA (pink strand). The new strands have two functions: ❹ They serve as mRNA for the synthesis of new viral proteins, and ❺ they serve as templates for synthesizing new viral genome RNA. ❻ The new coat proteins assemble around the new viral RNA. ❼ Finally, the viruses leave the cell by cloaking themselves in plasma membrane. Thus, the virus obtains its envelope from the host cell, leaving the cell without necessarily lysing it.

Not all animal viruses reproduce in the cytoplasm. For example, herpesviruses, which you read about in the chapter introduction, are enveloped DNA viruses that reproduce in the host cell's nucleus; they acquire their envelopes from the cell's nuclear membranes. Herpesviruses have another important characteristic: While inside the nuclei of certain nerve cells, herpesvirus DNA may remain latent, without destroying these cells. From time to time, physical stress, such as a cold or sunburn, or emotional stress may stimulate the herpesvirus DNA to begin production of the virus, which then infects cells at the body's surface and brings about cold sores or genital sores. Once acquired, herpes infections may flare up repeatedly throughout a person's life.

The amount of damage a virus causes our body depends partly on how quickly our immune system responds to fight the infection and partly on the ability of the infected tissue to repair itself. We usually recover completely from colds because our respiratory tract tissue can efficiently replace dam-

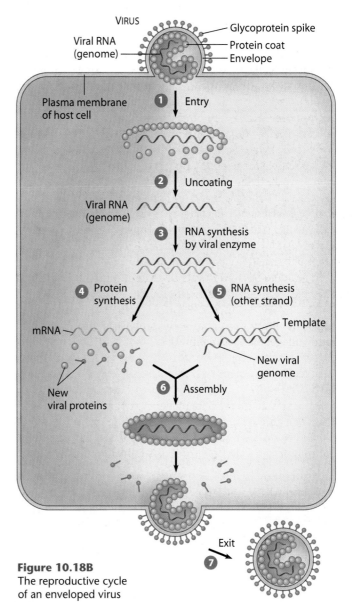

Figure 10.18B
The reproductive cycle of an enveloped virus

aged cells by mitosis. In contrast, the poliovirus attacks nerve cells, which do not divide. The damage to such cells by polio, unfortunately, is permanent. In such cases, we try to prevent the disease with vaccines (see Module 24.4). The antibiotic drugs that help us recover from bacterial infections are powerless against viruses. The development of antiviral drugs has been slow because it is difficult to find ways to kill a virus without killing its host cell.

Web/CD Activity 10J *Simplified Reproductive Cycle of a DNA Virus*

? Explain how some viruses replicate without having DNA.

■ The genetic material of these viruses is RNA, which is replicated inside the host cell by special enzymes encoded by the virus. The viral genome (or its complement) serves as mRNA for the synthesis of viral proteins.

10.19 Plant viruses are serious agricultural pests

Viruses that infect plant cells can stunt plant growth and diminish crop yields. Most plant viruses discovered to date are RNA viruses. Many of them, like the tobacco mosaic virus in **Figure 10.19**, are rod-shaped with a spiral arrangement of proteins surrounding the nucleic acid.

Figure 10.19 Tobacco mosaic disease (mottling of leaves) and the structure of the virus (right)

To infect a plant, a virus must first get past the plant's outer protective layer of cells. Thus, a plant damaged by wind, chilling, injury, or insects is more susceptible to infection than a completely healthy plant. Besides injuring plants, some insects also carry and transmit plant viruses. Farmers and gardeners, too, may spread plant viruses through the use of pruning shears and other tools. And infected plants may pass viruses to their offspring.

Once a virus enters a plant cell and begins reproducing, it can spread throughout the entire plant through plasmodesmata, the cytoplasmic connections that penetrate the walls between adjacent plant cells (see Figure 4.18A). As with animal viruses, there are no cures for most viral diseases of plants. Agricultural scientists focus instead on reducing the number of plants that become infected and on breeding genetic varieties of crop plants that resist viral infection.

? What are three ways that plant cells may be exposed to viruses?

■ Through lesions caused by injuries, through transfer by insects that feed on the plants, and through contaminated farming or gardening tools

10.20 Emerging viruses threaten human health

Viruses that have appeared suddenly or have recently come to the attention of medical scientists are referred to as **emerging viruses**. HIV, the AIDS virus, is a classic example: This virus appeared in San Francisco in the early 1980s, seemingly out of nowhere. The deadly Ebola virus (**Figure 10.20A**), recognized initially in 1976 in central Africa, is one of several emerging viruses that cause hemorrhagic fever, an often fatal syndrome characterized by fever, vomiting, massive bleeding, and circulatory system collapse. A number of other dangerous new viruses cause encephalitis, inflammation of the brain. One example is the West Nile virus, which appeared for the first time in North America in 1999 and had spread to all 48 contiguous U.S. states by the end of 2004.

A viral disease that has emerged even more recently is severe acute respiratory syndrome (SARS), first reported in China in February 2003. Within three months, about 8,450 people were known to be infected, of whom some 10% subsequently died. Researchers quickly identified the agent causing SARS as a coronavirus, so named for its halo-like "corona" of spikes (**Figure 10.20B**). This virus, with a single-stranded RNA genome, was not previously known to cause disease in humans.

From where and how do such viruses burst on the human scene, giving rise to rare or previously unknown diseases? Three processes contribute to the emergence of viral diseases: mutations, contact between species, and spread from isolated populations.

The mutation of existing viruses is a major source of new viral diseases. RNA viruses tend to have unusually high rates of mutation because errors in replicating their RNA genomes cannot be corrected by proofreading (unlike DNA replication; see Module 10.5). Some mutations enable existing viruses to evolve into new strains (genetic varieties) that can cause disease in individuals who had developed immunity to the ancestral virus. Flu epidemics, for instance, are caused by new influenza virus strains that are genetically different enough from earlier strains that people have little immunity to them.

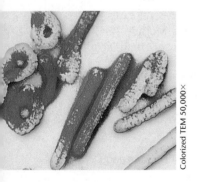

Colorized TEM 50,000×

Figure 10.20A Ebola virus, each an enveloped thread of protein-coated RNA

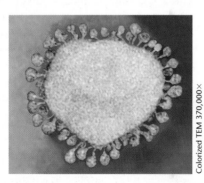

Colorized TEM 370,000×

Figure 10.20B A coronavirus like the one that causes SARS

New viral diseases often arise from the spread of existing viruses from one host species to another. Scientists estimate that about three-quarters of new human diseases have originated in other animals. For example, hantavirus is common in rodents, especially deer mice. The population of deer mice in the southwestern United States exploded in 1993 after unusually wet weather increased the rodents' food supply. Many people who inhaled dust containing traces of urine and feces from infected mice became infected with hantavirus, and dozens died. The source of the SARS-causing virus is still undetermined, although candidates include exotic animals found in livestock markets in China. And early 2004 brought reports of the first cases of people in Southeast Asia infected with a flu virus previously seen only in birds. If this virus evolves so that it can spread easily from person to person, the potential for a major human outbreak is significant. Indeed, evidence is strong that the flu pandemic of 1918–1919, which killed about 40 million people, originated in birds.

The spread of a viral disease from a small, isolated population can lead to widespread epidemics. For instance, AIDS went unnamed and virtually unnoticed for decades before it began to spread around the world. In this case, technological and social factors, including affordable international travel, blood transfusions, sexual promiscuity, and the abuse of intravenous drugs, allowed a previously rare human disease to become a global scourge.

Web/CD Thinking as a Scientist *Connection: Why Do AIDS Rates Differ Across the U.S.?*

? Why doesn't a bout of flu give us immunity to flu in subsequent years?

■ Influenza viruses evolve rapidly by frequent mutation; thus, the strains that infect us later will most likely be different from the ones to which we've developed immunity.

10.21 The AIDS virus makes DNA on an RNA template

The devastating disease AIDS is caused by a type of RNA virus with some special twists. In outward appearance, the AIDS virus (HIV) resembles the flu or mumps virus. As illustrated in **Figure 10.21A**, HIV has a membranous envelope and glycoprotein spikes. These components enable HIV to enter and leave a host cell much the way the mumps virus does (see Figure 10.18B). Notice, however, that HIV contains two identical copies of its RNA instead of one. HIV also has a different mode of reproduction. HIV carries molecules of an enzyme called **reverse transcriptase**, which catalyzes reverse transcription, the synthesis of DNA on an RNA template. This unusual phenomenon, which is opposite the usual DNA → RNA flow of genetic information, characterizes **retroviruses** ("retro" means "backward").

Figure 10.21B illustrates what happens after HIV RNA is uncoated in the cytoplasm of a host cell. The reverse transcriptase (yellow) ❶ uses the RNA as a template to make a DNA strand and then ❷ adds a second, complementary DNA strand. ❸ The resulting double-stranded DNA then enters the cell's nucleus and inserts itself into the chromosomal DNA, becoming a *provirus* (analogous to a prophage). Occa-

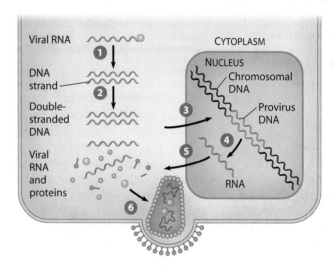

Figure 10.21B The behavior of HIV nucleic acid in a host cell

sionally, the provirus is ❹ transcribed into RNA and ❺ translated into viral proteins. ❻ New viruses assembled from these components leave the cell and can then infect other cells. This is the standard reproductive cycle for retroviruses.

AIDS stands for acquired immune deficiency syndrome, and **HIV** for human immunodeficiency virus; these terms describe the main effect of the virus on the body. HIV infects and eventually kills several kinds of white blood cells that are important in immunity. We discuss AIDS in more detail when we take up the immune system in Chapter 24.

Web/CD Activity 10K *Retrovirus (HIV) Reproductive Cycle*

Web/CD Thinking as a Scientist *Connection: What Causes Infections in AIDS Patients?*

? Why is HIV classified as a retrovirus?

■ Because it synthesizes DNA from its RNA genome. This is the reverse ("retro") of the usual DNA → RNA information flow.

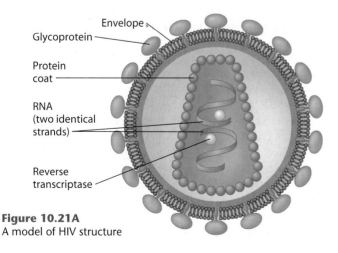

Figure 10.21A
A model of HIV structure

Envelope
Glycoprotein
Protein coat
RNA (two identical strands)
Reverse transcriptase

10.22 Bacteria can transfer DNA in three ways

By studying viral reproduction, researchers also learn about the mechanisms that regulate DNA replication and gene expression in cells. Bacteria are equally valuable as microbial models in genetics research, but for different reasons. As prokaryotic cells, bacteria allow researchers to investigate molecular genetics in the simplest living organisms.

Most of a bacterium's DNA is found in a single bacterial (prokaryotic) chromosome, which is a closed loop of DNA with associated proteins. In the diagrams here, we show the chromosome much smaller than it actually is relative to the cell. A bacterial chromosome is hundreds of times longer than its cell; it fits inside the cell because it is highly folded.

Bacterial cells divide by binary fission, which is preceded by replication of the bacterial chromosome (see Module 8.3). Because binary fission is an asexual process involving only a single parent, the bacteria in a colony are genetically identical to the parental cell. But this does not mean that bacteria lack for ways to produce new combinations of genes. In fact, in the bacterial world, there are three mechanisms by which genes can move from one cell to another: transformation, transduction, and conjugation.

Figure 10.22A illustrates **transformation**, the uptake of foreign DNA from the surrounding environment. In Frederick Griffith's "transforming factor" experiment (see Module 10.1), a harmless strain of bacteria took up pieces of DNA left from the dead cells of a disease-causing strain. The DNA from the pathogenic bacteria carried a gene that made the cells resistant to an animal's defenses, and the previously harmless bacteria could cause pneumonia in infected animals when they acquired this gene.

Bacteriophages provide the second means of bringing together genes of different bacteria. The transfer of bacterial genes by a phage is called **transduction**. In this process, a fragment of DNA belonging to a phage's previous host cell is accidentally packaged within the phage's coat instead of phage DNA. When the phage infects a new bacterial cell, the DNA stowaway from the former host cell is injected into the new host (**Figure 10.22B**).

Figure 10.22C shows what happens at the DNA level when two bacterial cells mate. This union of cells and the DNA transfer between them is called **conjugation** (from the Latin *conjugatus*, united). No-

tice that the "male" donor cell has sex pili, one of which is attached to the "female" recipient cell. The outside layers of the cells have fused, and a cytoplasmic bridge has formed between them. Through this mating bridge, donor cell DNA (bright blue in the figure) passes to the recipient cell. The donor cell replicates its DNA as it transfers it, so the cell doesn't end up lacking any genes. The DNA replication is a special type that allows one copy to peel off and transfer into the recipient cell.

Once new DNA gets into a bacterial cell, by whatever mechanism, part of it may then integrate into the recipient's chromosome. As **Figure 10.22D** indicates, integration occurs by crossing over between the two DNA molecules, a process similar to crossing over between eukaryotic chromosomes (see Module 8.18). Here we see that two crossovers result in a piece of the donated DNA replacing part of the recipient cell's original DNA. The leftover pieces of DNA are broken down, leaving the recipient bacterium with a recombinant chromosome. We look more closely at gene transfer by conjugation in the next module.

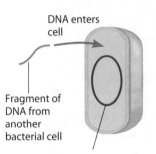

Figure 10.22A
Transformation

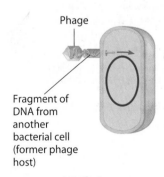

Figure 10.22B   Transduction

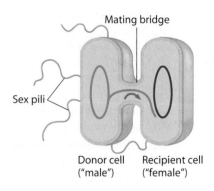

Figure 10.22C   Conjugation

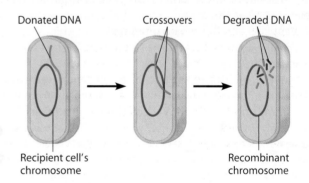

Figure 10.22D   Integration of donated DNA into the recipient cell's chromosome

? The three modes of gene transfer between bacteria are _____, which is transfer via a virus; _____, which is the uptake of DNA from the surrounding environment; and _____, which is the bacterial version of mating.

transduction . . . transformation . . . conjugation

Bacterial plasmids can serve as carriers for gene transfer

The ability of a donor *E. coli* cell to carry out conjugation is usually due to a specific piece of DNA called the **F factor** (F for fertility). The F factor carries genes for making sex pili and other things needed for conjugation; it also contains an origin of replication, where DNA replication can start.

Let's see how the F factor behaves during conjugation. In **Figure 10.23A**, the F factor (bright blue) is integrated into the male bacterium's chromosome. When this cell conjugates with a recipient cell, the male chromosome starts replicating at the F factor's origin of replication, indicated by the blue dot on the DNA. The growing copy of the DNA peels off the chromosome and heads into the recipient cell. Thus, part of the F factor serves as the leading end of the transferred DNA, but right behind it are genes from the donor's original chromosome. The rest of the F factor stays in the donor cell. Once inside the recipient cell, the transferred donor genes can recombine with the corresponding part of the recipient chromosome by crossing over. If crossing over occurs, the recipient cell may be genetically changed, but it usually remains female because the two cells break apart before the rest of the F factor transfers.

Alternatively, as **Figure 10.23B** shows, an F factor can exist as a **plasmid**, a small, circular DNA molecule separate from the bacterial chromosome. Every plasmid has an origin of replication, required for its replication within the cell, and some plasmids, including the F-factor plasmid, can bring about conjugation and move to another cell. When the male cell in Figure 10.23B mates with a recipient cell, the F factor replicates and at the same time transfers one whole copy of itself, in linear rather than circular form, to the recipient cell. The transferred plasmid re-forms a circle in the recipient cell, and the cell becomes male.

E. coli and other bacteria have many different kinds of plasmids. You can see several from one cell in **Figure 10.23C**, along with part of the bacterial chromosome, which extends in loops from the ruptured cell. Some plasmids carry genes that can affect the survival of the cell. Plasmids of one class, called **R plasmids**, pose serious problems for human medicine. Transferable R plasmids carry genes for enzymes that destroy antibiotics such as penicillin and tetracycline. Bacteria containing R plasmids are resistant (hence the designation R) to antibiotics that would otherwise kill them. The widespread use of antibiotics in medicine and agriculture has tended to kill off bacteria that lack R plasmids, while those with R plasmids have multiplied. As a result, an increasing number of bacteria that cause human diseases, such as food poisoning and gonorrhea, are becoming resistant to antibiotics (see Module 13.13).

We will return to the topic of plasmids in Chapter 12. But first, we continue our study of molecular genetics in Chapter 11, where we explore what is known about how genes themselves are controlled.

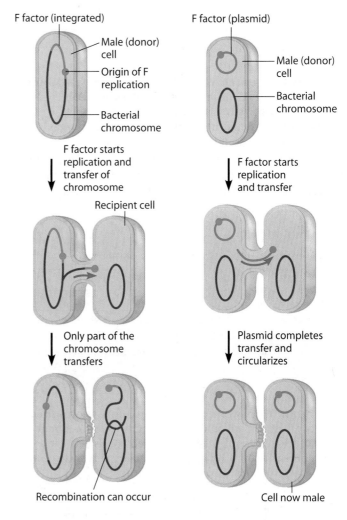

Figure 10.23A Transfer of chromosomal DNA by an integrated F factor

Figure 10.23B Transfer of an F-factor plasmid

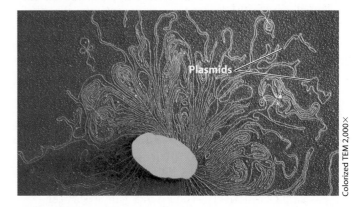

Colorized TEM 2,000×

Figure 10.23C Plasmids and part of a bacterial chromosome released from a ruptured *E. coli* cell

? What is a plasmid?

Web/CD Thinking as a Scientist *How Can Antibiotic-Resistant Plasmids Transform* E. coli?

Reviewing the Concepts

The Structure of the Genetic Material (Introduction–10.3)

DNA as the genetic material. Viruses provided some of the earliest evidence that genes are made of DNA. One key experiment showed that certain bacterial viruses (phages) reprogram host cells to produce more phages by injecting their DNA **(Introduction–10.1).**

DNA is a nucleic acid, made of long chains of nucleotide monomers **(10.2):**

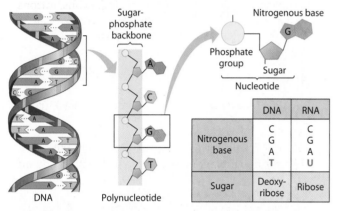

Watson and Crick worked out the three-dimensional structure of DNA: two polynucleotide strands wrapped around each other in a double helix. Hydrogen bonds between bases hold the strands together. Each base pairs with a complementary partner: A with T, and G with C **(10.3).**

DNA Replication (10.4–10.5)

DNA replication starts with the separation of DNA strands. Then enzymes use each strand as a template to assemble new nucleotides into a complementary strand. Using the enzyme DNA polymerase, the cell synthesizes one daughter strand as a continuous piece, the other as a series of short pieces, which are then connected by the enzyme DNA ligase **(10.4–10.5).**

The Flow of Genetic Information from DNA to RNA to Protein (10.6–10.16)

From genotype to phenotype. The information constituting an organism's genotype is carried in the sequence of its DNA bases. A particular gene—a linear sequence of many nucleotides—specifies a polypeptide. The DNA of the gene is transcribed into RNA, which is translated into the polypeptide **(10.6).** The "words" of the DNA "language" are triplets of bases called codons. The codons in a gene specify the amino acid sequence of a polypeptide. Nearly all organisms use exactly the same genetic code **(10.7–10.8).**

Transcription. In the nucleus, the DNA helix unzips, and RNA nucleotides line up and hydrogen-bond along one strand of the DNA, following the base-pairing rules. As the single-stranded messenger RNA (mRNA) peels away from the gene, the DNA strands rejoin **(10.9).** Eukaryotic RNA is processed before leaving the nucleus as mRNA. Noncoding segments called introns are spliced out, and a cap and tail are added to the ends **(10.10).**

Translation takes place in the cytoplasm. A ribosome attaches to the mRNA and translates its message into a specific polypeptide, aided by transfer RNAs (tRNAs) that act as interpreters. Each tRNA is a folded molecule bearing a base triplet called an anticodon on one end; a specific amino acid is added to the other end **(10.11).** The mRNA moves a codon at a time relative to the ribosome, and a tRNA with a complementary anticodon pairs with each codon, adding its amino acid to the peptide chain **(10.12–10.14):**

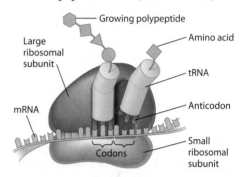

The sequence of codons in DNA, via the sequence of codons in mRNA, spells out the primary structure of a polypeptide **(10.15).**

Mutations are changes in the DNA base sequence, caused by errors in DNA replication or recombination, or by mutagens. Substituting, inserting, or deleting nucleotides alters a gene, with varying effects on the organism **(10.16).**

Microbial Genetics (10.17–10.23)

Viruses can be regarded as genes packaged in protein. When phage DNA enters a lytic cycle inside a bacterium, it is replicated, transcribed, and translated; the new viral DNA and protein molecules then assemble into new phages, which burst from the host cell. In the lysogenic cycle, phage DNA inserts into the host chromosome and is passed on to generations of daughter cells. Much later, it may initiate phage production **(10.17).** Many viruses cause disease when they invade animal or plant cells. Many, such as flu viruses, have RNA, rather than DNA, as their genetic material. Some animal viruses steal a bit of host cell membrane as a protective envelope. Some can remain latent in the host's body for long periods **(10.18).** Most plant viruses have RNA genomes. They enter their hosts via wounds in the plant's outer layers **(10.19).** Emerging viral diseases pose a threat to human health **(10.20).** HIV, the AIDS virus, is a retrovirus; inside a cell it uses its RNA as a template for making DNA, which then inserts into a host chromosome **(10.21).**

Bacteria can transfer genes from cell to cell by one of three processes: transformation, transduction, or conjugation **(10.22).** Plasmids, small circular DNA molecules separate from the bacterial chromosome, can serve as carriers for the transfer of genes **(10.23).**

Connecting the Concepts

1. Check your understanding of the flow of genetic information through a cell by filling in the blanks.

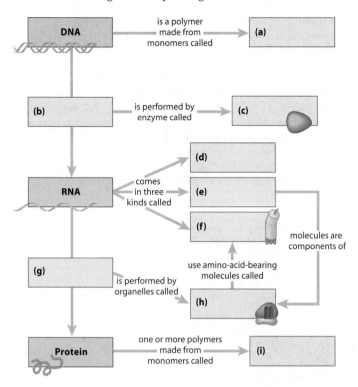

Testing Your Knowledge

2. Scientists have discovered how to put together a bacteriophage with the protein coat of phage T2 and the DNA of phage lambda. If this composite phage were allowed to infect a bacterium, the phages produced in the host cell would have _____. (*Explain your answer.*)
   a. the protein of T2 and the DNA of lambda
   b. the protein of lambda and the DNA of T2
   c. a mixture of the DNA and proteins of both phages
   d. the protein and DNA of T2
   e. the protein and DNA of lambda

3. A geneticist found that a particular mutation had no effect on the polypeptide coded by a gene. This mutation probably involved
   a. deletion of one nucleotide.
   b. alteration of the start codon.
   c. insertion of one nucleotide.
   d. deletion of the entire gene.
   e. substitution of one nucleotide.

4. Which of the following correctly ranks the structures in order of size, from largest to smallest?
   a. gene-chromosome-nucleotide-codon
   b. chromosome-gene-codon-nucleotide
   c. nucleotide-chromosome-gene-codon
   d. chromosome-nucleotide-gene-codon
   e. gene-chromosome-codon-nucleotide

5. The nucleotide sequence of a DNA codon is GTA. A messenger RNA molecule with a complementary codon is transcribed from the DNA. In the process of protein synthesis, a transfer RNA pairs with the mRNA codon. What is the nucleotide sequence of the tRNA anticodon?
   a. CAT   d. CAU
   b. CUT   e. GT
   c. GUA

Describing, Comparing, and Explaining

6. Describe the process of DNA replication: the ingredients needed, the steps in the process, and the final product.

7. Describe the process by which the information in a gene is transcribed and translated into a protein. Correctly use these words in your description: tRNA, amino acid, start codon, transcription, RNA splicing, exons, introns, mRNA, gene, codon, RNA polymerase, ribosome, translation, anticodon, peptide bond, stop codon.

Applying the Concepts

8. A cell containing a single chromosome is placed in a medium containing radioactive phosphate, so that any new DNA strands formed by DNA replication will be radioactive. The cell replicates its DNA and divides. Then the daughter cells (still in the radioactive medium) replicate their DNA and divide, so that a total of four cells are present. Sketch the DNA molecules in all four cells, showing a normal (nonradioactive) DNA strand as a solid line and a radioactive DNA strand as a dashed line.

9. The base sequence of the gene coding for a short polypeptide is CTACGCTAGGCGATTGACT. What would be the base sequence of the mRNA transcribed from this gene? Using the genetic code in Figure 10.8A, give the amino acid sequence of the polypeptide translated from this mRNA. (*Hint:* What is the start codon?)

10. Researchers on the Human Genome Project have determined the nucleotide sequences of human genes and in many cases identified the proteins encoded by the genes. Knowledge of the nucleotide sequences of genes might be used to develop lifesaving medicines or treatments for genetic defects. In the United States, both government agencies and biotechnology companies have applied for patents on their discoveries of genes. In Britain, the courts have ruled that a naturally occurring gene cannot be patented. Do you think individuals and companies should be able to patent genes and gene products? Before answering, consider the following: What are the purposes of a patent? How might the discoverer of a gene benefit from a patent? How might the public benefit? What might be some positive and negative results of patenting genes?

CHAPTER 11

GENE REGULATION

11.1 Proteins interacting with DNA turn prokaryotic genes on or off in response to environmental changes

11.2 Differentiation yields a variety of cell types, each expressing a different combination of genes

11.3 Differentiated cells may retain all of their genetic potential

11.4 DNA packing in eukaryotic chromosomes helps regulate gene expression

11.5 In female mammals, one X chromosome is inactive in each cell

11.6 Complex assemblies of proteins control eukaryotic transcription

11.7 Eukaryotic RNA may be spliced in more than one way

11.8 Translation and later stages of gene expression are also subject to regulation

11.9 Review: Multiple mechanisms regulate gene expression in eukaryotes

ANIMAL CLONING

11.10 Nuclear transplantation can be used to clone animals

11.11 Reproductive cloning has valuable applications, but human reproductive cloning raises ethical issues

11.12 Therapeutic cloning can produce stem cells with great medical potential

THE GENETIC CONTROL OF EMBRYONIC DEVELOPMENT

11.13 Cascades of gene expression and cell-to-cell signaling direct the development of an animal

11.14 Signal transduction pathways convert messages received at the cell surface to responses within the cell

11.15 Key developmental genes are very ancient

THE GENETIC BASIS OF CANCER

11.16 Cancer results from mutations in genes that control cell division

11.17 Oncogene proteins and faulty tumor-suppressor proteins can interfere with normal signal transduction pathways

11.18 Multiple genetic changes underlie the development of cancer

11.19 Mary-Claire King discusses mutations that cause breast cancer

11.20 Avoiding carcinogens can reduce the risk of cancer

To Clone or Not to Clone?

ALL THE ANIMALS SHOWN ON THESE PAGES have what is—for the time being, anyway—a highly unusual origin: They are all clones. In this usage, a **clone** is an individual created by asexual reproduction and thus genetically identical to a single parent. First demonstrated in the 1950s with frogs, animal cloning became much more commonplace after 1997, when Scottish researchers announced the first successful cloning of a mammal—the celebrated Dolly the sheep (above), using a mammary cell from an adult ewe. In the years since Dolly's landmark birth, researchers have cloned a variety of other mammals, such as mules (upper left), horses, cats, cows, goats, pigs, monkeys, and mice.

Some researchers have concentrated their efforts on cloning members of endangered species. Among the rare animals that have been cloned are a gaur (an Asian ox) and a mouflon (a small European sheep). One remarkable case was the 2003 cloning of a banteng (a Javanese bovine, shown at far right) using frozen skin cells from a zoo-raised individual that died in 1980. Nuclei from the donor were inserted into enucleated cow eggs, which were implanted into surrogate Angus cows. This case proves the possibility of producing new animals even when a female of the donor species is unavailable. Scientists may

Cloned animals are less healthy than those arising from a fertilized egg

The Control of Gene Expression

someday even be able to use similar cross-species methods to clone an animal from a recently extinct species.

The potential use of cloning to slow or reverse extinction and repopulate endangered species holds tremendous promise. However, cloning may create new problems. Conservationists object that cloning may trivialize the tragedy of extinction and detract from efforts to preserve natural habitats. They correctly point out that cloning, like other forms of asexual reproduction, does not increase genetic diversity and is therefore not as beneficial to endangered species as natural reproduction.

Researchers are also concerned that cloned animals are less healthy than those arising from a fertilized egg. In 2003, at age 6, Dolly was euthanized after suffering complications from a lung disease normally seen only in much older sheep. Dolly's premature death, as well as her arthritic condition, led to speculation

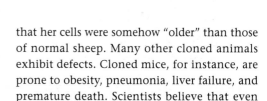

that her cells were somehow "older" than those of normal sheep. Many other cloned animals exhibit defects. Cloned mice, for instance, are prone to obesity, pneumonia, liver failure, and premature death. Scientists believe that even cloned animals that appear normal are likely to have subtle defects. Recent work traces these problems to physical changes within the chromosomes of the cloned organism that derail the normal regulation of genes.

The ability to clone an animal using a transplanted nucleus demonstrates an important point: The nuclei of adult somatic cells contain a complete genome capable of directing the production of all the cell types in an organism. The development of a multicellular organism, with many different kinds of cells, thus depends on the turning on and off of genes—the control of gene expression. The health problems associated with cloned animals underscore the fact that proper gene regulation is important to the well-being of all organisms.

This chapter focuses first on how genes are regulated, beginning with simple prokaryotic systems and then describing gene regulation in eukaryotes. Next we look at the methods and applications of animal cloning. Then we examine the connection between gene regulation and embryonic development. Finally, we discuss cancer, a disease that results from genetic defects in cell regulation. ■ ■ ■

11.1 Proteins interacting with DNA turn prokaryotic genes on or off in response to environmental changes

Gene regulation—the turning on and off of genes—can help organisms respond to environmental changes. But what do we actually mean when we say that genes are turned on or off? As we discussed in Chapter 10, genes determine the nucleotide sequences of specific RNA molecules; if this RNA is mRNA, it in turn determines the sequences of amino acids in protein molecules. Thus, a gene that is turned on is being transcribed into RNA, and that message is being translated into specific protein molecules. The overall process by which genetic information flows from genes to proteins—that is, from genotype to phenotype—is called **gene expression.** The control of gene expression makes it possible for cells to produce specific kinds of proteins when and where they are needed. The turning on and off of transcription is the main way that gene expression is regulated in all organisms.

Our earliest understanding of gene control came from studies of the bacterium *Escherichia coli* (**Figure 11.1A**). *E. coli* can change its metabolic activities in response to changes in its environment. For example, when a certain nutrient is plentiful, *E. coli* does not squander valuable resources to make it from scratch. Let's look at how the regulation of gene transcription helps *E. coli* accomplish such control.

The lac *Operon* Picture an *E. coli* cell living in your intestine. It is dependent on your dietary whims for its nutrients. If you eat a sweet roll for breakfast, the bacterium will soon be bathed in sugars and broken-down fats. Later on, if you have a glass of fat-free milk and a salad for lunch, *E. coli*'s environment will change drastically.

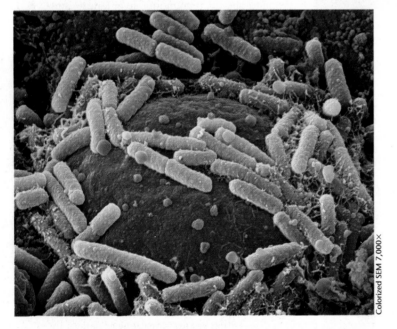

Figure 11.1A Cells of *E. coli* bacteria

Colorized SEM 7,000×

Let's focus on your glass of milk for a moment. One of the main nutrients in milk is the disaccharide sugar lactose. When lactose is plentiful in the intestine, *E. coli* makes the enzymes necessary to absorb the sugar and use it as an energy source. Conversely, when lactose is not plentiful, *E. coli* does not waste its energy producing these enzymes.

Remember that enzymes are proteins; their production is an outcome of gene expression. *E. coli* can make lactose-utilization enzymes because it has genes that code for these enzymes. **Figure 11.1B** presents a model (first proposed in 1961 by French biologists François Jacob and Jacques Monod) to explain how an *E. coli* cell can turn genes coding for lactose-utilization enzymes off or on, depending on whether lactose is available.

E. coli uses three enzymes to take up and start metabolizing lactose, and the genes coding for these three enzymes are regulated as a single unit. The DNA at the top of Figure 11.1B represents a small segment of the bacterium's chromosome. Notice that the three genes that code for the lactose-utilization enzymes (light blue) are next to each other in the DNA.

Adjacent to the group of lactose enzyme genes are two *control sequences*, short sections of DNA that help control the enzyme genes. One stretch of nucleotides is a **promoter** (green), a site where the transcription enzyme, RNA polymerase, attaches and initiates transcription—in this case, transcription of all three lactose enzyme genes (as depicted in the bottom panel of Figure 11.1B). Between the promoter and the enzyme genes, a DNA segment called an **operator** (yellow) acts as a switch. The operator determines whether RNA polymerase can attach to the promoter and start transcribing the genes.

Such a cluster of genes with related functions, along with a promoter and an operator, is called an **operon;** with rare exceptions, operons exist only in prokaryotes. The key advantage to the grouping of related genes into operons is that a single "on-off switch" can control the whole cluster. The operon discussed here is called the *lac* operon, short for lactose operon. When an *E. coli* encounters lactose, all the enzymes needed for its use are made at once because the operon's genes are all controlled by a single switch, the operator. But what determines whether the operator switch is on or off?

The top panel of Figure 11.1B shows the *lac* operon in "off" mode, its status when there is no lactose in the cell's environment. Transcription is turned off by a molecule called a **repressor** (red), a protein that functions by binding to the operator and physically blocking the attachment of RNA polymerase to the promoter. On the left side of the figure, you can see where the repressor comes from. A gene called a **regulatory gene** (dark blue), located outside the operon, codes for the repressor. The regulatory gene is expressed continually, so the cell always has a small supply of repressor molecules.

How can an operon be turned on if its repressor is always present? As the bottom panel of Figure 11.1B indicates, lactose interferes with the attachment of the *lac* repressor to the operator by binding to the repressor and changing its shape. With its new shape, the repressor cannot bind to the operator, and the operator switch remains on. RNA polymerase can now bind to the promoter (since it is no longer being blocked) and from there transcribe the genes of the operon. The resulting mRNA carries coding sequences for all three enzymes (purple) needed for lactose use. The cell can translate the single mRNA message into three separate polypeptides because the mRNA has multiple codons signaling the start and stop of translation.

The newly produced mRNA and protein molecules will remain intact for only a short time before cellular enzymes break them down. When the synthesis of mRNA and protein stops because lactose is no longer present, the molecules quickly disappear.

Other Kinds of Operons The *lac* operon is only one type of operon in bacteria. Other types also have a promoter, an operator, and several adjacent genes, but they differ in the way the operator switch is controlled. **Figure 11.1C** shows two types of repressor-controlled operons. The *lac* operon's repressor is active when alone and inactive when bound to lactose. A second type of operon, represented here by the *trp* operon, is controlled by a repressor that is *inactive* alone. To be active, this type of repressor must combine with a specific small molecule. In our example, the small molecule is tryptophan, an amino acid essential for protein synthesis. *E. coli* can make tryptophan from scratch, using enzymes encoded in the *trp* operon. But it will stop making tryptophan and simply absorb it in prefabricated form from the surroundings whenever possible. When *E. coli* is swimming in tryptophan, which occurs in large amounts in such foods as milk and poultry, tryptophan binds to the repressor of the *trp* operon. This activates the *trp* repressor, enabling it to switch off the operon. Thus, this type of operon allows bacteria to stop making certain essential molecules when the molecules are already present in the environment, saving materials and energy for the cells.

A third type of operon uses **activators**, proteins that turn operons *on* by binding to DNA. These proteins act by making it easier for RNA polymerase to bind to the promoter, rather than by blocking RNA polymerase, as repressors do. Armed with a variety of operons, regulated by repressors and activators, *E. coli* and other prokaryotes can thrive in frequently changing environments.

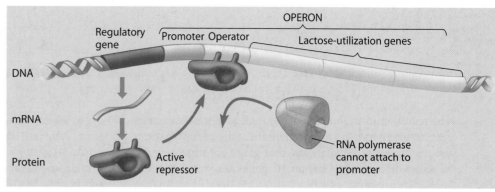

Operon turned off (lactose absent)

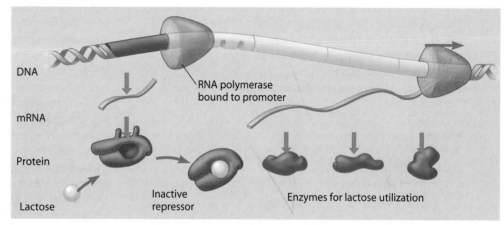

Operon turned on (lactose inactivates repressor)

Figure 11.1B The *lac* operon

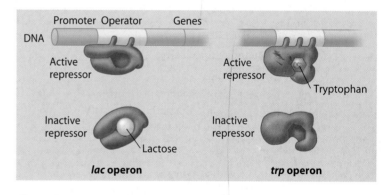

Figure 11.1C Two types of repressor-controlled operons

Next we'll examine how more complex eukaryotes achieve gene regulation.

Web/CD Activity 11A *The* lac *operon in* E. coli

? A certain mutation in *E. coli* impairs the ability of the *lac* operator to bind to the repressor. How would this affect the cell?

■ The cell would wastefully produce the enzymes for lactose metabolism continuously, even in the absence of lactose.

11.2 Differentiation yields a variety of cell types, each expressing a different combination of genes

Compared with bacteria, eukaryotic organisms, especially multicellular ones, face elaborate gene regulation challenges. During the repeated cell divisions that lead from a zygote to a multicellular adult, individual cells must undergo **differentiation**—that is, become specialized in structure and function. Differentiation results from selective gene expression, the turning on and off of genes.

The light micrographs **in Figure 11.2** show several types of human cells. Each sample was stained with dyes to bring out important cellular details, and all three micrographs are printed at about the same magnification (750×). The micrograph on the left shows a short segment of a muscle cell, which is a long fiber with multiple nuclei (dark horizontal rods). The vertical stripes result from the arrangement of the proteins that bring about muscle contraction. As you'd ex-

pect, the genes encoding these proteins are active in muscle cells.

The center micrograph shows cells from the pancreas, the organ that produces the hormones glucagon and insulin, which regulate blood sugar. The gene for glucagon is turned on only in the alpha cells (pink) and the gene for insulin only in the beta cells (light purple). The dark blue spots are cell nuclei.

In the micrograph on the right, we see a single white blood cell (purple with a dark, two-part nucleus) surrounded by red blood cells (pink). In immature red blood cells, the genes for the oxygen-carrying protein hemoglobin are turned on full blast. Later in differentiation, mammalian red blood cells lose their nuclei and other organelles. Mature red blood cells (shown in the micrograph) are packed full with hemoglobin.

The genes for "housekeeping" enzymes, such as those involved in glycolysis, are active in all metabolizing cells. In contrast, the genes for specialized proteins are turned on only in particular types of cells.

In summary, the particular genes that are active in each type of differentiated cell are the source of its particular function and structure. Next we look more closely at some of the evidence that differentiated cells carry the entire genome.

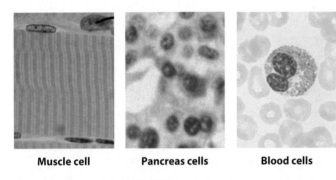

Muscle cell　　**Pancreas cells**　　**Blood cells**

Figure 11.2 Some types of human cells (all shown at 750×)

? If a nerve cell and a skin cell in your body have the same genes, how can the cells be so different?

■ Each cell type must be expressing certain genes that are present in, but not expressed in, the other cell types.

11.3 Differentiated cells may retain all of their genetic potential

Differentiated cells express only a small percentage of their genes. But are all the genes still present? And if they are, do the differentiated cells retain the potential to express them?

One way to approach these questions is to see if a differentiated cell can generate a whole new organism. In plants, this ability is common, as was first demonstrated during the 1950s by F. C. Steward and his students at Cornell University. As shown in **Figure 11.3**, they found that when they transferred cells from a carrot to culture medium, a single cell could begin dividing and eventually grow into an adult plant, a genetic replica of the parent plant. This method can be used to produce thousands of genetically identical organisms—clones—from the somatic cells of a single plant. In this way, it is possible to propagate large numbers of crop plants with desirable traits such as high fruit yield or resistance to disease. The fact that a mature plant cell can dedifferentiate (reverse its differentiation) and

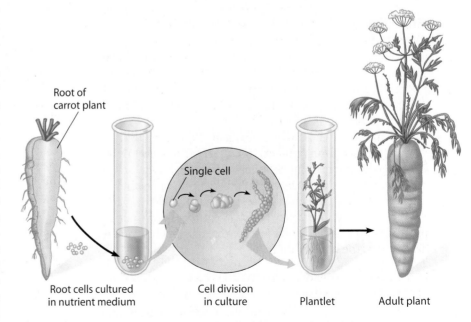

Root of carrot plant

Root cells cultured in nutrient medium

Single cell

Cell division in culture

Plantlet

Adult plant

Figure 11.3 Growth of a carrot plant from a differentiated root cell

then give rise to all the different kinds of specialized cells of a new plant shows that differentiation does not necessarily involve irreversible changes in the DNA.

But is this sort of cloning possible in animals? An indication that differentiation need not impair an animal cell's genetic potential is the natural process of **regeneration**, the regrowth of lost body parts. When a salamander loses a leg, for example, certain cells in the leg stump dedifferentiate, divide, and then redifferentiate, giving rise to a new leg. Many animals can regenerate lost parts, especially among the invertebrates, and in a few relatively simple animals, isolated differentiated cells can dedifferentiate and then develop into an organism. Further evidence for the complete genetic potential of animal cells comes from cloning experiments. We will focus on that topic in detail later. First, let's examine how genes are regulated in eukaryotes.

? What evidence supports the view that differentiation is based on the control of gene expression rather than on irreversible changes in the genome?

■ The cloning of plants and animals from differentiated cells

11.4 DNA packing in eukaryotic chromosomes helps regulate gene expression

Let's begin our exploration of gene regulation in eukaryotes by looking at the chromosomes, where almost all of the cell's genes are located. The DNA in just a single human chromosome averages 4 cm in length, thousands of times more than the diameter of the nucleus. All this DNA can fit within the nucleus because of an elaborate, multilevel system of coiling and folding, or *packing*, of the DNA in each chromosome. A crucial aspect of DNA packing is the association of the DNA with small proteins called **histones**. In fact, histone proteins account for about half the mass of eukaryotic chromosomes. (Prokaryotes have analogous proteins, but lack the degree of DNA packing in eukaryotes.)

Figure 11.4 shows a model for the main levels of DNA packing. At the top, notice that the double helix has a diameter of 2 nm, which is not altered by packing. At the first level of packing, histones attach to the DNA. In electron micrographs, the DNA-histone complex has the appearance of beads on a string. Each "bead," called a **nucleosome**, consists of DNA wound around a protein core of eight histone molecules. Short stretches of DNA, called linkers, join consecutive nucleosomes (the "string" between the "beads").

At the next level of packing, the beaded string is wrapped into a tight helical fiber. Then this fiber coils further into a thick supercoil with a diameter of about 300 nm. Looping and folding can further compact the DNA, as you can see in the metaphase chromosome at the bottom of the figure. Viewed as a whole, Figure 11.4 gives a sense of how successive levels of coiling and folding enable a huge amount of DNA to fit into a cell nucleus.

DNA packing tends to prevent gene expression by preventing transcription proteins from contacting the DNA. Cells seem to use such higher levels of packing for long-term inactivation of genes. Highly compacted chromatin, which is found not only in metaphase chromosomes but also in varying regions of interphase chromosomes, is generally not expressed at all. One intriguing instance of this phenomenon is described in the next module.

? How does dense packing of DNA in chromosomes prevent gene expression?

■ RNA polymerase and other proteins required for transcription do not have access to the DNA in tightly packed regions of a chromosome.

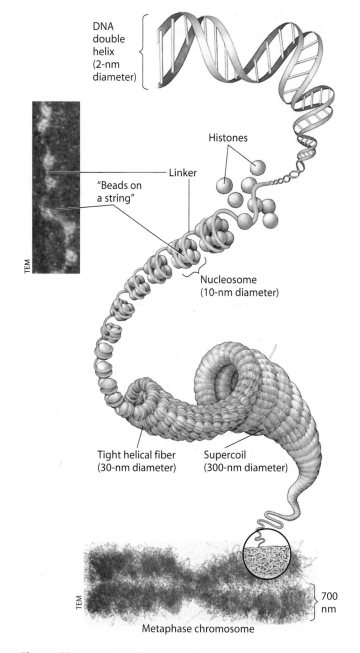

Figure 11.4 DNA packing in a eukaryotic chromosome

DNA double helix (2-nm diameter)

Histones

Linker

"Beads on a string"

Nucleosome (10-nm diameter)

Tight helical fiber (30-nm diameter)

Supercoil (300-nm diameter)

700 nm

Metaphase chromosome

11.5 In female mammals, one X chromosome is inactive in each cell

In female mammals, one X chromosome in each somatic cell exists in a highly compacted and almost entirely inactive form. This **X chromosome inactivation** is initiated early in embryonic development, when one of the two X chromosomes in each cell is inactivated at random. Which X chromosome is inactivated is a matter of chance in each embryonic cell, but once an X chromosome in inactivated, all descendant cells have the same copy turned off. Consequently, adult females have a patchwork mosaic of cells, with populations of cells that express different X chromosomes.

A striking example of this phenomenon is the tortoiseshell cat, which has orange and black patches of fur (**Figure 11.5**). The relevant fur-color gene is on the X chromosome, and the tortoiseshell phenotype requires the presence of two different alleles, one for orange fur and one for black fur. Normally, only females can have both alleles because only they have two X chromosomes. If a female is heterozygous for the tortoiseshell gene, she will have the tortoiseshell phenotype. Orange patches are formed by populations of cells in which

the X chromosome with the orange allele is active; black patches have cells in which the X chromosome with the nonorange allele is active.

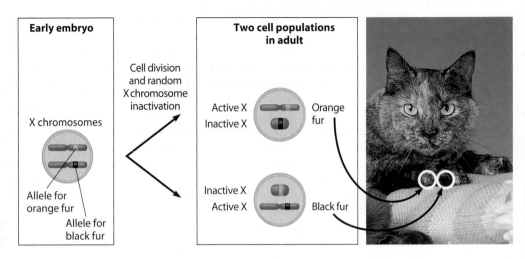

Figure 11.5 Tortoiseshell pattern on a cat, a result of X chromosome inactivation

? Why are tortoiseshell cats usually female?

■ In general, only females have two X chromosomes (but see Module 8.22).

11.6 Complex assemblies of proteins control eukaryotic transcription

The packing and unpacking of chromosomal DNA provide a coarse adjustment for eukaryotic gene expression by making a region of DNA either more or less available for transcription. The fine-tuning begins with the initiation of RNA synthesis—transcription. In both prokaryotes and eukaryotes, the initiation of transcription (whether transcription starts or not) is the most important stage for regulating gene expression.

Like prokaryotes (see Module 11.1), eukaryotes employ regulatory proteins that bind to DNA and turn the transcription of genes on and off. The eukaryotic control mechanisms involve proteins that, like prokaryotic repressors and activators, bind to specific segments of DNA. However, eukaryotic cells have more regulatory proteins and more control sequences in their DNA. The current model for the initiation of eukaryotic transcription features an intricate array of regulatory proteins that interact with DNA and with one another to turn genes on or off.

In contrast to the clustered genes of bacterial operons, each eukaryotic gene usually has its own promoter and other control sequences. Moreover, activator proteins seem to be more important in eukaryotes than repressors. In multicellular eukaryotes, the "default" state for most genes seems to be "off." A typical animal or plant cell needs to turn on (transcribe) only a small percentage of its genes,

those required for the cell's specialized structure and function. Housekeeping genes, those continually active in virtually all cells for routine activities such as glycolysis, may be in an "on" state by default.

Transcription Factors In order to function, eukaryotic RNA polymerase requires the assistance of proteins called **transcription factors**. Activators are one type of transcription factor. In the model depicted in **Figure 11.6**, the first step in initiating gene transcription is the binding of activators (green) to DNA sequences called **enhancers** (yellow). In contrast to the operators of prokaryotic operons, enhancers are usually far away from the gene they help regulate and may be on either side of the gene. The binding of activators to enhancers leads to bending of the DNA. Once the DNA is bent, the bound activators interact with other transcription factor proteins (purple), which then bind as a complex at the gene's promoter. This large assembly of proteins facilitates the correct attachment of RNA polymerase to the promoter and the initiation of transcription. As shown in the figure, several enhancers and activators may be involved. Not shown are *repressor* proteins that may bind to DNA sequences called **silencers** and function analogously to *inhibit* the start of transcription.

In summary, both eukaryotes and prokaryotes control transcription by using proteins that bind to DNA. However, many more regulatory proteins are involved in eukaryotes, and the interactions among these proteins are far more complex.

Coordinating Eukaryotic Gene Expression If eukaryotic genomes only rarely have operons, how does the eukaryotic cell deal with genes of related function that need to be turned on or off at the same time? Genes coding for the enzymes of a metabolic pathway, for example, are often scattered over different chromosomes. Coordinated gene expression in eukaryotes seems to depend on the association of a specific enhancer (or collection of enhancers) with every gene of a dispersed group. Copies of the transcription factors that recognize these DNA sequences bind to them all at once, promoting simultaneous transcription of the genes, no matter where they are in the genome.

Web/CD Thinking as a Scientist *How Do You Design a Gene Expression System?*

> **?** In stimulating transcription of a specific eukaryotic gene, an enhancer does not act directly on the gene's promoter, but has its effect via DNA-binding proteins called _____ _____.

■ transcription factors

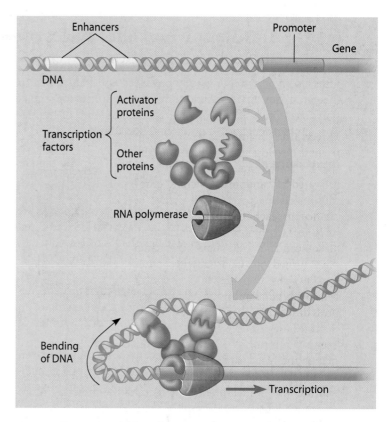

Figure 11.6 A model for the turning on of a eukaryotic gene

11.7 Eukaryotic RNA may be spliced in more than one way

Once transcription of a eukaryotic RNA molecule is completed, the noncoding segments called introns are removed by splicing (see Module 10.10). RNA splicing provides several possible ways for regulating gene expression. Some scientists think that the splicing process itself may help control the flow of mRNA from nucleus to cytoplasm because until splicing is completed, the RNA is attached to the molecules of the splicing machinery and cannot pass through the nuclear pores. Moreover, in some cases, the cell can carry out splicing in more than one way, generating different mRNA molecules from the same RNA transcript. Notice in Figure 11.7, for example, that one mRNA molecule ends up with the green exon and the other with the brown exon. (The light-colored segments are the introns.) With this sort of **alternative RNA splicing**, an organism can get more than one type of polypeptide from a single gene.

One interesting example of two-way splicing is found in the fruit fly. In this animal, the differences between males and females are largely due to different patterns of RNA splicing. And as you will learn in Chapter 12, recent results from the Human Genome Project suggest that alternative splicing is very common in humans. The 100 or so instances already known include one gene whose transcript can be spliced to encode *seven* alternative versions of a certain protein involved in cellular contraction. Each of the seven is made in a different type of cell.

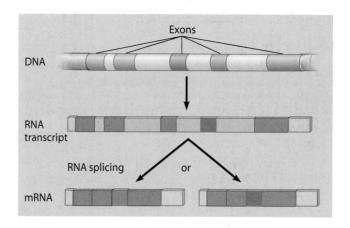

Figure 11.7 Production of two different mRNAs from the same gene

> **?** How does alternative RNA splicing enable a single gene to encode more than one kind of polypeptide?

■ Each kind of polypeptide is encoded by an mRNA molecule containing a different combination of exons.

Translation and later stages of gene expression are also subject to regulation

After eukaryotic mRNA is fully processed and transported to the cytoplasm, there are additional opportunities for regulation. These include mRNA breakdown, initiation of translation, protein activation, and protein breakdown.

Breakdown of mRNA   Molecules of mRNA do not remain intact forever. Enzymes in the cytoplasm eventually break them down, and the timing of this event is an important factor regulating the amounts of various proteins that are produced in the cell. Long-lived mRNAs can be translated into many more protein molecules than short-lived ones. Prokaryotic mRNAs have very short lifetimes; they are degraded by enzymes within a few minutes after their synthesis. This is one reason bacteria can change their proteins relatively quickly in response to environmental changes. In contrast, the mRNA of eukaryotes can have lifetimes of hours or even weeks.

A striking example of long-lived mRNA is found in vertebrate red blood cells, which act like factories for manufacturing the protein hemoglobin. In most species of vertebrates, the mRNAs for hemoglobin are unusually stable. They probably last as long as the red blood cells that contain them—about a month or a bit longer in reptiles, amphibians, and fishes—and are translated again and again. Mammals are an exception, as you learned in Module 11.2. When their red blood cells mature, they lose their ribosomes (along with their other organelles) and thus cease to make new hemoglobin. However, mammalian hemoglobin itself lasts about as long as the red blood cells last, around four months.

Initiation of Translation   The process of translating mRNA into polypeptide also offers opportunities for regulation. Among the molecules involved in translation are a great many proteins that control the start of polypeptide synthesis. Red blood cells, for instance, have an inhibitory protein that prevents translation of hemoglobin mRNA unless the cell has a supply of heme, the iron-containing chemical group essential for hemoglobin function. (It is the iron atom of the heme group to which oxygen molecules actually attach.)

Protein Activation   After translation is complete, polypeptides may require alteration to become functional. Posttranslational control mechanisms in eukaryotes often involve the cleavage (cutting) of a polypeptide to yield a smaller final product that is the active protein, able to carry out a specific function in the organism. In **Figure 11.8** we see the example of the hormone insulin, which is a protein. Insulin is synthesized in the cells of the pancreas as one long polypeptide that has no hormonal activity. After translation is completed, the polypeptide folds up, and covalent bonds form between the sulfur (S) atoms of sulfur-containing amino acids (see Figure 3.12B). (Two H atoms are lost as each S—S bond forms.) The result is that parts of the polypeptide are linked together in a specific way. Finally, a large center portion is cut away, leaving two shorter chains held together by the sulfur linkages. This combination of two shorter polypeptides is the form of insulin that functions as a hormone.

Protein Breakdown   The final control mechanism operating after translation is the selective breakdown of proteins. Though mammalian hemoglobin may last as long as the red blood cell housing it, the lifetimes of many other proteins are closely regulated. Some of the proteins that trigger metabolic changes in cells are broken down within a few minutes or hours. This regulation allows a cell to adjust the kinds and amounts of its proteins in response to changes in its environment. It also enables the cell to maintain its proteins in prime working order. Indeed, when proteins are damaged, they are usually broken down right away and replaced by new ones that function properly.

Web/CD Activity 11B *Gene Regulation in Eukaryotes*

? Once mRNA encoding a particular protein reaches the cytoplasm, what are four mechanisms that can regulate the amount of the active protein in the cell?

■ Breakdown of the mRNA; regulation of translation initiation; activation of the protein; and breakdown of the protein

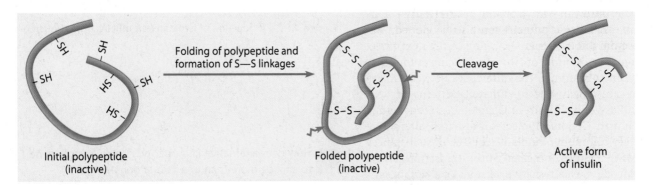

Figure 11.8   Protein activation: The role of polypeptide cleavage in producing the active insulin protein

11.9 Review: Multiple mechanisms regulate gene expression in eukaryotes

Figure 11.9 provides a review of eukaryotic gene expression and highlights the multiple control points where the process can be turned on or off, speeded up, or slowed down. Picture the series of pipes that carry water from your local water supply, perhaps a reservoir, to a faucet in your home. At various points, valves control the flow of water. We use this model in the figure to illustrate the flow of genetic information from a chromosome—a reservoir of genetic information—to an active protein that has been synthesized in the cell's cytoplasm. The multiple mechanisms that control gene expression are analogous to the control valves in water pipes. In the figure, each gene expression "valve" is indicated by a control knob. Note that these knobs represent *possible* control points; for most proteins, only a few control points are probably important. As we have seen, the most important control point, in both eukaryotes and prokaryotes, is usually the start of transcription. In the diagram, the large yellow knob represents the mechanisms that regulate the start of transcription.

After transcription, RNA processing in the nucleus adds nucleotides to the ends of the RNA (cap and tail) and splices out introns. As we discussed in Module 11.7, a growing body of evidence suggests the importance of control at this stage. Once mRNA reaches the cytoplasm, additional stages subject to regulation include mRNA translation and eventual breakdown, possible alteration of the polypeptide to give the active protein, and the eventual breakdown of the protein.

Despite its numerous steps, Figure 11.9 oversimplifies the control of gene expression. What it does not show is the web of control that connects different genes, often through their products. We have seen examples in both prokaryotes and eukaryotes of the actions of gene products (usually proteins) on other genes or on other gene products within the same cell. The genes of operons in *E. coli*, for instance, are controlled by repressor or activator proteins encoded by regulatory genes on the same DNA molecule. In eukaryotes, many genes are controlled by proteins encoded by regulatory genes on different chromosomes.

In eukaryotes, cellular differentiation results from the selective turning on and off of genes. But each cell in a eukaryote still retains its full genetic potential. We saw evidence for this in Module 11.3: the cloning of plants from differentiated plant cells. Do animals, like plants, retain their full set of genes in every somatic cell? The answer is yes, and this implies that animals can also be cloned. We turn to an examination of animal cloning in the next three modules.

Web/CD Activity 11C *Review: Gene Regulation in Eukaryotes*

? Of the nine regulatory "valves" in Figure 11.9, which five can also operate in a prokaryotic cell?

■ (1) Control of transcription; (2) control of mRNA breakdown; (3) control of translation; (4) control of protein activation; and (5) control of protein breakdown

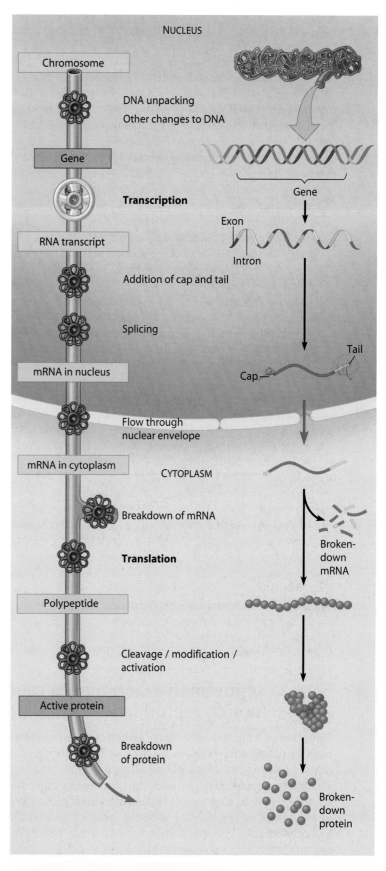

Figure 11.9 The gene expression "pipeline" in a eukaryotic cell

11.10 Nuclear transplantation can be used to clone animals

As mentioned in Module 11.3, cloning provides strong evidence that differentiated cells retain their full genetic potential. Animal cloning, including the cases described in the chapter introduction, is achieved through a procedure called **nuclear transplantation (Figure 11.10)**. First performed in the 1950s on frog embryos, nuclear transplantation involves replacing the nucleus of an egg cell or a zygote with the nucleus of an adult somatic cell. The egg cell may then begin to divide. About 5 days later, repeated cell divisions form a blastocyst, a ball of cells. At this point, the blastocyst may be used for different purposes, as indicated by the branching in Figure 11.10.

If the animal to be cloned is a mammal, further development requires implanting the blastocyst into the uterus of a surrogate mother (Figure 11.10, upper branch). The resulting animal will be genetically identical to the donor of the nucleus—a "clone" of the donor. This type of cloning, which results in birth of a new individual, is called **reproductive cloning**. Scottish researcher Ian Wilmut and his colleagues used reproductive cloning to produce the celebrated sheep Dolly (see introduction), using a fully differentiated mammary cell of an adult ewe as the source of the nucleus.

In a different cloning procedure (Figure 11.10, lower branch), **embryonic stem cells (ES cells)** are harvested from the blastocyst. In nature, embryonic stem cells give rise to all the different kinds of specialized cells of the body. In the laboratory, embryonic stem cells are easily grown in culture, where, given the right conditions, they can perpetuate themselves indefinitely. When the major aim is to produce embryonic stem cells for therapeutic treatments, the process is called **therapeutic cloning**. In the next two modules, we will discuss applications of reproductive and therapeutic cloning.

? What are the intended products of reproductive cloning and therapeutic cloning?

■ Reproductive cloning seeks the production of new individuals. Therapeutic cloning seeks the harvest of embryonic stem cells.

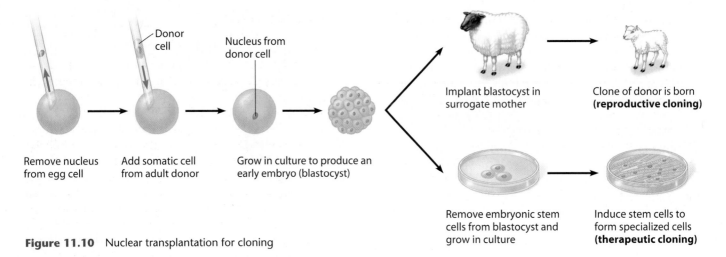

Figure 11.10 Nuclear transplantation for cloning

Remove nucleus from egg cell

Add somatic cell from adult donor

Grow in culture to produce an early embryo (blastocyst)

Implant blastocyst in surrogate mother

Clone of donor is born **(reproductive cloning)**

Remove embryonic stem cells from blastocyst and grow in culture

Induce stem cells to form specialized cells **(therapeutic cloning)**

Donor cell

Nucleus from donor cell

11.11 Reproductive cloning has valuable applications, but human reproductive cloning raises ethical issues

Since Dolly's birth, researchers have cloned many other mammals, including mice, cats, horses, cows, and pigs. Such reproductive cloning has many potential applications. By using genetically engineered donor nuclei, geneticists can study the effects of changing single genes or combinations of genes. And in the future, biologists may routinely produce genetically identical animals for experimentation, a potential benefit to genetics research.

On an experimental basis, agricultural scientists are already cloning farm animals with specific sets of desirable traits in the hope of creating high-yielding, genetically identical herds. The pharmaceutical industry is experimentally cloning mammals for the production of potentially valuable drugs. Figure 12.16B shows cloned sheep that secrete a human blood protein into their milk, one that could prove useful in treating cystic fibrosis. The piglets in **Figure 11.11** are clones that lack one of two copies of a gene for a protein that can cause immune system rejection in humans. Such pigs may one day provide organs for transplant into humans. Some wildlife biologists hope that reproductive cloning can be used to restock the populations of

endangered animals (although others caution against such an approach, as discussed in the chapter introduction).

The successful cloning of various mammals has heightened speculation that humans could be cloned. In 2001, a biotechnology company called Advanced Cell Technology provoked an uproar with the announcement that it had created the first human embryos by cloning. Their most advanced embryo, however, had stopped growing at about six cells, and their intention was solely to harvest embryonic stem cells. Although these embryos were not allowed to grow further, this achievement brings us one step closer to the possibility of human reproductive cloning.

Critics point out that there are many obstacles—both practical and ethical—to human cloning. Practically, animal cloning is extremely difficult and inefficient. Only a small percentage of cloned embryos develop normally. Ethically, creating human embryos for such purposes raises many troubling questions. An ethical consensus on the status of the human blastocyst is unlikely any time soon. Meanwhile, the research and the debate continue.

Figure 11.11   Piglet clones

? Describe the method used to clone mammals.

The nucleus of an egg cell is replaced with a somatic cell nucleus; after development starts, the early embryo is implanted in the uterus of a surrogate mother.

11.12   Therapeutic cloning can produce stem cells with great medical potential

Therapeutic cloning produces embryonic stem cells, cells that in the early animal embryo differentiate to give rise to all the cell types in the body. When grown in laboratory culture, embryonic stem cells can divide indefinitely (like cancer cells; see Module 8.10). But the right conditions—such as the presence of certain growth factors—can induce changes in gene expression that cause differentiation into a particular cell type (Figure 11.12). If scientists can discover the right conditions, they will be able to grow cells for the repair of injured or diseased organs. Such cells could be made by inserting a cell nu-

cleus from a patient into an ES cell from which the nucleus has been removed. When implanted in the patient, these cells would not be rejected by the immune system because they would be genetically identical to the patient's own cells.

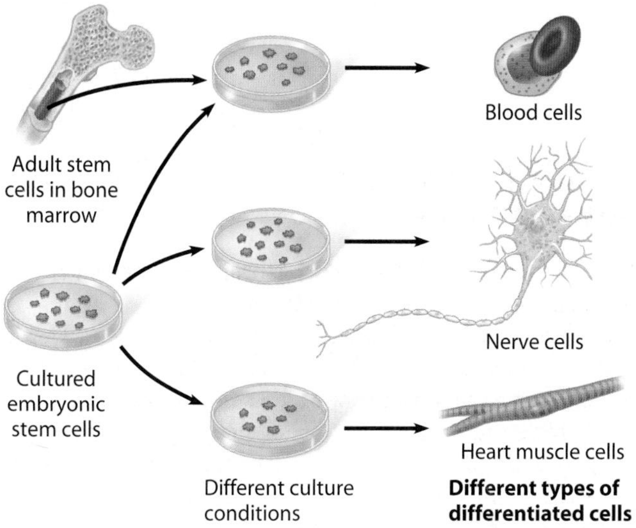

But the use of ES cells raises both ethical and technical problems. Human ES cells must be obtained, at least initially, by destroying human embryos (such as ones donated by patients undergoing infertility treatment). This might be avoided by using **adult stem cells**, cells present in adult tissues that generate replacements for nondividing differentiated cells. Unlike ES cells, adult stem cells are partway along the road to differentiation. They can often give rise to multiple types of specialized cells, but it is not clear whether they can give rise to *all* types of cells.

Like ES cells, adult stem cells can be grown in culture and induced to differentiate into a range of cell types. For example, adult stem cells in bone marrow generate all types of blood cells. Perhaps adult stem cells, ethically less problematic to obtain than ES cells, may someday provide the answer to human tissue and organ replacement. However, embryonic stem cells are currently more promising than adult stem cells for such applications. While most people believe that reproductive cloning is unethical, opinions vary more widely about the morality of therapeutic cloning using embryonic stem cells.

? In nature, how do ES cells differ from adult stem cells?

ES cells in the embryo give rise to all the different kinds of cells in the body. Adult stem cells generate only a few related types of cells.

Figure 11.12   Differentiation of stem cells in culture

Adult stem cells in bone marrow

Cultured embryonic stem cells

Different culture conditions

Blood cells

Nerve cells

Heart muscle cells

Different types of differentiated cells

11.13 Cascades of gene expression and cell-to-cell signaling direct the development of an animal

Some of the first glimpses into the relationship between gene expression and embryonic development came from studies of mutants of the fruit fly *Drosophila melanogaster* (see Module 9.20). **Figure 11.13A**, showing front views of the heads of two fruit flies, includes a mutant that developed in a strikingly abnormal way: It has two legs where its antennae should be! Research on this and other developmental mutants has led to the identification of many of the genes that program development in the normal fly. This genetic approach has revolutionized developmental biology.

Among the earliest events in fruit fly development are ones that determine which end of the egg cell will become the head and which end will become the tail. These events occur in the ovaries of the mother fly and involve communication between an unfertilized egg cell and cells adjacent to it in its follicle (egg chamber). As **Figure 11.13B** indicates, ❶ one of the first genes activated in the egg cell codes for a protein that leaves the egg cell and signals adjacent follicle cells. ❷ These follicle cells are stimulated to turn on genes for other proteins, which signal back to the egg cell. ❸ One of the egg cell's responses is to localize a specific type of mRNA (red) at one end of the cell. This mRNA marks the end of the egg where the fly's head will develop and thus defines the animal's head-to-tail axis. (One piece of evidence from mutant studies is that a defective gene for the "head" mRNA causes the embryo to develop tails at both ends!) In a similar way, other egg cell genes direct the positioning of the top-to-bottom and side-to-side axes.

After the egg is fertilized and laid, repeated mitoses transform the zygote into an embryo. ❹ Translation of the "head" mRNA in the early embryo produces a regulatory protein (green dots) that diffuses through the embryo but remains most concentrated at the head end. In turn, this protein gradient triggers a corresponding gradient of transcription in the embryo's nuclei. ❺ The proteins (purple

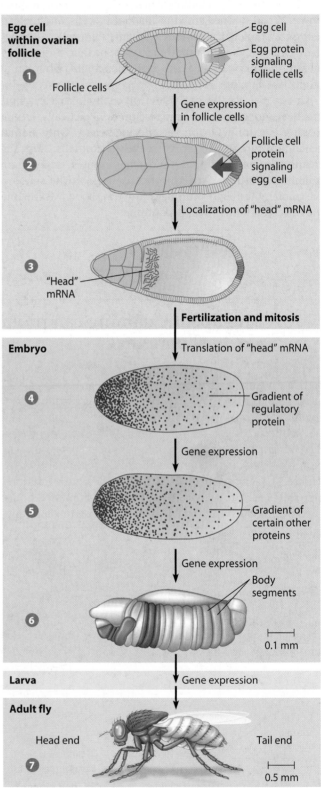

Figure 11.13B Key steps in the early development of head-tail polarity in a fruit fly

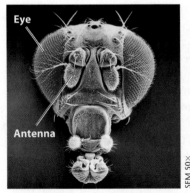

Head of a normal fruit fly   Head of a developmental mutant

Figure 11.13A A mutant fruit fly with legs coming out of its head, compared with a normal fruit fly

dots) resulting from translation of this RNA initiate more rounds of gene expression. Cell signaling—now among the cells of the embryo—helps drive the process. ❻ The result is the subdivision of the embryo's body into segments.

Now finer details of the fly can take shape. Protein products of some of the axis-specifying genes and segment-forming genes activate yet another set of genes. These genes, called homeotic genes, determine what body parts will develop from each segment. A **homeotic gene** is a master control gene that regulates batteries of other genes that actually determine the anatomy of parts of the body. For example, one set of homeotic genes in fruit flies instructs cells in the segments of the head and thorax (midbody) to form antennae and legs, respectively. Elsewhere, these homeotic genes remain turned off, while others are turned on. ❼ The eventual outcome is an adult fly. Notice that the adult's body segments correspond to those of the embryo in step 6. It was mutation of a homeotic gene that was responsible for the abnormal fly in Figure 11.13A.

Cascades of gene expression, with the protein products of one set of genes activating another set of genes, and so on, are a common theme in development. Many of the proteins, such as the one in step 1 of Figure 11.13B, do not act on genes directly. Instead, they act as signals that indirectly trigger expression of a gene in a neighboring cell. Next we'll take a closer look at how this happens.

Web/CD Activity 11D *Development of Head-Tail Polarity*

Web/CD Thinking as a Scientist *How Can the "Head" Gene Be Regulated to Alter Development?*

? What determines which end of a developing fruit fly will become the head?

■ A specific kind of mRNA localizes at the end of the unfertilized egg that will become the head. After fertilization of the egg, this mRNA is translated into a regulatory protein that forms a gradient starting at the head end.

11.14 Signal transduction pathways convert messages received at the cell surface to responses within the cell

Cell-to-cell signaling, with proteins or other kinds of molecules carrying messages from signaling cells to receiving (target) cells, is a key mechanism in development, as well as in the coordination of cellular activities in the mature organism. In most cases, a signal molecule acts by binding to a receptor protein in the plasma membrane of the target cell and initiating a signal transduction pathway in the cell. A **signal transduction pathway** is a series of molecular changes that converts a signal on a target cell's surface to a specific response inside the cell.

Figure 11.14 shows the main elements of a signal transduction pathway in which the target cell's response is the transcription of a gene. ❶ The signaling cell secretes the signal molecule. ❷ This molecule binds to a receptor protein embedded in the target cell's plasma membrane. ❸ The binding activates the first in a series of relay proteins within the target cell. Each relay molecule activates another. ❹ The last relay molecule in the series activates a transcription factor that ❺ triggers transcription of a specific gene. ❻ Translation of the mRNA produces a protein.

Signal transduction pathways are crucial to many cellular functions. We encountered them when we studied the cell cycle control system earlier in this unit, and we'll revisit them when we discuss cancer shortly and when we study hormone function in Chapters 26 and 33.

Web/CD Activity 11E *Signal Transduction Pathway*

? How can a signal molecule from one cell alter gene expression in a target cell without even entering the target cell?

■ By binding to a receptor protein in the membrane of the target cell and triggering a signal transduction pathway that activates transcription factors

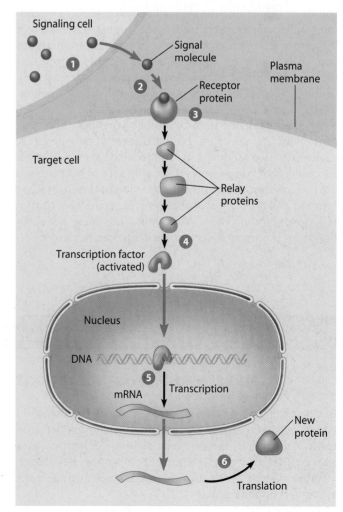

Figure 11.14 A signal transduction pathway that turns on a gene

11.15 Key developmental genes are very ancient

Among the most exciting biological discoveries in recent years is that a class of similar genes—homeotic genes—help direct embryonic development in a wide variety of organisms. Researchers studying homeotic genes in fruit flies found a common structural feature: Every homeotic gene they looked at contained a common sequence of 180 nucleotides. Very similar sequences have since been found in virtually every eukaryotic organism examined so far, including yeasts, plants, and humans—and even some prokaryotes. These nucleotide sequences are called **homeoboxes**, and each is translated into a segment (60 amino acids long) of the protein product of the homeotic gene. The homeobox polypeptide segment binds to specific sequences in DNA, enabling homeotic proteins that contain it to turn groups of genes on or off during development.

Figure 11.15 highlights some striking similarities in the chromosomal locations and the developmental roles of some homeobox-containing homeotic genes in two quite different animals. The figure shows portions of chromosomes that carry homeotic genes in the fruit fly and the mouse. The colored boxes represent homeotic genes that are very similar in flies and mice. Notice that the order of genes on the fly chromosome is the same as on the four mouse chromosomes and that the gene order on the chromosomes corresponds to analogous body regions in both animals. These similarities suggest that the original version of these homeotic genes arose very early in the history of life and that the genes have remained remarkably unchanged for eons of animal evolution.

By their presence in such diverse creatures, homeotic genes illustrate one of the central themes of biology: unity in diversity (see Module 1.4). The fact that these key genes are *control* genes underscores the importance of regulation in the lives of organisms. In the next section of the chapter, we turn to regulatory genes of another type: genes that control cell growth and division. When mutations make the functioning of these genes go awry, the result can be cancer.

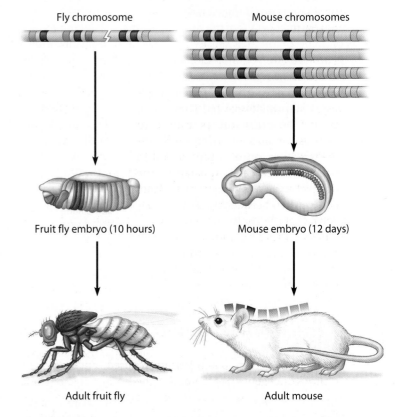

Figure 11.15 Comparison of fruit fly and mouse homeotic genes

> **?** If the DNA sequences called homeoboxes, which help homeotic genes direct development, are common to flies and mice, then why aren't flies and mice more alike?

■ Homeotic genes have much DNA besides their homeoboxes, so there can be much variation in homeotic genes. Moreover, the genes regulated by the protein products of the homeotic genes can be very different in different organisms.

THE GENETIC BASIS OF CANCER

11.16 Cancer results from mutations in genes that control cell division

In Chapter 8, we introduced cancer as a variety of diseases in which cells escape from the control mechanisms that normally limit their growth and division. Scientists have learned that this escape from normal controls is due to changes in some of the cells' genes, changes that affect the expression of other genes.

The abnormal behavior of cancer cells was observed years before anything was known about the cell cycle, its control, or the role genes play in making cells cancerous. One of the earliest clues to the cancer puzzle was the discovery, in 1911, of a virus that causes cancer in chickens. Recall that viruses are simply molecules of DNA or RNA coated with protein and in some cases a membranous envelope. Viruses that cause cancer can become permanent residents in host cells by inserting their nucleic acid into the DNA of host chromosomes. It is now known that a number of viruses that can cause cancer carry specific cancer-causing genes in their nucleic acid. When inserted into a host cell, these genes can

make the cell cancerous. Such a gene, which can cause cancer when present in a single copy in the cell, is called an **oncogene** (from the Greek *onkos*, tumor).

Proto-Oncogenes In 1976, American molecular biologists J. Michael Bishop, Harold Varmus, and their colleagues made a startling discovery. They found that a virus that causes cancer in chickens contains an oncogene that is an altered version of a gene found in normal chicken cells. Apparently, the virus picked up the gene from a former host cell. Subsequent research has shown that the chromosomes of many animals, including humans, contain genes that can be converted to oncogenes. A normal gene that has the potential to become an oncogene is called a **proto-oncogene.** Thus, a cell can acquire an oncogene either from a virus or from the conversion of one of its own genes.

The work by Bishop and Varmus focused cancer research on proto-oncogenes. Searching for the normal role of these genes, researchers found that many proto-oncogenes code for growth factors—proteins that stimulate cell division—or for other proteins that somehow affect growth factor function or some other aspect of the cell cycle. When all these proteins are functioning normally, in the right amounts at the right times, they help control cell division and cellular differentiation.

For a proto-oncogene to become an oncogene, a mutation must occur in the cell's DNA. Mutations that produce most types of cancer occur in somatic cells, those not involved in gamete formation. **Figure 11.16A** illustrates three kinds of changes in somatic cell DNA that can produce active oncogenes. Let's assume that the starting proto-oncogene codes for a protein that stimulates cell division. On the left in the figure, a mutation (green) in the proto-oncogene itself creates an oncogene that codes for a hyperactive protein, one whose stimulating effect is stronger than normal. In the center, an error in DNA replication or recombination generates multiple copies of the gene, which are all transcribed and translated; the result is an excess of the normal stimulatory protein. On the right, the proto-oncogene has been moved from its normal location in the cell's DNA to another location. At its new site, the gene is under the control of a different promoter, one that causes it to be transcribed more often than normal; and the normal protein is again made in excess. So in all three cases, normal gene expression is changed, and the cell is stimulated to divide excessively.

Tumor-Suppressor Genes Changes in genes whose products *inhibit* cell division are also involved in cancer.

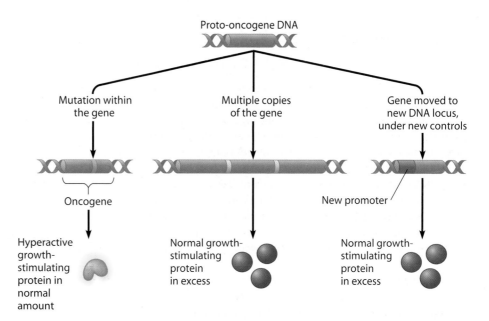

Figure 11.16A Alternative ways to make oncogenes from a proto-oncogene (all leading to excessive cell growth)

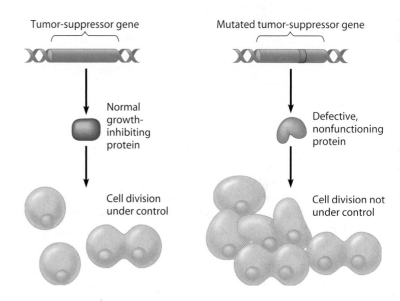

Figure 11.16B The effect of a mutation in a tumor-suppressor gene

These genes are called **tumor-suppressor genes** because the proteins they encode normally help prevent uncontrolled cell division. When a mutation in a tumor-suppressor gene results in a defective protein, as in **Figure 11.16B**, cells that are usually under the control of the normal protein may divide excessively, eventually forming a cancerous tumor.

? In what respect is "proto-oncogene" a misnomer?

■ The term does not describe the normal function of the gene, which is generally the regulation of the cell cycle. The conversion of a proto-oncogene to an oncogene is an aberration, not a natural progression.

11.17 Oncogene proteins and faulty tumor-suppressor proteins can interfere with normal signal transduction pathways

To understand how oncogenes and defective tumor-suppressor genes can contribute to uncontrolled cell growth, we need to look more closely at the normal functions of proto-oncogenes and tumor-suppressor genes. Genes in both categories often code for proteins involved in signal transduction pathways leading to gene expression, pathways similar to the one described in Module 11.14.

The figures below (excluding, for the moment, the white boxes) illustrate two types of signal transduction pathways leading to the synthesis of proteins that influence the cell cycle. In **Figure 11.17A**, the pathway leads to the stimulation of cell division. The initial signal is a growth factor, and the target cell's ultimate response is the production of a protein that stimulates the cell to divide. By contrast, **Figure 11.17B** shows an inhibitory pathway, in which a growth-*inhibiting* factor causes the target cell to make a protein that inhibits cell division. In both cases, the newly made proteins

function by interacting with components of the cell cycle control system (see Module 8.9). The figures here do not show these interactions.

Now let's see what can happen when the target cell undergoes a cancer-causing mutation. The white box in Figure 11.17A shows the protein product of an oncogene resulting from mutation of a proto-oncogene called *ras*. The normal product of *ras* is a relay protein. Ordinarily, a stimulatory pathway like this will not operate unless the growth factor is available. However, an oncogene protein that is a hyperactive version of a protein in the pathway may trigger the pathway even in the absence of a growth factor. In this example, the oncogene protein is a hyperactive version of the *ras* relay protein that issues signals on its own. In fact, abnormal versions or amounts of any of the pathway's components—from the growth factor itself to the transcription factor—could have the same final effect: overstimulation of cell division.

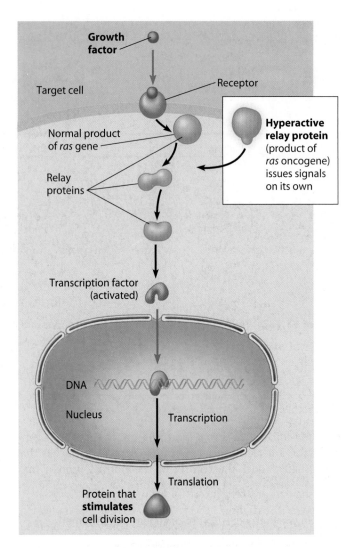

Figure 11.17A A stimulatory signal transduction pathway and the effect of an oncogene protein

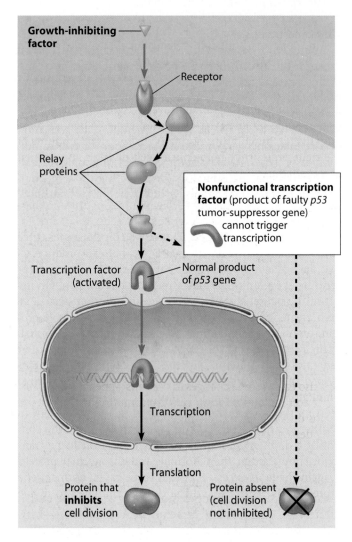

Figure 11.17B An inhibitory signal transduction pathway and the effect of a faulty tumor-suppressor protein

The white box in Figure 11.16B indicates how a mutant tumor-suppressor protein can affect cell division. In this case, the mutation affects a gene called *p53*, which codes for the transcription factor. This mutation leads to the production of a faulty transcription factor, one that the signal transduction pathway cannot activate. As a result, the gene for the inhibitory protein at the bottom of the figure remains turned off, and excessive cell division may occur.

Mutations of the *ras* and *p53* genes have been implicated in many kinds of cancer. In fact, mutations in *ras* occur in about 30% of human cancers, and mutations in *p53* occur in more than 50%. As we see next, most forms of cancer probably result from a series of changes in multiple genes.

? Contrast the action of an oncogene with that of a cancer-causing mutation in a tumor-suppressor gene.

■ An oncogene produces an abnormal protein that stimulates cell division via a signal transduction pathway; a mutant tumor-suppressor gene produces a defective protein unable to function in a pathway that normally inhibits cell division.

11.18 Multiple genetic changes underlie the development of cancer

About 150,000 Americans will be stricken by cancer of the colon (large intestine) or rectum this year, perhaps including some of your own relatives or friends. One of the best-understood types of human cancer, colon cancer illustrates an important principle about how cancer develops: *More than one somatic mutation is needed to produce a full-fledged cancer cell.* As in many cancers, the development of a colon cancer that metastasizes is gradual. (See Module 8.10 to review cancer terms.)

As shown in **Figure 11.18A**, ❶ the first sign of a colon cancer is the unusually frequent division of apparently normal cells in the colon lining. ❷ Later, a small benign tumor (polyp) appears in the colon wall, ❸ eventually becoming a malignant tumor (a carcinoma). These cellular changes parallel the changes at the DNA level, including the activation of a cellular oncogene and the inactivation of two tumor-suppressor genes. These genetic changes (mutations) result in altered signal transduction pathways like the ones outlined in Module 11.17. The requirement for several mutations—the actual number is usually four or more—explains why cancers can take a long time to develop.

Figure 11.18B indicates how mutations that lead to cancer may accumulate in a lineage of somatic cells. Colors distinguish the normal cell (tan) from cells (shades of pink) with one or more mutations leading to increased cell division and cancer. Once a cancer-promoting mutation occurs (red band on chromosome), it is passed to all the descendants of the cell carrying it. In our example, the first two mutations make the cells divide more rapidly; otherwise, the cells appear normal. The third mutation further increases the rate of cell division and also causes some changes in the cells' appearance. Finally, a cell accumulates a fourth cancer-promoting mutation and begins dividing uncontrollably. The structure of this cell and of its descendants is grossly altered.

Our understanding of the genetic basis of cancer has grown by leaps and bounds in recent years. Among the discoveries is an additional category of tumor-suppressor genes. Rather than being components of pathways like the one in Figure 11.17B, the normal proteins encoded by these tumor-suppressor genes function in the repair of damaged DNA. When they are mutated, other cancer-causing mutations are more likely to accumulate.

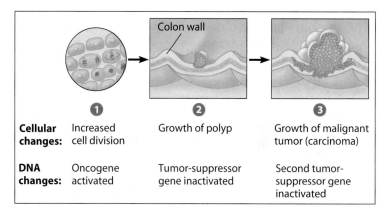

Figure 11.18A Stepwise development of a typical colon cancer

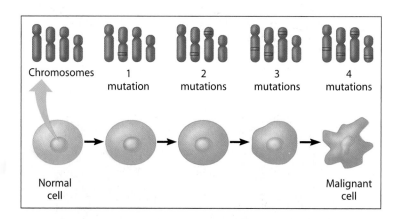

Figure 11.18B Accumulation of mutations in the development of a cancer cell

? One exposure to a particular carcinogenic (cancer-causing) chemical at a certain concentration for a certain length of time has a 10% chance of causing any one of the four mutations in Figure 11.18B. What is the chance (in %) that one exposure to the carcinogen will result in a malignant cell?

■ $0.1 \times 0.1 \times 0.1 \times 0.1 = 0.0001$, or 0.01%

11.19 Mary-Claire King discusses mutations that cause breast cancer

Mary-Claire King (Figure 11.19) has spent almost 30 years exploring the genetic basis of breast cancer, a disease that strikes one out of every ten American women. Dr. King was a math major at Carleton College and did not become interested in genetics until she arrived at UC Berkeley to study statistics in the mid-1960s. For her Ph.D. research, with the late Allan Wilson, she carried out a genetic comparison of humans and chimpanzees, finding extraordinary similarities (see Module 19.2). Dr. King turned to studying breast cancer in the 1970s, while at UC San Francisco, because she wanted "to do work that was important to people in an everyday sort of way." We interviewed her at the University of Washington, where she is now a professor of genome sciences.

Like all cancers, breast cancer is sometimes referred to as a "genetic disease." We asked Dr. King to explain:

It means two things. Cancer is always genetic in the sense that cancer is always the consequence of changes in DNA. Cells that have cancer-causing mutations no longer divide and develop as they should. The great majority of mutations that lead to cancer arise in the tissue where the cancer starts—the colon or the breast, for example. These mutations are somatic mutations; they are in the body but not in the germ line, the cells that give rise to eggs or sperm.

In some families, however, there are germ-line mutations in one or more of these same genes. Such a mutation is passed on from parent to child and predisposes the recipient to cancer. We call this cancer "inherited," even though it doesn't appear unless the person acquires additional, somatic mutations.

The vast majority of breast cancer cases seem to have nothing to do with inherited mutations. But there are many accounts, going back to the ancient Greeks, of families in which breast cancer appears frequently, suggesting that an inherited trait might be involved.

Dr. King was intrigued by familial breast cancer:

It occurred to me that if mutations were involved in familial breast cancer, we might be able to identify some of the genes that are mutated and then figure out the functions of the normal versions of these genes. The results might give us insight into the more common, nonhereditary forms of the disease.

After almost two decades of work, Dr. King and her colleagues succeeded in identifying a gene on chromosome 17 that is mutated in many families with familial breast cancer. Mutations in this gene, called *BRCA1,* put a woman at high risk of breast cancer (and ovarian cancer as well)—a more than 60% risk of developing cancer before age 50. Research from other laboratories suggests that the protein encoded by the normal version of *BRCA1* acts as a tumor suppressor, although its exact role in the cell remains unclear. Regardless of how mutation of *BRCA1* contributes to cancer, it is likely that certain environmental influences, by tending to cause new somatic mutations, also play a role. Dr. King is attempting to sort out these environmental factors:

We're now working with Jewish families in New York City and Israel, families with known mutations in *BRCA1* or *BRCA2,* a gene found soon after *BRCA1.* (Among breast cancer patients of Jewish ancestry, about 10% have mutations in one of these genes.) With the participants' permission, we test the DNA of breast cancer patients from these families. Then we ask, "At what age did cancer appear?" If some women developed breast cancer at 70 and some at 30, were there any differences in their environmental exposures?

Genetic testing of research participants has a clear goal: to learn more about the disease in the hope of helping prevent or cure the disease in future generations. But what about testing women in the general population for *BRCA* mutations? Should healthy women be tested, especially those who may be predisposed to breast cancer? Dr. King believes that genetic testing is useful only when an individual chooses it with a full understanding of its limitations:

For a woman today to know that she is predisposed to breast and ovarian cancer offers her a problem—but no solution except preventive surgical removal of her breasts and/or ovaries.

In addition, King believes, testing should be carried out only under the condition that the results remain confidential and do not affect the woman's access to jobs or health insurance.

Figure 11.19 Mary-Claire King

? (a) In what sense is breast cancer always "genetic"?
(b) Why is most breast cancer considered "nonhereditary"?

■ (a) DNA change is always involved. (b) Most breast cancers are associated with somatic mutations, not inherited mutations that are passed from parents to offspring via gametes.

11.20 Avoiding carcinogens can reduce the risk of cancer

Cancer is the second-leading cause of death in the United States, exceeded only by heart disease. Death rates due to certain forms of cancer (including stomach, cervical, and uterine cancers) have decreased in recent years, but the overall cancer death rate is still on the rise, currently increasing at about 1% per decade. Table 11.20 lists the most common cancers in the United States and associated risk factors for each.

Cancer-causing agents, factors that alter DNA and make cells cancerous, are called **carcinogens.** Most mutagens are carcinogens, agents capable of bringing about cancer-causing DNA changes (see Figure 11.18B). Two of the most potent carcinogens (and mutagens) are X-rays and ultraviolet radiation in sunlight. X-rays are a significant cause of leukemia and brain cancer. Exposure to UV radiation from the sun is known to cause skin cancer, including a deadly type called melanoma.

The largest group of carcinogens are mutagenic chemical compounds and substances containing them. Among the most important carcinogens, the one substance known to cause more cases and types of cancer than any other single agent is tobacco. In 1900, lung cancer was a rare disease. Since then, largely because of an increase in cigarette smoking that has only begun to reverse in the last few decades, lung cancer has greatly increased, and today more people die of lung cancer than any other form of cancer. Most tobacco-related cancers come from actually smoking, but the passive inhalation of secondhand smoke is also a risk. As the table here indicates, tobacco use, sometimes in combination with alcohol consumption, causes a number of other types of cancer in addition to lung cancer. In nearly all cases, cigarettes are the main culprit, but smokeless tobacco products (snuff and chewing tobacco) are linked to cancer of the mouth and throat.

How do carcinogens cause cancer? As we have seen, most cancers result from multiple genetic changes, including the activation of oncogenes and inactivation of tumor-suppressor genes. In many cases, these changes result from decades of exposure to the mutagenic effects of carcinogens. Carcinogens can also produce their effect by promoting cell division. Generally, the higher the rate of cell division, the greater the chance for mutations resulting from errors in DNA replication or recombination. Some carcinogens seem to have both effects. For instance, the hormones that cause breast and uterine cancers promote cell division and may also cause genetic changes that lead to cancer. In other cases, several different agents, such as viruses and one or more carcinogens, may together produce cancer.

We are still a long way from knowing all the factors that contribute to cancer. The effects of many environmental pollutants have yet to be evaluated, although some substances, such as asbestos, are definitely known to be carcinogenic. Exposure to certain carcinogens is often a matter of individual choice. Tobacco use, consumption of animal fat and alcohol, and time spent in the sun, for example, are all behavioral factors that affect our cancer risk.

TABLE 11.20 CANCER IN THE UNITED STATES

Cancer	Risk Factors	Estimated Number of Cases in 2004
Prostate	African heritage; possibly dietary fat	230,100
Breast	Estrogen	217,400
Lung	Tobacco smoke	173,800
Colon, rectum, and anus	High dietary fat; smoking; alcohol	151,000
Lymphomas	Viruses (for some types)	62,300
Urinary bladder	Cigarette smoke	60,200
Melanoma of skin	Ultraviolet light	55,100
Uterus	Estrogen	40,300
Kidney	Cigarette smoke	35,700
Leukemias	X-rays; benzene; virus (for one type)	33,400
Pancreas	Tobacco smoke; obesity	31,900
Mouth and throat	Tobacco in various forms; alcohol	28,300
Ovary	Obesity; Many ovulation cycles	25,600
Stomach	Table salt; cigarette smoke	22,700
Liver	Alcohol; hepatitis viruses	18,900
Brain and nerve	Trauma; X-rays	18,400
Cervix	Sexually transmitted viruses; tobacco	10,500
All others		152,400

Avoiding carcinogens is not the whole story, for there is growing evidence that some food choices significantly reduce cancer risk. For instance, eating 20–30 grams (g) of plant fiber daily (about twice the amount the average American consumes) and at the same time reducing animal fat intake may help prevent colon cancer. There is also evidence that other substances in fruits and vegetables, including vitamins C and E and certain compounds related to vitamin A, may offer protection against a variety of cancers. Cabbage and its relatives, such as broccoli and cauliflower (see Figure 13.2A), are thought to be especially rich in substances that help prevent cancer, although the identities of these substances are not yet established. Determining how diet influences cancer has become a major research goal.

The battle against cancer is being waged on many fronts, and there is reason for optimism in the progress being made. It is especially encouraging that we can help reduce our risk of acquiring some of the most common forms of cancer by the choices we make in our daily life.

Web/CD Activity 11F *Connection: Causes of Cancer*

? Of all known environmental factors, which one causes the most cancer cases?

■ Tobacco

CHAPTER REVIEW

Reviewing the Concepts

A clone is an individual created by asexual reproduction and thus genetically identical to a single parent **(Introduction)**.

Gene Regulation (11.1–11.9)

In prokaryotes, genes for related enzymes are often controlled together in units called operons. Regulatory proteins bind to control sequences in the DNA and turn operons on or off in response to environmental changes **(11.1)**.

In multicellular eukaryotes, cells become specialized, or differentiated, as a zygote develops into a mature organism. Different types of cells make different proteins because different combinations of genes are active in each type **(11.2)**. Most differentiated cells retain a complete set of genes **(11.3)**. A chromosome contains DNA wound around clusters of histone proteins, forming a string of bead-like nucleosomes. This beaded fiber is further wound and folded. DNA packing tends to block gene expression, presumably by preventing access of transcription proteins to the DNA **(11.4)**. An extreme example of DNA packing in interphase cells is X chromosome inactivation in the cells of female mammals **(11.5)**. A variety of regulatory proteins interact with DNA and with each other to turn the transcription of eukaryotic genes on or off **(11.6)**. After transcription, alternative RNA splicing may generate two or more types of mRNA from the same transcript **(11.7)**. The lifetime of an mRNA molecule helps determine how much protein is made, as do protein factors involved in translation. A protein may need to be activated in some way, and eventually the cell will break it down **(11.8)**. Figure 11.9 reviews the multiple stages of eukaryotic gene expression, each stage offering an opportunity for regulation **(11.9)**.

Animal Cloning (11.10–11.12)

Nuclear transplantation is used to clone animals **(11.10)**:

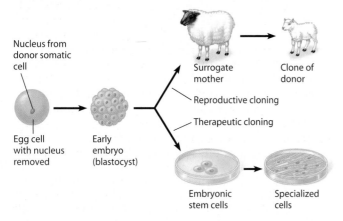

Nucleus from donor somatic cell

Egg cell with nucleus removed

Early embryo (blastocyst)

Surrogate mother

Clone of donor

Reproductive cloning

Therapeutic cloning

Embryonic stem cells

Specialized cells

Reproductive cloning of nonhuman mammals is useful in research, agriculture, and medicine **(11.11)**. Like embryonic stem cells, adult stem cells can both perpetuate themselves in culture and give rise to differentiated cells. Unlike embryonic stem cells, however, adult stem cells normally give rise to only a limited range of cell types **(11.12)**.

The Genetic Control of Embryonic Development (11.13–11.15)

Gene regulation and animal development. A cascade of gene expression controls the development of an animal from a fertilized egg. Homeotic genes, for example, control batteries of genes that shape anatomical parts such as antennae **(11.13)**. Signal transduction pathways convert molecular messages to cell responses **(11.14)**. Homeotic genes contain nucleotide sequences, called homeoboxes, that are very similar in many kinds of organisms **(11.15)**.

The Genetic Basis of Cancer (11.16–11.20)

Cancer cells, which divide uncontrollably, result from mutations in genes whose protein products affect the cell cycle. A mutation can change a proto-oncogene (a normal gene that promotes cell division) into an oncogene, which causes cells to divide excessively. Mutations that inactivate tumor-suppressor genes have similar effects **(11.16)**. Many proto-oncogenes and tumor-suppressor genes code for proteins active in signal transduction pathways regulating cell division **(11.17)**. Cancers result from a *series* of genetic changes in a cell lineage **(11.18)**. Researchers have gained insight into the genetic basis of breast cancer by studying families in which a disease-predisposing mutation is inherited **(11.19)**. Reducing exposure to carcinogens (which induce cancer-causing mutations) and making other lifestyle choices can help reduce cancer risk **(11.20)**.

Connecting the Concepts

1. Complete the following concept map to test your knowledge of gene regulation.

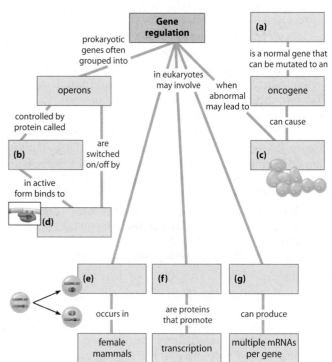

Gene regulation

(a)

prokaryotic genes often grouped into

in eukaryotes may involve

when abnormal may lead to

is a normal gene that can be mutated to an

operons

oncogene

controlled by protein called

are switched on/off by

can cause

(b)

(c)

in active form binds to

(d)

(e)

(f)

(g)

occurs in

are proteins that promote

can produce

female mammals

transcription

multiple mRNAs per gene

Testing Your Knowledge

Multiple Choice

2. The control of gene expression is more complex in multicellular eukaryotes than in prokaryotes because _____. (*Explain your answer.*)
   a. eukaryotic cells are much smaller
   b. in a multicellular eukaryote, different cells are specialized for different functions
   c. prokaryotes are restricted to stable environments
   d. eukaryotes have fewer genes, so each gene must do several jobs
   e. eukaryotic genes code for proteins

3. Your bone cells, muscle cells, and skin cells look different because
   a. each cell contains different kinds of genes.
   b. they are present in different organs.
   c. different genes are active in each kind of cell.
   d. they contain different numbers of genes.
   e. each cell has different mutations.

4. Which of the following methods of gene regulation do eukaryotes and prokaryotes have in common?
   a. elaborate packing of DNA in chromosomes
   b. activator and repressor proteins, which attach to DNA
   c. the addition of a cap and tail to mRNA after transcription
   d. *lac* and *trp* operons
   e. the removal of noncoding portions of RNA

5. A eukaryotic gene was inserted into the DNA of a bacterium. The bacterium then transcribed this gene into mRNA and translated the mRNA into protein. The protein produced was useless; it contained many more amino acids than the protein made by the eukaryotic cell. Why?
   a. The mRNA was not spliced, as it is in eukaryotes.
   b. Eukaryotes and prokaryotes use different genetic codes.
   c. Repressor proteins interfered with transcription and translation.
   d. The lifetime of the bacterial mRNA was too short.
   e. Ribosomes were not able to bind to the mRNA.

6. A homeotic gene does which of the following?
   a. It serves as the ultimate control for prokaryotic operons.
   b. It regulates the expression of groups of other genes during development.
   c. It represses the histone proteins that package eukaryotic DNA.
   d. It helps splice mRNA after transcription.
   e. It inactivates one of the X chromosomes in a female mammal.

7. All your cells contain proto-oncogenes, which can change into cancer-causing genes. Why do cells possess such potential time bombs?
   a. Viruses infect cells with proto-oncogenes.
   b. Proto-oncogenes are genetic "junk" with no known function.
   c. Proto-oncogenes are unavoidable environmental carcinogens.
   d. Cells produce proto-oncogenes as a by-product of mitosis.
   e. Proto-oncogenes are necessary for normal control of cell division.

8. Which of the following is a valid difference between embryonic stem cells and the stem cells found in adult tissues?
   a. In laboratory culture, only adult stem cells are immortal.
   b. In nature, only embryonic stem cells give rise to all the different types of cells in the organism.
   c. Only adult stem cells can differentiate in culture.
   d. Embryonic stem cells are generally more difficult to grow in culture than adult stem cells.
   e. Only embryonic stem cells are found in every tissue of the adult body.

Describing, Comparing, and Explaining

9. A mutation in a single gene may cause a major change in the body of a fruit fly, such as an extra pair of legs or wings. Yet it probably takes the combined action of hundreds or thousands of genes to produce a wing or leg. How can a change in just one gene cause such a big change in the body?

Applying the Concepts

10. Study the illustrations of the *lac* operon in Module 11.1. Normally, the genes are turned off when lactose is not present. Lactose activates the genes, which code for enzymes that enable the cell to use lactose. Mutations can alter the function of this operon; in fact, it was the effects of various mutations that enabled Jacob and Monod to figure out how the operon works. Predict how the following mutations would affect the function of the operon in the presence and absence of lactose:
    a. Mutation of regulatory gene; repressor will not bind to lactose.
    b. Mutation of operator; repressor will not bind to operator.
    c. Mutation of regulatory gene; repressor will not bind to operator.
    d. Mutation of promoter; RNA polymerase will not attach to promoter.

11. A chemical called dioxin is produced as a by-product of some chemical manufacturing processes. Trace amounts of this substance were present in Agent Orange, a defoliant sprayed on vegetation during the Vietnam War. There has been a continuing controversy over its effects on soldiers exposed to it during the war. Animal tests have suggested that dioxin can be lethal and can cause birth defects, cancer, liver and thymus damage, and immune system suppression. But its effects on humans are unclear, and even animal tests are inconclusive; a hamster is not affected by a dose that can kill a guinea pig. Researchers have discovered that dioxin enters a cell and binds to a protein that in turn attaches to the cell's DNA. How might this mechanism help explain the variety of dioxin's effects on different body systems and in different animals? How might you determine whether a particular individual became ill as a result of exposure to dioxin?

Answers to all questions can be found in Appendix 3.

For study help and Activities, go to campbellbiology.com or the student CD-ROM.

BACTERIAL PLASMIDS AND GENE CLONING

12.1   Plasmids are used to customize bacteria: An overview

12.2   Enzymes are used to "cut and paste" DNA

12.3   Genes can be cloned in recombinant plasmids: A closer look

12.4   Cloned genes can be stored in genomic libraries

12.5   Reverse transcriptase helps make genes for cloning

12.6   Recombinant cells and organisms can mass-produce gene products

12.7   DNA technology is changing the pharmaceutical industry and medicine

RESTRICTION FRAGMENT ANALYSIS AND DNA FINGERPRINTING

12.8   Nucleic acid probes identify clones carrying specific genes

12.9   DNA microarrays test for the expression of many genes at once

12.10   Gel electrophoresis sorts DNA molecules by size

12.11   Restriction fragment length polymorphisms can be used to detect differences in DNA sequences

12.12   DNA technology is used in courts of law

12.13   Gene therapy may someday help treat a variety of diseases

12.14   The PCR method is used to amplify DNA sequences

GENOMICS

12.15   The Human Genome Project is an ambitious application of DNA technology

12.16   Most of the human genome does not consist of genes

12.17   The science of genomics compares whole genomes

GENETICALLY MODIFIED ORGANISMS

12.18   Genetically modified organisms are transforming agriculture

12.19   Could GM organisms harm human health or the environment?

12.20   Genomics researcher Eric Lander discusses the Human Genome Project

CHAPTER 12

DNA and Crime Scene Investigations

ON NOVEMBER 22, 1983, the sleepy English village of Narborough awoke to news of a horrific crime: A 15-year-old-girl named Lynda Mann had been raped and murdered on a country lane near her home. The killer left behind few clues, except for semen on the victim's body and clothes. Despite extensive investigation, the trail of evidence ran cold and the crime went unsolved.

Three years later, the horror resurfaced when another 15-year-old-girl, Dawn Ashworth, was also raped and murdered less than a mile away from the first crime scene. When tests indicated that the 1983 and 1986 semen samples could be from the same man, police began to search for a double murderer. After another extensive investigation, a maintenance worker from a nearby hospital was arrested and charged with both crimes. Under considerable pressure from police, the worker confessed to the second murder, but denied committing the first.

In an attempt to pin both murders on the suspect, investigators turned to Alec Jeffreys, a professor at nearby Leicester University, who had recently developed the first DNA fingerprint identification system. Because the DNA sequence of every person is unique (except for identical twins), DNA fingerprinting can be used to determine with near certainty whether two samples of genetic material are from the same individual. Jeffreys compared DNA from the 1983 and 1986 semen samples. As the police suspected, the DNA analysis proved that the same person had committed both crimes. However, when Jeffreys analyzed the suspect's DNA, it did not match either crime scene sample, proving that the suspect must be innocent. The police quickly released the suspect, making him the first person in legal history to be exonerated by DNA evidence.

The detectives were back at square one. In an attempt to collect more evidence, they asked every young male from the surrounding area to donate blood for DNA testing. Although nearly 5,000 men were sampled, none had DNA that matched the evidence from the crime scenes. The police were once again stymied. The case finally broke when a pub-goer described how a local named Colin Pitchfork (inset photo, above) had bullied him into

DNA Technology and Genomics

submitting blood on Pitchfork's behalf. The police promptly arrested Pitchfork and took a sample of his blood. Indeed, his DNA matched the samples from the two crime scenes. Colin Pitchfork pleaded guilty to both crimes, closing the first murder case ever to be solved by DNA evidence.

The Narborough murders were the first of many criminal investigations that have relied on DNA evidence. **DNA technology**—methods for studying and manipulating genetic material—has rapidly revolutionized the field of forensics, the scientific analysis of evidence for legal investigations. Since its introduction, DNA fingerprinting has become a standard law enforcement tool and has provided crucial evidence (of both innocence and guilt) in many famous cases, including the O.J. Simpson murder trial and the impeachment of President Bill Clinton. As we will see, DNA technology has applications in many other fields, from cancer research to agriculture and even history. Perhaps

the most exciting use of DNA technology in basic research is the Human Genome Project, whose goal is to map all the human DNA down to the level of its nucleotide sequences. This project is uncovering the genetic basis of what it means to be human. On a more practical level, this ambitious endeavor is expected to help us better understand and treat many diseases.

In this chapter, we will examine several significant roles that DNA technology has assumed in society, including gene cloning to produce useful products, DNA fingerprinting and forensic science, human gene therapy for the treatment of disease, comparisons of genomes from different organisms, and the agricultural production of genetically modified organisms. Along the way, we'll consider the specific techniques involved, how they are applied, and some of the social, legal, and ethical issues that are raised by these new technologies. ■ ■ ■

Investigator at one of the crime scenes (above), Narborough, England (left)

DNA Technology and Genomics

submitting blood on Pitchfork's behalf. The police promptly arrested Pitchfork and took a sample of his blood. Indeed, his DNA matched the samples from the two crime scenes. Colin Pitchfork pleaded guilty to both crimes, closing the first murder case ever to be solved by DNA evidence.

The Narborough murders were the first of many criminal investigations that have relied on DNA evidence. **DNA technology**—methods for studying and manipulating genetic material—has rapidly revolutionized the field of forensics, the scientific analysis of evidence for legal investigations. Since its introduction, DNA fingerprinting has become a standard law enforcement tool and has provided crucial evidence (of both innocence and guilt) in many famous cases, including the O.J. Simpson murder trial and the impeachment of President Bill Clinton. As we will see, DNA technology has applications in many other fields, from cancer research to agriculture and even history. Perhaps the most exciting use of DNA technology in basic research is the Human Genome Project, whose goal is to map all the human DNA down to the level of its nucleotide sequences. This project is uncovering the genetic basis of what it means to be human. On a more practical level, this ambitious endeavor is expected to help us better understand and treat many diseases.

In this chapter, we will examine several significant roles that DNA technology has assumed in society, including gene cloning to produce useful products, DNA fingerprinting and forensic science, human gene therapy for the treatment of disease, comparisons of genomes from different organisms, and the agricultural production of genetically modified organisms. Along the way, we'll consider the specific techniques involved, how they are applied, and some of the social, legal, and ethical issues that are raised by these new technologies. ▪ ▪ ▪

Investigator at one of the crime scenes (above), Narborough, England (left)

12.1 Plasmids are used to customize bacteria: An overview

Most DNA technology methods depend on bacteria—in particular, *Eschericia coli*. In fact, it was research into the genetics of *E. coli* during the 1970s that led to the development of **recombinant DNA technology**, a set of laboratory techniques for combining genes from different sources— even different species—into a single DNA molecule. Recombinant DNA technology is now widely used to alter the genes of many types of cells for practical purposes. For example, scientists have genetically engineered bacteria to mass-produce many useful chemicals, from cancer drugs to pesticides. Furthermore, genes have been transferred from bacteria into plants and from humans to farm animals (see Table 12.6).

To manipulate genes in the laboratory, biologists often use bacterial **plasmids**, which are small, circular DNA molecules that replicate separately from the much larger bacterial chromosome (see Module 10.23). Because plasmids can carry virtually any gene and replicate in bacteria, they are key tools for **gene cloning**, the production of multiple identical copies of a gene-carrying piece of DNA.

Figure 12.1 presents an overview of gene cloning. The procedure begins when ❶ a plasmid is isolated from a bacterium and ❷ DNA carrying a gene of interest is obtained from another cell. The gene of interest could be, for instance, a human gene encoding a protein of medical value or a plant gene conferring resistance to pests. ❸ A piece of DNA containing the gene is inserted into the plasmid. The resulting plasmid now consists of **recombinant DNA**, DNA in which genes from two different sources are combined *in vitro* into the same DNA molecule. ❹ Next, a bacterial cell takes up the plasmid through transformation (see Module 10.22). ❺ This recombinant bacterium then reproduces to

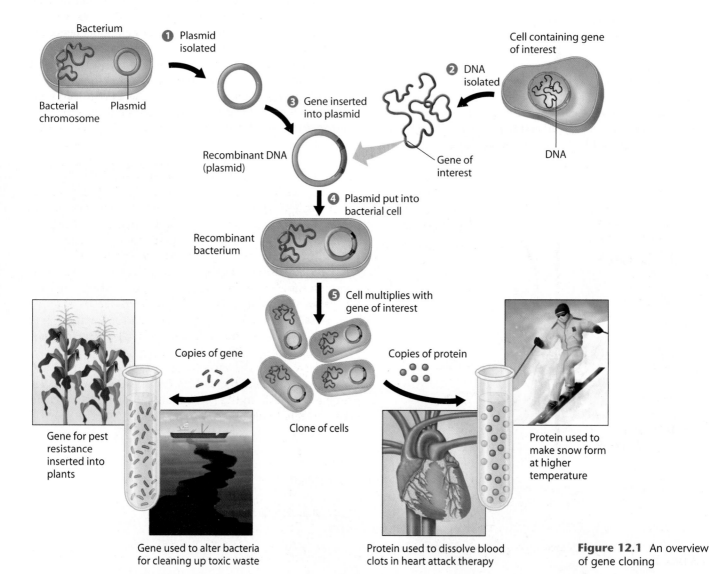

Figure 12.1 An overview of gene cloning

form a **clone** of cells (a group of identical cells descended from a single ancestral cell), each carrying a copy of the gene. As shown at the bottom of the figure, cloned genes can be used directly or to manufacture protein products.

Gene-cloning methods are central to **genetic engineering**, the direct manipulation of genes for practical purposes. Genetic engineering has launched a revolution in **biotechnology**, the use of organisms or their components to make useful products. In the next several modules, we examine some of the tools and procedures of this technology in more detail.

? How does the rapid reproduction of bacteria make them a good choice for cloning a foreign gene?

■ A foreign gene located within plasmid DNA inside a bacterium is replicated each time the cell divides, resulting in rapid accumulation of many copies of the gene.

12.2 Enzymes are used to "cut and paste" DNA

For the gene-cloning procedure outlined in Figure 12.1, a piece of DNA containing the gene of interest must be cut out of a chromosome and "pasted" into a bacterial plasmid. The cutting tools are bacterial enzymes called **restriction enzymes**. In nature, these enzymes protect bacterial cells against intruding DNA from other organisms or viruses. They work by chopping up the foreign DNA, a process that *restricts* foreign DNA from surviving in the cell. (The bacterial cell's own DNA is protected from restriction enzymes through chemical modification by other enzymes.)

Hundreds of different restriction enzymes have been identified and isolated. Each restriction enzyme is very specific, recognizing a particular short DNA sequence (usually four to eight nucleotides long). Once the DNA sequence is recognized, the restriction enzyme cuts both DNA strands at specific points within the sequence. In **Figure 12.2**, ❶ we start with a piece of DNA containing one recognition sequence for a particular restriction enzyme from *E. coli*. In this case, the restriction enzyme will cut the DNA strands between the bases A and G within the sequence, producing pieces of DNA called **restriction fragments**. ❷ The staggered cuts yield two double-stranded DNA fragments with single-stranded ends, called "sticky ends." Sticky ends are the key to joining DNA restriction fragments originating from different sources. These short extensions can form hydrogen-bonded base pairs with complementary single-stranded stretches of DNA.

❸ A piece of DNA (gray) from another source is now added. Notice that the gray DNA has single-stranded ends identical in base sequence to the sticky ends on the blue DNA. The gray, "foreign" DNA has ends with this particular base sequence because it was cut from a larger molecule by the same restriction enzyme used to cut the blue DNA. ❹ The complementary ends on the blue and gray fragments allow them to stick together by base-pairing. (The hydrogen bonds are not shown.)

The union between the blue and gray DNA fragments shown in step 4 is made permanent by the "pasting" enzyme **DNA ligase**. This enzyme, which the cell normally uses in DNA replication (see Module 10.5), catalyzes the formation of covalent bonds between adjacent nucleotides, sealing the breaks in the DNA strands. ❺ The final outcome is a stable molecule of recombinant DNA.

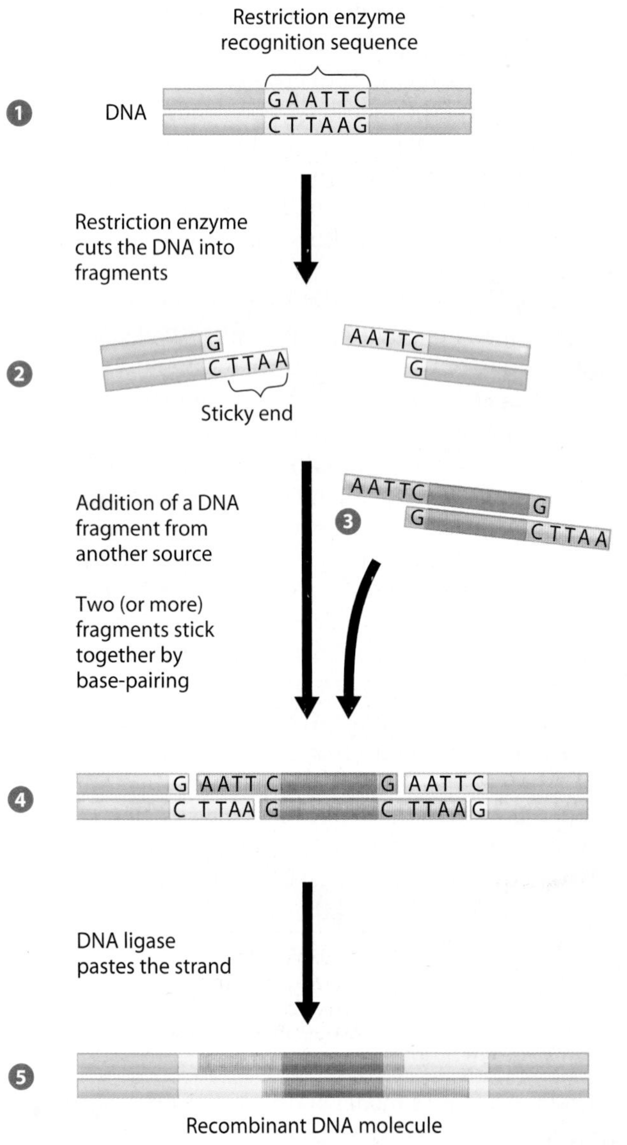

Web/CD Activity 12A *Restriction Enzymes*

Figure 12.2 Creating recombinant DNA using a restriction enzyme and DNA ligase

? What are "sticky ends"?

■ Single-stranded regions whose unpaired bases can hydrogen-bond to the complementary sticky ends of other fragments created by the same restriction enzyme

Genes can be cloned in recombinant plasmids: A closer look

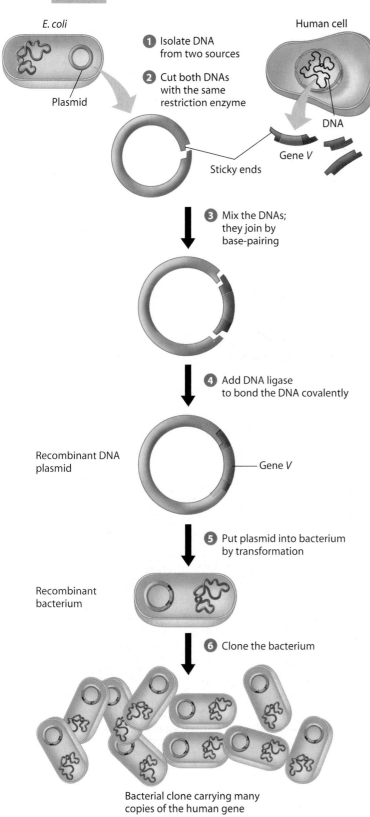

Figure 12.3 Cloning a gene in a bacterial plasmid

E. coli

① Isolate DNA from two sources

Human cell

② Cut both DNAs with the same restriction enzyme

Plasmid

DNA

Gene V

Sticky ends

③ Mix the DNAs; they join by base-pairing

④ Add DNA ligase to bond the DNA covalently

Recombinant DNA plasmid

Gene V

⑤ Put plasmid into bacterium by transformation

Recombinant bacterium

⑥ Clone the bacterium

Bacterial clone carrying many copies of the human gene

Consider a typical genetic engineering challenge: A molecular biologist at a pharmaceutical company has identified a human gene that codes for a valuable product: a hypothetical substance called protein V that kills certain human viruses. The biologist wants to set up a system for making large amounts of the gene so that the protein can be manufactured on a large scale. **Figure 12.3** illustrates a way to make many copies of the gene using the techniques we have been discussing.

❶ The biologist isolates two kinds of DNA: the bacterial plasmid that will serve as the **vector** (gene carrier), and the human DNA containing gene *V*. In this example, the DNA containing the gene of interest comes from human tissue cells that have been growing in laboratory culture. The plasmid comes from the bacterium *E. coli*.

❷ The researcher treats both the plasmid and the human DNA with the same restriction enzyme. An enzyme is chosen that cleaves the plasmid in only one place. The human DNA, with thousands of restriction sites, is cut into many fragments, one of which carries gene *V*. In making the cuts, the restriction enzyme creates sticky ends on both the human DNA fragments and the plasmid. The figure shows the processing of just one human DNA fragment and one plasmid, but actually millions of plasmids and human DNA fragments (most of which do not contain gene *V*) are treated simultaneously.

❸ The human DNA is mixed with the cut plasmid. The sticky ends of the plasmid base-pair with the complementary sticky ends of the human DNA fragment. ❹ The enzyme DNA ligase joins the two DNA molecules by covalent bonds, and the result is a recombinant DNA plasmid containing gene *V*. ❺ The recombinant plasmid is added to a bacterium. Under the right conditions, the bacterium takes up the plasmid DNA by transformation.

❻ This step is the actual gene cloning. The bacterium is allowed to reproduce, forming a clone of cells that all carry the recombinant plasmid. In our example, the biologist will grow a cell clone large enough to produce protein V in marketable quantities.

The cloning procedure described here, which uses a mixture of fragments from the entire genome of an organism, is called a "shotgun" approach. Thousands of different recombinant plasmids are produced in step 3, and a clone of each is made during steps 5 and 6. The complete set of plasmid clones, each carrying copies of a particular segment from the initial genome, is a type of library. The next two modules discuss such libraries in more detail.

Web/CD Activity 12B *Cloning a Gene in Bacteria*

Web/CD Thinking as a Scientist *How Can Antibiotic-Resistant Plasmids Transform* E. coli?

? Which of the three modes of gene transfer described in Module 10.22 is exploited by scientists to introduce a foreign gene into a bacterial cell?

■ Transformation, the uptake of foreign DNA from the surrounding environment

12.4 Cloned genes can be stored in genomic libraries

Each bacterial clone from the procedure in Module 12.3 consists of identical cells with recombinant plasmids carrying one particular fragment of human DNA. The entire collection of all the cloned DNA fragments from a genome is called a **genomic library**. On the left side of Figure 12.4, the red, yellow, and green DNA segments represent three of the thousands of different library "books" that are "shelved" in plasmids inside bacterial cells. A typical cloned DNA fragment is big enough to carry one or a few genes, and together the fragments include the entire genome of the organism from which the DNA was derived.

Bacterial plasmids are one type of vector that can be used in the cloning of genes, but not the only type. Phages can also serve as vectors. When a phage is used, the DNA fragments are inserted into phage DNA molecules. The recombinant phage DNA can then be introduced into a bacterial cell through the normal infection process. Inside the cell, the phage DNA replicates and produces new phage particles, each carrying the foreign DNA. A collection of phage clones can constitute a second type of genomic library (Figure 12.4, right). In Module 12.8, you will learn how to find one particular book from among all those found in either type of library. In the next module, we look at another source of DNA for cloning: eukaryotic mRNA.

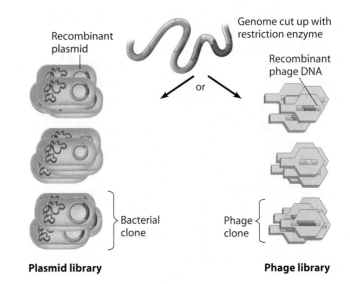

Figure 12.4 Genomic libraries

? In what sense does a genomic library have multiple copies of each "book"?

■ Each "book"—a piece of DNA from the genome that was the source of the library—is present in every recombinant bacterium or phage in a clone.

12.5 Reverse transcriptase helps make genes for cloning

Rather than starting with an entire eukaryotic genome (as described in the previous two modules), a researcher can focus on the genes *expressed* in a particular kind of cell by using its mRNA as the starting material. As shown in Figure 12.5, ❶ the chosen cells transcribe their genes and ❷ process the transcripts to produce mRNA. ❸ The researcher isolates the mRNA and makes single-stranded DNA transcripts from it using the enzyme **reverse transcriptase** (yellow in the figure), which is obtained from retroviruses (see Module 10.21). ❹ Enzymes are added to break down the mRNA, and ❺ DNA polymerase is used to synthesize a second DNA strand.

The DNA that results from such a procedure, called **complementary DNA (cDNA)**, represents only the subset of genes that were transcribed into mRNA in the starting cells. Among other purposes, a cDNA library is useful for studying the genes responsible for the specialized functions of a particular cell type, such as brain or liver cells. And because cDNAs lack introns, they are shorter than the full versions of the genes, and therefore easier to work with.

Now that we've examined some cloning methods, we'll focus on practical applications of this technology.

? Why is a cDNA gene made using reverse transcriptase often shorter than the natural form of the gene?

■ Because cDNAs are made from spliced mRNAs, which lack introns

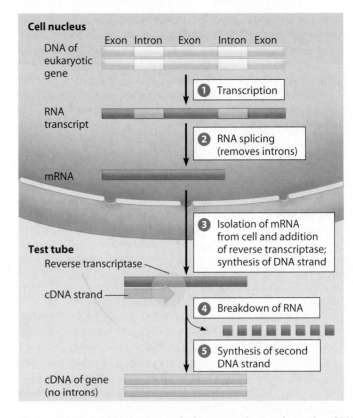

Figure 12.5 Making an intron-lacking gene from eukaryotic mRNA

12.6 Recombinant cells and organisms can mass-produce gene products

Recombinant cells and organisms constructed by DNA technology are used to manufacture many useful products, chiefly proteins (Table 12.6). Most of these products are made by cells grown in culture. By transferring the gene for a desired protein into a bacterium, yeast, or other kind of cell that is easy to grow, one can produce large quantities of proteins that are present naturally in only minute amounts.

Bacteria are often the best organisms for manufacturing a protein product. Major advantages of bacteria include the plasmids and phages available for use as gene-cloning vectors and the fact that bacteria can be grown rapidly and cheaply in large tanks. Furthermore, bacteria can be readily engineered to produce large amounts of particular proteins and in some cases to secrete the protein products into their growth medium. Secretion into the growth medium simplifies the task of collecting and purifying the products. As Table 12.6 shows, a number of proteins of importance in human medicine and agriculture are being produced in the bacterium *E. coli*.

Despite the advantages of bacteria, it is sometimes desirable or necessary to use eukaryotic cells to produce a protein product. Often, the first-choice eukaryotic organism for protein production is the yeast used in making bread and beer, *Saccharomyces cerevisiae*. As bakers and brewers have recognized for centuries, yeast cells are easy to grow. And like *E. coli,* yeast cells can take up foreign DNA and integrate it into their genomes. Yeast cells also have plasmids that can be used as gene vectors, and yeast is often better than bacteria at synthesizing and secreting eukaryotic proteins. *S. cerevisiae* is currently used to produce a number of proteins. In some cases,

Figure 12.6 "Pharm" animals that produce a human protein

the same product (for example, interferons used in cancer research) can be made in either yeast or bacteria. In other cases, such as the hepatitis B vaccine, yeast alone is used.

The cells of choice for making some gene products come from mammals. Genes for these products are often cloned in bacteria as a preliminary step. For example, the genes for two proteins that affect blood clotting, Factor VIII and TPA, are cloned in a bacterial plasmid before transfer to mammalian cells for large-scale production. Many proteins that mammalian cells normally secrete are glycoproteins, proteins with chains of sugars attached. Because only mammalian cells can attach the sugars correctly, mammalian cells must be used for making these products.

Recently, pharmaceutical researchers have been exploring the mass production of gene products by whole animals or plants rather than cultured cells. For example, using recombinant DNA technology, genetic engineers can add a gene for a desired human protein to the genome of a mammal in such a way that the gene's product is secreted in the animal's milk. The sheep in **Figure 12.6** carry a gene for a human blood protein that is a potential treatment for cystic fibrosis. We continue with medical applications of DNA technology in the next module.

? Why can't glycoproteins be mass-produced by engineered bacteria or yeast cells?

TABLE 12.6 SOME PROTEIN PRODUCTS OF RECOMBINANT DNA TECHNOLOGY

Product	Made In	Use
Human insulin	E. coli	Treatment for diabetes
Human growth hormone (HGH)	E. coli	Treatment for growth defects
Epidermal growth factor (EGF)	E. coli	Treatment for burns, ulcers
Interleukin-2 (IL-2)	E. coli	Possible treatment for cancer
Bovine growth hormone (BGH)	E. coli	Improving weight gain in cattle
Cellulase	E. coli	Breaking down cellulose for animal feeds
Taxol	E. coli	Treatment for ovarian cancer
Interferons (alpha and gamma)	S. cerevisiae; E. coli	Possible treatment for cancer and viral infections
Hepatitis B vaccine	S. cerevisiae	Prevention of viral hepatitis
Erythropoietin (EPO)	Mammalian cells	Treatment for anemia
Factor VIII	Mammalian cells	Treatment for hemophilia
Tissue plasminogen activator (TPA)	Mammalian cells	Treatment for heart attacks

■ Because bacteria and yeast cells cannot correctly attach the sugar groups to the protein of glycoproteins.

12.7 DNA technology is changing the pharmaceutical industry and medicine

DNA technology, and gene cloning in particular, is widely used to produce medicines and to diagnose diseases.

Therapeutic Hormones Consider the first two products in Table 12.6, human insulin and human growth hormone (HGH). In the United States alone, about 2 million people with diabetes depend on insulin treatment. Before 1982, the main sources of this hormone were pig and cattle tissues obtained from slaughterhouses. Insulin extracted from these animals is chemically similar, but not identical, to human insulin, and it causes harmful side effects in some people. Genetic engineering has largely solved this problem by developing bacteria that synthesize and secrete actual human insulin. In 1982, Humulin (Figure 12.7A)—human insulin produced by bacteria—became the first recombinant DNA drug approved by the Food and Drug Administration.

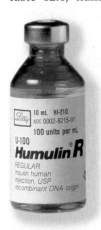

Figure 12.7A Human insulin produced by bacteria

Human growth hormone was harder to produce than insulin because the HGH molecule is about twice as big. Because growth hormones from other animals are not effective in humans HGH was urgently needed. In 1985, molecular biologists made an artificial gene for HGH by joining a human DNA fragment to a chemically synthesized piece of DNA; using this gene, they were able to produce HGH in *E. coli*. Before this genetically engineered hormone became available, children with a HGH deficiency had to rely on scarce supplies from human cadavers or else face dwarfism.

Diagnosis and Treatment of Disease DNA technology is being used increasingly in disease diagnosis. Of obvious value for identifying the alleles associated with genetic diseases, it can also pinpoint infections. For example, DNA analysis can help track down and identify elusive viruses such as HIV. An individual's gene expression profile may someday allow physicians to tailor treatments for many different disorders. (We'll learn how alleles are identified in the next module.)

Vaccines DNA technology is also helping medical researchers develop vaccines. A **vaccine** is a harmless variant or derivative of a pathogen (usually a bacterium or virus) that is used to prevent an infectious disease (see Module 24.4). When a person—a potential host for the pathogen—is inoculated, the vaccine stimulates the immune system to develop lasting defenses against the pathogen. For the many viral diseases for which there is no effective drug treatment, prevention by vaccination is virtually the only medical way to prevent illness.

Genetic engineering can be used in several ways to make vaccines. One approach is to use genetically engineered cells

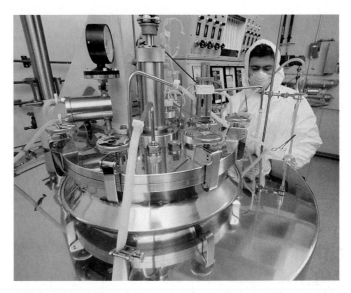

Figure 12.7B Equipment used in the production of a vaccine against hepatitis B

(or organisms) to produce large amounts of a protein molecule that is found on the pathogen's outside surface. This method has been used to make the vaccine against hepatitis B virus. Hepatitis is a disabling and sometimes fatal liver disease, and the hepatitis B virus may also cause liver cancer. **Figure 12.7B** shows a tank for growing yeast cells that have been engineered to carry hepatitis B genes.

Another way to use DNA technology in vaccine development is to make a harmless artificial mutant of the pathogen by altering one or more of its genes. When a harmless mutant is used as a vaccine, it multiplies in the body and may trigger a strong immune response. Artificial-mutant vaccines may cause fewer side effects than those that have traditionally been made from natural mutants.

Yet another scheme for making vaccines employs a virus related to the one that causes smallpox. Smallpox was once a dreaded human disease, but it was eradicated worldwide in the 1970s by widespread vaccination with a harmless variant (natural mutant) of the smallpox virus. Using this harmless virus, genetic engineers can replace some of the genes encoding proteins that induce immunity to smallpox with genes that induce immunity to other diseases. In fact, the virus could be engineered to carry the genes needed to vaccinate against several diseases simultaneously. In the future, one inoculation may prevent a dozen diseases.

? Human growth hormone and insulin produced by DNA technology are used in the treatment of _____ and _____, respectively.

■ dwarfism . . . diabetes

12.8 Nucleic acid probes identify clones carrying specific genes

Often the most difficult task in gene cloning is finding the right "shelf" in a genomic library—that is, identifying a bacterial or phage clone containing a desired gene from among all those created. If bacterial clones containing a specific gene actually translate the gene into protein, they can be identified by testing for the protein product. This is not always the case, however. Fortunately, researchers can also test directly for the gene itself.

Methods for detecting genes directly depend on base pairing between the gene and a complementary sequence on another nucleic acid molecule, either DNA or RNA. When at least part of the nucleotide sequence of a gene is already known or can be guessed, this information can be used to advantage. Taking a simplified example, if we know that a hypothetical gene contains the sequence TAGGCT, a biochemist can synthesize a short single strand of DNA with the complementary sequence (ATCCGA) and label it with a radioactive isotope or fluorescent dye. This labeled, complementary molecule is called a **nucleic acid probe** because it is used to find a specific gene or other nucleotide sequence within a mass of DNA. (In practice, a probe molecule would usually be considerably longer than six nucleotides.)

Figure 12.8 shows how a probe works. The DNA sample to be tested is treated with heat or alkali to separate the DNA strands. When the DNA probe is added to these strands, it tags the correct molecule—finds the correct shelf in the library—by hydrogen-bonding to the complementary sequence in the gene of interest. Such a probe can be simultaneously applied to many bacterial colonies to screen them all for a gene of interest. For example, a piece of filter paper can be pressed against the colonies growing on a solid

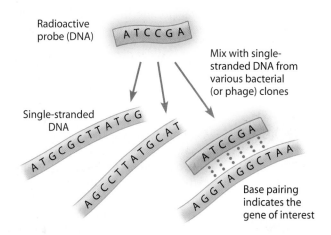

Figure 12.8 How a DNA probe tags a gene by base pairing

growth medium, picking up cells from each colony, and then soaked in probe solution. Any bacterial colonies carrying the gene of interest will be tagged on the filter paper, marking them for easy identification. Once the researcher identifies a colony carrying the desired gene, the cells can be grown further and the gene of interest (and/or its protein product) isolated in large amounts.

? How does a probe consisting of radioactive DNA or RNA enable a researcher to find the bacterial clones carrying a particular gene?

■ The probe molecules bind to and label DNA only in the cells containing the gene of interest, which contains a complementary DNA sequence.

CONNECTION

12.9 DNA microarrays test for the expression of many genes at once

Besides hunting for one specific gene (as shown in Figure 12.8), nucleic acid probes can be used to perform large-scale analyses that determine which of many genes are active (transcribed) in particular cells at particular times. This technique relies on DNA microarrays. A **DNA microarray** is a glass slide carrying thousands of different kinds of single-stranded DNA fragments arranged in an array (grid). Each DNA fragment is obtained from a particular gene; a single microarray thus carries DNA from thousands of genes.

Figure 12.9 outlines how microarrays are used. ❶ A researcher isolates all of the mRNA transcribed from genes in a particular type of cell. ❷ This collection of mRNAs is mixed with reverse transcriptase to produce a mixture of cDNAs (see Module 12.5), each one complementary to one of the mRNAs. The cDNAs are produced in the presence of nucleotides that

have been modified to fluoresce (glow). ❸ A small amount of the fluorescently labeled cDNA mixture is added to each of the DNA fragments in the microarray. If a molecule in the cDNA mixture is complementary to a DNA fragment at a particular location on the grid, the cDNA molecule binds to it, becoming fixed there. ❹ After nonbinding cDNA is rinsed away, the remaining cDNA produces a detectable glow in the microarray (see Figure 12.9, top). The pattern of glowing spots enables the researcher to quickly determine which genes are turned on or off in the starting cells.

Analyses using DNA microarrays may contribute to a better understanding of certain diseases and suggest new diagnostic techniques or therapies. For example, comparing patterns of gene expression in breast cancer tumors and noncancerous breast tissue has already resulted in more informed and effec-

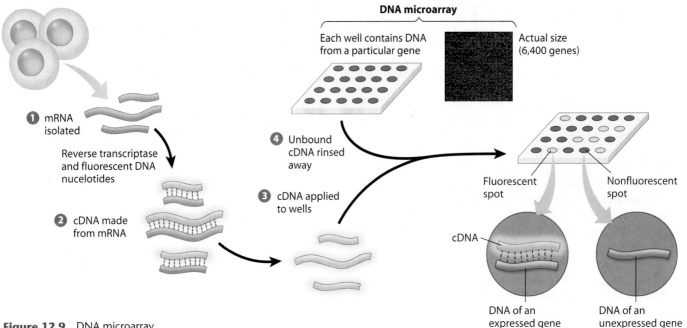

DNA microarray

Each well contains DNA from a particular gene

Actual size (6,400 genes)

① mRNA isolated

Reverse transcriptase and fluorescent DNA nucelotides

② cDNA made from mRNA

④ Unbound cDNA rinsed away

③ cDNA applied to wells

Fluorescent spot

Nonfluorescent spot

cDNA

DNA of an expressed gene

DNA of an unexpressed gene

Figure 12.9 DNA microarray

tive treatment protocols. Ultimately, information from DNA microarray experiments should provide us a grander view—how ensembles of genes interact to form a living organism.

? What is learned from a DNA microarray assay?

■ What genes are active (transcribed) in a particular sample of cells

12.10 Gel electrophoresis sorts DNA molecules by size

Many approaches for studying DNA molecules make use of **gel electrophoresis**. This technique uses a gel (a thin slab of jellylike material) as a molecular sieve to separate nucleic acids or proteins on the basis of size or electrical charge. Figure 12.10 shows how we would use gel electrophoresis to separate the various DNA molecules in three different mixtures. A sample of each mixture is placed in a well at one end of a flat, rectangular gel. A negatively charged electrode from a power supply is attached near the DNA-containing end of the gel, and a positive electrode is attached near the other end. Because DNA molecules have negative charge owing to

their phosphate groups, they all travel through the gel toward the positive pole. As they move, a thicket of polymer fibers within the gel impedes longer molecules more than it does shorter ones, separating them by length. Thus, gel electrophoresis separates a mixture of linear DNA molecules into bands, each consisting of DNA molecules of the same length, with shorter molecules toward the bottom.

Web/CD Activity 12C *Gel Electrophoresis of DNA*

? (a) What causes DNA molecules to move toward the positive pole during electrophoresis? (b) Why do large molecules move more slowly than smaller ones?

■ (a) The negatively charged phosphate groups of the DNA are attracted to the positive pole. (b) The gel resists their movement.

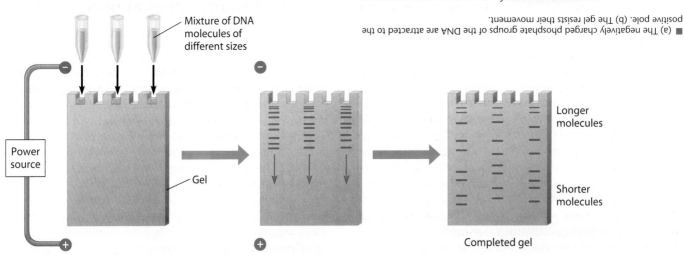

Mixture of DNA molecules of different sizes

Power source

Gel

Longer molecules

Shorter molecules

Completed gel

Figure 12.10 Gel electrophoresis of DNA

12.11 Restriction fragment length polymorphisms can be used to detect differences in DNA sequences

Unless you have an identical twin, your DNA is different from everyone else's; its total nucleotide sequence is unique. Some of your DNA consists of genes, and even more of it is composed of noncoding stretches of DNA. Whether a segment of DNA codes for amino acids or not, it is inherited just like any other part of a chromosome. For this reason, geneticists can use *any* DNA segment that varies from person to person as a **genetic marker**, a chromosomal landmark whose inheritance can be studied. And just like a gene, a noncoding segment of DNA is more likely to be an exact match to the comparable segment in a relative than to the segment in an unrelated individual.

Restriction fragment analysis is a method for detecting differences in nucleotide sequence between homologous samples of DNA, usually from two different individuals. In restriction fragment analysis, two of the methods you have learned about are used in succession: DNA fragments produced by restriction enzymes (see Module 12.2) are sorted by gel electrophoresis (see Module 12.10). *The number of restriction fragments and their sizes reflect the specific sequence of nucleotides in the starting DNA.* The differences in restriction fragments produced in this way are called **restriction fragment length polymorphisms** (**RFLPs**, pronounced "rif-lips"). To understand RFLPs, we need to examine Figures 12.11A and 12.11B.

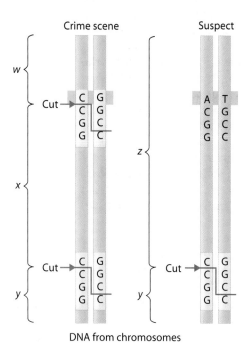

Figure 12.11A Restriction site differences between two homologous samples of DNA

How Restriction Fragments Reflect DNA Sequence In **Figure 12.11A**, we see corresponding segments of DNA from two DNA samples prepared from human tissue; let's imagine that the first was collected at a crime scene and the second from a suspect. Notice that the DNA sequences differ by a single base pair (highlighted in gold). In this case, the restriction enzyme cuts the DNA between two cytosine (C) bases in the sequence CCGG and in its complement, GGCC. Because the crime scene DNA has two recognition sequences for the restriction enzyme, it is cleaved in two places, yielding three restriction fragments (labeled *w, x,* and *y*). DNA from the suspect, however, has only one recognition sequence and yields only two restriction fragments (*z* and *y*). Notice that the lengths of restriction fragments, as well as their numbers, differ depending on the exact sequence of bases in the DNA.

To detect the differences between the collections of restriction fragments, we need to separate the restriction fragments in the two mixtures and compare their lengths. We can accomplish these things through gel electrophoresis. As shown in **Figure 12.11B**, the three kinds of restriction fragments from the crime scene separate into three bands in the gel, while those from the suspect separate into only two bands. Notice that the smallest fragment from the crime scene DNA (*y*) produces a band at the same location as the identical small fragment from the suspect DNA. So you can see that electrophoresis allows us to see similarities as well as differences between mixtures of restriction fragments—and similarities as well as differences between the base sequences

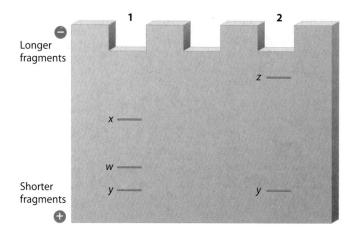

Figure 12.11B Gel electrophoresis of restriction fragments

in DNA from two individuals. The restriction fragment analysis in Figure 12.11B clearly shows that the DNA sample from the suspect does not match the DNA from the crime scene.

In real life, the samples of DNA used as starting material for preparing restriction fragments would not be pure preparations of a single DNA segment. More likely, the starting material would be bulk DNA from cells, which would yield huge numbers of bands on the gel. Fortunately, we can use a DNA probe to focus in on the bands coming solely from the DNA sequences we are interested in without having to purify them from the rest of the DNA. **Figure 12.11C** shows how this is done. The paper to which the DNA is transferred

(blotted) holds the DNA stationary while it is being exposed to the probe.

Using DNA Probes to Detect Harmful Alleles

An important application of restriction fragment analysis is the detection of potentially harmful alleles in heterozygous individuals who are free of symptoms (see Module 9.15). The heterozygotes may be carriers of a harmful recessive allele, such as for cystic fibrosis, or a dominant allele that is not expressed until later in life, such as for Huntington's disease (see Module 9.9). A key to detecting the harmful allele is that within a particular family, the allele for a disease is generally identical in all family members who carry it. That allele often contains one or more restriction sites that are different from the ones in the normal allele.

Once the restriction fragment patterns for the normal and harmful alleles are known, restriction fragment analysis can be used to test family members who are suspected carriers of the harmful allele. Figure 12.11C shows the use of this procedure on blood samples from three relatives: a known carrier and two suspected carriers. In this example, the banding pattern from the known carrier (I) matches one relative (II) but not another (III).

In the next two modules, we'll examine further the use of DNA technology in forensics, and we'll see how it might be employed in the treatment of human diseases.

Web/CD Activity 12D *Analyzing DNA Fragments Using Gel Electrophoresis*

Web/CD Thinking as a Scientist *How Can Gel Electrophoresis Be Used to Analyze DNA?*

? What does RFLPs stand for? Explain the name.

■ *Restriction fragment length polymorphisms; the length of restriction fragments differs (is polymorphic) in samples of homologous DNA from different sources.*

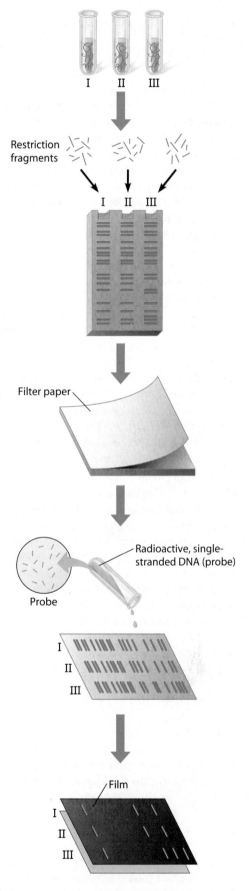

1 Restriction fragment preparation. DNA is extracted from white blood cells taken from individuals I, II, and III. Individual I is known to carry a disease allele. A restriction enzyme is added to the three samples of DNA to produce restriction fragments.

2 Gel electrophoresis. The mixture of restriction fragments from each sample is separated by electrophoresis. Each sample forms a characteristic pattern of bands. (Since all the DNA in the blood cells was used, there would be many more bands than shown here.)

3 Blotting. The DNA bands are treated to separate the strands of the molecules, and the single strands are transferred by simple blotting onto special filter paper.

4 Radioactive probe. The paper blot is immersed in a solution of a radioactive probe, a single-stranded DNA molecule that is complementary to the DNA sequence of the genetic marker of interest. The probe attaches by base-pairing to restriction fragments arising from the marker DNA. (Thus, it may stick to several bands in a lane.)

5 Detection of radioactivity (autoradiography). The unattached probe is rinsed off, and a sheet of photographic film is laid over the paper. The radioactivity in the probe exposes the film to form an image corresponding to specific bands—the bands containing DNA that base-pairs with the probe. In this example, the band pattern is the same for individuals I and II but different for individual III. Since we know that individual I carries the disease allele, the test shows that individual II is also a carrier but that individual III is not.

Figure 12.11C The use of restriction fragment analysis to detect a harmful allele

12.12 DNA technology is used in courts of law

As discussed in the chapter introduction, DNA technology now plays an important role in **forensic science**, the scientific analysis of evidence for crime scene and other legal investigations. In violent crimes, body fluids or small pieces of tissue may be left at the crime scene or on the clothes of the victim or assailant. If rape has occurred, semen may be recovered from the victim's body. With enough tissue or semen, forensic scientists can determine the blood type or tissue type using older methods that test for proteins. However, such tests require fresh samples in relative large amounts. Also, because many people have the same blood or tissue type, this approach can only exclude a suspect; it cannot provide strong evidence of guilt.

DNA testing, on the other hand, can identify the guilty individual with a high degree of certainty because the DNA sequence of every person is unique (except for identical twins). RFLP analysis is one major type of DNA testing (see Module 12.11). It is a powerful method for comparing DNA samples and requires only about 1,000 cells. In a murder case, for example, such analysis can be used to compare DNA samples from the suspect, the victim, and bloodstains on the suspect's clothes (**Figure 12.12A**). Radioactive probes mark the electrophoresis bands that contain certain markers. Usually about a dozen markers are tested; in other words, only a few selected portions of the DNA are compared. However, even such a small set of markers from an individual can provide a **DNA fingerprint**, or specific pattern of bands, that is of forensic use, because the probability that two people would have exactly the same set of markers is very small (**Figure 12.12B**). The autoradiograph in Figure 12.12A resembles the type of evidence presented to juries in murder trials. Notice that DNA from blood on the defendant's clothes matches the DNA of the victim but

Figure 12.12B
DNA data for forensic use

differs from that of the defendant, providing strong evidence of guilt.

DNA fingerprinting can also be used to establish family relationships. A comparison of the DNA of a mother, her child, and the purported father can conclusively settle a question of paternity. Sometimes paternity is of historical interest: DNA fingerprinting provided strong evidence that Thomas Jefferson or one of his close male relatives fathered at least one child with his slave Sally Hemings.

Today, the markers most often used in DNA fingerprinting are inherited variations in the lengths of repetitive DNA (see Module 12.16). These repetitive DNA sequences are highly variable from person to person, providing even more markers than RFLPs. For example, one person may have the nucleotides ACA repeated 65 times at one genome locus and 118 times at a second locus, whereas another person is likely to have different numbers of repeats at these loci.

Just how reliable is DNA fingerprinting? In most legal cases, the probability of two people having identical DNA fingerprints is between one chance in 100,000 and one in a billion. The exact figure depends on how many markers are compared and on how common those markers are in the population. For this reason, DNA fingerprints are now accepted as compelling evidence by legal experts and scientists alike. In fact, DNA analysis on stored forensic samples has provided the evidence needed to solve many "cold cases" in recent years. DNA fingerprinting has also exonerated many wrongly convicted people, some of whom were on death row.

Web/CD Activity 12E *Connection: DNA Fingerprinting*

? In what way is DNA fingerprinting valuable for determining innocence as well as guilt?

Defendant's blood | Blood from defendant's clothes | Victim's blood

4 µg 8 µg

D | jeans _shirt_ | V

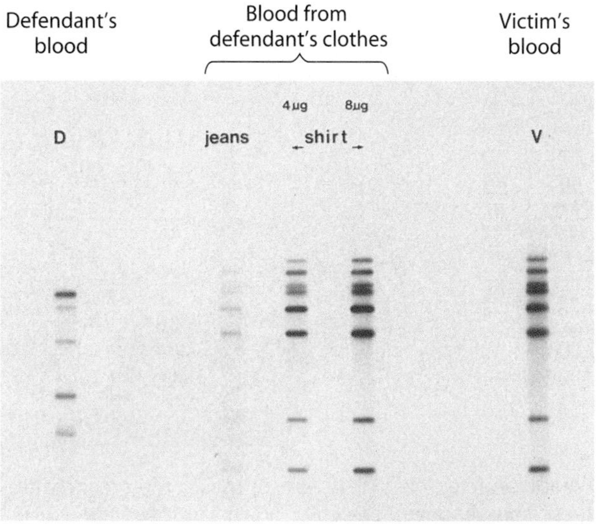

Figure 12.12A DNA fingerprints from a murder case

■ A DNA fingerprint can prove with near certainty that a sample of DNA does or does not come from a particular individual. DNA analysis therefore has an equal chance of providing evidence in support of guilt or innocence.

12.13 Gene therapy may someday help treat a variety of diseases

Techniques for manipulating DNA have the potential for treating a variety of diseases by **gene therapy**—alteration of an afflicted individual's genes. In people with disorders traceable to a single defective gene, it should theoretically be possible to replace or supplement the defective gene with a normal allele. The new allele could be inserted into somatic cells of the tissue affected by the disorder.

For gene therapy to be permanent, the normal allele would have to be transferred to cells that multiply throughout a person's life. Bone marrow cells, which include the stem cells that give rise to all the cells of the blood and immune system, are prime candidates (see Modules 11.5 and 23.17). Figure 12.13 outlines one possible procedure for gene therapy in an individual whose bone marrow cells do not produce a vital protein product because of a defective gene. ❶ The normal gene is cloned and then inserted into the nucleic acid of a retrovirus vector that has been rendered harmless. ❷ Bone marrow cells are taken from the patient and infected with the virus. ❸ The virus inserts its nucleic acid, including the human gene, into the cells' DNA (see Module 10.21). ❹ The engineered cells are then injected back into the patient. If the procedure succeeds, the cells will multiply throughout the patient's life and produce the missing protein. The patient will be cured.

A procedure like the one shown in Figure 12.13 was used in the first human gene therapy trial, which began in 1990. This trial sought to treat severe combined immunodeficiency disease (SCID), a disorder in which the patient lacks a funtional immune system (see Module 24.16). But the clinical results from this trial did not confirm the effectiveness of the treatment. In another trial beginning in 2000, ten young children with SCID were treated by the same procedure. Nine of these patients showed significant improvement after two years, providing the first definitive success of gene therapy. However, two of the patients subsequently developed leukemia, a cancer of the blood cells. Researchers discovered that in both cases, the virus used to carry the normal allele into bone marrow cells had inserted DNA near a gene involved in proliferation and development of blood cells. This insertion somehow caused the leukemia. Thus, although the concept of gene therapy remains promising, very little scientifically strong evidence of effective gene therapy has yet appeared. Active research into human gene therapy, with new, tougher safety guidelines, continues.

Human gene therapy raises both technical and ethical issues. One important technical question is, How can researchers build in gene control mechanisms to ensure that cells with the transferred gene make appropriate amounts of the gene product at the right time and in the right parts of the body? And how can they be sure that the gene's insertion does not harm some other necessary cell function?

Among the ethical questions posed by gene therapy is, Who will have access to it? The procedures now being tested are expensive and require expertise and equipment found only in major medical centers. A related question is, Should gene therapy be reserved for treating serious diseases? And what about its potential use for enhancing athletic ability, physical appearance, and even intelligence?

Technically easier than modifying genes in somatic cells is the genetic engineering of germ cells or zygotes—already accomplished in lab animals. But this possibility raises the most difficult ethical question of all: whether we should try to eliminate genetic defects in our children and their descendants. Should we interfere with evolution in this way? From a biological perspective, the elimination of unwanted alleles from the gene pool could backfire. Genetic variety is a necessary ingredient for the survival of a species as environmental conditions change with time. Genes that are damaging under some conditions may be advantageous under others (one example is the sickle-cell allele; see Module 9.14). Are we willing to risk making genetic changes that could be detrimental to our species in the future? We may have to face this question soon.

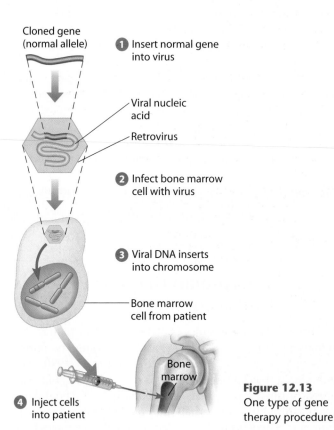

Cloned gene
(normal allele)

❶ Insert normal gene into virus

Viral nucleic acid

Retrovirus

❷ Infect bone marrow cell with virus

❸ Viral DNA inserts into chromosome

Bone marrow cell from patient

Bone marrow

❹ Inject cells into patient

Figure 12.13
One type of gene therapy procedure

❓ What characteristic of retroviruses makes them candidate vectors for gene therapy?

■ They integrate DNA into the DNA of host cells.

12.14 The PCR method is used to amplify DNA sequences

DNA cloning in cells (see Module 12.3) is often the best method for preparing large quantities of a particular gene. However, when the source of DNA is scanty or impure, the **polymerase chain reaction**, or **PCR**, is a much better method. In this technique, any specific target segment within a DNA molecule can be quickly amplified (copied many times) in a test tube. Starting with a single DNA molecule, automated PCR can generate 100 billion similar molecules in a few hours.

In principle, PCR is simple. A DNA sample is mixed with the DNA replication enzyme DNA polymerase, nucleotide monomers, and a few other ingredients. The solution is then exposed to cycles of heating (to separate the DNA strands) and cooling. During each cycle, the DNA is replicated, doubling the amount of DNA (Figure 12.14). The key to automating PCR was the discovery of an unusual heat-stable DNA polymerase, first isolated from prokaryotes living in hot springs. Unlike most proteins, this enzyme can withstand the heat at the start of each cycle.

For PCR to work, only minute amounts of DNA need be present in the starting material, and this DNA can be in a partially degraded state. From such a scant starting sample, PCR can produce enough DNA for restriction fragment analysis or other DNA technologies. However, occasional errors during PCR replication impose limits on the number of good copies that can be made by this method. So PCR cannot replace gene cloning in cells when large amounts of DNA are needed.

Devised in 1985, PCR has had a major impact on biological research and biotechnology. PCR has been used to amplify DNA from a wide variety of sources: fragments of ancient DNA from a 40,000-year-old frozen woolly mammoth; DNA from fingerprints or from tiny amounts of blood, tissue, or semen found at crime scenes; DNA from single embryonic cells for rapid prenatal diagnosis of genetic disorders; and DNA of viral genes from cells infected with such difficult-to-detect viruses as HIV.

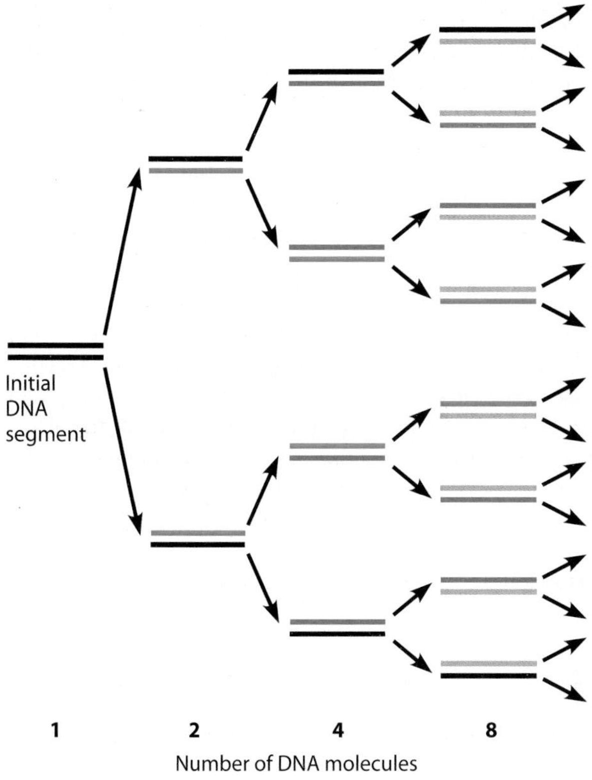

Initial DNA segment

| 1 | 2 | 4 | 8 |

Number of DNA molecules

Figure 12.14 DNA amplification by PCR

? Why must DNA polymerase from heat-stable prokaryotes be used during PCR?

■ The enzyme must be able to survive the heating stage of each cycle.

GENOMICS

CONNECTION

12.15 The Human Genome Project is an ambitious application of DNA technology

The **Human Genome Project** (HGP) is an effort to map the human genome in total detail by determining the entire nucleotide sequence of human DNA. Begun in 1990, this ambitious project was expected to take 15 years but was largely finished several years ahead of schedule. The project was organized by an international, publicly funded consortium of researchers and proceeded through three stages that provided progressively more detailed views of the human genome:

1. *Genetic (linkage) mapping.* In Module 9.17, you learned how geneticists use data from genetic crosses to map genes on a chromosome. For the HGP, geneticists combined pedigree analysis of large families with DNA technology to map over 5,000 genetic markers. The resulting low-resolution linkage map provided a framework for mapping other markers and for arranging later, more detailed maps of particular regions.

2. *Physical mapping.* To create a physical map, researchers determined the number of base pairs between markers. This is done by cutting the DNA of each chromosome into a number of restriction fragments, cloning them, and then figuring out the original order of the fragments. The key is to make fragments that overlap and then use probes or automated nucleotide sequencing of the ends

to find the overlaps. In this way, more and more fragments can be assigned to a sequential order that corresponds to their order in a chromosome.

3. *DNA sequencing.* The most arduous part of the project is determining the nucleotide sequences of a set of DNA fragments covering the entire genome, the fragments already mapped in stage 2. Advances in automatic DNA sequencing have been crucial to this endeavor (Figure 12.15). Sequencing machines can handle DNA molecules up to about 800 nucleotides in length.

This three-stage approach is logical and thorough. However, in the mid 1990s, J. Craig Venter, a former government scientist, proposed an alternative strategy and set up the company Celera Genomics to implement it. Venter's "whole genome shotgun" approach was essentially to proceed directly to the sequencing of small, random DNA fragments, relying on software to determine the order of the pieces. Celera actually made significant use of the consortium's data from stages 1 and 2, but the competition between the two groups hastened progress. In February 2001, Celera announced the sequencing of over 90% of the human genome. At the same time, HGP researchers made a similar announcement. Sequencing of the human genome is now virtually complete, although some gaps remain to be mapped because certain parts of the chromosomes resist mapping by the usual methods.

The potential benefits of having a complete map of the human genome are great. For basic science, the information is already providing insight into such fundamental mysteries as embryonic development and evolution. For human health, the identification of genes will aid in the diagnosis, treatment, and possibly prevention of many of our more common ailments, including heart disease, allergies, diabetes, schizophrenia, alcoholism, Alzheimer's disease, and

Figure 12.15 DNA sequencing

cancer. Hundreds of disease-associated genes have already been identified as a result of the project.

The DNA sequences from the HGP are deposited in a database available to researchers all over the world via the Internet. Scientists use software to analyze the sequences. Then comes the most exciting challenge: figuring out the functions of the genes and how they work together to direct the structure and function of a living organism. This challenge and the applications of the new knowledge should keep scientists busy well into the twenty-first century.

Web/CD Activity 12F *The Human Genome Project: Human Chromosome 17*

 What is meant by the human genome *sequence*?

■ The order of the nucleotides in the DNA of all the human chromosomes

---

12.16 Most of the human genome does not consist of genes

The biggest surprise from the HGP is the small number of human genes. The current estimate is about 25,000 genes—only one and a half to two times the number found in the fruit fly and nematode worm. How, then, to account for human complexity? Part of the answer may lie in alternative RNA splicing (see Module 11.9); scientists think that a typical human gene probably specifies several polypeptides.

In addition to genes, humans, like most complex eukaryotes, have a huge amount of noncoding DNA, about 97% of the total. Some noncoding DNA is made up of gene control sequences such as promoters and enhancers. The remaining DNA includes introns (whose total length may be ten times greater than the exons of a gene) and noncoding DNA located between genes.

Much of the DNA between genes consists of **repetitive DNA**, nucleotide sequences present in many copies in the genome. In one type of repetitive DNA, a unit of just a few nucleotide pairs is repeated many times in a row. Stretches of DNA with thousands of such repetitions are prominent at the centromeres and ends of chromosomes, suggesting that this

DNA plays a role in chromosome structure. Recent research supports the idea that the repetitive DNA at chromosome ends—called **telomeres**—also has a protective function; a significant loss of telomeric DNA quickly leads to cell death. Furthermore, abnormal lengthening of this DNA may help "immortal" cancer cells evade normal cell aging.

In the second main type of repetitive DNA, each repeated unit is hundreds of nucleotides long, and the copies are scattered around the genome. Most of these sequences seem to be associated with **transposons** ("jumping genes"), DNA segments that can move or be copied from one location to another in a chromosome and even between chromosomes. Transposons can land in the middle of other genes and disrupt them. Researchers believe that transposons, through their copy-and-paste mechanism, are responsible for the proliferation of dispersed repetitive DNA in the human genome.

 How many genes are in the human genome?

■ About 25,000

12.17 The science of genomics compares whole genomes

In addition to mapping human DNA, researchers are also working on the genomes of other species, focusing first on species that are important in biological research. As of 2005, the genomes of about 150 species have been (or are almost) sequenced.

Table 12.17 lists a few of the completed genomes. While the vast majority of genomes under study are from prokaryotes, there are 20 or so eukaryotic species in the group, including vertebrates, invertebrates, and plants. The first eukaryotic genome to be completed was that of the brewer's/baker's yeast *Saccharomyces cerevisiae*, a single-celled organism. The nematode *Caenorhabditis elegans*, a simple worm, was the first multicellular organism to be sequenced. Other important research organisms, such as *Arabidopsis thaliana* (mustard plant), *Drosophila melanogaster* (fruit fly), and *Mus musculus* (mouse), have also been sequenced. And other species whose genomes have been (or are currently being) sequenced include the mosquito, the dog, the rat, the chicken, and the frog.

Why map so many genomes? All genomes are of interest in their own right, but comparative analysis with the genes of other species also helps scientists interpret the human genome. For example, when scientists find a nucleotide sequence in the human genome similar to a yeast gene whose function is known, they have a valuable clue to the function of the human sequence. Indeed, several yeast protein-coding genes are so similar to certain human disease-causing genes that researchers have figured out the functions of the disease genes by studying their normal yeast counterparts. Many genes of disparate organisms are turning out to be astonishingly similar, to the point that one researcher has joked that he now views fruit flies as "little people with wings."

Now that the sequences of many entire genomes are available, scientists can study whole sets of genes and their interactions, an approach called **genomics**. Taking a systems approach (see Module 1.3), genomics is yielding new insights into fundamental questions about genome organization, regulation of gene expression, growth and development, and evolution.

Comparisons of genome sequences from different species also allow us to evaluate the evolutionary relationships between those species. The more similar in sequence the same gene is in two species, the more closely related those species are by their evolutionary history. Likewise, comparisons of multiple genes between species can shed light on higher groupings of species, which reflect their evolutionary history. Indeed, comparisons of the completed genome sequences of bacteria, archaea, and eukarya strongly support the theory that these are the three fundamental domains of life.

The success in sequencing genomes and studying whole genomes is encouraging scientists to attempt similar systematic study of the full protein sets (proteomes) encoded by genomes, an approach called **proteomics**. The number of proteins in humans far exceeds the number of genes. And since proteins, not genes, actually carry out the activities of the cell, scientists must study when and where proteins are produced in an organism and how they interact in order to understand the functioning of cells and organisms. Assembling and analyzing proteomes pose many experimental challenges, but ongoing advances are providing the tools to continue the investigation.

Genomics and proteomics are enabling biologists to approach the study of life from an increasingly global perspective. Biologists are now in a position to compile catalogs of genes and proteins—that is, a listing of all the "parts" that contribute to the operation of cells, tissues, and organisms. With such catalogs in hand, researchers are shifting their attention from the individual parts to how they function together in biological systems.

? Why is it useful to sequence nonhuman genomes?

■ Besides their intrinsic interest, comparative analysis of their genes helps scientists interpret the human data.

TABLE 12.17 SOME IMPORTANT COMPLETED GENOMES

Organism	Year completed	Size of genome (in base pairs)	Approximate number of genes
Haemophilus influenzae (bacterium)	1995	1.8 million	1,700
Saccharomyces cerevisiae (yeast)	1996	12 million	6,000
Escherichia coli (bacterium)	1997	4.6 million	4,400
Caenorhabditis elegans (nematode)	1998	97 million	19,000
Drosophila melanogaster (fruit fly)	2000	180 million	13,700
Arabidopsis thaliana (mustard plant)	2000	118 million	25,500
Mus musculus (mouse)	2001	2.6 billion	25,000
Oryza sativa (rice)	2002	430 million	60,000
Homo sapiens (humans)	2003	2.9 billion	25,000

12.18 Genetically modified organisms are transforming agriculture

Scientists concerned with feeding the growing human population are using DNA technology to make **genetically modified (GM) organisms** for use in agriculture. A GM organism (or GMO) is one that has acquired one or more genes by artificial means rather than by traditional breeding methods. (The new gene may or may not be from another species.)

To make genetically modified plants, researchers can manipulate the DNA of a single somatic cell and then grow a plant with a new trait from the engineered cell. Already in commercial use are a number of crop plants carrying new genes for desirable traits, such as delayed ripening and resistance to spoilage and disease.

The most common vector used to introduce new genes into plant cells is a plasmid from the soil bacterium *Agrobacterium tumefaciens* called the **Ti plasmid (Figure 12.18A)**. ❶ With the help of a restriction enzyme and DNA ligase, the gene for the desired trait (red) is inserted into a segment of the plasmid called T DNA. ❷ Then the recombinant plasmid is put into a plant cell, where the T DNA carrying the new gene integrates into a plant chromosome. ❸ Finally, the recombinant cell is cultured and grows into a whole plant. If the newly acquired gene is from another species, the recombinant organism is called a **transgenic organism.**

Genetic engineering is rapidly replacing traditional plant-breeding programs, especially in cases where useful traits are determined by one or only a few genes. For example, the majority of the American soybean and cotton crops are genetically modified. Many of these GM plants have received bacterial genes that make the plants resistant to herbicides or pests. Farmers can more easily grow these crops with far less tillage and reduced use of chemical insecticides.

Genetic engineering also has great potential for improving the nutritional value of crop plants. "Golden rice," a transgenic variety with a few daffodil genes, produces grains containing beta-carotene, which our body uses to make vitamin A. This rice could help prevent vitamin A deficiency—and resulting blindness—among the half of the world's people who depend on rice as their staple food.

Agricultural researchers are also making transgenic animals. To do this, scientists first remove egg cells from a female and fertilize them *in vitro*. They then inject a previously cloned gene directly into the nuclei of the fertilized eggs. Some of the cells integrate the foreign DNA into their genomes. The engineered embryos are then surgically implanted in a surrogate mother. If an embryo develops successfully, the result is a transgenic animal, containing a gene from a third "parent" that may even be of another species.

Figure 12.18B
Transgenic pigs

The goals of creating a transgenic animal are often the same as the goals of traditional breeding—for instance, to make a sheep with better quality wool or a cow that will mature in a shorter time. Scientists might, for example, identify and clone a gene that causes the development of larger muscles (muscles make up most of the meat we eat) in one variety of cattle and transfer it to other cattle or even to sheep.

Transgenic animals also have been engineered to be pharmaceutical "factories" that produce otherwise rare biological substance for medical use (**Figure 12.18B**; see also Module 12.6). Recently, researchers have engineered transgenic chickens that express large amounts of the foreign product in their eggs. This success suggests that transgenic chickens may emerge as relatively inexpensive pharmaceutical factories in the near future. Unfortunately, genetically modified organisms can pose potential hazards as well as rewards. We discuss some of these risks next.

Web/CD Activity 12G *Connection: Applications of DNA Technology*

? What is the function of the Ti plasmid in the creation of transgenic plants?

■ It is used as the vector for introducing foreign genes into a plant cell.

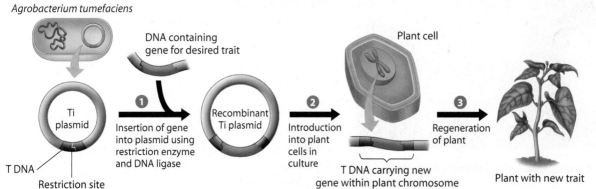

Agrobacterium tumefaciens

DNA containing gene for desired trait

Plant cell

Ti plasmid

T DNA

Restriction site

❶ Insertion of gene into plasmid using restriction enzyme and DNA ligase

Recombinant Ti plasmid

❷ Introduction into plant cells in culture

T DNA carrying new gene within plant chromosome

❸ Regeneration of plant

Plant with new trait

Figure 12.18A
Using the Ti plasmid as a vector for genetically engineering plants

12.19 Could GM organisms harm human health or the environment?

As soon as scientists realized the power of DNA technology, they began to worry about potential dangers. Early concerns focused on the possibility that recombinant DNA technology might create new pathogens. What might happen, for instance, if cancer cell genes were transferred into bacteria or viruses? To guard against such rogue microbes, scientists developed a set of guidelines that were adopted as formal government regulations in the United States and some other countries. One safety measure is a set of strict laboratory procedures designed to protect researchers from infection by engineered microbes and to prevent the microbes from accidentally leaving the laboratory (Figure 12.19A). In addition, strains of microorganisms to be used in recombinant DNA experiments are genetically crippled to ensure that they cannot survive outside the laboratory. Finally, certain obviously dangerous experiments have been banned.

Today, most public concern about possible hazards centers not on recombinant microbes but on genetically modified (GM) crop plants. Advocates of a cautious approach fear that some crops carrying genes from other species might be hazardous to human health or the environment.

One specific concern is that genetic engineering could transfer allergens, which are molecules to which some people are allergic (see Module 24.17), to plants people eat. Although there is some evidence that this could happen, advocates claim that these proteins could be tested for their ability to cause allergic reactions.

Nevertheless, because of health concerns, activists continue to lobby for the clear labeling of all foods containing products of GM organisms. Early in 2000, negotiators from 130 countries (including the United States) agreed on a Biosafety Protocol that requires all exporters to identify GM organisms present in bulk food shipments and allows importing countries to decide whether they pose environmental or health risks. Some biotechnology advocates, however, point out that similar demands were not made when "transgenic" crop plants produced by traditional breeding techniques were put on the market. One example of such a plant is triticale, which was synthesized decades ago by combining the genomes of wheat and rye—two plants that do not interbreed in nature. Triticale is grown worldwide.

Advocates of a cautious approach toward GM crops fear that transgenic plants might pass their new genes to close relatives in nearby wild areas (Figure 12.19B). We know that lawn and crop grasses, for example, commonly exchange genes with wild relatives via pollen transfer. If crop plants carrying genes for resistance to herbicides, diseases, or insect pests pollinated wild ones, the offspring might become "superweeds" that would be very difficult to control. However, researchers may be able to prevent the escape of such plant genes in various ways—for example, by engineering plants so that they cannot hybridize.

Today, governments and regulatory agencies throughout the world are grappling with how to facilitate the use of biotechnology in agriculture, industry, and medicine while ensuring that new products and procedures are safe. In the United States, all projects are evaluated for potential risks by regulatory agencies such as the Food and Drug Administration, Environmental Protection Agency, National Institutes of Health, and Department of Agriculture. These agencies are under increasing pressure from some consumer groups. In the case of GM plants and certain other applications of DNA technology, zero risk is probably unattainable. Scientists and the public need to weigh the possible benefits versus risks on a case-by-case basis. The best scenario would be for us to base our decisions on sound scientific information rather than on either irrational fear or blind optimism.

Figure 12.19B Pollen might transfer genes from genetically engineered crop plants to wild relatives nearby

Web/CD Activity 12H *Connection: DNA Technology and Golden Rice*

? What is one of the concerns about engineering crop plants by adding genes for herbicide resistance?

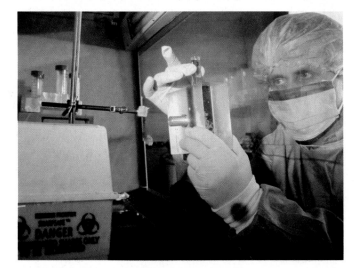

Figure 12.19A A maximum-security laboratory at the Pasteur Institute in Paris

■ The possibility that the genes could escape, via cross-pollination, to weeds that are closely related to the crop species

12.20 Genomics researcher Eric Lander discusses the Human Genome Project

Dr. Eric Lander (Figure 12.20) is the founding director of the Broad Institute of MIT and Harvard, which uses genomics to develop new tools and approaches to understanding and treating disease. Previously, he supervised a team that played a leading role in the Human Genome Project (HGP). Although he is now a leader in the field of genomics, Dr. Lander's study of biology began much later in life than for most scientists. In fact, he taught economics at Harvard Business School before turning to biology full-time in 1990. In a recent interview, Dr. Lander discussed his unusual career path:

Figure 12.20 Eric Lander

In high school, I took biology, but I loved math, and I was a math major in college. I went on to get my Ph.D. in mathematics but decided I didn't want to be a pure mathematician. One day, my brother suggested I might be interested in the coding theory of the brain and sent me some papers on mathematical neurobiology. I realized that to understand them, I had to learn something about cellular neurobiology. This required me to study cell biology. Next came molecular biology, and finally, I really had to know genetics. So one thing led to another—and here I am still learning genetics!

Dr. Lander's interest in applying math to biology led him to the field of genetic analysis at a time when the HGP was starting. Dr. Lander describes the importance of the project:

You can study the detailed properties of individual genes, as biologists did before the Human Genome Project and still do, or you can study how all the components of the system interact. Important discoveries are made at both levels, but there are some things you only see when you step back. Imagine looking at a pointillist painting. Up close, the dots are interesting, but when you step back you can see patterns that weren't evident at first. Until the Human Genome Project, it hadn't been possible to step back and get the big picture of the human genome.

One of the major surprises of the HGP was the number of human genes. Dr. Lander describes this finding:

My best guess today is that there are 20,000 to 25,000 protein-coding genes. One of the most surprising findings of the Human Genome Project was a gene count much lower than people had expected based on the total size of the human genome; not so many years ago, textbooks gave 100,000 as the likely number of human genes. When we got the rough draft of the sequence, we thought there might be as many as 40,000 genes, but we soon learned that many of these were actually pseudogenes, which are defective, nonfunctional copies of true genes. The gene count has been falling and falling.

In addition to the number of genes, the HGP has yielded many other unexpected results. Dr. Lander thus sees the completion of the HGP not as an end, but as the beginning of a deeper exploration:

The extent of our ignorance became clear from comparing the human genome with the mouse genome. When we lined up the two, we found that about 5% of the human genome showed strong similarity to that of the mouse, indicating strong evolutionary conservation since the last common ancestor of mouse and human. But only about a third of the 5% could be accounted for by known genes and regulatory sequences, leaving a lot more that evolution "cares about" than we can explain today. That's what I love about genomics. We're learning that there are vast tracts of biology we have missed. It's as if we suddenly could look at the whole Earth and see that, golly, there are several continents we hadn't known about! Genomics is revealing huge territories for the next generation of young scientists to explore.

Dr. Lander points out that the availability of genomic data is revolutionizing the field of evolutionary biology. We will consider how genetics and many biological fields apply to evolution in Unit III.

? How does the actual number of human genes compare to what was expected before the Human Genome Project?

■ There are far fewer, only about 25,000 instead of 100,000.

CHAPTER REVIEW

Reviewing the Concepts

Bacterial Plasmids and Gene Cloning (Introduction–12.7)

DNA fingerprinting is a set of laboratory procedures that determines with near certainty whether two samples of DNA are from the same individual **(Introduction).**

Gene cloning is one application of DNA technology, methods for studying and manipulating genetic material. Researchers can insert desired genes into plasmids, creating recombinant DNA, and insert those plasmids into bacteria. If the recombinant bacteria multiply into a clone, the foreign genes are also copied **(12.1).** The tools used to make recombinant DNA include restriction enzymes, which cut DNA at specific sequences, and DNA ligase, which "pastes" DNA fragments together **(12.2).** Bacteria can take up recombinant plasmids from their surroundings and reproduce, thereby cloning the plasmids and the genes they carry **(12.3).**

Genomic libraries, sets of DNA fragments containing all of an organism's genes, can be constructed and stored in

cloned bacterial plasmids or phages **(12.4)**:

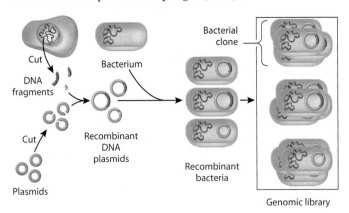

Reverse transcriptase can be used to make cDNA libraries containing only the genes that are transcribed by a particular type of cell **(12.5)**.

Applications of gene cloning include the mass production of gene products for medical and other uses. Different organisms, including bacteria, yeast, and mammals, can be used for this purpose **(12.6)**. Researchers use gene cloning to produce hormones, diagnose and treat diseases, and produce vaccines **(12.7)**.

Restriction Fragment Analysis and DNA Fingerprinting (12.8–12.14)

DNA technology methods can be used to identify specific pieces of DNA. A nucleic acid probe, a short, single-stranded molecule of radioactively or fluorescently labeled DNA or RNA, can tag a desired gene in a library **(12.8)**. DNA microarray assays can reveal patterns of gene expression in different kinds of cells **(12.9)**. Gel electrophoresis can sort DNA molecules by size **(12.10)**:

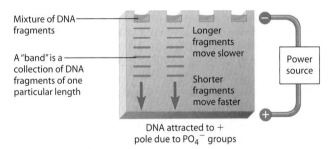

Restriction fragment length polymorphisms (RFLPs) reflect differences in the sequences of DNA samples. After digestion by restriction enzymes, the fragments are run through a gel, and then radioactive probes can reveal bands of interest **(12.11)**. DNA fingerprinting can help solve crimes and establish paternity **(12.12)**. Gene therapy may one day be used to treat both genetic diseases and nongenetic disorders; progress is slow, however **(12.13)**. The polymerase chain reaction (PCR) can be used to clone a small sample of DNA quickly, producing enough copies for analysis **(12.14)**.

Genomics (12.15–12.17)

The Human Genome Project (HGP), begun in 1990 and now largely completed, involved genetic and physical mapping of chromosomes followed by DNA sequencing. The data are providing insight into development, evolution, and many diseases **(12.15)**. The haploid human genome contains about 25,000 genes and a huge amount of noncoding DNA. Much of the noncoding DNA consists of repetitive nucleotide sequences and transposons that can move about within the genome **(12.16)**. The sequencing of many prokaryotic and eukaryotic genomes has produced data for genomics, the study of whole genomes. Besides being interesting themselves, nonhuman genomes can be compared with the human genome. Proteomics is the study of the full sets of proteins produced by organisms **(12.17)**.

Genetically Modified Organisms (12.18–12.20)

Recombinant DNA technology can be used to produce new genetic varieties of plants and animals; a number of important crop plants are genetically modified **(12.18)**. Genetic engineering involves risks, such as ecological damage from GM crops **(12.19)**. Genomics pioneer Eric Lander points out that much remains to be learned from the HGP **(12.20)**.

Connecting the Concepts

1. You have discovered a tiny drop of blood at a crime scene. Fill in the following diagram, which outlines the steps of a DNA technology-based investigation of the mystery drop.

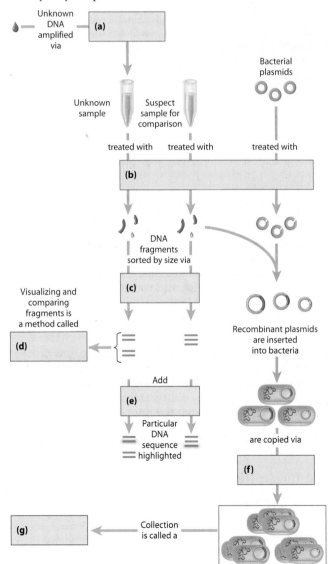

Testing Your Knowledge

Multiple Choice

2. Which of the following would be considered a transgenic organism?
   a. a bacterium that has received genes via conjugation
   b. a human given a corrected human blood-clotting gene
   c. a fern grown in cell culture from a single fern root cell
   d. a rat with rabbit hemoglobin genes
   e. a human treated with insulin produced by bacteria

3. When a typical restriction enzyme cuts a DNA molecule, the cuts are uneven, so that the DNA fragments have single-stranded ends. These ends are useful in recombinant DNA work because
   a. they enable a cell to recognize fragments produced by the enzyme.
   b. they serve as starting points for DNA replication.
   c. the fragments will bond to other fragments with complementary ends.
   d. they enable researchers to use the fragments as molecular probes.
   e. only single-stranded DNA segments can code for proteins.

4. DNA fingerprints used as evidence in a murder trial look something like supermarket bar codes. The pattern of bars in a DNA fingerprint shows
   a. the order of bases in a particular gene.
   b. the presence of various-sized fragments from chopped-up DNA.
   c. the presence of dominant or recessive alleles for particular traits.
   d. the order of genes along particular chromosomes.
   e. the exact location of a specific gene in a genomic library.

5. A biologist isolated a gene from a human cell, attached it to a plasmid, and inserted the plasmid into a bacterium. The bacterium made a new protein, but it was nothing like the protein normally produced in a human cell. Why? (*Explain your answer.*)
   a. The bacterium had undergone transformation.
   b. The gene did not have sticky ends.
   c. The gene contained introns.
   d. The gene did not come from a genomic library.
   e. The biologist should have cloned the gene first.

6. A paleontologist has recovered a bit of organic material from the 400-year-old preserved skin of an extinct dodo. She would like to compare DNA from the sample with DNA from living birds. Which of the following would be most useful for increasing the amount of DNA available for testing?
   a. restriction fragment analysis
   b. polymerase chain reaction
   c. molecular probe analysis
   d. electrophoresis
   e. Ti plasmid technology

7. How many genes are there in a human sperm cell?
   a. 23
   b. 46
   c. 5,000–10,000
   d. about 25,000
   e. about 3 billion

Describing, Comparing, and Explaining

8. Explain how you might engineer *E. coli* to produce human growth hormone (HGH) using the following: *E. coli* containing plasmids, DNA carrying a gene for HGH, DNA ligase, a restriction enzyme, equipment for manipulating and growing bacteria, a method for extracting and purifying the hormone.

9. Recombinant DNA techniques are used to custom-build bacteria for two main purposes: to obtain multiple copies of certain genes and to obtain useful proteins produced by certain genes. Give an example of each of these applications in medicine and agriculture.

Applying the Concepts

10. A biochemist hopes to find a gene in human liver cells that codes for an important blood-clotting protein. She knows that the nucleotide sequence of a small part of the gene is CTGGACTGACA. Briefly explain how to obtain the desired gene.

11. What is left for genetic researchers to do now that the Human Genome Project has determined the complete nucleotide sequences of nearly all of the human chromosomes? Explain.

12. Today, it is fairly easy to make transgenic plants and animals. What are some important safety and ethical issues raised by this use of recombinant DNA technology? What are some of the possible dangers of introducing genetically engineered organisms into the environment? What are some reasons for and against leaving decisions in these areas to scientists? To business owners and executives? What are some reasons for and against more public involvement? How might these decisions affect you? How do you think these decisions should be made?

13. In the not-too-distant future, gene therapy may be an option for the treatment and cure of some inherited disorders. What do you think are the most serious ethical issues that must be dealt with before human gene therapy is used on a large scale? Why do you think these issues are important?

14. The possibility of extensive genetic testing raises questions about how personal genetic information should be used. For example, should employers or potential employers have access to such information? Why or why not? Should the information be available to insurance companies? Why or why not? Is there any reason for the government to keep genetic files? Is there any obligation to warn relatives who might share a defective gene? Might some people avoid being tested for fear of being labeled genetic outcasts? Or might they be compelled to be tested against their wishes? Can you think of other reasons to proceed with caution?

Answers to all questions can be found in Appendix 3.

For study help and Activities, go to campbellbiology.com or the student CD-ROM.

UNIT

III

Concepts of Evolution

13 HOW POPULATIONS EVOLVE

14 THE ORIGIN OF SPECIES

15 TRACING EVOLUTIONARY HISTORY

CHAPTER 13

DARWIN'S THEORY OF EVOLUTION

13.1   A sea voyage helped Darwin frame his theory
       of evolution

13.2   Darwin proposed natural selection as the mechanism of evolution

13.3   The study of fossils provides strong evidence for evolution

13.4   A mass of other evidence reinforces the evolutionary view of life

13.5   Scientists can observe natural selection in action

POPULATION GENETICS AND
THE MODERN SYNTHESIS

13.6   Populations are the units of evolution

13.7   The gene pool of a nonevolving population remains constant over
       the generations

13.8   The Hardy-Weinberg equation is useful in public health science

13.9   In addition to natural selection, genetic drift and gene flow can
       contribute to evolution

13.10 Endangered species often have reduced variation

VARIATION AND NATURAL SELECTION

13.11 Variation is extensive in most populations

13.12 Mutation and sexual recombination generate variation

13.13 The evolution of antibiotic resistance in bacteria is a
       serious public health concern

13.14 Diploidy and balancing selection preserve variation

13.15 The perpetuation of genes defines evolutionary fitness

13.16 Natural selection can alter variation in a population in three ways

13.17 Sexual selection may produce sexual dimorphism

13.18 Natural selection cannot fashion perfect organisms

Clown, Fool, or Simply Well Adapted?

IS THE BLUE-FOOTED BOOBY really as awkward or foolish as its name implies? With its bright blue feet and trusting demeanor, boobies were certainly noticed by early travelers to the Galápagos Islands, a chain of relatively young volcanic islands located about 900 km off the Pacific coast of South America. Spanish sailors may have called them "clown" (*bobo* in Spanish). British seamen may have called them "booby" (slang for "stupid"), as the birds were so approachable that they were easily killed. Either way, their comical feet are hard to miss.

British seamen may have called them "booby" (slang for "stupid")

Charles Darwin, whose observations in the Galápagos contributed greatly to his theory of evolution, no doubt encountered this friendly bird. And like the other species he observed, boobies have physical features that help them succeed in their environment. For example, their large, webbed feet make great flippers, propelling the birds through the water at high speeds. In the clear Galápagos waters, you can watch boobies gracefully "fly" beneath the surface. Their feet are thus a huge advantage while hunting fish. On land, however, those same feet are awkward, making for klutzy walking and for an even clumsier flight takeoff; their wings often touch the ground before they become airborne.

How Populations Evolve

In addition to their infamous feet, boobies have other characteristics that serve them well in their seafaring environment. The booby's body and bill are streamlined, like a torpedo, minimizing friction when it dives from heights of up to 24 m into the shallow water below. To pull themselves out of this high-speed dive once they hit the water, boobies use their large tail. During their dives, their nostrils close, preventing water from being forced into the lungs.

A number of specialized glands help boobies stay afloat and manage salt intake while at sea. A gland at the base of the booby's large tail secretes oil that keeps the booby waterproof. Another gland, in the bird's eye socket, keeps the salt level in the bird's body from reaching dangerous levels. The gland accumulates salt from body fluids and produces a concentrated salt solution. This salty liquid then trickles into the nasal cavity and is expelled when the bird shakes its head from side to side, a characteristic clown-like movement.

As useful as adaptations may be, they often represent a trade-off between different needs. The booby's webbed feet are a case in point: extremely functional in water, but clumsy for walking on dry land. But because boobies spend a great deal of time in the water, and especially because they find most of their food there, this adaptation represents a net advantage, improving chances of an individual's surviving long enough to reproduce.

The booby's big webbed feet, streamlined shape, large tail, nostrils that close, and specialized salt-excreting glands are all examples of **evolutionary adaptations**, inherited traits that enhance an organism's ability to survive and reproduce in a particular environment. In this chapter, we examine some of the processes by which the inhabitants of the Galápagos, and species everywhere, develop evolutionary adaptations to their environment. We thus begin our study of **evolution**, the changes in organisms over time that have transformed life on Earth from its earliest forms to the vast diversity that we see today. ■ ■ ■

13.1 A sea voyage helped Darwin frame his theory of evolution

When we think of an animal, plant, or other organism, we tend to envision it in its natural surroundings. Nature films don't just show us interesting organisms; they also tell us how special structures or behaviors help the organisms survive and reproduce in a particular environment. If you visited the Galápagos Islands today, you would see many of the same sights that fascinated Darwin over a century ago: blue-footed boobies waddling around; lumbering, long-necked giant tortoises (whose Spanish name is *galápagos*); and marine iguanas basking on dark lava rocks (Figure 13.1A). Being aware of each organism's adaptations and how they fit the particular conditions of the environment would help you appreciate everything you saw and heard. An appreciation for the close ties between organisms and their environment is part of the legacy of Charles Darwin.

Like many concepts in science, the main ideas Darwin advanced—that living species have arisen from earlier life-forms and that species change over time—can be traced back to the ancient Greeks. About 2,500 years ago, the Greek philosopher Anaximander promoted the idea that life arose in water and that simpler forms of life preceded more complex ones. The road from Anaximander to Darwin, however, was long and tortuous. The Greek philosopher Aristotle, whose views had an enormous impact on Western culture, generally held that species are fixed, or permanent, and do not evolve. Judeo-Christian culture fortified this idea with a literal interpretation of the Book of Genesis, holding that all species were individually designed by a divine creator. The idea that all living species are static in form and inhabit an Earth that is at most about 6,000 years old dominated the intellectual and cultural climate of the Western world for centuries.

In the century prior to Darwin, only a few scientists questioned the belief that species are fixed, unchanging, and per-

fect. In the mid-1700s, the study of **fossils** (the imprints or remains of organisms that lived in the past) led French naturalist Georges Buffon to suggest that Earth might be much older than 6,000 years. In addition, scientists observed similarities between fossils and living organisms. Buffon proposed the possibility that certain fossil forms might be ancient versions of similar living species. In the early 1800s, French naturalist Jean Baptiste Lamarck suggested that the best explanation for this relationship of fossils to current organisms is that life evolves. Today, we remember Lamarck mainly for his erroneous view of *how* species evolve. He proposed that by using or not using its body parts, an individual may develop certain traits that it passes on to its offspring. Lamarck's idea is known as the inheritance of acquired characteristics. He suggested, for instance, that the ancestors of the giraffe had lengthened their necks by stretching higher and higher into the trees to reach leaves. This mistaken idea obscures the important fact that Lamarck helped set the stage for Darwin by strongly advocating evolution and by proposing that species evolve as a result of interactions with their environment.

Charles Darwin was born in 1809, the same year that Lamarck published some of his ideas on evolution. In December 1831, at the age of 22, Darwin began a round-the-world sea voyage (Figure 13.1B) that profoundly influenced his thinking and eventually the thinking of the entire world. He accompanied the captain of the HMS *Beagle,* a surveying ship, on a mission to chart poorly known stretches of the South American coastline. Darwin actually spent most of his time onshore, collecting thousands of specimens of fossils and living plants and animals. He noted the unique adaptations of these South American organisms in places as different as the Brazilian jungle, the grasslands of the pampas, and the frigid lands near Antarctica. Darwin asked himself why fossils of the South American continent were more similar to modern South American species than to fossils of other continents. Other questions arose from Darwin's visit to the Galápagos Islands. Darwin observed that these islands had many unique organisms, most of which were similar to, but different from, the plants and animals of the nearest mainland. Even the individual islands had some species that differed from those on other islands. Referring to the islands and their unique inhabitants, he later wrote, "Both in space and time, we seem to be brought somewhat near to that great fact—that mystery of mysteries—the first appearance of new beings on the earth."

While on the voyage, Darwin read and was strongly influenced by the recently published *Principles of Geology*, by Scottish geologist Charles Lyell. Lyell's work led Darwin to realize that natural forces gradually change Earth's surface and that these forces are still operating in modern times. Darwin had collected fossils of marine snails in the Andes. Having read Lyell's book and witnessed an earthquake that

Figure 13.1A A marine iguana, one of the unique inhabitants of the Galápagos Islands

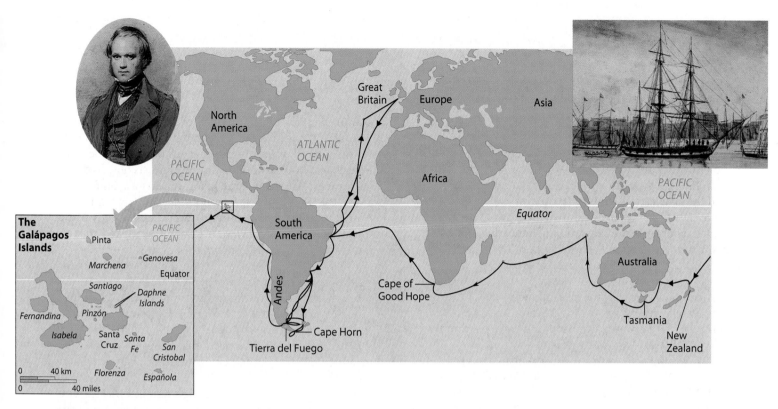

Figure 13.1B The voyage of the *Beagle* (1831–1836). The insets show a young Charles Darwin and his ship.

raised part of the coastine of Chile almost a meter, he came to believe that slow, natural processes, such as the growth of mountains as a result of earthquakes, could account for the presence of marine snails on mountaintops.

By the time Darwin returned to Great Britain five years after the *Beagle* first set sail, his experiences and reading had led him to doubt seriously that Earth and living organisms were unchangeable and had been specially created only a few thousand years earlier. By then he believed that Earth was very old and constantly changing. Upon his return, Darwin analyzed his collections, discussed them with colleagues, continued to read, and maintained extensive journals of his observations, studies, and thoughts.

By the early 1840s, Darwin had composed a long essay describing the major features of his theory of evolution. He realized that his ideas would cause a social furor, however, and he delayed publishing his essay. Then, in the mid-1850s, Alfred Wallace, a British naturalist doing fieldwork in Indonesia, conceived a theory almost identical to Darwin's. Wallace asked Darwin to evaluate the manuscript he had written about his theory to see if it merited publication. In 1858, two of Darwin's colleagues presented Wallace's paper and excerpts of Darwin's earlier essay together to the scientific community. With the publication in 1859 of his book, *On the Origin of Species by Means of Natural Selection*, Darwin presented the world with an avalanche of evidence and a strong, logical argument for evolution. He also described his theory of natural selection, an explanation of how evolution occurs.

In the first edition of his book, Darwin did not actually use the word *evolved* until the very end, referring instead to **descent with modification**. This phrase summarized Dar-

win's view of life. He perceived a unity among species, with all organisms related through descent from an ancestor that lived in the remote past. As the descendants of that ancestor spread into various habitats over millions of years, they accumulated diverse modifications, or adaptations, that accommodated them to diverse ways of life. The history of life seemed to resemble a tree, with multiple branchings from a common trunk to the tips of the twigs. At each fork of the evolutionary tree is an ancestor common to all lines of descent branching from that fork. Species that are closely related, such as the blue-footed booby and the red-footed booby, share many characteristics because their lineage of common descent extends to the smallest branches of the tree of life.

Evolution is the great unifying theme of biology, and *The Origin of Species* fueled an explosion in biological research and knowledge that continues today. As we will see in this chapter and the next two, evolutionary theory continues to expand beyond Darwin's basic ideas. Nonetheless, few contributions in all of science have explained as much, withstood as much repeated testing over the years, and stimulated as much other research as those of Darwin.

Web/CD Activity 13A *Darwin and the Galápagos Islands*

Web/CD Activity 13B *The Voyage of the* Beagle: *Darwin's Trip Around the World*

? What was Darwin's phrase for evolution? What does it mean?

■ Descent with modification. An ancestral species could diversify into many descendant species by the accumulation of adaptations to various environments.

Darwin devoted much of *The Origin of Species* to the ways that organisms become adapted to their environment. His theory of how this happens arose from several key observations and an inference (see Module 1.6). First of all, Darwin recognized that all species tend to produce excessive numbers of offspring. He had read an influential essay on human population written in 1798 by British economist Thomas Malthus. Malthus contended that much of human suffering—disease, famine, homelessness, and war—was the inescapable consequence of the human population's potential to grow much faster than the rate at which supplies of food and other resources could be produced. It was apparent to Darwin that Malthus's concepts applied to all species. Darwin deduced that because natural resources are limited, the production of more individuals than the environment can support leads to a struggle for existence among the individuals of a population, with only some offspring surviving in each generation. Many eggs are laid, young born, and seeds spread, but only a tiny fraction complete their development and leave offspring of their own. The rest are eaten, starved, diseased, unmated, or unable to reproduce for other reasons.

In addition to the overproduction of offspring, two other important observations by Darwin were that individuals of a population vary extensively in their characteristics and that many of the varying traits are inherited—that is, passed from one generation to the next.

What do overproduction of offspring, limited natural resources, and heritable variations have to do with organisms becoming adapted to their environment? Darwin saw that every environment has only a limited supply of resources and that survival in a limited environment depends in

part on the features the organisms inherit from their parents. He inferred that within a varied population, individuals whose characteristics adapt them best to their environment are most likely to survive and reproduce; these individuals thus tend to leave more offspring than less fit individuals do. Reproduction is central to what Darwin saw as the basic mechanism of evolution, the process he called **natural selection.** Darwin perceived that the essence of natural selection is differential, or unequal, success in reproduction.

Darwin's insight was both simple and profound: Reproduction is unequal, with the individuals that best meet specific environmental demands having the greatest reproductive success. Put another way, differential reproductive success (natural selection) is the means by which the environment filters variations, favoring some over others. As Darwin reasoned, natural selection results in favored traits being represented more and more and unfavored ones less and less in ensuing generations. Thus, the unequal ability of individuals to survive and reproduce leads to a gradual change in the characteristics of a population of organisms, with favored characteristics accumulating over the generations.

Darwin found convincing evidence for his ideas in the results of **artificial selection,** the selective breeding of domesticated plants and animals. He saw that by selecting individuals with the desired traits as breeding stock, humans were playing the role of the environment and bringing about differential reproduction. They were, in fact, modifying species. We see evidence of what Darwin was talking about in the vegetables we eat. For example, broccoli, cauliflower, cabbages, Brussels sprouts, and kale (**Figure 13.2A**) are all varieties of a single species of wild mustard, and all were produced by artificial selection. These and many other domesticated plants and animals bear little resemblance to the wild species from which they were derived. **Figure 13.2B** shows some of the enormous diversity that breeders have produced in a few thousand years within a single species, the domestic dog.

Darwin reasoned that if so much change could be achieved in a relatively short period of time by artificial selection, then over hundreds or thousands of generations, natural selection should be able to modify species considerably. With natural selection operating over vast spans of time, many heritable changes would accumulate. Such changes would account for the evolution of new species—for example, the five species of canines in **Figure 13.2C**—from an ancestral species.

Let's summarize the two main features of Darwin's theory: The diverse forms of life have arisen by descent with modification from ancestral species, and the mechanism of modification has been natural selection working over enormous spans of time. In the next two modules, we examine some of the evidence for evolution.

Figure 13.2A Artificial selection in plants: five vegetables (below) derived from wild mustard (left)

Web/CD Thinking as a Scientist How Do Environmental Changes Affect a Population?

Figure 13.2B Eight breeds of dogs (all members of the same species), the results of hundreds to thousands of years of artificial selection

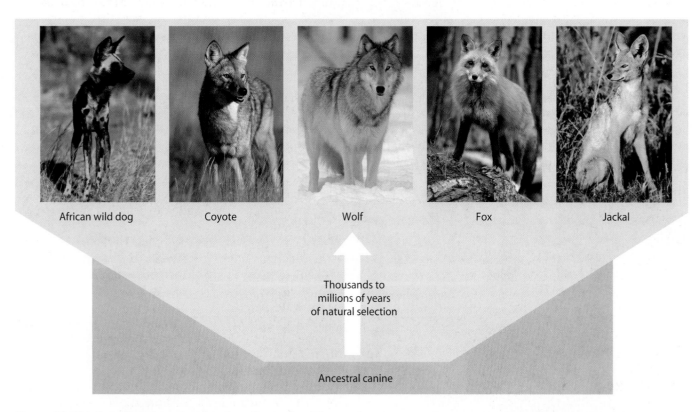

Figure 13.2C Five different species of canines, the results of thousands to millions of years of natural selection

❓ Differential reproductive success among a population's varying individuals in their natural environment is called _____ _____.

■ natural selection

Darwin developed his theory of descent with modification mainly with evidence from the geographic distribution of species, examples of artificial selection, and the fossil record. His careful documentation convinced many of the scientists of his day that organisms do indeed evolve. Subsequent discoveries in many fields support this great principle. Fossils document some of the drastic changes that life has undergone over time. The photographs on this page illustrate a number of fossils, each of which formed in a somewhat different way.

The organic substances of a dead organism usually decay rapidly, but hard parts of an animal that are rich in minerals, such as the bones and teeth of dinosaurs and the shells of clams and snails, may remain as fossils. The fossilized skull in **Figure 13.3A** is from one of our early relatives, *Homo erectus,* who lived some 1.5 million years ago in Africa.

Sometimes the remains of dead organisms are actually turned into stone by a process called petrification. Petrification occurs when minerals dissolved in groundwater seep into the tissues of dead organisms and replace organic matter. The petrified

tree in **Figure 13.3B** stood about 190 million years ago in what is now a desert in eastern Arizona.

Many of the fossils that **paleontologists** (scientists who study fossils) find in their digs are not the actual remnants of organisms at all, but rocks that are replicas of the organisms. Such fossils result when a dead organism captured in sediment decays and leaves an empty mold that is filled by minerals dissolved in water. The casts that form when the minerals harden are replicas of the organism, as seen in the **375-million-year-old** casts of ammonites (shelled marine organisms) shown in **Figure 13.3C**. Trace fossils are footprints, burrows, and other remnants of an ancient organism's behavior. The boy in **Figure 13.3D** is standing in a 150-million-year-old dinosaur track in Colorado.

Some fossils actually retain organic material. The leaf in **Figure 13.3E** is about 40 million years old. It is a thin film pressed in rock, still greenish with remnants of its chlorophyll and well enough preserved that biologists can analyze its molecular and cellular structure. In rare instances, an entire organism, including its soft parts, is fos-

Figure 13.3 A gallery of fossils

A Skull of *Homo erectus*

B Petrified tree

C Ammonite casts

D Dinosaur tracks

E Fossilized organic matter of a leaf

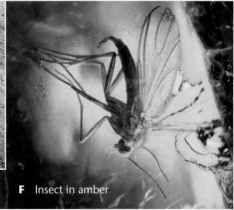

F Insect in amber

G "Ice Man"

Figure 13.3H Strata of sedimentary rock at the Grand Canyon

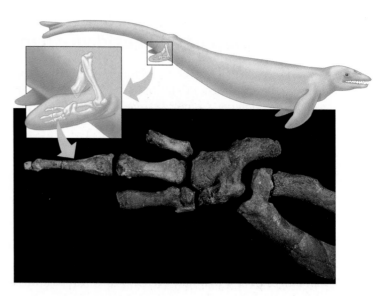

Figure 13.3I *Basilosaurus,* an extinct whale whose hind legs link living whales with their land-dwelling ancestors

silized. This can happen only if the individual is buried in a medium that prevents bacteria and fungi from decomposing the corpse. The insect in **Figure 13.3F** got stuck in the resin of a tree about 35 million years ago. The resin hardened into amber (fossilized resin), preserving the insect. Other media can preserve whole organisms. Explorers have discovered mammoths, bison, and even prehistoric humans frozen in ice or preserved in acid bogs. Such rare discoveries make the news, as did the 1991 discovery of the "Ice Man" in **Figure 13.3G**, who had died 5,000 years ago. However, biologists rely mainly on more common sedimentary fossils to reconstruct the history of life.

The **fossil record**—the ordered array in which fossils appear within layers of sedimentary rocks like those in **Figure 13.3H**—provides some of the strongest evidence of evolution. Sedimentary rocks form from layers of sand and mud that settle to the bottom of seas, lakes, and marshes. Over millions of years, deposits pile up and compress the older sediments below into rock. When aquatic organisms die, they settle along with the sediments and may be preserved as fossils. Many organisms living on land may also be swept into swamps and seas. Land organisms that remain in place when they die may first be covered by windblown silt and then buried in waterborne sediments when sea levels rise over them.

Changes in sea level and the drying and refilling of lakes and swamps affect sedimentation. The rate of sedimentation and the types of particles that settle also vary over time. As a result, the rock forms in **strata**, or layers. Younger strata are on top of older ones; thus, the relative ages of fossils can be determined by the layers in which they are found. Figure 13.3H shows strata of sedimentary rock at the Grand Canyon. The Colorado River has cut through over 2,000 m of rock, exposing sedimentary layers that can be read like huge pages from the book of life. Scan the canyon wall from rim to floor, and you look back through hundreds of millions of years. Each layer entombs fossils that represent some of the organisms from that period of Earth's history.

The fossil record testifies that organisms have evolved in a historical sequence. The oldest known fossils, dating from about 3.5 billion years ago, are prokaryotes. Molecular and cellular evidence also indicates that prokaryotes were the ancestors of all life. Fossils in younger layers of rock reveal the evolution of various groups of eukaryotic organisms. One example is the successive appearance of the different classes of vertebrates (animals with backbones). Fishlike fossils are the oldest vertebrates in the fossil record. Amphibians are next, followed by reptiles, then mammals and birds.

The evolutionary view of life predicts that we would find signs in the fossil record of linkages between ancient extinct organisms and species living today. Indeed, paleontologists have discovered many fossils that link past and present. For example, a series of fossils documents the changes in skull shape and size that occurred as mammals evolved from reptiles. Another series of fossils indicates that whales evolved from four-legged land mammals that lived some 55 million years ago. Whales living today have forelegs in the form of flippers but lack hind legs, although they do have small hind-leg and foot bones that do not extend from the body. Paleontologists digging in Egypt and Pakistan have identified fossils of extinct whales that actually had hind limbs. **Figure 13.3I** shows the fossilized leg bones of *Basilosaurus,* one of these ancient whales. The legs were about half a meter long and included bones similar to those of land mammals. These whales were already aquatic animals that no longer used their legs to support their weight. Other fossil whales have been found with larger or smaller limbs, and all are related to still older animals that spent only part of their time in water.

In the next module, we look at other sources of evidence of evolution.

? Why is amber a good source for recovering organic material from ancient organisms?

■ The fossilized resin prevents bacteria and fungi from decomposing the organic material.

13.4 A mass of other evidence reinforces the evolutionary view of life

As we discussed in Chapter 1, evolution is biology's core theme, and every aspect of life shows signs of evolutionary change. Let's now take a look at some of the evidence that reinforces the fossil record of evolution.

Biogeography It was the geographic distribution of species, known as **biogeography**, that first suggested to Darwin that organisms evolve from common ancestors. Darwin noted that Galápagos animals resembled species of the South American mainland more than they resembled animals on similar but distant islands. The logical explanation was that the Galápagos species evolved from South American immigrants. There are many other examples from biogeography that make sense only in the historical context of evolution. The unique collection of marsupials (pouched mammals) in Australia, such as kangaroos and koalas, diversified in isolation on that island continent after it became separated from the landmasses on which placental mammals diversified.

Comparative anatomy Also providing support for evolution and cited extensively by Darwin is **comparative anatomy**, the comparison of body structures in different species. Anatomical similarities between many species give signs of common descent. Similarity in characteristics that results from common ancestry is known as **homology**. As Figure 13.4A shows, the same skeletal elements make up the forelimbs of humans, cats, whales, and bats, all of which are mammals. The functions of these forelimbs differ. A whale's flipper does not do the same job as a bat's wing, so if these structures had been uniquely engineered, we would expect that their basic designs would be very different. However, their structural similarity would not be surprising if all mammals descended from a common ancestor with the same basic limb elements. The logical explanation is, in fact, that the arms, forelegs, flippers, and wings of different mammals are variations on a common anatomical plan that has become adapted to different functions. Biologists call such similarities in different organisms **homologous structures**—features that often have different functions but are structurally similar because of common ancestry.

Comparative anatomy testifies that evolution is a remodeling process in which ancestral structures that originally functioned in one capacity become modified as they take on new functions—the kind of process that Darwin called descent with modification. We see many signs that evolution remodels structures rather than creating them anew. For example, the human spine and knee joint were derived from ancestral structures that supported four-legged mammals. Almost none of us will reach old age without experiencing knee or back problems. If these structures had been designed specifically to support our bipedal posture, we would expect them to be less subject to painful ailments.

Some of the most interesting homologous structures are **vestigial organs**, structures of marginal, if any, importance to the organism. Vestigial organs are remnants of structures that served important functions in the organism's ancestors. The small hind-leg and foot bones of modern whales discussed in Module 13.3 are examples of vestigial organs. The skeletons of some snakes retain vestiges of the pelvis and leg bones of walking ancestors. Because limbs are a hindrance to a snake's way of life, natural selection favored snake ancestors with successively smaller limbs.

Comparative embryology The comparison of early stages of development, called **comparative embryology**, is another major source of evidence for the common descent of organisms. One sign that vertebrates evolved from a common ancestor is that all of them have an embryonic stage in which they have structures on the sides of the throat called pharyngeal (throat) pouches. At this stage, the embryos of fishes, frogs, snakes, birds, and mammals look relatively alike. They take on more and more distinctive features as development progresses. For example, pharyngeal pouches develop into gills in fishes, but into parts of the ears and throat in humans. Note the pharyngeal pouches and tails of both the bird embryo (left) and the human embryo (right) in **Figure 13.4B**.

Molecular biology Anatomical homology is not helpful in linking very distantly related organisms such as plants and animals and microorganisms. But in recent decades, advances in **molecular biology** have enabled biologists to read a molecu-

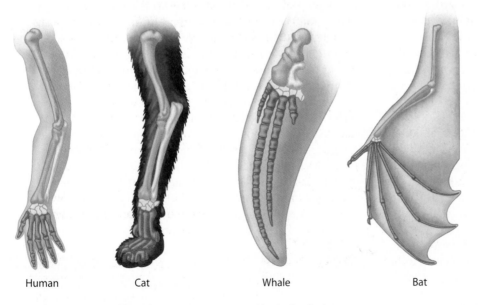

| Human | Cat | Whale | Bat |

Figure 13.4A Homologous structures: vertebrate forelimbs

lar history of evolution in the DNA sequences of organisms. As we saw in Chapter 10, the hereditary background of an organism is documented in its DNA and in the proteins encoded in the DNA. Siblings have greater similarity in their DNA and proteins than do unrelated individuals of the same species. And if two species have genes with sequences that match closely, biologists conclude that these sequences must have been inherited from a relatively recent common ancestor. In contrast, the greater the number of sequence differences between species, the less likely they share a close common ancestor. Molecular comparisons between diverse organisms have allowed biologists to develop hypotheses about the evolutionary divergence of major branches on the tree of life.

Studies of the amino acid sequences of similar (homologous) proteins in different species have been a rich source of data about evolutionary relationships. Table 13.4 compares the amino acid sequence of human hemoglobin, the protein that carries oxygen in the blood, with the sequences of the hemoglobin of five other vertebrate animals. You can see that 95% of the amino acids of the rhesus monkey polypeptide are identical to the human polypeptide, whereas only 14% of the lamprey amino acids are the same as those in human hemoglobin. These amino acid comparisons lead to a hypothesis about evolutionary relationships: Rhesus monkeys are much more closely related to humans than are lampreys; mice, chickens, and frogs fall in between. It turns out that this hypothesis agrees with earlier conclusions from comparative anatomy and embryology. Furthermore, the hypothesis agrees with fossil evidence. The lineage that led to humans diverged from the one leading to monkeys only about 25 million years ago, whereas lampreys (some of the most primitive living vertebrates) branched off the trunk of vertebrate evolution some 450 million years ago.

Another molecular indication of the relatedness of organisms comes from master control genes, which regulate groups of other genes during embryonic development. Research on the genetics of development has turned up very similar nucleotide sequences in these genes from organisms as different as fruit flies and mammals, as we saw when we examined the homeotic genes in Module 11.13. Biologists

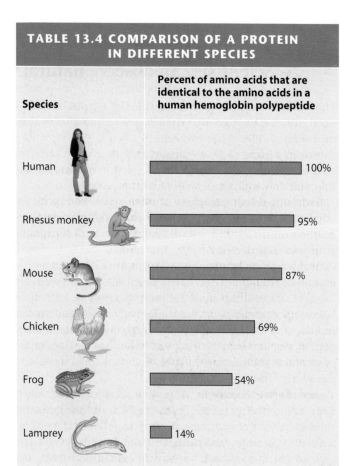

TABLE 13.4 COMPARISON OF A PROTEIN IN DIFFERENT SPECIES

Species	Percent of amino acids that are identical to the amino acids in a human hemoglobin polypeptide
Human	100%
Rhesus monkey	95%
Mouse	87%
Chicken	69%
Frog	54%
Lamprey	14%

predict that most, if not all, multicellular eukaryotes will be found to have similar genes regulating their early development. The logical conclusion is that these genes first arose in a common ancestor.

Darwin's boldest hypothesis is that *all* life-forms are related. About 100 years after Darwin made this claim, molecular biology has provided strong evidence for it: All forms of life use DNA and RNA, and the genetic code is essentially universal. This genetic language has been passed along through all the branches of evolution ever since its beginnings in an early form of life.

Having examined some of the evidence that supports evolution, we consider examples of natural selection in action in the next module.

Web/CD Activity 13C *Reconstructing Forelimbs*

? (a) What is homology? (b) How does the concept of homology relate to Table 13.4?

■ (a) Homology is similarity between different species due to their evolution from a common ancestor. (b) Proteins are gene products; thus, similarities in their amino acid sequences reflect the hereditary connection that is the basis of homology.

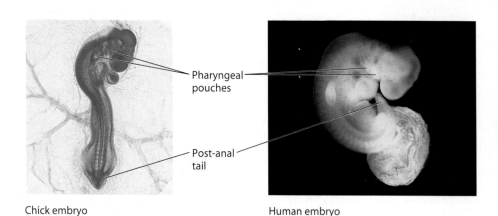

Pharyngeal pouches

Post-anal tail

Chick embryo   Human embryo

Figure 13.4B Homologous structures in vertebrate embryos

13.5 Scientists can observe natural selection in action

The blue-footed boobies described in the chapter introduction exhibit traits that are evolutionary adaptations to their ocean-based life. The exquisite camouflage adaptations shown in **Figure 13.5A** by insects that evolved in different environments are also examples of the results of natural selection. But do we have examples of natural selection in action?

Indeed, a great many cases of natural selection in nature have been documented. A classic example involves Peter and Rosemary Grant's work with finches in the Galápagos Islands over **a period of** 20 years (see Module 14.9). As part of their research, they measured changes in beak size in a population of ground finches. In dry years, when small seeds are in short supply, birds must eat more large seeds. Birds with larger, stronger beaks have a feeding advantage and greater reproductive success, and the average beak depth for the population increases. During wet years, smaller beaks are more efficient for eating the now abundant small seeds, and the average beak depth decreases.

An unsettling example of natural selection in action is the evolution of pesticide resistance in hundreds of insect species. Pesticides are poisons used to kill insect pests in farmlands, swamps, backyards, and homes. Whenever a new type of pesticide is used to control agricultural pests, the story is similar (**Figure 13.5B**): A relatively small amount of poison dusted onto a crop may kill 99% of the insects, but subsequent sprayings are less and less effective. The few survivors of the first pesticide wave are insects with genes that somehow enabled them to resist the chemical attack. So the poison kills most members of the population, leaving the resistant individuals to reproduce and pass the genes for pesticide resistance to their offspring. The proportion of pesticide-resistant individuals increases in each generation. Like the finches, the insect population has adapted to environmental change through natural selection.

These two examples of evolutionary adaptation highlight three key points about natural selection. First, natural selection is more of an editing process than a creative mechanism. A pesticide does not create resistant individuals, but selects for resistant insects that were already present in the population. Second, natural selection is contingent on time and place: It favors those characteristics in a varying population that fit the current, local environment. For instance, mutations that endow houseflies with resistance to the pesticide DDT also reduce their growth rate. Before DDT was introduced, such mutations were a handicap to the flies that had them. But once DDT was part of the environment, the mutant alleles were advantageous, and natural selection increased their frequency in fly populations. Finally, these examples show that significant evolutionary change can occur in a short time.

In the next module, we begin to examine the genetic basis for such evolutionary change.

? In what sense is natural selection more of an editing process than a creative process?

■ Natural selection cannot create beneficial traits on demand, but "edits" variation in a population by selecting for those traits that work best in the current environment.

Figure 13.5A Camouflage as an example of evolutionary adaptation

A flower mantid in Malaysia

A leaf mantid in Costa Rica

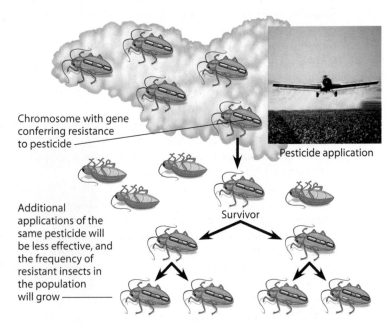

Chromosome with gene conferring resistance to pesticide

Pesticide application

Additional applications of the same pesticide will be less effective, and the frequency of resistant insects in the population will grow

Survivor

Figure 13.5B Evolution of pesticide resistance in an insect population

13.6 Populations are the units of evolution

A **population** is a group of individuals of the same species living in the same place at the same time. (For now, we can define a **species** as a group of populations whose members are capable of interbreeding and producing fertile offspring.) A population is the smallest unit that can evolve. We can, in fact, measure evolution as a change in the prevalence of certain heritable characteristics in a population over a span of generations. The increasing proportion of resistant insects in areas sprayed with pesticide is one example. Natural selection favored insects with genes for pesticide resistance; these insects left more offspring, and the population changed, or evolved. Note that the individual insects did not evolve. It is true that natural selection acts on individuals; their characteristics affect their chances of survival and their reproductive success within a local environment. But the evolutionary impact of natural selection is only apparent when tracking how a population changes over time.

Darwin understood that it is populations that evolve. He saw that natural selection, in favoring some heritable traits over others, changes populations over successive generations. What eluded him was the genetic basis of population change. Without this basis, he could not explain the cause of variation among the individuals making up a population, nor could he account for the perpetuation of parents' traits in their offspring. Today, we know that heritable characteristics are carried by genes on chromosomes and that mutations may produce new alleles. Also, we understand how Mendel's principles of segregation and independent assortment of alleles operate during meiosis to produce genetic variation in gametes and resulting offspring (see Module 9.18). Although Mendel and Darwin were contemporaries, the significance of Mendel's work was not recognized until after Darwin's death.

An important turning point for evolutionary theory was the development in the 1920s of **population genetics**, a field of science that combines Darwin's and Mendel's ideas by studying how populations change genetically over time. A comprehensive theory of evolution that incorporates population genetics and ideas from paleontology, taxonomy, and biogeography took form in the 1940s and continues to develop today. Known as the **modern synthesis**, this theory focuses on populations as the units of evolution as well as on the central role of natural selection.

Let's examine some key features of populations. Populations may be isolated from others of the same species, with little interbreeding and thus little exchange of genes between them. Such isolation is common for populations confined to widely separated islands or mountain ranges separated by lowlands. However, populations are not always isolated, nor do they necessarily have sharp boundaries. Figure 13.6 illustrates the tendency for humans to concentrate locally. This is a nighttime satellite photograph showing the lights of population centers in North America. We know that these popula-

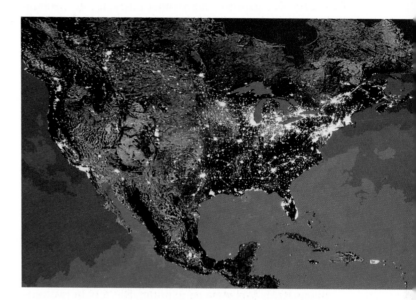

Figure 13.6 Human population centers in North America

tions are not really isolated; people move around, and there are suburban and rural communities between cities. Nevertheless, people are most likely to choose mates locally. As a result, for humans and other species, individuals in one population center are, on average, more closely related to one another than to members of other populations.

In studying evolution at the population level, geneticists focus on what is called the **gene pool**, the total collection of genes in a population at any one time. The gene pool consists of all alleles (alternative forms of genes) in all the individuals making up a population. For many gene loci, there are two or more alleles in the gene pool. For example, in an insect population, there may be two alleles relating to pesticide breakdown, one that codes for an enzyme that breaks down a certain pesticide and one for a version of the enzyme that does not. In fields sprayed with pesticide, the allele for the enzyme conferring resistance will increase in frequency and the other allele will decrease in frequency. When the relative frequencies of alleles in a population change like this over a number of generations, evolution is occurring on its smallest scale. Such a change in a gene pool is often called **microevolution**.

To understand how microevolution works, it helps to examine the genetics of a simple hypothetical population whose gene pool is *not* changing. Before you continue with the next module, you might wish to review Mendelian genetics in Chapter 9 (especially Modules 9.2 and 9.4).

? Why can't an individual evolve?

■ Evolution is only apparent as the changes in the genetic makeup of a population over time. An individual's genes don't change during its lifetime.

The gene pool of a nonevolving population remains constant over the generations

Let's consider an imaginary, nonevolving population of blue-footed boobies with individuals that differ in foot webbing (Figure 13.7A). Let's also imagine that foot webbing is controlled by a single gene. We'll assume that the allele for non-webbed feet (*W*) is completely dominant to the allele for webbed feet (*w*). The term *dominant* (see Module 9.3) may seem to suggest that over many generations of sexual reproduction, the *W* allele will somehow come to "dominate," becoming more and more common at the expense of the recessive *w* allele. In fact, this is not what happens. The shuffling of alleles that accompanies sexual reproduction does not alter the genetic makeup of the population. In other words, sexual reproduction alone does not lead to evolution. No matter how many times alleles are segregated into different gametes by meiosis and united in different combinations by fertilization, the frequency of each allele in the gene pool will remain constant unless acted on by other agents. This principle is known as **Hardy-Weinberg equilibrium**, named for the two scientists who derived it independently in 1908.

To test Hardy-Weinberg equilibrium, let's look at two generations of our booby population. **Figure 13.7B** shows the genetic makeup of the original population. We have a total of 500 boobies; of these, 320 birds have the genotype *WW* (nonwebbed feet), 160 have the heterozygous genotype, *Ww* (also nonwebbed feet, since *W* is dominant), and 20 have the genotype *ww* (webbed feet). The proportions of the three possible genotypes (the genotype frequencies) are 0.64 for *WW* ($\frac{320}{500}$ = 0.64), 0.32 for *Ww* ($\frac{160}{500}$ = 0.32), and 0.04 for *ww* ($\frac{20}{500}$ = 0.04).

From the genotype frequencies, we can calculate the frequency of each allele in this population. Each booby carries two alleles for foot type, so the population has 1,000 alleles for this characteristic. To find the number of *W* alleles, we add the number carried by the *WW* boobies, 2 × 320 = 640, to the number carried by the *Ww* boobies, 160. The total number of *W* alleles is thus 800. The frequency of the *W* allele, which we will call *p*, is $\frac{800}{1,000}$, or 0.8. We can calculate the frequency of the *w* allele in a similar way; this frequency, called *q*, is 0.2. (The letters *p* and *q* are often used to represent allele frequencies.)

What happens when the boobies of this parent population form gametes? At the end of meiosis, each gamete has one allele for foot type, either *W* or *w*. The frequency of the two alleles in the gametes will be the same as it is in the parental population, 0.8 for *W* and 0.2 for *w*.

Figure 13.7C (next page) shows a Punnett square that uses these gamete allele frequencies and the rule of multiplication (see Module 9.7) to calculate the frequencies of the three possible genotypes in the next generation. The probability of producing a *WW* individual (by combining two *W* alleles from the pool of gametes) is $p \times p = p^2$, or 0.8 × 0.8 = 0.64. Thus, the frequency of *WW* boobies in the next generation would be 0.64. Likewise, the frequency of *ww* individuals would be q^2 = 0.04. For heterozygous individuals, *Ww*, the genotype can form in two ways, depending on whether the sperm or egg supplies the dominant allele. In other words, the frequency of *Ww* would be $2pq = 2 \times 0.8 \times 0.2 = 0.32$. Thus, the three possible genotypes have the same frequency in the next generation as they did in the parent generation.

Finally, what about the frequencies of the alleles in this next generation? Since the genotype frequencies are the same as in the parent population, the allele frequencies, *p* and *q*, are also the same. In fact, we could follow the frequencies through many generations, and the results would continue to be the same. Thus, the gene pool of this population is in a state of equilibrium—Hardy-Weinberg equilibrium.

Now let's write a general formula for calculating the frequencies of genotypes in a gene pool from the frequencies of alleles. In our imaginary blue-footed booby population, the frequency of the *W* allele (*p*) is 0.8, and the frequency of the *w* allele (*q*) is 0.2. Note that *p* + *q* = 1; this is always true, since the combined frequencies of all alleles must be 100% of the alleles for that gene in the population. If there are only two alleles and we know the frequency of one, we can calculate the frequency of the other one:

$$p = 1 - q \quad \text{and} \quad q = 1 - p$$

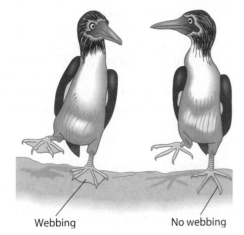

Figure 13.7A
Imaginary blue-footed boobies, with and without foot webbing

Webbing   No webbing

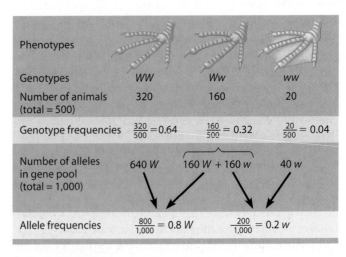

Phenotypes			
Genotypes	*WW*	*Ww*	*ww*
Number of animals (total = 500)	320	160	20
Genotype frequencies	$\frac{320}{500}$=0.64	$\frac{160}{500}$ = 0.32	$\frac{20}{500}$ = 0.04
Number of alleles in gene pool (total = 1,000)	640 *W*	160 *W* + 160 *w*	40 *w*
Allele frequencies	$\frac{800}{1,000}$ = 0.8 *W*		$\frac{200}{1,000}$ = 0.2 *w*

Figure 13.7B Gene pool of original population of boobies

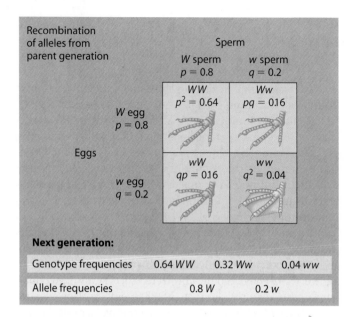

Recombination of alleles from parent generation	Sperm	
	W sperm $p = 0.8$	w sperm $q = 0.2$
W egg $p = 0.8$	WW $p^2 = 0.64$	Ww $pq = 0.16$
w egg $q = 0.2$	wW $qp = 0.16$	ww $q^2 = 0.04$

Next generation:

Genotype frequencies	0.64 WW	0.32 Ww	0.04 ww
Allele frequencies		0.8 W	0.2 w

Figure 13.7C Gene pool of next generation of boobies

Notice in Figures 13.7B and 13.7C that the frequencies of all possible genotypes in the populations also add up to 1 (that is, 0.64 + 0.32 + 0.04 = 1). We can represent these relationships symbolically with the Hardy-Weinberg equation:

$$p^2 \quad + \quad 2pq \quad + \quad q^2 \quad = \quad 1$$

Frequency of WW	Frequency of Ww	Frequency of ww

If a population is in Hardy-Weinberg equilibrium and its members continue to mate randomly generation after generation, allele and genotype frequencies will remain constant over time. Hardy-Weinberg equilibrium tells us that something other than the reshuffling processes of sexual repro-

duction is required to alter a gene pool—that is, to change allele frequencies in a population from one generation to the next. One way to find out what factors *can* change a gene pool is to identify the conditions that must be met if genetic equilibrium is to be maintained. For a population to be in Hardy-Weinberg equilibrium, it must satisfy five main conditions:

1. The population is very large.
2. There is no gene flow; that is, there is no migration of individuals or gametes into or out of the population.
3. Mutations (changes in genes) do not alter the gene pool.
4. Mating is random.
5. All individuals are equal in reproductive success; that is, natural selection does not occur.

These five conditions are rarely met, and thus, we do not really expect a natural population to be in Hardy-Weinberg equilibrium. In many populations, however, the rate of evolutionary change is so slow that the population appears close to equilibrium, allowing researchers to estimate allele and genotype frequencies. Hardy-Weinberg equilibrium also provides a benchmark for studying populations in which gene pools are changing.

Web/CD Thinking as a Scientist *How Can Frequency of Alleles Be Calculated?*

? In a different imaginary booby population, the frequency of the recessive allele for webbed feet is 0.4. What is the frequency of individuals that have nonwebbed feet?

■ 0.84 (If $q = 0.4$, $p = 0.6$; 0.36 for WW and 0.48 for Ww), assuming the population is in Hardy-Weinberg equilibrium

---

CONNECTION

13.8 The Hardy-Weinberg equation is useful in public health science

The Hardy-Weinberg equation has broad application. For instance, public health scientists use it to estimate how many people carry alleles for certain inherited diseases. Consider the case of phenylketonuria (PKU), which is an inherited inability to break down the amino acid phenylalanine. PKU occurs in about one out of 10,000 babies born in the United States and, if untreated, results in severe mental retardation. Newborn babies are now routinely tested for PKU, and symptoms can be prevented or lessened by following a strict diet. Packaged foods with ingredients such as aspartame, a common artificial sweetener that contains phenylalanine, must list warnings for people with PKU.

PKU is due to a recessive allele, so the frequency of individuals in the U.S. population born with PKU corresponds to the q^2 term in the Hardy-Weinberg equation. Given one PKU occurrence per 10,000 births, $q^2 = 0.0001$. Therefore,

the frequency of the recessive allele for PKU in the population, q, equals the square root of 0.0001, or 0.01. And the frequency of the dominant allele, p, equals $1 - q$, or 0.99. The frequency of carriers, heterozygous people who do not have PKU but may pass the PKU allele on to offspring, is $2pq$, which equals $2 \times 0.99 \times 0.01$, or 0.0198. Thus, the equation tells us that about 2% (actually 1.98%) of the U.S. population carries the PKU allele. Estimating the frequency of a harmful allele is part of any public health program dealing with genetic diseases.

? Which term in the Hardy-Weinberg equation—p^2, $2pq$, or q^2—corresponds to the frequency of individuals who have no alleles for the disease PKU?

■ p^2

13.9 In addition to natural selection, genetic drift and gene flow can contribute to evolution

Deviations from the conditions for Hardy-Weinberg equilibrium can cause changes in gene pools. The three main causes of such evolutionary change are genetic drift, gene flow, and natural selection. Of the three, only natural selection leads to adaptive evolution. Although new genes and new alleles originate by mutation, these random and rare events probably change allele frequencies little from one generation to the next.

Genetic drift is a change in the gene pool of a population due to chance. The smaller the population, the more impact genetic drift is likely to have. Flip a coin a thousand times, and a result of 700 heads and 300 tails would make you very suspicious about that coin. But flip a coin ten times, and an outcome of seven heads and three tails would seem within reason. The smaller the sample, the greater the chance of deviation from an idealized result—in this case, an equal number of heads and tails. Let's apply this logic to a population's gene pool. The frequencies of alleles in a gene pool will be more stable from one generation to the next when a population is large. If a population of organisms is small, its gene pool may not be accurately represented in the next generation. In fact, its allele frequencies may vary erratically from generation to generation. In many cases, an allele may be lost from a population by such chance fluctuations. Over time, genetic drift tends to reduce genetic variation through such losses.

Two situations that can shrink populations down to a small size—small enough for genetic drift to have a large impact on allele frequencies—are the bottleneck effect and the founder effect.

The **bottleneck effect** is an event that drastically reduces population size. Earthquakes, floods, or fires may kill large numbers of individuals, producing a small surviving population that is unlikely to have the same genetic makeup as the original population. The analogy of shaking just a few marbles through a bottleneck illustrates how a bottlenecking event works (**Figure 13.9A**). Certain alleles (purple marbles) may be present at higher frequency in the surviving population than in the original population, others (green marbles) may be present at lower frequency, and some (orange marbles) may not be present at all. Genetic drift may continue to change the gene pool for many generations until the population is again large enough for fluctuations due to chance to have less of an impact.

In a real example, human hunters in the 1890s reduced the population of northern elephant seals in California to about 20 individuals. Since then, this mammal (pictured in **Figure 13.9B**) has become a protected species, and the population has grown back to over 30,000 members. However, in examining 24 gene loci in a representative sample of the seals, researchers found *no* variation. For each of the 24 genes, they found only one allele, probably because of bottlenecking. In contrast, genetic variation abounds in populations of a closely related species, the southern elephant seal, which was not bottlenecked.

A second situation that can produce a small population is the colonization of a new location by a small number of individuals. The smaller the group, the less likely the genetic makeup of the colonists will represent the gene pool of the larger population they left. If the colony is successful, random changes in allele frequencies will continue until the population grows large enough for genetic drift to be minimal. Such change in the gene pool of a small colony is called the **founder effect.** The founder effect undoubtedly contributed to the evolutionary divergence of the finches and other South American organisms that arrived as strays on the remote Galápagos Islands.

The founder effect explains the relatively high frequency of certain inherited disorders among some human populations established by small numbers of colonists. In 1814, 15 people founded a British colony on Tristan da Cunha, a group of small islands in the middle of the Atlantic Ocean. Apparently, one of the colonists carried a recessive allele for retinitis pigmentosa, a progressive form of blindness. Of the 240 descendants who still lived on the islands in the 1960s, four had retinitis pigmentosa, and at least nine others were known to be heterozygous carriers of the allele. The frequency of this allele was much higher in this population than in the population from which the founders came.

Another source of evolutionary change is **gene flow.** A population may gain or lose alleles when fertile individuals move into or out of a population or when gametes (such as plant pollen) are transferred between populations. Gene flow tends to reduce differences between populations. For example, humans today move more freely about the world than in the past, and gene flow has become an important agent of evolutionary change in previously isolated human populations.

In natural selection, individuals with characteristics that best adapt them to their environment have the most repro-

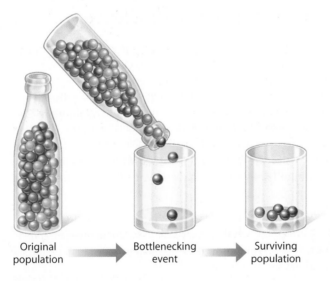

Original population → Bottlenecking event → Surviving population

Figure 13.9A The bottleneck effect

ductive success. The condition for Hardy-Weinberg equilibrium that there be no natural selection—that all individuals in a population be equal in their ability to reproduce—is probably never met in nature. Populations of sexually reproducing organisms consist of varied individuals, and some variants leave more offspring than others. In our imaginary blue-footed booby population, birds with webbed feet (genotype *ww*) might produce more offspring because they are more efficient at swimming and catching food than birds without webbed feet (genotype *Ww* or *WW*). Genetic equilibrium would be disturbed as the frequency of the *w* allele increased in the gene pool. Natural selection results in the accumulation and maintenance of traits that adapt a population to its environment. If the environment should change, natural selection would favor traits adapted to the new conditions. The degree of adaptation that can occur is limited by the amount and kind of genetic variation in the population—a potential problem for endangered species, as we see next.

Web/CD Activity 13D *Causes of Microevolution*

Figure 13.9B Elephant seals descended from bottleneck survivors

? List three causes of microevolution. Which one will adapt a population to its environment?

■ Genetic drift, gene flow, and natural selection. Natural selection is adaptive.

CONNECTION

13.10 Endangered species often have reduced variation

The loss of genetic variation as a result of bottlenecking events may reduce the ability of a population to adapt to environmental change. This issue has direct bearing on one of our most pressing problems, the unprecedented rate of the worldwide loss of species. Endangered species typically have low genetic variability. As their populations are severely reduced, the diversity of their gene pools also declines. One such species is the cheetah (*Acinonyx jubatus*), pictured in Figure 13.10. The fastest of all running animals, the cheetah can run down the swiftest antelope. These magnificent cats were formerly widespread in Africa and Asia. Their numbers fell drastically during the last ice age some 10,000 years ago. At that time, the species may have suffered a severe bottleneck, possibly as a result of disease, human hunting, and periodic droughts. Some researchers think that the South African cheetah population suffered a second bottleneck during the 19th century, when South African farmers hunted the animals to near extinction. Today, only a few small cheetah populations exist in the wild.

Studies of the African cheetah populations show very low genetic variation. In fact, their genetic variability is far less than average for mammals and is even less than that of some highly inbred varieties of laboratory mice. This lack of variability, coupled with an increasing loss of habitat, makes the cheetah's future precarious. The cheetahs remaining in Africa are being crowded into nature reserves and parks as human demands on the land increase. Along with crowding comes greater opportunity for predation by other large cats, such as lions, and increased potential for the spread of dis-

Figure 13.10 The cheetah, a species with low genetic variability

ease. With so little variability, the cheetah may have a reduced capacity to adapt to such environmental challenges. Captive breeding programs are already under way and may be required for the cheetah's long-term survival.

We look more closely at genetic variation and natural selection in the next modules.

? Why might new strains of pathogens pose a greater threat to cheetah populations than to mammalian populations having more genetic variation?

■ Because cheetah populations have so little genetic variation, there is potential for some new disease against which no individuals are resistant.

13.11 Variation is extensive in most populations

We have no trouble recognizing our friends in a crowd. Each person has a unique genome, reflected in individual variations in appearance, voice, and temperament. We are very conscious of human diversity, but individuality in populations of other animals and plants may escape our notice. Nonetheless, individual variation occurs in populations of all species that reproduce sexually. In addition to anatomical differences, most populations have a great deal of variation that can only be observed at the molecular level.

Not all variation in a population is heritable. The phenotype results from a combination of the genotype, which is inherited, and many environmental influences. For instance, a strength-training program can build up your muscle mass, but you would not pass this environmentally produced physique on to your offspring. Only the genetic component of variation is relevant to natural selection.

Many of the variable traits of the individuals in a population result from the combined effect of several genes. As we saw in Module 9.15, polygenic inheritance produces traits that vary more or less continuously—in human height, for instance, from very short individuals to very tall ones. By contrast, other features, such as human ABO blood groups (see Module 9.13), are determined by a single gene locus, with different alleles producing only distinct phenotypes.

When individuals differ in a discrete phenotypic character, the different forms are called *morphs*. A population is said to be **polymorphic** for a characteristic if two or more morphs are present in noticeable numbers. Figure 13.11 illustrates a striking example of polymorphism. These four garter snakes belong to the same species and were all captured in one Oregon field. Interestingly, the behavior of each different morph seems to correlate with its coloration. For example, when approached, the spotted snakes, which blend with their background, generally freeze. In contrast, the snakes with stripes, which make it difficult to judge the speed of motion, usually flee rapidly when approached.

In addition to variation within populations, most species exhibit geographic variation between populations. Because the environment is likely to differ from one place to another, natural selection will contribute to this geographic variation. Sometimes this variation occurs in what is called a **cline**, a graded change in an inherited characteristic along a geographic continuum. For example, the body size of many

Figure 13.11 Polymorphism in a population of garter snakes

birds and mammals tends to increase with increasing latitude in North America. Presumably, large size is adaptive in colder latitudes because it reduces the ratio of body surface area to volume and helps conserve body heat.

How is genetic variation measured? Population geneticists look at variation both at the level of whole genes (gene diversity) and at the molecular level of DNA (nucleotide diversity). Gene diversity is the average percent of gene loci that are heterozygous in a population. For example, the average heterozygosity of fruit flies (*Drosophila*) is 14%, meaning that a typical fruit fly is heterozygous at about 14% of its gene loci and homozygous at the rest. Nucleotide diversity is determined by comparing the nucleotide sequences of DNA samples. Interestingly, humans have less genetic variation than most species. Our average nucleotide diversity is only about 0.1%; you and your neighbor have the same nucleotide at 999 out of every 1,000 nucleotide sites in your DNA. We are far more genetically alike than we are different, but there is still enough variation to account for the genetic component of the enormous individuality we observe in people.

How do inherited variations arise? We address this question in the next module.

? Which of the following variations in a human population is the best example of polymorphism: height; blood group (A, B, AB, or O); number of fingers; math proficiency?

■ Blood group

13.12 Mutation and sexual recombination generate variation

Mutation is the ultimate source of the genetic variation that serves as raw material for evolution. As we saw in Module 10.16, **mutations**, or changes in the nucleotide sequence of DNA, can create new alleles. Most mutations occur in

body cells, however, and are lost when the individual dies. Only mutations in cells that produce gametes can be passed to offspring and potentially affect a population's gene pool.

A mutation that substitutes one nucleotide for another will be harmless if it does not affect the function of the protein the DNA encodes. However, if it does affect the protein's function, the mutation will probably be harmful. An organism is a refined product of thousands of generations of past selection, and a random change in its DNA is not likely to improve its genome any more than shooting a bullet through the hood of a car is likely to improve the engine performance.

On rare occasions, however, a mutant allele may actually improve the adaptation of its bearer to the environment and enhance reproductive success. This kind of effect is more likely when the environment is changing in such a way that mutations that were once disadvantageous are favorable under the new conditions. The evolution of DDT-resistant houseflies (see Module 13.5) illustrates this point. In another example, mutations that make the HIV virus resistant to antiviral drugs also slow the reproductive rate of the virus (see Module 10.21). However, once antiviral drugs are used to treat HIV-infected individuals, these mutant alleles are advantageous because they allow the virus to survive, and natural selection increases their frequency in the HIV population.

Chromosomal mutations that delete, disrupt, or rearrange many gene loci at once are almost certain to be harmful (see Figures 8.23A and 8.23B). Very rarely, a chromosomal rearrangement may be beneficial if it links together genes that produce some positive effect. And duplication of part of a chromosome can be an important source of new genetic variation. If a repeated segment of a chromosome can persist over the generations, it may provide a bigger genome with extra genes that may eventually take on new functions by mutation. For example, our remote ancestors carried a single gene for detecting odors. This gene has been duplicated repeatedly through mutation, resulting in hundreds of olfactory receptor genes that allow humans to detect a wide variety of odors.

In microorganisms with very short generation spans, mutation generates genetic variation very rapidly. Bacteria multiply so rapidly that a beneficial mutation can increase its frequency in descendant populations in a matter of hours or days. Because bacteria are generally haploid, with only a single gene for each inherited characteristic, a newly created allele can have an effect immediately. Its expression is not obscured by another allele for the same characteristic.

Mutation rates tend to be low in animals and plants, averaging about one in every 100,000 genes per generation. For most animals and plants, low mutation rates, long generation times, and diploid genomes prevent most mutations from significantly affecting genetic variation from one generation to the next. Consequently, animals and plants depend mainly on sexual recombination for the genetic variation that makes adaptation possible. (Of course, the origin of this allele variation is past mutations.) As we saw in Modules 8.14 and 8.15, fresh assortments of existing alleles arise every generation from three random components of sexual recombination: crossing over, independent assortment of homologous chromosomes, and random fertilization. During meiosis, each pair of homologous chromosomes, one inherited from

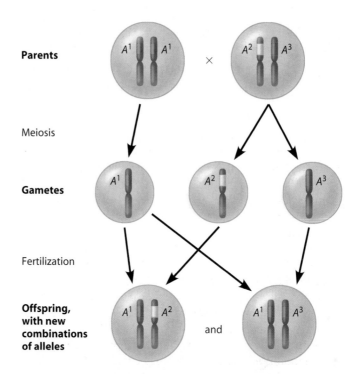

Figure 13.12 Shuffling alleles by sexual recombination

each parent, trade some of their genes by crossing over, and then the chromosomes separate into gametes independently of other chromosome pairs. Gametes from one individual vary extensively in their genetic makeup, and each zygote made by a mating pair has a unique assortment of alleles resulting from the random union of a sperm and an ovum. In Figure 13.12, we see the results of a mating of parents with the genotypes A^1A^1 and A^2A^3, where A^1, A^2, and A^3 are three different alleles for a gene. Even without considering independent assortment or crossing over, the offspring have different combinations of alleles than were present in their parents.

Bacteria and many viruses can also undergo recombination, but they do so less regularly than animals and plants and often in ways that allow them to cross species barriers (see Module 14.3). In the strain of *E. coli* bacterium that is the frequent culprit in food poisoning cases, many genes are actually "mosaics" of genes from different bacteria. The ability of pathogens to evolve rapidly through such recombination, together with their high mutation rates, makes them especially dangerous adversaries. We explore the problem of antibiotic resistance in bacteria in the next module.

Web/CD Activity 13E *Genetic Variation from Sexual Recombination*

? _____ is the ultimate source of genetic variation in a population, but in a sexual population with a relatively long generation span, most of the variation we observe is due to _____.

■ Mutation . . . sexual recombination

13.13 The evolution of antibiotic resistance in bacteria is a serious public health concern

Antibiotics are drugs that disable or kill infectious microorganisms. Most antibiotics are naturally occurring chemicals derived from other microorganisms. Penicillin, for example, was originally isolated from a mold and has been widely prescribed since the 1940s. A revolution in human health rapidly followed its introduction, rendering many previously fatal diseases easily curable (such as strep throat and surgical infections). During the 1950s, some doctors even predicted the end of human infectious disease.

Why hasn't this optimistic prediction come true? It did not take into account the force of evolution. In the same way that pesticides select for resistant insects (see Module 13.5), antibiotics select for resistant bacteria. The genes that confer such antibiotic resistance are often carried on R plasmids (see Module 10.23), which are passed on to bacterial offspring and may even be transferred to other bacteria. For nearly every antibiotic that has been developed, a resistant strain of bacterium has appeared within a few decades. For example, some strains of the tuberculosis-causing bacterium are now resistant to all three of the antibiotics commonly used to treat the disease. **Figure 13.13** shows an X-ray of a TB patient.

In what ways do we contribute to the problem of antibiotic resistance? Livestock producers add antibiotics to animal feed as a growth promoter. As a result, much of the packaged meat for sale in supermarkets contains bacteria that are resistant to standard antibiotics. Doctors contribute to the problem by overprescribing antibiotics—for example, to patients with viral infections, which do not respond to antibiotic treatment. And patients contribute to this problem through the misuse of prescribed antibiotics—for example, by prematurely stopping the medication because they feel better. This allows mutant bacteria that may be killed more slowly by the drug to survive and multiply. Subsequent mutations in such bacteria may lead to full-blown antibiotic resistance. During the anthrax crisis of 2001, public health officials urged panicked citizens to avoid unnecessarily taking ciprofloxacin (Cipro), the

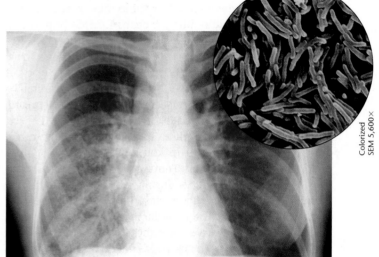

Colorized
SEM 5,600×

Figure 13.13 Colorized X-ray of the lungs of a tuberculosis patient (infection highlighted in red), with inset showing *Mycobacterium tuberculosis*, the infecting agent

drug used to treat the deadliest form of anthrax infection, because doing so could select for resistant bacteria.

Difficulty in treating common human infections is a serious public health concern. Penicillin was effective against nearly all bacterial infections in the 1940s but is virtually useless today in its original form. New drugs have since been developed, but they continue to be rendered ineffective as resistant bacteria evolve. The medical community and pharmaceutical companies are engaged in an ongoing race against the powerful force of bacterial evolution.

Web/CD Thinking as a Scientist *Connection: What Are the Patterns of Antibiotic Resistance?*

 Explain why the following statement is incorrect: "Antibiotics have created resistant bacteria."

■ The use of antibiotics has increased the frequency of alleles for resistance that were already naturally present in bacterial populations.

13.14 Diploidy and balancing selection preserve variation

Natural selection acting on the variations within a population adapts the population to its environment. But what prevents natural selection from eliminating this variation as it selects against unfavorable genotypes? Why aren't the less adaptive alleles eliminated as the best alleles are passed on to the next generation? The tendency for natural selection to reduce variation in a population is countered by mechanisms that maintain variation.

Most eukaryotes are diploid, and having two sets of chromosomes helps to prevent populations from becoming genetically uniform. The effects of recessive alleles are not

often displayed in diploid organisms. A recessive allele is subject to natural selection only when it influences the phenotype, and this occurs only when two copies of it appear in a homozygous individual. In a heterozygote, a recessive allele is, in effect, hidden, or protected, from natural selection. This hiding of recessive alleles in the presence of dominant ones can allow a large number of recessive alleles to remain in a gene pool, and these alleles may prove advantageous in later generations. Individuals with a trait resulting from two copies of a recessive allele might be eliminated by natural selection in one environment, but if the environ-

ment changes, such individuals might have greater reproductive success.

Genetic variability may also be preserved by natural selection, the very same force that generally reduces it. **Balancing selection** occurs when natural selection maintains stable frequencies of two or more phenotypic forms in a population. This balanced polymorphism may be the result of a **heterozygote advantage.** When heterozygous individuals have greater reproductive success than homozygotes, two or more alleles for a characteristic will be maintained by natural selection. One such example is the protection from malaria conferred by the sickle-cell allele (see Module 9.14). The frequency of the sickle-cell allele in Africa is generally highest in areas where malaria is a major cause of death. The environment in these areas favors heterozygotes, who are protected from the most severe effects of malaria. Less favored are individuals homozygous for the normal hemoglobin allele, who are susceptible to malaria, and individuals homozygous for the sickle-cell allele, who develop sickle-cell disease.

A second type of balancing selection that promotes balanced polymorphism is **frequency-dependent selection**, in which the survival and reproduction of any one morph declines if that phenotypic form becomes too common in the population. Perhaps a species of butterfly has several different coloration patterns. Birds may develop a "search image" that enables them to more easily locate and feed on any one morph that becomes too common. The frequency of the other color patterns would then increase.

Some genetic variations in populations seem to have little or no impact on reproductive success. The diversity of human fingerprints (**Figure 13.14**), for example, seems to be

Figure 13.14 Human fingerprints, probably an example of neutral variation

an example of **neutral variation**—genetic variation that provides no apparent selective advantage for some individuals over others. Some of these supposedly neutral alleles will increase their frequency in the gene pool and others will decrease by the chance effects of genetic drift, but natural selection will not affect them.

It is possible, of course, that variations appearing to be neutral may influence reproductive success in ways that are difficult to measure. Also, a variation may be neutral in one environment but not in another. Biologists continue to debate the extent of neutral variation. But we can be certain that even if only a fraction of the extensive variation in a gene pool significantly affects the organism, that is still an enormous resource of raw material for natural selection and the adaptive evolution it causes.

? Why would natural selection tend to reduce genetic variation more in populations of haploid organisms than in populations of diploid organisms?

■ All alleles in a haploid organism are phenotypically expressed and are hence screened by natural selection.

13.15 The perpetuation of genes defines evolutionary fitness

The phrases "struggle for existence" and "survival of the fittest" are misleading if we take them to mean direct competitive contests between individuals. There *are* animal species in which individuals lock horns or otherwise do combat to determine mating privilege. But reproductive success is generally more subtle and passive. In a varying population of moths, certain individuals may survive and produce more offspring than others because their wing colors hide them from predators better. Plants in a wildflower population may differ in reproductive success because some attract more pollinators, owing to slight variations in flower color, shape, or fragrance. A frog may produce more eggs than her neighbors because she is better at catching insects for food. These examples point to a biological definition of **fitness**, the contribution an individual makes to the gene pool of the next generation relative to the contributions of other individuals. Thus, the fittest individuals in the context of evolution are those that produce the largest number of viable, fertile offspring and thus pass on the most genes to the next generation.

Survival alone does not guarantee reproductive success. The biggest, fastest, toughest frog in the pond has a fitness of

zero if it is sterile. Production of fertile offspring is the only score that counts in natural selection.

Evolutionary fitness has to do with genes, but it is the phenotype of an organism—its physical traits, metabolism, and behavior—that is directly exposed to the environment. Acting on phenotypes, selection indirectly adapts a population to its environment by increasing or maintaining favorable genotypes in the gene pool. The fitness of any one allele, however, depends on the entire genetic context in which it works. For example, alleles that enhance the growth of the trunk of a tree may be useless or even detrimental in the absence of alleles at other loci that enhance the growth of roots required to support the tree. On the other hand, alleles that contribute nothing to an organism's success or are even maladaptive may be perpetuated because they are present in individuals whose overall fitness is high. The whole baseball team wins the World Series, including the players with the worst batting average or the most errors.

? What determines an organism's fitness?

■ The number of fertile offspring it leaves and thus its contribution to the gene pool of the next generation

Natural selection can alter variation in a population in three ways

Let's see how the culling effects of natural selection can affect the distribution of phenotypes using an imaginary deer mouse population. The bell-shaped curve in the top graph of **Figure 13.16** depicts the frequencies of individuals that could result from a polygenic inheritance pattern for variation in fur color. In this initial population, fur color varies along a continuum from very light (only a few individuals) through various intermediate shades (many individuals) to very dark (few individuals). The other three graphs show three different ways in which natural selection could alter the phenotypic variation in the idealized population. The blue downward arrows symbolize the pressure of natural selection working against certain phenotypes.

Stabilizing selection favors intermediate variants. It typically occurs in relatively stable environments, where conditions tend to reduce phenotypic variation. In the mouse population depicted in the graph on the bottom left, stabilizing selection has eliminated the extremely light and dark individuals, and the population has a greater number of intermediate phenotypes, which are best suited to a stable environment. Stabilizing selection probably prevails most of the time in most populations. For example, this type of selection keeps the majority of human birth weights in the range of 3–4 kg (6.5–9 pounds). For babies a lot smaller or larger than this size, infant mortality may be greater.

Directional selection shifts the overall makeup of the population by acting against individuals at *one* of the phenotypic extremes. For the mouse population in the bottom center graph, the trend is toward darker fur color, as might occur if the landscape has become shaded by the growth of trees. Directional selection is most common during periods of environmental change or when members of a species migrate to some new habitat with different environmental conditions. The changes we described in populations of insects exposed to insecticides and HIV exposed to antiviral drugs are examples of directional selection.

Disruptive selection typically occurs when environmental conditions are varied in a way that favors individuals at *both* extremes of a phenotypic range. For the mice in the graph on the bottom right, individuals with light and dark fur have increased their numbers relative to intermediate variants. Perhaps the mice colonized a patchy habitat where a background of light soil was studded with areas of dark rocks. Disruptive selection can lead to balanced polymorphism with two or more contrasting morphs, such as in the population of garter snakes in Figure 13.11.

Next we consider a special case of selection, one that leads to phenotypic differences between males and females.

? Of the three modes of natural selection, which is most common?

Stabilizing selection ■

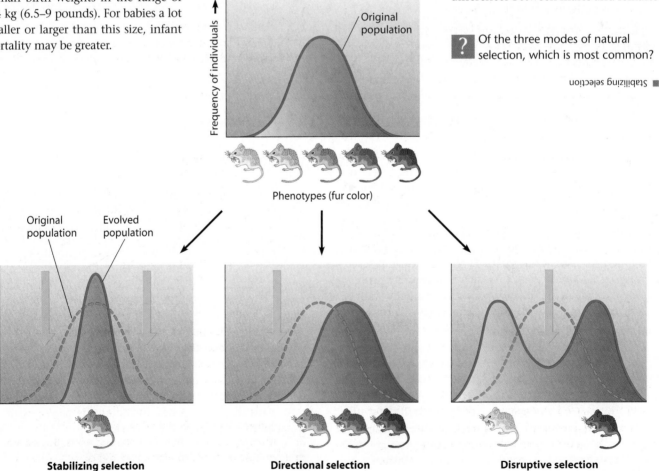

Figure 13.16 Three possible effects of natural selection on a phenotypic character

13.17 Sexual selection may produce sexual dimorphism

The males and females of an animal species obviously have different reproductive organs. But they may also have other noticeable differences, not directly associated with reproduction, called secondary sexual characteristics. This distinction in appearance is called **sexual dimorphism.** It is often manifested in a size difference, but can also be evident in the form of male adornment, such as colorful plumage in birds, manes on lions, or antlers on deer. Males are usually the showier sex, at least among vertebrates.

Darwin considered sexual selection, the determining of who mates with whom, to be a separate selection process that produces sexual dimorphism. In some species, secondary sex structures may be used to compete with members of the same sex for mates (Figure 13.17A). Contests may involve physical combat, but are more often ritualized displays (see Chapter 35). This so-called intrasexual selection (within the same sex) is common in species where the winning male garners a harem of females.

In a more common type of sexual selection, called intersexual selection or mate choice, individuals of one sex (usually females) are choosy in selecting their mates. Males with the largest or most colorful adornments are often the most attractive to females. The extraordinary feathers of a peacock's tail are an example of this sort of "choose me" statement (Figure 13.17B). What intrigued Darwin is that some of these mate-attracting features do not seem to be otherwise adaptive and may in fact pose some risks. For example, showy plumage may make male birds more visible to predators. But if such secondary sexual characteristics help a male gain a mate, then they will be reinforced over the generations for the most Darwinian

Figure 13.17A
A contest for access to mates

of reasons—because they enhance reproductive success. Every time a female chooses a mate based on a certain appearance or behavior, she perpetuates the alleles that caused her to make that choice and allows a male with a particular phenotype to perpetuate his alleles.

Figure 13.17B A peacock's advertisement for mates

? Males with the most elaborate ornamentation may garner the most mates. How might such a mate choice be advantageous to a female?

■ An elaborate display may signal good health and therefore good genes that would be provided to the female's offspring.

13.18 Natural selection cannot fashion perfect organisms

There are at least four reasons why natural selection cannot produce perfection:

1. *Organisms are limited by historical constraints.* Each species has a legacy of descent with modification from a long line of ancestral forms. Evolution does not scrap ancestral anatomy and build each new complex structure from scratch, but co-opts existing structures and adapts them to new situations. The wings of birds are fashioned from bones that supported the walking legs of their reptilian ancestors.

2. *Adaptations are often compromises.* Each organism must do many different things. A blue-footed booby uses its webbed feet to swim as it dives into the ocean for prey, but these same feet make for clumsy travel on land.

3. *Chance and natural selection interact.* Chance probably affects the genetic structure of populations to a greater extent than was once believed. For instance, when a storm blows insects hundreds of miles over an ocean to an island, the wind does not necessarily transport the

specimens that are best suited to the new environment. And not all alleles fixed by genetic drift in the gene pool of a small population are better suited to the environment than alleles that are lost.

4. *Selection can only edit existing variations.* Natural selection favors only the fittest variations from the phenotypes that are available, which may not be the ideal traits. New alleles do not arise on demand.

With all these constraints, we cannot expect evolution to craft perfect organisms. Natural selection operates on a "better than" basis. We can see evidence for evolution in the subtle imperfections of the organisms it produces.

? Humans owe much of their physical versatility and athleticism to their flexible limbs and joints. But we are prone to sprains, torn ligaments, and dislocations. Why?

■ Adaptations are compromises: Structural reinforcement has been compromised for agility.

Reviewing the Concepts

Darwin's Theory of Evolution (Introduction–13.5)

Evolutionary adaptations are inherited characteristics that enhance an organism's ability to survive and reproduce in a particular environment **(Introduction)**.

Darwin's context. Aristotle and the Judeo-Christian culture held that species are fixed. Fossils suggested that life-forms change; geologists proposed that a very old Earth is changed by gradual processes. Darwin's experiences during the voyage of the *Beagle* helped frame his ideas **(13.1)**.

Descent with modification. Darwin proposed that living species are descended from earlier life-forms and that natural selection is the mechanism of evolution **(13.2)**.

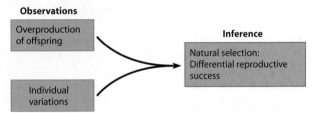

Evidence of evolution. The fossil record reveals that organisms have evolved in a historical sequence **(13.3)**. Further evidence for evolution comes from biogeography, comparative anatomy, comparative embryology, and molecular biology **(13.4)**. Evolutionary adaptations resulting from natural selection have been observed in populations of birds, insects, and many other organisms **(13.5)**.

Population Genetics and the Modern Synthesis (13.6–13.10)

Populations evolve. The modern synthesis connects Darwin's theory with population genetics. Microevolution is a change in the relative frequencies of alleles in a population's gene pool **(13.6)**.

Hardy-Weinberg equilibrium. The shuffling of alleles during sexual reproduction does not alter their proportions in a gene pool. The Hardy-Weinberg equation calculates allele and genotype frequencies in a population **(13.7)**.

Allele frequencies	$p + q = 1$
Genotype frequencies	$p^2 + 2pq + q^2 = 1$

Dominant homozygotes · Heterozygotes · Recessive homozygotes

Public health scientists use the Hardy-Weinberg equation to estimate frequencies of disease-causing alleles **(13.8)**.

Causes of evolution. Allele frequencies may change because of genetic drift (change in a gene pool due to chance), gene flow (movement of individuals or gametes between populations), and natural selection (differential reproductive success). The bottleneck effect and founder effect lead to genetic drift **(13.9)**. Low genetic variability may reduce the capacity of endangered species to survive **(13.10)**.

Variation and Natural Selection (13.11–13.18)

Variation in populations. Many populations exhibit polymorphism—different forms of phenotypic characteristics—and geographic variation **(13.11)**. Genetic variation is generated by mutation and by sexual recombination **(13.12)**. The excessive use of antibiotics is leading to the evolution of antibiotic resistance in bacteria **(13.13)**. Diploidy preserves variation by "hiding" recessive alleles. Balanced polymorphism may result from heterozygote advantage or from frequency-dependent selection **(13.14)**.

Fitness is the relative contribution an individual makes to the gene pool of the next generation **(13.15)**.

Outcomes of selection (13.16):

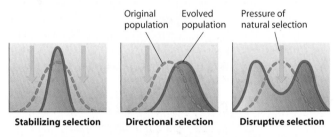

Sexual selection leads to the evolution of secondary sex characteristics, which can give individuals an advantage in mating **(13.17)**.

Perfect organisms? Natural selection is limited by historical constraints, adaptive compromises, chance events, and availability of variations **(13.18)**.

Connecting the Concepts

1. Darwin described his theory of evolution as "descent with modification." Summarize the key points of his theory, including his proposed mechanism of evolution.

2. Complete this map describing potential causes of evolutionary change within populations.

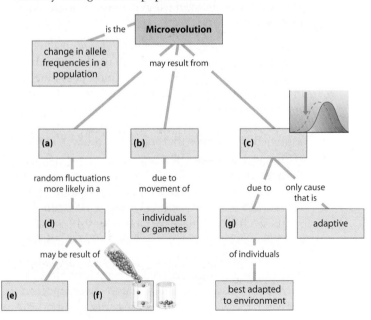

Testing Your Knowledge

Multiple Choice

3. The processes of ____ and ____ generate variation, and ____ produces adaptation to the environment.
   a. sexual recombination . . . natural selection . . . mutation
   b. mutation . . . sexual recombination . . . genetic drift
   c. genetic drift . . . mutation . . . sexual recombination
   d. mutation . . . natural selection . . . sexual recombination
   e. mutation . . . sexual recombination . . . natural selection

4. Natural selection is sometimes described as "survival of the fittest." Which of the following most accurately measures an organism's fitness?
   a. how strong it is when pitted against others of its species
   b. its mutation rate
   c. how many fertile offspring it produces
   d. its ability to withstand environmental extremes
   e. how much food it is able to make or obtain

5. Mutations are rarely the cause of evolution in populations of plants and animals because
   a. they are often harmful and do not get passed on.
   b. they do not directly produce most of the genetic variation present in a diploid population.
   c. they occur very rarely.
   d. they are only passed on when they occur in cells that lead to gametes.
   e. of all of the above.

6. A geneticist studied a grass population growing in an area of erratic rainfall. She found that plants with alleles for curled leaves reproduced better in dry years, and plants with alleles for flat leaves reproduced better in wet years. This situation would tend to
   (*Explain your answer*)
   a. cause genetic drift in the grass population.
   b. preserve the variability in the grass population.
   c. lead to stabilizing selection in the grass population.
   d. lead to uniformity in the grass population.
   e. cause gene flow in the grass population.

7. Birds with average-sized wings survived a severe storm more successfully than other birds in the same population with longer or shorter wings. This illustrates
   a. the founder effect.
   b. stabilizing selection.
   c. artificial selection.
   d. gene flow.
   e. disruptive selection.

8. If an allele is recessive and lethal in homozygotes before they reproduce,
   a. the allele is present in the population at a frequency of 0.001.
   b. the allele will be removed from the population by natural selection in approximately 1,000 years.
   c. the allele will most likely remain in the population at a low frequency because it cannot be selected against in heterozygotes.
   d. the fitness of the homozygous recessive genotype is 0.
   e. both c and d are correct.

9. In a population with two alleles, *B* and *b,* the allele frequency of *b* is 0.4. *B* is dominant to *b*. What is the frequency of individuals with the dominant phenotype if the population is in Hardy-Weinberg equilibrium?
   a. 0.16
   b. 0.36
   c. 0.48
   d. 0.84
   e. You cannot tell from this information.

10. Darwin's claim that all life is descended from a common ancestor is best supported with evidence from
    a. the fossil record.
    b. molecular biology.
    c. taxonomy.
    d. comparative anatomy.
    e. comparative embryology.

Describing, Comparing, and Explaining

11. Write a paragraph briefly describing the kinds of evidence for evolution.

12. Define fitness from an evolutionary perspective.

13. Sickle-cell disease is caused by a recessive allele. Roughly one out of every 400 African-Americans (0.25%) is afflicted with sickle-cell disease. Use the Hardy-Weinberg equation to calculate the percentage of African-Americans who are carriers of the sickle-cell allele.
    (*Hint:* $q^2 = 0.0025$.)

14. It seems logical that natural selection would work toward genetic uniformity; the genotypes that are most fit produce the most offspring, increasing the frequency of adaptive alleles and eliminating less adaptive alleles. Yet there remains a great deal of variability within populations. Describe some of the factors that contribute to this genetic variability.

Applying the Concepts

15. A population of snails is preyed upon by birds that break the snails open on rocks, eat the soft bodies, and leave the shells. The snails occur in both striped and unstriped forms. In one area, researchers counted both live snails and broken shells. Their data are summarized below:

	Striped	Unstriped	Total	Percent Striped
Living	264	296	560	47.1
Broken	486	377	863	56.3

    Which snail form seems to be better adapted to this environment? Why? Predict how the frequencies of striped and unstriped snails might change in the future.

16. School districts in several states have been criticized by groups demanding that science classes give "equal time" to alternative, usually fundamentalist Christian, interpretations of the origin and history of life. They argue that it is only fair to let students evaluate both evolution and the idea that all species were created by God as the Bible relates. Do you think religious views about the origin of species should receive the same emphasis as evolution in science courses? Why or why not?

Answers to all questions can be found in Appendix 3.

For study help and Activities, go to campbellbiology.com or the student CD-ROM.

14.1 The origin of species is the source of biological diversity

CONCEPTS OF SPECIES

14.2 What is a species?
14.3 Reproductive barriers keep species separate

MECHANISMS OF SPECIATION

14.4 Geographic isolation can lead to speciation
14.5 Reproductive barriers may evolve as populations diverge
14.6 New species can arise within the same geographic area as the parent species
14.7 Polyploid plants clothe and feed us
14.8 Adaptive radiation may occur in new or newly vacated habitats
14.9 Peter and Rosemary Grant study the evolution of Darwin's finches
14.10 The tempo of speciation can appear steady or jumpy

MACROEVOLUTION

14.11 Evolutionary novelties may arise in several ways
14.12 Genes that control development are important in evolution
14.13 Evolutionary trends do not mean that evolution is goal directed

Mosquito Mystery

IN THE DANK TUNNELS of the London Underground and the backyards of North America hums a tiny but important mystery that baffles biologists—and makes thousands of people sick each year. It's a mosquito.

These small, everyday bloodsuckers are biologically perplexing. Which of these mosquitoes belong together in the same species, and which do not? How does their wily avoidance of easy grouping affect everything from our understanding of species science to public health?

Mosquitoes buzz through London's subway system. Underground riders may swat impatiently and wonder how so many got in from the stations above. But these mosquitoes didn't have to travel down from the top. They live and breed in the Underground system. They may look the same as mosquitoes flying above ground, but the likeness largely ends there. London's above-ground mosquitoes hibernate in winter and prefer to bite birds. Their subway counterparts are active year-round and feed on rodents and humans.

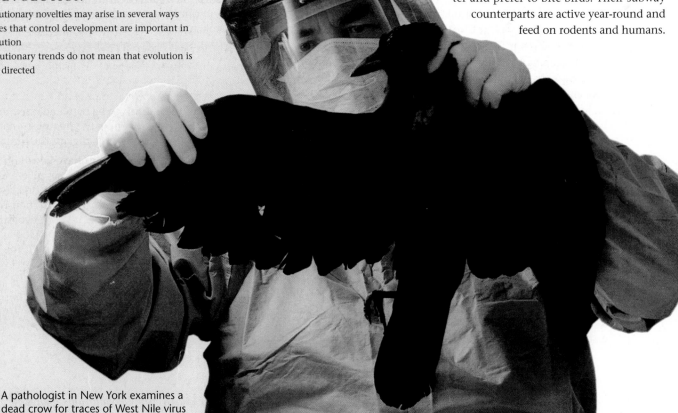

A pathologist in New York examines a dead crow for traces of West Nile virus

The Origin of Species

London's above-ground mosquitoes belong to a species found around the world called *Culex pipiens*. Some scientists lumped the Underground mosquitoes into the same species. Others disagreed, saying that the subterranean mosquitoes act so differently that they belong to a separate species, which they named *Culex molestus* for the insects' pesty behavior.

In 1999, two University of London researchers thought they had the answer, announcing that London's *C. pipiens* and *C. molestus* were indeed different species. They based their findings on the inability of the mosquitoes from the two locations to interbreed. As you'll learn in this chapter, one definition of a species is a group of organisms that can breed and produce fertile offspring. The British team said that *C. molestus* represented a rare example of observable **speciation,** or the emergence of a new species. This process can occur if members of a species are isolated from each other for so long that they evolve differently. But this gradual process rarely occurs quickly enough for humans to observe. Mosquitoes, however, with their rapid breeding and ability to adapt to their environment, might move on a quicker evolutionary clock. The London Underground opened in the 1860s. The researchers maintained that *C. molestus* had evolved from *C. pipiens* in less than 150 years.

This speciation hypothesis seemed to settle the squabbles over London's mosquitoes until 2004, when a different research project turned it upside down. This time, a team of U.S. scientists decided to look at mosquitoes around the world to determine which mosquitoes transmit West Nile virus to humans. The virus can be deadly: Since appearing in the United States in 1999, it has spread quickly and killed hundreds of people.

The U.S. team analyzed the genetic material of mosquitoes from seven countries. In Britain, they found important genetic differences between London's above-ground and subterranean mosquitoes. Those differences were so pronounced, they said, that the subway mosquitoes did not appear to have evolved from their counterparts above. Instead, they seemed to be related to mosquitoes found living in more southern climates. Forget speciation, their findings said—*C. molestus* mosquitoes had just traveled to London and found a warm home underground.

That analysis, while refuting the speciation hypothesis, still bolstered the argument that the *C. pipiens* and *C. molestus* mosquitoes in London are two separate species. With their inability to breed, different behaviors, and divergent genetic material, how could they be the same species? But then the West Nile researchers turned to North America.

When the researchers looked at the genes of U.S. mosquitoes, more than 40 percent had genes from both bird-biting *C. pipiens* and people-biting *C. molestus*. On the North American continent, the mosquitoes were swapping genetic material and creating hybrids. If these hybrids bite both birds and people, they are thus capable of spreading West Nile virus between the two. The researchers concluded that this might explain why human West Nile virus has spread so quickly in the United States and done so much harm. But the research further muddied the question of mosquito species. How could *C. pipiens* and *C. molestus* behave like two species on one continent and one species on another?

As you will see in this chapter, such geographic, behavioral, and genetic issues are just some of the factors that separate species and influence the emergence of new ones. Let's begin our study of the origin of species—a fascinating and complex topic. ■ ■ ■

14.3 Reproductive barriers keep species separate

Clearly, a fly will not mate with a frog or a fern. But what prevents species that are closely related from interbreeding? While geographic barriers may prevent similar species from interbreeding, geography is not intrinsic to organisms. It takes a **reproductive barrier**—a biological feature of the organisms themselves—to prevent populations belonging to closely related species from interbreeding when their ranges overlap. As shown in Table 14.3, the various types of reproductive barriers that isolate the gene pools of species can be categorized as either prezygotic or postzygotic, depending on whether they function before or after zygotes (fertilized eggs) form.

Prezygotic Barriers Prezygotic barriers actually prevent mating or fertilization between species. There are five main types of prezygotic barriers. The first type, called **temporal isolation**, occurs when two species breed at different times— during different seasons, at different times of the day, or even in different years. For example, **Figure 14.3A** shows the eastern spotted skunk and the very similar-looking western spotted skunk. The geographic ranges of these two species overlap in the Great Plains, but the western species breeds in the fall and the eastern species in late winter. Many plants also exhibit seasonal differences in breeding time. Two species of pine trees inhabit some of the same areas of central California, but the Monterey pine (*Pinus radiata*) releases pollen in February, while the Bishop's pine (*P. muricata*) does so in April. Some plants are temporally isolated because their flowers open at different times of the day, so pollen cannot be transferred from one to another.

Figure 14.3A Eastern spotted skunk (top) and western spotted skunk (bottom), reproductively isolated by a temporal barrier

TABLE 14.3	REPRODUCTIVE BARRIERS BETWEEN SPECIES
PREZYGOTIC BARRIERS: **Prevent Mating or Fertilization**	
Temporal isolation:	Mating or flowering occurs at different seasons or times of day.
Habitat isolation:	Populations live in different habitats and do not meet.
Behavioral isolation:	There is little or no sexual attraction between different species.
Mechanical isolation:	Structural differences in genitalia or flowers prevent copulation or pollen transfer.
Gametic isolation:	Male and/or female gametes die before uniting or fail to unite.
POSTZYGOTIC BARRIERS: **Prevent the Development of Fertile Adults**	
Hybrid inviability:	Hybrids fail to develop or to reach sexual maturity.
Hybrid sterility:	Hybrids fail to produce functional gametes.
Hybrid breakdown:	Offspring of hybrids are weak or infertile.

In a second type of prezygotic barrier, **habitat isolation**, two species live in the same general area but not in the same kinds of places. Two closely related species of garter snake are found in western North America, but one lives mainly in water and the other on land. Habitat isolation also affects parasites that are confined to certain plant or animal host species. Two species of parasites living in different hosts will not have the opportunity to interbreed.

In **behavioral isolation**, a third type of prezygotic barrier, there is little or no sexual attraction between females and males of different species. Special signals that attract mates and elaborate mating behaviors that are unique to a species are probably the most important reproductive barriers between closely related animals. For example, male fireflies of various species signal to females of their kind by blinking their lights in particular rhythms. Females respond only to signals of their own species, flashing back and attracting the males.

Figure 14.3B shows a form of behavioral isolation called a courtship ritual. Many species will not mate until the male and female have performed an elaborate ritual that is unlike that of any other species. These blue-footed boobies are involved in a courtship dance in which the male points his beak, tail, and wing tips to the sky. Part of the "script" calls for the male to high-step, a dance that advertises his bright blue feet.

Mechanical isolation, a fourth type of prezygotic barrier, occurs when female and male sex organs are not compatible; for instance, the male copulatory organs of many insect species have a unique and complex structure that fits the female parts of only one species. In the plant kingdom,

The Origin of Species

London's above-ground mosquitoes belong to a species found around the world called *Culex pipiens*. Some scientists lumped the Underground mosquitoes into the same species. Others disagreed, saying that the subterranean mosquitoes act so differently that they belong to a separate species, which they named *Culex molestus* for the insects' pesty behavior.

In 1999, two University of London researchers thought they had the answer, announcing that London's *C. pipiens* and *C. molestus* were indeed different species. They based their findings on the inability of the mosquitoes from the two locations to interbreed. As you'll learn in this chapter, one definition of a species is a group of organisms that can breed and produce fertile offspring. The British team said that *C. molestus* represented a rare example of observable **speciation,** or the emergence of a new species. This process can occur if members of a species are isolated from each other for so long that they evolve differently. But this gradual process rarely occurs quickly enough for humans to observe. Mosquitoes, however, with their rapid breeding and ability to adapt to their environment, might move on a quicker evolutionary clock. The London Underground opened in the 1860s. The researchers maintained that *C. molestus* had evolved from *C. pipiens* in less than 150 years.

This speciation hypothesis seemed to settle the squabbles over London's mosquitoes until 2004, when a different research project turned it upside down. This time, a team of U.S. scientists decided to look at mosquitoes around the world to determine which mosquitoes transmit West Nile virus to humans. The virus can be deadly: Since appearing in the United States in 1999, it has spread quickly and killed hundreds of people.

The U.S. team analyzed the genetic material of mosquitoes from seven countries. In Britain, they found important genetic differences between London's above-ground and subterranean mosquitoes. Those differences were so pronounced, they said, that the subway mosquitoes did not appear to have evolved from their counterparts above. Instead, they seemed to be related to mosquitoes found living in more southern climates. Forget speciation, their findings said—*C. molestus* mosquitoes had just traveled to London and found a warm home underground.

That analysis, while refuting the speciation hypothesis, still bolstered the argument that the *C. pipiens* and *C. molestus* mosquitoes in London are two separate species. With their inability to breed, different behaviors, and divergent genetic material, how could they be the same species? But then the West Nile researchers turned to North America.

When the researchers looked at the genes of U.S. mosquitoes, more than 40 percent had genes from both bird-biting *C. pipiens* and people-biting *C. molestus*. On the North American continent, the mosquitoes were swapping genetic material and creating hybrids. If these hybrids bite both birds and people, they are thus capable of spreading West Nile virus between the two. The researchers concluded that this might explain why human West Nile virus has spread so quickly in the United States and done so much harm. But the research further muddied the question of mosquito species. How could *C. pipiens* and *C. molestus* behave like two species on one continent and one species on another?

As you will see in this chapter, such geographic, behavioral, and genetic issues are just some of the factors that separate species and influence the emergence of new ones. Let's begin our study of the origin of species—a fascinating and complex topic. ■ ■ ■

14.1 The origin of species is the source of biological diversity

Even though Darwin titled his seminal work *On the Origin of Species,* he had relatively little to say about how this process occurred. Most of Darwin's theory of evolution focused on natural selection and the gradual adaptation of a population to its environment, a process sometimes called microevolution. But if microevolution were *all* that happened, then Earth would be inhabited only by a highly adapted version of the first form of life.

Speciation—the origin of new species—is at the focal point of evolution. **Figure 14.1** shows one species evolving into two. Each time speciation occurs, biological diversity increases. Over the course of about 3.6 billion years, an ancestral life-form first gave rise to two or more different types of cells, which then branched to form new lineages, which branched again, until we arrive at the millions of life-forms that have existed and now exist on Earth. How could such an amazing diversification of organisms occur?

Macroevolution encompasses these dramatic biological changes, which begin with the origin of new species. Sometimes a new species has properties novel enough to define a major branch in the tree of life, such as the flowers of flowering plants or the legs of land vertebrates. In this chapter,

we consider how new species arise and how novel properties may evolve. But first, let's define what we mean by species.

? What is the difference between microevolution and macroevolution?

■ Microevolution deals with changes in the gene pool of a single population. Macroevolution includes the origin of new species and higher taxonomic groupings.

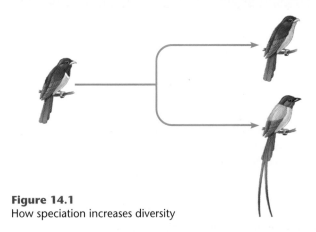

Figure 14.1
How speciation increases diversity

CONCEPTS OF SPECIES

14.2 What is a species?

The word *species* is from the Latin for "kind" or "appearance," and indeed, even young children learn to distinguish between kinds of plants or animals—between dogs and cats, for instance—from differences in their appearance. Although the basic idea of species as distinct life-forms seems intuitive, devising a more formal definition is not so easy.

Taxonomy is the branch of biology concerned with naming and classifying the diverse forms of life. In the 18th century, Swedish physician and botanist Carolus Linnaeus introduced the two-part, or binomial, system of naming organisms, which we still use. For our own species, the binomial designation is *Homo sapiens* (Latin for "wise human being"). Linnaeus named over 11,000 species based on each one's physical characteristics.

As you read in the chapter introduction about London's mosquitoes, appearance alone does not always define a species. The two birds in **Figure 14.2A** look much the same, even though they represent two species; the one on the left is an eastern meadowlark (*Sturnella magna*), and the one on the right is a western meadowlark (*Sturnella neglecta*). Though their body shapes and colorations are very similar, the songs of the two species are different. Distinct songs help these birds choose mates of their own species.

Whereas the individuals of these meadowlark species seem to exhibit fairly limited variation in physical appearance, cer-

Figure 14.2A   Similarity between two species of songbirds, the eastern meadowlark, *Sturnella magna* (left), and the western meadowlark, *Sturnella neglecta*

tain other species—our own, for example—seem extremely varied. If we did not know that humans all belong to one species, *Homo sapiens,* the physical diversity within our species (partly illustrated in **Figure 14.2B**) might lead us to guess that there are several human species. We see that populations of the same species, as well as individuals of the same population, may be very similar or may vary greatly in appearance.

The Biological Species Concept   How do biologists define a species? What keeps one species distinct from others? One view of species, called the **biological species concept,**

Figure 14.2B Diversity within one species

lineages resulting from reproduction. Even with most living, sexually reproducing species, we lack sufficient information about interbreeding to use reproductive isolation as the sole criterion for species assignment.

Other Species Concepts Biologists have developed several other ways to define species. The use of these different concepts depends on the organisms involved and the questions being asked. In practice, for most organisms—sexual, asexual, and fossils alike—classification is based mainly on observable and measurable phenotypic traits. This commonly used method, based on the **morphological species concept**, is how scientists have identified most of the 1.8 million species that have been named to date. The disadvantage of this method, however, is that it relies on subjective critera, and researchers may disagree on the structural features that should be used to distinguish a species.

defines a species as a population or group of populations whose members have the potential to interbreed and produce fertile offspring (offspring who themselves can reproduce). The diverse men and women in Figure 14.2B have the potential to interbreed and produce viable babies that develop into fertile adults. All humans belong to the same species.

Members of different species do not usually mate with each other. And, if members of one species do mate with members of another species, the offspring (or those of later generations) will not be fertile. In effect, this reproductive isolation prevents genetic exchange (gene flow) and maintains the gap between species.

But what about those pesky mosquitoes you learned about? In London, even if *Culex pipiens* mosquitoes do enter the Underground, they are reproductively isolated from *C. molestus*; individuals of these species do not interbreed and produce fertile hybrids. In the United States, however, it appears that there has been genetic exchange between the two types of mosquitoes. And the resulting hybrids may be the culprits in the rapid spread of the West Nile virus. Determining a clear-cut species identification on the basis of reproductive isolation from other species is sometimes more complex than it may seem.

There are other instances in which applying the biological species concept is problematic. For example, there is no way to determine whether organisms that are now fossils were once able to interbreed. Also, this criterion is useless for organisms that are completely asexual in their reproduction, as are prokaryotes and many single-celled protists. As one of these organisms (a single cell) divides, it produces a lineage of genetically identical cells. Some asexual organisms can exchange genes (for instance, the transfer of DNA between even distantly related bacteria, described in Chapter 12), but otherwise there is little or no gene flow between the various

Another approach to species definition, called the **ecological species concept**, identifies species in terms of their ecological niches, focusing on unique adaptations to particular roles in a biological community. (We will examine the concept of ecological niche in more detail in Chapter 37.) For example, two species of Galápagos finches may be similar in appearance but distinguishable on the basis of what they eat.

Finally, the **phylogenetic species concept** defines a species as a set of organisms with a unique genetic history—that is, as one tip on the branching tree of life. Biologists trace the history of a species by comparing its physical characteristics or its DNA sequences to the corresponding data for other species. Such analysis can distinguish groups of individuals that are sufficiently different to be considered separate species. Of course, agreeing on the amount of difference required to distinguish separate species remains a problem.

Each species concept is useful, depending on the situation and the questions being asked. As you saw with the mosquitoes, identifying species is a challenging enterprise. The biological species concept, however, is very useful when focusing on how these discrete groups of organisms may arise and be maintained by reproductive isolation. Because reproductive isolation is an essential factor in the evolution of many species, we look at it more closely next.

? By defining a species by its reproductive _____ from other populations, the biological species concept can only be applied to organisms that reproduce _____.

isolation · · · sexually ∎

14.3 Reproductive barriers keep species separate

Clearly, a fly will not mate with a frog or a fern. But what prevents species that are closely related from interbreeding? While geographic barriers may prevent similar species from interbreeding, geography is not intrinsic to organisms. It takes a **reproductive barrier**—a biological feature of the organisms themselves—to prevent populations belonging to closely related species from interbreeding when their ranges overlap. As shown in Table 14.3, the various types of reproductive barriers that isolate the gene pools of species can be categorized as either prezygotic or postzygotic, depending on whether they function before or after zygotes (fertilized eggs) form.

Prezygotic Barriers Prezygotic barriers actually prevent mating or fertilization between species. There are five main types of prezygotic barriers. The first type, called **temporal isolation**, occurs when two species breed at different times—during different seasons, at different times of the day, or even in different years. For example, Figure 14.3A shows the eastern spotted skunk and the very similar-looking western spotted skunk. The geographic ranges of these two species overlap in the Great Plains, but the western species breeds in the fall and the eastern species in late winter. Many plants also exhibit seasonal differences in breeding time. Two species of pine trees inhabit some of the same areas of central California, but the Monterey pine (*Pinus radiata*) releases pollen in February, while the Bishop's pine (*P. muricata*) does so in April. Some plants are temporally isolated because their flowers open at different times of the day, so pollen cannot be transferred from one to another.

Figure 14.3A Eastern spotted skunk (top) and western spotted skunk (bottom), reproductively isolated by a temporal barrier

TABLE 14.3	REPRODUCTIVE BARRIERS BETWEEN SPECIES
PREZYGOTIC BARRIERS: Prevent Mating or Fertilization	
Temporal isolation:	Mating or flowering occurs at different seasons or times of day.
Habitat isolation:	Populations live in different habitats and do not meet.
Behavioral isolation:	There is little or no sexual attraction between different species.
Mechanical isolation:	Structural differences in genitalia or flowers prevent copulation or pollen transfer.
Gametic isolation:	Male and/or female gametes die before uniting or fail to unite.
POSTZYGOTIC BARRIERS: Prevent the Development of Fertile Adults	
Hybrid inviability:	Hybrids fail to develop or to reach sexual maturity.
Hybrid sterility:	Hybrids fail to produce functional gametes.
Hybrid breakdown:	Offspring of hybrids are weak or infertile.

In a second type of prezygotic barrier, **habitat isolation**, two species live in the same general area but not in the same kinds of places. Two closely related species of garter snake are found in western North America, but one lives mainly in water and the other on land. Habitat isolation also affects parasites that are confined to certain plant or animal host species. Two species of parasites living in different hosts will not have the opportunity to interbreed.

In **behavioral isolation**, a third type of prezygotic barrier, there is little or no sexual attraction between females and males of different species. Special signals that attract mates and elaborate mating behaviors that are unique to a species are probably the most important reproductive barriers between closely related animals. For example, male fireflies of various species signal to females of their kind by blinking their lights in particular rhythms. Females respond only to signals of their own species, flashing back and attracting the males.

Figure 14.3B shows a form of behavioral isolation called a courtship ritual. Many species will not mate until the male and female have performed an elaborate ritual that is unlike that of any other species. These blue-footed boobies are involved in a courtship dance in which the male points his beak, tail, and wing tips to the sky. Part of the "script" calls for the male to high-step, a dance that advertises his bright blue feet.

Mechanical isolation, a fourth type of prezygotic barrier, occurs when female and male sex organs are not compatible; for instance, the male copulatory organs of many insect species have a unique and complex structure that fits the female parts of only one species. In the plant kingdom,

Figure 14.3B Courtship ritual in blue-footed boobies as a behavioral barrier between species

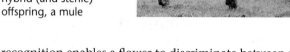

Figure 14.3D Hybrid sterility: a horse and a donkey may produce a hybrid (and sterile) offspring, a mule

mechanical barriers contribute to reproductive isolation of flowering plants. Many species have flower structures that are adapted to specific insect or animal pollinators that transfer pollen only between plants of the same species. Figure 14.3C shows a hummingbird obtaining nectar as its head is dusted with pollen, which the bird will transfer to the next flower it visits. In some cases, the beak of a particular species of hummingbird is just the right length for the flower tube of the one plant species it pollinates, ensuring that pollen is transferred only between plants of that species.

Gametic isolation is a fifth type of prezygotic barrier. A male and a female from two different species may copulate, but the gametes do not unite to form a zygote. In many mammals, for example, the sperm cannot survive in the reproductive tract of a female of a different species. Gametic isolation can operate even if fertilization is external. Male and female sea urchins of many different species release eggs and sperm into the sea, but fertilization occurs only if species-specific molecules on the surfaces of egg and sperm attach to each other. A similar mechanism of molecular

recognition enables a flower to discriminate between pollen of the same species and pollen of different species.

Postzygotic Barriers In contrast to prezygotic barriers, postzygotic barriers operate after hybrid zygotes are formed. (Hybrid zygotes result from the union of gametes of two different species.) In some cases, there is **hybrid inviability;** that is, the hybrids do not survive. Hybrid inviability may occur, for example, in certain frogs of the genus *Rana* that live in the same habitats. Occasional hybrids are produced, but they do not complete development or are very frail.

Another type of postzygotic barrier is **hybrid sterility,** in which hybrid offspring of two different species reach maturity and are vigorous but sterile and therefore unable to bring about gene flow between the parent species. A mule, for example, is the robust offspring of a female horse and a male donkey (Figure 14.3D). The horse and donkey remain separate species because a mule virtually never interbreeds with a horse or a donkey. Therefore, the gene pools of the horse and donkey remain isolated.

In a third type of postzygotic barrier, called **hybrid breakdown,** the first-generation hybrids are viable and fertile, but when these hybrids mate with one another or with either parent species, the offspring are feeble or sterile. For example, different species of cotton plants can produce fertile hybrids, but the offspring of the hybrids do not survive.

In summary, reproductive barriers form the boundaries around many closely related species. The process of speciation may depend on the formation of these barriers that prevent the sharing of genes. Next we examine situations that make reproductive isolation and speciation possible.

> **?** Two closely related tropical bird species live in the same forest, but one feeds and mates in the forest canopy and the other on the forest floor. This is an example of _____ isolation, which is a _____zygotic reproductive barrier.

Figure 14.3C Mechanical isolation of a flowering plant

■ habitat . . . pre

14.4 Geographic isolation can lead to speciation

A key event in the origin of many species is the separation of a population—with its gene pool—from other populations of the same species. With its gene pool isolated, the splinter population can follow its own evolutionary course. Changes in its allele frequencies caused by natural selection, genetic drift, and mutations are unaffected by gene flow from other populations. In the formation of many species living today, the initial block to gene flow seems to have been a geographic barrier that isolated a population. This mode of speciation is called **allopatric speciation** (from the Greek *allos,* other, and *patra,* fatherland). Populations separated by a geographic barrier are known as allopatric populations.

Several geologic processes can fragment a population into two or more isolated populations. A mountain range may emerge and gradually split a population of organisms that can inhabit only lowlands. A large lake may subside until there are several smaller lakes, isolating certain fish populations. A land bridge such as the Isthmus of Panama may form and separate the marine life on either side.

How large must a geographic barrier be to keep allopatric populations apart? The answer depends on the ability of the organisms to move about. Birds, mountain lions, and coyotes can easily cross mountain ranges, rivers, and canyons. The windblown pollen of pine trees is also not hindered by such barriers, and the seeds of many plants may be carried back and forth by animals. In contrast, small rodents may find a deep canyon or a wide river a formidable barrier. The Grand Canyon and Colorado River (Figure 14.4) separate two species of antelope squirrels. Harris's antelope squirrel (*Ammospermophilus harrisi*) inhabits the south rim. Just a few miles away on the north rim is the closely related white-tailed antelope squirrel (*A. leucurus*).

The likelihood of allopatric speciation increases when a population is both small and isolated. A small population is more likely to have its gene pool changed substantially by factors such as genetic drift and natural selection. For example, in less than 2 million years, the few animals and plants from the South American mainland that colonized the Galápagos Islands gave rise to all the new species now found there.

Geographic isolation creates opportunities for speciation, but it does not necessarily lead to new species. Indeed, even when gene pool changes result in the adaptation of an isolated population to a local environment, speciation may or may not occur. Speciation occurs only when the gene pool undergoes changes that establish *reproductive* barriers between the isolated population and its parent population.

Next we look at ways in which biologists study allopatric speciation.

? A new species will not arise just because a population becomes geographically isolated. For _____ speciation to occur, changes in the gene pool must produce _____ _____.

■ allopatric . . . reproductive isolation

Figure 14.4 Geographically isolated species of antelope squirrels

A. harrisi

A. leucurus

Reproductive barriers may evolve as populations diverge

How can biologists confirm that two isolated populations have changed enough that speciation has occurred? In some cases, researchers bring together members of separated populations to see if they can interbreed in the laboratory. The initial claim that the above-ground and Underground mosquitoes had evolved into separate species was based on such an experiment. (You may also remember from the chapter introduction that genetic data indicate that these species are not reproductively isolated in the United States.)

Researchers also attempt to document the evolution of reproductive isolation with more extensive laboratory experiments. While at Yale University, Diane Dodd tested this hypothesis: Reproductive barriers can evolve as a by-product of the adaptive divergence of populations in different environments. She divided a sample of fruit flies into laboratory populations that were bred for several generations on different food sources (Figure 14.5A). Some populations were fed with starch, while others were fed with maltose (malt sugar). Acting over several generations, natural selection favored those individuals that were best suited to using the available nutrient; the populations improved in their digestion of either starch or maltose.

Dodd then combined flies from various populations in mate-choice experiments. As you can see from the results in Figure 14.5A, female "maltose flies" were more likely to mate with male maltose flies than with male "starch flies," even when the maltose flies came from a different maltose fly population. The female starch flies also discriminated in favor of starch flies as mates. (In the control groups, flies did not show a mating preference between flies from the same population and flies from different populations cultured on the same food source.) The breeding preferences shown in the experimental group is an example of a prezygotic barrier, a behavioral obstacle to interbreeding. The reproductive

Figure 14.5B Geographic isolation and speciation in Death Valley

barrier was not absolute—some mating between maltose flies and starch flies did occur—but reproductive isolation was apparently well under way after several generations of evolutionary divergence.

Biologists have also documented cases of natural populations in the process of becoming reproductively isolated. For example, females of the Galápagos ground finch *Geospiza difficilis* respond to the songs of males from the same island, but ignore the songs of males of the same species from other islands. This finding indicates that different behavioral (prezygotic) barriers have begun to accumulate in these separated ground finch populations.

Figure 14.5B illustrates a geographic isolation that has led to allopatric speciation. About 50,000 years ago, what is now the Death Valley region of California and Nevada had a wet climate and an extensive system of interconnected lakes and rivers. A drying trend began about 10,000 years ago, and by 4,000 years ago, the region had become a desert. Today, all that is left of the network of lakes and rivers are isolated springs. In certain cases, a single spring is home to a species of pupfish found nowhere else in the world (see inset). Apparently, these desert pool fishes evolved from one ancestral species whose range was broken up when the region became arid. By either genetic drift or natural selection, and in just a few thousand years, the isolated populations evolved into separate species.

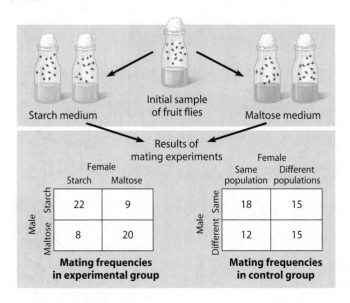

Figure 14.5A Evolution of reproductive barriers in lab populations of fruit flies

	Female			Female	
	Starch	Maltose		Same population	Different populations
Male Starch	22	9	**Male** Same	18	15
Maltose	8	20	Different	12	15
Mating frequencies in experimental group			**Mating frequencies in control group**		

? As separated populations adapt to different environments, changes in their gene pools may coincidentally lead to the evolution of _____ _____.

reproductive barriers ■

14.6 New species can arise within the same geographic area as the parent species

Not all species arise as a result of geographic isolation. In **sympatric speciation** (from the Greek *syn,* together, and *patra,* fatherland), reproductive isolation develops and new species arise without geographic separation. Sympatric speciation does not seem to be widespread among animals but has been important in plant evolution.

Many plant species have originated from accidents during cell division that result in extra sets of chromosomes. In this type of sympatric speciation, the new species is a **polyploid,** meaning that its cells have more than two complete sets of chromosomes. Figure 14.6A shows how a polyploid zygote can result from a single parent species that is diploid. The key abnormality is that meiosis fails to occur properly during gamete formation, and the chromosome number is not reduced. Thus, diploid, rather than haploid, gametes are produced. If self-fertilization occurs (as it commonly does in plants), the resulting zygote is tetraploid; that is, it has four of each type of chromosome. This zygote may develop into a mature plant that can reproduce by self-fertilization.

These new tetraploid plants will also be able to breed with diploid plants of the parental type, but the resulting hybrids will be triploid (3*n*). The triploid zygote comes from the fusion of a diploid (2*n*) gamete from the tetraploid parent and a haploid (*n*) gamete from the diploid parent. Triploid individuals are sterile; they cannot produce normal gametes because the odd number of chromosomes cannot form homologous pairs and separate normally during meiosis. Thus, the creation of the tetraploid (4*n*) plant is an instantaneous speciation event: A new species, reproductively isolated from its parent species, is produced in just one generation.

Sympatric speciation by polyploidy was first discovered by Dutch botanist Hugo de Vries. Working in the early 1900s, de Vries studied genetic diversity in evening primroses (Figure 14.6B). In his breeding experiments with *Oenothera lamarckiana* (upper left), a diploid species of primrose with 14 chromosomes, he noticed an unusual variant. Microscopic inspection revealed that it was a tetraploid with 28 chromosomes. De Vries named this new primrose species, which could not interbreed with its parent species, *Oenothera gigas,* for its large size (lower left).

Most polyploid species, however, do not arise from a single parent species, but from the hybridization of two parent species. The creation of a polyploid species in this way requires the coupling of two accidental events: the hybridization of the two parent species and a cell division error in the resultant hybrid. A hybrid offspring is normally sterile because its chromosomes are not homologous and cannot pair during meiosis. However, the hybrid may reproduce asexually. A mitotic error that doubles the chromosome number or a meiotic error in gamete production may produce functional gametes that can then unite and give rise to a new, fertile species, reproductively isolated from both parent species. We look at examples of this common type of hybrid speciation next.

Web/CD Thinking as a Scientist *How Do New Species Arise by Genetic Isolation?*

? Return to the reproductive barriers in Table 14.3 and choose the one that reproductively isolates a viable polyploid plant from its parental species.

■ Hybrid sterility

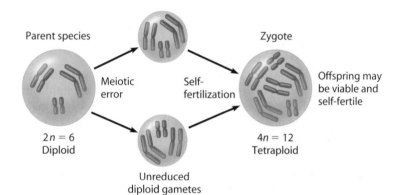

Figure 14.6A Sympatric speciation by polyploid formation

Parent species — Meiotic error — Self-fertilization — Zygote — Offspring may be viable and self-fertile

2*n* = 6 Diploid — Unreduced diploid gametes — 4*n* = 12 Tetraploid

O. lamarckiana

O. gigas

Figure 14.6B Botanist Hugo de Vries with two species of evening primrose

14.7 Polyploid plants clothe and feed us

Plant biologists estimate that 25–50% of all plant species are polyploids. Hybridization between two species accounts for most of this polyploidy, perhaps because the unusually diverse assortment of genes a hybrid inherits from parents of different species can be advantageous.

Many of the plants we grow for food are polyploids, including oats, potatoes, bananas, peanuts, barley, plums, apples, sugarcane, coffee, and wheat. Cotton, also a polyploid, remains the source of one of the world's most popular clothing fibers. Cotton cloth is made from the long white plumes that extend from the seeds of the plant.

Wheat, the most widely cultivated plant in the world, occurs as 20 different species of *Triticum*. We know that humans domesticated diploid species of wheat at least 11,000 years ago because wheat grains of *Triticum monococcum* (with $2n = 14$) have been found in the remains of Middle Eastern farming villages about this old. These diploids have small seed heads and are not highly productive, but some grow wild and others are still cultivated in the Middle East.

Our most important wheat species today is bread wheat (*Triticum aestivum*, Figure 14.7A), a polyploid with 42 chromosomes. Figure 14.7B illustrates how this species may have evolved; the uppercase letters represent not genes but *sets of chromosomes* that have been traced through the lineage. The process began with hybridization between two wheats, one the domesticated species *T. monococcum* (AA), the other one of several wild species that probably grew as weeds at the edges of cultivated fields (BB). Chromosome sets A and B of the two species would not have been able to pair at meiosis, making the AB hybrid sterile. However, an error in cell division and self-fertilization produced a new species (AABB) with 28 chromosomes. Today, we know this species as emmer wheat (*T. turgidum*), varieties of which are grown widely in Eurasia and western North America. It is used mainly for making macaroni and other noodle products because its proteins hold their shape better than bread-wheat proteins.

The final step in the evolution of bread wheat is believed to have occurred in early farming villages on the shores of European lakes over 8,000 years ago. At that time, the cultivated emmer wheat, with its 28 chromosomes, hybridized spontaneously with the closely related wild species *T. tauschii* (DD), which has 14 chromosomes. The hybrid (ABD, with 21 chromosomes) was sterile, but a cell division error in this hybrid and self-fertilization doubled the chromosome number to 42. The result was bread wheat, with two each of the three ancestral sets of chromosomes (AABBDD).

Today, plant geneticists create new polyploids in the laboratory by using chemicals that induce meiotic and mitotic errors. They can produce new hybrids with special qualities, such as a hybrid combining the high yield of wheat with the hardiness of rye.

Web/CD Activity 14A *Polyploid Plants*

Figure 14.7A
Bread wheat

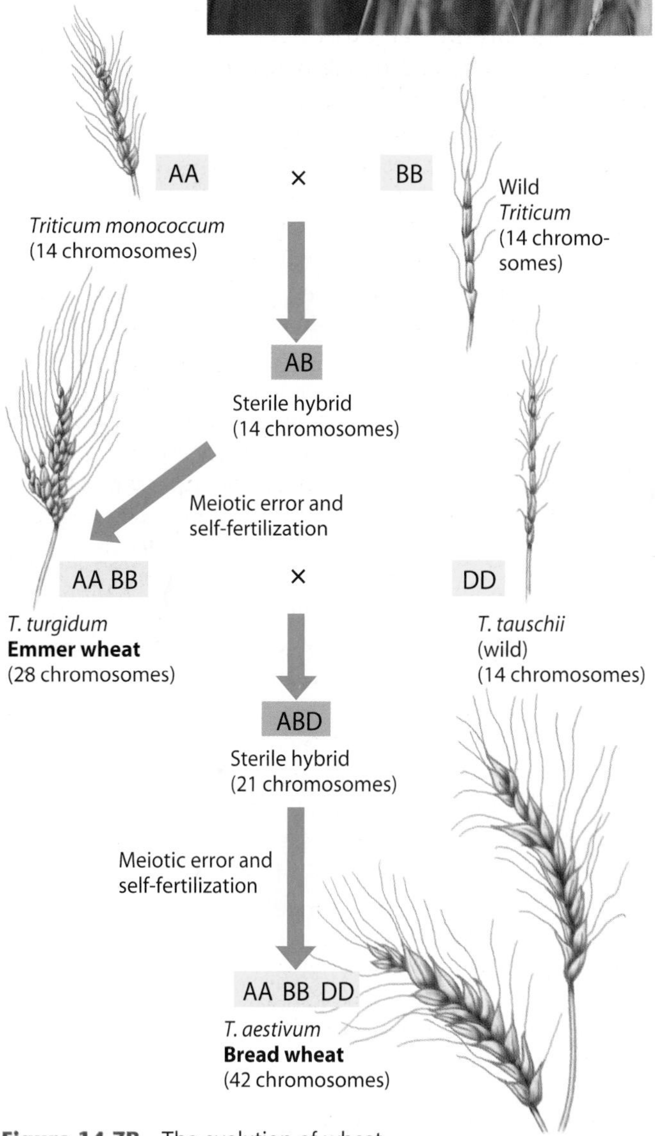

AA × BB

Triticum monococcum (14 chromosomes)

Wild *Triticum* (14 chromosomes)

AB

Sterile hybrid (14 chromosomes)

Meiotic error and self-fertilization

AA BB × DD

T. turgidum **Emmer wheat** (28 chromosomes)

T. tauschii (wild) (14 chromosomes)

ABD

Sterile hybrid (21 chromosomes)

Meiotic error and self-fertilization

AA BB DD

T. aestivum **Bread wheat** (42 chromosomes)

Figure 14.7B The evolution of wheat

? Each speciation episode in the evolution of bread wheat is an example of _____ speciation, which is the origin of a new species without geographic isolation from the parent species.

sympatric ■

14.8 Adaptive radiation may occur in new or newly vacated habitats

The evolution of many new species from a common ancestor introduced to a new and diverse environment is called **adaptive radiation**. Adaptive radiation typically occurs when a few organisms make their way to new, unexploited areas or when environmental changes cause numerous extinctions, opening up various opportunities for the survivors. For example, fossil evidence indicates that mammals underwent a dramatic adaptive radiation after the extinction of the dinosaurs 65 million years ago.

Isolated island chains with physically diverse habitats are often the sites of explosive adaptive radiations. Colonizers may undergo multiple allopatric and sympatric speciation events, producing species that are found nowhere else on Earth. The Galápagos Archipelago, located west of Ecuador, is one of the world's great showcases of adaptive radiation. Each island was born naked from underwater volcanoes and was gradually clothed by plants, animals, and microorganisms derived from strays that rode the ocean currents and winds from other islands and the South American mainland.

The Galápagos island chain has a total of 14 species of closely related birds called Galápagos finches (also known as Darwin's finches because Darwin wrote about them). These birds have many similarities but differ in their feeding habits and their beak type, which is correlated with what they eat. The three examples shown in **Figure 14.8A** illustrate the range of beak shape and size, each adapted for a specific diet.

Evidence accumulated since Darwin's time indicates that all 14 finch species evolved from a single small population of ancestral birds that colonized one of the islands. **Figure 14.8B** illustrates how this might have happened. ❶ Completely isolated on the island after migrating from the mainland, the founder population (species A) may have undergone significant changes in its gene pool and ❷ become a new species, which we'll call species B. Later, a few individuals of species B may have been blown by storms to a neighboring island, where, under different conditions, ❸ they evolved into species C. Some of these birds may have recolonized the first island and coexisted there with the ancestral species if reproductive barriers kept the species distinct. ❹ Species C may also have colonized a new island and ❺ evolved into species D. Species D may then have dispersed to the two islands of its ancestors.

Actually, the Galápagos finches colonized and speciated repeatedly on many separate islands of the Galápagos. Today,

Cactus-seed-eater (cactus finch)

Tool-using insect-eater (woodpecker finch)

Seed-eater (medium ground finch)

Figure 14.8A Examples of differences in beak shape and size in Galápagos finches, each adapted for specific diets

each of the Galápagos Islands has several species of finches, with as many as ten on some islands. The effects of the adaptive radiation of Darwin's finches are evident in the many types of beaks, specialized for different foods (see Figure 14.8A). In the next module, we meet some of the researchers who study Darwin's finches.

Web/CD Activity 14B *Exploring Speciation on Islands*

? Why would allopatric speciation be less common on an island close to a mainland than on a more isolated island of the same size?

■ Continued gene flow between mainland populations and those on nearby islands reduces the chance of enough genetic divergence for speciation.

Figure 14.8B
Adaptive radiation
on an island chain

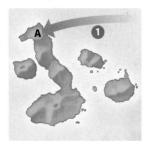
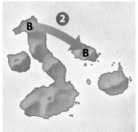
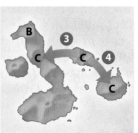
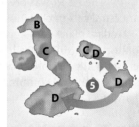

14.9 Peter and Rosemary Grant study the evolution of Darwin's finches

Some theories wait a long time to be tested. Such was the case with Darwin's 150-year-old hypothesis that the beaks of the diverse Galápagos finch species had adapted to different food sources through natural selection. Then came the classic research of Peter and Rosemary Grant (see Module 13.5). For almost 30 years, the Grants have been documenting natural selection acting on finches.

How did the Grants come to work with Darwin's finches? They were looking for a pristine, undisturbed place to study variation within populations. As we saw in the previous module, islands, with their isolated populations, make ideal laboratories in which to study evolution. And the Grants knew from other researchers that the Galápagos were promising. In 1973, Peter banded about 60 medium ground finches on Daphne Major, an isolated, uninhabited island in the Galápagos. When he returned 8 months later with Rosemary and their young daughters, they were able to find all but two of the banded birds. With such an opportunity to work with a small, isolated population, they decided to study these birds for 3 years. One evolutionary question led to another, and for 30 years they have spent up to 3 months a year on Daphne Major, an island about the size of a football stadium. Each year, the Grants (Figure 14.9) and their students have captured, marked, measured, and studied every finch on this rather inhospitable island. Here is how Peter Grant describes their rugged and isolated research site:

> There is no beach on Daphne. There are just steep rocks. To land on the island, you have to find some little platform that the waves have cut out of the rock and then climb on from the boat when there are no waves. Then you climb up until you reach a slope where you can actually stand up and walk. And you have to get supplies up there too —something on the order of 30 5-gallon water jugs, cans of food, packets of rice, sugar . . . plus a stove and cylinder of gas for cooking, as well as other camping supplies.

What were some of the evolutionary questions that kept the Grants on this rocky island for so many years? In addition to their study of beak size and natural selection, the Grants focused on another evolutionary mystery: What keeps two finch species distinct despite their ability to interbreed? They found that the occasional interbreeding between the medium ground finch and the cactus finch (see Figure 14.8A) happens when a male learns to sing the song of the other species. Nestlings whose father died or did not sing much may learn a neighbor's song, even if the neighbor is a different finch species. Thus, a medium ground finch might breed with a cactus finch because he sings her song.

To find out whether these interspecies couples would create a new hybrid species, the Grants followed the survival of their offspring. They found that the hybrids have intermediate bill sizes and thus can only survive during wet years when there are plenty of soft, small seeds. During dry years, the hybrids can't crack the larger, harder seeds that the medium ground finches can eat and can't compete with cactus finches for cactus seeds. As Rosemary Grant explains:

> There is this occasional hybridization through a breakdown of a learned cultural trait, the song. And so you get this balance between an input of genes and then selection, during drought years, keeping the populations on divergent trajectories in spite of the episodes of hybridization.

In other words, when hybrids breed with members of the parent species, they introduce new genes on which natural selection can act. But the severe selection during drought years (when the populations of both finch species are greatly reduced and the hybrids die off) keeps the medium ground finch and the cactus finch on separate evolutionary paths.

Peter Grant conjectures about hybrid finches and their adaptive radiation, which was first documented by Darwin:

> Perhaps hybrids occasionally disperse . . . to another island that has neither the hybrids nor the parent species. The hybrids could start a new population with a range of genetic variation different from the parent species. . . . I see no reason why hybridization hasn't been important right from the beginning, from the first divergence of the ancestral finch stock that reached the islands. We don't have the early stages, but that's the big challenge of evolutionary biology—trying to infer from modern clues what happened in the past.

Another challenge of evolutionary biology, at least as practiced by the Grants, is to enjoy field research, even when it means camping on the rocks.

? Despite the rocks, what were the advantages of Daphne Major as a research site?

■ The resident finch populations were small and isolated, and individual birds and their offspring could be followed over several years.

Figure 14.9 Rosemary and Peter Grant

14.10 The tempo of speciation can appear steady or jumpy

Although biologists continue to gather examples of evolution in progress, much of the evidence of evolution comes from the fossil record, the chronicle of extinct organisms engraved in layers of rock over millions of years of geologic time. Let's take a look at two models that have proved useful in interpreting the evolutionary patterns suggested by the fossil record.

Figure 14.10A illustrates the evolution of two lineages of hypothetical butterflies by what has been called the **gradualism model**. This model fits Darwin's view of the origin of species: Differences gradually evolve in populations as they become adapted to their local environments; and new species (represented by the two butterflies at the top) evolve gradually from the ancestral population. According to the gradualism model, big changes (speciations) occur by the steady accumulation of many small changes.

Many evolutionary biologists since Darwin's time, and even Darwin himself, have been struck by how few sequences of fossils have ever been found that clearly show a gradual, steady accumulation of small changes in evolutionary lineages. Instead, most fossil species appear suddenly in a layer of rock and persist essentially unchanged through several layers (strata) until disappearing from the record of the rocks as suddenly as they appeared. Paleontologists Niles Eldredge and Stephen Jay Gould coined the term **punctuated equilibrium** to describe these abrupt episodes of speciation punctuating long periods of little change, or equilibrium. Illustrated in Figure 14.10B, this model suggests that the evolution of our hypothetical butterflies actually occurs in spurts. Notice that, in contrast to the gradualism model, Figure 14.10B shows no transitional stages in the lineages. The butterflies look the same at the bottom and top of each lineage; the species change little, if at all, once they appear.

The fossil record suggests that successful species last for a few million years, on average. Suppose that a particular species survived for 5 million years, but that most of the changes in its body features occurred during the first 50,000 years of its existence, just 1% of the overall history of the species. Because time periods this short often cannot be distinguished in fossil strata, the species would seem to have appeared abruptly and continued, apparently unchanged, in rocks spanning millions of years before disappearing. Even though the emergence of this species actually took thousands of years, the overall history of the lineage as depicted in the fossil record may seem to fit the punctuated equilibrium model.

The periods of equilibrium in the fossil record may also be explained in a manner consistent with the gradualism model. All species continue to adapt after they come into existence,

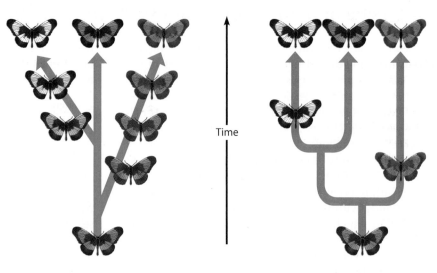

Figure 14.10A Gradualism model

Figure 14.10B Punctuated equilibrium model

Time

but often in ways that cannot be detected from fossils. By necessity, paleontologists base evolutionary hypotheses almost entirely on external anatomy and skeletons. During periods of apparent equilibrium, changes in behavior, internal anatomy, and physiology may go undetected.

Is it likely that most species evolve abruptly and then remain essentially unchanged for most of their existence? Rapid speciation certainly occurs in some cases. As we saw earlier, abrupt speciation can occur by polyploidy in plants and even in a few animals. It appears that genetic drift and natural selection can significantly alter the gene pool of a small population isolated in a challenging new environment in a few hundred to a few thousand generations. Also, mutation of just a few of the genes that regulate embryonic development may produce radically new body features, a topic we will explore in Module 14.12. Debates about the tempo of evolution continue to catalyze research and ultimately lead to a better understanding of the process.

Speciation may begin with small differences as one species gives rise to another similar species. However, as species diverge and speciate again and again, these differences accumulate and become more pronounced. Thus, speciation constitutes the beginning of macroevolutionary change. The more dramatic transformations associated with macroevolution are the topics we consider in the final modules of this chapter.

> ? How does the punctuated equilibrium model account for the relative rarity of transitional fossils linking newer species to older ones?

■ If speciation takes place in a relatively short time compared to the overall time the species exists, the transition of one species to another may be difficult to find in the fossil record.

14.11 Evolutionary novelties may arise in several ways

The Darwinian theory of gradual change can account for the evolution of intricate structures such as eyes or of new body structures such as wings. In most cases, complex structures have evolved in increments from simpler versions having the same basic function—a process of refinement. In others, we can trace the origin of evolutionary novelties to the gradual adaptation of existing structures to new functions.

As an example of the process of gradual refinement, consider the amazing camera-like eyes of vertebrates and squids. Evidence today supports the hypothesis that all complex eyes evolved from a simple ancestral patch of photoreceptor cells through a series of incremental modifications that benefited their owners at each stage. Figure 14.11 illustrates the range of complexity in the structure of eyes among molluscs living today. Some molluscs have very simple eyes. Others have eye cups that have no lenses or other means of focusing images. In those molluscs that do have complex eyes, the organs probably evolved in small steps of adaptation. Examples of such small steps may be seen in Figure 14.11.

Throughout their evolutionary history, eyes retained their basic function of vision. But evolutionary novelty can also arise through the gradual acquisition of new functions. The term **exaptation** refers to a structure that evolved in one context and later was adapted for another function. This term suggests that a structure can become adapted to alternative functions; it does not mean that a structure somehow evolves in anticipation of future use. Indeed, natural selection can only result in the improvement of a structure in the context of its current utility.

Consider the probable evolution of birds from a dinosaur ancestor. Feathers could not have evolved as an adaptation for upcoming flights. Their first utility may have been for insulation. Likewise, the lightweight, honeycombed bones of birds are homologous to the bones of their earthbound ancestors. If light bones predated flight, as is clearly indicated by the fossil record, then they must have had some function on the ground. The ancestors of birds were probably relatively small, agile, bipedal dinosaurs that also would have benefited from a light frame. It is possible that longer, wing-like forelimbs and feathers, which increased the surface area of these forelimbs, were co-opted for flight after functioning in some other capacity, such as mating displays, thermoregulation, and camouflage (functions that feathers still serve today). The first flights may have been only extended hops to pursue prey or escape from a predator. Once flight itself became an advantage, natural selection would have remodeled feathers and wings to fit their additional function.

The flippers of penguins are another example of the modification of existing structures for different functions. Penguins cannot fly, but their modified wings are powerful flippers that make them strong, fast, underwater swimmers.

Web/CD Activity 14C *Mechanisms of Macroevolution*

? Explain why the concept of exaptation does not imply that a structure evolves *in anticipation of* some future environmental change.

■ Although a structure is co-opted for new or additional functions in a new environment, it existed because it worked as an adaptation to the old environment.

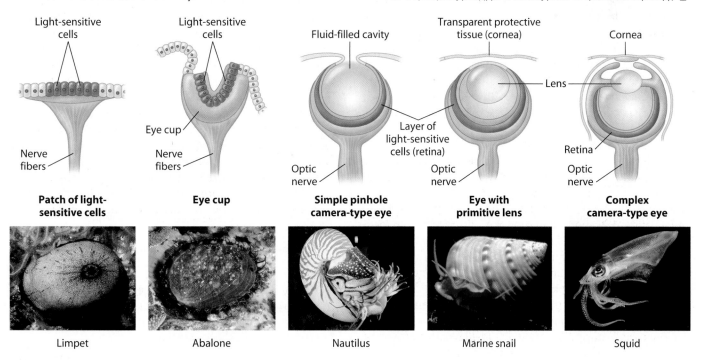

Patch of light-sensitive cells	**Eye cup**	**Simple pinhole camera-type eye**	**Eye with primitive lens**	**Complex camera-type eye**
Limpet	Abalone	Nautilus	Marine snail	Squid

Figure 14.11 A range of eye complexity among molluscs

14.12 Genes that control development are important in evolution

Scientists working at the interface of evolutionary biology and developmental biology—the research field abbreviated **"evo-devo"**—are studying how slight genetic divergences can become magnified into major morphological differences between species. Genes that program development control the rate, timing, and spatial pattern of changes in an organism's form as it is transfigured from a zygote into an adult.

Homeotic genes, described in Chapter 11, determine such basic features as where a pair of wings or legs will develop on a fruit fly or a bird. Changes in such genes can have a profound impact on morphology. Consider, for example, the evolution of terrestrial vertebrates from fishes. The location within a limb bud where certain homeotic genes are expressed determines how far developing bones extend into the limb. Changes in the expression of these genes may have led to the evolution of walking legs from the paired fins of fishes.

Many striking evolutionary transformations are the result of a change in the rate or timing of developmental events. **Figure 14.12A** is a photograph of an axolotl, a salamander that illustrates a phenomenon called **paedomorphosis** (from the Greek *paedos*, child, and *morphosis*, shaping), the retention in the adult of features that were juvenile in its ancestors. The axolotl grows to full size and reproduces without losing its external gills, a juvenile feature in most species of salamanders.

Changes in the timing and rate of growth have also been important in human evolution. As the skulls in **Figure 14.12B** show, humans and chimpanzees are much more alike as fetuses than they are as adults. In the fetuses of both species, the skulls are rounded and the jaws are small, making the face rather flat and rounded. As development proceeds, uneven bone growth makes the chimpanzee skull sharply angular, with heavy browridges and massive jaws. In contrast, the adult human has a skull with decidedly rounded, more fetus-like contours.

Our large skull is one of our most distinctive features. Our large, complex brain is another. Compared to the brain of chimpanzees, our brain continues to grow for several more years, which can be interpreted as the prolonging of a juvenile process. Our brain develops to its unparalleled size and complexity during childhood, a period of development unique to humans. All mammals have a period of infant dependency, when the young are fed milk and require parental protection. After weaning, most mammals mature rapidly to adulthood. Apes, including chimpanzees, have a longer period of infancy than most mammals, but only humans have a true *childhood*, a prolonged period when we remain dependent on parental care. The main function of childhood may be to provide more time to learn from adults.

The evolutionary biologist Stephen Jay Gould contended that there is a connection between our physical traits and our unusually long period of dependency. Gould suggested that our juvenile physical traits—rounded head with large forehead, bulging cheeks, and small chin—may be visual clues that make adults feel affectionate and thus caring and protective during our childhood. Gould used the cartoon character Mickey Mouse to illustrate his point. Early Mickey Mouse renditions were not nearly as popular as later ones (**Figure 14.12C**). Gould contended that the present-day Mickey is successful because he elicits affectionate, parental responses. He certainly *is* more youthful looking: His head and eyes are larger relative to his body, and his legs are shorter.

Whether or not the Mickey Mouse analogy is applicable, changes in genes that control development were probably important events in the divergence of our ancestors from the lineage we share with the great apes.

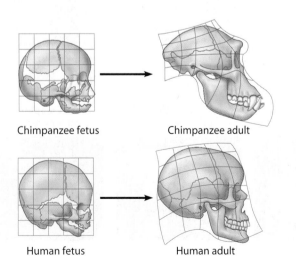

Figure 14.12A An axolotl, a paedomorphic salamander

Chimpanzee fetus   Chimpanzee adult

Human fetus   Human adult

Figure 14.12B Chimpanzee and human skulls compared

Figure 14.12C The "evolution" of Mickey Mouse

Web/CD Activity 14D *Paedomorphosis: Morphing Chimps and Humans*

? Most invertebrates have one cluster of homeotic genes. A mutation that involved duplication of this cluster may have been instrumental in the evolution of vertebrates. What research field tackles such questions?

■ Evo-devo, in which researchers combine evolutionary and developmental biology

Evolutionary trends do not mean that evolution is goal directed

The fossil record seems to reveal trends in the evolution of many species. For instance, two trends in the human lineage were toward a larger skull and a more complex brain. Some lineages may show a trend toward larger or smaller body size. The modern horse is a descendant of an ancestor about the size of a large dog that lived some 40 million years ago. Named *Hyracotherium,* this ancestor had four toes on its front feet, three toes on its hind feet, and teeth adapted to browsing on shrubs and trees. In contrast, modern horses (*Equus*) are larger, have only one toe on each foot, and have teeth modified for grazing on grasses.

As you can see in **Figure 14.13**, the fossil record of horses includes many species that descended from *Hyracotherium.* The yellow-highlighted names track a sequence of fossil horses that were intermediate in form between *Hyracotherium* and *Equus.* If these were the only fossils known, they could create the illusion of a single trend in an unbranched lineage, progressing toward larger size, reduced number of toes, and teeth modified for grazing. However, this would ignore other fossil horses in lineages that died out (the terminal arrows in the figure); an example is the lineage represented by the browser *Anchitherium.* Actually, *Equus* represents the only surviving twig of a multibranched evolutionary bush with several divergent trends.

What accounts for the continuance of some evolutionary trends and the cessation of others? One model of long-term trends considers species to be analogous to individuals: Speciation is their birth, extinction their death, and daughter species their offspring. According to this model of species selection, unequal generation of new species and unequal survival of species play a role in macroevolution similar to the role of differential reproduction in microevolution (see Module 13.6). In other words, the species that endure the longest and generate the greatest number of new species determine the direction of major evolutionary trends. Evolutionary biologists continue to explore this model and other possible sources of the trends observed in the fossil record.

It is important to recognize that the appearance of an evolutionary trend does not imply that there is some intrinsic drive toward a particular goal. Evolution is the result of the interactions between organisms and their current environments. If environmental conditions change, an apparent evolutionary trend may cease or even reverse itself.

In the next chapter, we will continue our study of speciation and macroevolution with a more detailed look at the fossil record and how biologists develop classification systems that reflect the evolutionary history of organisms.

? A general trend in the evolution of mammals was toward larger brain size relative to body size. Use the species selection model to explain how such a trend could occur.

■ This could occur if, on average, those species with larger brains persisted longer before extinction and gave rise to more "daughter" species than did species with smaller brains.

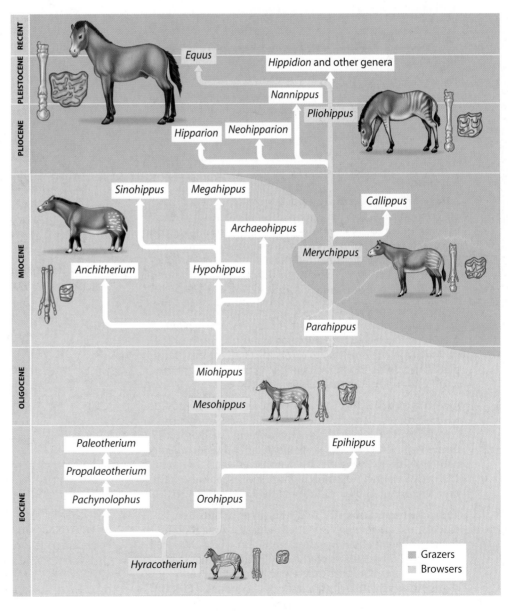

Figure 14.13 The branched evolution of horses

CHAPTER REVIEW

Reviewing the Concepts

Speciation. Earth's incredible biological diversity is the result of macroevolution, which begins with the origin of new species **(Introduction–14.1)**.

Concepts of Species (14.2–14.3)

Defining species. Linnaeus used physical characteristics to distinguish species. His binomial system is the basis of taxonomy, the naming and classification of life's forms.

The biological species concept defines a species as a population or group of populations whose members can interbreed and produce fertile offspring. Most organisms are classified based on observable phenotypic traits, a method using the morphological species concept. The ecological species concept defines a species by its ecological niche. The phlyogenetic species concept defines a species as a set of organisms representing a specific evolutionary lineage **(14.2)**.

Reproductive barriers serve to isolate a species' gene pool and prevent interbreeding **(14.3)**.

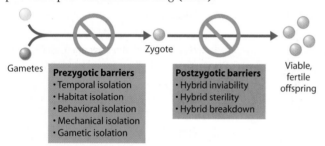

Gametes

Zygote

Prezygotic barriers
• Temporal isolation
• Habitat isolation
• Behavioral isolation
• Mechanical isolation
• Gametic isolation

Postzygotic barriers
• Hybrid inviability
• Hybrid sterility
• Hybrid breakdown

Viable, fertile offspring

Mechanisms of Speciation (14.4–14.10)

Allopatric speciation. Geographically separated from other populations, a small population may become genetically unique as its gene pool is changed by natural selection or genetic drift. Laboratory studies and examples in natural populations have documented the evolution of reproductive barriers during a population's adaptive evolution under a new set of environmental conditions **(14.4–14.5)**.

Sympatric speciation. A new species may arise without geographic isolation from its parent species. Many plant species have evolved by polyploidy (multiplications of the chromosome number due to errors in cell division). Many plants, including food plants such as bread wheat, are the result of hybridization and polyploidy **(14.6–14.7)**.

Adaptive radiation can occur among survivors provided with expanded opportunities following mass extinctions or colonization of a diverse new environment, such as newly formed islands. Peter and Rosemary Grant have documented natural selection acting on populations of Galápagos finches. The occasional hybridization of finch species may also have been important in their adaptive radiation **(14.8–14.9)**.

Tempo of speciation. According to the gradualism model, new species evolve by the gradual accumulation of changes brought about by natural selection. The punctuated equilibrium model draws on the fossil record, where species change most as they arise from an ancestral species and then change relatively little for the rest of their existence **(14.10)**.

Macroevolution (14.11–14.13)

Evolutionary novelties. Many complex structures evolve in many stages from simpler versions having the same basic function. Other novel structures result from the gradual adaptation of existing structures to new functions **(14.11)**.

Evolution of genes controlling development. "Evo-devo" is a field that combines evolutionary and developmental biology. Large evolutionary changes may have resulted from changes in genes that control the spatial organization of body parts or the rate and timing of development **(14.12)**.

Evolutionary trends do not imply that evolution is goal directed. They may reflect species selection—the unequal speciation or unequal survival of species on a branching evolutionary tree **(14.13)**.

Connecting the Concepts

1. Name the two types of speciation represented by this diagram. For each type, describe how reproductive barriers may develop between the new species.

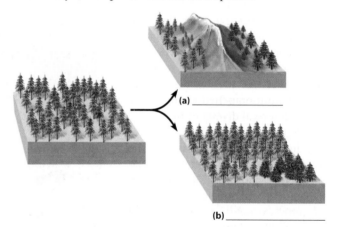

(a) _____

(b) _____

2. Fill in the blanks in the following concept map, which presents some of the mechanisms of macroevolution.

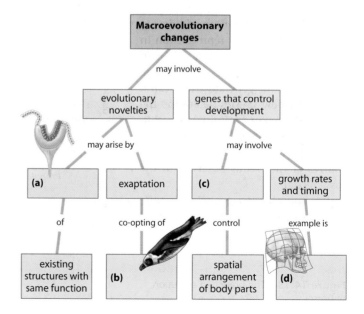

Macroevolutionary changes

may involve

evolutionary novelties

genes that control development

may arise by

may involve

(a)

exaptation

(c)

growth rates and timing

of

co-opting of

control

example is

existing structures with same function

(b)

spatial arrangement of body parts

(d)

Testing Your Knowledge

Multiple Choice

3. Biologists have found more than 500 species of fruit flies on the various Hawaiian Islands, all apparently descended from a single ancestor species. This example illustrates
   a. polyploidy.
   b. temporal isolation.
   c. adaptive radiation.
   d. sympatric speciation.
   e. postzygotic barriers.

4. Bird guides once listed the myrtle warbler and Audubon's warbler as distinct species that lived side by side in parts of their ranges. However, recent books show them as eastern and western forms of a single species, the yellow-rumped warbler. Apparently, it has been found that the two kinds of warblers
   a. live in the same areas.
   b. successfully interbreed.
   c. are almost identical in appearance.
   d. are merging to form a single species.
   e. have the same number of chromosomes.

5. Which of the following is an example of a postzygotic reproductive barrier?
   a. One *Ceanothus* shrub lives on acid soil, another on basic soil.
   b. Mallard and pintail ducks mate at different times of year.
   c. Two species of leopard frogs have different mating calls.
   d. Hybrid offspring of two species of jimsonweeds always die before reproducing.
   e. Pollen of one kind of tobacco cannot fertilize another kind.

6. A small, isolated population is more likely to undergo speciation than a large one because a small population
   a. contains a greater amount of genetic diversity.
   b. is more susceptible to gene flow.
   c. is more affected by genetic drift.
   d. is more subject to errors during meiosis.
   e. is more likely to survive in a new environment.

7. A new plant species C, which formed from hybridization of species A ($2n = 16$) with species B ($2n = 12$), would probably produce gametes with a chromosome number of
   a. 12.
   b. 14.
   c. 16.
   d. 28.
   e. 56.

8. A horse ($2n = 64$) and a donkey ($2n = 62$) can mate and produce a mule. How many chromosomes would there be in a mule's cells?
   a. 31
   b. 62
   c. 63
   d. 126
   e. 252

9. What prevents horses and donkeys from hybridizing to form a new species?
   a. hybrid sterility
   b. hybrid inviability
   c. hybrid breakdown
   d. gametic isolation
   e. prezygotic barrier

10. A swim bladder is a gas-filled sac that helps fish maintain buoyancy. The evolution of the swim bladder from lungs of an ancestral fish is an example of
    a. an evolutionary trend.
    b. paedomorphosis.
    c. changes in homeotic gene expression.
    d. punctuated equilibrium.
    e. exaptation.

11. Which concept of species would be most useful to a field biologist identifying new species in a tropical forest?
    a. biological
    b. morphological
    c. phylogenetic
    d. ecological
    e. None of these would be useful; specimens would have to go back to the lab for DNA testing.

Describing, Comparing, and Explaining

12. Explain how each of the following makes it difficult to clearly define a species: variation within a species, geographically isolated populations, asexual species, fossil organisms.

13. Explain why allopatric speciation would be less likely on an island close to a mainland than on a more isolated island.

14. Compare the gradualism and punctuated equilibrium models of evolution.

15. Differentiate between microevolution and macroevolution.

Applying the Concepts

16. Cultivated American cotton plants have a total of 52 chromosomes ($2n = 52$). In each cell, there are 13 pairs of large chromosomes and 13 pairs of smaller chromosomes. Old World cotton plants have 26 chromosomes ($2n = 26$), all large. Wild American cotton plants have 26 chromosomes, all small. Propose a testable hypothesis to explain how cultivated American cotton probably originated.

17. The red wolf, *Canis rufus,* once widespread in the southeastern and southcentral United States, nearly became extinct in the late 1970s. Saved by a captive breeding program under the authority of the Endangered Species Act (ESA), it has been reintroduced in areas such as the Great Smoky Mountains National Park. Recent genetic evidence indicates that the red wolf may not be a separate species, but a hybrid of the coyote, *Canis latrans,* and the gray wolf, *Canis lupus.* Though the original intent of the ESA was to protect all endangered groups—whether species, subspecies, or hybrids—the costs may be prohibitive. What criteria should be applied if we must decide which organisms to protect? Are there reasons to preserve hybrids, subspecies, or local populations when the species as a whole is not at risk?

Answers to all questions can be found in Appendix 3.

For study help and Activities, go to campbellbiology.com or the student CD-ROM.

Archaeopteryx fossil

CHAPTER 15

MACROEVOLUTION AND EARTH'S HISTORY

15.1 The fossil record chronicles macroevolution
15.2 The actual ages of rocks and fossils mark geologic time
15.3 Continental drift has played a major role in macroevolution
15.4 Tectonic trauma imperils local life
15.5 Mass extinctions were followed by diversification of life-forms

PHYLOGENY AND SYSTEMATICS

15.6 Phylogenies are based on homologies in fossils and living organisms
15.7 Systematics connects classification with evolutionary history
15.8 Cladograms are diagrams based on shared characters among species
15.9 Molecular biology is a powerful tool in systematics
15.10 Arranging life into kingdoms is a work in progress

Are Birds Really Dinosaurs with Feathers?

DID BIRDS EVOLVE FROM DINOSAURS? Evolutionary biologists have been pondering this question for decades. If birds evolved from dinosaurs, then, in a sense, dinosaurs are not really extinct, but rather are flying around today. Some biologists find this idea absurd and accept the older view that birds evolved from an earlier group of reptiles, the pseudosuchians, which are more closely related to today's crocodiles and alligators than to dinosaurs. How can biologists investigate a question like this?

A fossil of the earliest known bird, called *Archaeopteryx* (shown in the photo above), was discovered in 1861, just two years after Darwin's publication of *The Origin of Species*. As you can see, it is a partial skeleton with impressions of feathers. Although some early scientists speculated on the kinship between dinosaurs and birds, it wasn't until the 1970s that John Ostrom, of Yale University, published a series of papers that ignited the controversy over the connection between birds and dinosaurs. Using evidence

Few experts agreed with Ostrom at first because they were convinced that birds evolved from other reptiles

Tracing Evolutionary History

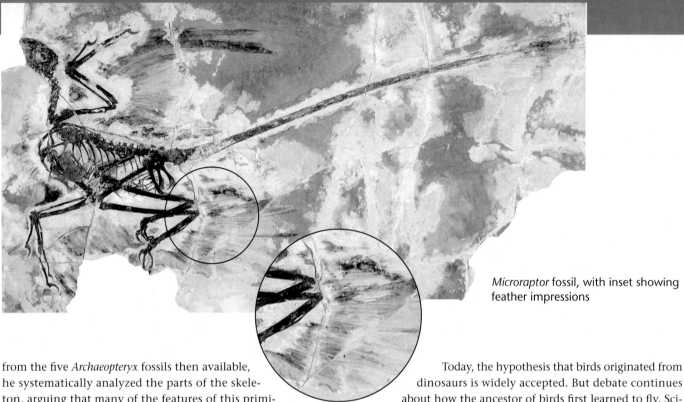

Microraptor fossil, with inset showing feather impressions

from the five *Archaeopteryx* fossils then available, he systematically analyzed the parts of the skeleton, arguing that many of the features of this primitive bird were dinosaur-like. He went so far as to contend that if it weren't for the feather imprints, an *Archaeopteryx* specimen could easily have been classified as a theropod dinosaur.

Few experts agreed with Ostrom at first because they were convinced that birds evolved from a reptile group very different from the theropods. But his arguments were interesting and sufficiently well developed to provoke a reaction. Others began to present opposing evidence. As is often the case in research, a controversial view is good for a field: It spurs inquiry on both sides of a question as researchers try to counter the arguments and evidence presented by the other side and find new ways to bolster their case.

Meanwhile, Ostrom and others sought more support for their bird-dinosaur hypothesis. A new method for evaluating evolutionary relationships, called cladistics, was emerging about the time that Ostrom was gathering his evidence. We will discuss cladistics later in this chapter and see how it is reshaping some evolutionary trees. In this case, cladistic analysis strongly supported Ostrom's conclusions.

As the weight of opinion among experts was swinging towards Ostrum's theory, the bird-dinosaur link was finally corroborated in the 1990s by the discovery of a number of fossils of feathered theropods.

Today, the hypothesis that birds originated from dinosaurs is widely accepted. But debate continues about how the ancestor of birds first learned to fly. Scientists have investigated two possibilities. In one scenario, small, ground-running dinosaurs chasing prey or escaping predators used feathers to gain extra lift as they jumped into the air. In another scenario, dinosaurs climbed trees and then glided back down to the ground. A purported 120-million-year-old fossil of a small theropod called *Microraptor* found in 2003 has been interpreted in favor of the second scenario. Look closely at the fossil shown above and you will see that, in addition to feathers on its arms, this therapod has feathers on its hind legs. Other scientists, however, question the conclusion that this was a four-winged gliding dinosaur indicative of an early stage in the origin of bird flight. Among other objections, they suggest that the anatomy of theropods (or of any known bird) would not have permitted the hind legs to extend sideways to such a gliding position. Thus, the evolution of birds and their ability to fly continues to be an intriguing and challenging scientific inquiry.

Reconstructing the history of life on Earth is a work in progress; tantalizing new clues emerge with each fossil that is unearthed. Fossils have much to tell us about extinct organisms and how they lived. They can also tell us about the major milestones in the evolutionary history of life, the main subject of this chapter. ■ ■ ■

15.1 The fossil record chronicles macroevolution

Chapter 13 considered microevolution, the generation-to-generation change in a population's allele frequencies, mainly due to genetic drift and natural selection. In Chapter 14, we discussed speciation, the origin of new species as local

TABLE 15.1 THE GEOLOGIC RECORD

Relative Duration of Eons	Era	Period	Epoch	Age (Millions of Years Ago)	Some Important Events in the History of Life
Phan-erozoic	Cenozoic	Neogene	Holocene		Historical time
				0.01	
			Pleistocene		Ice ages; humans appear
				1.8	
			Pliocene		Origin of genus *Homo*
				5.3	
			Miocene		Continued radiation of mammals and angiosperms; apelike ancestors of humans appear
				23	
		Paleogene	Oligocene		Origins of many primate groups, including apes
				33.9	
			Eocene		Angiosperm dominance increases; continued radiation of most modern mammalian orders
				55.8	
			Paleocene		Major radiation of mammals, birds, and pollinating insects
				65.5	
Proter-ozoic	Mesozoic	Cretaceous			Flowering plants (angiosperms) appear; many groups of organisms, including dinosaurs, become extinct at end of period (Cretaceous extinctions)
				145.5	
		Jurassic			Gymnosperms continue as dominant plants; dinosaurs abundant and diverse
				199.6	
		Triassic			Cone-bearing plants (gymnosperms) dominate landscape; radiation of dinosaurs; origin of mammal-like reptiles
				251	
	Paleozoic	Permian			Radiation of reptiles; origin of most present-day orders of insects; extinction of many marine and terrestrial organisms at end of period
				299	
		Carboniferous			Extensive forests of vascular plants; first seed plants; origin of reptiles; amphibians dominant
				359.2	
		Devonian			Diversification of bony fishes; first tetrapods and insects
				416	
		Silurian			Diversification of early vascular plants
				443.7	
		Ordovician			Marine algae abundant; colonization of land by plants and arthropods
				488.3	
		Cambrian			Sudden increase in diversity of many animal phyla (Cambrian explosion)
				542	
Archaean		Precambrian		600	Diverse algae and soft-bodied invertebrate animals
				2,200	Oldest fossils of eukaryotic cells
				2,500	
				2,700	Concentration of atmospheric oxygen begins to increase
				3,500	Oldest fossils of cells (prokaryotes)
				3,800	Oldest known rocks on Earth's surface
				Approx. 4,600	Origin of Earth

populations diverge enough to become reproductively isolated from a parent species. We also looked at macroevolution, the major changes in the history of life. Macroevolution results from the cumulative effects of many speciation events over vast tracts of time and encompasses the origin of evolutionary novelties, such as feathers.

The fossil record, the sequence in which fossils appear in rock strata, is an archive of evolutionary history. By studying fossils in rock strata in one area, we gain a local glimpse of long-term evolutionary change. Studying the order of fossils in strata from many sites enables us to trace macroevolution.

Based on this sequence of fossils, geologists have established a geologic record, as shown in Table 15.1 on the facing page. Earth's history is divided into three eons. The time line to the left of the table shows the relative lengths of these eons. The earliest two eons—the Archaean and the Proterozoic—lasted approximately 4 billion years and are collectively referred to as the Precambrian. The Phanerozoic eon, covering roughly the last half-billion years, is divided into three eras: the Paleozoic, Mesozoic, and Cenozoic. Most eras are subdivided into periods. The boundaries between eras are marked by mass extinctions, when many forms of life disappeared from the fossil record and were replaced by species that diversified from the survivors. Lesser extinctions often mark the boundaries of the periods that make up an era.

Rocks from the Precambrian (Archaean and Proterozoic eons) have undergone extensive change over time, and much of their fossil content is no longer visible. Nonetheless, paleontologists have found some fossil-rich Precambrian strata and have pieced together ancient events in life's history. The oldest known fossils, dating from 3.5 billion years ago, are of prokaryotes; the oldest known eukaryotes are from 2.2 billion years ago. Strata from the late Precambrian (600 million years ago) bear diverse fossils of algae and soft-bodied animals.

Dating from about 542 million years ago, rocks of the Paleozoic ("ancient animal") era contain fossils of lineages that gave rise to modern organisms, as well as many lineages that have become extinct. During the early Paleozoic, virtually all life was aquatic, but by about 400 million years ago, plants and animals were well established on land.

Following the Paleozoic era was the Mesozoic ("middle animal") era, also known as the age of reptiles because of its abundance of reptilian fossils, including those of dinosaurs. The Mesozoic also saw the beginnings of mammals and flowering plants (angiosperms). By the end of the Mesozoic, dinosaurs had become extinct, except, as discussed in the chapter introduction, for one lineage—the birds.

An explosive period of evolution of mammals, birds, and angiosperms began at the dawn of the Cenozoic ("recent animal") era, about 65 million years ago. Because much more is known about the Cenozoic than about earlier eras, our table subdivides the two Cenozoic periods into finer intervals called epochs. Our own species, *Homo sapiens,* appeared on the scene only about 100,000–200,000 years ago, during the Pleistocene (Ice Age) epoch. Thus, our tenure on the planet is only a tiny portion of the immense saga of geologic time. In fact, if a clock of Earth's history were scaled to represent 1 hour, humans appeared only 1 second ago.

Web/CD Activity 15A *A Scrolling Geologic Record*

? Use Table 15.1 to estimate how long prokaryotes inhabited Earth before eukaryotes evolved.

■ About 1,300 million years (1.3 billion years)

15.2 The actual ages of rocks and fossils mark geologic time

The record of the rocks chronicles the *relative* ages of fossils, the order in which species present in a succession of strata evolved. However, the sequence alone does not tell the *actual* ages in years of the embedded fossils.

Geologists use several techniques to determine the ages of rocks and the fossils they contain. The method most often used, called **radiometric dating**, is based on the measurement of certain radioactive isotopes (see Module 2.4). Fossils contain isotopes of elements that accumulated when the organisms were alive. For example, the carbon in a living organism includes both the most common isotope, carbon-12 (^{12}C), and a less common radioactive isotope, carbon-14 (^{14}C), in the same ratio as is present in the atmosphere. Once an organism dies, it stops accumulating carbon, and its carbon-14 starts slowly to decay to another element. Each radioactive isotope has a fixed rate of decay, known as its half-life. Carbon-14 has a half-life of 5,730 years, meaning that half the carbon-14 in a specimen decays in about 5,730 years, half the remaining carbon-14 decays in the next 5,730 years, and so on. Knowing both the half-life of a radioactive isotope and the ratio of radioactive to stable isotope in a fossil enables us to tell how old the fossil is. For instance, if a fossil has a ^{14}C-to-^{12}C ratio half that of the atmosphere, it is about 5,730 years old; a fossil with one-fourth the atmosphere's ratio is about 11,460 years old.

Carbon-14 is useful for dating relatively young fossils. To date older fossils, paleontologists use radioactive isotopes with longer half-lives. For instance, potassium-40, with a half-life of 1.3 billion years, can be used to date volcanic rocks hundreds of millions of years old. The age of fossils can then be inferred from the ages of surrounding rock layers. Radiometric dating has an error factor of plus or minus 10%.

The dates in the geologic record in Table 15.1 were established by dating rocks and fossils. In the next module, we examine some of the geologic processes that can produce the distinct changes marking the boundaries between periods.

? Your measurements indicate that a fossilized skull you unearthed has a ^{14}C-to-^{12}C ratio about one-sixteenth that of the atmosphere. What is the approximate age of the skull?

■ 22,920 years (four half-life reductions)

15.3 Continental drift has played a major role in macroevolution

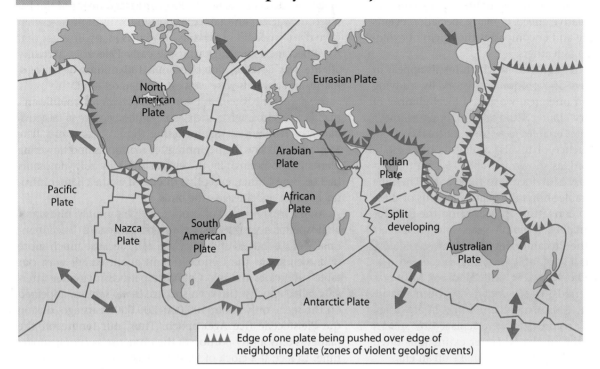

Figure 15.3A
Earth's crustal plates

▲▲▲▲ Edge of one plate being pushed over edge of neighboring plate (zones of violent geologic events)

In 1912, German meteorologist Alfred Wegener proposed the hypothesis of **continental drift.** Wegener postulated that all land on Earth was once one great mass, which broke up into continents that drifted like rafts to their present positions. Wegener suggested that the shapes of our modern continents, like pieces of a jigsaw puzzle, reflect their former positions in the original supercontinent. Like many ideas generated before their time, Wegener's were not taken seriously for decades. Until the 1960s, the general belief was that the continents have always been fixed in their present positions.

In recent decades, geologists, paleontologists, and biologists have accumulated overwhelming evidence for the concept of continental drift. We now know that the continents and seafloors form a thin outer layer of planet Earth, called the crust, and that the crust covers a mass of hot material called the mantle. Furthermore, the crust is divided into giant, irregularly shaped plates (outlined in red in **Figure 15.3A**). Because the mantle circulates constantly, the crustal plates move about slowly but incessantly—they literally float—on the underlying mantle. As a result of plate movements, world geography changes constantly, for unless landmasses are embedded in the same crustal plate, their positions relative to each other do not remain the same. North America and Europe, for example, are presently drifting apart at a rate of about 2 cm per year. Throughout geologic time, continental movements have greatly influenced the distribution of organisms around the world, and continental drift explains much of the history of life.

Two chapters in the continuing saga of continental drift seem to have been especially significant in their influence on life. The first occurred about 250 million years ago, near the

end of the Paleozoic era, when plate movements brought all the landmasses together into the supercontinent Wegener had originally proposed. Shown at the bottom in **Figure 15.3B** on the facing page, this supercontinent is called **Pangaea,** meaning "all land." Imagine some of the possible effects on life as massive continents joined. Species that had been evolving in isolation came together and competed. When the landmasses fused, the total amount of shoreline and shallow coastal areas was reduced. Ocean basins became deeper, lowering the sea level and draining the shallow coastal seas. Then, as now, most marine species inhabited shallow waters, and many of these organisms probably died out as their habitats shrank. The formation of Pangaea also would have altered terrestrial environments, as the changing ocean currents affected climates on land. The interior of the vast continent was cold and dry. Overall, the fossil record indicates that the formation of Pangaea reshaped biological diversity, causing great numbers of extinctions. These, in turn, provided new opportunities for organisms that survived the crisis.

The second dramatic chapter in continental drift began about 180 million years ago, during the Mesozoic era. Pangaea started to break apart again, causing a geographic isolation of colossal proportions. As the continents drifted apart, each became a separate evolutionary arena—a huge island on which organisms evolved in isolation from their previous neighbors. At first, Pangaea split into northern and southern landmasses, which we call Laurasia and Gondwana, respectively. This split was more or less complete by about 135 million years ago, as shown in Figure 15.3B. By the end of the Mesozoic era (and the Cretaceous period), some 65 million years ago, the modern continents were beginning to take shape. Then, just 10 million

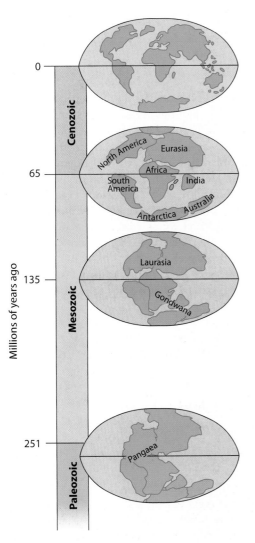

Figure 15.3B Continental drift

years ago, India collided with Eurasia, and the slow, steady crunching of the Indian and Eurasian plates formed the Himalayas, the tallest and youngest of Earth's major mountain ranges. (Regions where two plates crunch together are marked by red arrowheads in Figure 15.3A.)

The pattern of continental mergings and separations solves many puzzles, including Australia's great diversity of marsupials (pouched mammals). Marsupials probably originated in what is now North America and spread southward while the continents were still joined. The subsequent breakup of continents set Australia "afloat" like a great ark of marsupials. Isolated for over 50 million years, marsupials evolved and diversified on Australia, filling ecological roles analogous to those filled by placental mammals on other continents.

Continental drift also explains the distribution of a group of ancient vertebrates called lungfishes (**Figure 15.3C**). Today, there are six species of lungfishes in the world, four in Africa and one each in Australia and South America (yellow striped areas in **Figure 15.3D**). What is the evolutionary history of these animals? As the orange triangles in Figure 15.3D indicate, fossil lungfishes have been found on all continents except Antarctica. This widespread fossil record indicates that lungfishes evolved when Pangaea was intact.

In the next module, we consider some of the perils associated with the movements of Earth's crustal plates.

? Paleontologists have discovered matching fossils of Triassic reptiles in West Africa and Brazil, regions that are separated by 3,000 km of ocean. How could continental drift explain such finds?

■ West Africa and Brazil were connected during the early Mesozoic era, and the reptiles must have ranged across both areas.

Figure 15.3C
An African lungfish

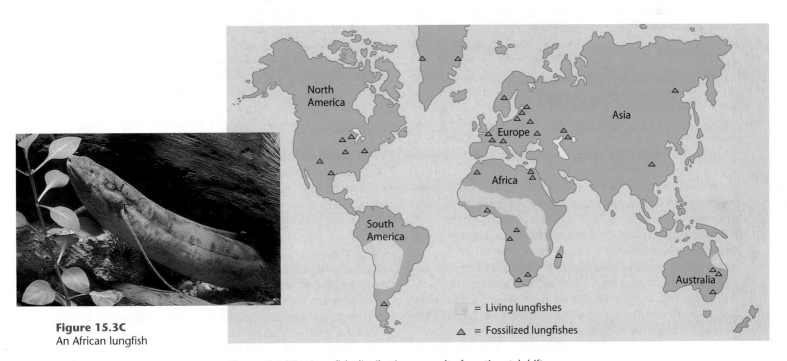

= Living lungfishes

△ = Fossilized lungfishes

Figure 15.3D Lungfish distribution, a result of continental drift

15.4 Tectonic trauma imperils local life

The theory of **plate tectonics** describes the forces involved in the movement of Earth's plates and the geologic processes that result. Not only do moving crustal plates cause continents to collide, pile up, and build mountain ranges, they also produce volcanoes and earthquakes. The boundaries of plates are hotspots of such geologic activity. California's frequent earthquakes result from movement along the infamous San Andreas Fault, part of the border where the Pacific and North American plates grind together and gradually slide past each other (Figure 15.4A). Undersea earthquakes can cause giant waves, such as the devastating 2004 tsunamis created when a large area of a fault in the Indian Ocean ruptured near the meeting point of the Indian, Eurasian, and Australian plates.

Erupting volcanoes, emitting hot, molten rock from beneath Earth's crust, can cause tremendous devastation, although they can also create opportunities for living organisms. For instance, volcanoes at sea can produce islands, such as the Galápagos and Hawaiian Islands. But the same volcanic activity that creates an oceanic island may destroy life that evolved there. In 1883, fiery pumice from a volcano covered the small volcanic island of Krakatau, near the boundary between the Australian and Eurasian plates (Figure 15.4B). Before the eruption, Krakatau was covered with a dense tropical rain forest. Afterward, it was virtually devoid of life. Despite the devastation, life from neighboring islands began recolonizing Krakatau soon after its surface cooled. Within 50 years after the eruption, tropical forest with a great diversity of plants and animals again covered the island.

Plate tectonics has played a role in many of the major changes that characterize evolutionary history. As we see in the next module, extraterrestrial objects may also be very important.

? The Andes Mountains in South America are associated with tectonic activity near which plate boundary? (See Figure 15.3A.)

■ Boundary between the Nazca and South American plates

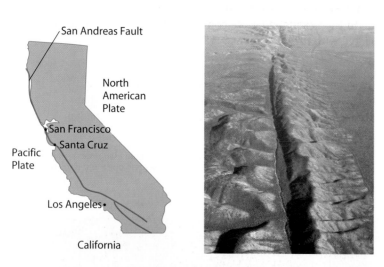

Figure 15.4A The San Andreas Fault (shown north of Los Angeles), a boundary between two crustal plates

San Andreas Fault
North American Plate
San Francisco
Santa Cruz
Pacific Plate
Los Angeles
California

Figure 15.4B The volcanic eruption that destroyed the island of Krakatau

15.5 Mass extinctions were followed by diversification of life-forms

Extinction is inevitable in a changing world. A species may become extinct because its habitat has been destroyed, because of unfavorable climatic changes, or because of changes in its biological community, such as the evolution of new predators or competitors. Extinctions occur all the time, but extinction rates have not been steady. The fossil record chronicles a number of occasions when global environmental changes were so rapid and disruptive that a majority of species were swept away. There have been at least six distinct periods of mass extinctions over the last 600 million years.

Of all the mass extinctions, the ones marking the ends of the Permian and Cretaceous periods have received the most attention. The Permian mass extinction, which occurred about 251 million years ago and defines the boundary between the Paleozoic and Mesozoic eras, claimed about 96% of marine animal species and took a tremendous toll on terrestrial life as well. At the end of the Cretaceous period, about 65 million years ago, the world again lost an enormous number of species—more than half of all marine species and many lineages of terrestrial plants and animals. At that point, dinosaurs had dominated the land and air for some 150 million years. Then, in less than 10 million years—a brief period in geologic time—all the dinosaurs were gone, leaving behind only the descendants of one lineage, the birds.

The cause of these mass extinctions is unclear, but we do know that the Permian mass extinction occurred at a time of enormous volcanic eruptions in what is now Siberia, constituting the most extreme volcanic activity in the past half-billion years. Besides spewing lava and ash into the atmosphere, the eruptions may have produced enough carbon dioxide to warm the global climate. Reduced temperature differences between the equator and the poles would have slowed the mixing of ocean water, which in turn would have reduced the amount of oxygen available to marine organisms. An oxygen deficit in the oceans may have played a large role in the Permian extinction.

One clue to a possible cause of the Cretaceous mass extinction is a thin layer of clay enriched in iridium that separates sediments from the Mesozoic and Cenozoic eras. Iridium is an element very rare on Earth but common in meteorites and other extraterrestrial material that occasionally fall to Earth. Many paleontologists conclude that the iridium layer is the result of fallout from a huge cloud of dust that billowed into the atmosphere when a large meteorite or asteroid hit Earth. The cloud would have blocked light and severely disturbed the global climate for months.

The so-called impact hypothesis has many supporters, and a large asteroid crater, the 65-million-year-old Chicxulub crater, has been found in the Caribbean Sea near the Yucatán Peninsula of Mexico (**Figure 15.5**). About 180 km in diameter, the crater is the right size to have been caused by an object with a diameter of 10 km. The horseshoe shape of the crater and the pattern of debris in sedimentary rocks indicate than an asteroid or comet struck at a low angle from the southeast. The artist's interpretation in Figure 15.5 represents the impact and its immediate effect—a cloud of hot vapor and debris that could have killed most of the plants and animals in North America within minutes.

The fossil record shows us that the climate cooled late in the Cretaceous and that shallow seas were receding from continental lowlands. Some researchers propose that climatic changes due to continental drift could have caused the Cretaceous extinctions, whether an asteroid collided with Earth or not. Still others point to evidence that a spike in volcanic activity took place in what is now India at the time of the Cretaceous mass extinction. Was this spike triggered by the Chicxulub impact? The various hypotheses are not mutually exclusive, and researchers continue to debate the extent to which each contributed to the extinctions.

Whatever their causes, mass extinctions affect biological diversity profoundly. But there is a creative side to the destruction. Each massive dip in species diversity has been followed by an explosive *increase* in diversity. Mass extinctions seem to have provided the surviving organisms with new environmental opportunities. For example, mammals existed for at least 75 million years before undergoing an explosive increase in diversity just after the Cretaceous. Their rise to prominence was undoubtedly associated with the void left by the extinction of the dinosaurs. The world might be a very different place today if many dinosaur lineages had escaped the Cretaceous extinction or if none of the mammals that lived in the Cretaceous had survived.

Web/CD Activity 15B *Mechanisms of Macroevolution*

? The Permian and Cretaceous mass extinctions mark the ends of the _____ and _____ eras, respectively. (*Hint:* Refer back to Table 15.1.)

■ Paleozoic . . . Mesozoic

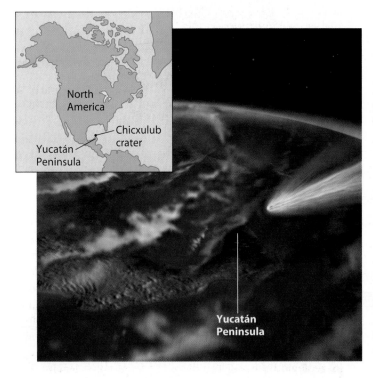

North America

Chicxulub crater

Yucatán Peninsula

Yucatán Peninsula

Figure 15.5 The impact hypothesis for the Cretaceous mass extinction

15.6 Phylogenies are based on homologies in fossils and living organisms

The evolutionary history of a group of organisms is called **phylogeny** (from the Greek *phylon,* tribe, and *genesis,* origin). The fossil record provides a substantial chronicle of evolutionary change that can help trace the phylogeny of many groups. It is, however, an incomplete record, as many of Earth's species probably never left fossils; many fossils that formed were probably destroyed by later geologic processes; and only a fraction of existing fossils have likely been discovered. Even with its limitations, however, the fossil record is a remarkably detailed account of biological change over the vast scale of geologic time.

In addition to evidence from the fossil record, phylogeny can also be inferred from morphological and molecular homologies among living organisms. As we discussed in Module 13.4, homologies are similarities due to shared ancestry. Homologous structures may look and function differently in different species, but they exhibit fundamental similarities because they evolved from the same structure in a common ancestor. Among vertebrates, for instance, the whale limb is adapted for steering in the water; the bat wing is adapted for flight. Nonetheless, there is a basic similarity in the bones supporting these two structures (see Figure 13.4A). Indeed, Ostrom used the homologous skeletal structures of the fossils of *Archaeopteryx* and a vicious, sickle-clawed theropod predator in his argument that birds were the direct descendants of theropod dinosaurs (see the chapter introduction).

Generally, organisms that share very similar morphologies or similar DNA sequences are likely to be closely related. The search for homologies is not without pitfalls, however, for not all likenesses are inherited from a common ancestor. In a process called **convergent evolution,** species from different evolutionary branches may come to resemble one another if they live in very similar environments. In such cases, natural selection may result in body structures and even whole organisms that look very similar. Similarity due to convergence is called **analogy,** not homology. For example, the two plants in **Figure 15.6** look remarkably similar, although they are not closely related. On the left is the

Figure 15.6 Analogous structures resulting from convergent evolution

ocotillo, which is common on the coastal desert of Baja California. On the right is the allauidia, which grows in desert areas in Madagascar. These two plants are on widely separated lineages and have evolved in isolation for millions of years. Nonetheless, as the plants became adapted to similar environments, analogous structures evolved. Analogous structures are also sometimes called homoplasies (from the Greek, meaning "to mold in the same way"). A critical task in reconstructing phylogenies is to distinguish homologies, which indicate common ancestry, from analogies, or homoplasies, which do not.

Reconstructing phylogeny is part of **systematics,** an analytical approach to the study of the diversity of life and the evolutionary relationships between organisms.

? Our forearms and a bat's wings (see Figure 13.4A) are _____. A bat's wings and a bee's wings are _____.

homologous · · · analogous ∎

15.7 Systematics connects classification with evolutionary history

Although the system of naming and classifying species that Linnaeus introduced in the 18th century was not based on geneology, or evolutionary relationship, many of its features remain useful in systematics. Two of these are the binomial designation of species and hierarchical classification.

Taxonomists (biologists who identify, name, and classify species) assign scientific names to each species. Common names, such as monkey, fruit fly, and pea, may work well in everyday communication, but they can be ambiguous be-

cause there are many species of each of these kinds of organisms. And some common names are misleading. Consider these three "fishes": jellyfish (a cnidarian), crayfish (a crustacean), and silverfish (an insect).

As we have seen, Linnaeus's system assigns to each species a two-part latinized name, or **binomial.** The first part of a binomial is the **genus** (plural, *genera*) to which a species belongs. The second part identifies one **species** within that genus. The two parts must be used together to name a

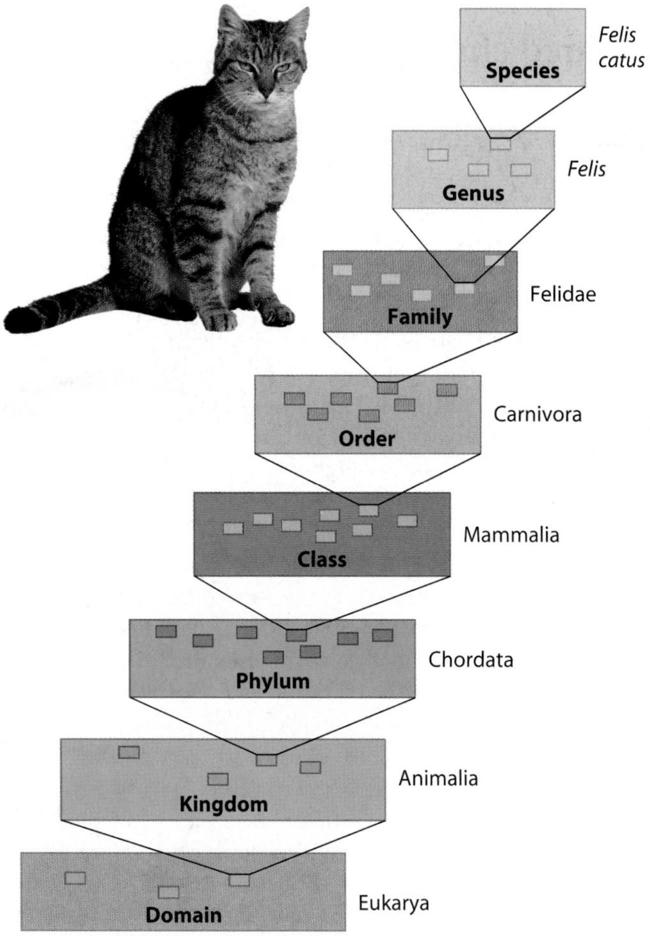
Species
Felis catus
Genus
Felis
Family
Felidae
Order
Carnivora
Class
Mammalia
Phylum
Chordata
Kingdom
Animalia
Domain
Eukarya

Figure 15.7A Hierarchical classification of the domestic cat

species. For example, the scientific name for the domestic cat is *Felis catus*. Notice that the first letter of the genus name is capitalized and that the binomial is italicized and latinized. (You can name a bug you discover after a friend, but you must add the appropriate Latin ending.)

In addition to naming species, Linnaeus also grouped species into a hierarchy of categories. Beyond the grouping of species within genera (as indicated by the binomial), the Linnaean system extends to progressively broader categories of classification. It places similar genera in the same **family**, puts families into **orders**, orders into **classes**, classes into **phyla** (singular, *phylum*), and phyla into **kingdoms.** Most taxonomists also group kingdoms into a higher taxonomic category called the **domain.**

Figure 15.7A illustrates the progressively more comprehensive classification of the domestic cat. The genus *Felis*, which includes the domestic cat (*Felis catus*) and several closely related wild cats, is grouped in the cat family, Felidae, along with the genus *Panthera*. (*Panthera* includes the tiger, leopard, jaguar, and African lion.) Family Felidae belongs to the order Carnivora, which also includes the family Canidae (for example, the wolf and domestic dog) and several other families. Order Carnivora is grouped with many other orders in the class Mammalia, the mammals. Class Mammalia is one of several classes belonging to the phylum Chordata in the kingdom Animalia, which is one of several kingdoms in the domain Eukarya. Each taxonomic unit at any level—family Felidae or class Mammalia, for instance—is called a **taxon** (plural, *taxa*).

Grouping organisms into more inclusive categories seems to come naturally to humans—it is a way to structure our world. Classifying species into higher taxa, however, is ultimately arbitrary. Higher classification levels are generally defined by various morphological characteristics chosen by taxonomists rather than by a quantitative measurement that could apply to the same taxon level across all lineages.

Ever since Darwin, systematics has had a goal beyond simple organization: to have classification reflect evolutionary relationships. Biologists traditionally use **phylogenetic trees** to depict hypotheses about the evolutionary history of species. These branching diagrams reflect the hierarchical classification of groups nested within more inclusive groups. As shown in Figure 15.7B, the "highest," or most inclusive, taxon is at the bottom, and each branch point represents the divergence of two lineages from a common ancestor. Figure 15.7B illustrates the connection between classification and phylogeny; it shows the classification of some of the taxa in the order Carnivora and the probable evolutionary relationships between these groups.

Our study of phylogenetic trees continues as we explore cladograms in the next module.

? How much of the classification in Figure 15.7A do we share with the domestic cat?

■ We are classified the same down to the class level: Both cats and humans are mammals. We do not belong to the same order.

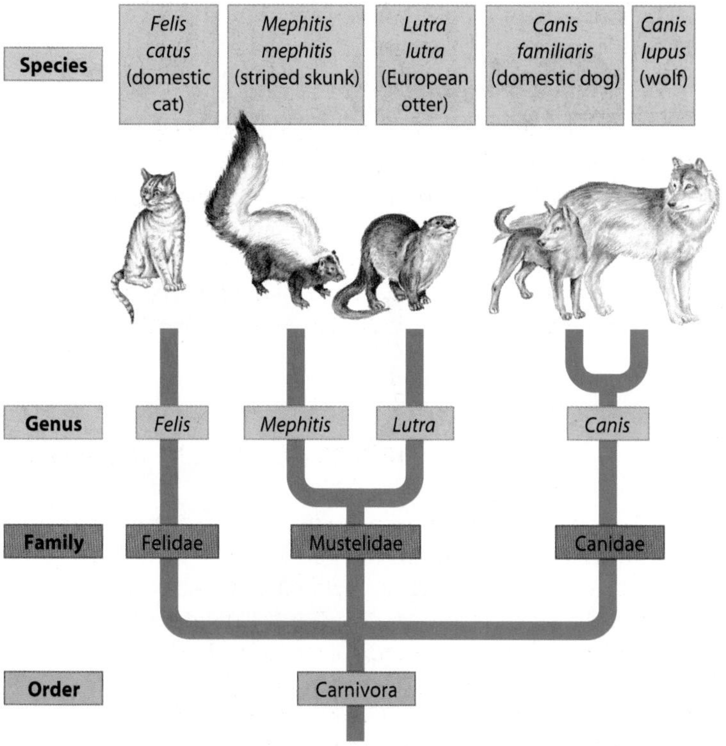
Species
Felis catus (domestic cat)
Mephitis mephitis (striped skunk)
Lutra lutra (European otter)
Canis familiaris (domestic dog)
Canis lupus (wolf)
Genus
Felis
Mephitis
Lutra
Canis
Family
Felidae
Mustelidae
Canidae
Order
Carnivora

Figure 15.7B The relationship between classification and phylogeny

15.8 Cladograms are diagrams based on shared characters among species

Charles Darwin envisioned the goals of modern systematics when he wrote in *The Origin of Species,* "Our classifications will come to be, as far as they can be so made, genealogies."

The science of phylogenetic systematics, with the goal of making classification as consistent as possible with evolutionary history, entered a vigorous new era in the 1960s. Just as molecular methods became readily available for comparing species, computer technology helped usher in a new approach called cladistics.

Cladistics involves the identification of **clades** (from the Greek *clados,* branch), evolutionary branches that consist of an ancestral species and all its descendants. Such a group of organisms, be it a genus, family, or some higher taxon, is said to be **monophyletic** (meaning "single tribe"). Identifying clades makes it possible to construct classification schemes that reflect the branching pattern of evolution.

Cladistics is based on the Darwinian concept that evolution proceeds when a new heritable trait develops in an organism and is passed on to its descendants. Groups of organisms that share such a new, or derived, trait are more closely related to each other than to groups that have only the original set of traits. The new traits are called **shared derived characters**, whereas the original traits present in the ancestral groups are called **shared primitive characters**.

The simplified example in **Figure 15.8A** illustrates the construction of a simple **cladogram**, a diagram depicting the pattern of shared characters. The figure compares four taxa (all vertebrate animal species) according to the presence or absence in these taxa of a set of four homologous traits, or characters. The color coding highlights how these four characters are shared among the four vertebrates.

An important part of cladistics is a comparison between a so-called ingroup and an outgroup. The **ingroup** (the three mammals in this simplified example) is the group of taxa that is actually being analyzed. The **outgroup** is closely related to the ingroup, but is not a member of it. In our example, the turtle (representing reptiles, the outgroup) and the mammals (collectively the ingroup) are all related in that they are vertebrates. The outgroup provides a reference point for distinguishing primitive characters from derived characters. In our example, a vertebral column (backbone) is a shared primitive character. It is present in all the taxa, including the outgroup. Derived characters are the evolutionary innovations that define the sequence of branch points in the phylogeny of the ingroup. Hair and mammary glands are derived characters that distinguish all mammals from reptiles.

Now let's see how having the outgroup for reference helps us formulate a hypothesis about the relationships between the taxa making up the ingroup. Consider the third character in the figure—gestation, the carrying of offspring in the uterus within the female parent. The outgroup does not exhibit gestation. Instead, turtles and most other reptiles lay eggs with a shell. One of the mammals, the duck-billed platypus, also lays eggs with a shell; and we might infer from this that egg-laying is a primitive feature of the vertebrates we are studying and that the duck-billed platypus is a primitive mammal. In fact, this hypothesis is strongly supported by structural and molecular evidence, and the duck-billed platypus is more like the ancestors of mammals (which were reptile-like) than the other mammals in the figure.

Our outgroup comparison tells us that egg-laying is primitive, so we can deduce that gestation (offspring carried in a uterus and nourished by a placenta) is a derived character. We can infer that gestation evolved in an ancestor common to the kangaroo and beaver that is more recent than the ancestor shared by all mammals, including the platypus.

A cladogram is constructed from a series of two-way branch points, each

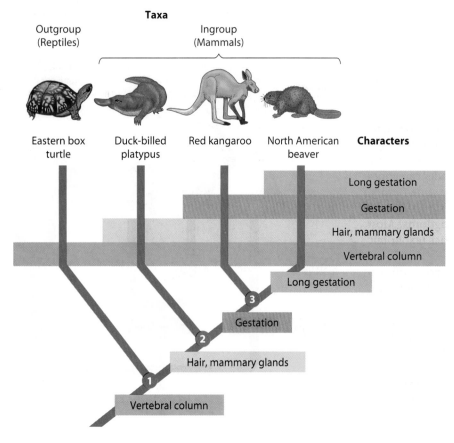

Figure 15.8A Constructing a cladogram

of which represents the divergence of two groups from a common ancestor with the emergence of a lineage possessing a new set of derived traits. The cladogram in Figure 15.8A is based on comparisons of the four homologous characters. Each node (branch point) represents an ancestor common to all taxa above that node. For instance, the tree indicates that the ancestor at node 2 gave rise to all the mammals. The sequence of branching represents the order in which new traits evolved and the historical chronology of when groups last shared a common ancestor. For instance, the yellow bar indicates that hair and mammary glands evolved in the lineage that led to all the mammals; gestation and long gestation were derived later in the course of mammalian evolution. The last common ancestor to all mammals lived longer ago than did the last ancestor shared by kangaroos and beavers. Remember that in a cladogram, the ancestor at each node and all of the taxa above that node should represent a monophyletic taxon. Thus, a cladogram can form the basis of a phylogenetic tree.

Another key aspect of cladistics is **parsimony**, the quest for the simplest (and thus probably the most likely) explanation for observed phenomena. Useful in many areas of science, parsimony in systematics means that the simplest hypotheses that are consistent with the comparative data are likely to be the correct ones. Systematists use the principle of parsimony to construct phylogenetic trees that represent the smallest number of evolutionary changes. For instance, parsimony leads to the hypothesis that a beaver is more closely related to a kangaroo than to a platypus because the beaver and the kangaroo both have gestation. It is possible that gestation evolved twice, once in the kangaroo lineage and independently in the beaver lineage, but this explanation is more complicated and less likely. Typical cladistic analyses involve much more complex data sets than we presented in Figure 15.8A and are usually handled by computer programs designed to construct parsimonious trees.

Cladistics has become the most widely used method in systematics, and it is shaking some phylogenetic trees. Cladistic analysis not only supports the once controversial view that birds evolved from theropod dinosaurs (see the chapter introduction), but also suggests that birds belong in the reptilian clade. In traditional vertebrate taxonomy, crocodiles, snakes, lizards, and other reptiles were classified in the class Reptilia, while birds were placed in the separate class Aves. In this traditional classification, however, the reptilian clade would not be monophyletic—it would not include its ancestral species and all of its descendants, one group of which includes the birds. Studies of shared derived characters and the fossil record indicate that birds evolved from a lineage of dinosaurs that is more closely related to crocodiles than to lizards and snakes. Thus, a classification scheme that reflects phylogenetic branchings would recognize that birds and crocodiles make up one clade and lizards and snakes make up another, as shown in **Figure 15.8B**. And if we go back as far as the ancestor that crocodiles share with lizards and snakes, we see that the reptilian clade must also include birds. Many classification schemes, including the one used in this textbook, adopt this view.

The more we know about an organism and its relatives, the more accurately we can portray its phylogeny. We can be less certain about the evolutionary history of extinct groups represented only by fossils. We can be more certain about phylogeny within a group that has living representatives that can be thoroughly studied. Systematists use many kinds of evidence, such as structural and developmental features, molecular data, and behavioral traits of organisms, to re-create evolutionary histories. However, even the best phylogenetic tree only represents the most likely hypothesis based on the available evidence. As new data accumulate, hypotheses are revised and new trees drawn. In the next module, we consider the source of much of this new evidence—molecular comparisons.

? To distinguish a particular clade of mammals within the larger clade that corresponds to class Mammalia, why is hair not a useful characteristic?

■ Hair is a shared primitive character common to all mammals and cannot be helpful in distinguishing different mammalian subgroups.

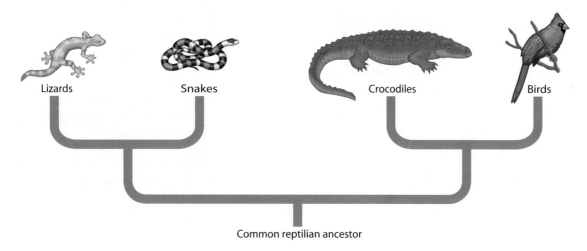

Figure 15.8B A phylogenetic tree of reptiles

Lizards   Snakes   Crocodiles   Birds

Common reptilian ancestor

15.9 Molecular biology is a powerful tool in systematics

Molecular systematics—comparing nucleic acids or other molecules to infer relatedness—is a valuable approach for tracing evolutionary histories. At the molecular level, the evolutionary divergence of species parallels the accumulation of differences in their genomes. The more recently two species have branched from a common ancestor, the more similar their DNA sequences should be.

Scientists have sequenced more than 20 billion bases of nucleic acid data from thousands of species. This enormous collection of data has fueled a boom in the study of phylogeny, clarifying many evolutionary relationships. The phylogenetic tree for the family Ursidae (bears) and the family Procyonidae (raccoons) shown in **Figure 15.9A** was constructed from comparisons of DNA and blood proteins. The molecular evidence indicates that the giant panda is more closely related to bears than to raccoons and that the lesser panda is a member of the raccoon family. Notice that this phylogenetic tree can include a time line because the fossil record has provided evidence for the timing of some of the branchings of ancestral groups.

Bears and raccoons are closely related mammals, but systematists can also use DNA analyses to assess relationships between groups of organisms that are so phylogenetically distant that structural similarities are absent. It is also possible to trace phylogenies among groups of present-day bacteria and other microorganisms for which we have no fossil record at all. And molecular systematics enables scientists to compare genetic divergence among individuals within a species. Molecular biology has helped to extend systematics to the extremes of evolutionary relationships far above and below the species level, ranging from the major branches of the tree of life to its finest twigs.

The ability of molecular trees to encompass both short and long periods of time is based on the observation that different genes evolve at different rates. The DNA coding for ribosomal RNA (rRNA) changes relatively slowly, so comparisons of DNA sequences in these genes are useful for investigating relationships between taxa that diverged hundreds of millions of years ago. Studies of rRNA sequences, for example, have shown that humans are more closely related to fungi than to green

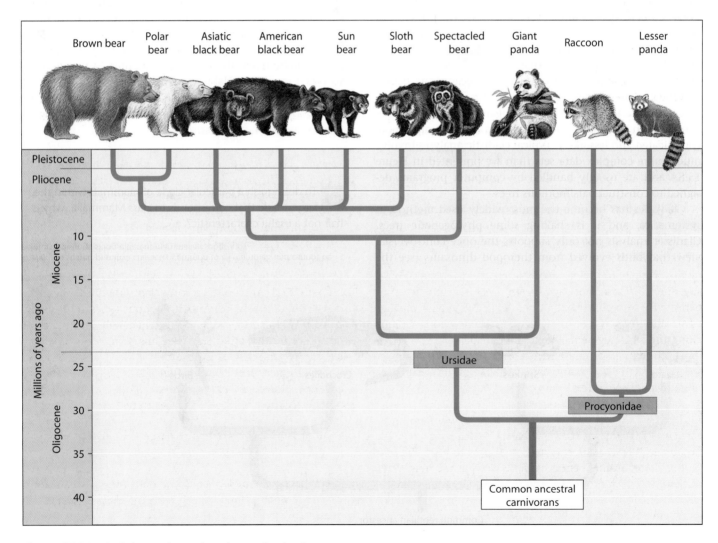

Figure 15.9A A phylogenetic tree based on molecular data

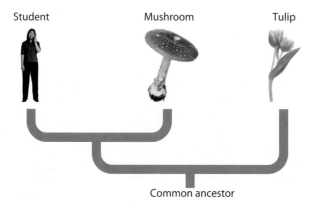

Figure 15.9B Phylogenetic tree of humans, plants, and fungi

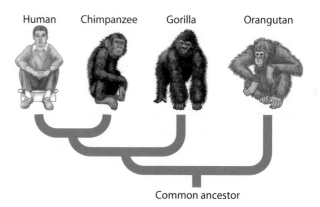

Figure 15.9C Phylogenetic tree of some hominoids

plants—something that morphological comparisons alone certainly could not do (**Figure 15.9B**).

In contrast, the DNA in mitochondria (mtDNA) evolves relatively rapidly and can be used to investigate more recent evolutionary events. For example, researchers have used mtDNA sequences to study the relationships between different Native American cultures. Their studies support earlier evidence that the Pima of Arizona, the Maya of Mexico, and the Yanomami of Venezuela are closely related, probably descending from the first wave of immigrants to cross the Bering Land Bridge from Asia to the Americas during the glaciation of the late Pleistocene epoch, about 13,000 years ago.

DNA Comparisons Molecular comparisons of nucleic acids often pose technical challenges. The first step is to align homologous nucleic acid sequences from the two species being studied. If the species are closely related, the sequences likely differ at only one or a few sites. In contrast, comparable nucleic acid sequences in distantly related species usually have different bases at many sites and may even have different lengths. This is the case because over longer periods of time, insertions and deletions accumulate, altering the lengths of the gene sequences (see Module 10.16). To address this problem, systematists use computer programs that search for similar sequences along DNA segments from both species. The programs then reestablish alignment by inserting gaps in one DNA segment to compensate for mutations that altered sequence length, making it possible to compare sequences in homologous regions along the DNA segments. Such comparisons can reveal the most fundamental similarities or differences between species—exactly how many bases are alike or different within equivalent sequences.

Molecular Clocks Some regions of genomes appear to accumulate changes at constant rates. Comparisons of certain homologous DNA sequences for taxa known to have diverged during a certain time period have shown that the number of nucleotide substitutions is proportional to the time that has elapsed since the lineages branched. For example, homologous genes of bats and dolphins are much more alike than are those of sharks and tuna. This is consistent with the fossil evidence that sharks and tuna have been on separate evolution-

ary paths much longer than have bats and dolphins. In this case, molecular divergence has kept better track of time than have changes in morphology.

For a gene shown to have a reliable average rate of change, a **molecular clock** can be calibrated in actual time by graphing the number of nucleotide differences against the dates of evolutionary branch points known from the fossil record. The graph line can then be used to estimate the dates of other evolutionary episodes not documented in the fossil record. Some biologists are skeptical about the accuracy of molecular clocks, but their judicial use may provide approximate markers of elapsed time. An abundant fossil record extends back only about 550 million years, and molecular clocks have been used to date evolutionary divergences that occurred a billion or more years ago. But the estimates assume that the clocks have been constant for all that time. Thus, such estimates are likely to have large errors.

Genome Evolution Now that we can compare entire genomes, including our own, some interesting facts have emerged. For example, the genomes of humans and chimpanzees are 99% identical. The phylogenetic tree of hominoids (great apes and humans) shown in **Figure 15.9C** reflects these molecular comparisons. An even more remarkable fact is that homologous genes are widespread and can extend over huge evolutionary distances. While 99% of the genes of humans and mice are certainly not identical, they are detectably homologous. And 50% of our genes are homologous with those of yeast. This remarkable commonality demonstrates that all living organisms share many biochemical and developmental pathways.

Evolutionary theory holds that all of life has a common ancestor. Molecular systematics is helping to link all living organisms into a vast tree of life, as we see next.

Web/CD Thinking as a Scientist *How Is Phylogeny Determined Using Protein Comparisons?*

> **?** What type of molecular comparisons would help to determine whether fungi are more closely related to plants or to animals?

■ Comparisons of molecules that change or evolve very slowly, such as ribosomal RNA (or the DNA that codes for rRNA)

15.10 Arranging life into kingdoms is a work in progress

As you have learned, phylogenetic trees are hypotheses about evolutionary history. Like all hypotheses, they are revised, or in some cases completely rejected, in accordance with new evidence. Molecular systematics and cladistics are combining to remodel phylogenetic trees and challenge conventional classifications, even at the kingdom level.

Over the years, many schemes have been proposed for classifying organisms into kingdoms. Historically, a two-kingdom system divided all organisms into plants and animals. But it was beset with problems. Where do prokaryotes fit? Where do unicellular, photosynthetic protists that move like animals belong? And what about the fungi?

In 1969, American ecologist Robert H. Whittaker argued effectively for a **five-kingdom system** that places prokaryotes in the kingdom Monera (Figure 15.10A). Organisms of the other four kingdoms all consist of eukaryotic cells. Kingdoms Plantae, Fungi, and Animalia contain multicellular eukaryotes that differ in structure, development, and modes of nutrition. Plants make their own food by photosynthesis. Fungi live by decomposing the remains of other organisms and absorbing small organic molecules. Most animals live by ingesting food and digesting it within their bodies. The kingdom Protista, containing all eukaryotes that do not fit the definition of plant, fungus, or animal, is a taxonomic grab bag in the five-kingdom system.

It is important to keep in mind that classification schemes are not facts of nature, but human constructions. The five-kingdom system was one attempt to classify the diversity of life into a scheme that is useful and reflective of evolutionary history. In the last decade, molecular studies have highlighted serious flaws in the five-kingdom system, and researchers have proposed new classifications ranging from six to dozens of kingdoms. But as debate continues at the kingdom level, there is more of a consensus that the kingdoms of life can now be grouped into three higher levels of classification called domains.

Figure 15.10B shows the **three-domain system.** Whereas the five-kingdom system reflects the two fundamentally different groups of living organisms—prokaryotes and eukaryotes—this newer scheme recognizes three basic groups: two domains of prokaryotes, Bacteria and Archaea, and one domain of eukaryotes, called Eukarya. Bacteria and archaea differ in a number of key ways, which we will discuss in Chapter 16.

Molecular and cellular evidence indicates that two lineages of prokaryotes (the bacteria and the archaea) diverged very early in the evolutionary history of life. Molecular evidence also suggests that archaea are more closely related to eukaryotes than to bacteria. Thus, Figure 15.10B indicates that the lineage of domain Eukarya diverged from domain Archaea after archaea and bacteria diverged. Comparisons of complete genomes from the three domains, however, show that, especially during the early history of life, there have been substantial interchanges of genes between organisms in the different domains. It is even possible that the first eukaryote arose

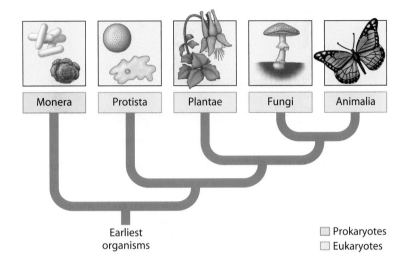

Figure 15.10A The five-kingdom classification scheme

□ Prokaryotes
□ Eukaryotes

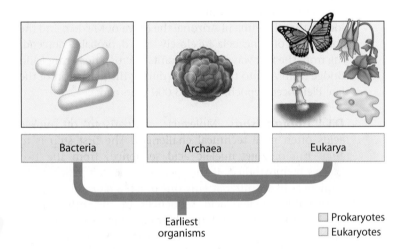

Figure 15.10B The three-domain classification scheme

□ Prokaryotes
□ Eukaryotes

through a fusion between a bacterial and an archaeal ancestor. As new data and new methods for analyzing that data emerge, constructing the universal tree of life will continue to engage and challenge systematists.

Thus, it is important to understand that defining the higher categories of classification (the kingdoms and domains of life) will always be a work in progress. In the next unit, we examine the enormous diversity of organisms that have populated Earth since life first arose over 3.5 billion years ago. As we do so, keep in mind that taxonomic groupings and phylogenetic trees are hypotheses, using the best available data to trace the evolutionary history of life.

Web/CD Activity 15C *Classification Schemes*

? In comparing the five-kingdom system with the three-domain system, how many of the kingdoms fall into domain Eukarya?

CHAPTER REVIEW

Reviewing the Concepts

Macroevolution and Earth's History (15.1–15.5)

The fossil record documents the main events in the history of life. In the geologic record, major transitions in life-forms separate eras; smaller changes divide eras into periods **(15.1)**. Radiometric dating, which measures the decay of radioactive isotopes, can gauge the actual ages of fossils and the rocks in which they are found **(15.2)**.

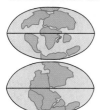 **Continental drift** is the slow, incessant movement of Earth's crustal plates on the hot mantle. The formation of Pangaea altered habitats and triggered extinctions. The separation of the continents affected the distribution and diversification of organisms **(15.3)**. Volcanoes and earthquakes result from plate tectonics, the movements of Earth's plates **(15.4)**.

Mass extinctions occurred at the end of the Permian and Cretaceous periods. The Cretaceaous extinction, which included the dinosaurs, may have been caused by an asteroid. A rebound in diversity follows mass extinctions **(15.5)**.

Phylogeny and Systematics (15.6–15.10)

Phylogeny, the evolutionary history of a group, is based on identifying homologous structures and molecular sequences that provide evidence of common ancestry. Analogous similarities result from convergent evolution in similar environments. Systematics involves the analytical study of diversity and phylogeny **(15.6)**.

Classification. Taxonomists assign a binomial, consisting of a genus and species name, to each species. A genus may include a group of related species. Genera are grouped into progressively larger categories: family, order, class, phylum, kingdom, and domain. A phylogenetic tree is a hypothesis of evolutionary relationships **(15.7)**.

Cladistics uses shared derived characters to define monophyletic taxa. Shared primitive characters are common to ancestral groups. The simplest (most parsimonious) hypothesis creates the most likely phylogenetic tree **(15.8)**.

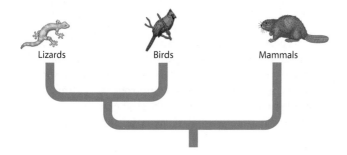

Molecular systematics develops phylogenetic hypotheses based on molecular comparisons. Some regions of DNA change at a rate consistent enough to serve as molecular clocks to date evolutionary events. Homologous genes are found in many diverse species **(15.9)**.

Kingdoms and domains. In the five-kingdom system, prokaryotes are in kingdom Monera. Eukaryotes (plants, animals, protists, and fungi) are grouped in separate kingdoms. The domain system recognizes the prokaryotic domains Bacteria and Archaea. Eukaryotes are placed in domain Eukarya **(15.10)**.

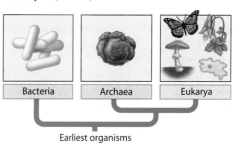

Connecting the Concepts

1. Fill in this concept map, which summarizes some of the key ideas about systematics.

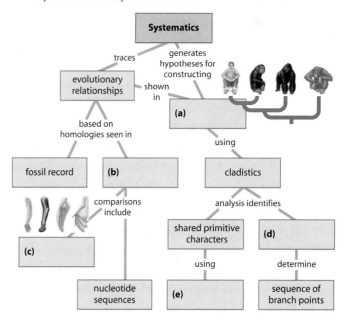

Testing Your Knowledge

Multiple Choice

2. The animals and plants of India are almost completely different from the species in nearby Southeast Asia. Why might this be true?
   a. They have become separated by convergent evolution.
   b. The climates of the two regions are different.
   c. India is in the process of separating from the rest of Asia.
   d. Life in India was wiped out by ancient volcanic eruptions.
   e. India was a separate continent until relatively recently.

CHAPTER 15 *Tracing Evolutionary History*   **311**

3. Mass extinctions that occurred in the past
   a. cut the number of species to the few survivors left today.
   b. resulted mainly from the separation of the continents.
   c. occurred regularly, about every million years.
   d. were followed by diversification of the survivors.
   e. wiped out land animals but had little effect on marine life.

4. If you were using cladistics to build a phylogenetic tree of cats, which of the following would be the best choice for an outgroup?
   a. wolf   d. turtle
   b. leopard   e. lion
   c. domestic cat

5. Which of the following would most likely provide the best data for determining the phylogeny of three very closely related species?
   a. the fossil record
   b. a comparison of embryological development
   c. an analysis of their morphological differences and similarities
   d. a comparison of nucleotide sequences in homologous genes and mitochondrial DNA
   e. a comparison of their ribosomal DNA sequences

6. Two worms in the same class must also be grouped in the same
   a. order.   c. genus.   e. species.
   b. phylum.   d. family.

7. Major divisions in the geologic record are marked by
   a. radioactive dating.
   b. distinct changes in the types of fossilized life.
   c. continental drift.
   d. regular time intervals measured in millions of years.
   e. the appearance, in order, of prokaryotes, eukaryotes, protists, plants, fungi, and animals.

Describing, Comparing, and Explaining

8. Which of these are more likely to be closely related: two species with similar appearance but very divergent gene sequences, or two species with very different appearances but nearly identical genes? Explain.

9. Discuss various hypotheses that have been suggested to explain the Cretaceous mass extinction that ended the reign of the dinosaurs. What kind of evidence supports each hypothesis?

10. Does each of the following pairs of structures more likely represent analogy or homology: (a) porcupine quills and cactus spines; (b) cat paw and human hand; (c) bird wing and dragonfly wing? Explain your answers.

11. What type of molecular comparisons would have been used to determine the very early branches in the universal tree of life? Explain.

12. What is a molecular clock? What assumptions underlie the use of a molecular clock?

Applying the Concepts

13. Your measurements indicate that a fossilized skull you unearthed has a carbon-14/carbon-12 ratio about one-sixteenth that of the skulls of present-day animals. What is the approximate age of the fossilized skull? (The half-life of carbon-14 is 5,730 years.)

14. Birds and mammals have four-chambered hearts. Most reptiles have three-chambered hearts. According to the principle of parsimony, birds and mammals should be placed in the same clade. Why do most phylogenetic trees show birds grouped with reptiles and mammals on a separate branch?

15. A paleontologist is comparing fossils from three dinosaurs and *Archaeopteryx*. The following table shows the distribution of characters for each species, where 1 means that the trait is present and 0 means it is not. The outgroup comparison (not shown in the table) had none of the traits. Arrange these species on the cladogram shown. Also indicate the derived character that defines each branch point on the cladogram.

Trait	Velociraptor	Coelophysis	Archaeopteryx	Allosaurus
Hollow bones	1	1	1	1
Three-fingered hand	1	0	1	1
Half-moon-shaped wristbone	1	0	1	0
Reversed first toe	0	0	1	0

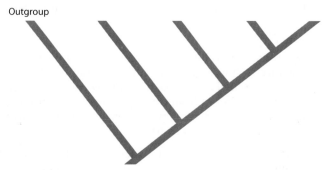

Outgroup

16. Experts estimate that human activities cause the extinction of hundreds of species every year. The natural "background" rate of extinction is thought to be a few species per year. As we continue to alter the environment, especially by destroying tropical rain forests and altering Earth's climate, the resulting wave of extinctions will probably rival the end of the Cretaceous period, when perhaps half the species on Earth disappeared. Most biologists are alarmed at this prospect, which they see as a catastrophe of unprecedented proportions. What are some of the reasons for their concern? Other people are not as worried, because life has endured numerous mass extinctions in the past and has always bounced back. Are the current mass extinctions different? Why or why not? What might be the consequences for the surviving species? What might be the consequences for humans?

Answers to all questions can be found in Appendix 3.

For study help and Activities, go to campbellbiology.com or the student CD-ROM.

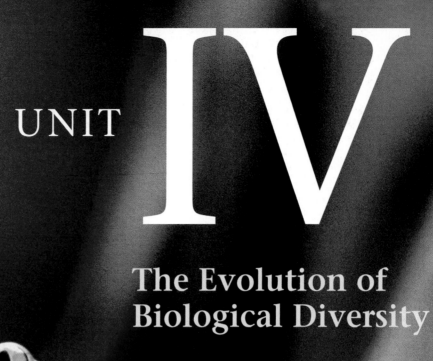

UNIT IV

The Evolution of Biological Diversity

16   THE ORIGIN AND EVOLUTION OF MICROBIAL LIFE: PROKARYOTES AND PROTISTS

17   PLANTS, FUNGI, AND THE COLONIZATION OF LAND

18   THE EVOLUTION OF ANIMAL DIVERSITY

19   HUMAN EVOLUTION

Layers of a bacterial mat

EARLY EARTH AND THE ORIGIN OF LIFE

16.1 Life began on a young Earth
16.2 How did life originate?
16.3 Stanley Miller's experiments showed that organic molecules could have arisen on a lifeless Earth
16.4 The first polymers may have formed on hot rocks or clay
16.5 The first genetic material and enzymes may both have been RNA
16.6 Membrane-enclosed molecular cooperatives may have preceded the first cells

PROKARYOTES

16.7 Prokaryotes have inhabited Earth for billions of years
16.8 Bacteria and archaea are the two main branches of prokaryotic evolution
16.9 Prokaryotes come in a variety of shapes
16.10 Various structural features contribute to the success of prokaryotes
16.11 Prokaryotes obtain nourishment in a variety of ways
16.12 Archaea thrive in extreme environments—and in other habitats
16.13 Bacteria include a diverse assemblage of prokaryotes
16.14 Some bacteria cause disease
16.15 Bacteria can be used as biological weapons
16.16 Prokaryotes help recycle chemicals and clean up the environment

PROTISTS

16.17 The eukaryotic cell probably originated as a community of prokaryotes
16.18 Protists are an extremely diverse assortment of eukaryotes
16.19 A tentative phylogeny of eukaryotes includes multiple clades of protists
16.20 Diplomonads and euglenozoans include some flagellated parasites
16.21 Alveolates have sacs beneath the plasma membrane and include dinoflagellates, apicomplexans, and ciliates
16.22 Stramenopiles are named for their "hairy" flagella and include the water molds, diatoms, and brown algae
16.23 Amoebozoans have pseudopodia and include amoebas and slime molds
16.24 Red algae and green algae are the closest relatives of land plants
16.25 Multicellularity evolved several times in eukaryotes

How Ancient Bacteria Changed the World

CRUISING IN SHALLOW WATER off an island in the Bahamas, an undersea research vessel (see the facing page) glides over an array of what looks like pockmarked boulders. Are they lava rocks from an ancient volcano? Mounds of coral? Massive sponges? Their surfaces are greenish and sticky. They seem rigid, and an edge of one reveals layers. Back in the laboratory, a look at a sample through a microscope shows that prokaryotic organisms coat the mounds. Most of the prokaryotes turn out to be cyanobacteria, photosynthetic bacteria.

The photograph at the top left of this page shows part of a cyanobacterial mat that was growing in a warm lagoon in Baja California. The reddish brown layers are bands of sediment built up by the bacteria. Coating the top of the mat where they are exposed to sunlight, sticky bacteria concentrate sand grains and other fine particles from the seawater. While a sediment layer accumulates, the bacteria keep migrating to the surface and growing over it. A layered mat builds up as this process is repeated over and over. The large Bahamian mounds formed in this way over thousands of years, hardening as their older sediment layers solidified. These rocklike structures, composed of many layers of bacteria and sediment, are called **stromatolites** (from the Greek *stroma*, bed, and *lithos*, rock).

Stromatolites, rocklike bacterial mats

The Origin and Evolution of Microbial Life: Prokaryotes and Protists

The cyanobacteria that produce such layered mats and stromatolites are descended from some of the oldest organisms known—photosynthetic prokaryotes that dominated Earth some 3 billion years ago. The fossil record indicates that at that time, greenish lawns of prokaryotes covered virtually every wet, sunlit surface on the planet. Animals that might have grazed the prokaryotic lawns had not evolved yet, and the unchecked growth of ancient prokaryotes built up huge expanses of stromatolites, many of which eventually became fossilized.

The oldest known fossils of life on Earth are stromatolites dating from 3.5 billion years ago. The prominence of fossilized stromatolites in rocks around 2.5 billion years old marks the time when ancient photosynthetic prokaryotes were so numerous that the gaseous oxygen (O_2) they produced was changing Earth's anaerobic atmosphere to an aerobic atmosphere. This change set the stage for the evolution of all other aerobic life.

Photosynthetic prokaryotes dominated Earth from 3 billion years ago to about 1 billion years ago and then declined significantly. Their direct descendants, the cyanobacteria and other photosynthetic bacteria of today, remain abundant in freshwater lakes and ponds and in shallow oceans, but they only form thick mats or mounds in environments inhospitable to most other life. Thus, great expanses of stromatolites are a thing of the past. Nevertheless, the grand, global legacy of ancient photosynthetic prokaryotes—the aerobic atmosphere—remains with us today. And the aerobic atmosphere is but one illustration of the profound effects that living organisms can have on the environment.

Tracing the roles played by various organisms in the history of life on Earth is one of two main objectives in this unit of chapters. Our other goal is to introduce you to the diversity of life. We start in this chapter with the prokaryotes and the protists. As mentioned in Modules 1.5 and 15.10, the prokaryotes are the members of the domains Bacteria and Archaea; the protists are a diverse assemblage of mostly unicellular eukaryotes. ■ ■ ■

16.1 Life began on a young Earth

Figure 16.1A An artist's conception of a common scene on Earth when mats of prokaryotes were the main form of life

Imagine visiting Earth some 3 billion years ago. The planet bristles with volcanoes spewing gases into the atmosphere and molten rock onto the surface. Greenish "stepping stones," actually thick mats of prokaryotes, dominate a shoreline. Oxygen released from photosynthetic prokaryotes in the mats will change the atmosphere forever. **Figure 16.1A** illuminates the inseparable histories of planet Earth and its organisms.

Geologic and biological history have been closely intertwined since life began. As we saw in Chapter 15, the formation and subsequent breakup of the supercontinent Pangaea had a tremendous effect on the diversity and geographic distribution of life. Conversely, life has changed the planet it inhabits, sometimes profoundly. The photosynthetic prokaryotes that first released oxygen to the air completely altered Earth's atmosphere. Much more recently, the emergence of *Homo sapiens* has changed the land, water, and air on a scale and at a rate unprecedented for a single species.

But Earth's story begins before life appeared. Earth is one of nine planets orbiting the sun, which is one of billions of stars in the Milky Way. The Milky Way, in turn, is one of billions of galaxies in the universe. Gazing at stars gives us a look back in time. The star closest to our sun is 4 light years—40 trillion kilometers—away; we see the light it emitted four years ago. Some stars you can see in the night sky are so far away that their light is just reaching us, though they burned out millions of years ago.

The universe has not always been so spread out. Physicists have evidence that before the universe existed in its present form, all of its matter was concentrated in one mass. The mass seems to have blown apart with a "big bang" sometime between 10 and 20 billion years ago and to have been expanding ever since. Our solar system probably arose from a swirling cloud of dust. Most of the dust condensed in the center to form the sun, but some matter was left orbiting around the infant sun in concentric rings. Within each

orbit, kernels of matter had enough gravity to draw nearby dust and ice particles together, forming the planets.

Planet Earth formed about 4.6 billion years ago as a cold world. Later on, heat generated by the impact of meteorites, radioactive decay, and compaction by gravity thawed Earth and eventually turned it into a molten mass. The mass then sorted into layers of varying densities, with most of the nickel and iron sinking to the center and forming a core. Less dense material became concentrated in a mantle surrounding the core, and the least dense material settled on the surface, solidifying into a thin crust.

The first atmosphere was probably composed mostly of hot hydrogen gas (H_2). Because the gravity of Earth was not strong enough to hold such small molecules, the H_2 soon escaped into space. Volcanoes and other vents through the crust belched gases that formed a new atmosphere. Analyses of gases vented by modern volcanoes have led scientists to speculate that this second early atmosphere consisted of carbon dioxide (CO_2), nitrogen (N_2), hydrogen sulfide (H_2S), and water vapor (H_2O), with possibly some methane (CH_4) and ammonia (NH_3). The first seas were created by torrential rains that began when the planet had cooled enough for water vapor in the atmosphere to condense. Not only was the atmosphere of young Earth very different from the one we know today, but lightning, volcanic activity, and ultraviolet radiation were much more intense. It was in such an environment that life began.

Fossil evidence indicates the presence of primitive cellular organisms within a few hundred million years after Earth's crust solidified. We would guess from the relatively simple structure of prokaryotic cells (compared with that of eukaryotic cells) that early organisms were primitive prokaryotes, and the fossil record supports this idea. Notice how similar the fossilized mat in **Figure 16.1B** is to the still-growing one shown in the chapter introduction. The fossilized stromatolite in Figure 16.1B is from western Australia. Dating from about 3.5 billion years ago, it is among the oldest signs of life found thus far.

Could other organisms have predated those that formed the ancient stromatolites? Photosynthetic prokaryotes are among the simplest kinds of organisms we know that are autotrophic (produce their own food). However, photosynthesis is not a simple metabolic process, and it is unlikely that photosynthetic prokaryotes were the first forms of life on Earth. In fact, the evidence that photosynthetic prokaryotes lived 3.5 billion years ago is strong support for the hypothesis that life in a simpler form, unable to make its own food, arose much earlier, perhaps as early as 3.9 billion years ago.

Figure 16.1C uses the analogy of a clock ticking down from the origin of Earth to the present to show some major events in the history of Earth and its life. The diagram puts these oldest of fossils in historical perspective. It also dates the atmospheric change caused by ancient prokaryotes and the appearance of single-celled eukaryotes, two events that are subjects of the last two sections of this chapter. Notice in

Figure 16.1B A cross section of a fossilized stromatolite

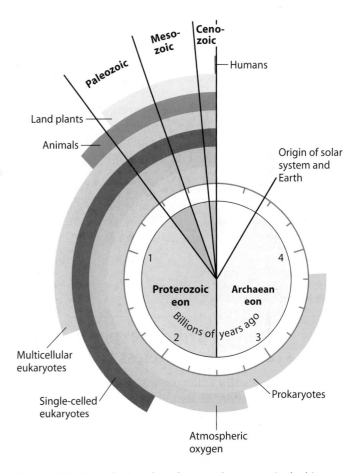

Figure 16.1C A clock analogy for some key events in the history of Earth and its life

Figure 16.1C how long prokaryotes inhabited Earth before the evolution of animals and land plants.

Web/CD Activity16A *The History of Life*

? What do fossilized stromatolites suggest about the evolution of prokaryotes?

■ Prokaryotes must have existed at least 3.5 billion years ago when the oldest fossilized stromatolites were formed.

16.2 How did life originate?

From the time of the ancient Greeks until well into the 19th century, it was common "knowledge" that life regularly arises from nonliving matter. Many people believed, for instance, that flies came from rotting meat, fish from ocean mud, and microorganisms from broth. Experiments performed in the 1600s showed that relatively large organisms, such as insects, cannot arise spontaneously from nonliving matter. However, debate about how microscopic organisms arise continued until the 1860s. In 1862, the French scientist Louis Pasteur confirmed what many others had suspected: All life today, including microbes, arises only by the reproduction of preexisting life.

Pasteur ended the argument over spontaneous generation of modern organisms, but he did not address the question of how life arose in the first place. The question remains with us today, although we have come a long way toward answering it in the last half century. We can be fairly certain that the first organisms came into being between about 3.9 billion years ago, when steam condensed to form seas, and 3.5 billion years ago, when the planet was inhabited by prokaryotes complex enough to build stromatolites. But what events led to the origin of life?

Most biologists subscribe to the hypothesis that the earliest form of life was much simpler than anything alive today and that life did first develop from nonliving materials. The earliest lifelike entities may have been aggregates of molecules with a particular arrangement that made simple metabolism and self-replication possible—two important properties that define life. Because living organisms consist of polymers formed from small organic molecules (monomers), the synthesis and accumulation of small organic molecules must have been the earliest chemical stage preceding the origin of life. Some scientists propose that meteorites and comets seeded Earth with organic molecules, but this contribution was likely to have been small. The prevailing opinion is that most of the first organic molecules arose from inorganic materials on the early planet. The formation of polymers from organic monomers would have been a second stage preceding the origin of life. Aggregates of abiotically produced molecules surrounded by a membrane may have provided the setting for the third stage: the origin of a mechanism of polymer replication, a primitive form of heredity. In the next four modules, we examine the stages through which organic molecules may have formed and assembled on early Earth, leading to the origin of the first forms of life.

? According to the hypothesis introduced in this module, the first cells were preceded by the abiotic formation of small _____ molecules, which then joined to form _____.

■ organic . . . polymers

16.3 Stanley Miller's experiments showed that organic molecules could have arisen on a lifeless Earth

In 1953, when Stanley Miller was a 23-year-old graduate student in the laboratory of Harold Urey at the University of Chicago, he performed some experiments that would attract global attention. Miller was the first to show that amino acids and other organic molecules could have been generated on a lifeless Earth (Figure 16.3A).

Miller's experiments were a test of a hypothesis about the origin of life developed in the 1920s by Russian biochemist A. I. Oparin and British geneticist J. B. S. Haldane. Oparin and Haldane independently proposed that the conditions on early Earth could have generated a collection of organic molecules that in turn could have given rise to the first living organisms. They reasoned that present-day conditions on Earth do not allow the spontaneous synthesis of organic compounds simply because the atmosphere is rich in O_2. Oxygen is corrosive: As a strong oxidizing agent, it tends to disrupt chemical bonds by extracting electrons from them. However, before the early prokaryotes added O_2 to the air, Earth probably had a reducing (electron-adding) atmosphere instead of an oxidizing one. An early reducing atmosphere could have caused simple molecules to combine, forming

Figure 16.3A Stanley Miller re-creating his 1953 experiment

more complex ones. As Miller, now a professor at the University of California, San Diego, told us several years ago,

> Oparin proposed that the primitive atmosphere contained the gases methane, ammonia, hydrogen, and water, and that chemical reactions in that primitive atmosphere produced the first organic molecules. That hypothesis had a good deal of appeal, but without the experiments, it was talked about but not very well accepted.

The construction of complex molecules from simple ones also requires energy, and Miller and Urey reasoned that there were abundant energy sources in the environment of early Earth. Besides lightning discharges, ultraviolet radiation probably reached Earth's surface with much greater intensity than it does today. Evidence indicates that young suns emit more ultraviolet (UV) radiation than older suns. Also, in our modern atmosphere, a layer of ozone (O_3) screens out most UV radiation. Ozone, which forms from ordinary oxygen (O_2), would not have been present before photosynthesis arose.

Miller and Urey predicted that organic molecules would form from inorganic ones under conditions like those on early Earth. The apparatus shown with Dr. Miller in Figure 16.3A is similar to the one he used to test the prediction. **Figure 16.3B** indicates how the apparatus simulated conditions on the early Earth. A flask of warmed water represented the primeval sea. The "atmosphere" consisted of a mixture of water vapor, H_2, CH_4, and NH_3—the gases that scientists in the 1950s thought prevailed in the ancient world. Electrodes discharged sparks into the gas mixture to mimic lightning. Below the spark chamber, a glass jacket called a condenser surrounded the apparatus. Filled with cold water,

the condenser cooled and condensed the water vapor in the gas mixture, causing "rain," along with any dissolved compounds, to fall back into the miniature sea. As material circulated through the apparatus, the solution in the flask slowly changed color. As Dr. Miller described it,

> The first time I did the experiment, it turned red. Very dramatic! And then, after it turned red, it got more yellow and then brown as the sparking went on.

After the experiment proceeded for a week, Miller found a variety of organic compounds in the solution, including some of the amino acids that make up the proteins of organisms:

> The surprise was that we . . . got mainly organic compounds of biological significance. And the amino acids were formed, not in trace quantities, but abundantly! The experiment went beyond our wildest hopes.

Miller's early experiments stimulated a great deal of interest and research on the prebiotic (before-life) origin of organic compounds. Since the 1950s, Miller and other researchers using modifications of Miller's apparatus have made most of the 20 amino acids commonly found in organisms, as well as sugars, lipids, the nitrogenous bases present in nucleotides of DNA and RNA, and even ATP. These laboratory studies support the idea that many of the organic molecules that make up living organisms could have formed before life itself arose on early Earth.

Scientists now think that the composition of the atmosphere of early Earth was somewhat different from what Miller assumed in his historic first experiment. Some scientists doubt that the early atmosphere played a direct role in early chemical reactions. Controversy exists over whether the atmosphere contained enough H_4 and NH_3 to be reducing. Growing evidence indicates that the early atmosphere as a whole was chemically neutral, made up primarily of N_2 and CO_2. Still, it is possible that small "pockets" of the early atmosphere—perhaps near volcanic openings—were reducing.

Alternatively, submerged volcanoes and deep-sea hydrothermal vents—gaps in the Earth's crust where hot water and minerals gush into deep oceans—may have provided the initial chemical resources for life.

We asked Dr. Miller in what sort of place he thought life actually began on Earth:

> The usual assumption is that it began in the ocean. But you can legitimately propose that some of the [earliest] processes occurred in different areas. For example, some of the polymerization reactions that made larger organic molecules probably occurred on beaches that had dried out and heated up.

We discuss this possibility in the next module.

Web/CD Thinking as a Scientist *How Might Conditions on Early Earth Have Created Life?*

? What hypothesis was Stanley Miller testing with his experiments?

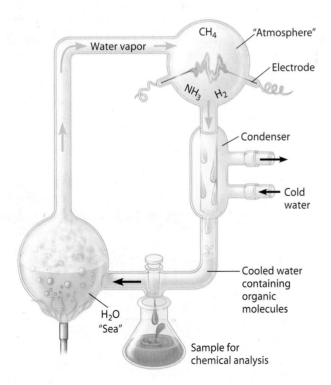

Figure 16.3B The synthesis of organic molecules in the Miller-Urey apparatus

■ The hypothesis that conditions on early Earth favored synthesis of organic molecules from inorganic ingredients

16.4 The first polymers may have formed on hot rocks or clay

After small organic molecules formed, the second major chemical step before life arose must have been polymerization—the formation of organic polymers, such as nucleic acids and proteins, from their monomers.

The polymers of life are synthesized by dehydration reactions that release a water molecule for each monomer added to the chain (see Module 3.3). In the living cell, specific enzymes catalyze these reactions. But polymerization also occurs in laboratory situations without enzymes—for example, when dilute solutions of organic monomers are dripped onto hot sand, clay, or rock. The heat vaporizes the water in the solutions and concentrates the monomers on the underlying substance. Some of the monomers then spontaneously bond together in chains, forming polymers.

Using this method, biochemist Sidney Fox, at the University of Miami, has succeeded in making polypeptides. On early Earth, in a similar fashion, raindrops or waves may have splashed dilute solutions of organic monomers onto fresh lava or other hot rocks and then rinsed polypeptides and other polymers back into the sea.

Clay surfaces may have been especially important as early polymerization sites. Even cool clay surfaces concentrate amino acids and other organic monomers from dilute solutions, because the monomers bind to electrically charged sites on the clay particles. Such binding sites could have brought monomers close together. Clay also contains metal atoms, such as iron and zinc, that can function as catalysts. On primitive Earth, metal atoms might have catalyzed the first dehydration reactions that linked monomers together, forming polymers.

? Before enzymes existed, clays with metal atoms might have catalyzed the _____ reactions that joined organic monomers, forming _____.

■ dehydration . . . polymers

16.5 The first genetic material and enzymes may both have been RNA

The formation of polymers on early Earth set the stage for the origin of early forms of life. But which polymers were most important? Here is a hint: An essential difference between life and nonlife is replication. Because nucleic acids are the biological polymers that replicate and store genetic information, they were most likely the essential first polymers.

Today's cells store their genetic information as DNA, transcribe the information into RNA, and then translate RNA messages into specific enzymes and other proteins. As we have seen in earlier chapters, this DNA → RNA → protein assembly system is extremely intricate. Most likely, it emerged gradually through a series of refinements to much simpler processes. What were the first genes like?

One hypothesis is that the first genes were short strands of RNA that replicated themselves without the assistance of proteins, perhaps on clay surfaces. Laboratory experiments support this idea. Short RNA molecules can assemble spontaneously from nucleotide monomers in the absence of enzymes. Furthermore, when RNA is added to a solution containing a supply of RNA monomers, new RNA molecules complementary to parts of the starting RNA sometimes assemble. So we can imagine a scenario on early Earth like the one in **Figure 16.5**: ❶ RNA monomers—nucleotides—spontaneously join, forming the first small "genes." ❷ Then an RNA chain complementary to one of these genes assembles. If the new chain, in turn, serves as a template for another round of RNA assembly, the result is a replica of the original gene.

This RNA replication process might have been aided by RNA molecules that acted as catalysts. Scientists have discovered that some RNAs, which they call **ribozymes**, can carry out a number of enzyme-like catalytic functions. For instance, modern ribozymes can catalyze RNA splicing (see Module 10.10).

Scientists use the term **RNA world** for the hypothetical period in the evolution of life when RNA served as both rudimentary genes and the sole catalytic molecules.

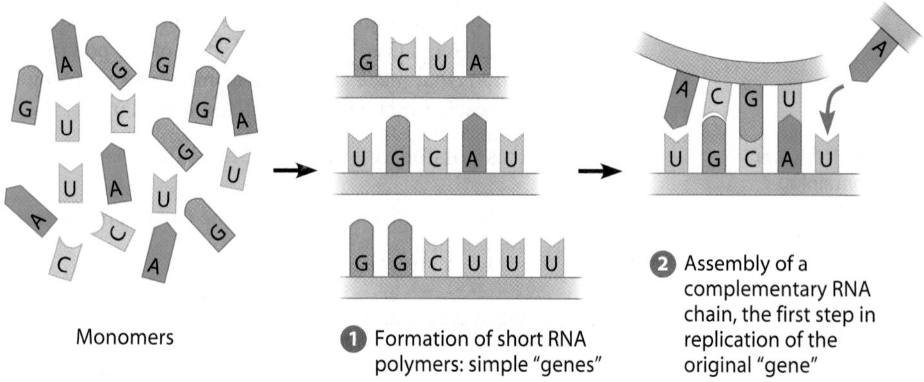

❶ Formation of short RNA polymers: simple "genes"

❷ Assembly of a complementary RNA chain, the first step in replication of the original "gene"

Monomers

Figure 16.5 A hypothesis for the origin of the first genes

? What is a ribozyme?

■ An RNA molecule that functions as a catalyst (enzyme)

16.6 Membrane-enclosed molecular cooperatives may have preceded the first cells

If self-replicating RNA molecules did arise as we have described, they were still a far cry from a living cell. Life as we know it requires a great number of complex organic molecules, and the molecules must interact and cooperate in precise ways. Put another way, life depends on intricate metabolic machinery that derives from the cooperation of many complex organic molecules. It is likely that some amount of molecular cooperation preceded life's origin.

The earliest form of molecular cooperation may have involved a primitive form of translation of simple RNA genes into polypeptides, translation that did not use ribosomes or tRNA. Suppose a strand of RNA acted as a rough template for polypeptide synthesis by weakly binding to amino acids and holding a few of them together long enough for them to be linked. If the sequence of nucleotides in the RNA influenced the sequence of amino acids, the resulting polypeptide would be a rough translation of the RNA gene. If, in turn, the polypeptide behaved as an enzyme that helped RNA molecules replicate (Figure 16.6A), then reciprocal molecular cooperation between nucleic acids and polypeptides would have begun.

A crucial step in the origin of life would have been the isolation of a collection of abiotically created molecules within a membrane, an entity called a **protobiont**. Laboratory experiments demonstate that protobionts could have formed spontaneously from abiotically produced organic compounds. For example, small membrane-bounded droplets can form when lipids are added to water (Figure 16.6B). These spheres are not alive, but they display some of the properties of living cells. They have a selectively permeable membrane-like surface, can grow by absorbing molecules from their surroundings, divide when they reach a certain size, and swell or shrink osmotically when placed in solutions of different salt concentrations. Similar spheres may have formed in the waters of a young Earth.

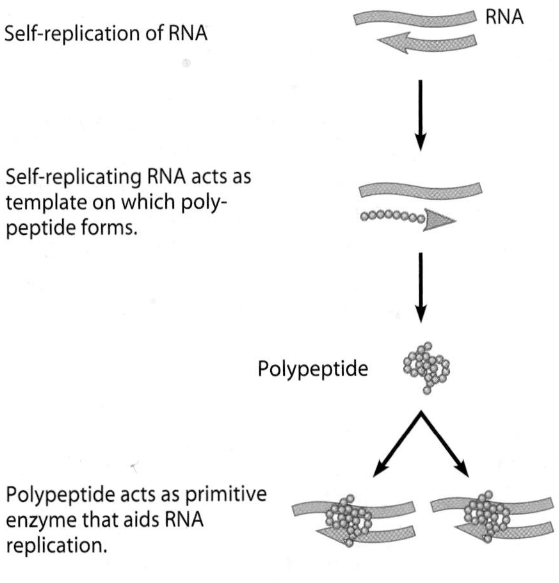

It is easy to imagine that certain of these cell-like entities on early Earth might have contained some self-replicating RNA molecules and RNA-polypeptide co-ops (Figure 16.6C). Once that happened, natural selection would have begun to shape the properties of these protobionts. Those that grew and replicated more efficiently than others would have increased in number, passing their abilities on to subsequent generations. In other words, protobionts could have come closer to evolving in a Darwinian sense.

With their cooperative associations of molecules now enclosed by membranes, protobionts would likely have developed the ability to replicate and to carry out essential chemical reactions—a primitive metabolism. But millions of years probably passed before the first living cells appeared on Earth. During this long period, the protobionts would have developed into complex metabolic machines containing DNA and capable of efficiently using a variety of raw materials from the environment.

The early protobionts, which relied on molecules present in the primitive soup of organic molecules surrounding them, were gradually replaced by organisms that could produce their own needed compounds. These organisms would have used sunlight or energy-rich molecules in the environment for energy. The diversification of these autotrophs encouraged the emergence of organisms that could live on products the autotrophs excreted or on the autotrophs themselves. These first autotrophs and the heterotrophs that came to rely on them were the first prokaryotes, Earth's sole living inhabitants from 3.5 to about 2 billion years ago. During this time, prokaryotes transformed the biosphere. For instance, atmospheric oxygen began to appear 2.7 billion years ago (see Figure 16.1D) as a result of prokaryotic photosynthesis. Prokaryotes continue to thrive today in a wide variety of environments, as we will see next.

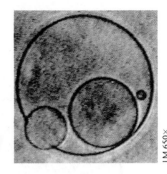

Figure 16.6B Microscopic spheres made of phospholipids

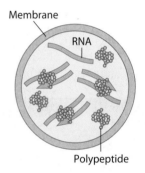

Membrane

RNA

Polypeptide

Figure 16.6C Cooperation among membrane-enclosed macromolecules

? According to the hypothesis presented in this module, why was the enclosure of molecular cooperatives by membranes a key requirement for the process of natural selection?

RNA

Self-replication of RNA

Self-replicating RNA acts as template on which polypeptide forms.

Polypeptide

Polypeptide acts as primitive enzyme that aids RNA replication.

Figure 16.6A The beginnings of molecular cooperation

■ Segregation within membrane-enclosed compartments allowed selection for the self-replicating molecular systems that grew the fastest and passed their successful properties to the next generation.

16.7 Prokaryotes have inhabited Earth for billions of years

The fossil record shows that prokaryotes were abundant 3.5 billion years ago. They continued to evolve all alone on Earth for the next 1.5 billion years. Today, prokaryotes are found wherever there is life. Their collective biological mass (biomass) is at least ten times that of all eukaryotes! More prokaryotes inhabit a handful of fertile soil than the total number of people who have ever lived. Prokaryotes also thrive in habitats too cold, too hot, too salty, too acidic, or too alkaline for any eukaryote. Biologists are discovering that these organisms have an astonishing genetic diversity. For example, comparing ribosomal RNA (rRNA) reveals that two strains of the bacterial species *Escherichia coli* are genetically more different than a human and a platypus.

You can get an idea of the size of most prokaryotes from **Figure 16.7**, a colorized scanning electron micrograph of the point of a pin (purple) covered with numerous bacteria (orange). Most prokaryotic cells have diameters in the range of 1–5 μm, much smaller than most eukaryotic cells (typically 10–100 μm).

Despite their small size, prokaryotes have an immense impact on our world. We hear most about a few species that cause serious illnesses. During the 14th century, Black Death—bubonic plague, a bacterial disease—spread across Europe, killing an estimated 25% of the human population at the time. Tuberculosis, cholera, and certain types of food poisoning are also caused by bacteria. We focus on bacterial **pathogens**, disease-causing agents, in Module 16.14.

Far more common than harmful bacteria are those that are benign or beneficial. We have bacteria in our intestines that provide us with important vitamins, and others living in our mouth help prevent harmful fungi from growing there. Prokaryotes that decompose dead organisms are essential to all life on Earth. They return vital chemical elements to the environment that can be used by plants, which in turn feed animals. If prokaryotic decomposers were to disappear, the chemical cycles that sustain life would halt, and all forms of eukaryotic life would also be doomed. In contrast, prokaryotic life would undoubtedly persist in the absence of eukaryotes, as it once did for billions of years.

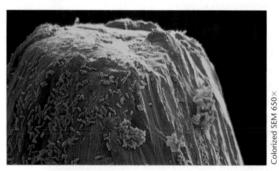

Figure 16.7 Bacteria on the point of a pin

Colorized SEM 650×

? Why might two strains of *E. coli* be genetically more different than a human and a platypus?

■ Bacteria have been evolving and accumulating differences for billions of years; placental mammals (such as humans) and egg-laying mammals (such as a platypus) diverged only about 200 million years ago.

16.8 Bacteria and archaea are the two main branches of prokaryotic evolution

Prokaryotes have a cellular organization fundamentally different from that of eukaryotes, as we saw in Modules 4.3 and 4.4. Whereas eukaryotic cells have a membrane-enclosed nucleus and numerous other membrane-enclosed organelles, most prokaryotic cells lack these structural features.

As we discussed in earlier chapters, however, there are two very different kinds of prokaryotes. When researchers first began comparing sequences of prokaryotic genes in the 1970s, they used one type of rRNA, a type found in all prokaryotes and eukaryotes, as a marker for evolutionary relationships. They concluded that many prokaryotes once classified as bacteria are actually more closely related to eukaryotes and belong in a domain of their own. Thus, prokaryotes are now classified in the domains **Bacteria** and **Archaea.** The name Archaea comes from the Greek *archaios* ("ancient"). Most biologists think that bacteria and archaea diverged from each other in very ancient times.

More recently, researchers have focused on DNA and have completely sequenced a number of bacterial and archaeal genomes (see Module 12.15). When compared with each other and with the genomes of eukaryotes such as yeast, these genome sequences strongly support the three-domain view of life. Some genes of archaea are similar to bacterial genes and others to eukaryotic genes; still others seem unique to archaea.

A current hypothesis is that present-day archaea and eukaryotes evolved from a common ancestor (see Figure 15.10B). But as described in Module 15.10, tracing their evolution is complicated by horizontal gene transfer between prokaryotic lineages. Over hundreds of millions of years, prokaryotes have acquired genes from distantly related species, and they continue to do so today. As a result, significant portions of the genomes of many prokaryotes are actually mosaics of genes imported from other species.

Table 16.8 summarizes some of the main differences between bacteria and archaea. In addition to the rRNA sequences, several other differences involve the cellular machinery for gene expression. These include differences in RNA polymerases (enzymes catalyzing the synthesis of RNA),

in the presence of introns within genes, and in sensitivity to certain antibiotics that inhibit protein synthesis. Subtle differences between bacterial and archaeal ribosomes—in both rRNA and proteins—undoubtedly account for the insensitivity of archaea to these antibiotics.

Other differences between bacteria and archaea show up in their cell walls and membranes. Nearly all prokaryotes have a cell wall outside their plasma membrane. As in plants, the wall maintains cell shape and provides physical protection. Bacterial cell walls contain a unique material called **peptidoglycan**, a polymer of sugars cross-linked by short polypeptides. No archaea have true peptidoglycan. Furthermore, the lipids forming the backbone of plasma membranes differ between the two domains.

For most of the features listed in the table, archaea are actually more like eukaryotes than like bacteria. In fact, archaea have at least as much in common with eukaryotes as they do with bacteria.

The main point here is to realize that there are two very different kinds of prokaryotic organisms. We will discuss the diversity within these two groups after a look at some more general features of prokaryotes.

TABLE 16.8 DIFFERENCES BETWEEN BACTERIA AND ARCHAEA		
Main Features	Bacteria	Archaea
rRNA sequences	Some unique to bacteria	Some unique to archaea; some match eukaryotic ones
RNA polymerase	One kind; relatively small and simple	Several kinds; complex; similar to eukaryotic ones
Introns (noncoding parts of genes)	Absent	Present in some genes
Antibiotic sensitivity (to streptomycin, chloramphenicol)	Growth inhibited	Growth not Inhibited
Peptidoglycan in cell wall	Present	Absent
Membrane lipids	Carbon chains unbranched	Some carbon chains branched
Histones associated with DNA	Absent	Present

? As different as bacteria and archaea are, both groups are characterized by _____ cells, which lack nuclei and other membrane-enclosed organelles.

■ prokaryotic

16.9 Prokaryotes come in a variety of shapes

Determining cell shape by microscopic examination is an important step in identifying prokaryotes. The micrographs below show three of the most common prokaryotic cell shapes. Spherical prokaryotic cells are called **cocci** (from the Greek word for "berries"). Cocci (singular, *coccus*) that occur in chains, like the ones in **Figure 16.9A**, are called streptococci (from the Greek *streptos,* twisted). The bacterium that causes strep throat in humans is a streptococcus. Other cocci occur in clusters; they are called staphylococci (from the Greek *staphyle,* cluster of grapes).

Figure 16.9B shows rod-shaped prokaryotes, which are called **bacilli** (singular, *bacillus*). Most bacilli occur singly, but the cells of some species occur in pairs (diplobacilli) and in chains (streptobacilli). The species shown here, which is common in fertile soil, exists as solitary cells.

A third prokaryotic cell shape is curved or spiral. Some bacteria in this category resemble commas and are called *vibrios*. Other bacteria and archaea have a helical shape, like a corkscrew. Helical prokaryotes that are relatively short and rigid are called *spirilla;* those with longer, more flexible cells are called *spirochetes* (**Figure 16.9C**). The bacterium that causes syphilis, for example, is a spirochete. Spirochetes include some giants by prokaryotic standards—cells 0.5 mm long (though very thin).

? How could a microscope help you distinguish the cocci that cause "staph" infections from those that cause "strep" throat?

■ It would show clusters of cells for staphylococcus and chains of cells for streptococcus.

Figure 16.9A Cocci

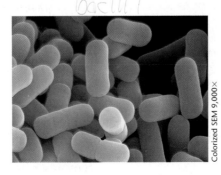

Figure 16.9B Bacilli

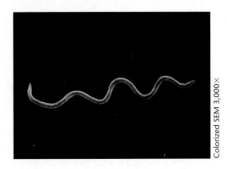

Figure 16.9C Spirochete

16.10 Various structural features contribute to the success of prokaryotes

In this module, we discuss some of the structural and functional adaptations that help prokaryotes thrive in a great variety of habitats.

External Structures   One of the most important features of nearly all prokaryotic cells is their cell wall. The wall maintains cell shape, provides physical protection, and prevents the cell from bursting in a hypotonic environment (see Module 5.17). In a hypertonic environment, most prokaryotes lose water and shrink away from their wall. Severe water loss inhibits reproduction, which explains why salt can be used to preserve certain foods, such as pork and fish.

As mentioned in Module 16.8, the cell walls of archaea differ from those of bacteria. The cell walls of bacteria fall into two general types, which scientists can identify with a technique called the **gram stain**. Gram-positive bacteria have simpler walls with a relatively thick layer of peptidoglycan. The walls of gram-negative bacteria stain differently. They have less peptidoglycan and are more complex, with an outer membrane that contains lipids bonded to carbohydrates.

Gram staining is a valuable identification tool in medicine. Among disease-causing bacteria, gram-negative species are generally more threatening than gram-positive species. The lipid molecules of the outer membrane of these bacteria are often toxic. The membrane also protects them against the body's defenses and impedes the entry of antibiotic drugs. The effectiveness of many antibiotics is due to their inhibition of cross-linking in the peptidoglycan cell wall. Such drugs can cripple many types of bacteria without adversely affecting human cells, which do not contain peptidoglycan.

The cell wall of many prokaryotes is covered by a capsule, a sticky layer of polysaccharides or protein. The capsule enables prokaryotes to adhere to their substrate or to other individuals in a colony. Capsules can also shield pathogenic prokaryotes from attacks by their host's immune system. The capsule surrounding the *Streptococcus* bacterium shown in Figure 16.10A enables it to attach to cells that line the human respiratory tract—in this image, a tonsil cell.

Some prokaryotes stick to their substrate or to one another by means of hairlike appendages called **pili** (singular, *pilus*). Pili show up clearly in the highly magnified micrograph in Figure 16.10B. Pili help prokaryotes stick to each other and to surfaces, such as rocks in flowing streams or the lining of human intestines. Specialized pili, called sex pili, link prokaryotes during conjugation, a process in which one cell transfers DNA to another (see Module 10.22).

Motility   Many bacteria and archaea are equipped with flagella, which enable them to move about. In response to chemical or physical signals in their environment, prokaryotes can move toward nutrients or other members of their species or away from a toxic substance. Flagella may be scattered over the entire cell surface or concentrated at one or both ends of the cell. Entirely different in structure from the flagellum of eukaryotic cells (described in Module 4.17), the prokaryotic flagellum is a naked protein structure that lacks microtubules. It is attached to the cell surface by a system of rotating rings anchored in the plasma membrane and cell wall. The rings give the flagellum a propeller-like rotary movement, as shown in Figure 16.10C, at the top of the next page. The flagellated organism in the photo is the bacterium *Proteus,* an especially fast swimmer.

Reproduction and Adaptation   Certainly a large part of the success of prokaryotes is their potential to reproduce quickly in a favorable environment. Dividing by binary fission, a single cell becomes 2 cells, which then become 4, 8, 16, and so on. While most prokaryotes produce a new generation within 1–3 hours, some species can produce a new generation in only 20 minutes under optimal conditions. If reproduction continued unchecked at this rate, a single prokaryote could give rise to a colony outweighing Earth in only three days! In reality, of course, prokaryotic reproduction is limited, as the cells eventually exhaust their nutrient supply, poison themselves with metabolic wastes, or are consumed by other organisms. Prokaryotes in nature also face competition from other microorganisms, many of which produce antibiotic chemicals that slow prokaryotic reproduction.

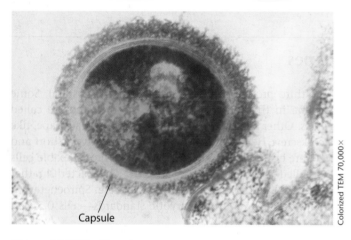

Colorized TEM 70,000×

Capsule

Figure 16.10A   Capsule

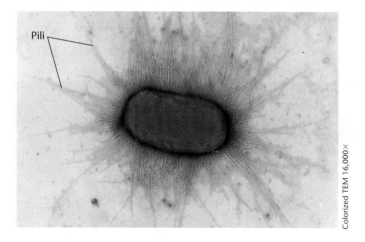

Pili

Colorized TEM 16,000×

Figure 16.10B   Pili

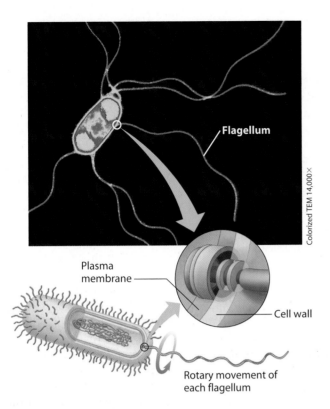

Plasma membrane

Cell wall

Rotary movement of each flagellum

Flagellum

Colorized TEM 14,000×

Figure 16.10C Prokaryotic flagella

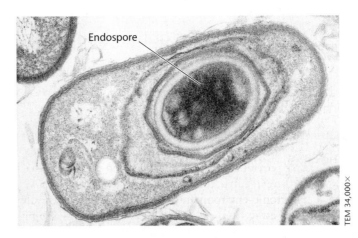

Endospore

TEM 34,000×

Figure 16.10D An endospore within a cell of the anthrax bacterium

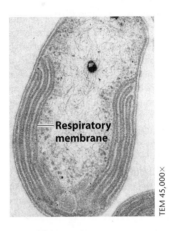

Respiratory membrane

TEM 45,000×

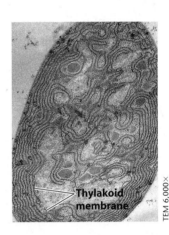

Thylakoid membrane

TEM 6,000×

Figure 16.10E Specialized membranes of an aerobic prokaryote (left) and a photosynthetic prokaryote (right)

Some prokaryotes are able to withstand harsh conditions. Certain bacteria, for example, can form specialized resistant cells when food is depleted or the environment has otherwise become inhospitable. **Figure 16.10D** shows an example of such an organism, *Bacillus anthracis,* the bacterium that produces the deadly disease called anthrax in cattle, sheep, and humans. There are actually two cells here, one inside the other. The outer cell, which will later disintegrate, produced the specialized inner cell, called an **endospore.** The endospore, which has a thick, protective coat, dehydrates and becomes dormant. It can survive all sorts of trauma, including extreme heat or cold. When the endospore receives cues that the environmental conditions have improved, it absorbs water and resumes growth.

Some endospores can remain dormant for centuries. Not even boiling water kills most of these resistant cells. To sterilize laboratory equipment, microbiologists use an autoclave, a pressure cooker that kills endospores by heating to a temperature of 121°C (250°F) with high-pressure steam. The food-canning industry uses similar methods to kill endospores of dangerous bacteria such as *Clostridium botulinum,* the source of the potentially fatal disease botulism.

Internal Organization As seen in Module 4.3, the cells of prokaryotes are simpler than those of eukaryotes in both their internal structure and organization of their genome. However, some prokaryotic cells do have specialized membranes that perform metabolic functions. The left side of **Figure 16.10E** shows an aerobic prokaryote with infoldings of the plasma membrane that function in cellular respiration. The right side of the figure shows a cyanobacterium with thylakoid membranes that function in photosynthesis.

The genome of a prokaryote has on average only about one-thousandth as much DNA as a eukaryotic genome. The DNA usually forms a circular chromosome. Smaller rings of DNA, called plasmids, carry genes that may provide resistance to antibiotics, direct the metabolism of rarely encountered nutrients, or have other "contingency" functions. The ability of many prokaryotes to transfer genes within and even between species contributes to the growing problem of antibiotic resistance in disease-causing bacteria.

Prokaryotic ribosomes are slightly smaller than eukaryotic ribosomes and differ in their protein and RNA content. These differences are great enough that certain antibiotics bind to prokaryotic ribosomes and block protein synthesis but do not affect eukaryotic ribosomes. As a result, we can use these antibiotics to kill bacteria without harming ourselves.

Next we consider some of the diverse ways in which prokaryotes obtain their nourishment.

Web/CD Activity 16B *Prokaryotic Cell Structure and Function*

? Why do microbiologists autoclave lab instruments and glassware, rather than simply boiling them?

■ To kill bacterial endospores, which can survive boiling water

16.11 Prokaryotes obtain nourishment in a variety of ways

When classifying diverse organisms, biologists often use the phrase "mode of nutrition" to describe how an organism obtains two main resources: carbon (for synthesizing organic compounds) and energy. As a group, prokaryotes exhibit much more nutritional diversity than eukaryotes.

Types of Nutrition Many prokaryotes are **autotrophs** ("self-feeders"), making their own organic compounds from inorganic sources. As shown in Table 16.11, autotrophs obtain their carbon atoms from carbon dioxide (CO_2). They get their energy from sunlight or from inorganic chemicals, such as hydrogen sulfide (H_2S), elemental sulfur (S), or compounds containing iron (Fe). Autotrophs that harness sunlight for energy and use CO_2 for carbon, such as the cyanobacteria, do so by photosynthesis; they are called **photoautotrophs**. Autotrophic organisms that obtain energy from inorganic chemicals instead of sunlight are called **chemoautotrophs**.

Most prokaryotes are **heterotrophs** ("other-feeders"), meaning they obtain their carbon atoms from organic compounds. Some heterotrophs, called **photoheterotrophs**, can obtain energy from sunlight. By far the largest group of prokaryotes, however, are nutritionally similar to animals in that they obtain both energy and carbon from organic molecules. Called **chemoheterotrophs**, these prokaryotes are so diverse that almost any organic molecule can serve as food for some species. Many species, such as *Escherichia coli,* a resident of the human intestine, can thrive on a variety of organic nutrients. The photograph in **Figure 16.11A** shows a culture of *E. coli* grown with only glucose as an organic nutrient. Each round spot in the culture dish is a colony, a clone of millions of bacterial cells.

Metabolic Cooperation Prokaryotes once were thought of as single-celled individualists, with each cell providing its own nutrition. Now, however, biologists recognize that some prokaryotes cooperate. The cyanobacterium *Anabaena* has genes for photosynthesis and for nitrogen fixation—the conversion of atmospheric nitrogen (N_2) to ammonia (NH_3). But because the O_2 produced by photosynthesis inactivates nitrogen-fixing enzymes, one cell cannot carry out both processes. *Anabaena* forms filamentous colonies (see Figure

TABLE 16.11 NUTRITIONAL CLASSIFICATION OF ORGANISMS		
Nutritional Type	**Energy Source**	**Carbon Source**
Photoautotroph (photosynthesizers)	Sunlight	CO_2
Chemoautotroph	Inorganic chemicals	CO_2
Photoheterotroph	Sunlight	Organic compounds
Chemoheterotroph	Organic compounds	Organic compounds

16.13D) in which most cells photosynthesize, while a few specialized ones fix nitrogen. Connections between cells allow them to exchange nitrogen for carbohydrates.

In some prokaryotes, metabolic cooperation occurs in surface-coating colonies called **biofilms** (**Figure 16.11B**). Cells in a colony secrete signaling molecules that recruit nearby cells. Once the colony becomes sufficiently large, the cells produce proteins that adhere them to the substrate and to each other. Channels in the biofilm allow nutrients to reach cells in the interior and wastes to be expelled.

In some instances, prokaryotes belonging to different species cooperate. For example, sulfate-consuming bacteria and methane-consuming archaea coexist in ball-shaped clumps in the mud of the ocean floor. The bacteria appear to use the archaea's organic waste products and, in turn, produce compounds that facilitate methane consumption by the archaea. This partnership has global ramifications: Each year, these archaea consume an estimated 300 billion kilograms of methane, a major greenhouse gas.

Web/CD Thinking as a Scientist *What Are the Modes of Nutrition in Prokaryotes?*

? Explain why a filament of *Anabaena* can be described as one of life's most self-sufficient organisms.

■ An *Anabaena* filament produces its food by photosynthesis and has specialized cells that supply it with the nitrogen needed for protein synthesis.

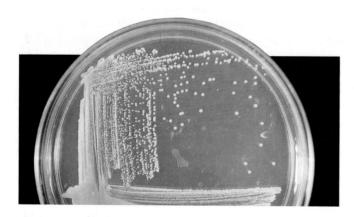

Figure 16.11A *E. coli* colonies grown on glucose

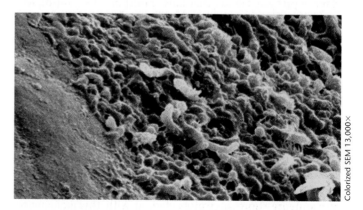

Colorized SEM 13,000×

Figure 16.11B Dental plaque, a biofilm that forms on teeth

16.12 Archaea thrive in extreme environments—and in other habitats

Archaea are abundant in many habitats, including places where few other organisms can survive. The archaeal inhabitants of extreme environments have unusual proteins and other molecular adaptations that enable them to metabolize and reproduce effectively. Scientists are only beginning to learn about these adaptations.

A group of archaea called the **extreme halophiles** ("salt lovers") thrive in very salty places, such as the Great Salt Lake in Utah, the Dead Sea, and seawater-evaporating ponds used to produce salt. **Figure 16.12A** shows an aerial view of some ponds of this sort next to San Francisco Bay. The colors of the ponds result from the dense growth of the archaea that thrive when the salinity of the water reaches 15–20%. (Before evaporation, seawater has a salt concentration of about 3%.) The purplish color of the ponds near the top of the photo is due to an archaean called *Halobacterium halobium*. A unique photosynthesizer, *H. halobium* lacks chlorophyll; it has a purple molecule called bacteriorhodopsin that traps solar energy.

Another group of archaea, the **extreme thermophiles** ("heat lovers"), thrive in very hot water; some even live near deep-ocean vents where temperatures are above 100°C, the boiling point of water at sea level! Other thermophiles thrive in acid. Many hot, acidic pools in Yellowstone National Park harbor such archaea, which give the pools a vivid greenish color **(Figure 16.12B)**. One of these organisms, *Sulfolobus*, can obtain energy by oxidizing sulfur or a compound of sulfur and iron; the mechanisms involved may be similar to those used billions of years ago by the first cells.

A third group of archaea, the **methanogens**, live in anaerobic environments and give off methane as a waste product. Many thrive in anaerobic mud at the bottom of lakes and swamps. You may have seen methane, also called marsh gas, bubbling up from a swamp. Great numbers of methanogens also inhabit the digestive tracts of animals. In humans, intestinal gas is largely the result of their metabolism. More importantly, methanogens aid digestion in cattle, deer, and other animals that depend heavily on cellulose for their nutrition. Unfortunately, these animals belch out large volumes of methane, contributing to the global release of this major greenhouse gas.

Accustomed to thinking of archaea as "extremophiles," scientists have been surprised to discover their abundance in more moderate environments, especially in the oceans. Archaea live at all depths, making up a substantial fraction of the prokaryotes in waters below 150 m and equaling the number of bacteria below 1,000 m. Archaea are thus one of the most abundant cell types in the Earth's largest habitat.

Because bacteria have been the subject of most prokaryotic research throughout the history of microbiology, much more is known about them than about archaea. Now that the evolutionary and ecological importance of archaea has come into focus, we can expect research on this domain to turn up many more surprises about the history of life and the roles of microbes in ecosystems.

Figure 16.12A "Salt-loving" archaea, extreme halophiles, growing in seawater-evaporating ponds near San Francisco Bay

Figure 16.12B "Heat-loving" archaea, extreme thermophiles, growing in a hot, highly acidic pool in Yellowstone National Park

? Some archaea are referred to as "extremophiles." Why?

■ Because they can thrive in extreme environments too hot, too salty, or too acidic for other organisms

16.13 Bacteria include a diverse assemblage of prokaryotes

In this module we sample some of the diversity of bacteria. Until the late 20th century, systematists based prokaryotic taxonomy on phenotypic criteria such as shape, motility, nutritional mode, and Gram staining. These criteria are still valuable in some contexts, such as the rapid identification of bacteria cultured from a patient's blood. But they do not reveal relationships. Although still a work in progress, molecular sytematics is starting to unravel bacterial phylogeny.

Domain Bacteria is currently divided into nine groups, five of which are considered subgroups of a clade of gram-negative bacteria called **proteobacteria**. Members of the subgroup alpha proteobacteria include *Rhizobium* species, which live in nodules in the roots of legumes, where they fix atmospheric N_2 (see Module 32.14), and *Agrobacterium* species, which produce tumors in plants and are used by genetic engineers to carry foreign DNA into the genomes of crop plants (see Module 12.18).

The photosynthetic members of the subgroup gamma proteobacteria include sulfur bacteria that oxidize H_2S. The small greenish globules you see in **Figure 16.13A** are sulfur wastes. Other gamma proteobacteria include many species that inhabit animal intestines, such as *Salmonella*, one cause of food poisoning; *Vibrio cholerae*, which causes cholera; and *Escherichia coli*, a common resident of the intestines of humans and other mammals and a favorite research organism.

The delta proteobacteria include the slime-secreting myxobacteria, which form elaborate colonies. When the soil dries out or food is scarce, the cells congregate into a fruiting body that releases resistant spores. *Bdellovibrio* bacteria are delta proteobacteria that attack other bacteria. They charge their prey at up to 100 μm/sec (comparable to a human running 600 km/hr) and bore into the prey by spinning at 100 revolutions per second (**Figure 16.13B**).

The **chlamydias** form a second bacterial group. *Chlamydia trachomatis* is a common cause of blindness in the world and also causes nongonococcal urethritis, the most common sexually transmitted disease in the United States.

Spirochetes are a group of helical bacteria (see Figure 16.9C) that spiral through their environment by means of rotating, internal filaments. Some spirochetes are notorious pathogens: *Treponema pallidum* causes syphilis, and *Borrelia burgdorferi* causes Lyme disease.

The group of **gram-positive bacteria** rivals the proteobacteria in diversity. One subgroup, the actinomycetes (from the Greek *mykes*, fungus, for which these bacteria were once mistaken), form colonies of branched chains of cells. Actinomycetes are very common in the soil, where they decompose organic matter. Soil-dwelling species in the genus *Streptomyces* (**Figure 16.13C**) are cultured by pharmaceutical companies as a source of many antibiotics, including streptomycin. Gram-positive bacteria also include many solitary species, such as *Bacillus anthracis* (see Figure 16.10D). *Staphylococcus* and *Streptococcus* are also gram-positive bacteria. Mycoplasmas are gram-positive bacteria that lack cell walls and are the tiniest of all known cells, with diameters as small as 0.1 μm, only about five times as large as a ribosome.

The **cyanobacteria** are the only prokaryotes with plant-like, oxygen-generating photosynthesis. Ancient cyanobacteria generated the stromatolites discussed in the chapter introduction and made the atmosphere aerobic. Today, both solitary and colonial cyanobacteria provide an enormous amount of food for freshwater and marine ecosystems. **Figure 16.13D** shows the cyanobacterium *Anabaena* (described in Module 16.11) with its specialized cells that fix nitrogen.

Web/CD Activity 16C *Diversity of Prokaryotes*

? Which group of bacteria uses H_2S as an electron source in its photosynthetic production of organic molecules?

■ Sulfur bacteria of the subgroup gamma proteobacteria

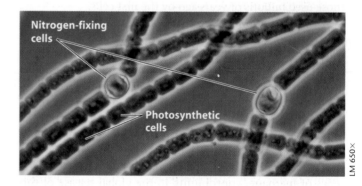

Nitrogen-fixing cells

Photosynthetic cells

LM 650×

Figure 16.13D *Anabaena*, a filamentous cyanobacterium

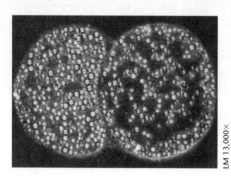

LM 13,000×

Figure 16.13A *Thiomargarita namibiensis,* showing globules of sulfur wastes

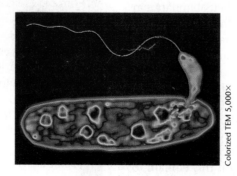

Colorized TEM 5,000×

Figure 16.13B *Bdellovibrio bacteriophorus* (flagellated cell) attacking a larger bacterium

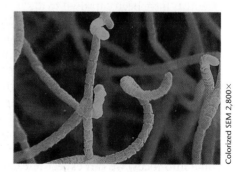

Colorized SEM 2,800×

Figure 16.13C *Streptomyces,* the source of many antibiotics

16.14 Some bacteria cause disease

All organisms, humans included, are almost constantly exposed to bacteria, some of which are potentially harmful. Most of us are well most of the time only because our body defenses check the growth of bacterial pathogens. Occasionally, the balance shifts in favor of a pathogen, and we become ill. Even some of the bacteria that are normal residents of the human body can make us ill when our defenses have been weakened by poor nutrition or by a viral infection.

Pathogenic bacteria cause about half of all human diseases. Between 2 and 3 million people a year die of the lung disease tuberculosis, which is caused by the actinomycete *Mycobacterium tuberculosis,* while another 2 million die from diarrheal diseases caused by various prokaryotes.

Most prokaryotes that cause illness do so by producing a poison—either an exotoxin or an endotoxin. **Exotoxins** are secreted by bacterial cells and include some of the most potent poisons known, such as the botulinum toxin that causes botulism.

Figure16.14A shows an electron micrograph of *Staphylococcus aureus*, with its characteristic grapelike clusters of cells. Although *S. aureus* may commonly be found on moist skin folds, if it enters the body through a wound, it can cause serious diseases. One of the exotoxins that *S. aureus* produces can cause layers of skin to slough off; another causes the potentially deadly toxic shock syndrome. Food contaminated with its exotoxins can cause vomiting and severe diarrhea.

Bacterial species that are generally harmless can also develop strains that cause illness. Since first identified in 1982, an exotoxin-producing strain of *E. coli* designated O157:H7 has caused a number of outbreaks of bloody diarrhea and hundreds of deaths. In the United States alone, there are an estimated 75,000 cases of O157:H7 infection per year and 60 deaths. Most cases are associated with contaminated beef. The recently sequenced genome of O157:H7 shows that more than 20% of its genes are not found in the genome of a harmless *E. coli* strain. These genes must have been incorporated into the genome through horizontal gene transfer from other species. For now, the best preventive measure is to avoid eating undercooked meat.

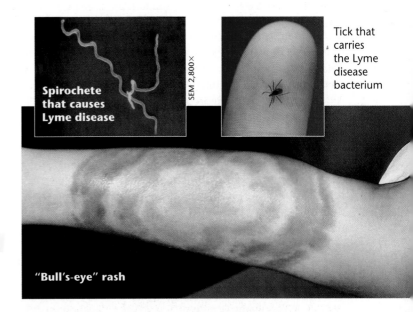

Spirochete that causes Lyme disease

SEM 2,800×

Tick that carries the Lyme disease bacterium

"Bull's-eye" rash

Figure 16.14B Lyme disease, a bacterial disease transmitted by ticks

Endotoxins are components of the outer membrane of gram-negative bacteria. All endotoxins induce the same general symptoms: fever, aches, and sometimes a dangerous drop in blood pressure (shock). Examples of endotoxin-producing bacteria include the species of *Salmonella* that cause food poisoning and typhoid fever.

During the last 100 years, improvements in sanitation in the developed world have greatly reduced the incidence of bacterial diseases. The installation of water treatment and sewage systems continues to be a public health priority throughout the world. Antibiotics can cure most bacterial diseases, although many pathogenic bacteria have evolved resistance to widely used antibiotics, becoming newly dangerous (see Module 13.13).

In addition to sanitation and antibiotics, a third defense against bacterial diseases is education. A case in point is **Lyme disease**, currently the most widespread pest-carried disease in the United States. The disease is caused by *Borrelia burgdorferi*, a spirochete carried by ticks that live on deer and field mice. As shown in **Figure 16.14B**, the disease usually starts as a red rash shaped like a bull's-eye around a tick bite. Antibiotics can cure the disease if administered within about a month after exposure. If untreated, Lyme disease can cause debilitating arthritis, heart disease, and nervous disorders. The best prevention is public education about avoiding tick bites and the importance of seeking treatment if a rash develops. Using insect repellent and wearing light-colored clothes can help reduce contact with ticks.

? Contrast exotoxins with endotoxins.

■ Exotoxins are poisons secreted by pathogenic bacteria; endotoxins are components of the cell walls of pathogenic bacteria.

Figure 16.14A *Staphylococcus aureus,* an exotoxin producer

SEM 12,000×

16.15 Bacteria can be used as biological weapons

In October 2001, endospores of *Bacillus anthracis,* the bacterium that causes anthrax, were found in envelopes mailed to members of the news media and the U.S. Senate (**Figure 16.15**). Eighteen people developed cases of anthrax, and five died. Unfortunately, while these attacks were shocking, they were not unique. There is a long and ugly history of using biological organisms as weapons. Animals, plants, fungi, and viruses have all served this purpose, but the most frequently employed agents have been bacteria.

During the Middle Ages, the bacterium *Yersinia pestis* caused bubonic plague throughout Europe. It also played a role in battle, when armies hurled the bodies of plague victims into enemy ranks. Early conquerors, settlers, and warring armies in South and North America gave native people items purposely contaminated with infectious bacteria, often wiping out whole tribes.

B. anthracis is a spore-forming bacterium (see Figure 16.10D) that lives in the soil of agricultural regions, where large grazing animals can become infected. People who work in agriculture, leather tanning, or wool processing may catch anthrax when exposed to spores from infected animal tissue. Anthrax is an obvious choice for biological weapons because *B. anthracis* is easy to obtain and easy to grow in the laboratory, and its hardy endospores can be stored for years. And anthrax can be deadly. In the bloodstream, the anthrax bacteria actively metabolize and multiply. As they metabolize, they release three proteins that combine to form a toxin that destroys body tissues and cells of the immune system.

Prognosis and treatment vary, depending on how anthrax spores enter the body. If spores enter through a break in the skin, they cause cutaneous (skin) lesions with a black center (the word *anthrax* is from the Greek word for "coal"); usually, these lesions can be readily cured with antibiotics such as penicillin. However, inhalation of a large number of spores can result in pulmonary (lung) anthrax, a deadly form of the disease. Antibiotics can be effective, but people often delay treatment because the early stages of the infection are indistinguishable from a common cold. Once the disease has advanced, antibiotics are ineffective against the accumulated toxin. In 2003, scientists published the complete genome of the strain of *B. anthracis* used in the October 2001 attack, in

Figure 16.15 Cleaning up a site where anthrax spores were released in October 2001

the hope of developing new vaccines and antibiotics. However, mass vaccinations are currently not an available solution to the frightening possibility of biowarfare and bioterrorism.

What is the status of bioweapons research in the United States? The first biological weapons research facility was opened in 1943 at Fort Detrick, Maryland. There, the military studied and bred new strains of bacteria that cause such illnesses as anthrax, botulism, and tularemia. To "weaponize" naturally occurring pathogens, researchers selected highly virulent strains, made them antibiotic resistant (see Module 10.23), and developed formulations for effective dispersion. But the practical difficulties of controlling such weapons—and a measure of ethical repugnance—led the United States to end this bioweapons program in 1969 and to order its products destroyed. In 1975, the United States signed the Biological Weapons Convention, pledging never to develop or store biological weapons. Eventually, 103 nations joined the ban, although not all signatories have honored it. The solution to the threat of biological weapons will ultimately have to come from cooperative efforts among the world's nations.

? Why is *Bacillus anthracis* an effective bioweapon?

■ It is easy to obtain, easy to grow in the lab, and forms potentially deadly endospores that resist destruction and can be easily dispersed.

16.16 Prokaryotes help recycle chemicals and clean up the environment

Despite the attention they demand, pathogenic prokaryotes are in the minority. Far more common are species that are vital to our well-being. All life depends on the cycling of chemical elements between organisms and the nonliving parts of our environment. Prokaryotes are indispensable components of chemical cycles. For example, in addition to

contributing to aquatic food chains and releasing oxygen to the atmosphere, some cyanobacteria also convert nitrogen in the atmosphere to nitrogen compounds that plants can take up and use. Other prokaryotes, such as *Rhizobium* living in nodules on the roots of beans and other legumes, contribute large amounts of nitrogen compounds to soil. Thus,

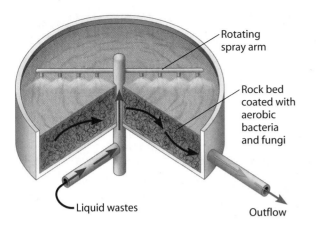

Figure 16.16A The trickling filter system at a sewage treatment plant

the nitrogen that plants use to make proteins and nucleic acids comes from prokaryotic metabolism in soil and water. In turn, animals get their nitrogen compounds from plants.

Another vital function of prokaryotes is the decomposition of organic wastes and dead organisms to inorganic chemicals. If it were not for such decomposers, carbon, nitrogen, and other elements essential to life would become locked in the organic molecules of corpses and waste products. We'll discuss chemical cycling more in Chapter 37.

The varied metabolic talents of prokaryotes also enable them to help solve environmental problems through **bioremediation**, the use of organisms to remove pollutants from soil, air, or water. Prokaryotic decomposers are the mainstays of our sewage treatment facilities. Raw sewage is first passed through a series of screens and shredders, and solid matter is allowed to settle out from the liquid waste. This solid matter, called sludge, is then gradually added to a culture of anaerobic prokaryotes, including both bacteria and archaea. The microbes decompose the organic matter in the sludge to material that can be used as landfill or fertilizer.

Liquid wastes are treated separately from the sludge. In Figure 16.16A, you can see a trickling filter system, one type of mechanism for treating liquid wastes. The long horizontal pipes rotate slowly, spraying liquid wastes through the air onto a thick bed of rocks, the filter. Aerobic bacteria and fungi growing on the rocks remove much of the organic material dissolved in the waste. Outflow from the rock bed is sterilized and then released, usually into a river or ocean.

In Figure 16.16B, workers are spraying fertilizers on an oil-polluted beach in Alaska. The fertilizers stimulate the growth of "oil-eating" bacteria occurring naturally in the soil and in some cases speed the natural breakdown process fivefold. Researchers are trying to genetically engineer bacteria to degrade oil even more efficiently.

Bacteria may also help us clean up old mining sites. The water that drains from mines is highly acidic and is also laced with poisons—often compounds of arsenic, copper, zinc, and the heavy metals lead, mercury, and cadmium. Contamination of our soils and groundwater by these toxic substances poses a widespread threat, and cleaning up the mess can be extremely expensive. Prokaryotes may be able to help. Bacteria called *Thiobacillus* thrive in the acidic waters that drain from mines, accumulating metals from the mine waters. Unfortunately, they also add sulfuric acid to the water. If this problem is solved, perhaps through genetic engineering, *Thiobacillus* and other prokaryotes may help us overcome some serious environmental problems.

Some mining companies use microbes to extract valuable metals from low-grade ores. Bacteria assist in extracting over 30 billion kilograms of copper from copper sulfides each year. Using prokaryotes that can extract gold from ore, one factory in the African nation of Ghana processes 1 million kilograms of gold concentrate a day.

? What is bioremediation?

■ The use of organisms to clean up pollution

■   ■   ■

In their various roles, prokaryotes have had a greater impact on the environment and on biological evolution than all other forms of life combined. Not only did ancient prokaryotes create Earth's aerobic atmosphere, but they also were the first organisms to tolerate the corrosive effects of atmospheric O_2 and the first to use O_2 in metabolizing organic molecules. Moreover, as just discussed, prokaryotes continue to be critical for chemical cycling on Earth today. In the next module, we describe the essential role of prokaryotes in another monumental evolutionary event—the formation of the first eukaryotic cells.

Figure 16.16B Treatment of an oil spill in Alaska

16.17 The eukaryotic cell probably originated as a community of prokaryotes

The fossil record indicates that eukaryotes evolved from pro-karyotes more than 2 billion years ago. One of biology's most basic questions is how this happened—in particular, how the membrane-enclosed organelles of eukaryotic cells arose. A widely accepted theory is that eukaryotic cells arose through a combination of two processes (Figure 16.17). In one process, **membrane infolding**, the eukaryotic cell's endomembrane system—all the membrane-enclosed organelles except mito-chondria and chloroplasts (see Chapter 4)—evolved from in-ward folds of the plasma membrane of a prokaryotic cell. The top portion of Figure 16.17 suggests how the nuclear envelope and endoplasmic reticulum (ER) may have devel-oped by infolding. The Golgi apparatus and other parts of the endomembrane system may then have evolved from the ER.

A second, very different process, called **endosymbiosis**, is thought to have generated mitochondria and chloroplasts. **Symbiosis** is a close association between organisms of two or more species. (The word is from the Greek for "living to-gether," and *endo*symbiosis refers to one species, called the endosymbiont, living *within* another, called the host.) Chloroplasts and mitochondria seem to have evolved from prokaryotes that established residence within other, larger prokaryotes (bottom portion of Figure 16.17). The ancestors of mitochondria may have been heterotrophic prokaryotes that were able to use oxygen and release large amounts of en-ergy from organic molecules by cellular respiration. An an-cestral host cell may have ingested some of these aerobic cells and packaged them in a vacuole. Some of the cells might have remained alive and continued to perform respiration in the host cell. In a similar way, photosynthetic prokaryotes ancestral to chloroplasts may have come to live inside a larger host cell. Because all eukaryotes have mitochondria (or remnants of mitochondria) but only some have chloro-plasts, scientists think that mitochondria evolved first.

It's not hard to imagine how a symbiosis between en-gulfed aerobic or photosynthetic cells and a larger host cell might have become mutually beneficial. In both cases, the engulfed cells may have grown increasingly dependent on the host cell for molecules and inorganic ions needed to carry out their biochemical activities. Likewise, the host cell may have derived increasing proportions of its ATP and or-ganic molecules from the engulfed cells. As the cells became more interdependent, they may have become truly a single organism, its parts inseparable.

Developed most extensively by Lynn Margulis, of the University of Massachusetts, the endosymbiosis model is supported by strong evidence. For instance, present-day mitochondria and chloroplasts are similar to prokaryotic cells in a number of ways. Both types of organelles contain small amounts of DNA, RNA, and ribosomes, all of which resemble their counterparts in prokaryotes more than

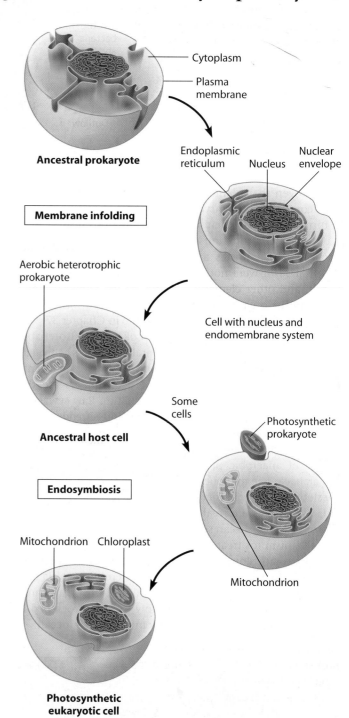

Figure 16.17 A model of the origin of eukaryotes

those in eukaryotes. Chloroplasts and mitochondria tran-scribe and translate their DNA into polypeptides, con-tributing to some of their own enzymes. They also replicate their own DNA and reproduce by a process similar to the binary fission of prokaryotes.

Endosymbiosis can also explain how chloroplasts and mitochondria came to be enclosed by two membranes. As Figure 16.17 indicates, the inner membranes of these organelles could have been derived from the plasma membranes of the engulfed prokaryotes, and their outer membranes could have come from the infolded plasma membranes of the original host cell. In fact, the inner membranes of mitochondria and chloroplasts have several enzymes and electron transport molecules that resemble those found in the plasma membranes of modern prokaryotes, presumably a result of their endosymbiotic origin.

What does molecular systematics tell us about the origin of the eukaryotic cell? Comparisons of RNA genes indicate that alpha proteobacteria are the closest relatives of mitochondria and that cyanobacteria are the closest relatives of chloroplasts (see Module 16.13). Some researchers have proposed that the nucleus evolved from an archaeal endosymbiont. The genome of eukaryotic cells appears to be the product of horizontal gene transfers between many different bacterial and archaeal lineages. According to the "you are what you eat" hypothesis, evolving eukaryotes consumed various bacteria and archaea and occasionally incorporated some of their genes into the nucleus.

The oldest fossils that most researchers agree are eukaryotic are about 2.1 billion years old. Let's now look at the most direct descendants of these first eukaryotes—the protists living on Earth today.

? Which organelles of eukaryotic cells probably descended from endosymbiotic bacteria?

■ Mitochondria and chloroplasts

16.18 Protists are an extremely diverse assortment of eukaryotes

The photograph in **Figure 16.18**—a drop of pond water viewed with the light microscope—illustrates a variety of **protists**, a diverse collection of mostly unicellular eukaryotes. Biologists used to classify all protists in a kingdom called Protista, but now most think that these organisms constitute several kingdoms within domain Eukarya. While our knowledge of the evolutionary relationships among these diverse groups remains incomplete, *protist* is still useful as a convenient term to refer to eukaryotes that are not plants, animals, or fungi.

Some protists synthesize their food by photosynthesis; these are called **algae** (another useful term that is not taxonomically meaningful). Others, informally called **protozoans** (from the Greek *protos,* first, and *zoion,* animal), are heterotrophic, eating bacteria and other protists. Some protists can be heterotrophic or autotrophic, depending on availability of light and nutrients. And some protists are fungus-like and obtain organic molecules by absorption.

Protist habitats are also diverse. Most protists are aquatic, and they are found almost anywhere there is water, including moist terrestrial habitats, such as damp soil and leaf litter, and the moist bodies of various host organisms.

As eukaryotes, protists are more complicated than any prokaryotes. Their cells have a membrane-enclosed nucleus (containing multiple chromosomes) and other organelles characteristic of eukaryotic cells. The flagella and cilia of protistan cells have a 9 + 2 pattern of microtubules, another typical eukaryotic trait (see Module 4.17).

Because most protists are unicellular, they are justifiably considered to be the simplest eukaryotes. However, the cells of many protists are among the most elaborate in the world. This level of cellular complexity is not really surprising, for each unicellular protist is a complete eukaryotic organism analogous to an entire animal or plant.

During the past 15 years, molecular and cellular studies have shaken the foundations of protistan taxonomy as much as they have that of the prokaryotes. The more intuitive groupings of protozoans and algae and fungus-like protists are phylogenetically meaningless, because the nutritional modes used to categorize them are spread across many different lineages. It is now clear that there are multiple clades of protists, with some lineages more closely related to plants, fungi, or animals than they are to other protists. Indeed, several protistan lineages appear to have arisen through secondary endosymbiosis, in which a eukaryotic green or red alga was engulfed by another eukaryote. In the rest of this chapter, we will use the tentative phylogeny illustrated in the next module to survey several protistan lineages.

? Why are protists especially important to biologists investigating the evolution of eukaryotic life?

■ Because the first eukaryotes were protists, and protists were ancestral to all other eukaryotes, including plants, fungi, animals, and modern protists

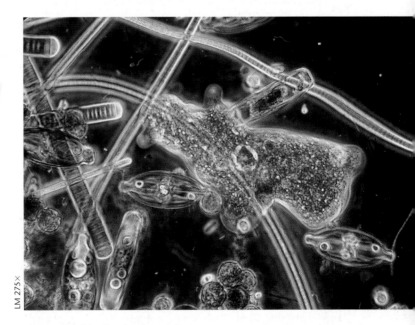

LM 275×

Figure 16.18 Protists in pond water

16.19 A tentative phylogeny of eukaryotes includes multiple clades of protists

The classification of protists, like all scientific inquiries, remains a work in progress. **Figure 16.19** illustrates an abbreviated version of one hypothesis of the phylogeny of kingdoms Fungi, Animalia, and Plantae and the groups that were traditionally placed in kingdom Protista. Don't be overwhelmed by this complex-looking scheme. We present it only as a means of providing an evolutionary organization for the protistan groups that we will survey. Certainly, the names, boundaries,

and placement of these clades will continue to change as the genomes of more protists are sequenced and compared. Next, let's examine a few present-day representatives of some of these diverse protistan lineages.

? According to the phylogenetic tree, are green algae more closely related to red algae or to brown algae?

■ Green algae share a more recent common ancestor with red algae.

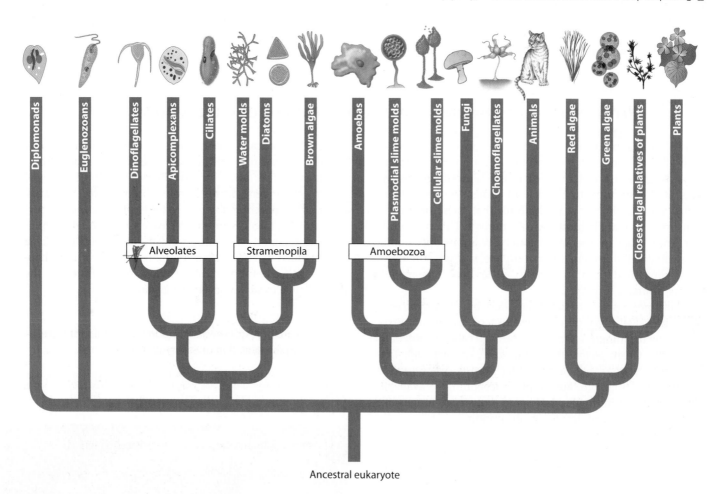

Figure 16.19 A tentative phylogeny of eukaryotes

16.20 Diplomonads and euglenozoans include some flagellated parasites

The **diplomonads** may represent the most ancient surviving lineage of eukaryotes. They have two nuclei and multiple flagella. Their modified mitochondria have no DNA or electron transport chains. Most diplomonads are anaerobic.

An infamous example of a diplomonad is *Giardia intestinalis* (**Figure 16.20A**), a parasite that lives in the intestines of mammals. People most often pick up *Giardia* by drinking water contaminated with feces containing the parasite. Drinking such contaminated water from a seemingly pris-

tine stream can cause severe diarrhea and ruin a camping trip. Boiling the water first will kill *Giardia*.

The **euglenozoans** are a diverse clade that includes heterotrophs, photosynthetic autotrophs, and pathogenic parasites. In **Figure 16.20B**, the squiggly "worms" are cells of *Trypanosoma;* the red cells are human red blood cells. This trypanosome causes sleeping sickness, which is spread by the African tsetse fly. This disease is invariably fatal if left untreated. Trypanosomes evade detection by their host's

immune system by altering the molecular structure of their coats frequently.

Euglena, a common inhabitant of pond water, is a member of a different euglenozoan lineage that is characterized by one or two flagella extending from one end of the cell (Figure 16.20C). Many species of photosynthetic *Euglena* can also absorb nutrients as heterotrophs in the dark.

? How do the nutritional modes of diplomonads and euglenozoans differ?

■ Diplomonads are anaerobic heterotrophs; euglenozoans exhibit all types of nutritional modes.

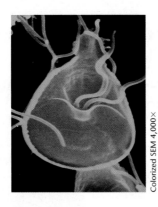

Figure 16.20A
A diplomonad: *Giardia intestinalis*

Colorized SEM 4,000×

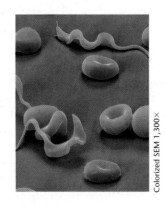

Figure 16.20B
A euglenozoan: *Trypanosoma* (with blood cells)

Colorized SEM 1,300×

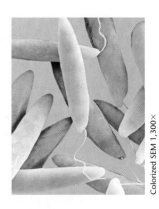

Figure 16.20C
A euglenozoan: *Euglena*

Colorized SEM 1,300×

16.21 Alveolates have sacs beneath the plasma membrane and include dinoflagellates, apicomplexans, and ciliates

Dinoflagellates, apicomplexans, and ciliates form another clade of protists, the **alveolates,** whose identity is emerging from molecular systematics. Alveolates are characterized by membrane-enclosed sacs just beneath the plasma membrane. Researchers hypothesize that these sacs, called alveoli, help stabilize the cell surface or regulate water and ion content.

Dinoflagellates (Figure 16.21A) are very common components of marine and freshwater phytoplankton (microscopic photosynthetic organisms). Some reside within coral animals, providing much of the food for coral reef communities. Other dinoflagellates are heterotrophic. Each species has a characteristic shape reinforced by plates made of cellulose. The beating of two flagella in grooves encircling the cell produces the spinning movement for which these organisms are named (from the Greek *dinos,* whirling). Dinoflagellate blooms—population explosions—sometimes cause warm coastal waters to turn pinkish orange, a phenomenon known as red tide. Toxins produced by some red-tide dinoflagellates have caused massive fish kills. Humans who consume molluscs that have accumulated the toxins are affected as well, sometimes fatally.

All **apicomplexans** are parasites of animals, and some cause serious human diseases. Apicomplexans are so named

because one end (the *apex*) of the infectious cell of these parasites contains a *complex* of organelles specialized for penetrating host cells. *Plasmodium* (Figure 16.21B), the apicomplexan that causes malaria, enters red blood cells, feeding on them from within and eventually destroying them. Spread by mosquitoes, malaria is one of the most debilitating and widespread human diseases. About 300 million people are now infected, and up to 2 million die each year.

Ciliates are a large, varied group of alveolates, named for their use of cilia to move and feed. Nearly all ciliates are free-living, including the common freshwater protist *Paramecium* (see Figure 4.12B). Ciliates are unique in having two types of nuclei: a single, large macronucleus, which controls everyday activities, and multiple tiny micronuclei, which function in sexual reproduction. You can see the macronucleus, which resembles a string of beads running the length of the cell, in the trumpet-shaped ciliate *Stentor* (Figure 16.21C).

? Where might autotrophic dinoflagellates be found?

■ Some are floating phytoplankton; others are symbionts within coral.

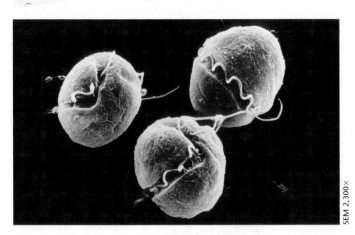

Figure 16.21A A dinoflagellate: *Gymnodium*

SEM 2,300×

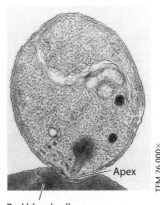

Apex

Red blood cell

TEM 26,000×

Figure 16.21B
An apicomplexan: *Plasmodium*

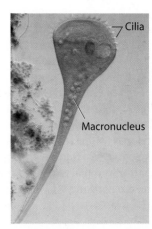

Cilia

Macronucleus

LM 60×

Figure 16.21C
A ciliate: *Stentor*

16.22 Stramenopiles are named for their "hairy" flagella and include the water molds, diatoms, and brown algae

Stramenopiles (from the Latin *stramen,* straw, and *pilos,* hair) are named for their "hairy" flagellum, which has numerous hairlike projections. This flagellum is usually paired with a "smooth" flagellum. The clade includes several groups of heterotrophs (such as water molds and downy mildews) as well as certain groups of algae.

Water molds are fungus-like protists that generally decompose dead plants and animals in freshwater habitats (Figure 16.22A). Parasitic water molds sometimes grow on the skin or gills of fish. Early morphological studies suggested that these organisms were fungi, but molecular systematics has confirmed that water molds are not closely related to fungi. Downy mildews are plant parasites that are relatives of water molds. One of these mildews causes late blight of potatoes, which contributed to the devastating Irish famine in the mid-1800s. A million people died and at least that many were forced to leave Ireland.

Diatoms (Figure 16.22B) are unicellular, photosynthetic algae with a unique, glassy cell wall containing silica, the mineral actually used to make glass. The cell wall consists of two halves that fit together like the bottom and lid of a shoe box. Both freshwater and marine environments are rich in diatoms, and the organic molecules these microscopic algae produce are a key source of food in all aquatic environments. Indeed, diatoms are as important a food source for many marine animals as grasses are for many land animals. Diatoms store their food reserves in the form of an oil, which also provides buoyancy, keeping diatoms floating near the surface in the sunlight. Massive accumulations of fossilized diatoms make up thick sediments known as diatomaceous earth, which is mined for use as both a filtering medium and a grinding and polishing agent.

Brown algae are the largest and most complex algae. They owe their characteristic brown color to some of the pigments in their chloroplasts. All are multicellular, and most are marine. Brown algae include many of the species commonly called seaweeds. (Some red and green algae are also referred to as seaweeds.) Seaweeds are sometimes used for food: A brown alga (kombu) is used in soups, and sheets of a red alga (nori) are used to wrap sushi.

We use the word *seaweeds* here to refer to marine algae that have large multicellular bodies. Some of these organisms even have specialized organs that resemble those of plants. However, seaweeds lack the true stems, leaves, and roots found in most plants. Figure 16.22C shows an underwater bed of brown algae called **kelp** off the coast of California. Anchored to the seafloor by rootlike structures called holdfasts, kelp may grow to heights of 100 m. Fish, sea lions, sea otters, and gray whales regularly use these kelp "forests" as their feeding grounds.

? Which stramenopiles are not photosynthetic?

■ Water molds and downy mildews

Figure 16.22A   A water mold breaking down a dead insect

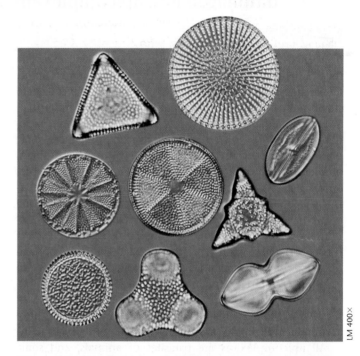

LM 400×

Figure 16.22B   Diatoms, unicellular algae

Figure 16.22C   Brown algae: a kelp "forest"

Amoebas move and feed by means of **pseudopodia** (singular, *pseudopodium*), which are temporary extensions of the cell. Molecular systematics now indicates that amoebas are dispersed across many taxonomic groups. We will consider only the **amoebozoans**, which have lobe-shaped pseudopodia. This clade includes many species of free-living amoebas, some parasitic amoebas, and the slime molds.

The amoeba in **Figure 16.23A** is ingesting a small protist. Its pseudopodia arch around the prey and enclose it in a food vacuole (see Module 4.10). Free-living amoebas creep over rocks, sticks, or mud at the bottom of a pond or ocean. One species of parasitic amoeba causes amebic dysentery, responsible for up to 100,000 deaths worldwide every year.

The yellow, branching growth on the dead log in **Figure 16.23B** is a **plasmodial slime mold**. These protists are common where there is moist, decaying organic matter and are often brightly pigmented. Large and branching as it is, the organism in Figure 16.23B is not multicellular. Containing many nuclei within one mass of cytoplasm undivided by plasma membranes, the whole thing is called a plasmodium. (Don't confuse this word with the apicomplexan *Plasmodium*, which causes malaria.) Because most nuclei go through mitosis at the same time, plasmodial slime molds can be used to study molecular details of the cell cycle.

The plasmodium extends pseudopodia through soil and rotting logs, engulfing food by phagocytosis as it grows. Its weblike form is an adaptation that enlarges the organism's surface area. Within the fine channels of the plasmodium, cytoplasm streams first one way and then the other in pulsing flows that are beautiful to watch with a microscope. The

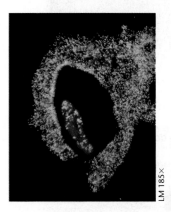

Figure 16.23A
An amoeba ingesting a smaller protist

LM 185×

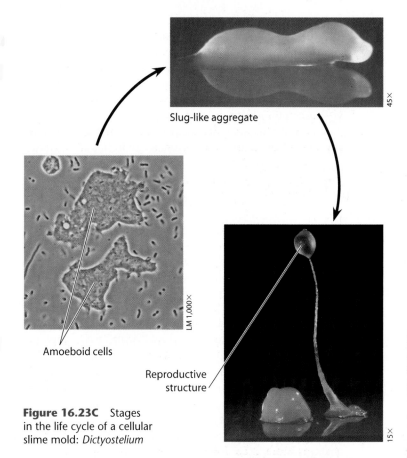

Slug-like aggregate

45×

Amoeboid cells

LM 1,000×

Reproductive structure

15×

Figure 16.23C   Stages in the life cycle of a cellular slime mold: *Dictyostelium*

cytoplasmic streaming probably helps distribute nutrients. When food and water are in short supply, the plasmodium stops growing and differentiates into reproductive structures (shown in the inset) that produce spores. When conditions become favorable, the spores release haploid cells that fuse to form a zygote, and the life cycle continues.

Cellular slime molds are also common on rotting logs and decaying organic matter. The micrographs in **Figure 16.23C** show three stages in the life cycle of a typical cellular slime mold, *Dictyostelium*. Most of the time, this organism exists as solitary amoeboid cells. The small, dark rods around the amoeboid cells are bacteria; the bacteria inside the cells are being digested in food vacuoles. When food is in short supply, the amoeboid cells swarm together, forming a slug-like aggregate that wanders around for a short time. Some of the cells then dry up and form a stalk supporting an asexual reproductive structure in which yet other cells develop into spores. *Dictyostelium* is a useful model for researchers studying the genetic mechanisms and chemical changes underlying cellular differentiation.

? Contrast the plasmodium of a plasmodial slime mold with the slug-like stage of a cellular slime mold.

■ A plasmodium is not multicellular, but is one cytoplasmic mass with many nuclei; the slug-like stage of a cellular slime mold consists of many cells.

Figure 16.23B   A plasmodial slime mold: *Physarum*

16.24　Red algae and green algae are the closest relatives of land plants

Molecular comparisons and other evidence support the following phylogenetic scenario: Over a billion years ago, a heterotrophic cell acquired a cyanobacterial endosymbiont. The descendants of this ancient protist evolved into red algae and green algae. Secondary endosymbiosis of a red alga led to the alveolates and stramenopiles (see Modules 16.21 and 16.22). At least 475 million years ago, the lineage that produced green algae gave rise to land plants, which we will cover in Chapter 17. In this module, we discuss both the red and green algae.

The warm coastal waters of the tropics are home to the majority of species of **red algae.** Their red color comes from an accessory pigment that masks the green of chlorophyll. Red algae are typically soft-bodied, but some have cell walls encrusted with hard, chalky deposits. Encrusted species, such as the one in **Figure 16.24A,** are common on coral reefs, and their hard parts are important in reef building.

Green algae are named for their grass-green chloroplasts. The micrographs in **Figure 16.24B** show two types of green algae. *Chlamydomonas* is a unicellular alga common in freshwater lakes and ponds. It is propelled through the water by two flagella. (Such cells are said to be biflagellated.) *Volvox* is a colonial green alga. Each *Volvox* colony is a hollow ball composed of hundreds or thousands of biflagellated cells. The large colonies shown here will eventually release the small daughter colonies within them. Some marine green algae are large and complex enough to qualify as seaweeds.

Most green algae have complex life cycles. The life cycle of *Ulva,* or sea lettuce (**Figure 16.24C**), follows a pattern called **alternation of generations,** in which a multicellular diploid (*2n*) form alternates with a multicellular haploid (*n*) form. Alternation of generations occurs in a number of multicellular algae and in all plants. Notice in the figure that the multicellular haploid forms are called **gametophytes.** The gametophyte generation alternates with a diploid generation that features a multicellular diploid form called a **sporophyte.** In *Ulva,* the gametophyte and sporophyte organisms are identical in appearance; both look like the one in the photograph, although they differ in chromosome number. The haploid gametophyte produces gametes by mitosis, and fusion of the gametes begins the sporophyte generation. In turn, cells in the sporophyte undergo meiosis and produce haploid, flagellated spores. The life cycle is completed when a spore settles to the bottom of the ocean and develops into a gametophyte.

Web/CD Thinking as a Scientist *What Kinds of Protists Are Found in Various Habitats?*

? What were the probable origins of the chloroplasts of green, red, and brown algae? (Remember that brown algae is a stramenopile.)

■ A cyanobacterial endosymbiont in both green and red algae; a red algal endosymbiont in brown algae

Figure 16.24A　A red alga: an encrusted type, on a coral reef

Volvox colonies　　　　　　　　　　　　*Chlamydomonas*

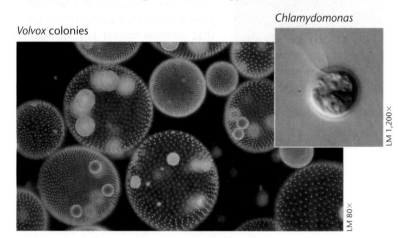

LM 1,200×

LM 80×

Figure 16.24B　Green algae, colonial and unicellular (inset)

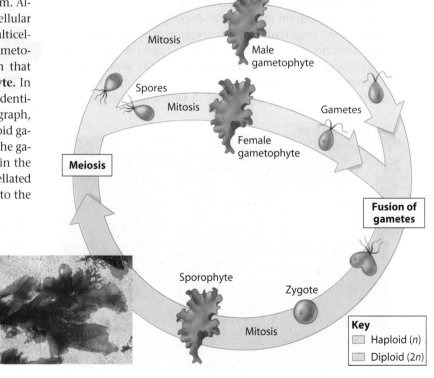

Mitosis

Male gametophyte

Spores

Mitosis

Gametes

Female gametophyte

Meiosis

Fusion of gametes

Sporophyte

Zygote

Mitosis

Key
　Haploid (*n*)
　Diploid (*2n*)

Figure 16.24C　A multicellular green alga: *Ulva* (sea lettuce) and its life cycle

Multicellularity evolved several times in eukaryotes

Increased complexity often makes more variations possible. Thus, the origin of the eukaryotic cell led to an evolutionary radiation of new forms of life. As you saw in this chapter, unicellular protists, which are structurally complex eukaryotic cells, are much more diverse in form than the simpler prokaryotes. The evolution of multicellular bodies broke through another threshold in structural organization.

Multicellular organisms—seaweeds, plants, animals, and most fungi—are fundamentally different from unicellular ones. In a unicellular organism, all of life's activities occur within a single cell. In contrast, a multicellular organism has various specialized cells, which perform different functions and are dependent on each other. For example, some cells give the organism its shape, while others make or procure food, transport materials, provide movement, or reproduce.

The most widely held view is that the organisms linking multicellular organisms to their unicellular ancestors were probably unicellular protists that lived as colonies, federations of independent cells sticking loosely together. **Figure 16.25** suggests how a unicellular protist with flagellated cells may have formed colonies that eventually gave rise to multicellular organisms. ❶ An ancestral colony may have formed, as colonial protists do today, when a cell divided and its offspring remained attached to one another. ❷ Next, the cells in the colony may have become somewhat specialized and interdependent, with different cell types becoming more and more efficient at performing specific, limited tasks. ❸ Later on, additional specialization among the cells in the colony may have led to distinctions between sex cells (gametes) and nonreproductive cells (somatic cells).

We see specialization and cooperation among cells today in several colonial protists, such as the green alga *Volvox* in Figure 16.24B. *Volvox* produces gametes, which depend on somatic cells while developing. Cells in truly multicellular organisms, as we know them today, are specialized for many more nonreproductive functions, including feeding, waste disposal, gas exchange, and protection, to name a few. Evolution of the division of labor to this extent involved many additional steps in somatic cell specialization.

As you saw in Figure 16.19, at least three major lineages from the ancestral eukaryote led to multicellular forms: one leading to brown algae, one to fungi and animals, and one to red algae, green algae, and plants. In Chapter 17, you will meet the charophyceans, the group of green algae most closely related to plants. And in Chapter 18, you will meet the choanoflagellates, the protists that are the closest relatives of animals.

The oldest known fossils of multicellular eukaryotes are small multicellular algae that date from 1.2 billion years ago. For the next 600 million years or so, the fossil record remains somewhat scanty, but by about 600 million years ago, a variety of multicellular algae had evolved, along with some soft-bodied animals resembling corals, jellies, and worms.

A period of mass extinctions separated the Precambrian from the Paleozoic era, but multicellular life again flourished soon thereafter. By about 500 million years ago, diverse animals, fungi, and multicellular algae populated aquatic environments. All life was still aquatic.

Around 500 million years ago, the move onto land began, probably as certain green algae living along the edges of lakes gave rise to primitive plants. In the next chapter, we trace the long evolutionary movement of plants onto land and their diversification there. After that, we pick up the threads of animal evolution in Chapter 18.

? About how long did life evolve in aquatic habitats before the algal ancestors of plants first began to colonize land? (*Hint:* See Module 16.1.)

■ About 3 billion years

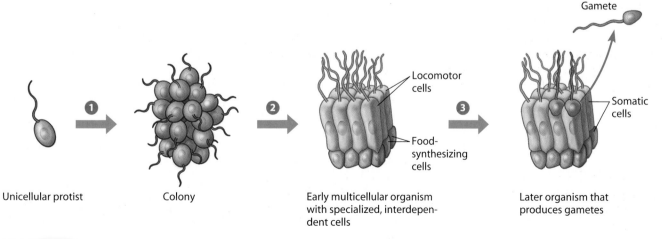

Figure 16.25 A model for the evolution of a multicellular organism from a unicellular protist

Unicellular protist · Colony · Early multicellular organism with specialized, interdependent cells · Locomotor cells · Food-synthesizing cells · Later organism that produces gametes · Gamete · Somatic cells

CHAPTER REVIEW

Reviewing the Concepts

Early Earth and the Origin of Life (Introduction–16.6)

Early Earth, which formed some 4.6 billion years ago, had an atmosphere that may have contained H_2O, CO_2, H_2S, and N_2 and some CH_4 and NH_3. Volcanic activity, lightning, and UV radiation were intense. Fossilized prokaryotes date back 3.5 billion years. Fossilized mats 2.5 billion years old mark a time when photosynthetic prokaryotes were producing enough O_2 to make the atmosphere aerobic **(Introduction–16.1).**

Organic molecules may have been formed abiotically in the conditions on early Earth **(16.2–16.3).** These molecules may then have polymerized on hot rocks **(16.4).**

RNA world. The first genes may have been RNA molecules that catalyzed their own replication **(16.5).** RNA might have acted as templates for the formation of polypeptides, which in turn assisted in RNA replication. Membranes may have separated various aggregates of self-replicating molecules, which could be acted on by natural selection **(16.6).**

Prokaryotes (16.7–16.16)

Prokaryotes are the oldest life-forms and remain the most numerous and widespread organisms **(16.7).**

Domains Bacteria and Archaea are distinguished on the basis of nucleotide sequences and other molecular and cellular features **(16.8).** Prokaryotes may be shaped as spheres (cocci), rods (bacilli), and curves or spirals **(16.9).**

Structural features that help prokaryotes thrive virtually everywhere are protective cell walls, sticky capsules, pili that cling to surfaces, flagella that provide motility, and resistant endospores. Rapid growth contributes to success **(16.10).**

Nutritional modes of prokaryotes are diverse **(16.11):**

Nutritional Mode	Energy Source	Carbon Source
Photoautotroph	Sunlight	CO_2
Chemoautotroph	Inorganic chemicals	CO_2
Photoheterotroph	Sunlight	Organic compounds
Chemoheterotroph	Organic compounds	Organic compounds

Archaea are common in extreme environments, such as salt lakes, acidic hot springs, and deep-sea hydrothermal vents. They are also a major life-form in the ocean **(16.12).**

Bacteria are currently organized into several subgroups of proteobacteria, chlamydias, spirochetes, gram-positive bacteria, and cyanobacteria. Cyanobacteria photosynthesize in a plantlike way **(16.13).**

Pathogenic bacteria cause disease by producing exotoxins or endotoxins **(16.14).** Bacteria such as the species that causes anthrax can be used as biological weapons **(16.15).**

Bioremediation is the use of organisms to clean up pollution. Prokaryotes are decomposers in sewage treatment and can clean up oil spills and toxic mine wastes. Prokaryotes are important in Earth's chemical cycles **(16.16).**

Protists (16.17–16.25)

Eukaryotic cells evolved from prokaryotic cells more than 2 billion years ago. The nucleus and endomembrane system probably evolved from infoldings of the plasma membrane. Mitochondria and chloroplasts probably evolved from aerobic and photosynthetic endosymbionts, respectively **(16.17).**

Protists are mostly unicellular eukaryotes. Molecular systematics is exploring eukaryotic phylogeny **(16.18–16.19).**

Diplomonads and euglenozoans are flagellated protists. The parasitic *Giardia* is a diplomonad with highly reduced mitochondria. Euglenozoans include trypanosomes and *Euglena* **(16.20).**

Alveolates have membrane sacs under the plasma membrane. Dinoflagellates are unicellular algae; apicomplexans are parasites, such as *Plasmodium,* which causes malaria; ciliates use cilia to move and feed **(16.21).**

Stramenopiles are named for their "hairy" flagella. This clade includes fungus-like water molds, photosynthetic, unicellular diatoms, and brown algae—large, complex seaweeds. Kelp forms important marine habitats **(16.22).**

Amoebozoans include amoebas with lobe-shaped pseudopodia and slime molds. A plasmodial slime mold is a multinucleate plasmodium that forms reproductive structures under adverse conditions. Cellular slime molds have unicellular and multicellular stages **(16.23).**

Red and green algae are in the lineage that includes plants. Red algae contribute to coral reefs. Green algae may be unicellular, colonial, or multicellular. The life cycles of many algae involve the alternation of haploid gametophyte and diploid sporophyte generations **(16.24).**

Multicellularity evolved in several different lineages, probably by specialization of the cells of colonial protists. Multicellular life arose over a billion years ago **(16.25).**

Connecting the Concepts

1. Using the figure below, describe the stages that may have led to the origin of life.

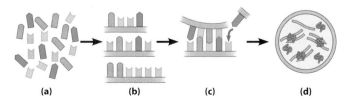

(a)　　　(b)　　　(c)　　　(d)

2. For each date, identify an important event in the history of Earth and its life.

Billions of years ago	Event
4.5	
3.5	
2.7	
2.1	
1.2	

Testing Your Knowledge

Multiple Choice

3. Ancient cyanobacteria, found in fossil stromatolites, were very important in the history of life because they
   a. were probably the first living things to exist on Earth.
   b. produced the oxygen in the atmosphere.
   c. are the oldest known archaea.
   d. were the first multicellular organisms.
   e. extracted heat from the atmosphere, cooling Earth.

4. You set your time machine for 3 billion years ago and push the start button. When the dust clears, you look out the window. Which of the following describes what you would probably see?
   a. plants and animals very different from those alive today
   b. a cloud of gas and dust in space
   c. green scum in the water
   d. land and water sterile and devoid of life
   e. an endless expanse of red-hot molten rock

5. In terms of nutrition, autotrophs are to heterotrophs as
   a. kelp are to diatoms.
   b. archaea are to bacteria.
   c. slime molds are to algae.
   d. algae are to slime molds.
   e. pathogenic bacteria are to harmless bacteria.

6. The bacteria that cause tetanus can be killed only by prolonged heating at temperatures considerably above boiling. This suggests that tetanus bacteria
   a. have cell walls containing peptidoglycan.
   b. protect themselves by secreting antibiotics.
   c. secrete endotoxins.
   d. are autotrophic.
   e. produce endospores.

7. Glycolysis is the only metabolic pathway common to nearly all organisms. To scientists, this suggests that it
   a. evolved many times during the history of life.
   b. was first seen in early eukaryotes.
   c. appeared early in the history of life.
   d. must be very complex.
   e. appeared rather recently in the evolution of life.

8. Of the following groups, which is thought to have most recently shared a common ancestor with animals?
   a. red algae   d. dinoflagellates
   b. cellular slime molds   e. green algae
   c. brown algae

9. Which of the following groups is incorrectly paired with an example of that group?
   a. euglenozoans—trypanosome causing sleeping sickness
   b. alveolates—apicomplexan such as *Plasmodium*
   c. stramenopiles—brown algae
   d. amoebozoans—water mold
   e. diplomonads—*Giardia*

10. Which of the following prokaryotes is not pathogenic?
    a. *Chlamydia*
    b. *Rhizobium*
    c. *Streptococcus*
    d. *Salmonella*
    e. *Bacillus anthracis*

Describing, Comparing, and Explaining

11. Describe how antibiotics may kill bacteria.

12. How do most biologists think that the mitochondria and chloroplasts of eukaryotic cells originated? What is the evidence for this idea?

13. *Chlamydomonas* is a unicellular green alga. How does it differ from a photosynthetic bacterium, which is also single-celled? How does it differ from a protozoan, such as an amoeba? How does it differ from larger green algae, such as sea lettuce (*Ulva*)?

14. What characteristic distinguishes true multicellularity from colonies of cells?

Applying the Concepts

15. Imagine you are on a team designing a moon base that will be self-contained and self-sustaining. Once supplied with building materials, equipment, and organisms from Earth, the base will be expected to function indefinitely. One of the team members has suggested that everything sent to the base be sterilized so that no bacteria of any kind are present. Do you think this is a good idea? Predict some of the consequences of eliminating all bacteria from an environment.

16. The buildup of CO_2 in the atmosphere resulting from the burning of fossil fuels is regarded as a major contributor to global warming (see Module 7.13). Diatoms and other microscopic algae in the oceans counter this buildup by using large quantities of atmospheric CO_2 in photosynthesis, which requires small quantities of iron. Experts suspect that a shortage of iron may limit algal growth in the oceans. Some scientists have suggested that one way to reduce CO_2 buildup might be to fertilize the oceans with iron. The iron would stimulate algal growth and thus the removal of more CO_2 from the air. A single supertanker of iron dust, spread over a wide enough area, might reduce the atmospheric CO_2 level significantly. Do you think this approach would be worth a try? Why or why not?

Answers to all questions can be found in Appendix 3.

For study help and Activities, go to campbellbiology.com or the student CD-ROM.

PLANT EVOLUTION AND DIVERSITY

17.1   Plants evolved from green algae
17.2   Plants have adaptations for life on land
17.3   Plant diversity reflects the evolutionary history of the plant kingdom

ALTERNATION OF GENERATIONS AND PLANT LIFE CYCLES

17.4   Haploid and diploid generations alternate in plant life cycles
17.5   Mosses have a dominant gametophyte
17.6   Ferns, like most plants, have a dominant sporophyte
17.7   Seedless plants dominated vast "coal forests"
17.8   A pine tree is a sporophyte with tiny gametophytes in its cones
17.9   The flower is the centerpiece of angiosperm reproduction
17.10 The angiosperm plant is a sporophyte with gametophytes in its flowers
17.11 The structure of a fruit reflects its function in seed dispersal
17.12 Agriculture is based almost entirely on angiosperms
17.13 Interactions with animals have profoundly influenced angiosperm evolution
17.14 Plant diversity is a nonrenewable resource

FUNGI

17.15 Fungi absorb food after digesting it outside their bodies
17.16 Fungi produce spores in both asexual and sexual life cycles
17.17 Fungi can be classified into five groups
17.18 Fungal groups differ in their life cycles and reproductive structures
17.19 Parasitic fungi harm plants and animals
17.20 Lichens consist of fungi living mutualistically with photosynthetic organisms
17.21 Fungi also form mutualistic relationships with animals
17.22 Fungi have enormous ecological benefits and practical uses

CHAPTER 17

Plants and Fungi—A Beneficial Partnership

WE TEND TO TAKE OUR ORANGE JUICE FOR GRANTED, but it is no small feat for citrus growers to produce it at a reasonable cost. Orange groves are found in Florida, Texas, and California. An enormous investment, trees like the ones pictured below take three to seven years to start producing fruit and require a rich supply of fertilizer. They are also vulnerable to freezing and to a long list of pathogenic bacteria, insects, and especially fungi.

Fungi are not always harmful to plants, however. There is another kind of association between fungi and plants in which the fungus plays the role of vital benefactor. You can see an example of this relationship in the photograph on the facing page, which shows a fungus growing around the roots of a red pine tree. The fungus is the dense network of white strands that totally ensheathes the roots. This fungus is called a mycorrhizal fungus; together, the root and the fungus form an intimate, mutually beneficial association called a **mycorrhiza** (meaning "fungus root"; plural, *mycorrhizae*). For its part, a mycorrhizal fungus absorbs phosphorus and other essential minerals as well as water from the soil, and these nutrients are then available to the plant. The sugars produced by the plant nourish the fungus.

Plants, Fungi, and the Colonization of Land

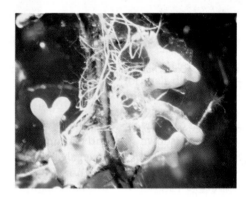

A mycorrhizal fungus enveloping roots of a red pine tree

Citrus trees can also have mycorrhizae, and the fungi may offer a way to cut the high economic and environmental costs of producing citrus fruits. Mycorrhizae can make a tree more resistant to disease, thereby reducing the need for pesticides that kill disease-causing organisms. Also, by enhancing a tree's uptake of nutrients, mycorrhizae can reduce, or even eliminate, the need for fertilizers. Unfortunately, the conditions in a typical citrus grove undermine the growth of mycorrhizae. Many citrus growers use fungus-killing chemicals (fungicides) to control fungi that cause disease, and the fungicides poison the mycorrhizal fungi as well. As a result, the grower loses the benefits of mycorrhizae and must apply expensive fertilizers. The environment also suffers because fungicides harm many kinds of organisms, and excess fertilizers pollute streams, lakes, and groundwater.

Some growers have tried to mitigate these problems by adding mycorrhizal fungi back into the soil or root zones

The colonization of land by plants was a major event in the history of life

of their groves, alternating between destroying and replenishing the fungi. Organic citrus growers, who avoid fungicides, often use such mycorrhiza-promoting applications, happy to find a natural boost for their trees.

Cultivated citrus groves treated with fungicides are unnatural in lacking mycorrhizae. In nature, nearly all plants have mycorrhizae. In fact, mycorrhizae appear in fossils of some of the oldest known plants, suggesting that these beneficial relationships with fungi may have played an important role as the first plants evolved from green algae and adapted to land.

The colonization of land by plants was a major event in the history of life. Plants transformed the landscape, creating new environmental opportunities for prokaryotes and protists and making it possible for herbivorous animals and their predators to evolve on land. This chapter continues our account of the evolution of life's diversity, focusing on plants and fungi. ■ ■ ■

17.3 Plant diversity reflects the evolutionary history of the plant kingdom

Figure 17.3A highlights some of the major events in the history of the plant kingdom and presents a widely held view of the relationships between surviving lineages of plants.

Approximately 475 million years ago, plants originated from an aquatic green algal ancestor. Early diversification gave rise to liverworts, hornworts, and mosses, plants that are informally called **bryophytes** (Figure 17.3B). Bryophytes resemble other plants in having apical meristems and embryos that are retained on the parent plant. Unlike other plants, however, bryophytes lack lignified vascular tissue and are therefore often called "nonvascular plants," although some do have tubular conducting cells. Without the strong cell walls of vascular tissue, bryophytes lack support. Thus, a mat of moss actually consists of many plants growing in a tight pack, holding one another up. The mat is spongy and retains water. The flagellated sperm of mosses and other bryophytes swim to the eggs, so fertilization requires the plant to be covered with a film of water.

Although the term *bryophyte* is still commonly used to refer to all nonvascular plants, debate continues over the relationships of liverworts, hornworts, and mosses to each other and to vascular plants. The broken lines in Figure 17.3A indicate the uncertainty about bryophyte phylogeny.

The **vascular plants** originated about 420 million years ago. Their lignin-hardened vascular tissues provide strong support, enabling stems to stand upright and grow tall on land. Two clades of vascular plants are informally called **seedless vascular plants** (Figure 17.3C, facing page): the lycophytes (such as club mosses) and the widespread pterophytes (ferns and their relatives).

A fern has well-developed roots and rigid stems. In many species, the leaves, commonly called fronds, sprout from stems that grow along the ground. Ferns are common in temperate forests, but they are most diverse in the tropics. In some tropical species, called tree ferns, upright stems can grow several meters tall. Like bryophytes, ferns have flagellated sperm that require a layer of water to reach the eggs, and they have spores enclosed in tough, protective walls.

As indicated in Figure 17.3A, the first vascular plants with seeds evolved about 360 million years ago. A **seed** consists of an embryo packaged with a food supply within a protective

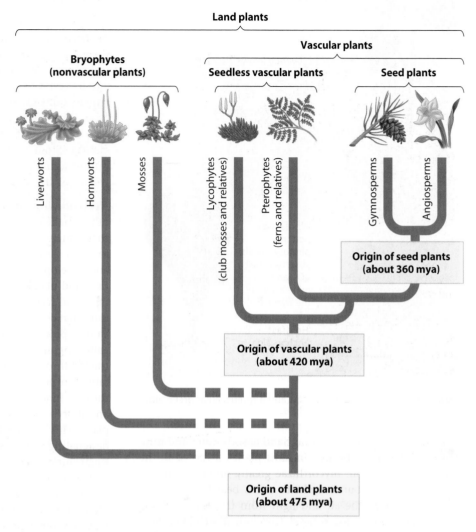

Figure 17.3A Some highlights of plant evolution

Figure 17.3B Bryophytes: liverwort (left), moss (center), hornwort (right)

Plants, Fungi, and the Colonization of Land

Citrus trees can also have mycorrhizae, and the fungi may offer a way to cut the high economic and environmental costs of producing citrus fruits. Mycorrhizae can make a tree more resistant to disease, thereby reducing the need for pesticides that kill disease-causing organisms. Also, by enhancing a tree's uptake of nutrients, mycorrhizae can reduce, or even eliminate, the need for fertilizers. Unfortunately, the conditions in a typical citrus grove undermine the growth of mycorrhizae. Many citrus growers use fungus-killing chemicals (fungicides) to control fungi that cause disease, and the fungicides poison the mycorrhizal fungi as well. As a result, the grower loses the benefits of mycorrhizae and must apply expensive fertilizers. The environment also suffers because fungicides harm many kinds of organisms, and excess fertilizers pollute streams, lakes, and groundwater.

Some growers have tried to mitigate these problems by adding mycorrhizal fungi back into the soil or root zones

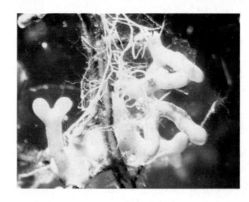

A mycorrhizal fungus enveloping roots of a red pine tree

The colonization of land by plants was a major event in the history of life

of their groves, alternating between destroying and replenishing the fungi. Organic citrus growers, who avoid fungicides, often use such mycorrhiza-promoting applications, happy to find a natural boost for their trees.

Cultivated citrus groves treated with fungicides are unnatural in lacking mycorrhizae. In nature, nearly all plants have mycorrhizae. In fact, mycorrhizae appear in fossils of some of the oldest known plants, suggesting that these beneficial relationships with fungi may have played an important role as the first plants evolved from green algae and adapted to land.

The colonization of land by plants was a major event in the history of life. Plants transformed the landscape, creating new environmental opportunities for prokaryotes and protists and making it possible for herbivorous animals and their predators to evolve on land. This chapter continues our account of the evolution of life's diversity, focusing on plants and fungi. ■ ■ ■

17.1 Plants evolved from green algae

The algal ancestors of plants may have carpeted moist fringes of lakes or coastal salt marshes over 500 million years ago. These shallow-water habitats were subject to occasional drying, and natural selection would have favored algae that could survive periodic droughts. Some species accumulated adaptations that enabled them to live permanently above the water line. The modern-day green alga *Coleochaete* (Figure 17.1A) may resemble an early plant ancestor.

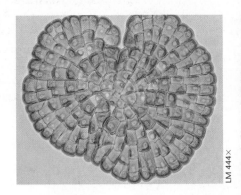

Figure 17.1A *Coleochaete,* a simple charophycean

Figure 17.1B *Chara,* an elaborate charophycean

It grows at the edges of lakes as disklike, multicellular colonies. *Coleochaete* and the more elaborate pond alga *Chara,* shown in Figure 17.1B, belong to a lineage of green algae called the **charophyceans.**

Plants and present-day charophyceans probably evolved from a common ancestor. Morphological, biochemical, and genetic evidence identify the charophyceans as the closest living relatives of land plants. Both groups have similar microscopic structures for making their cellulose cell walls and a similar mechanism for forming the cell plate that divides the cytoplasm during cell division (see Module 8.7). Their peroxisomes contain similar enzymes (see Figure 4.4B). Also, in plants that have flagellated sperm, the structure of the sperm closely resembles that of charophycean sperm. Molecular similarities of both nuclear and chloroplast genes also indicate that charophyceans, particularly *Coleochaete* and *Chara,* are the closest living relatives of land plants.

Adaptations enabling permanent life on dry land apparently accumulated in at least one population of an ancestral green alga by about 475 million years ago, the age of the oldest known plant fossils. The evolutionary novelties of these first land plants opened the new frontier of a terrestrial habitat. Early plant life would have thrived in the new environment. Bright sunlight was virtually limitless on land; the atmosphere had an abundance of carbon dioxide (CO_2); and at first there were relatively few pathogens and plant-eating animals. Next we examine the adaptations that allowed plants to access these environmental benefits by surviving and reproducing on land.

? What are some key homologies between plants and the green algae called charophyceans?

■ Mechanisms for construction of cellulose cell walls and cell plate formation during cell division; peroxisome enzymes; nuclear and chloroplast genes

17.2 Plants have adaptations for life on land

Plants are multicellular eukaryotes that make organic molecules by photosynthesis. Trees, shrubs, and grasses fit this definition and are clearly members of kingdom **Plantae.** But as we saw in Modules 16.18 and 16.24, multicellular algae also fit this definition. However, a set of derived characters distinguishes land plants as a clade (a monophyletic group), setting land plants apart from their closest algal relatives: growth-producing regions called apical meristems, alternation of haploid and diploid generations, walled spores produced in sporangia, male and female gametangia, and multicellular, dependent embryos. As we discuss further in this section, some of these traits reflect adaptations for terrestrial life.

Figure 17.2A on the facing page contrasts how plants and multicellular green algae interact with their environments.

Many algae are anchored by a holdfast, but generally they have no rigid tissues and are supported by the surrounding water. The whole algal body obtains CO_2 and minerals directly from the water. And almost all of the organism receives light and can perform photosynthesis. The algal ancestors of plants changed drastically as they became adapted to living on land. We briefly describe the key adaptations of plants to four challenges of terrestrial life.

Obtaining Resources from Two Locations A typical plant must obtain chemicals from both soil and air, two very different media. Water and mineral nutrients are mainly found in the soil; light and CO_2 are mainly available aboveground. Most plants have discrete organs—roots, stems, and leaves—that help meet this resource challenge.

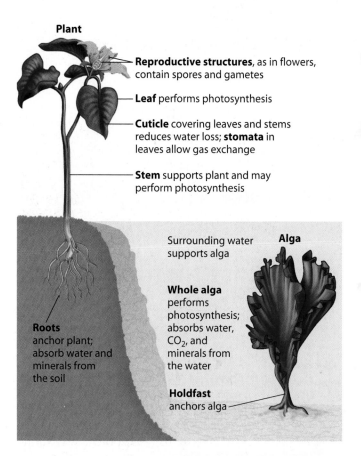

Plant

Reproductive structures, as in flowers, contain spores and gametes

Leaf performs photosynthesis

Cuticle covering leaves and stems reduces water loss; **stomata** in leaves allow gas exchange

Stem supports plant and may perform photosynthesis

Roots anchor plant; absorb water and minerals from the soil

Surrounding water supports alga

Alga

Whole alga performs photosynthesis; absorbs water, CO_2, and minerals from the water

Holdfast anchors alga

Figure 17.2A Comparing a plant and a multicellular green alga

Plant roots provide anchorage and absorb water and mineral nutrients from the soil. In most plants, as noted in the chapter introduction, mycorrhizae greatly enhance this absorption. Above-ground, a plant's stems bear leaves, which obtain CO_2 from the air and light from the sun, enabling them to perform photosynthesis. The elongation and branching of a plant's roots and stems maximize exposure to the resources in the soil and air. Growth-producing regions of cell division, called **apical meristems**, are found at the tips of stems and roots.

A plant must be able to connect its subterranean and aerial parts, conducting water and minerals upward from its roots to its leaves and distributing sugars produced in the leaves throughout its body. Most plants have **vascular tissue**, a network of thick-walled cells joined into narrow tubes that extend throughout the plant body. The photograph of part of an aspen leaf in **Figure 17.2B** shows the leaf's network of veins, which are fine branches of the vascular tissue. There are two types of vascular tissue. One, called **xylem**, includes dead cells that form microscopic pipes conveying water and minerals up from the roots. The other, called **phloem**, consists of living cells and distributes sugars throughout the plant.

Supporting the Plant Body Because air provides no support, a plant must be able to support itself against the pull

of gravity. The cell walls of some plant tissues, including vascular tissue, are thickened and strengthened by a chemical called **lignin**. A tree would collapse were it not for its lignin-rich cell walls.

Maintaining Moisture Another challenge of terrestrial life is the loss of water to the air. Helping most plants retain water is a waxy **cuticle** that covers their aerial parts (stems and leaves). Gas exchange cannot occur directly through the cuticle, but CO_2 and oxygen (O_2) diffuse across the leaf surfaces through tiny pores called **stomata** (singular, *stoma*). Two surrounding cells regulate each stoma's opening and closing. Stomata are open to allow gas exchange, usually during sunlight hours, and closed at other times to prevent water loss by evaporation.

Reproducing on Land Reproduction on land presents other challenges. For an alga, the surrounding water ensures that gametes and offspring stay moist and provides the means for their dispersal. Plants, however, must keep their gametes and developing embryos from drying out in the air. Like the first land plants, many living plants still produce gametes in male and female **gametangia** (singular, *gametangium*), structures that consist of protective jackets of cells surrounding the gamete-producing cells. The egg remains in the female gametangium and is fertilized there. Either the sperm swim to the egg through a film of water, or sperm-producing cells contained in pollen are conveyed, by wind or animals, close to the egg. In all plants, the fertilized egg (zygote) develops into an embryo while attached to and nourished by the parent plant. This multicellular, dependent embryo is the basis for designating plants as **embryophytes** (*phyte* means plant), distinguishing them from algae.

The life cycles of all plants involve an alternation of haploid and diploid generations, which we will describe in Module 17.4. Seed plants, whose offspring are protected in seeds, may rely on wind or animals (such as fruit-eating birds or mammals) to disperse their offspring. Plant reproduction also includes the production of haploid spores in protective structures called **sporangia** (singular, *sporangium*). A **spore** is a cell that can develop into a new organism without fusing with another cell. Plants that do not produce seeds often rely on these tough-walled, resistant spores for dispersal.

In summary, we can define a plant as a multicellular photosynthetic eukaryote with various adaptations to life on land. Next we trace the evolution of the major plant groups.

Web/CD Activity 17A *Terrestrial Adaptations of Plants*

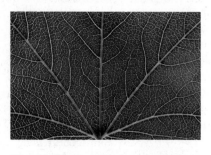

Figure 17.2B The network of veins in a leaf

? Which adaptations enable plants to reproduce on land?

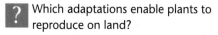
■ Multicellular gametangia that protect gametes from drying out, protection of embryo within female gametangia, and tough-walled spores

Plant diversity reflects the evolutionary history of the plant kingdom

Figure 17.3A highlights some of the major events in the history of the plant kingdom and presents a widely held view of the relationships between surviving lineages of plants.

Approximately 475 million years ago, plants originated from an aquatic green algal ancestor. Early diversification gave rise to liverworts, hornworts, and mosses, plants that are informally called **bryophytes** (Figure 17.3B). Bryophytes resemble other plants in having apical meristems and embryos that are retained on the parent plant. Unlike other plants, however, bryophytes lack lignified vascular tissue and are therefore often called "nonvascular plants," although some do have tubular conducting cells. Without the strong cell walls of vascular tissue, bryophytes lack support. Thus, a mat of moss actually consists of many plants growing in a tight pack, holding one another up. The mat is spongy and retains water. The flagellated sperm of mosses and other bryophytes swim to the eggs, so fertilization requires the plant to be covered with a film of water.

Although the term *bryophyte* is still commonly used to refer to all nonvascular plants, debate continues over the relationships of liverworts, hornworts, and mosses to each other and to vascular plants. The broken lines in Figure 17.3A indicate the uncertainty about bryophyte phylogeny.

The **vascular plants** originated about 420 million years ago. Their lignin-hardened vascular tissues provide strong support, enabling stems to stand upright and grow tall on land. Two clades of vascular plants are informally called **seedless vascular plants** (Figure 17.3C, facing page): the lycophytes (such as club mosses) and the widespread pterophytes (ferns and their relatives).

A fern has well-developed roots and rigid stems. In many species, the leaves, commonly called fronds, sprout from stems that grow along the ground. Ferns are common in temperate forests, but they are most diverse in the tropics. In some tropical species, called tree ferns, upright stems can grow several meters tall. Like bryophytes, ferns have flagellated sperm that require a layer of water to reach the eggs, and they have spores enclosed in tough, protective walls.

As indicated in Figure 17.3A, the first vascular plants with seeds evolved about 360 million years ago. A **seed** consists of an embryo packaged with a food supply within a protective

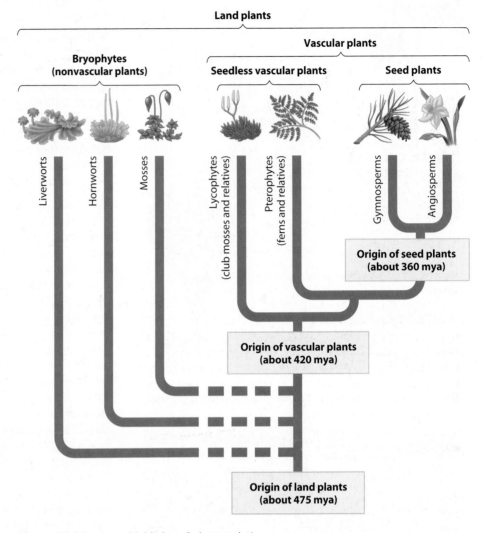

Figure 17.3A Some highlights of plant evolution

Figure 17.3B Bryophytes: liverwort (left), moss (center), hornwort (right)

Figure 17.3C Seedless vascular plants: fern (left) and club moss (right)

Figure 17.3D Gymnosperms: (clockwise, from lower left) ginkgo, cycad, blue spruce (a conifer), *Ephedra*

covering. Today, the seed plant lineage accounts for over 90% of the approximately 290,000 species of living plants. Several key adaptations underlie the enormous success of seed plants. First, seeds are survival packets for life on land, contributing to the spread of plants to diverse habitats by allowing plant embryos to be dispersed more widely. Second, seed plants do not require a water layer for fertilization. They produce **pollen**, which transfers sperm to egg-producing parts of the plant. Pollen is carried passively by wind or animals.

Among the earliest seed plants were the **gymnosperms** (from the Greek *gymnos*, naked, and *sperma*, seed). Seeds of gymnosperms are said to be "naked" because they are not produced in specialized chambers. The largest clade of gymnosperms are the conifers, consisting mainly of cone-bearing trees, such as pine, spruce, and fir. (The term *conifer* means "cone-bearing.") Some examples of gymnosperms that are less common are the ornamental ginkgo tree, the palmlike cycads, and the desert shrubs of the genus *Ephedra* (Figure 17.3D).

The most recent major episode in plant evolution was the appearance of flowering plants, or **angiosperms** (from the Greek *angion*, container, and *sperma*, seed), at least 140 million years ago. Flowers are complex reproductive structures that develop seeds within protective chambers. The great majority of living plants—some 250,000 species—are angiosperms and include a wide variety of plants, such as grasses, flowering shrubs, and flowering trees (Figure 17.3E).

In summary, four key adaptations for life on land distinguish the main lineages of the plant kingdom. (1) Dependent embryos are present in all plants. (2) Lignified vascular tissues mark a lineage that gave rise to most living plants. (3) Seeds are found in a lineage that includes all living gymnosperms and angiosperms and that dominates the plant kingdom today. (4) Flowers mark the angiosperm lineage. As we will see in the next several modules, the life cycles of living plants reveal additional details about plant evolution.

Figure 17.3E Angiosperms: barley grass (top) and jacaranda trees (bottom)

Web/CD Activity 17B *Highlights of Plant Evolution*

? Identify which of the following traits is shared by all plants: flowers, seeds, retained embryo, vascular tissue.

■ Embryo retained on parent plant

17.4 Haploid and diploid generations alternate in plant life cycles

Plants have life cycles that are very different from ours. Humans are diploid individuals; that is, each of us has two sets of chromosomes, one from each parent. The only haploid stages in the human life cycle are sperm and eggs. By contrast, plants have an **alternation of generations**: the diploid and haploid stages are distinct, multicellular generations.

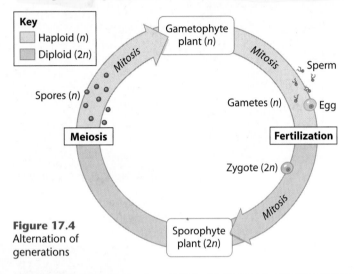

Key
Haploid (*n*)
Diploid (2*n*)

Mitosis
Gametophyte plant (*n*)
Mitosis
Sperm
Gametes (*n*)
Egg
Fertilization
Zygote (2*n*)
Mitosis
Sporophyte plant (2*n*)
Meiosis
Spores (*n*)
Mitosis

Figure 17.4
Alternation of generations

The haploid generation of a plant produces gametes and is called the **gametophyte.** The diploid generation produces spores and is called the **sporophyte.** In a plant's life cycle, these two generations alternate in producing each other.

In **Figure 17.4,** you can see that haploid gametophyte plants produce gametes (sperm and eggs) by mitosis. Fertilization results in a diploid zygote. The zygote divides by mitosis and develops into the multicellular diploid sporophyte plant. The sporophyte produces haploid spores by meiosis. A spore then develops by mitosis into a multicellular haploid gametophyte.

Although some algae exhibit alternation of generations, the closest algal relatives of plants, the charophyceans, do not. Thus, this life cycle appears to have evolved independently in plants; it is a derived character of land plants.

The next module highlights the life cycle of mosses. In mosses, as in all bryophytes, the gametophyte is the dominant generation—the larger, more obvious stage of the life cycle.

? Gametophyte is to _____ as _____ is to diploid.

■ haploid · · · sporophyte

17.5 Mosses have a dominant gametophyte (See text at top of facing page.)

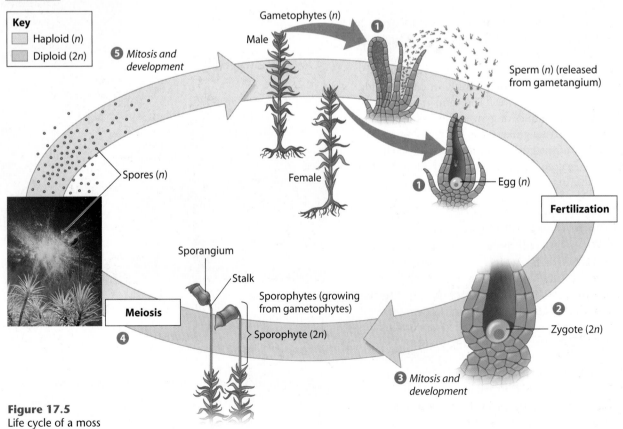

Key
Haploid (*n*)
Diploid (2*n*)

❺ *Mitosis and development*

Gametophytes (*n*)
Male

❶
Sperm (*n*) (released from gametangium)

Female

❶
Egg (*n*)

Fertilization

Spores (*n*)

Sporangium
Stalk
Meiosis
❹
Sporophytes (growing from gametophytes)
Sporophyte (2*n*)

❷
Zygote (2*n*)

❸ *Mitosis and development*

Figure 17.5
Life cycle of a moss

In a moss, most of the green, cushiony growth we see consists of gametophytes. Follow the moss life cycle in **Figure 17.5**. ❶ Gametes develop in male and female gametangia, usually on separate plants. The flagellated sperm swim through a film of water to the egg in the female gametangium. After fertilization, ❷ the zygote remains in the gametangium. ❸ There it divides by mitosis, developing into a sporophyte embryo and then a mature sporophyte, which remains attached to the gametophyte. ❹ Meiosis occurs in the sporangia at the tips of the sporophyte stalks. As you can see in the photograph, after meiosis, haploid spores are released from the sporangium. ❺ The spores undergo mitosis and develop into gametophyte plants.

Web/CD Activity 17C *Moss Life Cycle*

> **?** How do moss sperm travel from male gametangia to female gametangia, where fertilization of eggs occurs?

▪ The flagellated sperm swim through a film of water.

17.6 Ferns, like most plants, have a dominant sporophyte

All we usually see of a fern is the sporophyte. But let's start the fern life cycle with the gametophyte. ❶ Fern gametophytes often have a distinctive heart-like shape (shown at the top of **Figure 17.6**), but they are quite small (about 0.5 cm across) and inconspicuous. Like mosses, ferns have flagellated sperm that require moisture to reach an egg. ❷ The zygote remains on the gametophyte, where ❸ it develops into an independent sporophyte. ❹ The yellow dots in the photograph are clusters of sporangia, in which cells undergo meiosis, producing haploid spores. The spores are released and ❺ develop into gametophytes by mitosis.

Today, about 95% of all plants, including all seed plants, have a dominant sporophyte in their life cycle. As seed plants evolved, their sporophyte became adapted to house the gametophyte and all reproductive stages (including spores, eggs, sperm, zygotes, and embryos). Before we resume this story, let's glance back to a time in plant history, before seed plants rose to dominance, when ferns and other seedless plants covered much of the land surface.

Web/CD Activity 17D *Fern Life Cycle*

Web/CD Thinking as a Scientist *What Are the Different Stages of a Fern Life Cycle?*

> **?** How is it possible for the fern gametophyte to produce haploid gametes without meiosis?

▪ All the gametophyte's cells are haploid, so there is no need to reduce chromosome number by meiosis to produce haploid gametes.

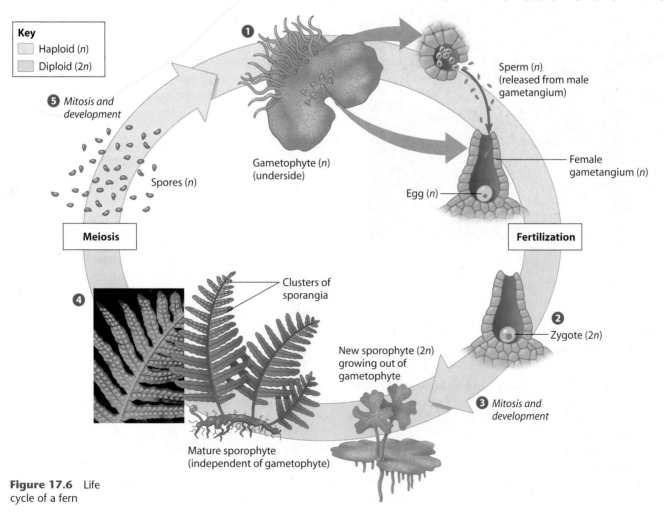

Key
- Haploid (*n*)
- Diploid (2*n*)

❶ Gametophyte (*n*) (underside)

Sperm (*n*) (released from male gametangium)

Female gametangium (*n*)

Egg (*n*)

Fertilization

Zygote (2*n*)

❷

❸ *Mitosis and development*

New sporophyte (2*n*) growing out of gametophyte

Mature sporophyte (independent of gametophyte)

❹

Clusters of sporangia

Meiosis

Spores (*n*)

❺ *Mitosis and development*

Figure 17.6 Life cycle of a fern

Seedless plants dominated vast "coal forests"

The two clades of seedless vascular plants, the lycophytes (introduced in Module 17.3 as club mosses) and the ferns and their relatives, were once the dominant plant forms on Earth. During the Carboniferous period (about 360–299 million years ago), vast forests of these plants grew in swampy areas that covered much of what is now Eurasia and North America. At that time, these continents were close to the equator and had tropical climates. Figure 17.7 shows a reconstruction of one of these forests based on fossil evidence. Most of the large trees with straight trunks are lycophytes. These giant woody trees had diameters of more than 2 m and heights of more than 40 m. On the far left, the tree with feathery branches is a horsetail, a giant ancestor of the few living species of horsetails, which are relatives of ferns. Tree ferns were also prominent, although they are not featured in this artistic reconstruction. Animals, including the giant dragonfly you see, also thrived in the swamp forests.

The tropical swamp forests of the Carboniferous period generated great quantities of organic matter. As the plants died, they fell into stagnant wetlands and did not decay completely. Their remains formed thick organic deposits called peat. Later, seawater covered the swamps, marine sediments covered the peat, and pressure and heat gradually converted the peat to coal. Coal is black sedimentary rock made up of fossilized plant material. Carboniferous coal deposits are the most extensive ever formed. (The name Carboniferous comes from the Latin *carbo*, coal, and *fer*, bearing.)

Coal, oil, and natural gas are **fossil fuels**—fuels formed from the remains of ancient organisms. Coal was crucial to the Industrial Revolution, and globally today people burn 6 billion tons of coal a year. When these vast Carboniferous forests were living, their photosynthesis removed so much CO_2 from the atmosphere that they may have contributed to global cooling. Ironically, burning these nonrenewable fossil fuels returns CO_2 and other greenhouse gases to the atmosphere and is now contributing to global warming.

Near the end of the Carboniferous period, the world climate turned drier, and the vast swamps and forests began to disappear. The climatic change provided an opportunity for the early seed plants, which grew along with the seedless plants in the Carboniferous swamps. With their wind-dispersed pollen and protective seeds, they could complete their life cycles on dry land.

Throughout the Mesozoic era, which began about 250 million years ago, gymnosperms dominated terrestrial ecosystems and served as the food supply for giant herbivorous dinosaurs. One group, the cycads, was particularly abundant, and so the "Age of Dinosaurs" is also known as the "Age of Cycads." Today, conifers are the dominant gymnosperms, being widespread in regions with dry, cool climates and short growing seasons. Their needlelike leaves have little surface area for evaporation, and a thick cuticle also helps retain water. Most conifers do not lose their leaves and can photosynthesize throughout the year.

Let's see how the pine tree life cycle fits with the major trends in plant history: dominance of the sporophyte generation and protection of the delicate reproductive stages.

Figure 17.7 A reconstruction of an extinct forest dominated by seedless plants

? Why are coal, oil, and natural gas called "fossil" fuels?

■ Because they are derived from ancient organisms that did not decay completely after dying

A pine tree is a sporophyte with tiny gametophytes in its cones

Pines and other conifers illustrate how drastically the relative roles of the haploid and diploid generations changed as plants evolved on land. A pine tree itself is a sporophyte; the gametophyte generation consists of very small stages that grow inside the tree's cones.

Cones are a significant adaptation to land, for they harbor all of a conifer's reproductive structures. These are the same structures that we described earlier for the mosses and ferns: diploid sporangia, which produce haploid spores by meiosis; haploid gametophytes; gametes (eggs and sperm); zygotes, resulting from fertilization; and embryos.

A pine tree bears two types of cones, which produce spores that develop into female and male gametophytes. The hard, woody ones are ovulate cones (❶ in Figure 17.8). An ovulate cone has many hard, radiating scales, each bearing a pair of **ovules**. An ovule starts out as a sporangium

with a covering, or integument. ❷ Pollen cones are generally much smaller than ovulate cones; they contain many sporangia, each of which makes numerous haploid spores by meiosis. Male gametophytes, or **pollen grains**, develop from the spores. Mature pollen cones release a cloud of pollen (millions of microscopic grains). You may have seen yellowish conifer pollen covering cars or floating on ponds in the spring.

Carried by the wind and not dependent on water, pollen grains house cells that develop into sperm. ❸ **Pollination** occurs when pollen grains land on and enter an ovule. Meanwhile, meiosis occurs in a spore mother cell in the ovule, and ❹ one surviving haploid spore cell develops into the female gametophyte. Not until months later do eggs appear within the female gametophyte. ❺ A tiny tube grows out of each pollen grain and eventually releases sperm near an egg. Fertilization usually occurs more than a year after pollination.

❻ Usually all eggs in an ovule are fertilized, but typically only one zygote develops fully into a sporophyte embryo. The ovule transforms into the seed, which contains the embryo's food supply (the remains of the female gametophyte) and has a tough seed coat (the ovule's integument). In a typical pine, seeds are shed about two years after pollina-

tion. ❼ The seed is dispersed by wind, and when conditions are favorable, it germinates, and its embryo grows into a pine seedling.

In summary, all the reproductive stages of conifers are housed in cones borne on sporophytes. The ovule is a key adaptation—a protective device for all the female stages in the life cycle, as well as the site of pollination, fertilization, and embryonic development. The ovule becomes the seed, an important terrestrial adaptation and a major factor in the success of the conifers and flowering plants.

Next we consider flowering plants, the most diverse and geographically widespread of all plants. Angiosperms dominate most landscapes today, and it is their flowers that account for their unparalleled success.

Web/CD Activity 17E *Pine Life Cycle*

? What key adaptations have contributed to the overwhelming success of seed plants?

■ Pollen transfers sperm to eggs without the need for water. Seeds protect, nourish, and help disperse plant embryos. The reduced gametophytes of seed plants are protected on the sporophyte plant.

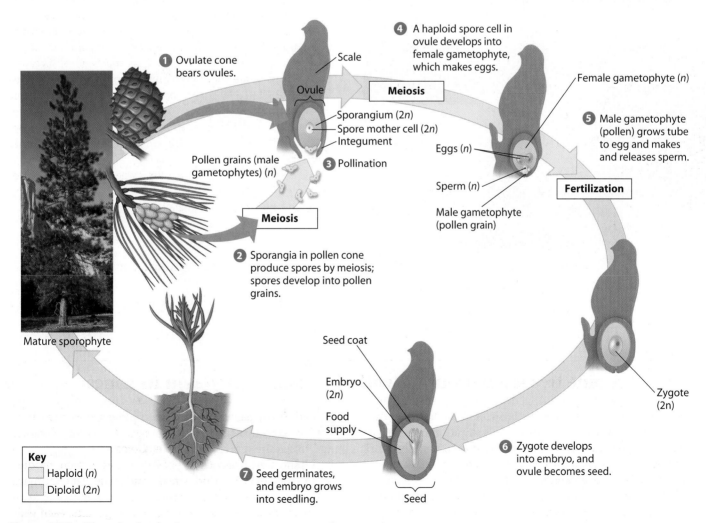

Figure 17.8 Life cycle of a pine tree

17.9 The flower is the centerpiece of angiosperm reproduction

No organisms make a showier display of their sex life than angiosperms (Figure 17.9A). From roses to cherry blossoms, flowers are the sites of pollination and fertilization. They also generate fruits, which contain the angiosperm's seeds.

Figure 17.9B shows the anatomy of a flower. Although the generalized flower drawn here looks different from the flowers in Figure 17.9A, these and all other kinds of flowers have a common anatomy. A flower is actually a short stem bearing up to four kinds of modified leaves attached at a region called the receptacle. At the base of the flower are the **sepals**, which are usually green. They enclose the flower before it opens (think of a rosebud). Above the sepals are the **petals**, which are usually the most striking part of the flower and are often important in attracting insects and other pollinators. (Wind-pollinated flowers generally lack brightly colored parts.) The actual reproductive structures are multiple stamens and one or more carpels. Each **stamen** consists of a stalk (filament) bearing a sac called an **anther**, in which pollen grains develop. The **carpel** consists of a stalk (style) with an ovary at the base and a sticky tip known as the **stigma**, which traps pollen. The **ovary** is a protective chamber containing one or more ovules, in which the eggs develop. As we see in the next module, a seed develops from each ovule, and the fruit develops from the ovary.

Figure 17.9A Some examples of floral diversity: lupine (left, a cluster of flowers), orchid (center), water lily (right)

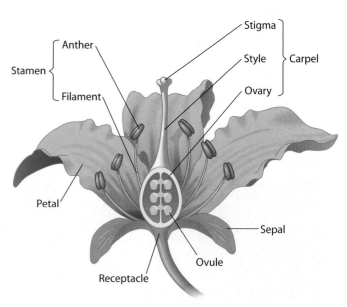

Figure 17.9B The parts of a flower

17.10 The angiosperm plant is a sporophyte with gametophytes in its flowers

In broad outline, the angiosperm life cycle resembles that of a gymnosperm. The plant we see is a sporophyte, and the tiny gametophyte generation lives on it. In contrast to a gymnosperm, in which gametophytes grow in cones and seeds are naked (not produced in specialized chambers), angiosperm gametophytes develop in flowers, and seeds are produced in an ovary and packaged inside a fruit.

Figure 17.10, illustrating the life cycle of a flowering plant, highlights features that have been especially important in angiosperm evolution. (We will discuss these, as well as double fertilization in angiosperms, in more detail in Modules 31.9–31.13.) Starting at the "Meiosis" box on the top of Figure 17.10, ❶ meiosis occurring in the anthers of the flower produces haploid spores that undergo mitosis and form the male gametophytes, or pollen grains. ❷ Meiosis in the ovule produces a haploid spore that undergoes mitosis and forms the few cells of the female gametophyte, one of which becomes an egg. ❸ Pollination occurs when a pollen grain, carried by the wind or an animal, lands on the stigma. As in gymnosperms, a tube grows from the pollen grain to the ovule, and a sperm fertilizes the egg, forming ❹ a zygote. Also as in gymnosperms, ❺ a seed develops from each

ovule. Each seed consists of an embryo (a new sporophyte) surrounded by a food supply and a seed coat. While the seeds develop, ❻ the ovary's wall thickens, forming the fruit that encloses the seeds. When conditions are favorable, ❼ a seed germinates and the embryo grows and develops into a mature sporophyte, completing the life cycle.

Several other features of angiosperm life have enhanced the success of these plants. One is the evolution of mutually dependent relationships with animals, which carry pollen more reliably than the wind. Another is the ability to reproduce rapidly. Fertilization in angiosperms usually occurs about 12 hours after pollination, making it possible for the plant to produce seeds in only a few days or weeks. As we mentioned in Module 17.8, a typical gymnosperm usually

takes several years to produce seeds. Rapid seed production is advantageous, particularly in harsh environments such as deserts, where growing seasons are extremely short.

Another feature contributing to the success of angiosperms is the development of fruits, which protect and help disperse the seeds, as we see in the next module.

Web/CD Activity 17F *Angiosperm Life Cycle*

? What is the difference between pollination and fertilization?

■ Pollination is the transfer of pollen by wind or animals from stamens to the tips of carpels. Fertilization is the union of egg and sperm, which are released from the pollen tube after the tube grows and makes contact with an ovule.

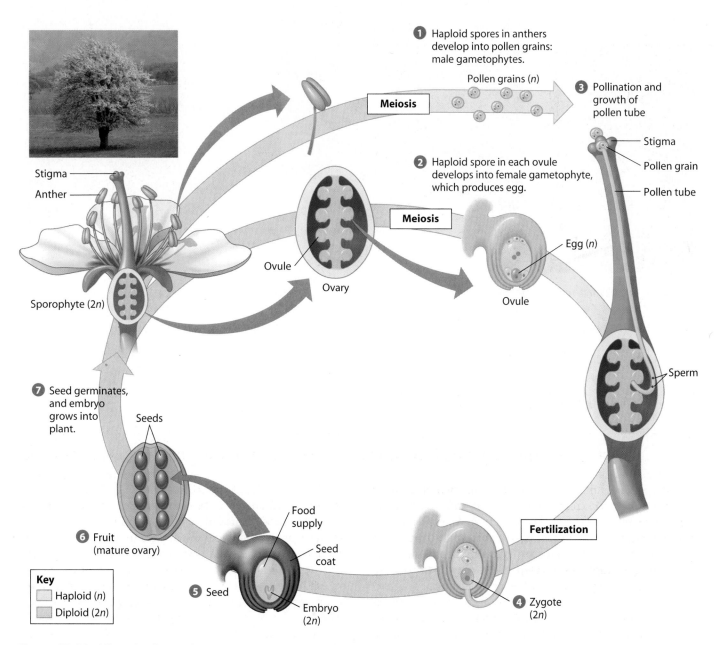

Figure 17.10 Life cycle of an angiosperm

17.11 The structure of a fruit reflects its function in seed dispersal

A **fruit**, the ripened ovary of a flower, is an adaptation that helps disperse seeds. Some angiosperms depend on wind for seed dispersal. For example, the fruit of a maple tree (**Figure 17.11A**) acts like a propeller, spinning a seed away from the parent tree on wind currents. Some angiosperms produce fruits that hitch a ride on animals. The barbs of cockleburs hook to the fur of animals (**Figure 17.11B**). These fruits may be carried for miles before they open and release their seeds.

Many angiosperms produce fleshy, edible fruits that are attractive to animals as food. When the mouse in **Figure 17.11C** eats a berry, it digests the fleshy part of the fruit, but most of the tough seeds pass unharmed through its digestive tract. The mouse may then deposit the seeds, along with a supply of natural fertilizer, some distance from where it ate the fruit.

The dispersal of seeds in fruits is one of the main reasons angiosperms are so widespread and successful. Angiosperms also contribute to the success of humans, as we see next.

? **What is a fruit?**

■ A ripened ovary of a flower, which contains, protects, and aids in the dispersal of seeds

Figure 17.11A Maple fruit can be dispersed by the wind

Figure 17.11B Cockleburs (fruit) may be carried by animal fur

Figure 17.11C Seeds within edible fruits are often dispersed in animal feces

CONNECTION

17.12 Agriculture is based almost entirely on angiosperms

Whereas gymnosperms supply most of our lumber and paper, flowering plants provide nearly all our food. Corn, rice, wheat, and the other grains are dry fruits, the main food source for most of the world's people and their domesticated animals. Many food crops are fleshy fruits, such as strawberries, apples, cherries, oranges, tomatoes, squash, and cucumbers. Others are modified roots, such as carrots and sweet potatoes, or modified stems, such as onions and potatoes.

We also grow angiosperms for spices, fiber, medications, perfumes, and decoration. Hardwoods, such as oak, cherry, and walnut, are flowering plants. Two of the world's most popular beverages come from coffee beans and tea leaves, and you can thank the tropical cacao tree for cocoa and chocolate.

Early humans probably collected wild seeds and fruits. Agriculture developed as humans began cultivating plants to have a more dependable food source. As they domesticated plants, humans began to intervene in plant evolution by selectively breeding to improve the quantity and quality of crops. New and improved plant species are now being genetically engineered. Agriculture is a unique kind of evolutionary relationship between plants and humans.

? **How have humans influenced plant evolution?**

■ By selective breeding and genetic engineering, humans intervene in the evolution of plants to maximize the harvest of products for human use.

17.13 Interactions with animals have profoundly influenced angiosperm evolution

Flowering plants and land animals have had mutually beneficial relationships throughout their evolutionary history. Most angiosperms depend on insects, birds, or mammals for pollination and seed dispersal. And most land animals depend on angiosperms for food. These mutual dependencies tend to improve the reproductive success of both the plants and the animals and thus are favored by natural selection.

In many cases, flowers and their pollinators are quite specialized in their mutual adaptations. This type of mutual evolutionary influence between two species is called **coevolution**, a process we will explore further in Chapter 37.

Many angiosperms produce flowers that attract pollinators that rely entirely on the flowers' nectar and pollen for food. Nectar is a high-energy fluid that is of use to the plant only for attracting pollinators. The color and fragrance of a flower are usually keyed to a pollinator's sense of sight and smell. Many flowers also have markings that attract pollinators, leading them past pollen-bearing organs on the way to nectar. For example, flowers that are pollinated by bees often have markings that reflect ultraviolet light. Such markings are invisible to us, but vivid to bees. The bee in **Figure 17.13A** is harvesting nectar and pollen from a scotch broom flower. The flower parts arched over the insect are the pollen-bearing stamens. Some of the pollen the bee picks up here will rub off onto the stigmas of other flowers it visits.

Many flowers pollinated by birds are red or pink, colors to which bird eyes are especially sensitive. The shape of the flower may also be important. Flowers that depend largely on hummingbirds, for example, typically have their nectar located deep in a floral tube, where only the long, thin beak and tongue of the bird are likely to reach. As a hummingbird (**Figure 17.13B**) flies among flowers in search of nectar, its feathers and beak pick up pollen, which it will deposit in other flowers of the same shape, and so probably of the same species, as it continues to feed.

Insects and birds are active mainly during the day. Some flowering plants, however, depend on nocturnal pollinators, such as bats. These plants typically have large, light-colored, highly scented flowers that can easily be found at night. The photograph in **Figure 17.13C** shows a bat approaching the large, white flower of a cactus. As the bat eats part of the flower, its body becomes dusted with pollen, which it passes on as it visits other flowers.

To a large extent, flowering plants are diverse and successful because they have close connections with other organisms. As we discussed in the chapter introduction, these organisms include fungi, which we discuss in Modules 17.15–17.22. But first let's consider how the amazing diversity of plants we have just surveyed is threatened by some connections with humans.

> **?** How are an angiosperm and its pollinators mutually rewarded by their relationship?

> ■ The pollinators obtain food, and the plant has its pollen transferred much more efficiently than if it were carried by the wind.

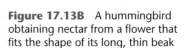

Figure 17.13B A hummingbird obtaining nectar from a flower that fits the shape of its long, thin beak

Figure 17.13A A bee picking up pollen as it feeds on nectar

Figure 17.13C A bat, a nighttime pollinator

17.14 Plant diversity is a nonrenewable resource

The exploding human population, with its demand for space and natural resources, is extinguishing plant species and other vital species at an unprecedented rate.

The threat to biodiversity is especially visible in the world's forests. From the stands of conifers in North America to swampy rain forest groves in the tropics, clear-cutting, burning, and other human-caused environmental damage have relentlessly reduced the globe's forest cover. As the trees fall, plants and wildlife that live alongside them also perish. What is lost is irreplaceable—entire ecosystems that provide medicinal plants, food, timber, and clean water and air. More than 25% of prescription drugs are extracted from plants, and researchers have investigated fewer than 5,000 of the world's approximately 290,000 known plant species as sources of medicine. Table 17.14 lists only a few of the unique medicinal compounds derived from plants.

People have been pushing into forestlands for thousands of years, but in the last century, scientists say, the rate of global forest reduction has reached alarming levels. About 50 million acres of forest—an area about the size of the state of Washington—are cleared every year. Much of Europe's original forests are gone. The forests of North America, which once dominated the landscape, have shrunk by almost 40% in the last two centuries to make room for people and meet the demand for lumber and paper (Figure 17.14). Not only have many of the animals that depend on these ecosystems—bears, wolves, birds of prey—disappeared, but hardy species of trees have also been replaced by species that grow quickly but are less resilient. Timber farms on land that once sustained natural forests have little of the biodiversity of the original forests, with pesticides and other chemicals allowing the land to support only a few kinds of life.

TABLE 17.14 A SAMPLING OF MEDICINES DERIVED FROM PLANTS

Compound	Example of Source	Example of Use
Atropine	Belladonna plant	Pupil dilator in eye exams
Digitalin	Foxglove	Heart medication
Menthol	Eucalyptus tree	Ingredient in cough medicines
Morphine	Opium poppy	Pain reliever
Quinine	Cinchona tree	Malaria preventive
Taxol	Pacific yew	Ovarian cancer drug
Tubocurarine	Curare tree	Muscle relaxant during surgery
Vinblastine	Periwinkle	Leukemia drug

About 20% of tropical forests were destroyed in the last third of the 20th century. About half of all the world's forests are found in the tropics, and they contain the vast majority of the world's plant and animal genetic resources. The diversity of life in these forests is astonishing, and its loss has heralded a biodiversity crisis of unprecedented scale.

Scientists are now rallying to stem the loss of genetic diversity and to offer less destructive ways for humans to work with forests. One ambitious effort, the All Species Foundation, is seeking to catalog every species on Earth within the next 25 years, from the smallest insect to the largest forest tree. The United Nations is also working to conserve the vast majority of plant species and to develop better ways of managing forests and plants by the year 2010. The goal of such efforts is to encourage forest management practices that are sustainable.

There is little doubt that forests will continue to be cut. But the search is on for ways of harvesting forests while still preserving tree cover and protecting plants and wildlife, allowing them to be studied and used for medicines, food, and tourism. If we begin to see forests and other ecosystems as living treasures that regenerate slowly, we may learn to work with them in more sustainable ways.

In the final section of this chapter, we look at another essential component of forests and many other ecosystems—the fungi.

Web/CD Activity 17G *Connection: Madagascar and the Biodiversity Crisis*

Web/CD Thinking as a Scientist *How Are Trees Identified by Their Leaves?*

? In what ways are forests nonrenewable resources? In what ways can they be renewable resources?

■ When the destruction of forest habitats results in a loss of species, that diversity can never be reclaimed. When forests are harvested at sustainable rates, regrowth will replace what was cut and habitats may not be changed as drastically.

Figure 17.14 Clear-cutting in Alaska

17.15 Fungi absorb food after digesting it outside their bodies

Members of the kingdom **Fungi** have body structures and modes of reproduction unlike those of any other organism. Although they are heterotrophs like animals, fungi don't eat (ingest) their food, but rather they acquire their nutrients by **absorption**. Fungi secrete powerful enzymes that digest their food outside their bodies and then absorb the small nutrient molecules into their cells.

Discussing fungi in the same chapter as plants may seem to indicate that these two kingdoms are close relatives. Actually, fungi are more closely related to animals than to plants. But the success of plants on land and the great diversity of fungi are interconnected. Mycorrhizae, the associations between plant roots and fungi that we discussed in the chapter introduction, helped make the colonization of land possible for both plants and fungi. Fungi cannot make their own food and must obtain organic molecules from other organisms. Before plants colonized land and began stocking soil with organic molecules, fungi may have thrived only in aquatic environments. And without fungi as mycorrhizal partners and as decomposers that return vital nutrients to the soil, plants could not have survived on land.

Fungi are found virtually everywhere, both in soil and in water, and are essential decomposers in most ecosystems. The orange umbrellas in **Figure 17.15A** are the reproductive structures of a fungus that is breaking down a dead log. Not all fungi are beneficial, however. Some are parasites, obtaining their nutrients at the expense of plants or animals.

A typical fungus consists of threadlike filaments that are called **hyphae** (singular, *hypha*). Hyphae branch repeatedly, forming a feeding network known as a **mycelium** (plural, *mycelia*), illustrated in **Figure 17.15B**. The photograph in **Figure 17.15C** shows the mycelium of a fungus growing on decaying leaves. The mushrooms in Figure 17.15A, solid as they seem, are actually made of tightly packed hyphae. A mushroom is just an above-ground reproductive structure growing from a much more extensive underground mycelium.

Fungal hyphae are surrounded by a cell wall. Unlike plants, which have cellulose cell walls, most fungi have cell walls made of chitin, a strong, flexible nitrogen-containing polysaccharide, identical to the chitin found in the external skeletons of insects. In most fungi, the hyphae consist of chains of cells separated by cross-walls that have pores large enough to allow ribosomes, mitochondria, and even nuclei to flow from cell to cell. Some fungi lack cross-walls and have many nuclei within a single mass of cytoplasm.

Fungi cannot run or fly in search of food. But their mycelium makes up for the lack of mobility by being able to grow at a phenomenal rate, branching throughout a food source and extending its hyphae into new territory. Because its hyphae grow longer without getting thicker, the fungus develops a huge surface area from which it can secrete diges-

Figure 17.15A Fungi decomposing a log

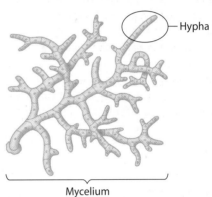

Hypha

Mycelium

Figure 17.15B A mycelium, made of numerous hyphae

Figure 17.15C The mycelium of a fungus growing on decaying leaves

tive enzymes and through which it can absorb food. A mycelium can add as much as a kilometer of hyphae each day. The mycelium of one giant fungus in Oregon spreads through 890 hectares (2,200 acres) of forest (equivalent to over 1,600 football fields). Scientists estimate that this fungus has been growing for 2,600 years, expanding outward in search of food.

Next we consider how fungi reproduce.

? Contrast the heterotrophic nutrition of a fungus with your own heterotrophic nutrition.

■ A fungus digests its food externally by secreting enzymes onto the food and then absorbing the small nutrients that result from digestion. In contrast, humans and most other animals "eat" relatively large pieces of food and digest the food within their bodies.

17.16 Fungi produce spores in both asexual and sexual life cycles

Many fungal species can reproduce either sexually or asexually (Figure 17.16). Reproduction typically involves the release of vast numbers of haploid spores, which are transported easily over great distances by wind or water. Fungal spores have been found more than 160 km above Earth. Spores that land in a moist place with food germinate and produce a new fungus.

In the sexual reproduction of many fungi, two haploid mycelia of different mating types release sexual signaling molecules, grow toward each other, and fuse. But this cytoplasmic fusion is often not followed immediately by the fusion of "parental" nuclei. Thus, many fungi have what is called a **heterokaryotic stage** (from the Greek, meaning "different nuclei"), in which cells contain two genetically distinct haploid nuclei. Hours, days, or even centuries may pass before the parental nuclei fuse, forming the usually short-lived diploid phase. Zygotes undergo meiosis inside specialized reproductive structures, from which haploid spores are dispersed.

Many fungi, including those commonly called molds and yeasts, typically reproduce asexually. The term **mold** refers to any rapidly growing fungus that reproduces asexually by producing spores, often at the tips of specialized hyphae. These familiar furry carpets often appear on aging fruit, bread, and other foods. The term **yeast** refers to any single-celled fungus that reproduces asexually by cell division or by budding—pinching off small "buds" from a parent cell. Yeasts inhabit liquid or moist habitats, such as plant sap and animal tissues.

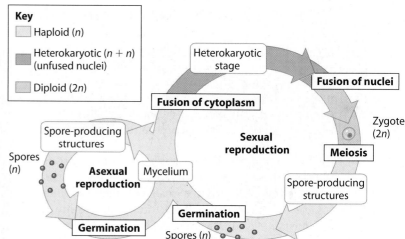

Key
- Haploid (*n*)
- Heterokaryotic (*n* + *n*) (unfused nuclei)
- Diploid (2*n*)

Figure 17.16 Generalized life cycle of a fungus

We will examine fungal reproduction more closely in Module 17.18 after we look at the classification of fungi.

Web/CD Activity 17H *Fungal Reproduction and Nutrition*

? What is the heterokaryotic stage of a fungus?

■ The stage in which each cell has two different nuclei (from two different parents), with the nuclei not yet fused

17.17 Fungi can be classified into five groups

Biologists who study fungi have described over 100,000 species, and there may be as many as 1.5 million species. Sexual reproductive structures are often used to classify species. Fungi that have no known sexual stage, including many molds and yeasts, are informally known as **imperfect fungi.** Increasingly, scientists use molecular data to try to recreate the phylogeny of fungi.

All but one group of fungi lack flagella, a condition that was once a criterion for placement in kingdom Fungi. Phylogenetic systematics, however, suggests that fungi evolved from a flagellated ancestor shared with animals. Based on molecular clock analysis (see Module 15.9), scientists estimate that the ancestors of animals and fungi diverged into separate lineages 1.5 billion years ago. The oldest undisputed fossils of fungi, however, are only about 460 million years old, perhaps because the ancestors of terrestrial fungi were microscopic and fossilized poorly.

Figure 17.17A shows a current hypothesis of fungal phylogeny. The dashed branches indicate groups that are probably not monophyletic. Although there are still areas of uncertainty and disagreement, most biologists recognize five groups of fungi.

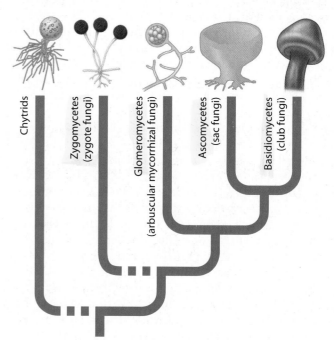

Figure 17.17A A proposed phylogenetic tree of fungi

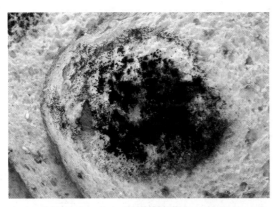

Figure 17.17B
Zygomycete:
Rhizopus stolonifer, the black bread mold

SEM 6,500×

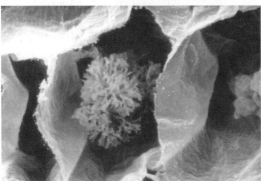

Figure 17.17C
Glomeromycete: arbuscule in a root cell

The **chytrids**, the only fungi with flagellated spores, are thought to represent the earliest lineage of fungi. They are common in lakes, ponds, and soil. Some species are decomposers; others parasitize protists, plants, or animals.

The **zygomycetes**, or **zygote fungi**, are characterized by their resistant zygosporangium, in which haploid spores form by meiosis. This diverse group includes fast-growing molds, such as black bread mold (Figure 17.17B) and molds that rot produce such as peaches, strawberries, and sweet potatoes. Some zygote fungi are parasites on animals.

The **glomeromycetes**, or **arbuscular mycorrhizal fungi**, were formerly grouped with zygomycetes, but genetic analyses indicate that they are a separate clade. They form a distinct type of mycorrhiza in which hyphae that invade plant roots branch into tiny treelike structures known as arbuscules (Figure 17.17C). About 90% of all plants have such symbiotic partnerships with glomeromycetes, which deliver phosphate and other minerals to plants while receiving organic nutrients in exchange.

The **ascomycetes**, or **sac fungi**, are named for saclike structures called asci that produce spores in sexual reproduction. They live in a variety of marine, freshwater, and terrestrial habitats and range in size from unicellular yeasts to elaborate morels and cup fungi (Figure 17.17D). Ascomycetes include some of the most devastating plant pathogens; other species live with green algae or cyanobacteria in symbiotic associations called lichens, which we discuss in Module 17.20.

The **basidiomycetes**, or **club fungi**, are probably the most familiar fungi—the mushrooms, the puffballs, and the shelf fungi (Figure 17.17E). They are named for their club-shaped, spore-producing structure called a basidium (plural, *basidia;* see Figure 17.18B). Many species excel at breaking down the lignin found in wood and play a key role as decomposers. For example, shelf fungi often break down the wood of weak or damaged trees and continue to decompose the wood after the tree dies. The basidiomycetes also include two groups of particularly destructive plant parasites, the rusts and smuts, which we discuss in Module 17.19.

In the next module, we explore the life cycles of some representative fungi.

Figure 17.17D Ascomycetes: edible morels (left) and cup fungus (right)

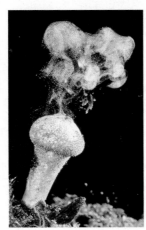

Figure 17.17E Basidiomycetes (club fungi): mushrooms (left), puffball (center), and shelf fungi (right)

? What is one reason that chytrids are thought to have diverged earliest in fungal evolution?

■ Chytrids are the only fungi that have flagellated spores, a characteristic of the ancestor of fungi.

Fungal groups differ in their life cycles and reproductive structures

The life cycle of black bread mold (see Figure 17.17B) is typical of zygomycetes. As hyphae expand through its food, the fungus reproduces asexually, forming spores in sporangia at the tips of upright hyphae. When the food is depleted, the fungus reproduces sexually. As shown in **Figure 17.18A**, mycelia of different mating types ❶ join and produce ❷ a cell containing nuclei from two parents. This young zygosporangium develops into a thick-walled structure ❸ that can tolerate dry or harsh environments. When conditions are favorable, the parental nuclei fuse, and the diploid nucleus undergoes meiosis, ❹ forming haploid spores.

Like the zygomycetes, ascomycetes, such as those shown in Figure 17.17D, reproduce asexually when conditions are suitable and sexually when conditions become harsh.

Now let's follow the life cycle of a mushroom, a basidiomycete, starting at the center bottom of **Figure 17.18B**. ❶ The heterokaryotic stage begins when hyphae of two different mating types fuse, forming ❷ a heterokaryotic mycelium, which grows and produces the mushroom. In the club-shaped cells called basidia, which line the gills of a mushroom, haploid nuclei fuse, forming ❸ diploid nuclei. Each diploid nucleus undergoes meiosis, and ❹ haploid spores are formed. A mushroom can release as many as a billion spores. If spores land on moist matter that can serve as food, ❺ they germinate and grow into haploid mycelia.

Much of the success of fungi is due to their reproductive capacity, both in the asexual production of spores and in sexual reproduction, which retains genetic variability and facilitates adaptation to harsh or changing conditions.

Web/CD Activity 17I *Fungal Life Cycles*

Web/CD Thinking as a Scientist *How Does the Fungus Pilobolus Succeed as a Decomposer?*

? Where would you find diploid cells in a mushroom?

■ In club-shaped cells called basidia lining the gills

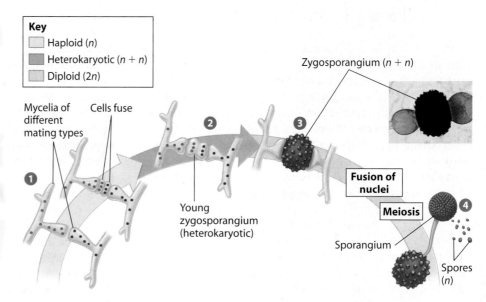

Key
- Haploid (*n*)
- Heterokaryotic (*n* + *n*)
- Diploid (2*n*)

Zygosporangium (*n* + *n*)

Mycelia of different mating types

Cells fuse

Young zygosporangium (heterokaryotic)

Fusion of nuclei

Meiosis

Sporangium

Spores (*n*)

Figure 17.18A Sexual reproduction of a zygote fungus

Figure 17.18B Life cycle of a mushroom

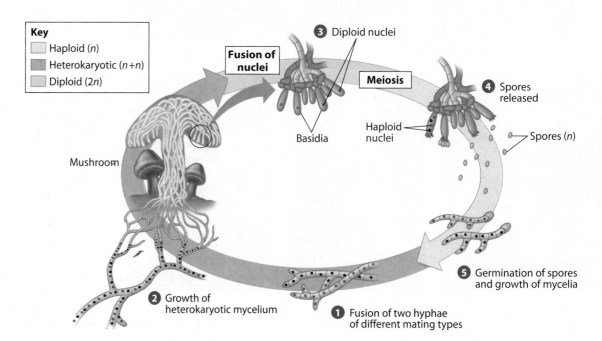

Key
- Haploid (*n*)
- Heterokaryotic (*n*+*n*)
- Diploid (2*n*)

❸ Diploid nuclei

Fusion of nuclei

Meiosis

❹ Spores released

Basidia

Haploid nuclei

Spores (*n*)

Mushroom

❷ Growth of heterokaryotic mycelium

❶ Fusion of two hyphae of different mating types

❺ Germination of spores and growth of mycelia

17.19 Parasitic fungi harm plants and animals

Of the 100,000 known species of fungi, about 30% make their living as parasites, mostly in or on plants. In some cases, fungi have literally changed landscapes. The dead tree in **Figure 17.19A** is an American elm killed by the parasitic fungus that causes Dutch elm disease. The fungus evolved with European species of elm trees, and it is relatively harmless to them. But it is deadly to American elms. Accidentally introduced into the United States on logs sent from Europe to pay World War I debts, the fungus has destroyed elm trees all across the northeastern United States.

Fungi are a serious problem as agricultural pests. About 80% of plant diseases are caused by fungi. Between 10% and 50% of the world's fruit harvest is lost each year to fungal attack. Species of club fungi called smuts and rusts are common on grain crops and cause tremendous economic losses each year. The ear of corn shown in **Figure 17.19B** is infected with a widespread fungal pathogen called corn smut. The grayish growths are called galls. Analogous to a mushroom, a gall is made up of heterokaryotic hyphae that invade a developing corn kernel and eventually displace it. When a gall matures, it breaks open and releases thousands of blackish spores. In parts of Central America, the smutted ears are cooked and eaten as a delicacy, but generally corn smut is regarded as a scourge. Fortunately, certain genetic strains of corn are resistant to it.

Some of the fungi that attack food crops are toxic to humans. The seed heads of many kinds of grain, including rye,

wheat, and oats, are sometimes infected with fungal growths called ergots, the dark structures on the seed head of rye shown in **Figure 17.19C**. Consumption of flour made from ergot-infested grain can cause gangrene, nervous spasms, burning sensations, hallucinations, temporary insanity, and death. One epidemic in Europe in the year 944 killed more than 40,000 people. Several kinds of toxins have been isolated from ergots. One, called lysergic acid, is the raw material from which the hallucinogenic drug LSD is made. Certain others are medicinal in small doses. An ergot compound is useful in treating high blood pressure and stopping maternal bleeding after childbirth.

Figure 17.19C
Ergots on rye

Animals are much less susceptible to parasitic fungi than are plants. Only about 50 species of fungi are known to be parasitic in humans and other animals. However, their effects are significant enough to make us take them seriously. In humans, fungi cause infections ranging from annoyances such as athlete's foot to deadly lung diseases.

The general term for a fungal infection is **mycosis.** Skin mycoses include the disease called ringworm, so named because it appears as circular red areas on the skin. The ringworm fungus can infect virtually any skin surface. Most commonly, it attacks the feet, causing the intense itching and sometimes blistering known as athlete's foot. Though highly contagious, athlete's foot and other ringworm infections can be treated with various fungicidal lotions and powders. Systemic mycoses are fungal infections that spread throughout the body, usually from spores that are inhaled. These can be very serious diseases. Coccidioidomycosis is a systemic mycosis that produces tuberculosis-like symptoms in the lungs. It is so deadly that it is now considered a potential biological weapon.

The yeast that causes vaginal yeast infections is an example of an opportunistic pathogen—a normal inhabitant of the body that causes problems only when some change in the body's microbiology, chemistry, or immunology allows the yeast to grow unchecked. Many other opportunistic mycoses have increased in recent decades, in part because of AIDS, which compromises the immune system.

In the next two modules, we look at mutualistic associations with fungi.

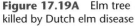

Figure 17.19A   Elm tree killed by Dutch elm disease

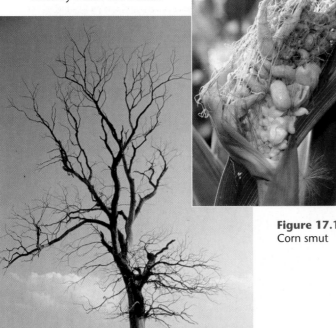

Figure 17.19B
Corn smut

? **What is a mycosis? What is an opportunistic pathogen?**

■ A mycosis is a fungal infection. An opportunistic pathogen is a normal inhabitant of the body that grows out of control when there is a change in the body's microbiology, chemistry, or immunology.

17.20 Lichens consist of fungi living mutualistically with photosynthetic organisms

Mutually beneficial connections are a recurring theme in both the plant and fungal kingdoms. The different types of lichens seen in **Figure 17.20A** are another example of a mutual relationship involving fungi. **Lichens** are associations of millions of green algae or cyanobacteria held in a mass of fungal hyphae (**Figure 17.20B**). The mutualistic merger of fungus and alga is so complete that lichens are actually named as species, as though they were single organisms.

In most lichens that have been studied, each partner provides something the other could not obtain on its own. The fungus receives food from its photosynthetic partner. The fungal mycelium, in turn, provides a suitable habitat for the alga, helping it to absorb and retain water and minerals.

Lichens are rugged and able to live where there is little or no soil. As a result, they are important pioneers on new land. Lichens grow into tiny rock crevices, adding to the forces that erode hard surfaces and paving the way for future plant growth. Some lichens can tolerate severe cold, and carpets of

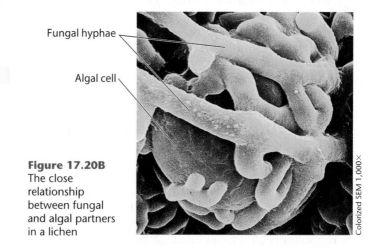

Fungal hyphae

Algal cell

Figure 17.20B The close relationship between fungal and algal partners in a lichen

Colorized SEM 1,000×

them cover the arctic tundra. Caribou feed on lichens in their winter feeding grounds in Alaska.

Lichens can also withstand severe drought. They are opportunists, growing in spurts when the conditions are favorable. When it rains, a lichen quickly absorbs water and photosynthesizes at a rapid rate. In dry air, it dehydrates and photosynthesis may stop, but the lichen remains alive more or less indefinitely. Some lichens are thousands of years old, rivaling the oldest plants and fungi as the oldest organisms on Earth.

As tough as lichens are, many do not withstand air pollution. Because they get most of their minerals from the air, in the form of dust or compounds dissolved in raindrops, lichens are very sensitive to airborne pollutants such as sulfur dioxide. The death of lichens may be a sign that air quality in an area is deteriorating.

? What benefits do algae in lichens receive from their fungal partners?

Figure 17.20A Lichens growing on rock

■ A suitable habitat for growth; absorption and retention of water and minerals

17.21 Fungi also form mutalistic relationships with animals

Fungi share their services not only with plants but also with some animals. Fungi help break down plant material in the guts of cows and other grazing mammals. Several species of ants and termites take advantage of the digestive power of fungi by raising them in "farms." For example, Central American leaf-cutting ants construct huge underground nests with chambers in which they cultivate fungal gardens. The ants scour tropical forests in search of leaves. A colony of ants, often well over a million individuals, can strip a large tree of its foliage in a single night. They carry their bounty back to the nest to feed to the fungi (**Figure 17.21**). The fungi feed on the leaves, using their enzymes to break down cellulose, which ants cannot digest. The ants harvest

Figure 17.21 Leaf-cutting ants carrying leaves to feed their fungi

the swollen tips of hyphae as their food. Some members of the colony "weed" undesirable fungi out of the fungal gardens. When a queen ant establishes a new colony, she takes fungal hyphae along in a pouch in her mouth.

Such farmer insects and their fungal "crops" have been evolving together for well over 50 million years. The fungi have become so dependent on their caretakers that in many cases they can no longer survive without the ants.

? Why do we consider leaf-cutter ants and their fungi to be a mutualistic association?

■ Both partners benefit from their interaction.

17.22 Fungi have enormous ecological benefits and practical uses

As you have read, fungi have been major players in terrestrial communities ever since they moved onto land in the company of plants. As symbiotic partners in mycorrhizae, fungi supply essential nutrients to plants and are enormously important in natural ecosystems and agriculture.

Fungi, along with prokaryotes, are essential decomposers in ecosystems, breaking down organic matter and restocking the environment with vital nutrients essential for plant growth. The air is so full of fungal spores that as soon as a leaf falls or an insect dies, it is covered with spores and is soon infiltrated by fungal hyphae. If fungi and prokaryotes in a forest suddenly stopped decomposing, leaves, logs, feces, and dead animals would pile up on the forest floor. Plants and the animals they feed would starve because elements taken from the soil would not be returned.

Almost any carbon-containing substance can be consumed by fungi. Some fungi that decompose wood, for example, also decompose many toxic pollutants, including the pesticide DDT and certain chemicals that cause cancer. Scientists are investigating the use of petroleum-digesting fungi to clean up oil spills and other chemical messes.

Fungi have a number of practical uses for humans. Most of us have eaten mushrooms, although we may not have realized that we were ingesting reproductive structures of subterranean fungi. The distinctive flavors of certain cheeses, including Roquefort and blue cheese (Figure 17.22A), come from fungi used to ripen them. Highly prized by gourmets are truffles, produced by certain mycorrhizal fungi associated with tree roots. Humans have used yeasts to produce alcoholic beverages and cause bread to rise for thousands of years. Under anaerobic conditions, yeasts ferment sugars to alcohol and CO_2 (see Module 6.13).

Fungi are medically valuable as well. Like the bacteria called actinomycetes (see Module 16.13), some fungi produce antibiotics that are used to treat bacterial diseases. In fact, the first antibiotic discovered was penicillin, which is made by the common mold called *Penicillium*. In Figure 17.22B, the clear area between the mold and the bacterial colony is where the antibiotic produced by *Penicillium* inhibits the growth of the bacteria (*Staphylococcus aureus*).

Fungi also figure prominently in research in molecular biology and in biotechnology. Researchers use yeasts to study the molecular genetics of eukaryotes because they are easy to culture and manipulate. Yeasts have been genetically modified to produce human proteins. And researchers are sequencing the genome of a wood-digesting fungus with the goal of deciphering the metabolic pathways by which it breaks down wood and using it to produce paper pulp.

As decomposers, as producers of antibiotics and food (for humans as well as ants), and as mutualistic partners in mycorrhizae and lichens, fungi are vital to the living world.

Fungi are the third group of eukaryotes we have surveyed so far. Strong evidence suggests that they evolved from protistan ancestors that also gave rise to the fourth and most diverse group of eukaryotes, the animals, which we study next.

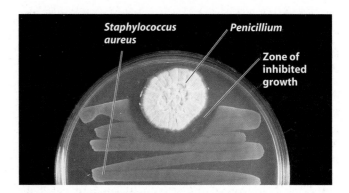

Figure 17.22B A culture of *Penicillium* and bacteria

? What do you think is the function of the antibiotics that fungi produce in their natural environments?

■ The antibiotics probably block the growth of microorganisms, especially prokaryotes, that compete with the fungi for nutrients and other resources.

Figure 17.22A Blue cheese

CHAPTER REVIEW

Reviewing the Concepts

Mycorrhizae, mutually beneficial associations of plant roots and fungi, are common and may have helped the first plants adapt to land **(Introduction).**

Plant Evolution and Diversity (17.1–17.3)

Green algal ancestors. Molecular, physical, and chemical evidence indicate that the green algae called charophyceans are the closest living relatives of plants **(17.1).**

Terrestrial adaptations of plants. Plants are multicellular photosynthetic eukaryotes, and most have the following adaptations for living on land **(17.2):**

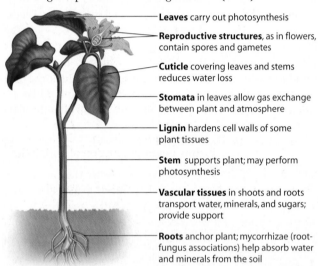

Leaves carry out photosynthesis

Reproductive structures, as in flowers, contain spores and gametes

Cuticle covering leaves and stems reduces water loss

Stomata in leaves allow gas exchange between plant and atmosphere

Lignin hardens cell walls of some plant tissues

Stem supports plant; may perform photosynthesis

Vascular tissues in shoots and roots transport water, minerals, and sugars; provide support

Roots anchor plant; mycorrhizae (root-fungus associations) help absorb water and minerals from the soil

Plant diversity. Bryophytes include the mosses, hornworts, and liverworts. Vascular plants have supportive vascular tissues. Ferns are seedless vascular plants with flagellated sperm. Seed plants have pollen grains that transport sperm and protect embryos in seeds. Gymnosperms, such as pines, produce seeds in cones. The seeds of angiosperms develop within protective ovaries **(17.3).**

Alternation of Generations and Plant Life Cycles (17.4–17.14)

Alternation of generations. The haploid gametophyte produces eggs and sperm by mitosis. The zygote develops into the diploid sporophyte, in which meiosis produces haploid spores. Spores grow into gametophytes **(17.4).**

Moss life cycle. A mat of moss is mostly gametophytes, which produce eggs and swimming sperm. The zygote develops on the gametophyte into the smaller sporophyte **(17.5).**

Fern life cycle. The sporophyte is the dominant generation. Sperm, produced by the gametophye, swim to the egg. **(17.6).** Ferns and other seedless plants once dominated ancient forests; their remains formed coal **(17.7).**

Conifer life cycle. A pine tree is a sporophyte; tiny gametophytes grow in its cones. A sperm from a pollen grain fertilizes an egg in the female gametophyte. The zygote develops into a sporophyte embryo, and the ovule becomes a seed, with stored food and a protective coat **(17.8).**

Angiosperms have independent sporophytes with tiny gametophytes protected in flowers, which usually consist of sepals, petals, stamens (produce pollen), and carpels (produce ovules). Ovules become seeds, and ovaries become fruits. Angiosperms provide most of our food **(17.9–17.12).**

Plant diversity has been influenced by interactions with animals and is a nonrenewable resource **(17.13–17.14).**

Fungi (17.15–17.22)

Fungal nutrition and body structure. Fungi are heterotrophic eukaryotes that digest their food externally and absorb the nutrients. A fungus usually consists of a mass of threadlike hyphae, called a mycelium **(17.15).**

Fungal reproduction usually involves the production of spores. In some fungi, fusion of haploid hyphae produces a heterokaryotic stage containing nuclei from two parents. After the nuclei fuse, meiosis produces haploid spores **(17.16).**

Fungal phylogeny. Fungi evolved from an aquatic, flagellated ancestor. Fungal groups include chytrids, zygomycetes, glomeromycetes, ascomycetes, and basidiomycetes **(17.17).**

Fungal life cycles often include asexual and sexual stages. Groups have characteristic reproductive structures **(17.18).**

Parasitic fungi cause 80% of plant diseases and some serious human mycoses **(17.19).**

Mutualistic associations. Lichens consist of algae or cyanobacteria within a fungal network. Some animals benefit from the digestive abilities of fungi **(17.20–17.21).**

Ecological and other benefits. Fungi are essential decomposers and provide food and antibiotics **(17.22).**

Connecting the Concepts

1. In this abbreviated diagram, identify the four major plant groups and the key terrestrial adaptation associated with each of the three major branch points.

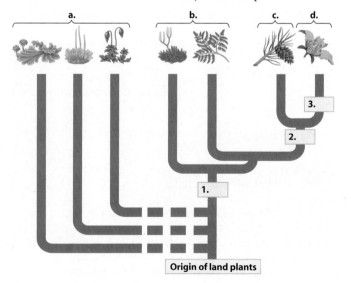

Origin of land plants

2. Identify the cloud seen in each photograph. Describe the life cycle events associated with each cloud.

   A. Pine tree, a gymnosperm

   B. Puffball, a club fungus

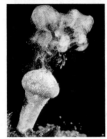

Testing Your Knowledge

Multiple Choice

3. Angiosperms are different from all other plants because only they have
   a. a vascular system.
   b. flowers.
   c. a life cycle that involves alternation of generations.
   d. seeds.
   e. a dominant sporophyte phase.

4. Which of the following produce eggs and sperm? (*Explain your answer.*)
   a. the sexual reproductive structures of a fungus
   b. fern sporophytes
   c. moss gametophytes
   d. the anthers of a flower
   e. moss sporangia

5. The eggs of seed plants are fertilized within ovules, and the ovules then develop into
   a. seeds.
   b. spores.
   c. gametophytes.
   d. fruit.
   e. sporophytes.

6. The diploid sporophyte stage is dominant in the life cycles of all of the following except
   a. a pine tree.
   b. a dandelion.
   c. a rose bush.
   d. a fern.
   e. a moss.

7. Under a microscope, a piece of a mushroom would look most like
   a. jelly.
   b. a tangle of string.
   c. grains of sugar or salt.
   d. a piece of glass.
   e. foam.

8. Which of the following is an opportunistic pathogen that can cause a mycosis?
   a. HIV, the AIDS virus
   b. the fungus that produces ergots on rye, which can cause serious symptoms if milled into flour
   c. smuts, serious pathogens of grain crops
   d. coccidioidomycosis, a deadly systemic mycosis
   e. the yeast that causes vaginal yeast infections

9. Which of the following terms includes all the others?
   a. angiosperm
   b. gymnosperm
   c. vascular plant
   d. fern
   e. seed plant

10. Which of the following is a plant with flagellated sperm and a sporophyte-dominated life cycle?
    a. chytrid
    b. moss
    c. charophycean
    d. fern
    e. liverwort

Describing, Comparing, and Explaining

11. Compare a seed plant with an alga in terms of adaptations for life on land versus life in the water.

12. How do animals help flowering plants reproduce? What do the animals get in return?

13. Why have fungi and plants been classified in different kingdoms?

Applying the Concepts

14. Many fungi produce antibiotics, such as penicillin, which are valuable in medicine. But of what value might the antibiotics be to the fungi? Similarly, fungi often produce compounds with unpleasant tastes and odors as they digest their food. What might be the value of these chemicals to the fungi? How might production of antibiotics and odors have evolved?

15. In April 1986, an accident at a nuclear power plant in Chernobyl, Ukraine, scattered radioactive fallout for hundreds of miles. In assessing the biological effects of the radiation, researchers found mosses to be especially valuable as organisms for monitoring the damage. As mentioned in Module 10.16, radiation damages organisms by causing mutations. Explain why it is faster to observe the genetic effects of radiation on mosses than on plants from other groups. Imagine that you are conducting tests shortly after a nuclear accident. Using potted moss plants as your experimental organisms, design an experiment to test the hypothesis that the frequency of mutations decreases with the organism's distance from the source of radiation.

16. Much of the conifer forest in the U.S. Pacific Northwest has been clear-cut; less than 10% of the original ancient forest, dominated by giant firs and hemlocks, remains. There is no law protecting endangered habitats, so to protect the northern spotted owl, which lives only in old-growth conifers, conservationists sued to stop logging under the Endangered Species Act. The lawsuits halted logging in many national forest areas. Lumber companies buy trees from national forests, loggers work there, and the economies of many small communities depend on logging. The reduction in timber supply has driven up the cost of lumber. Imagine that it is your task to deal with this situation. What are the opposing issues? What would you suggest to resolve this conflict, and how would you defend your policy?

Answers to all questions can be found in Appendix 3.

For study help and Activities, go to campbellbiology.com or the student CD-ROM.

ANIMAL EVOLUTION AND DIVERSITY

18.1   What is an animal?

18.2   The ancestor of animals was probably a colonial, flagellated protist

18.3   Animals can be characterized by basic features of their "body plan"

18.4   The body plans of animals can be used to build phylogenetic trees

INVERTEBRATES

18.5   Sponges have a relatively simple, porous body

18.6   Cnidarians are radial animals with tentacles and stinging cells

18.7   Flatworms are the simplest bilateral animals

18.8   Nematodes have a pseudocoelom and a complete digestive tract

18.9   Diverse molluscs are variations on a common body plan

18.10 Annelids are segmented worms

18.11 Arthropods are segmented animals with jointed appendages and an exoskeleton

18.12 Insects are the most diverse group of organisms

18.13 Echinoderms have spiny skin, an endoskeleton, and a water vascular system for movement

18.14 Our own phylum, Chordata, is distinguished by four features

VERTEBRATES

18.15 Derived characters define the major clades of chordates

18.16 Lampreys are vertebrates that lack hinged jaws

18.17 Jawed vertebrates with gills and paired fins include sharks, ray-finned fishes, and lobe-fins

18.18 Amphibians were the first tetrapods—vertebrates with two pairs of limbs

18.19 Reptiles are amniotes—tetrapods with a terrestrially adapted egg

18.20 Birds are feathered reptiles with adaptations for flight

18.21 Mammals are amniotes that have hair and produce milk

ANIMAL PHYLOGENY AND DIVERSITY REVISITED

18.22 An animal phylogenetic tree is a work in progress

18.23 Humans threaten animal diversity by introducing non-native species

What Am I?

OF SOME 1.7 MILLION SPECIES OF ORGANISMS known to scientists, 1.3 million are animals. This large number of identified animal species not only indicates how successful animals are as a group, but also reflects our keen interest in the other members of our kingdom. Humans have a long history of studying, appreciating, and using animal diversity. But classifying a new animal isn't always easy. Imagine you were the first European zoologist to encounter the animal at left in its native Australia. What would you make of it? It has a bill and webbed feet similar to a duck's, but the rest of its furry body looks very much like that of a muskrat or other aquatic rodent. To make the case even more confusing, this animal lays eggs. So how would you classify it? Is it a bird or a mammal? The decision is easier once you study it a little more closely. This animal, called a duck-billed platypus, has mammary glands that produce milk for its young. That trait, along with its hair, is enough to place it in the mammalian class of animals.

Scientists investigating the platypus bill, which does look similar to that of a duck, found that it is not a hard, inert bird's bill but rather is covered by soft skin filled with sensitive nerve endings. While the duck and the platypus both use their bills to dig for food in muddy waters, the platypus's bill serves an additional purpose as a sensory organ to help it locate food and avoid obstacles underwater. When a platypus dives,

The Evolution of Animal Diversity

A Tasmanian tiger, 1928

it closes its eyes, so it relies heavily on its bill to "see" its surroundings. Indeed, biologists have found that a large portion of the platypus brain is devoted to processing sensory information from its bill.

Part of the interest in studying animals is exploring their many fascinating adaptations. The incredible diversity of animal life arose through hundreds of millions of years of evolution as natural selection shaped animal adaptations to Earth's diverse and changing environments.

The duck-billed platypus belongs to a small group of egg-laying mammals, called monotremes, that are found only in Australia and New Guinea. Unlike the rest of the world, Australia has relatively few placental mammals (mammals that bear fully developed live young). Most Australian mammals are marsupials, such as the kangaroos pictured here, whose young complete their development in the mother's pouch.

Why do marsupials represent the majority of mammalian species in Australia, but are rare in other parts of the world? Marsupials are an ancient group of mammals that used to be common on other continents. But it appears that they did not compete well with placental mammals, and most became extinct. However, when Australia broke off from Pangaea over 60 million years ago, its isolated marsupials could flourish, filling the roles, or niches, that placental mammals fill on other continents. For example, the now-extinct Tasmanian tiger (or thylacine), shown above, was a marsupial that once filled the large-predator niche. The quoll, a small marsupial catlike animal that you will encounter in the last module of this chapter, fills a small-predator niche. And the kangaroo, which grazes on grass and other plants, fills the niche occupied by horses or antelopes on other continents. These are just a few examples of convergent evolution, in which unrelated species have adaptations that enable them to fill the same ecological niche in different parts of the world. In some cases, the species look different—nothing else looks quite like a kangaroo or a platypus. But in other cases, similar burrowing or gliding or even long-snouted ant-eating forms have evolved in unrelated animal species.

In this chapter, we embark on a tour of the vast diversity found in the animal kingdom—a diversity that has been evolving for perhaps a billion years. We will encounter a spectacular range of forms ranging from corals to cockroaches to crocodiles. We will look at 9 major phyla of the roughly 35 phyla in the kingdom **Animalia.** And we will see that identifying, classifying, and arranging this diversity remain a work in progress. But first let's define what an animal is! ■ ■ ■

18.1 What is an animal?

Animals are multicellular, heterotrophic eukaryotes that obtain nutrients by ingestion. Now that's a mouthful—and speaking of mouthfuls, **Figure 18.1A** shows a rock python just beginning to ingest a gazelle. **Ingestion** means eating food. This mode of nutrition contrasts animals with fungi, which absorb nutrients after digesting food outside their body. Animals digest food within their body after ingesting other organisms, dead or alive, whole or by the piece.

Animals also have other distinctive features. Animal cells lack the cell walls that provide strong support in the bodies of plants and fungi. Animal cells are held together by extracellular structural proteins and by unique types of intercellular junctions (see Module 4.18). And most animals have muscle cells for movement and nerve cells for conducting impulses.

Other unique features are seen in animal life cycles. Most animals are diploid and reproduce sexually; eggs and sperm are the only haploid cells, as shown in the life cycle of a sea star in **Figure 18.1B**. ❶ Male and female adult animals make haploid gametes by meiosis, and ❷ an egg and a sperm fuse, producing a zygote. ❸ The zygote divides by mitosis, forming ❹ an early embryonic stage called a **blastula**, which is usually a hollow ball of cells. ❺ In the sea star and most other animals, one side of the blastula folds inward, forming a stage called a **gastrula**. ❻ The internal sac formed by gastrulation becomes the digestive tract, lined by a cell layer called the **endoderm**. The embryo also has an **ectoderm**, an outer cell layer that gives rise to the outer covering of the animal and, in some phyla, to the central nervous system. Most animals have a third embryonic layer, known as the **mesoderm**, that forms the muscles and most other internal organs.

After the gastrula stage, many animals develop directly into adults. Others, such as the sea star, develop into one or more larval stages first. ❼ A **larva** is an immature individual that looks different from the adult animal. The larva under-

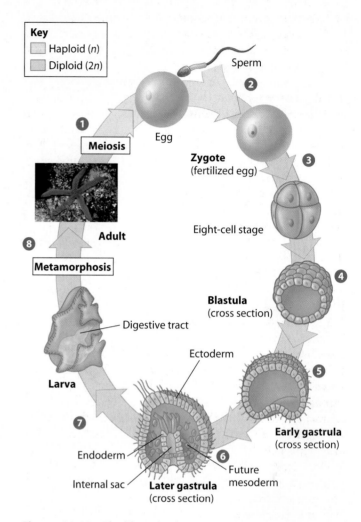

Key
- ☐ Haploid (n)
- ☐ Diploid (2n)

Sperm ❷

Egg

❶ Meiosis

Zygote (fertilized egg) ❸

Eight-cell stage

Adult ❽

Metamorphosis

Blastula (cross section) ❹

Digestive tract

Ectoderm

Larva

Early gastrula (cross section) ❺

❼

Endoderm

Internal sac

Later gastrula (cross section) ❻ Future mesoderm

Figure 18.1B The life cycle of a sea star

goes a major change of body form, called **metamorphosis**, ❽ in becoming an adult capable of reproducing sexually.

This transformation of a zygote into an adult animal is controlled by specific regulatory genes. All eukaryotes have regulatory genes that contain DNA sequences called homeoboxes (see Module 11.15). But animals share a unique homeobox-containing family of genes called *Hox* genes that play important roles in the development of animal embryos. Thus, molecular biology helps us distinguish animals from other life-forms and, as you will see later in this chapter, also helps us investigate the phylogenetic relationships among the highly diverse animal forms we are about to survey.

❓ List the distinguishing characteristics of animals.

■ Multicellular, eukaryotic heterotrophs that ingest their food; no cell walls; unique cell junctions; nerve and muscle cells; sexual reproduction and life cycles with unique embryonic stages; unique developmental genes

Figure 18.1A Ingestion, the animal way of life

18.2 The ancestor of animals was probably a colonial, flagellated protist

Biologists have long speculated about the origins of animals. Some molecular clock calculations (see Module 15.9) estimate that the common ancestor of living animals lived about a billion years ago. This ancestor may have resembled modern choanoflagellates, colonial protists that are the closest living relatives of animals.

Figure 18.2A shows one hypothesis for how a colonial flagellated protist may have evolved into a simple animal with specialized cells arranged in two layers. ❶ The earliest colonial aggregates may have been only a few cells. ❷ Some larger colonies may have formed hollow spheres. ❸ Eventually, cells in the colony may have become specialized for functions such as reproduction, locomotion, and feeding. ❹ A simple multicellular organism with cell layers might have evolved as cells on one side of the colony folded inward. ❺ Eventually, a gastrula-like "proto-animal" may have evolved.

There are no fossils of this early evolutionary event. The oldest known animal fossils date from the late Precambrian time, about 575 million years ago. These fossils represent several different soft-bodied forms that were too complex to have been the first animals.

Beginning at the dawn of the Cambrian period, about 542 million years ago, the fossil record marks a dramatic increase in animal diversity. So many animal body plans and new phyla appear in such an evolutionarily short time span (about 15 million years) that biologists call this episode the **Cambrian explosion.** Many Cambrian animals may seem bizarre (Figure 18.2B), but most biologists classify the Cambrian fossils as ancient representatives or at least relatives of animal phyla living today.

What ignited the Cambrian explosion? One hypothesis emphasizes ecological causes: The evolution of hard body coverings led to increasingly complex predator-prey relationships and diverse adaptations for feeding, motility, and protection. A second hypothesis focuses on geologic changes: Perhaps atmospheric oxygen had finally reached a high enough concentration to support the metabolism of more active, mobile animals. Another hypothesis looks to genetic causes—the evolution of the *Hox* complex of regulatory

Figure 18.2B A drawing based on fossils from the early Cambrian period

genes. Much of the diversity in body form among the animal phyla is associated with variations in the spatial and temporal expression of these genes within developing embryos. These three hypotheses are not mutually exclusive—all three causes may have contributed to the relatively rapid radiation of animal phyla over half a billion years ago.

Most animals are **invertebrates,** so called because they lack a vertebral column (backbone): Of the 35 or so animal phyla (systematists disagree on the precise number), the animals in all but one are invertebrates. We'll begin with the far more numerous and diverse invertebrates and end the chapter with a survey of the vertebrates. But first we look at some of the anatomical features biologists use to classify this vast animal diversity.

? What is the main difference between a colonial organism and an organism that is truly multicellular?

■ The cells of a multicellular organism are more extensively specialized and interdependent than are the cells of a colonial organism.

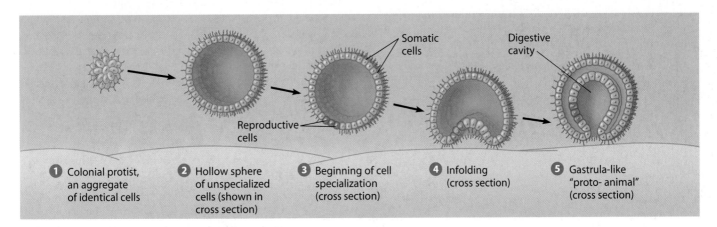

Figure 18.2A A hypothesis for the evolution of animals from a colonial flagellated protist (The arrows symbolize evolutionary time.)

❶ Colonial protist, an aggregate of identical cells

❷ Hollow sphere of unspecialized cells (shown in cross section)

❸ Beginning of cell specialization (cross section)

❹ Infolding (cross section)

❺ Gastrula-like "proto-animal" (cross section)

Somatic cells

Reproductive cells

Digestive cavity

CHAPTER 18 *The Evolution of Animal Diversity* **369**

18.3 Animals can be characterized by basic features of their "body plan"

One way that biologists categorize the diversity of animals is by certain general features of body structure, which together describe what is referred to as an animal's "body plan." These distinctions are used to help infer the phylogenetic relationships between animal groups.

One such feature is symmetry. Some animals have **radial symmetry**: As illustrated by the sea anemone on the left in **Figure 18.3A**, the body parts radiate from the center. Any imaginary slice through the central axis divides a radially symmetrical animal into mirror images. Thus, the animal has a top and a bottom, but not right and left sides. As shown by the lobster in the figure, an animal with **bilateral symmetry** has mirror-image right and left sides; a distinct head, or **anterior**, end; a tail, or **posterior**, end; a back, or **dorsal**, surface; and a bottom, or **ventral**, surface. The brain, sense organs, and mouth are usually located in the head.

The symmetry of an animal fits its lifestyle. A radial animal is typically sedentary or passively drifting, meeting its environment equally on all sides. In contrast, most bilaterally symmetrical animals are active and travel headfirst through the environment, with their eyes and other sense organs contacting the environment first.

Body plans also vary in the organization of tissues. True tissues are collections of specialized cells, usually isolated from other tissues by membrane layers, that perform specific functions (an example is the nervous tissue of your brain and spinal cord). Sponges lack true tissues (see Module 18.5), but in other animals, the cell layers formed in the process of gastrulation (see Figure 18.1B) give rise to true tissues and to organs. Some animals have only ectoderm and endoderm; most animals also have mesoderm, making a body with three tissue layers.

Animals with three tissue layers may be characterized by the presence or absence of a **body cavity**. This fluid-filled space between the digestive tract and body wall cushions the internal organs and enables them to grow and move independently. In soft-bodied animals, a noncompressible fluid in the body cavity forms a **hydrostatic skeleton** that provides a rigid structure against which muscles contract, moving the animal.

In the figures to the right, colors indicate the tissue layers: ectoderm (blue), mesoderm (red), and endoderm (yellow). The cross section through a flatworm (**Figure 18.3B**) reveals a body that is solid except for the cavity of the digestive sac.

A roundworm (**Figure 18.3C**) has a body cavity called a **pseudocoelom** (from the Greek *pseudes*, false, and *koilos*, hollow). A pseudocoelom is not completely lined by tissue derived from mesoderm. A segmented worm (**Figure 18.3D**) has a body cavity called a true **coelom**, which is completely lined by tissue derived from mesoderm.

Animals with three tissue layers can be separated into two groups based on details of their embryonic development, such as the fate of the opening formed during gastrulation that leads to the developing digestive tract. In **protostomes** (from the Greek *protos*, first, and *stoma*, mouth), this opening becomes the mouth; in **deuterostomes** (from the Greek *deutero*, second), this opening becomes the anus, and the mouth forms from a second opening. Other differences between protostomes and deuterostomes include the pattern of early cell divisions and the way the coelom forms.

Next we see how these general features of body plan are used to infer relationships between animal groups.

? List four features that can describe an animal's body plan.

■ Symmetry, number of tissue layers, body cavity type, embryonic development

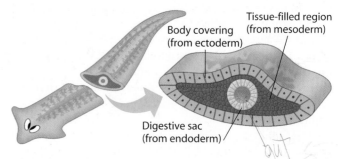

Figure 18.3B No body cavity (a flatworm)

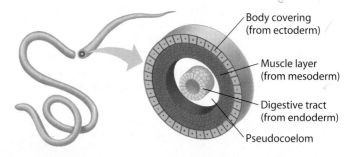

Figure 18.3C Pseudocoelom (a roundworm)

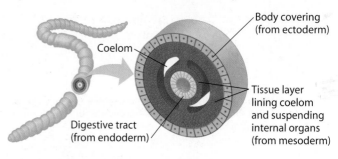

Figure 18.3D True coelom (a segmented worm)

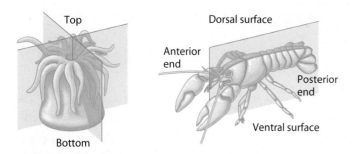

Figure 18.3A Radial (left) and bilateral (right) symmetry

18.4 The body plans of animals can be used to build phylogenetic trees

Because animals diversified so rapidly on the scale of geologic time, it is difficult, using only the fossil record, to sort out the evolutionary relationships between the various phyla. Traditionally, biologists have proposed hypotheses about animal phylogeny based on morphological studies, often using the characteristics of body plan and embryonic development described in the preceding module.

Figure 18.4 presents a morphology-based phylogenetic tree of the major phyla of the animal kingdom. At the bottom is the ancestral colonial protist. The tree has a series of branch points. The first branch splits the sponges from the clade of **eumetazoans** ("true animals"), the animals with true tissues. The next branch point separates the animals with radial symmetry from those with bilateral symmetry. Most animal phyla belong to the clade of **bilaterians.** This morphology-based tree then divides the bilaterians into two clades based on embryology: deuterostomes and protostomes.

The type of body cavity has historically been used to infer phylogeny, but research suggests that coeloms and pseudocoeloms have been gained or lost multiple times in the course of evolution and thus cannot be used to define clades.

All phylogenetic trees are hypotheses for the key events in the evolutionary history that led to the animal phyla now living on Earth. Increasingly, researchers are adding molecular comparisons to their data sets for identifying clades. In Module 18.22, we will see that such data are leading to new hypotheses for grouping animal phyla.

? Why are phylogenetic trees considered hypotheses?

■ The fossil record is incomplete. A tree is based on the best available data and will be revised based on new research and molecular comparisons.

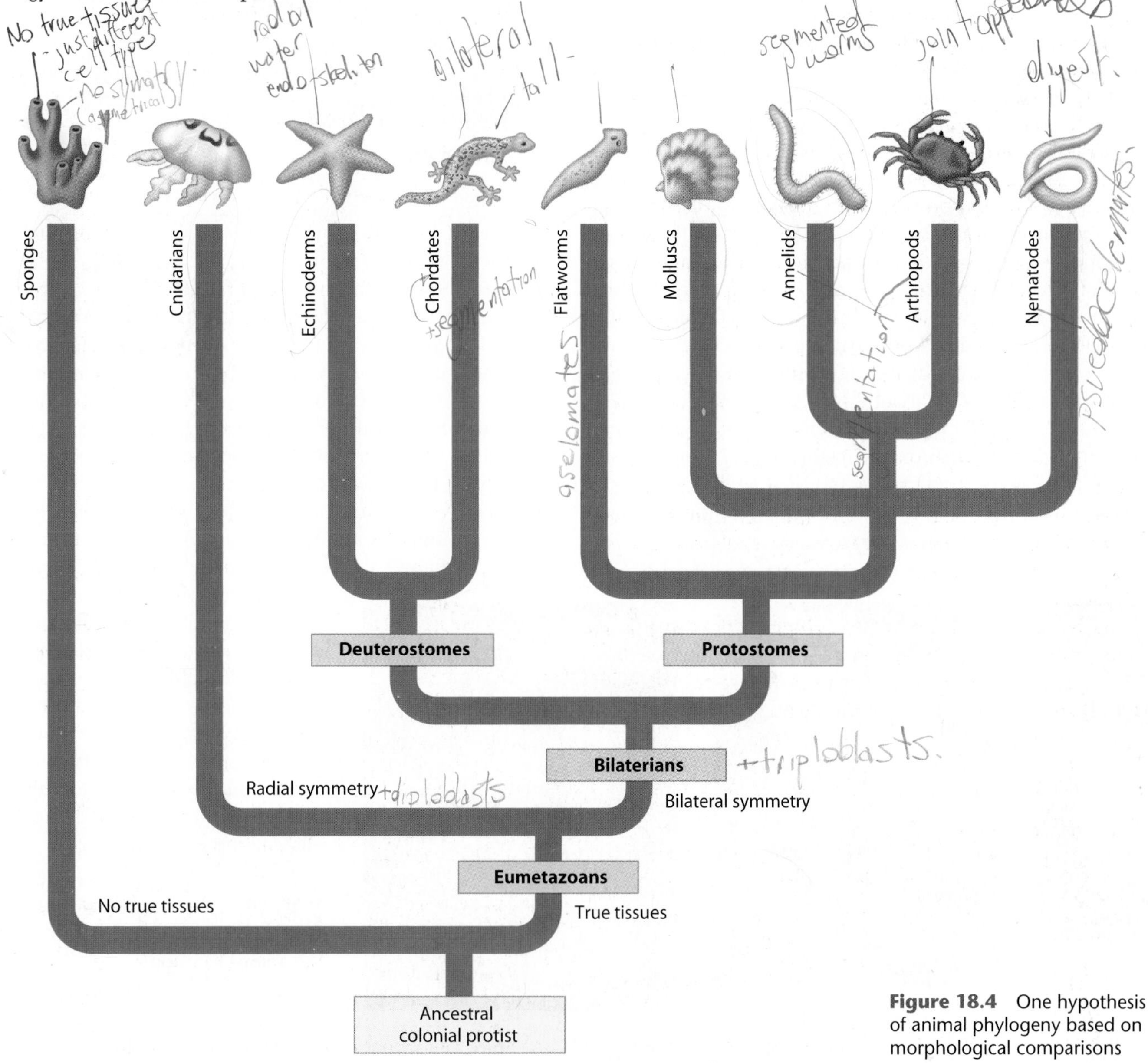

Figure 18.4 One hypothesis of animal phylogeny based on morphological comparisons

18.5 Sponges have a relatively simple, porous body

Sponges (phylum Porifera) are stationary animals that are so sedentary that the ancient Greeks believed them to be plants. The majority of species are marine, although some are found in fresh water. **Figure 18.5A** shows two individuals of the genus *Scypha*, a small sponge only about 1–3 cm high. The purple tube sponge in **Figure 18.5B**, on the other hand, can reach heights of 1.5 m. Some sponges, such as *Scypha*, resemble simple sacs. Others, such as the azure vase sponge in **Figure 18.5C**, have folded body walls and irregular shapes. Sponges lack body symmetry.

A simple sponge resembles a thick-walled sac perforated with holes. (*Porifera* means "pore-bearer" in Latin.) Water is drawn through the pores into a central cavity, then flows out through a larger opening (**Figure 18.5D**). More complex sponges have branching water canals.

The body of a sponge consists of two layers of cells separated by a gelatinous region. The inner layer of flagellated cells called **choanocytes** (purple in Figure 18.5D) help to sweep water through the sponge's body. Wandering through the middle body region are **amoebocytes** (blue), which produce skeletal fibers (yellow) composed of either mineral-containing particles or a flexible protein called spongin. We use the flexible, honeycombed skeletons of some sponges as bath sponges.

Sponges are examples of **suspension feeders** (also known as filter feeders), animals that collect food particles from water passed through some type of food-trapping equipment. Sponges feed by collecting food particles from water that streams through their porous bodies. To obtain enough food to grow by 100 g (about 3 ounces), a sponge must filter 1,000 kg (about 275 gallons) of seawater. Choanocytes trap food particles in mucus on the membranous collars that surround the base of their flagella and then engulf the food by phagocytosis (see Module 5.19). Amoebocytes pick up food

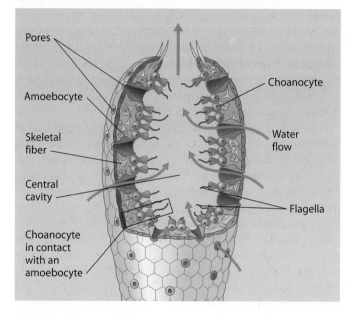

Figure 18.5D Structure of a simple sponge

packaged in food vacuoles from choanocytes, digest it, and carry the nutrients to other cells.

Adult sponges are **sessile,** meaning they are anchored in place; therefore, they cannot escape from predators. Researchers have found that sponges produce defensive compounds such as toxins and antibiotics that deter pathogens, parasites, and predators. Some of these compounds may prove useful to humans as new drugs. Also, as sponges trap particles, they perform an important ecological service by purifying the water around them.

Sponges are the simplest of all animals. They have no nerves or muscles, though their individual cells can sense and react to changes in the environment. Since the cell layers are loose federations of cells, they are not considered true tissues. Biologists hypothesize that the sponge lineage arose very early from the multicellular organisms that gave rise to the animal kingdom. The choanocytes of sponges and the cells of living choanoflagellates are similar, supporting the molecular evidence that animals evolved from a flagellated protist ancestor.

> ? Why is it thought that sponges represent the earliest branch of the animal kingdom?

Figure 18.5A
Scypha

Figure 18.5B A purple tube sponge

Figure 18.5C An azure vase sponge

■ Sponges lack symmetry and true tissues, and their choanocytes resemble certain flagellated protists.

Cnidarians are radial animals with tentacles and stinging cells

All animals except sponges have true tissues and belong to the clade Eumetazoa ("true animals"). Among eumetazoans, one of the oldest groups is phylum Cnidaria, which includes the hydras, jellies (also called "jellyfish"), sea anemones, and corals. **Cnidarians** are characterized by radial symmetry and only two tissue layers. The simple body of most cnidarians has an outer epidermis and an inner cell layer that lines the digestive cavity. A jelly-filled middle region may have scattered amoeboid cells. Contractile tissues and nerves occur in their simplest forms in cnidarians.

Cnidarians exhibit two kinds of radially symmetrical body forms. Hydras, common in freshwater ponds and lakes, have a cylindrical body with tentacles projecting from one end. This body form is a **polyp** (Figure 18.6A). The other type of cnidarian body is the **medusa**, exemplified by the marine jelly in **Figure 18.6B**. While polyps are mostly stationary, medusae move freely about in the water. They are shaped like an umbrella with a fringe of tentacles around the lower edge. Some jellies have tentacles over 100 m long dangling from an umbrella up to 2 m in diameter.

Some cnidarians pass sequentially through both a polyp stage and a medusa stage in their life cycle. Others exist only as medusae; still others, such as hydras and sea anemones (Figure 18.6C), exist only as polyps.

Cnidarians are carnivores that use their tentacles to capture small animals and protists and to push the prey into their mouths. In a polyp, the mouth is on the top of the body, at the hub of the radiating tentacles (see Figure 21.3A). In a medusa, the mouth is in the center of the undersurface. The mouth leads into a digestive compartment called a **gastrovascular cavity** (from the Greek *gaster,* belly, and Latin *vas,* vessel). The mouth is the only opening in the body, so undigested food and other wastes exit through it. The gastrovascular cavity also circulates fluid that services internal cells (hence the "vascular" in gastrovascular; see Figure 23.2A). Acting as a hydrostatic skeleton, fluid in the cavity provides body support and helps give a cnidarian its shape, much like water in a balloon. When the animal closes its mouth, the volume of the cavity is fixed. Then contraction of selected cells changes the shape of the animal and produces movement.

Phylum Cnidaria (from the Greek *cnide,* nettle) is named for its unique stinging cells, called **cnidocytes,** that function in defense and in capturing prey. Each cnidocyte contains a fine thread coiled within a capsule (Figure 18.6D). When it is discharged, the thread can sting or entangle prey. Some large marine cnidarians use their stinging threads to catch fish. A group of cnidarians called cubozoans have highly toxic cnidocytes. The sea wasp, a cubozoan found off the coast of northern Australia, is the deadliest organism on Earth: One animal has enough poison to kill 60 people.

Coral animals are cnidarians that secrete a hard external skeleton. Each generation builds on top of previous generations, constructing characteristic shapes of "rocks" we call coral. Many coral animals harbor symbiotic algae that produce food for the corals. Tropical coral reefs are also home to an enormous variety of invertebrates and fishes.

? What are three functions of a cnidarian's gastrovascular cavity?

■ (1) Digestion, (2) circulation, and (3) physical support and movement

Figure 18.6A Polyp body form: a hydra (about 2–25 mm high)

Figure 18.6B Medusa body form: a marine jelly called a sea nettle (about 5 cm in diameter)

Figure 18.6C Sea anemones, such as this *Anthopleura* (about 6 cm in diameter), exist only as polyps

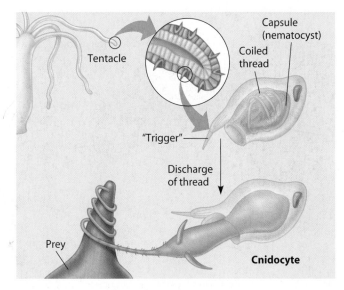

Figure 18.6D Cnidocyte action

18.7 Flatworms are the simplest bilateral animals

The vast majority of animal species belong to the clade Bilateria, which consists of animals with bilateral symmetry and three embryonic tissue layers. **Flatworms**, phylum Platyhelminthes (from the Greek *platys,* flat, and *helmis,* worm), are the simplest of the bilaterians. These thin, often ribbonlike animals range in length from about 1 mm to 20 m and live in marine, freshwater, and damp terrestrial habitats. In addition to free-living forms, there are many parasitic species. In common with cnidarians, most flatworms have a gastrovascular cavity with only one opening. The fine branches of the gastrovascular cavity distribute food throughout the animal.

There are three major groups of flatworms. Worms called planarians (Figure 18.7A) represent a group known as the **free-living flatworms.** A planarian has a head with a pair of light-sensitive eyespots and a flap at each side that detects chemicals. Dense clusters of nerve cells form a simple brain, and a pair of nerve cords connect with small nerves that branch throughout the body.

The gastrovascular cavity of a planarian is highly branched. When the animal feeds, it sucks food in through a mouth at the tip of a muscular tube that projects from the mid-ventral surface of the body (as shown in the figure). Planarians live on the undersurfaces of rocks in freshwater ponds and streams. Using cilia on their ventral surface, they crawl about in search of food. They also have muscles that enable them to twist and turn.

A second group of flatworms, the **flukes,** live as parasites in other animals. Many flukes have suckers that attach to their host and a tough protective covering. Reproductive organs occupy nearly the entire interior of these worms.

Many flukes have complex life cycles with an intermediate host in which larvae develop. The larvae then infect the final host in which they live as adults. For example, blood flukes that parasitize humans spend part of their life cycle in snails. These flukes cause a long-lasting disease called schistosomiasis that affects 200 million people around the world, causing severe abdominal pain, anemia, and dysentery.

Tapeworms are another parasitic group of flatworms. Adult tapeworms inhabit the digestive tracts of vertebrates, including humans. In contrast with planarians and flukes, most tapeworms have a very long, ribbonlike body with repeated units. They also differ from other flatworms in not having a digestive tract. Living in partially digested food in the intestines of their hosts, they simply absorb nutrients across their body surface. As Figure 18.7B shows, the anterior end, called the scolex, is armed with hooks and suckers that grasp the host. Behind the scolex is a long ribbon of repeated units filled with both male and female reproductive structures. Full of ripe eggs, those at the posterior end break off and pass out of the host's body in feces.

Like parasitic flukes, tapeworms have a complex life cycle, usually involving more than one host. Most species benefit from the predator-prey relationships of their hosts. A prey species—a sheep or a rabbit, for example—may become infected by eating grass contaminated with tapeworm eggs. Larval tapeworms develop in these hosts, and a predator—a coyote or a dog, for instance—becomes infected when it eats an infected prey animal. The adult tapeworms develop in the predator's intestines.

Humans can be infected with tapeworms by eating undercooked pork or beef infected with tapeworm larvae. The larvae are microscopic, but the adults can reach lengths of 6 m in the human intestine. An orally administered drug called niclosamide kills the adult worms.

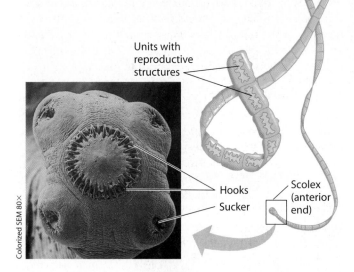

Units with reproductive structures

Hooks

Sucker

Scolex (anterior end)

Figure 18.7B   A tapeworm, a parasitic flatworm

? Flatworms and cnidarians differ in symmetry, with flatworms being _____ and cnidarians being _____, but the animals of both phyla have a _____ .

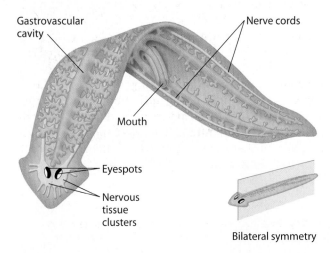

Gastrovascular cavity

Nerve cords

Mouth

Eyespots

Nervous tissue clusters

Bilateral symmetry

Figure 18.7A   A free-living flatworm, the planarian (most are about 5–10 mm long)

■ bilateral . . . radial . . . gastrovascular cavity

18.8 Nematodes have a pseudocoelom and a complete digestive tract

Nematodes, also called roundworms, make up the phylum Nematoda. As bilaterians, these animals have bilateral symmetry and a three-tissue layer construction. In contrast with flatworms, roundworms have a fluid-filled body cavity (a pseudocoelom, not completely lined with mesoderm) and a digestive tract with two openings.

Nematodes are cylindrical with a blunt head and tapered tail. They range in size from less than 1 mm to more than a meter. The body is covered by a tough, nonliving covering, or **cuticle,** that resists drying and crushing. When the worm grows, it periodically sheds its cuticle (molts) and secretes a new, larger one. What looks like a corduroy coat on the nematode in **Figure 18.8A** is its cuticle.

You can also see the mouth at the tip of the blunt anterior end of the nematode in Figure 18.8A. Nematodes have a **complete digestive tract,** extending as a tube from the mouth to the anus near the tip of the tail. Food travels only one way through the system and is processed as it moves along. In animals with a complete digestive tract, the anterior regions of the tract churn and mix food with enzymes, while the posterior regions absorb nutrients and then dispose of wastes. This division of labor allows each part of the digestive tract to be specialized for its particular function.

Fluid in the pseudocoelom of a nematode distributes nutrients absorbed from the digestive tract throughout the body. The pseudocoelom also functions as a hydroskeleton, and contraction of longitudinal muscles produces the characteristic thrashing motion of nematodes.

Nematodes are among the most numerous of all animals in both number of species and number of individuals. Nematodes live virtually everywhere there is rotting organic matter. These worms are important decomposers in soil and on the bottom of lakes and oceans. Other nematodes thrive as parasites in the moist tissues of plants and in the body fluids and tissues of animals. (The largest known nematodes are parasites of whales and measure more than 7 m.)

Little is known about most free-living nematodes. A notable exception is the soil-dwelling species *Caenorhabditis elegans,* an important research organism. A *C. elegans* adult consists of only about 1,000 cells—in contrast with the human body, which consists of some 60 trillion cells. By following every cell division in the developing embryo, biologists have been able to trace the lineage of every cell in the adult worm. The genome of *C. elegans* has been sequenced, and the ongoing research contributes to our understanding of how genes control animal development—and even to our understanding of some of the mechanisms involved in aging in humans.

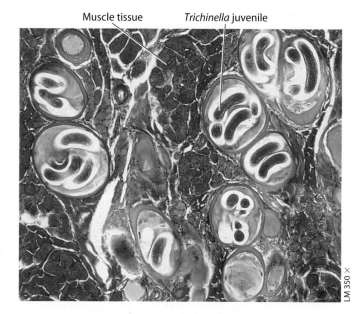

Muscle tissue   *Trichinella* juvenile

LM 350×

Figure 18.8B   A parasitic nematode (*Trichinella spiralis*); juvenile worms encysted in human muscle tissue

Many species of nematodes are serious agricultural pests that attack the roots of plants or parasitize animals. Humans are host to at least 50 species of roundworms, including a number of disease-causing organisms. Among these are hookworms, which attach to the intestinal wall and suck blood. Dogs, cats, and many other mammals are also susceptible to hookworms. Other nematodes called heartworms are deadly to dogs. Spread by mosquitoes and also infectious to humans, heartworms seem to be on the increase in the United States.

One of the most notorious nematodes is *Trichinella spiralis,* which causes trichinosis in a wide variety of mammals, including humans. People usually acquire the worms by eating undercooked pork containing the juvenile worms. You can see some of these encysted juveniles in the section of human muscle shown in **Figure 18.8B.** Trichinosis causes severe nausea and sometimes death when large numbers of the worms penetrate heart muscle. Cooking meat until it is no longer pink kills the worms.

Estimates of nematode species range as high as 500,000. We might expect that an animal group so numerous and widespread would include a great diversity of body form. In fact, the opposite is true. Most species look very much alike. In sharp contrast, animals in the phylum Mollusca, which we examine next, exhibit enormous diversity in body form.

? Why does a nematode have to shed its cuticle when it grows?

Mouth

Colorized SEM 400×

Figure 18.8A   A free-living nematode

■ This nonliving cuticle does not expand as the animal grows. The nematode must molt its old cuticle and secrete a new, larger one.

Snails, slugs, oysters, clams, octopuses, and squids are just a few of the great variety of animals known as **molluscs** (phylum Mollusca). Molluscs are soft-bodied animals (from the Latin *molluscus*, soft), but most are protected by a hard shell.

It may seem that animals as different as squids and clams could not belong in the same phylum, but these and other molluscs have inherited several common features from their ancestors. **Figure 18.9A** illustrates the basic body plan of a mollusc, consisting of three main parts: a muscular **foot** (gray in the drawing), which functions in locomotion; a **visceral mass** (orange) containing most of the internal organs; and a **mantle** (purple), a fold of tissue that drapes over the visceral mass and secretes a shell in molluscs such as clams and snails. In many molluscs, the mantle extends beyond the visceral mass, producing a water-filled chamber called the mantle cavity, which houses the gills (left side in Figure 18.9A).

Figure 18.9A shows yet another body feature found in many molluscs—a unique rasping organ called a **radula**, which is used to scrape up food. In a snail, for example, the radula extends from the mouth and slides back and forth like a backhoe, scraping and scooping algae off rocks.

Most molluscs have separate sexes, with reproductive organs located in the visceral mass. The life cycle of many marine molluscs includes a ciliated larva called a trochophore, which is also characteristic of some other invertebrate phyla.

In contrast with flatworms, which have no body cavity, and nematodes, which have a pseudocoelom, molluscs have a true coelom (blue in Figure 18.9A). Also unlike flatworms and nematodes, molluscs have a **circulatory system**—an organ system that pumps blood and distributes nutrients and oxygen throughout the body.

These basic body features have evolved in markedly different ways in different groups of molluscs. The three most diverse groups (classes) are the gastropods (including snails and slugs), bivalves (including clams, scallops, and oysters), and cephalopods (including squids and octopuses).

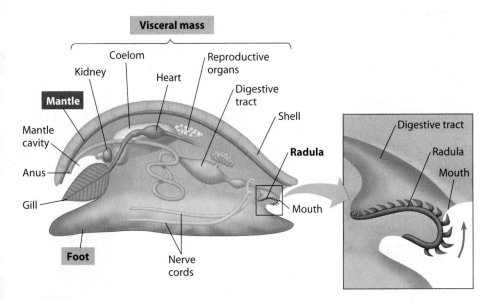

Figure 18.9A The general body plan of a mollusc

Gastropods The largest group of molluscs are the **gastropods** (from the Greek *gaster*, belly, and *pous*, foot), found in fresh water, salt water, and terrestrial environments. In fact, they include the only molluscs that live on land. Most gastropods are protected by a single, spiraled shell into which

the animal can retreat when threatened. Many gastropods have a distinct head with eyes at the tips of tentacles, like the land snail in **Figure 18.9B**. Terrestrial snails lack the gills typical of aquatic molluscs; instead, the lining of the mantle cavity functions as a lung, exchanging gases with the air.

Most gastropods are marine, and shell collectors delight in their variety. Slugs, however, are unusual molluscs in that they have lost their mantle and shell during their evolution. The long, colorful projections on the sea slug shown in **Figure 18.9C** (about 5 cm long) function as gills.

Bivalves The **bivalves** (from the Latin *bi*, double, and *valva*, leaf of a folding door) include numerous species of clams, oysters, mussels, and scallops. They have shells divided into two halves that are hinged together. Most bivalves are suspension feeders. The mantle cavity contains gills that are used for feeding as well as gas exchange. The mucus-coated gills trap fine food particles suspended in the water, and cilia sweep the particles to the mouth. Most bivalves are sedentary, living in sand or mud. They may use their muscular foot for digging and anchoring. Mussels are sessile, secreting strong threads that attach them to rocks, docks, and boats. The scallop in **Figure 18.9D** (about 10 cm in diameter) can skitter along the seafloor by flapping its shell, rather like the mechanical false teeth sold in novelty shops. Notice the many bluish eyes peering out between the two halves of its hinged shell. The eyes are set into the fringed edges of the animal's mantle.

Cephalopods The **cephalopods** (from the Greek *kephale*, head, and *pous*, foot) differ from gastropods and bivalves in being adapted to the lifestyle of fast, agile predators.

Figure 18.9B A terrestrial gastropod: a land snail

Figure 18.9C A marine gastropod: a sea slug

Figure 18.9E A cephalopod with an internal shell: a squid

Figure 18.9D A bivalve: a scallop

Figure 18.9F A cephalopod without a shell: an octopus

The chambered nautilus is a descendant of ancient groups with external shells, but in other cephalopods, the shell is small and internal (as in squids) or missing altogether (as in octopuses). Cephalopods use beak-like jaws and a radula to crush or rip prey apart. The mouth is at the base of the foot, which is drawn out into several long tentacles for catching and holding prey.

The squid in **Figure 18.9E** (about 20 cm long) ranks with fishes as a fast, streamlined predator. It darts about by drawing water into its mantle cavity and then forcing a jet of water out through a muscular siphon. It steers by pointing the siphon in different directions. The octopus in **Figure 18.9F** (about 30 cm long) lives on the seafloor, where it creeps about in search of crabs and other food.

All cephalopods have large brains and sophisticated sense organs, and these contribute to their being successful, mobile predators. Cephalopod eyes are among the most complex sense organs in the animal kingdom. Each eye contains a lens that focuses light and a retina on which clear images form. Octopuses are considered among the most intelligent invertebrates and have shown remarkable learning abilities in laboratory experiments.

The giant squid is the largest of all invertebrates. Until recently, the biggest specimen on record was 18 m long and weighed about 2 tons. In 2003, however, an even larger specimen of another species, dubbed the colossal squid, was caught near Antarctica. Some biologists think that this specimen was a juvenile and estimate that adults could be twice as large. It is likely that these two species of squid spend most of their time in the deep ocean, and scientists have never observed either species in its natural habitat. Thus, these marine giants remain one of the great mysteries of invertebrate life.

The next animals we consider look much different from molluscs; their bodies all show segmentation.

? As representatives of classes of molluscs, a garden snail is an example of a _____; a clam is an example of a _____; and a squid is an example of a _____.

■ gastropod . . . bivalve . . . cephalopod

18.10 Annelids are segmented worms

A segmented body resembling a series of fused rings is the hallmark of phylum Annelida (from the Latin *anellus,* ring). **Segmentation,** the subdivision of the body along its length into a series of repeated parts (segments), played a central role in the evolution of many complex animals. A segmented body allows for greater flexibility and mobility, and it probably evolved as an adaptation facilitating movement. An earthworm, a typical **annelid,** uses its flexible, segmented body to crawl and burrow rapidly into the soil.

Annelids range in length from less than 1 mm to 3m, the length of some giant Australian earthworms. They are found in damp soil, in the sea, and in most freshwater habitats. Some aquatic annelids swim in pursuit of food, but most are bottom-dwelling scavengers that burrow in sand and mud. There are three main groups of annelids: earthworms and their relatives, polychaetes, and leeches.

Earthworms and Their Relatives Figure 18.10A illustrates the segmented anatomy of an earthworm. Internally, the coelom is partitioned by membrane walls (only a few are fully shown here). Many of the internal body structures are repeated within each segment. The nervous system (yellow) includes a simple brain and a ventral nerve cord with a cluster of nerve cells in each segment. Excretory organs (green), which dispose of fluid wastes, are also repeated in each segment (only a few are shown in this diagram). The digestive tract, however, is not segmented; it passes through the segment walls from the mouth to the anus.

Many invertebrates, including most molluscs and all arthropods (which we will meet next), have what is called an **open circulatory system,** in which blood is pumped through vessels that open into body cavities where organs are bathed directly in blood. Annelids and vertebrates, in contrast, have a **closed circulatory system,** in which blood remains enclosed in vessels as it distributes nutrients and oxygen throughout the body. As you can see in the diagram at the lower left, the main vessels of the earthworm circulatory system—a dorsal blood vessel and a ventral blood vessel—are connected by segmental vessels. The pumping organ, or "heart," is simply an enlarged region of the dorsal blood vessel plus five pairs of segmental vessels near the anterior end.

Each segment is surrounded by longitudinal and circular muscles. Earthworms move by coordinating the contraction of these two sets of muscles (see Figure 30.1D). These muscles work against the coelomic fluid in each segment, which acts as a hydrostatic skeleton. Each segment also has four pairs of stiff bristles that provide traction for burrowing.

Earthworms are hermaphrodites; that is, they have both male and female reproductive structures. But they mate and cross-fertilize by exchanging sperm. A specialized organ, visible as the thickened region of the worm in Figure 18.10A, secretes a cocoon made of mucus. The cocoon slides along the worm, picking up the eggs and the received sperm. The cocoon slips off the worm into the soil, where the embryos develop.

Earthworms eat their way through the soil, extracting nutrients as soil passes through their digestive tube. Undigested

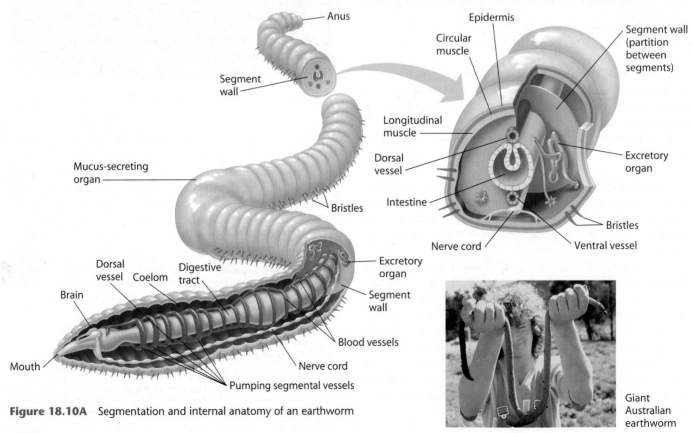

Figure 18.10A Segmentation and internal anatomy of an earthworm

Giant Australian earthworm

material, mixed with mucus secreted into the digestive tract, is eliminated as castings (feces) through the anus. Farmers value earthworms because the animals aerate the soil and their castings improve the soil's texture. Darwin estimated that a single acre of British farmland had about 50,000 earthworms, producing 18 tons of castings per year.

Polychaetes The **polychaetes** (from the Greek *polys,* many, and *chaeta,* hair) form the largest group of annelids. Figure 18.10B shows a polychaete called a sandworm, which lives on the seafloor. Each segment of a polychaete has a pair of fleshy, paddle-like appendages with many stiff bristles (called chaetae) that help the worm wriggle about in search of small invertebrates to eat. In many polychaetes, the appendages are richly supplied with blood vessels and function in gas exchange.

Most polychaetes are marine. Many live in tubes and extend feathery appendages coated with mucus that trap suspended food particles. Tube-dwellers usually build their tubes by mixing mucus with bits of sand and broken shells. Some, such as the Christmas tree worm (Figure 18.10C), bore their tubes into the limestone of coral reefs. The two brightly colored spires you see in the photograph are feeding appendages extending from the worm's head. The white projection on the bottom is a stopper that seals off the tube when the worm withdraws.

Leeches The third main group of annelids is the **leeches,** which are notorious for their bloodsucking habits. However, most species are free-living carnivores that eat small invertebrates such as snails and insects. The majority of leeches inhabit fresh water, but there are also marine species and a few terrestrial species that inhabit moist vegetation in the tropics. They range in length from 1 to 30 cm.

Some bloodsucking leeches use razor-like jaws to slit the skin of an animal. The host is usually oblivious to this attack because the leech secretes an anesthetic as well as an anticoagulant into the wound. The leech then sucks as much blood as it can hold, often more than ten times its own weight. After this gorging, a leech can last for months without another meal.

Until this century, physicians frequently used leeches for bloodletting, removing what was considered "bad blood" from sick patients. While this medical application has fallen out of favor, many researchers are investigating the potential use of leech anticoagulant to dissolve blood clots that form during surgery or as a result of heart disease. This chemical is now being produced by genetic engineering and is in clinical trials.

Leeches are still occasionally used to remove blood from bruised tissues (Figure 18.10D) and to help relieve swelling in fingers or toes that have been sewn back on after accidents. Blood tends to accumulate in a reattached finger or toe until small veins have a chance to grow back into it. Leeches are applied to remove this blood.

The segments of an annelid are all very similar. In the next group we explore, the arthropods, body segments and their appendages have become specialized, serving a variety of functions.

Figure 18.10B
A sandworm (about 15 cm long), a polychaete that lives on the seafloor

Figure 18.10C   A Christmas tree worm, a tube-building polychaete (5–10 cm long)

Figure 18.10D
A medicinal leech applied to drain blood from a patient

? Tapeworms and leeches are both parasites. What are the key differences between these two?

■ Whereas both are composed of repeated segments, the segments of a tapeworm are filled mostly with reproductive organs and are shed from the posterior end of the animal. Tapeworms are flatworms with no body cavity and, in their parasitic lifestyle, not even a gastrovascular cavity. Leeches have a true coelom and a complete digestive tract.

18.11 Arthropods are segmented animals with jointed appendages and an exoskeleton

Over a million species of **arthropods**—including crayfish, lobsters, crabs, barnacles, spiders, ticks, and insects—have been identified. It is estimated that the arthropod population of the world numbers about a billion billion (10^{18}) individuals! In terms of species diversity, geographic distribution, and sheer numbers, phylum Arthropoda must be regarded as the most successful animal phylum.

The diversity and success of arthropods are largely related to their segmentation, their hard exoskeleton, and their jointed appendages, for which the phylum is named (from the Greek *arthron*, joint, and *pous*, foot). As indicated in the drawing of a lobster in **Figure 18.11A**, the appendages are variously adapted for sensory reception, defense, feeding, walking, and swimming. The arthropod body, including the appendages, is covered by an **exoskeleton**, an external skeleton that protects the animal and provides points of attachment for the muscles that move the appendages. This nonliving covering, or cuticle, is constructed from layers of protein and chitin, a polysaccharide. The exoskeleton is thick around the head, where its main function is to house and protect the brain. It is paper-thin and flexible in other locations, such as the joints of the legs. As it grows, an arthropod must periodically shed its old exoskeleton and secrete a larger one, a complex process called **molting**.

In contrast with annelids, which have similar segments throughout their body, the body of most arthropods is formed of several distinct groups of segments: the head, thorax, and abdomen. (In some arthropods, including the lobster, the exoskeleton of the head and thorax is partly fused, forming a body region called the cephalothorax.) Each of the segment groups is specialized for a different function. In a lobster, the head bears sensory antennae, eyes, and jointed mouthparts on the ventral side. The thorax bears a pair of defensive appendages (the pincers) and four pairs of legs for walking. The abdomen has swimming appendages.

Also unlike annelids, arthropods have an open circulatory system in which a tube-like heart pumps blood through short arteries into spaces surrounding the organs. A variety of gas exchange organs have evolved. Most aquatic species have gills. Terrestrial insects have internal air ducts that branch throughout the body (see Module 22.4).

Fossils and molecular evidence suggest that living arthropods represent four major lineages that diverged early in the evolution of arthropods. The figures in this module illustrate representatives of three of these lineages. The fourth, the insects, will be discussed in Module 18.12.

Chelicerates **Figure 18.11B** shows a number of **horseshoe crabs**, members of a species that has survived with little change for hundreds of millions of years. This "living fossil" is the only surviving member of a group of marine chelicerates that were abundant in the sea some 300 million years ago. One member of this group, the water scorpion, could grow up to 3 m long. Horseshoe crabs are common on the Atlantic and Gulf coasts of the United States.

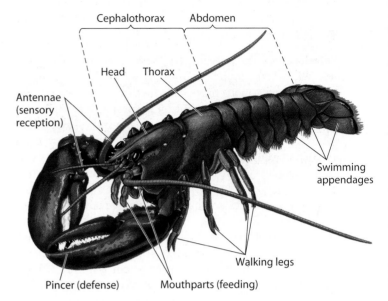

Figure 18.11A The structure of a lobster, an arthropod

Living **chelicerates** also include the scorpions, spiders, ticks, and mites, collectively called **arachnids**. Most arachnids live on land. Scorpions (**Figure 18.11C**, left) are nocturnal hunters. Their ancestors were among the first terrestrial carnivores, preying on other arthropods that fed on early land plants. Scorpions have a large pair of pincers for defense and capturing prey. The tip of the tail bears a poisonous stinger. Scorpions eat mainly insects and spiders and attack people only when prodded or stepped on. Only a few species are dangerous to humans.

Spiders, a diverse group of arachnids, are usually active during the daytime, hunting insects or trapping them in webs of silk that they spin from specialized glands on their abdomen (see Figure 18.11C, center). Mites make up another large group of arachnids. On the right in Figure

Figure 18.11B Horseshoe crabs (up to about 30 cm wide)

A black widow spider (about 1 cm wide)

A scorpion (about 8 cm long)

A dust mite (about 420 μm long)

Colorized SEM 900×

Figure 18.11C Arachnids

18.11C is a micrograph of a dust mite, a ubiquitous scavenger in our homes. Thousands of these microscopic animals can thrive in a few square centimeters of carpet or in one of the dust balls that form under a bed. Dust mites do not carry infectious diseases, but many people are allergic to them.

Millipedes and Centipedes The animals in this lineage have similar segments over most of their body and superficially resemble annelids; however, their jointed legs identify them as arthropods. **Millipedes (Figure 18.11D)** are worm-like terrestrial creatures that eat decaying plant matter. They have two pairs of short legs per body segment. **Centipedes** are terrestrial carnivores with a pair of poison claws used in defense and to paralyze prey, such as cockroaches and flies. Each of their body segments bears a single pair of long legs.

Crustaceans The **crustaceans** are nearly all aquatic. Lobsters and crayfish are in this group, along with numerous crabs, shrimps, and barnacles. Barnacles **(Figure 18.11E)** are marine crustaceans with a cuticle that is hardened into a shell containing calcium carbonate, which may explain why they were once classified as molluscs. Their jointed appendages project from their shell to strain food from the water. Most barnacles anchor themselves to rocks, boat hulls, pilings, or even whales. The adhesive they produce is as strong as any glue ever invented.

Other crustaceans are small copepods and krill, which serve as food sources for many fishes and whales.

We turn next to the fourth lineage of arthropods, the insects, whose numbers dwarf all other groups combined.

? The phylum Arthropoda is named for its members'
_____ _____.

jointed appendages ∎

Figure 18.11D A millipede (about 7 cm long)

Figure 18.11E Crustaceans: goose barnacles (about 2 cm high)

Insects are the most diverse group of organisms

The total number of insect species is greater than the total of all other species combined. **Entomology** is a branch of biology that specializes in the study of insects. Insects live in almost every terrestrial habitat and in fresh water, and flying insects fill the air. Insects are rare in the seas, where crustaceans are the dominant arthropods.

Insects have a number of common features. Like the grasshopper in **Figure 18.12A**, most have a three-part body, consisting of a head, a thorax, and an abdomen. The head usually bears a pair of sensory antennae and a pair of eyes. Several pairs of mouthparts are adapted for particular kinds of eating—for example, for chewing plant material in grasshoppers; for lapping up fluids in houseflies; and for piercing skin and sucking blood in mosquitoes. Most adult insects have three pairs of legs and one or two pairs of wings. The wings are extensions of the cuticle, so insects have acquired the ability to fly without sacrificing any legs. By contrast, birds and bats have one of their two pairs of walking legs modified as wings and are generally clumsy on ground. The ability to fly is a major factor in the success of insects.

Many insects undergo metamorphosis in their development. For instance, insects in the three groups on this page undergo **incomplete metamorphosis;** the young resemble adults but are smaller with different body proportions. By contrast, the groups on the facing page undergo **complete metamorphosis;** their larval stages (such as caterpillars, which are the larvae of moths and butterflies, and maggots, which are fly larvae) are specialized for eating and growing and look very different from the adults, which are specialized for dispersal and reproduction. In these groups, metamorphosis from larva to adult occurs during a pupal stage.

Systematists classify insects into about 26 orders, mostly on the basis of wing and mouthpart structure. The drawings in **Figures 18.12A–18.12G** illustrate representative insects in seven of the most common orders.

Web/CD Thinking as a Scientist *How Are Insect Species Identified?*

? Contrast incomplete and complete metamorphosis.

■ In complete metamorphosis, adults look very different from larvae.

A. Order Orthoptera. The grasshopper represents this group, which contains about 13,000 species. Other orthopterans are the crickets, katydids, and locusts. These insects have biting and chewing mouthparts, and most species are herbivorous. They have large hind legs adapted for jumping and two pairs of wings (one leathery, one membranous). Males commonly make courtship sounds by rubbing together body parts, such as a ridge on the hind leg.

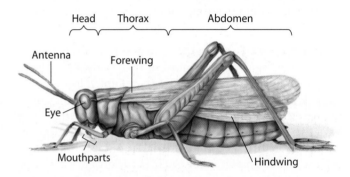

Figure 18.12A Insect anatomy, as seen in a grasshopper

B. Order Odonata. This order includes about 5,000 species of dragonflies and damselflies. These insects have two pairs of similar wings and large, compound eyes. They have chewing mouthparts and are carnivorous, often catching and eating other insects while flying. The larvae of larger species sometimes eat tadpoles and small fishes.

Figure 18.12B A damselfly

C. Order Hemiptera. Often called the "true bugs," hemipterans (about 85,000 species) include bedbugs, plant bugs, stinkbugs, and water striders. They have piercing, sucking mouthparts, and most species feed on plant sap. A few, such as bedbugs, feed on blood. The true bugs have two pairs of wings, and the front half of each forewing is thickened and leathery. Water striders walk on water by taking advantage of surface tension (see Module 2.11).

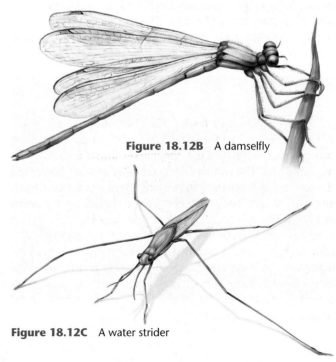

Figure 18.12C A water strider

D. Order Coleoptera. Beetles make up the largest order in the animal kingdom. There are about 350,000 species known worldwide. Beetles occur almost everywhere, from high mountains to the seashore, and their habitats are extremely varied. Forests, streams, ponds, soil, dung, carrion, and plant material each support a number of species. Beetles have biting and chewing mouthparts, and they include carnivores, herbivores, and omnivores. Beetles vary in length from less than 1 mm to about 12 cm. They have two pairs of wings, but only the hindwings function in flight. The forewings are hardened and thickened as protective covers for the hindwings.

Figure 18.12D A ground beetle

E. Order Lepidoptera. These are the moths and butterflies (about 120,000 species). They have two pairs of wings, with the hind pair being smaller. Typically, the wings and body are covered by scales—the dust you find on your fingers after holding a butterfly. The mouthparts form a long drinking tube that is adapted for drinking nectar from flowers. The tube is coiled under the head when not in use. When the insect drinks nectar, the tube is uncoiled and extended deep into the flower.

Figure 18.12E A hawk moth

F. Order Diptera. Dipterans (about 151,000 species) are the flies, including fruit flies, houseflies, gnats, and mosquitoes. Flies have a single pair of wings; instead of hindwings, they have small, club-shaped organs called halteres, which function in maintaining balance during flight. Mosquitoes have piercing, sucking mouthparts, and the females suck blood. Most other dipterans have lapping mouthparts and feed on nectar or other liquids. Mosquitoes and other bloodsuckers may transmit disease, such as malaria or African sleeping sickness, to their human hosts.

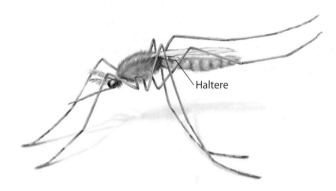

Haltere

Figure 18.12F A mosquito

G. Order Hymenoptera. The hymenopterans (about 125,000 species) are the ants, bees, and wasps. They have two pairs of wings (except for worker ants), and both are used in flight. The hindwings are smaller and are hooked to the rear of the forewings, improving flight efficiency. Generally, the thorax and abdomen are separated by a narrow waist, which, along with the four translucent wings, makes it easy to identify an insect as a member of this order. With chewing and sucking mouthparts, some hymenopterans are herbivorous; others are carnivorous, eating mainly other insects. Females have a posterior stinging organ. Many hymenopterans display complex behavior, including social organization (see Module 35.20).

Figure 18.12G A paper wasp

18.13 Echinoderms have spiny skin, an endoskeleton, and a water vascular system for movement

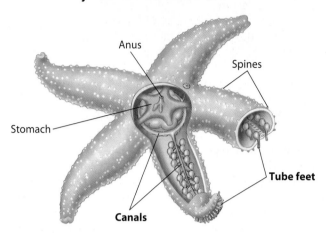

Figure 18.13A The water vascular system (canals and tube feet) of a sea star

Echinoderms, such as sea stars, sand dollars, and sea urchins, are slow-moving or sessile marine animals. Most are radially symmetrical as adults. Both the external and the internal parts of a sea star, for instance, radiate from the center like spokes of a wheel. The bilateral larval stage of echinoderms, however, tells us that echinoderms are not closely related to cnidarians or other animals that never show bilateral symmetry.

The phylum name Echinodermata (from the Greek *echin*, spiny, and *derma*, skin) refers to the prickly bumps or spines of a sea star or sea urchin. These are extensions of the hard calcareous plates that form the **endoskeleton**, or internal skeleton, under the thin skin of the animal.

Unique to echinoderms is the **water vascular system**, a network of water-filled canals that branch into extensions called tube feet. Tube feet function in locomotion, feeding, and gas exchange (Figure 18.13A). A sea star pulls itself slowly over the seafloor using its suction-cup-like tube feet. Its mouth is centrally located on its undersurface. When a sea star encounters an oyster or clam, its favorite food, it grips the mollusc with its tube feet and positions its mouth next to the narrow opening between the two valves of the shell. The sea star then turns its stomach inside out, pushing it through its mouth and into the opening of the mollusc's shell. The sea star's stomach proceeds to digest the soft parts of the mollusc inside the mollusc's shell (Figure 18.13B).

Sea stars and some other echinoderms are capable of regeneration. Arms that are lost are readily regrown.

In contrast with sea stars, sea urchins are spherical and have no arms. They do have five rows of tube feet that project through tiny holes in the animal's globe-like case. If you look carefully in Figure 18.13C, you can see the long, thread-like tube feet projecting among the spines of this purple sea urchin (the tube feet are darker than the spines). Sea urchins move by pulling with their tube feet. They also have muscles that pivot their spines, which can aid in locomotion. Unlike the carnivorous sea stars, most sea urchins eat algae.

Other echinoderm classes include brittle stars, which move by slashing their long, flexible arms; sea lilies, which live attached to the substrate by a stalk; and sea cucumbers, elongated animals that resemble their name rather than other echinoderms, but have five rows of tube feet.

Though echinoderms have many unique features, we see evidence of their relation to other animals in their embryonic development. As we discussed in Modules 18.3 and 18.4, differences in patterns of development have led biologists to identify echinoderms and chordates (which include vertebrates) as a clade of bilateral animals called deuterostomes. Thus, echinoderms are more closely related to our phylum, the chordates, than to the protostome animals, such as molluscs, annelids, and arthropods. We begin our examination of the chordates next.

Web/CD Activity 18A *Characteristics of Invertebrates*

? Contrast the skeleton of an echinoderm with that of an arthropod.

■ An echinoderm has an endoskeleton; an arthropod has an exoskeleton.

Figure 18.13B A sea star feeding on a clam

Figure 18.13C A sea urchin (about 12 cm in diameter)

18.14 Our own phylum, Chordata, is distinguished by four features

Four distinctive features appear in the embryos, and often in the adults, of animals in the phylum Chordata: (1) a **dorsal, hollow nerve cord**; (2) a **notochord**, a flexible, supportive, longitudinal rod located between the digestive tract and the nerve cord; (3) **pharyngeal slits** located in the pharynx, the region just behind the mouth; and (4) a muscular **post-anal tail** (a tail posterior to the anus). Together, these features identify a **chordate**. You can see these four features in the diagrams in Figures 18.14A and 18.14B. The two chordates shown, tunicates and lancelets, are called invertebrate chordates because they do not have a backbone.

Adult **tunicates** are stationary and look more like small sacs than anything we usually think of as a chordate (see Figure 18.14A). Tunicates often adhere to rocks and boats, and they are common on coral reefs. The adult has no trace of a notochord, nerve cord, or tail, but it does have prominent pharyngeal slits that function in feeding. Very different from the adult, the tunicate larva is a swimming, tadpole-like organism that exhibits all four chordate trademarks.

Tunicates are suspension feeders. Seawater enters the adult animal through an opening at the top, passes through the pharyngeal slits into a large cavity in the animal, and exits back into the ocean via an excurrent siphon on the side of the body (see the photo in Figure 18.14A). Food particles are trapped in a net made of mucus and then transported to the intestine, where they are digested. Because they shoot a jet of water through their excurrent siphon when threatened, tunicates are often called sea squirts.

Lancelets, another group of marine invertebrate chordates, also feed on suspended particles. Lancelets are small, bladelike chordates that live in marine sands (see Figure 18.14B). When feeding, a lancelet wriggles backward into the sand with its head sticking out. As in tunicates, a net of mucus secreted across the pharyngeal slits traps food particles. Water flowing through the slits exits via an opening in front of the anus.

Lancelets clearly illustrate the four chordate features. They also have segmental muscles that flex their body from side to side, producing slow swimming movements. These serial muscles are evidence of the lancelet's segmentation. The muscles develop from blocks of mesoderm called somites (see Module 27.12), which are found along each side of the notochord in all chordate embryos. Although not unique to chordates, body segmentation is another chordate characteristic.

What is the relationship between the invertebrate chordates and the vertebrates? The tunicates likely represent the deepest branch of the chordate lineage. Molecular evidence indicates that the lancelets are the closest living relatives of vertebrates. Research has shown that the same genes that organize the major regions of the vertebrate brain are expressed in the same pattern at the anterior end of the lancelet nerve cord.

The invertebrate chordates have helped us identify the four chordate hallmarks. In the next module, we look at a phylogenetic tree that highlights some of the important developments in chordate evolution.

? What four features do we share with invertebrate chordates, such as lancelets?

■ (1) Dorsal, hollow nerve cord; (2) notochord; (3) pharyngeal slits; (4) post-anal tail

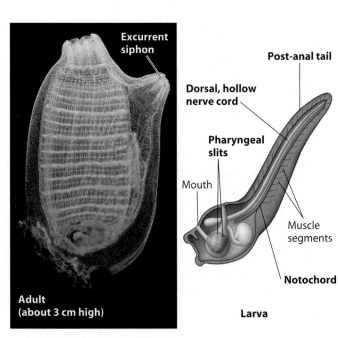

Figure 18.14A A tunicate

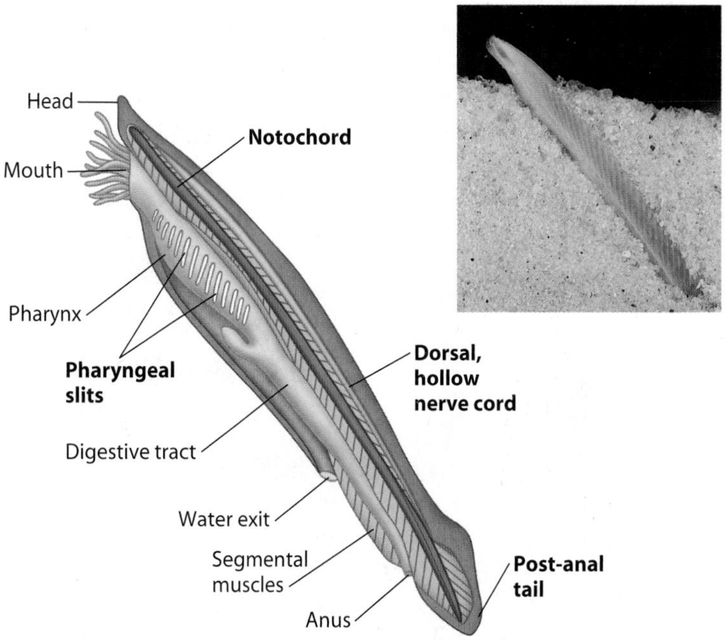

Figure 18.14B A lancelet (5–15 cm long)

18.15 Derived characters define the major clades of chordates

Using a combination of anatomical, molecular, and fossil evidence, biologists have developed hypotheses for the evolution of chordate groups. **Figure 18.15** illustrates a current view of the major clades of chordates and lists some of the derived characters that define the clades. You can see that the tunicates are thought to be the first group to branch from the chordate lineage. Unlike tunicates, all other chordates have a brain, albeit a small one in the lancelets (only a swollen tip of the nerve cord).

The next transition was the development of a head that consists of a brain at the anterior end of the dorsal nerve cord, eyes and other sensory organs, and a skull. These innovations opened up a completely new way of feeding for chordates: active predation. All chordates that have a head are called **craniates** (from the word *cranium,* meaning "skull"). Hagfishes, the most primitive surviving craniate lineage, have a skull composed of cartilage but no vertebrae. Their main body support is an adult version of the embryo's notochord.

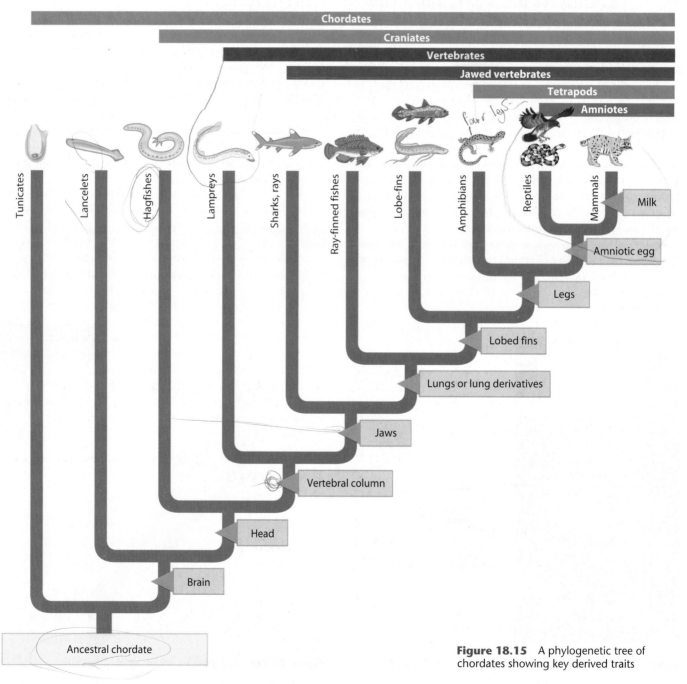

Figure 18.15 A phylogenetic tree of chordates showing key derived traits

The origin of a backbone came next. The **vertebrates** are distinguished by a more extensive skull and a backbone, or **vertebral column**, composed of a series of bones called **vertebrae** (singular, *vertebra*). These skeletal elements enclose the main parts of the nervous system. The skull forms a case for the brain, and the vertebrae enclose the nerve cord. The vertebrate skeleton is an endoskeleton, made of either flexible cartilage or a combination of hard bone and cartilage. Bone and cartilage are mostly nonliving material. But because there are living cells that secrete the nonliving material, the endoskeleton can grow with the animal.

The next major transition was the origin of jaws, which opened up new feeding opportunities. The evolution of lungs or lung derivatives, followed by muscular lobed fins with skeletal support, opened the possibility of life on land. **Tetrapods**, jawed vertebrates with two pairs of limbs, were the first vertebrates on land. The evolution of **amniotes**, tetrapods with a terrestrially adapted egg, enhanced reproduction on land.

In the rest of the chapter, we survey the vertebrates, from the jawless lampreys to the fishes to the tetrapods to the amniotes.

> **?** List the hierarchy of clades to which mammals belong.
>
> ■ Chordates, craniates, vertebrates, jawed vertebrates, tetrapods, amniotes

---

18.16   Lampreys are vertebrates that lack hinged jaws

Lampreys represent the oldest living lineage of vertebrates. Living in freshwater streams, the larvae are suspension feeders that resemble lancelets and spend much of the time buried in sediment. Most lampreys migrate to the sea or lakes as they mature into adults.

Just seeing the mouth of a sea lamprey (**Figure 18.16A**) suggests what it can do. It clamps onto the side of a fish and, using its rasping tongue to penetrate the skin, sucks its victim's blood. After invading the Great Lakes over the past 150 years, these voracious vertebrates multiplied rapidly, decimating fish populations as they spread. By the mid-1900s, sea lampreys, along with an increase in commercial fishing and water pollution, had destroyed virtually all the large fishes in the lakes. Since the 1960s, streams that flow into the lakes have been treated with a chemical that reduces lamprey numbers, and fish populations have been recovering.

In addition to the lineage that led to the lampreys, the fossil record chronicles other jawless vertebrates, from soft-bodied predators with barbed hooks and mineralized "teeth" to scavengers with paired fins and armored bodies. These extremely diverse vertebrates, however, were all extinct by 360 million years ago.

Jawed vertebrates appeared in the fossil record in the mid-Ordovician period, about 470 million years ago, and steadily became more diverse. Their success probably relates to their paired fins and tail, which allowed them to swim after prey, and to their jaws, which enabled them to catch and eat a wide variety of prey instead of feeding as mud-suckers or suspension feeders.

Sharks, fishes, amphibians, reptiles (including birds), and mammals—the vast majority of living vertebrates—have jaws supported by two skeletal parts held together by a hinge. Where did these hinged jaws come from? According to one hypothesis, they evolved by modification of skeletal supports of the anterior pharyngeal (gill) slits. The first part of **Figure 18.16B** shows the skeletal rods supporting the gill slits in a hypothetical ancestor. The main function of these gill slits was trapping suspended food particles. The other two parts show changes that may have occurred as jaws evolved. By following the red and green structures, you can see that the jaws and their supports evolved from two pairs of skeletal rods located between gill slits that were near the mouth. The remaining gill slits, no longer required for suspension feeding, remained as sites of gas exchange.

> **?** Vertebrate jaws probably evolved from the skeletal rods supporting _____.
>
> ■ gills

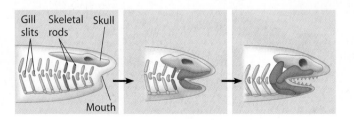

Figure 18.16A   A sea lamprey, with its rasping mouth (inset)

Figure 18.16B   A hypothesis for the origin of vertebrate jaws

18.17 Jawed vertebrates with gills and paired fins include sharks, ray-finned fishes, and lobe-fins

Three lineages of jawed vertebrates with gills and paired fins are commonly called fishes. The sharks and rays of the class Chondrichthyes, which means "cartilage fish," have changed little in over 300 million years. As shown in Figure 18.15, lungs or lung derivatives are the key derived character of the clade that includes the ray-finned fishes and the lobe-fins. Muscular fins supported by stout bones further characterize the lobe-fins.

Chondrichthyans Sharks and rays, the **chondrichthyans**, have a flexible skeleton made of cartilage. The largest sharks are suspension feeders that eat small, floating plankton. Most sharks, however, are adept predators—fast swimmers with a streamlined body and powerful jaws (**Figure 18.17A**). A shark has sharp vision and a keen sense of smell. On its head are electrosensors, organs that can detect minute electrical fields produced by muscle contractions in nearby animals. Sharks and most of the other aquatic vertebrates have a **lateral line system**, a row of sensory organs running along each side that are sensitive to changes in water pressure and can detect minor vibrations caused by animals swimming nearby.

Ray-finned Fishes The **ray-finned fishes**, which include the familiar tuna, trout, and goldfish, have a skeleton reinforced with a hard matrix of calcium phosphate. Their fins are supported by thin, flexible skeletal rays. Most have flattened scales covering their skin and secrete a coating of mucus that reduces drag during swimming. **Figure 18.17B** highlights key features of a ray-finned fish, such as the rainbow trout shown in the photograph. On each side of the head, a protective flap called an **operculum** covers a chamber housing the gills. Movement of the operculum allows the fish to breathe without swimming. (By contrast, sharks must generally swim to pass water over their gills.) Ray-finned fishes also have a lung derivative that helps keep them buoyant—the **swim bladder**, a gas-filled sac. Swim bladders evolved from balloon-like lungs, which the ancestral fishes may have used to supplement their gas exchange by gills in shallow water.

Lobe-fins Ray-finned fishes emerged during the Devonian period along with another major lineage—the **lobe-fins**. The

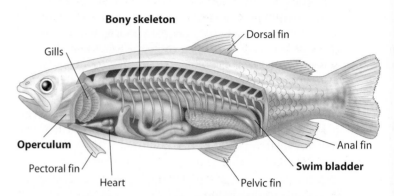

key derived character of this group is a series of rod-shaped bones in their muscular pectoral and pelvic fins. During the Devonian, they lived along coastal wetlands and may have used their lobed fins to "walk" underwater. Today, three lineages of lobe-fins survive: The coelacanth (**Figure 18.17C**) is a deep-sea dweller once thought extinct. The lungfishes are represented by a few Southern Hemisphere genera (see Figure 15.3C) that generally inhabit stagnant waters and gulp air into lungs connected to the pharynx. And the third lineage adapted to life on land during the mid-Devonian and gave rise to terrestrial vertebrates, as we see next.

Figure 18.17B Anatomical features of a ray-finned fish

? From what structure may the swim bladder of ray-finned fishes have evolved?

■ Simple lungs of an ancestral species

Figure 18.17A A sand bar shark: a chondrichthyan

Figure 18.17C A coelacanth, a lobe-fin

Amphibians were the first tetrapods—vertebrates with two pairs of limbs

Tetrapods, which means "four feet" in Greek, are jawed vertebrates with limbs and feet that can support their weight on land. During the Devonian period, a diversity of plants and arthropods already inhabited the land. Lobe-fins were supplementing gills for gas exchange, and sturdy, muscular fins supported by extensions of the skeleton were probably better than fins for paddling and crawling through the dense vegetation at the water's edge. Thus, lungs and appendages evolved in certain lobe-fins millions of years before the earliest amphibians crawled onto land.

The fossil record chronicles this transition to amphibians over the period from 400 to 350 million years ago. For example, fossils of *Acanthostega* (**Figure 18.18A**) have the bony supports of gills but also have four appendages with the same basic skeletal elements as the walking limbs of amphibians, reptiles, and mammals. Though *Acanthostega* was aquatic, it represents a period of vertebrate evolution when adaptations that equipped certain lobe-fins for shallow water preadapted one lineage of those lobe-fins for a gradual transition to spending more and more time walking and breathing on the terrestrial side of the water's edge.

The early amphibians probably feasted on insects and other invertebrates in the lush forests of the Carboniferous period (see Module 17.7). As a result, amphibians became so widespread and diverse that the Carboniferous period is sometimes called the age of amphibians.

Amphibians include salamanders, frogs, and caecilians. Some present-day salamanders are entirely aquatic, but those that live on land walk with a side-to-side bending of the body that may resemble the swagger of early terrestrial tetrapods (**Figure 18.18B**). Frogs are more specialized for moving on land, using their powerful hind legs to hop along the terrain. Caecilians are nearly blind and are legless, looking like earthworms. However, they evolved from a legged ancestor. Most species burrow in moist soil in tropical areas.

In Greek, the word *amphibios* means "living a double life," a reference to the metamorphosis of many frogs. A frog spends

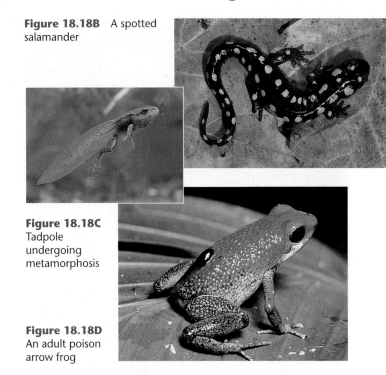

Figure 18.18B A spotted salamander

Figure 18.18C Tadpole undergoing metamorphosis

Figure 18.18D An adult poison arrow frog

much of its time on land, but it lays its eggs in water. The larval stage, called a tadpole, is a legless, aquatic algae-eater with gills, a lateral line system resembling that of fishes, and a long, finned tail. In changing into a frog, the tadpole undergoes a radical metamorphosis (**Figure 18.18C**). When a young frog crawls onto shore and continues life as a terrestrial insect-eater, it has four legs and air-breathing lungs instead of gills. Not all amphibians live such a double life, however. Some species are strictly terrestrial, and others are only aquatic. *Toad* is a term generally used to refer to frogs that have rough skin and live entirely in terrestrial habitats.

Most amphibians are found in damp habitats, where their moist skin supplements their lungs for gas exchange. Amphibian skin usually has poison glands that often play a role in defense. Poison arrow frogs have particularly deadly poisons, and their vivid coloration may warn away potential predators (**Figure 18.18D**). Frogs are usually quiet creatures, but during the breeding season, many species fill the air with their mating calls.

For the past 25 years, zoologists have been documenting a rapid and alarming decline in amphibian populations throughout the world. The causes may be multiple, including habitat degradation and the spread of a pathogenic fungus. The environmental assaults include acid rain, which is especially damaging to amphibians because of their dependence on wet places for completing their life cycles.

The next clade of vertebrates we discuss, the amniotes, are able to complete their life cycles entirely on land.

? How is the term *amphibian* misleading?

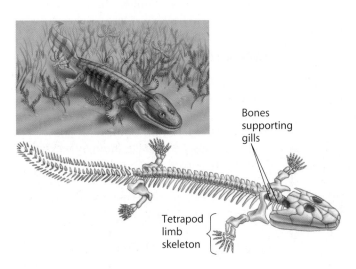

Bones supporting gills

Tetrapod limb skeleton

Figure 18.18A *Acanthostega*, a Devonian tetrapod relative

■ Not all amphibians lead a "double life." Some are terrestrial, and some are aquatic.

Reptiles are amniotes—tetrapods with a terrestrially adapted egg

Figure 18.19A A bull snake laying eggs

Reptiles (including birds) and mammals are **amniotes.** The major derived character of this clade is the **amniotic egg,** inside of which the embryo develops within a protective, fluid-filled sac called the amnion (see Chapter 27). The evolution of the amniotic egg, which functions as a "self-contained pond," enabled reptiles to complete their life cycles on land. Figure 18.19A shows a bull snake laying its eggs on dry land. As we saw in Module 17.3, the seed played a similar role in the evolution of plants.

The clade of amniotes called **reptiles** includes lizards, snakes, turtles, crocodilians, and birds, along with a number of extinct groups such as most of the dinosaurs. Lizards are the most numerous and diverse reptiles other than birds. Snakes, which are closely related to lizards, may have become limbless as their ancestors adapted to a burrowing lifestyle. Turtles have changed little since they evolved, although their ancestral lineage is still uncertain. Crocodiles and alligators (crocodilians) are the largest living reptiles, although some turtles are heavier. Crocodilians spend most of the time in water, breathing through upturned nostrils. As we discussed in the introduction to Chapter 15, birds are reptiles with feathers and various adaptations for flight.

In addition to an amniotic egg protected in a waterproof shell, reptiles have several other adaptations for terrestrial living not found in amphibians. Reptilian skin, covered with scales waterproofed with the tough protein keratin, keeps the body from drying out. Reptiles cannot breathe through their dry skin and obtain most of their oxygen with their lungs, using their rib cage to help ventilate their lungs.

Nonbird reptiles are sometimes said to be "cold-blooded" because they do not use their metabolism to produce body heat. Nonetheless, reptiles do have ways to regulate their temperature. The bearded dragon of the Australian outback (Figure 18.19B) commonly warms up in the morning by sitting on warm rocks and basking in the sun. If the lizard gets too hot, it seeks shade. Because nonbird reptiles absorb the external heat rather than generating much of their own, they are said to be **ectothermic** (from the Greek *ektos,* outside, and *therme,* heat), a term that is more appropriate than the term *cold-blooded.*

Figure 18.19B
A bearded dragon

Like the amphibians from which they evolved, reptiles were once much more prominent than they are today. Following the decline of amphibians, reptilian lineages expanded rapidly, creating a dynasty that lasted 200 million years. Dinosaurs, the most diverse group, included the largest animals ever to inhabit land. Some were "gentle giants" like the large dinosaur in Figure 18.19C, lumbering about while browsing on vegetation. Others, like the 3-m-long *Deinonychus* (Greek for "terrible claw"), were voracious carnivores that ran on two legs. Unlike living reptiles (excluding birds), *Deinonychus* and other small dinosaurs may have been **endothermic,** using heat generated by metabolism to maintain a warm, steady body temperature. Paleontologists have also discovered signs of parental care in some dinosaurs.

Most dinosaurs died out during the period of mass extinctions about 65 million years ago (see Module 15.5). Descendants of one dinosaur lineage, however, survive today as the reptilian group we know as birds.

? What is an amniotic egg?

Figure 18.19C Artist's reconstruction of a pack of *Deinonychus* attacking a tenontosaurus

■ An egg with a shell, housing an embryo in a fluid-filled sac, the amnion.

Birds are feathered reptiles with adaptations for flight

Strong fossil evidence indicates that **birds** evolved from a lineage of small, two-legged dinosaurs called theropods (see the introduction to Chapter 15). Chinese paleontologists have recently unearthed several fossils appearing to be feathered theropods. Such findings imply that feathers evolved long before powered flight. Among the possible functions of these early feathers were insulation and courtship displays.

Figure 18.20A is an artist's reconstruction based on a 150-million-year-old fossil of the oldest known, most primitive bird, called *Archaeopteryx* (from the Greek *archaios*, ancient, and *pteryx*, wing). Like living birds, it had feathered wings, but otherwise it was more like a small bipedal dinosaur of its era, with its teeth, wing claws, and tail with many vertebrae. Fossils of a number of other birds from the Cretaceous period show a gradual loss of certain features of dinosaurs, such as teeth, as well as the acquisition of traits shared by all birds today, including a short tail. Living birds seem to have evolved from a lineage of birds that survived the Cretaceous mass extinctions.

Nearly every part of the body of most birds reflects adaptations that enhance flight. Many features help reduce weight for flight: Present-day birds lack teeth; their tail is supported by only a few small vertebrae; their feathers have hollow shafts; and their bones have a honeycombed structure, making them strong but light. For example, the huge seagoing frigate bird (**Figure 18.20B**) has a wingspan of more than 2 m, but its whole skeleton weighs only about 113 g (4 oz).

Flight feathers shape bird wings into airfoils, providing lift and maneuverability in the air (see Figure 30.1E). Providing power for flight are large breast (flight) muscles, which are anchored to a keel-like breastbone. Most of what we call white meat on a turkey or chicken is the flight muscles.

Figure 18.20B   A soaring frigate bird, with a wingspan of 2 m

Flying requires a great amount of energy, and present-day birds have a high rate of metabolism and are endothermic. Insulating feathers help to maintain their warm body temperature. Supporting their high metabolic rate, birds have a highly efficient circulatory system, and their lungs are even more efficient at extracting oxygen from the air than are the lungs of mammals (see the introduction to Chapter 22).

Flying safely requires acute senses. Birds have excellent vision, perhaps the best of all vertebrates. They have relatively large brains and display very complex behaviors, particularly during breeding season. The amniotic eggs of birds are covered with a hard shell, and in many species, male and female birds take turns incubating the eggs and then feeding the young. Some birds migrate great distances each year to different feeding or breeding grounds.

With wings driven by powerful flight muscles, present-day birds are masterful flyers. Some species, such as the frigate bird, have wings adapted to soaring on air currents, and they flap their wings only occasionally. Others, such as hummingbirds, excel at maneuvering but must flap almost continuously to stay aloft. Flight ability is typical of birds, but there are a few flightless species, including penguins and the ostrich and emu of Australia (**Figure 18.20C**).

Web/CD Thinking as a Scientist *How Does Bone Structure Shed Light on the Origin of Birds?*

List some adaptations of birds that enhance flight.

■ Reduced weight, endothermic with high metabolism, efficient respiratory and circulatory systems, feathered wings shaped like airfoils, good eyesight

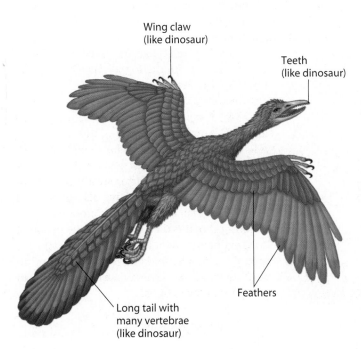

Wing claw
(like dinosaur)

Teeth
(like dinosaur)

Feathers

Long tail with
many vertebrae
(like dinosaur)

Figure 18.20A   *Archaeopteryx*, an extinct bird

Figure 18.20C
An emu, a
flightless bird

18.21 Mammals are amniotes that have hair and produce milk

There are two major lineages of amniotes: one that led to the reptiles (including birds) and one that produced the mammals. The first true **mammals** arose about 200 million years ago and were probably small, nocturnal insect-eaters. Of the three main groups of living mammals, monotremes are the oldest lineage. Marsupials diverged from eutherians (placental mammals) about 180 million years ago. During the Mesozoic era, mammals remained about the size of today's shrews, which are very small insectivores. After the extinction of large dinosaurs at the end of the Cretaceous period, however, mammals underwent an adaptive radiation, giving rise to large terrestrial herbivores and predators, as well as bats and aquatic porpoises and whales. The blue whale, an endangered species, grows to lengths of nearly 30 m and is the largest animal that has ever existed.

Two features—hair and mammary glands that produce milk to nourish young—are mammalian hallmarks. The main function of hair is insulation that helps maintain a warm body temperature. Like birds, mammals are endothermic. Efficient respiratory and circulatory systems (including a four-chambered heart) support a high rate of metabolism. A sheet of muscle called the diaphragm helps ventilate the lungs (see Module 22.5). Differentiation of teeth adapted for eating many kinds of foods is another mammalian trait.

Mammals generally have a larger brain than other vertebrates of the same size. The relatively long period of parental care provides time for offspring to learn from their parents.

Echidnas (spiny anteaters) and the duck-billed platypus, described in the chapter introduction, are the only existing **monotremes**, the egg-laying mammals. The female platypus usually lays two eggs and incubates them in a nest. After hatching, the young lick up milk secreted onto the mother's fur (**Figure 18.21A**).

Most mammals are born rather than hatched. During gestation, the embryos are nurtured inside the mother's body. The lining of the uterus and some membranes from the embryo (which are homologous to those of the amniotic egg of reptiles) form a **placenta**, a structure in which nutrients from the mother's blood diffuse into the embryo's blood.

Marsupials have a brief gestation and give birth to tiny, embryonic offspring that complete development while attached to the mother's nipples. The nursing young are usually housed in an external pouch, called the marsupium (**Figure 18.21B**). Nearly all marsupials live in Australia, New Zealand, and North and South America. As we discussed in the chapter introduction, Australia has been a marsupial sanctuary for much of the past 60 million years. Marsupials have diversified there, filling terrestrial habitats that on other continents are occupied by eutherians.

Eutherians are mammals that bear fully developed live young. They are commonly called **placental mammals** because their placentas are more complex than those of marsupials, and the young complete their embryonic development in the mother's uterus attached to the placenta. The large silvery membrane still clinging to the newborn zebra in **Figure 18.21C** is the amniotic sac, which held the developing embryo in a bath of protective amniotic fluid.

Elephants, rodents, rabbits, dogs, cows, whales, and bats are all examples of eutherians. Humans, also eutherians, belong to the order Primates, along with monkeys and apes. We examine human evolution in Chapter 19.

Web/CD Activity 18B *Characteristics of Chordates*

? What are two hallmarks of mammals?

■ Hair and mammary glands

Figure 18.21B Marsupials: a gray kangaroo with her young in her pouch

Figure 18.21A Monotremes: a duck-billed platypus with newly hatched young

Figure 18.21C Eutherians: a zebra with newborn

18.22   An animal phylogenetic tree is a work in progress

We have looked at examples of an immense diversity of animals in this chapter—representatives of about one-third of the known animal phyla. The diversity of the animal kingdom is spectacular, but it can also be overwhelming to a student. One way to help organize all this diversity is to arrange these groups on a phylogenetic tree, as you saw in Figure 18.4. Such trees summarize the proposed evolutionary threads that tie all this diversity together. Biologists continue to sort out the evolutionary history of the animals using evidence from the fossil record, morphology, embryology, and molecular data. Thus, phylogenetic trees are hypotheses that will continue to be tested and refined.

Figure 18.4 presented a traditional, morphology-based set of hypotheses about the relationships between the nine phyla we surveyed in this chapter. Figure 18.22 presents a slightly revised tree based on molecular comparisons. Let's see how the hypotheses based on certain sets of molecular data agree with and differ from the morphology-based tree.

At the bottom of both trees is the ancestral colonial protist that was the probable ancestor of animals. Both trees agree on the early divergence of phylum Porifera (the sponges), with their lack of true tissues and body symmetry.

The main trunk of the animal kingdom represents the clade of eumetazoans, animals with true tissues. The eumetazoans split into two distinct lineages that differ in body symmetry and the number of cell layers formed in gastrulation. The hydras, jellies, sea anemones, and corals of phylum Cnidaria are radially symmetrical and have two cell layers. The other lineage consists of bilateral animals, the bilaterians. Both trees agree on these early branchings.

The traditional phylogenetic tree in Figure 18.4 divides the bilaterians into deuterostomes and protostomes based on patterns of embryological development. The molecular-based tree also recognizes the deuterostomes, which include the echinoderms and chordates, as a clade.

How do the trees differ? As you can see in the branches highlighted in light blue, the molecular data distinguish two clades within the protostomes: lophotrochozoans and ecdysozoans. The lophotrochozoans, while grouped based on molecular similarities, are named for the feeding apparatus (called a lophophore) of some phyla in the group (which we did not discuss) and for the trochophore larva found in molluscs and annelids. This group includes the flatworms, molluscs, annelids, and many other phyla that we did not survey. The ecdysozoans include the nematodes and arthropods. Both of these phyla have exoskeletons that must be shed for the animal to grow. This molting process is called ecdysis, the basis for the name ecdysozoan.

Both of these trees represent hypotheses for the key events in the evolutionary history that led to the animal phyla now living on Earth. Like other phylogenetic trees, this molecular one serves to stimulate research and discussion, and it is subject to revision as new information is acquired. Even as zoologists continue to revise some branch points, however, both trees' overall message remains the same: The animal kingdom's great diversity arose through the process of evolution, and all animals exhibit features reminiscent of their evolutionary history.

In the last module in this chapter on animal diversity, we look at the unintended and often negative consequences when humans relocate some of that diversity.

Web/CD Activity 18C *Animal Phylogenetic Tree*

Web/CD Thinking as a Scientist *How Do Molecular Data Fit Traditional Phylogenies?*

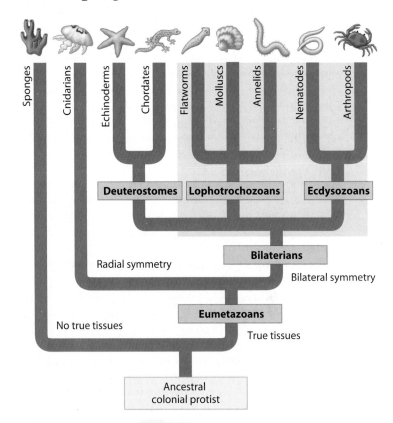

Figure 18.22   A molecular-based phylogenetic tree

?   Compare the placement of annelids, arthropods, and molluscs in Figures 18.4 and 18.22. How do they differ?

■ In the morphology-based tree, annelids and arthropods are shown to be more closely related to each other than to molluscs, probably based on their segmented bodies. In the molecular-based tree, arthropods are separated from both annelids and molluscs and are placed in the ecdysozoan clade.

18.23 Humans threaten animal diversity by introducing non-native species

The animal kingdom's amazing diversity arose through hundreds of millions of years of evolution. But this diversity can be quickly threatened by actions that throw slowly developed ecosystems into disarray—as when a small catlike creature meets an unstoppable toad from overseas.

That catlike creature is the quoll (Figure 18.23A), one of the Australian marsupials discussed in the chapter introduction. Like the majority of the more than 1 million species of plants and animals found in Australia, quolls are endemic to that geographically isolated continent; that is, they are found only there. And when such local species encounter invasive non-native species, the results are often disastrous.

Quolls are predators, hunting smaller animals, including Australia's many types of frogs. Their diet didn't present a problem until 1935, when sugarcane growers in northern Australia decided to import a boxful of foreign toads to fight beetles that were damaging sugar crops. The non-native amphibians from South America, known as cane toads (shown in Figure 18.23B), didn't do much to stop the beetles. Instead, the 102 toads in that box quickly turned into one of Australia's biggest wildlife disasters.

Australia had no native toad species of its own. The invaders bred and spread quickly, with female cane toads able to lay as many as 20,000 eggs in a single season. Cane toads now inhabit vast stretches of northern and eastern Australia.

A number of the cane toad's behaviors and characteristics spell doom for native inhabitants. Cane toad tadpoles develop more quickly than those of native Australian frogs, beating them in the search for food. Adult cane toads have voracious appetites, eating everything from dog food to mice and devouring many native Australian insects and small animals along the way. They can grow to weigh more than 4 pounds! And, more importantly, they are poisonous to almost all their predators, including quolls. Quolls that eat cane toads die. Together with habitat loss and other environmental pressures, the introduction of cane toads to Australia has driven native quolls into serious decline.

This scenario of invasive non-native versus native species has been repeated over and over in Australia, as a rush of human-introduced species in the last 200 years has threatened to turn the landscape upside down. In the mid-1800s, a would-be hunter brought over a few dozen rabbits, creating a population explosion of more than 200 million rabbits that have devoured native vegetation and turned grasslands into dusty deserts that offer little food for native animals (see Figure 37.6A). Non-native foxes, also brought in by sport hunters, have devoured populations of bilbies (Figure 18.23C), wallabies (Figure 18.23D), and other small marsupials not adapted to confront such predators.

Australia now spends millions of dollars each year combating non-native plants and animals, trying everything from mass hunting programs to biological control efforts. One effort involves moving species to safer territory. In 2003, 65 quolls were relocated from toad-infested lands in northern Australia to uninhabited offshore islands beyond the cane toad's reach. The transplanted quolls have thrived, and biologists hope that the island shelters will protect at least a few members of the species while the cane toad problem receives further study. Australia's ongoing efforts to undo the damage from invasive species have become a continent-wide example of how precarious diversity can be: Millions of years worth of evolutionary changes can be threatened by just a few years of human-caused changes to native habitats.

We consider other human-related threats to the environment in Chapter 38. But in the next chapter we look at the evolutionary history of the human species.

Figure 18.23A A quoll, an endangered cane toad predator

Figure 18.23B A cane toad, a poisonous non-native species

Figure 18.23C Bilbies, preyed on by non-native foxes

? Why are native species often threatened by non-native species?

■ Non-native species may compete with or prey on (or poison, in the case of cane toads) native species that did not evolve with the newcomers and thus have not been able to adapt to their presence.

Figure 18.23D A wallaby, preyed on by non-native foxes

CHAPTER REVIEW

Reviewing the Concepts

Animal Evolution and Diversity (18.1–18.4)

Animals ingest their food. Animal development may include a blastula, gastrula, and larval stage **(18.1)**.

Ancestor. Animals evolved from a colonial protist. Animal diversity exploded during the Cambrian period **(18.2)**.

Animal body plans may vary in symmetry (radial or bilateral), body cavity (none, pseudocoelom, or true coelom), and embryonic development (protostomes or deuterostomes). These differences may be used to infer phylogenetic relationships **(18.3–18.4)**.

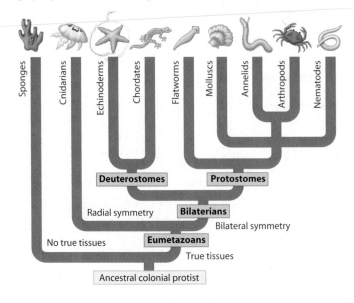

Invertebrates (18.5–18.14)

Sponges (phylum Porifera) are the simplest animals and have no true tissues. Their flagellated choanocytes filter food from water passing through the porous sponge body **(18.5)**.

Cnidarians (phylum Cnidaria) have true tissues and radial symmetry. Their two body forms are polyps (such as hydra) and medusae (jellies). They have cnidocytes on tentacles that sting prey and a gastrovascular cavity **(18.6)**.

Flatworms (phylum Platyhelminthes) are bilateral animals with no body cavity. A planarian has a gastrovascular cavity and a simple nervous system. Flukes and tapeworms are parasitic flatworms with complex life cycles **(18.7)**.

Nematodes (phylum Nematoda) have a pseudocoelom and a complete digestive tract and are covered by a protective cuticle. Many nematodes (roundworms) are free-living decomposers; others are plant or animal parasites **(18.8)**.

Molluscs (phylum Mollusca) include gastropods (such as snails and slugs), bivalves (such as clams), and cephalopods (such as squids). All have a muscular foot and a mantle, which may secrete a shell and which encloses the visceral mass. Many molluscs feed with a rasping radula **(18.9)**.

Annelids (phylum Annelida) are segmented worms and include earthworms, polychaetes, and leeches **(18.10)**.

Arthropods (phylum Arthropoda) are segmented animals with exoskeletons and jointed appendages. The four lineages are chelicerates (arachnids such as spiders), the aquatic crustaceans (lobsters and crabs), the lineage of millipedes and centipedes, and the terrestrial insects **(18.11)**.

Insects have a three-part body (head, thorax, and abdomen) and three pairs of legs; most have wings. These most successful arthropods are grouped in about 26 orders. Their development often includes metamorphosis **(18.12)**.

Echinoderms (phylum Echinodermata), such as sea stars, have spiny skins, an endoskeleton, and a water vascular system with tube feet that aids in respiration and locomotion. They are radially symmetrical as adults **(18.13)**.

Chordates (phylum Chordata) have a dorsal, hollow nerve cord, a stiff notochord, pharyngeal slits, and a muscular post-anal tail. The simplest chordates are lancelets and tunicates, marine invertebrates that use their pharyngeal slits for suspension feeding **(18.14)**.

Vertebrates (18.15–18.21)

A chordate phylogenetic tree is based on a sequence of derived characters. Most chordates are vertebrates, with a head and a backbone made of vertebrae **(18.15)**.

 Lampreys lack hinged jaws and paired fins. Most vertebrates have hinged jaws, which may have evolved from skeletal supports of the gill slits **(18.16)**.

 Jawed "fishes" include chondrichthyans (sharks and rays with skeletons made of cartilage), ray-finned fishes (with operculi that move water over the gills and a buoyant swim bladder), and lobe-fins (with muscular fins supported by bones) **(18.17)**.

 Amphibians, represented by frogs, salamanders, and caecilians, were the first tetrapods, with limbs allowing movement on land. Most amphibian embryos and larvae still must develop in water **(18.18)**.

 Reptiles are amniotes. The shelled amniotic egg and waterproof scales of reptiles (snakes, lizards, turtles, crocodilians, extinct dinosaurs, and birds) are terrestrial adaptations. Living reptiles other than birds are ectothermic **(18.19)**.

 Birds are reptiles that have wings, feathers, an endothermic metabolism, and many other adaptations related to flight **(18.20)**.

 Mammals are endothermic amniotes with hair, which insulates their bodies, and mammary glands, which produce milk. Monotremes lay eggs. The embryos of marsupials and eutherians are nurtured by the placenta within the uterus. Marsupial offspring complete development attached to the mother's nipples, usually in a pouch. Eutherians (placental mammals) complete development before birth **(18.21)**.

Animal Phylogeny and Diversity Revisited (18.22–18.23)

Molecular-based phylogenetic trees distinguish two protostome clades: the lophotrochozoans and the ecdysozoans **(18.22)**.

Introduced species threaten native animals **(18.23)**.

Connecting the Concepts

1. The table below lists the common names of the nine animal phyla surveyed in this chapter. For each phylum, identify the key characteristics and some representatives.

Phylum	Characteristics	Representatives
Sponges		
Cnidarians		
Flatworms		
Nematodes		
Molluscs		
Annelids		
Arthropods		
Echinoderms		
Chordates		

2. In this diagram of the major clades of chordates, fill in the key derived character that defines each clade.

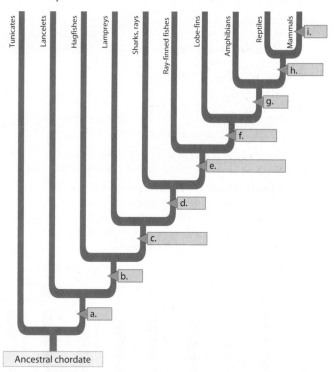

Testing Your Knowledge

Multiple Choice

3. Bilateral symmetry in animals is best correlated with
   a. an ability to see equally in all directions.
   b. the presence of a skeleton.
   c. motility and active predation and escape.
   d. development of a true coelom.
   e. adaptation to terrestrial environments.

4. Jon found an organism in a pond, and he thinks it's a freshwater sponge. His friend Liz thinks it looks more like an aquatic fungus. How can they decide whether it is an animal or a fungus?
   a. See if it can swim.
   b. Figure out whether it is autotrophic or heterotrophic.
   c. See if it is a eukaryote or a prokaryote.
   d. Look for cell walls under a microscope.
   e. Determine whether it is unicellular or multicellular.

5. Reptiles are much more extensively adapted to life on land than amphibians in that reptiles
   a. have a complete digestive tract.
   b. lay shelled eggs.
   c. are endothermic.
   d. have legs.
   e. do not go through a larval stage.

6. In Australia, marsupials fill the niches that placental mammals fill in other parts of the world because
   a. they are better adapted and have outcompeted placental mammals (eutherians).
   b. they originated in Australia.
   c. they evolved from monotremes that migrated to Australia about 50 million years ago.
   d. human-caused environmental changes have favored the success of marsupials.
   e. after Pangaea broke up, they diversified in physical isolation from placental mammals.

7. A lamprey, a shark, a lizard, and a rabbit share all the following characteristics except
   a. pharyngeal slits in the embryo or adult.
   b. vertebrae.
   c. hinged jaws.
   d. a dorsal, hollow nerve cord.
   e. a post-anal tail.

8. Which of the following groupings includes the largest number of species? (*Explain your answer.*)
   a. invertebrates
   b. chordates
   c. arthropods
   d. insects
   e. vertebrates

9. In which of the following pairs do both animals undergo metamorphosis that includes a larval stage that looks quite different from the adult?
   a. human, hydra
   b. grasshopper, frog
   c. sea star, tunicate
   d. earthworm, lancelet
   e. human, sea anemone

10. Which of the following animal groups does not have tissues derived from mesoderm?
    a. annelids   d. cnidarians
    b. amphibians   e. flatworms
    c. echinoderms

11. Molecular comparisons place nematodes and arthropods in clade Ecdysozoa. What characteristic do they share that is the basis for the name Ecdysozoa?
    a. a complete digestive tract   d. bilateral symmetry
    b. body segmentation   e. a true coelom
    c. molting of an exoskeleton

Matching

12. Include the vertebrates
13. Medusa and polyp body forms
14. The simplest bilateral animals
15. The most primitive animal group
16. Earthworms, polychaetes, and leeches
17. Largest phylum of all
18. Closest relatives of chordates
19. Body cavity is a pseudocoelom
20. Have a muscular foot and a mantle

a. annelids
b. nematodes
c. sponges
d. arthropods
e. flatworms
f. cnidarians
g. molluscs
h. echinoderms
i. chordates

Describing, Comparing, and Explaining

21. Compare the structure of a planarian (a flatworm) and an earthworm with regard to the following: digestive tract, body cavity, and segmentation.

22. Name two phyla of animals that are radially symmetrical and two that are bilaterally symmetrical. How do the overall lifestyles of radial and bilateral animals differ?

23. One of the key characteristics of arthropods is their jointed appendages. Describe four functions of these appendages in four different arthropods.

24. Birds and mammals are both endothermic, and both have four-chambered hearts. Most reptiles are ectothermic and have three-chambered hearts. Why don't biologists group birds with mammals? Why do most biologists now consider birds to be reptiles?

Applying the Concepts

25. A marine biologist has dredged up an unknown animal from the seafloor. Describe some of the characteristics that could be used to determine the animal phylum to which the creature should be assigned.

26. Coral reefs harbor a greater diversity of animals than any other environment in the sea. Australia's Great Barrier Reef has been protected as a marine reserve and is a mecca for scientists and nature enthusiasts. Elsewhere, such as Indonesia and the Philippines, coral reefs are in danger. Many reefs have been depleted of fish, and runoff from the shore has covered coral heads with sediment. Nearly all the changes in the reefs can be traced back to human activities. What kinds of activities do you think might be contributing to the decline of the reefs? What are some reasons to be concerned about this decline? Do you think the situation is likely to improve or worsen in the future? Why? What might the local people do to halt the decline? Should the developed countries help? Why or why not?

Answers to all questions can be found in Appendix 3.

For study help and Activities, go to campbellbiology.com or the student CD-ROM.

CHAPTER 19

Neanderthal skull

PRIMATE DIVERSITY

19.1 The human story begins with our primate heritage
19.2 Hominoids include humans and four other groups of apes

HOMINID EVOLUTION

19.3 The human branch of the primate tree is only a few million years old
19.4 Upright posture evolved well before an enlarged brain in hominids
19.5 Larger brains and reduced sexual dimorphism mark the evolution of *Homo*
19.6 When and where did *Homo sapiens* arise?
19.7 Human skin colors reflect adaptations to varying amounts of sunlight
19.8 A genetic difference helped humans start speaking

OUR CULTURAL HISTORY AND ITS CONSEQUENCES

19.9   Culture gives humans enormous power to change the environment
19.10 Scavenging, gathering, and hunting were the earliest human endeavors
19.11 Agriculture was a major development in human history
19.12 Development of complex tools affects human culture and the world

How Are We Related to Neanderthals?

NEANDERTHALS (see skull above) were an evolutionarily recent human relative who lived in Europe and became extinct about 30,000–40,000 years ago. But do Neanderthals survive, in some small measure, in our DNA? Are modern-day humans a distinct species from Neanderthals, or did our ancestors interbreed with this relative that died out long ago?

Scientists, examining fossils that are tens of thousands of years old, have long puzzled over certain similarities between Neanderthals and humans. Neanderthals were muscular and robust, with a brain similar to ours in size but slightly different in shape. They had large noses, as well as heavy brows and cheekbones, and made hunting tools from stone and wood. Analysis of skeletons has revealed that Neanderthals had hands as nimble as those of living humans, with a thumb and fingers that could touch to form a precise grip. Neanderthals lived

Human Evolution

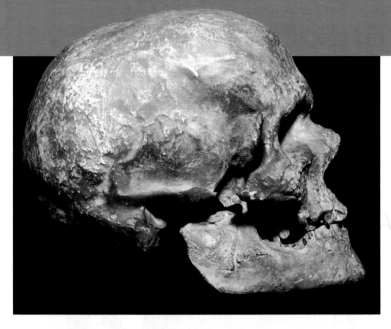

Cro-Magnon skull

alongside Cro-Magnons, a somewhat different-looking group (see skull above) who were the direct ancestors of present-day Europeans. Neanderthals and Cro-Magnons likely competed for territory, food, and other resources. Were the two groups biologically separate, or did they interbreed?

Researchers studying human origins debated this question for decades until recent genetic analyses indicated little, if any, interbreeding between the two groups. In 1997, researchers isolated some DNA from Neanderthal bones found in the Feldhofer cave in the Neander Valley in Germany, where the first Neanderthal remains were discovered about 150 years ago. Analysis showed that the DNA is different from that of all living humans. A greater similarity to living Europeans would be expected if Neanderthals had indeed contributed genes to Europeans. Instead, it seems that there was little or no genetic contribution by Neanderthals to present-day humans.

Neanderthals had hands as nimble as those of living humans

In 2001, analysis of DNA from the breastbone of a Neanderthal baby found in the northern Caucasus Mountains in Russia produced similar results. The baby's DNA was compared with that of the previously studied Feldhofer DNA, with DNA of living humans, and with chimpanzee DNA. The researchers found that the DNA from the child is most similar to that of the Feldhofer sample and is distinct from that of living humans. Scientists have also compared DNA from four Neanderthals with that of living humans from Europe, Africa, and Asia. If Neanderthals had given rise to Europeans, then both groups should share a common ancestor, with other living humans being more distantly related. Instead, all of the Neanderthals studied form their own distinct genetic group, while the Europeans are more closely related to the Africans and Asians.

This genetic evidence supports previously known fossil evidence that Neanderthals arose from a distinct species that arrived in Europe long before the ancestors of modern-day humans and that Neanderthals became extinct without contributing to the gene pool of humans. The relationship between humans and Neanderthals may best be described as sister species: They last shared a common ancestor long before Neanderthals spread to Europe and before our own species, *Homo sapiens* (Latin for "wise man"), emerged and spread around the world.

This type of exploration and inquiry underlies much of the study into our enigmatic origins. Relatively few skeletal remains have been discovered, and most of these are incomplete, so it is difficult to reconstruct our complete genealogy. In this chapter, we examine this compelling and sometimes controversial subject—the evolutionary history of our species. We begin with a look at our primate heritage. ■ ■ ■

Human Evolution

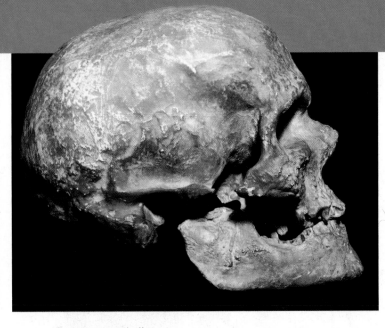

Cro-Magnon skull

alongside Cro-Magnons, a somewhat different-looking group (see skull above) who were the direct ancestors of present-day Europeans. Neanderthals and Cro-Magnons likely competed for territory, food, and other resources. Were the two groups biologically separate, or did they interbreed?

Researchers studying human origins debated this question for decades until recent genetic analyses indicated little, if any, interbreeding between the two groups. In 1997, researchers isolated some DNA from Neanderthal bones found in the Feldhofer cave in the Neander Valley in Germany, where the first Neanderthal remains were discovered about 150 years ago. Analysis showed that the DNA is different from that of all living humans. A greater similarity to living Europeans would be expected if Neanderthals had indeed contributed genes to Europeans. Instead, it seems that there was little or no genetic contribution by Neanderthals to present-day humans.

Neanderthals had hands as nimble as those of living humans

In 2001, analysis of DNA from the breastbone of a Neanderthal baby found in the northern Caucasus Mountains in Russia produced similar results. The baby's DNA was compared with that of the previously studied Feldhofer DNA, with DNA of living humans, and with chimpanzee DNA. The researchers found that the DNA from the child is most similar to that of the Feldhofer sample and is distinct from that of living humans. Scientists have also compared DNA from four Neanderthals with that of living humans from Europe, Africa, and Asia. If Neanderthals had given rise to Europeans, then both groups should share a common ancestor, with other living humans being more distantly related. Instead, all of the Neanderthals studied form their own distinct genetic group, while the Europeans are more closely related to the Africans and Asians.

This genetic evidence supports previously known fossil evidence that Neanderthals arose from a distinct species that arrived in Europe long before the ancestors of modern-day humans and that Neanderthals became extinct without contributing to the gene pool of humans. The relationship between humans and Neanderthals may best be described as sister species: They last shared a common ancestor long before Neanderthals spread to Europe and before our own species, *Homo sapiens* (Latin for "wise man"), emerged and spread around the world.

This type of exploration and inquiry underlies much of the study into our enigmatic origins. Relatively few skeletal remains have been discovered, and most of these are incomplete, so it is difficult to reconstruct our complete genealogy. In this chapter, we examine this compelling and sometimes controversial subject—the evolutionary history of our species. We begin with a look at our primate heritage. ■ ■ ■

19.1 The human story begins with our primate heritage

The mammalian order Primates includes the lemurs, tarsiers, monkeys, and apes. Humans are members of the ape group. The earliest primates were probably small arboreal (tree-dwelling) mammals that arose before 65 million years ago, when dinosaurs still dominated the planet. Most living primates are still arboreal, and the primate body has a number of features that were shaped, through natural selection, by the demands of living in trees. Although humans never lived in trees, the human body retains many of the traits that evolved in our arboreal ancestors.

The squirrel-sized slender loris in **Figure 19.1A** illustrates a number of primate features. It has limber shoulder and hip joints, enabling it to climb and brachiate (swing from one branch to another). The five digits of its grasping feet and hands are highly mobile; the separation of its big toe from the other toes and its flexible thumb give the lemur the ability to hang onto branches and manipulate food. The great sensitivity of the hands and feet to touch also aids in manipulation. Lorises have a short snout and eyes set close together on the front of the face. The position of the eyes makes their two fields of vision overlap, enhancing depth perception, an important trait for maneuvering in trees. We humans share all these basic primate traits with the slender loris except for the widely spaced big toe.

As shown in the phylogenetic tree in **Figure 19.1B**, the slender loris belongs to one of three main groups of living primates. The lorises and pottos of tropical Africa and southern Asia are placed in one group along with the lemurs. Ranging from the pygmy mouse lemur, which weighs 25 g (1 oz) to the sifakas (**Figure 19.1C**, facing page), which may weigh as much as 8 kg (17.6 lbs), lemurs are a diverse group. They are found only in Madagascar, a Texas-sized island in the Indian Ocean about 420 km off the eastern coast of Africa. Severed from the African continent by plate movements, the island has been an isolated hotbed of speciation for well over 100 million years. But of about 50 species of lemurs originally present, 18 have become extinct since humans first colonized Madagascar about 2,000 years ago. Most lemurs are agile climbers and leapers that spend nearly all their time in trees. Thus, they are threatened by the destruction of their tropical forest homes.

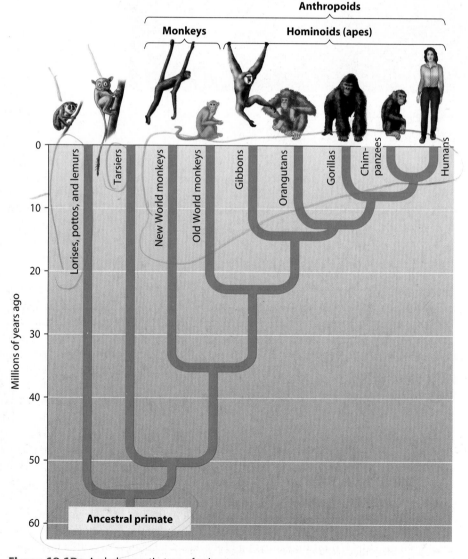

Figure 19.1A A slender loris

Figure 19.1B A phylogenetic tree of primates

The tarsiers form a second group of primates. Limited to Southeast Asia, these small, nocturnal tree-dwellers have flat faces with large eyes (Figure 19.1D). Fossil evidence indicates that tarsiers are more closely related to anthropoids, the third group of primates, than to the loris–potto group.

The **anthropoids** (from the Greek *anthropos,* man, and *eidos,* form) include monkeys and apes. The ape group, the **hominoids,** includes humans. Anthropoids generally have a larger brain relative to body size and rely more on eyesight and less on sense of smell than other mammals. Anthropoids have a fully **opposable thumb;** that is, they can touch the tip of all four fingers with their thumb. In monkeys and most apes, the opposable thumb functions in a grasping "power grip," but in humans, a distinctive bone structure at the base of the thumb allows it to be used for more precise manipulation.

The fossil record indicates that anthropoids began diverging from other primates about 50 million years ago. Notice that monkeys do not constitute a monophyletic group. The first monkeys probably evolved in the Old World (Africa and Asia) and may have reached South America by rafting on logs from Africa. The monkeys of the Old World and New World (the Americas) have been evolving separately for over 30 million years.

New World monkeys, found in Central and South America, are all arboreal. Their nostrils are wide open and far apart, and many, such as the woolly spider monkey, an inhabitant of rain forests in eastern Brazil (Figure 19.1E, left), have a long tail that is prehensile—specialized for grasping tree limbs. The squirrel-sized golden lion tamarin (Figure 19.1E, right) is a New World monkey that inhabits lowland rain forests of eastern Brazil. Most of its habitat having been replaced by housing developments, this species was reduced to fewer than 200 individuals in the wild in the 1970s. With an intense international conservation effort to save the species from extinction, its numbers have rebounded to about 1,200.

In contrast, Old World monkeys lack a prehensile tail, and their nostrils open downward. They include macaques (Figure 19.1F), mandrills, baboons, and rhesus monkeys. Many species are arboreal, but some, such as the baboons of the African savanna, are ground-dwelling. Both New and Old World monkeys are active during the day and usually live in bands held together by social behavior.

Monkeys differ from most apes in having forelimbs that are about equal in length to their hind limbs. Old World monkeys and apes (hominoids) diverged about 20–25 million years ago. The human lineage probably diverged from an ancestor shared with chimpanzees somewhere between 5 and 7 million years ago. We look at the apes next.

? (a) To which mammalian order do we belong? (b) What are the three main groups of this order?

■ (a) Primates; (b) lorises, pottos, and lemurs; tarsiers; and anthropoids

Figure 19.1C A sifaka, a type of lemur, with its young

Figure 19.1D A tarsier, member of a distinct primate group

Figure 19.1E New World monkeys: The woolly spider monkey (left; note prehensile tail), and the golden lion tamarin (right; note nostrils that open to the side)

Figure 19.1F Old World monkey: a macaque with its young

Hominoids (apes) include the gibbons, orangutans, gorillas, chimpanzees (and bonobos), and humans. Apart from the human lineage, the apes have a smaller geographic range than the monkeys; they evolved and diversified only in Africa and Southeast Asia and are confined to tropical regions (mainly rain forests). Apes lack a tail and have relatively long arms and short legs. They are chiefly vegetarians, although chimpanzees also eat insects and some vertebrates, such as young antelopes, pigs, and monkeys. Humans also are omnivorous. Apes have larger brains relative to body size than other primates, and their behavior is consequently more flexible. Gorillas, chimpanzees, and humans have a high degree of social organization.

Nine species of gibbons, all found in Southeast Asia, are the only entirely arboreal apes (**Figure 19.2A**). Gibbons are the smallest, lightest, and most acrobatic of the apes. They are also the only nonhuman apes that are monogamous, with mated pairs remaining together for life.

The orangutan is a shy, solitary species that lives in the rain forests of Sumatra and Borneo. The largest living arboreal mammal, it moves rather slowly through the trees, supporting its stocky body with all four limbs (**Figure 19.2B**). Orangutans may occasionally venture onto the forest floor.

The gorilla (**Figure 19.2C**) is the largest ape: Some males are almost 2 m tall and weigh about 200 kg (440 lb). Found only in African rain forests, gorillas usually live in groups of up to about 20 individuals. They spend nearly all their time on the ground. Gorillas can stand upright on their hind legs. But when they walk on all fours, their knuckles contact the ground.

Like the gorilla, the chimpanzee and a closely related species called the bonobo are knuckle walkers. These apes spend as much as a quarter of their time on the ground. Both species inhabit tropical Africa. Chimpanzees have been studied extensively, and many aspects of their behavior resemble human behavior. For example, chimpanzees make and use simple tools. The individual in **Figure 19.2D** is using a blade of grass to "fish" for termites. Chimpanzees also raid other social groups of their own species, exhibiting behavior formerly thought to be uniquely human. Researchers have demonstrated repeatedly that chimpanzees can learn human sign language. However, we do not yet know what role symbolic communication plays in the behavior of wild chimpanzees.

One of our most entrenched beliefs is that humans are the only thinking, self-aware beings. The behavior of chimpanzees in front of mirrors, however, challenges this belief. When first introduced to a mirror, a chimpanzee responds the way most other animals do—as if it were seeing another individual of its species. After several days, though, a chimp will begin using a mirror in ways that indicate it has a concept of self. It will inspect its face and other parts of its body

Figure 19.2A A gibbon

Figure 19.2B An orangutan

Figure 19.2C A gorilla and offspring

Figure 19.2D A chimpanzee

that it cannot see without the mirror. It will also make faces at the mirror, using expressions different from those used in communicating with others.

Molecular evidence indicates that chimpanzees and gorillas are more closely related to humans than they are to other apes. Humans and chimpanzees are especially closely related; their genomes are 99% identical. Primate researchers are acutely aware of the special significance of the great apes to us. In the words of chimpanzee authority Jane Goodall, "The most important spin-off of the chimp research is probably the humbling effect it has on us who do the research. We are not, after all, the only aware, reasoning beings on this planet."

Web/CD Activity 19A *Primate Diversity*

? Which primate groups are classified as anthropoids, and which groups are classified as hominoids?

■ Anthropoids include the monkeys and hominoids. Hominoids include the gibbons, orangutans, gorillas, chimpanzees, and humans.

19.3 The human branch of the primate tree is only a few million years old

Humans and chimpanzees diverged from a common ancestor, probably between 5 and 7 million years ago (see Figure 19.1B). **Paleoanthropology,** the study of human origins and evolution, focuses on this tiny slice of biological history. If we compress the history of life to a year, humans and chimpanzees diverged less than 18 hours ago.

A number of derived characters distinguish humans from other hominoids. Most obviously, humans are *bipedal*—they stand upright and walk on two legs. Humans have *shorter jaws* and flatter faces than other hominoids. Humans have a much *larger brain* and are capable of *language, symbolic thought,* and the manufacture and use of *complex tools*. Other traits include long-term pair-bonding between mates and a much longer period of parental care than is seen in other hominoids.

Paleoanthropologists have unearthed fossils of approximately 20 species of extinct **hominids,** species that are more closely related to humans than to chimpanzees and are therefore on the human branch of the evolutionary tree. (Remember that the term *hominoid* refers to all apes, including humans, and that *anthropoid* is an even broader term, since it includes monkeys.)

Figure 19.3 is a time line for some hominid species. The vertical bars indicate the approximate time period when each

species existed, as judged from the fossil record. The oldest hominid yet discovered in the fossil record, *Sahelanthropus tchadensis,* lived from about 7 to 6 million years ago. As Figure 19.3 shows, the fossil record suggests that hominid diversity increased dramatically in the period between 4 and 2 million years ago. Many of the groups from that time are collectively called **australopiths,** although their phylogeny is unresolved and they almost certainly do not represent a monophyletic group. Australopiths got their name from the 1924 discovery in South Africa of *Australopithecus africanus* ("southern ape of Africa"), which lived from 3 to 2.4 million years ago. Some australopiths overlapped with species of the genus *Homo.*

A common misconception is to think of human evolution as a parade of hominids leading directly from an ancestral hominoid to Homo sapiens. But if human evolution is a parade, it is a very disorderly one, with many groups breaking away from the march to wander down dead-end alleyways. At times several hominid species coexisted, but all except one—our own—ended in extinction.

? Based on the fossil evidence represented in Figure 19.3, how many hominid species existed 1.7 million years ago?

■ *Five: P. boisei, P. robustus, H. habilis, H. ergaster, H. erectus*

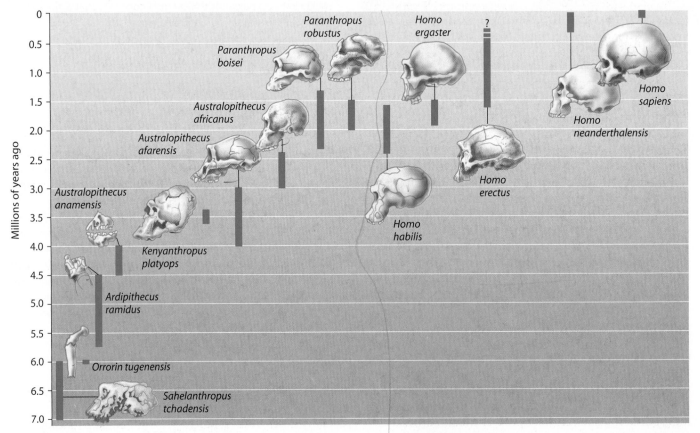

Figure 19.3 A time line for some hominid species

19.4 Upright posture evolved well before an enlarged brain in hominids

Bipedalism is a very old hominid trait. The 4-million-year-old *Australopithecus anamensis* had leg bones that suggest that early hominids were increasingly bipedal. And some 3.5 million years ago, several upright-walking hominids left footprints in damp volcanic ash in what is now Tanzania in East Africa (**Figure 19.4**, left). The prints fossilized and were discovered in 1978 by British anthropologist Mary Leakey.

 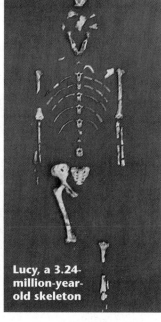

Ancient footprints

Lucy, a 3.24-million-year-old skeleton

Figure 19.4 Evidence of bipedalism in early hominids

Another clue to upright stance is the location of the opening in the base of the skull through which the spinal cord exits (see Module 30.3). In early hominids (and in humans), this opening is located underneath the skull, allowing the head to be held directly over the body.

In 1974, in the Afar region of Ethiopia, paleoanthropologists discovered a 3.24-million-year-old skeleton that was 40% complete (Figure 19.4, right). "Lucy," as the fossil was named, was petite—only about 1 m tall, with a brain the size of a softball. Lucy and similar fossils were assigned to the species *Australopithecus afarensis* (after the Afar region). Fossils show that *A. afarensis* walked on two legs and existed as a species for at least 1 million years.

Australopiths seem to have had various styles of locomotion, and some spent more time on the ground than others. Paleoanthropologists have long seen a link between the rise of bipedal hominids and the increase in savanna (grassland) habitat as a result of climate change. Although all recently discovered fossils of early hominids show signs of bipedalism, however, none of them appear to have lived in savannas. Whatever the selective pressure that lead to bipedalism, this trait freed the forelimbs of our ancestors for other uses, such as the manufacture and use of tools.

These early hominids show that the fundamental human trait of bipedalism evolved millions of years before the other major human trait—an enlarged brain—became evident.

? How can paleoanthropologists conclude that a species was bipedal based on only a fossil skull?

■ By the location of the opening where the spinal cord exits the skull

19.5 Larger brains and reduced sexual dimorphism mark the evolution of *Homo*

Enlargement of the hominid brain is first evident in fossils from East Africa dating to about 2.4 million years ago. These larger skulls also had shorter jaws. Sharp stone tools are sometimes found with these fossils, which have been dubbed *Homo habilis* ("handy man") (see Figure 19.3).

Fossils dating to 1.9 to 1.6 million years ago, considered by many paleoanthropologists to be those of a distinct species, *Homo ergaster*, mark a new stage in hominid evolution. *H. ergaster* had a substantially larger brain than *H. habilis*, as well as long, slender legs with hip joints well adapted for long-distance walking. *H. ergaster* has been associated with more sophisticated stone tools. Its smaller teeth also suggest that *H. ergaster* either ate different foods (more meat and less plant matter) or prepared some of its food before chewing, perhaps by cooking or mashing it.

H. ergaster marks an important shift in the relative sizes of the sexes. In primates, size difference is a major feature of sexual dimorphism (see Module 13.17). Male gorillas weigh about twice as much as females of their species. In *Australo-*

pithecus afarensis, males were 1.5 times as heavy as females. But in early *Homo*, sexual dimorphism was significantly reduced, and this trend continues in our own species: Human males average about 1.2 times the weight of females.

The reduced sexual dimorphism may offer some clues to the social systems of extinct hominids. In many mammal groups, more pronounced sexual dimorphism is associated with intense male-male competition for multiple females. In species that exhibit more pair-bonding, sexual dimorphism is less dramatic. Thus, *H. ergaster* males and females may have engaged in more pair-bonding than earlier hominids did. This shift may have been associated with long-term parental care of babies by both parents. Human babies depend on their parents for food and protection much longer than do the young of chimpanzees and other hominoids.

Fossils of *H. ergaster* were originally considered early members of another species, *Homo erectus*, and they are certainly closely related. *H. erectus* ("upright man") was the first hominid to migrate out of Africa. The oldest known fossils

of hominids outside Africa, discovered in 2000 in the former Soviet Republic of Georgia, date back 1.8 million years. *H. erectus* eventually migrated as far as Indonesia. What was the fate of these larger-brained hominids who traveled across Asia? Fossil evidence indicates that *H. erectus* became extinct at some point about 200,000 years ago.

19.6 When and where did *Homo sapiens* arise?

Evidence from fossils and DNA studies is coming together to support a compelling hypothesis about how our own species, *Homo sapiens*, emerged and spread around the world.

It is now clear that the ancestors of humans originated in Africa. Older species (perhaps *H. ergaster* or *H. erectus*) gave rise to newer species such as *H. heidelbergensis* (which originated in Africa about 600,000 years ago) and ultimately *H. sapiens*. The oldest known fossils of our own species have been discovered in Ethiopia, and include fossils that are 160,000 and 195,000 years old (Figure 19.6A). These

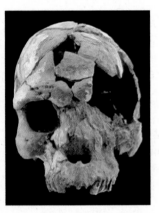

Figure 19.6A One of the oldest known *H. sapiens* fossils

early humans lacked the heavy brow ridges of *H. erectus* and Neanderthals (*H. neanderthalensis*) and were more slender. As discussed in the chapter introduction, evidence indicates that Neanderthals had appeared in Europe by 200,000 years ago. They were likely descendants of *H. heidelbergensis*.

The Ethiopian fossils support molecular evidence about the origin of humans. In addition to showing that living humans are more closely related to one another than to Neanderthals, DNA studies indicate that Europeans and Asians share a more recent common ancestor and that many African lineages represent earlier branches on the human tree. These findings strongly suggest that all living humans have ancestors that originated as *Homo sapiens* in Africa.

This conclusion is further supported by analyses of mitochondrial DNA, which is maternally inherited, and Y chromosomes, which are transmitted from fathers to sons. Such studies suggest that all living humans inherited their mitochondrial DNA from a common ancestral woman who lived 150,000–200,000 years ago. Mutations on the Y chromosomes can serve as markers for tracing the ancestry and relationships among males alive today. By comparing the Y chromosomes of males from various geographic regions, researchers were able to infer divergence from a common African ancestor.

Evidence suggests that our species emerged from Africa in one or more waves, spreading first into Asia and then to Europe and Australia. The oldest fossils of *H. sapiens* outside Africa date back about 50,000 years. The date of the arrival of humans in the New World is uncertain, although the oldest generally accepted evidence puts it at 15,000 years ago.

As humans migrated into Asia and then Europe, they encountered Neanderthals. This hominid species may have been driven to extinction by the combined stresses of the last ice age and competition from humans.

New findings continually update our understanding of the context of *H. sapiens* evolution. In 2004, researchers reported the discovery of an adult hominid skeleton dating from just 18,000 years ago. Found on the Indonesian island of Flores, this fossil represents a previously unknown hominid species, now named *Homo floresiensis*. The individual was much shorter and had a much smaller brain volume than *H. sapiens*. A question to be answered is whether this species ever encountered *H. sapiens*, which coexisted in Indonesia during that time period.

The rapid expansion of *H. sapiens* may have been spurred by the evolution of human cognition as our species evolved in Africa. Although Neanderthals and other hominids were able to produce sophisticated tools, they showed little creativity and not much capability for symbolic thought, as far as we can tell. In contrast, researchers are beginning to find evidence of more sophisticated thought as *H. sapiens* evolved. In 2002, researchers reported the discovery in South Africa of 77,000-year-old art—geometric markings on pieces of ochre (Figure 19.6B). By 36,000 years ago, humans were producing spectacular cave paintings.

As humans migrated across the globe, they encountered environments that shaped the evolution of various traits. One of these traits is skin color, which we explore next.

Web/CD Activity 19B *Human Evolution*

Figure 19.6B Engravings on a 77,000-year-old piece of ochre, among the earliest signs of symbolic thought in humans

19.7 Human skin colors reflect adaptations to varying amounts of sunlight

As early species of hominids began walking long distances, a reduction in body hair helped keep these more active individuals cool. But their bare skin was now exposed to the damaging ultraviolet (UV) rays of sunlight.

Dark pigmentation shields skin from cancer-causing UV radiation. But because skin cancers usually arise after reproductive age, this protection was not likely a factor in skin pigment evolution. An alternative hypothesis is that dark skin may have been selected for because it protects against the UV-induced loss of folate (folic acid). Folate is essential for sperm production and for proper development of a fetus.

So why aren't all humans dark-skinned? Although most effects of sunlight are harmful, UV radiation catalyzes the synthesis of vitamin D in the skin. This vitamin facilitates the absorption of calcium from food, and calcium is required for bone development. Dark-skinned humans evolving in equatorial Africa received sufficient UV radiation to make vitamin D. Northern latitudes receive less sunlight. The loss of skin pigmentation in humans that migrated north from Africa may have allowed their skin to receive sufficient UV radiation to produce vitamin D.

The evolution of differing skin pigmentations likely provided a balance between folate protection and vitamin D production. In more recent human migrations, many people may now live in areas where they receive either more or less UV radiation than is ideal for their skin pigmentation.

? How could natural selection have driven the evolution of dark skin pigmentation in hominids?

■ Individuals with dark skin pigmentation were protected from folate-destroying UV radiation. Folate is essential for successful production of offspring.

19.8 A genetic difference helped humans start speaking

Another adaptation that facilitated the spread of humans across the globe involved a mutation in a gene linked to one of our most human characteristics—speech (Figure 19.8).

Other primates vocalize sounds and can use them to communicate. Chimpanzees, for example, utter cries that warn others of impending danger. But many nonprimate animals also share this form of simple communication. Why are humans able not only to vocalize sounds but to conduct complex communication, learning to talk from elders and creating intricate societies that function through the use of shared language? Furthermore, humans use language in abstract ways—referring to objects that are not present, talking about the future, and generalizing.

In the early 1990s, researchers studying a family with a rare developmental disorder were able to link a specific gene to language ability. Many members of this family have severe difficulty in speaking. By analyzing the genes of this family, as well as the DNA of an unrelated boy with a similar disorder, the researchers zeroed in on a gene called *FOXP2*.

This gene has broad power to shape brain development. Rather than controlling just one function, *FOXP2* codes for a transcription factor (see Module 11.6) that controls the expression of many genes. In the family with the speech disorder, affected individuals carry a slightly different version of the gene. That small difference is powerful enough to interfere with development of brain areas linked to speech.

That finding prompted comparisons to other species. Other animals, including crocodiles, birds, mice, and monkeys, also have a *FOXP2* gene. But the human version is unique, perhaps explaining our unique linguistic abilities. The animals with a *FOXP2* gene most similar to humans are song-learning birds, such as canaries and parakeets. These birds also learn their vocal behavior from others of their species. Further research has determined that the human form of *FOXP2* likely arose within the last 100,000 years, a time period that matches the emergence of *Homo sapiens*.

FOXP2 is not the only gene that allows humans to speak; other genes shape other parts of the brain and anatomy critical for talking. But researchers think that *FOXP2* is a key to the unique linguistic abilities that enable the expression of human thought and make it possible for older members of the species to pass large amounts of knowledge on to younger members. Language is the foundation of culture, a concept we will examine in the remainder of this chapter.

Figure 19.8 Conversation—a uniquely human trait

? *FOXP2* shapes the development of which organ?

■ The brain

19.9 Culture gives humans enormous power to change the environment

Three major milestones highlight the evolutionary history leading to *Homo sapiens:* (1) evolution of an erect stance, which required major remodeling of the foot, pelvis, and vertebral column; (2) enlargement of the brain, which paralleled a lengthening of the period of growth of the skull and its contents after birth (see Module 14.12); and (3) evolution of a prolonged period of parental care.

Indeed, the challenge of human birth reflects both our upright posture, which limits the size of the pelvic opening through which the baby must pass, and the exceptionally large head of humans relative to body size. The brain continues to grow in size and complexity during childhood, a period of development unique to humans. Humans care for their offspring far longer than any other species, giving children the chance to learn from the experiences of earlier generations. This is the basis of **culture**—the accumulated knowledge, customs, beliefs, arts, and other human products that are socially transmitted over the generations. As de-

scribed in the preceding module, the major means of this transmission is language, written and spoken.

More than any other factor in human history, culture has made *Homo sapiens* a unique force in the history of life on Earth—a species that can defy its physical limitations and alter nature at a rate far exceeding that of biological evolution. Our culture enables us to change the environment to meet our needs. As we will see in Chapter 36, it has also allowed the human population to explode and, in effect, threaten the existence of many other species and challenge Earth's capacity to sustain us. In the next three modules, we examine three stages in the development of more sophisticated tools and culture, each of which gave humans increasing power to alter the environment.

? What role does a long period of parental care play in the development of culture?

■ It extends the opportunity for parents to transmit the lessons of the past to offspring.

19.10 Scavenging, gathering, and hunting were the earliest human endeavors

The earliest hominids in Africa "made their living" by scavenging, gathering, and hunting. These three central activities continued to be the way of life for the groups of hominids called australopiths and for the various species of *Homo*. These activities were the norm for *Homo sapiens* during most of the last 100,000 years. Actually, for most of our existence, humans have probably relied more on scavenging, especially stealing fresh kills from other predators, and gathering wild fruits, seeds, and roots than on hunting. Only in the last 50,000 years or so did toolmaking become sophisticated enough to allow hunting to become a major food-producing activity. Gathering and hunting continue successfully to this day in various societies. **Figure 19.10** shows a !Kung tribesman, a skilled bow hunter. The !Kung, who live in southwestern Africa, are one of several hunter-gatherer peoples of modern times. (The exclamation point represents a clicking sound used in the !Kung people's language.)

With the development of tools that could be thrown and cooperative hunting techniques, humans began to profoundly affect certain other species. Fossil records indicate that humans decimated populations of woolly rhinoceros and giant deer in Europe. About 50,000 years ago, humans reached Australia by boat and may have killed off that continent's giant kangaroos. Nomadic hunters migrated from Asia to North America via the Bering land bridge perhaps 15,000 years ago. These early migrants may also have pushed to extinction some of North America's large mammals.

Hunter-gatherers not only made tools, but also organized communal activities and divided labor. Many groups also

developed semipermanent residences near rich hunting grounds and along seacoasts and lakes. Some grew crops and traded with other groups. These activities ushered in a new stage of human history—the rise of agriculture.

? Humans became better hunters as they developed more sophisticated _____.

■ tools and cooperative hunting techniques

Figure 19.10 A modern-day !Kung hunter with bow and arrows

19.11 Agriculture was a major development in human history

The second major stage of cultural evolution came with the development of agriculture in Africa, Eurasia, and the Americas about 10,000 to 15,000 years ago.

In the Middle East, major agricultural centers sprang up in an area called the Fertile Crescent. Extending northeast from the Nile River in Africa and including the modern nations of Israel, Syria, Iraq, and Kuwait, the Fertile Crescent was once covered with forests and grasses and had rich soils. Farmers cleared the forests and farmed the soil for years, growing mostly wheat, without seriously reducing the soil's fertility.

Figure 19.11
A primitive plow

About 5,000 years ago, farmers in the area began using primitive plows, like the one shown in **Figure 19.11**. The plow broke up the thick sod of grasslands and opened vast new areas for farming. With more intensive agriculture, local populations increased, placing an ever-increasing burden on the soil. Herds of domestic animals overgrazed the land, leaving it exposed to eroding winds and rains, while the crops eventually depleted the soil of nutrients. These patterns of overuse combined with changes in climate led to increasing desertification. Today, much of the Fertile Crescent is a desert.

Despite these eventual environmental problems, by adopting farming as a way of life, *Homo sapiens* took the first big step toward becoming the dominant species on Earth. Along with agriculture came permanent settlements and the first cities. Fewer people were needed to grow a group's food, and many people could specialize in other activities.

Agriculture thus paved the way for the development of industry and technology—the next major change in human culture, which continues today.

> **?** What is the connection between agriculture and cities?

■ Agriculture made food sources more localized so that larger settlements were possible, and the increasing efficiency of agriculture provided time for people to engage in activities other than food gathering.

19.12 Development of complex tools affects human culture and the world

The Industrial Revolution, which began in England in the 18th century, brought a switch from small-scale hand production of tools and goods to large-scale machine production (**Figure 19.12**, top). With the machine age came increasing demands for energy sources to fuel the machines—mostly timber and coal at first, then oil, and recently also hydroelectric power and nuclear power. The invention of the tractor and other farm machinery reduced the need for farm laborers, and many people migrated to cities in search of work. As more food was produced and medical advances reduced the number of deaths from disease, the human population began to grow faster throughout the world. The growth of the human population and of technology continue at a phenomenal pace.

Through the cultural changes from scavenging-gathering-hunting to high-tech societies, humans have not changed biologically in any significant way. We are probably no more intelligent than our forebears who lived in caves. Toolmakers once chipped away at stones; now they fashion microchips for computers (Figure 19.12, bottom). The know-how to build computers and spaceships is stored not in our genes but in what is passed along by parents, teachers, and books.

We are now the most numerous and widespread of all large animals, and our activities cause change everywhere. As we saw when we discussed biological evolution in Unit III, there is nothing new about environmental change. What is new is the *speed* of change, for technological change outpaces biological change by orders of magnitude. As we will

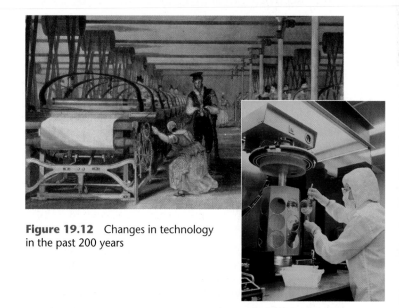

Figure 19.12   Changes in technology in the past 200 years

discuss in Chapter 38, we are changing the world so quickly that many species cannot adapt. In fact, we may be jeopardizing our own existence as well.

> **?** How has technology contributed to growth of the human population?

■ By making more food available and by reducing disease

CHAPTER REVIEW

Reviewing the Concepts

Primate Diversity (19.1–19.2)

Primates had evolved as small arboreal mammals by 65 million years ago. Primate characters include limber joints, grasping hands and feet with flexible digits, a short snout, and forward-pointing eyes that enhance depth perception. The three groups of living primates are the lorises, pottos, and lemurs; the tarsiers; and the anthropoids (monkeys and apes) **(19.1)**.

Hominoids include gibbons, orangutans, gorillas, chimpanzees (and bonobos), and humans. Hominoids have larger brains than other primates and lack tails. Humans share 99% of our genome with chimpanzees, our closest living relatives **(19.2)**.

Hominid Evolution (19.3–19.8)

Hominids, which include species on the human evolutionary branch, and chimpanzees probably diverged from a common ancestor between 5 and 7 million years ago. Major milestones in hominid evolution included the appearance of bipedalism, a larger brain, and longer parental care. Language, symbolic thought, the use of complex tools, and pair-bonding are also hominid traits. The oldest known hominid dates from 7 million years ago. Diverse species called australopiths lived from about 4 to 2 million years ago. *Homo sapiens* is the only hominid that has not become extinct **(19.3)**.

Bipedalism was evident in early hominids and preceded the evolution of the enlarged brain **(19.4)**.

The genus *Homo* includes hominids with larger brains and evidence of tool use. *Homo ergaster* had a larger brain than *H. habilis* and reduced sexual dimorphism, indicative of pair-bonding and perhaps biparental care. *H. erectus* was the first hominid to spread out of Africa **(19.5)**.

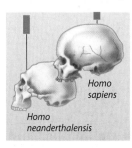

Homo sapiens

Homo neanderthalensis

Homo sapiens evolved in Africa and later migrated to Asia and then to Europe, where they probably competed with Neanderthals (*H. neanderthalensis*). The Neanderthals preceded *H. sapiens* into Europe and became extinct about 30,000 to 40,000 years ago, apparently without contributing to the gene pool of humans **(19.6)**.

Human skin color may vary as a result of natural selection balancing dark skin's ability to block UV radiation, which destroys folate, while still allowing enough radiation to synthesize vitamin D in different environmental conditions. Both of these vitamins are essential for fetal development **(19.7)**.

Language is a uniquely human trait that permits the creation of human cultures. Linguistic ability has been linked to the human version of the *FOXP2* gene **(19.8)**.

Our Cultural History and Its Consequences (19.9–19.12)

Human culture—our accumulated knowledge, customs, beliefs, arts, and other products—has changed enormously over the course of human evolution **(19.9)**. The first humans survived by scavenging, gathering, and hunting **(19.10)**. The rise of agriculture began about 10,000 to 15,000 years ago, when people formed more permanent settlements and began growing food and domesticating animals **(19.11)**. The Industrial Revolution began in the 1700s, initiating a change to energy-intensive, large-scale machine production. Mechanized farming and improved medicine accelerated the growth of the human population and our impact on the environment **(19.12)**.

Connecting the Concepts

1. Fill in the missing names of the groups on this phylogenetic tree of primates. Also indicate which groups are collectively called anthropoids and which are called hominoids.

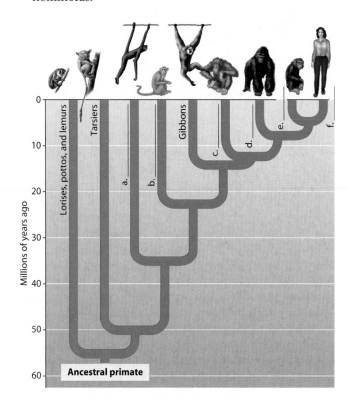

2. List some of the derived characters of these groups.

Group	Derived characters
a. Primates	
b. Hominoids	
c. Hominids	

Testing Your Knowledge

Multiple Choice

3. Fossils suggest that the first major trait distinguishing hominids from other primates was
   a. a larger brain.
   b. erect posture.
   c. forward-facing eyes with depth perception.
   d. grasping hands.
   e. toolmaking.

4. Which of the following correctly lists possible ancestors of humans from the oldest to the most recent?
   a. *Homo erectus, Australopithecus, Homo habilis*
   b. *Australopithecus, Homo habilis, Homo erectus*
   c. *Australopithecus, Homo erectus, Homo habilis*
   d. *Homo ergaster, Homo erectus, Homo neanderthalensis*
   e. *Homo habilis, Homo erectus, Australopithecus*

5. Which of these is not a member of the anthropoids?
   a. chimpanzee
   b. ape
   c. tarsier
   d. human
   e. New World monkey

6. Studies of DNA support which of the following?
   a. Members of the group called australopiths were the first to migrate from Africa.
   b. Neanderthals are more closely related to humans in Europe than to humans in Africa.
   c. *Homo sapiens* originated in Africa.
   d. *Sahelanthropus* was the earliest hominid.
   e. Chimpanzees are more similar to gorillas and orangutans than to humans.

7. According to Figure 19.3, which of these species seems to have survived the longest?
   a. *Homo erectus*
   b. *Australopithecus africanus*
   c. *Homo ergaster*
   d. *Homo sapiens*
   e. *Australopithecus afarensis*

8. The earliest members of the genus *Homo*
   a. had a large brain compared to other hominids.
   b. probably hunted dinosaurs.
   c. lived in Europe.
   d. lived about 4 million years ago.
   e. were the first hominids to be bipedal.

9. In which way are we significantly different from our 100,000-year-old ancestors?
   a. We use tools.
   b. We are much more intelligent.
   c. We have accumulated more information.
   d. We converse.
   e. We are more emotional.

10. To which of the following human traits has the *FOXP2* gene been linked?
    a. opposable thumb enabling a precision grip
    b. brain development associated with language
    c. an extended period of parental care
    d. sophisticated tool manufacture and use
    e. the development of dark skin pigmentation

Describing, Comparing, and Explaining

11. What adaptations inherited from our primate ancestors enable humans to make and use tools?

12. In what ways is chimpanzee behavior similar to human behavior?

13. Contrast hominoids with hominids.

14. Summarize one hypothesis that explains variation in human skin color as an adaptive compromise.

15. Describe the lifestyle of an early hominid.

Applying the Concepts

16. Explain some of the reasons why humans have been able to expand in number and distribution to a greater extent than most other animals.

17. Anthropologists are interested in locating areas in Africa where fossils 4–8 million years old might be found. Why?

18. Some researchers think that drying and cooling of the climate caused expansion of the African savanna and that this environment favored upright walking in early hominids. Why might bipedalism be advantageous in the savanna? How might an erect posture relate to the evolution of a larger brain?

19. The human body has not changed much in the last 100,000 years, but human culture has changed a great deal. As a result of our culture, we change the environment at a rate far greater than many species, including our own, can evolve. What evidence of rapid environmental change do you see regularly? What aspect(s) of human culture are responsible for these changes? Do you see any evidence of a decrease in the rate of human-caused environmental changes?

Answers to all questions can be found in Appendix 3.

For study help and Activities, go to campbellbiology.com or the student CD-ROM.

Testing Your Knowledge

Multiple Choice

3. Fossils suggest that the first major trait distinguishing hominids from other primates was
   a. a larger brain.
   b. erect posture.
   c. forward-facing eyes with depth perception.
   d. grasping hands.
   e. toolmaking.

4. Which of the following correctly lists possible ancestors of humans from the oldest to the most recent?
   a. *Homo erectus, Australopithecus, Homo habilis*
   b. *Australopithecus, Homo habilis, Homo erectus*
   c. *Australopithecus, Homo erectus, Homo habilis*
   d. *Homo ergaster, Homo erectus, Homo neanderthalensis*
   e. *Homo habilis, Homo erectus, Australopithecus*

5. Which of these is not a member of the anthropoids?
   a. chimpanzee
   b. ape
   c. tarsier
   d. human
   e. New World monkey

6. Studies of DNA support which of the following?
   a. Members of the group called australopiths were the first to migrate from Africa.
   b. Neanderthals are more closely related to humans in Europe than to humans in Africa.
   c. *Homo sapiens* originated in Africa.
   d. *Sahelanthropus* was the earliest hominid.
   e. Chimpanzees are more similar to gorillas and orang-utans than to humans.

7. According to Figure 19.3, which of these species seems to have survived the longest?
   a. *Homo erectus*
   b. *Australopithecus africanus*
   c. *Homo ergaster*
   d. *Homo sapiens*
   e. *Australopithecus afarensis*

8. The earliest members of the genus *Homo*
   a. had a large brain compared to other hominids.
   b. probably hunted dinosaurs.
   c. lived in Europe.
   d. lived about 4 million years ago.
   e. were the first hominids to be bipedal.

9. In which way are we significantly different from our 100,000-year-old ancestors?
   a. We use tools.
   b. We are much more intelligent.
   c. We have accumulated more information.
   d. We converse.
   e. We are more emotional.

10. To which of the following human traits has the *FOXP2* gene been linked?
    a. opposable thumb enabling a precision grip
    b. brain development associated with language
    c. an extended period of parental care
    d. sophisticated tool manufacture and use
    e. the development of dark skin pigmentation

Describing, Comparing, and Explaining

11. What adaptations inherited from our primate ancestors enable humans to make and use tools?

12. In what ways is chimpanzee behavior similar to human behavior?

13. Contrast hominoids with hominids.

14. Summarize one hypothesis that explains variation in human skin color as an adaptive compromise.

15. Describe the lifestyle of an early hominid.

Applying the Concepts

16. Explain some of the reasons why humans have been able to expand in number and distribution to a greater extent than most other animals.

17. Anthropologists are interested in locating areas in Africa where fossils 4–8 million years old might be found. Why?

18. Some researchers think that drying and cooling of the climate caused expansion of the African savanna and that this environment favored upright walking in early hominids. Why might bipedalism be advantageous in the savanna? How might an erect posture relate to the evolution of a larger brain?

19. The human body has not changed much in the last 100,000 years, but human culture has changed a great deal. As a result of our culture, we change the environment at a rate far greater than many species, including our own, can evolve. What evidence of rapid environmental change do you see regularly? What aspect(s) of human culture are responsible for these changes? Do you see any evidence of a decrease in the rate of human-caused environmental changes?

Answers to all questions can be found in Appendix 3.

For study help and Activities, go to campbellbiology.com or the student CD-ROM.

UNIT V

Animals: Form and Function

20 UNIFYING CONCEPTS OF ANIMAL STRUCTURE AND FUNCTION

21 NUTRITION AND DIGESTION

22 GAS EXCHANGE

23 CIRCULATION

24 THE IMMUNE SYSTEM

25 CONTROL OF THE INTERNAL ENVIRONMENT

26 CHEMICAL REGULATION

27 REPRODUCTION AND EMBRYONIC DEVELOPMENT

28 NERVOUS SYSTEMS

29 THE SENSES

30 HOW ANIMALS MOVE

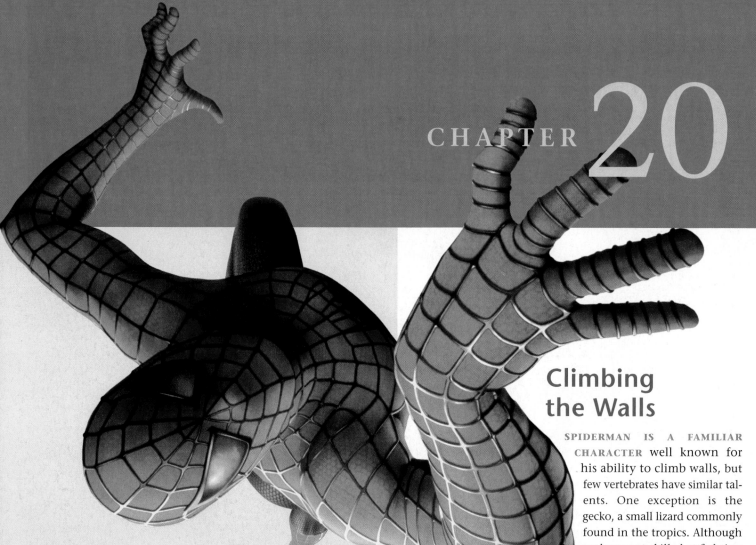

CHAPTER 20

Climbing the Walls

SPIDERMAN IS A FAMILIAR CHARACTER well known for his ability to climb walls, but few vertebrates have similar talents. One exception is the gecko, a small lizard commonly found in the tropics. Although perhaps not skilled at fighting crime or rescuing those in danger, geckos have no trouble walking up walls and even across ceilings. How do they do it? Several hypotheses, including either a sticky adhesive or suction cups on their toes, have turned out to be wrong. Instead, the explanation relates to hairs, called setae, on the gecko's toes. These hairs are made of the protein keratin, just like our own hair.

The micrographs on the opposite page reveal the microscopic structure of setae. They are arranged in rows, and each seta ends in many split ends called spatulae, which have flattened tips. It took a multidisciplinary team of biologists and engineers to work out how the setae stick to surfaces with enough strength to support the animal's weight. In a recent study using the Tokay gecko (*Gekko gecko*), engineers designed an apparatus to measure the force of attraction between individual setae and the surface they touched—a difficult task because of the microscopic size of setae. This force of attraction turned out to be ten times greater than had been predicted.

But what is causing this attraction? The researchers attribute it to attractions between molecules at the tips of the spatulae and molecules making up the surface. Even uncharged molecules have regions that temporarily carry charges, and a region of positive charge on one molecule will be attracted to a region of negative

THE HIERARCHY OF STRUCTURAL ORGANIZATION IN AN ANIMAL

20.1 Structure fits function in the animal body
20.2 Animal structure has a hierarchy
20.3 Tissues are groups of cells with a common structure and function
20.4 Epithelial tissue covers the body and lines its organs and cavities
20.5 Connective tissue binds and supports other tissues
20.6 Muscle tissue functions in movement
20.7 Nervous tissue forms a communication network
20.8 Artificial tissues have medical uses
20.9 Organs are made up of tissues
20.10 Organ systems work together to perform life's functions
20.11 New imaging technology reveals the inner body

EXCHANGES WITH THE EXTERNAL ENVIRONMENT

20.12 Structural adaptations enhance exchange between animals and their environment
20.13 Animals regulate their internal environment
20.14 Homeostasis depends on negative feedback

Unifying Concepts of Animal Structure and Function

Spatulae coming from a single seta

Rows of setae on a gecko's foot

charge on another. (These attractions, called van der Waals forces, also help hold individual protein and nucleic acid molecules in the characteristic shapes you saw in Chapter 3.) Each instance of attraction is fleeting and very weak, but there are so many setae—about half a million on each toe, each ending in hundreds of spatulae—that the combined strength of these forces becomes significant. In fact, a single seta could hold up an ant!

If the combined forces are so strong, why doesn't the gecko get stuck—its toes adhering so firmly to a surface that it can't move? The answer has to do with the angle at which setae make contact with a surface. The researchers discovered that slight changes in the angle of attachment cause large changes in the amount of force. This means that a slight change in the position of the toes makes it easy for the gecko to lift its foot.

The gecko's remarkable ability to walk on walls is thus a function resulting from special structural adaptations of its body, adaptations that extend to the microscopic level. Other structural features of the gecko's body correlate with their functions, from the scales (also made of the protein keratin) that protect its body from drying out to the arrangement of the muscles and bones that move its feet as it walks up walls.

The relationship between structure and function is an important overarching concept of biology. It also helps us understand animals. The chapters in this unit explore animal form and function in the context of the various problems animals must solve: how to nourish themselves, obtain oxygen from their environment and distribute it throughout their body, excrete wastes, sense and respond to the environment, move, and reproduce. These various adaptations have been fashioned by natural selection, fitting structure to function by selecting, over many generations, for what works best within a particular population in its environment.

This chapter opens the unit with an overview of animal structure and function. ■ ■ ■

20.1 Structure fits function in the animal body

Anatomy is the study of an organism's structures; **physiology** is the study of the functions of an organism's structures. An anatomist studying the gecko, for instance, might focus on the arrangement of muscles and bones in a gecko's legs or on the shape and number of setae on its toes that allow it to climb walls (see chapter introduction). A physiologist might study the functioning of muscles in the gecko's legs or the production of setae by epidermal cells. Despite their different approaches, both biologists are working toward a better understanding of the connection between structure and function—such as how structural adaptations give the gecko its remarkable ability to walk on walls.

Another elegant example of the correlation between structure and function is the flight apparatus of birds. Consider the feathers of the wing. Feathers give the wing its broad shape without adding much weight to the body. They remain dry because they are lightly coated with oil, and they also trap air, which provides insulation that helps a bird maintain its high body temperature and metabolism.

The functions of feathers result from their unique structure. Produced by special pits in the bird's skin, feathers consist entirely of nonliving material, mainly the protein keratin. A flight feather, like the one enlarged in **Figure 20.1**, has a hollow keratin shaft that provides a central support with minimum weight. Small flat rods called barbs extend from both sides of the shaft. Still finer rods called barbules extend from the sides of the barbs. Each barbule has tiny hooks that interlock with adjacent barbules. With all of its barbules interlocked, a feather has a particular shape and rigidity that support flight. When barbules become detached, these features are lost, and the bird's flight ability is impaired. When a bird preens, it draws its feathers through its beak, hooking the barbules back together, like zipping up a zipper.

The muscles and bones in a bird wing also show the relationship between structure and function. Muscles provide power, and the bones provide support for flight. The flight muscles are situated on the breast and around the base of the wings, keeping most of the weight off the wings and helping the bird maintain balance in flight.

The bones in a bird wing are homologous to those in the human arm, but the number of bones in the wing diminished as birds evolved. The bird wing has only three fingers (numbered in the figure), and only the middle one (finger 2) has a complete set of bones. The bird's wrist and palm also have fewer bones than ours do. This adaptation helps make the wing lighter but less flexible than the human wrist and hand. This reduced flexibility stabilizes the wing and helps it function as a unit in flight. The photograph of the dissected bone on the right side of the figure illustrates another adaptation in the bird skeleton. Many bird bones are hollow but reinforced internally with trusses similar to those used in airplane wings. This structure provides maximum strength with minimum weight, the ideal combination for flight.

The ability to fly or to walk or to climb walls emerges from the specific arrangement of specialized structures. As we will see throughout our study of the anatomy and physiology of animals, structure fits function.

Web/CD Activity 20A Correlating Structure and Function of Cells

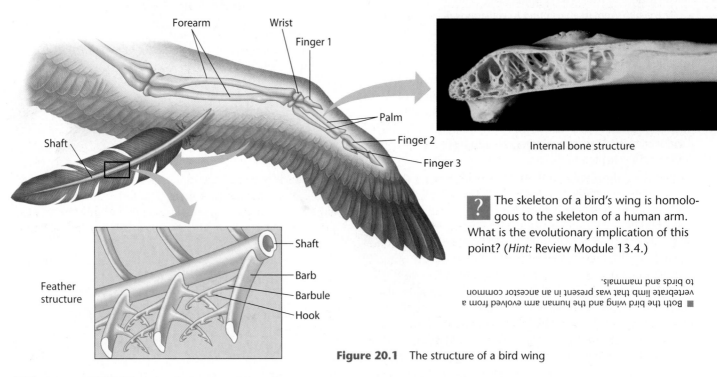

Forearm
Wrist
Finger 1
Shaft
Palm
Finger 2
Finger 3
Internal bone structure

? The skeleton of a bird's wing is homologous to the skeleton of a human arm. What is the evolutionary implication of this point? (*Hint:* Review Module 13.4.)

■ Both the bird wing and the human arm evolved from a vertebrate limb that was present in an ancestor common to birds and mammals.

Shaft
Barb
Barbule
Hook
Feather structure

Figure 20.1 The structure of a bird wing

20.2 Animal structure has a hierarchy

Structure in the living world is organized in a series of hier-archical levels. We followed the progression from atoms to molecules to cells in Unit I. Now let's trace the hierarchy from cells to organisms. (In Unit VII, we will pick up the trail again, moving from organisms to ecosystems.)

Figure 20.2 illustrates structural hierarchy in a pelican. Part A shows a single muscle cell in the bird's heart. This cell's main function is to contract, and the stripes in the cell result from the precise alignment of strands of proteins that perform that function. Each muscle cell is also branched, providing for multiple connections to other cells that ensure coordinated contractions of all the muscle cells in the heart. Together these cells make up a tissue (part B), the second level of structure and function. A **tissue** is an integrated group of similar cells that perform a specific function.

Part C, the heart itself, illustrates the organ level of the hi-erarchy. An **organ** is made up of two or more types of tissues that together perform a specific task. In addition to muscle tissue, the heart includes nervous tissue and connective tis-sue. Part D shows the circulatory system, the organ system of which the heart is a part. An **organ system** consists of multiple organs that together perform a vital body function. The other parts of the circulatory system are the blood ves-sels: arteries, veins, and capillaries.

In part E, the pelican itself forms the final level of this hi-erarchy. An **organism** contains a number of organ systems, each specialized for certain functions and all functioning to-gether as an integrated, coodinated unit. For example, the pelican's circulatory system cannot function without oxy-gen supplied by the respiratory system and nutrients sup-plied by the digestive system. And it takes the coordination of several other organ systems to enable this bird to fly.

At each level in this hierarchy, we find that structure and function are related. In the next several modules, we focus on the tissue level of this biological hierarchy.

Web/CD Activity 20B *The Levels of Life Card Game*

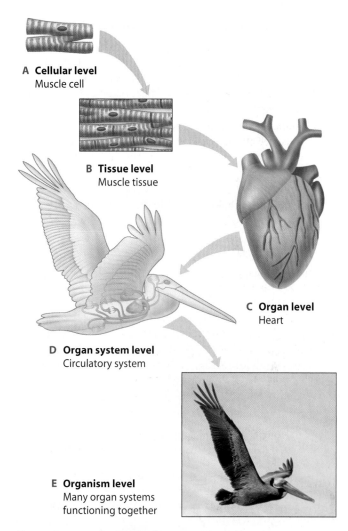

A Cellular level
Muscle cell

B Tissue level
Muscle tissue

C Organ level
Heart

D Organ system level
Circulatory system

E Organism level
Many organ systems
functioning together

Figure 20.2   An example of structural hierarchy in a pelican

> [?] What level of animal structure does the human brain represent?

■ Organ

20.3 Tissues are groups of cells with a common structure and function

In almost all animals, most of the cells of the body are organ-ized into tissues. The cells composing a tissue are specialized; they have a particular structure that enables them to perform a specific task. As we saw in the last module, for example, each of the heart's muscle cells contains strands of contrac-tile proteins and has branches that connect to other muscle cells, helping the cells contract in a coordinated manner.

The term *tissue* is from a Latin word meaning "weave," and some tissues resemble woven cloth in that they consist of a meshwork of nonliving fibers and other extracellular substances surrounding living cells. Other tissues are held together by a sticky glue that coats the cells or by special

junctions between adjacent plasma membranes (see Module 4.18). An animal has four major categories of tissue: epithe-lial tissue, connective tissue, muscle tissue, and nervous tis-sue. We examine each of these separately in the next four modules.

Web/CD Activity 20C *Overview of Animal Tissues*

> [?] An organ is usually constructed from two or more different _____ .

■ tissues

20.4 Epithelial tissue covers the body and lines its organs and cavities

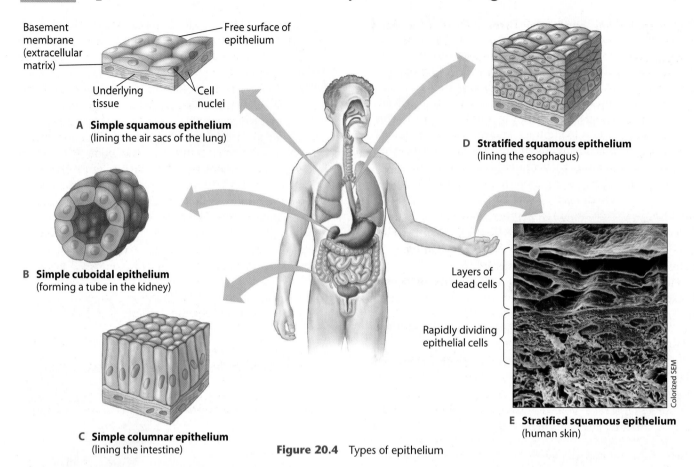

A **Simple squamous epithelium**
(lining the air sacs of the lung)

Basement membrane (extracellular matrix)

Free surface of epithelium

Underlying tissue

Cell nuclei

B **Simple cuboidal epithelium**
(forming a tube in the kidney)

C **Simple columnar epithelium**
(lining the intestine)

D **Stratified squamous epithelium**
(lining the esophagus)

Layers of dead cells

Rapidly dividing epithelial cells

Colorized SEM

E **Stratified squamous epithelium**
(human skin)

Figure 20.4 Types of epithelium

Epithelial tissue, also called **epithelium**, occurs as sheets of tightly packed cells that cover body surfaces and line internal organs and cavities. One side of an epithelium is anchored to underlying tissues by a **basement membrane**, a dense mat of extracellular matrix consisting of fibrous proteins and sticky polysaccharides. The other side—the free surface—faces the outside of an organ or the inside of a tube or passageway. Together, the tightly knit cells and basement membrane form a protective barrier and, in some cases, a surface for exchange with the fluid or air on the other side.

Epithelial tissues are named according to the number of cell layers they have and according to the shape of the cells on their free surface. A simple epithelium has a single layer of cells, whereas a stratified epithelium has multiple layers. The shape of the cells may be squamous (like floor tiles), cuboidal (like dice), or columnar (like bricks on end). Parts A, B, and C of Figure 20.4 show examples of simple epithelia with the three cell shapes; part D shows a stratified squamous epithelium. In each case, the pink color identifies the cells of the epithelium itself.

The structure of each type of epithelium fits its function. Simple squamous epithelium is thin and suitable for exchanging materials by diffusion. We find it lining our capillaries (smallest blood vessels) and the air sacs of our lungs. In contrast, stratified squamous epithelium is well suited for lining surfaces subject to abrasion, such as the esophagus, which

can be abraded by rough food. Stratified squamous epithelium regenerates rapidly by division of the cells near the basement membrane. New cells move toward the free surface as older cells slough off. Our epidermis, shown in the micrograph in part E, is a different type of stratified squamous epithelium, with a thick layer of dead cells at the free surface. The protein keratin is deposited in these epithelial cells and makes our skin waterproof. (Keratin also makes up the setae on a gecko's toes and the feathers of a bird.)

Both cuboidal and columnar epithelia have cells with a relatively large amount of cytoplasm, facilitating their role of secretion or absorption of substances. Epithelial tissue specialized to produce and secrete various chemicals is called glandular epithelium. In addition to making up various glands, glandular epithelia line the digestive tract and respiratory tubes. Here they form a **mucous membrane**, which secretes a slimy solution called mucus that lubricates the surface and keeps it moist. The mucous membrane of our air tubes helps keep our lungs clean by trapping dust, pollen, and other particles in its secretions. The beating of cilia on this mucous membrane then sweeps the mucus-trapped materials upward and out of the breathing passageways.

Web/CD Activity 20D *Epithelial Tissue*

? Epithelial tissues are classified according to the _____ of their cells and the number of cell _____.

shape . . . layers

20.5 Connective tissue binds and supports other tissues

Unlike epithelium, **connective tissue** consists of a sparse population of cells scattered through an extracellular matrix. The cells produce and secrete the matrix, which usually consists of a web of fibers embedded in a liquid, jelly, or solid. Connective tissues may be grouped into six major types. Figure 20.5 shows micrographs of each type and illustrates where each would be found in an arm, for example.

The most common connective tissue in the human body is called **loose connective tissue** (part A) because its matrix is a loose weave of fibers. Many of the fibers consist of the strong, ropelike protein collagen. Other fibers are more elastic, making the tissue resilient as well as strong. Loose connective tissue serves mainly as a binding and packing material, holding other tissues and organs in place. In the figure, we show the loose connective tissue that lies directly under the skin, where it helps bind the skin to underlying muscles.

Fibrous connective tissue (B) has a matrix of densely packed parallel bundles of collagen fibers, an arrangement that maximizes its nonelastic strength. Fibrous connective tissue forms tendons, which attach muscles to bone, and ligaments, which join bones together.

Adipose tissue (C) stores fat in large, closely packed adipose cells held in a sparse matrix of fibers. This tissue pads and insulates the body and stores energy. Each adipose cell contains a large fat droplet that swells when fat is stored and shrinks when fat is used as fuel.

The matrix of **cartilage** (D), connective tissue that forms a strong but flexible skeletal material, consists of an abundance of collagen fibers embedded in a rubbery substance. Cartilage commonly surrounds the ends of bones, where it forms a shock-absorbing surface. It also supports the nose and ears and forms the cushioning disks between our vertebrae.

Bone (E), a rigid connective tissue, has a matrix of collagen fibers embedded in calcium salts. This combination makes bone hard without being brittle. Compact bone, as shown in the micrograph, contains repeating circular units of matrix, each with a central canal containing blood vessels and nerves, which service the bone cells. Like other tissues, bone contains living cells and can therefore grow with the animal.

Blood (F) functions differently from other connective tissues. Its extensive extracellular matrix is a liquid called plasma that consists of water, salts, and dissolved proteins. Red and white blood cells are suspended in the plasma. Blood functions mainly in transporting substances from one part of the body to another and in immunity.

Web/CD Activity 20E *Connective Tissue*

> ? Why does blood qualify as a type of connective tissue?

■ Because it consists of a relatively sparse population of cells surrounded by a noncellular matrix, which is plasma in this case

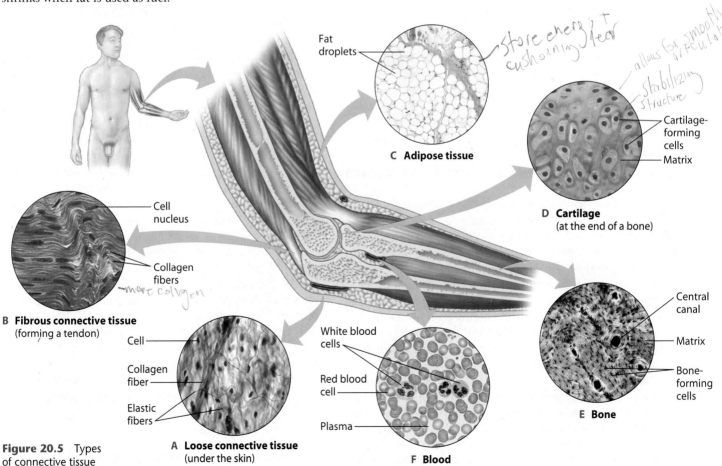

Fat droplets

C Adipose tissue

Cartilage-forming cells

Matrix

D Cartilage (at the end of a bone)

Cell nucleus

Collagen fibers

B Fibrous connective tissue (forming a tendon)

Cell

Collagen fiber

Elastic fibers

A Loose connective tissue (under the skin)

White blood cells

Red blood cell

Plasma

F Blood

Central canal

Matrix

Bone-forming cells

E Bone

Figure 20.5 Types of connective tissue

20.6 Muscle tissue functions in movement

Muscle tissue consists of bundles of long cells called muscle fibers and is the most abundant tissue in most animals. Within the cytoplasm of muscle fibers are large numbers of contractile proteins arranged in parallel. Geckos, birds, humans, and all other vertebrates have three types of muscle tissue. **Figure 20.6** shows micrographs of these three types: skeletal muscle, cardiac muscle, and smooth muscle.

Skeletal muscle is attached to bones by tendons and is responsible for voluntary movements of the body. The arrangement of the contractile units along the length of muscle cells gives them a striped or striated appearance, as you can see in part A below.

Cardiac muscle (B) forms the contractile tissue of the heart. It is striated like skeletal muscle, but its cells are branched, interconnecting at specialized junctions that rapidly relay the signal to contract from cell to cell during the heartbeat.

Smooth muscle (C) gets its name from its lack of striations. This type of muscle is found in the walls of the digestive tract, urinary bladder, arteries, and other internal organs. The cells (fibers) are shaped like spindles. They contract more slowly than skeletal muscles, but they can sustain contractions for a longer period of time.

Web/CD Activity 20F *Muscle Tissue*

? The muscles responsible for a gecko climbing a wall are _____ muscles.

■ skeletal

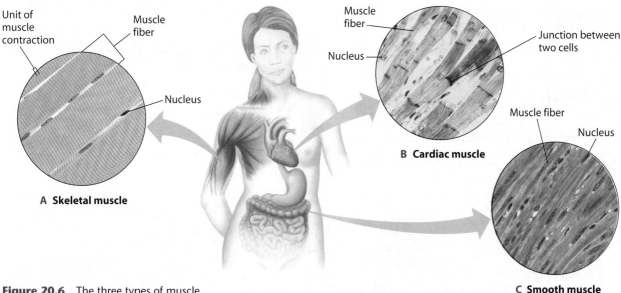

Figure 20.6 The three types of muscle

A **Skeletal muscle**
- Unit of muscle contraction
- Muscle fiber
- Nucleus

B **Cardiac muscle**
- Muscle fiber
- Nucleus
- Junction between two cells

C **Smooth muscle**
- Muscle fiber
- Nucleus

20.7 Nervous tissue forms a communication network

Nervous tissue senses stimuli and rapidly transmits information from one part of an animal to another. The structural and functional unit of nervous tissue is the nerve cell, or **neuron**, which is uniquely specialized to conduct electrical nerve impulses. The micrograph in **Figure 20.7** shows two neurons, each consisting of a cell body (containing the cell's nucleus) and a number of slender extensions. One type of extension, called a dendrite, generally conveys signals toward the cell body; another type, the axon, usually transmits signals away from the cell body, often to another neuron or to a muscle cell.

Nervous tissue is not made up entirely of neurons. It actually contains many more supporting cells than neurons. Some of these cells nourish the neurons. Others surround and insulate axons, promoting faster transmission of signals.

Web/CD Activity 20G *Nervous Tissue*

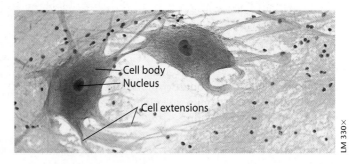

LM 330×

Figure 20.7 Neurons in the spinal cord
- Cell body
- Nucleus
- Cell extensions

? How does the long length of some axons (such as those that extend from your lower spine to your toes) relate to the function of a neuron?

■ It allows for the rapid transmission of a nerve signal over a long distance.

20.8 Artificial tissues have medical uses

Scientists are learning how to replace tissue when the body cannot. Many of the body's tissues are self-repairing, but a few types of tissue, such as cartilage, do not regenerate easily. In other cases, serious injury can overwhelm a tissue's healing process. Tissue loss can bring permanent disability. Badly damaged knee cartilage, for example, has hobbled many athletes. And a sharp blow to the face can knock out a tooth for good. In more severe cases, such as deep or widespread burns, tissue destruction can turn deadly as infection invades areas once shielded by skin.

By understanding how tissues develop and grow, scientists and doctors now have new ways to heal tissues that can no longer recover on their own. One of the most successful advances has come in the form of artificial skin (Figure 20.8), a type of human-engineered tissue designed for everyone from burn victims to diabetics with skin ulcers. The tissue is grown from human fibroblasts, tissue-generating cells often harvested from newborn foreskin tissue. These cells are applied along a tiny scaffolding, where they multiply and produce a three-dimensional skin substitute containing active living cells. The scaffolding, often made from collagen, is later absorbed by the body.

While artificial skin is costly and complicated to use, doctors have welcomed this medical advance. The success of artificial skin is spurring more research into artificial tissue. Some scientists are seeking ways to grow artificial cartilage. In London, dentists are trying to develop replacement teeth, relying on the biological processes that build teeth in children to urge a new tooth to grow in an adult mouth.

Figure 20.8 Dermagraft, an artificial tissue

? What types of new tissues would have to develop to heal an extensive, deep burn?

■ Epithelial and several types of connective tissues (loose, adipose, blood)

20.9 Organs are made up of tissues

In all but the simplest animals (sponges and some cnidarians), tissues are arranged into organs that perform specific functions. The heart, for example, while mostly muscle, also has epithelial, connective, and nervous tissues. Epithelial tissue lining the heart chambers prevents leakage and provides a smooth surface over which blood can flow with little friction. Connective tissue makes the heart elastic and strengthens its walls and valves. Neurons direct the rhythmic contractions of cardiac muscles.

In some organs, tissues are organized in layers, as you can see in the diagram of the small intestine in **Figure 20.9**. The lumen, or space, within the small intestine is lined by a columnar epithelium that absorbs nutrients and secretes mucus and digestive juices. Notice the fingerlike projections that increase the surface area of this lining. Surrounding this layer (and extending into the projections) is a zone of connective tissue that contains blood and lymph vessels. Two layers of smooth muscle (oriented in different directions) surround the connective tissue and move food through the digestive tract. The smooth muscle, in turn, is surrounded by another layer of connective tissue and epithelial tissue.

An organ represents a higher level of structure than the tissues composing it, and it performs functions that none of its component tissues can carry out alone. These functions emerge from the coordinated efforts of tissues. Coordinated interaction is a basic feature at all levels in an animal's structural hierarchy.

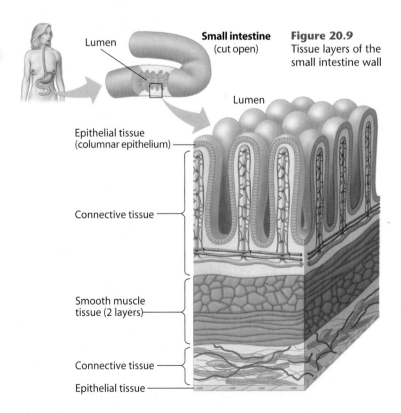

Figure 20.9 Tissue layers of the small intestine wall

Lumen

Small intestine (cut open)

Lumen

Epithelial tissue (columnar epithelium)

Connective tissue

Smooth muscle tissue (2 layers)

Connective tissue

Epithelial tissue

? Explain why a disease that damages connective tissue can impair most of the body's organs.

■ Connective tissue is a component of most organs.

20.10 Organ systems work together to perform life functions

Just as it takes many specialized cells to create a tissue and several different tissues to build an organ, it takes the integration of several organs into organ systems to carry out the major body functions. There are 12 major organ systems in vertebrate animals. Using the human as an example, **Figure 20.10** outlines the components that make up these systems. As you read through the brief descriptions of the functions of each system, remember that all of the body's organ systems are interdependent and work together to create a functional organism. We will examine these systems in detail in the chapters of this unit.

Figure 20.10 Summary of the functions of human organ systems

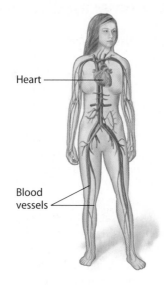

C Circulatory system
Delivers nutrients and O_2 to body cells; carries CO_2 to the lungs and metabolic wastes to the excretory organs, the kidneys.

Heart

Blood vessels

A Digestive system
Ingests and breaks down food into smaller chemical units to be used as fuel for cellular respiration; eliminates undigested material.

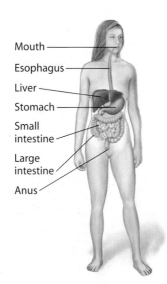

Mouth
Esophagus
Liver
Stomach
Small intestine
Large intestine
Anus

D Immune system
Defends the body against infections and cancer.

E Lymphatic system
Returns fluid that leaks from blood vessels to the circulatory system; functions as part of the immune system.

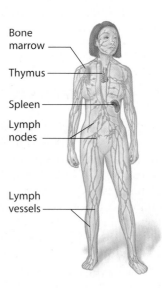

Bone marrow
Thymus
Spleen
Lymph nodes
Lymph vessels

Nasal cavity
Larynx
Trachea
Bronchus
Lung

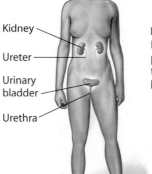

Kidney
Ureter
Urinary bladder
Urethra

F Excretory system
Removes nitrogen-containing waste products from the blood; regulates the chemical makeup and water balance of the blood.

B Respiratory system Exchanges gases with the environment; supplies the blood with oxygen (O_2) and disposes of carbon dioxide (CO_2).

G Endocrine system

Secretes chemicals, called hormones, that regulate body activities such as digestion, metabolism, growth, reproduction, heart rate, and water balance.

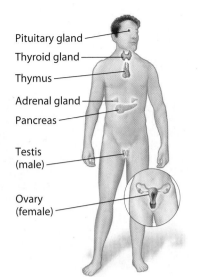

Pituitary gland
Thyroid gland
Thymus
Adrenal gland
Pancreas
Testis (male)
Ovary (female)

H Nervous system

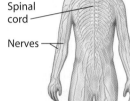

Brain
Sense organ
Spinal cord
Nerves

Coordinates body activities by detecting stimuli, integrating information, and directing the body's responses.

I Integumentary system

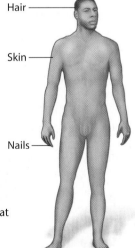

Hair
Skin
Nails

Protects against mechanical injury, infection, excessive heat or cold, and drying out.

J Skeletal system

Supports the body; protects certain internal organs, such as the brain and lungs; provides the framework for muscles to produce movement.

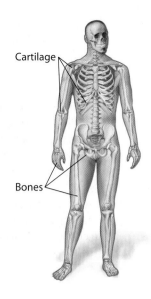

Cartilage
Bones

K Muscular system

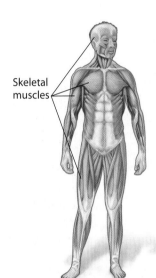

Skeletal muscles

Skeletal muscles produce movement, maintain posture, and produce heat.

L Reproductive systems

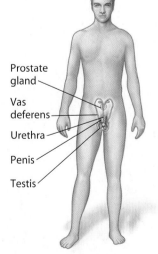

Male
Prostate gland
Vas deferens
Urethra
Penis
Testis

Female
Oviduct
Ovary
Uterus
Vagina

Produce gametes and sex hormones; female system provides organs to support a developing embryo and glands for producing milk.

? The two organ systems most directly involved in regulating all other systems are the _____ and _____ systems.

■ nervous . . . endocrine

20.11 New imaging technology reveals the inner body

Among the most exciting recent developments in medical technology are techniques that allow physicians to "see" the organs and organ systems we have just surveyed without resorting to surgery. We mentioned one of these techniques—ultrasound—in Module 9.10. Here we look at some others.

X-Rays X-rays, discovered in 1895, were the first means of producing a photographic image of internal organs and the only imaging method available until the 1950s. X-rays are a type of high-energy radiation (see Module 7.6). They pass readily through soft tissues, such as skin, nerves, and muscle. The photographic film on which an X-ray image is recorded is placed behind the body, and the features that show up most distinctly are the shadows of hard structures that block the rays—bones and dense tumors, for instance.

Conventional X-rays are used routinely to check for broken bones and tooth cavities. However, there are some problems with routine X-rays. One obvious shortcoming is their failure to make soft tissues clearly visible. In addition, the standard X-ray technique produces only a flat, two-dimensional image, with anatomical structures often confusingly overlapped. Finally, the X-rays themselves, in large enough doses, can cause cancer.

CT Modern X-ray techniques have overcome some of the disadvantages cited above. An exciting extension of the standard technique has conquered the overlapping-image

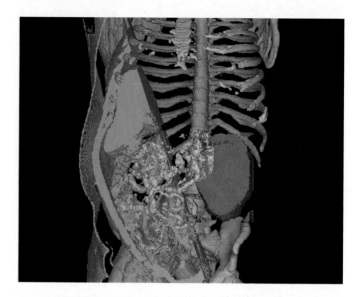

Figure 20.11B A three-dimensional CT image showing a large tumor (red object) surrounding a kidney

problem. This newer X-ray method is called computed tomography (CT), a computer-assisted technique that produces images of a series of thin cross sections through the body. The patient is often given a special liquid to improve the contrast of the images and is then slowly moved through a doughnut-shaped CT machine (**Figure 20.11A**) as the X-ray source circles around the body, illuminating successive sections from many angles. The CT scanner's computer then produces high-resolution video images of the cross sections, which can be studied individually or combined into various three-dimensional views.

CT scans are excellent diagnostic tools. They can detect small differences between normal and abnormal tissues in many organs, but are especially useful for evaluating problems that affect the abdomen and brain—areas where conventional X-ray procedures are of little help. In the CT image inset in Figure 20.11A, you can see a brain hemorrhage (pale oval at far right), the result of a ruptured blood vessel. This CT image is two-dimensional; it shows only a single, thin slice of the brain. **Figure 20.11B** shows a three-dimensional CT scan of a man's chest and abdomen. This routine CT scan revealed a large cancerous growth surrounding a kidney. The kidney and a 10-pound cancerous mass were successfully removed by surgery.

Another useful diagnostic technique is one that uses ultrafast CT scanners to show the actual movements and changes in volumes of body organs, such as the heart beating and blood flowing through vessels. Physicians use this technique to identify heart defects and constricted or blocked blood vessels and to monitor the status of coronary bypass grafts.

MRI A completely different technique, magnetic resonance imaging (MRI) uses no X-rays or any other high-energy radi-

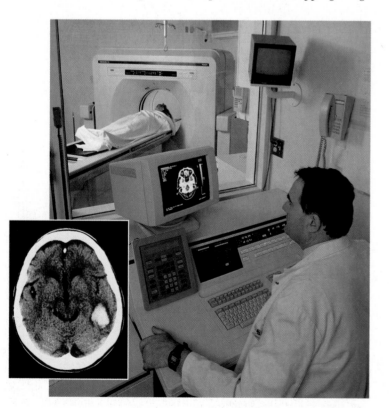

Figure 20.11A A technician monitors the output as a patient moves through a CT scanner

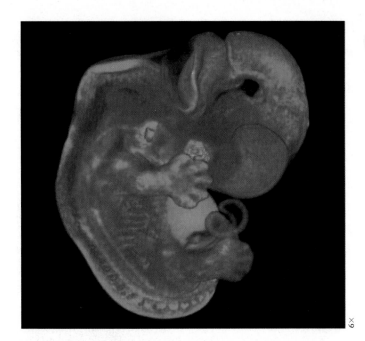

Figure 20.11C An MRM scan of a 47-day-old human embryo

ation. Instead, MRI takes advantage of the behavior of the hydrogen atoms in water molecules. The nuclei of hydrogen atoms are usually oriented in random directions, but in a magnetic field they align in the same direction. MRI uses powerful magnets to align the hydrogen nuclei, then knocks the nuclei out of alignment with a brief pulse of radio waves. In response, the hydrogen atoms give out faint radio signals of their own, which are picked up by the MRI scanner and translated by computer into an image.

Since water is a major component of all of our soft tissues, MRI visualizes them well. At the same time, dense structures such as bone, which contains little water, are nearly invisible to MRI. These qualities make MRI particularly good for detecting problems in nervous tissue that is surrounded by bone. For example, MRI allows physicians to see delicate nerve fibers in the spinal cord.

Three-dimensional CT and MRI scans are not only used diagnostically. They are often used before surgery to map out the surgical procedure or design artificial implants for reconstructive surgery. MRI scans are also used during surgery to guide delicate procedures.

MRM A more powerful type of MRI, called magnetic resonance microscopy (MRM), has revolutionized our ability to create detailed three-dimensional images of very small structures. **Figure 20.11C** shows a colorized MRM of an early human embryo. The developing eye is yellow-orange. The liver shows up bright green in the abdomen, and a bright-green ear is visible above the shoulder. With this imaging technique, researchers can study the development of our organ systems.

PET Positron-emission tomography (PET) is an imaging technology that differs from both CT and MRI in its ability to yield information about metabolic processes at specific locations in the body. In preparation for a PET scan, the patient is injected with a biological molecule—glucose, for example—labeled with a radioactive isotope (see Module 2.5). Used only in small quantities, the isotope is not dangerous. Metabolically active cells take up more of the labeled glucose than less active cells. The isotope emits positively charged subatomic particles called positrons. When the positrons collide with electrons inside the cells, enough high-energy radiation (gamma rays) is released to be detected by an instrument called a PET scanner. Thus, PET pinpoints metabolic hot spots by highlighting in vivid colors the sites of most intense radiation.

PET is proving most valuable for measuring the metabolic activity of various parts of the brain. This technique is providing insights into brain activity in people affected by illnesses such as schizophrenia, epilepsy, and Alzheimer's disease and in stroke patients. Equally exciting is the use of PET to learn about the healthy brain. **Figure 20.11D**, for example, shows PET scans of a person's brain during four different kinds of mental activity that involve language. Metabolic hot spots in the brain appear white, orange, and yellow in these scans, clearly revealing the regions of the brain that are most active during each activity. Researchers have also identified the areas of the brain most active during different types of problem solving and changes in the locations of brain activity associated with learning. Research into brain function has also benefited from a new technique called functional MRI, which can track changes in blood flow into small areas of the brain in real time.

These new imaging techniques are providing medical science with powerful diagnostic tools and researchers with incredibly detailed anatomical and physiological studies. They are significantly increasing our knowledge of both the structure and function of the human body.

? Why are the imaging techniques described in this module regarded as relatively noninvasive, in contrast to such diagnostic methods as exploratory surgery or biopsy?

■ Although they may involve injections, these imaging techniques require no penetration of the body with instruments such as scalpels.

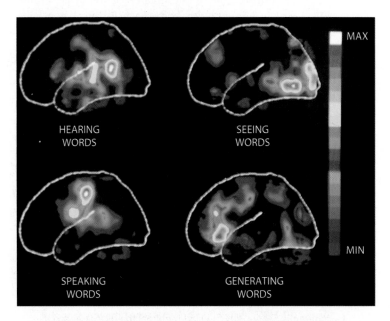

Figure 20.11D PET scans of a brain engaged in different mental activities

20.12 Structural adaptations enhance exchange between animals and their environment

Animals cannot survive unless they can exchange materials with their environment, and this exchange must extend to the cellular level. Oxygen and nutrients must enter the cell, and carbon dioxide and other metabolic wastes must exit. Because a living cell must be bathed in aqueous fluid for its plasma membrane to remain intact, only molecules dissolved in water can move across the plasma membrane.

A freshwater hydra has a body wall only two cell layers thick (Figure 20.12A). The outside layer is in contact with the environment; the inner layer is bathed by fluid in the saclike gastrovascular cavity. Water flushes in and out of the cavity, which opens to the outside via the mouth. As the arrows in the figure indicate, materials diffuse back and forth between the cells, the hydra's surroundings, and the gastrovascular cavity. With this body structure, each cell in a hydra has enough surface area exposed to an aqueous environment to service its entire volume of cytoplasm by direct diffusion and active transport.

The saclike body of a hydra or the paper-thin one of the flatworms we discussed in Module 18.7 works well for animals with a simple body structure. However, most animals have an outer surface that is relatively small compared with the animal's overall volume. As an extreme example, the surface-to-volume ratio (see Module 4.2) of a whale is hundreds of thousands of times smaller than that of a hydra. Still, every cell in the whale's body must be bathed in fluid, have access to oxygen and nutrients, and be able to dispose of its wastes. How is all this accomplished?

Most animals have specialized surfaces for exchanging materials with the environment. Figure 20.12B is a schematic model illustrating four of the organ systems of an animal with a structurally complex body. Each system has a large, specialized internal exchange surface. We have placed the circulatory system in the middle because of its central role in transporting substances between the other three systems. The blue arrows indicate exchange of materials between the circulatory system and the other systems.

Actually, direct exchange does not occur between the blood and the cells making up tissues and organs. The cells of the body are bathed in a solution called **interstitial fluid** (see the circular enlargement). Materials are exchanged between the blood and the interstitial fluid and between the interstitial fluid and the body cells. In other words, to get from the blood to body cells or vice versa, materials must pass through interstitial fluid.

The digestive system, especially the intestine, has an expanded surface area resulting from folds and projections of its epithelial tissues (see Figure 20.9). Nutrients from digested food are absorbed from the lumen into the cells of this large surface area. They then pass into the interstitial fluid and from there into capillaries, tiny branched blood vessels that form an exchange network with the digestive surfaces. This system of exchange from the cells lining the intestine to the surrounding interstitial fluid to the blood is so effective that enough nutrients move into the circulatory system to support the rest of the cells in the body.

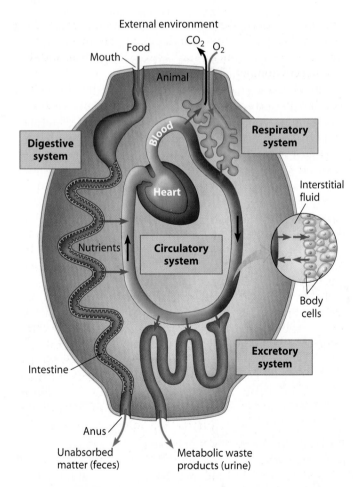

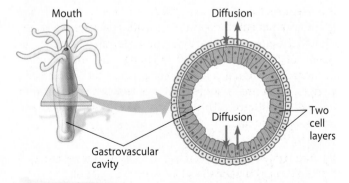

Figure 20.12A Direct exchange between the environment and the cells of a structurally simple animal (a hydra)

Figure 20.12B Indirect exchange between the environment and the cells of a complex animal

The extensive, epithelium-lined tubes of the excretory system are equally effective at increasing the surface area for exchange. Enmeshed in capillaries, excretory tubes extract metabolic wastes that the blood brings from body cells throughout the body. The wastes move out of the blood into the excretory tubes and pass out of the body in urine.

The respiratory system also has an enormous internal surface area associated with a vast number of capillaries. Blood vessels that convey blood from the heart to the lungs divide into tiny capillaries that radiate throughout the lungs, shown in Figure 20.12C as the red banches. The white branches represent tiny air tubes that end in multilobed sacs lined with thin squamous epithelium. Oxygen readily moves from the air in the lungs across this epithelium and into the blood in the capillaries. The blood returns to the heart and is then pumped throughout the body to supply all cells with O_2.

Figure 20.12B highlights two basic concepts in animal biology: First, any animal with a complex body—one with most of its cells not in direct contact with the outside environment—must have internal structures that provide enough surface area to service those cells. Second, the organ

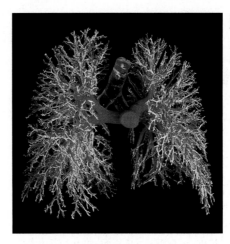

Figure 20.12C
The finely branched surfaces of the human lungs

systems of the body are interdependent, working together to produce a functional organism.

> ? The lungs, small intestine, and kidneys all have an extensive _____ _____ of epithelium. This structure is related to their function of _____ with the environment.

■ surface area … exchange

20.13 Animals regulate their internal environment

Over a century ago, French physiologist Claude Bernard recognized that *two* environments are important to an animal: the external environment, surrounding the animal, and the internal environment, where its cells actually live. The internal environment of a vertebrate is the interstitial fluid that fills the spaces around the cells. Many animals maintain relatively constant conditions in their internal environment. Our own bodies maintain the salt and water balance of our internal fluids and also keep the fluids at about 37°C (98.6°F). A bird, such as the ptarmigan, also maintains salt and water balance and temperature (about 40°C), even in winter (Figure 20.13A). The bird uses energy from its food to generate body heat, and it has a thick, insulating coat of down feathers. A gecko does not generate its own body heat, but it can maintain a fairly constant body temperature by basking in the sun or resting in the shade. And it does regulate the salt and water balance of its internal fluids.

Today, Bernard's concept of the constant internal environment is included in the broader principle of **homeostasis**, which means "a steady state." **Figure 20.13B** illustrates this principle. Conditions may fluctuate widely in the animal's external environment, but homeostatic mechanisms regulate internal conditions, resulting in much smaller fluctuations in the animal's internal environment. For example,

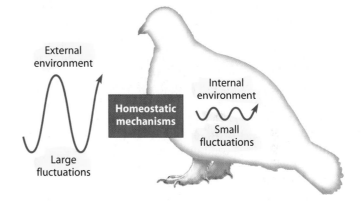

Figure 20.13B A model of homeostasis

birds and mammals have a control system that keeps body temperature within a narrow range, despite wide fluctuations in the temperature of the external environment.

The internal environment of an animal always fluctuates slightly. Homeostasis is a dynamic state, an interplay between outside forces that tend to change the internal environment and internal control mechanisms that oppose such changes. An animal's homeostatic control systems maintain internal conditions within a range where life's metabolic processes can occur.

> ? The ability of your body to regulate its internal pH and the sugar concentration of its blood are examples of _____.

■ homeostasis

Figure 20.13A A white-tailed ptarmigan in its snowy habitat

Homeostasis depends on negative feedback

Most of the control mechanisms of homeostasis are based on **negative feedback**, in which a change in a variable triggers mechanisms that reverse that change. To identify the components of a negative-feedback system, consider the simple example of the regulation of room temperature. You set the thermostat at a comfortable temperature—call this its *set point*. When a sensor in the thermostat detects that the temperature has dropped below this set point, the thermostat turns on the furnace. The response (heat) reverses the drop in temperature. Then, when the temperature rises to the set point, the thermostat turns the furnace off. Physiologists would call the sensor a *receptor* that is triggered by a *stimulus* (room temperature below the set point) and the furnace an *effector*, which produces a *response* (heat). The thermostat represents a *control center,* which processes information from the receptor and directs the response by the effector.

Many of the control centers that maintain homeostasis in animals are located in the brain. For example, your "thermostat" operates by negative feedback to switch on and off mechanisms that maintain body temperature around 37°C. As shown in the upper part of **Figure 20.14**, when the thermostat senses a rise in temperature above the set point, it activates cooling mechanisms, such as the dilation of blood vessels in the skin and sweating. Once body temperature returns to normal, the thermostat shuts off these cooling mechanisms. When body temperature falls below the set point (lower part of the figure), the thermostat activates warming mechanisms, such as constriction of blood vessels to reduce heat loss and shivering to generate heat. Again, a return to normal temperature shuts off these mechanisms.

As we examine the body's organ systems in detail in the chapters of this unit, we will see many examples of homeostatic control and negative feedback, as well as constant reminders of the relationship between structure and function.

Web/CD Activity 20H *Regulation: Negative and Positive Feedback*

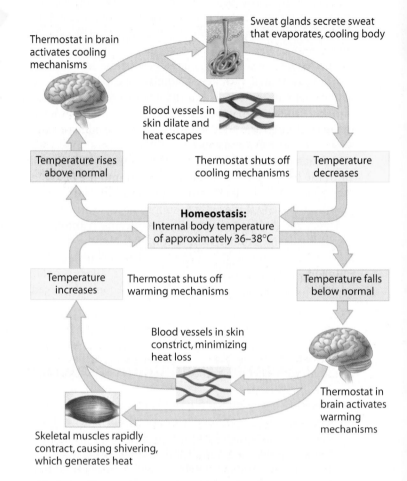

Figure 20.14 Control of body temperature

❓ Explain how homeostatic control is achieved by negative-feedback mechanisms.

■ Negative feedback maintains an internal balance by triggering mechanisms that reverse the movement of variables away from a set point.

CHAPTER REVIEW

Reviewing the Concepts

The Hierarchy of Structural Organization in an Animal (Introduction–20.11)

Correlation between structure and function is one of biology's most fundamental concepts. Anatomy is the study of structure; physiology studies how structures function **(Introduction–20.1).**

Structural hierarchy of the body: cell, tissue, organ, organ system, and organism **(20.2).**

Tissues are groups of many similar cells that perform a specific function **(20.3).**

Tissue	Epithelial (20.4)	Connective (20.5)	Muscle (20.6)	Nervous (20.7)
Structure	Sheets of closely packed cells	Sparse cells in extracellular matrix	Long cells (fibers) with contractile proteins	Neurons with branching extensions
Function	Protection, exchange, secretion	Binding and support of other tissues	Movement of body parts	Transmission of nerve signals

Artificial tissues have medical applications **(20.8)**.

Organs are made of several tissues that collectively perform specific functions **(20.9)**.

Organ systems consist of several related organs. The coordination of organ systems produces a functional organism. The integumentary system covers and protects the body. Skeletal and muscular systems support and move it. The digestive and respiratory systems obtain food and oxygen, and the circulatory system transports these materials. The excretory system disposes of certain wastes, while the immune and lymphatic systems protect the body from infection and cancer. The nervous and endocrine systems control and coordinate body functions. The reproductive system produces offspring **(20.10)**. New technologies, such as CT, MRI, MRM, and PET, are used in medical diagnosis and research **(20.11)**.

Exchanges with the External Environment (20.12–20.14)

Surface areas for exchange. Large, complex animals have specialized structures that increase surface area. Exchange of materials between blood and body cells takes place through the interstitial fluid **(20.12)**.

Homeostasis. Animals regulate their internal environment to achieve homeostasis, an internal steady state **(20.13)**. Control systems detect change and direct responses. Negative-feedback mechanisms keep internal variables fairly constant, with small fluctuations around set points **(20.14)**.

Connecting the Concepts

1. There are several key concepts introduced in this chapter: Structure correlates with function; an animal body has a hierarchy of organization; and complex bodies have structural adaptations to increase surface areas for exchange. Label the tissue layers shown in this section of the small intestine and describe how this diagram illustrates these key concepts.

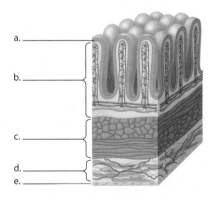

a. _____
b. _____
c. _____
d. _____
e. _____

Testing Your Knowledge

Multiple Choice

2. Which of the following pairs of body systems primarily regulates the activities of the other systems?
   a. circulatory and muscular systems
   b. nervous and endocrine systems
   c. lymphatic and integumentary systems
   d. endocrine and lymphatic systems
   e. integumentary and nervous systems

3. The cells in the human body are in contact with an internal environment consisting of
   a. blood.   d. matrix.
   b. connective tissue.   e. mucous membranes.
   c. interstitial fluid.

4. Negative-feedback mechanisms are
   a. most often involved in maintaining homeostasis.
   b. activated only when a variable rises above a set point.
   c. analogous to a furnace that produces heat.
   d. involved in contractions during childbirth.
   e. found only in birds and mammals.

5. Which of the following best illustrates homeostasis? (*Explain your answer.*)
   a. Most adult humans are between 5 and 6 feet tall.
   b. The lungs and intestines have large surface areas.
   c. When blood salt concentration goes up, the kidney expels more salt.
   d. All the cells of the body are about the same size.
   e. When oxygen in the blood decreases, you feel dizzy.

Matching (*Terms in the right-hand column may be used more than once.*)

6. Closely packed cells covering a surface   a. connective tissue

7. Neurons   b. muscle tissue

8. Adipose tissue, blood, and cartilage   c. nervous tissue

9. May be simple or stratified   d. epithelial tissue

10. Scattered cells embedded in matrix

11. Senses stimuli and transmits signals

12. Cells are called fibers

13. Cells may be squamous, cuboidal, or columnar

14. Skeletal, cardiac, or smooth

Describing, Comparing, and Explaining

15. Briefly explain how the structure of each of these tissues is well suited to its function: stratified squamous epithelium in the skin, neurons in the brain, simple squamous epithelium lining the lung, bone in the skull.

16. Describe ways in which the bodies of large, complex animals are structured for exchanging materials with the environment. Why can some smaller creatures get along without such structural features?

Applying the Concepts

17. Some companies are promoting whole-body CT scans as general health screens that consumers may purchase. Many parents now request CT scans when they take an injured child to the ER. There is growing concern that widespread and repeated use of this technology is exposing adults and children to potentially high levels of radiation for procedures that may not be medically necessary. How can consumers evaluate the health risks versus benefits of these procedures? What is the government's role in regulating these commercial ventures?

Answers to all questions can be found in Appendix 3.

For study help and Activities, go to campbellbiology.com or the student CD-ROM.

OBTAINING AND PROCESSING FOOD

21.1   Animals ingest their food in a variety of ways
21.2   Overview: Food processing occurs in four stages
21.3   Digestion occurs in specialized compartments

HUMAN DIGESTIVE SYSTEM

21.4   The human digestive system consists of an alimentary canal and accessory glands
21.5   Digestion begins in the oral cavity
21.6   The food and breathing passages both open into the pharynx
21.7   The Heimlich maneuver can save lives
21.8   The esophagus squeezes food along to the stomach by peristalsis
21.9   The stomach stores food and breaks it down with acid and enzymes
21.10 Bacterial infections can cause ulcers
21.11 The small intestine is the major organ of chemical digestion and nutrient absorption
21.12 The large intestine reclaims water and compacts the feces

DIETS AND DIGESTIVE ADAPTATIONS

21.13 Adaptations of vertebrate digestive systems reflect diet

NUTRITION

21.14 Overview: A healthy diet satisfies three needs
21.15 Chemical energy powers the body
21.16 An animal's diet must supply essential nutrients
21.17 Vegetarians must be sure to obtain all eight essential amino acids
21.18 A healthy diet includes 13 vitamins
21.19 Essential minerals are required for many body functions
21.20 Do you need to take vitamin and mineral supplements?
21.21 What do food labels tell us?
21.22 Obesity is a human health problem
21.23 What are the health risks and benefits of fad diets?
21.24 Diet can influence cardiovascular disease and cancer

CHAPTER 21

Getting Their Fill of Krill

WHALES ARE THE LARGEST ANIMALS in the world. Few other species, living or extinct, even approach their great size. The humpback whale, shown in the pictures on the facing page, is a medium-sized member of the whale clan. It can be 16 m long and weigh up to 65,000 kg (72 tons), about as much as 70 mid-size cars.

It takes an enormous amount of food to support a 72-ton animal. Humpback whales eat small fishes and crustaceans called krill, shown above. The painting on the next page shows a remarkable technique humpbacks often use to corral food organisms before gulping them in. Beginning about 20 m below the ocean surface, a humpback swims slowly in an upward spiral, blowing air bubbles as it goes. The rising bubbles form a cylindrical screen, or "bubble net." Krill and fish inside the bubble net swim away from the bubbles and become concentrated in the center of the cylinder. The whale then surges up through the center of the net with its mouth open, harvesting the catch in one giant gulp.

Humpback whales strain their food from seawater. Instead of teeth, these giants have an array of comblike plates called baleen on each side of their upper jaw. Notice the white baleen in the open mouth of the whale in the photograph on the facing page. To start feeding, a humpback whale opens its mouth, expands its throat, and takes a huge gulp of seawater. When its mouth closes, the baleen acts as a sieve: Water is forced back out through spaces in the baleen, trapping a mass of food in the mouth. The food is then swallowed whole, passing into the stomach, where digestion begins. The humpback's stomach can hold about half a ton of food at a time, and in a typical day, the animal's digestive system will process as much as 2 tons of krill and fish.

In about four months, a humpback whale eats and digests over 200 tons of food

The humpback and most other large whales are endangered species, having been hunted almost to extinction for meat and whale oil by the 1960s. Today, most nations honor an international ban on whaling, and some species are showing signs of recovery. Humpbacks still roam the Atlantic and Pacific oceans. They feed in polar regions during summer months and migrate

Nutrition and Digestion

to warmer oceans to breed when temperatures begin to fall. The photograph on this page was taken during summer in the Pacific Northwest. Food is so abundant there that humpbacks harvest much more energy than they burn each day. Much of the excess is stored as a thick layer of fat, or blubber, just under their skin. After a summer of feasting, humpback whales leave Glacier Bay and head south to breeding and calving grounds off the Hawaiian Islands, some 6,000 km away. Living off body fat, they eat little, if at all, until they return to Alaskan waters eight months later.

In about four months, a humpback whale eats and digests over 200 tons of food and stores enough fat to keep its 72-ton body active for another eight months—a remarkable feat and a fitting introduction to this chapter on animal nutrition and digestion. ■ ■ ■

21.1 Animals ingest their food in a variety of ways

All animals eat other organisms—dead or alive, whole or by the piece. In general, animals fall into one of three dietary categories. **Herbivores** (from the Latin *herba*, green crop, and *vorus*, devouring), such as cattle, gorillas, snails, and sea urchins, eat mainly autotrophs (plants and algae). **Carnivores** (from the Latin *carne*, flesh), such as lions, hawks, spiders, and snakes, eat other animals. Animals that ingest *both* plants and animals are called **omnivores** (from the Latin *omnis*, all). Omnivores include crows, cockroaches, raccoons, and humans, who evolved as hunters, gatherers, and scavengers.

How do animals obtain and ingest their food? There are a variety of ways. **Suspension feeders** extract food particles suspended in the surrounding water. For example, the humpback whale described in the chapter introduction uses its baleen to sift krill and small fish from the water. Clams and oysters are also suspension feeders. A film of mucus on their gills traps tiny morsels suspended in the water, and beating cilia on the gills sweep the food along to the mouth. Tube worms (**Figure 21.1A**) filter food particles with their feathery tentacles.

Substrate feeders live in or on their food source and eat their way through it. **Figure 21.1B** shows a leaf miner caterpillar, the larva of a moth, eating its way through the soft green tissue inside an oak leaf. The dark spots are a trail of feces that the caterpillar leaves in its wake. Earthworms are also substrate feeders. They eat their way through the soil, digesting partially decayed organic material as they go. In doing so, they help aerate the soil, making it more suitable for plants.

Fluid feeders obtain food by sucking nutrient-rich fluids from a living host, either a plant or an animal.

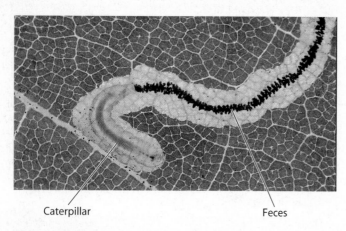

Figure 21.1A A suspension feeder (tube worm)

Aphids, for example, tap into the sugary sap in plants. Bloodsuckers, such as mosquitoes and ticks, pierce animals with needlelike mouthparts. The female mosquito in **Figure 21.1C** has just filled her abdomen with a meal of human blood. (Only female mosquitoes suck blood; males live on plant nectar.) In contrast to such parasitic fluid feeders, which harm their hosts, some fluid feeders actually benefit their hosts. For example, hummingbirds and bees move pollen between flowers as they fluid-feed on nectar.

Rather than filtering food from water, eating their way through a substrate, or sucking fluids, most animals are **bulk feeders**, meaning they ingest relatively large pieces of food. **Figure 21.1D** shows a great blue heron preparing to swallow its prey. A bulk feeder uses such diverse utensils as tentacles, pincers, claws, poisonous fangs, or jaws and teeth to kill its prey, to tear off pieces of meat or vegetation, or to take mouthfuls of animal or plant products.

Whatever the type of food or feeding mechanism, the processing of food involves four stages, as we see next.

? Blue whales, the largest animals ever to live, feed on krill. Blue whales are _____ feeders.

■ suspension

Figure 21.1C A fluid feeder (mosquito)

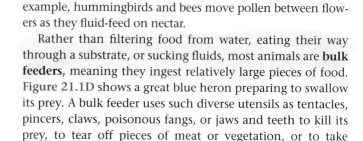

Caterpillar   Feces

Figure 21.1B A substrate feeder (caterpillar)

Figure 21.1D A bulk feeder (great blue heron)

21.2 Overview: Food processing occurs in four stages

So far we have discussed what animals eat and how they feed. As shown in **Figure 21.2A**, ❶ **ingestion**, the act of eating, is only the first of four main stages of food processing. ❷ The second stage, **digestion**, is the breaking down of food into molecules small enough for the body to absorb. Digestion typically occurs in two phases. First, food may be mechanically broken into smaller pieces. In animals with teeth, the process of chewing or tearing breaks large chunks of food into smaller ones. The second phase of digestion is the chemical breakdown process called hydrolysis. Catalyzed by specific enzymes, hydrolysis breaks chemical bonds in food molecules by adding water to them (see Module 3.3).

Most of the organic matter in food consists of proteins, fats, and carbohydrates—all large polymers (multi-unit molecules made up of small units called monomers). Animals cannot use these materials directly for two reasons. First, as macromolecules, these polymers are too large to pass through plasma membranes and enter the cells. Second, an animal needs monomers to make the polymers of its own body. Most of the polymers in food (for instance, the proteins in beans) are different from those that make up an animal's body.

All organisms use the same monomers to make their polymers. For instance, cats, humans, and bean plants all make their proteins from the same 20 kinds of amino acids. Digestion in an animal breaks the macromolecules in food into their component monomers. As shown in **Figure 21.2B**, proteins are split into amino acids, polysaccharides and disaccharides are split into monosaccharides, nucleic acids are split into nucleotides, and fats are split into glycerol and fatty acids. The animal can then use these monomers to make the specific polymers it needs (see Modules 6.14 and 6.15).

The last two stages of food processing occur after digestion. ❸ In the third stage, **absorption**, the cells lining the digestive tract take up (absorb) the products of digestion—small molecules such as amino acids and simple sugars.

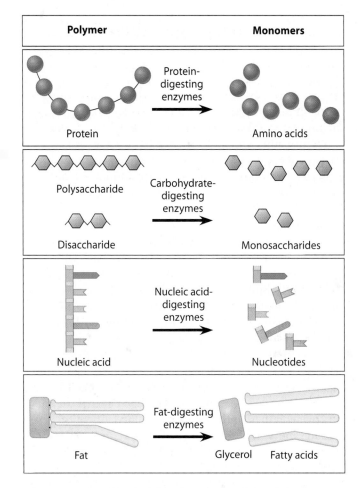

Figure 21.2B Chemical digestion: the breakdown of polymers to monomers

From the digestive tract, these nutrients travel in the blood to body cells, where they are joined together to make the macromolecules of the cells or broken down further to provide energy. In an animal that eats much more than its body immediately uses, many of the nutrient molecules are converted to fat for storage. ❹ In the fourth and last stage of food processing, **elimination**, undigested material passes out of the digestive tract.

How can an animal digest food without digesting its own cells and tissues? After all, digestive enzymes hydrolyze the same biological materials (such as proteins, carbohydrates, and fats) that animals are made of—and it is obviously important to avoid digesting oneself! Animals avoid the risk of self-digestion by processing food in specialized compartments, as we discuss in the next module.

> ? What are the two main digestive processes?

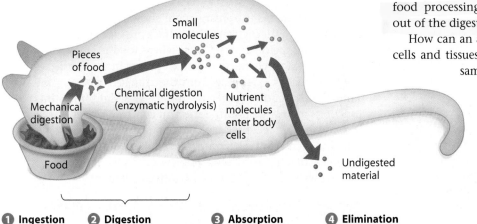

❶ Ingestion ❷ Digestion ❸ Absorption ❹ Elimination

Figure 21.2A The four main stages of food processing

■ Mechanical breakdown and chemical breakdown (enzymatic hydrolysis)

Food vacuoles are the simplest digestive compartments. A cell engulfs food by phagocytosis, and the newly formed food vacuole fuses with a lysosome containing hydrolytic enzymes (see Module 4.10). Sponges (see Module 18.5) digest their food entirely in food vacuoles. In contrast, most animals have an internal compartment in which digestion occurs outside of cells, enabling an animal to devour much larger food than could be ingested by phagocytosis alone.

As we saw in Chapter 18, cnidarians and flatworms have a **gastrovascular cavity**, a digestive compartment with a single opening, the **mouth**. **Figure 21.3A** shows a hydra digesting a small crustacean called *Daphnia*. ❶ Gland cells lining the gastrovascular cavity secrete digestive enzymes that ❷ break down the soft tissues of the prey. ❸ Other cells engulf small food particles, which ❹ are broken down in food vacuoles. Undigested materials are expelled through the mouth.

Most animals have an **alimentary canal**, a digestive tube with two openings, a mouth and an anus. Because food moves in one direction, specialized regions of the tube can carry out digestion and absorption of nutrients in sequence.

Food entering the mouth usually passes into a **pharynx**, or throat. Depending on the species, the **esophagus** may channel food to a crop, gizzard, or stomach. A **crop** is a pouch-like organ in which food is softened and stored. **Stomachs** and **gizzards** may also store food temporarily, but they are more muscular and they churn and grind the food. Chemical digestion and nutrient absorption occur mainly in the **intestine**. Undigested materials are expelled through the **anus**.

Figure 21.3B illustrates three examples of alimentary canals. The digestive tract of an earthworm includes a muscular pharynx that sucks food in through the mouth. Food passes through the esophagus and is stored in the crop. The muscular gizzard, which contains small bits of sand and gravel, pulverizes the food. Digestion and absorption occur

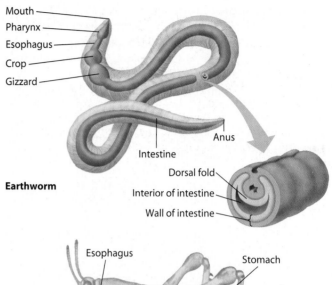

Earthworm

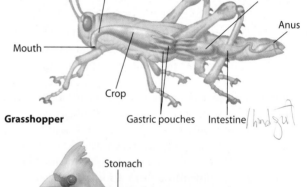

Grasshopper

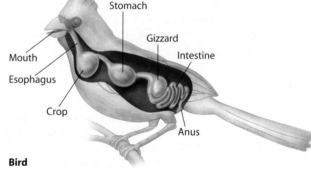

Bird

Figure 21.3B Three examples of alimentary canals

in the intestine. As the enlargement shows, a dorsal fold of the intestinal wall increases the surface area for absorption.

A grasshopper also has a crop where food is stored and moistened. Most chemical digestion occurs in the stomach. Gastric pouches extending from the stomach increase the surface area for nutrient absorption. The short intestine functions mainly to absorb water and compact wastes.

Many birds have three separate chambers: a crop, a stomach, and a gravel-filled gizzard, in which food is pulverized. Chemical digestion and absorption occur in the intestine.

Next we look at the human digestive system.

? What is an advantage of an alimentary canal, compared to a gastrovascular cavity?

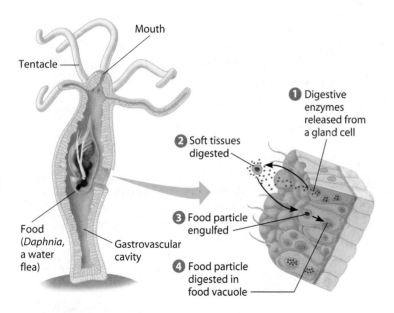

Figure 21.3A Digestion in the gastrovascular cavity of a hydra

■ Specialized regions can carry out digestion and absorption sequentially.

21.4 The human digestive system consists of an alimentary canal and accessory glands

As an introduction to our own digestive system, **Figure 21.4** provides an overview of the human alimentary canal and the digestive glands associated with it. The main parts of the canal are the mouth, oral cavity, tongue, pharynx, esophagus, stomach, small intestine, large intestine, rectum, and anus. The digestive glands—the salivary glands, pancreas, and liver—are labeled in blue on the figure. They secrete digestive juices that enter the alimentary canal through ducts. Secretions from the liver are stored in the gallbladder before they are released into the intestine.

Once food is swallowed, muscles propel it through the alimentary canal by **peristalsis**, rhythmic waves of contraction of smooth muscles in the walls of the digestive tract (see Module 21.8). In only 5–10 seconds, food passes from the pharynx down the esophagus and into the stomach. A constriction at the base of the esophagus keeps food in the stomach.

A muscular ringlike valve, called the **pyloric sphincter**, regulates the passage of food out of the stomach and into the small intestine. The sphincter works like a drawstring, closing off the tube and keeping food in the stomach long enough for stomach acids and enzymes to begin digestion. The final steps of digestion and nutrient absorption occur in the small intestine over a period of 5–6 hours. Undigested material moves slowly through the large intestine (taking 12–24 hours), and feces are expelled through the anus.

In the next several modules, we follow a snack—an apple and some crackers and cheese—through the alimentary canal to see in more detail what happens to the food in each of the processing stations along the way.

? What is peristalsis, and what is its function in our digestive system?

■ Wavelike contractions of smooth muscles that move food along the alimentary canal

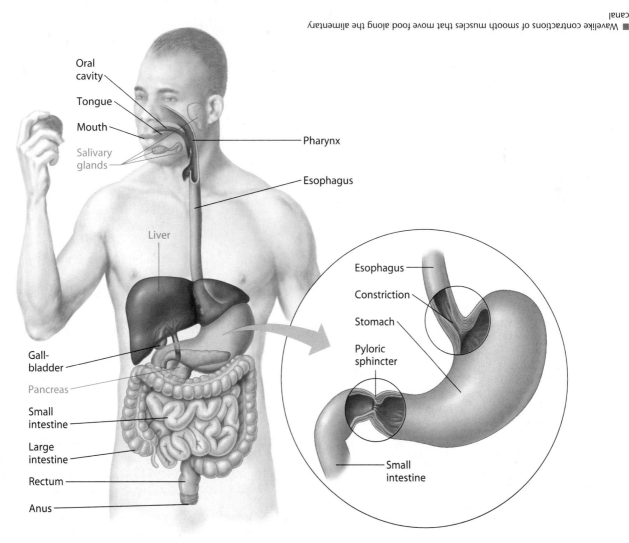

Figure 21.4 The human digestive tract

21.5 Digestion begins in the oral cavity

As you anticipate your apple, cheese, and crackers, your salivary glands may start delivering **saliva** through ducts to the oral cavity even before you take a bite. This is a response to the sight or smell (or even thought) of food. The presence of food in the oral cavity continues to stimulate salivation. In a typical day, your salivary glands secrete more than a liter of saliva.

Saliva contains several substances important in food processing. A slippery glycoprotein protects the soft lining of the mouth and lubricates food for easier swallowing. Buffers neutralize food acids, helping prevent tooth decay. Antibacterial agents kill many of the bacteria that enter the mouth with food. Saliva also contains salivary amylase, a digestive enzyme that begins hydrolyzing the starch in your cracker.

Mechanical and chemical digestion begin in the oral cavity. Chewing cuts, smashes, and grinds food, making it easier to swallow and exposing more food surface to digestive enzymes. As **Figure 21.5** shows, you have four kinds of teeth. Starting at the front on one side of the upper or lower jaw, there are two bladelike incisors. These you use for biting into your apple. Behind the incisors is a single pointed canine tooth. (Canine teeth are much bigger in carnivores, which use them to kill and rip apart prey.) Next come two premolars and three molars, which grind and crush the food. (The third molar, a "wisdom" tooth, does not appear in some people.)

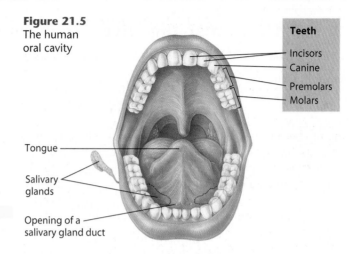

Figure 21.5
The human oral cavity

Teeth
- Incisors
- Canine
- Premolars
- Molars

Tongue

Salivary glands

Opening of a salivary gland duct

Also prominent in the oral cavity is the tongue, a muscular organ covered with taste buds. Besides enabling you to taste your meal, the tongue manipulates food and helps shape it into a ball called a bolus. In swallowing, the tongue pushes the bolus to the back of the oral cavity and into the pharynx.

? Chewing functions in _____ digestion, and salivary amylase initiates the chemical digestion of _____.

■ mechanical . . . starch

21.6 The food and breathing passages both open into the pharynx

Openings into both the esophagus and the **trachea** (windpipe) are in the pharynx. Most of the time, as shown on the left in **Figure 21.6**, the esophageal opening is closed off by a sphincter (blue arrows), and air enters the trachea and proceeds to the lungs (black arrows). This situation changes when you start to swallow some of the apple you've just finished chewing. A bolus of food enters the pharynx, triggering the swallowing reflex (center drawing); the esophageal sphincter relaxes and allows the bolus to enter the esophagus (green arrow). At the same time, the larynx (voice box) moves upward and tips the epiglottis (a flap of cartilage and fibrous connective tissue) over the tracheal opening. In this position, the epiglottis prevents food from passing into the trachea. You can see this motion in the bobbing of your larynx (also called your Adam's apple) during swallowing. After the bolus enters the esophagus, the larynx moves downward, the epiglottis moves up again, and the breathing passage reopens (right drawing). The esophageal sphincter contracts above the bolus.

? What prevents food from going down the wrong tube?

■ The epiglottis tips down over the opening to the trachea during swallowing.

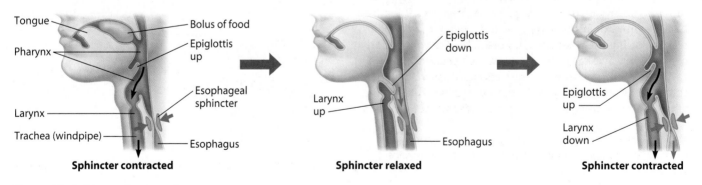

Tongue · Bolus of food · Epiglottis up

Pharynx

Esophageal sphincter

Larynx

Trachea (windpipe) · Esophagus

Sphincter contracted

Epiglottis down · Larynx up · Esophagus

Sphincter relaxed

Epiglottis up · Larynx down

Sphincter contracted

Figure 21.6 The swallowing reflex

21.7 The Heimlich maneuver can save lives

Sometimes our swallowing mechanism goes awry. A person may eat too quickly or fail to chew food thoroughly. Or an infant may swallow a toy or other object too big to pass through the esophagus. Such mishaps can lead to a blocked pharynx or trachea. Air cannot flow into the trachea, causing the person to choke. If breathing is not restored within minutes, brain damage or death may result.

To save someone who is choking, it is essential to dislodge any foreign objects in the throat and get air flowing. This quick assistance often comes through the use of the Heimlich maneuver. The procedure, invented by Dr. Henry Heimlich in the 1970s, allows people with little medical training to step in and aid a choking victim.

The maneuver is often performed on someone who is seated or standing up. Stand behind the victim and place your arms around the victim's waist. Make a fist with one hand, and place it against the victim's upper abdomen, well below the rib cage. Then place the other hand over the fist and press into the victim's abdomen with a quick upward thrust. When done correctly, the diaphragm is forcibly elevated, pushing air into the trachea. This procedure should be repeated until the lodged object is forced up and out of the victim's airway (Figure 21.7). The maneuver can also be performed on drowning victims to first clear the lungs of water before beginning CPR. And individuals can use their own fists or the back of a chair to force air upward and dislodge a foreign object from their own pharynx or trachea.

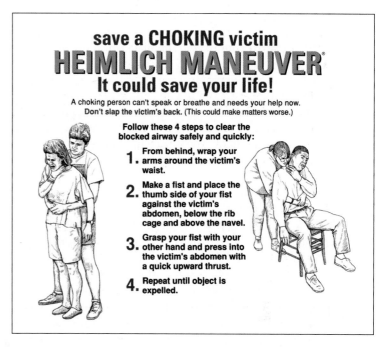

Figure 21.7 The Heimlich maneuver for helping choking victims

? If a piece of food is stuck in the pharynx, what effect could it have on nearby structures? (*Hint:* See Figure 21.6.)

■ The epiglottis may be tipped down, blocking the opening to the trachea.

21.8 The esophagus squeezes food along to the stomach by peristalsis

The esophagus is a muscular tube that conveys food boluses from the pharynx to the stomach. The muscles at the very top of the esophagus are under voluntary control; thus, the act of swallowing begins voluntarily. But then involuntary waves of contraction by smooth muscles in the rest of the esophagus take over. Figure 21.8 shows how waves of muscle contraction—peristalsis—squeeze a bolus toward the stomach. As food—such as your chewed up bite of apple—is swallowed, muscles above the bolus contract (blue arrows), pushing the bolus downward. At the same time, muscles around the bolus relax, allowing the passageway to open. Muscle contractions continue in waves until the bolus enters the stomach.

Waves of smooth muscle contraction also move materials through the small and large intestine. But first we explore what happens when your snack reaches the stomach.

? How does food get from the pharynx to the stomach of an astronaut in the weightless environment of a space station?

■ By peristalsis.

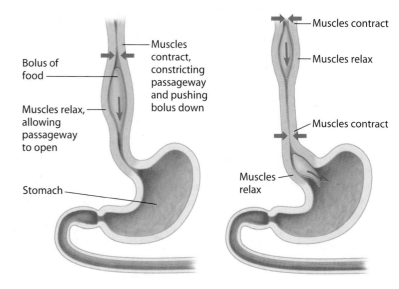

Figure 21.8 Peristalsis moving a food bolus down the esophagus

21.9 The stomach stores food and breaks it down with acid and enzymes

Having a stomach is the main reason we do not need to eat constantly. Our stomach is highly elastic and can stretch to accommodate about 2 liters (L) of food and drink, usually enough to satisfy our body's needs for many hours.

Some chemical digestion occurs in the stomach. The stomach secretes **gastric juice**, which is made up of mucus, enzymes, and strong acid. One function of the acid is to break apart the cells in food. The acid also kills most bacteria and other microbes that are swallowed with food.

The interior surface of the stomach wall is highly folded, and as both the micrograph and diagram in **Figure 21.9** show, it is dotted with pits leading down into tubular gastric glands. The gastric glands have three types of cells that secrete different components of the gastric juice. Mucous cells (dark pink) secrete mucus, which lubricates and protects the cells lining the stomach; parietal cells (yellow) secrete hydrochloric acid (HCl); and chief cells (tan) secrete pepsinogen, an inactive form of the enzyme pepsin.

The diagram on the far right of the figure indicates how pepsinogen, HCl, and pepsin interact in the stomach. ❶ Pepsinogen and HCl are secreted into the lumen (cavity) of the stomach. ❷ Next, the HCl converts pepsinogen to pepsin. ❸ Pepsin then activates more pepsinogen, starting a chain reaction. Pepsin begins the chemical digestion of proteins—those in your cheese snack, for instance. It splits the polypeptide chains of the proteins into smaller polypeptides. This action primes the proteins for further digestion, which will occur in the small intestine.

What prevents gastric juice from digesting away the stomach lining? Secreting pepsin in the inactive form of pepsinogen helps protect the cells of the gastric glands, and mucus helps protect the stomach lining from both pepsin and acid. Still, the epithelium is constantly eroded. Enough new cells are generated by mitosis to replace the stomach lining completely about every three days.

Cells in our gastric glands do not secrete gastric juice constantly. Their activity is regulated by a combination of nerve signals and hormones. When you see, smell, or taste food, a signal from your brain to your stomach stimulates your gastric glands to secrete gastric juice. Once you have food in your stomach, substances in the food stimulate cells in the stomach wall to release the hormone **gastrin** into the circulatory system. Gastrin circulates in the bloodstream, returning to the stomach wall. When it arrives there, it stimulates further secretion of gastric juice. As much as 3 L of gastric juice may be secreted a day. A negative-feedback mechanism like the one we described in Module 20.14 inhibits the secretion of gastric juice when the stomach contents become too acidic. The acid inhibits the release of gastrin, and with less gastrin in the blood, the gastric glands secrete less gastric juice.

About every 20 seconds, the stomach contents are mixed by the churning action of muscles in the stomach wall. You

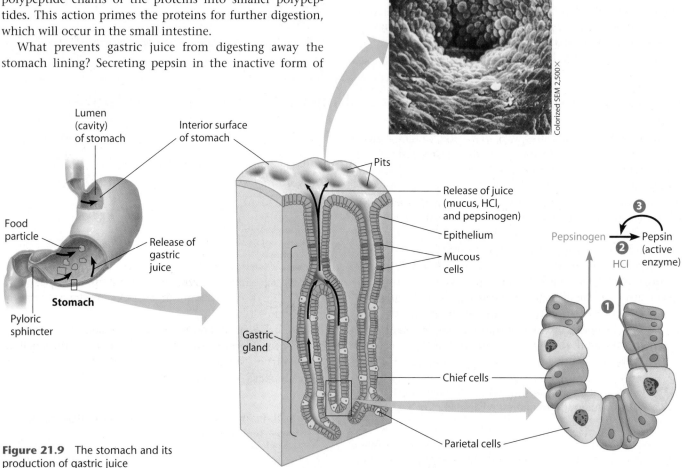

Figure 21.9 The stomach and its production of gastric juice

may feel hunger pangs when your empty stomach churns. (Other sensations of hunger result from appetite-controlling hormones that we will discuss in Module 21.22.) As a result of mixing and enzyme action, what begins in the stomach as a recently swallowed apple, cracker, and cheese snack becomes a nutrient-rich broth known as **acid chyme.**

The opening between the esophagus and the stomach is usually closed until a bolus arrives. Occasional backflow of acid chyme into the lower end of the esophagus causes the feeling we call heartburn (which should more accurately be called esophagus-burn). Some people suffer this backflow frequently and severely enough to harm the lining of the esophagus, a condition called GERD (gastroesophageal reflux disease).

Between the stomach and the small intestine, the pyloric sphincter helps regulate the passage of acid chyme from the stomach into the small intestine. With the acid chyme leaving the stomach only a squirt at a time, the stomach takes about 2–6 hours to empty after a meal. An acid chyme rich in fats stimulates the small intestine to release a hormone that slows the emptying of the stomach, providing more time for the digestion of fats in the intestine. Other hormones secreted by the small intestine influence the release of digestive juices from the pancreas and gallbladder.

We'll continue with the digestion of your snack in Module 21.11. But first, we'll consider the digestive problem of gastric ulcers.

? If you add pepsinogen to a test tube containing protein dissolved in distilled water, not much protein will be digested. What inorganic substance could you add to the tube to accelerate protein digestion? What effect will it have?

■ Hydrochloric acid or some other acid will convert inactive pepsinogen to active pepsin, which will begin the digestion of proteins.

21.10 Bacterial infections can cause ulcers

A stomachful of digestive juice laced with strong acid breaks apart the cells in our food, kills bacteria, and begins the digestion of proteins. At the same time, these chemicals, acidic enough to dissolve iron nails, can be harmful. A gel-like coat of mucus normally protects the stomach wall from the corrosive effect of digestive juice, but this is not foolproof protection. When it fails, open sores called **gastric ulcers** can develop in the stomach wall. The symptoms are usually a gnawing pain in the upper abdomen, which may occur a few hours after eating.

Gastric ulcers were formerly thought to result from the production of too much pepsin and/or acid or too little mucus. For years, the blame was put on factors that may cause these effects, such as aspirin, ibuprofen, smoking, alcohol, coffee, and stress. However, strong evidence now points to a spiral-shaped bacterium called *Helicobacter pylori* as the primary culprit (**Figure 21.10**). The low pH of the stomach kills most microbes, but not the acid-tolerant *H. pylori*. This bacterium burrows beneath the mucus and releases harmful chemicals. Growth of *H. pylori* seems to result in a localized loss of protective mucus and damage to the cells lining the stomach. Numerous white blood cells move into the stomach wall to fight the infection, and their presence is associated with mild inflammation of the stomach, called gastritis. Gastric ulcers develop when pepsin and hydrochloric acid destroy cells faster than the cells can regenerate. Eventually, the stomach wall may erode to the point that it actually has a hole in it. This hole can lead to life-threatening infection within the abdomen or internal bleeding. *H. pylori* is found in 70–90% of ulcer and gastritis sufferers. It is also found, however, in more than 33% of healthy people. Some studies link *H. pylori* to certain kinds of stomach cancer.

Gastric ulcers usually respond to a combination of antibiotics and bismuth (the active ingredient in Pepto-Bismol),

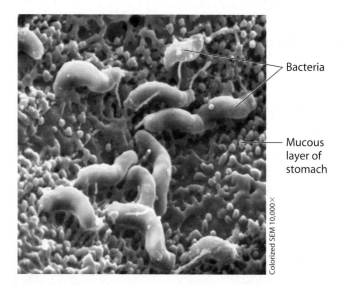

Bacteria

Mucous layer of stomach

Colorized SEM 10,000×

Figure 21.10 Ulcer-causing bacteria on the mucous layer of stomach

which eliminates the bacteria and promotes healing. Drugs that reduce stomach acidity may also help, and researchers are making progress toward developing a vaccine to prevent infection by *H. pylori*.

When digesting food leaves the stomach, it is accompanied by gastric juices, and so the first section of the small intestine—the duodenum—is also susceptible to ulcers, as is the esophagus in cases of severe GERD.

? In contrast to most microbes, the species that causes ulcers thrives in an environment with a very low _____.

■ pH

21.11 The small intestine is the major organ of chemical digestion and nutrient absorption

Returning to our journey through the digestive tract, what is the status of your snack as it passes out of the stomach into the small intestine? The food has been mechanically reduced to smaller pieces and mixed with juices; it now resembles a thick soup. Chemically, starch digestion began in the mouth, and protein breakdown began in the stomach. All other chemical digestion of the original macromolecules in the apple, cheese, and crackers occurs in the **small intestine**. Nutrients are also absorbed into the blood from the small intestine. With a length of over 6 m, the small intestine is the longest organ of the alimentary canal. (Its name is based not on its length but on its diameter, which is only about 2.5 cm; the large intestine is much shorter but has twice the diameter.)

Two large glandular organs, the pancreas and the liver, contribute to digestion in the small intestine (Figure 21.11A). The **pancreas** produces pancreatic juice, a mixture of digestive enzymes and an alkaline solution rich in bicarbonate. The alkaline solution neutralizes acid chyme as it enters the small intestine. The **liver** performs a wide variety of functions, including the production of bile. **Bile** contains bile salts that emulsify fats, making them more susceptible to attack by digestive enzymes. The **gallbladder** stores bile until it is needed in the small intestine. The first 25 cm or so of the small intestine is called the **duodenum.** This is where the acid chyme squirted from the stomach mixes with bile from the gallbladder, pancreatic juice from the pancreas, and digestive enzymes from gland cells in the intestinal wall itself.

Table 21.11 summarizes the processes of enzymatic digestion that occur in the small intestine. All four types of macromolecules (carbohydrates, proteins, nucleic acids, and fats) are digested. As we discuss the digestion of each, the table will help you keep track of the enzymes involved (shown in blue).

The digestion of carbohydrates that began in the oral cavity is completed in the small intestine. An enzyme called

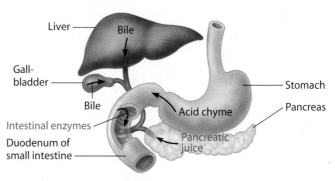

Figure 21.11A The small intestine and related digestive organs

pancreatic amylase hydrolyzes starch (a polysaccharide) into the disaccharide maltose. The enzyme maltase then splits maltose into the monosaccharide glucose. Maltase is one of a family of enzymes, each specific for the hydrolysis of a different disaccharide. Another enzyme, sucrase, hydrolyzes table sugar (sucrose), and lactase digests milk sugar (lactose, common in milk and cheese). Children generally have much more lactase than adults. Some adults lack lactase altogether, and ingesting milk products can give them cramps and diarrhea because they cannot digest the lactose (see the introduction to Chapter 3).

The small intestine also completes the digestion of proteins that was begun in the stomach. The pancreas and the duodenum secrete hydrolytic enzymes that completely dismantle polypeptides into amino acids. The enzymes trypsin and chymotrypsin break polypeptides into smaller polypeptides. Two other enzymes, aminopeptidase and carboxypeptidase, split off one amino acid at a time, working from both ends of a polypeptide. Another type of enzyme, dipeptidase, hydrolyzes fragments only two or three amino acids long.

TABLE 21.11 ENZYMATIC DIGESTION IN THE SMALL INTESTINE

Carbohydrates

Starch —— Pancreatic amylase ——→ Maltose (and other disaccharides) —— Maltase, sucrase, lactase, etc. ——→ Monosaccharides

Proteins

Polypeptides —— Trypsin, _or pepsin_ chymotrypsin ——→ Smaller polypeptides —— Aminopeptidase, carboxypeptidase, dipeptidase ——→ Amino acids

Nucleic acids

DNA and RNA —— Nucleases ——→ Nucleotides —— Other enzymes ——→ Nitrogenous bases, sugars, and phosphates

Fats

Fat globules —— Bile salts ——→ Fat droplets (emulsified) —— Lipase ——→ Fatty acids and glycerol

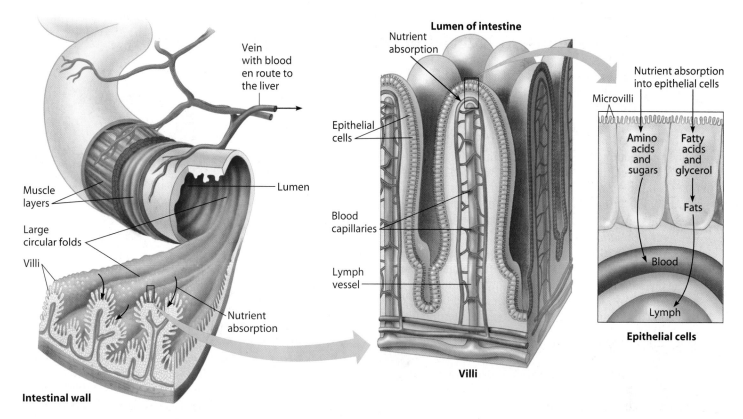

Figure 21.11B Structure of the small intestine

Working together, this enzyme team digests proteins much faster than any single enzyme could.

Yet another team of enzymes, the nucleases, hydrolyzes the nucleic acids in food. Nucleases from the pancreas split DNA and RNA (which are present in the cells of food items) into their component nucleotides. The nucleotides are then broken down into nitrogenous bases, sugars, and phosphates by other enzymes produced by the duodenal cells.

In contrast to starch and proteins, nearly all the fat in your cheese remains completely undigested until it reaches the duodenum. Hydrolysis of fats is a special problem because fats are insoluble in water. How is this problem solved? First, bile salts in bile cause fat globules to be physically broken up into smaller fat droplets, a process called emulsification. When there are many small droplets, a larger surface area of fat is exposed to lipase, an enzyme that breaks fat molecules down into fatty acids and glycerol.

By the time peristalsis has moved the mixture of chyme and digestive juices through the duodenum, chemical digestion of your meal is just about complete. The main function of the rest of the small intestine is the absorption of nutrients and water.

Structurally, the small intestine is well suited for its task of absorbing nutrients. Its lining has a huge surface area—roughly 300 m² , about the size of a tennis court. As **Figure 21.11B** shows, the extensive surface area results from several kinds of folds and projections. Around the inner wall of the small intestine are large circular folds with numerous small, fingerlike projections called **villi** (singular, *villus*). Each of the epithelial cells lining a villus has many tiny surface projections, called **microvilli**. The microvilli extend into the lumen of the intestine and greatly increase the surface area across which nutrients are absorbed.

Some nutrients are absorbed by simple diffusion; other nutrients are pumped against concentration gradients into the epithelial cells. Notice that a small lymph vessel (yellow) and a network of capillaries (red, purple, and blue) penetrate the core of each villus. After fatty acids and glycerol are absorbed by an epithelial cell, these building blocks are recombined into fats that are then transported into the lymph vessel. Other absorbed nutrients, such as amino acids and sugars, pass out of the intestinal epithelium and then across the thin walls of the capillaries into the blood.

The capillaries that drain nutrients away from the villi converge into larger veins and eventually into a main vessel, the hepatic portal vein, that leads directly to the liver. The liver thus gets first access to nutrients absorbed from a meal. The liver converts many of the nutrients into new substances that the body needs. One of its main functions is to remove excess glucose from the blood and convert it to glycogen (a polysaccharide), which is stored in liver cells. From the liver, blood travels to the heart, which pumps the blood and the nutrients it contains to all parts of the body. The nutrients from your apple, cheese, and crackers are now on their way to being incorporated into your body.

Web/CD Activity 21A *Digestive System Function*

Web/CD Thinking as a Scientist *What Role Does Amylase Play in Digestion?*

> [?] Amylase is to _____ as _____ is to DNA.

■ starch . . . nuclease

21.12 The large intestine reclaims water and compacts the feces

The **large intestine**, or **colon**, is about 1.5 m long and 5 cm in diameter. As the enlargement in **Figure 21.12** shows, it joins the small intestine at a T-shaped junction, where a sphincter controls the passage of unabsorbed food material out of the small intestine. One arm of the T is a blind pouch called the **cecum**. The **appendix**, a small, fingerlike extension of the cecum, contains a mass of white blood cells that make a minor contribution to immunity. Despite this role, the appendix itself is prone to infection (appendicitis). If this occurs, the appendix can be surgically removed without weakening the immune system.

The colon's main function is to absorb water from the alimentary canal. Altogether, about 7 L of fluid enters the lumen of the digestive tract each day as the solvent of the various digestive juices. About 90% of this water is absorbed back into the blood and tissue fluids, with the small intestine reclaiming most of it and the colon finishing the job. As the water is absorbed, the remains of the digested food become more solid as they are conveyed along the colon by peristalsis. These waste products of digestion, the **feces**, consist mainly of indigestible plant fibers (cellulose from your apple, for instance) and prokaryotes that normally live in the colon. Some of our colon bacteria, such as *E. coli*, produce important vitamins, including biotin, folic acid, several B vitamins, and vitamin K. These vitamins are absorbed into the bloodstream through the colon.

The terminal portion of the colon is the **rectum**, where the feces are stored until they can be eliminated. Strong contractions of the colon create the urge to defecate. Two rectal sphincters, one voluntary and the other involuntary, regulate the opening of the anus.

If the lining of the colon is irritated—by a viral or bacterial infection, for instance—the colon is less effective in re-

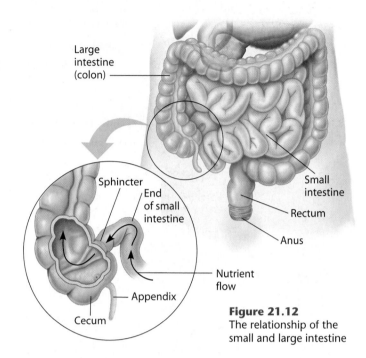

Figure 21.12
The relationship of the small and large intestine

claiming water, and diarrhea may result. The opposite problem, constipation, occurs when peristalsis moves the feces along too slowly; the colon reabsorbs too much water, and the feces become too compacted. Constipation often results from a diet that does not include enough plant fiber or from a lack of exercise.

? Explain why treatment of a chronic infection with antibiotics for an extended period of time may cause a vitamin K deficiency.

■ The antibiotics may kill the bacteria that synthesize vitamin K in the colon.

DIETS AND DIGESTIVE ADAPTATIONS

21.13 Adaptations of vertebrate digestive systems reflect diet

We have used our own alimentary canal to illustrate the basic plan of the vertebrate digestive system. However, vertebrate groups exhibit many variations on this plan. In each case, the structure and function of the animal's digestive system are adapted to the kind of food the animal eats.

The length of an animal's digestive tract tells us something about its diet. In general, herbivores and omnivores have longer alimentary canals, relative to their body size, than carnivores. A longer canal provides the extra time it takes to extract nutrients from vegetation, which is more difficult to digest than meat because of the cell walls in plant material. A longer canal also provides more surface area for absorbing nutrients, which are usually less concentrated in vegetation than in meat. A model case is the frog, which is carnivorous as an adult but mainly herbivorous as a tadpole, eating mostly

algae. A tadpole's intestine is long relative to its body size. When a tadpole transforms into an adult, the rest of its body grows more than its intestine, leaving the adult frog with an intestine that is shorter relative to its overall size.

In addition to a long alimentary canal, most herbivorous animals also have special chambers that house great numbers of microbes—bacteria and protists. The animals themselves cannot digest the cellulose in plants. The microbes break down the cellulose to simple sugars and other nutrients, which the animals then absorb directly or obtain by digesting the microbes.

Many herbivorous mammals—horses, elephants, rabbits and some rodents, for example—house cellulose-digesting microbes in the colon and in a large cecum, the pouch where the small and large intestines connect. Some of the nutrients

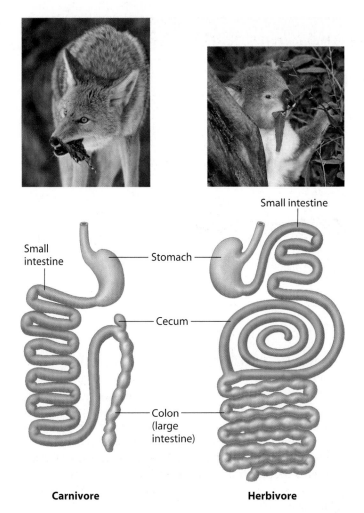

cecum digest plant material, making it possible for the koala to get almost all its food and water from the leaves of eucalyptus trees.

Ruminant mammals, such as cattle, sheep, and deer, have a more elaborate system for cellulose digestion. The stomach of a ruminant has four chambers, as shown in **Figure 21.13B.** The arrows in this figure indicate the pathway of food. When a cow first chews and swallows a mouthful of grass, the food enters ❶ the rumen and then ❷ the reticulum (green arrows). Bacteria and protists in the rumen and reticulum immediately go to work on the cellulose-rich meal, and the cow helps by periodically regurgitating and rechewing her food (red arrows). This rumination, or "chewing the cud," softens and helps break down plant fibers, making them more accessible to digestion by the microbes.

As the blue arrows indicate, the cow swallows her cud into ❸ the omasum, where water is absorbed. The cud finally passes to ❹ the abomasum, where the cow's own enzymes complete digestion. Here, a cow obtains many of its nutrients by digesting the microbes along with the nutrients they produce. The microbes reproduce so rapidly that their numbers remain stable despite this constant loss. With its microbes and multistage food-processing system, a ruminant harvests more energy and nutrients from the cellulose in hay or grass than a nonruminant herbivore such as a horse or an elephant.

? When a tadpole becomes an adult frog during metamorphosis, there is little growth of the intestine relative to the rest of the body. Thus, compared with the adult frog, a tadpole has a longer intestine relative to its body size. What does this suggest about the diets of these two stages in the frog's life history?

■ The tadpole is mainly herbivorous (eats mostly algae), while the adult frog is carnivorous (eats insects, for example).

Figure 21.13A The alimentary canal in a carnivore (coyote) and an herbivore (koala)

produced by the microbes are absorbed in the cecum and colon. Most of the nutrients, however, are lost in the feces because they do not go through the small intestine, the main site of nutrient absorption. Rabbits and some rodents obtain these nutrients by eating some of their feces, thus passing the food through the alimentary canal a second time. The feces from the second round of digestion are more compact and are not reingested. Many desert rodents conserve water by eating their first round of feces.

Figure 21.13A compares the digestive tract of a carnivore, the coyote, with that of an herbivore, the koala. The koala is an Australian marsupial (see Module 18.21). These two mammals are about the same size, but the koala's intestine is much longer and includes the longest cecum (about 2 m) of any animal of its size. Bacteria in the

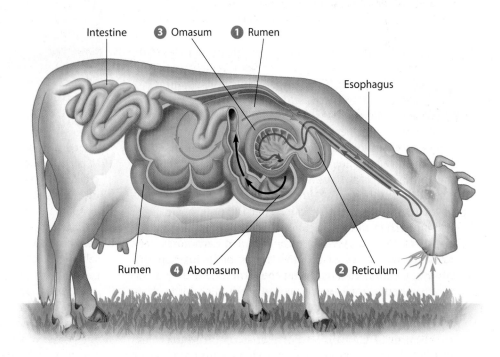

Figure 21.13B The digestive system of a ruminant mammal

21.14 Overview: A healthy diet satisfies three needs

All animals—whether herbivores like cows, carnivores like cats, or omnivores like humans—have the same basic nutritional needs. All animals must obtain (1) fuel to power all body activities, (2) organic raw materials to make the animal's own molecules, and (3) essential nutrients, or substances the animal cannot make for itself from raw materials but must obtain from food, with "no assembly required."

We have seen that digestion breaks the organic polymers in food into monomers. Cells can then oxidize these monomers for energy or assemble them into their own polymers—the proteins, carbohydrates, lipids, and nucleic acids needed to build and maintain cell structure and function. Starting with the need for fuel and paying particular attention to humans, we discuss basic nutritional needs in the rest of this chapter.

? **What is an "essential nutrient"?**

■ A substance that an organism requires but cannot make

21.15 Chemical energy powers the body

Reading a book, walking to class, eating and digesting food, and every other activity your body performs require fuel in the form of chemical energy. Cellular metabolism produces the body's energy currency, ATP, by oxidizing organic molecules digested from food (see Module 6.12). Usually, cells use carbohydrates and fats as fuel sources. But when carbohydrates or fats are in short supply, cells will use proteins as an energy source. Fats are especially rich in energy: The oxidation of a gram of fat liberates more than twice the energy liberated from a gram of carbohydrate or protein. The energy content of food is measured in **kilocalories** (1 **kcal** = 1,000 calories). The calories listed on food labels or referred to elsewhere regarding nutrition are actually kilocalories and are written as *Calories* (with a capital C).

Cellular metabolism must continuously drive several processes for an animal to remain alive. These include cell maintenance, breathing, the beating of the heart, and, in birds and mammals, the maintenance of body temperature. The number of kilocalories (Calories) a resting animal requires to fuel these essential processes for a given time is called the **basal metabolic rate (BMR).** The BMR for humans averages 1,300–1,500 kcal per day for adult females and about 1,600–1,800 kcal per day for adult males. This is about equivalent to the energy requirement of a 75-watt light bulb. But this is only a basal (base) rate—the amount of energy we "burn" lying motionless. Any activity, even working quietly at your desk, consumes kilocalories in addition to the BMR. The more strenuous the activity, the greater the energy demand. Table 21.15 gives you an idea of the amount of activity it takes for a 68-kg (150-lb) person to use up the kilocalories contained in several common foods.

What happens when we take in more kilocalories than we use up? Rather than discarding the extra energy, our cells store it in various forms. Our liver and muscles store energy in the form of glycogen, a polymer of glucose molecules. Most of us can store enough glycogen to supply about a day's worth of basal metabolism. Our cells also store excess energy as fat. This happens even if our diet contains little fat because the liver converts excess carbohydrates and proteins to fat. The average human's energy needs can be fueled by the oxidation of only 0.3 kg of fat per day. Most healthy people have enough stored fat to sustain them through several weeks of starvation. We discuss fat storage and its consequences in Module 21.22. But first we consider the essential nutrients that must be supplied in the diet.

? **What is the basal metabolic rate?**

■ The minimum number of kilocalories that a resting animal needs to maintain life's basic processes

TABLE 21.15 EXERCISE REQUIRED TO "BURN" THE CALORIES (KCAL) IN COMMON FOODS

	Jogging	Swimming	Walking
Speed of exercise	9 min/mi	30 min/mi	20 min/mi
kcal "burned"/hour	700	540	160
Cheeseburger ($\frac{1}{4}$ lb) 560 kcal	48 min	1 hr, 2 min	3 hr, 30 min
Cheese pizza (1 slice) 450 kcal	38 min	50 min	2 hr, 49 min
Soft drink (12 oz) 173 kcal	15 min	19 min	65 min
Whole wheat bread (1 slice) 100 kcal	9 min	11 min	38 min

These data are for a person weighing 68 kg (150 lb).

21.16 An animal's diet must supply essential nutrients

Besides providing fuel and organic raw materials, an animal's diet must also supply **essential nutrients**. These are materials that must be obtained in preassembled form because the animal's cells cannot make them from *any* raw material. An individual whose diet is chronically deficient in calories is said to be *undernourished*. An individual whose diet is missing one or more essential nutrients is said to be *malnourished*. Because a diet of a single staple such as rice or corn can often provide sufficient calories, undernourishment is generally common only where drought, war, or some other crisis has severely disrupted the food supply. Another cause of undernourishment is anorexia nervosa, an eating disorder in which a person does not eat enough because of an intense fear of becoming fat. In human populations, malnutrition is much more common than undernutrition, and it is even possible for an overnourished (obese) individual to be malnourished. There are four classes of essential nutrients: essential fatty acids, essential amino acids, vitamins, and minerals.

Our cells make fats and other lipids by combining fatty acids with other molecules, such as glycerol (see Module 3.8). We can make most of the fatty acids we need. Those we cannot make, called **essential fatty acids**, we must obtain in our diet. One essential fatty acid, linoleic acid, is especially important because it is needed to make some of the phospholipids of cell membranes. Most diets furnish ample amounts of essential fatty acids, and deficiencies are rare.

Adult humans cannot make eight of the 20 kinds of amino acids needed to synthesize proteins. These eight, known as the **essential amino acids**, must be obtained from the diet. (Infants also require a ninth, histidine.) Because the body cannot store excess amino acids, a deficiency of a single essential amino acid limits the use of other amino acids, impairs protein synthesis, and can lead to protein deficiency, a serious type of malnutrition. This is the most common type of malnutrition among humans. The victims are usually children, who, if they survive infancy, are likely to be retarded mentally and underdeveloped physically.

The simplest way to get all the essential amino acids is to eat meat and animal by-products such as eggs, milk, and cheese. The proteins in these products are said to be "complete," meaning they provide all the essential amino acids in the proportions needed by the body. In contrast, most plant proteins are "incomplete," or deficient in one or more essential amino acids. In the next module, we discuss how vegetarians can obtain all the essential amino acids in their diet.

? **What is the difference between undernutrition and malnutrition?**

■ Undernutrition is a diet deficient in calories; malnutrition is a diet deficient in an essential nutrient.

21.17 Vegetarians must be sure to obtain all eight essential amino acids

Vegetarian diets may range from avoiding meat to the vegan diet of avoiding animal products of any kind, including eggs, milk, and cheese. Is it possible for a person to be a vegetarian and obtain adequate nutrition? The answer is yes, but vegetarians have to know how to get all the essential nutrients.

People may become vegetarians by choice or, more commonly, because they simply cannot afford animal protein. Animal protein is more expensive to produce, and usually to buy, than plant protein, and most of the human population is primarily vegetarian. Nutritional problems can result when people have to rely on a single type of plant food—just corn, rice, or wheat, for instance. On such a limited diet, people are likely to become protein-deficient because they lack certain essential amino acids.

The key to being a healthy vegetarian is to eat a variety of plant foods that together supply sufficient quantities of all the essential amino acids. Simply by eating a combination of beans and corn, for example, vegetarians can get all the essential amino acids (**Figure 21.17**). Most societies have, by trial and error, developed balanced diets that prevent protein deficiency. The Mexican staple diet of corn tortillas and beans is one example.

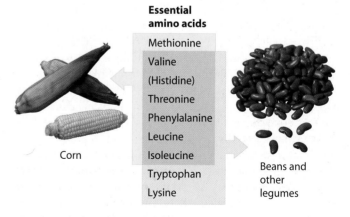

Essential amino acids

Methionine
Valine
(Histidine)
Threonine
Phenylalanine
Leucine
Isoleucine
Tryptophan
Lysine

Corn

Beans and other legumes

Figure 21.17 Essential amino acids from a vegetarian diet

? **Look carefully at Figure 21.17. A diet consisting strictly of beans would probably result in a deficiency of the essential amino acid _____.**

■ methionine

21.18 A healthy diet includes 13 vitamins

A **vitamin** is an organic nutrient that we must obtain from our diet, but is required in much smaller quantities than the essential amino acids. Most vitamins serve as coenzymes or parts of coenzymes; they have catalytic functions and are used over and over in metabolic reactions (see Module 5.7). Table 21.18 lists 13 essential vitamins and their major dietary sources. Though needed in only tiny amounts, vitamins are absolutely necessary, as you can see from the functions and symptoms of deficiencies listed in the table. Extreme excesses, however, can also be dangerous.

Water-soluble vitamins include the B complex and vitamin C. Fat-soluble vitamins include vitamins A, D, E, and K. In general, excess water-soluble vitamins will be eliminated in urine. Excessive amounts of fat-soluble vitamins, however, build up in body fat. Thus, overdoses may have toxic effects.

? Why are vitamins required in such small doses, compared with other essential organic nutrients?

■ Because vitamins generally have catalytic functions as coenzymes, and thus each vitamin molecule can repeat its function many times

TABLE 21.18 VITAMIN REQUIREMENTS OF HUMANS

Vitamin	Major Dietary Sources	Functions in the Body	Symptoms of Deficiency / Symptoms of Extreme Excess
Water-Soluble Vitamins			
Vitamin B-1 (thiamine)	Pork, legumes, peanuts, whole grains	Coenzyme used in removing CO_2 from organic compounds	Beriberi (nerve disorders, emaciation, anemia)
Vitamin B-2 (riboflavin)	Dairy products, meats, enriched grains, vegetables	Component of coenzyme FAD	Skin lesions such as cracks at corners of mouth
Niacin	Nuts, meats, grains	Component of coenzymes NAD^+ and $NADP^+$	Skin and gastrointestinal lesions, nervous disorders Liver damage
Vitamin B-6 (pyridoxine)	Meats, vegetables, whole grains	Coenzyme used in amino acid metabolism	Irritability, convulsions, muscular twitching, anemia Unstable gait, numb feet, poor coordination
Pantothenic acid	Most foods: meats, dairy products, whole grains, etc.	Component of coenzyme A	Fatigue, numbness, tingling of hands and feet
Folic acid (folacin)	Green vegetables, oranges, nuts, legumes, whole grains	Coenzyme in nucleic acid and amino acid metabolism; neural tube development in embryo	Anemia, gastrointestinal problems May mask deficiency of vitamin B-12
Vitamin B-12	Meats, eggs, dairy products	Coenzyme in nucleic acid metabolism; maturation of red blood cells	Anemia, nervous system disorders
Biotin	Legumes, other vegetables, meats	Coenzyme in synthesis of fat, glycogen, and amino acids	Scaly skin inflammation, neuromuscular disorders
Vitamin C (ascorbic acid)	Fruits and vegetables, especially citrus fruits, broccoli, cabbage, tomatoes, green peppers	Used in collagen synthesis (e.g., for bone, cartilage, gums); antioxidant; aids in detoxification; improves iron absorption	Scurvy (degeneration of skin, teeth, blood vessels), weakness, delayed wound healing, impaired immunity Gastrointestinal upset
Fat-Soluble Vitamins			
Vitamin A (retinol)	Dark green and orange vegetables and fruits, dairy products	Component of visual pigments; maintenance of epithelial tissues; antioxidant; helps prevent damage to cell membranes	Vision problems; dry, scaly skin Headache, irritability, vomiting, hair loss, blurred vision, liver and bone damage
Vitamin D	Dairy products, egg yolk (also made in human skin in presence of sunlight)	Aids in absorption and use of calcium and phosphorus; promotes bone growth	Rickets (bone deformities) in children; bone softening in adults Brain, cardiovascular, and kidney damage
Vitamin E (tocopherol)	Vegetable oils, nuts, seeds	Antioxidant; helps prevent damage to cell membranes	None well documented; possibly anemia
Vitamin K	Green vegetables, tea (also made by colon bacteria)	Important in blood clotting	Defective blood clotting Liver damage and anemia

21.19 Essential minerals are required for many body functions

Minerals are simple inorganic nutrients, usually required in small amounts. We must acquire the essential minerals listed in Table 21.19 from our diet; some of the major dietary sources are listed. The table also lists the functions and symptoms of deficiency for each mineral.

Along with other vertebrates, we humans require relatively large amounts of calcium and phosphorus to construct and maintain our skeleton. Too little calcium can result in the degenerative bone disease osteoporosis. Calcium is also necessary for the normal functioning of nerves and muscles, and phosphorus is an ingredient of ATP and nucleic acids. Iron is a component of hemoglobin, the oxygen-carrying protein of red blood cells, and of several electron carrier molecules that function in cellular respiration. Vertebrates need iodine to make the hormone thyroxine, which regulates metabolic rate. Many minerals are components of various enzymes.

Sodium, potassium, and chlorine are important in nerve function and help maintain the osmotic balance of cells. Most people ingest far more salt (sodium chloride) than they need. The average U.S. citizen eats enough salt to provide about 20 times the required amount of sodium. Ingesting too much sodium may be associated with high blood pressure.

? Which of the minerals listed in the table are involved with the formation or maintenance of teeth?

■ Calcium, phosphorus, and fluorine

TABLE 21.19 MINERAL REQUIREMENTS OF HUMANS

Mineral (chemical symbol)	Dietary Sources	Functions in the Body	Symptoms of Deficiency*
Calcium (Ca)	Dairy products, dark green vegetables, legumes	Bone and tooth formation, blood clotting, nerve and muscle function	Stunted growth, possibly loss of bone mass
Phosphorus (P)	Dairy products, meats, grains	Bone and tooth formation, acid-base balance, nucleotide synthesis	Weakness, loss of minerals from bone, calcium loss
Sulfur (S)	Proteins from many sources	Component of certain amino acids	Symptoms of protein deficiency
Potassium (K)	Meats, dairy products, many fruits and vegetables, grains	Acid-base balance, water balance, nerve function	Muscular weakness, paralysis, nausea, heart failure
Chlorine (Cl)	Table salt	Acid-base balance, water balance, nerve function, formation of gastric juice	Muscle cramps, reduced appetite
Sodium (Na)	Table salt	Acid-base balance, water balance, nerve function	Muscle cramps, reduced appetite
Magnesium (Mg)	Whole grains, green leafy vegetables	Component of certain enzymes	Nervous system disturbances
Iron (Fe)	Meats, eggs, legumes, whole grains, green leafy vegetables	Component of hemoglobin, of certain enzymes, and of electron carriers in energy metabolism	Iron-deficiency anemia, weakness, impaired immunity
Fluorine (F)	Fluoridated drinking water, tea, seafood	Maintenance of tooth (and probably bone) structure	Higher frequency of tooth decay
Zinc (Zn)	Meats, seafood, grains	Component of certain digestive enzymes and other proteins	Growth failure, scaly skin inflammation, reproductive failure, impaired immunity
Copper (Cu)	Seafood, nuts, legumes, organ meats	Component of enzymes in iron metabolism, electron transport	Anemia, bone and cardiovascular changes
Manganese (Mn)	Nuts, grains, vegetables, fruits, tea	Component of certain enzymes	Abnormal bone and cartilage
Iodine (I)	Seafood, dairy products, iodized salt	Component of thyroid hormones	Goiter (enlarged thyroid)
Cobalt (Co)	Meats, dairy products	Component of vitamin B-12	None, except as B-12 deficiency
Selenium (Se)	Seafood, meats, whole grains	Component of enzymes; functions in association with vitamin E	Muscle pain, maybe heart muscle deterioration
Chromium (Cr)	Brewer's yeast, liver, seafood, meats, some vegetables	Involved in glucose and energy metabolism	Impaired glucose metabolism
Molybdenum (Mo)	Legumes, grains, some vegetables	Component of certain enzymes	Disorder in excretion of nitrogen-containing compounds

*All of these minerals can be harmful when consumed in extreme excess.

21.20 Do you need to take vitamin and mineral supplements?

A healthy diet usually includes enough vitamins and minerals for most people and is considered the best source of these nutrition mainstays. Such diets meet the **Recommended Dietary Allowances (RDAs)**, minimum amounts of nutrients that are needed each day by healthy people, as determined by a national scientific panel. The subject of vitamin dosage, however, can cause heated scientific and popular debate. Some people argue that RDAs are set too low for some vitamins, and some of these people believe, probably mistakenly, that *massive* doses of vitamins confer health benefits. Research is far from complete, and debate continues, especially over optimal doses of vitamins C and E. Evidence indicates, however, that excessive amounts of some vitamins and minerals, such as vitamin A and iron, can definitely be harmful.

Vitamin and mineral supplements may be advisable for people who do not follow a healthy diet, who cannot consume certain foods, who are vegetarians, or who are dieting. Pregnant women and older people may have specific nutritional needs that require a supplement. However, unless recommended by a doctor, people should generally avoid megavitamins—supplements that far exceed daily recommended doses.

? **What are RDAs?**

■ Recommended Dietary Allowances: minimal daily amounts of nutrients determined to be required for health

21.21 What do food labels tell us?

Have you ever found yourself sitting at the breakfast table reading the label on a cereal box? What does it all mean?

The Food and Drug Administration (FDA) requires that various types of information be given on packaged-food labels, as shown in Figure 21.21. You'll find the ingredients listed in order from the greatest amount (by weight) to the least. There are also several kinds of "nutrition facts" found on food labels. First, a serving size of the food is defined according to standards set by the FDA. The energy content in "Calories" (that is, kilocalories) is listed per serving. Selected nutrients are also listed as amounts per serving and as percentages of a daily value. The daily values are based on a diet containing 2,000 kcal per day. For example, the 1.5 g of fat in a slice of this bread provides 2% of the daily fat allowance for a person needing 2,000 kcal per day.

Food labels emphasize nutrients believed to be associated with disease risks (fats, cholesterol, and sodium) and with a healthy diet (such as dietary fiber, protein, and certain vitamins and minerals). From the data shown, you can tell that each serving of this bread contains 19 g of total carbohydrate. Dietary fiber consists of indigestible complex carbohydrates, mainly cellulose. Subtracting 3 g of dietary fiber and 3 g of sugars (simple carbohydrates) from the 19 g of total carbohydrate tells you that each serving of this bread contains 13 g of digestible complex carbohydrate. This is chiefly starch.

To help consumers compare nutrient amounts in a particular food with their total daily needs, food labels also provide some general nutritional information. For example, the lower part of the label recommends less than 20 g of saturated fat and at least 25 g of dietary fiber for those following a 2,000-kcal daily diet.

Ingredients: whole wheat flour, water, high fructose corn syrup, wheat gluten, soybean or canola oil, molasses, yeast, salt, cultured whey, vinegar, soy flour, calcium sulfate (source of calcium).

Figure 21.21
Whole wheat bread label

Nutrition Facts

Serving Size 1 slice (43g)
Servings Per Container 16

Amount Per Serving

Calories 100   Calories from Fat 10

	% Daily Value*
Total Fat 1.5g	**2%**
Saturated Fat 0g	**0%**
Cholesterol 0mg	**0%**
Sodium 190mg	**8%**
Total Carbohydrate 19g	**6%**
Dietary Fiber 3g	**12%**
Sugars 3g	
Protein 4g	

Vitamin A 0%	•	Vitamin C 0%
Calcium 2%	•	Iron 4%
Thiamine 6%	•	Riboflavin 2%
Niacin 6%	•	Folic Acid 0%

* Percent Daily Values are based on a 2,000 calorie diet. Your daily values may be higher or lower depending on your calorie needs:

		Calories:	2,000	2,500
Total Fat	Less than		65g	80g
Sat. Fat	Less than		20g	25g
Cholesterol	Less than		300mg	300mg
Sodium	Less than		2,400mg	2,400mg
Total Carbohydrate			300g	375g
Dietary Fiber			25g	30g

Calories per gram:
Fat 9 • Carbohydrate 4 • Protein 4

? **What percent of the daily requirements for the fat-soluble vitamins is provided by a slice of the bread in Figure 21.21?** (*Hint:* Review Table 21.18.)

■ 0%

21.22 Obesity is a human health problem

The World Health Organization now recognizes obesity as a major global health problem. The increased availability of fattening foods and large portions in many countries combined with more sedentary lifestyles puts excess weight on bodies. In the United States, the percentage of obese (very overweight) people has doubled to 30% in the past two decades, and another 35% are overweight. Weight problems often begin at a young age; 15% of children and adolescents in the United States are overweight.

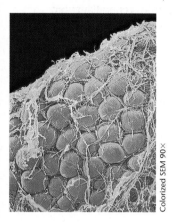

Colorized SEM 90×

Figure 21.22A Human fat cells

Obesity contributes to a number of health problems, including the most common type of diabetes, cancer of the colon and breasts, and cardiovascular disease. Obesity is estimated to be a factor in 300,000 deaths per year in the United States.

The obesity epidemic has stimulated an increase in scientific research on the causes and possible treatments for weight-control problems. Inheritance is one factor in obesity, which helps explain why certain people have to struggle so hard to control their weight, while others can eat and eat without gaining weight. Researchers have identified dozens of the genes that code for weight-regulating hormones. As researchers continue to study the signaling pathways that regulate both long-term and short-term appetite and the body's storage of fat, there is reason to be somewhat optimistic that obese people who have inherited defects in these weight-controlling mechanisms may someday be treated with a new generation of drugs. But so far, the diversity of defects in these complex systems has made it difficult to develop drugs that are effective and free from serious side effects.

The complexity of weight control in humans is evident from studies of the hormone leptin, one of the key long-term appetite regulators in mammals. Leptin is produced by adipose (fat) cells (**Figure 21.22A**). As adipose tissue increases, leptin levels in the blood rise, which normally cues the brain to suppress appetite. This is one of the feedback mechanisms that usually keep people from becoming obese in spite of access to an abundance of food. Conversely, loss of body fat decreases leptin levels, signaling the brain to increase appetite. Mice that inherit a defect in the gene for leptin become very obese (**Figure 21.22B**). Researchers found that they could treat these obese mice by injecting leptin.

The discovery of the leptin-deficiency mutation in mice made front-page news and initially generated much excitement because humans also have a leptin gene. And indeed, obese children who have inherited a mutant form of the leptin gene do lose weight after leptin treatments. However, relatively few obese people have such defects in leptin production. In fact, most obese humans have an abnormally high level of leptin, which, after all, is produced by adipose tissue. For some reason, the brain's satiety center does not respond to the high leptin levels in many obese people. One hypothesis is that in humans, perhaps in contrast to many other mammals, the function of the leptin system is to prevent weight loss, not protect against weight gain. Thus, the decline in leptin when body fat is lost stimulates appetite; but the high levels of leptin produced by large stores of body fat do not function to depress appetite. This physiological nuance may be a consequence of our evolutionary history.

Most of us crave foods that are fatty: fries, chips, burgers, cheese, and ice cream. Though fat hoarding can be a health liability today, it may actually have been advantageous in our evolutionary past. Only in the past few centuries have large numbers of people had access to a reliable supply of high-calorie food. Our ancestors on the African savanna were hunter-gatherers who probably survived mainly on seeds and other plant products, a diet only occasionally supplemented by hunting game or scavenging meat from animals killed by other predators. In such a feast-and-famine existence, natural selection may have favored those individuals with a physiology that induced them to gorge on rich, fatty foods on those rare occasions when such treats were available. Individuals with genes promoting the storage of fat during feasts may have been more likely than their thinner friends to survive famines. So perhaps our modern taste for fats, which contributes to the obesity epidemic, is partly an evolutionary vestige of our past, when food was less plentiful.

Next we explore some popular weight loss diets.

Figure 21.22B A mouse with a defect in a gene for leptin, an appetite-suppressing hormone (left); a normal mouse (right)

? What are two roles of the hormone leptin? Which of these roles does leptin appear not to play in obese humans?

■ A drop in leptin due to a loss of adipose tissue stimulates appetite; a high level of leptin, produced by increased body fat, depresses appetite. The second mechanism does not seem to function in some humans.

21.23 What are the health risks and benefits of fad diets?

As the numbers of people who are overweight or obese rise, so does interest in ways to shed body fat. The obesity epidemic has fueled explosive growth in the weight loss industry. Most of the trendy diets promoted by weight loss businesses claim to improve a person's health and appearance quickly and dramatically, and judging from the growing amounts of money flowing to these enterprises, dieters believe them. According to some estimates, the U.S. market for weight loss products and services, worth about $60 million in 1999, could reach more than $150 billion in 2007.

But the popularity of weight loss programs does not mean that all diet regimens are effective or medically sound. Weight loss can improve a person's health and, in a culture that values slender physiques, improve a person's self-image. But the health benefits and staying power of any diet are closely tied to how weight loss occurs. If a fad diet sheds pounds through unhealthy methods, any benefits may be short-lived, and long-term health problems may follow.

In recent years, many of the most popular—and controversial—weight loss schemes have focused on reduced intake of carbohydrates. Backers of these diets say that with fewer carbohydrates, the body must burn stored fat instead. Such diets dramatically restrict carbohydrate consumption, some allowing as little as 20 g of carbohydrates a day—less than 10% of the current RDA. Dieters on such plans are often told to replace carbohydrates with proteins and fats. People following "low-carb" diets often drop sugar, bread, fruits, and potatoes from their diets, swapping in cheese, nuts, and meat instead.

Some people have lost weight quickly on the regimen of a low-carbohydrate diet, and because of these success stories and the fatty foods the diet allows, the approach has surged in popularity. According to surveys, as many as one in five Americans have tried some type of low-carbohydrate diet. Americans spend as much as $15 billion a year on "low-carb" diet aids and foods. Although some studies have found these diets to be effective, others have found that they offer only short-lived benefits. Much of the initial weight reduction in

such a diet comes through water loss. Once the diet is stopped, the lost weight is quickly regained. The fatty foods encouraged in such diets may contribute to heart disease and kidney problems. Reductions in fruits and vegetables cut a person's intake of vitamins, minerals, and fiber, increasing the risk of diseases such as cancer. As a result, few doctors recommend low-carbohydrate diets as a healthy way to long-term weight loss.

Low-carbohydrate diets unseated low-fat diets, an earlier dieting trend with its own flood of low-fat (but often high-sugar) processed foods. Such low-fat regimens dramatically cut dietary fat, often reducing consumption of dairy products, meat, nuts, and oils. But a healthy body composition often requires a certain amount of fat. Low-fat diets often lack adequate amounts of fatty acids and protein, and may make it difficult for the body to absorb fat-soluble vitamins. A reasonable amount of fats and related lipids are essential components of the human body and seem to correlate with a healthy immune system.

Healthy women may have as much as 20–25% of their body weight in fat; for healthy men, the amount is typically 15–19%. Extremely thin people tend to have lower levels of vitamin A and beta-carotene in their blood, which may make them more susceptible to certain forms of cancer.

As the problems with fad diets reveal, the body requires a balance of nutrients for good health and long-term weight control. The best approach to weight control is a combination of exercise and a restricted but balanced diet that provides at least 1,200 kcal per day and adequate amounts of all essential nutrients (Table 21.23). A restricted, balanced diet, along with regular aerobic exercise, can trim the body gradually and keep extra fat off without harmful side effects.

? In what sense is maintaining a stable body weight a matter of caloric bookkeeping?

■ When we burn as many calories a day through BMR and activities as we take in with our food, a stable body weight will result.

TABLE 21.23 TYPES OF WEIGHT LOSS DIETS

Diet Type	Health Effects and Potential Problems
Extremely low-carbohydrate diets Less than 100 g of carbohydrates per day	Initial loss of weight is primarily water; problems may include fatigue and headaches; long-term adherence to diet may be associated with muscle loss
Extremely low-fat diets Less than 20% of kilocalories from fat	May be inadequate in essential fatty acids, protein, and certain minerals; may decrease absorption of fat-soluble vitamins; may result in irregular menstrual periods in women
Formula diets Based on formulated or packaged products; many are very low in kilocalories	If very low in kilocalories (less than 800 kcal per day), may result in loss of body protein and may cause dry skin, thinning hair, constipation, and salt imbalance
Balanced diet of 1,200 kcal or more	If carefully chosen, such a diet can meet all nutrient needs; weight loss is usually 1–2 pounds per week; dieter may become discouraged by slow progress

21.24 Diet can influence cardiovascular disease and cancer

Food influences far more than size and appearance. Diet also plays an important role in a person's risk of developing serious illnesses, including cardiovascular disease and cancer. **Figure 21.24** shows some of the risk factors associated with cardiovascular disease. Though certain factors are unavoidable, we can influence others through our behavior. Diet is an example of a behavioral factor that may affect cardiovascular health. For instance, a diet rich in saturated fats is linked to high blood cholesterol levels, which in turn may be linked to cardiovascular disease. Saturated fats are found in eggs, butter, lard, and most other animal products. Saturated fats are also found in artificially saturated ("hydrogenated") vegetable oils. The hydrogenation process, which solidifies vegetable oils, also produces a type of fatty acid called *trans* fats. These dangerous fats have also been linked to heart disease, and many doctors now recommend that people consume as few *trans* fats as possible. This requires careful reading of food ingredient labels, since hydrogenated oils are widely used in many processed foods, including crackers, store-bought cookies, and cereals. A food that contains hydrogenated or partially hydrogenated oils is likely to have *trans* fats.

Cholesterol is another factor in cardiovascular health. This molecule travels through the body in blood lipoproteins, which are particles made up of thousands of molecules of cholesterol and other lipids and one or more protein molecules (see Figure 5.20). High blood levels of a family of lipoproteins called **low-density lipoproteins (LDLs)** generally correlate with a tendency to develop blocked blood vessels, high blood pressure, and consequent heart attacks. In contrast to LDLs, cholesterol carriers called **high-density lipoproteins (HDLs)** may decrease the risk of vessel blockage, perhaps because some HDLs convey blood cholesterol to the liver, where it is broken down. Many researchers believe that reducing LDLs while maintaining or increasing HDLs lowers the risk of cardiovascular disease. Exercise tends to increase HDL levels, while smoking lowers them.

A diet high in saturated fats tends to increase LDL levels. *Trans* fats tend not only to increase LDL levels, but also to lower HDL levels, a two-pronged attack on cardiovascular

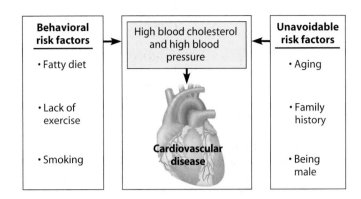

Figure 21.24 Risk factors associated with cardiovascular disease

health. By contrast, eating mainly unsaturated fats, such as fish oil and most liquid vegetable oils (including corn, soybean, and olive oils), tends to lower LDL levels and raise HDL levels. These oils are also important sources of vitamin E, whose antioxidant effect may help prevent blood vessel blockage.

As discussed in Module 11.20, diet also seems to be involved in some forms of cancer. Some research suggests a link between diets heavy in fats or carbohydrates and the incidence of breast cancer. The incidence of colon cancer and prostate cancer may be linked to a diet rich in saturated fat or red meat. Other foods may help fight cancer. For example, some fruits and other plant-based foods are rich in antioxidants, chemicals that help protect cells from damaging molecules known as free radicals. Antioxidants may help prevent cancer, although this link is still debated by scientists.

The relationship between foods and cancer is complex, and much remains to be learned. The American Cancer Society suggests that following the dietary guidelines listed in **Table 21.24**, in combination with physical activity, can help lower cancer risk. The ACS's main recommendation is to "eat a variety of healthful foods, with an emphasis on plant sources."

■   ■   ■

In this chapter, we have seen that a balanced diet supplies enough raw materials to make all the macromolecules we need, the proper amounts of prefabricated essential nutrients, and enough kilocalories to satisfy our energy needs. Energy remains a theme in Chapter 22, as we look at how the body obtains the oxygen it needs to harvest energy from food molecules.

TABLE 21.24 DIETARY GUIDELINES FOR REDUCING CANCER RISK
Eat 5 or more servings of a variety of fruits and vegetables daily.
Choose whole grain rice, bread, pasta, and cereals.
Limit consumption of red meats, especially those high in fat.
To avoid possible carcinogens, minimize consumption of cured and smoked foods, such as hot dogs, salami, and bacon. Avoid moldy foods. Avoid charred foods.
If you drink alcoholic beverages, limit yourself to a maximum of one or two drinks a day (a drink = 12 oz beer, 5 oz wine, 1.5 oz 80% distilled spirits).

? If you are trying to minimize the damaging effects of blood cholesterol on your cardiovascular system, your goal is to increase/decrease (*choose one*) your LDLs and increase/decrease (*choose one*) your HDLs.

■ decrease . . . increase

Reviewing the Concepts

Obtaining and Processing Food (Introduction–21.3)

Animal feeding mechanisms include suspension, substrate, fluid, and bulk feeding. Animals may be herbivores, carnivores, or omnivores **(Introduction–21.1)**.

Food processing includes four stages: ingestion, digestion, absorption, and elimination **(21.2)**.

Digestive compartments may be food vacuoles (sponges), a gastrovascular cavity (cnidarians and flatworms), or, in most animals, an alimentary canal running from mouth to anus with specialized regions **(21.3)**.

Human Digestive System (21.4–21.12)

The human digestive system consists of an alimentary canal and accessory glands. The rhythmic muscle contractions of peristalsis squeeze food along the alimentary canal. The pyloric sphincter regulates passage of food from the stomach to the small intestine **(21.4)**.

Oral cavity. The teeth break up food, saliva moistens it, and salivary enzymes begin the hydrolysis of starch. The tongue pushes the bolus of food into the pharynx **(21.5)**.

Pharynx and esophagus. The swallowing reflex moves food from the pharynx into the esophagus, while keeping it out of the trachea. The Heimlich maneuver can dislodge food from the pharynx or trachea during choking **(21.6–21.7)**. Peristalsis in the esophagus moves food into the stomach **(21.8)**.

The stomach stores food and mixes it with acidic gastric juice. Pepsin in gastric juice begins the hydrolysis of protein **(21.9)**. Bacterial infections in the stomach and duodenum are associated with ulcers **(21.10)**.

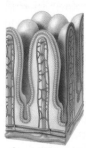

The small intestine is the site of most digestion and absorption. Alkaline pancreatic juice neutralizes the acid chyme, and its enzymes digest food polymers. Bile, made in the liver and stored in the gallbladder, emulsifies fat for attack by enzymes. Enzymes from cells of the intestine complete the digestion of many nutrients. Folds of the intestinal lining and tiny, fingerlike villi (with microscopic microvilli) increase the absorptive surface. Nutrients pass across the epithelium and into the blood, which flows to the liver, where nutrients are processed and stored **(21.11)**.

The large intestine, or colon, reabsorbs water from undigested material. Feces are stored in the rectum **(21.12)**.

Diets and Digestive Adaptations (21.13)

Dietary adaptations of herbivores include longer alimentary canals and cellulose-digesting microbes housed in special chambers. Ruminants such as cows process food with the aid of microbes in four chambers **(21.13)**.

Nutrition (21.14–21.24)

A healthy diet provides fuel for activities, raw materials for biosynthesis, and essential nutrients **(21.14)**. The basal metabolic rate (BMR) is the energy a resting animal requires each day. Excess energy is stored as glycogen or fat **(21.15)**.

Essential nutrients are those that an animal must obtain from its diet. The eight essential amino acids can be obtained from animal protein or from the proper combination of plant foods **(21.16–21.17)**.

Vitamins and minerals are essential in the human diet. Most vitamins function as coenzymes **(21.18)**. Minerals are inorganic nutrients that play a variety of roles **(21.19)**. Supplements ensure a sufficient quantity of vitamins and nutrients; megadoses may be dangerous **(21.20)**. Food labels provide important nutritional information **(21.21)**.

Obesity is a serious health problem, caused by lack of exercise and abundance of fattening foods, and may partly stem from an evolutionary advantage of fat hoarding **(21.22)**.

Weight loss diets may help individuals lose weight but may have health risks. A healthy diet may reduce the risk of cardiovascular disease and cancer **(21.23–21.24)**.

Connecting the Concepts

1. Label the parts of the human digestive system below and indicate the functions of these organs and glands.

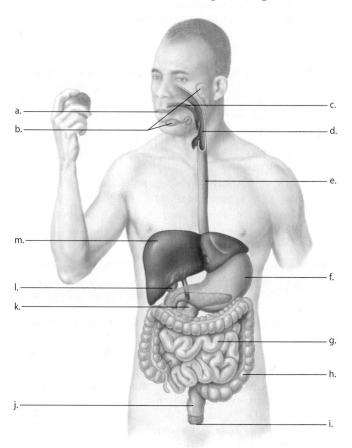

a.

b.

c.

d.

e.

f.

g.

h.

i.

j.

k.

l.

m.

2. Complete the following map summarizing the nutritional needs of animals that are met by a healthy diet.

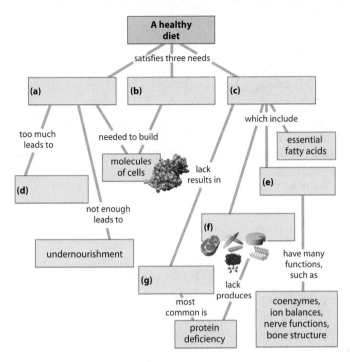

Testing Your Knowledge

Multiple Choice

3. Earthworms, which are substrate feeders,
   a. feed mostly on mineral substrates.
   b. filter small organisms from the soil.
   c. are bulk feeders.
   d. are herbivores that eat autotrophs.
   e. eat their way through the soil, feeding on partially decayed organic matter.

4. The energy content of fats
   a. is released by bile salts.
   b. may be lost unless an herbivore eats some of its feces.
   c. is more than two times that of carbohydrates or proteins.
   d. can reverse the effects of malnutrition.
   e. Both c and d are correct.

5. Which of the following statements is false?
   a. The average human has enough stored fat to supply calories for several weeks.
   b. An increase in leptin levels leads to an increase in appetite and weight gain.
   c. The interconversion of glucose and glycogen takes place in the liver.
   d. After glycogen stores are filled, excessive calories are stored as fat, regardless of their original food source.
   e. Carbohydrates and fats are preferentially used as fuel before proteins are used.

6. Which of the following is mismatched with its function?
   a. most B vitamins—coenzymes
   b. vitamin E—antioxidant
   c. vitamin K—blood clotting
   d. iron—component of thyroid hormones
   e. phosphorus—bone formation, nucleotide synthesis

7. Why do many vegetarians combine different protein sources or eat some eggs or milk products?
   a. to make sure they obtain sufficient calories
   b. to provide sufficient vitamins
   c. to make sure they ingest all essential fatty acids
   d. to make their diet more interesting
   e. to provide all essential amino acids at the same time

Describing, Comparing, and Explaining

8. A peanut butter and jelly sandwich contains carbohydrates, proteins, and fats. Describe what happens to the sandwich when you eat it. Discuss ingestion, digestion, absorption, and elimination.

Applying the Concepts

9. How might our craving for fatty foods, which is helping to fuel the obesity crisis, have evolved through natural selection?

10. Use this Nutrition Facts label to answer these questions:
    a. What percentage of the total Calories in this product is from fat?
    b. Is this product a good source of vitamin A and calcium? Explain.
    c. Each gram of fat supplies 9 Calories. Based on the grams of saturated fat and its % Daily Value, calculate the upper limit of saturated fat (in grams and Calories) that an individual on a 2,000-Calorie/day diet should consume.

Nutrition Facts

Serving Size 1/2 Cup (83g)
Servings Per Container 8

Amount Per Serving

Calories 190 Calories from Fat 110

	% Daily Value*
Total Fat 12g	**18%**
Saturated Fat 8g	**40%**
Cholesterol 45mg	**15%**
Sodium 75mg	**3%**
Total Carbohydrate 18g	**6%**
Dietary Fiber 0g	**0%**
Sugars 17g	
Protein 3g	

Vitamin A 10%	•	Vitamin C 8%
Calcium 10%	•	Iron 0%

*Percent Daily Values (DV) are based on a 2,000 calorie diet.

11. The media report numerous claims and counterclaims about the benefits and dangers of certain foods, dietary supplements, and diets. Have you modified your eating habits on the basis of nutritional information disseminated by the media? Why or why not? How should one evaluate whether such nutritional claims are valid?

12. It is estimated that 10% of Americans don't get enough to eat on a regular basis. Worldwide, at least 840 million people go to bed hungry most nights, and millions of people have starved to death in recent decades. In some cases, war, poor crop yields, and disease epidemics strip people of food. Many also blame global food distribution systems, saying it is not inadequate food production but unequal food distribution that causes food shortages. What responsibility do nations have for feeding their citizens? For feeding the people of other countries? What do you think you can do to lessen world hunger?

Answers to all questions can be found in Appendix 3.
For study help and Activities, go to campbellbiology.com or the student CD-ROM.

Surviving in Thin Air

THE HIGH MOUNTAINS of the Himalayas have claimed the lives of even the world's top mountain climbers; the journey into thin air can weaken their muscles, cloud their minds, and sometimes fill their lungs with fluid. The air at the height of the world's highest peak, 9,700-m Mount Everest, is so low in oxygen (O_2) that most people would pass out instantly if exposed to it.

But if you were ever to make it to the top of Mount Everest, you might see birds flying by. Twice a year, flocks of geese migrate over the Himalayas, traveling between winter quarters in India and summer breeding grounds in Russia. These geese, along with other species of migratory birds, can travel easily at heights that would leave most people drowsy, lethargic, or dead.

How do geese and ducks manage to fly at such heights? One factor is the efficiency of their lungs, which can draw far more oxygen from the air than our own lungs can. These birds also have blood containing hemoglobin with a very high affinity for oxygen, picking it up in the lungs and carrying it to tissues throughout the body. Their circulatory system has large number of capillaries (tiny blood vessels) that carry oxygen-rich blood to their flight muscles, and the muscles themselves pack a protein that stores a ready supply of oxygen. All these adaptations allow the high-flying birds you see in these photos to travel even where the air is very thin.

> **It is the continuous supply of oxygen to body cells that makes the difference between life and death in the thin air of the Himalayas**

Humans can try to adapt to higher elevations, but success is less certain. Most people function well only below 3,300 m and are helpless at higher elevations without an oxygen mask. There are permanent villages at extremely high elevations in the Himalayas and the Andes, but the people living there have adapted in ways that allow them to function with relatively little oxygen, including large lungs, a large heart, and blood that carries additional hemoglobin and red blood cells. Even then, thin air can take its toll. Local inhabitants such as the Nepalese Sherpas have a reputation for their strength as porters and guides in the

MECHANISMS OF GAS EXCHANGE

22.1   Overview: Gas exchange involves breathing, transport of gases, and exchange of gases with tissue cells
22.2   Animals exchange O_2 and CO_2 across moist body surfaces
22.3   Gills are adapted for gas exchange in aquatic environments
22.4   The tracheal system of insects provides direct exchange between the air and body cells
22.5   Terrestrial vertebrates have lungs
22.6   Smoking is a deadly assault on our respiratory system
22.7   Breathing ventilates the lungs
22.8   Breathing is automatically controlled

TRANSPORT OF GASES IN THE BODY

22.9   Blood transports respiratory gases
22.10 Hemoglobin carries O_2 and helps transport CO_2 and buffer the blood
22.11 The human fetus exchanges gases with the mother's bloodstream

Gas Exchange

high Himalayas. But many die doing such work, their bodies succumbing to altitude-related illnesses under the burdens of long travel and heavy loads.

Altitude-caused disorders include everything from mild headaches, dizziness, and nausea to life-threatening fluid buildup in the lungs and swelling of the brain. Avoiding these disorders requires careful conditioning for high altitudes, and the higher one goes, the longer the adjustment takes. As you move from sea level up into the mountains, your body starts adjusting immediately. Your heart pumps faster, and some blood vessels may increase in diameter if you stay in the mountains more than a few days. Within weeks, the rate and depth of your breathing increase to bring more air into your lungs. At the same time, your body may develop more capillaries, and your red blood cell count may go up, allowing your blood to carry more oxygen. After long-term training, some Everest climbers have been able to survive for a short time at the top of the world's highest peak without oxygen masks. Runners and cyclists may also use this type of training, moving to high altitudes to gain stronger lungs and more oxygen-rich blood and then returning to sea level to blow past competitors who trained at lower elevations.

Our study of cellular respiration in Chapter 6 showed why animals require oxygen. Without O_2, the metabolic machinery that releases energy from food molecules shuts down. It is the continuous supply of O_2 to body cells that makes the difference between life and death in the thin air of the Himalayas.

The process of **gas exchange**, often called **respiration**, is the interchange of O_2 and the waste product CO_2 between an animal and its environment. In this chapter, we will explore the respiratory systems of animals. ■ ■ ■

22.1 Overview: Gas exchange involves breathing, transport of gases, and exchange of gases with tissue cells

Gas exchange makes it possible for animals to put to work the food molecules the digestive system provides. Figure 22.1 presents an overview of the three phases of gas exchange in an animal with lungs. ❶ Breathing is the first phase of the gas exchange process. When an animal breathes, a large, moist internal surface is exposed to air. O_2 diffuses across the cells lining the lungs and into surrounding blood vessels. At the same time, CO_2 diffuses out of the blood and into the lungs. As the animal exhales, CO_2 is removed from the body.

❷ A second phase of gas exchange is the transport of gases by the circulatory system. The O_2 that has diffused into the blood attaches to hemoglobin in red blood cells and is carried from the lungs to the body's tissues. CO_2 is also transported in blood from the tissues back to the lungs.

❸ In the third phase of gas exchange, body cells take up O_2 from the blood and release CO_2 to the blood. This O_2 is required for cells to obtain energy from the food molecules the body has digested and absorbed. As we learned in Module 6.4, O_2 functions in cellular respiration as the final electron acceptor in the stepwise breakdown of fuel molecules. H_2O and CO_2 are waste products, and ATP is produced to power cellular work.

Thus, our cells require a continuous supply of O_2 and must dispose of CO_2. Gas exchange involves the respiratory system and the circulatory system in servicing the cells of the body.

? Humans cannot survive for more than a few minutes without O_2. Why?

■ Cells require a steady supply of ATP in order to function. Cellular respiration requires O_2 to produce this ATP. Without ATP, cells and the organism die.

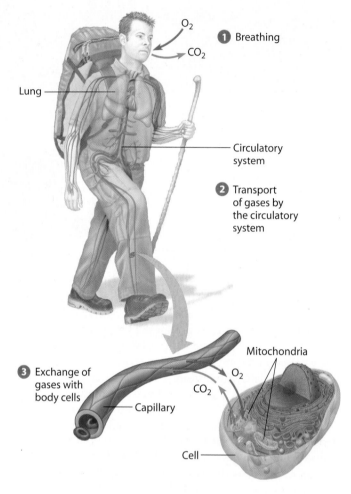

Figure 22.1 The three phases of gas exchange

22.2 Animals exchange O_2 and CO_2 across moist body surfaces

The part of an animal where gases are exchanged with the environment is called the **respiratory surface.** (In this context, the word *respiratory* refers to the process of breathing, not to cellular respiration.) Respiratory surfaces are made up of living cells, whose plasma membranes must be wet to function properly. Thus, the respiratory surfaces of terrestrial as well as aquatic animals must be moist, and gases must be dissolved in water before they can diffuse across them. The surface area of the respiratory surface must be extensive enough to take up sufficient O_2 for every cell in the body and to dispose of all waste CO_2. Usually, a single layer of cells covers or lines the entire respiratory surface. Being thin and moist, the layer allows O_2 to diffuse rapidly into the circulatory system or directly into body tissues and allows CO_2 to diffuse out.

The four figures on the facing page illustrate, in simplified form, four types of respiratory organs, structures where gas exchange with the external environment occurs. In each case, the circle represents a cross section of the animal's body through the respiratory surface. The yellow areas represent the respiratory surfaces; the green circles represent body surfaces with little or no role in respiration. The boxed enlargements show a portion of the respiratory surface in the process of exchanging O_2 and CO_2.

Some animals use their entire outer skin as a gas exchange organ. The earthworm in Figure 22.2A is an example. Notice in the cross-sectional diagram that its whole body surface is yellow; there are no specialized gas exchange surfaces. Oxygen diffuses into a dense net of thin-walled capillaries lying just beneath the skin. Earthworms and other "skin-breathers" must live in damp places or in water because their whole body surface has to stay moist. Animals that breathe only through their skin and lack specialized gas

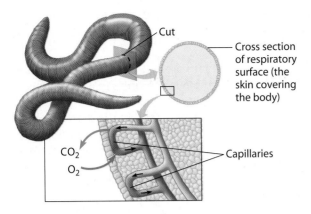

Figure 22.2A The entire outer skin

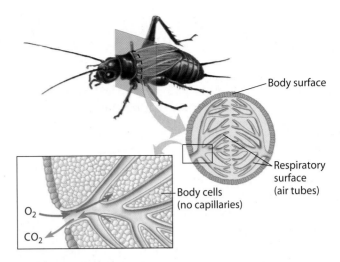

Figure 22.2C Tracheal system

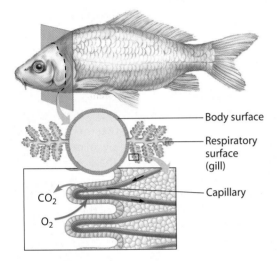

Figure 22.2B Gills

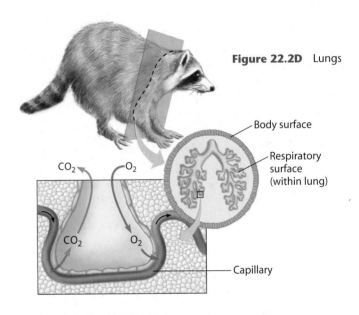

Figure 22.2D Lungs

exchange organs are generally small, and many are long and thin or flattened. Small size or flatness provides a high ratio of respiratory surface to body volume, allowing for sufficient gas exchange for the entire body.

In most animals, the skin surface is not extensive enough to exchange gases for the whole body. Consequently, certain parts of the body have become adapted as respiratory surfaces. Gills have evolved in most aquatic animals. Lungs or an internal system of gas exchange tubes called tracheae have evolved in most terrestrial animals. Gills, lungs, and tracheae all have extensive surfaces for gas exchange, as shown in Figures 22.2B–22.2D.

Gills are extensions, or outfoldings, of the body surface specialized for gas exchange. Many marine worms have flap-like gills that extend from each body segment. The gills of clams and crayfish are clustered in one body location. A fish **(Figure 22.2B)** has a set of feather-like gills on each side of its head. As indicated in the enlargement, O_2 diffuses across the gill surfaces into capillaries, and CO_2 diffuses in the opposite direction, out of the capillaries and into the external environment. Since the respiratory surfaces of aquatic animals extend into the surrounding water, keeping the surface moist is not a problem.

In most terrestrial animals, the respiratory surfaces are folded into the body rather than projecting from it. The in-

folded surfaces open to the air only through narrow tubes, an arrangement that helps retain the moisture that is essential for the cells of the respiratory surfaces to function.

The **tracheal system** of insects is an extensive system of branching internal tubes **(Figure 22.2C)**. As we will see in Module 22.4, the smallest branches exchange gases directly with body cells. Thus, gas exchange in insects requires no assistance from the circulatory system.

Most terrestrial vertebrates have **lungs (Figure 22.2D)**, which are internal sacs lined with moist epithelium. As the diagram indicates, the inner surfaces of the lungs branch extensively, forming a large respiratory surface. Gases are carried between the lungs and the body cells by the circulatory system.

We examine gills, tracheae, and lungs more closely in the next several modules.

? What is the main difference between gills and lungs in terms of their spatial relationship to the rest of an animal's body?

■ The extensive respiratory surface of gills extends outward from the body into the surrounding environment (water); in contrast, lungs are internal sacs with respiratory surfaces.

Gills are adapted for gas exchange in aquatic environments

Oceans, lakes, and other bodies of water contain O_2 in the form of dissolved gas. The gills of fishes and many invertebrate animals, including lobsters and clams, tap this source of O_2. The total surface area of the gills is often much greater than that of the rest of the body.

An advantage of exchanging gases in water is that there is no problem keeping the respiratory surface wet. On the other hand, the amount of available oxygen (dissolved O_2) in water is only about 3–5% of what it is in the air, and the warmer and saltier the water, the less dissolved O_2 it holds. Thus, gills—especially those of large, active animals in warm oceans—must be very efficient to obtain enough oxygen from water.

The drawings in **Figure 22.3** show the architecture of fish gills, which are among the most efficient gas exchange organs in the aquatic world. There are four supporting gill arches on each side of the body. Two rows of gill filaments project from each gill arch. Each filament bears many plate-like structures called lamellae (singular, lamella), which are the actual respiratory surfaces. A lamella is full of tiny capillaries that are separated from the outside by only one or a few layers of cells. Capillaries are so narrow that red blood cells must pass through them in single file. As a result, every red blood cell comes in close contact with oxygen dissolved in the surrounding water.

What you can't see in the drawings are the movements that ventilate the gills. We use the term **ventilation** to refer to any mechanism that increases the flow of the surrounding water or air over the respiratory surface (gills, tracheae, or lungs). Increasing this flow ensures a fresh supply of O_2 and the removal of CO_2. The blue arrows in the drawings represent the one-way flow of water into the mouth, across the gills, and out the side of the fish's body. Swimming fish simply open their mouths and let water flow past the gills. Fish also pump water across the gills by the coordinated opening and closing of the mouth and operculum, the stiff flap that covers and protects the gills. Because water is dense and contains so little oxygen, most fish must expend considerable energy in ventilating their gills.

The arrangement of capillaries in a fish gill greatly enhances gas exchange. Blood flows in the direction opposite to the movement of water past the gills. This makes it possible to transfer oxygen to the blood by a very efficient process called countercurrent exchange. **Countercurrent exchange** is the transfer of something from a fluid moving in one direction to another fluid moving in the opposite direction. The name comes from the fact that the two fluids are moving *counter* to each other. Their opposite flow maintains a diffusion gradient that enhances transfer of the substance. Let's see how this principle works in a fish gill.

In the circular enlargement on the right of Figure 22.3, notice that the direction of water flow over the surface of a lamella (blue arrow) is opposite that of the blood flow within the lamella (arrow turning from blue to red). The changing in-

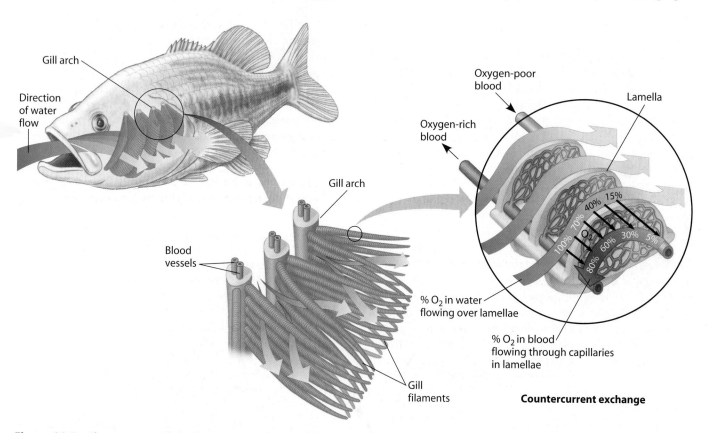

Figure 22.3 The structure of fish gills

tensities of these arrows and the numbers on them indicate the changing amount of O_2 dissolved in each fluid: the darker the color, the more O_2. Notice that as blood flows through a lamella and picks up more and more O_2, the blood comes in contact with water that has even more O_2 available because it is just beginning its passage over the gills. As a result, a diffusion gradient is maintained that favors the transfer of O_2 from the water to the blood along the entire length of the capillary.

This countercurrent exchange mechanism is so efficient that fish gills can remove more than 80% of the oxygen dissolved in the water flowing through them. The basic mechanism of countercurrent exchange is also important in temperature regulation, as you will see in Chapter 25.

Gills are unsuitable for an animal living on land. An expansive surface of wet membrane extending out from the body and exposed to air would lose too much water to evaporation. Most terrestrial animals house their respiratory surfaces within the body, opening to the atmosphere through narrow tubes, as we see next.

? (This is a tough one!) What would be the maximum percentage of the water's oxygen a gill could extract if its blood flowed in the same direction as the water instead of counter to it?

■ 50%. As O_2 diffused from the water into the blood as they flowed in the same direction, the concentration gradient would become less and less steep until there was the same amount of O_2 dissolved in both, and O_2 could no longer diffuse from water to blood.

22.4 The tracheal system of insects provides direct exchange between the air and body cells

There are two big advantages to exchanging gases by breathing air: Air contains a much higher concentration of O_2, and air is much lighter and easier to move than water. Thus, a terrestrial animal expends much less energy than an aquatic animal ventilating its respiratory surface. The main problem facing any air-breathing animal is the loss of water to the air by evaporation. With respiratory surfaces occurring as tiny tubes deep in the body of an insect, evaporation is reduced, and the respiratory system loses very little water.

The tracheal system of insects is made up of air tubes that branch throughout the body (Figure 22.4A). The largest tubes, called tracheae, open to the outside and are reinforced by rings of chitin, as shown in the blowup on the bottom right of the figure. Enlarged portions of tracheae form air sacs near organs that require a large supply of O_2.

The micrograph in Figure 22.4A shows how these tubes branch repeatedly. The smallest branches, called tracheoles, extend to nearly every cell in the insect's body. The tiny tips of the tracheoles are closed and contain fluid (dark blue in the figure). Gas is exchanged with body cells by diffusion across the moist epithelium that lines these tips. Thus, the circulatory system of insects is not involved in transporting oxygen.

For a small insect, diffusion through the tracheae brings in enough O_2 and removes enough CO_2 to support cellular respiration. Larger insects may ventilate their tracheal systems with rhythmic body movements that compress and expand the air tubes like bellows. An insect in flight (Figure 22.4B) has a very high metabolic rate and consumes 10 to 200 times more O_2 than it does at rest. In many insects, alternating contraction and relaxation of the flight muscles rapidly pumps air through the tracheal system.

? In what basic way does the process of gas exchange in insects differ from that in both fish and humans?

■ The circulatory system of insects is not involved in transporting gases to and from the body cells.

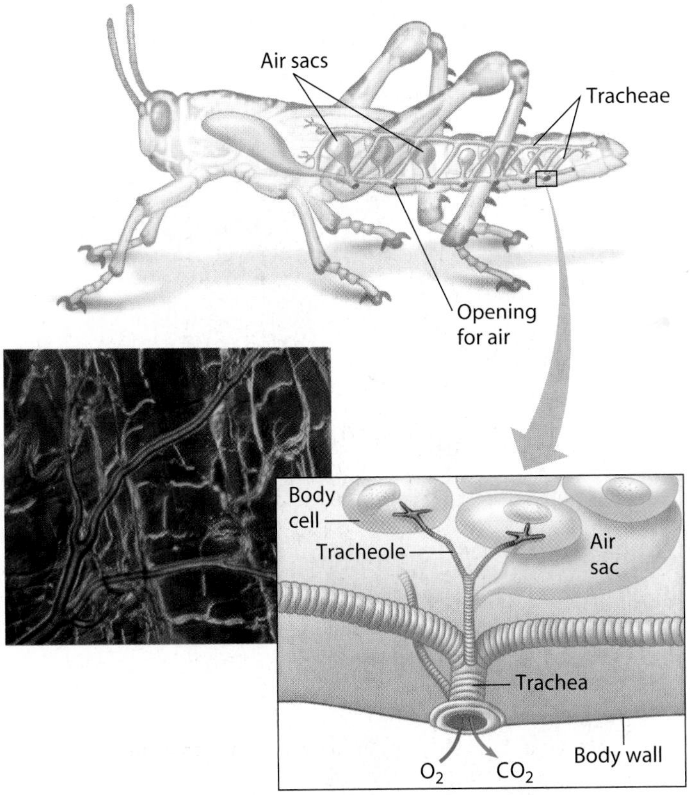

Figure 22.4A The tracheal system of an insect

LM 250×

Air sacs

Tracheae

Opening for air

Body cell

Tracheole

Air sac

Trachea

O_2 CO_2

Body wall

Figure 22.4B
A grasshopper in flight

22.5 Terrestrial vertebrates have lungs

Reptiles (including birds), mammals, and most amphibians exchange gases in lungs. In contrast to the tracheae of insects, lungs are restricted to one location in the body. Therefore, the circulatory system must transport gases between the lungs and the rest of the body.

Amphibians have small lungs (some salamanders lack lungs altogether) and rely heavily on the diffusion of gases across body surfaces. The skin of frogs, for example, supplements gas exchange in the lungs. Most reptiles (including all birds) and mammals rely entirely on lungs for gas exchange. In general, the size and complexity of lungs are correlated with an animal's metabolic rate (and thus oxygen need). For example, the lungs of endotherms (birds and mammals) have a greater area of exchange surface than the lungs of similar-sized ectotherms (amphibians and nonbird reptiles). The total respiratory surface of human lungs is about 100 m^2, equal to the surface area of a racquetball court.

Figure 22.5A shows the human respiratory system (along with the esophagus and heart, for orientation). Our lungs are in the chest cavity, which is bounded at the bottom by a sheet of muscle called the **diaphragm.** Air passes to our lungs via a system of branching narrow tubes.

Air usually enters our respiratory system through the nostrils. It is filtered by hairs and warmed, humidified, and sampled for odors as it flows through a maze of spaces in the nasal cavity. We can also draw in air through the mouth, but mouth breathing does not allow the air to be processed by the nasal cavity. From the nasal cavity or mouth, air passes to the **pharynx**, where the paths for air and food cross. As we saw in Module 21.6, the air passage in the pharynx is open for breathing except when we swallow.

From the pharynx, air is inhaled into the **larynx** (voice box). When we exhale, the outgoing air rushes by a pair of **vocal cords** in the larynx, and we can produce sounds by voluntarily tensing muscles in the voice box, stretching the cords and making them vibrate. We produce high-pitched sounds when our vocal cords are tense and therefore vibrating very fast. When the cords are less tense, they vibrate slowly and produce low-pitched sounds.

From the larynx, inhaled air passes toward the lungs through the **trachea**, or windpipe. Rings of cartilage maintain the shape of the trachea, much as metal rings keep the hose of a vacuum cleaner from collapsing. The trachea forks into two **bronchi** (singular, *bronchus*), one leading to each lung. Within the lung, the bronchus branches repeatedly into finer and finer tubes called **bronchioles.** Bronchitis is a condition in which these small tubes become inflamed and constricted, making breathing difficult.

As **Figure 22.5B** shows, the bronchioles dead-end in grapelike clusters of air sacs called **alveoli** (singular, *alveolus*). Each

Figure 22.5B The structure of alveoli

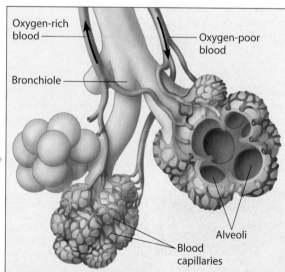

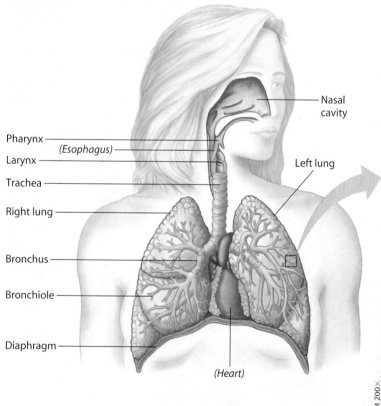

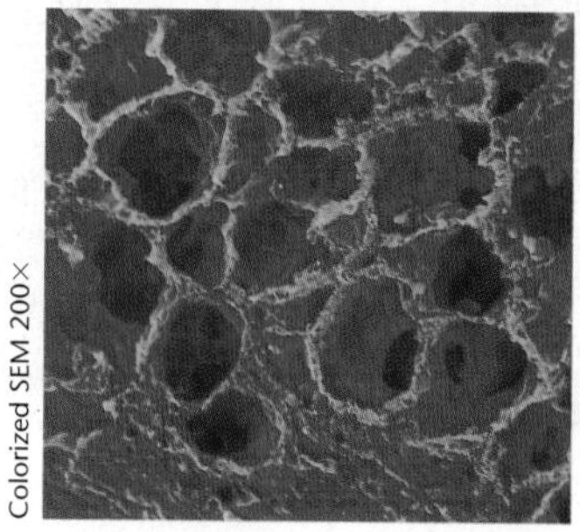

Figure 22.5A The human respiratory system

Colorized SEM 200×

Figure 22.5C
Air spaces in alveoli

of our lungs contains millions of these tiny sacs. Figure 22.5C, produced by a scanning electron microscope, is a cutaway view of alveoli showing the air spaces. The inner surface of each alveolus is lined with a thin layer of epithelial cells that form the respiratory surface. The O_2 in inhaled air dissolves in a film of moisture on the epithelial cells. It then diffuses across the epithelium and into a web of blood capillaries that surrounds each alveolus. The CO_2 diffuses the opposite way—from the capillaries, across the epithelium of the alveolus, into the air space of the alveolus, and finally out in the exhaled air.

The trachea and major branches of the respiratory system are lined by a moist epithelium covered by cilia and a thin film of mucus. The cilia and mucus are the system's cleaning elements. The mucus traps dust, pollen, and other contaminants, and the beating cilia move the mucus upward to the pharynx, where it is usually swallowed. In the next module, we explore one of the most serious threats to this delicate epithelium and to our lungs.

Web/CD Activity 22A *The Human Respiratory System*

? List the parts of the respiratory system in the order that an inhaled breath of air would encounter them.

■ Nasal cavity → pharynx → larynx → trachea → bronchus → bronchiole → alveolus

---

CONNECTION

22.6 Smoking is a deadly assault on our respiratory system

Virtually everywhere today, the air we breathe exposes the cells in our respiratory system to chemicals that they are not adapted to tolerate. Air pollutants such as sulfur dioxide, carbon monoxide, and ozone can all cause respiratory problems. One of the worst sources of air pollutants is tobacco smoke. The visible smoke from a cigarette, cigar, or pipe is mainly microscopic particles of carbon. Sticking to the carbon particles are many toxic chemicals. A single drag on a cigarette exposes a person to over 4,000 chemicals, more than 50 of which are carcinogens (cancer-causing agents).

Tobacco smoke irritates the cells lining the bronchi, inhibiting or destroying their cilia. Frequent coughing—common in heavy smokers—becomes the respiratory system's attempt to clear the mucus no longer moved by the cilia. Smoke's noxious particles also kill macrophages, defensive cells that reside in the respiratory tract and engulf fine particles and microorganisms. Thus, smoking disables the normal cleansing and protective mechanisms of the respiratory system, allowing even more toxin-laden particles to reach the lung's delicate alveoli.

Some of the toxins in tobacco smoke cause lung cancer. The photographs in Figure 22.6 show a cutaway view of a pair of healthy human lungs (left) and the lungs of a smoker with cancer. The lungs on the right are black from the long-term buildup of smoke particles, except where pale cancerous tumors appear. Smokers account for 90% of all lung cancer cases. Most victims die within one year of diagnosis. Smokers also have a markedly greater risk than nonsmokers of developing cancers of the bladder, pancreas, mouth, throat, and several other organs.

Smoking can also cause **emphysema**, a disease in which the walls of the alveoli lose their elasticity and deteriorate, reducing the lungs' capacity for gas exchange. Breathlessness and constant fatigue result, as the body is forced to spend more and more energy just breathing.

The second highest number of smoking-related deaths come from cardiovascular disease. Smokers have a higher rate of heart attacks and stroke. Smoking raises blood pressure and increases harmful cholesterol levels in the blood.

Every year in the United States, smoking kills about 440,000 people, more than all the deaths caused by traffic accidents, alcohol and drug abuse, HIV, and murders combined. On average, adults who smoke cigarettes die 13 to 14 years earlier than nonsmokers. Moreover, studies show that nonsmokers exposed to secondary cigarette smoke are also at risk. Young children are particularly susceptible, with increased risk of asthma, bronchitis, and pneumonia.

Clearly, efforts to reduce smoking and secondary exposure to smoke are important to public and personal health. No lifestyle choice has a more positive impact on long-term health than the decision not to smoke. This is true even if you already smoke. Quitting smoking has immediate health benefits, and after about 15 years, the risk of lung cancer and heart disease is similar to that of people who have never smoked.

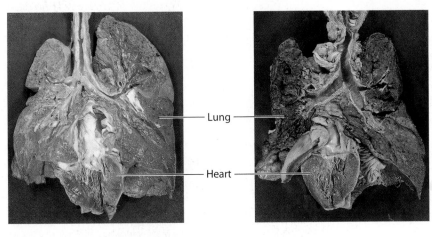

Lung

Heart

? What causes smoker's cough?

■ Tobacco smoke damages cilia, inhibiting their ability to sweep mucus and trapped particles out of the respiratory tract. The body tries to compensate by coughing.

Figure 22.6 Healthy lungs (left) and cancerous lungs (right)

22.7 Breathing ventilates the lungs

Breathing is the alternation of inhalation and exhalation. This ventilation of our lungs maintains high O_2 and low CO_2 concentrations at the respiratory surface. Like all mammals, we breathe by pulling air into the lungs and then pushing it back out.

Figure 22.7A shows the changes that occur in our rib cage, chest cavity, and lungs during breathing. During inhalation (left diagram), both the rib cage and chest cavity expand, and the lungs follow suit. The ribs move upward and the rib cage expands as muscles between the ribs contract. At the same time, the diaphragm contracts, moving downward and expanding the chest cavity as it goes.

The increase in the volume of the lungs during inhalation lowers the air pressure in the alveoli to less than atmospheric pressure. Flowing from a region of higher pressure to one of lower pressure, air rushes through the nostrils and down the breathing tubes to the alveoli. This type of ventilation is called **negative pressure breathing**.

The diagram on the right in Figure 22.7A shows exhalation. The rib muscles and diaphragm both relax, decreasing the volume of the rib cage and chest cavity and forcing air out of the lungs. Notice that the diaphragm curves upward into the chest cavity when relaxed.

Each year, a human adult may take between 4 million and 10 million breaths. The volume of air in each breath is about 500 mL when we breathe quietly. The maximum volume of air that we can inhale and exhale during forced breathing is called **vital capacity.** It averages about 3.4 L and 4.8 L for college-age females and males, respectively. (Women tend to have smaller rib cages and lungs.) The lungs actually hold more air than the vital capacity. Because the alveoli do not completely collapse, a residual volume of "dead" air remains in the lungs even after we blow out as much air as we can. As lungs lose resilience (springiness) with age or as the result of disease, such as emphysema, our residual volume increases at the expense of vital capacity.

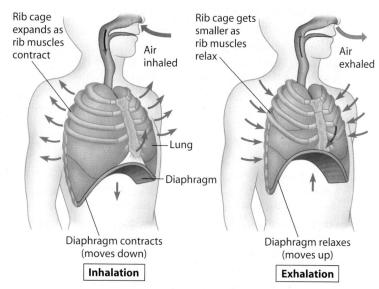

Figure 22.7A How a human breathes

As we mentioned in the introduction, the gas exchange system of birds is different from ours. Let's now take a look at some of the adaptations that make a bird's respiratory system so efficient.

Unlike the in-and-out flow of air in the human alveoli, birds have a one-way flow of air through the lungs. Birds have several large air sacs in addition to their lungs. These do not function directly in gas exchange, but act as bellows that keep air flowing through the lungs. As the simplified diagrams in Figure 22.7B indicate, both sets of air sacs expand during inhalation. The posterior sacs fill with fresh air (red) from the outside, while the anterior sacs fill with stale air (blue) from the lungs. During exhalation, both sets of air sacs deflate, forcing air from the posterior sacs into the lungs, and air from the anterior sacs out of the system via the trachea.

Instead of alveoli, bird lungs contain tiny parallel tubes (shown in the electron micrograph in the circular inset). Gas exchange occurs across the walls of these tubes as air passes one-way through them (red arrows). Because of the one-way flow of air, there is no dead air (residual volume) in the bird lung, so lung oxygen concentrations are higher in birds than in mammals. Birds can extract about 5% more oxygen from a volume of inhaled air than we can.

? Compare the pathway of air flow in the lungs of mammals and birds.

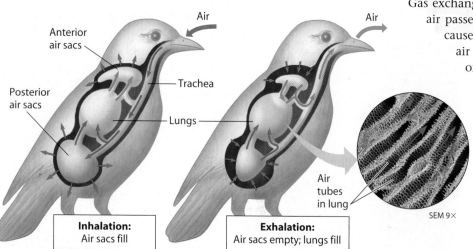

Figure 22.7B How a bird breathes

■ In mammals, air enters and leaves the lungs by the same pathway, and newly inhaled air mixes with oxygen-depleted residual air. In birds, air flows unidirectionally through the lungs.

22.8 Breathing is automatically controlled

What controls our breathing? We obviously have some conscious control over it because we can voluntarily hold our breath for a short while or breathe faster and deeper. Most of the time, however, automatic control centers in our brain regulate our breathing movements. Automatic control is essential, for it ensures coordination between the respiratory and circulatory systems and the body's metabolic needs for gas exchange.

Our **breathing control centers** (represented by the gold circles in **Figure 22.8**) are located in parts of the brain called the pons and medulla oblongata (medulla, for short). Nerves from the medulla's control center signal the diaphragm and rib muscles to contract, making us inhale. These nerves send out signals that result in about 10 to 14 inhalations per minute when we are at rest. Between inhalations, the muscles relax, and we exhale. The control center in the pons smooths out the basic rhythm of breathing set by the medulla.

How does the medulla's control center adjust our breathing rate in response to the body's varying needs? The control center monitors the CO_2 level of the blood and regulates breathing rate in response. Its main cues about CO_2 concentration come from slight changes in the pH of the blood and in the fluid bathing the brain (cerebrospinal fluid). The pH starts to drop when the amount of CO_2 in-

creases in the blood. When we exercise vigorously, for instance, our metabolism speeds up and our body cells generate more CO_2 as a waste product. The CO_2 goes into the blood, where it reacts with water to form carbonic acid. The acid lowers the pH of the blood and cerebrospinal fluid slightly. When the medulla senses this pH drop, its breathing control center increases the breathing rate and depth. As a result, more CO_2 is eliminated in the exhaled air, and the pH returns to normal.

When you were a kid, did you ever make yourself dizzy by **hyperventilating**, excessively taking rapid, deep breaths? Hyperventilating demonstrates the action of your breathing control center, but it's hard on your body. Deep, rapid breathing purges the blood of so much CO_2 that the control center temporarily ceases to send signals to the rib muscles and diaphragm. Breathing stops until the CO_2 level increases enough to switch the breathing center back on.

Our breathing control center responds directly to CO_2 levels, but it usually does not respond directly to oxygen levels. Since the same process that consumes O_2, cellular respiration, also produces CO_2, a rise in CO_2 (drop in pH) is generally a good indication of a drop in blood oxygen. Thus, by responding to lowered pH, the breathing control center also controls blood oxygen level.

Secondary control over breathing is exerted by sensors in the aorta and carotid arteries that monitor concentrations of O_2 as well as CO_2. When the O_2 level in the blood is severely depressed, these sensors signal the control center via nerves to increase the rate and depth of breathing. This response may occur, for example, at high altitudes, where the air is so thin that we cannot get enough O_2 by breathing normally.

The breathing control center responds to a variety of nervous and chemical signals to keep the breathing rate and depth in tune with the changing demands of the body. Breathing rate must also be coordinated with the activity of the circulatory system. During exercise, the rate at which our heart beats and the amount of blood it pumps with each beat must be matched with the increased breathing rate. We examine the role of the circulatory system in gas exchange more closely in the next module.

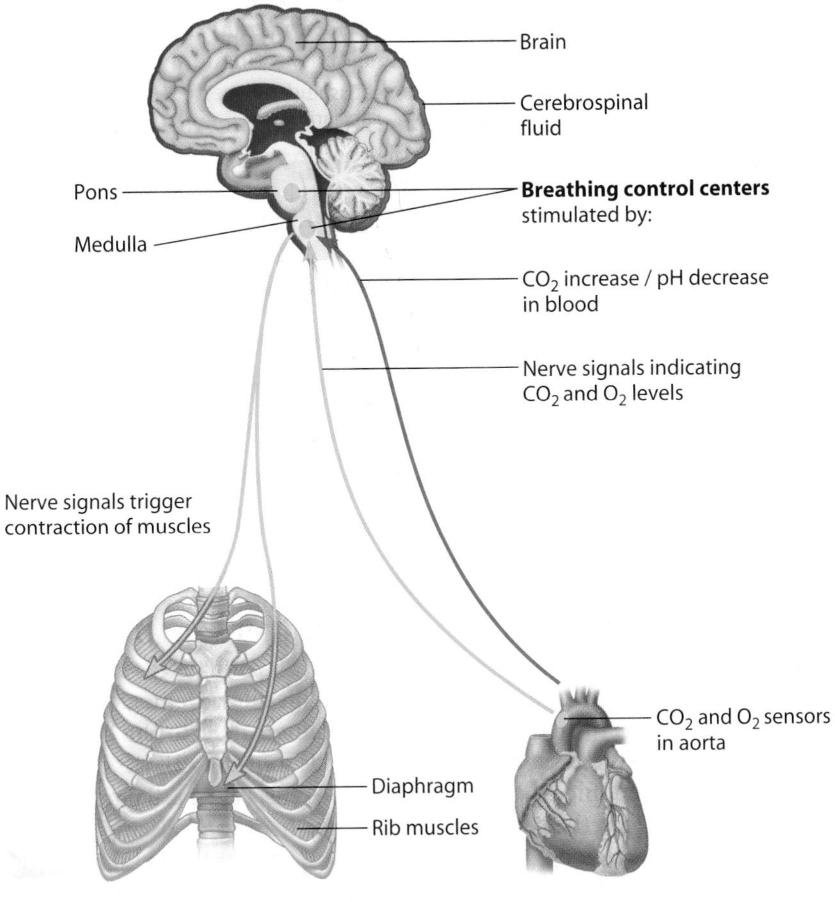

Figure 22.8 Control centers that regulate breathing

? Explain how hyperventilation disrupts the control of breathing.

■ By purging the blood of CO_2 (and hence carbonic acid), which indirectly stimulates inhalation via its action on the breathing control center, hyperventilation temporarily suspends breathing.

CHAPTER REVIEW

Reviewing the Concepts

Mechanisms of Gas Exchange (Introduction–22.8)

Gas exchange, the interchange of O_2 and CO_2 between an organism and its environment, provides O_2 for cellular respiration and removes its waste product, CO_2. Gas exchange often involves breathing, transport of gases, and exchange of gases with body cells **(Introduction–22.1).**

Respiratory surfaces must be thin and moist for diffusion of O_2 and CO_2 to occur. Some animals use their entire skin as a gas exchange organ. In most animals, specialized body parts—such as gills, tracheal systems, or lungs—provide large respiratory surfaces for gas exchange **(22.2).**

Gills are extensions of the body that absorb O_2 dissolved in water. In a fish, gas exchange is enhanced by ventilation and the countercurrent flow of water and blood **(22.3).**

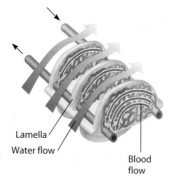

Lamella
Water flow
Blood flow

Tracheal systems in insects transport O_2 directly to body cells through a network of finely branched tubes **(22.4).**

Lungs. Most terrestrial vertebrates have lungs. In mammals, air inhaled through the nostrils passes through the pharynx and larynx into the trachea, bronchi, and bronchioles to the alveoli, where gas exchange occurs. Mucus and cilia in the respiratory passages protect the lungs, but smoking can destroy these protections. Smoking causes lung cancer, heart disease, and emphysema **(22.5–22.6).**

Breathing is the alternation of inhalation and exhalation. The contraction of rib muscles and diaphragm expands the chest cavity and reduces air pressure in the alveoli (negative pressure breathing). Vital capacity is the maximum volume of air that can be inhaled and exhaled, but the lungs still hold a residual volume. Air flows in one direction through the more efficient lungs of birds **(22.7).**

Breathing control centers in the brain keep breathing in tune with body needs, sensing and responding to the CO_2 level in the blood. A drop in blood pH triggers an increase in the rate and depth of breathing **(22.8).**

Transport of Gases in the Body (22.9–22.11)

Circulation. The heart pumps oxygen-poor blood to the lungs, where it picks up O_2 and drops off CO_2. Then the heart pumps the oxygen-rich blood to body cells, where it drops off O_2 and picks up CO_2. Gases diffuse down partial-pressure gradients in lungs and body tissues **(22.9).**

Hemoglobin in red blood cells transports oxygen, helps buffer the blood, and carries some CO_2. Most CO_2 is transported as bicarbonate ions in the plasma **(22.10).** A human fetus exchanges gases with maternal blood in the placenta. Fetal hemoglobin enhances oxygen transfer from maternal blood. At birth, rising CO_2 in fetal blood stimulates the breathing control centers to initiate breathing **(22.11).**

Connecting the Concepts

1. Complete this map to review some of the concepts of gas exchange.

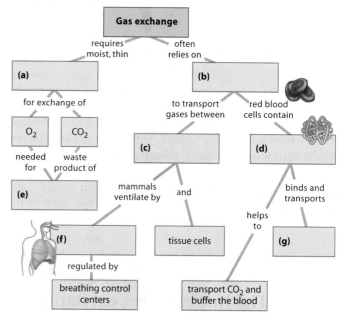

2. Label the parts of the human respiratory system.

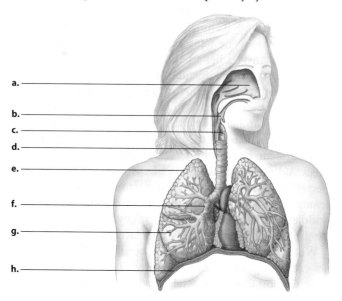

a.
b.
c.
d.
e.
f.
g.
h.

Testing Your Knowledge

Multiple Choice

3. When you hold your breath, which of the following first leads to the urge to breathe?
   a. falling CO_2
   b. falling O_2
   c. rising CO_2
   d. rising pH of the blood
   e. both c and d

22.8 Breathing is automatically controlled

What controls our breathing? We obviously have some conscious control over it because we can voluntarily hold our breath for a short while or breathe faster and deeper. Most of the time, however, automatic control centers in our brain regulate our breathing movements. Automatic control is essential, for it ensures coordination between the respiratory and circulatory systems and the body's metabolic needs for gas exchange.

Our **breathing control centers** (represented by the gold circles in Figure 22.8) are located in parts of the brain called the pons and medulla oblongata (medulla, for short). Nerves from the medulla's control center signal the diaphragm and rib muscles to contract, making us inhale. These nerves send out signals that result in about 10 to 14 inhalations per minute when we are at rest. Between inhalations, the muscles relax, and we exhale. The control center in the pons smooths out the basic rhythm of breathing set by the medulla.

How does the medulla's control center adjust our breathing rate in response to the body's varying needs? The control center monitors the CO_2 level of the blood and regulates breathing rate in response. Its main cues about CO_2 concentration come from slight changes in the pH of the blood and in the fluid bathing the brain (cerebrospinal fluid). The pH starts to drop when the amount of CO_2 in-creases in the blood. When we exercise vigorously, for instance, our metabolism speeds up and our body cells generate more CO_2 as a waste product. The CO_2 goes into the blood, where it reacts with water to form carbonic acid. The acid lowers the pH of the blood and cerebrospinal fluid slightly. When the medulla senses this pH drop, its breathing control center increases the breathing rate and depth. As a result, more CO_2 is eliminated in the exhaled air, and the pH returns to normal.

When you were a kid, did you ever make yourself dizzy by **hyperventilating,** excessively taking rapid, deep breaths? Hyperventilating demonstrates the action of your breathing control center, but it's hard on your body. Deep, rapid breathing purges the blood of so much CO_2 that the control center temporarily ceases to send signals to the rib muscles and diaphragm. Breathing stops until the CO_2 level increases enough to switch the breathing center back on.

Our breathing control center responds directly to CO_2 levels, but it usually does not respond directly to oxygen levels. Since the same process that consumes O_2, cellular respiration, also produces CO_2, a rise in CO_2 (drop in pH) is generally a good indication of a drop in blood oxygen. Thus, by responding to lowered pH, the breathing control center also controls blood oxygen level.

Secondary control over breathing is exerted by sensors in the aorta and carotid arteries that monitor concentrations of O_2 as well as CO_2. When the O_2 level in the blood is severely depressed, these sensors signal the control center via nerves to increase the rate and depth of breathing. This response may occur, for example, at high altitudes, where the air is so thin that we cannot get enough O_2 by breathing normally.

The breathing control center responds to a variety of nervous and chemical signals to keep the breathing rate and depth in tune with the changing demands of the body. Breathing rate must also be coordinated with the activity of the circulatory system. During exercise, the rate at which our heart beats and the amount of blood it pumps with each beat must be matched with the increased breathing rate. We examine the role of the circulatory system in gas exchange more closely in the next module.

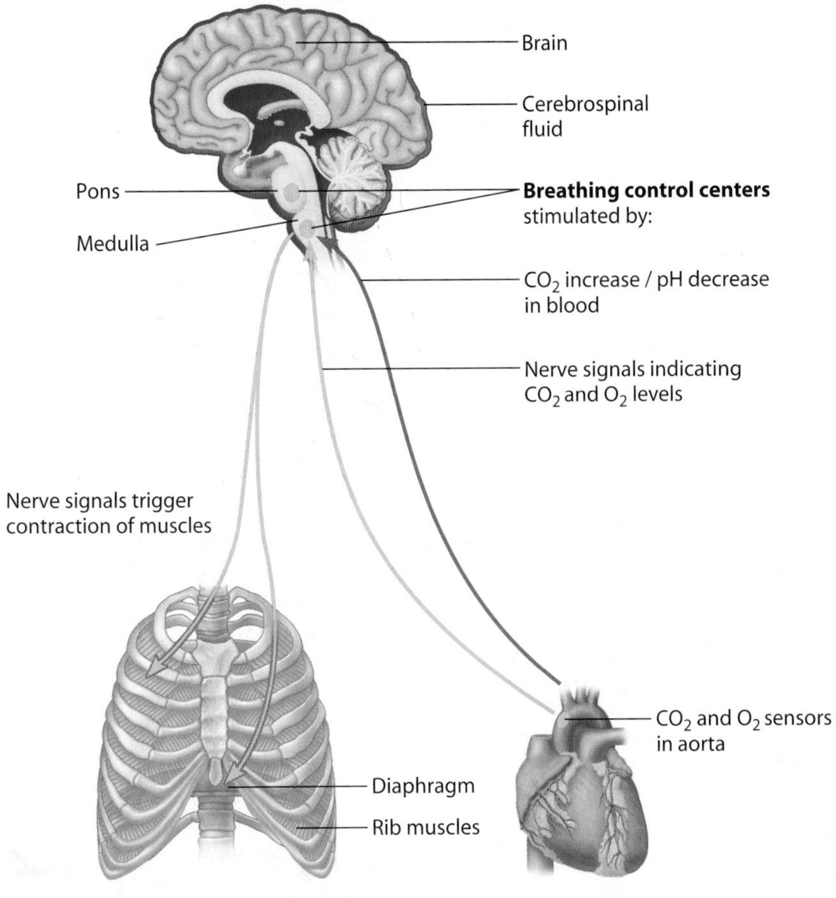

Brain

Cerebrospinal fluid

Pons

Medulla

Breathing control centers stimulated by:

CO_2 increase / pH decrease in blood

Nerve signals indicating CO_2 and O_2 levels

Nerve signals trigger contraction of muscles

CO_2 and O_2 sensors in aorta

Diaphragm

Rib muscles

Figure 22.8 Control centers that regulate breathing

? Explain how hyperventilation disrupts the control of breathing.

■ By purging the blood of CO_2 (and hence carbonic acid), which indirectly stimulates inhalation via its action on the breathing control center, hyperventilation temporarily suspends breathing.

22.9 Blood transports respiratory gases

How does O_2 get from our lungs to all the other tissues in our body, and how does CO_2 travel from the tissues to the alveoli? To answer these questions, we must jump ahead a bit to the subject of Chapter 23 and look at the basic organization of our circulatory system.

Figure 22.9 is a schematic diagram showing the main components of the human circulatory system and their role in gas exchange. Let's start with the heart, in the middle of the diagram. One side of the heart handles oxygen-poor blood (colored blue). The other side handles oxygen-rich blood (red). As indicated in the lower portion of the diagram, oxygen-poor blood returns to the heart from capillaries in body tissues. The heart pumps this blood to the alveolar capillaries in the lungs. At the top of the diagram, gases are exchanged between air in the alveolar spaces and blood in the alveolar capillaries. Blood leaves the alveolar capillaries, having lost CO_2 and gained O_2. This oxygen-rich blood returns to the heart and is pumped out to body tissues.

The exchange of gases between capillaries and the cells around them occurs by the diffusion of gases down gradients of pressure. A mixture of gases, such as air, exerts pressure. (You see evidence of gas pressure whenever you open a can of soda, releasing the pressure of the CO_2 it contains.) Each kind of gas in a mixture accounts for a portion, called the **partial pressure**, of the mixture's total pressure. Molecules of each kind of gas will diffuse down a gradient of its own partial pressure independent of the other gases. At the bottom of the figure, for instance, O_2 moves from oxygen-rich blood, through the interstitial fluid, and into tissue cells because it diffuses from a region of higher partial pressure to a region of lower partial pressure. The tissue cells maintain this gradient as they consume O_2 in cellular respiration. The CO_2 produced as a waste product of cellular respiration diffuses down its own partial-pressure gradient out of the cells and into the capillaries. Diffusion also accounts for gas exchange in the alveoli.

? What is the physical process underlying gas exchange?

■ Diffusion of each gas down its partial-pressure gradient

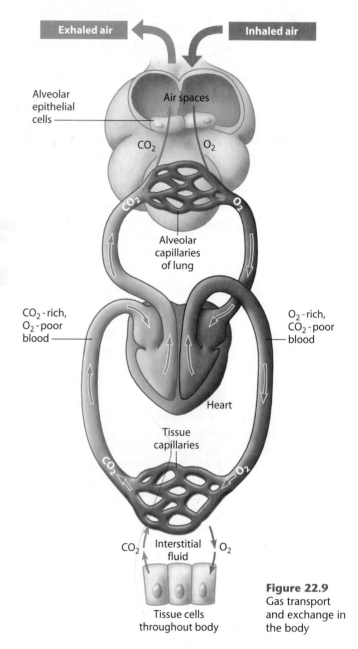

Figure 22.9
Gas transport and exchange in the body

Exhaled air
Inhaled air
Alveolar epithelial cells
Air spaces
CO_2
O_2
Alveolar capillaries of lung
CO_2-rich, O_2-poor blood
O_2-rich, CO_2-poor blood
Heart
Tissue capillaries
CO_2
Interstitial fluid
O_2
Tissue cells throughout body

22.10 Hemoglobin carries O_2 and helps transport CO_2 and buffer the blood

Oxygen is not very soluble in water, and most of the O_2 in blood is carried by **hemoglobin** in the red blood cells. A hemoglobin molecule consists of four polypeptide chains of two different types, distinguished by the two shades of purple in Figure 22.10 on the next page. Attached to each polypeptide is a chemical group called a heme (green), at the center of which is an iron atom (black). Each iron atom can carry one O_2 molecule. Thus, every hemoglobin molecule can carry up to four oxygen molecules. Hemoglobin loads up with O_2 in the lungs and transports it to the body's tissues. There, hemoglobin unloads some or all of its cargo, depending on the O_2 needs of the cells. (The partial pressure of O_2 in the tissue reflects how much O_2 the cells are using.)

Hemoglobin is a multipurpose molecule. It also helps the blood transport CO_2 and assists in buffering the blood—that is, preventing harmful changes in pH.

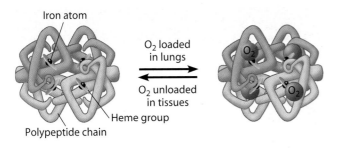

Iron atom

O_2 loaded in lungs

O_2 unloaded in tissues

Heme group

Polypeptide chain

Figure 22.10 Hemoglobin loading and unloading of O_2

When CO_2 leaves a tissue cell, it diffuses through the interstitial fluid, across the wall of a capillary, and into the blood fluid (plasma). Most of the CO_2 enters the red blood cells, where some of it combines with hemoglobin. The rest reacts with water molecules, forming carbonic acid (H_2CO_3). Red blood cells contain an enzyme that hastens this reaction. H_2CO_3 then breaks apart into a hydrogen ion (H^+) and a bicarbonate ion (HCO_3^-). Hemoglobin binds most of the H^+, minimizing the change in blood pH. The bicarbonate ions diffuse into the plasma, where they are carried to the lungs. This reversible reaction is shown here:

$$CO_2 + H_2O \longleftrightarrow H_2CO_3 \longleftrightarrow H^+ + HCO_3^-$$

Carbon dioxide, Water, Carbonic acid, Hydrogen ions, Bicarbonate

As blood flows through capillaries in the lungs, this process is reversed. Carbonic acid forms when bicarbonate combines with H^+. The carbonic acid is then converted back to CO_2 and water. Finally, the CO_2 diffuses from the blood into the alveoli and out of the body in exhaled air.

We have seen how O_2 and CO_2 are transported between the lungs and body tissue cells via the bloodstream. In the next module, we consider a special case of gas exchange between two circulatory systems.

Web/CD Activity 22B *Transport of Respiratory Gases*

? O_2 in the blood is transported bound to _____ within _____ _____ cells, while CO_2 is mainly transported as _____ ions within the plasma.

■ hemoglobin . . . red blood . . . bicarbonate

22.11 The human fetus exchanges gases with the mother's bloodstream

Figure 22.11 shows a human fetus inside the mother's uterus. The fetus literally swims in a protective watery bath, the amniotic fluid. Its lungs are full of fluid and are nonfunctional. How does the fetus exchange gases with the outside world? The answer lies in the function of the placenta, a composite organ that includes tissues from both the fetus and the mother. A large net of capillaries fans out into the placenta from blood vessels in the umbilical cord of the fetus. These fetal capillaries exchange gases with the maternal blood that circulates in the placenta, and the maternal circulatory system carries the gases to and from the mother's lungs. Aiding O_2 uptake by the fetus is fetal hemoglobin, a special type that attracts O_2 more strongly than does adult hemoglobin. Among the many health risks of smoking (see Module 22.6) is a reduction, perhaps by as much as 25%, in the supply of oxygen reaching the placenta.

What happens when a baby is born? Suddenly placental gas exchange ceases, and the baby's lungs must begin to work. Carbon dioxide in fetal blood acts as a signal. As soon as CO_2 stops diffusing from the fetus into the placenta, a CO_2 rise in fetal blood causes blood pH to fall, stimulating the breathing control centers in the infant's brain, and the newborn takes its first breath.

A human birth and the radical changes in gas exchange mechanisms that accompany it are extraordinary events. Resulting from millions of years of evolutionary adaptation, these events are on a par with the remarkable flying ability of the geese we discussed in the chapter introduction. For a goose to breathe the thin air and fly great distances high above Earth, or for a human baby to switch almost instantly from living in water and exchanging gases with maternal blood to breathing air directly, requires truly remarkable adaptations in the organism's respiratory system. Also required are adaptations of the circulatory system, which, as we have seen, supports the respiratory system in its gas exchange function. We turn to the circulatory system in Chapter 23.

? How does fetal hemoglobin enhance oxygen transfer from mother to fetus across the placenta?

■ Fetal hemoglobin has a greater affinity for O_2 than does adult hemoglobin, which helps "pull" the O_2 from maternal blood to fetal blood.

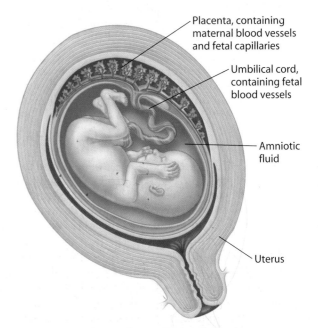

Placenta, containing maternal blood vessels and fetal capillaries

Umbilical cord, containing fetal blood vessels

Amniotic fluid

Uterus

Figure 22.11 A human fetus and placenta in the uterus

Reviewing the Concepts

Mechanisms of Gas Exchange (Introduction–22.8)

Gas exchange, the interchange of O_2 and CO_2 between an organism and its environment, provides O_2 for cellular respiration and removes its waste product, CO_2. Gas exchange often involves breathing, transport of gases, and exchange of gases with body cells **(Introduction–22.1).**

Respiratory surfaces must be thin and moist for diffusion of O_2 and CO_2 to occur. Some animals use their entire skin as a gas exchange organ. In most animals, specialized body parts—such as gills, tracheal systems, or lungs—provide large respiratory surfaces for gas exchange **(22.2).**

Gills are extensions of the body that absorb O_2 dissolved in water. In a fish, gas exchange is enhanced by ventilation and the countercurrent flow of water and blood **(22.3).**

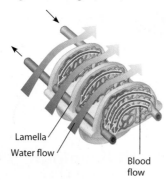

Lamella
Water flow
Blood flow

Tracheal systems in insects transport O_2 directly to body cells through a network of finely branched tubes **(22.4).**

Lungs. Most terrestrial vertebrates have lungs. In mammals, air inhaled through the nostrils passes through the pharynx and larynx into the trachea, bronchi, and bronchioles to the alveoli, where gas exchange occurs. Mucus and cilia in the respiratory passages protect the lungs, but smoking can destroy these protections. Smoking causes lung cancer, heart disease, and emphysema **(22.5–22.6).**

Breathing is the alternation of inhalation and exhalation. The contraction of rib muscles and diaphragm expands the chest cavity and reduces air pressure in the alveoli (negative pressure breathing). Vital capacity is the maximum volume of air that can be inhaled and exhaled, but the lungs still hold a residual volume. Air flows in one direction through the more efficient lungs of birds **(22.7).**

Breathing control centers in the brain keep breathing in tune with body needs, sensing and responding to the CO_2 level in the blood. A drop in blood pH triggers an increase in the rate and depth of breathing **(22.8).**

Transport of Gases in the Body (22.9–22.11)

Circulation. The heart pumps oxygen-poor blood to the lungs, where it picks up O_2 and drops off CO_2. Then the heart pumps the oxygen-rich blood to body cells, where it drops off O_2 and picks up CO_2. Gases diffuse down partial-pressure gradients in lungs and body tissues **(22.9).**

Hemoglobin in red blood cells transports oxygen, helps buffer the blood, and carries some CO_2. Most CO_2 is transported as bicarbonate ions in the plasma **(22.10).** A human fetus exchanges gases with maternal blood in the placenta. Fetal hemoglobin enhances oxygen transfer from maternal blood. At birth, rising CO_2 in fetal blood stimulates the breathing control centers to initiate breathing **(22.11).**

Connecting the Concepts

1. Complete this map to review some of the concepts of gas exchange.

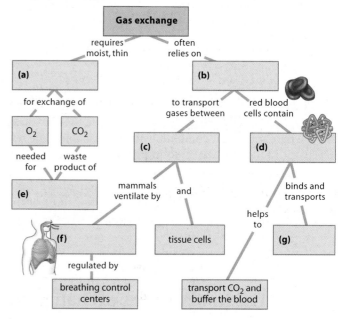

2. Label the parts of the human respiratory system.

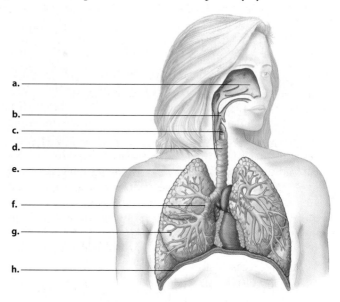

a.
b.
c.
d.
e.
f.
g.
h.

Testing Your Knowledge

Multiple Choice

3. When you hold your breath, which of the following first leads to the urge to breathe?
   a. falling CO_2
   b. falling O_2
   c. rising CO_2
   d. rising pH of the blood
   e. both c and d

4. Countercurrent gas exchange in the gills of a fish
   a. speeds up the flow of water through the gills.
   b. maintains a gradient that enhances diffusion.
   c. enables the fish to obtain oxygen without swimming.
   d. means that blood and water flow at different rates.
   e. allows O_2 to diffuse against its partial-pressure gradient.

5. When you inhale, the diaphragm
   a. relaxes and moves upward.
   b. relaxes and moves downward.
   c. contracts and moves upward.
   d. contracts and moves downward.
   e. is not involved in the breathing movements.

6. In which of the following organisms does oxygen diffuse directly across a respiratory surface to cells, without being carried by the blood?
   a. a grasshopper   d. a sparrow
   b. a whale   e. a mouse
   c. an earthworm

7. What is the function of the cilia in the trachea and bronchi?
   a. to sweep air into and out of the lungs
   b. to increase the surface area for gas exchange
   c. to vibrate when air is exhaled to produce sounds
   d. to dislodge food that may have slipped past the epiglottis
   e. to sweep mucus with trapped particles up and out of the respiratory tract

8. What do the alveoli of mammalian lungs, the gill filaments of fish, and the tracheal tubes of insects have in common?
   a. use of a circulatory system to transport gases
   b. respiratory surfaces that are invaginations (infoldings) of the body wall
   c. countercurrent exchange
   d. a large, moist surface area for gas exchange
   e. all of the above

9. Which of the following is the best explanation for why birds can fly over the Himalayas while most humans require oxygen masks to climb these mountains?
   a. Birds are much smaller and require less oxygen.
   b. Birds use positive pressure breathing, whereas humans use negative pressure breathing.
   c. With their one-way flow of air and efficient ventilation, the lungs of birds extract more O_2 from the air.
   d. The circulatory system of birds is much more efficient at delivering oxygen to tissues than is that of humans.
   e. Humans are endotherms and thus require more oxygen than do birds, which are ectotherms.

Describing, Comparing, and Explaining

10. What are two advantages of breathing air, compared to obtaining dissolved oxygen from water? What is a comparative disadvantage of breathing air?

11. Trace the path of an oxygen molecule from the air to a muscle cell in your arm, naming all the structures involved along the way.

Applying the Concepts

12. Partial pressure reflects the relative amount of gas in a mixture and is measured in millimeters of mercury (mm Hg). Llamas are native to the Andes Mountains in South America. The partial pressure of O_2 (abbreviated P_{O_2}) in the atmosphere where llamas live is about half of the P_{O_2} at sea level. As a result, the P_{O_2} in the lungs of llamas is about 50 mm Hg, whereas it is about 100 mm Hg in human lungs at sea level. A dissociation curve shows the % saturation (the amount of O_2 bound to hemoglobin) at increasing P_{O_2}. As you see in this graph, the dissociation curves for llama and human hemoglobin differ. Compare these two curves and explain how the hemoglobin of llamas is an adaptation to living where the air is "thin."

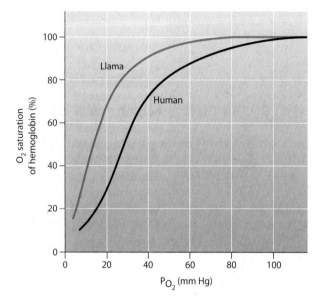

13. One of the many mutant opponents that the movie monster Godzilla contends with is Mothra, a giant moth-like creature with a wingspan of 7 to 8 m. Science fiction creatures like these can be critiqued on the grounds of biomechanical and physiological principles. Focusing on the principles of gas exchange that you learned about in this chapter, what problems would Mothra face? Why do you think truly giant insects are improbable?

14. Hundreds of studies have linked smoking with cardiovascular and lung disease. According to most health authorities, smoking is the leading cause of preventable, premature death in the United States. Antismoking and health groups have proposed that cigarette advertising in all media be banned entirely. What are some arguments in favor of a total ban on cigarette advertising? What are arguments in opposition? Do you favor or oppose such a ban? Defend your position.

Answers to all questions can be found in Appendix 3.

For study help and Activities, go to campbellbiology.com or the student CD-ROM.

23.1 The circulatory system connects with all body tissues

MECHANISMS OF INTERNAL TRANSPORT

23.2 Several types of internal transport have evolved in animals

23.3 Vertebrate cardiovascular systems reflect evolution

THE MAMMALIAN CARDIOVASCULAR SYSTEM

23.4 The human heart and cardiovascular system are typical of mammals

23.5 The structure of blood vessels fits their functions

23.6 The heart contracts and relaxes rhythmically

23.7 The pacemaker sets the tempo of the heartbeat

23.8 What is a heart attack?

23.9 Blood exerts pressure on vessel walls

23.10 Measuring blood pressure can reveal cardiovascular problems

23.11 Smooth muscle controls the distribution of blood

23.12 Capillaries allow the transfer of substances through their walls

STRUCTURE AND FUNCTION OF BLOOD

23.13 Blood consists of red and white blood cells suspended in plasma

23.14 Too few or too many red blood cells can be unhealthy

23.15 Blood clots plug leaks when blood vessels are injured

23.16 Stem cells offer a potential cure for blood cell diseases

How Does Gravity Affect Blood Circulation?

FEW ANIMALS SEEM LESS ALIKE than the ones you see here. On this page is a corn snake, found throughout much of the United States. This nonpoisonous predator eats mostly rats and mice, but it can also climb trees and dine on bird eggs. The giraffes on the right live in central Africa. They are herbivores, using their long necks to browse trees.

Despite their differences, snakes and giraffes have many features in common. As land vertebrates, they both have a backbone, lungs, and a circulatory system. They also have something in common with *all* animals that live on land: Every part of their body is subject to the persistent, unwavering force of gravity.

Gravity does not greatly affect aquatic animals because their body is supported by water, but it has profound effects on terrestrial species. Our own body shows signs of constant, long-term exposure to gravity, such as the tendency of the skin on our face to sag with age. Our circulatory system is strongly affected by gravity, which tends to pull blood downward into the lower parts of the body. The pull of gravity is a problem for us because we stand upright on two legs; the giraffe is similarly affected because of its long neck and legs; the corn snake faces the problem of gravity when it climbs a tree looking for bird eggs.

What are the solutions to these problems of gravity? Mammals, including humans and giraffes, have a very strong heart that keeps blood circulating despite gravity's pull. When we are standing, our heart must pump blood against gravity from our heart to our brain. The challenge is even greater for a giraffe. A standing giraffe requires a great deal more pressure to pump

Circulation

blood the 2.5 m from its heart to its head. Indeed, the blood pressure of a giraffe is about twice that of a human. But when a giraffe bends down to drink, as shown below, the pull of gravity almost doubles the blood pressure in the arteries leading to its head. Special check valves, sinuses, and other mechanisms protect the giraffe's brain from this potentially dangerous spike in blood pressure.

How does blood travel uphill in the veins of a giraffe's long legs, or in our own legs, to return to the heart? Skeletal muscle contractions help blood move along its way. As we walk or run, our leg muscles squeeze the veins and force the blood upward toward the heart. Our veins also have valves that allow the blood to flow in only one direction, preventing it from flowing back down the legs.

How does gravity affect a corn snake when it is climbing a tree? Because its heart is located close to its head, the snake's brain receives enough blood even when it is vertical. Also, blood vessels in its tail constrict when it climbs, helping to maintain blood flow to the head. After a climb, a corn snake

wriggles vigorously. This motion contracts muscles all over its body, squeezing veins and increasing circulation.

Like these land vertebrates, all organisms must exchange materials with their environment and distribute materials within their body. Most animals have a system of internal transport—a **circulatory system**—that transports oxygen and carbon dioxide, distributes nutrients to body cells, and conveys the waste products of metabolism to specific sites for disposal. This chapter focuses on the evolution, structure, and function of circulatory systems. ■ ■ ■

23.1 The circulatory system connects with all body tissues

A circulatory system is necessary in any animal whose body is too large or too complex for vital chemicals to reach all its parts by diffusion alone. Diffusion is inadequate for transporting materials over distances greater than a few cell widths—far less than the distance oxygen must travel between our lungs and brain or the distance nutrients must go between our small intestine and the muscles in our arms and legs. The circulatory system provides an efficient long-distance internal transport system that brings resources close enough to cells for diffusion to occur.

To be effective, the circulatory system must have an intimate connection with body tissues. In the human body, for instance, the heart pumps blood that has just been oxygenated in the lungs through a system of blood vessels that lead to microscopic vessels called **capillaries.** The capillaries form an intricate network among the cells of a tissue, such that no substance has to diffuse far to enter or leave a cell. The micrograph in **Figure 23.1A** shows the relationship between capillaries and our body tissues. This particular capillary supplies oxygenated, nutrient-rich blood to smooth muscle cells. Notice that red blood cells pass single-file through the capillary. Each red blood cell comes close enough to the surrounding tissue that O_2 can diffuse out of it into the muscle cells.

In **Figure 23.1B**, the downward arrows show the route that molecules take in diffusing from blood into tissue cells. As we discussed in Module 20.12, materials are not exchanged directly between blood and body cells. Each body cell is immersed in a watery interstitial fluid. Molecules such as O_2 (red dots in the figure) and nutrients (green dots) diffuse first out of a capillary into the interstitial fluid and then from the interstitial fluid into a tissue cell.

The circulatory system has several other major functions in addition to transporting O_2 and nutrients. It conveys metabolic wastes to waste disposal organs: CO_2 to the lungs and a variety of other metabolic wastes to the kidneys. The upward arrows in Figure 23.1B represent the diffusion of waste molecules (gray dots) out of a tissue cell, through the interstitial fluid, and into the capillary.

The circulatory system also plays a key role in maintaining a constant internal environment (homeostasis). By ex-

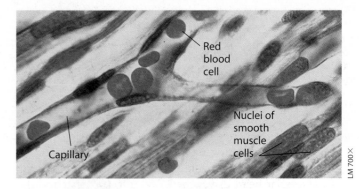

Figure 23.1A A capillary in muscle tissue

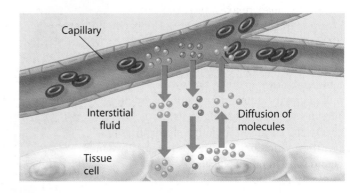

Figure 23.1B Diffusion between blood and tissue cells

changing molecules with the interstitial fluid, the circulatory system helps control the makeup of the environment in which the tissue cells live. And the circulatory system helps control the makeup of the blood by continuously moving it through organs, such as the liver and kidneys, that regulate the blood's contents. As we will see in later chapters, the circulatory system also is involved in body defense, temperature regulation, and hormone distribution.

? The "internal pond" that bathes your cells is _____ _____.

■ interstitial fluid

MECHANISMS OF INTERNAL TRANSPORT

23.2 Several types of internal transport have evolved in animals

The body plan of a hydra and other cnidarians makes a circulatory system unnecessary. As we saw in Module 21.3, the body wall of a hydra is only two or three cells thick, so all the cells can exchange materials directly with the water surrounding the animal or with the water in its gastrovascular cavity. Water is drawn into the gastrovascular cavity through the mouth and exits back out through the mouth. Digestion

occurs in the gastrovascular cavity and in the cells lining it. Only these cells have direct access to nutrients, but nutrients have only a short distance to diffuse to cells of the outer layer.

The jelly in **Figure 23.2A** (facing page) is a cnidarian with a more elaborate gastrovascular cavity, with branches radiating from the mouth to a circular canal. Ciliated cells lining these branches and canals circulate the gastrovascular fluid.

Planarians and most other flatworms also have a gastrovascular cavity that exchanges materials with the environment through a single opening. A gastrovascular cavity provides adequate internal transport for such flat, thin animals, but it is not adequate for animals with thick, multiple layers of cells. Such animals require a true circulatory system containing a specialized circulatory fluid, **blood.**

Two basic types of circulatory systems have evolved in animals. Many invertebrates, including most molluscs and all arthropods, have what is called an **open circulatory system.** The system is termed "open" because blood is pumped through open-ended vessels and flows out among the cells; there is no distinction between blood and interstitial fluid. In an insect, such as the grasshopper (**Figure 23.2B**), pumping of the tubular heart drives the blood into the head and the rest of the body (black arrows). Body movements help circulate the blood as chemical exchange occurs with body cells. When the heart relaxes, blood returns to it through several pores. Each pore has a valve that closes when the heart contracts, preventing backflow of the blood. As we saw in Module 22.4, respiratory gases are conveyed to and from the insect's body cells by the tracheal system (not shown here), not by the circulatory system.

Earthworms, squids, octopuses, and vertebrates have **closed circulatory systems.** The vertebrate circulatory system is often called a **cardiovascular system** (from the Greek *kardia,* heart, and Latin *vas,* vessel). The blood is confined to the vessels, which keep it distinct from the interstitial fluid. There are three kinds of vessels in a closed circulatory system. **Arteries** carry blood away from the heart to organs and tissues throughout the body; **veins** return blood to the heart; and capillaries convey blood between arteries and veins within each tissue. Arteries and veins are distinguished by the *direction* in which they carry blood, not by the quality of the blood they contain. Although most arteries convey oxygen-rich blood and most veins transport blood depleted of oxygen (oxygen-poor blood), there are important exceptions. For example, we have two arteries, called pulmonary arteries, that carry oxygen-poor blood from our heart to our lungs; and we have four pulmonary veins that carry freshly oxygenated blood from the lungs to the heart.

The cardiovascular system of a fish (**Figure 23.2C**) illustrates key features of a closed circulatory system. The heart of a fish has two main chambers. The **atrium** (plural, *atria*) receives blood from the veins, and the **ventricle** pumps blood to the gills via large arteries. As in all figures depicting closed circulatory systems in this chapter, red represents oxygen-rich blood and blue represents oxygen-poor blood. After passing through the gill capillaries, oxygen-rich blood flows into large arteries that carry it to all other parts of the body. The large arteries branch into **arterioles,** small vessels that give rise to capillaries. Networks of capillaries called **capillary beds** infiltrate every organ and tissue in the body. The thin walls of the capillaries allow chemical exchange between the blood and the interstitial fluid. The capillaries converge into **venules,** which converge into veins that return blood to the heart.

In the next module, we explore vertebrate cardiovascular systems in more detail.

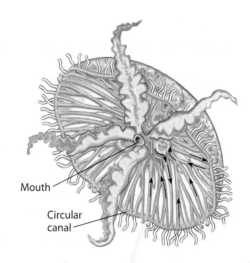

Figure 23.2A The gastrovascular cavity (salmon color) in a jelly

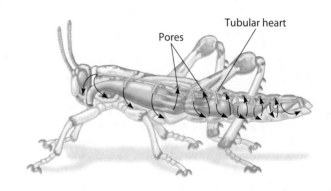

Figure 23.2B The open circulatory system (vessels in gold) in a grasshopper

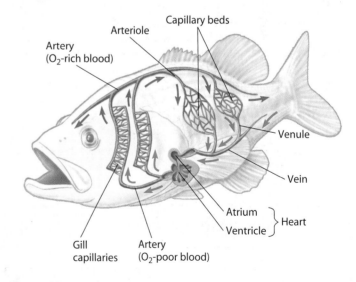

Figure 23.2C The closed circulatory system in a fish

? A(n) _____ is a blood vessel that carries blood toward the heart, to a heart chamber called a(n) _____. A(n) _____ is a blood vessel that carries blood away from the heart, from a heart chamber called a(n) _____.

■ vein · · · atrium · · · artery · · · ventricle

23.3 Vertebrate cardiovascular systems reflect evolution

The colonization of land by vertebrates was a momentous episode in the history of life, opening vast new opportunities. As aquatic vertebrates became adapted for terrestrial life, nearly all of their organ systems underwent major changes. One of the most drastic was the switch from gill breathing to lung breathing, and this switch was accompanied by important changes in the cardiovascular system.

As diagrammed in Figure 23.3A, a fish has a single circuit of blood flow and two heart chambers. Blood pumped from the ventricle travels first to the gill capillaries, where it picks up oxygen. The blood delivers oxygen to the tissues of the body as it passes through a second set of capillaries, known as systemic capillaries, before returning to the atrium of the heart. The blood slows down considerably while passing through the gill capillaries, although it is helped on its way to the other organs by the animal's swimming movements.

In terrestrial vertebrates, a more vigorous flow of blood is supplied to body organs by means of a **double circulation**, in which blood is pumped a second time after it slows down in the capillary beds of the lungs. The **pulmonary circuit** carries blood between the heart and the gas exchange tissues in the lungs, and the **systemic circuit** carries blood between the heart and the rest of the body.

You can see an example of these two circuits in Figure 23.3B. Notice that the right side of the animal's heart is on the left in the diagram, and the left side of the heart is on the right. (It is customary to draw the system this way, as though the heart is in a body facing you from the page.) Frogs and other amphibians have a three-chambered heart. The right atrium receives blood returning from the systemic capillaries. The ventricle pumps blood to capillary beds in the lungs and skin. Because gas exchange occurs both in the lungs and across the thin, moist skin, this is called a pulmocutaneous

circuit. Oxygen-rich blood returns to the left atrium, and most of it is pumped by the ventricle into the systemic circuit. Although blood mixes in the single ventricle, most of the oxygen-poor blood (blue) is diverted to the pulmocutaneous circuit and most of the oxygen-rich blood (red) goes to the systemic circuit.

Reptiles (except birds) also have a three-chambered heart, but the ventricle is partially divided, and less mixing of blood occurs. The ventricle is completely divided in crocodilians.

In all birds and mammals, the ventricle is completely divided and the heart has four chambers: two atria (A) and two ventricles (V) (Figure 23.3C). The left side of the heart receives and pumps only oxygen-rich blood (red), while the right side handles only oxygen-poor blood (blue). The evolution of a powerful four-chambered heart was an essential adaptation to support the high metabolic rate characteristic of birds and mammals, which are endothermic. Endotherms use about ten times as much energy as equal-sized ectotherms, such as reptiles (see Module 18.19); therefore, their circulatory system needs to deliver about ten times as much fuel and oxygen to body tissues. This requirement is met by a large and powerful heart that is able to pump a large volume of blood and by separate systemic and pulmonary circulations. Birds and mammals descended from different reptilian ancestors, and their four-chambered hearts evolved independently.

Next we look at the mammalian system in more detail.

> **?** What is the difference between the single circuit of a fish cardiovascular system and double circulation?

■ In a fish, blood travels through the gill capillaries and then the systemic capillaries before returning to the heart. In the double circulation of land vertebrates, blood returns to the heart and is pumped a second time between the pulmonary and systemic circuits.

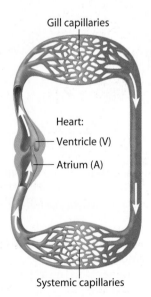

Figure 23.3A Diagram of the cardiovascular system of a fish

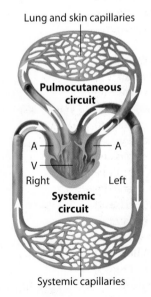

Figure 23.3B Diagram of the cardiovascular system of an amphibian

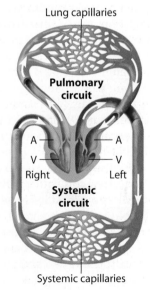

Figure 23.3C Diagram of the cardiovascular system of a bird or mammal

23.4 The human heart and cardiovascular system are typical of mammals

About the size of a clenched fist, the human heart (Figure 23.4A) is enclosed in a sac just under the breastbone. The heart is formed mostly of cardiac muscle tissue. Its thin-walled atria collect blood returning to the heart, most of which then flows into the ventricles. Contraction of the atria complete the filling of the ventricles. The thicker-walled ventricles pump blood to the lungs and to all other body tissues. Flap-like valves in the heart regulate the direction of blood flow, as we will see in Module 23.6.

Let's follow the blood through the entire circulatory system, as shown in Figure 23.4B. Beginning with the pulmonary (lung) circuit, ❶ the right ventricle pumps blood to the lungs via ❷ two **pulmonary arteries.** As the blood flows through ❸ capillaries in the lungs, it takes up oxygen and unloads carbon dioxide. Oxygen-rich blood then flows back through ❹ the **pulmonary veins** to ❺ the left atrium. Next, the oxygen-rich blood flows from the left atrium into ❻ the left ventricle.

In Figure 23.4A, you can see that the walls of the left ventricle are thicker than those of the right ventricle. The powerful muscles in the left ventricle pump blood to all body tissues through the systemic circuit. As Figure 23.4B shows, oxygen-rich blood leaves the left ventricle through ❼ the **aorta.** The aorta is our largest blood vessel, with a diameter of roughly 2.5 cm, about the same diameter as a quarter. The first branches from the aorta are the coronary arteries (not shown), which supply blood to the heart muscle itself. Several large arteries branch from the aorta, leading to ❽ the head, chest, and arms (top) and to the abdominal region and legs (bottom). For simplicity, Figure 23.4B does not show the individual organs, but within each organ, arteries lead to arterioles that branch to capillaries. The capillaries rejoin as venules, which convey the blood back into veins. ❾ Oxygen-poor blood from the upper body is channeled into a large vein called the **superior vena cava,** and another large vein, the **inferior vena cava,** drains blood from the lower body. The two venae cavae empty their blood into ❿ the right atrium. As the blood flows from the right atrium into the right ventricle, we complete our journey.

Remember that the path of any single blood cell is always heart to lung capillaries to heart to body tissue capillaries and back to heart. In one systemic circuit, a blood cell may travel to the brain; in the next (after a pulmonary circuit), it may travel to the legs. It never travels from the brain to the legs without first returning to the heart and being pumped to the lungs to be recharged with oxygen.

Now that we have surveyed the cardiovascular system as a whole, let's take a closer look at the structure and function of its parts, first the vessels and then the heart.

Web/CD Activity 23A *Mammalian Cardiovascular System Structure*

Web/CD Activity 23B *Path of Blood Flow in Mammals*

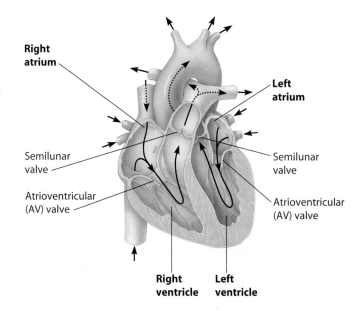

Figure 23.4A Blood flow through the human heart

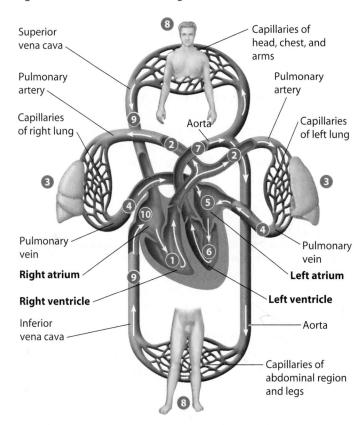

Figure 23.4B Blood flow through the human cardiovascular system

? Vena cava is to right _____ as _____ vein is to left atrium.

■ atrium . . . pulmonary

23.5 The structure of blood vessels fits their functions

Figure 23.5 illustrates the structures of the different kinds of blood vessels and how the vessels are connected. Look first at the capillaries (center), which form fine branching networks where materials are exchanged between the blood and the interstitial fluid that bathes the cells. Appropriate to this function, capillaries have very thin walls formed of a single layer of epithelial cells, which is wrapped in a thin basement membrane (see Module 20.4). The inner surface of the capillary is smooth and keeps the blood cells from being abraded as they tumble along.

Arteries, arterioles, veins, and venules have thicker walls than those of capillaries. The walls have the same epithelium as capillaries, but they are reinforced by two other tissue layers. An outer layer of connective tissue with elastic fibers enables the vessels to stretch and recoil. The middle layer consists mainly of smooth muscle. Both these layers are thicker and sturdier in arteries, providing the strength and elasticity to accommodate the rapid flow and high pressure of blood pumped by the heart. Arteries are also able to regulate blood flow by constricting or relaxing their smooth muscle layer. The thinner-walled veins convey blood back to the heart at low velocity and pressure. Within large veins, flaps of tissue act as one-way valves. As you learned in the chapter introduction, the valves prevent backflow, permitting blood to flow only toward the heart.

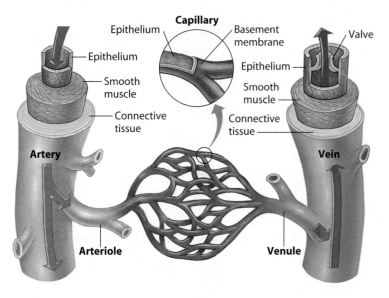

Figure 23.5 Structural relationships of blood vessels

? Which tissue layer is common to all blood vessels?

■ Epithelium

23.6 The heart contracts and relaxes rhythmically

The heart is the hub of the circulatory system. In a continuous cycle, it passively fills with blood and then actively contracts. A complete sequence of filling and pumping is called the **cardiac cycle.**

Figure 23.6 shows a cardiac cycle that takes about 0.8 second, corresponding to a heart rate of 75 beats per minute. When the entire heart is relaxed, in the phase called ❶ **diastole,** blood flows into all four of its chambers. Blood enters the right atrium from the venae cavae and the left atrium from the pulmonary veins. The valves between the atria and the ventricles (atrioventricular, or AV, valves) are open. Diastole lasts about 0.4 second, during which the ventricles nearly fill with blood.

The other main phase of the cardiac cycle is called **systole.** ❷ Systole begins with a very brief (0.1-second) contraction of the atria that completely fills the ventricles with blood (atrial systole). ❸ Then the ventricles contract for about 0.3 second (ventricular systole). The force of their contraction closes the AV valves, opens the semilunar valves located at the exit from each ventricle, and pumps blood into the large arteries. Blood flows into the atria during the second part of systole, as the green arrows in step 3 indicate.

The volume of blood per minute that the left ventricle pumps into the systemic circuit is called **cardiac output.** This volume is equal to the amount of blood pumped by the

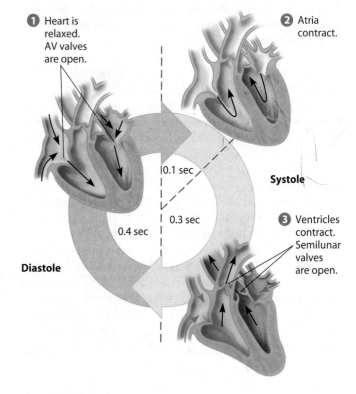

❶ Heart is relaxed. AV valves are open.

❷ Atria contract.

Systole

❸ Ventricles contract. Semilunar valves are open.

Diastole

0.1 sec

0.3 sec

0.4 sec

Figure 23.6 A cardiac cycle

left ventricle each time it contracts (about 75 mL per beat for the average person) times the heart rate. An average person at rest might have a heart rate of 70 beats per minute. At this rate, cardiac output would be 75 mL/beat × 70 beats/min = 5,250 mL/min = 5.25 L/min, roughly equivalent to the total volume of blood. Thus, a drop of blood travels the entire circuit in just 1 minute. Heart rate and cardiac output vary, depending on age, fitness, and other factors. Both increase, for instance, during heavy exercise, in which cardiac output can increase fivefold.

The heart valves, made of flaps of connective tissue, prevent backflow and keep blood moving in the correct direction. The closing of the AV valves when the ventricles contract keeps blood from flowing back into the atria. When the ventricles relax in diastole, blood in the arteries starts to flow back toward the heart, causing the flaps of the semilunar valves to close and preventing blood from flowing back into the ventricles. The heart sounds we can hear with a stethoscope—"lub-dup, lub-dup"—are caused by these closings of the heart valves. The "lub" sound comes from the recoil of blood against the closed AV valves. The "dup" comes as the semilunar valves snap shut.

A trained ear can also detect the hissing sound of a heart murmur, which may indicate a defect in one or more of the heart valves. A murmur occurs when a stream of blood squirts backward through a valve. Some people are born with murmurs, while others have their valves damaged by infection (from rheumatic fever, for instance). Most valve defects do not reduce the efficiency of blood flow enough to warrant surgery. Those that do can be corrected by replacing the damaged valves with artificial ones or with valves taken from an organ donor (human or other animal, usually a pig).

? During a cardiac cycle of 0.8 second, the atria are generally relaxed for ____ second.

■ 0.7

23.7 The pacemaker sets the tempo of the heartbeat

A specialized region of cardiac muscle called the **pacemaker,** or **SA (sinoatrial) node,** maintains the heart's pumping rhythm by setting the rate at which all the muscle cells of the heart contract.

The pacemaker is situated in the wall of the right atrium (Figure 23.7). ❶ The pacemaker generates electrical signals (black arrows) much like those produced by nerve cells. ❷ Because cardiac muscle cells are electrically connected by specialized junctions between cells, signals (yellow color) spread quickly through both atria, making them contract in unison. The signals also pass to a relay point called the **AV (atrioventricular) node,** in the wall between the right atrium and right ventricle. Here the signals are delayed about 0.1 second. The delay ensures that the atria contract and empty before the ventricles contract. ❸ Specialized cardiac muscle fibers (orange) then relay the signals to the apex of the heart and ❹ up through the walls of the ventricles, triggering the strong contractions that drive the blood out of the heart.

The electrical signals in the heart generate electrical changes in the skin, which can be detected by electrodes and recorded as an electrocardiogram (ECG or EKG). The yellow color in the graphs under the hearts indicates the part of an ECG that matches the electrical event shown in yellow in the heart. In step 4, the portion of the ECG to the right of the yellow "spike" is the electrical activity of the ventricles becoming primed to conduct the next contraction signals.

In certain kinds of heart disease, the heart's self-pacing system fails to maintain a normal heart rhythm. The remedy is an **artificial pacemaker,** a tiny electronic device surgically implanted near the AV node. Artificial pacemakers emit electrical signals that trigger normal heartbeats.

Two sets of nerves with opposite effects can direct the pacemaker to speed up or slow down, depending on physiological and emotional cues. Heart rate is also influenced by hormones, such as epinephrine, the "fight-or-flight" hormone released at times of stress. Heart rate also increases with exercise, enabling the circulatory system to provide the additional oxygen needed by hardworking muscles. Thus, the heart responds to stimuli from our surroundings—something the pacemaker cannot do on its own.

? A slight decrease in blood pH causes the pacemaker to speed up. What is the function of this control mechanism? (*Hint:* See Module 22.8.)

■ More CO_2 in the blood causes pH to drop. The increased heart rate enhances delivery of CO_2-rich blood to the lungs for removal of the CO_2.

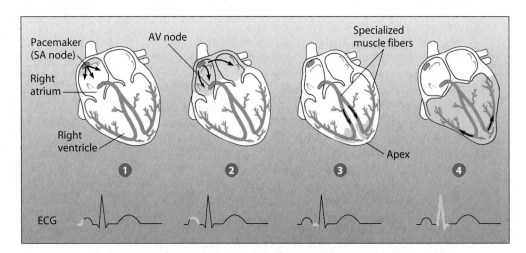

Figure 23.7 Control of the heart's rhythm (top) and an electrocardiogram (bottom). (Yellow color represents electrical signals.)

23.8   What is a heart attack?

Like all of our cells, heart muscle cells require nutrients and oxygen-rich blood to survive. Indeed, their needs are high, as our hearts contract over 100,000 times per day. When blood exits the heart via the aorta, several coronary arteries (shown in red in **Figure 23.8A**) immediately branch off to feed the heart muscle. If one or more of these blood vessels become blocked, heart muscle cells will quickly die (gray area in Figure 23.8A). Such an event, and the subsequent failure of the heart to function properly, is called a **heart attack.** Approximately one-third of heart attack victims die almost immediately. For those who survive, the ability of the damaged heart to pump blood may be seriously impaired.

Many people die each year of diseases of the heart and blood vessels, known as **cardiovascular disease.** Heart attacks and strokes, the death of brain tissue resulting from blockage of arteries in the head, are leading causes of death, ranking first and third, respectively, in the United States. Cardiovascular disease accounts for almost 50% of all deaths in the United States, killing over 1 million people each year—about one every 30 seconds.

The suddenness of a heart attack or stroke belies the fact that the arteries of most victims became impaired gradually, by a chronic cardiovascular disease known as **atherosclerosis**

(from the Greek *athero*, paste, and *sclerosis*, hardness). During the course of this disease, growths called plaques develop in the inner walls of arteries, narrowing the passages through which blood can flow (**Figure 23.8B**). The smooth muscle layer of an artery becomes thickened and infiltrated with cholesterol and fibrous connective tissue. A blood clot is more likely to become trapped in a vessel that has been narrowed by plaques. Furthermore, plaques are common sites of blood clot formation.

To some extent, the tendency to develop cardiovascular disease appears to be inherited. There are three everyday behaviors, however, that significantly impact the risk. Smoking doubles the risk of heart attack and harms the circulatory system in several other ways. Exercise can cut the risk of heart disease in half, but most adults fail to achieve recommended amounts of physical activity. Eating a heart-healthy diet, low in cholesterol and saturated fat, can reduce the risk of atherosclerosis (see Module 21.24).

There are treatments available for cardiovascular disease, and more continue to be developed. Heart attack victims are treated with clot-dissolving drugs, which stop many heart attacks and help prevent damage. Diagnostic tests ranging from cholesterol and blood pressure measurements to sophisticated imaging techniques such as CT and MRI help identify those at risk. Drugs can lower cholesterol and blood pressure, a risk factor we discuss in Module 23.10. Angioplasty (inserting a tiny catheter with a balloon that is inflated to compress plaques and widen clogged arteries) and stents (small wire mesh tubes that prop arteries open) can also help. Bypass surgery is a more drastic remedy. In this procedure, blood vessels removed from a patient's legs are sewn into the heart to shunt blood around clogged arteries. In extreme cases, a heart transplant may be necessary. With the severe shortage of donor hearts, various artificial pumping devices are under development.

Fortunately, the U.S. death rate from cardiovascular disease has been cut in half over the past 50 years. Health education, early diagnosis, and reduction of risk factors, particularly smoking, are responsible. The availability of automatic external defibrillators (AEDs) has also saved thousands of lives. These devices deliver electric shocks that can reverse a short circuit of the heart's pacemaker and reestablish normal electrical rhythms in the heart. Unlike hospital defibrillators, AEDs are designed to be used by laypeople and are placed in emergency vehicles and in public places (such as airports and shopping malls) where they are quickly accessible.

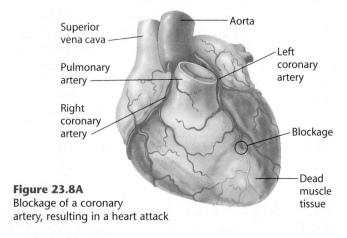

Figure 23.8A
Blockage of a coronary artery, resulting in a heart attack

Superior vena cava — Aorta — Left coronary artery — Pulmonary artery — Right coronary artery — Blockage — Dead muscle tissue

? List three things you can do to lower your risk of cardiovascular disease.

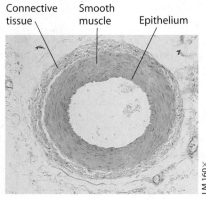

Connective tissue   Smooth muscle   Epithelium

LM 160×

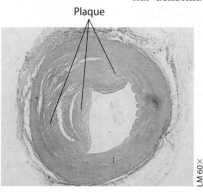

Plaque

LM 60×

Figure 23.8B   Atherosclerosis: a normal artery (left) and an artery partially closed by plaque (right)

■ Exercise regularly, reduce dietary fat and cholesterol, and don't smoke.

23.9 Blood exerts pressure on vessel walls

Blood pressure is the force that blood exerts against the walls of our blood vessels. Created by the pumping of the heart, blood pressure is the force driving blood from the heart through arteries and arterioles to capillary beds.

When the ventricles contract, blood is forced into the arteries faster than it can flow into the arterioles. This stretches the elastic walls of the arteries. You can feel this effect of blood pressure when you measure your heart rate by taking your pulse. The **pulse** is the rhythmic stretching of the arteries. At the top of Figure 23.9A, you can see the surge in pressure caused by ventricular contraction (systolic pressure). The elastic arteries snap back during diastole, maintaining pressure on the blood (diastolic pressure) and a continuous flow of blood into arterioles and capillaries.

Blood pressure depends partly on cardiac output (the volume of blood pumped into the aorta) and partly on the resistance to blood flow imposed by the narrow openings of the arterioles. These openings are controlled by smooth muscles. When the muscles relax, the arterioles dilate, and blood flows through them more readily, causing a fall in blood pressure. Physical and emotional stress can raise blood pressure by triggering nervous and hormonal signals that constrict these blood vessels. Regulatory mechanisms coordinate cardiac output and changes in the arteriole resistance to maintain adequate blood pressure as demands on the circulatory system change. Thus, a giraffe can eat leaves high in a tree and then bend to get a drink (see chapter introduction).

As Figure 23.9A indicates, blood pressure (expressed in millimeters of mercury, mm Hg) and the blood's velocity (rate of flow, expressed in centimeters per second, cm/sec) are highest in the aorta and arteries. Blood pressure and velocity both decline abruptly as the blood enters the arterioles. The pressure drop results mainly from the resistance to blood flow caused by friction between the blood and the large surface area it contacts in the walls of the numerous tiny arterioles.

The velocity decline in the arterioles is mainly a result of the structural arrangement indicated in the middle of Figure 23.9A. The total combined cross-sectional area of all the arterioles is much greater than the diameter of the artery that feeds blood into them. If there were only one small arteriole per artery, the blood would flow faster through the arteriole, the way water does when you add a narrow nozzle to a garden hose. However, there are many arterioles per artery, so the effect is like taking the nozzle off the hose: As you increase the diameter of a pipe, the flow rate goes down.

The cross-sectional area is greatest in the capillaries, and, as you can see, the velocity of blood is slowest through them. The overall result of this decline in velocity and pressure is a steady, leisurely flow of blood in the capillaries, enhancing the exchange of substances between blood and interstitial fluid.

By the time blood reaches the veins, its pressure has dropped to near zero. The blood has encountered so much resistance as it passes through the millions of tiny arterioles and capillaries that the force from the pumping heart

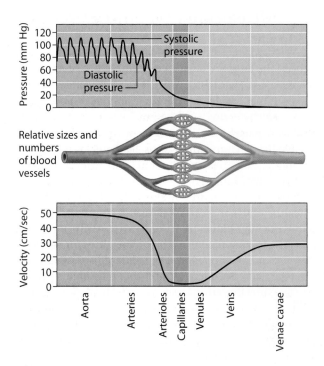

Figure 23.9A Blood pressure and velocity in the blood vessels

no longer propels it. How, then, does blood return to the heart, especially when it must travel up the legs against gravity? As we discussed in the chapter introduction, the veins of mammals such as humans and giraffes are sandwiched between skeletal muscles (Figure 23.9B). Consequently, whenever the body moves, the muscles pinch the veins and squeeze blood along toward the heart. The large veins of mammals have valves that allow the blood to flow only toward the heart. Breathing also helps return blood to the heart. When we inhale, the change in pressure within our chest cavity causes the large veins near our heart to expand and fill.

Web/CD Activity 23C *Mammalian Cardiovascular System Function*

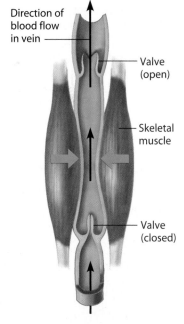

Figure 23.9B Blood flow in a vein

❓ If blood pressure in the veins drops to zero, how is it that blood velocity increases as blood flows from venules to veins?

■ The combined diameter of the vessels through which the blood flows decreases. The velocity of the flow increases, just as water flows faster when a nozzle narrows the opening of a hose.

23.10 Measuring blood pressure can reveal cardiovascular problems

Blood exerts a force on the inside walls of blood vessels throughout the entire circulatory system, but the term *blood pressure* usually refers to the force pushing against arterial walls. As indicated in **Figure 23.10**, ❶ a typical blood pressure for a healthy young adult is 110/70. The units are millimeters of mercury (mm Hg), indicating how tall a column of mercury the pressure could support. The first number is the systolic pressure; the second number is the diastolic pressure (see Module 23.9).

Blood pressure is an important indicator of cardiovascular health, and abnormal readings can indicate serious problems. Luckily, blood pressure can be easily measured using a sphygmomanometer, or blood pressure cuff. ❷ Once wrapped around the upper arm, where large arteries are accessible, the cuff is inflated until the pressure is strong enough to close the artery and cut off blood flow to the lower arm. ❸ A stethoscope is used to listen for sounds of blood flow below the cuff, and systolic blood pressure is the first measurement taken as the cuff is gradually deflated. The first sound of blood spurting through the constricted artery indicates that the blood pressure is stronger than the pressure exerted by the cuff. The pressure at this point is the systolic pressure. ❹ The sound of blood flowing unevenly through the artery continues until the pressure of the cuff falls below the pressure of the artery during diastole. Blood now flows continuously through the artery, and the sound of blood flow ceases. The reading on the pressure gauge at this point is the diastolic pressure.

Optimal blood pressure for adults is below 120 mm Hg for systolic pressure and below 80 mm Hg for diastolic pressure. Lower values are generally considered better, except in rare cases where low blood pressure may indicate a serious underlying condition (such as endocrine disorders, malnutrition, or internal bleeding). Blood pressure higher than the normal range, however, may indicate a serious cardiovascular disorder.

High blood pressure, or **hypertension**, is persistent systolic blood pressure at or higher than 140 mm Hg and/or diastolic blood pressure at or higher than 90 mm Hg. Hypertension affects almost one-third of the adult population in the United States. It is sometimes called a "silent killer" because high blood pressure often displays no outward symptoms for years, but may be leading to severe health problems.

High blood pressure harms the cardiovascular system in several ways. Elevated pressure requires the heart to work harder to pump blood throughout the body, and over time the left ventricle may enlarge. When the coronary blood supply does not keep up with the demands of this increase in muscle mass, the heart muscle weakens. In addition, the increased force on arterial walls causes tiny ruptures that promote plaque formation, aggravating atherosclerosis (see Module 23.8) and increasing the risk of blood clot formation. Prolonged hypertension is the major cause of heart attack, heart disease, stroke, and kidney failure.

In the vast majority of patients, the exact cause of hypertension cannot be firmly established. Some predispositions

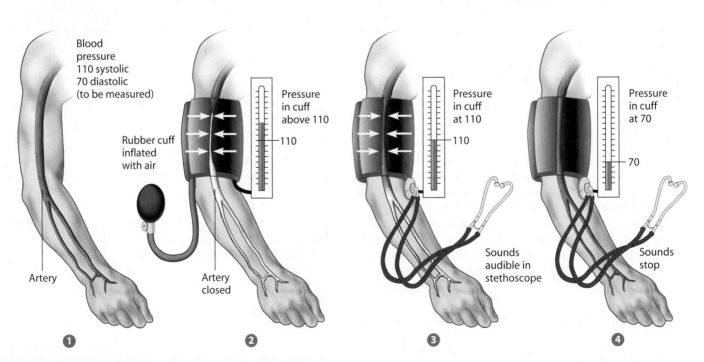

Figure 23.10 Measuring blood pressure

to hypertension cannot be avoided, such as gender, ethnicity, age, and heredity. Males have a greater risk of high blood pressure up to age 55, but females have a greater risk over age 75. African-Americans are more prone to hypertension than Caucasians. Blood pressure generally increases with age. Heredity also plays a role, since children of parents with hypertension are twice as likely to develop the condition.

However, no matter how many unavoidable predispositions a person may have, there are lifestyle changes that can prevent or control hypertension in just about everybody: eating a heart-healthy diet, not smoking, avoiding excess alcohol (more than two drinks per day), exercising regularly (30 minutes of moderate activity on most days), and maintaining a healthy weight. Many people associate salt with high blood pressure, but it is a contributing factor only in a small percentage of people. If lifestyle changes don't lower blood pressure, there are several effective antihypertensive medications.

Web/CD Thinking as a Scientist *Connection: How Is Cardiovascular Fitness Measured?*

? Listening with a stethoscope below a sphygmomanometer cuff, you hear sounds that begin at 140 mm Hg and cease at 90 mm Hg. What are the systolic and diastolic blood pressures for this person? What is this person's blood pressure?

■ Systolic = 140; diastolic = 90; blood pressure = 140/90

23.11 Smooth muscle controls the distribution of blood

You learned in Module 23.9 that smooth muscles in arteriole walls can influence blood pressure by changing the resistance to blood flow out of the arteries and into arterioles. The smooth muscles in the arteriole walls also regulate the distribution of blood to the capillaries of the various organs. At any given time, only about 5–10% of the body's capillaries have blood flowing through them. However, each tissue has many capillaries, so every part of the body is supplied with blood at all times. Capillaries in a few organs, such as the brain, heart, kidneys, and liver, usually carry a full load of blood, but in many other sites, the blood supply varies as blood is diverted from one destination to another, depending on need.

In addition to the smooth muscles that can constrict or dilate an arteriole leading into a capillary bed, a second mechanism, illustrated in **Figure 23.11**, regulates the distribution of blood. Notice that in both parts of this figure there is a capillary called a thoroughfare channel, through which blood streams directly from arteriole to venule. This channel is always open. Capillaries branching off from thoroughfare channels form the bulk of the capillary bed. Passage of blood into these branching capillaries is regulated by rings of smooth muscle, called precapillary sphincters because they are located at the entrance to capillary beds. As you can see in the figure, ❶ blood flows through a capillary bed when its precapillary sphincters are relaxed. ❷ It bypasses the capillary bed when the sphincters are contracted. After a meal, for instance, precapillary sphincters in the wall of the digestive tract let a larger quantity of blood pass through the capillary beds than when food is not being digested. During strenuous exercise, many of the capillaries in the digestive tract are closed off, and blood is supplied more generously to skeletal muscles.

Nerves and hormones influence the contraction of the smooth muscles in both these mechanisms that regulate the flow of blood to capillary beds. Next we consider how substances are exchanged when these smooth muscles relax and blood is allowed to flow through a capillary.

? What two mechanisms control the distribution of blood to the capillary beds of the body?

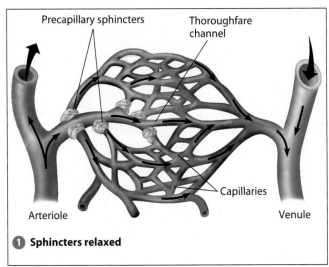

Precapillary sphincters **Thoroughfare channel**

Capillaries

Arteriole Venule

❶ **Sphincters relaxed**

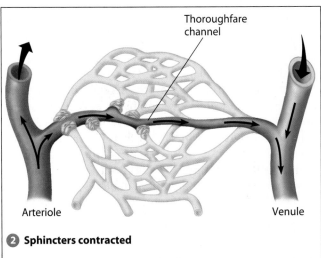

Thoroughfare channel

Arteriole Venule

❷ **Sphincters contracted**

Figure 23.11 The control of capillary blood flow by precapillary sphincters

■ Constriction of an arteriole, so that less blood reaches a capillary bed, and contraction of precapillary sphincters, so that blood flows through thoroughfare channels only, not capillary beds

Capillaries allow the transfer of substances through their walls

Capillaries are the only blood vessels with walls thin enough for substances to cross between the blood and the interstitial fluid that bathes the body cells. This transfer of materials is the most important function of the circulatory system, so let's take a closer look at it.

Figure 23.12A shows a cross section of a capillary that serves skeletal muscle cells, along with a drawing to help you interpret the micrograph. The capillary wall consists of adjoining epithelial cells that enclose a lumen, or space, that is just large enough for red blood cells to tumble through in single file. The nucleus you see here belongs to one of the two cells making up this portion of the capillary wall. (The other cell's nucleus does not appear in this particular cross section.) The blue area around the capillary is a space containing interstitial fluid.

The exchange of substances between the blood and the interstitial fluid occurs in several ways. Some substances, such as oxygen and carbon dioxide, simply diffuse through the epithelial cells of the capillary wall. Some larger molecules may be carried across an epithelial cell in vesicles that form by endocytosis on one side of the cell and then release their contents by exocytosis on the other side (see Module 5.19).

In addition, the capillary wall is leaky; there are narrow clefts between the epithelial cells making up the wall (see Figure 23.12A). Water and small solutes, such as sugars and salts, move freely through these clefts. Blood cells and dissolved proteins remain inside the capillary because they are too large to pass through these passageways. Much of the exchange between blood and interstitial fluid is the result of the pressure-driven flow of fluid (consisting of water and dissolved solutes) through these clefts.

The diagram in Figure 23.12B shows part of a capillary with blood flowing from its arterial end (near an arteriole) to its venous end (near a venule). What are the active forces that drive fluid into or out of the capillary? One of these forces is blood pressure, which tends to push fluid outward. The other is osmotic pressure, a force that tends to draw fluid into the capillary because the blood has a higher concentration of solutes than the interstitial fluid. Proteins dissolved in the blood account for much of this high solute concentration. (To review the principles of osmosis, see Module 5.16.)

The direction of fluid movement into or out of the capillary at any point depends on the difference between blood pressure and osmotic pressure. At the upstream (arterial) end of the capillary, the blood pressure exceeds the osmotic pressure, and there is a net movement of fluid out of the capillary. At the downstream (venous) end of the capillary, the situation is reversed. The blood pressure drops so much in

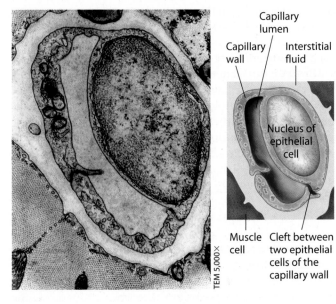

Figure 23.12A A capillary in cross section

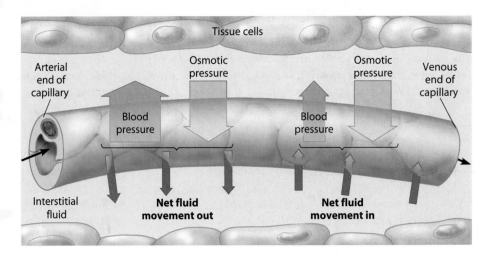

Figure 23.12B The movement of fluid into and out of a capillary

the capillary bed that the osmotic pressure outweighs it, and fluid reenters the capillary.

Most of the fluid that leaves the blood at the arterial end of a capillary bed reenters the capillaries at the venous end. The remaining fluid is returned to the blood by the vessels of the lymphatic system, which we discuss in Module 24.3.

Now that we have examined the structure and function of the heart and blood vessels, we turn our focus in the next several modules to the composition of blood itself.

? Explain how edema, the accumulation of fluid in body tissues, can result from a severe protein deficiency in the diet that decreases the concentration of blood plasma proteins.

■ Decreased blood protein concentration reduces the osmotic gradient across the capillary, thus reducing the amount of fluid that moves back into the capillary.

23.13 Blood consists of red and white blood cells suspended in plasma

Blood consists of several types of cells suspended in a liquid called **plasma.** When a blood sample is taken, the cells can be separated from the plasma by spinning the sample in a centrifuge (a chemical must be added to prevent the blood from clotting). The cellular elements (cells and cell fragments), which make up about 45% of the volume of blood, settle to the bottom of the centrifuge tube, underneath the transparent, straw-colored plasma (Figure 23.13).

Plasma is about 90% water. Among its many solutes are inorganic salts in the form of dissolved ions. These ions have several functions, such as maintaining osmotic balance, keeping the pH of blood at about 7.4, and contributing to the proper environment needed for nerve and muscle function.

Plasma also includes proteins, which help maintain the osmotic balance between blood and interstitial fluid. Some proteins act as buffers. Others have specific functions. For example, fibrinogen functions in blood clotting, and immunoglobulins are important in body defense (immunity).

Plasma also contains a wide variety of substances in transit from one part of the body to another, such as nutrients, waste products, O_2, CO_2, and hormones.

Two classes of cells are suspended in blood plasma: **red blood cells** and **white blood cells.** A third cellular element, **platelets,** are cell fragments involved in clotting.

Red blood cells are also called **erythrocytes.** There are about 25 trillion of these cells in the average person's 5 L of blood. The structure of a red blood cell suits its main function, which is to carry oxygen. Human red blood cells are small biconcave disks, thinner in the center than at the sides. Their small size and shape create a large surface area across which oxygen can diffuse. Each tiny red blood cell contains about 250 million molecules of hemoglobin and, thus, can transport about a billion oxygen molecules. It lacks a nucleus, which allows more room to pack in hemoglobin.

There are five major types of white blood cells, or **leukocytes,** as pictured in Figure 23.13: monocytes, neutrophils, basophils, eosinophils, and lymphocytes. Their collective function is to fight infections and cancer. For example, monocytes and neutrophils are **phagocytes,** which engulf and digest bacteria and debris from our own dead cells. White blood cells actually spend much of their time moving through interstitial fluid, where most of the battles against infection are waged. There are also great numbers of white cells in the lymphatic system. You will learn more about the functions of leukocytes in body defense in the next chapter.

? For every white blood cell in normal human blood, there are about ____ to ____ red blood cells (see figure).

■ 500 . . . 1,000

Plasma (55%)	
Constituent	**Major functions**
Water	Solvent for carrying other substances
Salts (ions) Sodium Potassium Calcium Magnesium Chloride Bicarbonate	Osmotic balance, pH buffering, and nerve and muscle function
Plasma proteins	Osmotic balance and pH buffering
Fibrinogen	Clotting
Immunoglobulins (antibodies)	Immunity
Substances transported by blood Nutrients (e.g., glucose, fatty acids, vitamins) Waste products of metabolism Respiratory gases (O_2 and CO_2) Hormones	

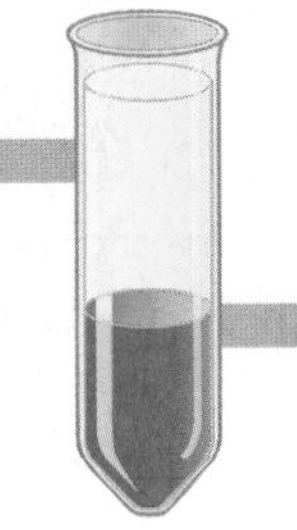

Centrifuged blood sample

Cellular elements (45%)		
Cell type	**Number** per µL (mm³) of blood	**Functions**
Erythrocytes (red blood cells)	5–6 million	Transport of oxygen (and carbon dioxide)
Leukocytes (white blood cells)	5,000–10,000	Defense and immunity
Basophil, Eosinophil, Neutrophil, Lymphocyte, Monocyte		
Platelets	250,000–400,000	Blood clotting

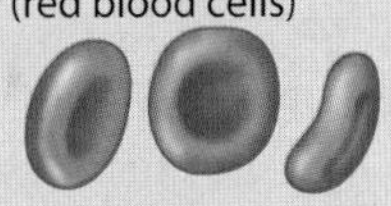

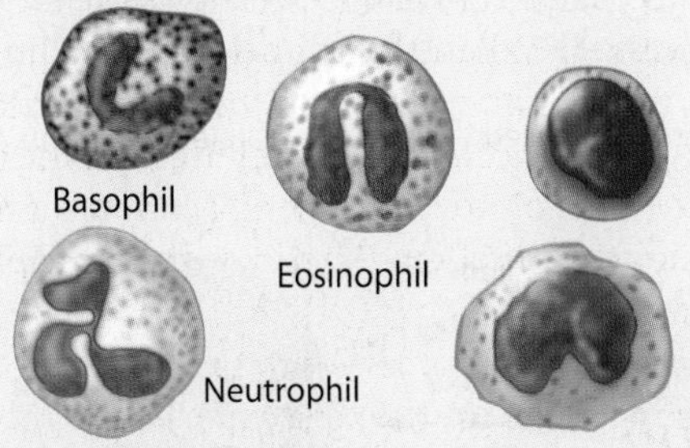

Figure 23.13 The composition of blood

23.14 Too few or too many red blood cells can be unhealthy

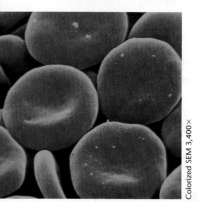

Figure 23.14
Human red blood cells

Colorized SEM 3,400×

Adequate numbers of red blood cells (Figure 23.14) are essential for healthy body function. After circulating in blood for 3 or 4 months, red blood cells are broken down and their molecules recycled. Much of the iron removed from the hemoglobin is returned to the bone marrow, where new red blood cells are formed at the amazing rate of 2 million per second.

An abnormally low amount of hemoglobin or a low number of red blood cells is a condition called **anemia.** An anemic person feels constantly tired and is often susceptible to infections because the body cells do not get enough oxygen. Anemia can result from a variety of factors, including excessive blood loss, vitamin or mineral deficiencies, and certain cancers. Iron deficiency is the most common cause. Women are more likely to develop iron deficiency than men because of blood loss during menstruation. Pregnant women generally benefit from iron supplements to support the developing fetus and placenta.

The production of red blood cells in the bone marrow is controlled by a negative-feedback mechanism that is sensitive to the amount of oxygen reaching the tissues via the blood. If the tissues are not receiving enough oxygen, the kidneys produce a hormone called **erythropoietin (EPO)** that stimulates the bone marrow to produce more red blood cells. Patients on kidney dialysis often have very low red blood cell counts because their kidneys do not produce enough erythropoietin. Genetically engineered EPO has significantly helped these patients.

One of the physiological adaptations of individuals who live at high altitudes, where oxygen levels are low, is the production of more red blood cells. Many athletes train at high altitudes to benefit from this effect. But other athletes take more drastic and illegal measures to increase the oxygen-carrying capacity of their blood and improve their performance. Injecting synthetic EPO can increase normal red blood cell volume from 45% to as much as 65%. Other athletes seek an unfair advantage by blood doping—withdrawing and storing their red blood cells and then reinjecting them before a competition. One way that athletic commissions test for cheaters is by measuring the percentage of red blood cells in the blood volume. Athletic organizations also test for EPO-like chemicals; several athletes who tested positive in the 2002 Winter Olympics were stripped of their medals.

But there can be even more serious consequences. In some athletes, a combination of dehydration from a long race and blood already thickened by an increased number of red blood cells has led to severe medical problems, such as clotting, stroke, heart failure, and even death.

? Why might increasing the number of red blood cells result in greater endurance and speed?

■ The additional red blood cells increase the oxygen-carrying capacity of blood and thus the oxygen supply to working muscles.

23.15 Blood clots plug leaks when blood vessels are injured

We all get cuts and scrapes from time to time, yet we don't bleed to death because our blood contains self-sealing materials that are activated when blood vessels are injured. These sealants are platelets and the plasma protein **fibrinogen.**

The immediate response to an injury is constriction of the damaged blood vessel, reducing blood loss and allowing time for repairs to begin. **Figure 23.15A** shows the stages of the clotting process. ❶ When the epithelium (tan) lining a blood vessel is damaged, connective tissue in the vessel wall is exposed to blood. Platelets rapidly adhere to the exposed tissue and release chemicals that make nearby platelets sticky. ❷ Soon a cluster of sticky platelets forms a plug that provides fast protection against further blood loss. Clotting factors released from the clumped platelets interact with clotting factors in the plasma, setting off a chain of reactions that culminates in the formation of a reinforced patch,

❶ Platelets adhere to exposed connective tissue

❷ Platelet plug forms

❸ Fibrin clot traps blood cells

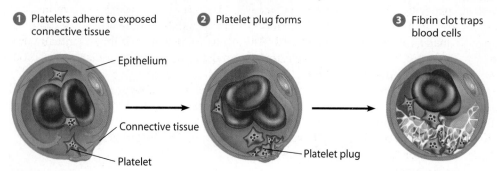

Figure 23.15A The blood-clotting process

Colorized SEM 3,400×

Figure 23.15B A fibrin clot

called a scab when it's on the skin. In this complex process, which involves more than a dozen different clotting factors, an enzyme is activated that converts fibrinogen to a thread-like protein called **fibrin**. ❸ Threads of fibrin (white) trap blood cells and more platelets. **Figure 23.15B** is a micrograph of a fibrin clot. Within an hour after a fibrin clot forms, the platelets contract, pulling the torn edges closer together and reducing the size of the area in need of repair. Chemicals released by platelets also stimulate cell division in smooth muscle and connective tissue, initiating the healing process.

The clotting mechanism is so important that any defect in it can be life-threatening. In the inherited disease hemophilia, excessive, sometimes fatal bleeding occurs from even minor cuts and bruises. Another type of defect can lead to a blood clot in the *absence* of injury; such a clot, called a thrombus, can be dangerous if it blocks a key blood vessel. An embolus is a thrombus that breaks free and is carried in the bloodstream. An embolus that lodges in and blocks an artery of the heart, brain, or lung can cause a heart attack, a stroke, or a pulmonary embolism that damages the lungs.

? What is the role of platelets in blood clot formation?

■ Platelets adhere to exposed connective tissue and release various chemicals that help a platelet plug form, activate the pathway leading to a fibrin clot, and promote healing.

23.16 Stem cells offer a potential cure for blood cell diseases

The red marrow of bones such as the ribs, vertebrae, breastbone, and pelvis all contain a spongy tissue in which unspecialized cells called **stem cells** differentiate into blood cells. As shown in **Figure 23.16**, one type of stem cell gives rise to lymphocytes, which function in the immune system (discussed in Chapter 24). A second type of stem cell can produce erythrocytes, other white blood cells, and the cells that produce platelets. After forming in the early embryo, these stem cells continually produce all the blood cells needed throughout life. Recently, researchers have been able to isolate stem cells and grow them in the laboratory.

Leukemia is cancer of the white blood cells, or leukocytes. Leukocytes protect the body against infections and cancer cells. However, sometimes leukocytes become cancerous themselves. Because cancerous cells grow uncontrollably, a person with leukemia has an unusually high number of leukocytes, most of which do not function normally. The overabundant white blood cells crowd out red blood cells and platelets, causing severe anemia and impaired clotting.

Leukemia is usually fatal unless treated, and not all cases respond to the standard cancer treatments—radiation and chemotherapy. An alternative treatment is transplanting healthy bone marrow tissue from a suitable donor, often a sibling, into a patient whose own cancerous marrow has been purposely destroyed. Such a patient requires lifelong treatment with drugs that suppress the tendency of some of the transplanted marrow cells to "reject" the cells of the recipient. To avoid the rejection problem, patients may be treated with their own bone marrow: Marrow from the patient is removed, processed to remove as many cancerous cells as possible, and then reinjected.

No matter the source (from a donor or from the patient), stem cells can be obtained for transplantation by three methods. In the oldest method, whole bone marrow is harvested by inserting a large-bore needle into the pelvic bone. A more recent technique uses drugs to draw stem cells out of the marrow and into the blood. Then the donor is connected to a refrigerated centrifuge that separates blood components, removes the ones needed for transplantation, and

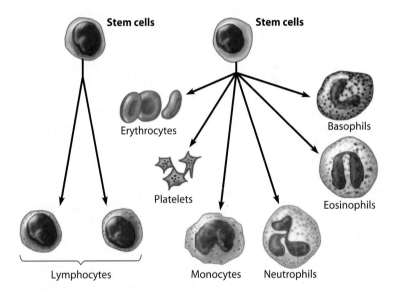

Figure 23.16 Differentiation of blood cells from stem cells

returns the rest. The newest method gathers stem cells from umbilical cord blood. These cells can be stored for possible later use by the child or donated to a compatible recipient in need of a stem cell transplant. Injection of as few as 30 stem cells can repopulate the blood and immune system.

Stem cell research holds great promise, and leukemia is just one of several blood diseases that may be treated by bone marrow stem cells. In a few cases, researchers have induced bone marrow stem cells to differentiate into more than just blood cells. Thus, these adult stem cells may eventually provide cells for human tissue and organ transplants. (See Module 11.12 for more on stem cells.)

In Chapter 24, we explore the diverse roles of white blood cells in the immune system.

? Name three possible sources of stem cells for transplantation.

■ Stem cells can be harvested from whole bone marrow, concentrated from the blood, or obtained from umbilical cord blood.

CHAPTER REVIEW

Reviewing the Concepts

A circulatory system transports O_2 and nutrients to cells and takes away CO_2 and other wastes. Capillaries, the sites of exchange between blood and interstitial fluid, form intricate networks among tissue cells **(Introduction–23.1).**

Mechanisms of Internal Transport (23.2–23.3)

Types of internal transport. Gastrovascular cavities function in both digestion and internal transport in cnidarians and flatworms. In the open circulatory systems of arthropods and many molluscs, a heart pumps blood through open-ended vessels to bathe tissue cells directly. In closed circulatory systems, a heart pumps blood through arteries to capillaries; veins return blood to the heart **(23.2).**

Cardiovascular systems of vertebrates. The two-chambered heart of a fish pumps blood in a single circuit from gill capillaries to systemic capillaries and back to the heart. Land vertebrates have double circulation with separate pulmonary and systemic circuits. Amphibians and reptiles (except birds) have three-chambered hearts; birds and mammals have four-chambered hearts **(23.3).**

The Mammalian Cardiovascular System (23.4–23.12)

The mammalian heart has two thin-walled atria that pump blood into the ventricles and two thick-walled ventricles that pump blood to all other body organs **(23.4).**

Blood vessel structure correlates with function (see diagram) **(23.5).**

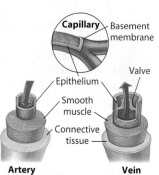

Cardiac cycle. During diastole (relaxation), blood flows from the veins into the heart chambers; during systole, contractions of the atria push blood into the ventricles, and then stronger contractions of the ventricles propel blood into the large arteries. Cardiac output is the amount of blood/minute pumped into the systemic circuit. Heart valves prevent the backflow of blood **(23.6).**

The pacemaker (SA node) generates electrical signals that trigger contraction of the atria. The AV node relays these signals to the ventricles. An electrocardiogram records the electrical changes in the heart. Heart rate adjusts to body needs **(23.7).** A heart attack is damage to cardiac muscle, usually resulting from a blocked coronary artery **(23.8).**

Blood pressure, the force blood exerts on vessel walls, depends on cardiac output and the resistance of vessels. Pressure is highest in the arteries and lowest in the veins. Blood velocity is slowest in the capillaries. Muscle contractions and one-way valves keep blood moving through veins to the heart **(23.9).** Blood pressure is measured as systolic and diastolic pressures. Hypertension is a serious cardiovascular problem **(23.10).**

Capillary Exchange. Constriction of arterioles and pre-capillary sphincters controls blood flow through capillary beds **(23.11).** The transfer of materials between blood and interstitial fluid occurs by diffusion and by pressure flow through clefts between epithelial cells. Blood pressure forces fluid out of the capillary at the arterial end, and osmotic pressure draws fluid in at the venous end **(23.12).**

Structure and Function of Blood (23.13–23.16)

Blood consists of cells in a fluid plasma, which contains various inorganic ions, proteins, nutrients, wastes, gases, and hormones. Red blood cells (erythrocytes) transport O_2 bound to hemoglobin. White blood cells (leukocytes) function both inside and outside the circulatory system to fight infections and cancer **(23.13).** The hormone erythropoietin regulates red blood cell production. Some athletes artificially increase their red blood cell number, a dangerous practice **(23.14).**

Blood clotting When a blood vessel is damaged, platelets help trigger the conversion of fibrinogen to fibrin, forming a clot that plugs the leak **(23.15).**

Stem cells divide in bone marrow to produce all blood cells and may be used to treat some blood disorders **(23.16).**

Connecting the Concepts

1. Use this diagram to review the flow of blood through a human cardiovascular system. Label the indicated parts, color the vessels that carry oxygen-rich blood red, and then trace the flow of blood by numbering the circles from 1 to 10, starting with 1 in the right ventricle.

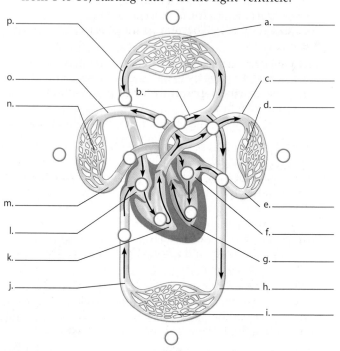

Testing Your Knowledge

Multiple Choice

2. Blood pressure is highest in ____, and blood moves most slowly in ____.
   - a. veins . . . capillaries
   - b. arteries . . . capillaries
   - c. veins . . . arteries
   - d. capillaries . . . arteries
   - e. arteries . . . veins

3. When the doctor listened to Janet's heart, he heard "lub-hisss, lub-hiss" instead of the normal "lub-dup" sounds. The hiss is most likely due to _____. (*Explain your answer.*)
   a. a clogged coronary artery
   b. a defective atrioventricular (AV) valve
   c. a damaged pacemaker
   d. a defective semilunar valve
   e. high blood pressure

4. Which of the following is the biggest difference between your cardiovascular system and the cardiovascular system of a fish?
   a. In a fish, blood is oxygenated by passing through a capillary bed.
   b. Your heart has two chambers; a fish heart has four.
   c. Your circulation has two circuits; fish circulation has one circuit.
   d. Your heart chambers are called atria and ventricles.
   e. Yours is a closed system; the fish's is an open system.

5. Paul's blood pressure is 150/90. The 150 indicates _____, and the 90 indicates _____.
   a. pressure in the left ventricle . . . pressure in the right ventricle
   b. arterial pressure . . . heart rate
   c. pressure during ventricular contraction . . . pressure during heart relaxation
   d. systemic circuit pressure . . . pulmonary circuit pressure
   e. pressure in the arteries . . . pressure in the veins

6. Which of the following *initiates* the process of blood clotting?
   a. damage to the lining of a blood vessel
   b. exposure of blood to the air
   c. conversion of fibrinogen to fibrin
   d. attraction of leukocytes to a site of infection
   e. conversion of fibrin to fibrinogen

7. Blood flows more slowly in the arterioles than in the artery that supplies them because the arterioles
   a. must provide opportunity for exchange with the interstitial fluid.
   b. have thoroughfare channels to venules that are often closed off, slowing the flow of blood.
   c. have sphincters that restrict flow to capillary beds.
   d. are narrower than the artery.
   e. collectively have a larger cross-sectional area than does the artery.

Describing, Comparing, and Explaining

8. Trace the path of blood starting in a pulmonary vein, through the heart, and around the body, returning to the pulmonary vein. Name, in order, the heart chambers and types of vessels through which the blood passes.

9. Explain how the structure of capillaries relates to their function of exchanging substances with the surrounding interstitial fluid. Describe how that exchange occurs.

10. Here is a blood sample that has been spun in a centrifuge. List, as completely as you can, the components you would find in the straw-colored fluid at the top of this tube and in the dense red portion at the bottom.

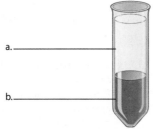

Applying the Concepts

11. Some babies are born with a small hole in the wall between the left and right ventricles. How might this affect the oxygen content of the blood pumped out of the heart into the systemic circuit?

12. Juan has a disease in which damaged kidneys allow some of his normal plasma proteins to be removed from the blood. How might this condition affect the osmotic pressure of blood in capillaries, compared to that of the surrounding interstitial fluid? One of the symptoms of this kidney malfunction is an accumulation of excess interstitial fluid, which causes Juan's arms and legs to swell. Can you explain why this occurs?

13. Recently, a 19-year-old woman received a bone marrow transplant from her 1-year-old sister. The woman was suffering from a deadly form of leukemia and was almost certain to die without a transplant. Their parents had decided to have another child in a final attempt to provide their daughter with a matching donor. Although the ethics of the parents' decision were criticized, doctors report that this situation is not uncommon. In your opinion, is it acceptable to have a child in order to provide an organ or tissue donation? Why or why not?

14. As we discussed in the chapter introduction, gravity affects blood circulation. Physiologists speculate about blood pressure and cardiovascular adaptations in dinosaurs—some of which had necks almost 10 m long, which would have required a systolic pressure of nearly 760 mm Hg to pump blood to the brain when the head was fully raised. Some analyses suggest that dinosaurs' hearts were not powerful enough to generate such pressures, leading to the speculation that the long-necked dinosaurs fed close to the ground rather than raising their heads to feed on high foliage. Scientists also debate whether dinosaurs had a "reptile-like" or "bird-like" heart. Most modern reptiles have a three-chambered heart with just one ventricle. Birds, which evolved from a lineage of dinosaurs, have a four-chambered heart with two ventricles. Some scientists believe that the circulatory needs of these long-necked dinosaurs provide evidence that dinosaurs must have had a four-chambered heart. Can you think of several reasons why they might conclude this?

Answers to all questions can be found in Appendix 3.

For study help and Activities, go to campbellbiology.com or the student CD-ROM.

CHAPTER 24

INNATE DEFENSES AGAINST INFECTION

24.1 Innate defenses against infection include the skin and mucous membranes, phagocytic cells, and antimicrobial proteins

24.2 The inflammatory response mobilizes nonspecific defense forces

24.3 The lymphatic system becomes a crucial battleground during infection

ACQUIRED IMMUNITY

24.4 The immune response counters specific invaders

24.5 Lymphocytes mount a dual defense

24.6 Antigens have specific regions where antibodies bind to them

24.7 Clonal selection musters defensive forces against specific antigens

24.8 Antibodies are the weapons of humoral immunity

24.9 Antibodies mark antigens for elimination

24.10 Monoclonal antibodies are powerful tools in the lab and clinic

24.11 Helper T cells stimulate humoral and cell-mediated immunity

24.12 HIV destroys helper T cells, compromising the body's defenses

24.13 Cytotoxic T cells destroy infected body cells

24.14 Cytotoxic T cells may help prevent cancer

24.15 The immune system depends on our molecular fingerprints

DISORDERS OF THE IMMUNE SYSTEM

24.16 Malfunction or failure of the immune system causes disease

24.17 Allergies are overreactions to certain environmental antigens

An AIDS Uproar

IN 1996, THE U.S. GOVERNMENT BEGAN FUNDING a series of studies intended to reduce the number of AIDS babies in poor countries. Despite this seemingly noble goal, the experiments soon sparked intense controversy. To understand the debate, one must consider the intersection of science, ethics, and public policy.

Since first identified in 1981, AIDS (acquired immune deficiency syndrome) has killed more than 20 million people. More than 40 million people are currently infected with HIV, the human immunodeficiency virus, which causes AIDS. In some developing countries, particularly in southern Africa, almost 40% of adults are infected.

Given these grim statistics, prevention is paramount in slowing the spread of this deadly disease. In particular, stopping the transmission of HIV from mother to newborn during childbirth is a major goal in the battle against AIDS. A large study of pregnant women in the

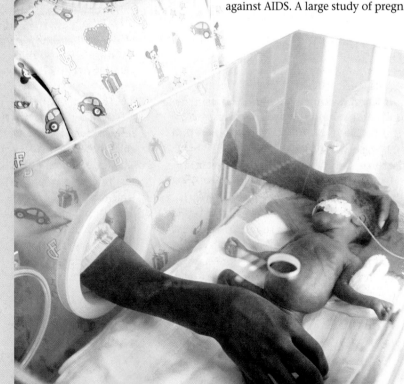

The Immune System

40 million
people living with AIDS

3 million
died in 2003

5 million
newly infected in 2003

500,000
children under the age of 15 who died in 2003

United States and France, completed in 1994, offered some hope. The study showed that intensive treatment with the antiviral drug AZT—which can block HIV's ability to multiply within the human body—cut mother-to-child HIV transmission by nearly 70%. Although effective, the treatment regimen costs about $800, well beyond the means of HIV-infected women in many developing nations, where it is most desperately needed.

Seeking cheaper treatment alternatives, the U.S. government funded a series of studies in 1996, using volunteer patients from Africa, Thailand, and the Caribbean. The patients, over 6,000 women who were HIV-positive and pregnant, were divided into two groups. The first group received a short course of AZT treatment that costs about $80. The other group served as a control group and received only placebo medication (a sugar pill that contained no HIV-fighting drugs). By comparing the rates of mother-to-child transmission between the two groups, researchers hoped to establish a dosage that was both effective and within the financial means of poorer countries.

Despite the scientists' good intentions, these trials caused an uproar. Critics charged that it is unethical to provide patients with placebos when a proven (albeit expensive) treatment is available. Some doctors decried the missed opportunity to prevent hundreds of babies from being born HIV-positive. Opponents testified before Congress that the government, by sponsoring such trials, was abusing vulnerable patients and allowing babies to die.

Proponents of the trials countered that the methodology had been scrutinized and approved by independent review boards within both the United States and the host countries. The participating researchers pointed to a well-accepted ethical precept that participants in clinical trials must receive at least the standard care of their home nation; since virtually no efforts were being made to prevent mother-to-child transmission in the participating countries, no patients were receiving worse care than they would have in the absence of the study. From a strictly scientific standpoint, some researchers argued that it was impossible to reliably measure the effect of the drug treatment without the placebo. The World Health Organization (WHO) of the United Nations agreed when it stated that "placebo-controlled trials offer the best option" for finding reliable alternatives to the too-expensive AZT regimen. The critics countered that when it comes to questionable ethical conduct, the ends cannot justify the means.

While the ethical debate raged, the research continued and data were collected. In 1998, one of the Thai studies showed that oral administration of AZT during the final four weeks of pregnancy (a relatively cheap course of medication) lowered HIV transmission by 50%. A year later, another trial in three African nations found a similar reduction from a slightly longer course of medication. Indeed, the cheaper short-term AZT treatment worked so well that researchers quickly modified the protocols for all of the remaining trials, substituting the newly established regimen for the controversial placebos. Although the change in protocol ended this particular debate, similar kinds of clinical experiments involving AIDS and other infectious diseases—and similar sorts of controversies—are ongoing.

The ravaging effects of AIDS demonstrates how much we depend on our body's built-in defenses. HIV is deadly because it disables these defenses, the immune system. As we will see in this chapter, our immune system is a very specific defense system in that it recognizes an invader, such as a virus, and then produces large numbers of cells and molecules that combat that particular agent. In a healthy individual, this acquired defense backs up several mechanisms of innate resistance, which provide inborn protection against a variety of invaders. Following a discussion of our innate defenses in the first three modules, we concentrate on the mechanisms of acquired immune defense. At the end of the chapter, we examine some of the problems—from seasonal nuisances to fatal illnesses—that may arise when the intricate interplay of our body's defenses goes awry. ■ ■ ■

24.1 Innate defenses against infection include the skin and mucous membranes, phagocytic cells, and antimicrobial proteins

Innate immunity, the human body's first line of defense against bacteria, viruses, and other pathogens, is present and effective long before exposure to pathogens. Innate defenses are also largely nonspecific: They do not distinguish one infectious microbe from another. For example, an outer layer of intact skin is a tough barrier of dead cells that most bacteria and viruses cannot penetrate, although even tiny abrasions may allow infection. Moreover, acids secreted by skin glands inhibit the growth of many microbes. Sweat, saliva, and tears also contain lysozyme, an enzyme that digests the cell walls of many bacteria. Organ systems that open to the external environment—such as the digestive, respiratory, and genitourinary systems—are guarded by the mucous membranes that line them (see Module 20.4) and have other innate defenses as well. Stomach acid kills most bacteria swallowed with food or saliva. Guarding the respiratory route, hairs in our nostrils filter incoming air, and mucus in our respiratory tubes traps most microbes and dirt that get past the nasal filter. Cilia on cells lining the tubes sweep mucus upward and out of the system.

Microbes that breach the body's external defenses, such as those that enter through a cut, are confronted by innate defensive cells. These are all classified as white blood cells (see Module 23.15), although they are found in interstitial fluid as well as blood vessels. Most, such as abundant **neutrophils**, are phagocytic; they engulf microbes in infected tissues. **Macrophages** ("big eaters") are large phagocytic cells that wander actively in the interstitial fluid, "eating" any bacteria and virus-infected cells they encounter. In **Figure 24.1A**, a large macrophage is shown using multiple extensions (pseudopodia) to snare bacteria. **Natural killer cells**, another type of white blood cell, are not phagocytes.

They attack cancer cells and virus-infected cells by releasing chemicals that promote apoptosis, also called programmed cell death (see Module 27.13).

Other innate defenses include proteins that either attack microbes directly or impede their reproduction. Especially important are interferons and the complement proteins.

Interferons are proteins produced by virus-infected cells that help other cells resist viruses. **Figure 24.1B** shows how the interferon mechanism works. ❶ The virus infects a cell, which causes ❷ interferon genes in the cell's nucleus to be turned on. ❸ The cell makes interferon. The infected cell then dies, but ❹ its interferon molecules may diffuse to neighboring healthy cells, ❺ stimulating them to produce other proteins that inhibit viral reproduction. This is a good example of a nonspecific defense, since interferon made in response to one virus confers resistance to unrelated viruses.

Additional innate immunity is provided by the **complement system**, a group of about 30 proteins that circulate in the blood plasma and act cooperatively (in complement) with other defense mechanisms. In the absence of infection, these proteins are inactive. Substances on the surfaces of many microbes, however, can trigger a cascade of steps that activate the complement system, leading to the lysis (bursting) of the invaders. Certain complement proteins also help trigger another innate defense, the inflammatory response, the subject of the next module.

? Which innate defenses (a) actually help prevent infection and (b) come into play only after infection has occurred?

■ (a) Skin, gland secretions, mucous membranes; (b) phagocytes (macrophages, neutrophils), natural killer cells, interferons, complement system

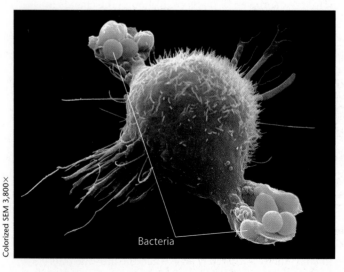

Colorized SEM 3,800×

Bacteria

Figure 24.1A Phagocytosis of bacteria by a macrophage

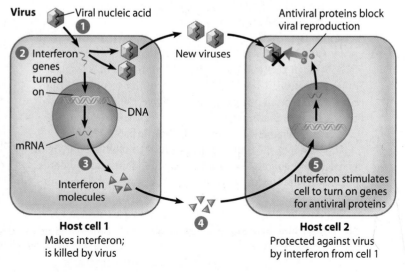

Virus — Viral nucleic acid
❶
❷ Interferon genes turned on
DNA
mRNA
❸
Interferon molecules
New viruses
Antiviral proteins block viral reproduction
❺ Interferon stimulates cell to turn on genes for antiviral proteins
❹

Host cell 1
Makes interferon; is killed by virus

Host cell 2
Protected against virus by interferon from cell 1

Figure 24.1B The interferon mechanism against viruses

24.2 The inflammatory response mobilizes nonspecific defense forces

The **inflammatory response** is a major component of our innate defense system. Any damage to tissue, whether caused by microorganisms or by physical injury—even just a scratch or insect bite—triggers this response. You can see signs of the inflammatory response if you fail to properly treat a cut. The area becomes red, swollen, and warmer than the surrounding area. This reaction is inflammation, which literally means "setting on fire."

Figure 24.2 shows the chain of events that make up the inflammatory response in a case where a pin has broken the skin, allowing infection by bacteria. ❶ The damaged cells soon release chemical alarm signals, such as **histamine.** ❷ The chemicals spark the mobilization of various defenses. Histamine, for instance, induces neighboring blood vessels to dilate and become leakier. Blood flow to the damaged area increases, and blood plasma passes out of the leaky vessels into the interstitial fluid of the affected tissues. Other chemicals (some that are part of the complement system) attract phagocytes to the area. Squeezing between the cells of the blood vessel wall, these white cells (yellow in the figure) migrate out of the blood into the tissue spaces. The local increase in blood flow, fluid, and cells produces the redness, heat, and swelling characteristic of inflammation.

The major results of the inflammatory response are to disinfect and clean injured tissues. ❸ The white blood cells mustered into the area engulf bacteria and the remains of any body cells killed by them or by the physical injury. Many of the white cells die in the process, and their remains are also engulfed and digested. The pus that often accumulates at the site of an injury or infection consists mainly of dead white cells and fluid that has leaked from the capillaries during the inflammatory response.

The inflammatory response also helps prevent the spread of infection to surrounding tissues. Clotting proteins (see Module 23.16) present in blood plasma pass into the interstitial fluid during inflammation. Along with platelets, these substances form local clots that help seal off the infected region and allow healing to begin.

The inflammatory response may be localized, as we have just described, or widespread (systemic). Sometimes microorganisms such as bacteria or protozoans get into the blood or release toxins that are carried throughout the body in the bloodstream. The body may react with several inflammatory weapons. For instance, the number of white blood cells circulating in the blood may increase. Severe infections like meningitis or appendicitis may cause the number of circulating defensive cells to increase severalfold within just a few hours. Another response to systemic infection is fever, an abnormally high body temperature. Toxins themselves may trigger the fever, or macrophages may release compounds that set the body's thermostat at a higher temperature. A very high fever is dangerous, but a moderate one may stimulate phagocytosis and hasten tissue repair.

Sometimes bacterial infections bring about an overwhelming systemic inflammatory response leading to a condition known as *septic shock*. Characterized by very high fever and low blood pressure, septic shock is a common cause of death in hospital critical care units. Clearly, while local inflammation is an essential step toward healing, widespread inflammation can be devastating.

❓ Why is the inflammatory response considered *nonspecific*?

■ Because the response is the same regardless of the cause of tissue damage

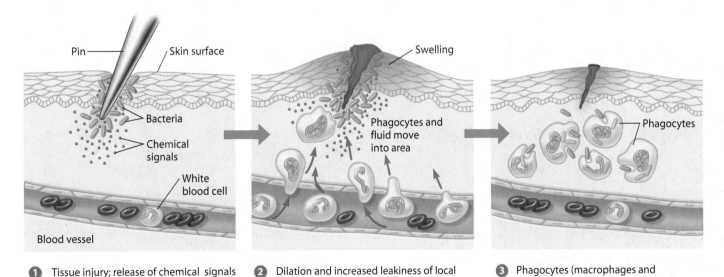

❶ Tissue injury; release of chemical signals such as histamine

❷ Dilation and increased leakiness of local blood vessels; migration of phagocytes to the area

❸ Phagocytes (macrophages and neutrophils) consume bacteria and cell debris; tissue heals

Figure 24.2 The inflammatory response

24.3 The lymphatic system becomes a crucial battleground during infection

Involved in both innate and acquired immunity, the **lymphatic system (Figure 24.3)** consists of a branching network of vessels, numerous lymph nodes (rounded organs packed with macrophages and white blood cells called lymphocytes), the tonsils and adenoids, the appendix, and the spleen. It also includes the bone marrow and the thymus (green labels), which are the sites where white blood cells develop. The lymphatic vessels carry a fluid called **lymph**, which is similar to interstitial fluid but contains less oxygen and fewer nutrients. The lymphatic system has two main functions: to return tissue fluid to the circulatory system and to fight infection.

As we noted in Module 23.12, a small amount of the fluid that enters the tissue spaces from the blood in a capillary bed does not reenter the blood capillaries. Instead, this fluid is returned to the blood via lymphatic vessels. The enlargement in Figure 24.3 (bottom right) shows a branched lymphatic vessel in the process of taking up fluid from tissue spaces in the skin. As shown here, fluid enters the lymphatic system by diffusing into tiny, dead-end lymphatic capillaries that are intermingled among the blood capillaries.

Lymph drains from the lymphatic capillaries into larger and larger lymphatic vessels. It reenters the circulatory system via two large lymphatic vessels, the thoracic duct and the right lymphatic duct, which fuse with veins in the chest. As the close-up indicates, the lymphatic vessels resemble veins in having valves that prevent the backflow of fluid toward the capillaries (see Module 23.5). Also, like veins, lymphatic vessels depend mainly on the movement of skeletal muscles to squeeze their fluid along. The black arrows in the close-ups indicate the flow of lymph.

As lymph circulates through the lymphatic organs (such as the lymph nodes shown in Figure 24.3, top right), it carries microbes from infection sites throughout the body. Once inside lymphatic organs (pink labels), macrophages that reside there permanently may engulf the invaders in a nonspecific fashion. Additionally, lymphocytes may be activated to mount a specific immune response against them. When your body is fighting an infection, the nodes and other organs of the lymphatic system become a major battleground. Lymph nodes fill with huge numbers of defensive cells, causing the tender "swollen glands" in your neck and armpits that your doctor looks for as a likely sign of infection.

? What are the two main functions of the lymphatic system?

■ To return fluid from interstitial spaces to the circulatory system and to combat infection

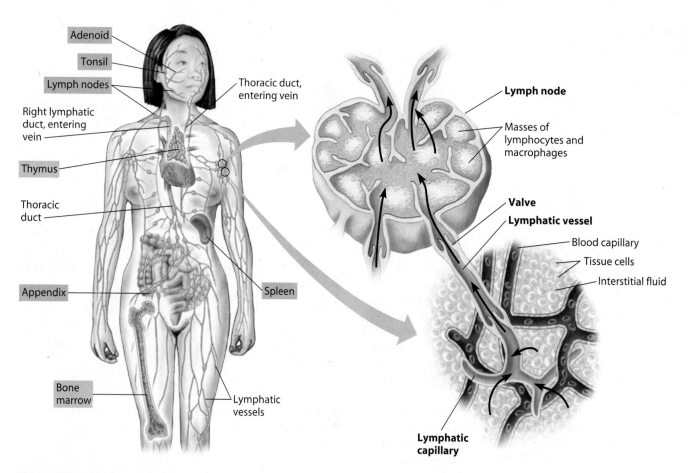

Figure 24.3 The human lymphatic system

24.4 The immune response counters specific invaders

When our innate defenses fail to ward off an infectious agent, the immune system provides another line of defense. The immune system recognizes and defends against invading microbes and against cancer cells (which the body usually identifies as foreign). Made up of over 2 trillion individual cells spread throughout the bloodstream and lymphatic system, the **immune system** often acts more effectively than innate resistance. This is because the immune system is highly specific—that is, it distinguishes one infectious agent from another. Moreover, it can amplify certain nonspecific responses, such as inflammation and the complement system.

The immunity confered by the immune system is called acquired immunity. Whereas our innate defenses are always ready to fight a variety of infections, **acquired immunity** fully develops *only after* exposure to pathogens. The immune response must be primed by the presence of a foreign substance, called an antigen. An **antigen** is any foreign molecule that elicits an immune response. (The word *antigen* is a contraction of "*anti*body-*gen*erating," a reference to the fact that the foreign agent provokes the immune response.) Antigens include certain molecules on the surfaces of viruses, bacteria, mold spores, cancer cells, pollen, and house dust, as well as molecules on the cell surfaces of transplanted organs. When the immune system detects an antigen, it responds with an increase in the number of cells that either attack the invader directly or produce defensive proteins called antibodies. An **antibody** is a protein found in blood plasma that attaches to one particular kind of antigen and helps counter its effects. The defensive cells and antibodies produced against that antigen are usually specific to that antigen; they are ineffective against any other foreign substance.

Besides being extremely specific, the immune system has a remarkable "memory." It can "remember" antigens it has encountered before and react against them more promptly and vigorously on subsequent exposures. For example, if a person gets chicken pox, the immune system remembers certain molecules on the virus that causes this disease. Should the virus enter the body again, the immune system mounts a quick and decisive attack that usually destroys the virus before symptoms appear. Thus, the immune response, unlike innate defenses, is adaptive; exposure to a particular foreign agent enhances future response to that same agent.

The term **immunity** means resistance to *specific* invaders. Immunity is usually acquired by natural exposure to antigens, but it can also be achieved by **vaccination** (also known as immunization). In this procedure, the immune system is confronted with a **vaccine** composed of a harmless variant or component of a disease-causing microbe, such as an inactivated bacterial toxin, a dead or weakened microbe, or part of a microbe. The vaccine stimulates the immune system to mount defenses against this antigen, defenses that will also be effective against the actual pathogen because it has similar antigens. Once we have been successfully vaccinated, our immune system will respond quickly if it is exposed to the microbe.

In the United States, widespread vaccination of children has virtually eliminated viral diseases such as polio, mumps, and measles. Researchers are trying to develop a vaccine for AIDS, but so far success has been elusive. One of the major success stories of modern vaccination involves smallpox, a potentially fatal viral infection that affected over 50 million people per year worldwide in the 1950s. A subsequent vaccination effort was so effective that there have been no cases of smallpox since 1977. Recently, however, the U.S. government has begun to stockpile hundreds of millions of doses of smallpox vaccine and to vaccinate high-risk health-care and military workers in case the smallpox virus is used in a bioterrorist attack (**Figure 24.4**).

Whether antigens enter the body naturally (if you catch the flu) or artificially (if you get a flu shot), the resulting immunity is called **active immunity** because the person's own immune system actively produces antibodies. It is also possible to acquire **passive immunity** by receiving premade antibodies. For example, a fetus obtains antibodies from its mother's bloodstream; babies receive antibodies from breast milk; and travelers sometimes get a shot containing antibodies to pathogens they are likely to encounter. In yet another example, the effects of a poisonous snakebite may be counteracted by injecting the victim with antivenin, which contains antibodies extracted from animals previously exposed to the venom. Passive immunity is temporary because the recipient's immune system is not stimulated by antigens. Immunity lasts only as long as the antibodies do—usually a few weeks or months.

Figure 24.4 A soldier receiving a smallpox vaccination

? Why is protection resulting from a vaccination considered *active* immunity rather than *passive* immunity?

■ Because the body itself produces the immunity by mounting an immune response and generating antibodies, even though the stimulus consists of artificially introduced antigens

Lymphocytes mount a dual defense

Lymphocytes, white blood cells that spend most of their time in the tissues and organs of the lymphatic system, produce the immune response. Like all blood cells, lymphocytes originate from stem cells in the bone marrow (see Module 23.17). As shown in **Figure 24.5A**, some immature lymphocytes continue developing in the bone marrow; these become specialized as B lymphocytes, or **B cells**. Other immature lymphocytes are carried by the blood from the bone marrow to the thymus, a gland in the upper chest region. There the lymphocytes become specialized as T lymphocytes, or **T cells** (**Figure 24.5B**, facing page). Both B cells and T cells eventually make their way via the blood to the lymph nodes, spleen, and other lymphatic organs.

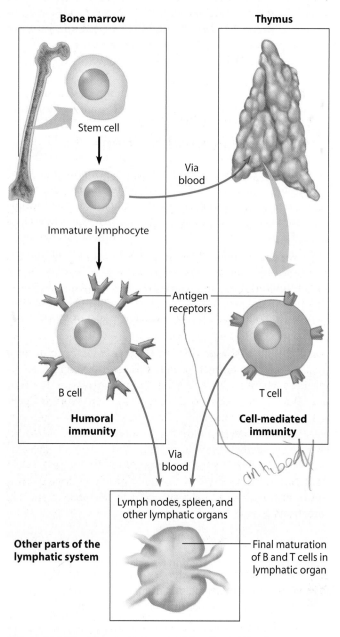

Figure 24.5A The development of B cells and T cells

The B cells and T cells of our immune system mount a dual defense. The B cells secrete antibodies, and because antibodies circulate in the blood and lymph, immunity conferred by B cells is called **humoral immunity.** (The body fluids were formerly called "humors.") The humoral system defends primarily against bacteria and viruses present in body fluids. Humoral immunity can be passively transferred by injecting blood plasma (containing antibodies) from an immune individual into a nonimmune individual.

In humoral immunity, antibodies are carried through body fluids to sites of infection wherever they occur. As we will see in Module 24.9, antibodies mark invaders by binding to them. The resulting antigen-antibody complexes are easily recognized for destruction and disposal by phagocytic cells.

The second type of immunity, produced by T cells, is called **cell-mediated immunity.** T cells circulate in the blood and lymph, attacking body cells that have been infected with bacteria or viruses. T cells also work against infections caused by fungi and protozoans and are thought to be important in protecting the body from its own cells if they become cancerous. In addition, T cells function indirectly by promoting phagocytosis by other white blood cells and by stimulating B cells to produce antibodies. Thus, T cells are involved in both cell-mediated and humoral immunity.

When a T cell develops in the thymus or a B cell develops in bone marrow, certain genes in the cell are turned on. This causes the cell to synthesize molecules of a specific protein, which are then incorporated into the plasma membrane. As indicated in Figure 24.5A, these protein molecules (shown in dark purple) stick out from the cell's surface. The molecules are **antigen receptors,** capable of binding one specific type of antigen. Each T or B cell has about 100,000 antigen receptors, and all the receptors on a single cell are identical—they all recognize the same antigen. In the case of a B cell, the receptors are almost identical to the particular antibody that the B cell will secrete. Once a B cell or T cell has its surface proteins in place, it can recognize a specific antigen and mount an immune response against it. One cell may recognize an antigen on the mumps virus, for instance, while another detects a particular antigen on a tetanus-causing bacterium.

We see in Figure 24.5A that after the B cells and T cells have developed their antigen receptors, these lymphocytes leave the bone marrow and thymus and move via the bloodstream to the lymph nodes, spleen, and other parts of the lymphatic system. In these organs, many B and T cells take up residence and encounter infectious agents that have penetrated the body's outer defenses. Because lymphatic capillaries extend into virtually all the body's tissues, bacteria or viruses infecting nearly any part of the body eventually enter the lymph and are carried to the lymphatic organs. As we will describe in Module 24.7, when a B or T cell within a lymphatic organ first confronts the specific antigen it is programmed to recognize, it differentiates further and becomes a fully mature component of the immune system.

An enormous diversity of B cells and T cells develops in each individual. Researchers estimate that each of us has millions of different kinds—enough to recognize and bind virtually every possible antigen. A small population of each kind of lymphocyte lies in wait in our body, genetically programmed to recognize and respond to a specific antigen. Only a tiny fraction of the immune system's lymphocytes will ever be used, but they are all available if needed. It is as if the immune system maintains a huge standing army of soldiers, each made to recognize one particular kind of invader. The majority of soldiers never encounter their target and remain idle. But when an invader does appear, chances are good that some lymphocytes will be able to recognize it, bind to it, and call in reinforcements.

? Contrast the targets of humoral immunity with those of cell-mediated immunity.

■ Humoral immunity works against pathogens in the body fluids; cell-mediated immunity attacks infected cells or abnormal cells (cancer cells, for example).

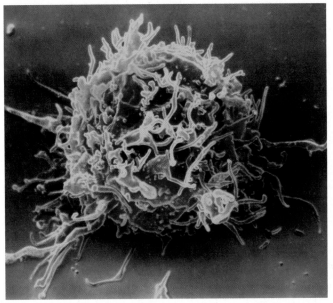

Colorized TEM 4,500×

Figure 24.5B A T cell

24.6 Antigens have specific regions where antibodies bind to them

As molecules that elicit the immune response, antigens usually do not belong to the host animal. Most antigens are proteins or large polysaccharides on the surfaces of viruses or foreign cells. Common examples are protein-coat molecules of viruses, parts of the capsules and cell walls of bacteria, and macromolecules on the surface cells of other kinds of organisms, such as protozoans and parasitic worms. (Sometimes a particular microbe is called an antigen, but this usage is misleading because the microbe will almost always have several kinds of antigenic molecules.) Other sources of antigenic molecules include blood cells or tissue cells from other individuals (of the same species or a different species). Antigenic molecules are also found dissolved in body fluids; foreign molecules of this type include bacterial toxins and bee venom.

As shown in **Figure 24.6**, an antibody usually recognizes and binds to a small surface-exposed region of an antigen, called an **antigenic determinant** (also known as an epitope). An antigen-binding site, a specific region on the antibody molecule, recognizes an antigenic determinant by the fact that the binding site and antigenic determinant have complementary shapes, like an enzyme and substrate or a lock and key. An antigen usually has several different determinants (there are three in the diagram here), so different antibodies (two, in this case) can bind to the same antigen. A single antigen molecule may stimulate the immune system to make several distinct antibodies against it. Notice that each antibody molecule has two identical antigen-binding sites. We'll return to antibody structure in Module 24.8. But first, let's see how the body produces large quantities of antibodies and defensive cells in response to specific infections.

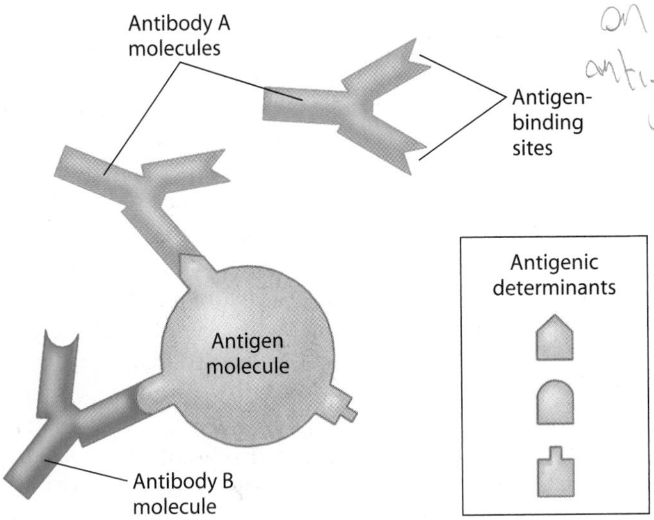

? Why is it inaccurate to refer to a pathogen, such as a virus, as an antigen?

■ It is inaccurate because antigens are not whole pathogens; they are molecules, which may be chemical components of a pathogen's surface. One pathogen may have several antigens.

Antibody A molecules

Antigen-binding sites

on antigenic

Antigen molecule

Antibody B molecule

Antigenic determinants

Figure 24.6 The binding of antibodies to antigenic determinants

24.7 Clonal selection musters defensive forces against specific antigens

The immune system's ability to defend against an almost infinite variety of antigens depends on a process called **clonal selection**. Once inside the body, an antigen encounters a diverse pool of B and T lymphocytes. However, one particular antigen interacts only with the tiny fraction of lymphocytes bearing receptors specific to that antigen. Once activated by the antigen, these few "selected" cells proliferate, forming a clone (genetically identical population) of thousands of cells all specific for the stimulating antigen. This antigen-driven cloning of lymphocytes—clonal selection—is a vital step in the immune system's acquired response to infection.

The Steps of Clonal Selection Figure 24.7A indicates how clonal selection works for the B cells of humoral immunity. ❶ The row of cells at the top of the figure represents a vast repertoire of B cells in a lymph node. Notice that each lymphocyte has its own specific type of antigen

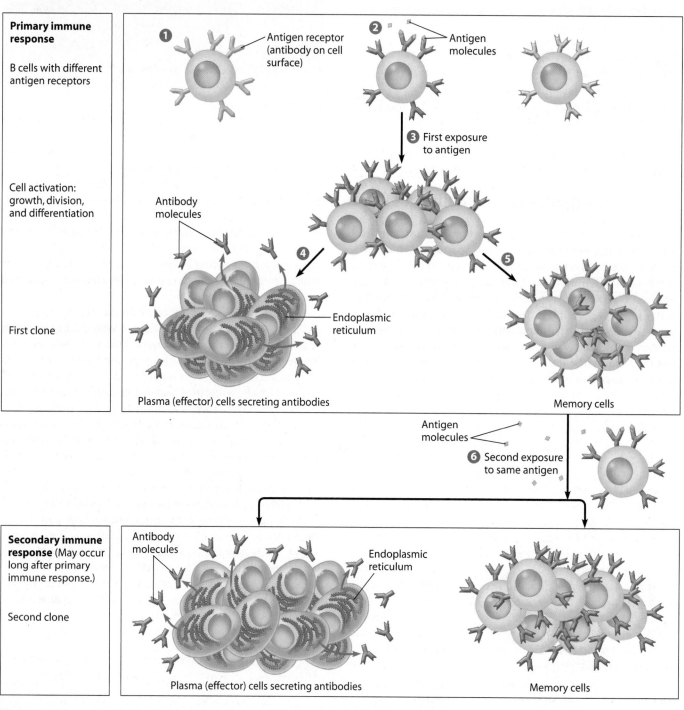

Figure 24.7A Clonal selection of B cells in the primary and secondary immune responses

receptor embedded in its surface. The cells have their receptors in place before they ever encounter an antigen.

The first time an antigen enters the body and is swept into a lymph node, ❷ antigenic determinants on its surface bind with the few B cells that happen to have complementary receptors. Other lymphocytes, without the appropriate binding sites, are not affected. Primed by the interaction with the antigen, ❸ the selected cell is activated: It grows, divides, and differentiates into two genetically identical yet physically distinct types of cells. Both newly produced types of cells are specialized for defending against the antigen that triggered the response. ❹ The first group of newly produced cells are **effector cells** that combat the antigen. Because this example involves B cells, the effector cells produced are **plasma cells.** Each plasma cell secretes antibody molecules, all of the same specific type. Each plasma cell makes as many as 2,000 copies of its antibody per second. (These plasma cells thus require large amounts of endoplasmic reticulum, a characteristic of cells actively synthesizing and secreting proteins.) These antibodies circulate in the blood and lymphatic fluid, contributing to humoral immunity. Although highly effective at combating infection, each effector cell lasts only 4 or 5 days.

❺ The second group of cells produced by the activated B cells are a smaller number of **memory cells,** which differ from effector cells in both appearance and function. In contrast to short-lived effector cells, memory cells may last for decades. They remain in the lymph nodes, poised to be activated by a second exposure to the antigen. In fact, in some cases, memory cells seem to confer lifetime immunity, as they may in such childhood diseases as mumps and measles. Steps 1–5 show the initial phase of acquired immunity, called the **primary immune response.** This occurs when lymphocytes are exposed to an antigen for the first time.

❻ When memory cells produced during the primary response are activated by a second exposure to the same antigen, they initiate the **secondary immune response.** This response is faster and stronger than the first. Another round of clonal selection ensues. The selected memory cells multiply quickly, producing a large second clone of lymphocytes that mount the secondary response. Like the first clone, the second clone includes effector cells that produce antibodies and memory cells capable of responding to future exposures to the antigen. In our example here with B cells, the secondary response produces very high levels of antibodies that, though they are short-lived, are often more effective against the antigen than those produced during the primary response.

The concept of clonal selection is so fundamental to understanding acquired immunity that it is worth restating: Each antigen, by binding to specific receptors, selectively activates a tiny fraction of lymphocytes; these few selected cells then give rise to many cells, all specific for and dedicated to eliminating the antigen that started the response. Thus, we see that the versatility of the immune system depends on a great diversity of preexisting lymphocytes with different antigen receptors.

Primary vs. Secondary Immune Response Now that we have seen how clonal selection works, let's take a look at

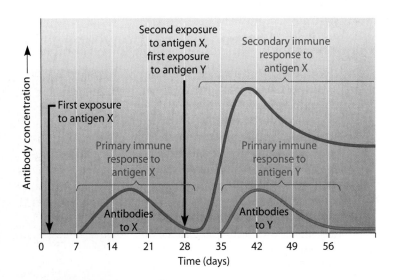

Figure 24.7B The two phases of the immune response

how two exposures to an identical antigen trigger the two phases of the acquired immune response in an individual. Figure 24.7B illustrates the difference between the two phases. On the far left of the graph, you can see that the primary response does not start right away; it usually takes several days for the lymphocytes to become activated by an antigen (called X here) and form clones of effector cells. When the effector cell clone forms, antibodies start showing up in the blood, as the graph shows. During this delay, a stricken individual may become ill. The antibody level reaches its peak about 2 weeks after initial exposure. As the antibody levels in the blood and lymph rise, the symptoms of the illness typically diminish and disappear. The primary response subsides as the effector cells die out.

The second exposure to antigen X (day 28 in the graph) triggers the secondary immune response. Notice that this secondary response occurs faster than the primary response (typically in 2 to 7 days, versus 10 to 17 days). As mentioned, the secondary response is also of greater magnitude (produces higher levels of antibodies) and is more prolonged.

The red curve in Figure 24.7B illustrates the specificity of the immune response. If the body is exposed to a different antigen (Y), even after it has already responded to antigen X, it responds with another primary response, this one directed against antigen Y. The response to Y is not enhanced by the response to X; that is, acquired immunity is specific.

Although we have focused on humoral immunity (produced by B cells) in this module, clonal selection, effector cells, and memory cells are features of cell-mediated immunity (produced by T cells) as well. In the next several modules, we discuss humoral immunity further. After that, we focus on how the cell-mediated arm of the immune system helps defend the body against pathogens.

? What is the immunological basis for referring to certain diseases, such as mumps, as *childhood* diseases?

■ One bout with the pathogen, which most often occurs during childhood, is usually enough to confer immunity for the rest of that individual's life.

24.8 Antibodies are the weapons of humoral immunity

B cells are the "frontline warriors" of the "defensive machine" we call humoral immunity. Plasma cells—the effector cells produced during clonal selection—make and secrete antibodies, the proteins that serve as molecular weapons of defense.

We have been using Y-shaped symbols to represent antibodies, and the Y actually does resemble their shape, as the antibody molecule in Figure 24.8A illustrates. Figure 24.8B is a simplified diagram explaining antibody structure. Each antibody molecule is made up of four polypeptide chains, two identical "heavy" chains and two identical "light" chains. In both figures, the parts colored in shades of pink represent the fairly long, heavy chains of amino acids that give the molecule its Y shape. Bonds (black lines in Figure 24.8B) at the fork of the Y hold these chains together. The two green regions in each figure are shorter chains of amino acids, the light chains. Each of the light chains is bonded to one of the heavy chains. As Figure 28.8A indicates, the bonded chains actually intertwine.

An antibody molecule has two related functions in humoral immunity: to recognize and bind to a certain antigen, and to assist in neutralizing the antigen it recognizes. The structure of an antibody allows it to perform these functions. Notice in Figure 24.8B that each of the four chains of the molecule has a C (constant) region and a V (variable) region. At the tip of each arm of the Y, a pair of V regions forms an **antigen-binding site**, a region of the molecule responsible for the antibody's recognition-and-binding function. A huge variety in the three-dimensional shapes of the binding sites of different antibody molecules arises from a similarly large variety in the amino acid sequences in the V regions; hence the term *variable*. This structural variety accounts for the diversity of lymphocytes and gives the humoral immune system the ability to react to virtually any kind of antigen.

The tail of the antibody molecule, formed by the constant regions of the heavy chains, helps mediate the disposal of the bound antigen. Antibodies with different kinds of heavy-chain C regions are grouped into different classes. Humans and other mammals have five major classes of antibodies, called IgA, IgD, IgE, IgG, and IgM (where Ig stands for immunoglobulin, another name for antibody). Each of the five classes differs in where it's found in the body and

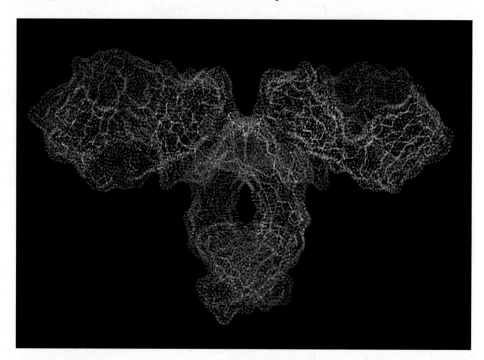

Figure 24.8A   A computer graphic of an antibody molecule

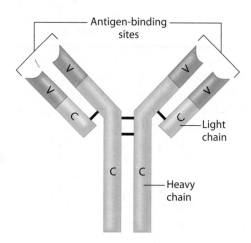

Figure 24.8B   Antibody structure

how it works. However, all five classes of antibodies perform the same basic function: to mark invaders for elimination. We take a closer look at this process next.

? How is the specificity of an antibody for an antigen analogous to an enzyme's specificity for its substrate?

■ Both antibodies and enzymes are proteins with binding sites of specific shape that recognize and bind to other molecules (antigens for antibodies, substrates for enzymes) with complementary shapes.

Antibodies mark antigens for elimination

The main role of antibodies in eliminating invading microbes or molecules is to mark the invaders. An antibody marks an antigen by combining with it to form an antigen-antibody complex. Weak chemical bonds between antigen molecules and the antigen-binding sites on antibody molecules hold the complex together.

As **Figure 24.9** illustrates, it is the binding of antibodies to antigens that actually triggers mechanisms to neutralize or destroy an invader. Such a mechanism is called an effector mechanism. Several effector mechanisms are depicted in the figure. In viral neutralization, antibodies bind to certain surface proteins on a virus, thereby blocking its ability to infect a host cell. Similarly, antibodies may bind to surface molecules on bacterial cells. The bound antibodies enhance macrophage destruction of the bacteria.

Another effector mechanism is the agglutination (clumping together) of viruses, bacteria, or foreign eukaryotic cells. Because each antibody molecule has at least two binding sites, antibodies can hold a clump of invading cells together. Agglutination makes the cells easy for phagocytes to capture.

A third effector mechanism, precipitation, is similar to agglutination, except that the antibody molecules link *dissolved* antigen molecules together. This makes the antigen molecules precipitate out of solution as solids. Precipitation, like the other effector mechanisms discussed so far, enhances engulfment by phagocytes.

One of the most important effector mechanisms in humoral immunity is the activation of the complement system (see Module 24.1) by antigen-antibody complexes. Activated complement system proteins (green in the diagram) can attach to a foreign cell and poke holes in its plasma membrane, causing cell lysis (rupture).

Taken as a whole, this figure illustrates a fundamental concept of immunity: All effector mechanisms involve a *specific* recognition-and-attack phase followed by a *nonspecific* destruction phase. Thus, the antibodies of humoral immunity, which identify and bind to foreign invaders, work with innate defenses, such as phagocytes and complement, to form a complete defense system.

? How does acquired humoral immunity interface with the body's innate defense system?

■ Antibodies mark specific antigens for destruction, but it is usually the complement system, phagocytes, or other components of innate defense that destroy the antigens.

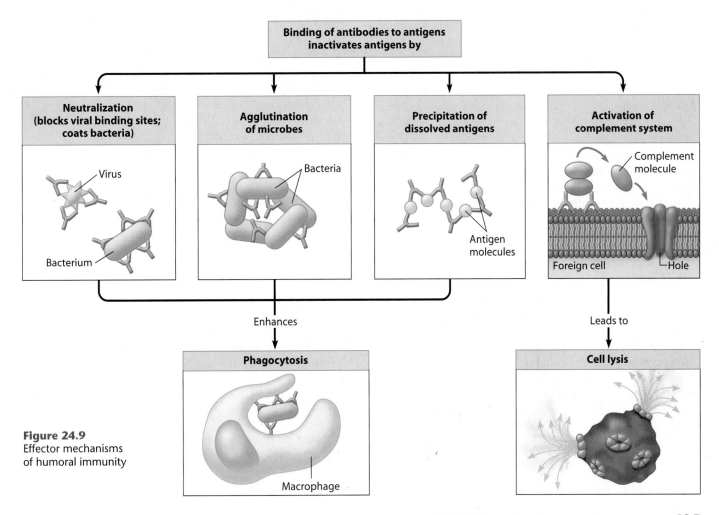

Binding of antibodies to antigens inactivates antigens by

Neutralization (blocks viral binding sites; coats bacteria)
Virus
Bacterium

Agglutination of microbes
Bacteria

Precipitation of dissolved antigens
Antigen molecules

Activation of complement system
Complement molecule
Foreign cell
Hole

Enhances

Leads to

Phagocytosis
Macrophage

Cell lysis

Figure 24.9
Effector mechanisms of humoral immunity

24.10 Monoclonal antibodies are powerful tools in the lab and clinic

Because of their ability to tag specific molecules or cells, antibodies are widely used in laboratory research, clinical diagnosis, and the treatment of disease. In the original procedure for preparing antibodies, a small sample of antigen is injected into a rabbit or mouse. In response to the antigen, the animal produces antibodies, which can be collected directly from its blood. However, because the antigen usually has many different antigenic determinants, the result is polyclonal: a mixture of different antibodies produced by different clones of cells.

In the late 1970s, a technique was developed for making **monoclonal antibodies**. The term *monoclonal* means that all the cells producing the antibodies are descendants of a single cell; thus, they all produce identical antibody molecules that are specific for the same antigenic determinant. Monoclonal antibodies are harvested from cell cultures rather than from animals.

As shown in **Figure 24.10A**, the trick to making monoclonal antibodies is the fusion of two cells to form a hybrid cell with a combination of desirable properties. First, an animal is injected with an antigen that will stimulate B cells to make the desired antibody. At the same time, cancerous tumor cells, which can multiply indefinitely, are grown in a culture. The scientist then fuses a tumor cell with a normal antibody-producing B cell from the animal. The hybrid cell makes antibody molecules specific for a single antigenic determinant and is able to multiply indefinitely in a laboratory dish. Thus, large amounts of identical antibody molecules can be prepared.

As mentioned earlier, monoclonal antibodies are particularly useful for diagnosis. For example, monoclonal antibodies can be prepared that bind to bacteria that cause a sexually transmissible disease or to a hormone that indicates pregnancy. If the antibodies have been labeled for easy detection (by a dye, for instance), they will reveal the presence of the bacteria or hormone. One type of home pregnancy test detects a hormone called human chorionic gonadotropin (hCG; see Module 27.16) that is present in the urine of pregnant women. When a testing strip is dipped into urine, monoclonal antibodies on the strip bind to any hCG that is present, causing a color change on the strip (Figure 24.10B).

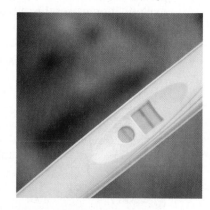

Figure 24.10B
A home pregnancy test

Monoclonal antibodies also have great promise for use in the treatment of certain diseases, including cancer. For example, Herceptin, a genetically engineered monoclonal antibody, is used to treat a common form of aggressive breast cancer. The Herceptin antibody molecules act by binding to growth factor receptors that are present in excess on the cancer cells. They thus prevent the receptors from transmitting "grow" signals to the cells. Furthermore, certain types of cancer are treated with tumor-specific monoclonal antibodies bound to toxin molecules. The toxin-linked antibodies carry out a precise search-and-destroy mission, selectively attaching to and killing tumor cells.

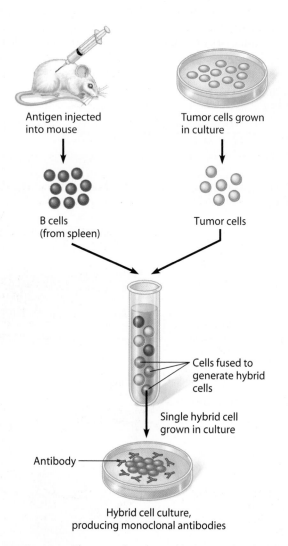

Antigen injected
into mouse

Tumor cells grown
in culture

B cells
(from spleen)

Tumor cells

Cells fused to
generate hybrid
cells

Single hybrid cell
grown in culture

Antibody

Hybrid cell culture,
producing monoclonal antibodies

Figure 24.10A   The procedure for making monoclonal antibodies

? How do home pregnancy tests based on monoclonal antibodies work?

■ The monoclonal antibodies produce a color change in the test solution when they react with a hormone present in the female's urine during early pregnancy.

24.11 Helper T cells stimulate humoral and cell-mediated immunity

The antibody-producing B cells of humoral immunity make up one army of the body's defense network. The humoral defense system identifies and helps destroy invaders that are in our blood, lymph, or interstitial fluid—in other words, outside our body cells. But many invaders, including all viruses, enter cells and reproduce there. It is the cell-mediated immunity produced by T cells that battles pathogens that have already entered body cells.

Whereas B cells respond to free antigens present in body fluids, T cells respond only to antigens present on the surfaces of the body's own cells. There are two main kinds of T cells. **Cytotoxic T cells** attack body cells that are infected with pathogens; we'll discuss these T cells in Module 24.13. **Helper T cells** play a role in many aspects of immunity. They help activate cytotoxic T cells and macrophages and even help stimulate B cells to produce antibodies.

Helper T cells interact with other white blood cells—including macrophages and B cells—that function as **antigen-presenting cells**. All of cell-mediated immunity and much of humoral immunity depend on the precise interaction of antigen-presenting cells and helper T cells. This interaction activates the helper T cells, which can then go on to activate other cells of the immune system.

As its name implies, an antigen-presenting cell *presents* a foreign antigen to a helper T cell. Consider a typical antigen-presenting cell, a macrophage. As shown in **Figure 24.11**, ❶ the macrophage ingests a microbe (or other foreign particle) and breaks it into fragments—foreign antigens (orange in the figure). Then molecules of a special protein (green) belonging to the macrophage, which we will call **self protein** (because it belongs to the body itself), ❷ bind the foreign antigens—**nonself molecules**—and ❸ display them on the cell's surface. Each of us has a unique set of self proteins, which serve as identity markers for our body cells. ❹ Helper T cells recognize and bind to the *combination* of a self protein and a foreign antigen—called a self-nonself complex—displayed on an antigen-presenting cell. This double-recognition system is like the system banks use for safe-deposit boxes: Opening your box requires the banker's key along with your specific key.

The ability of a helper T cell to recognize a unique self-nonself complex on an antigen-presenting cell depends on the receptors (purple) embedded in the T cell's plasma membrane. As indicated in the figure, a T cell receptor actually has two binding sites: one for antigen and one for self protein. The two binding sites enable a T cell receptor to recognize the overall shape of a self-nonself complex on an antigen-presenting cell. The immune response is highly specific because the receptors on each helper T cell can bind only one kind of self-nonself complex on an antigen-presenting cell.

The binding of a T cell receptor to a self-nonself complex activates the helper T cell. Several other kinds of signals can enhance this activation. For example, certain proteins secreted by the antigen-presenting cell, such as interleukin-1 (green arrow), diffuse to the helper T cell and stimulate it.

Activated helper T cells promote the immune response in several ways, with a major mechanism being the secretion of additional stimulatory proteins. One such protein, interleukin-2 (blue arrows), has three major effects. ❺ First, it makes the helper T cell itself grow and divide, producing both memory cells and additional active helper T cells. This positive-feedback loop amplifies the cell-mediated defenses against the antigen at hand. Second, interleukin-2 ❻ helps activate B cells, thus stimulating humoral immunity. And third, ❼ it stimulates the activity of cytotoxic T cells, the topic of Module 24.13.

> **?** How can one helper T cell stimulate both humoral and active immunity?

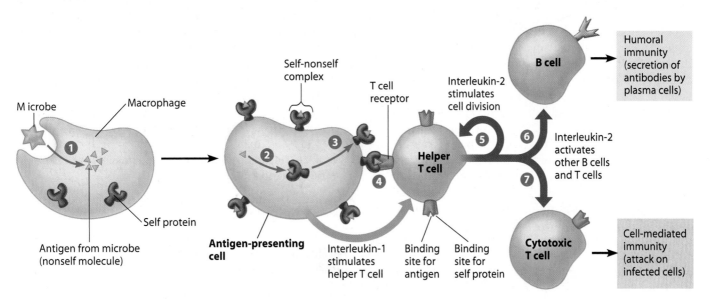

Figure 24.11 The activation of a helper T cell and its roles in immunity

24.12 HIV destroys helper T cells, compromising the body's defenses

AIDS (acquired immune deficiency syndrome) results from infection by HIV, the human immunodeficiency virus. HIV can infect a variety of cells, but it most often attacks helper T cells (Figure 24.12). As HIV depletes the body of helper T cells, both cell-mediated and humoral immunity are severely impaired. This drastically compromises the body's ability to fight infections.

How does HIV destroy helper T cells? First it must be transmitted into a host organism. HIV is transmitted by body fluids carrying infected cells. Most often, HIV enters the body through small cuts or sores during sexual contact or via needles contaminated with infected blood. Once HIV is in the bloodstream, proteins on the surface of the virus can bind to proteins on the surface of a helper T cell. Attached to the T cell, HIV may enter and begin to reproduce.

Inside the T cell, the RNA genome of HIV is reverse-transcribed, and the newly produced DNA is integrated into the host T cell's genome (see Module 10.21). This viral genome can now direct the production of new viruses from inside the T cell, generating up to a thousand or more per day. Eventually, the host helper T cell dies, from the damaging effects of virus reproduction and/or virus-triggered apoptosis (programmed cell death).

HIV in the bloodstream may infect and kill other helper T cells. As the number of T cells decreases, the body's ability to fight even the mildest infection is hampered, and AIDS eventually develops. It may take ten years or more for full-blown AIDS symptoms to appear after the initial HIV infection.

Immune system impairment makes AIDS patients susceptible to **opportunistic infections** and cancers that are rebuffed by a healthy individual but can harm those with weak immunity. For example, infection by a ubiquitous fungus called *Pneumocystis carinii* rarely occurs among people with healthy immune systems. In a person with AIDS, however, infection by *P. carinni* can cause severe pneumonia and death. Likewise, Kaposi's sarcoma, a very rare skin cancer, used to be seen exclusively among the elderly or patients receiving chemotherapy treatments for cancer. It is now most frequently seen among AIDS patients. In fact, it was an unusual cluster of Kaposi's sarcoma and pneumocystis pneumonia cases in 1981 that first brought AIDS to the attention of the medical community.

Since its discovery, AIDS has been the focus of considerable research. AIDS remains incurable, although certain drugs can slow HIV reproduction and the progress of AIDS. As discussed in the chapter introduction, there are treatments that can drastically lower transmission rates of HIV from mother to child. However, these drugs are expensive and not available to all who need them. In addition, mutational changes that occur as the virus reproduces can generate strains of HIV that are drug resistant. Combinations of drugs can help counter this effect, since viruses newly resistant to one drug may be

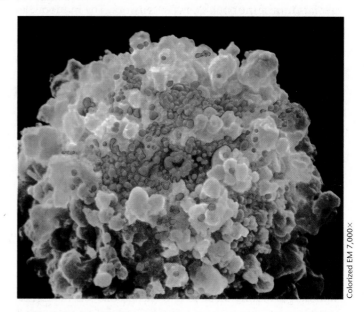

Figure 24.12 A human helper T cell (green) under attack by HIV (red spheres)

Colorized EM 7,000×

defeated by another. However, some HIV virus strains are resistant even to multiple-drug "cocktails," reducing the effectiveness of these drugs in some patients. Additionally, the multidrug regimens are complicated, expensive, and may have debilitating side effects. The tendency of HIV surface antigens to undergo frequent mutational changes has also hampered efforts to develop an effective vaccine.

Until there is a vaccine or a cure, the best way to stop AIDS is to prevent the spread of HIV. This can only happen by educating people about how the virus is transmitted—through unprotected sex, through direct contact with blood, or from mother to baby. Although using a latex condom or other barrier to body fluids during sex does not completely eliminate the risk of transmitting HIV (or other similarly transmitted viruses, such as the hepatitis B virus), it does significantly reduce it. Practicing safe sex and avoiding intravenous drugs could save millions of lives.

Web/CD Activity 24A *HIV Reproductive Cycle*

Web/CD Thinking as a Scientist *Connection: What Causes Infections in AIDS Patients?*

Web/CD Thinking as a Scientist *Connection: Why Do AIDS Rates Differ Across the U.S.?*

? Why is it so difficult to develop an AIDS vaccine?

■ Because the HIV virus evolves so rapidly

24.13 Cytotoxic T cells destroy infected body cells

Two types of T cells participate in cell-mediated immunity: helper T cells and cytotoxic T cells. In Module 24.11, we learned how helper T cells activate cytotoxic T cells, the only T cells that actually kill infected cells.

Once activated, cytotoxic T cells identify infected cells in the same way that helper T cells identify antigen-presenting cells. An infected cell has foreign antigens—molecules belonging to the viruses or bacteria infecting it—attached to self proteins on its surface (Figure 24.13). Like a helper T cell, a cytotoxic T cell carries receptors (purple) that can bind with a self-nonself complex on the infected cell.

The self-nonself complex on an infected body cell is like a red flag to cytotoxic T cells that have matching receptors. As shown in the figure, ❶ the cytotoxic T cell binds to the infected cell. The binding activates the T cell, which then synthesizes several new proteins that act on the bound cell, including one called **perforin** (red dots in the figure). ❷ Perforin is discharged and attaches to the infected cell's membrane, making holes in it. T cell enzymes (blue dots) then enter the infected cell and promote its death by apoptosis. ❸ The infected cell dies and is destroyed.

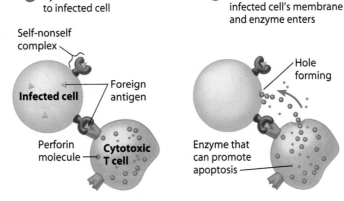

❶ Cytotoxic T cell binds to infected cell

❷ Perforin makes holes in infected cell's membrane and enzyme enters

❸ Infected cell is destroyed

Self-nonself complex

Foreign antigen

Infected cell

Perforin molecule

Cytotoxic T cell

Hole forming

Enzyme that can promote apoptosis

Figure 24.13 How a cytotoxic T cell kills an infected cell

? Compare and contrast the T cell receptor with the antigen receptor on the surface of a B cell.

■ Both receptors bind to a specific antigen, but the T cell receptor only recognizes that antigen when it is presented along with a "self" marker on the surface of one of the body's own cells.

24.14 Cytotoxic T cells may help prevent cancer

People with immune deficiencies are often unusually susceptible to cancer, a fact suggesting that the immune system plays a watchdog role against at least some forms of cancer. Researchers are therefore actively engaged in studying the interactions between the immune system and cancer cells.

As we discussed in Module 8.10 and in Chapter 11, the genetic changes that lead to cancer can result in the production of new proteins not found in normal body cells. Some of these will be displayed on the surfaces of the cancerous cells, marking them as abnormal. These distinctive molecules, called tumor antigens, can be identified as foreign by cytotoxic T cells, leading to the destruction of affected cells. Thus, cytotoxic T cells can defend against malignant tumors the same way they defend against invading microbes.

Figure 24.14 shows a troop of cytotoxic T cells (light blue) attacking a tumor cell (gold) in a laboratory culture. Why this built-in surveillance system sometimes fails in the body, allowing tumors to develop, is still largely a mystery. Tumors may develop if cancer cells shed the surface molecules that mark them as foreign, or if cancer cells otherwise evade the immune system. Studying how the immune system normally helps prevent cancer is an important part of the cancer research effort.

Figure 24.14 Cytotoxic T cells attacking a cancer cell

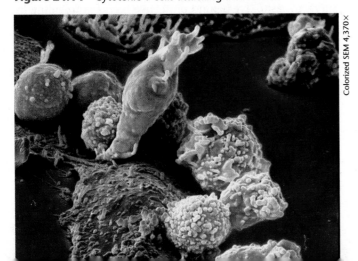

Colorized SEM 4,370×

? Stress causes the adrenal glands to secrete steroid hormones that, among many other functions, suppress the immune response. How might this link prolonged and severe stress to a greater probability of developing cancer?

■ Suppressing the immune response might make it possible for new cancer cells to avoid destruction by cytotoxic T cells.

24.15 The immune system depends on our molecular fingerprints

As we have seen, our immune system's ability to recognize the body's own molecules—that is, to distinguish *self* from *nonself*—enables it to battle foreign invaders without harming healthy body cells. Self proteins on cell surfaces are the keys to this ability. Each person's cells have a particular collection of self proteins that provide the molecular "fingerprints" recognized by the immune system.

Each of us has two sets of self proteins on the surfaces of our cells. Class I proteins occur on almost all nucleated cells in the body. Class II proteins are found only on a few types of cells, including B cells, activated T cells, and macrophages. The particular collections of proteins in the two sets are specific to the individual in whom they are found, marking the body cells as "off-limits" to the immune system; our lymphocytes do not attack these molecules.

The immune system not only distinguishes body cells from microbes, but also can tell your cells from those of other people. Genes at multiple chromosomal loci code for the main self proteins. The group of self-protein genes is called the **major histocompatibility complex**, or **MHC**. Because there are hundreds of alleles in the human population for each MHC gene, it is virtually impossible for any two people (except identical twins) to have completely matching sets of self proteins.

The immune system's ability to recognize foreign antigens does not always work in our favor. For example, when a person receives an organ transplant or tissue graft, the person's immune system recognizes the MHC markers on the donor's cells as foreign and attacks them. To minimize rejection, doctors look for a donor with self proteins matching the recipient's as closely as possible. The best match is to transplant the patient's own tissue, as when a burn victim receives skin grafts removed from other parts of his or her body. Otherwise, identical twins provide the closest match, followed by nonidentical siblings. Additionally, doctors use drugs to suppress the immune response against the transplant. Unfortunately, these drugs may also reduce the ability to fight infections and cancer. A few, however, such as cyclosporine, can suppress cell-mediated responses without crippling humoral immunity.

? In what sense is a cell's set of MHC surface markers analogous to a fingerprint?

■ The set of MHC ("self") markers is unique to each individual.

DISORDERS OF THE IMMUNE SYSTEM

CONNECTION

24.16 Malfunction or failure of the immune system causes disease

Overall, our immune system is highly effective, protecting us against a vast array of potentially harmful invaders. When the immune system doesn't function properly, serious disease can result.

Autoimmune diseases result when the immune system goes awry and turns against some of the body's own molecules. In *systemic lupus erythematosus (lupus)* for example, B cells make antibodies against a wide range of self molecules, such as histones and DNA released by the normal breakdown of body cells. Lupus is characterized by skin rashes, fever, arthritis, and kidney malfunction. *Rheumatoid arthritis* is another antibody-mediated autoimmune disease; it leads to damage and painful inflammation of the cartilage and bone of joints. In *insulin-dependent diabetes mellitus* (see Module 26.9), the insulin-producing cells of the pancreas are the targets of cytotoxic T cells. In *multiple sclerosis (MS)*, T cells react against the myelin sheath that surrounds many neurons (see Figure 28.2). People with MS experience a number of serious neurological abnormalities.

Most medicines currently available for treating autoimmune diseases either suppress immunity in general or are limited to the alleviation of specific symptoms. However, as research scientists learn more about these diseases and about the normal operation of the immune system, they hope to develop more effective therapies.

In contrast to autoimmune diseases are a variety of defects called **immunodeficiency diseases.** Immunodeficient people lack one or more of the components of the immune system. This makes them susceptible to frequent and recurrent infections and cancer. In the rare congenital disease *severe combined immunodeficiency (SCID)*, both T cells and B cells are absent or inactive. People with SCID are extremely sensitive to even minor infections. Until recently, their only hope for survival was to live behind protective barriers (providing inspiration for "bubble boy" stories in the popular media) or to receive a successful bone marrow transplant that would continue to supply functional lymphocytes. Since the early 1990s, medical researchers have been testing a gene therapy for this disease with some success (see Module 12.19).

Immunodeficiency is not always an inborn condition; it may be acquired later in life. For instance, *Hodgkin's disease,* a type of cancer that affects the lymphocytes, can depress the immune system. Radiation therapy and the drug treatments used against many cancers can have the same effect. Another well-known acquired immune deficiency is AIDS (see Module 24.12).

There is growing evidence that physical and emotional stress can harm immunity. Hormones secreted by the adrenal glands during stress affect the numbers of white blood cells and may suppress the immune system in other ways.

The association between emotional stress and immune function also involves the nervous system. Some neurotransmitters secreted when we are relaxed and happy may enhance immunity. In one study, college students were examined just after a vacation and again during final exams.

Their immune systems were impaired in various ways during exam week; for example, interferon levels were lower. These and other observations indicate that general health and state of mind affect immunity.

? What is a probable side effect of autoimmune disease treatments that suppress the immune system?

■ Lowered resistance to infections

24.17 Allergies are overreactions to certain environmental antigens

Allergies are abnormal sensitivities to antigens in our surroundings. Antigens that cause allergies are called **allergens.** Protein molecules on pollen grains, on the feces of tiny mites that live in house dust, and in animal dander (shed skin cells) are common allergens. Many people who are allergic to cats and dogs are actually allergic to proteins in the animal's saliva that get deposited on the fur when the animal licks itself. Allergic reactions typically occur very rapidly and in response to tiny amounts of an allergen. A person allergic to cat or dog saliva, for instance, may react to a few molecules of the allergen in a matter of minutes. Allergic reactions can occur in many parts of the body, including the nasal passages, bronchi, and skin. Symptoms may include sneezing, runny nose, coughing, wheezing, and itching.

The symptoms of an allergy result from a two-stage reaction sequence outlined in **Figure 24.17.** The first stage, called sensitization, occurs when a person is first exposed to an allergen—pollen, for example. ❶ After an allergen enters the bloodstream, it binds to B cells with complementary receptors. ❷ The B cells then proliferate through clonal selection and secrete large amounts of antibodies to this allergen. ❸ Some of these antibodies attach to receptor proteins on the surfaces of **mast cells,** body cells that produce histamine and other chemicals that trigger the inflammatory response (see Module 24.2).

The second stage of an allergic response begins when the person is exposed to the same allergen later. The allergen enters the body and ❹ binds to the antibodies attached to mast cells. ❺ This causes the mast cells to release histamine, which triggers the allergic symptoms. As in inflammation, histamine causes blood vessels to dilate and leak fluid and so causes nasal irritation, itchy skin, and tears. **Antihistamines** are drugs that interfere with histamine's action and give temporary relief from an allergy.

Allergies range from seasonal nuisances to severe, life-threatening responses. **Anaphylactic shock** is an especially dangerous type of allergic reaction. Some people are extremely sensitive to certain allergens, such as bee venom, penicillin, or allergens in peanuts or shellfish. Any contact with these allergens makes their mast cells release inflammatory chemicals very suddenly. As a result, their blood vessels dilate abruptly, causing a rapid, potentially fatal drop in blood pressure. Fortunately, anaphylactic shock can be counteracted with injections of the hormone epinephrine.

? How do antihistamines relieve allergy symptoms?

■ By blocking the action of histamine, which produces the symtomps of the inflammatory response.

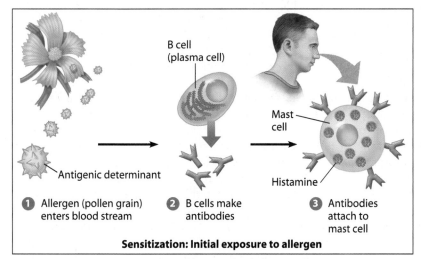
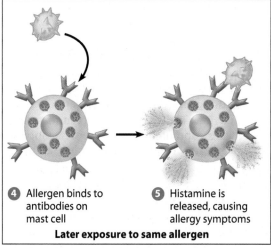

❶ Allergen (pollen grain) enters blood stream
❷ B cells make antibodies
❸ Antibodies attach to mast cell

Sensitization: Initial exposure to allergen

❹ Allergen binds to antibodies on mast cell
❺ Histamine is released, causing allergy symptoms

Later exposure to same allergen

B cell (plasma cell)
Mast cell
Histamine
Antigenic determinant

Figure 24.17 The two stages of an allergic reaction

CHAPTER REVIEW

Reviewing the Concepts

Innate Defenses Against Infection (24.1–24.3)

Innate defenses against infection include the skin, mucous membranes, phagocytic cells, and antimicrobial proteins **(24.1).** Tissue damage triggers the inflammatory response, which can disinfect tissues and limit further infection **(24.2).** The lymphatic system is a network of vessels and organs. The vessels collect fluid from body tissues and return it as lymph to the blood. Lymph organs, such as the spleen and lymph nodes, are packed with white blood cells that fight infections **(24.3).**

Acquired Immunity (24.4–24.15)

The immune system counters specific invaders by responding to antigens. Infection or vaccination triggers active immunity, which allows the immune system to "remember" an invader. We can also temporarily acquire passive immunity **(24.4).** Two kinds of lymphocytes carry out the immune response:

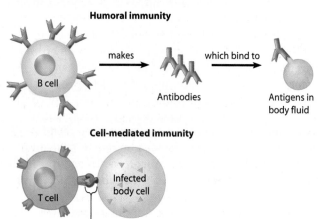

Humoral immunity

B cell — makes → Antibodies — which bind to → Antigens in body fluid

Cell-mediated immunity

T cell / Infected body cell

Self-nonself complex

Millions of kinds of B cells and T cells, each with different membrane receptors, wait in the lymphatic system, where they may respond to invaders **(24.5).** Antigen molecules have specific regions, called antigenic determinants, to which antibodies bind **(24.6).**

Clonal selection. When an antigen enters the body, it activates only a small subset of lymphocytes, those with complementary receptors. The selected cells multiply into clones of short-lived effector cells specialized for defending against the antigen that triggered the response and memory cells, which confer long-term immunity. Activated by subsequent exposure to the antigen, memory cells mount a rapid and massive secondary immune response **(24.7).**

Humoral immunity. Secreted by plasma (effector) B cells, an antibody molecule has antigen-binding sites specific to the antigenic determinants that elicited its secretion **(24.8).** Antibodies promote antigen elimination through several mechanisms **(24.9).** Monoclonal antibodies are useful in research, diagnosis, and treatment of certain cancers **(24.10).**

Cell-mediated immunity. Helper T cells and cytotoxic T cells are the main effectors of cell-mediated immunity, and helper T cells also stimulate the humoral responses. In cell-mediated immunity, an antigen-presenting cell displays a foreign antigen (a nonself molecule) and one of the body's own self proteins to a helper T cell. The helper T cell's receptors recognize the self-nonself complexes, and the interaction activates the helper T cell. In turn, the helper T cell can activate cytotoxic T cells and B cells **(24.11).** The AIDS virus attacks helper T cells, opening the way for opportunistic infections **(24.12).** Cytotoxic T cells bind to infected body cells and destroy them **(24.13).** Cytotoxic T cells may also attack cancer cells, which have abnormal surface molecules **(24.14).** The immune system normally reacts only against nonself substances, not against self. Transplanted organs may be rejected because these cells lack the unique "fingerprint" of the recipient's self proteins **(24.15).**

Disorders of the Immune System (24.16–24.17)

When immunity malfunctions. In autoimmune diseases, the system turns against the body's own molecules. In immunodeficiency diseases, immune components are lacking, and frequent infections occur **(24.16).** Allergies are abnormal sensitivities to antigens (allergens) in the surroundings **(24.17).**

Connecting the Concepts

1. Complete this concept map to summarize the key concepts concerning the body's defenses.

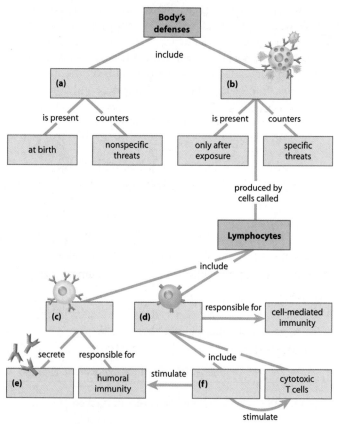

Testing Your Knowledge

Multiple Choice

2. Foreign molecules that evoke an immune response are called
   a. pathogens.
   b. antibodies.
   c. lymphocytes.
   d. histamines.
   e. antigens.

3. Which of the following is *not* part of the body's innate defense system?
   a. natural killer cells
   b. antibodies
   c. interferons
   d. complement system
   e. inflammation

4. Which of the following best describes the difference in the way B cells and cytotoxic T cells deal with invaders?
   a. B cells confer active immunity; T cells confer passive immunity.
   b. B cells send out antibodies to attack; T cells themselves do the attacking.
   c. T cells handle the primary immune response; B cells handle the secondary response.
   d. B cells are responsible for cell-mediated immunity; T cells are responsible for humoral immunity.
   e. B cells attack the first time the invader is present; T cells attack subsequent times.

5. The antigen-binding sites of an antibody molecule are formed from the molecule's variable regions. Why are these regions called variable?
   a. They can change their shapes on command to fit different antigens.
   b. They change their shapes when they bind to an antigen.
   c. Their specific shapes are unimportant.
   d. They can be different shapes on different antibody molecules.
   e. Their sizes vary considerably from one antibody to another.

6. Cytotoxic T cells are able to recognize and attack cancer cells because
   a. cancer changes the surfaces of cancerous cells.
   b. B cells help them.
   c. cancer is a bacterial infection.
   d. cancer cells release antibodies into the blood.
   e. cancer is an autoimmune disease.

Matching

7. Attacks infected body cells
8. Carries out humoral immunity
9. Causes allergy symptoms
10. Phagocytic white blood cell
11. General name for a B or T cell
12. Carries out the secondary immune response
13. Cell most commonly attacked by HIV

   a. lymphocyte
   b. cytotoxic T cell
   c. helper T cell
   d. mast cell
   e. macrophage
   f. B cell
   g. memory cell

Describing, Comparing, and Explaining

14. Describe (a) how AIDS is transmitted and (b) how immune system cells in an infected person are affected by HIV. Why is AIDS particularly deadly compared to other viral diseases? What are the most effective means of preventing HIV transmission?

15. What is inflammation? How does it protect the body? Why is inflammation considered a nonspecific defense?

16. Your roommate is rushed to the hospital after suffering a severe allergic reaction to a bee sting. After she is treated and released, she asks you (the local biology expert!) to explain what happened. She says, "I don't understand how this could have happened. I've been stung by bees before and didn't have a reaction." Suggest a hypothesis to explain what has happened to cause her severe allergic reaction and why she did not have the reaction after previous bee stings.

Applying the Concepts

17. Organ donation saves many lives each year. Even though some transplanted organs are derived from living donors, the majority come from patients who die but still have healthy organs that can be of value to a transplant recipient. Potential organ donors can fill out an organ donation card to specify their wishes. If the donor is in critical condition and dying, the donor's family is usually consulted to discuss the donation process. Generally, the next of kin must approve before donation can occur, regardless of whether the patient has completed an organ donation card. In some cases, the donor's wishes are overridden by a family member. Do you think that family members should be able to deny the stated intentions of the potential donor? Why or why not? Have you signed up to be an organ donor? Why or why not?

18. There is great concern about the rate of HIV infection among teenagers. Schools in some large cities have instituted programs to make condoms available to students, along with counseling about safer sex. These plans have divided school boards and communities. Some citizens and church groups are opposed to giving condoms to students on the grounds that it might appear to encourage sexual activity. By contrast, many school and public health officials view the situation as a health issue rather than a moral issue. The heart of the controversy seems to be whether the schools should take such a direct role in this part of student life. What are the reasons for and against distribution of condoms? What do you think the school's role should be?

Answers to all questions can be found in Appendix 3.

For study help and Activities, go to campbellbiology.com or the student CD-ROM.

CHAPTER 25

Let Sleeping Bears Lie

WHEN WE THINK OF HIBERNATION, we usually think of bears curled up in dens for the winter. Ironically, bears don't really hibernate, at least not in the scientific sense of the word. Hibernation is a state in which the body temperature of an animal goes well below normal, and the animal goes into a state of suspended animation from which it is not easily aroused. The body temperatures of small hibernating mammals, such as chipmunks and ground squirrels, may drop as much as 30°C. In contrast, a bear's body temperature goes down only a few degrees, and a bear is easily awakened. Biologists refer to this state as dormancy rather than hibernation. Field biologists who have visited bears in their winter dens to study dormancy have been surprised—and distressed—to find how easily bears are roused and how annoyed they are at being awakened.

In their dormant state, bears undergo significant physiological changes while maintaining homeostasis. For example, a dormant black bear's body temperature fluctuates between 31°C and 34°C after dropping from the bear's normal temperature of 37°C. This relative stability of the lower temperature indicates that **thermoregulation**, the maintenance of internal temperature within narrow limits, continues even during dormancy. Bears and other mammals are **endotherms**, animals that derive most of their body heat from their metabolism. Other endotherms include birds and a few other reptiles, some fishes, and numerous insect species. **Ectotherms**, which include most invertebrates, fishes, amphibians, lizards, snakes, and turtles, warm themselves mainly by absorbing heat from their surroundings.

How does a bear maintain its body temperature during dormancy? A number of adaptations help keep it warm. In anticipation of dormancy, the bear's appetite mechanisms are reset and the animal goes through a period of heavy eating (see photo on the facing page). During this time a black bear may gain up to 14 kg (31 lb) a week to provide it with the energy (stored as fat) that it needs to survive the winter. A bear's body fat and

A grizzly bear raking leaves into a den for the winter

THERMOREGULATION

25.1   Heat is gained or lost in four ways
25.2   Thermoregulation involves adaptations that balance heat gain and loss
25.3   Reducing metabolic rate and body temperature saves energy

OSMOREGULATION AND EXCRETION

25.4   Osmoregulation: Animals balance the gain and loss of water and solutes
25.5   Do we need to drink eight glasses of water each day?
25.6   Animals must dispose of nitrogenous wastes
25.7   The liver performs many functions, including the production of urea
25.8   Alcohol consumption can damage the liver
25.9   The excretory system plays several major roles in homeostasis
25.10 Overview: The key processes of the excretory system are filtration, reabsorption, secretion, and excretion
25.11 From blood filtrate to urine: A closer look
25.12 Kidney dialysis can be a lifesaver

Control of the Internal Environment

A black bear stocking up on juniper berries

dense fur provide superb insulation. Its habit of curling up in its den also helps keep heat loss to a minimum. Reduced blood flow to the bear's extremities decreases heat loss and maintains higher temperatures in its head and torso.

Besides a reduction in body temperature, bears undergo other physiological changes during dormancy. For example, dormant bears do not eat, expel solid waste, or urinate. This last point is of interest to researchers who study **excretion**, the disposal of nitrogen-containing wastes, and **osmoregulation**, the control of the gain and loss of water and solutes. Physicians who treat patients suffering from kidney failure wonder why the bears are not poisoned by the nitrogenous compound urea. As you will learn in this chapter, urea is produced by the breakdown of protein and is normally excreted in the urine. It turns out that while bears do break down proteins during dormancy, there is a change in their metabolism whereby the urea is recycled back to form new amino acids. These amino acids are used to make new proteins, helping to reduce muscle loss during dormancy. This recycling is very different from what happens in a fasting nondormant mammal. In a fasting mammal, the urea level rises as proteins are broken down to provide energy, and urea is removed in the urine. The changes in metabolism that take place in dormant bears are good examples of how homeostatic mechanisms can adjust to an animal's changing needs.

Dormant bears do not eat, expel solid waste, or urinate

In this chapter, we explore the homeostatic control mechanisms of thermoregulation, osmoregulation, and excretion. We will see that most animals, like the bear, can survive fluctuations in the external environment because these control mechanisms allow only a narrow range of fluctuation in the fluid environment that bathes their cells. Let's take a closer look at thermoregulation first. ■ ■ ■

25.1 Heat is gained or lost in four ways

An animal exchanges heat with the environment by four physical processes. *Conduction* is the transfer of thermal motion (heat) between molecules of objects that are in direct contact, as when an animal is physically touching an object in its environment. Heat is always conducted from an object of higher temperature to one of lower temperature. In Figure 25.1, heat conducted from the warm rock is elevating the lizard's body temperature (red arrows).

Convection is the transfer of heat by the movement of air or liquid past a surface. In the figure, a breeze removes heat (orange arrows) by convection from the lizard's tail.

Radiation is the emission of electromagnetic waves. Radiation can transfer heat between objects that are not in direct contact, as when an animal absorbs heat radiating from the sun (yellow arrows). The lizard also radiates some of its own heat into the external environment.

Evaporation is the loss of heat from the surface of a liquid that is losing some of its molecules as a gas. A lizard loses heat as moisture evaporates from its nostrils (blue arrow).

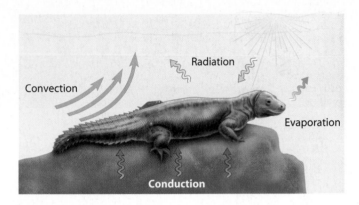

Figure 25.1   Mechanisms of heat exchange

? If you are sweating on a hot day and turn a fan on yourself, what mechanisms contribute to your cooling?

■ Evaporation (sweating) and convection (fan moving air)

25.2 Thermoregulation involves adaptations that balance heat gain and loss

Different animals are adapted to different environmental temperatures, and each species has an optimal temperature range. Within that range, endotherms and many ectotherms maintain a fairly constant internal temperature as the external temperature fluctuates. Five general categories of adaptations help animals thermoregulate.

Metabolic Heat Production   In cold weather, hormonal changes tend to boost the metabolic rate of birds and mammals, increasing their heat production. Simply moving around more or shivering also produces heat as a metabolic by-product of the contraction of skeletal muscles. Honeybees survive cold winters by clustering together and shivering in their hive (Figure 25.2A). The metabolic activity of all the bees together generates enough heat to keep the cluster alive. Many endothermic insects, such as bees and moths, also increase heat production through contraction of their powerful flight muscles before taking off in cold weather.

Figure 25.2A   A cluster of shivering honeybees generating heat

Insulation   A major thermoregulatory adaptation in mammals and birds is insulation—hair (often called fur), feathers, or fat layers. Most land mammals and birds react to cold by raising their fur or feathers, which traps a layer of air next to the warm skin, improving the insulation. (Although humans have little hair, muscles raise our hair in the cold, causing goose bumps, a vestige from our furry ancestors.) Aquatic mammals are insulated by a thick layer of fat.

Circulatory Adaptations   Heat loss can be altered by changing the amount of blood flowing to the skin. In a bird or mammal (and some ectotherms), nerves signal surface blood vessels to constrict or dilate, depending on the external temperature (see Module 20.14). When the vessels are constricted, less blood flows from the warm body core to the body surface, reducing the rate of heat loss. Conversely, dilation of surface blood vessels increases the rate of heat loss.

Figure 25.2B, on the facing page, illustrates a circulatory adaptation called a **countercurrent heat exchanger**, in which warm and cold blood flow in opposite (countercurrent) directions in two adjacent blood vessels. Warm blood (red) from the body core cools as it flows down the goose's leg or the dolphin's flipper. But the arteries carrying the warm blood are in close contact with veins (blue) conveying cool blood back toward the body core. As shown by the black arrows, heat passes from the warmer blood to the cooler blood along the whole length of these side-by-side vessels. By the time returning blood leaves the leg or the flipper, it is almost as warm as the body core. Thus, heat loss is minimal, even when the animal is standing on ice or swimming in frigid water.

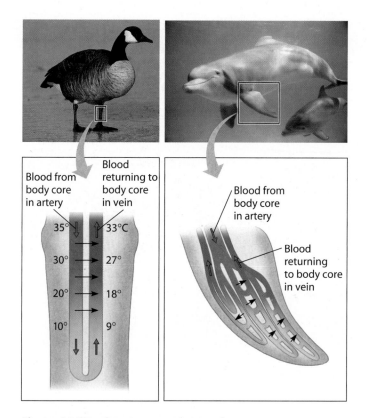

Figure 25.2B Countercurrent heat exchangers

cold blood returning from the gills is transported in large vessels lying just under the skin. Small branches of these vessels deliver oxygenated to the deep muscles. Each branch runs side by side with a vessel carrying warm blood outward from the inner body. The resulting countercurrent heat exchange retains heat around the main swimming muscles and enables the vigorous, sustained activity of these endothermic animals.

Evaporative Cooling Many animals live in places where thermoregulation requires cooling as well as warming. Some animals have adaptations that greatly increase evaporative cooling, such as panting, sweating, or even spreading saliva on body surfaces. Honeybees cool their hive during hot weather by transporting water into it and fanning with their wings, promoting heat loss by evaporation and convection.

Behavioral Responses Both endotherms and ectotherms use behavioral responses to control body temperature. Some animals, including some birds and butterflies, migrate seasonally to more suitable climates. Others, such as desert lizards, bask in the sun when it is cold and find cool, damp areas or burrows when it is hot. Many animals seek relief from heat by bathing, which brings immediate cooling by convection and continues to cool for some time by evaporation.

Web/CD Thinking as a Scientist *How Does Temperature Affect Metabolic Rate in* **Daphnia?**

? Compare countercurrent heat exchange with the countercurrent exchange of oxygen in fish gills. (See Module 22.3.)

■ In both cases, countercurrent exchange enhances transfer all along the length of a blood vessel—transfer of heat from one vessel to another in the case of a heat exchanger and transfer of oxygen between water and vessels in the case of gills.

Some endothermic fishes and sharks also have countercurrent exchange mechanisms. All fishes and sharks lose heat as blood passes through the gills. In large, powerful swimmers such as bluefin tuna, swordfish, and the great white shark,

---

25.3 Reducing metabolic rate and body temperature saves energy

The North American gray tree frog can spend much of the winter frozen (Figure 25.3). A frozen frog has almost no metabolic activity and burns almost no energy. A solution of "antifreeze" compounds (cryoprotectants) keeps ice crystals from rupturing its cells, protecting this overwintering ectotherm.

Endotherms are generally able to remain active in cold weather, although they use a great deal of food energy for heat production. When food supplies are low and environmental temperatures extreme, some endotherms are also able to reduce their metabolic rates to save energy.

Torpor is a state of reduced activity in which metabolic rate and body temperature decrease. **Hibernation** (Latin *hibernus*, winter) is a type of long-term torpor in cold weather. Some ground squirrels hibernate for as much as eight months of the year, living on energy that was stored in body fat when food was abundant.

Some animals have a summer torpor, or **estivation** (Latin *aestas*, summer). Their slowed metabolism and inactivity allow them to survive long periods of high temperatures and reduced water supplies.

Many small mammals and birds with extremely high metabolic rates exhibit a daily torpor. Most bats feed at night and go into torpor during daylight. Chickadees and hummingbirds feed during the day and often go into torpor on cold nights.

While some animals can tolerate fluctuations in body temperature, few can withstand changes in the balance of water and solutes in body fluids. We consider osmoregulation next.

? Why doesn't the dormancy of bears qualify as hibernation? (See chapter introduction.)

■ A bear's temperature drops only slightly, and a bear is awakened easily

Figure 25.3 The gray tree frog: active (top) and frozen (bottom)

25.4 Osmoregulation: Animals balance the gain and loss of water and solutes

Just as thermoregulation depends on balancing heat loss and gain, osmoregulation depends on balancing the uptake and loss of water and solutes, such as salt (NaCl) and other ions. Whether an animal inhabits land, fresh water, or salt water, its cells cannot survive a *net* water gain or loss. Animal cells swell and burst if there is a net uptake of water; they shrivel and die if there is a substantial net loss of water.

Osmoregulation is based largely on regulating solutes because water follows the movement of solutes by osmosis. Osmosis occurs whenever two solutions separated by a membrane differ in total solute concentration (see Module 5.16). There is a net movement of water from the hypotonic solution (the one with lower solute concentration) to the hypertonic solution (the one with higher solute concentration).

Some aquatic animals that live in the sea have body fluids with a solute concentration equal to that of seawater. Called **osmoconformers**, such animals do not undergo a net gain or loss of water. The sea, where animals first evolved, is the only environment that supports osmoconformers. The total solute concentration of most marine invertebrates conforms to that of seawater, and thus these animals do not expend energy regulating their water content. However, the concentration of certain ions in their body fluids is different from that of seawater. For example, for cell membranes to function properly, the concentration of potassium ions (K^+) must be higher within cells than in either the interstitial fluid or seawater. Because it takes energy to actively transport ions into or out of cells, even an osmoconformer expends some energy to maintain its ion concentrations.

Freshwater animals, land animals, and marine vertebrates (such as porpoises, seabirds, and fishes) have body fluids whose solute concentration differs from that of their environment. Therefore, they must use energy to regulate water loss or gain. Such animals are called **osmoregulators**.

The freshwater fish in Figure 25.4A has a much higher solute concentration in its internal fluids than that of fresh water, creating an osmotic problem for the animal. The fish constantly gains water by osmosis through its body surface, especially through its gills. It also loses salt by diffusion to its more dilute environment.

How does a freshwater fish maintain the proper water and solute balance? It does not drink water. Its food helps supply some ions, and its gills actively take up salt. The fish's kidneys work constantly to produce large amounts of dilute **urine**, the waste material produced by its excretory system. By excreting dilute urine, the fish disposes of excess water and conserves solutes.

The saltwater fish in Figure 25.4B has the opposite osmoregulatory problem. Because its internal fluids are lower in total solutes than seawater, a saltwater fish loses water by osmosis across its body surfaces. It also gains salt both by diffusion and from the food it eats. The fish balances the water

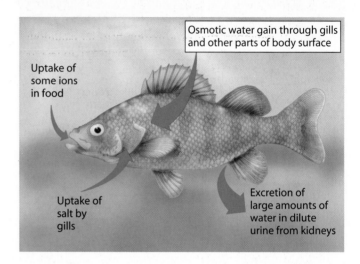

Figure 25.4A Osmoregulation in a freshwater fish, a perch

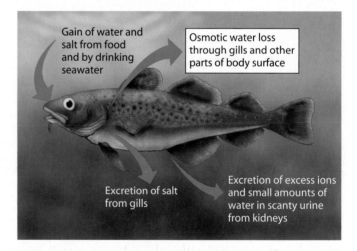

Figure 25.4B Osmoregulation in a saltwater fish, a cod

loss by drinking large amounts of seawater, and it pumps excess salt out through its gills. It also saves water by producing only small amounts of urine, in which it disposes of some excess ions.

Most fishes have little tolerance for changes in the solute concentration of their surroundings. The perch and the cod shown here, for instance, are restricted to fresh water and the ocean, respectively. However, a few fishes, such as salmon, can migrate between seawater and fresh water. Salmon have remarkable osmoregulatory adaptations. While in the ocean, they drink seawater and excrete excess salt from their gills, osmoregulating like the cod. When salmon move into fresh water to spawn, their osmoregulatory mechanism switches to the freshwater mode of the perch; they cease drinking, and their gills take up salt from the dilute environment.

What about land animals? They are osmoregulators, but they are not surrounded by water; therefore, they cannot directly exchange water with the environment by osmosis. Most terrestrial animals gain water by drinking and by eating moist foods, while constantly losing water from moist surfaces in their lungs or respiratory tubes, in urine and feces, and by evaporation across the skin. The osmoregulatory situation of a land animal resembles that of a marine fish; its paramount problem is losing water and becoming dehydrated. In fact, dehydration is such a severe problem on land that it may largely explain why only two groups of animals, arthropods and vertebrates, have colonized land with great success.

What is it about arthropods and vertebrates that gives them an edge in the battle against dehydration? Adaptations that reduce water loss are key. Insects, the prevalent arthropods on land, have tough exoskeletons impregnated with waterproof wax. Most terrestrial vertebrates, including humans, have an outer skin formed of multiple layers of dead, water-resistant cells. Also essential to survival on land are adaptations that protect fertilized eggs and developing embryos from drying out. Many insects lay their eggs in moist areas, and the eggs of many species are surrounded by a tough, watertight shell. Likewise, the embryos of amniotes—reptiles (including birds) and mammals—develop in a water-filled amniotic sac surrounded by protective membranes. Behavioral adaptations can also save water. For instance, many desert mammals and arthropods spend much of their time in moist burrows and venture out only at night. And, as we will see in Module 25.11, the kidney plays a major role in conserving water.

As terrestrial animals, we obtain our water from food and drink. Just how much water do we need to drink each day? We examine that question in the next module.

? Why are no freshwater animals osmoconformers?

■ Osmoconformers have solute concentrations equal to that of the environment. The body fluid of a freshwater osmoconformer would be too dilute to support life's processes.

25.5 Do we need to drink eight glasses of water each day?

The traditional wisdom has been that in addition to the water we get from food and other beverages, we need eight glasses of water a day for adequate hydration. This idea has fueled the popularity of bottled water, which now pulls in more than $8 billion a year in the United States. And indeed, students with their water bottles are a common sight on most campuses (Figure 25.5).

But the "eight glasses of water a day" adage is not based on scientific evidence. The Institute of Medicine, a national research panel, examined this notion and came to very different conclusions on water requirements. The panel determined that most people do not deliberately need to drink additional water. The average healthy person gets enough fluid from a mix of beverages—even those with caffeine—as well as the water content of food. For men, an average of 3.8 liters (L) a day is adequate; women need about 2.6 L. In the United States, men now get an average of 3.1 L a day from beverages, while women average 2.1 L. Most people absorb another 0.5–0.7 L from their food, with water-rich foods such as fruits and vegetables the best source. A tossed salad, for example, may contain a cup of water; a sandwich with lettuce and tomato provides almost one-half cup of water.

The scientific panel also refuted the common perception that caffeinated beverages remove water from the body. If these substances are consumed in moderate amounts, the body compensates to maintain adequate hydration. Thus, a cup of coffee can contribute to a person's total daily water intake. Small amounts of alcohol consumption also may not deprive the body of water, although larger quantities interfere with the hormones regulating water balance in the body, as we will discuss in Module 25.11.

Some people, such as athletes or those who live in hot climates, may need more water than others. As we saw in

Figure 25.5 Students maintaining hydration

Module 25.2, sweating is an important thermoregulatory mechanism, but it can also cause osmoregulatory problems. During heavy exercise, for example, fluid loss from sweating can exceed 2 L/hr. Drinking water before, during, and after exercise is important to maintaining proper hydration.

But most people, by letting their thirst guide their drinking and eating, absorb enough water. This instinct to drink is part of the body's system of osmoregulation. We examine these water-balancing processes in more detail next.

? How do most people meet their hydration needs?

■ From the beverages they drink during the day and from their food

Animals must dispose of nitrogenous wastes

Waste disposal is as important to homeostasis as water and solute balance. And because most wastes must be removed from the body dissolved in water, the type and quantity of wastes may have a large impact on water balance. Metabolism produces a number of toxic by-products, particularly the nitrogenous (nitrogen-containing) wastes that result from the breakdown of proteins and nucleic acids. An animal must dispose of (excrete) these metabolic wastes.

The form of an animal's nitrogenous wastes depends on its evolutionary history and its habitat. As **Figure 25.6** indicates, most aquatic animals dispose of their nitrogenous wastes as **ammonia**. Among the most toxic of all metabolic by-products, ammonia (NH_3) is formed when amino groups ($-NH_2$) are removed from amino acids and nucleic acids. It is too toxic to be stored in the body, but it is highly soluble in water and diffuses rapidly across cell membranes. If an animal is surrounded by water, ammonia readily diffuses out of its cells and body. Small, soft-bodied invertebrates, such as planarians (flatworms), excrete ammonia across their whole body surface. Fishes excrete it mainly across the gills.

Ammonia excretion does not work well for land animals. Ammonia does not diffuse readily into the air. Because it is so toxic, it must be transported and excreted in large volumes of very dilute solutions, and most terrestrial animals simply do not have access to that much water. Land animals convert ammonia into less toxic compounds, either urea or uric acid, that can be safely transported and stored in the body and released periodically by the excretory system. The disadvantage of excreting urea or uric acid is that the animal must use energy to produce these compounds from ammonia.

As shown in the figure, mammals, most amphibians, sharks, and some bony fishes excrete **urea**. Urea is produced in the vertebrate liver and transported by the circulatory system to the excretory organs, the kidneys. Urea is highly soluble in water. It is also some 100,000 times less toxic than ammonia, so it can be held in a concentrated solution in the body and disposed of with relatively little water loss. Some animals can switch between excreting ammonia and urea, depending on environmental conditions. Certain toads, for example, excrete ammonia (thus saving energy) when in water, but they excrete mainly urea (reducing excretory water loss) when on land.

Urea can be stored in a concentrated solution, but it still takes water to dispose of it. By contrast, land animals that excrete **uric acid** (birds and many other reptiles, insects, land snails, and a few amphibians living in deserts) avoid the water loss problem almost completely. As you can see in the figure, uric acid is a considerably more complex molecule than either urea or ammonia. Like urea, uric acid is relatively nontoxic. But unlike either ammonia or urea, uric acid is largely insoluble in water. In most cases,

it is excreted as a semisolid paste. (The white material in bird droppings is mostly uric acid.) An animal must expend more energy to excrete uric acid than urea, but the higher energy cost is balanced by the great savings in body water.

An animal's type of reproduction also influences whether it excretes urea or uric acid. Urea can diffuse out of a shell-less amphibian egg or be carried away from a mammalian embryo in the mother's blood. However, the shelled eggs produced by birds and other reptiles are not permeable to liquids. In these animals, natural selection apparently favored the use of uric acid, which can be stored in the egg as a harmless solid left behind when the animal hatches.

In the next module we will look more closely at the liver, the urea-synthesizing organ of vertebrates.

? Aquatic turtles excrete both urea and ammonia; land turtles excrete mainly uric acid. What could account for this difference?

■ Although uric acid evolved as a waste product in terrestrial reptiles with their shelled eggs, natural selection favored the energy savings of ammonia and urea for aquatic turtles.

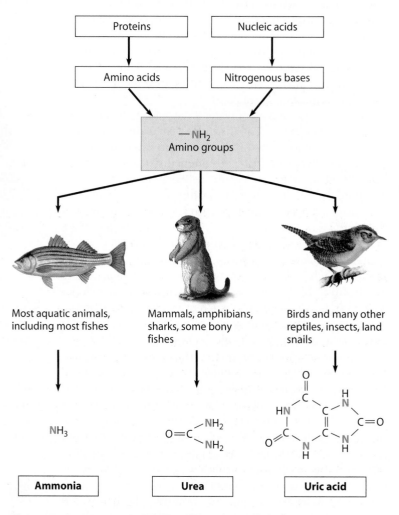

Figure 25.6 Nitrogen-containing metabolic waste products

25.7 The liver performs many functions, including the production of urea

A discussion of excretion and, indeed, of homeostasis in general would not be complete without mentioning the liver, which performs more functions than any other organ in the body. After breaking down amino acids for energy or for recycling into other molecules, the liver prepares the nitrogenous wastes (ammonia) for disposal by synthesizing urea. It also converts toxins such as alcohol and other drugs into inactive products that can be excreted in urine.

The liver can perform many functions because the metabolic machinery of its cells is so versatile. Liver cells synthesize bile for the digestion of fats (see Module 21.11), as well as proteins essential to many of the body's functions. Among these are plasma proteins important in blood clotting and in maintaining the osmotic balance of the blood, as well as lipoproteins that transport fats and cholesterol to body tissues. The liver also regulates the amount of glucose in the blood. One of its most important functions is to convert glucose into glycogen, which it stores for later use. In balancing the amount of glycogen it stores with the amount of glucose it releases to the blood, the liver plays a key role in regulating body metabolism. You will read about the hormonal control of this important function in Chapter 26.

The liver has a strategic location in the body—between the intestines and the heart. As indicated in Figure 25.7, capillaries from the small and large intestines converge into vessels that lead into the **hepatic portal vein.** This large vein transports nutrients absorbed by the intestines directly to the liver. Thus, the liver has a chance to modify and detoxify substances absorbed by the digestive tract before the blood carries these materials to the heart for distribution to the rest of the body.

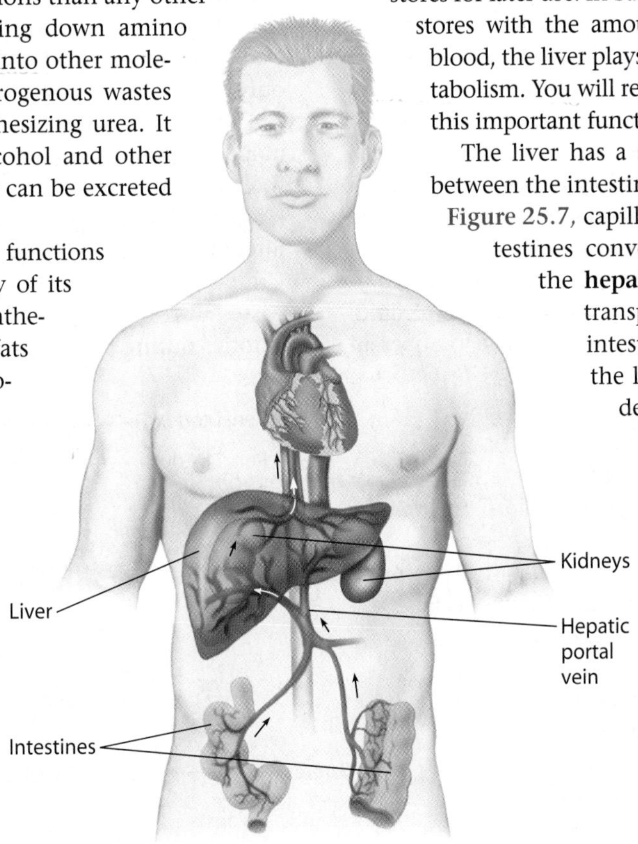

Kidneys

Liver

Hepatic portal vein

Intestines

Figure 25.7
The hepatic portal system

? What role does the liver play in the body's processing of nitrogenous waste?

■ The liver breaks down excess amino acids for energy or recycling and produces urea as the nitrogenous waste product.

25.8 Alcohol consumption can damage the liver

The liver's cleansing abilities do not protect it from toxins, such as alcohol, that it processes. Some breakdown products of alcohol are more toxic than alcohol itself, leading to liver cell damage, cell death, and tissue inflammation. Under heavy, frequent exposure to alcohol, this damage may become so severe that abnormal scar tissue forms in the liver. This condition, called cirrhosis, distorts the liver's internal structure. The scar tissue slowly replaces functioning liver tissue, gradually diminishing blood flow through the liver. This damage impairs the liver's many functions.

A person with cirrhosis of the liver often suffers from a variety of health problems, including a build-up of toxins in the blood, excessive bruising or bleeding, diabetes, and cancer. About 25,000 Americans die each year from alcohol-related cirrhosis of the liver. Other causes of cirrhosis include hepatitis, a viral disease that harms the liver. Liver damage from cirrhosis cannot be reversed, but treatment can stop or delay further progression. For people suffering from alcohol-related cirrhosis of the liver, the most effective treatment is to stop drinking. Some people with severe cirrhosis may require a liver transplant. The best way to prevent cirrhosis of the liver is to limit alcohol consumption. Current U.S. guidelines advise no more than two alcoholic drinks per day for men and one per day for women. (One drink is equal to 12 ounces of beer, 5 ounces of wine, or 1.5 ounces of distilled spirits.)

A healthy liver functions as a key part of the excretory system by producing urea and other wastes that the kidneys can remove from the blood and excrete in the urine. We will examine the excretory system in more detail next.

? What is cirrhosis of the liver?

■ Widespread scar tissue that disrupts the liver's function

25.9 The excretory system plays several major roles in homeostasis

Survival in any environment requires a precise balance between waste disposal and the animal's need for water. The human excretory system plays a central role in homeostasis, forming and excreting urine while regulating the amount of water and ions in the body fluids.

The main processing centers of our excretory system are the two kidneys. Each is a compact organ, about the size of your fist, nearly filled with about 80 km of fine tubes (tubules) and an intricate network of blood capillaries. The human body contains only about 5 L of blood, but since this blood circulates repeatedly, about 1,100–2,000 L pass through the capillaries in our kidneys every day. From this enormous traffic of blood, our kidneys extract daily about 180 L of fluid, called **filtrate**, consisting of water, urea, and a number of valuable solutes, including glucose, amino acids, ions, and vitamins. If we excreted all the filtrate as urine, we would lose vital nutrients and dehydrate rapidly. But our kidneys refine the filtrate, concentrating the urea and returning most of the water and solutes to the blood. In a typical day, we excrete only about 1.5 L of urine.

Figure 25.9 illustrates the "plumbing" plan and the blood supply of the human excretory system. Starting with the whole system in part A, blood to be filtered enters each kidney via a renal artery, shown in red; blood that has been filtered leaves the kidney in the renal vein, shown in blue. Urine leaves each kidney through a duct called a **ureter** and passes into the **urinary bladder**. Periodically, the bladder empties during urination. Urine leaves the body through a

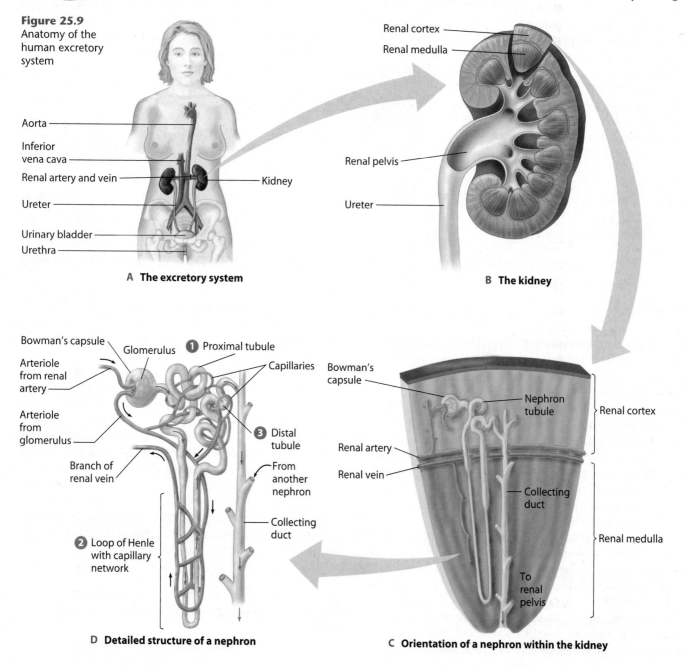

Figure 25.9
Anatomy of the human excretory system

Aorta

Inferior vena cava

Renal artery and vein

Ureter

Urinary bladder

Urethra

Kidney

A The excretory system

Renal cortex

Renal medulla

Renal pelvis

Ureter

B The kidney

Bowman's capsule

Glomerulus

Arteriole from renal artery

Arteriole from glomerulus

Branch of renal vein

❶ Proximal tubule

Capillaries

❸ Distal tubule

From another nephron

Collecting duct

❷ Loop of Henle with capillary network

D Detailed structure of a nephron

Bowman's capsule

Nephron tubule

Renal artery

Renal vein

Renal cortex

Collecting duct

Renal medulla

To renal pelvis

C Orientation of a nephron within the kidney

tube called the **urethra**, which empties near the female vagina or through the male penis.

As shown in part B, the kidney has two main regions, the **renal cortex** (outer layer) and the **renal medulla** (inner region). From the medulla, urine flows into a chamber called the renal pelvis, and from there into the ureter.

Each kidney contains about a million tiny functional units called **nephrons**, one of which is shown in part C. A nephron consists of a nephron tubule and its associated blood vessels. Performing the kidney's functions in miniature, the nephron extracts a tiny amount of filtrate from the blood and then refines the filtrate into a much smaller quantity of urine. Each nephron starts and ends in the kidney's cortex; some extend into the medulla, as in part C. The receiving end of the nephron is a cup-shaped swelling, called **Bowman's capsule**. At the other end of the nephron is the **collecting duct**, which carries urine to the renal pelvis.

Part D shows a nephron in more detail, along with its blood vessels. Bowman's capsule envelops a ball of capillaries called the **glomerulus** (plural, *glomeruli*). The glomerulus and Bowman's capsule make up the blood-filtering unit of the nephron. Here, blood pressure forces water and solutes from the blood in the glomerular capillaries across the wall of Bowman's capsule and into the nephron tubule. This process creates the filtrate, leaving blood cells and large molecules such as plasma proteins behind in the capillaries.

The rest of the nephron refines the filtrate. The tubule has three sections: ❶ the **proximal tubule** (in the cortex); ❷ the **loop of Henle**, a hairpin loop carrying filtrate toward—in some cases, into—the medulla and then back toward the cortex; and ❸ the **distal tubule** (called distal because it is the most distant from Bowman's capsule). The distal tubule drains into a collecting duct, which receives filtrate from many nephrons. From the kidney's many collecting ducts, the processed filtrate, or urine, passes into the renal pelvis and then into the ureter.

The intricate association between blood vessels and tubules is the key to nephron function. As shown in part D, the nephron has two distinct networks of capillaries. One network is the glomerulus, a finely divided portion of an arteriole that branches from the renal artery. Leaving the glomerulus, the arteriole re-forms and carries blood to the second capillary network, which surrounds the proximal and distal tubules. This second network functions with the tubule in refining the filtrate. Some of the vessels in this network parallel the loop of Henle, with blood flowing downward in one vessel and back up through another. Leaving the nephron, the capillaries converge to form a venule leading toward the renal vein.

With the structure of a nephron in mind, we focus next on what actually happens as our excretory system filters blood, refines the filtrate, and excretes urine.

Web/CD Activity 25A *Structure of the Human Excretory System*

[?] Place these parts of a nephron in the order in which filtrate moves through them: proximal tubule, Bowman's capsule, distal tubule, loop of Henle.

■ Bowman's capsule, proximal tubule, loop of Henle, distal tubule

---

25.10 Overview: The key processes of the excretory system are filtration, reabsorption, secretion, and excretion

Our excretory system produces and disposes of urine in four major processes, shown in Figure 25.10. First of all, during **filtration**, water and virtually all other molecules small enough to be forced through the capillary wall enter the nephron tubule from the glomerulus.

Two processes then refine the filtrate. In **reabsorption**, water and valuable solutes, including glucose, salt, other ions, and amino acids, are returned to the blood from the filtrate. In **secretion**, substances in the blood are transported into the filtrate. When there is an excess of H^+ in the blood, for example, these ions are secreted into the filtrate, thus keeping the blood from becoming acidic. Secretion also eliminates certain drugs and other toxic substances from the blood. In both reabsorption and secretion, water and solutes move between the tubule and capillaries by passing through the interstitial fluid (see Module 23.1).

Finally, in **excretion**, urine—the product of filtration, reabsorption, and secretion—passes from the kidneys to the outside via the ureters, urinary bladder, and urethra.

[?] Urine differs in composition from the fluid that enters a nephron tubule by filtration because of the processes of _____ and _____.

■ reabsorption . . . secretion

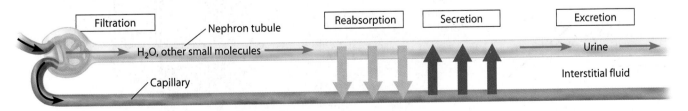

Figure 25.10 Major processes of the excretory system

25.11 From blood filtrate to urine: A closer look

Let's take a closer look at how a single nephron and collecting duct in the kidney produce urine from a blood filtrate that initially consists of a large amount of water and a number of valuable solutes, along with waste molecules.

The broad arrows in **Figure 25.11** indicate where reabsorption and secretion occur along the nephron tubule. The pink arrows pointing out of the tubule represent reabsorption, which may occur by active transport, passive diffusion, or osmosis. The blue arrows pointing into the tubule represent secretion. For simplicity, this figure omits the capillary network that surrounds the tubule.

The colored area of the figure represents the interstitial fluid, through which solutes and water move between the tubule and capillaries. The intensity of the color corresponds to the concentration of solutes in the interstitial fluid: The concentration is lowest in the cortex of the kidney and becomes progressively higher toward the inner medulla. We will see that it is by maintaining this solute gradient that the kidney can extract and save most of the water from the filtrate. All along the tubule, wherever you see water passing out of the filtrate into the interstitial fluid, the water is moving by osmosis. It does so because the solute concentration of the interstitial fluid exceeds that of the filtrate.

Let us first discuss the activities of the proximal and distal tubules. The proximal tubule actively transports nutrients such as glucose and amino acids from the filtrate into the interstitial fluid, to be reabsorbed into the capillaries. NaCl (salt) is reabsorbed from both the proximal and distal tubules. As NaCl is transported from the filtrate to the interstitial fluid, water follows by osmosis. Secretion of excess H^+ and reabsorption of HCO_3^- also occur in the proximal and distal tubules, helping to regulate the blood's pH. Potassium concentration in the blood is regulated by secretion of excess K^+ into the distal tubule. Drugs and poisons that were processed in the liver are secreted into the proximal tubule.

The loop of Henle and the collecting duct have one major function: water reabsorption. The long loop of Henle in the figure carries the filtrate deep into the medulla and then back to the cortex. The presence of NaCl and some urea in the interstitial fluid in the medulla maintains the high concentration gradient that increases water reabsorption by osmosis. Water leaves the tubule because the interstitial fluid in the medulla has a higher solute concentration than the filtrate. As soon as the water passes into the interstitial fluid, it moves into nearby blood capillaries and is carried away. This prompt removal is essential because the water would otherwise dilute the interstitial fluid surrounding the loop and destroy the concentration gradient necessary for water reabsorption.

Just after the filtrate rounds the hairpin turn in the loop of Henle, water reabsorption stops because the tubule there is impermeable to water. As the filtrate moves back toward the cortex, NaCl leaves the filtrate, first passively and then actively as the cells of the tubule pump NaCl into the interstitial fluid. It is primarily this movement of salt that creates the solute gradient in the interstitial fluid of the medulla.

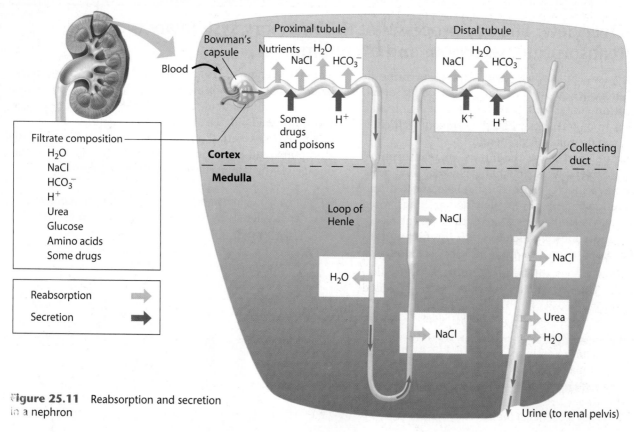

Figure 25.11 Reabsorption and secretion in a nephron

Final refining of the filtrate occurs in the collecting duct. By actively reabsorbing NaCl, the collecting duct is important in determining how much salt is excreted in the urine. In the inner medulla, the collecting duct becomes permeable to urea and some leaks out, adding to the high concentration gradient in the interstitial fluid. As the filtrate moves through the medulla, more water is reabsorbed before the urine passes into the renal pelvis. In sum, the nephron returns much of the water that filters into it from the blood. This water conservation is one of the major functions of the kidneys.

Our kidneys also maintain a precise and essential balance between water and solutes in our body fluids. When the solute concentration rises above a set point, a control center in the brain increases the blood level of **antidiuretic hormone (ADH)**, which signals the nephrons to step up water reabsorption. When the solute concentration is diluted below the set point, as when we drink a lot of water, blood levels of ADH drop and water reabsorption is reduced, resulting in an increased discharge of dilute urine. (Increased urination is called diuresis, and it is because ADH opposes this state that it is called *anti*diuretic hormone.) Alcohol inhibits the release of ADH and can cause excessive urinary water loss and dehydration, which may account for some of the symptoms of a hangover.

Still other hormones are involved in the kidney's regulation of blood volume and pressure. By adjusting both the flow of blood to the nephrons and the nephrons' reabsorption of Na^+ and water, these hormones alter the volume (and thus the pressure) of blood throughout the body.

We see that our kidneys' regulatory functions are controlled by an elaborate system of checks and balances. The coordination of all the body's regulatory systems by hormones is the subject of Chapter 26.

Web/CD Activity 25B *Nephron Function*

Web/CD Activity 25C *Control of Water Reabsorption*

Web/CD Thinking as a Scientist *What Affects Urine Production?*

? Some of the drugs classified as diuretics make the epithelium of the collecting duct less permeable to water. How would this affect kidney function?

■ The collecting ducts would reabsorb less water, and thus the diuretic would increase water loss in the urine.

25.12 Kidney dialysis can be a lifesaver

A person can survive with only one functioning kidney, but if both kidneys fail, the buildup of toxic wastes and the lack of regulation of blood pressure, pH, and ion concentrations will lead to certain death if untreated. Over 60% of kidney disease cases are caused by hypertension and diabetes, but the prolonged use of pain relievers, alcohol, and other drugs and medicines are also possible causes.

Knowing how the nephron works helps us understand how some of its functions can be performed artificially when the kidneys are damaged. **Figure 25.12** illustrates a type of artificial kidney, called a dialysis machine. **Dialysis** means "separation" in Greek. Like the nephrons of the kidney, the machine sorts small molecules of the blood, keeping some and discarding others. The patient's blood is pumped from an artery through a series of tubes made of a selectively permeable membrane. The tubes are immersed in a dialyzing solution much like the interstitial fluid that bathes the nephrons. As the blood circulates through the tubing, urea and excess ions diffuse out. Needed substances, such as bicarbonate ions, diffuse from the dialyzing solution into the blood. The machine continually discards the used dialyzing solution as wastes build up.

While dialysis is life sustaining, it is costly and time consuming (three times a week for 4–6 hours at a time). It also requires severe dietary and lifestyle restrictions. Many individuals benefit from a kidney transplant, from either a living donor (often a relative) or an organ donor who has just died. The waiting list for kidney transplants, unfortunately, is quite long.

? How would the composition of dialyzing solution compare with that of the patient's blood plasma?

■ Dialyzing solution has a solute concentration similar to interstitial fluid. The solution contains no urea, which allows urea from the patient's blood to diffuse into it.

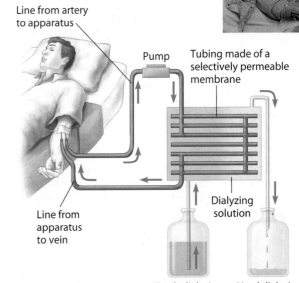

Figure 25.12 Kidney dialysis

Line from artery to apparatus

Pump

Tubing made of a selectively permeable membrane

Line from apparatus to vein

Dialyzing solution

Fresh dialyzing solution

Used dialyzing solution (with urea and excess ions)

Reviewing the Concepts

Homeostatic mechanisms. Thermoregulation maintains body temperature within a tolerable range. Endotherms derive body heat mainly from their metabolism; ectotherms absorb heat from their surroundings. Osmoregulation balances the gain and loss of water and solutes. Excretion disposes of nitrogenous wastes **(Introduction)**.

Thermoregulation (25.1–25.3)

Heat exchange with the environment may occur by conduction, convection, radiation, and evaporation **(25.1)**.

Adaptations for thermoregulation include increased metabolic heat production; insulation such as fur, feathers, and fat layers; circulatory adaptations such as adjusting blood flow to the skin or countercurrent heat exchange; evaporative cooling by sweating or panting; and behavioral responses such as moving to the sun or shade, migrating, or bathing **(25.2)**. Torpor, which includes hibernation in cold weather, is an inactive state in which a lowered metabolism saves energy **(25.3)**.

Osmoregulation and Excretion (25.4–25.12)

Osmoregulation. Osmoconformers, such as many marine invertebrates, have the same internal solute concentration as seawater. Osmoregulators control their solute concentrations. Freshwater fishes gain water by osmosis, excrete excess water, and pump in salt across their gills. Marine fishes lose water by osmosis, drink seawater, and excrete and pump out excess salt. Land animals gain water by drinking and eating; they lose it by evaporation and waste disposal. Their kidneys, waterproof skin, and reproductive and behavioral adaptations conserve water **(25.4)**. Thirst is usually an adequate guide to water intake **(25.5)**.

Nitrogenous wastes are toxic breakdown products of protein. Ammonia (NH_3) is poisonous but soluble and is easily disposed of by aquatic animals. Urea is less toxic and easier to store. Some land animals save water by excreting uric acid, a virtually dry waste. Urea and uric acid take energy to produce **(25.6)**.

Urea

The liver produces urea, breaks down toxins, produces bile, plasma proteins, and lipoproteins, and adjusts the nutrient content of the blood **(25.7)**. Liver function can be impaired by excessive alcohol consumption or hepatitis **(25.8)**.

The excretory system expels wastes and regulates water and ion balance. Nephrons, the functional units of kidneys, extract a filtrate from the blood and refine it to urine. Urine leaves the kidneys via the ureters, is stored in the urinary bladder, and is expelled through the urethra **(25.9)**. The key processes of urine formation are filtration (blood pressure forces water and many small solutes into the nephron); reabsorption (valuable solutes are re-

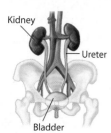
Kidney
Ureter
Bladder

claimed from the filtrate); secretion (excess H^+ and toxins are added to the filtrate); and excretion of urine **(25.10)**.

Filtrate to urine. Nutrients, salt, and water are reabsorbed from the proximal and distal tubules. Secretion of H^+ and reabsorption of HCO_3^- help regulate pH. High NaCl concentration in the medulla promotes reabsorption of water. Antidiuretic hormone (ADH) regulates the amount of water the kidneys excrete **(25.11)**. Compensating for kidney failure, a dialysis machine removes wastes from the blood and maintains its solute concentration **(25.12)**.

Connecting the Concepts

1. Complete this map, which presents the three main topics of this chapter.

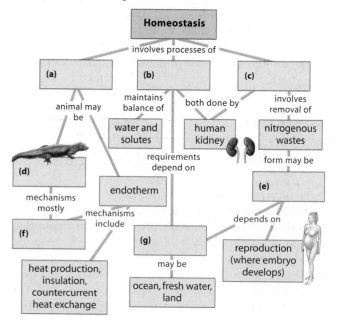

2. In this schematic of urine production in a nephron, label the four processes involved and list some of the substances that are moved in each process.

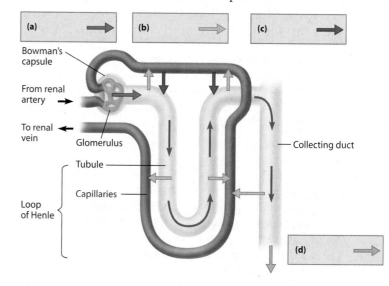

Testing Your Knowledge

Multiple Choice

3. The main difference between endotherms and ectotherms is
   a. how they conserve water.
   b. whether they are warm or cold.
   c. the source of most of their body heat.
   d. whether they live in a warm or cold environment.
   e. whether they maintain a fairly stable body temperature.

4. In each nephron of the kidney, the glomerulus and Bowman's capsule
   a. filter the blood and capture the filtrate.
   b. reabsorb water into the blood.
   c. break down harmful toxins and poisons.
   d. reabsorb ions and nutrients.
   e. refine and concentrate the urine for excretion.

5. As filtrate passes through the loop of Henle, salt is removed and concentrated in the interstitial fluid of the medulla. This high concentration enables nephrons to
   a. excrete the maximum amount of salt.
   b. neutralize toxins that might be found in the kidney.
   c. control the pH of the interstitial fluid.
   d. excrete a large amount of water.
   e. reabsorb water very efficiently.

6. Birds and insects excrete uric acid, while mammals and most amphibians excrete mainly urea. What is the chief advantage of uric acid over urea as a waste product?
   a. Uric acid is more soluble in water.
   b. Uric acid is a much simpler molecule.
   c. It takes less energy to make uric acid.
   d. Less water is required to excrete uric acid.
   e. More solutes are removed excreting uric acid.

7. Which process in the nephron is least selective?
   a. secretion
   b. reabsorption
   c. filtration
   d. active transport of salt
   e. passive diffusion of salt

8. All of the following are functions of the liver except
   a. detoxification of drugs and toxins.
   b. synthesis of plasma (blood) proteins.
   c. interconversion of glucose and glycogen.
   d. synthesis of urine.
   e. synthesis of bile (which is stored in the gallbladder).

9. A freshwater fish would be expected to
   a. pump salt out through its gills.
   b. produce copious quantities of dilute urine.
   c. diffuse urea out through the gills.
   d. have scales and a covering of mucus that reduce water loss to the environment.
   e. do all of the above.

10. Which of the following is not an adaptation for reducing the rate of heat loss to the environment?
    a. feathers or fur
    b. reducing blood flow to surface blood vessels
    c. contraction of flight muscles before a moth takes off
    d. countercurrent heat exchanger
    e. thick layer of fat

Matching

Match each of the following components of blood with what happens to it as the blood is processed by the kidney.

11. Water
12. Glucose
13. Plasma protein
14. Toxins or drugs
15. Red blood cell
16. Urea

a. passes into filtrate; almost all excreted in urine
b. remains in blood
c. passes into filtrate; mostly reabsorbed
d. secreted and excreted

Describing, Comparing, and Explaining

17. Compare the problems of water and salt regulation a salmon faces when it is swimming in the ocean and when it migrates into fresh water to spawn.

18. Can ectotherms have stable body temperatures? Explain.

Applying the Concepts

19. Assuming equal size, which of these organisms would produce the greatest amount of nitrogenous wastes? Explain.
    a. An endotherm or an ectotherm?
    b. A carnivore or an herbivore (assume both are endotherms)?

20. You are studying a large tropical reptile that has a high and relatively stable body temperature. How would you determine whether this animal is an endotherm or an ectotherm?

21. Riding by a lake in midwinter, you notice a small flock of geese standing on the ice. Imagine what it would be like for you to stand there with no boots or warm pants. Propose a hypothesis to explain why the birds' legs do not freeze. You may assume you have equipment for measuring temperatures in the birds' legs. What results would you expect if your hypothesis is correct?

22. The kidneys remove many drugs from the blood, and these substances show up in the urine. Some employers require a urine drug test at the time of hiring and/or at intervals during the term of employment. Why do some employers feel that drug testing is necessary? Do you think that passing a drug test is a valid criterion for employment? If so, for what types of jobs? Would you take a drug test to get or keep a job? Why or why not?

23. Kidneys were the first organs to be successfully transplanted. A donor can live a normal life with a single kidney, making it possible for individuals to donate a kidney to an ailing relative or even an unrelated individual. In some countries, poor people sell kidneys to transplant recipients through organ brokers. What are some of the ethical issues associated with organ commerce?

Answers to all questions can be found in Appendix 3.

For study help and Activities, go to campbellbiology.com or the student CD-ROM.

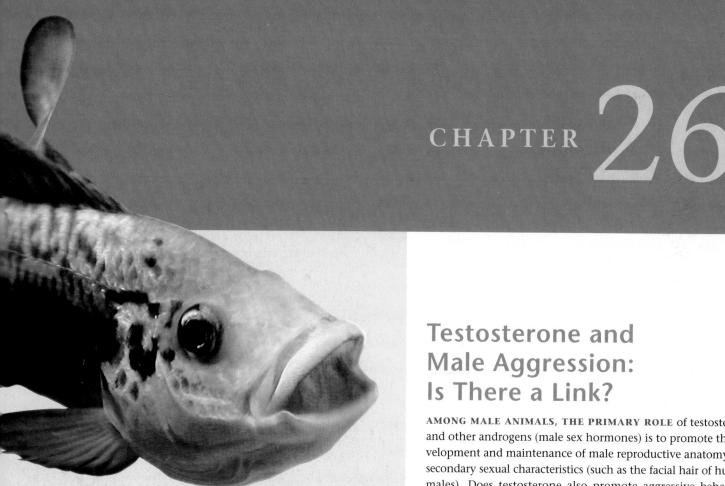

Testosterone and Male Aggression: Is There a Link?

AMONG MALE ANIMALS, THE PRIMARY ROLE of testosterone and other androgens (male sex hormones) is to promote the development and maintenance of male reproductive anatomy and secondary sexual characteristics (such as the facial hair of human males). Does testosterone also promote aggressive behavior? While the link between testosterone levels and aggression in humans remains questionable, such a connection has been conclusively established in other animal species.

Consider the cichlid fish *Oreochromis mossambicus,* pictured at left. Also called the Mozambique tilapia, this native of eastern Africa now lives in warm rivers in many parts of the world, as well as in hobbyists' aquariums. During the breeding season, the males fight fiercely with other males that enter their mating territories. Researchers have measured androgens in male cichlids in the laboratory and have found elevated levels in males engaged in territorial battles; the victors tend to be those with higher levels. A recent study has shown that androgens surge even in male *spectators* of these fights. This finding, in combination with results from other studies, suggests that the increased androgens sharpen the bystanders' alertness and readiness for fighting.

Cichlids are not the only animal in which an individual's androgen production responds to certain social interactions with other group members. However, it is difficult to generalize across species about a link between androgens and aggression. The correlation is most obvious in species where males defend territories (as cichlids do) or fight for access to females.

Establishing connections between androgens and human male aggression is especially difficult. How should we define aggression in humans? Should we base it on psychological testing or reserve the term for actual fighting? Even the latter criterion is problematic because of the many variables involved in human interactions. As a case in point, researchers have found a correlation between high androgen levels and fighting among male prison inmates, but incarceration is such an abnormal situation, with so many other factors coming into play, that the

THE NATURE OF CHEMICAL REGULATION

26.1 Chemical signals coordinate body functions
26.2 Hormones affect target cells by two main signaling mechanisms

THE VERTEBRATE ENDOCRINE SYSTEM

26.3 Overview: The vertebrate endocrine system
26.4 The hypothalamus, closely tied to the pituitary, connects the nervous and endocrine systems

HORMONES AND HOMEOSTASIS

26.5 The thyroid regulates development and metabolism
26.6 Hormones from the thyroid and parathyroids maintain calcium homeostasis
26.7 Pancreatic hormones regulate blood glucose levels
26.8 Diabetes is a common endocrine disorder
26.9 The adrenal glands mobilize responses to stress
26.10 Glucocorticoids offer relief from pain, but not without serious risks
26.11 The gonads secrete sex hormones

Chemical Regulation

applicability of these results to males in general is questionable. Moreover, several other hormones are also implicated in aggression, such as epinephrine (adrenaline). All these complications help explain why the results of research on androgens and human male aggression vary widely from study to study. If a consensus among researchers exists, it is that within the wide normal range of human testosterone levels, higher levels of the hormone do not lead directly to higher levels of aggression.

What do we know about the effects of androgens in humans? Besides their roles in the development of the male reproductive anatomy, they also influence sexual arousal in both males and females. (Women's ovaries produce small amounts of testosterone.) There is also evidence correlating androgen levels with spatial ability (visualizing objects accurately in three dimensions). While demonstrating clear links to violent aggression is prob-

lematic, researchers have made connections to competition in a broader sense. Testosterone levels in men usually rise before a competition, whether boxing or chess. Afterward, the testosterone level remains high in the winner and falls in the loser.

The hormone testosterone is one of a number of chemical signals that have multiple effects in the body, as we will see in this chapter. The overarching role of hormones is to coordinate activities in different parts of the body, enabling the organ systems to function cooperatively. Hormones regulate our most basic bodily functions, such as energy use, metabolism, and growth.

This chapter is specifically about hormones and other kinds of chemical signals. However, our general theme is homeostasis, with a focus on how chemical signals maintain an animal body's dynamic steady state. We begin on the next page with a look at the main kinds of chemicals that regulate body functions and the cells that secrete them. ■ ■ ■

Does testosterone play a role in aggressive behavior?

26.1 Chemical signals coordinate body functions

Animals rely on many kinds of chemical signals to regulate their body activities. Testosterone is an example of one kind, a hormone (from the Greek *hormon,* to excite). An animal **hormone** is a chemical signal that is carried by the circulatory system (usually in the blood) and that communicates regulatory messages within the body. Hormones are made and secreted mainly by organs called **endocrine glands**. Collectively, all of an animal's hormone-secreting cells constitute its **endocrine system**, the body's main chemical-regulating system.

Because most hormones are carried in the blood, they reach all parts of the body, and the endocrine system is thus especially important in controlling whole-body activities. For example, hormones coordinate responses to stimuli such as stress, dehydration, or low blood glucose levels. Hormones also regulate long-term developmental processes, such as growth, maturation, and reproduction.

Figure 26.1A sketches the activity of a hormone-secreting cell. Secretory vesicles in the endocrine cell are full of molecules of the hormone (blue). The endocrine cell secretes the molecules directly into the circulatory system. From there, hormones may travel to all parts of the body, but only certain types of cells, called **target cells**, are equipped to respond. A single hormone molecule may dramatically alter a target cell's metabolism by turning on or off the production of a number of enzymes. A tiny amount of a hormone can govern the activities of enormous numbers of target cells in a variety of organs. (In the next module, we'll look at *how* hormones trigger responses in their target cells.) A hormone is ignored by other types of cells (nontarget cells).

Chemical signals play a major role in coordinating the functioning of all animals. Hormones are the body's long-distance chemical regulators and convey information via the bloodstream to target cells throughout the body. Other chemical signals—**local regulators**—are secreted into the interstitial fluid and affect only nearby target cells. Still other chemical signals, called **pheromones**, carry messages between different individuals of a species, as in mate attraction.

The endocrine system often collaborates with the body's other system of internal communication and regulation: the nervous system (the subject of Chapter 28). The nervous system transmits electrical signals via nerve cells. These rapid messages control split-second responses. The flick of a frog's tongue catching a fly and the jerk of your hand away from a flame result from high-speed nerve signals. The endocrine system, in contrast, coordinates slower but longer-lasting responses. In many cases, the endocrine system takes hours or even days to act, partly because of the time it takes for hormones to be made and transported to all their target organs.

While it is convenient to distinguish between the endocrine and nervous systems, in reality the lines between these two regulatory systems are blurred. In particular, certain specialized nerve cells called **neurosecretory cells** perform functions of both systems (Figure 26.1B). Like all nerve cells, neurosecretory cells conduct nerve signals, but they also make and secrete hormones into the blood.

A few chemicals serve both as hormones in the endocrine system and as chemical signals in the nervous system. Epinephrine (adrenaline), for example, functions in vertebrates as the "fight-or-flight" hormone (so called because it prepares the body for sudden action) and as a neurotransmitter. A **neurotransmitter** is a chemical that carries information from one nerve cell to another or from a nerve cell to another kind of cell that will react, such as a muscle cell or an endocrine cell. When a nerve signal reaches the end of a nerve cell, it triggers the secretion of neurotransmitter molecules (Figure 26.1C). Unlike most hormones, however, neurotransmitters usually do not travel in the bloodstream. We discuss neurotransmitters further in Chapter 28. For the rest of this chapter, we'll explore the endocrine system and its hormones.

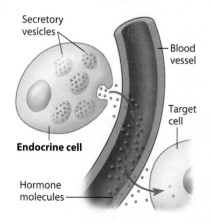

Figure 26.1A Hormone from an endocrine cell

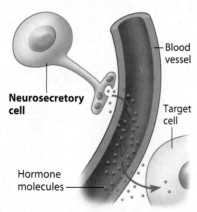

Figure 26.1B Hormone from a neurosecretory cell

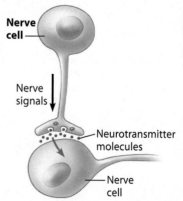

Figure 26.1C Neurotransmitter

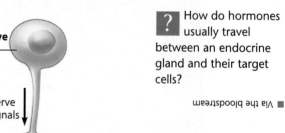

? How do hormones usually travel between an endocrine gland and their target cells?

■ Via the bloodstream

26.2 Hormones affect target cells by two main signaling mechanisms

Three major classes of molecules function as hormones in vertebrates: proteins and peptides, amines, and steroids. Proteins and peptides (small polypeptides containing 3 to 30 amino acids) and amines (molecules derived from amino acids) are water-soluble. Steroid hormones are lipid-soluble. (For examples of each type, see Table 26.3.)

Regardless of their chemical structure, however, signaling by any of these molecules involves three key events: reception, signal transduction, and response. *Reception* of the signal occurs when a hormone binds to a specific receptor protein on or in the target cell. Each signal molecule has a specific shape that can be recognized by its target cell receptors. The binding of a signal molecule to a receptor protein triggers events within the target cell—*signal transduction*—that result in a *response*, a change in the cell's behavior. Cells that lack receptors for a particular chemical signal do not respond to that signal.

While both water- and lipid-soluble hormones carry out the three key steps outlined above, they do so by different mechanisms. We'll now take a closer look at how each type of hormone elicits cellular responses.

The receptors for most water-soluble hormones are embedded in the plasma membrane of target cells and project outward from the cell surface (**Figure 26.2A**). ❶ A hormone molecule binds to the receptor protein, activating it. ❷ This initiates a signal transduction pathway, a series of changes in cellular proteins (relay molecules) that converts an extra-cellular chemical signal to a specific intracellular response. ❸ The final relay molecule activates a protein that carries out the cell's response, either in the cytoplasm or in the nucleus. One hormone may trigger a variety of responses in target cells because the cells may contain different receptors for that hormone, diverse signal transduction pathways, or several proteins that can carry out the response.

While water-soluble hormones bind to receptors in the plasma membrane, steroid hormones bind to receptors *inside* the cell. **Steroid hormones,** such as the sex hormones testosterone and estrogen, are lipids made from cholesterol (see Module 3.9). Steroid hormones (as well as a few other hormones we'll discuss later) are small, nonpolar molecules that can diffuse through the phospholipid membranes of cells. As shown in **Figure 26.2B,** ❶ a steroid hormone enters a cell in this way. If the cell is a target cell, the hormone ❷ binds to a receptor protein in the cytoplasm or nucleus. Rather than triggering a signal transduction pathway, the hormone-receptor complex itself usually carries out the transduction of the hormonal signal: The complex acts as a transcription factor—a gene activator (see Module 11.6). ❸ The hormone-receptor complex attaches to specific sites on the cell's DNA in the nucleus. (These sites are enhancers; see Module 11.6.) ❹ The binding of the hormone-receptor complex to DNA stimulates transcription of certain genes into RNA, which is translated into new proteins. All steroid hormones act by turning genes on or off.

Because a hormone can bind to a variety of receptors in various kinds of target cells, different kinds of cells can respond differently to the same hormone. The main effect of epinephrine on heart muscle cells, for example, is cellular contraction, which speeds up the heartbeat; its main effect on liver and muscle cells, however, is glycogen breakdown, providing glucose (an energy source) to body cells.

Web/CD Activity 26A *Overview of Cell Signaling*

Web/CD Activity 26B *Nonsteroid Hormone Action*

Web/CD Activity 26C *Steroid Hormone Action*

? What are two major differences between the mechanisms of action of steroid and nonsteroid hormones?

■ (1) Steroid hormones bind to receptors inside the cell; most other hormones bind to plasma membrane receptors. (2) Steroid hormones always affect gene expression; other hormones have this or other effects.

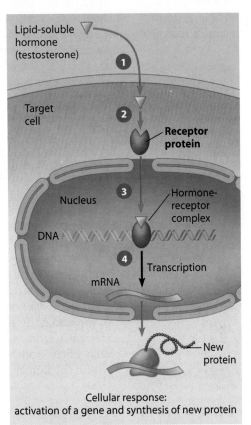

Figure 26.2A A hormone that binds a plasma-membrane receptor

Figure 26.2B A hormone that binds an intracellular receptor

CHAPTER 26 *Chemical Regulation* **521**

26.3 Overview: The vertebrate endocrine system

The vertebrate endocrine system consists of more than a dozen major glands. Some of these, such as the thyroid and the pituitary gland, are endocrine specialists; their sole or main function is secreting hormones into the blood. Several other glands have both endocrine and nonendocrine functions. The pancreas, for example, secretes hormones that influence the level of glucose in the blood and also secretes digestive enzymes into the intestine via ducts (see Module 21.11). Still other organs, such as the stomach, are primarily nonendocrine but have some cells that secrete hormones.

Figure 26.3 shows the locations of the major human endocrine glands. Table 26.3 summarizes the actions of the main hormones they produce, as well as how the glands themselves are regulated. The table provides an overview of the human endocrine system, and you may wish to refer to it as we focus on the individual glands and their hormones in later modules. (Keep in mind, however, that this chapter only covers the major endocrine glands and their hormones; there are other hormone-secreting structures—the heart, liver, and stomach, for example—and other hormones that will not be discussed in this chapter.)

Notice the distribution of the three chemical classes of hormone (proteins/peptides, amines, and steroids) in Table 26.3. Only the sex organs and the cortex of the adrenal gland produce steroid hormones, the main type of hormone that actually enters target cells. Most of the endocrine glands produce water-soluble hormones, which generally bind to plasma membrane receptors and act via signal transduction.

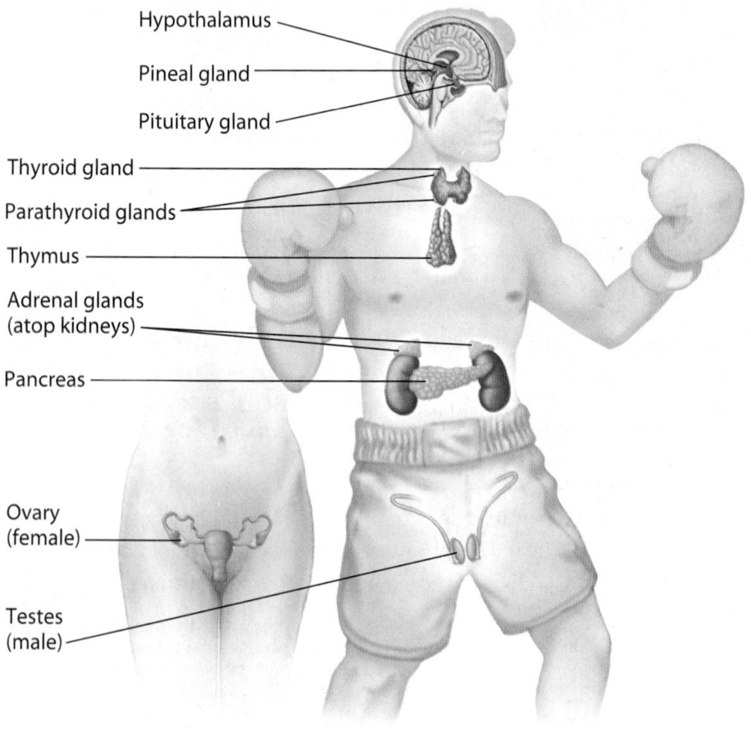

Hormones have a wide range of targets. Some, like the sex hormones, which promote male and female characteristics, affect most of the tissues of the body. Other hormones, such as glucagon from the pancreas, have only a few kinds of target cells (liver and fat cells for glucagon). Some hormones have other endocrine glands as their targets. For example, the pituitary gland produces thyroid-stimulating hormone, which promotes activity of the thyroid gland.

The close association between the endocrine system and the nervous system is apparent in both Figure 26.3 and Table 26.3. For example, the hypothalamus, which is part of the brain, secretes many hormones that regulate other endocrine glands, especially the pituitary. We'll explore structural and functional connections between the endocrine system and the nervous system further in Module 26.4.

Endocrine glands that we will not be discussing in later modules include the pineal gland and the thymus. Much remains to be learned about both of these organs. The **pineal gland** is an outgrowth of the brain that secretes melatonin, a hormone that links environmental light conditions with daily or seasonal rhythms. (In mammals, the cells that detect light for this purpose are in the eye, intermingled with the cells used for vision.) We know the most about melatonin's function in mammals that breed during certain seasons. For example, in sheep and deer that breed in the fall, when days are short, high levels of melatonin in the blood stimulate reproductive activity. In contrast, in mammals that breed in the spring, longer days and less melatonin in the blood promote reproductive activity. We do not yet know exactly what effects melatonin has on the body cells that produce these rhythms.

The **thymus gland** lies under the breastbone in humans and is quite large during childhood. Not until the 1960s was the important role of the thymus in the immune system discovered. Thymus cells secrete several important hormones, including a peptide that stimulates the development of T cells (see Module 24.5). Beginning at puberty, when the immune system has been well established, the thymus shrinks drastically. However, it continues to secrete its T-cell–stimulating hormones throughout life.

In the remainder of this chapter, we will explore several of the endocrine glands listed in Figure 26.3 and Table 26.3. Our discussion will focus on the hormones produced by each organ and how they help to maintain homeostasis within the human body.

Hypothalamus
Pineal gland
Pituitary gland
Thyroid gland
Parathyroid glands
Thymus
Adrenal glands (atop kidneys)
Pancreas
Ovary (female)
Testes (male)

? Of the glands listed in Table 26.3, which ones secrete lipid-soluble hormones?

Figure 26.3 The major endocrine glands in humans

■ The testes, ovaries, and adrenal cortex

TABLE 26.3 MAJOR HUMAN ENDOCRINE GLANDS AND SOME OF THEIR HORMONES

Gland	Hormone	Chemical Class	Representative Actions	Regulated by
Hypothalamus	Hormones released by the posterior pituitary and hormones that regulate the anterior pituitary (see below)			
Pituitary gland Posterior lobe (releases hormones made by hypothalamus)	Oxytocin	Peptide	Stimulates contraction of uterus and mammary gland cells	Nervous system
	Antidiuretic hormone (ADH)	Peptide	Promotes retention of water by kidneys	Water/salt balance
Anterior lobe	Growth hormone (GH)	Protein	Stimulates growth (especially bones) and metabolic functions	Hypothalamic hormones
	Prolactin (PRL)	Protein	Stimulates milk production	Hypothalamic hormones
	Follicle-stimulating hormone (FSH)	Protein	Stimulates production of ova and sperm	Hypothalamic hormones
	Luteinizing hormone (LH)	Protein	Stimulates ovaries and testes	Hypothalamic hormones
	Thyroid-stimulating hormone (TSH)	Protein	Stimulates thyroid gland	Thyroxine in blood; hypothalamic hormones
	Adrenocorticotropic hormone (ACTH)	Protein	Stimulates adrenal cortex to secrete glucocorticoids	Glucocorticoids; hypothalamic hormones
Pineal gland	Melatonin	Amine	Involved in rhythmic activities (daily and seasonal)	Light/dark cycles
Thyroid gland	Thyroxine (T_4) and triiodothyronine (T_3)	Amine	Stimulate and maintain metabolic processes	TSH
	Calcitonin	Peptide	Lowers blood calcium level	Calcium in blood
Parathyroid glands	Parathyroid hormone (PTH)	Peptide	Raises blood calcium level	Calcium in blood
Thymus	Thymosin	Peptide	Stimulates T cell development	Not known
Adrenal glands Adrenal medulla	Epinephrine and norepinephrine	Amine	Increase blood glucose; increase metabolic activities; constrict certain blood vessels	Nervous system
Adrenal cortex	Glucocorticoids	Steroid	Increase blood glucose	ACTH
	Mineralocorticoids	Steroid	Promote reabsorption of Na^+ and excretion of K^+ in kidneys	K^+ in blood
Pancreas	Insulin	Protein	Lowers blood glucose	Glucose in blood
	Glucagon	Protein	Raises blood glucose	Glucose in blood
Testes	Androgens	Steroid	Support sperm formation; promote development and maintenance of male secondary sex characteristics	FSH and LH
Ovaries	Estrogens	Steroid	Stimulate uterine lining growth; promote development and maintenance of female secondary sex characteristics	FSH and LH
	Progesterone	Steroid	Promotes uterine lining growth	FSH and LH

The hypothalamus, closely tied to the pituitary, connects the nervous and endocrine systems

The distinction between the endocrine system and the nervous system often blurs, especially when we consider the diverse roles of the hypothalamus and its intricate association with the pituitary gland. As part of the brain (**Figure 26.4A**), the **hypothalamus** receives information from nerves about the internal condition of the body and about the external environment. It then responds to these conditions by sending out appropriate nervous or endocrine signals. Its hormonal signals directly control the pituitary gland, which in turn secretes hormones that influence numerous body functions. The hypothalamus thus exerts master control over the endocrine system by using the pituitary to relay directives to other glands.

As Figure 26.4A shows, the **pituitary gland** consists of two distinct parts: a posterior lobe and an anterior lobe, both situated in a pocket of skull bone at the base of the hypothalamus. The **posterior pituitary** is composed of nervous tissue and is actually an extension of the hypothalamus. It stores and secretes two hormones made in the hypothalamus. In contrast, the **anterior pituitary** is composed of endocrine cells that synthesize and secrete numerous hormones directly into the blood. Several of these hormones control the activity of other endocrine glands. The hypothalamus exerts control over the anterior pituitary by secreting two kinds of hormones—releasing hormones and inhibiting hormones—into short blood vessels that connect the two organs. **Releasing hormones** stimulate the anterior pituitary to secrete hormones, while **inhibiting hormones** induce the anterior pituitary to stop secreting hormones.

Figures 26.4B and 26.4C, on the facing page, emphasize the structural and functional connections between the hypothalamus and the pituitary. As **Figure 26.4B** indicates, a set of neurosecretory cells extends from the hypothalamus into the posterior pituitary. These cells synthesize the peptide hormones **oxytocin** and **antidiuretic hormone (ADH)**. These hormones (blue triangles) are channeled along the neurosecretory cells into the posterior pituitary. When released into the blood from the posterior pituitary, oxytocin causes uterine muscles to contract during childbirth and mammary glands to eject milk during nursing. ADH helps cells of the kidney tubules reabsorb water, thus decreasing urine volume when the body needs to retain water (see Module 25.11). When the body has too much water, the hypothalamus responds to negative feedback, slowing the release of ADH from the posterior pituitary.

Figure 26.4C shows a second set of neurosecretory cells in the hypothalamus. These cells secrete releasing and inhibiting hormones (red dots) that control the anterior pituitary. A system of small blood vessels carries these hormones from the hypothalamus to the anterior pituitary. In response to hypothalamic releasing hormones, the anterior pituitary synthesizes and releases many different peptide and protein hormones (blue triangles), which influence a broad range of body activities. **Thyroid-stimulating hormone (TSH)**, **adrenocorticotropic hormone (ACTH)**, **follicle-stimulating hormone (FSH)**, and **luteinizing hormone (LH)** all activate other endocrine glands. Feedback mechanisms control the secretion of these hormones by the anterior pituitary.

Figure 26.4D shows how the hypothalamus operates through the anterior pituitary to direct the activity of another endocrine organ, the thyroid gland. The hypothalamus secretes a releasing hormone known as **TRH (TSH-releasing hormone)**. In turn, TRH stimulates the anterior pituitary to produce TSH. Under the influence of TSH, the thyroid secretes the hormone thyroxine into the blood. **Thyroxine** increases the metabolic rate of most body cells, warming the body as a result.

Precise regulation of the TRH-TSH-thyroxine system keeps the hormones at levels that maintain homeostasis. The hypothalamus takes some cues from the environment; for instance, cold temperatures tend to increase its secretion of TRH. In addition, as the red arrows in Figure 26.4D indicate, negative-feedback mechanisms control the secretion of thyroxine. When thyroxine increases in the blood, it acts on the hypothalamus and anterior pituitary, inhibiting TRH and TSH secretion and consequently thyroxine synthesis.

Of all the pituitary secretions, none has a broader effect than the protein called **growth hormone (GH)**. GH promotes protein synthesis and the use of body fat for energy metabolism in a wide variety of target cells. In young mammals, GH promotes the development and enlargement of all parts of the body. Abnormal production of GH can result in several human disorders. Too much GH during childhood, usually due to a pituitary tumor, can lead to gigantism. Excessive production of GH in adulthood, a condition known as acromegaly, stimulates bony growth in the face, hands, and feet. In contrast, too little GH in childhood can lead to pituitary dwarfism. Administer-

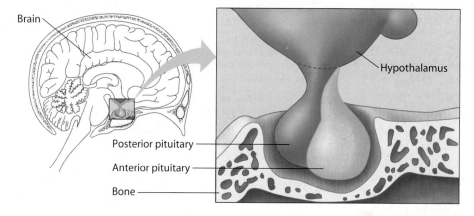

Figure 26.4A Location of the hypothalamus and pituitary

Brain
Hypothalamus
Posterior pituitary
Anterior pituitary
Bone

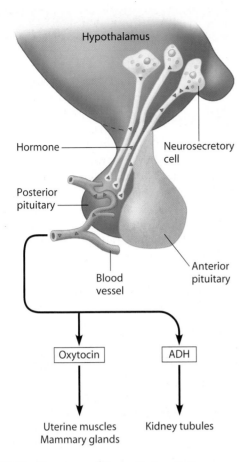

Figure 26.4B Hormones of the posterior pituitary

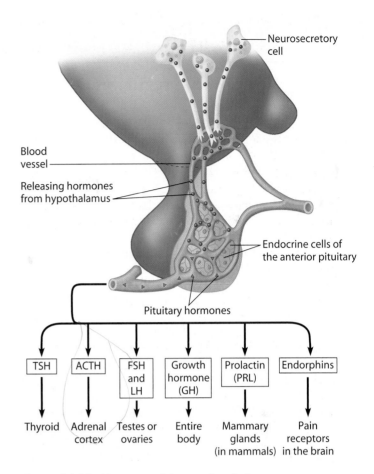

Figure 26.4C Hormones of the anterior pituitary

ing growth hormone to children with GH deficiency can successfully prevent dwarfism. Once extracted only in minute quantities from the pituitary glands of cadavers, human GH is now produced in large amounts by bacteria modified to carry the human GH gene. Unfortunately, its increased availability has caused some athletes to abuse human GH to bulk up their muscles. Such abuse is extremely dangerous and can lead to disfigurement, heart failure, and multiple cancers.

Another anterior pituitary hormone, **prolactin (PRL)**, produces very different effects in different species. In mammals, it stimulates mammary glands to produce milk; in birds, it controls fat metabolism and reproduction; in amphibians, it regulates larval development and delays metamorphosis; and in freshwater fishes, it regulates salt and water balance. These diverse effects suggest that prolactin is an ancient hormone whose functions diversified during vertebrate evolution.

The **endorphins**, hormones produced by the anterior pituitary as well as the brain, are the body's natural painkillers. These chemical signals bind to receptors in the brain and dull the perception of pain. The effect on the nervous system is similar to that of the drug morphine, earning endorphins the nickname "natural opiates" (although it might be more accurate to call opiates "artificial endorphins"!). Some researchers speculate that the so-called "runner's high" results partly from the release of endorphins when stress and

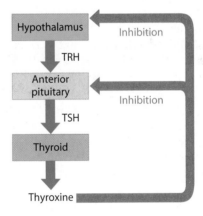

Figure 26.4D Control of thyroxine secretion

pain in the body reach critical levels. It has also been suggested that endorphins may be released during deep meditation, by acupuncture treatments, or even by eating very spicy foods.

? Alcohol inhibits secretion of ADH by the anterior pituitary. Predict how this action of alcohol would affect urination.

■ Alcohol increases the volume of urine produced.

26.5 The thyroid regulates development and metabolism

Your **thyroid gland** is located just under your larynx (voice box). Thyroid hormones affect virtually all the tissues of vertebrate animals.

The thyroid produces two very similar amine hormones, both of which contain the element iodine. One of these, **thyroxine**, is often called T_4 because it contains four iodine atoms; the other, **triiodothyronine**, is called T_3 because it contains three iodine atoms.

T_3 and T_4 have essentially the same effects on their target cells. One of their crucial roles is in development and maturation. In a bullfrog, for example, they trigger the profound reorganization of body tissues that occurs as a tadpole—a strictly aquatic organism—transforms into an adult frog, which may spend much of its time on land. Thyroid hormones are equally important in mammals, especially in bone and nerve cell development. In humans, a thyroid deficiency known as cretinism results in retarded skeletal growth and poor mental development.

The thyroid gland has several important homeostatic functions. For example, T_3 and T_4 help maintain normal blood pressure, heart rate, muscle tone, digestion, and reproductive function. Throughout the body, these hormones tend to increase the rate of oxygen consumption and cellular metabolism. Too much or too little of these hormones in the blood can result in serious metabolic disorders. An excess of T_3 and T_4 in the blood (*hyper*thyroidism) can make a person overheat, sweat profusely, become irritable, develop high blood pressure, and lose weight. The most common form of hyperthyroidism is Graves disease; protruding eyes are a typical symptom (Figure 26.5A). Conversely, insufficient amounts of T_3 and T_4 (*hypo*thyroidism) can cause weight gain, lethargy, and intolerance to cold.

Hypothyroidism can result from a defective thyroid gland or from dietary

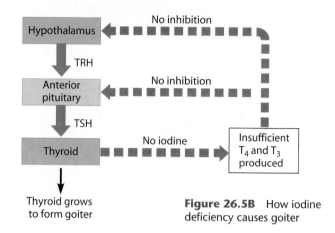

Figure 26.5B How iodine deficiency causes goiter

Hypothalamus → TRH → Anterior pituitary → TSH → Thyroid → Thyroid grows to form goiter

No inhibition — No inhibition — No iodine — Insufficient T_4 and T_3 produced

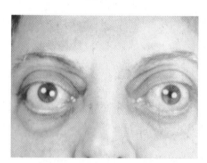

Figure 26.5A Graves disease, a form of hyperthyroidism

disorders. For example, severe iodine deficiency during childhood can cause cretinism. And in adults, insufficient iodine in the diet can cause **goiter**, an enlargement of the thyroid (see Figure 2.2A). In such cases, the thyroid gland cannot synthesize adequate amounts of its T_3 and T_4 hormones. The lack of T_3 and T_4 interrupts the feedback loops that control thyroid activity (**Figure 26.5B**). The blood never carries enough of the T_3 and T_4 hormones to shut off the secretion of TRH (TSH-releasing hormone) or TSH. The thyroid enlarges because TSH continues to stimulate it.

Fortunately, both hypo- and hyperthyroidism can be successfully treated. For example, many cases of goiter can be improved simply by adding iodine to the diet. Seawater is a rich source of iodine, and goiter rarely occurs in people living near the seacoast, where the soil is iodine-rich and a lot of seafood is consumed. Goiter is less common today than in the past thanks to the incorporation of iodine into table salt, but it still affects thousands of people in developing nations.

Web/CD Thinking as a Scientist *How Do Thyroxine and TSH Affect Metabolism?*

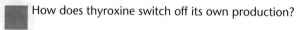

How does thyroxine switch off its own production?

■ By negative feedback: It inhibits the secretion of TRH from the hypothalamus and TSH from the pituitary.

26.6 Hormones from the thyroid and parathyroids maintain calcium homeostasis

An appropriate level of calcium in the blood and interstitial fluid is essential for many body functions. Without calcium, nerve signals cannot be transmitted from cell to cell, muscles cannot function properly, blood cannot clot, and cells cannot transport molecules across their membranes. The thyroid and parathyroid glands function in the homeostasis of calcium

ions (Ca^{2+}), keeping the concentration of the ions within a narrow range (about 10 mg of Ca^{2+} per 100 mL of blood).

There are four **parathyroid glands**, all embedded in the surface of the thyroid. Two peptide hormones, **calcitonin** from the thyroid gland and **parathyroid hormone (PTH)**, secreted by the parathyroids, regulate blood calcium levels.

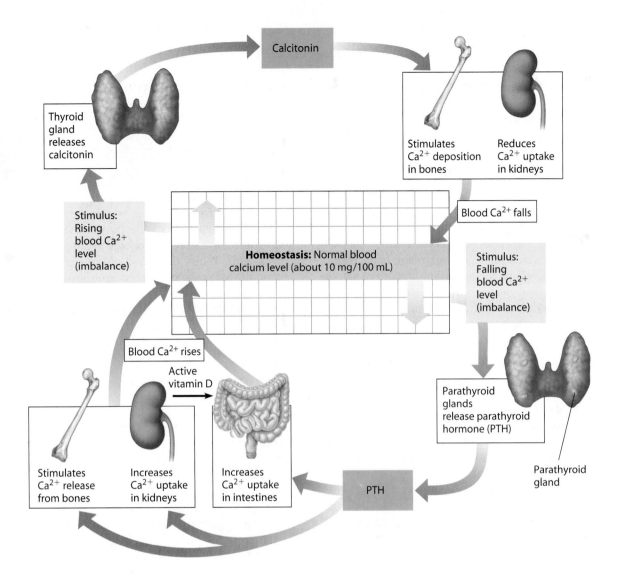

Figure 26.6 Calcium homeostasis

Calcium and PTH are said to be **antagonistic hormones** because they have opposite effects: Calcitonin lowers the calcium level in the blood, whereas PTH raises it. As **Figure 26.6** indicates, these two antagonistic hormones operate by means of feedback systems that keep the calcium level near the homeostatic set point of 10 mg/100 mL. To read the diagram, start with the tan box on the right and follow the arrows to the bottom part of the figure.

When the blood Ca^{2+} level drops below 10 mg/100 mL, the parathyroids release PTH into the blood. PTH stimulates the release of calcium ions from bones and increases Ca^{2+} reabsorption by the kidneys. The kidneys also play an indirect role in calcium homeostasis, which involves vitamin D. We obtain this vitamin in inactive form from food and also from chemical reactions in our skin when it is exposed to sunlight. Transported in the blood, inactive vitamin D undergoes sequential steps of activation in the liver and kidneys. The active form of vitamin D, secreted by the kidneys, acts as a hormone. It stimulates the intestines to increase uptake of Ca^{2+} from food. The result is a higher Ca^{2+} level in the blood.

Starting from the tan box on the left, now follow the top part of the diagram to see how calcitonin from the thyroid gland reverses the effects of PTH. A rise in blood Ca^{2+} above the set point induces the thyroid gland to secrete calcitonin. Calcitonin, in turn, has two main effects: It causes more Ca^{2+} to be deposited in the bones, and it makes the kidneys reabsorb less Ca^{2+} as they form urine. The result is a lower Ca^{2+} level in the blood.

In summary, a sensitive balancing system maintains calcium homeostasis. The system depends on feedback control by two antagonistic hormones. Failure of the system can have far-reaching effects in the body. For example, a shortage of PTH causes the blood calcium level to drop dramatically, leading to convulsive contractions of the skeletal muscles. This condition, known as tetany, can be fatal.

? In the control of calcium ion levels in the blood by calcitonin and PTH, what are the two main target organs of the hormones?

■ Bones and kidneys

26.7 Pancreatic hormones regulate blood glucose levels

The **pancreas** produces two hormones that play a large role in managing the body's energy supplies. Clusters of endocrine cells, called islets of Langerhans, are scattered throughout the pancreas. Each islet has a population of beta cells, which produce the hormone **insulin**, and a population of alpha cells, which produce another hormone, **glucagon**. Insulin and glucagon—both protein hormones—are secreted directly into the blood.

As shown in **Figure 26.7**, insulin and glucagon are antagonistic hormones that regulate the concentration of glucose in the blood. The two hormones counter each other in a feedback circuit that precisely manages the amount of circulating glucose available as cellular fuel versus the amount of glucose stored as the polymer glycogen in body cells. By negative feedback, the concentration of glucose in the blood determines the relative amounts of insulin and glucagon secreted by the islet cells. In the top half of the diagram, you see what

happens when the glucose concentration of the blood rises above the set point of about 90 mg of glucose per 100 mL of blood, as it does shortly after we eat a carbohydrate-rich meal. The rising blood glucose level (tan box on the left) stimulates the beta cells in the pancreas to secrete more insulin. The insulin stimulates nearly all body cells to take up glucose from the blood, thereby decreasing the blood glucose level. Liver cells (and skeletal muscle cells) take up much of the glucose and use it to form glycogen, which they store. Insulin also stimulates cells to metabolize the glucose for immediate energy use, for the storage of energy in fats, or for the synthesis of proteins. When the blood glucose level falls to the set point, the beta cells lose their stimulus to secrete insulin.

Following the bottom half of the diagram, you see what happens when the blood glucose level starts to dip below the set point (tan box on the right), as it may between meals or during strenuous exercise. The pancreatic alpha cells respond

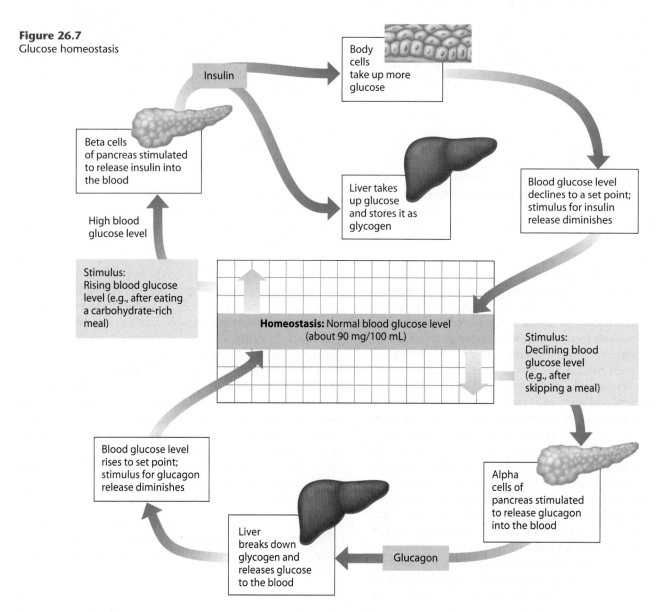

Figure 26.7
Glucose homeostasis

Insulin

Body cells take up more glucose

Beta cells of pancreas stimulated to release insulin into the blood

High blood glucose level

Liver takes up glucose and stores it as glycogen

Blood glucose level declines to a set point; stimulus for insulin release diminishes

Stimulus: Rising blood glucose level (e.g., after eating a carbohydrate-rich meal)

Homeostasis: Normal blood glucose level (about 90 mg/100 mL)

Stimulus: Declining blood glucose level (e.g., after skipping a meal)

Blood glucose level rises to set point; stimulus for glucagon release diminishes

Liver breaks down glycogen and releases glucose to the blood

Glucagon

Alpha cells of pancreas stimulated to release glucagon into the blood

by secreting more glucagon. Glucagon is a fuel mobilizer, signaling liver cells to break glycogen down into glucose, convert amino acids and fat-derived glycerol to glucose, and release the glucose into the blood. Then, when the blood glucose level returns to the set point, the alpha cells slow their secretion of glucagon.

In the next module, we see what can happen when this delicately balanced system breaks down.

? How is the insulin-glucagon relationship similar to the calcitonin-PTH relationship?

■ In both cases, the two hormones are antagonists that help maintain homeostasis by counteracting one another's effects. Their actions keep the blood concentration of a key chemical (glucose for insulin-glucagon; calcium ions for calcitonin-PTH) near the set point.

CONNECTION

26.8 Diabetes is a common endocrine disorder

Diabetes mellitus is a serious hormonal disease in which the body cells are unable to absorb glucose from the blood. It affects about 18 million Americans—6% of the total population—and an estimated 5 million of them do not even know they are ill. Diabetes develops when there is not enough insulin in the blood or when body cells do not respond normally to blood insulin. In either case, the cells cannot obtain enough glucose from the blood, and thus, starved for fuel, they are forced to burn the body's supply of fats and proteins. Meanwhile, since the digestive system continues to absorb glucose from the diet, the glucose concentration in the blood can become extremely high—so high, in fact, that measurable amounts of glucose are excreted in the urine. (Normally, the kidney leaves no glucose in the urine.)

There are treatments for diabetes mellitus—insulin supplements and/or special diets—but no cure. Untreated diabetes can cause dehydration, blindness, cardiovascular and kidney disease, and nerve damage. Every year, almost 300,000 Americans die from the disease or its complications.

There are two types of diabetes mellitus. Type 1 (insulin-dependent) diabetes is an autoimmune disease, in which white blood cells (T cells) of the body's own immune system attack and destroy the pancreatic beta cells. As a result, the pancreas does not produce enough insulin, and glucose builds up in the blood. Type 1 diabetes generally develops during childhood. Treatment consists of injections of human insulin—produced by genetically engineered bacteria—several times daily.

Type 2 (non-insulin-dependent) diabetes is characterized either by a deficiency of insulin or, more commonly, by reduced responsiveness of target cells due to some change in insulin receptors. Type 2 diabetes is almost always associated with being overweight and underactive, although whether obesity causes diabetes (and, if so, how) remains unknown. This form of diabetes generally appears after age 40, but even young people who are overweight and inactive can develop the disease. In the U.S., more than 90% of diabetics are type 2. Many of them can manage their blood glucose with regular exercise and a healthy diet high in soluble fiber and low in fat and sodium; some require medications.

How is diabetes detected? Early signs of either type include a lack of energy, a craving for sweets, frequent urination, and

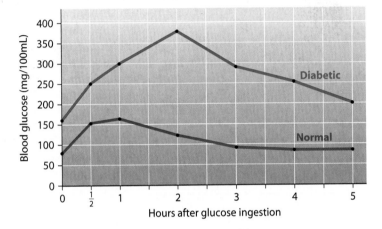

Figure 26.8 Results of glucose tolerance tests

persistent thirst. The diagnostic test for diabetes is a glucose tolerance test: The person swallows a sugar solution and then has blood drawn at prescribed time intervals. Each blood sample is tested for glucose. In **Figure 26.8**, you can compare the glucose tolerance of a person with diabetes with that of a healthy individual. A healthy body can maintain a more constant concentration of blood glucose.

Diabetes is not the only disease that can result from problems with insulin. Some people have hyperactive beta cells that put too much insulin into the blood when sugar is eaten. As a result, their blood glucose level can drop well below normal. This condition, called **hypoglycemia**, usually occurs 2–4 hours after a meal and may be accompanied by hunger, weakness, sweating, and nervousness. In severe cases, when the brain receives inadequate amounts of glucose, a person may develop convulsions, become unconscious, and even die. Hypoglycemia is not common, and most forms of it can be controlled by reducing sugar intake and eating more frequently, in smaller amounts.

? Three hours after glucose ingestion, the person with diabetes whose test is shown in the graph above has a blood glucose concentration about __ times that of the normal individual.

■ 3

The adrenal glands mobilize responses to stress

The human body has two **adrenal glands** sitting atop the kidneys. As you can see in the inset at the far left of Figure 26.9, each adrenal gland is actually made up of two glands fused together: a central portion called the **adrenal medulla** and an outer portion called the **adrenal cortex**. Though the cells they contain and the hormones they produce are different, both the medulla and the cortex secrete hormones that enable the body to respond to stress.

The adrenal medulla produces the "fight-or-flight" hormones, which ensure a rapid, short-term response to stress. You've probably felt your heart beat faster and your skin develop goose bumps when sensing danger or approaching a stressful situation, such as entering a final exam. Positive emotions—extreme pleasure, for instance—can produce the same effects. These reactions are triggered by two amine hormones secreted by the adrenal medulla, **epinephrine** (adrenaline) and **norepinephrine** (noradrenaline).

Stressful stimuli, whether negative or positive, activate certain nerve cells in the hypothalamus. As indicated on the left side of the diagram, these cells send nerve signals via the spinal cord to the adrenal medulla, stimulating it to secrete epinephrine and norepinephrine into the blood. Norepinephrine and epinephrine have somewhat different effects

on tissues, but both contribute to the short-term stress response. Both hormones stimulate liver cells to release glucose, thus making more fuel available for cellular work. They also prepare the body for action by raising the blood pressure, breathing rate, and metabolic rate. In addition, epinephrine and norepinephrine change blood-flow patterns, making some organs more active and others less so. For example, epinephrine dilates blood vessels in the brain and skeletal muscles, thus increasing alertness and the muscles' ability to react to stress. At the same time, epinephrine and norepinephrine constrict blood vessels elsewhere, thereby reducing activities that are not immediately involved in the stress response, such as digestion. The short-term stress response occurs and subsides rapidly.

In contrast to hormones from the adrenal medulla, those secreted by the adrenal cortex can provide a slower, longer-lasting response to stress. The adrenal cortex responds to endocrine signals—chemical signals in the blood—rather than to nerve cell signals. As the right side of the diagram indicates, the hypothalamus secretes a releasing hormone that stimulates target cells in the anterior pituitary to secrete the hormone adrenocorticotropic hormone (ACTH). In turn, ACTH stimulates cells of the adrenal cortex to synthesize

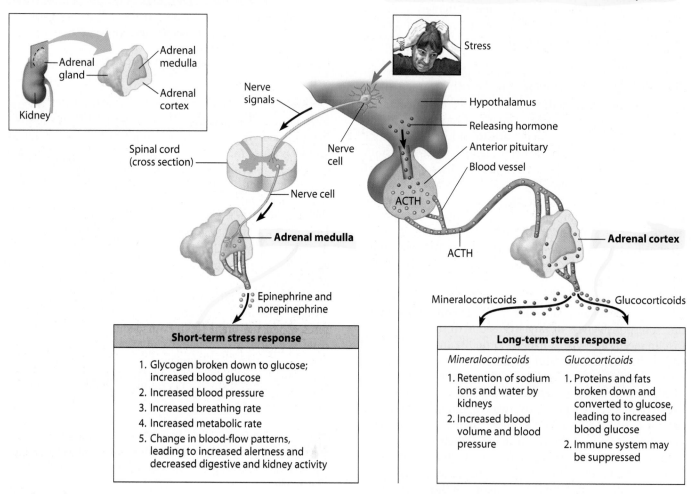

Figure 26.9 How the adrenal glands control our responses to stress

and secrete a family of steroid hormones called **cortico-steroids.** The two main types in humans are the mineralo-corticoids and the glucocorticoids. Both help maintain homeostasis when the body experiences long-term stress.

Mineralocorticoids act mainly on salt and water bal-ance. One of these hormones (aldosterone) stimulates the kidneys to reabsorb sodium ions and water, with the overall effect of increasing the volume of the blood and raising blood pressure as a response to prolonged stress.

Glucocorticoids function mainly in mobilizing cellular fuel, thus reinforcing the effects of glucagon. Glucocorticoids promote the synthesis of glucose from noncarbohydrates such as proteins and fats. When the body cells consume more glucose than the liver can provide from glycogen stores, glucocorticoids stimulate the breakdown of muscle proteins, making amino acids available for conversion to glu-cose by the liver. This makes more glucose available in the blood as cellular fuel in response to stress.

Very high levels of glucocorticoids can suppress the body's defense system, including the inflammatory response that occurs at infection sites (see Module 24.2). For this reason, physicians may use glucocorticoids to treat diseases in which excessive inflammation is a problem. The glucocorticoid cor-tisone, for example, was once regarded as a miracle drug for treating serious inflammatory conditions such as arthritis. Cortisone and other glucocorticoids can relieve swelling and pain from inflammation; but by suppressing immunity, they can also make a person highly susceptible to infection. We discuss some other dangers of glucocorticoids next.

[?] How would a deficiency of receptors in the hypothalamus for adrenal steroids affect levels of those hormones in the blood? (*Hint:* Apply to adrenal steroids what you learned in Module 26.5 about the regulation of thyroxine.)

■ This deficiency would cause abnormally high levels of adrenal steroids.

26.10 Glucocorticoids offer relief from pain, but not without serious risks

Pain is often part of a professional athlete's life, and few are better acquainted with it than Bill Walton, former basketball superstar for UCLA and the NBA's Portland Trail Blazers and Boston Celtics and current TV sportscaster (**Figure 26.10**). Walton was born with high arches and a malformed left foot. Running or jumping usually hurt, but for years he ac-cepted the pain as part of his heavy workouts.

In 1977, Walton, a 6-foot 11-inch, 225-pound center, had been with the Blazers three years and led them to the National Basketball Association championship. But Walton's stardom with the Blazers was all too brief. Midway through the next season, he was sidelined with painful injuries. Following the team physician's advice, he started taking gluco-corticoids and other painkillers so he could stay active. During the 1978 play-offs, primed with oral doses of dexa-methasone, a glucocorticoid, and several other drugs, he played in two games. Though limping badly, Walton scored 27 points and got 22 rebounds. The morning after the second game, X-rays showed he had been playing with a frac-tured bone in the arch of his left foot.

Walton's superstar days were over. Amid a storm of media at-tention, he made several comebacks in the 1980s, but he was never his former self on the basketball court. Nevertheless, in recognition of his early triumphs, Walton was elected to the Basketball Hall of Fame in 1993.

Physicians often prescribe oral glucocorticoids to relieve pain from athletic injuries. However, glucocorticoids are po-tentially very dangerous; taking them orally for more than five days can depress the activity of the adrenal glands and may cause psychological changes. Taking them for more than a few weeks can increase the risk of diabetes, eye problems, and bone fractures. It is safer, but still potentially dangerous, to inject a glucocorticoid at the site of an injury. With this treatment, the pain usually subsides, but its underlying cause remains. Masking the pain covers up the pain's message—that tissue is damaged and may get worse if not allowed to heal. If an athlete exercises an injured site before the tissue has recov-ered, the added stress can cause more serious damage.

Bill Walton's case was complicated. It never was firmly es-tablished that glucocorticoids worsened his condition be-cause he had been playing with foot pain and may have seriously injured his foot before he started using painkillers. One physician contends that Walton fractured the same bone in his left foot four times during his basketball career. An important outcome of Walton's plight was the wide-spread attention it drew, making more people aware of the potential dangers of painkilling drugs.

Figure 26.10 Bill Walton as a professional athlete (in green), and today as a sportscaster

[?] Some patients who are treated with high doses of gluco-corticoids to reduce inflammation of joints have difficulty fighting infections. Explain this side effect.

■ The very same action that reduces inflammation also suppresses the body's defenses against disease.

26.11 The gonads secrete sex hormones

The sex hormones are steroid hormones that affect growth and development and also regulate reproductive cycles and sexual behavior. The **gonads**, or sex glands (ovaries in the female and testes in the male), secrete sex hormones, in addition to producing gametes.

The gonads of mammals produce three major categories of sex hormones: estrogens, progestins, and androgens. Both females and males have all three types, but in different proportions. Females have a high ratio of estrogens to androgens. **Estrogens** maintain the female reproductive system and promote the development of such female features as the generally smaller body size, higher-pitched voice, breasts, and wider hips. **Progestins**, such as progesterone, are primarily involved in preparing and maintaining the uterus to support the embryo, at least in mammals.

In general, **androgens** stimulate the development and maintenance of the male reproductive system. Males have a high ratio of androgens to estrogens, their main androgen being **testosterone**. In humans, androgens produced by male embryos during the seventh week of development stimulate the embryo to develop into a male rather than a female. During puberty, high concentrations of androgens trigger the development of male characteristics, such as a lower-pitched voice, facial hair, and large skeletal muscles. Androgens have somewhat different effects in different animals. In the chapter introduction, we discussed the role of androgens in the behavior of male cichlid fish. In elephant seals (**Figure 26.11**), male androgens produce bodies weighing 2 tons and more, an inflatable enlargement of the nasal cavity, a thick hide that can withstand bloody conflicts, and aggressive behavior toward other males. The two males in the photo are fighting. One will establish dominance over the other and the right to mate with many females.

The synthesis of sex hormones by the gonads is regulated by the hypothalamus and anterior pituitary. In response to a releasing factor from the hypothalamus, the anterior pituitary secretes follicle-stimulating hormone (FSH) and luteinizing hormone (LH). These stimulate the ovaries or testes to synthesize and secrete the sex hormones, among other effects. We examine the complex effects of these hormones when we focus on human reproduction in the next chapter.

Figure 26.11 Male elephant seals in combat

Web/CD Activity 26D *Human Endocrine Glands and Hormones*

? Estrogens are to _____ as _____ are to testes.

ovaries . . . androgens ■

CHAPTER REVIEW

Reviewing the Concepts

The Nature of Chemical Regulation (26.1–26.2)

Hormones are chemical signals, usually carried in the blood, that cause specific changes in target cells. Endocrine glands and neurosecretory cells secrete hormones. All hormone-secreting cells make up the endocrine system, which works with the nervous system in regulating body activities **(26.1)**. Hormones trigger changes in target cells by two general mechanisms **(26.2)**:

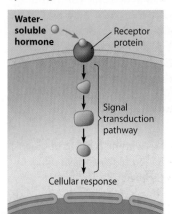
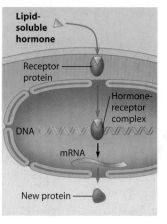

The Vertebrate Endocrine System (26.3–26.4)

The vertebrate endocrine system consists of more than a dozen glands secreting more than 50 hormones. Some glands are specialized for hormone secretion only; some also do other jobs. Some hormones have a very narrow range of targets and effects; others have numerous effects on many kinds of target cells **(26.3)**.

The hypothalamus and pituitary:

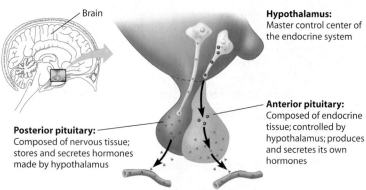

Hypothalamus: Master control center of the endocrine system

Anterior pituitary: Composed of endocrine tissue; controlled by hypothalamus; produces and secretes its own hormones

Posterior pituitary: Composed of nervous tissue; stores and secretes hormones made by hypothalamus

Releasing and inhibiting hormones from the hypothalamus control the secretion of several other hormones **(26.4)**.

Hormones and Homeostasis (26.5–26.11)

Vertebrate hormones. Two amine hormones from the thyroid gland, T_4 and T_3, regulate an animal's development and metabolism. Negative feedback maintains homeostatic levels of T_4 and T_3 in the blood. Thyroid imbalance can cause disease **(26.5)**. Blood calcium level is regulated by a tightly balanced antagonism between calcitonin from the thyroid and parathyroid hormone from the parathyroid glands **(26.6)**. The pancreas secretes two hormones, insulin and glucagon, that control blood glucose levels. Insulin signals cells to use and store glucose. Glucagon causes cells to release stored glucose into the blood **(26.7)**. Diabetes mellitus results from a lack of insulin or a failure of cells to respond to it **(26.8)**. Hormones from the adrenal glands help maintain homeostasis when the body is stressed. Nerve signals from the hypothalamus stimulate the adrenal medulla to secrete epinephrine and norepinephrine, which quickly trigger the fight-or-flight response. ACTH from the pituitary causes the adrenal cortex to secrete glucocorticoids and mineralocorticoids, which boost blood pressure and energy in response to long-term stress **(26.9)**. Glucocorticoids relieve inflammation and pain, but they can mask injury and suppress immunity **(26.10)**. Estrogens, progestins, and androgens are steroid sex hormones produced by the gonads in response to signals from the hypothalamus and pituitary. Estrogens and progestins stimulate the development of female characteristics and maintain the female reproductive system. Androgens, such as testosterone, trigger the development of male characteristics **(26.11)**.

Connecting the Concepts

Match each hormone (top) with the gland where it is produced (center) and its effect on target cells (bottom).

1. thyroxine
2. epinephrine
3. androgens
4. insulin
5. melatonin
6. FSH
7. PTH
8. ADH

 Pineal gland

Testes

 Parathyroid gland

Adrenal medulla

 Hypothalamus

Pancreas

 Anterior pituitary

Thyroid gland

a. lowers blood glucose
b. stimulates ovaries
c. triggers fight-or-flight
d. promotes male traits
e. regulates metabolism
f. related to daily rhythm
g. raises blood calcium level
h. boosts water retention

Testing Your Knowledge

Multiple Choice

9. Which of the following controls the activity of all the others?
   a. thyroid gland
   b. pituitary gland
   c. adrenal cortex
   d. hypothalamus
   e. ovaries

10. The pancreas increases its output of insulin in response to
    a. an increase in body temperature.
    b. changing cycles of light and dark.
    c. a decrease in blood glucose.
    d. a hormone secreted by the anterior pituitary.
    e. an increase in blood glucose.

11. Which of the following hormones have antagonistic (opposing) effects?
    a. parathyroid hormone and calcitonin
    b. glucagon and thyroxine
    c. growth hormone and epinephrine
    d. ACTH and cortisone
    e. epinephrine and norepinephrine

12. The body is able to maintain a relatively constant level of thyroxine in the blood because
    a. thyroxine stimulates the pituitary to secrete thyroid-stimulating hormone (TSH).
    b. thyroxine inhibits the secretion of TSH-releasing hormone (TRH) from the hypothalamus.
    c. TRH inhibits the secretion of thyroxine by the thyroid gland.
    d. thyroxine stimulates the hypothalamus to secrete TRH.
    e. thyroxine stimulates the pituitary to secrete TRH.

13. Which of the following hormones has the broadest range of targets?
    a. ADH
    b. oxytocin
    c. TSH
    d. epinephrine
    e. ACTH

Describing, Comparing, and Explaining

14. Explain how the hypothalamus controls body functions through its action on the pituitary gland. How does control of the anterior and posterior pituitary differ?

15. Explain how the same hormone might have different effects on two different target cells and no effect on a third type of cell.

Applying the Concepts

16. A strain of transgenic mice remains healthy as long as you feed them regularly and do not let them exercise. After they eat, their blood glucose level rises slightly and then declines to a homeostatic level. However, if these mice fast or exercise at all, their blood glucose drops dangerously. Which hypothesis best explains their problem? (*Explain your choice.*)
    a. The mice have insulin-dependent diabetes.
    b. The mice lack insulin receptors on their cells.
    c. The mice lack glucagon receptors on their cells.
    d. The mice cannot synthesize glycogen from glucose.

17. How could a hormonal imbalance result in a person who is genetically male but physically female?

Answers to all questions can be found in Appendix 3.

For study help and Activities, go to campbellbiology.com or the student CD-ROM.

CHAPTER 27

ASEXUAL AND SEXUAL REPRODUCTION

27.1 Sexual and asexual reproduction are both common among animals

HUMAN REPRODUCTION

27.2 Reproductive anatomy of the human female
27.3 Reproductive anatomy of the human male
27.4 The formation of sperm and ova requires meiosis
27.5 Hormones synchronize cyclic changes in the ovary and uterus
27.6 The human sexual response occurs in four phases
27.7 Sexual activity can transmit disease
27.8 Contraception can prevent unwanted pregnancy

PRINCIPLES OF EMBRYONIC DEVELOPMENT

27.9 Fertilization results in a zygote and triggers embryonic development
27.10 Cleavage produces a ball of cells from the zygote
27.11 Gastrulation produces a three-layered embryo
27.12 Organs start to form after gastrulation
27.13 Changes in cell shape, cell migration, and programmed cell death give form to the developing animal
27.14 Embryonic induction initiates organ formation
27.15 Pattern formation organizes the animal body

HUMAN DEVELOPMENT

27.16 The embryo and placenta take shape during the first month of pregnancy
27.17 Human development from conception to birth is divided into three trimesters
27.18 Childbirth is hormonally induced and occurs in three stages
27.19 Reproductive technology increases our reproductive options

Baby Bonanza

IF YOU WALK THROUGH A BUSY PARK ON A SUNNY DAY, chances are you'll have to share the footpath with proud parents pushing two babies in a double-wide stroller. That's because in 2002, over 125,000 twins were born in the United States, more than in any previous year. Adding to that, over 7,000 "supertwins"—three or more children born at one time—also entered the world. Indeed, multiple birth rates have been on the rise for decades. Between 1980 and 2002, the rate of twin births in the United States rose by 83%. The rate of supertwin births rose even more dramatically during this period: over 550%. Multiple pregnancies, multiple births, and hence multiple strollers have suddenly become much more common.

What is the cause of this remarkable baby bonanza? One answer: the increased use of fertility drugs. **Infertility**, defined as the inability to bear children after one year of trying, affects about one in seven American couples. Many have sought medical assistance to conceive children. Although infertility is frequently traced to problems in the male reproductive system, such problems are rarely treated with medications. There are, however, several drugs that can effectively treat female infertility, which is often caused by a failure to ovulate. Ovulation, the release of a mature egg during the monthly ovarian cycle, is essential for conception and is controlled by several hormones. By altering the levels of one or more of these hormones in the female, fertility drugs have allowed thousands of infertile couples to have babies.

Fertility drugs have been prescribed in the United States for over 30 years and are relatively safe and effective. However, they are sometimes too effective: They can promote the release of multiple eggs during a single ovulation. Each one of these eggs may then be fertilized (that is, fuse with a sperm to form a zygote), resulting in multiple embryos. Over 10% of women taking fertility drugs become pregnant with more than one embryo—sometimes quite a few more. Indeed, fertility drugs are responsible for some extraordinary multiple births, such as the "Iowa septuplets" (shown at right)—four boys and three girls, born in 1997—the world's first surviving septuplets.

Reproduction and Embryonic Development

Despite the successful births now enjoyed by previously infertile couples, multiple births carry some risk. Newborns from multiple births are premature more often than babies from single births. They also have lower birth weights, are less likely to survive (the mortality rate is 5 times higher among twins and 12 times higher among supertwins), and are more likely to suffer life-long disabilities if they do survive. The Iowa septuplets, for example, were born more than two months premature and each weighed less than half of the national average birth weight. Thus, while modern medicine offers infertile couples new options, these options carry risks that require careful consideration.

Fertility drugs are sometimes too effective

Fertility drugs are one example of how reproductive technologies can alter the normal reproductive cycle. We investigate these reproductive options and their risks and implications at the end of this chapter. First, we survey animal reproduction and development. Following a brief introduction to the diverse ways that animals reproduce, we focus on the reproductive system of our own species. We examine how human eggs and sperm form, and then we return to the subject of hormones, to see how they affect our reproductive activities. In the second half of the chapter, we discuss the processes of fertilization and embryonic development in vertebrates and then focus on human embryonic development and birth. ■ ■ ■

The Iowa septuplets, born in 1997

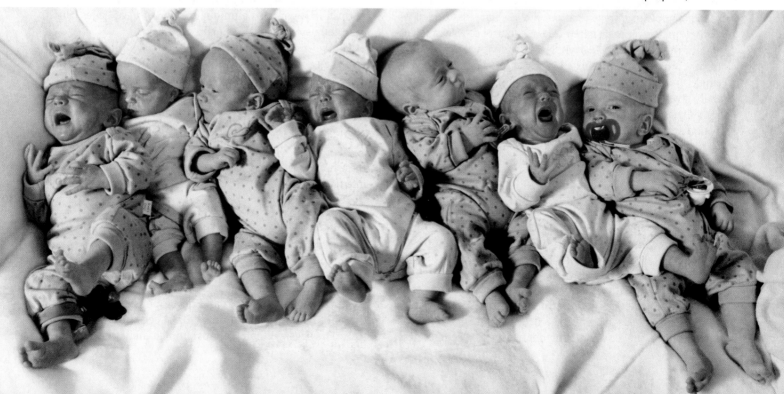

27.1 Sexual and asexual reproduction are both common among animals

Individuals have a finite life span. A population transcends the limit of finite life spans only by **reproduction**, the creation of new individuals from existing ones. Animals reproduce in a great variety of ways, but there are two principal modes: asexual and sexual.

Asexual reproduction (reproduction without sex) is the creation of offspring whose genes all come from one parent, without the fusion of egg and sperm. The offspring of asexual reproduction are genetic copies of the lone parent.

Several types of asexual reproduction are found among animals. Many invertebrates reproduce asexually by **budding**, splitting off new individuals from existing ones (Figure 27.1A; also see Figure 8.11C). The sea anemone in the center of Figure 27.1B is undergoing **fission**, the separation of a parent into two or more individuals of about equal size. Asexual reproduction can also start with the process of **fragmentation**, the breaking of the parent body into several pieces, some or all of which develop into complete adults. For an animal to reproduce this way, fragmentation must be accompanied by **regeneration**, the regrowth of lost body parts. Sea stars, for example, have remarkable powers of regeneration. If a sea star loses one of its arms, it may regenerate a new one in a matter of weeks. In sea stars of the genus *Linckia*, a whole new individual can develop from a broken-off arm plus a bit of the central body. Thus, a single animal with five arms, if broken apart, could potentially give rise to five offspring via asexual reproduction.

Asexual reproduction has several potential advantages. For one thing, it allows animals that do not move from place to place or that live in isolation to produce offspring without finding mates. Another advantage is that it enables an animal to produce many offspring quickly; no time or energy is lost in gamete production or fertilization. Asexual reproduction perpetuates a particular genotype precisely and rapidly. Therefore, it can be an effective way for animals that are genetically well suited to a particular environment to quickly expand their populations and exploit available resources.

A potential disadvantage of asexual reproduction is that it produces genetically uniform populations. Genetically similar individuals may thrive in one particular environment, but if the environment changes and becomes less

Figure 27.1A Budding in a hydra

LM 25×

Figure 27.1B Asexual reproduction of a sea anemone by fission

favorable, all individuals may be affected equally, and the entire population may die out.

In contrast to asexual reproduction, **sexual reproduction** is the creation of offspring by the fusion of two haploid (n) sex cells, or **gametes**, to form a diploid ($2n$) **zygote** (fertilized egg). The male gamete, the **sperm**, is a relatively small cell that moves by means of a flagellum. The female gamete, the unfertilized **egg**, or **ovum** (plural, **ova**), is a much larger cell that is not self-propelled. The zygote—and the new individual it develops into—contains a unique combination of genes carried from the parents via the egg and sperm.

Unlike asexual reproduction, sexual reproduction increases genetic variability among offspring. As we discussed in Modules 8.16 and 8.18, meiosis and random fertilization can generate enormous genetic variation. The variability produced by the reshuffling of genes in sexual reproduction may provide greater adaptability to changing environments. In theory, when an environment changes suddenly or drastically, there is a better chance that some of the offspring will survive and reproduce if they aren't all genetically very similar.

Many animals can reproduce both sexually and asexually, benefiting from both modes. The microscopic animal in Figure 27.1C is a rotifer. Rotifers abound in freshwater ponds and lakes, where their diet consists mainly of algae, bacteria, and protozoans. Most rotifers reproduce asexually when there is ample food and when water temperatures are favorable for rapid growth and development. The female in the photograph is laying unfertilized eggs produced by mitosis. The eggs will hatch almost immediately into a clone of

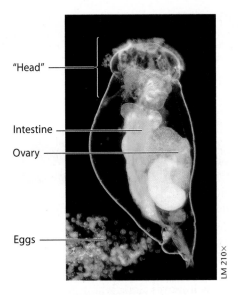

"Head"

Intestine

Ovary

Eggs

LM 210×

Figure 27.1C A rotifer laying eggs

Eggs

Figure 27.1E Frogs in an embrace that triggers the release of eggs and sperm (the sperm are too small to be seen)

new females. Asexual reproduction usually continues until cold temperatures signal the approach of winter or until the food supply dwindles or the habitat starts to dry up. The rotifers then reproduce sexually, producing a generation of genetically varied individuals. The fertilized eggs of the sexual generation have a thick shell and can withstand harsh conditions, such as freezing and drying.

Although sexual reproduction has advantages, it presents a problem for nonmobile animals and for those that live solitary lives: how to find a mate. One solution that has evolved is **hermaphroditism**, in which each individual has both female and male reproductive systems. (The term comes from the Greek myth in which Hermaphroditus, son of the gods Hermes and Aphrodite, fused with a woman to form a single individual of both sexes.) Although some hermaphrodites, such as tapeworms, can fertilize their own eggs, most must mate with another member of the same species. When hermaphrodites mate (for example, the two earthworms in Figure 27.1D), each animal serves as both male and female, donating and receiving sperm. For hermaphrodites, every individual encountered is a potential

mate, and mating can result in twice as many offspring than if only one individual's eggs were fertilized.

The mechanics of fertilization play an important part in sexual reproduction. Many aquatic invertebrates and most fishes and amphibians exhibit **external fertilization:** The parents discharge their gametes into the water, where fertilization then occurs, often without the male and female even making physical contact. Timing is crucial because the eggs must be ripe for fertilization when sperm contact them. For many species—certain clams that live in freshwater rivers and lakes, for instance—environmental cues such as temperature and day length cause a whole population to release gametes all at once. Males or females may also emit a chemical signal as they release their gametes. The signal triggers gamete release in members of the opposite sex. Most fishes and amphibians with external fertilization have specific courtship rituals that trigger simultaneous gamete release in the same vicinity by the female and male. An example of such a mating ritual is the clasping of a female frog by a male (Figure 27.1E).

In contrast to external fertilization, **internal fertilization** occurs when sperm are deposited in or close to the female reproductive tract, and gametes unite within the tract. Nearly all terrestrial animals exhibit internal fertilization, which is an adaptation that protects developing eggs in a dry environment. Internal fertilization usually requires **copulation**, or sexual intercourse. It also requires complex reproductive systems, including organs for gamete storage and transport and organs that facilitate intercourse. For examples of these complex structures, we turn next to the human female and male.

Figure 27.1D Earthworms mating

? In terms of genetic makeup, what is the most important difference between the outcome of sexual reproduction and that of asexual reproduction?

■ The offspring of sexual reproduction are genetically diverse.

27.2 Reproductive anatomy of the human female

The drawings in this and the next module illustrate the structures of the human female and male reproductive systems. Both sexes have a pair of gonads (ovaries or testes) where the gametes are produced, a system of ducts that house and conduct the gametes, and structures that facilitate copulation.

A woman's **ovaries** are each about an inch long, with a bumpy surface (**Figure 27.2A**). The bumps are **follicles,** each consisting of a single developing egg cell surrounded by one or more layers of follicle cells that nourish and protect the developing egg cell. In addition to producing egg cells, the ovaries produce hormones, as we saw in Chapter 26. Specifically, the follicle cells produce the female sex hormone estrogen. (In this chapter, we use the word *estrogen* to refer collectively to several closely related chemicals that affect the body similarly.)

Most or all of the 400,000 follicles a woman will ever have are thought to be formed before her birth, but only several hundred will release egg cells during her reproductive years. Starting at puberty and continuing until menopause, one follicle (or rarely two or more) matures and releases its egg cell about every 28 days. An egg cell is ejected from the follicle in a process called **ovulation,** shown in **Figure 27.2B.**

After ovulation, the remaining follicular tissue grows within the ovary to form a solid mass called the **corpus luteum** (Latin for "yellow body"); you can see one in the ovary on the left in Figure 27.2A. The corpus luteum secretes progesterone, a hormone that helps maintain the uterine lining during pregnancy, and additional estrogen. If the egg is not fertilized, the corpus luteum degenerates, and a new

follicle matures during the next cycle. We discuss ovulation and female hormonal cycles further in later modules.

Notice in Figure 27.2A that each ovary lies next to the opening of an **oviduct,** also called a fallopian tube. The oviduct opening resembles a funnel fringed with fingerlike projections. The projections touch the surface of the ovary, but the ovary is actually separated from the opening of the oviduct by a tiny space. When ovulation occurs, the egg cell passes across the space and into the oviduct, where cilia sweep it toward the uterus. Fertilization usually occurs in the upper third of the oviduct. The resulting zygote starts to divide, thus becoming an embryo, as it moves along within the oviduct.

The **uterus,** also known as the womb, is the actual site of pregnancy. The uterus is only about 3 inches long in a woman who has never been pregnant, but during pregnancy it expands considerably to accommodate a baby. The uterus has a thick muscular wall, and its inner lining, the **endometrium,** is richly supplied with blood vessels. The embryo implants (digests a place for itself) in the endometrium, and development is completed there. The term **embryo** is used for the stage in development from the first division of the zygote until body structures begin to appear, about the ninth week in humans. From the ninth week until birth, a developing human is called a **fetus.**

The uterus is the *normal* site of pregnancy. However, in about 1% of pregnancies, the embryo implants somewhere else, resulting in an **ectopic pregnancy** (from the Greek *ektopos,* out of place). Most ectopic pregnancies occur in the oviduct and are called tubal pregnancies. Ectopic pregnancies require surgical removal; otherwise, they can rupture

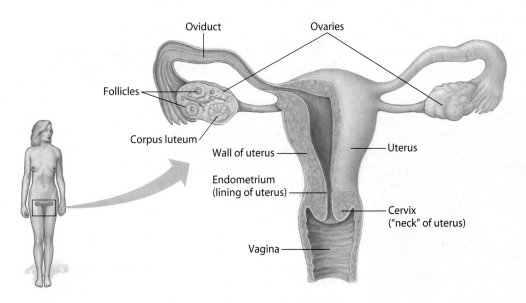

Figure 27.2A Front view of female reproductive anatomy (upper portion)

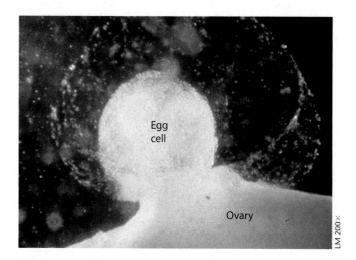

Figure 27.2B Ovulation

surrounding tissues, causing severe bleeding and even death of the mother.

The narrow neck of the uterus is the **cervix**, which opens into the vagina. The **vagina** is a thin-walled, but strong, muscular chamber that serves as the birth canal through which the baby is born. The vagina is also the repository for sperm during copulation.

You can see more features of female reproductive anatomy in Figure 27.2C, a side view. Notice that the vagina opens to the outside just behind the opening of the urethra, the tube through which urine is excreted. A pair of slender skin folds,

the **labia minora**, border the openings, and a pair of thick, fatty ridges, the **labia majora**, protect the vaginal opening. Until sexual intercourse or vigorous physical activity ruptures it, a thin piece of tissue called the **hymen** partly covers the vaginal opening. **Bartholin's glands**, near the vaginal opening, secrete mucus during sexual arousal, lubricating the vagina and facilitating intercourse.

Several female reproductive structures are important in sexual arousal, and stimulation of them can produce highly pleasurable sensations. The vagina, labia minora, and a structure called the **clitoris** all engorge with blood and enlarge during sexual activity. The sole function of the clitoris is sexual arousal. It consists of a short shaft supporting a rounded **glans**, or head, covered by a small hood of skin called the **prepuce**. In Figure 27.2C, blue highlights the spongy tissue within the clitoris that fills with blood during arousal. The clitoris, especially the glans, has an enormous number of nerve endings and is very sensitive to touch. Accompanied by other arousing stimuli, gentle stimulation of the glans can often trigger orgasm. We discuss the human sexual response in more detail in Module 27.6.

? In which organ of the human female does the fetus develop?

■ The uterus

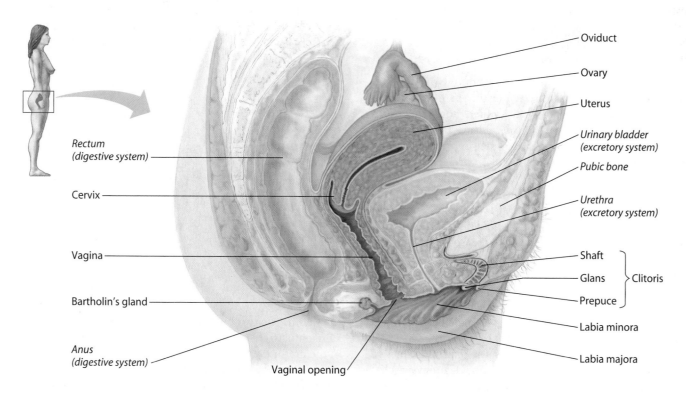

Figure 27.2C   Side view of female reproductive anatomy (with nonreproductive structures in italic)

27.3 Reproductive anatomy of the human male

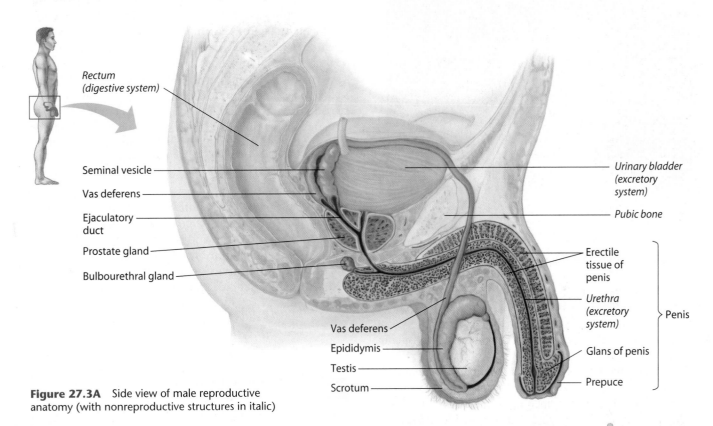

Figure 27.3A Side view of male reproductive anatomy (with nonreproductive structures in italic)

Figures 27.3A and 27.3B present two views of the male reproductive system. The male gonads, or **testes** (singular, *testis*), are each housed outside the abdominal cavity in a sac called the **scrotum.** Sperm cannot develop at human core body temperature, but the scrotum keeps the sperm-forming cells cool enough to function normally.

Now let's track the path of sperm from one of the testes out of the male's body. From each testis, sperm pass into a coiled tube called the **epididymis,** which stores the sperm while they continue to develop. Sperm leave the epididymis during **ejaculation,** the expulsion of sperm-containing fluid from the penis. At that time, muscular contractions propel the sperm from the epididymis through another duct called the **vas deferens.** The vas deferens passes upward into the abdomen and loops around the urinary bladder. Next to the bladder, the vas deferens joins a short duct from a gland, the seminal vesicle (see Figure 27.3A). The two ducts unite to form a short **ejaculatory duct,** which joins its counterpart conveying sperm from the other testis. The union of the two ejaculatory ducts forms the urethra, which conveys both urine and sperm out through the penis, although not at the same time. Thus, unlike the female, the male has a connection between the reproductive and excretory systems.

In addition to the testes and ducts, the male reproductive system contains three sets of glands: the seminal vesicles, the prostate gland, and the bulbourethral glands. The two **seminal vesicles** secrete a thick fluid that contains fructose, which provides most of the energy used by the sperm. The

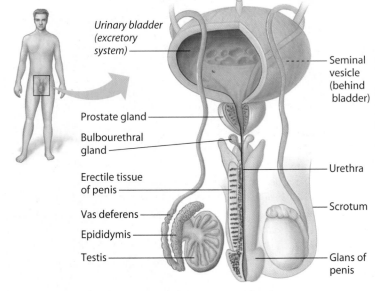

Figure 27.3B Front view of male reproductive anatomy

prostate gland secretes a thin fluid that further nourishes the sperm. The two **bulbourethral glands** secrete a clear, alkaline mucus that balances the acidity of any traces of urine in the urethra.

Together, the sperm and the glandular secretions make up **semen,** the fluid discharged (ejaculated) from the penis during orgasm. About 2–5 mL (1 teaspoonful) of semen are discharged during a typical ejaculation. About 95% of the fluid

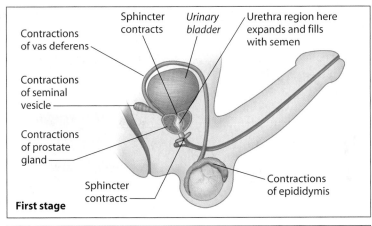

First stage

Contractions of vas deferens

Contractions of seminal vesicle

Contractions of prostate gland

Sphincter contracts

Sphincter contracts

Urinary bladder

Urethra region here expands and fills with semen

Contractions of epididymis

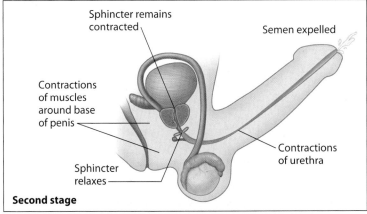

Second stage

Sphincter remains contracted

Contractions of muscles around base of penis

Sphincter relaxes

Semen expelled

Contractions of urethra

Figure 27.3C The two stages of ejaculation

consists of glandular secretions. The other 5% is made up of 50–130 million sperm, only one of which may eventually fertilize a single egg. The alkalinity of the semen helps neutralize the acidic environment of the vagina, protecting the sperm and increasing their motility.

The human **penis** consists mainly of tissue that can fill with blood to cause an erection during sexual arousal. The erectile tissue is shown in blue in Figures 27.3A and 27.3B. Erection is essential for insertion of the penis into the vagina. (See Module 28.8 for a discussion of a signal molecule important for erection and how drugs can affect its action.) Like the clitoris, the penis consists of a shaft that supports the glans, or head. The glans is richly supplied with nerve endings and is highly sensitive to stimulation. As in the female, a fold of skin called the prepuce, or foreskin, covers the glans. Circumcision, the surgical removal of the prepuce, arose from religious traditions. Scientific studies have not verified that circumcision has an overall positive or negative impact on a man's health or hygiene.

Figure 27.3C illustrates the process of ejaculation and summarizes the production of semen and its expulsion. Ejaculation occurs in two stages. At the peak of sexual arousal, muscles in the epididymis, seminal vesicles, prostate gland, and vas deferens contract (upper drawing). These contractions force secretions from the glands into the vas deferens and propel sperm from the epididymis. At the same time, a sphincter muscle at the base of the bladder contracts, preventing urine from leaking into the urethra from the bladder.

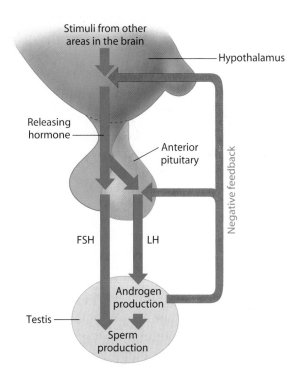

Stimuli from other areas in the brain

Hypothalamus

Releasing hormone

Anterior pituitary

Negative feedback

FSH

LH

Testis

Androgen production

Sperm production

Figure 27.3D Hormonal control of the testis

Another sphincter also contracts, closing off the entrance of the urethra into the penis. The section of the urethra between the two sphincters fills with semen and expands. In the second stage of ejaculation, the expulsion stage (lower drawing), the sphincter at the base of the penis relaxes, admitting semen into the penis. Simultaneously, a series of strong muscle contractions around the base of the penis and along the urethra expels the semen from the body.

Figure 27.3D shows how hormones control sperm production by the testes. Influenced by signals from other parts of the brain, the hypothalamus secretes a releasing hormone that regulates release of follicle-stimulating hormone (FSH) and luteinizing hormone (LH) by the anterior pituitary (see Module 26.4). FSH increases sperm production by the testes, while LH promotes the secretion of androgens, mainly testosterone. Androgens stimulate sperm production. In addition, androgens carried in the blood help maintain homeostasis by a negative-feedback mechanism (red arrows), inhibiting secretion of both the releasing hormone and LH. Under the control of this chemical regulating system, the testes produce hundreds of millions of sperm every day, from puberty well into old age. Next we'll see how sperm and eggs are made.

Web/CD Activity 27A *Reproductive System of the Human Male*

Web/CD Thinking as a Scientist *Connection: What Might Obstruct the Male Urethra?*

? Arrange the following organs in the correct sequence for the travel of sperm: epididymis, testis, urethra, vas deferens.

■ Testis, epididymis, vas deferens, urethra

27.4 The formation of sperm and ova requires meiosis

Both sperm and ova are haploid cells that develop by meiosis from diploid cells in the gonads. Before we turn to the formation of gametes, **gametogenesis**, you may want to review Modules 8.12–8.14 as background for our discussion.

Spermatogenesis, the formation of sperm cells, takes about 65–75 days in the human male. **Figure 27.4A** outlines spermatogenesis. Recall that the diploid chromosome number in humans is 46; that is, $2n = 46$.

Sperm develop in the testes in coiled tubes called the **seminiferous tubules.** Diploid cells that begin the process are located near the outer wall of the tubules (at the top of the enlarged wedge of tissue in Figure 27.4A). These cells multiply constantly by mitosis, and each day about 3 million of them differentiate into **primary spermatocytes**, the cells that undergo meiosis. Meiosis I of a primary spermatocyte produces two **secondary spermatocytes**, each with the

Figure 27.4A Spermatogenesis

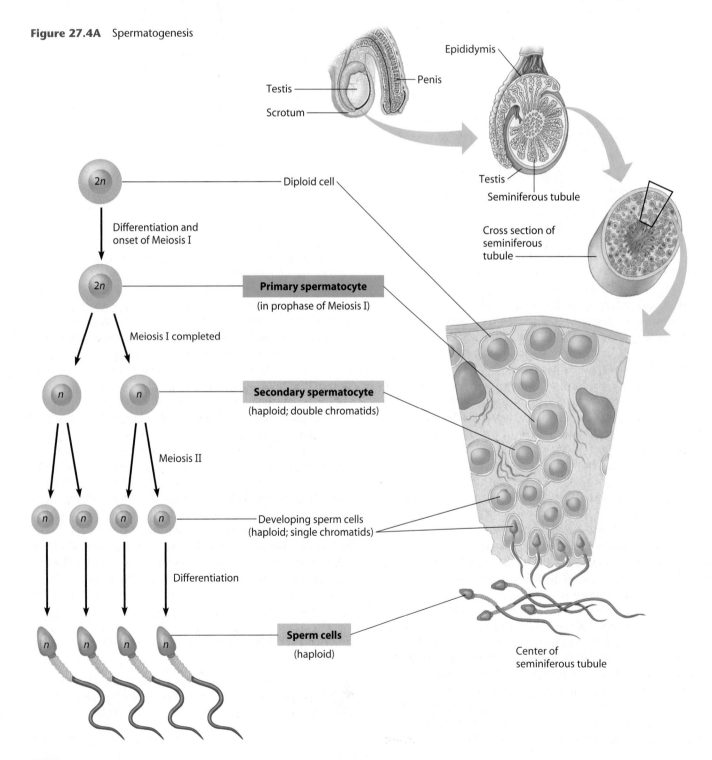

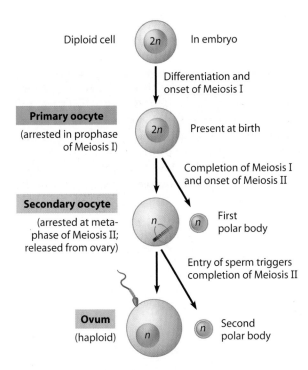

Figure 27.4B Meiosis in oogenesis

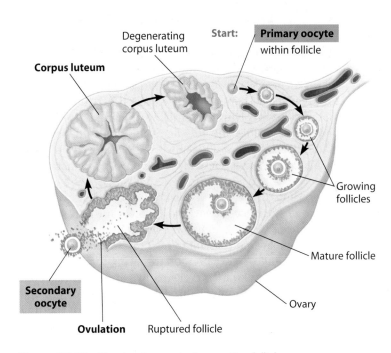

Figure 27.4C The development of an ovarian follicle

haploid number of chromosomes (*n* = 23). The chromosomes are still in their duplicated state, each consisting of two identical chromatids. Meiosis II then forms four cells, each with the haploid number of single-chromatid chromosomes. A sperm cell develops by differentiation of each of these haploid cells and is gradually pushed toward the center of the seminiferous tubule. From there it passes into the epididymis, where it matures, becomes motile, and is stored until ejaculation.

Figures **27.4B** and **27.4C** show **oogenesis**, the development of mature ova (egg cells). Most of the process occurs in the ovary. Oogenesis actually begins prior to birth, when a diploid cell in each developing follicle begins meiosis. At birth, each follicle contains a dormant **primary oocyte**, a diploid cell that is resting in prophase of meiosis I. A primary oocyte can be hormonally triggered to develop further. After puberty, about every 28 days, FSH (follicle-stimulating hormone) from the pituitary stimulates one of the dormant follicles to develop. The follicle enlarges, and the primary oocyte completes meiosis I and begins meiosis II. Meiosis then halts again at metaphase II. In the female, the division of the cytoplasm in meiosis I is unequal, with a single **secondary oocyte** receiving almost all of it. The smaller of the two daughter cells, called the first polar body, receives almost no cytoplasm.

The secondary oocyte is the stage released by the ovary during ovulation. It enters the oviduct, and if a sperm cell penetrates it, the secondary oocyte completes meiosis II. Meiosis II yields a second polar body and the actual ovum. The haploid nucleus of the ovum can then fuse with the haploid nucleus of the sperm cell, producing a zygote.

Although not shown in Figure 27.4B, the first polar body may also undergo meiosis II, forming two cells. These and the second polar body receive virtually no cytoplasm and quickly degenerate. Polar body formation leaves the ovum

with nearly all the cytoplasm and thus the bulk of the nutrients contained in the original diploid cell.

Figure 27.4C is a cutaway view of an ovary. The series of follicles here represents the changes one follicle undergoes over time; the arrows indicate the sequence. An actual ovary would have thousands of dormant follicles, each containing a primary oocyte. Usually, only one follicle has a dividing oocyte at any one time, and as it develops, that follicle stays in one place in the ovary. Meiosis I occurs as the follicle matures. About the time the secondary oocyte forms, the pituitary hormone LH (luteinizing hormone) triggers ovulation, the rupture of the follicle and expulsion of the secondary oocyte. The ruptured follicle then develops into a corpus luteum. Unless fertilization occurs, the corpus luteum degenerates before another follicle starts to develop.

Oogenesis and spermatogenesis are alike in that they both produce haploid gametes. However, these two processes differ in three important ways. First, only one ovum results from each diploid cell that undergoes meiosis. The other products of oogenesis, the polar bodies, degenerate. By contrast, in spermatogenesis, all four products of meiosis develop into mature gametes. Second, although the cells from which sperm develop continue to divide by mitosis throughout the male's life, this is thought not to be the case for the comparable cells in the human female (although recent research has brought this point into question). Third, oogenesis has long "resting" periods, whereas spermatogenesis produces mature sperm in an uninterrupted sequence.

? Which process in the development of sperm and ova is responsible for the genetic variation among gametes? (*Hint:* Review Module 8.17.)

■ Meiosis, specifically meiosis I

27.5 Hormones synchronize cyclic changes in the ovary and uterus

Oogenesis is one part of a female mammal's reproductive cycle, a recurring sequence of events that produces gametes, makes them available for fertilization, and prepares the body for pregnancy. The reproductive cycle is actually one integrated cycle involving cycles in two different reproductive organs: the ovaries and the uterus. In discussing oogenesis in the last module, we described the **ovarian cycle**, cyclic events that occur about every 28 days in the human ovary. Hormonal messages synchronize the ovarian cycle with related events in the uterus called the **menstrual cycle**. The hormone story is complex and involves intricate feedback mechanisms. Table 27.5 lists the major hormones and their roles. You may find it useful to refer to this table while reading this module. Figure 27.5, on the facing page, shows how the events of the ovarian cycle (part C) and menstrual cycle (part E) are synchronized through the actions of multiple hormones (shown in parts A, B, and D). Notice the time scale at the bottom of part E; it also applies to parts B–D. Follow Figure 27.5 carefully as you read the descriptions of the events in the ovarian and menstrual cycles.

An Overview of the Ovarian and Menstrual Cycles

Let's begin with the structural events of the ovarian and menstrual cycles. For simplicity, we have divided the ovarian cycle (part C of the figure) into two phases separated by ovulation: the pre-ovulatory phase, when a follicle is growing and a secondary oocyte is developing, and the post-ovulatory phase, after the follicle has become a corpus luteum.

Events in the menstrual (or uterine) cycle (part E) are synchronized with the ovarian cycle. By convention, the first day of a woman's "period" is designated day 1 of the menstrual cycle. Uterine bleeding, called **menstruation**, usually persists for 3–5 days. Notice that this corresponds to the beginning of the pre-ovulatory phase of the ovarian cycle. During menstruation, the endometrium (inner lining of the uterus) breaks down and leaves the body through the vagina. The menstrual discharge consists of blood, small clusters of endometrial cells, and mucus. After menstruation, the endometrium regrows. It continues to thicken through the time of ovulation, reaching a maximum at about 20–25 days. If an embryo has not implanted in the uterine lining by this time, menstruation begins again, marking the start of the next ovarian and menstrual cycles.

Now let's consider the hormones that regulate the ovarian and menstrual cycles. The ebb and flow of the five hormones listed in Table 27.5 synchronize events in the ovarian cycle (the growth of the follicle and ovulation) with events in the menstrual cycle (preparation of the uterine lining for possible implantation of an embryo). A releasing hormone from the hypothalamus in the brain regulates secretion of the two pituitary hormones FSH and LH. The blood levels of FSH, LH, and two other hormones—estrogen and progesterone—coincide with specific events in the ovarian and menstrual cycles.

TABLE 27.5	HORMONES OF THE OVARIAN AND MENSTRUAL CYCLES	
Hormone	**Secreted by**	**Major Roles**
Releasing hormone	Hypothalamus	Regulates secretion of LH and FSH by pituitary
FSH	Pituitary	Stimulates growth of ovarian follicle
LH	Pituitary	Stimulates growth of ovarian follicle and production of secondary oocyte; promotes ovulation; promotes development of corpus luteum and secretion of hormones
Estrogen	Ovarian follicle	Low levels inhibit pituitary; high levels stimulate hypothalamus; promotes endometrium
Estrogen and progesterone	Corpus luteum	Maintain endometrium; high levels inhibit hypothalamus and pituitary; sharp drops promote menstruation

Hormonal Events Before Ovulation

Focusing on part A of Figure 27.5, we see that the releasing hormone from the hypothalamus stimulates the anterior pituitary to ❶ increase its output of FSH and LH. True to its name, ❷ FSH stimulates the growth of an ovarian follicle, in effect starting the ovarian cycle. In turn, the follicle secretes estrogen. Early in the pre-ovulatory phase, the follicle is small (part C) and secretes relatively little estrogen (part D). As the follicle grows, ❸ it secretes more and more estrogen, and the rising but still relatively low level of estrogen exerts negative feedback on the pituitary. This keeps the blood levels of FSH and LH low for most of the pre-ovulatory phase (part B). As the time of ovulation approaches, hormone levels change drastically, with estrogen reaching a critical peak (part D) just before ovulation. This high level of estrogen exerts positive feedback on the hypothalamus (green arrow in part A), which then ❹ makes the pituitary secrete bursts of FSH and LH. By comparing parts B and D of the figure, you can see that the peaks in FSH and LH occur just after the estrogen peak. It may help to place a piece of paper over the figure and slide it slowly to the right. As you uncover the figure, you will see the follicle getting bigger and the estrogen level rising to its peak, followed almost immediately by the LH and FSH peaks. ❺ Then, just to the right of the peaks, comes the dashed line representing ovulation.

Hormonal Events at Ovulation and After

LH stimulates the completion of meiosis, transforming the primary

oocyte in the follicle into a secondary oocyte. It also signals enzymes to rupture the follicle, allowing ovulation to occur, and triggers the development of the corpus luteum from the ruptured follicle (hence its name, luteinizing hormone). LH also promotes the secretion of progesterone and estrogen by the corpus luteum. In part D of the figure, you can see the progesterone peak and the second (lower and wider) estrogen peak after ovulation.

High levels of estrogen and progesterone in the blood following ovulation have a strong influence on both ovary and uterus. The combination of the two hormones exerts negative feedback on the hypothalamus and pituitary, producing ❻ falling FSH and LH levels. The drops in FSH and LH prevent follicles from developing and ovulation from occurring during the post-ovulatory phase. Also, the LH drop is followed by the gradual degeneration of the corpus luteum. Near the end of the post-ovulatory phase, unless an embryo has implanted in the uterus, the corpus luteum stops secreting estrogen and progesterone. ❽ As blood levels of these hormones decline, the hypothalamus once again can stimulate the pituitary to secrete more FSH and LH, and a new cycle begins.

Control of the Menstrual Cycle Hormonal regulation of the menstrual cycle is simpler than that of the ovarian cycle. The menstrual cycle (part E) is directly controlled by estrogen and progesterone alone. You can see the effects of these hormones by comparing parts D and E of the figure. Starting around day 5 of the cycle, the endometrium thickens in response to the rising levels of estrogen and, later, progesterone. ❼ When the levels of these hormones drop, the endometrium begins to slough off. Menstrual bleeding begins soon thereafter, on day 1 of a new cycle.

We have now described what happens in the human ovary and uterus in the absence of fertilization. As we'll see later, the ovarian and menstrual cycles are put on hold if fertilization and pregnancy occur. Early in pregnancy, the developing embryo, implanted in the endometrium, releases a hormone (human chorionic gonadotropin, or HCG) that acts like LH. The hormone maintains the corpus luteum, which continues to secrete progesterone and estrogen, keeping the endometrium intact. We'll return to the events of pregnancy in Modules 27.16 and 27.17.

Web/CD Activity 27B *Reproductive System of the Human Female*

❓ Which hormonal change triggers the onset of menstruation?

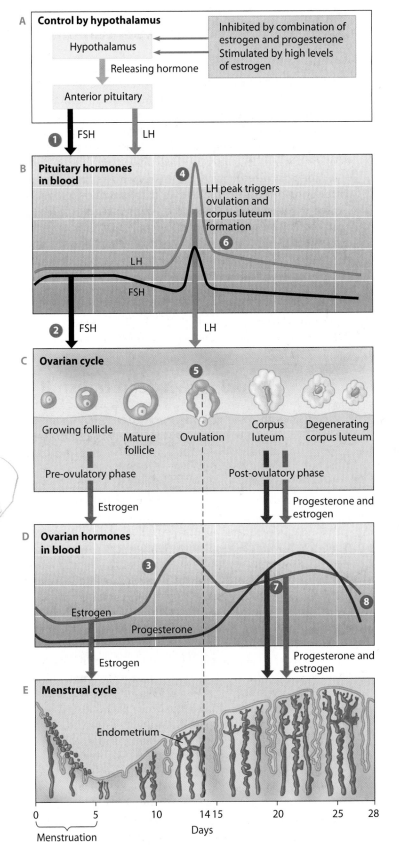

Figure 27.5 The reproductive cycle of the human female

[Handwritten annotation, inverted at bottom of text column:] The drop in the levels of estrogen and progesterone. These changes are caused by negative feedback of these hormones on the hypothalamus and pituitary after ovulation.

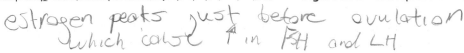

[Handwritten annotation: estrogen peaks just before ovulation which cause ↑ in FSH and LH]

27.6 The human sexual response occurs in four phases

Most female mammals are receptive to males only on certain days—in many species, for only a brief period once or a few times a year. A female deer or bear, for example, will mate only during a few weeks in the autumn. During specific mating times, a female is said to be in estrus, meaning she is at her peak of sexual readiness. This is the only time she ovulates and the only time her uterus is primed for implantation.

Humans and several other primates are unusual in having no distinct mating periods; females are potentially receptive to males throughout the year. The arousal of sexual interest in humans involves a variety of psychological as well as physical factors. Nevertheless, human sexual response is characterized by a common physiological pattern.

The physical events of the human sexual response occur in a sequence of four phases. The **excitement phase** prepares the sexual organs for coitus (sexual intercourse): Sexual passion builds, the penis and clitoris become erect, the testes, labia, and nipples may swell, the vagina secretes lubricating fluid, and muscles tighten in the arms and legs. These responses continue during the **plateau phase**, which is marked by increases in breathing and heart rate. **Orgasm** follows, characterized by rhythmic involuntary contractions of the reproductive structures, extreme pleasure for both partners, and ejaculation by the male. The **resolution phase** completes the cycle and reverses the previous responses: Organs return to normal size, muscles relax, and passion subsides.

? How does the timing of mating in humans contrast with that of most other mammals?

■ Human females are potentially receptive to mating throughout the year, in contrast to the seasonal mating of most other mammals.

27.7 Sexual activity can transmit disease

Sexually transmitted diseases (STDs) are contagious diseases spread by sexual contact. AIDS is caused by HIV (see Modules 10.21 and 24.12); genital herpes and genital warts are also caused by viruses. Viral STDs are not curable. They can be controlled by medications, but symptoms and the ability to infect others remain a possibility through a person's lifetime. Other STDs are usually curable with drugs, especially if diagnosed early.

Many STDs can cause long-term problems or even death if left untreated. Anyone who is sexually active should have regular medical exams, be tested for STDs, and seek immediate help if any suspicious symptoms appear—even if they are mild (Table 27.7).

STDs are most prevalent among teenagers and young adults; nearly two-thirds of infections occur among people under 25. The best way to avoid the spread of STDs is, of course, abstinence. Alternatively, latex condoms provide the best protection for "safe sex."

TABLE 27.7 STDS COMMON IN THE UNITED STATES

Disease	Microbial Agent	Major Symptom and Effects	Treatment
Bacterial			
Chlamydial infections	*Chlamydia trachomatis*	Genital discharge, itching, and/or painful urination; often no symptoms in women; pelvic inflammatory disease (PID)	Antibiotics
Gonorrhea	*Neisseria gonorrhoeae*	Genital discharge; painful urination; sometimes no symptoms in women; PID	Antibiotics
Syphilis	*Treponema pallidum*	Ulcer (chancre) on genitalia in early stages; spreads throughout body and can be fatal if not treated	Antibiotics can cure in early stages
Viral			
Genital herpes (see Chapter 10 introduction and Module 10.18)	Herpes simplex virus type 2, occasionally type 1	Recurring symptoms: small blisters on genitalia, painful urination, skin inflammation; linked to cervical cancer, miscarriage, birth defects	Valacyclovir can prevent recurrences
Genital warts	Papilloma-viruses	Painless growths on genitalia; some of the viruses linked to cancer	Removal by freezing
AIDS and HIV infection	HIV	See Module 24.12	Combination of drugs
Protozoan			
Trichomoniasis	*Trichomonas vaginalis*	Vaginal irritation, itching, and discharge; usually no symptoms in men	Antiprotozoal drugs
Fungal			
Candidiasis (yeast infections)	*Candida albicans*	Similar to symptoms of trichomoniasis; frequently acquired nonsexually	Antifungal drugs

? Besides abstinence from sexual contact, what can prevent the spread of STDs?

■ Latex condoms

27.8 Contraception can prevent unwanted pregnancy

Contraception is the deliberate prevention of pregnancy. Table 27.8 lists common methods of contraception, with their failure rates when used correctly and when used typically. Note that these two rates are often quite different, emphasizing the importance of learning to use contraception correctly. It is also important to note that the "safe sex" provided by condoms can prevent both unwanted pregnancy and sexually transmitted diseases; this is not true of other contraceptive methods.

Complete abstinence (avoiding intercourse) is the only totally effective method of birth control, but other methods are effective to varying degrees. Sterilization, surgery that prevents sperm from reaching an ovum, is very reliable. A woman may have a **tubal ligation**, in which a doctor removes a short section from each oviduct (and may tie, or ligate, the remaining ends). A man may undergo a **vasectomy**, in which a doctor cuts a section out of each vas deferens to prevent sperm from reaching the urethra. Both forms of sterilization are relatively safe and free from side effects but are permanent.

The effectiveness of other methods of contraception depends on how they are used. Temporary abstinence, also called the **rhythm method** or **natural family planning**, depends on refraining from intercourse during the days around ovulation, when fertilization is most likely. It is difficult, however, to predict ovulation accurately. Thus, while the rhythm method is reliable in theory, in practice it is among the least reliable methods of contraception. **Withdrawal** of the penis from the vagina before ejaculation is also ineffective because some sperm may be released before climax.

If used correctly, **barrier methods** can be quite effective at physically preventing the union of sperm and egg. Condoms are sheaths, usually made of latex, that fit over the penis or within the vagina (**Figure 27.8**). A diaphragm is a dome-shaped rubber cap that covers the cervix; a cervical cap is similar but thimble-shaped and smaller. Both require a doctor's visit for proper fitting. Barrier devices (including condoms) are more effective when used in combination with

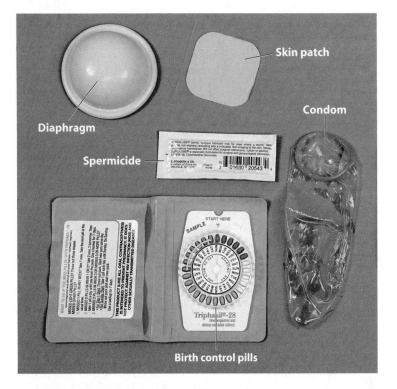

Figure 27.8 Some contraceptive devices

spermicides, sperm-killing foam or jelly; spermicides used alone are not particularly reliable.

Some of the most effective methods of contraception prevent the release of gametes. **Oral contraceptives**, or **birth control pills**, come in several different forms that vary in their combinations of hormones. The most widely used pills contain a combination of synthetic estrogen and a synthetic progesterone-like hormone called progestin. "The pill" prevents ovulation and keeps follicles from developing. Combined hormone contraceptives are also available as an injection, a ring inserted into the vagina, or a skin patch. The hormone progestin by itself is available as tablets (the "minipill") or as injections that last for three months (Depo-Provera).

Certain drugs can prevent fertilization or implantation even after intercourse has occurred. Combination birth control pills can be prescribed in high doses for emergency contraception; they function as **morning after pills** (**MAPs**). If taken within three days after unprotected intercourse, MAPs are about 75% effective. Such treatments should only be used in emergencies because they have significant side effects. If pregnancy has already occurred, the drug mifepristone can induce an abortion during the first seven weeks of pregnancy. Mifepristone requires a doctor's prescription and several visits to a medical facility.

TABLE 27.8	CONTRACEPTIVE METHODS	
	Pregnancies/100 Women/Year*	
Method	**Used Perfectly**	**Typically**
Birth control pill (combination)	0.1	5
Vasectomy	0.1	0.15
Tubal ligation	0.2	0.5
Progestin minipill	0.5	5
Rhythm	1–9	20
Withdrawal	4	19
Condom (male)	3	14
Diaphragm and spermicide	6	20
Spermicide alone	6	26

*Without contraception, about 85 pregnancies would occur.

? _____ is to males as tubal ligation is to _____.

Vasectomy . . . females

27.9 Fertilization results in a zygote and triggers embryonic development

The last seven modules have focused on the anatomy and physiology of the human reproductive system. In the next seven modules, we will examine the results of reproduction: the formation and development of an embryo. The concepts presented in this section apply to most vertebrates. We will return to the human story in the final section of this chapter.

Embryonic development begins with **fertilization**, the union of a sperm and an egg to form a diploid zygote. Fertilization combines haploid sets of chromosomes from two individuals and also activates the egg by triggering metabolic changes that start embryonic development.

The Properties of Sperm Cells Figure 27.9A is a micrograph of an unfertilized human egg almost covered by sperm. Of all these sperm, only one will enter and fertilize the egg. All the other sperm—the ones shown here and millions more that were ejaculated with them—will die. The one sperm that penetrates the egg adds its unique set of genes to those of the egg and contributes to the next generation.

Figure 27.9B illustrates the structure of a mature human sperm. Here is another case of form fitting function. The sperm's streamlined shape is an adaptation for swimming through fluids in the vagina, uterus, and oviduct of the female. The sperm cell's thick head contains a haploid nucleus and is tipped with a vesicle, the **acrosome**, which lies just inside the plasma membrane. The acrosome contains enzymes that help the sperm penetrate the egg. The neck and middle piece of the sperm contain a long, spiral mitochondrion. The sperm absorbs high-energy nutrients, especially the sugar fructose, from the semen. Thus fueled, its mitochondrion provides ATP for movement of the tail, which is actually a flagellum. By the time a sperm has reached the egg, it has consumed much of the energy available to it. But a successful sperm will have enough energy left to penetrate the egg and deposit its nucleus in the egg's cytoplasm.

The Process of Fertilization Figure 27.9C on the facing page illustrates the sequence of events in fertilization. This diagram is based on fertilization in sea urchins (phylum Echinodermata—see Module 18.13), on which a great deal of research has been done. Similar processes occur in other animals, including humans. The diagram traces one sperm through the successive activities of fertilization. Notice that to reach the egg nucleus, the sperm nucleus must pass through three barriers: the egg's jelly coat (yellow), a middle region of glycoproteins called the vitelline layer (pink), and the egg cell's plasma membrane.

Let's follow the steps shown in the figure. As a sperm ❶ approaches and then ❷ contacts the jelly coat of the egg, the acrosome in the sperm head releases a cloud of enzyme molecules that digest a cavity into the jelly. When the sperm

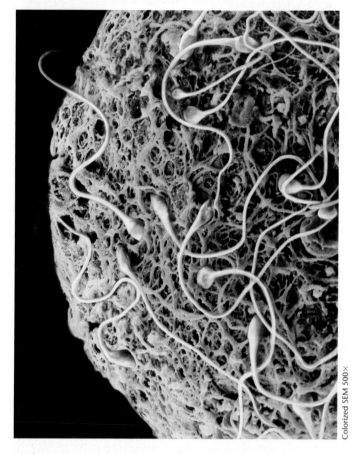

Colorized SEM 500×

Figure 27.9A A human egg cell surrounded by sperm

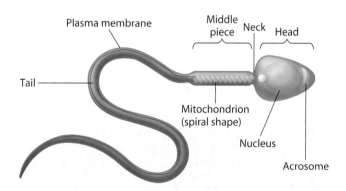

Plasma membrane — Middle piece — Neck — Head

Tail —

Mitochondrion (spiral shape)

Nucleus

Acrosome

Figure 27.9B The structure of a human sperm cell

head reaches the vitelline layer, ❸ species-specific protein molecules on its surface bind with specific receptor proteins on the vitelline layer. The binding between these proteins ensures that sperm of other species cannot fertilize the egg. This specificity is especially important when fertilization is

external, because the sperm of other species may be present in the water. After the specific binding occurs, the sperm proceeds through the vitelline layer, and ❹ the sperm's plasma membrane fuses with that of the egg. Fusion of the two membranes makes it possible for ❺ the sperm nucleus to enter the egg.

Fusion of the sperm and egg plasma membranes triggers a number of important changes in the egg. Two such changes prevent other sperm from entering the egg. About 1 second after the membranes fuse, the entire egg plasma membrane becomes impenetrable to other sperm cells. Shortly thereafter, ❻ the vitelline layer hardens and separates from the plasma membrane. The space quickly fills with water, and the vitelline layer becomes the so-called **fertilization envelope**, another barrier impenetrable to sperm. If these events did not occur and an egg were fertilized by more than one sperm, the resulting zygote nucleus would contain too many chromosomes, and the zygote could not develop normally.

Membrane fusion also triggers a burst of metabolic activity in the egg. In preparation for the enormous growth and development that will follow fertilization, the egg gears up from near dormancy, increasing cellular respiration and protein synthesis. Next, ❼ the egg and sperm nuclei fuse, producing the diploid nucleus of the zygote. In the next module, we begin to trace the development of the zygote into a new animal.

? What is the function of the fertilization envelope?

■ It helps prevent the entry of more than one sperm into the egg.

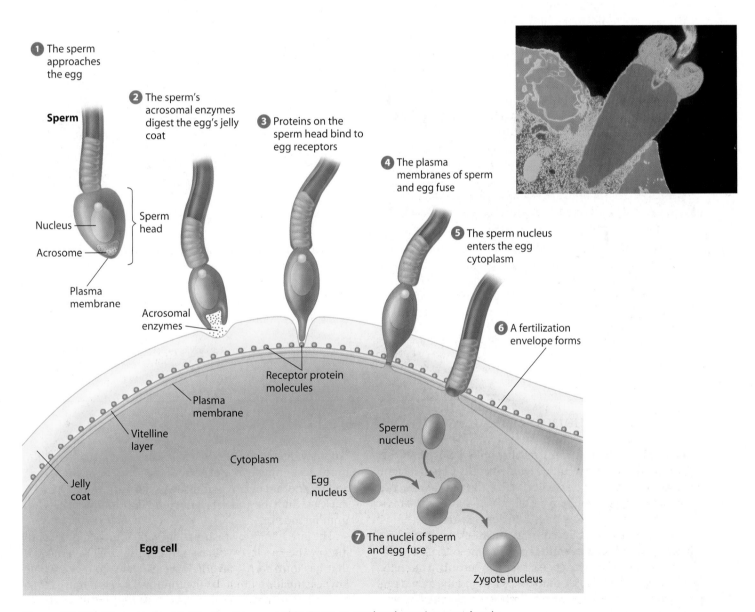

Figure 27.9C   The process of fertilization in a sea urchin (inset: sperm head entering cytoplasm)

27.10 Cleavage produces a ball of cells from the zygote

An animal consists of many thousands, millions, even trillions of cells organized into complex tissues and organs. The transformation from a zygote to this multicellular state is truly phenomenal. Order and precision are required at every step, and both are clearly displayed in the first two major phases of embryonic development: cleavage and gastrulation. We focus on cleavage in this module and gastrulation in the next.

Cleavage is a rapid succession of cell divisions that produces a ball of cells—a multicellular embryo—from the zygote. DNA replication, mitosis, and cytokinesis occur rapidly, but gene transcription virtually shuts down, and few new proteins are synthesized. As a result, the embryo of most animals does not grow larger during cleavage. Nutrients stored in the egg nourish the dividing cells, and the cell divisions partition the zygote into many smaller cells.

Figure 27.10 illustrates cleavage in a sea urchin. As the first three steps show, the number of cells doubles with each cleavage division. In a sea urchin, a doubling occurs about every 20 minutes, and the whole cleavage process takes about 3 hours to produce a solid ball of cells. Notice that each cell in the ball is much smaller than the zygote. As cleavage continues, a fluid-filled cavity called the **blastocoel** forms in the center of the embryo. At the completion of cleavage, there is a large cavity surrounded by one or more layers of cells. This hollow ball of cells is called the **blastula**.

Cleavage makes two very important contributions to early development. It creates a multicellular embryo, the blastula, from a single-celled zygote. Cleavage is also an organizing process, partitioning the multicellular embryo into developmental regions. As we discussed in Module 11.13, the cytoplasm of the zygote contains a variety of chemicals that control gene expression during early development. During cleavage, regulatory chemicals become localized in particular groups of cells, where they activate the genes that direct the formation of specific parts of the animal. Gastrulation, the next phase of development, further refines the embryo's cellular organization.

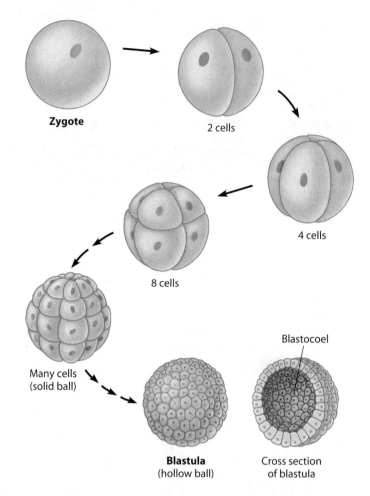

Figure 27.10 Cleavage in a sea urchin

? How does the reduction of cell size during cleavage increase oxygen supply to the cells' mitochondria? (*Hint:* Review Module 4.2.)

■ Smaller cells have a greater plasma membrane surface area relative to cellular volume, and this facilitates diffusion of oxygen from the environment to the cell's cytoplasm.

27.11 Gastrulation produces a three-layered embryo

Gastrulation, the second major phase of embryonic development, adds more cells to the embryo; more importantly, it sorts all the cells into distinct cell layers. In the process, the embryo is transformed from a hollow ball of cells—the blastula—into a three-layered stage called the **gastrula**.

The three layers produced in gastrulation are embryonic tissues called **ectoderm**, **endoderm**, and **mesoderm**. The ectoderm forms the outer layer (skin) of the gastrula. The endoderm forms an embryonic digestive tract. And the mesoderm partly fills the space between the ectoderm and endoderm. Eventually, these three cell layers develop into all the parts of the adult animal. For instance, our nervous system and the

outer layer (epidermis) of our skin come from ectoderm; the innermost lining of our digestive tract arises from endoderm; and most other organs and tissues, such as the kidney, heart, muscles, and the inner layer of our skin (dermis), develop from mesoderm.

The mechanics of gastrulation vary somewhat, depending on the species. We have chosen the frog, a vertebrate that has long been a favorite of researchers, to demonstrate how gastrulation produces the three cell layers. **Figure 27.11** takes us from a blastula at the top to a three-layered gastrula at the bottom. The diagrams in the left column show an external view of gastrulation; each drawing represents a multi-

cellular embryo, and the black arrows indicate movements of cell layers. The cutaway drawings on the right reveal the internal structures that develop as gastrulation occurs. The timing of these events varies with the species and the temperature in which the frog develops. In many frogs, cleavage and gastrulation together take about 15–20 hours.

❓ Gastrulation forms a new cavity, the _____, which is lined by _____ and which develops into the animal's _____ tract.

■ archenteron . . . endoderm . . . digestive

① The blastula. Formed by cleavage, the frog blastula is a partially hollow ball of unequally sized cells. As the cross section shows, the cells toward one end, called the animal pole, are smaller than those near the opposite end, the vegetal pole. The cells near the vegetal pole are larger because they contain yolk granules, which make them divide at a slower rate than those at the animal pole. The three colors on the blastula indicate regions of cells that will give rise to the primary cell layers: ectoderm (blue), endoderm (yellow), and mesoderm (pink). (Notice that each layer may be more than one cell thick.) In real embryos, these regions have been identified by dyeing the cells with harmless stains and observing where the cells go as development proceeds. A glance ahead at parts 2–4 will show you that the cells that will form endoderm move from the surface to the inside of the embryo.

② Blastopore formation. Gastrulation begins when a small groove, called the **blastopore,** appears on one side of the blastula. The blastopore is the place where cells of the future endoderm move inward from the surface (the dashed part of the arrow indicates inward movement). Meanwhile, the cells that will form ectoderm spread over more of the surface of the embryo, and the cells that will form mesoderm begin to spread inside the embryo.

③ Cell migration to form layers. The beginnings of the three layers can now be seen in the cross section. Migrating endodermal cells (yellow) have produced a simple digestive cavity called the **archenteron.** The advancing endoderm and the archenteron have filled some of the space formerly occupied by the blastocoel. Cells that will form the mesoderm (pink) are located between the endoderm and the ectoderm (blue).

④ Completion of gastrulation. Gastrulation is completed when the embryo is three-layered. Ectoderm covers the surface except for a cluster of endodermal cells called the **yolk plug.** The yolk plug marks the site of the blastopore and of the future anus. At this stage, the endoderm and its archenteron have replaced the blastocoel. Mesoderm forms a layer between the ectoderm and the endoderm.

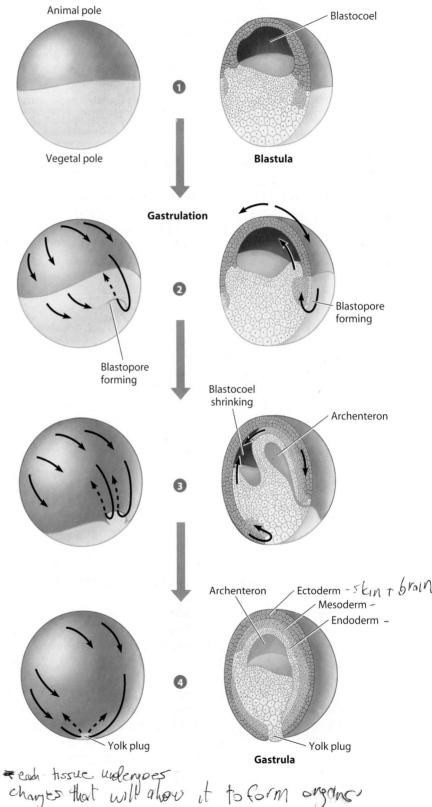

Animal pole

Vegetal pole

Blastocoel

Blastula

Gastrulation

Blastopore forming

Blastopore forming

Blastocoel shrinking

Archenteron

Archenteron

Ectoderm – skin + brain
Mesoderm –
Endoderm –

Yolk plug

Yolk plug

Gastrula

Figure 27.11 Development of the frog gastrula

each tissue undergoes changes that will allow it to form organs

Organs start to form after gastrulation

In organizing the embryo into three layers, gastrulation sets the stage for the shaping of an animal. Once the ectoderm, endoderm, and mesoderm form, cells in each layer begin to differentiate into tissues and embryonic organs. The cutaway drawing in **Figure 27.12A** shows the developmental structures that appear in a frog embryo a few hours after the completion of gastrulation. The orientation drawing at the upper left of the figure indicates a corresponding cut through an adult frog.

We see two structures in the embryo in Figure 27.12A that were not present at the gastrula stage described in the last module. An organ called the notochord has developed in the mesoderm, and a structure that will become the hollow nerve cord is beginning to form in the ectoderm. Recall that the notochord and dorsal, hollow nerve cord are hallmarks of the chordates (see Module 18.14).

The notochord is visible in cross section in the drawing in Figure 27.12A. It forms from mesoderm just above the archenteron. Made of a substance similar to cartilage, the **notochord** extends for most of the embryo's length and provides support for other developing tissues. Later in development, the notochord will function as a core around which mesodermal cells gather and form the backbone.

You can also see in Figure 27.12A the beginnings of the frog's hollow nerve cord, formed from a portion of the ectoderm. The area shown in green in the cutaway drawing is a thickened region of ectoderm called the neural plate. From it arises a pair of pronounced ectodermal ridges, called neural folds, visible in both the drawing and the micrograph below it. If you now look at the series of diagrams in **Figure 27.12B**, you will see what happens as the neural folds and neural plate develop further. The neural plate rolls up and forms the neural tube, which then sinks beneath the surface of the embryo and is covered by an outer layer of ectoderm. The **neural tube** is destined to become the brain and spinal cord.

Figure 27.12C shows a later frog embryo (about 12 hours older than the one in Figure 27.12A), in which the neural tube has formed. Notice in the drawing that the neural tube lies directly above the notochord. The relative positions of the neural tube, notochord, and archenteron give us a preview of the basic body plan of a frog. The spinal cord will lie within extensions of the dorsal surface of the backbone (which will replace the notochord), and the digestive tract will be ventral to the backbone. We see this same arrangement of organs in all vertebrates.

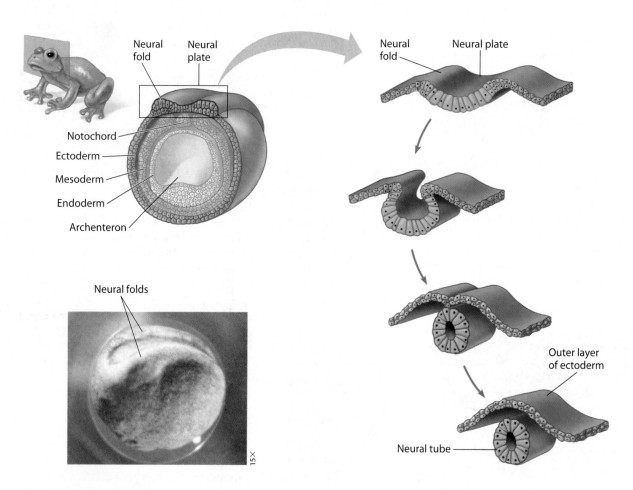

Figure 27.12A The beginning of organ development in a frog: the notochord, neural folds, and neural plate

Figure 27.12B
Formation of the neural tube

Figure 27.12C shows several other fundamental changes. In the micrograph, which is a side view, you can see that the embryo is more elongated than the one in Figure 27.12A. You can also see the beginnings of an eye and a tail (called the tail bud). Part of the ectoderm has been removed to reveal a series of internal ridges called somites. The **somites** are blocks of mesoderm that will give rise to segmental structures (constructed of repeating units), such as the vertebrae and associated muscles of the backbone. In the cross-sectional drawing, notice that the mesoderm next to the somites is developing a hollow space—the body cavity, or **coelom.** Segmented body parts and a coelom are basic features of chordates (see Module 18.14).

In this and the previous two modules, we have observed the sequence of changes that occur as an animal begins to take shape. To summarize, the key phases in embryonic development are cleavage (which creates a multicellular animal from a zygote), gastrulation (which organizes the embryo into three discrete layers), and organ formation (which generates embryonic organs from the three embryonic tissue layers). These same three phases occur in nearly all animals.

If we followed a frog's development beyond the stage represented in Figure 27.12C, within a few hours we would be able to monitor muscular responses and a heartbeat and see a set of gills with blood circulating in them. A long tail fin would grow

TABLE 27.12	DERIVATIVES OF THE THREE EMBRYONIC TISSUE LAYERS
Embryonic Layer	**Organs and Tissues in the Adult**
Ectoderm	Epidermis of skin and its derivatives; epithelial lining of mouth and rectum; sense receptors in epidermis; cornea and lens of eye; nervous system; adrenal medulla; tooth enamel
Endoderm	Epithelial lining of digestive tract (except mouth and rectum); epithelial lining of respiratory system; liver; pancreas; thyroid; parathyroids; thymus; lining of urethra, urinary bladder, and reproductive system
Mesoderm	Notochord (in animals retaining it as adults); skeletal system; muscular system; circulatory system; excretory system; reproductive system (except gamete-forming [germ] cells, which differentiate during cleavage); dermis of skin; lining of body cavity; adrenal cortex

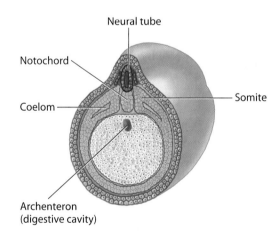

Neural tube

Notochord

Coelom

Somite

Archenteron
(digestive cavity)

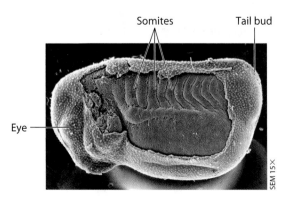

Somites   Tail bud

Eye

SEM 15×

Figure 27.12C An embryo with completed neural tube, somites, and coelom

from the tail bud. The timing of the later stages in frog development varies enormously, but in many species, by 5–8 days after development begins, we would see all the body tissues and organs of a tadpole emerge from cells of the ectoderm, mesoderm, and endoderm. Eventually, the structures of the tadpole (Figure 27.12D) would transform into the tissues and organs of an adult frog. Table 27.12 lists the major organs and tissues that arise in frogs (and other vertebrates) from each of the three main embryonic tissue layers.

Figure 27.12D A tadpole

Watching embryos develop helps us appreciate the enormous changes that occur as one tiny cell, the zygote, gives rise to a highly structured, many-celled animal. Your own body, for instance, is a complex organization of some 60 trillion cells, all of which arose from a zygote smaller than the period at the end of this sentence. Discovering how this incredibly intricate arrangement is achieved is one of biology's greatest challenges. Through research that combines the experimental manipulation of embryos with molecular biology and genetics, developmental biologists have begun to work out the mechanisms that underlie development. We examine several of these mechanisms in the next three modules.

Web/CD Activity 27D *Frog Development Video*

? What is the embryonic basis for the dorsal, hollow nerve cord that is common to all members of our phylum?

■ The nerve cord, which becomes the brain and spinal cord, develops from a dorsal ectodermal plate that folds to form an interior tube.

Figure 27.12C shows several other fundamental changes. In the micrograph, which is a side view, you can see that the embryo is more elongated than the one in Figure 27.12A. You can also see the beginnings of an eye and a tail (called the tail bud). Part of the ectoderm has been removed to reveal a series of internal ridges called somites. The **somites** are blocks of mesoderm that will give rise to segmental structures (constructed of repeating units), such as the vertebrae and associated muscles of the backbone. In the cross-sectional drawing, notice that the mesoderm next to the somites is developing a hollow space—the body cavity, or **coelom.** Segmented body parts and a coelom are basic features of chordates (see Module 18.14).

In this and the previous two modules, we have observed the sequence of changes that occur as an animal begins to take shape. To summarize, the key phases in embryonic development are cleavage (which creates a multicellular animal from a zygote), gastrulation (which organizes the embryo into three discrete layers), and organ formation (which generates embryonic organs from the three embryonic tissue layers). These same three phases occur in nearly all animals.

If we followed a frog's development beyond the stage represented in Figure 27.12C, within a few hours we would be able to monitor muscular responses and a heartbeat and see a set of gills with blood circulating in them. A long tail fin would grow

TABLE 27.12	DERIVATIVES OF THE THREE EMBRYONIC TISSUE LAYERS
Embryonic Layer	**Organs and Tissues in the Adult**
Ectoderm	Epidermis of skin and its derivatives; epithelial lining of mouth and rectum; sense receptors in epidermis; cornea and lens of eye; nervous system; adrenal medulla; tooth enamel
Endoderm	Epithelial lining of digestive tract (except mouth and rectum); epithelial lining of respiratory system; liver; pancreas; thyroid; parathyroids; thymus; lining of urethra, urinary bladder, and reproductive system
Mesoderm	Notochord (in animals retaining it as adults); skeletal system; muscular system; circulatory system; excretory system; reproductive system (except gamete-forming [germ] cells, which differentiate during cleavage); dermis of skin; lining of body cavity; adrenal cortex

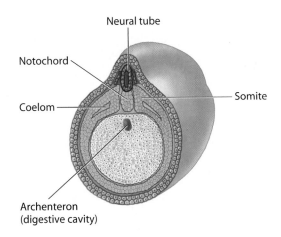

Neural tube

Notochord

Coelom

Somite

Archenteron (digestive cavity)

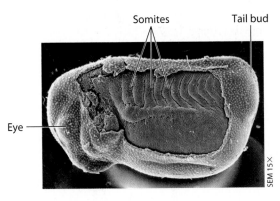

Somites

Tail bud

Eye

SEM 15×

Figure 27.12C An embryo with completed neural tube, somites, and coelom

from the tail bud. The timing of the later stages in frog development varies enormously, but in many species, by 5–8 days after development begins, we would see all the body tissues and organs of a tadpole emerge from cells of the ectoderm, mesoderm, and endoderm. Eventually, the structures of the tadpole (Figure 27.12D) would transform into the tissues and organs of an adult frog. Table 27.12 lists the major organs and tissues that arise in frogs (and other vertebrates) from each of the three main embryonic tissue layers.

Figure 27.12D A tadpole

Watching embryos develop helps us appreciate the enormous changes that occur as one tiny cell, the zygote, gives rise to a highly structured, many-celled animal. Your own body, for instance, is a complex organization of some 60 trillion cells, all of which arose from a zygote smaller than the period at the end of this sentence. Discovering how this incredibly intricate arrangement is achieved is one of biology's greatest challenges. Through research that combines the experimental manipulation of embryos with molecular biology and genetics, developmental biologists have begun to work out the mechanisms that underlie development. We examine several of these mechanisms in the next three modules.

Web/CD Activity 27D *Frog Development Video*

? What is the embryonic basis for the dorsal, hollow nerve cord that is common to all members of our phylum?

■ The nerve cord, which becomes the brain and spinal cord, develops from a dorsal ectodermal plate that folds to form an interior tube.

27.13 Changes in cell shape, cell migration, and programmed cell death give form to the developing animal

Many events in development depend on a combination of several kinds of cellular processes. **Figure 27.13A** shows how two changes in cell shape bring about the formation of the neural tube (see Figure 27.12B). Cells of the ectoderm fold inward by first elongating and then becoming wedge-shaped. The result is a tube of ectoderm—the start of the brain and spinal cord.

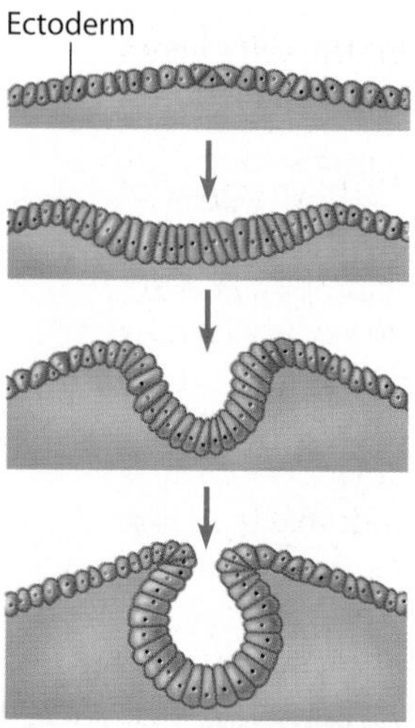

Cell migration is also essential in development. For example, during gastrulation, ectodermal cells use protrusions similar to the pseudopodia of amoeboid cells to "crawl" to the embryo's surface. Migrating cells may follow chemical trails secreted by cells near their specific destination. Once a migrating cell reaches its destination, surface proteins enable it to recognize similar cells. The cells join together and secrete glycoproteins that glue them in place. Finally, they differentiate, taking on the characteristics of a particular tissue.

Another key developmental process is **programmed cell death**, or **apoptosis**, the timely and tidy suicide of cells. Animals have genes coding for proteins that kill the cell that produces them. In humans, the timely death of specific cells in developing hands and feet creates the spaces between fingers and toes. Cell death is also essential for the normal development of our nervous and immune systems. In **Figure 27.13B**, the cell on the left shrinks and dies because a suicide gene has been turned on. Meanwhile, signals from the dying cell make an adjacent cell phagocytic. This cell engulfs and digests the dead cell, keeping the embryo free of harmful debris.

? Which usually comes first in the developmental history of a cell, its migration within the embryo or its differentiation into a specialized cell?

■ Migration

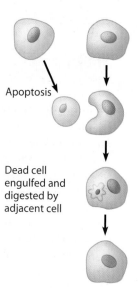

Figure 27.13B
Programmed cell death

Figure 27.13A Changes in cell shape during neural tube formation

Ectoderm

27.14 Embryonic induction initiates organ formation

All developmental processes depend on signals passed between neighboring cells and cell layers, telling embryonic cells precisely what to do when. The mechanism by which one group of cells influences the development of an adjacent group of cells is called **induction**. Induction may be mediated by diffusible signals or, if the cells are in direct contact, by cell-surface interactions. Induction plays a major role in the early development of virtually all tissues and organs from ectoderm, endoderm, and mesoderm. Its effect is to switch on a set of genes whose expression makes the receiving cells differentiate into a specific tissue. A sequence of inductive signals leads to increasingly greater specialization of cells as organs begin to take shape.

Figure 27.14 illustrates the differentiation of cells that form the vertebrate eye and two of the inductions that occur during this process. Cells destined to give rise to the eye actually begin receiving inductive signals during gastrulation. ❶ The eye begins to take shape from an outgrowth of the developing brain (the optic vesicle) and an adjacent cluster of cells on the body surface (the lens ectoderm). ❷ As a result of earlier inductions, some of the cells of the optic vesicle and lens ectoderm undergo shape changes that cause them to fold inward. The optic vesicle transforms into the optic cup, which will become the retina, and the optic stalk, which will become the optic nerve. ❸ Cells of the optic cup induce the lens ectoderm to start forming the lens of the

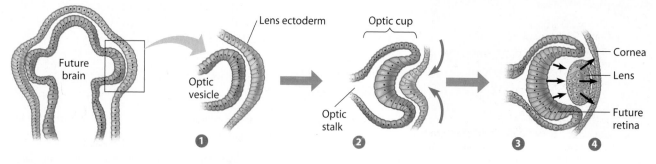

Figure 27.14 Induction during eye development

eye. (These inductive signals from the optic cup are shown as black arrows.) Finally, ❹ cells of the developing lens induce development of the cornea, which covers the lens.

Web/CD Thinking as a Scientist *What Determines Cell Differentiation in the Sea Urchin?*

? **How do signal transduction pathways function in induction?**

■ They mediate between the chemical signal received by the cell and the resulting changes in gene expression and other responses by the cell.

27.15 Pattern formation organizes the animal body

Forming the parts of an eye is one thing, but what about the development of an entire region of the body? An arm and a leg, for instance, share the same kinds of tissues—muscle, connective tissue, cartilage, and skin—but these tissues are arranged somewhat differently in the two limbs. What directs the formation of major body parts?

The shaping of an animal's major parts involves **pattern formation**, the emergence of a body form with specialized organs and tissues in the right places. Research indicates that master control genes respond to chemical signals that tell a cell where it is relative to other cells in the embryo (see Module 11.13). These positional signals determine which master control genes will be expressed and, consequently, which body parts will form.

Figure 27.15A indicates how positional signals affect the development of vertebrate limbs, which develop from embryonic structures called limb buds. Bird wings, for example, develop from the two anterior limb buds. For a wing to form properly, each embryonic wing cell must receive signals specifying its position in three dimensions: how close it is to the main axis of the embryo, to the anterior or posterior edge of the developing wing, and to the dorsal or ventral surface of

the embryo. Only with this information will the cell's genes direct the synthesis of the proteins needed for normal differentiation in that cell's specific location.

Experiments reveal that vertebrate limbs have zones of cells that provide positional information to other cells via chemical signals. Researchers have located one such pattern-forming zone on the posterior surface of the wing-forming limb buds of birds. Cells nearest the zone—presumably those exposed to the highest concentration of chemical signals from it—develop into posterior wing structures; cells farthest from the zone form anterior structures. As indicated in Figure 27.15B, if a block of cells from this zone is removed from one bird embryo (the donor) and grafted onto the anterior part of the limb bud of another embryo (the host), the host will develop additional wing structures—almost a double wing.

A major goal of developmental research is to learn how the one-dimensional information encoded in the nucleotide sequence of a zygote's DNA directs the development of the three-dimensional form of an animal. Pattern formation requires cells to receive and interpret environmental cues that vary from one location to another. These cues, acting together along three axes, tell cells where they are in the three-dimensional realm of a developing organ. In the next two modules, we'll see the results of this process as we watch an individual of our own species take shape.

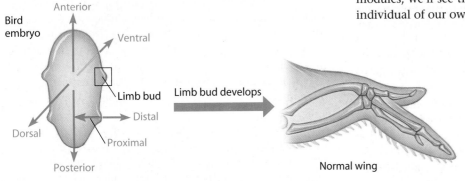

Figure 27.15A The normal development of a wing

? **How is pattern formation already apparent at the gastrula stage?** (*Hint:* Review Figure 27.11.)

■ The major axes of the animal—anterior/posterior, dorsal/ventral, and left/right—are already set at the gastrula stage.

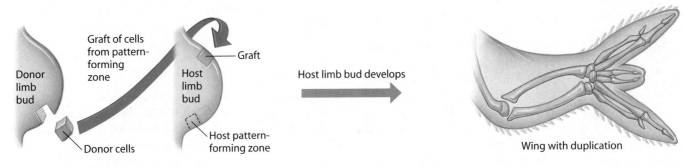

Figure 27.15B Experimental evidence for a pattern-forming zone

CHAPTER 27 *Reproduction and Embryonic Development*   **555**

27.16 The embryo and placenta take shape during the first month of pregnancy

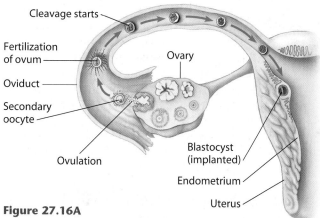

Figure 27.16A
From ovulation to implantation

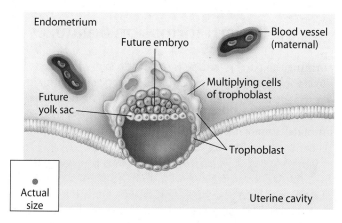

Figure 27.16C   Implantation under way (about 7 days)

Pregnancy, or **gestation**, is the carrying of developing young within the female reproductive tract. It begins at conception, the fertilization of the egg by a sperm, and continues until birth. Duration of pregnancy varies considerably among animal species; gestation in mice lasts about 21 days, while elephants carry their young for 600 days. Human pregnancy averages 266 days (38 weeks) from conception, or 40 weeks from the start of the last menstrual cycle.

An Overview of Developmental Events   The figures in this module illustrate, in cross section, the changes that occur during the first month of human development. The insets at the lower left of Figures 27.16C–27.16F show the embryo's actual size at each stage.

Conception occurs in the oviduct (**Figure 27.16A**). Cleavage starts about 24 hours after fertilization and continues as the embryo moves down the oviduct toward the uterus. By the sixth or seventh day after fertilization, the embryo has reached the uterus, and cleavage has produced about 100 cells. The embryo is now a hollow sphere of cells called a **blastocyst** (the mammalian equivalent of the sea urchin blastula we saw in Figure 27.10).

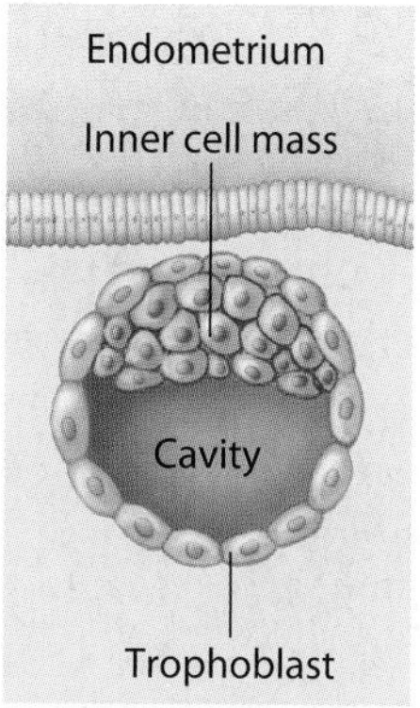

The human blastocyst (**Figure 27.16B**) has a fluid-filled cavity, an inner cell mass that will actually form the baby, and an outer layer of cells called the **trophoblast**. The trophoblast secretes enzymes that enable the blastocyst to implant in the endometrium, the uterine lining (gray in all the figures).

The blastocyst starts to implant in the uterus about a week after conception. In **Figure 27.16C**, you can see extensions of the trophoblast

Figure 27.16B
Blastocyst (6 days after conception)

spreading into the endometrium; these extensions consist of multiplying cells. The trophoblast cells eventually form part of the **placenta**, the organ that provides nourishment and oxygen to the embryo and helps dispose of its metabolic wastes. As we'll see, the placenta consists of both embryonic and maternal tissues.

In Figure 27.16C, the cells colored purple and yellow are derived from the inner cell mass. Most of the purple cells will give rise to the embryo. The yellow cells, some purple cells, and some trophoblast cells will give rise to four structures called the **extraembryonic membranes**, which develop as attachments to the embryo and help support it. You can see three of these membranes—the amnion (from purple cells), the yolk sac (from yellow cells), and the chorion (partly from trophoblast)—starting to take shape in **Figure 27.16D**. A later stage (**Figure 27.16E**) shows the fourth extraembryonic membrane, the allantois, developing as an extension of the yolk sac.

Gastrulation, the stage shown in Figure 27.16D, is under way by 9 days after conception. There is already evidence of the three embryonic layers—ectoderm (blue), endoderm (yellow), and mesoderm (pink). The embryo itself (not including the membranes) develops from the three inner cell layers shown in Figure 27.16E. The ectoderm layer will form the outer part of the embryo's skin. As indicated in the drawing, the ectoderm layer is continuous with the amnion. Similarly, the embryo's digestive tract will develop from the endoderm layer, which is continuous with the yolk sac. The bulk of most other organs will develop from the central layer of mesoderm.

Roles of the Extraembryonic Membranes   Figure 27.16F shows the embryo about a month after conception, along with its life-support system, made up largely of the four extraembryonic membranes. By this time, the **amnion** has grown to enclose the embryo. The amniotic cavity is filled with fluid, which protects the embryo. The amnion usually

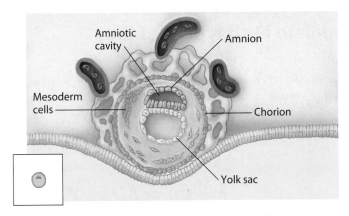

Figure 27.16D Embryonic layers and extraembryonic membranes starting to form (9 days)

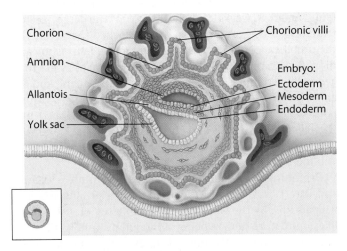

Figure 27.16E Three-layered embryo and four extraembryonic membranes (16 days)

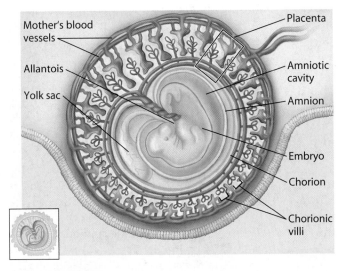

Figure 27.16F Placenta formed (31 days)

breaks just before childbirth, and the amniotic fluid ("water") leaves the mother's body through her vagina.

In humans and most other mammals, the **yolk sac** contains no yolk, but is given the same name as the homologous structure in other vertebrates. In a bird egg, the yolk sac contains a large mass of yolk. Isolated within a shelled egg outside the mother's body, a developing bird obtains nourishment from the yolk rather than from a placenta. In mammals, the yolk sac, which remains small, has other important functions. It produces the embryo's first blood cells and its first germ cells, the cells that will give rise to the gamete-forming cells in the gonads.

The **allantois** also remains small in mammals. It forms part of the umbilical cord—the lifeline between the embryo and the placenta. It also forms part of the embryo's urinary bladder. In birds and other reptiles, the allantois expands around the embryo and functions in waste disposal.

The outermost extraembryonic membrane, the **chorion**, develops from the trophoblast and from mesoderm cells derived from the yolk sac (see Figure 27.16D). The chorion becomes the embryo's part of the placenta. Cells in the chorion secrete a hormone called **human chorionic gonadotropin (HCG)**, which maintains production of estrogen and progesterone by the corpus luteum of the ovary during the first few months of pregnancy. Without these hormones, menstruation would occur, and the embryo would abort spontaneously. Levels of HCG in maternal blood are so high that some is excreted in the urine, where it can be detected by pregnancy tests.

The Placenta Notice in Figure 27.16D the knobby outgrowths on the outside of the chorion. In Figure 27.16E, these outgrowths, now called **chorionic villi**, are larger and contain mesoderm. In Figure 27.16F, the chorionic villi contain embryonic blood vessels formed from the mesoderm. By this stage, the placenta is fully developed. Starting with the chorion and extending outward, the placenta is a composite organ consisting of chorionic villi closely associated with the blood vessels of the mother's endometrium. The villi are actually bathed in tiny pools of maternal blood (purple in the figure). The mother's blood and the embryo's blood are not in direct contact. However, the chorionic villi absorb nutrients and oxygen from the mother's blood and pass these substances to the embryo via the chorionic blood vessels colored red. The chorionic vessels shown in blue carry wastes away from the embryo. The wastes diffuse into the mother's bloodstream and are excreted by her kidneys.

The placenta takes care of many of the embryo's needs. For example, it allows protective antibodies to pass from the mother to the fetus. However, the placenta cannot always protect the embryo from substances circulating in the mother's blood. A number of viruses—the German measles virus and HIV, for example—can cross the placenta. German measles can cause serious birth defects; HIV-infected babies usually die of AIDS within a few years. Most drugs, both prescription and not, also cross the placenta, and many can harm the developing embryo. Alcohol and the chemicals in tobacco smoke, for instance, raise the risk of miscarriage and birth defects. Alcohol can cause a set of birth defects called fetal alcohol syndrome, which includes mental retardation.

? Why does testing for HCG in a woman's urine or blood work as an early test of pregnancy?

■ Because this hormone is secreted by the chorion of an embryo

27.17 Human development from conception to birth is divided into three trimesters

In the previous module, we followed human development through the first four weeks. In this module, we use photographs to illustrate the rest of human development in the uterus. For convenience, we divide the period of human development from conception to birth into three **trimesters** of about 3 months each.

First Trimester

The first trimester is the time of most radical change for both mother and embryo. **Figure 27.17A** shows a human embryo about 5 weeks after fertilization. In that brief time, this highly organized multicellular embryo has developed from a single cell. Not shown here are the extraembryonic membranes that surround the embryo or most of the umbilical cord that attaches it to the placenta. This embryo is about 7 mm (0.28 in.) long and has a number of features in common with the somite stage of a frog embryo (see Figure 27.12C). The embryo has a notochord and a coelom, both formed from mesoderm. Its brain and spinal cord have begun to take shape from a tube of ectoderm, as in the frog. The human embryo also has four stumpy limb buds, a short tail, and elements of gill pouches. The gill pouches appear during embryonic development in all chordates; in land vertebrates, they eventually develop into parts of the throat and middle ear. Overall, a month-old human embryo is similar to other vertebrates at the somite stage of development.

Figure 27.17B shows a developing human, now called a fetus, about 9 weeks after fertilization. The large pinkish structure on the left is the placenta, attached to the fetus by the umbilical cord. The clear sac around the fetus is the amnion. By this time, the fetus is decidedly human, rather than generally vertebrate. It is about 5.5 cm (2.2 in.) long and has all of its organs and major body parts, including a disproportionately large head. The somites have developed into the segmental muscles and the bones of the back and ribs. The limb buds have become tiny arms and legs with fingers and toes.

Beginning at about 9 weeks, the fetus can move its arms and legs, turn its head, frown, and make sucking motions with its lips. By the end of the first trimester, the fetus looks like a miniature human being, although its head is still oversized for the rest of the body. The sex of the fetus is usually evident by this time, and its heartbeat can be detected with a stethoscope.

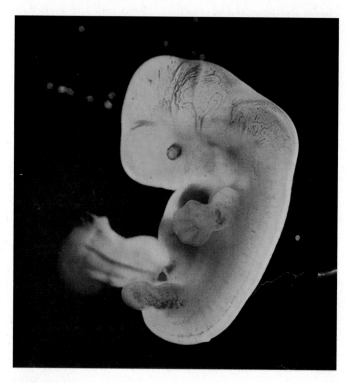

Figure 27.17A 5 weeks

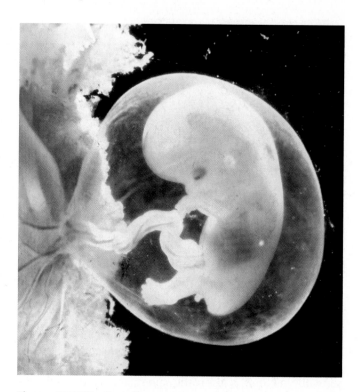

Figure 27.17B 9 weeks

Human development from conception to birth is divided into three trimesters

In the previous module, we followed human development through the first four weeks. In this module, we use photographs to illustrate the rest of human development in the uterus. For convenience, we divide the period of human development from conception to birth into three **trimesters** of about 3 months each.

First Trimester

The first trimester is the time of most radical change for both mother and embryo. **Figure 27.17A** shows a human embryo about 5 weeks after fertilization. In that brief time, this highly organized multicellular embryo has developed from a single cell. Not shown here are the extraembryonic membranes that surround the embryo or most of the umbilical cord that attaches it to the placenta. This embryo is about 7 mm (0.28 in.) long and has a number of features in common with the somite stage of a frog embryo (see Figure 27.12C). The embryo has a notochord and a coelom, both formed from mesoderm. Its brain and spinal cord have begun to take shape from a tube of ectoderm, as in the frog. The human embryo also has four stumpy limb buds, a short tail, and elements of gill pouches. The gill pouches appear during embryonic development in all chordates; in land vertebrates, they eventually develop into parts of the throat and middle ear. Overall, a month-old human embryo is similar to other vertebrates at the somite stage of development.

Figure 27.17B shows a developing human, now called a fetus, about 9 weeks after fertilization. The large pinkish structure on the left is the placenta, attached to the fetus by the umbilical cord. The clear sac around the fetus is the amnion. By this time, the fetus is decidedly human, rather than generally vertebrate. It is about 5.5 cm (2.2 in.) long and has all of its organs and major body parts, including a disproportionately large head. The somites have developed into the segmental muscles and the bones of the back and ribs. The limb buds have become tiny arms and legs with fingers and toes.

Beginning at about 9 weeks, the fetus can move its arms and legs, turn its head, frown, and make sucking motions with its lips. By the end of the first trimester, the fetus looks like a miniature human being, although its head is still oversized for the rest of the body. The sex of the fetus is usually evident by this time, and its heartbeat can be detected with a stethoscope.

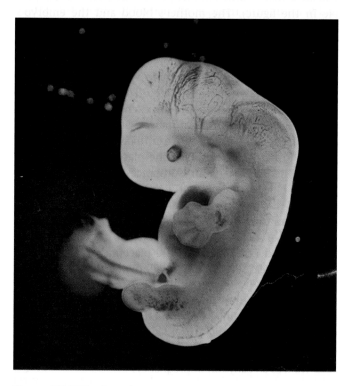

Figure 27.17A 5 weeks

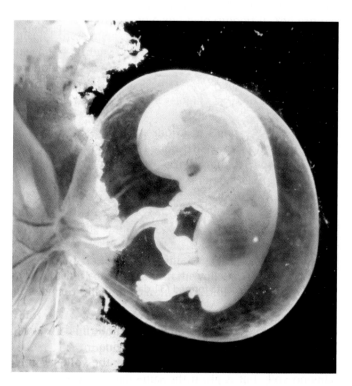

Figure 27.17B 9 weeks

Second Trimester

The main developmental changes during the second and third trimesters involve an increase in size and general refinement of the human features—nothing as dramatic as the changes of the first trimester. **Figure 27.17C** shows a fetus at 14 weeks, 2 weeks into the second trimester. The fetus is now about 6 cm (2.4 in.) long. During the second trimester, the placenta takes over the task of maintaining itself by secreting progesterone, rather than receiving it from the corpus luteum. At the same time, the placenta stops secreting HCG, and the corpus luteum, no longer needed to maintain pregnancy, degenerates.

At 20 weeks (**Figure 27.17D**), well into the second trimester, the fetus is about 19 cm (7.6 in.) long, weighs about half a kilogram (1 lb), and has the face of an infant, complete with eyebrows and eyelashes. Its arms, legs, fingers, and toes have lengthened. It also has fingernails and toenails and is covered with fine hair. Fetuses of that age are usually quite active. The mother's abdomen has become markedly enlarged, and she may often feel her baby move. Because of the limited space in the uterus, the fetus flexes forward into the so-called fetal position. By the end of the second trimester, the fetus's eyes are open and its teeth are forming.

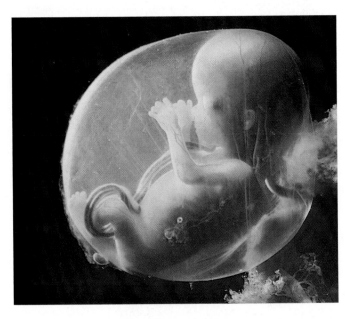

Figure 27.17C   14 weeks

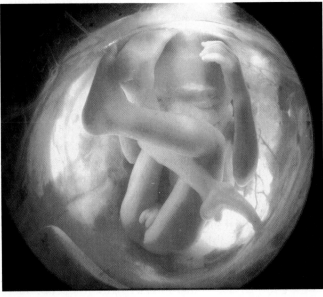

Figure 27.17D   20 weeks

Third Trimester

The third trimester (28 weeks to birth, **Figure 27.17E**) is a time of rapid growth as the fetus gains the strength it will need to survive outside the protective environment of the uterus. Babies born prematurely—as early as 24 weeks—may survive, but they require special medical care after birth. During the third trimester, the fetus's circulatory system and respiratory system undergo changes that will allow the switch to air breathing (see Module 22.12). The fetus gains the ability to maintain its own temperature, and its bones begin to harden and its muscles thicken. It also loses much of its fine body hair, except on its head. The head itself changes its proportions. The fetus becomes less active as it fills the space in the uterus. At birth, babies average about 50 cm (20 in.) in length and weigh 3–4 kg (6–8 lb).

? Certain drugs cause their most serious damage to an embryo very early in pregnancy, often before the mother even realizes she is pregnant. Why?

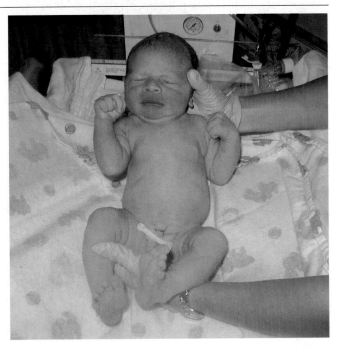

Because organ systems begin to develop early in the first trimester

Figure 27.17E   At birth

Childbirth is hormonally induced and occurs in three stages

The birth of a child is brought about by a series of strong, rhythmic contractions of the uterus, called **labor.** Several hormones are thought to play key roles (Figure 27.18A). One hormone, estrogen, reaches its highest level in the mother's blood during the last weeks of pregnancy. An important effect of this estrogen is to trigger the formation of numerous oxytocin receptors on the uterus. Cells of the fetus produce the hormone oxytocin, and late in pregnancy, the mother's pituitary gland secretes it in increasing amounts. Oxytocin stimulates powerful contractions of smooth muscles in the wall of the uterus. It also stimulates the placenta to make prostaglandins, local tissue regulators that also stimulate uterine muscle cells, making the muscles contract even more.

The hormonal induction of labor involves positive-feedback control. In this case, oxytocin and prostaglandins cause uterine contractions that in turn stimulate the release of more and more oxytocin and prostaglandins. The result is climactic—the intense muscle contractions that propel a baby from the uterus (womb).

Figure 27.18B shows the three stages of labor. As the process begins, the cervix (neck of the uterus) gradually opens, or dilates. ❶ The first stage, dilation, is the time from the onset of labor until the cervix reaches its full dilation of about 10 cm. Dilation is the longest stage of labor, lasting 6–12 hours or even considerably longer.

❷ The period from full dilation of the cervix to delivery of the infant is called the expulsion stage. Strong uterine contractions, lasting about 1 minute each, occur every 2–3 minutes, and the mother feels an increasing urge to push or bear down with her abdominal muscles. Within a period of 20 minutes to an hour or so, the infant is forced down and out of the uterus and vagina. An attending physician or midwife clamps and cuts the umbilical cord after the baby is expelled.

❸ The final stage is the delivery of the placenta, usually within 15 minutes after the birth of the baby.

Hormones continue to be important after the baby and placenta are delivered. Decreasing levels of progesterone and estrogen allow the uterus to start returning to its prepregnancy state. Less progesterone in the maternal blood also

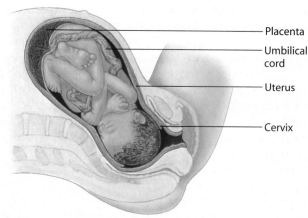

❶ Dilation of the cervix

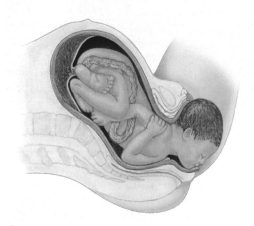

❷ Expulsion: delivery of the infant

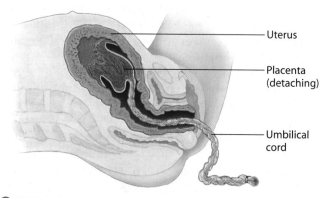

❸ Delivery of the placenta

Figure 27.18A The hormonal induction of labor

Figure 27.18B The three stages of labor

allows the pituitary hormone prolactin to promote milk production by the mammary glands. About 2–3 days after birth, the mother begins to secrete milk under the direct influence of both oxytocin and prolactin.

? The onset of labor is marked by dilation of the _____.

■ cervix

27.19 Reproductive technology increases our reproductive options

About 15% of couples who want children are unable to conceive, even after a year of unprotected sex. In most cases, such infertility can be traced to problems with the man. His testes may not produce enough sperm (a "low sperm count"), or those that are produced may be defective. Underproduction of sperm is frequently caused by the man's scrotum being too warm, so a switch of underwear from briefs (which hold the scrotum close to the body) to boxers may help. In other cases, infertility is caused by **impotence**, also called erectile dysfunction, the inability to maintain an erection. Female infertility can result from a lack of ova, a failure to ovulate, or blocked oviducts (often caused by scarring due to sexually transmitted diseases). Other women are able to conceive, but cannot support a growing embryo in the uterus.

Reproductive technologies can help many cases of infertility. Drug therapies (including Viagra) and penile implants can be used to treat impotence. If a man produces no functioning sperm, the couple may elect to use another man's sperm that had been anonymously donated to a sperm bank.

If a woman has normal ova that are not being released properly, hormone injections can induce ovulation. As discussed at the start of the chapter, such treatments frequently result in multiple pregnancies. If a woman has no ova of her own, they, too, can be obtained from a donor for fertilization and injection into the uterus. While sperm can be collected without any danger to the donor, collection of ova involves surgery and therefore pain and risk for the donating woman.

If a woman produces ova but is unable to support a growing fetus, she and her partner may hire a surrogate mother. In such cases, the couple enters into a legal contract with a woman who agrees to be implanted with the couple's embryo and carry it to birth. However, a number of states have laws restricting surrogate motherhood owing to the serious ethical and legal problems that can arise.

Many infertile couples turn to fertilization procedures called **assisted reproductive technology (ART)**. These procedures involve surgically removing eggs (secondary oocytes) from a woman's ovaries, fertilizing the eggs, and returning them to the woman's body. Eggs, sperm, and embryos from such procedures can be frozen for later pregnancy attempts.

With **in vitro fertilization (IVF)**, the most common ART procedure, a woman's eggs are mixed with sperm in culture dishes (in vitro) and incubated for several days to allow fertilized eggs to start developing (**Figure 27.19**). When they have developed into embryos of at least 8 cells each, the embryos are carefully inserted into the woman's uterus. In ZIFT (zygote intrafallopian transfer), eggs are also fertilized in vitro,

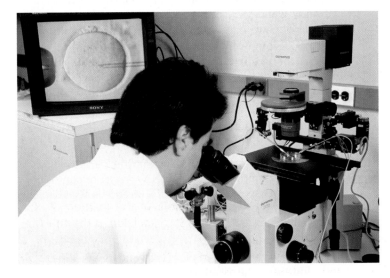

Figure 27.19 A technician performing in vitro fertilization

but zygotes are then transferred immediately to the woman's oviducts (fallopian tubes). In GIFT (gamete intrafallopian transfer), the eggs are not fertilized in vitro. Instead, the eggs and sperm are placed in the woman's oviducts, in the hope that fertilization will occur there.

Recent research has shown increased risks (lower birth weights and higher rates of birth defects) for babies born from IVF (compared with "natural" conception). Despite such risks and the high cost (typically $10,000 per attempt, whether it succeeds or not), IVF techniques are now performed at medical centers throughout the world and result in the birth of thousands of babies each year.

? Why is the term "test-tube baby" an inaccurate reference to the product of in vitro fertilization?

■ Because the embryo is transferred very early to a human container, the uterus of a woman, where the rest of its development occurs

■   ■   ■

In this chapter, we have watched a single-celled product of sexual reproduction, the zygote, become transformed into a new organism, complete with all organ systems. One of the first of those organ systems to develop is the nervous system. In the next chapter, we see how the nervous system functions together with the endocrine system to regulate virtually all body activities.

CHAPTER REVIEW

Reviewing the Concepts

Asexual and Sexual Reproduction (27.1)

Animal reproduction. In asexual reproduction, one parent produces genetically identical offspring by budding, fission, or fragmentation/regeneration. Asexual reproduction enables an individual to produce many offspring rapidly. Sexual reproduction involves the fusion of gametes from two parents, resulting in genetic variation among offspring. This may enhance reproductive success in changing environments **(27.1)**.

Human Reproduction (27.2–27.8)

The human reproductive system consists of a pair of ovaries (in females) or testes (in males), ducts that carry gametes, and structures for copulation. A woman's ovaries contain follicles that nurture eggs and produce sex hormones. Oviducts convey eggs to the uterus, where a fertilized egg develops. The uterus opens into the vagina, which receives the penis during intercourse and forms the birth canal **(27.2)**. A man's testes produce sperm, which are expelled through ducts during ejaculation. Several glands contribute to the formation of fluid that nourishes and protects sperm. This fluid and the sperm constitute semen **(27.3)**.

Spermatogenesis and oogenesis produce sperm and ova, respectively. Primary spermatocytes are made continuously in the testes; these diploid cells undergo meiosis to form four haploid sperm. Each month, one primary oocyte matures to form a secondary oocyte, which, if fertilized, completes meiosis and becomes a haploid ovum **(27.4)**:

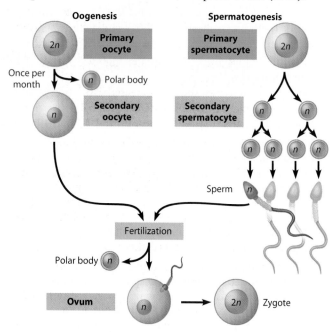

The female reproductive cycle. Hormones synchronize cyclic changes in the ovaries and uterus. Approximately every 28 days, the hypothalamus signals the anterior pituitary to secrete FSH and LH, which trigger the growth of a follicle and ovulation, the release of an egg. The follicle becomes the corpus luteum, which secretes both estrogen and progesterone. These two hormones stimulate the endometrium (the uterine lining) to thicken, preparing the uterus for implantation. They also inhibit the hypothalamus, reducing FSH and LH secretion. If the egg is not fertilized, the drop in LH shuts down the corpus luteum and its hormones. This triggers menstruation, the breakdown of the endometrium. The hypothalamus and pituitary then stimulate another follicle, starting a new cycle. If fertilization occurs, a hormone from the embryo maintains the uterine lining and prevents menstruation **(27.5)**.

Reproductive health. The human sexual response occurs in four phases: excitement, plateau, orgasm, and resolution **(27.6)**. Sexual intercourse carries the risk of exposure to sexually transmitted diseases **(27.7)**. Contraception can prevent pregnancy **(27.8)**.

Principles of Embryonic Development (27.9–27.15)

Fertilization and early embryonic development. In fertilization, a sperm releases enzymes that pierce the egg's coat. Sperm surface proteins bind to egg receptor proteins, sperm and egg plasma membranes fuse, and the two nuclei unite. Changes in the egg membrane prevent entry of additional sperm, and the fertilized egg (zygote) develops into an embryo **(27.9)**. Cleavage is a rapid series of cell divisions that results in a blastula, a ball of cells **(27.10)**. In gastrulation, cells migrate inward and form a rudimentary digestive cavity. The resulting gastrula has three layers of cells **(27.11)**:

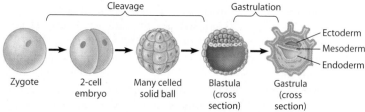

After gastrulation, the three embryonic tissue layers give rise to specific organ systems **(27.12)**. Tissues and organs take shape in a developing embryo as a result of cell shape changes, cell migration, and programmed cell death **(27.13)**. In a process called induction, adjacent cells and cell layers influence each other's differentiation via chemical signals **(27.14)**. Pattern formation, the emergence of the parts of a structure in their correct relative positions, involves the response of genes to spatial variations of chemicals in the embryo **(27.15)**.

Human Development (27.16–27.19)

Human development begins with fertilization in the oviduct. Cleavage produces a blastocyst, whose inner cell mass becomes the embryo. The blastocyst's outer layer, the trophoblast, implants in the uterine wall. Gastrulation occurs, and organs develop from the three embryonic layers. Meanwhile, the four extraembryonic membranes develop: the amnion, the chorion, the yolk sac, and the allantois. The embryo floats in the fluid-filled amniotic cavity, while the chorion and embryonic mesoderm form the embryo's part of the placenta. The placenta's chorionic villi absorb food and oxygen from the mother's blood **(27.16)**. Human embryonic development is divided into three trimesters of

about 3 months each. The most rapid changes occur during the first trimester. By 9 weeks, the embryo is called a fetus. The second and third trimesters are times of growth and preparation for birth **(27.17).** Hormonal changes induce birth. Estrogen makes the uterus more sensitive to oxytocin, which acts with prostaglandins to initiate labor. The cervix dilates, the baby is expelled by strong muscular contractions, and the placenta follows **(27.18).**

Reproductive technologies. New techniques can provide help to many infertile couples, but some of these methods raise important ethical and legal questions **(27.19).**

Connecting the Concepts

1. This graph plots the rise and fall of pituitary and ovarian hormones during the human ovarian cycle.

   Identify each hormone (A–D) and the reproductive events with which each one is associated (P–S). For A–D, choose from: estrogen, LH, FSH, and progesterone. For P–S, choose from: ovulation, growth of follicle, menstruation, and development of corpus luteum. How would the right-hand side of this graph be altered if pregnancy occurred? What other hormone is responsible for triggering this change?

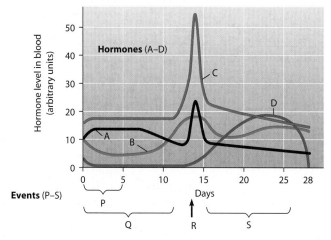

Testing Your Knowledge

Multiple Choice

2. After a sperm penetrates an egg, the fertilization envelope
   a. secretes important hormones.
   b. enables the fertilized egg to implant in the uterus.
   c. prevents more than one sperm from entering the egg.
   d. attracts additional sperm to the egg.
   e. activates the egg for embryonic development.

3. In an experiment, a researcher colored a bit of tissue on the outside of a frog gastrula with an orange fluorescent dye. The embryo developed normally. When the tadpole was placed under an ultraviolet light, which of the following glowed bright orange? (*Explain your answer.*)
   a. the heart   c. the brain   e. the liver
   b. the pancreas   d. the stomach

4. How does a zygote differ from an ovum?
   a. A zygote has more chromosomes.
   b. A zygote is smaller.
   c. A zygote consists of more than one cell.
   d. A zygote is much larger.
   e. A zygote divides by meiosis.

5. A woman had several miscarriages. Her doctor suspected that a hormonal insufficiency was causing the lining of the uterus to break down, as it does during menstruation, terminating her pregnancies. Treatment with which of the following might help her remain pregnant?
   a. oxytocin
   b. follicle-stimulating hormone
   c. testosterone
   d. luteinizing hormone
   e. prolactin

Matching

6. Turns into the corpus luteum   a. vas deferens
7. Female gonad   b. prostate gland
8. Site of spermatogenesis   c. endometrium
9. Site of fertilization in humans   d. testis
10. Human gestation occurs here   e. follicle
11. Sperm duct   f. uterus
12. Secretes seminal fluid   g. ovary
13. Lining of uterus   h. oviduct

Describing, Comparing, and Explaining

14. Compare sperm formation with egg formation. In what ways are the processes similar? In what ways are they different?

15. The embryos of reptiles (including birds) and mammals have systems of extraembryonic membranes. What are the functions of these membranes, and how do fish and frog embryos survive without them?

16. In an embryo, nerve cells grow out from the spinal cord and form connections with the muscles they will eventually control. What mechanisms described in this chapter might explain how these cells "know" where to go and which cells to connect with?

Applying the Concepts

17. As a frog embryo develops, the neural tube forms from ectoderm along what will be the frog's back, directly above the notochord. To study this process, a researcher carefully extracted a bit of notochord tissue and inserted it underneath the ectoderm where the belly of the frog would normally develop. What can the researcher hope to learn from this experiment? Predict the possible outcomes. What experimental control would you suggest?

18. Should parents undergoing in vitro fertilization have the right to choose which embryos to implant based on genetic criteria, such as the presence or absence of disease-causing genes? How about based on the sex of the embryo? How could you distinguish acceptable criteria from unacceptable ones? Do you think such options should be legislated? How would you formulate the law?

Answers to all questions can be found in Appendix 3.

For study help and Activities, go to campbellbiology.com or the student CD-ROM.

The late actor Christopher Reeve, an influential advocate for spinal cord research

NERVOUS SYSTEM STRUCTURE AND FUNCTION

28.1 Nervous systems receive sensory input, interpret it, and send out appropriate commands
28.2 Neurons are the functional units of nervous systems

NERVE SIGNALS AND THEIR TRANSMISSION

28.3 A neuron maintains a membrane potential across its membrane
28.4 A nerve signal begins as a change in the membrane potential
28.5 The action potential propagates itself along the neuron
28.6 Neurons communicate at synapses
28.7 Chemical synapses make complex information processing possible
28.8 A variety of small molecules function as neurotransmitters
28.9 Many drugs act at chemical synapses

AN OVERVIEW OF ANIMAL NERVOUS SYSTEMS

28.10 Nervous system organization usually correlates with body symmetry
28.11 Vertebrate nervous systems are highly centralized and cephalized
28.12 The peripheral nervous system of vertebrates is a functional hierarchy
28.13 Opposing actions of sympathetic and parasympathetic neurons regulate the internal environment
28.14 The vertebrate brain develops from three anterior bulges of the neural tube

THE HUMAN BRAIN

28.15 The structure of a living supercomputer: The human brain
28.16 The cerebral cortex is a mosaic of specialized, interactive regions
28.17 Injuries and brain operations provide insight into brain function
28.18 Several parts of the brain regulate sleep and arousal
28.19 The limbic system is involved in emotions, memory, and learning
28.20 Changes in brain physiology can produce neurological disorders

CHAPTER 28

Can an Injured Spinal Cord Be Fixed?

PROTECTED INSIDE THE BONY VERTEBRAE of the spine is an inch-thick gelatinous bundle of nervous tissue called the spinal cord. Shown in the image on the facing page, the spinal cord acts as the central communication conduit between the brain and the rest of the body. Millions of nerve fibers carry motor information from the brain to the muscles, while other fibers bring sensory information (about touch, pain, and body position, for example) from the body to the brain. The spinal cord acts like a transcontinental telephone cable jam-packed with wires, each of which carries messages to or from the central hub and an outlying area.

But what happens if that cable is cut? Signals cannot get through, communication is lost, and the cable must be repaired or replaced. In humans, the spinal cord is rarely severed because the vertebrae provide rigid protection. However, a traumatic blow to the spinal column and subsequent bleeding, swelling, and scarring can crush the delicate nerve bundles and prevent signals from passing. The result may be a debilitating injury. Such trauma along the back can cause paraplegia—paralysis

Millions of nerve fibers carry motor information from the brain to the muscles

of the lower half of the body. Trauma higher up in the neck can cause quadriplegia—paralysis from the neck down, which may necessitate permanent breathing assistance from an artificial respirator. Such injuries are usually permanent because the spinal cord, unlike other body tissues, cannot repair itself.

The late actor Christopher Reeve (best known for playing Superman in several movies) suffered a spinal cord injury during an equestrian competition. He was thrown from his horse and landed headfirst. Two vertebrae in his neck were fractured, crushing the spinal cord at the base of his skull and causing quadriplegia. Reeve died of complications related to his injury in 2004.

Over 10,000 Americans suffer spinal cord injuries each year. The most common causes are car crashes, violence (usually from gunshots), falls, and sports. Because the majority of spinal cord

Nervous Systems

injuries happen to people younger than 30, the subsequent disabilities often last for decades at great monetary and emotional cost.

Historically, spinal cord injuries have been considered untreatable. In fact, a 3,700-year-old Egyptian papyrus describes them as "an ailment not to be treated." Recently, however, there has been some minor progress: In 1988, researchers discovered that administering a powerful steroid drug within hours of a spinal cord injury limits its severity. But reversing spinal cord damage is a formidable challenge.

Some researchers are coaxing damaged nerve cells to regenerate by administering growth factor proteins (see Module 8.8) or transplanting the cells that produce growth factors to the site of the injury. Other researchers are attempting to block proteins that inhibit growth. Still others, believing that damaged nerve cells cannot be fixed, are trying to find ways to replace them with either mature nerve cells from elsewhere in the body or fetal tissue. A recent version of this approach is the attempt to use

Spinal cord

embryonic stem cells—progenitor cells capable of developing into all other cell types (see Module 11.5)—or partially differentiated neural stem cells to grow new nerve connections. Several recent studies involving combinations of these proposed therapies have shown promise in rats, even partially restoring motor function below the injury. While none of these strategies may ever "cure" a damaged spinal cord, they may still offer benefits of great value, such as regained control of the bladder, bowels, respiration, or a limb. Thanks to the efforts of Reeve and others to raise public awareness, spinal damage is now the subject of much research. The years ahead hold great promise for improving the prognosis after spinal cord injuries.

In this chapter, we explore the structure, function, and evolution of nervous systems. We focus in particular on the vertebrate nervous system and the structure and function of the human brain. We'll begin with an introduction to the central nervous system (the brain and spinal cord) and peripheral nervous system (the nerves that carry information to and from the rest of the body). ■ ■ ■

28.1 Nervous systems receive sensory input, interpret it, and send out appropriate commands

Nervous systems are the most intricately organized data-processing systems on Earth. Your brain, for instance, contains an estimated 100 billion neurons (nerve cells), which are specialized for carrying signals from one location in the body to another. A **neuron** consists of a cell body, containing the nucleus and cell organelles, and long, thin extensions called neuron fibers that convey signals. Each neuron may communicate with thousands of others, forming networks that enable us to learn, remember, perceive our surroundings, and move.

With few exceptions, nervous systems have two main divisions. The division called the **central nervous system (CNS)** consists of the brain and, in vertebrates, the spinal cord. The other division, the **peripheral nervous system (PNS)**, is made up mostly of communication lines called nerves that carry signals into and out of the CNS. A **nerve** is a cable-like bundle of neuron extensions tightly wrapped in connective tissue. In addition to nerves, the PNS also has **ganglia** (singular, *ganglion*), clusters of neuron cell bodies.

A nervous system has three interconnected functions (**Figure 28.1A**). **Sensory input** is the conduction of signals from sensory receptors, such as light-detecting cells of the eye, to integration centers. **Integration** is the interpretation of the sensory signals and the formulation of responses. **Motor output** is the conduction of signals from the integration centers to **effector cells**, such as muscle cells or gland cells, which perform the body's responses. The integration of sensory input and motor output is not usually rigid and linear, but involves the continuous background activity symbolized by the circular arrow in Figure 28.1A.

The relationship between neurons and nervous system structure and function is easiest to see in the relatively simple circuits that produce **reflexes**, or automatic responses to stimuli (**Figure 28.1B**). The small colored balls in the figure represent neuron cell bodies; the thin lines represent neuron fibers. Three functional types of neurons correspond to a nervous system's three main functions: **Sensory neurons** convey signals (information) from sensory receptors into the CNS. **Interneurons** are located entirely within the CNS. They integrate data and then relay appropriate signals to other interneurons or to motor neurons. **Motor neurons** convey signals from the CNS to effector cells. (For simplicity, this figure shows only one neuron of each functional type, but virtually any body activity actually involves many of each.)

When the knee is tapped, ❶ a sensory receptor detects a stretch in the tendon, and ❷ a sensory neuron conveys this information into the CNS (spinal cord). In the CNS, the information goes to ❸ a motor neuron and to ❹ an interneuron. One set of muscles (quadriceps) responds to motor signals conveyed by a motor neuron by contracting, jerking the lower leg forward. At the same time, another motor neuron, responding to signals from the interneuron, inhibits the flexor muscles, making them relax and not resist the action of the quadriceps.

As you can see from this example, the nervous system depends on the ability of neurons to receive and convey signals. Next, we'll examine how that is done.

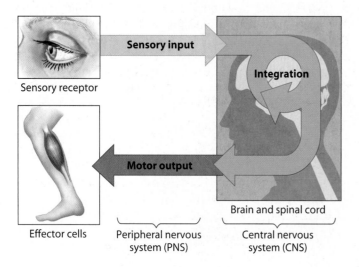

Figure 28.1A Organization of a nervous system

Sensory receptor — Sensory input → Integration
Effector cells — Motor output

Effector cells | Peripheral nervous system (PNS) | Central nervous system (CNS)
Brain and spinal cord

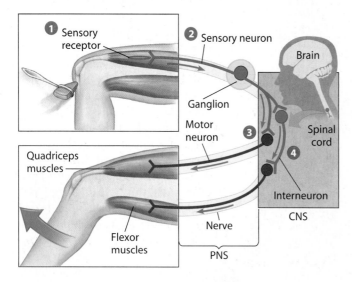

Figure 28.1B The knee-jerk reflex

❶ Sensory receptor
❷ Sensory neuron
Brain
Ganglion
Motor neuron ❸
Spinal cord
❹
Quadriceps muscles
Interneuron
CNS
Nerve
Flexor muscles
PNS

? Arrange the following neurons into the correct sequence for information flow during the knee-jerk reflex: interneuron, sensory neuron, motor neuron.

■ Sensory neuron → interneuron → motor neuron

Neurons are the functional units of nervous systems

The ability of neurons to receive and transmit information depends on their structure. **Figure 28.2** depicts a motor neuron, like those that carry command signals from your spinal cord to your skeletal muscles. The inset shows an SEM of a neuron.

Most of a neuron's organelles, including its nucleus, are located in the **cell body**. Arising from the cell body are two types of extensions: numerous dendrites and a single axon. **Dendrites** (from the Greek *dendron*, tree) are highly branched extensions that *receive* signals from other neurons and convey this information toward the cell body. Dendrites are often short. In contrast, the **axon** is typically a much longer extension that *transmits* signals to other cells, which may be other neurons or effector cells. Some axons, such as the ones that reach from your spinal cord to muscle cells in your feet, can be over a meter long.

Neurons actually make up only part of a nervous system. Outnumbering neurons by as many as 50 to 1 are **supporting cells,** or **glia,** that are essential for the structural integrity and normal functioning of the nervous system. Figure 28.2 shows one kind of supporting cell called a Schwann cell, which is found in the PNS. (Similar cells, called oligodendrocytes, are found in the CNS.) In many animals, axons that convey signals very rapidly are enclosed along most of their length by a thick insulating material. In vertebrates, the insulating material, called the **myelin sheath,** resembles a chain of oblong beads. Each bead is actually a Schwann cell, and the myelin sheath is essentially a chain of Schwann cells, each wrapped many times around the axon. The spaces between Schwann cells are called **nodes of Ranvier,** and they are the only points on the axon where signals can be transmitted. Everywhere else, the myelin sheath insulates the axon, preventing signals from passing along it. When a signal travels along a myelinated axon, it jumps from node to node, as indicated in the figure. By jumping along the axon, the signal travels much faster than it could if it had to take the long route along the whole length of the axon. In the human nervous system, signals can travel along a myelinated axon about 150 m/sec (over 330 mi/hr!), which means that a command from your brain can make your fingers move in just a few milliseconds. Without myelin sheaths, the signals could go only about 5 m/sec.

The debilitating disease multiple sclerosis (MS) demonstrates the importance of myelin sheaths. MS leads to a gradual destruction of myelin sheaths by the individual's own immune system. The result is a progressive loss of signal conduction, muscle control, and brain function.

Notice in Figure 28.2 that the axon ends in a cluster of branches. A typical axon has hundreds or thousands of these branches, each with a **synaptic terminal** at the very end. The site of communication between a synaptic terminal and another cell is called a **synapse.** As we will see later, this is where information is passed between neurons. With the basic structure of a neuron in mind, let's take a closer look at the signals that neurons convey.

Web/CD Activity 28A *Neuron Structure*

? What is the function of the myelin sheath?

■ It speeds up conduction of signals along axons.

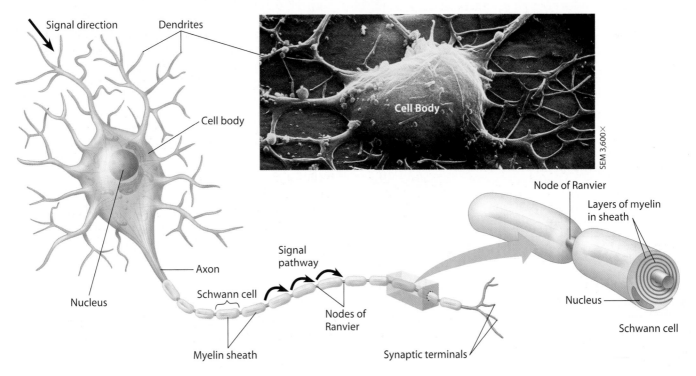

Figure 28.2 Structure of a myelinated motor neuron

NERVE SIGNALS AND THEIR TRANSMISSION

28.3 A neuron maintains a membrane potential across its membrane

To understand nerve signals, we must first study a resting neuron, one that is not transmitting a signal. A resting neuron has potential energy, energy that can be put to work to send signals from one part of the body to another. This potential energy, called the **membrane potential**, resides in an electrical charge difference across the neuron's plasma membrane. The cytoplasm just inside the membrane is negative in charge, and the fluid just outside the membrane is positive. Since opposite charges tend to move toward each other, a membrane stores energy by holding opposite charges apart, like a battery. The strength (voltage) of a neuron's stored energy can be measured with microelectrodes connected to a voltmeter (**Figure 28.3A**). The voltage across the plasma membrane of a resting neuron is called the **resting potential.** A neuron's resting potential is about −70 millivolts (mV)—about 5% of the voltage in a flashlight battery. (The minus sign indicates that the inside of the cell is negative relative to the outside.)

The resting potential exists because of differences in ionic composition of the fluids inside and outside the cell. The plasma membrane surrounding the neuron has channels and pumps, made of proteins, that regulate the passage of inorganic ions. One result is that a resting membrane allows much more potassium (K⁺) than sodium (Na⁺) to diffuse across the membrane. Notice in Figure 28.3B that Na⁺ (blue)

is more concentrated outside the cell than inside; the Na⁺ channels allow very little to diffuse in. But K⁺ (green), which is more concentrated inside, can freely flow out. As the positively charged potassium ions diffuse out, they leave behind an excess of negative charge. Also helping maintain the resting potential are membrane proteins called **sodium-potassium (Na⁺-K⁺) pumps.** These pumps actively transport Na⁺ out of the cell and K⁺ in, thereby helping keep the concentration of Na⁺ low in the cell and K⁺ high.

This ionic gradient (high K⁺/low Na⁺ concentrations inside coupled with low K⁺/high Na⁺ concentrations outside) produces an electrical potential difference, or voltage, across the membrane—the resting potential. Notice an important point: The membrane potential can change from its resting value if the membrane's permeability to particular ions changes. As we will see in the next module, this is the basis of nearly all electrical signals in the nervous system.

? If a neuron's membrane suddenly becomes more permeable to sodium ions, there is a rapid net movement of Na⁺ into the cell. What are the two forces that drive the ions inward?

■ The greater concentration of Na⁺ outside the cell than inside and the membrane potential (negatively charged inside vs. outside) favor the inward diffusion of Na⁺.

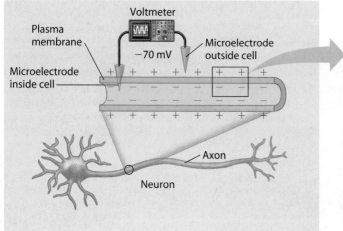

Figure 28.3A Measuring a neuron's resting potential

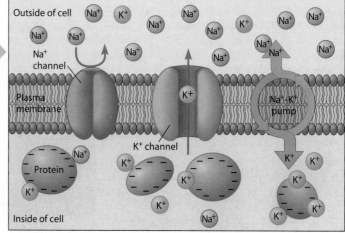

Figure 28.3B How the resting potential is generated

28.4 A nerve signal begins as a change in the membrane potential

Turning on a flashlight uses the potential energy stored in a battery to create light. In a similar way, stimulating a neuron's plasma membrane can trigger the release and use of the membrane's potential energy to generate a nerve signal. A **stimulus** is any factor that causes a nerve signal to be generated. Examples of stimuli include light, sound, a tap on the knee, or a chemical signal from another neuron.

The discovery of giant axons in squids (up to 1 mm in diameter) gave researchers their first chance to study how stimuli trigger signals in a living neuron. From microelectrode studies with squid neurons, British biologists A. L. Hodgkin and A. F. Huxley worked out the details of nerve signal transmission in the 1940s. Their findings, summarized in **Figure 28.4**, apply to neurons in all animals.

Graphing electrical changes in neuron membranes was the first step in discovering how nerve signals are generated. The multicolored line on the graph below traces the electrical changes that make up the **action potential**, a nerve signal that carries information along an axon. The graph records electrical events over time (in milliseconds) at a particular place on the membrane where a stimulus is applied. ❶ The graph starts out at the membrane's resting potential (-70 mV). ❷ The stimulus is applied. If it is strong enough, the voltage rises to what is called the **threshold** (-50 mV, in this case). The difference between the threshold and the resting potential is the minimum change in the membrane's voltage that must occur to generate the action potential. ❸ Once the threshold is reached, the action potential is triggered. The membrane polarity reverses abruptly, with the interior of the cell becoming positive with respect to the outside. ❹ The membrane then rapidly repolarizes as the voltage drops back down, ❺ undershoots the resting potential, and ❶ finally returns to it.

What actually causes the electrical changes of the action potential? The rapid flip-flop of the membrane potential results from the rapid movements of ions across the membrane at Na$^+$ and K$^+$ channels (called *voltage-gated channels* because they have special gates that open and close, depending on changes in membrane potential). The diagrams surrounding the graph show the ion movements. Starting at lower left,

❶ the resting membrane is positively charged on the outside, and the cytoplasm just inside the membrane is negatively charged. ❷ A stimulus triggers the opening of a few Na$^+$ channels in the membrane, and a tiny amount of Na$^+$ enters the axon. This makes the inside surface of the membrane slightly less negative. If the stimulus is strong enough, a sufficient number of Na$^+$ channels open to raise the voltage to the threshold. ❸ Once the threshold is reached, more Na$^+$ channels open, allowing even more Na$^+$ to diffuse into the cell. As more and more Na$^+$ moves in, the voltage soars to its peak. ❹ The peak voltage triggers closing and inactivation of the Na$^+$ channels. Meanwhile, the K$^+$ channels open, allowing K$^+$ to diffuse rapidly out. These changes produce the downswing on the graph. ❺ A very brief undershoot of the resting potential results because the K$^+$ channels close slowly. ❶ The membrane then returns to its resting potential. In a typical neuron, this entire process is very brief—only about 1–2 msec in duration.

Web/CD Thinking as a Scientist *What Triggers Nerve Impulses?*

? In what way is an action potential a self-boosted response—that is, an example of positive feedback?

■ The opening of Na$^+$ gates caused by stimulation of the neuron changes the membrane potential, and this change causes more of the voltage-gated Na$^+$ channels to open.

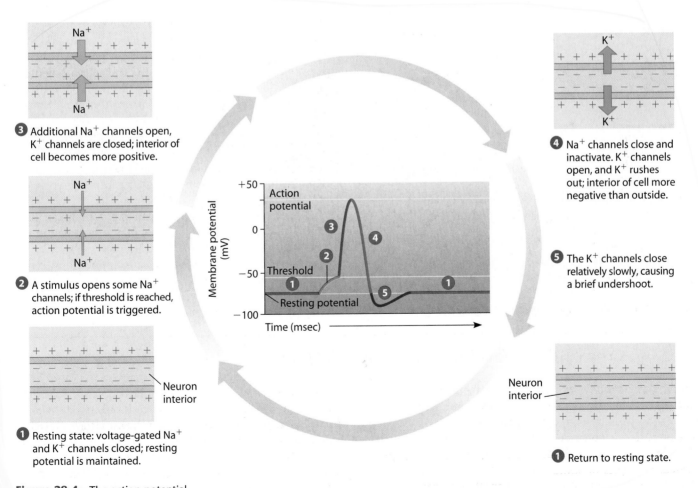

3 Additional Na$^+$ channels open, K$^+$ channels are closed; interior of cell becomes more positive.

2 A stimulus opens some Na$^+$ channels; if threshold is reached, action potential is triggered.

1 Resting state: voltage-gated Na$^+$ and K$^+$ channels closed; resting potential is maintained.

4 Na$^+$ channels close and inactivate. K$^+$ channels open, and K$^+$ rushes out; interior of cell more negative than outside.

5 The K$^+$ channels close relatively slowly, causing a brief undershoot.

1 Return to resting state.

Figure 28.4   The action potential

The action potential propagates itself along the neuron

For an action potential to function as a long-distance signal, it must travel along the axon from the cell body to the synaptic terminals. It does so by regenerating itself along the axon. You can follow this process in **Figure 28.5**. A nerve signal starts out as an action potential generated in the axon near the cell body of the neuron. The effect of this action potential is like tipping the first of a row of standing dominoes: The first domino does not travel along the row, but its fall is relayed to the end of the row, one domino at a time.

The three steps in Figure 28.5 show the changes that occur in an axon segment as a nerve signal passes from left to right. Let's first focus on the axon region on the far left. ❶ When this region of the axon (blue) has its Na⁺ channels open, Na⁺ rushes inward (blue arrows), and an action potential is generated. This corresponds to the upswing of the curve (step 2) in Figure 28.4. ❷ Soon, the K⁺ channels in that same region open, allowing K⁺ to diffuse out of the axon (green arrows); at this time, its Na⁺ channels are closed and inactivated, and we would see the downswing of the action potential. ❸ A short time later, we would see no signs of an action potential at this (far-left) spot because the axon membrane here has returned to its resting potential.

Now let's see how these events lead to the "domino effect" of a nerve signal. In step 1 of the figure, the blue arrows pointing sideways within the axon indicate local spreading of the electrical changes caused by the inflowing Na⁺ associated with the first action potential. These changes trigger the opening of Na⁺ channels in the membrane just to the right of the action potential. As a result, a second action potential is generated, as indicated by the blue region in step 2. In the same way, a third action potential is generated in step 3, and each action potential generates another all the way down the axon.

So why are action potentials propagated in only one direction along the axon (left to right in the figure)? As the blue arrows indicate, local electrical changes do spread in both directions in the axon. However, these changes cannot open Na⁺ channels and generate an action potential when the Na⁺ channels are inactivated (step 4 in Figure 28.4). Thus, an action potential cannot be generated in the regions where K⁺ is leaving the axon (green in the figure) and Na⁺ channels are still inactivated. Consequently, the inward flow of Na⁺ that depolarizes the axon membrane ahead of the action potential cannot produce another action potential behind it. Once an action potential starts, it moves only in the one direction, toward the synaptic terminals.

So we see that a nerve signal propagates itself in one direction by the electrical changes it produces in the neuron membrane. If you rap your finger on a desk, for instance, the contact is a stimulus that triggers action potentials in the tips of sensory neurons in your skin. The action potentials propagate along the axons, carrying the information (that your finger has hit a hard object) into your central nervous system.

Action potentials are *all-or-none* events; that is, they are the same no matter how strong or weak the stimulus that triggers them. How, then, do action potentials relay different intensities of information to your central nervous system? It is the *frequency* of action potentials that changes with the intensity of stimuli. If you rap your finger hard against the desk, your CNS receives many more action potentials per millisecond than after a soft tap.

In the past three modules, we've examined how nerve signals are conducted along a single neuron. In the next four modules, we focus on how signals pass from one neuron to another.

Web/CD Activity 28B *Nerve Signals: Action Potentials*

? During an action potential, ions move across the neuron membrane in a direction perpendicular to the direction of the impulse along the neuron. What is it that actually travels along the neuron as the signal?

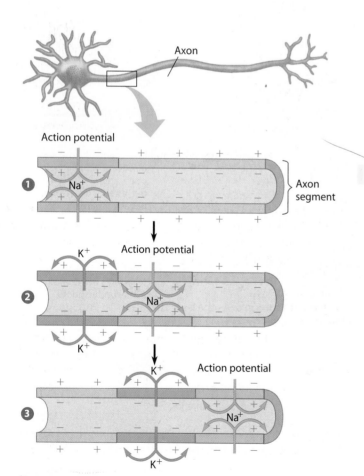

Figure 28.5 Propagation of the action potential along an axon

■ The signal is the wavelike change in membrane potential; the self-perpetuated action potential that regenerates sequentially at points farther and farther away from the site of stimulation.

Neurons communicate at synapses

When an action potential reaches the end of an axon, it generally stops there. In most cases, action potentials are not transmitted from cell to cell. However, *information* is transmitted, and this transmission occurs at synapses—the regions of communication between a synaptic terminal and another cell.

Synapses come in two varieties: electrical and chemical. In an electrical synapse, electrical current passes directly from one neuron to the next. The receiving neuron is stimulated quickly and at the same frequency of action potentials as the sending neuron. Lobsters and many fishes can flip their tails with lightning speed because the neurons that carry signals for these movements communicate by electrical synapses. In the human body, electrical synapses are found in the heart and digestive tract, where nerve signals maintain steady, rhythmic muscle contractions. In contrast, chemical synapses are prevalent in most other organs, including the central nervous system, where signaling among neurons is more complex and varied.

Unlike electrical synapses, chemical synapses have a narrow gap, called the **synaptic cleft**, separating a synaptic terminal of the sending neuron from the receiving neuron. The cleft prevents the action potential in the sending neuron from spreading directly to the receiving neuron. Instead, the action potential (an electrical signal) is first converted to a chemical signal. The chemical signal, consisting of molecules of **neurotransmitter** (see Module 26.1), may then generate an action potential in the receiving cell.

Now let's follow the events that occur at a chemical synapse in Figure 28.6. The neurotransmitter is contained in **synaptic vesicles** in the sending neuron's synaptic terminals. ❶ An action potential (red arrow) arrives at the synaptic terminal. ❷ The action potential triggers chemical changes that cause some synaptic vesicles to fuse with the plasma membrane of the sending cell. ❸ The fused vesicles release their neurotransmitter molecules (brown) by exocytosis into the synaptic cleft, and the neurotransmitter diffuses across the cleft.

What happens next varies among different types of chemical synapses. ❹ In one common type, the released neurotransmitter binds to receptor molecules on the receiving cell's plasma membrane. ❺ The binding of neurotransmitter to receptor opens *chemically gated ion channels* in the receiving cell's membrane. With the channels open, ions can diffuse into the receiving cell and trigger new action potentials. ❻ The neurotransmitter is broken down by an enzyme or transported back into the signaling cell, and the ion channels close. Step 6 ensures that the neurotransmitter's effect is brief and precise.

You can review what we have covered so far by thinking about what is happening right now in your own nervous

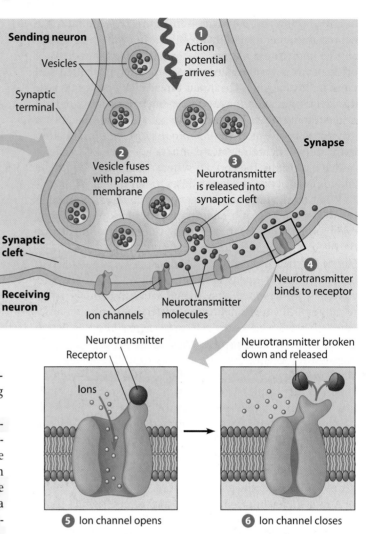

Figure 28.6 Neuron communication

system. Action potentials carrying information about the words on this page are streaming along sensory neurons from your eyes to your brain. Arriving at synapses with receiving cells (interneurons in the brain), the action potentials are triggering the release of neurotransmitters at the ends of the sensory neurons. The neurotransmitters are diffusing across synaptic clefts and triggering changes in some of your interneurons—changes that lead to integration of the signals and ultimately to what the signals actually mean (in this case, the meaning of words and sentences). Next, motor neurons in your brain will send out action potentials to muscle cells in your fingers, telling them to contract in just the right way to turn the page.

Web/CD Activity 28C *Neuron Communication*

> **?** How does a synapse ensure that signals pass *only* in one direction, from a sending neuron to a receiving cell?

■ The signal can go only one way at any one synapse because the sending neuron releases neurotransmitter, and only the receiving cell has receptors for the neurotransmitter.

28.7 Chemical synapses make complex information processing possible

As the drawing and micrograph in **Figure 28.7** indicate, one neuron may interact with many others. In fact, a neuron may receive information via neurotransmitters from hundreds of other neurons via thousands of synaptic terminals (red and green in the drawing). The inputs can be highly varied because each sending neuron may secrete a different quantity or kind of neurotransmitter. The membrane of a neuron resembles a tiny circuit board, receiving and processing bits of information in the form of neurotransmitter molecules. These living circuit boards account for the nervous system's ability to process data and formulate appropriate responses to stimuli.

What do neurotransmitters actually do to receiving neurons? The binding of a neurotransmitter to a receptor may open ion channels in the receiving cell's plasma membrane or trigger a signal transduction pathway that does so. The effect of the neurotransmitter depends on the kind of membrane channel it opens. Neurotransmitters that open Na$^+$ channels, for instance, may trigger action potentials in the receiving cell. Such neurotransmitters and the synapses where they are released are referred to as excitatory (green in the drawing). In contrast, many neurotransmitters open membrane channels for ions that *decrease* the tendency to develop action potentials in the receiving cell—such as channels that admit Cl$^-$ or release K$^+$. These neurotransmitters and their synapses are called inhibitory (red). The effects of both excitatory and inhibitory neurotransmitters can vary in magnitude. In general, the more neurotransmitter molecules that bind to receptors on the receiving cell and the closer the synapse is to the base of the receiving cell's axon, the stronger the effect.

A receiving neuron's membrane may receive signals—both excitatory and inhibitory—from many different presynaptic neurons. If the excitatory signals are collectively strong enough to raise the membrane potential to threshold, an action potential will be generated in the receiving cell. That neuron then passes signals along its axon to other cells at a rate that represents a summation of all the information it has received. (Signal frequency is key because action potentials are all-or-none events.) Each new receiving cell, in turn, processes this information along with all its other inputs.

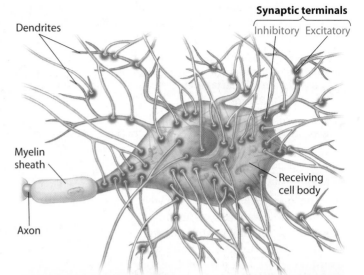

Figure 28.7 A neuron's multiple synaptic inputs

? Contrast excitatory and inhibitory synapses in how they change a receiving cell's membrane potential relative to triggering an action potential.

■ Neurotransmitters from an excitatory synapse open ion channels that move the receiving cell's membrane potential closer to threshold; neurotransmitters from an inhibitory synapse open ion channels that move the cell's membrane potential farther from threshold.

28.8 A variety of small molecules function as neurotransmitters

As we've discussed, the propagation of nerve signals across chemical synapses depends on neurotransmitters. A variety of small molecules serve this function.

Many neurotransmitters are small, nitrogen-containing organic molecules. One, called **acetylcholine**, is important in the brain and at synapses between motor neurons and muscle cells. Depending on the kind of receptors on receiving cells, acetylcholine may be excitatory or inhibitory. For instance, acetylcholine makes our skeletal muscles contract but slows the rate of contraction of cardiac muscles.

Nitrogen-containing neurotransmitters called **biogenic amines** are derived from amino acids. The biogenic amines include epinephrine, norepinephrine, serotonin, and dopamine, all of which also function as hormones (see Chapter 26). Biogenic amines are important neurotransmitters in the central nervous system. Serotonin and dopamine affect sleep, mood, attention, and learning. Imbalances of biogenic amines are associated with various disorders. For example, the degenerative illness Parkinson's disease is associated with a lack of dopamine in the brain, and an excess of

dopamine is linked to schizophrenia. Reduced levels of nor-epinephrine and serotonin seem to be linked with some types of depression. Some psychoactive drugs, including LSD, apparently produce their hallucinatory effects by binding to serotonin and dopamine receptors in the brain.

Four other neurotransmitters—aspartate, glutamate, glycine, and GABA (gamma aminobutyric acid)—are actually amino acids. All are known to be important in the central nervous system. Aspartate and glutamate are excitatory; glycine and GABA are inhibitory.

Several peptides, relatively short chains of amino acids, also serve as neurotransmitters. One, called substance P, is an excitatory neurotransmitter that mediates our perception of pain. The endorphins are peptides that function as both neurotransmitters and hormones, decreasing our perception of pain during times of physical or emotional stress.

Neurons also use some dissolved gases, notably nitric oxide (NO), as chemical signals. During sexual arousal in human males, certain neurons release NO into the erectile tissue of the penis, and the NO triggers an erection. (The erectile dysfunction drug Viagra promotes this effect of NO.) Neurons produce NO molecules on demand, rather than storing them in synaptic vesicles. The dissolved gas diffuses into neighboring cells, produces a change, and is quickly broken down.

? What determines whether a neuron is affected by a specific neurotransmitter?

■ To be affected by a particular neurotransmitter, a neuron must have specific receptors for that neurotransmitter.

CONNECTION

28.9 Many drugs act at chemical synapses

Many psychoactive drugs, even common ones such as caffeine, nicotine, and alcohol, affect the action of neurotransmitters in the brain's billions of synapses (**Figure 28.9**). Caffeine, found in coffee, tea, chocolate, and many soft drinks, keeps us awake by countering the effects of inhibitory neurotransmitters. Nicotine acts as a stimulant by binding to and activating acetylcholine receptors. Alcohol is a strong depressant. Its precise effect on the nervous system is not yet known, but it seems to increase the inhibitory effects of the neurotransmitter GABA.

Many prescription drugs used to treat psychological disorders alter the effects of neurotransmitters (see Module 28.20). The most popular class of antidepressant medication works by affecting the action of serotonin. Called selective serotonin reuptake inhibitors (SSRIs), these medications block the removal of serotonin from synapses, increasing the amount of time this mood-altering neurotransmitter is available to receiving cells. Tranquilizers such as diazepam (Valium) and alprazolam (Xanax) activate the receptors for GABA, increasing the effect of this inhibitory neurotransmitter. In other cases, a drug may bind to and block a receptor, reducing a neurotransmitter's effect. For instance, some antipsychotic drugs used to treat schizophrenia block dopamine receptors. Some drugs used to treat attention deficit hyperactivity disorder (ADHD), such as methylphenidate (Ritalin), are chemically similar to dopamine and norepinephrine. ADHD medications are believed to block the reuptake of these neurotransmitters, but their precise actions are poorly understood.

What about illegal drugs? Stimulants such as amphetamines and cocaine increase the release and availability of norepinephrine and dopamine at synapses. Abuse of these drugs can produce symptoms resembling schizophrenia. The active ingredient in marijuana (tetrahydrocannabinol, or THC) binds to brain receptors normally used by another neurotransmitter (anandamide) that seems to play a role in pain, depression, appetite, memory, and fertility. Opiates—morphine, codeine,

and heroin—bind to endorphin receptors, reducing pain and producing euphoria. Not surprisingly, opiates can be used medicinally for pain relief. However, abuse of opiates may permanently change the brain's chemical synapses and reduce normal synthesis of neurotransmitters. As explained in Module 28.15, these drugs are also highly addictive.

The drugs discussed here are used for a variety of purposes, both medicinal and recreational. While they have the ability to increase alertness and sense of well-being or to reduce physical and emotional pain, they also have the potential to disrupt the brain's finely tuned neural pathways, altering the chemical balances that are the product of millions of years of evolution.

Figure 28.9 Alcohol, nicotine, and caffeine—drugs that alter the effects of neurotransmitters

? When people say that "alcohol lowers a person's inhibitions," it is a behavioral description. At the neurological level, it is probably more accurate to say that "alcohol raises inhibitions." Why?

■ Alcohol probably depresses the brain by enhancing the inhibitory effects of GABA.

28.10 Nervous system organization usually correlates with body symmetry

To this point, we have concentrated on the cellular mechanisms that are fundamental to the nervous system. There is remarkable uniformity throughout the animal kingdom in the way nerve cells function. However, there is great variety in how nervous systems as a whole are organized. In this module, we will briefly survey animal nervous systems, starting with the simplest.

Some animals lack a nervous system altogether; sponges, for instance, have no cells that are specialized for generating and transmitting nerve signals. Hydras (cnidarians) have one of the simplest types of nervous systems (Figure 28.10A). A hydra has a **nerve net,** a diffuse, weblike system of neurons extending throughout the body. A hydra has no head or brain, and its nerve net has no central or peripheral divisions. The neurons of the nerve net control the contractions of the digestive cavity.

Hydras and other cnidarians and adult echinoderms are radially symmetrical, and so are their nervous systems. Radially symmetrical nervous systems tend to be uncentralized, but this does not necessarily imply that they are structurally or functionally simple. Sea stars and many other echinoderms have a nerve net in each arm, connected by radial nerves to a central nerve ring. This organization is better suited than a diffuse nerve net for controlling complex motion, such as the coordinated movements of hundreds of tube feet that allow a sea star to move in one direction.

Most animals are bilaterally symmetrical, with a head and tail and a tendency to move headfirst through the environment. The head—most often the first part of the animal to encounter new stimuli—is usually equipped with sense organs and a brain. Flatworms are the simplest animals that show two evolutionary hallmarks of bilateral symmetry: **cephalization,** or concentration of the nervous system at the head end, and **centralization,** the presence of a central nervous system (CNS) distinct from a peripheral nervous system (PNS). The planarian worm in Figure 28.10B has a small brain composed of ganglia (clusters of nerve cell bodies) and two parallel **nerve cords** that control the animal's movements. These elements constitute the simplest clearly defined central nervous system in the animal kingdom. Nerves that connect the CNS with the rest of the body make up the peripheral nervous system.

More complex bilaterally symmetrical invertebrates exhibit the same basic pattern of a CNS composed of a brain and one or more nerve cords, but the CNS tends to be more complex than in flatworms. For instance, the brain of a leech (Figure 28.10C) contains a greater concentration of neurons than that of a flatworm, and the ventral nerve cord contains segmentally arranged ganglia. The insect shown in Figure 28.10D has a brain composed of several fused ganglia, and its ventral nerve cord also has a ganglion in each body segment. Each of these ganglia directs the activity of muscles in its segment of the body.

Molluscs are good examples of how the structure of a nervous system correlates with how animals interact with the environment. Sessile or slow-moving molluscs such as clams have little or no cephalization and relatively simple sense organs. In contrast, squids and octopuses have the most sophisticated invertebrate nervous systems, rivaling those of some vertebrates. As shown in Figure 28.10E, the large brain of a squid, accompanied by complex eyes and rapid signaling along giant axons, correlates well with the active predatory life of these animals.

In the next several modules, we explore the complex nervous systems of our own subphylum, the vertebrates.

? Why is it advantageous for the brain of most bilateral animals to be located at the head end?

■ Cephalization places the brain near major sense organs that are concentrated on the end of the animal that leads the way as the animal moves through its environment.

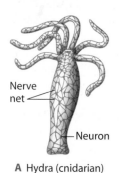

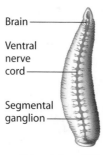

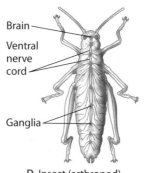

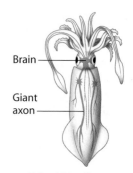

A Hydra (cnidarian)   **B** Flatworm (planarian)   **C** Leech (annelid)   **D** Insect (arthropod)   **E** Squid (mollusc)

Figure 28.10   Invertebrate nervous systems (not to scale)

Vertebrate nervous systems are highly centralized and cephalized

Vertebrate nervous systems are diverse in both structure and level of sophistication. For instance, the brains of dolphins and humans are much more complex structurally than the brains of frogs or fishes; they are also much more powerful integrators. However, all vertebrate nervous systems have some fundamental similarities. All have distinct central and peripheral elements and are highly centralized and cephalized. In all vertebrates, the brain and spinal cord make up the CNS, while the PNS comprises the rest of the nervous system (Figure 28.11A). The **spinal cord**, which runs lengthwise inside the vertebral column, or spine, conveys information to and from the brain and integrates simple responses to certain kinds of stimuli (such as the knee-jerk reflex). The master control center, the **brain**, includes homeostatic centers that keep the body functioning smoothly; sensory centers that integrate data from the sense organs; and (in humans, at least) centers of emotion and intellect. The brain also sends out motor commands to muscles.

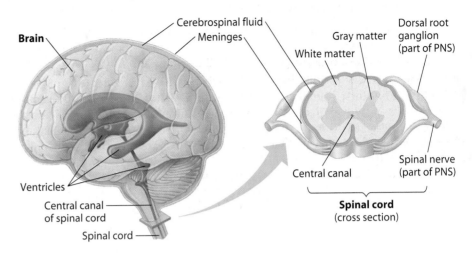

Figure 28.11B Fluid-filled spaces of the vertebrate CNS

A vast network of blood vessels services the CNS. Brain capillaries are more selective than those elsewhere in the body. They allow essential nutrients and oxygen to pass freely into the brain, but keep out many other chemicals, such as metabolic wastes from other parts of the body. This selective mechanism, called the **blood-brain barrier**, maintains a stable chemical environment for the brain.

Both the brain and spinal cord contain fluid-filled spaces (Figure 28.11B). Fluid-filled spaces called **ventricles** in the brain are continuous with the narrow **central canal** of the spinal cord. These cavities are filled with **cerebrospinal fluid**, which is formed in the brain by the filtration of blood. Circulating slowly through the central canal and ventricles (and then draining back into veins), the cerebrospinal fluid cushions the CNS and assists in the supply of nutrients and hormones and the removal of wastes. Also protecting the brain and spinal cord are layers of connective tissue, called **meninges**. In mammals, cerebrospinal fluid circulates between two of the meninges, providing an additional protective cushion for the CNS. As shown in the spinal cord diagram in Figure 28.11B, the CNS has two distinct areas. **White matter** is mainly composed of axons (with their whitish myelin sheaths); **gray matter** consists mainly of nerve cell bodies and dendrites.

The ganglia and nerves of the vertebrate PNS are a vast communication network. **Cranial nerves** originate in the brain and terminate mostly in structures of the head and upper body (your eyes, nose, and ears, for instance). **Spinal nerves** originate in the spinal cord and extend to parts of the body below the head. All spinal nerves and most cranial nerves contain both sensory neurons and motor neurons. Thousands of incoming and outgoing signals pass each other within the same nerves all the time.

Next, let's take a closer look at the vertebrate PNS.

? A vertebrate's central nervous system consists of the _____ and the _____.

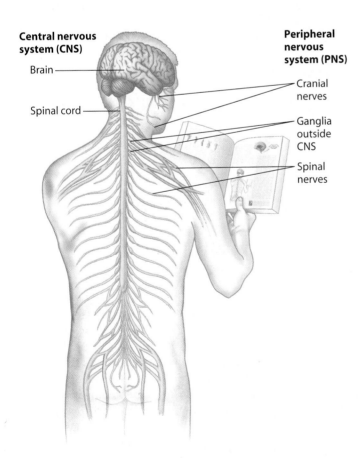

Figure 28.11A A vertebrate nervous system (back view)

■ brain . . . spinal cord

28.12 The peripheral nervous system of vertebrates is a functional hierarchy

The PNS can be divided into two functional components: the somatic nervous system and the autonomic nervous system (**Figure 28.12**). The **somatic nervous system** carries signals to and from skeletal muscles, mainly in response to *external* stimuli. When you touch a hot stove, for instance, these neurons carry commands to your arm to pull away. The somatic nervous system is said to be voluntary because many of its actions (which we take up in Chapter 30) are under conscious control. But much skeletal muscle activity is actually controlled by reflexes mediated by the spinal cord or brainstem.

The **autonomic nervous system** regulates the *internal* environment by controlling smooth and cardiac muscles and the organs of the digestive, cardiovascular, excretory, and endocrine systems. This control is generally involuntary.

As you can see in Figure 28.12, the autonomic nervous system is composed of three divisions: sympathetic, parasympathetic, and enteric. In the next module, we discuss the varied functions of the autonomic nervous system in more detail.

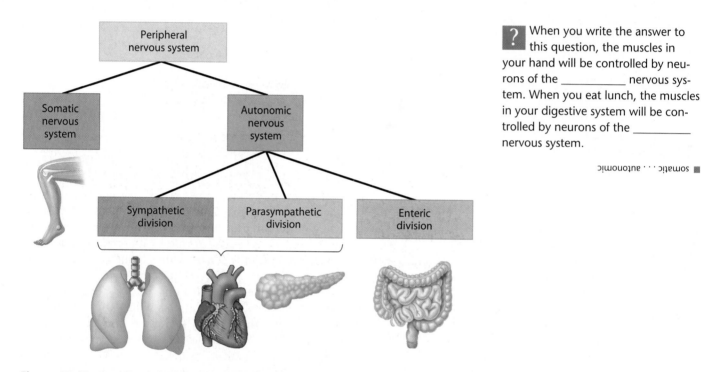

? When you write the answer to this question, the muscles in your hand will be controlled by neurons of the _____ nervous system. When you eat lunch, the muscles in your digestive system will be controlled by neurons of the _____ nervous system.

■ somatic . . . autonomic

Figure 28.12 Functional divisions of the vertebrate PNS

28.13 Opposing actions of sympathetic and parasympathetic neurons regulate the internal environment

Your autonomic nervous system contains two sets of neurons with opposing effects on most body organs. One set, called the **parasympathetic division**, primes the body for activities that gain and conserve energy for the body ("rest and digest"). A sample of the effects of parasympathetic signals appears on the left in **Figure 28.13**. These include stimulating the digestive organs, such as the salivary glands, stomach, and pancreas; decreasing the heart rate; and increasing glycogen production.

The other set of neurons, the **sympathetic division**, tends to have the opposite effect, preparing the body for intense, energy-consuming activities, such as fighting, fleeing,

or competing in a strenuous game (the "fight-or-flight" response). You see some of the effects of sympathetic signals on the right side of the figure. The digestive organs are inhibited, the bronchi dilate so that more air can pass through them, the heart rate increases, the liver releases the energy compound glucose into the blood, and the adrenal glands secrete the hormones epinephrine and norepinephrine.

Fight-or-flight and relaxation are opposite extremes. Your body usually operates at intermediate levels, with most of your organs receiving both sympathetic and parasympathetic signals (the liver and adrenal medulla are exceptions). The opposing signals adjust an organ's activity to a suitable level.

The **enteric division** (not shown in Figure 28.13) of the autonomic nervous system consists of complex networks of neurons in the digestive tract, pancreas, and gallbladder. The enteric networks control the secretions of these organs as well as activity in the smooth muscles that produce peristalsis. Although the enteric division can function independently, it is normally regulated by the sympathetic and parasympathetic divisions.

Sympathetic and parasympathetic neurons emerge from different regions of the CNS, and they use different neurotransmitters. As the green dots in the figure indicate, neurons of the parasympathetic system emerge from the brain and the lower part of the spinal cord. Most parasympathetic neurons produce their effects by releasing the neurotransmitter acetylcholine at synapses within target organs. In contrast, neurons of the sympathetic system emerge from the middle regions of the spinal cord (red dots). Most sympathetic neurons release the neurotransmitter norepinephrine at target organs.

While it is convenient to divide the PNS into somatic and autonomic components, it is important to realize that these two divisions cooperate to maintain homeostasis. In response to a drop in body temperature, for example, the brain signals the autonomic nervous system to constrict surface blood vessels, which reduces heat loss. At the same time, the brain also signals the somatic nervous system to cause shivering, which increases heat production.

In the next several modules, we take a closer look at the highest level of the nervous system's structural hierarchy: the brain.

? How would a drug that inhibits the parasympathetic nervous system affect a person's pulse?

■ The pulse, or heart rate, would probably increase.

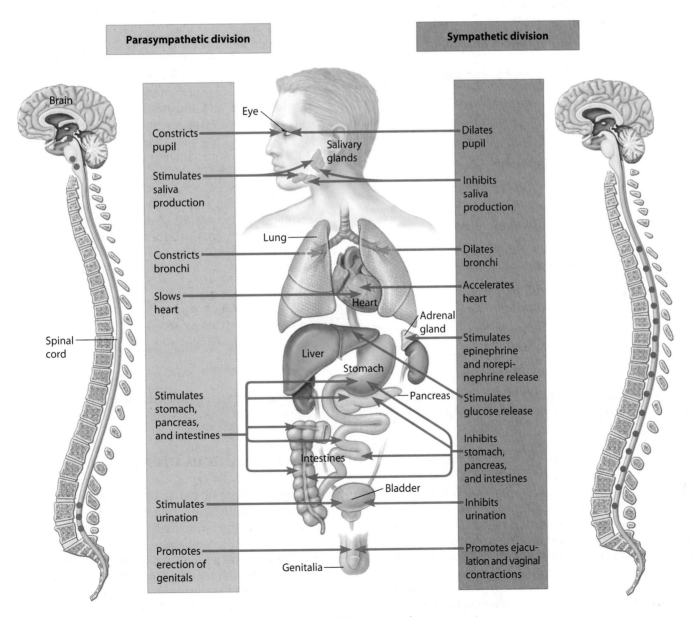

Parasympathetic division	Sympathetic division

Brain

Eye — Constricts pupil / Dilates pupil

Salivary glands

Stimulates saliva production / Inhibits saliva production

Lung

Constricts bronchi / Dilates bronchi

Slows heart / Accelerates heart

Heart

Adrenal gland

Stimulates epinephrine and norepinephrine release

Spinal cord

Liver

Stomach

Stimulates stomach, pancreas, and intestines

Pancreas

Stimulates glucose release

Inhibits stomach, pancreas, and intestines

Intestines

Bladder

Stimulates urination / Inhibits urination

Promotes erection of genitals / Promotes ejaculation and vaginal contractions

Genitalia

Figure 28.13 The parasympathetic and sympathetic divisions of the autonomic nervous system

28.14 The vertebrate brain develops from three anterior bulges of the neural tube

We begin our study of the vertebrate brain by examining its embryonic development from the dorsal hollow nerve cord, one of the four distinguishing features of chordates (see Module 18.14). During early embryonic development in all vertebrates, three bilaterally symmetrical bulges—the **forebrain**, **midbrain**, and **hindbrain**—appear at the anterior end of the neural tube (**Figure 28.14**, left). In the course of vertebrate evolution, the forebrain and hindbrain gradually became subdivided—both structually and functionally—into regions that assume specific responsibilities.

Another trend in brain evolution was the increasing integrative power of the forebrain. Evolution of the most complex vertebrate behavior paralleled the evolution of the **cerebrum**, an outgrowth of the forebrain that is the most sophisticated center of homeostatic control and integration. The cerebrum of birds and mammals is much larger relative to other parts of the brain than the cerebrum of fishes, amphibians, and other reptiles. And the sophisticated behavior of birds and mammals is directly correlated with their large cerebrum.

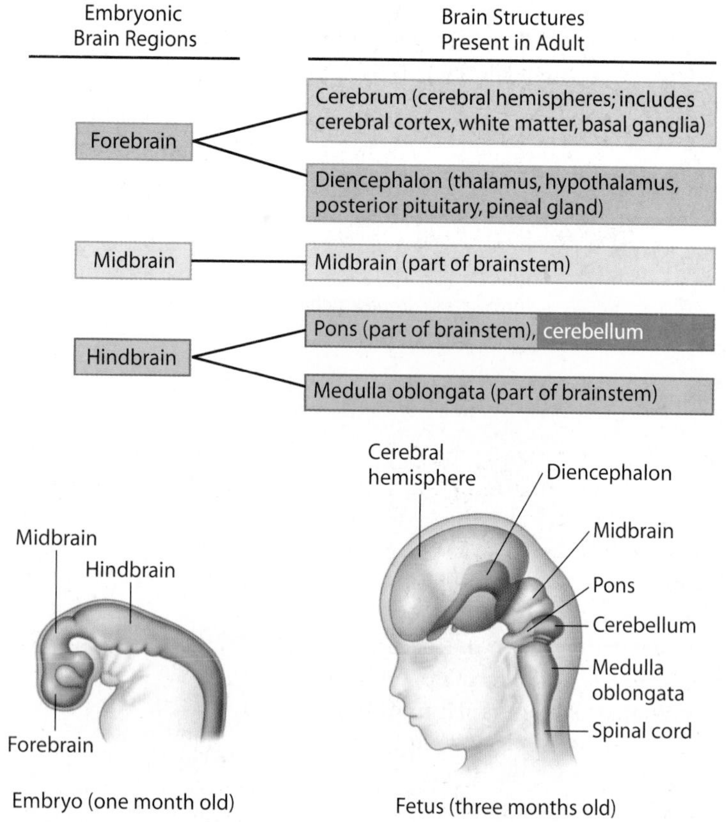

During the embryonic development of the human brain, the most profound changes occur in the region of the forebrain. Rapid, expansive growth of this region during the second and third months creates the two halves of the cerebrum, called the left and right cerebral hemispheres, which extend over and around much of the rest of the brain (Figure 28.14, right). By the sixth month of development, foldings increase the surface area of the outer cerebral hemispheres. This extensively convoluted outer region is called the cerebral cortex.

Porpoises, whales, and primates also have large and complex cerebral cortices. Porpoises communicate using a large repertory of sounds. They also have the ability to locate objects, such as prey, using sound echoes. Much of a porpoise's cerebral cortex may be devoted to processing information about its sound-oriented world. In contrast, the brain of humans (one species of primate) is strongly oriented toward visual perceptions. Humans have the largest brain surface area, relative to body size, of all animals. Next we take a look at the main components of the human brain.

Figure 28.14 Embryonic development of the human brain

? Which region of the brain has changed the most during the course of vertebrate evolution?

■ The cerebrum, in particular the cerebral cortex

THE HUMAN BRAIN

28.15 The structure of a living supercomputer: The human brain

Composed of perhaps as many as 100 billion intricately organized neurons, with a much larger number of supporting cells, the human brain is more powerful than the most sophisticated computer. In **Figure 28.15A** and **Table 28.15**, we see that the three ancestral brain regions have evolved considerably from their original state.

Looking first at the lower brain centers, two sections of the hindbrain (blue), the **medulla oblongata** and **pons**, and the midbrain (purple) make up a functional unit called the **brainstem**. Consisting of a stalk with cap-like swellings at the anterior end of the spinal cord, the brainstem is, evolutionarily speaking, one of the older parts of the vertebrate brain. The brainstem coordinates and filters the conduction of information from sensory and motor neurons to the higher brain regions. It also regulates sleep and arousal and helps coordinate body movements, such as walking. Table 28.15 lists some of the individual functions of the medulla oblongata, pons, and midbrain.

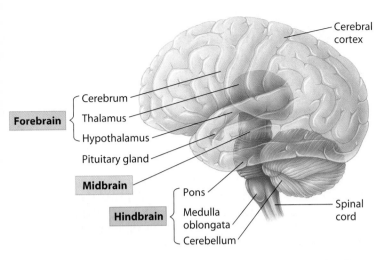

Figure 28.15A The main parts of the human brain

TABLE 28.15 MAJOR STRUCTURES OF THE HUMAN BRAIN

Brain Structure	Major Functions
Brainstem	Conducts data to and from other brain centers; helps maintain homeostasis; coordinates body movement
Medulla oblongata	Controls breathing, circulation, swallowing, digestion
Pons	Controls breathing
Midbrain	Receives and integrates auditory data; coordinates visual reflexes; sends sensory data to higher brain centers
Cerebellum	Coordinates body movement; plays role in learning and in remembering motor responses
Thalamus	Serves as input center for sensory data going to the cerebrum, output center for motor responses leaving the cerebrum; sorts data
Hypothalamus	Functions as homeostatic control center; controls pituitary gland; serves as biological clock
Cerebrum	Performs sophisticated integration; plays major role in memory, learning, speech, emotions; formulates complex behavioral responses

Another part of the hindbrain, the **cerebellum** (light blue), is a planning center for body movements. Evidence indicates that it also plays a role in learning, decision making, and remembering motor responses. The cerebellum receives sensory information about the position of joints and the length of muscles, as well as information from the auditory and visual systems. It also receives input concerning motor commands issued by the cerebrum. The cerebellum uses this information to coordinate movement and balance. Hand-eye coordination is an example of such control. If the cerebellum is damaged, the eyes can follow a moving object, but they will not stop at the same place as the object.

The most sophisticated integrating centers are those derived from the forebrain (orange and gold)—the thalamus, the hypothalamus, and the cerebrum. The **thalamus** contains most of the cell bodies of neurons that relay information to the cerebral cortex. The thalamus first sorts data into categories (all the touch signals from a hand, for instance). It also suppresses some signals and enhances others. The thalamus then sends information on to the appropriate higher brain centers for further interpretation and integration.

In Module 26.4, we saw that the hypothalamus controls the pituitary gland and the secretion of many hormones. The hypothalamus also regulates body temperature, blood pressure, hunger, thirst, the sex drive, and fight-or-flight responses, and it helps us experience emotions such as rage and pleasure. A "pleasure center" in the hypothalamus could also be called an addiction center, for it is strongly affected by certain addictive drugs, such as amphetamines and cocaine. As described in Module 28.9, these drugs increase the effects of norepinephrine and dopamine at synapses in the pleasure center, producing a short-term high, often followed by depression. Cocaine addiction may involve chemical changes in the pleasure center and elsewhere in the hypothalamus.

In another part of the hypothalamus, clusters of neurons known as suprachiasmatic nuclei function as an internal timekeeper, our **biological clock.** Receiving visual input from the eyes (light/dark cycles, in particular), the clock maintains our **circadian rhythms**—patterns that are repeated daily—such as the sleep/wake cycle. Research with many different species has shown that without environmental cues, biological clocks keep time in a free-running way. For example, when humans are placed in artificial settings that lack environmental cues, our biological clocks and circadian rhythms maintain a cycle of approximately 24 hours 11 minutes.

The cerebrum, the largest and most complex part of our brain, consists of right and left **cerebral hemispheres** (Figure 28.15B), each of which is responsible for the opposite side of the body. A thick band of nerve fibers called the **corpus callosum** facilitates communication between the cerebral hemispheres, enabling them to process information together. Under the corpus callosum, groups of neurons called the **basal ganglia** (or basal nuclei) are important in motor coordination. If they are damaged, a person may be immobilized. Degeneration of the basal ganglia occurs in Parkinson's disease (see Module 28.20). The most extensive portion of our cerebrum, the cerebral cortex, is the subject of the next module.

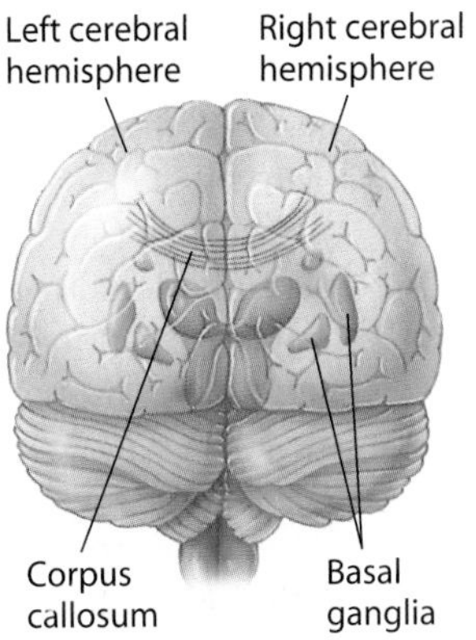

Figure 28.15B A rear view of the brain

? Choosing from the structures in Table 28.15, identify the brain part most important in solving an algebra problem.

■ Cerebrum

28.16 The cerebral cortex is a mosaic of specialized, interactive regions

Although less than 5 mm thick, the highly folded **cerebral cortex** accounts for about 80% of the total human brain mass. It contains some 10 billion neurons and hundreds of billions of synapses. Its intricate neural circuitry produces our most distinctive human traits: reasoning and mathematical abilities, language skills, imagination, artistic talent, and personality traits. Assembling information it receives from our eyes, ears, nose, taste buds, and touch sensors, the cortex also creates our sensory perceptions—what we are actually aware of when we see, hear, smell, taste, or touch. The cerebral cortex also regulates our voluntary movements.

Like the rest of the cerebrum, the cerebral cortex is divided into right and left sides connected by the corpus callosum. Each side of the cerebral cortex has four lobes (represented by different colors in Figure 28.16, which shows the left cerebral hemisphere). Researchers have identified a number of functional areas within each lobe.

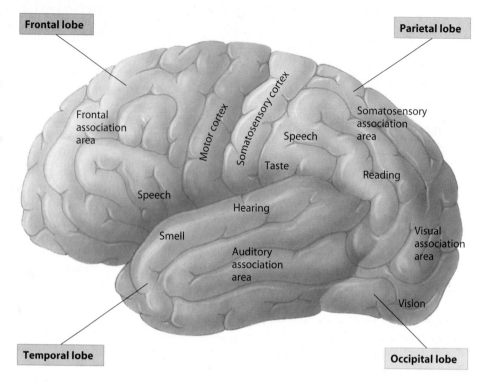

Figure 28.16 Functional areas of the cerebrum's left hemisphere

Two areas of known function form the boundary between the frontal and parietal lobes. One, called the motor cortex, functions mainly in sending commands to skeletal muscles, signaling appropriate responses to sensory stimuli. Next to the motor cortex, the somatosensory cortex receives and partially integrates signals from touch, pain, pressure, and temperature receptors throughout the body. The cerebral cortex also has centers that receive and begin processing sensory information concerned with vision, hearing, taste, and smell. Each of these centers, as well as the somatosensory cortex, cooperates with an adjacent area, called an association area. Imaging techniques, such as the PET scans described in Module 20.11, are beginning to show how a complicated interchange of signals among receiving centers and association areas produces our sensory perceptions.

Making up most of our cerebral cortex, the association areas are the sites of higher mental activities—roughly, what we call thinking. In humans, a large association area in the frontal lobe uses varied inputs from many other areas of the brain to evaluate consequences, make considered judgments, and plan for the future. Language results from extremely complex interactions among several association areas. For instance, the parietal lobe of the cortex has association areas used for reading and speech. These areas obtain visual information (the appearance of words on a page) from the vision centers. Then, if the words are to be spoken aloud, they arrange the information into speech patterns and tell another speech center, in the frontal lobe, how to make the motor cortex move the tongue, lips, and other muscles to form words. When we hear words, the parietal areas perform similar functions using information from auditory centers of the cortex. You can see the locations of the various language centers in the PET scans in Figure 20.11D.

You may have heard people comment that they are "left-brained" or "right-brained." In a phenomenon known as **lateralization**, areas in the two hemispheres become specialized for different functions during infant and child brain development. In most people, the left hemisphere becomes adept at language, logic, and mathematical operations, as well as detailed skeletal motor control and processing of fine visual and auditory details. The right hemisphere is stronger at spatial relations, pattern and face recognition, and emotional processing in general. (In about 10% of us, these roles of the left and right hemispheres are reversed or the hemispheres are less specialized.)

How have researchers identified the functions of different parts of the brain? We explore this question next.

? A stroke that causes loss of speech and numbness of the right side of the body has probably damaged brain tissue in the _____ lobe of the _____ hemisphere.

■ parietal . . . left

28.17   Injuries and brain operations provide insight into brain function

The physiology of the human brain is exquisitely complex, making it one of the most difficult structures to study in all of biology. No animal model or computer simulation can accurately predict its complicated functions. New techniques, such as PET scans and MRIs (see Module 20.11), are allowing researchers to associate specific parts of the brain with various activities. Much of what has been learned about the brain, however, has come from rare individuals whose brains were altered through injury, illness, or surgery. By studying such "broken brains," researchers have gained insight into how healthy brains operate.

The first well-publicized case of this type involved a man named Phineas Gage. In 1848, while working as a railroad construction foreman, Gage accidentally exploded a dynamite charge that propelled a 3-foot-long spike through his head. The 13-pound steel rod entered his left cheek and traveled upward behind his left eye and out the top of his skull, landing several yards away. Incredibly, Gage walked away from the accident and appeared to have an intact intellect. However, his associates soon noticed drastic changes in his personality, with new propensities toward meanness, vulgarity, irresponsibility, and an inability to control his behavior.

At the time, Gage's doctor was able to note these changes, but understanding of the brain was insufficient to explain them. Luckily, the doctor preserved Gage's skull and the spike, allowing a group of researchers in 1994 to produce a computer model of the injury (Figure 28.17A). The modern analysis offered an explanation for Gage's bizarre behavior: The rod had pierced both frontal lobes of his brain. People with these sorts of injuries often exhibit irrational decision making and difficulty processing emotions. As you will learn in Module 28.19, the frontal lobes are part of the limbic system, a group of brain structures involved with emotions.

Beginning with the work of several neurosurgeons in the 1950s, many of the functional areas of the cerebral cortex have been identified during brain surgery. The cortex lacks cells that detect pain; thus, after anesthetizing the scalp, a neurosurgeon can operate on the cerebrum with the patient awake. Parts of the cortex can be stimulated with a harmless electrical current. Stimulation of specific areas can cause someone to experience different sensations or recall memories. Researchers can obtain information about the effects simply by questioning the conscious patient.

Neurophysiologists have also gained insight into the interrelatedness of the brain's two hemispheres. As discussed in Module 28.16, association areas in the left and right sides become specialized for different functions. Much of what we know about this lateralization stems from the work of Roger Sperry with patients whose corpus callosum (communicating fibers between the two hemispheres; see Module 28.15) had been surgically cut to treat severe epileptic seizures. In a series of ingenious experiments, Sperry demonstrated that his patients were unable to verbalize sensory information that was received by only the right hemisphere.

One of the most radical surgical alterations of the brain is a hemispherectomy (Figure 28.17B)—the removal of most of one half of the brain, excluding deep structures such as the thalamus, brainstem, and basal ganglia. This procedure is performed to alleviate severe seizure disorders that originate from one of the hemispheres as a result of illness, abnormal development, or stroke. Incredibly, with just half a brain, hemispherectomy patients recover quickly, often leaving the hospital within a few weeks. Their intellectual capacities are undiminished, although the side of the body opposite the surgery remains partially paralyzed. Higher brain functions that previously originated from the missing half of the brain begin to be controlled by the opposite side. The younger the patient is, ideally less than 5 or 6, the faster and more complete the recovery. Development after hemispherectomy is a striking example of the remarkable plasticity of the brain.

In the upcoming modules, we examine several other topics of active brain research: sleep and arousal, emotions, memory, learning, and neurological disorders.

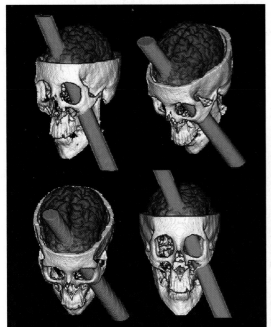

Figure 28.17A Computer model of Phineas Gage's injury

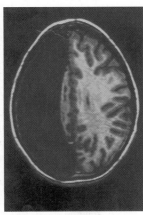

Figure 28.17B X-ray of hemispherectomy patient after surgery

? How are researchers able to investigate brain function during brain surgery?

■ The cortex lacks pain receptors. Regions of the brain can be stimulated during surgery, and the conscious patient can report sensations or memories.

Several parts of the brain regulate sleep and arousal

As anyone who has sat through a lecture on a warm day knows, attentiveness and mental alertness vary from moment to moment. *Arousal* is a state of awareness of the outside world. Its counterpart is *sleep,* a state in which we continue to receive stimuli but are not conscious of them.

Acting with other brain regions, the hypothalamus helps regulate our sleep/wake cycles. The pons and medulla oblongata contain centers that promote sleep when stimulated, and the midbrain has a center that causes arousal. Serotonin may be the neurotransmitter of the sleep-producing centers. Drinking milk before bedtime may induce sleep because milk contains large amounts of tryptophan, the amino acid from which serotonin is synthesized.

Also important in regulating sleep and arousal is a diffuse functional system of neurons that extends through the core of the brainstem. Called the **reticular formation** (Figure 28.18A), this system receives data from sensory receptors (blue arrows in the figure). It filters out some familiar and repetitive information that constantly enters the nervous system—the feel of your clothes against your skin, for example—and sends useful data to the cerebral cortex (green arrows). Generally, the more input the cortex receives from the reticular formation, the more alert and aware we are.

Researchers can study the electrical activity in the brain during arousal and sleep. Electrical contacts are placed on the scalp (Figure 28.18B), with wires leading to a device that records the patterns of electrical activity, called brain waves, on an **electroencephalogram,** or **EEG** (Figure 28.18C). In general, the less mental activity taking place, the more regular are the brain waves recorded on the EEG. The top part of Figure 28.18C shows the fairly regular, slow brain waves recorded when a person is lying quietly with closed eyes. The middle recording, seen when a person is solving a complex mental problem, consists of faster, more irregular waves.

The bottom part, showing a portion of a sleep cycle, reveals two alternating types of deep sleep. Slow-wave (SW) sleep is characterized by delta waves, which are fairly regular.

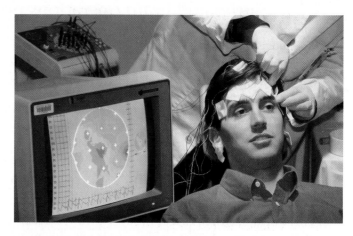

Figure 28.18B   Electrodes placed on scalp for an EEG

Awake but quiet (alpha waves)

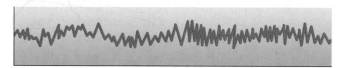

Awake during intense mental activity (beta waves)

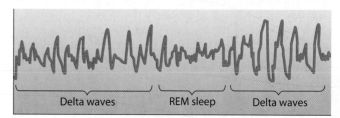

Delta waves   REM sleep   Delta waves

Asleep

Figure 28.18C   Brain waves recorded by an EEG

In **REM sleep,** the brain waves are rapid and less regular, more like those of the awake state. During REM (rapid-eye-movement) sleep, the eyes move quickly under the closed lids. The brain itself is highly active and may consume more oxygen than it does when awake. We have most of our dreams during REM sleep, which typically occurs about six times a night for periods of 5 to 50 minutes each.

Understanding the function of sleeping and dreaming remains a compelling research problem. One hypothesis is that sleep is involved in the consolidation of learning and memory, and experiments show that regions of the brain activated during a learning task may become active again during sleep.

? What prevents the cerebral cortex from being overwhelmed by all the sensory stimuli arriving from sensory receptors?

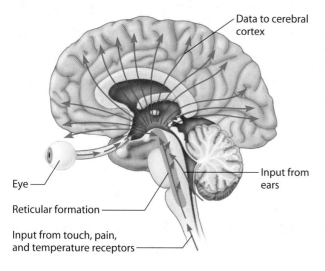

Data to cerebral cortex

Input from ears

Input from touch, pain, and temperature receptors

Eye

Reticular formation

Figure 28.18A   The reticular formation

■ The reticular formation filters out unimportant stimuli.

28.19 The limbic system is involved in emotions, memory, and learning

Mapping the parts of the brain involved in human emotions, learning, and memory and studying the interactions between these parts are among the great challenges in biology today. Much of human emotion, learning, and memory depends on our **limbic system.** This functional unit (**Figure 28.19**) includes parts of the thalamus and hypothalamus and two partial rings around them formed by portions of the cerebral cortex. Two cerebral structures, the amygdala and the hippocampus, play key roles in memory, learning, and emotions.

The limbic system is central to such behaviors as nurturing infants and bonding emotionally to other people. Primary emotions that produce laughing and crying are mediated by the limbic system, and it also attaches emotional "feelings" to basic survival mechanisms of the brainstem, such as feeding, aggression, and sexuality. The intimate relationship between our feelings and our thoughts results from interactions between the limbic system and the prefrontal cortex, which is involved in complex learning, reasoning, and personality.

Memory, which is essential for learning, is the ability to store and retrieve information derived from experience. The **amygdala** is central in recognizing the emotional content of facial expressions and laying down emotional memories. Sensory data converge in the amygdala, which seems to act as a memory filter, somehow labeling information to be remembered by tying it to an event or emotion of the moment. The **hippocampus** is involved in both the formation of memories and their recall. Portions of the frontal lobes are involved in associating primary emotions with different situations.

We sense our limbic system's role in both emotion and memory when certain odors bring back "scent memories." Have you ever had a particular smell suddenly make you nostalgic for something that happened when you were a child? As indicated in Figure 28.19, signals from your nose enter your brain through the olfactory bulb, which connects with the limbic system. Thus, a specific scent can immediately trigger emotional reactions and memories.

Short-term memory, as the name implies, lasts only a short time—usually only a few minutes. It is short-term memory that allows you to dial a phone number just after looking it up. You may, however, store the number in **long-term memory** and be able to recall it weeks after you originally looked it up, or even longer. The transfer of information from short-term to long-term memory is enhanced by rehearsal, positive or negative emotional states mediated by the amygdala, and the association of new data with data previously learned and stored in long-term memory. For example, it's easier to learn a new card game if you already have "card sense" from playing other games.

Factual memories, involving names, faces, words, and places, are different from skill or procedural memories. Skill memories usually involve motor activities that are learned by repetition without consciously remembering specific information. You perform skills, such as tying your shoes, rid-

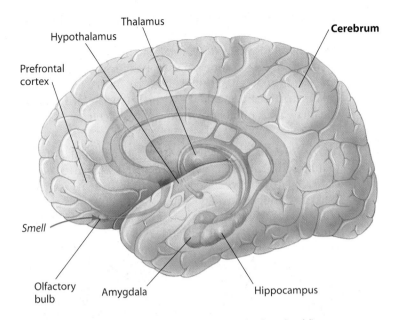

Figure 28.19 The limbic system (shown in shades of gold)

ing a bicycle, or hitting a baseball, without consciously recalling the individual steps required to do these tasks correctly. Once a skill memory is learned, it is difficult to unlearn. For example, a person who has played tennis with a self-taught, awkward backhand has a tougher time learning the correct form than a beginner just learning the game. Bad habits, as we know, are hard to break.

Information processing by the brain generally seems to involve a complex interplay of several integrating centers. By experimenting with animals, studying amnesia (memory loss) in humans, and using brain-imaging techniques, scientists have begun to map some of the major brain pathways involved in memory. Their proposed pathway involves the hippocampus and amygdala, which receive sensory information from the cortex and convey it to other parts of the limbic system and to the prefrontal cortex. The memory storage is completed when signals return to the area in the cortex where the sensory perception originated.

In the final module of this chapter, we'll discuss four major disorders of the nervous system, including their symptoms and treatments.

? Which three factors help transfer information from short-term to long-term memory?

■ Rehearsal, emotional associations, and connection with previously learned data

28.20 Changes in brain physiology can produce neurological disorders

Neurological disorders (diseases of the nervous system) take an enormous toll on society. Examples of neurological disorders are schizophrenia, depression, Alzheimer's disease, and Parkinson's disease. While these conditions are not yet curable, there are a number of treatments available.

Schizophrenia   About 1% of the American population suffer from **schizophrenia**, a severe mental disturbance characterized by psychotic episodes in which patients lose the ability to distinguish reality. Symptoms of schizophrenia typically include hallucinations (most often "voices" telling patients that they are worthless and evil); delusions (generally paranoid); blunted emotions; distractibility; lack of initiative; and difficulty with verbal expression. Contrary to commonly held belief, schizophrenics do not have a "split personality." There seem to be several different forms of schizophrenia, and it is unclear whether they represent different disorders or variations of the same underlying disease.

The causes of schizophrenia are unknown, although the disease has a strong genetic component. Studies of identical twins show that if one twin has schizophrenia, there is a 50% chance that the other twin will have it, too. Since identical twins share identical genes, this indicates that schizophrenia has an equally strong environmental component, the nature of which has not been identified.

Current treatments for schizophrenia focus on brain pathways that use dopamine as a neurotransmitter. Despite their ability to alleviate symptoms, many of the drugs used to treat schizophrenia have such negative side effects that patients frequently stop taking them. Now that the human genome has been sequenced, there is a vigorous effort under way to find the mutant genes that predispose a person to the disease. This effort includes sequencing DNA from families with a high incidence of schizophrenia. Identification of genetic mutations that contribute to schizophrenia may yield new insight into the causes of the disease, which may in turn lead to new therapies.

Depression   In recent years, researchers have begun to learn how brain physiology is involved in depression. Two broad forms of depressive illness have been identified: major depression and bipolar disorder.

People with **major depression** may experience persistent sadness, loss of interest in pleasurable activities, changes in body weight and sleeping patterns, loss of energy, and suicidal thoughts. While all of us experience feelings of sadness from time to time, major depression is extreme and more persistent, leaving a person unable to live a normal life. Major depression occurs in about 5% of the population; if left untreated, symptoms may become more frequent and severe over time.

Bipolar disorder, or manic–depressive disorder, involves extreme mood swings and affects about 1% of the population. The manic phase is characterized by high self-esteem, increased energy, a flow of ideas, and extreme talkativeness, as well as behaviors that often bring disaster, such as increased risk taking, promiscuity, and reckless spending. In its milder forms, this phase is sometimes associated with great creativity, and some well-known artists, musicians, and literary figures (including Keats, Tolstoy, and Hemingway) have had periods of intense output during their manic phases. The depressive phase is marked by sleep disturbances, feelings of worthlessness, and decreased ability to experience interest and pleasure.

Both bipolar disorder and major depression have a genetic component. As in schizophrenia, there is also a strong environmental influence; stress, especially severe stress in childhood, may be an important factor.

Several treatments for depression are available. Many depressed people have an imbalance of neurotransmitters, particularly the biogenic amine serotonin. Some medications are intended to correct such imbalances. The most commonly prescribed class of antidepressant drugs are selective serotonin reuptake inhibitors (SSRIs). As the name implies, these medications increase the amount of time that serotonin is available to stimulate certain neurons in the brain, which appears to relieve some symptoms. In 1987, fluoxetine (best known as Prozac) became the first SSRI approved to treat depression. Other frequently prescribed SSRIs include paroxetine (Paxil) and sertraline (Zoloft). As shown in **Figure 28.20A**, the number of prescriptions for SSRIs in the United States has tripled over the last decade. SSRIs now represent one of the largest classes of medication in the United States, in terms of both dollars spent and number of prescriptions.

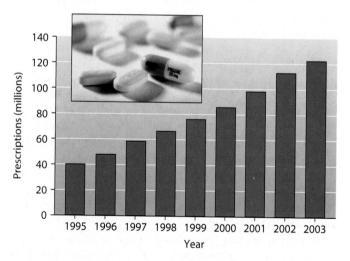

Figure 28.20A   SSRI (inset) prescriptions in the United States by year

Alzheimer's disease A form of mental deterioration, or dementia, **Alzheimer's disease (AD)** is characterized by confusion, memory loss, and a variety of other symptoms. Its incidence is usually age related, rising from about 10% at age 65 to about 35% at age 85. Thus, by helping humans live longer, modern medicine is increasing the proportion of AD patients in the population. The disease is progressive; patients gradually become less able to function and eventually need to be dressed, bathed, and fed by others. There are also personality changes, and patients often lose their ability to recognize people, even family members.

At present, a firm diagnosis of AD is difficult to make while the patient is alive because it is one of several forms of dementia. However, AD results in a characteristic pathology of the brain: Neurons die in huge areas of the brain, and brain tissue often shrinks. This shrinkage is visible with brain imaging but is not enough to positively identify AD. What is diagnostic is the postmortem finding of two features—neurofibrillary tangles and senile plaques—in the remaining brain tissue (**Figure 28.20B**). Neurofibrillary tangles are bundles of degenerated brain cells. Senile plaques are aggregates of beta-amyloid, an insoluble peptide containing 40 to 42 amino acids that is cleaved from a membrane protein normally present in neurons. The plaques appear to trigger the death of surrounding neurons, but whether amyloid plaques cause Alzheimer's or are a symptom of it remains unclear.

Parkinson's Disease Approximately 1 million people in the United States suffer from **Parkinson's disease (Figure 28.20C)**, a motor disorder characterized by difficulty in initiating movements, slowness of movement, and rigidity. Patients often have a masked facial expression, muscle tremors, poor balance, a flexed posture, and a shuffling gait. Like Alzheimer's disease, Parkinson's is progressive, and the risk increases with age. The incidence of Parkinson's disease is about 1% at age 65 and about 5% at age 85.

The symptoms of Parkinson's disease result from the death of neurons in the midbrain. These neurons normally release dopamine from their synaptic terminals.

Parkinson's disease appears to result from a combination of environmental and genetic factors. Evidence for a genetic role includes the fact that some families with an increased incidence of Parkinson's disease carry a mutated form of the gene for a protein important in normal brain function.

At present, there is no cure for Parkinson's disease, although various treatments can help control the symptoms. Treatments include drugs such as L-dopa, a dopamine precursor, and surgery. One potential cure is to develop embryonic stem cells into dopamine-secreting neurons in the laboratory and implant them in patients' brains. In laboratory experiments, transplantation of such cells into rats with a Parkinson's-like condition improved the animals' motor control. It remains to be seen whether this approach will work in humans.

Unraveling the biological basis of neurological disorders remains one of the most challenging tasks of modern biology. Recent advances in locating genes that correlate with

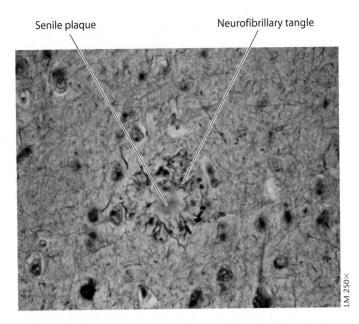

Senile plaque   Neurofibrillary tangle

LM 250×

Figure 28.20B Neurofibrillary tangles and senile plaques in the brain of a patient with Alzheimer's disease

Figure 28.20C Actor Michael J. Fox (right) and boxer Muhammad Ali, both of whom suffer from Parkinson's disease, testifying before the Senate about funding for the disorder

CNS disorders and creating animal models for these diseases have provided new insights, but many aspects of our nervous system remain mysterious. In the next chapter, we examine another aspect of nervous systems—how sense organs gather information about the environment.

? What do the initials SSRI stand for? Relate the name to its mechanism of action.

■ SSRIs (selective serotonin reuptake inhibitors) are antidepressant drugs that specifically (*selectively*) prevent (*inhibit*) the reabsorption (*reuptake*) of the neurotransmitter serotonin in the brain.

Reviewing the Concepts

Injuries to the spinal cord disrupt communication between the central nervous system (brain and spinal cord) and the rest of the body **(Introduction).**

Nervous System Structure and Function (28.1–28.2)

The nervous system obtains and processes sensory information and sends commands to effector cells (such as muscles) that carry out appropriate responses:

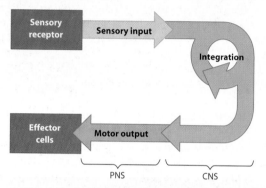

Sensory neurons conduct signals from sensory receptors to the central nervous system (CNS), which consists of the brain and, in vertebrates, the spinal cord. Interneurons in the CNS integrate information and send it to motor neurons. Motor neurons, in turn, convey signals to effector cells. Located outside the CNS, the peripheral nervous system (PNS) consists of nerves (bundles of fibers of sensory and motor neurons) and ganglia (clusters of cell bodies of the neurons) **(28.1).** The functional units of the nervous system are neurons, cells specialized for carrying signals. A neuron consists of a cell body and two types of extensions (fibers) that conduct signals, dendrites and axons. Many axons are enclosed by cellular insulation called the myelin sheath, which speeds up signal transmission **(28.2).**

Nerve Signals and Their Transmission (28.3–28.9)

The conduction of nerve signals. At rest, a neuron's plasma membrane has an electrical voltage called the resting potential. The resting potential is caused by the membrane's ability to maintain a positive charge on its outer surface opposing a negative charge on its inner (cytoplasmic) surface **(28.3).** A stimulus alters the permeability of a portion of the membrane, allowing ions to pass through and changing the membrane's voltage. A nerve signal, called an action potential, is a change in the membrane voltage from the resting potential to a maximum level and back to the resting potential **(28.4).** Action potentials are self-propagated in a one-way chain reaction along a neuron. An action potential is an all-or-none event. The frequency of action potentials (but not their strength) changes with the strength of the stimulus **(28.5).**

Synapses. The transmission of signals between neurons or between neurons and effector cells occurs at junctions called synapses. Electrical signals pass between cells at electrical synapses. At chemical synapses, the sending cell secretes a chemical signal, a neurotransmitter, which crosses the synaptic cleft and binds to a specific receptor on the surface of the receiving cell **(28.6).** Some neurotransmitters excite the receiving cell; others inhibit the receiving cell's activity by decreasing its ability to develop action potentials. A cell may receive differing signals from many neurons; the summation of excitation and inhibition determines whether or not it will transmit a nerve signal **(28.7).** Many small, nitrogen-containing molecules serve as neurotransmitters **(28.8).** Many psychoactive drugs act at synapses and affect neurotransmitter action **(28.9).**

An Overview of Animal Nervous Systems (28.10–28.14)

Animal nervous systems. Radially symmetrical animals have a nervous system arranged in a weblike system of neurons called a nerve net. Most bilaterally symmetrical animals exhibit cephalization, the concentration of the nervous system in the head end, and centralization, the presence of a central nervous system **(28.10).** Vertebrate nervous systems are highly centralized and cephalized. The brain and spinal cord contain fluid-filled spaces. Cranial and spinal nerves make up the peripheral nervous system **(28.11).**

Organization of the vertebrate nervous system (28.12–28.13):

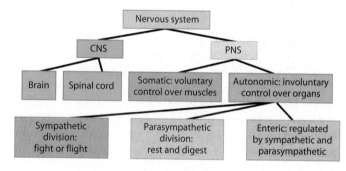

The vertebrate brain evolved by the enlargement and subdivision of three anterior bulges of the neural tube: the hindbrain, midbrain, and forebrain. The size and complexity of the cerebrum in birds and mammals correlates with their sophisticated behavior **(28.14).**

The Human Brain (28.15–28.20)

Anatomy of the human brain. The midbrain and subdivisions of the hindbrain, together with the thalamus and hypothalamus of the forebrain, function mainly in conducting information to and from higher brain centers. They regulate homeostatic functions, keep track of body position, and sort sensory information. The forebrain's cerebrum is the largest and most complex part of the brain. Most of the cerebrum's integrative power resides in the cerebral cortex of the two cerebral hemispheres **(28.15).** Specialized integrative regions of the cerebral cortex include the somatosensory cortex and centers for vision, hearing, taste, and smell. The motor cortex directs responses. Association areas, concerned with higher mental activities such as reasoning and language, make up most of the cerebrum. The right and left cerebral hemispheres tend to specialize in different mental tasks **(28.16).**

Higher brain functions. Brain injuries and surgeries have been used to study brain function **(28.17).** Sleep and arousal involve activity by the hypothalamus, medulla oblongata, pons, and neurons of the reticular formation. An EEG measures brain waves during sleep and arousal **(28.18).** The limbic system, a functional group of integrating centers in the cerebral cortex, thalamus, and hypothalamus, is involved in emotions, memory, and learning **(28.19).** Many neurological disorders can be linked to changes in brain physiology. Treatment can often improve patients' lives, but cures remain elusive **(28.20).**

Connecting the Concepts

1. Test your understanding of the nervous system by matching the following labels with their corresponding letters: CNS, effector cells, interneuron, motor neuron, PNS, sensory neuron, sensory receptor, spinal cord, synapse.

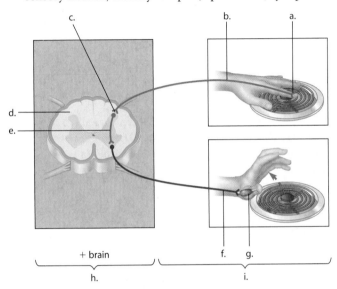

Testing Your Knowledge

Multiple Choice

2. Joe accidentally touched a hot pan. His arm jerked back, and an instant later, he felt a burning pain. How would you explain that his arm moved before he felt the pain?
   a. His limbic system blocked the pain momentarily, but the important pain signals eventually got through.
   b. His response was a spinal cord reflex that occurred before the pain signals got to the brain.
   c. It took a while for his brain to search long-term memory and figure out what was going on.
   d. Motor neurons are myelinated; sensory neurons are not. The signals traveled faster to his muscles.
   e. This scenario is not actually possible. The brain must register pain before a person can react.

3. Which of the following mediates sleep and arousal?
   a. the reticular formation, along with the hypothalamus and thalamus
   b. the limbic system, which includes the amygdala and hippocampus
   c. the left hemisphere of the cerebral cortex
   d. the midbrain and cerebellum
   e. the parasympathetic and sympathetic divisions of the nervous system

4. Anesthetics block pain by blocking the transmission of nerve signals. Which of these three chemicals might work as anesthetics? (*Circle all that apply and explain your selections.*)
   a. a chemical that prevents the opening of sodium channels in membranes
   b. a chemical that inhibits the enzymes that degrade neurotransmitters
   c. a chemical that blocks neurotransmitter receptors

Describing, Comparing, and Explaining

5. As you hold this book, nerve signals are generated in nerve endings in your fingertips and sent to your brain. Once a touch has caused an action potential at one end of a neuron, what causes the nerve signal to move from that point along the length of the neuron to the other end? What is the nerve signal, exactly? Why can't it go backward? How is the nerve signal transmitted from one neuron to the next across a synapse? Write a short paragraph that answers these questions.

Applying the Concepts

6. Using microelectrodes, a researcher recorded nerve signals in four neurons in the brain of a snail. The neurons are called A, B, C, and D in the table below. A, B, and C all can transmit signals to D. In three experiments, the animal was stimulated in different ways. The number of nerve signals transmitted per second by each of the cells is recorded in the table. Write a short paragraph explaining the different results of the three experiments.

	Signals/sec			
	A	**B**	**C**	**D**
Experiment #1	50	0	40	30
Experiment #2	50	0	60	45
Experiment #3	50	30	60	0

7. Brain injuries tend to be severe because most neurons, once damaged, will not regenerate. The use of embryonic stem cells has been proposed as a potential treatment for many neurological diseases. As mentioned in the text, neurons developed from such stem cells in laboratory culture might be implanted in the brain of a person with Alzheimer's or Parkinson's disease. These neurons might be able to replace those damaged by the disease. Do you favor or oppose research along those lines? Explain your answer.

8. Alcohol's depressant effects on the nervous system cloud judgment and slow reflexes. Alcohol consumption is a factor in most fatal traffic accidents in the United States. What are some other impacts of alcohol abuse on society? What are some of the responses of people and society to alcohol abuse? Do you think this is primarily an individual or societal problem? Do you think our responses to alcohol abuse are appropriate and proportional to the seriousness of the problem?

Answers to all questions can be found in Appendix 3.

For study help and Activities, go to campbellbiology.com or the student CD-ROM.

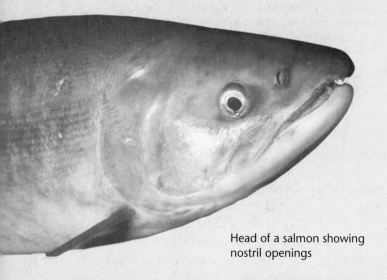

Migrating salmon

29.1   Sensory inputs become sensations and perceptions in the brain

SENSORY RECEPTION

29.2   Sensory receptors convert stimulus energy to action potentials
29.3   Specialized sensory receptors detect five categories of stimuli

VISION

29.4   Several types of eyes have evolved among invertebrates
29.5   Vertebrates have single-lens eyes
29.6   To focus, a lens changes position or shape
29.7   Artificial lenses or surgery can correct focusing problems
29.8   Our photoreceptors are rods and cones

HEARING AND BALANCE

29.9   The ear converts air pressure waves to action potentials that are perceived as sound
29.10   The inner ear houses our organs of balance
29.11   What causes motion sickness?

TASTE AND SMELL

29.12   Taste and odor receptors detect chemicals present in solution or air
29.13   Our sense of taste may change as we age
29.14   Review: The central nervous system couples stimulus with response

Head of a salmon showing nostril openings

An Animal's Senses Guide Its Movements

A FATAL ENCOUNTER between a brown bear and a migrating salmon in a glacial stream in Alaska: Salmon are fast and agile, but easy prey for a bear when in shallow water or when leaping up a waterfall, as shown on the next page. Despite the bear's rather poor eyesight and great size (up to about 500 kg, or over 1,000 pounds), it has lightning reflexes and can snatch up a large fish with little effort.

What brought these animals together at this particular moment? The bear learned as a cub to head for the salmon streams in late autumn. It followed its mother there and learned how to fish by watching her. The bear's acute sense of smell helped it find the stream. For the salmon, becoming a meal for the bear suddenly ends a remarkable odyssey. After nearly eight years at sea, it was returning to reproduce in the stream where its life began. The salmon has a keen sense of smell, and odors led it to this particular stream among thousands of possibilities.

In the fall, a female salmon lays her eggs in a shallow depression on the bottom of a small, fast-flowing stream. A male covers the eggs with sperm, and both parents die soon thereafter. The following spring, larval salmon hatch and begin drifting downstream toward the ocean, starting a journey that will take them far out to sea. While at sea, they feed in large schools with salmon from other river systems. When they are 4–6 years old and sexually mature, the salmon segregate into groups of common geographic origin and start migrating back toward the river from which they emerged as juveniles.

Research on how salmon locate their home stream has been going on since the early 1950s. The fish seem to navigate to the mouth of their home river system by using the angle of the sun for reference. Once there, however, their sense of smell takes over.

The water that flows from each stream into a river seems to carry a unique scent, a mixture of chemicals from the plants and soils in the area. The scent of its home stream apparently becomes fixed in the memory of a young salmon before it migrates to the sea. Years later, when the mature salmon arrives in the vicinity of

The Senses

its home river system, it swims along the coast until it detects the faint odors matching the scent memory in its brain. In response to perhaps only a few molecules of its "home chemicals," it enters that river and begins its upstream journey.

A salmon has a nostril on each side of its head. In the photograph at the bottom of the opposite page, you can see that a nostril has two openings. As the fish swims, water flows in one opening and out the other, and highly sensitive cells in the nostril detect chemicals dissolved in the water. When researchers blocked the nostrils of migrating salmon with cotton, the fish lost their homing ability. They still migrated, but could not make the correct choices at forks of streams.

As a salmon swims upstream, its nostrils follow a scent trail in the water. The closer it gets to its home stream, the stronger the

The scent of its home stream apparently becomes fixed in the memory of a young salmon

scent becomes. More and more of the scent molecules stimulate sensory cells in its nostrils, and the cells send more and more signals to the fish's brain, telling it that it's on the right track.

The cells in the salmon's nostrils aren't the only source of sensory information helping the salmon make it upstream. Like most fishes, a salmon has special receptor cells that respond to the movement of water over its skin. These cells are located in a lateral line system running along the sides of its body. This system of openings and tubes (visible as the dark line along the sides of the fish in the photo at the top of the opposite page) enables the salmon to sense the direction and velocity of water currents, so it can distinguish which direction is upstream. The lateral line system also enables the salmon to sense other water movements, including those generated by prey and predators. It cannot, however, perceive a bear's paw descending from above.

Sensory information gathered by sensory receptors and processed by the brain guide salmon and brown bears to specific stream sites. This chapter focuses on sensory structures and how they gather information. To begin, we examine the distinction between information gathering and information processing. ■ ■ ■

29.3 Specialized sensory receptors detect five categories of stimuli

Based on the type of signals to which they respond, we can group sensory receptors into five general categories: pain receptors, thermoreceptors, mechanoreceptors, chemoreceptors, and electromagnetic receptors.

Figure 29.3A, showing a section of human skin, reveals why the surface of our body is sensitive to such a variety of stimuli. Our skin contains pain receptors, thermoreceptors (sensors for both heat and cold), and mechanoreceptors (sensors for touch and pressure). Each of these receptors is a modified dendrite of a sensory neuron (see Module 28.2). The neuron both transduces stimuli and sends action potentials to the central nervous system. In other words, each receptor serves as both a receptor cell and a sensory neuron. Most of the dendrites in the dermis (the underlying region of the skin) are wrapped in one or more layers of connective tissue (purple areas in the figure); however, the pain and touch receptors in the epidermis (outer skin layer) and the touch receptors around the base of hairs are naked dendrites.

Pain Receptors Probably all animals have **pain receptors**, although we cannot say what nonhuman perceptions of pain are like. Pain often indicates danger and usually makes an animal withdraw to safety. All parts of the human body except the brain have pain receptors. Pain can make us aware of injury or disease. Pain receptors may respond to excess heat or pressure or to chemicals released from damaged or inflamed tissues. Histamines and acids are some of the chemicals that trigger pain. Local regulators (see Module 26.1) increase pain by sensitizing pain receptors. Aspirin and ibuprofen reduce pain by inhibiting prostaglandin synthesis.

Thermoreceptors **Thermoreceptors** in the skin detect either heat or cold. Other temperature sensors located deep in the body monitor the temperature of the blood. The hypothalamus in the brain is the body's major thermostat. Receiving action potentials from both surface and deep sensors, the hypothalamus keeps a mammal's body temperature within a narrow range (see Module 20.14).

Mechanoreceptors Different types of **mechanoreceptors** are stimulated by various forms of mechanical energy, such as touch and pressure, stretching, motion, and sound. All these forces produce their effects by bending or stretching the plasma membrane of a receptor cell. When the membrane changes shape, it becomes more permeable to sodium or potassium ions, and the mechanical energy of the stimulus is transduced into a receptor potential.

At the top of Figure 29.3A are two types of mechanoreceptors that detect light touch. Both types transduce very slight inputs of mechanical energy into action potentials. A third type of pressure sensor, lying deeper in the skin, is stimulated by strong pressure. A fourth type of mechanoreceptor, the touch receptor around the base of the hair, detects hair movements. Touch receptors at the base of the stout whiskers on a cat are extremely sensitive and enable the animal to detect close objects by touch in the dark. Another type of

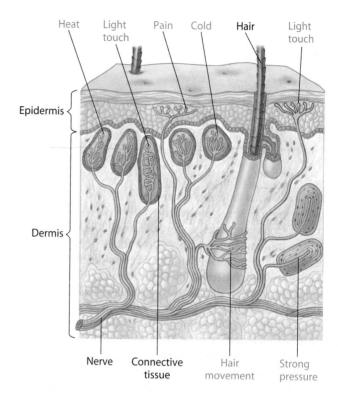

Figure 29.3A Sensory receptors (labels in blue) in the human skin

mechanoreceptor (not shown) is found in our skeletal muscles. Sensitive to changes in muscle length, **stretch receptors** monitor the position of body parts (see Figure 28.1B).

A variety of mechanoreceptors collectively called **hair cells** detect sound waves and other forms of movement in water. The "hairs" on these sensors are either specialized types of cilia or cellular projections called microvilli. The lateral line system of a salmon (see chapter introduction) contains hair cells that sense water currents. The sensory hairs project from the surface of a receptor cell into either the external environment, such as the water surrounding a fish, or an internal fluid-filled compartment, such as our inner ear. Figure 29.3B indicates how hair cells work, ❶ starting with a receptor cell at rest. ❷ When fluid movement bends the hairs in one direction, the hairs stretch the cell membrane, increasing its permeability to certain ions. This makes the hair cell secrete more neurotransmitter molecules and increases the rate of action potential production by a sensory neuron. ❸ When the hairs bend in the opposite direction, ion permeability decreases, the hair cell releases fewer neurotransmitter molecules, and the rate of action potential generation decreases. We'll see later that hair cells are involved in both hearing and balance.

Chemoreceptors **Chemoreceptors** include the sensory receptors in our nose and taste buds, which are attuned to chemicals in the external environment, as well as some receptors that detect chemicals in the body's internal environ-

The Senses

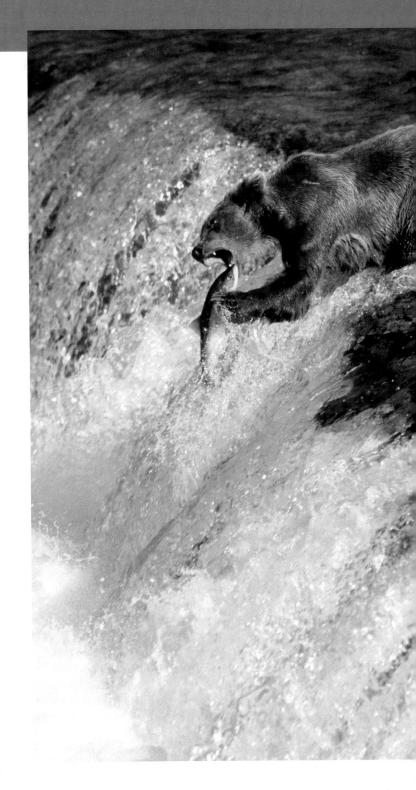

its home river system, it swims along the coast until it detects the faint odors matching the scent memory in its brain. In response to perhaps only a few molecules of its "home chemicals," it enters that river and begins its upstream journey.

A salmon has a nostril on each side of its head. In the photograph at the bottom of the opposite page, you can see that a nostril has two openings. As the fish swims, water flows in one opening and out the other, and highly sensitive cells in the nostril detect chemicals dissolved in the water. When researchers blocked the nostrils of migrating salmon with cotton, the fish lost their homing ability. They still migrated, but could not make the correct choices at forks of streams.

As a salmon swims upstream, its nostrils follow a scent trail in the water. The closer it gets to its home stream, the stronger the scent becomes. More and more of the scent molecules stimulate sensory cells in its nostrils, and the cells send more and more signals to the fish's brain, telling it that it's on the right track.

The scent of its home stream apparently becomes fixed in the memory of a young salmon

The cells in the salmon's nostrils aren't the only source of sensory information helping the salmon make it upstream. Like most fishes, a salmon has special receptor cells that respond to the movement of water over its skin. These cells are located in a lateral line system running along the sides of its body. This system of openings and tubes (visible as the dark line along the sides of the fish in the photo at the top of the opposite page) enables the salmon to sense the direction and velocity of water currents, so it can distinguish which direction is upstream. The lateral line system also enables the salmon to sense other water movements, including those generated by prey and predators. It cannot, however, perceive a bear's paw descending from above.

Sensory information gathered by sensory receptors and processed by the brain guide salmon and brown bears to specific stream sites. This chapter focuses on sensory structures and how they gather information. To begin, we examine the distinction between information gathering and information processing. ■ ■ ■

29.1 Sensory inputs become sensations and perceptions in the brain

Sensory receptor cells are tuned to the conditions of the external world and the internal organs. They detect stimuli, such as chemicals, light, tension in a muscle, sounds, electricity, cold, heat, and touch, and send information to the central nervous system. The sensory cells in a bear's nose, for instance, detect chemicals in the air and send reports about the chemicals to the brain. The reports take the form of action potentials, the same signals used throughout the nervous system (see Module 28.4). The sensory receptor's job is completed when it triggers action potentials that go to the central nervous system.

An action potential triggered in response to the smell of a salmon is the same as an action potential triggered in response to the sight of a salmon. The ability to distinguish the type of stimulus, such as smell or sight, depends on the part of the brain that receives the action potential. Once the brain is aware of **sensations**, the action potentials it receives from sense receptors, the brain interprets them to produce a perception. **Perceptions**, such as colors, smells, sounds, and tastes, are constructed by the brain as it processes sensations and integrates them with other information, forming a meaningful interpretation or conscious understanding of sensory data.

Figure 29.1 may further demonstrate the difference between sensation and perception for you. What do you see when you first look at the figure? If you see only some black splotches on a blue background, you have developed a sensation—an awareness of this sensory information. What if we say that the figure shows a person riding a horse? With this

clue, the brain forms a perception; it converts the sensation into a meaningful image. Your brain integrated the new information with some of its stored data.

What does the brain actually do with sensory information in creating a perception? As we discussed in Chapter 28, researchers using brain-imaging techniques are beginning to find out. Perceptions appear to re-

Figure 29.1 Black splotches or a person riding on a horse?

sult from communication among neurons arranged in extremely complex circuits involving multiple areas of the cerebrum. These areas include sensory and association areas and the limbic system, which may access stored memories.

Perceptions are the product of a continuum of information processing, beginning with the detection of stimuli by sensory receptors. In this chapter, we focus on sensory receptors and how they function.

? In comparing sensations and perceptions of a specific environmental stimulus, humans probably vary more in their _____ than in their _____.

■ perceptions . . . sensations

29.2 Sensory receptors convert stimulus energy to action potentials

Sensory organs, such as your eyes or your taste buds, contain **sensory receptors**, specialized cells or neurons that detect stimuli. All stimuli represent forms of energy. The sensory receptors in your eyes detect light energy; those in your taste buds detect chemicals, such as salt or sugar.

What exactly do we mean when we say that a sensory receptor detects a stimulus—a photon of light or a molecule of sugar, for instance? Stimulus detection means that the receptor cell converts one type of signal (the stimulus) to an electrical signal. This conversion, called **sensory transduction**, produces a change in the membrane potential (the potential energy stored by the membrane; see Module 28.3) of the receptor cell. This change in the membrane potential is a result of the opening or closing of ion channels in the sensory receptor's plasma membrane.

Figure 29.2A shows sensory transduction occurring when sensory receptor cells in a taste bud detect sugar molecules. ❶ The sugar molecules first enter the taste bud, where

❷ they bind to specific protein molecules on a taste receptor membrane. The binding initiates a signal transduction pathway (see Module 11.14) that causes ❸ some ion channels in the membrane to close and others to open. Changes in the flow of ions create a graded change in membrane potential called a **receptor potential.** In contrast to action potentials, which are all-or-none phenomena, receptor potentials vary; the stronger the stimulus, the larger the receptor potential.

Once a stimulus is converted to a receptor potential, the receptor potential usually results in signals entering the central nervous system. In our taste bud example, ❹ each receptor cell forms a synapse with a sensory neuron. This is a chemical synapse just like the one between neurons described in Module 28.6. In many cases, a receptor cell constantly secretes neurotransmitter into this synapse at a set rate, triggering a steady stream of action potentials in the sensory neuron. ❺ The graph on the left shows the rate at which the sensory neuron sends action potentials when the taste receptor is not detect-

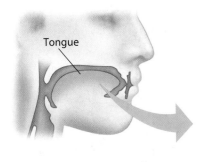

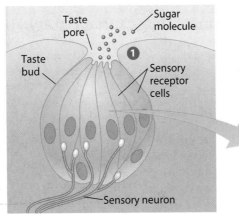

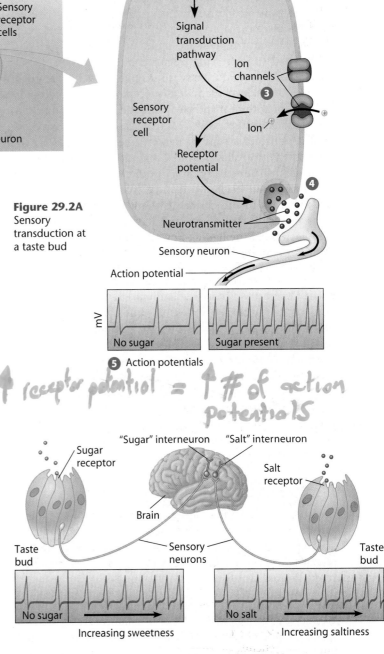

Figure 29.2A Sensory transduction at a taste bud

ing any sugar. The graph on the right shows what happens when there are enough sugar molecules to trigger a strong receptor potential, causing the receptor cell to release more neurotransmitter than usual. This additional neurotransmitter increases the rate of action potential generation in the sensory neuron. It is the change in the rate of action potentials that signals the brain that the sensory receptor detects a stimulus.

Thus we see that sensory receptors transduce (convert) stimuli into receptor potentials, which trigger action potentials that enter the central nervous system for processing. Since action potentials are the same no matter where or how they are produced, how do they communicate a sweet taste instead of a salty one? In **Figure 29.2B**, the taste bud on the left has sensory receptors that respond to sugar, and the taste bud on the right responds to salt. The sensory neurons from the sugar-detecting taste bud synapse with different interneurons in the brain than those contacted by neurons from the salt-detecting taste bud. The brain distinguishes stimulus types (in this case, sugar from salt) by the patterns in which interneurons are stimulated.

The graphs in Figure 29.2B also indicate how action potentials communicate information about the *intensity* of stimuli (for example, very sweet or less sweet). In each case, the left part of the graph represents the rate at which the sensory neurons in the taste bud transmit action potentials when the taste receptors are not stimulated. The right side of each graph shows that the rate of transmission depends on the intensity of the stimulus. The stronger the stimulus, the more neurotransmitter released by the receptor cell and the more frequently the sensory neuron transmits action potentials to the brain. The brain interprets the intensity of the stimulus from the rate at which it receives action potentials.

There is an important qualification to what we have just said about stimulus intensity. Have you ever noticed how an odor that is strong at first seems to fade with time, even when you know the substance is still there? The same effect helps you adjust to a hot or cold shower and enables you to wear clothes without being constantly aware of them. The effect is called **sensory adaptation**, the tendency of sensory receptors to become less sensitive when they are stimulated repeatedly. When receptors become less sensitive, they trigger fewer action potentials, and the brain may lose its awareness of stimuli as a result. Sensory adaptation keeps the body from reacting to

Figure 29.2B How action potentials transmit different taste sensations

normal background stimuli. Without it, our nervous system would become overloaded with useless information.

This overview of sensory transduction, transmission, and adaptation explains how sensory receptors work in general. Now let's look at the receptors themselves.

? What is meant by sensory transduction?

■ The conversion of a stimulus signal to an electrical signal (a receptor potential) by a sensory receptor cell

Specialized sensory receptors detect five categories of stimuli

Based on the type of signals to which they respond, we can group sensory receptors into five general categories: pain receptors, thermoreceptors, mechanoreceptors, chemoreceptors, and electromagnetic receptors.

Figure 29.3A, showing a section of human skin, reveals why the surface of our body is sensitive to such a variety of stimuli. Our skin contains pain receptors, thermoreceptors (sensors for both heat and cold), and mechanoreceptors (sensors for touch and pressure). Each of these receptors is a modified dendrite of a sensory neuron (see Module 28.2). The neuron both transduces stimuli and sends action potentials to the central nervous system. In other words, each receptor serves as both a receptor cell and a sensory neuron. Most of the dendrites in the dermis (the underlying region of the skin) are wrapped in one or more layers of connective tissue (purple areas in the figure); however, the pain and touch receptors in the epidermis (outer skin layer) and the touch receptors around the base of hairs are naked dendrites.

Pain Receptors Probably all animals have **pain receptors,** although we cannot say what nonhuman perceptions of pain are like. Pain often indicates danger and usually makes an animal withdraw to safety. All parts of the human body except the brain have pain receptors. Pain can make us aware of injury or disease. Pain receptors may respond to excess heat or pressure or to chemicals released from damaged or inflamed tissues. Histamines and acids are some of the chemicals that trigger pain. Local regulators (see Module 26.1) increase pain by sensitizing pain receptors. Aspirin and ibuprofen reduce pain by inhibiting prostaglandin synthesis.

Thermoreceptors **Thermoreceptors** in the skin detect either heat or cold. Other temperature sensors located deep in the body monitor the temperature of the blood. The hypothalamus in the brain is the body's major thermostat. Receiving action potentials from both surface and deep sensors, the hypothalamus keeps a mammal's body temperature within a narrow range (see Module 20.14).

Mechanoreceptors Different types of **mechanoreceptors** are stimulated by various forms of mechanical energy, such as touch and pressure, stretching, motion, and sound. All these forces produce their effects by bending or stretching the plasma membrane of a receptor cell. When the membrane changes shape, it becomes more permeable to sodium or potassium ions, and the mechanical energy of the stimulus is transduced into a receptor potential.

At the top of Figure 29.3A are two types of mechanoreceptors that detect light touch. Both types transduce very slight inputs of mechanical energy into action potentials. A third type of pressure sensor, lying deeper in the skin, is stimulated by strong pressure. A fourth type of mechanoreceptor, the touch receptor around the base of the hair, detects hair movements. Touch receptors at the base of the stout whiskers on a cat are extremely sensitive and enable the animal to detect close objects by touch in the dark. Another type of

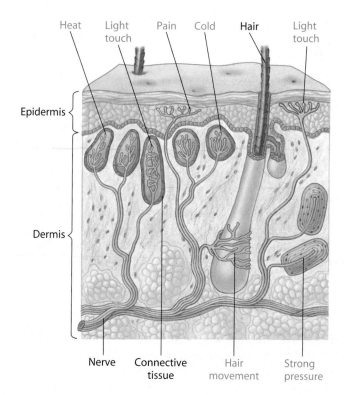

Figure 29.3A Sensory receptors (labels in blue) in the human skin

mechanoreceptor (not shown) is found in our skeletal muscles. Sensitive to changes in muscle length, **stretch receptors** monitor the position of body parts (see Figure 28.1B).

A variety of mechanoreceptors collectively called **hair cells** detect sound waves and other forms of movement in water. The "hairs" on these sensors are either specialized types of cilia or cellular projections called microvilli. The lateral line system of a salmon (see chapter introduction) contains hair cells that sense water currents. The sensory hairs project from the surface of a receptor cell into either the external environment, such as the water surrounding a fish, or an internal fluid-filled compartment, such as our inner ear. Figure 29.3B indicates how hair cells work, ❶ starting with a receptor cell at rest. ❷ When fluid movement bends the hairs in one direction, the hairs stretch the cell membrane, increasing its permeability to certain ions. This makes the hair cell secrete more neurotransmitter molecules and increases the rate of action potential production by a sensory neuron. ❸ When the hairs bend in the opposite direction, ion permeability decreases, the hair cell releases fewer neurotransmitter molecules, and the rate of action potential generation decreases. We'll see later that hair cells are involved in both hearing and balance.

Chemoreceptors **Chemoreceptors** include the sensory receptors in our nose and taste buds, which are attuned to chemicals in the external environment, as well as some receptors that detect chemicals in the body's internal environ-

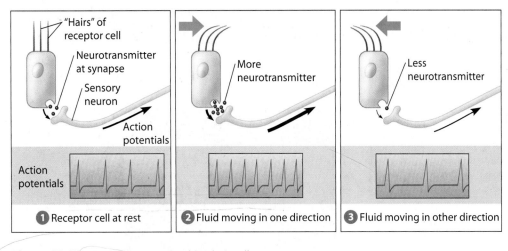

Figure 29.3B Mechanoreception by a hair cell

The three panels are labeled:
1. Receptor cell at rest
2. Fluid moving in one direction
3. Fluid moving in other direction

Labels in the figure: "Hairs" of receptor cell; Neurotransmitter at synapse; Sensory neuron; Action potentials; More neurotransmitter; Less neurotransmitter

ment. Internal chemoreceptors include sensors in some of our arteries that can detect changes in the amount of O_2 in the blood. Osmoreceptors in the brain detect changes in the total solute concentration of the blood and stimulate thirst when blood osmolarity increases (see Module 25.11).

One of the most sensitive chemoreceptors is found on the antennae of the male silkworm moth *Bombyx mori* (Figure 29.3C). The antennae are covered with thousands of sensory hairs (visible in the micrograph). The hairs have chemoreceptors that detect a sex pheromone released by the female.

Electromagnetic Receptors Energy of various wavelengths, occurring as electricity, magnetism, or light, may be detected by **electromagnetic receptors**. Certain fishes discharge electrical currents into the water and use electroreceptors to detect nearby obstacles and prey. The platypus (see Chapter 18 intro-

duction) has electroreceptors on its bill that can detect electrical fields generated by the muscles of prey, such as crustaceans, frogs, and small fishes.

Many animals appear to use Earth's magnetic field to orient themselves as they migrate. The iron-containing mineral magnetite is found in the abdomen of bees, in the skulls of a variety of vertebrates (including salmon, sea turtles, pigeons, and humans), and in certain protists and prokaryotes that orient with respect to Earth's magnetic field. Once used by sailors as a primitive compass, magnetite may be part of an orienting mechanism in many organisms.

Photoreceptors, including eyes, are probably the most common type of electromagnetic receptor. Photoreceptors detect the electromagnetic energy we call light, which may be in the visible or ultraviolet part of the electromagnetic spectrum (see Module 7.6). The rattlesnake in **Figure 29.3D** has prominent eyes that detect visible light. Below its eyes are two infrared receptors. These specialized electromagnetic receptors detect the body heat of its preferred prey, small mammals and birds. The receptors can detect the infrared radiation emitted by a mouse a meter away.

A variety of light detectors have evolved in the animal kingdom. Despite their differences, however, all photoreceptors contain similar pigment molecules that absorb light, and evidence indicates that all photoreceptors may be homologous (having a common ancestry). In the next module, we consider some of the different types of eyes found in invertebrates.

? For each of the following senses in humans, identify the type of receptor: seeing, tasting, hearing, smelling.

■ Photoreceptor; chemoreceptor; mechanoreceptor; chemoreceptor

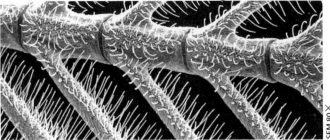

SEM 80×

Figure 29.3C Chemoreceptors on insect antennae

Labels: Eye; Infrared receptor

Figure 29.3D Electromagnetic receptors in a snake

29.4 Several types of eyes have evolved among invertebrates

Most invertebrates have some kind of light-detecting organ. One of the simplest is the **eye cup** of planarians, which provides information about light intensity and direction but does not form an image. The eye cup contains photoreceptor cells that are partially shielded by darkly pigmented cells (**Figure 29.4A**). Light can enter the eye cup only where there are no pigmented cells, and the openings of the two eye cups face opposite directions. The brain compares the rate of nerve impulses coming from the two eye cups, and the animal turns until the sensations are equal and minimal. The result is that the animal moves directly away from the light source and reaches a dark hiding place.

Two major types of image-forming eyes have evolved in invertebrates. Compound eyes are found in insects and crustaceans. A **compound eye** consists of many tiny light-detecting units, called ommatidia. You can see some of the thousands of ommatidia (the tiny dots) making up the two compound eyes of a fly in **Figure 29.4B**. Each ommatidium has its own light-focusing lens and several photoreceptor cells. Every ommatidium picks up light from a tiny portion of the field of view. The animal's brain then forms a mosaic visual image by assembling the data from all the ommatidia.

Compound eyes are extremely acute motion detectors, an important advantage for flying insects and other small ani-

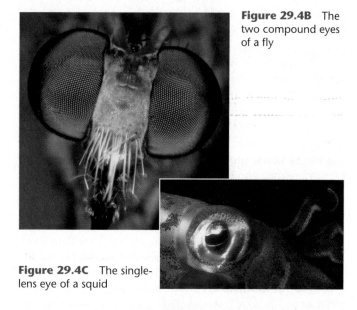

Figure 29.4B The two compound eyes of a fly

Figure 29.4C The single-lens eye of a squid

mals that are often threatened by predators. The compound eyes of most insects also provide excellent color vision. Some species, such as honeybees, can see ultraviolet light (invisible to humans), which helps them locate certain nectar-bearing flowers.

The second type of image-forming eye, the **single-lens eye**, works on a principle similar to that of a camera. For example, the eye of a squid (**Figure 29.4C**) has a small opening, the pupil, through which light enters. Analogous to a camera's shutter, an adjustable iris changes the diameter of the pupil. Behind the pupil, a single lens focuses light onto the retina, which consists of many photoreceptor cells. Muscles in the eye move the lens forward or backward, focusing on objects at different distances. We look at the single-lens eyes of vertebrates next.

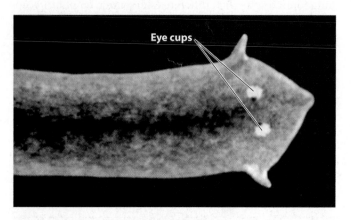

Eye cups

Figure 29.4A The eye cups of a planarian

? What key optical feature is found in the eyes of both insects and squids but is not present in planarians?

■ Lenses, which focus light onto photoreceptor cells

29.5 Vertebrates have single-lens eyes

The vertebrate eye is like the eye of a squid in that it has a single lens and is camera-like, but it evolved independently and differs in several details from the single-lens eyes of invertebrates. Our eyes are remarkable sense organs, able to detect a multitude of colors, form images of objects miles away, and respond to minute amounts of light energy.

The outer surface of the human eyeball is a tough, whitish layer of connective tissue called the **sclera** (**Figure 29.5**). At the front of the eye, the sclera becomes the transparent

cornea, which lets light into the eye and also helps focus light. The sclera surrounds a pigmented layer called the **choroid**. The anterior choroid forms the doughnut-shaped **iris**, which gives the eye its color. By changing size, the iris regulates the size of the **pupil**, the opening in the center of the iris that lets light into the interior of the eye. After going through the pupil, light passes through the disklike **lens**, which is held in position by ligaments. The lens focuses images onto the **retina**, a layer just inside the choroid. Photo-

receptor cells of the retina transduce light energy, and action potentials pass via sensory neurons in the optic nerve to the visual centers of the brain. Photoreceptor cells are highly concentrated at the retina's center of focus, called the **fovea**. There are no photoreceptor cells in the part of the retina where the optic nerve passes through the back of the eye. We cannot detect light that is focused on this **blind spot**, but having two eyes with overlapping fields of view enables us to perceive uninterrupted images.

Two chambers make up the bulk of the eye. The large chamber behind the lens is filled with jellylike **vitreous humor.** The much smaller chamber in front of the lens contains the thinner **aqueous humor**. The humors help maintain the shape of the eyeball. In addition, the aqueous humor circulates through its chamber. This fluid, secreted by the **ciliary body,** supplies nutrients and oxygen to the lens, iris, and cornea and carries off wastes. Blockage of the ducts that drain this fluid can cause glaucoma, increased pressure inside the eye that may lead to blindness. If diagnosed early, glaucoma can be treated with medications that increase the circulation of aqueous humor.

A thin mucous membrane (see Module 20.4) helps keep the outside of the eye moist. This membrane, called the conjunctiva, lines the inner surface of the eyelids and folds back over the white of the eye (but not the cornea). A gland above the eye secretes a dilute salt solution that is spread across the eyeball by blinking and drains into ducts that lead into the nasal cavities. This fluid cleanses and moistens the eye surface. Excess secretion, in response to eye irritation or emotional dis-

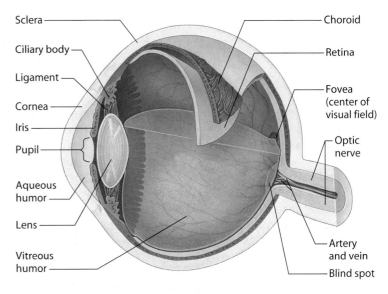

Figure 29.5 The single-lens eye of a vertebrate

tress (or happiness), causes tears to spill over the eyelid and fill the nasal cavities, producing sniffles. Only humans shed emotional tears, which may play a role in reducing stress.

? Arrange the following eye parts into the correct sequence encountered by photons of light traveling into the eye: pupil, retina, cornea, lens, vitreous humor, aqueous humor.

■ Cornea → aqueous humor → pupil → lens → vitreous humor → retina

sclera   cornea   cornea thor   choroid

29.6   To focus, a lens changes position or shape

A lens focuses light onto a retina by bending light rays. Focusing can occur in two ways. The lens may be rigid, as in squids and many fishes, and focusing occurs as muscles move it back or forth, as you might focus on an object using a magnifying glass. Or, as in the mammalian eye, focusing is accomplished by changing the shape of the lens. The thicker the lens, the more sharply it bends light.

The shape of the lens is controlled by the ciliary muscles, which are attached to the choroid and the ligaments that suspend the lens (**Figure 29.6**). When the eye focuses on a nearby object, these muscles contract, pulling the choroid toward the lens, which reduces the tension on the ligaments. As these ligaments slacken, the elastic lens becomes thicker and rounder, as shown in the top diagram of Figure 29.6. This change, called **accommodation,** allows the diverging light rays from a close object to be bent and focused.

Light from distant objects approaches in parallel rays that require less bending for proper focusing on the retina. When the eye focuses on a distant object, the ciliary muscles relax, and the choroid moves away from the lens. This puts tension on the ligaments and flattens the elastic lens, as shown in the bottom diagram.

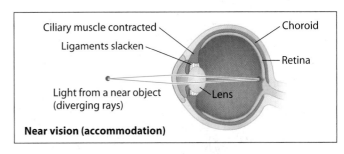

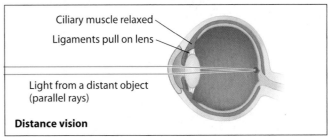

Figure 29.6 How lenses focus light

? As you read this text, are your lenses relatively thick or thin?

Web/CD Activity 29A Structure and Function of the Eye

■ Thick

29.7 Artificial lenses or surgery can correct focusing problems

When you have your vision tested, you are asked to read letters on a special chart. The chart measures your **visual acuity**, the ability of your eyes to distinguish fine detail. The examiner asks you to read a line of letters sized for legibility at a distance of 20 feet, using one eye at a time. If you can do this, you have so-called normal (20/20) acuity in each eye. This means that from a distance of 20 feet, each of your eyes can read the chart's line of letters designated for 20 feet.

Suppose you find out that your visual acuity is 20/10. This is actually better than normal; it means that you can read letters from a distance of 20 feet that a person with 20/20 vision can only read at 10 feet. On the other hand, someone with 20/50 acuity has worse than normal vision. He or she must stand at a distance of 20 feet to read what a person with normal acuity can read at 50 feet.

Three of the most common visual problems are nearsightedness, farsightedness, and astigmatism. All three are focusing problems, easily corrected with artificial lenses. Nearsighted people cannot focus well on distant objects, although they can see well at short distances (the condition is named for the type of vision that is *unimpaired*). A nearsighted eyeball (Figure 29.7A) is longer than normal. The lens cannot flatten enough to compensate, and it focuses distant objects in front of the retina, instead of on it. As shown by the drawing on the right in Figure 29.7A, **nearsightedness** (also known as myopia) is corrected by glasses or contact lenses that are thinner in the middle than at the outside edge. The corrective lenses make the light rays from distant objects diverge slightly as they enter the eye. The focal point formed by the lens in the eye then falls directly on the retina. Other treatment options include surgery that cuts slits in the cornea or laser surgery that removes corneal tissue. Both techniques reshape the cornea, thus reducing the bending of light rays.

Farsightedness (also known as hyperopia) is the opposite of nearsightedness. It occurs when the eyeball is shorter than normal, and the focal point of the lens is behind the retina (Figure 29.7B). Farsighted people see distant objects normally but cannot focus on close objects. Corrective lenses that are thicker in the middle than at the outside edge compensate for farsightedness by making light rays from nearby objects converge slightly before they enter the eye. A type of farsightedness called presbyopia (from the Greek *presbys*, old man) develops with age. Beginning around the mid-40s, the lens of the eye becomes less elastic. As a result, the lens gradually loses its ability to focus on nearby objects (accomodation), and reading without glasses becomes difficult.

Astigmatism is blurred vision caused by a misshapen lens or cornea. Any such distortion makes light rays converge unevenly and not focus at any one point on the retina. Lenses that correct astigmatism are asymmetrical in a way that compensates for the asymmetry in the eye.

? A person with 20/100 vision in both eyes must stand at _____ feet to read what someone with normal vision can read at _____ feet.

■ 20 . . . 100

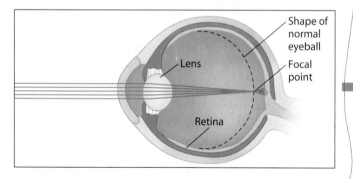

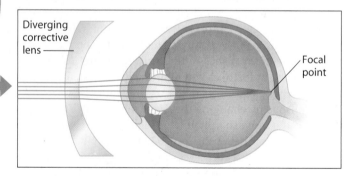

Figure 29.7A A nearsighted eye (eyeball too long)

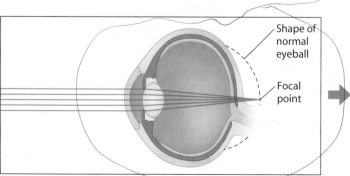

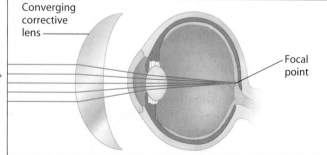

Figure 29.7B A farsighted eye (eyeball too short)

Our photoreceptors are rods and cones

Built into the human retina are about 125 million rod cells and 6 million cone cells, two types of photoreceptors named for their shapes (**Figure 29.8A**). **Cones** are stimulated by bright light and can distinguish color, but they contribute little to night vision. **Rods** are extremely sensitive to light and enable us to see in dim light at night, though only in shades of gray. The relative numbers of rods and cones correlate with whether an animal is most active during the day or night.

In humans, rods are found in greatest density at the outer edges of the retina and are completely absent from the fovea, the retina's center of focus. If you face directly toward a dim star in the night sky, the star is hard to see. Viewing it at an angle, however, makes your lens focus the starlight onto the parts of the retina with the most rods, and you can see the star. By contrast, you achieve your sharpest day vision by looking straight at the object of interest. This is because cones are densest (about 150,000 per square millimeter) in the fovea. Some birds, such as hawks, have more than a million cones per square millimeter, which enables them to spot small prey from high in the air.

How do rods and cones detect light? As Figure 29.8A shows, each rod and cone includes an array of membranous disks containing light-absorbing visual pigments. Rods contain a visual pigment called **rhodopsin**, which can absorb dim light. Cones contain visual pigments called **photopsins**, which absorb bright, colored light. We have three types of cones, each containing a different type of photopsin. These cells are called blue cones, green cones, and red cones, referring to the colors absorbed best by their photopsin. We can perceive a great number of colors because the light from each particular color triggers a unique pattern of stimulation among the three types of cones. Color blindness, more common in males than females because it is inherited as a sex-linked trait (see Module 9.24), results from a deficiency in one or more types of cones. The most common type is red-green color blindness, in which red and green are seen as the same color—either red or green—depending on which type of cone is deficient.

Figure 29.8B shows the pathway of light into the eye and through the cell layers of the retina. Notice that the rods and cones have their tips embedded in the back of the retina (pink cells). Light must pass through several relatively transparent layers of neurons before reaching the pigments in the rods and cones. Like all sensory receptors, rods and cones are stimulus transducers. When rhodopsin and photopsin absorb light, they change chemically, and the change alters the permeability of the cell's membrane. The resulting receptor potential triggers a change in the release of neurotransmitter, which initiates a complex integration process in the retina. As shown in Figure 29.8B, visual information transduced by the rods and cones passes from the photoreceptor cells through a network of neurons (black arrows). Notice the numerous synapses between the photoreceptor cells and the neurons and among the neurons themselves. Integration in this maze of synapses helps sharpen images and increases the contrast between light and dark areas. Action potentials carry the partly integrated information into the brain via the optic nerve. Three-dimensional perceptions (what we actually see) result when visual input coming from the two eyes is integrated further in several processing centers of the cerebral cortex.

? Explain why our night vision is mostly in black and white rather than in color.

■ Rods are more sensitive to light than cones, and thus the low light intensity at night stimulates far more rods than cones.

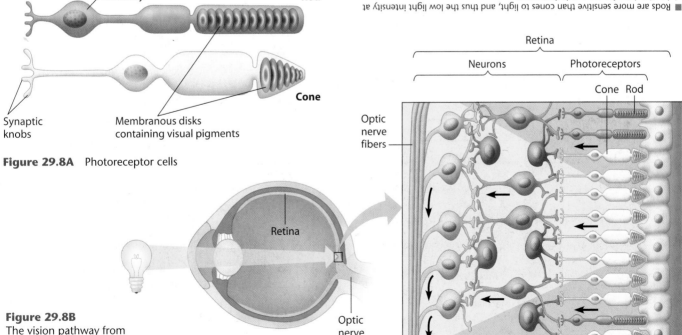

Figure 29.8A Photoreceptor cells

Cell body — Rod
Synaptic knobs
Membranous disks containing visual pigments
Cone

Figure 29.8B
The vision pathway from light source to optic nerve

Retina
Optic nerve

Retina
Neurons — Photoreceptors
Cone Rod
Optic nerve fibers

29.9 The ear converts air pressure waves to action potentials that are perceived as sound

The human ear is really two separate organs, one for hearing and the other for maintaining balance. We look at the structure and function of our hearing organ in this module and then turn to our sense of balance in Module 29.10. Both organs operate on the same basic principle, the stimulation of long microvilli-like projections on hair cells (mechanoreceptors) in fluid-filled canals.

The ear is complex, and it helps to learn its basic structure before studying how it functions. The ear is composed of three regions: the outer ear, the middle ear, and the inner ear (Figure 29.9A). The **outer ear** consists of the flap-like **pinna**—the structure we commonly refer to as our "ear"—and the **auditory canal**. The pinna and the auditory canal collect sound waves and channel them to the **eardrum**, a sheet of tissue that separates the outer ear from the **middle ear** (Figure 29.9B). When sound pressure waves strike the eardrum, it vibrates and passes the vibrations to three small bones: the hammer, anvil, and stirrup. The stirrup is connected to the **oval window**, a membrane-covered hole in the skull bone, through which vibrations pass into the inner ear. The middle ear also opens into the **Eustachian tube,** which connects with the pharynx (back of the throat) and equalizes pressure between the middle ear and the atmosphere. This tube is what enables you to move air in or out to equalize pressure on either side of the eardrum ("pop" your ears) when changing altitude rapidly in an airplane or car.

The **inner ear** consists of fluid-filled channels in the bones of the skull. Sound vibrations or movements of the head set the fluid in motion. One of the channels, the **cochlea** (Latin for "snail"), is a long, coiled tube. The cross-sectional view of the cochlea in **Figure 29.9C**, on the bottom of the next page, shows that inside it are three fluid-filled canals. Our hearing organ, the **organ of Corti**, is located within the middle canal. The organ of Corti consists of an array of hair cells embedded in a **basilar membrane** (the floor of the middle canal). The hair cells are the sensory receptors of the ear. As you can see in the enlargement, a shelflike projection called the tectorial membrane extends from the wall of the middle canal. Notice that the tips of the hair cells are in contact with the tectorial membrane. Sensory neurons synapse with the base of the hair cells and carry action potentials to the brain via the auditory nerve.

Now let's see how the parts of the ear function in hearing. Sound waves are pressure waves in the air that are collected by the pinna and auditory canal of the outer ear. Represented in **Figure 29.9D**, these pressure waves make your eardrum vibrate with the same frequency as the sound. The frequency, measured in hertz (Hz), is the number of vibrations per second (1 Hz is equal to one vibration per second).

From the eardrum, the vibrations are amplified as they are transferred through the hammer, anvil, and stirrup in the middle ear to the oval window. Vibrations of the oval window then produce pressure waves in the fluid within the cochlea. The vibrations first pass from the oval window into the fluid in the upper canal of the cochlea. Pressure waves travel through the upper canal to the tip of the cochlea, at the coil's center. The pressure waves then enter the lower canal and gradually fade away.

As a pressure wave passes through the upper canal of the cochlea, it pushes downward on the middle canal, making

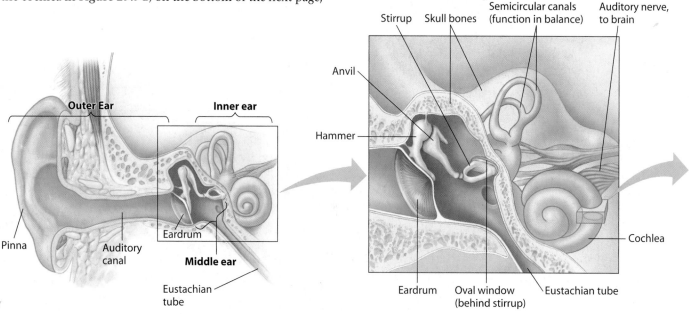

Figure 29.9A An overview of the human ear

Figure 29.9B The middle ear and the inner ear

the basilar membrane vibrate. Vibration of the basilar membrane makes the hairlike projections on the hair cells alternately brush against and draw away from the overlying tectorial membrane. When a hair cell's projections are bent, ion channels in its plasma membrane open, and positive ions enter the cell. As a result, the hair cell develops a receptor potential and releases more neurotransmitter molecules at its synapse with a sensory neuron. In turn, the sensory neuron sends more action potentials to the brain through the auditory nerve.

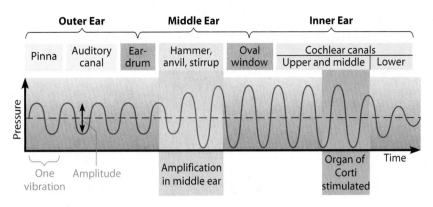

Figure 29.9D The route of sound wave vibrations through the ear

Volume and Pitch The brain senses a sound as an increase in the frequency of action potentials it obtains from the auditory nerve. But how is the quality of the sound determined? The higher the volume (loudness) of sound, the higher the amplitude of the pressure wave it generates. In the ear, a higher amplitude of pressure waves produces more vigorous vibrations of fluid in the cochlea, more pronounced bending of the hair cells, and thus, more action potentials generated in the sensory neurons. The loudness of sound is measured in decibels (dB). The decibel scale for human hearing ranges from 0 to 120 dB, the loudest we can hear without intolerable pain.

The pitch of a sound depends on the frequency of the sound waves. High-pitched sounds, such as high notes sung by a soprano, generate high-frequency waves. Low-pitched sounds, like low notes sung by a bass, generate low-frequency waves. How does the cochlea distinguish sounds of different pitch? The key is that the basilar membrane is not uniform along its length. The end near the oval window is relatively narrow and stiff, while the other end, near the tip of the cochlea, is wider and more flexible. Each region of the basilar membrane is most sensitive to a particular frequency of vibration, and the region vibrating most vigorously at any instant sends the most action potentials to auditory centers in the brain. The brain interprets the information and gives us a perception of pitch. Young people with healthy ears can hear pitches in the range of 20–20,000 Hz. Dogs can hear sounds as high as 40,000 Hz, and bats can emit and hear clicking sounds as high-pitched as 100,000 Hz.

Deafness, the loss of hearing, can be caused by the inability to conduct sounds, resulting from middle-ear infections, a ruptured eardrum, or stiffening of the middle-ear bones (a common age-related problem). Deafness can also result from damage to sensory receptors or neurons. Few parts of our anatomy are more delicate than the organ of Corti. Frequent or prolonged exposure to sounds over 90 dB can damage or destroy hair cells. In the United States, employees exposed to occupational noise above that level must wear ear protection. Amplified rock music often reaches 120 dB. Ear plugs can provide protection to both rock musicians and their fans.

? How does the ear convert sound waves in the air to pressure waves of the fluid in the cochlea?

■ Sound waves in air cause the eardrum to vibrate. The small bones attached to the inside of the eardrum transmit the movement to the oval window on the wall of the inner ear. Vibrations of the oval window set in motion the fluid in the inner ear, which includes the cochlea.

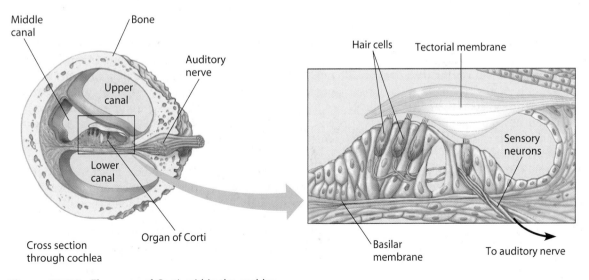

Figure 29.9C The organ of Corti, within the cochlea

29.10 The inner ear houses our organs of balance

Several organs in the inner ear detect body position and movement. These fluid-filled structures lie next to the cochlea (Figure 29.10) and include three semicircular canals and two chambers, the utricle and the saccule. All the equilibrium structures operate on the same principle, by the bending of hairs on hair cells.

The three **semicircular canals** detect changes in the head's rate of rotation or angular movement. As shown in the figure, the canals are arranged in three perpendicular planes and can therefore detect movement in all directions. A swelling at the base of each semicircular canal contains a cluster of hair cells with their hairs projecting into a gelatinous mass called a cupula (shown in the enlargements). When you rotate your head in any direction, the thick, sticky fluid in the canals moves more slowly than your head. Consequently, the fluid presses against the cupula, bending the hairs. The faster you rotate your head, the greater the pressure and the higher the frequency of action potentials sent to the brain. If you rotate your head at a constant speed, the fluid in the canals begins moving with the head, and the pressure on the cupula is reduced. But if you stop suddenly, the fluid continues to move and again stimulates the hair cells, which may make you feel dizzy.

Clusters of hair cells in the **utricle** and **saccule** detect the position of the head with respect to gravity. The hairs of these cells project into a gelatinous material containing many small calcium carbonate particles. When the position of the head changes, this heavy material bends the hairs in a different direction, causing an increase or decrease in the rate at which action potentials are sent to the brain. The brain determines the new position of the head by interpreting the altered flow of action potentials.

The equilibrium receptors provide data the brain needs to determine the position and movement of the head. Using this information, the brain develops and sends out commands that make the skeletal muscles balance the body.

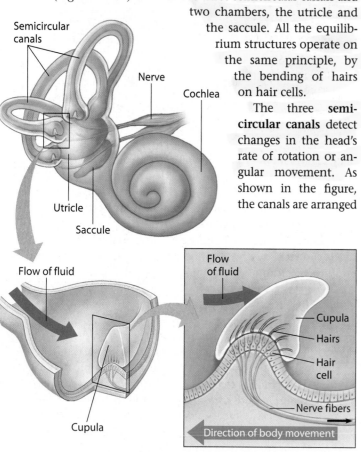

Figure 29.10 Equilibrium structures in the inner ear

? What type of receptor cell is common to our senses of hearing and equilibrium?

■ Hair cells, which are mechanoreceptors

CONNECTION

29.11 What causes motion sickness?

Boating, flying, or even riding in a car can make us dizzy and nauseated, a condition called motion sickness. Some people start feeling ill just from thinking about getting on a boat or plane. Many others get sick only during storms at sea or during turbulence in flight. Motion sickness is thought to result from the brain's receiving signals from equilibrium receptors in the inner ear that conflict with visual signals from the eyes. When a susceptible person is inside a moving ship, for instance, signals from the equilibrium receptors in the inner ear indicate, correctly, that the body is moving (in relation to the environment outside the ship). In conflict with these signals, the eyes may tell the brain that the body is in a stationary environment, the cabin. Somehow the conflicting signals make the person feel ill. Symptoms may be relieved by closing the eyes, limiting head movements, or focusing on a stable horizon. Many sufferers of motion sickness take a sedative such as Dramamine or Bonine to relieve their symptoms. Long-lasting, drug-containing skin patches prevent motion sickness by inhibiting input from the equilibrium sensors. Ginger tablets and pressure-point wristbands may also help.

Motion sickness can be a severe problem for astronauts, and the National Aeronautics and Space Administration conducts research on the problem. One of NASA's most interesting findings is that some people can learn to consciously control body functions, such as the vomiting reflex. Astronauts receive intensive training in how to exert "mind over body" when zero gravity starts to induce motion sickness.

? Explain how someone could suffer motion sickness when watching a film shot from the front of a roller coaster.

■ There would be conflicting information between vision ("I'm moving") and the equilibrium sense ("I'm sitting still in my theater seat").

29.12 Taste and odor receptors detect chemicals present in solution or air

Our senses of taste and smell depend on receptor cells that detect chemicals in the environment. Chemoreceptors in our taste buds detect molecules in solution; chemoreceptors in our nose detect airborne molecules.

Our sense of taste depends on taste receptors organized into taste buds on the tongue. In addition to the four familiar taste perceptions—sweet, sour, salty, and bitter—a fifth, called umami (Japanese for "delicious"), is elicited by glutamate, an amino acid in the flavor enhancer monosodium glutamate (MSG) and in meat. Module 29.2 describes the functioning of a receptor stimulated by sugar. Each type of taste receptor is most responsive to a particular type of substance. The brain integrates a variety of inputs from different receptors to create the flavors you perceive.

The olfactory (smell) receptors are sensory neurons that line the upper portion of the nasal cavity and send impulses along their axons directly to the olfactory bulb of the brain (Figure 29.12). Notice the cilia extending from the tips of these chemoreceptors into the mucus that coats the nasal cavity. When an odorous substance diffuses into this region (blue dots), it binds to specific receptor proteins on the cilia. The binding triggers a membrane depolarization and generates action potentials. Integration of the signals in the brain results in an odor perception. Humans can distinguish thousands of different odors.

Many animals, such as the salmon and brown bear in the chapter introduction, rely heavily on their sense of smell for survival. Most mammals have a much more discriminating sense of smell than humans. Odors often provide more information than visual images about food, the presence of mates,

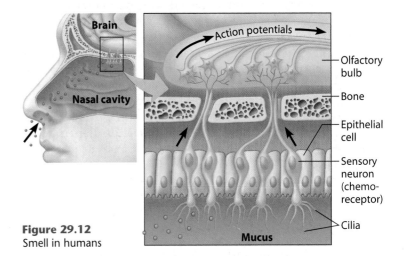

Figure 29.12
Smell in humans

or danger. In contrast, humans often pay more attention to sights and sounds than to smells.

Although the receptors and brain pathways for taste and smell are independent, the two senses do interact. Indeed, much of what we call taste is really smell, as you have undoubtedly noticed when a stuffy head cold dulls your perception of taste. How else might your sense of taste change?

? What is the key structural difference between taste receptors and olfactory receptors?

■ Taste receptors release neurotransmitter, triggering action potentials carried to the brain by sensory neurons. Olfactory receptors are modified sensory neurons.

29.13 Our sense of taste may change as we age

As you grow older, you may find yourself choosing foods you once refused. A meal of a spinach salad and grilled fish, for example, is often more appealing to an adult than a young child. This shift may reflect age-related changes in our taste perceptions.

Scientific studies have found that small children usually possess greater taste sensitivity than adults. This sensitivity has been documented in children as young as 3 years old, and appears to hold steady through the teen years. In adulthood, however, this sensitivity begins to lessen and can diminish rapidly with age. Adults in their 60s or 70s, for example, have shown reduced sensitivity to a range of tastes, especially salty and umami tastes.

Taste perception can vary from person to person, often because of genetic traits. Some individuals, for example, have a

hypersensitivity to bitter tastes, which may correlate with their dislike for spinach, broccoli, and other vegetables rich in bitter, but beneficial, phytochemicals. Thus, taste-related food choices can have health implications.

Researchers theorize that age-related loss of taste sensitivity is likely tied to the blend of senses that allows us to perceive flavors in our food. Smell is known to decline with age even more than taste. Hence, a reduced sense of smell is also likely to diminish perceived flavors—and may make certain foods more palatable as we mature.

? What are some potentially harmful effects of the decline in taste sensitivity in older adults?

■ They may not consume enough nutrients if food no longer tastes appealing.

29.14 Review: The central nervous system couples stimulus with response

In this chapter and the previous one, we focused on information gathering and processing. Sensory receptors provide an animal's nervous system with vital data that enable the animal to avoid danger, find food and mates, and maintain homeostasis—in short, to survive.

A brown bear catching a salmon helps us summarize the sequence of information flow in an animal. A bear sees a flash in the stream. Within milliseconds, photoreceptor cells in the bear's retinas transduce the light energy focused on them by the lens, and action potentials representing a glimpse of the salmon enter the brain. Before the salmon can swim out of reach, a vast network of neurons in the bear's brain, with millions of synapses, integrates the information and sends out command signals, again in the form of action potentials. The commands go out via motor neurons to muscles, and the bear lunges and grabs its meal.

In the next chapter, we see how muscles carry out the commands they receive from the nervous system.

? What three general types of neurons are involved when a bear catches a salmon? (*Hint:* Review Module 28.1.)

■ Sensory neurons, interneurons, and motor neurons

CHAPTER REVIEW

Reviewing the Concepts

Animal senses gather information that guides feeding, migrating, and other behaviors **(Introduction)**. Perception is the brain's integration of sensations **(29.1)**.

Sensory Reception (29.2–29.3)

Sensory receptors are specialized cells or neurons that detect stimuli. Sensory transduction converts stimulus energy into receptor potentials, which trigger action potentials that are transmitted to the brain. Action potential frequency reflects stimulus strength. Repeated stimuli may lead to adaptation, a decrease in sensitivity **(29.2)**.

Types of sensory receptors. Pain receptors sense dangerous stimuli. Thermoreceptors detect heat or cold. Mechanoreceptors respond to mechanical energy (such as touch, pressure, and sound), and chemoreceptors to chemicals. Electromagnetic receptors respond to electricity, magnetism, and light (sensed by photoreceptors) **(29.3)**.

Vision (29.4–29.8)

Invertebrate eyes range from simple eye cups that sense light intensity and direction to the many-lensed compound eyes of insects to the single-lens eyes of squids **(29.4)**.

Vertebrate eyes are single-lens eyes. In the human eye, the cornea and flexible lens focus light on photoreceptor cells in the retina **(29.5)**.

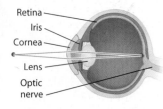

Focusing involves changing the shape of the lens **(29.6)**. Nearsightedness and farsightedness result when the focal point is not on the retina. Corrective lenses bend the light rays to compensate **(29.7)**.

Rods and cones allow us to see shades of gray in dim light and to see color in bright light, respectively **(29.8)**.

Hearing and Balance (29.9–29.11)

The human ear channels sound waves through the outer ear to the eardrum to a chain of bones in the middle ear to the fluid in the coiled cochlea in the inner ear. Pressure waves in the fluid bend hair cells of the organ of Corti against a membrane, triggering nerve signals to the brain. Louder sounds generate more action potentials; pitches stimulate different regions of the organ of Corti **(29.9)**.

The organs of balance, the semicircular canals and utricle and saccule located in the inner ear, sense body position and movement **(29.10)**. Conflicting signals from the inner ear and eyes may cause motion sickness **(29.11)**.

Taste and Smell (29.12–29.14)

Taste and smell depend on chemoreceptors that bind specific molecules. Taste receptors, located in taste buds on the tongue, produce five taste sensations. Olfactory (smell) sensory neurons line the nasal cavity. Various odors and tastes result from the integration of input from many receptors **(29.12)**. Taste sensitivity declines with age **(29.13)**.

Stimulus and response. The nervous system receives sensory information, integrates it, and commands appropriate muscle responses **(29.14)**.

Connecting the Concepts

1. Complete this map summarizing sensory receptors.

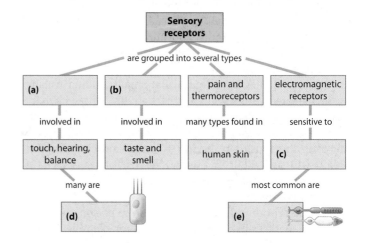

2. Label the parts in this diagram of the human ear. Describe the passage of sound waves through the ear to the perception of sound in the brain.

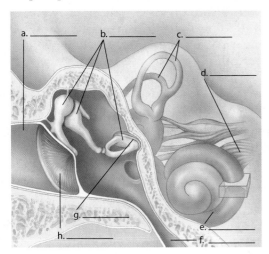

Testing Your Knowledge

Multiple Choice

3. Eighty-year-old Mr. Johnson was becoming slightly deaf. To test his hearing, his doctor held a vibrating tuning fork tightly against the back of Mr. Johnson's skull. This sent vibrations through the bones of the skull, setting the fluid in the cochlea in motion. Mr. Johnson could hear the tuning fork this way, but not when it was held away from the skull a few inches from his ear. The problem was probably in the *(Explain your answer.)*
   a. auditory center in Mr. Johnson's brain.
   b. auditory nerve leading to the brain.
   c. hair cells in the cochlea.
   d. bones of the middle ear.
   e. fluid of the cochlea.

4. Which of the following correctly traces the path of light into your eye?
   a. lens, cornea, pupil, retina
   b. cornea, pupil, lens, retina
   c. cornea, lens, pupil, retina
   d. lens, pupil, cornea, retina
   e. pupil, cornea, lens, retina

5. If you look away from this book and focus your eyes on a distant object, the eye muscles _____ and the lenses _____ to focus images on the retinas.
   a. relax . . . flatten
   b. relax . . . become more rounded
   c. contract . . . flatten
   d. contract . . . become more rounded
   e. contract . . . relax

6. Which of the following are *not* present in human skin?
   a. thermoreceptors
   b. electromagnetic receptors
   c. touch receptors
   d. pressure receptors
   e. pain receptors

7. Jim had his eyes tested and found that he has 20/40 vision. This means that
   a. the muscles in his iris accommodate too slowly.
   b. he is farsighted.
   c. the vision in his left eye is normal, but his right eye is defective.
   d. he can see at 40 feet what a person with normal vision can see at 20 feet.
   e. he can see at 20 feet what a person with normal vision can see at 40 feet.

8. What do the lateral line of fish and the cochlea of your ear have in common?
   a. They use hair cells to sense sound or pressure waves.
   b. They are organs of equilibrium.
   c. They use electromagnetic receptors to sense pressure waves in fluid.
   d. They use granules that signal a change in position and stimulate their receptor cells.
   e. They are homologous structures that share a common evolutionary origin with all organs of hearing.

Describing, Comparing, and Explaining

9. How does your brain determine the volume and pitch of sounds?

10. As you read these words, the lenses of your eyes project patterns of light representing the letters onto your retinas. There the photoreceptors respond to the patterns of light and dark and transmit nerve signals to the brain. The brain then interprets the words. In this example, which processes relate to sensation and which to perception?

11. For what purposes do animals use their senses of taste and smell?

Applying the Concepts

12. Sensory organs tend to come in pairs. We have two eyes and two ears. Similarly, a planarian worm has two eye cups, a rattlesnake has two infrared receptors, and a butterfly has two antennae. Propose a testable hypothesis that could explain the advantage of having two ears or eyes instead of one.

13. Sea turtles bury their eggs on the beach above the high-tide line. When the baby turtles hatch, they dig their way to the surface of the sand and quickly head straight for the water. How do you think the turtles know which way to go? Outline an experiment to test your hypothesis.

14. Have you ever felt your ears ringing after listening to loud music from a stereo or at a concert? Can this music be loud enough to permanently impair your hearing? Do you think people are aware of the possible danger of prolonged exposure to loud music? Should anything be done to warn or protect them? If you think so, what action would you suggest? What effect might warnings have?

Answers to all questions can be found in Appendix 3.

For study help and Activities, go to campbellbiology.com or the student CD-ROM.

John Hutchinson (far left) of Stanford University paints spots on Tika, an Asian elephant

MOVEMENT AND LOCOMOTION

30.1   Diverse means of animal locomotion have evolved

SKELETAL SUPPORT

30.2   Skeletons function in support, movement, and protection
30.3   The human skeleton is a unique variation on an ancient theme
30.4   Bones are complex living organs
30.5   Broken bones can heal themselves
30.6   Weak, brittle bones are a serious health problem, even in young people

MUSCLE CONTRACTION AND MOVEMENT

30.7   The skeleton and muscles interact in movement
30.8   Each muscle cell has its own contractile apparatus
30.9   A muscle contracts when thin filaments slide across thick filaments
30.10   Motor neurons stimulate muscle contraction
30.11   Athletic training increases strength and endurance
30.12   The structure-function theme underlies all the parts and activities of an animal

CHAPTER 30

Elephants Do the "Groucho Gait"

ONE IS THE SIZE OF A SMALL TRUCK, with skin like leather and feet bigger than a pie plate. The other, all bushy eyebrows and wobbling cigar, is an icon of vintage American comedy. Elephants and Groucho Marx don't seem to have much in common—unless you take a look at how the two hustle and what that reveals about how animals move.

Movement is a distinctive feature of animals. Some motions are critical to survival, as when a gopher darts back down its hole to escape a swooping hawk. Others are more for convenience, helping you turn the page of this book or reach for the glass of water at your bedside. In all its necessity and variation, movement is a wide field of scientific inquiry. How do tiny insects pick up and move relatively huge objects? Can disabled people be helped with mobility problems? Why is one creature's motion so different from another's?

Scientists who study such questions focus on the field of biomechanics, which analyzes animal motion one movement at a time. The 19th-century photographer Eadweard Muybridge helped open the door to the study of animal movement by pioneering the use of stop-action photography. In 1872 Leland Stanford, the founder of Stanford University, hired Muybridge to settle a bet: Does a trotting horse ever have all four hooves off the

How Animals Move

Groucho Marx

ground simultaneously? After five years of work, Muybridge was able to create a series of stop-action photographs that settled the wager with a definitive "Yes." More than a century later, John Hutchinson and his colleagues at Stanford tried a similar approach—but on a far bigger four-footed creature.

The Stanford researchers, wanting to better understand how large animals move, focused on one hefty creature, the Asian elephant, and an old question in biomechanics—Do elephants run? They are certainly known to be speedy, and other creatures in their way usually head elsewhere fast. But does their imposing gait add up to running? Gathering some paint and video equipment, the researchers decided to find out.

The scientists had already clocked the swiftness of Asian elephants kept in the United States and found they stuck to a speed of about 16 km/hr. Thinking that seemed a little slow, the team headed to Thailand and an elephant center that cared for pachyderms raised in the wild. There, they studied 42 elephants, feeling along their leg and hip muscles for the joints between bones and then marking those spots with dots of white paint that could be easily tracked on video.

Then they put each elephant at the start of a 30-m course, plied the would-be track star with cheers and food rewards,

For their size, elephants have relatively small muscles

turned on the video camera, and had the creature hustle along.

The Thai elephants reached speeds of up to 25 km/hr, the fastest ever recorded for a pachyderm. They also yielded biomechanical images that hint at running. When many four-footed animals reach their top speed, all their feet are off the ground simultaneously at some point in their gait. Elephants, however, keep three feet on the ground when they stride, even at high speeds. But their bodies, especially their centers of gravity, pick up more of a bounce as they speed up. That bouncing gait is another feature of running. It is also similar to the crouched stride of Groucho Marx.

The researchers are now planning more studies to get a definitive answer on whether elephants truly meet the biomechanical definition of running. But their work has already pointed to important findings. For their size, elephants have relatively small muscles. Understanding how they move could give insights into the movements of other large animals, such as dinosaurs. The research could also help obese people with mobility problems or people with weak or paralyzed muscles.

The mechanics of elephant motion have this potential to explain the movements of other animals and people because of a common structure all such creatures share. The elements of the elephant's body examined in the study—muscles and bones—are essential to the way many animals move. Whether an animal walks, runs, swims, crawls, flies, or sits in one place and only moves its mouth, the interplay of three organ systems provides its motion. The nervous system plays the key role in issuing commands to the muscular system. The muscular system exerts the force that actually makes an animal or its parts move. The skeletal system is essential because the force exerted by muscles produces movement only when the force is applied against a firm structure, the skeleton. These systems keep any creature in motion, from the biggest elephant to the goofiest comedian. In this chapter, we'll focus on skeletons, muscles, and the movements their interactions produce. ■ ■ ■

30.1 Diverse means of animal locomotion have evolved

Animal movement is extremely diverse. Many animals stay in one place and move only certain parts of their body. For example, an adult sponge's only movements are the opening and closing of cellular pores on its surface and the beating of flagella, drawing suspended food particles in through the pores. Most animals, however, spend much of their time and energy moving about in search of food or mates or escaping from danger. Active travel from place to place, also called **locomotion**, requires that an animal expend energy to overcome two forces that tend to keep it stationary: friction and gravity. The relative importance of these two forces varies, depending on the environment.

Swimming Gravity is not much of a problem for a swimming animal, because water supports much or all of the animal's weight. On the other hand, overcoming friction is more difficult for a swimmer, because water is dense and offers considerable resistance to a body moving through it.

Animals swim in diverse ways. Many insects, for example, swim the way we do, using their legs as oars to push against the water. Squids, scallops, and some jellies are jet-propelled, taking in water and squirting it out in bursts. Fishes swim by moving their body and tail from side to side (Figure 30.1A). Whales and other aquatic mammals move by undulating their body and tail from top to bottom. A sleek, streamlined shape, like that of seals, porpoises, penguins, and many fishes, is an adaptation that aids rapid swimming.

Figure 30.1A A fish swimming

Locomotion on Land: Hopping, Walking, Running, and Crawling The problems of locomotion on land are more or less the opposite of those in the water. Air offers very little resistance to an animal moving through it, at least at moderate speeds. However, air provides little support for an animal's body, and a land animal must be able to support itself and overcome the force of gravity. When a land animal walks, runs, or hops, its leg muscles expend energy both to propel it and to keep it from falling down. To move on land, powerful muscles and strong skeletal support are more important than a streamlined shape.

The kangaroo travels mainly by hopping (Figure 30.1B). Large muscles in its hind legs generate a lot of power. Tendons (which connect muscle to bone) in the legs also momentarily store energy when the kangaroo lands—somewhat like the spring on a pogo stick. The higher the jump, the tighter the spring coils when a pogo stick lands and the greater the tension in the tendons when a kangaroo lands. In both cases, the stored energy is available for the next jump. For the kangaroo, the tension in its legs is a cost-free energy

Figure 30.1B Kangaroos hopping

Figure 30.1C Kangaroos sitting

boost that reduces the total amount of energy the animal expends to travel. The pogo stick analogy applies to many land animals. The legs of an insect, a horse, or a human, for instance, retain some spring during walking or running, although less than those of a hopping kangaroo.

The kangaroo in Figure 30.1C illustrates a solution to another problem of terrestrial life: maintaining balance. The animal can sit upright with relatively little energy cost because it has three parts of its body touching the ground simultaneously, its two hind legs and the base of its tail. Similar to a camera tripod, this arrangement stabilizes the upright body. The same principle applies to four-legged animals. When walking slowly, they usually keep three feet on the ground at all times. And, as you read in the chapter introduction, elephants keep three feet on the ground even when moving quite fast. Bipedal (two-footed) animals, such as birds and humans, are less stable on land and keep part of at least one foot on the ground when walking. When an animal is running, all of its feet may be off the ground simultaneously. At running speeds, momentum, more than foot contact, stabilizes the body's position, just as a moving bicycle stays upright.

A crawling animal, such as a snake or an earthworm, faces very different problems. Because much of the animal's body is in contact with the ground, friction offers considerable resistance to movement. Many snakes crawl rapidly by undu-

lating the entire body from side to side. Aided by large, movable scales on their underside, a snake's body pushes against the ground, driving the animal forward. Boa constrictors and pythons creep forward in a straight line, driven by muscles that lift belly scales off the ground, tilt them forward, and then push them backward against the ground.

Earthworms crawl by peristalsis, a type of movement produced by rhythmic waves of muscle contractions passing from head to tail. (In Module 21.8, we saw how peristalsis squeezes food through our digestive tract.) To move by peristalsis, an animal needs a set of muscles that elongates the body and another set that shortens it. Also required are a way to anchor its body to the ground and a hydrostatic skeleton, which we will discuss more in Module 30.2. As illustrated in **Figure 30.1D**, the contraction of circular muscles constricts and elongates some regions of the fluid-filled body segments of a crawling earthworm, while longitudinal muscles shorten and thicken other regions. In position ❶, segments at the head and tail end of the worm are short and thick (longitudinal muscles contracted) and anchored to the ground by bristles. Just behind the head, a group of segments is thin and elongated (circular muscles contracted), with bristles held away from the ground. In position ❷, the head has moved forward because circular muscles in the head segments have contracted. Segments just behind the head and near the tail are now thick and anchored, thus preventing the head from slipping backward. In position ❸, the head segments are thick again and anchored to the ground in their new position, well ahead of their starting point. The rear segments of the worm now release their hold on the ground and are pulled forward.

Flying Many phyla of animals include species that crawl, walk, or run, and almost all phyla include swimmers. But

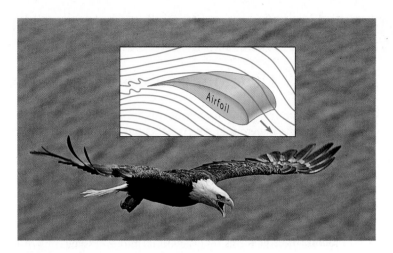

Figure 30.1E A bald eagle in flight

flying has evolved in only a few animal groups: insects, reptiles (including birds), and, among the mammals, bats. A group of large flying reptiles died out millions of years ago, leaving birds and bats as the only flying vertebrates.

For an animal to become airborne, its wings must develop enough lift to completely overcome the pull of gravity. The key to flight is the shape of wings. All types of wings, including those of airplanes, are airfoils—structures whose shape alters air currents in a way that creates lift. As **Figure 30.1E** shows, an airfoil has a leading edge that is thicker than the trailing edge. It also has an upper surface that is somewhat convex and a lower surface that is flattened or concave. This shape makes the air passing over the wing travel farther than the air passing under the wing. As a result, air molecules are spaced farther apart above the wing than below it, and the air pressure underneath the wing is greater. This pressure difference provides the "lift" for flight.

Birds can reach great speeds and cover enormous distances. Swifts, which can fly 170 km/hr (105 mph), are the fastest. The bird that migrates the farthest is the arctic tern, which flies round-trip between the North and South Poles each year.

All types of animal movement have certain underlying similarities. At the cellular level, every form of movement is based on one of two basic contractile systems: microtubules or microfilaments. Both of these systems consume energy moving protein strands against one another. The movements of cilia and flagella result from the bending of microtubules, as we discussed in Module 4.17. Microfilaments play a major role in amoeboid movement and are the contractile elements of muscle cells. Later in this chapter, we will look at the contraction of muscles and how muscle contraction translates into movement when the muscles work against a firm skeleton. First, let's look at skeletons.

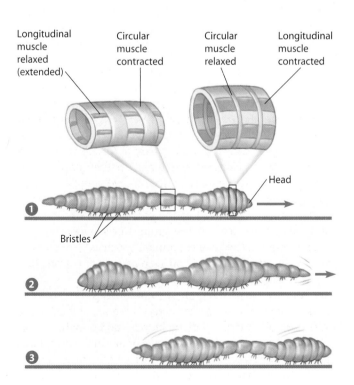

Longitudinal muscle relaxed (extended)

Circular muscle contracted

Circular muscle relaxed

Longitudinal muscle contracted

Head

Bristles

❶

❷

❸

Figure 30.1D An earthworm crawling by peristalsis

? Contrast swimming with walking in terms of the forces an animal must overcome to move.

■ Friction resists an animal moving through water, but gravity has little effect because of the animal's buoyancy; air poses little resistance to an animal walking on land, but the animal must support itself against the force of gravity.

30.2 Skeletons function in support, movement, and protection

A skeleton has many functions. An animal could not move without its skeleton, and most land animals would sag from their own weight if they had no skeleton to support them. Even an animal in water would be a formless mass without a skeletal framework to maintain its shape. Skeletons also may protect an animal's soft parts. For example, the vertebrate skull protects the brain, and the ribs form a cage around the heart and lungs. There are three main types of skeletons: hydrostatic skeletons, exoskeletons, and endoskeletons. All three types have multiple functions.

Hydrostatic Skeletons A **hydrostatic skeleton** consists of fluid held under pressure in a closed body compartment. This is very different from the more familiar skeletons made of hard materials. Nonetheless, a hydrostatic skeleton helps protect other body parts, cushioning them from shocks. It also gives the body shape and provides support for muscle action.

Earthworms have a fluid-filled internal body cavity, or coelom (see Module 18.3). As a segmented animal, the earthworm has its coelom divided into separate compartments. The fluid in these segments functions as a hydrostatic skeleton, and the action of circular and longitudinal muscles working against the hydrostatic skeleton produces the peristaltic movement described in Module 30.1.

Cnidarians, such as hydras and jellies, also have a hydrostatic skeleton. A hydra, for example, holds fluid in its gastrovascular cavity and can alter its body shape drastically using contractile cells in its body wall. When a hydra closes its mouth and the cells encircling its body wall constrict, it elongates and its tentacles extend, as shown on the left in **Figure 30.2A**. Because water is not compressible, constricting the cavity forces the animal to elongate, somewhat like squeezing a water-filled balloon. A hydra often sits in this position for hours, waiting for prey to swim by. If it is disturbed, its mouth opens, allowing water to flow out, and longitudinal

cells in its body wall contract, shortening the body (Figure 30.2A, right).

Most animals with hydrostatic skeletons are soft and flexible. A hydra can extend its body and tentacles and also expand its body around ingested prey. An earthworm can burrow through soil because it is flexible and has a hydrostatic skeleton. Similarly, having an expandable body and a hydrostatic skeleton enables many tube-dwelling animals to extend out of their tube for feeding and gas exchange and then quickly squeeze back into their tube when threatened (see Figure 18.10C). Hydrostatic skeletons work well for many aquatic animals and for terrestrial animals that crawl or burrow by peristalsis. However, a hydrostatic skeleton cannot support the forms of terrestrial locomotion in which an animal's body is held off the ground, such as walking.

Exoskeletons A great variety of aquatic and terrestrial animals have a rigid external skeleton, or **exoskeleton.** In insects and other arthropods, muscles attached to knobs and plates on the inner surfaces of the exoskeleton move the jointed body parts (see Module 18.11). At the joints of legs, the skeleton is thin and flexible, making possible a wide variety of body movements. You may have experienced the hard exoskeleton, flexible joints, and interior muscles of an arthropod when eating crab legs.

The armor-like protection, support, and flexibility of the exoskeleton have helped to ensure the evolutionary success of the arthropods. The arthropod exoskeleton is secreted by living cells, but it is nonliving material that does not grow with the animal. It must be shed (molted) and replaced by a larger exoskeleton at intervals to allow for the animal's growth (**Figure 30.2B**). Depending on the species, most

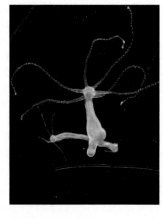

Figure 30.2A
The hydrostatic skeleton of a hydra in two states

Figure 30.2B The exoskeleton of an arthropod: a crab molting

Figure 30.2C The exoskeleton of a mollusc: a clam

Figure 30.2D A sea urchin (above) and its endoskeleton (right)

insects molt from four to eight times before reaching adult size. A few insect species and certain other arthropods, such as lobsters and crabs, molt at intervals throughout life.

In Figure 30.2B, on the previous page, you saw a crab in the process of molting. Its old shell (the dark one on the right) split open when the crab outgrew it. An arthropod is never without an exoskeleton of some sort. A newly molted crab, for instance, has a new, soft, elastic exoskeleton, which formed under the old one. Soon after molting, the crab expands its body by gulping air or water. Its new exoskeleton then hardens in the expanded position, and the animal has room for further growth. Until its exoskeleton hardens, an arthropod is very susceptible to predation; besides being weakly armored, it is usually less mobile, because the soft exoskeleton cannot support the full action of its muscles. The soft-shelled crabs found seasonally on many menus are newly molted crabs.

The shells of molluscs, such as clams (**Figure 30.2C**), are also exoskeletons, but unlike the chitinous arthropod exoskeleton, mollusc shells are made of a mineral, calcium carbonate. The mantle, a sheetlike extension of the animal's body wall, secretes the shell (see Module 18.9). As a mollusc grows, it does not molt; rather, it enlarges the diameter of its shell by adding to its outer edge.

Endoskeletons An **endoskeleton** consists of hard or leathery supporting elements situated among the soft tissues of an animal. Sponges, for example, are reinforced by a framework of tough protein fibers or by hard structures called spicules (see Module 18.5). Usually microscopic and sharp-pointed, a spicule consists of inorganic material such as calcium salts or silica. Sea stars, sea urchins, and most other echinoderms have an endoskeleton of hard plates beneath their skin (see Module 18.13). In living sea urchins,

about all you see are movable spines, which are attached to the endoskeleton by muscles (**Figure 30.2D**, left). A dead urchin with its spines removed reveals the plates that form a rigid skeletal case (right photo).

Vertebrates have endoskeletons consisting of cartilage or a combination of cartilage and bone (see Module 20.5). One major lineage of vertebrates, the sharks, have skeletons of cartilage reinforced with calcium. **Figure 30.2E** shows the more common condition for vertebrates. Bone makes up most of a frog's skeleton, as it does in bony fishes and land vertebrates. The frog skeleton and the skeletons of most other vertebrates also include some cartilage (blue in the figure), mainly in areas where flexibility is needed. Next, let's look at our own skeleton.

? What are the advantages and disadvantages of an exoskeleton as compared to an endoskeleton?

■ An exoskeleton may offer greater protection to body parts, but must usually be molted for the animal to grow.

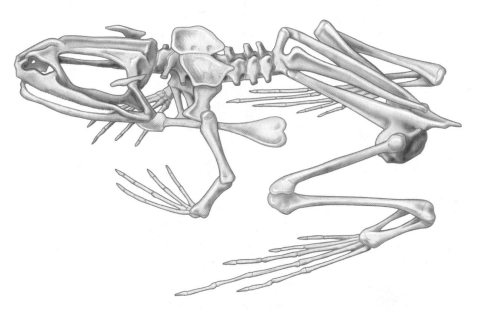

Figure 30.2E Bone (tan) and cartilage (blue) in the endoskeleton of a vertebrate: a frog

30.3 The human skeleton is a unique variation on an ancient theme

In contrast to the frog skeleton, which supports an animal that sits on all fours and hops with its hind legs, the human skeleton supports an upright body that sits on its hindquarters and walks or runs on two legs. Despite the differences, the skeletons of these and other vertebrates have a number of similarities. For instance, all vertebrates have an axial skeleton (green in **Figure 30.3A**) supporting the axis, or trunk, of the body. The **axial skeleton** consists of the skull, enclosing and protecting the brain; the vertebral column (backbone), enclosing the spinal cord; and, in most vertebrates, a rib cage around the lungs and heart.

Most vertebrates also have an **appendicular skeleton** (tan in Figure 30.3A), which is made up of the bones of the appendages (arms, legs, wings, fins) and the bones that anchor the appendages to the axial skeleton. In a land vertebrate, the shoulder girdle and the pelvic girdle provide a base of support for the bones of the forelimbs and hind limbs. The skeletal features humans have in common with other vertebrates stem from a basic skeletal model that originated in a group of fishes ancestral to land vertebrates. Hundreds of millions of years of adaptation have reshaped the model, but the main components remain.

What are some of the distinctive features of the human skeleton? Our distant ancestors were quadrupedal (four-footed), and virtually every part of the skeleton changed as upright posture and bipedalism evolved in the human lineage. **Figure 30.3B** contrasts the human skeleton (side view) with that of a modern quadrupedal primate, a baboon. Housing our large brain, our skull is large and flat-faced; its rounded part is the largest braincase relative to body size in the animal kingdom. In humans, the skull is balanced atop the backbone with the spinal cord exiting directly underneath. In quadrupeds, the spinal cord attaches toward the back of the skull, which is slung from the leading edge of the backbone. Our backbone is S-shaped, which helps balance the body in the vertical plane. In contrast, the baboon's is arched horizontally. Our pelvic girdle is shorter and rounder and oriented more vertically than the quadruped's. The bones of our hands and feet are also different from the baboon's. Free of locomotor functions, the human hand is adapted for strong gripping and precise manipulation. Our feet, with ground-touching heel, straight-facing big toe for propulsion, and shock-absorbing arches, are specialized for supporting the entire body and for bipedal walking.

Nothing produced by evolution's restructuring process is perfect, and the human skeleton is no exception. Our vertical backbone bears weight unevenly, and our lower back carries much of the load.

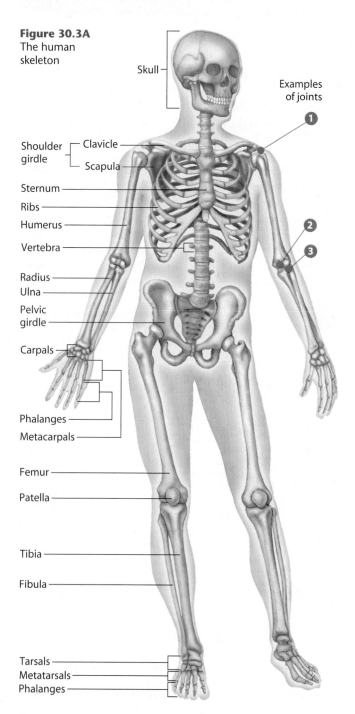

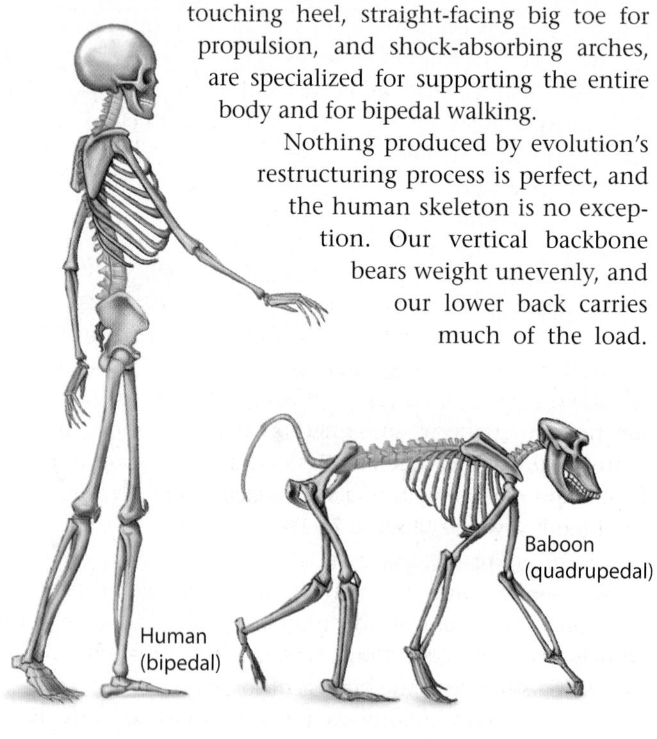

Figure 30.3A
The human skeleton

Skull

Examples of joints

Shoulder girdle — Clavicle

Scapula

Sternum

Ribs

Humerus

Vertebra

Radius

Ulna

Pelvic girdle

Carpals

Phalanges

Metacarpals

Femur

Patella

Tibia

Fibula

Tarsals
Metatarsals
Phalanges

Human (bipedal)

Baboon (quadrupedal)

Figure 30.3B Bipedal and quadrupedal primate skeletons compared

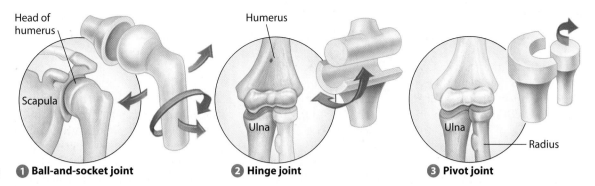

Figure 30.3C
Three kinds of joints

1 Ball-and-socket joint   **2 Hinge joint**   **3 Pivot joint**

The lower back is easily strained, especially when we bend over or lift heavy objects.

Much of the versatility of the vertebrate skeleton comes from its joints. Strong fibrous connective tissues called **ligaments** hold together the bones of movable joints. As shown in **Figure** 30.3C and at the numbers in Figure 30.3A, **1 ball-and-socket joints**, such as where the humerus joins the shoulder girdle, enable us to rotate our arms and legs and move them in several planes. **2** Shown here in the arm but also found in the knee, a **hinge joint** permits movement in a single plane. **3** A **pivot joint** enables us to rotate the forearm at the elbow. Hinge and pivot joints in our wrists and hands enable us to make precise manipulations.

Web/CD Activity 30A *The Human Skeleton*

? What type of joint joins the femur to the pelvic girdle?

■ A ball-and-socket joint

30.4 Bones are complex living organs

The expression "dry as a bone" should not be taken literally. Bones are actually complex organs consisting of several kinds of moist, living tissues. **Figure 30.4** shows a human humerus (upper arm bone). A sheet of fibrous connective tissue, shown in pink (most visible in the enlargement on the lower right), covers most of the outside surface. This tissue helps form new bone in the event of a fracture. A thin sheet of cartilage (blue) forms a cushion-like surface for joints, protecting the ends of bones as they move against one another. The bone itself contains living cells that secrete a surrounding material, or matrix. Bone matrix consists of flexible fibers of the protein collagen embedded in a hard mineral made of calcium and phosphate (see Figure 20.5E). The collagen keeps the bone flexible and nonbrittle, while the hard mineral matrix resists compression.

The shaft of this long bone is made of compact bone, so named because it has a dense structure. Notice that the compact bone surrounds a central cavity. The central cavity contains **yellow bone marrow**, which is mostly stored fat brought into the bone by the blood. The ends, or heads, of the bone have an outer layer of compact bone and an inner layer of spongy bone, so named because it is honeycombed with small cavities. The cavities contain **red bone marrow** (not shown in the figure), a specialized tissue that produces our blood cells (see Module 23.16).

Like all living tissues, bone cells require servicing. Blood vessels course through channels in the bone, transporting nutrients and regulatory hormones to its cells. Nerves (not shown) paralleling the blood vessels help regulate the traffic of materials between the bone and the blood.

Minor changes in bone architecture occur continually in a process known as bone remodeling. These dynamic changes are key to bone self-repair, which we explore next.

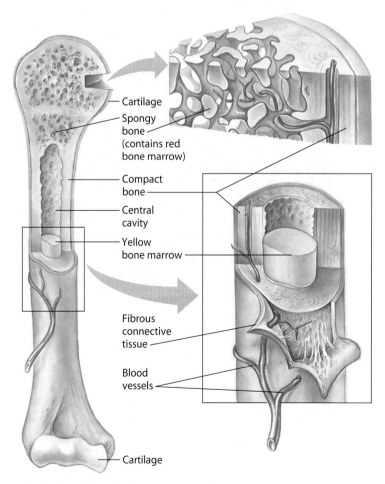

Cartilage
Spongy bone (contains red bone marrow)
Compact bone
Central cavity
Yellow bone marrow
Fibrous connective tissue
Blood vessels
Cartilage

Figure 30.4 The structure of an arm bone

? What causes the colors of yellow and red bone marrow?

■ Stored fat and developing red blood cells, respectively

30.5 Broken bones can heal themselves

Bones are rigid but not inflexible; they will bend in response to external forces. For example, if you fall forward with your hands outstretched, the bones of your hands, wrists, and arms will flex to absorb the shock and then return to their original shape. However, as many of us know from personal experience, the skeletal system has its limits. If a force is applied that exceeds a bone's resiliency, the result is a broken bone, or fracture. The average American will break two bones during his or her lifetime, most commonly the forearm or, for people over 75, the hip.

There are two important factors that determine whether a bone might break: the strength of the skeleton (which is affected by factors discussed in Module 30.6) and the angle and amount of force applied. Usually, a fracture occurs from a sudden impact such as a fall or car accident. Wearing protective gear, such as seat belts, helmets, or padding, can protect bones from high-energy trauma. Less often, so-called stress fractures occur from long-term repeated forces, such as those applied during jogging.

Human bones are living, dynamic tissues that are continuously broken down and rebuilt. This natural process is also quite effective at healing fractures. Treatment involves two steps: putting the bone back into its natural shape and then immobilizing it until the body's normal bone-building cells can repair the break. A splint (which may immobilize a limb or allow limited movement) or cast is usually sufficient to protect the area, prevent movement, and promote healing. Sometimes, external pressure must be applied to align the broken parts, a process called traction. In more severe cases, a fracture can be repaired surgically by inserting plates, rods, and/or screws that hold the broken pieces together (Figure 30.5A).

Once a bone heals, it is actually thicker and stronger than before, making it unlikely that another fracture will occur in the same spot. In certain cases, however, severely injured or diseased bone is beyond repair and must be replaced. For several decades, broken hip joints have been replaced with artificial ones made of titanium or cobalt alloys (Figure 30.5B). More recently, bone grafts (from the patient or from a cadaver) or synthetic polymers have been used to replace defective bone.

 What two steps are necessary to heal a broken bone?

■ Return the bone to its natural position and then immobilize it

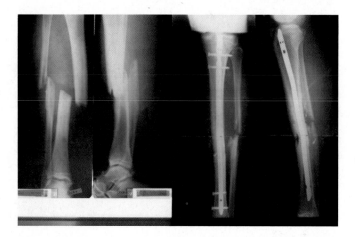

Figure 30.5A X-rays of a broken leg (left, two views) and the same leg after the bones were set with a plate and screws (right)

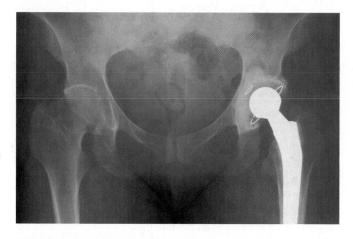

Figure 30.5B X-ray of an artificial hip joint (right) and its natural counterpart (left)

30.6 Weak, brittle bones are a serious health problem, even in young people

The risk of bone fracture increases if bones are porous and weak. Figure 30.6 contrasts healthy bone tissue (left) and bone diseased by osteoporosis (right). **Osteoporosis** is a growing health concern in the United States. It is characterized by low bone mass and structural deterioration of bone tissue. This weakness emerges from disruptions in the skeleton's continuous process of self-repair. Throughout a person's life, older bone tissue is removed and replaced by new tissue, a renewal that must happen for bones to remain strong. If bone is replaced too slowly, osteoporosis may result.

Until recently, this condition was mostly considered a problem for women after menopause. Estrogen contributes to normal bone maintenance, and with lowered production of the hormone, bones may become more porous and easily

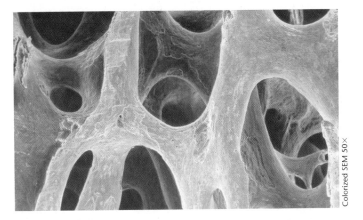
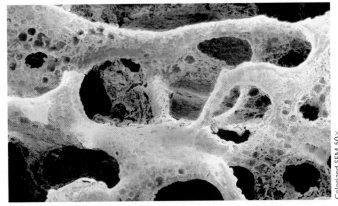

Figure 30.6 Healthy spongy bone tissue (left) and bone diseased by osteoporosis (right)

broken. But while osteoporosis remains a serious health problem for older women, it is also emerging as a concern for men and younger people. Of the 44 million Americans threatened by osteoporosis, scientists estimate that 32% are men. Doctors have also noted a dramatic increase in bone fractures in children and teenagers in recent years, a rise many link to weaker bones in young people.

Many scientists believe that these trends reflect unhealthy changes in the American lifestyle. Healthy bones require weight-bearing exercise, such as jogging or lifting weights, because mild stress makes bone denser. Strong bones also require an adequate intake of dietary calcium. As one of bone's main components, this mineral is essential to bone replacement. With many Americans of all ages getting less exercise and more children consuming beverages such as soda instead of calcium-rich milk, bone health is likely to suffer.

Other lifestyle habits, such as smoking, and diseases such as diabetes mellitus may also contribute to osteoporosis. Treatments include calcium and Vitamin D supplements and drugs that slow bone loss. Prevention of osteoporosis begins with sufficient calcium intake while bones are still increasing in density (up until about age 30). Weight-bearing exercise is also beneficial, both while young and later in life.

? What two things can a young person do to help prevent osteoporosis?

■ Ingest plenty of calcium and exercise regularly

MUSCLE CONTRACTION AND MOVEMENT

30.7 The skeleton and muscles interact in movement

Figure 30.7 shows how an animal's muscles interact with its bones, which act as levers, to produce movement. Muscles are connected to bones by **tendons** (see Module 20.5). For instance, one end of the biceps muscle shown in the figure is attached by tendons to bones of the shoulder; the other end is attached across the hinge joint of the elbow—which acts as the fulcrum—to one of the bones in the forearm.

The action of a muscle is always to contract, or shorten. A muscle's relaxation to an extended position is a passive process. The ability to move an arm in opposite directions requires that muscles be attached to the arm bones in antagonistic pairs. As shown in the figure, contraction of the biceps muscle raises the forearm. The triceps muscle is the biceps's antagonist. The upper end of the triceps attaches to the shoulder, while its lower end attaches to the elbow. At the right, you see that contraction of the triceps lowers the forearm, extending the biceps in the process.

All animals—very small ones like ants and giant ones like elephants—have antagonistic pairs of muscles that apply opposite forces to move parts of their skeleton. Next, we see how a muscle's structure explains its ability to contract.

? When exercising to strengthen muscles, why is it important to impose resistance while both flexing and extending the limbs?

■ This exercises both muscles of antagonistic pairs, which only do work when they are contracting.

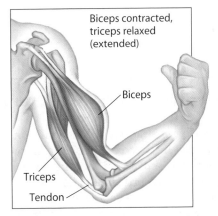
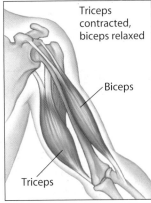

Biceps contracted, triceps relaxed (extended)

Biceps

Triceps

Tendon

Triceps contracted, biceps relaxed

Biceps

Triceps

Figure 30.7 Antagonistic action of muscles in the human arm

30.8 Each muscle cell has its own contractile apparatus

We looked briefly at the various types of muscle tissue in Module 20.6. **Skeletal muscle**, which is attached to the skeleton and produces body movements, is made up of a hierarchy of smaller and smaller parallel strands. As indicated at the top in **Figure 30.8**, a muscle consists of bundles of parallel muscle fibers. Each muscle fiber is a single cell with many nuclei.

Farther down in the drawing, notice that each muscle fiber is itself a bundle of smaller **myofibrils**. A myofibril consists of repeating units called **sarcomeres**. Skeletal muscle is also called striated (striped) muscle because the sarcomeres produce alternating light and dark bands when viewed with a microscope. Structurally, a sarcomere is the region between two dark, narrow lines, called Z lines, in the myofibril. Functionally, the sarcomere is the contractile apparatus in a myofibril—the muscle fiber's fundamental unit of action.

The micrograph and the diagram below it reveal the structure of a sarcomere in more detail. They show that a sarcomere is composed of regular arrangements of two kinds of filaments: thin filaments, colored blue in the diagram, and thick filaments, colored pink. A **thin filament** consists of two strands of the protein actin and two strands of a regulatory protein, coiled around each other. Each **thick filament** contains a staggered array of multiple strands of the protein myosin. The broad, dark band centered in the sarcomere is where the thick filaments are (bottom diagram); they are interspersed with thin filaments that project toward the center of the sarcomere. The light bands at both edges of the sarcomere have only thin filaments. Within the light bands, the Z lines consist of proteins that connect adjacent thin filaments.

This specific arrangement of repeating units of thin and thick filaments is directly related to the mechanics of muscle contraction. In the next module, we see how this structure contributes to the functioning of a muscle cell and hence how an entire muscle contracts.

Web/CD Activity 30B *Skeletal Muscle Structure*

? The two most abundant proteins of a myofibril are
_____ and _____.

actin . . . myosin

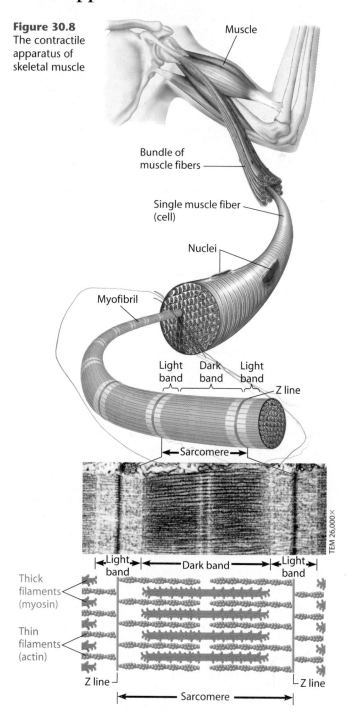

Figure 30.8
The contractile apparatus of skeletal muscle

Muscle

Bundle of muscle fibers

Single muscle fiber (cell)

Nuclei

Myofibril

Light band — Dark band — Light band

Z line

Sarcomere

TEM 26,000×

Light band — Dark band — Light band

Thick filaments (myosin)

Thin filaments (actin)

Z line

Z line

Sarcomere

30.9 A muscle contracts when thin filaments slide across thick filaments

Let's now relate the structure of a sarcomere to its function. According to the **sliding-filament model** of muscle contraction, a sarcomere contracts (shortens) when its thin filaments slide across its thick filaments. **Figure 30.9A**, on the next page, is a simplified diagram that shows a sarcomere in a relaxed muscle, in a contracting one, and in a fully contracted one. Notice in the contracting sarcomere that the Z lines and

the thin filaments (blue) have moved toward the middle of the sarcomere. And when the muscle is fully contracted, the thin filaments overlap in the middle of the sarcomere. Contraction only shortens the sarcomere; it does not change the lengths of the thick and thin filaments. A whole muscle can shorten about 35% of its resting length when all its sarcomeres contract.

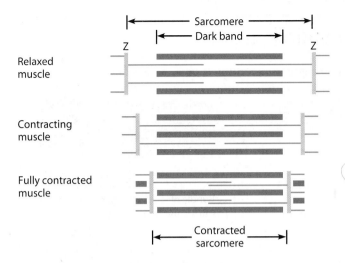

Figure 30.9A The sliding-filament model of muscle contraction

Relaxed muscle

Contracting muscle

Fully contracted muscle

Sarcomere
Dark band
Contracted sarcomere

What makes the thin filaments slide when a sarcomere contracts? The key events are energy-consuming interactions between the myosin molecules of the thick filaments and the actin of the thin filaments. Electron micrographs show that parts of myosin molecules, called heads, bind with specific sites on the thin filaments. In a muscle fiber at rest, these sites are covered by a regulatory protein. Stimulation by a motor neuron causes the binding sites to be exposed so that actin and myosin can interact. Energy for sliding comes from ATP.

Figure 30.9B indicates how sliding seems to work. ❶ ATP binds to a myosin head, causing the head to detach from a binding site on actin. (Binding sites on actin are indicated by purple spots.) ❷ Next, the myosin head hydrolyzes ATP to ADP and ℗ (phosphate), releasing energy that changes the shape of the myosin head, extending it to its high-energy position. ❸ This energized myosin head binds to an exposed binding site on actin. ❹ The molecular event that actually causes sliding is called the power stroke. ADP and ℗ are released from the myosin head, and the head bends back to its low-energy position, pulling the thin filament toward the center of the sarcomere, in the direction of the green arrow. After the power stroke, another ATP binds with the myosin head, and the whole process repeats. On the next power stroke, the myosin head attaches to a binding site ahead of the previous one on the thin filament (closer to the Z line).

This sequence—detach, extend, attach, pull—occurs again and again in a contracting muscle. Though we show only one myosin head in the figure, a typical thick filament has about 350 heads, each of which can bind and unbind to a thin filament about five times per second. Preventing the filaments from backsliding during contraction, some myosin heads hold the thin filaments in position, while others are reaching for new binding sites. As long as sufficient ATP is present, the process continues until the muscle is fully contracted or until the signal to contract stops.

The sliding-filament model explains how the sarcomeres of skeletal muscles contract. Next we consider how the nervous system controls muscle contraction.

Web/CD Activity 30C *Muscle Contraction*

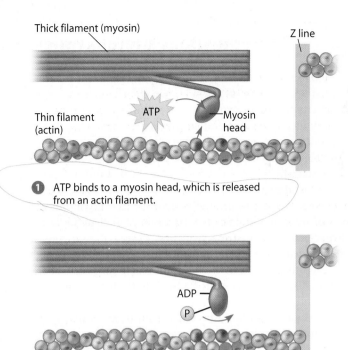

Thick filament (myosin)

Thin filament (actin)

ATP

Myosin head

Z line

❶ ATP binds to a myosin head, which is released from an actin filament.

ADP
℗

❷ Hydrolysis of ATP extends the myosin head.

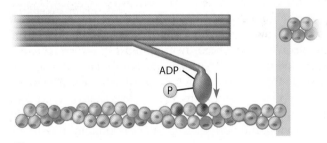

ADP
℗

❸ The myosin head attaches to an actin binding site.

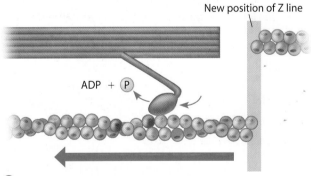

New position of Z line

ADP + ℗

❹ The power stroke slides the actin (thin) filament toward the center of the sarcomere.

Figure 30.9B The mechanism of filament sliding. (The regulatory protein strand of the thin filament is not shown in these drawings.)

? Which region of a sarcomere becomes shorter during contraction of a muscle: the dark band, the light bands, or both dark and light bands?

■ The light bands shorten and even disappear as the thin filaments slide (are pulled) toward the center of the sarcomere.

30.10 Motor neurons stimulate muscle contraction

The sarcomeres of a muscle fiber do not contract on their own. They must be stimulated to contract by motor neurons. A typical motor neuron can stimulate more than one muscle fiber because each neuron has many branches. In the example shown in **Figure 30.10A**, you see two so-called **motor units,** each consisting of a neuron and all the muscle fibers it controls (two or three, in this case). A motor neuron has its dendrites and cell bodies in the central nervous system (here, the spinal cord). Its axon forms synapses, called **neuromuscular junctions,** with the muscle fibers. When a motor neuron sends out an action potential, its synaptic knobs release the neurotransmitter acetylcholine. Acetylcholine diffuses across the neuromuscular junctions to the muscle fibers, making all the fibers of the motor unit contract simultaneously.

The organization of individual neurons and muscle cells into motor units is the key to the action of whole muscles. We know that we can vary the amount of force our muscles develop; an arm wrestler, for example, may change the amount of force developed by the biceps and triceps several times in the course of a match. The ability to do so depends mainly on the nature of motor units. Each motor neuron in a large muscle like the biceps may serve several hundred fibers scattered throughout the muscle. Stimulation of the muscle by a single motor neuron, however, would produce only a weak contraction. More forceful contractions result when additional motor units are activated. Thus, depending on how many motor units your brain commands to contract, you can apply a small amount of force to lift a fork or considerably more to lift, say, this textbook. In muscles re-

quiring precise control, such as those controlling eye movements, a motor neuron may control only one muscle fiber.

How does a motor neuron make a muscle fiber contract? The initial events of stimulation are the same as those that occur at a synapse between two neurons in the nervous system (see Module 28.6): The acetylcholine that diffuses across the neuromuscular junction changes the permeability of the muscle fiber's plasma membrane. The change triggers action potentials that sweep across the muscle cell membrane. As indicated in **Figure 30.10B**, the action potentials (black arrows) then pass deep into the muscle cell along membranous tubules that fold inward from the plasma membrane. Inside the cell, the action potentials make the endoplasmic reticulum (ER) release Ca^{2+} (green dots) into the cytoplasm. Remember that thin filaments consist of two strands of actin and two strands of a regulatory protein, which wrap around and block the binding sites on actin. Ca^{2+} moves this regulatory protein so that the myosin heads can attach to actin, initiating filament sliding, as we saw in the previous module. When motor neurons stop sending action potentials to the muscle fibers, the ER pumps Ca^{2+} back out of the cytoplasm, actin binding sites become blocked again, the sarcomeres stop contracting, and the muscle relaxes.

Web/CD Thinking as a Scientist *How Do Electrical Stimuli Affect Muscle Contraction?*

? How does the endoplasmic reticulum help regulate muscle contraction?

■ By reversibly storing and releasing Ca2+, the ER regulates the cytoplasmic concentration of this ion, which is required in the cytoplasm for the binding of myosin to actin.

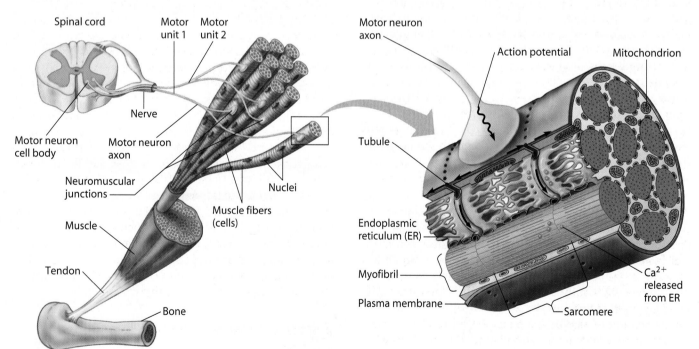

Figure 30.10A The relation between motor neurons and muscle fibers

Figure 30.10B Part of a muscle fiber (cell) at a neuromuscular junction

30.11 Athletic training increases strength and endurance

Motor neurons stimulate muscles to contract, but does the muscle always listen? If you have ever worked your muscles so hard that you couldn't move, you know that there is a point at which your muscles will no longer contract. Most of us never push ourselves to total muscle fatigue; we are much more likely to give in to psychological fatigue first. The drive to win in the face of mental fatigue is often what defines elite athletes. Nonetheless, athletes train to increase their resistance to physical fatigue.

What causes muscles to tire? Recall that ATP is required for muscle contraction (see Module 30.9). If ATP supply does not keep up with demand, muscle contraction ceases. This can happen in a number of different ways. If the circulatory system cannot supply enough oxygen to a muscle, muscle fibers are not able to generate ATP via cellular respiration (aerobically). As you learned in Chapter 6, muscle cells can also generate ATP anaerobically, but this can only continue for a minute or so before lactic acid builds up and causes muscle fatigue. Muscle fatigue also occurs if the muscle runs out of fuel. Muscle fibers receive glucose from the blood and also store glycogen, a polymer of glucose. These supplies can be exhausted during long periods of exertion.

So how are elite athletes able to compete in physically strenuous sports without their muscles quickly going on strike? Training programs increase both the stamina and the strength of their muscles. **Aerobic exercise** increases the efficiency and fatigue resistance of muscles. It is therefore the backbone of all endurance sports, such as distance running, biking, and cross-country skiing. Over time, aerobic training increases the size and number of muscle mitochondria, improves blood flow to muscles, and strengthens the heart and circulatory system. There are other benefits as well: Bones become stronger and gas exchange in the lungs becomes more efficient.

While overall muscle strength is also increased through aerobic exercise, building larger muscles that can generate greater power requires **anaerobic exercise.** Sprinters, weight lifters, and other "power" athletes rely on this high-intensity strength conditioning. Anaerobic training pushes muscles, especially the body's fast muscle fibers (see Chapter 6, introduction), to a point where they are contracting so forcefully that they no longer receive adequate oxygen and must switch to anaerobic respiration. Top-level athletes work on raising this "anaerobic threshold," the point at which muscles switch to anaerobic respiration. Anaerobic training increases the size of muscle fibers, enabling them to generate more force. Vigorously exercised muscle fibers also store more glycogen as fuel reserves.

The amount of aerobic and anaerobic exercise in an athlete's training program must be balanced. Weight lifters who only think about bulk will not build the aerobic capacity to help muscles resist fatigue. And distance runners who don't pay attention to strength training are likely to favor some

Figure 30.11 Heptathlete Carolina Kluft excels at many events

muscle groups over others. Over time, this imbalance can limit performance or lead to injury. Most athletes follow a workout regime called "cross-training," routines that alternate aerobic and anaerobic activities and improve both endurance and strength. Muscles need to be exercised in balance as well. Because most muscles function in antagonistic pairs, such as the biceps and triceps shown in Figure 30.7, they must be equally strong to work well together. Thus, most weight-training programs exercise both equally.

World-class athletes understand the need for balance in their training and work to maintain it, even as they compete in events that may require more of one capability (such as speed or endurance or strength) than another. But how about the heptathlete in **Figure 30.11**? This athlete performs in seven different events and must excel in everything from the shot put and sprints (anaerobic) to distance runs (aerobic). Such athletes and their trainers know that the body can only reach its potential if all its abilities are challenged and developed in a balanced manner.

The same basic principles underlie any workout program. Consistent, balanced training will increase blood flow and oxygen to muscle fibers, giving them more power and resistance to fatigue. Not everyone can get to the Olympics just by working out, but most people can reach a healthy level of both cardiovascular fitness and physical strength.

? Should a marathoner engage in aerobic or anaerobic training?

■ Both, a balance of strength and endurance will improve performance. Anaerobic training will increase muscle strength; aerobic training will develop the resistance to fatigue essential for marathon running.

30.12 The structure-function theme underlies all the parts and activities of an animal

Figure 30.12 Integration of the sensory, nervous, and motor systems

The batter steps up to the plate. He drives the ball hard toward an infield gap. The shortstop dives and robs the batter of a base hit.

■   ■   ■

Though far removed from the arena of natural selection, a baseball game demonstrates some of the remarkable evolutionary adaptations of the human body. A hit and a catch are both spectacular displays of our nervous system's ability to almost instantly link environmental stimuli with appropriate muscular activity (Figure 30.12). Sensory receptors provide both batter and infielder with information about the path of the ball. The batter can perceive a ball traveling over 90 mph. Photoreceptors in the infielder's eyes detect the ball as soon as it leaves the bat. His ears hear the crack of the bat hitting the ball. Sensory neurons convey this information as a series of action potentials. In both cases, a supercomputer we call the human brain integrates information about the angle and speed of the ball and then signals multiple muscles to perform a very specific action—to swing a bat at precisely the right moment or to dive and thrust out a hand to meet the ball, again at just the right moment. The fans see a spectacular play by a talented athlete. These visible activities result from a nervous system that integrates information gathered by sensory receptors and then directs muscles to respond. These muscles contract against a skeletal system to produce rapid and precise movements. These remarkable abilities result from adaptations that have been refined through natural selection and contribute to our survival as a species.

? In a sentence, describe how Figure 30.12 illustrates the popular sports term *hand-eye coordination*.

We have now completed our unit on the organ systems of animals. One overriding theme has been homeostasis: the coordination of organ systems in maintaining a stable internal environment. As we've seen, these systems include a digestive system that provides fuel and building materials, a respiratory system that provides oxygen for aerobic respiration, a circulatory system that distributes materials, an immune system for body defense, an excretory system that balances body fluids and removes nitrogenous wastes, an endocrine system for chemical regulation, a reproductive system to produce the next generation, and a nervous system that integrates sensory input and motor output.

Another key theme has been that structure underlies function—that the structural adaptations of a cell, tissue, organ, or organ system determine the job it can perform. We have seen numerous examples of this basic principle, from our early study of the digestive system to our discussion of the immune system and our later discussions on nervous control, sensory receptors, and movement. Watching the precise movements of a baseball player or the outward activity of any animal gives us an overview of the structure-function relationship and a reminder of the evolutionary process that generated it.

We'll see the structure-function theme emerge again in the context of adaptations to the environment when we take up the study of plants in the next unit.

■ One possible answer: Ball meets glove because the brain coordinates visual input about the path and speed of the ball with motor output that places the hand at the right place at the right time.

CHAPTER REVIEW

Reviewing the Concepts

Movement and Locomotion (Introduction–30.1)

Movement is one of the most distinctive features of animals **(Introduction).**

Locomotion, active travel from place to place, requires that an animal use energy to overcome friction and gravity.

 Animals that swim are supported by water but are slowed by friction. Animals that walk, hop, or run on land are less affected by friction, but must support themselves against gravity. Burrowing or crawling animals must overcome friction. They may move by side-to-side undulation or by peristalsis. The wings of birds, bats, and flying insects are airfoils, which generate lift **(30.1).**

Skeletal Support (30.2–30.6)

Skeletons function in body support, movement as muscles pull against them, and protection of internal organs. Worms and cnidarians have hydrostatic skeletons—fluid held under pressure in closed body compartments. Exoskeletons are hard external cases, such as the chitinous, jointed skeletons of arthropods. The vertebrate endoskeleton is composed of cartilage and bone **(30.2).**

The human skeleton consists of an axial skeleton (skull, vertebrae, and ribs) and an appendicular skeleton (shoulder girdle, upper limbs, pelvic girdle, and lower limbs). Movable joints provide flexibility. The human skeleton changed dramatically as upright posture and bipedalism evolved **(30.3).**

Bones are living organs. Cartilage at the ends of bones cushions the joints. Bone cells, serviced by blood vessels and nerves, live in a matrix of flexible protein fibers and hard calcium salts. Long bones have a fat-storing central cavity and spongy bone at their ends. Spongy bone contains red marrow, where blood cells are made **(30.4).** Broken bones are realigned and immobilized, and bone cells build new bone, repairing the break **(30.5).**

Osteoporosis, a bone disease characterized by weak, porous bones, is a growing health concern **(30.6).**

Muscle Contraction and Movement (30.7–30.12)

Muscles and bones interact to produce movement. Antagonistic pairs of muscles produce opposite movements. Muscles perform work only when contracting **(30.7).**

Muscle fibers, or cells, consist of bundles of myofibrils, which contain bundles of overlapping thick (myosin) and thin (actin) protein filaments. Sarcomeres, repeating groups of thick and thin filaments, are the contractile units **(30.8).**

The sliding-filament model explains muscle contraction. The myosin heads of the thick filaments bind ATP and extend to high-energy states. The heads then attach to bind-

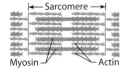

ing sites on the actin molecules and pull the thin filaments toward the center of the sarcomere **(30.9).**

Motor neurons carry action potentials that initiate muscle contraction. A neuron and the muscle fibers it controls constitute a motor unit. Acetylcholine released at a neuromuscular junction triggers an action potential that passes along tubules into the center of the muscle cell. Calcium released from the endoplasmic reticulum initiates muscle contraction **(30.10).** A balance of aerobic and anaerobic exercise increases endurance and strength **(30.11).**

Structure and function. Animal movement is a visible reminder that function emerges from structure. An animal's nervous system connects sensations derived from environmental stimuli to responses carried out by its muscles **(30.12).**

Connecting the Concepts

1. Complete this concept map on animal movement.

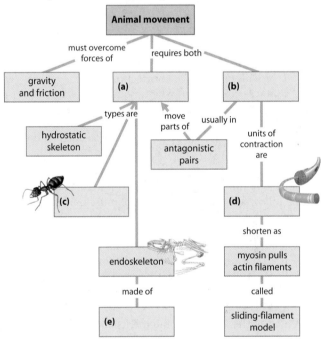

Testing Your Knowledge

Multiple Choice

2. A human's internal organs are protected mainly by the
   a. hydrostatic skeleton.
   b. motor unit.
   c. axial skeleton.
   d. exoskeleton.
   e. appendicular skeleton.

3. Arm and leg muscles are arranged in antagonistic pairs. How does this affect their functioning?
   a. It provides a backup if one of the muscles is injured.
   b. One muscle of the pair pushes while the other pulls.
   c. A single motor neuron can control both of them.
   d. It allows the muscles to produce opposing movements.
   e. It doubles the strength of contraction.

4. Gravity would have the least effect on the movement of which of the following? (*Explain your answer.*)
   a. a salmon   d. a sparrow
   b. a human   e. a grasshopper
   c. a snake

5. Which of the following bones in the human arm would correspond to the femur in the leg?
   a. radius   d. metacarpal
   b. tibia   e. ulna
   c. humerus

6. Which of the following animals is correctly matched with its type of skeleton?
   a. fly—endoskeleton
   b. earthworm—exoskeleton
   c. dog—exoskeleton
   d. lobster—exoskeleton
   e. bee—hydrostatic skeleton

7. When a horse is running fast, its body position is stabilized by
   a. side-to-side undulation.
   b. energy stored in tendons.
   c. the lift generated by its movement through the air.
   d. foot contact with the ground.
   e. its momentum.

8. What is the role of acetylcholine in muscle contraction?
   a. It moves the regulatory protein, exposing actin binding sites to the myosin heads.
   b. It provides energy for contraction.
   c. It blocks contraction when the muscle relaxes.
   d. It is the neurotransmitter released by a motor neuron, and it initiates an action potential in a muscle fiber.
   e. It forms the heads of the myosin molecules in the thick filaments inside a muscle fiber.

9. Muscle A and muscle B are the same size, but muscle A is capable of much finer control than muscle B. Which of the following is likely to be true of muscle A? (*Explain your answer.*)
   a. It is controlled by more neurons than muscle B.
   b. It contains fewer motor units than muscle B.
   c. It is controlled by fewer neurons than muscle B.
   d. It has larger sarcomeres than muscle B.
   e. Each of its motor units consists of more cells than the motor units of muscle B.

10. Which of the following statements about skeletons is true?
    a. Hydrostatic skeletons are soft and do not protect body parts.
    b. Chitin is a major component of vertebrate skeletons.
    c. Evolution of bipedalism involved little change in the axial skeleton.
    d. Most cnidarians must shed their skeleton periodically in order to grow.
    e. Vertebrate bones contain living cells.

Describing, Comparing, and Explaining

11. In terms of both numbers of species and numbers of individuals, insects are the most successful land animals. Write a paragraph explaining how their exoskeletons help them live on land. Are there any disadvantages to having an exoskeleton?

12. A hawk swoops down, seizes a mouse in its talons, and flies back to its perch. In a few sentences, explain how its wings enable it to overcome the downward pull of gravity as it flies upward.

13. Describe how you bend your arm, starting with action potentials and ending with the contraction of a muscle. How does a strong contraction differ from a weak one?

14. In what ways did the human skeleton change as an upright posture and bipedalism evolved? Describe the changes by comparing the human skeleton with the skeleton of a quadruped such as a baboon.

Applying the Concepts

15. Drugs are often used to relax muscles during surgery. Which of the following chemicals do you think would make the best muscle relaxant, and why? Chemical A: Blocks acetylcholine receptors on muscle cells. Chemical B: Floods the cytoplasm of muscle cells with calcium.

16. An earthworm's body consists of a number of fluid-filled compartments, each with its own set of longitudinal and circular muscles. In a different kind of worm, the roundworm, a single fluid-filled cavity occupies the body, and there are only longitudinal muscles that run its entire length. Predict how the movement of a roundworm would differ from the movement of an earthworm.

17. When a person dies, muscles become rigid and fixed in position—a condition known as rigor mortis, which often figures importantly in mystery novels. Rigor mortis occurs because muscle cells are no longer supplied with ATP (when breathing stops, ATP synthesis ceases). Calcium also flows freely into dying cells. The rigor eventually disappears because the biological molecules break down. Explain, in terms of the mechanism of contraction described in Module 30.9, why the presence of calcium and the lack of ATP would cause muscles to become rigid, rather than limp, after death.

18. A goal of the Americans with Disabilities Act is to allow people with physical limitations to fully participate in and contribute to society. Perhaps you have a disability or know someone with a disability. Imagine that a neuromuscular disease or injury makes it impossible for you to walk. Think about your activities during the last 24 hours. How would your life be different if you had to get around in a wheelchair? What kinds of barriers or obstacles would you encounter? What kinds of changes would have to be made in your activities and surroundings to accommodate your change in mobility?

19. Some athletes take anabolic steroids illegally to increase the buildup of muscle that occurs with training. In young people, whose bones are still growing, steroids also speed up the conversion of cartilage to bone. With regard to bone growth, what might be some negative consequences of steroid use? Do you think that steroid use is worth the risks? Why or why not? As an athlete, would you use anabolic steroids if you knew you could get away with it? Why or why not?

Answers to all questions can be found in Appendix 3.

For study help and Activities, go to campbellbiology.com or the student CD-ROM.

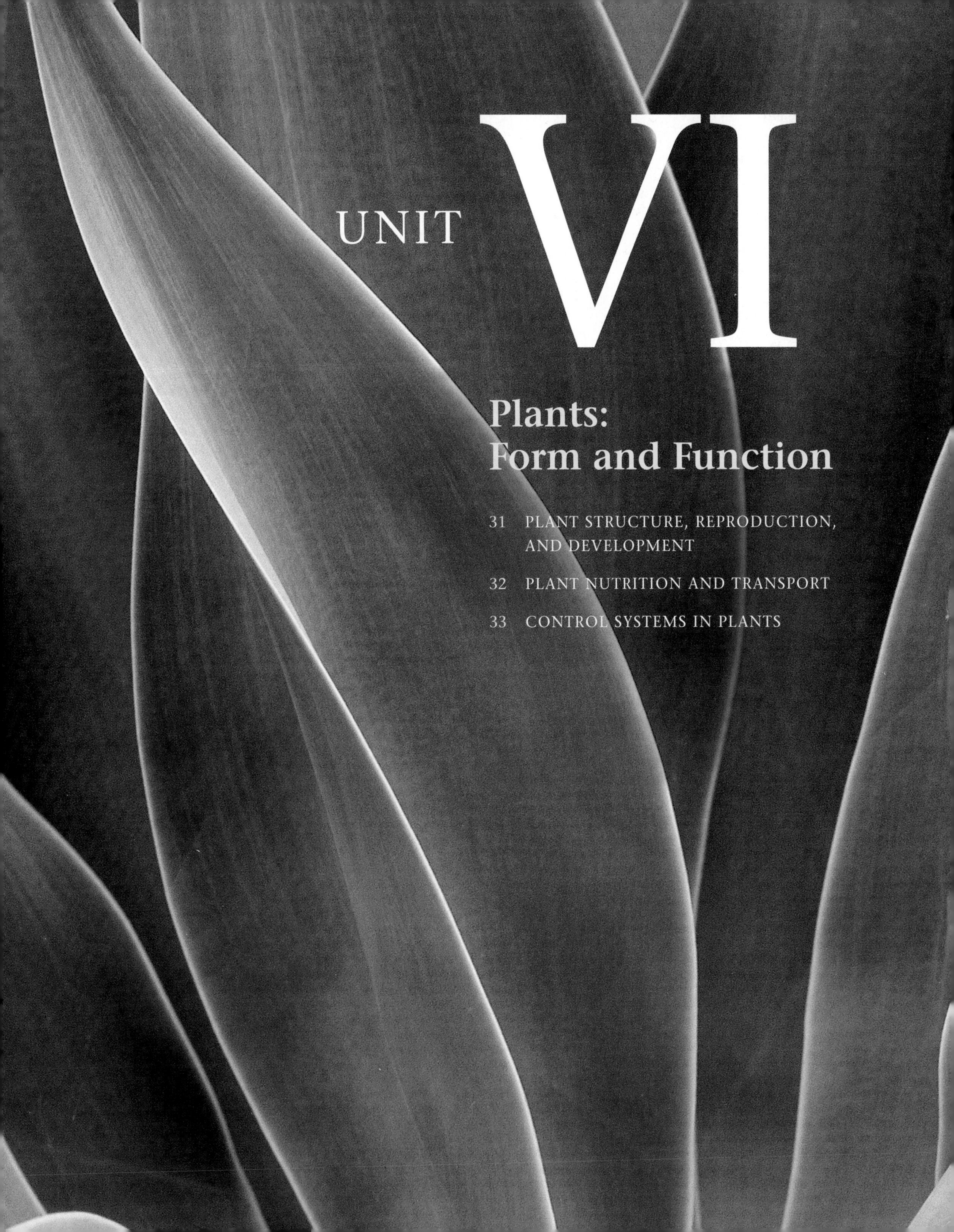

UNIT **VI**

Plants:
Form and Function

31 PLANT STRUCTURE, REPRODUCTION,
    AND DEVELOPMENT

32 PLANT NUTRITION AND TRANSPORT

33 CONTROL SYSTEMS IN PLANTS

CHAPTER 31

31.1 Plant scientist Natasha Raikhel studies the *Arabidopsis* plant as a model biological system

PLANT STRUCTURE AND FUNCTION

31.2 The two main groups of angiosperms are the monocots and the dicots
31.3 A typical plant body consists of roots and shoots
31.4 Many plants have modified roots, stems, and leaves
31.5 Plant cells and tissues are diverse in structure and function
31.6 Three tissue systems make up the plant body

PLANT GROWTH

31.7 Primary growth lengthens roots and shoots
31.8 Secondary growth increases the girth of woody plants

REPRODUCTION OF FLOWERING PLANTS

31.9 Overview: The sexual life cycle of a flowering plant
31.10 The development of pollen and ovules culminates in fertilization
31.11 The ovule develops into a seed
31.12 The ovary develops into a fruit
31.13 Seed germination continues the life cycle
31.14 Asexual reproduction produces plant clones
31.15 Asexual reproduction is a mainstay of modern agriculture

A Gentle Giant

IN 1951, AUTHOR AND CONSERVATIONIST Freeman Tilden wrote: "Not a single *Sequoiadendron gigantea* [giant sequoia] should be cut. They represent a unique survival; they belong to our remote and future histories." The giant sequoia pictured at the bottom of the facing page, named the General Sherman for the controversial Union commander in the Civil War, is a case in point: It has been growing for about 2,500 years, and plant biologists believe it could live for many more.

Plant Structure, Reproduction, and Development

A cross section of the trunk of a giant sequoia

A unique survivor and a truly awesome representative of the plant kingdom, the General Sherman resides in the Giant Forest in Sequoia–Kings Canyon National Park, in the Sierra Nevada Mountains of central California. At 84 m (275 feet) tall, it is over 26 m taller than Niagara Falls. Its massive trunk, weighing nearly 1,400 tons, is about 10 m in diameter at the base; its lowest limb, about 40 m above the ground, is over 2 m in diameter. This is the largest plant on Earth.

The photograph above shows the ancient trunk of another giant sequoia. The rings you see in this cross section are called growth rings because each one marks a year in the tree's life. The rings vary in thickness, depending on weather conditions during the growing seasons. About 1,700 years old when it was felled in 1917, this tree was a seedling at the time of the Roman Empire. It was over 200 years old when the Saxons invaded England in A.D. 449. By 1492, when Columbus landed in the New World, it had laid down 80% of its growth rings. Having survived earthquakes, numerous fires, droughts, and storms, most of the giant sequoias were cut down between 1850 and 1900, a time of virtually uncontrolled lumbering. A tree like this one or the General Sherman would yield enough prime boards for nearly 50 four- or five-room houses, and many giant sequoias were cut down for just that purpose. Today, only a few groves remain, nearly all in national and state parks, where they are viewed as living treasures.

As we become more aware of the importance of environmental conservation, our population continues to grow and our demands for plant products increase; lumber, fabric, paper, most of our food, and numerous industrial chemicals all come from plants. Beyond our immediate needs, plants are vital to the planet's well-being. Virtually all land animals—including humans—depend on them for food, either eating plants directly or eating other animals that eat plants. Above and below the ground, plants provide food, shelter, and breeding areas for animals, fungi, and microorganisms. Plant roots prevent soil erosion, and photosynthesis in plant leaves helps reduce carbon dioxide levels in the atmosphere and adds oxygen to the air.

The giant sequoia is one of about 600 living species of cone-bearing plants known as conifers. As we saw in Chapter 17, conifers are gymnosperms, one of two groups of seed plants. Gymnosperms bear seeds in cones, whereas plants of the other group, the angiosperms (flowering plants), produce seeds enclosed in fruits. Because angiosperms make up over 90% of the plant kingdom, we concentrate on them in this unit. (The structures of gymnosperms and seedless plants such as mosses and ferns are covered in Chapter 17.) By studying angiosperm structure in relation to growth and reproduction, we can gain insight into the structure-function relationship of all plants and, by extension, all biological systems. ■ ■ ■

The General Sherman tree

31.1 Plant scientist Natasha Raikhel studies the *Arabidopsis* plant as a model biological system

Dr. Natasha Raikhel, Distinguished Professor of Plant Biology at the University of California, Riverside, is one of America's most prominent plant biologists (Figure 31.1A). Born in Russia, she traded a promising musical career for the study of zoology. After emigrating to the U.S. in the 1970s, Dr. Raikhel became interested in plant biology. Today, Dr. Raikhel performs research on the mustard plant *Arabidopsis* (Figure 31.1B). In a recent interview, she explained how such research can provide insight into other biological systems:

Arabidopsis provides such a convenient model organism for studying processes that are important in all plants, including crop plants. For example, resistance to cold and resistance to pathogens are two traits that are very useful in crop plants. But such traits are much easier to analyze in *Arabidopsis*. So, if we first identify certain genes and regulatory pathways that can help *Arabidopsis* resist environmental stress, we can then transfer some of that knowledge to improve crop plants.

Dr. Raikhel applies her research findings to the field of systems biology, which is an approach to studying biology that aims to model the dynamic behavior of whole biological systems. Dr. Raikhel explains that systems biology seeks "to predict how various pathways in the organism interact and how a change in one pathway will affect the whole organism." She points out that it is a good time to conduct such research on plants:

One reason is that we now know a lot about . . . the plant *Arabidopsis*. Its genome has been sequenced, and many of its proteins have been identified. Once you understand the functions of these proteins, how the proteins interact, and how

pathways in the cells interact, then you can really start to answer questions about how cells function and how the whole plant works. Systems biology becomes possible, and I think this makes plant biology much more exciting.

In addition to her work with *Arabidopsis* and her leadership role in the international community of plant cell biologists, Dr. Raikhel participates in a program that provides summer research opportunities to undergraduate community college students. Dr. Raikhel explains why she enjoys having young students in her lab:

I love to be around these young people because they keep me young. At first, when these students join the lab, they are completely naive about research and just don't know how to get started. But once you put a fire under them and get them excited, it's just the most rewarding thing in the world. For example, this past summer, we had a student from San Bernardino Valley College, a community college nearby. I got him involved with chemical genetics, and he just loved his project. He will be on a paper that we are writing. He has transferred to UCR, and he still works in my lab for many hours each week, even though he's taking lots of courses. This student now wants to become a plant researcher—he's a convert! And that gives me great satisfaction.

Dr. Raikhel's enthusiasm for studying the structure-function relationship in plants—and the joy she receives from sharing it with students—leads us into this chapter.

? What is systems biology?

■ A branch of biology that seeks to model the behavior of whole organisms

Figure 31.1A Dr. Natasha Raikhel (right) with students in her lab

Figure 31.1B *Arabidopsis thaliana*, a model plant organism

31.2 | The two main groups of angiosperms are the monocots and the dicots

Angiosperms have dominated the land for over 100 million years, and there are about 250,000 known species of flowering plants living today. Most of our foods come from a few hundred domesticated species of flowering plants. Among these foods are roots, such as beets and carrots; the fruits of trees and vines, such as apples, nuts, berries, and squashes; the fruits and seeds of legumes, such as peas and beans; and grains, the fruits of grasses such as rice, wheat, and corn (maize).

On the basis of several structural features, we can divide angiosperms into two groups, called monocots and dicots (Figure 31.2). The names *monocot* and *dicot* refer to the first leaves that appear on the plant embryo. These embryonic leaves are called seed leaves, or **cotyledons.** A **monocot** embryo has one seed leaf; a **dicot** embryo has two seed leaves.

Monocots are a large group of related plants that include the orchids, bamboos, palms, and lilies, as well as the grains and other grasses. You can see the single cotyledon inside the seed on the top left in Figure 31.2. The leaves, stems, flowers, and roots of monocots are also distinctive. Most monocots have leaves with parallel veins. Monocot stems have vascular tissues (tissues that transport water and nutrients) arranged in a complex array of bundles. The flowers of most monocots have their petals and other parts in multiples of three. The roots of monocots form a fibrous system—a mat of threads—that spreads out below the soil surface. With most of their roots in the top few centimeters of soil, monocots, especially

grasses, make excellent ground cover that reduces erosion. The roots of a rye grass plant, for instance, have more than 20 times the surface area of its stems and leaves.

Most angiosperms are dicots. The great majority of dicots, called the eudicots, or "true dicots," are related; a few smaller groups have evolved the dicot-type anatomy independently. The true dicots include most shrubs and trees (except for the conifers, which are gymnosperms), as well as the majority of our ornamental plants and many of our food crops. You can see the two cotyledons of a typical dicot in the seed on the lower left in Figure 31.2. Dicot leaves have a multibranched network of veins, and dicot stems have vascular bundles arranged in a ring. The dicot flower usually has petals and other parts in multiples of four or five. The large, vertical root of a dicot, known as a taproot, goes deep into the soil, as you know if you've ever tried to pull up a dandelion.

As we saw in the preceding unit on animals, a close look at a structure often reveals its function. Conversely, function provides insight into the "logic" of a structure. In the modules that follow, we'll take a detailed look at the correlation between plant structure and function.

? The terms *monocot* and *dicot* refer to the number of _____ in the seed.

■ cotyledons (seed leaves)

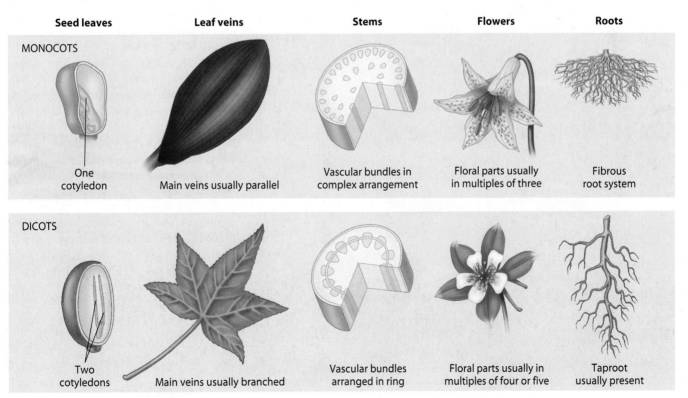

| Seed leaves | Leaf veins | Stems | Flowers | Roots |

MONOCOTS

One cotyledon | Main veins usually parallel | Vascular bundles in complex arrangement | Floral parts usually in multiples of three | Fibrous root system

DICOTS

Two cotyledons | Main veins usually branched | Vascular bundles arranged in ring | Floral parts usually in multiples of four or five | Taproot usually present

Figure 31.2 A comparison of monocots and dicots

A typical plant body consists of roots and shoots

Plants, like multicellular animals, have organs composed of various tissues, and these tissues are composed of cells. In looking at the hierarchy of plant structure—organs, tissues, and cells—we will focus first on organs as the most readily observable features of plant structure.

The basic structure of plants reflects their evolutionary history as land-dwelling organisms. Most plants must draw resources from two very different environments: They must absorb water and minerals from soil, and they must obtain CO_2 and light from air. The subterranean roots and aerial shoots of a typical land plant, such as the generalized flowering plant shown in **Figure 31.3**, perform these vital functions. Neither roots nor shoots can survive without each other. Lacking chloroplasts and living in the dark, most roots would starve without sugar and other organic nutrients transported from the photosynthetic leaves of the shoot system. Conversely, stems and leaves depend on the water and minerals absorbed by roots.

A plant's **root system** anchors it in the soil, absorbs and transports minerals and water, and stores food. The fibrous root system of a monocot consists of a mat of generally thin roots spread out shallowly in the soil. In contrast, dicots have one main vertical taproot with many small secondary lateral roots growing outward. Near the root tips in both dicots and monocots, a vast number of tiny projections called **root hairs** enormously increase the root surface area for absorption of water and minerals. As shown on the far right of the figure, each root hair is an extension of an epidermal cell (a cell in the outer layer of the root).

The **shoot system** of a plant is made up of stems, leaves, and adaptations for reproduction—flowers, in angiosperms. (We'll return to the angiosperm flower in Module 31.9.) As indicated in Figure 31.3, the **stems** are the parts of the plant that are generally above the ground and that support the leaves and flowers. In the case of a tree, the stems are the trunk and all the branches, including the smallest twigs. A stem has **nodes**, the points at which leaves are attached, and **internodes**, the portions of the stem between nodes. The **leaves** are the main photosynthetic organs in most plants, although green stems also perform photosynthesis. Most leaves consist of a flattened blade and a stalk, or petiole, which joins the leaf to a node of the stem.

The two types of buds you see in the figure are undeveloped shoots. When a plant stem is growing in length, the **terminal bud,** at the apex (tip) of the stem, has developing leaves and a compact series of nodes and internodes. The **axillary buds,** one in each of the angles formed by a leaf and the stem, are usually dormant. In many plants, the terminal bud produces hormones that inhibit growth of the axillary buds, a phenomenon called **apical dominance.** By concentrating resources on growing taller, apical dominance is an evolutionary adaptation that increases the plant's exposure to light. This is especially important where vegetation is dense. However, branching is also important for increasing the exposure of the shoot system to the environment, and under certain conditions, the axillary buds begin growing. Some develop into shoots bearing flowers, and others become nonreproductive branches complete with their own terminal buds, leaves, and axillary buds. Removing the terminal bud usually stimulates the growth of axillary buds. This is why pruning fruit trees and "pinching back" houseplants makes them bushier.

The drawing in Figure 31.3 gives us an overview of plant structure, but it by no means represents the enormous diversity of angiosperms. Next, let's look briefly at some variations on the basic themes of root and stem structure.

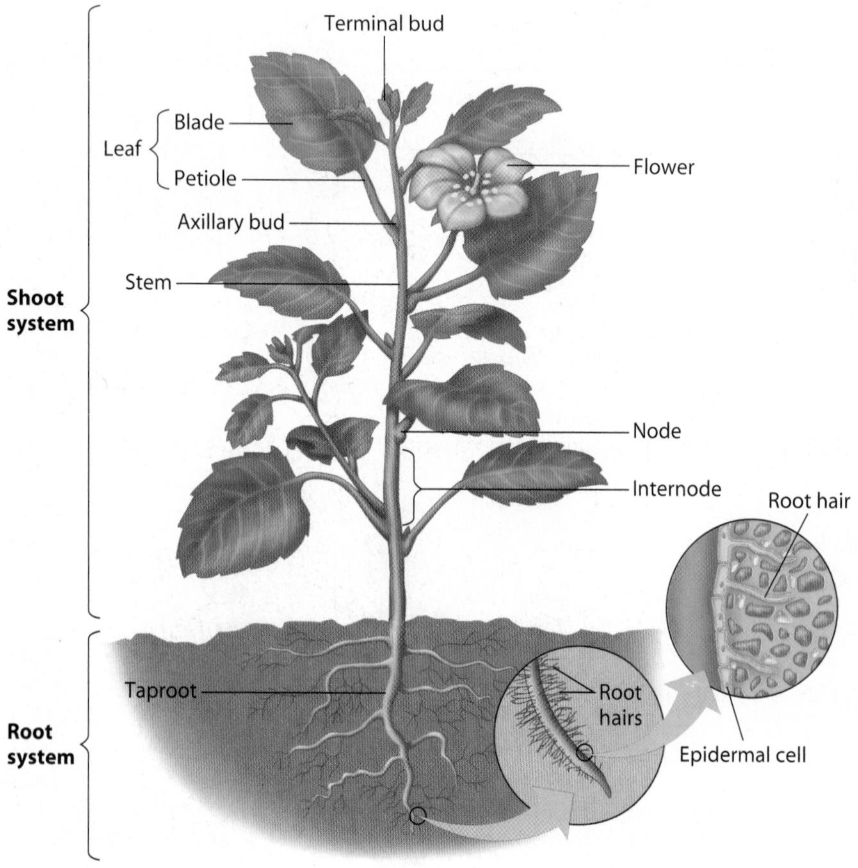

Figure 31.3 The body plan of a flowering plant (a dicot)

? Explain why pruning certain types of fruit trees increases future fruit harvest.

■ Removal of terminal buds from major branches results in more branching by reducing inhibition of axillary buds. More branches produce more flowers and hence more fruit.

31.4 Many plants have modified roots, stems, and leaves

Roots, stems, and leaves come in a variety of sizes and shapes, and they are adapted for a variety of functions. Many are adapted for storage. Carrots, turnips, sugar beets, and sweet potatoes, for instance, all have unusually large taproots that store food in the form of carbohydrates such as starch (Figure 31.4A). The plants consume the stored sugars during flowering and fruit production. For this reason, root crops are harvested before flowering.

Figure 31.4B shows three examples of modified stems. The strawberry plant has a horizontal stem called a **stolon** (or "runner") that grows along the ground surface. Stolons enable a plant to reproduce asexually, as plantlets form at nodes along their length. You've seen a different stem modification if you have ever dug up an iris plant or cooked with fresh ginger; the large, brownish, rootlike structures are actually **rhizomes**, horizontal stems that grow just below or along the soil surface. The rhizomes store food and, having buds, they can also spread and form new plants. The potato plant has rhizomes that end in

Figure 31.4A The modified root of a sugar beet plant

enlarged structures specialized for storage called **tubers** (the potatoes we eat). Potato "eyes" are clusters of axillary buds on the tubers that mark the nodes. Plant bulbs are underground shoots containing swollen leaves that store food. As you peel an onion, you are removing layers of leaves attached to a short stem.

Plant leaves, too, are highly varied. Grasses and many other monocots, for instance, have long leaves without petioles. Some dicots, such as celery, have enormous petioles—the stalks we eat—which contain a lot of water and stored food. The upper photograph in **Figure 31.4C** shows a modified leaf called a **tendril**, with its tips coiled around a stem. Tendrils help plants climb. (Some tendrils, as in grapevines, are modified stems.) The spines of the barrel cactus (lower photo) are modified leaves that protect the plant from animals. The main part of the cactus is the stem, which is adapted for photosynthesis and water storage.

So far we have examined plants as we see them with the unaided eye. Next, we move further down the structural hierarchy to explore plant tissues and cells microscopically.

? If a potato "sprouts" from an eye, are the resulting appendages part of the root system or shoot system?

■ The potato tuber is a modified stem, so the eyes are part of the shoot system.

Strawberry plant

Potato plant

Stolon (runner)

Taproot

Ginger plant

Rhizome

Rhizome

Tuber

Root

Figure 31.4B Three kinds of modified stems

Figure 31.4C Modified leaves: the tendrils of a pea plant (top) and cactus spines (bottom)

31.5 Plant cells and tissues are diverse in structure and function

In addition to features shared with other eukaryotic cells (see Module 4.4), most plant cells have three unique structures (**Figure 31.5A**): chloroplasts, the sites of photosynthesis; a central vacuole containing fluid that helps maintain cell turgor (firmness); and a cell wall that surrounds the plasma membrane.

The enlargement on the right in Figure 31.5A highlights the adjoining cell walls of two cells. Plant cell walls are made mainly of the structural carbohydrate cellulose. Many plant cells, especially those that provide structural support, have a two-part cell wall; a primary cell wall is laid down first, and then a more rigid secondary cell wall is secreted between the plasma membrane and the primary wall. The primary walls of adjacent cells in plant tissues are held together by a sticky layer called the middle lamella. Pits, where the cell wall is relatively thin, allow the contents of adjacent cells to lie close together. Plasmodesmata are channels of communication and circulation between adjacent plant cells.

The structure of a plant cell and the nature of its wall often correlate with the cell's main functions. As you consider the five major types of plant cells shown in Figures 31.5B–31.5F, notice the structural adaptations that make their specific functions possible.

Parenchyma cells (Figure 31.5B) are the most abundant type of cell in most plants. They remain alive when mature and usually have only primary cell walls, which are thin and flexible. Parenchyma cells perform most of the metabolic functions of a plant, such as photosynthesis, aerobic respiration and food storage. Most parenchyma cells can divide and differentiate into other types of plant cells under certain conditions, such as during the repair of an injury. In the laboratory, it is even possible to regenerate an entire plant from a single parenchyma cell (as we will see in Module 31.15).

Collenchyma cells (Figure 31.5C) resemble parenchyma cells in lacking secondary walls, but they have unevenly thickened primary walls. Their main function is to provide flexible support in parts of the plant that are still growing; young stems and petioles often have collenchyma cells just below their surface (the "string" of a celery stalk, for example). These living cells elongate as stems and leaves grow.

Sclerenchyma cells have thick secondary cell walls (yellow in **Figure 31.5D**) usually strengthened with lignin, which is the main chemical component of wood. Mature sclerenchyma cells cannot elongate, and they occur in regions of the plant that have stopped growing in length. When mature, most sclerenchyma cells are dead, their cell walls forming a rigid "skeleton" that supports the plant.

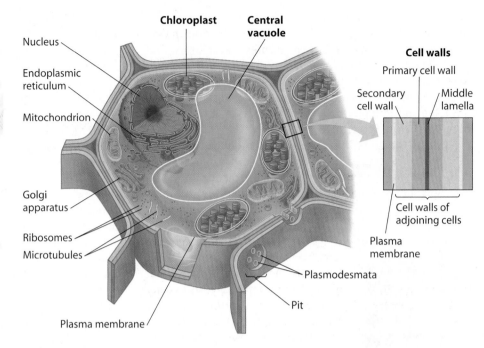

Figure 31.5A The structure of a plant cell

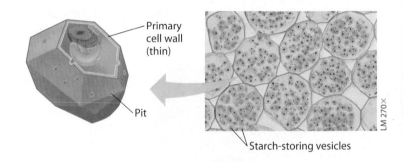

Figure 31.5B Parenchyma cell

Figure 31.5C Collenchyma cell

Figure 31.5D shows two types of sclerenchyma cells. One, called a **fiber**, is long and slender and is usually arranged in bundles. Some plant tissues with abundant fiber cells are commercially important; hemp fibers, for example, are used to make rope and clothing. **Sclereids**, which are shorter than fiber cells, have thick, irregular, and very hard secondary walls. Sclereids impart the hardness to nutshells and seed coats, and the gritty texture to the soft tissue of a pear.

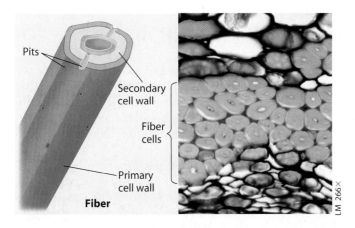

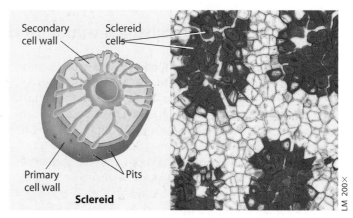

Figure 31.5D Sclerenchyma cells: fiber (left) and sclereid (right)

Angiosperms have two types of **water-conducting cells:** tracheids and vessel elements. Both have rigid, lignin-containing secondary cell walls. As **Figure 31.5E** shows, **tracheids** are long, thin cells with tapered ends. **Vessel elements** are wider, shorter, and less tapered. Chains of tracheids or vessel elements with overlapping ends form a system of tubes that conveys water from the roots to the stems and leaves. The tubes are hollow because both tracheids and vessel elements are dead when mature, with only their cell walls remaining. Water passes through pits in the walls of tracheids and vessel elements and through openings in the end walls of vessel elements. Because of their thick, rigid walls, these cells also function in support.

Food-conducting cells, known as **sieve-tube members**, are also arranged end to end, forming tubes (**Figure 31.5F**). Unlike water-conducting cells, however, sieve-tube members remain alive at maturity, though they lose most organelles, including the nucleus and ribosomes. This reduction in cell contents enables nutrients to pass more easily through the cell. The end walls between sieve-tube members, called **sieve plates**, have pores that facilitate the flow of fluid from cell to cell along the sieve tube. Alongside each sieve-tube member is at least one **companion cell**, which is connected to the sieve-tube member by numerous plasmodesmata. One companion cell may serve multiple sieve-tube members by producing and transporting proteins to all of them.

As in animals, the cells of plants are grouped into tissues with characteristic functions. For example, there are two types of vascular tissues, one conducting water, the other, food. Vascular tissue called **xylem** contains water-conducting cells that convey water and dissolved minerals upward from the roots. Vascular tissue called **phloem** contains sieve-tube members that transport sugars from leaves or storage tissues to other parts of the plant. In addition to conducting cells, phloem and xylem contain sclerenchyma cells, which provide support, and parenchyma cells, which store various materials.

In this module, we have examined plant cells and tissues. In the next module, we examine the level above plant tissues, called tissue systems.

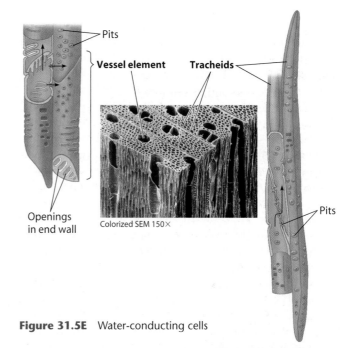

Figure 31.5E Water-conducting cells

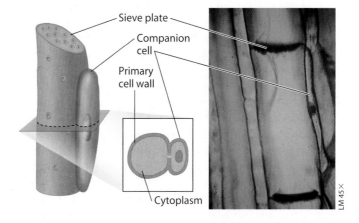

Figure 31.5F Food-conducting cell (sieve-tube member)

? Which of the following cell types has the potential to give rise to all others in the list: collenchyma, sclereid, parenchyma, vessel element, companion cell?

■ Parenchyma

A **tissue system** consists of one or more tissues organized into a functional unit within a plant. Each plant organ—root, stem, or leaf—is made up of three tissue systems: the dermal, vascular, and ground tissue systems. Each tissue system is continuous throughout the entire plant body, but the systems are arranged differently in leaves, stems, and roots (**Figure 31.6**). We examine the tissue systems of young roots and shoots in this module. Later, we will see that the tissue systems are somewhat different in older roots and stems.

The **dermal tissue system** (brown in the figure) forms an outer protective covering. Like our own skin, it acts as the first line of defense against physical damage and infectious organisms. In many plants, the dermal tissue system consists of a single layer of tightly packed cells called the **epidermis**. The epidermis of leaves and most stems has a waxy coating called the **cuticle**, which helps prevent water loss. The **vascular tissue system** (purple), made up of xylem and phloem tissues, provides support and long-distance transport. Tissues that are neither dermal nor vascular make up the **ground tissue system** (yellow). It accounts for most of the bulk of a young plant, filling the spaces between the epidermis and vascular tissue system. Ground tissue internal to the vascular tissue is called **pith**, and ground tissue external to the vascular tissue is called **cortex**. The ground tissue system has diverse functions, including photosynthesis, storage, and support.

The close-up views on the right side of Figure 31.6 show how these three tissues systems are organized in typical plant roots, stems, and leaves. The micrograph at the bottom shows in cross section the three tissue systems in a young dicot root. Water and minerals absorbed from the soil must enter through the epidermis. In the center of the root, the vascular tissue system forms a **vascular cylinder**, with xylem cells radiating from the center, like spokes of a wheel, and phloem cells filling in the wedges between the spokes. The ground tissue system of the root, the region between the vascular cylinder and epidermis, consists entirely of cortex. The cortex cells store food as starch and take up minerals that have entered the root through the epidermis. The innermost layer of the cortex is the **endodermis**, a cylinder one cell thick. The endodermis is a selective barrier, determining which substances pass between the rest of the cortex and the vascular tissue (see Module 32.2).

As the center of Figure 31.6 indicates, the young stem of a dicot looks quite different from that of a monocot. Both stems have their vascular tissue system arranged in numerous **vascular bundles**. However, the monocot stem has vascular bundles scattered throughout the ground tissue, whereas the dicot stem has vascular bundles arranged in a ring. Unlike

roots, the ground tissue of dicot stems consists of both a cortex region and a pith region. The cortex fills the space between the vascular ring and the epidermis. The pith fills the center of the stem and is often important in food storage. The ground tissue of a monocot stem is not divided into cortex and pith.

The top of Figure 31.6 illustrates the arrangement of the three tissue systems in a typical dicot leaf. The epidermis is interrupted by pores called **stomata** (singular, *stoma*), which allow CO_2 exchange between the surrounding air and the photosynthetic cells inside the leaf. Each stoma is flanked by two **guard cells**, which regulate the size of the stoma.

The ground tissue system of a leaf, called the **mesophyll**, is sandwiched between the upper and lower epidermis. Mesophyll consists mainly of photosynthetic parenchyma cells. The green structures in the diagram are their chloroplasts. In this dicot leaf, notice that cells in the lower mesophyll are loosely arranged, with a labyrinth of air spaces through which CO_2 and O_2 circulate. The air spaces are particularly large in the vicinity of stomata, where gas exchange with the outside air occurs. In many monocot leaves and in some dicot leaves, the mesophyll is not arranged in distinct upper and lower layers.

In both monocots and dicots, the leaf's vascular tissue system is made up of a network of veins. As you can see in Figure 31.6, each **vein** is a vascular bundle composed of xylem and phloem surrounded by a sheath of parenchyma cells. The veins' xylem and phloem, continuous with the vascular bundles of the stem, are in close contact with the leaf's photosynthetic tissues. This ensures that those tissues are supplied with water and mineral nutrients from the soil and that sugars made in the leaves are transported throughout the plant. The vascular structure also functions as a skeleton that reinforces the shape of the leaf.

This completes our survey of basic plant anatomy. Next, we examine how plants grow.

Web/CD Activity 31A *Root, Stem, and Leaf Sections*

Web/CD Thinking as a Scientist *What Are the Functions of Monocot Tissues?*

? Biologists generally define an "animal tissue" as a group of cells with common structure and function. How does this definition of a tissue contrast with what biologists call a "tissue system" in plants?

■ A plant tissue system may consist of several types of specialized cells, such as the different cell types of the vascular tissue system.

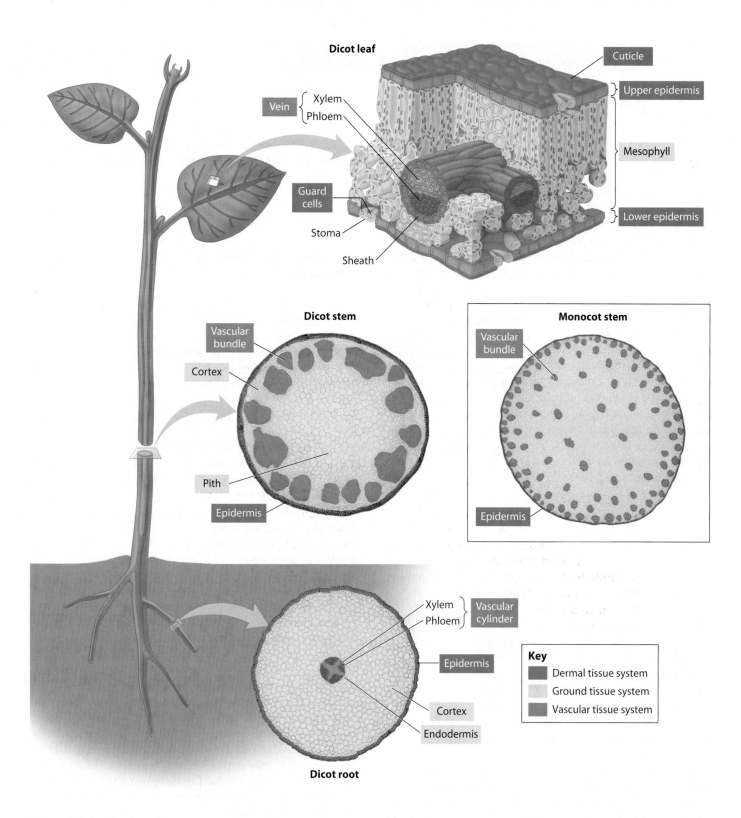

Figure 31.6 The three tissue systems

31.7 Primary growth lengthens roots and shoots

So far, we have looked at the structure of plant tissues and cells in mature organs. We will now examine how such organization arises through plant growth.

Most species of plants continue to grow as long as they live, a condition known as **indeterminate growth**. Most animals, in contrast, are characterized by **determinate growth**; that is, they cease growing after reaching a certain size. These differences underlie a broader distinction between plants and animals. Most animals *move* through their environment. Plants, in contrast, *grow* through their environment. Indeterminate growth can thus enable plants to increase their exposure to sunlight, air, and soil throughout life.

Indeterminate growth does not mean that plants are immortal; they do, of course, die. Flowering plants are categorized as annuals, biennials, or perennials, based on the length of their life cycle (the time from germination through flowering and seed production to death). **Annuals** complete their life cycle in a single year or less. Our most important food crops (including grains and legumes) are annuals, as are a great number of wildflowers. **Biennials** complete their life cycle in two years; flowering usually occurs during the second year. Beets and carrots are biennials, but we usually harvest them in their first year and so miss seeing their flowers. **Perennials** are plants that live and reproduce for many years. They include trees, shrubs, and some grasses.

Some plants are among the oldest organisms alive. The giant sequoias we discussed in the chapter's introduction are ancient (the oldest living one began growing about 3,000 years ago), and another conifer, the bristlecone pine, can live thousands of years longer than that. Some buffalo grass of the North American plains is believed to have been growing for 10,000 years from seeds that sprouted at the close of the last ice age. When a plant dies, it is not usually from old age, but from an infection or some environmental trauma, such as fire or severe drought.

Growth in plants is made possible by tissues called meristems. A **meristem** consists of cells that divide frequently, generating additional cells. Some products of this division remain in the meristem and produce still more cells, while others differentiate and are incorporated into tissues and organs of the growing plant. Meristems at the tips of roots and in the buds of shoots are called **apical meristems** (Figure 31.7A). Cell division in the apical meristems produces the new cells that enable a plant to grow in length, a process known as **primary growth.** Primary growth enables roots to push through the soil and allows shoots to increase exposure to light and CO_2. Although apical meristems lengthen both roots and shoots, there are important differences in the mechanisms of primary growth in each system. We will examine them separately, starting at the bottom.

Figure 31.7B illustrates primary growth in a longitudinal section through a growing onion root. The root tip is covered by a thimble-like **root cap** that protects the delicate, actively dividing cells of the apical meristem. Growth in length occurs just behind the root tip, where three zones of cells at successive stages of primary growth are located. Moving away from the root tip, they are called the zone of cell division, the zone of elongation, and the zone of maturation. The three zones of cells grade together, with no sharp boundaries between them.

The zone of cell division includes the root apical meristem and cells that derive from it. New root cells are produced in this region, including the cells of the root cap. In the zone of elongation, root cells elongate, sometimes to more than ten times their original length. It is cell elongation that is mainly responsible for pushing the root tip farther into the soil. The cells lengthen, rather than expand equally in all directions, because of the circular arrangement of cellulose fibers in parallel bands in their cell walls. The enlargement diagrams at the left of the root indicate how this works. The cells elongate by taking up water, and as they do, the cellulose fibers (shown in red) separate, somewhat like an expanding accordion. The cells cannot expand greatly in width because the cellulose fibers do not stretch much.

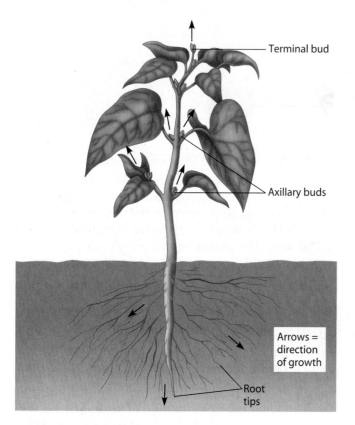

Terminal bud

Axillary buds

Arrows = direction of growth

Root tips

Figure 31.7A Locations of apical meristems, which are responsible for primary growth

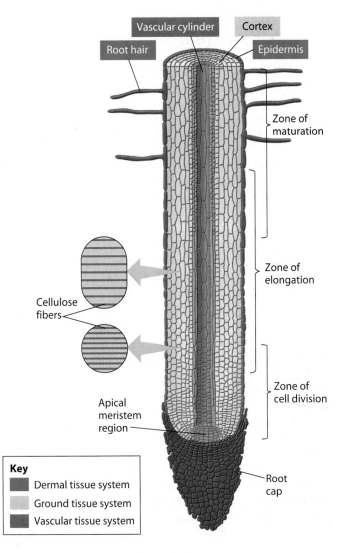

Figure 31.7B Primary growth of a root

Labels on figure 31.7B:
- Vascular cylinder
- Cortex
- Root hair
- Epidermis
- Zone of maturation
- Zone of elongation
- Cellulose fibers
- Zone of cell division
- Apical meristem region
- Root cap

Key
- Dermal tissue system
- Ground tissue system
- Vascular tissue system

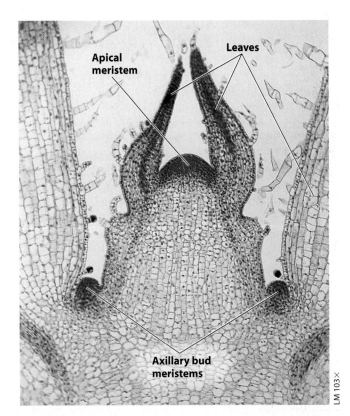

Labels on micrograph:
- Apical meristem
- Leaves
- Axillary bud meristems
- LM 103×

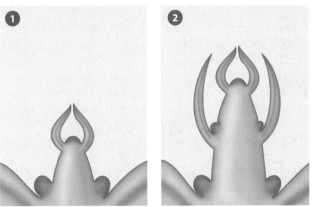

Figure 31.7C Primary growth of a shoot

The three tissue systems of a mature plant (dermal, ground, and vascular) take shape in the zone of maturation. Cells of the vascular cylinder differentiate into **primary xylem** and **primary phloem.** Differentiation of cells (specialization of their structure and function) results from differential gene expression (see Module 11.2). Cells in the vascular cylinder, for instance, develop into primary xylem or phloem cells because certain genes are turned on and are expressed as specific proteins, whereas other genes in these cells are turned off. Genetic control of differentiation is one of the most active areas of plant research.

The micrograph in **Figure 31.7C** shows a section through the end of a growing shoot that was cut lengthwise from its tip to just below its uppermost pair of axillary buds. You can see the apical meristem, which is a dome-shaped mass of dividing cells at the tip of the terminal bud. Elongation occurs just below this meristem, and the elongating cells push the apical meristem upward, instead of downward as in the root. As the apical meristem advances upward, some of its cells remain behind, and these become new axillary bud meristems at the base of the leaves.

The drawings in Figure 31.7C show two stages in the growth of a shoot. Stage ❶ is just like the micrograph. At the later stage shown in sketch ❷, the apical meristem has been pushed upward by elongating cells underneath.

Primary growth accounts for a plant's lengthwise growth. The stems and roots of many plants increase in thickness too, and in the next module, we see how this usually happens.

? You have cells in the lower layers of your skin that continue dividing, replacing dead cells that slough from the surface. Why is it inaccurate to compare such regions of active cell division in your body to a plant meristem? (*Hint:* Consider the types of cells produced.)

■ Your dividing cells normally are limited in the types of cells they can form. In contrast, the products of cell division in a plant meristem differentiate into all the diverse cell types of a plant.

31.8 Secondary growth increases the girth of woody plants

In the previous module, we saw how the apical meristems of plants produce primary growth. Woody plants (such as trees, shrubs, and vines) continue to grow in girth, thickening after primary growth has ceased. This increase in thickness of stems and roots, called **secondary growth**, is caused by the activity of two cylinders of dividing cells called **lateral meristems**. These two cylinders, known as the vascular cambium and the cork cambium, extend along the length of roots and stems.

The **vascular cambium** first appears as a cylinder of actively dividing cells between the primary xylem and primary phloem, as you can see in the pie section of a dicot stem at the left in **Figure 31.8A**. This region of the stem is just beginning secondary growth. Except for the vascular cambium, the stem at this stage of growth is virtually the same as a young stem undergoing primary growth (compare this figure with the dicot stem in Figure 31.6A). Secondary growth adds layers of vascular tissue on either side of the vascular cambium, as indicated by the green arrows.

The drawings at the center and the right show the results of secondary growth. In the center drawing, the vascular cambium has given rise to two new tissues: **secondary xylem** to its interior and **secondary phloem** to its exterior. Each year, the vascular cambium gives rise to layers of secondary xylem and secondary phloem that are larger in diameter than the previous layer (see the drawing at the right). In this way, the vascular cambium brings about thickening of a root or stem.

Secondary xylem makes up the **wood** of a tree, shrub, or vine. Over the years, a woody stem gets thicker and thicker as its vascular cambium produces layer upon layer of secondary xylem. The cells of the secondary xylem have thick

walls rich in lignin, giving wood its characteristic hardness and strength. The annual growth rings, such as those in the trunk of a giant sequoia or the locust tree in **Figure 31.8B**, result from the layering of secondary xylem. The layers are visible as rings because of uneven activity of the vascular cambium during the year. In woody plants that live in temperate regions, such as most of the United States, the vascular cambium becomes dormant each year during winter, and secondary growth is interrupted. When secondary growth resumes in the spring, a cylinder of early wood forms. Made up of the first new xylem cells to develop, early wood cells are usually larger in diameter and thinner-walled than those produced later in summer. The boundary between the large cells of early wood and the smaller cells of the late wood produced during the previous growing season is usually a distinct ring visible in cross sections of tree trunks and roots. Therefore, a tree's age can be estimated by counting its annual rings. The rings may have varying thicknesses, reflecting the amount of seasonal growth in a given year.

Now let's return to Figure 31.8A and see what happens to the parts of the stem that are *external* to the vascular cambium. Unlike xylem, the external tissues do not accumulate over the years. Instead, they are sloughed off at about the same rate they are produced.

Notice at the left of the figure that the epidermis and cortex, both the result of primary growth, make up the young stem's external covering. When secondary growth begins (center drawing), the epidermis is sloughed off and replaced with a new outer layer called **cork** (brown). Mature cork cells are dead and have thick, waxy walls, which protect the underlying tissues of the stem from water loss, physical dam-

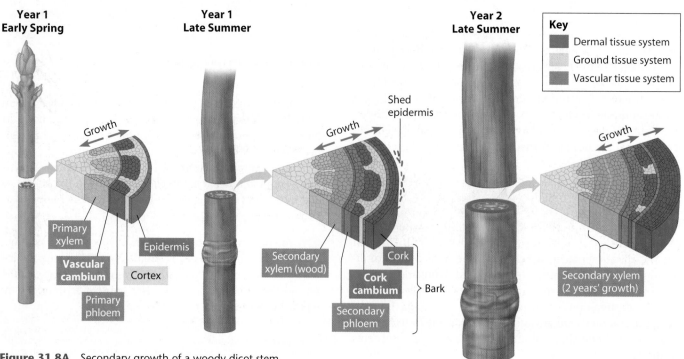

Figure 31.8A Secondary growth of a woody dicot stem

age, and pathogens. Cork is produced by meristematic tissue called the **cork cambium** (light brown), which first forms from parenchyma cells in the cortex. As the stem thickens and the secondary xylem expands, the original cork and cork cambium are pushed outward and fall off, as is evident in the cracked, peeling bark of many tree trunks. A new cork cambium forms to the inside. When no cortex is left, it forms from parenchyma cells in the phloem.

Everything external to the vascular cambium is called **bark.** As indicated in the diagram at the right in Figure 31.8A, its main components are the secondary phloem, the cork cambium, and the cork. The youngest secondary phloem (next to the vascular cambium) functions in sugar transport. The older secondary phloem dies, as does the cork cambium you see here. Pushed outward, these tissues and cork produced by the cork cambium help protect the stem until they, too, are sloughed off as part of the bark. Keeping pace with secondary growth, cork cambium keeps regenerating from the secondary phloem and keeps producing a steady supply of cork.

The log on the left in Figure 31.8B, from a locust tree, shows the results of several decades of secondary growth. The bulk of a trunk like this is dead tissue. The living tissues in it are the vascular cambium, the youngest secondary phloem, the cork cambium, and cells in the wood rays, which you can see radiating from the center of the log in the drawing on the right. The **wood rays** consist of parenchyma cells that transport water to the outer living tissues in the trunk. The **heartwood,** in the center of the trunk, consists of older layers of secondary xylem. These cells no longer transport water and minerals (xylem sap); they are clogged with resins and other metabolic by-products that make the heartwood resistant to rotting. The lighter-colored **sapwood**

consists of younger secondary xylem that does conduct xylem sap.

Thousands of useful products are made from wood—from construction lumber to fine furniture, musical instruments, paper, insulation, and a long list of chemicals, including turpentine, alcohols, artificial vanilla flavoring, and preservatives. Among the qualities that make wood so useful are a unique combination of strength, hardness, lightness, high insulating properties, durability, and workability. In many cases, there is simply no good substitute for wood. A wooden oboe, for instance, produces far richer sounds than a plastic one. Fence posts made of locust tree wood actually last much longer in the ground than metal ones. Ball bearings are sometimes made of a very hard wood called lignum vitae. Unlike metal bearings, they require no lubrication because a natural oil completely penetrates the wood.

In a sense, wood is analogous to the hard endoskeletons of many land animals. It is an evolutionary adaptation that enables a shrub or tree to remain upright and keep growing year after year on land—sometimes to attain enormous masses and heights, as we saw in the chapter's introduction. In the next few modules, we examine some additional adaptations that enable plants to live on land—those that facilitate reproduction.

Web/CD Activity 31B *Primary and Secondary Growth*

? (a) What type of plant tissue makes up wood? (b) What is bark?

■ (a) Secondary xylem; (b) all tissues exterior to the vascular cambium—secondary phloem, cork cambium, and cork.

Sapwood

Heartwood

Figure 31.8B Anatomy of a locust log

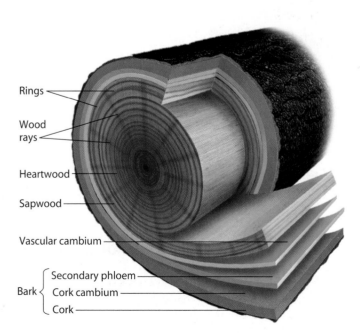

Rings

Wood rays

Heartwood

Sapwood

Vascular cambium

Bark { Secondary phloem
Cork cambium
Cork

31.9 Overview: The sexual life cycle of a flowering plant

It has been said that an oak tree is merely an acorn's way of making more acorns. Indeed, evolutionary fitness for any organism is measured only by its ability to replace itself with healthy, fertile offspring. Thus, from an evolutionary viewpoint, all the structures and functions of a plant can be interpreted as mechanisms contributing to reproduction. In the remaining modules, we explore the reproductive biology of angiosperms, beginning here with a brief overview.

Flowers, the reproductive shoots of angiosperms, typically contain four types of modified leaves called floral organs: sepals, petals, stamens, and carpels (Figure 31.9A). **Sepals**, which enclose and protect the flower bud, are usually green and more leaflike than the other floral organs. The **petals** are often colorful and advertise the flower to pollinators. Stamens and carpels are the reproductive organs, containing sperm and eggs, respectively.

A **stamen** consists of a stalk (filament) tipped by an anther. Within the **anther** are sacs in which meiosis occurs and in which pollen is produced. Pollen grains house the cells that develop into sperm.

A **carpel** has a long slender neck (style) with a sticky stigma at its tip. The **stigma** is the landing platform for pollen. The base of the carpel is the **ovary**, within which are **ovules**, each containing a developing egg and supporting cells.

Figure 31.9B shows the life cycle of a generalized angiosperm. Fertilization occurs in the ovule, which then develops into a seed containing the embryo. Meanwhile, the ovary develops into a fruit, which protects the seed and aids in dispersing it. Completing the life cycle, the seed **germinates** (begins to grow) in a suitable habitat; the embryo develops into a seedling; and the seedling grows into a mature plant.

In the next four modules, we examine key stages in the angiosperm sexual life cycle in more detail. We will see that there are a number of variations in the basic themes presented here.

? Pollen develops within the _____ of _____.
Ovules develop within the _____ of _____.

■ anthers · · · stamens · · · ovaries · · · carpels

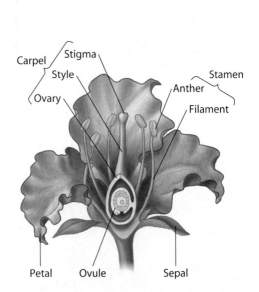

Figure 31.9A The structure of a flower

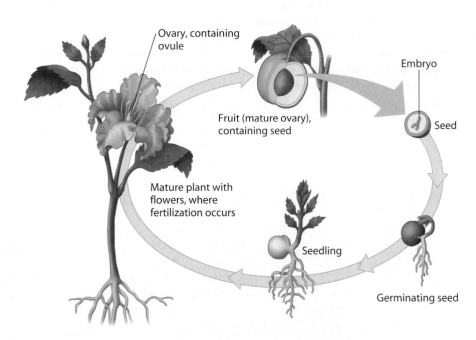

Figure 31.9B Life cycle of a generalized angiosperm

31.10 The development of pollen and ovules culminates in fertilization

Recall from Chapter 17 that the life cycles of all plants include alternation of haploid (*n*) and diploid (*2n*) generations. The roots, stems, leaves, and most of the reproductive structures of a rose plant, an oak tree, or a grass—in fact, all angiosperms— are diploid. The diploid plant body is called the **sporophyte**. A sporophyte produces special structures, the anthers and ovules, in which cells undergo meiosis. Haploid spores result. Each of these then divides mitotically and becomes a multicellular **gametophyte**, the plant's haploid generation. The gametophyte produces gametes by mitosis. At fertilization, the male

and female gametes unite, producing a diploid zygote. The life cycle is completed when the zygote divides by mitosis and develops into a new sporophyte. Without a microscope, all we can see of the angiosperm life cycle is the sporophytes and occasionally the pollen (male gametophytes) produced by them. In this module, we take a microscopic look at the gametophytes of a flowering plant.

We begin with the development of the male gametophyte, the pollen grain. The cells that develop into pollen grains are found within a flower's anthers (top left in **Figure 31.10**). Each cell first undergoes meiosis, forming four haploid cells called spores. Each spore then divides mitotically, forming two haploid cells, called the tube cell and the generative cell. A thick wall forms around these cells, and the resulting pollen grain is ready for release from the anther.

Moving to the top right of the figure, we can follow the development of the flower parts that form the female gametophyte and eventually the egg. In most species, the ovary of a flower contains several ovules, but only one is shown here. An ovule contains a central cell (gold) surrounded by a protective covering of smaller cells (yellow). The central cell enlarges and undergoes meiosis, producing four haploid spores. Three of the spores usually degenerate, but the surviving one enlarges and divides mitotically, producing a multicellular structure called the **embryo sac** (gold area). Housed in several layers of protective cells (yellow) produced by the sporophyte plant, the embryo sac is the female gametophyte. The sac contains a large central cell with two haploid nuclei. One of its other cells is the haploid egg, ready to be fertilized.

The first step leading to fertilization is **pollination** (at the center of the figure), the transfer of pollen from anther to stigma. Most angiosperms are dependent on insects, birds, or other animals to transfer their pollen. But the pollen of some plants—such as grasses and many trees—is windborne, as anyone bothered by seasonal pollen allergies knows.

After pollination, the pollen grain germinates on the stigma. Its tube cell gives rise to the pollen tube, which grows downward into the ovary. Meanwhile, the generative cell divides mitotically, forming two sperm. When the pollen tube reaches the base of the ovule, it enters the ovary and discharges its two sperm near the embryo sac. One sperm fertilizes the egg, forming the zygote. The other contributes its haploid nucleus to the large diploid central cell of the embryo sac. This cell, now with a triploid (3*n*) nucleus, will give rise to a food-storing tissue called **endosperm.**

The union of two sperm with two different nuclei of the embryo sac is called **double fertilization,** and the resulting production of endosperm is unique to angiosperms. Endosperm will develop only in ovules containing a fertilized egg, thereby preventing angiosperms from squandering nutrients.

Web/CD Activity 31C *Angiosperm Life Cycle*

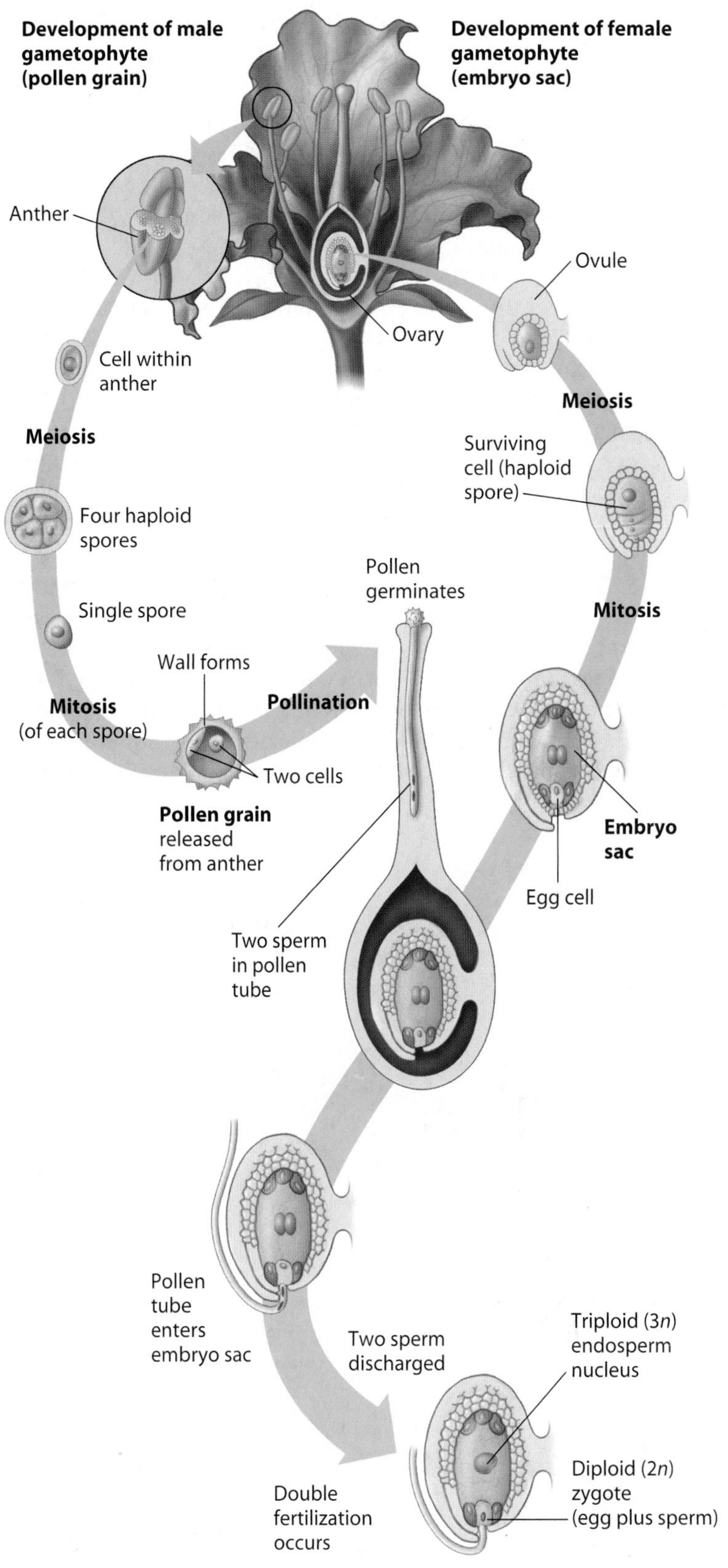

? Fertilization unites the sperm cell, which develops within the male gametophyte (or ____ ____), with the egg cell, which develops within the female gametophyte (or ____ ____).

■ pollen grain . . . embryo sac.

Figure 31.10 Gametophyte development and fertilization in an angiosperm

31.11 The ovule develops into a seed

After fertilization, the ovule, containing the triploid central cell and the zygote, begins developing into a seed. The triploid cell divides and develops into the nutrient-rich tissue called endosperm. The endosperm nourishes the embryo until it becomes a self-supporting seedling.

As shown in **Figure 31.11A**, embryonic development begins when the zygote divides into two cells. Repeated division of one of the cells then produces a ball of cells that becomes the embryo. The other cell divides to form a thread of cells that pushes the embryo into the endosperm. The bulges you see on the embryo are the developing cotyledons. You can tell that the plant in this drawing is a dicot since it has two cotyledons.

The result of embryonic development in the ovule is a mature seed, which you see on the bottom right of Figure 31.11A. Near the end of its maturation, the seed loses most of its water and forms a hard, resistant **seed coat** (brown). The embryo, surrounded by its endosperm food supply (gold), becomes dormant; it will not develop further until the seed germinates. **Seed dormancy**, a condition in which growth and development are suspended temporarily, is an important evolutionary adaptation. It allows time for a plant to disperse its seeds and increases the chance that a new generation of plants will begin growing only when environmental conditions, such as temperature and moisture, favor survival.

The dormant embryo contains a miniature root and shoot, each equipped with an apical meristem. After the seed germinates, the apical meristems will sustain primary growth as long as the plant lives. Also present in the embryo are the three tissue cylinders that will form the epidermis, cortex, and primary vascular tissues.

Figure 31.11B contrasts the internal structures of dicot and monocot seeds. In the dicot (a bean), the embryo is an elongated structure with two fleshy cotyledons (tan). The embryonic root develops just below the point at which the cotyledons are attached to the rest of the embryo. The embryonic shoot, tipped by a pair of miniature embryonic leaves, develops just above the point of attachment. The bean seed contains no endosperm because its cotyledons absorb the endosperm nutrients as the seed forms. The nutrients start passing from the cotyledons to the embryo when it germinates.

A kernel of corn, an example of a monocot, is actually a fruit containing one seed. Everything you see in the drawing is the seed, except the kernel's outermost covering. The covering is the dried tissue of the fruit, tightly bonded to the seed coat. Different from the bean, the corn seed contains a large endosperm and a single cotyledon. The cotyledon absorbs the endosperm's nutrients during germination. Also unlike the bean, the embryonic root and shoot in corn each have a protective sheath.

? What is the role of the endosperm in a seed?

■ The endosperm provides nutrients to the developing embryo.

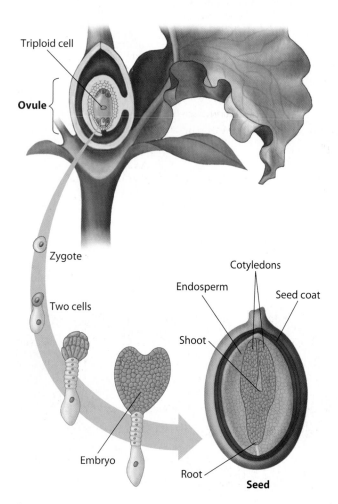

Figure 31.11A Development of a dicot plant embryo

Triploid cell

Ovule

Zygote

Two cells

Embryo

Endosperm

Cotyledons

Seed coat

Shoot

Root

Seed

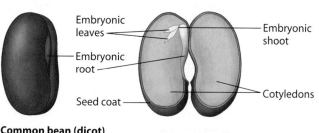

Embryonic leaves

Embryonic shoot

Embryonic root

Seed coat

Cotyledons

Common bean (dicot)

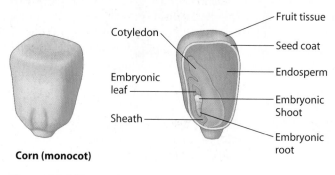

Cotyledon

Embryonic leaf

Sheath

Fruit tissue

Seed coat

Endosperm

Embryonic Shoot

Embryonic root

Corn (monocot)

Figure 31.11B Seed structure

31.12 The ovary develops into a fruit

In the previous two modules, we followed the angiosperm life cycle from the flower on the sporophyte plant through the transformation of an ovule into a seed. Simultaneous with seed development, fertilization triggers hormonal changes that cause a flower's ovary to mature into a **fruit**, a specialized vessel that houses and protects seeds and helps disperse them from the parent plant. A pea pod is a fruit, as is a peach, orange, tomato, cherry, or corn kernel.

The photographs in **Figure 31.12A** illustrate the changes in a pea plant leading to pod formation. ❶ Soon after pollination, ❷ the flower drops its petals, and the ovary starts to grow. The ovary expands tremendously, and its wall thickens, ❸ forming the pod, or fruit.

Figure 31.12B matches the parts of a pea flower with what they become in the pod. The wall of the ovary becomes the pod. The ovules, within the ovary, develop into the seeds. The small, threadlike structure at the end of the pod is what remains of the upper part of the flower's carpel. The sepals of the flower often stay attached to the base of the green pod. Peas are usually harvested at this stage of fruit development. If the pods are allowed to develop further, they become dry and brownish and will split open, releasing the seeds.

Fruits are highly varied, as **Figure 31.12C** illustrates. In some cases, what we commonly call fruits are actually more than developed ovaries. In an apple, for instance, the part we discard, the core, is the thickened ovary and therefore the true fruit. The soft, fleshy part we eat develops from parts of the flower base that fuse with the ovary. In contrast to the pea pod, which dehydrates when it is fully developed, a fleshy fruit like an apple becomes softer as enzymes weaken the cell walls. This is usually accompanied by a color change from green to red, orange, or yellow. Additionally, the fruit becomes sweeter as organic acids or starch molecules are converted to sugar, which may reach a concentration of as much as 20% when the fruit is fully ripe. Fleshy, edible fruits such as apples serve as enticements to animals that help spread seeds.

A typical fruit is derived from a flower with a single carpel and is called a **simple fruit**. Some simple fruits are fleshy, such as a peach, whereas others are dry, such as a pea pod or a

nut. An **aggregate fruit** results from a single flower that has more than one carpel, each forming a small fruit. These "fruitlets" are clustered together, as in a raspberry. A **multiple fruit** develops from a group of separate flowers tightly clustered together. When the walls of the many ovaries start to thicken, they fuse together and become incorporated into one fruit, as in a pineapple.

Web/CD Activity 31D *Seed and Fruit Development*

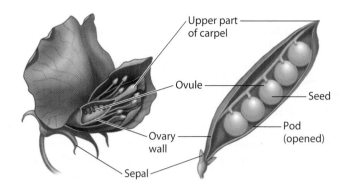

Figure 31.12B The correspondence between flower and fruit in the pea plant

Figure 31.12C
A variety of fruits, including simple, aggregate, and multiple fruit

? Seed is to _____ as _____ is to ovary.

■ ovule · · · fruit

Figure 31.12A Development of a simple fruit, a pea pod

31.13 Seed germination continues the life cycle

The germination of a seed is often used to symbolize the beginning of life, but as we have seen, the seed already contains a miniature plant, complete with embryonic root and shoot. Thus, at germination, the plant does not begin life but rather resumes the growth and development that was temporarily suspended during seed dormancy.

Germination usually begins when the seed takes up water. The hydrated seed expands, rupturing its coat. The inflow of water triggers metabolic changes in the embryo that make it start growing again. Enzymes begin digesting stored nutrients in the endosperm or cotyledons, and these nutrients are transported to the growing regions of the embryo.

The figures below trace germination in a dicot (a garden pea) and a monocot (corn). In Figure 31.13A, notice that the embryonic root of a pea emerges first and grows downward from the germinating seed. Next, the embryonic shoot emerges, and a hook forms near its tip. The hook protects the delicate shoot tip by holding it downward, rather than pushing it up through the abrasive soil. As the shoot breaks through the soil surface, its tip is lifted gently out of the soil as exposure to light stimulates the hook to straighten. The first foliage leaves then expand from the shoot tip and begin making food by photosynthesis. The cotyledons, their food reserves used by the germinating embryo, remain behind in the soil and decompose.

Corn and other monocots use a different mechanism for breaking ground at germination (Figure 31.13B). A protective sheath surrounding the shoot pushes upward and breaks through the soil. The shoot tip then grows up through the tunnel provided by the sheath. As in the pea, the corn cotyledon remains in the soil and decomposes.

In the wild, only a small fraction of fragile seedlings endure long enough to reproduce. Production of enormous numbers of seeds compensates for the odds against individual survival. Asexual reproduction, generally simpler and less hazardous for offspring than sexual reproduction, is an alternative means of plant propagation, as we see next.

Web/CD Thinking as a Scientist *What Tells Desert Seeds When to Germinate?*

? Which meristems provide additional cells for early growth of a seedling after germination?

■ The apical meristems of the shoot and root

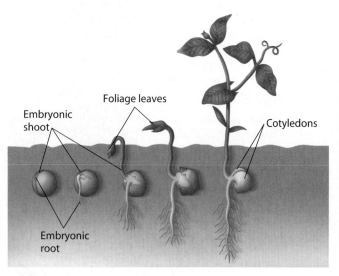

Figure 31.13A Pea germination (a dicot)

Foliage leaves

Embryonic shoot

Cotyledons

Embryonic root

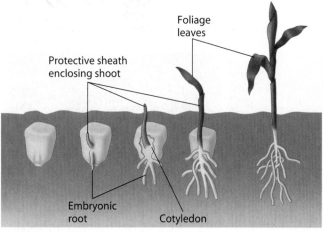

Foliage leaves

Protective sheath enclosing shoot

Embryonic root

Cotyledon

Figure 31.13B Corn germination (a monocot)

31.14 Asexual reproduction produces plant clones

Imagine chopping off your finger and watching it grow and develop into an exact copy of you. This would be an example of asexual reproduction (also called vegetative propagation in plants), in which offspring are derived from a single parent. The resulting offspring, an asexually produced, genetically identical organism, is often called a **clone**.

Asexual reproduction in angiosperms and other plants is an extension of their capacity to grow throughout life. A plant's meristematic tissues can sustain growth indefinitely. In addition, parenchyma cells throughout a plant can divide and differentiate into the various types of cells.

The photographs on this page show four examples of plant cloning in nature. Asexual reproduction in plants often involves **fragmentation**, the separation of a parent plant into parts that develop into whole plants. The garlic bulb (**Figure 31.14A**) is actually an underground stem that functions in storage. A single large bulb fragments into several parts, called cloves. Each clove can give rise to a separate plant, as indicated by the green shoots emerging from some of them. The white sheaths are leaves attached to the stem.

Each of the small trees you see in **Figure 31.14B** is a sprout from the roots of a coast redwood tree, a close relative of the giant sequoia we discussed in the chapter's introduction. Eventually, one or more of these root sprouts may take the place of its parent in the forest.

The ring of plants in **Figure 31.14C** is a clone of creosote bushes growing in the Mojave Desert in southern California.

In this case, the word *clone* refers to a group of genetically identical organisms. All these bushes came from generations of asexual reproduction by roots. Making the oldest sequoias seem youthful, this clone apparently began with a single plant that germinated from a seed about 12,000 years ago. The original plant probably occupied the center of the ring.

Figure 31.14D shows a patch of dune grass in Massachusetts. Most grasses can propagate asexually by sprouting shoots and roots from runners. A small patch of grass can spread until it covers an acre or more of surface.

Many plants can reproduce both sexually and asexually. What advantages can asexual reproduction offer? For one thing, a parent plant well suited to its environment can clone many copies of itself, all of which would be equally well suited to current conditions. Also, the offspring may not be as fragile as seedlings produced by sexual reproduction. Both asexual reproduction and sexual reproduction have played important roles in the evolutionary adaptation of plant populations to their environments.

? Which mode of reproduction (sexual or asexual) would generally be more advantageous in a location where the composition of the soil is constantly changing? Why?

■ Sexual, because it generates genetic variation among the offspring, which enhances the potential for adaptation to a changing environment

Figure 31.14A Cloves of a garlic bulb

Figure 31.14B Sprouts from the roots of coast redwood trees

Figure 31.14C A ring of creosote bushes

Figure 31.14D Dune grass

31.15 Asexual reproduction is a mainstay of modern agriculture

Figure 31.15 Test-tube cloning

The ability of plants to reproduce asexually provides many opportunities for producing large numbers of plants with minimal effort and expense. For example, most of our fruit trees and houseplants are asexually propagated from cuttings. Several other plants are propagated from root sprouts (raspberries, for example) or pieces of underground stems (such as potatoes).

Plants can also be propagated by test-tube methods. The laboratory tube in **Figure 31.15** contains an apple plantlet that was grown from a few meristem cells cut from a large tree and cultured on a chemical medium. Using this method, a single plant can be cloned into thousands of copies. Orchids and certain pine trees used for mass plantings are commonly propagated this way.

As discussed in Module 12.18, plant cell culture methods also enable researchers to grow plants from genetically engi-neered plant cells. Foreign genes are incorporated into a single parenchyma cell, and the cell is then cultured so that it multiplies and develops into a new plantlet. The resulting "GM" (genetically modified) plant may then be able to grow and reproduce normally. The commercial adoption by farmers of GM crops has been one of the most rapid cases of technology transfer in the history of agriculture. Many people are concerned about possible risks to the environment and human health associated with the use of GM crop plants (see Module 12.19).

Aside from the issues raised by GM plants, modern agriculture faces some potentially serious problems. Nearly all of today's crop plants have very little genetic variability. Furthermore, we grow most crops in **monocultures**, large areas of land with a single plant variety. Given these conditions, plant scientists fear that a small number of diseases could devastate large crop areas. In response, plant breeders are working to maintain "gene banks," storage sites for seeds of many different plant varieties that can be used to breed new hybrids.

? What is the most serious threat to an agriculture based on monoculture?

■ New or newly arrived pathogens to which the clone lacks resistance

---

CHAPTER REVIEW

Reviewing the Concepts

Angiosperms, or flowering plants, are the most familiar and diverse group of plants **(Introduction).** One angiosperm, *Arabidopsis*, is a popular model organism **(31.1).**

Plant Structure and Function (31.2–31.6)

Plant structure. The two main types of angiosperms, monocots and dicots, differ in the number of seed leaves and the structure of roots, stems, leaves, and flowers **(31.2).** Here is the body plan of a typical flowering plant **(31.3):**

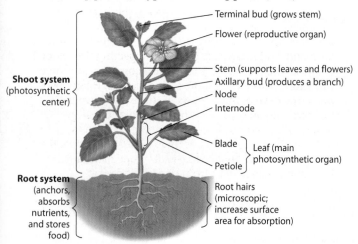

Roots and stems may function in storing food, asexual reproduction, and protection **(31.4).**

Tissues and cells. The five major types of plant cells are parenchyma, collenchyma, sclerenchyma (including fiber and sclereid cells), water-conducting cells (including tracheids and vessel elements), and food-conducting cells (sieve-tube members). Two kinds of vascular tissue are xylem, which conveys water and dissolved minerals, and phloem, which transports sugars **(31.5).** Roots, stems, and leaves are made up of three tissue systems: the dermal, vascular, and ground tissue systems. Dermal tissue covers and protects the plant. The vascular tissue system contains xylem and phloem. The ground tissue system consists of parenchyma cells and supportive collenchyma and sclerenchyma cells **(31.6).**

Plant Growth (31.7–31.8)

Meristems, areas of unspecialized, dividing cells, are where plant growth originates. Apical meristems at the tips of roots and in the terminal buds and axillary buds of shoots initiate primary (lengthwise) growth by producing new cells. A root or shoot lengthens further as the cells elongate and differentiate **(31.7).** An increase in a plant's girth, called secondary growth, arises from cell division in a cylindrical meristem called the vascular cambium. The vascular cambium thickens a stem by adding layers of secondary xylem, or wood, next to its inner surface. Outside the

vascular cambium, the bark consists mainly of secondary phloem, cork cambium, and protective cork cells that are produced by the cork cambium (31.8).

Reproduction of Flowering Plants (31.9–31.15)

The angiosperm flower consists of sepals, petals, stamens, and carpels. Pollen grains develop in anthers, at the tips of stamens. The tip of the carpel, the stigma, receives pollen grains. The ovary, at the base of the carpel, houses the egg-producing structure, the ovule (31.9).

The angiosperm life cycle. Haploid spores are formed within ovules and anthers. The spores in the anthers give rise to male gametophytes—pollen grains—which produce sperm. A spore in an ovule produces the embryo sac, the female gametophyte. Each embryo sac contains an egg cell. Pollination is the arrival of pollen grains onto a stigma. A pollen tube grows into the ovule, and sperm pass through it and fertilize both the egg and a second cell. This process is called double fertilization (31.10). After fertilization, the ovule becomes a seed, and the fertilized egg within it divides and becomes an embryo. The other fertilized cell develops into the endosperm, which stores food for the embryo (31.11). The ovary develops into a fruit, which helps protect and disperse the seeds (31.12). A seed starts to germinate when it takes up water and expands. The embryo resumes growth and absorbs nutrients from the endosperm. An embryonic root emerges, and a shoot pushes upward and expands its leaves (31.13).

Asexual reproduction can be achieved via bulbs, sprouts, or runners (31.14). Propagating plants asexually from cuttings or bits of tissue can increase agricultural productivity but can also reduce genetic diversity (31.15).

Connecting the Concepts

1. Create a concept map or a detailed sketch that shows the relationships between the following parts of an angiosperm body: root system, root hairs, shoot system, leaf, petiole, blade, stem, node, internode, flower, stamen, carpel, sepal, petal, stigma, style, filament, ovary, ovule.

Testing Your Knowledge

Multiple Choice

2. Which of the following is closest to the center of a woody stem? (*Explain your answer.*)
   a. vascular cambium   d. primary xylem
   b. primary phloem   e. secondary xylem
   c. secondary phloem

3. A pea pod is formed from ____. A pea inside the pod is formed from ____.
   a. an ovule . . . a carpel   d. an anther . . . an ovule
   b. an ovary . . . an ovule   e. endosperm . . . an ovary
   c. an ovary . . . a pollen grain

4. While walking in the woods, you encounter an unfamiliar nonwoody flowering plant. If you want to know whether it is a monocot or dicot, it would *not* help to look at the

   a. number of seed leaves, or cotyledons, present in its seeds.
   b. shape of its root system.
   c. number of petals in its flowers.
   d. arrangement of vascular bundles in its stem.
   e. size of the plant.

5. In angiosperms, each pollen grain produces two sperm. What do these sperm do?
   a. Each one fertilizes a separate egg cell.
   b. One fertilizes an egg, and the other fertilizes the fruit.
   c. One fertilizes an egg, and the other is kept in reserve.
   d. Both fertilize a single egg cell.
   e. One fertilizes an egg, and the other fertilizes a cell that develops into stored food.

Matching

6. Attracts pollinator   a. pollen grain
7. Develops into seed   b. ovule
8. Protects flower before it opens   c. anther
9. Produces sperm   d. ovary
10. Produces pollen   e. sepal
11. Houses ovules   f. petal

Describing, Comparing, and Explaining

12. The scent of apple blossoms and the buzzing of bees fill an orchard on a warm spring day. Describe the processes by which the pollen carried from flower to flower by the bees results in the apple you might pick in the fall.

13. Name three kinds of asexual reproduction. Explain two advantages of asexual reproduction over sexual reproduction. What is the primary drawback?

14. What part of a plant are you eating when you consume each of the following: tomato, celery stalk, peanut, strawberry, lettuce, artichoke, beet?

Applying the Concepts

15. Plant scientists are searching Peru, Mexico, and the Middle East for the wild ancestors of potatoes, corn, and wheat. Why is this search important?

16. Tropical forests contain a wealth of plants that are potential new sources of food, as well as sources of medicine and other useful products. The developing nations of the tropics cannot develop these resources themselves. Under pressure from growing populations and debt, they are cutting their forests for lumber and farmland, and many species are disappearing. Developed countries are pressuring the tropical countries to protect the forests before even more species are lost. Many people in the developing nations see little incentive to preserve the forests only to have corporations from industrialized countries profit from new products obtained from the forests. Is there a way to preserve the tropical forests so that both the developed and developing nations will benefit from their abundance?

Answers to all questions can be found in Appendix 3.

For study help and Activities, go to campbellbiology.com or the student CD-ROM.

CHAPTER 32

THE UPTAKE AND TRANSPORT OF PLANT NUTRIENTS

32.1   Plants acquire their nutrients from soil and air
32.2   The plasma membranes of root cells control solute uptake
32.3   Transpiration pulls water up xylem vessels
32.4   Guard cells control transpiration
32.5   Phloem transports sugars

PLANT NUTRIENTS AND THE SOIL

32.6   Plant health depends on a complete diet of essential inorganic nutrients
32.7   You can diagnose some nutrient deficiencies in your own plants
32.8   Fertile soil supports plant growth
32.9   Soil conservation is essential to human life
32.10 Organic farmers must follow ecological principles
32.11 Agricultural research is improving the yields and nutritional values of crops

PLANT NUTRITION AND SYMBIOSIS

32.12 Fungi help most plants absorb nutrients from the soil
32.13 Most plants depend on bacteria to supply nitrogen
32.14 Legumes and certain other plants house nitrogen-fixing bacteria
32.15 The plant kingdom includes parasites and carnivores

Plants That Clean Up Poisons

THE HEALTHY FERNS pictured below and at right with University of Florida researcher Dr. Lena Ma have a secret: Their leaves contain high levels of arsenic, a chemical element toxic to most plants and animals. Amazingly, this fern species, called the brake fern (*Pteris vittata*), actually thrives on the poison!

Dr. Ma and her colleagues discovered the brake fern's unusual affinity for arsenic in a lumberyard. There they found this plant growing in soil heavily contaminated with arsenic, which is a major component of a common wood preservative. The arsenic concentrations in its leaves turned out to be 200 times higher than that of the soil. And this hardy and fast-growing fern not only absorbed arsenic but also grew 40% bigger in arsenic-laden soil. While over 400 other plant species are known "metal accumulators"—absorbing large amounts of lead, zinc, and other heavy metals—the brake fern is the first arsenic-loving plant identified.

Because of the brake fern's special properties, biologists are hoping to add it to their toolkit for *phytoremediation*, the use of plants to help clean up polluted soil and groundwater. Phytoremediation is one type of bioremediation, which includes the use of prokaryotes and protists to detoxify polluted sites. With conventional methods, the cost of cleaning up the toxins that

Plant Nutrition and Transport

have accumulated at thousands of factories, farms, and military sites in the United States could top $700 billion, so finding alternative cleanup methods is essential.

A number of successful phytoremediation projects have already been completed. In one project, sunflowers and Indian mustard, two species that absorb lead, were planted on contaminated land near an automobile factory in Detroit. The plants reduced the soil's lead contamination by 43%, down to acceptable levels. The lead-rich crops were then harvested and hauled away to a hazardous-waste landfill. This proved to be a cost-effective and efficient way to clean up the toxic soil. Before phytoremediation, cleanup crews would have had to haul away 4,400 m³ (cubic meters) of soil, instead of a few cubic meters of plant material, and the cost would have been more than double.

Sunflowers have helped clean up one of the most dangerous types of toxic substances: radioactive metals. In a contaminated pond near the destroyed nuclear power plant in Chernobyl (see Module 2.5), sunflower plants were set adrift on foam rafts (see photo at right). Within days, the concentrations of two radioactive metals, strontium-90 and cesium-137, reached levels in the sunflower roots several thousand times higher than in the water.

As it turns out, toxic metals are just one plant absorption specialty. Poplar and willow trees are distinguished by their ability to pump large amounts of water up through their roots—along with solvents and other organic pollutants that are present. The organic pollutants may then be broken down. From Hawaii to Montana, oil refineries and other polluting industries are using such trees to treat groundwater polluted with organic contaminants. Poplars are such effective water pumps that one environmental engineer calls them "a self-assembling solar-powered pump-and-treat system."

Despite its successes, phytoremediation is not without problems. Plants that accumulate toxic substances still have to be disposed of. And researchers are concerned that some toxins absorbed by plants can evaporate from leaves and contaminate the air. They also worry about animals eating toxin-laden plants. However, they are encouraged by studies indicating that at least some animals avoid plants containing high concentrations of toxins. (This work suggests why toxin accumulation may have evolved in the first place—as protection from hungry animals.) The biggest limitation of phytoremediation is its slowness. While soil can be hauled away in a matter of days, albeit at high cost, plants can take months to

Sunflower plants absorbing radioactive metals from a contaminated pond.

grow. Furthermore, remediating crops must often be planted for multiple seasons to reduce soil toxicity to acceptable levels.

In using plants to clean up toxic wastes, we are benefiting from millions of years of plant evolution. Unable to move about in search of food, plants have evolved amazing abilities to pull water and nutrients out of the soil and air. While arsenic and other toxins are not normally considered plant nutrients, it's obvious that some species have adapted to survive, and even thrive, in their presence. Researchers are still studying the brake fern's taste for arsenic. In this chapter, we'll see how plants obtain essential nutrients and how they transport them throughout their roots, stems, and leaves. ■ ■ ■

32.1 Plants acquire their nutrients from soil and air

Watch a plant grow from a tiny seed, and you can't help wondering where all the mass comes from. Aristotle thought that soil provided all the substance for plant growth, and 17th-century physician Jan Baptista van Helmont performed an experiment to test the hypothesis that plants grow by absorbing material from soil. He planted a willow seedling in a pot containing 91 kg of soil. After five years, the willow had grown into a tree weighing 76.8 kg, but only 0.06 kg of soil had disappeared from the pot. Van Helmont concluded that the willow had grown mainly from added water. A century later, an English physiologist named Stephen Hales postulated that plants are nourished mostly by air.

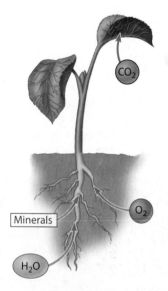

As it turns out, there is some truth in all these early ideas about plant nutrition; air, water, and soil all contribute to plant growth (Figure 32.1A). A plant's leaves absorb carbon dioxide (CO_2) from the air; in fact, about 96% of a plant's dry weight is organic (carbon-containing) material built mainly from CO_2. The figure also points out that a plant gets water (H_2O), minerals, and some oxygen (O_2) from the soil.

Figure 32.1A
The uptake of nutrients by a plant

What happens to the materials a plant takes up from the air and soil? The sugars a plant makes by photosynthesis are composed of the elements carbon, oxygen, and hydrogen. In Chapter 7, we saw that the carbon and oxygen used in photosynthesis come from atmospheric CO_2 and that the hydrogen comes from water molecules. Plant cells use the sugars made by photosynthesis in constructing all the other organic materials they need, primarily carbohydrates. The giant trunks of the redwood trees in **Figure 32.1B**, for instance, consist mainly of sugar derivatives, such as the cellulose of cell walls.

Plants use cellular respiration to break down some of the sugars they make, obtaining energy from them in a process that consumes O_2. A plant's leaves take up some O_2 from the air, but we do not show this in Figure 32.1A because plants are actually net producers of O_2, giving off more of this gas than they use. When water is split during photosynthesis, O_2 gas is produced and released through the leaves. The O_2 being taken up from the soil by the plant's roots in Figure 32.1A is actually atmospheric O_2 that has diffused into the soil; it is used in cellular respiration in the roots themselves.

What does a plant do with the minerals it absorbs from the soil? A look at three elements that plant roots take up as inorganic ions provides a partial answer. Nitrogen is a com-ponent of all nucleic acids, all proteins, ATP, and many plant hormones and coenzymes. Nitrogen and magnesium are both components of chlorophyll, the plant's key light-absorbing molecule. Phosphorus is a major component of nucleic acids, phospholipids, and ATP.

A plant's ability to move water from its roots to its leaves and its ability to deliver sugars to specific areas of its body are staggering feats of evolutionary engineering. Figure 32.1B highlights the distance between the bottom of a tree and its leaves; the roots of a redwood can be over 100 m (330 feet) below the topmost leaves! In the next four modules, we follow the movement of water, dissolved mineral nutrients, and sugar throughout the plant body.

? Plants require inorganic nutrients, which they acquire from the atmosphere in the form of _____ and from the soil in the form of _____ and _____.

■ carbon dioxide . . . water . . . minerals

Figure 32.1B Redwood trees, giant products of photosynthesis

32.2 The plasma membranes of root cells control solute uptake

With its surface area enormously expanded by thousands of root hairs (**Figure 32.2A**), a plant root has a remarkable ability to extract water and other materials from soil. Recall from Module 31.3 that root hairs are extensions of epidermal cells that cover the root. Root hairs form a huge surface area; for example, the root hairs of a single sunflower plant, if laid end to end, could stretch for miles. All this surface area in contact with nutrient-containing soil allows a plant to absorb the water and minerals it needs for growth.

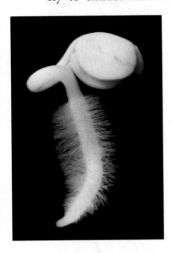

Figure 32.2A Root hairs of a radish seedling

All substances that enter a plant root are in solution. For water and solutes to be transported from the soil throughout the plant, they must move through the epidermis and cortex of the root and then into the water-conducting xylem tissue in the central cylinder of the root. Any route the water and solutes take from the soil to the xylem requires that they pass through some of the plasma membranes of the root cells. Because plasma membranes are selectively permeable, only certain solutes can enter the xylem. This selectivity helps regulate the mineral composition of a plant's vascular system.

You can see two possible routes to the xylem in the bottom part of **Figure 32.2B**. The blue arrows indicate an *intracellular* route. Water and selected solutes cross the cell wall and plasma membrane of an epidermal cell (usually at a root hair). The cells within the root are all interconnected by plasmodesmata (channels through the walls of adjacent cells); there is a continuum of living cytoplasm among the root cells. Therefore, once inside the epidermal cell, the solution can move inward from cell to cell without crossing any other plasma membranes, diffusing through the interconnected cytoplasm all the way into the root's endodermis. An endodermal cell then discharges the solution into the xylem (purple).

The red arrows indicate an alternative route. This route is *extracellular;* the solution moves inward within the hydrophilic walls and extracellular spaces of the root cells but does not enter the cytoplasm of the epidermis or cortex cells. The solution crosses no plasma membranes, and there is no selection of solutes until they reach the endodermis. Here, a continuous waxy barrier called the **Casparian strip** stops water and solutes from entering the xylem via

cell walls. The Casparian strip forces water and ions that travel the extracellular (red) route to cross a plasma membrane into an endodermal cell. Ion selection occurs at this membrane instead of in the epidermis, and once the selected solutes and water are in the endodermal cell, they can be discharged into the xylem.

Actually, water and solutes rarely follow just the two kinds of routes in Figure 32.2B. In a real plant, they may take any combination of these routes, and they may pass through numerous plasma membranes and cell walls en route to the xylem. Because of the Casparian strip, however, there are no nonselective routes; the water and solutes must cross a plasma membrane at some point. Next, we see how water and minerals move upward within the xylem from the roots to the shoots.

? What is the function of the Casparian strip?

■ It regulates the passage of minerals (inorganic ions) into the xylem by blocking access via cell walls and requiring all minerals to cross a selectively permeable plasma membrane.

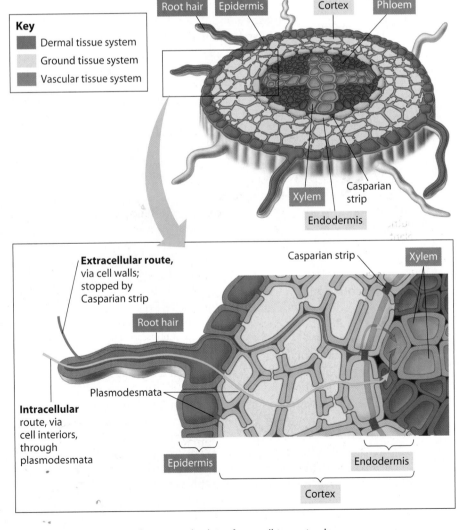

Key
■ Dermal tissue system
□ Ground tissue system
■ Vascular tissue system

Root hair Epidermis Cortex Phloem

Xylem Casparian strip
Endodermis

Extracellular route, via cell walls; stopped by Casparian strip

Casparian strip Xylem

Root hair

Plasmodesmata

Intracellular route, via cell interiors, through plasmodesmata

Epidermis Endodermis

Cortex

Figure 32.2B Routes of water and solutes from soil to root xylem

32.3 Transpiration pulls water up xylem vessels

As a plant grows upward toward sunlight, it needs to extract an increasing supply of water and dissolved mineral ions from the soil. To thrive, a plant must be able to transport these resources from its roots to the rest of the plant.

We saw in Figure 31.5E that xylem tissue of angiosperms includes two types of conducting cells: tracheids and vessel elements. When mature, both types of cells are dead, consisting only of cell walls, and both are in the form of very thin tubes that are arranged end to end. Because the cells have openings in their ends, a solution of water and inorganic nutrients, called **xylem sap**, can flow through these tubes. Xylem sap flows all the way up from the plant's roots through the shoot system to the tips of the leaves.

What force moves xylem sap up against the downward pull of gravity? Is it pushed or pulled upward? Plant biologists have found that the roots of some plants do exert a slight upward push on xylem sap. The root cells actively pump inorganic ions into the xylem, and the root's endodermis holds the ions there. As ions accumulate in the xylem, water tends to enter by osmosis, pushing xylem sap upward ahead of it. This force, called **root pressure**, can push xylem sap up a few meters.

For the most part, however, xylem sap is not pushed from below by root pressure but pulled upward by the leaves. Plant biologists have determined that the pulling force is **transpiration**, which is the loss of water from the leaves and other aerial parts of a plant.

Figure 32.3 illustrates transpiration and its effect on water movement in a tree. At the top right of the figure, water molecules (blue arrows) are shown leaving the leaf through a stoma, a microscopic pore on the surface of a leaf (see also Figures 31.6 and 32.4). When the stoma is open, water diffuses out of the leaves because the concentration of water molecules is higher in the spaces between cells inside the leaf than in the surrounding air. Under certain atmospheric conditions, this exiting moisture may condense to form dew.

Transpiration can pull xylem sap up the tree because of two special properties of water: cohesion and adhesion. Both of these properties arise from the polarity of water molecules (see Modules 2.9–2.11). **Cohesion** is the sticking together of molecules of the same kind. In the case of water, hydrogen bonds make the H_2O molecules stick to one another, as the circular enlargement in the figure shows. The cohering water molecules in the xylem tubes form continuous strings, extending all the way from the leaves down to the roots. In contrast, **adhesion** is the sticking together of molecules of different kinds. Water molecules tend to adhere via hydrogen bonds to hydrophilic cellulose molecules in the walls of xylem cells.

What effect does transpiration have on a string of water molecules that tend to adhere to the walls of xylem tubes? Before a water molecule can leave the leaf, it must break off from the end of the string. In effect, it is pulled off by a steep diffusion gradient between the moist interior of the leaf and the drier surrounding air. Cohesion resists the pulling force of the diffusion gradient, but it is not strong enough to overcome it. The molecule breaks off, and the opposing forces of cohesion and transpiration put tension on the remainder of the string of water molecules. As long as transpiration continues, the string is kept tense and is pulled upward as one molecule exits the leaf and the one right behind it is tugged up into its place. Adhesion of the water molecules to the walls of the xylem cells assists the upward movement of the xylem sap by counteracting the downward pull of gravity.

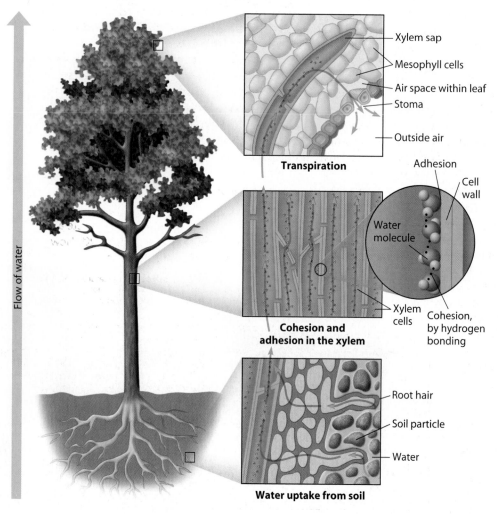

Flow of water

Xylem sap
Mesophyll cells
Air space within leaf
Stoma
Outside air
Transpiration

Adhesion
Cell wall
Water molecule
Xylem cells
Cohesion, by hydrogen bonding
Cohesion and adhesion in the xylem

Root hair
Soil particle
Water
Water uptake from soil

Figure 32.3 The flow of water up a tree

Plant biologists call this explanation for the ascent of xylem sap the **transpiration-cohesion-tension mechanism.** We can summarize it as follows: Transpiration exerts a pull that is relayed downward along a string of water molecules held together by cohesion and helped upward by adhesion. Transpiration is an efficient means of moving large volumes of water upward from roots to shoots. A single corn plant, for example, transpires 125 L of water in a growing season. And yet, all this transport of xylem sap requires no energy expenditure by the plant. Physical properties—cohesion, adhesion, and evaporation—move water and dissolved minerals from a plant's roots to its shoots.

Web/CD Thinking as a Scientist *How Is the Rate of Transpiration Calculated?*

? (a) Contrast cohesion and adhesion and (b) describe the role of each in the ascent of xylem sap.

■ (a) Cohesion is the sticking together of identical molecules—water molecules in the case of xylem sap. Adhesion is the sticking together of different kinds of molecules, as in the adhesion of water to the cellulose of xylem walls. (b) Cohesion maintains a continuous string of water during transpiration; adhesion helps to support the downward pull of gravity.

32.4 Guard cells control transpiration

Adaptations that increase photosynthesis—such as large leaf surface areas—have the serious drawback of increasing water loss by transpiration. Viewed this way, a plant's tremendous requirement for water is part of the cost of making food by photosynthesis. As long as water moves up from the soil fast enough to replace the water that is lost, transpiration presents no problem. But if the soil dries out and transpiration exceeds the delivery of water to the leaves, the leaves will wilt. Unless the soil and leaves are rehydrated, the plant will eventually die.

The leaf stomata, which can open and close, are adaptations that help plants regulate their water content and adjust to changing environmental conditions. As shown in **Figure 32.4**, a pair of guard cells flank each stoma. The guard cells control the opening of a stoma by changing shape, widening or narrowing the gap between the two cells.

What actually causes guard cells to change shape and thereby open or close stomata? Figure 32.4 illustrates the principle. A stoma opens (left) when its guard cells gain potassium ions (K^+, red dots) and water (blue arrows) from neighboring cells (shown in light gray). The cells actively take up K^+, and water then enters by osmosis. (For a review of osmosis, see Module 5.16.) When the vacuoles in the guard cells gain water, the cells become more turgid and bowed. The cell wall of a guard cell is not uniformly thick, as you can see in the drawing, and the cellulose molecules are oriented in such a way that the cell buckles away from its companion guard cell when it is turgid. The result is an increase in the size of the gap (stoma) between the two cells. Conversely, when the guard cells lose K^+, they also lose water by osmosis and become flaccid and less bowed, closing the space between them (right).

In general, guard cells keep the stomata open during the day and closed at night. During the day,

CO_2 can enter the leaf from the atmosphere and thus keep photosynthesis going when sunlight is available. At night, when there is no light for photosynthesis and therefore no need to take up CO_2, the closed stomata save water.

At least three cues contribute to stomatal opening at dawn. One is sunlight, which stimulates guard cells to accumulate K^+ and become turgid. A low level of CO_2 in the leaf can have the same effect. A third cue is an internal timing mechanism—a biological clock—found in the guard cells; even if you keep a plant in a dark closet, stomata will continue their daily rhythm of opening and closing. (We'll return to biological clocks in plants in Module 33.10.)

Even during the day, the guard cells may close the stomata if the plant is losing water too fast. This response reduces further water loss and may prevent wilting, but it also slows down CO_2 uptake and photosynthesis—one reason that droughts reduce crop yields. In summary, guard cells arbitrate the photosynthesis-transpiration compromise on a moment-to-moment basis by integrating a variety of stimuli.

Web/CD Activity 32A *Transpiration*

? Some leaf molds, fungi that parasitize plants, secrete a chemical that causes guard cells to accumulate K^+. How does this adaptation help the mold infect the plant?

■ Accumulation of K^+ by guard cells results in osmotic water uptake, and the turgid condition of the cells keeps the stomata open. The mold can then grow into the leaf interior via the stomata.

Figure 32.4 How guard cells control stomata

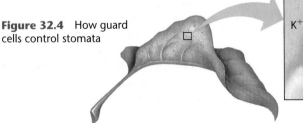

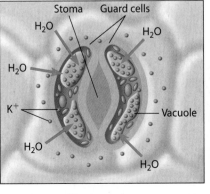

Stoma opening

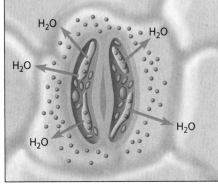

Stoma closing

32.5 Phloem transports sugars

A plant has two separate transport systems: xylem (the topic of Module 32.3) and phloem. Xylem transports xylem sap (water and dissolved minerals), while the main function of phloem is to transport the products of photosynthesis. In angiosperms, phloem contains food-conducting cells called sieve-tube members arranged end to end as tubes (see Figure 31.5F). The micrograph in **Figure 32.5A** shows two sieve-tube members and the sieve plate between them. Through the perforations in sieve plates, a sugary solution called **phloem sap** moves freely from one cell to the next (the cytoplasms of these living cells are continuous). Phloem sap may contain inorganic ions, amino acids, and hormones, but its main solute is usually the disaccharide sucrose. When highly concentrated by boiling, the sucrose-rich phloem sap produced by certain varieties of North American maple trees is better known as maple syrup.

In contrast to xylem sap, which only flows upward from the roots, phloem sap moves throughout the plant in various directions. However, sieve tubes always carry sugars from a sugar source to a sugar sink. A **sugar source** is a plant organ that is a net producer of sugar, by photosynthesis or by breakdown of starch. Mature leaves are the primary sugar sources. A **sugar sink** is an organ that is a net consumer or storer of sugar. Growing roots, buds, stems, and fruits are

sugar sinks. A storage organ, such as a tuber or a bulb, may be a source or sink, depending on the season. When stockpiling carbohydrates in the summer, it is a sugar sink. After breaking dormancy in the spring, it is a source as its starch is broken down to sugar, which is carried to the growing tips of the plant. Thus, each food-conducting tube in phloem tissue has a source end and a sink end, but these may change with the season or the developmental stage of the plant.

What causes phloem sap to flow from a sugar source to a sugar sink? Flow rates may be as high as 1 m/hr, which is much too fast to be accounted for by diffusion. (It would take phloem sap 8 years to travel a meter if it moved by diffusion alone.) Plant biologists have tested a number of hypotheses for phloem sap movement. A hypothesis called the **pressure flow mechanism** is now widely accepted for angiosperms. **Figure 32.5B**, on the facing page, illustrates how this works, using a beet plant as an example. The pink dots in the phloem tube represent sugar molecules; notice their concentration gradient from top to bottom. The blue color represents a parallel gradient of water (hydrostatic) pressure in the phloem sap.

At the sugar source (leaves, in this example), ❶ sugar is loaded into a phloem tube by active transport. Sugar loading at the source end raises the solute concentration inside the phloem tube. ❷ The high solute concentration draws water into the tube by osmosis. The inward flow of water raises the water pressure at the source end of the tube.

At the sugar sink (the beet root, in this case), both sugar and water leave the phloem tube. ❸ As sugar departs from the phloem, ❹ water follows by osmosis. The exit of sugar lowers the sugar concentration in the sink end; the exit of water lowers the hydrostatic pressure in the tube. The building of water pressure at the source end and the reduction of that pressure at the sink end cause water to flow from source to sink—down a gradient of hydrostatic pressure. Since the sugar is dissolved in the water and the sieve plates allow free movement of solutes as well as water, the sugar is carried along from source to sink at the same rate as the water. As indicated on the right side of Figure 32.5B, xylem tubes recycle the water back from sink to source.

The pressure flow mechanism explains why phloem sap always flows from a sugar source to a sugar sink, regardless of their locations in the plant. However, the mechanism is somewhat difficult to test because most experimental procedures disrupt the structure and function of the phloem tubes. Some of the most interesting studies have taken advantage of natural phloem probes: insects called aphids, which feed on phloem sap.

The three photographs in **Figure 32.5C** show how plant biologists have used aphids to study phloem sap. On the left, an aphid feeds by inserting its needlelike mouthpart, called a stylet, into the phloem of a tree branch. The aphid is releasing from its anus a drop of so-called honeydew—actually, a tiny amount of phloem sap lacking some solutes that the insect's digestive tract has removed for food. The

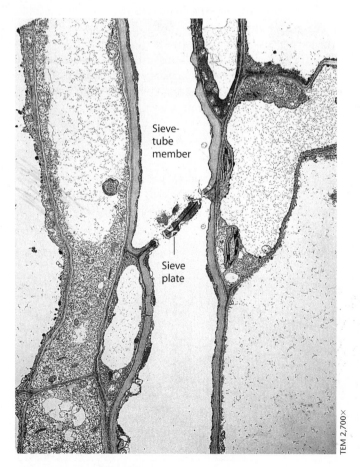

TEM 2,700×

Figure 32.5A Food-conducting cells of phloem

micrograph in the center shows an aphid's stylet inserted into one of the plant's food-conducting cells. The pressure within the phloem force-feeds the aphid, swelling it to several times its original size. While the aphid is feeding, it can be anesthetized and severed from its stylet. The stylet then serves the researcher as a miniature tap that drips phloem sap for hours. The photograph on the right shows a droplet of phloem sap on the cut end of a stylet. Studies using this technique support the pressure flow model: The closer the stylet is to a sugar source, the faster the sap flows out and the greater its sugar concentration. This is what we would expect if pressure is generated at the source end of the phloem tube by the active pumping of sugar into the tube.

We now have a broad picture of how a plant transports materials from one part of its body to another. Water and inorganic ions enter from the soil and are distributed by xylem. The xylem sap is pulled upward by transpiration. Carbon dioxide enters the plant through leaf stomata and is converted into sugars in the leaves. A second transport system, phloem, distributes the sugars. Pressure flow drives the phloem sap from leaves and storage sites to other parts of the plant, where the sugars are used or stored.

In Chapter 7, we discussed how plants convert raw materials into organic molecules by photosynthesis. We have yet to say much about the kinds of inorganic nutrients a plant needs and what it does with them. This is the subject of plant nutrition, which we discuss in the next section.

Web/CD Activity 32B *Transport in Phloem*

? Contrast the forces that move phloem sap with the forces that move xylem sap.

■ Pressure is generated at the source end of a sieve tube by the loading of sugar and the resulting osmotic flow of water into the phloem. This pressure pushes phloem sap from the source end to the sink end of the tube. In contrast, transpiration generates a pulling force that drives the ascent of xylem sap.

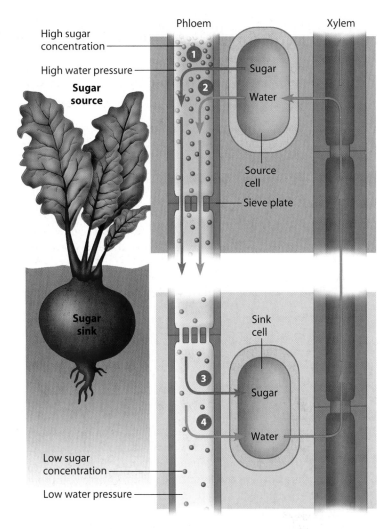

Figure 32.5B Pressure flow in plant phloem from a sugar source to a sugar sink (and the return of water to the source via xylem)

Aphid feeding on a small branch

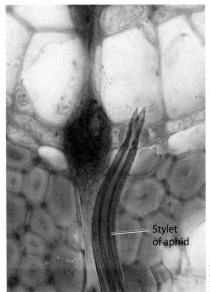

Aphid's stylet inserted into a phloem cell

Severed stylet dripping phloem sap

Figure 32.5C Tapping phloem sap with the help of an aphid

32.6 Plant health depends on a complete diet of essential inorganic nutrients

In contrast to animals, which require a complex diet of organic (carbon-containing) foods, plants survive and grow solely on inorganic substances (that is, plants are autotrophs; see Module 7.1). The ability of plants to assimilate CO_2 from the air, extract water and inorganic ions from the soil, and synthesize organic compounds is essential not only to the survival of plants but also to the survival of humans and other animals.

A chemical element is considered an **essential element** if a plant must obtain it to complete its life cycle—that is, to grow from a seed and produce another generation of seeds. A method called hydroponic culture can be used to determine which chemical elements are essential nutrients. As shown in **Figure 32.6**, a researcher grows plants without soil by bathing the roots in mineral solutions. Air is bubbled into the water to give the roots oxygen for cellular respiration. By omitting a particular element, such as potassium, from the medium, a researcher can test whether that element is essential to the plant.

If the element left out of the solution is an essential nutrient, then the incomplete medium will make the plant abnormal in appearance compared to control plants grown on a complete nutrient medium. The most common symptoms of a nutrient deficiency are stunted growth and discolored leaves. Hydroponic culture studies have helped identify 17

essential elements needed by all plants. Most research has involved crop plants and houseplants; little is known about the nutritional needs of uncultivated plants.

Of the 17 essential elements, nine are called **macronutrients** because plants require relatively large amounts of them. Six of the nine macronutrients—carbon, oxygen, hydrogen, nitrogen, sulfur, and phosphorus—are the major ingredients of organic compounds forming the structure of a plant. These six elements make up almost 98% of a plant's dry weight. The other three macronutrients—potassium, calcium, and magnesium—make up another 1.5%.

How does a plant use calcium, potassium, and magnesium? Calcium has several functions. For example, it is important in the formation of cell walls, and it combines with certain proteins to form a glue that holds plant cells together in tissues. Calcium also helps maintain the structure of cell membranes and helps regulate their selective permeability. Potassium is crucial as a cofactor required for the activity of several enzymes. (Recall from Module 5.7 that a cofactor is an atom or molecule that cooperates with an enzyme in catalyzing a reaction.) Potassium is also the main solute for osmotic regulation in plants; we saw in Module 32.4 how potassium ion movements regulate the opening and closing of stomata. Magnesium is a component of chlorophyll and thus essential for photosynthesis. Magnesium is also a cofactor for several enzymes.

Elements that plants need in very small amounts are called **micronutrients.** The eight known micronutrients are chlorine, iron, manganese, boron, zinc, copper, nickel, and molybdenum. Micronutrients function in plants mainly as cofactors. Iron, for example, is a component of cytochromes, proteins in the electron transport chains of chloroplasts and mitochondria. Because micronutrients function mainly in catalysis (and are therefore used over and over), plants need only minute quantities of these elements. The requirement for molybdenum, for example, is so modest that there is only one atom of this rare element for every 60 million atoms of hydrogen in dried plant material. Yet a deficiency of molybdenum or any other micronutrient can weaken or kill a plant.

Complete solution containing all minerals (control)

Solution lacking potassium (experimental)

Figure 32.6 A hydroponic culture experiment

? You conduct an experiment like the one in Figure 32.6 to test whether a certain plant species requires a particular chemical element as a micronutrient. Why is it important that the glassware be completely clean?

■ Because micronutrients are required in only minuscule amounts, even the smallest amount of dirt in the experimental flask may contain enough of the element you are testing to allow normal growth and invalidate your results.

32.7 You can diagnose some nutrient deficiencies in your own plants

The quality of soil—especially the availability of nutrients—determines the health of a growing plant and the quality of our own nutrition. Figure 32.7A shows two experimental corn crops. The plants on the left are growing in soil rich in nitrogen. The small, lighter-colored plants on the right are growing in nitrogen-deficient soil. Even if the nitrogen-deficient plants produce grain, the crop will have a lower nutritional value, and its nutrient deficiences will then be passed on to livestock or human consumers.

Fortunately, the symptoms of nutrient deficiency are often distinctive enough for a grower to diagnose its cause. Many growers make visual diagnoses and then check their conclusions by having soil and plant samples chemically analyzed at a state or local laboratory.

Figure 32.7B shows a healthy tomato plant for comparison with genetically identical plants (Figures 32.7C–32.7E) suffering from macronutrient deficiencies. Nitrogen shortage is the most common nutritional problem for plants. Soils are usually not deficient in nitrogen, but they are often deficient in the nitrogen compounds that plants can use: dissolved nitrate ions (NO_3^-) and ammonium ions (NH_4^-). Stunted growth and yellow-green leaves, starting at the tips of older leaves (Figure 32.7C), are signs of nitrogen deficiency.

Phosphorus deficiency is the second most common nutritional ailment in plants. As is the case with nitrogen, soils usually contain plenty of phosphorus, but not always in the ionic, water-soluble forms ($H_2PO_4^-$ or HPO_4^{2-}) that plants can use. A phosphorus-deficient plant may have green leaves, but its growth rate is markedly reduced, and its new growth is often spindly and brittle. Also, in some plants, such as the one in Figure 32.7D, phosphorus deficiency produces a purplish color on the undersides of the leaves.

Figure 32.7E shows potassium deficiency. Plants take up potassium as K^+ dissolved in soil water. Again, most potassium compounds in the soil are only slightly soluble in water and therefore not available to plants. The signs of a potassium shortage are generally more localized than those of nitrogen and phosphorus deficiencies. The older leaves usually show the most obvious signs; they often turn yellow and develop dead, brownish tissue at the edges or in spots. Stems and roots are also weakened, leading to stunting.

Once a diagnosis of a nutrient deficiency is made, treating the problem is usually simple. You can choose from a number of fertilizers, natural or manufactured, to enrich the soil. Many of these consist of the inorganic compounds plants can use directly, such as nitrates and phosphates; others contain organic materials that are broken down to the usable inorganic compounds by microbes in the soil.

? (a) What is the most common nutrient deficiency in plants? (b) What are the signs?

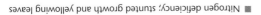
■ Nitrogen deficiency; stunted growth and yellowing leaves

Figure 32.7A The effect of nitrogen availability on corn growth: corn grown in nitrogen-rich soil (left) and nitrogen-poor soil (right)

Figure 32.7B
A healthy plant

Figure 32.7C
Nitrogen deficiency

Figure 32.7D
Phosphorus deficiency

Figure 32.7E
Potassium deficiency

32.8 Fertile soil supports plant growth

Along with climate, the major factor determining whether a plant can grow well in a particular location is the quality of the soil. Fertile soil can support abundant plant growth by providing conditions that enable plant roots to absorb water and dissolved nutrients.

Different layers of soil are visible in a road cut or deep hole, such as the cross section shown in **Figure 32.8A**. You can see three distinct soil layers, called horizons, in the cut. The A horizon, or **topsoil**, is subject to extensive weathering (freezing, drying, and erosion, for example). Topsoil is a mixture of rock particles of various sizes, living organisms, and **humus**, the remains of partially decayed organic material. The rock particles provide a large surface area that retains water and minerals while also forming air spaces containing oxygen that can diffuse into plant roots. Fertile topsoil is home to an astonishing number and variety of bacteria, protozoans, fungi, and small animals such as earthworms, roundworms, and burrowing insects. Along with plant roots, these organisms loosen and aerate the soil and contribute organic matter to the soil as they live and die. Nearly all plants depend on bacteria and fungi in the soil to break down organic matter into inorganic molecules that roots can absorb. Besides providing nutrients, humus also tends to retain water while keeping the topsoil porous enough for good aeration of the plant roots. Topsoil is rich in organic materials and is therefore most important for plant growth. Plant roots branch out in the A horizon and usually extend into the next layer, the B horizon.

The soil's B horizon contains many fewer organisms and much less organic matter than the topsoil and is less subject to weathering. Fine clay particles and nutrients dissolved in soil water drain down from the topsoil and often accumulate in the B horizon. Below the B horizon, the C horizon is composed mainly of partially broken-down rock that serves at the "parent" material for the upper layers of soil.

Figure 32.8B illustrates the intimate association between a plant's root hairs, soil water, and the tiny particles of topsoil. The root hairs are in direct contact with the water that surrounds the particles. The soil water is not pure but a solution containing dissolved inorganic ions. Oxygen diffuses

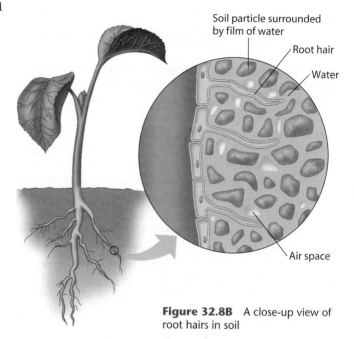

Figure 32.8B A close-up view of root hairs in soil

into the water from small air spaces in the soil. Roots absorb this soil solution.

Cation exchange is a mechanism by which root hairs take up certain positively charged ions (cations). Inorganic cations—such as calcium (Ca^{2+}), magnesium (Mg^{2+}), and potassium (K^+)—adhere by electrical attraction to the negatively charged surfaces of soil particles. This adherence helps prevent these positively charged nutrients from draining away during heavy rain or irrigation. In cation exchange (**Figure 32.8C**), root hairs release hydrogen ions (H^+) into the soil solution. The hydrogen ions displace cations on the clay particle surfaces, and root hairs can then absorb them.

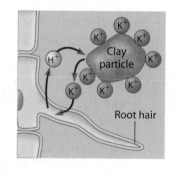

Figure 32.8C Cation exchange

In contrast to cations, negative ions (anions)—such as nitrate (NO_3^-)—are usually not bound tightly by soil particles. Unbound ions are readily available to plants, but they tend to drain out of the soil quickly. If they do, the soil may become deficient in nitrogen.

It may take centuries for a soil to become fertile through the breakdown of rock and the accumulation of organic material. The loss of soil fertility is one of our most pressing environmental problems, as we discuss next.

Web/CD Activity 32C *Absorption of Nutrients from Soil*

Web/CD Thinking as a Scientist *Connection: How Does Acid Precipitation Affect Mineral Deficiency?*

? How do roots actively increase the availability of mineral nutrients that are cations?

Figure 32.8A Three soil horizons in a Tennessee cotton field

■ By secreting hydrogen ions, which displace cations from soil particles

32.9 Soil conservation is essential to human life

Our survival as a species depends on soil, and yet erosion and chemical pollution threaten this vital resource throughout the world. As the human population continues to grow and more and more land is cultivated, farming practices that conserve soil fertility will become essential to our survival. Three critical goals of soil conservation are proper irrigation, prevention of erosion, and prudent fertilization.

Irrigation can turn a desert into a garden, but farming in dry regions is a huge drain on water resources. Additionally, irrigation can gradually make the soil salty. The whitish deposits on the soil in the photograph in **Figure 32.9A** are salts that were dissolved in irrigation water flooded onto a field. Left behind when the excess water evaporated, the deposits will eventually make the soil too salty for crop plants to tolerate. Instead of flooding fields, modern irrigation often employs perforated pipes that drip water slowly into the soil close to plant roots. This drip irrigation uses less water, allows the plants to absorb most of the water, and reduces water loss from evaporation and drainage.

Preventing erosion—the blowing or washing away of soil—is one of the most important challenges of modern agriculture. Thousands of acres of farmland are lost to water and wind erosion each year in the United States alone. Precautions such as planting rows of trees as windbreaks, terracing hillside crops, and cultivating in a contour pattern can prevent loss of topsoil. The crops in **Figure 32.9B** are planted in rows that go around, rather than up and down, the hill in the field. This contour tillage helps slow the runoff of water and topsoil after heavy rains. Crops such as alfalfa and wheat provide good ground cover and protect the soil better than other crops that are usually planted in more widely spaced rows.

Prehistoric farmers may have started fertilizing their fields after noticing that grass grew faster and greener where animals had defecated. In developed nations today, most farmers use commercially produced fertilizers containing minerals that are either mined or prepared by industrial processes. These fertilizers are usually enriched in nitrogen, phosphorus, and potassium, the macronutrients most commonly deficient in farm and garden soils.

Figure 32.9B Planting to prevent soil erosion in a hilly area

Manure, fishmeal, and compost (decaying plant matter) are called "organic" fertilizers because they are of biological origin and contain decomposing organic material. Before the nutrients in these substances can be used by plants, the organic material must be broken down by bacteria and fungi to inorganic nutrients that roots can absorb. Whether from organic fertilizer or a chemical factory, the minerals a plant extracts from the soil are in the same form. The difference is that organic fertilizers release nutrients gradually, whereas minerals in inorganic commercial fertilizers are available immediately. However, because they are soluble in water, minerals from inorganic fertilizers may not be retained in the soil for long. Problems arise when fields are overfertilized with inorganic products and excess nutrients are not taken up by plants. The excess minerals are often leached from the soil by rainwater or irrigation. This mineral runoff may pollute groundwater, streams, and lakes.

Agricultural researchers are developing ways to maintain crop yields while reducing fertilizer use. One approach is to genetically engineer "smart" plants that inform the grower when a nutrient deficiency is imminent but *before* damage has occurred. One such plant contains a reporter gene that leads to the production of a blue pigment in leaf cells when the phosphorus content of plant tissues declines. When leaves of these smart plants develop a blue tinge, the farmer knows it is time to add phosphorus-containing fertilizer.

? Why do organic fertilizers generally contaminate water resources less than inorganic fertilizers?

■ Organic fertilizers release mineral nutrients gradually as they decompose, so there is less likelihood of the minerals leaching into the groundwater or running off into streams and lakes.

Figure 32.9A Flood irrigation

32.10 Organic farmers must follow ecological principles

If you find tomatoes at a local farmers' market labeled "organic" and also find tomatoes at a giant grocery store chain marked the same way, can you be sure they were both *grown* the same way? What does an "organic" label mean? As of 2002, all products sold as "organic" in the United States must be grown and processed according to strict guidelines. These are intended to promote goals such as protecting biological diversity, maintaining and replenishing soil fertility (as by crop rotation), managing pests without pesticides (for example, providing habitat for predators and parasites of crop pests), avoiding genetically modified organisms, and using few or no synthetic fertilizers and pesticides.

To ensure compliance with these guidelines, all organic farmers must have their operations certified by groups accredited by the U.S. Department of Agriculture. Yearly inspections ensure proper organic farming practices, accurate record keeping, and a buffer of land between organic farms and neighboring conventional farms.

Farmers in the United States have dedicated over 2 million acres to organic farming (Figure 32.10). Organic farming is one of the fastest growing segments of agriculture; U.S. organic farmland acreage quadrupled during the 1990s alone.

The ultimate aim of many organic farmers is to restore as much to the soil as is drawn from it, creating fields that are bountiful and self-sustaining. Many have chosen organic farming to both protect the environment and answer the growing demand for more naturally produced foods. The benefits of organic farming are clear: fewer synthetic chemicals in the environment and less risk of exposing farmworkers and wildlife to potential toxins. And since organic fruits and vegetables are often picked when ripe and sold locally, rather than treated with preservatives and shipped long distances, they can be fresher and better tasting than conventional produce.

But while organic farming is spreading, it hasn't replaced conventional agriculture: Only about 0.3% of U.S. cropland is certified organic, and only about 2% of the U.S. food supply

Figure 32.10 An organic farmer harvesting peppers

is grown using organic methods. Furthermore, an organic label is no guarantee of safety or extra health benefits. Scientists disagree, for example, about the nutritional differences, if any, between organic and conventional produce.

Still, organic farmers continue to improve their practices. Some are trying new growing methods that promote greater biological diversity among the plants and wildlife in their fields. Others are looking for better natural fertilizers that increase crop yields. The future of farming, they say, lies in working toward two goals simultaneously: feeding the world's people and promoting a healthy environment.

? If you buy some "organic" apples, does that tell you anything about how they were grown?

■ An organic label indicates that the grower has been certified to be following standards meant to promote long-term ecological principles.

32.11 Agricultural research is improving the yields and nutritional values of crops

Either by choice or by economic necessity, the majority of people in the world have a predominantly vegetarian diet. Thus, particularly in developing countries, people depend mainly on plants for protein. Unfortunately, many plants have a low protein content, and the proteins they have may be deficient in one or more amino acids that humans need from their diet. Protein deficiency is thus the most common form of malnutrition among humans.

Improving the quality and quantity of proteins in crops is a major goal of agricultural researchers. Plant breeding has

resulted in new varieties of corn, wheat, and rice that are enriched in protein (Figure 32.11). However, many of these "super" varieties have an extraordinary demand for nitrogen, usually supplied in the form of commercial fertilizer. Unfortunately, these fertilizers are expensive to produce. Thus, the countries that most need high-protein crops are usually the ones least able to afford to grow them.

As discussed in Chapter 12, genetic engineering is already helping improve crops. New genetically modified (GM) varieties include cotton plants engineered to resist viruses and

Figure 32.11 Plant researchers with "super" rice

potato plants engineered to produce their own insecticide, making them resistant to attack by beetles that can destroy whole crops. In addition, tomato plants have been genetically modified to produce fruit that is slow to spoil. A major goal of agricultural scientists working in this area is to create plants that provide more nutritious food, such as the vitamin A–rich "golden rice" described in Module 12.18. Eventually,

researchers hope to develop grains that have a full complement of the amino acids humans need to make proteins.

Genetic engineering holds great potential for increasing agricultural production. There are potential problems, however. Crop plants containing genes that help the plant resist herbicides or diseases might, for example, escape into the wild and overgrow native species. Governments around the world are grappling with how to proceed: whether to promote a technological agricultural revolution or slow its progress until more information is available about the potential hazards. We will cover the subject of agricultural productivity again in Chapter 33, where we discuss plant hormones.

Another strategy that could potentially increase protein yields in crops is to increase the activity of organisms that form symbiotic relationships with plants, the focus of the next section.

Web/CD Activity 32D *Connection: Genetic Engineering of Golden Rice*

? Why is research on the protein content of crop plants so important to human health worldwide?

■ Because the most common form of malnutrition is protein deficiency, and most people in the world get most of their protein from plants

---

32.12 Fungi help most plants absorb nutrients from the soil

Reliance on the soil for nutrients that may be in short supply makes it imperative that the roots of plants have a large absorptive surface area. As we have seen, root hairs add a great deal of surface to plant roots. Most plants gain even more absorptive surface by teaming up with fungi.

The micrograph in Figure 32.12 shows a root of a eucalyptus tree. The root is covered with a twisted mat of fungal filaments. Together, the roots and the fungi form a mutually beneficial structure called a **mycorrhiza** ("fungus root"). The fungi benefit from a steady supply of sugar supplied by the host plant. In return, the fungi increase the surface area for water uptake and selectively absorb phosphate and other minerals from the soil and supply them to the plant. The fungi of mycorrhizae also secrete growth factors that stimulate roots to grow and branch, as well as antibiotics that may protect the plant from pathogens in the soil.

The plant-fungus symbiosis of mycorrhizae might have been one of the evolutionary adaptations that made it possible for plants to colonize land. Indeed, fossilized roots from some of the earliest plants include mycorrhizae. When terrestrial ecosystems were young, the soil was probably not very rich in nutrients. The fungi of mycorrhizae, which are more efficient at absorbing minerals than the roots, would have helped nourish the pioneering plants. Even today, the plants that first become established on nutrient-poor soils, such as abandoned farmland or eroded hillsides, are usually heavily colonized with mycorrhizae.

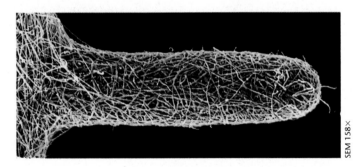

SEM 158×

Figure 32.12 A mycorrhiza on a eucalyptus root

However, roots can be transformed into mycorrhizae only if they are exposed to the appropriate species of fungus. For example, if seeds are collected in one environment and planted in foreign soil, the plants may show signs of malnutrition resulting from the absence of the plants' natural mycorrhizal partners. Farmers may avoid this problem by inoculating seeds with spores of mycorrhizal fungi. Plants also form symbiotic relationships with other organisms, as we see next.

? What are mycorrhizae?

■ Symbiotic associations of roots and fungi

32.13 Most plants depend on bacteria to supply nitrogen

As discussed in Module 32.7, nitrogen deficiency is the most common nutritional problem in plants. The atmosphere is nearly 80% nitrogen. But atmospheric nitrogen is gaseous N_2, a form that plants cannot use. For plants to absorb nitrogen, it must first be converted to ammonium (NH_4^+) or nitrate (NO_3^-).

In contrast to other minerals, the NH_4^+ and NO_3^- in soil are not derived from the breakdown of rock. Instead, ammonium and nitrate in the soil are produced from atmospheric N_2 or from organic matter by bacteria. As shown in **Figure 32.13**, certain soil bacteria, called nitrogen-fixing bacteria, convert atmospheric N_2 to ammonia (NH_3), a metabolic process called **nitrogen fixation**. In soil, ammonia picks up another H^+ to form an ammonium ion (NH_4^+). A second group of bacteria, called ammonifying bacteria, adds to the soil's supply of ammonium by decomposing organic matter (humus).

Plant roots can absorb nitrogen as ammonium. However, plants acquire their nitrogen mainly in the form of nitrate (NO_3^-), which is produced in the soil by a third group of soil bacteria called nitrifying bacteria. After nitrate is absorbed by roots, plant enzymes convert the nitrate back into ammonium, which is then incorporated into amino acids. Amino acids are used to make proteins and other nitrogen-containing organic molecules.

? What is the danger in applying a bactericide to the soil around plants?

■ The bactericide might kill soil bacteria that make nitrogen available to plants, causing nitrogen deficiency.

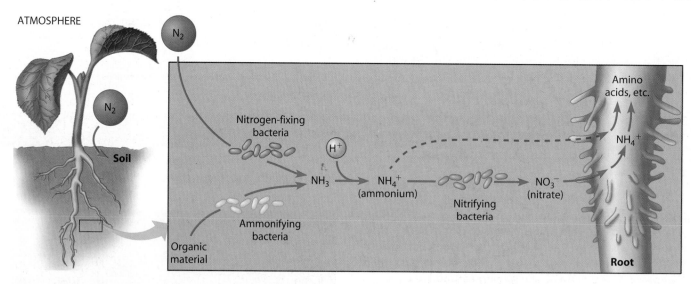

Figure 32.13 The roles of bacteria in supplying nitrogen to plants

32.14 Legumes and certain other plants house nitrogen-fixing bacteria

Symbiotic relationships with nitrogen-fixing bacteria provide some plant species with a built-in source of ammonium. For example, the roots of plants in the legume family—including peas, beans, peanuts, alfalfa, and many other plants that produce their seeds in pods—have swellings called **nodules** (Figure 32.14A). Within these nodules, plant cells have been "infected" by nitrogen-fixing bacteria of the genus *Rhizobium* ("root living"). *Rhizobium* bacteria reside in cytoplasmic vesicles formed by the root cell.

The micrograph at the top of the facing page (Figure 32.14B) shows a cross section of one such cell; notice the vesicles full of bacteria. Each

legume is associated with a particular strain of *Rhizobium*. Other nitrogen-fixing bacteria (actinomycetes) are found in the root nodules of some plants that are not legumes, such as alders.

The relationship between a plant and its nitrogen-fixing bacteria is mutually beneficial. The plant provides the bacteria with carbohydrates and other organic compounds. The bacteria have enzymes that catalyze the conversion of atmospheric N_2 to ammonium ions (NH_4^+), a form readily used by the plant. When conditions are favorable, root nodule bacteria fix so much nitrogen that the nodules secrete excess NH_4^+, which increases the fertility of the soil. This is one

Figure 32.14A Root nodules on a pea plant

— Shoot

— Nodules

— Roots

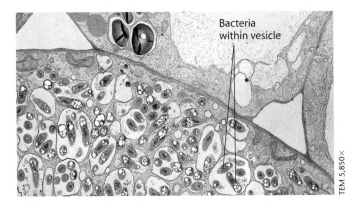

Bacteria
within vesicle

TEM 5,850×

Figure 32.14B Bacteria within a root nodule cell

reason farmers practice crop rotation, one year planting a nonlegume, such as corn, and the next year planting a legume, such as alfalfa. The legume crop may be plowed under so that it will decompose as "green manure," reducing the need for manufactured fertilizer.

> **?** How do the nitrogen-fixing bacteria of root nodules benefit from their symbiotic relationship with plants?

■ The bacteria, which do not photosynthesize, depend on the host plant for a supply of certain organic compounds produced by the photosynthesizing plant.

32.15 The plant kingdom includes parasites and carnivores

The mutualistic associations of plants with fungi (in mycorrhizae) and bacteria (in nodules) are not the only kinds of symbiosis that can increase a plant's nutrient supply and chances of survival. Some plants have evolved ways of obtaining food from other plants or animals. Figure 32.15A shows a parasitic plant called dodder (the yellow-orange threads wound around the green plant). Dodder cannot photosynthesize; it obtains organic molecules from other plant species, using specialized roots that tap into the host's vascular tissue.

Figure 32.15B shows part of an oak tree parasitized by mistletoe, the plant traditionally tacked above doorways during the Christmas season. All the leaves you see here are mistletoe; the oak has lost its leaves for winter. Mistletoe is photosynthetic, but it supplements its diet by siphoning sap from the vascular tissue of the host tree.

Certain plants are carnivorous, obtaining some of their nutrients, especially nitrogen, by killing and digesting insects and other small animals. Carnivorous plants grow in habitats (such as acid bogs) where soils are poor in nitrogen and other minerals. Organic matter decays so slowly in acidic soils that there is little inorganic nitrogen available for plant roots to take up. Though they are photosynthetic, the sundew and Venus' flytrap (Figures 32.15C and 32.15D) thrive by obtaining their nitrogen from insects.

Few species illustrate the correlation of structure and function better than carnivorous plants. The sundew plant (Figure 32.15C) has modified leaves, each bearing many club-shaped hairs. A sticky, sugary secretion at the tips of the hairs attracts insects and traps them. The presence of an insect triggers the hairs to bend and the leaf to enfold its prey. The hairs then secrete digestive enzymes, and the plant absorbs nutrients released as the insect is digested.

The Venus' flytrap (Figure 32.15D) has hinged leaves that close around small insects, usually ants and grasshoppers. As insects walk on the insides of these leaves, they touch sensory hairs that trigger closure of the trap. The leaf then secretes digestive enzymes and absorbs nutrients from the prey.

Using insects as a source of nitrogen is a nutritional adaptation that enables carnivorous plants to thrive in soils

Figure 32.15A Dodder growing on a pickleweed

Figure 32.15B Mistletoe growing on an oak

Figure 32.15C A sundew plant trapping a damselfly

Figure 32.15D A Venus' flytrap digesting a katydid

where most other plants cannot. Fortunately for animals, such predator-prey turnabouts are rare!

> **?** Carnivorous plants are most common in locales where the soil is deficient in _____.

■ nitrogen

CHAPTER REVIEW

Reviewing the Concepts

The Uptake and Transport of Plant Nutrients (Introduction–32.5)

Water and nutrient uptake. In phytoremediation, certain plant species that absorb toxic substances are used to help clean up polluted soil and groundwater **(Introduction)**. As a plant grows, its roots absorb water, minerals (inorganic ions), and some O_2 from the soil. Its leaves take carbon dioxide from the air **(32.1)**. Root hairs greatly increase a root's absorptive surface. Water and solutes can move through the root's epidermis and cortex by going either through cells or between them. However, all water and solutes must pass through the selectively permeable plasma membranes of cells of the endodermis to enter the xylem (water-conducting tissue) for transport upward **(32.2)**.

Transport of water, minerals, and sugar. Transpiration can move xylem sap, consisting of water and dissolved inorganic nutrients, to the top of the tallest tree **(32.3–32.4)**:

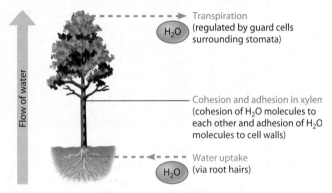

Transpiration (regulated by guard cells surrounding stomata)

Cohesion and adhesion in xylem (cohesion of H_2O molecules to each other and adhesion of H_2O molecules to cell walls)

Water uptake (via root hairs)

Phloem transports food molecules made by photosynthesis by a pressure flow mechanism. At a sugar source, sugar is loaded into a phloem tube. The sugar raises the solute concentration in the tube, and water follows, raising the pressure in the tube. As sugar is removed at a sugar sink, water follows. The increase in pressure at the sugar source and decrease at the sugar sink causes phloem sap to flow from source to sink **(32.5)**.

Plant Nutrients and the Soil (32.6–32.11)

Plant nutrition. A plant must obtain usable sources of the chemical elements—"nutrients"—it requires from its surroundings. Macronutrients, such as carbon and nitrogen, are needed in large amounts, mostly to build organic molecules. Micronutrients, including iron and zinc, act mainly as cofactors of enzymes **(32.6)**. Stunting, wilting, and color changes indicate nutrient deficiencies **(32.7)**.

Fertile soil contains a mixture of small rock and clay particles that hold water and ions and also allow O_2 to diffuse into plant roots. Humus (decaying organic material) provides nutrients and supports the growth of organisms that enhance soil fertility. Anions (negatively charged ions), such as nitrate (NO_3^-), are readily available to plants because they are not bound to soil particles. However, anions tend to drain out of soil rapidly. Cations (positively charged

ions), such as K^+, adhere to soil particles. In cation exchange, root hairs release H^+ ions, which displace cations from soil particles; the root hairs then absorb the free cations **(32.8)**. Water-conserving irrigation, erosion control, and the prudent use of herbicides and fertilizers are aspects of good soil management **(32.9)**.

Nutrition and agriculture. Organic farmers are certified to follow ecologically sound practices **(32.10)**. Through traditional plant breeding and DNA technologies, researchers are developing new varieties of crop plants with improved yields and nutritional value **(32.11)**.

Plant Nutrition and Symbiosis (32.12-32.15)

Relationships with other organisms help plants obtain nutrients. Many plants form mycorrhizae, mutually beneficial associations with fungi. A network of fungal threads increases a plant's absorption of nutrients and water, and the fungus receives some nutrients from the plant **(32.12)**. Bacteria in the soil convert atmospheric N_2 to forms that can be used by plants **(32.13)**:

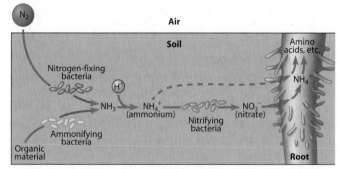

Legumes and certain other plants have nodules in their roots that house nitrogen-fixing bacteria **(32.14)**. Parasitic plants siphon sap from host plants. Carnivorous plants can obtain nitrogen by digesting insects **(32.15)**.

Connecting the Concepts

1. Fill in the blanks in this concept map to help you tie together key concepts concerning transport in plants.

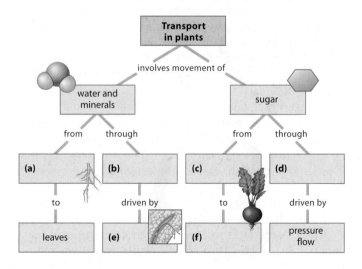

Testing Your Knowledge

Multiple Choice

2. Plants require the smallest amount of which of the following nutrients?
   a. oxygen
   d. iron
   b. phosphorus
   e. hydrogen
   c. carbon

3. Which of the following activities of soil bacteria does *not* contribute to creating usable nitrogen supplies for plant use?
   a. the fixation of atmospheric nitrogen
   b. the conversion of ammonium ions to nitrate ions
   c. the decomposition of dead animals
   d. the assembly of amino acids into proteins
   e. the generation of ammonium from proteins in dead leaves

4. By trapping insects, carnivorous plants obtain ____, which they need ____. (*Pick the best answer.*)
   a. water . . . because they live in dry soil
   b. nitrogen . . . to make sugar
   c. phosphorus . . . to make protein
   d. sugars . . . because they can't make enough by photosynthesis
   e. nitrogen . . . to make protein

5. A major long-term problem resulting from flood irrigation is the
   a. drowning of crop plants
   b. accumulation of salts in the soil
   c. erosion of fine soil particles
   d. encroachment of water-consuming weeds
   e. excessive cooling of the soil

Describing, Comparing, and Explaining

6 Explain how guard cells limit water loss from a plant on a hot, dry day. How can this be harmful to the plant?

Applying the Concepts

7. Acid rain contains an excess of hydrogen ions (H^+). One effect of acid rain is to deplete the soil of plant nutrients such as calcium (Ca^{2+}), potassium (K^+), and magnesium (Mg^{2+}). Offer a hypothesis to explain why acid rain washes these nutrients from the soil. How might you test your hypothesis?

8. In some situations, the application of nitrogen fertilizer to crops has to be increased each year because the fertilizer decreases the rate of nitrogen fixation in the soil. Propose a hypothesis to explain this phenomenon. Describe a test for your hypothesis. What results would you expect from your test?

9. Transpiration is fastest when humidity is low and temperature is high, but in some plants it seems to increase in response to light as well. During one 12-hour period when cloud cover and light intensity varied frequently, a scientist studying a certain crop plant recorded the data in the table below. (The transpiration rates are grams of water per square meter of leaf area per hour.)

Time (hr)	Temperature (°C)	Humidity (%)	Light (% of full sun)	Transpiration Rate (g/m²/hr)
8 A.M.	14	88	22	57
9	14	82	27	72
10	21	86	58	83
11	26	78	35	125
12 P.M.	27	78	88	161
1	33	65	75	199
2	31	61	50	186
3	30	70	24	107
4	29	69	50	137
5	22	75	45	87
6	18	80	24	78
7	13	91	8	45

Do these data support the hypothesis that the plants transpire more when the light is more intense? If so, is the effect independent of temperature and humidity? Explain your answer. (*Hint:* Look for overall trends in each column, and then compare pairs of data within each column and between columns.)

10. Agriculture is by far the biggest user of water in arid western states, including Colorado, Arizona, and California. The populations of these states are growing, and there is an ongoing conflict between cities and farm regions over water. To ensure water supplies for urban growth, cities are purchasing water rights from farmers. This is often the least expensive way for a city to obtain more water, and some farmers can make more money selling water than growing crops. Discuss the possible consequences of this trend. Is this the best way to allocate water for all concerned? Why or why not?

Answers to all questions can be found in Appendix 3.

For study help and Activities, go to campbellbiology.com or the student CD-ROM.

What Are the Health Benefits of Soy?

PLANT HORMONES

33.1 Experiments on how plants turn toward light led to the discovery of a plant hormone
33.2 Five major types of hormones regulate plant growth and development
33.3 Auxin stimulates the elongation of cells in young shoots
33.4 Cytokinins stimulate cell division
33.5 Gibberellins affect stem elongation and have numerous other effects
33.6 Abscisic acid inhibits many plant processes
33.7 Ethylene triggers fruit ripening and other aging processes
33.8 Plant hormones have many agricultural uses

GROWTH RESPONSES AND BIOLOGICAL RHYTHMS IN PLANTS

33.9 Tropisms orient plant growth toward or away from environmental stimuli
33.10 Plants have internal clocks
33.11 Plants mark the seasons by measuring photoperiod
33.12 Phytochrome is a light detector that may help set the biological clock
33.13 Joanne Chory studies the effects of light and hormones in the model plant *Arabidopsis*

PLANT DEFENSES

33.14 Defenses against herbivores and infectious microbes have evolved in plants
33.15 Plant biochemist Eloy Rodriguez studies how animals use defensive chemicals made by plants

Estrogen (Estradiol)   Phytoestrogen (Genistein)

Chemical structures of a human estrogen and a plant phytoestrogen

AMERICANS ARE DISCOVERING SOY. In fact, soy product sales have almost quadrupled in the last decade. Since soy offers a number of dietary benefits, this is a positive trend. Soy protein is one of the few plant proteins that contain all the essential amino acids, making it a healthy meat substitute. In addition, the FDA now allows soy food labels to claim that 25 g of soy protein per day (for example, 2.5 glasses of soy milk or 8 ounces of tofu) "may reduce the risk of heart disease." Soy is rich in antioxidants and fiber, low in fat, and has been shown to lower levels of LDL ("bad cholesterol") and triglycerides while maintaining HDL ("good cholesterol"). Not bad for a plant that until recently was mostly used for animal feed! Many supermarkets now carry soy foods such as soy milk (ground soybeans mixed with water and flavorings), tofu (cooked pureed soybeans formed into cakes), soy flour (which lacks gluten and so must be mixed with wheat flour for baking), and miso (fermented soybean paste used for seasoning).

When we eat plants, we eat plant hormones

While the nutritional benefits of soybeans are not surprising, people are usually less aware that soybeans also contain non-nutritive phytochemicals (literally, "plant chemicals") that may have significant metabolic effects on the human body. As you will learn in this chapter, plants, like humans, use hormones as chemical signals that control growth and development. It is only natural, then, that when we eat plants we consume plant hormones.

Phytoestrogens, a class of plant hormones, are found in soy. Their chemical structure is similar to the human female sex hormone estrogen (figure at left), allowing them to bind to estrogen receptors on human cells. One type of phytoestrogen, the isoflavones, help regulate the soy plant's growth and appear to exert a weak hormonal effect on the human body as well. In menopausal women (whose ovaries greatly curtail estrogen production), isoflavones may help reduce the negative effects of lower estrogen production, such as hot flashes and the risk of osteoporosis. So some women choose dietary supplements with isoflavones instead of hormone replacement

Control Systems in Plants

therapy (HRT), which often contains estrogen isolated from the urine of pregnant mares.

However, the health benefits of isoflavones are still being investigated. The strongest evidence comes from epidemiological studies (studies of the incidence and distribution of health-related problems in various populations). For example, women in China, who generally consume high levels of soy, have lower incidence of hot flashes and fewer hip fractures compared with Chinese women living in the West, who consume less soy. Controlled clinical trials, however, have shown only small effects.

While most health professionals agree that soy is a good addition to any diet, all the benefits and risks of isolated isoflavones have not been established. The metabolism of estrogen and the related phytoestrogens involves risks and benefits that are dose related. For example, while moderate levels of estrogen relieve menopausal symptoms, high levels appear to increase the risk of breast cancer. It is also hard for women to know if they are receiving beneficial quantities of isoflavones because amounts may vary widely among soy products. Soy flour, for example, contains over 15 times more isoflavones per weight than soy milk. In addition, dietary supplements are not subject to strict federal regulation, so women may be placing themselves at risk of too high a dose. Another area of potential concern is the use of soy protein–based infant formulas. Too much soy consumed by such a small person

might introduce unhealthy levels of phytoestrogens. Scientists will continue to explore the potential health risks and benefits of isoflavones and other plant hormones.

The effects of plant hormones on humans may not be fully known, but we do know that hormones are crucial to the life of plants. In this chapter, we explore the diverse roles of plant hormones: how they affect plant movement, growth, flowering, fruit development, and even defense. ■ ■ ■

Soybeans

Control Systems in Plants

therapy (HRT), which often contains estrogen isolated from the urine of pregnant mares.

However, the health benefits of isoflavones are still being investigated. The strongest evidence comes from epidemiological studies (studies of the incidence and distribution of health-related problems in various populations). For example, women in China, who generally consume high levels of soy, have lower incidence of hot flashes and fewer hip fractures compared with Chinese women living in the West, who consume less soy. Controlled clinical trials, however, have shown only small effects.

While most health professionals agree that soy is a good addition to any diet, all the benefits and risks of isolated isoflavones have not been established. The metabolism of estrogen and the related phytoestrogens involves risks and benefits that are dose related. For example, while moderate levels of estrogen relieve menopausal symptoms, high levels appear to increase the risk of breast cancer. It is also hard for women to know if they are receiving beneficial quantities of isoflavones because amounts may vary widely among soy products. Soy flour, for example, contains over 15 times more isoflavones per weight than soy milk. In addition, dietary supplements are not subject to strict federal regulation, so women may be placing themselves at risk of too high a dose. Another area of potential concern is the use of soy protein–based infant formulas. Too much soy consumed by such a small person

might introduce unhealthy levels of phytoestrogens. Scientists will continue to explore the potential health risks and benefits of isoflavones and other plant hormones.

The effects of plant hormones on humans may not be fully known, but we do know that hormones are crucial to the life of plants. In this chapter, we explore the diverse roles of plant hormones: how they affect plant movement, growth, flowering, fruit development, and even defense. ■ ■ ■

Soybeans

33.1 Experiments on how plants turn toward light led to the discovery of a plant hormone

The shoot of a houseplant on a windowsill grows toward light (Figure 33.1A). If you rotate the plant, it soon reorients its growth until the leaves again face the window. The growth of a shoot in response to light is called **phototropism** (from the Greek *photos*, light, and *tropos*, turn). Phototropism is an adaptive response, directing growing seedlings and the shoots of mature plants toward the sunlight that drives photosynthesis.

Microscopic observations of plants growing toward light indicate the cellular mechanism that underlies phototropism. **Figure 33.1B** shows a grass seedling curving toward light coming from one side. As the enlargement shows, cells on the darker side of the seedling are larger—actually, they have elongated faster—than those on the brighter side. The different cellular growth rates made the shoot bend toward the light. If a

Figure 33.1A A houseplant growing toward light

seedling is illuminated uniformly from all sides or if it is kept in the dark, the cells all elongate at a similar rate. In these situations, the seedling grows straight upward.

What causes plant cells in a shoot to grow at different rates? We found in Chapter 26 that hormones help coordi-

nate internal activities, such as growth rates, in animals. We might predict, therefore, that plants also have hormones that regulate growth. Actually, the idea that plants have hormones emerged from a series of classic experiments on how shoots respond to light.

Showing That Light Is Detected by the Shoot Tip In the late 19th century, Charles Darwin and his son Francis conducted some of the earliest experiments on phototropism. They observed that grass seedlings could bend toward light only if the tips of their shoots were present. The first five grass plants in **Figure 33.1C** summarize the Darwins' findings. When they removed the tip of a grass shoot, the shoot did not curve toward the light. The shoot also remained straight when the Darwins placed an opaque cap on its tip. However, the shoot curved normally when they placed a transparent cap on its tip or an opaque shield around its base. The Darwins concluded that the tip of the shoot was responsible for sensing light. They also recognized that the growth response, the bending of the shoot, occurs below the tip. Therefore, they speculated that some signal was transmitted downward from the tip to the growth region of the shoot.

A few decades later, Danish botanist Peter Boysen-Jensen further tested the chemical signal idea of the Darwins; the last two plants in Figure 33.1C summarize his findings. In one group of seedlings, Boysen-Jensen inserted a block of gelatin between the tip and the lower part of the shoot. The gelatin block prevented cellular contact but allowed chemicals to diffuse through. The seedlings with gelatin blocks behaved normally, bending toward light. In a second set of seedlings, Boysen-Jensen inserted a thin piece of the mineral mica under the shoot tip. Mica is an impermeable barrier, and the seedlings with mica had no phototropic response. These experiments supported the hypothesis that the signal for phototropism is a mobile chemical.

Isolating the Chemical Signal In 1926, Frits Went, a Dutch graduate student, modified Boysen-Jensen's techniques and extracted the chemical messenger for phototropism in grasses. As shown in **Figure 33.1D**, Went first removed the tips of grass seedlings and placed them on blocks of agar, a gelatin-like material. He reasoned that the chemical messenger (pink in the figure) from the shoot tips should diffuse into the agar and that the blocks should then be able to substitute for the shoot tips. Went tested the effects of the agar blocks on tipless seedlings, which he kept in the dark to eliminate the effect of light. First, he centered the treated agar blocks on the cut tips of a batch of seedlings. These plants grew straight upward. They also grew faster than the decapitated control seedlings, which hardly grew at all. Went concluded that the agar had

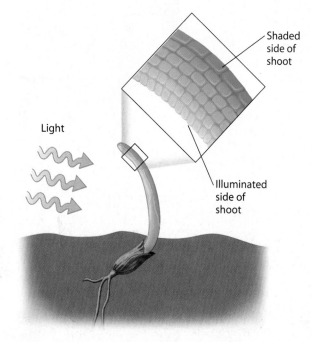

Light

Shaded side of shoot

Illuminated side of shoot

Figure 33.1B Phototropism in a grass seedling

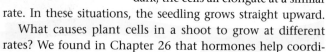

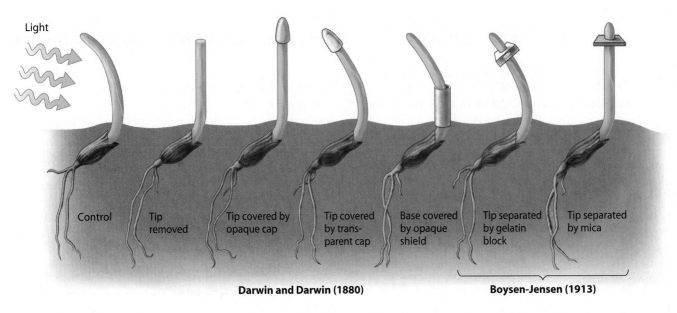

Light

Control | Tip removed | Tip covered by opaque cap | Tip covered by transparent cap | Base covered by opaque shield | Tip separated by gelatin block | Tip separated by mica

Darwin and Darwin (1880)　　　　　　　　**Boysen-Jensen (1913)**

Figure 33.1C Early experiments on phototropism: detection of light by shoot tips and evidence for a chemical signal

absorbed the chemical messenger produced in the shoot tip and that the chemical had passed into the shoot and stimulated it to grow. He then placed agar blocks off-center on another batch of tipless seedlings. These plants bent away from the side with the chemical-laden agar block, as though growing toward light. Control seedlings with *blank* agar blocks (whether offset or not) grew no more than the first control. Went concluded that the agar block contained a chemical produced in the shoot tip, that this chemical stimulated growth as it passed down the shoot, and that a shoot curved toward light because of a higher concentration of the growth-promoting chemical on the darker side. For this chemical messenger, or hormone, Went chose the name auxin, from the Greek *auxein* ("to increase"). In the 1930s, biochemists determined the chemical structure of Went's auxin.

The classical hypothesis for what causes grass shoots to grow toward light, based on these early experiments, is that an uneven distribution of auxin moving down from the shoot tip causes cells on the darker side to elongate faster than cells on the brighter side. Studies with plants other than grass shoots, however, do not always support this hypothesis. For example, there is no evidence that light from one side causes an uneven distribution of auxin in the stems of sunflowers, radishes, and other dicots. There is, however, a greater concentration of substances that may act as growth inhibitors on the lighted side

of a stem. Still, auxin's role in the phototropism of grass shoots opened up the field of research on plant hormones.

> **?** How do the experiments illustrated in Figures 33.1C and 33.1D provide evidence that phototropism depends on a chemical signal—that is, a hormone?

■ Light is detected by the shoot tip, but the bending response occurs farther down the shoot. The fact that the signal can pass through a barrier that prevents cell contact but allows chemicals to pass suggests that the signal is a chemical.

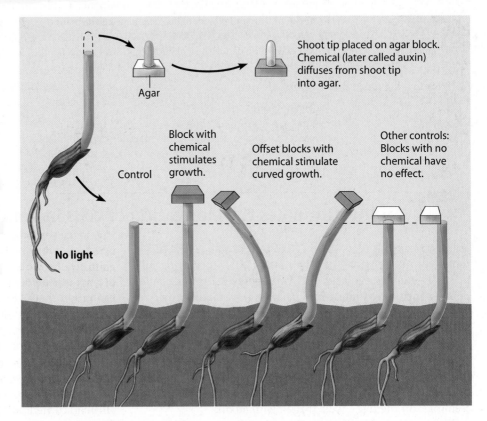

Agar

Shoot tip placed on agar block. Chemical (later called auxin) diffuses from shoot tip into agar.

No light

Control

Block with chemical stimulates growth.

Offset blocks with chemical stimulate curved growth.

Other controls: Blocks with no chemical have no effect.

Figure 33.1D Went's experiments: isolation of the chemical signal

33.2 Five major types of hormones regulate plant growth and development

Found in all plants and animals, **hormones** are chemical signals that coordinate the activities in an organism. A hormone is a compound that is produced by one part of the body and then transported to other parts, where it binds to a specific receptor and triggers a signal transduction pathway that amplifies the hormone's signal (see Module 11.3). Only a minute quantity of a hormone is required to have a profound effect on target cells and tissues.

Plant biologists have identified five major types of plant hormones, which are previewed in Table 33.2. (It is important to note that each "hormone" listed in the table represents a group of related hormones.) In general, hormones exert control by affecting the division, elongation, and differentiation of cells. As Table 33.2 indicates, each type of hormone can produce a variety of effects. Notice that all five types of hormones influence growth (stimulating or inhibiting cell division and elongation), and four of them affect development (cell differentiation).

Each hormone has multiple effects, depending on its site of action, its concentration, and the developmental stage of the plant. In most situations, no single hormone acts alone. Instead, it is usually the balance of several plant hormones—their relative concentrations—that controls the growth and development of a plant. These interactions will become apparent during this chapter's survey of hormone functions.

Not listed in Table 33.2 are the brassinosteroids (see Module 33.13). The brassinosteroids are chemicals similar to the steroids cholesterol and the animal sex hormones. Their effects are very similar to those of auxin. Additionally, many of the plant defense molecules we will discuss in Module 33.14 are probably plant hormones as well. The next five modules focus on the five major types of plant hormones.

Web/CD Thinking as a Scientist *What Plant Hormones Affect Organ Formation?*

? Hormones elicit cellular responses by binding to extracellular receptors and triggering _____.

■ signal transduction pathways

TABLE 33.2 MAJOR TYPES OF PLANT HORMONES

Hormone	Major Functions	Where Produced or Found in Plant
Auxins	Stimulate stem elongation; affect root growth, differentiation, and branching; development of fruit; apical dominance; phototropism and gravitropism (response to gravity)	Meristems of apical buds; young leaves; embryos within seeds
Cytokinins	Affect root growth and differentiation; stimulate cell division and growth; stimulate germination; delay aging	Made in roots and transported to other organs
Gibberellins	Promote seed germination, bud development, stem elongation, and leaf growth; stimulate flowering and fruit development; affect root growth and differentiation	Meristems of apical buds and roots; young leaves; embryos
Abscisic acid (ABA)	Inhibits growth; closes stomata during water stress; helps maintain dormancy	Leaves, stems, roots, green fruits
Ethylene	Promotes fruit ripening; opposes some auxin effects; promotes or inhibits growth and development of roots, leaves, and flowers, depending on species	Ripening fruits, nodes of stems, aging leaves and flowers

---

33.3 Auxin stimulates the elongation of cells in young shoots

The term **auxin** is used for any chemical substance that promotes seedling elongation. The natural auxin occurring in plants is indoleacetic acid, or IAA, but several other compounds, including some synthetic ones, have auxin activity. When we use the term *auxin* in this text, however, we are referring specifically to IAA.

Figure 33.3A (on the facing page) shows the effect of auxin on growing pea plants. All the seedlings in the photograph were grown under controlled conditions for the same length of time. The only difference is that the taller seedlings, on the right, were treated with auxin.

The apical meristem at the tip of a shoot (see Module 31.7) is a major site of auxin synthesis. As auxin moves downward, it stimulates growth of the stem by making cells elongate. As the blue curve in Figure 33.3B shows, auxin promotes cell elongation in stems only within a certain concentration range. Above a certain level (0.9 g of auxin per liter of solution, in this case), it usually inhibits cell elongation in stems, probably by inducing the production of ethylene, a hormone that generally counters the effects of auxin.

The red curve on the graph shows the effect of auxin on root growth. An auxin concentration too low to stimulate shoot cells will cause root cells to elongate. On the other hand, an auxin concentration high enough to make stem cells elongate is in the concentration range that inhibits root cell elongation. These effects of auxin on cell elongation reinforce two

Figure 33.3A The effect of auxin (IAA) on pea plants (right)

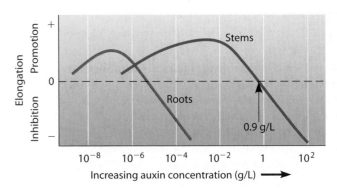

Figure 33.3B The effect of auxin concentration on cell elongation

points: (1) the same chemical messenger may have different effects at different concentrations in one target cell, and (2) a given concentration of the hormone may have different effects on different target cells.

How does auxin make plant cells elongate? One hypothesis is that auxin initiates elongation by weakening cell walls. As shown in **Figure 33.3C**, ❶ auxin may stimulate certain proteins in a plant cell's plasma membrane to pump hydrogen ions into the cell wall, lowering the pH within the wall. The low pH activates enzymes that separate cross-linking molecules from larger cellulose molecules ❷ in the wall (see Figure 3.7). ❸ The cell then swells with water and elongates because its weakened wall no longer resists the cell's tendency to take up water osmotically. After this initial elongation caused by the uptake of water, the cell sustains the growth by synthesizing more cell wall material and cytoplasm. These processes are also stimulated by auxin.

Auxin produces a number of other effects in addition to stimulating cell elongation and causing stems and roots to lengthen. Auxin induces cell division in the vascular cambium, thus promoting growth in stem diameter (see Module 31.8). Furthermore, auxin is produced by developing seeds and promotes the growth of fruit. Under greenhouse conditions, a lack of insect pollinators can lead to poorly developed tomato fruits. To address this problem, synthetic auxins sprayed on greenhouse-grown tomato vines induce fruit development without a need for pollination, making it possible to grow seedless tomatoes. The use of synthetic auxins as herbicides is described in Module 33.8.

? Suppose you had a tiny pH electrode that could measure the pH of a plant cell's wall. How could you use it to test the hypothesis presented in Figure 33.3C for how auxin stimulates cell elongation?

■ The hypothesis predicts that addition of auxin to the cell should stimulate H⁺ pumps and lower the pH of the wall (make it more acidic). You could test this prediction with your pH electrode by measuring the wall pH in the presence of (experimental group) or absence of (control group) auxin.

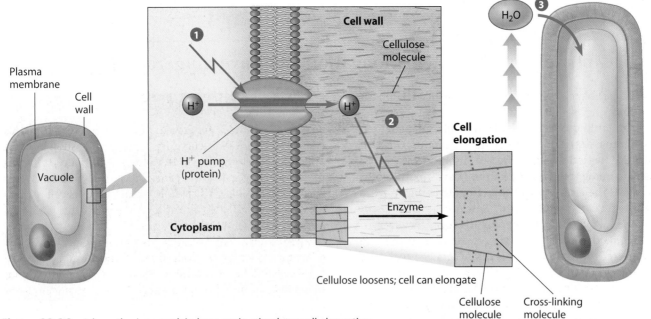

Figure 33.3C A hypothesis to explain how auxin stimulates cell elongation

33.4 Cytokinins stimulate cell division

Cytokinins are growth regulators that promote cell division, or cytokinesis. A number of cytokinins have been extracted from plants—the most common is zeatin, so named because it was first discovered in corn (*Zea mays*). Also, several synthetic ones have been made. Natural cytokinins are produced in actively growing tissues, particularly in roots, embryos, and fruits. Cytokinins made in the roots reach target tissues in stems by moving upward in xylem sap.

Plant biologists have found that cytokinins enhance the division, growth, and development of plant cells grown in culture. Cytokinins also retard the aging of flowers and leaves, and florists use cytokinin sprays to keep cut flowers fresh.

Cytokinins and auxin interact in the control of apical dominance, the ability of the terminal bud to suppress the growth of axillary buds (see Modules 31.3 and 31.7). The photographs in **Figure 33.4** show the results of a simple experiment that demonstrates this effect. Both basil plants pictured are the same age. The one on the left has an intact terminal bud; the one on the right had its terminal bud removed several weeks earlier. In the plant on the left, apical dominance resulted in lengthwise growth but inhibited growth of the axillary buds (the buds that produce side branches). As a result, the shoot grew in height but did not branch out to the sides very much. In the plant on the right, the lack of a terminal bud resulted in the activation of the axillary buds, making the plant grow more branches and become bushy.

The leading hypothesis to explain the hormonal regulation of apical dominance—the direct inhibition hypothesis—proposes that auxin and cytokinin act antagonistically in regulating axillary bud growth. According to this view, auxin transported down the shoot from the terminal bud directly inhibits axillary buds from growing, causing a shoot to lengthen at the expense of lateral branching. Meanwhile, cytokinins entering the shoot system from roots counter the action of auxin by signaling axillary buds to begin growing. Thus, the ratio of auxin and cytokinin is viewed as the critical factor in controlling axillary bud inhibition. Many observations and experiments are consistent with the direct inhibition hypothesis. However, there are also some data that contradict it. The direct inhibition hypothesis, therefore, does not account for all experimental findings: It is likely that plant biologists have not uncovered all the pieces of this puzzle.

— Terminal bud

No terminal bud

Figure 33.4 The effects of naturally occurring auxin and cytokinins on plant growth

? According to the direct inhibition hypothesis, the status of axillary buds—dormant or growing—depends on the relative concentrations of _____ moving down from the shoot tip and _____ moving up from the roots.

■ auxin . . . cytokinins

33.5 Gibberellins affect stem elongation and have numerous other effects

Farmers in Asia have long noticed that some rice seedlings in their paddies grew so tall and spindly that they toppled over before they could produce grain. In the 1920s, Japanese scientists discovered that a fungus of the genus *Gibberella* caused this "foolish seedling disease." By the 1930s, Japanese scientists had determined that the fungus produced hyperelongation of rice stems by secreting a chemical, which was given the name **gibberellin**. Researchers later discovered that gibberellin exists naturally in plants, where it is a growth regulator. Foolish seedling disease occurs when rice plants infected with the *Gibberella* fungus get an overdose of gibberellin.

More than 100 different gibberellins have been identified in plants. Roots and young leaves are major sites of gibberellin production. One of the main effects of gibberellins

Figure 33.5A Reversing dwarfism in pea plants with gibberellin

Figure 33.5B An effect of gibberellin treatment on grapes (right)

is to stimulate cell elongation and cell division in stems and leaves. This action generally enhances that of auxin and can be clearly demonstrated when certain dwarf plant varieties grow to normal height after treatment with gibberellins (Figure 35.5A). Also in combination with auxin, gibberellins can influence fruit development, and gibberellin-auxin sprays can make apples, currants, and eggplants develop without fertilization. The most important commercial application of gibberellins is in the production of the Thompson variety of seedless grapes. Gibberellins make the grapes grow larger and farther apart in a cluster. The left cluster of grapes in

Figure 33.5B is untreated; the right cluster shows the effect of gibberellin treatment.

Gibberellins are also important in seed germination in many plants. Many seeds that require special environmental conditions to germinate, such as exposure to light or cold temperatures, will germinate when sprayed with gibberellins. In nature, gibberellins in seeds are probably the link between environmental cues and the metabolic processes that renew growth of the embryo. For example, when water becomes available to a grass seed, it causes the embryo in the seed to release gibberellins, which promote germination by mobilizing nutrients stored in the seed. In some plants, gibberellins seem to interact antagonistically with another hormone, abscisic acid, which we discuss next.

? A gibberellin deficiency probably caused the dwarf variety of pea plants studied by Gregor Mendel (see Figure 9.2D). Given the role of gibberellin as a growth hormone, why does it make sense that the dwarf pea plant allele is recessive?

■ Because minute amounts of hormone are usually enough to elicit a strong response, a heterozygous plant would still produce enough gibberellin to result in normal growth.

33.6 Abscisic acid inhibits many plant processes

In the 1960s, one research group studying bud dormancy and another team investigating leaf abscission (the dropping of autumn leaves) isolated the same compound, **abscisic acid (ABA).** Ironically, ABA is no longer thought to play a primary role in either bud dormancy or leaf abscission (for which it was named), but it is a plant hormone of great importance in other functions. Unlike the growth-stimulating hormones we have studied so far—auxin, cytokinins, and gibberellins—ABA slows down growth.

One of the times in the life of a plant when it is advantageous to suspend growth is the onset of seed dormancy. Seed dormancy has great survival value because it ensures that the seed will germinate only when there are favorable conditions of light, temperature, and moisture. What prevents a

Figure 33.6
The Mojave Desert in California blooming after a rain

seed dispersed in autumn from germinating immediately only to be killed by winter? For that matter, what prevents seeds from germinating in the dark, moist interior of the fruit? ABA is the answer. Levels of ABA may increase 100-fold during seed maturation. Many types of dormant seeds will only germinate when ABA is removed or inactivated in some way. For example, some seeds require prolonged exposure to cold to trigger ABA inactivation. In these plants, the breakdown of ABA in the winter is required for seed germination in the spring.

The seeds of some desert plants remain dormant in parched soil until a downpour washes ABA out of the seeds, allowing them to germinate. For example, the evening primroses and purple sand verbena in Figure 33.6 grew from seeds that germinated just after a hard rain.

As we saw in the previous module, gibberellins promote seed germination. For many plants, the ratio of ABA to gibberellins determines whether the seed will remain dormant or germinate.

In addition to its role in dormancy, ABA is the primary internal signal that enables plants to withstand drought. When a plant begins to wilt, ABA accumulates in its leaves and causes stomata to close (see Module 32.4). This reduces transpiration and prevents further water loss. In some cases, water shortage can stress the root system before the shoot system. ABA transported from roots to leaves may function as an "early warning system."

? Which two hormones regulate seed dormancy and germination? What are their opposing effects?

■ Abscisic acid maintains seed dormancy; gibberellins promote germination.

Ethylene triggers fruit ripening and other aging processes

Early in the 20th century, oranges and grapefruits were ripened for market in sheds equipped with kerosene stoves. Fruit growers thought it was the heat that ripened the fruit, but when they tried newer, cleaner-burning stoves, the fruit did not ripen fast enough. Plant biologists learned later that ripening in the sheds was actually due to **ethylene**, a gaseous by-product of kerosene combustion. We now know that plants produce their own ethylene, which functions as a hormone that triggers a variety of aging responses, including fruit ripening and programmed cell death. Ethylene is also produced in response to stresses such as drought, flooding, mechanical pressure, injury, and infection.

Fruit Ripening A burst of ethylene production in a fruit triggers its ripening. Because ethylene is a gas, the signal to ripen spreads from fruit to fruit: One bad apple really does spoil the lot! The ripening process includes the enzymatic breakdown of cell walls, which softens the fruit, and the conversion of starches and acids to sugars, which makes the fruit sweet. The production of new scents and colors attracts animals, which eat the fruits and disperse the seeds.

You can make some fruits ripen faster if you store them in a plastic bag so that the ethylene gas can accumulate. **Figure 33.7A** shows the results of a fruit-ripening demonstration. Three unripe bananas were stored for the same time period in plastic bags: (1) with an ethylene-releasing orange, (2) with a beaker of an ethylene-releasing chemical, and (3) alone. As you can see, the more ethylene present, the riper (darker) the banana. On a commercial scale, many kinds of fruit—tomatoes, for instance—are often picked green and then ripened in huge storage bins into which ethylene gas is piped—a modern variation on the old storage shed.

In other cases, growers take measures to *retard* the ripening action of natural ethylene. Stored apples are often flushed with CO_2, which inhibits ethylene synthesis. Also, gas is circulated around the apples to prevent ethylene from accumulating. In this way, apples picked in autumn can be stored for sale the following summer.

The Falling of Leaves Like fruit ripening, the changes that occur in deciduous trees each autumn—color changes, drying, and the loss of leaves—are also aging processes. Leaves lose their green color because chlorophyll is broken down during autumn. Fall colors result from a combination of new pigments made in autumn and pigments that were already present in the leaf but masked by the green chlorophyll. Autumn leaf drop is an adaptation that helps keep the tree from drying out in winter. Without its leaves, a tree loses less water by evaporation when its roots cannot take up water from the frozen ground. Before leaves fall, many essential elements are salvaged from them and stored in the stem.

When an autumn leaf falls, the base of the leaf stalk separates from the stem. The separation region is called the abscission layer. As indicated in **Figure 33.7B**, the abscission layer consists of a narrow band of small parenchyma cells with thin walls that are further weakened when enzymes digest the cell walls. The leaf drops off when its weight, often helped by wind, splits the abscission layer apart. Notice the layer of protective cells adjacent to the abscission layer. Even before the leaf falls, these cells form a leaf scar on the

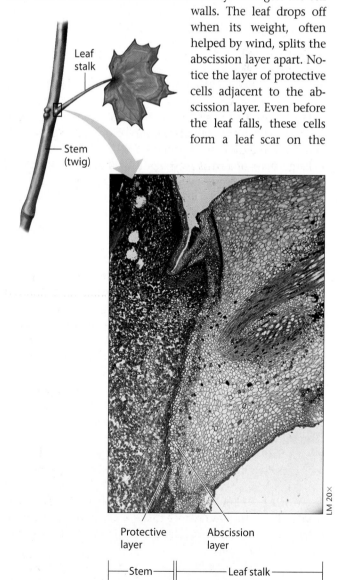

Leaf stalk

Stem (twig)

LM 20×

Protective layer

Abscission layer

Stem — Leaf stalk

Figure 33.7B Abscission layer at the base of a leaf

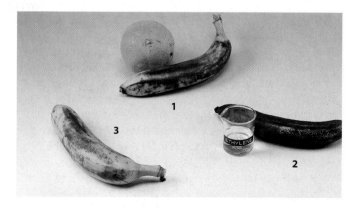

Figure 33.7A The effect of ethylene on the ripening of bananas

stem. Dead cells covering the scar help protect the plant from infectious organisms.

Leaf drop is triggered by environmental stimuli, including the shortening days of autumn and, to a lesser extent, cooler temperatures. These stimuli apparently cause a change in the balance of ethylene and auxin. The auxin prevents abscission and helps maintain the leaf's metabolism, but as a leaf ages, it produces less auxin. Meanwhile, cells begin producing ethylene, which stimulates formation of the abscission layer. The ethylene primes the abscission layer to split by promoting the synthesis of enzymes that digest cell walls in the layer.

We have now completed our survey of the five major types of plant hormones. Before moving on to the topic of plant behavior, let's look at some agricultural uses of these chemical regulators.

Web/CD Activity 33A *Leaf Abscission*

? Botanists sometimes refer to ethylene as the "senescence hormone." What is the basis for this term?

■ Many of ethylene's functions, including fruit ripening and leaf abscission, are associated with aging-like changes in cells.

CONNECTION

33.8 Plant hormones have many agricultural uses

Much of what we know about plant hormones—and there is much more to be learned—has a direct application to agriculture. As already mentioned, the control of fruit ripening and the production of seedless fruits are two of several major uses of these chemicals. Plant hormones also enable farmers to control when plants will drop their fruit. For instance, synthetic auxins are often used to prevent orange and grapefruit trees from dropping their fruit before they can be picked. **Figure 33.8** shows a citrus grower in Florida spraying an orange grove with auxin. The quantity of auxin must be carefully monitored because too much of the hormone may stimulate the plant to release more ethylene, making the fruit ripen and drop off sooner.

Large doses of auxins are often used intentionally to promote premature fruit drop. For example, auxin may be sprayed on apple and olive trees to thin the developing fruits; the remaining fruits will grow larger. Ethylene is used to thin peaches and prunes, and it is sometimes sprayed on berries, grapes, and cherries to loosen the fruit so it can be picked by machines.

In combination with auxin, gibberellins are used to produce seedless fruits, as mentioned in Module 33.5. Sprayed on other kinds of plants, at an earlier stage, gibberellins can have the opposite effect: the *promotion* of seed production. A large dose of gibberellins will induce many biennial plants, such as carrots, beets, and cabbage, to flower and produce seeds during their first year of growth. Ordinarily, biennials do not produce seeds until their second year of growth.

Research on plant hormones has had other spin-offs. One of the most widely used herbicides, or weed killers, is 2,4-D, a synthetic auxin that disrupts the normal balance of hormones that regulate plant growth. Because dicots are more sensitive than monocots to this herbicide, 2,4-D can be used to selectively remove dandelions and other broadleaf dicot weeds from a lawn or grainfield. By applying herbicides to cropland, a farmer can reduce the amount of tillage required to control weeds, thus reducing soil erosion, fuel consumption, and labor costs.

Modern agriculture relies heavily on the use of synthetic chemicals. Without chemically synthesized herbicides to control weeds and synthetic plant hormones to help grow and

Figure 33.8 Using auxins to prevent early fruit drop

preserve fruits, less food would be produced, and food prices could increase considerably. At the same time, there is growing concern that the heavy use of artificial chemicals in food production may pose environmental and health hazards. A chemical called dioxin, for example, is a by-product of 2,4-D synthesis. Though 2,4-D itself does not appear to be toxic to mammals, dioxin causes birth defects, liver disease, and leukemia in laboratory animals. Therefore, dioxin is a serious hazard when it leaks into the environment. Also, many consumers are concerned that foods produced with artificial help may not be as tasty or nutritious as those raised more naturally. At present, however, organic foods are relatively expensive to produce. As we discussed in Module 32.10, these issues involve both economics and ethics: Should we continue to produce cheap, plentiful food using artificial chemicals and tolerate the potential problems, or should we put more of our agricultural effort into farming without these potentially harmful substances, recognizing that foods may be less plentiful and more expensive as a result?

? (a) What is the main commercial incentive for agribusiness to treat its products with plant hormones? (b) What behavior of consumers helps drive this use of hormone sprays in agriculture?

■ (a) The need to compete in the market by increasing production and lowering cost; (b) shopping for the lowest price on produce

33.9 Tropisms orient plant growth toward or away from environmental stimuli

Having surveyed the hormones that carry signals within a plant, we now shift our focus to the responses of plants to physical stimuli from the environment. A plant cannot migrate to water or a sunny spot, and a seed cannot maneuver itself into an upright position if it lands upside down in the soil. Because of their immobility, plants must respond to environmental stimuli through developmental and physiological mechanisms. **Tropisms** are directed growth responses that cause parts of a plant to grow toward or away from a stimulus. Phototropism, the growth of a plant shoot in response to light, is one type of tropism. Two other types are gravitropism, a response to gravity, and thigmotropism, a response to touch.

Response to Light As we saw in Module 33.1, the mechanism for phototropism is a greater rate of cell elongation on the darker side of a stem. In grass seedlings, the signal linking the light stimulus to the cell elongation response is auxin. Researchers have shown that illuminating a grass shoot from one side causes auxin to migrate across the tip from the bright side to the dark side. The shoot tips contain a protein pigment that detects the light and somehow passes the "message" to molecules that affect auxin transport. (We discuss protein light receptors in Module 33.12.)

Response to Gravity A plant's growth response to gravity, **gravitropism**, is illustrated by the corn seedlings in **Figure 33.9A**. These seedlings were both germinated in the dark. The one on the left was left untouched; notice that its shoot grew straight up and its root straight down. The seedling on the right was germinated in the same way, but two days later it was turned on its side so that the shoot and root were horizontal. By the time the photo was taken, the shoot had turned back upward, exhibiting a negative response to gravity, and the root had turned down, exhibiting positive gravitropism.

One hypothesis for how plants tell up from down is that gravity pulls special organelles containing dense starch grains to the low points of cells. The uneven distribution of organelles may in turn signal the cells to redistribute auxin. This effect has been documented in roots. A higher auxin concentration on the lower side of a root inhibits cell elongation. As cells on the upper side continue to elongate, the root curves downward. This tropism continues until the root is growing straight down. Gravitropism is an important adaptation, making the shoot of a germinating plant grow upward into the light and the root grow into the soil, no matter how the seed lands or is planted in the soil.

Response to Touch Directional growth in response to touch, **thigmotropism** (from the Greek *thigma*, touch), is illustrated in **Figure 33.9B** by the tendril of a pea plant coiling around a wire fence. The tendril (actually a modified leaf) grew straight until it touched the support. Contact then stimulated the cells to grow at different rates on opposite sides of the tendril (slower in the contact area), making the tendril coil around the wire. Most climbing plants have tendrils that respond by coiling and grasping when they touch rigid objects. Thigmotropism enables these plants to use other objects for support while growing toward sunlight.

Tropisms all have one function in common: They help plant growth stay in tune with the environment. In the next module, we see that plants also have a way of keeping time with their environment.

? Why are tropisms called "growth responses"?

■ Because the movement of a plant organ toward or away from an environmental stimulus takes place by growing. An organ bends when cells on one side grow faster than cells on the other side, and it elongates in one direction when cells grow evenly.

Figure 33.9A Gravitropism

Figure 33.9B Thigmotropism

33.10 Plants have internal clocks

Your pulse rate, blood pressure, body temperature, rate of cell division, blood cell count, alertness, urine composition, metabolic rate, sex drive, and responsiveness to medications all fluctuate rhythmically with the time of day. Plants also display rhythmic behavior; examples include the opening and closing of stomata (see Module 32.4) and the "sleep movements" of many species that fold their leaves or flowers in the evening and unfold them in the morning.

Innate Biological Rhythms and Their Fine-Tuning by Environmental Cues An innate biological cycle of about 24 hours is called a **circadian rhythm** (from the Latin *circa,* about, and *dies,* day). A circadian rhythm persists even when an organism is sheltered from environmental cues. A bean plant, for example, exhibits sleep movements at about the same intervals even if kept in constant light or darkness. Thus, circadian rhythms occur with or without external stimuli such as sunrise and sunset. Research on many organisms indicates that circadian rhythms are controlled by internal timekeepers called **biological clocks.**

Although a biological clock continues to mark time in the absence of environmental cues, to remain tuned to a period of *exactly* 24 hours, it requires daily signals from the environment. This is because innate circadian rhythms generally differ somewhat from a 24-hour period. Consider bean plants, for instance. As shown in **Figure 33.10**, the leaves of a bean plant are normally horizontal at noon and folded downward at midnight. When the plant is held in darkness, however, its sleep movements change to a cycle of about 26 hours.

The light/dark cycle of day and night provides the cues that usually keep biological clocks precisely synchronized with the outside world. But a biological clock cannot immediately adjust to a sudden major change in the light/dark cycle. We observe this problem ourselves when we cross several time zones in an airplane: When we reach our destination, we have "jet lag"; our internal clock is not synchronized with the clock on the wall. Moving a plant across several time zones produces a similar lag. In the case of either the plant or the human traveler, resetting the clock usually takes several days.

The Nature of Biological Clocks Just what is a biological clock? Researchers are actively investigating this question. In humans and other mammals, the clock is located within a cluster of nerve cells in the hypothalamus of the brain (see Module 28.15). But for most other organisms, including plants, we know little about where the clocks are located or what kinds of cells are involved. A leading hypothesis is that biological timekeeping in plants may depend on the synthesis of a protein that regulates its own production through feedback control. Once the protein accumulates to a sufficient concentration, it turns off its own gene. When the concentration of the protein falls, transcription restarts. The result would be a cycling of the protein's concentration over a roughly 24-hour period—

Noon

Midnight

Figure 33.10   Sleep movements of a bean plant

a clock! Some research indicates that such a molecular mechanism may be common to all eukaryotes. However, much research remains to be done in this area.

Unlike most metabolic processes, biological clocks and the circadian rhythms they control are affected little by temperature. Somehow, a biological clock compensates for temperature shifts. This adjustment is essential, for a clock that speeds up or slows down with the rise and fall of outside temperature would be an unreliable timepiece.

In attempting to answer questions about biological clocks, it is essential to distinguish between the clock and the processes it controls. You could think of the sleep movements of leaves as the "hands" of a biological clock, but they are not the essence of the clockwork itself. You can restrain the leaves of a bean plant for several hours so that they cannot move. But on release, they will rush to the position appropriate for the time of day. Thus, we can interfere with an organism's rhythmic activity, but its biological clock goes right on ticking. In the next module, we'll learn more about the interface between a plant's internal biological clock and the external environment.

? If a bean plant is kept in constant darkness, its sleep movements change to a cycle of about 26 hours. How many days will it take for its leaves to be in the "noon position" when the time is actually midnight? Explain your answer.

■ About 6 days. The 26-hour cycle in a constant environment is 2 hours longer than a real day. Thus, it would take 6 days for the plant's leaf position to be 12 hours out of step with the actual time of day.

A biological clock not only times a plant's everyday activities, but may also influence seasonal events that are important in a plant's life cycle. Flowering, seed germination, and the onset and ending of dormancy are all stages in plant development that usually occur at specific times of the year. The environmental stimulus plants most often use to detect the time of year is called **photoperiod**, the relative lengths of day and night.

Plants whose flowering is triggered by photoperiod fall into two groups. One group, the **short-day plants**, generally flower in late summer, fall, or winter, when light periods shorten. Chrysanthemums and poinsettias are examples of short-day plants. In contrast, **long-day plants**, such as spinach, lettuce, iris, and many cereal grains, usually flower in late spring or early summer, when light periods lengthen. Spinach, for instance, flowers only when daylight lasts at least 14 hours. (Some plants, such as dandelions, are day-neutral; their flowering is unaffected by photoperiod.)

In the 1940s, researchers discovered that flowering and other responses to photoperiod are actually controlled by *night* length, not day length. In fact, the so-called short-day plants are actually long-night plants, and the so-called long-day plants are actually short-night plants. However, the day-length terms are embedded firmly in the literature of plant biology.

Figure 33.11 illustrates the evidence for the night-length effect and also shows the difference between the flowering response of a short-day plant and a long-day plant. The left side of the figure represents short-day plants. The first two bars and corresponding plant drawings show that a short-day plant will not flower until it is exposed to a *continuous* dark period exceeding a critical length (about 10 hours, in this case). The continuity of darkness is important. The short-day plant will not blossom if the nighttime part of the photoperiod is interrupted by even a brief flash of light (third bar). (There is no effect if the daytime portion of the photoperiod is broken by a brief exposure to darkness.)

Florists apply this information about short-day plants to bring us flowers out of season. Chrysanthemums, for instance, are short-day plants that normally bloom in the autumn, but their blooming can be stalled until Mother's Day in May by punctuating each long night with a flash of light, thus turning one long night into two short nights.

The right side of the figure demonstrates the effect of night length on a long-day plant. In this case, flowering occurs when the night length is *shorter* than a critical length (less than 10 hours, in this example). As the third bar shows, flowering can be induced in a long-day plant by a flash of light during the night.

Notice that we distinguish long-day from short-day plants not by an *absolute* night length but by whether the critical night length sets a maximum (long-day plants) or minimum (short-day plants) number of hours of darkness required for flowering. In both cases, the actual number of hours in the critical night length is specific to each species of plant.

? A particular short-day plant won't flower in the spring. We try to induce flowering by using a short dark interruption to split the long-light period of spring into two short-light periods. What result do you predict?

■ The plants still won't flower because it is actually night length, not day length, that counts in the photoperiodic control of flowering.

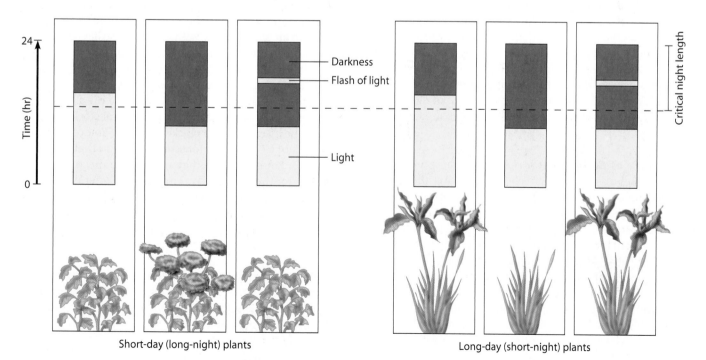

Short-day (long-night) plants   Long-day (short-night) plants

Darkness
Flash of light
Light
Critical night length
Time (hr)
24
0

Figure 33.11 Photoperiodic control of flowering

Phytochrome is a light detector that may help set the biological clock

The discovery that photoperiod (specifically night length) determines the seasonal responses of plants poses another question: How does a plant actually measure photoperiod? Much remains to be learned about this, but photoreceptive pigments called phytochromes are part of the answer. **Phytochromes** are proteins with a light-absorbing component. Because the light absorbed is at the red end of the spectrum, the molecules appear blue or bluish green (see Module 7.6).

Phytochromes were discovered during studies on how different wavelengths of light affect seed germination. Red light, with a wavelength of 660 nanometers (nm), was found to be most effective at increasing germination. Light of a longer wavelength, called far-red light (730 nm, near the edge of human visibility), both inhibited germination and reversed the effect of red light.

Researchers also learned that red light (which is one component of white daylight) is the most effective wavelength for interrupting night length and affecting flowering in short-day and long-day plants. Bar 1 in **Figure 33.12A** shows the results we saw in the previous module for both short-day and long-day plants that receive a flash of light during their critical dark period. The letter R on the light flash stands for red light.

The other three bars in Figure 33.12A show how flashes of far-red (FR) light affect flowering. As bar 2 shows, the effect of a flash of red light that interrupts a period of darkness can be reversed by a subsequent flash of far-red light: Both types of

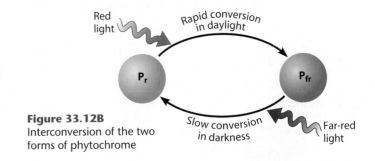

Figure 33.12B
Interconversion of the two forms of phytochrome

plants behave as though there is no interruption in the night length. Bars 3 and 4 indicate that no matter how many flashes of light a plant receives, only the wavelength of the *last* flash affects the plant's measurement of night length.

Now we're ready to consider the role of phytochromes. A phytochrome, researchers have learned, reverts back and forth between two forms that differ only slightly in structure. One form absorbs red light and the other absorbs far-red light. The two forms of a phytochrome are designated P_r (red absorbing) and P_{fr} (far-red absorbing). As diagrammed in **Figure 33.12B**, when the P_r form absorbs red light (660 nm), it is quickly converted to P_{fr}, and when P_{fr} absorbs far-red light (730 nm), it is converted back to P_r. These interconversions help account for the results shown in Figure 33.12A.

In nature, of course, the night is not punctuated by flashes of red and far-red light. Instead, the P_r form slowly accumulates in the continuous darkness that follows sunset. Each night, new phytochrome is synthesized in the P_r form, and P_{fr} is broken down by enzymes more readily than P_r. Also, P_{fr} in some plant species reverts to P_r in the dark. After sunrise, the red wavelengths of sunlight cause much of the phytochrome to be rapidly converted from the P_r form to P_{fr}. It is this sudden increase in P_{fr} each day at dawn that resets a plant's biological clock. Interactions between phytochrome and the biological clock enable plants to measure the passage of night and day. In doing so, the clock monitors photoperiod and cues appropriate seasonal physiological responses, such as seed germination, flowering, and the beginning and ending of bud dormancy.

Plants also have a group of blue-light photoreceptors that control such light-sensitive plant responses as phototropism and the opening of stomata at daybreak. Light is an especially important environmental factor in the lives of plants, and diverse receptors and signaling pathways have evolved that mediate a plant's responses to light.

Web/CD Activity 33B *Flowering Lab*

? How do phytochrome molecules help the plant recognize dawn each day?

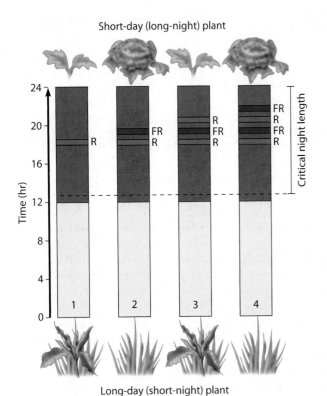

Short-day (long-night) plant

Long-day (short-night) plant

Figure 33.12A   The reversible effects of red and far-red light

■ Phytochrome molecules are mainly in the P_r form during the night. The sudden conversion of P_r to P_{fr} due to the absorption of the red wavelengths of sunlight, signals dawn.

33.13 Joanne Chory studies the effects of light and hormones in the model plant *Arabidopsis*

Figure 33.13
Joanne Chory

Joanne Chory (**Figure 33.13**) is a professor of biology and Howard Hughes Medical Institute Investigator at the Salk Institute for Biological Studies in La Jolla, California. Dr. Chory's research has revealed key steps in the signal transduction pathways by which light regulates the development of plants. As is often the case in biology, Dr. Chory's success has depended in part on selecting an appropriate organism as a research model—in this case, *Arabidopsis*, the "laboratory mouse" of modern plant biology. (See Module 31.1.) In a recent interview, Dr. Chory commented on how the study of this tiny mustard plant has had agricultural applications:

One example is what we've learned from *Arabidopsis* mutants about how the hormone ethylene functions in fruit ripening. The same genes responsible for the ethylene pathway in *Arabidopsis* are found in such fruits as tomatoes, and understanding how these genes work enables us to control the ripening process. Another application is that identifying genes in *Arabidopsis* can help breeders of crop plants. . . . For instance, sorghum would not normally grow in Texas, but breeders have selected for a mutation affecting a photoreceptor in the plant that we know, based on *Arabidopsis* research, would allow sorghum to complete its life cycle in Texas fields.

Plants use various photoreceptors to detect light. These photoreceptors then trigger complex signal transduction pathways that mediate a plant's many important responses to light. Dr. Chory's research with *Arabidopsis* mutants that are affected abnormally by light has led to her discovery of the role of steroid hormones in this process and the identification of a plant steroid hormone receptor. What is she learning about such hormones?

Plant steroids, which are called brassinosteroids, do a lot of the same kinds of things as sex steroids do in humans. The more steroid a plant has, the bigger and tougher and more robust it is. Steroids also regulate sexual reproduction in plants. I think it's interesting how a certain group of molecules began functioning in diverse organisms as signaling molecules. Many of the enzymes a plant uses to make its steroids are also found in animals that make their own types of steroids. So some of the genes for these enzymes have probably been conserved since plants and animals diverged from a common ancestor over a billion years ago.

Dr. Chory's research relates to the signal transduction pathways you learned about in Chapters 11 and 26, to the importance of light in regulating the lives of plants, which you read about in this chapter, and to the central theme of biology—evolution.

? Why is *Arabidopsis* called the "laboratory mouse" of plant biology?

■ This plant serves as a model system in which researchers can study many complex processes.

33.14 Defenses against herbivores and infectious microbes have evolved in plants

Plants are at the base of most food webs and are therefore subject to attack by a wide range of **herbivores** (plant-eating animals). There are also a number of pathogens that have the potential to damage or kill plants. Plants counter these threats with a variety of defense systems.

Defenses Against Herbivores Plants counter herbivores with both physical defenses, such as thorns, and chemical defenses, such as distasteful or toxic compounds. For example, some plants produce an unusual amino acid called canavanine. Canavanine resembles arginine, one of the 20 amino acids normally used to make proteins. If an insect eats a plant containing canavanine, the molecule is incorporated into the

insect's proteins in place of arginine. Because canavanine is different enough from arginine to change the shape and hence the function of proteins, the insect is harmed.

Some plants even recruit predatory animals that help defend the plants against certain herbivores. Such helpful predators include wasps that kill caterpillars feeding on plants. The recruitment process is outlined in **Figure 33.14A**. ❶ When a caterpillar bites into the plant, the physical damage to the plant and a chemical in the caterpillar's saliva together trigger ❷ a signal transduction pathway within the plant cells. The pathway leads to a specific cellular response: ❸ the synthesis and release of volatile (gaseous) chemicals that ❹ attract ("recruit") the wasp. ❺ The wasp injects its eggs into the

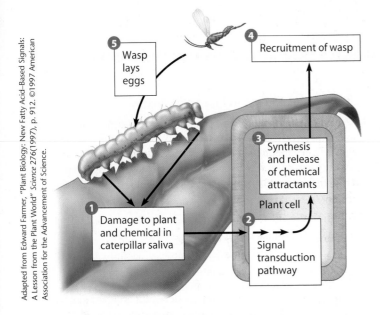

Figure 33.14A Recruitment of a wasp in response to an herbivore

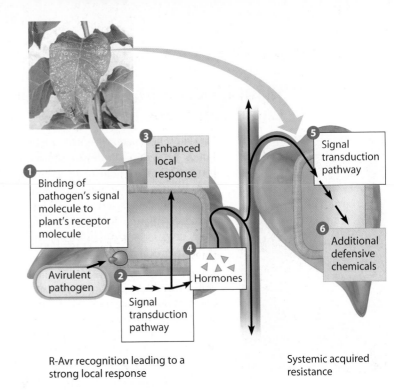

R-Avr recognition leading to a strong local response

Systemic acquired resistance

Figure 33.14B Defense responses against an avirulent pathogen

caterpillar. When the eggs hatch, the wasp larvae eat their way out of the caterpillar, killing it.

Defenses Against Pathogens A plant, like an animal, is subject to infection by pathogenic microbes: viruses, bacteria, and fungi. And like an animal, a plant has defense systems that hinder infection and counter pathogens that do manage to infect the plant.

A plant's first line of defense against infection is the physical barrier of the plant's "skin," the epidermis (see Chapter 31). However, microbes can cross this barrier through wounds or through natural openings such as stomata. Once infected, the plant uses chemicals as a second line of defense. Plant cells damaged by the infection release microbe-killing molecules and, in addition, chemicals that signal nearby cells to mount a similar chemical defense. Infection also stimulates chemical changes in the plant cell walls, which toughen the walls and thus slow the spread of the microbe within the plant.

The plant's chemical defense system is enhanced by the plant's inherited ability to recognize certain pathogens. A kind of "compromise" has coevolved between plants and most of their pathogens: The pathogen gains enough access to its host to perpetuate itself without severely harming the plant. The plant is said to be resistant to that pathogen, and the pathogen is said to be avirulent to the plant.

This resistance to destruction by a specific pathogen is based on the ability of the plant and the microbe to make a complementary pair of molecules. A plant has many *R* genes (for *resistance*), and each pathogen has a set of *Avr* genes (for *avirulence*). Researchers hypothesize that an *R* gene encodes a receptor protein on the plant's cells and that the complementary *Avr* gene leads to the production of some "signal" molecule of the pathogen that binds specifically to that receptor. Recognition of pathogen-derived molecules by R proteins

triggers a signal transduction pathway that leads to a defense response in the infected plant tissue.

Figure 33.14B shows this interaction and subsequent events in the plant. ❶ The binding of the pathogen's signal molecule (magenta) to the plant's receptor (purple) triggers ❷ a signal transduction pathway, which leads to ❸ a defense response that is much stronger than would occur without the R-Avr matchup. The cells at the site of infection mount a vigorous chemical defense, tightly seal off the area, and then kill themselves. The spots on one of the leaves in the photo in Figure 33.14B result from this sort of local response. As sick as such a leaf appears, it will survive.

The defense response at the site of infection helps protect the rest of the plant in yet another way. Among the signal molecules produced there are ❹ hormones that sound an alarm throughout the plant. ❺ At destinations distant from the original site, these hormones trigger signal transduction pathways leading to ❻ the production of additional defensive chemicals. This defense response, called **systemic acquired resistance**, is actually nonspecific, providing protection against a diversity of pathogens for days.

Researchers suspect that one of the alarm hormones is salicylic acid, a compound whose pain-relieving effects led early cultures to use the salicylic-acid–rich bark of willows (*Salix*) as a medicine. Aspirin is a chemical derivative of this compound. With the discovery of systemic acquired resistance, biologists have learned one function of salicylic acid in plants.

? What is released at a site of infection that triggers the development of general resistance to pathogens elsewhere in the plant?

■ A hormone

Adapted from Edward Farmer, "Plant Biology: New Fatty Acid–Based Signals: A Lesson from the Plant World" *Science* 276(1997), p. 912. ©1997 American Association for the Advancement of Science.

33.15   Plant biochemist Eloy Rodriguez studies how animals use defensive chemicals made by plants

Figure 33.15
Eloy Rodriguez in his research laboratory

Dr. Eloy Rodriguez (**Figure 33.15**), James A. Perkins Professor of Environmental Biology at Cornell University, is one of the world's leading experts on defensive chemicals produced by plants. In a recent interview, he described the significance of these chemicals this way:

A plant's ability to survive is really due to chemistry. A large array of organisms eat plants, and a plant can't just get up and run. Natural selection favors those plants with the right kinds of chemical compounds that ward off fungi, bacteria, viruses, insects, and large herbivores.

Dr. Rodriguez spends much of his time in his laboratory, but he is also a field biologist. His studies in tropical rain forests have had far-reaching influence:

One project that I got involved with has now developed into a discipline. It's called zoopharmacognosy, the study of how animals possibly medicate themselves with plants. I got involved with Richard Wrangham, a primatologist who was studying primates in the Kibale Forest of Uganda. In one observation, researchers followed a particular chimp for several days, and for two whole days this animal concentrated on one plant species. It wouldn't eat the whole plant. It would take off the leaves, crack the stalk, and then suck out the juice.

Richard pointed out to me that sometimes chimps seem to select young leaves from certain plant species. These animals get up in the morning, make a beeline toward these plants, and take a certain amount of the young leaves. We calculated that they were more or less getting a set dosage of the drug or drugs in the leaf.

Some of the compounds that Rodriguez and other researchers have discovered by following apes and other animals have potential use in human medicine:

Some of the compounds we've identified kill parasitic worms, and some may be useful against tumors. . . . A lot of plant collection is done randomly—you go out and collect a bunch of bark, for example. It is a very tedious way to go about getting drugs. It's nice when you have animals basically telling you, "Here, try these leaves."

Additionally, as Dr. Rodriguez relates, modern scientists also benefit from observing native peoples:

Indigenous people have been extremely successful [at using medicinal plants]. Anthropologists have documented that they extract plant materials, grind them up, filter them, and treat them with burned leaves (which act as a base)—almost the same process that I use. They get out materials that are relatively pure for use as medicine. A lot of the drug companies started out thanks to those folks.

Though his work focuses on the medicinal potential of rain forest chemicals, Rodriguez is quick to emphasize that the importance of rain forests goes beyond human concerns:

The more we search the rain forest, the more in awe of it we are. We see how important it is. In the tropical rain forest, we are talking about the ultimate diversity in plants, animals, fungi—life! The arguments that have been made for preserving the forests are excellent. We are talking about the health of the planet.

Rodriguez's words lead us to our next and final unit, Ecology. In these chapters, we will explore connections between organisms and see how organisms interact with their environments.

? What do researchers hope to learn by observing the feeding behavior of sick chimpanzees?

■ The selective feeding of the ill animals on certain parts of specific plants might lead researchers to medicines that could help humans, too.

CHAPTER REVIEW

Reviewing the Concepts

Plant Hormones (Introduction–33.8)

Plant hormones, such as isoflavones from soy, may provide human health benefits **(Introduction)**. Hormones coordinate the activities of plant cells and tissues. Experiments carried out by Darwin and others showed that the tip of a grass seedling detects light and transmits a signal down to the growing region of the shoot. This signal is a hormone named auxin. Auxin promotes faster cell elongation on the shaded site of the root **(33.1)**. By triggering signal transduction pathways, very small amounts of plant hormones regulate plant growth and development **(33.2)**.

Major types of hormones. Plants produce auxin (IAA) in the apical meristems at the tips of shoots. At different concentrations, auxin stimulates or inhibits the elongation of shoots and roots. It may act by weakening cell walls, allowing them to stretch when cells take up water. Auxin also stimulates the development of vascular tissues and cell division in the vascular cambium, promoting growth in stem diameter **(33.3)**. Cytokinins, produced by growing roots, embryos, and fruits, are hormones that promote cell division. Cytokinins from roots may balance the effects of auxin from apical meristems, causing lower buds to develop into branches **(33.4)**. Gibberellins stimulate the elongation of stems and leaves and the development of fruit. Gibberellins released from embryos function in some of the early events

of seed germination **(33.5).** Abscisic acid (ABA) inhibits the germination of seeds. The ratio of ABA to gibberellins often determines whether a seed will remain dormant or germinate. Seeds of many plants remain dormant until their ABA is inactivated or washed away. ABA also acts as a "stress hormone," causing stomata to close when a plant is dehydrated **(33.6).** As fruit cells age, they give off ethylene gas, which hastens ripening. A changing ratio of auxin to ethylene, triggered mainly by shorter days, probably causes autumn color changes and the loss of leaves from deciduous trees **(33.7).** Plant hormones have a variety of agricultural uses. Farmers use auxin to delay or promote fruit drop. Auxin and gibberellins are used to produce seedless fruits. A synthetic auxin called 2,4-D is used to kill weeds. There are questions about the safety of using such chemicals **(33.8).**

Growth Responses and Biological Rhythms in Plants (33.9–33.13)

Tropisms. Plants sense and respond to environmental changes in a variety of ways. Tropisms are growth responses that change the shape of a plant or make it grow toward or away from a stimulus:

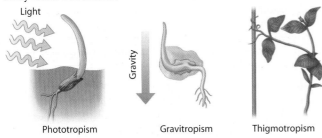

Phototropism   Gravitropism   Thigmotropism

Phototropism, bending in response to light, may result from auxin moving from the light side to the dark side of a stem. A response to gravity, or gravitropism, may be caused by the settling of special organelles on the low sides of shoots and roots, which may trigger a change in the distribution of hormones. Thigmotropism, a response to touch, is responsible for the coiling of tendrils and vines around objects **(33.9).**

Biological clocks and circadian rhythms. An internal biological clock controls sleep movements and other daily cycles in plants. These cycles, called circadian rhythms, persist with periods of about 24 hours even in the absence of environmental cues, but such cues are needed to keep them synchronized with day and night **(33.10).** Plants mark the seasons by measuring photoperiod, the relative lengths of night and day. The timing of flowering is one of the seasonal responses to photoperiod **(33.11):**

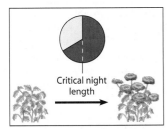 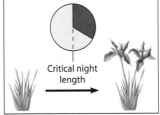

Short-day (long-night) plants   Long-day (short-night) plants

Light-absorbing proteins called phytochromes may help plants set their biological clock and monitor photoperiod **(33.12).** A small, wild mustard called *Arabidopsis* is a popular model organism for plant molecular biologists **(33.13).**

Plant Defenses (33.14–33.15)

Defensive chemicals. Plants use chemicals to defend themselves against both herbivores and pathogens. So-called avirulent plant pathogens interact with host plants in a specific way that stimulates both local and systemic defenses in the plant. Local defenses include microbe-killing chemicals and sealing off of the infected area. Hormones trigger generalized defense responses in other organs (systemic acquired resistance) **(33.14).** Some animals may medicate themselves by eating plants containing certain defensive chemicals **(33.15).**

Connecting the Concepts

1. Test your knowledge of the five major classes of plant hormones (auxins, cytokinins, gibberellins, abscisic acid, ethylene) by matching one hormone to each lettered box. (Note that some hormones will match up to more than one box.)

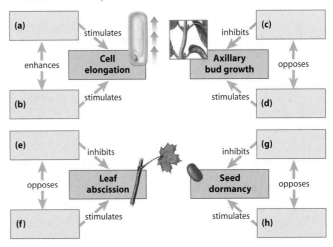

Testing Your Knowledge

Multiple Choice

2. During winter or periods of drought, which one of the following plant hormones inhibits growth and seed germination?
   a. ethylene   d. auxin
   b. abscisic acid   e. cytokinin
   c. gibberellin

3. A certain short-day plant flowers only when days are less than 12 hours long. Which of the following would cause it to flower? (*Explain your answer.*)
   a. a 9-hour night and 15-hour day with 1 minute of darkness after 7 hours
   b. an 8-hour day and 16-hour night with a flash of white light after 8 hours
   c. a 13-hour night and 11-hour day with 1 minute of darkness after 6 hours
   d. a 12-hour day and 12-hour night with a flash of red light after 6 hours

4. Auxin causes a shoot to bend toward light by
   a. causing cells to shrink on the dark side of the shoot.
   b. stimulating growth on the dark side of the shoot.
   c. causing cells to shrink on the lighted side of the shoot.
   d. stimulating growth on the lighted side of the shoot.
   e. inhibiting growth on the dark side of the shoot.

CHAPTER 33 *Control Systems in Plants*   **679**

5. In the autumn, the amount of ____ increases and ____ decreases in fruit and leaf stalks, causing a plant to drop fruit and leaves.
   a. ethylene . . . auxin
   b. gibberellin . . . abscisic acid
   c. cytokinin . . . abscisic acid
   d. auxin . . . ethylene
   e. gibberellin . . . auxin

6. Plant hormones act by affecting the activities of
   a. genes.
   b. membranes.
   c. enzymes.
   d. genes, membranes, and enzymes.
   e. genes and enzymes.

7. Buds and sprouts often form on tree stumps. Which of the following hormones would you expect to stimulate their formation?
   a. auxin
   b. cytokinin
   c. abscisic acid
   d. ethylene
   e. gibberellin

8. A plant's defense response at the site of initial infection by a pathogen will be especially strong if
   a. the pathogen is virulent.
   b. the plant makes a receptor protein that recognizes a signal molecule from the microbe.
   c. the pathogen is a fungus.
   d. the plant has an *Avr* gene that is the right match for one of the microbe's *R* genes.
   e. the right combination of hormones travel from the infection site to other parts of the plant.

Matching

9. Bending of a shoot toward light
10. Growth response to touch
11. A cycle with a period of about 24 hours
12. Pigment that helps control flowering
13. Relative lengths of night and day
14. Growth response to gravity
15. Folding of plant leaves at night

a. phytochrome
b. photoperiod
c. sleep movement
d. circadian rhythm
e. thigmotropism
f. phototropism
g. gravitropism

Describing, Comparing, and Explaining

16. If apples are to be stored for long periods, it is best to keep them in a place with good air circulation. Explain why.

17. Write a short paragraph explaining why a houseplant becomes bushier if you pinch off its terminal buds.

Applying the Concepts

18. Jon just started a new job as night watchman at a plant nursery. His boss told him to stay out of a room where chrysanthemums (which are short-day plants) were about to flower. Around midnight, looking for the restroom, Jon accidentally opened the door to the chrysanthemum room and turned on the lights for a moment. How might this affect the chrysanthemums? What could Jon do to correct his mistake?

19. A plant biologist observed a peculiar pattern when a tropical shrub was attacked by caterpillars. The biologist noticed that after a caterpillar ate a leaf, it would skip over nearby leaves and attack a leaf some distance away. The researcher found that when a leaf was eaten, nearby leaves started making a chemical that deterred the caterpillars. Simply removing a leaf did not trigger the same change nearby. The biologist suspected that a damaged leaf sent out a chemical that signaled other leaves. How could this hypothesis be tested?

20. In the 1950s, scientists discovered that many plants emit a gaseous compound called isoprene into the atmosphere. Isoprene is a precursor of the cytokinin hormones, but why plants emit it remains an open question. Researchers have found that plants emit isoprene when they are photosynthesizing, and the amount emitted increases as the temperature of a plant's leaves increases. Plants also synthesize more isoprene when they cannot obtain enough water. Darkness, substances that inhibit photosynthesis, and pure nitrogen gas (N_2) slow or stop isoprene emission. One hypothesis is that isoprene emissions help prevent damage to plants caused by high temperatures. Chlorophyll fluorescence has been used as a measure of irreversible leaf damage; the more fluorescence, the greater the damage. The graph below displays the results of some experimental tests of this hypothesis.

   The red line on the graph shows the effect of increasing temperature on leaves in air with no isoprene. The green line shows the effect on leaves from the same species of plant in air containing isoprene. Do these results support the hypothesis? Explain your answer. The leaves were exposed to N_2 while the recordings were made. Why do you suppose the experimenters did this?

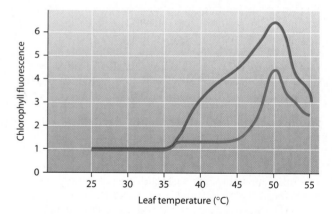

21. Imagine the following scenario: A plant scientist has developed a synthetic chemical that mimics the effects of a plant hormone. The chemical can be sprayed on apples before harvest to prevent flaking of the natural wax that is formed on the skin. This makes the apples shinier and gives them a deeper red color. What kinds of questions do you think should be answered before farmers start using this chemical on apples? How might the scientist go about finding answers to these questions?

Answers to all questions can be found in Appendix 3.

For study help and Activities, go to campbellbiology.com or the student CD-ROM.

5. In the autumn, the amount of ____ increases and ____ decreases in fruit and leaf stalks, causing a plant to drop fruit and leaves.
   a. ethylene . . . auxin
   b. gibberellin . . . abscisic acid
   c. cytokinin . . . abscisic acid
   d. auxin . . . ethylene
   e. gibberellin . . . auxin

6. Plant hormones act by affecting the activities of
   a. genes.
   b. membranes.
   c. enzymes.
   d. genes, membranes, and enzymes.
   e. genes and enzymes.

7. Buds and sprouts often form on tree stumps. Which of the following hormones would you expect to stimulate their formation?
   a. auxin
   b. cytokinin
   c. abscisic acid
   d. ethylene
   e. gibberellin

8. A plant's defense response at the site of initial infection by a pathogen will be especially strong if
   a. the pathogen is virulent.
   b. the plant makes a receptor protein that recognizes a signal molecule from the microbe.
   c. the pathogen is a fungus.
   d. the plant has an *Avr* gene that is the right match for one of the microbe's *R* genes.
   e. the right combination of hormones travel from the infection site to other parts of the plant.

Matching

9. Bending of a shoot toward light
10. Growth response to touch
11. A cycle with a period of about 24 hours
12. Pigment that helps control flowering
13. Relative lengths of night and day
14. Growth response to gravity
15. Folding of plant leaves at night

a. phytochrome
b. photoperiod
c. sleep movement
d. circadian rhythm
e. thigmotropism
f. phototropism
g. gravitropism

Describing, Comparing, and Explaining

16. If apples are to be stored for long periods, it is best to keep them in a place with good air circulation. Explain why.

17. Write a short paragraph explaining why a houseplant becomes bushier if you pinch off its terminal buds.

Applying the Concepts

18. Jon just started a new job as night watchman at a plant nursery. His boss told him to stay out of a room where chrysanthemums (which are short-day plants) were about to flower. Around midnight, looking for the restroom, Jon accidentally opened the door to the chrysanthemum room and turned on the lights for a moment. How might this affect the chrysanthemums? What could Jon do to correct his mistake?

19. A plant biologist observed a peculiar pattern when a tropical shrub was attacked by caterpillars. The biologist noticed that after a caterpillar ate a leaf, it would skip over nearby leaves and attack a leaf some distance away. The researcher found that when a leaf was eaten, nearby leaves started making a chemical that deterred the caterpillars. Simply removing a leaf did not trigger the same change nearby. The biologist suspected that a damaged leaf sent out a chemical that signaled other leaves. How could this hypothesis be tested?

20. In the 1950s, scientists discovered that many plants emit a gaseous compound called isoprene into the atmosphere. Isoprene is a precursor of the cytokinin hormones, but why plants emit it remains an open question. Researchers have found that plants emit isoprene when they are photosynthesizing, and the amount emitted increases as the temperature of a plant's leaves increases. Plants also synthesize more isoprene when they cannot obtain enough water. Darkness, substances that inhibit photosynthesis, and pure nitrogen gas (N_2) slow or stop isoprene emission. One hypothesis is that isoprene emissions help prevent damage to plants caused by high temperatures. Chlorophyll fluorescence has been used as a measure of irreversible leaf damage; the more fluorescence, the greater the damage. The graph below displays the results of some experimental tests of this hypothesis.

    The red line on the graph shows the effect of increasing temperature on leaves in air with no isoprene. The green line shows the effect on leaves from the same species of plant in air containing isoprene. Do these results support the hypothesis? Explain your answer. The leaves were exposed to N_2 while the recordings were made. Why do you suppose the experimenters did this?

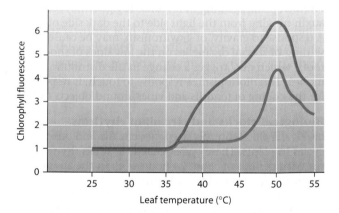

21. Imagine the following scenario: A plant scientist has developed a synthetic chemical that mimics the effects of a plant hormone. The chemical can be sprayed on apples before harvest to prevent flaking of the natural wax that is formed on the skin. This makes the apples shinier and gives them a deeper red color. What kinds of questions do you think should be answered before farmers start using this chemical on apples? How might the scientist go about finding answers to these questions?

Answers to all questions can be found in Appendix 3.

For study help and Activities, go to campbellbiology.com or the student CD-ROM.

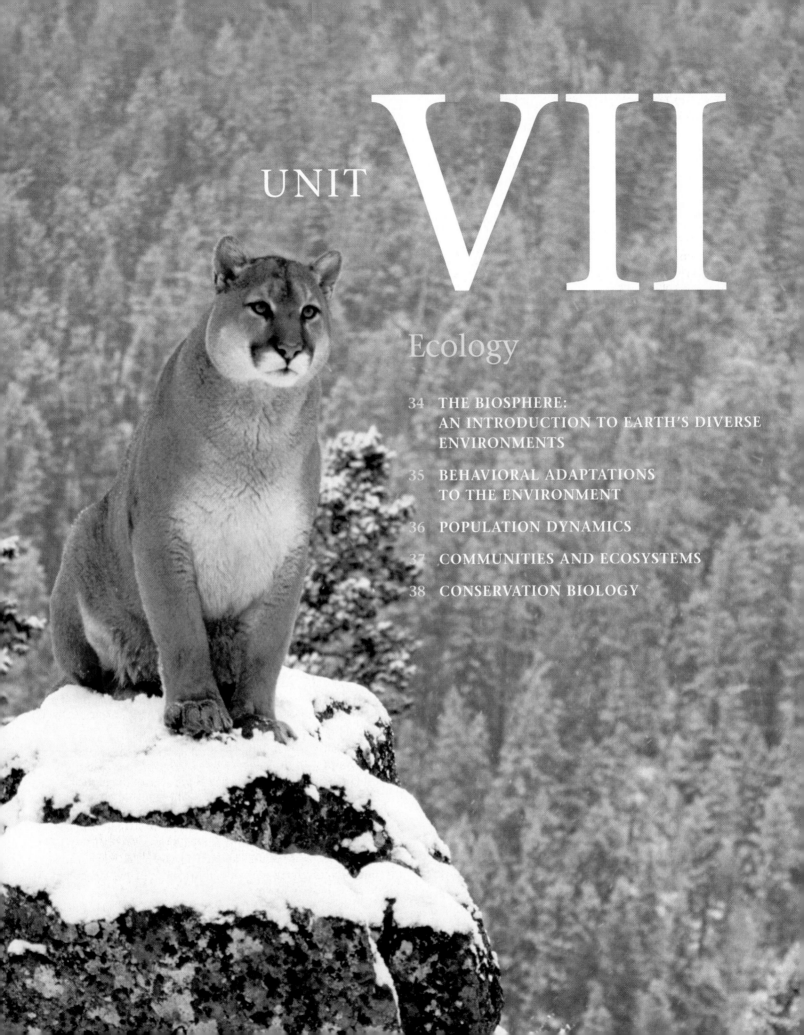

UNIT **VII**

Ecology

34   THE BIOSPHERE:
     AN INTRODUCTION TO EARTH'S DIVERSE
     ENVIRONMENTS

35   BEHAVIORAL ADAPTATIONS
     TO THE ENVIRONMENT

36   POPULATION DYNAMICS

37   COMMUNITIES AND ECOSYSTEMS

38   CONSERVATION BIOLOGY

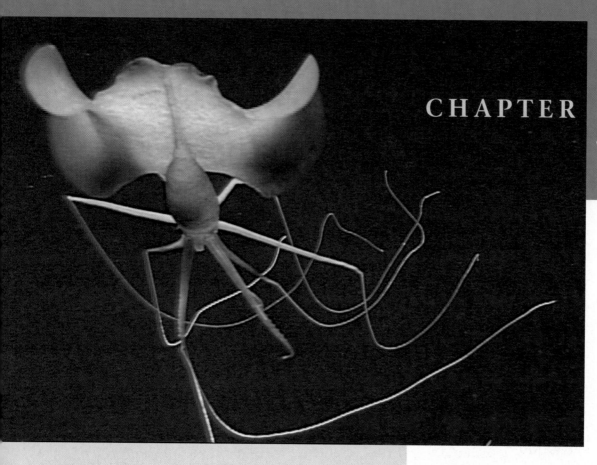

34.1   Ecologists study how organisms interact with their environment at several levels

THE BIOSPHERE

34.2   The biosphere is the total of all of Earth's ecosystems
34.3   Environmental problems reveal the limits of the biosphere
34.4   Physical and chemical factors influence life in the biosphere
34.5   Organisms are adapted to abiotic and biotic factors by natural selection
34.6   Regional climate influences the distribution of biological communities

AQUATIC BIOMES

34.7   Oceans occupy most of Earth's surface
34.8   Freshwater biomes include lakes, ponds, rivers, streams, and wetlands

TERRESTRIAL BIOMES

34.9   Terrestrial biomes reflect regional variations in climate
34.10   Tropical forests cluster near the equator
34.11   Savannas are grasslands with scattered trees
34.12   Deserts are defined by their dryness
34.13   Spiny shrubs dominate the chaparral
34.14   Temperate grasslands include the North American prairie
34.15   Broadleaf trees dominate temperate forests
34.16   Coniferous forests are often dominated by a few species of trees
34.17   Long, bitter-cold winters characterize the tundra
34.18   Ecologist Ariel Lugo studies tropical forests in Puerto Rico

A Mysterious Giant of the Deep

NEARLY 3,350 M UNDER THE PACIFIC, the deep-sea submersible *Tiburon* maneuvers through the dark. No light from above penetrates here. Controlled by marine biologists at the surface, *Tiburon*'s camera suddenly catches sight of something strange: a huge creature, 6-m-long with thin 3-m tentacles and large wing-like mantles. The mantles, like wings of wide pink ribbon undulating in a breeze, propel the mysterious giant along the seafloor. The squid hovers in what scientists think is its "fishing position," tentacles spread wide and dangling down "like a living spider web." When its long arms contact the vessel, it quickly swims away. Another squid is not so lucky, getting itself entangled with the submersible and only dislodging its arms and tentacles with great difficulty.

There are about 1,000 known species of squids, but until recently, no one had reported on this "mystery squid." It may represent not just a new species, but an entirely new family of squids. Scientists won't be sure until one is captured, and this may take some time. Although the squid has now been sighted perhaps a dozen times around the world, the ocean is Earth's largest and least explored ecosystem.

Submersibles such as the remotely operated *Tiburon* are helping us learn more about ocean life. Another submersible that has helped researchers find previously unknown deep-sea species is *Alvin* (Figure A). Accommodating a pilot and two other people and equipped with spotlights and sampling arms,

The Biosphere: An Introduction to Earth's Diverse Environments

Figure A Deep-sea submersible *Alvin*

it has carried scientists down some 2,500 m. Off the southern tip of Baja California in Mexico, scientists in *Alvin* discovered seafloor life whose ultimate energy source is not sunlight, but energy that comes from the interior of the planet. Scientists found these new organisms inhabiting the unique world of hydrothermal vents, sites near the adjoining edges of giant plates of Earth's crust, where molten rock and hot gases surge upward from Earth's interior. **Figure B** shows one such site, where a chimneylike vent, perhaps as much as 30 m high, emits scalding water and hot gases such as hydrogen sulfide (H_2S). The water temperature is 350°C, but boiling is prevented by immense water pressure. Just a meter from the vents, the water temperature is only about 2°C. This is truly an environment of extremes.

A variety of animals, including sea anemones, giant clams, shrimps, crabs, a few fishes, and tube worms (**Figure C**), thrive near hydrothermal vents. These animals live on energy extracted from chemicals by bacteria, rather than on light energy trapped by plants, algae, or photosynthetic bacteria, as other animals do.

The bacteria inhabiting hydrothermal vents are chemoautotrophs, most of which obtain energy by oxidizing hydrogen sulfide. These so-called sulfur bacteria use the energy to convert CO_2 from seawater into organic food molecules. Many of the animals in vent communities, such as the giant tube worms (as large as 3 m long), harbor sulfur bacteria within their body. The worms absorb sulfur compounds from the water; the bacteria use the compounds as an energy source and make organic food molecules.

The unusual organisms living at hydrothermal vents and near the ocean floor highlight how diverse life is on our planet. Until deep-sea submersible expeditions, no one knew that entire communities of organisms could live on energy from inorganic compounds or that mystery squids were anything more than tall tales. The scientific study of the interactions of organisms with their environments—from hydrothermal vents to solar-powered terrestrial systems—is called **ecology** (from the Greek *oikos*, home). As we will see in this chapter and those that follow, ecology is a critically important field of biology, with implications for all forms of life on Earth. ■ ■ ■

Figure C Tube worms

Figure B Hydrothermal vents

683

34.1 Ecologists study how organisms interact with their environment at several levels

Figure 34.1 Research on giant clams near an ocean vent

The definition of ecology just given sounds straightforward, but it covers an enormously complex, exciting, and important area of biology. The study of ecology reveals the great richness of our planet and provides the basic understanding we need to guide our decisions about how to conserve and sustain that richness.

The interactions between organisms and their environment are two-way. Organisms are affected by their environment, but by their very presence and activities, they also change the environment. In transforming energy, bacteria change the ocean vent environment and make it habitable for animals. Likewise, plants drastically alter their environment, extracting CO_2 from the atmosphere and adding O_2 to it, in the process providing themselves and animals with food.

Ecologists study environmental interactions at several levels. At the **organism** level, they may examine how one kind of organism meets the challenges of its environment, either through its physiology or behavior. An ecologist working at this level might study, for instance, the adaptations of clams to the extreme temperatures around hydrothermal vents (**Figure 34.1**).

Another level of study in ecology is the **population**, a group of individuals of the same species living in a particular geographic area. The clams of one species living near a particular ocean vent would constitute a population. An ecologist might study what factors limit their population size.

A third level, the **community**, consists of all the populations of different species that inhabit a particular area. All the organisms supported by a particular hydrothermal vent would constitute a community. An ecologist working at this level might focus on interspecies interactions, such as the effect of predation by crabs on tube worms.

The fourth level of ecological study, the **ecosystem**, includes all the forms of life in a certain area and all the nonliving factors as well. The latter, called **abiotic components**, include temperature, forms of energy, water, inorganic nutrients, and other chemicals. The organisms making up the community are called **biotic components**. Some critical questions at the ecosystem level concern how chemicals cycle and how energy flows between organisms and their surroundings. For a vent community, one ecosystem-level question would be, How much of the energy available to them do giant clams and other animals actually use?

Ecological research at any level uses the elements of the scientific process: the posing of hypotheses and the use of observations and experiments to test those hypotheses. Many ecologists perform experiments not only in the laboratory, where conditions can be simplified and controlled, but also in the field. As we discussed in Module 1.8, testing ecological hypotheses must take into account the multiple variables of the environment. Ecologists also devise mathematical and computer models that enable them to simulate large-scale experiments that are impossible to conduct in the field.

? A (an) _____ consists of a biological _____ (all the biotic components of an area), along with the nonliving environmental factors, or _____ components.

■ ecosystem · · · community · · · abiotic

THE BIOSPHERE

34.2 The biosphere is the total of all of Earth's ecosystems

Gazing at a view like that in **Figure 34.2A**, Apollo astronaut Rusty Schweickart once remarked: "On that small blue-and-white planet below is everything that means anything to you. National boundaries and human artifacts no longer seem real. Only the biosphere, whole and home of life."

The **biosphere** is the global ecosystem—the sum of all the planet's ecosystems. The most complex level in ecology, the biosphere includes the atmosphere to an altitude of several kilometers, the land down to and including water-bearing rocks about 3,000 m under Earth's surface, lakes and streams, caves, and the oceans to a depth of several kilometers. Isolated in space, the biosphere is self-contained, or closed, except that its photosynthesizers derive energy from sunlight, and it loses heat to space.

Figure 34.2A Earth as seen from the moon

Another feature of the biosphere is its patchiness, and we can see this on several levels. On a global scale, we see it in the distribution of continents and oceans. On a regional scale, patchiness occurs in the distribution of deserts, grasslands, forests, lakes, and streams, for example. The aerial view of a wilderness area in **Figure 34.2B** shows patchiness on a local scale. Here we see a mixture of forest, small lakes, a meandering river, and open meadows. If we moved even closer, into any one of these different environments, we would find patchiness on yet a smaller scale. For example, we would find that each lake has several different **habitats** (specific environments in which organisms live). Each habitat has characteristic abiotic factors—such as water depth, temperature, and dissolved O_2 in a lake habitat—that largely determine the kinds of organisms that live in that habitat.

Standing in a wilderness can be misleading; the lakes and streams appear untouched, and the forest seems almost boundless. Views from space are more sobering, for they remind us that our planet is a finite home in the vastness of

Figure 34.2B Patchiness of the environment in the Alaskan wilderness

space, not an unlimited frontier that we humans can abuse indefinitely.

? Why is it more accurate to define the biosphere as the global ecosystem rather than the global community?

■ Because the biosphere includes both abiotic and biotic components

34.3 Environmental problems reveal the limits of the biosphere

Figure 34.3 Rachel Carson

> The "control of nature" is a phrase conceived in arrogance, born of the Neanderthal age of biology and philosophy, when it was supposed that nature exists for the convenience of man.
>
> —Rachel Carson, *Silent Spring*

Our current awareness of the biosphere's limits stems mainly from the 1960s, a time of growing disillusionment with environmental practices of the past. In the 1950s, technology seemed poised to free humankind from several age-old bonds. New chemical fertilizers and pesticides, for example, showed great promise for increasing agricultural productivity and eliminating insect-borne diseases. Fertilizers were applied extensively, and pests were attacked by massive aerial spraying. The immediate results were astonishing: Increases in farm productivity enabled developed nations such as the United States to grow surplus food and market it overseas, and the worldwide incidence of malaria and several other insect-borne diseases was markedly reduced. DDT and other chemicals were hailed as miracle weapons with potential use anywhere insects caused problems.

Our enthusiasm for chemical fertilizers and pesticides began to wane as some of the side effects of DDT and other widely used poisons began appearing in the late 1950s. One of the first to perceive the global dangers of pesticide abuse

was Rachel Carson (**Figure 34.3**). Much of our current environmental awareness stems from her book *Silent Spring,* published in 1962. Her warnings were underscored when, shortly thereafter, scientists reported that DDT was threatening the survival of predatory birds and was showing up in human milk. Another serious problem to arise was genetic resistance to pesticides, evolving in an increasing number of pest populations. By the early 1970s, disillusionment with the overuse of chemicals and a realization that our finite biosphere could not tolerate unlimited exploitation had developed into widespread concern about environmental problems.

Today, it's clear that no part of the biosphere is untouched by human activities, and many people are concerned about the abusive impact of human populations and technology. The growing list of species that are extinct or endangered by loss of habitat; localized famines aggravated by land misuse and an expanding population; and the poisoning of soil and streams with toxic wastes—these are just a few of the problems that we have created and must solve. We will examine some of our environmental problems in Chapter 38. But analyzing environmental issues and planning for better practices begin with an understanding of the basic concepts of ecology, which we start to explore now.

Web/CD Activity 34A *Connection: DDT and the Environment*

? As a call for environmental consciousness, Rachel Carson's *Silent Spring* focused on the destructive consequences of toxic pollutants, especially _____.

■ the pesticide DDT

Physical and chemical factors influence life in the biosphere

The biosphere is finite, but it is also extremely diverse. Understanding its structure and dynamics can help us understand our environmental dilemmas. Of the variety of physical and chemical factors affecting the organisms living in a particular ecosystem, solar energy, water, temperature, and wind are among the most important.

Solar energy powers nearly all terrestrial and shallow-water ecosystems. Other ecosystems, such as the hydrothermal vents and ecosystems in dark caves and in groundwater deep beneath Earth's surface, are powered by energy that bacteria extract from inorganic chemicals. In aquatic environments that sunlight reaches, the availability of light has a significant effect on the growth and distribution of photosynthetic bacteria and algae. Because water itself and the microorganisms in it absorb light and keep it from penetrating very far, most photosynthesis occurs near the surface of a body of water. In terrestrial environments, light is often not the most important factor limiting plant growth. In many forests, however, shading by trees creates intense competition for light at ground level.

Water, a second abiotic factor, is essential to all life. Aquatic organisms have a seemingly unlimited supply of water, but they face problems of water balance if their own solute concentration does not match that of their surroundings. As we saw in Module 25.4, aquatic organisms confront solute concentrations in freshwater lakes and streams that are very different from those in the sea. For a terrestrial organism, the main water problem is the threat of drying out. Many land species have watertight coverings that reduce water loss. Some terrestrial animals have kidneys that save water by excreting very concentrated urine.

Temperature is an important abiotic factor because of its effect on metabolism. Few organisms can maintain a sufficiently active metabolism at temperatures close to 0°C, and temperatures above 45°C destroy the enzymes of most organisms. Extraordinary adaptations enable some species to live outside this temperature range. For example, some of the frogs and turtles living in the northern United States and Canada can freeze during winter months and still survive, and bacteria living in hydrothermal vents and hot springs have enzymes that function optimally at extremely high temperatures. Mammals and birds can remain considerably warmer than their surroundings and can be active in a fairly wide range of temperatures, but even these animals function best at certain temperatures.

Wind is an important abiotic factor for several reasons. Local wind damage often creates openings in forests, contributing to patchiness in ecosystems. Wind also increases an organism's rate of water loss by evaporation. The resulting increase in evaporative cooling can be advantageous on a hot summer day, but it can cause dangerous wind chill in the winter.

These abiotic factors combine with others to produce the physical and chemical components of an organism's environment. For example, most soils are complex combinations of inorganic nutrients, organic materials in various stages of decomposition, water, and air. Such variables as soil structure, pH, and nutrient content often play major roles in determining the distribution of organisms.

Other abiotic factors include such unpredictable disturbances as fires, hurricanes, tornadoes, tsunamis, and volcanic eruptions. Fire occurs frequently enough in some communities, however, that many plants have adapted to this periodic disturbance. The photographs in **Figure 34.4** were taken during and a few months after an extensive fire in Yellowstone National Park. Small plants rapidly colonized the area, taking advantage of the increased sunlight and nutrients released from the trees that burned.

In the next module, we examine the interaction between one animal species and the abiotic factors of its environment. We also look at some of the influencing biotic factors in the animal's ecosystem.

Figure 34.4 Fire and recovery in Yellowstone National Park

Web/CD Thinking as a Scientist *How Do Abiotic Factors Affect Distribution of Organisms?*

? _____ energy is such an important abiotic factor because _____ provides the organic fuel and building material for the organisms of almost all ecosystems.

■ Solar . . . photosynthesis

Organisms are adapted to abiotic and biotic factors by natural selection

An organism's ability to survive and reproduce in a particular environment is a result of natural selection, as we discussed in Chapter 13. By eliminating the least fit individuals in populations, environmental forces help adapt species to the mix of abiotic and biotic factors that they encounter.

The presence of a species in a particular place can come about in two ways: The species may evolve in that location, or it may disperse to that location and be able to survive once it is there. The pronghorn "antelope," pictured in Figure 34.5, evolved on the open plains and shrub deserts of North America over 1 million years ago. It is found nowhere else and is only distantly related to the numerous species of antelopes in Africa. Taking the pronghorn as an example, let's see how some of its unique adaptations fit the environmental conditions in which it evolved.

First, what about the major abiotic factors? The pronghorn's habitat is arid, windswept, and subject to extreme temperature fluctuations both daily and seasonally. The pronghorn is superbly adapted to these conditions. If you drive through Wyoming or parts of Colorado in the winter, you will see herds of these animals foraging in the open when temperatures are well below 0°C. The pronghorn has a thick coat made of hollow hairs that trap air and use it as insulation. In hot weather, the pronghorn can raise patches of this stiff hair to release body heat. Water is rarely a problem for a pronghorn because it obtains a great deal of moisture from the vegetation it eats.

What about the pronghorn's adaptations to the biotic components of its habitat? The pronghorn's main foods are forbs (small broadleaf plants), grasses, and woody shrubs, and its teeth are adapted for biting and chewing these plants. Also, like a cow, it has a stomach containing cellulose-digesting bacteria. As the pronghorn eats plants, the bacteria digest cellulose, and the animal obtains most of its nutrients from the bacteria. As the pronghorn evolved, it became adapted to predation by wolves, coyotes, and cougars. The pronghorn's main adaptations to escape predators are great speed and endurance. Capable of sprinting about 95 km/hr (60 mph) on flat ground, it is one of the fastest mammals. An adult pronghorn can also keep up a pace of about 65 km/hr for at least 30 minutes—a definite benefit when being chased by long-distance runners such as wolves. Other adaptations that help the pronghorn foil predators include its tan and white coat, which often camouflages the animal on the open plains, and its keen eyes, which can detect movement at great distances. The pronghorn also derives protection from living in herds. When one pronghorn starts to run, its white rump patch seems to alert other herd members to danger.

Organisms vary a great deal in their ability to tolerate fluctuations and long-term changes in their environment. The pronghorn has survived significant environmental changes. During the pioneering days of the American West, its numbers were seriously reduced by human hunting. Unlike the bison, elk, wolf, and cougar (all of which were extinguished

Figure 34.5 Pronghorns (*Antilocapra americana*)

from the plains), wild populations of the pronghorn survived. Today, the species is common in several western states, where it competes mainly with domestic cattle and sheep for food and is hunted mainly by humans and coyotes.

The pronghorn is a highly successful herbivorous running mammal of open country. In a very different environment, such as a wooded area where predators would be more easily hidden by vegetation and could stalk the pronghorn at close range, the pronghorn's adaptations for escaping predators might not be as effective. This suggests that an organism can usually tolerate environmental fluctuations only within the set of conditions to which it is adapted. Outside that set, the organism may not survive long enough to reproduce. Thus, in adapting populations to local environmental conditions, natural selection may limit the distribution of organisms. The absence of the pronghorn outside North America, however, does not necessarily imply that the species could not survive elsewhere; it may only mean that it was never able to disperse beyond this region.

Web/CD Activity 34B *Adaptations to Biotic and Abiotic Factors*

? Why does the set of evolutionary adaptations characterizing a species tend to limit the geographic distribution of that species?

■ A species well adapted to a particular set of environmental conditions may not be as well equipped to survive and reproduce where the environment poses different challenges.

34.6 Regional climate influences the distribution of biological communities

When we ask what determines whether a particular organism or community of organisms lives in a certain area, the climate of the region—especially temperature and rainfall—is often a large part of the answer. Earth's global climate patterns are largely determined by the input of solar energy and the planet's movement in space.

Figure 34.6A shows that because of its curvature, Earth receives an uneven distribution of solar energy. The sun's rays strike equatorial areas most directly (perpendicularly). Away from the equator, the rays strike Earth's surface at oblique angles. As a result, the same amount of solar energy is spread over a larger area. Thus, any particular area of land or ocean near the equator absorbs more heat than comparable areas in the more northern or southern latitudes.

The seasons of the year result from the permanent tilt of the planet on its axis as it orbits the sun. As **Figure 34.6B** shows, the globe's position relative to the sun changes through the year. The Northern Hemisphere, for instance, is tipped most toward the sun in June, creating the long days of summer in that hemisphere; at the same time, days are short and it is winter in the Southern Hemisphere. Conversely, the Southern Hemisphere is tipped farthest toward the sun in December, creating summer there and winter in the Northern Hemisphere. The **tropics**, the region between latitudes 23.5° north (the Tropic of Cancer) and 23.5° south (the Tropic of Capricorn), experience the greatest annual input and least seasonal variation in solar radiation.

Figure 34.6C shows some of the effects of the intense solar radiation near the equator on global patterns of rainfall and winds. Arrows indicate air movements. High temperatures in the tropics evaporate water from Earth's surface. Heated by the direct rays of the sun, moist air at the equator rises, creating an area of calm or of very light winds known as the **doldrums.** As warm equatorial air rises, it cools and releases much of its water content, creating the abundant precipitation typical of most tropical regions. High temperatures

throughout the year and ample rainfall largely explain why rain forests are concentrated near the equator.

After losing their moisture over equatorial zones, high-altitude air masses spread away from the equator until they cool and descend again at latitudes of about 30° north and south. This descending dry air absorbs moisture from the land. Thus, many of the world's great deserts—the Sahara in North Africa and the Arabian on the Arabian Peninsula, for example—are centered at these latitudes. As the dry air descends, some of it spreads back toward the equator. This movement creates the cooling **trade winds,** which dominate the tropics. As the air moves back toward the equator, it warms and picks up moisture until it ascends again.

The latitudes between the tropics and the Arctic Circle in the north and the Antarctic Circle in the south are called **temperate zones.** Generally, these regions have seasonal variations in climate and more moderate temperatures than the tropics or the polar zones. Notice in Figure 34.6C that some of the descending air heads into the latitudes above 30°. At first these air masses pick up moisture, but they tend to drop it as they cool at higher latitudes. This is why the north and south temperate zones, especially latitudes around 60°, tend to be moist. Broad expanses of coniferous forest dominate the landscape at these fairly wet but cool latitudes.

Figure 34.6D shows the major global air movements, called the **prevailing winds.** Prevailing winds (pink arrows) result from the combined effects of the rising and falling of air masses (blue and brown arrows) and Earth's rotation (gray arrows). Because Earth is spherical, its surface moves faster at the equator (where its diameter is greatest) than at other latitudes. In the tropics, Earth's rapidly moving surface deflects vertically circulating air, making the trade winds blow from east to west. In temperate zones, the slower-moving surface produces the **westerlies,** winds that blow from west to east.

A combination of the prevailing winds, the planet's rotation, unequal heating of surface waters, and the locations

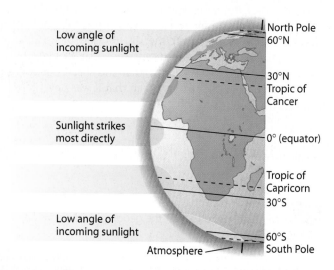

Figure 34.6A How solar radiation varies with latitude

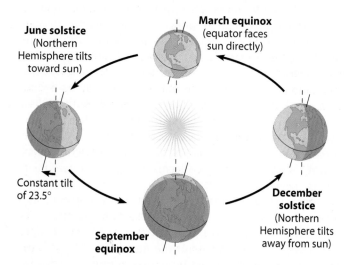

Figure 34.6B How Earth's tilt causes the seasons

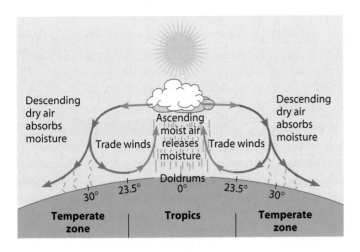

Figure 34.6C How uneven heating causes rain and winds

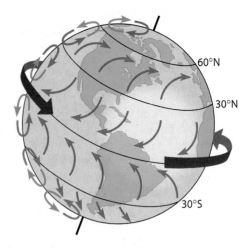

Figure 34.6D Prevailing wind patterns

and shapes of the continents creates **ocean currents**, river-like flow patterns in the oceans. Ocean currents have a profound effect on regional climates. For instance, the Gulf Stream circulates warm water northward from the Gulf of Mexico and makes the climate on the west coast of Great Britain warmer during winter than the coast of New England, which is actually farther south but is cooled by a current flowing south from the coast of Greenland.

Oceans generally moderate the climate of nearby land. On a good March day in Southern California, it is possible to spend the morning snowboarding in freezing temperatures in the mountains and then go to the beach in the afternoon to surf and bask under 18°C (65°F) skies. And as **Figure 34.6E** shows, in August the beach offers welcome relief to those who live just 40 or so miles inland.

Landforms can also affect local climate. Air temperature declines by about 6°C with every 1,000-m increase in elevation, an effect you've probably experienced if you've hiked up a mountain. **Figure 34.6F** illustrates the effect of mountains on rainfall. This drawing represents major landforms across the state of California, but mountain ranges cause similar effects elsewhere. California is a temperate area in which the prevailing winds are westerlies. As moist air moves in off the Pacific Ocean and encounters the westernmost mountains (the Coast Range), it flows upward, cools at higher altitudes, and drops a large amount of water. The world's tallest trees, the coastal redwoods, thrive here. Farther inland, precipitation increases again as the air moves up and over higher mountains (the Sierra Nevada). Some of the world's deepest snow packs occur here. On the eastern side of the Sierra, there is little precipitation, and the descending air also absorbs moisture. As a result of this rain shadow, much of central Nevada is desert.

Rain forests and deserts are among the world's major ecosystems, called **biomes.** Just as rain forests appear where there is abundant precipitation, and deserts where dry air descends over land, the appearance of other types of biological communities can be explained by regional climates. We'll see this clearly when we survey the biosphere's major terrestrial biomes. First, let's take a brief look at aquatic biomes.

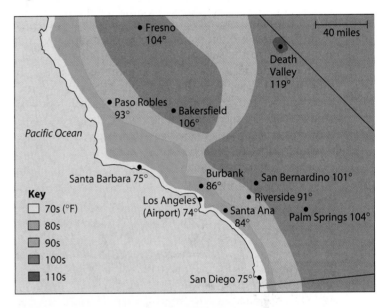

Figure 34.6E Local high temperatures for August 12, 2004, in Southern California

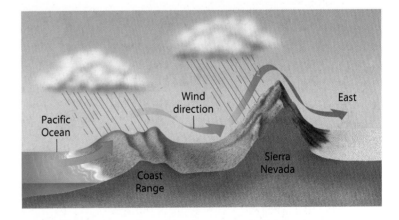

Figure 34.6F How mountains affect rainfall (California)

? What causes summer in the Northern Hemisphere?

■ Because of the fixed angle of Earth's axis relative to the orbital plane around the sun, the Northern Hemisphere is tilted toward the sun during the portion of the annual orbit that corresponds to the summer months.

34.7 Oceans occupy most of Earth's surface

Life originated in the sea and evolved there for almost 3 billion years before plants and animals began moving onto land. Covering about 75% of the planet's surface, oceans have always had an enormous impact on the biosphere. As you just learned in Module 34.6, their evaporation provides most of Earth's rainfall, and ocean temperatures have a major effect on climate and wind patterns. Photosynthesis by marine algae and cyanobacteria supplies a substantial portion of the biosphere's oxygen.

Abiotic factors such as amount of light and distance from shore influence the distribution of aquatic biomes. The area of shore where land meets ocean is called the **intertidal zone.** This area is often pounded by waves during high tide and exposed to the sun and drying winds during low tide. The rocky intertidal zone is home to many sedentary organisms, such as algae, barnacles, and mussels, which attach to rocks and are thus prevented from being washed away. On sandy beaches, suspension-feeding worms, clams, and predatory crustaceans bury themselves in the ground. Surface-dwelling crabs and shorebirds feed along the shore. Other common intertidal animals are sponges, sea anemones, echinoderms, and small fishes. **Figure 34.7A** shows some of the diverse organisms in a tide pool on the coast of central California. A tide pool is a small body of water that remains in a rock or sand depression during low tide.

The intertidal zone is one of several oceanic zones (**Figure 34.7B**). The open ocean itself, called the **pelagic zone** (from the Greek *pelagos,* sea), supports communities dominated by highly motile animals such as fishes, squids, and marine mammals, including whales and dolphins. Microscopic algae and cyanobacteria, collectively called **phytoplankton** (from the Greek *phyton,* plant, and *plankton,* wandering), drift passively in the pelagic zone. Phytoplankton are the ocean's main photosynthesizers. **Zooplankton** are small, drifting animals that usually have morphological features that keep them afloat. They eat phytoplankton and in turn are consumed by larger animals.

The seafloor is called the **benthic zone** (from the Greek *benthos,* bottom of the sea). Depending on depth and light penetration, the benthic community consists of attached algae, fungi, bacteria, sponges, burrowing worms, clams, sea anemones, crabs, and fishes.

As Figure 34.7B indicates, marine biologists often group the illuminated regions of the benthic and pelagic communities together, calling them the photic zone. The **photic zone** is a relatively small portion of ocean water and bottom into which light penetrates and in which photosynthesis occurs. Underlying the photic zone is a vast, dark region called the **aphotic zone.** This is the most extensive part of the biosphere. Even without light, life is still diverse in the aphotic zone. Many kinds of invertebrates, such as sea urchins and polychaete worms, and some fishes scavenge organic matter that sinks from the lighted waters above. Fish may have enlarged eyes, enabling them to see in the very dim light, and

Figure 34.7A Intertidal zone organisms

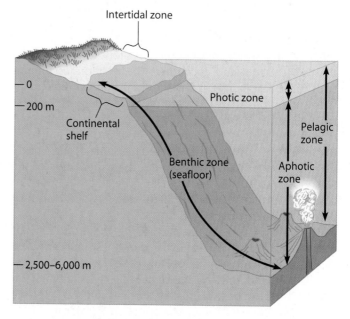

Figure 34.7B Oceanic zones

luminescent organs that attract mates and prey. And as we discussed in the chapter introduction, the hydrothermal vent communities, powered by chemical energy rather than sunlight, are densely populated with diverse and unique organisms.

On the submerged parts of continents called **continental shelves**, the benthic communities usually receive some light, and nutrients from the seafloor circulate in the shallow water. **Coral reefs**, a visually spectacular and biologically diverse ecosystem, are found in warm tropical waters above the continental shelf. The reef is built up slowly by successive generations of coral animals—a diverse group of cnidarians (see Module 18.6) that secrete a hard external skeleton—and by multicellular algae encrusted with limestone (see Module 16.24). Unicellular mutualistic algae live within the corals, providing the coral with food. Many of the branching forms in **Figure 34.7C** are skeletons of coral animals. Coral reefs support a huge diversity of invertebrates and fishes. These highly productive biomes, however, are easily degraded by pollution, native and introduced predators, and human souvenir hunters.

Figure 34.7D shows an **estuary**, an area where a freshwater stream or river merges with the ocean. The saltiness of estuaries ranges from nearly that of fresh water (less than 1% salt) to that of the ocean (3% salt). This particular estuary is part of the Chesapeake Bay in Maryland. Several rivers empty into this bay, which then merges with the Atlantic Ocean. With their waters enriched by nutrients from the river, estuaries are among the most productive biomes on Earth. Oysters, crabs, and many fishes live in estuaries or reproduce in them. Estuaries are also crucial nesting and feeding areas for waterfowl.

Mudflats and salt marshes are extensive coastal wetlands that often border estuaries. A **wetland** is an area that is transitional between an aquatic ecosystem and a terrestrial one. Covered with water either permanently or periodically, wetlands support the growth of aquatic plants.

Until fairly recently, many people viewed the ocean as a bountiful, virtually limitless resource, and we have harvested the ocean heavily and used it as a dumping ground for wastes. Estuaries and coastal wetlands have been especially abused, with few undisturbed areas remaining and many totally replaced by commercial and residential developments on landfill. We are now seeing the effects of our disregard for marine communities: Seafood is becoming less plentiful, the result of overharvesting and pollution; many whales and other species are in danger of extinction, mainly from overhunting; and oil and other pollutants foul coastal areas. Laws in many countries, including the United States,

Figure 34.7C A coral reef with its immense variety of invertebrates and fishes

now prohibit whaling and the disposal of sewage and other wastes at sea. Many countries are also taking steps to restore and conserve estuaries and other wetlands.

? The _____, small photosynthetic organisms inhabiting the _____ zone of the pelagic zone, provide most of the food for oceanic life.

■ phytoplankton . . . photic

Figure 34.7D An estuary in Maryland

Freshwater biomes include lakes, ponds, rivers, streams, and wetlands

Figure 34.8A A stream in the Great Smoky Mountains, Tennessee

Light has a significant impact on freshwater biomes, just as it does on ocean ecosystems. In all but the smallest lakes and ponds, there is usually a distinct photic (lighted) zone and an aphotic zone. Phytoplankton grow in the photic zone, and rooted plants often inhabit shallow waters. Large populations of microorganisms in the benthic (bottom-dwelling) community decompose dead organisms that sink to the bottom. Respiration by microbes also removes oxygen from water near the bottom, and in some lakes, benthic areas are unsuitable for any organisms except anaerobic microbes.

Temperature may also have a profound effect on freshwater communities. During the summer, lakes often have a distinct upper layer of water that has been warmed by the sun and does not mix with underlying, cooler water. Fish often spend much of their time in the deep, cool waters of a lake unless oxygen levels there become depleted by decomposers.

Nitrogen and phosphorus are the nutrients that usually limit the amount of phytoplankton growth in a lake or pond. When there are temperature layers in a lake, for instance, nutrients released by decomposers can become trapped near the bottom, out of reach of the phytoplankton. During the summer months, this may limit the growth of algae (and thus photosynthesis) in the photic zone. As winter approaches, the surface water becomes denser as it cools; it then tends to mix with the deeper water, allowing nutrients to return to the surface, where phytoplankton can again use them. Seasonal mixing also restores oxygen to the depths.

Today, many lakes and ponds are affected by large inputs of nitrogen and phosphorus from sewage and from runoff from fertilized lawns and agricultural fields. These nutrients often produce blooms, or population explosions of algae. Heavy algal growth reduces light penetration into the water, and when the algae die and decompose, a pond or lake can suffer serious oxygen depletion.

Rivers and streams generally support communities of organisms quite different from those of lakes and ponds. A river or a stream changes greatly between its source (perhaps a spring or snowmelt) and the point at which it empties into a lake or the ocean. Near the source, the water is usually cold, low in nutrients, and clear (**Figure 34.8A**). The channel is often narrow, with a swift current that does not allow much silt to accumulate on the bottom. The current also inhibits the growth of phytoplankton; most of the organisms found here are supported by the photosynthesis of algae attached to rocks or by organic material (such as leaves) carried into the stream from the surrounding land. The most abundant benthic animals are usually arthropods such as small crustaceans and insect larvae that have physical and behavioral adaptations that enable them to resist being swept away. Trout are often the predominant fishes, locating their food, including insects, mainly by sight in the clear water.

Downstream, a river or stream generally widens and slows. The water is usually warmer and may be murkier because of sediments and phytoplankton suspended in it. Worms and insects that burrow into mud are often abundant, as are waterfowl, frogs, and catfish and other fishes that find food more by scent and taste than by sight.

Freshwater wetlands range from swamps, as shown in **Figure 34.8B**, to marshes and bogs. They may form in shallow basins or along the banks of rivers or lakes. Wetlands are among the richest of biomes in terms of species diversity. They provide water storage areas that reduce flooding and improve water quality by filtering pollutants. The recognition of their ecological and economic value has led to governmental and private efforts to protect and restore wetlands.

Web/CD Activity 34C *Aquatic Biomes*

Figure 34.8B The Okefenokee National Wetland Reserve, Georgia

? Why does sewage cause algal blooms in lakes?

■ The sewage adds nutrients, such as nitrates and phosphates, that stimulate growth of the algae.

34.9 Terrestrial biomes reflect regional variations in climate

Figure 34.9 introduces the eight major types of terrestrial biomes. Many of the biomes are named for climatic features and for their predominant vegetation, but each is also characterized by microorganisms, fungi, and animals adapted to that particular environment. A grassland, for instance, is more likely than a forest biome to be populated by grazing animals, such as the pronghorn described in Module 34.5.

The distribution of the biomes largely depends on climate, with temperature and rainfall often the key factors determining the kind of biome that exists in a particular region. If the climate in two geographically separate areas is similar, the same type of biome may occur in both places; notice on the map that each biome type occurs on at least two continents. Each biome is characterized by a *type* of biological community, not a specific assemblage of species. For example, the groups of species living in the deserts of the American Southwest and in the Sahara Desert of Africa are different, but the species in both are adapted to desert conditions. Widely separated biomes may look alike because of convergence, the evolution of similar traits in independently evolved species living in similar environments (see Module 15.6).

Within each biome there is local variation, giving the vegetation a patchy, rather than a uniform, appearance. For example, in northern coniferous forests, snowfall may break branches and small trees, causing openings where broadleaf trees such as aspen and birch can grow. Local storms and fires also create openings in many biomes.

Fire has a very important effect in some biomes. Without periodic burning, many grasslands would be replaced by forests. Most grasses survive burning because the growing points of their shoots are below-ground and are not killed. Some species of pines that dominate coniferous forests produce cones that release their seeds only when subjected to extreme heat. Germinating soon after a forest fire has blackened an area, they have access to open space and a rich supply of nutrients from burned plants.

Today, many natural biomes are broken up and altered by human activity. In fact, as we discuss in Chapter 38, a high rate of biome alteration by humans is correlated with an unusually high rate of species loss throughout the globe.

We now begin a more detailed survey of the major biomes. To help you locate the biomes, we include with each module an orientation map that is color-coded to match the map here.

Web/CD Activity 34D *Terrestrial Biomes*

? Test your knowledge of world geography: Which biome is most closely associated with a "Mediterranean climate"?

■ Chaparral

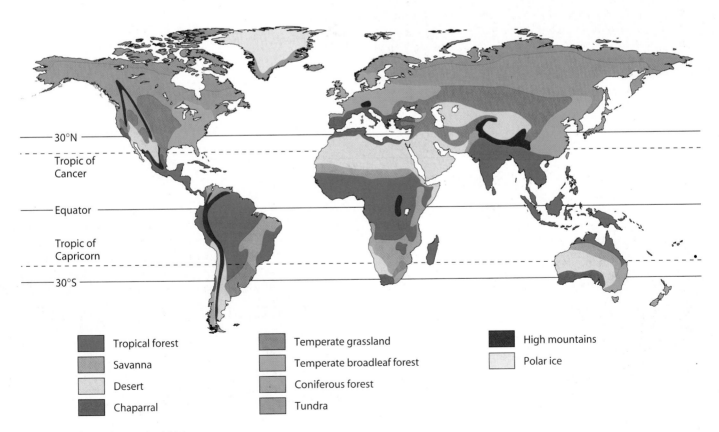

Tropical forest	Temperate grassland	High mountains
Savanna	Temperate broadleaf forest	Polar ice
Desert	Coniferous forest	
Chaparral	Tundra	

Figure 34.9 Major terrestrial biomes

34.10 Tropical forests cluster near the equator

Tropical forests occur in equatorial areas where the temperature is warm and days are 11–12 hours long year-round. Rainfall in these areas is quite variable, and this, rather than temperature or photoperiod, generally determines the kind of vegetation that grows in a particular kind of tropical forest. In areas where rainfall is scarce or there is a prolonged dry season, tropical dry forests predominate. The plants found there are a mixture of thorny shrubs and deciduous trees and succulents. Tropical rain forests are found in very humid equatorial areas where rainfall is abundant (200–400 cm per year).

The tropical rain forest, such as the lush area in Borneo shown in Figure 34.10, is among the most complex of all biomes, harboring enormous numbers of different species. Up to 300 species of trees can be found in a single hectare (2.5 acres). Vertical stratification, providing many different habitats, includes a layer of emergent trees growing above a closed upper canopy, one or two layers of lower trees, the shrub understory, and the ground layer of herbaceous plants. Many trees are covered by woody vines known as lianas and by epiphytes (plants that grow on other plants rather than in soil) such as bromeliads and orchids. Many of the animals are also tree-dwellers; monkeys, birds, insects, snakes, bats, and frogs find food and shelter many meters above the ground.

The soils of tropical rain forests are typically poor, because high temperatures and rainfall lead to rapid decomposition and recycling rather than to a buildup of organic material.

Human impact on the world's tropical rain forests is currently a source of great concern. It is a common practice to clear the forest for lumber or simply burn it, farm the land for a few years, and then abandon it. Mining has also devastated large tracts of rain forest. Once stripped, the tropical rain forest recovers very slowly because the soil is so nutrient-poor. We discuss the potential consequences of destroying the tropical forests in Chapter 38, including the impact on world climate. In the last module of this chapter, you will learn more about ecological studies in a tropical forest in Puerto Rico.

Figure 34.10 Tropical rain forest

? Why are the soils in most tropical rain forests so poor in nutrients that they can only support farming for a few years after the forest is cleared?

■ The tropical conditions favor rapid decomposition of organic litter in the forest soil, and most of the ecosystem's nutrients are tied up in the vegetation that is cleared away before farming.

34.11 Savannas are grasslands with scattered trees

Figure 34.11, a photograph taken in Kenya, shows a typical **savanna**, a biome dominated by grasses and scattered trees. Rainfall averages 30–50 cm per year, and the temperature is warm year-round. Savannas (from the Spanish *sabana*, meadow) are simple in structure compared with tropical forests. Frequent fires, caused by lightning or human activity, and grazing animals inhibit further invasion by trees. The dominant plants are fire adapted.

Figure 34.11 Savanna

Grasses and forbs (small broadleaf plants) grow rapidly during the rainy season, providing a good food source for many animal species. Large grazing mammals must migrate to greener pastures and scattered watering holes during regular periods of seasonal drought. The dominant herbivores in savannas are actually insects, especially ants and termites. Also common are many burrowing animals, including mice, moles, gophers, snakes, ground squirrels, worms, and numerous arthropods.

Many of the world's large herbivores and their predators inhabit savannas. African savannas are home to giraffes, zebras, and many species of antelopes, as well as to lions and cheetahs. Several species of kangaroos are the dominant large herbivores of Australian savannas.

? How do fires help maintain savannas as grassland ecosystems?

■ By repeatedly preventing the spread of trees and other woody plants

---

34.12 Deserts are defined by their dryness

Deserts are the driest of all terrestrial biomes, characterized by low and unpredictable rainfall (less than 30 cm per year). Some deserts can be very hot, with daytime soil surface temperatures above 60°C (140°F) and large daily temperature fluctuations. Other deserts, such as those west of the Rocky Mountains, are relatively cold. Air temperatures in cold deserts may fall below −30°C. The large deserts in central Australia and central Africa have average annual rainfall of less than 2 cm, and the Atacama Desert in Chile, the driest place on Earth, averages less than 0.1 mm! There is often no rain at all for decades at a time. But not all desert air is dry. Coastal sections of the Atacama and of the Namib Desert in Africa are often shrouded in fog, although the ground remains extremely dry.

Figure 34.12 Desert

As we discussed in Module 34.6, large tracts of desert occur in two regions of descending dry air, centered around the 30° north and 30° south latitudes. At higher latitudes, large deserts may occur in the rain shadows of mountains; these include much of central Asia east of the Caucasus Mountains and much of California and Nevada east of the Sierra Nevada.

The cycles of growth and reproduction in the desert are keyed to rainfall. The driest deserts have no perennial vegetation at all, but in less arid regions the dominant plants are scattered deep-rooted shrubs, often interspersed with cacti, as in the Sonoran Desert in southern Arizona, shown in **Figure 34.12**. The pleated structure of the saguaro cacti you see in the photo enables these plants to expand when they absorb water during wet periods. Desert plants typically produce great numbers of seeds, which may remain dormant until a heavy rain triggers germination. Periods of rainfall (often in late winter) may produce spectacular blooms of annual plants.

Like desert plants, desert animals are adapted to drought and extreme temperatures. Many live in burrows and are active only during the cooler nights, and most have special adaptations that conserve water. Seed-eaters such as ants, many birds, and rodents are common in deserts. Lizards, snakes, and hawks eat the seed-eaters.

The process of **desertification**, the conversion of semiarid regions to desert, is a significant environmental problem. In central Africa, for example, a burgeoning human population, overgrazing, and dry land farming are converting large areas of savanna to desert.

? Why isn't "cold desert" an oxymoron?

■ Because deserts are defined by relatively little precipitation and dry soil, not mainly by temperature

34.13 Spiny shrubs dominate the chaparral

Chaparral (the Spanish word for "place of evergreen scrub oaks") is a region of dense, spiny shrubs with tough, evergreen leaves. Occurring in midlatitude coastal areas, the chaparral climate results mainly from cool ocean currents circulating offshore, which tend to produce mild, rainy winters and long, hot, dry summers.

First described in the Mediterranean region, chaparral vegetation is also found in other areas, including California. The photograph in **Figure 34.13** was taken in the Los Padres National Forest in California. In addition to the perennial shrubs that dominate chaparral, annual plants are also commonly seen, especially during the wet winter and spring months.

Chaparral vegetation is adapted to periodic fires, most often caused by lightning; in fact, the vegetation requires occasional fires for long-term maintenance. Shrubs usually regenerate quickly from their fire-resistant roots, using stored food reserves and mineral nutrients released by the fires. In addition, many chaparral plant species produce seeds that will germinate only after a hot fire. Houses do not fare as well in the firestorms that race through these areas, such as the devastating fires in the fall of 2003 in the densely populated canyons of Southern California.

Animals characteristic of the chaparral are browsers such as deer, fruit-eating birds, and seed-eating rodents, as well as lizards and snakes.

? What is the main reason homeowner's insurance is relatively expensive for people who choose to build in the chaparral?

■ High fire risk

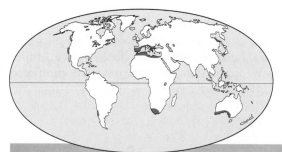

Figure 34.13 Chaparral

34.14 Temperate grasslands include the North American prairie

Figure 34.14 Temperate grassland

Temperate grasslands have some of the characteristics of tropical savannas, but they are mostly treeless, except along rivers or streams, and are found in regions of relatively cold winter temperatures. Most grasslands persist because of periodic drought, fires, and grazing by large mammals, all of which inhibit growth of woody plants but do not harm the below-ground grass shoots.

Grasslands expanded in range following the retreat of the glaciers after the last ice age. Coupled with this expansion was the proliferation of large grazing mammals. The bison and pronghorn of North America, the gazelles and zebras of the African veldt, and the wild horses and sheep

of the Asian steppes are some examples. Enriched by glacial deposits and mulch from decaying plant material, the soil of grasslands supports a great diversity of microorganisms and small animals, especially annelids, arthropods, and burrowing mammals such as prairie dogs.

The amount of annual rainfall influences the height of grassland vegetation. The photograph in Figure 34.14 (previous page) was taken on the relatively dry, short-grass prairie of South Dakota. Tallgrass prairie occurs in wetter areas, such as eastern Kansas. Little remains of North American prairies today. Most of the region is intensively farmed, and it is one of the most productive agricultural regions in the world.

> **?** How do humans now use most of the North American land that was once temperate grasslands?

■ For farming

34.15 Broadleaf trees dominate temperate forests

Temperate broadleaf forests grow throughout midlatitude regions, where there is sufficient moisture to support the growth of large trees. This includes most of the eastern United States, most of central Europe, and parts of eastern Asia and Australia. In the Northern Hemisphere, deciduous trees characterize temperate broadleaf forests. Some of the dominant trees include species of oak, hickory, birch, beech, and maple. The mix of tree species varies widely, depending on such factors as climate in different latitudes and local soil conditions. Figure 34.15 shows a photograph taken in the autumn in Great Smoky Mountains National Park in North Carolina.

Temperatures in temperate broadleaf forests range from very cold in the winter to hot in the summer (−30°C to +30°C). Precipitation is relatively high and usually evenly distributed throughout the year. These forests typically have a growing season of 5 to 6 months and a distinct annual rhythm, in which the trees drop leaves and become dormant in late autumn, then produce new leaves each spring. The loss of leaves in winter prevents evaporation of water from the tree at a time when freezing reduces the available water.

Temperate broadleaf forests are more open than tropical rain forests and are not as tall or as diverse. However, their soils are rich in inorganic and organic nutrients. Rates of decomposition are lower in temperate forests than in the tropics, and a thick layer of leaf litter accumulates on forest floors, which conserves many of the biome's nutrients. Many invertebrates live in the soil and leaf litter. Many vertebrates, such as mice, shrews, and ground squirrels, burrow for shelter and food. Temperate broadleaf forests are also home

Figure 34.15 Temperate broadleaf forest

to many species of birds and—where not eliminated by humans—bobcats, foxes, black bears, and mountain lions.

Virtually all the original broadleaf forests in North America were destroyed by logging and by the clearing of land for agriculture and urban development. In contrast to drier ecosystems, these forests tend to recover after disturbance, and today we see broadleaf trees dominating undeveloped areas over much of their former range.

> **?** How does the soil of a temperate broadleaf forest differ from that of a tropical rain forest?

■ The soil in temperate broadleaf forests is rich in inorganic and organic nutrients.

CHAPTER 34 *The Biosphere: An Introduction to Earth's Diverse Environments*   **697**

34.16 Coniferous forests are often dominated by a few species of trees

Cone-bearing evergreen trees, such as spruce, pine, fir, and hemlock, dominate **coniferous forests**. The northern coniferous forest, or taiga, is the largest terrestrial biome on Earth, stretching in a broad band across North America and Eurasia to the southern border of the arctic tundra. Taiga (from the Russian word for "mountain") is also found at cool, high elevations in more temperate latitudes, as in much of the mountainous region of western North America. Figure 34.16 shows part of a coniferous forest in Oregon.

The taiga is characterized by long, cold winters and short, wet summers that are sometimes warm. The soil is nutrient-poor, thin, and acidic, forming slowly because of the slow decomposition of conifer needles. There may be considerable precipitation, mostly in the form of snow. The snow usually falls before the coldest temperatures occur, and it insulates the soil, keeping it from freezing to such depths that it would never thaw out during the short summers. Animals of the taiga include moose, elk, hares, bears, wolves, grouse, and migratory birds.

Coniferous forests of coastal North America (Northern California to Alaska) are actually temperate rain forests. Warm, moist air from the Pacific Ocean supports this unique biome, which, like most coniferous forests, is dominated by a few tree species, such as hemlock, Douglas fir, and redwood. Coniferous forests are being logged at an alarming rate, and the old-growth stands of these trees may soon disappear.

? How does the soil of the northern coniferous forests differ from that of a broadleaf forest?

Figure 34.16 Coniferous forest

■ The soil is thinner, nutrient-poor, and acidic because conifer needles decompose slowly in the low temperatures.

34.17 Long, bitter-cold winters characterize the tundra

At the northernmost limits of plant growth and at high altitudes is the **tundra** (from the Russian word for "marshy plain"). Vegetation in the tundra includes dwarf shrubs, grasses, mosses, and lichens. The arctic tundra encircles the North Pole, extending southward to the coniferous forests. Alpine tundras are found above the treeline on high mountains, even in the tropics.

Figure 34.17 shows the arctic tundra in central Alaska in the autumn. The climate here is often extremely cold, with little light for long periods of time. During the brief, warm summers, when there is nearly constant daylight, plants grow quickly and flower in a rapid burst.

The arctic tundra is characterized by **permafrost**, continuously frozen subsoil. Only the upper part of tundra soil thaws in the summer. The permafrost prevents the roots of plants from penetrating very far into the soil. Extremely cold winter air temperatures, high winds, and permafrost explain the absence of trees. The arctic tundra may receive as little precipitation as some deserts. But poor drainage, due to the permafrost, and slow evaporation, because of the low temperatures, keep the soil continually saturated.

Animals of the tundra withstand the cold by having good insulation that retains heat. Large herbivores of the tundra include musk oxen and caribou. The principal smaller animals are rodents called lemmings and a few predators, such as the arctic fox and snowy owl. Many animals, especially birds, are migratory, using the tundra as a summer breeding ground. During the brief growing season, clouds of mosquitoes often fill the tundra air.

? What three abiotic factors account for the rarity of trees in arctic tundra?

■ Long, very cold winters (short growing season), high winds, and permafrost

Figure 34.17 Tundra

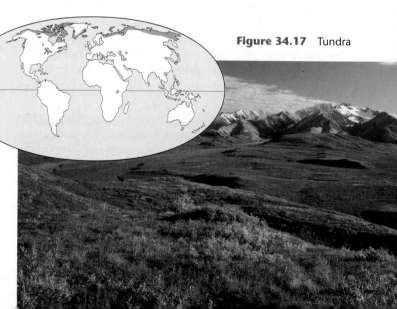

34.18 Ecologist Ariel Lugo studies tropical forests in Puerto Rico

Figure 34.18A The Luquillo Experimental Forest

Figure 34.18B Dr. Ariel Lugo

High in the mountains of Puerto Rico, birds and bats swoop through the air in search of insects. Small streams roll down rocky slopes past dense dwarf forest, through patches of palms, and under leafy canopies sheltered by tall groves. This is the Luquillo Experimental Forest (Figure 34.18A), and what is being studied here may provide answers to the problems caused by wide-scale cutting of trees in the tropics.

For scientists, the Luquillo forest is a fortunate accident. The 28,000-acre area fell into the hands of researchers only after the tropical forests of Puerto Rico had been abused for centuries. By the early 1900s, the kind of deforestation described in Module 34.10 had already ruined much of Puerto Rico's forestland. Widespread abandonment of ruined farmland allowed a patchwork of cleared and still-forested areas to be gathered into a single reserve, and the Luquillo Experimental Forest project was formally launched in 1988. The research effort is run by the University of Puerto Rico and the U.S. Forest Service, but biologists throughout the world study the forest and its recovery. By studying this forest, scientists have a window into one of the world's most diverse wonders.

The ecology of Luquillo mirrors that of other tropical forests. About half of all the world's forests are in the tropics, covering less than 6% of the world's land area but containing the vast majority of its plant and animal genetic resources. They have faced unrelenting pressure from humans in the last century: Scientists working for the United Nations estimate that about 53,000 acres of tropical forests are destroyed each year, an area the size of North Carolina.

Dr. Ariel Lugo (Figure 34.18B), a forest ecologist, has been one of the key scientists at Luquillo in his native Puerto Rico. In an interview, he described the consequences of deforestation:

When deforestation occurs in an unplanned, uncontrolled way, then you lose species and productivity and degrade your soil and water resources. These regional effects fragment the landscape. And when you fragment the landscape, you disconnect nature—you lose the value of ecosystems working in synchrony. When you add up all the deforestation, you get into the global impact. Tropical forests are so huge that they help regulate climate, they help regulate the cycles of nutrients and water and gases.

At Luquillo, work by scientists such as Dr. Lugo has offered valuable insights into the peril these forests face and the promise they hold. Researchers have watched cleared parts of the forest slowly recover, getting a better understanding of how such damage affects the entire ecosystem and how much time nature needs to heal from human impact. Hurricanes that have ripped through the island have given scientists a chance to watch the forest recover from a natural disaster—and confirm the troubling conclusion that ecosystems bounce back from natural catastrophes much more readily than they do human ones. Dr. Lugo explained:

Hurricane Hugo in 1989 was such a huge disturbance that it took out all the leaves in large areas of tropical forests. But when we studied hurricane damage in Luquillo, we found that the mortality of trees was relatively low—no more than 20%. What we're learning is that hurricanes are the main organizing force of that forest. The forest goes through a cycle that averages 60 years, starting with great impact by winds and rain of a hurricane and then about 60 years of regrowth. In those 60 years, we see the species change, the growth rates change, the size of the trees change, and the density of the wood change. In other words, the hurricane might appear destructive but it's actually constructive; it makes the forest more productive; it rejuvenates the forest.

Now you ask me, how does this impact of natural disturbance compare with human impact on the rain forest? Well, there's no comparison! Life can adapt to these regular disturbances. For example, animal populations have actually exploded after a hurricane, which means they have adaptations that allow them to respond in a positive way. But human activity is unpredictable because humans act on impulse, in the short term, or without reason. So it's much harder for nature to adapt to human activity.

Luquillo is giving scientists the chance to understand not only what humans have done wrong, but also what can be done to correct those errors. As Dr. Lugo said:

The challenge is to harmonize or integrate all the legitimate claims on the forest. Who's going to organize development such that you have indigenous people, and you have conservation of the resource, and you maintain climatic balances, and you develop some areas, and you have agriculture, and you get fuel wood? That requires management in my view, and it requires a lot of compromise, and a lot of hard thinking, and a lot of good will and political will.

? Why can widespread destruction of tropical forests affect ecosystems far from the forests?

■ Because tropical forests help regulate climate and chemical cycling on a global scale

Reviewing the Concepts

Ecology. Ecologists study the interactions of organisms with their environments at the organismal, population, community, and ecosystem levels. Ecosystem interactions involve living (biotic) communities and nonliving (abiotic) physical and chemical factors **(Introduction–34.1).**

The Biosphere (34.2–34.6)

The biosphere is the global ecosystem. Patchiness characterizes the biosphere. Each habitat has a unique community of species **(34.2).** Human activities, including the widespread use of chemicals, affect all parts of the biosphere **(34.3).**

Abiotic and biotic factors. Solar energy, water, temperature, wind, and disturbances are abiotic factors determining the biosphere's structure and dynamics **(34.4).** Natural selection adapts organisms to both abiotic factors and biotic ones, such as predation and competition **(34.5).**

Climate determines the distribution of communities. Most climatic variations are due to the uneven heating of Earth's surface as it orbits the sun, setting up patterns of precipitation and prevailing winds. Ocean currents influence coastal climate. Landforms such as mountains affect rainfall **(34.6).**

Aquatic Biomes (34.7–34.8)

Oceans. Light, distance from shore, and the availability of nutrients shape ocean communities. Oceanic zones include the intertidal, pelagic, and benthic zones. Coral reefs are found in wam waters above continental shelves. Estuaries are productive areas where rivers flow into the ocean **(34.7).**

Freshwater ecosystems. Light, temperature, and the availability of nutrients and dissolved oxygen shape lake and pond communities. Abiotic factors change from the source of a river to its mouth, and communities vary accordingly. Wetlands include marshes and swamps **(34.8).**

Terrestrial Biomes (34.9–34.18)

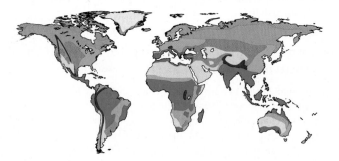

Eight terrestrial biomes. Temperature and rainfall mainly determine the terrestrial biomes **(34.9).** Tropical forests occur along the equator. Tropical rain forests are the most diverse ecosystem. Savannas are grasslands with scattered trees. Deserts are the driest biomes. The chaparral is a shrubland with cool, rainy winters and dry, hot summers. Temperate grasslands are found where winters are cold. Forests of broadleaf trees grow in some temperate areas. The northern coniferous forest, or taiga, is found where there

are short summers and long, snowy winters. Arctic tundra is a treeless biome characterized by extreme cold, wind, and permafrost. Alpine tundra occurs above the treeline on high mountains **(34.10–34.17).** The Luquillo Experimental Forest allows ecologists to study the effects of disruption on tropical forests **(34.18).**

Connecting the Concepts

1. You have seen that Earth's terrestrial biomes reflect regional variations in climate. But what determines these climatic variations? Interpret the following diagrams in reference to global patterns of temperature, rainfall, and winds.

   a. Solar radiation and latitude:

   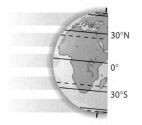

   b. Earth's orbit around the sun:

   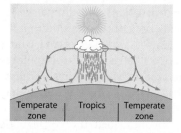

   c. Global patterns of air circulation and rainfall:

   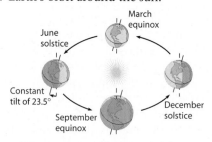

2. Aquatic biomes differ in levels of light, nutrients, oxygen, and water movement. These abiotic factors influence the productivity and diversity of the ecosystem.

   a. Productivity, roughly defined as photosynthetic output, is greater in estuaries, coral reefs, and shallow ponds than it is in the headwaters (source) of streams, the oceanic photic zone, and clear mountain lakes. Describe the abiotic factors that contribute to high productivity in the first three ecosystems.

   b. Do the same abiotic factors limit productivity in a stream's source, the oceanic photic zone, and a clear mountain lake? Explain.

Testing Your Knowledge

Matching

3. The most complex and diverse biome
4. Ground permanently frozen
5. Broadleaf trees such as hickory and birch
6. Mediterranean climate
7. Spruce, fir, pine, and hemlock trees
8. Home of ants, antelopes, and lions
9. The steppes, African veldts, and North American plains

a. chaparral
b. savanna
c. taiga
d. temperate forest
e. temperate grassland
f. tropical rain forest
g. tundra

Multiple Choice

10. Changes in the seasons are caused by
    a. the tilt of Earth's axis toward or away from the sun.
    b. annual cycles of temperature and rainfall.
    c. variation in the distance between Earth and the sun.
    d. an annual cycle in the sun's energy output.
    e. the periodic buildup of heat energy at the equator.

11. What makes the Gobi Desert of Asia a desert?
    a. The growing season there is very short.
    b. Its vegetation is sparse.
    c. It is hot.
    d. Temperatures vary little from summer to winter.
    e. It is dry.

12. Which of the following sea creatures might be described as a pelagic animal of the aphotic zone?
    a. a coral reef fish
    b. a giant clam near a deep-sea hydrothermal vent
    c. an intertidal snail
    d. a deep-sea squid
    e. a harbor seal

13. Why do the tropics and the windward side of mountains receive more rainfall than areas around latitudes 30° north and south and the leeward side of mountains?
    a. Rising warm, moist air cools and drops its moisture as rain.
    b. Descending air condenses, creating clouds and rain.
    c. The tropics and the windward side of mountains are closer to the ocean.
    d. There is more solar radiation in the tropics and on the windward side of mountains.
    e. Earth's rotation creates seasonal differences in rainfall.

14. Phytoplankton are the major photosynthesizers in
    a. swamps.
    b. streams.
    c. the ocean photic zone.
    d. the intertidal zone.
    e. coral reefs.

15. An ecologist monitoring the number of great apes in a wildlife refuge over a five-year period is studying ecology at which level?
    a. organism
    b. population
    c. community
    d. ecosystem
    e. biosphere

16. Many plant species have adaptations for dealing with periodic fires. Such fires are typical of a
    a. chaparral.
    b. savanna.
    c. temperate grassland.
    d. temperate broadleaf forest.
    e. a, b, or c

Describing, Comparing, and Explaining

17. Tropical rain forests are the most diverse biomes. What factors contribute to this diversity?

18. What biome do you live in? Describe your climate and the factors that have produced that climate. What plants and animals are typical of this biome? If you live in an urban or agricultural area, how have human interventions changed the natural biome?

Applying the Concepts

19. In the climograph below, biomes are plotted by their range of annual mean temperature and annual mean precipitation. Identify the following biomes: arctic tundra, coniferous forest, desert, grassland, temperate forest, and tropical forest. Explain why there are areas in which biomes overlap on this graph.

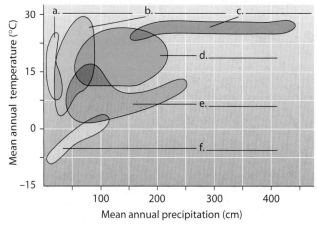

20. The North American pronghorn looks and acts like the antelopes of Africa. But the pronghorn is the only survivor of a family of mammals restricted to North America. Propose a hypothesis to explain how these widely separated animals came to be so much alike.

21. Near Lawrence, Kansas, there was a rare patch of the original North American temperate grassland that had never been plowed. It was home to numerous native grasses and annual plants, including two endangered plants. Environmental activists thought that the area should be set aside as a nature reserve, and they started to raise money to save it. In 1990, the owner of the land plowed it, stating that there are no federal laws protecting endangered plants on private grasslands and that he did not want to be told what to do with his property. What issues and values are in conflict in this situation? How could this story have had a more satisfactory ending for all concerned?

Answers to all questions can be found in Appendix 3.

For study help and Activities, go to campbellbiology.com or the student CD-ROM.

CHAPTER 35

THE SCIENTIFIC STUDY
OF BEHAVIOR

35.1   Behavioral ecologists ask both
       proximate and ultimate
       questions
35.2   Early behaviorists used
       experiments to study fixed
       action patterns
35.3   Behavior is the result of both
       genes and environmental factors

LEARNING

35.4   Learning ranges from simple behavioral changes to complex
       problem solving
35.5   Imprinting is learning that involves innate behavior and experience
35.6   Imprinting poses problems and opportunities for conservation
       programs
35.7   Animal movement may be a simple response to stimuli or involve
       spatial learning
35.8   Movements of animals may depend on internal maps
35.9   Animals may learn to associate a stimulus or behavior with a
       response
35.10 Social learning involves observation and imitation of others
35.11 Problem-solving behavior relies on cognition

FORAGING AND MATING BEHAVIORS

35.12 Behavioral ecologists use cost-benefit analysis in studying foraging
35.13 Mating behaviors enhance reproductive success
35.14 Mating behavior often involves elaborate courtship rituals

SOCIAL BEHAVIOR AND SOCIOBIOLOGY

35.15 Sociobiology places social behavior in an evolutionary context
35.16 Territorial behavior parcels space and resources
35.17 Rituals involving agonistic behavior often resolve confrontations
       between competitors
35.18 Dominance hierarchies are maintained by agonistic behavior
35.19 Behavioral biologist Jane Goodall discusses dominance hierarchies
       and reconciliation behavior in chimpanzees
35.20 Social behavior requires communication between animals
35.21 Altruistic acts can often be explained by the concept of inclusive
       fitness
35.22 Both genes and culture contribute to human social behavior
35.23 Edward O. Wilson promoted the field of sociobiology and is a
       leading conservation activist

Leaping Herds of Herbivores

IN THE AFRICAN GRASSLANDS, where hungry leopards and hyenas lunge for their prey with only the briefest of warnings, a twitchy antelope with a piercing snort shows that its survival is related to its behavior. In the ecosystem of the African savanna, the impala (*Aepyceros melampus*) is both one of the most commonly eaten creatures and one of the most successful.

Numerous scientific studies have shown that leopards, hyenas, wild dogs, and other carnivores devour large numbers of impalas. Despite this heavy predation pressure, these antelopes are actually growing in number and broadening their range. Even more surprising, fossil records indicate that the impala has remained almost unchanged since modern antelope lineages diverged from a common ancestor about 8 million years ago. No new species have arisen in the impala lineage, while the lineage that includes the wildebeests and hartebeests has split into new species at least 18 times. Impalas seem to have remained well adapted to their environment throughout their history and continue to be so today, even as many of them wind up as other animals' meals.

What characteristics enable the impala's success in its environment? Impalas can turn from peaceful foragers to airborne reddish-brown streaks in an instant. They are smaller than many other African antelopes, standing only about 1 m tall. But, impalas are impressive jumpers, covering as much as 11 m in a single leap (see photo on the opposite page).

Like many creatures preyed on by carnivores, impalas find strength in numbers. Alert and high-strung, they live and graze in herds, enabling plenty of watchful eyes and twitching ears to be alert for predators. Hints of danger often bring a loud snort that gets the attention of every animal in the herd. Or an explosion of legs and hooves may occur as one fleeing impala sets the entire group bounding and zigzagging in different directions, confusing their hunters. As they leap, impalas kick up their back legs, releasing a scent from glands on their heels that helps the scattering members of the herd to keep track of each other.

Smaller threats get equal attention. Impalas may watch for big hunters, but they also have a unique grooming system that helps remove bloodsucking ticks that dig into their skin. Such ticks are plentiful in the grassy, shrubby areas impalas prefer, and large numbers of them on an animal can weaken it or transmit disease.

Behavioral Adaptations to the Environment

Impalas use their teeth to pull these bloodsuckers off themselves and each other. Often they allow a bird called an oxpecker to land on their head and neck and feed on ticks (see photo on previous page). It isn't unusual to see an impala in the grass chewing its food while watching for large predators and wearing a bevy of birds behind its ears.

Impala herds are far from fixed. Impalas do not appear to form long-lasting relationships between individuals; mothers and their offspring, for example, rarely show any sign of recognition after the lambs are weaned. In the rainy season, when food is plentiful, herds of several hundred impalas can form, only to break into smaller groups as water and food dry up. Dominant males establish their own territories and try to control groups of females who wander into them, but they also tolerate the presence of "bachelor herds" of other males.

Impalas are not picky eaters or drinkers—a definite advantage in the harsh African climate. Impalas eat a wide variety of plant species, nibbling everything from short grasses to the leaves and twigs of shrubs. Their diet allows them to move into transitional habitats where open grasslands and forest meet. While they prefer areas with water, they can go for days without an actual drink, getting moisture from their forage, from the morning dew on the stems of grasses, and even from the dew that clings to their own coats.

Impalas, who thrive amid regular attacks from fierce predators and fill the African savanna with their graceful gymnastics, reflect this chapter's main objective—to illustrate the connections between animal behavior, evolution, and ecology. ▪ ▪ ▪

How can this species thrive when so many skilled hunters favor it? The answer lies largely in the impala's behavior.

35.1   Behavioral ecologists ask both proximate and ultimate questions

We humans have studied animal activities for as long as we have lived on Earth. Because our early ancestors were hunters—and sometimes the hunted—knowledge of animal behavior was essential to their survival. But the other animal species that share our planet have been a source of interest beyond the need for practical information. Our fascination with animal behavior is reflected in the many nature programs on television that highlight the vast diversity of animal life, from penguins at the South Pole to impalas on the African grasslands.

We commonly think of behavior as the observable actions an animal performs. But it also includes unobservable activities such as the chemical communication between animals and the neural processes of learning. Thus, we can think of **behavior** as everything an animal does and how it does it.

The foundation for the scientific study of behavior was established in the mid-20th century with the work of Karl von Frisch and Konrad Lorenz, of Austria, and Niko Tinbergen, of the Netherlands. These researchers shared a Nobel Prize in 1973 for their discoveries in animal behavior. In the early 1900s, Karl von Frisch, who pioneered the use of experimental methods in behavior, studied honeybee behavior in detail. Naturalist Konrad Lorenz, often regarded as the founder of behavioral biology, emphasized the importance of studying and comparing the behavior of various animals in response to different stimuli. Niko Tinbergen worked closely with Lorenz, concentrating on experimental studies of innate behavior and on simple forms of learning. We will see several examples of the studies of these pioneering behavioral biologists throughout this chapter.

One of Tinbergen's most important contributions came in a paper he wrote in 1963, where he proposed that to fully understand any behavior, one must consider both the mechanisms underlying the behavior and its evolutionary significance—how the behavior contributes to survival and reproduction (fitness). This research approach forms the basis of **behavioral ecology**, the field that studies behavior in an evolutionary context.

Behavioral ecologists ask both proximate and ultimate questions when studying a behavior. **Proximate questions** focus on the proximate (immediate) causes of a behavior. These include the environmental stimuli that trigger a specific behavior and the genetic, physiological, and anatomical mechanisms that bring it about. Proximate questions are often referred to as "how" questions. For example, impalas use their incisors to pull ticks off each other, as described in the chapter introduction and shown in **Figure 35.1**. This tit-for-tat grooming behavior occurs even between adult males, who at other times engage in aggressive behavior toward one another. A proximate question might be how this grooming behavior is initiated. Researchers observed male impalas to identify the signals that communicate that a particular encounter will result in grooming rather than in aggression. After an initial approach and nose-to-nose contact, a pair of impalas stand facing each other with heads turned. The grooming session is initiated as one impala scrapes the other 4 to 8 times, then stops and waits for a response in kind. After about 6 to 12 exchanges, the pair separate. The environmental stimuli that trigger an impala's grooming behavior appear to be both the presence of ticks and the behavioral signals of potential grooming partners.

But behavioral ecologists also pursue **ultimate questions**, those that focus on the ultimate, or evolutionary, cause of a behavior. These are "why" questions. A reasonable hypothesis for why natural selection would favor this grooming behavior is that the fitness (reproductive success) of impalas that engage in mutual grooming may be enhanced because they are not weakened by loss of blood to ticks.

The complementary nature of proximate and ultimate perspectives can be demonstrated with the behaviors frequently studied by the early behavioral biologists. We explore one of these types of behavior—the fixed action pattern—next.

Figure 35.1   Impalas grooming each other for ticks

?   When you touch a hot plate, your arm automatically recoils. What might be the proximate and ultimate causes of this behavior?

■ The proximate cause is a simple reflex, a neural pathway linking stimulation of receptors in your finger to motor response by muscles of your arm and hand; the ultimate cause is the natural selection for a behavior that minimizes damage to the body, thereby contributing to survival and reproductive success.

Lorenz and Tinbergen were among the first to demonstrate the importance of **innate behavior,** behavior that appears to be performed in virtually the same way by all individuals of a species. Many of Lorenz's and Tinbergen's studies were concerned with essentially unchangeable behavioral sequences called **fixed action patterns (FAPs).** Like someone who has memorized a poem or a piece of music but must start over at the beginning if interrupted, an animal can only perform the FAP as a whole. Once an animal initiates a FAP, it usually carries the sequence to completion, even if it receives different stimuli before it finishes.

Figure 35.2 illustrates one of the FAPs that Lorenz and Tinbergen studied in detail. The bird is the graylag goose, a common European species that nests in shallow depressions on the ground. If the goose happens to bump one of her eggs out of the nest, she always retrieves it in the same manner. As indicated in the figure, she stands up, extends her neck, uses a side-to-side head motion to nudge the egg back

with her beak, and then sits down on the nest again. If the egg slips away (or is pulled away by an experimenter) while the goose is retrieving it, she continues as though the egg were still there. Only after she sits back down on her eggs does she seem to notice that the egg is still outside the nest. Then she begins another retrieval sequence. If the egg is again pulled away, the goose again completes the retrieval motion without the egg. A goose would even perform the sequence when Lorenz and Tinbergen placed a foreign object, such as a small toy or a ball, near her nest.

Fixed action patterns are important in the lives of many species. When a hatchling senses that an adult bird is near, it responds with a FAP: It begs for food by raising its head, opening its mouth, and cheeping. In turn, the parent responds with another FAP: It stuffs food in the gaping mouth.

In its simplest form, a FAP is an innate response to a certain stimulus. A stimulus that triggers a FAP is called a **sign stimulus.** For the graylag goose, the sign stimulus for egg retrieval is the presence of an egg (or other object) near the nest. For a parent bird feeding a chick, it is the chick's gaping mouth. A sign stimulus is often a simple cue in an animal's environment, leading the animal to respond quickly and appropriately without integrating a lot of sensory information.

These relatively simple, innate behaviors seem to occur in all animals, including humans. Human infants grasp strongly with their hands in response to a touch stimulus on the hand. They smile in response to a face or even something that vaguely resembles a face, such as two dark spots in a circle.

How might we explain FAPs in the context of behavioral ecology? Automatically performing certain behaviors may maximize fitness to the point that genes for variants of that behavior are lost. For example, there are some things that a young animal has to get right on the first try if it is to stay alive. Kittiwakes are gulls that nest on cliff ledges. Unlike other gull species, kittiwakes show an innate aversion to cliff edges; they turn away from the edge. Chicks in earlier generations that did not show this edge-aversion response would not have lived to pass the genes for their risk-taking behavior on to the next generation.

In many cases, we can see a clear advantage of FAPs over learned behavior. A graylag goose protects its offspring by automatically retrieving its eggs, and a chick does not have to learn how to obtain food from its parents. Similarly, a newborn impala starts nursing without any previous experience in obtaining milk from its mother.

Innate behaviors are under strong genetic control. Most innate behaviors, however, improve with experience. In the next module, we look at this intersection of genes and environment in producing a behavior.

? How would you explain FAPs in the context of proximate and ultimate causes?

■ The proximate cause of a FAP is often a simple environmental cue (sign stimulus). The ultimate cause is that natural selection would favor behaviors that enable animals to perform tasks essential to survival without any previous experience.

Figure 35.2 A graylag goose retrieving an egg—a FAP

Behavior is the result of both genes and environmental factors

How much is our ability to play a musical instrument programmed by genes and how much by experience? A virtuoso pianist must have extraordinary hand coordination and the ability to make the mechanical instrument resound with the dynamics and emotions of the music. How much of an artist's virtuosity results from genes (what we call innate talent), and how much comes from study and training? And is the capacity for disciplined training itself mainly learned or inherited?

A myth that is still perpetuated to some extent by popular media is that behavior is due either to genes (nature) or to environmental influences (nurture). In biology, however, the nature-versus-nurture debate is not about either/or; it is about how both genes and the environment influence the development of phenotypic traits, including behavioral ones. As we discussed in Chapter 9, phenotype depends on both genes and the environment; behavioral traits have genetic and environmental components, as do all of an animal's structural and functional features.

Behavioral research has uncovered a variety of mammalian behaviors that are under relatively strong genetic control. One of the most striking lines of research concerns the mating behavior of prairie voles, *Microtus ochrogaster*. Prairie voles are monogamous, a social trait found in only about 3% of mammalian species. (We will discuss the mating patterns of animals in Module 35.13.) A male prairie vole associates closely with his mate, engaging in grooming and huddling behaviors (**Figure 35.3A**). An unmated male prairie vole shows little aggression toward other voles, whether male or female. A mated male, on the other hand, becomes in-

tensely aggressive toward any strange male or female vole, while remaining nonaggressive toward his mate. Male prairie voles also help their mates care for the young, another relatively uncommon trait among male mammals. A few days following the birth of pups, a male prairie vole will spend a great deal of time hovering over them, licking them, and carrying them around.

Over the past decade, researchers at Emory University have investigated the genetic and physiological controls of this complex mating and parental behavior. Prairie voles form a monogamous pair after a night of repeated matings. Two hormones, vasopressin and oxytocin, are released in the brain during such matings and are implicated in forming mate preferences. Studies have shown differences in the number and distribution of oxytocin receptors in female prairie voles compared to their distribution in female montane (mountain-region) voles (**Figure 35.3B**). Unlike prairie voles, montane voles are promiscuous, meaning that they do not form lasting pair-bonds. Vasopressin appears to be the chemical that mediates mate preference in males, and differences in vasopressin receptor locations have been demonstrated between male prairie voles and male montane voles.

To test whether the distribution of vasopressin receptors is a key factor controlling the parental and mating behavior of male prairie voles, researchers inserted the prairie vole gene for the vasopressin receptor into a line of laboratory mice. The transgenic mice developed a vasopressin receptor distribution pattern remarkably similar to that of prairie voles, and they showed many of the mating behaviors of male prairie voles. Wild-type mice have neither that distribution of receptors nor similar mating behaviors. Thus, it appears that a single gene may mediate a considerable amount of the complex mating behavior that occurs in prairie voles.

Parallel studies of another monogamous rodent, the California mouse (*Peromyscus californicus*), provide evidence for the influence of social environment on behavior. Like prairie voles, male California mice provide extensive parental care and are highly aggressive toward other mice. In contrast to prairie voles, however, male California mice that have not mated are already aggressive.

Figure 35.3A A pair of mate-for-life prairie voles huddling

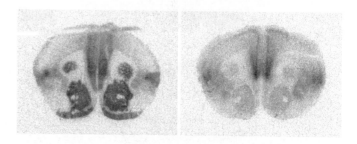

Figure 35.3B Distribution and density of oxytocin receptors (indicated by dark stain) in the brain of a female monogamous prairie vole (left) and a promiscuous montane vole (right)

Researchers at the University of Wisconsin at Madison studied what influence the social environment might have on the behavior of male California mice. To manipulate the environment, they placed newborn California mice in the nests of white-footed mice (*Peromyscus leucopus*), a species in which males are not monogamous and engage in little parental care. They also placed white-footed mice in the nests of California mice. This "cross-fostering" altered the behavior of both species. For instance, male white-footed mice raised by California mice were more aggressive than those raised in nests of their own species. In contrast, male California mice that were cross-fostered were less aggressive toward intruders than those reared normally. When young California mice crawled away from the nest of their foster parents, they were not retrieved as often as those raised by their own species; when these cross-fostered California mice became parents, they, too, spent less time retrieving their pups. In one study, the brains of California mice raised by white-footed mice were found to contain reduced levels of vasopressin, the brain chemical associated with elevated aggression and parental care in male prairie voles.

The fact that cross-fostered California and white-footed mice adopted some of the behaviors of their foster parents suggests that experience during development can lead to changes in parental and aggressive behaviors. And these changes can be passed to future generations, not through genes but through the social environment.

Together, these complementary studies of voles and mice provide strong evidence that the mating and parenting behaviors of these rodents are the product of both genes and environmental factors. Indeed, the interaction of genes and the environment appears to determine most animal behaviors.

One of the most powerful ways that the environment can influence behavior is through learning, the topic we consider next.

? Without doing any experiments, how might you distinguish between a behavior that is mostly controlled by genes and one that is mostly determined by the environment?

■ A genetically controlled behavior would not differ much between populations, regardless of the environment, while the environmentally controlled behavior would differ widely across populations located in different environments.

35.4 Learning ranges from simple behavioral changes to complex problem solving

Learning is modification of behavior as a result of specific experiences. Learning enables animals to change their behaviors in response to changing environmental conditions. As Table 35.4 indicates, there are various forms of learning, ranging from a simple behavioral change in response to a single stimulus, to complex problem solving involving entirely new behaviors.

One of the simplest forms of learning is **habituation**, in which an animal learns not to respond to a repeated stimulus that conveys little or no information. There are many examples of habituation in both invertebrate and vertebrate animals. The cnidarian *Hydra*, for example, contracts when disturbed by a slight touch; it stops responding, however, if disturbed repeatedly by such a stimulus. Similarly, a scarecrow stimulus will usually make birds avoid a tree with ripe fruit for a few days. But the birds soon become habituated to the scarecrow and may even land on it on their way to the fruit tree. Once habituated to a particular stimulus, an animal still senses the stimulus—its sensory organs detect it—but the animal has learned not to respond to it.

In terms of ultimate causation, habituation may increase fitness by allowing an animal's nervous system to focus on stimuli that signal food, mates, or real danger and not waste time or energy on a vast number of other stimuli that are irrelevant to survival and reproduction.

Proximate and ultimate causes of behavior are also evident in imprinting—the type of learning we look at next.

? What type of learning enables your brain to ignore the constant sensations of touch from the clothes you're wearing?

■ Habituation

TABLE 35.4 TYPES OF LEARNING

Learning Type	Defining Characteristic
Habituation	Loss of response to a stimulus after repeated exposure
Imprinting	Learning that is irreversible and limited to a sensitive time period in an animal's life; often results in a strong bond between offspring and parents
Spatial learning	Using landmarks to learn the spatial structure of the environment
Cognitive mapping	An internal representation of the spatial relationships among objects in the environment
Associative learning	Behavioral change based on linking a stimulus or behavior with a reward or punishment; includes trial-and-error learning
Social learning	Learning by observing and mimicking others
Problem solving	Inventive behavior that arises in response to a new situation

Imprinting is learning that involves innate behavior and experience

Learning often interacts closely with innate behavior. Some of the most interesting examples of this interaction involve the phenomenon known as imprinting. **Imprinting** is learning that is limited to a specific time period in an animal's life and that is generally irreversible. The limited phase in an animal's development when it can learn certain behaviors is called the **sensitive period.**

In classic experiments done in the 1930s, Konrad Lorenz used the graylag goose to demonstrate imprinting. When incubator-hatched goslings spent their first few hours with Lorenz, rather than with their mother, they steadfastly followed Lorenz (**Figure 35.5A**) and showed no recognition of their mother or other adults of their species. Even as adults, the birds continued to prefer the company of Lorenz and other humans to that of geese.

Lorenz demonstrated that the most important imprinting stimulus for graylag geese was movement of an object (normally the parent bird) away from the hatchlings. The effect of movement was increased if the moving object emitted some sound. The sound did not have to be that of a goose, however; Lorenz found that a box with a ticking clock in it was readily and permanently accepted as a "mother."

Just as a young bird requires imprinting to know its parents, the adults must also imprint to recognize their young. For a day or two after their own chicks hatch, adult herring gulls will accept and even defend a strange chick introduced into their nesting territory. However, once imprinted on their offspring, adults will kill and eat any strange chicks.

Not all examples of imprinting involve parent-offspring bonding. Newly hatched salmon, for instance, do not receive parental care but seem to imprint on the complex mixture of odors unique to their stream. As we saw in the introduction to Chapter 29, this imprinting enables adult salmon to find their way back to their home stream to spawn after spending a year or more at sea.

For many kinds of birds, imprinting plays a role in song development. For example, researchers studying song development in white-crowned sparrows (**Figure 35.5B**) found that male birds memorize the song of their species during a sensitive period (the first 50 days of life). They do not sing during this phase, but several months later they begin to practice this song, eventually learning to reproduce it correctly. The birds do not need to hear the adult song during their practice phase; isolated males raised in soundproof chambers learned to sing normally as long as they heard a recorded song of their species during the sensitive period. In contrast, isolated males that did not hear their species' song until after 50 days sang an abnormal song. Researchers also discovered a purely genetic component of white-crowned sparrow song development: Isolated males exposed to recorded songs of other species during the sensitive period did not adopt these foreign songs. When they later learned to sing, these birds sang an abnormal song similar to that of males that heard no recorded bird songs.

The ability of parents and offspring to keep track of each other and the ability of male songbirds to attract mates are examples of behaviors that have direct and immediate effects on survival and reproduction. Imprinting provides a way for such behavior to become more or less fixed in an animal's nervous system.

? Explain why we say that imprinting has both innate and learned components.

■ Its innate component is the tendency to imprint on a stimulus during a sensitive period. The imprinting itself is a form of learning.

Figure 35.5A Konrad Lorenz with geese imprinted on him

Figure 35.5B A male white-crowned sparrow

35.6 Imprinting poses problems and opportunities for conservation programs

Imprinting on a parent can help ensure that young animals thrive, but it sometimes presents a new set of survival challenges. When scientists are trying to save endangered species, providing newborn animals with proper objects for imprinting takes on increased importance.

In many cases where a species is at the edge of extinction, biologists have tried to increase its numbers in captivity. These captive breeding programs have sheltered animals such as the giant pandas of China and the California condor trying to create safe conditions in which threatened animals can multiply and improve their chances for survival.

In some captive breeding programs, adult animals are caught and kept in conditions thought beneficial for breeding. Offspring are usually raised by the parents and may be kept for breeding or released back to the wild. In other programs, parents are absent, as when eggs are removed from a nest. Artificial incubation is often successful, but as demonstrated by Lorenz's gosling experiments, parental presence becomes more important once the bird is out of its shell.

When there are no parents to rear new chicks, scientists have tried using birds of other species, or even bird-shaped hand puppets, as foster parents. And for one species of endangered bird, surrogate parenting has come in the unlikely form of another winged object—a small airplane.

People have taken to the skies in ultralight aircraft to help save the whooping crane (*Grus americana*). This migratory waterfowl, which reaches a height of about 1.5 m, has a white body with black-tipped wings that spread out over 2 m. Its name comes from its distinctive call. Once common in North American skies during their north-south migrations, whooping cranes were almost killed off by habitat loss and hunting. By the 1940s, only 16 wild birds returned from their summer breeding ground in Canada to their wintering area on the Texas coast.

Protections for whooping cranes were established, and in 1967, U.S. wildlife officials launched long-term recovery and captive breeding efforts. A young whooping crane doesn't reach breeding maturity until it is 4 years old, preventing new generations of birds from reproducing quickly. And although whooping cranes often lay two eggs, parents usually successfully rear only one chick.

Breeding program managers tried taking eggs from whooping crane nests and placing them with foster parents (sandhill cranes). Whooping crane chicks were accepted, raised, and taught to migrate by the sandhill crane parents, aptly learning behaviors needed for survival. But as the whooping cranes reached maturity, the problem with foster parenting became clear. The whooping cranes had bonded with sandhill cranes and showed no interest in breeding with their own kind.

Program managers realized that they needed another approach. With plans to set up a separate breeding colony in Wisconsin that would migrate to Florida for the winter, they turned to Operation Migration. This bird advocacy group,

Figure 35.6 Whooping cranes following a surrogate parent

based in the United States and Canada, had already succeeded in getting geese and sandhill cranes to follow a small, lightweight plane. Beginning in the mid-1980s, the group had developed ways to hatch birds, teach them to recognize the plane as a parent figure, and persuade them to follow the plane along migratory routes.

In 2001, Operation Migration applied its techniques to whooping cranes. The plane parenting effort started while the cranes were still in their shells. Even before they hatched, incubating crane eggs were serenaded with recorded sounds of the plane's engine. When the chicks emerged from their shells, the first thing they saw was a hand puppet, shaped and painted in the form of an adult whooping crane.

As the chicks grew, they were taken out for exercise and training, guided by the same type of puppet. But now the puppet was attached to the plane, which coaxed the chicks to follow it by rolling along the ground. Pilots and other humans were shielded in hooded suits to make sure the birds bonded with the puppet and plane, not people. Eventually, the birds started following the plane on short flights (**Figure 35.6**).

October 2001 brought the real test. Would the young whooping cranes follow the plane from their protected grounds in Wisconsin along a migratory route to Florida? The trip took 48 days but ultimately proved successful. Each flight day, the young cranes lined up eagerly behind the plane, wings raised, ready to follow their "parent" to the next stop. And the next spring, five of the eight young cranes retraced the route to Wisconsin on their own. Operation Migration has since taught three more generations of whooping cranes to migrate, boosting the species' chances of survival. There are now about 300 whooping cranes in the wild, a steady comeback for the slow-recovering species.

? What features of whooping cranes have made their recovery so difficult?

■ They do not breed until 4 years of age and usually only raise one chick a breeding season. As migratory birds, they require resources and protection in two habitats.

35.7 Animal movement may be a simple response to stimuli or involve spatial learning

Moving in a directed way enables animals to avoid predators, migrate to a more favorable environment, obtain food, and find mates and nest sites. The simplest kinds of movement do not involve learning. A random movement in response to a stimulus is called a **kinesis** (plural, *kineses;* from the Greek word for "movement"). A kinesis may be merely starting or stopping, changing speed, or turning more or less frequently. Sow bugs are small woodland crustaceans that exhibit kinesis in dry areas, becoming more active. The more they move, the greater the chance they will find a moist area. Once they are there, their decreased activity tends to keep them in that more favorable environment.

In contrast, a **taxis** (plural, *taxes;* from the Greek *tasso,* put in order) is a more or less automatic movement directed toward (positive) or away from (negative) some stimulus. For example, many stream fish, such as trout, exhibit positive rheotaxis (from the Greek *rheos,* current); they automatically swim or orient in an upstream direction (**Figure 35.7A**). This orientation keeps them from being swept away by the current and keeps them facing in the direction food is likely to come from.

The capacity for **spatial learning** may enhance the fitness of an organism by helping it move through its environment. Spatial learning involves the use of **landmarks** and is a more complex mechanism than a kinesis or taxis. Figure 35.7B illustrates a classic experiment done by Tinbergen in 1932. It deals with nesting behavior in an insect called the digger wasp, which builds its nest in a small burrow in the ground. A female wasp will often excavate four or five separate nests and fly to each one daily, cleaning them and bringing food to the larvae in the nests. To test his hypothesis that the digger wasp uses landmarks to keep track of her nests, Tinbergen ❶ placed a circle of pinecones around a nest opening and waited for the mother wasp to return and tend the nest. After the wasp flew away, Tinbergen ❷ moved the pinecones a few feet to one side of the nest opening. The next time the wasp returned, she flew to the center of the pinecone circle instead of to the actual nest opening.

This experiment indicated that the wasp did use landmarks and that she could learn new ones to keep track of her nests. Next, to test whether the wasp was responding to the pinecones themselves or to their circular arrangement, Tinbergen ❸ arranged the pinecones in a triangle around the nest and made a circle of small stones to one side of the nest

opening. This time, the wasp flew to the stones, indicating that her cue was the *arrangement* of the landmarks rather than their appearance.

Many animals learn the particular set of landmarks in their area and use them to find their way. Some animals appear to use more sophisticated navigation mechanisms, as we see next.

Web/CD Thinking as a Scientist *How Can Pillbug Responses to Environments Be Tested?*

? Planarians (see Figure 29.4A) move directly away from light into dark places. What type of movement is this?

■ Taxis away from light (negative phototaxis)

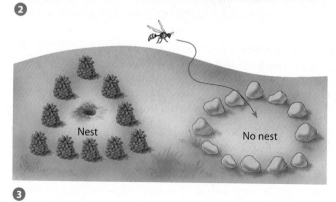

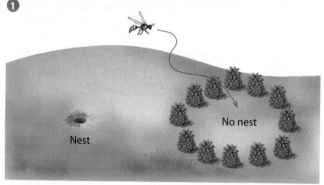

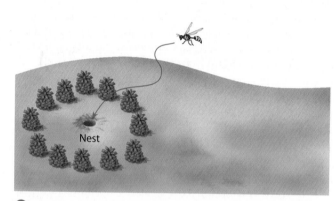

Figure 35.7B Nest-locating behavior of the digger wasp

Figure 35.7A Positive rheotaxis of a trout

35.8 Movements of animals may depend on internal maps

An animal can move around its environment using landmarks alone. Honeybees, for instance, might learn ten or so landmarks and locate their hive and flowers in relation to those features. A more powerful mechanism is a **cognitive map**, an internal representation, or code, of the spatial relationships among objects in an animal's surroundings.

It is actually very difficult to distinguish experimentally between an animal that is simply using landmarks and one that is using a true cognitive map. The best evidence for cognitive maps comes from research on the family of birds that includes jays, crows, and nutcrackers. Many of these birds store food in caches. A single bird may store nuts in thousands of caches that may be widely dispersed. The bird not only relocates each cache, but also keeps track of food quality, bypassing caches in which the food was relatively perishable and would have decayed. It would seem that these birds use cognitive maps to memorize the location of their food stores.

The most extensive studies of internal maps have involved animals that exhibit **migration**, the regular back-and-forth movement of animals between two geographic areas. Migration enables many species, such as the whooping cranes discussed in Module 35.6, to access food resources throughout the year and to breed or winter in areas that favor survival.

Another long-distance traveler is the gray whale. During the summer, these giant mammals feast on small, bottom-dwelling invertebrates that abound in northern oceans. In the autumn, they leave their feeding grounds north of Alaska and begin the long trip south along the North American coastline to winter in the warm lagoons of Baja California (Mexico). Females give birth there, before migrating back north with their young. The yearly round-trip, some 20,000 km, is the longest made by any mammal.

Researchers have found that migrating animals stay on course by using a variety of cues. Gray whales seem to use the coastline to pilot their way north and south. Whale watchers sometimes see gray whales stick their heads straight up out of the water, perhaps to obtain a visual fix on land.

Many birds migrate at night, navigating by the stars the way ancient human sailors did. Navigating by the sun or stars requires an internal timing device to compensate for the continuous daily movement of celestial objects. Consider what would happen if you started walking one day, orienting yourself by keeping the sun on your left. In the morning, you would be heading south, but by evening you would be heading back north, having made a circle and gotten nowhere. A calibration mechanism must also allow for the apparent change in position of celestial objects as the animal moves over its migration route.

At least one night-migrating bird, the indigo bunting, seems to avoid the need for a timing mechanism by fixing on the North Star, the one bright star in northern skies that appears almost stationary. **Figure 35.8** illustrates an experimental setup that was used to study the bunting's navigational mechanism. During the migratory season, wild and laboratory-reared birds were placed in funnel-like cages in a planetarium (see photograph). Each funnel had an ink pad at its base and was lined with blotting paper. When a bird stepped on the ink pad and then tried to fly in a certain direction, it tracked ink on the paper. The researchers found that wild buntings tracked ink in the direction of the North Star, as did those raised in the lab and introduced to the northern sky in a planetarium. Birds raised under a planetarium sky with a different fixed-location star oriented to that star. Apparently, buntings learn a star map and fix on a stationary star when navigating at night.

Some animals appear to migrate using only innate responses to environmental cues. Each fall, a new generation of monarch butterflies flies about 4,000 km over a route they've never flown before to very specific wintering sites. Studies of other animals show the interaction of genes and experience in migration. Research efforts continue to reveal the complex mechanisms by which animals traverse Earth.

? Why is a timekeeping mechanism essential for stellar navigation?

■ Because the positions of the stars change with time of night and season

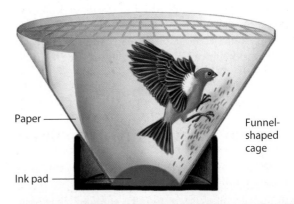

Paper

Funnel-shaped cage

Ink pad

Figure 35.8 An experiment demonstrating star navigation

35.9 Animals may learn to associate a stimulus or behavior with a response

In **associative learning**, an animal learns that a particular stimulus or a particular response is linked to a reward or punishment. If you keep a pet, you probably have observed one type of associative learning firsthand. A dog or cat can learn to associate a particular sound, word, or gesture with some type of punishment or reward. For example, the sound of a can being opened may bring a cat running for food.

The best-known laboratory studies involving associative learning date from the work of American psychologist B. F. Skinner in the 1930s. A rat placed in a "Skinner box" finds and manipulates a lever in the box, usually by accident, and is rewarded by the release of food. The animal quickly learns to associate manipulation of the lever with a food reward. Most animal trainers use a reward to reinforce a desired behavior. Eventually, the animal performs the behavior on command, without always receiving a reward.

In natural settings, a common form of associative learning is called **trial-and-error learning**. In this case, an animal learns to associate one of its own behavioral acts with a positive or negative effect. The animal then tends to repeat the response if it is rewarded or avoid the response if it is harmed. For example, predators quickly learn to associate certain kinds of prey with painful experiences. A porcupine's sharp quills and ability to roll into a quill-covered ball are strong deterrents against many predators. Coyotes, mountain lions, and domestic dogs often learn the hard way to avoid attacking porcupines nose-first (Figure 35.9). Trial-and-error learning often provides responses that are important to survival.

? How might the fact that many bad-tasting or stinging insect species have similar color patterns benefit both the insects and the animals that may prey on them?

■ Potential predators associate insects with that coloration with a negative effect; all these insects gain protection when predators learn to avoid them.

Figure 35.9 Trial-and-error learning by a coyote

35.10 Social learning involves observation and imitation of others

Another form of learning is **social learning**—learning by observing the behavior of others. Many predators, including cats, coyotes, and wolves, seem to learn some of their basic hunting tactics by observing and imitating their mother.

Studies of the alarm calls of vervet monkeys in Amboseli National Park in Kenya provide an interesting example of how performance of a behavior can improve through social learning. Vervet monkeys (*Cercopithecus aethiops*) are about the size of a domestic cat. They give distinct alarm calls when they see leopards, eagles, or snakes, all of which prey on vervets. When a vervet sees a leopard, it gives a loud barking sound; when it sees an eagle, it gives a short two-syllable cough; and the snake alarm call is a "chutter." Upon hearing a particular alarm call, other vervets in the group behave in an appropriate way: They run up a tree on hearing the alarm for a leopard (vervets are nimbler than leopards in trees); look up on hearing the alarm for an eagle; and look down on hearing the alarm for a snake (Figure 35.10).

Infant vervet monkeys give alarm calls, but in a relatively undiscriminating way. For example, they give the "eagle" alarm on seeing any bird, including harmless birds such as bee-eaters. With age, the monkeys improve their accuracy.

Figure 35.10 On seeing a python (foreground), a vervet monkey gives a distinct "snake" alarm call (inset).

In fact, adult vervet monkeys give the eagle alarm only on seeing an eagle belonging to either of the two species that eat vervets. Infants probably learn how to give the right call by observing other members of the group and receiving social confirmation. For instance, if the infant gives the call on the right occasion—an eagle alarm when there is an eagle overhead—another member of the group will also give the eagle call. But if the infant gives the call when a bee-eater flies by, the adults in the group are silent. Thus, vervet monkeys have an initial, unlearned tendency to give calls on seeing potentially threatening objects in the environment. Learning fine-tunes the calls so that by adulthood, vervets give calls only in response to genuine danger and are prepared to fine-tune the alarm calls of the next generation. However, neither vervets nor any other species comes close to matching the social learning and cultural transmission that occurs among humans, a topic we'll explore later in the chapter.

> ? What type of learning in humans is exemplified by identification with a role model?
>
> ■ Social learning (observation and imitation)

35.11 Problem-solving behavior relies on cognition

A broad definition of **cognition**—and the way we use the term in this book—is the ability of an animal's nervous system to perceive, store, process, and use information gathered by sensory receptors. One area of research in the study of animal cognition is how an animal's brain represents physical objects in the environment. For instance, some researchers have discovered that many animals, including insects, are capable of categorizing objects in their environment according to concepts such as "same" and "different." One research team has trained honeybees to match colors and black-and-white patterns. Other researchers have developed innovative experiments, in the tradition of Lorenz and Tinbergen, for demonstrating pattern recognition in birds called nuthatches. These studies suggest that nuthatches apply simple geometric rules to locate their many seed caches.

Some animals have complex cognitive abilities that include problem solving—the ability to apply past experience to novel situations. Problem-solving behavior is highly developed in some mammals, especially dolphins and primates. If a chimpanzee is placed in a room with a banana hung high above its head and several boxes on the floor, the chimp will gradually "size up" the situation and then stack the boxes in order to reach the food. In Figure 35.11A, a chimp is solving the problem of cracking oil palm nuts by using two stones as a hammer and anvil.

Problem-solving behavior has also been observed in some bird species. For example, researchers placed ravens in situations in which they had to obtain food hanging from a string. Interestingly, the researchers observed a great deal of variation in the ravens' solutions. The raven in Figure 35.11B used one foot to pull up the string incrementally and the other foot to secure the string so the food didn't drop.

An excellent test of human problem-solving behavior is in the construction of experiments that allow us to explore the cognition and problem-solving behavior of other animals.

In the next section of the chapter, we explore feeding and mating behaviors, both of which directly affect fitness.

> ? Besides problem solving, what other type of learning is illustrated by Figure 35.11A?
>
> ■ Social learning (observation)

Figure 35.11A A chimpanzee solving a problem

Figure 35.11B
A raven solving a problem

35.12 Behavioral ecologists use cost-benefit analysis in studying foraging

Because adequate nutrition is essential to an animal's survival and reproductive success, we should expect natural selection to refine behaviors that enhance the efficiency of feeding. Food-obtaining behavior, or **foraging**, includes not only eating, but also any mechanism an animal uses to recognize, search for, and capture food items.

Animals forage in a great many ways. Some animals are "generalists," whereas others are "specialists." The gull in **Figure 35.12A** is an extreme generalist; it will eat just about anything that is readily available—plant or animal, alive or dead. In sharp contrast, the koala of Australia, an extreme feeding specialist, eats only the leaves of a few species of eucalyptus trees (**Figure 35.12B**). Most animals are somewhere in between gulls and koalas in the range of their diet. Impalas, as you read in the chapter introduction, are both grazers and browsers, feeding on grass when it is plentiful, but able to eat shrubs and even dry leaves and herbs unpalatable to most other herbivores. The success of impalas in many regions may be attributed to their adaptable foraging. Koalas, on the other hand, are restricted to areas with eucalyptus trees and are thus extremely vulnerable to habitat loss.

Often, even a generalist will concentrate on a particular item of food when it is readily available. The mechanism that enables an animal to find particular foods efficiently is called a **search image**. If the favored food item becomes scarce, the animal may develop a search image for a different food item. (Humans often use search images; for example, when you look for something on a kitchen shelf, you proba-

bly scan rapidly to find a package of a certain size and color rather than reading all the labels.)

Whenever an animal has food choices, there are trade-offs involved in the selection. Some behavioral ecologists apply a cost-benefit analysis, to evaluate the efficiency of foraging behaviors. According to the predictions of **optimal foraging theory**, an animal's feeding behavior should provide maximal energy gain with minimal energy expense and minimal risk of being eaten while foraging.

Consider, for example, the bass in **Figure 35.12C**, which can readily consume both minnows and crayfish. The minnow will provide more usable energy per unit of weight (a crayfish has a lot of hard-to-digest exoskeleton), but less total energy (the minnow is smaller). The minnow may be harder to catch than the crayfish, but the crayfish may take longer to eat. Complicating the picture even more, the bass must be alert to other predators while feeding. Which will expose it more to a predatory turtle or a larger fish, chasing a minnow or mouthing a thrashing crayfish?

In most natural environments, there are so many variables that it's hard to imagine any animal could forage in an absolutely optimal manner. Nonetheless, numerous studies indicate that when prey is plentiful, animals of many species forage in such a way that their overall energy intake-to-expenditure ratio is high. A bass forages efficiently, if not exactly optimally, by switching between different prey—minnows, crayfish, aquatic insects, and other invertebrates—as conditions such as a prey size, density, and ease of capture change.

Predation is one of the most significant potential costs of foraging. Studies have shown that foraging in groups such as

Figure 35.12A A gull: a feeding generalist

Figure 35.12B A koala: a feeding specialist

Figure 35.12C A bass eating a crayfish: an optimal forager?

herds of impalas, flocks of birds, or schools of fish reduces the individual's risk of predation. And for some predators, hunting in groups improves their success. Thus, group behavior may improve foraging efficiency by both reducing the costs and increasing the benefits of foraging.

? Early humans were hunter-gatherers, but evidence suggests that nutrition was based more on gathering than on hunting. How does this relate to optimal foraging theory?

■ Meat is very nutritious, but hunting also poses relatively high costs in effort and risk compared to the gathering of plant products and dead animals.

35.13 Mating behaviors enhance reproductive success

Natural selection also favors mating behaviors that enhance reproductive success. Several kinds of mating systems are found among animals. When mating is **promiscuous**, no strong pair-bonds or lasting relationships exist between males and females. In species where the mates remain together for a longer period, the relationship may be **monogamous** (one male with one female) or **polygamous** (an individual of one sex mating with several of the other). Polygamous relationships most often involve a single male and many females, although in some species, this is reversed and a single female mates with several males.

The needs of young are an important factor in the evolution of mating systems. Most newly hatched birds, for example, cannot care for themselves and require a large, continuous food supply that a single parent may not be able to provide. This may explain why most birds are monogamous: A male may leave more viable offspring by helping a single mate than by going off to seek more mates. Birds whose young can feed and care for themselves almost immediately after hatching, such as pheasants and quail, have less need for parents to stay together. In these species, males can maximize their reproductive success by seeking other mates, and polygamy is common.

In the case of mammals, the lactating female is often the only food source for the young, and males usually play no role in caring for their offspring. In species, such as lions, where males protect the females and young, a male or small group of males typically guard many females at once in a harem. The prairie voles described in Module 35.3 are unusual mammals in regard to both monogamy and parental care.

Certainty of paternity may also be a factor in the evolution of mating behavior and parental care. By helping their offspring survive to maturity, parents help ensure the perpetuation of their own genes. Young born or eggs laid definitely contain the mother's genes. But even in the case of a normally monogamous relationship, the young may have been fathered by a male other than the female's usual mate. The certainty of paternity is relatively low in most species with internal fertilization because the acts of mating and birth (or mating and egg laying) are separated over time. This could help explain why species in which males are the sole parental caregiver are rare in birds and mammals. However, the males of many species with internal fertilization do engage in behaviors that appear to increase their certainty of paternity, such as guarding females from other males.

Certainty of paternity is much higher when egg laying and mating occur together, as occurs when fertilization is

Figure 35.13 A male jawfish with his mouth full of eggs

external. This may explain why parental care in aquatic invertebrates, fishes, and amphibians, when care occurs at all, is at least as likely to be by males as by females. **Figure 35.13** shows a male jawfish exhibiting paternal care of eggs. Jawfish, which are found in tropical marine habitats, hold the eggs they have fertilized in their mouth, keeping them aerated (by spitting them out and sucking them back in) and protected from predators until they hatch.

It is important to realize that when behavioral ecologists use the term *certainty of paternity*, they do not mean that animals are "aware" of paternity when they behave a certain way, any more than they mean that animals consciously "choose" their food. Foraging that maximizes benefit over cost or parental behaviors that correlate with certainty of paternity exist because they have been reinforced over generations by natural selection. Animals with genes for such behaviors reproduced more successfully and passed those genes on to the next generation.

Mating behaviors may include courtship rituals and "choosing" among potential mates, as we see in the next module.

? A male lion is polygamous but not promiscuous. Explain.

■ A male lion mates with several females and is thus polygamous. But he forms long-lasting bonds with these females and thus is not considered promiscuous.

35.14 Mating behavior often involves elaborate courtship rituals

Animals of many species tend to view members of their own species as competitors, to be driven away. Even animals that forage and travel in groups often maintain a distance from

their companions. How, then, is mating accomplished? In many species, prospective mates must perform an elaborate courtship ritual, which confirms that individuals are of the same species, of the opposite sex, physically primed for mating, and not threats to each other.

Figure 35.14A shows the courtship and mating of the common loon, which breeds on secluded lakes in the northern United States and Canada. The courting male and female swim side by side, performing a series of displays, ❶ frequently turning their heads away from each other. (In contrast, a male loon defending his territory charges at an intruder with his beak pointed straight ahead.) ❷ The birds then dip their beaks in the water and submerge their heads. ❸ The male invites the female onto land by turning his head backward with his beak down. There, ❹ they copulate.

In some species, courtship is a group activity in which members of one or both sexes choose mates from a group of candidates. (See Module 13.17 to review mate choice and sexual selection.) Sage grouse are chicken-like birds that inhabit high sagebrush plateaus in the western United States. Each day in early spring, 50 or more males congregate in an open area where they strut about, erecting their tail feathers in a bright, fanlike display (Figure 35.14B). Dominant males usually defend a small territory near the center of the area. Females arrive several weeks after the males. After watching the males perform, a female selects one, and the pair copulates. Usually, all the females choose dominant males, so only about 10% of the males actually mate. Is there an advantage to this group mating ritual? In choosing a dominant male, a female sage grouse may be giving her offspring, and thus her own genes, the best chance for survival. Research studies on several species of animals have shown a connection between a male's physical characteristics and the quality of his genes.

? How is a female's fitness improved by choosing a mate based on his mating displays or physical adornments?

■ She is more likely to have healthy offspring by mating with a healthy male than with a sickly one.

Figure 35.14A Courtship and mating of the common loon

Figure 35.14B Courtship display by a male sage grouse

716   **UNIT VII** *Ecology*

35.15 Sociobiology places social behavior in an evolutionary context

The mating behaviors of loons and sage grouse are examples of social behavior. Biologists define **social behavior** as any kind of interaction between two or more animals, usually of the same species. Courtship, aggression, and cooperation are all examples of social behavior. An essential ingredient of all types of social behavior is communication—some means of transferring information between individuals.

Many animals migrate and feed in large groups (flocks, packs, herds, or schools). As discussed in the chapter introduction, impalas feed in herds that provide protection from predators through their early warning and escape behavior. Wolves usually hunt in a pack consisting of a tightly knit group of family members. Hunting in packs enables them to kill large animals, such as moose or elk, that would be unavailable to an individual wolf.

The discipline of **sociobiology** applies evolutionary theory to the study and interpretation of social behavior—how social behaviors are adaptive and how they could have evolved by natural selection. We discuss several aspects of social behavior in the next several modules.

? What key process is required for social behavior within a population?

■ Communication

35.16 Territorial behavior parcels space and resources

Many animals exhibit territorial behavior. A **territory** is an area, usually fixed in location, that individuals defend and from which other members of the same species are usually excluded. The size of the territory varies with the species, the function of the territory, and the resources available. Territories are typically used for feeding, mating, rearing young, or combinations of these activities.

Figure 35.16A shows a nesting colony of gannets in Quebec, Canada. Space is at a premium, and the birds defend territories just large enough for their nests by calling out and pecking at other birds. As you can see, each gannet is literally only a peck away from its closest neighbors. Such small nesting territories are characteristic of many colonial seabirds. In contrast, most cats, including jaguars, leopards, cheetahs, and even domestic cats, defend much larger territories, which they use for foraging as well as breeding.

Individuals that have established a territory usually proclaim their territorial rights continually; this is the function of most bird songs, the noisy bellowing of sea lions, and the chattering of squirrels. Scent markers are frequently used to signal a territory's boundaries. The male cheetah in Figure 35.16B, a resident of Africa's Serengeti National Park, is spraying urine on a tree. The odor will serve as a chemical "No Trespassing" sign. Other males that approach the area will sniff the marked tree and recognize that the urine is not their own. Usually, the intruder will avoid the marked territory and a potentially deadly confrontation with its proprietor.

Not all species are territorial. However, for those that are, the territory can provide exclusive access to food supplies, breeding areas, and places to raise young. Familiarity with a specific area may help individuals avoid predators or forage more efficiently. In a territorial species, such benefits increase fitness and outweigh the energy costs of defending a territory.

? Why is the territory of a gannet so much smaller than the territory of a cheetah?

■ The gannet only uses its territory for raising young, not for foraging. Cheetahs use their territories for foraging as well as for breeding.

Figure 35.16A Gannet territories

Figure 35.16B A cheetah marking its territory

Social behavior requires communication between animals

Social behavior depends on some form of signaling between the participating animals. In behavioral ecology, a **signal** is a behavior that causes a change in behavior in another animal. As we have seen, animals use many kinds of signals. These include sounds, such as the snort that signals danger to a herd of impalas; odors, such as the urine signs left by many cats to mark their territories; visual displays, such as beak pointing in loons; and touches, such as the grooming behavior of chimpanzees. The sending of, reception of, and response to signals constitute animal **communication**, an essential element of interactions between individuals.

What determines the type of signal animals use to communicate? Most terrestrial mammals are nocturnal, which makes visual displays relatively ineffective. Many mammals use odor and sound signals, which work well in the dark. Birds, by contrast, are mostly diurnal (active in daytime) and use visual and sound signals. Humans are also diurnal and likewise use mainly visual and sound signals. Therefore, we can detect the bright colors and songs birds use to communicate. If we had the well-developed olfactory abilities of most mammals and could detect their rich world of odor cues, mammal-sniffing might be as popular as bird-watching.

What types of signals are effective in aquatic environments? A common visual signal used by territorial fishes is to erect their fins, which is generally enough to drive off intruders. Electrical signals produced by certain fishes communicate hierarchy or status. And chemical alarm substances released from an injured fish may signal nearby fish to form a tightly packed school, often near the bottom, where they are safer from attack. Some aquatic animals use auditory signals to communicate. Humpback whales have complex songs that can be heard hundreds of kilometers away, helping the whales locate each other over vast areas of ocean.

In general, the more complex the social organization of a species, the more complex the signaling required to sustain it. Animals often use more than one type of signal simultaneously. **Figure 35.20A** shows a ring-tailed lemur, a tree-dwelling primate of Madagascar that lives in social groups averaging 15 individuals. Lemurs use visual displays, scent communication, and vocalizations to maintain the dominance hierarchy in the group. The animal shown here is communicating aggression with its prominent tail. Prior to this display, it smeared its tail with odorous secretions from glands on its forelegs. By waving its scented tail over its head, the lemur transmits both visual and chemical signals.

In addition to primates, many other vertebrates exhibit intricate social behavior that involves various types of signals. Some invertebrates also have complex social systems, such as that of honeybees. Often numbering over 50,000 individuals, a honeybee colony has complicated communication needs. Pheromones released by the queen bee help maintain the complex social order of the colony. Alarm pheromones call for aggressive or defensive behavior.

Intrigued with the question of how bees communicate about food resources, Karl von Frisch performed experiments in which he put dishes of scented sugar water at varying distances and directions from observation hives. Von Frisch observed and described a very complex signaling system. A worker bee returning from a food source is surrounded by bees (**Figure 35.20B**). The worker regurgitates some nectar that the others taste and smell and then performs a buzzing "dance" that seems to signal the location of the food. The other workers leave the hive and begin foraging. This innovative research continues as biologists investigate spatial learning in bees and seek to distinguish the behavioral signals and the visual and scent cues bees use to forage.

Web/CD Activity 35A
Honeybee Waggle Dance

? What types of signals do bees use?

Figure 35.20A A lemur communicating aggression

Sound, touch, and chemical ■

Figure 35.20B Bee communication

35.15 Sociobiology places social behavior in an evolutionary context

The mating behaviors of loons and sage grouse are examples of social behavior. Biologists define **social behavior** as any kind of interaction between two or more animals, usually of the same species. Courtship, aggression, and cooperation are all examples of social behavior. An essential ingredient of all types of social behavior is communication—some means of transferring information between individuals.

Many animals migrate and feed in large groups (flocks, packs, herds, or schools). As discussed in the chapter introduction, impalas feed in herds that provide protection from predators through their early warning and escape behavior. Wolves usually hunt in a pack consisting of a tightly knit group of family members. Hunting in packs enables them to kill large animals, such as moose or elk, that would be unavailable to an individual wolf.

The discipline of **sociobiology** applies evolutionary theory to the study and interpretation of social behavior—how social behaviors are adaptive and how they could have evolved by natural selection. We discuss several aspects of social behavior in the next several modules.

? What key process is required for social behavior within a population?

■ Communication

35.16 Territorial behavior parcels space and resources

Many animals exhibit territorial behavior. A **territory** is an area, usually fixed in location, that individuals defend and from which other members of the same species are usually excluded. The size of the territory varies with the species, the function of the territory, and the resources available. Territories are typically used for feeding, mating, rearing young, or combinations of these activities.

Figure 35.16A shows a nesting colony of gannets in Quebec, Canada. Space is at a premium, and the birds defend territories just large enough for their nests by calling out and pecking at other birds. As you can see, each gannet is literally only a peck away from its closest neighbors. Such small nesting territories are characteristic of many colonial seabirds. In contrast, most cats, including jaguars, leopards, cheetahs, and even domestic cats, defend much larger territories, which they use for foraging as well as breeding.

Individuals that have established a territory usually proclaim their territorial rights continually; this is the function of most bird songs, the noisy bellowing of sea lions, and the chattering of squirrels. Scent markers are frequently used to signal a territory's boundaries. The male cheetah in **Figure 35.16B**, a resident of Africa's Serengeti National Park, is spraying urine on a tree. The odor will serve as a chemical "No Trespassing" sign. Other males that approach the area will sniff the marked tree and recognize that the urine is not their own. Usually, the intruder will avoid the marked territory and a potentially deadly confrontation with its proprietor.

Not all species are territorial. However, for those that are, the territory can provide exclusive access to food supplies, breeding areas, and places to raise young. Familiarity with a specific area may help individuals avoid predators or forage more efficiently. In a territorial species, such benefits increase fitness and outweigh the energy costs of defending a territory.

? Why is the territory of a gannet so much smaller than the territory of a cheetah?

■ The gannet only uses its territory for raising young, not for foraging. Cheetahs use their territories for foraging as well as for breeding.

Figure 35.16A Gannet territories

Figure 35.16B A cheetah marking its territory

35.17 Rituals involving agonistic behavior often resolve confrontations between competitors

Agonistic behavior (from the Greek *agon*, struggle) includes a variety of threats or actual combat that settles disputes between individuals in a population. Conflicts often arise over limited resources, such as food, mates, or territories. An agonistic encounter may involve a test of strength or, more commonly, exaggerated posturing and other symbolic displays, or rituals, that make the individuals look large or aggressive. Eventually, one individual stops threatening and becomes submissive, exhibiting some type of appeasement display—in effect, surrendering.

Figure 35.17 Ritual wrestling by rattlesnakes

Because violent combat may injure the victor as well as the vanquished in a way that reduces reproductive fitness, we would predict that natural selection would favor ritualized contests. And, in fact, this is what usually happens in nature.

The rattlesnakes pictured in **Figure 35.17**, for example, are rival males wrestling over access to a mate. If they bit each other, both would die from the toxin in their fangs, but they are engaged in a pushing, rather than a biting, match. One snake usually tires before the other, and the winner pins the loser's head to the ground. In a way, the snakes are like two people who settle a serious argument by arm wrestling instead of resorting to fists or guns. In a typical case, the agonistic ritual inhibits further aggressive activity. Once two individuals have settled a dispute by agonistic behavior, future encounters between them usually involve less dispute, with the original loser giving way to the original victor. Often the victor of an agonistic ritual gains first or exclusive access to mates, and so this form of social behavior can directly affect an individual's evolutionary fitness.

? Why is "fighting to the death" an unusual form of agonistic behavior among animals?

■ Because ritualized posturing or nonlethal combat can usually produce a winner without injuries that would lower reproductive fitness for the winner and eliminate it altogether for the loser.

35.18 Dominance hierarchies are maintained by agonistic behavior

Many animals live in social groups maintained by agonistic behavior. Chickens are an example. If several hens unfamiliar to one another are put together, they respond by chasing and pecking each other. Eventually, they establish a clear "pecking order." The alpha, or top-ranked, hen in the pecking order (the one on the left in **Figure 35.18**) is dominant; she is not pecked by any other hens and can usually drive off all the others by threats rather than actual pecking. The alpha hen also has first access to resources such as food, water, and roosting sites. The beta, or second-ranked, hen similarly subdues all others except the alpha, and so on down the line to the omega, or lowest, animal.

Pecking order in chickens is an example of a **dominance hierarchy**, a ranking of individuals based on social interactions. Once a hierarchy is established, each animal's status in the group is fixed, often for several months or even years. Consequently, rather than fighting with others, group members can concentrate on finding food, watching for predators, locating a mate, or caring for young.

Dominance hierarchies are common, especially in vertebrate populations. In a wolf pack, for example, there is a dominance hierarchy among the females, and the hierarchy may control the pack's size. When food is abundant, the alpha female mates and also allows others to do so. When food is scarce, she usually monopolizes males for herself and keeps other females from mating.

Next we hear from a scientist who has conducted long-term studies of dominance hierarchies in nature.

? How may a dominance hierarchy enhance the reproductive success for all individuals of an animal population?

■ By reducing the amount of time and energy that agonistic behavior diverts from feeding and other survival activities, all individuals may have more reproductive success.

Figure 35.18 Chickens exhibiting pecking order

35.19 Behavioral biologist Jane Goodall discusses dominance hierarchies and reconciliation behavior in chimpanzees

Chimpanzees are humans' closest relatives. Dr. Jane Goodall (Figure 35.19A), one of the world's best-known biologists, has studied these remarkable primates in their natural habitat in East Africa since the early 1960s. She has described her discoveries in her many books, her appearances on National Geographic Society television specials, and her frequent lecture tours. In all of her writings and interviews, Dr. Goodall promotes a better understanding of animal behavior, especially that of primates. She also works tirelessly to encourage better living conditions for animals in medical research labs and zoos. She founded the Roots & Shoots program to educate young people to take action to improve the environment and their local communities.

In an interview, Goodall describes dominance hierarchies:

Some male chimpanzees devote much time and effort to improving or maintaining their position in the hierarchy. For the most part, the male uses the impressive charging display, during which he races across the ground, hurls rocks, drags branches, leaps up and shakes the vegetation. In other words, he makes himself look larger and more dangerous than he may actually be. In this way he can often intimidate a rival without having to risk an actual fight, which could be dangerous for him as well as for his rival. The more frequent, the more vigorous, and the more imaginative his charging display, the more likely it is that he will attain a high social position.

And what about female chimpanzees?

Females have a hierarchy too. . . . The reproductive advantage to the high-ranking female is clear. She can better appropriate choice food items and thus make her milk richer. In addition, her offspring are likely to become high-ranked since she will support them. In the supportive family group situation, all have a better chance of survival.

Chimpanzees live in fairly permanent social groups, and there are benefits to maintaining friendly relations within the group. Following a conflict, there is usually some kind of reconciliation behavior. For example, a chimpanzee that has threatened another member of its group may invite reconciliation by a hand gesture (Figure 35.19B), leading to a bout of friendly grooming. In Goodall's words:

Social grooming is the single most important social activity in the chimp community. It improves bad relationships and maintains good ones. A few brief grooming movements serve to reassure, to appease a higher-ranking individual, or to calm a subordinate. A mother pacifies her child by embracing and then grooming him or her. Adult males enjoy particularly long grooming sessions—this is important. Males do sometimes compete quite vigorously for dominance rank, and their relationship may then become tense. Yet it is crucial that they be able to cooperate in order to jointly protect the territory of their community.

Social primates seem to spend substantial time in reconciliation and pacification-type behavior.

Studies of chimpanzee behavior can make us more aware of what we have in common with other species. Jane Goodall's years of chimpanzee research have convinced her that chimpanzees are truly conscious beings. As she explains:

Science has been very quick to recognize the incredible similarity in [the anatomy] of the chimpanzee brain and human brain . . . and all the other amazing physiological similarities. So it stands to reason that you would find similarities in the emotions . . . and in certain kinds of behavior and intellect.

? Why is the study of chimpanzee behavior relevant to understanding the origins of certain human behaviors?

■ Because chimpanzees and humans share a common ancestor

Figure 35.19A Jane Goodall with Goblin, an alpha male

Figure 35.19B Reconciliation in chimpanzees

35.20 Social behavior requires communication between animals

Social behavior depends on some form of signaling between the participating animals. In behavioral ecology, a **signal** is a behavior that causes a change in behavior in another animal. As we have seen, animals use many kinds of signals. These include sounds, such as the snort that signals danger to a herd of impalas; odors, such as the urine signs left by many cats to mark their territories; visual displays, such as beak pointing in loons; and touches, such as the grooming behavior of chimpanzees. The sending of, reception of, and response to signals constitute animal **communication**, an essential element of interactions between individuals.

What determines the type of signal animals use to communicate? Most terrestrial mammals are nocturnal, which makes visual displays relatively ineffective. Many mammals use odor and sound signals, which work well in the dark. Birds, by contrast, are mostly diurnal (active in daytime) and use visual and sound signals. Humans are also diurnal and likewise use mainly visual and sound signals. Therefore, we can detect the bright colors and songs birds use to communicate. If we had the well-developed olfactory abilities of most mammals and could detect their rich world of odor cues, mammal-sniffing might be as popular as bird-watching.

What types of signals are effective in aquatic environments? A common visual signal used by territorial fishes is to erect their fins, which is generally enough to drive off intruders. Electrical signals produced by certain fishes communicate hierarchy or status. And chemical alarm substances released from an injured fish may signal nearby fish to form a tightly packed school, often near the bottom, where they are safer from attack. Some aquatic animals use auditory signals to communicate. Humpback whales have complex songs that can be heard hundreds of kilometers away, helping the whales locate each other over vast areas of ocean.

In general, the more complex the social organization of a species, the more complex the signaling required to sustain it. Animals often use more than one type of signal simultaneously. **Figure 35.20A** shows a ring-tailed lemur, a tree-dwelling primate of Madagascar that lives in social groups averaging 15 individuals. Lemurs use visual displays, scent communication, and vocalizations to maintain the dominance hierarchy in the group. The animal shown here is communicating aggression with its prominent tail. Prior to this display, it smeared its tail with odorous secretions from glands on its forelegs. By waving its scented tail over its head, the lemur transmits both visual and chemical signals.

In addition to primates, many other vertebrates exhibit intricate social behavior that involves various types of signals. Some invertebrates also have complex social systems, such as that of honeybees. Often numbering over 50,000 individuals, a honeybee colony has complicated communication needs. Pheromones released by the queen bee help maintain the complex social order of the colony. Alarm pheromones call for aggressive or defensive behavior.

Intrigued with the question of how bees communicate about food resources, Karl von Frisch performed experiments in which he put dishes of scented sugar water at varying distances and directions from observation hives. Von Frisch observed and described a very complex signaling system. A worker bee returning from a food source is surrounded by bees (**Figure 35.20B**). The worker regurgitates some nectar that the others taste and smell and then performs a buzzing "dance" that seems to signal the location of the food. The other workers leave the hive and begin foraging. This innovative research continues as biologists investigate spatial learning in bees and seek to distinguish the behavioral signals and the visual and scent cues bees use to forage.

Web/CD Activity 35A
Honeybee Waggle Dance

? What types of signals do bees use?

Figure 35.20A A lemur communicating aggression

■ Sound, touch, and chemical

Figure 35.20B Bee communication

35.21 Altruistic acts can often be explained by the concept of inclusive fitness

Many social behaviors are selfish. Behavior that maximizes an individual's survival and reproductive success is favored by selection, regardless of how much the behavior may harm others. For example, superior foraging ability by one individual may leave less food for others. **Altruism,** on the other hand, is defined as behavior that reduces an individual's fitness while increasing the fitness of others in the population. How can we explain examples of what appears to be atruistic, or selfless, behavior?

Consider the Belding's ground squirrel, which lives in regions of the western United States and is vulnerable to predators such as coyotes and hawks. Upon seeing a predator approach, a squirrel often gives a high-pitched alarm call (Figure 35.21A), which alerts nearby squirrels, who then retreat to their burrows. Field observations have confirmed that the conspicuous alarm call identifies the caller's location and increases the risk of being killed.

Altruistic behavior is often evident in animals that live in cooperative colonies. For example, workers in a honeybee hive are sterile females who labor all their lives on behalf of the queen. When a worker stings an intruder in defense of the hive, the worker usually dies.

Figure 35.21A Belding's ground squirrel giving an alarm call

The animals in Figure 35.21B are highly social rodents called naked mole rats. Almost hairless and nearly blind, they live in colonies in underground chambers and tunnels in southern and northeastern Africa. With a social structure resembling that of honeybees, each colony has only one reproducing female, called the queen. The queen mates with one to three males, called kings. The rest of the colony consists of nonreproductive females and males who forage for roots and care for the queen, her young, and the kings. While trying to protect the queen or kings from a snake that invades the colony, a nonreproductive naked mole rat may sacrifice its own life.

You might wonder how altruistic behavior can evolve if it reduces the reproductive success of self-sacrificing individuals. It is easy to see how selfless behavior might be selected for when it involves parents and offspring. When parents sacrifice their own well-being to ensure the survival of their young, they are maximizing the survival of their own genes. But reproducing is only one way to pass along genes; helping a close relative reproduce is another. Siblings, like parents and offspring, have half their genes in common, and an individual shares one-fourth of its genes with offspring of a sibling. The concept of **inclusive fitness** describes an individual's success at perpetuating its genes by producing its own offspring and by helping close relatives, who likely share many of those genes, to produce offspring.

Figure 35.21B The queen of a naked mole rat colony nursing offspring while surrounded by other individuals of the colony

Altruism increases inclusive fitness when it maximizes the reproduction of close relatives. The natural selection favoring altruistic behavior that benefits relatives is called **kin selection.** Thus, the genes for altruism may be propagated if individuals that benefit from altruistic acts are themselves carrying those genes.

If kin selection explains altruism, then the examples of unselfish behavior we observe should involve close relatives. This is in fact the case. Most alarm calls are given by female Belding's squirrels, whose close relatives live nearby. Researchers have also found that all the individuals in a naked mole rat colony are closely related. The nonreproductive members are the queen's descendants or siblings; by enhancing a queen's chances of reproducing, they increase the chances that some genes identical to their own will be propagated. Likewise, bees in a hive all share genes with the queen. Their work (or even death) in support of the queen helps ensure that many of their genes will survive.

Kin selection does not explain all types of altruism. Chimpanzees sometimes save the lives of nonrelatives. Similarly, female dolphins without young will often help unrelated mothers care for their young. In these cases, there can be no immediate enhancement of the altruists' fitness. However, the beneficiary may reciprocate—that is, "return the favor" someday. Thus, we can explain altruism toward nonrelatives as **reciprocal altruism:** an altruistic act that may be repaid at a later time by the beneficiary. Reciprocal altruism appears to be fairly rare, limited largely to species with social groups stable enough that individuals have many chances to exchange aid. It is often used to explain altruism in humans. In the last two modules of this chapter, we take a look at human social behavior.

? What is the ultimate cause for altruism between kin?

■ Natural selection reinforces altruistic behavior through the reproductive success of closely related individuals that have many genes in common with the altruist, including genes for altruism.

35.22 Both genes and culture contribute to human social behavior

We have discussed how genes and environment combine to influence individual behavior. In human individuals, such as the talented pianist mentioned in Module 35.3, genetic and environmental influences must blend to draw beautiful music from the player's fingers. But how do these forces shape behavior not just in one person, but in many?

Group behaviors, like individual behaviors, reflect a complex mix of genetic and environmental influences. Consider the example of reciprocal altruism. As you just learned, genes for this behavior may have been perpetuated in some species because the altruists benefit in the long run. But in the case of humans, compelling environmental forces are also at work. For thousands of years, human cultures have emphasized the values of compassion and giving. People teach their children to share with others, and they honor those who help the needy. Humans help strangers who may never be able to return the favor. Today, with the advent of modern communications and transport, millions of individuals donate food and money to people they will never meet on the other side of the world.

We see this combination of innate and learned influences in numerous human social behaviors. Behavioral scientists have tracked these influences in everything from how people stake out space and territory to how humans choose a mate.

One recent study, published in 2003 by the National Academy of Sciences, looked at the factors that influence mate choice in Western societies. The researchers devised a list of ten attributes desirable in a long-term partner, based on four criteria with reproductive and therefore evolutionary relevance: wealth and status; family commitment; physical appearance; and sexual fidelity. They then surveyed almost 1,000 heterosexual young men and women associated with Cornell University, asking them to rank the ten attributes in terms of their importance when choosing a long-term partner. Study partici-

pants were then asked to rate themselves as a potential mate using those same attributes.

If we expect genetic influences to dominate the process of mate selection, we might predict that participants would choose wealth or physical attractiveness in a mate above all else, as ample resources mean a greater chance of rearing successful offspring, and physical attractiveness is often associated with good health and fertility. However, study participants consistently valued the same attributes in potential mates that they saw in themselves. Instead of always wanting a rich or beautiful partner, study participants most wanted someone like themselves. The results of this study suggest that contrary to the old saying "opposites attract," the process of mate selection could be described as "likes attract."

The researchers surmised that this behavior could be partially explained by evolution. Other research has suggested that partners who are similar to each other are more likely to have long, stable relationships—the sorts of unions more likely to produce and provide for children. But environmental forces clearly play a role. Western societies, along with many other human societies, have long urged people to value more than looks or money when choosing a mate. People are encouraged to look for other qualities, such as similar values, personality, and temperament. Thus, social forces can help steer people toward long-term partners with greater compatibility and more staying power (Figure 35.22).

This environmental influence, in which societies transmit behavioral guidelines to younger members, is an example of the social learning that is the root of culture. **Culture** can be defined as a system of information transfer through social learning or teaching that influences the behavior of individuals in a population. While humans are not the only creatures on the planet to exhibit social learning and culture, these processes are essential to maintaining human societies.

As scientists study human social learning and culture, they continue to explore the influences of genes and environment. Sociobiology, the area of research introduced in Module 35.15, centers on the idea that social behavior evolves, like anatomical traits, as an expression of genes that have been perpetuated by natural selection. When applied to humans, this idea might make life sound predetermined. But its backers say that human behavior could never be explained as simply the result of genes. Perhaps more than any other example, human social behavior reflects the critical combination of genes and environment.

Next we look at one contemporary scientist who has promoted sociobiology and who explains that the concept ultimately holds great hope for human behavior and the future of life on Earth.

Figure 35.22 Similar interests as part of human mate choice

? How does the very long childhood of humans, compared with that of most other animals, contribute to culture?

■ It prolongs the opportunity for the young to learn from parents, relatives, teachers, and other mentors.

35.23 Edward O. Wilson promoted the field of sociobiology and is a leading conservation activist

Figure 35.23 Evolutionary biologist and conservation activist Edward O. Wilson

In 1975, E. O. Wilson (**Figure 35.23**), of Harvard University, published a book entitled *Sociobiology: The New Synthesis*. Drawing from his work with ants and from numerous studies on vertebrate and invertebrate animals, Wilson promoted the concept of sociobiology, arguing that genetic influences had become undervalued in the scientific study of social behavior. In an interview, Professor Wilson recalls this book's social effects:

In *Sociobiology: The New Synthesis*, I primarily intended to cover the social insects and vertebrate animals. I then saw that I could not leave out the most familiar vertebrate animal, *Homo sapiens*. I included two chapters on human beings primarily for completeness. I expected the book to have an impact, but I didn't expect it to stir up a hornet's nest of controversy in the social science community, as it did.

Sociobiology involves a search for evidence that social activities improve an animal's fitness. The concepts of kin selection and reciprocal altruism, with their focus on how the survival of genes underlies social behavior, are key ideas in sociobiology. Professor Wilson elaborates on why these ideas were so controversial:

Sociobiology came out at a time when most scholars in the social sciences believed not only that heredity has no importance in human social behavior, but that it is dangerous to speak of heredity, because to do so might imply that human destiny is fixed, and that there is nothing we can do about social ills. This was a primary reason for resistance both from social scientists, who had already settled on a sociocultural explanation for social behavior, and from some biologists.

In his introduction to the twenty-fifth-anniversary edition of *Sociobiology*, Wilson explains that recent research in human genetics and neuroscience strengthens the case for a biological understanding of human behavior. A new field of study, called evolutionary psychology, has grown out of sociobiology and draws theory and data from both biology and the social sciences.

We asked Wilson about current directions in biological research:

Modern biology consists of two major fronts of advance. One of them addresses the physical and chemical basis of life's operation and the development of organisms. The other addresses the behavior and the living together of organisms, as studied in behavioral biology and ecology, increasingly with a new emphasis on biodiversity. These latter areas—behavioral biology and ecology, or evolutionary biology for short—are attaining new importance to society.

Wilson elaborated on the impact of science on society:

Science is no longer just a fun thing, like landing on the moon or discovering a new species of bird. It is vital—and people know it. They see science as a major part of modern ethics and legislative action. They also see the environment as something that they have got to know about.

Why? As Wilson explains in his 1992 book, *The Diversity of Life*, intricately interconnected ecosystems are threatened by a human-made biodiversity crisis—an extinction crisis that rivals the extinction event that wiped out the dinosaurs and other species 65 million years ago. And why does biodiversity matter?

Biologists define biodiversity in the broadest sense as all of the variety of life—from the different genes [alleles] at the same chromosome position within populations, up through different species of organisms, on up to different aggregations of species in ecosystems. We should never knowingly allow a species to go extinct if appropriate measures can save it. That, in essence, is the biodiversity ethic.

In his most recent book, *The Future of Life*, Wilson elaborates on this ethic and the value of biodiversity. What should bother us most, he claims, is that we have no idea what we are destroying, what vital treasures vanish each day. He explains that it makes economic as well as ethical sense to preserve all forms of life.

The causes of our biodiversity crisis are twofold: the exploding human population and the increasing use of energy and resources. Wilson points to the biological basis for some of the human behaviors that have created this crisis, such as caring only about a small piece of land, a limited band of kinfolk, and a time span of a few generations. But he also cites another biologically based human behavior that may help to reduce global population growth. As women become socially and economically empowered, they have fewer children. "Reduced reproduction by female choice can be thought a fortunate gift of human nature to future generations."

In Wilson's many books, he documents the need for a universal environmental ethic that can save Earth's biodiversity and, in the process, save our own species.

? What is sociobiology?

■ The study of the evolutionary basis of social behavior

CHAPTER REVIEW

Reviewing the Concepts

The Scientific Study of Behavior (35.1–35.3)

Behavioral ecology studies behavior in an evolutionary context, considering both proximate (immediate) and ultimate (evolutionary) causes of an animal's actions. Natural selection preserves behaviors that enhance fitness **(35.1)**.

Innate behavior is performed the same way by all members of a species. Sign stimuli are simple cues that trigger fixed action patterns (FAPs). FAPs ensure that activities essential to survival are performed correctly without practice **(35.2)**.

Nature and nurture. Behavior usually involves both genetic and environmental influences **(35.3)**.

Learning (35.4–35.11)

Learning is a change in behavior resulting from experience. Habituation is learning to ignore a repeated, unimportant stimulus **(35.4)**.

Imprinting is irreversible learning limited to a sensitive period. Captive breeding programs for endangered species must provide proper imprinting models **(35.5–35.6)**.

Spatial learning involves using landmarks to move through the environment. Kineses and taxes are simple movements in response to a stimulus **(35.7)**.

Cognitive maps are internal representations of spatial relationships of objects in the surroundings. Migratory animals may move between areas using the sun, stars, landmarks, or other cues **(35.8)**.

Associative learning. Many animals can learn by associating external stimuli or their own behavior with positive or negative effects **(35.9)**.

Social learning involves changes in behavior that result from observation and imitation of others **(35.10)**.

Cognition is the ability of an animal's nervous system to perceive, store, process, and use information. Some animals exhibit problem-solving behavior, which involves complex cognitive processes **(35.11)**.

Foraging and Mating Behaviors (35.12–35.14)

Foraging includes identifying, obtaining, and eating food. Optimal foraging theory predicts that feeding behavior will maximize energy gain and minimize energy expenditure and risk **(35.12)**.

Mating systems may be promiscuous, monogamous, or polygamous. The needs of offspring and certainty of paternity help explain differences in mating systems and parental care by males **(35.13)**. Courtship rituals advertise the species, sex, and physical condition of potential mates **(35.14)**.

Social Behavior and Sociobiology (35.15–35.23)

Sociobiology studies social behavior, the interactions of two or more animals, in the context of evolution **(35.15)**.

Territorial behavior allocates space and resources. Animals exhibiting this behavior defend their territories **(35.16)**.

Agonistic behavior, including threats, rituals, and sometimes combat, settles disputes over resources **(35.17)**.

Dominance hierarchies partition resources among members of a social group **(35.18)**. Chimpanzees exhibit dominance hierarchies and reconciliation behaviors **(35.19)**.

Signaling in the form of sounds, scents, displays, or touches provides communication needed for social behavior **(35.20)**.

Altruism can usually be explained by the concepts of inclusive fitness and kin selection: An animal can propagate its own genes by helping relatives reproduce. In reciprocal altruism, individuals do favors that may later be repaid **(35.21)**.

Human behavior has a genetic basis but is quite variable, being strongly influenced by learning and culture **(35.22)**. According to sociobiologist E. O. Wilson, natural selection underlies many human behaviors, including behaviors that have led to our current biodiversity crisis **(35.23)**.

Connecting the Concepts

1. Complete this map, which reviews behavioral ecology.

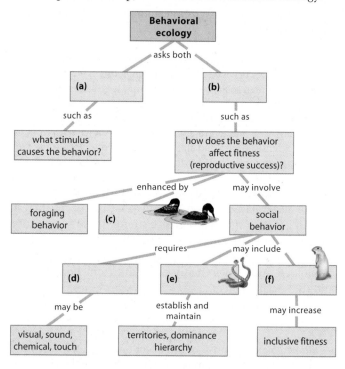

2. Create your own concept map to organize your understanding of the genetic and environmental components of animal behavior and their relationship to learning. Include examples where possible.

Testing Your Knowledge

Multiple Choice

3. Although many chimpanzee populations live in environments containing oil-palm nuts, members of only a few populations use stones to crack open the nuts. The most likely explanation for this behavioral difference between populations is that
   a. members of different populations differ in manual dexterity.
   b. members of different populations have different nutritional requirements.
   c. members of different populations differ in learning ability.
   d. the cultural tradition of using stones to crack nuts has arisen in only some populations.
   e. the behavioral difference is caused by genetic differences between populations.

4. Pheasants do not feed their chicks. Immediately after hatching, a pheasant chick starts pecking at seeds and insects on the ground. How might a behavioral ecologist explain the ultimate cause of this behavior?
   a. Pecking is a fixed action pattern.
   b. Pheasants learned to peck, and their offspring inherited this behavior.
   c. Pheasants that pecked survived and reproduced best.
   d. Pecking is a result of imprinting during a sensitive period.
   e. Pecking is an example of habituation.

5. A blue jay that aids its parents in raising its siblings is increasing its
   a. reproductive success.
   b. status in a dominance hierarchy.
   c. altruistic behavior.
   d. inclusive fitness.
   e. certainty of paternity.

6. Ants carry dead ants out of the anthill and dump them on a "trash pile." If a live ant is painted with a chemical from dead ants, other ants repeatedly carry it, kicking and struggling, to the trash pile, until the substance wears off. Which of the following best explains this behavior?
   a. The chemical is a sign stimulus for a fixed action pattern.
   b. The ants have become imprinted on the chemical.
   c. The ants continue the behavior until they become habituated.
   d. The ants can learn only by trial and error.
   e. The chemical triggers a negative taxis.

Describing, Comparing, and Explaining

7. Almost all the behaviors of a housefly are innate. What are some advantages and disadvantages to the fly of innate behaviors compared with behaviors that are mainly learned?

8. A wolf pack has both a dominance hierarchy and a territory defended against other wolf packs. What are the benefits to the pack of a dominance hierarchy? What are the benefits of holding a territory?

9. A chorus of frogs fills the air on a spring evening. The frog calls are courtship signals. What are the functions of courtship behaviors? How might a behavioral ecologist explain the proximate cause of this behavior? The ultimate cause?

Applying the Concepts

10. Crows break the shells of certain molluscs before eating them by dropping them onto rocks. Hypothesizing that crows drop the molluscs from a height that gives the most food for the least effort (optimal foraging), a researcher dropped shells from different heights and counted the drops it took to break them.

Height of drop (m)	Average number of drops required to break shell	Total flight height (number of drops × height per drop)
2	55	110
3	13	39
5	6	30
7	5	35
15	4	60

   a. The researcher measured the average drop height for crows and found it was 5.23 m. Does this support the researcher's hypothesis? Explain.
   b. Describe an experiment to determine whether this feeding behavior of crows is learned or innate.

11. Scientists studying scrub jays found that it is common for "helpers" to assist mated pairs of birds in raising their young. The helpers lack territories and mates of their own. Instead, they help the territory owners gather food for their offspring. Propose a hypothesis to explain what advantage there might be for the helpers to engage in this behavior instead of seeking their own territories and mates. How would you test your hypothesis? If your hypothesis is correct, what kind of results would you expect your tests to yield?

12. Researchers are very interested in studying identical twins who were raised apart. Among other things, they hope to answer questions about the roles of inheritance and upbringing in human behavior. So far, data suggest that identical twins raised apart are much more alike than researchers would have predicted based on their different environments. They have similar personalities, mannerisms, habits, and interests. Why do identical twins make such good subjects for this kind of research? What do the results suggest to you? What are the potential pitfalls of this research? What abuses might occur in the use of these data if the studies are not evaluated critically?

13. Do animals think and feel the same kinds of things we do? These questions bear on animal rights, a subject much in the news. Many important biological discoveries have come from experiments performed on animals, yet some animal rights activists believe that all animal experimentation is cruel and should be stopped. They have harassed researchers, and even vandalized laboratories and set animals free. Why are animals used in experiments? Are there uses of animals that should be discontinued? What kinds of guidelines should researchers follow in using animals in experiments, and who should establish and enforce the guidelines?

Answers to all questions can be found in Appendix 3.

For study help and Activities, go to campbellbiology.com or the student CD-ROM.

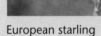

European starling

36.1 Population ecology studies how and why populations change

POPULATION STRUCTURE AND DYNAMICS

36.2 Density and dispersion patterns are important population variables
36.3 Life tables track mortality and survivorship in populations
36.4 Idealized models help us understand population growth
36.5 Multiple factors may limit population growth
36.6 Some populations have "boom-and-bust" cycles

LIFE HISTORIES AND THEIR EVOLUTION

36.7 Evolution shapes life histories
36.8 Principles of population ecology have practical applications

THE HUMAN POPULATION

36.9 Human population growth has started to slow after centuries of exponential increase
36.10 Birth and death rates and age structure affect population growth

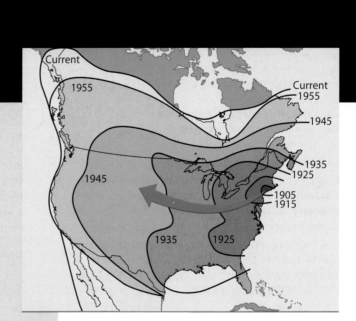

CHAPTER 36

The spread of starlings across North America

Population Dynamics

The Spread of Shakespeare's Starlings

CITY-DWELLERS OFTEN DETOUR around buildings and trees where starlings roost. Ranchers and farmers watch dense flocks of these birds devour grain from fields and feedlots. Yet if you lived in North America a little over a century ago, you would not have seen this bird. Once restricted to Europe and Asia, the European starling is now an abundant and destructive pest in North America, eastern Australia, New Zealand, and South Africa. Omnivorous, aggressive, and tenacious, starlings often replace native species. They oust woodpeckers, bluebirds, and swallows from nesting sites and may pull nestlings of these and other species out of nests to make room for their own offspring.

During the 1800s and early 1900s, introducing foreign species of animals and plants to North America was a popular, unregulated activity. Many people belonged to "acclimatization societies," whose purpose was to bring in species from other countries. Private groups and state game agencies imported the ring-necked pheasant, an Asian native, for sport hunting. Civic authorities in over 100 cities introduced the now-widespread house sparrow for "aesthetic reasons" and to control pest insects. A citizens' group introduced the starling as part of a campaign to bring all the birds mentioned in Shakespeare's works to the New World. In *Henry the Fourth*, a character named Hotspur alludes to the starling's ability to mimic human speech: "Nay, I'll have a starling shall be taught to speak nothing but 'Mortimer.'" Who would have thought this line from a play would trigger an environmental calamity 300 years after it was written?

Shakespeare enthusiasts released about 120 starlings in New York's Central Park in 1890. New Yorkers cheered as a breeding pair built a nest under the eaves of the American Museum of Natural History. From that foothold, starlings spread rapidly throughout the United States and Canada, as shown on the map on the facing page. Their range now extends from Mexico to Alaska. In less than a century, the North American starling population increased to about 100 million. Population estimates now are well over that number. As many as 5 million birds have been counted in a single roost. Local, state, and federal agencies in the United States spend millions of dollars each year trying to control starling populations. Mass trappings, hunting, electrified wires on buildings, fireworks,

Omnivorous, aggressive, and tenacious, starlings often replace native species

chemical repellents, and poisons have all proved ineffective in controlling the birds. California attempted a full-scale eradication, killing 9 million birds in three years. But even if every bird in the state had been killed, more would have moved right in.

The starling population in North America has some features in common with the global human population. Both are expanding and are virtually uncontrolled. Both are also harming other species: The starling threatens other bird species; the human population threatens much of the biosphere. In this chapter, we discuss why neither of these populations can grow indefinitely. But when will they stop, and what will stop them?

Population ecology, the subject of this chapter, is concerned with changes in population size and the factors that regulate populations over time. In the first few modules, we look at how ecologists study populations and some of the major factors that control populations in nature. Then we return to the topic of rapidly expanding populations and see how the principles of population ecology apply to the global human population. ■ ■ ■

36.1 Population ecology is the study of how and why populations change

Ecologists generally define a **population** as a group of individuals of a single species that occupy the same general area. These individuals rely on the same resources, are influenced by the same environmental factors, and have a high likelihood of interacting and breeding with one another.

The starling population of North America and the global human population are both very large. Most of our knowledge of population dynamics comes from studies of much smaller groups that are confined by more restricted geographic boundaries—for instance, a population of birds on an island, or fish in a lake, or protists in a laboratory culture.

A researcher must define a population by geographic boundaries appropriate to the questions being asked. For example, a population biologist studying the effects of hunting on deer might define a population as all the deer within a particular state. Another researcher, studying the effects of the AIDS epidemic on the human population, might focus on the HIV infection rate in one nation or throughout the world. Regardless of the scale, two important characteristics of any population are its density and its dispersion patterns. We discuss these characteristics next.

? What is the relationship between a population and a species?

■ A population is a localized group of individuals of a species.

POPULATION STRUCTURE AND DYNAMICS

36.2 Density and dispersion patterns are important population variables

Population density is the number of individuals of a species per unit area or volume—the number of oak trees per square kilometer (km^2) in a forest, for example, or the number of earthworms per cubic meter (m^3) in forest soil.

How do we measure population density? In rare cases, it is possible to count all individuals within the boundaries of the population. For example, it is possible to count the number of sea stars in a tide pool. And herds of large mammals such as elephants may be counted from airplanes.

In most cases, it is impractical or impossible to count all individuals in a population. Instead, ecologists use a variety of sampling techniques to estimate population densities. For example, they might base an estimate of the density of alligators in the Florida Everglades on a count of individuals in a few sample plots of 1 km^2 each. The larger the number and size of sample plots, the more accurate the estimates. In some cases, population densities are estimated not by counts of or-

ganisms but by indirect indicators, such as number of bird nests or rodent burrows or even animal droppings or tracks.

Within a population's geographic range, local densities may vary greatly. The **dispersion pattern** of a population refers to the way individuals are spaced within their area. These patterns are important characteristics for an ecologist to study, since they provide insights into the environmental effects and social interactions in the population.

A **clumped** pattern, in which individuals are aggregated in patches, is the most common in nature. Clumping often results from an unequal distribution of resources in the environment. For instance, plants or fungi may be clumped in areas where soil conditions and other factors favor germination and growth. Clumping of animals is often associated with uneven food distribution or with mating or other social behavior. For instance, fish are often clumped in schools (Figure 36.2A), which may reduce predation risks and in-

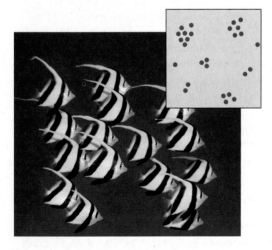

Figure 36.2A Clumped dispersion of schooling fish

Figure 36.2B Uniform dispersion of nesting king penguins (left) and human habitations (right)

crease feeding efficiency. Mosquitoes often swarm in great numbers, increasing their chances for mating.

A **uniform**, or even, pattern of dispersion often results from interactions between the individuals of a population. For instance, some plants secrete chemicals that inhibit the germination and growth of nearby plants that could compete for resources. And animals may exhibit uniform dispersion as a result of territorial behavior. **Figure 36.2B** (previous page) shows uniform dispersion in king penguins and humans.

In a **random** type of dispersion, individuals in a population are spaced in a patternless, unpredictable way. Clams living in a coastal mudflat, for instance, might be randomly dispersed at times of the year when they are not breeding and when resources are plentiful and do not affect their dis-tribution. However, varying habitat conditions and social interactions make random dispersion rare.

Estimates of population density and dispersion patterns are both important in analyzing populations. They enable researchers to monitor changes in a population and to compare and contrast the growth and stability of populations in different areas. The next module describes another tool that ecologists use to study population dynamics.

Web/CD Activity 36A *Techniques for Estimating Population Density and Size*

? What dispersion pattern would you predict in a forest population of sowbugs (arthropods that require high humidity)?

■ Clumped (under leaves or rotting logs)

36.3 Life tables track mortality and survivorship in populations

When the life insurance industry was established about a century ago, insurance companies developed an interest in the mathematics of survival. Needing to determine how long, on average, an individual of a given age could be expected to live, they invented what are called **life tables.** Table 36.3 was compiled using U.S. population data from 1999 and based on 100,000 births. According to this table, what is the chance of a 20-year-old surviving to age 30? In which age decade does the mortality rate increase significantly? Population ecologists have adopted this technique, constructing life tables for various plant and animal species to study the dynamics of population growth.

A graph like the one in **Figure 36.3** makes the data in a life table easy to comprehend. These **survivorship curves** plot the proportion of individuals alive at each age. By using a percentage scale instead of actual ages on the horizontal axis, we can compare species with widely varying life spans on the same graph. The curve for the human population tells us that most people die in older age intervals, as we saw in the life table. Species that exhibit this Type I curve—humans and many other large mammals—usually produce few offspring but give them good care, increasing the likelihood that they will survive to maturity.

In contrast, a Type III curve indicates high death rates for the very young and then a period when death rates are much lower for those few individuals who survive to a certain age. Species with this type of survivorship curve usually produce very large numbers of offspring but provide little or no care for them. An oyster, for instance, may release millions of eggs, but most offspring die as larvae from predation or other causes. A Type II curve is intermediate, with mortality more constant over the life span. This type of survivorship has been observed in some invertebrates, lizards, and rodents, such as squirrels.

Population ecologists also use mathematical models that describe and predict population growth, as we see next.

TABLE 36.3 LIFE TABLE FOR THE U.S. POPULATION IN 1999

Age Interval	Number Living at Start of Age Interval (*N*)	Number Dying During Interval (*D*)	Mortality (Death Rate) During Interval (*D/N*)	Chance of Surviving Interval (1 − *D/N*)
0–10	100,000	930	0.009	0.991
10–20	99,070	448	0.005	0.995
20–30	98,621	937	0.010	0.990
30–40	97,684	1,362	0.014	0.986
40–50	96,321	2,800	0.029	0.971
50–60	93,521	6,068	0.065	0.935
60–70	87,453	13,169	0.151	0.849
70–80	74,284	23,757	0.320	0.680
80–90	50,526	32,082	0.635	0.365
90 +	18,445	18,445	1.000	0.000

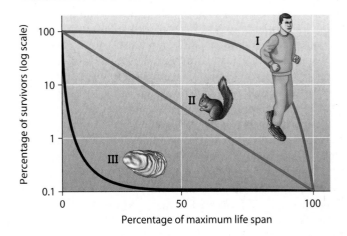

Figure 36.3 Three types of survivorship curves

? What is the key feature of the "mortality" column in the life table for a population with a Type II survivorship curve?

■ The mortality is about the same for every age interval.

Web/CD Activity 36B *Investigating Survivorship Curves*

To appreciate the explosive potential for population increase, consider a single bacterium that can reproduce by fission every 20 minutes under ideal laboratory conditions. There would be two bacteria after 20 minutes, four after 40 minutes, eight after 60 minutes, and so on. If this continued for a day and a half—a mere 36 hours—there would be bacteria enough to form a layer a foot deep over the entire Earth. At the other extreme, elephants may produce only six young in a 100-year life span. Still, Darwin estimated that it would take only 750 years for a single pair of elephants to give rise to a population of 19 million. And the starling population described in this chapter's introduction increased from 100 to a million in less than a century.

Time	Number of Cells	
0 minutes	1	$= 2^0$
20	2	$= 2^1$
40	4	$= 2^2$
60	8	$= 2^3$
80	16	$= 2^4$
100	32	$= 2^5$
120 (= 2 hours)	64	$= 2^6$
3 hours	512	$= 2^9$
4 hours	4,096	$= 2^{12}$
8 hours	16,777,216	$= 2^{24}$
12 hours	68,719,476,736	$= 2^{36}$

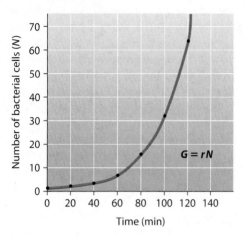

Figure 36.4A Exponential growth of bacteria

The Exponential Growth Model The rate of population increase under ideal conditions is called exponential growth. The whole population multiplies by a constant factor during each time interval. For example, the factor for the bacterial population represented in **Figure 36.4A** is 2 (because each parent cell splits to produce two daughter cells), and the time interval is 20 minutes. The progression for bacterial growth—2, 4, 8, 16, and so on—is the number 2 raised to a successively higher power (exponent) each generation. Graphing these data produces a J-shaped curve, typical of exponential growth. Notice that the curve gets steeper over time: The larger the population of bacteria, the faster the population grows.

The simple equation $G = rN$ describes this J-shaped curve. The G stands for the growth rate of the population (the number of new individuals added per time interval); N stands for the population size (the number of individuals in the population at a particular time); and r stands for the per capita rate of increase (the average contribution of each individual to population growth). How do we estimate this per capita rate? Population growth reflects the number of individuals added to the population minus the number removed from the population. If a population of 1,000 members has 500 births and 300 deaths in a time period, the net increase per individual is $(500 - 300)/1,000$, or 0.2. In a population growing in an ideal environment with unlimited space and resources, r is called the **intrinsic rate of increase**—the maximum capacity of members of that population to reproduce. The value of r depends on the kind of organism. For example, bacteria have a much higher r than do elephants. For any population expanding without limits, r remains constant.

To determine population growth, we multiply the reproductive contribution an average member is expected to make (r) by the number of individuals in the population. The exponential equation tells us that if r is constant, the rate at which a population grows depends on the number of individuals already in the population. It's like compound interest on a savings account. At 7% interest per year (analogous to the constant r in our equation), you make only $70 the first year on a $1,000 deposit. But leave the principal and interest in the bank until retirement age, and you'll be supplementing your retirement with $2,200 per year on an account that has grown to over $30,000. Similarly, in our equation for exponential population growth, the bigger the value of N, the faster the population increases. On a graph, the lower part of the J results from the relatively slow growth when the population is small. The steep, upper part of the J results from N being large.

The **exponential growth model** gives an idealized picture of unregulated population growth. For bacteria, unregulated growth means there is no restriction on the abilities of the cells to live, grow, and reproduce. Given a few days of unregulated growth, bacteria would smother every other living thing. Obviously, no population—neither bacteria nor elephants nor humans—can grow exponentially indefinitely.

Limiting Factors and the Logistic Growth Model In nature, a population that is introduced to a new environment or is rebounding from a catastrophic decline in numbers may grow exponentially for a while, but eventually, one or more environmental factors will limit its growth. Population size then stops increasing or may even crash. Environmental factors that restrict population growth are called **limiting factors.**

You can see the effect of population-limiting factors in the graph in **Figure 36.4B**, which illustrates the growth of a population of fur seals on St. Paul Island, off the coast of Alaska. (For simplicity, only the mated bulls were counted. Each has a harem of a number of females, as shown in the photograph.) Before 1925, the seal population on the island remained low because of uncontrolled hunting, although it changed from year to year. After hunting was controlled, the population increased rapidly until about 1935, when it began to level off and started fluctuating around a population size of about 10,000 bull seals. At this point, a number of

Figure 36.4B Growth of a population of fur seals

limiting factors, including some hunting and the amount of space suitable for breeding, restricted population growth.

The fur seal growth curve resembles the **logistic growth model,** a description of idealized population growth that is slowed by limiting factors as the population size increases. **Figure 36.4C** compares the logistic growth model (red) with the exponential growth model (blue). As you can see, the logistic curve is J-shaped at first, but gradually levels off to resemble an S shape.

The equation for logistic growth is more complicated than the exponential equation because it describes the effect of limiting factors on an increasing population size:

$$G = rN \frac{(K - N)}{K}$$

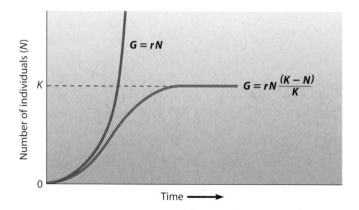

Figure 36.4C Logistic growth and exponential growth compared

This equation is actually simpler than it may appear. As you can see, the logistic equation is the exponential equation modified by the term $(K - N)/K$. The only new letter in the equation is K, which stands for carrying capacity. **Carrying capacity** is the maximum population size that a particular environment can support ("carry"). For the fur seal population on St. Paul Island, for instance, K is about 10,000 mated males. The value of K varies, depending on the species and the resources available in the habitat. K might be considerably less than 10,000 for a fur seal population on a smaller island with fewer breeding sites.

Let's see how the term $(K - N)/K$ works in producing the S-shaped logistic curve. When the population first starts growing, N is very small compared to the carrying capacity K. Thus, the term $(K - N)/K$ nearly equals K/K, or 1, and population growth G is close to rN—that is, exponential growth. However, as the population increases and N gets closer to carrying capacity, the term $(K - N)/K$ becomes an increasingly smaller fraction. The population growth rate slows as rN is multiplied by that fraction. At carrying capacity, the population is as big as it can theoretically get in its environment; at this point, $N = K$, and $(K - N)/K = 0$. The population growth rate (G) becomes zero.

What does the logistic growth model suggest to us about real populations in nature? The model predicts that a population's growth rate will be small when the population size is *either* small or large, and highest when the population is at an intermediate level relative to the carrying capacity. At a low population level, resources are abundant, and the population is able to grow nearly exponentially. At this point, however, the increase is small because N is small. In contrast, at a high population level, limiting factors strongly oppose the population's potential to increase. There might be less food available per individual or fewer breeding territories, nest sites, or shelters. These limiting factors cause the birth rate to decrease, the death rate to increase, or both. Eventually, the population stabilizes at the carrying capacity (K), when the birth rate equals the death rate.

It is important to realize that the logistic growth model presents a mathematical ideal. No natural populations fit it perfectly. In the fur seal example in Figure 36.4B, you see that the population first overshot its carrying capacity before stabilizing at a lower level. Some natural populations fluctuate greatly, making it difficult even to define carrying capacity. Overall, the logistic model is a useful starting point for studying population growth and for constructing more complex models. In Module 36.8 we will discuss some of the ways these models are useful in conservation biology and pest management. And like any good starting hypothesis, the logistic model has stimulated research, leading to a better understanding of the factors affecting population growth. We take a closer look at some of these factors next.

> **?** Why is population growth greatest when N is $\frac{1}{2} K$, when population size is half the carrying capacity?

■ At this population size, there are more reproducing individuals than at lower population sizes and still lots of space or other resources available for growth.

36.5 Multiple factors may limit population growth

The logistic growth model predicts that population growth slows and eventually ceases as population density increases. In other words, increasing population density results in a decrease in birth rate, an increase in death rate, or both. What could cause these **density-dependent** rates—declining birth rates and rising death rates in response to increasing population density?

Several factors appear to regulate growth in natural populations. The most obvious is competition among members of a growing population for limited resources. As a limited food supply is divided among more and more individuals, birth rates may decline. Field studies of songbirds have demonstrated this effect. **Figure 36.5A** shows one such study of a song sparrow population on a small island in British Columbia. As the density of females increases, the clutch size (number of eggs laid) decreases. A shortage of food appears to cause this decrease. In an experiment in which females were given extra food when population densities were high, they did not show this decrease in clutch size.

A limited resource may be something other than food or nutrients. In many vertebrates that defend a territory, the availability of space may limit reproduction. For instance, the number of nesting sites on rocky islands may limit the population size of oceanic birds such as gannets.

Population density also influences the health and thus the survival of organisms. Plants grown under crowded conditions tend to be smaller and less likely to survive. And those that do survive produce fewer flowers, fruits, and seeds. Gardeners who understand this density-dependent result thin their seedlings to produce the best possible yield. Animals, too, experience increased mortality at high population densities. These deaths may be a result of increased disease transmission under crowded conditions or the accumulation of toxic waste products. Predation may also be an important cause of density-dependent mortality. A predator may concentrate on and capture more of a particular kind of prey as that prey becomes abundant, thus limiting further growth of the population.

For some animal species, physiological factors appear to regulate population size. White-footed mice in a small field enclosure will multiply from a few to a colony of 30 to 40 individuals, but reproduction then declines until the population ceases to grow, even when additional food and shelter are provided. High population densities in mice appear to induce a stress syndrome in which hormonal changes can delay sexual maturation, cause reproductive organs to shrink, and depress the immune system. In this case, high densities cause both an increase in mortality and a decrease in birth rate. Similar effects of crowding have been observed in wild populations of other rodents.

In these examples of population regulation, we have seen how increased density causes population growth rate to decline by reducing birth rate and/or increasing death rate. In many natural populations, however, abiotic factors such as weather may limit or reduce population size well before other limiting factors become important. If we look at the

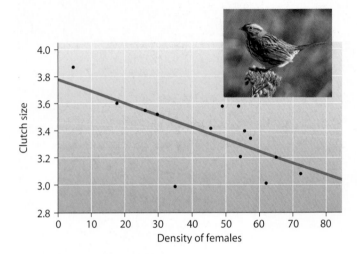

Figure 36.5A Decrease in song sparrow clutch size as population density increases

growth curve of such a population, we see something like exponential growth followed by a rapid decline, rather than a leveling off. **Figure 36.5B** shows this effect for a population of aphids, insects that feed on the phloem sap of plants. These and many other insects often show virtually exponential growth in the spring and then rapid die-offs when the weather turns hot and dry in the summer. A few individuals may remain, and these may allow population growth to resume again if favorable conditions return. In some populations of insects—many mosquitoes and grasshoppers, for instance—adults die off entirely, leaving only eggs, which initiate population growth the following year. In addition to seasonal changes in the weather, environmental factors, such as fire, floods, storms, and habitat disruption by human activity, can affect population densities.

Over the long term, most populations are probably regulated by a mixture of factors. Some populations remain fairly

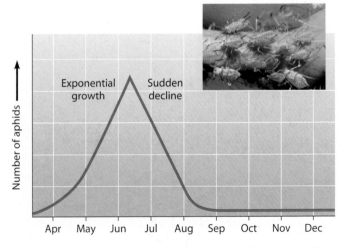

Figure 36.5B The effect of an abiotic factor (climate) on aphid population size

stable in size and are presumably close to a carrying capacity that is determined by biotic factors such as competition or predation. Most populations for which we have long-term data, however, show fluctuations in numbers. For example, though density has a population-regulating influence on clutch size in a song sparrow population (see Figure 36.5A), over a 25-year period this natural population shows alternating spurts of growth and drastic decline (in periods of severe winter weather) (Figure 36.5C). Thus, the dynamics of many populations result from a complex interaction of both density-dependent birth and death rates and abiotic factors such as climate and disturbances.

? List some of the factors that may reduce birth rate or increase death rate as population density increases.

■ Food and nutrient limitations, insufficient territories, increase in disease and predation, accumulation of toxins

Figure 36.5C Fluctuations in a song sparrow population, with periodic catastrophic reductions due to severe winter weather

36.6 Some populations have "boom-and-bust" cycles

Some populations of insects, birds, and mammals fluctuate in density with remarkable regularity. A striking example is the dramatic "boom-and-bust" growth cycles of lemming populations, which can occur every three to four years. Some researchers hypothesize that natural changes in the lemmings' food supply may be the underlying cause. Another hypothesis, as discussed in Module 36.5, is that stress from crowding during the "boom" may reduce reproduction, causing a "bust."

Figure 36.6 illustrates another well-known example—the cycles of snowshoe hare and lynx. The lynx is one of the main predators of the snowshoe hare in the far northern forests of Canada and Alaska. About every ten years, both hare and lynx populations show a rapid increase followed by a sharp decline.

What causes these boom-and-bust cycles? Since ups and downs in the two populations seem to almost match each other on the graph, does this mean that changes in one directly affect the other? For the hare cycles, there are three main hypotheses. First, cycles may be caused by increasing food shortages during winter resulting from overgrazing. Second, cycles may be due to predator-prey interactions. Many predators other than lynx, such as coyotes, foxes, and great-horned owls, eat hares, and the combination of predators might overexploit their prey. Third, cycles could be affected by a combination of food resource limitation and excessive predation. Recent experimental studies performed in the field support the hypothesis that the ten-year cycles of the snowshoe hare are largely driven by excessive predation, but also influenced by fluctuations in the hare's food supplies.

For the lynx and many other predators that depend heavily on a single species of prey, the availability of prey can influence population. Thus, the ten-year cycles in the lynx population probably do result at least in part from the ten-year cycles in the hare population. And when prey become scarce, predators often turn on one another, accelerating the

collapse of predator populations. Prey populations, released from predator pressure, can then climb again.

Long-term studies are the key to unraveling the complex causes of such population cycles. Next we look at some life history traits that influence the population dynamics of different organisms.

? In one experiment, increasing food supply to hares increased their population density, but the population continued to show cyclic collapses. What might you conclude from these results?

■ Hare population cycles are not primarily caused by food shortage.

Figure 36.6 Population cycles of the snowshoe hare and the lynx

36.7 Evolution shapes life histories

An organism's **life history** is the series of events from birth through reproduction to death. Some key life history traits are the age at which reproduction first occurs, the frequency of reproduction, the number of offspring, and the amount of parental care given. For a given population in a particular environment, natural selection will favor the combination of life history traits that maximizes an individual's output of viable, fertile offspring. Life history traits, like anatomical features, are shaped by adaptive evolution.

Figure 36.7A illustrates a type of life history called "big-bang" reproduction. The agave, or century plant, grows in arid climates with sparse and unpredictable rainfall. It may grow for decades without flowering or reproducing. Then one rainy spring, it grows a floral stalk that may be as tall as a telephone pole, produces many seeds, and withers and dies. By growing and storing nutrients until an unusually wet year and then putting all its resources into reproduction, the agave's big-bang strategy is a life history adaptation to erratic climate. In contrast, a more predictable environment may favor repeated reproduction throughout the life span.

Some ecologists hypothesize that different life history patterns are favored under different population densities and conditions. Selection for life history traits that maximize reproductive success in uncrowded, unpredictable environments can be called *r*-selection; such populations maximize *r*, the intrinsic rate of increase (see Module 36.4). Individuals in these populations mature early and/or produce a large number of offspring at a time. Many insect and weed species show such life histories, maximizing reproductive output whenever environmental opportunity knocks.

In contrast, some species, typically larger-bodied and longer-lived, exhibit life history traits that are said to represent *K*-selection. Common in populations that live at densities close to the carrying capacity (*K*) of their environment, such traits include maturity and reproduction at a later age and the production of a few, well-cared-for offspring. The life histories of many large terrestrial vertebrates fit this model. For example, a female polar bear has only one or two offspring every three years, but the cubs remain in her protective custody for over two years.

The concept of *r*- and *K*-selection has been criticized as an oversimplification, and the characteristics of most species place them in between the extremes of the *r*- and *K*-selection dichotomy. This hypothesis, however, has stimulated research on factors affecting the evolution of life histories.

Let's look at one particular environmental factor, predation, that has been shown to shape life history traits. For years, researchers have been studying guppy populations living in small, relatively isolated pools on the Caribbean island of Trinidad. Guppies are small fish you probably recognize as popular aquarium pets. As you can see in **Figure 36.7B**, certain guppy populations live in pools with predators called killifish, which eat mainly small, immature guppies. Other

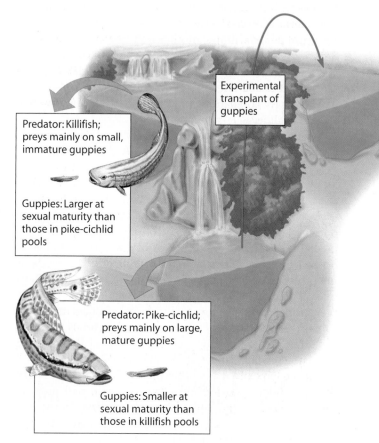

Experimental transplant of guppies

Predator: Killifish; preys mainly on small, immature guppies

Guppies: Larger at sexual maturity than those in pike-cichlid pools

Predator: Pike-cichlid; preys mainly on large, mature guppies

Guppies: Smaller at sexual maturity than those in killifish pools

Figure 36.7A The "big-bang" reproduction of the agave

Figure 36.7B Effect of predation on life history traits of guppies

guppy populations live where larger fish, called pike-cichlids, eat mostly mature, large-bodied guppies. Guppies in populations exposed to these pike-cichlids tend to be smaller, mature earlier, and produce more offspring at a time than those in areas with killifish. Thus, guppy populations differ in certain life history traits, depending on the kind of predator in their environment. For these differences to be the result of natural selection, the traits should be heritable. And indeed, guppies from both populations raised in the laboratory without predators retained their life history differences.

To test whether the feeding preferences of different predators caused these differences in life histories by natural selection, researchers introduced guppies from a pike-cichlid habitat into a guppy-free pool inhabited by killifish. The scientists tracked the weight and age at sexual maturity in the experimental guppy populations for 11 years, comparing these guppies with control guppies that remained in the pike-cichlid pools. The average weight and age at sexual maturity of the transplanted populations increased significantly as compared with the control populations. These studies demonstrate not only that life history traits are heritable and shaped by natural selection, but also that questions about evolution can be tested by field experiments.

As we have seen, population ecology involves theoretical model building as well as observations and experiments in the field. Next we look at how the principles of population ecology may be applied in conservation and management.

? Refer back to Module 36.3. Which type of survivorship curve would you expect to find in a population experiencing *r*-selection? *K* selection?

■ Type III for a population experiencing *r*-selection; type I for *K*-selection

CONNECTION

36.8 Principles of population ecology have practical applications

We often attempt to manage natural resources—trying to increase populations we wish to harvest or save from extinction and trying to decrease populations we consider pests. Principles of population ecology help guide us toward these various resource management goals.

Wildlife managers, fishery biologists, and foresters try to practice **sustainable resource management:** harvesting crops without damaging the resource. According to the concept of **maximum sustained yield,** harvesting should be done at a level that produces a consistent yield without forcing a population into decline. A population growing according to the logistic growth model increases the fastest when its density is at an intermediate level relative to its carrying capacity (see Module 36.4). One approach has been to harvest populations down to this level to ensure high growth rates.

Human economic and political pressures, however, often outweigh ecological concerns, and scientific information is frequently insufficient. For example, in the collapse of the northern cod fishery, shown in **Figure 36.8,** estimates of cod stocks were too high, and the practice of discarding young cod (not of legal size) at sea caused a higher mortality rate than was predicted. Following the collapse of many other fish and whale populations, resource managers are trying to minimize the risk of resource collapse by setting minimum population sizes or imposing protected, harvest-free areas.

For species that are in decline or facing extinction, resource managers may try to provide additional habitat or improve the quality of existing habitat to raise the carrying capacity, *K,* and thus increase population growth.

Reducing the size of a population is also a challenging task. As we saw with the starlings, simply killing many individuals will not usually decrease the size of pest populations. Many insect and weed species have life history traits that are *r*-selective (see Module 36.7), adapted to produce rapid population growth. Also, most pesticides kill both the pest and their natural predators. Because prey species often have a higher reproductive rate than predators, pest populations rapidly rebound before their predators can reproduce.

Integrated pest management (IPM) uses a combination of biological, chemical, and cultural methods to control agricultural pests. IPM relies on knowledge of the population ecology of the pest and its associated predators and parasites, as well as crop growth dynamics.

As we've learned, there are many factors that influence a population's size. To effectively manage any population, we must identify those variables, account for the unpredictability of the environment, consider interactions with other species, and weigh the economic and political issues. These same issues apply to the growth of the human population, which we explore next.

? What is the rationale behind hunting seasons for deer and other wildlife?

■ To protect wildlife from overharvest yet maintain lower population levels so that growth rate is high and mortality from resource limitation is reduced.

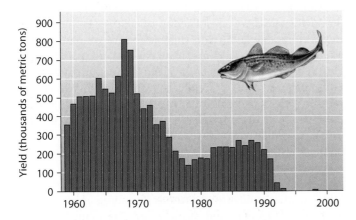

Figure 36.8 Collapse of northern cod fishery off Newfoundland

36.9 Human population growth has started to slow after centuries of exponential increase

On October 12, 1999, the world population reached 6 billion. By 2004, we numbered 6.4 billion. Over 200,000 people are added each day, and it takes only 4 years for world population growth to add the equivalent of another United States. As Figure 36.9A shows, the human population increased relatively slowly until about 1650, when approximately 500 million people inhabited the world. It took 200 years for the global population to double to 1 billion. In the next 80 years it doubled again to 2 billion and doubled still again in the next 45 years to more than 4 billion. Put another way, it took from the beginning of human history until the 1800s for the world's population to reach 1 billion, yet it took less than 12 years for the population to expand from 5 to 6 billion. If you compare the graph of human population growth with the graph in Figure 36.4A, it looks as if we have been multiplying like bacteria, proliferating in the biosphere as though it were an enormous petri dish.

Throughout most of human history, parents had many children. But only two, on average, survived to adulthood. So the population merely replaced itself. But with the advent of better nutrition, sanitation, and medical care, enough children are surviving to reproductive age to boost the population to unprecedented numbers. And as you saw in Module 36.4, for a population growing exponentially, the greater the numbers, the faster the growth.

Actually, the annual rate of population growth has finally begun to slow, from a peak rate of 2.19% in 1962 to 1.16%

in 2003. The rate is projected to continue its decline, slowing but not stopping population growth over the next decades. Even at these lowered growth rates, the world's population is projected to reach 7.3 to 8.4 billion by 2025.

The full weight of human activity no longer fits comfortably within the confines of Earth. To accommodate all the people expected on earth by 2025 and improve their diets, the world will have to double food production. Already, agricultural lands are under pressure. Overgrazing by the world's growing herds of livestock is turning vast areas of grassland into desert. In Africa, the number of livestock often exceeds grassland carrying capacity by half or more. Water use has risen sixfold over the past 70 years, causing rivers to run dry, water for irrigation to be depleted, and levels of groundwater to drop. And because so much open space will be needed to support this expanding human population, many thousands of other species are expected to become extinct.

How large a population of humans can Earth hold? In Module 36.4, we defined carrying capacity as the maximum population size that an environment can support. Some scientists maintain that we have already exceeded Earth's carrying capacity; others believe that Earth can support 10–15 billion people or even more. Researchers have used equations such as the one for the logistic growth model and estimates of limiting factors such as food or habitable land to arrive at estimates of carrying capacity.

An especially promising approach to estimating Earth's carrying capacity, based on the "ecological footprint" concept, considers multiple constraints, including food, fuel, water, housing, and waste disposal. The human **ecological footprint** is an estimate of the amount of land needed to support our multiple demands on Earth's resources. Researchers calculates, in hectares of land per person (1 ha = 2.47 acres), the current demand on resources made by each country and by the human population as a whole. Six types of ecologically productive areas are considered in calculating the ecological footprint: arable land (land suitable for crops), pasture, forest, ocean, built-up land, and fossil energy land. Fossil energy land is calculated on the basis of the land required for vegetation to absorb the CO_2 produced by burning fossil fuels. All measures are converted to land area per person. Worldwide, there is a huge disparity in the size of these ecological footprints. In 1997, when this study was done, a U.S. resident had a footprint estimated to be 8.4 ha, while Indians and Bangladeshis got by on 0.8 and 0.5 ha, respectively.

Now let's consider the available ecological capacity of various countries and Earth as a whole. By adding up all the ecologically productive land available on the planet, we can determine that the world had an ecological capacity of about

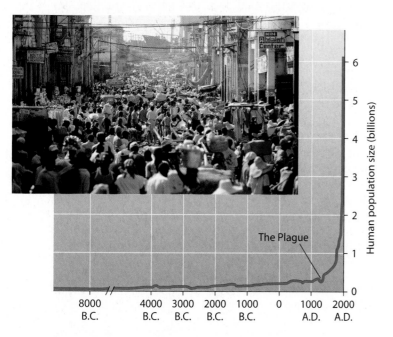

Figure 36.9A The history of human population growth

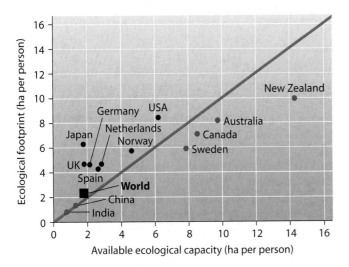

Figure 36.9B Ecological footprint in relation to ecological capacity

2 ha per person alive in 1997. If we wish to reserve land for parks and conservation, we must reduce this to 1.7 ha. Figure 36.9B graphs the ecological footprint for 13 countries and the whole world in relation to their available ecological capacities. The countries above the red diagonal line are in ecological deficit; countries with blue dots still have resource surpluses. The graph indicates two things: First, the world in general was already in ecological deficit in 1997. Second, countries vary greatly in their individual footprint size and in their available ecological capacity. The United States has a bigger ecological footprint (8.4 ha per person) than its own land and resources can support (6.2 ha per person). By this measure, the U.S. population is already above carrying capacity. And compared with the world's available ecological capacity, we're using four times our share. The overall analysis of human impacts via ecological footprints suggests that the world is already at or slightly above its carrying capacity.

So the problem is not just overpopulation, but overconsumption. The world's richest countries, with 20% of the global population, use 86% of the world's resources, leaving just 14% of global resources—energy, food, water, and other essentials—for the other 80% of the world's population to share. Indeed, the poorest 20% of the population accounts for just 1.3% of resource consumption. Ecologists warn that Earth isn't big enough to allow everyone living on it even now to indulge in the U.S. standard of living. Some researchers estimate that to do so would require the resources of three more planet Earths.

We can only speculate about Earth's ultimate carrying capacity for the human population or about what factors will eventually limit our growth. Perhaps food will be the main factor. Malnutrition and famines are common in some countries, but they result mainly from unequal distribution rather than inadequate production of food. So far, technological improvements in agriculture have allowed food supplies to keep up with global population growth. However, we also know, based on principles of energy flow through ecosystems, that environments can support a larger number of herbivores than carnivores (see Modules 37.13 and 37.14).

If everyone ate as much meat as the wealthiest people in the world, less than half of the present world population could be fed on current food harvests. But it seems unlikely that people in wealthier countries will abandon the consumption of meat. Also, as economic conditions improve in other countries, meat consumption will probably increase.

Perhaps we will eventually be limited by suitable space, like nesting birds on ocean islands. Certainly, as our population grows, the conflict over space utilization will intensify. As the photographs in **Figure 36.9C** indicate, however, there seem to be few limits on how closely humans can be crowded together.

Barring some worldwide calamity, it is likely that the human population will continue to grow well into this century. Whatever happens, we know that the population must eventually stop growing. Unlike other organisms, we have the ability to decide whether this occurs mainly by decreased reproduction (a result of social changes involving individual choice and/or government intervention) or by increased mortality (a result of resource limitation, plagues, war, and environmental degradation). The expanding human population is probably the greatest crisis ever faced by life on Earth.

Web/CD Activity 36C *Human Population Growth*

? Looking at Figure 36.9B, which country has the greatest ecological deficit?

◼ Japan, because its ecological footprint is over three times larger than its available ecological capacity.

Figure 36.9C
Some very crowded places

Traffic in downton Cairo, Egypt

Manhattan, New York City

Refugee camp in Zaire

36.10 Birth and death rates and age structure affect population growth

Worldwide, human population growth is a mosaic of the rates of growth in different countries. Some developed countries, such as Sweden, have virtually stable populations, whereas the populations of most developing nations are still growing rapidly. How can countries achieve population stability? Zero population growth (ZPG) is when birth rates equal death rates. There are two possible ways to reach ZPG:

ZPG = High birth rates − high death rates

or

ZPG = Low birth rates − low death rates

The movement from the first configuration to the second is called the **demographic transition.** Most developed countries have made that transition, whereas most developing nations are still in the process. **Figure 36.10A** illustrates the demographic transition for Mexico, with projections of birth and death rates to 2050. You can see that Mexico's death rate dropped well before its birth rate began to decline, but the gap between the two is now narrowing. (The spike in the death rate corresponds to the worldwide flu epidemic of 1918–1919.) Population size will continue to grow until birth rate has dropped down to equal death rate.

After 1950, mortality (death) rates declined rapidly in most developing countries, but birth rates have declined in a more variable manner. Birth rate decline has been most dramatic in China, a result of government restrictions on family size. In 1970, the birth rate in China predicted an average of 5.9 children per woman; by 2004, the expected number per woman was 1.7. In India, the second most populous country, birth rates have fallen more slowly. Some African countries have had a dramatic transition to lower birth rates, though birth rates remain high in most of sub-Saharan Africa. About 80% of the world's people now live in developing countries, and almost all of the current population growth is occurring

there. Indeed, of the next billion people to be added to the global population, 98% of them will be born in developing nations.

In the developed countries, populations are nearing equilibrium, with birth rates at or below the replacement level of 2.1 children per female. In Canada, Japan, Italy, and a few other nations, birth rates are in fact below replacement, and in some eastern and central European countries, populations are declining.

The **age structure** of a population is the proportion of individuals in different age-groups. The diagrams in **Figure 36.10B** show age structure for Afghanistan, the United States, and Italy. In these diagrams, each of the three different colors represents the fraction of the population for one of three age-groups (0–14, 15–44, and 45 and older). Within each of these broader groups, each horizontal bar represents the population in a 5-year age-group. The area to the left of each vertical center line represents the percentage of males in each age-group; females are represented on the right side of the line.

The tan bands in Figure 36.10B correspond to people who are in their reproductive years; the green bands represent those who will contribute to population growth in the future. Afghanistan, whose population is growing at 2.4% a year, has an age structure that is bottom-heavy, skewed toward young individuals who are now or will soon be reproducing. Almost 42% of Afghanistan's population is under 15. The United States is growing at 0.6% per year, and its age structure is relatively even except for a bulge that corresponds to the "baby boom" that lasted for about two decades after the end of World War II. Even though couples born during those years have had an average of fewer than two children, the nation's overall birth rate still exceeds the death rate because there are still so many "boomers" and their offspring of reproductive age. The current U.S. birth rate is 2.1 children per woman. The smaller base in the age-structure diagram for Italy, which has negative population growth, indicates that individuals younger than reproductive age are relatively underrepresented in the population. The 14% of Italy's population that is under 15 and its birth rate of 1.2 children per woman both contribute to negative population growth.

Age-structure diagrams not only reveal a population's growth trends; they also indicate social conditions. For instance, we can predict that most developing nations will have an increasing need for employment and schools as their populations continue to expand, not to mention the need to increase food production, sanitation, and infrastructure. In contrast, many developed nations must plan for more health-care facilities for their aging populations. For many developed nations, the age-structure diagrams tell us that a decreasing number of working-age people will soon be supporting an increasing number of retired people. In the United States, it is this demographic feature that has made the future of Social Security and Medicare such major political issues.

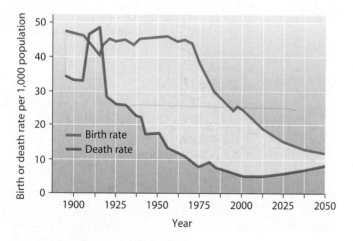

Figure 36.10A   Demographic transition in Mexico
From Population Reference Bureau, 2000 and U.S. Census Bureau International Data Base, 2003

You can now see why the developing countries account for virtually all global population growth. Uneven age structures, with a high percentage of young adults and children who are likely to reproduce in the near future, sustain their explosive population growth. Other demographic differences between developing and developed countries include *infant mortality,* the number of infant deaths per 1,000 live births, and *life expectancy at birth,* the predicted average length of life at the time of birth. For example, in 2003, Afghanistan had an infant mortality rate of 143 and a life expectancy at birth of 47 years. By contrast, in Japan only 3 children out of each 1,000 born died in infancy, and life expectancy was 81 years. While global life expectancy has been increasing since about 1950, more recently it has dropped in a number of regions, including sub-Saharan Africa. In these regions, the combination of social upheaval and infectious diseases such as AIDS and tuberculosis is reducing life expectancy. In the east African country of Rwanda, for instance, life expectancy in 2003 was approximately 39 years. These factors reflect the quality of life faced by children at birth. High infant mortality also tends to maintain high birth rates in a population.

A unique feature of human population growth is our ability to control it with voluntary contraception and government-sponsored family planning. Social change and the rising educational and career aspirations of women in many cultures encourage them to delay reproduction. Delayed reproduction dramatically decreases population growth rates. You learned in Module 36.7 that reproduction at an early age is an *r*-selected life history trait, a trait that supports a high growth rate. Family sizes are smaller when women begin having children in their early 30s instead of in their teens, and the time span between generations is longer.

Reduced family size is the key to the demographic transition. As women's status and education increase, they choose to have fewer children. This phenomenon has been observed in both developed and developing countries, anywhere that the lives of women have improved. Given access to affordable contraceptive methods, women generally practice birth control, and many countries now subsidize family planning services and have official population policies. In many other countries, however, issues of family planning remain socially and politically charged, with heated disagreement over how much support should be provided for family planning.

It seems more desirable for population control to result from a decrease in the birth rate by individual choice and social changes than by an increase in the death rate. As we discuss in Chapter 38, our rapidly expanding global population and the consumption of our planet's limited resources pose serious threats to our future and that of most other species. For better or worse, we have the unique responsibility to decide the fate of our species and the rest of the biosphere.

Web/CD Activity 36D *Analyzing Age-Structure Diagrams*

? During the demographic transition from high birth and death rates to low birth and death rates, countries usually undergo rapid population growth. Explain why.

■ The death rate declines before the birth rate declines, creating a period when births greatly outnumber deaths. This also sets up a skewed age structure, so that population increase continues for a while even after birth rates decline.

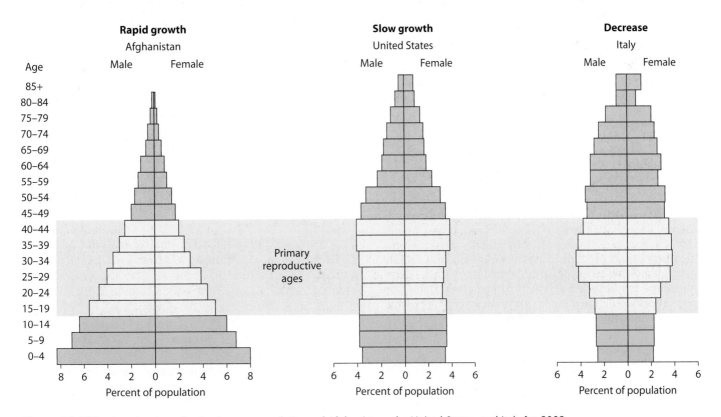

Figure 36.10B Age structures for the human populations of Afghanistan, the United States, and Italy for 2003

Reviewing the Concepts

Population ecology is concerned with changes in population size and the factors that regulate populations over time. A population consists of members of a species living in the same place at the same time **(Introduction–36.1).**

Population Structure and Dynamics (36.2–36.6)

Population characteristics.   Population density is the number of individuals in a given area or volume. Environmental and social factors influence the spacing of individuals in various dispersion patterns: clumped (most common), uniform, or random **(36.2).**

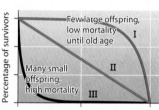

Life tables and survivorship curves predict an individual's statistical chance of dying or surviving during each interval in its life. The three types of survivorship curves reflect species' differences in reproduction and mortality **(36.3).**

Exponential growth is the accelerating increase that occurs when growth is unlimited. The equation $G = rN$ describes this J-shaped growth curve; G = the population growth rate, r = an organism's inherent capacity to reproduce, and N = the population size.

Logistic growth is the model that represents the slowing of population growth as a result of limiting factors and the leveling off at carrying capacity, which is the number of individuals the environment can support. The equation $G = rN(K − N)/K$ describes a logistic growth curve, where K = carrying capacity and the term $(K − N)/K$ accounts for the leveling off of the curve **(36.4).**

Limiting factors.   As a population's density increases, factors such as limited food supply and increased disease or predation may increase the death rate, decrease the birth rate, or both. Abiotic factors such as severe weather may limit many natural populations. Most populations are probably regulated by a mixture of factors and fluctuations in numbers are common **(36.5).** Some populations undergo regular boom-and-bust cycles of growth and decline **(36.6).**

Life Histories and Their Evolution (36.7–36.8)

Diversity of life histories.   Natural selection shapes a species' life history, the series of events from birth through reproduction to death. Populations with so-called *r*-selected life history traits produce many offspring and grow rapidly in unpredictable environments. Populations with *K*-selected traits raise few offspring and maintain relatively stable populations **(36.7).** Principles of population ecology are useful in managing natural resources **(36.8).**

The Human Population (36.9–36.10)

Earth's carrying capacity.   The human population has been growing almost exponentially for centuries, standing now at about 6.4 billion. The ecological footprint represents the amount of land per person needed to support a nation's resource needs. The ecological capacity of the world may already be smaller than the population's ecological footprint **(36.9).**

The demographic transition is the shift from high birth and death rates to low birth and death rates. The age structure of a population—the proportion of individuals in different age-groups—affects its future growth. Increasing the status of women may help to reduce family size **(36.10).**

Connecting the Concepts

1. Use this graph of the idealized exponential and logistic growth curves to complete the following.
   a. Label the axes and curves on the graph.
   b. Give the formula that describes the blue curve.
   c. What does the dotted line represent?
   d. For each curve, indicate and explain where population growth is the most rapid.
   e. Which of these curves best represents global human population growth?

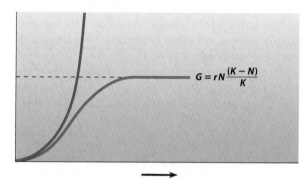

2. The graph below shows the demographic transition for a hypothetical country. Many developed countries that have achieved a stable population size have undergone a transition similar to this. Answer the following questions concerning this graph.
   a. What does the blue line represent? The red line?
   b. This diagram has been divided into four sections. Describe what is happening in each section.
   c. In which section(s) is the population size stable?
   d. In which section is the population growth rate the highest?

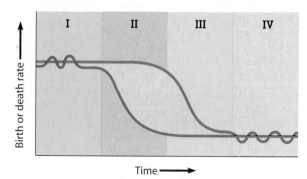

Testing Your Knowledge

Multiple Choice

3. To figure out the human population density of your community, you would need to know the number of people living there and
   a. the land area in which they live.
   b. the birth rate of the population.
   c. whether population growth is logistic or exponential.
   d. the dispersion pattern of the population.
   e. the carrying capacity.

4. The term $(K - N)/K$
   a. is the carrying capacity for a population.
   b. is greatest when K is very large.
   c. is zero when population size equals carrying capacity.
   d. increases in value as N approaches K.
   e. accounts for the overshoot of carrying capacity.

5. Which of the following represents a demographic transition?
   a. A population switches from exponential to logistic growth.
   b. A population reaches zero population growth when the birth rate drops to zero.
   c. There are equal numbers of individuals in all age-groups.
   d. A population exhibits boom-and-bust cycles.
   e. A population switches from high birth and death rates to low birth and death rates.

6. With regard to its rate of growth, a population that is growing logistically
   a. grows fastest when density is lowest.
   b. has a high intrinsic rate of increase.
   c. grows fastest at an intermediate population density.
   d. grows fastest as it approaches carrying capacity.
   e. is always slowed by abiotic factors.

7. Skyrocketing growth of the human population since the beginning of the Industrial Revolution appears to be mainly a result of
   a. migration to thinly settled regions of the globe.
   b. a drop in death rate due to sanitation and health care.
   c. better nutrition boosting the birth rate.
   d. the concentration of humans in cities.
   e. social changes that make it desirable to have more children.

8. According the the ecological footprint study produced in 1997,
   a. the carrying capacity of the world is 10 billion.
   b. the carrying capacity of the world would increase if all people ate more meat.
   c. the current demand on global resources by industrialized countries is below the ecological capacity of those countries.
   d. the United States has a larger ecological footprint than the ecological capacity of its own land.
   e. nations with the largest ecological footprints have the fastest population growth rates.

Describing, Comparing, and Explaining

9. What are some factors that might have a density-dependent limiting effect on population growth?

10. What is survivorship? What does a survivorship curve show? Explain what the three survivorship curves tell us about humans, squirrels, and oysters.

11. Describe the factors that might produce the following three types of dispersion patterns in populations.

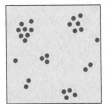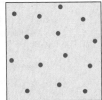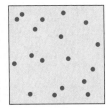

Applying the Concepts

12. A sampling technique commonly used to estimate the sizes of wildlife populations is the mark-recapture method. The researcher traps animals in the study area and marks them with tags, bands, or spots of dye. The marked animals are released, and traps are set a second time. This second capture will yield both marked and unmarked individuals. The proportion of marked (recaptured) animals to unmarked animals in the second trapping is assumed to be equivalent to the proportion of the animals marked in the first catch to the total population. For example, suppose a researcher captures, tags, and releases 50 voles. Two weeks later, 100 voles are captured, of which 10 are marked. What percent of the total population is estimated to have been marked? Using this information, how could you estimate the total population size?

13. Considering the learning behavior of animals, what might be some limits to the accuracy of the mark-recapture method?

14. The mountain gorilla, spotted owl, giant panda, snow leopard, and grizzly bear are all endangered by human encroachment on their environments. Another thing these animals have in common is that they all have K-selected life history traits. Why might they be more easily endangered than animals with r-selected life history traits? What general type of survivorship curve would you expect these species to exhibit? Explain your answer.

15. How does the age structure of the U.S. population explain the current surplus in the Social Security fund? If the system is not changed, why will the surplus give way to a deficit sometime in the next few decades?

16. Many people regard the rapid population growth of developing countries as our most serious environmental problem. Others feel that the growth of developed countries, though slower, is actually a greater threat to the environment. What kinds of environmental problems result from population growth in (a) developing countries and (b) developed countries? Which do you think is the greater threat? Why?

Answers to all questions can be found in Appendix 3.

For study help and Activities, go to campbellbiology.com or the student CD-ROM.

Apanteles glomeratus

A cabbage white butterfly (*Pieris rapae*)

STRUCTURAL FEATURES OF COMMUNITIES

37.1   A community includes all the organisms inhabiting a particular area
37.2   Competition may occur when a shared resource is limited
37.3   Predation leads to diverse adaptations in both predator and prey
37.4   Predation can maintain diversity in a community
37.5   Herbivores and the plants they eat have various adaptations
37.6   Symbiotic relationships help structure communities
37.7   Disturbance is a prominent feature of most communities
37.8   Fire specialist Max Moritz discusses the role of fire in ecosystems
37.9   Trophic structure is a key factor in community dynamics
37.10   Food chains interconnect, forming food webs

ECOSYSTEM STRUCTURE AND DYNAMICS

37.11   Ecosystem ecology emphasizes energy flow and chemical cycling
37.12   Primary production sets the energy budget for ecosystems
37.13   Energy supply limits the length of food chains
37.14   A production pyramid explains why meat is a luxury for humans
37.15   Chemicals are recycled between organic matter and abiotic reservoirs
37.16   Water moves through the biosphere in a global cycle
37.17   The carbon cycle depends on photosynthesis and respiration
37.18   The nitrogen cycle relies heavily on bacteria
37.19   The phosphorus cycle depends on the weathering of rock

ECOSYSTEM ALTERATION

37.20   Ecosystem alteration can upset chemical cycling
37.21   David Schindler talks about the effects of nutrients on freshwater
        ecosystems

CHAPTER 37

Dining In

A 4-MM-LONG WASP called *Apanteles glomeratus* stabs through the skin of a caterpillar and lays her eggs (see photo at upper left). The caterpillar, a larva of the cabbage white butterfly (*Pieris rapae,* above), is doomed. It will be destroyed from within as the wasp larvae hatch and nourish themselves on its internal organs. We benefit from the wasp's behavior, for *Pieris* caterpillars eat cabbages and broccoli and are abundant agricultural pests.

The hapless caterpillar houses a three-step food chain

Apanteles wasps help control cabbage butterfly populations, but they never come close to eliminating them. There are always fewer *Apanteles* wasps than *Pieris* caterpillars, for *Apanteles* has problems of its own. Other wasps, called ichneumons (opposite page, left), can detect when a *Pieris* caterpillar contains *Apanteles* larvae. When a female ichneumon finds one of these caterpillars, she pierces it and deposits her eggs inside the *Apanteles* larvae. And the story may go on. Yet another wasp, a chalcid (opposite page, right), may lay its eggs inside the ichneumon larvae. When this happens, the hapless caterpillar houses a three-step food chain:

Communities and Ecosystems

A chalcid wasp

An ichneumon wasp

chalcid larvae eating ichneumon larvae eating *Apanteles* larvae eating the caterpillar larva. Usually, only the chalcids will emerge from the dead husk of the caterpillar.

Though he was unaware of this unusual food chain, 18th-century English satirist Jonathan Swift wrote:

> So, Nat'ralists observe, a Flea
> Hath smaller Fleas that on him prey,
> And these have smaller Fleas to bite 'em,
> And so proceed *ad infinitum*.

Swift's flea analogy was meant to ridicule picky literary critics, but his lines paint a good picture of the complex relationships between *Pieris* caterpillars and wasps.

Connections between organisms is the major topic of this chapter. In the first half of the chapter, we see that a biological community derives its structure from the interactions and interdependence of the organisms living within it. Later, we find that the hapless caterpillar houses a three-step food chain ecosystem functioning involves complex interactions between the community and its physical environment. When we get to ecosystems, we take another look at Jonathan Swift's poem; although it works as satire, its last line can't be taken literally, because no food chain, including the extensive one in a *Pieris* caterpillar, can be infinite. ■ ■ ■

37.1 A community includes all the organisms inhabiting a particular area

In the previous chapter, we saw that a population is a group of interacting individuals of a particular species. We now move one step up the hierarchy of nature to the level of the community. A biological **community** is an assemblage of all the populations of organisms living close enough together for potential interaction. For example, the butterflies, grasses, wildflowers, and shrubs in Figure 37.1 are components of a community in an open field. Butterflies interact with many species, pollinating flowers, laying eggs on plants, and serving as food for birds.

Just as a population has certain characteristics, such as density and dispersion pattern, a community has its own set of properties. The key characteristics of a community are species diversity, dominant species, response to disturbances, and trophic structure.

The **species diversity** of a community—the variety of different kinds of organisms that make it up—has two components. One is species richness, or the total number of different species in the community. The other is the relative abundance of the different species. For example, imagine two communities, each with 100 individuals distributed among four different species (A, B, C, and D) as follows:

Community 1:   25 A   25 B   25 C   25 D
Community 2:   97 A   1 B   1 C   1 D

The species richness is the same for both communities, because they both contain four species, but the relative abundance is very different. Suppose these were communities of wildflowers in neighboring fields. You would easily notice the four different types of flowers in community 1,

but without looking carefully, you might see only the abundant species A in the second field. Most observers would intuitively describe community 1 as the more diverse of the two communities. Indeed, ecologists consider both richness and relative abundance in determining species diversity.

The second property of a community is its dominant species. In general, a small number of species exert strong control over a community's composition and diversity. In terrestrial situations, the dominant species is usually the most prevalent form of vegetation. For example, wildflowers are the dominant species of the community shown in Figure 37.1. The types and structural features of plants largely determine the kinds of animals that live in a community. As you'll see in Module 37.4, however, some species may have an impact on community structure because of their pivotal roles in community dynamics.

The third property of a community is its response to disturbances, such as storms, fire, drought, floods, or human disruptions. For example, a forest dominated by cedar and hemlock trees may last for thousands of years with little change in species diversity. Large cedars and hemlocks even withstand most lightning-caused fires, which kill small trees and shrubs growing in the forest. However, if a fire does kill the dominant trees, it will take a long time for the forest to return to its original species composition. On the other hand, some communities depend on fire to maintain their species diversity, as you'll see in Module 37.8. Almost all communities change in response to disturbances. Thus, disturbance plays a vital role in determining community structure and composition.

The fourth property of a community is **trophic structure** (from the Greek *trophe*, nourishment), the feeding relationships among the various species making up the community. A community's trophic structure determines the passage of energy and nutrients from plants and other photosynthetic organisms to herbivores and then to carnivores. For example, as we saw in the chapter introduction, the energy in a cabbage leaf eaten by a moth caterpillar may end up supporting one, two, or three kinds of carnivorous wasp larvae.

With the four main properties of a community in mind, we turn next to the various kinds of interactions between species that contribute to the structure of a community. These interspecific interactions are of four main types: competition, predation, herbivory (herbivores eating plants or algae), and symbiosis. In discussing these interactions, we'll see that they are all influenced by evolution through natural selection.

Web/CD Thinking as a Scientist *How Are Impacts on Community Diversity Measured?*

? How could a community appear to have relatively little diversity even though it is rich in species?

Figure 37.1 Butterflies, grasses, and flowers in a field community

■ One or a few of the diverse species could account for almost all the organisms in the community, with the other species being rare.

37.2 Competition may occur when a shared resource is limited

As you learned in Module 36.5, as a population increases in density, competition for limited resources may slow that population's growth. If two different species are competing for the same resource, called **interspecific competition**, the growth of one or both populations may be inhibited. Weeds growing in a garden compete with garden plants for nutrients and water. Lynx and foxes compete for prey such as snowshoe hares in northern forests. Interspecific competition may play a major role in structuring a community.

In 1934, Russian ecologist G. F. Gause studied the effects of interspecific competition in laboratory experiments with two closely related species of *Paramecium*. Gause cultured these protists under stable conditions with a constant amount of food added every day. When he grew the two species in separate cultures, each population grew rapidly and then leveled off at what was apparently the carrying capacity of the culture (see Module 36.4). But when Gause cultured the two species together, one species was driven to extinction. Gause concluded that two species so similar that they compete for the same limiting resources cannot coexist in the same place. One will use the resources more efficiently and thus reproduce more rapidly than the other. Even a slight reproductive advantage will eventually lead to local elimination of the inferior competitor. Ecologists call Gause's ideas the **competitive exclusion principle.**

Tests of the competitive exclusion principle include some classic field experiments on the interactions between two species of barnacles (sessile crustaceans) that attach to intertidal rocks on the North Atlantic coast (**Figure 37.2A**). Both types of barnacles, *Balanus* and *Chthamalus,* grow on rocks that are exposed during low tide; when immersed at high tide, they feed on organic particles suspended in the water. Both types have free-swimming larvae that may settle and begin to develop on virtually any rock surface.

Chthamalus (brown in Figure 37.2A) occupies the upper parts of the rocks, which are out of the water longer during low tides. *Balanus* (tan) fails to survive as high on the rocks, apparently because it cannot survive prolonged exposure to air during low tides. When experimenters removed *Balanus* from the lower rocks, *Chthamalus* spread lower, colonizing the unoccupied rocks. Researchers concluded that the lower limit of *Chthamalus*'s distribution is set by competition with *Balanus.*

The competitive exclusion principle applies to what is called a species' niche. In ecology, a **niche** is a species' role in its community, or the sum total of its use of the biotic and abiotic resources of its habitat. For example, the attachment sites on intertidal rocks, the amount of exposure to seawater and air, and the food it consumes are some of the aspects of each barnacle's niche. Combining the niche

Figure 37.2A A test of competitive exclusion: two species of barnacles on intertidal rocks

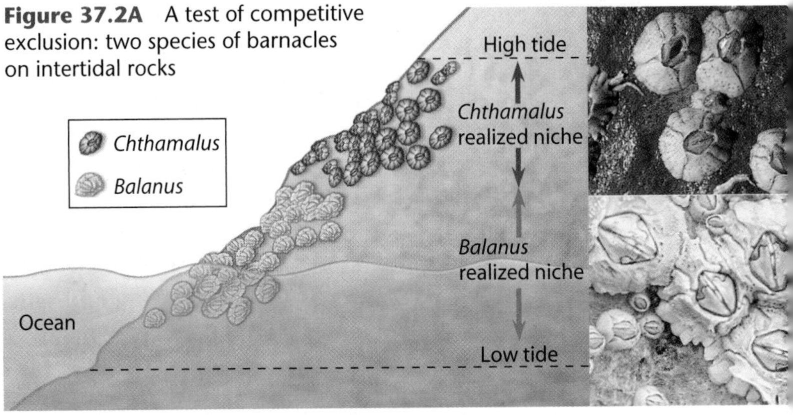

concept with the competitive exclusion principle, we might predict that two species cannot coexist in a community if their niches are identical.

There are two possible outcomes of competition between species having identical niches: Either the less competitive species will be driven to local extinction, or one of the species may evolve enough through natural selection to use a different set of resources. This differentiation of niches that enables similar species to coexist in a community is called **resource partitioning.** **Figure 37.2B** shows the different feeding habitats of seven species of *Anolis* lizards that live near each other in the Dominican Republic. This resource partitioning may be "the ghost of competition past"—circumstantial evidence of earlier interspecific competition resolved when differences in niches evolved.

Competition is a major factor in adaptive evolution, as is predation, an interaction that we explore next.

? How could you test the hypothesis that it is susceptibility to drying, not competitive exclusion, that keeps *Balanus* from populating the upper rocks shown in Figure 37.2A?

■ You could remove *Chthamalus* from the upper parts of the rocks to see whether *Balanus* failed to spread upward even in the absence of a competitor.

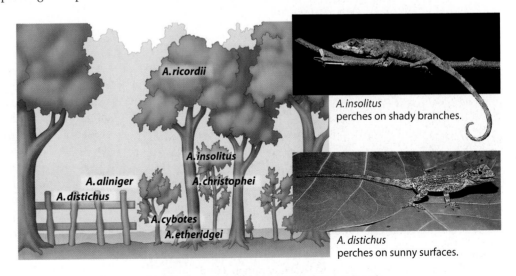

Figure 37.2B Resource partitioning among seven species of *Anolis* lizards

37.3 Predation leads to diverse adaptations in both predator and prey

Predation is an interaction between species in which one species, the **predator**, kills and eats another, the **prey.** Because eating and avoiding being eaten are prerequisites to reproductive success, the adaptations of both predators and prey tend to be refined through natural selection.

Many feeding adaptations of predators are both obvious and familiar. Most predators have acute senses that enable them to locate prey. In addition, adaptations such as claws, teeth, fangs, stingers, or poison, help catch and subdue prey. Predators that pursue their prey are generally fast and agile, whereas those that lie in ambush are often camouflaged in their environments.

Camouflage is also a very common type of defense against predation. As **Figure 37.3A** shows, the gray tree frog (*Hyla arenicolor*), common in the southwestern United States, becomes almost invisible on a gray tree trunk.

Other animal defenses include mechanical defenses, such as the porcupine's sharp quills or the hard shells of clams and oysters. Chemical defenses are also widespread. Animals with effective chemical defenses are often brightly colored, a warning to predators. The vivid markings of the poison-arrow frog (**Figure 37.3B**), an inhabitant of rain forests in Costa Rica, warn of deadly alkaloids in the frog's skin. In some parts of South America, human hunters in the rain forest tip their arrows with poisons from similar frogs to bring down large mammals.

A species of prey may gain significant protection through mimicry, a "copycat" adaptation in which one species mimics the appearance of another. In **Batesian mimicry**, a palatable or harmless species mimics an unpalatable or harmful model. Recall the king snake in Chapter 1, which mimics the poisonous coral snake. In another intriguing example, the hawkmoth larva puffs up its head and thorax when disturbed, which then resemble the head of a small poisonous snake (**Figure 37.3C**). The mimicry even involves behavior; the larva weaves its head back and forth and hisses like a snake.

Figure 37.3D illustrates another kind of mimicry, called **Müllerian mimicry**, in which two unpalatable species that inhabit the same community mimic each other. On the top is a species of bee called a cuckoo bee; on the bottom is a

Figure 37.3A Camouflage: a gray tree frog on bark

Figure 37.3B Chemical defenses: the poison-arrow frog

Figure 37.3C Batesian mimicry: a snake (above) and a hawkmoth larva (right)

Figure 37.3D
Müllerian mimicry: a cuckoo bee (above) and a yellow jacket (right)

wasp called a yellow jacket. Both species have stingers that release toxic chemicals. Presumably, both gain an adaptive advantage beyond their own defenses because predators will learn more quickly to avoid any prey with this appearance.

Predators may also use mimicry. Some snapping turtles have a tongue that resembles a wriggling worm, thus luring small fish. Any fish that tries to eat the "bait" is itself quickly consumed as the turtle's strong jaws snap closed.

? Explain why predation is a potent factor in adaptive evolution.

■ Both the predators with adaptations that increase their success at obtaining food and the prey that are best able to avoid being eaten will be most likely to survive and reproduce and pass those adaptations on to their offspring.

37.4 Predation can maintain diversity in a community

You might think that organisms eating other organisms would always reduce species diversity, but field studies have shown that predator-prey relationships can actually help maintain community diversity. Experiments by American ecologist Robert Paine in the 1960s were among the first to provide such evidence. Paine manually removed the dominant predator, a sea star of the genus *Pisaster* (shown in **Figure 37.4A** eating its favorite food, a mussel), from experimental areas within the intertidal zone of the Washington coast. The result was that *Pisaster*'s main prey, a mussel of the genus *Mytilus,* outcompeted many of the other shoreline organisms (algae, barnacles, and snails, for instance) for the important resource of space on the rocks. The number of different organisms present in experimental areas dropped from over 15 species to under 5 species.

The experiments of Paine and others gave rise to the concept of a **keystone species**, a species that exerts strong control on community structure because of its ecological role, or niche. *Pisaster* is a keystone predator that reduces the density of the strongest competitors in the community, thus preventing the competitive exclusion of weaker competitors.

Sea otters are a keystone predator in the North Pacific. Once relatively abundant, they were reduced to near extinction by the fur trade during the 19th century. An interna-tional treaty provided protection in the 20th century, enabling popula-tions to recover to very high densi-ties. Sea otters feed on sea urchins, and sea urchins feed mainly on kelp, a large seaweed. In areas where sea otters are abundant, sea urchins are rare and kelp forests are well devel-oped. Where sea otters are rare, sea urchins are common and kelp is al-most absent.

During the last 20 years, sea ot-ters have declined dramatically in large areas off the coast of western Alaska. The loss of this keystone species has allowed sea urchin popu-lations to increase, resulting in the destruction of kelp forests. Ecolo-gists suspect that killer whales are the cause of the sea otter decline (**Figure 37.4B**). Killer whales have probably been eating sea otters for the past two decades because the previous prey of the whales, mainly seals and sea lions, have declined in density. The decline of these prey species reflects a decline in the pop-ulations of fish species that the seals and sea lions eat. And all of these changes in the Alaskan marine com-munities are probably related to human overfishing in the North Pa-cific. From case studies such as these, ecologists are beginning to outline the key interactions that help struc-ture communities and to document the impact of human populations on all communities.

Figure 37.4B
Predation of killer whales on sea otters has allowed sea urchins to overgraze kelp

? What is the advantage to a keystone predator of being specialized to feed mainly on those prey species that are otherwise the most successful among potential prey species?

■ The most competitive prey species probably represent the most abundant and dependable food source for the predator.

Figure 37.4A A *Pisaster* sea star, a keystone species, eating a mussel

Herbivores and the plants they eat have various adaptations

The sea urchins that feed on kelp beds are **herbivores**, animals that eat plants or algae. Other aquatic herbivores include snails and some fishes. In terrestrial ecosystems, large mammalian herbivores such as cattle, sheep, and deer may be most familiar, but the majority of herbivores are small insects, such as grasshoppers and beetles.

Herbivorous insects may locate food by using chemical sensors on their feet, and their mouthparts are adapted for shredding tough vegetation or sucking plant juices. Herbivorous vertebrates may have specialized teeth or digestive systems adapted for processing vegetation (see Module 21.13). They may also use their sense of smell to identify food plants.

Because plants cannot run away from herbivores, chemical toxins, often in combination with various kinds of antipredator spines and thorns, are their main weapons against being eaten. Among such chemical weapons are the poison strychnine, produced by a tropical vine called *Strychnos toxifera;* morphine, from the opium poppy; nicotine, produced by the tobacco plant; mescaline, from peyote cactus; tannins, from a variety of plant species; and many substances that we use as flavorings but that are distasteful to some herbivores (cinnamon, cloves, and mint, for instance). Some plants even produce chemicals that cause abnormal development in insects that eat them.

Some herbivore-plant interactions illustrate the concept of **coevolution**, a series of reciprocal evolutionary adaptations in two species. Coevolution occurs when a change in one species acts as a new selective force on another species, and counteradaptation of the second species in turn affects the selection of individuals in the first species. **Figure 37.5** illustrates a classic example of coevolution of an herbivorous insect (the caterpillar of the butterfly *Heliconius,* top left) and a plant (the passionflower *Passiflora,* a tropical vine). *Passiflora* produces toxic chemicals that protect its leaves from most insects, but *Heliconius* caterpillars have digestive enzymes that break down the toxins. As a result, *Heliconius* gains access to a food source that few other insects can eat.

These poison-resistant caterpillars seem to be a strong selective force for *Passiflora* plants, some of which have evolved defenses against them. For instance, the leaves of some *Passiflora* species produce yellow sugar deposits that look like *Heliconius* eggs. You can see two eggs in the top right photograph of Figure 37.5 and two egg-like sugar deposits in the bottom photo. Female butterflies avoid laying their eggs on leaves that already have eggs, presumably ensuring that only a few caterpillars will hatch and feed on any one leaf. Because the butterfly often mistakes the yellow sugar deposits for eggs, *Passiflora* species with the yellow deposits are less likely to be eaten.

The story is even more complicated, however. The egg-like sugar deposits, as well as smaller ones scattered over the leaf, attract ants and wasps that prey on *Heliconius* eggs and larvae. Thus, adaptations that appear to be coevolutionary responses between just two species may in fact involve interactions among many species in a community.

Eggs

Sugar deposits

Figure 37.5 Coevolution: *Heliconius* and the passionflower vine

? Is Batesian mimicry, in which a harmless species mimics an unpalatable or poisonous species, an example of coevolution? Explain your answer.

■ No. Coevolution involves adaptive responses of two species to each other. In this case, the "model" species is not adapting to changes in the mimic.

Symbiotic relationships help structure communities

A **symbiotic relationship** is an interaction between two or more species that live together in direct contact. There are three main types of symbiotic relationships: parasitism, commensalism, and mutualism. Parasitism and mutualism can be key factors in community structure.

In **parasitism** (from the Greek *para,* near, and *sitos,* food), a parasite lives on or in its host and obtains its nourishment from the host. A tapeworm is an internal parasite that lives inside the intestines of a larger animal and absorbs nutrients from its host. Ticks, which suck blood from animals, and

aphids, which tap into the sap of plants, are examples of external parasites. The parasitic wasps described in the chapter introduction lay their eggs on or in living hosts, and the larvae that hatch then feed on the host's body.

Natural selection favors the parasites that are best able to find and feed on hosts. Natural selection also favors the evolution of host defenses. For example, the immune system of vertebrates (see Chapter 24) provides a multipronged defense against specific internal parasites. With natural selection working on both host and parasite, the eventual outcome is often a relatively stable relationship in which the host is not usually killed.

Pathogens are disease-causing bacteria, viruses, or protists that could be thought of as microscopic parasites. Unlike most parasites, however, many pathogens inflict lethal harm on their hosts. At times, it is possible to watch the effects of natural selection in host-pathogen relationships. Scenes like the one shown in **Figure 37.6A** were common in Australia during the 1940s. The continent was overrun by hundreds of millions of European rabbits, whose population had exploded from just 12 pairs imported onto an estate for sport hunting a century earlier. The rabbits destroyed huge expanses of Australian farmland and threatened the sheep and cattle industries. In 1950, myxoma virus, a pathogen that infects rabbits, was deliberately introduced into Australia to control the rabbit population. Spread rapidly by mosquitoes, the virus devastated the rabbit population. The virus was less deadly to the offspring of surviving rabbits, however, and it has caused less and less harm over the years. Apparently, genotypes in the rabbit population were selected for that were better able to resist the pathogen. Meanwhile, the deadliest strains of the virus perished with their hosts as natural selection favored strains that could infect hosts but not kill them. Thus, natural selection stabilized this host-pathogen relationship, and today, the virus that was originally introduced has only a mild effect on the rabbit population. The Australian government has had more success with a different virus introduced in 1995.

Figure 37.6A Rabbits in Australia before a pathogenic virus was introduced

Figure 37.6B Mutualism between acacia trees and ants

In contrast to parasitism, in **commensalism** (from the Latin *com,* together, and *mensa,* table), one partner benefits without significantly affecting the other. Few cases of absolute commensalism have been documented, because it is unlikely that one partner will be completely unaffected. Algae that grow on the shells of sea turtles, barnacles that attach to whales, and birds that feed on insects flushed out of the grass by grazing cattle are sometimes considered commensal.

The third type of symbiosis, **mutualism** (from the Latin *mutualis,* reciprocal) benefits both partners in the relationship. We have discussed several mutualistic associations in previous chapters—for instance, the association of legume plants and nitrogen-fixing bacteria and the association of mycorrhizal fungi and plant roots. **Figure 37.6B** illustrates another case of mutualism that is somewhat similar to that of passionflower vines and the predaceous ants they attract (see Module 37.5). This is part of a branch of a bull's horn acacia tree, which grows in Central and South America. The tree provides room and board for ants of the genus *Pseudomyrmex.* The ants live in large, hollow thorns and eat sugar secreted by the tree. As the photograph suggests, they also eat the yellow structures at the tips of leaflets; these are protein-rich swellings that seem to have no function for the tree except to attract ants. The ants benefit the host tree by attacking virtually anything that touches it. They sting other insects and large herbivores and even clip surrounding vegetation that grows near the tree. When the ants are removed, the trees usually die, probably because herbivores damage them so much that they are unable to compete with surrounding vegetation for light and growing space.

The complex interplay of species in symbiotic relationships highlights an important point about communities: Their structure depends on a web of diverse connections between organisms. In the next module, we explore how disturbances affect community structure.

Web/CD Activity 37A *Interspecific Interactions*

? Which type of symbiotic relationship is represented by mycorrhizae, the associations of plant roots and fungi described in Module 32.12?

■ Mutualism

aphids, which tap into the sap of plants, are examples of external parasites. The parasitic wasps described in the chapter introduction lay their eggs on or in living hosts, and the larvae that hatch then feed on the host's body.

Natural selection favors the parasites that are best able to find and feed on hosts. Natural selection also favors the evolution of host defenses. For example, the immune system of vertebrates (see Chapter 24) provides a multipronged defense against specific internal parasites. With natural selection working on both host and parasite, the eventual outcome is often a relatively stable relationship in which the host is not usually killed.

Pathogens are disease-causing bacteria, viruses, or protists that could be thought of as microscopic parasites. Unlike most parasites, however, many pathogens inflict lethal harm on their hosts. At times, it is possible to watch the effects of natural selection in host-pathogen relationships. Scenes like the one shown in **Figure 37.6A** were common in Australia during the 1940s. The continent was overrun by hundreds of millions of European rabbits, whose population had exploded from just 12 pairs imported onto an estate for sport hunting a century earlier. The rabbits destroyed huge expanses of Australian farmland and threatened the sheep and cattle industries. In 1950, myxoma virus, a pathogen that infects rabbits, was deliberately introduced into Australia to control the rabbit population. Spread rapidly by mosquitoes, the virus devastated the rabbit population. The virus was less deadly to the offspring of surviving rabbits, however, and it has caused less and less harm over the years. Apparently, genotypes in the rabbit population were selected for that were better able to resist the pathogen. Meanwhile, the deadliest strains of the virus perished with their hosts as natural selection favored strains that could infect hosts but not kill them. Thus, natural selection stabilized this host-pathogen relationship, and today, the virus that was originally introduced has only a mild effect on the rabbit population. The Australian government has had more success with a different virus introduced in 1995.

Figure 37.6A Rabbits in Australia before a pathogenic virus was introduced

Figure 37.6B Mutualism between acacia trees and ants

In contrast to parasitism, in **commensalism** (from the Latin *com,* together, and *mensa,* table), one partner benefits without significantly affecting the other. Few cases of absolute commensalism have been documented, because it is unlikely that one partner will be completely unaffected. Algae that grow on the shells of sea turtles, barnacles that attach to whales, and birds that feed on insects flushed out of the grass by grazing cattle are sometimes considered commensal.

The third type of symbiosis, **mutualism** (from the Latin *mutualis,* reciprocal) benefits both partners in the relationship. We have discussed several mutualistic associations in previous chapters—for instance, the association of legume plants and nitrogen-fixing bacteria and the association of mycorrhizal fungi and plant roots. **Figure 37.6B** illustrates another case of mutualism that is somewhat similar to that of passionflower vines and the predaceous ants they attract (see Module 37.5). This is part of a branch of a bull's horn acacia tree, which grows in Central and South America. The tree provides room and board for ants of the genus *Pseudomyrmex.* The ants live in large, hollow thorns and eat sugar secreted by the tree. As the photograph suggests, they also eat the yellow structures at the tips of leaflets; these are protein-rich swellings that seem to have no function for the tree except to attract ants. The ants benefit the host tree by attacking virtually anything that touches it. They sting other insects and large herbivores and even clip surrounding vegetation that grows near the tree. When the ants are removed, the trees usually die, probably because herbivores damage them so much that they are unable to compete with surrounding vegetation for light and growing space.

The complex interplay of species in symbiotic relationships highlights an important point about communities: Their structure depends on a web of diverse connections between organisms. In the next module, we explore how disturbances affect community structure.

Web/CD Activity 37A *Interspecific Interactions*

? Which type of symbiotic relationship is represented by mycorrhizae, the associations of plant roots and fungi described in Module 32.12?

■ Mutualism

Disturbance is a prominent feature of most communities

A traditional view of biological communities is that they are characterized by a more or less stable structure and composition of species. But in many communities, at least on a local scale, change seems to be more common than stability. A more recent model describes communities as constantly changing in response to disturbances.

Disturbances are events such as storms, fire, floods, droughts, overgrazing, or human activity that damage biological communities, remove organisms from them, and alter the availability of resources. The types of disturbances and their frequency and severity vary from community to community. By gathering data from specific communities over many years, ecologists are beginning to appreciate and understand the impact of disturbances.

We tend to think of disturbances in negative terms, but this is not always the case. Small-scale disturbances often have positive effects. For example, when a large tree falls in a windstorm, it disturbs the immediate surroundings, but it also creates new habitats. For instance, more light may now reach the forest floor, giving small seedlings the opportunity to grow; or the depression left by its roots may fill with water and be used as egg-laying sites by frogs, salamanders, and numerous insects. Small-scale disturbances may enhance environmental patchiness, which can contribute to species diversity in a community.

Communities change drastically following a severe disturbance that strips away vegetation and even soil. The disturbed area may be colonized by a variety of species, which are gradually replaced by a succession of other species, in a process called **ecological succession.**

When ecological succession begins in a virtually lifeless area with no soil, it is called **primary succession.** Examples of such areas are new volcanic islands or the rubble left by a retreating glacier. Often the only life-forms initially pres-ent are autotrophic bacteria. Lichens and mosses, which grow from windblown spores, are commonly the first large photosynthesizers to colonize the area. Soil develops gradually as rocks weather and organic matter accumulates from the decomposed remains of the early colonizers. Lichens and mosses are gradually overgrown by grasses and shrubs that sprout from seeds blown in from nearby areas or carried in by animals. Eventually, the area is colonized by plants that become the community's prevalent form of vegetation. Primary succession can take hundreds or thousands of years.

Ecologists studying primary succession on the rocky moraines left by retreating glaciers around Glacier Bay, Alaska, have identified a predictable pattern of changes in vegetation and soil characteristics. Some of these successional stages are shown in **Figure 37.7**. Early pioneering species include *Dryas*, a mat-forming shrub, which is later replaced by thick stands of alder. Both these species have symbiotic nitrogen-fixing bacteria, which improve the soil for later species. Later, spruce trees use the soil nitrogen and lower the soil pH as the trees' acidic needles decompose. By altering soil properties, pioneer plant species permit new species to grow, and the new plants in turn alter the environment. By about 300 years after glacial retreat, spruce-hemlock forest dominates on well-drained slopes in this area.

Secondary succession occurs where a disturbance has destroyed an existing community but left the soil intact. For instance, forested areas in the eastern United States that are cleared for farming will, if abandoned, undergo secondary succession and may eventually return to forest. Secondary succession also occurs as areas recover from fires or floods. Understanding the diverse effects of disturbance in communities is especially important today; as we discuss in Chapter 38, humans are the most widespread and significant agents of disturbance.

Web/CD Activity 37B *Primary Succession*

? What is the main abiotic factor that distinguishes primary from secondary succession?

■ Absence (primary) versus presence (secondary) of soil at the onset of succession

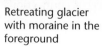
Retreating glacier with moraine in the foreground

Dryas stage

Spruce starting to appear in the alder and cottonwood forest

Spruce and hemlock forest

Figure 37.7 Primary succession after glacier retreat in Glacier Bay, Alaska

37.8   Fire specialist Max Moritz discusses the role of fire in ecosystems

Figure 37.8A   Fire specialist Max Moritz

When wildfires tear through a mountain forest or dry stretch of chaparral, people are quick to call them disasters. But while fire can take lives and destroy homes, it is also a natural part of some landscapes, deeply tied to cycles of climate and vegetation. One researcher who is improving our knowledge of this ecological force is Dr. Max Moritz, a wildland fire specialist with the College of Natural Resources at the University of California, Berkeley (Figure 37.8A). One of his special areas of interest is fire in the chaparral ecosystems that mark much of the American West.

Chaparral wildfires drove walls of flame through large parts of southern California in recent years (Figure 37.8B), raising new questions about the best ways to manage wildfires. For Moritz, that starts with understanding the factors that give rise to fire anywhere—and how variable those factors can be:

> For wildfires to occur naturally in an ecosystem, certain conditions must be met. You must have enough fuel to carry a fire. You have to have a dry season or some period in which fire is possible. And you have to have ignition. You have to have all three come together. And at any given location, there can be a lot of variation in those three things. That variation really creates a diversity of fire regimes (or patterns).

Until recent decades, the benefits of fire were poorly understood or overlooked entirely. People have long tried to suppress wildfires, a practice that has proved more harmful in some places than others, Moritz says, because of how a particular ecosystem works:

> Many dry forest ecosystems of the West, such as those dominated by ponderosa pines, were once characterized by frequent low-intensity surface fires. We know that from examining fire scars in tree ring data. Fire suppression has changed this pattern, letting smaller plants and trees take hold and increasing the potential for high-intensity fires.

Fires can devastate areas unaccustomed to flames. But in ecosystems in which wildfire is a natural occurrence, Moritz has seen how some forms of life depend on it:

> Where fire has played a natural role, either in structuring the ecosystem or as an evolutionary force, it can have a really important role to play. Wildfires are important in nutrient cycling and in creating the necessary habitat conditions for many organisms. Fire creates the conditions necessary for the regeneration of many plants. Fire will remove older plants and expose soil for seeds to grow and mature. There are species in chaparral systems, such as manzanita, that actually require the heating and chemical effects of fire for their seedlings to germinate. That is how important fire is to them. They have evolved to live with fire.

Figure 37.8B   Fire burning near Simi Valley, California in October, 2003

Now, as human understanding of fire increases, people need to be aware of all the complex—and simple—ways they influence wildfire. As Moritz says:

> In some big fires, past human fire suppression is a likely culprit. But other activities, such as logging, landscape fragmentation, and grazing, can also play a role. In places with larger human populations, there are more chances that a person will start a fire, and we're more likely to see a fire during periods of extreme fire weather. Recent fires in southern California are a great example of that—we just had a whole bunch of ignitions at a terrible, terrible time.

And understanding wildfire means living with it, not just managing it. Through his research, Moritz hopes to help people coexist more easily with incendiary cycles of weather, vegetation growth, and flame.

> Long-term, it comes down to where and how we build. We need to start thinking more about how we design neighborhoods and how we design homes, if they are built in places where they are likely to burn. There are lots of fire-resistant materials available and guidelines for clearing vegetation. We need to start thinking of fires as natural events that have an incredible amount of variation built into them—and then build accordingly. We can design our communities with the possibility of fire in mind and rethink the way we live in fire-prone ecosystems.

In his multifaceted view of how fire is linked to plants, climate, and people, Moritz reflects an important principle of ecology—the connections that shape an ecosystem. Next, we look at trophic structure, another characteristic of communities.

> **?** Why is a fire in a pine forest likely to destroy mature trees if it occurs in an area that has been "protected" from fire for many decades?

■ Because fuel accumulates in the form of undecomposed materials on the forest floor, resulting in a fire that is hot enough to ignite and destroy mature trees

37.9 Trophic structure is a key factor in community dynamics

Every community has a trophic structure, a pattern of feeding relationships consisting of several different levels. The sequence of food transfer up the trophic levels is known as a **food chain.** In **Figure 37.9**, the trophic levels are arranged vertically, and the names of the levels appear in colored boxes. The arrows connecting the organisms point from the food to the consumer. This transfer of food moves chemical nutrients and energy from the producers up through the trophic levels in a community.

Figure 37.9 compares a terrestrial food chain and an aquatic food chain. Starting at the bottom, the trophic level that supports all others consists of autotrophs, which ecologists call **producers.** Photosynthetic producers use light energy to power the synthesis of organic compounds. Plants are the main producers on land. In water, the producers are mainly photosynthetic protists and cyanobacteria, collectively called phytoplankton. Multicellular algae and aquatic plants are also important producers in shallow waters.

All organisms in trophic levels above the producers are heterotrophs, or consumers, and all consumers are directly or indirectly dependent on the output of producers. Herbivores, which eat plants, algae, or phytoplankton, are the **primary consumers.** Primary consumers on land include grasshoppers and many other insects, snails, and certain vertebrates, such as grazing mammals and birds that eat seeds and fruits. In aquatic environments, primary consumers include a variety of zooplankton (mainly protists and microscopic animals such as small shrimps) that eat phytoplankton.

Above the primary consumers, the trophic levels are made up of carnivores, which eat the consumers from the level below. On land, **secondary consumers** include many small mammals, such as the mouse shown here eating an herbivorous insect, and a great variety of small birds, frogs, and spiders, as well as lions and other large carnivores that eat grazers. In aquatic ecosystems, secondary consumers are mainly small fishes that eat zooplankton.

Higher trophic levels include **tertiary consumers,** such as snakes that eat mice and other secondary consumers. Most ecosystems have secondary and tertiary consumers. As the figure indicates, some also have a higher level, **quaternary consumers.** These include hawks in terrestrial ecosystems and killer whales in the marine environment.

Not shown in Figure 37.9 is another trophic level of consumers called **detritivores,** or **decomposers,** which derive their energy from **detritus,** the dead material produced at all the trophic levels. Detritus includes animal wastes, plant litter, and all sorts of dead organisms. Most organic matter eventually becomes detritus and is consumed by detritivores. A great variety of animals, often called scavengers, eat detritus. For instance, earthworms, many rodents, and insects eat fallen leaves and other detritus. Other scavengers include crayfish, catfish, crows, and vultures.

A community's main detritivores are the prokaryotes and fungi, which secrete enzymes that digest organic material and then absorb the breakdown products. Enormous numbers of microscopic fungi and prokaryotes in the soil and in mud at the bottom of lakes and oceans convert (recycle) most of the community's organic materials to inorganic compounds that plants or phytoplankton can use. The breakdown of organic materials to inorganic ones is called **decomposition.** In a sense, all organisms perform decomposition. In cellular metabolism, they all break down organic material and release inorganic products, such as carbon dioxide and ammonia, to the environment. But the decomposition by prokaryotes and fungi links all trophic levels and is essential for all communities and, indeed, for the continuation of life on Earth.

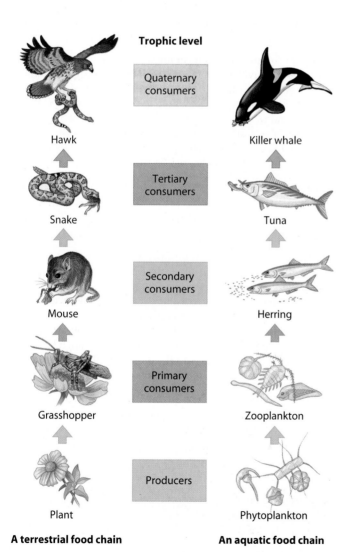

Trophic level

Quaternary consumers

Tertiary consumers

Secondary consumers

Primary consumers

Producers

Hawk · Snake · Mouse · Grasshopper · Plant

A terrestrial food chain

Killer whale · Tuna · Herring · Zooplankton · Phytoplankton

An aquatic food chain

Figure 37.9 Two food chains

? I'm eating a cheese pizza. At which trophic level(s) am I feeding?

■ Primary consumer (flour and tomato sauce) and secondary consumer (cheese, a product from cows, which are primary consumers)

37.10 Food chains interconnect, forming food webs

A more realistic view of the trophic structure of a community is a **food web**, a network of interconnecting food chains. Figure 37.10 shows a simplified example of a food web in a Sonoran desert community. As in the food chains of Figure 37.9, the arrows indicate the direction of nutrient transfer ("who eats whom") and are color-coded by trophic levels.

Notice that a consumer may eat more than one type of producer, and several species of primary consumers may feed on the same species of producer. Some animals weave into the web at more than one trophic level. The lizard and mantid are strictly secondary consumers, eating insects. The woodpecker on the left, however, is a primary consumer when it eats cactus seeds and a secondary consumer when it eats ants or grasshoppers. The hawk at the top of the web is a secondary, tertiary or quaternary consumer, depending on its prey. Food webs, like food chains, do not typically show detritivores, which consume dead organic material from all trophic levels.

We have now looked at some of the characteristics of a community: species diversity, interspecific interactions, dominant species, disturbances, and trophic structure. In the next modules, we broaden our scope to look at ecosystems, the highest level of ecological complexity.

Web/CD Activity 37C *Food Webs*

? Even consumers at the highest level of a food web eventually become food for _____.

■ detritivores

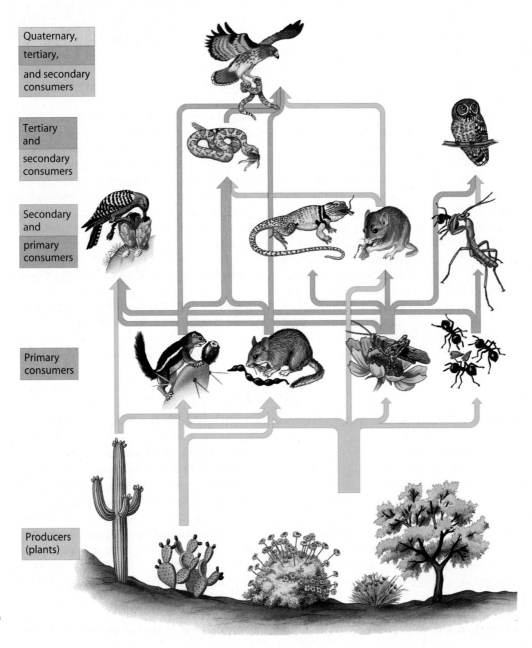

Figure 37.10
A food web

37.11 Ecosystem ecology emphasizes energy flow and chemical cycling

An **ecosystem** consists of all the organisms in a community as well as the abiotic environment with which the organisms interact. Ecosystems can range from a microcosm, such as the terrarium in **Figure 37.11**, to a large area such as a forest. Farms and cities are human-dominated ecosystems. The entire biosphere may be regarded as a global ecosystem.

Regardless of an ecosystem's size, its dynamics involve two processes—energy flow and chemical cycling. These processes are highlighted by the arrows in Figure 37.11. The yellow, orange, and red arrows represent **energy flow**, the passage of energy *through* the components of the ecosystem. Energy enters the terrarium in the form of sunlight. Plants (producers) convert the light energy to chemical energy. Animals (consumers) take in some of this chemical energy in the form of organic compounds when they eat the plants. Detritivores, such as bacteria and fungi in the soil, obtain chemical energy when they decompose the dead remains of plants and animals. Every use of chemical energy by organisms involves a loss of some energy to the surroundings in the form of heat (see Module 5.2). Eventually, therefore, the ecosystem would run out of energy if it were not powered by a continuous inflow of energy from an outside source. For most ecosystems, the sun is the energy source, but exceptions include several unusual kinds of ecosystems powered by chemical energy obtained from inorganic compounds (see Chapter 34's introduction).

In contrast to energy flow, **chemical cycling** (blue arrows) involves the transfer of materials *within* the ecosystem. An ecosystem, especially an artificial one like a terrarium, is more or less self-contained in terms of matter. Chemical elements such as carbon and nitrogen are cycled between abiotic components (air, water, and soil) and biotic components of the ecosystem. The plants acquire these elements in inorganic form from the air and soil and fix them into organic molecules, some of which animals consume. Detritivores

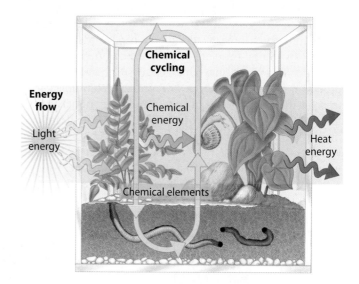

Figure 37.11 A terrarium ecosystem

return most of the elements in inorganic form to the soil and air. Some elements are also returned as the by-products of plant and animal metabolism.

In summary, both energy flow and chemical cycling involve the transfer of substances through the trophic levels of the ecosystem. However, energy flows through ecosystems, whereas chemicals are recycled within them. We explore these fundamental ecosystem dynamics in the rest of the chapter.

Web/CD Activity 37D *Energy Flow and Chemical Cycling*

❓ Why is the transfer of energy in an ecosystem referred to as energy *flow*, not energy *cycling*?

■ Because energy passes through an ecosystem, entering as sunlight and leaving as heat. It is not recycled within the ecosystem.

37.12 Primary production sets the energy budget for ecosystems

Each day, planet Earth receives about 10^{19} kcal of solar energy, the energy equivalent of about 100 million atomic bombs. Most of this energy is absorbed, scattered, or reflected by the atmosphere or by Earth's surface. Of the visible light that reaches plants, algae, and cyanobacteria, only about 1% is converted to chemical energy by photosynthesis. But on a global scale, this is enough to produce about 170 billion tons of organic material per year in the biosphere.

Ecologists call the amount, or mass, of living organic material in an ecosystem the **biomass**. The amount of solar energy converted to chemical energy (organic compounds) by an ecosystem's producers for a given area and during a given

time period is called **primary production**. It can be expressed in units of energy or of mass. The primary production of the entire biosphere is 170 billion tons of biomass per year.

Different ecosystems vary considerably in their primary production as well as in their contribution to the total production of the biosphere. **Figure 37.12**, at the top of the next page, contrasts the net primary production of a number of different ecosystems. (Net primary production refers to the amount of biomass produced minus the amount used by producers as fuel for their own cellular respiration.) Tropical rain forests are among the most productive terrestrial ecosystems and contribute a large portion of the planet's overall produc-

tion of biomass. Coral reefs also have very high production, but their contribution to global production is small because they cover such a small area. Interestingly, even though the open ocean has very low production, it contributes the most to Earth's total net primary production because of its huge size—it covers 65% of Earth's surface area.

Next, we consider how the energy budget set by producers gets divided among the trophic levels of an ecosystem.

Web/CD Thinking as a Scientist *How Do Temperature and Light Affect Primary Production?*

? Deserts and semidesert scrub cover about the same amount of surface area as tropical forests but contribute less than 1% of Earth's net primary production, while rain forests contribute 22%. Explain this difference.

■ The primary production of tropical rain forests is over 20 times greater than that of deserts and semidesert scrub ecosystems.

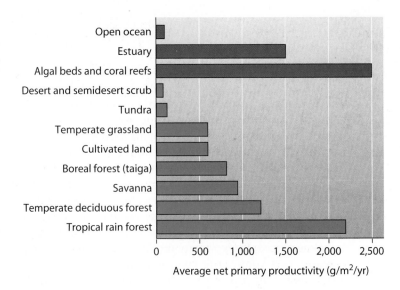

Figure 37.12 Net primary production of various ecosystems

37.13 Energy supply limits the length of food chains

When energy flows as organic matter through the trophic levels of an ecosystem, much of it is lost at each link in a food chain. Consider the transfer of organic matter from producers to primary consumers (herbivores). In most ecosystems, herbivores manage to eat only a fraction of the plant material produced, and they can't digest all of what they do consume. For example, a grasshopper feeding on a flower might digest and absorb only about half the organic material it eats, passing the indigestible wastes as feces. Of the organic compounds it does absorb from its food, the grasshopper typically uses two-thirds as fuel for cellular respiration. Thus, only the chemical energy left over after respiration can be converted to grasshopper biomass. And only this biomass or amount of energy is available to the next trophic level.

Figure 37.13, called a pyramid of production, illustrates the cumulative loss of energy with each transfer in a food chain. Each tier of the pyramid represents one trophic level, and the width of each tier indicates how much of the chemical energy of the tier below is actually incorporated into the organic matter of that trophic level. Note that producers convert only about 1% of the energy in the sunlight available to them to primary production. In this idealized pyramid, 10% of the energy available at each trophic level becomes incorporated into the next higher level. The efficiencies of energy transfer usually range from 5 to 20%. In other words, 80 to 95% of the energy at one trophic level never transfers to the next.

An important implication of this stepwise decline of energy in a trophic structure is that the amount of energy available to top-level consumers is small compared with that available to lower-level consumers. Only a tiny fraction of the energy stored by photosynthesis flows through a food chain to a tertiary consumer, such as a snake feeding on a mouse. This explains why top-level consumers such as lions and hawks require so much geographic territory; it takes a lot of

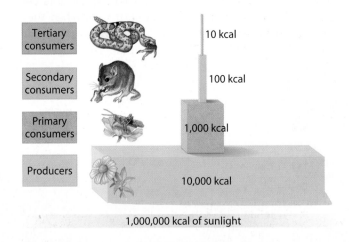

Figure 37.13 An idealized pyramid of production

vegetation to support trophic levels so many steps removed from photosynthetic production.

Pyramids of production help you understand why most food chains are limited to three to five levels; there is simply not enough energy at the very top of an ecological pyramid to support another trophic level. We can now see why the last line of Jonathan Swift's verse quoted in the chapter introduction—"And so proceed *ad infinitum*"—cannot be taken literally for an ecosystem. The number of levels in a trophic structure is anything but infinite. Instead, as the production pyramid indicates, trophic structures and their component food chains are severely limited by the availability of energy.

? Approximately what proportion of the energy produced by photosynthesis makes it to the snake in Figure 37.13?

37.14 A production pyramid explains why meat is a luxury for humans

The dynamics of energy flow apply to the human population as much as to other organisms. Like most other consumers, we depend entirely on production by plants for our food. As omnivores, we eat both plant material and meat. When we eat grain or fruit, we are primary consumers; when we eat beef or other meat from herbivores, we are secondary consumers. When we eat fish like trout and salmon (which eat insects and other small animals), we are tertiary or quaternary consumers.

The production pyramid on the left in **Figure 37.14** indicates energy flow from primary producers to humans as vegetarians (herbivores). The energy in the producer trophic level comes from a corn crop. The pyramid on the right illustrates energy flow from the same corn crop, with humans as secondary consumers, eating beef. These two pyramids are generalized models, based on the rough estimate that about 10% of the energy available in a trophic level appears at the next higher trophic level. Thus, the pyramids indicate that the human population has about ten times more energy available to it when people eat corn than when they process the same amount of corn through another trophic level and eat corn-fed beef.

Eating meat of any kind is an expensive luxury, both economically and environmentally. In many countries, people cannot afford to buy much meat and are vegetarians by necessity. Of course, some herbivores graze on vegetation that humans cannot process for food. But producing meat for human consumption usually requires that more land be cultivated, more water be used for irrigation, and more chemical fertilizers and pesticides be applied to croplands used for growing grain. It is likely that as the human population expands, meat consumption will become even more of a luxury than it is today (see Module 36.9).

We turn next to the subject of chemical nutrients, all of which differ from energy in that they cycle within ecosystems.

Web/CD Activity 37E Energy Pyramids

? Why does demand for meat also tend to drive up prices of grains such as wheat and rice, fruits, and vegetables?

■ Potential supply of plants for direct consumption as food for humans is diminished by the use of agricultural land to grow feed for cattle, chickens, and other meat sources.

Trophic level

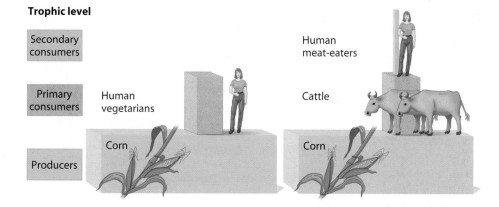

Secondary consumers

Primary consumers

Producers

Human vegetarians

Corn

Human meat-eaters

Cattle

Corn

Figure 37.14 Food energy available to the human population at different trophic levels

37.15 Chemicals are recycled between organic matter and abiotic reservoirs

The sun (or in some cases Earth's interior) supplies ecosystems with a continual influx of energy, but aside from the occasional meteorite, there are no extraterrestrial sources of chemical elements. Life, therefore, depends on the recycling of chemicals. While an organism is alive, much of its chemical stock changes continuously, as nutrients are acquired and waste products are released. Atoms present in the complex molecules of an organism at the time of its death are returned to the environment by the action of detritivores, replenishing the pool of inorganic nutrients that plants and other producers use to build new organic matter.

Because chemical cycles in an ecosystem involve both biotic and abiotic components, they are called **biogeochemical cycles. Figure 37.15,** at the top of the next page, is a general

scheme for the cycling of a nutrient within an ecosystem. Note that the cycle has an **abiotic reservoir** where a chemical accumulates or is stockpiled outside of living organisms. The atmosphere, for example, is an abiotic reservoir for carbon.

Let's trace our way around a biogeochemical cycle. ❶ Producers incorporate chemicals from the abiotic reservoir into organic compounds. ❷ Consumers feed on the producers, incorporating some of the chemicals into their own bodies. ❸ Both producers and consumers release some chemicals back to the environment in waste products (CO_2 and nitrogen wastes of animals). ❹ Detritivores play a central role. As organisms die, these decomposers return chemicals in inorganic form to the soil, water, and air. The producers gain a renewed supply of raw materials, and the cycle continues.

Biogeochemical cycles can be local or global. The chemical phosphorus, for example, cycles almost entirely within local areas, at least over the short term. Soil is the main reservoir for nutrients in a local cycle. In contrast, for those chemicals that spend part of their time in gaseous form—carbon, nitrogen, and water are examples—the cycling is essentially global. For instance, some of the carbon a plant acquires from the air may have been released into the atmosphere by the respiration of a plant or animal on another continent.

In the next four modules, we look at the cyclic movement of water, carbon, nitrogen, and phosphorus. As you study the cycles, look for the four basic steps we have cited, as well as the geologic processes that may move chemicals around and between ecosystems. In each cycle diagram, the arrow widths indicate the relative contribution of a process to the movement of substances in that cycle.

? How does the movement of chemicals in an ecosystem differ from the movement of energy?

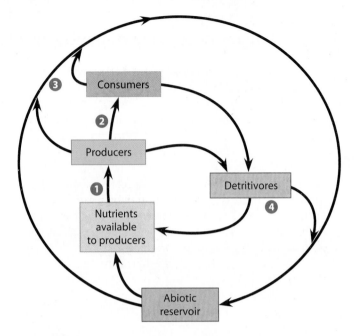

Figure 37.15 A general model of nutrient cycling

37.16 Water moves through the biosphere in a global cycle

Water is essential to life, both because organisms are made mostly of water and because water's unique properties make environments on Earth habitable (see Modules 2.11–2.14).

Figure 37.16 illustrates the global water cycle. Three major processes driven by solar energy—precipitation, evaporation, and transpiration from plants—continuously move water between the land, oceans, and the atmosphere. Over the oceans (left side of the figure), evaporation exceeds precipitation. The result is a net movement of water vapor to clouds that are carried by winds from the oceans across the land. On land (right side of the figure), precipitation exceeds evaporation and transpiration. Excess precipitation forms systems of surface water (such as lakes and rivers) and groundwater, all of which flow back to the sea, completing the water cycle. The water cycle has a global character because there is a large reservoir of water in the atmosphere, as well as in the oceans. Thus, water molecules that have evaporated from the Pacific Ocean, for instance, may appear in a lake or in an animal's body far inland in North America.

Human activity affects the global water cycle in a number of important ways. One of the main sources of atmospheric water is transpiration from the dense vegetation making up tropical rain forests. The destruction of these forests, which is occurring rapidly, will change the amount of water vapor in the air. This, in turn, will likely alter local, and perhaps global, weather patterns.

Another human acivity that affects the water cycle is the pumping of large amounts of groundwater to the surface for irrigation. This practice can increase the rate of evaporation over land and may deplete groundwater supplies.

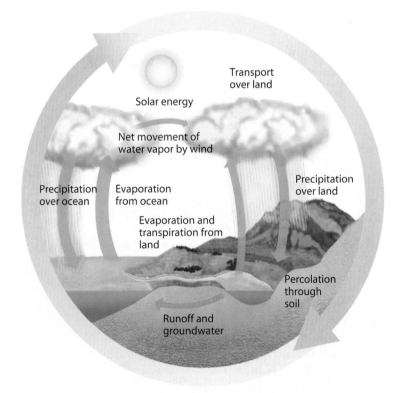

Figure 37.16 The global water cycle

? What is the main way that living organisms contribute to the water cycle?

37.17 The carbon cycle depends on photosynthesis and respiration

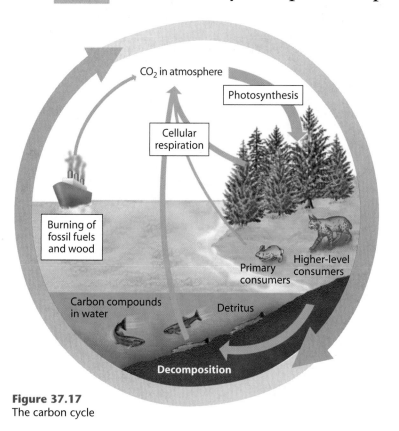

Figure 37.17
The carbon cycle

Carbon is the major ingredient of all organic molecules. Like water, the element carbon has an atmospheric reservoir and cycles globally. Other abiotic reservoirs of carbon include fossil fuels, dissolved carbon compounds in the oceans, and sedimentary rocks such as limestone.

As shown in **Figure 37.17**, the reciprocal metabolic processes of photosynthesis and cellular respiration are mainly responsible for the cycling of carbon between the biotic and abiotic worlds. Carbon compounds in detritus—animal wastes, plant litter, and dead organisms of all kinds—are consumed and decomposed by detritivores, and detritivore respiration, along with that of plants, animals, and other organisms, returns CO_2 to the atmosphere.

On a global scale, the return of CO_2 to the atmosphere by respiration closely balances its removal by photosynthesis. However, the increased burning of wood and fossil fuels (coal and petroleum) is raising the level of CO_2 in the atmosphere. This appears to be leading to the significant problem of global warming, as we will discuss in Module 38.5.

Web/CD Activity 37F *The Carbon Cycle*

? What would happen to the carbon cycle if all the detritivores suddenly went on "strike" and stopped working?

■ Carbon would accumulate in organic mass, the atmospheric reservoir of carbon would decline, and plants would eventually be starved for CO_2.

37.18 The nitrogen cycle relies heavily on bacteria

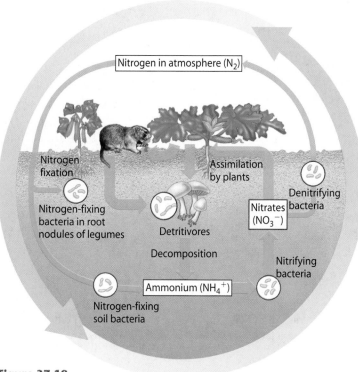

Figure 37.18
The nitrogen cycle

As an ingredient of proteins and nucleic acids, nitrogen is essential to the structure and functioning of all organisms. In particular, it is a crucial and often limiting plant nutrient. The atmosphere contains a huge reservoir of nitrogen; almost 80% of the atmosphere is N_2. However, N_2 is a form of nitrogen that is unavailable to plants. Plants can only use nitrogen in the form of nitrate ions (NO_3^-) or ammonium ions (NH_4^+). These compounds can be made from N_2 by soil bacteria.

Notice the two groups of nitrogen-fixing bacteria on the left side of **Figure 37.18**. Nitrogen fixers in the soil and those living symbiotically in the roots of beans and other legumes (see Module 32.13) convert atmospheric N_2 to ammonium (NH_4^+). (In aquatic ecosystems, cyanobacteria are important nitrogen fixers.) Following the arrows to the lower right, you can see that nitrifying bacteria convert NH_4^+ to NO_3^-, the main source of nitrogen for plants. Used by plants to make amino acids and proteins, the nitrogen is then available to consumers.

Bacteria and fungi acting as detritivores decompose nitrogen-containing detritus back into ammonium, thus keeping the nitrogen available for plants. Another group of soil bacteria, the denitrifiers (far right in the figure), complete the nitrogen cycle by converting soil nitrates to atmospheric N_2.

Most of the nitrogen cycling in many ecosystems involves the inner cycle in the diagram, the thicker arrows linking

plants and animals, detritivores, and nitrifying bacteria. The processes indicated by the thinner arrows make a relatively smaller contribution to the nitrogen cycle in many natural ecosystems.

Although not shown in the figure, some NH_4^+ and NO_3^- are made in the atmosphere by chemical reactions involving N_2 and NH_3 gas (ammonia). These ions reach the soil in precipitation and dust and are a crucial source of nitrogen for plants in some ecosystems.

Human activity has altered the nitrogen cycle balance in many areas. Sewage treatment facilities often empty large amounts of dissolved inorganic nitrogen compounds into rivers or streams. Large amounts of inorganic nitrogen fertilizers are routinely applied to agricultural croplands. Lawns and golf courses also receive sizable doses of fertilizers. Crop and lawn plants take up some of the nitrogen compounds, and denitrifying bacteria convert some into atmospheric N_2, but chemical fertilizers usually exceed the soil's natural recycling capacity. The excess nitrogen compounds often enter streams, lakes, and groundwater.

In lakes and streams, these nitrogen compounds continue to fertilize, causing heavy growth of algae. Groundwater pollution by nitrogen fertilizers is a serious problem in many agricultural areas. Nitrates in drinking water are converted to nitrites (NO_2^-), which can be toxic, in the human digestive tract. In Module 32.10, we discussed some alternatives to the extensive use of agricultural fertilizers.

Web/CD Activity 37G *The Nitrogen Cycle*

Web/CD Activity 37H *Water Pollution from Nitrates*

? What is the main abiotic reservoir of nitrogen?

■ The atmospheric supply of N_2.

37.19   The phosphorus cycle depends on the weathering of rock

Organisms require phosphorus as an ingredient of nucleic acids, phospholipids, and ATP and (in vertebrates) as a mineral component of bones and teeth. In contrast to nitrogen and carbon, phosphorus has its main abiotic reservoirs in rocks rather than in the atmosphere. (This is also true of the elements potassium and calcium.)

At the center of **Figure 37.19**, the weathering of rock gradually adds inorganic phosphate (PO_4^{3-}) to the soil. Plants absorb the dissolved phosphate ions in the soil and build them into organic compounds. Consumers obtain phosphorus in organic form from plants. Phosphates are returned to the soil by animal excretion and the action of detritivores. As indicated in the central part of Figure 37.19, some phosphorus drains from terrestrial ecosystems into the sea, where it may settle and eventually become part of new rocks. This phosphorus will not cycle back into living organisms until geologic processes uplift the rocks and expose them to weathering.

Because weathering is generally a slow process, the amount of phosphates available to plants in natural ecosystems is often quite low. Plant growth can, in fact, be limited by the small amount of soluble phosphates in the soil.

In lakes that have not been altered by human activity, a low level of dissolved phosphates often keeps algal growth to a minimum, thereby helping keep the water clear. In many areas, however, excess phosphates are a problem. Like nitrogen compounds, phosphates are a major component of sewage outflow. They are also used extensively in agricultural fertilizers and are a common ingredient in pesticides. Phosphate pollution of lakes and rivers, like nitrate pollution, leads to heavy growth of algae and cyanobacteria.

In the next two modules, we discuss some long-term ecological studies of nutrient cycling that demonstrate some of the effects of human alteration of nutrient cycles.

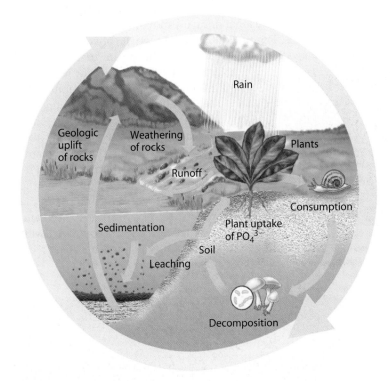

Figure 37.19   The phosphorus cycle

? Over the short term, why does phosphorus cycling tend to be more localized than either carbon or nitrogen cycling?

■ Because phosphorus is cycled almost entirely within the soil rather than transferred over long distances via the atmosphere

37.20 Ecosystem alteration can upset chemical cycling

The cycling of any chemical element in an ecosystem depends on the web of feeding relationships between plants, animals, and detritivores and on geologic processes. Obtaining an accurate picture of chemical cycling requires long-term study, and a number of research groups have been conducting such studies for several decades. In one of the longest-running examples of long-term ecological research, a team has been monitoring nutrient cycling in a forest ecosystem since 1963. The study site, the Hubbard Brook Experimental Forest in the White Mountains of New Hampshire, is a deciduous forest with several valleys, each drained by a small creek that is a tributary of Hubbard Brook. Bedrock impenetrable to water is close to the surface of the soil, and each valley constitutes a watershed that can drain only through its creek.

The Hubbard Brook team set out to study water and nutrient dynamics under natural conditions and after severe human intrusion. They first determined the amounts of water and key nutrients that normally move in and out of six of the watersheds. Figure 37.20A shows a small concrete dam with a V-shaped spillway built across the stream at the bottom

of a watershed to monitor water and nutrient losses. When monitoring began, about 60% of the water that fell as rain and snow exited through the streams, and the remaining 40% was lost by transpiration from plants and evaporation from the soil. Preliminary data also indicated that the flow of nutrients into and out of the watersheds was nearly balanced and was relatively small compared with the quantity of nutrients being recycled within the forest itself.

Figure 37.20A A dam at the Hubbard Brook study site

In 1966, one of the valleys was completely logged and then sprayed with herbicides for three years to prevent regrowth of plants. All the original plant material was left in place to decompose. This severely altered watershed is the white (snow-covered) area in the center of Figure 37.20B. The inflow and outflow of water and minerals for this watershed were compared with a control (unaltered) watershed for three years. Water runoff from the altered system increased by 30–40%, apparently because there were no plants to absorb and transpire water from the soil. Net losses of nutrients were huge, as shown for nitrate in Figure 37.20C. Following deforestation, the nitrate concentration in the creek was 60 times greater in the altered watershed than in the control. Not only was this vital nutrient drained from the ecosystem, but nitrate in the stream reached a level considered unsafe for drinking water.

After 40 years, data from Hubbard Brook point to some other long-term trends. For instance, since the 1950s, acid rain and snow have dissolved most of the Ca^{2+} from the forest soil, and the streams have carried it away. By the 1990s, forest plants at Hubbard Brook had virtually stopped adding new growth, apparently because of a lack of Ca^{2+} in the soil. In 1998, ecologists began a massive experiment to test this hypothesis. They monitored control and experimental watersheds for two years and then added Ca^{2+} by helicopter. Initial results show significant increases in pH, nitrate concentrations, and soil respiration in the experimental watershed. Changes in tree growth may take longer to record.

In the next module, we discuss another ecologist's work studying the effects of nutrients in freshwater ecosystems.

? How can clear-cutting a forest (removing all trees) damage the water quality of nearby lakes?

■ Without the growing trees to assimilate minerals from the soil, more of the minerals run off and end up polluting water resources.

Figure 37.20B Logged watershed in the Hubbard Brook Forest

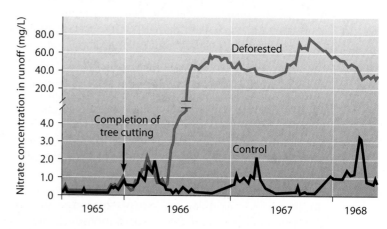

Figure 37.20C The loss of nitrate from a deforested watershed

37.20 Ecosystem alteration can upset chemical cycling

The cycling of any chemical element in an ecosystem depends on the web of feeding relationships between plants, animals, and detritivores and on geologic processes. Obtaining an accurate picture of chemical cycling requires long-term study, and a number of research groups have been conducting such studies for several decades. In one of the longest-running examples of long-term ecological research, a team has been monitoring nutrient cycling in a forest ecosystem since 1963. The study site, the Hubbard Brook Experimental Forest in the White Mountains of New Hampshire, is a deciduous forest with several valleys, each drained by a small creek that is a tributary of Hubbard Brook. Bedrock impenetrable to water is close to the surface of the soil, and each valley constitutes a watershed that can drain only through its creek.

The Hubbard Brook team set out to study water and nutrient dynamics under natural conditions and after severe human intrusion. They first determined the amounts of water and key nutrients that normally move in and out of six of the watersheds. Figure 37.20A shows a small concrete dam with a V-shaped spillway built across the stream at the bottom

Figure 37.20A A dam at the Hubbard Brook study site

of a watershed to monitor water and nutrient losses. When monitoring began, about 60% of the water that fell as rain and snow exited through the streams, and the remaining 40% was lost by transpiration from plants and evaporation from the soil. Preliminary data also indicated that the flow of nutrients into and out of the watersheds was nearly balanced and was relatively small compared with the quantity of nutrients being recycled within the forest itself.

In 1966, one of the valleys was completely logged and then sprayed with herbicides for three years to prevent regrowth of plants. All the original plant material was left in place to decompose. This severely altered watershed is the white (snow-covered) area in the center of Figure 37.20B. The inflow and outflow of water and minerals for this watershed were compared with a control (unaltered) watershed for three years. Water runoff from the altered system increased by 30–40%, apparently because there were no plants to absorb and transpire water from the soil. Net losses of nutrients were huge, as shown for nitrate in Figure 37.20C. Following deforestation, the nitrate concentration in the creek was 60 times greater in the altered watershed than in the control. Not only was this vital nutrient drained from the ecosystem, but nitrate in the stream reached a level considered unsafe for drinking water.

After 40 years, data from Hubbard Brook point to some other long-term trends. For instance, since the 1950s, acid rain and snow have dissolved most of the Ca^{2+} from the forest soil, and the streams have carried it away. By the 1990s, forest plants at Hubbard Brook had virtually stopped adding new growth, apparently because of a lack of Ca^{2+} in the soil. In 1998, ecologists began a massive experiment to test this hypothesis. They monitored control and experimental watersheds for two years and then added Ca^{2+} by helicopter. Initial results show significant increases in pH, nitrate concentrations, and soil respiration in the experimental watershed. Changes in tree growth may take longer to record.

In the next module, we discuss another ecologist's work studying the effects of nutrients in freshwater ecosystems.

? How can clear-cutting a forest (removing all trees) damage the water quality of nearby lakes?

■ Without the growing trees to assimilate minerals from the soil, more of the minerals run off and end up polluting water resources.

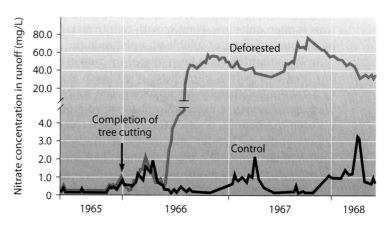

Figure 37.20B Logged watershed in the Hubbard Brook Forest

Figure 37.20C The loss of nitrate from a deforested watershed

37.21 David Schindler talks about the effects of nutrients on freshwater ecosystems

Figure 37.21A
David Schindler

The Hubbard Brook experiment shows that major change in a terrestrial ecosystem disrupts chemical cycling, moving nutrients to other areas, such as streams and lakes. How do nutrients from deforested lands and agricultural areas affect aquatic ecosystems? Nutrient levels may naturally increase over time, making a lake more productive. However, in a process known as *cultural eutrophication,* added nutrients cause algae and cyanobacteria to multiply rapidly, resulting in a "bloom." Heavy bacterial and algal growth greatly reduces oxygen levels at night, when the photosynthesizers respire. As these organisms die and accumulate at the bottom of the lake, decomposers can use up much of the oxygen dissolved in deep waters. As a result, the pond or lake may lose much of its species diversity. Human-caused eutrophication, for example, wiped out commercially important fishes in Lake Erie by the 1960s.

Dr. David Schindler (**Figure 37.21A**) is a professor of ecology at the University of Alberta. Before becoming a professor, he was a government scientist at the Experimental Lakes Project in northern Ontario. There he performed classic experiments that led to the banning of phosphates in detergents. He describes his research in this recent interview:

> When the project started in 1968, our mandate was to test water management issues at the level of whole-lake ecosystems. The main objective was the study of eutrophication, the overfertilization of lakes with mineral nutrients. The big issue at the time was whether phosphorus was the main culprit in eutrophication. Laboratory experiments implicated phosphorus, but my bosses at the time had a hard time convincing politicians and managers to invest millions or billions of dollars in phosphorus management schemes based solely on these small-scale experiments.

He explains why there was resistance to managing the input of phosphorus into freshwater ecosystems:

> One of the big sources of phosphorus in those days was phosphate detergents, and there was a big political lobby defending the use of these detergents. The companies that produced detergents were pointing the finger at carbon as the main problem.

Dr. Schindler tested whether it was carbon, phosphorus, or some other nutrient that was causing the eutrophication and algal blooms:

> In our best-known experiment, we divided a lake into two basins. We added just carbon and nitrogen to one basin. We included phosphorus along with carbon and nitrogen in the other basin. We got a tremendous algal bloom within weeks

Figure 37.21B Experimental eutrophication of part of a lake (phosphorus added to left basin)

after adding phosphorus, but no change with just the carbon and nitrogen.

The results of that experiment, shown in **Figure 37.21B**, prompted government regulation of phosphate detergents:

> We published our results in *Science* in 1974, and that pretty well set off a cascade of phosphorus regulations for detergents and sewage effluents. The response in Eastern Canada was quick. But it took 17 years to get all the U.S. states in the Great Lakes region on board.

That helped solve one problem. Dr. Schindler describes the most serious current threats to freshwater ecosystems:

> Acid precipitation is one. Warming of the climate is another. Changes in land use can affect lakes. For example, if you bulldoze a forest to pasture cows on the land, you increase the runoff of nutrients into the water four- or fivefold at least. And if you plow that land, plant crops, and add fertilizer, you increase the yield of nutrients to the water even more. And now we have these huge, intensive livestock operations—up to 30,000 cattle or 80,000 hogs. Very few of these big livestock operations have sewage treatment. The animals have a lot of the same intestinal microbes that humans have. We would no longer think of discharging raw sewage from a city into a river without treating it. But we do it all the time with intensive livestock operations. We're not handling our fresh water very well. And I think a big problem is that we've tended to look at the water issues, such as land use and acid precipitation, in isolation, when it's usually a combination of factors damaging the ecosystems.

Human disruption of both aquatic and terrestrial ecosystems is a global problem. In the next chapter, we look at conservation biology, efforts to conserve the diversity of the ecosystems on which humans depend.

? How does excessive addition of mineral nutrients to a lake eventually result in the loss of most fish species?

■ The eutrophication initially causes population explosions of algae and cyanobacteria. The respiration of so much life, including the decomposers of all the organic refuse, consumes most of the lake's oxygen, which the fish require.

Reviewing the Concepts

Structural Features of Communities (37.1–37.10)

A community includes all the organisms in a particular area. Communities are characterized by species diversity, dominant species, response to disturbance, and trophic structure **(37.1)**.

Interspecific competition occurs between two species if they both require the same limited resource. A species' niche includes its total use of biotic and abiotic resources. Species with identical niches may not be able to coexist, or natural selection may lead to resource partitioning **(37.2)**.

Predation. The adaptations of both predators and prey tend to be refined through natural selection. Some prey gain protection through camouflage and mimicry **(37.3)**. A keystone predator may maintain community diversity by reducing the numbers of the strongest competitors **(37.4)**.

Herbivores have adaptations for eating plants; many plants produce protective toxic chemicals. Some herbivore-plant interactions illustrate coevolution, or reciprocal evolutionary adaptations **(37.5)**.

Symbiotic relationships are interactions in which species live together in direct contact. In parasitism, a parasite obtains food at the expense of its host. In commensalism, one species benefits while the other is unaffected. In mutualism, both partners benefit **(37.6)**.

Disturbance, whether human caused or not, is characteristic of most communities. Ecological succession is a transition in species composition of a community. Primary succession is the gradual colonization of barren rocks. Secondary succession occurs after a disturbance has destroyed a community but left the soil intact **(37.7)**. Fire is a key abiotic factor in many ecosystems **(37.8)**.

Trophic structure of a community can be represented by a food chain—the stepwise flow of energy and nutrients from plants (producers), to herbivores (primary consumers), to carnivores (secondary and higher-level consumers). Detritivores (animal scavengers, fungi, and prokaryotes) decompose waste matter and recycle nutrients **(37.9)**. A food web is a network of interconnecting food chains **(37.10)**.

Ecosystem Structure and Dynamics (37.11–37.19)

An ecosystem includes a community and the abiotic factors with which it interacts **(37.11)**.

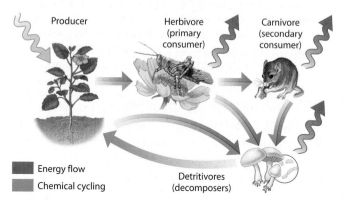

Energy flow
Chemical cycling

Producer
Herbivore (primary consumer)
Carnivore (secondary consumer)
Detritivores (decomposers)

Primary production is the rate at which producers convert sunlight to chemical energy in organic matter (biomass). A pyramid of production shows the flow of energy from producers to primary consumers and to higher trophic levels. Only about 10% of the energy stored at each trophic level is available to the next level. Thus, a field of corn can support many more human vegetarians than meat-eaters **(37.12–37.14)**.

Biogeochemical cycles. Nutrients recycle between organisms and abiotic reservoirs **(37.15)**. Solar heat drives the global water cycle of precipitation, evaporation, and transpiration **(37.16)**. Carbon is taken from the atmosphere by photosynthesis, used to make organic molecules, and returned to the atmosphere by cellular respiration **(37.17)**. Various bacteria in soil (and legume root nodules) convert gaseous N_2 to compounds that plants can use: ammonium (NH_4^+) and nitrate (NO_3^-). Detritivores decompose organic matter and recycle nitrogen to plants **(37.18)**. Phosphorus and other soil minerals are recycled locally; they are in long-term storage in rocks **(37.19)**.

Ecosystem Alteration (37.20–37.21)

Ecosystem studies show that drastic alterations, such as the total removal of vegetation, can increase the runoff of water and loss of soil nutrients. Environmental changes caused by humans, such as acid rain, can unbalance nutrient cycling over the long term **(37.20)**. Nutrient runoff from agricultural lands and large livestock operations may cause excessive algal growth. This cultural eutrophication reduces species diversity and harms water quality **(37.21)**.

Connecting the Concepts

1. Fill in the blanks in this concept map summarizing the interspecific interactions in a community.

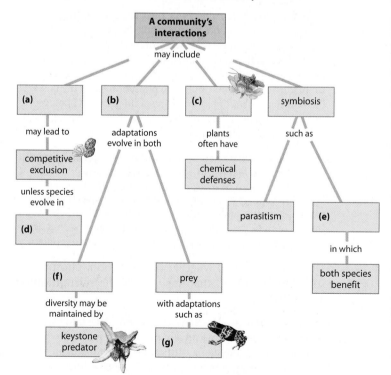

A community's interactions

may include

(a)   (b)   (c)   symbiosis

may lead to   adaptations evolve in both   plants often have   such as

competitive exclusion   chemical defenses

unless species evolve in   parasitism   (e)

(d)   in which

both species benefit

(f)   prey

diversity may be maintained by   with adaptations such as

keystone predator   (g)

2. Fill in this concept map summarizing ecosystem dynamics.

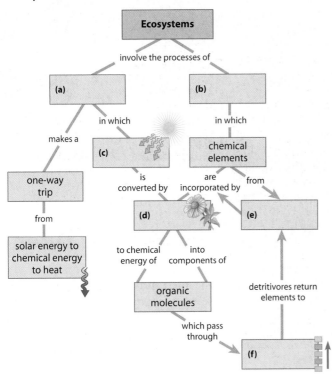

Testing Your Knowledge

Multiple Choice

3. Which of the following groups is absolutely essential to the functioning of an ecosystem?
   a. producers
   b. producers and herbivores
   c. producers, herbivores, and carnivores
   d. detritivores
   e. producers and detritivores

4. Which of these is a result of resource partitioning?
   a. Two species can compete for the same prey item.
   b. Slight variations in niche allow closely related or very similar species to coexist in the same habitat.
   c. Two species can share identical niches in a habitat.
   d. Competitive exclusion results in the success of the superior species.
   e. When two species share the same niche, one species migrates to a new habitat.

5. The open ocean and tropical rain forests contribute the most to Earth's net primary production because
   a. both have high rates of net primary production.
   b. both cover huge surface areas of the Earth.
   c. nutrients cycle fastest in these two ecosystems.
   d. the ocean covers a huge surface area and the tropical rain forest has a high rate of production.
   e. Both a and b are correct.

6. Which of the following organisms is mismatched with its trophic level?
   a. algae—producer
   b. fungi—detritivore
   c. phytoplankton—primary consumer
   d. carnivorous fish larvae—secondary consumer
   e. eagle—tertiary or quaternary consumer

7. Which of the following best illustrates ecological succession?
   a. A mouse eats seeds, and an owl eats the mouse.
   b. Decomposition in soil releases nitrogen that plants can use.
   c. Grasses grow in a deserted field, followed by shrubs and then trees.
   d. Imported pheasants increase in numbers, while local quail disappear.
   e. Overgrazing causes a loss of nutrients from soil.

Describing, Comparing, and Explaining

8. What is cultural eutrophication? What steps might be taken to slow this process?

9. Local conditions, such as heavy rainfall or the removal of plants, may limit the amount of nitrogen, phosphorus, or calcium available, but the amount of carbon available in an ecosystem is seldom a problem. Explain.

10. In Southeast Asia, there's an old saying: "There is only one tiger to a hill." In terms of energy flow in ecosystems, explain why big predatory animals such as tigers and sharks are relatively rare.

11. For which chemicals are biogeochemical cycles global? Explain.

12. What roles do bacteria play in the nitrogen cycle?

Applying the Concepts

13. An ecologist studying plants in the desert performed the following experiment. She staked out two identical plots, which included a few sagebrush plants and numerous small, annual wildflowers. She found the same five wildflower species in roughly equal numbers on both plots. She then enclosed one of the plots with a fence to keep out kangaroo rats, the most common grain-eaters of the area. After two years, to her surprise, four of the wildflower species were no longer present in the fenced plot, but one species had increased dramatically. The control plot had not changed. Using the principles of ecology, propose a hypothesis to explain her results. What additional evidence would support your hypothesis?

14. Sometime in 1986, near Detroit, a freighter pumped out water ballast containing larvae of European zebra mussels. The molluscs multiplied wildly, spreading through Lake Erie and entering Lake Ontario. In some places, they have become so numerous that they have blocked the intake pipes of power plants and water treatment plants, fouled boat hulls, and sunk buoys. We have seen similar population explosions before, with rabbits in Australia and starlings in North America, for example. What makes this kind of population explosion occur? What might happen to native organisms that suddenly must share the Great Lakes ecosystem with zebra mussels? How would you suggest trying to solve the mussel population problem?

Answers to all questions can be found in Appendix 3.

For study help and Activities, go to campbellbiology.com or the student CD-ROM.

CHAPTER 38

Myanmar tiger
photographed
by a remote
"camera trap"

THE BIODIVERSITY CRISIS: AN OVERVIEW

38.1 Human activities threaten Earth's biodiversity
38.2 Biodiversity is vital to human welfare
38.3 Habitat destruction, introduced species, and overexploitation are the major threats to biodiversity
38.4 Pollution of the environment compounds our impact on other species
38.5 Rapid global warming could alter the entire biosphere

CONSERVATION OF POPULATIONS AND SPECIES

38.6 Two ways to study endangered populations are the small-population approach and the declining-population approach
38.7 Identifying critical habitat factors can guide conservation efforts

MANAGING AND RESTORING ECOSYSTEMS

38.8 Sustaining ecosystems and landscapes is a conservation priority
38.9 Protected areas are established to slow the loss of biodiversity
38.10 The Yellowstone to Yukon Conservation Initiative seeks to preserve biodiversity by connecting protected areas
38.11 The study of how to restore degraded habitats is a developing science
38.12 The Kissimmee River project is a case study in restoration ecology
38.13 Zoned reserves are an attempt to reverse ecosystem disruption
38.14 Sustainable development is an ultimate goal

Saving the Tiger

THE MASSIVE ORANGE CAT PROWLS the forest stealthily, sniffing the air as it searches for prey. The tiger is free to hunt and roam in this dark valley, a lowland stretch of trees where people are still relatively few. But just because humans don't live near this forest in Myanmar (formerly called Burma) doesn't mean the tiger goes unnoticed. As the cat glides toward a large tree, a hidden camera senses its body heat and snaps the photo shown above. And so, another endangered tiger has been tracked in this Southeast Asian country—part of the effort to protect tigers in Myanmar and keep them from joining the legions of species sliding toward extinction around the world.

Tigers (*Panthera tigris*) once roamed across Asia from Turkey to the Russian Far East. A hundred years ago, scientists estimate, about 100,000 tigers could be found in the wild. Now that number has plummeted to as low as 5,000. Three of the world's eight tiger subspecies have disappeared entirely.

Human arrival also brought the tiger's greatest nemesis—hunting

But the big cats persist in 14 Asian countries today, and some of these nations, such as Myanmar, are making an effort to protect this majestic part of their natural heritage. For Myanmar, that has meant focusing on the Hukawng Valley, an isolated forest not very hospitable to people but ideal habitat for tigers. The cats ruled here for thousands of years, feeding on deer and other wildlife without much interference from humans.

The modern era shook that serenity. Better roads brought loggers and miners who destroyed parts of the forest and ruined habitat for tigers and their prey. Human arrival also brought the tiger's greatest nemesis—hunting. While such killing may be illegal, a dead tiger still brings big money. Tiger skins, heads, and claws are prized as trophies, and the cats' bones and internal organs are considered a potent source of some traditional Asian medicines. In recent years, tiger bones have fetched as much as $200 per kilogram on the black market—a tempting price in a country as poor as Myanmar. Between people hoping to make a living off the for-

Conservation Biology

est and people hoping to make a living off tiger parts, the big cats' future looked grim.

In 2001, officials in Myanmar decided to protect tigers by turning 6,500 km^2 of the Hukawng Valley into a reserve with strict prohibitions against mining, logging, and hunting. International wildlife biologists, working with the Myanmar Forest Department, helped determine how many tigers remained. They set up remote cameras to locate tigers, and the results weren't encouraging. Using photos to identify individual cats by their stripes, the scientists determined that only about 100 tigers remained.

Myanmar responded by giving the cats even more protection. In 2004, government officials tripled the size of the reserve, expanding its size to more than 20,000 km^2, nearly as large as the state of Vermont. Plans are under way to provide training and jobs to people living in the area, either making them part of the tiger protection efforts or giving them other alternatives to hunting tigers. Scientists estimate that if properly protected, the tiger population in the newly expanded reserve could grow about tenfold.

The effort to save Myanmar's tigers is part of a worldwide struggle to preserve **biodiversity**, the diversity of living things. We are now presiding over an alarming **biodiversity crisis**, a rapid decrease in Earth's great variety of organisms. Biology, the science of life, is a critical part of modern attempts to reverse this trend and conserve biodiversity.

In this final chapter, it is fitting to discuss **conservation biology**, a goal-oriented science that seeks to counter the biodiversity crisis. Conservation biology can focus on a single species, such as the Myanmar tigers. Or it can reach more broadly, trying to protect many species at once by preserving habitats and ecosystems. In this chapter, we examine some of the major factors behind the biodiversity crisis. We also look at some of the research and conservation strategies biologists are using to try to slow the rate of species loss. In our look at the fight to save our biological heritage, we will see that conservation biology touches all levels of ecology, from a single tiger to the forest it roams. ■ ■ ■

Hukawng Valley, Myanmar

38.1 Human activities threaten Earth's biodiversity

In many ways, the plight of the tiger in the chapter introduction illustrates the effect of modern human culture worldwide. Throughout the biosphere, human activities are altering trophic structure, energy flow, chemical cycling, and natural disturbances—ecosystem processes on which we and other species depend. By some estimates, we are doing more damage to the biosphere and pushing more species toward extinction than the changes that triggered the mass extinctions of dinosaurs about 65 million years ago. The current mass extinction is due to the evolution of a single species—a big-brained, manually dexterous, environment-manipulating toolmaker that has named itself *Homo sapiens*.

To date, scientists have described and formally named about 1.8 million species. Some biologists believe that the total number of species is about 10 million, but others estimate it to be as high as 200 million. Because we do not know the number of species currently in existence, we cannot determine the actual rate of species loss. But by some estimates, the global extinction rate may be as much as 1,000 times higher than at any time in the past 100,000 years. Several researchers estimate that at the current rate of destruction, over half of all currently living plant and animal species will be gone by the end of the 21st century.

Biodiversity has three levels: genetic diversity, species diversity, and ecosystem diversity. The *genetic diversity* within and between populations of a species is the raw material that makes microevolution and adaptation to the environment possible. If local populations are lost and the number of individuals in a species declines, so, too, do the genetic resources for that species. As you learned in Module 13.10, a severe reduction in genetic variation theatens the survival of a species.

Much of the popular and political discussion about the biodiversity crisis centers on *species diversity*—the variety of species in an ecosystem or throughout the biosphere. The U.S. Endangered Species Act (ESA) defines an **endangered species** as one that is "in danger of extinction throughout all or a significant portion of its range." Also defined for protection by the ESA, **threatened species** are those that are likely to become endangered in the foreseeable future. Here are just a few examples of why conservation biologists are so concerned about species loss:

- About 12% of the 9,946 known bird species and 24% of the 4,763 known mammalian species in the world are threatened with extinction.

- About 20% of the known freshwater fishes in the world have either become extinct during historical times or are seriously threatened.

- Of the approximately 20,000 known plant species in the United States, 200 species have become extinct since good records have been kept and 730 species are endangered or threatened.

Ecosystem diversity is the third component of biological diversity. Because of the network of community interactions among populations of different species within an ecosystem, the local extinction of one species—say, a keystone predator—can have a negative impact on the overall species richness of the ecosystem (see Module 37.4). And each ecosystem has characteristic patterns of energy flow and chemical cycling that can affect the whole biosphere. For example, the productive "pastures" of phytoplankton in the oceans may help moderate global warming by consuming massive quantities of CO_2 for photosynthesis and for building bicarbonate shells.

Figure 38.1A shows an all-too-common example of ecosystem damage and loss—the destruction of tropical

Figure 38.1A Clearing a rain forest in Borneo

Figure 38.1B A coral reef, one of the most diverse ecosystems

forests to make room for and support an expanding human population. The coral reef shown in **Figure 38.1B** (previous page) is another ecosystem known for its species richness and productivity. About 93% of coral reefs have been damaged by human activities. At the current rate of destruction, 40–50% of the reefs, which are home to one-third of marine fish species, could be lost in the next 30 to 40 years.

As we discuss in the next module, the biodiversity crisis threatens our species as well.

? Why is it too narrow to define the biodiversity crisis as a loss of species?

■ In addition to species loss, the biodiversity crisis includes the loss of genetic diversity within populations and species and the degradation of entire ecosystems.

38.2 Biodiversity is vital to human welfare

Why should we care about the loss of biodiversity? Perhaps the purest reason is what Harvard biologist E. O. Wilson (see Module 35.23) calls *biophilia*, our sense of connection to nature and other forms of life. Many people also share a moral belief that other species have an inherent right to life.

But in addition to ethical and aesthetic reasons for preserving biodiversity, there are practical ones as well. We depend on other species for food, clothing, shelter, oxygen, soil fertility—the list goes on and on. In the United States, 25% of all prescriptions dispensed from pharmacies contain substances derived from plants. For instance, two substances effective against Hodgkin's disease and certain other forms of cancer come from the rosy periwinkle, a flowering plant native to the island of Madagascar (**Figure 38.2**). Madagascar alone harbors some 8,000 species of flowering plants, 80% of which occur only there. With an estimated 200,000 species of plants and animals, Madagascar is among the top five most biologically diverse countries in the world. Unfortunately, most of Madagascar's species are in serious trouble.

In the 2,000 years that humans have lived on the island, Madagascar has lost 80% of its forests and about 50% of its native species.

The loss of species means the loss of genes. The enormous genetic diversity of all the organisms on Earth has great potential benefit. Consider PCR (see Module 12.14), the DNA-replicating technology based on an enzyme extracted from thermophilic prokaryotes. Currently, biotechnology companies are searching for commercially useful enzymes in the prokaryotes found in the numerous hot springs in Yellowstone National Park. Many researchers and biotechnology leaders are enthusiastic about the potential that such "bioprospecting" holds for future development of new medicines, industrial chemicals, and other products.

Another reason to be concerned about the changes that underlie the biodiversity crisis is the threat of large-scale alterations in the biosphere to the human population itself. Like all species, we evolved in Earth's ecosystems and are dependent on their living and nonliving components.

In an attempt to counter what they see as a tendency of policymakers and governments to undervalue the biosphere's life-sustaining features, a team of ecologists estimated the economic value of "services" provided by ecosystems, such as the contribution of wetlands to reducing the severity of floods or the control of agricultural pests by natural enemies. Other ecosystem services include purification of air and water, decomposition of wastes, pollination of crops, nutrient cycling, and protection from UV rays, to name just a few. For the year 1997, these scientists estimated the average annual value of ecosystem services at 33 trillion U.S. dollars. In contrast, the global gross national product for the same year was $18 trillion. Although rough, these estimates make the important point that we cannot afford to take ecosystems for granted.

In the next module, we explore the human activities that pose the greatest threats to biodiversity.

Web/CD Activity 38A Connection: Madagascar and the Biodiversity Crisis

? What are two reasons to be concerned about the relationship of the biodiversity crisis to human welfare?

■ (1) The environmental degradation threatening other species may also harm us; (2) we are dependent on biodiversity, both directly through use of organisms and their products and indirectly through ecosystem services.

Figure 38.2 The rosy periwinkle (*Catharanthus roseus*), a source of anticancer drugs

38.3 Habitat destruction, introduced species, and overexploitation are the major threats to biodiversity

Human alteration of habitats poses the single greatest threat to biodiversity throughout the biosphere. Agriculture, urban development, forestry, mining, and environmental pollution have brought about massive destruction and fragmentation of habitats. Deforestation has been occurring at an alarming rate in tropical forests (see Figure 38.1A).

The amount of human-altered land surface is approaching 50%, and we use over half of all accessible surface fresh water. Some of the most productive aquatic habitats in estuaries and intertidal wetlands are also prime locations for commercial and residential development. The loss of marine habitat is severe, especially in coastal areas and coral reefs. According to the International Union for Conservation of Nature and Natural Resources, habitat destruction is implicated in the decline of 73% of the species in modern history that have become extinct, endangered, vulnerable, or rare.

Ranking second behind habitat loss as a cause of the biodiversity crisis are introduced (sometimes called exotic) species, which disrupt communities by competing with or preying on native species. Introduced species are imported in various ways. People inadvertently carry hitchhiking seeds or insects with them when they travel, and shipments of goods throughout the world have often included unintended passengers, such as the zebra mussels released in ballast water from a cargo ship in the Great Lakes. Many foreign plants and animals have been intentionally introduced for agricultural or ornamental purposes. If your campus is in an urban setting, there is a good chance that the birds you see most often are starlings (see Chapter 36 introduction), rock doves (often called "pigeons"), and house sparrows—all introduced species that have replaced native birds in many areas of North America. Another intentional introduction with disastrous effects is the kudzu vine, which was brought into the southern United States in the 1930s to help control erosion but now covers vast expanses of the landscape (Figure 38.3A). The United States has at least 50,000 introduced species, which have cost over $130 billion in damage and

Figure 38.3B North Atlantic bluefin tuna being auctioned in a Japanese fish market

control efforts. And that figure does not include the priceless loss of native species.

One of the largest rapid-extinction events yet recorded is the loss of freshwater fishes in Lake Victoria, in East Africa. About 200 species of native fishes found nowhere else but in this lake have been lost, mainly due to the introduction of the predatory Nile perch by Europeans in the 1960s. An unusually big freshwater fish (up to 2 m long), the Nile perch was introduced to provide high-protein food for the growing human population. Unfortunately, the perch's main effect has been to wipe out the smaller native species, reducing its own food supply in the process.

The third major threat to biodiversity is overexploitation of wildlife by harvesting at rates exceeding the ability of populations to rebound. Such overharvesting has threatened some rare trees that produce valuable wood. Animal species whose numbers have been drastically reduced by excessive commercial harvest, poaching, or sport hunting include tigers, whales, the American bison, Galápagos tortoises, and numerous fishes. Many fish stocks in the ocean have been reduced to levels that cannot sustain further exploitation. The North Atlantic bluefin tuna is one example. In the 1980s, wholesalers began airfreighting fresh, iced bluefin to Japan for sushi and sashimi. The fish, which used to bring just a few cents per pound for cat food, now brings up to $100 per pound (Figure 38.3B). With that kind of demand, it took just ten years to reduce the bluefin population to less than 20% of its 1980 size. An expanding, often illegal world trade in wildlife products (such as rhinoceros horns, elephant tusks, and grizzly bear gallbladders) also threatens many species.

Web/CD Activity 38B *Connection: Fire Ants as an Exotic Species*

? What is an introduced species?

Figure 38.3A Kudzu in South Carolina

■ A species that has been accidentally or purposefully transferred from one location to another, where it did not occur naturally

38.4 Pollution of the environment compounds our impact on other species

Another way in which the human population contributes to the biodiversity crisis is through the release of pollutants into the environment. These substances can have local, regional, and global effects. Some pollutants, such as oil spills, contaminate local areas. Pollutants emitted into the atmosphere, however, may be carried aloft and foul the air thousands of miles away. Acid precipitation, for instance, can harm lakes and forests far from where the sulfur and nitrogen pollutants are released (see Module 2.16).

As you learned in Module 7.14, the **ozone layer** in the upper atmosphere protects Earth from the harmful ultraviolet rays in sunlight. Measurements by atmospheric scientists document that the ozone layer has been gradually thinning since 1975, probably as a result of the accumulation of chlorofluorocarbons (CFCs), chemicals used as refrigerants, as propellants in aerosol cans, and in certain manufacturing processes. The consequences of ozone depletion for life on Earth may be quite severe, not only increasing skin cancers and cataracts among humans, but harming crops and natural communities, especially the phytoplankton that are responsible for a large proportion of Earth's primary production.

In Module 37.21, we discussed the accelerated eutrophication of lakes and streams caused by nutrient pollution and algal blooms. In a far-reaching example of this problem, nitrogen runoff from farm fields as far north as Minnesota and Montana has been linked to a large, recurring "dead zone" in the Gulf of Mexico. Extending outward from where the Mississippi River deposits its nutrient-laden waters, these deep, oxygen-starved waters disrupt benthic communities, displacing fish and invertebrates that can move and killing those left behind. More than 150 recurring and permanent coastal dead zones have been documented in seas worldwide.

As we discussed in Module 34.3, chemical pesticides have helped us grow more food and fight infectious diseases. Many of these toxins, however, persist in the environment and are transported to areas far from where they are applied. For instance, in the 1960s, researchers began finding traces of DDT in marine mammals in the Arctic, far from any places DDT had been used. The chemical had been transported and concentrated as it passed through food webs. This concentration, or **biological magnification**, occurs because the biomass at any given trophic level is produced from a much larger toxin-containing biomass ingested from the level below (Module 37.13). Thus, top-level predators are usually the organisms most severely damaged by toxic compounds that have been released into the environment. In the Great Lakes food chain shown in **Figure 38.4**, the concentration of industrial chemicals called PCBs measured in herring gull eggs was almost 5,000 times higher than that measured in phytoplankton. The concentration increased at each successive trophic level. Current research implicates many of the chlorinated hydrocarbons such as DDT and PCBs in endocrine disruption in a large number of animal species, including humans.

Human activities produce an immense variety of chemicals, including thousands of synthetics previously unknown in nature. Many of these chemicals cannot be degraded by microorganisms and consequently persist in the environment. In other cases, chemicals may be converted to more toxic products by reaction with other substances or by the metabolism of microorganisms. For example, mercury, a by-product of plastic production and coal-fired power generation, was once routinely expelled into rivers and the sea. Bacteria in the bottom mud converted the waste to methyl mercury, an extremely toxic soluble compound that then accumulated in the tissues of organisms, including humans who consumed fish from the contaminated waters.

Perhaps no alteration of the biosphere threatens to cause more harm than global warming, which we explore next.

Web/CD Activity 38C Connection: DDT and the Environment

? How does accelerated eutrophication lead to the development of coastal dead zones?

■ Excess nutrients produce algal blooms. As these algae die, vast numbers of bacteria use up the O_2 from the bottom water as they decompose the algae.

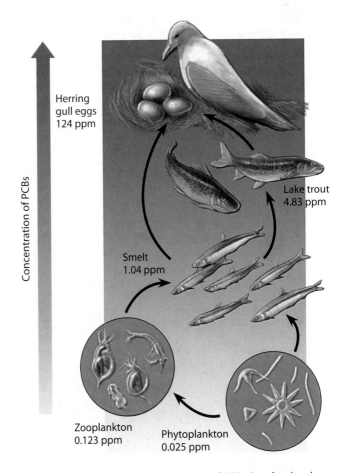

Figure 38.4 Biological magnification of PCBs in a food web

Herring gull eggs 124 ppm

Lake trout 4.83 ppm

Smelt 1.04 ppm

Concentration of PCBs

Zooplankton 0.123 ppm

Phytoplankton 0.025 ppm

38.5 Rapid global warming could alter the entire biosphere

The contribution of so-called greenhouse gases in the atmosphere to the warming of Earth has been understood for some time (see Module 7.13). The potential for global climate change as a result of increases in these gases has taken longer to emerge. But as more scientists agree on the severity of the problem of **global warming**, almost everyone sees the need to take steps to reduce climate change and its potential to worsen the biodiversity crisis.

Fossil fuels power most of our industry, food production, homes, and transportation. Since the Industrial Revolution, the concentration of CO_2 in the atmosphere has been increasing as a result of the combustion of fossil fuels and the burning of enormous quantities of wood from deforestation. Various measurement methods have estimated that the average CO_2 concentration in the atmosphere before 1850 was about 274 ppm. Today, it is more than 370 ppm. The CO_2 concentration has increased about 17% in the 45 years since a monitoring station in Hawaii began making very accurate measurements. The black line in **Figure 38.5A** shows this increase. Its zigzag shape is due to seasonal changes in CO_2 concentration during the summer growing season. The red line plots the warming trend in temperatures during this same period of measured increases in CO_2 concentration. Climatologists have also found a positive correlation between CO_2 levels and temperature in prehistoric times.

Carbon dioxide is one of several greenhouse gases—molecules that absorb heat and slow its escape from Earth. The effect of CO_2 on global warming is illustrated in **Figure 38.5B**, on the next page. Other greenhouse gases include methane

and nitrous oxide, both of which are also increasing as a result of fossil fuel consumption, industry, and agriculture.

As a natural phenomenon, the so-called greenhouse effect is essential for virtually all life on Earth; without it, the average air temperature at Earth's surface would be about $-18°C$. But the CO_2 that has flooded the atmosphere in the industrial age is turning this natural process from a protection to a danger. A number of studies predict that by the end of the 21st century, atmospheric CO_2 concentration will double, and average global temperature will increase by about 2°C.

What are the possible consequences of global warming? An increase of less than 2°C would melt polar ice and raise sea levels significantly. Scientists, using aerial photography, have already recorded noticeable reductions in polar ice. Over time, heavily populated coastal areas could be flooded, leaving cities such as New York and Los Angeles under water. A warming trend would also likely alter patterns of global rainfall and plant life. The grain belts of central United States, for example, might become much drier and unable to produce the wheat and corn now grown there.

Vast amounts of habitat, from tundra to tropical forest, would be altered by global warming, putting many species at risk. A growing body of research has already found numerous species struggling to adapt to Earth's rising temperature. Along Canada's Hudson Bay, careful long-term studies of polar bears have found these animals losing weight and body fat. These massive, meat-eating bears (**Figure 38.5C**) stalk their prey on ice and need to store up body fat for the warmer months, when ice melts and leaves them without a hunting ground. With warmer temperatures shortening the hunting season,

bears go longer without food. Biologists have found many bears with lower weight and less body fat than bears had decades ago, and some are starving.

Studies have documented effects on numerous other species. More than 20 species of butterflies in the northern hemisphere, from the orange-and-black dappled Edith's Checkerspot in California to the tawny Sooty Copper in Europe, have shifted their range northward as the climate warms. Most plants, however, disperse so slowly that they may not be able to migrate to new ranges in time to keep up with temperature changes.

The effects of a warmer Earth on human health are also beginning to be felt. The deadly European heat wave of 2003 claimed more than 30,000 lives. Climate projections suggest that such heat waves will become

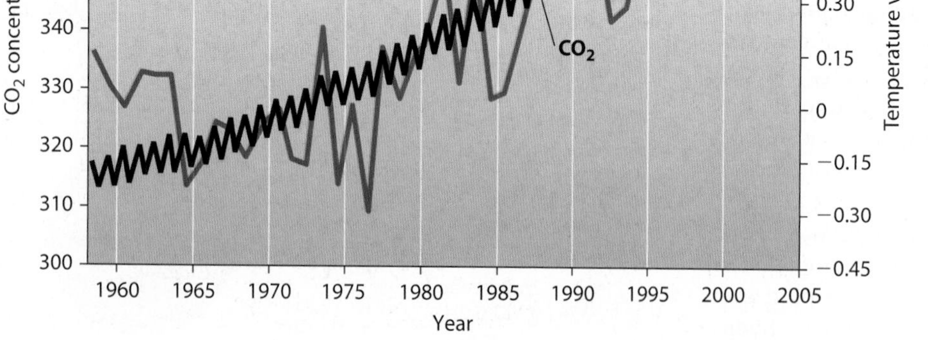

Figure 38.5A The increase of atmospheric CO_2 and temperature variation at Mauna Loa, Hawaii, since 1958

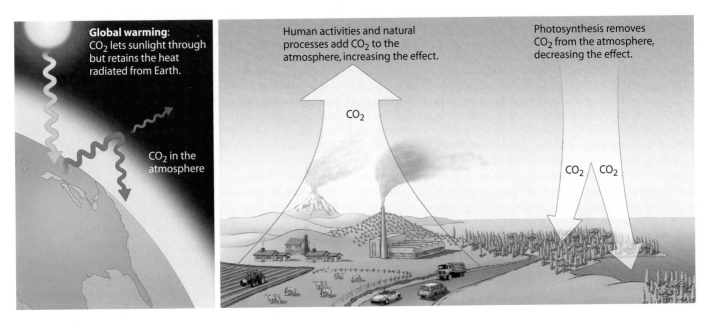

Figure 38.5B Global warming and factors influencing it

more common. What's more, tropical diseases may broaden their ranges into temperate regions. Also, less productive farms could increase food prices and decrease supply.

At present, the growing research on global warming has led most scientists to acknowledge the rapid onset of climate change. There is far less agreement, however, on what a warmer Earth will mean. The implications of global warming are now some of the largest and most contentious areas of scientific research. Some researchers have called for drastic cuts in the use of fossil fuels to slow temperature increases. Other researchers, business leaders, and government officials say that the effects of global warming are not yet known. Skeptics point out that Earth has a long history of severe climate shifts, none of which was caused by humans. Others raise the possibility that increasing smoke and cloud cover from fossil fuel combustion could actually *decrease* warming by reducing the amount of solar heat that reaches Earth.

Despite these debates, however, a majority of scientists and world leaders have concluded that immediate steps to slow the warming trend are needed to prevent catastrophic environmental change. The steps likely to reduce the chances of a climate-change disaster include everything from individual efforts to global cooperation. More than 189 countries have signed an international agreement known as the Kyoto Protocol that intends to reduce greenhouse gas emissions worldwide. The United States, the world's largest producer of greenhouse gases, had not accepted the agreement as of 2004, mostly because of opposition from business leaders and global warming skeptics. Other large industrial nations, however, are already rethinking their own economies to help fend off global warming. China, an economic giant with a huge appetite for cars and other fossil fuel–powered machines, is working to create new power sources and transit systems that create less carbon dioxide. A few auto makers are offering new types of "hybrid" gas-electric cars that use less fossil fuel and release fewer harmful emissions.

These cars have been snapped up by U.S. consumers, a sign that the public welcomes a shift away from fossil fuels.

But such actions by governments and industry are not enough to solve the problem. Fighting global warming will require billions of people around the world to make major lifestyle changes. Some measures may be less convenient (such as changing our driving patterns); others may be more costly (such as building more public transit systems). Failure to act, however, may have far more serious consequences for our health and economic welfare. And the costs in species loss, as warmer temperatures worsen the biodiversity crisis, are immeasurable.

Web/CD Activity 38D *Connection: The Greenhouse Effect*

? How has the deforestation of large areas contributed to an increase in atmospheric CO_2?

■ The burning of enormous quantities of wood adds CO_2 to the atmosphere, and the loss of vegetation reduces the amount of CO_2 removed by photosynthesis.

Figure 38.5C A polar bear, whose hunting season has been shortened by warmer temperatures

38.6 Two ways to study endangered populations are the small-population approach and the declining-population approach

Now that we have outlined some of the causes underlying the current decline in biodiversity, let's consider the ways in which conservation biologists attempt to protect the often small and fragmented populations of endangered or threatened species. Severe population fragmentation, the splitting and consequent isolation of portions of populations by habitat degradation, is one of the most harmful effects of habitat loss due to human activities.

The aerial photograph in **Figure 38.6A** illustrates fragmentation of a coniferous forest ecosystem in the Mount Hood National Forest in northwestern Oregon. The forest was originally continuous. The open, snow-covered areas in the photo were logged, creating forest fragments, some of which are islands within clear-cut areas. This kind of habitat alteration has reduced and fragmented populations of many species. One of the most controversial endangered species, the northern spotted owl, inhabits coniferous forests of the U.S. Pacific Northwest (**Figure 38.6B**). Owl populations were fragmented and declined markedly when these forests were logged.

In studying endangered populations, some conservation biologists adopt a *small-population approach*, which focuses on the problem of smallness itself. These biologists are concerned with the processes that ultimately drive a population to extinction after such factors as habitat loss have taken their toll on population size. A small population may enter an "extinction vortex"—a downward spiral toward smaller and smaller population size, leading to extinction. The key factor driving the extinction vortex is the loss, due to inbreeding and genetic drift, of the genetic variation on which populations depend for adaptive evolutionary responses to environmental change (see Module 13.9).

How small does a population have to be before it starts down the extinction vortex? The answer depends on the type of organism and must be evaluated case by case. Ecologists use various factors and computer models to estimate the minimum size a population needs to be to remain viable. According to the small-population approach, conservation measures should focus on maintaining sufficient habitat to support the *minimum viable population size* and, when possible, on importing individuals from other populations to increase the genetic variation of small, isolated populations.

Another approach to understanding the biology of extinction is the *declining-population approach*. This is a proactive conservation strategy for detecting, diagnosing, and halting population declines, even if the population is still far greater than its minimum viable size. The declining-population approach requires that researchers carefully dissect the causes of a decline before recommending or trying corrective measures. This approach often involves a series of logical steps: (1) Confirm that the species is presently in decline. (2) Study the species' natural history to determine its environmental requirements. (3) Develop hypotheses for all the possible causes of the decline, and list the predictions of each hypothesis. (4) Test the most likely hypothesis first, designing an experiment to determine if this factor is the main cause of the decline. In the ideal experiment, researchers remove the suspected agent of decline to see if the experimental population rebounds relative to a control population. (5) Apply the results of the diagnosis to the management of the threatened

Figure 38.6A Fragmentation of a forest ecosystem

Figure 38.6B
The northern spotted owl (*Strix occidentalis caurina*)

species. This requires monitoring recovery until the problem of decline is resolved. The next module presents an example of diagnosing and treating the decline of an endangered species, the red-cockaded woodpecker.

? How do the small-population and declining-population approaches to endangered species differ?

■ The first approach emphasizes a population's smallness and lack of genetic diversity as a cause of extinction; the second approach focuses on the environmental factors that are causing the population to decline.

38.7 Identifying critical habitat factors can guide conservation efforts

Identifying the specific combination of habitat factors that is critical for a species is pivotal in conservation biology. Figures 38.7A–38.7C illustrate several critical factors for the red-cockaded woodpecker (*Picoides borealis*), an endangered species originally found throughout the southeastern United States. This species requires mature pine forests, preferably ones dominated by the longleaf pine. Most woodpeckers nest in dead trees, but the red-cockaded woodpecker drills its nest holes in mature, living pine trees (**Figure 38.7A**). It also drills small holes around the nest hole entrance, making resin from the tree ooze down the trunk. The resin seems to repel some predators, such as corn snakes, which eat bird eggs and nestlings (see Chapter 23 introduction).

Figure 38.7B shows another critical factor for this woodpecker—low growth of plants among the mature pine trees. Historically, periodic fires swept through longleaf pine forests, keeping the undergrowth low. Fire suppression in more recent times, however, has altered the forest composition. Breeding birds tend to abandon nests when vegetation among the pines is thick and higher than about 4.5 meters. Apparently, the birds require a clear flight path between their home trees and the neighboring feeding grounds.

Leading to the decline of the red-cockaded woodpecker is the destruction or fragmentation of suitable habitats by logging and agriculture. The recent recovery of this woodpecker species from near-extinction to sustainable populations is largely due to recognizing and providing its key habitat factors: the protection of some longleaf pine forests and the use of controlled fires to reduce forest undergrowth (**Figure 38.7C**). Researchers also found that creating nest cavities in pine trees in unoccupied areas of suitable habitat helped young birds disperse to new territories.

Management aimed at conserving a single species carries with it the possibility of negatively affecting populations of other species. To test how management of pine forests for the red-cockaded wookpecker might impact migratory birds, ecologists compared bird communities in managed and unmanaged forests. They found that managed sites supported higher numbers and a higher diversity of birds than did control forests. In this case, managing for one species enhanced the diversity of an entire community of birds.

Determining habitat needs is only one aspect in the effort to save species. It is usually necessary to weigh other conflicting demands. For example, an ongoing, sometimes bitter debate in the U.S. Pacific Northwest pits saving habitat for populations of the timber wolf, grizzly bear, and bull trout against demands for jobs in timber, mining, and agriculture. Programs to restock wolves and to bolster populations of grizzly bears and other large carnivores are opposed by some recreationists concerned about safety and by many ranchers concerned with potential losses of livestock. Similar problems confront decision makers around the globe.

Because we will not be able to save every endangered species, we must determine which are most important for conserving biodiversity as a whole. For instance, identifying and protecting keystone species (Module 37.4) may help preserve whole communities. In many situations, conservation must look beyond individual species to communities.

? In what way does the recovery of the red-cockaded woodpecker illustrate the declining-population approach to conservation?

■ Researchers identified and then provided the key habitat factors of mature pine forests with low understory. They also created nesting cavities in unoccupied suitable habitat.

Figure 38.7A The red-cockaded woodpecker at its nest site

Figure 38.7B Forest habitat that can sustain red-cockaded woodpeckers

Figure 38.7C A controlled burn to restore the woodpeckers' habitat

38.8 Sustaining ecosystems and landscapes is a conservation priority

Most conservation efforts in the past have focused on saving individual species. But increasingly, conservation biology aims to sustain the biodiversity of entire ecosystems and landscapes. Ecologically, a landscape is a regional assemblage of interacting ecosystems, such as a forest, adjacent fields, wetlands, streams, and streamside habitats. **Landscape ecology** is the application of ecological principles to the study of the structure and dynamics of a collection of ecosystems. One goal of landscape ecology is to study human land-use patterns in the past, present, and foreseeable future and to make biodiversity conservation a priority.

Edges, or boundaries, between ecosystems are prominent features of landscapes. The photograph in **Figure 38.8A** shows a landscape area in Yellowstone National Park that includes grassland and forest. Human activities, such as logging and road building, often create edges that are more abrupt than those delineating natural landscapes.

Edges have their own sets of physical conditions and thus their own communities of organisms. Some organisms thrive in edges because they require resources of the two adjacent areas. For instance, whitetail deer browse on woody shrubs found in edge areas between woods and fields, and their populations often expand when forests are logged or interrupted with housing developments.

Communities where human activities have generated many edges often have less diversity and are dominated by a few species that are adapted to edges. In one example, populations of the brown-headed cowbird (**Figure 38.8B**), an edge-adapted species that lays its eggs in the nests of other birds, are currently expanding in many areas of North America. Cowbirds forage in open fields on insects disturbed by or attracted to cattle and other large herbivores; the cowbirds also need forests, where they can parasitize the nests of other birds. Increasing cowbird parasitism and loss of habitats are correlated with declining populations of several songbird species.

Where habitats have been severely fragmented, a **movement corridor**, a narrow strip or series of small clumps of high-quality habitat connecting otherwise isolated patches, can be a deciding factor in conserving biodiversity. Streamside habitats often serve as corridors, and government policy in some nations prohibits destruction of these areas. In areas of heavy human use, artificial corridors are sometimes constructed. In many areas, bridges or tunnels have reduced the number of animals killed as they try to cross highways (**Figure 38.8C**).

Figure 38.8B A male brown-headed cowbird (*Molothrus ater*)

Corridors can also promote dispersal and reduce inbreeding in declining populations. Corridors are especially important to species that migrate between different habitats seasonally. In some European countries, amphibian tunnels have been constructed to help frogs, toads, and salamanders cross roads to access their breeding territories. On the other hand, a corridor can be harmful—as, for example, in the spread of diseases, especially among small subpopulations in closely situated habitat patches. The effects of movement corridors between habitats in a landscape are not yet well understood, and researchers continue to study them.

? How can "living on the edge" be a good thing for some species, such as whitetail deer and cowbirds?

■ Because they use a combination of resources from the two ecosystems on either side of the edge

Figure 38.8A A landscape with distinct edges

Figure 38.8C Animal bridge in Banff National Park, Canada

Protected areas are established to slow the loss of biodiversity

Conservation biologists are applying their understanding of population, community, ecosystem, and landscape dynamics in establishing parks, wilderness areas, and other legally protected nature reserves. Choosing locations for protection often focuses on **biodiversity hot spots.** These relatively small areas have a large number of endangered and threatened species and an exceptional concentration of **endemic species**, those that are found nowhere else. Together, the "hottest" of Earth's biodiversity hot spots, shown in **Figure 38.9A**, total less than 1.5% of Earth's land but are home to a third of all species of plants and vertebrates. There are also hot spots in aquatic ecosystems, such as certain river systems and coral reefs.

Because endemic species are limited to specific areas, they are highly sensitive to habitat degradation. At the current rate of human development, some biologists estimate that loss of habitat will cause the extinction of about half of the species in terrestrial biodiversity hot spots in the next 10 to 15 years. Thus, biodiversity hot spots can also be hot spots of extinction. They rank high on the list of areas demanding strong global conservation efforts.

Concentrations of species provide an opportunity to protect many species in very limited areas. However, species endangerment is a truly global problem, and focusing on hot spots should not detract from efforts to conserve habitats and species diversity in other areas.

Another example of the uneven distribution of species is the localized concentration of species that migrate seasonally. As you learned in Module 35.8, monarch butterflies occupy much of the United States and Canada during the summer months, but migrate in the autumn to specific local sites in Mexico and California, where they congregate in huge numbers. Overwintering populations are particularly susceptible to habitat disturbances because they are concentrated in small areas. Thus, habitat preservation in Canada, the United States, or Mexico alone will not remove the threats to the monarch, and the situation is similar for many species of migratory songbirds, waterfowl, marine mammals, and sea turtles.

Sea turtles, such as the loggerhead turtle (**Figure 38.9B**), are threatened in their ocean feeding grounds and on land. Loggerheads take about 20 years to reach sexual maturity, and great numbers of juveniles and adults are drowned at sea when caught in fishing nets. The adults mate at sea, and the females migrate to specific sites on sandy beaches to lay their eggs. Buried in shallow depressions, the eggs are susceptible to predators, especially raccoons. And many egg-laying sites have become housing developments and beachside resorts. An ongoing international effort to conserve sea turtles focuses on protecting egg-laying sites and minimizing the death rate of adults and juveniles at sea.

Currently, governments have set aside about 7% of the world's land in various forms of reserves. One major conservation question is whether it is better to create one large reserve or a group of smaller ones. Far-ranging animals with

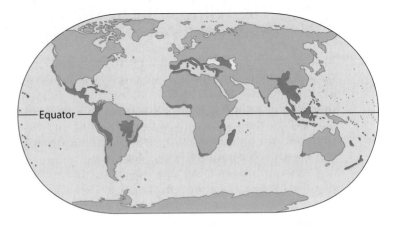

Figure 38.9.A Earth's terrestrial biodiversity hot spots (purple)

low-density populations, such as tigers, require extensive habitats. As conservation biologists learn more about the requirements for achieving minimum viable population sizes for endangered species, it is becoming clear that most national parks and other reserves are far too small. Given political and economic realities, it is unlikely that many existing parks will be enlarged, and most new reserves will also be too small. In Module 38.13, we will look at one approach to this problem, called a zoned reserve, that sets up buffer zones of private and public land surrounding a reserve. Another large-scale effort in North America is attempting to link existing reserves. We discuss this Yellowstone to Yukon Conservation Initiative next.

? **What is a biodiversity hot spot?**

■ A relatively small area with a disproportionate number of species, many of which are endemic

Figure 38.9B An adult loggerhead turtle (*Caretta caretta*) swimming in the Caribbean Sea

38.10 The Yellowstone to Yukon Conservation Initiative seeks to preserve biodiversity by connecting protected areas

If many existing reserves are too small to sustain a large number of threatened species, how can biologists draw the land around reserves into conservation efforts? In North America, one ambitious biodiversity plan is creating innovative ways to give creatures more room, building on lessons from research on a howling predator that once roamed a vast stretch of the northern Rocky Mountains.

This was a single animal, a gray wolf dubbed Pluie that was captured by scientists in western Canada in 1991. The biologists fitted the five-year-old female with a radio tracking collar and released her—routine work in studies of threatened animals. But the scientists were stunned by what they learned from Pluie. Over the next two years, this wolf roamed over an area more than 100,000 km² (38,600 mi²) in size, crossing between Canada and the United States and traveling between protected reserves and lands where she was fair game for killing. In 1995, her story came to a bloody end. While moving through lands outside the boundary of a nearby national park, Pluie, her mate, and one of her pups were shot (legally) by a hunter.

Biologists who had studied Pluie realized that the wolf's life captured all the promise—and all the pitfalls—of efforts to protect her. She had thrived for years within the sporadic shelter of parks and other protected territory. But such lands were never big enough to hold her. Like others of her species (*Canis lupus*), Pluie needed more room. Reserves could shield animals briefly, the scientists realized. True protection would have to include paths of safe passage between reserves.

This conclusion inspired the creation of the Yellowstone to Yukon Conservation Initiative (Y2Y), one of the world's most ambitious conservation biology efforts. The initiative aims at nothing less than preserving the web of life that has long defined the Rocky Mountains of Canada and the northern United States. This area is dotted with famous parks—Canada's Banff National Park and Wyoming's Yellowstone National Park among them—but scientists behind Y2Y now say that those areas alone cannot protect native species from human threats.

Y2Y seeks to knit together a string of parks and reserves, creating a vast 3,200-km wildlife corridor stretching down from Alaska across Canada to northern Wyoming (Figure 38.10A). The idea is not to create one giant park, but rather to connect parks with protected corridors where wildlife can travel safely.

Many of the signature species that live in this vast region, such as grizzly bears (Figure 38.10B), lynx, moose, and elk, don't confine themselves to human boundaries. But few have as great a range as the wolf. If Y2Y can provide safe passage for gray wolves, it will have also created secure zones for other animals in the Rockies.

Gray wolves once roamed all of North America. These carnivorous hunters live in packs that protect pups and search cooperatively for food. A pack may have a territory of about 130 km² or range much farther to find prey. The wolf's hunting prowess kept it the top predator of Northern American ecosystems as long as the human population was small. Things changed when large numbers of people migrated from Europe and pushed into the continent.

Deeming wolves a dangerous predator and competitor that threatened people and livestock, settlers in the United States launched widespread campaigns to wipe out wolves. By the early 20th century, gray wolves were nearly extinct in the lower 48 states, with only a few hundred surviving in northern Minnesota. More managed to stay alive in the wilds of less populated western Canada and Alaska.

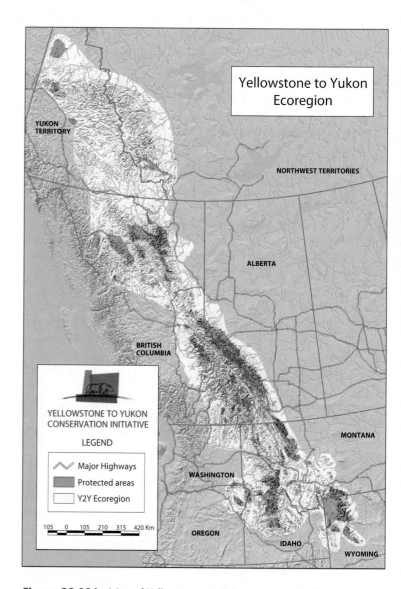

Figure 38.10A Map of Yellowstone to Yukon ecoregion with protected areas shown in green

Figure 38.10B Grizzly bear with cubs in Yellowstone Park

Figure 38.10C Gray wolf

Scientists gradually realized that widespread damage rippled through habitats after wolves had been removed. Without a predator to control their numbers, populations of elk and deer grew unchecked. As these increasing numbers of herbivores foraged for food, vegetation that sheltered smaller animals was damaged. Other animals, such as ravens and foxes, had once fed on the carcasses of wildlife killed by wolves and were now left without an important source of food. Wolves, biologists determined, were a keystone species—a species critical to the balance and maintenance of an ecosystem.

That understanding led to one of the most important and controversial conservation biology efforts in the Yellowstone to Yukon area. In 1991, the U.S. Fish and Wildlife Service launched a campaign to bring wolves back to Yellowstone National Park, a reserve that hadn't sheltered the animals in at least 50 years. After careful planning, which included compensation for angry ranchers who feared losing their cattle and sheep, about 60 wolves from Canada were released in the park in 1995 and 1996.

As wolf howls once again echoed through the Wyoming darkness, the wolf quickly became a hopeful symbol for Y2Y backers (Figure 38.10C). Yellowstone's wolves formed new packs and raised pups. By 2004, scientists counted 12 wolf packs inside the park, totaling about 300 wolves. And the wolf's return has brought more than howling. Park officials noted significant environmental improvements as wolves once again roamed Yellowstone. As wolves killed elk, moose, and deer, streambeds and other lands near waterways started to shelter more plants and animals. Fewer hoofed animals meant more grasses and taller trees. Those plants brought more birds, along with more water-dwelling beaver. In all, park biologists report that the wolf's return has affected at least 25 different species.

True to their nature, Yellowstone's wolves haven't followed human borders; six packs have been found just outside the park. Meanwhile, the migrations of Canadian wolves, along with smaller release programs, have brought the animals back to Idaho and Montana. In June 2004, a Yellowstone wolf was found hundreds of kilometers away in Colorado—a reminder that travels like Pluie's are common to wolves throughout the Y2Y region.

Such successes also bring risks—and reminders from scientists that wolves need safe corridors. The Colorado wolf was discovered dead by the side of a highway, most likely the victim of a car. Its appearance sparked angry protests from ranchers in the state, who said that any new wolves that appear should either be shot or shipped back to Yellowstone. Meanwhile, wildlife advocates maintain that wolves should be allowed to migrate naturally.

That argument reflects a broader debate about how to treat wolves as they return to their old ranges. Federal laws have protected them in the northern United States, but as wolves thrive, federal officials want to remove those protections. They have asked states where wolves are found to submit management plans—and then have delayed since some plans seemed designed to kill off wolves all over again. As of mid-2004, agreement between the two sides appears far off.

The biologists involved in the Yellowstone to Yukon Conservation Initiative are studying wildlife population dynamics on a landscape scale to help support regional conservation planning. The initiative is backing a range of research projects to determine the requirements for maintaining terrestrial and aquatic ecosystems in the Y2Y region. Their efforts to connect habitats include wildlife bridges, such as the one in Banff National Park (Figure 38.8C). This cross-border initiative is also providing broader lessons for global conservation. In Myanmar, for example, plans for the tiger reserve discussed in the chapter introduction include proposals to link the Myanmar reserve with others in neighboring countries.

In addition to creating reserves to protect species and their habitats from human disruptions, conservation efforts also attempt to restore ecosystems degraded by human activities. We look at the field of restoration ecology next.

? In what ways are wolves considered a keystone species?

■ Wolves regulate the populations of their prey, preventing these herbivores from damaging vegetation and degrading habitat for other members of the community; Wolf kills also provide food for other animals.

38.11 The study of how to restore degraded habitats is a developing science

Eventually, some areas that we alter and degrade are abandoned. For instance, mining activities may go on for several decades, but afterwards the lands may be abandoned in a degraded state. And many ecosystems are damaged by industrial practices such as dumping of toxic chemicals or by such mishaps as oil spills. A new and expanding field, called **restoration ecology**, uses ecological principles to develop ways to return degraded ecosystems to conditions as similar as possible to their natural, predegraded state.

One of the major strategies in restoration ecology is **bioremediation**, the use of living organisms—usually prokaryotes, fungi, or plants—to detoxify polluted ecosystems. For instance, the bacterium *Pseudomonas* has been used with some success to clean up oil spills on beaches (see Module 16.16). More common still is the use of certain prokaryotes to metabolize toxins in dump sites.

Some other promising uses of bioremediation involve the potential use of lichens and plants to concentrate mining wastes. Researchers in the United Kingdom discovered a lichen species that grows on soil polluted with uranium dust left over from mining. The lichen actually concentrates the uranium in a dark pigment similar to the melanin in human skin, as seen in the dark brown structures in the photograph in **Figure 38.11A**. Other researchers are experimenting with plants capable of extracting potentially toxic metals such as zinc, nickel, lead, and cadmium from contaminated soil (see Chapter 32 introduction). Prokaryotes, lichens, and plants may be useful in restoring degraded sites, and studies indicate that some species can concentrate metals in commercially marketable quantities.

In contrast to bioremediation, which is a strategy for *removing* harmful substances, **biological augmentation** uses organisms to *add* essential materials to a degraded ecosystem. A researcher must first find out what factors, such as chemical nutrients, have been removed from an area and are limiting its recovery rate. For instance, the soils of many tropical areas become nutrient-deficient and unproductive less than five years after being cleared for farming. Encouraging the growth of plants that thrive on nutrient-poor soils can hasten the rate of recovery of some tropical areas. In the photograph in **Figure 38.11B**, Dr. Ariel Lugo of the U.S. Forest Service stands next to an introduced plant, a legume called *Albizzia* that has colonized roadsides and deforested sites with nitrogen-poor soil in Puerto Rico (see Module 34.18). Apparently, the rapid buildup of organic material from dense stands of *Albizzia* helps set the stage for recolonization by native species, which can then overgrow the exotic plant.

Because of the complexity of ecosystems and the unique features of each situation, restoration ecologists usually must learn as they go. Many advocate experimenting with several promising types of management to find out what works best. The key seems to be to consider alternative plans and learn from mistakes as well as successes.

A great number of restoration projects are in progress around the world. In Japan, attempts to restore coastal seagrass beds reduced by urban development include construc-

Figure 38.11A Metal-concentrating lichens (grayish) growing on uranium mineral (green)

Figure 38.11B Dr. Ariel Lugo with an *Albizzia* plant

tion of seafloor habitat, transplantation from natural beds using artificial substrates, and even underwater hand seeding. Another project is restoring tropical dry forest in Costa Rica by feeding fruits of native trees to domestic livestock and using them to disperse the seeds to open grasslands. This project is a model for joining restoration ecology and the local economy. In the next module, we describe a particularly large and ambitious wetland restoration project.

Web/CD Thinking as a Scientist *Connection: How Are Potential Restoration Sites Analyzed?*

? Contrast the use of organisms in bioremediation with augmentation for improving a degraded ecosystem.

■ In bioremediation, certain organisms are used to remove harmful chemicals from the environment; in augmentation, certain organisms are used to add essential chemicals to the environment.

CONNECTION

38.12 The Kissimmee River project is a case study in restoration ecology

The Kissimmee River in south central Florida was once a meandering shallow river that wound its way through diverse wetlands from Lake Kissimmee southward into Lake Okeechobee. The periodic flooding of the river covered a 1.5–3.0 km wide floodplain during about half of the year, creating wetlands that provided critical habitat for vast numbers of birds, fishes, and invertebrates. And as the flood pulse deposited silt on the floodplain, it improved nutrient cycling and maintained the water quality of the river.

In the 1940s, a rapidly growing population and development of the area created pressure for flood control. Between 1962 and 1971, the U.S. Army Corps of Engineers converted the 166-km wandering river into a straight canal 9 m deep, 100 m wide, and 90 km long. This project drained approximately 31,000 acres of wetlands, with significant negative impacts on fish and wetland bird populations. Spawning and foraging habitats for fishes were eliminated, and important sport fishes, such as largemouth bass, were replaced by nongame species more tolerant of the lower O_2 concentration in the deeper canal. The populations of waterfowl declined by 92%, and the number of bald eagle nesting territories decreased by 70%. Without the marshes to help filter and reduce agricultural runoff, phosphorus and other excess nutrients were transported through Lake Okeechobee into the Everglades ecosystem to the south.

As these negative ecological effects began to be recognized, public pressure grew to restore the river. In 1992, Congress authorized the Kissimmee River Restoration Project, one of the largest landscape restoration projects and ecological experiments in the world. Scheduled for completion in 2012, the project will refill about 35 km of the artificial canal and restore flow to 75 km of the original meandering channels. The goal is to reclaim 27,000 acres of wetlands. Numerous research efforts will monitor and evaluate the responses of the populations, communities, and ecosystems of the Kissimmee River landscape.

The first phase of the project was finished in 2004. **Figure 38.12A** shows a section of the Kissimmee canal that has been plugged, diverting flow into the remnant channels in the center of the photo. Birds and ducks are returning in unexpected numbers to the 11,000 acres of wetlands that have been restored. The marshes have filled with native vegetation, and game fishes again swim in the river channels.

Former canal

Figure 38.12A Restoring the natural water flow patterns in the Kissimmee River

Economic considerations are part of all conservation and restoration projects. The cost of the Kissimmee River Restoration Project is estimated at $578 million. The project is expected to improve water quality in the surrounding areas and to benefit over 320 fish and wildlife species, including several endangered species, such as the wood stork shown in **Figure 38.12B**. Increased recreational usage (hunting and fishing) and ecotourism are expected to contribute to local and regional economies.

In the next module, we look at how the economic welfare of the local community is part of the zoned reserve approach to biodiversity preservation and restoration.

Figure 38.12B Wood stork feeding in a marsh

? How will the Kissimmee River Restoration Project improve water quality in the Everglades ecosystem?

■ The wetlands filter agricultural runoff and prevent excess nutrients from entering the Everglades.

Zoned reserves are an attempt to reverse ecosystem disruption

Today, few, if any, ecosystems remain unaltered by human activities. Accelerated eutrophication reduces species diversity in lakes and rivers because many organisms cannot tolerate the rapid changes in water quality. Large tracts of forest in taiga biomes are still being clear-cut in the United States, Canada, and Siberia to satisfy demands for lumber and urban development. The current rate of conversion of tropical forests to farmland threatens the survival of thousands of species and may contribute to global climate change.

In an attempt to slow the disruption of ecosystems, a number of countries are setting up what they call zoned reserves. A **zoned reserve** is an extensive region of land that includes one or more areas undisturbed by humans. The lands surrounding these areas continue to be used to support the human population, but they are protected from extensive alteration. As a result, they serve as a buffer zone, or shield, against further intrusion into the undisturbed areas.

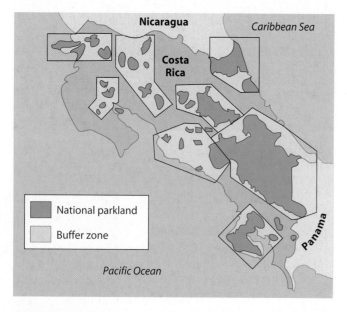

Figure 38.13A Zoned reserves in Costa Rica

Figure 38.13B Local schoolchildren marvel at the diversity of life in one of Costa Rica's reserves

A primary goal of the zoned reserve approach is to develop a social and economic climate in the buffer zone that is compatible with the long-term viability of the protected area.

The small Central American nation of Costa Rica has become a world leader in establishing zoned reserves. In exchange for a reduction in its international debt, the Costa Rican government established eight zoned reserves, called "conservation areas," outlined in black in **Figure 38.13A**. The green areas on the map are designated national parklands, which remain relatively unchanged by human activity; the yellow buffer zones are privately owned areas where people live and work.

Zoned reserves contribute to **sustainable development**, the long-term prosperity of human societies and the ecosystems that support them. Costa Rica is making progress in managing its reserves so that the buffer zones provide a steady, lasting supply of forest products, water, and hydroelectric power and also support sustainable agriculture. Costa Rica looks to its zoned reserve system to maintain at least 80% of its native species and to make this rich resource accessible to tourists and to its own citizens (**Figure 38.13B**). An important goal is providing a stable economic base for people living there. Destructive practices that are not compatible with long-term ecosystem stability and from which there is often little local profit are gradually being discouraged. Such destructive practices include massive logging, large-scale single-crop agriculture, and extensive mining.

The success of conservation in Costa Rica has involved leadership by the national government as well as essential partnerships between government, nongovernment organizations, and private citizens. How has the focus on conservation affected the Costa Rican people? As we discussed in Module 36.10, two indicators of living conditions are infant mortality rate and life expectancy. By both measures, living conditions have improved in Costa Rica over the past half century. In 2003, the literacy rate was 96%. Thus, we can infer that Costa Rica's conservation initiatives have not compromised human welfare. Nevertheless, many problems remain. One of the largest challenges will be maintaining a commitment to conservation in the face of a growing population. Although Costa Rica is in the middle of a rapid demographic transition (see Module 36.10) and birth rates are dropping, the population is predicted to grow from its current 4 million until it levels off at 6 million by the middle of this century. As always, a growing human population increases the demand on resources. Given its recent history, it seems probable that Costa Rica will continue its move to sustainable development, the topic of our final module.

? In zoned reserves, regulations prevent large-scale alterations of habitat in the buffer zones but do support sustainable development for the people living there. Why?

■ Large-scale disruptions could impact the nearby undisturbed areas, and preservation is a realistic goal only if it is compatible with an acceptable standard of living for the local people.

38.14 Sustainable development is an ultimate goal

In numbers, geographic range, and capacity to alter the biosphere, our species is clearly one of the most successful ones ever to inhabit planet Earth. Yet we seem to have set ourselves and the rest of the biosphere on a precarious path into the future. Facing increasing degradation of ecosystems, fragmentation of habitats, and loss of biodiversity, how can we best manage Earth's resources? Among the limited choices, which habitat areas are most practical to protect and manage if we are to save endangered species or the greatest number of species?

We must understand the complex interconnections of the biosphere in order to make sensible decisions about how to conserve these networks. To this end, many nations, scientific societies, and private foundations have embraced the concept of sustainable development. The Ecological Society of America, the world's largest organization of ecologists, endorses a research agenda called the Sustainable Biosphere Initiative. The goal of this initiative is to acquire the basic ecological information necessary for the intelligent and responsible development, management, and conservation of Earth's resources. The research agenda includes ways to sustain the productivity of natural and artificial ecosystems and studies of the relationship between biological diversity, global climate change, and ecological processes.

Sustainable development will depend not only on continued research and application of ecological knowledge. It will also require us to connect the life sciences with the social sciences, economics, and humanities. Conservation and restoration of biodiversity is only one side of sustainable development; the other key facet is improving the human condition. Public education and the political commitment and cooperation of nations are essential to the success of this endeavor.

It is unlikely that human nature will suddenly change drastically—that we will abruptly lose our environmental manipulativeness. What we must seek instead are ways to be more accommodating with other species and with the biosphere. Those of us living in affluent developed nations are responsible for the greatest amount of environmental degradation. Our long-term welfare and that of future generations demand that we work toward changing some of our values, learning to revere the natural processes that sustain us and reducing our orientation toward short-term personal gain. The current state of the biosphere demonstrates that we are treading precariously on uncharted ecological ground. The importance of our scientific and personal efforts cannot be overstated.

The striking image of a brown pelican on this book's cover and in **Figure 38.14** symbolizes the biodiversity crisis. In the 1960s, the biological magnification of DDT and other pesticides through the food chain reduced the populations of many predatory birds by causing their eggshells to thin and their reproductive efforts to fail. In 1970, the U.S. Fish and Wildlife Service declared the once-plentiful brown pelican an endangered species. After DDT was banned in 1972, the fish that had been poisoning the pelicans became less

Figure 38.14 The brown pelican (*Pelecanus occidentalis*)

toxic and pelican populations rebounded. Once again, the great birds are a common sight as they fly along the coastlines of the Americas. In their decline and recovery, brown pelicans show how a solid understanding of biology can help to preserve creatures and protect life.

An awareness of our unique ability to alter the biosphere and jeopardize the existence of other species, as well as our own, may help us choose a path toward a sustainable future. The goal is a world in which each generation inherits an adequate supply of natural and economic resources and a relatively stable environment. Despite the uncertainties, now is not a time for gloom and doom, but a time to aggressively pursue more knowledge about life and to work toward long-term sustainability.

Biology is the scientific expression of the human desire to know nature. We are most likely to save what we appreciate, and we are most likely to appreciate what we understand. By learning about the processes and diversity of life, we also become more aware of ourselves and our place in the biosphere.

Web/CD Activity 38E *Conservation Biology Review*

? Why is a concern for the well-being of future generations essential for progress toward sustainable development?

■ Sustainable development is a long-term goal—longer than a human lifetime. Preoccupation with personal gain in the here-and-now is an obstacle to sustainable development because it discourages behavior that benefits future generations.

Reviewing the Concepts

The Biodiversity Crisis: An Overview (Introduction–38.5)

Conservation biology is a goal-driven science that seeks to counter the biodiversity crisis, the rapid decrease in the number of species on Earth (**Introduction**).

Biodiversity includes genetic diversity, within and between populations; species diversity; and ecosystem diversity. Human activities threaten diversity at all levels (**38.1**). Biodiversity, while valuable for its own sake, also provides food, fiber, medicines, and ecosystem services (**38.2**).

Threats to biodiversity

Habitat destruction   Introduced species   Overexploitation

Another threat is pollution, which harms the environment. Effects include acid rain, ozone depletion, eutrophication, and dead zones. Chemical pesticides may be concentrated by biological magnification (**38.3–38.4**).

Global warming. Burning of fossil fuels is increasing the amount of CO_2 and other greenhouse gases in the air, which may warm Earth enough to change climate patterns, melt polar ice, and flood coastal regions. Global warming may increase the rate of species loss (**38.5**).

Conservation of Populations and Species (38.6–38.7)

Endangered populations. Habitat degradation often fragments populations. The small-population approach identifies the minimum viable population size and focuses on preserving genetic variation. The declining-population approach diagnoses and treats the causes of a population's decline (**38.6**). Preserving critical habitat may help endangered species recover. Conflicts may arise between habitat preservation and resource use by humans (**38.7**).

Managing and Restoring Ecosystems (38.8–38.14)

Landscape structure and biodiversity. Conservation efforts are increasingly aimed at sustaining ecosystems and landscapes. Edges between ecosystems have distinct sets of features and species. The increased frequency and abruptness of edges caused by human activities can increase species loss. Movement corridors connecting isolated habitats may be helpful to fragmented populations (**38.8**).

Nature reserves. Biodiversity hot spots have large concentrations of endemic species. Migratory species, both terrestrial and aquatic, may require international protection (**38.9**). The Yellowstone to Yukon Conservation Initiative is an international research and conservation effort that seeks to connect reserves and protect species and ecosystems (**38.10**).

Restoration ecology includes study of bioremediation to detoxify polluted areas and biological augmentation to

restore nutrients. Large-scale restoration projects attempt to restore damaged landscapes. The Kissimmee River Restoration Project is restoring river flow and wetlands, thus improving water quality and wildlife habitat (**38.11–38.12**).

Zoned reserves are undisturbed wildlands surrounded by buffer zones of compatible economic development. Costa Rica has established many zoned reserves (**38.13**).

Sustainable development seeks to improve the human condition while conserving biodiversity. It depends on increasing and applying ecological knowledge as well as valuing our linkages to the biosphere (**38.14**).

Connecting the Concepts

1. Complete the following map, which organizes some of the key concepts of conservation biology.

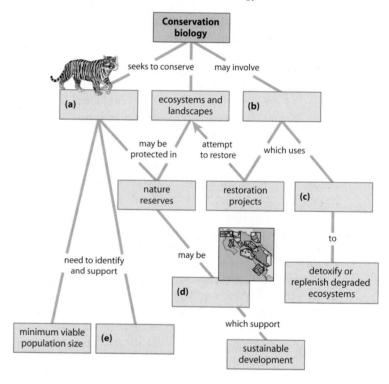

Testing Your Knowledge

Multiple Choice

2. Which of these statements best describes what conservation biologists mean by the "biodiversity crisis"?
   a. Introduced species, such as starlings and zebra mussels, have rapidly expanded their ranges.
   b. Harvests of marine fishes, such as cod and bluefin tuna, are declining.
   c. The current species extinction rate is as much as 1,000 times greater than at any time in the last 100,000 years.
   d. Many potential medicines are being lost as plant species become extinct.
   e. The number of hot spots worldwide is rapidly declining.

3. Which of the following poses the single greatest threat to biodiversity?
   a. introduced species
   b. overhunting
   c. movement corridors
   d. habitat loss
   e. global warming

4. Which of the following is characteristic of endemic species?
   a. They are often found in biodiversity hot spots.
   b. They are distributed widely in the biosphere.
   c. They require edges between ecosystems.
   d. Their trophic position makes them very susceptible to the effects of biological magnification.
   e. They are often keystone species whose presence helps to structure a community.

5. Ospreys and other top predators are most severely affected by pesticides such as DDT because they
   a. are especially sensitive to chemicals.
   b. have rapid reproductive rates.
   c. have very long life spans.
   d. store the pesticides in their tissues.
   e. consume prey in which pesticides are concentrated.

6. Movement corridors are
   a. the routes taken by migratory animals.
   b. strips or clumps of habitat that connect isolated fragments of habitat.
   c. landscapes that include several different ecosystems.
   d. edges or boundaries between ecosystems.
   e. buffer zones that promote the long-term viability of protected areas.

7. With limited resources, conservation biologists need to prioritize their efforts. Of the following choices, which should receive the greatest attention for the goal of conserving biodiversity?
   a. the northern spotted owl
   b. a commercially important species
   c. threatened and endangered vertebrate species
   d. a declining keystone species in a community
   e. all endangered species

8. Which of the following statements about protected areas is *not* correct?
   a. We now protect 25% of the land areas of the planet.
   b. National parks are only one type of protected area.
   c. Most reserves are smaller in size than the ranges of some of the species they are meant to protect.
   d. Management of protected areas must coordinate with the management of lands outside the protected zone.
   e. Biodiversity hot spots are important areas to protect.

Describing, Comparing, and Explaining

9. What are the three levels of biological diversity? Explain how human activities threaten each of these levels.

10. What is the so-called greenhouse effect? How is it important to life on Earth?

11. What are the possible causes and consequences of global warming? Why is international cooperation necessary if we are to solve this problem?

Applying the Concepts

12. Biologists in the United States are concerned that populations of many migratory songbirds are declining. Evidence suggests that some of these birds might be victims of pesticides. Most of the pesticides implicated in songbird mortality have not been used in the United States since the 1970s. Suggest a hypothesis to explain the current decline in songbird numbers.

13. You may have heard that human activities cause the extinction of one species every hour. Such estimates vary widely because we do not know how many species exist or how fast their habitats are being destroyed. You can make your own estimate of the rate of extinction. Start with the number of species that have been identified. To keep things simple, ignore extinction in the temperate latitudes and focus on the 80% of plants and animals that live in the tropical rain forest. Assume that destruction of the forest continues at a rate of 1% per year, so the forest will be gone in 100 years. Assume (optimistically) that half the rain forest species will survive in preserves, forest remnants, and zoos. How many species will disappear in the next century? How many species is that per year? Per day? Recent studies of the rain forest canopy have led some experts to predict that there may be as many as 30 million species on Earth. How does starting with this figure change your estimates?

14. One of the reasons the developed countries consume so much energy is that the price of energy does not reflect its real costs. What kinds of hidden environmental costs are not reflected in the price of fossil fuels? How are these costs paid, and by whom? Do you think these costs could or should be figured into the price of oil? How might that be done?

15. Many scientific studies indicate that global warming is under way and is linked to increased CO_2. Some believe we should gather more data before we act. What are the advantages and disadvantages of doing something now to slow global warming? What are the advantages and disadvantages of waiting until more data are available?

16. Until recently, response to environmental problems has been fragmented—an antipollution law here, incentives for recycling there. Meanwhile, the problems of the gap between the rich and poor nations, diminishing resources, and pollution continue to grow. Now people and governments are starting to envision a sustainable society. The Worldwatch Institute, a respected environmental monitoring organization, estimates that we must reach sustainability by the year 2030 to avoid economic and environmental disaster. To get there, we must begin shaping a sustainable society during this decade. In what ways is our present system not sustainable? What might a sustainable society be like? Do you think a sustainable society is an achievable goal? Why or why not? What is the alternative? What might we do to work toward sustainability? What are the major roadblocks to achieving sustainability? How would your life be different in a sustainable society?

Answers to all questions can be found in Appendix 3.

For study help and Activities, go to campbellbiology.com or the student CD-ROM.

Appendix 1: Metric Conversion Table

Measurement	Unit and Abbreviation	Metric Equivalent	Approximate Metric-to-English Conversion Factor	Approximate English-to-Metric Conversion Factor
Length	1 kilometer (km)	= 1000 (10^3) meters	1 km = 0.6 mile	1 mile = 1.6 km
	1 meter (m)	= 100 (10^2) centimeters	1 m = 1.1 yards	1 yard = 0.9 m
		= 1000 millimeters	1 m = 3.3 feet	1 foot = 0.3 m
			1 m = 39.4 inches	
	1 centimeter (cm)	= 0.01 (10^{-2}) meter	1 cm = 0.4 inch	1 foot = 30.5 cm
				1 inch = 2.5 cm
	1 millimeter (mm)	= 0.001 (10^{-3}) meter	1 mm = 0.04 inch	
	1 micrometer (μm)	= 10^{-6} meter (10^{-3} mm)		
	1 nanometer (nm)	= 10^{-9} meter (10^{-3} μm)		
	1 angstrom (Å)	= 10^{-10} meter (10^{-4} μm)		
Area	1 hectare (ha)	= 10,000 square meters	1 ha = 2.5 acres	1 acre = 0.4 ha
	1 square meter (m^2)	= 10,000 square centimeters	1 m^2 = 1.2 square yards	1 square yard = 0.8 m^2
			1 m^2 = 10.8 square feet	1 square foot = 0.09 m^2
	1 square centimeter (cm^2)	= 100 square millimeters	1 cm^2 = 0.16 square inch	1 square inch = 6.5 cm^2
Mass	1 metric ton (t)	= 1000 kilograms	1 t = 1.1 tons	1 ton = 0.91 t
	1 kilogram (kg)	= 1000 grams	1 kg = 2.2 pounds	1 pound = 0.45 kg
	1 gram (g)	= 1000 milligrams	1 g = 0.04 ounce	1 ounce = 28.35 g
			1 g = 15.4 grains	
	1 milligram (mg)	= 10^{-3} gram	1 mg = 0.02 grain	
	1 microgram (mg)	= 10^{-6} gram		
Volume (Solids)	1 cubic meter (m^3)	= 1,000,000 cubic centimeters	1 m^3 = 1.3 cubic yards	1 cubic yard = 0.8 m^3
			1 m^3 = 35.3 cubic feet	1 cubic foot = 0.03 m^3
	1 cubic centimeter (cm^3 or cc)	= 10^{-6} cubic meter	1 cm^3 = 0.06 cubic inch	1 cubic inch = 16.4 cm^3
	1 cubic millimeter (mm^3)	= 10^{-9} cubic meter (10^{-3} cubic centimeter)		
Volume (Liquids and Gases)	1 kiloliter (kL or kl)	= 1000 liters	1 kL = 264.2 gallons	1 gallon = 3.79 L
	1 liter (L)	= 1000 milliliters	1 L = 0.26 gallon	1 quart = 0.95 L
			1 L = 1.06 quarts	
	1 milliliter (mL or ml)	= 10^{-3} liter	1 mL = 0.03 fluid ounce	1 quart = 946 mL
		= 1 cubic centimeter	1 mL = approx. $\frac{1}{4}$ teaspoon	1 pint = 473 mL
			1 mL = approx. 15–16 drops	1 fluid ounce = 29.6 mL
				1 teaspoon = approx. 5 mL
Volume (Liquids and Gases)	1 microliter (ml or mL)	= 10^{-6} liter (10^{-3} milliliters)		
Time	1 second (s)	= $\frac{1}{60}$ minute		
	1 millisecond (ms)	= 10^{-3} second		
Temperature	Degrees Celsius (°C)		°F = $\frac{9}{5}$ °C - 32	°C = $\frac{5}{9}$ (°F − 32)

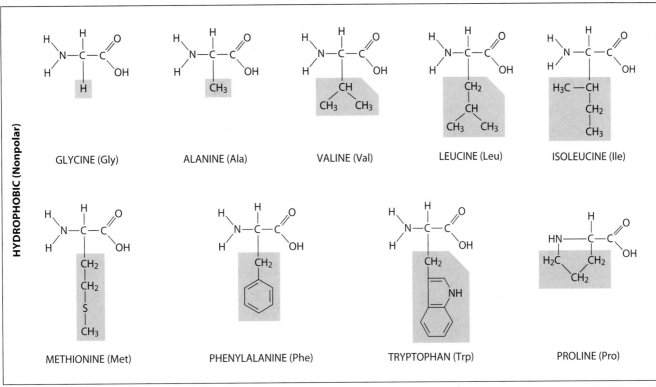

HYDROPHOBIC (Nonpolar)

GLYCINE (Gly)

ALANINE (Ala)

VALINE (Val)

LEUCINE (Leu)

ISOLEUCINE (Ile)

METHIONINE (Met)

PHENYLALANINE (Phe)

TRYPTOPHAN (Trp)

PROLINE (Pro)

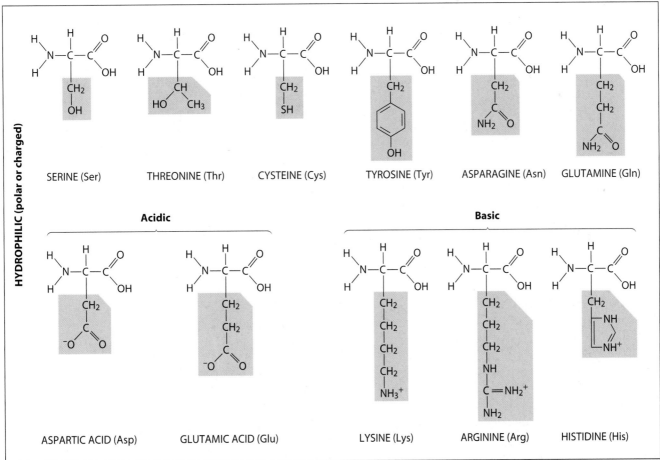

HYDROPHILIC (polar or charged)

SERINE (Ser)

THREONINE (Thr)

CYSTEINE (Cys)

TYROSINE (Tyr)

ASPARAGINE (Asn)

GLUTAMINE (Gln)

Acidic

Basic

ASPARTIC ACID (Asp)

GLUTAMIC ACID (Glu)

LYSINE (Lys)

ARGININE (Arg)

HISTIDINE (His)

Chapter 1

1. The vertical scale of biology refers to the hierarchical organization: from molecules to organelles, cells, tissues, organs, organ systems, organisms, populations, communities, ecosystems, and biosphere. At each level, emergent properties arise from the interaction and organization of component parts. The horizontal scale of biology refers to the incredible diversity of living organisms, past and present, including the 1.8 million species that have been named so far that can be grouped into three domains—Bacteria, Archaea, and Eukarya—and divided among numerous kingdoms.

2. a. life; b. evolution; c. natural selection; d. unity of life; e. three domains (or numerous kingdoms; 1.8 million species)

3. d **4.** c **5.** e **6.** d **7.** c **8.** b **9.** d

10. Both energy and chemical nutrients are passed through an ecosystem from producers to consumers to decomposers. But energy enters an ecosystem as sunlight and leaves as heat. Chemical nutrients are recycled from the abiotic soil or atmosphere through plants, consumers, and decomposers and returned to the soil and water.

11. Darwin hypothesized that natural selection operates in populations whose individuals have varied traits that are inherited. When natural selection favors the reproductive success of certain individuals in a population more than others, it changes the proportions of heritable variations in the population, gradually adapting a population to its environment.

12. In pursuit of answers to questions about nature, a scientist uses a logical thought process involving the key elements: observations about natural phenomena, questions derived from observations, hypotheses posed as tentative answers to questions, logical predictions of the outcome of tests if the hypotheses are true, and actual tests of hypotheses. Scientific research is not a rigid method because a scientist must adapt these key components to the set of conditions particular to each study. Intuition, chance, and luck are also part of science.

13. Technology is the application of scientific knowledge. For example, the use of solar power to run a calculator or heat a home is an application of our knowledge, derived by the scientific process, of the nature of light as a type of energy and how light energy can be converted to other forms of energy. Another example is the use of pieces of DNA removed from bacteria to insert new genes into crop plants. This process, often called genetic engineering, stems from decades of scientific research on the structure and function of DNA from many kinds of organisms.

14. Natural selection screens heritable variations by favoring the reproductive success of some individuals over others. These individuals pass more genes to the next generation than individuals that are not favored. As a result, the genetic makeup of a population changes. The change results from a screening (editing) of individuals (and consequently their genes), not from the creation of new genes or new individuals.

15. a. *Hypothesis:* Giving rewards to mice will improve their learning. *Prediction:* If mice are rewarded with food, they will learn to run a maze faster.

b. The control group was the mice that were not rewarded. Without them, it would be impossible to know if the mice who were rewarded decreased their time running the maze only because of practice.

c. Both groups of mice should be about the same age. Both experiments should be run at the same time of day and under the same conditions.

d. Yes, the data show that the rewarded mice began to run the maze faster by day 3, and improved their performance (ran faster than the control mice) each day thereafter.

16. The researcher needed to compare the number of attacks on artificial king snakes with attacks on artificial brown snakes. It may be that there were simply more predators in the coral snake areas or that they were hungrier than the snakes in the other areas. The experiment needed a control and proper data analysis.

17. Virtually any news report or magazine contains stories that are mainly about biology or at least have biological connections. How about biological connections in advertisements?

Chapter 2

1. a. protons; b. neutrons; c. electrons; d. different isotopes; e. covalent bonds; f. ionic bonds; g. polar covalent bonds; h. hydrogen bonding

2.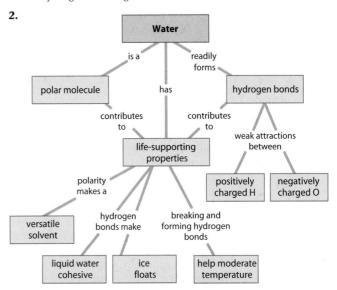

3. b **4.** c **5.** c **6.** e **7.** a (It needs to share 2 more electrons for a full outer shell of 8.) **8.** c **9.** d **10.** F (Only salt and water are compounds.) **11.** F (The smallest particle is an atom.) **12.** T **13.** T **14.** F (Water molecules in ice are farther apart.) **15.** T **16.** F (Most acid precipitation results from burning fossil fuels.) **17.** T

18. For diagram, see Figure 2.10. Water molecules form hydrogen bonds because they are polar. Their slightly negative O atoms are attracted to the slightly positive H atoms of neighboring molecules. The unique properties of water that result from hydrogen bonding are cohesion, surface tension, the ability to absorb and store large amounts of heat, a high boiling point, a solid form (ice) that is less dense than liquid water, and solvent properties.

19. First: Because increasing the temperature of water (the average speed of its molecules) requires breaking hydrogen bonds, a process that uses heat, a large amount of heat can be added to water before the water's temperature starts to rise. Conversely, when the surrounding temperature falls, new hydrogen bonds form in water, with the release of heat that slows the cooling process. Second: When the body becomes overheated, water evaporating from its surface decreases the body's temperature (evaporative cooling) because the hotter water molecules leave.

20. A covalent bond forms when atoms complete their outer shells by sharing electrons. Atoms can also complete their outer shells by gaining or losing electrons. This leaves the atoms as ions, with − and + charges. The oppositely charged ions are attracted to each other, forming an ionic bond.

21. An acid is a compound that donates hydrogen ions (H^+) to a solution. A base is a compound that accepts hydrogn ions and removes them from solution. Acidity is described by the pH scale, which measures H^+ concentration on a scale of 0 (most acidic) to 14 (most basic).

22. Fluorine needs 1 electron for a full outer shell of 8, and if potassium loses 1 electron, its outer shell will have 8. Potassium will lose an electron (becoming a + ion), and fluorine will pick it up (becoming a − ion). The ions will form an ionic bond.

23. Give a mouse sugar or oxygen gas containing a radioactive isotope of oxygen. Then see whether the carbon dioxide it exhales is radioactive.

24. *Some issues and questions to consider:* Which is less expensive, power from nuclear power plants or power from fossil fuel plants? Does the price of electricity reflect its actual cost, including environmental costs? Which would be more harmful: the environmental effects of acid rain and global warming from fossil fuel power plants or the effects of nuclear wastes and potential nuclear accidents? Do you favor development of nuclear energy or fossil fuel power plants? Which would you prefer to have near you? Do your answers to these last two questions differ? If so, why?

Chapter 3

1. Carbon's ability to form four covalent bonds, either with other carbon atoms, producing chains or rings of various lengths and shapes, or with other atoms, such as functional groups that confer specific properties on a molecule, is the basis for the incredible diversity of organic compounds. Organisms can link a small number of monomers into different arrangements to produce a huge variety of polymers.

2. a. Glucose; b. Energy storage; c. Cellulose; d. Fats; e. Cell membrane component; f. Steroids; g. Amino group; h. Carboxyl group; i. R group; j. Enzyme; k. Hair, tendons; l. Movement; m. Hemoglobin; n. Defense; o. Phosphate group; p. Nitrogenous base; q. Ribose sugar; r. DNA; s. Code for enzymes

3. d (The second kind of molecule is a polymer of the first.)   4. c
5. d 6. c 7. a 8. b 9. e 10. d

11. Triglycerides—store energy. Phospholipids—are major components of membranes. Waxes—make up waterproof coatings. Steroids—one kind, cholesterol, is a component of cell membranes; other kinds function as hormones.

12. Weak bonds that stabilize the three-dimensional structure of a protein are disrupted, and the protein unfolds. Function depends on shape, so if the protein is the wrong shape, it won't function properly.

13. Proteins are made of 20 amino acids arranged in many different sequences into chains of many different lengths. Genes, defined stretches of DNA, dictate the amino acid sequences of proteins in the cell.

14. Proteins function as enzymes to catalyze chemical reactions. They also function in structure, contraction, transport, defense, signaling, and storage of amino acids (see Module 3.11).

15. The sequence of nucleotides in DNA is transcribed into a sequence of nucleotides in RNA, which determines the sequence of amino acids that will be used to build a protein. Each set of three nucleotides is a code word for a particular amino acid. Proteins mediate all the activities of a cell; thus, by coding for proteins, DNA controls the functions of a cell.

16. This is a hydrolysis reaction, which consumes water. It is essentially the reverse of the diagram in Figure 3.5, except that fructose is a different shape than glucose.

17. Circle NH_2, an amino group; COOH, a carboxyl group; and OH, a hydroxyl group on the R group. This is an amino acid, a monomer of proteins.

18. *Some issues and questions to consider:* How will you choose your test subjects? How many subjects should you have? Will you give them all vitamin C, or just some of them? What criteria will you use to divide the test subjects into groups? What is a control group? Should the subjects know whether they are getting vitamin C or not? Should the experimenters who are giving out the drug and measuring the severity of cold symptoms know which of the subjects are getting vitamin C? What is a "double-blind" study? If there is a difference between your groups, how can you be sure it is due to vitamin C?

19. a. A: at about 37°C; B: at about 78°C.
    b. A from humans; B from thermophilic bacteria.
    c. Above 40°C, the human enzyme denatures and loses its shape and thus its function. The increased thermal energy disrupts the weak bonds that maintain secondary and tertiary structure in an enzyme.

Chapter 4

1. a. nucleus; b. ribosomes; c. Golgi apparatus; d. plasma membrane; e. mitochondrion; f. cytoskeleton; g. peroxisome; h. centriole; i. lysosome; j. flagellum; k. rough endoplasmic reticulum; l. smooth endoplasmic reticulum. For functions, see Table 4.19.

2. flagellum, lysosome, centriole (involved with microtubule formation in cell division)

3. chloroplast, central vacuole, cell wall

4. d 5. c (Small cells have a greater ratio of surface area to volume.)
6. b 7. d 8. e 9. a 10. c 11. b 12. d 13. c

14. DNA as genetic material, ribosomes, plasma membrane

15. Tight junctions form leakproof bonds. Anchoring junctions link cells to each other but allow materials to pass between them; they form strong sheets of cells. Communicating junctions are channels that allow flow from cell to cell.

16. Both process energy. A chloroplast converts light energy to chemical energy (sugar molecules). A mitochondrion converts chemical energy (food molecules) to another form of chemical energy (ATP).

17. Different conditions and conflicting processes can occur simultaneously within separate, membrane-enclosed compartments. Also, there is increased area for membrane-bounded enzymes that carry out metabolic processes.

18. Cilia may propel a cell through its environment or sweep a fluid environment past the cell.

19. A protein inside the ER is packaged inside transport vesicles that bud off the ER and then join to the Golgi apparatus. A transport vesicle containing the finished protein product then buds off the Golgi and travels to and joins with the plasma membrane, expelling the protein from the cell.

20. Part true, part false. Both animal and plant cells have mitochondria.

21. Cell 1: $S = 1,256$ μm^2; $V = 4,187$ μm^3; $S/V = 0.3$. Cell 2: $S = 5,024$ μm^2; $V = 33,493$ μm^3; $S/V = 0.15$. The smaller cell has a larger surface area relative to volume for absorbing food and oxygen and excreting waste. Small cells thus perform these activities more efficiently.

22. Both cilia and flagella rely on the same structural 9 + 2 organization of microtubules and dynein motor proteins. A reasonable explanation is that a hereditary defect in this mechanism results in sperm that do not swim and cilia that do not sweep mucus out of lungs.

23. *Some issues and questions to consider:* Were the cells Moore's property, a gift, or just surplus? Was Moore asked to donate the cells? Was he informed about how the cells might be used? Is it important to ask permission or inform the patient in such a case? How much did the researchers modify the cells? What did they have to do to them to sell the product? Do the researchers and the university have a right to make money from Moore's cells? Is the fact that they saved Moore's life a factor here? Does Moore have the right to sell his cells? Would Moore have been able to sell the cells without the researchers' help?

Chapter 5

1. a. enzyme molecule; b. active site of enzyme; c. substrate molecule; d. substrate in active site; an induced fit strains substrate bonds; e. substrate converted to products; f. product molecules; enzyme is ready for next catalytic cycle

2. a. active transport; b. concentration gradient; c. small nonpolar molecules; d. facilitated diffusion; e. energy (ATP)

3. d 4. b 5. d 6. c (Only active transport can move solute against a concentration gradient.) 7. d

8. Energy is neither created nor destroyed, but can be transferred and transformed. Plants transform the energy of sunlight into chemical energy stored in organic molecules. Almost all organisms rely on the products of photosynthesis for the source of their energy. In every energy transfer or transformation, disorder increases as some energy is lost to the random motion of heat.

9. The work of cells falls into three main categories: mechanical, chemical, and transport. ATP provides the energy for cellular work, usually by transferring a phosphate group to a protein (movement and transport) or a substrate (chemical).

10. Energy is stored in the chemical bonds of organic molecules. The barrier of activation energy prevents these molecules from spontaneously breaking down and releasing that energy. When a substrate fits into an enzyme's active site with an induced fit, its bonds may be strained and thus easier to break, or the active

site may orient two substrates in such a way as to facilitate the reaction.

11. Cell membranes are composed of a lipid bilayer with embedded proteins. The bilayer creates the hydrophobic boundary between cells and their surroundings (or between organelles and the cytoplasm). The proteins perform the many functions of membranes, such as enzyme action, transport, attachment, and signaling.

12. Heating, pickling, and salting denature enzymes, changing their shapes so they do not fit substrates. Freezing decreases the kinetic energy of molecules, so they lack energy of activation, even in the presence of enzymes.

13. a. The more enzyme present, the faster the rate of reaction, because it is more likely that enzyme and substrate molecules will meet.
    b. The more substrate present, the faster the reaction, for the same reason, but only up to a point. An enzyme molecule can work only so fast; once it is saturated (working at top speed), more substrate does not increase the rate.

14. *Some issues and questions to consider:* Does a woman have a right to work in an unsafe environment, even if it may put her child at risk? What are the rights of the child? Does the company have the right to "protect" women who are not pregnant? Is the company trying to protect the mother and child or protect itself from a potential lawsuit? Who is responsible for protecting employees and their children? The employees? The company? The government?

Chapter 6

1. a. glycolysis; b. citric acid cycle; c. oxidative phosphorylation; d. oxygen; e. electron transport chain; f. CO_2 and H_2O

2. e 3. c (NAD^+ and FAD, which are recycled by electron transport, are limited.) 4. e 5. d 6. c 7. b (and NADH is oxidized to NAD^+)

8. Glycolysis is considered the most ancient because it occurs in all living cells and doesn't require oxygen or membrane-enclosed organelles.

9. Oxygen picks up electrons from the oxidation of glucose at the end of the electron transport chain. Carbon dioxide results from the oxidation of glucose. It is released in the grooming of pyruvate and the citric acid cycle.

10. Fermentation recycles NADH back to NAD^+ so that ATP can continue to be produced by glycolysis. Thus, fermentation allows a cell to break down fuel molecules to produce ATP in the absence of oxygen.

11. In lactic acid fermentation (in muscle cells), pyruvate is reduced by NADH to form lactate, and NAD^+ is recycled. In alcohol fermentation, pyruvate is broken down to CO_2 and ethanol as NADH is oxidized to NAD^+.

12. As carbohydrates are broken down in glycolysis and the grooming of pyruvate, glycerol can be made from G3P and fatty acids from acetyl CoA. Amino groups, containing N atoms, must be supplied to various intermediates of glycolysis and the citric acid cycle to produce amino acids.

13. 100 kcal per day is 700 kcal per week. On the basis of Table 6.4, walking 4 mph would require $\frac{700}{231}$ = about 3 hr; swimming, 1.3 hr; running, 0.8 hr.

14. 10 NAD^+ and 2 FAD are needed to pick up the electrons and hydrogen atoms from a glucose molecule. NAD^+ and FAD are

recycled between electron transport and glycolysis and the citric acid cycle. We need a small additional supply to replace those that are damaged.

15. a. No, this shows the blue color getting more intense. The reaction decolorizes the blue dye.

b. No, this shows the dye being decolorized, but it also shows the three mixtures with different initial color intensities. The intensities should have started out the same, since all mixtures used the same concentration of dye.

c. Correct. The mixtures all start out the same, and then the ones with more succinate (reactant) decolorize faster. The mixture with the highest concentration of succinate decolorizes the fastest. (If we included mixtures with even greater concentrations of succinate, we would predict that eventually the mitochondria would be working as fast as possible, and no additional increase in the rate of decolorization would result from further additions of succinate.)

16. *Some issues and questions to consider:* Is your customer aware of the danger? Do you have an obligation to protect the customer, even against her wishes? Does your employer have the right to dismiss you for informing the customer? For refusing to serve the customer? Could you or the restaurant later be held liable for injury to the fetus? Or is the mother responsible for willfully disregarding warnings about drinking?

Chapter 7

1. a. electron transport chain; b. ATP synthase; c. thylakoid space (higher H^+ concentration); d. stroma; e. ATP

2. In chemiosmosis in mitochondria: a. Electrons come from food molecules. b. Electrons get their energy from the chemical energy of bonds in organic molecules. c. Electrons are passed to oxygen, which picks up H^+ and forms water.

In chemiosmosis in chloroplasts: a. Electrons come from splitting of water. b. Light energy excites the electrons. c. Electrons flow from water to the reaction center chlorophyll in Photosystem II to the reaction center chlorophyll in Photosystem I to NADPH.

In both processes: d. Energy is used to transport H^+ through a membrane. As H^+ flows back through ATP synthase, the energy is used to make ATP.

3. a. light energy; b. light reactions; c. Calvin cycle; d. O_2 released; e. electron transport chain; f. NADPH; g. ATP; h. 3-PGA is reduced

4. d **5.** c **6.** c **7.** b **8.** a **9.** e **10.** c (NADPH and ATP from light-dependent reactions are used by the Calvin cycle.) **11.** d

12. Light reactions: light and water are inputs; ATP, NADPH, and O_2 are outputs. Calvin cycle: CO_2, ATP, and NADPH are inputs; G3P, ADP, and $NADP^+$ are outputs.

13. Plants can break down the sugar for energy in cellular respiration or use the sugar as a raw material for making other organic molecules. Excess sugar is stored as starch.

14. *Some issues and questions to consider:* What are the risks that we take and costs we must pay if greenhouse warming continues? How certain do we have to be that warming is caused by human activities before we act? What can we do to reduce CO_2 emissions? What are the risks and costs if we reduce CO_2 emissions and greenhouse warming turns out not to be a real threat? Is it possible that the costs and sacrifices of reducing CO_2 emissions might actually improve our lifestyle?

15. *Some issues and questions to consider:* How much land would be required for large-scale conversion to biomass energy? Would different power plants have to be built? How do the benefits of energy plantations (habitat for wildlife, reduced erosion, diversification for farmers) compare with the benefits of fossil fuel harvesting? Is there a difference in net output of CO_2 between production and use of fossil fuels and production and use of biomass energy? Which one offers a more long-term solution to energy needs?

Chapter 8

1.

	Mitosis	Meiosis
Number of chromosomal duplications	1	1
Number of cell divisions	1	2
Number of daughter cells produced	2	4
Number of chromosomes in daughter cells	Diploid (2n)	Haploid (n)
How chromosomes line up during metaphase	Singly	In tetrads (metaphase I), then singly (metaphase II)
Genetic relationship of daughter cells to parent cell	Genetically identical	Genetically unique
Functions performed in the human body	Growth, development, repair	Production of gametes

2. c **3.** a **4.** b **5.** c **6.** e (A diploid cell would have an even number of chromosomes; the odd number suggests that meiosis I has been completed. Sister chromatids are together only in prophase and metaphase of meiosis II.) **7.** c **8.** b **9.** c **10.** b **11.** d

12. Various orientations of chromosomes at metaphase I of meiosis lead to different combinations of chromosomes in gametes. Crossing over during prophase I results in an exchange of chromosome segments and new combinations of genes. Random fertilization of eggs by sperm further increases possibilities for variation in offspring.

13. Interphase (e.g., third column from left in micrograph, third cell from top): Growth; metabolic activity; DNA synthesis. Prophase (e.g., second column, cell at bottom): Chromosomes shorten and thicken; mitotic spindle forms. Metaphase (e.g., first column, middle cell): Chromosomes line up on a plane going through the cell's equator. Anaphase (e.g., third column, second cell from top): Sister chromatids separate and move to the poles of the cell. Telophase (e.g., fourth column, fourth complete cell from top): Daughter nuclei form around chromosomes; cytokinesis usually occurs.

14. In culture, normal cells usually divide only when they are in contact with a surface but not touching other cells on all sides (the cells usually grow to form only a single layer). The density-dependent inhibition of cell division apparently results from local depletion of substances called growth factors. Growth factors are proteins secreted by certain cells that stimulate other cells to divide; they act via signal transduction pathways to signal the cell cycle control system of the affected cell to proceed past its checkpoints. The cell cycle control systems of cancer cells do not function properly. Cancer cells generally do not require externally supplied growth factors to complete the cell cycle, and

they divide indefinitely (in contrast to normal mammalian cells, which stop dividing after 20–50 generations)—two reasons why they are relatively easy to grow in the lab. Furthermore, cancer cells can often grow without contacting a solid surface, making it possible to culture them in suspension in a liquid medium.

15. A ring of microfilaments pinches an animal cell in two, a process called cleavage. In a plant cell, membranous vesicles form a disk called the cell plate at the midline of the parent cell, cell plate membranes fuse with the plasma membrane, and a cell wall grows in the space, separating the daughter cells.

16. See Figures 8.21A and 8.21B.

17. a. No. For this to happen, the chromosomes of the two gametes that fused would have to represent, together, a complete set of the donor's maternal chromosomes (the ones that originally came from the donor's mother) and a complete set of the donor's paternal chromosomes (from the donor's father). It is much more likely that the zygote would be missing one or more maternal chromosomes and would have an excess of paternal chromosomes, or vice versa.

b. Correct. Consider what would have to happen to produce a zygote genetically identical to the gamete donor: The zygote would have to have a complete set of the donor's maternal chromosomes and a complete set of the donor's paternal chromosomes. The first gamete in this union could contain any mixture of maternal and paternal chromosomes, but once that first gamete was "chosen," the second one would have to have one particular combination of chromosomes—the combination that supplies whatever the first gamete did not supply. So, for example, if the first three chromosomes of the first gamete were maternal, maternal, and paternal, the first three of the second gamete would have to be paternal, paternal, and maternal. The chance that all 23 chromosome pairs would be complementary in this way is only one in 2^{23} (i.e., one in 8,388,608). Because of independent assortment, it is much more likely that the zygote would have an unpredictable combination of chromosomes from the donor's father and mother.

c. No. First, the zygote could not be genetically identical to the gamete donor (see b). Second, the zygote could not be identical to either of the gamete donor's parents because the donor only has half the genetic material of each of his or her parents. For example, even if the zygote were formed by two gametes containing only paternal chromosomes, the combined set of chromosomes could not be identical to that of the donor's father because it would still be missing half of the father's chromosomes.

d. No. See answer c.

18. *Some possible hypotheses:* The replication of the DNA of the bacterial chromosome takes less time than the replication of the DNA in a eukaryotic cell. The time required for a growing bacterium to roughly double its cytoplasm is much less than for a eukaryotic cell. Bacteria have a cell cycle control system much simpler than that of eukaryotes.

19. 1 cm³ = 1,000 mm³, so 5,000 mm³ of blood contain 5,000 × 1,000 × 5,000,000 = 25,000,000,000,000, or 2.5×10^{13}, red blood cells. The $\frac{1}{120}$ of the cells that are replaced each day = $2.5 \times 10^{13}/120 = 2.1 \times 10^{11}$ cells. There are 24 × 60 × 60 = 86,400 seconds in a day. Therefore, the number of cells replaced each second = $2.1 \times 10^{11}/86,400$ = about 2×10^6, or 2 million. Thus, about 2 million cell divisions must occur each second to replace red blood cells that are lost.

20. Each chromosome is on its own in mitosis; chromosome replication and the separation of sister chromatids occur independently for each horse or donkey chromosome. Therefore, mitotic divisions, starting with the zygote, are not impaired. In meiosis, however, homologous chromosomes must pair in prophase I. This process of synapsis cannot occur properly because horse and donkey chromosomes do not match in number or content.

21. *Some issues and questions to consider:* Could it be that less money is spent on prevention because effective prevention is so much cheaper? Or because prevention has been tried, and it does not work well? Are lifestyle changes the kind of measures that could benefit from a shift in resources? Is prevention an individual matter of avoiding exposure or a social matter of preventing exposure? How might the answer to this question shape prevention policy? If more money were devoted to prevention, how could it be used to encourage you or others to make lifestyle changes? Would prevention work better for younger or older people? Might older people, already exposed to cancer-causing agents, actually be harmed by a shift of resources to prevention?

Chapter 9

1. a. alleles; b. loci; c. homozygous; d. dominant; e. recessive; f. incomplete dominance

2. c **3.** d **4.** d (Neither parent is ruby-eyed, but some offspring are, so it is recessive. Different ratios among male and female offspring show that it is sex-linked.) **5.** a **6.** e

7. Genes on the single X chromosome in males are always expressed because there are no corresponding genes on the Y chromosome to mask them. A male needs only one recessive color blindness allele (from his mother) to show the trait; a female must inherit the allele from both parents, which is less likely.

8. See Figure 9.20A. The parental gametes are *WS* and *ws*. Recombinant gametes are *Ws* and *wS*, produced by crossing over.

9. Height appears to be a quantitative trait, resulting from polygenic inheritance, like human skin color. See Module 9.15.

10. The character of freckles is dominant, so Tim and Jan must both be heterozygous. There is a $\frac{3}{4}$ chance that they will produce a child with freckles and a $\frac{1}{4}$ chance that they will produce a child without freckles. The probability that the next two children will have freckles is $\frac{3}{4} \times \frac{3}{4} = \frac{9}{16}$.

11. As in problem 10, both Tim and Jan are heterozygous, and Mike is homozygous recessive. The probability of the next child having freckles is $\frac{3}{4}$. The probability of the next child having a straight hairline is $\frac{1}{4}$. The probability that the next child will have freckles and a straight hairline is $\frac{3}{4} \times \frac{1}{4} = \frac{3}{16}$.

12. The genotype of the black short-haired parent rabbit is *BBSS*. The genotype of the brown long-haired parent rabbit is *bbss*. The F$_1$ rabbits will all be black and short-haired, *BbSs*. The F$_2$ rabbits will be $\frac{9}{16}$ black short-haired, $\frac{3}{16}$ black long-haired, $\frac{3}{16}$ brown short-haired, and $\frac{1}{16}$ brown long-haired.

13. If the genes are not linked, the proportions among the offspring will be 25% gray red, 25% gray purple, 25% black red, 25% black purple. The actual percentages show that the genes are linked. The recombination frequency is 6%.

14. The recombination frequencies are: black dumpy 36%, purple dumpy 41%, and black purple 6% (see problem 13). Since these

recombination frequencies reflect distances between the genes, the sequence must be purple-black-dumpy (or dumpy-black-purple).

15. $\frac{1}{4}$ will be boys suffering from hemophilia. $\frac{1}{4}$ will be female carriers. (The mother is a heterozygous carrier, and the father is normal.)

16. For a woman to be color-blind, she must inherit X chromosomes bearing the color blindness allele from both parents. Her father has only one X chromosome, which he passes on to all his daughters, so he must be color-blind. A male only needs to inherit the color blindness allele from a carrier mother; both his parents are usually phenotypically normal.

17. Start out by breeding the cat to get a population to work with. If the curl allele is recessive, two curl cats can have only curl kittens. If the allele is dominant, curl cats can have "normal" kittens. If the curl allele is sex-linked, ratios will differ in male and female offspring of some crosses. If the curl allele is autosomal, the same ratios will be seen among males and females. Once you have established that the curl allele is dominant and autosomal, you can determine if a particular curl cat is true-breeding (homozygous) by doing a testcross with a normal cat. If the curl cat is homozygous, all offspring of the testcross will be curl; if heterozygous, half of the offspring will be curl and half normal.

Chapter 10

1. a. nucleotides; b. transcription; c. RNA polymerase; d. mRNA; e. rRNA; f. tRNA; g. translation; h. ribosomes; i. amino acids

2. e (Only the phage DNA enters a host cell; lambda DNA determines both DNA and protein.) **3.** e **4.** b **5.** c

6. *Ingredients:* original DNA, nucleotides, several enzymes and other proteins, including DNA polymerases and DNA ligase. *Steps:* Original DNA strands separate at a specific site (origin of replication), nucleotides line up along each strand according to base-pairing rules, DNA polymerase links the nucleotides to form new strands. New nucleotides are added only to the 3′ end of a growing strand. One new strand is made in one continuous piece; the other new strand is made in a series of short pieces that are then joined by DNA ligase. *Product:* two identical DNA molecules, each with one old strand and one new strand.

7. A gene is the polynucleotide sequence with information for making one polypeptide. Each codon—a triplet of bases in DNA or RNA—codes for one amino acid. Transcription occurs when RNA polymerase produces RNA using one strand of DNA as a template. In prokaryotic cells, the RNA transcript may immediately serve as mRNA. In eukaryotic cells, the RNA is processed: A cap and tail are added, and RNA splicing removes introns and links exons together to form a continuous coding sequence. A ribosome is the site of translation, or polypeptide synthesis, and tRNA molecules serve as interpreters of the genetic code. Each tRNA molecule has an amino acid attached at one end and a three-base anticodon at the other end. Beginning at the start codon, mRNA moves relative to the ribosome a codon at a time. A tRNA with a complementary anticodon pairs with each codon, adding its amino acid to the polypeptide chain. The amino acids are linked by peptide bonds. Translation stops at a stop codon, and the finished polypeptide is released. The polypeptide folds to form a functional protein, sometimes in combination with other polypeptides.

8.

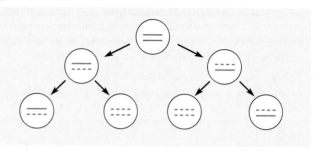

9. mRNA: GAUGCGAUCCGCUAACUGA. Amino acids: Met-Arg-Ser-Ala-Asn.

10. *Some issues and questions to consider:* Is it fair to issue a patent for a gene or gene product that occurs naturally in every human being? Or should a patent be issued only for something new that is invented rather than found? Suppose another scientist slightly modifies the gene or protein. How different does the gene or protein have to be in order that the patent is not infringed? Might patents encourage secrecy and interfere with the free flow of scientific information? What are the benefits to the holder of a patent? When research discoveries cannot be patented, what are the scientists' incentives for doing the research? What are the incentives for the institution or company that is providing financial support?

Chapter 11

1. a. proto-oncogene; b. repressor (or activator); c. cancer; d. operator; e. X inactivation; f. transcription factors; g. alternative RNA splicing

2. b (Different genes are active in different kinds of cells.) **3.** c
4. b **5.** a **6.** b **7.** e **8.** b

9. A mutation in a single gene can influence the actions of many other genes if the mutated gene is a control gene, such as a homeotic gene. A single control gene may encode a protein that affects (activates or represses) the expression of a number of other genes. In addition, some of the affected genes may themselves be control genes that in turn affect other batteries of genes. Cascades of gene expression are common in embryonic development.

10. a. If the mutated repressor could still bind to the operator on the DNA, it would continuously repress the operon; enzymes for lactose utilization would not be made, whether or not lactose was present.
b. The *lac* genes would continue to be transcribed and the enzymes made, whether or not lactose was present.
c. Same predicted result as for b.
d. RNA polymerase would not be able to transcribe the genes; no proteins would be made, whether or not lactose was present.

11. The protein to which dioxin binds in the cell is probably a transcription factor that regulates multiple genes (see Module 11.6). If the binding of dioxin influences the activity of this transcription factor (either activating or inactivating it), dioxin could thereby affect multiple genes and thus have a variety of effects on the body. The differing effects in different animals might be explained by differing genetic details in the different species. It would be extremely difficult to demonstrate conclusively that dioxin exposure was the cause of illness in a particular individual, even if dioxin had been shown to be present in

the person's tissues. However, if you had detailed information about how dioxin affects patterns of gene expression in humans and were able to show dioxin-specific abnormal patterns in the patient (perhaps using DNA microarrays; see Module 12.9), you might be able to establish a strong link between dioxin and the illness.

Chapter 12

1. a. PCR; b. restriction enzymes; c. gel electrophoresis; d. DNA fingerprinting (or restriction analysis or restriction fragment length polymorphisms); e. nucleic acid probe; f. cloning; g. genomic library

2. d 3. c 4. b 5. c (Bacteria lack the RNA-splicing machinery needed to delete eukaryotic introns.) 6. b 7. d

8. Isolate plasmids from *E. coli*. Cut the plasmids and the human DNA containing the HGH gene with restriction enzyme to produce molecules with sticky ends. Join the plasmids and the fragments of human DNA with ligase. Allow *E. coli* to take up recombinant plasmids. Bacteria will then replicate plasmids and multiply, producing clones of bacterial cells. Identify a clone carrying and expressing the HGH gene. Grow large amounts of the bacteria and extract and purify HGH from the culture. For the purposes of this question, we assume that the HGH gene provided does not have introns.

9. *Medicine:* Genes can be used to engineer viruses for vaccines, to produce transgenic lab animals for AIDS research, or for research related to human gene therapy. Proteins can be hormones, enzymes, blood-clotting factor, or the starting material for making vaccines. *Agriculture:* Foreign genes can be inserted into plant cells or animal eggs to produce transgenic crop plants or farm animals or inserted into viruses for making animal vaccines. Animal growth hormones are examples of agriculturally useful proteins that can be made using recombinant DNA technology.

10. She could start with DNA isolated from liver cells (the entire genome) and carry out the procedure outlined in Module 12.3 to produce a collection of recombinant bacterial clones, each carrying a small piece of liver cell DNA. To find the clone with the desired gene, she could then make a probe of radioactive RNA with a nucleotide sequence complementary to part of the gene: GACCUGACUGU. This probe would bind to the gene, labeling it and identifying the clone that carries it. Alternatively, the biochemist could start with mRNA isolated from liver cells and use it as a template to make DNA (using reverse transcriptase). Cloning this DNA rather than the entire genome would yield a smaller library of genes to be screened—only those active in liver cells. Furthermore, the genes would lack introns, making the desired gene easier to manipulate after isolation.

11. Determining the nucleotide sequences is just the first step. Once researchers have written out the DNA "book," they will have to try to figure out what it means—what the nucleotide sequences code for and how they work.

12. *Some issues and questions to consider:* What are some of the unknowns in recombinant DNA experiments? Do we know enough to anticipate and deal with possible unforeseen and negative consequences? Do we want this kind of power over evolution? Who should make these decisions? If scientists doing the research were to make the decisions about guidelines, what factors might shape their judgment? What might shape the judgment of business executives in the decision-making process? Does the public have a right to a voice in the

direction of scientific research? Does the public know enough about biology to get involved in this decision-making process? Who represents "the public," anyway?

13. *Some issues and questions to consider:* What kinds of impact will gene therapy have on the individuals who are treated? On society? Who will decide what patients and diseases will be treated? What costs will be involved, and who will pay them? How do we draw the line between treating disorders and "improving" the human species?

14. *Some issues and questions to consider:* Should genetic testing be mandatory or voluntary? Under what circumstances? Why might employers and insurance companies be interested in genetic data? Since genetic characteristics differ among ethnic groups and between the sexes, might such information be used to discriminate? Which of these questions do you think is most important? Which issues are likely to be the most serious in the future?

Chapter 13

1. The two major components of Darwin's theory are that all life has descended from a common ancestral form and that this "descent with modification" has been the result of natural selection. Individuals in a population have hereditary variations. The overproduction of offspring in the face of limited resources leads to a struggle for existence. The differential reproductive success of the most fit individuals in a population leads to the gradual accumulation of evolutionary adaptations to the local environment.

2. a. genetic drift; b. gene flow; c. natural selection; d. small population; e. founder effect; f. bottleneck effect; g. differential reproductive success

3. e 4. c 5. e 6. b (Erratic rainfall and differential reproductive success would ensure that a mixture of both forms remained in the population.) 7. b 8. e 9. d 10. b

11. Your paragraph should include such evidence as fossils and the fossil record, biogeography, comparative anatomy, comparative embryology, DNA and protein comparisons, artificial selection, and examples of natural selection.

12. The fitness of an individual (or of a particular genotype) is measured by the relative number of genes that it contributes to the gene pool of the next generation compared with the contribution of others. Thus, the number of fertile offspring produced determines an individual's fitness.

13. If $q^2 = 0.0025$, then $q = 0.05$. Since $p + q = 1$, $p = 1 - q = 0.95$. The proportion of heterozygotes is $2pq = 2 \times 0.95 \times 0.05 = 0.095$. About 9.5% of African-Americans are carriers.

14. Genetic variation is retained in a population by diploidy and balanced polymorphism. Recessive alleles are hidden from selection when in the heterozygote; thus, less adaptive or even harmful alleles are maintained in the gene pool and are available should selective pressures change. Both heterozygote advantage and frequency-dependent selection tend to produce balanced polymorphism and maintain alternate alleles in a population.

15. The unstriped snails appear to be better adapted. Striped snails make up 47% of the living population but 56% of the broken shells. Assuming that all the broken shells result from the meals of birds, we would predict that bird predation would reduce the frequency of striped snails, and the frequency of unstriped individuals would increase.

16. *Some issues and questions to consider:* Who should decide curriculum, scientific experts in a field or members of the community? Are the two alternatives both scientific ideas? Who judges what is scientific? If it is fairer to consider alternatives, should the door be open to all alternatives? Are constitutional issues (separation of church and state) involved here? Can a teacher be compelled to teach an idea he or she thinks is wrong? Should a student be required to learn an idea he or she thinks is wrong?

Chapter 14

1. a. Allopatric speciation: The barrier is habitat isolation.
   b. Sympatric speciation: The barrier is temporal isolation if the species flower at different times or gametic isolation if new species arose as a result of polyploidy.

2. a. gradual refinement; b. existing structure for new function; c. homeotic genes; d. paedomorphosis or extended growth period of human skull and brain

3. c **4.** b **5.** d **6.** c **7.** b **8.** c **9.** a **10.** e **11.** b

12. Different physical appearance may indicate different species or just differences within a species. Isolated populations may or may not be able to interbreed. Organisms that reproduce only asexually and fossil organisms do not have the potential to interbreed and produce fertile offspring; therefore, the biological species concept does not apply to them.

13. There is more chance for gene flow between populations on a mainland and nearby island. This interbreeding would make it more difficult for reproductive isolation to separate the two populations.

14. The gradualism model proposes that it is the gradual accumulation of small differences that eventually leads to speciation. According to the punctuated equilibrium model, most species diverge the most as they arise from an ancestral species and then remain fairly unchanged for the rest of their existence as a species.

15. Microevolution is the change in the gene pool of a population from one generation to the next. Macroevolution involves the origin of new species and evolutionary novelties.

16. A broad hypothesis would be that cultivated American cotton arose from a sequence of hybridization, meiotic failure, and self-fertilization. We can divide this broad statement into at least three testable hypotheses. *Hypothesis 1:* The first step in the origin of cultivated American cotton was hybridization between a wild American cotton plant (with 13 pairs of small chromosomes) and an Old World cotton plant (with 13 pairs of large chromosomes). If this hypothesis is correct, we would predict that the hybrid offspring would have 13 small chromosomes and 13 large chromosomes. *Hypothesis 2:* The second step in the origin of cultivated American cotton was a failure of meiosis in the hybrid offspring. If this hypothesis is true, we would expect the gametes resulting from meiotic failure to each have 13 large chromosomes and 13 small chromosomes. *Hypothesis 3:* The third step in the origin of cultivated American cotton was self-fertilization of the gametes resulting from meiotic failure. If this hypothesis is true, we would expect the outcome of self-fertilization to be a polyploid plant with 52 chromosomes—13 pairs of large ones and 13 pairs of small ones. This is the genetic makeup of cultivated American cotton.

17. *Some issues and questions to consider:* The rationale behind protecting all endangered groups is the desire to preserve genetic diversity. Each species, subspecies, and hybrid group may represent a unique mix of genes. Studies of the degree of genetic distinctiveness of a subspecies or hybrid group may help decision makers if cost is an issue. If the species as a whole is not at risk, it seems appropriate to determine how distinctive the gene pool of a subspecies or hybrid group is before assigning it a lower priority for saving. A question for society in general is: What is the value of any particular species and its genetically distinct subgroups? And how far are we willing (should we be willing) to go to preserve a genetically distinct group of organisms? How should the costs of preserving genetic diversity compare with the costs of other public projects?

Chapter 15

1. a. phylogenetic tree or cladogram; b. living organisms; c. morphology; d. shared derived characters; e. outgroup comparison

2. e **3.** d **4.** a **5.** d **6.** b **7.** b

8. The latter is more likely, since even small genetic changes can produce divergent physical appearances. But if genes have diverged greatly, it implies that lineages have been separate for some time, and the similar appearances may be analogous, not homologous.

9. An asteroid or comet may have hit Earth, raising dust, blocking sunlight, cooling the climate, and killing plants. Evidence includes a layer of iridium in rocks and a large crater under the Caribbean Sea. Fossil evidence suggests that increased volcanic activity in India may have released dust and cooled the climate. Continental drift may have altered climate and shorelines at the end of the Paleozoic.

10. (a) Analogy, because most animals and plants do not have homologous structures; (b) homology, since cats and humans are both mammals and have homologous forelimbs; (c) analogy, since birds and dragonflies are not closely related and the structure of their wings is very different.

11. The ribosomal RNA genes, which code for the RNA parts of ribosomes. They have evolved so slowly that homologies between even distantly related organisms can still be detected.

12. A molecular clock is a method of estimating the actual time of evolutionary events based on the numbers of molecular changes in homologous genes or regions of the genome. It is based on the assumption that the sequences being compared have evolved or changed at a constant rate.

13. 22,920 years old, a result of four half-life reductions

14. The four-chambered heart of birds and mammals is analogous, not homologous. It evolved independently in the two groups. An abundance of evidence supports the hypothesis that birds and reptiles are in the same clade and that birds and mammals evolved from different ancestors.

15.

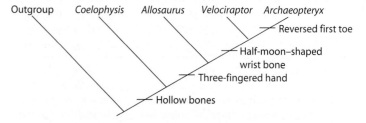

16. *Some issues and questions to consider:* Whereas previous mass extinctions have resulted from catastrophic events, such as asteroid collisions or volcanism, the current mass extinction is the result of human-caused environmental alteration. The rate of this mass extinction appears to be much greater than that of the others. Just because life bounces back, does that mean we will? Do we have any ethical responsibility to preserve other species?

Chapter 16

1. a. Abiotic synthesis of organic molecules using chemicals in atmosphere and lightning or other source of energy
b. Polymerization of monomers, perhaps on hot rocks
c. Beginnings of heredity and metabolism as RNA molecules may have replicated themselves and served as templates for polypeptides, which then may have catalyzed the formation of nucleotides and aided replication
d. Enclosure within a lipid membrane, which regulated passage of materials and allowed natural selection to act on the self-replicating protobionts

2.

Billions of years ago	Event
4.5	Origin of earth
3.5	First fossil evidence of life
2.7	Oxygen accumulates in atmosphere
2.1	First fossil evidence of eukaryotes
1.2	First evidence of multicellular eukaryotes

3. b **4.** c **5.** d (Algae are autotrophs; slime molds are heterotrophs.) **6.** e **7.** c **8.** b **9.** d **10.** b

11. Some antibiotics interfere with cross-linking in peptidoglycan and are most effective against gram-positive bacteria. Other antibiotics inhibit prokaryotic ribosomes.

12. Small, free-living prokaryotes were probably engulfed by a larger cell and took up residence inside. A symbiotic relationship developed between the host cell and engulfed cells. DNA, RNA, ribosomes, and inner membranes of mitochondria and chloroplasts are similar to those of bacteria. These organelles make some of their own proteins, replicate their own DNA, and reproduce by a fission-like process.

13. *Chlamydomonas* is a eukaryotic cell, much more complex than a prokaryotic bacterium. It is autotrophic, while amoebas are heterotrophic. It is unicellular, unlike multicellular sea lettuce.

14. Multicellular organisms have a greater extent of cellular specialization and more interdependence of cells. New organisms are produced from a single cell, either an egg or an asexual spore.

15. Not a good idea; all life depends on bacteria. You could predict that eliminating all bacteria from an environment would result in a buildup of toxic wastes and dead organisms (both of which bacteria decompose), a shutdown of all chemical cycling, and the consequent death of all organisms.

16. *Some issues and questions to consider:* Could we determine beforehand whether the iron would really have the desired effect? How? Would the "fertilization" need to be repeated? Could it be a cure for the problem, or would it merely treat the symptoms? Might the iron treatment have side effects? What might they be?

Chapter 17

1. a. bryophytes; b. ferns and relatives; c. gymnosperms; d. angiosperms; 1. lignin-hardened vascular tissue; 2. seeds that protect and disperse embryos; 3. flowers that enclose gametophyte generation; fruits

2. A. This is a cloud of pollen being released from a pollen cone of a pine tree. The pine tree is the sporophyte generation. Sporangia in ovulate and pollen cones produce spores by meiosis. In ovulate cones, spores develop into the female gametophyte in the ovule. In pollen cones, spores produce millions of the male gametophytes—the pollen grains. Pollination followed by fertilization produces a sporophyte embryo that is protected within the seed.

B. This is a cloud of haploid spores produced by a puffball fungus. Each spore may germinate to produce a haploid mycelium. Haploid hyphae of different mating types will fuse, creating a heterokaryotic mycelium, which is the feeding structure. Eventually, another puffball may develop, in which nuclei fuse in reproductive cells called basidia. Meiosis will produce spores that will be released, as shown in the photo.

3. b **4.** c (It is the only gametophyte among the possible answers.)
5. a **6.** e **7.** b **8.** e **9.** c **10.** d

11. The alga is surrounded and supported by water, and it has no supporting tissues, vascular system, or special adaptations for obtaining or conserving water. Its whole body is photosynthetic, and its gametes and embryos are dispersed into the water. The seed plant has lignified vascular tissues that support it against gravity and carry food and water; and organs that absorb water and minerals (roots), provide support (stems and roots), and photosynthesize (leaves and stems). It is covered by a waterproof cuticle and has stomata for gas exchange. Its sperm are carried by pollen grains, and embryos develop on the parent plant and are then protected and provided for by seeds.

12. Animals carry pollen from flower to flower and thus help fertilize eggs. They also disperse seeds by consuming fruit or carrying fruit that clings to their fur. In return, they get food (nectar, pollen, fruit).

13. Plants are autotrophs; they have chlorophyll and make their own food by photosynthesis. Fungi are heterotrophs that digest food externally and absorb nutrient molecules. There are also many structural differences; for example, the threadlike fungal mycelium is different from the plant body, and their cell walls are made of different substances. Plants evolved from green algae, fungi from a unicellular, flagellated, protistan ancestor. Evidence indicates that fungi are more closely related to animals than to plants.

14. Antibiotics probably kill off bacteria that compete with fungi for food. Similarly, bad tastes and odors deter animals that eat or compete with fungi. They are valuable warnings to animals that might eat spoiled food. Those fungi that produce antibiotics and bad-smelling and bad-tasting chemicals would survive and reproduce better than fungi unable to inhibit competitors. Animals that could recognize the smells and tastes also would survive and reproduce better than their competitors. Thus, natural selection would favor fungi that produce the chemicals and, to some extent, the competitors deterred by them.

15. Moss gametophytes, the dominant stage in the moss life cycle, are haploid plants. The diploid (sporophyte) generation is dominant in most other plants. Mutations are more likely to show up in haploid organisms because haploid organisms have only one set of genes. Recessive mutations may not be expressed in a diploid organism. Some factors to consider in designing your experiment: What are the advantages/disadvantages of

performing the experiment in the laboratory? In the field? What variables would be important to control? How many potted plants should you use? At what distances from the radiation source should you place them? What would serve as a control group for the experiment? What age of plants should you use?

16. *Some issues and questions to consider:* What are the other functions of forestland? How are other uses affected by logging? Must all the trees in an area be clear-cut? Should a particular area have multiple uses, or should different areas be used for different purposes? How much timber do we need? Could we conserve and recycle more? Are government-managed forests subsidizing private industry? Should we protect habitats as well as species? Aren't trees a renewable resource? Does the rate of regrowth match the rate of harvest? Will the ancient forests grow back? Are jobs and the economy at least as important as the owl?

Chapter 18

1. Sponges: sessile, saclike body with pores, suspension feeder; sponges

   Cnidarians: radial symmetry, gastrovascular cavity, cnidocytes, polyp or medusa body form; hydras, sea anemones, jellies, corals

   Flatworms: bilateral symmetry, gastrovascular cavity, no body cavity; free-living planarians, flukes, tapeworms

   Nematodes: pseudocoelom, covered with cuticle, complete digestive tract, ubiquitous, free-living and parasitic; roundworms, heartworms, hookworms, trichinosis worms

   Molluscs: muscular foot, mantle, visceral mass, circulatory system, many with shells, radula in some; snails and slugs, bivalves, cephalopods (squids and octopuses)

   Annelids: segmented worms, closed circulatory system, many organs repeated in each segment; earthworms, polychaetes, leeches

   Arthropods: exoskeleton, jointed appendages, segmentation, open circulatory system; chelicerates (spiders), crustaceans (lobsters, crabs), millipedes and centipedes, insects

   Echinoderms: radial symmetry as adult, water vascular system with tube feet, endoskeleton, spiny skin; sea stars, sea urchins

   Chordates: notochord; dorsal, hollow nerve cord; pharyngeal slits; post-anal tail; tunicates, lancelets, hagfish, and all the vertebrates (lampreys, sharks, ray-finned fishes, lobe-fins, amphibians, reptiles (including birds), mammals

2. a. brain; b. head; c. vertebral column; d. jaws; e. lungs or lung derivatives; f. lobed fins; g. legs; h. amniotic egg; i. milk

3. c 4. d 5. b 6. e 7. c 8. a (The invertebrates include all animals except the vertebrates.) 9. c 10. d 11. c 12. i 13. f 14. e 15. c 16. a 17. d 18. h 19. b 20. g

21. The gastrovascular cavity of a flatworm is an incomplete digestive tract; the worm takes in food and expels waste through the same opening. An earthworm has a complete digestive tract; food travels one way, and different areas are specialized for different functions. The flatworm's body is solid and unsegmented. The earthworm has a coelom, allowing its internal organs to grow and move independently of its outer body wall. Fluid in the coelom cushions internal organs, acts as a skeleton, and aids circulation. Segmentation of the earthworm, including its coelom, allows for greater flexibility and mobility.

22. Cnidarians and most adult echinoderms are radially symmetrical, while most other animals, such as arthropods and chordates, are bilaterally symmetrical. Most radially symmetrical animals stay in one spot or float passively. Most bilateral animals are more active and move headfirst through their environment.

23. For example, the legs of a horseshoe crab are used for walking, while the antennae of a grasshopper have a sensory function. Some appendages on the abdomen of a lobster are used for swimming, while the scorpion catches prey with its pincers. (Note that the scorpion stinger and insect wings are not considered jointed appendages.)

24. Fossil evidence supports the evolution of birds from a small, bipedal, feathered dinosaur, which was probably endothermic. The last common ancestor that birds and mammals shared was the ancestral amniote. The four-chambered hearts of birds and mammals must have evolved independently.

25. Important characteristics include symmetry, the presence and type of body cavity, segmentation, type of digestive tract, type of skeleton, and appendages.

26. *Some issues and questions to consider:* How does the decline of the reefs relate to agriculture? Deforestation? Overfishing? Rapid population growth? What value are the reefs to the local people? What is their value as a world biological resource? What might be the consequences if the reefs disappear? What is likely to make the situation worse? What is likely to improve the situation? In what ways might developed countries contribute to this problem? In what ways might developed countries be able to help the local people preserve the reefs? How might developed countries benefit from helping?

Chapter 19

1. a. New World monkeys; b. Old World monkeys; c. Orangutans; d. Gorillas; e. Chimpanzees; f. Humans

   Anthropoids include monkeys and apes (which include humans). Hominoids, or apes, include gibbons, orangutans, gorillas, chimpanzees, and humans.

2. a. Primates: limber shoulder and hip joints, mobile digits of hands and feet, flexible thumb and big toe, short snout, and eyes set close together
   b. Hominoids: larger brains relative to body size than other primates, lack tails, more flexible behavior. (Along with other anthropoids, hominoids have an opposable thumb.)
   c. Hominids: bipedalism, larger brains, reduced sexual dimorphism, shorter jaws, flatter faces, language, symbolic thought, manufacture and use of complex tools, longer period of parental care. (Note that these traits appeared at various times in hominid evolution.)

3. b 4. b 5. c 6. c 7. a 8. a 9. c 10. b

11. Several primate characteristics make it easy for us to make and use tools—mobile digits, opposable fingers and thumb, and great sensitivity of touch. Primates also have forward-facing eyes, which enhances depth perception and eye-hand coordination, and a relatively large brain.

12. Chimpanzees make and use simple tools, raid other social groups of their own species, can learn sign language and may use symbolic communication in the wild, and seem self-aware and able to use some form of reasoning to solve problems.

13. Hominoids include all apes (gibbons, orangutans, gorillas, chimpanzees, and humans). Hominids include the species on

the human evolutionary branch, more closely related to humans than to chimpanzees.

14. Skin pigmentation must be dark enough to protect against degradation of folate, a vitamin essential to normal embryonic development, but light enough to permit sufficient exposure to UV radiation (particularly in higher latitudes) to catalyze the production of vitamin D, a vitamin that permits the calcium absorption needed for both maternal and fetal bones.

15. The early hominids likely roamed the open forests and savanna of Africa, probably in small groups. They may have subsisted on nuts and seeds, birds' eggs, and whatever animals they could catch or scavenge from larger predators. They lived mostly on the ground but could climb trees, probably to pick fruits, escape predators, or steal carcasses left by leopards and other large cats.

16. Our intelligence and culture—accumulated and transmitted knowledge, beliefs, arts, and products—have enabled us to overcome our physical limitations and alter the environment to fit our needs and desires.

17. Most anthropologists think that humans and chimpanzees diverged from a common ancestor 5–7 million years ago. Primate fossils 4–8 million years old might help us understand how the human lineage first evolved.

18. Bipedalism would have allowed early hominids to see farther and run faster in the more open savanna environment, enabling them to find food and escape predators more easily. Bipedalism would have freed the hands to carry objects and use tools. Some anthropologists think that tool users with larger brains would have been favored by natural selection.

19. *Some issues and questions to consider:* If you live in an urban area, you probably see commercial and domestic development expanding into areas at the city's edge. In rural areas, agricultural and forestry practices alter land and habitats. News reports often mention global warming, acid precipitation, and other widespread changes associated with human activity. Evidence of a decrease in human-caused environmental changes is difficult to find.

Chapter 20

1. a. epithelial tissue; b. connective tissue; c. smooth muscle tissue; d. connective tissue; e. epithelial tissue

   The structure of the specialized cells in each type of tissue fits their function. For example, columnar epithelial cells are specialized for absorption and secretion; the fibers and cells of connective tissue provide support and connect the tissues. The hierarchy from cell to tissue to organ is evident in this diagram. The many projections of the lining of the small intestine greatly increase the surface area for absorption of nutrients.

2. b 3. c 4. a 5. c (Expelling salt opposes the increase in blood salt concentration, thereby maintaining a constant internal environment.) 6. d 7. c 8. a 9. d 10. a 11. c 12. b 13. d 14. b

15. Stratified squamous epithelium consists of many cell layers, which protect the body. Neurons are cells with long branches that conduct signals to other cells, making connections in the brain. Simple squamous epithelium is a single, thin layer of cells that allows for diffusion of gases across the lining of the lung. Bone cells are surrounded by a matrix containing fibers and calcium salts, forming a hard protective covering around the brain.

16. The surfaces of the intestine, excretory system, and lungs are highly folded and divided, with many blood vessels, increasing their surface area for exchange. Smaller creatures have a greater surface-to-volume ratio, and their cells are closer to the surface, enabling direct exchange between cells and the outside environment.

17. *Some issues and questions to consider:* Should a doctor's prescription be required for a whole-body CT scan? Should such scans only be available to those who can pay for them? Are CT scan machines calibrated so that they expose children or small adults to less radiation? Have there been research studies to test the effect of repeated exposures to radiation from CT scans? Whose responsibility is it to perform such studies and then publicize the results?

Chapter 21

1. a. mouth—ingests food; b. salivary glands—produce saliva; c. oral cavity—where food is chewed, moistened, and formed into bolus; d. pharynx—site of openings into esophagus and trachea; e. esophagus—transports bolus to stomach by peristalsis; f. stomach—food storage, begins digestion of proteins; g. small intestine—digestion and absorption; h. large intestine—absorption of water, compacts feces; i. anus—elimination of feces; j. rectum—storage of feces before elimination; k. pancreas—produces digestive enzymes and bicarbonate; l. gall bladder—stores bile; m. liver—produces bile

2. a. fuel, chemical energy; b raw materials, monomers; c. essential nutrients; d. obesity; e. vitamins and minerals; f. essential amino acids; g. malnourishment

3. e 4. c 5. b 6. d 7. e

8. You ingest the sandwich one bite at a time. In the oral cavity, chewing begins mechanical digestion, and salivary enzyme action on starch begins chemical digestion. When you swallow, food passes through the pharynx and esophagus to the stomach. Mechanical and chemical digestion continue in the stomach, where HCl in gastric juice breaks apart food cells and pepsin begins protein digestion. In the small intestine, enzymes from the pancreas and intestinal wall break down starch, protein, and fat to monomers. Bile from the liver and gallbladder emulsifies fat droplets for attack by enzymes. Most nutrients are absorbed into the bloodstream through the walls of the small intestine. In the large intestine, water is absorbed from undigested material, and feces are produced and eliminated.

9. Our craving for fatty foods may have evolved from the feast-and-famine existence of our ancestors. Natural selection may have favored individuals who gorged on and stored high-energy molecules, as they were more likely to survive famines.

10. a. 58%
    b. Based on a 2,000-Calorie diet, this product would supply about 9.5% of daily Calories, and it supplies 10% of vitamin A and calcium. If all food consumed supplied a similar quantity, the daily requirement for these two nutrients would be met.
    c. The 8 g of saturated fat in this product represent 40% of the daily value. Thus, the daily value must be 20 g (8/0.4 = 20). This represents 180 Calories from saturated fat.

11. *Some issues and questions to consider:* What are the roles of family, school, advertising, media, and government in providing nutritional information? How might the available information be improved? What types of scientific studies form the foundation of various nutritional claims?

12. *Some issues and questions to consider:* In wealthy countries, what are the factors that make it difficult for some people to get enough food? In your community, what types of help exist to feed hungry people? Think of two recent food crises in other countries and what caused them. Did other countries or international organizations provide aid? Which ones, and how did they help? Did that aid address the underlying causes of malnutrition and starvation in the stricken area or only provide temporary relief? How might that aid be changed to offer more permanent solutions to food shortages?

Chapter 22

1. a. respiratory surface; b. circulatory system; c. lungs; d. hemoglobin; e. cellular respiration; f. negative pressure breathing; g. O_2

2. a. nasal cavity; b. pharynx; c. larynx; d. trachea; e. lung; f. bronchus; g. bronchiole; h. diaphragm

3. c 4. b 5. d 6. a 7. e 8. d 9. c

10. Advantages of breathing air: It has a higher concentration of O_2 than water and is easier to move over the respiratory surface. Disadvantage of breathing air: It dries out the respiratory surface; living cells on the respiratory surface must remain moist.

11. Nasal cavity, pharynx, larynx, trachea, bronchus, bronchiole, alveolus, through wall of alveolus into blood vessel, blood plasma, into red blood cell, attaches to hemoglobin, carried by blood through heart, blood vessel in muscle, dropped off by hemoglobin, out of red blood cell, into blood plasma, through capillary wall, through interstitial fluid, and into muscle cell.

12. Llama hemoglobin has a higher affinity for O_2 than does human hemoglobin. The dissociation curve shows that its hemoglobin becomes saturated with O_2 at the lower P_{O_2} of the high altitudes to which llamas are adapted.

13. Insects have a tracheal system for gas exchange. In order to provide O_2 to all the body cells in such a huge moth, the tracheal tubes would have to be wider (to provide enough ventilation across larger distances) and very extensive (to service large flight muscles and other tissues), thus presenting problems of water loss and increased weight.

14. *Some issues and questions to consider:* Would a total ban on advertising decrease the number of cigarette smokers? If cigarettes are legal, can the right of cigarette manufacturers to advertise their product be restricted in this manner? Have similar bans on advertising of other legal, but potentially deadly products (such as alcohol) been tried? How have they worked? Do health concerns outweigh commercial concerns? If cigarettes are so bad, should they be declared illegal?

Chapter 23

1. a. capillaries of head, chest, and arms; b. aorta; c. pulmonary artery; d. capillaries of left lung; e. pulmonary vein; f. left atrium; g. left ventricle; h. aorta; i. capillaries of abdominal region; j. inferior vena cava; k. right ventricle; l. right atrium; m. pulmonary vein; n. capillaries of right lung; o. pulmonary artery; p. superior vena cava

   See text Figure 23.4B for numbers and red vessels.

2. b 3. d (The second sound is the closing of the semilunar valves as the ventricles empty.) 4. c 5. c 6. a 7. e

8. Pulmonary vein, left atrium, left ventricle, aorta, artery, arteriole, body tissue capillary bed, venule, vein, vena cava, right atrium, right ventricle, pulmonary artery, capillary bed in lung, pulmonary vein

9. Capillaries are very numerous, producing a large surface area for exchange close to body cells. The capillary wall is only one epithelial cell thick. Clefts between epithelial cells allow fluid with small solutes to move by bulk flow out at the arteriole end and back in at the venous end of the capillary.

10. a. Plasma (the straw-colored fluid) would contain water, inorganic salts (ions such as sodium, potassium, calcium, magnesium, chloride, and bicarbonate), plasma proteins such as fibrinogen and immunoglobulins (antibodies), and substances transported by blood, such as nutrients (e.g., glucose, fatty acids, vitamins), waste products of metabolism, respiratory gases (O_2 and CO_2), and hormones. b. The red portion would contain erythrocytes (red blood cells), leukocytes (white blood cells—basophils, eosinophils, neutrophils, lymphocytes, and monocytes), and platelets.

11. Oxygen content is reduced as oxygen-poor blood returning to the right ventricle from the systemic circuit mixes with oxygen-rich blood of the left ventricle.

12. Proteins are important solutes in blood, accounting for much of the osmotic pressure that draws fluid into the blood. If protein concentration is reduced, the inward pull of osmotic pressure will fail to balance the outward push of blood pressure. The net pressure at the arterial end of a capillary forces fluid out into the interstitial fluid. Net pressure at the venous end is not great enough to draw fluid back in.

13. *Some issues and questions to consider:* Is it ethical to have a child in order to save the life of another? Is it right to conceive a child as a means to an end—to produce a tissue or organ? Is this a less acceptable reason than most reasons parents have for bearing children? Do parents even need a reason for conceiving a child? Do parents have the right to make decisions like this for their young children? How will the donor (and recipient) feel about this when the donor is old enough to know what happened?

14. With a three-chambered heart, there is some mixing of oxygen-rich blood returning from the lungs with oxygen-poor blood returning from the systemic circulation. Thus, the blood of a dinosaur might not have supplied enough O_2 to support the higher metabolism and strong cardiac muscle contractions needed to generate such a high systolic blood pressure. Also, with a single ventricle pumping simultaneously to both pulmonary and systemic circuits, the blood pumped to the lungs would be at such a high pressure that it would damage the lungs.

Chapter 24

1. a. innate defenses (innate immunity); b. acquired defenses (acquired immunity); c. B cells; d. T cells; e. antibodies; f. helper T cells

2. e 3. b 4. b 5. d 6. a 7. b 8. f 9. d 10. e 11. a 12. g 13. c

14. AIDS is mainly transmitted in blood and semen. It enters the body through slight wounds during sexual contact or via needles contaminated with infected blood. AIDS is deadly because it infects helper T cells, crippling immunity and leaving the body vulnerable to other infections and cancer. The most effective way to avoid HIV transmission is to prevent contact with bodily fluids by practicing safe sex and avoiding intravenous drugs.

15. Inflammation is triggered by tissue injury. Damaged cells release histamine and other chemicals, which cause nearby blood vessels to dilate and become leakier. Blood plasma leaves vessels, and phagocytes are attracted to the site of injury. An increase in blood flow, fluid accumulation, and increased cell population cause redness, heat, and swelling. Inflammation disinfects and cleans the area and curtails the spread of infection from the injured area. Inflammation is considered nonspecific because similar defenses are presented in response to any infection.

16. One hypothesis is that your roommate's previous bee stings caused her to become sensitized to the allergens in bee venom. During sensitization, antibodies to allergens attach to receptor proteins on mast cells. During this sensitization stage, she would not have experienced allergy symptoms. When she was exposed to the bee venom again at a later time, the bee venom allergens bound to the mast cells, which triggered her allergic reaction.

17. *Some issues and questions to consider:* There is no correct answer to this question. Possible directions include the idea that if the donor felt strongly about the process, then his or her wishes should be respected. The opposite direction would be that the next of kin should be able to approve or deny the procedure. Other considerations may be appropriate, including religious beliefs.

18. *Some issues and questions to consider:* How important is it to protect students from AIDS? Is this a function of schools? Do schools serve other such "noneducational" purposes? Should parents or citizens' and church groups—on either side of the issue—have a say in this, or is it a matter between the school and the student? Does the distribution of condoms condone or sanction sexual activity or promiscuity? Is a school legally liable if a school-issued condom fails to protect a student? Are there alternative measures, such as education, that might be as effective for slowing the spread of AIDS?

Chapter 25

1. a. thermoregulation; b. osmoregulation; c. excretion; d. ectotherm; e. ammonia, urea, uric acid; f. behavioral responses; g. environment

2. a. filtration: water; NaCl, HCO_3^-, H^+, urea, glucose, amino acids, some drugs; b. reabsorption: nutrients, NaCl, water, HCO_3^-, urea; c. secretion: some drugs and toxins, H^+, K^+; d. excretion: urine containing water, urea, and excess ions

3. c **4.** a **5.** e **6.** d **7.** c **8.** d **9.** b **10.** c **11.** a **12.** c **13.** b **14.** d **15.** b **16.** a

17. In salt water, the fish loses water by osmosis. It drinks salt water and disposes of salts through its gills. Its kidneys conserve water and excrete excess ions. In fresh water, it gains water by osmosis. Its kidneys excrete a lot of dilute urine. Its gills take up salt, and some ions are ingested with food.

18. Yes. Ectotherms that live in very stable environments, such as tropical seas or deep oceans, have stable body temperatures. And terrestrial ectotherms can maintain relatively stable temperatures by behavioral means.

19. a. An endotherm would produce more nitrogenous wastes because it must eat more food to maintain its higher metabolic rate.
 b. A carnivore, because it eats more protein and thus produces more breakdown products of protein digestion—nitrogenous wastes.

20. You could take it back to the laboratory and measure its body temperature under different ambient temperatures.

21. A countercurrent heat exchanger in the birds' legs reduces the loss of heat from the body. You would expect the temperature of blood flowing back to the body from the legs to be only slightly cooler than the blood flowing from the body to the legs.

22. *Some issues and questions to consider:* Could drug use endanger the safety of the employee or others? Is drug testing relevant to jobs where safety is not a factor? Is drug testing an invasion of privacy, interfering in the private life of an employee? Is an employer justified in banning drug use off the job if it does not affect safety or ability to do the job? Do the same criteria apply to employers requiring the test? Could an employer use a drug test to regulate other employee behavior that is legal, such as smoking?

23. *Some issues and questions to consider:* Should human organs be sold? If the donor is poor, he or she may also not be in the best of health. What are the health risks to the donor? Will the recipient be assured of buying a healthy organ? How much does the organ broker make from this transaction? Both parties in such a transaction are probably fairly desperate. Should regulations be in place when people may not be able to make reasoned decisions? Or do people have a right to sell parts of their bodies, just as they can now sell other possessions or their labor?

Chapter 26

1–8. Pineal gland: 5, f
 Testes: 3, d
 Parathyroid gland: 7, g
 Adrenal medulla: 2, c
 Hypothalamus: 8, h
 Pancreas: 4, a
 Anterior pituitary: 6, b
 Thyroid gland: 1, e

9. d **10.** e **11.** a **12.** b (Negative feedback: When thyroxine increases, it inhibits TSH, which reduces thyroxine secretion.) **13.** d

14. The hypothalamus secretes releasing hormones and inhibiting hormones, which are carried by the blood to the anterior pituitary. In response to these signals from the hypothalamus, the anterior pituitary increases or decreases its secretion of a variety of hormones that directly affect body activities or influence other glands. Neurosecretory cells that extend from the hypothalamus into the posterior pituitary secrete hormones that are stored in the posterior pituitary until they are released into the blood.

15. Only cells with the proper receptors will respond to a hormone. For a steroid hormone, the presence (or absence) and types of receptor proteins inside the cell determine the hormone's effect. For a nonsteroid hormone, the types of receptors on the cell's plasma membrane are key, and the proteins of the signal transduction pathway may have different effects inside different cells.

16. a. No. Blood sugar level goes too low. Diabetes would tend to make the blood sugar level go too high after a meal.
 b. No. Insulin is working, as seen by the homeostatic blood sugar response to feeding.
 c. Correct. Without glucagon, exercise and fasting lower blood sugar, the cells cannot mobilize any sugar reserves, and their blood sugar level drops. Insulin (which lowers blood sugar) has no effect.
 d. No. If this were true, blood sugar level would increase too much after a meal.

17. If cells within a male embryo do not secrete testosterone at the proper time during development, the embryo will develop into a female despite being genetically male.

Chapter 27

1. A. FSH; B. estrogen; C. LH; D. progesterone; P. menstruation; Q. growth of follicle; R. ovulation; S. development of corpus luteum.

   If pregnancy occurs, the embryo produces human chorionic gonadotrophin (HCG), which maintains the corpus luteum, keeping levels of estrogen and progesterone high.

2. c   3. c (The outer layer in a gastrula is the ectoderm; of the choices given, only the brain develops from ectoderm.)   4. a   5. d   6. e   7. g   8. d   9. h   10. f   11. a   12. b   13. c

14. Both produce haploid gametes. Spermatogenesis produces four small sperm; oogenesis produces one large ovum. In humans, the ovary contains all the primary oocytes at birth, while testes can keep making primary spermatocytes throughout life. Oogenesis is not complete until fertilization, but sperm mature without eggs.

15. The extraembryonic membranes provide a moist environment for the embryos of terrestrial vertebrates and enable the embryos to absorb food and oxygen and dispose of wastes. Such membranes are not needed when an embryo is surrounded by water, as are those of fishes and amphibians.

16. The nerve cells may follow chemical trails to the muscle cells and identify and attach to them by means of specific surface proteins.

17. The researcher might find out whether chemicals from the notochord stimulate the nearby ectoderm to become the neural tube, a process called induction. Transplanted notochord tissue might cause ectoderm anywhere in the embryo to become neural tissue. Control: Transplant non-notochord tissue under the ectoderm of the belly area.

18. *Some issues and questions to consider:* What characteristics might parents like to select for? If parents had the right to choose embryos based on these characteristics, what are some of the possible benefits? What are potential pitfalls? Could an imbalance of the population result?

Chapter 28

1. a. sensory receptor; b. sensory neuron; c. synapse; d. spinal cord; e. interneuron; f. motor neuron; g. effector cell; h. CNS; i. PNS

2. b   3. a   4. Both a and c would prevent action potentials from occurring; b could actually increase the generation of action potentials.

5. At the point where the action potential is triggered, sodium ions rush into the neuron. They diffuse laterally and cause sodium gates to open in the adjacent part of the membrane, triggering another action potential. The moving wave of action potentials, each triggering the next, is a moving nerve signal. Behind the action potential, sodium gates are temporarily inactivated, so the action potential can only go forward. At a synapse, the transmitting cell releases a chemical neurotransmitter, which binds to receptors on the receiving cell and may trigger a nerve signal in the receiving cell.

6. The results show the cumulative effect of all incoming signals on neuron D. Comparing experiments 1 and 2, we see that the more nerve signals D receives from C, the more it sends; C is excitatory. Because neuron A is not varied here, its action is unknown; it may be either excitatory or mildly inhibitory.

Comparing experiments 2 and 3, we see that neuron B must release a strongly inhibitory neurotransmitter, because when B is transmitting, D stops.

7. *Some issues and questions to consider:* Some people might be against the use of embryonic stem cells for any disorder. This may be related to religious or moral beliefs. Potentially, some people may not be aware of the source of these stem cells and may be against their use because they believe that the cells come from elective abortions. Other people may agree with the use of stem cell research because of the potential to cure diseases that are currently fatal. Some people who may have a neutral opinion on the issue might be swayed by the thought of a loved one who might be helped by stem cell therapy.

8. *Some issues and questions to consider:* What is the role of alcohol in crime? What are its effects on families and in the workplace? In what ways is the individual responsible for alcohol abuse? The family? Society? Who is affected by alcohol abuse? How effective are treatment and punishment in curbing alcohol abuse? Who pays for alcohol abuse and consequent treatment or punishment? Is it possible to enjoy alcohol without abusing it?

Chapter 29

1. a. mechanoreceptors; b. chemoreceptors; c. electricity, magnetism, light; d. hair cells; e. photoreceptors

2. a. auditory canal (part of the outer ear); b. middle ear bones (hammer, anvil, stirrup); c. semicircular canals; d. auditory nerve; e. cochlea; f. Eustachian tube; g. oval window (behind stirrup); h. eardrum

   Sound waves strike the eardrum, which moves the bones of the middle ear. The bones vibrate fluid in the cochlea, which moves the basilar membrane. Hair cells on the basilar membrane bend against the tectorial membrane, causing changes in the membrane's permeability to ions and the generation of receptor potentials in the hair cells. The hair cells release neurotransmitter, stimulating sensory neurons, which transmit action potentials to the brain.

3. d (He could hear the tuning fork against his skull, so the cochlea, nerve, and brain are OK. Apparently, sounds are not being transmitted to the cochlea; therefore, the bones are the problem.)   4. b   5. a   6. b   7. e   8. a

9. Louder sounds create pressure waves with greater amplitude, moving hair cells more and generating a greater frequency of action potentials. Different pitches affect different parts of the basilar membrane; different hair cells stimulate different sensory neurons that transmit action potentials to different parts of the brain.

10. Sensation is the detection of stimuli (light) by the photoreceptors of the retina and transmission of action potentials to the brain. Perception is the interpretation of these nerve signals—sorting out the patterns of light and dark and determining their meaning.

11. Taste is used to sample food and determine its quality. Smell is used for many functions—communication of territories (scent marking), navigation (salmon), mate location (moths), sensing danger (predators, fires), and finding food.

12. *Some possible hypotheses:* Paired sensory receptors enable an animal to determine the direction from which stimuli come. Paired receptors enable comparison of the intensity of stimuli on either side. Paired receptors enable comparison of slightly

different images seen by the eyes or sounds heard by the ears (thus enabling the brain to perceive depth and distance).

13. Do the turtles hear the surf? Plug the ears of some turtles and not others. If turtles without earplugs head for the water and turtles with earplugs get lost, they probably hear the ocean. Or do they smell the water? Plug their nostrils and follow the same process.

14. *Some issues and questions to consider:* Assuming that the sound is loud enough to impair hearing, how long an exposure is necessary for this to occur? Does exposure have to occur all at once, or is damage cumulative? Who is responsible, concert promoters or listeners? Should there be regulations regarding sound exposure at concerts (as there are for job-related noise)? Are young people sufficiently mature and aware to heed such warnings?

Chapter 30

1. a. skeleton; b. muscles; c. exoskeleton; d. sarcomeres; e. bone and cartilage

2. c 3. d 4. a (Water supports aquatic animals, reducing the effects of gravity.) 5. c 6. d 7. e 8. d 9. a (Each neuron controls a smaller number of muscle fibers.) 10. e

11. Advantages of an insect exoskeleton include strength, good protection for the body, flexibility at joints, and protection from water loss. The major disadvantage is that the exoskeleton must be shed periodically as the insect grows, leaving the insect temporarily weak and vulnerable.

12. The bird's wings are airfoils, with convex upper surfaces and flat or concave lower surfaces. As the wings beat, air passing over them travels farther than air beneath. Air molecules above the wings are more spread out, lowering pressure. Higher pressure beneath the wings pushes them up.

13. Action potentials from the brain travel down the spinal cord and along a motor neuron to the muscle. The neuron releases a neurotransmitter, which triggers action potentials in a muscle fiber membrane. These action potentials initiate the release of calcium from the ER of the cell. Calcium enables myosin heads of the thick filaments to bind with actin of the thin filaments. ATP provides energy for the movement of myosin heads, which causes the thick and thin filaments to slide along one another, shortening the muscle fiber. The shortening of muscle fibers pulls on bones, bending the arm. If more motor units are activated, the contraction is stronger.

14. The human skull is balanced on top of (instead of in front of) the backbone. The human backbone is S-shaped, not arched like the baboon's. The human pelvic girdle is shorter, rounder, and oriented more vertically. The human hand is adapted for gripping, and our feet are adapted for support of the entire body and bipedal locomotion.

15. Chemical A would work better, because acetylcholine triggers contraction. Blocking it would prevent contraction. Chemical B would actually increase contraction, because Ca^{2+} signals filaments to interact and slide.

16. Circular muscles in the earthworm body wall decrease the diameter of each segment, squeezing internal fluid and lengthening the segment. Longitudinal muscles shorten and thicken each segment. Different parts of the earthworm can lengthen while others shorten, producing a crawling motion. The whole roundworm body moves at once because of a lack of segmentation. The body can only shorten or bend, not lengthen,

because of a lack of circular muscles. Roundworms simply thrash from side to side.

17. Calcium ions move a regulatory protein out of the way so myosin heads can attach to actin, resulting in muscle contraction. ATP causes the myosin heads of the thick filaments to detach from the thin filaments (Figure 30.9B, step 1). If there is no ATP present, the myosin heads remain attached to the thin filaments, and the muscle fiber remains fixed in position.

18. *Some issues and questions to consider:* Are the places where you live, work, or attend class accessible to a person in a wheelchair? If you were in a wheelchair, would you have trouble with doors, stairs, drinking fountains, toilet facilities, and eating facilities? What kinds of transportation would be available to you, and how convenient would they be? What activities would you have to forgo? How might your disability alter your relationships with your friends and family? How well would you manage on your own?

19. *Some issues and questions to consider:* With regard to bone growth, the major negative consequence in young people is that the bones will become completely hardened before they have grown to their full size. This means that the steroid user will be shorter than he or she would otherwise have been. Do steroid users know this? Do they think that the effects of steroids are temporary or reversible? Why is such emphasis placed on appearance and athletic prowess? Will steroid users later regret their use? If steroid use were not against the rules, would it be acceptable?

Chapter 31

1. Here is one possible concept map:

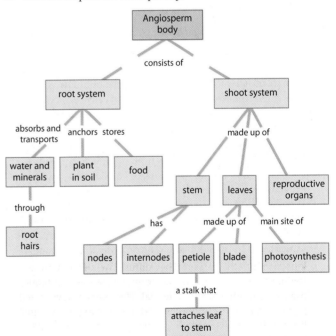

2. d (The vascular cambium forms to the outside of the primary xylem. The secondary xylem forms between primary xylem and the vascular cambium. The secondary phloem and primary phloem are outside the vascular cambium.) 3. b 4. e 5. e 6. f 7. b
8. e 9. a 10. c 11. d

12. Bees deposit the pollen on the stigma of a carpel, and a pollen tube grows to the ovary at the base of the carpel. Sperm travel down the pollen tube and fertilize egg cells in ovules. The

ovules grow into seeds, and the ovary grows into the flesh of the fruit (actually, in an apple, just the core). As the seeds mature, the fruit ripens and falls (or is picked).

13. Bulbs, root sprouts, and runners are all examples of asexual reproduction. Asexual reproduction is less wasteful and costly than sexual reproduction and less hazardous for young plants. The primary disadvantage of asexual reproduction is that it produces genetically identical offspring, decreasing genetic variability that can help a species survive times of environmental change.

14. Tomato: fruit (ripened ovary); celery stalk: leaf stalk (petiole); peanut: seed (ovule); strawberry: fruit (ripened ovary); lettuce: leaf blades; artichoke: terminal bud; beet: root

15. Modern methods of plant breeding and propagation have increased crop yields but have decreased genetic variability, so plants become more vulnerable to epidemics. Primitive varieties of crop plants could contribute to gene banks and be used for breeding new strains.

16. *Some issues and questions to consider:* Why are the forests being cut when preserving them is more beneficial in the long run? Why can't the developing nations create new forest products themselves? Should developing and developed nations share in the profits from the forest? What incentive is there for a company to create a new product if it can't keep the profits? What about companies paying an exploration fee for exclusive rights to take plants from a certain area? What about a contract for the developing nation to get a percentage of the profits from new products? Are there reasons to preserve the forest other than profits and beneficial new products?

Chapter 32

1. a. roots; b. xylem; c. sugar source; d. phloem; e. transpiration; f. sugar sink

2. d 3. d 4. e 5. b

6. If the plant starts to dry out, K^+ is pumped out of the guard cells. Water follows by osmosis, the guard cells become flaccid, and the stomata close. This prevents wilting, but it keeps leaves from taking in carbon dioxide, needed for photosynthesis.

7. *Hypothesis:* The hydrogen ions in acid precipitation displace positively charged nutrient ions from negatively charged clay particles. *Test:* In the laboratory, place equal amounts and types of soil in separate filters. The pore size of the filter must not allow any undissolved soil particles to pass through. Spray (to simulate rain) soil samples in the filters with solutions of different pH (e.g., pH 5, 6, 7, 8, 9). Determine the concentration of nutrient ions in the solutions. (The only variable in the solutions should be the hydrogen ion concentration. Ideally, the solutions would contain no dissolved nutrient ions.) Collect fluid that drips through soil samples and filters. Determine the hydrogen ion concentration and the nutrient ion concentration in each sample of fluid. *Prediction:* If the hypothesis is correct, the fluid collected from the soil samples exposed to pH lower than 5.6 (acid rain) will contain the highest concentration of positively charged nutrient ions.

8. *Hypothesis:* When fixed nitrogen levels increase to a certain level in the soil, it becomes harmful to the nitrogen-fixing bacteria that provide usable nitrogen to the crops. *Test:* Expose cultures of nitrogen-fixing bacteria (symbiotic and nonsymbiotic ones found in soil) to solutions of different concentrations of fixed nitrogen (i.e., NO_3^- and NH_4^+). Determine the concentrations

of nitrogen-fixing enzymes produced by the surviving bacteria in each sample. *Prediction:* If your hypothesis is correct, and the fixed nitrogen concentration is high enough to cause harm in some of the samples, you would expect the enzyme concentration to be measurably lower in samples whose fixed nitrogen concentration is above the level that causes harm.

9. The hypothesis is supported if transpiration varies with light intensity when humidity and temperature are about the same. These conditions are seen at two places in the table; at hours 11 and 12, recordings for temperature and humidity are about the same, but light intensity increased markedly from 11 to 12, as did the transpiration rate. The recordings made at hours 3 and 4 show the same effects. Also, the recordings made at hours 1 and 2 generally support the hypothesis. Here, both temperature and humidity decreased, so you might expect the transpiration rate to stay about the same or perhaps increase because the temperature decrease is small; however, the transpiration rate dropped, as did the light intensity.

10. *Some issues and questions to consider:* How were the farmers assigned or sold "rights" to the water? How is the price established when a farmer buys or sells water rights? Is there enough water for everyone who "owns" it? What kinds of crops are these farmers growing? What will the water be used for in the city? Are there other users with no rights, such as wildlife? Is any effort being made to curb urban growth and conserve water? Should millions of people be living in what is essentially a desert? What are the reasons for farming desert land?

Chapter 33

1. a. auxin; b. gibberellin; c. auxin; d. cytokinin; e. auxin; f. ethylene; g. gibberellin; h. abscisic acid

2. b 3. c (A short-day plant requires a long night. The 13-hour night of answer c is the only uninterrupted night longer than 12 hours.) 4. b 5. a 6. d 7. b 8. b 9. f 10. e 11. d 12. a 13. b 14. g 15. c

16. Fruit produces ethylene gas, which triggers the ripening and aging of the fruit. Ventilation prevents a buildup of ethylene and delays its effects.

17. The terminal bud produces auxins, which counter the effects of cytokinins from the roots and inhibit the growth of axillary buds. If the terminal bud is removed, the cytokinins predominate, and lateral growth occurs at the axillary buds.

18. The red wavelengths in the room's lights quickly convert the phytochrome in the mums to the P_{fr} form, which inhibits flowering in a long-night plant. The mums will not flower unless Jon can rig up some far-red lights. Exposure to a burst of far-red light would convert the phytochrome to the P_r form, allowing flowering to occur.

19. The biologist could remove leaves at different stages of being eaten to see how long it takes for changes to occur in nearby leaves. The "hormone" could be captured in an agar block, as in the phototropism experiments in Module 33.1, and applied to an undamaged plant. Another experiment would be to block "hormone" movement out of a damaged leaf or into a nearby leaf.

20. The results illustrated in the graph generally support the hypothesis; measured by chlorophyll fluorescence, greater leaf damage occurred in air with no isoprene than in air with isoprene. The experimenters exposed leaves to N_2 to prevent the leaves themselves from emitting isoprene. By doing so, they

were able to control the amount of isoprene in the air to which the leaves were exposed.

21. *Some issues and questions to consider:* Is the hormone safe for human consumption? What are its effects in the environment? Could its production produce impurities or wastes that might be harmful? What kinds of tests need to be done to demonstrate its safety? How much does it cost to make and use? Are the benefits worth the costs and risks? Is it worth using an artificial chemical on food simply to improve its appearance? A scientist could seek answers by studying the stability of the hormone in a variety of laboratory simulations of natural conditions. The toxicity of the hormone, of the materials used to produce it, and of its breakdown products could be determined in laboratory tests.

Chapter 34

1. a. The shape of Earth results in uneven heating, such that the tropics are warm and polar regions are cold.
b. The seasonal differences of winter and summer in temperate and polar regions are produced as Earth tips toward or away from the sun during its orbit around the sun.
c. Intense solar radiation in the tropics evaporates moisture; warm air rises, cools, and drops its moisture as rain; air circulates, cools, and drops around 30°N and S, warming as it descends and evaporating moisture from land, which creates arid regions.

2. a. All three areas have plenty of sunlight and nutrients.
b. Productivity in a stream's source is limited by lack of nutrients but also by a swift current, which doesn't allow for phytoplankton populations. Also, many streams may begin in wooded areas, where a lack of sunlight would be limiting. The oceanic photic zone is limited by low nutrients, except in areas of upwellings or temperate oceans where mixing occurs. A clear mountain lake's productivity could be limited by both lack of nutrients and low temperatures that slow growth and photosynthesis.

3. f **4.** g **5.** d **6.** a **7.** c **8.** b **9.** e **10.** a **11.** e **12.** d **13.** a **14.** c **15.** b **16.** e

17. Tropical rain forests have a warm, moist climate, with favorable growing conditions year-round. The diverse plant growth provides various habitats for other organisms.

18. Consider the main abiotic factors discussed in Module 34.4. Module 34.5 describes an example of one species' adaptations to the biome it lives in.

19. a. desert; b. grassland; c. tropical forest; d. temperate forest; e. coniferous forest; f. arctic tundra. Areas of overlap have to do with seasonal variations in temperature and precipitation.

20. Through convergent evolution, these unrelated animals adapted in similar ways to similar environments—temperate grasslands and savanna.

21. *Some issues and questions to consider:* What reasons are there for protecting areas like this? Is a public refuge or reserve the only way to protect species and habitats? Is there some way to protect organisms without confiscating land? Does protection take precedence over property rights? Would the prairie have been better off without the publicity? Should there be state or federal laws to protect endangered plants on private land? Should there be laws to protect endangered habitats, like this patch of prairie?

Chapter 35

1. a. proximate questions; b. ultimate questions; c. mating behavior; d. communication/signals; e. agonistic behaviors; f. altruism

2.

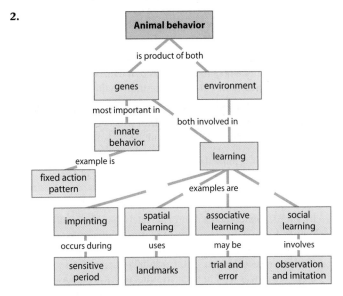

3. d **4.** c **5.** d **6.** a

7. *Main advantage:* Flies do not live long. Innate behaviors can be performed the first time without learning, enabling flies to find food, mates, etc., without practice. *Main disadvantage:* Innate behaviors are rigid; flies cannot learn to adapt to specific situations.

8. A dominance hierarchy minimizes energy wasted in fighting, maximizes efficiency in hunting, and ensures that at least some in the pack will get adequate resources. A territory ensures adequate resources for the pack.

9. Courtship behaviors reduce aggression between potential mates and confirm their species, sex, and physical condition. Environmental changes such as rainfall, temperature, and day length probably lead frogs to start calling, so these would be the proximate causes. The ultimate cause relates to evolution. Fitness (reproductive success) is enhanced for frogs that engage in courtship behaviors.

10. a. Yes. The experimenter found that 5 m provided the most food for the least energy because the total flight height (number of drops × height per drop) was the lowest. Crows appear to be using an optimal foraging strategy.
b. An experiment could measure the average drop height for juvenile and adult birds, or it could trace individual birds during a time span from juvenile to adult and see if their drop height changed.

11. One likely hypothesis is that the helper is closely related to one or both of the birds in the mated pair. Because closely related birds share relatively many genes, the helper bird is indirectly enhancing its own fitness by helping its relatives raise their young. (In other words, this behavior evolved by kin selection.) The easiest way to test the hypothesis would be to determine the relatedness of the birds by comparing tissue samples using molecular methods such as those described in Module 15.9. If birds are closely related, their DNA and proteins should be more similar than those of more distantly related or unrelated birds.

12. Identical twins are genetically the same, so any differences between them are due to environment. Thus, the study of

identical twins enables researchers to sort out the effects of "nature" and "nurture" on human behavior. The data suggest that many aspects of human behavior are inborn. Some people find these studies disturbing because they seem to leave less room for free will and self-improvement than we would like. Results of such studies may be carelessly cited in support of a particular social agenda.

13. Animals are used in experiments for various reasons. A particular species of animal may have features that make it well suited to answer an important biological question. Squids, for example, have a giant nerve fiber that made possible the discovery of how all nerve cells function. Animal experiments play a major role in medical research. Many vaccines that protect humans against deadly diseases, as well as drugs that can cure diseases, have been developed using animal experiments. Animals also benefit, as vaccines and drugs are developed for combating their own pathogens. Some researchers point out that the number of animals used in research is a small fraction of those killed as strays by animal shelters and a minuscule fraction of those killed for human food. They also maintain that modern research facilities are models of responsible and considerate treatment of animals. Whether or not this is true, the possibility that at least some kinds of animals used in research suffer physical pain as a result, and perhaps mental anguish as well, raises serious ethical issues. *Some questions to consider:* What are some medical treatments or products that have undergone testing in animals? Have you benefited from any of them? Are there alternatives to using animals in experiments? Would alternatives put humans at risk? Are all kinds of animal experiments equally valuable? In your opinion, what kinds of experiments are acceptable, and what kinds are unnecessary? What kinds of treatment are humane, and what kinds are inhumane?

Chapter 36

1. a. The *x*-axis is time; the *y*-axis is the number of individuals (*N*). The blue curve represents exponential growth; red is logistic growth.
   b. $G = rN$
   c. Carrying capacity of the environment (*K*).
   d. In exponential growth, population growth continues to increase as the population size increases. In logistic growth, the population grows fastest when the population is about $\frac{1}{2}$ the carrying capacity—when *N* is large enough so that *rN* produces a large increase, but the term $(K − N)/K$ has not yet slowed growth as much as it will as *N* gets closer to *K*.
   e. Exponential growth curve, although the human population has finally slowed from its exponential growth.

2. a. Birth rate is blue line; death rate is red line.
   b. I. Both birth and death rates are high. II. Birth rate remains high; death rate decreases, perhaps as a result of increased sanitation and health care. III. Birth rate declines, often coupled with increased opportunities for women and access to birth control; death rates are low. IV. Both birth and death rates are low.
   c. I and IV
   d. II, when death rate has fallen but birth rate remains high

3. a 4. c 5. e 6. c 7. b 8. d

9. Food and resource limitation, such as food or nesting sites; accumulation of toxic wastes; disease; increase in predation; stress responses, such as seen in some rodents

10. Survivorship is the fraction of individuals in a given age interval that survive to the next interval. It is a measure of the probability of surviving at any given age. A survivorship curve shows the fraction of individuals in a population surviving at each age interval during the life span. Oysters produce large numbers of offspring, most of which die young, with a few living a full life span. Few humans die young; most live out a full life span and die of old age. Squirrels have approximately constant mortality and about an equal chance of surviving at all ages.

11. Clumped is the most common dispersion pattern, usually associated with unevenly distributed resources or social grouping. Uniform dispersion may be related to territories or inhibitory interactions between plants. A random dispersion is least common and may occur when other factors do not influence the distribution of organisms.

12. 10% of the animals in the second capture were marked. Knowing that 50 animals were initially marked, and assuming that 10% of the whole population is marked, then the population is estimated to be 500. The equation for the mark-recapture method is:

$$N = \frac{\text{Number marked in first catch} \times \text{total second catch}}{\text{recaptured marked individuals}}$$

13. An animal that has been trapped before may be wary of the traps or, having learned that traps contain food, may have deliberately returned for more. This would cause estimates to be either higher or lower, respectively, than actual population numbers.

14. Populations with *K*-selected life history traits tend to live in fairly stable environments, held near carrying capacity by density-dependent limiting factors. They reproduce later and have fewer offspring than species with *r*-selected traits. Their lower reproductive rate makes it hard for them to recover from human-caused disruption of their habitat. We would expect species with *K*-selected life histories to have a Type I survivorship curve (see Module 36.3).

15. The largest population segment, the baby boomers, are currently in the workforce in their peak earning years, paying into the Social Security system. However, they are approaching the end of their contributing years, and the smaller working numbers following them will provide less money, driving the fund into a deficit.

16. *Some issues and questions to consider:* How does population growth in developing countries relate to food supply, pollution, and the use of natural resources such as fossil fuels? How are these things affected by population growth in developed countries? Which of these factors are most critical to our survival? Are they affected more by the growth of developing or developed countries? What will happen as developing countries become more developed? Will it be possible for everyone to live at the level of the developed world?

Chapter 37

1. a. competition; b. predation; c. herbivory; d. slightly different niches; e. mutualism; f. predator; g. warning coloration (poisonous), camouflage, mimicry

2. a. energy flow; b. chemical cycling; c. solar energy; d. primary producers; e. abiotic reservoirs; f. trophic levels (food chain or web)

3. e **4.** b **5.** d **6.** c **7.** c

8. Cultural eutrophication is excessive fertilization of bodies of water from agricultural or suburban runoff that results in blooms of cyanobacteria and algae. Respiration from these organisms and their decomposers depletes oxygen levels, leading to fish kills.

9. The abiotic reservoir of the first three nutrients is the soil. Carbon is available as carbon dioxide in the atmosphere.

10. These animals are secondary or tertiary consumers, at the top of the production pyramid. Stepwise energy loss means not much energy is left for them; thus, they are rare and require large territories in which to hunt.

11. Chemicals with a gaseous form in the atmosphere, such as carbon and nitrogen, have a global biogeochemical cycle.

12. Nitrogen fixation of atmospheric N_2 into ammonium; decomposition of detritus into ammonium; nitrification of ammonium into nitrate; denitrification of nitrates into N_2

13. *Hypothesis:* The kangaroo rat is a keystone herbivore in the desert. (Apparently, herbivory by the rats kept the one plant from outcompeting the others; removing the rats reduced plant diversity.) *Additional supporting evidence:* Observations of the rats preferentially eating dominant plants; finding that the dominant plant recovers from herbivore damage faster

14. *Some issues and questions to consider:* What relationships (predators, competitors, parasites) might exist in the mussels' native habitat that are altered in the Great Lakes? How might the mussels compete with Great Lakes organisms? Might competitive exclusion occur? What happens to displaced competitors? Might the Great Lakes species adapt in some way? Might the mussels adapt? Could possible solutions present problems of their own?

Chapter 38

1. endangered and threatened species; b. restoration ecology; c. bioremediation and augmentation; d. zoned reserves; e. critical habitat needs

2. c **3.** d **4.** a **5.** e **6.** b **7.** d **8.** a

9. Genetic, species, and ecosystem diversity. As populations become smaller, genetic diversity is usually reduced. Genetic biodiversity is also threatened when local populations of a species become extinct owing to habitat destruction or other assaults or when entire species are lost. Many human activities have led to the extinction of species. The greatest threats include habitat loss, introduced species, and overexploitation. Species extinction may alter the structure of whole communities. Pollution and other widespread disruptions may lead to the loss of entire ecosystems.

10. Carbon dioxide and several other gases in the atmosphere absorb infrared radiation and thus slow the escape of heat from Earth. This is called the greenhouse effect. The greenhouse effect is beneficial to life on Earth; without CO_2 in the atmosphere, the temperature at the surface of Earth would be much colder and less hospitable for life.

11. Fossil fuel consumption, industry, and agriculture are increasing the quantity of greenhouse gases—such as CO_2, methane, and nitrous oxide—in the atmosphere. This could trap more heat and raise atmospheric temperatures 2–5°C over the next century. Logging and the clearing of forests for farming contribute to global warming by reducing the uptake of CO_2 by plants (and adding CO_2 to the air when trees are burned). Global warming could shift patterns of precipitation, turning farmland into deserts. It could also cause polar ice caps to melt, raising ocean levels and flooding coastal areas. Global warming is an international problem; air and climate do not recognize international boundaries. Greenhouse gases produced by industrialized nations add to those resulting from deforestation in less developed tropical nations. Cooperation and commitment to a less consumerist lifestyle will be needed to slow global warming.

12. These birds might be affected by pesticides while in their wintering grounds in Central and South America, where such chemicals may still be in use. The birds are also affected by deforestation throughout their range.

13. About 1.5 million species have been named and described. Assume that 80% of all living things (not just plants and animals) live in tropical rain forests. This means that there are $1.5 \times 0.8 = 1.2$ million species there. If half the species survive, this means that 0.6 million species will be extinct in 100 years, or 0.6 million/100 = 6,000 per year. This means that 6,000/365 = 16 species will disappear per day, or almost one per hour. If there are 30 million species on Earth, 24 million live in the tropics, and 12 million will disappear in the next century. This is 120,000 per year, 329 per day, or 14 per hour.

14. *Some issues and questions to consider:* How does our use of fossil fuels affect the environment? What about oil spills? Disruption of wildlife habitat for construction of oil fields and pipelines? Burning of fossil fuels and possible climate change and flooding from global warming? Pollution of lakes and destruction of property by acid precipitation? Health effects of polluted air on humans? How are we paying for these "side effects" of fossil fuel use? In taxes? In health insurance premiums? Do we pay a nonfinancial price in terms of poorer health and quality of life? Could oil companies be required to pick up the tab for environmental effects of fossil fuel use? Could these costs be covered by an oil tax? How would this change the price of oil? How would a change in the price of oil change our pattern of energy use, our lifestyle, and our environment?

15. *Some issues and questions to consider:* Does the current evidence constitute "proof" that global warming is occurring? Why is it difficult to prove conclusively that human activities have an effect on global warming? How much information do we need before we act? Which of the following would be more serious, and why? Acting now to curb global warming, and later finding out that we were wrong—that global warming is not a threat? Or failing to take action now and later finding out that global warming is a real and serious problem?

16. *Some issues and questions to consider:* How do population growth, resource consumption, pollution, and reduction in biodiversity relate to sustainability? How do poverty, economic growth and development, and political issues relate to sustainability? Why might developed and developing nations take different views of a sustainable society? What would life be like in a sustainable society? Have any steps toward sustainability been taken in your community? What are the obstacles to sustainability in your community? What steps have you taken toward a sustainable lifestyle? How old will you be in 2030? What do you think life will be like then?

Appendix 4: Credits

Photo Credits

Unit Openers: Unit I p 15 Nancy Kedersha/SPL/Photo Researchers; **Unit II p 123** Jennifer Waters Shuler/Photo Researchers; **Unit III p 255** Colin Keates/Dorling Kindersley; **Unit IV p 313** Michael Sewell/Peter Arnold; **Unit V p 411** Michael & Patricia Fogden/Minden Pictures; **Unit VI p 621** Chris Jones/Photolibrary.com; **Unit VII p 681** Joseph Van Os/The Image Bank

Frontmatter: p xix Tui De Roy/Minden Pictures; **p xix** RCSB; **p xx** Mark Cassino/SuperStock; **p xxi** Getty Images; **p xxi** Getty Images; **p xxii** Andrew Olney/Stone; **p xxiii** Advanced Cell Technology, Inc; **p xxiv** Wolfgang Kaehler/CORBIS; **p xxv** Getty Images; **p xxvi** Dana Richter/Visuals Unlimited; **p xxvii** Dave Watts/Tom Stack & Assoc; **p xxvii** Columbia/Marvel/Kobal Collection; **p xxviii** Richard Schlecht/National Geographic Image Collection; **p xxxix** Chris Taylor, Cordaiy Photo Library Ltd./CORBIS; **p xxx** Brooks Kraft/Sygma; **p xxxi** Richard Lair/Thai Elephant Conservation Center; **p xxxii** Robert Holmes/CORBIS; **p xxxiii** The Image Bank; **p xxxiv** John Serrao/Visuals Unlimited; **p xxxv** Steve Winter/National Geographic Image Collection

Chapter 1: Opening photos left and center Tui De Roy/Minden Pictures; **right** Konrad Wothe/Minden Pictures; **1.1 top to bottom** World-Sat International/Photo Researchers; Dan Guravich/CORBIS; CORBIS; **1.4B** John Foxx/Imagestate; **1.4C** Joe McDonald/CORBIS; **1.4D** Jeff Lepore/Photo Researchers; **1.4E** Gerry Ellis/Minden Pictures; **1.5A** Oliver Meckes/Nicole Ottawa/Photo Researchers; **1.5B** Ralph Robinson/Visuals Unlimited; **1.5C top left** D. P. Wilson/Photo Researchers; **bottom left** Michael & Patricia Fogden/Minden Pictures; **middle** Frank Frank Young/CORBIS; **right** Michael and Patricia Fogden/CORBIS; **1.6A** Department of Library Services/American Museum of Natural History; **1.6C top** Stuart Westmorland/The Image Bank; **bottom** N. J. Dennis/Photo Researchers; **1.8B** E. R. Degginger/Photo Resarchers; **1.8C** Breck P. Kent; **1.8D both** David Pfennig; **1.9** Benjamin Cummings

Chapter 2: Opening photos top left and center Thomas Eisner, Cornell University; **top right** Robert Barker, Cornell University; **bottom** Photodisc; **2.2A** Ivan Polunin/Bruce Coleman, Inc; **2.2B** Benjamin Cummings; **2.3 left** Chip Clark; **center and right** Benjamin Cummings; **2.5A and B** CTI, Inc; **2.11** George Bernard/Animals Animals; **2.12** Greg Fiume/NewSport/CORBIS; **2.16A** Oliver Strewe/Stone; **2.16B** ArsNatura; **2.17B** Dorling Kindersley

Chapter 3: Opening photos left RCSB; **right** Michael S. Yamashita/CORBIS; **3.2** Digital Vision; **3.4A** Scott Camazine/Photo Researchers; **3.7 top and bottom** Biophoto Associates/Photo Researchers; **3.7 center** L. M. Beidler, Florida State University; **3.8A** Tony Hamblin, Frank Lane Picture Agency/CORBIS; **3.10** Mike Neveux Photography; **3.11** Eric Feferberg/AFP/Getty Images; **p 44** Wolfgang Kaehler/Liaison; **3.15** Archives, California Institute of Technology

Chapter 4: Opening photos left CNAC/MNAM/Dist. Réunion des Musées Nationaux/Art Resource, NY; **center** ArsNatura; **4.1A** Leica Microsystems Inc.; **4.1B** Michael Abbey/Visuals Unlimited; **4.1C** Andrew Syred/Photo Researchers; **4.1D** Dennis Kunkel/Visuals Unlimited; **4.1E** J. D. Eisenback; **4.1F** Torsten Wittman, The Nikon Small World Competition; **4.3A left** S. C. Holt, University of Texas Health Center/Biological Photo Service; **right** CNRI/Science Photo Library/Photo Researchers; **4.7** R. Bolender, D. Fawcett/Photo Researchers; **4.9** Don Fawcett/Visuals Unlimited; **4.10B** R. Rodewald, University of Virginia/Biological Photo Service; **4.10C** Daniel S. Friend, Harvard Medical School; **4.12A** Dr. Henry C. Aldrich/Visuals Unlimited; **4.12B** Roland Birke/Peter Arnold, Inc; **4.14** W.P. Wergin and E.H. Newcomb, University of Wisconsin/Biological Photo Service; **4.15** Nicolae Simionescu; **4.16 left to right** Mary Osborn, Max Planck Institute; Frank Solomon and J. Dinsmore, Massachusetts Institute of Technology; Mark S. Ladinsky and J. Richard McIntosh, University of Colorado;

4.17A P. Motta/Univ. "La Sapienza"/Photo Researchers; **4.17B** Biophoto Associates/Photo Researchers; **4.17C both** W.L. Dentler, Universtiy of Kansas/Biological Photo Service

Chapter 5: Opening photos left Mark Cassino/SuperStock; **right** James E. Lloyd; **5.1A** Dennis Curran/Index Stock Imagery; **5.1B** Benjamin Cummings; **5.1C** Taxi; **5.2A** Wally McNamee/CORBIS; **5.10** David Robertson; **5.19C left** Michael Abbey/Visuals Unlimited; **center** D.W Fawcett/Science Source/Photo Researchers; **right** M.M. Perry and A.B. Gilbert, J. Cell Sci. 39(1979): 257. Copyright 1979 The Company of Biologists Ltd.

Chapter 6: Opening photos left Getty Images; **right** Jose Azel/Aurora & Quanta Productions; **6.2** Getty Images; **6.13C** Charles O'Rear/CORBIS; **6.14** Dorling Kindersley; **6.15** Dorling Kindersley; **6.16** Gerry Ellis/Minden Pictures

Chapter 7: Opening photos both Timothy Volk, Biomass Program, SUNY-ESF; **7.1A** Theresa DeSalis/Peter Arnold, Inc; **7.1B** EyeWire; **7.1C** Ralph A. Glevenger/CORBIS; **7.1D** T. E. Adams/Visuals Unlimited; **7.2 bottom left** W. P. Wergin and E.H. Newcomb, University of Wisconsin/Biological Photo Service; **7.2 top right** Graham Kent, Benjamin Cummings; **7.3A** Runk/Schoenberger/Grant Heilman Inc; **7.6B** Adam Smith/Taxi; **7.7A** Christine L. Case; **7.12 left** David Muench/CORBIS; **right** David Bartruff/CORBIS; **7.13A** Raymond Gehman/CORBIS; **7.14A** Stella Johnson, Benjamin Cummings; **7.14B** NASA/Goddard Space Flight Center

Chapter 8: Opening photos left Richard Neville/Transtock Inc./Alamy Images; **top right** Courtesy of the National Tropical Botanical Garden, Kauai, Hawaii; **bottom right** Michael T. Sedam/CORBIS; **8.1A** Biophoto Associates/Photo Researchers; **8.1B** Gary Buss/Taxi; **8.3B** Lee D. Simon/Photo Researchers; **8.4A** Andrew Bajer, University of Oregon; **8.4B** Biophoto/Photo Researchers; **8.6 all** Conly Rieder; **8.7A** David M. Phillips/Visuals Unlimited; **8.7B** B.A. Palevitz, courtesy of E.H. Newcomb, University of Wisconsin; **8.11A** Brian Capon; **8.11B** Biophoto/Science Source/Photo Researchers; **8.11C** Biophoto Associates/Science Source/Photo Researchers; **8.17B left** F. Schussler/PhotoDisc; **right** Chris Collins/Bettmann/CORBIS; **8.18A** Cabisco/Visuals Unlimited; **8.19 left** Veronique Burger/Photo Researchers; **right** CNRI/Science Photo Library/Photo Researchers; **8.20A** CNRI/Science Photo Library/Photo Researchers; **8.20B** Lauren Shear/Photo Researchers; **p 151** Carolina Biological Supply/Phototake NYC

Chapter 9: Opening photos left Anthony Edgeworth/CORBIS; **right** Andrew Olney/Stone; **9.2A** The Bettmann Archive/CORBIS; **9.8A top left** CORBIS; **top right** Eyewire; **center left and right** PhotoDisc; **bottom left and right** Anthony Loveday, Benjamin Cummings; **9.9B** Dick Zimmerman/Shooting Star; **9.10A** CNRI/Science Photo Library/Photo Researchers; **9.10B** Eric J. Simon; **inset** Stone; **9.14** Bill Longcore/Photo Researchers; **9.16** Carl Schneider/Taxi; **9.21A** From Thomas Hunt Morgan: The Man and His Science, Garland Allen (Princeton University Press, 1978). Photo by Dr. Tove Mohr, Fredrikstad, Norway, provided by Garland Allen; **9.23A** Carolina Biological/Visuals Unlimited; **9.24B** FPG International **p 179** Breeder/owner: Patricia Speciale; photographer: Norma Jubinville

Chapter 10: Opening photos left Peter Hince/The Image Bank; **right** Eye of Scienct/Photo Researchers; **10.1A** Robley C. Williams, University of California Berkeley/Biological Photo Services; **10.3A** Elliott & Fry, National Portrait Gallery; **10.3B** National Institutes of Health; **10.3D** Richard Wagner, UCSF Graphics; **10.6B** Arthur M. Siegelman/Visuals Unlimited; **10.11C** M. A. Rould, J. J. Perona, P. Vogt, and T. A. Steitz, Science 246 (1 December 1989):cover. Copyright 1989 by the American Association for the Advancement of Science; **10.12A** Joachim Frank, Howard Hughes Medical Institute; **10.19** N. Thomas/Photo Researchers;

Illustration and Text Credits

The following figures are adapted from C. K. Mathews and K. E. van Holde, *Biochemistry,* 2nd ed. (Menlo Park, CA: Benjamin/Cummings, 1996), © 1996 The Benjamin/Cummings Publishing Company: **6.10**, **6.11**, and **11.4**.

The following figures are adapted from Elaine N. Marieb, *Human Anatomy and Physiology,* 5th ed. (San Francisco, Ca: Benjamin Cummings, 2001), © Benjamin Cummings: **4.4A**, **4.5**, **4.7**, and **EOC 4.2**.

The following figures are adapted from Elaine N. Marieb, *Human Anatomy and Physiology,* 4th ed. (Menlo Park, Ca: Benjamin/Cummings, 1998), © 1998 The Benjamin/Cummings Publishing Company: **22.9**, **22.11**, **25.11**, **27.2A**, **27.2C**, **27.3A**, **27.18B**, **29.7A**, **29.7B**, and **30.4**

The following figures are adapted from Gerard J. Tortora, Berdell R. Funke, and Christine L. Case, *Microbiology: An Introduction,* 6th ed. (Menlo Park, CA: Benjamin/Cummings, 1998), © 1998 Benjamin/Cummings: **12.1**, **16.16A**, **24.8B**, and **Table 27.7**.

The following figures are adapted from Lawrence G. Mitchell, John A. Mutchmor, and Warren D. Dolphin, *Zoology* (Menlo Park, CA: Benjamin/Cummings, 1988), © 1988 The Benjamin/Cummings Publishing Company: **2.5**, **18.1B**, **18.14B**, **21.11B**, **25.6**, **30.2E**, and **35.7B**.

Figure 1.8E: Data in graph based on D. W. Pfennig et al. 2001. Frequency-dependent Batesian mimicry. *Nature* 410: 323.

Chapter 2 introduction: Adapted from interview in Neil Campbell and Jane Reece, *Biology,* 6th ed. (San Francisco, CA: Benjamin Cummings, 2002). © 2002 Pearson Education, Inc., publishing as Benjamin Cummings.

Figures 3.13A and **B:** Adapted from D. W. Heinz, W. A. Baase, F. W. Dahlquist, B. W. Matthews, 1993. How amino-acid insertions are allowed in an alpha-helix of T4 lysozyme. *Nature* 361:561.

Unnumbered figure, page 44: Illustration adapted from Irving Geis. Rights owned by Howard Hughes Medical Institute. Not to be reproduced without permission.

Module 3.15: Talking About Science: Adapted from an interview in Neil Campbell *Biology,* 1st ed. (Menlo Park, CA: Benjamin/Cummings, 1987), © 1987 The Benjamin/ Cummings Publishing Company.

Table 6.4: Data from C. M. Taylor and G. M. McLeod, *Rose's Laboratory Handbook for Dietetics,* 5th ed. (New York: Macmillan, 1949), p. 18; J. V. G. A. Durnin and R. Passmore, *Energy and Protein Requirements in FAO/ WHO Technical Report* No. 522, 1973; W. D. McArdle, F. I. Katch, and V. L. Katch, *Exercise Physiology* (Philadelphia, PA: Lea & Feibiger, 1981); R. Passmore and J. V. G. A. Durnin, *Physiological Reviews* Vol. 35, pp. 801–840 (1955).

7.8B: Adapted from Richard and David Walker, ENERGY, PLANTS and MAN, fig. 4.1, p. 69. Oxygraphics <http://www.oxygraphics.co.uk> Copyright © Richard Walker. Reprinted by permission.

Module 7.14: Talking About Science: Adapted from an interview in Neil Campbell, Jane Reece, and Lawrence Mitchell, *Biology,* 5th ed. (Menlo Park, CA: Benjamin/Cummings, 1999), © 1999 Benjamin/Cummings.

Figure 8.22A, B: Adapted from F. Vogel and A. G. Motulski, HUMAN GENETICS. Copyright © 1982 Springer-Verlag.

Figure 9.8B: Adapted from *Everyone Here Spoke Sign Language* by Nora Ellen Grocer. Copyright © 1985 by Nora Ellen Grocer. Reprinted by permission of Harvard University Press.

Figure 9.14: Adapted from *Introduction to Genetic Analysis,* 4th ed. by Suzuki, Griffiths, Miller, and Leonine. Copyright © 1976, 1981, 1986, 1989, 1993, 1996 by W. H. Freeman and Company. Used with permission.

Module 10.20: Text quotation from Barbara J. Colleton, "Emerging Viruses, Emerging Threats," *Science,* Vol. 247, p. 279 (19 January 1990).

Figure 11.3: Adapted from W. H. Becker, *The World of the Cell,* p. 592 (Redwood City, CA: Benjamin/Cummings, 1986), © 1986 by The Benjamin/Cummings Publishing Company.

Figure 11.15: Adapted from an illustration by William McGinnis, UCSD.

Module 11.19: Talking About Science: Adapted from an interview in Neil Campbell, Jane Reece, and Lawrence Mitchell, *Biology,* 5th ed. (Menlo Park, CA: Benjamin/Cummings, 1999), © 1999 Benjamin/Cummings.

Table 11.20: Data from *2002 Facts & Figures: Estimated New Cancer Cases and Deaths,* published by the American Cancer Society.

Module 12.20: Talking About Science: Adapted from an interview in Neil Campbell and Jane Reece, *Biology,* 7th ed. (San Francisco, CA: 2005), © 2005 Pearson Education, publishing as Benjamin Cummings.

Figure 14.5A: Fig. 14.5A adapted from D. M. B. Dodd, *Evolution,* Vol. 11, pp. 1308-1311. Reprinted by permission of the Society for the Study of Evolution.

Module 14.9: Talking About Science: Adapted from an interview in Neil Campbell and Jane Reece, *Biology,* 6th ed. (San Francisco, CA: 2002), © 2002 Pearson Education, publishing as Benjamin Cummings.

Figure 14.11: Adapted from M. Strickberger. 1990. *Evolution.* Boston: Jones & Bartlett.

Figure 15.3D: From W. K. Purves and G. H. Orians, *Life: The Science of Biology,* 2nd ed., fig. 16.5c, p. 1180 (New York: Sinauer Associates, 1987.) Used with permission.

Figure 16.1A: Artist: Peter Sawyer © NMNH Smithsonian Institution.

Module 16.3: Talking About Science: Adapted from an interview in Neil Campbell, *Biology,* 2nd ed. (Redwood City, CA: Benjamin/Cummings, 1990), © 1990 The Benjamin/Cummings Publishing Company.

Table 17.14: Adapted from Randy Moore et al., *Botany,* 2nd ed. Dubuque, IA: Brown, 1998, Table 2.2, p. 37.

Figure 18.18A: Adapted from C. Zimmer. 1999. *At the Water's Edge.* Free Press, Simon & Schuster p. 90.

Figure 18.22: Adapted from André Adouette, et al., *Proceedings of the National Academy of Science,* Vol. 97, Issue 9, p. 4454, April 25, 2000. Copyright © 2000 National Academy of Sciences, U.S.A. Reprinted by permission.

Figure 19.3: Drawn from photos of fossils: *O. tugenensis* photo in Michael Balter, Early hominid sows division, *ScienceNow,* Feb. 22, 2001, © 2001 American Association for the Advancement of Science. *A. ramidus kadabba* photo by Timothy White, 1999/Brill Atlanta. *A. anamensis, A. garhi,* and *H. neanderthalensis* adapted from *The Human Evolution Coloring Book. K platyops* drawn from photo in Meave Leakey et al., New hominid genus from eastern Africa shows diverse middle Pliocene lineages, *Nature,* March 22, 2001, 410:433. *P. boisei* drawn from a photo by David Bill. *H. ergaster* drawn from a photo at www.inhandmuseum.com. *S. tchadensis* drawn from a photo in Michel Brunet et al., A new hominid from the Upper Miocene of Chad, Central Africa, *Nature,* July 11, 2002, 418:147, fig. 1b.

Figure 21.17: Adapted from Murray W. Nabors, *Introduction to Botany,* San Francisco, CA: Benjamin Cummings. © 2004 Pearson Education, Inc., Upper Saddle River, New Jersey.

Table 21.18: Data from RDA Subcommittee, *Recommended Dietary Allowances* (Washington, D. C.: National Academy Press, 1989); M. E. Shils and V. R. Young, *Modern Nutrition in Health and Disease* (Philadelphia, PA: Lea & Feibiger, 1988).

Glossary

A

A site One of two of a ribosome's binding sites for tRNA during translation. The A site holds the tRNA carrying the next amino acid to be added to the polypeptide chain. (A stands for aminoacyl tRNA.)

abiotic component (ā´-bī-ot´-ik) A nonliving component of an ecosystem, such as air, water, or temperature.

abiotic reservoir The part of an ecosystem where a chemical, such as carbon or nitrogen, accumulates or is stockpiled outside of living organisms.

ABO blood groups Genetically determined classes of human blood that are based on the presence or absence of carbohydrates A and B on the surface of red blood cells. The ABO blood group phenotypes, also called blood types, are A, B, AB, and O.

abscisic acid (ABA) (ab-sis´-ik) A plant hormone that inhibits cell division and promotes dormancy; interacts with gibberellins in regulating seed germination.

absorption The uptake of small nutrient molecules by an organism's own body; the third main stage of food processing, following digestion.

accommodation The automatic changes made by the eye as it focuses on nearby objects.

acetyl CoA (acetyl coenzyme A) The entry compound for the citric acid cycle in cellular respiration; formed from a fragment of pyruvate attached to a coenzyme.

acetylcholine (as´-uh-til-kō´-lēn) A nitrogen-containing neurotransmitter. Among other effects, it slows the heart rate and makes skeletal muscles contract.

achondroplasia (uh-kon´-druh-plā´-zhuh) A form of human dwarfism caused by a single dominant allele; the homozygous condition is lethal.

acid A substance that increases the hydrogen ion concentration in a solution.

acid chyme (kīm) A mixture of recently swallowed food and gastric juice.

acid precipitation Rain, snow, sleet, hail, drizzle, and so on, with a pH below 5.6. Acid precipitation can damage or destroy organisms by acidifying lakes, streams, and possibly land habitats.

acquired immunity The kind of defense that is mediated by B lymphocytes (B cells) and T lymphocytes (T cells). It exhibits specificity, memory, and self-nonself recognition. Also called adaptive immunity.

acrosome (ak´-ruh-sōm) A membrane-enclosed sac at the tip of a sperm; contains enzymes that help the sperm penetrate an egg.

action potential A self-propagating change in the voltage across the plasma membrane of a neuron; a nerve signal.

activator A protein that switches on a gene or group of genes.

active immunity Immunity conferred by recovering from an infectious disease.

active site The part of an enzyme molecule where a substrate molecule attaches (by means of weak chemical bonds); typically, a pocket or groove on the enzyme's surface.

active transport The movement of a substance across a biological membrane against its concentration gradient, aided by specific transport proteins and requiring input of energy (often as ATP).

adaptive immunity *See* acquired immunity.

adaptive radiation The emergence of numerous species from a common ancestor introduced to new and diverse environments.

adenine (A) (ad´-uh-nēn) A double-ring nitrogenous base found in DNA and RNA.

adhesion The attraction between different kinds of molecules.

adipose tissue A type of connective tissue whose cells contain fat.

adrenal cortex (uh-drē´-nul) The outer portion of an adrenal gland, controlled by ACTH from the anterior pituitary; secretes hormones called glucocorticoids and mineralocorticoids.

adrenal gland One of a pair of endocrine glands, located atop each kidney in mammals, composed of an outer cortex and a central medulla.

adrenal medulla (uh-drē´-nul muh-dul´-uh) The central portion of an adrenal gland, controlled by nerve signals; secretes the fight-or-flight hormones epinephrine and norepinephrine.

adrenocorticotropic hormone (ACTH) (uh-drē´-nō-cōr´-ti-kō-trop´-ik) A protein hormone secreted by the anterior pituitary that stimulates the adrenal cortex to secrete corticosteroids.

adult stem cell A cell present in adult tissues that generates replacements for nondividing differentiated cells.

aerobic exercise Exercising muscles at a level at which sufficient oxygen remains available for aerobic respiration.

age structure The relative number of individuals of each age in a population.

aggregate fruit A fruit, such as a blackberry, that develops from a single flower with many carpels.

agonistic behavior (a´-gō-nis´-tik) Confrontational behavior involving a contest waged by threats, displays, or actual combat, which settles disputes over limited resources, such as food or mates.

AIDS Acquired immune deficiency syndrome; the late stages of HIV infection, characterized by a reduced number of T cells and the appearance of characteristic secondary infections.

alcohol fermentation The conversion of the acid produced by glycolysis to carbon dioxide and ethyl alcohol.

alga (al´-guh) (plural, **algae**) One of a great variety of protists, most of which are unicellular or colonial photosynthetic autotrophs with chloroplasts containing the pigment chlorophyll *a*. Heterotrophic and multicellular protists closely related to unicellular autotrophs are also regarded as algae.

alimentary canal (al´-uh-men´-tuh-rē) A digestive tract consisting of a tube running between a mouth and an anus.

allantois (al´-an-tō´-is) In animals, an extraembryonic membrane that develops from the yolk sac; helps dispose of the embryo's nitrogenous wastes and forms part of the umbilical cord in mammals.

allele (uh-lē´-ul) An alternative form of a gene.

allergen (al´-er-jen) An antigen that causes an allergy.

allergy A disorder of the immune system caused by an abnormal sensitivity to an antigen. Symptoms are triggered by histamines released from mast cells.

allopatric speciation The formation of a new species as a result of an ancestral population's becoming isolated by a geographic barrier.

alpha helix (al´-fuh hē´-liks) The spiral shape resulting from the coiling of a polypeptide in a protein's secondary structure.

alternation of generations A life cycle in which there is both a multicellular diploid form, the sporophyte, and a multicellular haploid form, the gametophyte; a characteristic of plants and multicellular green algae.

alternative RNA splicing A type of regulation at the RNA-processing level in which different mRNA molecules are produced from the same primary transcript, depending on which RNA segments are treated as exons and which as introns.

altruism (al´-trū-iz-um) Behavior that reduces an individual's fitness while increasing the fitness of another individual.

alveolate A clade of protists that includes dinoflagellates, apicomplexans, and ciliates.

alveolus (al-vē´-oh-lus) (plural, **alveoli**) One of millions of tiny sacs within the vertebrate lungs where gas exchange occurs.

Alzheimer's disease (AD) An age-related dementia (mental deterioration) characterized by confusion, memory loss, and other symptoms.

amine (uh-mēn´) An organic compound with one or more amino groups.

amino acid (uh-mēn´-ō) An organic molecule containing a carboxyl group and an amino group; serves as the monomer of proteins.

amino group In an organic molecule, a functional group consisting of a nitrogen atom bonded to two hydrogen atoms.

ammonia A small and very toxic nitrogenous waste produced by metabolism.

amniocentesis (am´-nē-ō-sen-tē´-sis) A technique for diagnosing genetic defects while a fetus is in the uterus. A sample of amniotic fluid, obtained via a needle inserted into the amnion, is analyzed for telltale chemicals and defective fetal cells.

amnion (am´-nē-on) In vertebrate animals, the extraembryonic membrane that encloses the fluid-filled amniotic sac containing the embryo.

amniote Member of a clade of tetrapods that have an amniotic egg containing specialized membranes that protect the embryo. Amniotes include mammals and birds and other reptiles.

amniotic egg (am´-nē-ot´-ik) A shelled egg in which an embryo develops within a fluid-filled amniotic sac and is nourished by yolk; produced by reptiles, birds, and egg-laying mammals, it enables them to complete their life cycles on dry land.

amoeba (uh-mē´-buh) A type of protist characterized by great flexibility and the presence of pseudopodia.

amoebocyte (uh-mē´-buh-sīt) An amoeba-like cell that moves by pseudopodia, found in most animals; depending on the species, may

digest and distribute food, dispose of wastes, form skeletal fibers, fight infections, and change into other cell types.

amoebozoan A clade of protists that includes amoebas and slime molds and is characterized by lobe-shaped pseudopodia.

amphibian Member of the tetrapode class Amphibia. Amphibians include frogs, toads, and salamanders.

amygdala (uh-mig´-duh-la) An integrative center of the cerebrum; functionally the part of the limbic system that seems to label information to be remembered.

anabolic steroid (an´-uh-bol´-ik stār´-oyd) A synthetic variant of the male hormone testosterone that mimics some of its effects.

anaerobic exercise Exercising muscles until they no longer receive adequate oxygen and must switch to anaerobic respiration.

analogy The similarity of structure between two species that are not closely related; attributable to convergent evolution.

anaphase The fourth stage of mitosis, beginning when sister chromatids separate from each other and ending when a complete set of daughter chromosomes have arrived at each of the two poles of the cell.

anaphylactic shock (an´-uh-fi-lak´-tik) A potentially fatal allergic reaction caused by extreme sensitivity to an allergen; involves an abrupt dilation of blood vessels and a sharp drop in blood pressure.

anatomy The study of the structure of an organism.

anchorage dependence The requirement that to divide, a cell must be attached to a solid surface.

anchoring junction A junction that connects tissue cells to each other (or to an extracellular matrix) and allows materials to pass from cell to cell.

androgen (an´-drō-jen) A steroid sex hormone secreted by the gonads that promotes the development and maintenance of the male reproductive system and male body features.

anemia (uh-nē´-me-ah) A condition in which an abnormally low amount of hemoglobin or a low number of red blood cells results in the body cells not receiving enough oxygen.

angiosperm (an´-jē-ō-sperm) A flowering plant, which forms seeds inside a protective chamber called an ovary.

Animalia (an-eh-mal´-ē-uh) The kingdom that contains the animals.

annelid (uh-nel´-id) A segmented worm. Annelids include earthworms, polychaetes, and leeches.

annual A plant that completes its life cycle in a single year or growing season.

antagonistic hormones Two hormones that have opposite effects.

anterior Pertaining to the front, or head, of a bilaterally symmetrical animal.

anterior pituitary (puh-tū´-uh-tār-ē) An endocrine gland, adjacent to the hypothalamus and the posterior pituitary, that synthesizes several hormones, including some that control the activity of other endocrine glands.

anther A sac in which pollen grains develop, located at the tip of a flower's stamen.

anthropoid (an´-thruh-poyd) A member of a primate group made up of the apes (gibbons, orangutans, gorillas, chimpanzees, bonobos, and humans) and monkeys.

antibody (an´-tih-bod´-ē) A protein dissolved in blood plasma that attaches to a specific kind of antigen and helps counter its effects.

anticodon (an´-tī-kō´-don) On a tRNA molecule, a specific sequence of three nucleotides that is complementary to a codon triplet on mRNA.

antidiuretic hormone (ADH) (an´-tē-dī´-yū-ret´-ik) A hormone made by the hypothalamus and secreted by the posterior pituitary that promotes water retention by the kidneys.

antigen (an´-tuh-jen) A foreign (nonself) molecule that elicits an immune response.

antigen receptor A transmembrane version of an antibody molecule that B cells and T cells use to recognize specific antigens; also called a membrane antibody.

antigen-binding site A region of the antibody molecule responsible for the antibody's recognition and binding function.

antigen-presenting cell (APC) One of a family of white blood cells (e.g., a macrophage) that ingests a foreign substance or a microbe and attaches antigenic portions of the ingested material to its own surface, thereby displaying the antigens to a helper T cell.

antigenic determinant A region on the surface of an antigen molecule to which an antibody binds.

antihistamine (an´-tē-his´-tuh-mēn) A drug that interferes with the action of histamine, providing temporary relief from an allergic reaction.

anus The opening through which undigested materials are expelled.

aorta (ā-or´-tuh) An artery that conveys blood directly from the heart to other arteries.

aphotic zone (ā-fō´-tik) The region of an aquatic ecosystem beneath the photic zone, where light does not penetrate enough for photosynthesis to take place.

apical dominance (ā´-pik-ul) In a plant, the hormonal inhibition of axillary buds by a terminal bud.

apical meristem (ā´-pik-ul mer´-uh-stem) A meristem at the tip of a plant root or in the terminal or axillary bud of a shoot.

apicomplexan (ap´-ē-kom-pleks´-un) Any of a group of parasitic protozoans, some of which cause human diseases.

apoptosis (ap´-op-tō´-sis) Programmed cell death brought about by signals that trigger the activation of a cascade of "suicide" proteins in the cells destined to die.

appendicular skeleton (ap´-en-dik´-yū-ler) Components of the skeletal system that support the fins of a fish or the arms and legs of a land vertebrate; cartilage and bones of the shoulder girdle, pelvic girdle, and forelimbs and hind limbs. *See also* axial skeleton.

appendix (uh-pen´-dix) A small, fingerlike extension of the vertebrate cecum; contains a mass of white blood cells that contribute to immunity.

aqueous humor (ā´-kwē-us hyū´-mer) Plasma-like liquid in the space between the lens and the cornea in the vertebrate eye; helps maintain the shape of the eye, supplies nutrients and oxygen to its tissues, and disposes of its wastes.

aqueous solution (ā´-kwē-us) A solution in which water is the solvent.

arachnid A member of a major arthropod group (chelicerates) that includes spiders, scorpions, ticks, and mites.

arbuscular mycorrhizal fungus *See* glomeromycete.

Archaea (ar´-kē-uh) One of two prokaryotic domains of life, the other being Bacteria.

archenteron (ar-ken´-tuh-ron) In a developing animal, the endoderm-lined cavity formed during gastrulation; the digestive cavity of a gastrula.

arteriole (ar-ter´-ē-ōl) A vessel that conveys blood between an artery and a capillary bed.

artery A vessel that carries blood away from the heart to other parts of the body.

arthropod (ar-thrō-pod´) A member of the most diverse phylum in the animal kingdom. Arthropods include the horseshoe crab, arachnids (e.g., spiders, ticks, scorpions, and mites), crustaceans (e.g., crayfish, lobsters, crabs, and barnacles), millipedes, centipedes, and insects. Arthropods are characterized by a chitinous exoskeleton, molting, jointed appendages, and a body formed of distinct groups of segments.

artifical pacemaker A tiny electronic device surgically implanted near the AV node that emits electronic signals that trigger normal heartbeats.

artificial selection Selective breeding of domesticated plants and animals to promote the occurrence of desirable inherited traits in offspring.

ascomycete *See* sac fungus.

asexual reproduction The creation of offspring by a single parent, without the participation of sperm and egg.

assisted reproductive technology (ART) Procedure that involves surgically removing eggs from a woman's ovaries, fertilizing them, and then returning them to the woman's body. *See also* in vitro fertilization.

associative learning Learning that a particular stimulus or response is linked to a reward or punishment; includes classical conditioning and trial-and-error learning.

astigmatism (uh-stig´-muh-tizm) Blurred vision caused by a misshapen lens or cornea.

atherosclerosis (ath´-uh-rō´-skluh-rō´-sis) A cardiovascular disease in which growths called plaques develop on the inner walls of the arteries, narrowing their inner diameters.

atom The smallest unit of matter that retains the properties of an element.

atomic number The number of protons in each atom of a particular element.

atomic mass The approximate total mass of an atom; also called atomic weight. Given as a whole number, the atomic mass approximately equals the mass number.

ATP Adenosine triphosphate, the main energy source for cells.

ATP synthase A complex (cluster) of several proteins found in a cellular membrane (including the inner membrane of mitochondria, the thylakoid membrane of chloroplasts, and the plasma membrane of prokaryotes) that functions in chemiosmosis with adjacent electron transport chains, using the energy of a hydrogen ion concentration gradient to make ATP. An ATP synthase provides a port through which hydrogen ions (H^+) diffuse.

atrium (ā´-trē-um) (plural, **atria**) A heart chamber that receives blood from the veins.

auditory canal Part of the vertebrate outer ear that channels sound waves from the pinna or outer body surface to the eardrum.

australopith (os-trā´-lō-pith) The first hominids; scavenger-gatherer-hunters who lived on African savannas between about 4.4 million years ago and 1.5 million years ago.

autoimmune disease An immunological disorder in which the immune system attacks the body's own molecules.

autonomic nervous system (ot´-ō-nom´-ik) The component of the vertebrate peripheral nervous system that regulates the internal environment; made up of sympathetic and parasympathetic subdivisions.

autosome A chromosome not directly involved in determining the sex of an organism; in mammals, for example, any chromosome other than X or Y.

autotroph (ot´-ō-trōf) An organism that makes its own food, thereby sustaining itself without eating other organisms or their molecules. Plants, algae, and photosynthetic bacteria are autotrophs.

auxin (ok´-sin) A plant hormone, indoleacetic acid or a related compound, whose chief effect is to promote the growth and development of shoots.

AV (atrioventricular) node A region of specialized muscle tissue between the right atrium and right ventricle. It generates electrical impulses that primarily cause the ventricles to contract.

axial skeleton (ak´-sē-ul) Components of the skeletal system that support the central trunk of the body; the skull, backbone, and rib cage in a vertebrate. *See also* appendicular skeleton.

axillary bud (ak´-sil-ār-ē) An embryonic shoot present in the angle formed by a leaf and stem.

axon (ak´-son) A neuron fiber that conducts signals to another neuron or to an effector cell.

B

B cell A type of lymphocyte that matures in the bone marrow and later produces antibodies; responsible for humoral immunity. *See also* T cell.

bacillus (buh-sil´-us) (plural, **bacilli**) A rod-shaped prokaryotic cell.

Bacteria One of two prokaryotic domains of life, the other being Archaea.

bacteriophage (bak-tēr´-ē-ō-fāj) A virus that infects bacteria; also called a phage.

balancing selection Natural selection that maintains stable frequencies of two or more phenotypic forms in a population (balanced polymorphism).

ball-and-socket joint A joint that allows rotation and movement in several planes. Examples in humans are the hip and shoulder joints.

bark All the tissues external to the vascular cambium in a plant that is growing in thickness. Bark is made up of secondary phloem, cork cambium, and cork.

barrier method Contraception that relies on a physical barrier to block the passage of sperm. Examples include condoms and diaphragms.

Bartholin's glands (bar´-tō-linz) Glands near the vaginal opening in a human female that secrete lubricating fluid during sexual arousal.

basal body (bā´-sul) A eukaryotic cell organelle consisting of a 9 + 0 arrangement of microtubule triplets; may organize the microtubule assembly of a cilium or flagellum; structurally identical to a centriole.

basal ganglia (gang´-lē-uh) Clusters of nerve cell bodies located deep within the cerebrum that are important in motor coordination.

basal metabolic rate (BMR) The number of kilocalories a resting animal requires to fuel its essential body processes for a given time.

base A substance that decreases the hydrogen ion (H^+) concentration in a solution.

basement membrane The extracellular matrix, consisting of a dense mat of proteins and sticky polysaccharides, that anchors an epithelium to underlying tissues.

basidiomycete *See* club fungus.

basilar membrane The floor of the middle canal of the inner ear.

Batesian mimicry (bāt´-zē-un mim´-uh-krē) A type of mimicry in which a species that a predator can eat looks like a different species that is poisonous or otherwise harmful to the predator.

behavior Everything an animal does and how it does it, including muscular activities such as chasing prey, certain nonmuscular processes such as secreting a hormone that attracts a mate, and learning.

behavioral ecology The scientific field that studies behavior in an evolutionary context.

behavioral isolation A type of prezygotic barrier between species; two species remain isolated because individuals of neither species are sexually attracted to individuals of the other species.

benign tumor An abnormal mass of cells that remains at its original site in the body.

benthic zone A seafloor, or the bottom of a freshwater lake, pond, river, or stream.

biennial A plant that completes its life cycle in two years.

bilateral symmetry An arrangement of body parts such that an organism can be divided equally by a single cut passing longitudinally through it. A bilaterally symmetrical organism has mirror-image right and left sides.

bilaterian Member of the clade of animals Bilateria exhibiting bilateral symmetry.

bile A solution of bile salts that is secreted by the liver and that emulsifies fats and aids in their digestion.

binary fission A means of asexual reproduction in which a parent organism, often a single cell, divides into two individuals of about equal size.

binomial A two-part, latinized name of a species; for example, *Homo sapiens*.

biodiversity *See* species diversity.

biodiversity crisis The current rapid decline in the variety of life on Earth, largely due to the effects of human culture.

biodiversity hot spot A small geographic area with an exceptional concentration of endangered and threatened species, especially endemic species (those found nowhere else).

biofilm A surface-coating colony of prokaryotes that engage in metabolic cooperation.

biogenic amine A neurotransmitter derived from an amino acid.

biogeochemical cycle Any of the various chemical circuits which involve both biotic and abiotic components of an ecosystem.

biogeography The study of past and present distribution of species.

biological augmentation An approach to restoration ecology that uses organisms to add essential materials to a degraded ecosystem.

biological clock An internal timekeeper that controls an organism's biological rhythms; marks time with or without environmental cues but often requires signals from the environment to remain tuned to an appropriate period. *See also* circadian rhythm.

biological magnification The accumulation of persistent chemicals in the living tissues of consumers in food chains.

biological species concept The definition of a species as a population or group of populations whose members have the potential in nature to interbreed and produce fertile offspring. A biological species is also called a sexual species.

biology The scientific study of life.

biomass The amount, or mass, of organic material in an ecosystem.

biome (bī´-ōm) A terrestrial ecosystem, largely determined by climate, usually classified according to the predominant vegetation and characterized by organisms adapted to the particular environments.

bioremediation The use of living organisms to detoxify and restore polluted and degraded ecosystems.

biosphere The entire portion of Earth inhabited by life; the sum of all the planet's ecosystems.

biotechnology The use of living organisms (often microbes) to perform useful tasks; today, usually involves DNA technology.

biotic component (bī-o´-tik) A living component of a biological community; an organism, or a factor pertaining to one or more organisms.

bird A group of reptiles with feathers and adaptations for flight.

bipolar disorder Depressive mental illness characterized by swings of mood from high to low; also called manic-depressive disorder.

birth control pill A chemical contraceptive that inhibits ovulation, retards follicular development, or alters a woman's cervical mucus to prevent sperm from entering the uterus.

bivalve A member of a group of molluscs that includes clams, mussels, scallops, and oysters.

blastocoel (blas´-tuh-sēl) In a developing animal, a central, fluid-filled cavity in a blastula.

blastocyst (blas´-tō-sist) A mammalian embryo (equivalent to an amphibian blastula) made up of a hollow ball of cells that results from cleavage and that implants in the mother's endometrium.

blastopore (blas´-tō-por) A small indentation on one side of a blastula where cells that will form endoderm and mesoderm leave the surface and move inward.

blastula (blas´-tyū-luh) An embryonic stage that marks the end of cleavage during animal development; a hollow ball of cells in many species.

blind spot The place on the retina of the vertebrate eye where the optic nerve passes through the eyeball and where there are no photoreceptor cells.

blood A type of connective tissue with a fluid matrix called plasma in which blood cells are suspended.

blood pressure The force that blood exerts against the walls of blood vessels.

blood-brain barrier A system of capillaries in the brain that restricts passage of most substances into the brain, thereby preventing large fluctuations in the brain's environment.

body cavity A fluid-containing space between the digestive tract and the body wall.

bone A type of connective tissue consisting of living cells held in a rigid matrix of collagen fibers embedded in calcium salts.

bottleneck effect Genetic drift resulting from a drastic reduction in population size.

Bowman's capsule A cup-shaped swelling at the receiving end of a nephron in the vertebrate kidney; collects the filtrate from the blood.

brain The part of the central nervous system involved in regulating and controlling bodily activity and interpreting information from the senses transmitted through the nervous system.

brainstem A functional unit of the vertebrate brain, composed of the midbrain, the medulla oblongata, and the pons; serves mainly as a sensory filter, selecting which information reaches higher brain centers.

breathing The alternation of inhalation and exhalation, supplying a lung or gill with O_2-rich air or water and expelling CO_2-rich air or water.

breathing control center A brain center that directs the activity of organs involved in breathing.

bronchiole (bron´-kē-ōl) A thin breathing tube that branches from a bronchus within a lung.

bronchus (bron´-kus) (plural, **bronchi**) One of a pair of breathing tubes that branch from the trachea into the lungs.

brown alga One of a group of marine, multicellular, autotrophic protists, the most common and largest type of seaweed. Brown algae include the kelps.

bryophyte (brī´-uh-fīt) One of a group of plants that lack xylem and phloem; a nonvascular plant. Bryophytes include mosses and their close relatives.

budding A means of asexual reproduction whereby a new individual developed from an outgrowth of a parent splits off and lives independently.

buffer A chemical substance that resists changes in pH by accepting hydrogen ions from or donating hydrogen ions to solutions.

bulbourethral gland (bul´-bō-yū-rē´-thrul) One of a pair of glands near the base of the penis in the human male that secrete fluid that lubricates and neutralizes acids in the urethra during sexual arousal.

bulk feeder An animal that eats relatively large pieces of food.

C

C_3 plant A plant that uses the Calvin cycle for the initial steps that incorporate CO_2 into organic material, forming a three-carbon compound as the first stable intermediate.

C_4 plant A plant that prefaces the Calvin cycle with reactions that incorporate CO_2 into four-carbon compounds, the end product of which supplies CO_2 for the Calvin cycle.

calcitonin (kal´-sih-tōn´-in) A peptide hormone secreted by the thyroid gland that lowers the blood calcium level.

Calvin cycle The second of two stages of photosynthesis; a cyclic series of chemical reactions that occur in the stroma of a chloroplast, using the carbon in CO_2 and the ATP and NADPH produced by the light reactions to make the energy-rich sugar molecule G3P.

CAM plant A plant that uses crassulacean acid metabolism, an adaptation for photosynthesis in arid conditions. Carbon dioxide entering open stomata during the night is converted to organic acids, which release CO_2 for the Calvin cycle during the day, when stomata are closed.

cancer cell A cell that is not subject to normal cell cycle control mechanisms and that will therefore divide continuously.

capillary (kap´-il-er-ē) A microscopic blood vessel that conveys blood between an artery and a vein or between an arteriole and a venule; enables the exchange of nutrients and dissolved gases between the blood and interstitial fluid.

capillary bed One of the networks of capillaries that infiltrate every organ and tissue in the body.

capsid The protein shell that encloses a viral genome. It may be rod-shaped, polyhedral, or more complex in shape.

capsule A sticky layer that surrounds the bacterial cell wall, protects the cell surface, and sometimes helps glue the cell to surfaces.

carbohydrate (kar´-bō-hī´-drāt) Member of the class of biological molecules consisting of simple single-monomer sugars (monosaccharides), two-monomer sugars (disaccharides), and other multi-unit sugars (polysaccharides).

carbon fixation The incorporation of carbon from atmospheric CO_2 into the carbon in organic compounds. During photosynthesis in a C_3 plant, carbon is fixed into a three-carbon sugar as it enters the Calvin cycle. In C_4 and CAM plants, carbon is fixed into a four-carbon sugar.

carbon skeleton The chain of carbon atoms that forms the structural backbone of an organic molecule.

carbonyl group (kar´-buh-nēl´) In an organic molecule, a functional group consisting of a carbon atom linked by a double bond to an oxygen atom.

carboxyl group (kar-bok´-sil) In an organic molecule, a functional group consisting of an oxygen atom double-bonded to a carbon atom that is also bonded to a hydroxyl group.

carboxylic acid An organic compound containing a carboxyl group.

carcinogen (kar-sin´-uh-jin) A cancer-causing agent, either high-energy radiation (such as X-rays or UV light) or a chemical.

carcinoma (kar´-sih-nō´-muh) Cancer that originates in the coverings of the body, such as skin or the lining of the intestinal tract.

cardiac cycle (kar´-dē-ak) The alternating contractions and relaxations of the heart.

cardiac muscle Striated muscle that forms the contractile tissue of the heart.

cardiac output The volume of blood per minute that the left ventricle pumps into the aorta.

cardiovascular disease (kar´-dē-ō-vas´-kyū-ler) Diseases of the heart and blood vessels.

cardiovascular system A closed circulatory system with a heart and branching network of arteries, capillaries, and veins.

carnivore An animal that eats other animals. *See also* herbivore; omnivore.

carpel (kar´-pul) The female part of a flower, consisting of a stalk with an ovary at the base and a stigma, which traps pollen, at the tip.

carrier An individual who is heterozygous for a recessively inherited disorder and who therefore does not show symptoms of that disorder.

carrying capacity In a population, the number of individuals that an environment can sustain.

cartilage (kar´-ti-lij) A type of connective tissue consisting of living cells embedded in a rubbery matrix with collagenous fibers.

Casparian strip (kas-par´-ē-un) A waxy barrier in the walls of endodermal cells in a root that prevents water and ions from entering the xylem without crossing one or more cell membranes.

cation exchange A process in which positively charged minerals are made available to a plant when hydrogen ions in the soil displace mineral ions from the clay particles.

cecum (sē´-kum) (plural, **ceca**) A blind outpocket of a hollow organ such as an intestine.

cell A basic unit of living matter separated from its environment by a plasma membrane; the fundamental structural unit of life.

cell body The part of a cell, such as a neuron, that houses the nucleus.

cell cycle An ordered sequence of events (including interphase and the mitotic phase) that extends from the time a eukaryotic cell is first formed from a dividing parent cell until its own division into two cells.

cell cycle control system A cyclically operating set of proteins that triggers and coordinates events in the eukaryotic cell cycle.

cell division The reproduction of a cell.

cell plate A double membrane across the midline of a dividing plant cell, between which the new cell wall forms during cytokinesis.

cell theory The theory that all living things are composed of cells and that all cells come from other cells.

cell wall A protective layer exernal to the plasma membrane in plant cells, bacteria, fungi, and some protists; protects the cell and helps maintain its shape.

cell-mediated immunity The type of specific immunity brought about by T cells; fights body cells infected with pathogens. *See also* humoral immunity.

cellular differentiation The specialization in the structure and function of cells that occurs during the development of an organism; results from selective activation and deactivation of the cells' genes.

cellular metabolism (muh-tab´-uh-lizm) The chemical activities of cells.

cellular respiration The aerobic harvesting of energy from food molecules; the energy-releasing chemical breakdown of food molecules, such as glucose, and the storage of potential energy in a form that cells can use to perform work; involves glycolysis, the citric acid cycle, and oxidative phosphorylation (the electron transport chain and chemiosmosis).

cellular slime mold A type of protist that has unicellular amoeboid cells and aggregated reproductive bodies in its life cycle.

cellulose (sel´-yū-lōs) A large polysaccharide composed of many glucose monomers linked into cable-like fibrils that provide structural support in plant cell walls.

centipede A carnivorous terrestrial arthropod that has one pair of long legs for each of its numerous body segments, with the front pair modified as poison claws.

central canal The narrow cavity in the center of the spinal cord that is continuous with the fluid-filled ventricles of the brain.

central nervous system (CNS) The integration and command center of the nervous system; the brain and, in vertebrates, the spinal cord.

central vacuole (vak´-yū-ōl) A membrane-enclosed sac occupying most of the interior of a mature plant cell, having diverse roles in reproduction, growth, and development.

centralization The presence of a central nervous system (CNS) distinct from a peripheral nervous system.

centriole (sen´-trē-ōl) A structure in an animal cell composed of microtubule triplets arranged in a 9 + 0 pattern. An animal cell usually has a pair of centrioles within each of its centrosomes.

centromere (sen´-trō-mēr) The region of a chromosome where two sister chromatids are joined and where spindle microtubules attach during mitosis and meiosis. The centromere divides at the onset of anaphase during mitosis and anaphase II during meiosis.

centrosome (sen´-trō-sōm) Material in the cytoplasm of a eukaryotic cell that gives rise to microtubules; important in mitosis and meiosis; also called microtubule-organizing center.

cephalization (sef´-uh-luh-zā´-shun) The concentration of a nervous system at the anterior end.

cephalopod A member of a group of molluscs that includes squids and octopuses.

cerebellum (sār´-ruh-bel´-um) Part of the vertebrate hindbrain; mainly a planning center that interacts closely with the cerebrum in coordinating body movement.

cerebral cortex (suh-rē´-brul kor´-teks) A folded sheet of gray matter forming the surface of the cerebrum. In humans, it contains integrating centers for higher brain functions such as reasoning, speech, language, and imagination.

cerebral hemisphere The right or left half of the vertebrate cerebrum.

cerebrospinal fluid (suh-rē´-brō-spī´-nul) Blood-derived fluid that surrounds, protects against infection, nourishes, and cushions the brain and spinal cord.

cerebrum (suh-rē´-brum) The largest, most sophisticated, and most dominant part of the vertebrate forebrain, made up of right and left cerebral hemispheres.

cervix (ser´-viks) The neck of the uterus, which opens into the vagina.

chaparral (shap´-uh-ral´) A biome dominated by spiny evergreen shrubs adapted to periodic drought and fires; found where cold ocean currents circulate offshore, creating mild, rainy winters and long, hot, dry summers.

charophycean (kār´-uh-fī´-sē-un) A member of the green algal group that shares two ultrastructural features with land plants. They are considered the closest relatives of land plants.

chelicerate A lineage of arthropods that includes horseshoe crabs, scorpions, ticks, and spiders.

chemical bond An attraction between two atoms resulting from a sharing of outer-shell electrons or the presence of opposite charges on the atoms. The bonded atoms gain complete outer electron shells.

chemical cycling The use and reuse of chemical elements such as carbon within an ecosystem.

chemical energy Energy stored in the chemical bonds of molecules; a form of potential energy.

chemical reaction A process leading to changes in the composition of matter; involves the making and/or breaking of chemical bonds.

chemiosmosis (kem´-ē-oz-mō´-sis) The production of ATP using the energy of hydrogen ion (H^+) gradients across membranes to phosphorylate ADP; powers most ATP synthesis in cells.

chemoautotroph An organism that obtains both energy and carbon from inorganic chemicals. A chemoautotroph makes its own organic compounds from CO_2 without using light energy.

chemoheterotroph (kē´-mō-het´-er-ō-trō f) An organism that obtains energy and carbon from organic molecules.

chemoreceptor (kē´-mō-rē-sep´-ter) A sensory receptor that detects chemical changes within the body or a specific kind of molecule in the external environment.

chiasma (kī-az´-muh) (plural, **chiasmata**) The microscopically visible site where crossing over has occurred between chromatids of homologous chromosomes during prophase I of meiosis.

chlamydia A group of bacteria that includes a parasite that causes a common sexually transmitted disease.

chloroplast (klō´-rō-plast) An organelle found in plants and photosynthetic protists. Enclosed by two concentric membranes, a chloroplast absorbs sunlight and uses it to power the synthesis of organic food molecules (sugars).

choanocyte (kō-an´-uh-sīt) A flagellated feeding cell found in sponges. Also called a collar cell, it has a collar-like ring that traps food particles around the base of its flagellum.

choanoflagellate An ancestral colonial protist from which sponges, and possibly all animals, probably arose.

chondrichthyan Member of the class Chondrichthyes, vertebrates with skeletons made mostly of cartilage, such as sharks and rays.

Chordate (kōr´-dāte) Member of the phylum Chordata, animals that at some point during their development have a dorsal hollow nerve cord, a notochord, pharyngeal slits, and a post-anal tail. Chordates include lancelets, tunicates, and vertebrates.

chorion (kō´r-ē-on) In animals, the outermost extraembryonic membrane, which becomes the mammalian embryo's part of the placenta.

chorionic villus (kōr´-ē-on´-ik vil´-us) An outgrowth of the chorion, containing embryonic blood vessels. As part of the placenta, chorionic villi absorb nutrients and oxygen from, and pass wastes into, the mother's bloodstream.

chorionic villus sampling (CVS) A technique for diagnosing genetic defects while the fetus is in the uterus. A small sample of the fetal portion of the placenta is removed and analyzed.

choroid (kōr´-oyd) A thin, pigmented layer in the vertebrate eye, surrounded by the sclera. The iris is part of the choroid.

chromatin (krō´-muh-tin) The combination of DNA and proteins that constitutes chromosomes; often used to refer to the diffuse, very extended form taken by the chromosomes when a eukaryotic cell is not dividing.

chromosome (krō´-muh-sōm) A threadlike, gene-carrying structure found in the nucleus of a eukaryotic cell and most visible during mitosis and meiosis; also, the main gene-carrying structure of a prokaryotic cell. Chromosomes consist of chromatin.

chromosome theory of inheritance A basic principle in biology stating that genes are located on chromosomes and that the behavior of chromosomes during meiosis accounts for inheritance patterns.

chytrid Member of the fungal phylum Chytridiomycota, mostly aquatic fungi with flagellated spores that probably represent the most primitive fungal lineage.

ciliary body A portion of the vertebrate eye associated with the lens. It produces the clear, watery aqueous humor that fills the anterior cavity of the eye.

ciliate (sil´-ē-it) A type of protozoan that moves by means of cilia.

cilium (sil´-ē-um) (plural, **cilia**) A short appendage that propels some protists through the water and moves fluids across the surface of many tissue cells in animals. In common with eukaryotic flagella, cilia have a 9 + 2 arrangement of microtubules covered by the cell's plasma membrane.

circadian rhythm (ser-kā´-dē-un) In an organism, a biological cycle of about 24 hours that is controlled by a biological clock, usually under the influence of environmental cues; a pattern of activity that is repeated daily. *See also* biological clock.

circulatory system The organ system that transports materials such as nutrients, O_2, and hormones to body cells and transports CO_2 and other wastes from body cells.

citric acid cycle The metabolic cycle fueled by acetyl CoA formed after glycolysis in cellular respiration. Chemical reactions in the citric acid cycle complete the metabolic breakdown of glucose molecules to carbon dioxide. The cycle occurs in the matrix of mitochondria and supplies most of the NADH molecules that carry energy to the electron transport chains.

clades Evolutionary branches that consist of an ancestor and all its descendants.

cladistics (kluh-dis´-tiks) The study of evolutionary history; specifically, the scientific search for monophyletic taxa (clades), taxonomic groups composed of an ancestor and all its descendants.

cladogram A diagram depicting patterns of shared characteristics among species.

class In classification, the taxonomic category above order.

cleavage (klē-vij) (1) Cytokinesis in animal cells and in some protists, characterized by pinching in of the plasma membrane. (2) In animal development, the succession of rapid cell divisions without cell growth that converts the animal zygote into a ball of cells.

cleavage furrow The first sign of cytokinesis during cell division in an animal cell; a shallow groove in the cell surface near the old metaphase plate.

cline A gradation in an inherited trait along a geographic continuum; variation in a population's phenotypic features that parallels an environmental gradient.

clitoris (klit´-uh-ris) An organ in the female that engorges with blood and becomes erect during sexual arousal.

clonal selection (klōn´-ul) The production of a lineage of genetically identical cells that recognize and attack the specific antigen that stimulated their proliferation. Clonal selection is the mechanism that underlies the immune system's specificity and memory of antigens.

clone As a verb, to produce genetically identical copies of a cell, organism, or DNA molecule. As a noun, the collection of cells, organisms, or molecules resulting from cloning; also (colloquially), a single organism that is genetically identical to another because it arose from the cloning of a somatic cell.

closed circulatory system A circulatory system in which blood is confined to vessels and is kept separate from the interstitial fluid.

club fungus Member of the phylum Basidiomycota, characterized by a club-shaped, spore-producing structure called a basidium.

clumped Describing a dispersion pattern in which individuals are aggregated in patches.

cnidarian (nī-dār´-ē-un) An animal characterized by cnidocytes, radial symmetry, a gastrovascular cavity, and a polyp and medusa body form. Cnidarians include the hydras, jellyfishes, sea anemones, corals, and related animals.

cnidocyte (nī´-duh-sīt) A specialized cell for which the phylum Cnidaria is named; consists of a capsule containing a fine coiled thread, which, when discharged, functions in defense and prey capture.

CNS *See* central nervous system.

coccus (kok´-us) (plural, **cocci**) A spherical prokaryotic cell.

cochlea (kok´-lē-uh) A coiled tube in the inner ear of birds and mammals that contains the hearing organ, the organ of Corti.

codominant Inheritance pattern in which a heterozygote expresses the distinct trait of both alleles.

codon (kō´-don) A three-nucleotide sequence in mRNA that specifies a particular amino acid or polypeptide termination signal; the basic unit of the genetic code.

coelom (sē´-lōm) A body cavity completely lined with mesoderm.

coenzyme (kō-en´-zīm) An organic molecule (usually a vitamin or a compound synthesized from a vitamin) that acts as a cofactor, helping an enzyme catalyze a metabolic reaction.

coevolution Evolutionary change in which adaptations in one species act as a selective force on a second species, inducing adaptations that in turn act as a selective force on the first species; mutual influence on the evolution of two different interacting species.

cofactor A nonprotein substance (such as a copper, iron, or zinc atom or an organic molecule) that helps an enzyme catalyze a metabolic reaction. *See also* coenzyme.

cognition The ability of an animal's nervous system to perceive, store, process, and use information obtained by its sensory receptors.

cognitive map A representation within the nervous system of spatial relations among objects in an animal's environment.

cohesion (kō-hē´-zhun) The attraction between molecules of the same kind.

collecting duct A tube in the vertebrate kidney that concentrates urine while conveying it to the renal pelvis.

collenchyma cell (kō-len´-kim-uh) In plants, a cell with a thick primary wall and no secondary wall, functioning mainly in supporting growing parts.

colon (kō´-lun) Large intestine; the tubular portion of the vertebrate alimentary tract between the small intestine and the anus; functions mainly in water absorption and the formation of feces.

commensalism (kuh-men´-suh-lizm) A symbiotic relationship in which one partner benefits without significantly affecting the other.

communication Animal behavior involving transmission of, reception of, and response to signals.

community An assemblage of all the organisms living together and potentially interacting in a particular area.

companion cell In a plant, a cell connected to a sieve-tube member whose nucleus and ribosomes provide proteins for the sieve-tube member.

comparative anatomy The study of the body structures in different organisms.

comparative embryology The study of the formation, early growth, and development of different organisms.

competitive exclusion principle The concept that populations of two species cannot coexist in a community if their niches are nearly identical. Using resources more efficiently and having a reproductive advantage, one of the populations will eventually outcompete and eliminate the other.

competitive inhibitor A substance that reduces the activity of an enzyme by binding to the enzyme's active site in place of the substrate. A competitive inhibitor's structure mimics that of the enzyme's substrate.

complement system A family of nonspecific defensive blood proteins that cooperate with other components of the vertebrate defense system to protect against microbes; can enhance phagocytosis, directly lyse pathogens, and amplify the inflammatory response.

complementary DNA (cDNA) A DNA molecule made *in vitro* using mRNA as a template and the enzyme reverse transcriptase. A cDNA molecule therefore corresponds to a gene but lacks the introns present in the DNA of the genome.

complete digestive tract A digestive tube with two openings, a mouth and an anus.

complete dominance A type of inheritance in which the phenotypes of the heterozygote and dominant homozygote are indistinguishable.

complete metamorphosis (met´-uh-mōr´-fuh-sis) The transformation of a larva into an adult that looks very different than the larva and often functions very differently in its environment.

compound A substance containing two or more elements in a fixed ratio. For example, table salt (NaCl) consists of one atom of the element sodium (Na) for every atom of chlorine (Cl).

compound eye The photoreceptor in many invertebrates; made up of many tiny light detectors, each of which detects light from a tiny portion of the field of view.

concentration gradient An increase or decrease in the density of a chemical substance in an area. Cells often maintain concentration gradients of ions across their membranes.

conduction The direct transfer of thermal motion (heat) between molecules of objects in direct contact with each other.

cone (1) In vertebrates, a photoreceptor cell in the retina, stimulated by bright light and enabling color vision. (2) In conifers, a reproductive structure bearing pollen or ovules.

conifer A gymnosperm, or naked-seed plant, that produces cones.

coniferous forest A biome characterized by conifers, cone-bearing evergreen trees.

conjugation The union (mating) of two bacterial cells or protist cells and the transfer of DNA between the two cells.

connective tissue Tissue consisting of cells held in an abundant extracellular matrix, which they produce.

conservation biology The science of species preservation; the scientific study of ways to slow the current high rate of species loss.

consumer An organism that obtains its food by eating plants or by eating animals that have eaten plants.

continental drift A change in the position of continents resulting from the incessant slow movement (floating) of the plates of Earth's crust on the underlying molten mantle. It has caused continents to fuse and break apart periodically throughout geologic history.

continental shelf The submerged part of a continent.

contraception The deliberate prevention of pregnancy.

controlled experiment A component of the process of science whereby a scientist carries out two parallel tests, an experimental test and a control test. The experimental test differs from the control by one factor, the variable.

convection The mass movement of warmed air or liquid to or from the surface of a body or object.

convergent evolution Adaptive change resulting in nonhomologous (analogous) similarities among organisms. Species from different evolutionary lineages come to resemble each other (evolve analogous structures) as a result of living in very similar environments.

copulation Sexual intercourse, usually necessary for internal fertilization to occur.

coral reef A warm-water, tropical ecosystem dominated by the hard skeletal structures secreted primarily by the resident cnidarians.

cork The outermost protective layer of a plant's bark, produced by the cork cambium.

cork cambium Meristematic tissue that produces cork cells during secondary growth of a plant.

cornea (kor´-nē-uh) The transparent frontal portion of the sclera, which admits light into the vertebrate eye.

corpus callosum (kor´-pus kuh-lō´-sum) The thick band of nerve fibers that connect the right and left cerebral hemispheres in placental mammals, enabling the hemispheres to process information together.

corpus luteum (kor´-pus lū´-tē-um) A small body of endocrine tissue that develops from an ovarian follicle after ovulation; secretes progesterone and estrogen during pregnancy.

cortex In plants, the ground tissue system of a root, made up mostly of parenchyma cells, which store food and absorb minerals that have passed through the epidermis. *See also* adrenal cortex; cerebral cortex; renal cortex.

corticosteroid A family of hormones synthesized and secreted by the adrenal cortex, consisting of the mineralocorticoids and glucocorticoids.

cotyledon (kot´-uh-lē´-don) The first leaf that appears on an embryo of a flowering plant; a seed leaf. Monocot embryos have one cotyledon; dicot embryos have two.

countercurrent exchange The transfer of a substance from a fluid or volume of air moving in one direction to another fluid or volume of air moving in the opposite direction.

countercurrent heat exchanger Parallel blood vessels that convey warm and cold blood in opposite directions, maximizing heat transfer to the cold blood.

covalent bond (kō-vā´-lent) An attraction between atoms that share one or more pairs of outer-shell electrons; symbolized by a single line between the atoms.

cranial nerve A nerve that leaves the brain and innervates an organ of the head or upper body.

craniate A chordate with a head.

crista (kris´-tuh) (plural, **cristae**) A fold of the inner membrane of a mitochondrion. Enzyme molecules embedded in cristae make ATP.

crop A pouch-like organ in a digestive tract where food is softened and may be stored temporarily.

cross *See* hybrid.

cross-fertilization The fusion of sperm and egg derived from two different individuals.

crossing over The exchange of segments between chromatids of homologous chromosomes during synapsis in prophase I of meiosis; also, the exchange of segments between DNA molecules in prokaryotes.

crustacean A member of a major arthropod group that includes lobsters, crayfish, crabs, shrimps, and barnacles.

cultural eutrophication (yū-trō-fuh-kā´-shun) A process by which nutrients become highly concentrated in a body of water, leading to increased growth in organisms such as algae. Cultural eutrophication refers to situations where the nutrietns added to the water body originate mainly from human sources.

culture The ideas, customs, skills, rituals, and similar activities of a people or group that are passed along to succeeding generations.

cuticle (kyū-tuh-kul) (1) In animals, a tough, nonliving outer layer of the skin. (2) In plants, a waxy coating on the surface of stems and leaves that helps retain water.

cyanobacteria (sī-an´-ō-bak-tēr´-ē-uh) Photosynthetic, oxygen-producing bacteria, formerly called blue-green algae.

cystic fibrosis (sis´-tik fī-brō´-sis) A genetic disease that occurs in people with two copies of a certain recessive allele; characterized by an excessive secretion of mucus and vulnerability to infection; fatal if untreated.

cytokinesis (sī-tō-kuh-nē´-sis) The division of the cytoplasm to form two separate daughter cells. Cytokinesis usually occurs during telophase of mitosis, and mitosis and cytokinesis make up the mitotic (M) phase of the cell cycle.

cytokinin (sī-tō-kī´-nin) One of a family of plant hormones that promotes cell division, retards aging in flowers and fruits, and may interact antagonistically with auxins in regulating plant growth and development.

cytoplasm (sī-tō-plaz´-um) Everything inside a cell between the plasma membrane and the nucleus; consists of a semifluid medium and organelles.

cytosine (C) (sī´-tuh-sin) A single-ring nitrogenous base found in DNA and RNA.

cytoskeleton A meshwork of fine fibers in the cytoplasm of a eukaryotic cell; includes microfilaments, intermediate filaments, and microtubules.

cytotoxic T cell (sī´-tō-tok´-sik) A type of lymphocyte that attacks body cells infected with pathogens.

D

decomposer *See* detritivore.

decomposition The breakdown of organic materials into inorganic ones.

dehydration reaction (dē-hī-drā´-shun) A chemical process in which a polymer forms as monomers are linked by the removal of water molecules. One molecule of water is removed for each pair of monomers linked. Also called condensation.

dehydrogenase (dē´-hī-droj´-uh-nās) An enzyme that catalyzes a chemical reaction during which one or more hydrogen atoms are removed from a molecule.

deletion The loss of one or more nucleotides from a gene by mutation; the loss of a fragment of a chromosome.

demographic transition A shift from zero population growth in which birth rates and death rates are high to zero population growth characterized instead by low birth and death rates.

denaturation (dē-nā´-chur-ā´-shun) A process in which a protein unravels, losing its specific conformation and hence function; can be caused by changes in pH or salt concentration or by high temperature; also refers to the separation of the two strands of the DNA double helix, caused by similar factors.

dendrite (den´-drīt) A neuron fiber that conveys signals from its tip inward, toward the rest of the neuron; in a motor neuron, one of several short, branched extensions that convey nerve signals toward the cell body.

density-dependent Referring to any characteristic that varies according to an increase in population density.

density-dependent inhibition The arrest of cell division that occurs when cells grown in a laboratory dish touch one another; generally due to an inadequate supply of growth factors.

deoxyribonucleic acid (DNA) (dē-ok´-sē-rī´-bō-nū-klā´-ik) The genetic material that organisms inherit from their parents; a double-stranded helical macromolecule consisting of nucleotide monomers with deoxyribose sugar and the nitrogenous bases adenine (A), cytosine (C), guanine (G), and thymine (T). *See also* gene.

dermal tissue system The outer protective covering of plants.

descent with modification Darwin's initial phrase for the general process of evolution.

desert A biome characterized by organisms adapted to sparse rainfall (less than 30 cm per year) and rapid evaporation.

desertification The conversion of semi-arid regions to desert.

determinate growth Termination of growth after reaching a certain size, as in most animals. *See also* indeterminate growth.

detritivore (duh-trī´-tuh-vor) An organism that derives its energy from organic wastes and dead organisms.

detritus (duh-trī´-tus) Dead organic matter.

deuterostome (dū-ter´-ō-stōm) An animal with a coelom that forms from hollow outgrowths of the digestive tube of the early embryo. The deuterostomes include the echinoderms and the chordates.

diabetes mellitus (dī´-uh-bē´-tis me-lī´-tis) A human hormonal disease in which body cells cannot absorb enough glucose from the blood and become energy starved; body fats and proteins are then consumed for their energy. Insulin-dependent diabetes results when the pancreas does not produce insulin; non-insulin-dependent diabetes results when body cells fail to respond to insulin.

dialysis (dī-al´-uh-sis) Separation and disposal of metabolic wastes from the blood by mechanical means; an artificial method of performing the functions of the kidneys.

diaphragm (dī´-uh-fram) The sheet of muscle separating the chest cavity from the abdominal cavity in mammals. Its contraction expands the chest cavity, and its relaxation reduces it.

diastole (dī-as´-tō-lē) The stage of the heart cycle in which the heart muscle is relaxed, allowing the chambers to fill with blood. *See also* systole.

diatom (dī´-uh-tom) A unicellular photosynthetic alga with a unique, glassy cell wall containing silica.

dicot (dī´-kot) A flowering plant whose embryo has two seed leaves, or cotyledons.

differentiation *See* cellular differentiation.

diffusion The spontaneous movement of particles of any kind from where they are more concentrated to where they are less concentrated.

digestion The mechanical and chemical breakdown of food into molecules small enough for the body to absorb; the second main stage of food processing, following ingestion.

digestive system The organ system that ingests food, breaks it down into smaller chemical units, and absorbs the nutrient molecules.

dihybrid cross (dī´-hī´-brid) An experimental mating of individuals differing at two genetic loci.

dinoflagellate (dī´-nō-flaj´-uh-let) A unicellular photosynthetic alga with two flagella situated in perpendicular grooves in cellulose plates covering the cell.

diploid cell In an organism that reproduces sexually, a cell containing two homologous sets of chromosomes, one set inherited from each parent; a *2n* cell.

diplomonad A protist that has modified mitochondria, two equal-sized nuclei, and multiple flagella.

directional selection Natural selection that acts against the relatively rare individuals at one end of a phenotypic range.

disaccharide (dī-sak´-uh-rīd) A sugar molecule consisting of two monosaccharides linked by a dehydration reaction.

dispersion pattern The manner in which individuals in a population are spaced within their area. Three types of dispersion patterns are clumped (individuals are aggregated in patches), uniform (individuals are evenly distributed), and random (unpredictable distribution).

disruptive selection Natural selection that favors extreme over intermediate phenotypes.

distal tubule In the vertebrate kidney, the portion of a nephron that helps refine filtrate and empties it into a collecting duct.

disturbance In an ecological sense, a force that changes a biological community and usually removes organisms from it. Disturbances, such as fire and storms, play a pivotal role in structuring many biological communities.

DNA *See* deoxyribonucleic acid.

DNA fingerprinting A procedure that analyzes an individual's unique collection of DNA restriction fragments, detected by electrophoresis and nucleic acid probes. DNA fingerprinting can be used to determine whether two samples of genetic material are from the same individual.

DNA ligase (līˊ-gās) An enzyme, essential for DNA replication, that catalyzes the covalent bonding of adjacent DNA nucleotides; used in genetic engineering to paste a specific piece of DNA containing a gene of interest into a bacterial plasmid or other vector.

DNA microarray A glass slide carrying thousands of different kinds of single-stranded DNA fragments arranged in an array (grid). A DNA microarray is used to detect and measure the expression of thousands of genes at one time. Tiny amounts of a large number of single-stranded DNA fragments representing different genes are fixed to the glass slide. These fragments, ideally representing all the genes of an organism, are tested for hybridization with various samples of cDNA molecules.

DNA polymerase (puh-limˊ-er-ās) An enzyme that assembles DNA nucleotides into polynucleotides using a preexisting strand of DNA as a template.

DNA technology Methods used to study and/or manipulate DNA, including recombinant DNA technology.

doldrums (dolˊ-drums) An area of calm or very light winds near the equator, caused by rising warm air.

domain A taxonomic category above the kingdom level. The three domains of life are Archaea, Bacteria, and Eukarya.

dominance hierarchy The ranking of individuals based on social interactions; usually maintained by agonistic behavior.

dominant allele In a heterozygote, the allele that determines the phenotype with respect to a particular gene.

dorsal Pertaining to the back of a bilaterally symmetrical animal.

dorsal, hollow nerve cord One of the four hallmarks of chordates.

double bond A type of covalent bond in which two atoms share two pairs of electrons; symbolized by a pair of lines between the bonded atoms.

double circulation Circulation with separate pulmonary and systemic circuits, which ensures vigorous blood flow to all organs.

double fertilization In flowering plants, the formation of both a zygote and a cell with a triploid nucleus, which develops into the endosperm.

double helix The form of native DNA, referring to its two adjacent polynucleotide strands wound into a spiral shape.

Down syndrome A human genetic disorder resulting from the presence of an extra chromosome 21; characterized by heart and respiratory defects and varying degrees of mental retardation.

Duchenne muscular dystrophy (duh-shenˊ disˊ-truh-fē) A human genetic disease caused by a sex-linked recessive allele; characterized by progressive weakening and a loss of muscle tissue.

duodenum (dū-ō-dēˊ-num) The first portion of the vertebrate small intestine after the stomach, where acid chyme from the stomach is mixed with bile and digestive enzymes.

duplication Repetition of part of a chromosome resulting from fusion with a fragment from a homologous chromosome; can result from an error in meiosis or from mutagenesis.

E

eardrum A sheet of connective tissue separating the outer ear from the middle ear that vibrates when stimulated by sound waves and passes the waves to the middle ear.

earthworm A member of one of the three large groups of annelids. *See* annelid.

echinoderm (uh-kīˊ-nō-derm) Member of a phylum of slow-moving or sessile marine animals characterized by a rough or spiny skin, a water vascular system, an endoskeleton, and radial symmetry in adults. Echinoderms include sea stars, sea urchins, and sand dollars.

ecological footprint A method of using multiple constraints, including food, fuel, water, housing, and waste deposits, to estimate the human carrying capacity of the Earth.

ecological species concept The idea that ecological roles (niches) define species.

ecological succession The process of biological community change resulting from disturbance; transition in the species composition of a biological community, often following a flood, fire, or volcanic eruption. *See also* primary succession; secondary succession.

ecology The scientific study of how organisms interact with their environments.

ecosystem (ē-kō-sis-tem) All the organisms in a given area, along with the nonliving (abiotic) factors with which they interact; a biological community and its physical environment.

ectoderm (ekˊ-tō-derm) The outer layer of three embryonic cell layers in a gastrula; forms the skin of the gastrula and gives rise to the epidermis and nervous system in the adult.

ectopic pregnancy (ek-topˊ-ik) The implantation and development of an embryo outside the uterus.

ectotherm (ekˊ-tō-therm) An animal that warms itself mainly by absorbing heat from its surroundings.

ecotothermic Referring to organisms that do not produce enough metabolic heat to have much effect on body temperature.

EEG *See* electroencephalogram.

effector cell A cell capable of carrying out some action in response to a command from the nervous system.

egg The female gamete; when unfertilized, also called an ovum.

ejaculation (ih-jakˊ-yū-lāˊ-shun) Discharge of semen from the penis.

ejaculatory duct The short section of the ejaculatory route in mammals formed by the convergence of the vas deferens and a duct from the seminal vesicle. The ejaculatory duct transports sperm from the vas deferens to the urethra.

electroencephalogram (EEG) (ih-lekˊ-trō-en-sefˊ-uh-lōˊ-gram) A graph that shows the patterns of electrical activity in the brain during arousal and sleep.

electromagnetic energy Solar energy, or radiation, which travels in space as rhythmic waves and can be measured in photons.

electromagnetic receptor A sensory receptor that detects energy of different wavelengths, such as electricity, magnetism, and light.

electron A subatomic particle with a single negative electrical charge. One or more electrons move around the nucleus of an atom.

electron microscope (EM) An instrument that focuses an electron beam through, or onto the surface of, a specimen. An electron microscope achieves a thousandfold greater resolution than a light microscope; the most powerful EM can distinguish objects as small as 0.2 nm.

electron shell An energy level representing the distance of an electron from the nucleus of an atom.

electron transport chain A series of electron carrier molecules that shuttle electrons during the redox reactions that release energy used to make ATP; located in the inner membrane of mitochondria, the thylakoid membranes of chloroplasts, and the plasma membranes of prokaryotes.

electronegativity The tendency for an atom to pull electrons toward itself.

element A substance that cannot be broken down to other substances by chemical means. Scientists recognize 92 chemical elements occurring in nature.

elimination The passing of undigested material out of the digestive compartment.

embryo (emˊ-brē-ō) A developing stage of a multicellular organism. In humans, the stage in the development of offspring from the first division of the zygote until body structures begin to appear, about the ninth week of gestation.

embryo sac The female gametophyte contained in the ovule of a flowering plant.

embryonic stem cell (ES cell) Cell in the early animal embryo that differentiates during development to give rise to all the different kinds of specialized cells in the body.

embryophyte Another name for land plants, recognizing that land plants share the common derived trait of multicellular, dependent embryos.

emergent properties New properties that emerge with each step upward in the hierarchy of life, owing to the arrangement and interactions of parts as complexity increases.

emerging virus A virus that has appeared suddenly or has recently come to the attention of medical scientists.

emphysema (emˊ-fuh-sēˊ-muh) A respiratory disease caused by smoking, in which the alveoli lose their elasticity and deteriorate, reducing the lungs' capacity for gas exchange.

endangered species As defined in the U.S. Endangered Species Act, a species that is in danger of extinction throughout all or a significant portion of its range.

endemic species A species of organism whose distribution is limited to a specific geographic area.

endergonic reaction (en´-der-gon´-ik) An energy-requiring chemical reaction, which yields products with more potential energy than the reactants. The amount of energy stored in the products equals the difference between the potential energy in the reactants and that in the products.

endocrine gland (en´-dō-krin) A ductless gland that synthesizes hormone molecules and secretes them directly into the bloodstream.

endocrine system The organ system consisting of ductless glands that secrete hormones and the molecular receptors on or in target cells that respond to the hormones; cooperates with the nervous system in regulating body functions and maintaining homeostasis.

endocytosis (en´-dō-sī-tō´-sis) The movement of materials into the cytoplasm of a cell via membranous vesicles or vacuoles.

endoderm (en´-dō-derm) The innermost of three embryonic cell layers in a gastrula; forms the archenteron in the gastrula and gives rise to the innermost linings of the digestive tract and other hollow organs in the adult.

endodermis The innermost layer (a one-cell-thick cylinder) of the cortex of a plant root; forms a selective barrier determining which substances pass from the cortex into the vascular tissue.

endomembrane system A network of membranous organelles that partition the cytoplasm of eukaryotic cells into functional compartments. Some of the organelles are structurally connected to each other, whereas others are structurally separate but functionally connected by the traffic of membranous vesicles between them.

endometrium (en´-dō-mē´-trē-um) The inner lining of the uterus in mammals, richly supplied with blood vessels that provide the maternal part of the placenta and nourish the developing embryo.

endoplasmic reticulum (ER) An extensive membranous network in a eukaryotic cell, continuous with the outer nuclear membrane and composed of ribosome-studded (rough) and ribosome-free (smooth) regions. *See also* rough ER; smooth ER.

endorphin (en-dōr´-fin) A pain-inhibiting hormone produced by the brain and anterior pituitary; also serves as a neurotransmitter.

endoskeleton A hard skeleton located within the soft tissues of an animal; includes spicules of sponges, the hard plates of echinoderms, and the cartilage and bony skeletons of many vertebrates.

endosperm In flowering plants, a nutrient-rich mass formed by the union of a sperm cell with two polar nuclei during double fertilization; provides nourishment to the developing embryo in the seed.

endospore A thick-coated, protective cell produced within a bacterial cell exposed to harsh conditions.

endosymbiosis (en´-dō-sim´-bē-ō-sis) A process by which the mitochondria and chloroplasts of eukaryotic cells probably evolved from symbiotic associations between small prokaryotic cells living inside larger ones.

endotherm An animal that derives most of its body heat from its own metabolism.

endothermic Referring to animals that use heat generated by metabolism to maintain a warm, steady body temperature.

endotoxin A poisonous component of the outer membrane of certain bacteria that is released only when the bacteria die.

energy The capacity to perform work, or to rearrange matter.

energy coupling In cellular metabolism, the use of energy released from an exergonic reaction to drive an endergonic reaction.

energy flow The passage of energy through the components of an ecosystem.

energy of activation (E$_A$) The amount of energy that reactants must absorb before a chemical reaction will start.

enhancer A eukaryotic DNA sequence that helps stimulate the transcription of a gene at some distance from it. An enhancer functions by means of a transcription factor called an activator, which binds to it and then to the rest of the transcription apparatus. *See* silencer.

enteric division Part of the autonomic nervous sytem consisting of complex networks of neurons in the digestive tract, pancreas, and gallbladder.

entomology The study of insects.

entropy (en´-truh-pē) A measure of disorder. One form of disorder is heat, which is random molecular motion.

enzyme (en´-zīm) A protein that serves as a biological catalyst, changing the rate of a chemical reaction without itself being changed into a different molecule in the process.

epidermis (ep´-uh-der´-mis) (1) In animals, the living layer or layers of cells forming the protective covering, or outer skin. (2) In plants, the tissue system forming the protective outer covering of leaves, young stems, and young roots.

epididymis (ep´-uh-did´-uh-mus) A long coiled tube into which sperm pass from the testis and are stored until mature and ejaculated.

epinephrine (ep´-uh-nef´-rin) An amine hormone (also called adrenaline) secreted by the adrenal medulla that prepares body organs for "fight or flight"; also serves as a neurotransmitter.

epithelial tissue (ep´-uh-thē´-lē-ul) A sheet of tightly packed cells lining organs and cavities; also called epithelium.

epithelium (plural, **epithelia**) Epithelial tissue.

erythrocyte (eh-rith´-rō-sīt) *See* red blood cell.

erythropoietin (EPO) A hormone produced in the kidney when tissues of the body do not receive enough oxygen. This hormone stimulates the production of erythrocytes.

esophagus (eh-sof´-uh-gus) The channel through which food passes in a digestive tract; usually receives food from the pharynx.

essential amino acid An amino acid that an animal cannot synthesize itself and must obtain from food. Eight amino acids are essential for the human adult.

essential element In plants, a chemical element required for the plant to complete its life cycle (to grow from a seed and produce another generation of seeds).

essential fatty acid An unsaturated fatty acid that animals need but cannot make.

essential nutrient A substance that an organism must absorb in pre-assembled form because it cannot be synthesized from any other material. In humans, there are essential vitamins, minerals, amino acids, and fatty acids.

estivation (es´-tuh-vā´-shun) An animal's state of reduced activity (torpor) during periods of high environmental temperatures and reduced food and water supplies.

estrogen (es´-trō-jen) One of several chemically similar steroid hormones secreted by the gonads; maintains the female reproductive system and promotes the development of female body features.

estuary (es´-chū-ār-ē) An area where fresh water merges with seawater.

ethylene A gas that functions as a hormone in plants, triggering aging responses such as fruit ripening and leaf drop.

euglenozoan A diverse clade of flagellated protists that includes trypanosones and Euglena.

Eukarya (yū-kār-ē-uh) The domain of eukaryotes, organisms made of eukaryotic cells; includes all of the protists, plants, fungi, and animals.

eukaryote An organism with eukaryotic cells.

eukaryotic cell (yū-kār-ē-ot´-ik) A type of cell that has a membrane-enclosed nucleus and other membrane-enclosed organelles. All organisms except bacteria and archaea are composed of eukaryotic cells.

eumetazoan Member of the clade of "true animals," the animals with true tissues (all animals except sponges).

Eustachian tube (yū-stā´-shun) An air passage between the middle ear and throat of vertebrates that equalizes air pressure on either side of the eardrum.

eutherian (yū-thēr´-ē-un) Placental mammal; mammal whose young complete their embryonic development within the uterus, joined to the mother by the placenta.

evaporation The loss of heat from the surface of a liquid that is losing some of its molecules as a gas.

evo-devo The research field that combines evolutionary biology with developmental biology.

evolution Genetic change in a population or species over generations; all the changes that transform life on Earth; the heritable changes that have produced Earth's diversity of organisms.

evolutionary adaptation An inherited characteristic that enhances an organism's ability to survive and reproduce in a particular environment.

exaptation (ek´-sap-tā´-shun) A structure that has evolved in one environmental context and later becomes adapted for a different function in a different environmental context.

excitement phase The first phase of the human sexual response cycle, where the sexual passion builds.

excretion (ek-skrē´-shun) The disposal of nitrogen-containing metabolic wastes.

excretory system (ek´-skruh-tōr-ē) The organ system that disposes of nitrogen-containing waste products of cellular metabolism.

exergonic reaction (ek´-ser-gon´-ik) An energy-releasing chemical reaction in which the reactants contain more potential energy than the products. The reaction releases an amount of energy equal to the difference in potential energy between the reactants and the products.

exocytosis (ek´-sō-sī-tō´-sis) The movement of materials out of the cytoplasm of a cell via membranous vesicles or vacuoles.

exon (ek´-son) In eukaryotes, a coding portion of a gene. *See* intron.

exoskeleton A hard, external skeleton that protects an animal and provides points of attachment for muscles.

exotoxin A poisonous protein secreted by certain bacteria.

exponential growth model A mathematical description of idealized, unregulated population growth.

external fertilization The fusion of gametes that parents have discharged into the environment.

extracellular matrix A substance in which the cells of an animal tissue are embedded; consists of protein and polysaccharides.

extraembryonic membranes Four membranes (the yolk sac, amnion, chorion, and allantois) that form a life-support system for the developing embryo of a reptile, bird, or mammal.

extreme halophile A microorganism that lives in a highly saline environment, such as the Great Salt Lake or the Dead Sea.

extreme thermophile A microorganism that thrives in a hot environment (often 60–80°C).

eye cup The simplest type of photoreceptor, a cluster of photoreceptor cells shaded by a cuplike cluster of pigmented cells; detects light intensity and direction.

F

F factor A piece of DNA that can exist as a bacterial plasmid. The F factor carries genes for making sex pili and other structures needed for conjugation, as well as a site where DNA replication can start. F stands for fertility.

F₁ generation The offspring of two parental (P generation) individuals; F_1 stands for first filial.

F₂ generation The offspring of the F_1 generation; F_2 stands for second filial.

facilitated diffusion The passage of a substance across a biological membrane down its concentration gradient, aided by specific transport proteins.

facultative anaerobe (fak´-ul-tā´-tiv an´-uh-rōb) A microorganism that makes ATP by aerobic respiration if oxygen is present, but that switches to fermentation when oxygen is absent.

family In classification, the taxonomic category above genus.

farsightedness An inability to focus on close objects; occurs when the eyeball is shorter than normal and the focal point of the lens is behind the retina. Also called hyperopia.

fat A large lipid molecule made from an alcohol called glycerol and three fatty acids; a triglyceride. Most fats function as energy-storage molecules.

feces The wastes of the digestive tract.

feedback inhibition A method of metabolic control in which a product of a metabolic pathway acts as an inhibitor of an enzyme within that pathway, thereby blocking the metabolic reaction.

fertilization The union of the nucleus of a sperm cell with the nucleus of an egg cell, producing a zygote.

fertilization envelope A barrier that forms around an egg seconds after fertilization, preventing penetration of additional sperm.

fetus (fē´-tus) A developing human from the ninth week of gestation until birth; has all the major structures of an adult.

fiber (1) In animals, an elongate, supportive thread in the matrix of connective tissue; an extension of a neuron; a muscle cell. (2) In plants, a long, slender sclerenchyma cell that usually occurs in a bundle.

fibrin (fī´-brin) The activated form of the blood-clotting protein fibrinogen, which aggregates into threads that form the fabric of a blood clot.

fibrinogen (fī´-brin´-uh-jen) The plasma protein that is activated to form a clot when a blood vessel is injured.

fibrous connective tissue A dense tissue with large numbers of collagenous fibers organized into parallel bundles. This is the dominant tissue in tendons and ligaments.

filtrate Fluid extracted by the excretory system from the blood or body cavity. The excretory system produces urine from the filtrate after extracting valuable solutes from it and concentrating it.

filtration In the vertebrate kidney, the extraction of water and small solutes, including metabolic wastes, from the blood by the nephrons.

first law of thermodynamics The natural law stating that the total amount of energy in the universe is constant and that energy can be transferred and transformed, but never destroyed; also called the principle of energy conservation.

fission A means of asexual reproduction whereby a parent separates into two or more genetically identical individuals of about equal size.

fitness The contribution an individual makes to the gene pool of the next generation, relative to the contribution of other individuals in the population.

five-kingdom system The system of taxonomic classification based on five basic groups: Monera, Protista, Plantae, Fungi, and Animalia.

fixed action pattern (FAP) A genetically programmed, virtually unchangeable behavioral sequence performed in response to a certain stimulus.

flagellate (flaj´-uh-lit) A protist (protozoan) that moves by means of one or more flagella.

flatworm A member of the phylum Platyhelminthes.

fluid feeder An animal that lives by sucking nutrient-rich fluids from another living organism.

fluid mosaic A description of membrane structure, depicting a cellular membrane as a mosaic of diverse protein molecules embedded in a fluid bilayer made of phospholipid molecules.

fluke One of a group of parasitic flatworms.

follicle (fol´-uh-kul) A cluster of cells that surround, protect, and nourish a developing egg cell in the ovary; also secretes estrogen.

follicle-stimulating hormone (FSH) A protein hormone secreted by the anterior pituitary that stimulates the production of eggs by the ovaries and sperm by the testes.

food chain A sequence of food transfers from producers through several levels of consumers in an ecosystem.

food web A network of interconnecting food chains.

food-conducting cell A specialized, living plant cell with thin primary walls; arranged end to end, such cells collectively form phloem tissue. Also called sieve-tube member.

foot In an invertebrate animal, a structure used for locomotion or attachment, such as the muscular organ extending from the ventral side of a mollusc.

foraging Behavior necessary to recognize, search for, capture, and consume food.

forebrain One of three ancestral and embryonic regions of the vertebrate brain; develops into the thalamus, hypothalamus, and cerebrum.

forensic science The scientific analysis of evidence for crime scene and other investigations.

fossil A preserved remnant or impression of an organism that lived in the past.

fossil fuel An energy deposit formed from the remains of extinct organisms.

fossil record The chronicle of evolution over millions of years of geologic time engraved in the order in which fossils appear in rock strata.

founder effect Random change in the gene pool that occurs in a small colony of a population.

fovea (fō´-vē-uh) An eye's center of focus and the place on the retina where photoreceptors are highly concentrated.

fragmentation A means of asexual reproduction whereby a single parent breaks into parts that regenerate into whole new individuals.

free-living flatworm One of a group of nonparasitic flatworms.

frequency-dependent selection A decline in the reproductive success of a morph's phenotype resulting from the morphiphenotype becoming too common in a population; a cause of balanced polymorphism in populations.

fruit A ripened, thickened ovary of a flower, which protects dormant seeds and aids in their dispersal.

functional group An assemblage of atoms that forms the chemically reactive part of an organic molecule.

Fungi (fun´-jē) The kingdom that contains the fungi.

fungus (plural, **fungi**) A heterotrophic eukaryote that digests its food externally and absorbs the resulting small nutrient molecules. Most fungi consist of a netlike mass of filaments called hyphae. Molds, mushrooms, and yeasts are examples of fungi.

G

gallbladder An organ that stores bile and releases it as needed into the small intestine.

gametangium (gam´-uh-tan´-jē-um) (plural, **gametangia**) A reproductive organ that houses and protects the gametes of a plant.

gamete (gam´-ēt) A sex cell; a haploid egg or sperm. The union of two gametes of opposite sex (fertilization) produces a zygote.

gametic isolation (guh-mē´-tik) A type of prezygotic barrier between species; the species remain isolated because male and female gametes of the different species cannot fuse or they die before they unite.

gametogenesis The creation of gametes within the gonads.

gametophyte (guh-mē´-tō-fīt) The multicellular haploid form in the life cycle of organisms undergoing alternation of generations; mitotically produces haploid gametes that unite and grow into the sporophyte generation.

ganglion (gang´-glē-un) (plural, **ganglia**) A cluster (functional group) of nerve cell bodies in a centralized nervous system.

gap junction A channel between adjacent tissue cells through which water and other small molecules pass freely.

gas exchange *See* respiration.

gastric juice The collection of fluids (mucus, enzymes, and acid) secreted by the epithelium lining the stomach.

gastric ulcer An open sore in the lining of the stomach, resulting when pepsin and hydrochloric acid destroy the lining tissues faster than they can regenerate.

gastrin A digestive hormone that stimulates the secretion of gastric juice.

gastropod A member of the largest group of molluscs, including snails and slugs.

gastrovascular cavity A digestive compartment with a single opening, the mouth; may function in circulation, body support, waste disposal, and gas exchange, as well as digestion.

gastrula (gas´-trū-luh) The embryonic stage resulting from gastrulation in animal development. Most animals have a gastrula made up of three layers of cells: ectoderm, endoderm, and mesoderm.

gastrulation (gas´-trū-lā´-shun) The phase of embryonic development that transforms the blastula into a gastrula. Gastrulation adds more cells to the embryo and sorts the cells into distinct cell layers.

gel electrophoresis (jel´ ē-lek´-trō-fōr-ē´-sis) A technique for separating and purifying macromolecules. A mixture of molecules is placed on a gel between a positively charged electrode and a negatively charged one. Negative charges on the molecules are attracted to the positive electrode, and the molecules migrate toward that electrode. The molecules separate in the gel according to their rates of migration.

gene A discrete unit of hereditary information consisting of a specific nucleotide sequence in DNA (or RNA, in some viruses). Most of the genes of a eukaryote are located in its chromosomal DNA; a few are carried by the DNA of mitochondria and chloroplasts.

gene cloning The production of multiple copies of a gene.

gene expression The process whereby genetic information flows from genes to proteins; the flow of genetic information from the genotype to the phenotype.

gene flow The gain or loss of alleles from a population by the movement of individuals or gametes into or out of the population.

gene pool All the genes in a population at any one time.

gene therapy A treatment for a disease in which the patient's defective gene is altered.

genetic code The set of rules giving the correspondence between nucleotide triplets (codons) in mRNA and amino acids in protein.

genetic drift A change in the gene pool of a population due to chance.

genetic engineering The direct manipulation of genes for practical purposes.

genetic marker (1) An allele tracked in a genetic study. (2) A specific section of DNA that earmarks a particular allele; may contain specific restriction sites (points where restriction enzymes cut the DNA) that occur only in DNA that contains the allele.

genetic recombination The production, by crossing over and/or independent assortment of chromosomes during meiosis, of offspring with allele combinations different from those in the parents. The term may also be used more specifically to mean the production by crossing over of eukaryotic or prokaryotic chromosomes with gene combinations different from those in the original chromosomes.

genetically modified (GM) organism An organism that has acquired one or more genes by artificial means. If the gene is from another species, the organism is also known as a transgenic organism.

genetics The scientific study of heredity and hereditary variations.

genome (jē´-nōm) A complete (haploid) set of an organism's genes; an organism's genetic material.

genomic library (juh-nō´-mik) A set of DNA segments from an organism's genome; each segment is usually carried by a plasmid or phage.

genomics The study of whole sets of genes and their interactions.

genotype (jē´-nō-tīp) The genetic makeup of an organism.

genus (jē´-nus) (plural, **genera**) In classification, the taxonomic category above species; the first part of a species' binomial; for example, *Homo*.

geologic record A time scale established by geologists that reflects a consistent sequence of geologic periods, grouped into three eons.

germinate To start developing or growing.

gestation (jes-tā´-shun) Pregnancy; the state of carrying developing young within the female reproductive tract.

gibberellin (jib´-uh-rel´-in) One of a family of plant hormones that triggers the germination of seeds and interacts with auxins in regulating growth and fruit development.

gill An extension of the body surface of an animal, specialized for gas exchange and/or suspension feeding.

gizzard A pouch-like organ in a digestive tract where food is mechanically ground.

glans The rounded, highly sensitive head of the clitoris in females and penis in males.

glia A supporting cell that is essential for the structural integrity and normal functioning of the nervous system.

global warming A slow but steady rise in Earth's surface temperature, caused by increasing concentrations of greenhouse gases (such as CO_2 and CH_4) in the atmosphere.

glomeromycete Member of the fungal phylum Glomeromycota, characterized by a distinct branching form of endomycorrhizae (symbiotic relationships with plant roots) called arbuscular mycorrhizae.

glomerulus (glō-mer´-ū-lus) (plural, **glomeruli**) In the vertebrate kidney, the part of a nephron consisting of the capillaries that are surrounded by Bowman's capsule; together, a glomerulus and Bowman's capsule produce the filtrate from the blood.

glucagon (glū´-kuh-gon) A peptide hormone, secreted by the islets of Langerhaus in the pancreas, that raises the level of glucose in the blood.

glucocorticoid (glū´-kuh-kor´-tih-koyd) A corticosteroid hormone secreted by the adrenal cortex that increases the blood glucose level and helps maintain the body's response to long-term stress.

glycogen (glī´-kō-jen) A complex, extensively branched polysaccharide of many glucose monomers; serves as an energy-storage molecule in liver and muscle cells.

glycolysis (glī-kol´-uh-sis) The multistep chemical breakdown of a molecule of glucose into two molecules of pyruvate; the first stage of cellular respiration in all organisms; occurs in the cytoplasmic fluid.

glycoprotein (glī-kō-prō´-tēn) A macromolecule consisting of one or more polypeptides linked to short chains of sugars.

goiter An enlargement of the thyroid gland resulting from a dietary iodine deficiency.

Golgi apparatus (gol´-jē) An organelle in eukaryotic cells consisting of stacks of membranous sacs that modify, store, and ship products of the endoplasmic reticulum.

gonad A sex organ in an animal; an ovary or testis.

gradualism model The view that evolution occurs as a result of populations becoming isolated from common ancestral stock and gradually becoming genetically unique as they are adapted by natural selection to their local environments; Darwin's view of the origin of species.

gram-positive bacteria Bacteria with a cell wall that is structurally less complex and contains more peptidoglycan than that of gram-negative bacteria. Gram-positive bacteria are usually less toxic than gram-negative bacteria.

granum (gran´-um) (plural, **grana**) A stack of hollow disks formed of thylakoid membrane in a chloroplast. Grana are the sites where light energy is trapped by chlorophyll and converted to chemical energy during the light reactions of photosynthesis.

gravitropism (grav´-uh-trō´-pizm) A plant's growth response to gravity.

gray matter Regions of dendrites and clusters of nerve cell bodies within the CNS.

green alga One of a group of photosynthetic protists that includes unicellular, colonial, and multicellular species. Green algae are plant-like in having biflagellated cells (gametes in colonial and multicellular species), chloroplasts with chlorophyll *a*, cellulose cell walls, and starch.

greenhouse effect The warming of the atmosphere caused by CO_2, CH_4, and other gases that absorb infrared radiation and slow its escape from Earth's surface.

ground tissue system A tissue of mostly parenchyma cells that makes up the bulk of a young plant and is continuous throughout its body. The ground tissue system fills the space between the epidermis and the vascular tissue system.

growth factor A protein secreted by certain body cells that stimulates other cells to divide.

growth hormone (GH) A protein hormone secreted by the anterior pituitary that promotes development and growth and stimulates metabolism.

guanine (G) (gwa´-nēn) A double-ring nitrogenous base found in DNA and RNA.

guard cell A specialized epidermal cell in plants that regulates the size of a stoma, allowing gas exchange between the surrounding air and the photosynthetic cells in the leaf.

gymnosperm (jim´-nō-sperm) A naked-seed plant. Its seed is said to be naked because it is not enclosed in a fruit.

H

habitat A place where an organism lives; an environmental situation in which an organism lives.

habitat isolation A type of prezygotic barrier between species; the species remain isolated because they breed in different habitats.

habituation Learning not to respond to a repeated stimulus that conveys little or no information.

hair cell A type of mechanoreceptor that detects sound waves and other forms of movement in air or water.

haploid cell In the life cycle of an organism that reproduces sexually, a cell containing a single set of chromosomes; an *n* cell.

Hardy-Weinberg equilibrium The principle that the shuffling of genes that occurs during sexual reproduction, by itself, cannot change the overall genetic makeup of a population.

HDL *See* high-density lipoprotein.

heart attack Death of cardiac muscle cells and the resulting failure of the heart to deliver enough blood to the body.

heartwood In the center of trees, the darkened, older layers of secondary xylem made up of cells that no longer transport water and are clogged with resins. *See also* sapwood.

heat Thermal energy; the amount of energy associated with the movement of the atoms and molecules in a body of matter. Heat is energy in its most random form.

helper T cell A type of lymphocyte that helps activate other types of T cells and may help stimulate B cells to produce antibodies.

hemoglobin (hē´-mō-glō-bin) An iron-containing protein in red blood cells that reversibly binds O_2 and transports it to body tissues.

hemophilia (hē´-mō-fil´-ē-uh) A human genetic disease caused by a sex-linked recessive allele; characterized by excessive bleeding following injury.

hepatic portal vein A blood vessel that conveys blood from capillaries surrounding the intestine directly to the liver.

herbivore An animal that eats only plants or algae. *See also* carnivore; omnivore.

hermaphroditism (her-maf´-rō-dī-tizm) A condition in which an individual has both female and male gonads and functions as both a male and female in sexual reproduction by producing both sperm and eggs.

heterokaryotic stage A fungal life cycle stage that contains two genetically different nuclei in the same cell.

heterotroph (het´-er-ō-trōf) An organism that cannot make its own organic food molecules and must obtain them by consuming other organisms or their organic products; a consumer or a decomposer in a food chain.

heterozygote advantage Greater reproductive success of heterozygous individuals compared to homozygotes; tends to preserve variation in gene pools.

heterozygous (het´-er-ō-zī´-gus) Having two different alleles for a given gene.

hibernation Long-term torpor during cold weather, during which an animal's metabolic rate is reduced and it survives on energy stored in body fat.

high-density lipoprotein (HDL) A cholesterol-carrying particle in the blood, made up of cholesterol and other lipids surrounded by a single layer of phospholipids in which proteins are embedded. An HDL particle carries less cholesterol than a related lipoprotein, LDL, and may be correlated with a decreased risk of blood vessel blockage.

hindbrain One of three ancestral and embryonic regions of the vertebrate brain; develops into the medulla oblongata, pons, and cerebellum.

hinge joint A joint that allows movement in only one plane. In humans, examples include the elbow and knee.

hippocampus (hip´-uh-kam´-pus) An integrative center of the cerebrum; functionally, the part of the limbic system that plays a central role in memory and learning.

histamine (his´-tuh-mēn) A chemical alarm signal released by injured cells that causes blood vessels to dilate during an inflammatory response.

histone (his´-tōn) A small protein molecule associated with DNA and important in DNA packing in the eukaryotic chromosome.

HIV Human immunodeficiency virus, the retrovirus that attacks the human immune system and causes AIDS.

homeobox (hō´-mē-ō-boks´) A 180-nucleotide sequence within a homeotic gene and some other developmental genes.

homeostasis (hō´-mē-ō-stā´-sis) The steady state of body functioning; a state of equilibrium characterized by a dynamic interplay between outside forces that tend to change an organism's internal environment and the internal control mechanisms that oppose such changes.

homeotic gene (hō´-mē-ot´-ik) A master control gene that determines the identity of a body structure of a developing organism, presumably by controlling the developmental fate of groups of cells. (In plants, such genes are called organ identity genes.)

hominid (hah´-mi-nid) A species on the human branch of the evolutionary tree; a member of the family Hominidae, including *Homo sapiens* and our ancestors.

hominoid A term that refers to great apes, including humans.

homologous chromosomes (hō-mol´-uh-gus) The two chromosomes that make up a matched pair in a diploid cell. Homologous chromosomes are of the same length, centromere position, and staining pattern and possess alleles for the same genes at corresponding loci. One homologous chromosome is inherited from the organism's father, the other from the mother.

homologous structures Structures that are similar in different species of common ancestry.

homology Anatomical similarity due to common ancestry.

homozygous (hō´-mō-zī´-gus) Having two identical alleles for a given gene.

hormone (1) In plants, a chemical that is produced in one part of the plant and that travels to another part, where it coordinates certain plant activities. (2) In animals, a regulatory chemical that travels in the blood from its production site, usually an endocrine gland, to other sites, where target cells respond to the regulatory signal.

horseshoe crab A bottom-dwelling marine chelicerate, a member of the phylum Arthropoda.

human chorionic gonadotropin (HCG) (kōr´-ē-on´-ik gon´-uh-dō-trō´-pin) A hormone secreted by the chorion that maintains the corpus luteum of the ovary during the first three months of pregnancy.

Human Genome Project (HGP) An international collaborative effort to map and sequence the DNA of the entire human genome.

humoral immunity The type of specific immunity brought about by antibody-producing B cells; fights bacteria and viruses in body fluids. *See also* cell-mediated immunity.

humus (hyū´-mus) Decomposing organic material found in topsoil.

Huntington's disease A human genetic disease caused by a dominant allele; characterized by uncontrollable body movements and degeneration of the nervous system; usually fatal 10 to 20 years after the onset of symptoms.

hybrid The offspring of parents of two different species or of two different varieties of one species; the offspring of two parents that differ in one or more inherited traits; an individual that is heterozygous for one or more pairs of genes.

hybrid breakdown A type of postzygotic barrier between species; the species remain isolated because the offspring of hybrids are weak or infertile.

hybrid inviability A type of postzygotic barrier between species; the species remain isolated because hybrid zygotes do not develop or hybrids do not become sexually mature.

hybrid sterility A type of postzygotic barrier between species; the species remain isolated because hybrids fail to produce functional gametes.

hydrocarbon A chemical compound composed only of the elements carbon and hydrogen.

hydrogen bond A type of weak chemical bond formed when the partially positive hydrogen atom participating in a polar covalent bond in one molecule is attracted to the partially negative atom participating in a polar covalent bond in another molecule (or in another part of the same macromolecule).

hydrolysis (hī-drol´-uh-sis) A chemical process in which macromolecules are broken down by the chemical addition of water molecules to the bonds linking their monomers; an essential part of digestion.

hydrophilic (hī´-drō-fil´-ik) "Water-loving"; pertaining to polar, or charged, molecules (or parts of molecules) that are soluble in water.

hydrophobic (hī´-drō-fō´-bik) "Water-fearing"; pertaining to nonpolar molecules (or parts of molecules) that do not dissolve in water.

hydrostatic skeleton A skeletal system composed of fluid held under pressure in a closed body compartment; the main skeleton of most cnidarians, flatworms, nematodes, and annelids.

hydroxyl group (hī-drok´-sil) In an organic molecule, a functional group consisting of a hydrogen atom bonded to an oxygen atom.

hymen A thin membrane that partly covers the vaginal opening in the human female and is ruptured by sexual intercourse or other vigorous activity.

hypercholesterolemia (hī´-per-kō-les´-tur-ah-lēm´-ē-uh) An inherited human disease characterized by an excessively high level of cholesterol in the blood.

hypertension Abnormally high blood pressure; a persistent blood pressure of 140/90 or higher.

hypertonic solution In comparing two solutions, the one with the greater concentration of solutes.

hyperventilating Taking several deep breaths so rapidly that the CO_2 level in the blood is reduced, causing the breathing control centers to temporarily shut down breathing movements.

hypha (hī´-fuh) (plural, **hyphae**) One of many filaments making up the body of a fungus.

hypoglycemia (hī´-pō-glī-sē´-mē-uh) An abnormally low level of glucose in the blood that results when the pancreas secretes too much insulin into the blood.

hypothalamus (hī´-pō-thal´-uh-mus) The master control center of the endocrine system, located in the ventral portion of the vertebrate forebrain. The hypothalamus functions in maintaining homeostasis, especially in coordinating the endocrine and nervous systems; secretes hormones of the posterior pituitary and releasing hormones that regulate the anterior pituitary.

hypothesis (hī-poth´-uh-sis) (plural, **hypotheses**) A tentative explanation a scientist proposes for a specific phenomenon that has been observed.

hypotonic solution In comparing two solutions, the one with the lower concentration of solutes.

I

immune system The organ system that protects the body by recognizing and attacking specific kinds of pathogens and cancer cells.

immunity Resistance to specific body invaders.

immunodeficiency disease An immunological disorder in which the immune system lacks one or more components, making the body susceptible to infectious agents that would ordinarily not be pathogenic.

imperfect fungus A fungus with no known sexual stage.

impotence The inability to maintain an erection; also called erectile dysfunction.

imprinting Learning that is limited to a specific critical period in an animal's life and that is generally irreversible.

in vitro fertilization (IVF) (vē´-tro) Uniting sperm and egg in a laboratory container, followed by the placement of a resulting early embryo in the mother's uterus.

inbreeding Mating between close relatives

inclusive fitness The total effect an individual has on proliferating its genes by producing its own offspring and by providing aid that enables close relatives to increase the production of their offspring.

incomplete dominance A type of inheritance in which the phenotype of a heterozygote (*Aa*) is intermediate between the phenotypes of the two types of homozygotes (*AA* and *aa*).

incomplete metamorphosis A type of development in certain insects, such as grasshoppers, in which the larvae resemble adults but are smaller and have different body proportions. The animal goes through a series of molts, each time looking more like an adult, until it reaches full size.

indeterminate growth Growth that continues throughout life, as in most plants. *See also* determinate growth.

induced fit The interaction between a substrate molecule and the active site of an enzyme, which changes shape slightly to embrace the substrate and catalyze the reaction.

induction During embryonic development, the influence of one group of cells on another group of cells.

inferior vena cava (vē´-nuh kā´-vuh) A large vein that returns O_2-poor blood to the heart from the lower, or posterior, part of the body. *See also* superior vena cava.

infertility The inability to conceive after one year of regular, unprotected intercourse.

inflammatory response A nonspecific body defense caused by a release of histamine and other chemical alarm signals, which trigger increased blood flow, a local increase in white blood cells, and fluid leakage from the blood. The results include redness, heat, and swelling in the affected tissues.

ingestion The act of eating; the first main stage of food processing.

ingroup In a cladistic study of evolutionary relationships among taxa of organisms, the group of taxa that is actually being analyzed. *See also* out-group.

inhibiting hormone A kind of hormone released from the hypothalamus that makes the anterior pituitary stop secreting hormone.

innate behavior Behavior that appears to be performed in virtually the same way by all members of a species.

innate immunity The kind of defense that is mediated by phagocytic cells, antimicrobial proteins, the inflammatory response, and natural killer cells. It is present before exposure to pathogens and is effective from the time of birth.

inner ear One of three main regions of the vertebrate ear; includes the cochlea, organ of Corti, and semicircular canals.

insulin A protein hormone, secreted by the islets of Langerhans in the pancreas, that lowers the level of glucose in the blood.

integration The interpretation of sensory signals within neural processing centers of the central nervous system.

integumentary system (in-teg´-yū-men´-ter-ē) The organ system consisting of the skin and its derivatives, such as hair and nails in mammals; helps protect the body from drying out, mechanical injury, and infection.

interferon (in´-ter-fē r´-on) A nonspecific defensive protein produced by virus-infected cells and capable of helping other cells resist viruses.

intermediate One of the compounds that form between the initial reactant and the final product in a metabolic pathway, such as between glucose and pyruvate in glycolysis.

intermediate filament An intermediate-sized protein fiber that is one of the three main kinds of fibers making up the cytoskeleton of eukaryotic cells. Intermediate filaments are ropelike, made of fibrous proteins.

intermembrane space One of the two fluid-filled internal compartments of the mitochondrion. The intermembrane space is the narrow region between the inner and outer membranes.

internal fertilization Reproduction in which sperm are typically deposited in or near the female reproductive tract and fertilization occurs within the tract.

interneuron (in´-ter-nūr´-on) A nerve cell, entirely within the central nervous system, that integrates sensory signals and may relay command signals to motor neurons.

internode The portion of a plant stem between two nodes.

interphase The period in the eukaryotic cell cycle when the cell is not actually dividing. *See also* mitotic phase.

interspecific competition Competition between individuals or populations of two or more species requiring a limited resource; may inhibit population growth and help structure communities.

interstitial fluid (in´-ter-stish´-ul) An aqueous solution that surrounds body cells and through which materials pass back and forth between the blood and the body tissues.

intertidal zone (in´-ter-tīd´-ul) A shallow zone where the waters of an estuary or ocean meet land.

intestine The region of a digestive tract located between the gizzard or stomach and the anus and where chemical digestion and nutrient absorption usually occur.

intrinsic rate of increase An organism's inherent capacity to reproduce.

intron (in´-tron) In eukaryotes, a nonexpressed (noncoding) portion of a gene that is excised from the RNA transcript. *See* exon.

inversion A change in a chromosome resulting from reattachment of a chromosome fragment to the original chromosome, but in a reverse direction. Mutagens and errors during meiosis can cause inversions.

invertebrate An animal that lacks a backbone.

ion (ī´-on) An atom or molecule that has gained or lost one or more electrons, thus acquiring an electrical charge.

ionic bond (ī-on´-ik) An attraction between two ions with opposite electrical charges. The electrical attraction of the opposite charges holds the ions together.

iris The colored part of the vertebrate eye, formed by the anterior portion of the choroid.

islets of Langerhans (ī´-lits) Clusters of endocrine cells in the pancreas that produce insulin and glucagon.

isomers (ī´-sō-mers) Organic compounds with the same molecular formula but different structures and, therefore, different properties.

isotonic solution (ī-sō-ton´-ik) A solution having the same solute concentration as another solution.

isotope (ī´-sō-tōp) A variant form of an atom. Isotopes of an element have the same number of protons but different numbers of neutrons.

K

karyotype (kār´-ē-ō-tīp) A display of micrographs of the metaphase chromosomes of a cell, arranged by size and centromere position.

kelp A giant brown alga, up to 100 m long, that forms extensive undersea forests.

keystone predator A predator species that reduces the density of the strongest competitors in a community, thereby helping maintain species diversity.

keystone species A species that is not usually abundant in a community yet exerts strong control on community structure by the nature of its ecological role or niche.

kilocalorie (kcal) a quantity of heat equal to 1,000 calories. Used to measure the energy content of food, it is usually called a "calorie."

kin selection A phenomenom of inclusive fitness, used to explain altruistic behavior between related individuals.

kinesis (kuh-nē´-sis) Random movement in response to a stimulus.

kinetochore (kuh-net´-ō-kor) A specialized protein structure at the centromere region on a sister chromatid. Spindle microtubules attach to the kinetochore during mitosis and meiosis.

kinetic energy (kuh-net´-ik) The energy of motion; the energy of a mass of matter that is moving. Moving matter performs work by transferring its motion to other matter, such as leg muscles pushing bicycle pedals.

kingdom In classification, the broad taxonomic category above phylum or division.

Koch's postulates A set of criteria used to establish that a particular infectious agent causes a disease.

K-selection The concept that in certain (K-selected) populations, life history is centered around producing relatively few offspring that have a good chance of survival.

L

labia majora (lā´-bē-uh muh-jor´-uh) A pair of outer thickened folds of skin that protect the female genital region.

labia minora (lā´-bē-uh mi-nor´-uh) A pair of inner folds of skin, bordering and protecting the female genital region.

labor A series of strong, rhythmic contractions of the uterus that expel a baby out of the uterus and vagina during childbirth.

lactic acid fermentation The conversion of pyruvate to lactate with no release of carbon dioxide.

lancelet One of a group of invertebrate chordates.

landmark A point of reference for orientation during navigation.

landscape ecology The application of ecological principles to the study of the structure and dynamics of a collection of ecosystems; the scientific study of the biodiversity of interacting ecosystems.

large intestine See colon.

larva (lar´-vuh) (plural, **larvae**) A free-living, sexually immature form in some animal life cycles that may differ from the adult in morphology, nutrition, and habitat.

larynx (lār´-inks) The voice box, containing the vocal cords.

lateral meristem A meristem that thickens the roots and shoots of woody plants. The vascular cambium and cork cambium are lateral meristems.

lateral line system A row of sensory organs along each side of a fish's body. Sensitive to changes in water pressure, it enables a fish to detect minor vibrations in the water.

lateralization The phenomenon in which the two hemispheres of the brain become specialized for different functions.

law of independent assortment A general rule in inheritance that when gametes form during meiosis, each pair of alleles for a particular characteristic segregate independently; also known as Mendel's second law of inheritance.

law of segregation A general rule in inheritance that individuals have two alleles for each gene and that when gametes form by meiosis, the two alleles separate, and each resulting gamete ends up with only one allele of each gene; also known as Mendel's first law of inheritance.

LDL See low-density lipoprotein.

leaf The main site of photosynthesis in a plant; consists of a flattened blade and a stalk (petiole) that joins the leaf to the stem.

learning Modification of behavior as a result of specific experiences.

leech A member of one of the three large groups of annelids. See annelid.

lens The structure in an eye that focuses light rays onto the retina.

leukemia (lū-kī´-mē-ah) A type of cancer of the blood-forming tissues, characterized by an excessive production of white blood cells and an abnormally high number of them in the blood; cancer of the bone marrow cells that produce leukocytes.

leukocyte (lū´-kō-sīt) See white blood cell.

lichen (lī´-ken) A mutualistic association between a fungus and an alga or between a fungus and a cyanobacterium.

life cycle The entire sequence of stages in the life of an organism, from the adults of one generation to the adults of the next.

life history The series of events from birth through reproduction to death.

life table A listing of survivals and deaths in a population in a particular time period and predictions of how long, on average, an individual of a given age will live.

ligament A type of fibrous connective tissue that joins bones together at joints.

light microscope (LM) An optical instrument with lenses that refract (bend) visible light to magnify images and project them into a viewer's eye or onto photographic film.

light reactions The first of two stages in photosynthesis; the steps in which solar energy is absorbed and converted to chemical energy in the form of ATP and NADPH. The light reactions power the sugar-producing Calvin cycle but produce no sugar themselves.

lignin A chemical that hardens the cell walls of plants.

limbic system (lim´-bik) A functional unit of several integrating and relay centers located deep in the human forebrain; interacts with the cerebral cortex in creating emotions and storing memories.

limiting factors Environmental factors that restrict population growth.

linked genes Genes located on the same chromosome that tend to be inherited together.

lipid An organic compound consisting mainly of carbon and hydrogen atoms linked by nonpolar convalent bonds, making the compound mostly hydrophobic. Lipids include fats, waxes, phospholipids, and steroids and are insoluble in water.

liver The largest organ in the vertebrate body. The liver performs diverse functions, such as producing bile, preparing nitrogenous wastes for disposal, and detoxifying poisonous chemicals in the blood.

lobe-fin A bony fish with strong, muscular fins supported by bones. Lobe-fins are extinct except for one species, the coelacanth.

local regulator A chemical messenger that is secreted into the interstitial fluid and causes changes in cells very near the point of secretion. Neurotransmitters are local regulators.

locomotion Active movement from place to place.

locus (plural, **loci**) The particular site where a gene is found on a chromosome. Homologous chromosomes have corresponding gene loci.

logistic growth model A mathematical description of idealized population growth that is restricted by limiting factors.

long-day plant A plant that flowers in late spring or early summer, when day length is increasing.

long-term depression (LTD) A reduced responsiveness to an action potential (nerve signal) by a receiving neuron.

long-term memory The ability to hold, associate, and recall information over one's life.

long-term potentiation (LTP) An enhanced responsiveness to an action potential (nerve signal) by a receiving neuron.

loop of Henle (hen´-lē) In the vertebrate kidney, the portion of a nephron that helps concentrate the filtrate while conveying it between a proximal tubule and a distal tubule.

loose connective tissue The most widespread connective tissue in the vertebrate body. It binds epithelia to underlying tissues and functions as packing material, holding organs in place.

low-density lipoprotein (LDL) A cholesterol-carrying particle in the blood, made up of cholesterol and other lipids surrounded by a single layer of phospholipids in which proteins are embedded. An LDL particle carries more cholesterol than a related lipoprotein, HDL, and high LDL levels in the blood correlate with a tendency to develop blocked blood vessels and heart disease.

lung An internal sac, lined with moist epithelium, where gases are exchanged between inhaled air and the blood.

luteinizing hormone (LH) (lū-tē-uh-nī´-zing) A protein hormone secreted by the anterior pituitary that stimulates ovulation in females and androgen production in males.

Lyme disease A debilitating human disease caused by the bacterium *Borrelia burgdorferi*; characterized at first by a red rash at the site of a tick bite and, if not treated, by heart disease, arthritis, and nervous disorders.

lymph A fluid similar to interstitial fluid that circulates in the lymphatic system.

lymphatic system (lim-fat´-ik) The organ system through which lymph circulates; includes lymph vessels, lymph nodes, and the spleen. The lymphatic system helps remove toxins and pathogens from the blood and interstitial fluid and returns fluid and solutes from the interstitial fluid to the circulatory system.

lymphocyte (lim´-fuh-sīt) A type of white blood cell that is chiefly responsible for the immune response; found mostly in the lymphatic system. See B cell; T cell.

lymphoma (lim-fō´-muh) Cancer of the tissues that form white blood cells.

lysogenic cycle (lī-sō-jen´-ik) A type of bacteriophage replication cycle in which the viral genome is incorporated into the bacterial host chromosome as a prophage. New phages are not produced, and the host cell is not killed or lysed unless the viral genome leaves the host chromosome.

lysosome (lī´-sō-sōm) A digestive organelle in eukaryotic cells; contains hydrolytic enzymes that digest the cell's food and wastes.

lytic cycle (lit´-ik) A type of viral replication cycle resulting in the release of new viruses by lysis (breaking open) of the host cell.

M

macroevolution Evolutionary change on a grand scale, encompassing the origin of new taxonomic groups, evolutionary trends, adaptive radiation, and mass extinction.

macromolecule A giant molecule in a living organism: a protein, carbohydrate, lipid, or nucleic acid.

macronutrient A chemical substance that an organism must obtain in relatively large amounts. *See also* micronutrient.

macrophage (mak´-rō-fāj) A large, amoeboid, phagocytic white blood cell that functions in innate immunity by destroying microbes and in acquired immunity as an antigen-presenting cell.

magnification An increase in the apparent size of an object.

major depression Depressive mental illness characterized by experiencing a low mood most of the time.

major histocompatibility complex (MHC) *See* self protein.

malignant tumor An abnormal tissue mass that can spread into neighboring tissue and to other parts of the body; a cancerous tumor.

mammal Member of the class Mammalia, amniotes that possess mammary glands and hair.

mantle In a mollusc, the outgrowth of the body surface that drapes over the animal. The mantle produces the shell and forms the mantle cavity.

marsupial (mar-sū´-pē-ul) A pouched mammal, such as a kangaroo, opossum, or koala. Marsupials give birth to embryonic offspring that complete development while housed in a pouch and attached to nipples on the mother's abdomen.

mass number The sum of the number of protons and neutrons in an atom's nucleus.

mast cell A vertebrate body cell that produces histamine and other molecules that trigger the inflammatory response.

matter Anything that occupies space and has mass.

maximum sustained yield The level of harvest that produces a consistent yield without forcing a population into decline.

mechanical isolation A type of prezygotic barrier between species; the species remain isolated because structural differences between them prevent fertilization.

mechanoreceptor (mek´-uh-nō-ri-sep´-ter) A sensory receptor that detects physical deformations in the environment associated with pressure, touch, stretch, motion, or sound.

medulla oblongata (meh-duh´-luh ob´-long-got´-uh) Part of the vertebrate hindbrain continuous with the spinal cord; passes data between the spinal cord and forebrain and controls autonomic, homeostatic functions, including breathing, heart rate, swallowing, and digestion.

medusa (med-ū´-suh) (plural, **medusae**) One of two types of cnidarian body forms; an umbrella-like body form. Also called a jellyfish.

meiosis (mī-ō´-sis) In a sexually reproducing organism, the division of a single diploid nucleus into four haploid daughter nuclei. Meiosis and cytokinesis produce haploid gametes from diploid cells in the reproductive organs of the parents.

membrane infolding A process by which the eukaryotic cell's endomembrane system evolved from inward folds of the plasma membrane of a prokaryotic cell.

membrane potential The charge difference between a cell's cytoplasm and extracellular fluid due to the differential distribution of ions.

memory The ability to store and retrieve information. *See also* long-term memory, short-term memory.

memory cell One of a clone of long-lived lymphocytes formed during the primary immune response; remains in a lymph node until activated by exposure to the same antigen that triggered its formation. When activated, a memory cell forms a large clone that mounts the secondary immune response.

meninges (muh-nin´-jēz) Layers of connective tissue that enwrap and protect the brain and spinal cord.

menstrual cycle (men´-strū-ul) The hormonally synchronized cyclic buildup and breakdown of the endometrium of some primates, including humans.

menstruation (men´-strū-ā´-shun) Uterine bleeding resulting from shedding of the endometrium during a menstrual cycle.

meristem (mār´-eh-stem) Plant tissue consisting of undifferentiated cells that divide and generate new cells and tissues.

mesoderm (mez´-ō-derm) The middle layer of the three embryonic cell layers in a gastrula; gives rise to muscles, bones, the dermis of the skin, and most other organs in the adult.

mesophyll (mes´-ō-fil) The green tissue in the interior of a leaf; a leaf's ground tissue system; the main site of photosynthesis.

messenger RNA (mRNA) The type of ribonucleic acid that encodes genetic information from DNA and conveys it to ribosomes, where the information is translated into amino acid sequences.

metamorphosis (met´-uh-mōr´-fuh-sis) The transformation of a larva into an adult.

metaphase (met´-eh-fāz) The third stage of mitosis, during which all the cell's duplicated chromosomes are lined up at an imaginary plane equidistant between the poles of the mitotic spindle.

metastasis (muh-tas´-tuh-sis) The spread of cancer cells beyond their original site.

methanogen A microorganism that obtains energy by using carbon dioxide to oxidize hydrogen, producing methane as a waste product.

MHC Major histocompatibility complex. *See* self protein.

microevolution A change in a population's gene pool over a succession of generations; evolutionary changes in species over relatively brief periods of geologic time.

microfilament The thinnest of the three main kinds of protein fibers making up the cytoskeleton of a eukaryotic cell; a solid, helical rod composed of the globular protein actin.

micrograph A photograph taken through a microscope.

micronutrient An element that an organism needs in very small amounts and that functions as a component or cofactor of enzymes. *See also* macronutrient.

microtubule The thickest of the three main kinds of fibers making up the cytoskeleton of a eukaryotic cell; a straight, hollow tube made of globular proteins called tubulins. Microtubules form the basis of the structure and movement of cilia and flagella.

microvillus (plural, **microvilli**) A microscopic projection on the surface of a cell. Microvilli increase a cell's surface area.

midbrain One of three ancestral and embryonic regions of the vertebrate brain; develops into sensory integrating and relay centers that send sensory information to the cerebrum.

middle ear One of three main regions of the vertebrate ear; a chamber containing three small bones (the hammer, anvil, and stirrup) that convey vibrations from the eardrum to the oval window.

migration The regular back-and-forth movement of animals between two geographic areas at particular times of the year.

millipede A terrestrial arthropod that has two pairs of short legs for each of its numerous body segments and that eats decaying plant matter.

mineral In nutrition, a simple inorganic nutrient that an organism requires for proper body functioning.

mineralocorticoid (min´-er-uh-lō-kort´-uh-koyd) A corticosteroid hormone secreted by the adrenal cortex that helps maintain salt and water homeostasis and may increase blood pressure in response to long-term stress.

mitochondrial matrix (mī´-tō-kon´-drē-ul mā´-triks) The fluid contained within the inner membrane of a mitochondrion.

mitochondrion (mī´-tō-kon´-drē-on) (plural, **mitochondria**) An organelle in eukaryotic cells where cellular respiration occurs. Enclosed by two concentric membranes, it is where most of the cell's ATP is made.

mitosis (mī-tō-sis) The division of a single nucleus into two genetically identical daughter nuclei. Mitosis and cytokinesis make up the mitotic (M) phase of the cell cycle.

mitotic phase (M phase) The part of the cell cycle when mitosis divides the nucleus and distributes its chromosomes to the daughter nuclei and cytokinesis divides the cytoplasm, producing two daughter cells.

mitotic spindle A spindle-shaped structure formed of microtubules and associated proteins that is involved in the movements of chromosomes during mitosis and meiosis. (A spindle is shaped roughly like a football.)

modern synthesis A comprehensive theory of evolution that incorporates genetics and includes most of Darwin's ideas, focusing on populations as the fundamental units of evolution.

mold A rapidly growing fungus that reproduces asexually by producing spores.

molecular biology The study of the molecular basis of genes and gene expression; molecular genetics.

molecular clock Evolutionary timing method based on the observation that at least some regions of genomes evolve at constant rates.

molecular systematics Comparing nucleic acids or other molecules to infer relatedness.

molecule A group of two or more atoms held together by covalent bonds.

mollusc (mol´-lusk) A soft-bodied animal characterized by a muscular foot, mantle, mantle cavity, and radula; includes gastropods (snails and slugs), bivalves (clams, oysters, and scallops), and cephalopods (squids and octopuses).

molting In arthropods, the process of shedding an old exoskeleton and secreting a new, larger one.

monoclonal antibody (mon´-ō-klōn´-ul) An antibody secreted by a clone of cells and, consequently, specific for the one antigen that triggered the development of the clone.

monocot (mon´-ō-kot) A flowering plant whose embryos have a single seed leaf, or cotyledon.

monoculture The cultivation of a single plant variety in a large land area.

monogamous Referring to a type of relationship in which one male mates with just one female.

monohybrid cross An experimental mating of individuals differing at one genetic locus.

monomer (mon´-uh-mer) A chemical subunit that serves as a building block of a polymer.

monophyletic (mon´-ō-fī-let´-ik) Pertaining to a taxon derived from a single ancestral species that gave rise to no species in any other taxa.

monosaccharide (mon´-ō-sak´-uh-rīd) The smallest kind of sugar molecule; a single-unit sugar. Monosaccharides are the building blocks of more complex sugars and polysaccharides.

monotreme (mon´-uh-trēm) An egg-laying mammal, such as the duck-billed platypus.

morning after pill (MAP) Birth control pill taken within three days of unprotected intercourse to prevent fertilization or implantation.

morphological species concept The idea that species are defined by measurable anatomical criteria.

motor neuron A nerve cell that conveys command signals from the central nervous system to effector cells, such as muscle cells or gland cells.

motor output The conduction of signals from a processing center in a central nervous system to effector cells.

motor unit A motor neuron and all the muscle fibers it controls.

mouth An opening through which food is taken into an animal's body.

movement corridor A series of small clumps or a narrow strip of quality habitat (usable by organisms) that connects otherwise isolated patches of quality habitat.

mRNA See messenger RNA.

mucous membrane (myū-kus) Smooth, moist epithelium that lines the digestive tract and air tubes leading to the lungs.

Müllerian mimicry (myū-lār´-ē-un mim´-uh-krē) A mutual mimicry by two species, both of which are poisonous or otherwise harmful to a predator.

multiple fruit A fruit, such as pineapple, that develops from a group of flowers tightly clustered together. When the walls of the many ovaries start to thicken, they fuse together and become incorporated into one fruit.

muscle tissue Tissue consisting of long muscle cells that are capable of contracting when stimulated by nerve impulses; the most abundant tissue in a typical animal. See skeletal muscle; cardiac muscle; smooth muscle.

muscular system All the skeletal muscles in the body. (Cardiac muscle and smooth muscle are components of other organ systems.)

mutagen (myū-tuh-jen) A chemical or physical agent that interacts with DNA and causes a mutation.

mutagenesis (myū-tuh-jen´-uh-sis) The creation of a mutation.

mutation A change in the nucleotide sequence of DNA; the ultimate source of genetic diversity.

mutualism A symbiotic relationship in which both partners benefit.

mycelium (mī-sē´-lē-um) (plural, **mycelia**) The densely branched network of hyphae in a fungus.

mycorrhiza (mī´-kō-rī´-zuh) (plural, **mycorrhizae**) A mutualistic association of plant roots and fungi.

mycosis The general term for a fungal infection.

myelin sheath (mī´-uh-lin) A series of cells, each wound around, and thus insulating, the axon of a nerve cell in vertebrates. Each pair of cells in the sheath is separated by a space called a node of Ranvier.

myofibril (mī´-ō-fī´-bril) A contractile thread in a muscle cell (fiber) made up of many sarcomeres. Longitudinal bundles of myofibrils make up a muscle fiber.

N

NAD⁺ Nicotinamide adenine dinucleotide; a coenzyme that assists enzymes by conveying electrons (from hydrogen atoms) during the redox reactions of cellular metabolism. The plus sign indicates that the molecule is oxidized and ready to pick up hydrogens; the reduced, hydrogen (electron)-carrying form is NADH.

natural family planning A form of contraception that relies on refraining from sexual intercourse when conception is most likely to occur; also called the rhythm method.

natural killer cell A nonspecific defensive cell that attacks cancer cells and infected body cells, especially those harboring viruses.

natural selection Differential success in reproduction by different phenotypes resulting from interactions with the environment. Evolution occurs when natural selection produces changes in the relative frequencies of alleles in a population's gene pool.

nearsightedness An inability to focus on distant objects; occurs when the eyeball is longer than normal and the lens focuses distant objects in front of the retina. Also called myopia.

negative feedback A control mechanism in which a chemical reaction, metabolic pathway, or hormone-secreting gland is inhibited by the products of the reaction, pathway, or gland. As the concentration of the products builds up, the product molecules themselves inhibit the process that produced them.

negative pressure breathing A breathing system in which air is pulled into the lungs.

nematode (nem´-uh-tōde) A roundworm, characterized by a pseudocoelom, a cylindrical, wormlike body form, and a tough cuticle.

nephron The tubular excretory unit and associated blood vessels of the vertebrate kidney; extracts filtrate from the blood and refines it into urine.

nerve A cable-like bundle of neuron fibers (axons and dendrites) tightly wrapped in connective tissue.

nerve cord An elongated bundle of axons and dendrites, usually extending longitudinally from the brain or anterior ganglia. One or more nerve cords and the brain make up the central nervous system in many animals.

nerve net A weblike system of neurons, characteristic of radially symmetrical animals such as a hydra.

nervous system The organ system that forms a communication and coordination network between all parts of an animal's body.

nervous tissue Tissue made up of neurons and supportive cells.

neural tube (nyūr´-ul) An embryonic cylinder that develops from ectoderm after gastrulation; gives rise to the brain and spinal cord.

neuromuscular junction A synapse between an axon of a motor neuron and a muscle fiber.

neuron (nyūr´-on) A nerve cell; the fundamental structural and functional unit of the nervous system, specialized for carrying signals from one location in the body to another.

neurosecretory cell A nerve cell that synthesizes hormones and secretes them into the blood, as well as conducting nerve signals.

neurotransmitter A chemical messenger that carries information from a transmitting neuron to a receiving cell, either another neuron or an effector cell.

neutral variation Genetic variation that provides no apparent selective advantage for some individuals over others.

neutron An electrically neutral particle (a particle having no electrical charge), found in the nucleus of an atom.

neutrophil (nyū´-truh-fil) A nonspecific, defensive, phagocytic white blood cell that can engulf bacteria and viruses in infected tissue; has a multilobed nucleus.

niche (nich) A population's role in its community; the sum total of a population's use of the biotic and abiotic resources of its habitat.

nitrogen fixation The conversion of atmospheric nitrogen (N_2) into nitrogen compounds (NH_4^+, NO_3^-) that plants can absorb and use.

nitrogenous base (nī-troj´-en-us) An organic molecule that is a base containing the element nitrogen.

node The point of attachment of a leaf on a stem.

node of Ranvier (ron´-vē-ā) An unmyelinated region on a myelinated axon of a nerve cell, where signal transmission occurs.

nodule A swelling on a plant root consisting of plant cells that contain nitrogen-fixing bacteria.

noncompetitive inhibitor A substance that impedes the activity of an enzyme without entering an active site and thus without competing directly with the normal substrate. By binding elsewhere on the enzyme, a noncompetitive inhibitor changes the shape of the enzyme so that the active site no longer functions.

nondisjunction An accident of meiosis or mitosis in which a pair of homologous chromosomes or a pair of sister chromatids fail to separate at anaphase.

nonpolar covalent bond An attraction between atoms that share one or more pairs of electrons equally because the atoms have similar electronegativity.

nonself molecule A foreign antigen; a protein or other macromolecule that is not part of an organism's body. See also self protein.

norepinephrine (nor´-ep-uh-nef´-rin) An amine hormone (also called noradrenaline) secreted by the adrenal medulla that prepares body organs for fight or flight; also serves as a neurotransmitter.

notochord (nō´-tuh-kord) A flexible, cartilage-like, longitudinal rod located between the digestive tract and nerve cord in chordate animals; present only in embryos in many species.

nuclear envelope A double membrane, perforated with pores, that encloses the nucleus and separates it from the rest of the eukaryotic cell.

nuclear transplantation A technique in which the nucleus of one cell is placed into another cell that already has a nucleus or in which the nucleus has been previously destroyed.

nucleic acid (nū-klā´-ik) A polymer consisting of many nucleotide monomers; serves as a blueprint for proteins and, through the actions of proteins, for all cellular structures and activities. The two types of nucleic acids are DNA and RNA.

nucleic acid probe In DNA technology, a labeled single-stranded nucleic acid molecule used to find a specific gene or other nucleotide sequence within a mass of DNA. The probe hydrogen-bonds to the complementary sequence in the targeted DNA.

nucleoid region (nū´-klē-oyd) The region in a prokaryotic cell consisting of a concentrated mass of DNA.

nucleolus (nū-klē´-ō-lus) A structure within the nucleus of a eukaryotic cell where ribosomal RNA is made and assembled with proteins to make ribosomal subunits; consists of parts of the chromatin DNA, RNA transcribed from the DNA, and proteins imported from the cytoplasm.

nucleosome (nū´-klē-ō-sōm) The bead-like unit of DNA packaging in a eukaryotic cell; consists of DNA wound around a protein core made up of eight histone molecules.

nucleotide (nū-klē-ō-tīd) An organic monomer consisting of a five-carbon sugar covalently bonded to a nitrogenous base and a phosphate group. Nucleotides are the building blocks of nucleic acids.

nucleus (plural, **nuclei**) (1) An atom's central core, containing protons and neutrons. (2) The genetic control center of a eukaryotic cell.

O

ocean current One of the river-like flow patterns in the oceans.

omnivore An animal that eats both plants and animals. *See also* carnivore; herbivore.

oncogene (on´-kō-jēn) A cancer-causing gene; usually contributes to malignancy by abnormally enhancing the amount or activity of a growth factor made by the cell.

oogenesis (ō´-uh-jen´-uh-sis) The formation of ova (egg cells).

open circulatory system A circulatory system in which blood is pumped through open-ended vessels and out among the body cells. In an animal with an open circulatory system, blood and interstitial fluid are one and the same.

operator In prokaryotic DNA, a sequence of nucleotides near the start of an operon to which an active repressor can attach. The binding of repressor prevents RNA polymerase from attaching to the promoter and transcribing the genes of the operon.

operculum (ō-per´-kyū-lum) (plural, **opercula**) A protective flap on each side of a fish's head that covers a chamber housing the gills.

operon (op´-er-on) A unit of genetic regulation common in prokaryotes; a cluster of genes with related functions, along with the promoter and operator that control their transcription.

opportunistic infection An infection by a microorganism that is normally rebuffed by a healthy immune system but can harm those with weakened immunity.

opposable thumb An arrangement of the fingers such that the thumb can touch the ventral surface of the fingertips of all four fingers.

optimal foraging theory The basis for analyzing behavior as a compromise of feeding costs versus feeding benefits.

oral contraceptive *See* birth control pills.

order In classification, the taxonomic category above family.

organ A structure consisting of several tissues adapted as a group to perform specific functions.

organ of Corti (kor´-tē) The hearing organ in birds and mammals, located within the cochlea.

organ system A group of organs that work together in performing vital body functions.

organelle (ōr-guh-nel´) A structure with a specialized function within a cell.

organic compound A chemical compound containing the element carbon and usually synthesized by cells.

organism An individual living thing, such as a bacterium, fungus, protist, plant, or animal.

orgasm Rhythmic contractions of the reproductive structures, accompanied by extreme pleasure, at the peak of sexual excitement in both sexes; includes ejaculation by the male.

osmoconformer (oz´-mō-con-form´-er) An organism whose body fluids have a solute concentration equal to that of its surroundings. Osmoconformers do not have a net gain or loss of water by osmosis.

osmoregulation The control of the gain and loss of water and dissolved solutes in an organism.

osmoregulator An organism whose body fluids have a solute concentration different from that of its environment and that must use energy in controlling water loss or gain.

osmosis (oz-mō´-sis) The diffusion of water across a selectively permeable membrane.

osteoporosis (os´-tē-ō-puh-rō´-sis) A skeletal disorder characterized by thinning, porous, and easily broken bones; most common among women after menopause and often related to low estrogen levels.

outer ear One of three main regions of the ear in reptiles, birds, and mammals; made up of the auditory canal and, in many birds and mammals, the pinna.

outgroup In a cladistic study of evolutionary relationships among taxa of organisms, a taxon or group of taxa with a known relationship to, but not a member of, the taxa being studied. *See also* in-group.

oval window In the vertebrate ear, a membrane-covered gap in the skull bone, through which sound waves pass from the middle ear into the inner ear.

ovarian cycle (ō-vār´-ē-un) Hormonally synchronized cyclic events in the mammalian ovary, culminating in ovulation.

ovary (1) In animals, the female gonad, which produces egg cells and reproductive hormones. (2) In flowering plants, the basal portion of a carpel in which the egg-containing ovules develop.

oviduct (ō´-vuh-dukt) The tube that conveys egg cells away from an ovary; also called a fallopian tube.

ovulation (ah´-vyū-lā´-shun) The release of an egg cell from an ovarian follicle.

ovule (ō´-vyūl) A reproductive structure in a seed plant; contains the female gametophyte and the developing egg. An ovule develops into a seed.

ovum (ō´-vum) (plural, **ova**) An unfertilized egg, or female gamete.

oxidation The loss of electrons from a substance involved in a redox reaction; always accompanies reduction.

oxidative phosphorylation (fos´-fōr-uh-lā´-shun) The production of ATP using energy derived from the redox reactions of an electron transport chain.

oxytocin (ok´-si-tō´-sin) A peptide hormone, made by the hypothalamus and secreted by the posterior pituitary, that stimulates contraction of the uterus and mammary gland cells.

ozone layer The layer of O_3 in the upper atmosphere that protects life on Earth from the harmful ultraviolet rays in sunlight.

P

P generation The parent individuals from which offspring are derived in studies of inheritance. (P stands for parental.)

P site one of two of a ribosome's binding sites for tRNA during translation. The P site holds the tRNA carrying the growing polypeptide chain. (P stands for peptidyl tRNA.)

pacemaker The SA (sinoatrial) node; a specialized region of cardiac muscle that maintains the heart's pumping rhythm (heartbeat) by setting the rate at which the heart contracts.

paedomorphosis (pē´-duh-mōr´-fuh-sis) The retention of juvenile body features in an adult.

pain receptor A sensory receptor that detects pain.

paleoanthropology (pā-lē-ō-an´-thruh-pol´-uh-jē) The study of human origins and evolution.

paleontologist (pā´-lē-on-tol´-uh-jist) A scientist who studies fossils.

pancreas (pan´-krē-us) A gland with dual functions: The nonendocrine portion secretes digestive enzymes and an alkaline solution into the small intestine via a duct; the endocrine portion secretes the hormones insulin and glucagon into the blood.

Pangaea (pan-jē´-uh) The supercontinent consisting of all the major landmasses of Earth fused together. Continental drift formed Pangaea near the end of the Paleozoic era.

parasitism (pār´-uh-sit-izm) A symbiotic relationship in which the parasite, a type of predator, lives within or on the surface of a host, from which it derives its food.

parasympathetic division A set of neurons in the autonomic nervous system that generally promotes body activities that gain and conserve energy, such as digestion and reduced heart rate. *See also* sympathetic division.

parathyroid glands (pār´-uh-thī´-royd) Four endocrine glands embedded in the surface of the thyroid gland that secrete parathyroid hormone.

parathyroid hormone (PTH) A peptide hormone secreted by the parathyroid glands that raises blood calcium level.

parenchyma cell (puh-ren´-kim-uh) In plants, a relatively unspecialized cell with a thin primary wall and no secondary wall; functions in photosynthesis, food storage, and aerobic respiration and may differentiate into other cell types.

Parkinson's disease A motor disorder caused by a progressive brain disease and characterized by difficulty in initiating movements, slowness of movement, and rigidity.

parsimony (par´-suh-mō´-nē) In scientific studies, the search for the least complex explanation for an observed phenomenon.

partial pressure A measure of the relative amount of gas in a mixture.

passive immunity Temporary immunity obtained by acquiring ready-made antibodies; lasts only a few weeks or months because the immune system has not been stimulated by antigens.

passive transport The diffusion of a substance across a biological membrane, without any input of energy.

pathogen A disease-causing organism.

pattern formation During embryonic development, the emergence of a spatial organization in which the tissues and organs of the organism are all in their correct places.

PCR *See* polymerase chain reaction (PCR).

pedigree A family tree representing the occurrence of heritable traits in parents and offspring across a number of generations.

pelagic zone (puh-laj´-ik) The region of an ocean occupied by seawater.

penis The copulatory structure of male mammals.

peptide bond The covalent linkage between two amino acid units in a polypeptide; formed by a dehydration reaction.

peptidoglycan (pep´-tid-ō-glī´-kan) A polymer of complex sugars cross-linked by short polypeptides; a material unique to bacterial cell walls.

perception The brain's meaningful interpretation, or conscious understanding, of sensory information.

perennial (puh-ren´-ē-ul) A plant that lives for many years.

perforin (per´-fuh-rin) A protein secreted by a cytotoxic T cell that lyses (ruptures) an infected cell by perforating its membrane.

peripheral nervous system (PNS) The network of nerves and ganglia carrying signals into and out of the central nervous system.

peristalsis (per´-uh-stal´-sis) Rhythmic waves of contraction of smooth muscles. Peristalsis propels food through a digestive tract and also enables many animals, such as earthworms, to crawl.

permafrost Continuously frozen ground found in the tundra.

petal A modified leaf of a flowering plant. Petals are the often colorful parts of a flower that advertise it to insects and other pollinators.

pH scale A measure of the relative acidity of a solution, ranging in value from 0 (most acidic) to 14 (most basic). The letters pH stand for potential hydrogen and refer to the concentration of hydrogen ions (H^+).

phage (fāj) *See* bacteriophage.

phagocyte (fag´-ō-sīt) A white blood cell (e.g., a neutrophil or a monocyte) that engulfs bacteria, foreign proteins, and the remains of dead body cells.

phagocytosis (fag´-ō-sī-tō´-sis) Cellular "eating"; a type of endocytosis whereby a cell engulfs macromolecules, other cells, or particles into its cytoplasm.

pharyngeal slit (fā-rin´-jē-ul) A gill structure in the pharynx; found in chordate embryos and some adult chordates.

pharynx (fār´-inks) The organ in a digestive tract that receives food from the oral cavity; in terrestrial vertebrates, the throat region where the air and food passages cross.

phenotype (fē´-nō-tīp) The expressed traits of an organism.

pheromone A small, volatile chemical signal that functions in communication between animals and acts much like a hormone in influencing physiology and behavior.

phloem (flō´-um) The portion of a plant's vascular system that conveys phloem sap throughout a plant. Phloem is made up of sieve-tube members.

phloem sap The solution of sugars, other nutrients, and hormones conveyed throughout a plant via phloem tissue.

phosphate group (fos´-fāt) A functional group consisting of a phosphorus atom covalently bonded to four oxygen atoms.

phospholipid (fos´-fō-lip´-id) A molecule that is a constituent of the inner bilayer of biological membranes, having a polar, hydrophilic head and a nonpolar, hydrophobic tail.

phosphorylation (fos´-fōr-uh-lā´-shun) The transfer of a phosphate group, usually from ATP, to a molecule. Nearly all cellular work depends on ATP energizing other molecules by phosphorylation.

photic zone (fō´-tik) The region of an aquatic ecosystem into which light penetrates and where photosynthesis occurs.

photoautotroph An organism that obtains energy from sunlight and carbon from CO_2 by photosynthesis.

photoheterotroph An organism that obtains energy from sunlight and carbon from organic sources.

photon (fō´-ton) A fixed quantity of light energy. The shorter the wavelength of light, the greater the energy of a photon.

photoperiod The length of the day relative to the length of the night; an environmental stimulus that plants use to detect the time of year.

photophosphorylation (fō´-tō-fos´-fōr-uh-lā´-shun) The production of ATP by chemiosmosis during the light reactions of photosynthesis.

photopsin (fō-top´-sin) One of a family of visual pigments in the cones of the vertebrate eye that absorb bright, colored light.

photoreceptor A type of electromagnetic receptor that detects light.

photorespiration In a plant cell, the breakdown of a two-carbon compound produced by the Calvin cycle. The Calvin cycle produces the two-carbon compound, instead of its usual three-carbon product G3P, when leaf cells fix O_2, instead of CO_2. Photorespiration produces no sugar molecules or ATP.

photosynthesis (fō´-tō-sin´-thuh-sis) The process by which plants, autotrophic protists, and some bacteria use light energy to make sugars and other organic food molecules from carbon dioxide and water.

photosystem A light-capturing unit of a chloroplast's thylakoid membrane, consisting of a reaction center surrounded by numerous light-harvesting complexes.

phototropism (fō´-tō-trō´-pizm) The growth of a plant shoot toward or away from light.

phylogenetic species concept The idea that species are defined as a set of organisms with a unique genetic history.

phylogenetic tree (fī-lō-juh-net´-ik) A branching diagram that represents a hypothesis about evolutionary relationships between organisms.

phylogeny (fī-loj´-uh-nē) The evolutionary history of a group of organisms.

phylum (fī´-lum) (plural, **phyla**) In classification, the taxonomic category above class and below kingdom. Members of a phylum all have a similar general body plan.

physiology (fi´-zi-ol´-uh-ji) The study of the functions of an organism's structures.

phytochrome (fī´-tuh-krōm) A colored protein in plants that contains a special set of atoms that absorbs light.

phytoplankton (fī´-tō-plank´-ton) Algae and photosynthetic bacteria that drift passively in aquatic environments.

pilus (pī´-lus) (plural, **pili**) A short projection on the surface of a prokaryotic cell that helps the prokaryote attach to other surfaces. Specialized sex pili are used in conjugation to hold the mating cells together.

pineal gland (pin´-ē-ul) An outgrowth of the vertebrate brain that secretes the hormone melatonin, which coordinates daily and seasonal body activities, such as reproductive activity, with environmental light conditions.

pinna (pin´-uh) The flap-like part of the outer ear, projecting from the body surface of many birds and mammals; collects sound waves and channels them to the auditory canal.

pinocytosis (pī´-nō-sī-tō´-sis) Cellular "drinking"; a type of endocytosis in which the cell takes fluid and dissolved solutes into small membranous vesicles.

pith Part of the ground tissue system of a dicot plant. Pith fills the center of a stem and may store food.

pituitary gland An endocrine gland at the base of the hypothalamus; consists of a posterior lobe, which stores and releases two hormones produced by the hypothalamus, and an anterior lobe, which produces and secretes many homones that regulate diverse body functions.

pivot joint A joint that allows precise rotations in multiple planes. An example in humans is the wrist.

placenta (pluh-sen´-tuh) In most mammals, the organ that provides nutrients and oxygen to the embryo and helps dispose of its metabolic wastes; formed of the embryo's chorion and the mother's endometrial blood vessels.

placental mammal (pluh-sen´-tul) Mammal whose young complete their embryonic development in the uterus, nourished via the mother's blood vessels in the placenta; also called a eutherian.

Plantae (plan´-tā) The kingdom that contains the plants.

plasma The liquid matrix of the blood in which the blood cells are suspended.

plasma cell An antibody-secreting B cell.

plasma membrane The thin layer of lipids and proteins that sets a cell off from its surroundings and acts as a selective barrier to the passage of ions and molecules into and out of the cell; consists of a phospholipid bilayer in which are embedded molecules of protein and cholesterol.

plasmid A small ring of DNA separate from the chromosome(s). Plasmids are found in prokaryotes and yeast.

plasmodesma (plaz´-mō-dez´-muh) (plural, **plasmodesmata**) An open channel in a plant cell wall, through which strands of cytoplasm connect from adjacent walls.

plasmodial slime mold (plaz-mō´-dē-ul) A type of protist that has amoeboid cells, flagellated cells, and an amoeboid plasmodial feeding stage in its life cycle.

plasmodium (1) A single mass of cytoplasm containing many nuclei. (2) The amoeboid feeding stage in the life cycle of a plasmodial slime mold.

plasmolysis A phenomenon that occurs in plants in an hypertonic solution. The cell loses water and shrivels, and its plasma membrane pulls away from the cell wall, usually killing the cell.

plate tectonics Geologic processes, such as continental drift, volcanoes, and earthquakes, resulting from plate movement.

plateau phase The phase of the human sexual response cycle that follows the excitement phase and is marked by increases in breathing and heart rate.

platelet A piece of membrane-enclosed cytoplasm from a large cell in the bone marrow of a mammal; a blood-clotting element.

pleated sheet The folded arrangement of a polypeptide in a protein's secondary structure.

pleiotropy (plī´-uh-trō-pē) The control of more than one phenotypic characteristic by a single gene.

polar covalent bond An attraction between atoms that share electrons unequally because the atoms differ in electronegativity. The shared electrons are pulled closer to the more electronegative atom, making it partially negative and the other atom partially positive.

polar molecule A molecule containing polar covalent bonds.

pollen *See* pollen grain.

pollen grain The structures that contain the male gametophyte in seed plants.

pollination In seed plants, the delivery, by wind or animals, of pollen from the male parts of a plant to the stigma of a carpel on the female.

polychaete (pol´-ē-kēt) A member of the largest group of annelids. *See* annelid.

polygamous Referring to a type of relationship in which an indvidual of one sex mates with several of the other.

polygenic inheritance (pol´-ē-jen´-ik) The additive effect of two or more gene loci on a single phenotypic characteristic.

polymer (pol´-uh-mer) A large molecule consisting of many identical or similar molecular units, called monomers, covalently joined together in a chain.

polymerase chain reaction (PCR) (puh-lim´-uh-rās) A techique used to obtain many copies of a DNA molecule or part of a DNA molecule. A small amount of DNA mixed with the enzyme DNA polymerase, DNA nucleotides, and a few other ingredients replicates repeatedly in a test tube.

polymorphic (pol´-ē-mōr´-fik) Referring to a population in which two or more physical forms are present in readily noticeable frequencies.

polynucleotide (pol´-ē-nū´-klē-ō-tīd) A polymer made up of many nucleotides covalently bonded together.

polyp (pol´-ip) One of two types of cnidarian body forms; a columnar, hydra-like body.

polypeptide A chain of amino acids linked by peptide bonds.

polyploid (pol´-ē-ployd) An organism whose cells have more than two complete sets of chromosomes.

polysaccharide (pol´-ē-sak´-uh-rīd) A carbohydrate polymer consisting of hundreds to thousands of monosaccharides (sugars) linked by covalent bonds.

pons (pahnz) Part of the vertebrate hindbrain that functions with the medulla oblongata in passing data between the spinal cord and forebrain and in controlling autonomic, homeostatic functions.

population A group of interacting individuals belonging to one species and living in the same geographic area.

population density The number of individuals of a species per unit area or volume.

population ecology The study of how members of a population interact with their environment, focusing on factors that influence population density and growth.

population genetics The study of genetic changes in populations; the science of microevolutionary changes in populations.

post-anal tail A tail posterior to the anus; found in chordate embryos and most adult chordates.

posterior Pertaining to the rear, or tail, of a bilaterally symmetrical animal.

posterior pituitary An extension of the hypothalamus composed of nervous tissue that secretes hormones made in the hypothalamus; a temporary storage site for hypothalamic hormones.

potential energy Stored energy; the capacity to perform work that matter possesses because of its location or arrangement. Water behind a dam and chemical bonds possess potential energy.

predation An interaction between species in which one species, the predator, eats the other, the prey.

predator A consumer in a biological community.

prepuce (prē´-pyūs) A fold of skin covering the head of the clitoris and penis.

pressure flow mechanism The method by which phloem sap is transported through a plant from a sugar source, where sugars are produced, to a sugar sink, where sugars are used.

prevailing winds Winds that result from the combined effects of Earth's rotation and the rising and falling of air masses.

prey An organism eaten by a predator.

primary consumer An organism in the trophic level of an ecosystem that eats plants or algae.

primary growth Growth in the length of a plant root or shoot, produced by an apical meristem.

primary immune response The initial immune response to an antigen, which appears after a lag of several days.

primary oocyte (ō´-uh-sīt) A diploid cell, in prophase I of meiosis, that can be hormonally triggered to develop into an ovum.

primary phloem *See* phloem.

primary production The amount of solar energy converted to chemical energy (organic compounds) by autotrophs in an ecosystem during a given time period.

primary spermatocyte (sper-mat´-eh-sīt´) A diploid cell in the testis that undergoes meiosis I.

primary structure The first level of protein structure; the specific sequence of amino acids making up a polypeptide chain.

primary succession A type of ecological succession in which a biological community arises in an area without soil. *See also* secondary succession.

primary xylem *See* xylem.

probe In DNA technology, a labeled single-stranded nucleic acid molecule used to find a specific gene, or other nucleotide sequence, within a mass of DNA. The probe hydrogen-bonds to the complementary sequence in the targeted DNA.

producer An organism that makes organic food molecules from CO_2, H_2O, and other inorganic raw materials: a plant, alga, or autotrophic bacterium.

product An ending material in a chemical reaction.

progesterone (prō-jes´-teh-rōn) A steroid hormone secreted by the corpus luteum of the ovary; maintains the uterine lining during pregnancy.

progestin (prō-jes´-tin) One of a family of steroid hormones, including progesterone, produced by the mammalian ovary. Progestins prepare the uterus for pregnancy.

programmed cell death The timely death (and disposal of the remains) of certain cells, triggered by certain genes; an essential process in normal development; also called apoptosis.

prokaryote An organism with prokaryotic cells.

prokaryotic cell (prō-kār´-ē-ot´-ik) A type of cell lacking a membrane-enclosed nucleus and other membrane-enclosed organelles; found only in the domains Bacteria and Archaea.

prokaryotic flagellum (plural, **flagella**) A long surface projection that propels a prokaryotic cell through its liquid environment; totally different from the flagellum of a eukaryotic cell.

prolactin (PRL) (pro-lak´-tin) A protein hormone secreted by the anterior pituitary that stimulates milk production in mammals.

prometaphase The second stage of mitosis, during which the nuclear envelope fragments and the spindle microtubules attach to the kinetochores of the sister chromatids.

promiscuous Referring to a type of relationship in which mating occurs with no strong pair-bonds or lasting relationships.

promoter A specific nucleotide sequence in DNA, located at the start of a gene, that is the binding site for RNA polymerase and the place where transcription begins.

prophage (prō´-fāj) Phage DNA that has inserted by genetic recombination into the DNA of a prokaryotic chromosome.

prophase The first stage of mitosis, during which the chromatin condenses to form structures (sister chromatids) visible with a light microscope and the mitotic spindle begins to form, but the nucleus is still intact.

prostate gland (pros´-tāt) A gland in human males that secretes an acid-neutralizing component of semen.

protein A biological polymer constructed from amino acid monomers.

proteobacteria A diverse clade of gram-negative bacteria that includes five subgroups.

proteomics The study of whole sets of proteins and their interactions.

protist (prō´-tist) A member of the kingdom Protista. Most protists are unicellular, though some are colonial or multicellular.

Protista (prō-tis´-tuh) In the five-kingdom classification system, the kingdom that contains the unicellular eukaryotes (and closely related multicellular organisms) called protists.

protobiont An aggregate of abiotically produced molecules surrounded by a membrane or membrane-like structure.

proton A subatomic particle with a single positive electrical charge, found in the nucleus of an atom.

proto-oncogene (prō´-tō-on´-kō-jēn) A normal gene that can be converted to a cancer-causing gene.

protoplast fusion The fusing of two protoplasts from different plant species that would otherwise be reproductively incompatible.

protostome An animal with a coelom that develops from solid masses of cells that arise between the digestive tube and the body wall of the embryo. The protostomes include the molluscs, annelids, and arthropods.

protozoan (prō´-tō-zō´-un) (plural, **protozoa**) A protist that lives primarily by ingesting food; a heterotrophic, animal-like protist.

proximal tubule In the vertebrate kidney, the portion of a nephron immediately downstream from Bowman's capsule that conveys and helps refine filtrate.

proximate question In animal behavior, an inquiry that focuses on the environmental stimuli, if any, that trigger a particular behavioral act, as well as the genetic, physiological, and anatomical mechanisms underlying it.

pseudocoelom (sū´-dō-sē´-lōm) A body cavity that is in direct contact with the wall of the digestive tract.

pseudopodium (sū-dō-pō´-dē-um) (plural, **pseudopodia**) A temporary extension of an amoeboid cell. Pseudopodia function in moving cells and engulfing food.

pulmonary artery A large blood vessel that conveys blood from the heart to a lung.

pulmonary circuit One of two main blood circuits in terrestrial vertebrates; conveys blood between the heart and the lungs. *See also* systemic circuit.

pulmonary vein A blood vessel that conveys blood from a lung to the heart.

pulse The rhythmic stretching of the arteries caused by the pressure of blood forced through the arteries by contractions of the ventricles during systole.

punctuated equilibrium The idea that speciation occurs in spurts followed by long periods of little change.

Punnett square A diagram used in the study of inheritance to show the results of random fertilization.

pupil The opening in the iris that admits light into the interior of the vertebrate eye. Muscles in the iris regulate the pupil's size.

purine (pyū´-rēn) One of two families of nitrogenous bases found in nucleotides. Adenine (A) and guanine (G) are purines.

pyloric sphincter (pī-lōr´-ik sfink´-ter) In the vertebrate digestive tract, a muscular ring that regulates the passage of food out of the stomach and into the small intestine.

pyrimidine (puh-rim´-uh-dēn) One of two families of nitrogenous bases found in nucleotides. Cytosine (C), thymine (T), and uracil (U) are pyrimidines.

Q

quaternary consumer (kwot´-er-ner-ē) An organism that eats tertiary consumers.

quaternary structure The fourth level of protein structure; the shape resulting from the association of two or more polypeptide subunits.

R

R plasmid A bacterial plasmid that carries genes for enzymes that destroy particular antibiotics, thus making the bacterium resistant to the antibiotics.

radial symmetry An arrangement of the body parts of an organism like pieces of a pie around an imaginary central axis. Any slice passing longitudinally through a radially symmetrical organism's central axis divides it into mirror-image halves.

radiation The emission of electromagnetic waves by all objects warmer than absolute zero.

radioactive isotope An isotope whose nucleus decays spontaneously, giving off particles and energy.

radiometric dating A method for determining the age of fossils and rocks from the ratio of a radioactive isotope to the nonradioactive isotope(s) of the same element in the sample.

radula (rad´-yū-luh) A toothed, rasping organ used to scrape up or shred food; found in many molluscs.

random Describing a dispersion pattern in which individuals are spaced in a patternless, unpredictable way.

ray-finned fish A bony fish having fins supported by thin, flexible skeletal rays. All but one living species of bony fishes are ray-fins. *See* lobe-fin.

reabsorption In the vertebrate kidney, the reclaiming of water and valuable solutes from the filtrate.

reactant A starting material in a chemical reaction.

reaction center In a photosystem in a chloroplast, the chlorophyll *a* molecule and the primary electron acceptor that trigger the light reactions of photosynthesis. The chlorophyll donates an electron excited by light energy to the primary electron acceptor, which passes an electron to an electron transport chain.

reading frame The way in which a cell's mRNA-translating machinery groups the mRNA nucleotides into codons.

receptor On or in a cell, a specific protein molecule whose shape fits that of a specific molecular messenger, such as a hormone.

receptor potential The electrical signal produced by sensory transduction.

receptor-mediated endocytosis (en´-dō-sī-tō´-sis) The movement of specific molecules into a cell by the inward budding of membranous vesicles. The vesicles contain proteins with receptor sites specific to the molecules being taken in.

recessive allele In a heterozygous individual, the allele that has no noticeable effect on the phenotype.

reciprocal altruism (al´-trū-izm) In animal behavior, a selfless act repaid at a later time by the beneficiary or by another member of the beneficiary's social system.

recombinant DNA A DNA molecule carrying genes derived from two or more sources.

recombinant DNA technology Techniques for synthesizing recombinant DNA *in vitro* and transferring it into cells, where it can be replicated and may be expressed; also known as genetic engineering.

recombination frequency With respect to two given genes, the number of recombinant progeny from a mating divided by the total number of progeny. Recombinant progeny carry combinations of alleles different from those in either of the parents as a result of independent assortment of chromosomes or crossing over.

Recommended Dietary Allowance (RDA) A recommendation for daily nutrient intake established by nutritionists.

rectum The terminal portion of the large intestine where the feces are stored until they are eliminated.

red alga One of a group of marine, mostly multicellular, autotrophic protists, which includes the reef-building coralline algae.

red blood cell A blood cell containing hemoglobin, which transports O_2. Also called erythrocyte.

red bone marrow A specialized tissue found in cavities in the ends of bones that produces blood cells.

red-green color blindness A category of common, sex-linked human disorders involving several genes on the X chromosome; characterized by a malfunction of light-sensitive cells in the eyes; affects mostly males but also homozygous females.

redox reaction Short for oxidation-reduction; a chemical reaction in which electrons are lost from one substance (oxidation) and added to another (reduction). Oxidation and reduction always occur together.

reduction The gain of electrons by a substance involved in a redox reaction; always accompanies oxidation.

reflex An automatic reaction to a stimulus, mediated by the spinal cord or lower brain.

regeneration The regrowth of body parts from pieces of an organism.

regulatory gene A gene that codes for a protein, such as a repressor, that controls the transcription of another gene or group of genes.

releasing hormone A hormone, secreted by the hypothalamus, that makes the anterior pituitary secrete hormones.

REM sleep Rapid-eye-movement sleep; a period of sleep when the brain is highly active, brain waves are fairly rapid and regular, and the eyes move rapidly under the closed eyelids. Most dreams occur during REM sleep.

renal cortex The outer portion of the vertebrate kidney.

renal medulla The inner portion of the vertebrate kidney, beneath the renal cortex.

repetitive DNA Nucleotide sequences that are present in many copies in the DNA of a genome. The repeated sequences may be long or short and may be located next to each other or dispersed in the DNA.

repressor A protein that blocks the transcription of a gene or operon.

reproduction The creation of new individuals from existing ones.

reproductive barrier A biological feature of a species that prevents it from interbreeding with other species even when populations of the two species live together.

reproductive cloning Using a somatic cell from a multicellular organism to make one or more genetically identical individuals.

reproductive system The body organ system responsible for reproduction.

reptile Member of the clade of amniotes that includes including snakes, lizards, turtles, crocodilians, and birds, along with a number of extinct groups such as dinosaurs.

resolution A measure of the clarity of an image; the ability of an optical instrument to show two objects as separate.

resolution phase The final phase of the human sexual response cycle, where the structures return to normal size, muscles relax, and passion subsides.

resource partitioning The division of environmental resources by coexisting species such that the niche of each species differs by one or more significant factors from the niches of all coexisting species.

respiration (1) Gas exchange, or breathing; the exchange of O_2 and CO_2 between an organism and its environment. An aerobic organism takes up O_2 and gives off CO_2. (2) Cellular respiration; the aerobic harvest of energy from food molecules by cells.

respiratory surface The part of an animal where gases are exchanged with the environment.

respiratory system The organ system that functions in exchanging gases with the environment. It supplies the blood with O_2 and disposes of CO_2.

resting potential The voltage across the plasma membrane of a resting neuron.

restoration ecology The use of ecological principles to develop ways to return degraded ecosystems to conditions as similar as possible to their natural, predegraded state.

restriction enzyme A bacterial enzyme that cuts up foreign DNA, thus protecting bacteria against intruding DNA from phages and other organisms. Restriction enzymes are used in DNA technology to cut DNA molecules in reproducible ways.

Restriction fragment length polymorphisms (RFLPs) (rif´-lips) The differences in homologous DNA sequences that are reflected in different lengths of restriction fragments produced when the DNA is cut up with restriction enzymes.

restriction fragments Molecules of DNA produced from a longer DNA molecule cut up by a restriction enzyme; used in genome mapping and other applications.

restriction site A specific sequence on a DNA strand that is recognized as a "cut site" by a restriction enzyme.

reticular formation A system of neurons, containing over 90 separate nuclei, that passes through the core of the brain stem.

retina (ret´-uh-nuh) The light-sensitive layer in an eye, made up of photoreceptor cells and sensory neurons.

retrovirus An RNA virus that reproduces by means of a DNA molecule. It reverse-transcribes its RNA into DNA, inserts the DNA into a cellular chromosome, and then transcribes more copies of the RNA from the viral DNA. HIV and a number of cancer-causing viruses are retroviruses.

reverse transcriptase (tran-skrip´-tās) An enzyme that catalyzes the synthesis of DNA on an RNA template.

rhizome (rī´-zōm) A horizontal stem that grows below the ground.

rhodopsin (ro-dop´-sin) A visual pigment that is located in the rods of the vertebrate eye and that absorbs dim light.

rhythm method A form of contraception that relies on refraining from sexual intercourse when conception is most likely to occur; also called natural family planning.

ribonucleic acid (RNA) (rī-bō-nū-klā´-ik) A type of nucleic acid consisting of nucleotide monomers with a ribose sugar and the nitrogenous bases adenine (A), cytosine (C), guanine (G), and uracil (U); usually single-stranded; functions in protein synthesis and as the genome of some viruses.

ribosomal RNA (rRNA) (rī´-buh-sōm´-ul) The type of ribonucleic acid that, together with proteins, makes up ribosomes; the most abundant type of RNA.

ribosomal RNA (rRNA) sequence analysis Determination of the nucleotide sequence of ribosomal RNA molecules.

ribosome (rī´-buh-sōm) A cell organelle consisting of RNA and protein organized into two subunits and functioning as the site of protein synthesis in the cytoplasm. The ribosomal subunits are constructed in the nucleolus.

ribozyme (rī´-bō-zīm) An enzymatic RNA molecule that catalyzes chemical reactions.

RNA See ribonucleic acid.

RNA polymerase (puh-lim´-uh-rās) An enzyme that links together the growing chain of RNA nucleotides during transcription, using a DNA strand as a template.

RNA splicing The removal of introns and joining of exons in eukaryotic RNA, forming an mRNA molecule with a continuous coding sequence; occurs before mRNA leaves the nucleus.

RNA world A hypothetical period in the evolution of life when RNA served as rudimentary genes and the sole catalytic molecules.

rod A photoreceptor cell in the vertebrate retina, enabling vision in dim light.

root cap A cone of cells at the tip of a plant root that protects the root's apical meristem.

root hair An outgrowth of an epidermal cell on a root, which increases the root's absorptive surface area.

root pressure The upward push of xylem sap in a vascular plant, caused by the active pumping of minerals into the xylem by root cells.

root system All of a plant's roots, which anchor it in the soil, absorb and transport minerals and water, and store food.

rough endoplasmic reticulum (rough ER) (reh-tik´-yuh-lum) A network of interconnected membranous sacs in a eukaryotic cell's cytoplasm. Rough ER membranes are studded with ribosomes that make membrane proteins and secretory proteins. The rough ER constructs membrane from phospholipids and proteins.

rough ER See rough endoplasmic reticulum.

rRNA See ribosomal RNA.

r-selection The concept that in certain (r-selected) populations, a high reproductive rate is the chief determinant of life history.

rule of addition A rule stating that the probability that an event can occur in two or more alternative ways is the sum of the separate probabilities of the different ways.

rule of multiplication A rule stating that the probability of a compound event is the product of the separate probabilities of the independent events.

ruminant mammal (rū-min-ent) A mammal with a four-chambered stomach housing microorganisms that can digest cellulose; examples are cattle, deer, and sheep.

S

SA (sinoatrial) node (sī´-nō-ā´-trē-ul) The pacemaker of the heart, located in the wall of the right atrium. At the base of the wall separating the two atria is another patch of nodal tissue called the AV (atrioventricular) node. See pacemaker.

sac fungus Member of the phylum Ascomycota, characterized by saclike structures called asci that produce spores in sexual reproduction.

saccule (sak´-yūl) A fluid-filled inner-ear chamber containing hair cells that detect the position of the head relative to gravity.

saliva Salivary gland secretion that contains substances to lubricate food, buffers, antibacterial agents, and the digestive enzyme amylase.

salt A compound resulting from the formation of ionic bonds; also called an ionic compound.

sapwood Light-colored, water-conducting secondary xylem in a tree. See also heartwood.

sarcoma (sar-kō´-muh) Cancer of the supportive tissues, such as bone, cartilage, and muscle.

sarcomere (sar´-kō-mē r) The fundamental unit of muscle contraction, composed of thin filaments of actin and thick filaments of myosin; the region between two narrow, dark lines, called Z lines, in the myofibril.

saturated Pertaining to fats and fatty acids whose hydrocarbon chains contain the maximum number of hydrogens and therefore have no double covalent bonds. Saturated fats and fatty acids solidify at room temperature.

savanna A biome dominated by grasses and scattered trees.

scanning electron microscope (SEM) A microscope that uses an electron beam to study the surface architecture of a cell or other specimen.

schizophrenia Severe mental disturbance characterized by psychotic episodes in which patients lose the ability to distinguish reality from hallucination.

sclera (sklār´-uh) A layer of connective tissue forming the outer surface of the vertebrate eye. The cornea is the frontal part of the sclera.

sclereid (sklār´-ē-id) In plants, a very hard, dead sclerenchyma cell found in nutshells and seed coats; a stone cell.

sclerenchyma cell (skluh-ren´-kē-muh) In plants, a supportive cell with rigid secondary walls hardened with lignin.

scrotum A pouch of skin outside the abdomen that houses a testis; functions in cooling sperm, thereby keeping them viable.

search image The mechanism that enables an animal to find a particular kind of food efficiently.

second law of thermodynamics The natural law stating that energy conversions reduce the order of the universe, increasing its entropy.

secondary consumer An organism that eats primary consumers.

secondary growth An increase in a plant's girth, involving cell division in the vascular cambium and cork cambium.

secondary immune response The immune response elicited when an animal encounters the same antigen at some later time. The secondary immune response is more rapid, of greater magnitude, and of longer duration than the primary immune response.

secondary oocyte (ō´-uh-sīt´) A haploid cell resulting from meiosis I in oogenesis, which will become an ovum after meiosis II.

secondary phloem *See* phloem.

secondary spermatocyte (sper-mat´-uh-sīt´) A haploid cell resulting from meiosis I in spermatogenesis, which will become a sperm cell after meiosis II.

secondary structure The second level of protein structure; the regular patterns of coils or folds of a polypeptide chain.

secondary succession A type of ecological succession that occurs where a disturbance has destroyed an existing biological community but left the soil intact. *See also* primary succession.

secondary xylem *See* xylem.

secretion (1) The discharge of molecules synthesized by a cell. (2) In the vertebrate kidney, the discharge of wastes from the blood into the filtrate from the nephron tubules.

secretory protein A protein (such as an antibody) that is secreted by a cell.

seed A plant embryo packaged with a food supply within a protective covering.

seed coat A tough outer covering of a seed, formed from the outer coat of an ovule. In a flowering plant, the seed coat encloses and protects the embryo and endosperm.

seed dormancy The temporary suspension of growth and development of a seed.

seedless vascular plants The informal collective name for the phyla Lycophyta (club mosses and their relatives) and Pteridophyta (ferns and their relatives).

segmentation Subdivision along the length of an animal body into a series of repeated parts called segments.

selective permeability (per´-mē-uh-bil´-uh-tē) A property of biological membranes that allows some substances to cross more easily than others and blocks the passage of other substances altogether.

self protein A protein on the surface of an antigen-presenting cell that can hold a foreign antigen and display it to helper T cells. Each individual has a unique set of self proteins that serve as molecular markers for the body. Lymphocytes do not attack self proteins unless the proteins are displaying foreign antigens; therefore, self proteins mark normal body cells as off-limits to the immune system. The technical name for self proteins is *major histocompatibility complex (MHC) proteins. See also* nonself molecule.

self-fertilize The fusion of sperm and egg produced by the same individual organism.

semen (sē´-mun) The sperm-containing fluid that is ejaculated by the male during orgasm.

semicircular canals Fluid-filled channels in the inner ear that detect changes in the head's rate of rotation or angular movement.

semiconservative model Type of DNA replication in which the replicated double helix consists of one old strand, derived from the old molecule, and one newly made strand.

seminal vesicle (sem´-uh-nul ves´-uh-kul) A gland in males that secretes a fluid component of semen that lubricates and nourishes sperm.

seminiferous tubule (sem´-uh-nif´-uh-rus) A coiled sperm-producing tube in a testis.

sensation A feeling, or general awareness, of stimuli resulting from sensory information reaching the central nervous system.

sensitive period A limited phase in an individual animal's development when learning of particular behaviors can take place.

sensory adaptation The tendency of sensory neurons to become less sensitive when they are stimulated repeatedly.

sensory input The conduction of signals from sensory receptors to processing centers in the central nervous system.

sensory neuron A nerve cell that receives information from sensory receptors and conveys signals into the central nervous system.

sensory receptor A specialized cell or neuron that detects stimuli and sends information to the central nervous system.

sensory transduction The conversion of a stimulus signal to an electrical signal by a sensory receptor.

sepal (sē´-pul) A modified leaf of a flowering plant. A whorl of sepals encloses and protects the flower bud before it opens.

sessile An organism that is anchored to its substrate.

sex chromosome A chromosome that determines whether an individual is male or female.

sex-linked gene A gene located on a sex chromosome.

sexual dimorphism (dī-mōr´-fizm) A special case of polymorphism based on the distinction between the secondary sex characteristics of males and females.

sexual reproduction The creation of offspring by the fusion of two haploid sex cells (gametes), forming a diploid zygote.

sexually transmitted disease (STD) A contagious disease spread by sexual contact.

shared derived characters Homologous features that have changed from a primitive (ancestral) condition and that are unique to an evolutionary lineage; features found in members of a lineage but not found in ancestors of the lineage.

shared primitive characters Homologous features found in members of a lineage and also in the ancestors of the lineage; ancestral features.

shoot system All of a plant's stems, leaves, and reproductive structures.

short-day plant A plant that flowers in late summer, fall, or winter, when day length is shortening.

short-term memory The ability to hold information, anticipations, or goals for a time and then release them if they become irrelevant.

sieve plate An end wall in a sieve-tube member which facilitates the flow of phloem sap.

sieve-tube member A food-conducting cell in a plant. Chains of sieve-tube members make up phloem tissue.

sign stimulus In animal behavior, a stimulus that triggers a fixed action pattern.

signal A behavior that causes a change in behavior in another animal.

signal transduction In cell biology, a series of molecular changes that converts a signal on a target cell's surface to a specific response inside the cell.

silencer A eukaryotic DNA sequence that functions to inhibit the start of gene transcription; may act analogously to an enhancer by binding a repressor.

simple fruit A fruit, such as an apple, that develops from a flower with a single carpel and ovary.

single-lens eye The camera-like eye found in some jellies, polychaetes, spiders, many molluscs, and in vertebrates.

sister chromatid (krō´-muh-tid) One of the two identical parts of a duplicated chromosome in a eukaryotic cell.

skeletal muscle Striated muscle attached to the skeleton. The contraction of striated muscles produces voluntary movements of the body.

skeletal system The organ system that provides body support and protects body organs such as the brain, heart, and lungs.

skull The bony framework of the head.

sliding-filament model The theory explaining how muscle contracts, based on change within a sarcomere, the basic unit of muscle organization, stating that thin (actin) filaments slide across thick (myosin) filaments, shortening the sarcomere; the shortening of all sarcomeres in a myofibril shortens the entire myofibril.

small intestine The longest section of the alimentary canal. It is the principal site of the enzymatic hydrolysis of food macromolecules and the absorption of nutrients.

smooth endoplasmic reticulum (smooth ER) A network of interconnected membranous tubules in a eukaryotic cell's cytoplasm. Smooth ER lacks ribosomes. Enzymes embedded in the smooth ER membrane function in the synthesis of certain kinds of molecules, such as lipids.

smooth ER *See* smooth endoplasmic reticulum.

smooth muscle Muscle made up of cells without striations, found in the walls of organs such as the digestive tract, urinary bladder, and arteries.

social behavior Any kind of interaction between two or more animals, usually of the same species.

social learning Modification of behavior through the observation of other individuals.

sociobiology The study of the evolutionary basis of social behavior.

sodium-potassium (Na^+-K^+) pump A membrane protein that transports sodium ions out of, and potassium ions into, a cell against their concentration gradients. The process is powered by ATP.

solar energy Energy obtained from the sun.

solute (sol´-yūt) A substance that is dissolved in a solution.

solution A liquid consisting of a homogeneous mixture of two or more substances, consisting of a dissolving agent, called the solvent, and a substance that is dissolved, called the solute.

solvent The dissolving agent in a solution. Water is the most versatile known solvent.

somatic cell (sō-mat´-ik) Any cell in a multicellular organism except a sperm or egg cell or a cell that develops into a sperm or egg.

somatic nervous system The component of the vertebrate peripheral nervous system that carries signals to and from skeletal muscles.

somite (sō´-mīt) A block of mesoderm in a chordate embryo that gives rise to vertebrae and other segmental structures.

spatial learning Modification of behavior based on experience of the spatial structure of the environment.

speciation (spē´-sē-ā´-shun) The evolution of new species.

species A group whose members possess similar anatomical characteristics and have the ability to interbreed. *See* biological species concept.

species diversity The variety of species that make up a community; concerns both species richness (the total number of different species) and the relative abundance of the different species.

sperm A male gamete.

spermatogenesis (sper-mat´-ō-jen´-uh-sis) The formation of sperm cells.

spermicidal Referring to a sperm-killing chemical (cream, jelly, or foam) that works with a barrier device as a method of contraception.

spinal cord The dorsal hollow nerve cord in vertebrates, located within the vertebral column; with the brain, makes up the central nervous system.

spinal nerve In the vertebrate peripheral nervous system, a nerve that carries signals to or from the spinal cord.

spirochete A large spiral-shaped (curved) prokaryotic cell.

sponge An aquatic animal characterized by a highly porous body.

sporangium (spuh-ranj´-ē-um´) (plural, **sporangia**) A capsule in fungi and plants in which meiosis occurs and haploid spores develop.

spore (1) In plants and algae, a haploid cell that can develop into a multicellular individual without fusing with another cell. (2) In prokaryotes, protists, and fungi, any of a variety of thick-walled life cycle stages capable of surviving unfavorable environmental conditions.

sporophyte (spōr´-uh-fīt) The multicellular diploid form in the life cycle of organisms undergoing alternation of generations; results from a union of gametes and meiotically produces haploid spores that grow into the gametophyte generation.

stabilizing selection Natural selection that favors intermediate variants by acting against extreme phenotypes.

stamen (stā´-men) A pollen-producing male reproductive part of a flower, consisting of a stalk and an anther.

starch A storage polysaccharide found in the roots of plants and certain other cells; a polymer of glucose.

start codon (kō´-don) On mRNA, the specific three-nucleotide sequence (AUG) to which an initiator tRNA molecule binds, starting translation of genetic information.

stem The part of a plant's shoot system that supports the leaves and reproductive structures.

stem cell A relatively unspecialized cell that can give rise to one or more types of specialized cells. *See* embryonic stem cell (ES cell); adult stem cell.

steroid (ster´-oyd) A type of lipid whose carbon skeleton is in the form of four fused rings: three 6-sided rings and one 5-sided ring; examples are cholesterol, testosterone, and estrogen.

steroid hormone A lipid made from cholesterol that acts as a regulatory chemical, activating the transcription of specific genes in target cells.

stigma (stig´-muh) (plural, **stigmata**) The sticky tip of a flower's carpel, which traps pollen grains.

stimulus (plural, **stimuli**) (1) In the context of a nervous system, a factor that triggers sensory transduction. (2) In behavioral biology, a factor that triggers a specific response.

stoma (stō´-muh) (plural, **stomata**) A pore surrounded by guard cells in the epidermis of a leaf. When stomata are open, CO_2 enters a leaf, and water and O_2 exit. A plant conserves water when its stomata are closed.

stomach A pouch-like organ in a digestive tract that grinds and churns food and may store it temporarily.

stop codon In mRNA, one of three triplets (UAG, UAA, UGA) that signal gene translation to stop.

stramenopile A clade of protists that includes water mold, diatoms, and brown algae and is characterized by a "hairy" flagellum.

stratum (plural, **strata**) A layer of sedimentary rock that was formed as sediments of sand and mud were compressed into rock by overlying deposits.

stretch receptor A type of mechanoreceptor sensitive to changes in muscle length; detects the position of body parts.

strict anaerobe An organism that cannot survive in an atmosphere of oxygen.

stroma (strō´-muh) A thick fluid enclosed by the inner membrane of a chloroplast. Sugars are made in the stroma by the enzymes of the Calvin cycle.

stromatolite (strō-mat´-uh-līt) Rock formed of layered, fossilized bacterial mats.

substrate (1) A specific substance (reactant) on which an enzyme acts. Each enzyme recognizes only the specific substrate or substrates of the reaction it catalyzes. (2) A surface in or on which an organism lives.

substrate feeder An organism that lives in or on its food source, eating its way through the food.

substrate-level phosphorylation The formation of ATP occurring when an enzyme transfers a phosphate group from an organic molecule (e.g., one of the intermediates in glycolysis or the citric acid cycle) to ADP.

succession *See* ecological succession; primary succession; secondary succession.

sugar sink A plant organ that is a net consumer or storer of sugar. Growing roots, shoot tips, stems, and fruits are sugar sinks supplied by phloem.

sugar source A plant organ in which sugar is being produced by either photosynthesis or the breakdown of starch. Mature leaves are the primary sugar sources of plants.

sugar-phosphate backbone The alternating chain of sugar and phosphate to which the DNA and RNA nitrogenous bases are attached.

superior vena cava (vē´-nuh kā´-vuh) A large vein that returns O_2-poor blood to the heart from the upper body and head. *See also* inferior vena cava.

supporting cell A cell that is essential for the structural integrity and normal functioning of the nervous system; also called a glia.

surface tension A measure of how difficult it is to stretch or break the surface of a liquid.

survivorship curve A plot of the number of members of a cohort that are still alive at each age; one way to represent age-specific mortality.

suspension feeder An animal that extracts food particles suspended in the surrounding water.

sustainable development The long-term prosperity of human societies and the ecosystems that support them.

sustainable resource management Management of a natural resource so as not to damage the resource.

swim bladder A gas-filled internal sac that helps bony fishes maintain buoyancy.

symbiosis An interspecific interaction in which one species, the symbiont, lives in or on another species, the host.

symbiotic relationship A close association between organisms of two or more species.

sympathetic division A set of neurons in the autonomic nervous system that generally prepares the body for energy-consuming activities, such as fleeing or fighting. *See also* parasympathetic division.

sympatric speciation The formation of a new species as a result of a genetic change that produces a reproductive barrier between the changed population (mutants) and the parent population.

synapse (sin´-aps) A junction between two neurons, or between a neuron and an effector cell. Electrical or chemical signals are relayed from one cell to another at a synapse.

synaptic cleft (sin-ap´-tik) In a chemical synapse, a narrow gap separating the synaptic terminal of a transmitting neuron from a receiving neuron or an effector cell.

synaptic terminal The tip of a transmitting neuron's axon, where signals are sent to another neuron or to an effector cell.

synaptic vesicle A membranous sac containing neurotransmitter molecules at the tip of the presynaptic axon.

system A more complex organization formed from a combination of components.

systematics An analytical approach to the study of the diversity of life and the evolutionary relationships between organisms.

systemic acquired resistance A defensive response in plants infected with a pathogenic microbe; helps protect healthy tissue from the microbe.

systemic circuit One of two main blood circuits in terrestrial vertebrates; conveys blood between the heart and the rest of the body. *See also* pulmonary circuit.

systole (sis´-tō-lē) The contraction stage of the heart cycle, when the heart chambers actively pump blood. *See also* diastole.

T

T cell A type of lymphocyte that matures in the thymus and is responsible for cell-mediated immunity; also involved in humoral immunity. *See also* B cell.

T₃ *See* triiodothyronine.

T₄ *See* thyroxine.

tapeworm A parasitic flatworm characterized by the absence of a digestive tract.

target cell A cell that responds to a regulatory signal, such as a hormone.

taxis (tak´-sis) (plural, **taxes**) Virtually automatic orientation toward or away from a stimulus.

taxon (tak´-son) (plural, **taxa**) A proper name, such as phylum Chordata, class Mammalia, or *Homo sapiens,* in the taxonomic hierarchy used to classify organisms.

taxonomy The branch of biology concerned with identifying, naming, and classifying species.

technology The practical application of scientific knowledge.

telomere (tel´-uh-mēr) The repetitive DNA at each end of a eukaryotic chromosome.

telophase The fifth and final stage of mitosis, during which daughter nuclei form at the two poles of a cell. Telophase usually occurs together with cytokinesis.

temperate broadleaf forest A biome located throughout midlatitude regions where there is sufficient moisture to support the growth of large, broadleaf deciduous trees.

temperate grassland A grassland region maintained by seasonal drought, occasional fires, and grazing by large mammals.

temperate zones Latitudes between the tropics and the Arctic Circle in the north and the Antarctic Circle in the south; regions with milder climates than the tropics or polar regions.

temperature A measure of the intensity of heat, reflecting the average kinetic energy or speed of molecules.

temporal isolation A type of prezygotic barrier between species; the species remain isolated because they breed at different times.

tendon Fibrous connective tissue connecting a muscle to a bone.

terminal bud Embryonic tissue at the tip of a shoot, made up of developing leaves and a compact series of nodes and internodes.

terminator A special sequence of nucleotides in DNA that marks the end of a gene. It signals RNA polymerase to release the newly made RNA molecule, and then to depart from the gene.

territory An area that one or more individuals defend and from which other members of the same species are usually excluded.

tertiary consumer (ter´-shē-ār-ē) An organism that eats secondary consumers.

tertiary structure The third level of protein structure; the overall, three-dimensional shape of a polypeptide in a protein.

testcross The mating between an individual of unknown genotype for a particular characteristic and an individual that is homozygous recessive for that same characteristic.

testis (plural, **testes**) The male gonad in an animal; produces sperm and, in many species, reproductive hormones.

testosterone (tes-tos´-tuh-rōn) An androgen hormone that stimulates an embryo to develop into a male and promotes male body features.

tetrapod A vertebrate with two pairs of limbs. Tetrapods include mammals, amphibians, and birds and other reptiles.

thalamus (thal´-uh-mus) An integrating and relay center of the vertebrate forebrain; sorts and relays selected information to specific areas in the cerebral cortex.

theory A widely accepted explanatory idea that is broad in scope and supported by a large body of evidence.

therapeutic cloning The cloning of human cells by nuclear transplantation for therapeutic purposes, such as the replacement of body cells that have been irreversibly damaged by disease or injury. *See* nuclear transplantation; reproductive cloning.

thermodynamics (ther´-mō-dī-nam´-iks) The study of energy transformations that occur in a collection of matter. *See* first law of thermodynamics; second law of thermodynamics.

thermoreceptor A sensor (sensory receptor) that detects heat or cold.

thermoregulation The maintenance of internal temperature within a range that allows cells to function efficiently.

thick filament A filament composed of staggered arrays of myosin molecules; a component of myofibrils in muscle fibers.

thigmotropism (thig´-mō-trō´-pizm) Growth movement of a plant in response to touch.

thin filament The smaller of the two myofilaments consisting of two strands of actin and one strand of regulatory protein coiled around each other.

threatened species As defined in the U.S. Endangered Species Act, a species that is likely to become endangered in the foreseeable future throughout all or a significant portion of its range.

three-domain system A system of taxonomic classification based on three basic groups: Bacteria, Archaea, and Eukarya.

threshold The minimum change in a membrane's voltage that must occur to generate a nerve signal (action potential).

thylakoid (thī´-luh-koyd) One of a number of disk-shaped membranous sacs inside a chloroplast. Thylakoid membranes contain chlorophyll and the enzymes of the light reactions of photosynthesis. A stack of thylakoids is called a granum.

thymine (T) (thī´-min) A single-ring nitrogenous base found in DNA.

thymus gland (thī´-mus) An endocrine gland in the neck region of mammals that is active in establishing the immune system; secretes several hormones that promote the development and differentiation of T cells.

thyroid gland (thī´-royd) An endocrine gland that secretes thyroxine (T₄), triiodothyronine (T₃), and calcitonin.

thyroid-stimulating hormone (TSH) A protein hormone secreted by the anterior pituitary that stimulates the thyroid gland to secrete its hormones.

thyroxine (T₄) (thī-rok´-sin) An amine hormone secreted by the thyroid gland that stimulates metabolism in virtually all body tissues.

Ti plasmid A bacterial plasmid that induces tumors in plant cells that it infects; often used as a vector to introduce new genes into plant cells. Ti stands for tumor-inducing.

tight junction A junction that binds tissue cells together in a leakproof sheet.

tissue A cooperative unit of many similar cells that perform a specific function within a multicellular organism.

tissue system One or more tissues organized into a functional unit wihtin a plant or animal.

tonicity The ability of a solution to cause a cell within it to gain or lose water.

topsoil A mixture of particles derived from rock, living organisms, and humus.

torpor (tor´-per) A state of reduced activity by an endotherm. Torpor reduces energy consumption because the metabolic rate, body temperature, heart rate, and breathing rate decrease.

trace element An element that is essential for the survival of an organism but only in minute quantities.

trachea (trā-kē-uh) (plural, **tracheae**) The windpipe; the portion of the respiratory tube that has C-shaped cartilagenous rings and passes from the larynx to two bronchi.

tracheal system In insects, the extensive system of tiny tubes that branch throughout the insect's body, enabling gas exchange between outside air and body cells.

tracheid (trā-kē-id) A tapered, porous, water-conducting and supportive cell in plants. Chains of tracheids or vessel elements make up the water-conducting, supportive tubes in xylem.

trade winds The movement of air in the tropics (those regions that lie between 23.5° north latitude and 23.5° south latitude).

transcription The synthesis of RNA on a DNA template.

transcription factor In the eukaryotic cell, a protein that functions in initiating or regulating transcription. Transcription factors bind to DNA or to other proteins that bind to DNA.

transduction (1) The transfer of bacterial genes from one bacterial cell to another by a phage. (2) *See* sensory transduction. (3) *See* signal transduction.

transfer RNA (tRNA) A type of ribonucleic acid that functions as an interpreter in translation. Each tRNA molecule has a specific anticodon, picks up a specific amino acid, and conveys the amino acid to the appropriate codon on mRNA.

transformation The incorporation of new genes into a cell from DNA that the cell takes up from the surrounding environment.

transgenic organism An organism that contains genes from another species.

translation The synthesis of a polypeptide using the genetic information encoded in an mRNA molecule. There is a change of "language" from nucleotides to amino acids.

translocation (1) During protein synthesis, the movement of a tRNA molecule carrying a growing polypeptide chain from the A site to the P site on a ribosome.(The mRNA travels with it.) (2) A change in a chromosome resulting from a chromosomal fragment attaching to a nonhomologous chromosome; can occur as a result of an error in meiosis or from mutagenesis.

transmission electron microscope (TEM) A microscope that uses an electron beam to study the internal structure of thinly sectioned specimens.

transpiration The evaporative loss of water from a plant.

transpiration-cohesion-tension mechanism The transport mechanism whereby transpiration exerts a pull that is relayed downward along a string of molecules held together by cohesion and helped upward by adhesion.

transport vesicle A tiny membranous sac in a cell's cytoplasm carrying molecules produced by the cell. The vesicle buds from the endoplasmic reticulum or Golgi and eventually fuses with another membranous organelle or the plasma membranes, releasing its contents.

transposon (tranz-pō´-zon) A transposable genetic element, or "jumping gene"; a segment of DNA that can move from one site to another within a cell and serve as an agent of genetic change.

TRH-releasing hormone A peptide hormone that triggers the release of TSH (thyroid stimulating hormone) which in turn stimulates the thyroid gland.

trial-and-error learning Learning to associate a particular behavioral act with a positive or negative effect.

triglyceride (trī-glis´-uh-rīd) A fat, which consists of a molecule of glycerol linked to three molecules of fatty acid.

triiodothyronine (T₃) (trī´-ī-ō-dō-thī´-rō-nēn) An amine hormone secreted by the thyroid gland that stimulates metabolism in virtually all body tissues.

trimester In human development, one of three 3-month-long periods of pregnancy.

triplet code A set of three-nucleotide-long words that specify the amino acids for polypeptide chains. *See* genetic code.

trisomy 21 *See* Down syndrome.

tRNA *See* transfer RNA.

trophic structure (trō´-fik) The feeding relationships in a community; determines the route of energy flow and the pattern of chemical cycling in an ecosystem.

trophoblast (trō´f-ō-blast) In mammalian development, the outer portion of a blastocyst. Cells of the trophoblast secrete enzymes that enable the blastocyst to implant in the endometrium of the mother's uterus.

tropical forest A terestrial biome characterized by warm temperatures year-round.

tropics Latitudes between 23.5° north and south.

tropism (trō´-pizm) A growth response that makes a plant grow toward or away from a stimulus.

true-breeding Referring to organisms for which sexual reproduction produces offspring with inherited traits identical to those of the parents; the organisms are homozygous for the characteristics under consideration.

tubal ligation A means of sterilization in which a woman's two oviducts (fallopian tubes) are tied closed to prevent eggs from reaching the uterus; a segment of each oviduct is removed.

tuber An enlargement at the end of a rhizome, in which food is stored.

tumor An abnormal mass of cells that forms within otherwise normal tissue.

tumor-suppressor gene A gene whose product inhibits cell division, thereby preventing uncontrolled cell growth.

tundra A biome at the northernmost limits of plant growth and at high altitudes, characterized by dwarf woody shrubs, grasses, mosses, and lichens.

tunicate One of a group of invertebrate chordates.

U

ultimate question In animal behavior, an inquiry that focuses on the evolutionary significance of a behavioral act.

ultrasound imaging A technique for examining a fetus in the uterus. High-frequency sound waves echoing off the fetus are used to produce an image of the fetus.

uniform Describing a dispersion pattern in which individuals are evenly distributed.

unsaturated Pertaining to fats and fatty acids whose hydrocarbon chains lack the maximum number of hydrogen atoms and therefore have one or more double covalent bonds. Unsaturated fats and fatty acids do not solidify at room temperature.

uracil (U) (yū´-ruh-sil) A single-ring nitrogenous base found in RNA.

urea (yū-rē´-ah) A soluble form of nitrogenous waste excreted by mammals and most adult amphibians.

ureter (yū-rē´-ter or yū´-reh-ter) A duct that conveys urine from the kidney to the urinary bladder.

urethra (yū-rē´-thruh) A duct that conveys urine from the urinary bladder to the outside. In the male, the urethra also conveys semen out of the body during ejaculation.

uric acid (yū´-rik) An insoluble precipitate of nitrogenous waste excreted by land snails, insects, birds, and some reptiles.

urinary bladder The pouch where urine is stored prior to elimination.

urine Concentrated filtrate produced by the kidneys and excreted via the bladder.

uterus (yū´-ter-us) In the reproductive system of a mammalian female, the organ where the development of young occurs; the womb.

utricle (yū´-truh-kul) A fluid-filled inner-ear chamber containing hair cells that detect the position of the head relative to gravity.

V

vaccination (vak´-suh-nā´-shun) A procedure that presents the immune system with a harmless variant or derivative of a pathogen, thereby stimulating the immune system to mount a long-term defense against the pathogen.

vaccine (vak-sēn´) A harmless variant or derivative of a pathogen used to stimulate a host organism's immune system to mount a long-term defense against the pathogen.

vacuole (vak´-ū-ōl) A membrane-enclosed sac, part of the endomembrane system of a eukaryotic cell, having diverse functions.

vagina (vuh-jī´-nuh) Part of the female reproductive system between the uterus and the outside opening; the birth canal in mammals; also accommodates the male's penis and receives sperm during copulation.

vas deferens (vas def´-er-enz) (plural, **vasa deferentia**) Part of the male reproductive system that conveys sperm away from the testis; the sperm duct; in humans, the tube that conveys sperm between the epididymis and the common duct that leads to the urethra.

vascular bundle (vas´-kyū-ler) A strand of vascular tissues (both xylem and phloem) in a plant stem.

vascular cambium (vas´-kyū-ler kam´-bē-um) During secondary growth of a plant, the cylinder of meristematic cells, surrounding the xylem and pith, that produces secondary xylem and phloem.

vascular plant A plant with xylem and phloem.

vascular tissue Plant tissue consisting of cells joined into tubes that transport water and nutrients throughout the plant body.

vascular tissue system A system formed by xylem and phloem throughout the plant, serving as a transport system for water and nutrients, respectively.

vasectomy (vuh-sek´-tuh-mē) Surgical removal of a section of the two sperm ducts (vasa deferentia) to prevent sperm from reaching the urethra; a means of sterilization in the male.

vector In molecular biology, a piece of DNA, usually a plasmid or a viral genome, that is used to move genes from one cell to another.

vegetative propagation Asexual reproduction by a plant.

vein (1) In animals, a vessel that returns blood to the heart. (2) In plants, a vascular bundle in a leaf, composed of xylem and phloem.

ventilation A mechanism that provides contact between an animal's respiratory surface and the air or water to which it is exposed. Contact between a respiratory surface and air or water enables gas exchange to occur.

ventral Pertaining to the underside, or bottom, of a bilaterally symmetrical animal.

ventricle (ven´-truh-kul) (1) A heart chamber that pumps blood out of a heart. (2) A space in the vertebrate brain, filled with cerebrospinal fluid.

venule (ven´-yūl) A vessel that conveys blood between a capillary bed and a vein.

vertebra (ver´-tuh-bruh) (plural, **vertebrae**) One of a series of segmented units making up the backbone of a vertebrate animal.

vertebral column backbone; composed of a series of segmented units called vertebrae.

vertebrate (ver´-tuh-brāt) A chordate animal with a backbone. Vertebrates include agnathans, cartilaginous fishes, bony fishes, amphibians, reptiles (including birds and mammals).

vessel element A short, open-ended, water-conducting and supportive cell in plants. Chains of vessel elements or tracheids make up the water-conducting, supportive tubes in xylem.

vestigial organ A strucure of marginal, if any, importance to an organism. Vestigial organs are historical remnants of structures that had important functions in ancestors.

villus (vil´-us) (plural, **villi**) (1) A fingerlike projection of the inner surface of the small intestine. (2) A fingerlike projection of the chorion of the mammalian placenta. Large numbers of villi increase the surface areas of these organs.

visceral mass (vis´-uh-rul) One of the three main parts of a mollusc, containing most of the internal organs.

visual acuity The ability of the eyes to distinguish fine detail.

vital capacity The maximum volume of air that a respiratory system can inhale and exhale.

vitamin An organic nutrient that an organism requires in very small quantities. Vitamins generally function as coenzymes.

vitreous humor (vit´-rē-us hyū´-mer) A jellylike substance filling the space behind the lens in the vertebrate eye; helps maintain the shape of the eye.

vocal cord One of a pair of string-like tissues in the larynx. Air rushing past the tensed vocal cords makes them vibrate, producing sounds.

W

water mold A fungus-like protists in the stramenopile clade.

water vascular system In echinoderms, a radially arranged system of water-filled canals that branch into extensions called tube feet. The system provides movement and circulates water, facilitating gas exchange and waste disposal.

water-conducting cell A specialized, dead plant cell with lignin-containing secondary walls, arranged end to end, forming xylem tissue. *See also* tracheid; vessel element.

wavelength The distance between crests of adjacent waves, such as those of the electromagnetic spectrum.

wax A type of lipid molecule consisting of one fatty acid linked to an alcohol; functions as a waterproof coating on many biological surfaces, such as apples and other fruits.

westerlies Winds that blow from west to east.

wetland An ecosystem intermediate between an aquatic ecosystem and a terrestrial ecosystem. Wetland soil is saturated with water permanently or periodically.

white blood cell A blood cell that functions in defending the body against infections and cancer cells. Also called leukocyte.

white matter Tracts of axons within the central nervous system.

wild type The phenotype most commonly found in nature.

withdrawal The withdrawal of the penis from the vagina before ejaculation, an unreliable method of contraception.

wood Secondary xylem of a plant. *See also* heartwood; sapwood.

wood ray A column of parenchyma cells that radiates from the center of a log and transports water to its outer living tissues.

X

X chromosome inactivation In female mammals, the inactivation of one X chromosome in each somatic cell.

xylem (zī´-lum) The nonliving portion of a plant's vascular system that provides support and conveys xylem sap from the roots to the rest of the plant. Xylem is made up of vessel elements and/or tracheids, water-conducting cells. Primary xylem is derived from the procambium. Secondary xylem is derived from the vascular cambium in plants exhibiting secondary growth.

xylem sap The solution of inorganic nutrients conveyed in xylem tissue from a plant's roots to its shoots.

Y

yeast A single-celled fungus that inhabits liquid or moist habitats and repoduces asexually by simple cell division or by the pinching of small buds off a parent cell.

yellow bone marrow A tissue found within the central cavities of long bones, consisting mostly of stored fat.

yolk plug A cluster of endodermal cells at the surface of an amphibian gastrula. The yolk plug marks the position of the blastopore and the site of the future anus.

yolk sac An extraembryonic membrane that develops from endoderm; produces the embryo's first blood cells and germ cells and gives rise to the allantois.

Z

zoned reserve An extensive region of land that includes one or more areas that are undisturbed by humans. The undisturbed areas are surrounded by lands that have been altered by human activity.

zooplankton (zō´-ō-plank´-tun) Animals that drift in aquatic environments.

zygomycete Member of the fungal phylum Zygomycota, characterized by a sturdy structure called a zygosporangium during sexual reproduction.

zygote (zī´-gōt) The fertilized egg, which is diploid, that results from the union of a sperm cell nucleus and an egg cell nucleus.

zygote fungus *See* zygomycete.

Index

A

Abiotic components of ecosystems, 684, 686, 754
  adaptation to, 687
  population size and, 732, 732 *fig.*
Abiotic reservoir, 756–57
ABO blood groups, 167, 270
  multiple alleles for, 167 *table*
Abortion, 165
Abscisic acid (ABA), 666 *table,* 669
Absorption, 357, 431
Acacia tree, mutualism between ants and, 749, 749 *fig.*
Acanthostega (fossilized four-legged fish), 389, 389 *fig.*
Accommodation of eye lens, 595 *fig.*
Acetic acid, 35
Acetylcholine, 572–73
Acetyl CoA (acetyl coenzyme A)
  citric acid cycle and, 97, 97 *fig.*
  conversion of pyruvate into, 96, 96 *fig.*
Achondroplasia, 162–63, 163 *fig.*
Acid, 27
Acid chyme, 437
Acid precipitation, 28, 28 *fig.*, 761
  amphibians and, 389
Acquired immunity, 489–500
Acromegaly, 524
Acrosome, 548, 548 *fig.*, 549 *fig.*
ACTH (adrenocorticotropic hormone), 524, 530–31
Actin filaments, 64
  muscle contraction and, 614–15, 614 *fig.*, 615 *fig.*
Actinomycetes, 328, 328 *fig.*, 363
Action potential, 569, 569 *fig.*
  propagation along axon of, 570, 570 *fig.*
  sensations represented by, 591 *fig.*
Activator protein, 211
Active immunity, 489
Active sites, enzyme, 77
Active transport, 84, 84 *fig.*
Adaptive change in populations, 302–3
Adaptive radiation, 288, 288 *fig.*
Adenine (A), 185
ADH (antidiuretic hormone), 515, 524
Adhesion, water transport in plant tissues and, 648–49
Adipose tissue, 417, 417 *fig.*
Adrenal cortex, 530
Adrenal glands, 530–31
  stress responses and, 530 *fig.*
Adrenal medulla, 530
Adrenocorticotropic hormone (ACTH), 524, 530–31
Adult stem cells, 219
Advanced Cell Technology (ACT), 219
Aerobic cellular respiration, 89–100
Aerobic exercise, 617
Aerobic prokaryotes, 325, 325 *fig.*
Afghanistan, population age structure, 738, 739 *fig.*
Africa, human evolution originating in, 405
Agave, "big-bang" reproduction of, 734, 734 *fig.*
Age of mother, Down syndrome and, 145 *fig.*
Age structure of populations, 738–39, 739 *fig.*
Agglutination of pathogens, 495, 495 *fig.*
Aggregate fruit, 639, 639 *fig.*
Aging, taste perception and, 601
Agonistic behaviors, 718

Agriculture. *See also* Genetically-modified organisms
  angiosperms and, 354
  applications of reproductive cloning to, 218
  DNA technology and, 247–48
  fungi and, 342–43
  human evolution and development of, 408 *fig.*
  organic, 656
  plant hormones used in, 671
  soil conservation and, 655, 655 *fig.*
  vegetative plant reproduction and, 642
Agrobacterium spp., 328
Agrobacterium tumefaciens, 247
AIDS (acquired immune deficiency syndrome), 202, 203, 484–85, 498, 500. *See also* HIV
Air, plant nutrients acquired from, 646
Air pollution
  lichen susceptibility to, 362
  smoking and, 459
Alcohol consumption
  birth defects and, 557
  dehydration and, 509
  liver damage and, 511
  nervous system and, 573
Alcohol fermentation, 101, 101 *fig.*
Alcohols, 35
Aldehydes, 35
Algae, 6
  brown (multicellular), 336, 336 *fig.*
  lichens and, 362, 362 *fig.*
  as photoautotrophs, 108 *fig.*
  photosynthetic, 333, 333 *fig.*
Ali, Muhammad, 585 *fig.*
Alimentary canal, 432
  of carnivore *versus* herbivore, 441 *fig.*
  earthworm, grasshopper, and bird, 432 *fig.*
  human, 433
Allantois, 556, 557, 557 *fig.*
Alleles, 156
  detecting harmful, with restriction fragment analysis, 240–41
  diploid organisms and recessive, 272–73
  dominant and recessive, 156–57
  frequencies of, and Hardy-Weinberg equilibrium, 266–67
  homologous chromosomes as carriers of, 157
  multiple, of single genes, 156, 167
  mutations and creation of, 199, 270–71
  new combinations of, produced by crossing over, 172–73
  sexual recombination of, 271 *fig.*
  testing for disease-causing, 170
Allergens, 501
Allergic reactions, two stages of, 501 *fig.*
Allergies, 501
Allopatric speciation, 284
All Species Foundation, 356
Alpha helix protein structure, 44, 45 *fig.*, 46
Alpha proteobacteria, 328
  mitochondria and, 333
Alternation of generations
  green algae *Ulva*, 338, 338 *fig.*
  plants, 348–56, 348 *fig.*
Alternative RNA splicing, 215
Altruism, 721
Alveolates, 335
Alveoli, 458–59
  structure of, 458 *fig.*
Alzheimer's disease (AD), 585
  brain tissue showing, 585 *fig.*

American Dental Association, 18–19
Amine hormones, 521
Amines, 35
  biogenic, 572
Amino acids, 42 *fig.*
  codons translated into sequences of, 191
  comparing sequences in vertebrate hemoglobin, 263
  as essential nutrients, 443 *fig.*, 444
  formation of, 319
  peptide bonds and linkage of, 42–43
Amino group, 35
Ammonia, excretion of, 510
Ammonite casts, 260 *fig.*
Amniocentesis, 164, 164 *fig.*
Amnion, 556–57
Amniotes, 390
  osmoregulation in, 509
Amniotic eggs, 390, 391
Amniotic fluid, 556
Amoebas, 126 *fig.*, 337, 337 *fig.*
Amoebocytes, 372
Amoebozoans, 337
Amphetamines, 573
Amphibians, 389
  cardiovascular system, 470, 470 *fig.*
  excretion in, 510 *fig.*
  respiratory system of, 458
  tadpole larva of, 553 *fig.*
Amygdala, 583
Anabaena (cyanobacterium), 326, 328, 328 *fig.*
Anabolic steroids, health risks using, 41
Anaerobic cellular respiration, 89, 101
Anaerobic exercise, 617
Analogy and analogous structures, 304, 304 *fig.*
Anaphase
  cell division I, of meiosis, 138 *fig.*
  cell division II, of meiosis, 139 *fig.*
Anaphase, mitosis, 130, 131 *fig.*
Anaphylactic shock, 501
Anatomy, 414
  evolution and comparative, 262
Anaximander, 256
Anchorage dependence of cell division, 133
Anchoring junctions, 66, 66 *fig.*
Androgens, 518–19, 532
Anemia, 480
Angiosperms, 347, 347 *fig.*
  agriculture and role of, 354
  animals and evolution of, 355
  body plan of, 626, 626 *fig.*
  dicot and monocot, 625, 625 *fig.*
  flower of, 352, 636 *fig.*
  fruit of, 354, 639
  life cycle of, 352–53, 353 *fig.*
  sexual reproduction, 636–40
  sporophyte and gametophyte of, 352–53
Animal(s). *See also names of individual animals*
  angiosperm evolution and, 355
  body plans of, 370, 373, 374, 376
  body segmentation in, 378–79
  chemical regulation in (*see* Chemical regulation in animals)
  circulation in (*see* Circulatory system)
  cloned, 208–9, 218–19
  cognition in, 713
  defenses of, 746–47, 746 *fig.*
  defining and describing, 368
  development in (*see* Embryonic development in animals)
  digestion in (*see* Digestive system)

Animal(s) (continued)
diseases of (see Disease; Human disease)
ectothermic and endothermic, 504
evolution and diversity of, 366–67, 368–71, 393–94
exchanges between external environment and, 424–26, 424 fig.
fungi and, 362–63, 362 fig.
gas exchange in (see Gas exchange in animals)
hermaphroditic, 537, 537 fig.
human threat to diversity of, 394
immune system (see Immune system)
invertebrates, 372–85
mammals, 366–67
medicinal plants used by, 678
movement in (see Movement)
mutation rates in, 271
nervous systems, 574–78
nutrition in (see Animal nutrition)
organs and organ systems of, 419–21
phylogenetic tree for, 371, 371 fig.
phylogeny of, 393–94
primates, 400–402
regulation of internal environment by (see Homeostasis; Regulatory systems, animal)
reproduction in (see Asexual reproduction; Human reproduction; Sexual reproduction)
respiration in (see Respiration)
seed dispersal by, 354 fig.
senses of (see Senses)
signaling behavior in, 720
speciation in (see Speciation)
structural organization in hierarchy of, 414–23
structure and function in body of, 618
transgenic, 247–48
vertebrates, 386–92
Animal body. See also Bilateral symmetry; Body cavity, of animals
body symmetry of, and nervous system coordination, 574
characteristics, 370
of mollusc, 376 fig.
organs of (see Organ(s))
pattern formation in, 555
radial, 373
segmented, 378–79, 378 fig.
senses of (see Senses)
structure and function in, 618
temperature of (see Body temperature)
tissues of (see Tissues, animal)
Animal cells, 56–57, 56 fig.
cleavage, 132 fig., 550
cytokinesis in, 132
meiosis with diploid number of 4, 138–39 fig.
organelles of (see Organelles)
osmoregulation of water balance in, 83, 83 fig.
plasma membrane of, 80 fig.
surfaces and junctions, 66 fig.
Animal development. See Embryonic development in animals
Animal disease. See Disease; Human disease
Animalia (kingdom), 7, 7 fig., 310, 367, 393–94. See also Animal(s)
Animal nutrition, 428–51
chemical conversions and, 29, 29 fig.
chemical energy for body and, 442
diets and digestive adaptations in, 440–41
digestion and, 431, 432
essential amino acids, vitamins, and minerals, 443–46
feeding, ingestion, and, 368, 430–31

food labels and, 446, 446 fig.
human body fat, fad diets and, 448
human digestive system, 433–40
human disease, diet, and, 449
human healthy diet, 443–49
obtaining and processing food and, 430–33
in whales, 428–29
Annelid (Annelida), 378–79, 574, 574 fig.
Annuals, 632
Ant(s), 383
mutualism between acacia tree and, 749, 749 fig.
Antagonistic behavior, 718
Antagonistic hormones, 527
Antagonistic muscles, 613 fig.
Anterior pituitary, 524
hormones of, 525 fig.
Anterior surface of animals, 370, 370 fig., 374
Anthers, 352, 636
Anthrax, 272, 325, 325 fig.
as biological weapon, 330
Anthropoids, 401
Antibiotics
actinomycetes and synthesis of, 328
bacterial resistance to, 205, 272, 325, 498
cellular respiration and effects of, 99
enzymes and effects of, 78
fungi and production of, 363
gram-negative bacteria insensitivity to, 324
Antibody(ies), 489
antigen-binding sites on, 491
antigens marked for elimination by, 495, 495 fig.
binding to antigenic determinants, 491 fig.
computer graphic of, 494 fig.
monoclonal, 496, 496 fig.
structure of, 494, 494 fig.
as weapons of humoral immunity, 494
Anticodons, 195
Antidiuretic hormone (ADH), 515, 524
Antigen(s)
allergies and environmental, 501
antibody binding to, 494 fig.
marking of, by antibodies, 495, 495 fig.
Antigen-binding sites, 491, 491 fig., 494
Antigenic determinants, 491, 491 fig.
Antigen-presenting cells (APCs), 497
Antigen receptors, 490
Antigens, 489
Antihistamines, 501
Anus, 432
Aorta, 471
Apes, relationship of humans to, 400–401, 402
Aphids
abiotic effects on population size in, 732 fig.
study of transport in phloem tissue using, 650–51, 651 fig.
Aphotic zone, 690–91
Apical dominance, 626
Apical meristems, 344, 345
primary growth and, 632, 632 fig., 633 fig.
Apicomplexans, 335, 335 fig.
Apoptosis, 554, 554 fig.
Appendicular skeleton, 610, 610 fig.
Appendix, 440
Aquaporins, 82
Aquatic animals, 690
excretion in, 510 fig.
at hydrothermal vents, 683 fig.
osmoregulation in, 508 fig.
respiration and gas exchange in, 455, 455 fig., 456–57, 456 fig.
Aquatic ecosystems, 690–92
coral reefs, 691 fig.

eutrophication in, 761
freshwater, 692
gills as adaptation to, 455, 456–57
hydrothermal vents in, 683
oceans, 682–83, 690–91
Aquatic food chain, 752 fig.
Aqueous humor, 595, 595 fig.
Aqueous solutions, 26
Arabidopsis thaliana (mustard plant), 246, 620, 676, 676 fig.
Arachnids, 380
Arbuscular mycorrhizal fungi, 358 fig., 359, 359 fig.
Archaea (domain), 6, 6 fig., 310
bacteria compared to, 322–23, 323 table
extreme habitats of select, 327
insensitivity of gram-negative bacteria to, 324–25
Archaean eon, 299
Archaeopteryx fossil, 296–97, 297 fig., 391 fig.
Archenteron, 551 fig.
Aristotle, 154, 256, 646
Arousal, brain regulation of sleep and, 582
Art, biology and, 50–51
Arteries, 469
atherosclerosis of, 474 fig.
coronary, 474 fig.
pulmonary, 471
pulse of blood in, 475
structure of, 472 fig.
Arterioles, 469, 472 fig.
Arthritis, rheumatoid, 500
Arthropods (Arthropoda), 380–81
exoskeleton of, 608, 608 fig., 609
nervous system of, 574 fig.
structure of, 380 fig.
Artificial pacemaker, 473
Artificial selection, 258
Ascomycetes, 358 fig., 359, 359 fig.
Asexual reproduction, 125
in animals, 536–37
in fungi, 358
mitosis and, 136
in plants, 641, 641 fig.
Aspartame, 38
Aspartic acid, 42
Aspirin, disruption of pain-inducing enzymes and, 78
Assisted reproductive technology (ART), 561
Associative learning, 712
Asteroids, 303, 303 fig.
Astigmatism, 596
Astronauts, motion sickness and, 600
Atherosclerosis, 474, 474 fig.
Athletes
growth hormone and, 525
red blood cell levels and, 480
training by, 617
Athlete's foot, 361
Atmosphere
greenhouse effect in, 119, 770–71
origin of life and Earth's early, 317, 318–19
ozone layer depletion in, 120, 120 fig., 769
Atomic mass, 20
Atomic number, 20
Atoms, 2 fig., 3
chemical reactions and rearrangements of, 29
covalent bonds and molecule formation from, 23
electron arrangement and properties of, 22
electronegativity of, 24
ionic bonds formed between, 22–23
radioisotopes of, 20
structure of, 20
ATP (adenosine triphosphate), 35
active transport and need for, 84

cellular respiration and synthesis of, 74, 89, 91 *fig.*, 93, 95 *fig.*, 98, 100 *fig.*
as chemical energy shuttle in cells, 75, 75 *fig.*
mitochondria and, 63
muscle contraction and, 615
photosynthesis and synthesis of, 111, 114 *fig.*, 115, 115 *fig.*
photosynthesis and synthesis of sugars using, 116
ATP cycle, 75 *fig.*
ATP synthases, 93
poisons as disruptors of, 99
Atria (atrium), 469
Attention deficit hyperactivity disorder (ADHD) medications, 573
Auditory canal, 598
Australia, animal diversity in, 367, 392, 394, 395 *fig.*
Australopithecus afarensis, 404
Australopithecus africanus, 403
Australopithecus anamensis, 403
Australopiths, 403
Autoimmune disease, 500
diabetes (Type I) as, 529
multiple sclerosis, 567
Automatic external defibrillators (AEDs), 474
Autonomic nervous system, 576, 577, 577 *fig.*
Autosomes, 137
Autotrophs, 108, 326
Auxins, 665, 666 *table,* 671
antagonistic interactions of, 668, 671
cell elongation in shoots regulated by, 666–67, 667 *fig.*
Avian flu, 203
AV (atrioventricular) node, 473
Axial skeleton, 610, 610 *fig.*
Axillary buds, 626
Axolotl, 292, 292 *fig.*
Axons, 418, 567
propagation of action potential along, 570, 570 *fig.*
AZT (antiviral drug), 485

B

Bacilli, 323, 323 *fig.*
Bacillus anthracis, 325, 325 *fig.,* 328, 330
Bacteria (domain), 6, 6 *fig.,* 310
antibiotic resistance in, 205, 272, 325, 498
archaea compared to, 322–23, 323 *table*
binary fission in, 127
bioterrorism using, 330
cell structure and size in, 54
chromosomes in, 127 *fig.*
customization using plasmids, 232–33
cyanobacteria, 314–15
as denitrifiers, 758 *fig.*
disease causing, 329, 330, 437, 437 *fig.,* 487 (*see also* Disease; Human disease)
diversity of, 328
exponential growth model applied to, 730 *fig.*
flagella and cilia, 65, 324–25
at hydrothermal vents, 683
mutation rates in, 271
natural means of DNA transfer in, 204
nitrogen cycle and, 758–59, 758 *fig.*
nitrogen-fixing, 658–59
oil spill remediation with, 778
plasmids of (*see* Plasmids)
recycling and cleaning environment with, 330–31
as tool for manipulating DNA, 232–37
Bacterial cell wall, 55
Bacteriophage(s), 182
genomic library of, 235, 235 *fig.*

Hershey and Chase experiments on, 182–83, 183 *fig.*
lambda, 200
reproductive cycle of, 183 *fig.,* 200 *fig.*
Bacteriorhodopsin, 327
Balance, inner ear and sensation of, 600
Balancing selection, 273
Ball-and-socket joints, 611, 611 *fig.*
Bamboo, 103
Banff National Park, Canada, animal bridge in, 774 *fig.,* 777
Bark, 635, 635 *fig.*
Barnacles, 381 *fig.,* 745 *fig.*
Barrier methods (contraception), 547
Bartholin's glands, 539
Basal body, 65
Basal ganglia, 579
Basal metabolic rate (BMR), 442
Base, 27
Basement membrane, 416
Basidiomycetes, 358 *fig.,* 359, 359 *fig.*
Basilar membrane, 598, 599, 599 *fig.*
Basophils, 479, 479 *fig.*
Bat(s), 508
Batesian mimicry, 746, 746 *fig.*
Bateson, William, 172
Bats, 392
B cells (B lymphocytes), 490–91, 490 *fig.*
Bdellovibrio bacteriophorus, 328, 328 *fig.*
Beadle, George, 190
Bears, 588, 589
global warming and polar, 770, 771 *fig.*
systematics of raccoons and, 308, 308 *fig.*
thermoregulation in, 504–5
Bees, 383. *See also* Honeybees
Beetles, 383, 383 *fig.*
Behavior, 702–25
foraging and mating, 714–16
of impalas, 702–3
innate, 705
learning, 707–13
scientific study of, 704–7
sociobiology and social, 717–23
thermoregulation and, 507
Behavioral ecology, 704
animal movement, 711
feeding behavior, 714–15
Behavioral isolation, 282
Benign tumors, 135
Benthic zone, 690
Bernard, Claude, 425
Beta-carotene, conversion to vitamin A, 29
Biennials, 632
Bilateral symmetry, 370, 370 *fig.,* 374, 574
Bilaterians, 371, 374, 375, 393
Bilbies, 394, 395 *fig.*
Bile, 438
Binary fission, 127
Binomial naming system, 304–5
Biodiversity
in communities, 744, 747, 766
conservation of, 772–73
endangered species protection and, 765
evolution of (*see* Evolution)
human activities as threat to, 766–67
human welfare and, 767
in plants as nonrenewable resource, 356
Biodiversity crisis
endangered species and, 428, 764–65
global warming and, 770–71
habitat destruction, exotic species, and overexploitation as causes of, 768
overview of, 766–71
ozone layer depletion and, 120, 120 *fig.,* 769
reasons for concern about, 767

technology and human overpopulation as factors in, 356
(E. O.) Wilson on, 723, 767
Biodiversity hotspots, 775, 775 *fig.*
Biofilms, 326, 326 *fig.*
Biogenic amines, 572
Biogeochemical cycles, 756–57, 757 *fig.*
Biogeography, evidence for evolution in, 262
Biological augmentation, 778–79
Biological clocks, 579, 673
phytochromes and, 675
plant circadian rhythms, 673
in plant guard cells, 649
seasonal, in plants, 674
Biological magnification, 769, 769 *fig.*
Biological species concept, 280–81
Biological weapons, 330
Biological Weapons Convention, 330
Biology, 2. *See also* Conservation biology; Ecology; Molecular biology
art and, 50–51
at atomic and molecular level, 18
everyday life and, 12
scope of, 2–3
systems biology, 624
Biomass, 754–55
Biomass energy, 107
Biomes, 689
chaparral, 696
coniferous forests, 698
deserts, 695
freshwater, 692
global terrestrial, 693 *fig.*
oceans, 690–91
savannas, 694–95
temperate broadleaf forests, 697
temperate grasslands, 696–97
tropical forest, 694
tundra, 698
Biophilia, 767
Bioremediation, 331, 778–79
Biosafety Protocol, 248
Biosphere, 2, 2 *fig.,* 684–89
abiotic and biotic factors in, 684, 687
aquatic ecosystems of, 682–83, 690–92
autotrophs and producers in, 108
chemical cycling in, 756–59
climate and distribution of communities in, 688–89
definition of, 684
ecology as study of, 683, 684
environmental problems and limits of, 685 (*see also* Environmental problems)
as global ecosystem, 684–85
physical and chemical factors in, 686
terrestrial ecosystems of, 693–99
Biotechnology, 233
biodiversity and bioprospecting for, 767
yeasts and, 363
Biotic components of ecosystems, 684, 754. *See also* Organisms
adaptation to, 687
Bipedalism, 403, 404
skeleton and, 610 *fig.*
Bipolar disorder, 584
Birds, 391 (*See also names of individual birds*)
alimentary canal of, 432 *fig.*
blue-footed booby, 254–55, 266–67, 283 *fig.*
bones of, 291, 414 *fig.*
breathing in, 460 *fig.*
cardiovascular system, 470, 470 *fig.*
chemical pesticides and, 769
courtship and mating behaviors of, 282, 716, 716 *fig.*
digestion in, 432
edge-adapted species of, 774

Birds (continued)
  endangered and threatened, 773 *fig.*
  endangered species of, 772 *fig.*
  evolution of, 291
  excretion in, 510 *fig.*
  extinct, 296–97 *fig.*, 391 *fig.*
  finches, 264, 285, 288 *fig.*, 289
  flight in, 607, 607 *fig.*
  foraging behavior of, 714, 714 *fig.*
  habituation in, 707
  in high altitudes, 452
  lungs of, 460 *fig.*
  migration and navigation in, 711
  osmoregulation in embryos of, 509
  phylogenetic tree of, 307, 307 *fig.*
  problem-solving behavior in, 713, 713 *fig.*
  ptarmigan, 425, 425 *fig.*
  respiration and gas exchange in, 460
  respiratory system of, 458
  songbird species, 280, 280 *fig.*, 708
  song sparrow populations, 732 *fig.*, 733, 733 *fig.*
  territorial behavior of, 717 *fig.*
  thermoregulation in, 506, 507
  wing formation in, 555 *fig.*
  wing structure and function in, 414 *fig.*
Bird song, imprinting and learning of, 708
Birth control pills, 547
Birth defects, 148
Birth rates, human population growth and, 738–39
Births, multiples, 534–35
Bishop, J. Michael, 223
Bivalves, 376, 377 *fig.*
Blastocoel, 550, 550 *fig.*
Blastocyst(s), 556
  embryonic stem cells in, 218
  implantation of, 556 *fig.*
Blastopore, 551 *fig.*
Blastula, 368, 550, 550 *fig.*, 551 *fig.*
Blind spot, 595, 595 *fig.*
Blood, 469, 479–81. *See also* Red blood cells; White blood cells
  ABO groups, 167, 270
  buffers, 462–63
  clotting, 480–81, 480 *fig.*, 487
  composition of, 479 *fig.*
  as connective tissue, 417, 417 *fig.*
  diffusion between tissues and, 468 *fig.*
  distribution of, controlled by smooth muscle, 477
  exchanges between interstitial fluid and, 478, 478 *fig.*
  fetal, 463
  flow path of, through cardiovascular system, 471 *fig.*
  flow path of, through heart, 471 *fig.*
  gravity and circulation of, 466–67
  hemoglobin in (*see* Hemoglobin)
  plasma, 479
  platelets, 479
  stem cells and diseases of, 481
  transport of respiratory gases by, 462–63
  velocity and pressure of, 475–77, 475 *fig.*
Blood-brain barrier, 575
Blood fluke (*Schistosoma*), 374
Blood pressure, 475, 475 *fig.*
  measuring, 476–77, 476 *fig.*
Blood tests for genetic disorders, 165, 170
Blood vessels. *See also* Capillaries
  blood pressure and velocity in, 475–77, 475 *fig.*
  cardiovascular system and, 469, 470 *fig.*, 471 *fig.*, 472
  control of blood flow through, 477
  structure and function of, 472

Blue-footed booby, 254–55, 283 *fig.*
  gene pool and microevolution in, 266–67
Body. *See* Animal body; Plant body
Body cavity, of animals, 370
  cnidarian, 373
  flatworm, 374
  nematodes, 375
Body segmentation, 378–79
Body temperature
  regulation of, 426, 426 *fig.*
  thermoreceptors and sensation of, 592
Bone(s), 611–13
  bird, 291, 414 *fig.*
  breaks and healing of, 612
  as connective tissue, 417, 417 *fig.*
  structure of arm, 611 *fig.*
Bone marrow, 481, 611
  cell replacement in, 136 *fig.*
Boom-and-bust population growth cycles, 733, 733 *fig.*
Borrelia burgdorferi, 328, 329
Bottleneck effect, 268, 268 *fig.*
Botulism, 329
Bowman's capsule, 512 *fig.*, 513, 514, 514 *fig.*
Boysen-Jensen, Peter, 664, 665 *fig.*
Brain, 575
  cerebral cortex of, 580
  development of vertebrate, 578, 578 *fig.*
  evolution of, 403, 404
  fluid-filled spaces of CNS and, 575 *fig.*
  human, 578–85
  lateralization, 580
  regulation of breathing by, 461
Brainstem, 578
Brake fern, 644–45
Brassinosteroids, 666
BRCA1 and *BRCA2* genes, 226
Bread mold, black (*Rhizopus stolonifer*), 359, 359 *fig.*
  life cycle, 360
Bread mold, *Neurospora crassa*, 190
Breast cancer, 135 *fig.*
  genetic testing for defective gene and, 170
  (M.) King on mutations responsible for, 226
Breathing, 454, 460
  automatic control of, 461
  in birds, 460 *fig.*
  cellular respiration and, 90, 90 *fig.*
  in humans, 460 *fig.*, 461, 461 *fig.*
  oxygen supplied and carbon dioxide removed by, 462–63
Breathing control centers, 461, 461 *fig.*
Bronchi, 458
Bronchioles, 458
Brown algae, 336, 336 *fig.*
Bryophytes, 346
Bubonic plague, 322, 330
Budding, 136, 536
Buffers, 27
  blood, 462–63
Buffon, Georges, 256
Bulbourethral glands, 540
Bulk feeders, 430, 430 *fig.*
Butterflies, 383
  global warming and, 770
  monarch, 711

C
C₃ plants, 118
C₄ plants, 118
Caenorhabditis elegans (nematodes), 246, 375
Caffeinated beverages, dehydration and, 509
Caffeine, 573
Calcitonin, 526–27

Calcium
  homeostasis of, 526–27, 527 *fig.*
  muscle contraction, motor neurons, and, 616
  as plant nutrient, 652
  storage of, in smooth ER, 59
California mouse, behavior in, 706–7
Calvin, Melvin, 111
Calvin cycle of photosynthesis, 111, 116, 116 *fig.*, 117, 117 *fig.*
Cambrian period, animal evolution and, 369
Camouflage, 264 *fig.*, 746, 746 *fig.*
CAM plants, 118
Cancer
  breast (*see* Breast cancer)
  chromosomal alterations as cause of, 225 *fig.*
  colon, 225 *fig.*
  cytotoxic T cells and prevention of, 499
  death rates from, in United States, 227 *table*
  faulty cell-cell control system leading to, 135
  leukemia(s), 135, 481
  lung, 459
  micrograph of, 53 *fig.*
  monoclonal antibodies as treatment for, 496
  multiple genetic changes underlying development of, 225
  mutations in genes controlling cell division and, 222–23
  oncogene proteins, faulty tumor-suppressor proteins and, 224–25
  radioactive isotopes for detection of, 21
  reducing risk of, by diet and lifestyle changes, 227, 449, 449 *table*
  viruses as cause of, 222–23
Cancer cells, 135
Cancer drugs, mechanism of action in, 78
Canines. *See also* Dogs
  natural selection in, 259 *fig.*
Capillaries, 468
  control of blood flow through, 477, 477 *fig.*
  lymphatic, 488, 488 *fig.*
  in muscle tissue, 468 *fig.*
  structure/function of blood vessels and, 472
  substance transfer through walls of, 468 *fig.*, 478, 478 *fig.*
Capillary beds, 469
Capsid, 200
Capsule, 55 *fig.*
Capsule, prokaryotic cells, 55, 324, 324 *fig.*
Carbohydrates
  biosynthesis of, 103 *fig.* (*see also* Photosynthesis)
  digestion of, 438, 438 *table*
  disaccharides, 38
  as fuel for cellular respiration, 102, 102 *fig.*
  human perception of sweetness and, 38
  monosaccharides as, 37
  polysaccharides, 39
  problems of diets extremely low in, 448, 448 *table*
Carbon
  atom, 20 *fig.*, 22 *fig.*
  in human body, 18, 18 *table*, 19
  isotopes of, 20 *table*
  organic compounds and properties of, 34
Carbon-14, for radiometric dating, 299
Carbon cycle, 758, 758 *fig.*
Carbon dioxide (CO₂)
  breathing and release of, 90, 454
  exchange of oxygen and, at respiratory surfaces, 454–55

global warming and atmospheric, 119, 119 *fig.*, 770
  as plant nutrient, 646
Carbon fixation, 111, 116, 116 *fig.*
Carboniferous period, 350, 389
Carbon monoxide (CO) as disruptor of cellular respiration, 99, 99 *fig.*
Carbon skeleton, 34
  variations, 34 *fig.*
Carbonyl group, 35
Carboxyl group, 35
Carboxylic acids, 35
Carcinogens, 227
Carcinomas, 135
Cardiac cycle, 472 *fig.*
Cardiac muscle, 418, 418 *fig.*
Cardiac output, 472–73
Cardiovascular disease, 474
  blood pressure and diagnosis of, 476–77
  diet and, 449, 449 *fig.*
Cardiovascular system, 469
  gas exchange and role of, 462–63
  mammalian, 470 *fig.*, 471–78
  of vertebrates, 470
Carnivores, 430
  alimentary canal of, 441 *fig.*
  relationship of classification and phylogeny for select, 305 *fig.*
Carnivorous plants, 659, 659 *fig.*
Carotenoids, 112, 113
Carpels, 352, 636
Carriers, genetic, 161
Carrying capacity (K), 731
Cars, hybrid gas-electric, 771
Carson, Rachel, 685
Cartilage, 417, 417 *fig.*
Cartilaginous fishes, 388
Casparian strip, 647, 647 *fig.*
Cat(s)
  classification of domestic, 305 *fig.*
  X chromosome and tortoiseshell pattern, 214 *fig.*
Cation exchange, root hair uptake of nutrients by, 654, 654 *fig.*
Cecum, 440
Celera Genomics, 245
Cell(s), 2, 2 *fig.*, 50–87. *See also* Chemistry of life
  art related to looking at, 50–51
  chemical signals and, 70–71 (*see also* Hormone(s), animal; Hormone(s), plant)
  cytoskeleton and related structures, 64–65
  differentiation of (*see* Cellular differentiation)
  embryonic development and changes in shape of, 554, 554 *fig.*
  energy and, 72–75 (*see also* Energy)
  energy transformations in, 73
  enzyme functions and, 76–79
  eukaryotic, 56–57, 66 (*see also* Eukaryotic cells)
  life structure, functions and, 4
  membrane structure and function in, 79–86
  microscopy and study of, 51–53
  molecules of, 33
  organelles, 58–62, 63, 67
  origin of first, 321
  prokaryotic, 55 (*see also* Prokaryotic cell(s))
  respiration (*see* Cellular respiration)
  size of, 54
  surfaces and junctions, 66
Cell body, neuron, 567, 567 *fig.*
Cell cycle, 129
  cytokinesis in, 131 *fig.*, 132
  growth factor as signal for, 133, 134
  phases of, 129 *fig.*

Cell cycle control system
  cancer and faulty, 135, 222–23
  growth factors and function of, 134
  mechanical model for, 134 *fig.*
Cell division, 125, 127
  anchorage dependence of, 133
  cancer caused by faulty, 135, 222–23
  cell cycle and, 129
  controlling factors in, 133
  cytokinesis and, 131 *fig.*, 132
  cytokinins and plant, 668
  in eukaryotes, 128–36
  growth factors and, 134
  mitosis in, 130–31
  in prokaryotes, 127
  stages of, 130–31 *fig.*
Cell junctions, 66
Cell-mediated immunity, 490, 497
  clonal selection and, 493
  T cells and, 497, 497 *fig.*
Cell migration, embryonic development and, 551 *fig.*, 554
Cell movement
  cytoskeleton fibers and, 64 *fig.*
  flagella and cilia and, 65, 67
Cell plate, 132, 132 *fig.*
Cell theory, 52
Cell-to-cell signaling, animal embryonic development and role of, 221
Cellular differentiation
  cloning and, 219
  definition of, 212
  genetic potential of differentiated cells, 212–13
  specialized cells produced by, 212
Cellular metabolism, 56
  chemical energy in body, nutrition, and, 442
  disposal of nitrogenous wastes from, 510
  endergonic and exergonic reactions of, 74
Cellular respiration, 88–105. *See also* Gas exchange in animals
  ATP production in, 91
  basic mechanisms of energy release and storage and, 92–94, 95 *fig.*
  breathing and, 90, 90 *fig.*
  carbon cycle and, 758, 758 *fig.*
  definition of, 89
  equation for, 91 *fig.*
  as exergonic reaction, 74
  glycolysis stage of, 94, 94 *fig.*, 100
  interconnection between molecular breakdown and synthesis in, 102–3
  introduction to, 90–93
  overview of, 93 *fig.*
  oxidative phosphorylation for ATP production in, 98
  photosynthesis as origin of fuels for, 90, 90 *fig.*, 103
  poisons and disruption of, 99
  as redox reaction, 92, 110 *fig.*
Cellular slime molds, 337, 337 *fig.*
Cellulose, 39, 39 *fig.*
Cell walls, 55 *fig.*
  plant, 628, 628 *fig.*
  prokaryotic, 55
Cenozoic era, 299
Centipedes, 381
Central canal, of spinal cord, 575
Centralization of nervous system, 574, 575
Central nervous system (CNS), 566, 574, 575, 575 *fig.*
  brain and, 578–85
  fluid-filled spaces in, 575 *fig.*
  sensory input into, and response of, 590, 602
Central vacuole, 62, 62 *fig.*

Centrioles, 65
Centromere, 128, 128 *fig.*
Centrosomes, 131
Cephalization, 574, 575
Cephalopods, 376–77, 377 *fig.*
Cerebellum, 579
Cerebral cortex, 580, 580 *fig.*
Cerebral hemispheres, 579
Cerebrospinal fluid, 575
Cerebrum, 578, 579 *fig.*
  cerebral cortex, 580, 580 *fig.*
  hemispheres of, 579
Cervix, 539
Chaparral, 696, 696 *fig.*, 751
Chara (green alga), 344, 344 *fig.*
Chargaff, Erwin, 186
Chargaff's rules on base pairing, 186
Charophyceans, 344
Chase, Martha, 182–83, 184
Cheese production, 363
Cheetah (*Acinonyx jubatus*)
  low genetic variability of, 269, 269 *fig.*
  territorial behavior of, 717 *fig.*
Chelicerates, 380–81
Chemical bonds, 22, 23 *fig.*
  breaking and making of, in chemical reactions, 29 *fig.*
  covalent, 23
  double, 23
  ionic, 22–23
  polar covalent, 24
Chemical cycling in ecosystems, 754, 756–57
  alteration of, 760, 760 *fig.*
Chemical defenses
  in animals, 746, 746 *fig.*
  in plants, 676–78, 748
Chemical ecology, 16–17
Chemical elements, 18
  compounds, 19
  differences in, 20
  emergent properties of, 19 *fig.*
  in human body, 18 *table*
  as plant nutrients, 652
Chemical energy, 72
  ATP as shuttle in cells for, 75
  nutrition, diet, and, 442
Chemically-gated ion channels, 571
Chemical mutagens, 199
Chemical reactions, 29
  energy stored or released by, 74
  matter and atoms rearranged by, 29
  membranes and cellular, 79
  redox, in cellular respiration, 110
  redox, in photosynthesis, 110–11
Chemical regulation in animals, 518–33
  nature of, 520–21
  testosterone and male aggression, 518–19
  vertebrate endocrine system and, 522–25
Chemical signals, 70–71. *See also* Hormone(s)
  coordination of body functions by, 520
Chemical synapses, 571
  action of stimulants and depressants on, 573
  complex information processing possible because of, 572
  neurotransmitter activity at, 571 *fig.*
Chemiosmosis, 93, 98 *fig.*
  ATP generation during photosynthesis from, 115
  oxidative phosphorylation and, 98
Chemistry of life, 16–29. *See also* Chemical reactions
  atoms and molecules, 18–24
  carbohydrates, 37–39
  Eisner on, 16–17
  lipids, 40–41
  nucleic acids, 47

Chemistry of life *(continued)*
  organic compounds, 34–36
  Pauling's contributions to study of, 46
  proteins, 42–44
  water, 25–28
Chemoautotrophs, 326
Chemoheterotrophs, 326
Chemoreceptors, 592–93, 601
Chemotherapy, 135
Chernobyl (Ukraine) nuclear disaster, 21, 645
Chiasma (chiasmata), 142, 142 *fig.*
Chicxulub asteroid crater, 303 *fig.*
Chief cells, 436, 436 *fig.*
Childbirth, 560–61, 560 *fig.*
Chimpanzees, 402 *fig.*
  human relationship to, 402
  problem solving by, 713 *fig.*
  skulls of, 292 *fig.*
  vocalizations of, 406
China
  global warming safeguards by, 771
  population growth in, 738
Chlamydias, 328
Chlamydia trachomatis, 328
Chlamydomonas (unicellular green alga), 338, 338 *fig.*
Chlorofluorocarbons (CFCs), 120, 769
Chlorophyll
  chlorophyll *a,* 112, 113
  chlorophyll *b,* 112, 113
  photosynthesis and, 113 *fig.*
Chloroplasts, 63
  energy flow and, 86
  origins and evolution of, 332, 332 *fig.*
  photosynthesis in, 109, 112–17
  structure and location of, 109 *fig.*
Choanocytes, 372
Choking, Heimlich maneuver and, 435, 435 *fig.*
Cholesterol, 41, 80
  cardiovascular disease and, 474
  faulty membranes and excess levels of, 85
  structure of, 41 *fig.*
Chondrichthyes (cartilaginous fishes), 388
Chordates (Chordata)
  four features of, 385
  invertebrate, 385
  phylogenetic tree, 386
  vertebrate, 386–92
Chorion, 556, 557, 557 *fig.*
Chorionic villi, 557
Chorionic villus sampling (CVS), 164 *fig.*, 165
Choroid, 594, 595 *fig.*
Chory, Joanne, 676
Christmas tree worm, 379, 379 *fig.*
Chromatids, 128
Chromatin, 58, 128
Chromatium, 328 *fig.*
Chromosomal DNA, transfer of, by plasmids, 205 *fig.*
Chromosomal theory of inheritance, 171–74, 171 *fig.*
  crossing over and new allele combinations, 172–73
  gene mapping and, 174, 174 *fig.*
  linked genes and, 172
  Mendelian laws explained by, 171
  sex chromosomes and, 175–77
Chromosome(s), 58, 126, 187. *See also* Sex chromosome(s); X chromosome; Y chromosome
  alteration in number and structure of, 144–48
  bacteria (prokaryotic), 127
  daughter, 131 *fig.*

deletions, duplications, inversions, and translocations in, 148
duplication and distribution of, 128 *fig.*
eukaryotic, 128
in gametes, 137
gene regulation aided by DNA packing in eukaryotic, 213
homologous, 136–37, 137 *fig.* (see also Homologous chromosomes)
independent orientation of, in meiosis, 141
karyotype of, 144, 144 *fig.*, 145 *fig.*
mapping, 244–45
meiosis and reduction of, to haploid number, 138–39
mitosis and duplication of, in eukaryotes, 129–31
mutations in, and creation of genetic variation, 271
sex determination by number of, 175, 175 *fig.*
in somatic cells, 136–37
Chronic myelogenous leukemia (CML), 148, 148 *fig.*
Chytrids, 358–59, 358 *fig.*
Cichlid fish, aggression in, 518
Cilia (cilium), 65, 65 *fig.*
Ciliary body, 595
Ciliates, 335, 335 *fig.*
Ciprofloxacin (Cipro), 272
Circadian rhythms, 579, 673
Circulatory system, 420 *fig.*, 466–83
  adaptations in, for thermoregulation, 506
  of arthropods, 380
  blood in, 479–81 (*see also* Blood)
  body tissues and, 468
  closed, 378, 469
  gas exchange and transport of gases by, 454, 462–63
  gravity and, 466–67
  in high altitudes, 452–53
  mammalian cardiovascular system and, 471–78
  mechanisms of internal transport and, 468–70
  of mollusc, 376
  open, 378, 469
Circumcision, 541
Cirrhosis, liver, 511
Citric acid cycle
  cellular respiration and process of, 93, 96–97, 100
  details of, 97 *fig.*
  overview of, 96 *fig.*
  pyruvate groomed for, 96, 96 *fig.*
Clades, 306
Cladistics, 297, 306–7
Cladograms, 306–7, 306 *fig.*
Classes (taxonomy), 305
Class I and Class II proteins, 500
Classification
  phylogeny and, 305 *fig.*
  three-domain *versus* five-kingdom systems of, 310 *fig.*
Clay surfaces, polymer formation on, 320
Cleavage, 550
Cleavage furrows, 132, 132 *fig.*
Climate
  distribution of biological communities linked to, 688–89
  global warming of, 119, 770–71
  terrestrial biomes and variations in, 693
Cline, 270
Clinton (Bill) impeachment, 231
Clitoris, 539
Clonal selection, 492–93, 492 *fig.*

Clones and cloning
  animal, 208–9
  gene, 232–37
  nuclear transplantation and, 218, 218 *fig.*
  plant, 212, 641, 641 *fig.*, 642, 642 *fig.*
  reproductive, 208, 218
  therapeutic, 218, 219
Closed circulatory system, 378, 469. *See also* Cardiovascular system
Clostridium botulinum, 325
Club fungi, 358 *fig.*, 359, 359 *fig.*
Club mosses, 346, 347 *fig.*
Clumped population-dispersion pattern, 728–29, 728 *fig.*
Cnidarians (Cnidaria), 373
  digestion and gastrovascular cavity of, 373, 432
  habituation in, 707
  hydrostatic skeleton of, 608, 608 *fig.*
  internal transport in, 468
  nerve net in, 574, 574 *fig.*
Cnidocytes, 373, 373 *fig.*
Coal, formation of, 350
Cocaine, 573
Cocci, 323, 323 *fig.*
Coccidioidomycosis, 361
Cochlea, 598, 599 *fig.*
Codominance, genetic, 167
Codons, 191
  anticodons, 195
  conversion of, by transfer RNA, 194–95
  start, and initiation of translation, 196–97
  stop, and termination of translation, 197
  translation and transcription of, 191 *fig.*
Coelacanth, 388, 388 *fig.*
Coelom, 553
  true, 370, 370 *fig.*
Coenzymes, 77
Coevolution, 355, 748
Cofactors, 77
Cognition, 713
Cognitive maps, 711
Cohesion of water, 25
  plant transpiration and, 648–49
Coleochaete (green alga), 344, 344 *fig.*
Coleoptera Order, 383
Collagen, quaternary structure of, 44, 44 *fig.*
Collecting duct, of kidney nephron, 512 *fig.*, 513, 514–15, 514 *fig.*
Collenchyma cells, 628, 628 *fig.*
Colon, 440
Colon cancer, 225, 225 *fig.*
Colonial algae, 338, 338 *fig.*
Color blindness, 177
Columnar epithelium, simple, 416, 416 *fig.*
Commensalism, 748, 749
Communication, among animals, 720, 720 *fig.*
Community(ies), 742–54. *See also* Biodiversity; Ecology; Ecosystem(s)
  competition in, 745, 745 *fig.*
  definition of, 2, 744
  disturbance as feature of, 750
  food chains in, 742–43, 753
  food webs in, 753
  interaction of, with environment, 684 (*see also* Ecology)
  predation in, 746–48
  properties of, 744
  regional climate and distribution of, 688–89
  symbiotic relationships in, 748–49
  trophic structure and dynamics of, 752
Companion cells, 629 *fig.*
Comparative anatomy, evidence for evolution in, 262

Comparative embryology, evidence for evolution in, 262
Competition in communities, 745, 745 *fig.*
   agonistic animal behaviors and, 718, 719
Competitive exclusion principle, 745, 745 *fig.*
Competitive inhibitors of enzymes, 78
Complementary DNA (cDNA), 235
Complement system
   activation of, by humoral immune system, 495, 495 *fig.*
   innate immunity and, 486
Complete digestive tract, in nematodes, 375
Complete metamorphosis, 382
Compound(s)
   formation of, 19
   hydrophilic, 35
   hydrophobic, 40
   organic, 34–36
Compound events, rule of probability and, 160
Compound eyes, 594, 594 *fig.*
Computed tomography (CT), 422, 422 *fig.*
Concentration gradient, 81
Conduction, heat gains or losses by, 506, 506 *fig.*
Cones, pine, 350, 351 *fig.*
Cones, retina, 597, 597 *fig.*
Coniferous forests, 698
Conifers, 350
   giant sequoias, 623
   pine life cycle, 350–51, 351 *fig.*
Conjugation, bacterial, 204, 205
Conjunctiva, 595
Connective tissue, 417, 417 *fig.*
Conservation biology, 764–83
   biodiversity crisis and, 766–71
   conservation of populations and species and, 772–73
   definition of, 765
   ecosystem management and restoration, 774–81
   Myanmar tigers and, 764–65
Consumers, in communities, 3, 3 *fig.*, 752
Continental drift, 300–301, 301 *fig.*
Continental shelves, 690
Contour farming, 655, 655 *fig.*
Contraception, 547, 547 *table*
Contractile proteins, 42
Control centers, homeostasis and, 426
Controlled experiments, 11
Control sequences, DNA, 210
Convection, heat gains or losses and, 506
Convergent evolution, 304
Copulation, 537
Coral reefs, 691, 691 *fig.*
Corals, 373
Cork, 634–35, 635 *fig.*
Cork cambium, 635, 635 *fig.*
Corn
   seed germination of, 640, 640 *fig.*
   seed structure of, 638, 638 *fig.*
Cornea, 594, 595 *fig.*
Corn smut, 361, 361 *fig.*
Coronary arteries, 474 *fig.*
Coronavirus, 202, 202 *fig.*
Corpus callosum, 579
Corpus luteum, 538
Cortex, plant tissue, 630
Corticosteroids, 531
Costa Rica
   habitat restoration in, 779
   zoned reserves in, 780, 780 *fig.*
Cotton cloth, 287
Cotyledons, 625
Countercurrent exchange, 456–57

Countercurrent heat exchanger, 506–7, 507 *fig.*
Courtship rituals, 282, 716, 716 *fig.*
Covalent bonds, 23
   polar, 24
Cowbirds as edge-adapted species, 774
Coyotes, trial-and-error learning by, 712, 712 *fig.*
Cranial nerves, 575
Craniates, 386
Crawling as locomotion, 606–7
Cretaceous mass extinction, 302, 303, 390
Crick, Francis, 12, 46, 186–87, 186 *fig.*
Cri du chat syndrome, 148
Crime scene investigations, DNA technology and, 230–31
Cristae, 63
Cro-Magnon humans, 398–99
Crop (animal), 432
Crop plants
   improving protein content of, 656–57
   monoculture of, 642
Crop rotation, 658–59
Cross (genetic), 155–59
   dihybrid, 158–59
   monohybrid, 156
Cross fertilization in plants, 155, 155 *fig.*
Crossing over, 142–43, 143 *fig.*
   mapping genes using data from, 174, 174 *fig.*
   new allele combinations produced by, 172–73
Crustaceans, 381, 381 *fig.*
Cryoprotectants, 508
CT (computed tomography), 422, 422 *fig.*
Cuboidal epithelium, simple, 416, 416 *fig.*
Cultural eutrophication, 761
Culture, human, 407–8, 722
Cuticle, 345, 375, 380, 630
Cyanea kuhihewa, 124–25, 125 *fig.*
Cyanide, 78, 99, 99 *fig.*
Cyanobacteria, 314–15, 328
   chemical cycling and, 330
   chloroplasts and, 333
   as photoautotrophs, 108 *fig.*
Cycads, 350
Cystic fibrosis, 162, 218
Cytokinesis, 129, 131 *fig.*
   in animal *versus* plant cells, 132
   telophase I and, 139 *fig.*
Cytokinins, 666 *table*, 668
Cytoplasm, 56
Cytosine (C), 185
Cytoskeleton, 64–65
Cytotoxic T cells, 497, 497 *fig.*
   cancer prevention and role of, 499
   destruction of infected cell by, 499 *fig.*

D

Damselfly, 382 *fig.*
Darwin, Charles R., 8, 8 *fig.*
   evolutionary theory of, 256–61, 263
   gradualism model of speciation and, 290
   modern systematics and, 306
   sea voyage and studies of, 256, 257 *fig.*
   studies of phototropism and hormones by, 664, 665 *fig.*
Darwin, Francis, 664, 665 *fig.*
Daughter cells
   in meiosis, 139
   in mitosis, 132, 132 *fig.*
Daughter DNA, 188, 188 *fig. See also* DNA polymerases
DDT (pesticide), 2, 769, 781
Deafness, 161, 599
Death rates, human population growth and lower, 738–39

Death Valley, 285, 285 *fig.*
Declining-population approach to endangered species, 772–73
Decomposers, 3, 3 *fig.*, 363, 752
Decomposition, 752
Deductive reasoning, 9
Deep-sea submersibles, 682–83
Defenses, animal, 746, 746 *fig.*
Defenses, plant, 676–78, 748
Defensive proteins, 42. *See also* Antibody(ies)
Deforestation, 356, 694, 699, 760, 768
Dehydration reactions, 36, 36 *fig.*, 40 *fig.*
Dehydrogenase, 92
Deinonychus (dinosaur), 390, 390 *fig.*
Deletion, chromosomal, 148, 148 *fig.*
Delta proteobacteria, 328
Demographic transition, 738–39
Denaturation, 43
Dendrites, 418, 567
Density-dependent factors limiting population growth, 732
Density-dependent inhibition of cell division, 133, 133 *fig.*
Dental plaque, 326 *fig.*
Deoxyribonucleic acid, 47. *See also* DNA
Depression, 584
Derived characters, shared, 306
Dermagraft (artificial tissue), 419 *fig.*
Dermal tissue system, plants, 630
Descent with modification, principle of, 257
Desertification, 695
Deserts, 695, 695 *fig.*
Determinate growth, 632
Detritivores, 752
   nitrogen cycle and, 758 *fig.*
Detritus, 752
Deuterostomes, 370, 371, 393
Devonian period, 388
de Vries, Hugo, 286
Dexamethasone, 531
Diabetes mellitus, 529
   insulin-dependent, 500
Dialysis, kidney, 480, 515, 515 *fig.*
Diaphragm, respiratory, 458
Diastole, 472, 476
Diatoms, 336, 336 *fig.*
Dicots, 625
   embryo development in, 638, 638 *fig.*
   monocots *versus*, 625 *fig.*
   seed germination in, 640, 640 *fig.*
   tissues in, 627 *fig.*, 630, 631 *fig.*
Dictyostelium (cellular slime mold), 337, 337 *fig.*
Diet, animal, 430–33
   adaptations of digestive system reflect, 440–41
   as chemical energy, 442
   needs satisfied by healthy, 442
Diet, human
   cancer and, 449
   cardiovascular disease and, 474
   meat in, 756
   nutrition and, 442–49
   obesity and, 447
   soy in, 662–63
   vitamin and mineral supplements in, 446
Differential interference-contrast microscopy, 53, 53 *fig.*
Differentiation. *See* Cellular differentiation
Diffusion, 81 *fig.*
   between blood and tissue cells, 468 *fig.*
   facilitated, 82
   passive, 81
Digestion, 431, 431 *fig. See also* Digestive system

Digestive system, 420 *fig.*
  diet and adaptations of vertebrate, 440, 441
  exchanges with environment and role of, 424–25, 424 *fig.*
  human, 433–40, 433 *fig.*
  in nematodes, 375
  nutrition and (*see* Nutrition)
  pharynx and, 458
  specialized compartments of animal, 432, 432 *fig.*
Dihybrid cross, 158–59
Dinitrophenol (DNP) as disruptor of cellular respiration, 99, 99 *fig.*
Dinoflagellates, 335, 335 *fig.*
Dinosaurs, 291, 390
  tracks, 260 *fig.*
Dipeptides, 43
Diploid cells, 137
  plant, 348, 350–51, 351 *fig.*, 636 (*see also* Sporophytes)
Diploid organisms, balancing selection and, 272–73
Diplomonads, 334, 335 *fig.*
Diptera Order, 383
Directional selection, 274, 274 *fig.*
Disaccharides, 38, 38 *fig.*
Discovery science, 9
Disease. *See also* Human disease; *names of specific diseases*
  amoebic dysentery, 337
  bacterial, 205, 272, 322, 324, 328, 329, 437, 437 *fig.*, 487
  bioterrorism, 330
  fungal, 361
  global warming and, 771
  immune system malfunctions and, 500–501
  viral, 202–3
Dispersion patterns, population, 728–29
Disruptive selection, 274, 274 *fig.*
Distal tubule, of kidney nephron, 512 *fig.*, 513, 514, 514 *fig.*
Disturbances, in biological communities, 744, 750
DNA (deoxyribonucleic acid), 47
  bacterial plasmids and transfer of, 233–36
  bacterial transfer of, in nature, 204
  complementary, 235
  double helix structure of, 186, 187 *fig.*
  experiments demonstrating, as genetic material, 182, 183 *fig.*
  gel electrophoresis of, 239, 240–41
  gene regulation aided by packing of, in eukaryotes, 213
  genotype of, expressed as proteins, 190
  human evolution and mitochondrial, 405
  microarrays, 238–39, 239 *fig.*
  molecular structure of, 4–5
  Neanderthal, 398–99
  nitrogenous bases of, 185 *fig.*
  as nucleotide polymer, 184–85
  polynucleotide, 184 *fig.*
  repetitive, 245
  replication of, and multiple "bubbles" and daughter strands, 189, 189 *fig.*
  replication of, dependent on base pairing, 188, 188 *fig.*
  replication of, in cell cycle, 129
  restriction enzymes used to create recombinant, 233, 233 *fig.*
  transcription of, into RNA, 190, 193, 198
  viruses, 201
DNA fingerprints, 242 *fig.*
DNA ligase, 189, 233
DNA polymerases, 189
DNA probes, 238, 240–41

DNA sequence(s)
  amplifying, by PCR method, 244
  analysis of, in systematics, 308–9
  analysis of, using restriction fragments, 240–41
  enhancer, 214, 215 *fig.*
  mutations in, 199
  operator, 210, 211 *fig.*
  promoter, 193, 210, 211 *fig.*
  silencer, 214, 215 *fig.*
DNA sequencing
  in Human Genome Project, 244–45
  by restriction fragment analysis, 240–41
DNA strands
  opposite orientation of, 189 *fig.*
  sticky ends on, 233
DNA technology, 230–51
  in agriculture, 247–48
  bacteria as tool for manipulating DNA, 232–37
  DNA microarrays test, 238–39, 239 *fig.*
  gel electrophoresis, 239
  gene cloning, 232–37
  gene therapy and, 243
  genomics and, 231
  human genome and, 231
  legal and forensic applications of, 242
  mass production of gene products using, 236
  medical uses of, 237, 243
  nucleic acid probes, 238
  PCR technique and DNA sequence amplification, 244
  pharmaceuticals and, 237
  restriction fragment analysis, 240
  reverse transcriptase and gene cloning, 235
  risks of, 248
Dodd, Diana, 285
Dogs. *See also* Canines
  artificial and natural selection in, 152–53
  independent assortment of two genes in, 159 *fig.*
  testcross to determine genotype of, 159 *fig.*
Doldrums, 688
Dolly (cloned sheep), 208, 209, 218
Domains (classification), 6, 305
Dominance hierarchies, 718, 719
Dominant alleles, 156, 161
Dominant species of communities, 744
Dominant traits
  codominance, 167
  genetic disorders as, 162–63, 163 *table*
  incomplete dominance, 166–67
Donkey, hybrid sterility in cross between horse and, 283 *fig.*
Dormancy, in bears, 504–5
Dorsal, hollow nerve cord, 385, 385 *fig.*
Dorsal surface of animals, 370, 370 *fig.*
Double bonds, 23
Double circulation, 470
Double fertilization, 637, 637 *fig.*
Double helix, 47
  DNA structure, 47 *fig.*, 186, 187 *fig.*
Down, John Langdon, 145
Down syndrome, 145
Drosophila melanogaster (fruit fly), 246
  crossing over and recombinant gametes of, 173 *fig.*
  evolution of reproductive barriers among, 285, 285 *fig.*
  genetic control of development in, 220–21
  head-tail polarity, 220 *fig.*
  heterozygosity of, 270
  mapping genes of, 174 *fig.*
  sex-linked traits in, 176 *fig.*

Drugs. *See also* Pharmaceuticals
  effects of, on nervous system, 573
  hypothalamus and addiction to, 579
  placenta-crossing, 557
  tolerance to, 59
Duchenne muscular dystrophy, 177
Duodenum, 438
Duplication, chromosomal, 148, 148 *fig.*
Dwarfism, pituitary, 524–25
Dynein arms, 65
Dysentery, amoebic, 337

E

E. coli. See Escherichia coli
Ear(s)
  balance and inner, 600
  hearing and, 598–99
  route of sound through, 599 *fig.*
  structure of, 598 *fig.*, 599 *fig.*
Eardrum, 598
Earth
  atmosphere of (*see* Atmosphere)
  biodiversity hotspots on, 775, 775 *fig.*
  biosphere of (*see* Biosphere)
  climate and distribution of biological communities on, 688–89
  continental drift and plate tectonics of, 300–302
  crustal plates of, 300 *fig.*
  fossil record and macroevolution on, 298–303
  major episodes in early history of, 317 *fig.*
  origin of life on, 316–17
Earthworms, 370, 370 *fig.*, 378–79
  alimentary canal of, 432 *fig.*
  body segmentation in, 378 *fig.*
  hydrostatic skeleton of, 608
  peristalsis locomotion in, 607, 607 *fig.*
  reproduction in, 537 *fig.*
Eastern coral snake, 11 *fig.*
Ebola virus, 202, 202 *fig.*
Ecdysis, 393
Ecdysozoans, 393
Echidnas (spiny anteaters), 392
Echinoderms (Echinodermata), 384, 574. See *also* Sea star
Ecological footprint, human, 736
  carrying capacity and, 737 *fig.*
Ecological Society of America, 781
Ecological species concepts, 281
Ecological succession, 750
Ecology, 683. *See also* Community(ies); Conservation biology; Population dynamics
  behavioral, 704, 711, 714–15
  biosphere and study of, 684
  fungus and, 363
  restoration, 778–79
  species' niches, 745
Ecosystem(s), 2, 2 *fig.*, 682
  alteration of, 760–61
  aquatic, 682–83, 690–92, 761 (*see also* Aquatic ecosystems)
  biological augmentation of, 778–79
  bioremediation of, 778
  biosphere as total of all, 684–85
  carbon cycle in, 758
  carrying capacity of, 731
  chemical cycling in, 756–59, 760
  definition of, 754
  diversity in, 766
  ecology and study of, 683, 684 (*see also* Ecology)
  economic value for services of, 767
  energy flow and chemical cycling in, 754
  energy supply and food chains in, 755
  fire and, 693, 751

INDEX

human meat-eating and production pyramid in, 756
interactions in, 3, 3 *fig.*
landscape ecology and, 774
managing and restoring, 774–81
net primary production of various, 755 *fig.*
nitrogen cycle in, 758–59
phosphorus cycle in, 759
primary production as energy budget for, 754–55
protection of biodiversity hotspots, 775, 775 *fig.*
structure and dynamics of, 754–59
terrestrial biomes, 693–99
water cycling in, 757
Yellowstone to Yukon Conservation Initiative (Y2Y), 776–77, 776 *fig.*
zoned reserves, 780, 780 *fig.*
Ectoderm, 368, 370, 370 *fig.*, 550
neural tube formation from, 552 *fig.*
organ and tissue derivatives of, 553 *table*
Ectopic pregnancy, 538–39
Ectothermic animals, 390, 504
thermoregulation in, 506–7
Edges of landscapes, 774, 774 *fig.*
EEG (electroencephalogram), 582, 582 *fig.*
Effector cells, 493
Effector mechanisms, humoral immunity, 495
Effectors, nervous system, 566
Egg(s) (ova/ovum), 536. *See also* Gametes
amniotic, 390, 391
fertilization of animal, 537
human, surrounded by sperm, 548 *fig.*
plant, 637 *fig.*
Egyptian farmer, 408
Eisner, Thomas, 16–17
Ejaculation, 540–41
two stages of male, 541 *fig.*
Ejaculatory duct, 540
Eldredge, Niles, 290
Electrical synapses, 571
Electrocardiogram, heart rhythm and, 473, 473 *fig.*
Electroencephalogram (EEG), 582, 582 *fig.*
Electromagnetic energy, 112
Electromagnetic receptors, 593
Electron(s), 20
cellular respiration and energy release from, 92
chemical properties of atoms determined by arrangement of, 22, 22 *fig.*
Electronegativity of atoms, 24
Electron microscopes (EMs), 52
Electron shells, 22, 22 *fig.*
Electron transport chain
in cellular respiration, 92, 93, 98, 98 *fig.*, 100
disruption of, by poison, 99
in photosynthesis, 114, 114 *fig.*
Elements, chemical. *See* Chemical elements
Elephants, 604, 605
Elimination, 431
Embryo(s), 538
development of (*see* Embryonic development in animals)
human, 556–57
plant (*see* Sporophytes)
Embryology, comparative, and evolution, 262
Embryonic development in animals, 548–55
brain, 578, 578 *fig.*
cell shape, cell migration, and programmed cell death in, 554
cleavage, 550
fertilization, zygote formation, and initiation of, 548–49
gastrulation, 550–51, 551 *fig.*, 556

genetic control of, 220–22
human, 556–61
induction and, 554–55, 554 *fig.*
organ formation, 552–55, 552–55 *fig.*
pattern formation in, 555
Embryonic development in plants, 638, 638 *fig.*, 639
Embryonic stem cells (ES cells), 218, 565
Embryophytes, 345
Embryo sac, plant, 637, 637 *fig.*
Emergent properties, 4
Emerging viruses, 202–3
Emotions, limbic system and, 583
Emphysema, 459
Encephalitis, 202
Endangered species
biodiversity crisis and, 764–65
cheetahs, 269
reduced genetic variation in, 269
study approaches for, 772–73
U.S. Endangered Species Act's definition of, 766
whales as, 428
Endemic species, 775
Endergonic reactions, 74, 74 *fig.*
Endocrine glands, 520
adrenal, 530–31
disorders of, 529
glucocorticoids and, 531
gonads, 532
human, 522 *fig.*
hypothalamus and pituitary, 524–25
major, and hormones of, 523 *table*
pancreas, 528–29
pineal, 522
thymus, 522
thyroid and parathyroid, 526–27
Endocrine system, 421 *fig.*, 520
hormones of, 520 *fig.*, 523 *table*
nervous system and, 520
vertebrate, 522–25
Endocytosis, 84–85, 84 *fig.*, 85 *fig.*
Endoderm, 368, 370, 370 *fig.*, 550
organ and tissue derivatives of, 553 *table*
Endodermis, 630
Endomembrane system, 58
organelles of, 58–62
Endometrium, 538
Endoplasmic reticulum (ER), 58–59
Endorphins, 525
Endoskeletons, 609
bones of, 611–13
echinoderm, 384
human, 610–11
Endosperm, 637, 638
Endospore, 325, 325 *fig.*
Endosymbiosis, 332–33, 332 *fig.*
Endothermic animals, 390, 391, 392, 504
thermoregulation in, 506–7
Endotoxins, 329
Energy. *See also* Solar energy
ATP as shuttle in cells for chemical, 75
cells and, 72–75
cellular harvesting of (*see* Cellular respiration)
chemical reactions and release or storage of, 74
definition of, 72
electromagnetic, 112
enzymes and, 76–78
forms of, 72, 72 *fig.*
human activities and consumption of, 91 *table*
laws on conversion of, 73
organelles and conversion of, 63
Energy conservation, laws on, 73
Energy coupling, 74

Energy flow in ecosystems, 754
Energy of activation (E_A), 76
Energy supply in ecosystems, 755
Energy utilization as property of all organisms, 5
Enhancer DNA sequence, 214
Enteric division, autonomic nervous system, 577
Entomology, 382–83
Entropy, 73
Environment. *See also* Biosphere; Ecology
abiotic factors in, 684
adaptations to changes in (*see* Behavior)
animal behavior resulting from genes and, 706–7
animal exchanges with external, 424–25
animal regulation of internal, 425–26 (*see also* Homeostasis; Regulatory systems, animal)
biotic factors in, 684
genetic influences of, 170
human culture and, 766–67
response to, as property of all organisms, 5, 684
risks of genetically modified organisms to, 248
Environmental problems. *See also* Biodiversity crisis
acid precipitation, 28, 761
deforestation, 694, 699, 768
desertification, 695
ecosystem alteration, 760–61
greenhouse effect and global warming, 119
human overpopulation and, 736–37
introduced (exotic) species as, 394, 726–27, 768
nitrogen cycle balance, 758
oil spills, 331, 331 *fig.*
ozone layer depletion, 120, 120 *fig.*, 769
technology and, 12
Enzyme(s), 76–78
as catalyst, 77, 77 *fig.*
chemical reactions controlled by, 71
cofactors of, 77
definition of, 42, 76
digestion in small intestine and, 438–39, 438 *table*
digestion in stomach and, 436, 436 *fig.*
energy and, 76–78
function of, 76, 77 *fig.*
inhibitors of, 78, 78 *fig.*
milk-digesting, 32 *fig.*
pesticides and antibiotics effects on, 78
restriction, 233
temperature, salt, and pH effects on, 77
Eosinophils, 479, 479 *fig.*
Epidermis
animal (*see* Skin)
plant, 630
Epididymis, 540
Epinephrine, 473, 501, 519, 520, 530
Epithelial tissue (epithelium), 416, 416 *fig.*
Equilibrium, sense of, 600
Ergots, 361, 361 *fig.*
Erythrocytes, 479, 479 *fig. See also* Red blood cells
Erythropoietin (EPO), 480
Escherichia coli (E. coli)
chromosome, 127
of gamma proteobacteria group, 328
infection of, by virus, 182–83, 200
nutrition, 326, 326 *fig.*
pathogenic, 329
recombinant DNA technology and genetics research in, 233
regulation of gene transcription in, 210–11

Esophagus, 432
  human, 435, 435 *fig.*
Essential amino acids, 443, 443 *fig.*
Essential elements, 652
Essential fatty acids, 443
Essential minerals, 445, 445 *table*
Essential nutrients, 443
Estivation, 508
Estrogen(s), 521, 532
  chemical structure of, 662 *fig.*
  childbirth and, 560 *fig.*
  osteoporosis and, 612–13
  ovarian and menstrual cycles and, 544–45, 545 *fig.*
Estuaries, 690, 691 *fig.*
Ethanol as product of alcohol fermentation, 101, 101 *fig.*
Ethics
  AIDS babies study by U.S. and, 484–85
  fetal imaging or testing and, 165
  gene therapy and, 243
  human reproductive cloning and, 218–19
  reproductive technology and, 561
Ethylene, 666 *table*, 670–71
  fruit ripening and, 670 *fig.*
  leaf fall and, 670 *fig.*, 671
Euglena spp., 52–53 *fig.*, 335, 335 *fig.*
Euglenozoans, 334–35, 335 *fig.*
Eukarya (domain), 6, 7 *fig.*, 310
Eukaryotes, 6, 310
  balancing selection in diploid, 272–73
  fungi, 357–63
  gene regulation in, 213–17, 263
  plants, 342–56
  tentative phylogeny of, 334, 334 *fig.*
Eukaryotic cells, 4, 4 *fig.*, 55, 55 *fig.*
  cell cycle and cell division of, 128–36
  cilia and flagella of, 65 *fig.*
  cytoskeleton and, 64–65
  energy-converting organelles in, 63
  flow of genetic information in, 190 *fig.*, 198
  functional categories of organelles in, 67 *table*
  gene regulation in, 213–17
  messenger RNA in, 194
  mitosis, cytokinesis, and chromosome duplication in, 129–31
  organelles as functional compartments in, 56–57
  organelles of endomembrane system in, 58–62
  origins of, as prokaryotic community, 332–33
  surfaces and junctions of, 66
Eumetazoans, 371, 393
Eustachian tubes, 598
Eutherians, 392, 392 *fig.*
Eutrophication in freshwater ecosystems, 761
Evaporative cooling, 25, 25 *fig.*, 506, 507
"Evo-devo" research field, 292
Evolution, 8–9, 253–312
  of angiosperm plants, 355
  of animal diversity, 366–67, 368–71, 393–94
  of birds, 291, 296–97
  of cardiovascular system, 470
  convergent, 304
  Darwin's theory of, 256–61
  definition of, 255
  domains of life and, 310
  evidence for, 260–64
  of fungi, 342–43, 357–63
  genetic variation, natural selection, and, 270–75
  of genomes, 309

of horses, 293 *fig.*
  human, 398–410
  of invertebrate animals, 372–85
  life history of organisms and, 734–35
  macroevolution and, 291–93
  macroevolution and Earth history, 298–303
  of mammals, 392
  microevolution (in gene pools), 266–69
  modern synthesis of Darwin's theory of, 265–69
  of multicellular organisms, 339, 339 *fig.*
  natural selection as mechanism of, 264 (*see also* Natural selection)
  origins of life and, 316–19
  phylogenetic trees as symbol of, 305, 305 *fig.*
  of plants, 344–45, 346 *fig.*, 347
  population as units of, 265
  of prokaryotes (bacteria), 316–17, 322–31
  as property of all organisms, 5
  of protists, 332–33, 334, 334 *fig.*
  of reproductive barriers, 285
  sociobiology and, 717
  speciation and, 278–79, 284–90
  species concepts and, 280–83
  systematics, phylogenetic biology, and, 304–10
  variation and natural selection, 270–75
  of vertebrate animals, 386–92
Evolutionary adaptations, 8, 255. *See also* Mass extinctions
  natural selection and, 264, 264 *fig.*
Evolutionary fitness, 273
Evolutionary history. *See* Phylogeny
Exaptation, 291
Excitement phase, of sexual response, 546
Excretion, 505
  as excretory system function, 513
  homeostasis and, 512–13
  liver's role in, 511
  nitrogenous waste disposal via, 510
  osmoregulation and, 508–15
Excretory system, 420 *fig.*, 512 *fig.*
  exchanges with external environment and role of, 424 *fig.*, 425
  homeostasis and, 512–13
  key processes of, 512–13, 513 *fig.*
  kidney function in, 514–15, 514 *fig.*
Exercise
  aerobic and anaerobic, 617
  cardiovascular disease and, 474
  energy consumed by various activities of, 91 *table*
  food equivalents of, 442 *table*
  genetic influences of, 170
  water loss in, 509
Exergonic reactions, 74, 74 *fig.*
Exocytosis, 84–85, 84 *fig.*
Exons, 194
Exoskeletons, 380, 608–9, 608 *fig.*
Exotic species, 394
  as cause of biodiversity crisis, 768
  dispersion of European starling, 726–27
Exotoxins, 329
Exponential growth model of population growth, 730
  logistic growth *versus*, 731 *fig.*
External fertilization, 537
Extinct species, 296–97 *fig.*, 391 *fig.*
Extracellular matrix, 66, 66 *fig.*
Extraembryonic membranes, 556–57, 557 *fig.*
Extreme halophiles, 327
Extreme thermophiles, 327
Eye(s)
  corrective lens and surgery for, 596
  induction during development of, 554–55, 554 *fig.*

  photoreceptor cells in vertebrate, 597, 597 *fig.*
  types of invertebrate, 594
  vertebrate single-lens, 594–95
Eye cups, 594, 594 *fig.*

F

F_1 generation, 155
F_2 generation, 155
Facilitated diffusion, 82
Facultative anaerobes, 101
Fad diets, 448
FAD, cellular respiration and, 96-99 *fig.*, 100
$FADH_2$ (flavin adenine dinucleotide), cellular respiration and, 93, 96, 97 *fig.*
Fallopian tubes, 538
Family(ies) (taxonomy), 305
Family size, human, 739
FAP (familial adenomatous polyposis), 170
Farming
  contour, 655, 655 *fig.*
  organic, 656
Farsightedness, 596, 596 *fig.*
Fat, body, 447
Fats, 40. *See also* Lipids
  biosynthesis of, 103 *fig.*
  digestion of, 438 *table*, 439
  as fuel for cellular respiration, 102, 102 *fig.*
  problems of diets extremely low in, 448, 448 *table*
Fat-soluble vitamins, 444 *table*
Fatty acids, 40
  essential, 443
Feathers, 391
  structure of, 414 *fig.*
Feces, 440
Feedback inhibition, 78. *See also* Negative feedback
Feeding
  animal behaviors associated with, 714–15
  animal ingestion for, 368, 368 *fig.*, 372, 430
  humpback whales, 428–29
  sponge, 372 *fig.*
Females, human
  childbirth in, 560–61
  contraception in, 547, 547 *table*
  hormonal regulation of reproductive cycle in, 544–45, 545 *fig.*
  ovulation in, 538
  pregnancy in, 538–39
  reproductive anatomy of, 538–39, 538 *fig.*, 539 *fig.*
  sexual response in, 539, 546
Fermentation, 101
Ferns, 346, 347 *fig.*, 350
  life cycle of, 349 *fig.*
Fertile Crescent, 408
Fertility, 534–35
Fertility drugs, 534–35
Fertilization, 124, 137
  after nondisjunction in the mother, 146 *fig.*
  animal, 548–49, 549 *fig.*
  external *versus* internal, 537
  genetically modified plants and, 655
  plant, 352–53, 353 *fig.*, 637 *fig.*
  random, as cause of genetic variety, 141, 171
Fertilization envelope, 549
Fertilizers
  inorganic, 685
  organic, 656
Fetus, human, 538
  development of, 558–59, 558–59 *fig.*
  diagnosis of genetic disorders by testing of, 164–65, 164 *fig.*, 165 *fig.*
  gas exchange in, 463

placental and embryonic development of, 556–57, 556–57 *fig.*
F factor, transfer of chromosomal DNA by, 205, 205 *fig.*
Fibers (sclerenchyma cells), 628, 628 *fig.*, 629 *fig.*
Fibrin, 480 *fig.*, 481
Fibrinogen, 480–81
Fibroblasts, human-engineered tissue from, 419
Fibrous connective tissue, 417, 417 *fig.*
Filtrate, 512, 514–15, 514 *fig.*
Filtration function of excretory system, 513, 513 *fig.*
Fire, ecosystem changes and, 693, 751
Fireflies, 70 *fig.*, 71 *fig.*
    chemical signals between, 70–71
First law of thermodynamics, 73
Fishes, 388
    cardiovascular system, 470, 470 *fig.*
    cartilaginous, 388
    circulatory system, 469, 469 *fig.*
    damage caused by exotic species of, 768
    foraging behavior of, 714, 714 *fig.*
    gills of, 455 *fig.*, 456–57, 456 *fig.*
    locomotion in, 606 *fig.*
    osmoregulation in saltwater and freshwater, 508, 508 *fig.*
    population collapse, 735 *fig.*
    salmon sensing and migration, 588–89, 708
Fission, 536
Fitness, evolutionary, 273
Five-kingdom classification scheme, 310, 310 *fig.*
Fixed-action patterns (FAPs), 705, 705 *fig.*
Flagella, 65
    cell movement by eukaryotic, 65 *fig.*
    fungal, 358–59
    prokaryotic, 55 *fig.*, 324–25, 324 *fig.*
Flatworms, 370, 370 *fig.*, 374
    digestion and gastrovascular cavity of, 432
    nervous system of, 574, 574 *fig.*
Flight, 391
    feathers and, 414
    as form of locomotion, 607, 607 *fig.*
    wing structure/function and, 414 *fig.*
Flower(s), 352
    attractiveness of, to animals, 355
    correspondence between fruit and, 639 *fig.*
    parts of, 352 *fig.*
    structure of, 636 *fig.*
Flowering, photoperiodic control of, 674 *fig.*
Flowering plants, 347. *See also* Angiosperms
Fluid feeders, 430, 430 *fig.*
Fluid mosaic, membranes as, 80
Flukes, 374
Fluorescence and confocal microscopy, 53, 53 *fig.*
Fluoride, 19
Flying. *See* Flight
Follicles, ovarian, 538, 543, 543 *fig.*
Follicle-stimulating hormone (FSH), 524, 532, 544, 545, 545 *fig.*
Food
    artificial chemicals in production of, 671
    cancer risk reduction by changes in, 227
    digestion of (*see* Digestive system)
    exercise equivalent of, 442 *table*
    four stages of food processing, 431, 431 *fig.*
    as fuel for chemical energy in body, 102–3
    fungi and production of, 363
    harvesting of chemical energy from, 92 (*see also* Cellular respiration)
    meat as, 756
    obtaining and processing, in animals, 430–33 (*see also* Feeding)
    overpopulation and shortages of, 737

photosynthesis and production of, 117
production of, using lactic acid fermentation, 101
soy products as, 662–63
Food chain(s), 752
    biological magnification of pesticides in, 769 *fig.*
    energy supply and limits to, 755
    insect, 742–43
    interconnecting, food webs' formation and, 753
    terrestrial, and aquatic, 752 *fig.*
Food-conducting cells, 629, 629 *fig.*
Food labels, 446, 446 *fig.*
Food plants
    improving protein content of, 656–57
    polyploid plants as, 287
Food webs
    biological magnification of PCBs in, 769, 769 *fig.*
    formation of, 753, 753 *fig.*
Foolish seedling disease, 668
Foot, mollusc, 376
Foraging, 714
Forebrain, 578, 579, 579 *fig.*
Forensic science
    DNA technology applied to, 231, 242
    restriction fragments analysis to detect differences in DNA sequences for, 240
Forest(s). *See also* Tree(s); Tropical rain forests
    acid precipitation effects on, 28 *fig.*
    chemical cycling and alterations in, 760
    coniferous (taiga), 698
    deforestation of, 356, 694, 699, 760, 768
    fires as ecological force in, 751
    medicinal plants in tropical rain, 678
    temperate broadleaf, 697
Formula diets, 448 *table*
Fortified foods, 19, 19 *fig.*
Fossil(s), 256. *See also* Fossil record
    age of rocks and dating of, 299
    as evidence for evolution, 260–61
    human, 404
    macroevolution chronicled in, 298–99
    prokaryote, 317 *fig.*
Fossil fuels, 350
    global warming and, 119, 770
Fossil record, 261
    of eukaryotes, 332
    evidence of evolution in, 290
    evolutionary divergence chronicled in, 308
    evolutionary temporal patterns in, 290
    macroevolution chronicled in, 298–99
    mass extinctions recorded in, 302–3
    of mycorrhizal fungus, 342
    of prokaryotes, 322
Founder effect, 268
Fovea, 595, 595 *fig.*
Fox, Michael J., 585 *fig.*
Fox, Sidney, 320
FOXP2 gene, speech development and, 406
Fragmentation, asexual reproduction by, 536, 641
Franklin, Rosalind, 186, 186 *fig.*, 187
Free-living flatworms, 374, 374 *fig.*
Frequency-dependent selection, 273
Freshwater biomes, 692
    eutrophication in, 761
Frogs, 389
    embryonic development in, 552–53, 552 *fig.*, 553 *fig.*
    endoskeleton of, 609, 609 *fig.*
    gastrulation in, 550–51, 551 *fig.*
    metamorphosis in, 389 *fig.*
    reproduction in, 537, 537 *fig.*
    thermoregulation in, 507, 507 *fig.*

Fructose, 37, 37 *fig.*
Fruit, 354
    auxins and development of, 667, 671
    correspondence between flower and, 639 *fig.*
    development of, 639 *fig.*
    ethylene and ripening of, 670 *fig.*
    gibberellins and development of, 669
    seed dispersal and, 354 *fig.*
    simple, multiple, and aggregate, 639
Fruit flies. *See Drosophila melanogaster*
FSH (follicle-stimulating hormone), 524, 532, 544, 545, 545 *fig.*
Functional groups, 35
    common, 35 *table*
    differences in sex hormones, 35 *fig.*
Fungi, 357–63
    animals and, 362–63, 362 *fig.*
    colonization of land by plants and, 357
    as detritivores, 752
    ecological and practical impacts of, 363
    feeding and nutrition of, 357
    humans and, 309 *fig.*
    lichens and, 362
    life cycle phases of, 358 *fig.*
    mycorrhiza, 342–43, 357, 657
    parasitic, 361
    plant disease caused by, 361, 668 *fig.*
    yeast, 101, 357
Fungi (kingdom), 7, 7 *fig.*, 310, 357

G
G3P (glyceraldehyde 3-phosphate)
    as intermediate in glycolysis, 102
    photosynthesis and, 116
Gage, Phineas, brain injury to, 581, 581 *fig.*
Galápagos Islands, 302
    Darwin's studies of life on, 254–55
    finches of, 264, 285, 288 *fig.*, 289
Gallbladder, 438
Galls, 361
Gametangia, 345
Gametes, 137. *See also* Egg(s) (ova/ovum)
    animal, 536, 542–43
    crossing over and production of recombinant, 143, 172, 173 *fig.*
    formation of, by meiosis, 542–43
    genetic variation in (*see* Genetic variation)
    plant, 348, 636–37, 637 *fig.*
Gametic isolation, 283
Gametogenesis, 542
Gametophytes, 338, 348
    in angiosperm, 352–53, 636–37, 637 *fig.*
    moss, 348 *fig.*, 349
    pine, 350–51, 351 *fig.*
Gamma proteobacteria, 328
Ganglia, 566
Gap junctions, 66, 66 *fig.*
Garrod, Archibald, 190
Gas exchange in animals, 420 *fig. See also* Respiration
    fetal, 463
    mechanisms of, 454–61
    three phases of, 454 *fig.*
    transport of gases in the body and, 462–63
Gastric glands, 436, 436 *fig.*
Gastric juice, 436, 436 *fig.*
Gastric ulcers, 437
Gastrin, 436
Gastroesophageal reflux disease (GERD), 437
Gastropods, 376, 377 *fig.*
Gastrovascular cavity, 432
    cnidarian, 373
    flatworm, 374
    internal transport provided by, 468, 469 *fig.*
Gastrula, 368, 550

Gastrulation, 550
  embryonic tissues formed in, 550–51, 551
    fig., 553 *table*, 556
  organ formation after, 552–53
Gause, G. F., 745
Gecko(s), 413 *fig.*
  locomotion in, 412–13
Gel electrophoresis
  analyzing restriction fragments by,
    240 *fig.*
  sorting DNA molecules by size with, 239
Gene(s), 187
  alleles of, 167 (*see also* Alleles)
  behavior resulting from environment and,
    706–7
  BRCA1 and *BRCA2* genes, 226
  cancer-causing (oncogene), 223–25
  cascades of expressed, in animal
    embryonic development, 220–21
  cloning of (*see* Gene cloning)
  controls on expression of (*see* Gene
    regulation)
  developmental, and evolution, 292
  DNA as genetic material in, 47, 182–83
    (*See also* DNA)
  DNA technology and mass-produced
    products of, 236
  evolutionary fitness and perpetuation of,
    273
  expression of, in proteins (*see* Gene
    expression)
  homeotic, 221, 222 *fig.*
  Hox, 368
  hypothesis for first, 320, 320 *fig.*
  "jumping" (transposons), 245
  linked, 172, 176
  locus, 137
  mapping, 174, 174 *fig.*, 244–45
  Mendelian laws applied to, 152–70
  mutations in, 199, 222–23, 223 *fig.*
  p53 tumor-suppressor gene, 225
  pleiotropy as influence of single, 168,
    168 *fig.*
  polygenic inheritance and influence of
    many, 169
  ras proto-oncogene, 224–25
  sex-linked, 175–77
  transcription of, 193 *fig.*
  transfer of, by plasmids, 205
  tumor-suppressor, 223
Gene clones
  genomic libraries of, 234, 235, 235 *fig.*
  identifying, with nucleic acid probes, 238
Gene cloning
  bacterial plasmids and, 232–37
  overview of, 232 *fig.*
  in recombinant plasmids, 234, 234 *fig.*
  reverse transcriptase and production of
    genes for, 235
Gene expression, 208–29
  adult stem cells and, 219
  animal cloning, 208–9
  cancer and faulty, 222–27
  cascade of, in animal embryonic
    development, 220–21
  cellular differentiation and, 212–13
  cloning of nonhuman animals and, 218
  definition of, 210
  embryonic development and, 220–22
  regulation of, in eukaryotes, 213–17
  regulation of, in prokaryotes, 210–11
  reproductive cloning and, 218–19
  summary of mechanisms regulating
    eukaryotic, 217, 217 *fig.*
  therapeutic cloning, embryonic stem cells,
    and, 219
Gene flow, 268

Gene pools
  Hardy-Weinberg equilibrium in, 266–67
  microevolution in, 265, 268–69
Gene regulation
  embryonic development and, 219
  in eukaryotes, 213–17
  in prokaryotes, 210–11
Genesis, Book of, 256
Gene therapy, 243
Genetically-modified (GM) organisms,
  247–48
  nutrients in soil and, 655
  plants, 642
  potential risks of, 248
  protein yield in crops and, 656–57
Genetic code, RNA codons as, 192
Genetic disorders, 153. *See also names of*
  specific disorders, e.g. Cystic fibrosis
  autosomal, 163 *table*
  breast cancer as, 226
  carriers of, 161
  detecting, through fetal testing, 164–65
  detecting, through genetic testing, 170
  detecting, with restriction fragment
    analysis, 240–41
  dominant, 162–63
  Down syndrome, 145
  gene therapy for, 243
  hypercholesterolemia, 85
  incomplete dominance and, 166–67
  lysosomal storage disease, 61
  recessive, 162
  sex-linked, 177
Genetic diversity, 766. *See also* Biodiversity
Genetic drift, 268
Genetic engineering, 233. *See also* Clones
  and cloning; DNA technology;
  Genetically-modified organisms
Genetic markers, 240
Genetic recombination, crossing over and,
  143
Genetics, 152. *See also* Chromosome(s);
  Gene(s); Inheritance
  ancient roots of, 154
  animal behavior as result of environment
    and, 706–7
  DNA (*see* DNA)
  human disorders in (*See* Genetic disorders)
  molecular (*see* Molecular biology)
  population, 265
Genetic testing
  for disease-causing alleles, 170, 240–41,
    241 *fig.*
  legal and forensic applications of, 242,
    242 *fig.*
Genetic variation
  crossing over between homologous
    chromosomes as cause of, 142–43
  differing genetic information of
    homologous chromosomes and, 142
  evolutionary fitness and, 273
  independent orientation of chromosomes
    in meiosis as cause of, 141
  mutations as cause of, 270–71 (*see also*
    Mutation(s))
  in populations, 270–75
  random fertilization as cause of, 141
  reduced, in endangered species, 269
  sexual recombination as cause of, 271 *fig.*
Genome(s), 124. *See also* Human genome;
  Human Genome Project
  completed, 246 *table*
  evolution of, 309
Genomic libraries, 234, 235, 235 *fig.*
Genomics, 231, 244–46
Genotype, 157
  expression of DNA, as proteins, 190

  relationship of, to phenotype, 157, 166
  using testcross to determine, 159
Genus, 304–5
Geographical isolation and speciation, 284,
  285 *fig.*
Geologic record (time scale), 298 *table*, 299
GERD (gastroesophageal reflux disease), 437
Germination, of seeds, 636, 640, 640 *fig.*
  abscisic acid and, 669
Gestation, 556
GH (growth hormone), human, 524–25
  DNA technology for synthesis of, 237
Giant sequoia (*Sequoiadendron gigantea*),
  622–23
Giardia intestinalis, 334, 335 *fig.*
Gibberellins, 666 *table*, 668–69, 671
Gibbons, 402, 402 *fig.*
Gigantism, 524
Gill(s), 455, 455 *fig.*, 456–57
  structure of, 456 *fig.*
Giraffes, 467 *fig.*
  gravity and circulatory system of, 466–67
Gizzards, 432
Glacier retreat, primary succession after, 750,
  750 *fig.*
Glans, 539, 541
Glia, 567
Global warming, 119, 770–71
  atmospheric CO_2 and, 119 *fig.*, 770 *fig.*,
    771 *fig.*
Glomeromycetes, 358 *fig.*, 359, 359 *fig.*
Glomeruli (glomerulus), 513
Glucagon, 528–29
Glucocorticoids, 531
Glucose, 37
  cellular respiration and metabolism of,
    89–102
  homeostasis in blood levels of, 511,
    528–29, 528 *fig.*
  structure of, 37 *fig.*
Glucose-tolerance test, 529 *fig.*
Glycogen, 39, 39 *fig.*
  epinephrine and breakdown of, 530,
    530 *fig.*
  homeostasis and, 511, 528 *fig.*
Glycolysis
  aerobic cellular respiration and process of,
    93, 94, 94 *fig.*, 95 *fig.*, 100
  anaerobic cellular respiration and process
    of, 101
Glycoprotein, 59
Goiter, 18, 18 *fig.*, 526
  iodine deficiency and, 526 *fig.*
Golgi, Camillo, 60
Golgi apparatus, 60, 60 *fig.*
Gonads, 532. *See also* Ovaries; Testes
Gondwana, 300
Goodall, Jane, 402, 719
Goose, graylag
  fixed action patterns in, 705 *fig.*
  imprinting in, 708, 708 *fig.*
Gorillas, 402, 402 *fig.*
Gould, Stephen Jay, 290, 292
Gradualism model of speciation, 290,
  290 *fig.*
Gram-positive bacteria, 328
Gram stain, bacterial identification using,
  324
Grana (granum), 63, 109. *See also* Thylakoids
Grand Canyon, 261 *fig.*, 284, 284 *fig.*
Grant, Peter and Rosemary, 264, 289, 289 *fig.*
Grapes, gibberellins effect on, 669
Grasshopper, 382 *fig.*
  alimentary canal, 432 *fig.*
  circulatory system, 469, 469 *fig.*
  digestion in, 432
Grasslands, 694–95, 696–97, 696 *fig.*

Graves disease, 526, 526 *fig.*
Gravitropism, 672, 672 *fig.*
Gravity
  blood circulation and effects of, 466–67
  plants and effects of, 672
Gray matter, central nervous system, 575
Gray wolves, Yellowstone to Yukon
    Conservation Initiative and, 776–77,
    777 *fig.*
Green algae, 338, 338 *fig.*
  comparison of plants to multicellular,
    345 *fig.*
Greenhouse effect, 119, 770–71
Griffith, Frederick, 182, 204
Ground tissue system, 630
Growth
  development and, as property of all
    organisms, 5
  mitosis, cell division and, 136
  plant, 632–35, 672
Growth factor(s)
  cell-cycle control system signaled by, 134
  cell division affected by, 133, 133 *fig.*
  proto-oncogenes and coding for, 224 *fig.*
Growth-factor proteins, 565
Growth hormone (GH), human, 524–25
  DNA technology for synthesis of, 237
Growth-inhibiting factor, 224, 224 *fig.*
Guanine (G), 185
Guard cells, 630
  transpiration control by, 649
Guppy populations, predation and evolution
    of, 734 *fig.*, 735
Gymnosperms, 347, 347 *fig.*
  coniferous forests, 350, 698
  pine, 350–51, 351 *fig.*

H
Habitat(s), 685. *See also* Global warming
  archaea living in extreme, 327
  fragmentation of, as threat to populations,
    772 *fig.*, 773
  identifying critical, 773
  loss of, biodiversity crisis and, 768
  restoring degraded, 778–79
Habitat isolation, 282
Habituation, 707
Hair cells (sensory receptors), 592, 592 *fig.*
  mechanoreception by, 593 *fig.*
Haldane, J. B. S., 318
Hales, Stephen, 646
Halobacterium halobium, 327
Halophiles, 327
Hantavirus, 203
Haploid cells, 137
  in humans, 348 (*see also* Egg(s); Sperm)
  in plants, 348, 350–51, 351 *fig.*, 636–37
    (*see also* Gametophytes)
Harcombe, William, 10, 11
Hardy-Weinberg equilibrium, 266–67
  conditions required for, 268–69
  public health science and use of, 267
Heart, 471–74
  cardiac cycle and contraction of, 472–73
  cardiac muscle, 418, 418 *fig.*
  cardiovascular system and, 470, 471
  diseases of, and attack in, 474, 474 *fig.*
  path of blood flow through human, 471 *fig.*
  radioactive isotopes for detection of
    disorders in, 21
  tempo and rhythm of heartbeat, 473
Heart attack, 474, 474 *fig.*
Heartwood, 635, 635 *fig.*
Heartworms, 375
Heat
  definition of, 25
  as form of energy, 72, 73

thermoregulation and gains or losses of,
    506–7
Heimlich, Henry, 435
Heimlich maneuver, 435, 435 *fig.*
Helicobacter pylori, 437
Helium, atomic structure of, 20 *fig.*
Helper T cells, 497, 497 *fig.*, 499
Hemings, Sally, 242
Hemiptera Order, 382
Hemispherectomy, 581, 581 *fig.*
Hemoglobin, 462–63
  comparisons of vertebrate, showing
    evolutionary relationships, 263 *fig.*
  gene regulation and synthesis of, 216
  mutation in, causing sickle-cell disease, 46
  oxygen loading/unloading by, 463 *fig.*
  Pauling's research on, 46
Hemophilia, 177, 481
Hemorrhagic fever, 202
Hepatic portal system, 511 *fig.*
Hepatic portal vein, 511
Herbicides, 671
Herbivores, 430
  alimentary canal, 440, 441 *fig.*
  digestive adaptations, 440–41
  plant defenses against, 676–77, 677 *fig.*,
    748
Herceptin, 496
Heredity, science of. *See* Genetics
Hermaphroditism, 537, 537 *fig.*
Herpesvirus, 180–81, 201
Hershey, Alfred, 182–83, 184
Heterokaryotic stage, fungal life cycle, 358
Heterotrophs, 326
Heterozygote advantage, 272–73
Heterozygous organisms, 156
Hibernation, 504–5, 508
High altitudes, respiration and circulation
    in, 452–53
High-density lipoproteins (HDLs), 449
Hindbrain, 578–79, 579 *fig.*
Hinge joints, 611, 611 *fig.*
Hippocampus, 583
Hippocrates, 154
Histamines
  allergies and release of, 501
  inflammatory response and, 487
Histones, 213
HIV (human immunodeficiency virus), 202,
    203, 484–85
  behavior and structure of, 203 *fig.*
  mutations and genetic variation in, 271
  placenta-crossing by, 557
  T cell destruction by, 498, 498 *fig.*
Hodgkin, A. L., 568
Hodgkin's disease, 500
Homeoboxes, 222
Homeostasis, 425
  animal regulatory systems and, 505
  brain and, 575
  calcium, 527 *fig.*
  chemical signals and, 520
  circulatory system and, 468
  endocrine system and, 522–25
  glucose, 528–29, 528 *fig.*
  liver's role in, 511
  model of, 425 *fig.*
  negative feedback mechanisms of, 426,
    426 *fig.*
  neurotransmitters in, 572–73
  osmoregulation, excretion, and, 508–15
  somatic and autonomic nervous systems
    and, 577
Homeotic gene(s)
  animal embryonic development and role
    of, 221, 368
  evolution and, 292

of fruit fly *versus* mouse, 222 *fig.*
  Hox, 368
Hominids, 403–5
  timeline for, 403 *fig.*
Hominoids, 401
  phylogenetic tree of some, 309 *fig.*
Homo erectus, 260 *fig.*, 404–5, 405
Homo ergaster, 404
Homo floresiensis, 405
Homo habilis, 404
Homo heidelbergensis, 405
Homologous chromosomes, 136–37, 137 *fig.*
  crossing over between, 142–43, 143 *fig.*
  gene alleles carried by, 157
  mutations and variations in, 271
  reduction of number from diploid to
    haploid, 138–39
Homology (homologous structure), 262
  analogy *versus,* 304
  phylogenetic relationships indicated by, 304
  vertebrate embryos, 263 *fig.*
  wings and limbs, 262 *fig.*
Homo sapiens. See also Human(s)
  appearance of, 299
  culture of, 407–8, 722
  evolution of, 403, 404, 405
  FOXP2 gene and speech development in,
    406
  oldest known fossil of, 405 *fig.*
Homozygous organisms, 156
Honeybees
  pattern recognition by, 713
  signaling behavior in, 720, 720 *fig.*
  thermoregulation in, 506, 506 *fig.*
Hooke, Robert, 51
Hookworms, 375
Hopping locomotion, 606
Hormone(s), animal, 520
  adrenal gland, 530–31
  anabolic steroids, 41
  antagonistic, 527
  DNA technology and production of
    therapeutic, 237
  from endocrine glands, 520 *fig.*
  of hypothalamus and pituitary, 524,
    525 *fig.*
  insulin, 216 *fig.*
  labor and childbirth regulated by, 560–61,
    560 *fig.*
  melatonin, 522
  from neurosecretory cell, 520 *fig.*
  ovarian and menstrual cycles regulated by,
    544–45, 544 *table,* 545 *fig.*
  for pain relief, 531
  of pancreas, 528–29
  prairie vole behavior and, 706
  sexual reproduction and, 538, 541, 541
    fig., 544–45
  signaling mechanisms of, 521
  steroid, 521, 522 (*see also* Sex hormones)
  of thyroid and parathyroid, 526–27
Hormone(s), insect, 16
Hormone(s), plant, 664–71
  abscisic acid (ABA), 669
  agricultural uses of, 671
  auxin, 665, 666–67, 667 *fig.*, 671
  cytokinins, 668
  discovery of, 664–65
  ethylene, 670–71
  five major, 666 *table*
  gibberellins, 668–69, 671
  phytoestrogens, 662–63
Hornworts, 346, 346 *fig.*
Horse(s)
  hybrid sterility in cross between donkey
    and, 283 *fig.*
  trends in evolution of, 293 *fig.*

Horseshoe crab, 380
Hox genes, zygote development controlled by, 368
Hubbard Brook Experimental Forest, New Hampshire, ecosystem alteration in, 760, 760 *fig.*
Human(s). *See also* Fetus, human; *Homo sapiens; specific aspect under* Human
  breathing in, 90, 460 *fig.*, 461, 461 *fig.*
  cloning, 219
  daily water requirements of, 509
  global warming and ozone depletion caused by activities of, 120
  osmoregulation in, 509
  relationship of plants and fungi to, 309 *fig.*
  smell and chemoreceptors in, 601 *fig.*
  species, 281 *fig.*
Human behavior
  biodiversity crisis and, 768
  culture and, 407–8, 722
  fixed-action patterns of behavior in, 705
  territorial, 717
Human body. *See also* Human health
  ATP use by, 91
  blood types, 167
  body fat, 448
  bones, 611–13
  cardiovascular system and blood, 471–78
  chemical composition of, 18 *table*
  development of (*see* Human development)
  digestive system, 433–40
  endocrine glands in, 522 *fig.*
  endocrine system and hormones, 523 *table*
  energy consumed by various activities of, 91 *table*
  gas exchange in, 458 *fig.*, 462–63
  imaging technology for viewing inner, 422–23
  immune response (*see* Immune response)
  liver and hepatic portal system, 511 *fig.*
  lungs, 425 *fig.*, 458–59, 458 *fig.*, 460 *fig.*
  lungs, healthy and cancerous, 459 *fig.*
  lymphatic system, 488, 488 *fig.*
  muscles and muscle contraction, 613–18
  organs and organ systems, 419–21
  poisons and disruption of cellular respiration in, 99
  respiratory system, 458 *fig.*
  senses (*see* Senses)
  skeleton, 610, 610 *fig.*
  skull, 292 *fig.*
  somatic cells, 136–37
  tissues, 415–19
  water content of, 25
Human brain, 578–85
  cerebral cortex, 580, 580 *fig.*
  cerebral hemispheres, 579 *fig.*
  embryonic development of, 578, 578 *fig.*
  evolution of, 404
  injuries in, and study of, 581
  lateralization, 580
  limbic system, 583, 583 *fig.*
  neurological disorders and physiology of, 584
  sleep and arousal regulation in, 582
  structure of, 578–79, 579 *fig.*
  structure of, linked to function, 579 *table*
Human chorionic gonadotropin (HCG), 557
Human development, 556–61
  childbirth and, 560–61
  embryo and placenta formation, 556–57
  three trimesters of, 558–59, 558–59 *fig.*
Human disease. *See also* Genetic disorders
  AIDS, 202, 203, 484–85, 498, 500
  allergies as, 501
  amoebic dysentery, 337
  bacterial, 205, 272, 322, 329, 487

bioterrorism, 330
blood, 481
cancer (*see* Cancer)
cardiovascular and heart disease, 474
diabetes mellitus, 529
diet effects on, 449
DNA technology for diagnosis and treatment of, 237
fungal, 361
gastric ulcers, 437
global warming and, 771
hypercholesterolemia, 85, 166–67
hypoglycemia, 529
immune system malfunctions and, 500–501
lysosomal storage disease, 61
malaria (*Plasmodium*), 168, 273, 335, 335 *fig.*, 685
multiple sclerosis, 567
osteoporosis, 612–13, 613 *fig.*
parasitic, 361, 374, 375
Parkinson's disease, 573
pesticides and, 685
sexually transmissible, 546, 546 *table*
sickle cell disease and malaria, 168
skeletal and bone, 612–13
viral, 202–3
Human evolution, 398–410
  agricultural developments and, 408
  bipedalism and, 403, 404
  cultural history and, 407–8
  evolutionary relationships of vertebrate hemoglobin, including humans, 263 *fig.*
  fossils, 404 *fig.*
  hominids and, 403–5
  primate diversity and, 400–402
  relationship of Neanderthals to, 398–99
  skin pigmentation, 406
  speech, 406
  theories on locations of, 405
Human genetics. *See also* Genetics; Inheritance
  environmental influences of, 170
  family pedigrees and, 161
  speech development and, 406
Human genome, 245
Human Genome Project (HGP), 215, 231, 244–45, 249
Human growth hormone, 524–25
  DNA technology for synthesis of, 237
Human health. *See also* Medicine
  anabolic steroids and, 41
  application of Hardy-Weinberg equation to, 267
  gene therapy for, 243
  threat of bacterial resistance to antibiotics to, 205, 272, 325, 498
  threat of poisons to, 99
  threat of smoking to, 459
  threat of viruses to, 202–3
Human population, 736–39
  biodiversity and, 767
  birth/death rates, age structure and growth of, 738–39, 739 *fig.*
  centers of, in North America, 265 *fig.*
  environmental problems caused by, 119
  exponential growth of, 736–37
  global warming and, 770–71
  history of growth of, 736 *fig.*
  life tables for U.S., 729, 729 *table*
  uniform dispersion of habitat for, 728 *fig.*
Human reproduction, 538–47. *See also* Embryonic development in animals
  contraception, 547, 547 *table*
  embryonic development and, 556–61
  female reproductive anatomy, 538–39, 538 *fig.*, 539 *fig.*

  fertilization and zygote formation, 137
  hormones and, 541 *fig.*, 544–45, 545 *fig.*
  male reproductive anatomy, 540–41, 540 *fig.*, 541 *fig.*
  sex determination in, 176
  sexually transmissible diseases, 546, 546 *table*
  sexual response, 539, 546
  sperm and ova formation, 542–43, 542 *fig.*, 543 *fig.*
  technology of, 561
Humoral immunity, 490
  antibodies as weapons of, 494
  antibody marking of antigens by, 495
  antigen sites for antibody binding, 491
  clonal selection and, 492–93
  effector mechanisms of, 495 *fig.*
  monoclonal antibodies and, 496, 496 *fig.*
  T cells and, 497, 497 *fig.*
Humus, 654
Hunting-gathering-scavenging culture, 407
Huntington's disease, 163
Hurricanes, impact of, on tropical forest, 699
Huxley, A. F., 568
Hybrid(s), 155
Hybrid breakdown, 283
Hybrid gas-electric cars, 771
Hybrid inviability, 283
Hybrid sterility, 283
Hydra, 373
  asexual reproduction of, 136 *fig.*, 536 *fig.*
  digestion in, 432, 432 *fig.*
  exchanges between environment and, 424 *fig.*
  hydrostatic skeleton, 608, 608 *fig.*
  nerve net, 574, 574 *fig.*
Hydrocarbons, 34
Hydrochloric acid (HCl), 436
Hydrogen
  atom, 22 *fig.*
  cellular respiration and electron release from, 92, 92 *fig.*, 98, 98 *fig.*
  in human body, 18, 18 *table*, 19
Hydrogen bonds, properties of water based on, 24, 24 *fig.*
Hydrolysis, 36, 36 *fig.*
  digestion as, 431
Hydrophilic amino acids, 42, 42 *fig.*
Hydrophilic compounds, 35
Hydrophobic amino acids, 42, 42 *fig.*
Hydrophobic compounds, 40
Hydroponic culture, 652 *fig.*
Hydrostatic skeletons, 370, 608, 608 *fig.*
Hydrothermal vents, 327, 683 *fig.*
Hydroxyl group, 35
Hymen, 539
Hymenoptera Order, 383
Hypercholesterolemia, 85, 166–67
Hypertension, 476–77
Hyperthyroidism, 526, 526 *fig.*
Hypertonic solutions, 83, 83 *fig.*
Hyperventilation, 461
Hyphae, 357
Hypoglycemia, 529
Hypothalamus, 524, 579 *fig.*
  biological clocks and, 579
  drug addiction and, 579
  location of pituitary and, 524 *fig.*
  nervous system, endocrine system and, 524–25
  sleep-wake cycles regulated by, 582
Hypothesis, definition of, 9
Hypothesis-based science, 9, 10–11, 10 *fig.*
Hypothyroidism, 526
Hypotonic solutions, 83, 83 *fig.*
Hyracotherium fossil, 293, 293 *fig.*

I

Ibuprofen, disruption of pain-inducing enzymes and, 78
"Ice Man," 260 *fig.*, 261
Iguana, marine, 256 *fig.*
Imaging technology, 422–23
Imitation, learning by, 712–13
Immune response, 489
 primary and secondary, 492 *fig.*, 493, 493 *fig.*
 specific immunity (*see* Immune system)
Immune system, 420 *fig.*, 484–503
 acquired immunity, 489–500
 antibodies and humoral immunity, 494
 antibody binding to antigens and, 495, 495 *fig.*
 antigens and, 491
 cell-mediated immunity, 490, 497
 clonal selection and, 492–93, 492 *fig.*
 cytotoxic T cells and, 499
 disorders of, 500–501
 HIV and AIDS effects on, 203, 484–85, 498
 humoral immunity, 490
 inflammatory response, 487, 487 *fig.*
 innate immunity, 486–88
 lymphatic system and, 488, 488 *fig.*
 lymphocyte B cells and T cells of, 490–91
 memory of, 489
 monoclonal antibodies and, 496
 self-recognition and nonself-recognition in, 497, 500
Immunity, 489
 acquired, 489–500
 active and passive, 489
 cell-mediated, 490, 497
 humoral, 490
 innate, 486–88
 stress and, 501
Immunodeficiency diseases, 500. *See also* AIDS; HIV
Immunoglobulins. *See also* Antibody(ies)
 classes of, 494
Impact hypothesis of mass extinction, 303
Impalas, behaviors of, 702–3, 702–3 *fig.*, 704 *fig.*
Imperfect fungi, 358
Implantation of blastocyst, 556, 556 *fig.*
Impotence, 561
Imprinting, 708–9
Inbreeding, 162
Inclusive fitness, altruism and, 721
Incomplete dominance, 166–67
Incomplete metamorphosis, 382
Independent assortment, law of, 158–59, 171
Independent events, rule of probability and, 160
Indeterminate growth, 632
India, population growth in, 738
Indigo bunting, star navigation by, 711, 711 *fig.*
Individual variation, 8
Indoleacetic acid (IAA), 666
Induced fit, of substrate binding to enzymes, 77
Induction, organ formation and embryonic, 554–55, 554 *fig.*
Inductive reasoning, 9
Industrial Revolution, 408
Infant mortality, 739
Infants, newborn
 diagnosis of genetic disorders by testing of, 165
 multiple, 534–35
Inferior vena cava, 471
Infertility, 534–35, 561
Inflammatory response, 487, 487 *fig.*

Influenza virus, 201 *fig.*, 202
Ingestion, 368, 368 *fig.*, 430–31
Ingroup, cladistic analysis and, 306
Inheritance, 152–79. *See also* Genetics
 of behavior patterns, 705–7
 chromosomal theory of, 171–74 (*see also* Chromosome(s); Gene(s))
 Mendel's laws of, 154–65
 polygenic, 169
 sex chromosomes and sex-linked genes related to, 175–77
 study of, 154
 variations on Mendel's laws, 166–70
Inhibiting hormones, 524
Injuries
 brain, 581
 spinal cord, 564–65
Innate behavior, 705
Innate immunity, 486–88
 inflammatory response, 487, 487 *fig.*
 lymphatic system and, 488, 488 *fig.*
Inner ear, 595 *fig.*, 598, 598 *fig.*
 balance and, 600
Insect(s), 380, 382–83
 aphids, 650–51, 651 *fig.*
 butterflies, 383, 711, 770
 camouflage in, 264 *fig.*
 chemoreceptors in, 593, 593 *fig.*
 circulatory system, 469, 469 *fig.*
 compound eye in, 594, 594 *fig.*
 excretion in, 510 *fig.*
 exoskeleton of, 608–9
 fossilized, 260 *fig.*, 261
 fruit flies (*see Drosophila melanogaster*)
 grasshopper, 382, 432 *fig.*, 457 *fig.*, 469
 honeybees (*see* Honeybees)
 mosquito, 278–79
 nervous system, 574 *fig.*
 orders of, 382–83
 osmoregulation in, 509
 pesticide resistance in, 264 *fig.*
 plant pollination and, 355, 355 *fig.*
 sex determination in, 175, 175 *fig.*
 tracheal system, 455, 455 *fig.*, 457, 457 *fig.*
 wasps (*see* Wasps)
Insulation, thermoregulation and, 506
Insulin
 diabetes and, 529
 DNA technology for synthesis of, 237
 glucose homeostasis and, 528 *fig.*
 production of, 528
 regulation of gene expression and synthesis of, 216 *fig.*
Insulin-dependent diabetes mellitus, 500, 529
Integrated pest management (IPM), 735
Integration, nervous system, 566
Integumentary system, 421 *fig.*
Interferons
 innate immunity and, 486
 mechanism of action against viruses, 486 *fig.*
Interleukin-2, 497, 497 *fig.*
Intermediate compounds in cellular respiration, 94, 95 *fig.*
 biosynthesis from, 103 *fig.*
Intermediate filaments, 64, 64 *fig.*
Intermembrane space, 63
Internal environment, animal regulation of, 504–17
 homeostatic functions of liver, 511
 osmoregulation and excretion, 505, 508–15
 sympathetic and parasympathetic neurons and, 576–77
 thermoregulation, 504
Internal fertilization, 537

Internal transport, mechanisms of animal, 468–70. *See also* Circulatory system
International Olympic Committee, 41
Interneurons, 566
Internodes, 626
Interphase
 meiosis, 138 *fig.*
 mitosis, 129 *fig.*, 130, 130 *fig.*
Interspecific competition, 745
Interstitial fluid, 424
 exchanges between blood and, 469, 478, 478 *fig.*
 urine production and, 514
Intertidal zones, 690
Intestines, 432
Intrinsic rate of increase, 730
Introduced (exotic) species, 394, 726–27, 768
Introns, 194
 making gene with no, 235 *fig.*
Inversion, chromosomal, 148, 148 *fig.*
Invertebrates, 372–85
 annelids, 378–79
 arthropods, 380–81
 body plans of, 373, 374, 376 *fig.*
 body segmentation in, 378–79, 378 *fig.*
 as chordates, 385
 circulatory system, open, 469
 cnidarians, 373
 echinoderms, 384
 evolution of, from protists, 369 *fig.*
 eyes of, 594
 flatworms, 374
 insects, 382–83
 molluscs, 376–77
 nervous system of, 574 *fig.*
 reproduction in, 534–35
 roundworms, 375
 sponges, 372
In vitro fertilization (IVF), 561
Iodine, as trace element, 18
Ion channels, chemically-gated, 571
Ionic bonds, 22–23
Iridium, 303
Iris (eye), 594, 595 *fig.*
Iron, as trace element, 18
Irrigation, 655, 655 *fig.*, 757
Islands, speciation on, 288, 288 *fig.*
Islets of Langerhans, 528
Isoflavones, 662–63
Isomers, 34
Isotonic solutions, 83, 83 *fig.*
Isotopes
 carbon, 20 *table*
 radioactive, 20, 21
Italy, population age structure for, 738, 739 *fig.*

J

Jacob, François, 210
Japan, habitat restoration in, 778–79
Jaws
 hinged, 387
 human evolution and, 403
Jefferson, Thomas, 242
Jeffreys, Alec, 230
Jellies (jellyfish), 373
 gastrovascular cavity and internal transport in, 468, 469 *fig.*
Joints, three types of, 611 *fig.*
Jumping bean analogy for energy of activation, 76 *fig.*
Jumping genes (transposons), 245

K

Kandinsky, Wassily, 50
Kangaroo, 367, 392 *fig.*
 locomotion in, 606, 606 *fig.*

Kaposi's sarcoma, 498
Karyotype, 144, 144 *fig.*, 145 *fig.*
Keystone species, 747
Kidneys
 conversion of blood filtrate to urine by, 513
 dialysis of, 480, 515, 515 *fig.*
 excretory system and, 512–13, 512 *fig.*
 nephron structure and orientation in, 512 *fig.*
 structure of, 512 *fig.*
Killer whales, 9, 9 *fig.*
Killifish, predation on guppy populations by, 734 *fig.*, 735
Kilocalories, 442
 expense of, in select activities, 91 *table*
Kinesis, 710
Kinetic energy, 72, 72 *fig.*
King, Mary-Claire, 170, 226, 226 *fig.*
Kingdoms (classification), 6, 305
Kin selection, altruism and, 721
Kissimmee River Restoration Project, 779, 779 *fig.*
Kittiwakes, fixed action patterns in, 705
Klinefelter syndrome, 147, 147 *table*
Kluft, Carolina, 617 *fig.*
Knee-jerk reflex, 566 *fig.*
Koalas, foraging behavior of, 714, 714 *fig.*
Krakatau, volcanic eruption on, 302, 302 *fig.*
Krebs, Hans, 96
Krebs cycle, cellular respiration and process of, 96–97. *See also* Citric acid cycle
Krill, 428–29
K-selection, 735
!Kung people, 407, 407 *fig.*
Kyoto Protocol on global warming, 771

L

Labia majora, 539
Labia minora, 539
Labor stages in childbirth, 560–61, 560 *fig.*
lac operon, 210, 211, 211 *fig.*
Lactase, 32–33, 42
Lactic acid fermentation, 101, 101 *fig.*
Lactose, metabolism of, in prokaryotes, 210–11
Lactose intolerance, 32–33
Lakes and streams
 acid precipitation and, 28, 761
 eutrophication in, 761
 as freshwater ecosystem, 692
Lamarck, Jean Baptiste, 256
Lamellae, 456–57
Lampreys, 387, 387 *fig.*
Lancelets, 385, 385 *fig.*
Land
 animal locomotion on, 606–7
 climate and landforms, 689
 plant colonization of, 344–45
 vertebrates and life on, 389
Lander, Eric, 249
Landmarks, animal behavior using, 710
Landscape ecology
 edges and corridors affecting, 774
 habitat restoration, 778–79
Language, human evolution and, 403
Large intestine, 440
Larva, 368
Larynx, 458
Lateralization, brain, 580
Lateral line system, 388
Laurasia, 300
Law of independent assortment, 158–59, 171. *See also* Mendelian inheritance, laws of
Law of segregation, 156–57, 171. *See also* Mendelian inheritance, laws of

Laws of thermodynamics, 73
Lead contamination, phytoremediation of, 645
Leaf (leaves), 626
 falling of, 669, 670–71, 670 *fig.*, 671
 modified, 627, 627 *fig.*
 stomata on (*see* Stomata)
 tissue systems in, 631, 631 *fig.*
 veins of, 630
Leakey, Mary, 404
Learning, 707–13
 associative, 712
 cognition and problem-solving behavior in, 713
 habituation, 707
 imprinting, 708–9
 limbic system role in, 583
 sleep and, 582
 social, 712–13
 trial-and-error, 712
 types of, 707 *table*
Leeches, 378, 379, 379 *fig.*
 nervous system, 574, 574 *fig.*
Legal applications of DNA technology, 242
Legumes
 biological augmentation using, 778, 778 *fig.*
 nitrogen-fixing bacteria on, 658–59
Lemurs, 400, 400 *fig.*, 720 *fig.*
Lens (eye), 594, 595, 595 *fig.*
 accommodation in, 595 *fig.*
 corrective, 596
Lepidoptera Order, 383
Leptin, obesity and, 447
Leucine, 42
Leukemia(s), 135, 481
Leukocytes, 479, 479 *fig. See also* White blood cells
LH (luteinizing hormone), 524, 532, 544–45, 545 *fig.*
Lichens, 362, 778 *fig.*
Life. *See also* Organisms
 arranging, into kingdoms, 310
 cells as structural and functional units of, 4
 chemistry of (*see* Chemistry of life)
 common features, 4–5
 diversity of, 6–7
 evolution of, 8–9 (*see also* Evolution)
 levels of organization in, 2–3, 2 *fig.*, 415
 major episodes in early history of, 317 *fig.*
 origins of, 316–21
 prokaryote evolution and, 322–31
 scope of biology and, 2–4
Life cycle, 124–25
 angiosperm, 352–53, 353 *fig.*, 636 *fig.*, 637 *fig.*
 fern, 349 *fig.*
 fungi, 358, 358 *fig.*
 moss, 348 *fig.*, 349
 mushrooms, 360, 360 *fig.*
 pine, 350–51, 351 *fig.*
 sea star, 368 *fig.*
 zygote fungi, 360, 360 *fig.*
Life expectancy at birth, 739
Life history of organisms, evolution and, 734–35
Life tables, population mortality and survivorship tracked in, 729, 729 *table*
Light
 Earth's tilt as cause of variations in, 688 *fig.*
 phytochromes as detector of, 675
 plant responses to, 664–65, 672, 676
 vision and, 597, 597 *fig.*
Light microscopes (LMs), 52, 52 *fig.*, 53
Light reactions of photosynthesis, 111, 112–15

ATP synthesis through chemiosmosis in, 115, 115 *fig.*
 mechanical analogy of, 114 *fig.*
Lignin, 345
Limbic system, 583, 583 *fig.*
Linked genes, 172
Linnaeus, Carolus, 280, 304, 305
Lipids, 40–41
 dehydration reaction and formation of, 40 *fig.*
 formation of, 319
 as fuel for cellular respiration, 102
 synthesis of, in smooth ER, 58
Liver
 alcohol consumption and damage to, 511
 digestion and role of, 438
 homeostatic functions of, 511
Liverworts, 346, 346 *fig.*
Lizards, 390, 390 *fig.*
 geckos, 412–13, 413 *fig.*
Lobe-fins, 388, 388 *fig.*
Lobster, 380 *fig.*
Local regulators, 520
Locomotion, 606–7. *See also* Movement
 flying as, 607
 on land, 606–7
 muscle contraction for, 613–18
 skeletal support for, 608–13
 swimming as, 606
Locus of gene, 137
Logistic growth model of population growth, 730–31, 731 *fig.*
London Underground, mosquito speciation in, 278–79
Long-day plants, 674
Long-term memory, 583
Loon, courtship and mating behaviors of, 716, 716 *fig.*
Loop of Henle, of kidney nephron, 512 *fig.*, 513, 514, 514 *fig.*
Loose connective tissue, 417, 417 *fig.*
Lophotrochozoans, 393
Lorenz, Konrad, 704, 705, 708, 708 *fig.*
Loris, 400 *fig.*
Low-density lipoproteins (LDLs), 85, 449
LSD, 361
Luciferase, 71
Luciferin, 71
Lucy, fossilized skeleton, 404, 404 *fig.*
Lugo, Ariel, 699, 778, 778 *fig.*
Lung(s), 387
 bird, 460 *fig.*
 breathing and ventilation of, 460
 human, 425 *fig.*, 458 *fig.*, 460 *fig.*, 461
 smoking effects on, 459
 in terrestrial vertebrates, 455, 455 *fig.*, 458–59
 vital capacity of, 460
Lung cancer, 459 *fig.*
Lungfishes, 388
 geographic distribution of, 301, 301 *fig.*
Lupus (systemic lupus erythematosus), 500
Luquillo Experimental Forest, Puerto Rico, 699, 699 *fig.*
Luteinizing hormone (LH), 524, 532, 544–45, 545 *fig.*
Lycophytes, 346, 350
Lyell, Charles, 256
Lyme disease, 329, 329 *fig.*
Lymph, 488
Lymphatic system, 420 *fig.*, 488, 488 *fig.*
Lymph nodes, 488, 488 *fig.*
Lymphocytes, 479, 479 *fig.*, 490–91
 clonal selection of, 492–93, 492 *fig.*
 development of B cell and T cell, 490 *fig.*
 immune response and, 488
Lymphomas, 135

Lysogenic cycle, viral, 200, 200 *fig.*
Lysosomal storage disease, 61
Lysosomes, 60–61
  formation and function of, 61 *fig.*
Lysozyme, shape and function of, 43, 43 *fig.*
Lytic cycle, viral, 200, 200 *fig.*

M

Ma, Lena, 644
Macroevolution, 280, 291–93, 298–303
  continental drift and, 300–301
  "evo-devo" and role of genes in, 292
  fossil record and, 298–99
  geologic record (time scale) and, 298–99, 298 *table*
  mass extinctions and, 302–3
  phylogenetic trees as representation of, 305–7
  plate tectonics and, 302
  trends *versus* goals in, 293
Macromolecules, 36
  biosynthesis of, as cellular respiration intermediates, 103 *fig.*
  origin of cells in cooperative associations of, 321, 321 *fig.*
  polymer chains and formation of, 36
Macronutrients, plant, 652
Macrophages
  acquired immunity and, 497, 497 *fig.*
  innate immunity and, 486
Madagascar, biodiversity in, 767
Magnesium as plant nutrient, 652
Magnetic resonance imaging (MRI), 422–23
Magnetic resonance microscopy (MRM), 423, 423 *fig.*
Magnification of microscopes, 52
Major depression, 584
Major histocompatibility complex (MHC), 500
Malaria. *See also Plasmodium*
  heterozygote advantage and, 273
  pesticides and, 685
  sickle-cell disease and, 168
Malathion, as pesticide, 78
Males, human
  contraception in, 547, 547 *table*
  formation of sperm in, 542 *fig.*
  hormonal control of testis, 541 *fig.*
  reproduction anatomy of, 540–41, 540 *fig.*, 541 *fig.*
  sex-linked genetic disorders in, 177
  sexual response in, 546
Malignant tumors, 135
Malnutrition, 443
Malthus, Thomas, 258
Maltose, 38, 38 *fig.*
Mammal(s) (Mammalia), 392
  cardiovascular system, 470, 470 *fig.*, 471–78
  cloning, 218–19
  excretion in, 510 *fig.*
  osmoregulation in embryos of, 509
  respiratory system, 458
  ruminant, 441 *fig.*
  thermoregulation in, 506
  X chromosome inactivation in female, 214
Mantle, mollusc, 376
Margulis, Lynn, 332
Marijuana, 573
Marsupials, 367, 392
Marx, Groucho, 604, 605, 605 *fig.*
Mass extinctions, 339, 390
  life-form diversification following, 302–3
Mass number, 20
Mast cells, 501
Mating
  among humans, 722
  courtship rituals, 716, 716 *fig.*

  in hermaphroditic animals, 537, 537 *fig.*
  reproductive barriers to, 282–83, 285
  sexual selection and sexual dimorphism in, 275
Mating types, fungal, 358
Matter, definition of, 18
Maximum sustained yield, 735
Measurement equivalents, 52 *table*
Mechanical isolation, 282–83
Mechanoreceptors, 592
Medicinal plants, 356 *table*, 678
Medicine(s)
  applications of reproductive cloning to, 218–19
  cancer treatment, 496
  fungi as source of, 363
  imaging technology, 422–23
  leeches used in, 379
  monoclonal antibodies and, 496
  plants as source of, 356 *table*, 678, 767
  problem of R plasmids and antibiotic resistance in bacteria to, 272
  use of radioactive isotopes in, 21
Medulla oblongata, 578, 579 *fig.*
Medusa, 373
Meiosis, 136–43
  in angiosperm plants, 352–53, 353 *fig.*
  errors in, and altered chromosome number, 146
  gamete formation and, 137
  genetic variety produced by, 141, 142–43
  homologous chromosomes and, 136–39
  human reproduction, 542–43
  mitosis *versus*, 140, 140 *fig.*
  nondisjunction in, 146 *fig.*
  in oogenesis, 543 *fig.*
  phase I and II, 138–39, 138 *fig.*
  reduction of chromosome number by, 138–39
  in spermatogenesis, 542 *fig.*
  stages of, 138–39 *fig.*
Melatonin, 522
Membrane(s), cellular, 79–86
  chemical activities organized by, 79
  of chloroplasts and mitochondria, energy harvesting and, 86
  disease caused by faulty, 85
  enzyme-controlled chemical reactions in, 71
  fluid mosaic model of plasma, 79, 80
  phospholipid bilayer of plasma, 79
  transport across, 81–85, 81 *fig.*
Membrane infolding, 332, 332 *fig.*
Membrane potential, 568
  nerve signal transmission and changes in, 568–69
Memory
  immunological, 489, 493
  limbic system and, 583
  sleep and, 582
Memory cells, 493
Mendel, Gregor, genetic studies conducted by, 152, 154–55
Mendelian inheritance, laws of, 154–65
  chromosomal basis of, 171–77, 171 *fig.*
  homologous chromosomes, alleles, and, 157
  human genetic disorders based on, 162–65
  law of independent assortment, 158–59, 171
  law of segregation, 156–57, 171
  Mendel's first experiments and, 154–55
  as revealed in family pedigrees, 161
  rule of probability and, 160
  use of testcross to determine genotype and, 159
  variations on, 166–70

Meninges, 575
Menstrual cycle, 544–45, 545 *fig.*
  hormones of, 544 *table*
Menstruation, 544, 545 *fig.*
Mercury, environmental degradation from, 769
Meristems
  primary growth in, 632, 632 *fig.*, 633 *fig.*
  secondary growth in, 634–35
Mescaline, 748
Mesoderm, 368, 370, 370 *fig.*, 550
  organ and tissue derivatives of, 553 *table*
Mesophyll, 109, 630
Mesozoic era, 299, 300
Messenger RNA (mRNA)
  initiation codon and, 196–97
  lac operon and production of, 211
  production of eukaryotic, 194
  RNA splicing and production of, from same gene, 215 *fig.*
  translation and, 194–95, 216
Metabolic heat production, 506
Metabolic rate, reducing, 508
Metabolism. *See* Cellular metabolism
Metal accumulators, plants as, 644–45
Metals, bioremediation of toxic, 778, 778 *fig.*
Metamorphosis, 368, 382
Metaphase
  cell division I of meiosis I, 138 *fig.*
  cell division II, of meiosis, 139 *fig.*
  of mitosis, 130–31, 131 *fig.*
Metastasis, 135
Meteorites, 303, 303 *fig.*
Methane, 34
  alternate ways to represent, 23 *fig.*
  structure of, 34 *fig.*
Methanogens, 327
Mexico, demographic transition in, 738 *fig.*
Mice, behavior of California *versus* white-footed, 706–7
Mickey Mouse, 292 *fig.*
Microarrays, DNA, 238–39, 239 *fig.*
Microevolution, 265
  gene pool and, 266–67
  Hardy-Weinberg equilibrium in gene pools and, 266–69
  potential causes of, 268–69
Microfilaments, 64, 64 *fig.*
Micrographia (Hooke), 51
Micrographs, 51, 52
Micronutrients, plant, 652
Microraptor fossil, 296 *fig.*, 297
Microscopes, 52–53
Microtubule-organizing centers, 131
Microtubules, 64, 64 *fig.*, 131
  bending of, in cilia and flagella, 65, 65 *fig.*
Microvilli, 439
Midbrain, 578, 579 *fig.*
Middle ear, 598–99, 598 *fig.*
Migration
  electromagnetic receptors and, 593
  salmon sensing and, 588–89
Migration, internal maps and animal, 711
Miller, Stanley, 318–19
Millipedes, 381, 381 *fig.*
Mimicry, 10–11, 11 *fig.*, 746–47
Mineralocorticoids, 531
Minerals, essential, 445, 445 *table*
Mining wastes, bioremediation of, 778, 778 *fig.*
Mitchell, Peter, 93
Mites, 380, 381, 381 *fig.*
Mitochondria, 63
  chemiosmosis and ATP synthesis in, 98 *fig.*
  energy flow and, 86
  origins and evolution of, 332, 332 *fig.*
  poisons and function of, 99

Mitochondrial matrix, 63
Mitosis, 129–31
  interphase and, 129
  meiosis *versus*, 140, 140 *fig.*
  review of, 136
  stages of, 130–31, 130–31 *fig.*
Mitotic phase (M phase) of cell cycle, 129, 130–31 *fig.*
Mitotic spindle, 130–31
Modern synthesis of evolutionary theory, 265–69
Molecular biology, 180–207
  definition of, 181
  DNA and RNA as nucleotide polymers, 184–85
  DNA as genetic material, 182–83
  DNA replication, 188–89
  DNA structure, 186–87
  evidence for evolution in, 262–63
  genetic information flow from DNA to RNA to protein, 190–99
  phylogenetic tree based on, 393 *fig.*
  as tool in systematics, 308–9
  of viruses, 200–205
  yeasts and research in, 363
Molecular clock, 309
Molecules, 2 *fig.*, 3
  alternative ways to represen, 23 *fig.*, 34 *fig.*
  covalent bonds and, 23
  origin of first cells from cooperatives of, 321, 321 *fig.*
  polar and nonpolar, 24
Mole rats, altruism among, 721, 721 *fig.*
Molina, Mario, 120
Molluscs (Mollusca), 376–77
  exoskeleton of, 609, 609 *fig.*
  nervous system, 574, 574 *fig.*
  range of eye complexity among, 291, 291 *fig.*
Molting, 380
Monera (kingdom), 310
Monkeys, 400, 401 *fig.*
  alarm calls of vervet, 712–13, 712 *fig.*
Monoclonal antibodies, 496, 496 *fig.*
Monocots, 625
  dicots *versus*, 625 *fig.*
  embryo development in, 638, 638 *fig.*
  seed germination in, 640, 640 *fig.*
  tissues in, 630, 631 *fig.*
Monocultures, 642
Monocytes, 479, 479 *fig.*
Monod, Jacques, 210
Monogamous mating behavior, 715
Monohybrid cross, 156
Monomers, 36
Monophyletic taxa, 306
Monosaccharides, 37, 37 *fig.*
Monotremes, 366–67, 392
Morgan, Thomas Hunt, 173, 174 *fig.*
Moritz, Max, 751
Morning-after pills (MAPs), 547
Morphine, 748
Morphological species concepts, 281
Morphs, 270
Mosquito(es), 383, 383 *fig.*
  speciation of, 278–79
Mosses, 346, 346 *fig.*
  life cycle of, 348 *fig.*, 349
Moths, 383, 383 *fig.*
  chemoreceptors in, 593 *fig.*
Motion sickness, 600
Motor neurons, 566
  muscle contraction stimulated by, 616
  relationship of muscle fibers to, 616 *fig.*
  structure of, 567 *fig.*
Motor output, nervous system, 566
Motor units, 616

Mountains, effect of, on rainfall, 689 *fig.*
Mouth, 432
  human, 434, 434 *fig.*
Movement, 604–20
  animal locomotion and, 606–7
  cell, 64, 65 *fig.*, 67, 551 *fig.*, 554
  in elephants, 604, 605
  internal maps in animal, 711
  muscles and animal, 613–18 (*see also* Muscle)
  skeletal support for animal, 608–13
  stimuli, landmarks, and animal, 710
Movement corridors, 774
MRI (magnetic resonance imaging), 422–23
MRM (magnetic resonance microscopy), 423, 423 *fig.*
Mucous cells (stomach), 436, 436 *fig.*
Mucous membrane, 416
  as defense against infection, 486
Müllerian mimicry, 746–47
Multicellular green algae, 338, 338 *fig.*
Multicellular organisms
  evolution of, from unicellular protists, 339, 339 *fig.*
  seaweed as, 336
Multiple fruit, 639, 639 *fig.*
Multiple sclerosis (MS), 500, 567
Muscle
  antagonistic pairs of, 613 *fig.*
  blood distribution controlled by smooth, 477
  contractile apparatus of, 614–15
  interaction of skeleton and, in movement, 613
  motor neurons and contraction of, 616
  types of, 418 *fig.*
Muscle cells
  aerobic function in slow muscle, 89
  anaerobic function in slow muscle, 89
  contractile apparatus in each, 614
Muscle contraction, 614–15
  motor neurons and, 616
  sliding filament model of, 614–15, 615 *fig.*
Muscle fibers
  neuromuscular junction and, 616 *fig.*
  relationship between motor neurons and, 616 *fig.*
  slow *versus* fast, 88–89
Muscle tissue, 418
  capillaries in, 468 *fig.*
  contraction of, 614–15
  motor neurons and, 616
  types of, 418 *fig.*
Muscular dystrophy, Duchenne, 177
Muscular system, 421 *fig.*
Mushrooms, 359, 359 *fig.*, 363
  life cycle, 360, 360 *fig.*
Mus musculus (mouse), 246
Mustard plant (*Arabidopsis thaliana*), 246, 624, 624 *fig.*
Mutagenesis, 199
Mutagens, 199
Mutation(s), 199
  cancer caused by, 222–23, 223 *fig.*, 225 *fig.*, 226
  emerging viruses due to, 202–3
  in gene expression in *Drosophila,* 220
  genetic variation due to, 199, 270–71
  by HIV virus, 498
  types of, and effects of, 199 *fig.*
Mutualism. *See also* Symbiosis
  fungi and animals, 362–63, 362 *fig.*
  lichens and algae, 362, 362 *fig.*
  mycorrhiza and plants association as, 342–43
  as symbiotic relationship, 748, 749
Muybridge, Eadweard, 604

Myanmar tigers, conservation of, 764–65
Mycelium, 357
Mycoplasmas, 328
Mycorrhiza, 342–43, 357, 657, 657 *fig.*
Mycosis, 361
Myelin sheath, 567
Myofibrils, 614
Myosin filaments, muscle contraction and, 614, 614 *fig.*, 615 *fig.*
Myxobacteria, 328

N

NAD⁺ (nicotinamide adenine dinucleotide), cellular respiration and, 92, 100, 101
NADH, cellular respiration and role of, 92, 92 *fig.*, 93, 93 *fig.*, 94, 96, 97 *fig.*
NADP⁺, photosynthesis and, 111, 115
NADPH, photosynthesis and, 111, 114, 115
Naked mole rats, altruism among, 721, 721 *fig.*
Native American cultures, relationships among, 309
Natural family planning, 547
Natural killer cells, 486
Natural selection, 8–9, 8 *fig.*
  adaptations to abiotic or biotic factors and, 687
  canine species as result of, 259 *fig.*
  evolutionary adaptation resulting from, 264, 264 *fig.*
  evolutionary fitness and, 273
  general outcomes of, 274, 274 *fig.*
  genetic variation in populations affected by, 272–73
  as mechanism of evolution, 258
  parasitism and, 749
  perfection not produced by, 275
  reproductive success and, 715
Nature, Cell of the Month Image Competition, 51
Nature *versus* nurture debate, 706
Neanderthals, 398–99, 405
Nearsightedness, 596, 596 *fig.*
Negative feedback. *See also* Feedback inhibition
  homeostasis and mechanisms of, 426, 426 *fig.*
Negative pressure breathing, 460
Nematodes (*Caenorhabditis elegans*), 246, 375
Nephrons, 513
  detailed structure of, 512 *fig.*
  orientation within kidney of, 512 *fig.*
  reabsorption and secretion in, 514–15, 514 *fig.*
Nerve(s), 566
  cranial and spinal, 575
Nerve cords, 574
  dorsal, hollow, 385, 385 *fig.*
Nerve gases, enzymes' effects on, 78
Nerve net, 574, 574 *fig.*
Nerve signals, transmission of, 568–73
Nervous systems, 421 *fig.*, 564–87. *See also* Central nervous system; Peripheral nervous system
  of animals, 574–78
  autonomic, 576, 577 *fig.*
  body symmetry and organization of, 574
  brain of vertebrate (human), 578–85
  disorders of, 584–85
  endocrine system and, 520
  immune system and, 501
  invertebrate, 574 *fig.*
  nerve signal transmission, 568–73
  neurons as functional units of, 567 (*see also* Neuron(s))
  organization of, 566 *fig.*

parasympathetic and sympathetic, 576 *fig.*, 577 *fig.*
sensory input and response of, 590, 602
somatic, 576
spinal cord, 564–65, 577 *fig.*
structural and functional overview of, 566–67
vertebrate, 575–85
Nervous tissue, 418
Neural folds, 552 *fig.*
Neural tube
development of brain from, 578
formation of, 552, 552 *fig.*
Neurological disorders, 584
Neuromuscular junctions, 616, 616 *fig.*
Neuron(s), 418, 418 *fig.*, 566
action potential propagation along, 570, 570 *fig.*
communication between, at synapses, 571–73, 571 *fig.*
electrical potential in plasma membrane of, 568
as functional units of nervous system, 567
nerve signal transmission and changes in membrane potential of, 568–69
structure and function of, 567 *fig.*
Neurosecretory cells, 520, 520 *fig.*
Neurotransmitter(s), 520, 520 *fig.*, 571
chemical synapses and, 571 *fig.*
small molecules functioning as, 572–73
Neutralization of pathogens, 495, 495 *fig.*
Neutral variation, 273
Neutrons, 20
Neutrophils, 479, 479 *fig.*, 486
New Guinea, animal diversity in, 367
New World monkeys, 401, 401 *fig.*
Niches, 745
Nicotine, 573, 748
Nikon Small World Competition, 51
Nirenberg, Marshall, 192
Nitrogen
atom, 22 *fig.*
in human body, 18, 18 *table*, 19
as plant nutrient, 646, 652, 653, 653 *fig.*
Nitrogen cycle, 758–59, 758 *fig.*
Nitrogen fixation, 658–59
Nitrogenous base(s), 47, 47 *fig.*
Chargaff's rules on pairing of, 186
of DNA and RNA, 185 *fig.*
DNA replication dependent upon pairing of complementary, 188
formation of, 319
mutations in, 199
sequences of (*see* DNA sequence(s))
triplet code of, 191 (*see also* Codons)
Watson and Crick's rules on, 187
Nodes, stem, 626
Nodes of Ranvier, 567
Noncompetitive inhibitors of enzymes, 78
Nondisjunction in meiosis, 146
Non-native (exotic) species, 394, 726–27, 768
Nonpolar covalent bonds, 24
Nonself molecules, 497
Norepinephrine, stress and release of, 530
North America, human population centers in, 265 *fig.*
Notochord, 385, 385 *fig.*, 552
Nuclear envelope, 58
Nuclear transplantation, 218, 218 *fig.*
Nucleic acid probes, 238
Nucleic acids, 36. *See also* DNA; RNA
in bacteria *versus* in archaea, 322–23
digestion of, 438 *table*, 439
as nucleotide polymers, 47
Nucleoid region, 55 *fig.*
of prokaryotic cells, 55
Nucleolus, 58

Nucleosomes, 213
Nucleotides, 47 *fig.*, 184. *See also* DNA; RNA
DNA and RNA as polymers of, 47, 184–85
homeobox sequences of, 222
nitrogenous bases of (*see* Nitrogenous base(s))
noncoding (introns, exons), 194
promoter sequence of, 193
reading frame, 199
restriction fragment length polymorphism analysis of, 240–41
sequences of (*see* DNA sequence(s))
Nucleus (atom), 20
Nucleus (cell), 58, 58 *fig.*
cloning techniques using, 218, 218 *fig.*
Nutrient pollution in ecosystems, 761
Nutrition
animal (*see* Animal nutrition)
classification of organisms by mode of, 326 *table*
fungal, 357
genetic influences of, 170
plant (*see* Plant nutrition)
reducing cancer risk by changes in, 227

O

Obesity, human, 447
Ocean(s), 690–91
climate and, 689
coral reefs, 691, 691 *fig.*
deep sea, 682–83
hydrothermal vents in, 327, 683 *fig.*
sulfate-consuming bacteria and methane-consuming archaea in, 326
zones of, 690–91, 690 *fig.*
Octopus, 377, 377 *fig.*
Odonata Order, 382
Oenothera gigas (evening primrose), 286, 286 *fig.*
Oenothera lamarckiana (evening primrose), 286, 286 *fig.*
Oil spills, 331, 331 *fig.*, 363, 778
Old World monkeys, 401, 401 *fig.*
Olfactory receptors, 601
Oligomycin, 99, 99 *fig.*
Omnivores, 430
Oncogenes, 223
alternative ways of making, from proto-oncogenes, 223 *fig.*
interference by, in signal-transduction pathways, 224–25
One gene–one polypeptide hypothesis, 190
Onion root, cell division in, 136 *fig.*
On the Origin of Species by Means of Natural Selection (Darwin), 8, 257, 280
Oocytes, primary and secondary, 543
Oogenesis, 543, 543 *fig.*, 544
Oparin, A. I., 318, 319
Open circulatory system, 378, 469
Operation Migration, 709, 709 *fig.*
Operator DNA segment, 210, 211 *fig.*
Operculum, 388
Operons, 210
activator controlled, 211
lac, 210, 211 *fig.*
repressor controlled, 210–11, 211 *fig.*
trp, 211
Opiates, 573
Opportunistic infections, 498
Opposable thumbs, 401
Optimal foraging theory, 714
Oral cavity, human, 434, 434 *fig.*
Oral contraceptives, 547
Orangutans, 402, 402 *fig.*
Order as property of all organisms, 5
Orders (taxonomy), 305
Ordovician period, 387

Organ(s), 2, 2 *fig.*, 419. *See also names of specific organs*
embryonic formation of, 552–55
structural hierarchy and, 415
systems of, 419–21
Organelles, 2, 2 *fig.*, 51, 56
of endomembrane system, 58–62
energy-converting, 63
functional categories of, 67, 67 *table*
Organic compounds, 34–36
carbohydrates, 37–39
carbon-based, 34
functional groups, 35, 35 *table*
lipids, 40–41
nucleic acids and, 47
polymers, of, and molecule formation, 36
proteins, 42–44
Organic farming, 656
Organic molecules, origin of, 318–19
Organisms, 2, 2 *fig. See also* Animal(s); Life; Plant(s)
adaptation of, to biospheric factors, 687
arranging, into kingdoms, 310
chemistry of (*see* Chemistry of life)
classification of, 304–5
communities of (See Community(ies))
evolution of (*see* Evolution)
genetically modified, 247–48
interaction of, with environment, 684 (*see also* Ecology)
levels of organization in, 2–3
life history of, 734–35
multicellular, 339, 339 *fig.*
nutritional classification of, 326 *table*
reproduction in (*see* Reproduction)
structural hierarchy of, 415
transgenic, 247
Organ of corti, 598, 599 *fig.*
Organ systems, 2, 2 *fig.*
human body, 419–21 *fig.*
structural hierarchy and, 415
Orgasm, 546
Orthoptera Order, 382
Osmoconformers, 508
Osmoregulation, 82–83, 83, 505, 508–9
Osmoregulators, 508–9
Osmosis, 82, 82 *fig.*
Osteoporosis, 612–13, 613 *fig.*
Ostrom, John, 296–97
Outer ear, 598
Outgroup, cladistic analysis and, 306
Ova. *See* Egg(s)
Oval window, 598
Ovarian cycle, 544–45, 545 *fig.*
hormones of, 544 *table*
Ovaries, 538
Ovary, plant, 352, 636
fruit development from, 639
ovule development in, and fertilization, 350–51, 637, 637 *fig.*
Overconsumption, 737
Overproduction and competition, 8
Oviduct (fallopian tube), 538
Ovulation, 538, 538 *fig.*, 539 *fig.*, 556 *fig.*
hormonal events at and after, 544–45
hormonal events before, 544
Ovules, plant, 350–51, 636
development of, and fertilization, 637, 637 *fig.*
pine tree, 350–51, 351 *fig.*
seed development from, 638, 638 *fig.*, 639
Ovum. *See* Egg(s)
Owl, endangered species of, 772 *fig.*
Oxidation, 92
Oxidative phosphorylation
in cellular respiration, 93, 98, 98 *fig.*, 100
photophosphorylation *versus*, 115

Oxygen
  atom, 22 *fig.*
  breathing and intake of, 90, 454
  double bonds in, 23
  exchange of carbon dioxide and, through
    respiratory surfaces, 454–55, 462–63
  hemoglobin and loading/unloading of,
    463 *fig.*
  in human body, 18, 18 *table*, 19
  photosynthetic light reactions and, 114
  plant production of, by splitting water,
    110
Oxytocin, 524
  childbirth and role of, 560 *fig.*
  prairie vole behavior and, 706, 706 *fig.*
Ozone layer depletion, 120, 120 *fig.*, 769

P
p53 tumor-suppressor gene, 225
Pacemaker, 473
Pacific yew, pharmaceuticals from, 135
Paedomorphosis, 292, 292 *fig.*
Pain
  glucocorticoids for relief of, 531
  nervous system and perception of, 592
Paine, Robert, 747
Pain receptors, 592
Paleoanthropology, 403
Paleontologists, 260
Paleozoic era, 299
  mass extinction and, 302, 339
Pancreas, 438, 528–29
Pancreatic hormones, 528–29
Pandas, 103
Pangea, 300
Pangolin, 8–9, 9 *fig.*
Paramecium, 62, 62 *fig.*, 745
Parasite(s)
  control of pests with, 749 *fig.*
  flukes (flatworms) as, 374
  fungi as, 361
  of plants, 361
  plants as, 659, 659 *fig.*
  roundworms as, 375 *fig.*
  tapeworms as, 374, 374 *fig.*
Parasitism, as symbiotic relationship, 748–49
Parasympathetic division, autonomic
    nervous system, 576 *fig.*
  regulation of internal environment and,
    576–77
Parathion, as pesticide, 78
Parathyroid glands, 526–27
  calcium homeostasis regulated by, 527 *fig.*
Parathyroid hormone (PTH), 526–27
Parenchyma cells, 628, 628 *fig.*
Parietal cells, 436, 436 *fig.*
Parkinson's disease, 572, 585
Parsimony, cladistics and, 307
Partial pressure, 462
Passive immunity, 489
Passive transport, 81
Pasteur, Louis, 318
Paternity, certainty of, 715
Pathogens, 322. *See also* Disease; Human
    disease
  bacterial, 328, 329, 330, 437, 487
  natural selection and, 749
  plant defenses against, 677
  viral, 202–3
Pattern formation, 555
Pauling, Linus, 46, 46 *fig.*, 186
PCBs (toxic compound), 769
PCR (polymerase chain reaction), amplifying
    DNA sequences using, 244
Peas, Mendel's studies on genetics of,
    154–60
Pecking order, 718

Pedigree, human genetic traits tracked
    through, 161
Pelagic zone, 690
Pelicans
  resurgence of brown, 1, 781
  structural hierarchy in, 415 *fig.*
Penicillin, 78, 272, 363 *fig.*
Penis, 541
Pepsin, 436, 436 *fig.*
Pepsinogen, 436, 436 *fig.*
Peptide bonds, 43
  formation of, 43 *fig.*, 197
Peptide hormones, 521
Peptidoglycans, 323
Perception, 590
Perennials, 632
Perforin, 499
Peripheral nervous system (PNS), 566, 574,
    575 *fig.*
  cranial nerves and spinal nerves of, 575
  functional hierarchy of, 576, 576 *fig.*
Peristalsis
  in alimentary canal, 433, 435, 435 *fig.*
  earthworm locomotion by, 607, 607 *fig.*
Periwinkle, pharmaceuticals from, 135,
    767 *fig.*
Permafrost, 698
Permian mass extinction, 302, 303
Pesticides, 685
  biological magnification and
    environmental damage caused by,
    769 *fig.*
  enzymes' effects on, 78
  pest populations resistant to, 264 *fig.*
Pests
  control of, with parasites, 749 *fig.*
  pesticide resistance in, 264 *fig.*
  wasps and control of, 673, 676–77, 742–43
Petals, 352, 636
PET (positron-emission tomography)
    imaging, 423, 423 *fig.*
  scanner, 21 *fig.*
Petrification, 260, 260 *fig.*
Pfennig, David and Karin, 10
P generation, 155
pH, 27
  enzymes' effects on, 77
  scale, 27 *fig.*
Phages, 182–83. *See also* Bacteriophage(s)
  T2, 182 *fig.*
Phagocytes, 479
  inflammatory response and, 487, 487 *fig.*
Phagocytosis, 85, 85 *fig.*, 486 *fig.*
Pharmaceuticals
  antibiotics, 328
  applications of reproductive cloning to,
    218–19
  for chemotherapy, 135
  DNA technology and production of, 237
  plants as source of, 356 *table*, 678, 767
Pharyngeal slits, 385, 385 *fig.*
Pharynx, 432, 458
  human, 434, 434 *fig.*
Phenotype, 157
  environmental influence on, 170
  evolutionary fitness, natural selection,
    and, 273
  genotype and, 157, 166
  incomplete dominance and intermediate,
    166–67
  influence of single genes on (pleiotropy),
    168, 168 *fig.*
  proteins as molecular basis for traits of,
    190
Phenylalanine (Phe), 192
Phenylketonuria (PKU), Hardy-Weinberg
    equation and incidence of, 267

Phenylketonuria screening, 165
Pheromones, 519
Philadelphia chromosome, 148, 148 *fig.*
Phloem, 345, 629
  food-conducting cells of, 650 *fig.*
  primary, 633
  secondary, 634–35
  transport of sugars in, 650–51
Phloem sap, 650
Phosphate group, 35, 47, 47 *fig.*
Phospholipid bilayer of membranes, 79,
    79 *fig.*
Phospholipids, 41
  as main components of membranes, 79,
    79 *fig.*
Phosphorus
  cultural eutrophication and, 761
  as plant nutrient, 652, 653, 653 *fig.*
Phosphorus cycle, 759, 759 *fig.*
Phosphorylation, 75
  active transport and, 84, 84 *fig.*
  oxidative, in cellular respiration, 93, 98, 98
    fig.
  substrate-level, 94, 94 *fig.*
Photic zone, 690
Photoautotrophs, 108, 108 *fig.*, 326
Photoheterotrophs, 326
Photons, 112
Photoperiod, 674
Photophosphorylation, 115
Photopsins, 597
Photoreceptors, 593, 593 *fig.*, 594–95
  rods and cones in vertebrate eye, 597, 597
    fig.
Photorespiration, 118
Photosynthesis, 106–22
  adaptations of C_3, C_4, and CAM plants,
    118
  ATP and NADPH production in, 114
  ATP production in, 111
  autotrophs as producers and, 108
  Calvin cycle in, 111, 116
  carbon cycling and, 758
  cellular respiration and, 90, 90 *fig.*
  chemical equation for, 110
  chloroplast as site of, 109
  cyanobacteria as generator of, 328
  definition of, 106
  as endergonic reaction, 74
  fuel for cellular respiration originating
    from, 103
  light reactions of, 111, 112–15
  overview of, 108–11
  oxygen production in, 110
  radioactive isotopes for research on, 21
  as redox process, 110–11
  review of, 117
  solar radiation, Earth's atmosphere, and,
    119–20
  stages of, linked by ATP and NADPH, 111
  summary of chemical processes of, 117 *fig.*
Photosynthetic bacteria, 314, 315. *See also*
    Cyanobacteria
Photosynthetic prokaryotes, 325, 325 *fig.*
Photosynthetic protists (algae), 333
Photosystem, 113
  I (P700), 113, 114 *fig.*, 117 *fig.*
  II (P680), 113, 114 *fig.*, 117 *fig.*
  components, 113 *fig.*
Phototropism, 664–65, 664 *fig.*, 665 *fig.*, 672
Phyla (taxonomy), 305
Phylogenetic species concepts, 281
Phylogenetic trees, 305
  of animals, 371, 371 *fig.*
  based on molecular data, of bears and
    raccoons, 308 *fig.*
  of chordates, 386

cladistic analysis used to construct, 306–7
of fungi, 358–59, 358 *fig.*
of humans, plants, and fungi, 309 *fig.*
molecular, of animal kingdom, 393 *fig.*
of plants, 346 *fig.*
of primates, 401 *fig.*
of reptiles, 307 *fig.*
Phylogeny, 304
   bacterial, 328
   classification of organisms and, 305 *fig.*
   of eukaryotes (tentative), 334, 334 *fig.*
   homologous structures and, 304
   systematics and, 304–5, 306–7
Physarum (plasmodial slime mold), 337 *fig.*
Physiology, 414
Phytochromes, as light detector, 675
Phytoestrogens, 662–63
   chemical structure of, 662 *fig.*
Phytoplankton, 690
Phytoremediation, 644–45
Pigments
   carotenoids, 112, 113
   chlorophyll (*see* Chlorophyll)
   vision-related, 597
Pili/pilus, 55, 324, 324 *fig.*
Pine, life cycle of, 350–51, 351 *fig.*
Pineal gland, 522
Pinna, 598
Pinocytosis, 85, 85 *fig.*
Pitch (sound), 599
Pith, 630
Pituitary gland, 524
   hormones of, 525 *fig.*
Pivot joints, 611, 611 *fig.*
Placebo-controlled trials, 485
Placenta, 463 *fig.*
   formation of, 556, 557, 557 *fig.*
Placental mammals, 392
Planarians, 374
   eye cups of, 594, 594 *fig.*
   nervous system, 574, 574 *fig.*
Plant(s), 342–56.
   adaptations for life on land, 344–45
   agave, "big-bang" reproduction of, 734, 734 *fig.*
   agriculture and, 354, 642, 655, 655 *fig.*, 656, 671
   analogous structures of, 304 *fig.*
   angiosperm (*see* Angiosperms)
   as autotrophs, 108
   biological rhythms in, 673–76
   C_3, 118
   C_4 and CAM, 118
   carnivorous, 659, 659 *fig.*
   cells of (*see* Plant cells)
   defenses of, 676–78, 748
   dicot and monocot, 625, 625 *fig.*
   diseases and pests of (*see* Plant diseases and pests)
   diversity of, 346–47, 356
   DNA technology and enhanced, 247–48
   evolution of, 344–45, 346 *fig.*, 347
   fungi association with, 342–43, 357
   genetically engineered, 247
   growth of, 632–35, 672
   growth responses (tropisms) in, 664–65, 672
   gymnosperms (*see* Gymnosperms)
   hormones of, 664–71
   humans, relationship of, 309 *fig.*
   life cycles of, 348–56
   mechanical isolation in, 283 *fig.*
   medicinal, 356 *table*, 678
   multicellular green alga compared to, 345 *fig.*
   mutation rates in, 271
   mycorrhizal fungus and, 342–43, 357
   nutrition (*see* Plant nutrition)

parasitic, 659, 659 *fig.*
photosynthesis in (*see* Photosynthesis)
phytoremediation by, 644–45
polyploidy and sympatric speciation in, 286 *fig.*, 287
Raikhel on, 624
red algae and green algae as relatives of, 338
reproduction in (*see* Plant reproduction)
seed, 346–47 (*see also* Seed(s))
self-fertilizing, 154
structure and function of, 625–31
transgenic, 247
trees (*see* Tree(s))
Plantae (kingdom), 6, 7 *fig.*, 310, 343. See *also* Plant(s)
Plant body, 626 *fig.*
   cells and tissues of, 628–29
   roots and shoots of, 626, 627
   tissue systems of, 630
Plant cells, 57, 57 *fig.*, 628–29
   auxin and elongation of, 667 *fig.*
   cell division in, 128 *fig.*
   cell plate formation in, 132 *fig.*
   chloroplast (*see* Chloroplasts)
   cytokinesis in, 132
   cytokinins and division of, 668
   osmoregulation of water balance in, 83 *fig.*
   structure of, 628 *fig.*
   types of, 628 *fig.*, 629 *fig.*
   vacuoles of, 62
   walls and junctions of, 66 *fig.*
Plant diseases and pests
   defenses against, 676–78, 748
   foolish seedling disease, 668
   fungal, 361
   viral, 202
Plant growth, 632–35
   hormone regulation of, 666
   primary, 632, 632 *fig.*, 633 *fig.*
   secondary, 634–35, 634 *fig.*
Plant nutrition, 644–61
   agriculture and, 656
   deficiencies in, 653
   fungi and, 342–43, 657
   metal accumulators, 644–45
   nitrogen-fixing bacteria and, 658–59
   nutrient uptake, 646–47
   phloem transport and, 650–51
   plants as parasites and carnivores and, 659
   soil and, 652–57
   symbiosis and, 657–59
   water transpiration and, 648–49
Plant reproduction, 636–42
   alternation of generation and cycles of, 348–56
   angiosperm, 352–55, 353 *fig.*, 636–40
   asexual, 641, 641 *fig.*, 642, 642 *fig.*
   flower in, 352, 636 *fig.*, 637, 637 *fig.*
   fruit and, 354, 639
   gymnosperm, 350–51, 351 *fig.*
   for modern agriculture, 642
   reproductive isolation in, 282–83
   seed, 346–47, 638, 638 *fig.*, 639
   seed germination, 636, 640, 640 *fig.*
   sexual life cycle, 636 *fig.*, 637 *fig.*
Plant tissues. *See* Tissue systems, plant
Plasma, blood, 479
   composition of, 479 *fig.*
Plasma membrane, 55 *fig.*, 66 *fig.*
   of animal cell, 80 *fig.*
   cellular chemical activity and, 79, 86
   cellular water balance and, 82–83
   in cross section, 79 *fig.*
   faulty, 85
   fluid mosaic model of, 80
   infolding of prokaryotic, and origins of endomembrane system, 332 *fig.*

phospholipid bilayer of, 79
of prokaryotic cell, 55
proteins of, 80–81
transport across, 81–85
Plasma-membrane receptor, hormones binding to, 521 *fig.*
Plasmids
   as carriers for gene transfer, 205
   cloning genes in recombinant, 234, 234 *fig.*
   customizing bacteria with, 232–33
   F-factor, 205 *fig.*
   genomic library of, 234, 235, 235 *fig.*
   Ti, 247, 247 *fig.*
Plasmodesmata, 66, 66 *fig.*
Plasmodial slime mold, 337, 337 *fig.*
Plasmodium, 335, 335 *fig.*
Plasmolysis, 83
Plateau phase, of sexual response, 546
Platelets, 479
   blood clotting and, 480–81, 480 *fig.*
Plate tectonics, 302
Platyhelminthes, 374
Platypus, 366–67, 392, 392 *fig.*, 593
Pleated sheet protein structure, 44, 45 *fig.*, 46
Pleiotropy, 168
Pneumocystis pneumonia, 498
Poison arrow frogs, 389, 389 *fig.*
Poisons as disruptors of cellular respiration, 99
Polar covalent bonds, 24
Polar molecules, 24 *fig.*
Poliovirus, 201
Politics, science and, 120
Pollen, 347
Pollen grains, 351
Pollination, 351, 637, 637 *fig.*
   animals and, 355, 355 *fig.*
Pollution, environmental, 769. *See also* Air pollution
   habitat destruction and, 768
   by nutrients, 761
Polychaetes, 378, 379
Polygamous mating behavior, 715
Polygenic inheritance, 169, 270
Polymerase chain reaction (PCR), amplifying DNA sequences using, 244
Polymers, 36
   first, 318
   of nucleotides, DNA and RNA as, 47
Polymorphic populations, 270
   balanced, 273
Polynucleotides, 184–85
Polyp(s), 373
Polypeptides, 43. *See also* Amino acids; Protein(s)
   synthesis of, 198, 198 *fig.*, 320
   translation and elongating chains of, 197 *fig.*
Polyploid cells, 286
Polysaccharides, 39, 39 *fig.*
Pompe's disease, 61
Pons, 578, 579 *fig.*
Population(s), 2, 2 *fig.*
   Darwin's theory of evolution in, 256–57
   declining population approach to, 772–73
   definition of, 265, 728
   genetic variation in, 270–75
   growth of (*see* Population growth)
   human (*see* Human population)
   interaction of, with environment, 684 (*see also* Ecology)
   life tables of mortality and survivorship in, 729, 729 *table*
   microevolution of, 268–69
   natural selection and evolution of, 258–61
   small-population approach to, 772
   as units of evolution, 265

Population density, 728–29
  factors limiting population growth
    dependent on, 732–33
Population dynamics, 726–41
  behavior and (see Social behavior)
  human, 736–39
  life histories and evolution in, 734–35
  non-native species and, 726–27
  population ecology principles and, 735
  structure of populations and, 728–33
Population ecology, 727, 735. See also
    Population dynamics
Population genetics, 265–69
  gene pool and, 266–67
  Hardy-Weinberg equilibrium and,
    268–69
  microevolution and, 268–69
Population growth
  boom-and-bust cycles of, 733, 733 fig.
  density-dependent factors limiting,
    732–33
  exponential growth model of, 730
  human, 736–39
  logistic growth model and limiting factors
    in, 730–31
Porifera, 372
Porpoises, 392
Positron-emission tomography (PET), 423,
    423 fig.
Post-anal tail, 385, 385 fig.
Posterior pituitary, 524
  hormones of, 525 fig.
Posterior surface of animals, 370, 370 fig.
Postzygotic barriers, 282 table, 283
Potassium-40, for radiometric dating, 299
Potassium as plant nutrient, 652, 653,
    653 fig.
Potential energy, 72, 72 fig.
Pottos, 400
Prairies, 696–97, 696 fig.
Prairie voles, behavior in, 706
Precambrian period, 299, 339
Precapillary sphincters, 477, 477 fig.
Precipitation
  acid, 28
  in deserts, 695
  of dissolved antigens, 495, 495 fig.
  mountains' effects on, 689 fig.
  uneven heating as cause of, 689 fig.
Predation
  adaptations of prey/predators for, 746–47
  coevolution and, 748
  community diversity maintained by, 747
  group foraging behaviors and, 714–15
  life history shaping through evolution
    and, 734–35, 734 fig.
  plant defenses against, 676–77, 677 fig.,
    748
  population density and, 732
Predators
  coevolution and prey interactions with,
    748
  group foraging behaviors of, 715
  keystone, 747
  plant recruitment of predatory animals,
    676–77, 677 fig.
  population cycles and, 733
  prey defenses against, 746–47
  scientific method and study of prey and,
    10–11
Pregnancy, human, 538–39
  childbirth and, 560–61
  development of embryo and placenta,
    556–57
  trimesters of, 558–59, 558–59 fig.
Pregnancy tests, home, 496
Prenatal testing, 164–65

Prepuce, 539, 541
Pressure-flow mechanism in plant phloem
    tissue, 650, 651 fig.
Prevailing winds, 688–89
Prey
  coevolution and predator interactions
    with, 748
  population cycles and, 733
  predator defenses by, 746–47
  scientific method and study of predators
    and, 10–11
Prezygotic barriers, 282–83, 282 table
Primary consumers, 752, 752 fig., 753 fig.
Primary immune response, 492 fig., 493
Primary oocytes, 543
Primary phloem, 633
Primary production, 754–55
Primary spermatocytes, 542–43
Primary structure of proteins, 44, 45 fig.
Primary succession, 750
Primary xylem, 633
Primates, 400–402
  apes as closest relatives of humans, 402
  hominid evolution and, 403–5
  human branch of, 403–5
  phylogenetic tree of, 401 fig.
  reasoning in, 713
  signaling behavior in, 720
  skeleton of, 610 fig.
Primitive characters, shared, 306
Principles of Geology (Lyell), 256
PRL (prolactin), 525
Probability, Mendelian inheritance and rules
    of, 160
Probes, nucleic acid, 238
Problem solving, 713
Producers, 3, 3 fig., 108, 752, 753 fig.
Production, pyramid of, 755 fig.
Products of chemical reactions, 29
Progesterone, ovarian and menstrual cycles
    and, 544–45, 545 fig.
Progestins, 532
Programmed cell death, 554, 554 fig.
Progressive retinal atrophy (PRA), 153
Prokaryotes, 6, 6 fig., 310, 322–31
  archaea and bacteria as main branches in
    evolution of, 322–23
  cyanobacteria, 314–15
  as detritivores, 752
  early origins of, 321, 322
  as first life forms, 316–17
  fossilized, 261, 317 fig.
  gene regulation in, 210–11
  nutritional modes of, 326
  recycling and cleaning environment using,
    330–31
  shapes of, 323, 323 fig.
  structural features of, 324–25
Prokaryotic cells, 4, 4 fig., 55 fig.
  binary fission of, 127 fig.
  size and structure of, 55, 324–25
Prokaryotic flagella, 55, 55 fig., 324–25,
    324 fig.
Prolactin (PRL), 525
Prometaphase, mitosis, 130–31, 130 fig.
Promiscuous mating behavior, 715
Promoter DNA sequences, 193, 210, 211 fig.
Pronghorn antelopes, 687, 687 fig.
Propagation, vegetative, 641
Prophages, 200
Prophase
  cell division I, of meiosis I, 138 fig.
  cell division II, of meiosis II, 139 fig.
  mitosis, 130, 130 fig.
Prostaglandins, 560, 560 fig.
Prostate gland, 540
Protease inhibitors, 78

Protein(s), 36, 42–44. See also Amino acids
  amino acids and formation of, 42–43
  biosynthesis of, 103 fig.
  as chemical compounds, 19
  classes of, 42
  Class I and II, 500
  comparison of, in systematics, 308–9
  denaturation of, 43
  digestion of, 438–39, 438 table
  DNA technology and mass production of,
    236 table
  essential amino acids and, 443
  as essential to life, 42
  as fuel for cellular respiration, 102, 102 fig.
  in membranes, 80–81
  nucleic acids as polymer blueprints for, 47
  peptide bonds in, 43 fig.
  post-translational breakdown of, 216
  post-translational controls on activation
    of, 216
  receptors, 80–81
  secretory, 59, 59 fig.
  self and nonself, 497, 500
  shape and function, 43
  signal transduction and role of, 81, 221
  structure of, 44
  translation of RNA information into, 190
  vegetarian diets and, 656–57
Protein hormones, 521
Proteobacteria, 328
Proteomics, 246
Proterozoic eon, 299
Proteus (bacterium), 324 fig.
Protista (kingdom), 310
Protists, 7 fig., 332–39
  alveolates, 335
  amoebozoans, 337
  classification of, 334–38, 334 fig.
  diplomonads, 334
  euglenozoans, 334–35
  evolution of animal kingdom from
    colonial, 369
  multicellular life originating in, 339
  origins and evolution of, 332–33
  Paramecium, 62, 62 fig., 745
  red algae and green algae, 338
  stramenopiles, 336
  as unicellular eukaryotes, 333, 333 fig.
Protobionts, 321
Protons, 20
Proto-oncogenes, 223
  ras, 224–25
Protostomes, 370, 371, 393
Protozoans, 6, 333
Proximal tubule, of kidney nephron, 512 fig.,
    513, 514, 514 fig.
Proximate questions, about behavior, 704
Pseudocoelom, 370, 370 fig.
Pseudopodia (pseudopodium), 337
Ptarmigan, 425, 425 fig.
Pterophytes, 346
Public health science, Hardy-Weinberg
    equation useful in, 267
Puerto Rico, tropical rainforest in, 699
Pulmocutaneous circuit, 470
Pulmonary arteries, 471
Pulmonary circuit, 470
Pulmonary veins, 471
Pulse, 475
Punctuated equilibrium model of speciation,
    290, 290 fig.
Punnett, Reginald, 172
Punnett square, 157
Pupfish, 285, 285 fig.
Pupil (eye), 594, 595 fig.
Purines, 185
Pyloric sphincter, 433, 436 fig.

Pyramid of production
  human meat-eating as luxury explained
    by, 756
  idealized, 755 *fig.*
Pyrimidines, 185
Pyruvate
  cellular respiration and production of, 93,
    94, 95 *fig.*
  chemical grooming for citric acid cycle,
    96, 96 *fig.*

Q

Quaternary consumers in communities, 752,
  752 *fig.*, 753 *fig.*
Quaternary structure of proteins, 44, 45 *fig.*
Quolls, 367, 394, 394 *fig.*

R

Rabbits, as non-native pest species, 394,
  749 *fig.*
Raccoons, systematics of bears and, 308,
  308 *fig.*
Radial symmetry, 370, 370 *fig.*, 373
Radiation
  heat gains or losses by, 506
  as mutagen, 199
Radiation therapy, 135
Radioactive isotopes, 20, 21
Radiometric dating, 299
Radon, 21
Radula, mollusc, 376
Raikhel, Natasha, 624, 624 *fig.*
Rainfall. *See* Precipitation
Rain forests. *See* Tropical rain forests
Ramón y Cajal, Santiago, 50, 51 *fig.*
Random population-dispersion patterns,
  728 *fig.*, 729
Ras proto-oncogene, 224–25
Rattlebox moth, 16, 16 *fig.*, 17
Rattlebox plant, 16
Ray-finned fishes, 388, 388 *fig.*
Reabsorption function of excretory system,
  513, 513 *fig.*, 514–15, 514 *fig.*
Reactants in chemical reactions, 29
Reaction center of photosynthetic pigments,
  113
Reading frame, 199
Receptor-mediated endocytosis, 85, 85 *fig.*
Receptor potential, 590–91
Receptors, protein, 80–81
Receptors for hormones, 521, 521 *fig.*
Recessive alleles, 156, 161
Recessive traits, genetic disorders as, 161,
  162, 163 *table*
Reciprocal altruism, 721
Recombinant cells and organisms, 236
Recombinant DNA, use of restriction
  enzymes and DNA ligase to create, 233,
    233 *fig.*
Recombinant DNA technology, 232. *See also*
  DNA technology
Recombinant gametes, production of, 143,
  172, 173 *fig.*
Recombination frequency, 173
Recommended Daily Allowances (RDAs), 446
Rectum, 440
Red algae, 338, 338 *fig.*
Red blood cells (erythrocytes), 479, 479 *fig.*,
  480 *fig.*
  healthy body function and levels of, 480
Red bone marrow, 611
Red-green color blindness, 177
Redox reactions, 92 *fig.*, 95 *fig.*, 97 *fig.*
  of cellular respiration, 92, 110 *fig.*
  of photosynthesis, 110–11, 116, 116 *fig.*
Red-tide dinoflagellates, 335
Reduction, 92

Reeve, Christopher, 564, 564 *fig.*, 565
Reflexes, 566
  knee-jerk, 566 *fig.*
Regeneration, 213, 536
  by sea stars and echinoderms, 384
Regulation, as property of all organisms, 5
Regulatory genes, 210, 211 *fig.*
Regulatory systems, animal, 504–17
  chemical regulation, overview of, 520–21
  endocrine system as, 522–25
  homeostasis and (*see* Homeostasis)
  hormones and, 523 *table*
  liver and, 511
  osmoregulation and excretion as, 505,
    508–15
  sympathetic and parasympathetic nervous
    systems and, 576–77
  thermoregulation as, 504, 506–8
Regulatory systems, plant, 662–80
  defenses, 676–78
  growth responses, 672
  hormones, 664–71
  internal clocks, 673–76
Releasing hormones, 524
REM sleep, 582
Renal cortex, 512 *fig.*, 513
Renal medulla, 512 *fig.*, 513
Repetitive DNA, 245
Repressor protein, 210
Reproduction, 534–63. *See also* Inheritance
  alterations of chromosome number and
    structure during, 144–48
  asexual, 125, 536–37
  "big-bang" of agave, 734, 734 *fig.*
  cellular (cell division), 126–27
  connections between cell division and,
    126–27
  in green algae, 337, 337 *fig.*
  human, 538–47
  meiosis and crossing over in, 136–43
  mitosis and eukaryotic cell cycle in,
    128–36
  natural selection and success in, 715
  plant, 636–42
  of prokaryotes, 325
  as property of all organisms, 5
  sexual, 124–25, 536–37 (*see also* Sexual
    reproduction)
  of viruses, 201 *fig.*
Reproductive barriers, 282–83, 282 *table*
  evolution of, 285
  speciation and, 284
Reproductive cloning, 218–19
  applications of, 218–19
Reproductive systems, 421 *fig.*
Reptiles (Reptilia), 390
  cardiovascular system, 470
  dinosaurs, 296–97, 390
  excretion in, 510 *fig.*
  gravity and circulatory system of, 466–67
  lizards, 390, 390 *fig.*
  osmoregulation in embryos of, 509
  phylogenetic tree of, 307 *fig.*
  respiratory system of, 458
  snakes, 390, 390 *fig.*
Research
  applications of reproductive cloning to,
    218–19
  placebo-controlled trials, 485
  radioactive isotopes used in, 21
Resolution, of microscopes, 52
Resolution phase, of sexual response, 546
Resource partitioning, 745, 745 *fig.*
Resources, human overexploitation of, 768
Respiration, 452–65. *See also* Cellular
    respiration
  in birds, 460

breathing and, 90, 460, 461
  cellular, 74
  gas exchange and, overview of, 454
  gas exchange through moist body surfaces
    as, 454–55, 455 *fig.*
  gills in aquatic animals and, 456–57,
    456 *fig.*
  in high altitudes, 452–53
  insect tracheal system and, 457, 457 *fig.*
  lungs in terrestrial animals and, 458–61
    (*see also* Lung(s))
  smoking effects on, 459
  transport of gases in body and, 462–63
Respiratory surfaces, 454–55, 455 *fig.*
Respiratory system, 420 *fig.*, 454–55
  exchanges with external environment and
    role of, 424 *fig.*, 425
  in high altitudes, 452–53
  human, 458 *fig.*
  smoking as assault on, 459
Resting potential, neuron membrane,
    568 *fig.*
  nerve signal and changes in, 568–69
Restoration ecology, 778
  Kissimmee River Project, 779, 779 *fig.*
Restriction enzymes, creating recombinant
    DNA with, 233
Restriction fragment length polymorphisms
    (RFLPs), 240
Restriction fragments, 240
  analyzing, to detect differences in DNA
    sequences, 240, 240 *fig.*
  detecting harmful alleles using, 240–41
  procedure for analysis using, 241 *fig.*
  sticky ends and, 233
Restriction sites, 233
Reticular formation, 582, 582 *fig.*
Retina, 594, 595 *fig.*
  nerve cells, 51 *fig.*
  photoreceptor cells (rods and cones) in,
    597 *fig.*
Retinitis pigmentosa, 268
Retrovirus, 203
Reverse transcriptase, 203
  gene cloning and role of, 235
Rheumatoid arthritis, 500
Rhizobium bacteria, 658, 658 *fig.*
Rhizobium spp., 328, 330–31
Rhizomes, 627, 627 *fig.*
Rhodopsin, 597
Rhythm method (contraception), 547
Ribonucleic acid (RNA), 47. *See also* RNA
Ribosomal RNA (rRNA), 196
  in bacteria *vs.* in archaea, 322
Ribosomal RNA (rRNA) sequence analysis,
    309
Ribosomes, 55, 55 *fig.*
  as factories for polypeptides, 196
Ribozymes, 320
Ringworm, 361
Rituals
  agonistic behaviors and, 718
  courtship, 282, 716, 716 *fig.*
Rivers. *See* Lakes and streams
RNA (ribonucleic acid), 47
  comparing DNA and, in systematics,
    308–9
  as first genetic material, 320
  messenger RNA (mRNA), 194
  as nucleotide polymer, 185 *fig.*
  polynucleotide, 185 *fig.*
  transcription of DNA into, 190, 193
  translation of, into protein, 190
  viruses, 201, 202–3
RNA polymerase, 193, 211, 322–23
RNA splicing, 194
  alternative, 215

RNA world, 320
Rocks
    age of, and geologic time, 299
    phosphorus cycle and weathering of, 759
Rodriguez, Eloy, 678
Rods, retina, 597, 597 *fig.*
Root(s)
    modified, 627, 627 *fig.*
    mycorrhiza on, 342–43, 357
    nutrient uptake by, 647, 647 *fig.*
    primary growth in, 632, 633 *fig.*
    tissue systems in, 630, 631 *fig.*
Root cap, 632, 633 *fig.*
Root hairs, 626, 626 *fig.*
    in soil, 654 *fig.*
    solute uptake and, 647, 647 *fig.*
Root nodules, nitrogen-fixing bacteria on, 658, 658 *fig.*, 659 *fig.*
Root pressure, 648
Root system, 626, 626 *fig.*, 627
Rotenone as disruptor of cellular respiration, 99, 99 *fig.*
Rotifers, reproduction in, 536, 537 *fig.*
Rough endoplasmic reticulum (rough ER), 59, 59 *fig.*
Roundworms, 370, 370 *fig.*, 375
R plasmids, 205
    bacterial resistance to antibiotics carried on, 272
r-selection, 735
Rubisco, 118
    Calvin cycle and, 116, 116 *fig.*
RuBP, Calvin cycle and regeneration of, 116, 116 *fig.*
Rule of addition, 160
Rule of multiplication, 160
Ruminant mammals, 441
    digestive system, 441 *fig.*
Runners, as modified stems, 627
Runners, sprinter *versus* marathoner, 89–90
Running as locomotion, 604–5, 606
Rusts, 361

S
Saccharomyces cerevisiae (yeast), 246
Saccule, 600, 600 *fig.*
Sac fungi, 358 *fig.*, 359
Safe sex, 498, 546, 547
Sage grouse, courtship and mating behaviors of, 716, 716 *fig.*
Sahelanthropus tchadensis, 403
Salamanders, 389, 389 *fig.*
    paedomorphic axolotl, 292, 292 *fig.*
Saliva, 434
Salmon, sense perception and migration in, 588–89, 708
Salmonella, of gamma proteobacteria group, 328
Salt
    enzymes' effects on, 77
    ionic bonds as, 23, 23 *fig.*
San Andreas fault, 302, 302 *fig.*
Sandworms, 379, 379 *fig.*
SA (sinoatrial) node, 473
Sapwood, 635, 635 *fig.*
Sarcomas, 135
Sarcomeres, 614–15
Sarin, paralytic effects of, 78
SARS (severe acute respiratory syndrome), 202, 203
Saturated fats, 40
Savannas, 694–95, 694 *fig.*
Scanning electron microscopes (SEMs), 53
Scarlet king snake, 11 *fig.*
Scavenging-gathering-hunting culture, 407
Schindler, David, 761
Schistosomiasis, 374

Schizophrenia, 572–73, 584
Schweickart, Rusty, 684
Science. *See also* Biology
    discovery, using inductive reasoning, 9
    hypothesis-based, 9
    politics and, 120
    theories and, 8
Sclera, 594, 595 *fig.*
Sclereids, 628, 629 *fig.*
Sclerenchyma cells, 628, 629 *fig.*
Scorpions, 380, 381 *fig.*
Scrotum, 540
Scypha (sponge), 372, 372 *fig.*
Sea anemones, 373
    fission in, 536 *fig.*
Sea cucumbers, 384
Seals
    bottleneck effect on populations of northern elephant, 268, 269 *fig.*
    combat between male elephant, 532 *fig.*
    population-limiting factors on fur, 730–31, 731 *fig.*
Sea otter, as keystone species, 747 *fig.*
Search image, 714
Season(s), Earth's tilt as cause of, 688 *fig.*
Seasonal cycles, plant, 674
Sea star, 384, 384 *fig.*
    endoskeleton of, 609
    as keystone predator, 747, 747 *fig.*
    life cycle of, 368 *fig.*
Sea turtles, conservation efforts for, 774
Sea urchins, 384, 384 *fig.*
    egg fertilization in, 548, 549 *fig.*
    endoskeleton of, 609, 609 *fig.*
    killer whales' predation on sea otters and, 747, 747 *fig.*
    zygote cleavage in, 550 *fig.*
Seaweeds, 336
Secondary consumers in communities, 752, 752 *fig.*, 753 *fig.*
Secondary growth in plants, 634–35, 634 *fig.*
Secondary immune response, 492 *fig.*, 493
Secondary oocytes, 543
Secondary phloem, 634–35
Secondary spermatocytes, 542–43
Secondary structure of proteins, 44, 45 *fig.*
Secondary succession, 750
Secondary xylem, 634–35
Second law of thermodynamics, 73
Secretion function of excretory system, 513, 513 *fig.*, 514, 514 *fig.*
Secretory proteins, 59, 59 *fig.*
Sedimentary rock, strata of, 261, 261 *fig.*
Seed(s), 346–47
    abscisic acid and dormancy of, 669
    development of, 638, 638 *fig.*, 639
    fruits and dispersal of, 354 *fig.*
    germination of, 640, 640 *fig.*, 669
    structure of, 638 *fig.*
Seed coat, 638
Seed dormancy, 638, 669
Seedless plants, 350
    carboniferous coal forests formed by, 350
Seedless vascular plants, 346–47, 347 *fig.*
Seed plants, 347, 347 *fig. See also* Angiosperms; Gymnosperms
Segmentation, animal body, 378–79, 378 *fig.*
Segregation, Mendel's law of, 156–57, 171
Selective breeding, 153
Selective permeability of plasma membrane, 79
Selective serotonin uptake inhibitors (SSRIs), 573, 584
    prescriptions in U.S. by year, 584 *fig.*
Self-fertilizing plants, 154–55
Self proteins, 497
    Class I and II, 500

Self-recognition and non-self recognition, immune system and, 497, 500
Semen, 540–41
Semicircular canals, 600, 600 *fig.*
Semiconservative model of DNA replication, 188
Seminal vesicles, 540
Seminiferous tubules, 542, 543
Sensations, 590
Senses, 588–603
    hearing and balance, 598–600
    movement and, 588–89
    sensory inputs into nervous system and, 590, 602
    sensory reception, 590–93
    taste and smell, 601
    vision, 594–97
Sensitive period, 708
Sensory adaptation, 591
Sensory input, central nervous system, brain, and, 566, 590
Sensory neurons, 566
Sensory organ, platypus bill as, 366–67
Sensory receptors, 590, 592 *fig.*
    for balance, 600
    for electromagnetism, 593
    for light (vision), 594–97
    for pain, 592
    for sound (hearing), 598–99
    specialized, for five categories of stimuli, 592–93
    stimuli converted into electrical energy by, 590–91
    for taste and smell, 591 *fig.*, 601
    for temperature and touch, 592
Sensory transduction, 590, 591 *fig.*
Sepals, 352, 636
Septic shock, 487
Septuplets, Iowa, 534–35
Sequoia–Kings Canyon National Park, 623
Serine, 42
Serotonin, 582
Sessile organisms, 372
Setae, 412, 413 *fig.*
Set point, homeostasis and, 426
Severe acute respiratory syndrome (SARS), 202, 203
Severe combined immunodeficiency disease (SCID), 500
    gene therapy for, 243
Sewage treatment, 331, 331 *fig.*
Sex, safe, 498, 546, 547
Sex chromosome(s), 137, 175–77
    abnormal numbers of, 147, 147 *table*
    sex determination due to, 175, 175 *fig.*
    sex-linked genes and unique inheritance patterns, 176
    sex-linked genetic disorders, 177
Sex determination, 175, 175 *fig.*
Sex hormones, 522, 532
    functional group differences in, 35 *fig.*
    steroids and, 41
Sex-linked genes, 176
Sex pili, 324
Sexual dimorphism, 275, 404
Sexually transmissible diseases (STDs), 546, 546 *table*
Sexual recombination, 271
Sexual reproduction, 124–25, 126
    in animals, 536–37
    description of, 536
    in fungi, 358
    in humans, 538–47
    meiosis and, 542–43, 542 *fig.*, 543 *fig.*
    in plants, 348, 636–40
    prezygotic and postzygotic barriers to, 282–83, 282 *table*

sexual selection and sexual dimorphism in, 275
zygote fungi, 360, 360 *fig.*
Sexual response, human, 546
Sexual selection, 275
Shared derived characters, 306
Shared primitive characters, 306
Sharks, 388, 388 *fig.*
Shoots
auxin stimulation of cell elongation in, 666–67, 667 *fig.*
modified, 627
primary growth in, 632, 633 *fig.*
Shoot system, 626, 626 *fig.*, 627
Short-day plants, 674
Short-term memory, 583
Sickle-cell disease, 46
base pair mutation causing, 199
as genetic pleiotropy, 168, 168 *fig.*
heterozygote advantage and, 273
malaria and, 273
molecular basis of, 199 *fig.*
Sieve plates, 629, 629 *fig.*, 650 *fig.*
Sieve-tube members, 629, 629 *fig.*, 650 *fig.*
Sifaka, 401 *fig.*
Signaling behavior in animals, 720, 720 *fig.*
Signal proteins, 42
Signal transduction, 81, 81 *fig.*
cell-cycle control system utilizing, 134
Signal transduction pathways
cell-to-cell signaling and initiation of, 221, 221 *fig.*
hormones and creation of, 521
interference of oncogene proteins and faulty tumor-suppressor proteins with normal, 224–25
light and hormones effects on plants and, 676
Sign stimulus, 705
Silencer DNA sequence, 214
Silent Spring (Carson), 685
Simple fruit, 639, 639 *fig.*
Simpson (O. J.) murder trial, 231
Single-lens eye, 594–97
corrective lenses and surgery for, 596
focusing of, 595, 595 *fig.*
photoreceptors in, 597
structure of, 595 *fig.*
Sister chromatids, 128, 128 *fig.*
Skeletal muscle, 418, 418 *fig.*, 614
contractile apparatus of, 614–16
training of, 617
Skeletal system, 421 *fig.*
Skeleton, 608–9
bones of, 611–13
endoskeleton, 609
exoskeleton, 380, 608–9
human, 610–11, 610 *fig.*
hydrostatic, 370, 608
muscle interaction with, in movement, 613
support, movement, and protection provided by, 608–9
Skin
artificial, 419
cell replacement in, 136
epithelial tissue of, 416
sensory receptors in, 592 *fig.*
Skinner, B. F., 712
Skinner box, 712
Skin pigmentation, adaptation to sunlight amounts in humans and, 406
Skull, 386, 387
human and chimpanzee, 292 *fig.*
Skunks, eastern spotted and western spotted, 282, 282 *fig.*

Sleep
brain regulation of arousal and, 582
plant sleep movements, 673 *fig.*
Sliding-filament model of muscle contraction, 614–15, 615 *fig.*
Slime molds, 337, 337 *fig.*
Small intestine, 438–39
enzymatic digestion in, 438 *table*
relationship of, to large intestine, 440 *fig.*
structure of, 439, 439 *fig.*
tissue layers in wall of, 419 *fig.*
Small-population approach to endangered species, 772
Smell sense
chemoreceptors for, 592–93, 601
in salmon, 588–89, 708
Smoking
birth defects and, 557
cardiovascular disease and, 474
damage to respiratory system caused by, 459
Smooth endoplasmic reticulum (smooth ER), 58–59, 59 *fig.*
Smooth muscle, 418, 418 *fig.*
blood distribution controlled by, 477
Smuts, 361
Snakes, 390 *fig.*
antagonistic behavior in, 718, 718 *fig.*
electromagnetic receptors in, 593, 593 *fig.*
gravity and circulatory system of, 466–67
locomotion in, 606–7
mimicry in, 10–11
Social behavior, 717–23. *See also* Rituals
agonistic, 718
altruistic, 721
antagonistic, 718
communication among animals, 720
culture and human, 722
dominance hierarchies, 718
Goodall on chimpanzee dominance hierarchies and reconciliation behaviors, 719
human, 722
mating, 716 (*see also* Mating)
territorial, 717
(E. O.) Wilson on sociobiology and, 723
Social learning, 712–13
Sociobiology, 717, 723
Sodium chloride, 23, 23 *fig.*
Sodium potassium pumps (Na+-K+), 84
in neuron membrane, 568, 569, 572
Soil
characteristics of, 654
conservation of, 655, 655 *fig.*
fungal absorption of nutrients from, 657
organic farming and, 656
plant nutrients acquired from, 646, 652–53
Soil conservation, 655, 655 *fig.*
Soil horizons, 654, 654 *fig.*
Solar energy
as biospheric abiotic factor, 686
as electromagnetic energy, 112
greenhouse effect in Earth's atmosphere and, 119
photosynthetic light reactions and, 113
Solutes, 26
animal osmoregulation and balance of dissolved, 508–9
plant uptake of, 647, 647 *fig.*
Solutions, 26, 26 *fig.*
acidic, basic, and neutral, 27 *fig.*
hypertonic, hypotonic, and isotonic, 83 *fig.*
Solvents, 26
Somatic cells, 136–37
mutations in, and development of cancer, 225

Somatic nervous system, 576, 577
Somites, 553
Sound, route through ear of, 589–99, 599 *fig.*
Soy foods, 662–63
Sparrows
imprinting in white-crowned, 708
population fluctuations, 732 *fig.*, 733, 733 *fig.*
Spatial learning, animal behavior and, 710
Spatulae, 412, 413 *fig.*
Speciation, 280, 284–90
definition of, 279
of Darwin's finches on Galápagos Islands, 288 *fig.*, 289
geographical isolation and allopatric, 284
islands as laboratories of, 288, 288 *fig.*
of mosquitoes in London Underground, 278–79
polyploidy and plant, 287
sympatric, 286
tempo (models) of, 290, 290 *fig.*
Species, 6, 278–90. *See also* Biodiversity
binomial classifications and, 305
classifying (*see* Classification; Systematics; Taxonomy)
concepts of, 280–81
of Darwin's finches on Galápagos Islands, 288 *fig.*, 289
deep-sea, 682, 683 *fig.*
definition of, 265
diversity, 744, 747, 766
endangered (*see* Endangered species)
endemic, 775
extinct, 296–97 *fig.*, 391 *fig.*
identifying critical habitat for, 773
keystone, 747
mass extinctions and diversification of, 302–3
mechanisms in development of, 284–90
naming, 304–5
niches, 745
non-native (exotic), 394, 726–27
reproductive barriers between, 282–83, 282 *table*
speciation and development of new, 278–79
threatened, 766
Speech, *FOXP2* gene and human, 406
Sperm. *See also* Gametes
animal, 536
fertilization and, 548, 549 *fig.*
human, 542–43, 542 *fig.*
plant, 637, 637 *fig.*
properties of, 548
structure of human, 548 *fig.*
Spermatocytes, primary and secondary, 542–43
Spermatogenesis, 542, 542 *fig.*, 543
Spermicides, 547
Sperry, Roger, 581
Spiderman, 412, 412 *fig.*
Spiders, 380–81, 381 *fig.*
Spider silk proteins, uses of, 44
Spinal cord, 575
fluid-filled spaces in, 575 *fig.*
injured, 564–65
neurons in, 418 *fig.*
Spinal nerves, 575
Spirilla, 323
Spirochetes, 323, 323 *fig.*
Sponges, 372, 432
Sporangia (sporangium), 345
Spores, 345, 358, 360, 360 *fig.*
Sporophytes, 338, 348, 348 *fig.*, 636–37
angiosperm plant as, 352–53
fern, 349
pine, 350–51, 351 *fig.*

Squamous epithelium, simple or stratified, 416, 416 *fig.*
Squid, 377, 377 *fig.*
  deep-sea species of, 682
  eye of, 594, 594 *fig.*
  giant axons of, 568–69
  nervous system of, 574, 574 *fig.*
Squirrels, geographically isolated species of antelope, 284, 284 *fig.*
SRY gene, 175
SSRIs (selective serotonin uptake inhibitors), 573, 584
  prescriptions in U.S. by year, 584 *fig.*
Stabilizing selection, 274, 274 *fig.*
Stamens, 352, 636
Staphylococci, 323, 323 *fig.*
Staphylococcus, 328, 329, 329 *fig.*
Starch, 39, 39 *fig.*
Starling, dispersion of European, in North America, 726–27
Star navigation in birds, 711, 711 *fig.*
Start codon, translation initiated by, 196–97
Stem cells, 565
  adult, 219
  blood cell diseases and, 481
  differentiation of blood cells from, 481 *fig.*
  differentiation of cultured, 219 *fig.*
  embryonic, 216–17, 216–17 *fig.*
Stems, 626
  gibberellins and elongation of, 668–69
  modified, 627, 627 *fig.*
  secondary growth in woody, 634 *fig.*
  tissue systems in, 629, 630, 631 *fig.*
Stentor, 335 *fig.*
Steroid hormones, 521, 522
Steroids, 41
  anabolic, as health risk, 41
  spinal injuries and, 565
Steward, F. C., 212
Stigmas, 352, 636
Stimulants, effects of, on nervous system, 573
Stimulus, nerve signal, 568–69
  intensity of, 591
  sensory, 590, 602
Stolons, 627, 627 *fig.*
Stomach(s), 432
  bacterial infection and ulcers in, 437
  gastric glands in, 436 *fig.*
  human, 436–37
Stomata, 109, 345, 630
  closing of, and reduced water loss, 118
  guard cell regulation of opening/closing of, 649 *fig.*
Stop codon, 197
Storage proteins, 42
Stramenopiles, 336
Strata, of sedimentary rock, 261, 261 *fig.*
Streptococcus, 328
Streptomyces, 328, 328 *fig.*
Stress
  adrenal glands and response to, 530–31
  immunity and, 501
Stretch receptors, 592
Strict anaerobes, 101
Stroma, 63, 109
  Calvin cycle and photosynthesis in, 111
Stromatolites, 315, 317, 317 *fig.*
Structural proteins, 42
Structure and function, relationship of, in animals, 414–15, 618
Strychnine, 748
Sturtevant, Alfred H., 174, 174 *fig.*
Subatomic particles, 20
Submersibles, deep-sea, 682–83
Substrate feeders, 430, 430 *fig.*
Substrate-level phosphorylation, 94, 94 *fig.*
Substrates, enzyme, 77

Sucrose, 38
  sweetness scale relative to, 38 *table*
Sugar-phosphate backbone of nucleotides, 184
Sugars, 35
  formation of, 319
  photosynthesis and production of, 116
  transport of, by phloem tissue in plants, 650–51
Sugar sinks, 650
Sugar sources, 650
Sulfur bacteria, 328, 328 *fig.*
Sunflowers, 645
Sun-tanning, genetic influences on, 170
Superior vena cava, 471
Supplements, vitamin and mineral, 446
Supporting cells, neuron, 567
Surface tension, 25, 25 *fig.*
Surface-to-volume ratio in cells, 54, 55 *fig.*
Surrogate motherhood, 561
Surroundings, scientific definition of, 73
Survivorship curves, population, 729, 729 *fig.*
Suspension feeders, 372, 385, 430, 430 *fig.*
Sustainable Biology Initiative, 781
Sustainable development, 780, 781
Sustainable resource management, 735
Swallowing reflex, 434, 434 *fig.*
  Heimlich maneuver and disrupted, 435, 435 *fig.*
Sweating. *See also* Evaporative cooling
  water loss and, 509
Sweetness, scale of perceived, 38 *table*
Swift, Jonathan, 743
Swim bladder, 388
Swimming as locomotion, 606
Symbiosis, 332. *See also* Mutualism
  of mycorrhiza and plants, 657, 657 *fig.*
  in nitrogen-fixing bacteria and plants, 658–59
  plant nutrition and, 657–59
Symbiotic relationships, 748–49
Sympathetic division, autonomic nervous system, 576 *fig.*
  regulation of internal environment and, 576–77
Sympatric speciation, 286
Synapses, 567, 571
Synaptic cleft, 571
Synaptic terminals, 567, 572 *fig.*
Systematics, 304–10
  cladistics and phylogenetic, 306–7
  classifying and naming species as, 304
  homologies and study of, 304
  molecular biology as tool in, 308–9
Systemic acquired resistance, 677
Systemic circuit, 470
Systemic lupus erythematosus (lupus), 500
Systems, 4
  definition of, 73
Systems biology, 624
Systole, 472, 476

T
T2 phage, 182–83, 182 *fig.*
T_3 (triiodothyronine), 526
T_4 (thyroxine), 524, 526
  secretions of, 525 *fig.*
Tadpole, 553 *fig.*
Taiga, 698
Tannin, 748
Tapeworms, 374, 374 *fig.*
Taproot, 627 *fig.*
Target cells, 520, 522
  hormone signaling mechanisms affecting, 521
Tarsiers, 400, 401, 401 *fig.*

Tasmanian tiger, 367
Taste
  action potentials and sensation of, 591 *fig.*
  chemoreceptors for, 592–93, 601
Taste buds, sensory transduction at, 591 *fig.*
Tatum, Edward, 190
Taxis, 710
Taxol (drug), 135
Taxonomy, 6, 280
Tay-Sachs disease, 61
T cells (T lymphocytes), 490–91, 491 *fig.*, 497
  cytotoxic, 499, 499 *fig.*
  development of, 490 *fig.*
  helper, 499
  HIV destruction of, 498, 498 *fig.*
Technology
  DNA (*see* DNA technology)
  human evolution and, 408, 408 *fig.*
  science applications using, 12
Teeth, human, 434, 434 *fig.*
  development of replacement, 419
Telomeres, 245
Telophase
  cell division I, cytokinesis and, 139 *fig.*
  mitosis, 130, 131 *fig.*
Temperate broadleaf forests, 697
Temperate grasslands, 696–97, 696 *fig.*
Temperate zones, 688
Temperature. *See also* Body temperature
  as biospheric abiotic factor, 686
  definition of, 25
  enzymes' effects on, 77
  local high, in southern California, 689 *fig.*
Temporal isolation, 282
Tendons, 613
Tendrils, 627, 627 *fig.*
Terminal bud, 626
Terminator nucleotide sequence, 193
Terrarium ecosystem, 754, 754 *fig.*
Terrestrial animals
  cardiovascular system of, 470, 470 *fig.*, 471–78
  gravity effects on, 466–67
  lungs of, 455 *fig.*, 458–59
  osmoregulation in, 509
Terrestrial biomes, 693–99
  human disruption of, 761
Terrestrial food chain, 752 *fig.*
Territorial behavior, 717
Tertiary consumers in communities, 752, 752 *fig.*, 753 *fig.*
Tertiary structure of proteins, 44, 45 *fig.*
Testcross, determining genotype using, 159
Testes, 540
Testosterone, 521
  anabolic steroid as synthetic, 41
  male aggression and, 518–19
Tetraploidy, 286
Tetrapods, 389
Thalamus, 579
Theories, definition of, 8
Therapeutic cloning, 218, 219
Thermal energy, 72
Thermodynamics, laws of, 73
Thermophiles, 327
Thermoreceptors, 592
Thermoregulation, 504
  heat gains or losses and, 506–7
  reducing metabolic rate for energy savings and, 508
Theropods, feathered, 391. *See also Archaeopteryx* fossil
THG (tetrahydrogestrinone), 41
Thick filaments, myosin, 614–15, 614 *fig.*, 615 *fig.*
Thigmotropism, 672, 672 *fig.*

Thin filaments, actin, 614–15, 614 *fig.*, 615 *fig.*
Thinking, brain and, 580. *See also* Thought, symbolic
Thiobacillus, 331
Thought, symbolic. *See also* Thinking
  human evolution and, 403, 405, 405 *fig.*
Threatened species, 766. *See also* Endangered species
Three-domain classification scheme, 310, 310 *fig.,* 322
Threshold, nerve signal, 569, 569 *fig.*
Thylakoids, 109
  light reactions of photosynthesis in, 111, 112–15, 117
Thymine (T), 185
Thymus gland, 522
Thyroid gland, 526
  calcium homeostasis regulated by, 526–27, 527 *fig.*
  development and metabolism regulation by, 526
Thyroid-stimulating hormone (TSH), 524
Thyroxine (T$_4$), 524, 526
  secretions of, 525 *fig.*
Ticks, 380
Tigers
  conservation of Myanmar, 764–65
  Tasmanian, 367
Tight junctions, 66, 66 *fig.*
Tilden, Freeman, 622
Tinbergen, Niko, 704, 705, 710
Ti plasmid, 247, 247 *fig.*
Tissues, 2, 2 *fig.,* 415
Tissues, animal, 415–19
  artificial, 419, 419 *fig.*
  circulatory system and, 468
  embryonic, 551 *fig.,* 552, 553 *table*
  gas exchange in (*see* Gas exchange in animals)
  organs formed from, 419
Tissue systems, plant, 629, 630, 631 *fig. See also* Phloem; Xylem
Toads, 389
  cane, 394, 394 *fig.*
Tobacco
  birth defects and, 557
  as carcinogen, 227, 459
Tobacco mosaic virus, 202, 202 *fig.*
Tokaimura (Japan) nuclear power plant accident, 21
Tongue, human, 434, 434 *fig.*
  perception of sweetness and, 38
Tonicity, 83
Tools, complex, human evolution and, 403, 404, 405, 408
Topsoil, 654
Torpor, 508
Touch
  plant response to, 672
  sensory receptors of, 592, 592 *fig.*
Toxins
  bioremediation of metallic, 778, 778 *fig.*
  chemical, as animal defense, 746, 746 *fig.*
  chemical, as plant defense, 676–78, 748
  endotoxins and exotoxins, 329
  environmental degradation from, 769
  food crops and, 671
  fungal, 361
  phytoremediation of, 644–45
Trace elements, 18–19, 18 *table*
Trachea (windpipe), 434, 434 *fig.,* 458
Tracheal system, insect, 455 *fig.,* 457, 457 *fig.*
Tracheids, 629, 629 *fig.,* 648
Tracheoles, 457
Trade winds, 688
Tranquilizers, 573

Transcription, 190, 193, 193 *fig.*
  codons and, 191 *fig.*
  codons and genetic code, 192
  messenger RNA and, 194
  regulating eukaryotic genes with regulatory proteins (transcription factors) during, 214, 215 *fig.*
  regulating prokaryotic genes by turning on and off, 210–11
  summary, 198, 198 *fig.*
Transcription factors, 214, 215 *fig.*
Transduction, bacterial, 204
trans fats, 449
Transfer RNA (tRNA)
  structure of, 195 *fig.*
  translation and, 194–95
Transformation, bacterial, 204
Transgenic organism, 247–48
Translation, 190
  of codons, 191 *fig.*
  initiation codon for, 196, 197 *fig.*
  regulating eukaryotic genes during, 216
  stop codon for termination of, 197
  summary, 198, 198 *fig.*
  transfer RNA and, 194–95
Translocation
  chromosomal, 148, 148 *fig.*
  polypeptide chain formation and, 197
Transmission electron microscope (TEM), 53
Transpiration, 648–49
  guard cells and control of, 649
  water cycling and, 757
Transpiration-cohesion-tension mechanism, 649
Transport across membranes, 81–85
  active, 84
  exocytosis, endocytosis and, 84–85
  osmosis, 82
  passive, 81
  transport proteins and, 82
  water balance and, 82–83
Transport in animal bodies, 468–70
Transport in plant tissues
  of sugars, 650–51
  of water, 648–49
Transport proteins, 42
  facilitated diffusion and, 82
Transport vesicles, 59
Transposons, 245
Transthyretin, molecular structure of, 45 *fig.*
Tree(s). *See also* Forest(s)
  citrus, 342–43
  Dutch elm disease, 361, 361 *fig.*
  giant sequoias, 622–23
  pine, 350–51, 351 *fig.*
  water flow up, 648 *fig.*
  as water pumps, 645
Treponema pallidum, 328
TRH (TSH-releasing hormones), 524
Trial-and-error learning, 712
Trichinella spiralis, 375, 375 *fig.*
Trichinosis, 375
Triglycerides, 40
Triiodothyronine (T$_3$), 526
Trimesters, pregnancy, 558–59, 558–59 *fig.*
Triplet code, 191
Trisomy 21, 145, 146
Triticale, 248
Trophic structure of communities, 744
  ecosystem dynamics and, 752
  food chains and, 752–53, 752 *fig.,* 753 *fig.*
  pyramid of production and, 756
Trophoblast, 556
Tropical rain forests, 694. *See also* Forest(s)
  deforestation of, 356, 694
  Lugo on, 699
  medicinal plants in, 678

Tropics, 688
Tropisms, plant, 664–65, 672
trp operon, 211
True-breeding varieties, 152, 155
Trypanosoma, 334–35, 335 *fig.*
Tryptophan, 195, 582
  repressor-controlled operons and, 211, 211 *fig.*
TSH (thyroid-stimulating hormone), 524
TSH-releasing hormone (TRH), 524
Tubal ligation, 547
Tuberculosis, 272, 329
Tubers, 627, 627 *fig.*
Tumors, 135
Tumor-suppressor genes, 223
  interference of faulty, in signal-transduction pathways, 224–25
  p53, 225
Tundra, 698
Tunicates, 385, 385 *fig.*
Turner syndrome, 147, 147 *table*
Turtles, sea, conservation efforts for, 774
Twins, 534–35

U

Ulcers, gastric, 437
Ultimate questions, about behavior, 704
Ultrasound imaging, 165, 165 *fig.*
Ultraviolet light, 227
Ulva (multicellular green alga), 338, 338 *fig.*
Uncouplers as poisons, 99
Undernourished individuals, 443
Unequal reproductive success, 8
Unicellular green algae, 338, 338 *fig.*
Uniform population-dispersion pattern, 729
United States
  AIDS babies study by, 484–85
  Biological Weapons Convention and, 330
  cardiovascular disease in, 474
  dispersion of starling population in, 726–27
  human population centers in, 265 *fig.*
  Kyoto Protocol on global warming and, 771
  population age structure for, 738, 739 *fig.*
  regulatory agencies, 41, 91, 248
Unsaturated fats, 40
Uracil (U), 185
Urea, excretion of, 510
Ureters, 512, 512 *fig.*
Urethra, 512 *fig.,* 513, 541
Urey, Harold, 318, 319
Uric acid, excretion of, 510
Urinary bladder, 512
Urine, 509
  production of, 512, 513–14, 513 *fig.*
Uterus, 463 *fig.,* 538
Utricle, 600, 600 *fig.*

V

Vaccination, 489
Vaccines, 489
  DNA technology and development of, 237
Vacuoles, 62
Vagina, 539
Van der Waals forces, 412–13
van Helmont, Jan Baptista, 646
van Leeuwenhoek, Anton, 51
Variation in populations, natural selection and, 270–75
Varmus, Harold, 223
Vascular bundles, 630
Vascular cambium, 634
Vascular cylinder, 630
Vascular plants, 346–47
Vascular tissue system, 345, 630
Vas deferens, 540

Vasectomy, 547
Vasopressin, prairie vole behavior and, 706
Vector, gene, 234
Vegetarians, human
    essential amino acids for, 443
    as primary consumers, 756, 756 *fig.*
    protein deficiencies and, 656–57
Vegetative propagation, 641, 642, 642 *fig.*
Veins, animal, 469
    blood flow in, 475 *fig.*
    pulmonary, 471
    structure of, 472 *fig.*
Veins, leaf, 630
Venter, J. Craig, 245
Ventilation, 456. *See also* Breathing
Ventral surface of animals, 370, 370 *fig.*, 374
Ventricles
    of brain, 575
    of heart, 469
Venules, 469, 472 *fig.*
Vertebrae, 387
Vertebral column, 387
Vertebrates, 386–92
    amphibians, 389
    birds, 391
    cardiovascular system, 470
    as chordates, 387
    comparison of hemoglobin in select, 263
    digestive system adaptations of, 440–41
    endocrine system of, 522–25
    fish, 388
    with hinged jaws, 387
    limb formation, 555, 555 *fig.*
    lungs of terrestrial, 458–59
    nervous system of, 575–85
    reptiles, 390
Vervet monkeys, alarm calls of, 712–13,
    712 *fig.*
Vessel elements, 629, 629 *fig.*, 648
Vestigial organs, 262
Vibrio cholerae, 328
Vibrios, 323
Villi, intestinal, 439
Vinblastin (drug), 135
Virchow, Rudolf, 127
Virus(es), 200–205
    AIDS and HIV, 202, 203, 484–85, 498, 500
    animal disease caused by, 201
    cancer-causing, 222–23
    DNA of, 200
    emerging, 202–3
    herpes, 180–81
    infection of bacteria by (bacteriophage),
        183, 200
    influenza, 201 *fig.*, 203
    lytic and lysogenic cycles of phage, 200 *fig.*
    placenta-crossing, 557
    plant disease caused by, 202
    reproductive cycle of enveloped, 201 *fig.*
Visceral mass, 376
Vision, 594–97
    corrections for focusing problems with,
        596
    invertebrate eye types and, 594
    photoreceptors for, 593, 597, 597 *fig.*
    single-lens eyes in vertebrates and, 594–95
Visual acuity, 596
Vital capacity, 460
Vitamin(s), human requirements for, 444,
    444 *table*
Vitamin A
    beta-carotene converting to, 29, 29 *fig.*
    as chemical compound, 19
Vitamin D synthesis, sunlight availability,
    human skin colors and, 406
Vitreous humor, 595, 595 *fig.*

Vocal cords, 458
Voltage-gated channels, 569
Volume (sound), 599
Volvox (colonial alga), 338, 338 *fig.*
von Frisch, Karl, 704, 720

W

Walking locomotion, 606
Wallabies, 394, 395 *fig.*
Wallace, Alfred, 257
Walton, Bill, 531, 531 *fig.*
Wasps, 383, 383 *fig.*
    behavior of digger, 710, 710 *fig.*
    food chain involving *Apanteles*, 742–43
    plant recruitment of predatory, 676–77,
        677 *fig.*
Water, 25–28
    acid precipitation, 28
    acids, bases and, 27
    alternate ways to represent, 23 *fig.*
    as biospheric abiotic factor, 686
    chemical reaction in formation of, 29
    cohesion of liquid, 25
    gas exchange in, 455, 456–57
    hydrogen bonds in, 25
    osmosis and passive transport of, 82,
        82 *fig.*
    plant adaptations for saving, 118
    plant nutrients acquired from, 646
    plant production of oxygen by splitting,
        110
    plant xylem uptake/transport of, 647,
        647 *fig.*, 648–49
    polar covalent bonds of, 24, 24 *fig.*
    as solvent, 26 *fig.*
    states of, 26, 26 *fig.*
    temperature moderation and, 25
Water balance (osmoregulation), 82–83,
    508–9
Water-conducting cells, 629, 629 *fig.*
Water cycle, global, 757, 757 *fig.*
Water molds, 336, 336 *fig.*
Water-soluble vitamins, 444 *table*
Water strider, 25 *fig.*, 382 *fig.*
Water vascular system, 384, 384 *fig.*
Watson, James D., 12, 46, 186–87, 186 *fig.*
Wavelength of electromagnetic energy,
    112
Waxes, 41
Wegener, Alfred, 300
Weight loss diets, 448, 448 *table*
Went, Fritz, 664–65, 665 *fig.*
Westerlies, 688
West Nile virus, 202
Wetlands, 690, 692
    restoration of, 778
Whales
    evolution of, 261, 392
    extinct, 261 *fig.*
    feeding and nutrition in humpback,
        428–29
    migration in gray, 711, 711 *fig.*
    sea urchin declines from sea otters'
        predation by, 747, 747 *fig.*
Wheat (*Triticum* sp.), 287 *fig.*
White blood cells (leukocytes), 479
    inflammatory response and, 487
    lymphocytes (*see* Lymphocytes)
    lysosomes in, 61
    types of, 479 *fig.*
White-crowned sparrows, imprinting in, 708
White-footed mouse, behavior in, 707
White matter, central nervous system, 575
Whittaker, Robert H., 310
Whooping cranes, imprinting migratory
    behavior in, 709, 709 *fig.*

Wildlife populations, overexploitation of, 768
Wilkins, Maurice, 186, 187
Willows, 107
Wilmut, Ian, 218
Wilson, Allan, 226
Wilson, Edward O., 723, 767
Winds
    as biospheric abiotic factor, 686
    prevailing patterns of, 689 *fig.*
    seed dispersal by, 354, 354 *fig.*
    trade and prevailing, 688–89
    uneven heating as cause of, 689 *fig.*
Winemaking, yeast and, 101
Wings
    pattern formation and development of,
        555 *fig.*
    structure and function of, 414 *fig.*
Withdrawal (contraception), 547
Wolves
    dominance hierarchies in, 718
    Yellowstone to Yukon Conservation
        Initiative and, 776–77
Womb, human, 538
Wood, 634–35, 635 *fig.*
Woodpecker, endangered species of, 773,
    773 *fig.*
Wood rays, 635, 635 *fig.*
World Health Organization, 485
Wrangham, Richard, 678

X

X chromosome, 137
    abnormal numbers of, 147
    inactive, 214
X-chromosome inactivation, 214
X-O chromosome system, sex determination
    and, 175, 175 *fig.*
X-rays, 227, 422
X-Y chromosome system, sex determination
    and, 175, 175 *fig.*
Xylem, 345, 629
    nutrient uptake to root, 647, 647 *fig.*
    primary, 633
    secondary, 634–35
    water transport in, 648–49
Xylem sap, 648

Y

Y chromosome, 137
    abnormal numbers of, 147
    human evolution and study of, 405
    sex determination and, 175, 175 *fig.*
Yeast, 357, 363
    Saccharomyces cerevisiae, 246
Yeast infections, vaginal, 361
Yellow bone marrow, 611
Yellowstone National Park, 686
Yellowstone to Yukon Conservation
    Initiative (Y2Y), 776–77, 776 *fig.*
Yolk plug, 551 *fig.*
Yolk sac, 556, 557 *fig.*

Z

Zebra, 392, 392 *fig.*
Zero population growth, 738
Zoned reserves, 780
Zoopharmacognosy, 678
Zooplankton, 690
Z-W chromosome system
    sex determination and, 175, 175 *fig.*
Zygomycetes, 358 *fig.*, 359, 359 *fig.*
Zygote(s), 137
    animal, 368, 536, 548–50
    plant, 351, 352–53, 637, 637 *fig.*, 638
Zygote fungi, 358 *fig.*, 359
    life cycle, 360, 360 *fig.*

STUDENT CD-ROM AND WEBSITE ACTIVITIES

ACTIVITIES

1.1 The Levels of Life Card Game
1.5 Classification Schemes
2.1 The Levels of Life Card Game
2.4 Structure of the Atomic Nucleus
2.6 Electron Arrangement
2.6 Build an Atom
2.7 Ionic Bonds
2.8 Covalent Bonds
2.9 Nonpolar and Polar Molecules
2.10 Water's Polarity and Hydrogen Bonding
2.11 Cohesion of Water
2.15 Acids, Bases, and pH
3.1 Diversity of Carbon-Based Molecules
3.2 Functional Groups
3.3 Making and Breaking Polymers
3.4 Models of Glucose
3.7 Carbohydrates
3.9 Lipids
3.11 Protein Functions
3.14 Protein Structure
3.16 Nucleic Acid Structure
4.1 Metric System Review
4.3 Prokaryotic Cell Structure and Function
4.4 Comparing Prokaryotic and Eukaryotic Cells
4.4 Build an Animal Cell and a Plant Cell
4.8 Overview of Protein Synthesis
4.13 The Endomembrane System
4.15 Build a Chloroplast and a Mitochondrion
4.17 Cilia and Flagella
4.18 Cell Junctions
4.19 Review: Animal Cell Structure and Function
4.19 Review: Plant Cell Structure and Function
5.2 Energy Transformations
5.3 Chemical Reactions and ATP
5.4 The Structure of ATP
5.6 How Enzymes Work
5.12 Membrane Structure
5.13 Signal Transduction
5.13 Selective Permeability of Membranes
5.14 Diffusion
5.15 Facilitated Diffusion
5.17 Osmosis and Water Balance in Cells
5.18 Active Transport
5.19 Exocytosis and Endocytosis
5.21 Build a Chemical Cycling System
6.6 Overview of Cellular Respiration
6.7 Glycolysis
6.9 The Citric Acid Cycle
6.10 Electron Transport and Chemiosmosis

6.13 Fermentation
7.2 The Sites of Photosynthesis
7.5 Overview of Photosynthesis
7.6 Light Energy and Pigments
7.9 The Light Reactions
7.10 The Calvin Cycle
7.12 Photosynthesis in Dry Climates
8.5 The Cell Cycle
8.7 Mitosis and Cytokinesis Animation
8.7 Mitosis and Cytokinesis Video
8.13 Asexual and Sexual Life Cycles
8.15 Meiosis Animation
8.18 Origins of Genetic Variation
9.3 Monohybrid Cross
9.6 Dihybrid Cross
9.7 Gregor's Garden
9.12 Incomplete Dominance
9.21 Linked Genes and Crossing Over
9.23 Sex-Linked Genes
10.1 The Hershey-Chase Experiment
10.1 Phage T2 Reproductive Cycle
10.3 DNA and RNA Structure
10.3 DNA Double Helix
10.5 DNA Replication
10.6 Overview of Protein Synthesis
10.9 Transcription
10.14 Translation
10.17 Phage Lysogenic and Lytic Cycles
10.18 Simplified Reproductive Cycle of a DNA Virus
10.21 Retrovirus (HIV) Reproductive Cycle
11.1 The *lac* Operon in *E. coli*
11.8 Gene Regulation in Eukaryotes
11.9 Review: Gene Regulation in Eukaryotes
11.13 Development of Head-Tail Polarity
11.14 Signal Transduction Pathway
11.19 Connection: Causes of Cancer
12.2 Restriction Enzymes
12.3 Cloning a Gene in Bacteria
12.10 Gel Electrophoresis of DNA
12.11 Analyzing DNA Fragments Using Gel Electrophoresis
12.12 Connection: DNA Fingerprinting
12.15 The Human Genome Project: Human Chromosome 17
12.18 Connection: Applications of DNA Technology
12.19 Connection: DNA Technology and Golden Rice
13.1 Darwin and the Galápagos Islands
13.1 The Voyage of the *Beagle*: Darwin's Trip Around the World
13.4 Reconstructing Forelimbs
13.9 Causes of Microevolution
13.12 Genetic Variation from Sexual Recombination
14.7 Polyploid Plants

(continued on next page)

(continued from previous page)

14.8   Exploring Speciation on Islands
14.11   Mechanisms of Macroevolution
14.12   Paedomorphosis: Morphing Chimps and Humans
15.1   A Scrolling Geologic Record
15.5   Mechanisms of Macroevolution
15.10   Classification Schemes
16.1   The History of Life
16.10   Prokaryotic Cell Structure and Function
16.13   Diversity of Prokaryotes
17.2   Terrestrial Adaptations of Plants
17.3   Highlights of Plant Phylogeny
17.5   Moss Life Cycle
17.6   Fern Life Cycle
17.8   Pine Life Cycle
17.10   Angiosperm Life Cycle
17.14   Connection: Madagascar and the Biodiversity Crisis
17.16   Fungal Reproduction and Nutrition
17.18   Fungal Life Cycles
18.13   Characteristics of Invertebrates
18.21   Characteristics of Chordates
18.22   Animal Phylogenetic Tree
19.2   Primate Diversity
19.6   Human Evolution
20.1   Correlating Structure and Function of Cells
20.2   The Levels of Life Card Game
20.3   Overview of Animal Tissues
20.4   Epithelial Tissue
20.5   Connective Tissue
20.6   Muscle Tissue
20.7   Nervous Tissue
20.14   Regulation: Negative and Positive Feedback
21.11   Digestive System Function
22.5   The Human Respiratory System
22.10   Transport of Respiratory Gases
23.4   Mammalian Cardiovascular System Structure
23.4   Path of Blood Flow in Mammals
23.9   Mammalian Cardiovascular System Function
24.12   HIV Reproductive Cycle
24.13   Immune Responses
25.9   Structure of the Human Excretory System
25.11   Nephron Function
25.11   Control of Water Reabsorption
26.2   Overview of Cell Signaling
26.2   Nonsteroid Hormone Action
26.2   Steroid Hormone Action
26.11   Human Endocrine Glands and Hormones
27.3   Reproductive System of the Human Male
27.5   Reproductive System of the Human Female
27.11   Sea Urchin Development Video
27.12   Frog Development Video
28.2   Neuron Structure
28.5   Nerve Signals: Action Potentials
28.6   Neuron Communication

29.6   Structure and Function of the Eye
30.3   The Human Skeleton
30.8   Skeletal Muscle Structure
30.9   Muscle Contraction
31.6   Root, Stem, and Leaf Sections
31.8   Primary and Secondary Growth
31.10   Angiosperm Life Cycle
31.12   Seed and Fruit Development
32.4   Transpiration
32.5   Transport in Phloem
32.8   Absorption of Nutrients from Soil
32.11   Connection: Genetic Engineering of Golden Rice
33.7   Leaf Abscission
33.12   Flowering Lab
34.3   Connection: DDT and the Environment
34.5   Adaptations to Biotic and Abiotic Factors
34.8   Aquatic Biomes
34.9   Terrestrial Biomes
35.20   Honeybee Waggle Dance
36.2   Techniques for Estimating Population Density and Size
36.3   Investigating Survivorship Curves
36.9   Human Population Growth
36.10   Analyzing Age-Structure Diagrams
37.6   Interspecific Interactions
37.7   Primary Succession
37.10   Food Webs
37.11   Energy Flow and Chemical Cycling
37.14   Energy Pyramids
37.17   The Carbon Cycle
37.18   The Nitrogen Cycle
37.18   Water Pollution from Nitrates
38.2   Connection: Madagascar and the Biodiversity Crisis
38.3   Connection: Fire Ants as an Exotic Species
38.4   Connection: DDT and the Environment
38.5   Connection: The Greenhouse Effect
38.14   Conservation Biology Review

THINKING AS A SCIENTIST

1.6   How Do Environmental Changes Affect a Population?
1.8   The Process of Science: How Does Acid Precipitation Affect Trees?
2.1   Connection: How Are Space Rocks Analyzed for Signs of Life?
2.16   Connection: How Does Acid Precipitation Affect Trees?
3.16   Connection: What Factors Determine the Effectiveness of Drugs?
4.2   Connection: What Is the Size and Scale of Our World?
4.19   Connection: How Are Space Rocks Analyzed for Signs of Life?
5.7   How Is the Rate of Enzyme Catalysis Measured?
5.13   How Do Cells Communicate with Each Other?
5.17   How Does Osmosis Affect Cells?
6.12   How Is the Rate of Cellular Respiration Measured?
7.6   How Does Paper Chromatography Separate Plant Pigments?
7.11   How Is the Rate of Photosynthesis Measured?